环境化学前沿

（第二辑）

江桂斌　郑明辉　孙红文　蔡勇　主编

科学出版社
北京

内 容 简 介

环境化学是环境科学的核心组成部分，环境化学主要基于化学理论和方法，结合地学、生物、医学等交叉学科技术，以污染物为研究对象，以解决相关环境问题为目标，是一门研究污染物的生成与释放、环境赋存与归宿、转化与代谢、毒性效应与健康影响以及污染物削减控制原理与技术的学科。

经过 40 多年的发展，我国环境化学在学科建设、人才培养、队伍规模、实现国家目标和提升国际影响力等方面均取得了长足进步。环境化学已成为化学的一个重要分支。

《环境化学前沿》连续出版物邀请我国环境化学领域著名的专家学者，包括 40 多位"国家杰出青年科学基金"获得者、"千人计划"和"长江学者"入选者撰稿。本书（第二辑）与第一辑内容互相补充、互不重复，能够反映和代表我国目前环境化学领域的工作特色和主流发展趋势。

本书内容丰富、前瞻性强，可供环境化学、环境科学与工程以及地学、生物、材料、食品、公共卫生、化学品安全等交叉学科领域从事研究的科研人员、研究生和政府部门管理人员阅读和参考。

图书在版编目（CIP）数据

环境化学前沿. 第二辑/江桂斌等主编. —北京：科学出版社，2019.8
ISBN 978-7-03-062007-1

Ⅰ. ①环… Ⅱ. ①江… Ⅲ. ①环境化学–研究 Ⅳ. ①X13

中国版本图书馆 CIP 数据核字(2019)第 158141 号

责任编辑：朱　丽 / 责任校对：何艳萍
责任印制：肖　兴 / 封面设计：耕者设计工作室

科学出版社 出版
北京东黄城根北街 16 号
邮政编码：100717
http://www.sciencep.com

三河市春园印刷有限公司 印刷
科学出版社发行　各地新华书店经销

*

2019 年 8 月第 一 版　开本：889×1194 1/16
2019 年 12 月第二次印刷　印张：55 1/4
字数：1 820 000

定价：280.00 元
（如有印装质量问题，我社负责调换）

《环境化学前沿（第二辑）》编辑委员会

主　　编　江桂斌　郑明辉　孙红文　蔡　勇

编 委 会（以下按姓氏汉语拼音排序）

蔡亚岐　蔡宗苇　曹宏斌　柴立元　陈　超　陈吉平

陈建民　陈景文　陈　威　陈勇生　戴家银　党　志

冯新斌　冯兆忠　葛茂发　关小红　郭红岩　胡　春

胡　敏　胡献刚　黄　霞　季　荣　阚海东　李芳柏

梁　勇　廖春阳　林道辉　林　璋　林志芬　刘稷燕

刘建国　刘　倩　刘思金　罗　义　宁　平　欧阳钢锋

潘丙才　全　燮　桑　楠　孙轶斐　汪美贞　王爱杰

王飞越　王　琳　王书肖　王亚韡　韦朝海　魏东斌

吴李君　闫　兵　尹大强　应光国　尤世界　俞汉青

曾永平　张爱华　张爱茜　张礼知　张庆竹　张淑贞

张伟贤　郑玉新　周炳升　周东美　周群芳　朱东强

朱永官　祝凌燕　庄树林

序 言

自 2002 年 10 月 24～27 日第一届全国环境化学大会在浙江大学召开以来，全国环境化学大会已经连续召开了 9 届，参会人数从当初的 200 人发展到 6000 多人规模，受到国内外学术界的高度重视。鉴于全国环境化学大会的规模与影响，根据专家建议并结合国内外学术会议的成功经验，中国化学会环境化学专业委员会决定自 2017 年起，编辑出版《环境化学前沿》，将此作为全国环境化学大会的一项成果，奉献给广大读者。每一辑的《环境化学前沿》均将总结上届会议以来我国环境化学领域所取得的部分重要成果、展望未来的发展趋势。由于环境化学涉猎范围非常广泛，研究内容极为丰富，所以每次会议出版的前沿成果或展望将不追求领域的全面覆盖和完整的系统性，而是通过一些方面的进展与前沿的总结，展示我国环境化学工作者的最新成果，以期进一步围绕国家环境保护与健康的重大需求，提高我国环境化学领域的创新能力与国际影响力。

《环境化学前沿》（第一辑），2017 年 10 月首印 900 册，广大读者响应热烈，出版后随即售罄。为满足广大读者的需求，科学出版社又再次加印 1000 册，也早已售罄。这本书的出版受到广大环境领域科研人员及在校学生的欢迎，特别在全国环境化学大会后，很多实验室组织了对书中 30 个研究专题的学习与讨论，撰写读后体会。该书的加购信息也通过邮件、电话等形式反馈给出版社和作者，大家普遍认为这本书的出版对读者把握环境化学学科发展方向及其最前沿进展情况意义重大。通过这本书，读者对环境化学研究的全貌有了更充分快捷的获取方式。这些反馈，增强了作者和出版社的信心——在推动环境化学学科进步，培养学科后续人才方面，这本书的出版无疑是一次有意义并且实践证明是成功的尝试。

与 2017 年相比，组织《环境化学前沿》（第二辑）在时间上要充裕一些。第二辑的撰写同样得到我国环境化学领域著名专家学者的积极支持，所选内容均由第十届全国环境化学大会各分会负责人和科研骨干牵头，本领域专家学者分工协作而成。作者队伍中包括了 40 多位"国家杰出青年科学基金"获得者、"长江学者"和"千人计划"入选者。这些作者都是在环境化学一线从事相关研究工作、能够准确把握国家学术前沿的专家，他们的论述充分反映了目前国内外环境化学领域的工作特色和主流方向。专家们的倾心支持和高度的责任感，使得书稿能够在短时间内完成，充分展现了我国环境化学学界空前的凝聚力、向上的精神风貌和充满生机的朝气。在此，谨向所有作者的辛勤工作表示诚挚的谢意。本着学术自由、学术民主和学术平等的原则，本书允许不同风格，鼓励各抒己见，强调文责自负。

当前我国环境保护与健康领域依旧面临着十分严峻和十分复杂的局面。首先是工业发展在一定时期内呈现出高能耗、高排放、低效率的特点。据 2018 年数据，我国钢铁、水泥的总产量分别占世界总产量的 51%和 57%，而中国的 GDP 仅占世界的 17%左右；我国煤炭消费高达 37 亿吨标煤，占全球的 47%，而发电量仅占全球的 25%；机动车保有量达 3.3 亿辆，年产销居世界第一；而家用小汽车的增长，也必将随着我国城乡经济的发展呈现出更快的增长速度。其次是全球化学品登记与使用的速度爆炸式增加。几年以前全球登记的化学品只有 7000 万，短短的几年时间就翻了一番，大约每天增长 4 万种，每年增加 20%。工业生产与日常生活中使用的化学品，存在于生产、储运、使用最后到环境排放整个链条中，各国对使用的化学品毒性知之甚少。未经严格的环境风险与健康风险评估而使用的化学品造成了环境管理的巨大被动。科学研究已经证明，环境污染与健康有着密不可分的关联与因果关系。根据世界银行和世界卫生组织（WHO）有关统计数据，世界上 70%的疾病和 40%的死亡人数与环境因素有关。此外，全球气候变化使得污染物蒸发、迁移、转化、降解、代谢、累积等速度加快，污染全方位传播及污染全球化的趋势加重，加速了污染物的环境地球化学循环过程。

随着经济的快速发展，发达国家百年发展过程中经历的不同污染阶段所产生的健康问题在我国集中显现，多种与环境污染密切相关的疾病发病率显著上升，先后在我国多地出现的肿瘤高发现象均与污染密切相关。我国独有的环境污染特点决定了其健康问题的特殊性，不能照搬国外研究模式与成果解析污染与相关疾病的因果关系。理论与方法创新是我国环境与健康研究的机遇和挑战。

国际环境保护公约的签订使得我国环境化学的研究结果直接为国际公约的谈判与实施提供了科学支撑。为了在全球范围内控制和削减持久性有机污染物，《关于持久性有机污染物的斯德哥尔摩公约》获得联合国环境规划署通过，并开放各国签署，该公约已经于2004年付诸实施，公约认定的持久性有机物是一个开放体系，首次名单中包含了二噁英、多氯联苯、部分有机氯农药等共12类化合物。随着科学研究的深入和经济社会的发展，不断有新成员加入。公约的正式生效标志着持久性有机污染物污染成为当今世界各国共同面临的全球性重大环境问题。

2013年1月19日，联合国环境规划署通过了旨在全球范围内控制和减少汞排放的国际公约——《关于汞的水俣公约》文本，就具体限排范围作出详细规定，以减少汞对环境和人类健康造成损害。2013年10月10日，《关于汞的水俣公约》在日本熊本通过，包括中国在内的87个国家和地区的代表共同签署公约，标志着全球减少汞污染迈出第一步。

2016年4月25日，十二届全国人大常委会第二十次会议通过我国加入《关于汞的水俣公约》。2017年5月18日，欧盟及其七个成员国批准了《关于汞的水俣公约》。2017年8月16日，《关于汞的水俣公约》正式生效。这是近十年来环境与健康领域新增的一项全球性公约。履行此项公约，我国面临着汞污染减排的巨大挑战，急需科学技术的支撑。

上述环境问题的研究赋予了环境化学发展的空前机遇。我国的环境化学学科日益成熟繁荣，研究的水平、深度和广度有了空前的提高，一些研究开始在国际学术界产生重要影响。当然，科学的发展永无止境，且与国际环境化学学科发展的先进水平和国内相关学科的快速发展相比，环境化学在学科积累、人才队伍和研究基础等方面差距仍然较大，尤其是在研究的原创性、系统综合性、应用性和产业化等方面差距更加明显。

21世纪初，中国化学会环境化学专业委员会的成立是我国环境化学快速发展的必然产物。专业委员会及环境化学学科赶上了国际环境科学迅速发展，国家改革开放、经济快速增长以及与此同时环境保护成为国家重大战略需求的机遇，直接见证和参与了我国环境化学学科发展的历程，为我国环境化学学科的发展提供了交流和研讨的平台，对推动我国环境化学研究的进步发挥了重要作用。

前九届全国环境化学大会的召开，对我国环境化学的跨越式发展起到了重要助推作用。"好雨知时节，当春乃发生"，第十届全国环境化学大会是在我国全面建成小康社会关键之年召开的又一次里程碑式的大会。本次大会由南开大学承办，孙红文教授担任组织委员会主席。大会将于2019年8月15~19日在南开大学隆重召开。大会的主题是"聚力污染治理攻坚战，引领环境科学发展新时代"，大会将充分体现"创新、参与、合作、前瞻"的会议宗旨，推动环境化学学科更加重视国家需求，更加聚焦国际前沿，更加重视民生目标。大会将设立63个分会场，100多位国际著名专家学者包括海外华裔学者将参会。截止到7月8日，大会在线注册7209人，其中教师与研究人员3319人，研究生代表3890人，收到会议正式摘要4449篇，会议规模、参会人数、会议水平、国际化程度等再创历史新高！

"众人拾柴火焰高"，本书聚集了专家学者对若干环境化学前沿问题的认识与展望，是集体努力的成果。我们应该客观清楚地看到，尽管我国环境化学学科取得了巨大的成就，但离国家生态文明建设的需求，离真正解决我国突出的环境污染问题，提高全民族的健康水平等远大目标尚有很大差距。环境化学学科的影响力还需要进一步提高。由于当前学术界在日常科研管理、人才基金评审等环节上存在的一些倾向，不利于青年人才的成长、国际视野的养成和团队精神的培养，我们也希望通过本书，为广大青年学者在选题方面提供参考，避免急功近利的思想，着眼国家目标需求的前沿环境问题，立足做长期系统的工作，形成自己的研究特色，成为环境化学发展真正有用的人才。

今年在出版《环境化学前沿》（第二辑）的同时，由香港理工大学的李向东教授等主编，邀请国际本领域著名专家学者撰稿，由 Springer 出版社出版了 *A New Paradigm for Environmental Chemistry and Toxicology*，此书代表了环境化学与毒理学领域当前的国际权威视角、专家们长期研究的心得及未来发展预见，也作为《环境化学前沿》（第二辑）的姊妹篇，呈献给广大读者。

受组织者学识水平之局限，在《环境化学前沿》（第二辑）撰稿和修订过程中难免失之偏颇，出现不同学术观点甚至缺点错误。环境化学学科一直是在不断学习中提高的，学科的发展是永无止境的，人们对环境问题的认识也总是随着时间的推移而不断发展和提高。我们希望本书能够对广大环境化学工作者、研究生及环境管理专家有所裨益，若能对读者了解并把握环境化学研究的热点和前沿领域起到抛砖引玉作用，引起广大读者的广泛兴趣、讨论、争论和批评指正，便是编者期待和深感欣慰之处。

本书成稿过程中，中国科学院生态环境研究中心郑明辉研究员、蔡勇教授，南开大学孙红文教授等在组织专家审稿、改稿过程中做出了突出贡献；科学出版社为本书的顺利出版提供了方便的条件；朱丽及其团队为本书的策划与出版发挥了重要作用。在此一并表示诚挚的感谢。

2019 年夏於北京

目 录

序言
第1章 原位样品前处理技术在环境分析和毒理学研究中的应用 1
  1 引言 2
  2 原位样品前处理技术 3
    2.1 SPME 技术与被动采样技术 3
    2.2 与原位分析技术的比较 4
  3 原位样品前处理技术在环境分析和毒理学研究中的应用 4
    3.1 原位样品前处理技术的定量方法 4
    3.2 原位样品前处理技术在非生命环境介质中的应用 6
    3.3 原位样品前处理技术在动植物活体组织中的应用 8
    3.4 原位样品前处理技术在内源性小分子有机物检测上的应用 10
  4 原位样品前处理技术的萃取相 11
  5 展望 12
  参考文献 13
第2章 环境稳定同位素技术研究进展 20
  1 引言 21
  2 稳定同位素技术基础理论 22
    2.1 同位素理论基本概念 22
    2.2 稳定同位素分馏 23
  3 环境稳定同位素分析方法进展 24
    3.1 MC-ICP-MS 分析性能的改进 24
    3.2 MC-ICP-MS 与各类进样系统的联用 26
    3.3 新同位素体系分析方法的开发 28
    3.4 小结 28
  4 稳定同位素技术在污染物示踪中的应用进展 28
    4.1 稳定同位素在大气细颗粒示踪中的应用 28
    4.2 稳定同位素在环境中金属示踪中的应用 30
    4.3 稳定同位素在环境纳米技术中的应用 33
  5 环境稳定同位素标记技术研究进展 33
  6 展望 35
  参考文献 36
第3章 有机污染物生物有效性研究的方法与应用 47
  1 引言 48
  2 有机污染物生物有效性的评价方法 49
    2.1 生物学方法 49
    2.2 化学方法 50

3 生物有效性在研究有机污染物在环境介质及界面间迁移行为中的应用 ·············· 53
 3.1 沉积物 ··· 53
 3.2 水体 ··· 54
 3.3 土壤 ··· 55
4 生物有效性在人体健康研究中的应用 ··· 57
 4.1 生物有效性的活体测试 ·· 57
 4.2 生物有效性的体外测试（经消化道途径）·· 59
 4.3 生物有效性的体外测试（经呼吸道途径）·· 60
5 生物有效性在生态毒性评价识别中的应用 ··· 63
 5.1 仿生萃取技术在毒性预测中的应用 ··· 63
 5.2 基于生物效应的复合毒性评价方法 ··· 65
参考文献 ··· 67

第4章 天然有机质的环境行为和效应 ·· 78

1 引言 ·· 79
2 天然有机质的结构表征 ·· 80
 2.1 天然有机质组成结构的认知发展 ··· 80
 2.2 样品前处理方法 ··· 81
 2.3 天然有机质的分子表征方法 ·· 82
3 天然有机质的环境效应 ·· 88
 3.1 天然有机质对有机污染物吸附行为的影响 ··· 89
 3.2 天然有机质对污染物转化行为的影响 ·· 91
4 展望 ·· 95
参考文献 ··· 96

第5章 全氟及多氟烷基类化合物 ··· 104

1 引言 ·· 105
2 PFASs 简介 ·· 105
 2.1 理化信息 ·· 105
 2.2 国内外管控信息 ··· 108
3 PFASs 的环境赋存水平与迁移转化规律 ··· 109
 3.1 PFASs 的环境赋存水平 ·· 109
 3.2 PFASs 的迁移转化规律 ·· 111
4 PFASs 毒性效应 ··· 113
 4.1 PFOS 和 PFOA ·· 113
 4.2 F-53B ·· 113
 4.3 HFPO-DA（GenX）和 HFPO-TA ·· 114
 4.4 其他结构类似物 ··· 114
5 PFASs 的人体接触与健康效应 ·· 115
 5.1 人体内 PFOS 和 PFOA 的赋存水平 ·· 115
 5.2 人体内 PFASs 的转化、消除规律 ··· 116
 5.3 流行病学研究结果 ··· 117
6 新型 PFASs 的相关研究 ·· 118

	6.1 新型PFASs简介	118
	6.2 新型PFASs的筛查与鉴定	119
	6.3 新型PFASs的环境赋存水平和迁移转化	120
	6.4 新型PFASs的生物累积和毒性效应	121
7	展望	122
	参考文献	123

第6章 硅氧烷类污染物的环境赋存与转化 ··· 131

1 引言 ··· 132
2 硅氧烷简介 ··· 132
   2.1 物化性质与应用 ··· 132
   2.2 生态环境效应 ··· 134
3 硅氧烷的环境行为 ··· 135
   3.1 生活区域 ··· 136
   3.2 工业区域 ··· 141
4 人群暴露 ··· 145
   4.1 硅氧烷生产工厂 ··· 145
   4.2 硅氧烷使用工厂 ··· 147
5 改性硅氧烷环境行为的初步研究 ··· 147
   5.1 氟硅氧烷 ··· 147
   5.2 苯基硅氧烷 ··· 148
6 展望 ··· 149
参考文献 ··· 150

第7章 汞的环境污染研究进展 ··· 155

1 人为源大气汞排放研究进展 ··· 156
   1.1 人为源汞排放特征研究 ··· 156
   1.2 人为源汞排放清单进展 ··· 157
   1.3 展望 ··· 159
2 水生生态系统汞的迁移转化 ··· 159
   2.1 水环境中汞的光化学转化 ··· 160
   2.2 水体环境硫化汞的形态转化及在汞循环中的作用 ··· 161
   2.3 颗粒形态汞参与的水环境汞迁移转化过程 ··· 162
   2.4 海洋环境汞迁移转化研究的热点问题 ··· 162
   2.5 生态系统演变对水环境汞生态健康风险的影响 ··· 163
3 汞在森林生态系统生物地球化学研究进展 ··· 164
   3.1 森林生态系统汞循环过程 ··· 165
   3.2 森林生态系统汞的质量平衡 ··· 165
   3.3 森林生态系统汞稳定同位素研究 ··· 167
   3.4 总结与展望 ··· 169
4 汞的甲基化研究进展 ··· 170
   4.1 甲基汞的性质和危害 ··· 170
   4.2 甲基汞的来源和分布 ··· 170

4.3	汞的微生物甲基化	170
4.4	甲基汞的微生物去甲基化	173
4.5	汞的非生物甲基化	173
4.6	结论与展望	173

参考文献 … 173

第8章 铁基材料及其污染控制技术研究进展 … 193

1 引言 … 194

2 纳米零价铁及其重金属废水治理技术 … 194
  2.1 纳米零价铁的结构与制备 … 194
  2.2 纳米零价铁在水环境转化 … 195
  2.3 纳米零价铁与重金属反应 … 197
  2.4 纳米零价铁工程应用 … 199

3 铁矿物及其污染控制技术研究进展 … 201
  3.1 铁矿物的形成、制备及表征 … 202
  3.2 基于铁矿物吸附性质的环境污染控制技术 … 205
  3.3 基于铁矿物氧化还原性质的环境污染控制技术 … 208

4 基于高铁酸盐的污染控制技术及原理 … 214
  4.1 Fe(VI)氧化特性及强化氧化研究 … 214
  4.2 Fe(VI)氧化过程中消毒副产物生成情况 … 218
  4.3 Fe(VI)还原产物特性及其在去除砷及重金属方面的研究 … 218

5 微生物介导亚铁氧化耦合硝酸盐还原机制与环境效应 … 220
  5.1 概述 … 220
  5.2 微生物耦合硝酸盐还原和亚铁氧化 … 221
  5.3 硝酸盐生物作用还原为亚硝酸盐和化学反硝化作用 … 223
  5.4 酶促亚铁氧化耦合硝酸盐还原 … 224
  5.5 环境意义 … 225

6 展望 … 226

参考文献 … 226

第9章 持久性有机污染物植物吸收、迁移与转化的分子机制 … 239

1 引言 … 240

2 POPs的植物吸收与迁移 … 240
  2.1 土壤中POPs的植物吸收和迁移 … 240
  2.2 空气中POPs的植物吸收和迁移 … 241
  2.3 POPs植物吸收的模型 … 242
  2.4 POPs对植物的影响 … 243

3 POPs的植物转化 … 243
  3.1 POPs植物转化过程概述 … 243
  3.2 POPs植物转化中酶的作用 … 243
  3.3 POPs在植物中典型的Ⅰ相反应 … 244
  3.4 POPs在植物中典型的Ⅱ相反应 … 246
  3.5 POPs在植物中典型的Ⅲ相反应 … 247

| 3.6 POPs 转化产物的检测与筛查方法 ································ 248
    4 POPs 手性物质的植物吸收和转化 ·· 248
    5 展望 ·· 249
    参考文献 ··· 249

第 10 章 中国大气环境化学研究进展 ·· 256
    1 引言 ·· 257
    2 大气光化学污染 ·· 257
    3 大气成核和新粒子形成机制 ·· 259
        3.1 近期实验室模拟研究进展 ·· 260
        3.2 近期外场观测研究进展 ·· 260
    4 二次有机气溶胶与大气非均相/多相反应 ·································· 262
        4.1 二次有机气溶胶 ·· 262
        4.2 HONO ·· 262
        4.3 多相反应 ·· 263
        4.4 环境污染控制 ·· 263
    5 展望 ·· 264
    参考文献 ··· 264

第 11 章 环境抗生素与耐药基因污染研究进展 ································ 267
    1 引言 ·· 268
    2 抗生素使用 ·· 268
    3 抗生素环境污染 ·· 269
        3.1 城市污水处理厂中抗生素污染 ···································· 269
        3.2 水环境中抗生素污染 ·· 270
        3.3 养殖环境中抗生素污染 ·· 270
        3.4 土壤中抗生素污染 ·· 270
    4 耐药基因环境污染 ·· 271
        4.1 城市污水处理系统中耐药基因的污染现状及来源 ···················· 271
        4.2 养殖环境耐药基因污染 ·· 273
        4.3 大气环境耐药基因污染 ·· 274
    5 展望 ·· 276
    参考文献 ··· 277

第 12 章 环境微塑料的形成、分析、行为及风险管控研究进展 ···················· 288
    1 微塑料的环境来源与分布 ·· 289
        1.1 环境微塑料的来源 ·· 289
        1.2 海洋环境中微塑料 ·· 291
        1.3 淡水环境中微塑料 ·· 296
        1.4 土壤环境中微塑料 ·· 298
        1.5 灰尘和大气中微塑料 ·· 298
    2 微塑料的检测及表征 ·· 299
        2.1 微米尺度采样与检测 ·· 299
        2.2 纳米尺度检测方法 ·· 300

	2.3 塑料种类、形态、表面特征分析	302
3	微塑料的生物毒性及人体暴露	304
	3.1 微塑料进入生物体的途径与赋存特征	304
	3.2 塑料的毒理学暴露方法	306
	3.3 微塑料的毒性效应	307
	3.4 微塑料对人体的暴露风险	307
4	微塑料的环境行为	308
	4.1 微塑料的形成、风化和迁移转运	308
	4.2 微塑料对环境污染物的吸附和解吸	309
	4.3 微塑料对污染物生物富集的影响	310
	4.4 微塑料对污染物环境迁移转化的影响	311
5	微塑料污染的环境风险和管控	312
	5.1 微塑料污染的社会、经济、政治效应	312
	5.2 微塑料及塑料污染的管控	313
6	展望	314
	参考文献	315

第13章 饮用水消毒及消毒副产物研究进展 330

1	引言	331
2	消毒方式	332
	2.1 氯化消毒	332
	2.2 氯胺消毒	333
	2.3 臭氧消毒	333
	2.4 紫外线消毒	333
	2.5 电化学消毒	333
	2.6 联合消毒	334
3	消毒副产物的识别及生成特征	334
	3.1 三卤甲烷（THMs）	334
	3.2 卤乙酸（HAAs）	336
	3.3 卤乙腈（HANs）	338
	3.4 卤代乙酰胺（HAcAms）	339
	3.5 卤代硝基甲烷（HNMs）	340
	3.6 亚硝胺（NMAs）	341
	3.7 其他消毒副产物	342
4	消毒副产物及其前体物的分析检测方法	342
	4.1 消毒副产物检测	342
	4.2 消毒副产物前体物解析方法	343
5	饮用水中消毒副产物的浓度分布	344
6	消毒副产物的毒理学特征	346
	6.1 细胞毒性	346
	6.2 遗传毒性	347
	6.3 其他毒性	347

	6.4 流行病学调查	348
7	消毒副产物的控制及削减	348
	7.1 消毒副产物前体物的去除	348
	7.2 优化消毒方法	350
	7.3 消毒副产物的去除	350
8	困难及展望	351
	参考文献	352

第14章 大气污染物毒理学效应研究进展 363

1 引言 364
2 大气污染物暴露对呼吸系统的影响 364
    2.1 大气污染与哮喘 365
    2.2 大气污染与慢性肺阻塞疾病 366
    2.3 大气污染与肺癌 367
3 大气污染物暴露对心血管系统的影响 369
    3.1 大气污染物与高血压 369
    3.2 大气污染物与动脉粥样硬化 370
    3.3 大气污染物与心力衰竭 371
    3.4 大气污染物与心肌纤维化 372
4 大气污染物暴露对中枢神经系统的影响 373
    4.1 大气污染物与脑缺血性损伤 374
    4.2 大气污染物与神经退行性损伤 375
5 展望 376
参考文献 377

第15章 环境毒理组学研究进展 386

1 引言 387
2 暴露组的概念形成及研究进展 388
    2.1 暴露组概念的形成 388
    2.2 暴露组学研究方法学 389
    2.3 外暴露组的测量 389
    2.4 体内外源污染物的分析 390
    2.5 生物组学技术在暴露组研究中的应用 391
    2.6 暴露组学的研究进展 391
3 转录组学分析技术及其研究进展 392
    3.1 转录组及转录组学的概念形成 392
    3.2 转录组学分析技术 392
    3.3 转录组学技术的研究进展 394
    3.4 转录组学在环境毒理学研究中的应用 395
4 蛋白组学分析技术及其研究进展 396
    4.1 蛋白组学概述 396
    4.2 蛋白组学分析技术 396
    4.3 蛋白组学的研究策略 397

4.4 蛋白组学在环境毒理学研究中的应用 ·············· 398
5 代谢组学分析技术及其研究进展 ·············· 400
5.1 代谢组学概念 ·············· 400
5.2 代谢组学分析技术 ·············· 400
5.3 代谢组学相关软件的应用 ·············· 401
5.4 代谢组学在典型有机污染物毒性评估中的应用 ·············· 402
6 展望 ·············· 403
参考文献 ·············· 404

第16章 环境计算化学与计算毒理学 ·············· 410
1 引言 ·············· 411
2 碳纳米材料吸附化学品的模拟预测研究 ·············· 413
2.1 C_{60} 吸附化学品的预测 ·············· 414
2.2 碳纳米管（CNTs）吸附化学品的预测 ·············· 414
2.3 石墨烯和氧化石墨烯吸附化学品的预测 ·············· 415
3 化学品环境降解转化行为的模拟预测研究 ·············· 416
3.1 大气中的化学品的降解转化行为 ·············· 416
3.2 水相中的化学品降解转化行为 ·············· 423
4 化学品毒性机制及效应的模拟预测研究 ·············· 428
4.1 酶代谢外源化合物的计算模拟 ·············· 428
4.2 环境内分泌干扰效应的毒性通路与模拟预测 ·············· 431
5 展望 ·············· 433
参考文献 ·············· 436

第17章 新型环境污染物的生态毒理研究进展 ·············· 444
1 全氟化合物的生态毒理研究 ·············· 445
1.1 全氟化合物的生物富集和人体暴露 ·············· 445
1.2 全氟化合物的毒性效应与机制 ·············· 447
1.3 展望 ·············· 449
2 新型全氟聚醚羧酸的生态毒理研究 ·············· 449
2.1 PFECAs 替代品的分析方法及环境分布 ·············· 450
2.2 新型全氟替代品在全球部分自然水体中的分布特征 ·············· 453
2.3 新型 PFECAs 替代品的肝脏毒性效应与机制 ·············· 454
2.4 新型 PFECAs 替代品同蛋白质相互作用 ·············· 455
2.5 展望 ·············· 456
3 氯化石蜡的生态毒理研究 ·············· 457
3.1 环境分布与迁移 ·············· 458
3.2 毒性效应与机制 ·············· 461
3.3 人体内暴露与健康风险 ·············· 464
3.4 展望 ·············· 466
4 典型有机污染物的内分泌干扰和神经发育毒性效应 ·············· 467
4.1 增塑剂类污染物对生殖内分泌系统的影响 ·············· 467
4.2 溴代阻燃剂对甲状腺内分泌系统的影响 ·············· 469

4.3　有机磷阻燃剂的神经发育毒性效应 ················· 471
　　　4.4　展望 ················· 472
　　参考文献 ················· 472

第18章　环境砷污染与健康研究进展 ················· 479
　1　引言 ················· 480
　2　环境砷污染来源 ················· 480
　　2.1　环境砷污染概况 ················· 480
　　2.2　环境砷污染来源 ················· 481
　3　环境砷污染健康危害 ················· 485
　　3.1　砷对皮肤系统的影响 ················· 486
　　3.2　砷对消化系统的影响 ················· 486
　　3.3　砷对呼吸系统的影响 ················· 486
　　3.4　砷对神经系统的影响 ················· 487
　　3.5　砷对免疫系统的影响 ················· 487
　　3.6　砷对心血管系统的影响 ················· 487
　　3.7　砷对内分泌系统的影响 ················· 487
　　3.8　砷对泌尿系统的影响 ················· 488
　　3.9　砷对生殖发育系统的影响 ················· 488
　4　环境砷污染致病机制 ················· 488
　　4.1　砷致氧化损伤 ················· 489
　　4.2　砷致遗传损伤 ················· 490
　　4.3　砷致信号通路异常 ················· 491
　　4.4　砷致表观遗传改变 ················· 492
　　4.5　砷与免疫功能损伤和炎症反应 ················· 494
　5　环境砷污染健康风险评估 ················· 495
　　5.1　环境砷污染的危害识别 ················· 495
　　5.2　环境砷污染剂量-反应关系评估 ················· 496
　　5.3　环境砷污染暴露评估 ················· 496
　　5.4　环境砷污染风险特征分析 ················· 497
　　5.5　环境砷污染健康风险评估的不确定性 ················· 498
　6　展望 ················· 499
　　6.1　砷与环境多因素交互作用对健康的影响研究有待加强 ················· 499
　　6.2　砷中毒机制与转化应用研究有待深入 ················· 499
　　6.3　建立或完善我国慢性砷中毒人群研究队列有待重视 ················· 499
　　6.4　加强易感人群和敏感生命阶段的识别与干预研究意义重大 ················· 499
　　6.5　环境砷污染健康风险评估研究须与时俱进 ················· 500
　　6.6　砷中毒靶向药物研发亟待加强 ················· 500
　　参考文献 ················· 500

第19章　邻苯二甲酸酯增塑剂的人体暴露与健康风险 ················· 510
　1　引言 ················· 511
　2　邻苯二甲酸酯增塑剂的环境污染及人体暴露现状 ················· 513

	2.1	水体及沉积物中的PAEs	513
	2.2	土壤中的PAEs	514
	2.3	室内空气及颗粒物中的PAEs	514
	2.4	PAEs在饮用水及食品中的污染现状	515
	2.5	个人护理品中及医疗用品中的PAEs	517
	2.6	PAEs人体暴露现状	518
3	邻苯二甲酸酯增塑剂的体内代谢及其与慢性疾病的关联		520
	3.1	PAEs体内代谢	520
	3.2	PAEs与生殖健康	522
	3.3	PAEs与肥胖	522
	3.4	PAEs与糖尿病	523
	3.5	PAEs与呼吸系统疾病	524
	3.6	PAEs与其他疾病的相关性调查	524
4	邻苯二甲酸酯增塑剂的健康效应与毒理机制		525
	4.1	生殖毒性	525
	4.2	肥胖、胰岛素抵抗及糖尿病	525
	4.3	内分泌干扰效应	526
	4.4	表观遗传毒性	527
	4.5	代谢组学研究	528
5	展望		530
参考文献			530

第20章 低浓度化学品Hormesis效应的机制与应用研究进展 540

1	引言		541
2	低浓度化学品Hormesis效应的广谱性		542
3	低浓度化学品Hormesis效应的发生条件		543
4	低浓度化学品Hormesis效应的机制研究		546
	4.1	个体水平的基于ROS和NO的氧化应激机制	547
	4.2	群体水平的基于群体感应QS信号分子的机制	548
	4.3	基于随时间变化的跷跷板理论的机制	548
5	低浓度化学品Hormesis效应的应用研究		550
	5.1	中药领域	551
	5.2	水华领域	552
	5.3	食品添加剂领域	553
	5.4	混合化合物联合毒性领域	555
6	展望		556
参考文献			557

第21章 环境纳米材料与技术研究进展 560

1	引言		561
2	催化臭氧氧化耦合膜分离技术原理与应用		561
	2.1	纳米反应器原理	562
	2.2	中试设备及其水处理性能	563

2.3 小结 ... 563
 3 多相催化高级氧化水中难降解有机物研究进展 ... 563
  3.1 表面络合促进金属单反应位芬顿催化过程 ... 564
  3.2 双反应中心芬顿催化体系构建与界面反应过程 ... 564
  3.3 氮化碳表面电子结构调控与强化过硫酸盐活化降解有机污染物 ... 565
 4 天然氢氧化镁纳米材料环境应用的基础理论研究进展 ... 566
  4.1 表面缺陷 $Mg(OH)_2$ 的调控合成及对低浓度重金属氧阴离子的选择性吸附 ... 567
  4.2 利用 CO_2 加压调控 $Mg(OH)_2$ 相循环提取回收含铅废水中 Pb^{2+} ... 568
 5 有机-无机协同作用对有机重金属微污染物的强化吸附机制研究进展 ... 569
 6 限域结构复合纳米材料的构效调控与水处理应用研究进展 ... 570
  6.1 树脂基纳米复合材料 ... 571
  6.2 纳米限域结构强化类 Fenton 催化反应及膜分离过程 ... 572
 7 展望 ... 572
 参考文献 ... 573

第22章 纳米材料的环境转化与归趋研究进展 ... 576

 1 引言 ... 577
 2 金属基纳米材料的环境转化与归趋 ... 578
  2.1 金属基纳米材料的胶体稳定性和多孔介质迁移性 ... 578
  2.2 金属基纳米材料的溶解与转化 ... 579
  2.3 小结 ... 581
 3 碳纳米材料的环境转化与归趋 ... 581
  3.1 碳纳米材料的胶体稳定性及多孔介质迁移 ... 581
  3.2 碳纳米材料在水环境中的光化学转化 ... 583
  3.3 小结 ... 584
 4 复合纳米材料及多元金属基纳米材料的环境转化与归趋 ... 584
 5 复杂环境中纳米材料转化与归趋研究的新方法 ... 585
  5.1 实验技术 ... 585
  5.2 模型方法 ... 586
 6 展望 ... 587
 参考文献 ... 587

第23章 环境中纳米材料的生物效应与分子机制研究进展 ... 597

 1 引言 ... 598
 2 环境冠对纳米材料生物效应的影响 ... 598
  2.1 纳米材料环境冠的形成与特征 ... 599
  2.2 环境冠对纳米材料毒性效应的影响 ... 601
  2.3 环境冠对纳米材料毒性机制的影响 ... 602
  2.4 小结 ... 603
 3 纳米材料不同层次的生物效应与关键调控因子 ... 604
  3.1 生物分子效应 ... 605
  3.2 离体细胞效应 ... 606
  3.3 个体效应 ... 607

3.4 生态系统效应 ·· 608
　　3.5 纳米-生物效应的构效关系及预测模型 ····································· 610
　　3.6 小结 ··· 610
4 纳米材料的植物毒性与分子机制 ·· 611
　　4.1 纳米材料的植物吸收路径 ··· 611
　　4.2 纳米材料的植物毒性效应及其机理 ·· 613
　　4.3 小结 ··· 614
5 纳米材料的水生毒理效应 ··· 614
　　5.1 金属纳米材料 ·· 614
　　5.2 金属氧化物纳米材料 ··· 615
　　5.3 碳纳米材料 ··· 616
　　5.4 小结 ··· 617
6 纳米材料对超重群体的毒性效应 ·· 617
7 纳米材料与污染物的复合生物效应 ··· 618
　　7.1 金属基纳米材料与污染物复合作用 ·· 619
　　7.2 碳纳米材料与污染物复合作用 ·· 621
　　7.3 小结 ··· 622
8 雾霾颗粒物关键致毒组分研究 ··· 622
　　8.1 细颗粒物核材料理化性质对其细胞毒性的影响 ························· 623
　　8.2 细颗粒物吸附污染物对其健康效应的影响 ······························· 623
　　8.3 小结 ··· 624
9 展望 ·· 625
参考文献 ··· 625

第24章　水污染与控制技术研究 ·· 642
1 引言 ·· 643
2 物化水处理方法与技术 ·· 644
　　2.1 高级氧化/还原处理技术 ··· 644
　　2.2 富集分离技术 ·· 655
3 生化水处理方法与技术 ·· 663
　　3.1 脱氮除磷技术 ·· 663
　　3.2 废水处理实践 ·· 669
4 废水处理新技术系统 ·· 676
　　4.1 生物电化学技术 ··· 676
　　4.2 膜生物反应器技术 ··· 678
　　4.3 化工废水处理技术 ··· 681
　　4.4 厌氧处理新技术 ··· 683
　　4.5 光催化-生物降解直接耦合技术 ·· 686
5 废水资源/能源新技术 ·· 688
　　5.1 有机碳回收技术 ··· 688
　　5.2 磷回用技术 ··· 690
　　5.3 重金属回收技术 ··· 692

参考文献...695

第25章　矿山环境污染控制研究进展..718
　1　引言...719
　2　源头控制——矿山尾矿中重金属释放抑制技术...720
　　2.1　物理屏蔽法...720
　　2.2　表面钝化法...721
　　2.3　生物抑制法...725
　3　过程控制——酸性矿山废水中重金属去除研究...726
　　3.1　酸性矿山废水中重金属的去除方法...726
　　3.2　农林废弃物的改性及其对重金属的吸附去除...729
　4　次生矿物——酸性矿山废水中重金属自然归宿...731
　　4.1　次生矿物的形成...731
　　4.2　次生矿物的稳定性...732
　　4.3　次生矿物的相转化...733
　　4.4　次生矿物与酸性矿山废水中重金属的归趋...735
　5　生态恢复——矿山废弃地治理技术研究及应用...736
　　5.1　矿山废弃地极端环境微生物生态...736
　　5.2　矿山废弃地生态恢复技术研究...736
　　5.3　矿山废弃地生态恢复技术应用...737
　6　展望...738
　　参考文献...739

第26章　危险废物处理与资源化研究进展..749
　1　引言...750
　2　危险废物中重金属的处置与资源化...751
　　2.1　基于物相调控的重金属提取技术...751
　　2.2　选冶联合分离新工艺...755
　　2.3　含铀多金属修复处置...759
　3　危险废物中持久性有机污染物的污染控制技术...762
　　3.1　催化分解技术...762
　　3.2　焚烧飞灰固化/稳定化技术..766
　4　危险废物能源利用与清洁工艺途径...770
　　4.1　废电池能源化...770
　　4.2　清洁工艺途径...776
　5　展望...780
　　参考文献...780

第27章　全球气候变化与环境健康研究进展..783
　1　引言...784
　2　全球气候变化及其发展趋势...785
　　2.1　全球气候变化的定义和缘起...785
　　2.2　全球气候变化的发展趋势...785
　3　典型污染物环境过程与气候变化...787

- 3.1 重金属的环境化学过程与气候变化 ... 787
- 3.2 持久性有机污染物的环境化学过程与气候变化 ... 789
- 3.3 纳米材料的环境化学过程与气候变化 ... 791
- 3.4 农药的环境化学过程与气候变化 ... 793
- 4 其他气候变化与环境健康问题 ... 794
  - 4.1 营养元素问题 ... 794
  - 4.2 花粉问题 ... 795
  - 4.3 天然毒素问题 ... 795
  - 4.4 病原微生物问题 ... 795
  - 4.5 大气臭氧浓度升高问题 ... 796
- 5 展望 ... 796
- 参考文献 ... 797

第28章 冰冻圈环境化学 ... 805

- 1 引言 ... 806
- 2 冰冻圈简介 ... 806
  - 2.1 冰冻圈 ... 806
  - 2.2 冰冻圈的消融和环境变化 ... 807
- 3 冰雪及其界面的化学过程 ... 808
  - 3.1 光化学反应 ... 808
  - 3.2 卤水溶液反应 ... 809
  - 3.3 气体溶解和挥发 ... 810
  - 3.4 表面化学反应 ... 810
  - 3.5 氧化还原反应 ... 810
- 4 北极地区冰冻圈环境化学 ... 810
  - 4.1 北冰洋海冰环境中的污染物 ... 811
  - 4.2 北极地区的汞污染 ... 812
  - 4.3 北极地区的持久性有机污染物 ... 814
- 5 南极地区冰冻圈环境化学 ... 815
  - 5.1 南极地区金属污染物 ... 815
  - 5.2 南极地区持久性有机污染物 ... 816
  - 5.3 南极海洋边界层有机气溶胶 ... 816
- 6 冰川、冻土和高山地区冰冻圈环境化学 ... 818
  - 6.1 冰川环境化学 ... 818
  - 6.2 冻土环境化学 ... 820
  - 6.3 其他冰冻圈要素环境化学 ... 821
  - 6.4 青藏高原冰冻圈环境化学 ... 821
- 7 冰冻圈环境化学的全球效应和展望 ... 822
- 参考文献 ... 822

第29章 食物-能源-水：从单一领域到系统关联 ... 831

- 1 引言 ... 833
- 2 FEWS基本概念 ... 834

  2.1 内部关联分析 837
  2.2 外部影响分析 837
  2.3 系统综合性能 838
3 FEWS 模型和方法论 839
  3.1 模型案例 839
  3.2 模型构建方法 841
  3.3 量化分析模型 842
  3.4 模型小结 845
4 FEWS 关联网络的工程技术 846
  4.1 污水中营养物质的回用 847
  4.2 可持续能源技术 849
  4.3 其他综合技术 850
5 FEWS 的相关政策 851
  5.1 世界各国和国际组织高度重视食物、水及能源安全 851
  5.2 在国际层级进行 FEW 管理和政策制定的复杂性 852
  5.3 夯实工程技术，推动 FEW 政策的革新和发展 853
  5.4 项目资助情况 853
6 展望 853
参考文献 854

第 1 章　原位样品前处理技术在环境分析和毒理学研究中的应用

- 1. 引言 /2
- 2. 原位样品前处理技术 /3
- 3. 原位样品前处理技术在环境分析和毒理学研究中的应用 /4
- 4. 原位样品前处理技术的萃取相 /11
- 5. 展望 /12

本章导读

原位样品前处理技术指的是直接在所研究体系内部对目标分析物进行萃取、富集和净化的技术。原位样品前处理技术消除了传统的采样过程，避免了样品储存和运输过程中样品的污染和分析物的损失，也极大地简化了样品前处理过程。同时，原位样品前处理技术通常能够有效减轻甚至避免对所研究体系中的环境过程或生命过程的扰动，从而揭示分析物在所研究体系中的分配和时空赋存情况。截至目前，原位样品前处理技术已经在非生命环境介质、动植物活体组织和单细胞中污染物和内源性小分子有机物的检测上有所应用。进一步推动该技术的定量方法的发展和新型萃取相的开发，并拓展该技术在环境分析和毒理学研究中的应用将是未来研究的方向。本章将对原位样品前处理技术在有机污染物和生物内源性小分子有机物检测上的应用现状进行简要介绍。

关键词

原位样品前处理，定量方法，萃取相，有机污染物，内源性小分子有机物

1 引 言

采样和样品前处理是经典化学分析方法中的重要步骤，它们直接影响着分析方法的灵敏度和分析结果的准确性[1]。然而，在传统的采样和样品前处理过程中，样品需要与所处的环境分离，样品的组成及样品在原环境中所起的功能会被破坏，而样品中分析物的分布、分配和形态等信息也都有可能丢失。近年来，"原位样品前处理（in situ sample pretreatment）"及其相似概念"原位萃取（in situ extraction）"在文献中已有出现。例如，将固相微萃取（solid-phase microextraction，SPME）探针置于水体沉积物和动植物活体组织中，用于其中目标分析物的原位萃取，能够揭示分析物的时空分布情况[2-8]。在这样的样品前处理过程中，样品（如沉积物和动植物组织）无需从其所处的环境中分离，萃取、富集和净化等样品前处理过程被集成为一步，样品前处理过程得以大大简化。另外，样品本身的原始状态（native state），即样品的组成和功能（如沉积物所承担的物质交换的环境功能[3]、动植物组织的代谢功能[4-8]）等受到的影响极小或能够被忽略。

实现原位样品前处理过程，有望赋予分析结果四个方面的优势：①降低甚至避免传统的采样和样品前处理过程对所研究体系造成的损伤或干扰，获得反映所研究体系原始状态的信息。例如，不破坏样品中污染物的分配以测得污染物的游离浓度[3]，不造成生物正常生理状态的变化来研究外源活性物质对生物体正常代谢过程的影响[5]。②规避一些易氧化、易挥发的分析物在传统的采样和样品前处理过程中可能损失的风险，获得更准确的分析结果[5, 6]。③获得分析物在样品中的动态变化信息[4, 7]。传统的采样和样品前处理过程对样品具有不可逆的损耗性，不利于考察样品中分析物的动态变化情况。④简化分析过程和降低分析成本。原位样品前处理技术可以避免样品的存储和运输，且将萃取、富集和净化步骤集成为一步[8]。

截至目前，原位样品前处理技术在检测非生命环境介质中污染物的分布情况[3]、考察污染物在动植物组织中的富集情况[8]、研究动植物组织中污染物的富集和消除动力学过程[7]以及研究生物内源性小分子有机物水平随外源物作用所发生的变化[5, 6]等方面展现出了巨大的应用前景，并有望为环境分析和毒理学研究提供新颖、经济且高效的研究手段。同时，推动原位样品前处理技术自身的发展是促进其应用的基石，这包括发展其定量理论及开发新的萃取相。

2 原位样品前处理技术

2.1 SPME 技术与被动采样技术

SPME 技术是一种典型的能够用于原位样品前处理的技术[1-8]。SPME 探针目前已广泛应用于活体动植物组织和水体沉积物等样品的原位前处理。相比之下，另外一种微萃取技术，液相微萃取技术（liquid-phase microextraction，LPME）在原位样品前处理方面的应用则鲜见报道[9]，这主要是由于 LPME 萃取相的机械强度不高，且萃取相易泄露到样品基质中[10]。

除 SPME 技术以外，微透析（microdialysis，MD）技术也可以实现原位样品前处理过程，并且透析液可以直接导入分析仪器进行检测。然而，微透析技术在有机污染物的检测方面只有少量的应用[11-14]，这主要是因为微透析技术需要用到输液泵和透析液传输管线，不利于对活体动物组织、水体沉积物等样品进行原位样品前处理。另外，由于很多有机污染物的极性不高，这些有机污染物容易被透析膜截留而不易在透析液中富集，导致微透析技术并不适用于这类低极性有机污染物的检测。但另一方面，微透析技术在神经化学研究上的应用已相当成熟[15, 16]，因此，该技术能够用于检测污染物作用下动物神经化学的变化情况，来研究和揭示污染物的神经毒性及毒性机理[17-20]。整体而言，微透析技术的设备较为复杂，使用成本较高。相比微透析技术，SPME 探针的体积小、制备简单、便于携带，也容易置于不同的环境介质和动植物组织中进行原位样品前处理[1-8]。目前，原位 SPME 技术除了用于污染物的检测，也用于一些药物的药效研究。SPME 技术的萃取相（extraction phase）种类丰富，能够实现不同极性的有机污染物和动植物内源性小分子有机物的有效富集[5, 6, 21, 22]。由此可见，SPME 技术在环境分析以及研究生物体对外源物作用的响应方面，相比于微透析技术更方便、经济，且有望实现对更多种类的目标分析物的检测。

另外，目前在环境领域广泛应用的被动采样（passive sampling）技术也能实现目标分析物的原位萃取[23-25]。该技术一般使用不同形状的聚乙烯、聚甲醛、聚氨酯等作为被动采样器。被动采样器从样品基质中取出以后，往往还需要使用有机溶剂回收其萃取到的污染物，然后再对洗脱溶液进行净化和浓缩，来完成整个样品前处理过程。尽管一般意义上的被动采样过程并不包含完整的样品前处理过程，后续仍需溶液浓缩和净化步骤，但被动采样过程本身也包含了一定程度的富集和净化功能。SPME 技术通常也被认为是一种被动采样技术，但 SPME 探针的小尺寸允许其能够直接在气相色谱仪的进样口中热脱附，因此 SPME 技术一般不需要单独的溶液浓缩和净化过程，样品前处理过程相对更加简便[26]。而即使采用溶剂洗脱的方法，SPME 探针所需的洗脱时间也要远远短于其他被动采样器[3, 27]。同样因为 SPME 探针的尺寸小，SPME 探针对动植物活体的损伤也较小，SPME 技术在动植物活体组织中的应用相比一般的被动采样技术要多得多[2, 28-30]。而另一方面，萃取相体积大的被动采样器则相比 SPME 探针具有更广的萃取量线性增长的萃取时间区间，适用于更长时间跨度的时间加权平衡（time weighted average，TWA）浓度的测定。

SPME 技术被归为被动采样技术的一种，是因为 SPME 过程与其他的被动采样过程一样由分析物在样品基质和萃取相之间的化学势差驱动。反过来，SPME 技术强调萃取过程对所研究体系造成的影响可以忽略，而原位被动采样技术也往往要求被动采样过程对所研究体系原有的分配平衡所造成的影响可以忽略[31]。由此可见，从不同的角度出发，两种技术存在着相互包含的关系。

本章所介绍的原位样品前处理技术包括被动采样技术和 SPME 技术。截至目前，原位样品前处理技术已经在非生命环境介质[32, 33]、活体动植物组织[34, 35]和单细胞[36]中得到应用（图 1-1），原位样品前处理技术将有望发展成为环境分析和毒理学研究的有效手段。

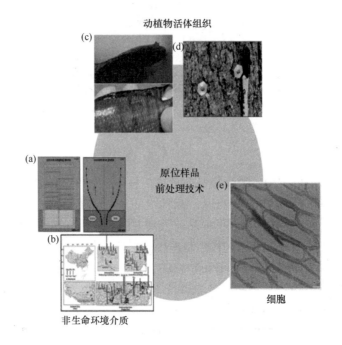

图 1-1 原位样品前处理技术的应用

被动采样技术检测：(a) 沉积物水间的污染物质量交换通量[32]；(b) 不同地域水体中污染物的游离浓度[33]

SPME 技术检测：(c) 鱼体内的药物[34]；(d) 树干中的含氯挥发性有机物[35]；(e) 细胞内的黄酮类物质[36]

2.2 与原位分析技术的比较

质谱成像技术[37]、光谱成像技术[38-40]和原位电化学分析技术[41]等原位分析技术能够获得动植物组织和单细胞中分析物的时空分布信息。然而，这些原位分析技术都容易受到样品基质的干扰。例如，质谱成像技术的离子化效率会受到样品基质的抑制[37]，而荧光成像技术、红外成像技术和拉曼成像技术都会因背景信号的干扰而损失检测灵敏度[38-40]，电化学分析技术则要求对电极进行特异性处理才能有效避免样品中其他成分的干扰[41]。另外，除质谱技术能够同时检测多种分析物外，其他的原位分析技术同时检测的分析物的种类有限，一般都只能特异性地检测某种分析物[38, 41]。相比之下，原位样品前处理技术将分析物从样品基质中分离开，能够有效降低基质效应对分析的干扰。并且，原位样品前处理技术也能够用于多种分析物的同时检测，甚至能够用于生物代谢组（metabolome）的检测[5, 6]。在空间分辨率方面，尽管成像技术的分辨率能够达到亚微米级别，原位样品前处理技术也已经能够对细胞质和细胞核进行分别检测[42]。

3 原位样品前处理技术在环境分析和毒理学研究中的应用

3.1 原位样品前处理技术的定量方法

不同于传统的样品制备技术有较高的绝对回收率，原位样品前处理技术通常只萃取样品中分析物总量的很小一部分。这一方面可以降低样品前处理过程对所研究体系的影响，另一方面也造成原位样品前处理技术的定量方法的建立相较完全萃取技术要复杂得多。在原位样品前处理技术中，传统的外标法、标准加入法等均难以保证准确度或不可用。

原位样品前处理技术的定量方法可以分为平衡校正方法和预平衡校正方法两种。当分析物在萃取相和样品基质间达到分配平衡，基于分析物在萃取相和样品基质间的分配系数以及萃取相的体积和达到平衡时分析物在萃取相中的量，就可以计算出样品中分析物的浓度［如式（1-1）所示］。式（1-1）要求原位样品前处理技术只萃取样品中分析物总量的很小一部分。如果分析物在萃取相和样品基质间尚未达到分配平衡，则需解析萃取相从样品基质中萃取分析物的动力学过程，并依托相应的动力学模型建立定量方法。平衡校正方法在环境被动采样技术中应用较多，通常需要耗费相当长的时间来达到萃取平衡状态[43, 44]。在预平衡校正方法中，人们普遍认为原位萃取过程都遵循一级动力学模型［如式（1-2）和图（1-2）中的萃取动力学曲线所示］[45]。现有的多种定量校正方法，包括动力学校正法（kinetic calibration）[3, 27]、采样速率校正法（sampling-rate calibration）[46, 47]和双探针法（dual-probe method）[48]等，均基于一级动力学模型发展而来。

$$C_s = \frac{n_e}{V_f K} \tag{1-1}$$

$$n = n_e\left(1 - e^{-at}\right) \tag{1-2}$$

式（1-1）和（1-2）中，C_s是样品基质中分析物的浓度，n_e是萃取相中分析物的平衡萃取量，V_f是萃取相体积，K是分析物在萃取相和样品间的分配系数，n是时间t内萃取相中分析物的萃取量，a是动力学常数。

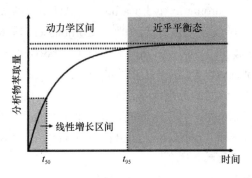

图 1-2　一级动力学模型描述的萃取量随时间的变化情况[45]

t_{50}和t_{95}分别指的是萃取量达到50%和95%的平衡萃取量的时间

在预平衡校正方法中，使用最多的是动力学校正法。该方法基于脱附过程与萃取过程间的各向同性（isotropy），即分析物和校正物在萃取相和样品基质中的传质速率的大小与传质方向无关。基于各向同性原则，该方法采用加载在萃取相中的校正物的脱附过程来预测萃取平衡状态[3, 27, 49-52]。使用的校正物也称"行为参照物（performance reference compound, PRC）"，理论上PRC的物化性质与分析物越接近，萃取平衡状态的预测越准确。另外，PRC也被用于原位校正萃取过程中不确定因素对萃取动力学过程的影响[53, 54]。需要特别指明的是，平衡校正方法可用于测定分析物的游离浓度，而动力学校正方法可以预测平衡状态，因此也可用于测定分析物的游离浓度。一般认为，游离浓度对于评估污染物的生物可利用性（bioavailability）和毒性相较总浓度更加合适。然而也有研究发现，在一些情况下PRC的脱附过程和分析物的萃取过程间的各向同性并不理想[55, 56]。

为保证定量结果的可信度，原位样品前处理技术的定量方法的建立有以下几个方面需要特别注意：①当采用平衡校正方法时，如何确认到达分配平衡状态，又如何获得可信的分配系数均至关重要[57, 58]；

②当采用预平衡校正方法时，需进一步确认利用 PRC 能否准确地预测平衡状态以及准确地校正采样速率[55, 56]；③复杂样品中的萃取动力学过程相较目前的理想一级动力学模型要复杂得多[59, 60]，分析物结合态的比例、传质的主导方式（对流或扩散）、样品的非均质性对萃取动力学的影响还有待深入探究。

3.2 原位样品前处理技术在非生命环境介质中的应用

被动采样技术被广泛应用于大气、水、土壤和水体沉积物等非生命环境介质（abiotic environmental medium）的原位样品前处理。被动采样技术在非生命环境介质中的应用对于揭示污染物在环境中的分布和迁移趋势，评估污染物的生物可利用度有重要促进作用。

3.2.1 测定污染物的游离浓度、评估污染物在环境介质间的迁移趋势

被动采样过程由分析物在样品基质和萃取相间的化学势差驱动，当被动采样达到平衡状态，分析物在萃取相中的化学势与分析物在样品基质中的化学势达到一致，分析物在各相间的分配达到平衡[61]。在采样达到平衡状态后，基于萃取相中分析物的浓度（由平衡萃取量和萃取相的体积计算得到）以及分析物在萃取相和水相之间的分配系数，能够计算水体或沉积物孔隙水中分析物的游离浓度[62, 63]。

同时，在被动采样达到平衡状态后，分析物在样品基质中的化学势（在一些文献中也称化学活度或逸度）就可以由分析物在萃取相中的化学势表征。而鉴于绝对的化学势不容易度量，因此可以通过比较相同材质的萃取相中污染物的平衡浓度，来比较污染物在不同环境介质间的化学势高低[61]。不同环境介质间的化学势差能够指明污染物在环境介质间被动迁移的趋势。例如 Cornelissen 等人发现，Oslo 港的沉积物孔隙水中多环芳烃（polycyclic aromatic hydrocarbons，PAHs）的游离浓度比沉积物上层水体要高，这也意味着沉积物有向上层水体释放 PAHs 的趋势[62]。Cornelissen 等人也比较了多氯代二苯并二噁英（polychlorinated dibenzo-p-dioxins，PCDDs）、多氯二苯并呋喃（polychlorinated dibenzo-p-furans，PCDFs）和多氯联苯（polychlorinated biphenyls，PCBs）在波罗的海（Baltic Sea）的沉积物和上层水体中的化学活度，发现污染物在沉积物和上层水体间的分配近乎达到平衡[64]。同样的，Mäenpää 等人和 Booij 等人都观察到沉积物孔隙水中 PCBs 的游离浓度比上层水体高，由此也推测沉积物会充当 PCBs 释放的源头[65, 66]。Lohmann 研究组则探究了多溴联苯醚（polybrominated diphenyl ethers，PBDEs）、PCBs 和 PAHs 在多个区域的水体、沉积物和空气之间的迁移趋势[67-69]。

由于在实际情况下达到采样平衡状态所需的时间通常很长，利用 PRC 来预测采样平衡状态可以有效缩短被动采样时间[3, 70, 71]。需要注意的是，无论使用平衡方法或是预平衡方法，要准确地测定游离浓度，都需要先获取可信的萃取相-水相分配系数值[72, 73]。然而，在一些情况下，分配系数仅由经验公式获得，而经验公式在不同类物质间的可迁移性还有待商榷[71]。另一方面，实验室测定分配系数的条件（如盐度和温度）又存在与环境真实条件不一致的问题[73]。分配系数的不准确将会影响游离浓度测定结果的准确度。另外，平衡状态的确认或准确预测对保证检测结果的准确度也非常重要。针对平衡采样方法，Mayer 等人提出了三种确认到达平衡状态的方法[61]。而对于预平衡采样方法，使用 PRC 预测平衡状态仍是最为常用的方法，尽管一些研究观察到 PRC 的脱附过程和污染物的被动采样过程间的各向同性并不好[55, 56]。

3.2.2 环境介质间污染物质量交换通量的检测

当污染物在相邻的两个环境介质间存在化学势差时，污染物将有从化学势高的环境介质向化学势低的环境介质发生被动迁移的趋势。因此，测定污染物在两个环境介质间的质量交换通量（mass exchange flux），对于准确揭示污染物在环境中的迁移规律有重要意义。

使用被动采样技术测定环境介质间的质量交换通量的最直接的方法是"两点法"[74]，即先用被动采

样器分别测得相邻两个环境介质中污染物的浓度，然后根据 Fick 第一扩散定律计算介质间的质量交换通量。这种方法被广泛地用于测定沉积物-水体[26, 75]、水体-大气[76-80]以及土壤-大气[81, 82]间的污染物质量交换通量。在计算沉积物-水体质量交换通量时，除了需要计算沉积物孔隙水和沉积物上层水体中污染物的游离浓度差，也需要通过一定的方式获得水体和沉积物之间的传质系数，该系数可由实验测得[75]，也可基于经验公式估算得到[26]。而在计算水体-大气间的质量交换通量时，则需要先基于污染物在水相和气相间的分配系数将被动采样器测得的气相浓度转化为水相浓度，同时，也需要通过一定的方式获得水体和大气之间的传质系数[76-80]。土壤-大气间的污染物质量交换通量的计算可以通过类似于沉积物-水体间的质量交换通量的计算方法得到[81]，也可由基于逸度差的公式计算得到[82]。

在沉积物—水体间的污染物质量交换通量的测定方面，我国曾永平教授研究组发明了由多块聚合物萃取相组装成的被动采样器[32, 83]。该被动采样器在距离沉积物—水体两相界面不同深度的沉积物中和不同高度的水体中同时原位萃取，由此测定污染物在不同深度的沉积物孔隙水中的浓度及在不同高度的水体中的浓度（图 1-3）。根据污染物在不同高度的水体中的浓度拟合得到泰勒系数，结合界面处污染物的浓度和污染物在水体中的扩散系数，就可以计算得到界面处的质量交换通量[32, 83]。Lin 等人也采用了类似的被动采样器，并对任意相邻的两块萃取相之间的质量交换通量进行了计算[84]。Fernandez 等人则是将整块聚合物膜竖直置于两相界面处进行被动采样，但他们仍然采用"两点法"的计算方法进行传质通量计算[26]。

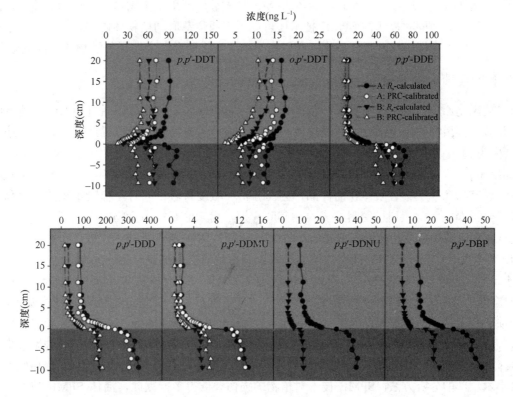

图 1-3 被动采样器测得的竖直方向上污染物浓度的分布情况[32]

下部深色代表沉积物，上部浅色代表水体，A、B 为两个不同的采样点，"R_s-calculated"表示由采样速率计算得到的结果，
"PRC-calibrated"代表由 PRC 校正得到的结果

3.2.3 污染物的区域分布情况研究

被动采样器价格低廉、便于携带，适用于较大数量采样点的原位应用。同时，被动采样技术能获得

其他技术难以获得污染物的水体游离浓度、污染物在沉积物孔隙水中的游离浓度以及污染物的气相浓度。因此，被动采样技术也被用于研究一定区域的污染物分布的情况，并研究污染物区域分布随时间的变化情况。根据污染物的区域分布情况及其随时间的变化情况，研究人员可以对污染的来源和迁移趋势进行深入分析[24, 44, 67-69, 71, 79, 85, 86]。例如，Lomann 研究组采用平衡被动采样技术，在 2011 年 4 月~10 月间分三个时间段，对北美苏必利尔湖（Lake Superior）区 19 个采样点的水体和大气中的 21 种 PAHs 和 11 种 PBDEs 进行了测定，较为系统地考察了污染物的分布情况及污染物在水体和大气间的质量交换情况[44]。他们也对 PBDEs 和 PCBs 在美国罗德岛州纳拉干西特（Narragansett）湾的分布情况以及分布随时间的变化情况进行了研究[67, 68]。另外，Lohmann 研究组还研究了 PBDEs 和 PHAs 跨大西洋的分布情况，同时研究了大西洋不同区域海水和大气间的污染物的质量交换情况[69, 79]。鲍恋君教授研究组则对中国的 42 个淡水区域（图 1-1b）和南极的几个内陆湖的 PAHs 的分布进行了考察，并讨论了人类活动对 PAHs 分布的影响[33, 71]。此外，一些研究组用大气被动采样器测定了大气中一些污染物的浓度，并对污染物在不同区域随时间的变化情况进行了研究[24, 85, 86]。

3.3 原位样品前处理技术在动植物活体组织中的应用

将活体（*in vivo*）SPME 技术用于动植物活体组织的原位样品前处理，能够避免对动植物活体造成严重损伤，并简化样品前处理过程。因此，活体 SPME 技术被广泛用于生物体内污染物的检测及污染物动态变化情况的研究[28]。同时，活体 SPME 技术还可以用于原位萃取动植物体内的内源性小分子有机物，以研究生物体在外源物作用下内源性小分子有机物的浓度变化情况，来揭示外源物的毒性机理[5, 6]。

3.3.1 地下污染情况的筛查和监测

在一些被氯化乙烷、氯化乙烯和甲基叔丁基醚等污染的地区，植物可以通过根系吸收地下的这些有机污染物并转移到植物的地上部分。通过检测植物地上部分污染物的含量以描绘某一时间点地下污染状况的技术称为植被筛查（phytoscreening），而长期跟踪地下污染情况变化的技术则称为植被监测（phytomonitoring）[87]。通常情况下，人们通过采集树干木质部或植物的其他组织来分析污染物的含量，但这种方法通常只允许采集有限体积的样品，因而分析灵敏度有限。同时，植物组织的保存和运输过程也容易引入分析误差[35]。将树枝密封起来检测其叶片挥发出来的有机污染物的方法则需要用到空气泵等复杂的装置，对于野外多点检测而言成本较大[88]。近年来，活体 SPME 技术也被用于植被筛查和植被监测。SPME 装置小巧，易于在野外使用，能够有效控制植被筛查和植被监测的成本。

Sheehan 等人将活体 SPME 技术应用于植物体内含氯挥发性有机物（volatile organic compounds, VOCs）的检测（图 1-1d），发现应用活体 SPME 技术比采集树干木质部进行顶空分析的灵敏度要高得多[35]。在另外一项研究中，Shetty 等人发现聚二甲基硅氧烷（polydimethylsiloxane, PDMS）和线性低密度聚乙烯对含氯 VOCs 的亲和作用好，且含氯 VOCs 在这两种聚合物中的扩散系数较大，两种聚合物均适用于含氯 VOCs 的活体萃取[30]。另外，SPME 技术可以在树干上钻取的单个孔内多次取样，相比每次检测都需重新采集木质部的分析方法，SPME 技术对植物的损伤要低得多。因此，活体 SPME 技术有望用于监测时间点更加密集、监测时间跨度更长的植被监测。Limmer 等人就采用平衡 SPME 方法对树木中含氯 VOCs 的变化进行了长达 4 年的监测。他们的监测结果显示，树木中含氯 VOCs 含量在春季开始增长，在夏季末期到达一年中的最高水平，然后又在秋季开始减少，在冬季达到最低水平（如图 1-4 所示）[89]。Reiche 等人则将 SPME 技术用于芦苇通气组织中甲基叔丁基酯的萃取，避免了传统的采样和样品前处理过程中通气组织中的甲基叔丁基酯的损失。Reiche 等人发现芦苇通气组织中的甲基叔丁基酯水平同样随季节发生变化[90]。

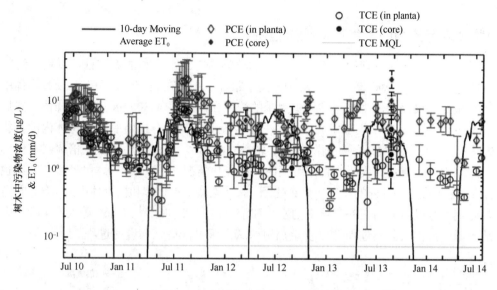

图 1-4 含氯 VOCs 在树木中的含量随时间的变化情况[89]

黑线表示植物蒸腾作用的强弱变化情况,"in planta"指的是基于 SPME 测得的浓度,"core"指的是采集木质部测得的浓度,"MQL"指的是方法定量限。PCE 和 TCE 分别指的是四氯乙烯和三氯乙烯。采集木质部获得的数据点的数量远少于应用 SPME 技术获得数据点的数量

3.3.2 动植物体内有机污染物的检测

活体 SPME 技术也被用于检测动植物活体内的有机污染物,以揭示有机污染物的生物富集特性。Zhou 等人首次将活体 SPME 技术用于野生鱼体背部肌肉中污染物的游离浓度检测[50]。结合溶剂萃取法测得的总浓度,Zhou 等人估算了污染物在鱼背部肌肉中的表观蛋白结合率[50]。鉴于 SPME 过程对被测动物造成的损伤小,Togunde 等人在野外将活体 SPME 技术用于加拿大珍稀鱼类大梭鱼(*Esox masquinongy*)体内富集的药物的检测[91]。Zhang 等人则发明了分段式的活体 SPME 探针,测定了分析物在脂鳍和背部肌肉中的游离浓度比(图 1-5a)[4]。另外,他们也使用长达 5 cm 的探针测得了探针插入不同深度的鱼背部肌肉中分析物的浓度(图 1-5b)[92]。欧阳钢锋教授研究组则首次在活鱼脑部萃取到其中富集的氟西汀和去甲氟西汀,并发现鱼脑部的药物的浓度显著高于鱼背部肌肉中的浓度(图 1-1c)[34]。截至 2019 年,活体 SPME 技术已经广泛用于检测活鱼背部肌肉中的农药[47]、药物[4, 34, 91, 92]和气味物质[93]。此外,Allan 等人也将硅胶植入活鱼体内,比较不同有机污染物在鱼体与其所处的水体之间的化学活度之比[29]。同样的,活体 SPME 技术在检测植物活体内蓄积的污染物方面也有许多应用[51, 94]。

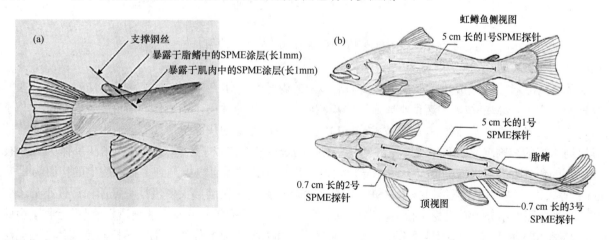

图 1-5 (a)分段式探针在鱼背部肌肉和脂鳍中活体萃取[4];(b)5 cm 长探针在鱼背部肌肉中活体萃取[92]

3.3.3 动植物体内污染物的富集和消除过程研究

SPME 技术也能够用于跟踪研究污染物在活体动植物体内的富集和消除过程，以揭示污染物的生物富集特性以及持久性。Win-Shwe 等人利用 SPME 技术跟踪了小鼠在腹腔注射甲苯后，海马体中甲苯的浓度变化情况[95]。跟踪结果显示，小鼠海马体中甲苯的浓度在注射甲苯 30 min 后达到峰值，然后在达到峰值后的 90 min 内迅速回落。欧阳钢锋研究组则研究了有机氯农药和有机磷农药在鱼体内的富集和消除过程，测得了这些农药的生物富集因子和消除动力学常数[7]。同时，研究组还研究了倍硫磷在鱼体内的代谢动力学过程，跟踪了倍硫磷及其代谢产物浓度随时间的变化情况（图 1-6）[7]。另外，研究组也研究了有机氯农药和有机磷农药在植物体内的富集和消除过程[96]，人造麝香在鱼体和植物体内的富集和消除过程[97]，以及碳纳米管对疏水有机污染物在芥菜体内的富集和消除过程的影响[98]。Zhou 等人和 Loi 等人则分别研究了翡翠木从土壤中富集农药的过程[51]以及番茄植株从土壤中富集化感素的过程[99]。

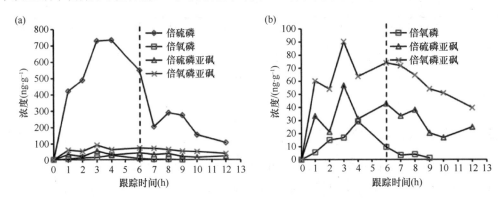

图 1-6 活体 SPME 技术研究鱼体内倍硫磷的代谢[7]

（b）是（a）的放大图，以清晰地显示代谢产物随时间的变化情况。竖直虚线左、右两侧分别表示暴露阶段和净化阶段

3.3.4 人体内有机污染物蓄积水平的评估

Allan 等人和 O'Connell 等人提出，人体内的硅胶植入体可以用于人体内持久性有机污染物的蓄积水平评估[100, 101]。Allan 等人认为持久性有机污染物会在人体与硅胶植入体间达到分配平衡，于是他们测定了的人体硅胶植入体中 PCBs 和 PBDEs 的浓度，再将硅胶植入体中 PCBs 和 PBDEs 的浓度换算成脂肪校正浓度，得到的结果与人乳汁及血液检测结果能够很好吻合[100]。O'Connell 等人则在人体硅胶植入体中鉴定了 1400 多种有机污染物，并估算了其中两种污染物在人体脂肪中的浓度[101]。

3.4 原位样品前处理技术在内源性小分子有机物检测上的应用

活体 SPME 技术也被成功用于检测动植物活体中的内源性小分子有机物[5, 6, 102, 103]。Vuckovic 等人的研究表明，应用 SPME 技术在活体小鼠血液中检测到的代谢物的种类比在离体的血液中检测到的小分子代谢物的种类多，特别是活体 SPME 技术能够检测到一些不稳定的代谢物，如谷胱甘肽、单磷酸腺苷等均难以用常规的溶剂沉淀法和超滤法检测到[6]。Bessonneau 等人比较了活体 SPME 技术和离体 SPME 技术检测到的鱼体暴露组（exposome）的差异，发现活体 SPME 技术可以检测到多种不稳定的代谢物[102]。Cudjoe 等人则发现，相比微透析技术，SPME 技术能够检测到大鼠脑部很多极性很低的内源性小分子有机物，这些内源性小分子有机物在使用微透析技术检测时易被透析膜截留而无法被透析液收集到[5]。由此可见，活体 SPME 技术在内源性小分子有机物的检测方面有着明显的优势：①能够检测到离体方法无法检测到的不稳定内源性小分子有机物，这是因为活体 SPME 技术消除了离体方法中样品储存的过程，且在组织原位使内源性小分子有机物与样品基质分离，从而避免内源性小分子有机物在样品前处理过程中发生酶解[6, 102]；②通过发展合适的萃取相，SPME 技术能够较为均衡地检测不同极性的代谢物[5, 22]。

目前，基于活体 SPME 技术的内源性小分子有机物检测方法已经被用于研究外源活性物质对生物体的作用。例如，Cudjoe 等人就用 SPME 技术研究了在大鼠腹腔注射氟西汀后，其脑部 5-羟色胺和多巴胺的浓度变化情况（图 1-7）[5]。Vuckovic 等人则发现在给小鼠一定剂量的卡马西平后，小鼠血液代谢组相较对照组有明显的变化[6]。这些通过监测内源性小分子有机物的变化成功揭示药物药效的研究，说明 SPME 活体技术也有望迁移到污染物毒性的研究上。事实上，Bessonneau 等人关于鱼体暴露组的研究已经表明活体 SPME 技术有潜力用于研究污染物暴露的生理效应[100]。

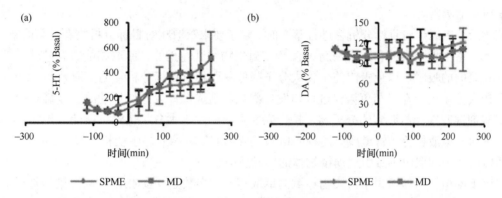

图 1-7　SPME 技术和微透析技术（MD）跟踪的大鼠脑部 5-羟色胺和多巴胺的变化情况[5]

时间零点为注射氟西汀的时间点

细胞是生命体的基本功能单元[104]，从细胞层面揭示污染物的毒性意义重大。相比其他单细胞组学（单细胞基因组学、单细胞转录组学和单细胞蛋白组学）信息，单细胞代谢组可以提供更即时、更动态的各类细胞表型的功能信息[104]。因此，检测单细胞代谢组将有望更即时、更动态地揭示污染物对细胞生理功能的影响。目前，检测单细胞代谢组的方法主要有质谱检测方法和核磁共振检测方法。其中，质谱检测方法因灵敏度更高而受到更加广泛的关注。然而，很多使用质谱检测单细胞代谢物的方法并不包含样品前处理过程[105-107]，培养基中的盐及细胞内液中的一些成分将影响分析物的离子化效率[108]。我国张新荣教授、栾天罡教授和加拿大 Janusz Pawliszyn 教授研究组都尝试用微型探针在单细胞中富集代谢物进行检测[36, 42, 109]。张新荣教授研究组用钨针比较了黄酮衍生物在洋葱细胞核和细胞质中的丰度[42]。Janusz Pawliszyn 教授研究组则制备了聚吡咯包覆的不锈钢丝探针，用于检测洋葱细胞中的槲皮黄酮和毛地黄黄酮（图 1-1e）[36]。而栾天罡教授研究组用十八烷基修饰的钨针对单细胞脂质进行了检测[109]。

4　原位样品前处理技术的萃取相

萃取相的选取主要基于分析物的极性[24, 46, 49, 57, 110]。而污染物在萃取相中的传质系数以及污染物在萃取相和样品基质间的分配系数则是重要的参考指标[30]。污染物在萃取相中的传质系数足够大，则有利于被动采样过程快速地到达平衡状态。而污染物在萃取相和样品基质间的分配系数越大，被动采样过程到达平衡状态所需的时间就越长。但另一方面，分配系数必须足够大才能够保证得到较高的分析灵敏度。另外，分配系数大也意味着被动采样器可以用来检测更长时间范围内的 TWA 浓度。

对于极性低的持久性有机污染物，如 PAHs、PCBs 和 PBDEs 等，常用的萃取相有 PDMS[57]、聚乙烯[49, 71, 78]、聚氨酯[109]和 XAD-2[24]等非极性的聚合物。聚甲醛相对这几种聚合物的极性更强，因此它能够作为有一定极性的有机污染物的萃取相[111]。针对药物等极性较强甚至部分电离的有机污染物，所采用的萃取相通常为商品化的亲水亲油平衡吸附剂[46, 112-114]或混合型离子交换吸附剂[115]。最近，也有研究组将离子液体作为药物和其他极性污染物的萃取相[116, 117]。

非生命环境介质中的被动采样所需的时间通常较长，特别是平衡被动采样所需的时间可达数月之久。被动采样器长期置于复杂的环境介质中，表面将有可能被微生物形成的生物膜或是吸附的腐殖质等屏蔽，这将影响被动采样的效率和重现性，也可能给检测结果带来误差[118-120]。因此，在实际应用中，研究人员应设法评估和校正微生物膜和其他屏蔽膜对被动采样过程的影响，或是采用一定的办法避免微生物膜的产生和腐殖质等的吸附。

在动植物活体组织中对目标分析物进行萃取时，除了要求萃取相对目标分析物有较高的萃取性能外，还要求萃取相具备生物相容性和耐生物腐蚀性[28]。耐生物腐蚀性指的是萃取相能够有效减少甚至避免生物大分子在其表面的吸附，以免吸附的生物大分子降低萃取相对目标分析物的萃取效率。SPME 探针表面吸附的生物大分子还会在气相色谱仪进样口发生碳化，造成探针表面性质的改变，影响探针寿命。另外，如果用溶剂洗脱探针萃取到的分析物，探针表面吸附的生物大分子也将可能随分析物一起洗脱到洗脱溶液中，进而造成液相色谱管道或分离柱的堵塞，或影响仪器响应。生物相容性则是保证 SPME 探针在动植物活体组织中萃取时不致引起动植物体的排异反应。

目前，用于动植物活体组织中污染物检测的萃取相绝大多数是 PDMS，目标污染物甚至也包括在生理 pH 下部分电离的药物[28]。PDMS 涂层能得到广泛应用，是因为它具有良好的生物相容性以及一定的耐生物腐蚀性[120, 121]。同时，PDMS 探针也很容易在实验室制备得到，且成本低廉[4, 7, 92]。另外，欧阳钢锋教授研究组还研制了一些新型的 SPME 探针涂层，用于鱼体内药物的高效萃取，如聚电解质-硅胶混合型涂层[34]和聚多巴胺鞘层包裹的纳米纤维涂层[122]等。

在检测动植物代谢组时，SPME 探针涂层则需要具备尽可能全面的代谢组覆盖率，即涂层需要对不同极性、不同电性的代谢物均具有较理想的萃取效率。Janusz Pawliszyn 教授研究组比较了多种商品化的固相萃取柱填料对代谢组的萃取效率，发现混合型吸附剂（修饰了烷基链和苯磺酸基的硅胶颗粒）、含苯硼酸基的吸附剂以及聚苯乙烯-二乙烯基苯能够获得较高的代谢组覆盖率[22]。Janusz Pawliszyn 教授研究组将混合型的吸附剂用生物相容的聚丙烯腈黏附在金属丝上制备成 SPME 探针，成功应用于小鼠血液和大鼠脑部的代谢组的检测[5, 6]。另外，十八烷基键合硅胶颗粒被用于研究鱼体的暴露组[100]，而商品化的 DVB/Carboxen/PDMS 探针也被用于检测苹果中的挥发性代谢物[101]。

用于单个细胞原位样品前处理的微型 SPME 探针并不多见，张新荣教授研究组对钨针进行硝酸处理制得了微型探针[42]，而 Janusz Pawliszyn 教授研究组则在不锈钢丝表面原位合成聚吡咯来制备微型探针[36]。

5 展　望

原位样品前处理技术应用于非生命环境介质、动植物活体组织和单细胞中有机污染物和内源性小分子有机物的检测，能够揭示污染物在环境中的分布和迁移规律以及污染物在生物体内的富集性和持久性，也能够探究外源物对生物体的作用及机理。目前，原位样品前处理技术在环境分析以及污染物毒理学研究的应用方面，依然有广阔的研究空间。第一，原位样品前处理技术在一些情况下的定量结果仍有一定的不确定性,这需要进一步探讨多个方面的问题,包括准确的分配系数的获取、PRC 方法可用性的验证等;第二,新型萃取相的研制是化学和材料科学的前沿热点[123-126],推动新型萃取相在原位样品前处理技术中的应用,对于拓展该技术的分析物的种类以及提高原位样品前处理效率都有重要的促进作用;第三,活体 SPME 技术能够检测到其他方法检测不到的不稳定的生物内源性小分子有机物[5, 6, 102],推动基于活体 SPME 技术的代谢组检测方法在污染物的毒理学研究上的应用,将有助于毒理学研究的进一步发展;第四,细胞是生命体的基本功能单元[104-107],从单细胞代谢组层面开展毒理学研究将提供污染物毒理学研究的新视角,而原位样品前处理技术在单细胞代谢组检测上具有极大的应用潜力[36, 42]。

参 考 文 献

[1] Pawliszyn J. Handbook of Solid Phase Microextraction. Beijing: Chemical Industry Press of China, 2009.

[2] Ouyang G, Vuckovic D, Pawliszyn J. Nondestructive sampling of living systems using in vivo solid-phase microextraction. Chemical Reviews, 2011, 111(4): 2784-2814.

[3] Cui X, Bao L, Gan J. Solid-phase microextraction (SPME) with stable isotope calibration for measuring bioavailability of hydrophobic organic contaminants. Environmental Science & Technology, 2013, 47(17): 9833-9840.

[4] Zhang X, Oakes K D, Cui S, et al. Tissue-specific in vivo bioconcentration of pharmaceuticals in rainbow trout (Oncorhynchus mykiss) using space-resolved solid-phase microextraction. Environmental Science & Technology, 2010, 44(9): 3417-3422.

[5] Cudjoe E, Bojko B, de Lannoy I, et al. Solid-phase microextraction: a complementary in vivo sampling method to microdialysis. Angewandte Chemie International Edition, 2013, 52(46): 12124-12126.

[6] Vuckovic D, de Lannoy I, Gien B, et al. In vivo solid-phase microextraction: capturing the elusive portion of metabolome. Angewandte Chemie International Edition, 2011, 50(23): 5344-5348.

[7] Xu J, Luo J, Ruan J, et al. In vivo tracing uptake and elimination of organic pesticides in fish muscle. Environmental Science & Technology, 2014, 48(14): 8012-8020.

[8] Zhang X, Oakes K D, Wang S, et al. In vivo sampling of environmental organic contaminants in fish by solid-phase microextraction. Trends in Analytical Chemistry, 2012, 32: 31-39.

[9] Donabella P, Rogers N, Levin R, et al. Development of supported liquid-phase microextraction probes for in vivo PK studies. Bioanalysis, 2015, 7(6): 661-670.

[10] Spietelun A, Marcinkowski Ł, de la Guardia M, et al. Green aspects, developments and perspectives of liquid phase microextraction techniques. Talanta, 2014, 119: 34-45.

[11] Solem L E, Kolanczyk R C, Mckim Ⅲ J M. An in vivo microdialysis method for the qualitative analysis of hepatic phase Ⅰ metabolites of phenol in rainbow trout (Oncorhynchus mykiss). Aquatic Toxicology, 2003, 62(4): 337-347.

[12] Nichols J W, Hoffman A D, Fitzsimmons P N, et al. Quantification of phenol, phenyl glucuronide, and phenyl sulfate in blood of unanesthetized rainbow trout by online microdialysis sampling. Toxicology Mechanisms and Methods, 2008, 18(5): 405-412.

[13] Nichols J W, Hoffman A D, Fitzsimmons P N, et al. Use of online microdialysis sampling to determine the in vivo rate of phenol glucuronidation in rainbow trout. Drug Metabolism and Disposition, 2008, 36(7): 1406-1413.

[14] Zhou S N, Oakes K D, Servos M R, et al. Use of simultaneous dual-probe microdialysis for the determination of pesticide residues in a jade plant (Crassula ovata). Analyst, 2009, 134(4): 748-754.

[15] Mizoguchi H, Katahira K, Inutsuka A, et al. Insular neural system controls decision-making in healthy and methamphetamine-treated rats. Proceedings of the National Academy of Sciences of the United States of America, 2015, 112(29): E3930-E3939.

[16] Kang X, Xu H, Teng S, et al. Dopamine release from transplanted neural stem cells in Parkinsonian rat striatum in vivo. Proceedings of the National Academy of Sciences of the United States of America, 2014, 111(44): 15804-15809.

[17] Stengård K, Tham R, O'Connor W T, et al. Acute toluene exposure increases extracellular GABA in the cerebellum of rat: a microdialysis study. Pharmacology and Toxicology, 1993, 73(6): 315-318.

[18] Faro L R F, Durán R, Do Nascimento J L M, et al. Effects of successive intrastriatal methylmercury administrations on dopaminergic system. Ecotoxicology and Environmental Safety, 2003, 55(2): 173-177.

[19] Rodríguez V M, Dufour L, Carrizales L, et al. Effects of oral exposure to mining waste on in vivo dopamine release from rat striatum. Environmental Health Perspective, 1998, 106(8): 487-491.

[20] Seegal R F, Okoniewski R J, Brosch K O, et al. Polychlorinated biphenyls alter extraneuronal but not tissue dopamine concentrations in adult rat striatum: an in vivo microdialysis study. Environmental Health Perspective, 2002, 110(11): 1113-1117.

[21] Xu J, Zheng J, Tian J, et al. New materials in solid-phase microextraction. Trends in Analytical Chemistry, 2013, 47: 68-83.

[22] Vuckovic D, Pawliszyn J. Systematic evaluation of solid-phase microextraction coatings for untargeted metabolomic profiling of biological fluids by liquid chromatography—mass spectrometry. Analytical Chemistry, 2011, 83(6): 1944-1954.

[23] Sorais M, Rezaei A, Okeme J O, et al. A miniature bird-borne passive air sampler for monitoring halogenated flame retardants. Science of the Total Environment, 2017, 599-600: 1903-1911.

[24] Gong P, Wang X, Liu X, et al. Field calibration of XAD-based passive air sampler on the Tibetan Plateau: wind influence and configuration improvement. Environmental Science & Technology, 2017, 51(10): 5642-5649.

[25] Liu Y, Wang S, McDonough C A, et al. Estimation of uncertainty in air—water exchange flux and gross volatilization loss of PCBs: a case study based on passive sampling in the lower Great Lakes. Environmental Science & Technology, 2016, 50(20): 10892-10902.

[26] Fernandez L A, Lao W, Maruya K A, et al. Calculating the diffusive flux of persistent organic pollutants between sediments and the water column on the Palos Verdes Shelf superfund site using polymeric passive samplers. Environmental Science & Technology, 2014, 48(7): 3925-3934.

[27] Lin K, Lao W, Lu Z, et al. Measuring freely dissolved DDT and metabolites in seawater using solid-phase microextraction with performance reference compounds. Science of the Total Environment, 2017, 599-600: 364-371.

[28] Xu J, Chen G, Huang S, et al. Application of in vivo solid-phase microextraction in environmental analysis. Trends in Analytical Chemistry, 2016, 85: 26-35.

[29] Allan I J, Bæk K, Haugen T O, et al. In Vivo Passive Sampling of Nonpolar Contaminants in Brown Trout (Salmo trutta). Environmental Science & Technology, 2013, 47(20): 11660-11667.

[30] Shetty M K, Limmer M A, Waltermire K, et al. In planta passive sampling devices for assessing subsurface chlorinated solvents. Chemosphere, 2014, 104: 149-154.

[31] Peijnenburg W J G M, Teasdale P R, Reible D, et al. Passive sampling methods for contaminated sediments: state of the science for organic contaminants. Integrated Environmental Assessment and Management, 2014, 10(2): 167-178.

[32] Liu H-H, Bao L-J, Zhang K, et al. Novel passive sampling device for measuring sediment—water diffusion fluxes of hydrophobic organic chemicals. Environmental Science & Technology, 2013, 47(17): 9866-9873.

[33] Yao Y, Huang C-L, Wang J-Z, et al. Significance of anthropogenic factors to freely dissolved polycyclic aromatic hydrocarbons in freshwater of China. Environmental Science & Technology, 2017, 51(15): 8304-8312.

[34] Xu J, Wu R, Huang S, et al. Polyelectrolyte microcapsules dispersed in silicone rubber for in vivo sampling in fish brains. Analytical Chemistry, 2015, 87(20): 10593-10599.

[35] Sheehan E M, Limmer M A, Mayer P, et al. Time-weighted average SPME analysis for in planta determination of cVOCs.

Environmental Science & Technology, 2012, 46(6): 3319-3325.

[36] Piri-Moghadam H, Ahmadi F, Gómez-Ríos G A, et al. Fast quantitation of target analytes in small volumes of complex samples by matrix-compatible solid-phase microextraction devices. Angewandte Chemie International Edition, 2016, 55(26): 7510-7514.

[37] Arts M, Soons Z, Ellis S R, et al. Detection of localized hepatocellular amino acid kinetics by using mass spectrometry imaging of stable isotopes. Angewandte Chemie International Edition, 2017, 56(25): 7146-7150.

[38] Cheng D, Xu W, Yuan L, et al. Investigation of drug-induced hepatotoxicity and its remediation pathway with reaction-based fluorescent probes. Analytical Chemistry, 2017, 89(14): 7693-7700.

[39] Harrison J P, Berry D. Vibrational spectroscopy for imaging single microbial cells in complex biological samples. Frontiers in Microbiology, 2017, 8: 675.

[40] Zhang X, Zong C, Xu M, et al. Raman imaging from microscopy to nanoscopy, and to macroscopy. Small, 2015, 11(28): 3395-3406.

[41] Li S, Zhu A, Zhu T, et al. Single biosensor for simultaneous quantification of glucose and pH in a rat brain of diabetic model using both current and potential outputs. Analytical Chemistry, 2017, 89(12): 6656-6662.

[42] Gong X, Zhao Y, Cai S, et al. Single cell analysis with probe ESI-mass spectrometry: detection of metabolites at cellular and subcellular levels. Analytical Chemistry, 2014, 86(8): 3840-3816.

[43] Tao Y, Xue B, Yao S. Using linoleic acid embedded cellulose acetate membranes to in situ monitor polycyclic aromatic hydrocarbons in lakes and predict their bioavailability to submerged macrophytes. Environmental Science & Technology, 2015, 49(10): 6077-6084.

[44] Ruge Z, Muir D, Helm P, et al. Concentrations, trends, and air–water exchange of PAHs and PBDEs derived from passive samplers in Lake Superior in 2011. Environmental Science & Technology, 2015, 49(23): 13777-13786.

[45] Ouyang G, Pawliszyn J. A critical review in calibration methods for solid-phase microextraction. Analytica Chimica Acta, 2008, 627(2): 184-197.

[46] Petrie B, Gravell A, Mills G A, et al. *In situ* calibration of a new chemcatcher configuration for the determination of polar organic micropollutants in wastewater effluent. Environmental Science & Technology, 2016, 50(17): 9469-9478.

[47] Ouyang G, Oakes K D, Bragg L, et al. Sampling-rate calibration for rapid and nonlethal monitoring of organic contaminants in fish muscle by solid-phase microextraction. Environmental Science & Technology, 2011, 45(18): 7792-7798.

[48] Musteata F M, de Lannoy I, Gien B, et al. Blood sampling without blood draws for in vivo pharmacokinetic studies in rats. Journal of Pharmaceutical and Biomedical Analysis, 2008, 47(4-5): 907-912.

[49] Apell J N, Gschwend P M. Validating the use of performance reference compounds in passive samplers to assess porewater concentrations in sediment beds. Environmental Science & Technology, 2014, 48(17): 10301-10307.

[50] Zhou S N, Oakes K D, Servos M R, et al. Application of solid-phase microextraction for in vivo laboratory and field sampling of pharmaceuticals in fish. Environmental Science & Technology, 2008, 42(16): 6073-6079.

[51] Zhou S N, Zhao W, Pawliszyn J. Kinetic calibration using dominant pre-equilibrium desorption for on-site and in vivo sampling by solid-phase microextraction. Analytical Chemistry, 2008, 80(2): 481-490.

[52] Zhang X, Oakes K D, Luong D, et al. Kinetically-calibrated solid-phase microextraction using label-free standards and its application for pharmaceutical analysis. Analytical Chemistry, 2011, 83(6): 2371-2377.

[53] Morrison S A, Belden J B. Characterization of performance reference compound kinetics and analyte sampling rate corrections under three flow regimes using nylon organic chemical integrative samplers. Journal of Chromatography A, 2016, 1466: 1-11.

[54] Mazzella N, Lissalde S, Moreira S, et al. Evaluation of the use of performance reference compounds in an Oasis-HLB

adsorbent based passive sampler for improving water concentration estimates of polar herbicides in freshwater. Environmental Science & Technology, 2010, 44(5): 1713-1719.

[55] Estoppey N, Schopfer A, Fong C, et al. An in-situ assessment of low-density polyethylene and silicone rubber passive samplers using methods with and without performance reference compounds in the context of investigation of polychlorinated biphenyl sources in rivers. Science of the Total Environment, 2016, 572: 794-803.

[56] Bao L-J, Wu X, Jia F, et al. Isotopic exchange on solid-phase micro extraction fiber in sediment under stagnant conditions: Implications for field application of performance reference compound calibration. Environmental Toxicology and Chemistry, 2016, 35(8): 1978-1985.

[57] Witt G, Lang S-C, Ullmann D, et al. Passive equilibrium sampler for in situ measurements of freely dissolved concentrations of hydrophobic organic chemicals in sediments. Environmental Science & Technology, 2013, 47(14): 7830-7839.

[58] Jalalizadeh M, Ghosh U. *In situ* passive sampling of sediment porewater enhanced by periodic vibration. Environmental Science & Technology, 2016, 50(16): 8741-8749.

[59] Xu J, Huang S, Jiang R, et al. Evaluation of the availability of bound analyte for passive sampling in the presence of mobile binding matrix. Analytica Chimica Acta, 2016, 917: 19-26.

[60] Xu J, Huang S, Wei S, et al. Study on the diffusion-dominated solid-phase microextraction kinetics in semisolid sample matrix. Analytical Chemistry, 2016, 88(18): 8921-8925.

[61] Mayer P, Tolls J, Hermens J L M, et al. Equilibrium sampling devices. Environmental Science & Technology, 2003, 37(9): 184A-191A.

[62] Cornelissen G, Pettersen A, Broman D, et al. Field testing of equilibrium passive samplers to determine freely dissolved native polycyclic aromatic hydrocarbon concentrations. Environmental Toxicology and Chemistry, 2008, 27(3): 499-508.

[63] Adams R G, Lohman R, Fernandez L A, et al. Polyethylene Devices: Passive Samplers for Measuring Dissolved Hydrophobic Organic Compounds in Aquatic Environments. Environmental Science & Technology, 2007, 41(4): 1317-1323.

[64] Cornelissen G, Wiberg K, Broman D, et al. Freely dissolved concentrations and sediment-water activity ratios of PCDD/Fs and PCBs in the open baltic sea. Environmental Science & Technology, 2008, 42(23): 8733-8739.

[65] Mäenpää K, Leppänen M T, Figueiredo K, et al. Fate of polychlorinated biphenyls in a contaminated lake ecosystem: Combining equilibrium passive sampling of sediment and water with total concentration measurements of biota. Environmental Toxicology and Chemistry, 2015, 34(11): 2463-2474.

[66] Booij K, Hoedemaker J R, Bakker J F. Dissolved PCBs, PAHs, and HCB in pore waters and overlying waters of contaminated harbor sediments. Environmental Science & Technology, 2003, 37(18): 4213-4220.

[67] Sacks V P, Lohmann R. Freely dissolved PBDEs in water and porewater of an urban estuary. Environmental Pollution, 2012, 162: 287-293.

[68] Mogan E J, Lohmann R. Detecting air-water and surface-deep water gradients of PCBs using polyethylene passive samplers. Environmental Science & Technology, 2008, 32(19): 7248-7253.

[69] Lohmann R, Klanova J, Pribylova P, et al. PAHs on a west-to-east transect across the tropical Atlantic Ocean. Environmental Science & Technology, 2013, 47(6): 2570-2578.

[70] Xue J, Liao C, Wang J, et al. Development of passive samplers for in situ measurement of pyrethroid insecticides in surface water. Environmental Pollution, 2017, 224: 516-523.

[71] Yao Y, Meng X-Z, Wu C-C, et al. Tracking human footprints in Antarctica through passive sampling of polycyclic aromatic hydrocarbons in inland lakes. Environmental Pollution, 2016, 213: 412-419.

[72] Allan I J, Booij K, Paschke A, et al. Field performance of seven passive sampling devices for monitoring of hydrophobic substances. Environmental Science & Technology, 2009, 43(14): 5383-5390.

[73] Eganhouse R P. Determination of polydimethylsiloxane–water partition coefficients for ten 1-chloro-4-[2, 2, 2-trichloro-1-(4-chlorophenyl)ethyl] benzene-related compounds and twelve polychlorinated biphenyls using gas chromatography/mass spectrometry. Journal of Chromatography A, 2016, 1438: 226-235.

[74] Liu H-H, Bao L-J, Zeng E Y. Recent advances in the field measurement of the diffusion flux of hydrophobic organic chemicals at the sediment-water interface. Trends in Analytical Chemistry, 2014, 54: 56-64.

[75] Koelmans A A, Poot A, de Lange H J, et al. Estimation of in situ sediment-to-water fluxes of polycyclic aromatic hydrocarbons, polychlorobiphenyls and polybrominated diphenylethers. Environmental Science & Technology, 2010, 44(8): 3014-3020.

[76] Tidwell L G, Allan S E, O'Connell S G, et al. PAH and OPAH flux during the deepwater horizon incident. Environmental Science & Technology, 2016, 50(14): 7489-7497.

[77] Khairy M, Muir D, Teixeira C, et al. Spatial trends, sources, and air-water exchange of organochlorine pesticides in the Great Lakes basin using low density polyethylene passive samplers. Environmental Science & Technology, 2014, 48(16): 9315-9325.

[78] Liu Y, Wang S, McDonough C A, et al. Gaseous and freely-dissolved PCBs in the Lower Great Lakes based on passive sampling: spatial trends and air–water exchange. Environmental Science & Technology, 2016, 50(10): 4932-4939.

[79] Lohmann R, Klanova J, Pribylova P, et al. Concentrations, fluxes, and residence time of PBDEs across the tropical Atlantic Ocean. Environmental Science & Technology, 2013, 47(24): 13967-13975.

[80] Apell J N, Gschwend P M. The atmosphere as a source/sink of polychlorinated biphenyls to/from the Lower Duwamish Waterway Superfund site. Environmental Pollution, 2017, 227: 263-270.

[81] Donald C E, Anderson K A. Assessing soil-air partitioning of PAHs and PCBs with a new fugacity passive sampler. Science of the Total Environment, 2017, 596-597: 293-302.

[82] Wang W, Simonich S, Giri B, et al. Atmospheric concentrations and air–soil gas exchange of polycyclic aromatic hydrocarbons(PAHs)in remote, rural village and urban areas of Beijing–Tianjin region, North China. Science of the Total Environment, 2011, 409: 2942-2950.

[83] Feng Y, Wu C-C, Bao L-J, et al. Examination of factors dominating the sediment-water diffusion flux of DDT-related compounds measured by passive sampling in an urbanized estuarine bay. Environmental Pollution, 2016, 219: 866-872.

[84] Lin D, Eek E, Oen A, et al. Novel Probe for in Situ Measurement of Freely Dissolved Aqueous Concentration Profiles of Hydrophobic Organic Contaminants at the Sediment-Water Interface. Environmental Science & Technology Letters, 2015, 2(11): 320-324.

[85] Niu S, Dong L, Zhang L, et al. Temporal and spatial distribution, sources, and potential health risks of ambient polycyclic aromatic hydrocarbons in the Yangtze River Delta(YRD)of eastern China. Chemosphere, 2017, 172: 72-79.

[86] Carratalá A, Moreno-González R, León V M. Occurrence and seasonal distribution of polycyclic aromatic hydrocarbons and legacy and current-use pesticides in air from a Mediterranean coastal lagoon(Mar Menor, SE Spain). Chemosphere, 2017, 167: 382-395.

[87] Burken J G, Vroblesky D A, Balouet J C. Phytoforensics, dendrochemistry, and phytoscreening: new green tools for delineating contaminants from past and present. Environmental Science & Technology, 2011, 45(15): 6218-6226.

[88] Doucette W, Klein H, Chard J, et al. Volatilization of trichloroethylene from trees and soil: measurement and scaling approaches. Environmental Science & Technology, 2013, 47(11): 5813-5820.

[89] Limmer M A, Holmes A J, Burken J G. Phytomonitoring of chlorinated ethenes in trees: a four-year study of seasonal chemodynamics in planta. Environmental Science & Technology, 2014, 48(18): 10634-10640.

[90] Reiche N, Mothes F, Fiedler P, et al. A solid-phase microextraction method for the in vivo sampling of MTBE in common

reed (Phragmites australis). Environmental Monitoring and Assessment, 2013, 185(9): 7133-7144.

[91] Togunde O P, Lord H, Oakes K D, et al. Development and evaluation of a new in vivo solid-phase microextraction sampler. Journal of Separation Science, 2013, 36(1): 219-223.

[92] Zhang X, Cai J, Oakes K D, et al. Development of the space-resolved solid-phase microextraction technique and its application to biological matrices. Analytical Chemistry, 2009, 81(17): 7349-7356.

[93] Bai Z, Pilote A, Sarker P K, et al. In vivo solid-phase microextraction with in vitro calibration: determination of off-flavor components in live fish. Analytical Chemistry, 2013, 85(4): 2328-2332.

[94] Zhou S N, Ouyang G, Pawliszyn J. Comparison of microdialysis with solid-phase microextraction for in vitro and in vivo studies. Journal of Chromatography A, 2008, 1196-1197: 46-56.

[95] Win-Shwe T-T, Mitsushima D, Nakajima D, et al. Toluene induces rapid and reversible rise of hippocampal glutamate and taurine neurotransmitter levels in mice. Toxicology Letters, 2007, 168(1): 75-82.

[96] Qiu J, Chen G, Xu J, et al. In vivo tracing of organochloride and organophosphorus pesticides in different organs of hydroponically grown malabar spinach (Basella alba L.). Journal of Hazardous Materials, 2016, 316: 52-59.

[97] Chen G, Jiang R, Qiu J, et al. Environmental fates of synthetic musks in animal and plant: An in vivo study. Chemosphere, 2015, 138: 584-591.

[98] Chen G, Qiu J, Liu Y, et al. Carbon nanotubes act as contaminant carriers and translocate within plants. Scientific Reports, 2015, 5: 15682.

[99] Loi R X, Solar M C, Weidenhamer J D. Solid-phase microextraction method for in vivo measurement of allelochemical uptake. Journal of Chemical Ecology, 2008, 34(1): 70-75.

[100] Allan I J, Bæk K, Kringstad A, et al. Should silicone prostheses be considered for specimen banking? A pilot study into their use for human biomonitoring. Environment International, 2013, 59: 462-468.

[101] O'Connell S G, Kerkvliet N I, Carozza S, et al. In vivo contaminant partitioning to silicone implants: Implications for use in biomonitoring and body burden. Environment International, 2015, 85: 182-188.

[102] Bessonneau V, Ings J, McMaster M, et al. In vivo microsampling to capture the elusive exposome. Scientific Reports, 2017, 7: 44038.

[103] Risticevic S, Souza-Silva E A, DeEll J R, et al. Capturing plant metabolome with direct-immersion in vivo solid phase microextraction of plant tissues. Analytical Chemistry, 2016, 88(2): 1266-1274.

[104] Zenobi R. Single-cell metabolomics: analytical and biological perspectives. Science, 2013, 342(6163): 1243259.

[105] Fessenden M. Metabolomics: small molecules, single cells. Nature, 2016, 540: 153-155.

[106] Comi T J, Do T D, Rubakhin S S, et al. Categorizing cells on the basis of their chemical profiles: progress in single-cell mass spectrometry. Journal of the American Chemical Society, 2017, 139(11): 3920-3929.

[107] Onjiko R M, Portero E P, Moody S A, et al. In situ microprobe single-cell capillary electrophoresis mass spectrometry: metabolic reorganization in single differentiating cells in the live vertebrate(Xenopus laevis)embryo. Analytical Chemistry, 2017, 89(13): 7069-7076.

[108] Zhang X C, Wei Z W, Gong X Y, et al. Integrated droplet-based microextraction with ESI-MS for removal of matrix interference in single-cell analysis. Scientific Reports, 2016, 6: 24730.

[109] Deng J, Li W, Yang Q, et al. Biocompatible surface-coated probe for in vivo, in situ, and microscale lipidomics of small biological organisms and cells using mass spectrometry. Analytical Chemistry, 2018, 90(11): 6936-6944.

[110] Peverly A A, Ma Y, Venier M, et al. Variations of flame retardant, polycyclic aromatic hydrocarbon, and pesticide concentrations in Chicago's atmosphere measured using passive sampling. Environmental Science & Technology, 2015, 49(9): 5371-5379.

[111] Endo S, Hale S E, Goss K-U, et al. Equilibrium partition coefficients of diverse polar and nonpolar organic compounds to polyoxymethylene (POM) passive sampling devices. Environmental Science & Technology, 2011, 45(23): 10124-10132.

[112] Guibal R R, Buzier R, Charriau A, et al. Passive sampling of anionic pesticides using the Diffusive Gradients in Thin films technique (DGT). Analytica Chimica Acta, 2017, 966: 1-10.

[113] Li Z, Sobek A, Radke M. Fate of pharmaceuticals and their transformation products in four small European rivers receiving treated wastewater. Environmental Science & Technology, 2016, 50(11): 5614-5621.

[114] Harman C, Reid M, Thomas K V. In situ calibration of a passive sampling device for selected illicit drugs and their metabolites in wastewater, and subsequent year-long assessment of community drug usage. Environmental Science & Technology, 2011, 45(23): 10124-10132.

[115] Fauvelle V, Mazzella N, Delmas F, et al. Use of mixed-mode ion exchange sorbent for the passive sampling of organic acids by Polar Organic Chemical Integrative Sampler(POCIS). Environmental Science & Technology, 2012, 46(24): 13344-13353.

[116] Męczykowska H, Kobylis P, Stepnowski P, et al. Ionic liquids for the passive sampling of sulfonamides from water—applicability and selectivity study. Analytical and Bioanalytical Chemistry, 2017, 409(16): 3951-3958.

[117] Caban M, Męczykowska H, Stepnowski P. Application of the PASSIL technique for the passive sampling of exemplary polar contaminants (pharmaceuticals and phenolic derivatives) from water. Talanta, 2016, 155: 185-192.

[118] Cristale J, Katsoyiannis A, Chen C, et al. Assessment of flame retardants in river water using a ceramic dosimeter passive sampler. Environmental Pollution, 2013, 172: 163-169.

[119] Harman C, Bøyum O, Thomas K V, et al. Small but different effect of fouling on the uptake rates of semipermeable membrane devices and Polar Organic Chemical Integrative Samplers. Environmental Toxicology and Chemistry, 2009, 28(11): 2324-2332.

[120] Souza Silva É A, Pawliszyn J. Optimization of fiber coating structure enables direct immersion solid phase microextraction and high-throughput determination of complex samples. Analytical Chemistry, 2012, 84(16): 6933-6938.

[121] Zheng J, Huang J, Xu F, et al. Powdery polymer and carbon aerogels with high surface areas for high-performance solid phase microextraction coatings. Nanoscale, 2017, 9(17): 5545-5550.

[122] Xu J, Huang S, Wu R, et al. Bioinspired polydopamine sheathed nanofibers for high-efficient in vivo solid-phase microextraction of pharmaceuticals in fish muscle. Analytical Chemistry, 2015, 87(6): 3453-3459.

[123] Alsbaiee A, Smith B J, Xiao L, et al. Rapid removal of organic micropollutants from water by a porous β-cyclodextrin polymer. Nature, 2016, 529(7585): 190-194.

[124] Chen Y, Chen F, Zhang S, et al. Facile fabrication of multifunctional metal–organic framework hollow tubes to trap pollutants. Journal of the American Chemical Society, 2017, 139(46): 16482-16485.

[125] Wang B, Lv X L, Feng D, et al. Highly stable Zr(IV)-based metal–organic frameworks for the detection and removal of antibiotics and organic explosives in water. Journal of the American Chemical Society, 2016, 138(19): 6204-6216.

[126] Desai A V, Manna B, Karmakar A, et al. A water-stable cationic metal–organic framework as a dual adsorbent of oxoanion pollutants. Angewandte Chemie International Edition, 2016, 55(27): 7811-7815.

作者：徐剑桥[1]，欧阳钢锋[1]

[1] 中山大学

第 2 章　环境稳定同位素技术研究进展

- 1. 引言 /21
- 2. 稳定同位素技术基础理论 /22
- 3. 环境稳定同位素分析方法进展 /24
- 4. 稳定同位素技术在污染物示踪中的应用进展 /28
- 5. 环境稳定同位素标记技术研究进展 /33
- 6. 展望 /35

本章导读

在自然界中，元素的生物地球化学循环包含氧化还原反应、络合反应、吸附、溶解、沉淀和生物循环等物理化学过程，通常伴随着元素的稳定同位素分馏效应，进而导致自然界中不同储库逐渐具有了不同的同位素组成特征。物质的同位素组成记录了物质的来源和经历的某些环境过程，是自然环境中物质的天然指纹。稳定同位素测量技术的发展，尤其是多接收器电感耦合等离子体质谱（MC-ICP-MS）的发明，使更多元素的同位素组成可以被准确测量，例如汞、铅、锶、铜、铁、锌、硅、镁、钙、银等，从而揭开了稳定同位素技术快速发展的新篇章。此外，得益于测量手段和分离纯化技术的进步，自然环境中越来越多的同位素分馏过程可以被详细研究，快速推动了稳定同位素分馏理论的完善，促使稳定同位素技术在环境地球化学领域进入了飞速发展时期。本章将针对稳定同位素（标记）技术的理论基础、分析方法及其在环境地球化学领域的典型应用进行较全面的综述，并对未来的研究方向进行展望。

关键词

稳定同位素，同位素标记，同位素分馏，环境污染物，来源示踪

1 引　言

"同位素"的概念起源 20 世纪初，是指质子数相同、中子数不同的一组原子。质量上的差别可以导致同位素之间出现轻微的物理化学性质差异，在宏观反应中表现为不同的反应或运动速率，导致在漫长的生物地球化学循环过程中，自然界中的不同物质逐渐具有了不同的同位素组成特征。稳定同位素技术就是基于物质的稳定同位素组成和同位素分馏机理去追溯物质的来源和经历的某些特定环境过程的技术。至今，稳定同位素技术经过一百多年的发展，从分析测量技术到同位素分馏理论，已经逐渐成熟。19 世纪末，Becquerel[1]和 Curie[2]首先发现了放射性元素，标志着"同位素地球化学"这一学科的诞生。几年后，Rutherford 等[3]创造性地提出了放射性元素的"半衰期"概念，为同位素年代学的研究奠定了基础。1913 年，Soddy 首次提出了"同位素"这一概念[4]，一直沿用至今。同一时期，Thomson 发明了第一台质谱仪，为今后稳定同位素的测量奠定了基础。从 20 世纪 30 年代开始，Urey、Murphy 和 Nier 等[5-7]引导了稳定同位素技术的发展，并对稳定同位素分馏理论进行了开拓性研究，例如 Urey 等[8]提出了"同位素古温度"的概念。到了近代，Walder 等[9]于 1993 年首次开发了基于多接收器电感耦合等离子体质谱（MC-ICP-MS）的同位素分析方法，自此，更多重质量元素的同位素组成可以被准确测量，使稳定同位素技术进入了一个新的发展阶段。目前，稳定同位素技术作为一种有力的示踪手段，已经广泛应用于陆地、大气和水体等自然环境体系的研究中。其中，碳、氮、氧和硫等传统稳定元素经过数十年发展，已经建立了相对成熟的分析方法和较为系统的理论体系。而像汞、锶、铜、铁、锌、硅、镁、钙、银等非传统稳定同位素的发展则相对滞后，直到 MC-ICP-MS 的发明应用，相应的同位素分析体系才开始建立。同时，得益于分离纯化技术的发展，不同环境过程中的同位素分馏机理和环境储库同位素组成的调查也不断获得新的进展，有力地推动了稳定同位素技术在各个环境地球化学领域的快速发展。此外，随着同位素分离富集技术的发展，商品化的富集同位素标准品越来越普遍，大大促进了稳定同位素标记技术的应用与发展。本章将依托非传统稳定同位素和富集稳定同位素重点介绍稳定同位素技术在环境地球化学领域的最新研究进展。

2 稳定同位素技术基础理论

2.1 同位素理论基本概念

同位素是质子数相同，中子数不同的一组原子，它们处于元素周期表的同一位置，但却具有不同的质量数。如图 2-1 所示，元素周期表中只有 21 种元素具有一种同位素，其他大部分元素都具有两种以上的同位素，因此大部分元素都适合进行同位素技术研究。由于具有相同的质子数和核外电子结构，同位素具有非常相似的物理化学性质，但它们的物理化学性质也因质量数上的不同而存在微小的差异，进而可以导致自然界中的各种物质出现同位素组成上的变化。同位素可分为两种基本类型：稳定同位素和放射性同位素，是根据是否存在放射性衰变行为进行区分的。放射性同位素经过放射性衰变会成为稳定同位素，进而可以引起特定元素的稳定同位素组成发生变化，例如放射性元素 ^{87}Rb 可以衰变为稳定元素 ^{87}Sr，进而影响到 Sr 同位素在自然界中的组成；U-Th-Pb 衰变体系可以改变 Pb 稳定同位素的组成。除了放射性元素发生放射性衰变这一因素，稳定同位素分馏效应也是导致自然界中稳定同位素组成发生变化的重要原因（详细介绍参见 2.2 节）。目前，稳定同位素又被分为传统稳定同位素和非传统稳定同位素。传统稳定同位素一般包含 5 种轻质量数元素：氢、碳、氮、氧、硫。对于传统稳定同位素，已经建立了成熟的同位素分析方法，并在环境地球化学领域应用了数十年。相比之下，非传统稳定同位素一般指的是传统稳定同位素之外的其他元素，例如过渡元素和重金属元素等，而这些元素的研究起步一般较晚。在过去的二十多年里，随着 MC-ICP-MS 的出现和相关分析技术的不断发展，非传统稳定同位素在地质学、生态学和环境化学等领域得到了快速发展。

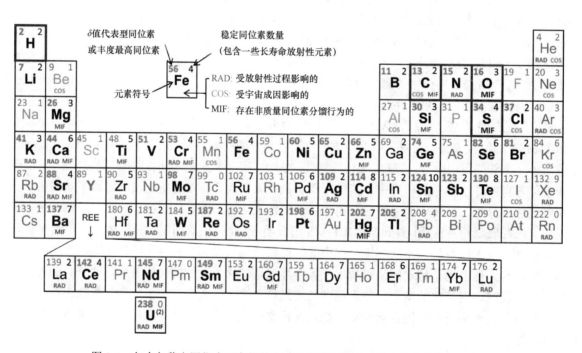

图 2-1 包含与稳定同位素研究相关内容的元素周期表（引自参考文献[202]）

2.2 稳定同位素分馏

稳定同位素分馏是指在反应过程中，不同质量数的同位素以不同比例分配到不同化合物或物相中的现象，其主要的驱动力是不同质量数的同位素之间存在微小的物理化学性质差异（热力学性质、扩散及反应速度上的差异等）。因此，同位素之间质量差别大小会影响到稳定同位素分馏的程度，如图 2-2 所示，原子序数越大的元素，通常其同位素之间质量的相对差异越小，因此普遍具有较小的同位素分馏范围。此外，稳定同位素分馏还受到相关元素地球化学行为影响，例如元素的氧化态数量、成键环境、反应活性以及气、液、固三相的存在状态等。在自然界中，稳定同位素分馏以一定的规律发生在环境地球化学的各个过程中，例如氧化还原反应、络合反应、吸附、溶解、沉淀和生物循环等，逐渐造成自然界中的不同储库具有了特定的同位素组成特征。因此，通过进行同位素分馏机理研究，可以很好地反推物质的来源和发生的地球化学过程。目前，稳定同位素分馏被视为一种高精准的示踪工具，已经被广泛应用于元素地球化学循环的研究中。

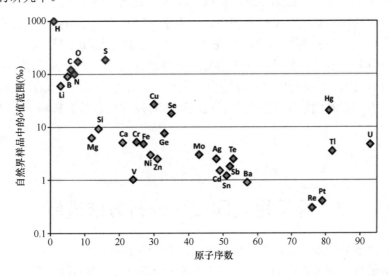

图 2-2　自然界中元素稳定同位素分布范围与原子序数的关系（引自参考文献[202]）

稳定同位素分馏可以分为质量依赖分馏（mass-dependent fractionation，MDF）和非质量依赖分馏（mass-independent fractionation，MIF），是以是否存在质量依赖效应进行区分的，其中核体积效应和磁效应是导致非质量分馏的主要原因。目前，自然界中发现的大部分同位素分馏现象都符合质量依赖分馏规律，仅有少数元素在实验室模拟（Ti、Cr、Zn、Sr 和 Mo 等）和自然界中（仅有 Hg、O 和 S）发现存在非质量分馏行为（图 2-1）。作为额外的示踪工具，Hg、O 和 S 三种元素在自然界中发生的非质量依赖分馏与稳定同位素分馏一起可以组成多维同位素示踪指纹，增加了溯源的维度和准确性。同位素分馏的程度可以通过比较反应前后物质间同位素组成变化的大小来反映，而物质的同位素组成通常是以参比于同位素标准物质的相对千分差（δ）来表示，即

$$\delta^x E = \left(\frac{\left(^xE/^yE \right)_{\text{样品}}}{\left(^xE/^yE \right)_{\text{标准物质}}} - 1 \right) \times 1000 \tag{2-1}$$

式中，E 代表某种化学元素，x 和 y 分别代表该元素两种同位素的质量数。而非质量依赖分馏的程度是以物质的同位素组成偏离质量分馏线的大小来表示，即

$$\Delta^y E = \delta^{x/y} E - \beta_{MDF} \times \delta^{y/z} E \qquad (2\text{-}2)$$

式中，Δ 表示该元素的质量依赖分馏和非质量依赖分馏之间的偏差，E 代表某种化学元素，δ 表示该元素相对于标准参考物质的同位素组成，y 代表该元素存在非质量分馏的同位素，x 和 z 代表该元素描述质量分馏线的两种同位素，β_{MDF} 代表该元素质量分馏线的斜率。

质量依赖分馏又可细分为热力学分馏和动力学分馏。热力学分馏又称为热力学平衡分馏，是指在平衡反应下，由热力学原因（熵、焓、内能等）导致互相接触的物质间出现同位素交换而产生的同位素分馏现象。发生热力学分馏时，重同位素通常更倾向于富集在"更强的成键环境"中，比如更高的氧化态、更低的配位数和更短的键长等。因此，在一定条件下达到热力学平衡时，系统中各组分的同位素组成往往出现一些规律性的变化，比如"矿物结晶序列"和"价态规律"等。在外部制约因素中，温度的影响最为显著。热力学分馏的程度随着温度的增加而减小（通常正比于 $1/T^2$），基于此，某些特定的同位素体系可以记录并反映古气候的变化，例如有孔虫的氧同位素组成。

动力学分馏是由于质量数不同的同位素具有不同的运动或反应速率造成的，通常发生在未达到浓度平衡状态的各种物理化学过程中，比如蒸发、扩散、冷凝、氧化还原反应及各种生物化学过程等。轻同位素反应速率更快，因此优先富集在反应产物中。动力学分馏程度主要由两个参数决定，分别是分馏系数和反应程度。通常，在反应的初始阶段，产物与反应物的同位素组成差异最大，但随着反应的进行，二者的同位素组成逐渐趋于一致。此外，许多动力学分馏过程符合瑞利分馏模型的特点（单向、不可逆且反应物混合均匀等），因此可以利用瑞利分馏模型来计算分馏系数或是定量反应程度。与热力学分馏不同的是，动力学分馏受多种因素（反应条件、反应速率和反应机理等）影响，所以很难进行理论模型计算。

3 环境稳定同位素分析方法进展

分析方法和相关仪器的进步是推动稳定同位素地球化学领域发展的一个至关重要的因素。质谱仪是目前测定稳定同位素最有效和最常用的方法。质谱仪利用不同质量和电荷的带电粒子（原子或分子）在磁场或电场中具有不同运动轨迹这一特性对不同的带电粒子进行分离，从而可以实现对单个同位素和同位素比值的高精度测定。早期的气体同位素质谱仪主要用于测定 C、H、O、N、S 等传统稳定同位素体系，而进入 21 世纪以来，热电离质谱仪（TIMS）、多接收电感耦合等离子体质谱仪（MC-ICP-MS）、二次离子质谱仪（SIMS）等新型同位素质谱技术的发展使元素周期表上以金属元素为代表的"非传统"稳定同位素体系能够得到精确的测定，极大地扩展了稳定同位素的环境应用，催生了非传统稳定同位素地球化学/环境地球化学等新兴研究领域。近年来，现代同位素质谱技术不断提高分析精度和质量分辨率，降低所需样品量，使同位素分析朝着精细化、微尺度和高分辨率的方向发展；并通过与色谱、激光剥蚀等技术联用，在在线同位素分析、元素不同形态的同位素分析、微区原位分析等前沿方向获得了不少进展。近期已有不少综述文章对稳定同位素的分析方法进行了详细总结[10-14]。由于篇幅限制，本节重点总结基于 MC-ICP-MS 的非传统稳定同位素分析方法的最新进展。

3.1 MC-ICP-MS 分析性能的改进

MC-ICP-MS 是目前应用最广泛的非传统稳定同位素质谱仪，与其他同位素质谱仪（如气体同位素质谱、TIMS 等）相比，具有离子化效率高、质量分辨率高、灵敏度高、分析时间短等优点，是进行高精度同位素比值测定的首选仪器[10]。但 MC-ICP-MS 的分析效果受到诸多因素的影响，例如信号灵敏度、干扰离子、仪器本身的同位素分馏效应等。近年的研究针对这些不足，从以下多个方面对 MC-ICP-MS 的分析效果进行了不少改进。

3.1.1 同位素信号灵敏度的提升

天然环境样品中非传统同位素组成的测定往往受限于元素含量。许多微量元素极低的含量及其部分同位素较低的天然丰度是其同位素组成测定的一大挑战。因此许多研究致力于如何提升同位素信号的灵敏度。其中一大进展是对 MC-ICP-MS 离子传输效率的改进。例如 Thermo Scientific 公司开发的"喷射式"离子传输界面（jet interface）使用新的样品锥和截取锥结构，配合效率更高的真空泵，显著增强了等离子体中的离子提取效率，使部分元素的灵敏度提升了两倍以上[15-18]。例如 Geng 等[17]利用喷射式离子传输界面配合冷蒸气发生系统使 Hg 同位素测定所需的 Hg 总量降低到了 0.7 ng。但喷射式界面也带来了一些新的问题。比如喷射式界面可能使仪器产生非质量分馏[19-20]，或加剧氧化物、氢化物等干扰离子的产生[21]。因此，离子传输界面的改进对高精度稳定同位素分析的利弊还有待进一步检验。

另一部分研究通过对质谱仪检测器的优化，增强了信号的灵敏度，提高了低元素含量样品的分析精度。例如在钒（V）同位素的测量中，Nielsen 等通过提高法拉第杯放大器电阻（10^{12} Ω）增强了丰度极低的 ^{50}V 的信号，使同位素分析所需的 V 总量降低了 90%，从而使更多低 V 含量样品能够进行高精度的同位素分析[13]。在铀（U）同位素测定中，Chen 等通过将较低电阻（10^{10} Ω）和较高电阻（10^{12} Ω）的法拉第杯放大器分别与高丰度的 ^{238}U 和低丰度的 ^{235}U 搭配，提高了 ^{235}U 的分析精度，同时使高浓度的 ^{238}U 能够直接测定，从而将 δ^{238}U/^{235}U 的分析精度提高到了 ± 0.02‰（2SD）[22]。类似的方法也被应用于 Pt、W 和 Hf 同位素测定[23]。

3.1.2 干扰离子的排除

干扰离子也是影响同位素组成高精度测定的一个主要因素。干扰离子是指与待测同位素质量数相同的多原子干扰（polyatomic interference）或同质异位素干扰（isobaric interference）。其信号往往与待测同位素信号部分重叠，严重影响同位素的准确测定。其中同质异位素干扰往往产生于样品的基质，可以通过样品的高度纯化来去除。而多原子干扰则往往产生于仪器分析过程中，例如 Ar、O、N、H 等来自仪器所使用的气体中的元素所形成的干扰离子（如 ^{40}Ar$^+$、^{40}Ar^{16}O$^+$ 等）。这些基于 Ar 的干扰离子对 K、Ca、Fe 等质量与 Ar 相近的元素同位素的影响尤为巨大。近年来的研究通过提高 MC-ICP-MS 的质量分辨率、加入碰撞反应池和冷等离子体等方法在干扰信号的分离或抑制方面获得了不少进展[24-27]。例如 Li 等[26]在钙同位素的分析中采用了近年来新开发的大型高分辨率 MC-ICP-MS（Nu Plasma 1700）。Nu Plasma 1700 采用了更大体积的磁铁和静电分析器（ESA），与其他主流 MC-ICP-MS 相比，在不牺牲灵敏度的情况下能达到更高的质量分辨率。Li 等[26]利用 Nu Plasma 1700 最大程度地分离了 ^{42}Ca、^{43}Ca 和 ^{44}Ca 与干扰离子信号，同时通过抑制 ^{40}Ar$^+$ 和 ^{40}Ca^{2+} 的散射干扰，使 δ^{44}Ca/^{42}Ca 分析精度提高到 < ± 0.07‰（2SD）。类似的方法也应用到了 Fe 同位素的分析中，较好地分离了干扰离子的信号，提高了分析精度[28-29]。但由于这种大型高分辨率 MC-ICP-MS 高昂的价格，其应用还较为局限。

在常规 MC-ICP-MS 中加入碰撞反应池单元（collision cell）是排除干扰离子的一个的新方法。碰撞反应池技术在质谱仪中应用广泛，通过引入 He 等特定气体与干扰离子发生碰撞或反应，能大幅度减少干扰离子的影响，是进行元素含量精确测定的首选方法。但在 MC-ICP-MS 中加入碰撞池进行同位素组成测定的研究还比较少。目前大部分的研究集中在钾（K）等受 ^{40}Ar$^+$ 及其衍生离子干扰影响强烈的同位素体系。例如 Wang 和 Jacobsen[30]利用 Ar 和 H$_2$ 作为反应气体引入碰撞反应池，使 ^{40}Ar$^+$ 和 ^{40}ArH$^+$ 等干扰信号降低到几个毫伏以下，从而可以在低分辨率模式下对低钾含量样品进行同位素分析。Li 等[25]采用 D$_2$ 替代

H_2 作为反应气体，进一步避免了碰撞反应池中 ArH^+ 的产生，使 $\delta^{41}K/^{39}K$ 的分析精度达到 < 0.2‰（2SD），足够区分环境样品中 >1‰ 的 $\delta^{41}K/^{39}K$ 变化。尽管碰撞反应池技术与 MC-ICP-MS 的联用显示出了较大的应用潜力，但该方法需要对现有的 MC-ICP-MS 硬件升级或者开发新的 MC-ICP-MS，而且该技术的稳定性和潜在负面影响（例如碰撞反应池中的同位素分馏）还需进一步验证。

冷等离子体法通过降低 ICP 的电离功率，抑制 Ar 的电离，从而减少基于 Ar 的干扰离子。这一方法也被用于提高 K、Ca、Fe 等同位素的分析精度[24, 27, 31-32]。但通常情况下降低 ICP 的电离效率也会降低待测同位素的信号，并且这一方法往往不能完全抑制干扰离子的产生，因此需要配合中、高分辨率模式使用，进一步影响信号强度。因此冷等离子体法更适合于待测元素含量较高的样品，或者搭配其他增强信号灵敏度的方法使用。

3.1.3 仪器同位素分馏效应的校正

与 TIMS 相比，MC-ICP-MS 会产生更大的仪器同位素分馏效应，因此对其进行有效的校正是开展高精度稳定同位素分析的关键之一。以往的研究认为仪器分馏效应主要产生于离子从等离子体中的提取和传输过程中的质量歧视效应，因此其校正方法主要针对质量分馏。然而近年来越来越多的研究发现，MC-ICP-MS 的仪器分馏效应不仅仅局限于质量分馏，非质量分馏（MIF）已在多种稳定同位素体系的测定中被观测到，例如 Nd、Sr、Ge、Pb、Hg、Si、Hf、Ba、Os 等[18, 33-37]。目前仪器 MIF 效应的产生原理还不清楚，但它无疑对校正方法提出了新的要求。目前主流的校正方法包括：内标法结合简单校正模型（如指数定律）、双稀释剂法、标准-样品间插法、标准-样品间插法与内标法或双稀释剂法的联用、线性回归法、同位素重量混合法。其中前两者仅能校正质量分馏，而后四种方法则能在不同程度上校正 MIF 效应。Yang 等在近期综述文章中对这 6 种校正方法的原理和各自的适用范围进行了详细介绍[20]。限于篇幅，本节不在此一一列出。

3.2 MC-ICP-MS 与各类进样系统的联用

除了 MC-ICP-MS 本身分析效果的改进，MC-ICP-MS 还常常与气相色谱、液相色谱、冷蒸气发生装置、激光剥蚀系统等各种进样系统联用来扩展其功能，开发了在线同位素分析、形态同位素分析、微区原位同位素分析等新方法。

3.2.1 在线同位素分析

在线同位素分析（online isotopic analysis）方法是指待测元素从样品基质中的分离、提纯、富集等前处理步骤在 MC-ICP-MS 的进样部分直接完成，与同位素分析步骤实时相连。与传统的离线前处理相比，在线方法往往具有省时省力、所需样品量更少、流程化程度更高等优点。例如 Yang 等[33]将 HPLC 与 MC-ICP-MS 串联测量 Si 稳定同位素，利用离子排斥色谱法对海水与河水中的 Si 进行了在线分离和实时同位素测定。该方法简化了天然水样中 Si 同位素测定的前处理流程，缩短了测量时间，达到了接近于离线方法的分析精度（$\delta^{30}Si$ 精度为 0.12‰，2SD）。在线同位素分析方法也被用于测定核燃料与核废料中的同位素，以降低样品前处理过程中的核污染和人体暴露风险。例如 Guéguen 等和 Martelat 等[38-40]利用 HPLC 和毛细管电泳法与 MC-ICP-MS 联用，对核燃料裂变产生的 U、Pu、Nd、Gd、Eu、Sm 等多种元素进行了在线分离并测定了这些元素的同位素比值。此外，在线方法对低浓度样品的同位素分析也具有优势。例如 Bérail 等[41]将汞同位素测定常用的在线冷蒸气发生系统与用于汞蒸气富集的双金汞齐系统结合，并与 MC-ICP-MS 联用，使汞浓度低至 5 ng/L 的样品（接近天然水体样品的汞浓度）也能够直接进行同位素测定。该方法在 5～50 ng/L 浓度范围的 $\delta^{202}Hg$ 分析精度为 0.29‰（2SD）。

目前在线同位素分析方法的主要挑战是该方法产生的同位素信号为瞬时信号,在计数统计、信号稳定性和分析精度方面都远远低于连续信号。并且瞬时信号产生过程中同位素比值会随时间变化,因此其同位素比值的计算需要采用特殊的数据处理方法,其中最常用的方法为峰面积积分法和线性回归斜率法[39, 42]。

3.2.2 形态同位素分析

形态同位素分析（compound specific isotopic analysis，CSIA）是通过液相或气相色谱对样品中同一元素的不同化学形态进行在线或离线分离和同位素测定的方法。CSIA 方法在 C、N 等传统稳定同位素和 Cl、Br、Hg 等非传统稳定同位素分析中都有广泛的应用。以汞同位素为例,在线 CSIA 方法主要通过气相色谱（GC）在线分离无机汞和甲基汞,从而分析这两种主要汞形态的同位素组成[43]。近年来的研究不断改进分离方法,提高汞 CSIA 的分析精度,降低所需样品浓度,使其更好地应用到低汞含量的天然样品。例如以往的在线 CSIA 方法中,由于注入 GC 的样品量有限（1～2 μL）,因此往往只能用于汞浓度较高的环境样品或实验样品。Bouchet 等[44]在 GC 之前加入了温控气化注射装置（programmed temperature vaporization injector，PTV）,该装置可使样品的注入体积提高达 100 倍（相对于直接注入）,从而实现低浓度样品中无机汞和甲基汞的在线富集,使在线 CSIA 方法能够用于测定形态汞含量 > 150 ng/g 的环境样品。该方法的分析精度为 0.19‰ 和 0.39‰（$\Delta^{199}Hg$ 和 $\delta^{202}Hg$，2SD）。Queipo-Abad 等[45]在最新的研究中详细报道了 GC-MC-ICP-MS 方法中各种参数（如峰宽度、积分时间、数据点数等）对汞 CSIA 的影响,并在不影响分析精度的前提下缩小了峰宽度（2～5 s）,有助于提高在线 CSIA 的分析效率。

除了在线 CSIA 以外,部分研究也开发了针对不同类型的环境样品的离线 CSIA 方法[46-49],相比于在线方法,其优势在于更灵活的样品处理和富集,以及更好的分析精度。除了 GC 以外,液相色谱（LC）也被用于分离形态汞,但目前还只用于离线 CSIA 分析[50]。与 GC 相比,LC 的主要优势在于可以对液体样品中的无机汞和甲基汞进行直接分离,不需要提前转化为气体形态,从而可以避免转化过程中的损失。

3.2.3 微区原位同位素分析

微区原位同位素分析（*in-situ* isotopic microanalysis）是同位素地球化学研究的一个前沿领域。其主要原理是利用激光剥蚀系统（laser ablation）与 MC-ICP-MS 联用（LA-MC-ICP-MS）或利用 SIMS 等技术,通过对环境样品不同部位进行原位分析,测定微米甚至纳米尺度上的同位素组成变化,从而可以揭示整体样品分析难以反映的精细环境地球化学过程。近年来微区原位技术被广泛应用于多种同位素体系和不同环境样品,尤其是 LA-MC-ICP-MS 技术获得不少进展之后。其中对原位分析方法促进最大的当属飞秒激光剥蚀系统（fs-LA）的发展。与传统的纳秒激光剥蚀系统（ns-LA）相比,fs-LA 在消除剥蚀过程的热效应、提高样品剥蚀效率和降低颗粒物粒径方面有显著进步[51-52]。例如 Zheng 等[53]详细比较了 fs-LA 与 ns-LA 对一些半导体材料的剥蚀效果（如产生的颗粒物形态与粒径分布）,以及对铁同位素分析的影响。发现 fs-LA 的剥蚀效果受到样品基质和激光剂量的影响更小,产生的颗粒粒径更均一,因此颗粒物传输效率和信号灵敏度更高,不同粒径的铁同位素组成变化也更小,更适合于高精度的原位铁同位素分析。fs-LA 也被用于 S、Si、Pb、Cu 等同位素体系的原位分析[54-58],与 ns-LA 相比能普遍提供更好的信号灵敏度、同位素比值精度更接近于溶液方法的同位素比值。

除了 fs-LA 技术以外,LA-MC-ICP-MS 技术在其他方面也获得了一些进展。例如部分研究针对 Pb 同位素的原位分析过程中 ^{204}Hg 的干扰,通过在 LA-MC-ICP-MS 系统中加入镀金除汞装置或基于多孔硅质膜的气体交换设备等,将气态汞从背景气体或激光剥蚀产生的气溶胶中移除,使 LA-MC-ICP-MS 能够直接分析富汞硫化物中的 Pb 同位素组成[59-60]。另外,一些研究使用分叉装置将激光剥蚀产生的气溶胶同时导入 MC-ICP-MS 和 Quadrupole-ICP-MS,实现了对同一个剥蚀点同时进行元素含量和同位素比值的测定,从而能更完整地反映样品原位的化学信息[61-62]。

由于 LA-MC-ICP-MS 技术也是一种在线同位素分析方法，因此它也有在线方法的普遍局限性。此外，激光剥蚀过程和等离子体离子化过程中同位素分馏效应的准确校正和基体匹配的标准物质缺乏等问题也是目前微区原位高精度同位素分析的主要瓶颈。

3.3 新同位素体系分析方法的开发

近年来基于 MC-ICP-MS 的分析技术的进展也促进了大量新的非传统同位素体系的开发，例如钒(V)、钡（Ba）、钾（K）、铂（Pt）、钯（Pd）、银（Ag）、铈（Ce）、铒（Er）、镓（Ga）等。这些新同位素体系与 Ca、Mg、Sr、Fe、Pb、Hg 等关注较多的非传统同位素体系相比，其分馏机理和环境应用的研究还处于起步阶段，但近年来在分析方法方面获得了不少进展。例如 Ba 同位素首先由 Von Allmen 等开发了基于 MC-ICP-MS 的高精度分析方法（2SD = 0.15 ‰）[63]。随后 Miyazaki 等和 Nan 等分别利用双稀释剂法和标准-样品间插法，通过对 ^{134}Xe 等干扰离子的仔细校正和样品与标准的高度匹配，使 $\delta^{137/134}$Ba 分析精度达到了 2SD < 0.05‰[37, 64]。又比如 Yuan 等率先开发了地质和生物样品中 Ga 的分离提纯方法以及基于 MC-ICP-MS 的 Ga 同位素高精度分析方法（2SD = 0.05 ‰），奠定了 Ga 同位素的应用基础[65]。

Lu 等在近期的综述中对上述新同位素体系的分析方法和相关应用有着较为系统的总结[12]。另外，本章 4.2 节也对部分新同位素体系进行了详细介绍。因此限于篇幅，本节对上述新同位素体系的分析方法不再一一详述。

3.4 小结

除以上进展以外，不少研究者也对同位素分析或前处理的自动化、非质谱的同位素分析方法等进行了探索[66-68]。尽管这些方面的研究还处于起步阶段，但它们揭示了环境稳定同位素分析方法的未来发展前景。总的来说，稳定同位素分析方法的发展趋势是更高精度、更微尺度、更高效率、更自动化和流程化，以及更多新同位素体系的开发。随着稳定同位素在环境与地球科学研究中的应用越来越广泛，也对其分析测试方法提出了新的挑战；而分析方法的进步，也将进一步激发新的应用研究。因此，分析方法的研究将永远是稳定同位素环境地球化学领域的前沿。

4 稳定同位素技术在污染物示踪中的应用进展

自然界中的生物地球化学循环过程通常包含复杂的物理化学反应，可以导致同位素分馏效应，从而改变物质的同位素组成。因此，基于物质的同位素组成和同位素分馏机理可以反推物质的来源和经历的某些环境过程。研究环境污染物的来源和迁移转化过程是判断其环境危害和进行污染防治的重要前提，而稳定同位素技术在来源示踪和过程示踪这两方面具有独特的优势，因此在环境污染物示踪研究中得到快速发展。本节主要介绍了稳定同位素技术在大气细颗粒物、环境重金属和纳米材料示踪研究中的典型应用。

4.1 稳定同位素在大气细颗粒示踪中的应用

1993 年，Walder 等[9]首次报道了基于 MC-ICP-MS 的同位素分析方法，揭开了非传统稳定同位素分析的新篇章。经过二十多年的发展,更多的非传统稳定同位素可以被精准测量,例如 Hg、Pb、Sr、Cu、Fe、Zn、Si、Nd、I 等。此外,得益于分离纯化技术的不断发展,逐渐克服了各种实际环境体系中复杂基质的干扰,使这些同位素体系在大气颗粒物溯源研究中得到广泛应用。其中,Hg、Pb 和 Sr 本身就是典型环境重金属污染物,它们在大气中的同位素溯源研究最先引起人们的关注,并得到迅速发展[69-77]。近些年,作为大气颗粒物中的高丰度元素,Cu、Fe 和 Zn 在大气颗粒物中的同位素研究也逐渐进入人们的视野,并获得了一些代表性研究成果[78-81]。而 Si、Nd 和 I 等元素在大气颗粒物溯源研究中应用最晚,仅有少量报道[82-85]。下面以 Hg、Cu 和 Si 作为这三个不同发展阶段的典型代表元素,简要介绍它们在大气颗粒物溯源方面的研究进展。

目前,Hg 同位素已成为示踪大气 Hg 污染来源和转化过程的重要手段。在自然界中,Hg 有 7 种稳定同位素:^{196}Hg(0.15%)、^{198}Hg(9.97%)、^{199}Hg(16.87%)、^{200}Hg(23.10%)、^{201}Hg(13.18%)、^{202}Hg(29.86%) 和 ^{204}Hg(6.87%),且同时存在两种分馏形式(MDF 和 MIF)。大气 Hg 主要有三种存在形式:气态单质汞(GEM)、活性气态汞(GOM)和颗粒态汞(PBM),它们在大气环境中可以通过氧化还原反应相互转化,同时伴随显著的同位素分馏效应。目前,GEM 和 GOM 的同位素组成和变化规律已经被广泛报道[74-76],而大气中 PBM 的同位素研究直到最近才逐渐引起人们的关注[86]。大气中 PBM 主要有两个来源:一是一次污染源的直接排放,二是经过 GEM 和 GOM 的吸附或转化而来,这两个来源都会影响 PBM 的同位素组成(δ^{202}Hg、Δ^{199}Hg 和 Δ^{200}Hg 分别代表汞元素的质量分馏、奇数非质量分馏和偶数非质量分馏)。其中,燃煤排放源作为重要的一次污染源,贡献了接近 40% 的人为源总 Hg 排放量,其烟气中的 Hg 同位素组成最先引起了人们的关注。Sun 等[87-89]和 Huang 等[90]发现燃煤燃烧排放过程可以导致显著的 Hg 同位素质量依赖分馏效应,进而影响大气 Hg 同位素的组成。具体地,燃煤燃烧释放的 GEM 在静电除尘器等大气污染防治装置中可以与 GOM 和 PBM 发生相互转化,进而导致显著的同位素分馏效应,最终造成烟气中 PBM 具有最小的 δ^{202}Hg 值。进一步,Das 等[91]发现在印度加尔各答省多个大气污染源附近(汽车尾气源、垃圾焚烧源和工业排放源)采集的 PM_{10} 中 Hg 同位素组成存在显著差异,反映了不同人为排放源可能具有不同的 Hg 同位素指纹特征。Huang 等[92]测量了北京地区一年四季 $PM_{2.5}$ 样品以及 30 个潜在污染源样品中的 Hg 同位素组成,通过分析 $PM_{2.5}$ 样品中 Hg 同位素组成的季节性变化特征,推测出 $PM_{2.5}$ 中 Hg 污染在冬季和夏季分别受到燃煤燃烧源和生物质燃烧源的显著影响,而春季和初夏更有可能受到长距离迁移的影响,从而证明了 Hg 同位素可以作为大气颗粒物中 Hg 污染溯源研究的有力工具。Fu 等[93]进一步将 Hg 同位素应用于大气 Hg 污染的区域性溯源研究,结果显示我国东北及东部地区的 PBM 污染主要来源于区域性人为排放,而西南和西北地区则主要受到来自南亚地区 PBM 的长距离迁移影响。需要特别指出的是,GEM 和 GOM 的吸附和转化过程也会改变 PBM 的同位素组成,进而增加 PBM 污染溯源的复杂性。Huang 等[94]发现北京地区 $PM_{2.5}$ 中 Hg 同位素组成存在较大的昼夜变化,揭示了大气光化学反应可以显著改变 $PM_{2.5}$ 中 Hg 元素的丰度和同位素组成。因此,在应用 Hg 同位素示踪大气 Hg 污染来源时,传输过程中光化学反应引起的 Hg 同位素组成变化不容忽视。综上,大气中不同形态 Hg 元素的同位素组成已经开展了较为全面的研究,且大气 Hg 污染源谱同位素组成和传输过程导致的 Hg 同位素分馏效应也有报道,由此可以看出 Hg 同位素在大气 Hg 污染溯源研究中具有很好的应用前景。

Cu 是大气颗粒物中高丰度重金属污染元素,其同位素组成在大气颗粒物示踪研究中逐渐引起人们关注。Cu 同位素有两种,分别是 ^{63}Cu(69.17%)和 ^{65}Cu(30.83%),它们在示踪大气颗粒物中 Cu 元素的来源方面具有了一定的优势。Gonzalez 等[78]发现巴塞罗那街道的大气细颗粒物中 Cu 同位素比值(δ^{65}Cu)在 +0.04‰±0.20‰ 和 +0.33‰±0.15‰ 之间,与汽车刹车片中 Cu 同位素组成一致,说明道路附近的大气颗粒物中 Cu 可能主要来自于汽车刹车导致的机械磨损,而非尾气排放。之后,Dong 等[95]采集了伦敦市不同道路站点的 PM_{10} 样品,经分析发现 PM_{10} 中 Cu 同位素组成与多种非尾气源(轮胎、刹车和道路粉尘)的同位素组成一致,进一步验证了城市环境大气颗粒物中 Cu 主要受到非尾气排放源的影响。进一步,Cu

同位素与其他多种元素同位素一起可以构建多维同位素示踪模型,定量估算各污染源对大气颗粒物的贡献大小。例如Oliveira等[79]基于Pb、Cu和Zn等多同位素示踪体系定量估算了巴西圣保罗市白天和夜间各个大气颗粒物一次污染源的贡献比重。综上,Cu同位素在大气颗粒物示踪研究中具有一定的应用潜力,下一步需要继续丰富污染源谱同位素数据,加强对各个一次排放源Cu同位素组成特征的研究。

Si广泛存在于大气颗粒物中,既有天然来源(例如土壤尘、建筑尘和扬尘中的矿物颗粒),也有人为排放(例如秸秆和燃煤燃烧后排放的无机颗粒),因此Si同位素适合作为大气颗粒物的示踪剂。Lu等[85]首次将Si稳定同位素应用于大气颗粒物的溯源研究中,发现大气颗粒物的不同一次排放源具有不同的Si同位素组成,且Si同位素在大气传输过程中不易发生同位素分馏(由于Si的高化学惰性),因此大气颗粒物中Si同位素的变化可以直接反映一次排放源的变化。此外,通过进一步分析北京地区2003年和2013年$PM_{2.5}$的浓度和其中Si同位素组成发现,相对于夏秋两季,春冬两季灰霾频发,且$PM_{2.5}$中显著富集轻同位素,结合污染源谱同位素数据推测春冬季节北京灰霾频发的主要原因可能是燃煤排放源的激增。此外,Lu等[96]进一步结合Si的化学惰性性质,建立了基于Si稀释效应定量评估二次气溶胶贡献比重的方法。综上,尽管Si同位素被证明可以用于示踪大气颗粒物的来源,但该方面的研究仍处于起步阶段。

稳定同位素在大气细颗粒物溯源研究中已经展示了广阔的应用前景。需要特别指出的是,大气颗粒物组分非常复杂(大量的有机和无机元素),容易导致严重的基质效应,这一难题已成为稳定同位素在大气细颗粒物溯源研究中应用的主要限制因素,因此还需要进一步加强相关同位素分析方法的开发与优化。此外,大气颗粒物中多同位素示踪体系的研究可以提供更加丰富的来源信息,亟待进一步发展。

4.2 稳定同位素在环境中金属示踪中的应用

随着金属同位素分馏理论的发展,作为一种新兴的地球化学工具,金属稳定同位素在过去的十几年中已经从天体化学、岩石地球化学等研究逐步拓展到与人类活动息息相关的环境地球化学领域。特别需要指出的是,环境中的重金属污染物不能够被降解去除,能够长久地存在于环境中,亟须运用新兴的金属稳定同位素这一强有力工具示踪重金属污染物质的源汇关系及迁移转化过程。

污染源谱同位素数据库的调查是稳定同位素示踪技术的重要前提。以锌元素为例,已经对自然地质储库和各类人为污染源中的锌同位素组成已经开展了广泛研究[97]。一般而言,沼泽中锌同位素组成最大(1.7‰±0.17‰)[98],闪锌矿中锌同位素组成最小(−0.804‰±0.015‰)[99],火成岩及其风化产物(土壤、沉积物和沉积岩等)的锌同位素组成较为均一。需要指出的是,某些来源具有较大的锌同位素变化范围。如在生命活动较强沼泽系统中,锌同位素组成($\delta^{66}Zn$)在沼泽剖面上的变化范围达到0.77‰[98];同样地,在表层海水(< 500 m)中,生物和有机物的协同作用可以造成较大的锌同位素分馏效应,进而增大锌同位素组成的分布范围(−1.1‰~0.9‰)[100]。对于典型的人为排放源而言,Pb-Zn矿冶炼产生的粉尘显著富集轻同位素,$\delta^{66}Zn$为−0.36‰[101]。综上,这些不同来源(闪锌矿、火成岩和铅锌矿冶炼等)具有不同的锌同位素组成,为基于锌同位素的溯源研究提供了重要依据。

此外,污染元素在从源到汇的过程中容易发生稳定同位素分馏效应,进而丢失污染源的同位素信息,影响溯源准确性。因此,研究从源到汇过程中伴生的稳定同位素分馏效应是稳定同位素示踪技术的重要保障。以锌元素为例,其同位素组成在自然水体中极易受到固相界面吸附的影响,因而引起广泛关注[101-107]。Pokrovsky等[107]首先对锌同位素在金属(氢)氧化物/溶液界面上的吸附分馏行为进行了研究,结果表明不同矿物的界面吸附效应可以造成不同的锌同位素分馏效应:赤铁矿、软锰矿、三水铝石和刚玉等矿物的界面倾向于吸附重的锌同位素,而针铁矿和水钠锰矿的界面倾向于吸附轻的锌同位素。不同的是,Bryan等[102]发现水钠锰矿界面倾向于吸附重的锌同位素,且分馏程度可以达到3‰。进一步,Gou等[104]将稳定同位素技术与扩展X射线吸收精细结构(EXAFS)光谱相结合,证明了锌元素在界面和液相中的局域结构（配位数和键长）差异是导致稳定同位素分馏效应的主要因素。具体来说，在 pH 较低或初始 Zn 浓度较低的情况下，锌原子以内圈络合的形态被氧化铝界面吸附，此时锌原子占据四面体的中心位置，和周围四个氧原子的距离为 1.97 Å，造成界面吸附相富集重的锌同位素；而在 pH 较高或初始 Zn 浓度较高的情况下，锌原子以表面沉淀的形态被氧化铝界面吸附，此时锌原子占据八面体的中心位置，和周围六个氧原子的距离拉长为 2.06 Å，缩小了与溶液相的差距，因此并未造成显著的同位素分馏效应。图 2-3 展示了键长差异程度与吸附分馏程度之间的关系，总体而言，吸附相和溶液相键长差异越大，同位素分馏程度越大[104]。

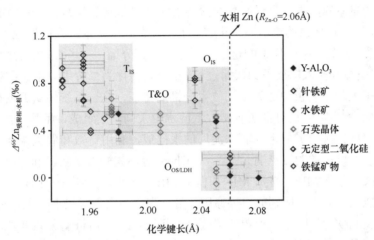

图 2-3　锌在不同矿物表面吸附产生的同位素分馏值与吸附产物 Zn-O 化学键长的关系图（引自参考文献[104]）

T_{IS} 表示四配位的内圈表面配合物，O_{IS} 表示六配位的内圈外面配合物，$O_{OS/LDH}$ 表示六配位的外圈表面配合物，T&O 表示六配位与四配位的混合

金属稳定同位素在污染物源解析方面具有事半功倍的强大功用。当前，金属稳定同位素已经逐渐应用在典型重金属污染的示踪方面。例如，①利用铊同位素示踪燃煤电厂排放[108]、水泥厂的粉尘排放[109]、对于周边土壤的污染[110]。②利用镉同位素示踪矿山废水对流域水体的影响[111]；发现铅锌冶炼厂排放烟尘和矿渣对土壤镉污染的影响，其中烟尘的贡献达 60%以上[112]；发现我国北江的三大镉来源为冶炼厂烟尘、冶炼厂炉渣以及当地背景和采矿活动[113]。③利用汞同位素示踪工业活动对其下游的土壤和沉积物的影响[114]，火电厂燃煤排放对大气干湿沉降的影响[115]，金属冶炼厂对湖泊沉积物的污染[116]，评估废弃汞矿周围生态环境危害[117]等。这些研究均表明金属同位素是示踪重金属污染的强有力工具。

金属同位素不仅能够应用在上述污染严重的点源研究中，随着分析方法及分析技术的提升，还能够解析人为干扰活动下流域内重金属贡献源及其迁移转化过程。Chen 等[118-121]率先在国际上开展了河流重金属的源解析工作，发现塞纳河流域水中锌同位素（$\delta^{66}Zn$）组成的变化范围较大（达 1‰，误差<0.04‰），且人为源和自然源具有不同的锌同位素组成；藉此，发现河水中的锌具有类似于 Na 或 Cl 一样的可保存性，奠定了锌同位素示踪污染源的理论基础；确定了塞纳河中锌的三个主要贡献端元（自然源即碳酸盐风化、偏远地区的家庭废水及城市污水）及其贡献比例；这些结论与法国十多个团队近三十年的研究结果一致，证明了金属稳定同位素是水体重金属污染源解析研究的强有力工具[120-121]。Chen 等[122]还结合溶解态锌同位素研究结果，发现塞纳河输送到大西洋的锌 90%以上来自于人类活动。这一系统性工作是国际上河流重金属同位素示踪研究的首个范例。Chen 等还率先利用多同位素体系（如锌和铁）给出了塞纳河悬浮颗粒物中锌和铁的同位素的指纹图谱，厘定了其污染源；并结合溶解态的铁同位素组成，判定了污染河流向海洋输入铁的同位素组成特征，证明海水中重的铁同位素的富集可能主要源自海洋内部生物过程。这项工作填补了受人类活动干扰的河流铁同位素研究为零的空白。后来，在另一项 Zn 同位素示源

工作中，Juillot 等[123]发现位于法国北部一个废弃 Zn 冶炼工厂附近的两个土壤剖面的 Zn 同位素组成从底部的 0.22‰±0.17‰～0.34‰±0.17‰到顶部逐渐增加到 0.76‰±0.14‰，说明顶部土壤受到了该工程排放的锌铁矿炉渣的影响（该炉渣的 Zn 同位素组成约 0.81‰）。

上述研究中主要是应用金属稳定同位素的质量分馏体系对其污染源和迁移转化过程进行有效示踪，除此之外，还存在金属稳定同位素的非质量分馏，如汞等多种金属同位素[124]，为环境中重金属的污染示踪增加了新的纬度。作为唯一具有国际公约的重点防控重金属，汞同位素是目前元素周期表中除氧和硫外唯一在自然界中具有明显同位素非质量分馏的重金属，且主要为奇数汞同位素非质量分馏（odd-MIF）[125]，此外还在与大气相关的自然样品中观察到明显的偶数汞同位素非质量分馏（even-MIF）[126-128]，使汞成为唯一具有"三维"同位素示踪体系的重金属，并展示了在金属同位素示踪领域的强大应用潜力。利用"三维"汞同位素分馏体系，尤其是汞同位素的奇数和偶数同位素分馏，一些国际学者团队如 Enrico、Obrist、Jiskra 和 Olson 发现正是植物摄入单质汞的代谢过程（单质汞转化为二价汞）引起了汞同位素的质量分馏，而植物本身仍然保留了大气单质汞中奇数汞同位素分馏的特征，这一系统性研究明确了大气单质汞干沉降进入植物体内的过程[75, 129-131]；通过分析我国大气降水中汞同位素组成，尤其是奇数同位素特征，解析了我国大气降水汞的主要来源为燃煤电厂和水泥厂的汞排放，利用奇数和偶数汞同位素特征发现长距离传输汞亦是大气降水汞不可忽视的贡献端元[132-135]。Chen 等率先开展了我国和加拿大湖泊汞同位素对比研究，发现我国湖泊汞同位素 MIF（奇数 MIF 和偶数 MIF）与加拿大明显不同，证明流域湿地光化学作用下甲基汞的输入可能是造成这一区别的主要原因，可能为我国高汞背景下野生鱼体汞（主要是甲基汞）为什么普遍低于北美及北欧背景区这一难题的解决提供新的思路。Yuan 等发现单次降雨过程中存在较大的汞同位素分馏，尤其是奇数汞同位素非质量分馏，降水中汞同位素组成随气象条件变化而变化，证明了汞同位素在气象及气候变化研究中的潜在应用价值[133, 135]。Zdanowicz 及其合作者分析了北极地区冰芯中汞浓度、形态和同位素组成，还能够评估北极地区大气汞的保存、积累和再释放通量，并重建该地区第四纪以来，特别是工业革命以来大气汞同位素变化规律及其沉降、污染历史，为全面认知北极圈内汞的循环及其对北极生态系统的危害提供了宝贵数据[136]。多位国际学者通过分析历史时期的地质样品，尤其是比较样品的奇数汞同位素分馏数据（$\Delta^{199}Hg$ 或 $\Delta^{201}Hg$），发现了三次生物大灭绝事件（白垩纪末恐龙大灭绝、二叠纪末大灭绝和奥陶纪末大灭绝）中火山诱因说的汞同位素新证据，展示了汞同位素在重建古气候或古环境方面的巨大应用潜力[137-139]。

除此之外，元素周期表中还有少数尚未被开发且极难开发的金属稳定同位素体系，如镓同位素。2016 年几乎同时发表的 3 篇关于地质样品中 Ga 同位素成的高精度分析方法文章[65, 140-142]，开辟了地球化学一个新的同位素研究体系，揭开了 Ga 同位素研究新篇章。初步结果显示 Ga 的同位素组成变化范围可达 1.83‰，有意思的是工业生产的标准物质与地质标准具有完全不同的 Ga 同位素组成，这预示着 Ga 同位素可对自然体系 Ga 进行有效源解析[65]。随后，Yuan 等在国际上首次开展镓在矿物表面吸附过程中同位素分馏研究，阐明了矿物表面吸附过程中镓同位素分馏机制，说明地表广泛存在的低温过程可能会产生明显的 Ga 同位素分馏，进一步预示 Ga 同位素可能是生物地球化学过程的有效示踪剂[143]。

金属稳定同位素不但可用来对金属进行准确的源解析，还由于其对环境条件改变极为敏感的特性，可作为追踪各种生物过程的有效指示剂。如将其应用到生命医学领域，可为准确诊断和预判疾病提供新视野。例如，Ca 稳定同位素可能成为骨骼疾病（如骨质疏松症）的直接诊断工具[144]；Fe 稳定同位素可能是遗传性疾病（如遗传性血色素沉着症）的有效示踪剂[145]；Cu 和 Zn 稳定同位素更可成为诊断潜在癌症（如乳腺癌）的重要标志物[146]。此外，即使在应用单一金属稳定同位素无法准确判定时，还可利用多个金属稳定同位素体系进行有效的联合示踪，可见，金属稳定同位素在生态环境健康效应方面具有潜在的应用价值。总之，新的分馏体系、分馏理论和分析方法的开发将极大地拓宽金属稳定同位素在环境方面的应用前景。

4.3 稳定同位素在环境纳米技术中的应用

目前,纳米材料使用量巨大,已广泛应用于食品、医药、化妆品和涂料等领域。但是,在提高人们生活质量的同时,纳米材料对人体健康和生态环境的潜在危害逐渐引起广泛关注。准确示踪环境中纳米颗粒物的来源和迁移转化过程是进行纳米材料毒理学评价的前提,然而由于受到传统分析方法的限制,纳米材料在自然界中的来源与归趋依然未能清晰阐明。稳定同位素技术主要是基于物质的同位素组成示踪物质的来源和经历的某些环境过程,在来源示踪和过程示踪方面具有独特的优势。在环境纳米技术领域,纳米材料的自然转化和工业合成过程通常包含复杂的物理化学反应,极易造成同位素分馏效应,进而改变纳米颗粒的同位素组成,基于此,可以准确示踪环境中纳米颗粒的来源和迁移转化过程。因此,稳定同位素技术在环境纳米技术领域具有广阔的应用前景。

近期,已有少量研究探索了稳定同位素技术在纳米颗粒溯源研究中的应用。Larner 等[147]和 Laycock 等[148]分别测量了工程 ZnO 和 CeO_2 纳米颗粒中 Zn 和 Ce 的同位素组成,发现它们的同位素组成与天然物质相比并无显著差异,因而判断稳定同位素技术无法识别环境中 ZnO 和 CeO_2 纳米颗粒的来源。Yang 等[149]测量了几种商业用品中银稳定同位素的组成,发现硫代癸烷修饰的纳米银、膳食补充剂和丝袜中的银同位素组成存在明显差异;进一步,Lu 等[150]发现人为源和天然源纳米银在转化过程中具有不同的银同位素分馏效应,为识别纳米银的来源提供了可能性;这些研究结果展示了稳定同位素技术在纳米银溯源研究中具有潜在的应用价值。值得一提的是,Yang 等[151]首次开展了基于多同位素示踪体系(硅和氧)的纳米颗粒溯源研究工作,发现人为源和天然源 SiO_2 纳米颗粒具有可分辨的硅/氧二维同位素指纹特征,并进一步结合机器学习模型(LDA 模型)给出了人为源和天然源 SiO_2 纳米颗粒的定量判别结果,准确率达到了 93.3%;此外,硅/氧二维同位素指纹还可以在一定程度上识别工程 SiO_2 纳米颗粒的合成方法和厂家来源,进一步证明了稳定同位素在环境纳米技术领域的强大应用潜力。

此外,稳定同位素技术还可以示踪纳米颗粒物的自然转化过程。第一,天然纳米颗粒物的生成过程可以导致显著的同位素分馏效应。例如在自然水体中,溶解性 Fe^{2+} 离子可以生成硫化铁类纳米颗粒(FeS_m),其生长过程伴随显著的铁同位素分馏效应[152];进一步,Wu 等[153]研究了 pH、温度和其他离子(HS^-)对该过程中铁同位素分馏因子的影响,结果显示,在 20℃的中性溶液中,Fe^{2+} 与硫化铁类纳米颗粒(FeS_m)之间的平衡分馏因子为−0.32‰,并受到溶液中特定离子(HS^-)的影响。第二,人为源纳米颗粒在自然水体中的转化过程也会导致同位素分馏效应,基于此,可以示踪人为源纳米颗粒在环境中的转化与归趋。Lu 等[150]首次报道了纳米银在自然水体中的转化过程可导致显著的银同位素分馏效应,同位素分馏因子达到 0.86‰;具体地,研究人员通过分析纳米银生成和溶解过程中伴随的银同位素分馏特征,揭示纳米银在自然环境中的生成和溶解机制,为纳米银的迁移转化研究提供了强有力的示踪工具。基于这一示踪工具,Zhang 等[154]结合纳米银在不同浓度腐殖酸水体中溶解过程伴随的银同位素分馏特征,进一步揭示了纳米银在实际水体中的稳定机制。

纳米材料的广泛使用导致了潜在的环境污染问题,引起了广泛关注。稳定同位素技术可以示踪自然环境中纳米颗粒的来源和迁移转化过程,因此在环境纳米技术领域具有广阔的应用前景。然而,目前这方面的研究尚处于起步阶段,还有广阔的发展空间,因此亟待更多的国内外同行加入该领域的研究中。

5 环境稳定同位素标记技术研究进展

富集同位素示踪是以富集的稳定同位素为示踪剂,追踪元素的环境与生物过程的技术。随着富集同位素的不断商品化，目前富集同位素示踪的应用日益广泛。相较于放射性同位素示踪，稳定同位素示踪避免了放射性同位素的使用及其可能的环境与健康危害[155]；在同位素分析中可采用高灵敏的 ICP-MS 或 MC-ICP-MS 检测，可将示踪同位素信号从内源性背景元素信号中分辨出来，从而降低金属的暴露浓度[156-157]；此外，多种稳定同位素的存在也使得多同位素示踪成为可能。

除了少数金属（如 Au、Ce、Co），大多数金属至少有两种稳定同位素。一般来说，采用自然丰度较低的富集同位素可为示踪提供最佳的灵敏度。在这种情况下，即使在具有天然金属背景的环境样品中掺入微量示踪同位素，也会导致金属同位素组成可分辨的变化，从而实现金属的高灵敏示踪。在选择示踪同位素时，还需考虑富集同位素的价格及其 ICP-MS 分析干扰[158]。

单稳定同位素示踪（single stable isotope tracing）广泛用于环境与生物体系中多种金属元素如铁、汞、锌等的行为研究。水相中溶解 Fe^{2+} 可显著促进铁矿物的再结晶过程。单一铁同位素可用于研究 Fe^{2+} 诱导铁矿物再结晶过程中的同位素交换行为。在针铁矿的再结晶过程中，示踪同位素 ^{57}Fe 开始时在针铁矿表面富集；随着再结晶的进行，示踪同位素 ^{57}Fe 在针铁矿中均有分布，其同位素组成与溶液相中的 Fe^{2+} 相同，即 Fe^{2+} 诱导针铁矿的再结晶经过表面交换—完全表面交换—整体渗透这一过程[159]。30～60d 同位素铁的摄入较 0～30d 下降，表明针铁矿的再结晶过程随时间逐渐变慢[160]。类似的过程在纤铁矿[161]、水铁矿[162-163]、赤铁矿[164]中同样存在，并受铁矿物结晶度与粒径、氧分压、水中有机碳含量等因素的影响[163-166]。利用 $^{202}Hg^{0}$ 单一同位素示踪农田作物对大气汞吸收的研究[167]表明，水稻与玉米叶片汞的沉降通量与大气汞的浓度呈正相关；两种农作物对大气汞的补偿点均低于国内背景地区大气汞浓度，表明农田植被是大气汞的重要汇。类似地，采用这一方法可研究土壤对汞的吸附动力学（$^{196}Hg^{2+}$）[168]、底泥/土壤中汞的甲基化与生物吸收（$^{199}Hg^{2+}/^{200}Hg^{2+}$）[169-170]、汞在硫化汞表面的再吸附（$^{202}Hg^{2+}$）[171]、二甲基汞的水稻吸收与分布［$(CH_3)_2^{199}Hg$］[172]。锌的单一同位素（$^{67}Zn^{2+}$）示踪广泛用于锌膳食吸收率的评价，发现老年人膳食平均吸收率为 27.9%，且男性高于女性[173]；牛奶可显著提高植酸含量较高大米对锌的吸收，超高温灭菌对牛奶促进锌吸收的作用没有影响[174]。采用类似的单一同位素示踪方法，可用于其他金属如铜（$^{65}Cu^{2+}$）[157, 175]、镉（$^{106}Cd^{2+}/^{110}Cd^{2+}/^{111}Cd^{2+}/^{113}Cd^{2+}$）[158, 176-178]、银（$^{107}Ag^{+}/^{109}AgNP$）[179-180]的形态转化与生物摄入。采用单一同位素添加，还可有效评估土壤、底泥等环境基质中金属的交换行为，定量其可交换量（exchangeable pool of M，M_E）。向污染土壤中加入 $^{199}Hg^{2+}$，经 72 h 平衡后，对滤液中各汞同位素进行分析，并用式（2-3）计算 Hg_E[181]。

$$Hg_E = \frac{M_{sl}C_{sp}V_{sp}\left(^{sp}IA_{sp} - {}^{rf}IA_{sp}R\right)}{M_{sp}W_{sl}\left(^{rf}IA_{sl}R - {}^{sp}IA_{sl}\right)} \quad (2\text{-}3)$$

式中，M_{sl} 和 M_{sp} 分别为土壤原有与加入的元素平均原子量，C_{sp} 为添加溶液中 $^{199}Hg^{2+}$ 的质量浓度（mg/L），V_{sp} 为添加溶液体积，W_{sl} 为土壤质量（kg），$^{sp}IA_{sp}$、$^{rf}IA_{sp}$ 分别为添加溶液中添加同位素与参比同位素的丰度，$^{sp}IA_{sl}$、$^{rf}IA_{sl}$ 分别为未加同位素富集溶液的对照土壤样品中的丰度，R 为 ICP-MS 测定的添加同位素与参照同位素的平衡比率[181]。研究表明，土壤中可交换态汞占总汞的 12%～25%，而传统的醋酸铵或氯化镁提取（分别小于 0.25%与 0.32%）显著低估了汞的可交换量[181]。采用类似的方法，可对铁（^{57}Fe）、镉（^{108}Cd）、铜（^{65}Cu）、镍（^{62}Ni）、铅（^{204}Pb）、锌（$^{67}Zn/^{70}Zn$）等金属的可交换量进行评估[182-184]。

在自然环境中，生物体通常处于多种途径的联合暴露之中。采用多稳定同位素示踪（multiple stable isotope tracing）方法可揭示不同暴露途径摄入的相对权重。采用三重稳定同位素示踪（^{110}Cd 示踪水相，^{111}Cd 示踪沉积物相，^{113}Cd 示踪膳物相）底栖生物诸氏鲻虾虎鱼的镉摄入途径[185]，结果显示虾虎鱼通过水相、沉积物相与膳食相的镉吸收速率常数分别为 3.1 L/(kg·d)、2.2×10^{-4} g/(g·d)和 3.3×10^{-3} g/(g·d)。沉积物中镉的生物利用率低于水相镉和膳食相镉，但当镉在沉积物和海水之间的分配系数（K_d）大于 6×10^4 L/kg 时，沉积物可能成为镉生物积累的主要来源。对虾虎鱼组织中镉同位素的分析[186]表明，鱼主要通过鳃和胃肠道吸收水相与沉积物相中的镉，饮食中镉主要通过胃肠道吸收。鳃吸收镉主要来自水相（77.2%～89.4%），而胃肠道镉主要来自饮食（81.3%～98.7%）。鱼体中摄取的镉主要来自饮食（47.1%～80.4%）和水（22.8%～1.6%）。多稳定同位素示踪可用于区分不同形态金属的环境分布与生物摄入[187]。^{68}ZnO 纳米颗粒与 ^{64}ZnCl$_2$ 加入土壤后，对土壤、间隙水、蚯蚓中锌同位素（^{68}Zn/^{64}Zn）的分析表明，ZnO 纳米颗粒与锌离子的环境分布与生物摄入无明显差异，这可能与 ZnO 的快速溶解有关[188]。双同位素示踪结合激光剥蚀(LA)-ICP-MS 分析显示，人视网膜色素上皮细胞对 ^{68}ZnSO$_4$ 与 ^{70}Zn-gluconate 的摄入无明显差异[187]。双同位素示踪也可区分不同给药途径锌摄入的差异。采用双同位素标记分别示踪锌的口服（^{67}Zn）与静脉注射（^{70}Zn）摄入，并以静脉注射为基准（定义为完全吸收），考察了口服谷物中锌的吸收效率，发现植酸裂解酶可显著促进儿童对谷物中锌的摄入[189]。类似的锌多稳定同位素示踪（^{67}Zn、^{68}Zn、^{70}Zn）发现，儿童环境肠道功能障碍可降低其对锌的吸收[190]。

金属形态分析技术结合多稳定同位素示踪，可揭示同一体系中不同形态金属行为的差异及其相互转化。采用同位素标记的无机汞与甲基汞（如 ^{199}Hg^{2+}、CH$_3^{201}$Hg$^+$），结合 GC-ICP-MS 联用技术，可区分同一体系中汞的甲基化与去甲基化过程[191-192]。双同位素示踪表明，盐沼中的植被可显著促进无机汞的微生物甲基化过程，抑制甲基汞的降解，从而有利于甲基汞的生成[191]。盐沼植物对汞的摄入分析表明，无机汞与甲基汞均可被植物摄入，大部分汞持留在根部，少部分汞可向茎叶部转运[193]。双同位素示踪（^{201}Hg^{2+}、^{199}Hg^0）揭示了湿地植物锯齿草对土壤与大气中汞的吸收特征[194]：土壤中添加的 ^{201}Hg^{2+} 可很快被根吸收；其中大部分 ^{201}Hg^{2+} 为根部所"捕集"（>77%），少部分 ^{201}Hg^{2+} 被运输到地上部分，但未检测到 ^{201}Hg 从叶片向空气的释放；同时，大气 ^{199}Hg^0 暴露显示，叶片吸收的 ^{199}Hg 大部分被固定化，再排放到大气中的 ^{199}Hg 比例极低（1.6%），叶片吸收的 ^{199}Hg 存在向根茎的传输过程（<24%）。水环境微宇宙中不同形态无机汞的多稳定同位素示踪[^{201}Hg(NO$_3$)$_2$、^{202}Hg^{2+}-OM（有机质）、β-^{198}HgS]分析表明，无机汞的甲基化、还原挥发速率与二价汞的化学形态密切相关[k_{meth}：^{201}Hg(NO$_3$)$_2$ > ^{202}Hg^{2+}-OM > β-^{198}HgS，k_{vol}：^{201}Hg(NO$_3$)$_2$ ≈ ^{202}Hg^{2+}-OM > β-^{198}HgS][195]，这与另一多稳定同位素示踪的结果一致[196]。采用双同位素标记银离子（^{109}Ag$^+$）与纳米银（^{107}AgNPs），研究了其在水中的光转化。在水溶液中，AgNPs 的氧化占主导，日光可显著促进 ^{107}AgNPs 的氧化溶解；但在溶解性有机质存在下，^{109}Ag$^+$ 存在显著的还原[197]。银离子与纳米银水稻摄入的双同位素示踪（^{107}Ag$^+$、^{109}AgNPs）分析显示，水稻对 ^{109}Ag 的摄入显著高于 ^{107}Ag，表明可能存在纳米银的直接生物摄入[198]。LC-ICP-MS 形态分析表明，银在水稻根部主要以纳米银形式存在，表明根部存在银的还原；而在茎叶部纳米银与银离子共存，提示茎叶部存在银的氧化[198]。

需要特别指出的是，除了 ICP-MS，塞曼原子吸收光谱检测也有望在同位素分析上发挥作用[199-200]。不同汞同位素具有较大的塞曼原子吸收信号响应差异，例如 ^{200}Hg 与 ^{202}Hg 的信号强度是 ^{198}Hg 与 ^{201}Hg 的 10 倍[201]。因此，塞曼原子吸收光谱可以用于不同汞同位素的定量以及富集同位素示踪研究。近期研究报道了采用塞曼原子吸收光谱分析单一或双富集同位素示踪汞的环境循环，如分配、离子交换、吸附/解吸附、甲基化/去甲基化等[202]。需要特别指出的是，该方法仍需对总汞（如采用原子荧光光谱）进行测定，以进行不同汞同位素的浓度校准。

6 展 望

稳定同位素技术在来源示踪和过程示踪方面具有独特优势，基于此，可以追溯污染物的来源、示踪污染物的迁移转化途径等。可以期待，稳定同位素技术将在环境地球化学领域得到越来越快的发展。但总体而言，目前该领域的研究尚处于起步阶段，亟需更多研究力量的投入。未来还需要在以下方面加强研究：

（1）发展更多元素的稳定同位素分析方法，开发更多实际环境体系的应用，进一步扩展环境稳定同位素技术的应用范围和提高解决实际环境问题的能力。

（2）进一步克服环境样品同位素分析的两大难点：复杂基质和痕量分析。因此，要继续加强对稳定同位素分析方法的开发与优化：一方面需开发更高效简便的消解和分离纯化方法，降低基质干扰；另一方面应进一步发展相关元素富集方法和提高仪器分析灵敏度，提高对痕量污染物的同位素分析能力。

（3）继续提高稳定同位素技术的示踪能力：一方面应加快对各个污染源谱同位素数据库的建立，提供更全面的污染源同位素指纹信息；另一方面应进一步加强对环境过程和同位素分馏机理的研究，增加对污染物从源到汇的过程中伴生的同位素分馏效应的认识，从而提高污染物溯源的准确性。

（4）继续扩大稳定同位素标记技术的应用潜力：一方面应提高富集稳定同位素标准品的制备技术，既要降低制备费用，又要兼顾应用领域，比如制备富集稳定同位素标记的不同粒径、不同表面修饰、不同组分的纳米颗粒，用以系统研究其在自然环境体系中的迁移转化过程；另一方面需提高相应应用及表征技术，比如富集同位素成像技术、多同位素联合标记技术和基于富集稳定同位素的污染物形态分析研究等。

参 考 文 献

[1] Becquerel A H. On the rays emitted by phosphorescent bodies. Comptes Rendus de Seances de l'academie de Sciences, 1896, 122: 501-503.

[2] Curie M. Rays emitted by compounds of uranium and thorium. Comptes Rendus de Seances de l'academie de Sciences, 1898, 126: 1101-1103.

[3] Rutherford E. A radio-active substance emitted from thorium compounds. Philosophical Magazine, 1900, 49: 1-14.

[4] Soddy F. Radioactivity. Annual Reports on the Progress of Chemistry, 1912, 9: 289-328.

[5] Urey H C, Brickwedde F G, Murphy G M. A hydrogen isotope of mass 2 and its concentration. Physical Review, 1932, 40: 1.

[6] Nier A O, Gulbransen E A. Variations in the relative abundance of the carbon isotopes. Journal of the American Chemical Society, 1939, 61: 697-698.

[7] Dole M, Slobod R L. Isotopic composition of oxygen in carbonate rocks and iron oxide ores. Journal of the American Chemical Society, 1940, 62: 471-479.

[8] Urey H C. Oxygen isotopes in nature and in the laboratory. Science, 1948, 108: 489-496.

[9] Walder A J, Platzner I, Freedman P A. Isotope ratio measurement of lead, neodymium and neodymium–samarium mixtures, hafnium and hafnium–lutetium mixtures with a double focusing multiple collector inductively coupled plasma mass spectrometer. Journal of Analytical Atomic Spectrometry, 1993, 8: 19-23.

[10] Yang L. Accurate and precise determination of isotopic ratios by MC‐ICP‐MS: A review, Mass Spectrometry Reviews, 2009, 28: 990-1011.

[11] 黄方, 田笙谕. 若干金属稳定同位素体系的研究进展: 以中国科大实验室为例. 矿物岩石地球化学通报, 2018, 37: 793-811.

[12] Lu D, Zhang T, Yang X, et al. Recent advances in the analysis of non-traditional stable isotopes by multi-collector inductively coupled plasma mass spectrometry. Journal of Analytical Atomic Spectrometry, 2017, 32: 1848-1861.

[13] Evans E H, Pisonero J, Smith C M, et al. Atomic spectrometry update: Review of advances in atomic spectrometry and related techniques. Journal of Analytical Atomic Spectrometry, 2018.

[14] Woodhead J D, Horstwood M S, Cottle J M. Advances in isotope ratio determination by LA–ICP–MS. Elements, 2016, 12: 317-322.

[15] Newman K. Effects of the sampling interface in MC-ICP-MS: Relative elemental sensitivities and non-linear mass dependent fractionation of Nd isotopes. Journal of Analytical Atomic Spectrometry, 2012, 27: 63-70.

[16] Gou L F, Jin Z D, Deng L, et al. Effects of different cone combinations on accurate and precise determination of Li isotopic composition by MC-ICP-MS. Spectrochimica Acta Part B: Atomic Spectroscopy, 2018, 146: 1-8.

[17] Geng H, Yin R, Li X. An optimized protocol for high precision measurement of Hg isotopic compositions in samples with low concentrations of Hg using MC-ICP-MS. Journal of Analytical Atomic Spectrometry, 2018, 33: 1932-1940.

[18] Xu L, Hu Z, Zhang W, et al. *In situ* Nd isotope analyses in geological materials with signal enhancement and non-linear mass dependent fractionation reduction using laser ablation MC-ICP-MS. Journal of Analytical Atomic Spectrometry, 2015, 30: 232-244.

[19] Newman K, Freedman P A, Williams J, et al. High sensitivity skimmers and non-linear mass dependent fractionation in ICP-MS. Journal of Analytical Atomic Spectrometry, 2009, 24: 742-751.

[20] Yang L, Tong S, Zhou L, et al. A critical review on isotopic fractionation correction methods for accurate isotope amount ratio measurements by MC-ICP-MS. Journal of Analytical Atomic Spectrometry, 2018, 33: 1849-1861.

[21] Georg R B, Newman K. The effect of hydride formation on instrumental mass discrimination in MC-ICP-MS: A case study of mercury (Hg) and thallium (Tl) isotopes. Journal of Analytical Atomic Spectrometry, 2015, 30: 1935-1944.

[22] Chen X, Romaniello S J, Herrmann A D, et al. Biological effects on uranium isotope fractionation (^{238}U/^{235}U) in primary biogenic carbonates. Geochimica et Cosmochimica Acta, 2018, 240: 1-10.

[23] Peters S T, Münker C, Wombacher F, et al. Precise determination of low abundance isotopes (^{174}Hf, ^{180}W and ^{190}Pt) in terrestrial materials and meteorites using multiple collector ICP-MS equipped with 1012 Ω Faraday amplifiers. Chemical Geology, 2015, 413: 132-145.

[24] Chen H, Tian Z, Tuller-Ross B, et al. High-precision potassium isotopic analysis by MC-ICP-MS: An inter-laboratory comparison and refined K atomic weight. Journal of Analytical Atomic Spectrometry, 2019, 34: 160-171.

[25] Li W, Beard B L, Li S. Precise measurement of stable potassium isotope ratios using a single focusing collision cell multi-collector ICP-MS. Journal of Analytical Atomic Spectrometry, 2016, 31: 1023-1029.

[26] Li M, Lei Y, Feng L, et al. High-precision Ca isotopic measurement using a large geometry high resolution MC-ICP-MS with a dummy bucket. Journal of Analytical Atomic Spectrometry, 2018, 33: 1707-1719.

[27] Morgan L E, Ramos D P S, Davidheiser-Kroll B, et al. High-precision ^{41}K/^{39}K measurements by MC-ICP-MS indicate terrestrial variability of δ^{41}K. Journal of Analytical Atomic Spectrometry, 2018, 33: 175-186.

[28] 梁鹏, 陈开运, 包志安, 等. 大型高分辨率多接收等离子体质谱准确测定地质标样的铁同位素组成. 矿物岩石地球化学通报, 2016: 473-478.

[29] Chen K Y, Yuan H L, Liang P, et al. Improved nickel-corrected isotopic analysis of iron using high-resolution multi-collector inductively coupled plasma mass spectrometry. International Journal of Mass Spectrometry, 2017, 421: 196-203.

[30] Wang K, Jacobsen S B. An estimate of the Bulk Silicate Earth potassium isotopic composition based on MC-ICPMS measurements of basalts. Geochimica et Cosmochimica Acta, 2016, 178: 223-232.

[31] Chernonozhkin S M, Costas-Rodríguez M, Claeys P, et al. Evaluation of the use of cold plasma conditions for Fe isotopic analysis via multi-collector ICP-mass spectrometry: Effect on spectral interferences and instrumental mass discrimination. Journal of Analytical Atomic Spectrometry, 2017, 32: 538-547.

[32] Hu Y, Chen X Y, Xu Y K, et al. High-precision analysis of potassium isotopes by HR-MC-ICPMS. Chemical Geology, 2018, 493: 100-108.

[33] Yang L, Zhou L, Hu Z, et al. Direct determination of Si isotope ratios in natural waters and commercial Si standards by ion exclusion chromatography multicollector inductively coupled plasma mass spectrometry. Analytical Chemistry, 2014, 86:

[34] Hu Z, Liu Y, Gao S, et al. Improved in situ Hf isotope ratio analysis of zircon using newly designed X skimmer cone and jet sample cone in combination with the addition of nitrogen by laser ablation multiple collector ICP-MS. Journal of Analytical Atomic Spectrometry, 2012, 27: 1391-1399.

[35] Yang L, Mester Z, Zhou L, et al. Observations of large mass-independent fractionation occurring in MC-ICPMS: Implications for determination of accurate isotope amount ratios. Analytical Chemistry, 2011, 83: 8999-9004.

[36] Zhu Z, Meija J, Tong S, et al. Determination of the isotopic composition of osmium using MC-ICPMS. Analytical Chemistry, 2018, 90: 9281-9288.

[37] Miyazaki T, Kimura J I, Chang Q. Analysis of stable isotope ratios of Ba by double-spike standard-sample bracketing using multiple-collector inductively coupled plasma mass spectrometry. Journal of Analytical Atomic Spectrometry, 2014, 29: 483-490.

[38] Guéguen F, Nonell A, Isnard H, et al. Multi-elemental Gd, Eu, Sm, Nd isotope ratio measurements by liquid chromatography coupled to MC-ICPMS with variable Faraday cup configurations during elution. Talanta, 2017, 162: 278-284.

[39] Guéguen F, Isnard H, Nonell A, et al. Neodymium isotope ratio measurements by LC-MC-ICPMS for nuclear applications: Investigation of isotopic fractionation and mass bias correction. Journal of Analytical Atomic Spectrometry, 2015, 30: 443-452.

[40] Martelat B, Isnard H, Vio L, et al. Precise U and Pu isotope ratio measurements in nuclear samples by hyphenating capillary electrophoresis and MC-ICPMS. Analytical Chemistry, 2018, 90: 8622-8628.

[41] Bérail S, Cavalheiro J, Tessier E, et al. Determination of total Hg isotopic composition at ultra-trace levels by on line cold vapor generation and dual gold-amalgamation coupled to MC-ICP-MS. Journal of Analytical Atomic Spectrometry, 2017, 32: 373-384.

[42] Rodríguez-Gonzàlez P, Epov V N, Pecheyran C, et al. Species-specific stable isotope analysis by the hyphenation of chromatographic techniques with MC-ICPMS. Mass Spectrometry Reviews, 2012, 31: 504-521.

[43] Epov V N, Rodriguez-Gonzalez P, Sonke J E, et al. Simultaneous determination of species-specific isotopic composition of Hg by gas chromatography coupled to multicollector ICPMS. Analytical Chemistry, 2008, 80: 3530-3538.

[44] Bouchet S, Bérail S, Amouroux D. Hg compound-specific isotope analysis at ultratrace levels using an on line gas chromatographic preconcentration and separation strategy coupled to multicollector-inductively coupled plasma mass spectrometry. Analytical Chemistry, 2018, 90: 7809-7816.

[45] Queipo-Abad S, Rodríguez-González P, Alonso J I G. Measurement of compound-specific Hg isotopic composition in narrow transient signals by gas chromatography coupled to multicollector ICP-MS. Journal of Analytical Atomic Spectrometry, 2019.

[46] Janssen S E, Johnson M W, Blum J D, et al. Separation of monomethylmercury from estuarine sediments for mercury isotope analysis. Chemical Geology, 2015, 411: 19-25.

[47] Qin C, Chen M, Yan H, et al. Compound specific stable isotope determination of methylmercury in contaminated soil. Science of the Total Environment, 2018, 644: 406-412.

[48] Li P, Du B, Maurice L, et al. Mercury isotope signatures of methylmercury in rice samples from the Wanshan mercury mining area, China: Environmental implications. Environmental science & technology, 2017, 51: 12321-12328.

[49] Masbou J, Point D, Sonke J E. Application of a selective extraction method for methylmercury compound specific stable isotope analysis (MeHg-CSIA) in biological materials. Journal of Analytical Atomic Spectrometry, 2013, 28: 1620-1628.

[50] Entwisle J, Malinovsky D, Dunn P J, et al. Hg isotope ratio measurements of methylmercury in fish tissues using HPLC with off line cold vapour generation MC-ICPMS. Journal of Analytical Atomic Spectrometry, 2018, 33: 1645-1654.

[51] 张文, 刘勇胜, 胡兆初. 微区原位 LA-MC-ICP-MS 铅同位素分析研究进展. 矿物岩石地球化学通报, 2018, 37: 812-826.

[52] Poitrasson F, d'Abzac F-X. Femtosecond laser ablation inductively coupled plasma source mass spectrometry for elemental and isotopic analysis: Are ultrafast lasers worthwhile? Journal of Analytical Atomic Spectrometry, 2017, 32: 1075-1091.

[53] Zheng X-Y, Beard B L, Lee S, et al. Contrasting particle size distributions and Fe isotope fractionations during nanosecond and femtosecond laser ablation of Fe minerals: Implications for LA-MC-ICP-MS analysis of stable isotopes. Chemical Geology, 2017, 450: 235-247.

[54] Schuessler J A, von Blanckenburg F. Testing the limits of micro-scale analyses of Si stable isotopes by femtosecond laser ablation multicollector inductively coupled plasma mass spectrometry with application to rock weathering. Spectrochimica Acta Part B: Atomic Spectroscopy, 2014, 98: 1-18.

[55] Lazarov M, Horn I. Matrix and energy effects during in-situ determination of Cu isotope ratios by ultraviolet-femtosecond laser ablation multicollector inductively coupled plasma mass spectrometry. Spectrochimica Acta Part B: Atomic Spectroscopy, 2015, 111: 64-73.

[56] Chen L, Chen K, Bao Z, et al. Preparation of standards for in situ sulfur isotope measurement in sulfides using femtosecond laser ablation MC-ICP-MS. Journal of Analytical Atomic Spectrometry, 2017, 32: 107-116.

[57] Kaiyun C, Honglin Y, Zhian B, et al. Precise and accurate in situ determination of lead isotope ratios in NIST, USGS, MPI - DING and CGSG glass reference materials using femtosecond laser ablation MC-ICP-MS. Geostandards and Geoanalytical Research, 2014, 38: 5-21.

[58] Shaheen M, Fryer B J. Improving the analytical capabilities of femtosecond laser ablation multicollector ICP-MS for high precision Pb isotopic analysis: The role of hydrogen and nitrogen. Journal of Analytical Atomic Spectrometry, 2010, 25: 1006-1013.

[59] Zhang W, Hu Z, Günther D, et al. Direct lead isotope analysis in Hg-rich sulfides by LA-MC-ICP-MS with a gas exchange device and matrix-matched calibration. Analytica Chimica Acta, 2016, 948: 9-18.

[60] Hu Z, Zhang W, Liu Y, et al. "Wave" signal-smoothing and mercury-removing device for laser ablation quadrupole and multiple collector ICPMS analysis: Application to lead isotope analysis. Analytical Chemistry, 2014, 87: 1152-1157.

[61] Prohaska T, Irrgeher J, Zitek A. Simultaneous multi-element and isotope ratio imaging of fish otoliths by laser ablation split stream ICP-MS/MC ICP-MS. Journal of Analytical Atomic Spectrometry, 2016, 31: 1612-1621.

[62] Aramendía M, Rello L, Bérail S, et al. Direct analysis of dried blood spots by femtosecond-laser ablation-inductively coupled plasma-mass spectrometry. Feasibility of split-flow laser ablation for simultaneous trace element and isotopic analysis. Journal of Analytical Atomic Spectrometry, 2014, 30: 296-309.

[63] Von Allmen K, Böttcher M E, Samankassou E, et al. Barium isotope fractionation in the global barium cycle: First evidence from barium minerals and precipitation experiments. Chemical Geology, 2010, 277: 70-77.

[64] Nan X, Wu F, Zhang Z, et al. High-precision barium isotope measurements by MC-ICP-MS. Journal of Analytical Atomic Spectrometry, 2015, 30: 2307-2315.

[65] Yuan W, Chen J B, Birck J-L, et al. Precise analysis of gallium isotopic composition by MC-ICP-MS. Analytical Chemistry, 2016, 88: 9606-9613.

[66] Enge T G, Field M P, Jolley D F, et al. An automated chromatography procedure optimized for analysis of stable Cu isotopes from biological materials. Journal of Analytical Atomic Spectrometry, 2016, 31: 2023-2030.

[67] Romaniello S, Field M, Smith H, et al. Fully automated chromatographic purification of Sr and Ca for isotopic analysis. Journal of Analytical Atomic Spectrometry, 2015, 30: 1906-1912.

[68] Lu X, Zhao J, Liang X, et al. The application and potential artifacts of zeeman cold vapor atomic absorption spectrometry in

mercury stable isotope analysis. Environmental Science & Technology Letters, 2019, 6.

[69] Li P, Duan X, Cheng H, et al. Application of lead stable isotopes to identification of environmental source. Environmental Science & Technology, 2013, 36: 63-67.

[70] Xianfang L I, Xiande L I U, Bing L I, et al. Isotopic determinations and source study of lead in ambient $PM_{2.5}$ in Beijing. Chinese Journal of Environmental Science, 2006, 27: 401-407.

[71] Simonetti A, Gariepy C, Carignan J. Pb and Sr isotopic evidence for sources of atmospheric heavy metals and their deposition budgets in northeastern North America. Geochimica et Cosmochimica Acta, 2000, 64: 3439-3452.

[72] Widory D, Liu X, Dong S. Isotopes as tracers of sources of lead and strontium in aerosols (TSP & $PM_{2.5}$) in Beijing. Atmospheric Environment, 2010, 44: 3679-3687.

[73] Hyeong K, Kim J, Pettke T, et al. Lead, Nd and Sr isotope records of pelagic dust: Source indication versus the effects of dust extraction procedures and authigenic mineral growth. Chemical Geology, 2011, 286: 240-251.

[74] Wang X, Bao Z, Lin C-J, et al. Assessment of global mercury deposition through litterfall. Environmental Science & Technology, 2016, 50: 8548-8557.

[75] Enrico M, Le Roux G, Marusczak N, et al. Atmospheric mercury transfer to peat bogs dominated by gaseous elemental mercury dry deposition. Environmental Science & Technology, 2016, 50: 2405-2412.

[76] Schleicher N J, Schäfer J, Blanc G, et al. Atmospheric particulate mercury in the megacity Beijing: Spatio-temporal variations and source apportionment. Atmospheric Environment, 2015, 109: 251-261.

[77] Rutter A P, Schauer J J, Shafer M M, et al. Dry deposition of gaseous elemental mercury to plants and soils using mercury stable isotopes in a controlled environment. Atmospheric Environment, 2011, 45: 848-855.

[78] Gonzalez R O, Strekopytov S, Amato F, et al. New insights from zinc and copper isotopic compositions into the sources of atmospheric particulate matter from two major European cities. Environmental Science & Technology, 2016, 50: 9816-9824.

[79] Souto-Oliveira C E, Babinski M, Araújo D F, et al. Multi-isotopic fingerprints(Pb, Zn, Cu)applied for urban aerosol source apportionment and discrimination. Science of The Total Environment, 2018, 626: 1350-1366.

[80] Beard B L, Johnson C M, Von Damm K L, et al. Iron isotope constraints on Fe cycling and mass balance in oxygenated Earth oceans. Geology, 2003, 31: 629-632.

[81] Flament P, Mattielli N, Aimoz L, et al. Iron isotopic fractionation in industrial emissions and urban aerosols. Chemosphere, 2008, 73: 1793-1798.

[82] Geagea M L, Stille P, Gauthier-Lafaye F, et al. Tracing of industrial aerosol sources in an urban environment using Pb, Sr, and Nd isotopes. Environmental Science & Technology, 2008, 42: 692-698.

[83] Grousset F, Biscaye P. Continental aerosols, isotopic fingerprints of sources and atmospheric transport: a review. Chemical Geology, 2005, 222: 149-167.

[84] Zhang L, Hou X, Xu S. Speciation analysis of ^{129}I and ^{127}I in aerosols using sequential extraction and mass spectrometry detection. Analytical Chemistry, 2015, 87: 6937-6944.

[85] Lu D, Liu Q, Yu M, et al. Natural silicon isotopic signatures reveal the sources of airborne fine particulate matter. Environmental Science & Technology, 2018, 52: 1088-1095.

[86] Xu H M, Sun R Y, Cao J J, et al. Mercury stable isotope compositions of Chinese urban fine particulates in winter haze days: Implications for Hg sources and transformations. Chemical Geology, 2019, 504: 267-275.

[87] Sun R, Streets D G, Horowitz H M, et al. Historical (1850—2010) mercury stable isotope inventory from anthropogenic sources to the atmosphere. Elementa-Science of the Anthropocene, 2016, 4: 1-15.

[88] Sun R, Sonke J E, Heimbuerger L E, et al. Mercury stable isotope signatures of world coal deposits and historical coal combustion emissions. Environmental Science & Technology, 2014, 48: 7660-7668.

[89] Sun R, Heimbuerger L E, Sonke J E, et al. Mercury stable isotope fractionation in six utility boilers of two large coal-fired power plants. Chemical Geology, 2013, 336: 103-111.

[90] Huang S, Yuan D, Lin H, et al. Fractionation of mercury stable isotopes during coal combustion and seawater flue gas desulfurization. Applied Geochemistry, 2017, 76: 159-167.

[91] Das R, Wang X, Khezri B, et al. Mercury isotopes of atmospheric particle bound mercury for source apportionment study in urban Kolkata, India. Elementa-Science of the Anthropocene, 2016, 4: 1-12.

[92] Huang Q, Chen J, Huang W, et al. Isotopic composition for source identification of mercury in atmospheric fine particles. Atmospheric Chemistry and Physics, 2016, 16: 11773-11786.

[93] Fu X, Zhang H, Feng X, et al. Domestic and transboundary sources of atmospheric particulate bound mercury in remote areas of China: Evidence from mercury isotopes. Environmental Science & Technology, 2019, 53: 1947-1957.

[94] Huang Q, Chen J, Huang W, et al. Diel variation in mercury stable isotope ratios records photoreduction of $PM_{2.5}$-bound mercury. Atmospheric Chemistry and Physics, 2019, 19: 315-325.

[95] Dong S, Gonzalez R O, Harrison R M, et al. Isotopic signatures suggest important contributions from recycled gasoline, road dust and non-exhaust traffic sources for copper, zinc and lead in PM_{10} in London, United Kingdom. Atmospheric Environment, 2017, 165: 88-98.

[96] Lu D, Tan J, Yang X, et al. Unraveling the role of silicon in atmospheric aerosol secondary formation: A new conservative tracer for aerosol chemistry. Atmospheric Chemistry and Physics, 2019, 19: 2861-2870.

[97] Moynier F, Vance D, Fujii T, et al. The isotope geochemistry of zinc and copper. Reviews in Mineralogy and Geochemistry, 2017, 82: 543-600.

[98] Weiss D J, Rausch N, Mason T F, et al. Atmospheric deposition and isotope biogeochemistry of zinc in ombrotrophic peat. Geochimica et Cosmochimica Acta, 2007, 71: 3498-3517.

[99] Pašava J, Tornos F, Chrastný V. Zinc and sulfur isotope variation in sphalerite from carbonate-hosted zinc deposits, Cantabria, Spain. Mineralium Deposita, 2014, 49: 797-807.

[100] Conway T M, John S G. The biogeochemical cycling of zinc and zinc isotopes in the North Atlantic Ocean. Global Biogeochemical Cycles, 2014, 28: 1111-1128.

[101] Juillot F, Maréchal C, Ponthieu M, et al. Zn isotopic fractionation caused by sorption on goethite and 2-Lines ferrihydrite. Geochimica et Cosmochimica Acta, 2008, 72: 4886-4900.

[102] Bryan A L, Dong S, Wilkes E B, et al. Zinc isotope fractionation during adsorption onto Mn oxyhydroxide at low and high ionic strength. Geochimica et Cosmochimica Acta, 2015, 157: 182-197.

[103] Dong S, Wasylenki L E. Zinc isotope fractionation during adsorption to calcite at high and low ionic strength. Chemical Geology, 2016, 447: 70-78.

[104] Gou W, Li W, Ji J, et al. Zinc isotope fractionation during sorption onto Al oxides: Atomic level understanding from EXAFS. Environmental Science & Technology, 2018, 52: 9087-9096.

[105] Guinoiseau D, Gélabert A, Moureau J, et al. Zn isotope fractionation during sorption onto kaolinite. Environmental Science & Technology, 2016, 50: 1844-1852.

[106] Nelson J, Wasylenki L, Bargar J R, et al. Effects of surface structural disorder and surface coverage on isotopic fractionation during Zn(II) adsorption onto quartz and amorphous silica surfaces. Geochimica et Cosmochimica Acta, 2017, 215: 354-376.

[107] Pokrovsky O, Viers J, Freydier R. Zinc stable isotope fractionation during its adsorption on oxides and hydroxides. Journal of Colloid and Interface Science, 2005, 291: 192-200.

[108] Vaněk A, Grösslová Z, Mihaljevič M, et al. Isotopic tracing of thallium contamination in soils affected by emissions from coal-fired power plants. Environmental Science & Technology, 2016, 50: 9864-9871.

[109] Kersten M, Xiao T, Kreissig K, et al. Tracing anthropogenic thallium in soil using stable isotope compositions. Environmental Science & Technology, 2014, 48: 9030-9036.

[110] Liu J, Wang J, Tsang D C W, et al. Emerging thallium pollution in China and source tracing by thallium isotopes. Environmental Science & Technology, 2018, 52: 11977-11979.

[111] Yang W J, Ding K B, Zhang P, et al. Cadmium stable isotope variation in a mountain area impacted by acid mine drainage. Science of the Total Environment, 2019, 646: 696-703.

[112] Cloquet C, Carignan J, Libourel G, et al. Tracing source pollution in soils using cadmium and lead isotopes. Environmental Science & Technology, 2006, 40: 2525-2530.

[113] Gao B, Zhou H, Liang X, et al. Cd isotopes as a potential source tracer of metal pollution in river sediments. Environmental Pollution, 2013, 181: 340-343.

[114] Grigg A R, Kretzschmar R, Gilli R S, et al. Mercury isotope signatures of digests and sequential extracts from industrially contaminated soils and sediments. Science of the Total Environment, 2018, 636: 1344-1354.

[115] Sherman L S, Blum J D, Keeler G J, et al. Investigation of local mercury deposition from a coal-fired power plant using mercury isotopes. Environmental Science & Technology, 2012, 46: 382-390.

[116] Ma J, Hintelmann H, Kirk J L, et al. Mercury concentrations and mercury isotope composition in lake sediment cores from the vicinity of a metal smelting facility in Flin Flon, Manitoba. Chemical Geology, 2013, 336: 96-102.

[117] Wiederhold J G, Smith R S, Siebner H, et al. Mercury isotope signatures as tracers for Hg cycling at the New Idria Hg mine. Environmental Science & Technology, 2013, 47: 6137-6145.

[118] Chen J, Bouchez J, Gaillardet J, et al. Behaviors of major and trace elements during single flood event in the Seine River, France. Procedia Earth and Planetary Science, 2014, 10: 343-348.

[119] Chen J B, Louvat P, Gaillardet J, et al. Direct separation of Zn from dilute aqueous solutions for isotope composition determination using multi-collector ICP-MS. Chemical Geology, 2009, 259: 120-130.

[120] Chen J, Gaillardet J, Louvat P, et al. Zn isotopes in the suspended load of the Seine River, France: Isotopic variations and source determination. Geochimica et Cosmochimica Acta, 2009, 73: 4060-4076.

[121] Chen J, Gaillardet J, Louvat P. Zinc isotopes in the Seine River waters, France: A probe of anthropogenic contamination. Environmental Science & Technology, 2008, 42: 6494-6501.

[122] Chen J B, Busigny V, Gaillardet J, et al. Iron isotopes in the Seine River(France): Natural versus anthropogenic sources. Geochimica et Cosmochimica Acta, 2014, 128: 128-143.

[123] Juillot F, Maréchal C, Morin G, et al. Contrasting isotopic signatures between anthropogenic and geogenic Zn and evidence for post-depositional fractionation processes in smelter-impacted soils from Northern France. Geochimica et Cosmochimica Acta, 2011, 75: 2295-2308.

[124] Malinovsky D, Vanhaecke F. Mass-independent isotope fractionation of heavy elements measured by MC-ICPMS: A unique probe in environmental sciences. Analytical and Bioanalytical Chemistry, 2011, 400: 1619-1624.

[125] Bergquist B A, Blum J D. Mass-dependent and-independent fractionation of Hg isotopes by photoreduction in aquatic systems. Science, 2007, 318: 417-420.

[126] Chen J, Hintelmann H, Feng X, et al. Unusual fractionation of both odd and even mercury isotopes in precipitation from Peterborough, ON, Canada. Geochimica et Cosmochimica Acta, 2012, 90: 33-46.

[127] Cai H, Chen J. Mass-independent fractionation of even mercury isotopes. Science Bulletin, 2016, 61: 116-124.

[128] Gratz L E, Keeler G J, Blum J D, et al. Isotopic composition and fractionation of mercury in Great Lakes precipitation and ambient air. Environmental Science & Technology, 2010, 44: 7764-7770.

[129] Obrist D, Agnan Y, Jiskra M, et al. Tundra uptake of atmospheric elemental mercury drives Arctic mercury pollution. Nature,

2017, 547: 201.

[130] Jiskra M, Sonke J E, Obrist D, et al. A vegetation control on seasonal variations in global atmospheric mercury concentrations. Nature Geoscience, 2018, 11: 244.

[131] Olson C L, Jiskra M, Sonke J E, et al. Mercury in tundra vegetation of Alaska: Spatial and temporal dynamics and stable isotope patterns. Science of the Total Environment, 2019, 660: 1502-1512.

[132] Wang Z, Chen J, Feng X, et al. Mass-dependent and mass-independent fractionation of mercury isotopes in precipitation from Guiyang, SW China. Comptes Rendus Geoscience, 2015, 347: 358-367.

[133] Yuan S, Zhang Y, Chen J, et al. Large variation of mercury isotope composition during a single precipitation event at Lhasa City, Tibetan Plateau, China. Procedia Earth and Planetary Science, 2015, 13: 282-286.

[134] Yuan S, Chen J, Cai H, et al. Sequential samples reveal significant variation of mercury isotope ratios during single rainfall events. Science of the Total Environment, 2018, 624: 133-144.

[135] Huang S, Sun L, Zhou T, et al. Natural stable isotopic compositions of mercury in aerosols and wet precipitations around a coal-fired power plant in Xiamen, southeast China. Atmospheric Environment, 2018, 173: 72-80.

[136] Zdanowicz C M, Krümmel E, Poulain A, et al. Historical variations of mercury stable isotope ratios in Arctic glacier firn and ice cores. Global Biogeochemical Cycles, 2016, 30: 1324-1347.

[137] Shen J, Chen J, Algeo T J, et al. Evidence for a prolonged Permian–Triassic extinction interval from global marine mercury records. Nature Communications, 2019, 10: 1563.

[138] Shen J, Algeo T J, Chen J, et al. Mercury in marine Ordovician/Silurian boundary sections of South China is sulfide-hosted and non-volcanic in origin. Earth and Planetary Science Letters, 2019, 511: 130-140.

[139] Sial A N, Chen J, Lacerda L, et al. Mukhopadhyay, Sucharita Pal, JP Shrivastava on the paper by Sial et al.(2016)Mercury enrichments and Hg isotopes in Cretaceous–Paleogene boundary successions: Links to volcanism and palaeoenvironmental impacts. Cretaceous Research 66, 60–81, Cretaceous Research, 2017, 76: 84-88.

[140] Kato C, Moynier F, Foriel J, et al. The gallium isotopic composition of the bulk silicate Earth. Chemical Geology, 2017, 448: 164-172.

[141] Kato C, Moynier F. Gallium isotopic evidence for the fate of moderately volatile elements in planetary bodies and refractory inclusions. Earth and Planetary Science Letters, 2017, 479: 330-339.

[142] Zhang T, Zhou L, Yang L, et al. High precision measurements of gallium isotopic compositions in geological materials by MC-ICP-MS. Journal of Analytical Atomic Spectrometry, 2016, 31: 1673-1679.

[143] Yuan W, Saldi G D, Chen J, et al. Gallium isotope fractionation during Ga adsorption on calcite and goethite. Geochimica et Cosmochimica Acta, 2018, 223: 350-363.

[144] Morgan J L, Skulan J L, Gordon G W, et al. Rapidly assessing changes in bone mineral balance using natural stable calcium isotopes. Proceedings of the National Academy of Sciences, 2012, 109: 9989-9994.

[145] Krayenbuehl P-A, Walczyk T, Schoenberg R, et al. Hereditary hemochromatosis is reflected in the iron isotope composition of blood. Blood, 2005, 105: 3812-3816.

[146] Télouk P, Puisieux A, Fujii T, et al. Copper isotope effect in serum of cancer patients. A pilot study. Metallomics, 2015, 7: 299-308.

[147] Larner F, Rehkaemper M. Evaluation of stable isotope tracing for ZnO nanomaterials: New constraints from high precision isotope analyses and modeling. Environmental Science & Technology, 2012, 46: 4149-4158.

[148] Laycock A, Coles B, Kreissig K, R et al. High precision $^{142}Ce/^{140}Ce$ stable isotope measurements of purified materials with a focus on CeO_2 nanoparticles. Journal of Analytical Atomic Spectrometry, 2016, 31: 297-302.

[149] Yang L, Dabek-Zlotorzynska E, Celo V. High precision determination of silver isotope ratios in commercial products by

MC-ICP-MS. Journal of Analytical Atomic Spectrometry, 2009, 24: 1564-1569.

[150] Lu D, Liu Q, Zhang T, et al. Stable silver isotope fractionation in the natural transformation process of silver nanoparticles. Nature Nanotechnology, 2016, 11: 682-686.

[151] Yang X, Liu X, Zhang A, et al. Distinguishing the sources of silica nanoparticles by dual isotopic fingerprinting and machine learning. Nature Communications, 2019, 10: doi: 10.1038/s41467-41019-09629-41465.

[152] Guilbaud R, Butler I B, Ellam R M, et al. Fe isotope exchange between Fe(II)(aq) and nanoparticulate mackinawite (FeSm) during nanoparticle growth. Earth and Planetary Science Letters, 2010, 300: 174-183.

[153] Wu L, Druschel G, Findlay A, et al. Experimental determination of iron isotope fractionations among Fe-aq(2+)-FeSaq-Mackinawite at low temperatures: Implications for the rock record. Geochimica et Cosmochimica Acta, 2012, 89: 46-61.

[154] Zhang T, Lu D, Zeng L, et al. Role of secondary particle formation in the persistence of silver nanoparticles in humic acid containing water under light irradiation. Environmental Science & Technology, 2017, 51: 14164-14172.

[155] Yin Y, Tan Z, Hu L, et al. Isotope tracers to study the environmental fate and bioaccumulation of metal-containing engineered nanoparticles: Techniques and applications. Chemical Reviews, 2017, 117: 4462-4487.

[156] Supiandi N I, Charron G, Tharaud M X, et al. Isotopically labelled nanoparticles at relevant concentrations: how low can we go?-The case of CdSe/ZnS QDs in surface waters. Environmental Science & Technology, 2019.

[157] Ubrihien R P, Taylor A M, Krikowa F, et al. Stable isotope analysis to detect copper(Cu)accumulation in species with high endogenous Cu concentrations: linking Cu accumulation with toxic effects in the gastropod Bembicium nanum. Marine and Freshwater Research, 2017, 68: 2087-2094.

[158] Cox A D, Noble A E, Saito M A. Cadmium enriched stable isotope uptake and addition experiments with natural phytoplankton assemblages in the Costa Rica Upwelling Dome. Marine Chemistry, 2014, 166: 70-81.

[159] Southall S C, Micklethwaite S, Wilson S A, et al. Changes in crystallinity and tracer-isotope distribution of goethite during Fe(II)-accelerated recrystallization. ACS Earth and Space Chemistry, 2018, 2: 1271-1282.

[160] Joshi P, Fantle M S, Larese-Casanova P, et al. Susceptibility of goethite to Fe^{2+}-catalyzed recrystallization over time. Environmental Science & Technology, 2017, 51: 11681-11691.

[161] Biswakarma J, Kang K, Borowski S C, et al. Fe(II)-catalyzed ligand-controlled dissolution of iron (hydr) oxides. Environmental Science & Technology, 2018, 53: 88-97.

[162] Zhou Z, Latta D E, Noor N, et al. Fe(II)-catalyzed transformation of organic matter–ferrihydrite coprecipitates: A closer look using Fe isotopes. Environmental Science & Technology, 2018, 52: 11142-11150.

[163] ThomasArrigo L K, Byrne J M, Kappler A, et al. Impact of organic matter on iron(II)-catalyzed mineral transformations in ferrihydrite–organic matter coprecipitates. Environmental Science & Technology, 2018, 52: 12316-12326.

[164] Frierdich A J, Nebel O, Beard B L, et al. Iron isotope exchange and fractionation between hematite(α-Fe_2O_3)and aqueous Fe(II): A combined three-isotope and reversal-approach to equilibrium study. Geochimica et Cosmochimica Acta, 2019, 245: 207-221.

[165] Chen C, Thompson A. Ferrous iron oxidation under varying pO_2 levels: the effect of Fe(III)/Al(III)oxide minerals and organic matter. Environmental Science & Technology, 2018, 52: 597-606.

[166] ThomasArrigo L K, Mikutta C, Byrne J, et al. Iron(II)-catalyzed iron atom exchange and mineralogical changes in iron-rich organic freshwater flocs: an iron isotope tracer study. Environmental Science & Technology, 2017, 51: 6897-6907.

[167] 朱宗强, 王训, 王衡, 等. 单一汞同位素示踪大气与农田作物汞的交换过程. 环境化学, 2018, 37: 419-427.

[168] Shetaya W H, Huang J H, Osterwalder S, et al. Sorption kinetics of isotopically labelled divalent mercury($^{196}Hg^{2+}$)in soil. Chemosphere, 2019, 221: 193-202.

[169] Strickman R, Mitchell C P. Accumulation and translocation of methylmercury and inorganic mercury in Oryza sativa: An

enriched isotope tracer study. Science of the Total Environment, 2017, 574: 1415-1423.

[170] Álvarez C R, Jiménez-Moreno M, Bernardo F G, et al. Using species-specific enriched stable isotopes to study the effect of fresh mercury inputs in soil-earthworm systems. Ecotoxicology and Environmental Safety, 2018, 147: 192-199.

[171] Jiang P, Li Y, Liu G, et al. Evaluating the role of re-adsorption of dissolved Hg^{2+} during cinnabar dissolution using isotope tracer technique. Journal of Hazardous Materials, 2016, 317: 466-475.

[172] Wang Z, Sun T, Driscoll C T, et al. Mechanism of accumulation of methylmercury in rice(*Oryza sativa* L.)in a mercury mining area. Environmental Science & Technology, 2018, 52: 9749-9757.

[173] LI Y J, Min L, Liu X B, et al. Zinc absorption from representative diet in a Chinese elderly population using stable isotope technique. Biomedical and Environmental Sciences, 2017, 30: 391-397.

[174] Talsma E F, Moretti D, Ly S C, et al. Zinc absorption from milk is affected by dilution but not by thermal processing, and milk enhances absorption of zinc from high-phytate rice in young Dutch women. The Journal of Nutrition, 2017, 147: 1086-1093.

[175] Fan W, Ren J, Wu C, et al. Using enriched stable isotope technique to study Cu bioaccumulation and bioavailability in *Corbicula fluminea* from Taihu Lake, China. Environmental Science and Pollution Research, 2014, 21: 14069-14077.

[176] Tang W, Zhong H, Xiao L, et al. Inhibitory effects of rice residues amendment on Cd phytoavailability: A matter of Cd-organic matter interactions? Chemosphere, 2017, 186: 227-234.

[177] Chen B, Nayuki K, Kuga Y, et al. Uptake and intraradical immobilization of cadmium by arbuscular mycorrhizal fungi as revealed by a stable isotope tracer and synchrotron radiation μX-ray fluorescence analysis. Microbes and Environments, 2018: ME18010.

[178] Tan Q G, Lu S, Chen R, et al. Making acute tests more ecologically relevant: Cadmium bioaccumulation and toxicity in an estuarine clam under various salinities modeled in a toxicokinetic-toxicodynamic framework. Environmental Science & Technology, 2019.

[179] Guo X, Yin Y, Tan Z, et al. Environmentally relevant freeze–thaw cycles enhance the redox-mediated morphological changes of silver nanoparticles. Environmental Science & Technology, 2018, 52: 6928-6935.

[180] Dang F, Chen Y, Huang Y, et al. Discerning the sources of silver nanoparticle in a terrestrial food chain by stable isotope tracer technique. Environmental Science & Technology, 2019.

[181] Shetaya W H, Osterwalder S, Bigalke M, et al. An isotopic dilution approach for quantifying mercury lability in soils. Environmental Science & Technology Letters, 2017, 4: 556-561.

[182] Mao L, Young S, Tye A, et al. Predicting trace metal solubility and fractionation in Urban soils from isotopic exchangeability. Environmental Pollution, 2017, 231: 1529-1542.

[183] Ren Z L, Sivry Y, Tharaud M, et al. Speciation and reactivity of lead and zinc in heavily and poorly contaminated soils: Stable isotope dilution. chemical extraction and model views, Environmental Pollution, 2017, 225: 654-662.

[184] Izquierdo M, Tye A, Chenery S. Using isotope dilution assays to understand speciation changes in Cd, Zn, Pb and Fe in a soil model system under simulated flooding conditions. Geoderma, 2017, 295: 41-52.

[185] Guo Z, Ye H, Xiao J, et al. Biokinetic modeling of Cd bioaccumulation from water, diet and sediment in a marine benthic goby: A triple stable isotope tracing technique. Environmental Science & Technology, 2018, 52: 8429-8437.

[186] Guo Z, Ni Z, Ye H, et al. Simultaneous uptake of Cd from sediment, water and diet in a demersal marine goby Mugilogobius chulae. Journal of Hazardous Materials, 2019, 364: 143-150.

[187] Rodríguez-Menéndez S, Fernández B, González-Iglesias H, et al. Isotopically-enriched tracers and ICP-MS methodologies to study zinc supplementation in single-cells of retinal pigment epithelium *in vitro*. Analytical Chemistry, 2019.

[188] Laycock A, Romero-Freire A, Najorka J, et al. Novel multi-isotope tracer approach to test ZnO nanoparticle and soluble Zn

bioavailability in joint soil exposures. Environmental Science & Technology, 2017, 51: 12756-12763.

[189] Brnić M, Hurrell R F, Songré-Ouattara L T, et al. Effect of phytase on zinc absorption from a millet-based porridge fed to young Burkinabe children. European Journal of Clinical Nutrition, 2017, 71: 137.

[190] Long J M, Mondal P, Westcott J E, et al. Zinc absorption from micronutrient powders Is low in bangladeshi toddlers at risk of environmental enteric dysfunction and may increase dietary zinc requirements. The Journal of Nutrition, 2019, 149: 98-105.

[191] Cesário R, Hintelmann H, Mendes R, et al. Evaluation of mercury methylation and methylmercury demethylation rates in vegetated and non-vegetated saltmarsh sediments from two Portuguese estuaries. Environmental Pollution, 2017, 226: 297-307.

[192] Figueiredo N, Serralheiro M, Canário J, et al. Evidence of mercury methylation and demethylation by the estuarine microbial communities obtained in stable Hg isotope studies. International Journal of Environmental Research and Public Health, 2018, 15: 2141.

[193] Cabrita M T, Duarte B, Cesário R, et al. Mercury mobility and effects in the salt-marsh plant Halimione portulacoides: Uptake, transport, and toxicity and tolerance mechanisms. Science of the Total Environment, 2019, 650: 111-120.

[194] Meng B, Li Y, Cui W, et al. Tracing the uptake, transport, and fate of mercury in sawgrass (*Cladium jamaicense*) in the Florida everglades using a multi-isotope technique. Environmental Science & Technology, 2018, 52: 3384-3391.

[195] Zhu W, Song Y, Adediran G A, et al. Mercury transformations in resuspended contaminated sediment controlled by redox conditions, chemical speciation and sources of organic matter. Geochimica et Cosmochimica Acta, 2018, 220: 158-179.

[196] Ndu U, Christensen G A, Rivera N A, et al. Quantification of mercury bioavailability for methylation using diffusive gradient in thin-film samplers. Environmental Science & Technology, 2018, 52: 8521-8529.

[197] Yu S, Yin Y, Zhou X, et al. Transformation kinetics of silver nanoparticles and silver ions in aquatic environments revealed by double stable isotope labeling. Environmental Science: Nano, 2016, 3: 883-893.

[198] Yang Q, Shan W, Hu L, et al. Uptake and transformation of silver nanoparticles and ions by rice plants revealed by dual stable isotope tracing. Environmental Science & Technology, 2018, 53: 625-633.

[199] Hadeishi T. Isotope-shift Zeeman effect for trace-element detection: an application of atomic physics to environmental problems. Applied Physics Letters, 1972, 21: 438-440.

[200] Batz L, Ganz S, Hermann G, et al. The measurement of stable isotope distribution using Zeeman atomic absorption spectroscopy. Spectrochimica Acta Part B: Atomic Spectroscopy, 1984, 39: 993-1003.

[201] Lu X, Zhao J, Liang X, et al. The application and potential artifacts of zeeman cold vapor atomic absorption spectrometry in mercury stable isotope analysis. Environmental Science & Technology Letters, 2019, 6: 165-170.

[202] Wiederhold J G. Metal stable isotope signatures as tracers in environmental geochemistry. Environmental Science & Technology, 2015, 49: 2606-2624.

作者：刘　倩[1]，陈玖斌[2]，李　伟[3]，阴永光[1]，郑　旺[2]，杨学志[1]
[1]中国科学院生态环境研究中心，[2]天津大学，[3]南京大学

第 3 章　有机污染物生物有效性研究的方法与应用

- 1. 引言 /48
- 2. 有机污染物生物有效性的评价方法 /49
- 3. 生物有效性在研究有机污染物在环境介质及界面间迁移行为中的应用 /53
- 4. 生物有效性在人体健康研究中的应用 /57
- 5. 生物有效性在生态毒性评价识别中的应用 /63

本章导读

进入21世纪以来，全球人口和社会经济急速增长。随着工农业生产的飞速发展及城市的扩张，大量的有机合成化学品被研发和使用，并通过各种途径排放到环境介质中，在水体、土壤和沉积物中富集，破坏生态系统稳态。它们还会通过生物积累或生物放大作用对环境生物和人类健康造成潜在危害。目前，环境中多种类型的有机污染物被广泛检出，主要包括多环芳烃（PAHs）、有机氯农药（OCPs）、多氯联苯（PCBs）、多溴二苯醚（PBDEs）、全氟化合物（PFOAs/PFOSs）以及药物和个人护理用品（PPCPs）等。这些有机污染物大都具有持久性、生物蓄积性和毒性，尤其部分化合物可能具有显著的内分泌干扰效应。因此，环境中有机污染物的水平现状和暴露风险评估已成为环境科学领域的重点研究内容。直接利用环境介质中有机污染物的总量评价其生物有效性和毒性效应，可能导致过高估计其环境风险。污染物进入环境介质后，与不同成分发生相互作用，其赋存状态随之发生变化，致使其移动性、生物有效性以及化学反应活性发生变化，因而对生态系统和人体健康的风险发生改变。环境中有机污染物的生物有效性评价对污染物环境行为、生物修复、生态毒性和健康风险研究有着重要的意义。本章将针对有机污染物生物有效性的评价方法及其在环境和健康领域的应用进行较全面的综述。

关键词

生物有效性，有机污染物，生态毒性，风险评估，评价方法

1 引　言

目前环境介质（大气、水体、沉积物、土壤等）中有机污染物风险评估大多采用有机污染物的总浓度。有机污染物总浓度可以反映其在环境中富集程度，但是并未考虑到污染物的赋存形态及环境因素影响，因此不能准确反映有机污染物对环境和生物体的生态毒性和健康危害，由此得到的结果可能会过高地估计了污染物的生态风险。

随着对污染物环境水平和生物效应的深入研究，科学家们提出了生物有效性（bioavailability）这一概念，但是大家对生物有效性的理解不同，因此在不同领域难以形成统一的概念。通常，针对污染物的生物有效性主要从环境科学和生物毒理学两个方面进行定义。环境科学中的生物有效性是指化学物质被生物吸收的程度和潜在的毒性，而生物毒理学中的生物有效性则侧重化学物质穿过生物膜进入细胞的量。在此基础之上，Alexander等提出了环境领域中生物有效性的定义，即用来表征污染物与生物接触或与之发生相互作用时被生物所吸收利用的程度和潜在毒性[1]。Semple等提议将生物有效性进一步划分为生物有效性和生物可及性（bioaccessibility）两个部分，其中生物有效性指环境中存在的化合物且可通过细胞膜直接进入生物体的部分，而生物可及性则包括生物有效性和潜在存在的可进入生物的两部分[2]。虽然对于生物有效性的概念还没有达成完全的统一，但大家认为其实质都是研究化学物质和生物体之间的潜在相关关系，这一过程都涉及物理、化学、生物等过程。

针对生物有效性定义上存在的分歧，2003年美国国家研究理事会（National Research Council，NRC）提出了"生物有效性过程"来描述化合物从环境介质进入生物体的过程[3]。该过程包括了污染物的形态之间相互转换、污染物的环境迁移、污染物通过生物膜被生物体吸收、污染源与靶标位点的结合四个过程。虽然对生物有效性的概念有了深入的认识，但在实际应用中不同的测定方法仍然无法获得准确统一的结果。为了更加具体化，Reichenberg和Mayer使用化学活度（chemical activity）来描述生物有效性，他们

认为可及性（accessibility）是潜在的生物有效性，可通过仿生萃取技术（温和溶剂萃取、超临界流体萃取、固相萃取等）进行提取和测定，而化学活度与逸度和自由溶解态浓度一样，是在平衡状态污染物进入生物体的过程，可以通过被动采样技术进行测定[4]。

近年来，在认识和完善生物有效性过程中，国内外科学家们已经发展了多种基于生物和化学的污染物生物有效性评价方法，但是这些方法在应用到不同环境介质和污染物种类时得到的评价结果也往往存在很大的差异。针对不同的环境介质和污染物，同时考虑到一系列环境影响因素，应建立统一的、标准的生物有效性预测和评价方法，从而提高环境中有机污染物风险诊断的准确性，也为生态环境和人类健康风险评价提供理论基础和依据。

2　有机污染物生物有效性的评价方法

目前，有机污染物生物有效性的评价方法主要包括生物学方法和化学方法两大类。生物学方法主要分为直接和间接两大类，化学方法主要有模型预测法和化学分析法。

2.1　生物学方法

利用生物评估有机污染物的生物有效性是比较直观的方法，可以直接得到污染物在生物体内的富集程度，常用的指示生物主要包括动物、植物和微生物。生物评估常用的测试方法主要有生物蓄积法、临界机体残留（critical body residue，CBR）法和离体生物标记物法。

生物蓄积法是将生物直接暴露于环境介质中，通过直接测量生物体内富集污染物含量或被微生物矿化的污染物含量。生活在环境介质中的生物体会通过不同途径受到化学物质暴露，而生物体在环境中的暴露方式也会影响其污染物的富集程度。生物体暴露除了受到自身形态、生理机能、代谢过程和生物行为的影响外，污染物的物理化学性质和周围环境条件（温度、气候等）也在一定程度上对富集结果产生影响。

由于不同污染物在生物体内的吸收特异性以及在不同生物体内的分布差异性，生物体内污染物的富集程度不能直观反映其生物有效性，因此引入靶位点浓度来评价污染物的生物有效性，即与生物体内靶位点发生相互作用并产生毒性作用的污染物含量，通常用临界机体残留浓度评估靶位点浓度。CBR法测定是以致死或半数致死效应为评价终点时生物体靶标内富集的污染物浓度，能将有机污染物的毒性和生物富集作用结合在一起[5]。

除了上述生物学直接评价方法外，间接评价法也得到了越来越广泛的应用。离体生物标记物法适用于环境中污染物浓度低于临界机体残留浓度或活体暴露生物实验较难完成时，通过测定暴露后生物体内大分子、细胞或生理的相应变化评估污染物的剂量-效应关系[6]。离体测试法中生物标记物的快速检测可高效、准确地提供受试污染物的生物有效性[7]。例如，Oikari等通过测定虹鳟鱼肝脏中CYP1A酶活程度评价PAHs的生物有效性[8]；Koganti等利用小鼠尿液中代谢水平和肺中化合物-DNA加合物水平来评价土壤中PAHs的生物有效性[9]。

生物学评估方法是最直观、准确的生物可利用性的评估方式，适用范围广，能涵盖大多数有机污染物，但是生物评估的周期较长、耗费昂贵、处理繁琐，而且生物个体的之间的差异性也会导致生物评估的重复性和平行性较差。

2.2 化学方法

2.2.1 模型预测法

长期以来，科研工作者在有机污染物生物有效性的预测方面开展了大量的研究工作，针对污染物的物理化学性质和生态系统参数发展建立了一系列评价预测模型。目前常用于预测生物效应和毒性终点的模型主要有定量结构-活性关系（quantitative structure-activity relationship，QSAR）模型[10]、定量结构-性质关系（quantitative structure-property relationship，QSPR）模型[11]和多介质数学模型（multimedia mathematical model，MMM）[12]。QSAR/QSPR是通过研究化合物的分子结构与其活性/性质之间的关系，达到预测化合生物活性或理化性质的目的，进而对化合物的性质、活性、毒性及环境行为进行预测和评价。QSAR/QSPR可评估污染物在多种环境介质中的迁移、分布、归趋、生物浓缩和生物富集过程。

尽管这些模型在环境介质中得到了比较广泛的应用，并且在一定程度上可以准确地预测污染物的生物有效性，但是这些模型大都建立在理想条件下，预测结果也会与实际情况有较大的偏差[13]。因此，在实际应用中可以将QSAR/QSPR模型与MMM耦合，提高模拟效率，同时将模型预测法与多种化学计量学方法结合，可以得到更准确的评价结果。

2.2.2 化学分析法

化学分析法评估环境介质中有机污染物生物有效性具有快捷简便、成本低廉的特点，近年来也得到了很大的发展，化学分析法主要包括传统的激烈的完全提取方法（如索氏提取、超声提取、加速溶剂提取、超临界流体萃取等）和温和的提取方法[如温和溶剂萃取、环糊精提取、聚2,6-二苯基对苯醚（Tenax）提取、被动采样等]。

早期的化学浸提法通常采用强烈的化学浸提剂和强有力的浸提程序，但后续研究发现这类方法会提取出大量不可被生物利用的化合物，过高地估计了这些污染物的生态风险，因此近些年来常用的化学分析法都是较为温和提取方法，或者是利用一些仿生材料来评价污染物的生物有效性。目前常用的两大类化学方法主要是提取法和被动采样法，提取法主要利用温和有机溶剂或第三相物质（如Tenax）提取快速解吸至水相的有机污染物，以此来评估其生物可利用性。被动采样方式主要是以非耗竭形式测定污染物从基质中自由溶出进入到采样器内的总量，当目标物选择性分配并吸附到替代相到达平衡后，即可获得给定采样周期内有机污染物的自由溶解态浓度，进而评估其生物有效性。目前应用较广泛的被动式采样装置主要包括固相微萃取（solid phase microextraction，SPME）、液相微萃取（liquid phase microextraction，LPME）以及半渗透膜被动式采样器（semi-permeable membrane device，SPMD）。以下将对常用的化学方法进行简述。

1）温和溶剂萃取

温和溶剂萃取通常指采用中等极性有机溶剂萃取与固相结合较弱的有机污染物，常用的温和溶剂主要有正丁醇、甲醇、乙腈、乙醇和异丙醇等。已有研究结果表明温和试剂萃取结果与生物富集系数或生物降解能力有很好的相关性。例如Liste等利用正丁醇提取土壤中PAHs的含量与蚯蚓体内富集量有很好的相关性[14]。Lei等用70%乙醇或异丙醇提取PAHs的结果与微生物降解量也具有显著相关性[15]。但温和溶剂萃取的结果有时也会与生物可利用性部分之间存在较大偏差。Tang等采用正丁醇、丙醇、甲醇及乙酸乙酯从老化土壤中提取PAHs，将提取结果与动植物富集量和微生物降解量对比，结果表明溶剂萃取量高于动植物富集吸收量，但与微生物降解量有较好相关性[16]。Macleod等通过研究也发现正丁醇-水混合提取结果高估了实际微生物利用量，但甲醇-水混合提取结果则低估了这一部分[17]。因此，针对不同环境介质和目标化合物选择适当的提取溶剂才能准确地评估其生物有效性，同时还应当考虑提取过程中环境条件、污染物理化性质及受试生物种属等影响因素。

2）环糊精提取法

环糊精是一种天然无毒的环状低聚糖，具有亲水外壳和非极性空腔，可以捕获基质中解吸的疏水性有机污染物分子形成水溶性的包合物，进而增加其溶解性和稳定性[18]。羟丙基-β-环糊精（hydroxypropyl-β-cyclodextrin，HPCD）是使用最为广泛的一类环糊精，常用于评价 PAHs、OCPs、PCBs 和脂肪烃等在土壤或沉积物中的生物有效性[19, 20]。HPCD 操作过程简便，无需附加装置，一般是直接将 HPCD 溶液加入到待测样品基质中，充分混合一定时间后离心取上清液即可分析测定。Reid 等利用 HPCD 对土壤中 ^{14}C 标记的菲进行萃取，结果发现 HPCD 萃取量和微生物对菲的矿化效果之间存在显著相关性，利用 HPCD 萃取土壤中有机污染物生物有效性的部分可以很好地模拟污染物在土壤和生物体之间的传质过程[21]。但对于蚯蚓的富集模拟却存在较大偏差，Hickman 等发现蚯蚓在土壤中积累菲与 HPCD 对菲萃取量之间没有相关性（$r^2=0.07$）[22]。同时，Barthe 等也发现 HPCD 萃取不能准确地模拟估算底栖无脊椎动物积累 PAHs 的量[23]。在实际应用中，尤其是对土壤样品的提取中，HPCD 的提取容量有限，可能会低估污染物的生物有效性，在此基础之上，Mayer 等进一步开发了吸附性生物有效性提取（sorptive bioaccessibility extraction，SBE）方法，通过引入高吸附容量的聚合物对 HPCD 溶液中的目标物连续吸附，进而避免 HPCD 吸附量达到饱和时造成的损失[24]。

3）Tenax 提取

Tenax 是一种以聚 2,6-二苯基对苯醚为颗粒的多孔膜材料，对疏水性有机污染物具有极强的亲和力，本身质地轻，易于在提取液表面过滤分离回收，因此被认为是较好的吸附剂，目前被广泛应用于不同环境介质中疏水性有机污染物（PAHs、PCBs、DDTs 等）的生物有效性评价[25]。Tenax 提取原理是基于生物体只能利用水溶性有机物，通过将 Tenax 树脂颗粒加入到样品匀浆液中，充分混合均匀后离心收集 Tenax 树脂颗粒并多次加入干净颗粒直至提取完全，收集的树脂颗粒利用有机溶剂进行萃取净化，即可得到基质中有机污染物解吸部分的浓度，进而预测该有机污染物的生物有效性。化学溶剂提取法中，Tenax 提取方法重现性好、试验周期短、易于再生，因此在沉积物、土壤和污泥中应用较为广泛。Cornelissen 等使用 Tenax 连续提取研究沉积物中氯苯（chlorobenzenes，CBs）、PAHs 和 PCBs 的解吸动力学，结果发现 Tenax 提取部分与快速解吸的 PAHs 部分具有很好地相关性，即 Tenax 提取土壤中 PAHs 可作为快速评价其生物有效性的方法[26]。孙红文等利用 Tenax 提取技术预测了土壤中芘的生物有效性，结果发现土壤中芘的解吸百分率与蚯蚓体内芘的浓度存在明显的相关性（$r^2=0.935$）[27]。Oleszczuk 等使用 Tenax 研究了污泥中 PAHs 对 *H. incongruens* 的潜在生物可利用性[28]。虽然 Tenax 适用于多种有机污染物的提取，但是对于某些化合物如复杂石油烃的模拟效果并不十分理想，因此需要对 Tenax 吸附剂表面进行改性，使之更好地适用于环境基质中有机污染物的提取，以期为环境有机污染物的生物有效性评价提供可靠手段。

4）固相微萃取法

SPME 是由加拿大 Pawliszyn 团队发明的一种新型样品前处理技术，其原理是将目标污染物富集于一根涂有薄层聚合物的玻璃棒或硅棒上，然后通过高温热解吸或溶剂洗脱将污染物解吸到分析仪器中进行测定[29]。SPME 作为一种平衡采样器，操作快速、简便，可模拟水体和土壤环境中无脊椎动物的富集行为，被广泛用于环境分析和生物分析等领域，可实现环境介质中有机污染物的生物有效性预测。传统的 SPME 测定的是环境介质中目标污染物的总浓度，目的是达到最大的萃取效率，Vaes 等将其改进并应用于自由溶解态浓度的测定，即微损耗固相微萃取（negligible depletion solid phase microextraction，nd-SPME）技术[30]。目前，nd-SPME 技术已经成为应用最广泛的测定自由溶解态浓度的技术之一，成功在多种环境介质和生物样品中实现了有机污染物的检测[31]。Poerschmann 等将 nd-SPME 应用于腐殖酸类溶解性有机质中有机锡化合物的吸附平衡研究[32]；Hermens 等利用 nd-SPME 测定了土壤孔隙水中 PAHs 的自由溶解态浓度，根据自由溶解态浓度计算得到的吸附系数略高于基于辛醇-水的分配系数，这一差异也是受土壤孔隙水中溶解有机质的影响[33]。虽然 SPME 在实际应用中备受欢迎，但对有机污染物生物有效性评估时仍有一些因素会对其产生影响，例如基质效应、纤维表面污损和样品的平衡时间，在评估过程中应当注意并且结合其他评价方法。

5）液相微萃取法

LPME 是一种微型化的液液萃取技术，同时也结合了 SPME 的优点，可以根据目标化合物的性质灵活地选择萃取溶剂从而实现对极性有机污染物的萃取[34]。根据萃取的表现形式，LPME 又可分为单滴液相微萃取（single-drop liquid phase microextraction，SD-LPME）和中空纤维膜支载的液相微萃取（hollow fiber based liquid phase microextraction，HF-LPME）。SD-LPME 是最早开发的液相微萃取法，通常适用于萃取分子量相对较大、熔点和沸点较高、在水中有适当溶解度的化合物，这类化合物辛醇-水分配系数较大（K_{OW} 约为 2000 左右），常用与样品相不互溶的沸点较低的有机萃取溶剂，如正辛醇、苯、甲苯、正己烷等[35]。由于 SD-LPME 萃取过程中萃取液滴易于滴落，不适合复杂基质的分析，因此 Pedersen 等提出了以多孔纤维膜为基础的 HF-LPME，中空的多孔纤维可增加溶剂与样品的接触面积，减小溶剂损失，降低复杂基质的影响[36]。HF-LPME 按萃取过程中相传质过程的不同可分为两相中空纤维液相微萃取和三相中空纤维液相微萃取。两相中空纤维液相微萃取形式采用有机溶剂作为萃取剂，将其填充在中空纤维的膜壁、膜孔及管腔内部，待测物由于萃取融入壁孔和管腔的有机溶剂中，实现稳定的富集作用。而三相中空纤维液相微萃取形式则先通过中空纤维膜壁膜孔中填充的有机溶剂萃取目标物，然后再通过置于纤维内部的接受相反萃取目标物，该方法比较适合于酸性或者碱性可解离物质的前处理。

6）半渗透膜被动式采样法

SPMD 作为一种模拟生物富集有机污染物的被动采样装置，被广泛应用于疏水性有机污染物的环境经分析和毒理效应评价。SPMD 通常利用低密度聚乙烯膜（low density polyethylene，LDPE）或其他微孔聚合物制成薄膜长袋，内部注入一层中性、大分子量脂类如三油酸甘油酯，三油酸甘油酯与生物体内一些种类的脂肪相似，进而使 SPMD 可以模拟生物对环境中污染物的吸收过程[37]。由于三油酸甘油酯具有很强的疏水性，因此 SPMD 更适合于测定强疏水性有机污染物（PCBs、PAHs 等）的自由溶解态浓度，且 LDPE 半透膜阻截的分子尺寸与生物膜类似，因此该方法测定的自由溶解态浓度可以评价污染物的生物有效性[38]。但传统的 SPMD 制备繁琐，样品富集平衡所需时间较长，采样相体积较大，在此基础之上，王子健等建立了一种新型被动采样装置——三油酸甘油酯-醋酸纤维素复合膜（triolein embedded cellulose acetate membrane，TECAM）[39,40]。TECAM 采样原理与 SPMD 类似，但构造有所不同，由三油酸甘油酯以脂滴的形式嵌于醋酸纤维素聚合物构造中，这种结构使得接触面积更大，易于快速达到萃取平衡，且制备过程简单，提取目标污染物的前处理过程也比较简洁，极大程度地提高了整体采样速率。SPMD 结合其他分析方法的技术手段为评估环境介质中有机污染物的生物有效性提供了新的思路和发展前景。

基于生物有效性对环境介质中有机污染物的环境风险评估一直是环境科学领域的重点研究方向，针对实验室内发展的现有生物有效性评估技术手段，需要经过长期实践和大量模拟比对，以期将其推广到实际环境检测中。目前现有的化学检测方法不能完全真实地反映污染物的生物有效性，而生物学方法又存在物种差异性，且实际操作繁琐，周期漫长，在实际应用中难以进行具体比较。因此，生物有效性评价方法的未来发展中应考虑到以下几点：①发展快速、简便、高效、成本低廉的检测方法和技术，用以测定有机污染物的自由溶解态浓度；②考虑环境介质的外在影响及污染物的环境行为和过程对其生物有效性的影响；③在现有生物方法的基础之上进行优化并建立统一的标准体系；④开发新型仿生技术手段，实现有机污染物生物有效性的准确预测。

3 生物有效性在研究有机污染物在环境介质及界面间迁移行为中的应用

有机污染物在各环境介质中的生物有效部分恰好是其最具迁移能力的部分。相对于在环境介质的其他迁移过程，如颗粒物再悬浮、大气的干湿沉降及土壤扬尘，有机污染物的生物有效部分的迁移主要以自由扩散为主，取决于其在各个环境介质中自由态浓度梯度，受环境影响因素较小。由于其疏水性，有机污染物在环境介质中易吸附在颗粒有机质上，影响其在生物有效态浓度，从而影响污染物的迁移行为。另一方面，污染物的迁移通量及方向是衡量其迁移行为的关键参数。在污染的初期与后期，有机污染物在两相环境介质的迁移方向有所不同。为此，如何有效定量及判定污染物在典型环境介质界面间，如沉积物-水、水-大气及土壤-大气间的迁移通量与方向，亦成为研究污染物的迁移行为的基础。

3.1 沉积物

作为水生生态系统的重要组成部分，沉积物-水界面是众多底栖生物的栖息场所，也是污染物地球化学循环的重要发生区域。在沉积物-水界面处发生的物质吸附/解吸、迁移/转化、扩散/掩埋以及生物活动因素等一系列的过程决定着污染物的迁移行为和最终归趋。因此，准确了解典型有机污染物在沉积物-水界面的迁移行为、估算污染物的迁移速率和释放通量，对于评估受污染区域的生态健康风险具有重要意义。目前，对于生物有效态有机污染物在沉积物内部以及沉积物-水界面之间的迁移释放有了比较系统的研究，包括在沉积物中的迁移速率，典型有机污染物在沉积物-水界面上的扩散通量以及影响有机污染物界面迁移行为的环境因素等。研究手段有传统主动采样法和新的被动采样技术的应用，研究方式也从实验室模拟到与野外原位实测相结合。

由于有机污染物一般具有疏水性，易吸附在沉积物颗粒。为此，生物有效态的有机污染物在沉积物内部迁移速率很慢。例如，Thibodeaux 等[41]报道了有机污染物的迁移速率为 0.00024~0.0024 cm/d，Valsaraj 等[42]指出 TCDD 在沉积物中的扩散速率为 0.019 cm/d。底栖生物活动可以显著影响有机污染物迁移扩散。例如 Josefsson 等[43]就发现含有底栖生物的沉积物中多氯联苯的扩散速率是对照组的 2~4 倍。

另一方面，研究有机污染物在界面处迁移行为的关键是准确定量其在两种环境介质中的浓度。在早期对沉积物-水迁移通量的研究中，主要是以主动采样方法在两个环境介质中各选取一个距离界面一定深度的点，利用动力源采集两相的样品并确定其疏水性有机污染物（hydrophobic organic contaminants, HOCs）的浓度，可称为两点采样法。通常选择沉积物面上 0.5 m 采集水体样品，并采集大约 0~5 cm 的沉积物，将所得到的数据代入到经典的质量传递模型确定其迁移通量及方向，即

$$F_{\text{sed-water}} = K_{\text{OL}}^{\text{sed-w}}(C_{\text{porewater}} - C_{\text{water}}) = K_{\text{OL}}^{\text{sed-w}}\left(\frac{C_{\text{sed}}}{K_{\text{P}}} - C_{\text{water}}\right) \tag{3-1}$$

式中，$C_{\text{porewater}}$、C_{water}、C_{sed} 分别是污染物在孔隙水、上层水体、沉积物中的浓度；$K_{\text{OL}}^{\text{sed-w}}$ 是目标污染物在沉积物孔隙水与上层水体间的质量传递系数，其与污染物在水体的扩散系数及两相间的扩散层厚度相关[44]；K_{P} 则是指污染物在沉积物与水体间的分配系数，在一定程度上需进行沉积物中有机碳含量的校正；如果界面迁移通量是正值，则表示目标物从沉积物向水体逃逸，若是负值则表示由水向沉积物富集的过程。有机污染物在孔隙水中的自由溶解态的部分是影响其迁移扩散的关键。不同区域沉积物的理化性质存在差异（如沉积物有机碳含量），目标物的 K_{P} 值在不同沉积物中并不是一个统一值，各界面的质量迁

移系数也存在很大的不确定性，这必然会有经验值估算通量造成偏差[45, 46]。

近年来，针对经典方法暴露出来的不足之处，越来越多的研究者利用被动采样技术对有机污染物在沉积物-水的界面迁移行为和影响因素进行了研究。被动采样是利用目标物在高分子材料吸附相与环境介质之间的高分配系数来采集环境介质中自由溶解态污染物的一种采样方法。这是一种建立在非耗竭条件下的采样技术，可以直接原位采集环境介质中的自由溶解态物质，对污染物本身的地球化学行为影响较小，能更好地反映环境中污染物的真实分布情况。鉴于此，研究者开始通过放置被动采样器来替代主动采样确定目标物的自由态浓度，再将其代入上述质量传输模型中进行估算目标物的界面通量。另外，区别于两点的测量可能存在的诸多问题，研究者研发出一系列整套的装置来研究有机污染物在沉积物-水界面的迁移释放情况。例如，Eek 等[47]设计的通量箱装置。该装置使用时，将空心的不锈钢圆柱体倒扣于沉积物上，由沉积物释放出的有机污染物被置于腔体顶部的吸附相捕集，从而计算出有机污染物在单位时间内的释放量。由此装置测量的挪威奥斯陆港口污染沉积物释放芘及 PCB-52 的释放通量分别为 $300\sim1600$ $ng/(m^2 \cdot d)$ 和 $2\sim8$ $ng/(m^2 \cdot d)$。

另一方面，多数研究者意识到有机污染物在环境界面处并不是从一个浓度到另一个浓度的突变，而是存在一个连续的浓度趋势，因此获得界面处分辨度更高的浓度梯度就显得更为重要。对此，Lin 等[48]将单个低密度聚乙烯膜以 0.5 cm 的间隔固定在一根不锈钢柱体上，测量了意大利马雷焦湖沉积物孔隙水（水面以下 40 cm）及上层水体（水面以上 25 cm）中滴滴涕及其代谢产物的分布趋势，并以菲克第一扩散定律计算了其迁移扩散通量为 3.9 $ng/(m^2 \cdot d)$。Belles 等[49]将一长条低密度聚乙烯膜固定在框架上，直接插入沉积物。放置结束后则将低密度聚乙烯膜取出分段剪出（每段膜宽度为 1 cm），用于萃取后确定沉积物孔隙水与上层水体中多环芳烃的浓度趋势。发现沉积物表层 1 cm 深处的多环芳烃浓度是深处的 500 倍，并计算出在沉积物 1 cm 深处的迁移通量为 $0.1\sim5$ $ng/(m^2 \cdot a)$。Fernandez 等[50]则用类似的方法测得滴滴涕类化合物在典型污染区域由沉积物扩散到水体的迁移扩散通量为 $240\sim1100$ $ng/(m^2 \cdot a)$。

但上述研究仍存在分辨率不够和忽视目标物在低密度聚乙烯膜上扩散的问题。在此基础上，Liu 等[51]研发了分辨度更高的沉积物-水界面被动采样装置。该装置由上层水体部分和下层沉积物部分组成。上层水体由采样单元呈水平纵向螺旋式排列，采样单元间的距离通过不同高度的不锈钢高度调节，共 37 层，接近于沉积物-水界面的最小高度可达 0.17 cm。下层沉积物部分则采样单元竖向设置，每个采样单元中不同高度镶嵌了相同厚度的低密度聚乙烯膜，用于定量深度达 10 cm 孔隙水中有机污染物的浓度分布趋势。该通量装置可获取有机污染物在上层水体高分辨的浓度分布趋势。此模式的优势在于用数学方法把界面扩散通量表达成被动采样装置的测量参数。该界面通量装置在广东省海陵湾中滴滴涕的界面通量的测定中进行了应用，并与通量箱测定的值进行相互验证。结果显示，界面通量采样装置测定的滴滴涕代谢产物的沉积物释放通量是 $5.9\sim150$ $ng/(m^2 \cdot d)$，与通量箱测定的值 [$5.5\sim85$ $ng/(m^2 \cdot d)$] 相当[51]。

对于影响有机污染物在沉积物-水界面行为的因素，主要是环境自身因素，包括沉积物有机碳含量及温度等。如 Feng 等的研究显示黑炭及干酪根在沉积物中的含量分别降低 6.7% 及 11%，将使得滴滴涕代谢产物（p,p'-DDE、p,p'-DDD 及 o,p'-DDD）的沉积物-水界面通量分别增长 $11\%\sim14\%$ 及 $12\%\sim23\%$[52]。在实际环境中影响污染物迁移行为是众多因素的共同作用的结果，不同的生态系统，如海洋、河流、湖泊等之间也有很大的差异。目前的研究多集中单因素的定性研究，还不能做到环境因素影响的复合化与定量化，研究的对象也多是河口、海湾，对于内陆的大型湖泊的研究还较少。后续对于有机污染物在复杂条件下界面迁移行为的研究仍在不断进行之中。

3.2 水体

水体是地球环境介质中有机污染物最大的流动载体。有机污染物在水体里存在两种形态：一个是颗粒物态；另一个是自由溶解态，而其生物有效态部分恰是自由溶解态。水体内自由溶解态的有机污染物向其他环境介质迁移的行为主要分三种：一是有水体沉降到沉积物；二是通过生物富集在食物链上的传递；三是由水体挥发至大气。第一种方式在本章 3.1 节已充分探讨。第二种在食物链上的传递，涉及有机污染物的生物放大及生物转化等行为，由于章节篇幅有限，暂不作讨论。本节只关注水体中自由溶解态有机污染物在不同深度水体的迁移及水体-大气间的交换。

浅层水体内有机污染物的扩散在水流的作用下较难分辨出具体的迁移机制。为此，研究者利用以低密度聚乙烯膜为吸附相的被动采样技术获得深海水体（最深达水深 5000m）自由溶解态有机污染物的浓度分布，从而探讨其迁移机制[53, 54]。前期研究者认为自由溶解态有机污染物可吸附在颗粒物上，随着颗粒沉降到海洋深部，再由于颗粒的矿化释放[55]。然而从直接测量自由溶解态污染物的不同深度水体的浓度分布来看，疏水性强的六氯代及七氯代多氯联苯的浓度峰值出现在水深 800m 以上，然后逐渐降低，而疏水性较弱的二氯代及三氯代多氯联苯反而在水深 800m 后，呈现出增加随后降低的趋势。这一结果说明颗粒物的沉降协同运输对于有机污染物在深海水体中的迁移的影响并不显著，反而海水侧向运输的作用更为重要。

水体中生物有效态的有机污染物在水体-大气间的交换行为通常以逸度大小来评估其迁移通量与方向。多数研究者利用被动采样技术获得有机污染物在一定高度与水深处的自由态浓度，再采用基于菲克第一扩散定律的经典两点法估算典型区域，如五大湖地区、墨西哥湾等有机物污染物在水体-大气间的迁移通量[56-58]，在五大湖地区，传统的多氯联苯已呈现出由水体扩散至大气趋势，而较为新型的多溴二苯醚则表现为由大气沉降到水体。然而，这一判定方法的不确定性在于不同高度处获得有机污染物的浓度不同，其用于判定交换方向可出现截然相反的结果。事实上，区别于污染物在沉积物-水界面间的分布，水体-大气界面间存在一个微表面层。微表面层富含有机质，易吸附在界面处的有机污染物[59]，或可影响有机污染物在两相间的交换。目前，室内模拟实验已经表明，当水体有机污染物自由溶解态浓度可维持在一个稳定状态时，微表面层的溶解性有机质反而可以促进污染物在水体-大气间的交换行为[60]。然而这目前仅限于室内测量，在原位自然环境中的研究甚少。在近期 Wu 等研究者[61]设置了一个能测定水体-大气近界面处有机污染物的浓度分布的以低密度聚乙烯膜为吸附相的被动采样装置。自然环境应用试验表明，采样装置测定的多环芳烃化合物在近水-大气水体层及水体主层的逸度分布呈现"C"或"L"形式，捕集了近水-大气界面处的多环芳烃化合物的逸度趋势，其递增的模式反映了有机污染物在近界面处的环境行为的复杂性。然而该水-大气被动采样装置目前仍难以定量污染物在微表面层的浓度，因此不能预估有机污染物在水-大气界面的交换通量。由此可见，水体中自由溶解态有机污染物在水体-大气间的迁移行为仍需更多的投入研究。

3.3 土壤

与沉积物类似，土壤是有机污染物的一个"储存库"。厘清和定量有机污染物在土壤中及土壤-大气间的迁移可以明晰土壤在污染物整个环境介质迁移的作用。土壤中有机污染物的形态相对于其在沉积物中更为复杂，存在固态、液态及气态形式，其中生物有效部分主要包括液态和气态。为此，土壤中污染物的生物有效态部分的迁移行为包括以滤液形式从土壤表面层迁移至下层甚至于地下水，或以气态形式从土壤挥发至大气及从大气沉降到土壤。目前数据显示，相对而言，多数研究者致力于探讨有机污染物在土壤与大气间的迁移，对有机污染物在土壤间的迁移研究较少。根据最近的研究结果表明，有机污染物从土壤表层迁移至下层的速率比预计的速度快。例如，Zheng 等研究者[62]通过加标 ^{13}C 标记的多氯联苯到土壤表面植物落叶，采用渗漏测定计结合聚氨酯泡沫收集目标物的生物有效态部分，原位测量了多氯联苯系列同系物在森林土壤中的迁移性。研究结果显示，在 120 天内，标记的多氯联苯已从表层植物落叶迁移在距离土壤表面 10 cm 深度的层，不同疏水性多氯联苯同系物从落叶到土壤表层及由表层到下层（10 cm 深度）向下迁移速率相差无几，大致范围是 0.0022~0.0034/d 及 0.00023~0.00047/d。此外，研究还发现这部分生物有效态多氯联苯与土壤中细颗粒及溶解性有机质协同迁移的现象。

生物有效态的有机污染物在土壤与大气间的迁移被认为主要以气态形式。通常情况下，研究者采用逸度形式统一生物有效态有机污染物在土壤与大气的浓度，从而评估其在两相间的迁移通量与方向。最经典的方法就是采用两点法，即通过常用的采样方式获得有机污染物在土壤与大气中的总浓度，再经过污染物在土壤-大气间的分配系数校正，得到其在两相间的逸度，从而计算得到土壤-大气的交换通量。交换方向的判定则通过污染物在土壤的逸度所占其在两相间逸度之和的比例与 0.5 的大小，如果大于 0.5，则认为污染物从土壤挥发至大气，反之，则从大气沉降到土壤。该方法的主要疑问在于有机污染物在土壤结构内部固相与气相间的分配是否已经达到平衡，能否采用平衡分配系数获得土壤中污染物的逸度，以及其在大气浓度能否代表近地面的值。例如，Zhang 等[63]以聚氨酯泡沫（PUF）作为吸附相，设计了一个由一系列逸度采样器镶嵌在不锈钢站台柱组成的装置，获取近于土壤表面上不同高度的气态多环芳烃垂直分布。结果显示，不同采样点处单个多环芳烃化合物在距离土壤表面不同高度中大气的浓度是不同的。研究者亦通过不同高度的浓度分布来判定目标物在土壤-大气间的迁移方向。这似乎更具有说服力。由此可见，采用两点法评估的结果具有一定的不确定性。

显而易见，采用逸度装置直接原位获取近土壤中生物有效态有机污染物，从而评估其在两相间的通量或许能减低评估的不确定性。为此，早在 2009 年，Cabrerizo 等[64]改进了土壤逸度采集装置（由一个表面面积 1m² 不锈钢平板置于土壤表面 3 cm 高度，在平板上中心处安装了一个空气采样器），将其空气采样器的流速从 200 L/min 调整为 8~10 L/min，使得有机污染物在土壤与气体间有充分接触时间，获得真正意义上土壤中有机污染物的逸度。后续多数研究者采用改进逸度装置[65-67]，例如，近期 Ren 等[66]研究者利用相同的装置，研究了青藏高原不同类型土壤（冻土、草地、森林）中持久性有机污染物在土壤-大气两相间的交换行为。研究结果显示，冻土可作为大气中持久性有机污染物的释放源，青藏高原土壤中 γ-HCH、HCB 及 PCB-28 的释放通量大约为 720 pg/(m²·d)、2935 pg/(m²·d) 及 538 pg/(m²·d)。在自然环境中，微小的空间模式（地表密集且矮小植物）显然会影响生物有效态污染物在土壤与大气间的迁移。Wang 等[68]采用多孔且小流量的主动采样方式（0.6~0.8 L/min）采集了水稻田大气样品，考察了水稻形成的微小空间对生物有效态的多环芳烃在土壤-大气间的迁移行为影响。研究结果表明水稻具有阻碍低环多环芳烃挥发的趋势；水稻下方高环多环芳烃（如蒽、芘等）含量要低于水稻上方，说明水稻具有阻碍高环多环芳烃沉降或向土壤迁移的趋势。

然而土壤逸度装置需要持续的电源支撑，限制了其应用范围。为此，Donald 与 Anderson[69]设计了以低密度聚乙烯膜为吸附相的土壤逸度被动采样装置，其装置由一个高度为 8cm 的腔体覆盖在土壤表面，腔体里放置了低密度聚乙烯膜用于确定有机污染物在土壤中的逸度。研究者将其试用于衡量点源污染场地多环芳烃的土壤-大气交换行为，其测试的结果与场地特征比较接近。纵观近期的研究进展可见有机污染物在土壤表层的生物有效态浓度是目前研究其在土壤-大气间迁移行为的关键参数。虽采用主动及被动的土壤逸度装置在近土壤表面可以获得"所谓"有机污染物在土壤的逸度，即有效态浓度，然而前提条件是采样时间内有机污染物在土壤固相-液相-气相达到平衡。这一条件需要充分验证。另一方面，目前测试的土壤表面都是无覆盖物的。在实际环境中，土壤表面具有植被或其他覆盖物，显然这类覆盖物可能影响有机污染物在土壤表层的生物有效态浓度，从而影响其迁移行为。前期研究采用有机污染物在土壤中的浓度与有机质含量的比值得到，森林土壤中不同种类的树（云杉、榉木及混合）产生的不同腐殖质类型影响多氯联苯的迁移性，例如，多氯联苯在阔叶场地（榉木）土壤中迁移性大于其在混合及云杉场地值[70]。总体而言，有机污染物在土壤中的生物有效态性及其在土壤内部与大气间的迁移行为仍需进一步研究。

4 生物有效性在人体健康研究中的应用

在人体健康研究领域，污染物的生物有效性是指穿过肠道上表皮细胞被人体吸收进入血液循环系统的部分。生物有效性可进一步划分为绝对生物有效性（absolute bioavailability，ABA）和相对生物有效性（relative bioavailability，RBA）。测定 ABA 分别采用口腔和静脉注射方式将污染物暴露于动物，两种暴露方式下的污染物的血药浓度曲线下面积（area under curve，AUC）的比值即为 ABA。但是静脉注射需要有经验的操作人员来进行，且取血过程繁琐，因此多采用 RBA 来表征污染物的生物有效性。RBA 是指通过口腔暴露测定介质和参照物质中污染物被人体吸收的部分之间的比值。

4.1 生物有效性的活体测试

活体动物实验是测定污染物生物有效性最直接和准确的方式。由于哺乳动物消化的相似性，目前主要的动物模型包括兔子、大鼠、小鼠、狗、猪、猴子等[71-74]。研究证明幼猪消化过程能够代表儿童胃肠道吸收污染物的生理过程[75]，由于解剖学上的相似性，幼猪成为用于检测人体暴露污染物生物有效性研究的理想动物模型。但幼猪模型成本高，实验周期长（约需一年），很大程度上限制了幼猪模型的应用。此外，从系统遗传学角度而言，猴子与人类最为接近，其胃肠道生理参数与人类类似。在研究药物肠道吸收方面，猴子被证明是最具有价值的动物模型，并能预测人体肠道吸收药物动力学过程[76]，但成本和道德伦理因素很大程度上制约了灵长类动物在测定污染物生物有效性方面的应用。因此，小鼠因其质量体积小、饲养成本低和实验操作过程简单等特点，是最常见的用于测定污染物生物有效性的动物模型。利用动物研究污染物生物有效性的暴露途径主要包括喂食[77-79]和灌胃[72, 80]。其中，长期喂食暴露是最简单也是最常用的暴露方式。

目前在活体测试领域的研究热点与难点在于生物靶点的筛选。根据生物有效性（即 ABA，RBA 的测定主要通过 AUC 获得）定义，最经典的生物有效性检测生物靶点是血液。血液中的污染物浓度会随着暴露时间延长而发生变化，可根据污染物在血液中的浓度随时间变化绘制出浓度-时间变化曲线，并计算浓度变化曲线下面积（AUC），根据 AUC 来计算污染物的生物有效性：

$$ABA = \frac{AUC_{oral}/Dose_{oral}}{AUC_{iv}/Dose_{iv}} \times 100\% \tag{3-2}$$

$$RBA = \frac{AUC_{sample}/Dose_{sample}}{AUC_{reference}/Dose_{reference}} \times 100\% \tag{3-3}$$

式中，AUC_{oral} 和 AUC_{iv} 是指通过口腔摄入和静脉注射的污染物在血液中的浓度随时间变化的 AUC；AUC_{sample} 和 $AUC_{reference}$ 是指通过口腔摄入的样品和参比物质中污染物在血液中的浓度随时间变化的 AUC；$Dose_{oral}$、$Dose_{iv}$、$Dose_{sample}$、$Dose_{reference}$ 分别指口腔暴露、静脉注射、样品、参比物质的污染物总剂量。

因为采用血液为生物标靶需要采集暴露不同时间后的血液样品，需要消耗大量的小鼠，且操作步骤繁琐。对于污染物，尤其是大多数的有机污染物来说，进入血液的污染物很快会分配到脂肪组织或其他靶器官中，致使血液中污染物浓度过低，不易被检测。因此一些具有连续累积污染物特征的器官或组织被用来作为生物学终点用于生物有效性的测试。由于小鼠和人都是通过尿液将砷排出体外的，所以尿液也常被用作评价砷生物有效性的生物靶点。例如 Bradham 等[81]采用长期（10 天）喂食的暴露方式，以小鼠尿液为生物靶点，测定了 9 种矿区或冶炼土壤中砷的 RBA。同样小鼠的肝脏、肾脏、骨头等也都被用作生物靶点测试重金属铅镉的生物有效性。近年来也有研究者发现，将不同器官中累积量相加作为生物靶点，可提高 RBA 的准确性和数据重复性。例如 Juhasz 等[82]通过测定小鼠肝脏和肾脏中 Cd 的富集浓度之和研究了 7 种土壤中 Cd 的 RBA。

通过测定污染物与参比物质中污染物在某一组织器官中的积累量比值计算相对生物有效性（RBA）的公式如式（3-4）：

$$\mathrm{RBA} = \frac{\mathrm{Organ}_{\mathrm{sample}}/\mathrm{Dose}_{\mathrm{reference}}}{\mathrm{Organ}_{\mathrm{reference}}/\mathrm{Dose}_{\mathrm{sample}}} \times 100\% \tag{3-4}$$

式中，$\mathrm{Organ}_{\mathrm{sample}}$ 和 $\mathrm{Organ}_{\mathrm{reference}}$ 是指暴露污染样品和参比物质后，污染物在组织器官中的积累量；$\mathrm{Dose}_{\mathrm{sample}}$ 和 $\mathrm{Dose}_{\mathrm{reference}}$ 是指污染样品和参比物质中污染物的总量。

对于有机污染物而言，生物靶点的选择更为复杂。表 3-1 列出测定持久性有机污染物相对生物有效性采用的动物模型、生物学终点及暴露方式。相对较为稳定的疏水性有机物多富集于脂肪组织，因此皮下脂肪组织被用来作为生物靶点测试土壤中滴滴涕、多氯联苯等有机氯物质的生物有效性，结果发现，喂食不同剂量的滴滴涕污染的石英砂后，脂肪中滴滴涕的累积量与暴露量呈显著的线性关系；且以脂肪为靶点的 RBA 的数据重复性较好，平均相对偏差为 16.7%，明显小于以肝脏、血液为靶点计算的 RBA 的平均标准偏差（25.6%和 29.3%）[83, 84]。而对于全氟化合物而言，则更倾向富集于肝脏。Li 等[85]以小鼠肝脏为靶器官测定了 17 种食物中的全氟辛酸的 RBA。然而对于一些易发生代谢的有机物而言，寻求适合的生物靶点器官组织仍旧是目前的难点。以多环芳烃为例，已经开展的研究采用粪便、尿液或血液中的多环芳烃富集量来计算 RBA，但是哪一种靶器官组织最佳，以及不同靶器官得到的 RBA 之间的对比研究并未开展。有幸的是，当代高分辨质谱 [如轨道阱（LTQ-Orbitrap）液质联用仪、飞行时间质谱等] 的长足发展可有针对性地进行有机代谢产物的分子结构鉴定工作，使分子结构解析和鉴定工作更加系统和简单，更直观准确地了解污染物在生物体内的运转及分布规律。与此同时也可结合基于生理学的药代动力学（physiologically based pharmacokinetic model，PBPK）等手段对结构类似的有机物的代谢转化进行模拟预测[86]。因此，利用现代质谱技术结合药代动力学模型有望明确有机污染物在动物体内的代谢途径，进而选择合适的生物学终点，更准确地测定其生物有效性。

表 3-1　测定持久性有机污染物相对生物有效性采用的动物模型、生物学终点及暴露方式

污染物	样品数	动物模型	暴露方式	生物学终点	相对生物有效性/%	参考文献
8 种多环芳烃	8	雌性 Landrace 幼猪（7～8 周）	喂食	血液	0.01～29	[78]
16 种多环芳烃	1	雌性 Balb/c 小鼠（22～24 g）	喂食	粪便	>65	[73]
菲	4	雄性 Sprague-Dawley 大鼠（275～350 g）	灌胃/静脉注射	血液	60.6～203.0	[80]
苯并芘	8	Landrace 幼猪（8～10 周，30～35 kg）	喂食	血液	22.1～108.1	[87]
苯并芘	12	雄性 Sprague-Dawley 大鼠（350±50 g）	喂食/静脉注射	血液/粪便	24.2～46.1	[88]
苯并芘	12	雄性 Sprague-Dawley 大鼠（350 g）	灌胃	血液	0～106.4	[89]
滴滴涕及代谢物	6	雌性 Balb/c 小鼠（22～25 g）	喂食	脂肪	17.9～65.4	[79]
滴滴涕及代谢物	8	雌性 Balb/c 小鼠（22～24 g）	喂食	肝脏和肾脏	18.7～60.8	[90]
滴滴涕及代谢物	7	雌性 Balb/c 小鼠（22～24 g）	喂食	脂肪，肝脏和肾脏	2.2～24.6	[91]
6 种多氯联苯	4	雄性幼猪（25 kg）	喂食	脂肪，肝脏和肌肉	3.0～101.8	[77]
6 种多氯联苯	3	雌性 Balb/c 小鼠（22～24 g）	喂食	脂肪，肝脏和肾脏	45～11	[84]

4.2 生物有效性的体外测试（经消化道途径）

尽管利用动物实验测定污染物的有效性直接且准确，但仍存在很多局限性，如成本高、实验周期较长、伦理道德因素制约等，因此急需建立简单经济准确的体外测试方法。过去二十年来欧美国家的一些研究机构开发了以模拟人体消化道为核心的体外测试方法。这些方法包括美国Ruby教授[92]建立的生理原理提取法（physiologically based extraction test，PBET）、Rodriguez教授[93]提出的体外胃肠（in vitro gastrointestinal，IVG）方法、德国标准化学会（Deutsches Institut für Normung e.V.）提出的DIN方法[94]、Kelly等[95]提出的溶解性生物可及性研究联合会（Solubility Bioaccessibility Research Consortium，SBRC）方法、欧洲生物可及性研究小组（BioAccessibility Research Group of Europe，BARGE）提出的UBM（Unified BARGE Method）[96]等。这些体外模拟方法的基本原理是根据人体胃液和小肠液的成分以及pH环境，配制模拟胃液和肠液，模拟胃相、肠相消化过程对环境介质中的污染物进行提取，可提取部分的污染物含量占总量的百分比即污染物通过口腔摄入这一暴露途径的生物可及性（in vitro bioaccessibility）。下面将从污染物类型（重金属、有机污染物）方面分别介绍体外方法的研究近况和局限：

1) 重金属

体外测试方法的应用前提是与活体实验具有显著线性相关关系。其中约定俗成的标准包括：线性相关系数r^2不小于0.60，斜率在0.8~1.2，截距接近于0，体外方法实验室内部数据的相对标准偏差<10%。上述体外方法的建立多针对土壤中的重金属，因此研究相对成熟。例如一些研究学者建立了PBET[92]、IVG[97]、UBM[75]方法与老鼠或幼猪模型之间的相关关系，相关系数分别达到了$r^2=0.93$、$r^2>0.74$、$r^2>0.7$和$r^2>0.92$。再例如澳大利亚的Juhasz教授[98]建立了活体实验与SBRC胃液提取之间的相关关系，与Juhasz等[99]对比发现，前后利用两批不同土壤样品建立的两个预测模型没有显著的差别，并利用多种交叉验证方法对基于SBRC方法的生物有效性预测模型进行了验证，进一步表明了SBRC方法的可靠性。SBRC胃液提取方法已成为美国环保局（USEPA）认定的评价污染土壤铅生物可获得性的标准方法[100]。但是这些体外方法也存在很多不足。例如针对同一样品，不同体外方法得到的结果会有较大的差异。例如Juhasz等[99]比较了四种体外测试方法（SBRC、IVG、DIN、PBET）与幼猪血液模型测定的砷相对生物有效性之间的相关关系，发现4种方法与活体实验之间都具有较强的线性相关关系，但以SBRC方法的结果最优。其次，建立体外方法时采用的土壤样本数量少且来源单一。如Smith等[101]以12个土壤样品为基础，研究了铅在SBRC方法胃液和肠液的生物可及性与小鼠活体实验测定的铅相对生物有效性之间的相关关系。

相比欧美国家，我国在重金属人体生物有效性体外测试方法的建立和标准化方面非常薄弱。例如Li等分别利用了五种体外实验方法（SBRC、IVG、DIN、PBET、UBM）和小鼠活体实验对采集自我国不同类型的砷、铅和镉污染土壤进行了人体生物有效性测试，发现IVG方法是替代动物活体实验测试土壤中砷人体生物有效性的最佳方法[102]，而UBM方法和PBET方法是测试土壤铅和镉人体生物有效性的最佳体外测试方法[83,103]。可见针对不同污染物，目前还缺乏一个统一的标准化测试方法。此外，以往研究多针对矿区/冶炼区污染土壤，且关注对象主要集中在砷、铅和镉，目前针对我国受冶炼、化工、焦化等行业影响的污染土壤中的多种污染物，还未建立标准体外测试方法。因此需要针对不同类型污染场地中的重金属，建立体外实验方法与动物活体实验测试人体生物有效性之间的线性相关关系，确立适合替代动物活体实验的体外实验方法，从而建立标准化的体外生物有效性测试方法。

2) 有机污染物

胃肠模拟法最早始于对食物中营养元素吸收的研究，后来被用于土壤重金属（如砷、铅等）的人体风险评价。直到近十年来，该类方法才开始应用于有机污染物，但是迄今为止并没有任何一种方法适用于有机污染物有效性的测定，极大程度上阻碍了有机污染的风险评估和修复治理。主要原因在于胃肠液体外方法最初是针对营养元素和重金属设计的,通过蛋白酶胆盐等组分的络合作用及 pH 的调节作用对重金属有很好的提取效果[104]。但是现有的胃肠模拟液中有机质组分偏少,对于有机污染物的吸附提取能力不强。尤其是疏水性很强的有机污染物与土壤中的有机质紧密结合,很难从固相解吸至胃肠液中。并且体外方法提取有机物主要是一种静态分配的过程,提取效果取决于污染物在胃肠液中的溶解度;而在实际的人体(动物)消化道中发生的是动态吸收过程,即肠道细胞吸收溶解的污染物后,会促使污染物进一步地解吸溶解。因此胃肠模拟液体外方法测定的疏水性有机物的生物有效性往往远低于动物活体实验的结果[73]。近年来在现有体外方法中加入吸附材料被认为是可以有效解决对有机物提取能力不足这一问题的措施,主要采用的吸附材料包括 C_{18} 膜材料[78]、硅胶[105]和 Tenax[74, 106]。这些吸附材料添加后可显著提高体外提取结果,例如 Gouliarmou 等[105]在 80 mL 模拟胃肠液中添加了 30cm 的硅胶管后,多环芳烃的生物可及性提高了 103~4567 倍。但仅有生物可及性的提高并不能完全说明体外方法可以用于有机污染物生物有效性的测定,同样也必须经过动物活体的验证。Li 等[74]采用小鼠测试和添加 Tenax 的 PBET 方法分别测定土壤中滴滴涕的生物有效性,结果发现添加 Tenax 可以显著提高体内-体外结果的线性相关性, $r^2 = 0.62$,斜率 = 1.17。之后来自澳大利亚的 Juhasz 教授也以小鼠活体测试为基准,验证了 PBET 方法中添加硅胶管后测得的生物可及性也与小鼠结果有显著的相关性[79]。

鉴于吸附材料对现有胃肠模拟方法的显著改善作用,一些研究者也尝试将吸附材料-胃肠模拟方法简化为单独吸附材料提取或吸附材料结合主要肠液成分(胆盐或蛋白酶)的提取方法。但是研究结果发现,这些简化方法测得的生物可及性与小鼠活体结果均无显著相关性[84],推测体外结果受吸附材料与模拟胃肠液成分之间的耦合作用共同影响。因此 Zhang 等[107]选择成分各异的五种体外胃肠模拟法,结合 Tenax 对多环芳烃污染土壤进行提取,并采用主成分分析和多元线性回归手段对测得的体外结果进行分析,发现除了吸附材料(Tenax)以外,模拟肠液中的胆盐成分和肠相培养时间是影响生物可及性的关键因子,该结果的研究思路为建立体外测试方法提供了重要的理论依据。总而言之,吸附材料-模拟胃肠法是适用于有机污染物生物有效性测定的体外方法。然而建立标准方法还需要更多更系统的研究,例如污染物的吸附解吸行为受到环境介质的理化性质的显著影响;吸附材料的选择、参数优化也都需要进一步研究。

4.3 生物有效性的体外测试(经呼吸道途径)

除消化道途径的暴露以外,经由呼吸道暴露于人体也是污染物产生健康危害的主要途径。近几年来,以模拟人体肺液为核心的测定污染物吸入生物有效性的体外方法受到越来越多的关注。经呼吸途径暴露的污染物通常存在于 pH 呈中性的细胞外环境和巨噬细胞内酸性较强的环境[108]。其中,细胞外环境以肺液为代表,主要包含钾、钙、钠、镁等阳离子和氯化物、碳酸氢盐、硫酸盐、有机酸、蛋白质等阴离子,以及以葡萄糖、氨基酸等营养素和蛋白质代谢产生的废物等为代表的非电解质。最早模拟细胞外环境的人造肺液(synthetic lung fluid, SLF)也被称为 Gamble 溶液。Gamble 溶液是肺部深处的细胞外环境的模拟液,在可吸入污染物的暴露评估中应用广泛,且经过多次改良[109, 110]。基于 Gamble 溶液改良的方法包括人造血清(synthetic serum)[111]、Gamble 血清模拟液(Gamble serum simulant)[112]、人造组织液(artificial interstitial fluid)[113]、改良 Gamble 溶液(modified Gamble solution)[114]、模拟肺液(simulated lung fluid)[115]等多种配方(表 3-2)。

尽管经过多次改良,但该类方法中溶液成分较为简单,鲜少含有抗氧化剂、有机酸等,并且不含肺部表面活性剂。研究认为肺部表面活性剂不仅是人类呼吸系统重要组成成分,同样也是强螯合剂。肺部表面活性剂是以单层形式存在于肺泡区域的气液界面之间的由脂质、蛋白质组成的复杂混合物[116],起到减少呼气时气液表面张力的作用。肺部表面活性剂可防止肺部气体完全排空,避免肺部充满液体,从而

表 3-2 不同 Gamble 溶液的成分及浓度

成分（mg/L）	原始 Gamble 溶液[130]	模拟肺液[115]	人造组织液[113]	Gamble 溶液[110]	改良 Gamble 溶液[114]	人造血清[111]	Gamble 溶液[109]
六水合氯化镁	203	212	203				
氯化铵				535	5300	535	118
氯化钠	6019	6400	6193	6786	6800	6786	6400
氯化钙				22		22	
二水合氯化钙	368	255	368		290		225
硫酸钠	71		71				
硫酸				45	510	45	
十水合硫酸钠		179					
磷酸氢二钠		148	142				
磷酸二氢钠	142			144		144	
一水合磷酸二氢钠					1700		
磷酸					1200		
碳酸氢钠	2604	2700	2604	2268	2300	2268	2700
碳酸钠					630		
二水合酒石酸氢钠		180					
柠檬酸二氢钠	97	153					
柠檬酸钠				52		52	160
一水合柠檬酸					420		
丙酮酸钠		172					
甘氨酸		118		375	450	450	190
L-半胱氨酸				121			
DPPC							200
三水合乙酸钠	953		952				
乙酸钠					580		
二水合柠檬酸铵			97		590		
乳酸钠		290					
氯化钾	298		298				
邻苯二甲酸氢钾					200		
过乙酸				79			
苯扎氯铵				50			
pH		7.4		7.3	7.4	7.3	

达到稳定肺部环境的效果[117]。目前加入模拟肺液的肺部表面活性剂主要是二棕榈酰卵磷脂和二棕榈酰磷脂酰胆碱（dipalmitoyl phosphatidyl choline，DPPC）。研究人员之后建立了含有谷胱甘肽、抗坏血酸、尿酸、二棕榈酰磷脂酰胆碱的模拟上皮肺液（simulated epithelial lung fluid，SELF）[118]等成分更为丰富的细胞外环境模拟肺液。颗粒物沉积数小时后，肺泡巨噬细胞会吞噬部分颗粒物。目前，模拟巨噬细胞内酸性较强环境的配方中应用较为广泛的是人工溶酶体液（artificial lysosomal fluid，ALF）[119]和模拟吞噬溶酶体液（simulated phagolysosomal fluid，PSF）[120]。

尽管当前有诸多模拟肺液配方，但并未建立标准配方，且主要针对颗粒物中重金属开展研究，利用体外模型对颗粒物中有机污染物生物有效性的研究较少。Borm 等[121]在37℃用100～10000 μg/mL 的 DPPC 溶液提取标准柴油尾气颗粒物和炭黑颗粒中的菲、芘、蒽、䓛、芴。发现生物可及性（即可提取部分）的菲少于 1.2%、芘少于 0.4%、蒽少于 1.0%、䓛少于 1.3%、芴少于 1.3%。Li 等[122]采用 SELF 提取 $PM_{2.5}$ 中 PAHs，发现 19 种 PAHs 的吸入生物可及性在 3.21%～44.2%，其中苯并［c］芴生物可及性最低，苊烯生物可及性最高，总体而言吸入生物可及性随着疏水性的增加而降低，并与 $PM_{2.5}$ 中的元素碳比例呈负相关。

由于呼吸途径颗粒物中污染物生物可及性的结果受到颗粒物性质、提取时间、模拟肺液配方、固液比等多方面影响。例如 Kastury 等[123]采用 1∶100、1∶500、1∶1000、1∶5000 四种固液比；1 h、8 h、24 h、48 h、72 h、96 h 和 120 h 七种振荡培养时间；磁力搅拌、偶尔搅拌、轨道搅拌、端到端搅拌等四种搅拌方法，利用 ALF 和 PSF 两种模拟肺液，测定 $PM_{2.5}$ 中金属的生物可及性，结果表明，模拟肺液配方、固液比、提取时间均会显著影响 $PM_{2.5}$ 中金属生物可及性。在各个影响因素中，研究者们针对不同类型的影响因素开展了研究。① 颗粒物类型：Julien 等[109]采用血清模拟 Gamble 溶液测定四种标准物质中金属的生物可及性，发现镉、铅和锌在汽车尾气颗粒物中生物可及性最高、在粉煤灰中最低。② 提取时间：传统认为通过黏膜纤毛作用从气管支气管清除耗时在数小时至一天，此外，颗粒物分布的区域和持续时间与人的个体和肺部大小有关[124]，而从肺泡区域的清除作用要更慢，从数月到数年不等[110, 125]。目前针对体外模拟吸入生物可及性的研究中，提取时间在 1 小时[126]到 630 小时[110]不等，但大多采用 24 小时[123]。③ 固液比：与颗粒物在人体内停留时间相似，大气颗粒物在肺部的负荷量也是不断变化的，主要取决于 PM_{10} 的浓度、呼吸系统中的沉积量以及肺部液体体积。Julien 等[109]通过计算采用每日呼吸量 10～20 m^3、颗粒物浓度 20～500 μg/m^3、肺液总量 10～20 mL，确定颗粒物在肺液中的固液比在 1∶1000～1∶100000，并且当固液比大于 1∶500 时金属生物可及性会下降，产生这一现象的主要原因是螯合剂饱和或者由于结块导致的表面积减少。在颗粒物中疏水性有机污染物吸入生物可及性的研究中，Li 等[122]采用 0.1 g/L 二棕榈酰磷脂酰胆碱（dipalmitoyl phosphatidyl choline，DPPC）水溶液、Gamble 溶液、人工溶酶体液（artificial lysosomal fluid，ALF）、SELF 四种配方的模拟肺液和 1∶100、1∶500、1∶1000、1∶5000、1∶10000 五种固液比条件下柴油尾气颗粒物中 PAHs 的吸入生物可及性。研究结果表明，采用 SELF 提取所得的 PAHs 吸入生物可及性（2.81%～92.9%）>Gamble 溶液（0.85%～63.0%）>ALF（0.92%～47.3%）>0.1 g/L DPPC 水溶液（1.94%～33.1%）。在四种模拟肺液方法中，PAHs 的生物可及性均会随着固液比的降低而上升。④ 颗粒大小：Xie 等[127]发现采用 Tenax 改良的 Gambler 溶液以及 ALF 提取颗粒物中 PAHs 时，颗粒物中疏水性有机物的生物可及性均随颗粒大小的增加而增加，这主要是由于细颗粒的表面积更大，更容易将污染物吸附于有机组分上[128, 129]。由上可见，体外方法中组成分、提取参数对污染物的吸入生物可及性均有影响，因此还需要更为系统的研究以确定关键影响因子，进而建立标准测试方法。

此外颗粒物中污染物的体外生物可及性对颗粒物健康风险评估以及污染管控都具有重要的启示作用。Li 等[122]发现考虑到生物可及性情况后，$PM_{2.5}$ 中 PAHs 的人体健康风险由 $4.44×10^{-4}$～$3.79×10^{-3}$ 降为 $2.90×10^{-5}$～$2.44×10^{-4}$，同时相比于 $PM_{2.5}$ 中 PAHs 的总浓度，PAHs 生物可及态浓度能更好预测由 $PM_{2.5}$ 中 PAHs 引起的类二噁英毒性（$r^2 = 0.40$～0.83，$p < 0.05$）。Xie 等[127]发现，未经过改良 Gambler 溶液、ALF 培养的电子垃圾燃烧产生的颗粒物引起的健康风险分别为 $6.2×10^{-5}$ 和 $2.56×10^{-4}$，但经改良 Gambler 溶液、ALF 培养后健康风险降为 $2.24×10^{-6}$ 和 $1.23×10^{-5}$。Li 等[83]研究了中国南京举办世界青年奥运会前和后的 $PM_{2.5}$ 中铅的生物可及性和形态变化，结果显示 2014 年南京青年奥运会期间南京市政府实施污染源头控制后，大气颗粒物中铅的来源由煤炭燃烧和金属冶炼改变为煤炭燃烧的单一来源，铅形态从 $PbSO_4$ 和 PbO 为主改变为 $PbSO_4$ 为主，因此铅的生物可及性（通过 Gamble 溶液和 ALF 溶液提取的部分）也随之降低。说明污染物的人体生物有效性研究不仅有利于准确地评价健康风险，还可以为制定合理经济的污染控制政策提供有力的理论支撑。

5 生物有效性在生态毒性评价识别中的应用

环境中有毒有害化学品种类繁多,形态各异,通常是多种污染物以不同形态、低剂量长期共存,毒性效应复杂。污染物的生物有效性与其迁移、降解等环境行为密切相关,同时也将直接影响其生物富集能力和毒性效应,因此准确测定环境介质中污染物的生物有效性是设定环境质量基准、有效开展生态风险评价的前提条件。近年来越来越多的研究提出在生态风险评价中,有必要考虑污染物的生物有效性对毒性的影响,通过建立不同环境介质中污染物的生物有效性的测量方法,并将其融入生物毒性评价和致毒物识别中,以便于降低生态风险评价的不确定性[131-135]。

生物有效性被认为是环境中污染物能被生物体吸收进入生物体而产生效应的部分,通常分为生物可利用性和生物可及性[2,4]。直接测定野外采集生物体内污染物的残留量可获取生物有效性信息,然而该数据受到可采集生物的种类区域差异大、生物样品量不足、高毒污染物检测灵敏度差等局限,方法可控性较弱,难以实现数据共享及标准化,加之于工作强度大等问题也限制该方法的规模化应用。加标受试污染物,开展实验室生物暴露,当污染物在生物体内达到稳态后,测试体内污染物残留的含量也常被用于表征生物有效性,但是由于生物测试实验周期长、分析手段复杂、个体差异大、不同物种之间难以推广等原因,限制了这些生物测试方法的应用范围。另一方面,基于化学分析的仿生萃取技术被用于估算污染物的生物有效性,其具有快捷简易、价格相对低廉、可比性较好的特点,近年来发展迅速。对应生物可利用性和生物可及性的两种概念,科研人员分别建立了测试污染物的化学活度(如被动采样)和可及性(如吸附剂辅助萃取)的仿生萃取方法,并将其与生物积累测试进行对比,验证了方法的适用性。在此基础上,开展方法标准化和规范化工作,使得生物有效性的概念和测试技术也从单纯的科学研究逐步发展到区域环境管理的应用中[136-138]。

尽管化学活性和生物可及性的测量手段和目标不一样,大量研究显示这两类仿生萃取技术的测定结果均与生物积累实验中的动物体内污染物的蓄积量密切相关,说明这些方法都可有效估算环境污染物的生物有效性[132,134]。通过加标沉积物所构建的 Tenax 树脂萃取效率和生物有效性之间的线性回归模型,不仅可用于测定从野外采集的沉积物中有机污染物的生物有效性,而且也成功地预测了从美国五大湖区采集的底栖动物体内的多环芳烃和多氯联苯的残留量[139-141]。与生物可及性测量主要依赖于经验公式不同,以化学活性为测定目标的被动采样技术一般通过量化环境介质中污染物的自由溶解态浓度来表达其生物有效性。通过污染物自由溶解态浓度及其在生物体内的浓缩因子进行计算,可预测污染物在生物体内的富集能力[142]。此外,对于疏水性较高,水体中自由溶解态浓度不易测量的污染物,可直接通过被动采样材料与生物脂肪等吸附相的分配系数来估算污染物在生物体内残留量[143-146]。除了对吸附和富集机制的描述更清晰之外,被动采样技术比吸附剂辅助萃取技术更适于原位采样测定,近年来得到广泛的关注[147,148]。2012 年被动采样技术被美国环保局推荐作为超级基金污染场址水体沉积物中有机污染物的监测手段,设定其标准化规范,为该技术的推广提供了有效的支持[138]。

5.1 仿生萃取技术在毒性预测中的应用

仿生萃取技术对环境污染物生物富集能力的测定,也被进一步运用到对毒性效应的估算中[132,134,135]。通常在开展沉积物与土壤样品的毒性评价时,因为有机碳含量是控制疏水性有机污染物分配的决定因素之一,有机碳标准化浓度常被用作毒性效应的剂量单位[149]。近年来的研究显示不同类型的有机碳可能具有不同的吸附能力，因此黑炭校正模型被提出用于生物有效性估算，然而由于对有机碳分类及其与污染物之间吸附机理的认知还很有限，该模型尚未能在毒性评价中得到实际应用[150-152]。另外，还有其他因素也可能影响环境污染物的生物有效性和毒性，例如土壤和沉积物性质（如沉积物粒径分布、芳香性和平面性等）[153, 154]、化合物的理化性质[139]和模式生物的生活习性等[155]。由于生物有效性和毒性受到多因素影响，基于总有机碳标准化浓度的传统毒性剂量单位往往可能导致环境样品毒性被高估[156, 157]。因此，进行毒性评价需要结合环境污染物的生物有效性测量方法。

仿生萃取技术的使用，可降低生态风险评价中使用在环境介质中污染物总浓度或有机碳标准化浓度产生的误差，可简单有效地将生物有效性的量化结合到毒性效应评价中，为提高生态风险评价的准确性奠定技术基础。Xu 等[158]对比了五种不同暴露表征方法，发现采用固相微萃取（SPME）测定的沉积物孔隙水中自由溶解态浓度最能准确地描述毒性效应。You 等[159]使用 Tenax 萃取的拟除虫菊酯类农药的生物有效浓度代替有机碳标准化农药浓度，提出生物有效毒性单位的概念，并以此预测野外沉积物对钩虾的毒性，显著提高了沉积物毒性评价的准确性。在此基础上，Li 等[154]比较了有机碳标准化浓度、Tenax 萃取浓度和 SPME 浓度作为沉积物毒性剂量指标的效果，结果显示以 SPME 浓度为暴露表达方式对毒性描述的不确定性最低，其次为 Tenax 萃取浓度，这也进一步反映了在沉积物毒性评价中考虑生物有效性是必要的。图 3-1 是基于被动采样（以 SPME 为例）的仿生萃取技术对污染物毒性效应测定的概念模型[160]。从图中可见，该方法通过被动采样技术测定环境污染物的自由溶解态浓度，然后估算生物体内污染物残留浓度，并以此来表征毒性效应[133, 161, 162]。此外，在毒性测试同时采用仿生萃取测定污染物对生物的暴露量，以此代替基于环境总浓度的暴露指标，也是在毒性评价中考虑生物有效性的一种常用方式，该方法不但用于单一化合物毒性评价，也被用于多种污染物共存的野外样品[157, 160, 163, 164]。为了有效地判断污染物对生物体的剂量-效应关系，构建了仿生萃取指标与生物毒性效应之间的关系，这也有利于推广仿生萃取技术在毒性预测中的应用[165]。

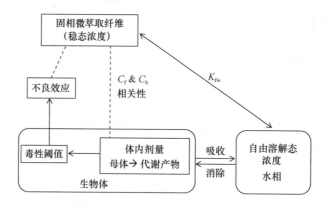

图 3-1　基于固相微萃取的仿生萃取技术测定污染物毒性效应的概念模型[160]

目前仿生萃取的研究主要集中于滴滴涕、多环芳烃和多氯联苯等传统污染物，致力于建立其测定指标和生物积累量（污染物在生物体内浓度）之间的关系，验证不同类型仿生萃取方法的可行性。此外，近年来有些研究也拓展性地将仿生萃取技术用于水体沉积物和土壤污染控制及修复方法的效果评估[166, 167]，但是应用仿生萃取技术测定环境污染物的生态毒性效应方面的研究仍然相对较少，而且主要集中在生物急性致死毒性效应的研究中[132, 135]。因此，考虑生物有效性的仿生萃取方法用于毒性评价的发展，应当朝着多生物物种、多类化合物和多毒性终点的方向不断完善，同时深入研究污染物在生物体内的转化过程对仿生萃取技术获取的毒性剂量指标的影响，以及仿生萃取技术测定指标与慢性亚致死毒性终点的关系等。

5.2 基于生物效应的复合毒性评价方法

对于复杂的环境样品，基于目标化合物的化学测定难以反映环境样品的毒性效应，而生物测试虽能直接检测毒害效应，但无法确定样品中主要致毒物质。此外，考虑不同类型污染物的生物有效性差异，对于准确评价多污染物共存的复合污染体系（例如水体沉积物和土壤）的生态风险尤为重要。随着人类活动需求的提升和科学技术的不断发展，新兴化学品不断被设计、合成、生产并进入市场。据美国化学会统计，目前世界上现有化学品种类多达上亿种（引自www.cas.org/index），且该数目仍然在高速递增中[168]。世界范围内化学品的使用量也随之快速增长，每年应用于工业和生活消费的合成化学品量约有 3 亿 t[169]，这些化学品在被使用后，可通过不同途径进入自然环境。迄今为止，进入环境的化学品已超过 10 万种[170]，这些化学品在环境中可进一步通过自然降解、生物代谢等过程形成新物质，导致了严峻的复合环境污染问题。面对这种与日俱增的化学品污染问题及其对生态系统的潜在危害，科学有效地监测污染状况、剖析生态效应、识别关键毒害因素至关重要。

传统的生态毒性评估方法有环境阈值判定、野外本土生物调查和实验室毒性测试等，但是在这些方法的应用中，缺少复合毒性效应及污染物生物有效性的考虑，难以反映污染物对生物体的真实暴露情况。目前最常见的生态风险评价方法仍主要基于化学分析手段，通过设立重点关注的污染物清单，分析环境介质中目标污染物浓度，将其与已有的生物毒性阈值数据进行比较，以此推断研究区域污染物的潜在生态风险。该方法在污染物来源相对明确的区域可能达到较好的评价效果，但是对于多类污染物共存的复杂体系，基于常规分析测试清单的评价手段往往难以应对数量日益剧增的有毒有害化学品带来的环境问题[171, 172]。主要表现在：①污染物监控范围受限。环境中污染物成千上万，而目标清单上分析物却寥寥无几，尽管不同国家/地区设定的清单存在一定差异（如中国水环境中"优控污染物"为 68 种、美国 129 种[173]），但是这些均与现实环境中污染物种类相比悬殊巨大，结果导致环境中关键致毒物可能不在清单上，尤其是面对化学品种类急增的局面，传统的基于化学分析的风险评价方法易忽略环境中新兴污染物的毒性贡献。虽然近年来随着环境学科的发展，优控污染物的列表不断扩充，但环境介质中的化学物质数量远远超出这些优控清单上的物质，将大多数物质排除在外，易出现遗漏和疏忽，尤其是环境中的不断涌现的新兴污染物和众多的降解代谢等过程产生的未知物。②化学分析过程中未能考虑污染物的生物有效性差异。即使传统方法能够检测出环境样品中所有致毒物质，但若忽略了不同物质的生物有效性差异，仅依据化学监测结果仍可能出现较大偏差。环境介质中污染物总浓度并非生物的真正暴露水平，可能给生态毒性评价和识别带来较大的误差。对于生物体来说，疏水性污染物都主要是以溶解态的形式进入有机体，被生物所利用，从而产生生态风险，该部分称为生物有效性。③基于单一化合物的分析结果，难以考虑各物质的联合毒性效应。环境介质中多种污染物共存，物质之间可能产生拮抗、协同等联合作用，仅仅从单一物质的环境浓度和毒性阈值的比较难以说明环境样品实际毒性的大小。总体而言，由于环境污染物的分析清单的局限、污染物的复合作用机制以及生物有效性的研究匮乏，使得单纯化学分析手段难以应对复杂样品的毒性效应的评价。

另一方面，生物测试虽可直接反映环境样品的毒性效应，但单纯的生物测试结果难以判定毒性产生的原因和主要致毒物质。单一的化学分析或单一的生物测试均具有片面性，亟须发展新的基于生物效应的复合毒性评价方法[174]。与常规基于化学分析的技术不同，基于生物效应的评价方法以环境样品的毒性效应为研究起点，运用样品分离、毒性测试、化学分析和统计建模等技术手段，逐步降低受试样品的复杂程度，缩小潜在致毒组分的范围，最终在复杂环境体系中筛出关键致毒污染物。综合运用生物测试与化学分析的测试方法——毒性鉴别评价（TIE）和效应导向分析（EDA），通过生物测试评价毒性效应，结合化学分析，筛查主要毒性物质，可有效避免传统单纯化学或生物监测方法的不足。以上两种基于生物效应的污染物筛查方法将环境样品的污染状况与生物效应相关联，既可评价环境样品的毒性效应，又可将毒性效应与具体的污染物相对应，利于判断主要致毒因素，已成为复杂环境体系毒性评价和致毒物识别的重要支撑技术[171, 172, 175]。

在复杂环境体系中，致毒物识别的研究过程中，考虑污染物生物有效性的影响同等重要，忽视生物有效性的影响可能导致关键致毒物被误判或漏判[172, 176]。例如，当运用全沉积物 TIE 技术评价美国伊利诺伊河流沉积物的生态风险时，发现多环芳烃类是该区域沉积物对底栖无脊椎动物的主要致毒因素[177]，而该结果与同样地段的采用孔隙水 TIE 方法进行鉴定的结果（氨氮是主要致毒因素[178]）不一致。进一步的研究显示由于孔隙水 TIE 的局限性，使其高估了极性污染物（氨氮）毒性，而低估了极性较弱的有机污染物的毒性，而全沉积物 TIE 更充分地考虑了不同污染物的生物有效性的差异，可更准确地反映沉积物对底栖动物的毒性效应[179]。另一方面，在应用 EDA 鉴别沉积物中主要致毒物质对绿藻的生长抑制效应时，在测试中将耗竭式萃取获得的污染物重新加标至水体中开展毒性测试，当采用常规溶剂加标方法，疏水性污染物生物有效性低的问题被忽略，多环芳烃的毒性被高估，使其被误判为主要致毒物，而真正的致毒物质却被忽视了。与此对应，在毒性测试中采用考虑生物有效性的基于硅橡胶的被动加标方法，则可提高致毒物识别的准确性[180]。

使用生物样品进行 EDA 的测试，是致毒物识别中结合生物有效性的最直接方式[181]。You 等[175]总结了使用生物样品开展 EDA 分析的研究，包括生物样品种类、毒性测试终点和测试方法，以及鉴别出的关键致毒物的信息。对于脊椎动物，血液和脂肪是常用的样品，而无脊椎动物的组织匀浆则被直接用于 EDA 分析，内分泌干扰效应是最常用的毒性终点。虽然生物样品的使用充分考虑了污染物的生物有效性，但是该方法使用受到样品量小、基质干扰高（内源性激素对内分泌毒性终点的干扰），以及污染物在生物体内转化过程对毒性识别影响机制不清的限制。

另一方面，在使用非生物样品开展 TIE 和 EDA，进行关键致毒物识别时可能遇到以下问题。①萃取方式：耗竭式溶剂萃取方式提取的是环境介质中污染物的总量，并非生物有效的部分，从而易导致化学鉴定和毒性测试结果出现较大偏差，甚至引起致毒物的错误判定。②加标方式：将沉积物萃取物直接加标于水相中进行测试时，由于生物有效性的差异，有机物在水相中的浓度与原沉积物孔隙水中的浓度可能出现较大偏差，导致错误结果。针对这两个问题，在毒性识别研究中考虑生物有效性，也主要从提取和加标两个方面来解决，即在环境样品中提取污染物时采用仿生萃取，而在开展萃取物的毒性测试时采用被动加标（passive dosing）的方式（图 3-2）。

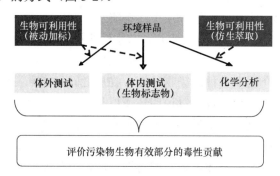

图 3-2　毒性评价识别中通过仿生萃取和被动加标考虑生物有效性的影响

当前 EDA 测试以及 TIE 第二步污染物鉴定多采用耗竭式萃取方法，如索氏抽提、加速溶剂萃取等，使得高疏水性污染物的生物有效性被高估，测定结果不能反映真实环境状况，从而导致毒性效应和致毒物的误判。已有研究表明，采用不同的萃取方法，化学鉴定和生物测试都可能出现不同结果[180, 182, 183]。为克服萃取方式导致的生物有效性偏差，研究尝试运用仿生萃取（基于生物可及性的提取方法、基于平衡分配的被动采样方法）替代耗竭式溶剂萃取[184, 185]。尽管被动采样技术可有效地获取生物有效性和毒性信息，但是由于其所获得样品量相对较小，难以满足 EDA 后续分离测试需求，故此在毒性识别研究中吸附剂辅助萃取的应用更为广泛。Yi 等[184]在 TIE 的毒性鉴定阶段采用 BCR 分级提取法和 Tenax 萃取分别分析了重金属和有机物的生物有效性，并将其用于估算生物有效性毒性单位，排除了低生物有效性的污染物对致毒物识别的干扰，缩减致毒物质清单，更准确地将广州城市河涌沉积物中毒性主要贡献重金属锁定为 Zn、Ni 和 Pb，主要致毒有机物为氯氰菊酯、氯氟氰菊酯和氟虫腈。Schwab 和 Brack[186]的研究表明大体积 Tenax 树脂萃取方法提取沉积物生物可及性部分可为基于体内测试的 EDA 方法提供足量样品。Schwab 等[187]运用 Tenax 树脂提取沉积物中生物可及性部分，并将其用于后续的 EDA 测试，结果表明绿藻的生长抑制效应主要由多环芳烃类导致。为降低吸附剂测试成本，Li 等[188]使用 XAD 代替 Tenax 树脂，提取沉积物中生物有效部分的污染物开展 EDA 分析，通过非目标污染物的筛查发现多环麝香类污染物对摇蚊幼虫毒性有重要贡献。

为降低传统溶剂加标方法导致的污染物在毒性测试样品和实际样品中的生物有效性偏差，Mayer 等[189]提出尝试使用被动加标代替传统的溶剂加标法。被动加标法利用污染物在有机相和水相间的分配平衡，维持测试水体中有机物的浓度恒定，同时模拟有机物在原环境介质（如沉积物）中的分配过程[190]。由于其可使污染物在生物暴露期间维持较稳定的水体浓度，也为毒性数据的准确性提供了重要保障。因具有生物相容性好、平衡分配机制比较明确等优点，聚二甲基硅氧烷（PDMS）在被动采样中应用广泛，而且为了适应不同测试容器的需要，多种形状的 PDMS 被使用，包括膜状[191]、棒状[182]、O 形环[192]等。

被动加标被大量用于疏水性污染物的毒性测试。Rojo-Nieto 等[193]指出被动加标的毒性实验重复性要明显优于溶剂加标的方法。马萍等[194]总结了近年来被动加标技术在离体和活体生物测试的应用。该方法在离体细胞实验中应用相对成熟，但是对大体系慢性活体测试的适用性尚待进一步研究。Xia 等[195]采用较大实验体系（1 L）、较长实验周期（16 d），成功使用被动加标方法研究了多环芳烃类化合物在食物链中的传递，这也为被动加标用于大体积慢性毒性测试提供了参考范例。被动加标技术也逐步用于 EDA 测试中[180]。Qi 等[196]以摇蚊幼虫综合生物标志物为亚致死毒性指标，借助于 PDMS 将沉积物萃取物的 EDA 分离的组分加入水体中开展摇蚊幼虫毒性测试，揭示了广州珠江段沉积物中目前使用农药通过氧化应激机制对摇蚊幼虫的毒性贡献。在 EDA 毒性测试过程中运用被动加标，利用 PDMS 模拟化合物在沉积物有机碳和水相中的分配，较好地考虑了沉积物中有机污染物的生物有效性，提高了致毒因素鉴别的准确性。将基于考虑生物有效性的提取模式与基于平衡分配的被动加标结合，进行致毒因素的鉴别，会成为 EDA 未来发展的主流。

在未来的研究中，需要进一步明确影响生物有效性的机制，完善化学分析技术测定指标与生物毒性效应的关系，并将相应测试手段标准化，以利于在环境管理中推广使用。

参 考 文 献

[1] Alexander M. Aging, bioavailability, and overestimation of risk from environmental pollutants. Environmental Science and Technology, 2000, 34(20): 4259-4265.

[2] Semple K T, Doick K J, Jones K C, et al. Defining bioavailability and bioac-cessibility of contaminated soil and sediment is complicated. Environmental Science and Technology, 2004, 38(12): 228-231.

[3] Ehlers L J, Luthy R G. Peer reviewed: Contaminant bioavailability in soil and sediment. Environmental Science and Technology, 2003, 37(15): 295-302.

[4] Reichenberg F, Mayer P. Two complementary sides of bioavailability: Accessibility and chemical activity of organic contaminations in sediments and soils. Environmental Toxicology and Chemistry, 2006, 25(5): 1239-1245.

[5] Verhaar H J M, Busser F J M, Hermens J L M. Surrogate parameter for the baseline toxicity content of contaminated water:

simulating the bioconcentration of mixtures of pollutants and counting molecules. Environmental Science and Technology, 1995, 29(3): 726-734.

[6] McCarthy J F, Shugart L R. Biomarkers of environmental contamination. Boca Raton: Lewis Publishers, 1990.

[7] 王海黎, 陶澍. 生物标志物在水环境研究中的应用. 中国环境科学, 1999, 19(5): 421-426.

[8] Oikari A, Fragoso N, Leppanen H, et al. Bioavailability to juvenile rainbow trout (*Oncorynchus mykiss*) of retene and other mixed-function oxygenase-active compounds from sediments. Environmental Toxicology and Chemistry, 2002, 21(1): 121-128.

[9] Koganti A, Spina D A, Rozett K, et al. Studies on the application of biomarkers in estimating the systemic bioavailability of polynuclear aromatic hydrocarbons from manufactured gas plant tarcontaminated soils. Environmental Science and Technology, 1998, 32(20): 3104-3112.

[10] Ramos E U, Vaes W H J, Verhaar H J M, et al. Quantitative structure-activity relationships for the aquatic toxicity of polar and nonpolar narcotic pollutants. Journal of Chemical Information and Modeling, 1998, 38(5): 845-852.

[11] Vaes W H J, Ramos E U, Verhaar H J M, et al. Understanding and estimating membrane/water partition coefficients approaches to derive quantitative structure property relationship. Chemical Research in Toxicology, 1998, 11(8): 847-854.

[12] Constantinou E, Seigneur C. A mathematical model for multimedia health risk assessment. Environmental Software, 1993, 8(4): 231-246.

[13] 胡霞林, 刘景富, 卢士燕, 等. 环境污染物的自由溶解态浓度与生物有效性. 化学进展, 2009, 21(2/3): 514-523.

[14] Liste H H, Alexander M. Butanol extraction to predict bioavailability of PAHs in soil. Chemosphere, 2002, 46: 1011-1017.

[15] Lei L, Bagchi R, Khodadoust A P, et al. Bioavailability prediction of polycyclic aromatic hydrocarbons in field-contaminated sediment by mild extractions. Journal of Environmental Engineering, 2006, 132(3): 384-391.

[16] Tang J, Alexander M. Mild extractability and bioavailability of polycyclic aromatic hydrocarbons in soil. Environmental Toxicology and Chemistry, 1999, 18: 2711-2714.

[17] Macleod C J A, Semple K T. Sequential extraction of low concentrations of pyrene and formation of non-extractable residues in sterile and non-sterile soils. Soil Biology and Biochemistry, 2003, 35: 1443-1450.

[18] Reid B J, Jones K C, Semple K T. Bioavailability of persistent organic pollutants in soils and sediments: A perspective on mechanisms, consequences and assessment. Environmental Pollution, 2000, 108(1): 103-112.

[19] Doick K J, Clasper P J, Urmann K, et al. Further validation of the HPCD-technique for the evaluátion of PAH microbial availability in soil. Environmental Pollution, 2006, 144(1): 345-354.

[20] Rhodes A H, McAllister L E, Semple K T. Linking desorption kinetics to phenanthrene biodegradation in soil. Environmental Pollution, 2010, 158(5): 1348-1353.

[21] Reid B J, Semple K T, Macleod C J, et al. Feasibility of using prokaryote biosensors to assess acute toxicity of polycyclic aromatic hydrocarbons. FEMS Microbiology Letters, 1998, 169(2): 227-233.

[22] Hickman Z, Reid B. Towards a more appropriate water based extraction for the assessment of organic contaminant availability. Environmental Pollution, 2005, 138(2): 299-306.

[23] Barthe M, Pelletier E. Comparing bulk extraction methods for chemically available polycyclic aromatic hydrocarbons with bioaccumulation in worms. Environmental Chemistry, 2007, 4(4): 271-283.

[24] Gouliarmou V, Mayer P. Sorptive bioaccessibility extraction (SBE) of soils: Combining a mobilization medium with an absorption sink. Environmental Science and Technology, 2012, 46(19): 10682-10689.

[25] White J C, Alexander M, Pignatello J J. Enhancing the bioavailability of organic compounds sequestered in soil and aquifer solids. Environmental Toxicology and Chemistry, 1999, 18: 182-187.

[26] Cornelissen G, Vannoort P C M, Govers H A J. Desorption kinetics of chlorobenzenes, polycyclic aromatic hydrocarbons,

and polychlorinated biphenyls: Sediment extraction with Tenax (R) and effects of contact time and solute hydrophobicity. Environmental Toxicology and Chemistry, 1997, 16(7): 1351-1357.

[27] Li B, Y T, Sun H. Meaning and progresses of studies on bioavailability of organic contaminants in soil. Science and Technology Review, 2016, 34(22): 48-55.

[28] Oleszczuk P. The Tenax fraction of PAHs relates to effects in sewage sludges. Ecotoxicology and Environmental Safety, 2009, 2(4): 1320-1325.

[29] Arthur C L, Pawliszyn J. Measurement of the free concentration using solid-phase microextraction: Binding to protein. Analytical Chemistry, 1990, 62(19): 2145-2148.

[30] Vaes W H J, Ramos E U, Verhaar H J M, et al. Measurement of the free concentration using solid-phase microextraction: Binding to protein. Analytical Chemistry, 1996, 68(24): 4463-4467.

[31] Heringa M B, Hermens J L M. Measurement of free concentrations using negligible depletion-solid phase microextraction (nd-SPME). Trends in Analytical Chemistry, 2003, 22(10): 575-587.

[32] Poerschmann J, Kopinke F D, Pawliszyn J. Solid phase microextraction to study the sorption of organotin compounds onto particulate and dissolved humic organic matter. Environmental Science and Technology, 1997, 31(12): 3629-3636.

[33] Ter Laak T L, Agbo S O, Barendregt A, et al. Freely dissolved concentrations of PAHs in soil pore water: Measurements via solid-phase extraction and consequences for soil tests. Environmental Science and Technology, 2006, 40(4): 1307-1313.

[34] Psillakis E, Kalogerakis N. Developments in liquid-phase microextraction. Trends in Analytical Chemistry, 2003, 22(9): 565-574.

[35] Jeannot M A, Cantwell F F. Solvent microextraction as a speciation tool: Determination of free progesterone in a protein solution. Analytical Chemistry, 1997, 69(15): 2935-2940.

[36] Pedersen-Bjergaard S, Rasussen K E. Liquid-liquid-liquid microextraction for sample preparation of biological fluids prior to capillary electrophoresis. Analytical Chemistry, 1999, 71: 2650-2656.

[37] Huckins J N, Tubergen M W, Manuweera G K. Semipermeable membrane devices containing model lipid: A new approach to monitoring the bioavailability of lipophilic contaminants and estimating their bioconcentration potential. Chemosphere, 1990, 20(5): 533-552.

[38] Buschini A, Giordani F, Pellacani C, et al. Cytotoxic and genotoxic potential of drinking water: A comparison between two different concentration methods. Water Research, 2008, 42(8/9): 1999-2006.

[39] Xu Y, Wang Z, Ke R, et al. Accumulation of organochlorine pesticides from water using triolein embedded cellulose acetate membranes. Environmental Science and Technology, 2005, 39(4): 1152-1157.

[40] Xu Y, Spurlock F, Wang Z, et al. Comparison of five methods for measuring sediment toxicity of hydrophobic contaminants. Environmental Science and Technology, 2007, 41(24): 8394-8399.

[41] Thibodeaux L J, Valsaraj K T, Reible D D. Bioturbation-driven transport of hydrophobic organic contaminants from bed sediment. Environmental Engineering Science, 2001, 18: 215-223.

[42] Valsaraj K T, Thibodeaux L J, Reible D D. A quasi-steady-state pollutant flux methodology for determining sediment quality criteria. Environmental Toxicology and Chemistry, 1997, 16: 391-396.

[43] Sarah J, Kjell L, Gunnarsson J S, et al. Bioturbation-driven release of buried PCBs and PBDEs from different depths in contaminated sediments. Environmental Science and Technology, 2010, 44: 7456-7464.

[44] Lick W. The sediment-water flux of HOCs due to "diffusion" or is there a well-mixed layer? If there is, does it matter? Environmental Science and Technology, 2006, 40: 5610-5617.

[45] Perlinger J A, Tobias D E, Morrow P S, et al. Evaluation of novel techniques for measurement of air-water exchange of persistent bioaccumulative toxicants in Lake Superior. Environmental Science and Technology, 2005, 39: 8411-8419.

[46] Bucheli T D, Gustafsson Ö. Ubiquitous observations of enhanced solid affinities for aromatic organo-chlorines in field situations: Are in situ dissolved exposures overestimated by existing partitioning models? Environmental Toxicology and Chemistry, 2001, 20: 1450-1456.

[47] Eek E, Cornelissen G, Breedveld G D. Field measurement of diffusional mass transfer of HOCs at the sediment-water interface. Environmental Science and Technology, 2010, 44: 6752-6759.

[48] Lin D, Eek E, Oen A, et al, Cornelissen G, Tommerdahl J, Luthy R G. Novel probe for in situ measurement of freely dissolved aqueous concentration profiles of hydrophobic organic contaminants at the sediment-water interface. Environmental Science and Technology Letters, 2015, 2: 320-324.

[49] Belles A, Alary C, Criquet J, et al. Assessing the transport of PAH in the surficial sediment layer by passive sampler approach. Science of the Total Environment, 2017, 579: 72-81.

[50] Fernandez L A, Wenjian L, Maruya K A, et al. Calculating the diffusive flux of persistent organic pollutants between sediments and the water column on the Palos Verdes shelf superfund site using polymeric passive samplers. Environmental Science and Technology, 2014, 48: 3925-3934.

[51] Liu H, Zhang K, Xu S P, et al. A novel passive sampling device for measuring sediment-water diffusion fluxes of hydrophobic organic chemicals. Environmental Science and Technology, 2013, 47: 9866-9873.

[52] Feng Y, Wu C C, Bao L J, et al. Examination of factors dominating the sediment-water diffusion flux of DDT-related compounds measured by passive sampling in an urbanized estuarine bay. Environmental Pollution, 2016, 219: 866-872.

[53] Booij K, van Bommel R, van Aken H M, et al. Passive sampling of nonpolar contaminants at three deep-ocean sites. Environmental Pollution, 2014, 195: 101-108.

[54] Sun C, Soltwedel T, Bauerfeind E, et al. Depth profiles of persistent organic pollutants in the North and Tropical Atlantic Ocean. Environmental Science and Technology, 2016, 50: 6172-6179.

[55] Sobek A, Gustafsson O. Deep water masses and sediments are main compartments for polychlorinated biphenyls in the Arctic Ocean. Environmental Science and Technology, 2014, 48: 6719-6725.

[56] Liu Y, Wang S, McDonough C A, et al. Gaseous and freely-dissolved PCBs in the Lower Great Lakes based on passive sampling: Spatial trends and air-water exchange. Environmental Science and Technology, 2015, 50: 4932-4939.

[57] Ruge Z, Muir D, Helm P A, et al. Concentrations, trends, and air-water exchange of PAHs and PBDEs derived from passive samplers in Lake Superior in 2011. Environmental Science and Technology, 2015, 49: 13777-13786.

[58] Tidwell L G, Allan S E, O'Connell S G, et al. Polycyclic aromatic hydrocarbon(PAH)and oxygenated PAH(OPAH)air-water exchange during the deepwater horizon oil spill. Environmental Science and Technology, 2015, 49: 141-149.

[59] Cunliffe M, Engel A, Frka S, et al. Sea surface microlayers: A unified physicochemical and biological perspective of the air-ocean interface. Prog Oceanogr, 2013, 109: 104-116.

[60] Ramus K, Kopinke F D, Georgi A. Sorption-induced effects of humic substances on mass transfer of organic pollutants through aqueous diffusion boundary layers: The example of water/air exchange. Environmental Science and Technology, 2012, 46: 2196-2203.

[61] Wu C C, Yao Y, Bao L J, et al. Fugacity gradients of hydrophobic organics across the air-water interface measured with a novel passive sampler. Environmental Pollution, 2016, 218: 1108-1115.

[62] Zheng Q, Nizzetto L, Liu X, et al. Elevated mobility of persistent organic pollutants in the soil of a tropical rainforest. Environmental Science and Technology, 2015, 49: 4302-4309.

[63] Zhang Y, Deng S, Liu Y, et al. A passive air sampler for characterizing the vertical concentration profile of gaseous phase polycyclic aromatic hydrocarbons in near soil surface air. Environmental Pollution, 2011, 159: 694-699.

[64] Cabrerizo A, Dachs J, Barcelo D. Development of a soil fugacity sampler for determination of air-soil partitioning of

persistent organic pollutants under field controlled conditions. Environmental Science and Technology, 2009, 43: 8257-8263.

[65] Wang Y, Luo C, Wang S, et al. The abandoned E-waste recycling site continued to act as a significant source of polychlorinated biphenyls: An *in situ* assessment using fugacity samplers. Environmental Science and Technology, 2016, 50: 8623-8630.

[66] Ren J, Wang X, Gong P, et al. Characterization of Tibetansoil as a source or sink of atmospheric persistent organic pollutants: Seasonal shift and impact of global warming. Environmental Science and Technology, 2019, 53: 3589-3598.

[67] Degrendele C, Audy O, Hofman J, et al. Diurnal variations of air-soil exchange of semivolatile organic compounds(PAHs, PCBs, OCPs, and PBDEs) in a central European receptor area. Environmental Science and Technology, 2016, 50: 4278-4288.

[68] Wang Y, Luo C, Wang S, et al. Assessment of the air-soil partitioning of polycyclic aromatic hydrocarbons in a paddy field using a modified fugacity sampler. Environmental Science and Technology, 2015, 49: 284-291.

[69] Donald C E, Anderson K A. Assessing soil-air partitioning of PAHs and PCBs with a new fugacity passive sampler. Science of the Total Environment, 2017, 596: 293-302.

[70] Komprdova K, Komprda J, Mensik L, et al. The influence of tree species composition on the storage and mobility of semivolatile organic compounds in forest soils. Science of the Total Environment, 2016, 553: 532-540.

[71] Chiou W L, Buehler P W. Comparison of oral absorption and bioavailability of drugs between monkey and human. Pharmaceutical Research, 2002, 19(6): 868-874.

[72] James K, Peters R E, Cave M R, et al. In vitro prediction of polycyclic aromatic hydrocarbon bioavailability of 14 different incidentally ingested soils in juvenile swine. Science of the Total Environment, 2018, 618: 682-689.

[73] Juhasz A L, Weber J, Stevenson G, et al. *In vivo* measurement, *in vitro* estimation and fugacity prediction of PAH bioavailability in post-remediated creosote-contaminated soil. Science of the Total Environment, 2014, 473-474: 147-154.

[74] Li C, Sun H, Juhasz A L, et al. Predicting the relative bioavailability of DDT and its metabolites in historically contaminated soils using a Tenax-improved physiologically based extraction test (TI-PBET). Environmental Science and Technology, 2015, 50: 1118-1125.

[75] Denys S, Caboche J, Tack K, et al. *In vivo* validation of the unified BARGE method to assess the bioaccessibility of arsenic, antimony, cadmium, and lead in soils. Environmental Science and Technology, 2012, 46(11): 6252-6260.

[76] Ikegami K, Tagawa K, Narisawa S, et al. Suitability of the cynomolgus monkey as an animal model for drug absorption studies of oral dosage forms from the viewpoint of gastrointestinal physiology. Biological and Pharmaceutical Bulletin, 2003, 26: 1442-1447.

[77] Delannoy M, Rychen G, Fournier A, et al. Effects of condensed organic matter on PCBs bioavailability in juvenile swine, an animal model for young children. Chemosphere, 2014, 104: 105-112.

[78] James K, Peters R E, Laird B D, et al. Human exposure assessment: A case study of 8 PAH contaminated soils using *in vitro* digestors and the juvenile swine model. Environmental Science and Technology, 2011, 45: 4586-4593.

[79] Juhasz A L, Herde P, Smith E. Oral relative bioavailability of dichlorodiphenyltrichloroethane(DDT)in contaminated soil and its prediction using in vitro strategies for exposure refinement. Environmental Research, 2016, 150: 482-488.

[80] Cave M R, Wragg J, Harrison I, et al. Comparison of batch mode and dynamic physiologically based bioaccessibility test for PAHs in soil samples. Environmental Science and Technology, 2010, 44: 3654-3660.

[81] Bradham K D, Scheckel K G, Nelson C M, et al. Relative bioavailability and bioaccessibility and speciation of arsenic in contaminated soils. Environmental Health Perspectives, 2011, 119(11): 1629-1634.

[82] Juhasz A L, Weber J, Naidu R, et al. Determination of cadmium relative bioavailability in contaminated soils and its prediction using *in vitro* methodologies. Environmental Science and Technology, 2010, 44: 5240-5247.

[83] Li S W, Sun H J, Li H B, et al. Assessment of cadmium bioaccessibility to predict its bioavailability in contaminated soils. Environment International, 2016, 94: 600-606.

[84] Li C, Zhang R, Li Y, et al. Relative bioavailability and bioaccessibility of PCBs in soils based on a mouse model and Tenax-improved physiologically-based extraction test. Chemosphere, 2017, 186: 709-715.

[85] Li K, Li C, Yu N Y, et al. *In vivo* bioavailability and in vitro bioaccessibility of perfluorooctanoic acid(PFOA)in food matrices: Correlation analysis and method development. Environmental Science and Technology, 2015, 49: 150-158.

[86] Chiu W A, Campbell J L Jr, Clewell H J 3rd, et al. Physiologically based pharmacokinetic(PBPK)modeling of interstrain variability in trichloroethylene metabolism in the mouse. Environmental Health Perspectives, 2014, 122(5): 456-463.

[87] Duan L, Palanisami T, Liu Y, et al. Effects of ageing and soil properties on the oral bioavailability of benzo[*a*]pyrene using a swine model. Environment International, 2014, 70: 192-202.

[88] Duan L, Naidu R, Liu Y, et al. Comparison of oral bioavailability of benzo[*a*]pyrene in soils using rat and swine and the implications for human health risk assessment. Environment International, 2016, 94: 95-102.

[89] Wu T, Taubel M, Holopainen R, et al. Infant and adult inhalation exposure to resuspended biological particulate matter. Environmental Science and Technology, 2018, 52: 237-247.

[90] Staskal D F, Scott L L F, Haws L C, et al. Assessment of polybrominated diphenyl ether exposures and health risks associated with consumption of southern Mississippi catfish. Environmental Science and Technology, 2008, 42: 6755-6761.

[91] Smith E, Weber J, Rofe A, et al. Assessment of DDT relative bioavailability and bioaccessibility in historically contaminated soils using an in vivo mouse model and fed and unfed batch *in vitro* assays. Environmental Science and Technology, 2012, 46: 2928-2934.

[92] Ruby M V, Davis A, Schoof R, et al. Estimation of lead and arsenic bioavailability using a physiologically based extraction test. Environmental Science and Technology, 1996, 30: 422-430.

[93] Rodriguez R R, Basta N T. An *in vitro* gastrointestinal method to estimate bioavailable arsenic in contaminated soils and solid media. Environmental Science and Technology, 1999, 33: 642-649.

[94] DIN, Deutsches Institut fur Normung e.V. Soil Quality−Absorption availability of organic and inorganic pollutants from contaminated soil material. DINE 19738, 2000.

[95] Kelley M E, Brauning S E, Schoof R A, et al. Assessing oral bioavailability of metals in soil. Columbus, OH: Battelle Press, 2002.

[96] Wragg J, Cave M, Basta N, et al. An inter-laboratory trial of the unified BARGE bioaccessibility method for arsenic, cadmium and lead in soil. Science of the Total Environment, 2011, 409: 4016-4030.

[97] Schroder J L, Basta N T, Casteel S W, et al. Validation of the *in vitro* gastrointestinal(IVG)method to estimate relative bioavailable lead in contaminated soils. Journal of Environmental Quality, 2004, 33: 513-521.

[98] Juhasz A L, Herde P, Herde C, et al. Validation of the predictive capabilities of the Sbrc-G *in vitro* assay for estimating arsenic relative bioavailability in contaminated soils. Environmental Science and Technology, 2014, 48: 12962-12969.

[99] Juhasz A L, Weber J, Smith E, et al. Assessment of four commonly employed *in vitro* arsenic bioaccessibility assays for predicting *in vivo* relative arsenic bioavailability in contaminated soils. Environmental Science and Technology, 2009, 43: 9487-9494.

[100] USEPA. *In Vitro* Bioaccessibility Assay for Lead in Soil. U.S. Environmental Protection Agency, Method. 1340, 2013.

[101] Smith E, Kempson I M, Juhasz A L, et al. *In vivo-in vitro* and XANES spectroscopy assessments of lead bioavailability in contaminated periurban soils. Environmental Science and Technology, 2011, 45: 6145-6152.

[102] Li J, Li K, Cui X Y, et al. *In vitro* bioaccessibility and *in vivo* relative bioavailability in 12 contaminated soils: Method comparison and method development. Science of the Total Environment, 2015, 532: 812-820.

[103] Li J, Li K, Cave M, et al. Lead bioaccessibility in 12 contaminated soils from China: Correlation to lead relative bioavailability and lead in different fractions. Journal of Hazardous Materials, 2015, 295: 55-62.

[104] 唐翔宇, 朱永官, 陈世宝. In-Vitro 法评估铅污染土壤对人体的生物有效性. 环境化学, 2003, 22: 503-506.

[105] Gouliarmou V, Collins C D, Christiansen E, et al. Sorptive physiologically based extraction of contami-nated solid matrices: Incorporating silicone rod as absorption sink for hydrophobic organic contaminants. Environmental Science and Technology, 2013, 47: 941-948.

[106] Fang M, Stapleton H M. Evaluating the bioaccessibility of flame retardants in house dust using an in vitro Tenax bead-assisted sorptive physiologically based method. Environmental Science and Technology, 2014, 48: 13323-13330.

[107] Zhang S, Li C, Li Y, et al. Bioaccessibility of PAHs in contaminated soils: Comparison of five in vitro methods with Tenax as a sorption sink. Science of the Total Environment, 2017, 601-602: 968-974.

[108] Zoitos B K, DeMeringo A, Rouyer E, et al. In vitro measurement of fiber dissolution rate relevant to biopersistence at neutral pH: An interlaboratory round robin. Inhalation Toxicology, 1997, 9(6): 525-540.

[109] Julien C, Esperanza P, Bruno M, et al. Development of an in vitro method to estimate lung bioaccessibility of metals from atmospheric particles. Journal of Environmental Monitoring, 2011, 13(3): 621-630.

[110] Wragg J, Klinck B. The bioaccessibility of lead from welsh mine waste using a respiratory uptake test. Journal of Environmental Science and Health Part a Toxic/Hazardous Substances and Environmental Engineering, 2007, 42(9): 1223-1231.

[111] Kanapilly G M, Raabe O G, Goh C H T, et al. Measurement of in vitro dissolution of aerosol particles for comparison to in vivo dissolution in the lower respiratory tract after inhalation. Health Physics, 1973, 24(5): 497-507.

[112] Ansoborlo E, Chalabreysse J, Escallon S, et al. In vitro solubility of uranium tetrafluoride with oxidizing medium compared with in vivo solubility in rats. International Journal of Radiation Biology, 1990, 58(4): 681-689.

[113] Stopford W, Turner J, Cappellini D, et al. Bioaccessibility testing of cobalt compounds. Journal of Environmental Monitoring, 2003, 5(4): 675-680.

[114] Gray J E, Plumlee G S, Morman S A, et al. In vitro studies evaluating leaching of mercury from mine waste calcine using simulated human body fluids. Environmental Science and Technology, 2010, 44(12): 4782-4788.

[115] Taunton A E, Gunter M E, Druschel G K, et al. Geochemistry in the lung: Reaction-path modeling and experimental examination of rock-forming minerals under physiologic conditions. American Mineralogist, 2010, 95(11-12): 1624-1635.

[116] Davies N M, Feddah M I R. A novel method for assessing dissolution of aerosol inhaler products. International Journal of Pharmaceutics, 2003, 255(1-2): 175-187.

[117] Dennis N A, Blauer H M, Kent J E. Dissolution fractions and half times of single source yellowcake in simulated lung fluids. Health Physics, 1982, 42: 469-477.

[118] Boisa N, Elom N, Dean J R, et al. Development and application of an inhalation bioaccessibility method (IBM) for lead in the PM_{10} size fraction of soil. Environment International, 2014, 70: 132-142.

[119] Midander K, Pan J, Wallinder I O, et al. Metal release from stainless steel particles in vitro influence of particle size. Journal of Environmental Monitoring, 2007, 9(1): 74-81.

[120] Stefaniak A B, Guilmette R A, Day G A, et al. Characterization of phagolysosomal simulant fluid for study of beryllium aerosol particle dissolution. Toxicology in Vitro, 2005, 19(1): 123-134.

[121] Borm P J A, Cakmak G, Jermann E, et al. Formation of PAH-DNA adducts after in vivo and vitro exposure of rats and lung cells to different commercial carbon blacks. Toxicology and Applied Pharmacology, 2005, 205(2): 157-167.

[122] Li Y Z, Juhasz A L, Ma L Q, et al. Inhalation bioaccessibility of PAHs in $PM_{2.5}$: Implications for risk assessment and toxicity prediction. Science of the Total Environment, 2019, 650: 56-64.

[123] Kastury F, Smith E, Juhasz A L. A critical review of approaches and limitations of inhalation bioavailability and bioaccessibility of metal (loid)s from ambient particulate matter or dust. Science of the Total Environment, 2017, 574: 1054-1074.

[124] Hofmann W, Asgharian B. The effect of lung structure on mucociliary clearance and particle retention in human and rat lungs. Toxicological Sciences, 2003, 73(2): 448-456.

[125] Twining J, McGlinn P, Loi E, et al. Risk ranking of bioaccessible metals from fly ash dissolved in simulated lung and gut fluids. Environmental Science and Technology, 2005, 39(19): 7749-7756.

[126] Voutsa D, Samara C. Labile and bioaccessible fractions of heavy metals in the airborne particulate matter from urban and industrial areas. Atmospheric Environment, 2002, 36(22): 3583-3590.

[127] Xie S Y, Lao J Y, Wu C C, et al. *In vitro* inhalation bioaccessibility for particle-bound hydrophobic organic chemicals: Method development, effects of particle size and hydrophobicity, and risk assessment. Environment International, 2018, 120: 295-303.

[128] Mehler W T, Li H Z, Pang J X, et al. Bioavailability of hydrophobic organic contaminants in sediment with different particle-size distributions. Archives of Environmental Contamination and Toxicology, 2011, 61: 74-82.

[129] Sun K, Ran Y, Yang Y, et al. Sorption of phenanthrene by nonhydrolyzable organic matter from different size sediments. Environmental Science and Technology, 2008, 42: 1961-1966.

[130] Moss O R. Simulant of lung interstitial fluid. Health Physics, 1979, 36(3): 447-448.

[131] Maruya K A, Dodder N G, Mehinto A C, et al. A tiered, integrated biological and chemical monitoring framework for contaminants of emerging concern in aquatic ecosystems. Integrated Environmental Assessment and Management, 2016, 12: 540-547.

[132] Mayer P, Parkerton T, Adams R, et al. Passive sampling methods for contaminated sediments: Scientific rationale supporting use of freely dissolved concentrations. Integrated Environmental Assessment and Management, 2014, 10: 197-209.

[133] Meador J. Rationale and procedure for using the tissue residue approach for toxicity assessment and determination of tissue, water, and sediment quality guideline for aquatic organisms. Human and Ecological Risk Assessment, 2006, 12: 1018-1073.

[134] You J, Harwood A D, Li H, et al. Chemical techniques for assessing bioavailability of sediment associated contaminants: SPME versus Tenax extraction. Journal of Environmental Monitoring, 2011, 13: 792-800.

[135] You J, Li H, Lydy M J. Assessment of sediment toxicity with SPME-based approaches. Comprehensive Analytical Chemistry, 2015, 67: 161-194.

[136] National Research Council. Bioavailability of contaminants in soils and sediments: Processes, tools and applications. Washington, DC: National Academies Press. 2003.

[137] International Standardization Organization. Soil quality—Requirements and guidance for the selection and application of methods for the assessment of bioavailability of contaminants in soil and soil materials. ISO 17402: 2008. 2008.

[138] USEPA, Guidelines for using passive samplers to monitor organic contaminants at superfund sediment sites. 9200.1-110 FS, Office of Superfund Remediation and Technology Innovation and Office of Research and Development, Washington, DC, USA, 2012.

[139] You J, Landrum P F, Lydy M J. Comparison of chemical approaches for assessing bioavailability of sediment-associated contaminants. Environmental Science and Technology, 2006, 40: 6348-6353.

[140] You J, Landrum P F, Trimble T A, et al. Availability of polychlorinated biphenyls in field-contaminated sediment. Environmental Toxicology and Chemistry, 2007, 26: 1940-1948.

[141] Trimble T A, You J, Lydy M J. Determining appropriate techniques for assessing the bioavailability of PCBs from field-collected sediment. Chemosphere, 2008, 71: 337-344.

[142] Kraaij R, Mayer P, Busser F J M, et al. Measured pore water concentrations make equilibrium partitioning work—A data analysis. Environmental Science and Technology, 2003, 37: 268-274.

[143] Jahnke A, McLachlan M S, Mayer P. Equilibrium sampling: Partitioning of organochlorine compounds from lipids into polydimethylsiloxane. Chemosphere, 2008, 73: 1575-1581.

[144] Mäenpää K, Leppänen M T, Figueiredo K, et al. Sorptive capacity of membrane lipids, storage lipids, and proteins: a preliminary study of partitioning of organochlorines in lean fish from a PCB-contaminated freshwater lake. Arch. Environ. Contam. Toxicol. 2015, 68: 193-203.

[145] Mäenpää K, Leppänen M T, Reichenberg F, et al. Equilibrium sampling of persistent and bioaccumulative compounds in soil and sediment: Comparison of two approaches to determine equilibrium partitioning concentrations in lipids. Environmental Science and Technology, 2011, 45: 1041-1047.

[146] Pei Y Y, Li H Z, You J. Determining equilibrium partition coefficients between lipid/protein and poly-dimethylsiloxane for highly hydrophobic organic contaminants using preloaded disks. Science of the Total Environment, 2017, 598: 385-392.

[147] Lydy M J, Landrum P F, Oen A, et al. Passive sampling methods for contaminated sediments: State of the science for organic contaminants. Integrated Environmental Assessment and Management, 2014, 10: 167-178.

[148] Lohmann R, Muir D, Zeng E Y, et al. Aquatic global passive sampling(AQUA-GAPS)revisited: First steps towards a network of networks for organic contaminants in the aquatic environment. Environmental Science and Technology, 2017, 51(3): 1060-1067.

[149] Di Toro D M, Mahony J D, Hansen D J, et al. Toxicity of cadmium in sediments: The role of acid volatile sulfide. Environmental Toxicology and Chemistry, 1990, 9: 1487-1502.

[150] Accardi-Dey A, Gschwend P M, Ralph M. Assessing the combined roles of natural organic, matter and black carbon as sorbents in sediments. Environmental Science and Technology, 2002, 36(1): 21-29.

[151] Pehkonen S, You J, Akkanen J, et al. Influence of black carbon and chemical planarity on bioavailability of sediment-associated contaminants. Environmental Toxicology and Chemistry, 2010, 29(9): 1976-1983.

[152] Maruya K A, Landrum P F, Burgess R M, et al. Incorporating contaminant bioavailability into sediment quality assessment frameworks. Integrated Environmental Assessment and Management, 2012, 8: 659-673.

[153] Lyytikainen M, Hirva P, Minkkinen P, et al. Bioavailability of sediment-associated PCDD/Fs and PCDEs: Relative importance of contaminant and sediment charateristics and biological factors. Environmental Science and Technology, 2003, 37: 3926-3934.

[154] Zhang J, You J, Li H-Z, et al. Particle-scale understanding of cypermethrin in sediment: Desorption, bioavailability, and bioaccumulation in benthic invertebrate Lumbriculus variegatus. Science Total of the Environment, 2018, 642: 638-645.

[155] Wang F, Goulet R R, Chapman P M. Testing sediment biological effects with the freshwater amphipod *Hyalella azteca*: The gap between laboratory and nature. Chemosphere, 2004, 57: 1713-1724.

[156] Kuivila K M, Hladik M L, Ingersoll C G, et al. Occurrence and potential sources of pyrethroid insecticides in stream sediments from seven U.S. metropolitan areas. Environmental Science and Technology, 2012, 46: 4297-4303.

[157] Li H Z, Sun B Q, Chen X, et al. Addition of contaminant bioavailability and species susceptibility to a sediment toxicity assessment: Application in an urban stream in China. Environmental Pollution, 2013, 178: 135-141.

[158] Xu Y P, Spurlock F, Wang Z J, et al. Comparison of five methods for measuring sediment toxicity of hydrophobic contaminants. Environmental Science and Technology, 2007, 41: 8394-8399.

[159] You J, Pehkonen S, Weston D P, et al. Chemical availability and sediment toxicity of pyrethroid insecticides to *Hyalella azteca*: Application to field sediment with unexpectedly low toxicity. Environmental Toxicology and Chemistry, 2008, 27(10): 2124-2130.

[160] Ding Y, Landrum P F, You J, et al. Use of solid phase microextraction to estimate toxicity: I relating fiber concentrations to toxicity. Environmental Toxicology and Chemistry, 2012, 31(9): 2159-2167.

[161] Meador J P, McCarty L S, Escher B I, et al. 10th Anniversary critical review: The tissue-residue approach for toxicity assessment: Concepts, issues, application, and recommendations. Journal of Environmental Monitoring, 2008, 10: 1486-1498.

[162] Escher B I, Hermens J L M. Internal exposure: linking bioavailability to effect. Environmental Science and Technology, 2004, 38: 455A-462A.

[163] Parkerton T F, Stone M A, Letinski D J. Assessing the aquatic toxicity of complex hydrocarbon mixtures using solid phase microextraction. Toxicology Letters, 2000, 112: 273-282.

[164] Harwood A D, Bunch A R, Flickinger D L, et al. Predicting the toxicity of permethrin to *Daphnia magna* in water using SPME fibers. Archives of Environmental Contamination and Toxicology, 2012, 62: 438-444.

[165] Du J, Pang J X, You J. Bioavailability-based chronic toxicity measurements of permethrin to *Chironomus dilutus*. Environmental Toxicology and Chemistry, 2013, 32(6): 1403-1410.

[166] Rakowska M I, Kupryianchyk D, Harmsen J, et al. *In situ* remediation of contaminated sediments using carbonaceous materials. Environmental Toxicology and Chemistry, 2012, 31: 693-704.

[167] Mackenbach E M, You J, Mills M A, et al. Application of a Tenax model to assess bioavailability of PCBs in field sediments. Environmental Toxicology and Chemistry, 2012, 31(10): 2210-2216.

[168] Karlsson M. TTIP and the environment: The case of chemicals policy. Global Affairs, 2015, 1: 21-31.

[169] Schwarzenbach R P, Escher B I, Fenner K, et al. The challenge of micropollutants in aquatic systems. Science, 2006, 313: 1972-1077.

[170] Pool R, Rusch E. Identifying and reducing environmental health risks of chemicals in our society: Workshop summary. National Academy of Sciences, 2014.

[171] Brack W, Ait-Aissa S, Burgess R M, et al. Effect-directed analysis supporting monitoring of aquatic environments: An in-depth overview. Science Total of the Environment, 2016, 544: 1073-1118.

[172] Li H-Z, Zhang J, You J. Diagnosis of complex mixture toxicity in sediment: Application of toxicity identification evaluation and effect-directed analysis. Environmental Pollution, 2018, 237: 944-954.

[173] Jin X, Wang Y, Giesy J P, et al. Development of aquatic life criteria in China: Viewpoint on the challenge. Environmental Science and Pollution Research, 2014, 21: 61-66.

[174] Kortenkamp A, Faust M. Regulate to reduce chemical mixture risk. Science, 2018, 361: 224-226.

[175] Burgess R M, Ho K T, Brack W, et al. Effects-directed analysis (EDA) and toxicity identification evaluation(TIE): Complementary but different approaches for diagnosing causes of environmental toxicity. Environmental Toxicology and Chemistry, 2013, 32: 1935-1945.

[176] You J, Li H-Z. Improving the accuracy of effect-directed analysis: The role of bioavailability. Environmental Science: Processes and Impacts, 2017, 19(12): 1484-1498.

[177] Mehler W T, Maul J D, You J, et al. Identifying the causes of sediment-associated contamination in the Illinois River (USA) using a whole-sediment toxicity identification evaluation. Environmental Toxicology and Chemistry, 2010, 29(1): 158-167.

[178] Sparks R E, Ross P E. Identification of toxic substances in the Upper Illinois River. ILENR/RE-WR-92/07. Illinois Department of Energy and Natural Resources, Springfield, IL, USA, 1992.

[179] Mehler W T, You J, Maul J D, et al. Comparative analysis of whole sediment and porewater toxicity identification evaluation techniques for ammonia and non-polar organic contaminants. Chemosphere, 2010, 78(7): 814-821.

[180] Bandow N, Altenburger R, Streck G, et al. Effect-directed analysis of contaminated sediments with partition-based dosing

using green algae cell multiplication inhibition. Environmental Science and Technology, 2009, 43: 7343-7349.

[181] Simon E, Bytingsvik J, Jonker W, et al. Jenssen B M, Lie E, Aars J, Hamers T, Lamoree M H. Blood plasma sample preparation method for the assessment of thyroid hormone-disrupting potency in effect-directed analysis. Environmental Science and Technology, 2011, 45: 7936-7944.

[182] Bandow N, Altenburger R, Varel U L V, et al. Partitioning-based dosing: An approach to include bioavailability in the effect-directed analysis of contaminated sediment samples. Environmental Science and Technology, 2009, 43: 3891-3896.

[183] Zielke H, Seiler T B, Niebergall S, et al. The impact of extraction methodologies on the toxicity of sediments in the zebrafish (*Danio rerio*) embryo test. Journal of Soils and Sediments, 2011, 11(2): 352-363.

[184] Yi X Y, Li H Z, Ma P, et al. Identifying the causes of sediment-associated toxicity in urban waterways in South China: Incorporating bioavailability-based measurements into whole-sediment toxicity identification evaluation. Environmental Toxicology and Chemistry, 2015, 34(8): 1744-1750.

[185] Brack W, Bandow N, Schwab K, et al. Bioavailability in effect-directed analysis of organic toxicants in sediments. Trends in Analytical Chemistry, 2009, 28: 543-549.

[186] Schwab K, Brack W. Large volume Tenax extraction of the bioaccessible fraction of sediment-associated organic compounds for a subsequent effect-directed analysis. Journal of Soils and Sediments, 2007, 7: 178-186.

[187] Schwab K, Altenburger R, Varel U L, et al. Effect-directed analysis of sediment-associated algal toxicants at selected hot spots in the river Elbe Basin with a special focus on bioaccessibility. Environmental Toxicology and Chemistry, 2009, 28: 1506-1517.

[188] Li H Z, Yi X Y, Cheng F, et al. Identifying organic toxicants in sediment using effect-directed analysis: A combination of bioaccessibility-based extraction and high-throughput midge toxicity testing. Environmental Science and Technology, 2019, 53: 996-1003.

[189] Mayer P, Wernsing J, Tolls J, et al. Establishing and controlling dissolved concentrations of hydrophobic organics by partitioning from a solid phase. Environmental Science and Technology, 1999, 33(13): 2284-2290.

[190] Smith K E C, Rein A, Trapp S, et al. Dynamic passive dosing for studying the biotransformation of hydrophobic organic chemicals: Microbial degradation as an example. Environmental Science and Technology, 2012, 46(9): 4852-4860.

[191] Perron M M, Burgess R M, Ho K T, et al. Development and evaluation of reverse polyethylene samplers for marine phase II whole-sediment toxicity identification evaluations. Environmental Toxicology and Chemistry, 2009, 28(4): 749-758.

[192] Smith K E C, Oostingh G J, Mayer P. Passive dosing for producing defined and constant exposure of hydrophobic organic compounds during *in vitro* toxicity tests. Chemical Research in Toxicology, 2010, 23(1): 55-65.

[193] Rojo-Nieto E, Smith K E C, Perales J A, et al. Recreating the seawater mixture composition of HOCs in toxicity tests with *Artemia franciscana* by passive dosing. Aquatic Toxicology, 2012, 120/121: 27-34.

[194] 马萍, 程飞, 李慧珍, 等. 聚二甲基硅氧烷被动加标法在水毒性测试中的应用: 现状与进展. 环境化学, 2017, 36: 1177-1188.

[195] Xia X, Li H, Yang Z, et al. How does predation affect the bioaccumulation of hydrophobic organic compounds in aquatic organisms? Environmental Science and Technology, 2015, 49(8): 4911-4920.

[196] Qi H X, Li H Z, Wei Y L, et al. Effect-directed analysis of toxicants in sediment with combined passive dosing and *in vivo* toxicity testing. Environmental Science and Technology, 2017, 51(11): 6414-6421.

作者：廖春阳[1]，鲍恋君[2]，崔昕毅[3]，游 静[2]

[1] 中国科学院生态环境研究中心，[2] 暨南大学，[3] 南京大学

第4章 天然有机质的环境行为和效应

- 1. 引言 /79
- 2. 天然有机质的结构表征 /80
- 3. 天然有机质的环境效应 /88
- 4. 展望 /95

本章导读

天然有机质广泛分布于土壤和水生生态系统中，在全球碳循环和污染物环境归趋过程中起着关键作用。天然有机质结构的多样性、复杂性和异质性是其环境过程研究的主要挑战之一。目前对天然有机质的结构表征仍然处于定性描述为主，半定量为辅的阶段，主要着眼于阐明来源及形成条件对其结构的影响，但对天然有机质分子结构在环境过程中的演变和关键作用了解还不深入。因此，需要借助现代分析技术的新发展，结合常规化学和谱学分析手段，全面解析天然有机质的结构信息。天然有机质具有丰富的极性基团和疏水微区，是地表环境污染物的重要吸附剂和载体，控制着环境污染物的分配传输和生物有效性，显著影响其环境效应。此外，天然有机质具有一定的氧化还原和光化学活性，从而介导金属/金属氧化物的氧化还原反应和有机污染物的降解反应。其介导的环境过程是决定污染物的赋存形态、环境转化和消减的重要过程，是地表水环境化学的研究重点之一。本章将针对天然有机质的表征方法和环境效应进行综述，主要聚焦于这两个领域近期的研究进展。

关键词

天然有机质，结构表征，吸附，转化，光化学，有机污染物，金属纳米材料

1 引 言

天然有机质（natural organic matter，NOM）主要是由动植物残体经微生物分解和转化及系列地球化学过程积累而形成，也有部分来自于火成过程。其广泛分布于土壤和水生生态系统中，在全球碳循环和污染物环境归趋过程中起着关键作用。NOM 具有丰富的极性基团和疏水微区，因此成为地表环境金属和有机污染物的重要吸附剂和载体，控制着环境污染物的分配传输和生物有效性。此外，NOM 具有氧化还原和光化学活性，可介导金属/金属氧化物的氧化还原反应和有机污染物的降解反应。NOM 与污染物之间的相互作用会显著影响其环境迁移、转化和归趋过程，继而影响其环境和健康效应。因此，NOM 的环境行为和效应研究成为环境化学和环境地球化学领域研究的热点问题。

目前，NOM 研究面临的首要问题是其结构的复杂性、多样性和异质性。长期以来，NOM 的来源、结构表征及特性描述一直是国内外环境与地球化学的研究热点。然而，地学上的表征数据较少用于研究 NOM 与污染物的相互作用。在传统表征手段 [如元素分析、X 射线光电子能谱（XPS）、紫外-可见光谱（UV-Vis）、荧光光谱、红外光谱（FTIR）等] 获取的信息基础上，傅里叶变换-离子回旋共振质谱（FT-ICR MS）、先进核磁（NMR）技术和同步辐射技术的发展和应用为解析 NOM 分子结构，探索其与污染物的微观作用机制提供了新的契机[1, 2]。FT-ICR MS 具有前所未有的质谱分辨率，是目前唯一能够准确检测 NOM 中绝大部分分子核质比的分析手段。不断发展的先进 NMR 技术不但可以定量分析 NOM 官能团含量，还可进一步解析不同官能团之间链接特性、结构域及异质特性。同步辐射技术具有高的灵敏度和空间分辨率（低至几纳米），因而能够用于环境/地质样品中有机质的原位形态和成像分析[3]。然而，由于前处理损失和分子离子化效率的差异，FT-ICR MS 只能作为定性或半定量分析手段；而 NMR 技术提供的官能团解析分辨率不高，无法提供 NOM 分子结构全信息。同步辐射技术的最大障碍在于需要大型的同步辐射装置，机时十分有限，限制了其在环境与地球化学领域的运用。因此，如何联合运用多种先进分析技术，并结合常规化学和谱学分析手段，全面表征 NOM 分子结构仍是目前研究的一个巨大挑战。

NOM 与有机污染物的相互作用是解决污染物多相分配和环境归趋问题的关键环节，也是环境化学研

究的重点[4]。非极性有机污染物在环境中主要的赋存介质为土壤、底泥和水体环境中的有机相,包括NOM和黑碳。NOM可通过自身吸附和改变黑碳吸附来影响污染物在环境中的吸附行为。其含量、组成和结构显著影响非极性有机污染物在环境介质中的分配规律。NOM介导的光化学反应是水环境化学研究的重点之一。NOM在太阳光照射下可产生多种光生活性中间体(PPRIs),PPRIs可继而与环境污染物发生反应,改变其赋存形态和降解速率。NOM中含有大量醌类、酚类以及芳香羧酸等氧化还原活性组分,可作为反应物或者电子传递载体,参与污染物在环境中的氧化还原转化过程。目前,对NOM氧化还原能力和产生PPRIs的量子产率的定量研究较为有限,尚不足以对NOM氧化还原和光化学特性及其对污染物归趋的影响机制进行系统定量讨论和预测。此外,环境纳米科技的兴起使得金属/金属氧化物纳米颗粒的环境归趋问题日益受到关注。该领域的研究近期取得了一些进展,但NOM对金属/金属氧化物纳米颗粒环境过程的影响机制认识尚不深入。研究NOM的吸附行为、光化学和氧化还原活性将有助于更准确地评估污染物的环境转化和归趋。因此,需厘清NOM与有机污染物相互作用和NOM光化学/氧化还原活性的微观机制,建立有效的定量预测方法。本章将针对NOM的表征方法和环境效应进行综述,主要聚焦于这两个领域近期的研究进展。

2 天然有机质的结构表征

天然有机质(NOM)存在于地表各圈层,有着十分重要的环境地球化学意义,但由于其来源和成因的复杂性,目前我们对其化学组成和结构的认知十分有限。如何从分子水平揭示NOM的组成与结构一直是环境科学和地球科学所面临的重大挑战。长期以来,科学家针对NOM的元素/同位素组成、高温裂解或化学分解的分子片段组成、分子量分布、酸碱特性、光学特性和氧化还原特性等各种物理化学特征发展了一系列的分析方法,这些方法对于认识NOM在环境中的含量及其周转速率和迁移转化及其对污染物迁移转化行为的影响等宏观过程有重要的意义,但这些方法往往得到的是有机质的整体信息或某一特征信息。本节将重点综述近年来在NOM分子水平表征方法方面的重要进展,这些方法同样适用于人工有机碳材料的分子表征。

2.1 天然有机质组成结构的认知发展

关于NOM的组成与结构的认知起源于人们对腐殖质(humic substance, HS)化学组成与结构的探索,早在18世纪土壤化学家就发现土壤中存在大量黑色难降解的有机物质并将其称作HS[5, 6],并逐步发展和完善了一套经典的按操作定义的HS组分的概念和提取方法。按照有机质在酸碱提取过程中的溶解性,HS被进一步分为酸碱均可溶的富里酸(fulvic acid, FA)、碱溶酸不溶的腐殖酸(humic acid, HA)以及酸碱均不溶的胡敏素(humin)。这一定义后来被扩展到水体有机质(如水体FA和HA)以及大气有机质(如腐殖类物质,humic-like substance, HULIS)的组成分析中。随着分析技术的发展和研究的深入,人们对于HS的认知也在不断发展,很长一段时间HS被认为是小分子经腐殖化过程(humification)形成的具有特殊结构的抗微生物降解的高分子聚合物(平均分子量20~50 kDa)[7, 8]。20世纪90年代,随着体积排阻色谱和核磁共振等技术的发展,一些学者开始对HS的聚合物(polymer)模型产生质疑,提出了超分子集合体(supramolecular association)的概念,认为HS是由分子量相对较小的有机分子(小于1000 Da)通过分子间氢键和疏水作用等形成的超分子集合体[7, 9],这一观点正在被越来越多的先进分析技术所支持。从"高分子聚合物"到"超分子集合体"的认知转变对NOM相关研究特别是对于认识有机碳的储存与周转过程具有重要的意义。最近，HS 的概念再次受到一些土壤化学家更为彻底的质疑，以 Johannes Lehmann 和 Markus Kleber 为代表的科学家认为自然环境中不存在有机质的腐殖化过程，NOM 是从生物残体到有机碎片到有机大分子再到 CO_2 不断分解过程中产生的有机物质的连续体（continuum），所谓的"HS"不是天然存在于土壤中的而是在强酸和强碱提取过程中产生的人为物质[10]。因此，他们强调原位表征土壤有机质的重要性，而反对将其进行分离和提取。这一观点足以颠覆整个 NOM 研究传统，因此遭到很多学者的强烈反对[5, 11]，到目前为止，关于 HS 证"实"与证"伪"的争论仍然十分激烈，但似乎都没有足够的科学证据驳倒对方的观点。但无论如何，这场学术争论对于 NOM 领域都将具有里程碑式的意义，而发展更先进的 NOM 组成与结构的表征方法，将是结束这场学术争论和揭开 NOM 神秘面纱的唯一途径。

2.2 样品前处理方法

虽然 Johannes Lehmann 和 Markus Kleber 等特别强调原位研究 NOM 的重要性[10]，但是原位分析 NOM 特别是土壤有机质仍然面临十分严峻的挑战，因为多数环境介质中的有机碳含量低而背景干扰严重，无法满足现有原位分析仪器和方法的要求，分离提纯仍然是大多数分析仪器必备的样品前处理方法。NOM 的定义十分宽泛，在实际操作中往往根据不同的样品提取方法衍生出各种子概念，包括前述提出的与 HS 相关的一系列概念。

按照其存在形式，NOM 可分为可溶性有机质（dissolved organic matter，DOM）和颗粒态有机质（particulate organic matter，POM）。二者的区分是根据操作定义的，DOM 是指能够通过一定孔径（0.1～1 μm）滤膜的有机物，相反，被截留的即为 POM[12]。通常自然水体中可溶性有机碳（DOC）浓度较低且含有无机杂质，难以对其进行分子水平的表征，需要对 DOM 进行分离提取和除盐纯化。常用的水体 DOM 分离纯化方法包括 XAD 树脂法、固相萃取法和反渗透/电渗析法（RO/ED）。XAD 树脂法是最早用于 DOM 分离提纯的方法，也是国际腐殖酸协会（IHSS）推荐的用于 FA 和 HA 提取纯化的方法。XAD 系列树脂是一类非离子的大孔型共聚化合物，它们通常有较大的比表面积，能够选择性吸附溶液中疏水性有机化合物。这类树脂吸附的有机质需要 NaOH 溶液才能洗脱，因此还需要 H 型阳离子交换树脂进一步去除洗脱液中的 Na^+ 离子，过程比较繁琐，亲水性有机物不能被树脂保留，因此有机碳的回收率较低。固相萃取法与 XAD 树脂法原理类似，但由于操作简单且有各种商品化的固相萃取柱，使其在 DOM 样品前处理中的应用更加广泛。Dittmar 等比较了不同固相萃取材料，包括改性苯乙烯二乙烯基苯聚合物型吸附剂（Varian PPL 和 ENV）以及不同碳链长度的硅基材料（Varian C_8，C_{18}，$C_{18}OH$ 和 $C_{18}EWP$）对河水和海水 DOM 的提取效率，结果表明 PPL 固相萃取柱具有最高的有机碳回收率（平均约 62%），其次是 C_{18} 固相萃取柱（约为 PPL 的 2/3），^{1}H-NMR 分析也表明 PPL-DOM 保留了更为丰富的化学基团[13]。但是，PPL 固相萃取材料对 DOM 的保留仍然是选择性的，亲水性组分比较容易损失。使用 PPL 吸附剂的另一优势在于用甲醇代替碱作为洗脱剂，能够直接用于质谱分析，因此，特别适合作为质谱分析的样品前处理方法。反渗透/电渗析法（RO/ED）是一种更为直接而高效的水体 DOM 分离提纯方法。它利用反渗透单元去除水而浓缩 DOM，然后再利用电渗析单元除去其中的盐类物质，从而实现水体 DOM 的分离提纯。这种方法最大的优势在于它对 DOM 的提取是非选择性的，因此具有更高的有机碳回收率（超过 75%），而且可以大规模的分离纯化水体 DOM[14]。

对于土壤 DOM 而言，情况更为复杂，准确地说应该按照提取方法进行操作定义，比如常用的水提取有机质（WEOM）、热水提取有机质（HWEOM）、碱提取有机质（AEOM）以及其他试剂提取有机质等。水提取是最温和的土壤 DOM 提取方法，因而不会对土壤 DOM 的化学结构造成任何影响，但由于提取不完全并不能全面代表土壤 DOM[15]。针对一些非极性土壤有机质组分，可以采用有机试剂如甲醇、正己烷或二氯甲烷等进行提取[16]。除了可提取的 DOM，土壤和沉积物中还存在大量不可提取的 POM，通常采用氢氟酸（HF）和强还原剂如连二亚硫酸盐去除其中的硅酸盐类和含铁矿物，并达到了浓缩有机质和去除磁性物质的目的[17, 18]，被广泛用作固体核磁共振分析的样品前处理方法。但以上化学试剂处理不但会导致大量和矿物结合的有机碳的损失，而且可能改变有机质的化学组成和结构[18]。因此，低有机碳含量的矿物-有机质样品前处理方法仍然是 NOM 分子表征的一大难题，既要考虑试剂和提取条件对有机质自身化学组成与结构的影响，又要考虑提取方法的选择性。

2.3 天然有机质的分子表征方法

2.3.1 核磁共振波谱

核磁共振波谱是目前运用最广泛的表征有机质中有机官能团的技术。^{1}H-NMR 是最早用于表征有机质化学结构的 NMR 技术，被广泛用于表征 DOM、FA 和 HA 等水溶或者有机溶剂可溶的有机质。^{1}H NMR 能够提供丰富的化合物指纹信息，可被用于分离提纯有机化合物的结构鉴定，但是对于有机质这类复杂的混合物，会因为谱峰重叠而大大降低分辨率。通过多维核磁特别是 ^{1}H-NMR-^{13}C-NMR 二维核磁能够有效提高分辨率，从而更加精准地鉴定出复杂有机质中存在的官能团结构。Hertkorn 等利用 ^{1}H-NMR-^{13}C-NMR 二维核磁鉴定出不同深度的海洋 DOM 中均存在大量羧基化脂环族分子（carboxyl-rich alicyclic molecules，CRAM），并认为这类物质可能是海洋中顽固有机质（refractory organic matter，ROM）的主要成分[19, 20]。对于难溶的样品也可以直接采用 ^{13}C-NMR 进行表征，由于固态物质具有较强偶极-偶极相互作用，固体 ^{13}C-NMR 的共振谱线通常较宽，因此分辨率远不如高分辨液体核磁，这一特点在研究复杂有机质时却有独到之处。一般而言，可以将有机质的固体 ^{13}C-NMR 谱图大致分为五个区域：①0～45 ppm 属于烷基碳(Alkyl C)区，②45～110 ppm 烷氧基碳(O-alkyl C)区，③110～160 ppm 芳香碳(Aromatic C)区，④160～190 ppm 羧基碳（Carboxyl C）区，⑤190～220 ppm 酮基碳（Ketone C）区。对每个区域进行积分即可计算出不同形态碳的相对含量，因此固体 ^{13}C-NMR 常被用来半定量分析有机碳的官能团组成。由于 ^{13}C 的检测灵敏度较低，通常采用交叉极化/魔角旋转（CP/MAS）技术来提高检测信号，但采用 CP/MAS 进行定量分析的一个主要缺陷就是存在旋转边带效应（spinning side bands，SBs），直接极化/魔角旋转（DP/MAS）技术可以有效避免旋转边带效应，大大提高了定量分析的可靠性[21]。除此之外，还有一些其他的技术如 DP/PASS（phase-adjusted sideband suppressing）、CP/TOSS（total side band suppressing）等也被用于消除旋转边带效应。但是 DP/MAS 技术在提高定量准确度的同时也需要付出时间的代价，其耗时通常比 CP/MAS 高出数倍[22]。由于碳含量低的样品采用固态 NMR 直接分析非常耗时，土壤/沉积物样品通常还含有顺磁性物质，需要利用一定的前处理方法去除无机矿物杂质，以消除样品磁性并提高有机碳浓度。目前仍缺乏一种既能够保证不破坏有机质的化学结构又能够高效去除矿物杂质的方法。尽管如此，^{13}C-NMR 仍然是目前唯一能够对复杂样品中不同形态有机碳进行定量分析的原位表征技术，在研究 NOM 形成演化等过程中的作用仍然不可取代。

2.3.2 同步辐射红外与软 X 射线吸收光谱

普通 FTIR 的灵敏度较低，而且受复杂环境介质的背景干扰严重，因此 FTIR 在原位表征有机质化学结构方面具有很大的局限性，通常只能用于表征有机质含量高或分离提取的有机质样品。同步辐射（SR）FTIR 具有更高的灵敏度和更高的空间分辨率（低至几 μm），因而能够用于环境/地质样品中有机质的原位形态和成像分析[3]。通过逐点或者整列式扫描土壤颗粒中不同空间位点的 FTIR 谱图，可以获得不同有机官能团和无机组分在土壤颗粒中的分布图，并能够进一步解析不同有机官能团之间以及有机官能团与无机组分之间在空间分布上的相关性，特别适合研究颗粒态有机质在环境中的分解转化过程以及土壤团聚体中有机质-矿物的相互作用。Lehmann 等利用 SR FTIR 研究了代表脂肪碳的 C—H（2922 cm^{-1}）、代表芳香碳的 C—C（1589 cm^{-1}）、代表多糖碳的 C—O（1035 cm^{-1}）与高岭土表面羟基（3695 cm^{-1}）在土壤团聚体中的分布，发现高岭土对脂肪碳在空间分布上的相关性最强，表明黏土矿物可能通过与脂肪碳的结合促进有机质的稳定[3]。由于一些有机污染物也具有特征红外吸收峰，因此能够用于土壤颗粒中污染物的原位分布及其与矿物-有机质的相互关系。Luo 等将 SR FTIR 用于研究一系列硝基化合物在土壤中的分布，发现有机质中的芳香组分对对硝基化合物在土壤颗粒表面的吸附贡献最大，黏土矿物也有一定贡献，但是其贡献随着污染物疏水性的增加而降低[23]，见图 4-1。

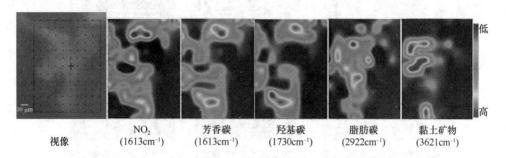

图 4-1　同步辐射 FTIR 原位表征土壤切片上 1,3-二硝基苯（1,3-dinitrobenzene，mDNB）、
不同形态碳及黏土矿物表面羟基的微区分布图（引自参考文献[23]）

图中方框区域大小为（100×120）μm^2

基于同步辐射软 X 射线光源的 C 1s 近边 X 射线吸收精细结构（C1s NEXAFS）在分析有机碳形态方面具有强大的能力，其所能获得的化学结构信息与 ^{13}C-NMR 类似，但是其灵敏度较 NMR 高而且不受磁性材料和其他元素的干扰，更适合于原位分析，已经被广泛用于腐殖质、土壤有机质、生物质和人工碳材料中有机碳形态的研究[24, 25]。不同有机碳官能团对应不同的 X 射线吸收能量吸收，因此可以根据 C1s NEXAFS 谱通过高斯拟合半定量地计算出不同有机碳官能团的含量（图 4-2）[26]。此外，结合软 X 射线扫描透射显微镜（STXM），可以在低至 10 nm 的横向分辨率上揭示不同形态 C 的空间分布[27]。利用这一技术，Lehmann 课题组开展了一系列研究，在纳米尺度上解析了有机碳在土壤团聚体中的异质分布

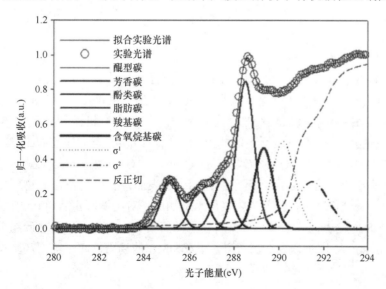

图 4-2　典型土壤有机质的 C1s 近边 X 射线吸收精细结构（C1s NEXAFS）谱图（引自参考文献[26]）

（图 4-3）[28, 29]，由于其他地球化学元素如 N、S、Ca、Fe、Al 和 Si 等的 K 边或 L 边吸收均在软 X 射线能量范围，也可以获得这些元素形态及其空间分布，与有机碳的形态分布结合，能够揭示土壤团聚体上有机质的转化以及矿物和有机质的相互作用，为土壤有机碳的循环和周转提供分子水平和纳米尺度上的证据[30]。当然，NEXAFS 和 STXM 技术也有其局限性，例如缺乏准确的定量方法，提高分辨率的同时灵敏度会降低，因此对于含碳量低的区域无法获得有效的化学信息。此外，这一技术的最大障碍在于需要大型的同步辐射装置，能够获得的机时十分有限，而环境、土壤和地球化学研究又需要大量的样品和机时作保障，因此，大大限制了其在环境地球化学领域的运用。

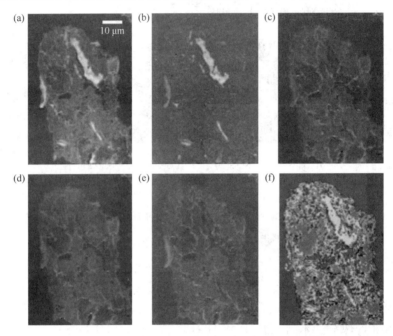

图 4-3　同步辐射 C1s NEXAFS 和 STXM 分析土壤团聚体中不同形态有机碳的空间分布（引自参考文献[28]）
（a）为总碳分布，（b）为芳香碳分布，（c）为脂肪碳分布，（d）为羧基官能团分布，（e）为酚官能团分布，（f）为不同形态碳分布的合成图

2.3.3　纳米离子探针

纳米离子探针又称为纳米二次离子质谱（NanoSIMS），以铯离子（Cs^+）或子氧离子（O^-）轰击样品表面产生的具有指纹特征的次级离子谱为检测基础，既可提供样品表面元素信息，也可提供化学组分信息，而且具有极高的空间分辨率（Cs^+ 源束斑小于 50nm，O^- 源束斑小于 200nm）。近年来这一技术被成功用于地质样品、土壤颗粒、大气颗粒物、生物和材料样品中有机碳和其他元素或同位素的原位成像分析[31-34]。Vogel 等利用 ^{15}N 和 ^{13}C 标记和非标记的凋落物培养土壤，通过 NanoSIMS 以铯离子作为离子源分析了土壤颗粒物表面 $^{12}C^-$、$^{13}C^-$、$^{12}C^{14}N^-$、$^{12}C^{15}N^-$、$^{16}O^-$ 的微区分布（图 4-4），发现新鲜有机质更容易富集在有机质—矿物复合体表面而不是裸露的矿物表面，表明包覆在矿物表面的老化有机质能够为新鲜有机质提供更多的结合位点，从土壤团聚体水平加深了对土壤固定有机质的认识[35]。Liu 等联合利用 STXM 和 NanoSIMS 原位表征了细菌胞外聚合物中的脂类、蛋白质类以及富硫组分更容易在针铁矿表面富集并呈 100～400 nm 大小的斑块式分布[36]。NanoSIMS 的局限性在于需要高真空的环境而且无法进行定量表征。

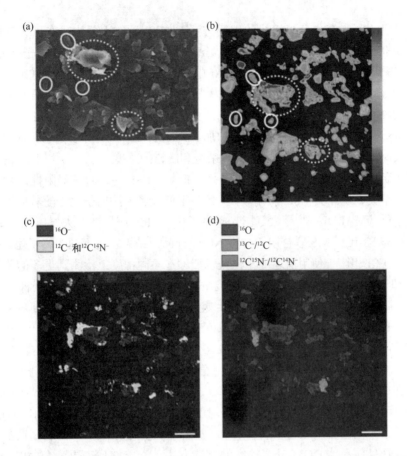

图 4-4 土壤颗粒 SEM 图片（a）及 NanoSIMS 表征土壤颗粒表面 $^{16}O^-$ 分布（b），$^{16}O^-$ 与 $^{12}C^-$ 和 $^{12}C^{14}N^-$ 分布（c），$^{16}O^-$、$^{13}C^-/^{12}C^-$ 与 $^{12}C^{15}N^-/^{12}C^{14}N^-$ 分布（d）（引自参考文献[35]）

2.3.4 傅里叶变换离子回旋共振质谱

在高分辨质谱技术出现之前，人们对于有机质的认知一直停留在元素组成和官能团水平。虽然高温热解-气相色谱质谱联用技术可以获得有机质的一些特征分子碎片，但是面对复杂的有机质样品，依靠色谱的分离手段以及传统质谱仪器的分辨率远远无法解析有机质的分子组成。近几十年来，以傅里叶变换离子回旋共振质谱（FT-ICR MS）为代表的超高分辨质谱技术的出现与快速发展为复杂有机体系的分子解析提供了强有力支持。1974 年 Melvin B. Comisarow 和 Alan G. Marshall 教授首次将离子回旋共振技术和傅里叶变换技术应用到质谱检测中。在 1981 年，Nicolet 仪器公司推出了世界首台商业化 FT-ICR MS，最初的磁场强度仅为 1.2T 的电磁体。次年，Bruker 公司开发出具有 4.7 T 超导磁体的 FT-ICR MS，时至今日，具有 21T 超导磁体的 FT-ICR MS 已经诞生[37]。与此同时，质谱质量分析器的技术日益发展，目前，一般的 FT-ICR MS 质量分辨率都能轻松超过 10^6，并具有超高质量准确性。此外，以大气压电喷雾离子化（ESI）技术为代表的软电离技术在 20 世纪 80 年代出现，并实现了与 FT-ICR MS 的联用[38]。这类软电离技术能够保证不破坏 DOM 分子中的共价键而只破坏分子间的非共价键，结合超高分辨的 FT-ICR MS 使得从单分子水平认识 NOM 的化学组成成为可能。

在 1996 年和 1997 年，William T. Cooper 教授团队和 Alan G. Marshall 教授合作首次将 9.6T 的 ESI FT-ICR MS 和 LDI（激光解离源）FT-ICR MS 用于 DOM（SRFA 和 SRHA）的分子解析[39]，从而为 DOM 研究开启了一扇新的大门。此后，科学家们从离子化模式（正离子与负离子）、离子化溶剂、仪器参数等各方面对 FT-ICR MS 在 DOM 的分析应用进行了研究[40-42]。通过一系列研究，基本确定了以负离子模式的 ESI FT-ICR MS 作为分析 DOM 分子组成的主要方法[43]。ESI 源的主要优势在于大气压电喷雾离子化技术能够离子化大部分带有酸性（如羧基）或碱性（如氨基）基团的极性或亲水性分子，而这些分子是 NOM 的主要成分。之所以采用负离子模式，是因为 NOM 中的大部分分子都带有酸性基团，容易在负离子模式下生成脱氢的分子离子（[M-H]⁻）。虽然在正离子模式下一些含有碱性基团的分子更容易形成加氢的分子离子[M+H]⁺，但是在 NOM 中占多数的含酸性基团的分子在正离子模式下的离子化效率很低。此外，正离子模式下一个分子可能同时产生加 H（[M+H]⁺）和加 Na（[M+Na]⁺）的分子离子，而且 NaH 与 C_2 的质量差仅为 2.4 mDa，因此大大增加了分子离子峰的复杂度和解谱的难度[38,44]。所以对于富含氮（氨基）的有机质样品同时选择正负两种离子化模式是更好的选择。但是，ESI 源也有其局限性，对于弱极性或非极性分子如多环芳烃类、糖类、烷烃类 ESI 源的离子化效率很低或者无法离子化。随着离子化技术的发展，基质辅助激光解吸离子化（MALDI）[45]、大气压光电离子化（APPI）[46,47]、大气压化学离子化（APCI）[47]等也相继作为离子源与 FT-ICR MS 联合用于分析 NOM 样品，结果表明不同的离子化方式对于 NOM 中的分子具有选择性，与 ESI 相比，使用前述离子化技术能够在不同程度上检测到更多的弱极性分子（如高芳香度和低氧化度），但是降低了高含氧量的极性分子的检出量。因此，选择其他一种或几种离子化技术与 ESI 结合能够更为全面地反映 NOM 的分子多样性。如果只选用一种离子化技术，则负离子模式的 ESI 离子源仍然是分析大多数 NOM 样品的首选，也是运用最多的方法，但如果样品具有明显的弱极性或非极组分，需要考虑其他的离子化方式。另一种增加 ESI FT-ICR MS 分析测得分子多样性的方法是与色谱技术联合，通过将 DOM 分成不同极性的组分从而降低同一样品分子间极性差异过大带来的离子化效率歧视效应。

金属离子的存在会干扰有机分子的离子化效率，因此除盐是 ESI FT-ICR MS 分析必须的样品前处理外，通常除盐的同时可以实现样品的浓缩。对于少量 DOM 样品的除盐和浓缩，固相萃取是目前常用的方法，一些研究也比较了不同固相萃取材料所得 DOM 样品中的分子组成[48-50]，表明商品化的 Agilent Bond Elut PPL 固相萃取柱是目前水体 DOM 分离提纯的首选方法[51]，能够代表更为全面的 DOM 分子多样性。理论上讲，几乎所有基于吸附的分离提纯方法都具有一定的选择性，PPL 固相萃取也不例外，相比而言 RO/ED 是一种非选择性的方法[52]，但是目前的 RO/ED 设备都需要较大的水样体积，因此更适合用于大量 DOM 样品的分离提纯，而不适合作为一种质谱分析 DOM 样品的前处理方法，如何发展小型化专门用于少量 DOM 样品前处理的 RO/ED 装置值得思考。

软电离模式加上 FT-ICR MS 的超高质量分辨率和质量准确度确保了 ESI FT-ICR MS 能够检测到 DOM 中成千上万个单个的分子离子，并通过精确的分子量和同位素信息匹配到精准的分子式。图 4-5 是一个典型的水体 DOM 样品（SRNOM）在负离子模式下的 ESI FT-ICR MS 谱图，从全谱可以看出 SRNOM 中检测到的分子离子的 m/z 在 200~1000 Da 范围内呈典型的正态分布，平均分子量在 450 Da 左右，局部放大可以看到分子离子峰的分布呈明显的簇状分布 [图 4-5（b）]，通过其中单一的一簇峰在 0.2 m/z 范围内就可以确定十几个分子式 [图 4-5（c）]。通过对 NOM 特别是 FA 和 HA 的分子解析，为腐殖质的超分子集合体假说提供了强有力的支持。

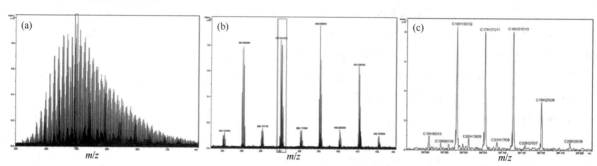

图 4-5 （a）负离子模式下 SRNOM 的 ESI FT-ICR MS 谱图，（b）m/z 394~402 段的局部放大图，（c）m/z 397 处的放大图

根据所得分子式的元素组成，可以获得一些经验参数并用于推测其化学结构和性质。典型的参数包括用于判断分子不饱和度的等价双键值（DBE）[53]，用于判断分子氧化度的平均碳氧化度（NOSC）[54]，用于判断分子芳香化度的芳香化指数（AI），该指数经过 Koch 和 Dittmar 修订之后能更加合理地反映分子的芳香化度（AI_{mod}）[53]。对于 $C_cH_hO_oN_nS_s$ 分子，上述参数计算公式如下：

$$DBE = 1 + 0.5(2c - h + n) \tag{4-1}$$

$$NOSC = 4 - (4c+h-3n-2o-2s)/c \tag{4-2}$$

$$AI_{mod} = (1 + c - 0.5o - s - 0.5h) / (c - 0.5o - s - n) \tag{4-3}$$

为了直观地分析所有测得分子式的特征，人们发展了各种根据不同参数间关系的图形化分析方法，其中运用最广泛的是 O/C 与 H/C 关系图，即 van Krevelen 图。图 4-6 是负离子模式 ESI FT-ICR MS 测得 SRNOM 中 $C_cH_hO_oN_n$ 分子式的 van Krevelen 图，根据经验可以将不同区域的分子划分成不同的化合物类型，通过统计不同区域分子的数量即可粗略估算不同组分分子所占的比例。从图中可知，SRNOM 以木质素类和单宁类分子为主，这是内陆水体 DOM 的一个重要特征，表明植物分解是 SRNOM 中分子的重要来源。当然，还可以根据具体的实验目的进行一些针对性的分析。

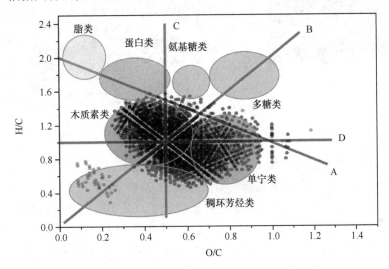

图 4-6　负离子模式 ESI FT-ICR MS 测得 SRNOM 中 $C_cH_hO_oN_n$ 分子式的 van Krevelen 图（引自参考文献[43]）
图中的圆圈代表经验的化合物分类，实线代表不同的化学反应过程，其中 A 代表甲基化和去甲基化过程，
B 代表加氢和脱氢过程，C 代表缩合和水解过程，D 代表氧化和还原过程

得益于 FT-ICR MS 对 DOM 分子组分强大的解析能力，近年来，越来越多的研究开始利用 FT-ICR MS 来揭示不同环境介质中 DOM 的分子多样性，并开始从分子水平探究 DOM 的环境迁移、转化、生物降解等环境生物地球化学行为。Kellerman 等利用 ESI FT-ICR MS 分析了瑞典境内 120 个湖泊的 DOM 的分子组成，共检测到 4032 个独特分子式，通过对 DOM 分子组成与水文和气候等参数进行统计分析发现，温度、降水和水体滞留时间等因素对于湖泊 DOM 分子多样性的影响十分重要[55]。Li 等分别利用 ESI FT-ICR MS 和 16S rRNA 基因测序技术分析了我国不同水稻种植区 88 份水稻土的 DOM 分子多样性和微生物多样性，发现二者之间存在显著的相关性[56]。矿物和有机质相互作用对土壤中有机质的固定有着重要的意义，Lv 等利用 ESI FT-ICR MS 研究了 DOM 在铁氧化物表面的吸附行为的分子机制，发现除了表观有机碳的吸附行为之外，DOM 中的分子在铁氧化物表面的选择性吸附，导致其在水-固界面发生了分子分馏。小分子量、不饱和度低、含氧量少的分子倾向于保留在溶液中，而具有相反特征的分子则更易与铁氧化矿物结合[57]。土壤中常见的几种铁（氢）氧化矿物中，无定形的水铁矿对于 DOM 的界面分子分馏作用更加显著，这一分馏行为可能进一步影响 DOM 的氧化还原特性[58]。为进一步揭示 DOM 在铁氧化物表面的

分子分馏机制，他们合成了具有可控表面的赤铁矿，发现 DOM 在赤铁矿表面的吸附分馏行为是受暴露晶面调控的，而铁氧化物表面的单配位羟基（—FeOH）丰度是其表面 DOM 分子分馏行为的决定因素（图 4-7）[59]。这些工作为土壤固碳及土壤有机碳循环研究提供了分子水平的视角。在工程应用领域，FT-ICR MS 也被用于揭示污水 DOM 的分子组成以及污水水处理、饮用水消毒等过程中 DOM 的转化降解过程以及消毒副产物的产生[60-62]，有助于进一步改进和完善污水处理和饮用水消毒措施。

图 4-7　DOM 在铁（氢）氧化物表面分子分馏行为示意图

总体而言，FT-ICR MS 在有机质解析方面的研究还处于起步阶段，目前来看它也表现出了一定的局限性，主要体现在以下几点，这也是未来研究需要突破的方向：①无论是样品前处理过程还是离子化方式都具有一定的选择性，因此利用 FT-ICR MS 检测到的分子式并不能完全代表 DOM 中的所有分子成分，未来需要发展非选择性的前处理方法和离子化技术；②FT-ICR MS 所得到的只是不带任何化学结构信息的分子式，所有与结构和化学性质相关的解释均出于经验性推测，未来需要将在线光/波谱技术与 FT-ICR MS 结合，实现 DOM 化学结构与分子式组成的同时解析；③由于化合物性质差异以及基质效应的存在，到目前为止，所有利用 FT-ICR MS 分析 DOM 分子式的方法均只是定性结果，无法获得分子式的定量信息，未来需要发展针对分子式的定量或者半定量的方法；④目前 FT-ICR MS 分析有机质样品所用到的离子化技术都仅限于分析水溶性或者有机溶剂可溶的有机质，未来需要发展可直接高效解离固体样品表面有机化合物的离子化技术并结合 FT-ICR MS 解析其分子组成。

3　天然有机质的环境效应

天然有机质（NOM）是水体和土壤系统中最为活跃的化学组分之一，在污染物的环境过程和归趋中扮演着极为重要的角色。NOM 富含芳香基团、羧基、羟基等活性功能基团，可作为天然吸附剂和载体，影响污染物在环境中的迁移性和生物可利用性。此外，NOM 也可以通过改变黑碳孔道结构来影响污染物的吸附量和可逆性。NOM 具有氧化还原和光化学活性，可作为反应物、促进剂或抑制剂，参与污染物的环境氧化还原和光转化过程，影响其环境归趋。研究 NOM 对污染物环境分配和转化的影响对全面认识 NOM 的生态环境效应、准确预测污染物的环境归趋都具有重要意义。本节将综述近期 NOM 对污染物吸附和转化行为影响机制的研究进展。

3.1 天然有机质对有机污染物吸附行为的影响

有机污染物在环境介质上的吸附行为控制着其环境分配、传输和反应活性，显著影响污染物的迁移、归趋和生物可利用性。非极性有机污染物在环境中主要的赋存介质为土壤、底泥和水体环境中的有机相，包括 NOM 和黑碳。NOM 的含量、组成和结构显著影响非极性有机污染物在环境介质中的赋存形态和分配规律。NOM 可通过自身吸附和改变黑碳吸附来影响污染物在环境介质中的吸附行为。

3.1.1 天然有机质吸附有机污染物的机制及主控因素

前期的工作已经报道了 NOM 与有机污染物之间的相互作用机制和构效关系[63-65]。NOM 对非极性有机污染物的吸附能力主要由其化学结构和空间构型所控制，但目前文献对其主控因素的认识仍有差异。一些研究指出 NOM 对非极性有机污染物（如多环芳烃和卤代化合物等）的吸附能力与其芳香性正相关，这主要归因于范德华力和 π-π 相互作用[63-69]。但也有研究中未观察到非极性有机物吸附与 NOM 芳香性之间的显著相关性[70, 71]。认为 NOM 组成中的脂肪族含量在非极性有机物的吸附中起更重要的作用[70, 71]。此外，非极性有机物的吸附亲和力通常与 NOM 的极性指标 [如(O+N)/C] 呈反比，表明疏水性分配的主导作用[65, 69, 72]。也有研究观察到 NOM 分子量的增大会促进其对非极性有机物的吸附[63, 72]。综合来看，NOM 对非极性有机污染物的吸附能力与其化学结构（芳香性、脂肪族、极性、分子量）密切相关，但主控因素尚不十分明确。

除了化学结构之外，NOM 对有机污染物的吸附行为也受其空间构型的影响。研究发现 NOM 分子可在水溶液中形成类胶束结构（pseudomicellar structure），即内部为疏水区而外部为亲水区，疏水内部区域被亲水外部区域屏蔽，为疏水性有机污染物的分配创造了疏水结构域[65, 73-77]。NOM 的空间构型受其化学结构（如芳香性/脂肪性、极性和分子量）以及溶液化学条件（如 pH、共存离子）的共同影响[74, 76, 78-81]。Tombácz 等发现具有较大烃含量的腐殖酸组分具有较高的临界胶束浓度[78]。Šmejkalová 和 Piccolo 基于扩散核磁共振谱的结果提出腐殖质的空间结构主要是由疏水芳香族组分的相互作用所控制的[81]。低 pH 可减少极性基团之间的排斥作用并促进疏水微区的结合，有利于类胶束结构的形成[76, 79, 80]。

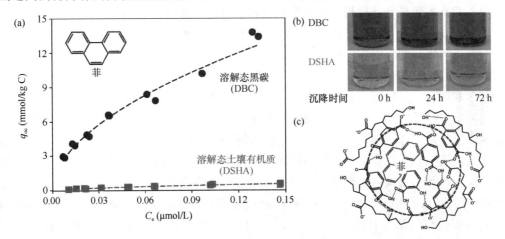

图 4-8 （a）菲在溶解态黑碳（DBC）和溶解态土壤腐殖酸（DSHA）上的吸附等温线；
（b）DBC 和 DSHA 聚合形成类胶束结构的性能差异；（c）菲在 DBC 类胶束结构分配的示意图

溶解态黑碳（dissolved black carbon，DBC）是黑碳连续体（black carbon continuum）中可溶解于水的组分，其含量约占淡水体系可溶性有机质库的10.6%[82]，是NOM的重要组分之一。Tang等率先证实了DBC对非极性有机污染物有较好的吸附能力[83]。他们考察了生物质黑碳溶出DBC对多环芳烃的吸附行为，发现多环芳烃在DBC上的分配系数（K_{OC}）与其正辛醇-水分配系数（K_{OW}，有机物的疏水性指标）具有良好的线性正相关性，这表明DBC吸附非极性有机污染物的方式与以往研究较多的腐殖类NOM相似，都是以疏水分配作用为主。虽然作用机制相似，但DBC吸附非极性有机污染物的能力远高于腐殖类NOM[84]，其吸附多环芳烃和氯苯的K_{OC}比腐殖类NOM高近一个数量级[图4-8（a）]。DBC对非极性有机污染物的强吸附能力与其独特的分子结构密不可分。与腐殖类NOM相比，DBC是一类相对简单、均一的有机混合物，其大部分组分为羧基取代的芳香化合物。借由芳香基团间的疏水作用和羧基间的分子间氢键作用，这些羧基芳香化合物容易聚合形成类胶束结构[图4-8（b）、（c）]，而类胶束结构内部较为致密的疏水域对有机污染物具有更高的分配能力[76, 85]。DBC另一个重要的结构特征是其含有丰富的羧基、羟基等极性官能团，因此其可通过氢键、静电作用等方式吸附极性有机污染物，也可与重金属污染物发生配位作用。

3.1.2 天然有机质对黑碳吸附污染物的影响机制

环境介质中的黑碳可有效吸附有机污染物，是土壤和底泥吸附非线性的重要来源。除自身吸附污染物外，NOM还会影响污染物在黑碳上的吸附，从而影响有机污染物的吸附量和吸附可逆性。黑碳孔结构对其吸附量和吸附可逆性具有显著影响。研究表明苯酚在模板合成的微孔碳上吸附是完全可逆的，这是由于微孔碳具有刚性、形状规则和互通的孔道结构[86]。环境中的黑碳均含有DBC成分，吸附有机物时具有一定的不可逆性。研究发现生物炭吸附嘧霉胺的可逆性随着孔径增加而增强，可能是由于大孔径提升了孔道的可及性[87]。水溶性有机质组分含量越低的黑碳对五氯酚的吸附能力越强[88]。相反，木炭上加载腐殖酸会导致多环芳烃的解吸动力学变慢，增加吸附不可逆性[89]。也有学者将有机物在黑碳上的不可逆吸附归因于吸附导致的黑碳孔道变形[90]。

与上述观点相吻合，Wang等发现黑碳中的少量DBC组分是其吸附不可逆性的重要来源。研究发现去除黑碳中的DBC组分后，黑碳对硝基苯、阿特拉津等有机污染物的吸附提高了11%~60%，且增强效果与污染物分子大小正相关[91]。同时，黑碳吸附污染物的速率明显变大、不可逆性降低。这表明DBC的存在是导致黑碳对有机污染物非理想吸附（吸附慢和不可逆）的重要原因之一。DBC对黑碳有机污染物吸附行为的影响主要源于其对黑碳孔道的堵塞效应降低了黑碳孔道内吸附位点的开放性和可利用性，进而影响黑碳对有机污染物的吸附速率和吸附能力[图4-9（a）]。为了进一步模拟NOM在黑碳非理想吸附中的作用，Wang等在有序中孔碳表面修饰了氨基官能团，并借此引入溶解性腐殖酸，考察了修饰前后中孔炭对硝基苯和阿特拉津的吸附性能。结果显示，经氨基和腐殖酸修饰后，中孔炭对污染物的吸附动力学变慢、吸附不可逆性增加，印证了NOM对黑碳表面非理想吸附的促成作用。NOM堵塞在黑碳孔道内侧时可形成"墨水瓶"孔道结构[图4-9（b）]；而有机污染物分子离开这种孔道时所需的活化能远高于进入时[92]，造成污染物在黑碳上吸附的不可逆性。此外，NOM具有较高的柔性，与有机污染物作用后容易发生形变，导致其与黑碳中刚性组分形成的孔道产生不可逆变形[图4-9（c）]，这也会引发黑碳对有机污染物的不可逆吸附。NOM的存在同样也可影响黑碳吸附重金属污染物。逐步释放至溶液相的NOM能与重金属通过配位作用形成有机-金属配合物，与黑碳产生竞争作用，抑制重金属在固相黑碳上的吸附。与此一致，已经有研究发现，利用弱酸、碱溶液去除黑碳中的NOM能有效提高黑碳对土壤中Cu(II)和Pb(II)的吸附固定能力[93]。

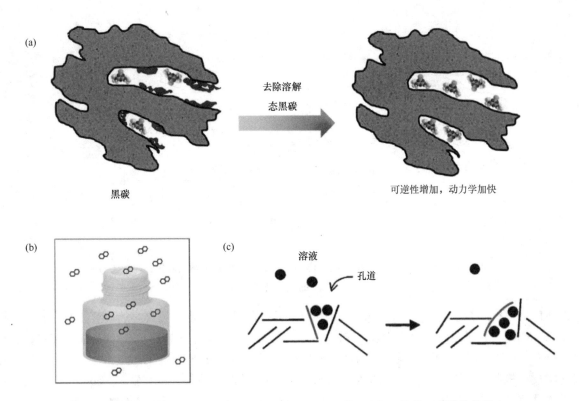

图4-9（a）溶解态黑碳（DBC）对黑碳（BC）孔道结构及有机污染物吸附性能的影响；
（b）、（c）"墨水瓶"孔道结构、孔道变形引发不可逆吸附的示意图

3.2 天然有机质对污染物转化行为的影响

NOM在太阳光照射下具有一定的光化学活性，可产生多种光生活性中间体（PPRIs）。PPRIs可与环境污染物发生反应，改变其赋存形态和降解速率。因此，NOM光化学过程参与的光解过程是有机污染物在天然环境中削减的重要途径。研究NOM的环境光化学活性及其影响因素有助于更准确地评估污染物的环境转化和归趋。

3.2.1 天然有机质光化学特性

NOM的吸收光谱随波长增加而指数性减小，没有明显的特征吸收峰。目前，NOM的光学特性可用两种模型解释：叠加模型或电子相互作用模型。叠加模型假设NOM组成中的发色团独立吸收和发射光子。因此，NOM的吸收和发射光谱代表了组分中所有发色团光谱的简单加和。该模型没有考虑发色团之间可能的相互作用，例如能量或电子转移过程。这些过程可能导致吸收和发射光谱的显著变化，如峰变宽或产生新峰。电子相互作用模型则提出NOM的光学性质至少部分来源于发色团之间的电子相互作用[94]。因此，观察到的NOM光谱将不再是单个发色团光谱的简单加和，还应包括发色团相互作用产生的吸收。电子相互作用模型的核心是木质素（或其他多酚前体）部分氧化形成的富电子供体和缺电子受体之间的短程电荷转移作用。该作用可以很好地解释NOM在近紫外和可见光波段的吸收和发射特性。这些近紫外和可见光的长波吸收和发射主要来源于羟基/甲氧基芳香电子供体和含羰基电子受体之间短程电荷转移作用能够产生新的较低能量的电荷跃迁。就目前的证据而言，电子相互作用模型可以更好地解释NOM的光学和光化学性质。

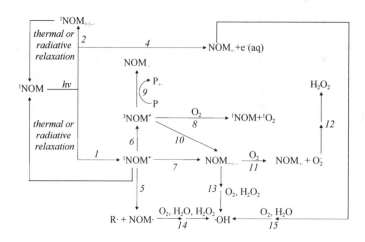

图 4-10 天然有机质的光化学反应机制（引自参考文献[94]）

P 代表外部的电子供体

NOM 的光化学过程是决定污染物环境归趋的重要过程，是地表水环境化学的研究重点之一。前期的研究指出 NOM 受光子激发后可产生多种 PPRIs，包括三线激发态（^{3}NOM*）、单线态氧（$^1O_2$）、超氧自由基（$O_2^{\cdot-}$）、双氧水（H_2O_2）、羟基自由基（·OH）、水合电子 [e^-（aq）] 以及有机自由基等[94]。NOM 的光化学反应机制可总结为图 4-10。NOM 在 250～400 nm 的激发波长下可产生 e^-（aq）[95]，其来源主要是芳香族基团，如酚类及其衍生物、芳香族羧酸和芳香族氨基酸等。水环境中 NOM 产生的量子产率和稳态浓度较高的 PPRIs 是 ^{3}NOM* 和 $^1O_2$。^{3}NOM* 的能量估计在 180～250kJ/mol，其单电子还原电位估计约为 1.7V，寿命在 1～100μs。目前尚无技术可定量 ^{3}NOM* 的量子产率 Φ_T。尽管如此，利用稳态技术和合理的动力学假设，可估计 Φ_T 在 0.01～0.1 之间，与 $\Phi_{^1O_2}$ 在同一个数量级，相对效率较高。目前的证据显示 ^{3}NOM* 来源于芳香酮和醛发色团。NOM 可通过 ^{3}NOM* 到 O_2 的能量转移过程产生 $^1O_2$。估算该过程的效应大概在 30%～50%。NOM 结构中 $^1O_2$ 的敏化剂主要是芳香酮和醛[96]。$^1O_2$ 的量子产率（$\Phi_{^1O_2}$）大约在 0.01～0.1 之间。具体量子产率主要取决于 NOM 结构、溶液 pH 和激发波长。通常微生物源的 NOM 比陆地源的 $\Phi_{^1O_2}$ 要高。这主要是由于微生物源的 NOM 具有较小的平均分子量。$\Phi_{^1O_2}$ 通常随着 pH 的上升而下降。太阳光照下，天然水体 O_2^- 和 H_2O_2 的产生速率约在 0.5～103 nmoL/(L·h) 之间。水体中 O_2^- 的稳态浓度在 0.1～10 nmoL/L，其中约一半可歧化为 H_2O_2。目前 NOM 结构对 $\Phi_{H_2O_2}$ 的影响机制尚不清晰。有报道指出较低分子量和较低芳香性的 NOM 可更有效地产生 H_2O_2[97]。然而，这种趋势在另一些样品中并不明确[98, 99]。NOM 在光照下也会形成类似·OH 活性的物质。这类物质（这里暂且称为·OH）在水环境中的生成速率在 10^{-11}～10^{-9} mol/(L·s) 之间[100-102]。在 300～400 nm 波长区域内，NOM 产生·OH 的量子产率（$\Phi_{\cdot OH}$）约在 10^{-5}～10^{-4} 之间。NOM 性质对 $\Phi_{\cdot OH}$ 的影响目前尚不明确。e^-（aq）的量子产率（Φ_{e^-}）在 300～400nm 的波长范围内大约在 10^{-5}～10^{-4} 之间。由于其极低的量子产率，NOM 产生的·OH 和 e^-（aq）不太可能对天然地表水环境中的污染物归趋产生显著影响。综上所述，水体中 ^{3}DOM* 和 $^1O_2$ 的稳态浓度和量子产率远高于其他 PPRIs，因此其对环境污染物的间接光解过程影响可能最为显著。

作为 NOM 的重要组分，DBC 中的芳香羧基、羟基和羰基也是常见的发色团，能够吸收特定波长的太阳光[65, 66]，这赋予了 DBC 光化学活性。现有研究表明，DBC 是 NOM 中光化学活性最活泼的组分之一。最近的工作利用分子探针技术定量研究了 DBC 产生 PPRIs 的能力，发现其可高效介导 $^1O_2$ 和 O_2^- 的产生[103, 104]。DBC 产生 $^1O_2$ 的量子效率约为 4.52 ± 0.98%，是 NOM 标准品的 2 倍；O_2^- 的产生速率为 1.2×10^{-9} mol/(L·s)，也高于天然海水中 O_2^- 的产生速率。其中，$^1O_2$ 生成过程与 DBC 的羰基结构密切相关，而 O_2^- 的生成则取决于 DBC 中有机质与矿物间的电子转移过程[104, 105]。

作为 PPRIs 中稳态浓度和量子产率最高的 ^{3}DOM*和 ^{1}O$_2$，其量子产率的定量预测有助于更好地评估环境污染物的间接光解过程动力学。研究发现尽管不同来源 NOM 或 NOM 各组分在形成过程、结构特征和光化学活性上差异很大，它们的 $\Phi_{singlet\ oxygen}$ 却可以通过同一组基于光谱学参数的定量模型进行预测[103]。其中预测效果最好的是基于 E_2/E_3 和 $S_{275-295}$ 的线性模型。因此，可认为 NOM 的 $\Phi_{singlet\ oxygen}$ 与 E_2/E_3 和 $S_{275-295}$ 之间的构效关系是建立在统一的物理化学理论基础上的。电荷转移理论认为 NOM 被光子激发后如发生分子内电荷转移则会削弱其通过能量传递过程产生 ^{1}O$_2$ 的能力。E_2/E_3 和 $S_{275-295}$ 越大的 NOM 分子量越小，分子内电荷转移越弱，因此其能量传递过程效率越高。此外，E_2/E_3 和 $S_{275-295}$ 也是 NOM 结构中类木质素成分的良好指针，而类木质素结构是 NOM 被光子激发产生三重激发态的活性中心。因此，E_2/E_3 和 $S_{275-295}$ 越大，^{3}NOM*（^{1}O$_2$ 前驱体）的产率越高。基于光谱学参数的 NOM 光化学活性预测模型具有良好的理论基础和准确性。由于紫外光谱设备可实现原位时间分辨连续监测，同时具有简易、经济和易维护的特点，未来可通过上述模型进行水体有机质光化学活性的原位实时监测和评估。

3.2.2 天然有机质对有机污染物光降解过程的影响

光降解是地表水中有机污染物环境消除的重要过程。理解有机污染物的直接和间接光降解机制和主控因素有助于更好地预测和评估其环境归宿和持久性。直接光降解过程中目标化合物会吸收光子的能量，发生键断裂或重排，从而形成新的稳定产物。这种途径仅适用于能吸收太阳光谱（即 $\lambda > 290$ nm）的化合物。间接光降解过程则更加普适，在该过程中光敏剂吸收光子并产生一系列可以降解目标化合物的活性物质（即 PPRIs）。地表水环境普遍存在的 NOM 可以同时影响有机污染物的直接和间接光降解过程。NOM 主要通过遮光效应（light screening）降低有机污染物的直接光降解速率。作为水生系统最为重要的敏化剂，NOM 可以诱导有机污染物的间接光降解；另一方面，NOM 也可以作为猝灭剂，抑制有机污染物的间接光降解过程。

NOM 可有效地介导酚类、农药、医药品和个人护理用品、内分泌干扰物等有机污染物的间接光降解过程[106-109]。不同的 PPRIs 在污染物降解过程中的相对贡献不尽相同。就稳态浓度和量子产率而言，^{3}NOM* 和 ^{1}O$_2$ 要远高于其他 PPRIs，是间接光解中需重考虑的 PPRIs。^{1}O$_2$ 是选择性氧化剂，它与一些特定结构的化合物（如酚类、呋喃、吲哚和咪唑）反应速率常数较大。研究表明，天然水体中 NOM 产生的 ^{1}O$_2$ 在酚类和部分氨基酸的光降解过程中起重要作用[108]。需要指出的是，在 NOM 结构内的微环境中 ^{1}O$_2$ 浓度要比在溶液中高出 1 个数量级[110]。因此，在 NOM 上有较强吸附的污染物可能会具有更高的反应速率。^{3}NOM* 在苯基脲类除草剂、磺胺类抗生素、阿莫西林、雌二醇和部分氨基酸光降解过程中起重要作用[108,109]。NOM 的结构特征与其介导间接光降解效能之间的构效关系研究较为有限。有学者尝试将污染物的降解速率常数与 NOM 的结构参数相关联，取得了一些初步的成果。Batista 等研究了不同 NOM 介导的磺胺甲基嘧啶光降解过程[111]，发现磺胺甲基嘧啶光降解速率与 ^{3}DOM*稳态浓度以及不同 NOM 生成 ^{3}NOM*的速率常数相关。其中 Aldrich 腐殖酸效率最高，而萨旺尼河富里酸反而会抑制磺胺甲基嘧啶的光降解过程。Zhou 等研究了不同 NOM（包括 DBC）介导的 17β-雌二醇光降解过程[109]，发现 17β-雌二醇光降解速率与其和 NOM 的结合态浓度正相关，与 ^{3}NOM*的稳态浓度呈线性正相关。从化学结构上看，分子量越小芳香性越高的 NOM 诱导 17β-雌二醇光降解的能力越强。与腐殖类 NOM 相比，DBC 具有更丰富的芳香羟基/羰基/羧基基团和较小的分子尺寸，这些结构特性有利于 ^{3}NOM*的产生，其促进效率是腐殖质类 NOM 的 6 倍以上。

3.2.3 天然有机质对污染物氧化还原反应的影响

近年来，关于 NOM 的氧化还原和电子传递特性逐渐引起人们的关注。Lovley 等最早在 Nature 上报道了 NOM 能够作为电子受体接受微生物厌氧呼吸过程产生的电子，继而将电子转移给其他电子受体如铁氧化物表面的 Fe(III)。其中 NOM 作为电子穿梭体，能够循环参与电子转移反应过程[112]。NOM 中含有大量醌类、酚类以及芳香羧酸等组分。研究表明，这些组分具有较高的氧化还原活性，可作为反应物或者电子传递载体，参与污染物在环境中的氧化还原转化过程[113-119]。例如，醌类、酚类、羧酸类组分既可作为还原剂，直接还原环境中的 Cr(VI)、As(V)等氧化性金属污染物[113, 119]；也可作为反应中间体，通过与污染物的氧化耦合（cross-coupling）促进多种有机污染物在铁、锰矿物表面的氧化性转化[114, 120]。醌类和酚类还能起电子穿梭体（electron shuttle）的作用，提高 Fe(III)、U(VI)、硝基芳香化合物和氯代烃等污染物在环境中的还原速率[115-118]。因此，NOM 可能会对污染物的环境氧化还原行为产生重要影响。

目前关于 NOM 中氧化还原组分的认识仍不十分明确。通常认为 NOM 的电子转移能力是主要源自其结构中包含的醌基官能团，通过醌、半醌和氢醌之间的可逆转化完成电子转移过程。但也有研究认为，醌基并不是 NOM 中唯一的电子转移体，还有更多非醌基基团如芳香结构、酚基、氨基、羧基等官能团同样具有电子转移能力[121]。预测和评估 NOM 介导污染物氧化还原反应活性的前提是准确定量其氧化还原能力[通常用 NOM 的电子接受能力（EAC）和电子提供能力（EDC）来衡量]。目前的主流方法包括化学法和电化学法两大类。化学法通过在添加氧化剂（柠檬酸铁或铁氰化物）和还原剂（H_2S 和 Zn^0）的条件下分别测定 NOM 的 EDC 和 EAC。该方法存在反应耗时、受体系 pH 变化影响、副反应多等问题，影响测量的准确度和效率[121, 122]。电化学法是近年来兴起的研究 NOM 氧化还原能力的方法，包括循环伏安法（CV）、直接电化学还原法（DER）等方法，它以电极作为电子供体或受体从而有效避免外加氧化/还原剂的影响[123, 124]。苏黎世大学的 Sander 课题组最近建立了活性有机分子介导的电化学测定新方法（ME）。该方法通过直接电化学还原法将 NOM 还原，然后利用计时电位法来间接测定 NOM 的供/得电子能力，与传统方法相比大大提高了测量的速度和精确度[125]。他们利用这种方法测定了一系列 NOM 的氧化还原能力，发现 NOM 还原能力与其中的酚羟基官能团呈显著正相关，而氧化能力与其中的芳香度呈显著正相关[125, 126]。最近，Zheng 等测定了不同来源 NOM（包括 DBC）的氧化还原能力，发现 DBC 普遍具有比标准腐殖质样品更高的 EDC 和更低的 EAC。DBC 更高的 EDC 主要来源于其更高含量的酚羟基官能团[127]。为进一步揭示 NOM 中不同分子组成的供电子能力，Lv 等通过对 NOM 进行极性分组，并结合 ME 方法和 FT-ICR MS 探究了不同 NOM 组分分子组成特征与其 EDC 之间的关系，发现分布在 van Krevelen 图中多酚区域的分子特别是中等含氧量的多酚类分子（O/C：0.4～0.67）与 EDC 呈显著的正相关，此外分子的等价双键值（DBE）也与 EDC 呈显著的正相关[58]。

3.2.4 天然有机质对金属纳米材料环境转化过程的影响

NOM 对金属纳米材料环境污染过程的影响研究是目前 NOM 环境效应研究的热点之一。NOM 可以作为络合剂、敏化剂和电子供体对金属纳米材料的环境污染过程产生截然不同的影响。NOM 可作为络合剂和敏化剂介导纳米银（nAg）在水环境中的赋存形态和毒性效应。nAg 的急性毒性主要来自其释放 Ag^+ 的能力[128, 129]，尽管也有报道指出其具有纳米颗粒特异性的毒性[130, 131]。银的形态对于理解 nAg 的环境归趋、生物可利用性和毒性效应至关重要。在环境中，nAg 会经历复杂的转化过程，包括络合、氧化还原和溶解[132]。nAg 颗粒较容易被氧化，形成 Ag_2O 壳层，随后可缓慢释放 Ag^+[133, 134]。释放出的 Ag^+ 可进一步与天然配体如 S^{2-} 和 NOM 络合。在一些情况下，NOM 会通过 PPRIs 介导的光氧化过程加速 nAg 的溶解[135, 136]。另一方面，Ag^+ 可以很容易地被 NOM 还原，在太阳光照下形成新的 nAg[135, 137-140]。光还原机理主要是配体金属电荷转移过程（LMCT）和 PPRIs（如 $O_2^{\cdot-}$）介导的还原过程[138, 139]。

NOM 介导的 Ag^+ 还原过程与水化学条件（如 pH 和共存离子）密切相关[137-139]。nAg 光生成速率随着 pH 的上升而增加，这主要归因于更强的 Ag^+ 吸附和 NOM 的还原电位。共存阳离子降低了 nAg 的光生成速率，可能是由于其与 Ag^+ 竞争 NOM 上的吸附位点[138]。不同的 NOM 或 NOM 组分介导 nAg 光生成速率差异较大，主要归因于 NOM 氧化还原电位和屏蔽光子能力的不同[137, 140]。近期研究显示 NOM 介导的

nAg 形成过程受 Ag⁺和氯离子浓度控制，并存在 Ag⁺临界诱导浓度[141]。特定 NOM 的临界诱导浓度受其介导下氧化还原反应平衡影响，NOM 产生的还原性 PPRIs 越多，氧化性 PPRIs 就越少，其临界诱导浓度越低。银离子浓度低于临界诱导浓度时，银纳米颗粒产生量极小；银离子浓度高于临界诱导浓度时，还原反应在与氧化反应及溶解作用竞争过程中占主导，促使银团簇快速聚集形成 nAg。水体中氯离子的存在可以通过形成晶核降低临界诱导浓度。

NOM 可主要作为电子供体抑制商业镉系颜料的环境光溶解过程。商业镉系颜料的主要成分为硫化镉、硒化镉和硫硒化镉，被广泛应用于陶瓷、塑料、玻璃、涂料等行业。最近的一项研究针对镉系颜料的半导体特性，揭示了商业镉系颜料由自然光照引发的镉释放过程及其反应机理[142]。研究结果显示，由于光生空穴和颜料晶格间的反应，水体中的镉系颜料在光照下发生快速溶解。24 小时光照下的镉释放比例可达 83%。镉系颜料的颗粒也相应减小，生成了小于 20 nm 的纳米颗粒。光溶解过程的主要产物是镉离子、硫酸根、氢离子和硒酸根。水体中普遍存在的 NOM 可通过三个途径影响镉系颜料的光溶解过程：①NOM 吸收光子，产生屏蔽作用，减少了颜料颗粒接收的光子量；②NOM 中的还原性基团可以与光生空穴反应，减轻镉系颜料的晶格氧化；③NOM 产生的 PPRIs 可与镉系颜料反应，加快其氧化溶解速率。NOM 对镉系颜料的光溶解过程的总体影响将取决于上述过程的相互作用。从现有结果来看，水体中的 NOM 可以显著抑制镉系颜料的光溶解过程。

另一方面，NOM 作为电子供体却可显著促进商业铬酸铅颜料的环境光溶解过程。铬酸铅颜料，包括铬黄和钼酸橙，是全球用量最大的低成本无机颜料之一。每年全球铬酸铅颜料市场约为 9 万 t，占全球铅消耗量的 3%[143]。由于其产量大，应用广泛，铬酸铅颜料在室内和室外环境中（如道路、游乐场、室内灰尘等）经常可以检出[144-147]。铬酸铅颜料中的 Cr(VI)具有高氧化还原电位（+1.33 V），易被还原[148]。近期的研究发现商业铬酸铅颜料可在 NOM 存在的情况下在水相发生光溶解，释放出 Pb(II)和 Cr(III)（未发表数据）。该过程是由光生电子介导颜料中 Cr(VI)的还原过程驱动的。铬酸铅颜料可被波长<514nm 的光子激发，产生电子和空穴。然而由于其电子空穴分离效率很低，所以光溶解过程非常缓慢。NOM 可以作为电子供体捕捉光生空穴，提高铬酸铅颜料的电子空穴分离效率。因此，NOM 存在时可以观察到显著的铬酸铅颜料光溶解过程。

上述两类无机颜料的环境污染过程有许多相似之处。这些颜料都是可以在太阳光谱中被激活的半导体。同时，它们都具有易于氧化还原反应的化学成分[如 S²⁻和 Cr(VI)]。当无机颜料同时具备这两个特征时，可能会通过自氧化还原反应（即颜料和光生电子/空穴之间的反应）发生光溶解。然而，这两类颜料的反应途径又有极大的不同之处。镉系颜料可以实现有效的电荷分离，并且它们的价带高于其阳极溶出标准电位。因此，它们的光溶解过程可以通过光生空穴和颜料晶格之间的氧化反应直接进行。在这种情况下，NOM 的存在主要通过竞争光子（即光屏蔽）和光生空穴来抑制光溶解过程。与之相反，铬酸铅颜料的电荷分离效率很低，抑制了光生电子和颜料晶格间的反应。它们的光溶解过程仅在存在电子供体（NOM）的情况下才显著。

4 展　望

近年来，天然有机质（NOM）的分析方法、概念模型和研究手段都取得了长足的发展。NOM 在环境和气候系统中的重要作用日益引起科学界的关注，吸引了许多优秀国际团队的加入。但目前 NOM 环境过程和效应的研究体系尚不完备，仍有许多问题亟待解决。

由于 NOM 组成结构的高度复杂性，目前关于 NOM 结构的认识还十分有限，成为了深入开展 NOM

研究的重要障碍。这就需要在表征方法和研究策略上有进一步的创新和突破。未来的研究需要重视以下方面：①NOM 的组成多样性决定了其化学多样性，不可能用一种方法全面表征其结构与组成，因此多手段结合与多参数表征是认识其化学多样性的有效途径；②NOM 的化学多样性取决于来源、分解转化过程、周围环境以及形成时间，需要结合多学科（包括化学、植物学、微生物学、地质学等）从多维度剖析 NOM 的组成多样性及其行为；③认识 NOM 的化学多样性是认识有机碳/氮循环的基础，需要将微观的有机质化学多样性耦合到宏观的碳/氮循环研究当中。

目前 NOM 与环境污染物相互作用的研究已有一定基础，初步揭示了 NOM 介导的污染物分配和转化过程机制。但该方向上的工作系统性不足，缺乏定量研究，对 NOM 结构多样性的兼容度较差。因此，不足以系统评估 NOM 对污染物分配、迁移和转化行为的影响。这首先归因于缺乏 NOM 结构的充分表征，导致目前的构效关系研究仍停留在定性描述为主、半定量研究为辅的水平。其次，NOM 环境效应的基础数据较为缺乏，不足以支撑定量预测模型的构建和优化。因此，未来研究首要在上述表征手段发展的基础上阐明 NOM 在时空尺度上的结构分异情况及其影响因素；其次要积累 NOM 与环境污染物相互作用的基础数据，解析主控因素，构建定量预测模型，并在环境条件下和较大空间尺度上进行模型验证和优化。

参 考 文 献

[1] Mopper K, Stubbins A, Ritchie J D, et al. Advanced instrumental approaches for haracterization of marine dissolved organic matter: Extraction techniques, mass spectrometry, and nuclear magnetic resonance spectroscopy. Chemical Reviews, 2007, 107(2): 419-442.

[2] Minor E C, Swenson M M, Mattson B M, et al. Structural characterization of dissolved organic matter: a review of current techniques for isolation and analysis. Environmental Science: Processes & Impacts, 2014, 16(9): 2064-2079.

[3] Lehmann J, Kinyangi J, Solomon D. Organic matter stabilization in soil microaggregates: implications from spatial heterogeneity of organic carbon contents and carbon forms. Biogeochemistry, 2007, 85(1): 45-57.

[4] 吴丰昌. 天然有机质及其与污染物的相互作用. 科学出版社, 2010.

[5] Hayes M H B, Swift R S. An appreciation of the contribution of Frank Stevenson to the advancement of studies of soil organic matter and humic substances. Journal of Soils and Sediments, 2018, 18(4): 1212-1231.

[6] Waksman S A. What is humus? Proceedings of the National Academy of Sciences of the United States of America, 1925. 11: 463-468.

[7] Sutton R, Sposito G. Molecular structure in soil humic substances: The new view. Environmental Science & Technology, 2005, 39(23): 9009-9015.

[8] Swift R S. Macromolecular properties of soil humic substances: Fact, fiction, and opinion. Soil Science, 1999, 164(11): 790-802.

[9] Piccolo A. The supramolecular structure of humic substances: A novel understanding of humus chemistry and implications in soil science. Advances in Agronomy, 2002, 75: 157-134.

[10] Lehmann J, Kleber M. The contentious nature of soil organic matter. Nature, 2015, 528(7580): 60-68.

[11] Piccolo A. In memoriam Prof. F. J. Stevenson and the Question of humic substances in soil. Chemical and Biological Technologies in Agriculture, 2016, 3: 1-3.

[12] Bolan N S, Adriano D C, Kunhikrishnan A, et al. Dissolved Organic Matter: Biogeochemistry, Dynamics, and Environmental Significance in Soils. Advances in Agronomy, 2011, 110: 1-75.

[13] Dittmar T, Koch B, Hertkorn N, et al. A simple and efficient method for the solid-phase extraction of dissolved organic matter (SPE-DOM) from seawater. Limnol Oceanogr-Meth, 2008, 6: 230-235.

[14] Koprivnjak J F, Pfromm P H, Ingall E, et al. Chemical and spectroscopic characterization of marine dissolved organic matter isolated using coupled reverse osmosis electrodialysis. Geochim Cosmochim Acta, 2009, 73: 4215-4231.

[15] Zsolnay A. Dissolved organic matter: artefacts, definitions, and functions. Geoderma, 2003, 113(3-4): 187-209.

[16] Tfaily M M, Chu R K, Tolic N, et al. Advanced solvent based methods for molecular characterization of soil organic matter by high-resolution mass spectrometry. Analytical Chemistry, 2015, 87(10): 5206-5215.

[17] Fox P M, Nico P S, Tfaily M M, et al. Characterization of natural organic matter in low-carbon sediments: Extraction and analytical approaches. Organic Geochemistry, 2017, 114: 12-22.

[18] Zhu L Y, Wang C Y, Zhang H, et al. Degradation and Mineralization of Bisphenol A by Mesoporous Bi(2)WO(6) under Simulated Solar Light Irradiation. Environmental Science & Technology, 2010, 44(17): 6843-6848.

[19] Hertkorn N, Benner R, Frommberger M, et al. Characterization of a major refractory component of marine dissolved organic matter. Geochim Cosmochim Acta, 2006, 70(12): 2990-3010.

[20] Hertkorn N, Harir M, Koch B P, et al. High-field NMR spectroscopy and FTICR mass spectrometry: powerful discovery tools for the molecular level characterization of marine dissolved organic matter. Biogeosciences, 2013, 10(3): 1583-1624.

[21] Ikeya K, Watanabe A. Application of 13C ramp CPMAS NMR with phase-adjusted spinning sidebands (PASS) for the quantitative estimation of carbon functional groups in natural organic matter. Analytical and Bioanalytical Chemistry, 2016, 408: 651–655.

[22] Simpson A J, Simpson M J, Soong R. Nuclear magnetic resonance spectroscopy and its key role in environmental research. Environmental Science & Technology, 2012, 46(21): 11488-11496.

[23] Luo L, Lv J, Chen Z. Synchrotron infrared microspectroscopy reveals the roles of aliphatic and aromatic moieties in sorption of nitroaromatic compounds to soils. Science of the Total Environment, 2017, 624: 210-214.

[24] Najafi E, Wang J A, Hitchcock AP, et al. Characterization of Single-Walled Carbon Nanotubes by Scanning Transmission X-ray Spectromicroscopy: Purification, Order and Dodecyl Functionalization. Journal of the American Chemical Society, 2010, 132(26): 9020-9029.

[25] Myneni S C B. Soft X-ray spectroscopy and spectromicroscopy studies of organic molecules in the environment. Applications of Synchrotron Radiation in Low-Temperature Geochemistry and Environmental Sciences, 2002, 49: 485-579.

[26] Solomon D, Lehmann J, Kinyangi J, et al. Long-term impacts of anthropogenic perturbations on dynamics and speciation of organic carbon in tropical forest and subtropical grassland ecosystems. Global Change Biology, 2007, 13(2): 511-530.

[27] Chao W, Fischer P, Tyliszczak T, et al.Real space soft x-ray imaging at 10 nm spatial resolution. Optics Express, 2012, 20(9): 9777-9783.

[28] Lehmann J, Solomon D, Kinyangi J, et al. Spatial complexity of soil organic matter forms at nanometre scales. Nat Geosci, 2008, 1(4): 238-242.

[29] Kinyangi J, Solomon D, Liang B I, et al. Nanoscale biogeocomplexity of the organomineral assemblage in soil: Application of STXM microscopy and C 1s-NEXAFS spectroscopy.Soil Sci Soc Am J, 2006, 70(5): 1708-1718.

[30] Solomon D, Lehmann J, Harden J, et al.Micro- and nano-environments of carbon sequestration: Multi-element STXM-NEXAFS spectromicroscopy assessment of microbial carbon and mineral associations. Chem Geol, 2012, 329: 53-73.

[31] Rennert T, Handel M, Hoschen C, et al. A NanoSIMS study on the distribution of soil organic matter, iron and manganese in a nodule from a Stagnosol. Eur J Soil Sci, 2014, 65(5): 684-692.

[32] Mueller C W, Kolbl A, Hoeschen C, et al.Submicron scale imaging of soil organic matter dynamics using NanoSIMS - From single particles to intact aggregates. Org Geochem, 2012, 42(12): 1476-1488.

[33] Ghosal S, Weber P K, Laskin A. Spatially resolved chemical imaging of individual atmospheric particles using nanoscale imaging mass spectrometry: insight into particle origin and chemistry. Anal Methods, 2014, 6(8): 2444-2451.

[34] Xiao J, Wen Y L, Li H, Hao J L, et al. In situ visualisation and characterisation of the capacity of highly reactive minerals to preserve soil organic matter (SOM) in colloids at submicron scale. Chemosphere, 2015, 138: 225-232.

[35] Vogel C, Mueller C W, Hoschen C, et al. Submicron structures provide preferential spots for carbon and nitrogen sequestration in soils. Nat Commun, 2014, 5.

[36] Liu X R, Eusterhues K, Thieme J, et al. STXM and NanoSIMS Investigations on EPS Fractions before and after Adsorption to Goethite. Environmental Science & Technology, 2013, 47(7): 3158-3166.

[37] Smith D F, Podgorski D C, Rodgers R P, et al. 21 Tesla FT-ICR Mass Spectrometer for Ultrahigh-Resolution Analysis of Complex Organic Mixtures. Anal Chem, 2018, 90(3): 2041-2047.

[38] Reemtsma T. Determination of molecular formulas of natural organic matter molecules by (ultra-) high-resolution mass spectrometry Status and needs. J Chromatogr A 2009, 1216(18): 3687-3701.

[39] Fievre A, Solouki T, Marshall A G, et al High-resolution Fourier transform ion cyclotron resonance mass spectrometry of humic and fulvic acids by laser desorption/ionization and electrospray ionization. Energ Fuel, 1997, 11(3): 554-560.

[40] Stenson A C, Landing W M, Marshall A G, et al. Ionization and fragmentation of humic substances in electrospray ionization Fourier transform-ion cyclotron resonance mass spectrometry. Anal Chem, 2002, 74(17): 4397-4409.

[41] Brown T L, Rice J A. Effect of experimental parameters on the ESI FT-ICR mass spectrum of fulvic acid. Anal Chem, 2000, 72(2): 384-390.

[42] Stenson A C, Marshall A G, Cooper W T. Exact masses and chemical formulas of individual Suwannee River fulvic acids from ultrahigh resolution electrospray ionization Fourier transform ion cyclotron resonance mass spectra. Anal Chem, 2003, 75(6): 1275-1284.

[43] Sleighter R L, Hatcher P G. The application of electrospray ionization coupled to ultrahigh resolution mass spectrometry for the molecular characterization of natural organic matter. J Mass Spectrom, 2007, 42(5): 559-574.

[44] Rostad C E, Leenheer J A. Factors that affect molecular weight distribution of Suwannee river fulvic acid as determined by electrospray ionization/mass spectrometry. Anal Chim Acta, 2004, 523(2): 269-278.

[45] Cao D, Huang H, Hu M, et al. Comprehensive characterization of natural organic matter by MALDI- and ESI-Fourier transform ion cyclotron resonance mass spectrometry. Anal Chim Acta, 2015, 866: 48-58.

[46] D'Andrilli J, Dittmar T, Koch B P, et al. Comprehensive characterization of marine dissolved organic matter by Fourier transform ion cyclotron resonance mass spectrometry with electrospray and atmospheric pressure photoionization. Rapid Commun Mass Spectrom, 2010, 24(5): 643-650.

[47] Hertkorn N, Frommberger M, Witt M, et al. Natural Organic Matter and the Event Horizon of Mass Spectrometry. Anal Chem, 2008, 80(23): 8908-8919.

[48] Perminova I V, Dubinenkov I V, Kononikhin A S, et al. Molecular Mapping of Sorbent Selectivities with Respect to Isolation of Arctic Dissolved Organic Matter as Measured by Fourier Transform Mass Spectrometry. Environmental Science & Technology, 2014, 48(13): 7461-7468.

[49] Waska H, Koschinsky A, Chancho M J R, et al. Investigating the potential of solid-phase extraction and Fourier-transform ion cyclotron resonance mass spectrometry (FT-ICR-MS) for the isolation and identification of dissolved metal-organic complexes from natural waters. Mar Chem, 2015, 173: 78-92.

[50] Raeke J, Lechtenfeld O J, Wagner M, et al. Selectivity of solid phase extraction of freshwater dissolved organic matter and its effect on ultrahigh resolution mass spectra. Environmental science Processes & impacts, 2016, 18(7): 918-927.

[51] Li Y, Harir M, Lucio M, et al. Proposed Guidelines for Solid Phase Extraction of Suwannee River Dissolved Organic Matter. Anal Chem, 2016, 88(13): 6680-6688.

[52] Green N W, Perdue E M, Aiken G R, et al. An intercomparison of three methods for the large-scale isolation of oceanic

dissolved organic matter. Mar Chem, 2014, 161: 14-19.

[53] Koch B P, Dittmar T. From mass to structure: an aromaticity index for high-resolution mass data of natural organic matter. Rapid Commun Mass Spectrom, 2006, 20(5): 926-932.

[54] Riedel T, Biester H, Dittmar T. Molecular Fractionation of Dissolved Organic Matter with Metal Salts. Environmental Science & Technology, 2012, 46(8): 4419-4426.

[55] Kellerman A M, Dittmar T, Kothawala D N, et al. Chemodiversity of dissolved organic matter in lakes driven by climate and hydrology. Nat Commun, 2014, 5: 3804.

[56] Li H Y, Wang H, Wang H T, et al. The chemodiversity of paddy soil dissolved organic matter correlates with microbial community at continental scales. Microbiome, 2018, 186: 2-16.

[57] Lv J, Zhang S, Wang S, et al. Molecular-Scale Investigation with ESI-FT-ICR-MS on Fractionation of Dissolved Organic Matter Induced by Adsorption on Iron Oxyhydroxides. Environ Sci Technol, 2016, 50(5): 2328-2336.

[58] Lv J T, Han R X, Huang Z Q, et al. Relationship between Molecular Components and Reducing Capacities of Humic Substances. Acs Earth and Space Chemistry, 2018, 2(4): 330-339.

[59] Lv J, Miao Y, Huang Z, et al. Facet-Mediated Adsorption and Molecular Fractionation of Humic Substances on Hematite Surfaces. Environ Sci Technol, 2018, 52 (20): 11660-11669.

[60] Zhang H F, Zhang Y H, Shi Q, et al. Characterization of Unknown Brominated Disinfection Byproducts during Chlorination Using Ultrahigh Resolution Mass Spectrometry. Environmental Science & Technology, 2014, 48(6): 3112-3119.

[61] Zhang H F, Zhang Y H, Shi Q, et al. Study on Transformation of Natural Organic Matter in Source Water during Chlorination and Its Chlorinated Products using Ultrahigh Resolution Mass Spectrometry. Environmental Science & Technology, 2012, 46(8): 4396-4402.

[62] Li Y Y, Fang Z, He C, et al. Molecular Characterization and Transformation of Dissolved Organic Matter in Refinery Wastewater from Water Treatment Processes: Characterization by Fourier Transform Ion Cyclotron Resonance Mass Spectrometry. Energ Fuel, 2015, 29(11): 6956-6963.

[63] Chin Y P, Aiken G R, Danielsen K M. Binding of pyrene to aquatic and commercial humic substances: The role of molecular weight and aromaticity. Environmental Science and Technology, 1997, 31(6): 1630-1635.

[64] Gauthier T D, Seitz W R, Grant C L. Effects of Structural and Compositional Variations of Dissolved Humic Materials on Pyrene Koc Values. Environmental Science and Technology, 1987, 21(3): 243-248.

[65] Kopinke F D, Georgi A, Mackenzie K. Sorption of pyrene to dissolved humic substances and related model polymers. 1. Structure - Property correlation. Environmental Science and Technology, 2001, 35(12): 2536-2542.

[66] Wang H B, Zhang Y J. Mechanisms of interaction between polycyclic aromatic hydrocarbons and dissolved organic matters. Journal of Environmental Science and Health - Part A Toxic/Hazardous Substances and Environmental Engineering, 2014, 49(1): 78-84.

[67] Uhle M E, Chin Y P, Aiken G R, et al. Binding of polychlorinated biphenyls to aquatic humic substances: The role of substrate and sorbate properties on partitioning. Environmental Science and Technology, 1999, 33(16): 2715-2718.

[68] Laor Y, Farmer W J, Aochi Y, et al. Phenanthrene binding and sorption to dissolved and to mineral-associated humic acid. Water Res, 1998, 32(6): 1923-1931.

[69] Jin J, Sun K, Wang Z, et al. Characterization and Phenanthrene Sorption of Natural and Pyrogenic Organic Matter Fractions. Environmental Science and Technology, 2017, 51(5): 2635-2642.

[70] Eriksson J, Frankki S, Shchukarev A, et al. Binding of 2, 4, 6-trinitrotoluene, aniline, and nitrobenzene to dissolved and particulate soil organic matter. Environmental Science and Technology, 2004, 38(11): 3074-3080.

[71] Ilani T, Schulz E, Chefetz B. Interactions of organic compounds with wastewater dissolved organic matter: Role of

hydrophobic fractions. J Environ Qual, 2005, 34(2): 552-562.

[72] Chiou C T, Malcolm R L, Brinton T I, et al. Water Solubility Enhancement of Some Organic Pollutants and Pesticides by Dissolved Humic and Fulvic Acids. Environmental Science and Technology, 1986, 20(5): 502-508.

[73] Conte P, Piccolo A. Conformational arrangement of dissolved humic substances. Influence of solution composition on association of humic molecules.Environmental Science and Technology, 1999, 33(10): 1682-1690.

[74] Sutton R, Sposito G.Molecular structure in soil humic substances: The new view. Environmental Science and Technology, 2005, 39(23): 9009-9015.

[75] Pan B, Ghosh S, Xing B. Nonideal binding between dissolved humic acids and polyaromatic hydrocarbons. Environmental Science and Technology, 2007, 41(18): 6472-6478.

[76] Pan B, Ghosh S, Xing B. Dissolved organic matter conformation and its interaction with pyrene as affected by water chemistry and concentration. Environmental Science and Technology, 2008, 42(5): 1594-1599.

[77] Marschner B, Winkler R, Jödemann D. Factors controlling the partitioning of pyrene to dissolved organic matter extracted from different soils. European Journal of Soil Science, 2005, 56(3): 299-306.

[78] Tombácz E. Colloidal properties of humic acids and spontaneous changes of their colloidal state under variable solution conditions. Soil Science, 1999, 164(11): 814-824.

[79] Guo J, Ma J. AFM study on the sorbed NOM and its fractions isolated from River Songhua.Water Res, 2006, 40(10): 1975-1984.

[80] Pace M L, Reche I, Cole J J, et al.pH change induces shifts in the size and light absorption of dissolved organic matter. Biogeochemistry, 2012, 108(1-3): 109-118.

[81] Šmejkalová D, Piccolo A. Aggregation and disaggregation of humic supramolecular assemblies by NMR diffusion ordered spectroscopy (DOSY-NMR). Environmental Science and Technology, 2008, 42(3): 699-706.

[82] Jaffe R, Ding Y, Niggemann J, et al. Global Charcoal Mobilization from Soils via Dissolution and Riverine Transport to the Oceans.Science, 2013, 340(6130): 345-347.

[83] Tang J, Li X, Luo Y, et al. Spectroscopic characterization of dissolved organic matter derived from different biochars and their polycylic aromatic hydrocarbons (PAHs) binding affinity. Chemosphere, 2016, 152: 399-406.

[84] Fu H, Wei C, Qu X, et al. Strong binding of apolar hydrophobic organic contaminants by dissolved black carbon released from biochar: A mechanism of pseudomicelle partition and environmental implications. Environmental Pollution, 2018, 232: 402-410.

[85] Wu J, Zhang H, He P J, et al.Insight into the heavy metal binding potential of dissolved organic matter in MSW leachate using EEM quenching combined with PARAFAC analysis. Water Res, 2011, 45(4): 1711-1719.

[86] Ji L, Liu F, Xu Z, et al. Zeolite-Templated Microporous Carbon As a Superior Adsorbent for Removal of Monoaromatic Compounds from Aqueous Solution. Environ Sci Technol, 2009, 43(20): 7870-7876.

[87] Yu X, Pan L, Ying G, et al. Enhanced and irreversible sorption of pesticide pyrimethanil by soil amended with biochars. Journal of Environmental Sciences, 2010, 22(4): 615-620.

[88] Peng P, Lang Y H, Wang X M. Adsorption behavior and mechanism of pentachlorophenol on reed biochars: pH effect, pyrolysis temperature, hydrochloric acid treatment and isotherms. Ecological Engineering, 2016, 90: 225-233.

[89] Zhou Z, Sun H, Zhang W.Desorption of polycyclic aromatic hydrocarbons from aged and unaged charcoals with and without modification of humic acids. Environmental Pollution, 2010, 158(5): 1916-1921.

[90] Braida W J, Pignatello J J, Lu Y, et al. Sorption Hysteresis of Benzene in Charcoal Particles. Environ Sci Technol, 2003, 37(2): 409-417.

[91] Wang B, Zhang W, Li H, et al. Micropore clogging by leachable pyrogenic organic carbon: A new perspective on sorption

irreversibility and kinetics of hydrophobic organic contaminants to black carbon. Environmental Pollution, 2017, 220: 1349-1358.

[92] Xing B, Pignatello J J, Gigliotti B.Competitive Sorption between Atrazine and Other Organic Compounds in Soils and Model Sorbents.Environ Sci Technol, 1996, 30(8): 2432-2440.

[93] Uchimiya M, Bannon D I. Solubility of Lead and Copper in Biochar-Amended Small Arms Range Soils: Influence of Soil Organic Carbon and pH. J Agric Food Chem, 2013, 61(32): 7679-7688.

[94] Sharpless C M, Blough N V. The importance of charge-transfer interactions in determining chromophoric dissolved organic matter (CDOM) optical and photochemical properties. Environ Sci-Process Impacts, 2014, 16(4): 654-671.

[95] Wang W, Zafiriou O C, Chan I Y, et al. Production of Hydrated Electrons from Photoionization of Dissolved Organic Matter in Natural Waters. Environ Sci Technol, 2007, 41(5): 1601-1607.

[96] Wilkinson F, Helman W P, Ross A B.Quantum Yields for the Photosensitized Formation of the Lowest Electronically Excited Singlet State of Molecular Oxygen in Solution. J Phys Chem Ref Data, 1993, 22(1): 113-262.

[97] Scully N, McQueen D, Lean D. Hydrogen peroxide formation: the interaction of ultraviolet radiation and dissolved organic carbon in lake waters along a 43–75 N gradient. Limnol Oceanogr, 1996, 41(3): 540-548.

[98] Dalrymple R M, Carfagno A K, Sharpless C M. Correlations between Dissolved Organic Matter Optical Properties and Quantum Yields of Singlet Oxygen and Hydrogen Peroxide.Environ Sci Technol, 2010, 44(15): 5824-5829.

[99] Zhang Y, Del Vecchio R, Blough N V. Investigating the Mechanism of Hydrogen Peroxide Photoproduction by Humic Substances. Environ Sci Technol, 2012, 46(21): 11836-11843.

[100] Vione D, Falletti G, Maurino V, et al. Sources and Sinks of Hydroxyl Radicals upon Irradiation of Natural Water Samples. Environ Sci Technol, 2006, 40(12): 3775-3781.

[101] Grannas A M, Martin C B, Chin Y P, et al. Hydroxyl Radical Production from Irradiated Arctic Dissolved Organic Matter. Biogeochemistry, 2006, 78(1): 51-66.

[102] White E M, Vaughan P P, Zepp R G. Role of the photo-Fenton reaction in the production of hydroxyl radicals and photobleaching of colored dissolved organic matter in a coastal river of the southeastern United States. Aquat Sci, 2003, 65(4): 402-414.

[103] Du Z, He Y, Fan J, et al. Predicting apparent singlet oxygen quantum yields of dissolved black carbon and humic substances using spectroscopic indices. Chemosphere, 2018, 194: 405-413.

[104] Fu H Y, Liu H T, Mao J D, et al.Photochemistry of Dissolved Black Carbon Released from Biochar: Reactive Oxygen Species Generation and Phototransformation. Environ Sci Technol, 2016, 50(3): 1218-1226.

[105] Fu H, Zhou Z, Zheng S, et al.Dissolved Mineral Ash Generated by Vegetation Fire Is Photoactive under the Solar Spectrum. Environ Sci Technol, 2018, 52(18): 10453-10461.

[106] Remucal C K.The role of indirect photochemical degradation in the environmental fate of pesticides: a review. Environ Sci-Process Impacts, 2014, 16(4): 628-653.

[107] Yan S W, Song W H.Photo-transformation of pharmaceutically active compounds in the aqueous environment: a review. Environ Sci-Process Impacts, 2014, 16(4): 697-720.

[108] 郤超, 李雁宾, 阴永光, 等. 天然水体中可溶性有机质的自由基光化学行为. 化学进展, 2012, 24(7): 1388-1397.

[109] Zhou Z, Chen B, Qu X, et al. Dissolved Black Carbon as an Efficient Sensitizer in the Photochemical Transformation of 17β-Estradiol in Aqueous Solution. Environ Sci Technol, 2018, 52(18): 10391-10399.

[110] Latch D E, McNeill K.Microheterogeneity of singlet oxygen distributions in irradiated humic acid solutions.Science, 2006, 311(5768): 1743-1747.

[111] Batista A P S, Teixeira A C S C, Cooper W J, et al. Correlating the chemical and spectroscopic characteristics of natural

organic matter with the photodegradation of sulfamerazine. Water Res, 2016, 93: 20-29.

[112] Lovley D R, Coates J D, BluntHarris E L, et al. Humic substances as electron acceptors for microbial respiration. Nature, 1996, 382(6590): 445-448.

[113] Jiang J, Bauer I, Paul A, et al Redox Changes by Microbially and Chemically Formed Semiquinone Radicals and Hydroquinones in a Humic Substance Model Quinone.Environ Sci Technol, 2009, 43(10): p. 3639-3645.

[114] Bialk H M, Simpson A J, Pedersen J A.Cross-Coupling of Sulfonamide Antimicrobial Agents with Model Humic Constituents.Environ Sci Technol, 2005, 39(12): 4463-4473.

[115] Gu B, Yan H, Zhou P, et al. Natural Humics Impact Uranium Bioreduction and Oxidation. Environ Sci Technol, 2005, 39(14): 5268-5275.

[116] Scott D T, McKnight D M, Blunt-Harris E L, et al. Quinone Moieties Act as Electron Acceptors in the Reduction of Humic Substances by Humics-Reducing Microorganisms. Environ Sci Technol, 1998, 32(19): 2984-2989.

[117] Curtis G P, Reinhard M. Reductive Dehalogenation of Hexachloroethane, Carbon Tetrachloride, and Bromoform by Anthrahydroquinone Disulfonate and Humic Acid.Environ Sci Technol, 1994, 28(13): 2393-2401.

[118] Schwarzenbach R P, Stierli R, Lanz K, et al. Quinone and iron porphyrin mediated reduction of nitroaromatic compounds in homogeneous aqueous solution.Environ Sci Technol, 1990, 24(10): 1566-1574.

[119] Sarkar B, Naidu R, Krishnamurti G S R, et al. Manganese(II)-Catalyzed and Clay-Minerals-Mediated Reduction of Chromium(VI) by Citrate. Environ Sci Technol, 2013, 47(23): 13629-13636.

[120] Polubesova T, Chefetz B.DOM-Affected Transformation of Contaminants on Mineral Surfaces: A Review. Critical Reviews in Environmental Science and Technology, 2014, 44(3): 223-254.

[121] Ratasuk N, Nanny M A. Characterization and quantification of reversible redox sites in humic substances.Environmental Science & Technology, 2007, 41(22): 7844-7850.

[122] Jiang J, Kappler A.Kinetics of microbial and chemical reduction of humic substances: Implications for electron shuttling. Environmental Science & Technology, 2008, 42(10): 3563-3569.

[123] Tian R, Yang G, Li H, et al. Activation energies of colloidal particle aggregation: towards a quantitative characterization of specific ion effects. Physical Chemistry Chemical Physics, 2014, 16(19): 8828-8836.

[124] Schwierz N, Horinek D, Sivan U, et al. Reversed Hofmeister series—The rule rather than the exception.Curr Opin Colloid Interface Sci, 2016, 23: 10-18.

[125] Aeschbacher M, Sander M, Schwarzenbach RP. Novel Electrochemical Approach to Assess the Redox Properties of Humic Substances. Environmental Science & Technology, 2010, 44(1): 87-93.

[126] Aeschbacher M, Graf C, Schwarzenbach RP, et al. Antioxidant Properties of Humic Substances. Environmental Science & Technology, 2012, 46(9): 4916-4925.

[127] Zheng Xiaojian, Liu Yafang, Fu Heyun, et al. Comparing electron donating/accepting capacities (EDC/EAC) between crop residue-derived dissolved black carbon and standard humic substances. The Science of the Total Environment, 2019, 673: 29-35.

[128] Xiu Z M, Zhang Q B, Puppala H L, et al. Negligible Particle-Specific Antibacterial Activity of Silver Nanoparticles. Nano Lett, 2012, 12(8): 4271-4275.

[129] Xiu Z M, Ma J, Alvarez P J J. Differential Effect of Common Ligands and Molecular Oxygen on Antimicrobial Activity of Silver Nanoparticles versus Silver Ions. Environ Sci Technol, 2011, 45(20): p. 9003-9008.

[130] Fabrega J, Fawcett S R, Renshaw J C, et al. Silver Nanoparticle Impact on Bacterial Growth: Effect of pH, Concentration, and Organic Matter. Environ Sci Technol, 2009, 43(19): 7285-7290.

[131] Yin L, Cheng Y, Espinasse B, et al. More than the Ions: The Effects of Silver Nanoparticles on Lolium multiflorum. Environ

Sci Technol, 2011, 45(6): 2360-2367.

[132] Levard C, Hotze E M, Lowry G V, et al. Environmental Transformations of Silver Nanoparticles: Impact on Stability and Toxicity. Environ Sci Technol, 2012, 46(13): 6900-6914.

[133] Cai W, Zhong H, Zhang L. Optical measurements of oxidation behavior of silver nanometer particle within pores of silica host. Journal of Applied Physics, 1998, 83(3): 1705-1710.

[134] Liu J, Hurt R H. Ion Release Kinetics and Particle Persistence in Aqueous Nano-Silver Colloids. Environ Sci Technol, 2010, 44(6): 2169-2175.

[135] Yu S J, Yin Y G, Chao J B, et al. Highly Dynamic PVP-Coated Silver Nanoparticles in Aquatic Environments: Chemical and Morphology Change Induced by Oxidation of Ag-0 and Reduction of Ag+. Environ Sci Technol, 2014, 48(1): 403-411.

[136] Grillet N, Manchon D, Cottancin E, et al. Photo-Oxidation of Individual Silver Nanoparticles: A Real-Time Tracking of Optical and Morphological Changes. The Journal of Physical Chemistry C, 2013, 117(5): 2274-2282.

[137] Yin Y G, Shen M H, Zhou X X, et al. Photoreduction and Stabilization Capability of Molecular Weight Fractionated Natural Organic Matter in Transformation of Silver Ion to Metallic Nanoparticle. Environ Sci Technol, 2014, 48(16): 9366-9373.

[138] Hou W C, Stuart B, Howes R, et al. Sunlight-Driven Reduction of Silver Ions by Natural Organic Matter: Formation and Transformation of Silver Nanoparticles. Environ Sci Technol, 2013, 47(14): 7713-7721.

[139] Yin Y G, Liu J F, Jiang G B. Sunlight-Induced Reduction of Ionic Ag and Au to Metallic Nanoparticles by Dissolved Organic Matter. ACS Nano, 2012, 6(9): 7910-7919.

[140] Adegboyega N F, Sharma V K, Siskova K, et al. Interactions of Aqueous Ag+ with Fulvic Acids: Mechanisms of Silver Nanoparticle Formation and Investigation of Stability. Environ Sci Technol, 2013, 47(2): 757-764.

[141] Liu H, Gu X, Wei C, et al. Threshold Concentrations of Silver Ions Exist for the Sunlight-Induced Formation of Silver Nanoparticles in the Presence of Natural Organic Matter. Environ Sci Technol, 2018, 52(7): 4040-4050.

[142] Liu H, Gao H, Long M, et al. Sunlight Promotes Fast Release of Hazardous Cadmium from Widely-Used Commercial Cadmium Pigment. Environ Sci Technol, 2017, 51(12): 6877-6886.

[143] Erkens L, Hamers H, Hermans R, et al. Lead chromates: A review of the state of the art in 2000. Surface Coatings International Part B: Coatings Transactions, 2001, 84(3): 169-176.

[144] MacLean L C W, Beauchemin S, Rasmussen P E. Lead Speciation in House Dust from Canadian Urban Homes Using EXAFS, Micro-XRF, and Micro-XRD. Environmental Science & Technology, 2011, 45(13): 5491-5497.

[145] Beauchemin S, MacLean L C W, Rasmussen P E. Lead speciation in indoor dust: a case study to assess old paint contribution in a Canadian urban house. Environmental Geochemistry and Health, 2011, 33(4): 343-352.

[146] Turner A, Kearl E R, Solman K R. Lead and other toxic metals in playground paints from South West England. Science of The Total Environment, 2016, 544: 460-466.

[147] Turner A, Solman K R. Lead in exterior paints from the urban and suburban environs of Plymouth, south west England. Science of The Total Environment, 2016, 547: 132-136.

[148] Ku Y, Jung I L. Photocatalytic reduction of Cr(VI) in aqueous solutions by UV irradiation with the presence of titanium dioxide. Water Research, 2001, 35(1): 135-142.

作者：瞿晓磊[1]，吕继涛[2]，张淑贞[2]，周东美[1]，朱东强[3]
[1]南京大学，[2]中国科学院生态环境研究中心，[3]北京大学

第 5 章　全氟及多氟烷基类化合物

- 1. 引言 /105
- 2. PFASs简介 /105
- 3. PFASs的环境赋存水平与迁移转化规律 /109
- 4. PFASs毒性效应 /113
- 5. PFASs的人体接触与健康效应 /115
- 6. 新型PFASs的相关研究 /118
- 7. 展望 /122

本章导读

全氟及多氟烷基类化合物（per- and polyfluoroalkyl substances，PFASs）在工业和生活领域应用广泛，在环境介质、生物组织及人体血液中被广泛检出。由于具有环境持久性、远距离迁移能力、生物累积能力和潜在的生物毒性，有关 PFASs 的研究已成为环境科学和毒理学领域的前沿课题和研究热点。本章首先对 PFASs 进行概述，包括其结构特点、物理化学性质和国内外管控信息等；其次，从前处理和仪器分析两个方面介绍了 PFASs 及其典型直链/支链异构体的分析方法，综述了 PFASs 在水体、沉积物、土壤和大气等环境介质中的赋存情况；再次，介绍 PFASs 的生物累积特点和毒性效应，以及人体对 PFASs 的暴露，包括暴露量与暴露途径、暴露水平、代谢和健康风险等；最后，简述新型全氟/多氟化合物相关的研究进展、PFASs 的污染控制技术和限制条款等。

关键词

全氟化合物，样品前处理，迁移转化，健康风险，限制条款

1 引言

自 1947 年美国 3M 公司（Minnesota Mining and Manufacturing Corporation）首次利用电化学氟化法生产出全氟化合物（perfluoroalkyl substances，PerFASs）以来，已有上百种含有磺酰基的全氟有机化合物系列产品被开发生产。PFASs 在工业和生活领域应用广泛，在环境介质、生物组织及人体血液中被广泛检出。由于具有环境持久性、远距离迁移能力、生物累积能力和潜在的生物毒性，2009 年全氟辛基磺酸及其盐类（perfluorooctane sulfonate，PFOS）和全氟辛基磺酰氟（perfluorooctane sulfonyl fluoride，PFOSF）被列入《斯德哥尔摩公约》附件 B 以限制其生产和使用；2015 年全氟辛酸（perfluorooctanoic acid，PFOA）被列入《斯德哥尔摩公约》的候选名单。2017 年全氟己基磺酸、其盐类（perfluorohexane sulfonate，PFHxS）及相关化合物被列入《斯德哥尔摩公约》的候选名单。有关 PFASs 的持久性、生物累积性及其生物毒性等问题的研究已成为环境科学和毒理学领域的前沿课题和研究热点。

2 PFASs 简介

全氟及多氟烷基类化合物（per- and polyfluoroalkyl substances，PFASs）是一类高持久性化合物，其所有或部分氢原子被氟原子取代，形成最稳定的碳氟键。该类物质具备优良的热稳定性、化学稳定性、生物稳定性、表面活性及疏水疏油性，被广泛应用于表面活性剂、消防泡沫（水成膜泡沫）、不粘锅、食物包装、防水衣物和防污剂等生产生活用品中[1]。

2.1 理化信息

PFASs 普遍被分成 3 类：全氟化合物（perfluoroalkyl substances，PerFASs）、多氟化合物（polyfluoroalkyl substances，PolyFASs）和氟聚合物（fluorinated polymers）。全氟化合物有全氟磺酸类（perfluoroalkane sulfonates，PFSAs，$C_nF_{2n+1}SO_3^-$)、全氟羧酸类（perfluoroalkyl carboxylates，PFCAs，$C_nF_{2n+1}COO^-$)、全氟磷酸类（perfluoroalkyl phosphonates，PFPAs，$C_nF_{2n+1}[O]P[OH]O^-$)、全氟磺酰胺类（perfluoroalkyl sulfonamides，FASAs，$C_nF_{2n+1}SO_2NH_2$)、全氟磺酰胺乙醇（perfluoroalkyl sulfonamidoethanols，FASEs，$C_nF_{2n+1}SO_2NHCH_2CH_2OH$) 和全氟磺酰胺乙酸（perfluoroalkyl sulfonamidoacetic acids，FASAAs，$C_nF_{2n+1}SO_2NHCH_2COOH$)，它们的烷基碳链上的氢原子全部被氟原子取代。多氟化合物为部分氢原子被氟原子取代，但是至少有一个氟原子，该类物质有多氟烷基磷酸酯（polyfluoroalkyl phosphoric acid esters，PAPs，$[O]P[OH]_{3-x}[OCH_2CH_2C_nF_{2n+1}]_x$)、氟调聚醇（fluorotelomer alcohols，FTOHs；$C_nF_{2n+1}CH_2CH_2OH$)、氟调聚磺酸类（x：2 fluorotelomer sulfonates，FTSAs，$C_nF_{2n+1}CH_2CH_2SO_3^-$)、氟调聚羧酸类（x：2 fluorotelomer carboxylates，FTCA，$C_nF_{2n+1}CH_2COO^-$)、氟调聚不饱和羧酸类（x：2 fluorotelomer unsaturated carboxylates，FTUCA，$C_{n-1}F_{2n-1}CF=CHCOO^-$)、氟调聚饱和醛类（n：2 fluorotelomer saturated aldehydes，FTALs，$C_nF_{2n+1}CH_2CHO$) 和氟调聚不饱和醛类（n：2 fluorotelomer unsaturated aldehydes FTUALs，$C_{n-1}F_{2n-1}CF=CHCHO$)。氟聚合物包括3种亚类：含氟聚合物（fluoropolymers）、全氟聚醚（fluoropolymers）和侧链含氟聚合物（side-chain fluorinated polymers）[2]。其中，全氟辛基磺酸（perfluorooctane sulfonate，PFOS）和全氟辛酸（PFOA）是环境中存在的典型PFASs，全氟己基磺酸（PFHxS）作为PFOS的一个替代品，是近几年环境介质中发现浓度较高的PFASs[3,4]，三种物质的分子结构分别见图5-1、图5-2、图5-3，主要理化性质见表5-1、表5-2和表5-3。

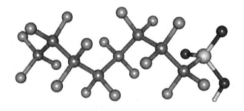

图 5-1　PFOS 分子结构示意图

表 5-1　PFOS 理化信息表[5]

	PFOS [$CF_3(CF_2)_7SO_3K$]
常温常压下存在状态	白色粉末
分子量	538
蒸气压	$3.31×10^{-4}$ Pa
水溶性	519 mg/L（20±0.5℃）
	680 mg/L（24~25℃）
熔点	≥400℃
沸点	—
气/水分配系数	$<2×10^{-6}$
亨利常数	$3.09×10^{-9}$ Pa·m^3/mol
pK_a	−3.27
pH（1 g/L）	—

图 5-2　PFOA 分子结构示意图

表 5-2 PFOA 理化信息表

属性	数值
20℃和 101.3 千帕条件下的物理状态	固体
熔点/凝固点	54.3℃
	44~56.5℃
沸点	188℃（1013.25 百帕）
	189℃（981 百帕）
蒸气压	全氟辛酸盐为 4.2 Pa（25℃）；从测量的数据外推
	全氟辛酸盐为 2.3 Pa（20℃）；从测量的数据外推
	全氟辛酸盐为 128 Pa（59.3℃）；测量数据
水溶性	9.5 g/L（25℃）
	4.14 g/L（22℃）
解离常数	<1.6，如 0.5
	1.5~2.8
pH	2.6（20 ℃时为 1g/L）
正辛醇/水分配系数 K_{ow}（对数值）	pH 7 和 25℃条件下为 6.3

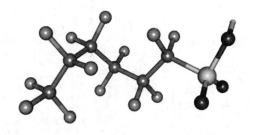

图 5-3 PFHxS 分子结构示意图

表 5-3 PFHxS 的理化性质

属性	数值
20 ℃和 101.3 千帕压力下的状态	固态白色粉末
熔点	320K
沸点	238~239℃
pKa 值	−3.45
	−3.3±0.5
	−5.8±1.3
蒸气压	58.9 Pa
水溶性	21.4 g/L（PFHxSK，20~25℃）
	2.3g/L（未解离）
正辛醇/水分配系数 K_{ow}（对数值）	5.17
空气/水分配系数 K_{aw}（对数值）	−2.38
辛醇-空气分配系数 K_{oa}（对数值）	7.55
有机碳/水分配系数，K_{oc}（对数值）（移动性）	2.05
	2.40

2.2 国内外管控信息

2000年，3M公司宣布停止PFOSF的生产。OECD在2002年的危险评估报告中指出，PFOS属PBT（persistent, bioaccumulative, toxic）性化学物质，并提出将其作为持久性有机污染物（persistent organic pollutants, POPs）候选者列入《斯德哥尔摩公约》。2005年，EPA发布了推进自主削减PFOA的"2010/2015 PFOA Stewardship Program"，提出把PFOA的工厂排放及产品含量以2000年为基准，到2010年削减95%，到2015年年底削减100%。欧盟议会于2006年12月12日发布限制PFOS销售和使用的法令，该法令于2007年12月27日前成为各成员国的国家法律，并于2008年6月27日起正式实施。2009年PFOS和全氟辛基磺酰氟（PFOSF被列入了《斯德哥尔摩公约》附件B以限制其生产和使用；2015年全氟辛酸其盐

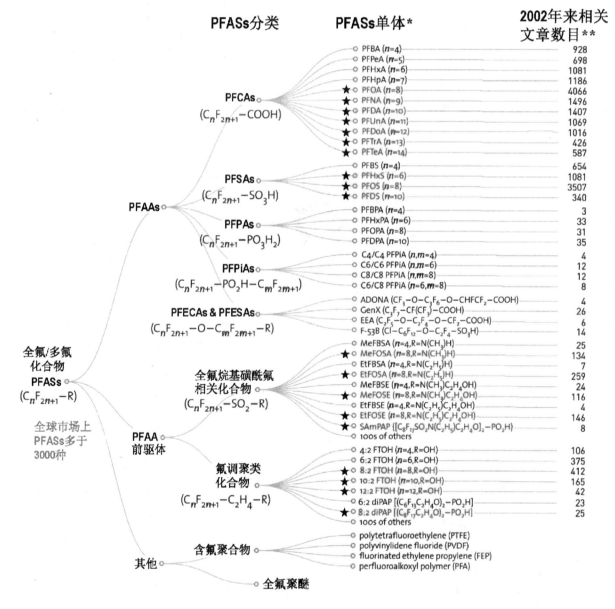

图 5-4　目前已有限制条款的PFASs种类

加粗字体的化合物受到国家/地区/全球条款限制

类及相关化合物被列入《斯德哥尔摩公约》的候选名单；2017年全氟己烷磺酸其盐类及相关化合物被列入《斯德哥尔摩公约》的候选名单[6-8]。鉴于PFASs可能存在的环境危害与健康风险，许多国家和地区都对PFASs的生产和使用进行了一定的限制[7]，目前已有限制条款的PFASs种类见图5-4（红色字体的化合物为目前已有限制条款的PFASs）[9]。

在我国，浙江省地方标准DB33/T 749—2009规定了纺织品、皮革中PFOS和PFOA的检测标准，中华人民共和国出入境检验检疫行业标准SN/T 2449—2010和SN/T 2842—2011分别规定了皮革与纺织品中PFOS的测定标准。我国环境保护部（现为生态环境部）于2008年将PFOA列入"高污染、高环境风险"化学品目录[10]。自2019年3月26日起，禁止PFOA和全氟辛基磺酰氟（perfluorooctane sulfonyl fluoride, PFOSF）除可接受用途外的生产、流通、使用和进出口[11]。

3　PFASs的环境赋存水平与迁移转化规律

3.1　PFASs的环境赋存水平

3.1.1　检测技术与方法（传统PFASs及其异构体）

1）样品前处理方法

目前监测环境中PFASs的浓度是环境科学研究的一大热点，而且被监测的环境介质多种多样，包括水、底泥、土壤以及生物样品等。不同基质中PFASs的前处理方法有较大差异。典型的PFASs比如PFOA和PFOS具有较强的极性和较高水溶性，水环境是其重要归趋。PFASs在天然水体中的浓度一般较低（pg/L～ng/L水平），为了能准确定量分析水样中的PFASs，在样品分析之前需要有一个浓缩富集和净化的过程。具有高富集倍数特点的固相萃取技术是目前PFASs水样前处理的主流方法，高回收率和高富集倍数是固相萃取技术的主要优势，同时有机溶剂的少量使用也使其成为一种环境友好的萃取分离技术，而技术的关键在于选择适合目标物的固相萃取柱。对于PFASs来说，HLB和WAX是两种比较常用的固相萃取柱。沉积物和土壤是长碳链PFASs重要的汇，测定PFASs在沉积物或者土壤样品中的赋存水平是研究PFASs的环境行为的重要组成部分。目前很多文献表明超声萃取法适用于提取沉积物、污泥、土壤等固体样品中的PFASs。Zhang等在超声提取的过程中采用了甲醇和水作为混合提取剂并添加NaOH溶液来提高提取效率[12]。超声提取方法的设备比较简单，而且容易获得理想的回收率。

与环境样品相比，生物样品的基质更为复杂，包含脂质、蛋白质、色素、纤维素等能与PFASs吸附和结合的成分。单一的溶剂萃取或者固相萃取往往不能有效地从生物样品中提取PFASs，离子对提取法是比较常用的提取生物样品的方法。离子对提取法主要是以NaOH为解离剂，用四丁基硫酸氢铵（TBA）与PFASs形成疏水离子对，然后通过较弱极性的溶剂甲基叔丁基醚（MTBE）萃取。目前离子对萃取法广泛应用于提取植物样本、动物样本和人体血液样本中PFASs的分析。

2）检测分析方法

液相质谱联用（LC-MS/MS）技术检测PFASs的方法已经趋于完善，可以满足同时测定多种PFASs的要求，而且具有灵敏度高、稳定性好等优点。而定量分析PFASs异构体是目前PFASs研究中的热点和难点。由于异构体之间物理化学性质接近，因此有效分离异构体的色谱峰是定量分析异构体的关键。同时，由于支链异构体PFASs在环境中的浓度较低，因此提高检测方法的灵敏度也是PFASs异构体研究面临的重要挑战。选择与 PFASs 存在特殊亲和作用的全氟辛基或五氟苯基修饰的反相液相色谱柱（例如 FluoroSep-RP Octyl 色谱柱）结合液相质谱联用（LC-MS/MS）技术的检测方法可以对 PFASs 异构体进行定性和定量。此外利用气相-质谱结合衍生化（四丁基硫酸氢铵）的方法也可以有效地分离 PFASs 异构体，而且这种方法具有较高的灵敏度和较短的分析时间，可以满足分析复杂环境样品的需求[13]。

3.1.2 各介质中赋存水平

由于 PFASs 较强的环境持久性和生物富集能力，因此广泛分布在各种环境介质（水体、大气、土壤）和生物体中（水生生物，陆生生物等）。目前关于水体中 PFASs 的研究较多，大多数研究发现 PFOA 和 PFOS 是水环境中主要的污染物。PFASs 在不同水体的一般规律是：点源污染水体（μg/L 水平）>河水（ng/L 水平）>海水（pg/L 水平）。例如，在大西洋赤道周围的表层海水中，PFOA 和 PFOS 的浓度范围为 10^2~10^3 pg/L，其他 PFASs 仅有少量检出[13, 14]，而在我国 19 条河流的入海口处，PFOA 的浓度范围 1.1~352.0 ng/L，PFOS 浓度范围为 0.4~14.8 ng/L[15]。在我国山东的小清河流域，由于该区域有大型氟化工厂，其 PFOA 的浓度甚至能够达到 579.0 μg/L[16]。

由于 PFASs 的中性前体物质具有半挥发性，能够进入到大气环境当中，很多研究也发现在大气中的颗粒物相和气相中都能够检测到多种 PFASs。挥发性前体物 FTOHs 是大气环境中主要的 PFASs，而且研究表明能够转化为 PFAAs 的前体物质（FTOH 和 FOSE）广泛存在于全球大气中的非颗粒物相中，从侧面证明了 PFASs 全球分布的广泛性[17]。大气中的 PFASs 污染具有明显的区域性，呈现出从城市到乡村递减的规律，而且污染情况与 PFASs 污染源密切相关。例如，Yao 等发现颗粒物相中 PFOA 在城市的浓度高达 29 ng/g，而在乡村地区的最高为 8.3 ng/g[18]。纺织品、不粘锅厨具等都有可能是室内环境中 PFASs 的释放源，因此室内环境中 PFASs 浓度一般要比室外大气环境中的高。室内空气中挥发性 PFASs 浓度一般在 pg/m^3~ng/m^3 水平，挥发性前体物质 FTOHs 特别是 8∶2 FTOH 是室内空气的主要污染物[19]。室内灰尘中的 PFASs 浓度一般在 ng/g 水平，与气相中的不同，灰尘中主要的 PFASs 为 PFOS 和 PFOA 等离子型 PFASs[20]。

目前关于 PFASs 的土壤监测研究大部分是围绕污染源进行的。Jin 等发现在江苏常熟的氟化工厂周围的土壤中全氟羧酸总浓度（ΣPFCAs）为 1.89~32.6 ng/g，全氟磺酸总浓度（ΣPFSAs）为 0.10~2.34 ng/g，PFOA 和 PFOS 是主要的 PFASs[21]。除了氟化工厂等 PFASs 的生产源头，改良土壤的污泥或者使用含有 PFASs 的消防泡沫的灭火剂也会导致土壤中 PFASs 浓度的增加。相对于水体来说，大部分研究发现长碳链的全氟羧酸（PFCAs）和全氟磺酸（PFSAs）在土壤中占比较高，这可能与长碳链 PFASs 更倾向于分配在土壤或者沉积物的分配行为有关。

由于生物富集作用，生物体内检测到的 PFASs 浓度相比环境中的浓度要高，甚至在极地动物样品中也能够检测到相当浓度的 PFASs。PFOS 在北极狐肝脏中的平均浓度为 80ng/g，在北极熊血浆中的浓度甚至达到了 237ng/g（中位数）[22]。水生环境中，Fang 等研究发现 PFOS 是太湖水生生物中主要的 PFASs，浓度范围在 13.4~468 ng/g，占总 PFASs 的 40.2%~90.3%[23]。Chen 等发现在太湖水生生物组织中 PFHxS 和 PFOS 浓度范围在 0.141~79.1 ng/g，同时发现 PFOS 的生物富集能力要强于 PFHxS[24]。对于分布在氟化工厂周围的两栖动物（黑斑蛙）而言，ΣPFASs 平均水平达到了 54.28 ng/g，而且新型替代品 2-氧-8-Cl-全氟辛基醚磺酸（商品名 F-53B）的富集能力要高于 PFOS[25]。蔬菜等植物样品中也能够检出 PFASs，我国汤逊湖周边的蔬菜中 PFASs 的浓度范围为 4.16~807 ng/g[26]，天津湿地地区白菜中 PFOS 和 PFOA 的浓度分别为 0.42 ng/g 和 0.84 ng/g[27]。由于可以从大气环境中吸收有机污染物，很多学者发现树皮树叶等样品可以用作监测大气中有机污染物的指示物。Jin 等经过研究发现，常熟氟工业园附近树皮和树叶中 PFASs 浓度达到了 250 ng/g 左右，明显高于背景区树木样品的浓度，而且树皮中 ΣPFASs 浓度呈现随着污染源距离增加而递减的趋势。PFOA、PFOS 和 PFHxS 异构体在树皮中所占比例与电化学氟化（ECF）法生产的产品中各异构体所占比例一致。这表明 PFAAs 主要由 ECF 法生产并可以经大气传输。Tian 等调查了在垃圾填埋场周围树叶中 ΣPFASs 的浓度（48 μg/g，最大值），树叶与空气中 ΣPFASs 的浓度具有较高的一致性[28]。这些研究说明树叶等植物样品可以作为指示物反映空气中 PFASs 的浓度水平。

3.2 PFASs 的迁移转化规律

3.2.1 环境介质中 PFASs 的短距离与长距离迁移规律

PFASs 因碳链长度不同、官能团种类不同理化性质差异较大，造成其在环境介质中的迁移规律存在明显的差异。碳链越长的 PFASs 疏水性越强，因此长碳链的 PFASs 较易吸附到灰尘、土壤和沉积物等固体颗粒相上，其迁移性较差。与长碳链 PFASs 相比，短碳链的 PFASs 在大气和水体等介质中具有更强的迁移性。较多的研究结果也表明长碳链 PFASs 主要富集在土壤、沉积物和生物体内，而离子型的短碳链 PFASs 主要富集在地表水、地下水等环境水体中。另外，一些中性全氟化合物 FTOH、PFOSF 等几乎不溶于水，但却具有较强的挥发性，大气是这些半挥发性的中性全氟化合物的主要归趋，它们最终会降解成 PFCAs 或 PFSAs 等离子型 PFASs。

大量的监测数据表明在偏远的高山湖泊、遥远的地球南北极环境中存在 PFOA、PFOS 等传统 PFCAs 和 PFSAs。关于 PFASs 的远距离迁移有四种可能的途径：一是半挥发性、挥发性的中性全氟化合物进入大气之后通过大气运动进行长距离迁移，然后在光照或者微生物的作用下降解转化生成 PFCAs 或 PFSAs。Benskin 等[29]通过对加拿大落基山脉高寒湖泊沉积物柱芯中的 PFASs 进行异构体分析，没有发现支链 PFCAs，推断大气中 FTOHs 的氧化是偏远地区 PFCAs 的主要来源。Wang 等[30]研究青藏高原大气中 PFASs 的分布，在雅鲁藏布大峡谷运输通道附近检测到较高水平的 FTOHs，表明可能存在长距离大气运输（LRAT）。二是直接排放到环境中的离子型 PFASs 由于水溶性较强，可通过海洋洋流或海浪迁移进行长距离迁移。Armitage 等[31]通过 GloboPOP 模型模拟，显示洋流的长距离传输方式对北极输送全氟辛酸盐量的贡献大于 FTOHs 经 LRAT 后降解的方式。Gomis 等[32]提出亲水性的 PFASs 替代物全氟聚醚羧酸和磺酸（PFESAs 和 PFECAs），如 4,8-二氧-3H-全氟壬酸铵（ADONA），2,3,3,3-四氟-2-(1,1,2,2,3,3,3-七氟丙氧基)丙酸铵（HFPO-DA），6：2 FTCA 和 F-53B 通过洋流作用而不是大气传输进行长距离迁移。三是虽然 PFOA、PFOS 等水溶解性强、挥发性差，难以通过直接挥发进入大气，但是可以通过飞沫、汽溶胶的形式从水中进入到大气，成为其进行大气传输的一个途径，Margot 等[33]发现 PFCAs 和 PFSAs 随着碳链长度的延长，其随汽溶胶进入大气的能力越强，但导致这种现象的原因尚不清楚。四是在生物体内富集的 PFASs 随食物链或生物迁徙向偏远地区传输。如 Bengtson Nash 等[34]发现南极半岛的贼鸥、白额鹱在南极寒季时迁徙到南美洲南部温暖地区暴露于 PFASs 污染环境中，回迁后通过生产、尸体分解、排泄物将其体内富集的 PFASs 带到南极地区。另外，高营养级的南象海豹觅食范围可由南极近海向北覆盖到几千千米以外的极锋带，它们也通过食物链将 PFASs 向南极传输。

3.2.2 PFASs 在环境中的转化规律

PFASs 的 C—F 键能高、稳定性极强，在自然环境条件下极难降解，在生物体内几乎不发生转化。而 PFASs 前体物是一大类在 PFASs 的羧基或磺酸基官能团结构基础上进一步衍生合成的化合物，通常含磷酸根、醚氧原子、非氟化碳原子等，它们在环境条件下较易发生降解，最终生成不同碳链长度的 PFCAs 和 PFSAs，成为环境中 PFASs 的间接来源，因此 PFASs 前体物在环境中的转化研究引起关注。目前应用广泛且研究较多的前体物主要是氟调聚酸类、全氟烷基磷酸酯类、全氟烷基磺酰胺及其 N 取代衍生物，其中 FTOHs 和 FOSAs 是环境介质中检出频率较高的前体物质。FTOHs 在污水处理厂好氧活性污泥中可以转化为PFCAs，且不同碳链长度的FTOHs微生物降解效率不同，碳链越短，被微生物转化的相对量就越多（4：2 FTOH>6：2 FTOH>8：2 FTOH）；另外8：2 FTOH在大气环境中主要氧化生成PFOA和较短链长的PFCA[35, 36]。FOSAs在污水处理厂好氧条件下也可从N-乙基全氟辛基磺酰氨基乙酸（N-Et FOSAA）向N-乙基全氟辛基磺酰胺（N-Et FOSA），再向FOSAs逐步降解转化生成PFOS[37]。FTOHs和FOSAs在环境中也可通过微生物、植物、动物生物途径转化生成PFAAs[38]。Zhao等[39, 40]发现N-EtFOSA和PFOSA在植物体内可代谢为PFOS和其他短链PFSAs，包括全氟丁基磺酸（PFBS）和PFHxS，但在土壤微生物、蚯蚓体内降解为稳定的PFOS，说明植物对PFOSA的转化途径不同于土壤微生物和动物；Zhao等[41]发现10：2 FTOH在小麦与蚯蚓体内也有完全不同的降解路径，在小麦中发现了一些更短碳链的PFCAs，表明植物可能会存在与动物不同的降解过程和机理。

此外，Mabury等[42]对多氟磷酸酯类的非生物和生物转化进行了系列研究，发现即使在pH 9、50℃的剧烈条件下，8：2 diPAP的水解仍然十分微弱，水解半衰期至少在26年以上。但是monoPAPs和diPAPs在大鼠体内肠道磷酸酶、肝肾中酶的作用下发生降解，生成氟调醇FTOHs，然后进一步氧化生成FTCA、FTUCA和不同碳链长度的PFCAs[43]。此外Lee等[44]发现6：2 diPAP可以被南瓜和苜蓿花吸收并发生进一步降解，产物以短链PFCAs（$C_{4\sim6}$）为主；Bizkarguenaga等[45]也发现胡萝卜和莴苣可以吸收土壤中的8：2 diPAP，但是植物中diPAP的降解路径和降解产率并不清楚。Mabury等[46]还研究了不同碳链长度（4：2, 6：2, 8：2, 10：2）的monoPAPs和6：2 diPAP在污泥和土壤中的降解，发现PAPs在微生物作用下同样能发生水解生成FTOHs，再进一步氧化，最终生成不同碳链长度的PFCAs，且diPAPs比monoPAPs更容易发生转化，可能是由于长碳链PAPs在污泥上有更强的吸附能力，从而降低了其生物可利用性，这一推测还有待于进一步研究。全氟烷基膦酸（PFPAs）与PFCAs和PFSAs相似，在环境中也十分稳定。但是全氟烷基膦酸酯（PFPiAs）的稳定性较差，在羟基自由基作用下很容易降解为PFPAs和PFCAs。在虹鳟鱼体内C_6/C_6, C_6/C_8和C_8/C_8-PFPiAs可以转化为C_6和C_8的PFPAs和PFCAs[47]。

PFOA和PFOS的异构体在环境介质、生物，甚至人体中广泛检出，由于生产工艺的特殊性，PFOA、PFOS（及其前体物）主要通过电化学氟化（ECF）法合成，一般含有稳定比例的直链（约70%～80%）和支链（约20%～30%）异构体组成，而PFOA的前体物，如氟调聚醇FTOHs等大多通过调聚合成法合成，只含纯直链（n-）异构体。由于CF_3基团的强拉电子能力，支链（Br-）异构化使磺酸或羧酸基团更稳定，因此在迁移转化过程中异构体的特征组成会产生变化。研究表明在不同环境介质中异构体的分配行为存在一定的差异，Chen等[48]对我国辽河和太湖湖水、悬浮颗粒物和底泥的研究结果显示，与支链异构体相比，直链的PFOA、PFOS和FOSAs更容易向颗粒相中分配，在悬浮颗粒物和底泥中直链的比例均高于水体。与工业品中直链PFOA（S）相比（n-PFOA 80%、n-PFOS 70%），动物体内n-PFOA（S）呈现富集特征（n-PFOA>90%、n-PFOS>70%），而在人体中n-PFOA同样呈现富集，并占绝对优势（96%～100%）。PFOA支链异构体优先通过尿液排泄或存在PFOA的氟调聚前体物降解，从而导致机体内直链异构体呈现富集，这一规律可以解释人与动物体中n-PFOA高比例（>90%高于工业品）的现象[49, 50]。然而，人体血液中n-PFOS的比例（48%～78%）普遍低于工业品的比例，这被视作存在前体物质间接暴露的证据，主要是由于支链前体物优先降解为支链PFOS，与相应的直链异构体相比，PFOSA的支链异构体Br-PFOSA消除得更快，优先转化为Br-PFOS。而当人体只暴露于PFOS，由于n-PFOS排泄相对较慢，血清n-PFOS比例一般也会>70%[50]。有研究表明长达十年以上的时间变化趋势显示，欧美人体血清中n-PFOS在逐年下降，而n-PFOA在逐年升高，说明前体物暴露及其转化对人体PFASs的贡献逐渐在增大[51]。另外，女性人体还可通过月经、怀孕、哺乳等方式排出PFASs，怀孕期特别是怀孕早期经胎盘传输，以及哺乳期的乳汁哺育是母体向胎儿或新生儿传递PFASs的重要方式。关于异构体母婴传递的研究发现对于PFOA和PFOS同分异构体，支链异构体比直链异构体更容易经胎盘传输，且异构体的胎盘转移效率随着支链点与官能团羧基或磺酸基距离的增加而降低。Beesoon等[52]将这种特殊的胎盘迁移归因于支链异构体比直链异构体更不容易与血清白蛋白结合；Thomsen 等[53]在挪威产妇的母乳中也发现了高比例的 n-PFOS，且异构体组成不随母亲个体和哺乳期而发生变化，该过程的具体机理有待进一步探究。

4 PFASs 毒性效应

PFASs 通过不同的工业用途，诸如灭火剂、铬雾抑制剂、纺织剂以及职业暴露和食物链传递途径，最终进入人体，并对人体健康造成潜在危害。大量研究证实，PFASs 造成的毒性效应主要表现在肝毒性、内分泌干扰毒性、发展和免疫毒性、神经毒性、致癌性。近年来，通过细胞、组织、动物以及理论模型，可以评估 PFASs 等毒物在体内的生物富集及对特定靶器官的毒性效应，借以预测 PFASs 对人类的健康危害。PFOS 和 PFOA 是环境中最有代表性的 PFASs，其他 PFASs 的毒性评价多以这两个化合物作为参照，进行毒性大小比较。对近三年来研究报道的典型 PFASs，例如 PFOS、PFOA 及其替代物的毒性效应进行总结归纳如下。

4.1 PFOS 和 PFOA

与其他典型的 POPs 具有较强的亲脂性相比，PFASs 对蛋白质具有很强的亲和力，因此 PFASs 主要分布在生物体的肝脏、血液等蛋白质含量相对更丰富的组织器官中。肝脏是动物体内的代谢中枢器官，糖、脂类和蛋白质作为生物体内三大营养物质，在肝脏中可相互进行转化与合成，并通过血液系统，输送到脂肪组织、肌肉组织、脑组织等肝外组织加以吸收和利用。肝脏作为 PFASs 的主要蓄积器官，受到 PFASs 的毒性作用显著。暴露于 PFOS 和 PFOA 的哺乳动物，随着暴露浓度的增加，受试动物的体重会显著下降，摄食量下降，体内脂肪组织萎缩，免疫器官萎缩，但肝脏组织明显肿大[54, 55]，肝脏中有大量的脂滴累积并伴随着肝糖原的下降，受试动物血液中葡萄糖、甘油三酯、游离脂肪酸和脂蛋白浓度均显著下降，这都说明肝脏作为代谢器官，其主要生物学功能，尤其是糖代谢和脂代谢受到干扰[56]。

作为环境中存在的典型内分泌干扰物，PFASs 可以通过与核受体结合而干扰生物体脂类代谢过程。目前关于其分子机理研究众多，最多的解释为 PFASs 作为脂肪酸类似物可激活过氧化物酶体激活受体 α（peroxisome proliferators-activated receptor α，PPARα），影响此核受体所调控的脂类代谢相关基因表达水平[57, 58]。此外，PFOS 可通过雌激素受体亚型 ERα 和 ERβ，诱导 MCF-10A 乳腺癌细胞的增殖，增加癌细胞转染、侵袭的能力[59, 60]。但相比于 PFOS，PFOA 诱导的乳腺癌细胞增殖不能被 ERα 的拮抗剂 ICI182,780 所抑制，却能被 PPARα 的拮抗剂 GW6471 抑制[61]。

另一方面，PPARα 的激活并非 PFASs 发挥其毒性效应的关键唯一靶标，PFASs 还可以直接作用于其他与脂类运输、合成和代谢有关的蛋白质，从而影响生物体的脂类代谢水平。如 PFOS 有可能影响肝脏中低密度脂蛋白 LDL 的分泌，干扰肝脏中脂类物质向脂肪、肌肉等周围组织器官的运输，从而造成小鼠肝脏中脂类物质的积累[55]；而 PFOA 则可通过扰乱脂类物质等在肝脏细胞内的转运和代谢，造成细胞核内脂滴的聚集[54]。

4.2 F-53B

先前的研究指出 PFOS/PFOA 急性暴露均可造成 BALB/c 小鼠或者斑马鱼的肝脏肿大、甘油三酯水平降低、脂肪变性等毒性效应特征[62]。而近年来有报道显示 6∶2 Cl-PFAES，一种 PFOS 替代物 F-53B 的主要组成物，能显著对人类肝脏细胞 HL-7702 表现出强于 PFOS 的细胞毒性。其针对模式动物斑马鱼的半数致死浓度为 15.5 mg/L（96 h），对比 PFOS 的 17 g/L 的半数致死浓度，6∶2 Cl-PFAES 的急性毒性相对稍强。

由于 F-53B 的结构与 PFOS 类似，研究者推测其也可能具有和 PFOS 类似的生物学功能和毒性效应。报道指出，F-53B 可以造成动物肝脏的肿大。并且通过和肝脏脂肪酸蛋白的高亲和力，F-53B 也可以在肝脏和 PFOS 保持相当的累积浓度，产生持久性的毒性效应输出[63]。例如对于核受体家族蛋白的 PPARs，F-53B 比 PFOS 更强地结合在该受体的内源性配体脂肪酸的结合口袋，诱导其介导的基因转录的激活，表现为配体依赖的激动效应[64]。小鼠的脂肪前体细胞 3T3-L1 暴露 F-53B 和 PFOS 后，油红染色结果显示 F-53B 可以造成细胞内脂类物质的累积[64]。另一方面，经长期暴露（56d）后检测小鼠体内标志性脂类化合物的浓度水平，结果显示 6∶2 Cl-PFAES 诱导血液中的甘油三酯和低密度脂蛋白胆固醇的含量升高，同时造成总胆固醇和高密度脂蛋白胆固醇的含量降低[65]。

此外，F-53B 可干扰甲状腺激素的体内平衡，表现出潜在的内分泌干扰效应。例如，F-53B 中的组成 6∶2 Cl-PFAES 相比于 PFOS 和 8∶2 Cl-PFAES，对甲状腺激素的内源性受体 TRα 和 TRβ 有着更强的结合亲和力。TRs 受体介导的基因激活的检测结果也验证了这一点：25 μmol/L 6∶2 Cl-PFAES 可显著诱导 TRs 的转录水平（近三倍），是上述三个化合物中最强的 TRs 激动剂。这种效应最终导致 6∶2 Cl-PFAES 可以诱导大鼠的 GH3 垂体肿瘤细胞的增殖[66]。而 6∶2 Cl-PFAES 的暴露则导致斑马鱼幼鱼显著的发育抑制，鱼体内 T4 的水平显著升高，同时伴随着甲状腺求蛋白 TG 以及涉及下丘脑的相关基因表达水平的降低[67]；6∶2 Cl-PFAES 也可引起斑马鱼的成鱼及其 F1 的子代 T4 水平的升高，以及甲状腺受体介导的基因激活[68]。体外竞争性结合实验和分子对接模拟计算指出，6∶2 Cl-PFAES 结合甲状腺激素转运蛋白 TTR 的亲和力与 PFOS 相当，这进一步证实了 6∶2 Cl-PFAES 可干扰生物体内甲状腺的生理平衡[66, 67, 69]。

另有研究证明，鸡蛋胚胎暴露 F-53B（20 d）后，胚胎心率显著降低，肝脏尺寸增加，表现出较强的发育毒性[70]。最新的研究报道指出，暴露 56d 6∶2 Cl-PFAES 可显著降低雄性 BALB/c 小鼠（1.0 mg/kg）睾丸和附睾的重量，但不影响血液中的雄激素、雌激素、促黄体激素等生殖激素的浓度水平，表现为相对 PFOS 较弱的生殖损害[71]。而 6∶2 Cl-PFAES 暴露斑马鱼成鱼和幼鱼后却表现出类似 PFOS 所诱导的生殖毒性。具体表现为：暴露 5 μg/L 的 6∶2 Cl-PFAES 可以抑制斑马鱼的性成熟指数、鱼卵产率、性腺形状，增加血液睾酮浓度，同时血液中的雌激素和卵黄蛋白原的水平也显著增长[72]。

4.3 HFPO-DA（GenX）和 HFPO-TA

PFOA 替代产品典型的分子结构为全氟链中引入不同数量的醚键，此类物质随着碳链长度的增长分别为有 HFPO-DA（也被称作 GenX）和 HFPO-TA。已有证据表明这些化合物具有与 PFOA 类似或者稍弱的毒性效应，能通过结合肝脏脂肪酸结合蛋白累积在肝脏细胞中[73]，诱导产生肝毒性[74, 75]。与 PFOA 相比，HFPO-TA 结合 PPARγ 受体的亲和力更强（是 PFOA 的 7.5 倍），诱导的转录激活效应也更显著（是 PFOA 的 2.5 倍）。HFPO-TA 暴露可促进人类的前体脂肪细胞中脂肪的变形以及脂质堆积。而链长较短的 HFPO-DA，对于 PPARγ 受体则无明显的结合能力和激活效应[76]。

4.4 其他结构类似物

6∶2 FTSA 和 6∶2 FTCA 作为 PFOS 和 PFOA 部分主链氢化的结构类似物，其毒性效应的研究相对较少。有报道指出，在小鼠模型的暴露实验中，6∶2 FTCA 表现为较小的毒性作用，6∶2 FTSA 的肝毒性相比 PFOS 较弱[73]。使用 4 碳短链化合物 PFBS 替代 PFOS，有研究表明，长期暴露于短碳链的 PFOS

替代品 PFBS，不仅 F0 代 Medaka 成鱼的肠道出现炎症反应，F1 代的子代成鱼也出现了明显的肠道损伤[77]。在一定内暴露剂量下，这些短碳链的 PFASs 对生物体的毒性效应不容忽视[78]。

综合以上毒性的报道，现有的 PFOS 和 PFOA 的替代品，其具有更严重的毒性效应和人体健康危害，对于它们的使用，更应谨慎。因此，为设计合成对人体健康安全的 PFOS 替代物，有研究重点考察了不同链长的全氟类化合物在大鼠体内血液和肝脏的累积规律，发现中长链的全氟类化合物易于在肝脏产生高浓度的累积。累积的浓度与其结合肝脏脂肪酸蛋白（liver fatty-acid binding protein, LFABP）的能力呈现强的相关。因此，LFABP 被鉴定作为重要的生物靶点，在全氟化合物的肝脏累积中起到主要贮藏作用[79]。

通过构建定量构效关系（quantitative structure-activity relationship, QSAR）模型来预测不同结构特征的 PFASs 与 LFABP 的结合强弱，以此反应化合物肝脏累积浓度的高低。模型所用到的数据集包括已知脂肪酸和全氟化合物结合人 LFABP 的解离常数。其中脂肪酸类化合物作为强的 LFABP 的配体为阳性对照，而低碳链的全氟化合物以及调聚醇类化合物为阴性对照。测试集和预测集包含的化合物活性范围从强结合到弱结合，尾端基团分别包含羧酸、磺酸、羟基等不同化学官能团结构。描述符的计算选用 ChemDes 软件得到，其可调用六种主流描述符计算引擎（Chemopy、CDK、RDKit、Pybel、BileDesc）分别计算一维、二维、三维以及相关量子化学参数等不同类型的超过三千种描述符。得到的 QSAR 预测方程有着良好的拟合性和稳健性，但该模型的应用区域只针对链状 PFASs。对于非链状分子的预测结果，可能会出现与实际值存在较大偏差的特殊情况。随后通过分子动力学模拟技术，研究者对上述 QSAR 预测方程进行了改进，将结合过程中的焓变、熵变、溶剂化能、吉布斯自由能等理论预测的热力学参数引入描述符的筛选过程，得到了拥有更广泛应用区域的 QSAR 方程。优化后的 QSAR 模型包含了焓变贡献（ΔH），可预测不同分子结构的全氟化合物的 LFABP 的解离常数，评估其结合该蛋白的亲和力强弱。因此，可以作为一种预警参数，间接地体现设计合成的 PFOS 替代物在肝脏累积浓度的高低。该研究进一步为了验证 QSAR 模型的预测能力，将两种与已知 PFOS 替代物结构完全不相似的全氟化合物 PFDecS 和 N-diPFBS 进行了 LFABP 结合能力的预测，发现这两种新型的 PFOS 替代化合物的理论 LFABP 结合能力与其在小鼠体内肝脏累积浓度呈现相一致的排序[79]。

进一步分析 PFOS、PFDecS 和 N-diPFBS 以及已知的两种替代物 6∶2 Cl-PFESA 和 6∶2 FTSA 的结合模式和关键结合氨基酸能量贡献可知：①环状替代品 PFDecS，增强了自身的分子的刚性，稳固了与 LFABP 的结合，增强了结合位点内 Arg122 和 Ser124 的氢键作用，因此提升了结合亲和力以及随后的累积量。②两性分子 PFBS 缩合形成的 N-diPFBS，隐藏了极性磺酸基团，减少了对应的氢键作用，因此结合能下降，累积也呈现轻微降低趋势。虽然可能会降低肝脏毒性，但是作为两性分子的特性也随之下降，工业用途作为灭火剂的添加剂和铬雾抑制剂的起泡效果也降低。综上所述，通过理论构建的预测方法，可以准确地评价其他新型 PFOS 替代品的 LFABP 的结合能力，以此可以推测这些化合物的肝脏累积强弱，以期设计合成肝脏累积浓度低的安全的 PFOS 替代品。

5　PFASs 的人体接触与健康效应

5.1　人体内 PFOS 和 PFOA 的赋存水平

人体通过呼吸、皮肤接触、饮食等途径暴露 PFASs。PFASs 在人体中的富集水平通常通过生物监测的方式获得，常用的生物监测介质有母乳、血液、尿液等。目前关于人体中 PFASs 暴露水平的数据比较充足。这部分数据主要集中在 PFAAs 类化合物。总的来讲，职业工人的暴露水平明显高于普通人群，工业区周边人群的暴露水平也高于普通人群的暴露水平[80-83]。

随着 FPOS 被纳入《斯德哥尔摩公约》的受控名单，有报道称近年来人体血清中 PFOS 呈先出下降的趋势。比如，Dong 等对美国 2003～2014 年 NHANES（the U.S. National Health and Nutrition Examination Survey）项目中 11895 个人体血清样本的 PFASs 含量进行了测定，发现相对于 2003～2004 年的数据，2013～2014 年的数据呈现下降趋势，表现为 PFOA、PFOS、PFHxS、PFNA 的浓度从 2003～2004 年的 3.7ng/mL、19.2 ng/mL、1.7 ng/mL、0.8 ng/mL 分别下降至 2013～2014 年的 1.8 ng/mL、4.7 ng/mL、1.3 ng/mL、0.6 ng/mL[84]。同时，澳大利亚、日本、挪威的研究也说明了这个趋势。值得注意的是，部分国家 PFASs 的含量在 2003 年之后是呈现上升趋势的。比如一项关于首尔 1994～2008 年人体 POFS 和 POFA 含量变化的报道显示，在这期间，PFOS 的浓度有轻微的变化，但是 PFOA 的浓度呈现出上升的趋势。同时研究也发现 2002 年后，中国沈阳市人体中的 PFOS 的浓度呈现上升的趋势。这些数据表明，2003 年 3M 停产 PFOS，加之《斯德哥尔摩公约》对生产 PFASs 方面的限制，PFASs 的生产地转移到了其他地区[81]。

随着 PFOS 及 PFOA 被逐渐禁用，国内外生产商开发出多种相应的替代品。所以，近几年人体中除了检出常见的 PFAS，比如：PFOS、PFOA 及 PFHxS 之外，还逐渐检出了一些常见 PFASs 的替代品，比如 Shi 等首次报道了人体中 F-53B 的含量。F-53B 是工业生产中的铬雾抑制剂，在中国常被当作 PFOS 的替代品所使用。该报道称 6∶2 Cl-PFESAs 和 8∶2 Cl-PFESAs 在 PFASs 高暴露地区食鱼者的体内含量的中值为 93.7 ng/mL 和 1.60 ng/mL，在金属电镀工人体内的浓度中值为 51.5 ng/mL 和 1.60 ng/mL[85]。Chen 等的研究发现，在人体脐带血中也发现了 F-53B 的存在[86]。其他的新型 PFASs 在人体中也有检出的报道。比如，Pan 等首次报道了 HFPO-TA（ammonium perfluoro-2-[（propoxy）propoxy]-1-propanoate）在环境中的含量。通过采集含氟聚合物生产商周边居民的血液样本，显示样本血清中 HFPO-TA 的含量为 2.93 ng/mL[87]。随着国内外对 PFOS 和 PFOA 管理得愈加严格，它们的替代品在市场上的应用越加广泛。目前关于替代品在人体中暴露水平的报道还较少，但可以预见关于这些替代品在人体中含量的报道会越来越多。

5.2 人体内 PFASs 的转化、消除规律

与其他化学物质类似，PFASs 进入人体后会有相应的代谢过程。目前关于 PFASs 在人体内的代谢规律主要集中在 PFASs 类物质在人体中的半衰期的测算方面。Olsen 等测算的 PFHxS 在人体中的半衰期算术均值和几何均值分别为 8.5 年和 7.3 年，PFHxS 和 PFOS 的半衰期比值为 1.5 年[87]。Fu 等的研究表明，PFHxS、PFOA 和 PFOS 在人体内的半衰期算术均值和几何均值分别为 14.7 年和 11.7 年、4.1 年和 4.0 年以及 32.6 年和 21.6 年[88]。Li 等的研究表明 PFHxS、PFOS、PFOA 在人体内的半衰期分别为 5.3 年、3.4 年和 2.7 年[89]。

除了普通人群之外，也有研究者研究 PFASs 在特殊人群中的赋存规律。同时，除了研究常见的 PFASs 之外，还研究了一些 PFOS 的替代品在人体内的赋存规律。比如，Chen 等研究了 PFOS 的替代品 6∶2 Cl-PFESA 和 8∶2 Cl-PFESA 在孕妇体内的赋存和母婴传输规律。研究发现在母体血清、脐带血以及胎盘中目标化合物的浓度顺序为 PFOS>6∶2 Cl-PFESA>8∶2 Cl-PFESA，并且 PFOS、6∶2 Cl-PFESA 和 8∶2 Cl-PFESA 在三类介质间的浓度存在显著相关。计算脐带血和母体血液中的目标化合物的比值，可以得到相应化合物的胎盘传输效率。结果显示，6∶2 Cl-PFESA 的传输效率较高（约为 40%），8∶2 Cl-PFESA 的传输效率更高。这可能是由于 8∶2 Cl-PFESA 的疏水性更高，并且 8∶2 Cl-PFESA 与血浆白蛋白的结合能力更低导致的[90]。

除此之外，也有研究报道了异构体在人体中的富集和消除规律。Gao 等的研究表明直链异构体在职

业工人血清中的富集水平比支链异构体高,血清中直链异构的比例高于尿液。通过计算肾脏消除速率,发现支链异构体的肾脏消除速率高于直链异构体。利用分子对接模型,发现直链异构体与人体血清白蛋白的结合能力更强,这一结果解释了上述直/支链异构体在人体中的差异富集、消除规律的原因[91]。除此之外,也有报道研究了异构体在母婴之间的传输规律。Chen 等的研究表明,直链和支链的 PFHxS、PFOS 以及 PFOA 均在母体血液、脐带血与胎盘中均有检出。直链和支链的 PFHxS、PFOS 以及 PFOA 可以通过胎盘屏障,可能对婴儿造成潜在的危险[92]。

5.3 流行病学研究结果

流行病学研究显示了 PFASs 水平与胆固醇、脂蛋白、甘油三酯和游离脂肪酸血清水平之间的关系[93-95]。在对 PFASs 流行病学研究的评估中,在 PFOA、PFOS 和 PFHxS 研究中观察到血清酶上升和血清胆红素增加,意味着肝损伤[96]。在对加拿大成年人进行的一项横向研究中,对抽样策略进行加权后观察到 PFHxS 与胆固醇结果(总胆固醇、低密度脂蛋白胆固醇、总胆固醇/高密度脂蛋白比率和非高密度脂蛋白胆固醇)之间存在着显著的相关性[93]。没有观察到支持 PFOA 和 PFOS 与胆固醇结果之间存在着相关性的证据。对居住在化工厂附近的人口进行调查的 C8 健康项目中的一项研究显示,随着 PFOA、PFOS 和 PFHxS 的浓度水平增加,胆固醇呈正单调递增[94]。不过,Nelson 等(2010 年)发现,在美国普通人群中 PFHxS 与总胆固醇、非高密度脂蛋白和低密度脂蛋白胆固醇呈负相关关系(全国健康和营养调查,2002~2003 年),而对 PFOA、PFOS 和 PFNA 观察到呈正相关关系。据观察,在挪威的孕妇中,5 种 PFASs 包括 PFOS、PFNA、PFDA、PFUnDA 和 PFHxS 与高密度脂蛋白-胆固醇呈正相关关系[95]。对西班牙孕妇进行的一项研究发现,PFOS 和 PFHxS 与糖耐量减低和妊娠糖尿病呈正相关关系[96]。

人类流行病学研究了血清中 PFASs 与儿童体内神经毒性作用或神经发育影响之间的相互关系。在 C8 健康项目中,Stein 和 Stavitz(2011 年)研究了儿童(5~18 岁;$n=10456$)血清中 PFOA、PFOS、PFNA 和 PFHxS 浓度与父母或自我报告并经医生诊断的注意力缺失多动症(不管目前是否正在使用治疗注意力缺失多动症的药物)之间的横向联系。尽管该人群接触到的 PFOA 水平最高,但据观察,接触与结果之间的最紧密联系是与 PFHxS 有关,就 PFOS、PFOA 和 PFNA 而言,没有观察到与注意力缺失多动症的强相关关系。在美国利用 1999~2000 年和 2003~2004 年全国健康和营养调查的数据进行的另一项研究中,在血清中全氟己烷磺酸水平上升 1 μg/mL 的情况下,也发现了注意力缺失多动症概率明显增加情况($n=571$)[97]。不过,在此项研究中,PFOS、PFOA 和 PFNA 都与父母报告的注意力缺失多动症呈正相关。在一项调查儿童血液中各种全氟化合物的水平及与行为抑制的关系的研究中,结果表明,PFHxS 的血液水平与儿童行为抑制缺陷高度相关($n=83$)[97]。在对格陵兰和乌克兰 5~9 岁儿童($n=1023$)进行的一项前瞻性研究中,使用"优点和困难问卷"评估了问题行为,与出生前接触 PFHxS 的程度低相比,接触程度高也与这些问题行为有关[98]。在一项对母亲脐带血中 PFASs 浓度和婴儿成长关系的研究中,研究者对新生儿和出生婴儿的生长发育进行了跟踪调查,结果发现,子宫内 PFASs 的暴露影响婴儿的生长发育,并且对于男孩和女孩的影响各异[99]。而近期的一项发现显示,出生前和儿童时期血液中 PFOA、PFOS 和 PFNA 的浓度与 5~8 岁儿童较强的阅读能力呈正相关,而 PFHxS 的浓度与阅读能力无关[100]。

流行病学研究指出了出生前和童年时期接触某些 PFASs 造成的免疫毒性效应或调节效应。日本对 1558 对母子调查了出生前接触某些 PFASs 与 4 岁前发生传染性疾病之间的关系。调查发现出生前接触 PFOS 和 PFHxS 与早年发生传染性疾病(例如,中耳炎、肺炎、RS 病毒和水痘)有关。PFOA 和 PFHxS 血清水平与 3 岁儿童的胃肠炎发作次数之间呈正相关关系($n=66$)[101]。在一项将 1997~2000 年和 2007~2009 年法罗岛的两个出生群组结合起来进行的后续研究中发现,5 岁时对破伤风疫苗产生的预增强血清抗体明显减少与出生时 PFOA 和 PFHxS 血清浓度加倍有关[102]。结构方程模型显示,就 5 种 PFASs(PFOA、

PFOS、PFNA、PFDA 和 PFHxS）而言，7 岁时 PFASs 接触加倍与 13 岁时白喉抗体浓度降低有关。目前的研究扩充了以前这一较年轻的群组中抗体反应不足的结论，因此进一步增强了这样一种认识：大力加强对 PFASs 接触的预防显得很有必要[103]。在一个未接种麻疹、流行性腮腺炎和风疹（MMR）三联疫苗儿童的小型亚群中，5 岁时接触 PFASs（PFOA、PFOS、PFNA、PFDA 和 PFHxS）与患哮喘概率加大有关[102]。一项对 1056 名妇女进行的前瞻性出生群组研究发现，出生前接触 PFOA、PFNA、PFDA、PFDoA 和 PFHxS 显著增大了女童在人生最初 24 个月里患儿童特应性皮炎的风险[100]。

一些研究表明，某些 PFASs 可能影响人的生殖。不论是流行病学研究还是体外研究都表明，全氟烷酸可能影响卵巢细胞信号传导以及总体生殖健康情况。在最近的一项研究中，从在美国接受过体外受精的 36 名研究对象那里收集了血液和滤泡液。结果显示，基线滤泡计数与血浆中 PFHxS 浓度呈负相关关系，滤泡液中 PFDA 和 PFuNA 的浓度与囊胚转化率呈负相关关系，表明在卵巢病理学背景下，这些 PFASs 有可能是一类值得关注的化合物[104]。在对中国 120 例原发性卵巢功能不全的妇女和 120 例对照人群的一项研究中，Zhang 等[100]发现，PFOA、PFOS 和 PFHxS 的高暴露增加了妇女罹患原发性卵巢功能不全的风险。在丹麦的一项病例对照研究中，观察到血清中 PFASs 的水平（PFNA 和 PFDA）与流产之间明显存在着强相关关系，据观察，流产与 PFHxS 几乎存在着明显的联系，不过，没有观察到 PFOA 和 PFOS 与流产之间的联系[105]。在加拿大的一项出生群组研究中（2008～2011 年，$n=1625$），PFOA 和 PFHxS 在妇女血浆中的浓度升高与用备孕时间更长和不孕概率提高来衡量的生育能力下降有关，不过，没有观察到与 PFOS 明显的联系[106]。

2009～2010 年全国健康和营养调查（$n=1566$）显示，血清中 PFOA，PFHxS 和 PFNA 的浓度与骨质疏松症的患病率呈正相关[107]。在工人中，溃疡性结肠炎和类风湿性关节炎与 PFOA 接触具有正相关性[108]。

6 新型 PFASs 的相关研究

6.1 新型 PFASs 简介

新型 PFASs（emerging PFASs）是相对于以 PFOS 和 PFOA 为代表的被充分研究和认识的库存 PFASs（legacy PFASs）而言的。目前认为，新型 PFASs 主要分为以下几种情况：一是为了应对国际公约对 PFOS 和 PFOA 的限制而开发的以短链或引入醚键或者杂原子等 PFASs 为代表的替代品；在 2012 年《斯德哥尔摩公约》缔约国大会上发布的 PFOS 等替代品的识别和评估技术报告中指出多种短链全氟化合物，例如全氟丁酸（PFBA）、全氟丁基磺酸（PFBS）、全氟己酸（PFHxA）和 PFHxS 等以及一些多氟化合物，例如 FTOHs 和调聚磺酸（FTSs）等已作为 PFOS 相关产品的替代物在电镀、石油、纺织、灭火等多个行业中使用，这种短链替代策略的理由是 PFASs 的生物累积性呈现随链长的减少而降低的趋势。然而已有文献指出 PFHxS 具有一定毒性和生物累积能力，因此被考虑作为 POPs 进行限制，目前已进入评估阶段[108,109]。除短链替代策略外，另一个策略是为了增加全氟烷基化合物在环境中的降解能力，在全氟烷基链中间引入一个或多个醚键，或者某个氟或多个氟原子被其他杂原子（例如氯）所取代得到全/多氟醚类物质（PFAESs），此类物质主要包括全/多氟醚羧酸（PFECAs）和全/多氟醚磺酸（PFESAs）。Wang 等[110,111]在总结多个工业和民用领域中 PFASs 替代品使用情况时指出，该类化合物是除短链 PFASs 以外另一类非常重要的全氟替代产品。HFPO-DA（商品名为 GenX）和 ADONA 分别是由全氟重要生产商杜邦公司和 3M 公司生产的 PFOA 或 APFO 的替代产品，目前已在聚四氟乙烯（TFE）的生产过程中作为分散剂使用[112]。

在2002年3M公司停止生产全氟辛基类水成膜泡沫灭火剂后,气溶胶态的氟化酮[$CF_3CF_2C(O)CF(CF_3)_2$, CAS:756-13-8]作为灭火剂的添加剂被生产和使用。而杜邦公司生产的水成膜泡沫灭火剂(AFFFs)添加的则是氟调聚磺胺甜菜碱(例如:6:2 FTAB, $C_6F_{13}C_2H_4SO_2\text{-}NHC_3H_6N^+(CH_3)_2CH_2COO^-$)或氟调聚磺胺氨基氧化物[$C_6F_{13}C_2H_4SO_2NHC_3H_6N(O)(CH_3)_2$, CAS No. 80475-32-7]。

据有关报道世界范围内登记有工业生产或者使用的属于 PFASs 的化合物种类多达 4000 多种,这些 PFASs 在全球被长期且大量地生产和使用,然而它们并未得到环境科学工作者的重视和研究,这类被忽视多年的 PFASs 也被归为新型 PFASs 的范畴。例如:F-53B 被作为铬雾抑制剂在我国电镀行业中与 PFOS 并行使用了近 30 年,在 PFOS 被限制后或成为电镀行业中 PFOS 的主要替代产品,然而其直到 2013 年才被首次报道存在于我国环境中[113]。据报道全氟壬烯氧基苯磺酸钠(OBS)早在 20 世纪 80 年代就被日本的 Neos 公司以 Neos Ftergent(NF)为商品名进行生产,该物质主要作为表面活性成分用于泡沫灭火剂和石油开采助剂中,目前在我国的生产量为 3500t/a,然而直到 2017 年才被首次报道广泛存在于油田周围环境中[114]。

除了上述替代产品或者被忽视的已生产和使用多年的新型 PFASs 外,由于化学合成工艺的限制而存在于库存 PFASs 或新型 PFASs 产品中的低含量的全氟或多氟化合物杂质,以及 PFASs 前体物及其在环境中的化学或生物转化产物,属于第三种类型的新型 PFASs。这些物质多数是库存 PFASs 的同系物等,它们的环境污染长期存在,只是以前的色谱-质谱仪器灵敏度等有限而未得到环境工作者的认识。例如 Place 和 Field[115]在多种 AFFFs 产品中识别出来的 10 种新型 PFASs,包括全氟链长为 4~12 的阴离子、阳离子或两性的表面活性剂,然而多数这类物质被认为并非来自于有意生产,其最大可能是来自于作为产品杂质、副产物或降解产物的 AFFF 的使用排放。Washington 等[116]在环境中检出不饱和多氟羧酸(uPFOA)和 2H 取代的多氟羧酸(2HPFCAs),并通过对基于氟调聚物的聚合物降解实验证明这两种物质的存在,说明 uPFOA 和 2HPFECAs 是作为氟聚合物降解后的产物被排入环境中。有研究显示水中 PFOS 和 PFOA 经过等离子体方法处理后会产生短链 PFASs、环形全氟烷(C_4F_8, C_5F_{10}, C_6F_{12}, C_7F_{14} 和 C_8F_{16})以及全氟烷基自由基、全氟醇或酮等[117]。Lin 等[118]认为环境中检出的 H-PFECAs 是 Cl-PFESA 的脱氯产物。Song 等在山东小清河一带识别出的多种新型 PFASs,例如:PFECAs、H-PFECAs、Cl-PFECAs 等均在 PFOA 的铵盐产品或其替代产品六氟环氧丙烷(HFPO)的三聚体产品中被检出,说明这些新型 PFASs 是以杂质的形式存在于库存 PFASs 或其替代产品中的,在该区域检出的 xH-PFCAs 则可能是生产 PVDF 过程中产生的一类副产物。作为航空液压油添加剂的全氟环己烷乙基磺酸(PFECHS)标准品中可检出其五元环异构体(PFPCPeS),该物质也会随 PFECHS 的使用排放而进入环境[119]。另外,一些可经过环境或生物过程生成库存 PFASs 的前体物,例如:全氟辛烷磺胺乙醇基磷酸二酯(SAmPAP)、全氟辛烷磺酰胺二乙醇(POSEs)、PAPs、FTOH、FTS 等也被作为新型 PFASs 进行研究。

6.2 新型 PFASs 的筛查与鉴定

高效液相色谱与质谱联用(HPLC-MS)技术在 PFASs 定量分析中具有不可替代的作用,然而这种分析依赖于 PFASs 标准品的使用,对于上述成百上千的无标准可用的新型 PFASs 的识别则由于质谱分辨率低而无法准确定性。随着科学仪器的飞速发展,飞行时间质谱(TOF-MS)和轨道离子阱质谱(Orbitrap MS)等高分辨质谱(HRMS)已经成为包括新型 PFASs 在内的有机物筛查的重要技术手段。HRMS 仪器具有超高的分辨率、质量准确度,可通过样品在 MS 中产生的精确分子量、特征碎片及同位素信息等对有机污染物进行识别确认。其中 HRMS 的非目标筛查技术在新型 PFASs 发挥着重要作用,在此过程中质量亏损过滤(mass defect filter,MDF)法是一个在大量质谱数据中筛查有机污染物的有效方法,MD 通常指元素、分子或离子的理论或测定的准确分子量与名义分子量之间的差值。对于 PFASs 来讲都含有一个或

多个 CF_2 基团,因此通常利用 Kendrick 质量亏损图进行 PFASs 的筛查,Kendrick 质量是用 CF_2 基团校正后的分子量(Kendrick Mass=Measured Mass*50/49.99681),KMD 值为名义的 Kendrick 分子量与 Kendrick 质量的差值,即 KMD= Nominal Mass–Kendrick Mass。以分子量为横坐标,KMD 为纵坐标作图即可得到质谱中出现的可疑 PFASs 碎片的 KMD 图。属于同一族的 PFASs 会出现在间隔 50Da 且平行于 X 轴的一条直线上,如图 5-5 所示为中国科学院生态环境研究中心蔡亚岐课题组分析样品时得到的 KMD 图,图中共有 11 类 PFASs,其中 9 类为新型 PFASs。

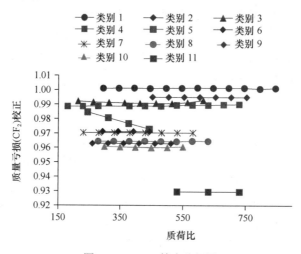

图 5-5　PFASs 筛查分组图

Jeniffer 等[120, 121]利用 KMD 技术不仅在 AFFFs 中识别出超短链的 PFASs(C_2 和 C_3),同时结合非目标的 R Script 在工业产品和 AFFFs 中识别出 40 类新型阴离子、阳离子和两性的氟化表面活性剂,其中 34 类源于电化学合成工艺,且具有相似的结构,同时在环境中检出的氟化表面活性剂有 11 种由电化学合成,2 类由氟调聚法合成,这些结果充分说明了 KMD 筛查技术的有效性。Liu 等[122]建立了一种大体积进样在线浓缩的液相色谱分离与 Orbitrap-HRMS 分析相结合的水体中新型 PFASs 的非目标筛查技术,该技术采用两次进样法最终识别环境样品中存在的新型 PFASs,第一次进样时采用三个平行的质谱事件循环扫描(两个源碎片标记扫描和一个 uHRMS 全扫描)来完成保留时间和经验分子式的匹配,其中一个源碎片标记扫描对能产生全氟烷基的诊断性碎片离子 $C_nF_{2n+1}^-$(例如 $C_2F_5^-$ m/z=118.993)和 $C_nF_{2n-1}^-$(例如:$C_3F_5^-$ m/z=139.992)及利于结构和元素组成确定的诊断性碎片离子 SO_3F^- m/z=98.956 和 SO_4H^- m/z=96.959 等化合物的保留时间进行标记;另一个源碎片标记扫描对能产生一些杂原子(例如:Cl^-)的化合物保留时间进行标记。第二次进样时通过对特定的分子离子进行扫描得到其碎片离子,对第一次进样所得到的经验分子式进行筛选,去掉由于不同元素或元素的不同位置而产生的具有相同 m/z 的碎片离子的分子式或结构,最终确定化合物的分子式和结构。他们采用该方法在工业废水、野生鱼体中识别出包括多氟磺酸、氯代全氟羧酸、氢代多氟羧酸在内的 8 类超过 160 种的新型 PFASs[122, 123]。除上述非目标筛查技术外,目标筛查和可疑目标筛查的方法同样可以在环境样品或工业产品中新型 PFASs 的识别分析中发挥作用,但这两种方法均是基于文献或数据库中已知的新型 PFASs 的相关信息实现对其识别和结构鉴定的。

6.3　新型 PFASs 的环境赋存水平和迁移转化

随着越来越多的新型 PFASs 在化工产品及环境中的识别,其环境赋存和归趋也逐渐成为研究热点。例如:F-53B 自 2013 年被在环境中报道起,就成为研究关注的对象。有研究表明,F-53B 在电镀园区的污水处理厂进出水中具有很高的浓度(进水:65~112 μg/L,出水:43~78 μg/L),其在受纳水体中的浓度在 10~50ng/L，且随着与排污口距离的增加逐渐下降[112]。Ruan 等[124]在我国多个城市采集的活性污泥中检出了 F-53B 的主要成分氯代多氟醚磺酸（Cl-PFESA）的两个主要同系物（C_8 和 C_{10}）的存在，其平均浓度分别为 2.15 ng/g dw 和 0.50 ng/g dw，二者在环境中的比值与它们在 F-53B 产品中比值之间的差异说明碳链更长的 C_{10} Cl-PFESA 由于更强的疏水性更易分配到污泥中；该研究结果充分说明了 F-53B 在我国电镀行业的广泛使用，因此 F-53B 等 Cl-PFESAs 在我国的污染普遍存在。此外，他们还发现 VB12 可在厌氧条件下将 Cl-PFESAs 结构中的 Cl 脱去并生成 H-PFESAs，并且在环境中检出了 H 取代的 PFESAs，从而在一定程度上证实 Cl-PFESAs 可能具有一定的环境可降解性。然而 Chen 等[125]的研究表明 Cl-PFESAs 在 42mmol/L 过硫酸氧化处理时并无显著变化，说明 Cl-PFESAs 与 PFOS 相比除了具有相似的环境持久性外，还具有更高的土壤吸附性。尽管 F-53B 被报道是具有中国特色的新型 PFAS，但是 Gebbink 等[126]却在北极熊、虎鲸和环斑海豹等极地生物中检出 F-53B 的存在，其浓度在 0.023~0.27 ng/g 之间，该结果表明，与 PFOS 类似，Cl-PFESA 可随洋流等途径进行远距离传输，并已成为全球性污染物。Pan 等[127]在全球七个国家采集的地表水中广泛检出 HFPO-DA、HFPO-TA 及 Cl-PFESA 的存在，其中位浓度分别为 0.95 ng/L、0.21 ng/L 和 0.31ng/L，表明作为 PFOA 替代产品使用的 PFECAs 也逐渐成为全球普遍存在的污染物。

Zhou 等较早地在我国武汉开展了短链 PFASs 的环境行为和归趋研究[128]，在汤逊湖水体中检出了较高浓度的 PFBS 和 PFBA，它们是水体中最主要的 PFASs，二者平均浓度分别为 3660 ng/L 和 4770 ng/L，研究表明二者的底泥水分配系数和生物富集因子较低，但因其随水流进行传输的能力较强，扩散范围较大，其环境影响仍然值得关注。氟聚合物（FP）生产厂被认为是 PFOA 的主要污染来源，在 PFOA 受到各国不同部门及国际公约的限制后，FP 厂会选用相应的替代产品进行氟聚合物的生产，因此将文献、产品信息、专利资料的调研和相关环境样品检测分析结合，是识别发现新型 PFASs 的有效方法。应用该方法，Song 等在我国最大的 FP 生产厂所在周边区域山东小清河流域识别鉴定出 42 种新型 PFASs 的存在，包括：Cl-PFCAs（$ClC_nF_{2n-2}HO_2$，$n=5$~12）、H-PFCAs（$C_nF_{2n-2}H_2O_2$，$n=6$~17）、x H-PFCAs（$C_{2n}F_{2n}H_2O_2$，$n=3,5,6,7$）、单醚 PFECAs（$C_nF_{2n-1}HO_3$，$n=3$~14）、多醚 PFECAs（$C_nF_{2n-1}HO_n$，$n=4$~7）和六氟环氧丙烷寡聚体（二聚、三聚和四聚，HFPO-DA，TrA 和 TeA）等六类化合物，对 HFPO-DA、TrA 和 TeA 进行了准确定量，HFPO-DA 和 TrA 在地表水中检出浓度分别在<LOQ~9.35×10^3 ng/L 和<LOQ~7.82×10^4 ng/L 之间；HFPO-TeA 是一种以前未曾报道过的新型 PFAS，尽管其在水体中未检出，但却在在底泥中检出了浓度在<LOQ~363ng/g dw 之间的 HFPO-TeA；根据新型 PFASs 在 HRMS 峰面积对其进行半定量后估算其在该区域的排放量，发现单醚 PFECAs 和多醚 PFECAs 的排放量可分别达到 2.84t/a 和 3.59t/a，其他新型 PFASs 的排放量低约一个数量级；新型和库存 PFASs 在水体和底泥中的相关性说明新型 PFASs 具有与库存 PFASs 相似的环境行为和归趋。Xu 等在油田周围湖泊水体中检出了一种含有苯环结构的全氟壬烯氧基苯磺酸（简称 OBS）的存在，其浓度最高可达 3200 ng/L，并且其在老旧油田区域环境的浓度显著高于背景地区和新油田，在背景地区、新油田和老油田区域湖水中的平均浓度分别为：6.9 ng/L、50 ng/L、560ng/L，说明 OBS 在油田周围环境中浓度水平主要受到采油行为和采油历史的影响[113]。此外，该团队围绕首都机场开展了作为航空液压油添加剂 PFECHS 的环境存在和行为研究[118]，在机场停机坪下游的湖泊中检出较高浓度的 PFECHS 及其五元环异构体 PFPCPeS，其总浓度在水体中在 1.04~324ng/L，在底泥中在<MLQ~2.23 ng/g dw 之间，并且浓度随着与机场距离的增加呈现降低的趋势。上述研究表明，许多新型 PFASs 由于具有与库存 PFASs 相似的结构，因此它们的环境行为和归趋具有一定的可预期性，但结构的差异也往往会引起其环境行为和归趋的分异现象，这应该引起研究者的重视。

6.4 新型 PFASs 的生物累积和毒性效应

生物富集能力是判断有机污染物是否属于持久性有机污染物（POPs）以及评价其环境风险的重要标

准之一。通常情况下，疏水性有机污染物的生物富集能力与其辛醇水分配系数（K_{ow}）具有直接关系，可选用理论模型对其理化性质和生物富集能力进行估算。然而与传统POPs不同的是，PFASs具有亲蛋白性而非亲脂性，这就使针对亲脂性有机污染物设计的估算模型并不能完全适用于PFASs的环境行为和生物富集能力的预测，目前人们最认可的评价PFASs生物富集的方式仍然是通过定量分析成对生物-水/底泥样品，得到浓度后计算生物累积因子BAF。总体来看，对于新型PFASs的生物累积和毒性效应的研究较少，无法支撑其环境风险研究。Shi等[129]在采自武汉汤逊湖和山东小清河的鲫鱼体的10种不同组织器官（血液、肝脏、性腺、鱼鳔、肾脏、鳃、脑、心脏、肌肉、胆汁等）中均检出Cl-PFESAs的存在，其浓度分布表现出与PFOS相似的器官特征，以血液浓度最高（TL：20.9 ng/g；XR：41.9 ng/g），其次为肝脏，肌肉中浓度最低，但是在计算总体负荷时肌肉所占负荷比例（～28%）仅次于血液负荷（～32%），比较表明F53-B的总体生物富集因子（log XR：4.124，TL：4.322）大于PFOS，同时超过了衡量生物富集能力的标准值5000 L/kg，说明F-53B具有较强的生物富集能力。此外，还有研究表明F-53B在水生食物网中具有一定的生物放大效应[130]。Shi等[85]对F-53B在人体内的暴露水平的研究发现在采集的98%以上人体血液样品中均可检出C_8 Cl-PFESA的存在，浓度最高达到5040 ng/mL，高食鱼人群和电镀工人血清中浓度（93.7 ng/mL，51.5 ng/mL）均显著高于普通暴露人群浓度，根据其在血液和尿液中浓度计算得到的肾清除半衰期也远大于PFOS，说明Cl-PFESA在人体内具有强于PFOS的累积性，其环境和人体健康风险值得我们进一步关注。除F-53B的相关研究外，PFOA的替代产品HFPO-TA的生物累积也有少量研究，Cui等[25]的研究表明HFPO-TA在鱼体内具有与PFOA相似的器官分布特征，其血液中的累积能力相对较高；他们还对Cl-PFESA和HFPO-TA在两栖动物-黑斑蛙体内的存在及累积能力进行了研究，结果表明Cl-PFESAs和HFPO-TA在黑斑蛙体内同样表现出强于PFOS或PFOA的生物富集能力。

Cl-PFESAs不仅具有以上所述的较强的生物累积能力，其各种毒性效应也被多项研究所证实，例如：Li等[64]的研究表明，Cl-PFESAs与过氧化物酶体增殖物激活受体（PPARs）的结合能力强于PFOS，C_8 Cl-PFESA具有与PFOS相似的PPARs信号通路结合活性，说明与PFOS一样，Cl-PFESAs同样会干扰PPARs的信号通路诱导细胞效应，从而诱发潜在的健康风险；小鼠干细胞的暴露研究表明F-53B对胚胎发育初期可能具有潜在的神经毒性[131]；成年鼠暴露不同浓度的F-53B[0.04 mg/(kg·d)，0.2 mg/(kg·d)，1mg/(kg·d)]时肝脏脂质聚集和肝脏质量均有显著增加，表现出明显的肝脏毒性效应。蛋黄暴露研究表明，F-53B暴露胚胎期20d时发现胚胎心率显著低于对照组，并且孵化出的高暴露小鸡的肝脏发生明显的肿大（增大8%）；虽然卵黄暴露F53-B不会影响胚胎的存活率以及孵化出小鸡的体重或氧化应激参数，但以上结果表明与PFOS类似，F-53B在环境相关浓度暴露的情况下会产生潜在的发育毒性。Sheng等[132]以小鼠为模型动物利用口服灌胃的方式考察PFOA替代品HFPO-TA的肝脏毒性，结果表明持续口服灌胃28d浓度为0.02 mg/(kg·d)、0.1 mg/(kg·d)、0.5 mg/(kg·d)HFPO-TA的情况下，会出现诸如肝肿大、肝坏死或氨基酸转氨酶活性增加等肝损伤的现象，并且HFPO-TA的肝脏毒性及潜在的生物富集能力明显强于文献报道的PFOA。以上研究说明，多数目前报道的新型PFASs作为长链PFASs的替代品使用的环境合理性，还需要更多更深入的生物累积、毒性效应方面的数据支撑，相关科学研究任重而道远。

7 展　　望

目前，PFOS和PFOA分别于2009年和2019年被正式增列为《斯德哥尔摩公约》新POPs物质名单。对于相应的物质管理措施和环境风险预防在我国面临比较严峻的挑战。我国需要在PFOS相关化合物信息和PFOA类物质的环境风险评估研究基础上制定相关审议议题的应对策略，制定国家管控对策。首先，摸清我国 PFOS 的生产、使用、贸易和污染情况是开展下一步风险防控的重要基础性工作。其次，我国涉及 PFOS 生产使用的行业较多，且多数在我国经济生活中占有较为重要的地位，应综合考虑社会、经济、技术的现实条件和不同行业的特殊情况，制定切实可行的有针对性的行业风险防控规划。最后，法规标准是对 PFOS 开展长效管理的重要依据，应对 PFOS 所涉及的法规标准进行梳理，将其纳入有关目录下，实现对其生产、使用、进出口、排放及废弃物处理的全过程管理。

参 考 文 献

[1] Lindstrom A B, Strynar M J, Libelo E L. Polyfluorinated compounds: past, present, and future. Environmental Science & Technology, 2011, 45: 7954-7961.

[2] Ahrens L, Bundschuh M. Fate and effects of poly- and perfluoroalkyl substances in the aquatic environment: a review. Environmental Toxicology and Chemistry, 2014, 33(9): 1921-1929.

[3] Cui Q, Pan Y, Zhang H, et al. Elevated concentrations of perfluorohexanesulfonate and other per- and polyfluoroalkyl substances in Baiyangdian Lake (China): Source characterization and exposure assessment. Environmental Pollution, 2018, 241: 684-691.

[4] Ma X, Shan G, Chen M, et al. Riverine inputs and source tracing of perfluoroalkyl substances (PFASs) in Taihu Lake, China. Science of Total Environment, 2018, 612: 18-25.

[5] OECD. Hazard assessment of perfluorooctanesulfonate (PFOS) and its salts. 2002.Unclassified ENV/JM/RD(2002)17/Final. Document No. JT00135607.

[6] UNEP.The new POPs under the Stockholm Convention [EB/O1]. 2009. http://chm.pops.int/TheConvention/ThePOPs/TheNewPOPs/tabid/2511/Default.aspx.

[7] UNEP.POPRC Recommendations for listing Chemicals [EB/O1]. 2015.http://chm.pops.int/Convention/POPsReviewCommittee/Chemicals/tabid/243/Default.aspx.

[8] UNEP. POPRC Recommendations for listing Chemicals [EB/O1].2017. http://chm.pops.int/Convention/POPsReviewCommittee/Chemicals/tabid/243/Default.aspx

[9] Wang Z, DeWitt J, Higgins C, et al. A Never-Ending Story of Per- and Polyfluoroalkyl Substances (PFASs). Environmental science & technology, 2017, 51: 2508-2518.

[10] 国家环保总局. 国家环保总局发布 2008 年第一批"高污染、高环境风险"产品名录 建议取消 39 种产品的出口退税和加工贸易 潘岳呼吁以绿色贸易推动产业调整保护公众健康. 2008. http://www.mee.gov.cn/gkml/sthjbgw/qt/200910/t20091023_180136.htm.

[11] 生态环境部. 关于禁止生产、流通、使用和进出口林丹等持久性有机污染物的公告.2019. http://www.mee.gov.cn/xxgk2018/xxgk/xxgk01/201903/t20190312_695462.html

[12] Zhang Z, Peng H, Wan Y, et al. Isomer-specific trophic transfer of perfluorocarboxylic acids in the marine food web of Liaodong Bay, North China. Environmental Science & Technology, 2015, 49: 1453-61.

[13] Nobuyoshi Y, Sachi T, Gert P, et al. Perfluorinated acids as novel chemical tracers of global circulation of ocean waters. Chemosphere, 2008, 70(7): 1247-1255.

[14] Wang T, Vestergren R, Herzke D, et al. Levels, isomer profiles, and estimated riverine mass discharges of perfluoroalkyl acids and fluorinated alternatives at the mouths of Chinese rivers. Environmental Science & Technology, 2016, 50(21): 11584-11592.

[15] Heydebreck F, Tang J, Xie Z, et al. Alternative and legacy perfluoroalkyl substances: differences between European and Chinese rver/estuary systems. Environmental Science & Technology, 2015, 49(14): 8386-8395.

[16] Annekatrin D, Ingo W, Christian T, et al. Polyfluorinated compounds in the atmosphere of the Atlantic and Southern Oceans:

evidence for a global distribution. Environmental Science & Technology, 2009, 43(17): 6507.

[17] Ellis D A, Martin J W, De Silva A O, et al. Degradation of fluorotelomer alcohols: a likely atmospheric source of perfluorinated carboxylic acids. Environ. Sci. Technol, 2004, 38: 3316-3321.

[18] Yao Y, Sun H, Gan Z, et al. Nationwide distribution of per- and polyfluoroalkyl substances in outdoor dust in mainland hina from eastern to western areas. Environmental Science & Technology, 2016, 50(7): 3676-85.

[19] Haug L S, Sandra H, Martin S, et al. Investigation on per- and polyfluorinated compounds in paired samples of house dust and indoor air from Norwegian homes. Environmental Science & Technology, 2011, 45(19): 7991-8.

[20] Mahiba S, Tom H, Glenys M W, et al. Indoor sources of poly- and perfluorinated compounds (PFCs) in Vancouver, Canada: implications for human exposure. Environmental Science & Technology, 2011, 45(19): 7999-8005.

[21] Jin H, Zhang Y, Zhu L, et al. Isomer profiles of perfluoroalkyl substances in water and soil surrounding a chinese fluorochemical manufacturing park. Environmental Science & Technology, 2015, 49(8): 4946-54.

[22] Routti H, Aars J, Fuglei E, et al. Emission changes dwarf the influence of feeding habits on temporal trends of per- and polyfluoroalkyl substances in two Arctic top predators. Environmental Science & Technology, 2017, 1:51.

[23] Shuhong F, Xinwei C, Shuyan Z, et al. Trophic magnification and isomer fractionation of perfluoroalkyl substances in the food web of Taihu Lake, China. Environmental Science & Technology, 2014, 48(4): 2173.

[24] Chen M, Wang Q, Shan G, et al. Occurrence, partitioning and bioaccumulation of emerging and legacy per- and polyfluoroalkyl substances in Taihu Lake, China. Science of the Total Environment, 2018, 634(25):1-9.

[25] Cui Q, Pan Y, Zhang H, et al. Occurrence and tissue distribution of novel perfluoroether carboxylic and sulfonic acids and legacy per/polyfluoroalkyl substances in black-spotted frog (Pelophylax nigromaculatus) [J]. Environmental Science & Technology, 2018, 52: 982-990.

[26] 胡哲. 不同生物介质中全氟化合物的分配特征和富集规律. 武汉: 华中农业大学博士学位论文, 2016

[27] 赵立杰, 周萌, 任新豪, 等. 全氟辛烷磺酸和全氟辛烷羧酸在天津大黄堡湿地地区鱼体和蔬菜中的分布研究. 农业环境科学学报, 2012, 31(12): 2321-7.

[28] Tian Y, Yao Y, Chang S, et al. Occurrence and phase distribution of neutral and ionizable per- and polyfluoroalkyl substances (PFASs) in the atmosphere and plant leaves around landfills: A case study in Tianjin, China. Environmental Science & Technology, 2018, 52(3): 1301-10.

[29] Benskin J P, Phillips V, St. Louis V L, et al. Source elucidation of perfluorinated carboxylic acids in remote alpine lake sediment cores. Environmental Science & Technology, 2011, 45(17): 7188-7194.

[30] Wang X P, Schuster J, Jones K C, et al. Occurrence and spatial distribution of neutral perfluoroalkyl substances and cyclic volatile methylsiloxanes in the atmosphere of the Tibetan Plateau. Atmospheric Chemistry and Physics, 2018, 18(12): 8745-8755.

[31] Armitage J, Cousins I T, Buck R C, et al. Modeling global-scale fate and transport of perfluorooctanoate emitted from direct sources. Environmental Science & Technology, 2006, 40: 6969-6975.

[32] Gomis M I, Wang Z, Scheringer M, et al. A modeling assessment of the physicochemical properties and environmental fate of emergingand novel per- and polyfluoroalkyl substances. Science of the Total Environment, 2015, 505: 981-991.

[33] Margot R A, Urs B A, Dag B A, et al. Water-to-air transfer of perfluorinated carboxylates and sulfonates in a sea spray simulator. Environmental Chemistry Letters, 2011, 8(4):381-388.

[34] Bengtson Nash S, Rintoul S R, Kawaguchi S, et al. Perfluorinated compounds in the Antarctic region: Ocean circulation provides prolonged protection from distant sources. Environmental Pollution, 2010, 158: 2985-2991.

[35] Wallington T J, Hurley M D, Xia J, et al. Formation of $C_7F_{15}COOH$ (PFOA) and other perfluorocarboxylic acids during the atmospheric oxidation of 8:2 fluorotelomer alcohol. Environmental Science & Technology, 2006, 40(3): 924-930.

[36] Yu X, Nishimura F, Hidaka T. Effects of microbial activity on perfluorinated carboxylic acids (PFCAs) generation during aerobic biotransformation of fluorotelomer alcohols in activated sludge. Science of the Total Environment, 2017, 610-611:776.

[37] Rhoads K R, Janssen E M L, Luthy R G, et al. Aerobic biotransformation and fate of N-ethyl perfluorooctane sulfonamidoethanol (N-EtFOSE) in activated sludge. Environmental Science & Technology, 2008, 42(8):2873-2878.

[38] Butt C M, Muir D C G, Mabury S A. Biotransformation pathways of fluorotelomer-based polyfluoroalkyl substances: A review. Environmental Toxicology and Chemistry, 2014, 33(2):243-267.

[39] Zhao S Y, Zhou T, Zhu L Y, et al. Uptake, translocation and biotransformation of N-ethyl perfluorooctanesulfonamide (N-EtFOSA) by hydroponically grown plants. Environmental Pollution, 2018, 235:404-410.

[40] Zhao S Y, Zhou T, Wang B H, et al. Different biotransformation behaviors of perfluorooctane sulfonamide in wheat (Triticum aestivum L.) from earthworms (Eisenia fetida). Journal of Hazardous Materials, 2018, 346:191.

[41] Zhao S Y, Zhu L Y. Uptake and metabolism of 10:2 fluorotelomer alcohol in soil-earthworm (Eisenia fetida) and soil-wheat (Triticum aestivum L.) systems. Environmental Pollution, 2017, 220:124-131.

[42] D'Eon J C, Mabury S A. Production of perfluorinated carboxylic Acids (PFCAs) from the biotransformation of polyfluoroalkyl phosphate surfactants (PAPS): Exploring routes of human contamination. Environmental Science & Technology, 2007, 41(13): 4799-4805.

[43] Jackson D A, Mabury S A. Enzymatic kinetic parameters for polyfluorinated alkyl phosphate hydrolysis by alkaline phosphatase. Environmental Toxicology and Chemistry, 2012, 31(9): 1966-1971.

[44] Lee H, Tevlin A G, Mabury S A, et al. Fate of polyfluoroalkyl phosphate diesters and their metabolites in biosolids-applied soil: biodegradation and plant uptake in greenhouse and field experiments. Environmental Science & Technology, 2014, 48(1): 340-349.

[45] Bizkarguenaga E, Zabaleta I, Prieto A, et al. Uptake of 8:2 perfluoroalkyl phosphate diester and its degradation products by carrot and lettuce from compost-amended soil. Chemosphere, 2016, 152: 309-317.

[46] Lee H, D'Eon, Jessica, Mabury S A. Biodegradation of polyfluoroalkyl phosphates as a source of perfluorinated acids to the environment. Environmental Science & Technology, 2010, 44(9): 3305-3310.

[47] Lee H, De Silva A O, Mabury S A. Dietary bioaccumulation of perfluorophosphonates and perfluorophosphinates in juvenile rainbow trout: Evidence of Metabolism of Perfluorophosphinates. Environmental Science & Technology, 2012, 46(6): 3489-3497.

[48] Chen X W, Zhu L Y, Pan X Y, et al. Isomeric specific partitioning behaviors of perfluoroalkyl substances in water dissolved phase, suspended particulate matters and sediments in Liao River Basin and Taihu Lake, China. Water Research, 2015, 80:235-244.

[49] Shan G Q, Wang Z, Zhou L Y, et al. Impacts of daily intakes on the isomeric profiles of perfluoroalkyl substances (PFASs) in human serum. Environment International, 2016, 89-90: 62-70.

[50] Zhang Y F, Beesoon S, Zhu L Y, et al. Biomonitoring of perfluoroalkyl acids in human urine and estimates of biological half-life. Environmental Science & Technology, 2013, 47(18): 10619-10627.

[51] Johansson J H, Berger U, Vestergren R, et al. Temporal trends (1999-2010) of perfluoroalkyl acids in commonly consumed food items. Environmental Pollution, 2014, 188: 102-108.

[52] Beesoon S, Martin J W. Isomer-specific binding affinity of perfluorooctanesulfonate (PFOS) and perfluorooctanoate (PFOA) to serum proteins. Environmental Science & Technology, 2015, 49(9): 5722-5731.

[53] Thomsen C, Haug L S, Stigum H, et al. Changes in concentrations of perfluorinated compounds, polybrominated diphenyl ethers, and polychlorinated biphenyls in norwegian breast-milk during twelve months of lactation. Environmental Science &

Technology, 2010, 44(24): 9550-9556.

[54] Wang L, Wang Y, Liang Y, et al. Specific accumulation of lipid droplets in hepatocyte nuclei of PFOA-exposed BALB/c mice. Scientific Reports, 2013, 3(6142): 2174.

[55] Wang L, Wang Y, Liang Y, et al. PFOS induced lipid metabolism disturbances in BALB/c mice through inhibition of low-density lipoproteins excretion. Scientific Reports, 2014, 4(1): 4582.

[56] Fabrega F, Kumar V, Schuhmacher M, et al. PBPK modeling for PFOS and PFOA: Validation with human experimental data. Toxicology Letters, 2014, 230(2): 244-251.

[57] Liu H, Wang J S, Sheng N, et al. Acot1 is a sensitive indicator for PPARα activation after perfluorooctanoic acid exposure in primary hepatocytes of Sprague-Dawley rats. Toxicology in Vitro, 2017, 42: 299-307.

[58] Ishibashi H, Hirano M, Kim E Y, et al. In vitro and in silico evaluations of binding affinities of perfluoroalkyl substances to baikal seal and human peroxisome proliferator-activated receptor α. Environmental Science & Technology, 2019, 53:2181-2188.

[59] Pierozan P, Karlsson O. PFOS induces proliferation, cell-cycle progression, and malignant phenotype in human breast epithelial cells. Arch Toxicol, 2018a, 92: 705-716.

[60] Cao H M, Wang L, Liang Y, et al. Protonation state effects of estrogen receptor α on the recognition mechanisms by perfluorooctanoic acid and perfluorooctane sulfonate: A computational study. Ecotoxicology and Environmental Safety, 2019a, 171: 647.

[61] Pierozan P, Jerneren F, Karlsson O, et al. Perfluorooctanoic acid (PFOA) exposure promotes proliferation, migration and invasion potential in human breast epithelial cells. Arch Toxicol, 2018b, 92: 1729-1739.

[62] Shi G H, Cui Q C, Wang J X, et al. Chronic exposure to 6:2 chlorinated polyfluorinated ether sulfonate acid (F-53B) induced hepatotoxic effects in adult zebrafish and disrupted the PPAR signaling pathway in their offspring. Environmental Pollution, 2019b, 249: 550-559.

[63] Sheng N, Cui R N, Wang J H, et al. Cytotoxicity of novel fluorinated alternatives to long-chain perfluoroalkyl substances to human liver cell line and their binding capacity to human liver fatty acid binding protein. Arch. Toxicol, 2018, 92:359-369.

[64] Li C H, Ren X M, Ruang T, et al. Chlorinated polyfluorinated ether sulfonates exhibit higher activity toward peroxisome proliferator-activated receptors signaling pathways than perfluorooctanesulfonate. Environmental Science & Technology, 2018, 52 (5): 3232-3239.

[65] Zhang H X, Zhou X J, Sheng N, et al. Subchronic hepatotoxicity effects of 6:2 chlorinated polyfluorinated ether sulfonate (6:2 Cl-PFESA), a novel perfluorooctanesulfonate (PFOS) alternative, on adult male mice. Environmental Science & Technology, 2018, 52: 12809-12818.

[66] Xin Y, Ren X M, Ruan T, et al. Chlorinated polyfluoroalkylether sulfonates exhibit similar binding potency and activity to thyroid hormone transport proteins and nuclear receptors as perfluorooctanesulfonate. Environmental Science & Technology, 2018, 52: 9412-9418

[67] Deng M, Wu Y, Xu C, et al. Multiple approaches to assess the effects of F-53B, a Chinese PFOS alternative, on thyroid endocrine disruption at environmentally relevant concentrations. Science of The Total Environment, 2018, 624: 215-224.

[68] Shi G H, Wang J X, Guo H, et al. Prental exposure to 6:2 chlorinated polyfluorinated ether sulfonate (F-53B) induced transgenerational thyroid hormone disruption in zebrafish. Science of The Total Environment, 2019, 665: 855-863.

[69] Ren X M, Qin W P, Cao L Y, et al. Binding interactions of perfluoroalkyl substances with thyroid hormone transport proteins and potential toxicological implications. Toxicology, 2016, 366: 32–42.

[70] Briels N, Ciesielski T M, Herzke D, et al. Developmental toxicity of perfluorooctanesulfonate (PFOS) and its chlorinated polyfluoroalkyl ether sulfonate alternative F-53B in the domestic chicken. Environmental Science & Technology, 2018, 52

(21): 12859–12867.

[71] Zhou X J, Wang J S, Sheng N, et al. Subchronic reproductive effects of 6：2 chlorinated polyfluorinated ether sulfonate (6：2 Cl-PFAES), an alternative to PFOS, on adult male mice. Journal of Hazardous Materials, 2018, 358: 256-264.

[72] Shi G H, Guo H, Sheng N, et al. Two-generational reproductive toxicity assessment of 6:2 chlorinated polyfluorinated ether sulfonate (F-53B, a novel alternative to perfluorooctane sulfonate) in zebrafish. Environmental Pollution, 2018, 243: 1517-1527.

[73] Sheng N, Zhou X J, Zheng F, et al. Comparative hepatotoxicity of 6:2 fluorotelomer carboxylic acid and 6:2 fluorotelomer sulfonic acid, two fluorinated alternatives to long-chain perfluoroalkyl acids, on adult male mice. Arch. Toxicol, 2017, 91: 2909-2919.

[74] Guo H, Wang J H, Yao J Z, et al. Comparative hepatotoxicity of novel PFOA alternatives (perfluoropolyether carboxylic acids) on male mice. Environmental Science & Technology, 2019, 53: 3929-3937.

[75] Sun S J, Guo H, Wang J S, et al. Hepatotoxicity of perfluorooctanoic acid and two emerging alternatives based on a 3D spheroid model. Environmental Environ. Pollution, 2019, 246: 955-962.

[76] Li C H, Ren X M, Guo L H, et al. Adipogenic activity of oligomeric hexafluoropropylene oxide (perfluorooctanoic acid alternative) through peroxisome proliferator-activated receptor γ pathway. Environmental Science & Technology, 2019, 53: 3287–3295.

[77] Chen L G, Lam J C W, Hu C Y, et al. Perfluorobutanesulfonate Exposure Causes Durable and Transgenerational Dysbiosis of Gut Microbiota in Marine Medaka. Environmental Science & Technology Letters, 2018, 5(12): 731–738.

[78] Chen F J, Wei C Y, Chen Q Y et al. Internal concentrations of perfluorobutane sulfonate (PFBS) comparable to those of perfluorooctane sulfonate (PFOS) induce reproductive toxicity in Caenorhabditis elegans. Ecotoxicology and Environmental Safety, 2018, 158: 223-229.

[79] Cao H M, Zhou Z, Wang L, et al. Screening of potential PFOS alternatives to decrease liver bioaccumulation: experimental and computational approaches. Environmental Science & Technology, 2019, 53: 2811-2819.

[80] Fromme H, Tittlemier S A, Völkel W, et al. Perfluorinated compounds – exposure assessment for the general population in western countries. International Journal of Hygiene and Environmental Health, 2009, 212(3): 239-270.

[81] Zhao Y G, Wong C K C, Wong M H. Environmental contamination, human exposure and body loadings of perfluorooctane sulfonate (PFOS), focusing on Asian countries. Chemosphere, 2012, 89(4): 355-368.

[82] Miralles-Marco A, Harrad S. Perfluorooctane sulfonate: A review of human exposure, biomonitoring and the environmental forensics utility of its chirality and isomer distribution. Environment International, 2015, 77: 148-159.

[83] Wang T, Wang P, Meng J, et al. A review of sources, multimedia distribution and health risks of perfluoroalkyl acids (PFAAs) in China. Chemosphere, 2015, 129: 87-99.

[84] Dong Z, Wang H, Yu Y Y, et al. Using 2003–2014 U.S. NHANES data to determine the associations between per- and polyfluoroalkyl substances and cholesterol: Trend and implications. Ecotoxicology and Environmental Safety, 2019, 173: 461-468.

[85] Shi Y, Vestergren R, Xu L, et al. Human exposure and elimination kinetics of chlorinated polyfluoroalkyl ether sulfonic acids (Cl-PFESAs). Environmental Science & Technology, 2016, 50(5): 2396-2404.

[86] Chen F, Yin S, Kelly B C, et al. Isomer-specific transplacental transfer of perfluoroalkyl acids: results from a survey of paired maternal, cord sera, and placentas. Environmental Science & Technology, 2017, 51(10): 5756-5763.

[87] Pan Y, Zhang H, Cui Q, et al. First report on the occurrence and bioaccumulation of hexafluoropropylene oxide trimer acid: an emerging concern. Environmental Science & Technology, 2017, 51(17): 9553-9560.

[88] Fu J, Gao Y, Cui L, et al. Occurrence, temporal trends, and half-lives of perfluoroalkyl acids (PFAAs) in occupational

workers in China. Scientific Reports, 2016, 6: 38039.

[89] Li Y, Fletcher T, Mucs D, et al. Half-lives of PFOS, PFHxS and PFOA after end of exposure to contaminated drinking water. Occupational and Environmental Medicine, 2018, 75(1): 46.

[90] Chen F, Yin S, Kelly B C, et al. Chlorinated polyfluoroalkyl ether sulfonic acids in matched maternal, cord, and placenta samples: a study of transplacental transfer. Environmental Science & Technology, 2017, 51(11): 6387-6394.

[91] Gao Y, Fu J, Cao H, et al. Differential accumulation and elimination behavior of perfluoroalkyl acid isomers in occupational workers in a manufactory in China. Environmental Science & Technology, 2015, 49(11): 6953-6962.

[92] Chen Q, Huang R, Hua L, et al. Prenatal exposure to perfluoroalkyl and polyfluoroalkyl substances and childhood atopic dermatitis: a prospective birth cohort study. Environmental Health, 2018b, 17:8.

[93] Fisher M, Arbuckle TE, Wade M, et al. Do perfluoroalkyl substances affect metabolic function and plasma lipids? -Analysis of the 2007-2009, Canadian Health Measures Survey (CHMS) Cycle 1. Environmental Research, 2013, 121:95-103.

[94] Steenland K, Tinker S, Frisbee S, et al. Association of perfluorooctanoic acid and perfluorooctane sulfonate with serum lipids among adults living near a chemical plant. American Journal of Epidemiology, 2009, 170(10):1268-1278.

[95] Starling A P, Engel S M, Whitworth K W, et al. Perfluoroalkyl substances and lipid concentrations in plasma during pregnancy among women in the Norwegian Mother and Child Cohort Study. Environment International, 2014, 62:104-112.

[96] ATSDR. Toxicological profile for perfluoroalkyls. draft for public comment, June 2018. U.S Department of Health and Human Services, Agency for Toxic Substances and Disease Registry (ATSDR) and the Environmental Protection Agency (EPA), 2018. https://www.atsdr.cdc.gov/toxprofiles/tp.asp?id=1117&tid=237.

[97] Hoffman K, Webster T F, Weisskopf M G, et al. Exposure to polyfluoroalkyl chemicals and attention deficit/hyperactivity disorder in U.S. children 12-15 years of age. Environmental Health Perspectives, 2010, 118(12): 1762-1767.

[98] Gump BB, Wu Q, Dumas AK, et al. Perfluorochemical (PFC) exposure in children: associations with impaired response inhibition. Environmental Science & Technology, 2011, 45(19): 8151-8159.

[99] Cao W, Liu X, Liu X, et al. Perfluoroalkyl substances in umbilical cord serum and gestational and postnatal growth in a Chinese birth cohort. Environment International, 2018, 116: 197-205.

[100] Zhang S, Tan R, Pan R, et al. Association of perfluoroalkyl and polyfluoroalkyl substances with premature ovarian insufficiency in Chinese women. Journal of Clinical Endocrinology & Metabolism, 2018, 103 (7): 2543.

[101] Granum B, Haug L S, Namork E, et al. Pre-natal exposure to perfluoroalkyl substances may be associated with altered vaccine antibody levels and immune-related health outcomes in early childhood. Journal of Immunotoxicology, 2013, 10(4): 373-379.

[102] Grandjean P, Heilmann C, Weihe P, et al. Estimated exposures to perfluorinated compounds in infancy predict attenuated vaccine antibody concentrations at age 5-years. Journal of Immunotoxicology, 2017a, 14(1): 188-195.

[103] Grandjean P, Heilmann C, Weihe P, et al. Serum vaccine antibody concentration in adolescents exposed to perfluorinated compounds. Environmental Health Perspectives, 2017b. https://doi.org/10.1289/EHP275.

[104] McCoy J A, Bangma J T, Reiner J L, et al. Associations between perfluorinated alkyl acids in blood and ovarian follicular fluid and ovarian function in women undergoing assisted reproductive treatment. Science of Total Environment, 2017, 605-606: 9-17.

[105] Jensen T K, Andersen L B, Kyhl H B, et al. Association between perfluorinated compound exposure and miscarriage in danish pregnant women. PLoS One., 2016, 10(4): e0123496. (and correction (2016). PLoS One. 11(2): e0149366).

[106] Velez M P, Arbuckle T E, Frazer W D. Maternal exposure to perfluorinated chemicals and reduced fecundity: the MIREC study. Human Reproduction, 2015, 30(3): 701–709.

[107] Khalil N, Chen A, Lee M, et al. Association of perfluoroalkyl substances, bone mineral density, and osteoporosis in the U.S.

population in NHANES 2009-2010. Environmental Health Perspectives, 2016, 124:81-87; http://dx.doi.org/10.1289/ehp.1307909.

[108] Steenland K, Zhao L, Winquist A. A cohort incidence study of workers exposed to perfluorooctanoic acid (PFOA). Occupational and Environmental Medicine, 2015, 72:373-80.

[109] Stockholm Convention on Persistent Organic Pollutants. Proposal to list perfluorohexane sulfonic acid (CAS No: 355-46-4, PFHxS), its salts and PFHxS-related compounds in annexes A, B and/or C to the Stockholm Convention on Persistent Organic Pollutants. 2017.

[110] Wang Z, Cousins I T, Scheringer M, et al. Fluorinated alternatives to long-chain perfluoroalkyl carboxylic acids (PFCAs), perfluoroalkane sulfonic acids (PFSAs) and their potential precursors. Environment International. 2013, 60: 242-248.

[111] Wang Z, Cousins I T, Scheringer M, et al. Hazard assessment of fluorinated alternatives to long-chain perfluoroalkyl acids (PFAAs) and their precursors: status quo, ongoing challenges and possible solutions. Environment International. 2015, 75:172-179.

[112] Gordon S C. Toxicological evaluation of ammonium 4, 8-dioxa-3H-perfluorononanoate, a new emulsifier to replace ammonium perfluorooctanoate in fluoropolymer manufacturing. Regulatory Toxicology and Pharmacology: RTP 2011, 59 (1): 64-80.

[113] Wang S, Huang J, Yang Y, et al. First report of a Chinese PFOS alternative overlooked for 30 years: its toxicity, persistence, and presence in the environment. Environmental Science & Technology 2013, 47 (18): 10163-10170.

[114] Xu L, Shi Y, Li C, et al. Discovery of a novel polyfluoroalkyl benzenesulfonic acid around oilfields in northern China. Environmental Science & Technology, 2017, 51(24): 14173-14181.

[115] Place B J, Field J A. Identification of novel fluorochemicals in aqueous film-forming foams used by the US military. Environmental Science & Technology, 2012, 46 (13) : 7120-7127.

[116] Washington, J. W.; Jenkins, T. M.; Weber, E. J. Identification of unsaturated and 2H polyfluorocarboxylate homologous series and their detection in environmental samples and as polymer degradation products. Environmental Science & Technology, 2015, 49 (22) :13256-13263.

[117] Singh R K, Fernando S, Fakouri Baygi S, et al. Breakdown products from perfluorinated alkyl substances (PFAS) degradation in a plasma-based water treatment process. Environmental Science & Technology, 2019, 53 (5) : 2731-2738.

[118] Lin Y, Ruan T, Liu A, et al. Identification of Novel Hydrogen-Substituted Polyfluoroalkyl Ether Sulfonates in Environmental Matrices near Metal-Plating Facilities. Environmental Science & Technology, 2017, 51 (20) : 11588-11596.

[119] Wang Y, Vestergren R, Shi Y, et al. Identification, tissue distribution, and bioaccumulation potential of cyclic perfluorinated sulfonic acids isomers in an airport impacted ecosystem. Environmental Science & Technology, 2016, 50 (20) : 10923-10932.

[120] Barzen-Hanson K A, Roberts S C, Choyke S, et al. Discovery of 40 classes of per- and polyfluoroalkyl substances in historical aqueous film-forming foams (AFFFs) and AFFF-impacted groundwater. Environmental Science & Technology, 2017, 51 (4) : 2047-2057.

[121] Barzen-Hanson K A, Field J A. Discovery and implications of C2 and C3 perfluoroalkyl sulfonates in aqueous film forming foams (AFFF) and groundwater. Environmental Science & Technology, 2015, 2(4) : 95-99

[122] Liu Y, Pereira Ados S, Martin J W. Discovery of C5-C17 poly- and perfluoroalkyl substances in water by in-line SPE-HPLC-Orbitrap with in-source fragmentation flagging. Analytical Chemistry, 2015, 87 (8) : 4260-4268.

[123] Liu Y, Qian M, Ma X, et al. Nontarget mass spectrometry reveals new perfluoroalkyl substances in fish from the yangtze river and Tangxun lake, China. Environmental Science & Technology, 2018, 52(10): 5830-5840.

[124] Ruan T, Lin Y, Wang T, et al. Identification of novel polyfluorinated ether sulfonates as PFOS alternatives in municipal

sewage sludge in China. Environmental Science & Technology, 2015, 49 (11): 6519-6527.

[125] Chen H, Choi Y J, et al. Sorption, aerobic biodegradation, and oxidation potential of PFOS alternatives chlorinated polyfluoroalkyl ether sulfonic acids. Environmental Science & Technology, 2018, 52 (17): 9827-9834.

[126] Gebbink W A, Bossi R, Rigét F F, et al. Observation of emerging per- and polyfluoroalkyl substances (PFASs) in Greenland marine mammals. Chemosphere, 2016, 144: 2384-2391.

[127] Pan Y, Zhang H, Cui Q, et al. Worldwide distribution of novel perfluoroether carboxylic and sulfonic acids in surface water. Environmental Science & Technology, 2018, 52 (14): 7621-7629.

[128] Zhou Z, Liang Y, Shi Y, et al. Occurrence and transport of perfluoroalkyl acids (PFAAs), including short-chain PFAAs in Tangxun Lake, China. Environmental Science & Technology, 2013, 47 (16): 9249-9257.

[129] Shi Y, Vestergren R, Zhou Z, et al. Tissue distribution and whole body burden of the chlorinated polyfluoroalkyl ether sulfonic acid F-53B in crucian carp (Carassius carassius): evidence for a highly bioaccumulative contaminant of emerging concern. Environmental Science & Technology, 2015, 49 (24): 14156-14165.

[130] Liu Y, Ruan T, Lin Y, et al. Chlorinated polyfluoroalkyl ether sulfonic acids in marine organisms from bohai sea, China: occurrence, temporal variations, and trophic transfer behavior. Environmental Science & Technology, 2017, 51(8): 4407-4414.

[131] Yin N, Yang R, Liang S, et al. Evaluation of the early developmental neural toxicity of F-53B, as compared to PFOS, with an in vitro mouse stem cell differentiation model. Chemosphere, 2018, 204: 109-118.

[132] Sheng N, Pan Y, Guo Y, et al. Hepatotoxic effects of hexafluoropropylene oxide trimer acid (HFPO-TA), a novel perfluorooctanoic acid (PFOA) alternative, on mice. Environmental Science & Technology, 2018, 52 (14): 8005-8015.

作者：王亚韡[1]，祝凌燕[2]，梁 勇[3]，蔡亚岐[1]
[1]中国科学院生态环境研究中心，[2]南开大学，[3]江汉大学

第 6 章　硅氧烷类污染物的环境赋存与转化

- 1. 引言 /132
- 2. 硅氧烷简介 /132
- 3. 硅氧烷的环境行为 /135
- 4. 人群暴露 /145
- 5. 改性硅氧烷环境行为的初步研究 /147
- 6. 展望 /149

本章导读

硅氧烷因其优良的性能，常作为润滑剂、疏水剂、助溶剂、消泡剂等被广泛应用于生活和工业领域。已有报道指出硅氧烷，特别是甲基硅氧烷，会对动物的神经、免疫和生殖系统产生毒害作用，具有致癌和致突变性。特别指出的是，鉴于甲基硅氧烷在全球的巨大产能和消费量，以及潜在的生态和人体健康影响，其环境问题已经成为研究热点。

为更好地了解硅氧烷的环境行为及人群暴露特性，本章将从以下三个方面进行较全面的综述：①甲基硅氧烷及其代谢产物在生活区域大气、水、土壤、沉积物等介质中的来源与归趋，如个人护理品的使用、生活污水的排放、固体废物的填埋处理等；基于两个工业案例（石油开采和造纸工业），介绍甲基硅氧烷氯转化产物的生成机制及其在污水处理过程和土壤中的归趋。②结合工厂周边空气、灰尘/土壤、工人血液中甲基硅氧烷残留水平，系统总结硅氧烷生产工厂与使用工厂的人群暴露的相关研究成果。③初步阐明多种改性硅氧烷的环境归趋与潜在的生态毒理效应。

关键词

甲基/改性硅氧烷，污染来源，迁移转化，人群暴露

1　引　言

由于甲基硅氧烷在众多领域得到广泛应用，并且其化学性质较为稳定，在人类生活和生产建设领域中可通过各种迁移途径进入大气、水、土壤等环境基质，形成持久存在。随着相关毒理学研究的开展，已有报道指出甲基硅氧烷能对动物的神经、免疫和生殖系统产生毒害作用，具有致癌和致突变性。加拿大环境保护部认为进入环境的甲基硅氧烷可能对人体健康或生物多样性产生急性或长期的危害，建议采取行政措施以尽可能减少或阻止其进入环境。鉴于甲基硅氧烷在全球的巨大产能和消费量，以及潜在的生态和人体健康影响，甲基硅氧烷的环境问题已经成为研究热点。

近二十年来，全球硅氧烷生产/使用规模迅速扩大，生产/使用过程中产生大量含硅氧烷的"三废"，其中低分子量环型硅氧烷（cyclic volatile methylsiloxane, cVMS）具有高挥发性，极易在生产流程中发生泄露排放，而一般的"三废"处理工艺并不能对硅氧烷进行有效的处理，因此硅氧烷残留很可能排入周边环境，并形成局部的硅氧烷重污染区域。因此，在硅氧烷典型污染区域，特别是硅氧烷生产及使用工业地区开展硅氧烷污染水平、迁移、转化等研究具有重要意义。另外，硅氧烷在生产、使用、排放过程中可通过多种暴露途径进入人体及其他生物体的体内循环系统，因此研究人体和生物体中该类化合物的负荷水平、富集/衰减速度、健康及生态效应也具有重要价值。

2　硅氧烷简介

2.1　物化性质与应用

硅是存在于自然界中最为广泛的元素之一，其含量约占地壳总质量的25.7%，仅次于氧（49.4%）。

在一般自然环境下，硅通常与氧结合以二氧化硅或硅酸盐等无机物的形式存在于尘土、沙砾、岩石之中，而有机硅化合物（organosilicone）则需要通过化学手段人工合成。

有机硅通常指含有 Si—O 键且硅原子直接与有机基团相连的化合物。有机硅兼具无机材料与有机材料的性能，具有表面张力低、黏度系数小、压缩性高、气体渗透性高等基本性质，也具有耐高温/低温、电气绝缘、耐氧化稳定性、耐候性、难燃、憎水、耐腐蚀等优异特性，广泛应用于航空航天、电子电气、建筑、运输、化工、纺织、食品、轻工、医疗等行业，它们大多应用于密封、黏合、润滑、涂层、表面活性、脱模、消泡、抑泡、防水、防潮、惰性填充等[1]。其中，以硅氧键（—Si—O—Si—）为骨架组成的硅氧烷类化合物，是有机硅化合物中数量最多，研究最深、应用最广的一类，约占有机硅化合物总用量的90%以上[2]。

支链全部由甲基构成的硅氧烷化合物称为甲基硅氧烷（methylsiloxane），按照主链结构，其可分为环型（以字母 D 命名）和线型（以字母 L 命名）两种，如图 6-1 所示。根据二甲基二氯硅烷单体（甲基硅氧烷生产原料）的产量换算，甲基硅氧烷的产量约占全球所有硅氧烷类化合物的80%[2]。甲基硅氧烷 Si—O 键能大于一般有机物中的 C—C 键能和 C—O 键能，分子间作用力小，并且甲基侧链具有高表面活性。因此，甲基硅氧烷能够适应各种形态变化，具有热稳定性、疏水性、润滑性等多种优良性能，在工业生产、日化及个人护理用品中作为原料、中间体、润滑剂、疏水剂、助溶剂、消泡剂等得到广泛应用[3-5]。据统计，我国目前硅氧烷主要用于电子产业（36%）和建筑业（25%），其次依次是纺织业（10%）、工业助剂（8%）、个人护理品（6%）[6]。自 20 世纪 80 年代后期开始，全球范围内的甲基硅氧烷的生产及应用得到迅速发展。八甲基环四硅氧烷（D4）、十甲基环五硅氧烷（D5）、十二甲基环六硅氧烷（D6）属于环型挥发性甲基硅氧烷，是生产 PDMS 的主要原料。目前全世界环硅氧烷单体的年产量约为 100 多万 t。美国八甲基环四硅氧烷（D4）和十甲基环五硅氧烷（D5）的年产量均达到 20 万 t，十二甲基环六硅氧烷（D6）的年产量为 2 万 t[7]。我国目前进入硅氧烷产业高速发展时期，2009 年报道我国的环硅氧烷总产量为 10 万 t，然而由于各地大批新建设的硅氧烷生产线相继投入运行，2017 年的产量迅速上升到 102 万 t，约占全球总产量的 50%[8]。

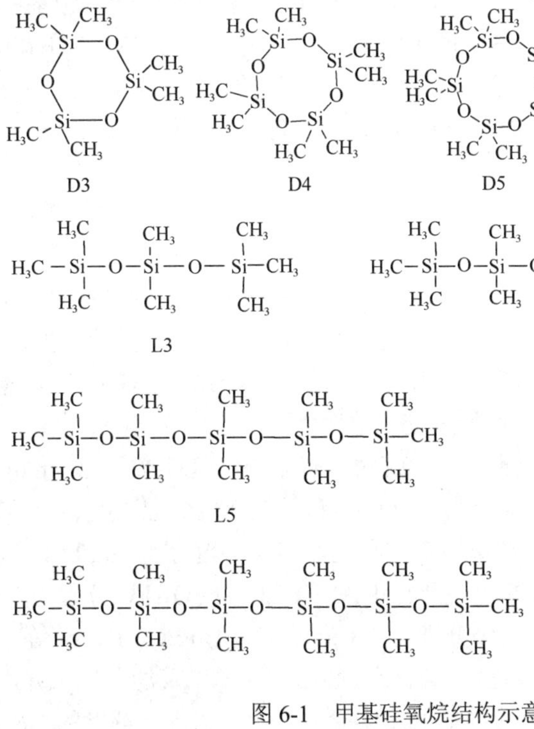

图 6-1 甲基硅氧烷结构示意图

2.2 生态环境效应

由于使用量巨大且应用范围较广，近年来甲基硅氧烷的生物累积及毒性效应已引起环境科学研究者的重视。近年来，甲基硅氧烷在水环境中的生物累积以及潜在的生态风险逐渐引起学术界的广泛关注。

2.2.1 生物富集/累积

Kierkegaard 等[9]报道了英国亨伯河口的沙蚕（hediste diversicolor）体内甲基硅氧烷 D4、D5、D6 的含量分别为 20 ng/g ww（湿重）、762 ng/g ww、27 ng/g ww。研究的比目鱼肌肉中环型硅氧烷 D4、D5、D6 的最高浓度分别为 10.4 ng/g ww、299 ng/g ww、4.7 ng/g ww。Kaj 等[10]报道了北欧地区海洋鱼类的肝脏中均有环型硅氧烷检出，其浓度范围为 5~100 ng/gww，而挪威地区采集的鳕鱼肝脏样品中 D5 的最高浓度达到了 2200 ng/g ww。

Warner 等[11]报道了北极区域鳕鱼和杜父鱼体样品中 D5 的平均浓度分别为 176 ng/g lw（脂重）和 531 ng/g lw。Kierkegaard[12]研究了在波罗的海海域中鲱鱼和灰海豹体内环型硅氧烷 D4、D5、D6 的含量，其中鲱鱼体内含量分别为 10 ng/g lw、200 ng/g lw、40 ng/g lw，灰海豹体内只有 D5 和 D6 检出，浓度范围为 9~24 ng/g lw 和 4.4~9.5 ng/g lw。在北欧地区的海洋哺乳动物中，海豹和巨头鲸体内的环型硅氧烷 D4、D5、D6 的最高含量分别为 12 ng/g ww、24 ng/g ww 和 7.9 ng/g ww。

生物浓缩因子（bioconcentration factor，BCF）是评价化合物在环境介质与生物体之间分配的重要参数。根据美国环保署（EPA）规定，当 BCF 值在 1000~5000，说明化合物具有潜在的生物富集性，当 BCF 大于 5000 时，则存在明显的生物富集性[13]。而目前，对于甲基硅氧烷的生物富集性和生物放大性，结论仍存在争议。之前的文献报道了 D4 和 D6 在鱼体内的 BCF 值分别为 12400 L/kg 和 1160 L/kg，而 D5 在鱼体内的 BCF 值变化较大，在 3362~13300 L/kg[14-17]。由于 BCF 并未涉及食物暴露，因此，为更准确地描述污染物的生物放大效应，引入营养级放大因子（TMF）来描述污染物在食物链和食物网间的传递。通常，TMF>1 时，表明化合物具有生物放大效应，TMF<1 时，则表明存在营养级稀释作用。Borgå 等[18]研究了 D4、D5 和 D6 在挪威 Mjøsa 湖食物网中的生物放大性，并与 PCB153、PCB180 以及一些 BDE 同系物比较，发现 D5 的 TMF 值明显大于 1，这表明 D5 具有明显的生物放大效应。而 Powell 等[19]的研究发现，在日本东京湾海洋食物网中，D4、D5 和 D6 则存在营养级稀释效应。

2.2.2 生物毒性

部分研究表明，甲基硅氧烷对动物的某些生理过程存在直接或间接的毒性作用，如一些环型甲基硅氧烷对免疫系统、呼吸系统以及肝脏系统存在不利影响[20-22]。体外试验表明高剂量的 D4 和 D5 可导致肝脏质量增大、卵巢质量减小和子宫肿大；小鼠注射一定浓度 D4 后表现出严重的肺损伤[23]；口腔暴露实验表明 D4 可通过与 α 受体结合产生雌激素效应[24, 25]，D5 则影响神经系统。将小鼠暴露在含有 D5 的空气中，在最高暴露浓度时，小鼠子宫癌的发病率显著上升[26]。

低浓度 D4 就会对部分水生生物产生危害，如对虹鳟鱼有影响，但对蚊虫类（midges）则无影响。对于小型虹鳟鱼，急性毒性实验中 D4 的半数致死浓度（LC_{50}）和最低不良效应浓度（LOAEC）分别为 10 μg/L 和 6.9 μg/L，而最大无影响浓度（NOEC）在急性实验和慢性实验中均为 4.4 μg/L[27, 28]。21 d 大型溞的暴露研究发现，当 D4 浓度达到 10 μg/L 时，开始出现死亡现象，NOEC 和 LOAEC 分别为 7.9 μg/L 和 15 μg/L。在急性毒性实验中[29]，红鲈鱼和糠虾的 NOEC 分别为 6.3 μg/L 和 9.1 μg/L。Parrot 等[30]在不同浓度的 D5 中进行黑头呆鱼的暴露实验，结果表明 D5 对鱼卵孵化率、胚胎存活率以及幼鱼生长并没有产生明显影响，其 NOEC 值为 8.7 g/kg。黑头呆鱼在 D6 溶液中的暴露实验也表明，接近溶解度的 D6 对鱼类也无明显不利现象。上述数据表明，对于一些较敏感的水生生物来说，D4 表现出比较明显的毒性效应，并且已经达到了欧盟制定的淡水及海洋水生生物的无效应浓度标准（小于 10 μg/L）。

也有部分研究关注甲基硅氧烷对沉积物及土壤的生态效应。Krueger 等[31]在一项沉积物毒性研究中发现，将蚊虫暴露于 D4 浓度为 355 μg/L 的沉积物中时，其数量明显减少，沉积物中 D4 的 NOEC 和 LOAEC 分别为 131 μg/L 和 355 μg/L。在慢性毒性实验中，以蚊虫为研究对象，D5 的 NOEC 和 LOAEC 分别为 69 μg/L 和 180 μg/L。在 Nordood 等[32]关于沉积物中底栖无脊椎动物端足虫的 D5 慢性毒性实验中发现，在有机碳含量分别为 0.5%和 11%的沉积物中，D5 的 LC_{50} 分别为 191 μg/g 和 857 μg/g。Velicogna 等[33]测试了土壤中 D5 的生态毒性，发现 D5 对植物大麦和陆生动物跳虫具有显著影响，其半数抑制浓度（IC_{50}）分别为 767 mg/kg 和 209 mg/kg。

2.2.3 环境风险评估与管理

虽然相关研究表明环型甲基硅氧烷对水生生物有明显的生物积累性，但由于目前实验和模拟研究得到的结论有矛盾，故不能断定其是否符合"持久性和生物积累性规定"中的标准。但有足够的证据表明，环型甲基硅氧烷可能具有对环境或生态多方面功能的有害作用。加拿大政府正在评估关于 D4、D5 生物积累性的资料，如果新的资料表明 D4、D5 具有生物积累性，则根据该国环保法 77（4）的规定，需要实质避免它们被释放到环境中去。

2009 年《加拿大政府公报》先后将 10 个有机硅化合物列为"有生物积累性、对非人类生物有毒"的化学物质。其中线型十甲基四硅氧烷和线型八甲基三硅氧是典型的低分子量二甲基硅氧烷，是某些低黏度甲基硅油的成分。这两种硅氧烷对鱼类的生物积累性（BCF、BAF）的数据都大于 5000；急、慢性毒理实验的数据（EC_{50}、LC_{50}）都小于 1.0 mg/L。目前这些化合物正处于征求意见阶段。

加拿大环境保护部认为进入环境的 D4 和 D5 的量/浓度可能对环境安全或生物多样性产生立即或长期的危害，属于典型的持久性有机污染物（persistent organic pollutants，POPs），建议对两者采用"生命周期法"进行管理，尽可能减少或阻止其进入环境。2009 年 1 月，加拿大环境保护部认为浓度相当低的环型硅氧烷 D4（小虹鳟鱼 LC_{50}，0.01 mg/L）和 D5 也能对敏感的深海水生生物有极高的短期和长期毒性。2010 年《加拿大政府公报》将线型硅氧烷 L3、L4 列入"有生物积累性、对非人类生物有毒的化学物质"。

2010 年 3 月，美国环保署宣布为了"化学品行动计划"，要将壬基酚/壬基酚乙氧基化物、六溴环十二烷、硅氧烷和二异氰酸酯等四类化学品列入下一批对象，要求有关公司向 EPA 提供证明该种化学品不具有不合理风险的信息。说明 EPA 有可能采取与加拿大环境保护部类似的做法[34-37]。2017 年，欧盟化学品管理局认定 D4 为 PBT；D5 和 D6 为 vPvB。列入 REACH 附录进行管控。2019 年，中国生态环境部开展"关于 D4、D5、D6 的环境风险及国内外行业和管控情况研究子项目"。

3 硅氧烷的环境行为

甲基硅氧烷在其生产和使用过程中可以通过各种途径迁移到环境中。近年来，随着分析方法的发展，作为一种新型污染物，甲基硅氧烷的环境存在已引起许多研究者和有关机构的重视。在全球范围内，甲基硅氧烷在不同介质中的浓度水平相继被报道。

3.1 生活区域

3.1.1 大气

甲基硅氧烷由于挥发性较强且在水中的溶解度较低，在相平衡条件下，这类化合物趋于在大气环境中分配。个人护理用品中，约有90%的硅氧烷直接挥发进入大气。据统计，全球日常用品（包括化妆品、护肤品等）大约平均每年向大气释放出 50~200t 的硅氧烷气体。文献报道，美国芝加哥家庭室内 cVMS（D4、D5、D6）总浓度的中位值为 2.2 μg/m³[38]；瑞典居民家庭室内空气样品中 D4、D5、D6 的浓度高达 100μg/m³[39]；英国和意大利两国居民家庭、学校等场所室内大气样品（n=91）挥发性甲基硅氧烷（D4~D6）总浓度最高分别为 820 μg/m³、940 μg/m³[40]。挥发到室内大气中的 cVMS 可吸附于大气颗粒中，然后通过干/湿沉降进入灰尘中。Tran 等报道美国、中国、日本、德国、印度等 12 个国家的室内灰尘样品中均可检测出 cVMS，总浓度为 33.5~42800 ng/g[41]。

Navea 等[42]研究发现，对流层中的 D4 和 D5 在气溶胶表面可以发生异相反应，这可能是大气中环型甲基硅氧烷消除的主要途径。在大气环境中，硅氧烷会与•NO₃、O₃ 和•OH 发生反应。其中，与•OH 发生反应的速度最快，脱去甲基而生成硅醇是大气中环型甲基硅氧烷的主要消除机制，该过程中每个甲基被•OH 替代的概率相同[43, 44]。环型甲基硅氧烷与•OH 反应的主要产物主要为硅醇类，其饱和蒸气压较低且溶解度较高，可通过湿沉降从大气中去除。在实验室模拟条件下，大气中 D4、D5 以及 D6 的一级动力学半衰期分别为 10.6 d，6.9 d 和 6.0 d；基于野外监测数据，通过模型拟合估算甲基硅氧烷在实际大气中的半衰期约为 10~30 d[45-47]。

虽然甲基硅氧烷在大气半衰期较短，但各国广泛而持续的使用和排放造成其在大气中具有"假持久性"特征，可随大气发生远距离迁移至人类活动影响较少的地区。Krogseth 等关于北极地区大气样品的研究中检出了环型甲基硅氧烷证实了这一推断，该地区夏季 D4、D5 平均浓度分别为 0.73 ng/m³、0.23 ng/m³，冬季平均浓度分别为 2.94 ng/m³、0.45 ng/m³[48]。大气中甲基硅氧烷浓度的季节性差异可能与大气羟基自由基浓度相关。Mclachlan 等[49]发现，夏季大气中•OH 在大气中的浓度通常高于冬季，这使得 D5 的浓度由 2~8 ng/m³ 降至 0.3~4 ng/m³。

3.1.2 水环境

除了通过挥发进入大气，日化及个人护理品中的小部分（10%）cVMS 可以随城市污水排放进入污水处理厂[45-47]。从生活污水中去除挥发性硅氧烷（cVMs）残留的第一步主要是在污水处理厂中进行。因此，研究挥发性甲基硅氧烷在污水处理厂（WWTP）的一系列迁移转化行为对于了解甲基硅氧烷的归趋具有重要意义。由于具有较高的有机碳-水分配系数（$\log K_{oc}$ = 4.2~5.86）且生物可降解性较差[50, 51]，cVMS 在污水处理系统中主要累积在活性污泥表面，进而随剩余污泥一起处置。我国松花江沿岸的 8 个污水处理厂的活性污泥中均检出了 cVMS，总浓度范围为 0.602~2.36 μg/g[52]；希腊 Athens 地区的一个污水处理厂污泥中 cVMS 总浓度为 21.1 μg/g[53]。Wang 等[54]研究了加拿大 11 个污水处理厂中环型甲基硅氧烷的浓度水平，其中进水 D4、D5 和 D6 的浓度分别为 0.282~6.69 μg/L、7.75~135 μg/L 和 1.53~26.9 μg/L，而出水浓度仅为 < 0.009~0.045 μg/L、< 0.027~1.56 μg/L 和 < 0.022~0.093，发现此类污染物在污水处理厂中的去除率高达 92%。

在水环境中，甲基硅氧烷由于水溶性较低且挥发性较强，在进入水相后会迅速分配到大气中。因此，在水体和底泥中，甲基硅氧烷的浓度通常较低且变化范围较大。水相中甲基硅氧烷可以被酸或者碱催化进而发生水解，其水解半衰期根据化合物的种类、温度以及 pH 的不同在几小时至 100 天之间变化。在

12℃且pH为7时，D4和D5在淡水中的水解半衰期分别为17d和315d，而在酸性或碱性条件下，其水解半衰期将会显著缩短[55, 56]。水环境中环型甲基硅氧烷的生物降解通常可以被忽略，这是因为这些化合物较低的水溶性限制了其生物可利用性。在污水处理厂中，甲基硅氧烷的主要去除方式为挥发、吸附和生物降解。在污水处理流程中，由于具有较强挥发性，大部分D4和D5可进入沼气中。同时，由于K_{ow}值较高，甲基硅氧烷容易吸附在污泥中，这是甲基硅氧烷在污水处理厂中主要的去除机制。Bletsou等[57]报道了希腊某污水处理厂中甲基硅氧烷的迁移行为，表明进水和出水中D5的比例分别为13%和50%，说明其在水环境中的停留时间较长，而D6在进水中为9%，在出水中则不足1%，这意味着D6在水环境中去除的比例相对较高，这些差异是由于甲基硅氧烷理化性质的不同而导致的。研究表明，污泥对甲基硅氧烷的吸附量随化合物疏水性的增加而升高，挥发量则随分子量的增大而降低[58]。

Xu等[59]比较系统地研究了北京某城市污水处理厂中A^2/O工艺过程中环型甲基硅氧烷的去除机制，结果表明，厌氧池中甲基硅氧烷主要通过生物降解去除，而好氧环境中环型甲基硅氧烷几乎不发生生物降解，在二沉池和好氧池主要通过挥发的方式去除。在该研究中，作者开发了固相微萃取提取以及气相色谱-质谱法联用的方法用于检测污水处理厂中水相和固体污泥样品中四种环型（D3~D6）和两种线型（L3~L4）甲基硅氧烷的浓度水平。在此方法的基础上，研究了污水处理厂不同处理单元的水样和污泥样品中挥发性硅氧烷的浓度，并通过模拟实验，研究了甲基硅氧烷在污水处理过程中的降解过程，系统分析了甲基硅氧烷在北京市某污水处理厂二级流程中的迁移和转化行为。

首先，通过对二级处理过程中活性污泥吸附对目标化合物去除的影响展开研究，在传统的厌氧-缺氧-好氧（A^2/O）工艺中D3、D4、D5和D6由剩余污泥吸附导致的质量损失的相对百分数（两次采样的平均值）分别为8.3%±1.8%、29.4%±2.8%、38.1%±7.2%、53.0%±13.5%，缺氧-厌氧-好氧（倒置A^2/O）工艺中分别为9.8%±1.4%、19.0%±1.3%、32.0%±2.3%、40.2%±6.0%。由于甲基硅氧烷具有比较高的辛醇-水分配系数（K_{ow}），环型甲基硅氧烷比较容易吸附在水环境中的污泥中。于是，作者引入分配系数（PC）来研究甲基硅氧烷在活性污泥的相关吸附特性，计算得到的活性污泥及其上层水中甲基硅氧烷的浓度比值比较大（D3、D4、D5、D6的PC值范围分别为142~1835、136~4869、162~4667、114~6814），这表明甲基硅氧烷在污泥中具有较强的吸附能力。

其次，研究了甲基硅氧烷在二级处理流程中每个单元通过挥发和降解的去除行为。在传统A^2/O过程的厌氧池中，D3、D4、D5、D6的质量损失相对百分数为44.4%~84.3%，而在缺氧池、好氧池、二沉池中的总质量损失相对百分数为6.3%~7.4%。在倒置的A^2/O过程厌氧池中，D3、D4、D5、D6的质量损失相对百分数为45.8%~77.1%，而在缺氧池、好氧池、二沉池中总质量损失相对百分数为12.7%~22.9%，如图6-2所示。结果表明，在传统和倒置的A^2/O过程中，挥发性硅氧烷大部分在厌氧池中消除。

最后，通过模拟实验研究了三种浓度水平下挥发性甲基硅氧烷在厌氧条件下的去除行为。在活性污泥系统中，D3和D6难以降解，培养60h降解质量比例为3.0%~18.1%，因此，可以推断D3和D6在厌氧池内的主要通过挥发作用去除。与D3和D6相比，D4和D5在厌氧环境下更容易降解，60h后降解比例约为44.4%~62.8%。于是，通过固相微萃取结合气相色谱-质谱方法在污泥培养系统检测甲基硅氧烷的水解产物-甲基硅二醇[$Me_2Si(OH)_2$]，通过硅元素的质量换算发现，培养60h后，D4和D5降解转化为甲基硅二醇的比例约为21.4%~30.6%（图6-3），水解过程除了甲基硅二醇，还包括一系列同系物。因此，可以认为微生物催化水解可能是厌氧池中环型硅氧烷降解的一种重要途径。

与日化及个人护理品不同，硅橡胶和密封剂等产品中的cVMS在使用过程中虽然也有一部分进入大气，但是剩余的另一部分极少进入污水而是持续残留在产品当中，进而随着废弃产品一起作为固体垃圾处置[45-47]。目前，垃圾处理主要有填埋、焚烧、堆肥三种方式，其中填埋法最为普及。全世界约有95%的城市垃圾通过填埋的方式处理，中国填埋处理的城市垃圾比例达90%以上[60]。文献报道美国1993年填埋的PDMS产品（70%为硅橡胶）约为12.6万t，欧盟2004年填埋的PDMS产品（89%为密封剂和硅橡

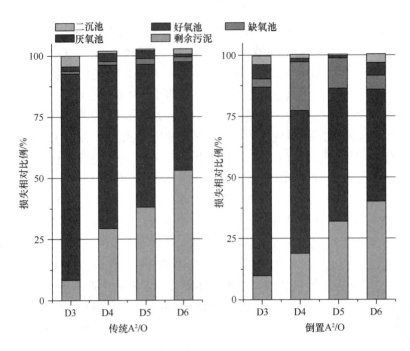

图 6-2 各流程甲基硅氧烷质量损失相对比例

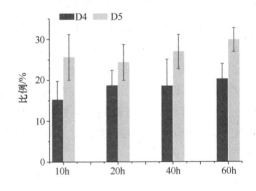

图 6-3 不同时间点降解的甲基硅氧烷转化为甲基硅二醇比例

胶）约为 39.2 万 t[45-47]。根据相关数据[61]，作者粗略估计中国 2015 年填埋的 PDMS 产品（90%为硅橡胶）为 80 万 t 左右。大量废弃的 PDMS 产品进入垃圾填埋场将有可能释放 cVMS 进入生物气和渗滤液，进而迁移到填埋场周边大气、土壤、水等环境介质中。Schweigkofler 和 Niessner 报道，德国两个垃圾填埋场生物气中 D4 和 D5 浓度范围分别为 4.2～8.8 mg/m^3 和 0.4～1.1 mg/m^3[62]。Paxéus 报道瑞典三个垃圾填埋场渗滤液中 D4、D5 浓度分别为 1～2 μg/L、0.1～0.4μg/L[63]。上述文献虽然报道了垃圾填埋场生物气和渗滤液中 cVMS 残留水平，然而由于采样周期短或者样品数量有限等原因，垃圾填埋场生物气和渗滤液中 cVMS 的长时间跨度（≥1 年）排放通量及其时间（或季节）变化规律和影响因素以及 cVMS 在填埋场周边环境中的环境行为和生态健康效应等问题并没有获得深入研究。另外，垃圾渗滤液在排放到环境水体之前需要经过污水处理流程。大量研究已表明传统的污水生物处理工艺不能有效降解 cVMS。然而，除了生物处理工艺之外，垃圾渗滤液处理过程中经常附加应用深度处理技术，通常是产生羟基自由基（·OH）的高级氧化技术。鉴于大气中·OH 可使 cVMS 发生羟基化反应[45-47]，高级氧化技术处理垃圾渗滤液过程中 cVMS 有可能发生羟基化反应。cVMS 羟基化产物的物理化学性能与母体不同，可能表现出与母体 cVMS 不同的环境行为和生态健康效应，因此研究垃圾渗滤液中 cVMS 羟基化产物在填埋场周边环境介质（特别是纳污水体）中的归趋也具有重要现实意义。

Xu 等[64]对中国某城市垃圾填埋场进行了全年采样，并对 cVMS 在垃圾填埋气和渗滤液中的赋存情况及其在渗滤液储存池中的去除机制进行系统研究。在垃圾填埋气样品中，D4 和 D5 的浓度为 0.105～2.33 mg/m³，而在渗滤液储存池进水、出水中，D4～D6 的浓度为<LOQ～30.5μg/L。在垃圾填埋气和渗滤液进水中环型甲基硅氧烷的质量负荷分别为 591～6575 mg/d 和 659～5760 mg/d，两种介质中环型甲基硅氧烷的质量负荷 1～7 月呈增长趋势，7～12 月则呈下降趋势，如图 6-4 所示。

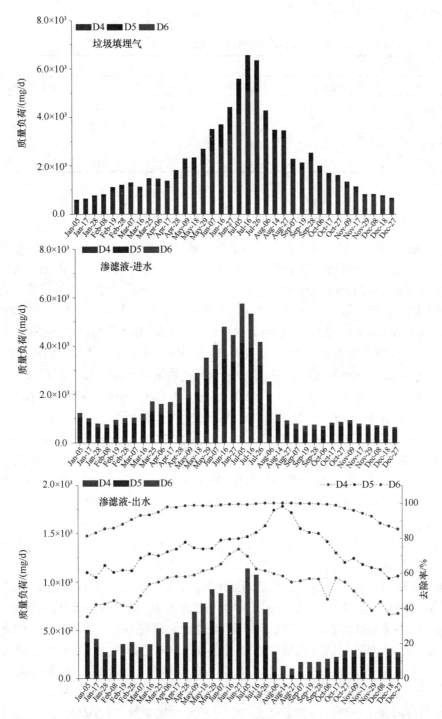

图 6-4　储存池渗滤液及垃圾填埋气中环型甲基硅氧烷的质量负荷及其去除比率

去除实验表明，对于 D4 和 D5，挥发是其在渗滤液储池中的主要去除方式，特别是在 8 月，去除比例可达 94.5%～100%。羟基自由基介导的间接光转化过程则为渗滤液中 D6 的主要去除途径，其去除效率在 6 月达到最大为 65.2%～73.7%。

D5 和 D6 的单羟基取代产物（D4TOH，D5TOH）在 7 月的渗滤液出水中均有检出。D4TOH 和 D5TOH 的挥发半衰期分别比 D5 和 D6 的挥发半衰期高 2.9 倍和 1.4 倍，而水解半衰期则比 D5 和 D6 分别低 6.1 倍和 10 倍，详见表 6-1。

表 6-1 模拟渗滤液中各目标物的一级去除速率常数

	总去除率		挥发		水解		光转化	
	速率常数/d	半衰期/d	速率常数/d	半衰期/d	速率常数/d	半衰期/d	速率常数/d	半衰期/d
D4TOH	4.49×10^{-2}	6.66	3.49×10^{-3}	86.3	4.01×10^{-2}	7.50	1.30×10^{-3}	231
D5TOH	1.96×10^{-2}	15.4	1.70×10^{-3}	177	1.40×10^{-2}	21.5	3.87×10^{-3}	77.8
D5	1.88×10^{-2}	16.0	1.02×10^{-2}	29.4	5.62×10^{-3}	53.6	2.98×10^{-3}	101
D6	1.37×10^{-2}	22.0	2.33×10^{-3}	129	1.41×10^{-3}	214	9.96×10^{-3}	30.2

3.1.3 土壤

由于污泥的土地利用越来越广泛，土壤成为甲基硅氧烷重要的环境受纳体。Wang 等[65]报道了加拿大安大略湖南部某处经过生物质改良的农业土壤中环型甲基硅氧烷 D4、D5 和 D6 的最高浓度分别为 17 ng/g dw、221 ng/g dw 和 711 ng/g dw（干重）。而西班牙的一项研究中报道了甲基硅氧烷在农业和工业土壤中的分布，结果表明工业土壤中甲基硅氧烷的浓度高于农业土壤，农业土壤中 D5 和 D6 的浓度分别为 9.2～56.9 ng/g dw 和 5.8～27.1 ng/g dw，线型硅氧烷 L5～L14 的总浓度为 191～292 ng/g dw；工业土壤中 D5 和 D6 的浓度分别为 22～184 ng/g dw 和 28～483 ng/g dw，线型硅氧烷 L5～L14 的总浓度为 1411～8532 ng/g dw。Sanchez-Brunete 等报道在西班牙的所有地区的工业土壤和农业土壤中均检测到 cVMS 的存在，总平均浓度范围分别在 15.0～84.0 ng/g dw 和 56.0～610.6 ng/g dw。在农业土壤中 D5 和 D6 的浓度范围为 9～57 ng/g dw 和 6～27 ng/g dw，在工业土壤中分别为 22～184 ng/g dw 和 28～483 ng/g dw，但只在一个工业土壤样本中检测到 D4 的存在，浓度为 58.6 ng/g dw。在混合了污泥的土壤中，D5 和 D6 的浓度分别为 30.8～37.5 ng/g dw 和 22.5～32.0 ng/g dw [66]。Sanchis 等检测到南极土壤样品中 cVMS 的存在，作者确定源于冰雪融化的环硅氧烷 D5 是主要组成部分，浓度的范围在< MLOD（最低检测下限）～最大值（110 ng/g dw），平均浓度 29.9 ng/gdw。D4 和 D6 的浓度分别在<LOD～23.9 ng/g dw（平均浓度= 14.3 ng/g dw）和<LOD～42.0 ng/g dw（平均浓度= 22.3 ng/g dw）[67]。但该研究使用的分析方法存在严重问题[68]，即该实验采集的样品被污染，因此该研究作为一种可行的南极污染机制的评估是有缺陷的，且研究没有检测南极冰雪和空气中 cVMS 的浓度。

通过对德国某处施用剩余污泥的土壤的长期监测发现，甲基硅氧烷并未在土壤中累积，这说明硅氧烷在土壤中可能发生降解和转化[69]。Lehamn 等[70, 71]研究了甲基硅氧烷在不同性质土壤中的降解情况，发现不同性质的土壤中甲基硅氧烷的半衰期范围为 4～28d。研究表明，土壤湿度、阳离子交换量、有机质含量及黏土含量等均对硅氧烷在土壤中的降解行为产生一定影响[72-75]。同时，硅氧烷在土壤中的降解是一个多步过程，首先可通过开环水解形成硅二醇低聚体，而后进一步水解成为二甲基硅二醇（DMSD）。随着分子量的增加，环型甲基硅氧烷的降解速率逐渐降低。cVMS 在土壤中降解的最终产物 DMSD 随后可以直接被生物降解或挥发至大气中进而发生光降解，而 DMSD 降解的最终产物为 SiO_2、CO_2 和 H_2O。

3.2 工业区域

相较于生活领域，甲基硅氧烷在工业生产领域中的使用量更为巨大。例如，美国 2010 年工业领域有机硅化合物（90%为硅氧烷）使用量约占其总使用量的 58%[76]。我国 2011 年电子产业、建筑业、纺织业等工业领域甲基硅氧烷的使用量约占其总使用量的 70%，居民生活中使用的个人护理品和日化用品消耗的甲基硅氧烷仅为其总使用量的 8.6%[77]。在工业生产过程中，挥发性甲基硅氧烷在整个生产流程中均可能向大气排放。另外，由于一般的工业废水处理工艺并不能对甲基硅氧烷进行有效的处理，废水中硅氧烷残留可能排入周边水环境，而吸附在污泥中的部分最终将进入土壤和大气。然而，相较于普通生活区域，目前国内外关于工业区域甲基硅氧烷的污染特征、源解析和迁移转化研究很少。

3.2.1 油田相关环境

聚二甲基硅氧烷作为破乳剂和消泡剂被大量应用于油田原油的生产中，其中的环型甲基硅氧烷会迁移至含油污水中，形成高浓度残留。同时，原油生产污水中又含有高浓度的氯离子，其在电氧化装置中会转化为氯自由基或游离氯化合物，这就为氯代甲基硅氧烷的生成提供了条件。Xu 等[78]检测了胜利油田采油污水处理站中氯代甲基硅氧烷的浓度，研究了目标化合物向周围环境中的迁移行为，测定了其在土壤中的挥发和降解半衰期［图 6-5（a）］。

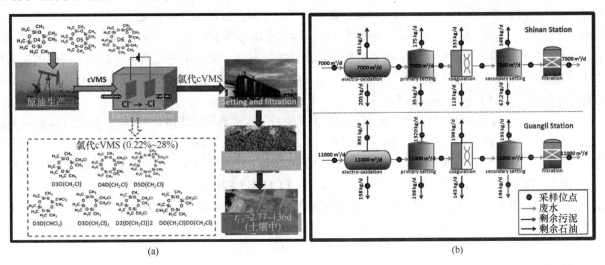

图 6-5 氯代甲基硅氧烷的生成过程（a）及含油污水站工艺流程（b）

首先，在处理站进水水相、固相和油相中检测到了甲基硅氧烷，而未检测到其氯代产物，这表明胜利油田原油生产中并没有应用氯代甲基硅氧烷。电化学氧化装置出水水相、固相和油相中 D4~D6 的一氯代产物的总浓度分别为 1.57~5.09 μg/L、52.3~95.1μg/g dw 和 136~312μg/g。同时，尽管广利和史南站电化学氧化装置进水中未检测到游离氯，但却在出水中检测到了高浓度的游离氯（4.55~9.44 g/L），表明游离氯是在该单元中由污水中的 Cl-经羟基自由基氧化而生成。氯原子作为氯离子氧化成游离氯的中间产物，可氯化非含硅有机化合物（磺胺甲噁唑和美托洛尔），因此可以推测氯原子同样可以氯化甲基硅氧烷。

采油污水处理站出水中未检测到氯代甲基硅氧烷的存在，这表明电化学氧化装置之后的处理单元对氯代甲基硅氧烷具有较好的去除效果。经调查发现，与其母体化合物的相似，剩余污泥/石油吸附对氯代甲基硅氧烷的去除起到主导作用。图 6-5（b）展示了含油污水处理站的工艺流程，含油污水处理站周围采集到的土壤样品中 D3D（CH$_2$Cl）的浓度为<LOD～273 ng/g dw（df=32.6%），D4D（CH$_2$Cl）为<LOD～524 ng/g dw（df = 45.6%），D5D（CH$_2$Cl）为<LOD～586 ng/g dw（df = 43.4%），D3D（CHCl$_2$）为<LOD～24.02 ng/g dw（df = 6.60%），D3D（CH$_2$Cl）$_2$为<LOD～78.0 ng/g dw（df = 16.6%），D2[D(CH$_2$Cl)]$_2$为<LOD～53.2 ng/g dw（df = 17.2%），DD（CH$_2$Cl）DD（CH$_2$Cl）为<LOD～35.1 ng/g dw（df = 17.2%）。只有采自距污水站较近的土壤样品（0.02～3 km）检测到了目标化合物，土壤中氯代甲基硅氧烷的浓度随采样点距污水站距离的增加呈指数下降趋势。此空间浓度变化趋势表明含油污水处理站是周围土壤环境中氯代甲基硅氧烷的污染源。

采样期间，土壤中氯代甲基硅氧烷的总平均浓度在最初的 6 年（2008～2013 年）里从 93.4 ng/g dw 增加到 184 ng/g dw，在之后的 4 年里（2014～2017 年）减小到 48.0 ng/g dw。而污水站脱水污泥中目标化合物的浓度并没有呈下降趋势，这表明电化学氧化装置中目标化合物的生成速率没有减慢。因此，土壤中氯代甲基硅氧烷浓度出现下降可能是其他原因导致的。一个可能的解释是污泥干燥作业方式的改变影响了污泥中目标化合物的浓度：自 2014 年起污泥干燥作业从露天干燥改为了机械脱水，与露天干燥相比，封闭式机械脱水工艺干燥速度快，可减少污泥中氯代甲基硅氧烷的挥发和污泥粉尘的产生。

土壤中氯代甲基硅氧烷的浓度是逐渐下降的，但这并不仅仅是污染源减少排放导致的，我们推测这与目标化合物在土壤中的挥发和降解有关。基于封闭和开放系统中的相关实验数据，较系统地研究了氯代甲基硅氧烷在土壤中的去除机制（挥发和降解）。

降解半衰期：在 TOC 浓度为 10mg/g 的封闭土壤中，氯代甲基硅氧烷在灭菌土壤中的半衰期与非灭菌土壤中的并无统计学差异（$p>0.05$，T 检验），这表明开环水解在氯代甲基硅氧烷的降解中起重要作用，此现象与其母体化合物类似。理论上说，在碱性（pH=7.4）土壤条件下，由于—CH$_2$Cl 比—CH$_3$ 的亲电性强，氯代甲基硅氧烷的水解速度应比甲基硅氧烷快。然而，作者发现一氯代甲基硅氧烷在封闭系统中的半衰期[D3D（CH$_2$Cl）为 6.92～26.8 d，D4D（CH$_2$Cl）为 14.7～65.9 d，D5D（CH$_2$Cl）为 40.1～234 d]，比其对应的母体化合物（5.64～151 d）长 1.2～2.4 倍。一个可能的解释是：由于氯代甲基硅氧烷具有较高的 K_{oc} 值，土壤有机质对其吸附性较强，相应的土壤中黏土矿物对其水解的催化作用会削弱。二氯代 D4 的降解速率比一氯代 D4 慢 1.12～4.66 倍，这同样是由于有机质对二氯代 D4 的吸附性较强导致的。二氯代 D4 的四种同分异构体中，半衰期最长的是 DD（CH$_2$Cl）DD（CH$_2$Cl）（$t_{1/2}$ = 12.0～99.9 d），之后分别是 D2[D(CH$_2$Cl)]$_2$（$t_{1/2}$ = 10.3～73.9 d）、D3D(CH$_2$Cl)$_2$（$t_{1/2}$ = 9.38～48.1 d）和 D3D（CHCl$_2$）（$t_{1/2}$ = 7.75～33.8 d）。此顺序与其 K_{oc} 值顺序一致，但作者推测分子极性差异同样是影响其水解速率的因素，例如：D3D(CHCl$_2$)或 D3D(CH$_2$Cl)$_2$ 的氯代基团都是连接在一个硅原子上的，而 D2[D(CH$_2$Cl)]$_2$ 或 DD（CH$_2$Cl）DD（CH$_2$Cl）的氯代基团则是连接在不同的硅原子上的，这导致了前者的极性较大，从而导致了其较快的水解速度。

挥发半衰期：D3D（CH$_2$Cl）、D4D（CH$_2$Cl）和 D5D（CH$_2$Cl）的挥发半衰期分别为 4.62～17.3 d、13.8～40.0 d 和 83.6～325 d，这比对应母体化合物要长 1.1～2.0 倍。此现象与理论经验一致，甲基硅氧烷在发生氯化反应后，分子质量会变大，这会导致其蒸气压下降，从而挥发性变差。与一氯代 D4 相比，二氯代 D4 的挥发速率要慢 15.7～41.9 倍。在四种同分异构体当中，D3D(CHCl$_2$)的挥发速率最快（$t_{1/2}$ = 135～271 d），其次分别是 D3D(CH$_2$Cl)$_2$（$t_{1/2}$ = 178～380 d）、D2[D(CH$_2$Cl)]$_2$（$t_{1/2}$ = 183～385 d）和 DD（CH$_2$Cl）DD（CH$_2$Cl）（$t_{1/2}$ = 194～461 d）。这与二氯代 D4 同分异构体沸点的顺序一致[D3D（CHCl$_2$）（118～120℃）<D3D(CH$_2$Cl)$_2$（124～126℃）< D2[D(CH$_2$Cl)]$_2$（128～129℃）和 DD（CH$_2$Cl）DD（CH$_2$Cl）（128～129℃）]。

3.2.2 造纸工业相关环境

聚二甲基硅氧烷（PDMS）由于具有低表面张力、高热稳定性和润滑性等优良性能，在工业生产建设领域和日常消费品行业中被大量广泛使用。例如，聚二甲基硅氧烷（PDMS）在造纸过程中作为消泡剂被大量应用于制浆、漂白和脱水等工艺中。目前，中国是全球范围内造纸产量最大的国家，每年的产量达

第 6 章 硅氧烷类污染物的环境赋存与转化

到了上亿吨。虽然传统的氯气漂白工艺在发达国家（如美国和加拿大等）造纸工厂已经被逐步淘汰，但是中国大多数造纸工厂造纸工艺中仍沿用该方法。意味着纸浆漂白工艺挥发性甲基硅氧烷（作为 PDMS 消泡剂的杂质存在）与氯气可能共存。

研究表明，挥发性环型甲基硅氧烷在气相中能够与羟基自由基、臭氧（O_3）、氯原子和 NO_3 自由基发生氧化作用，氯原子氧化甲基硅氧烷（六甲基二硅氧烷，硅醚）的双分子速率常数是羟基自由基的 10 倍。然而，目前国内外还没有相关文献报道在实际环境条件中甲基硅氧烷在水相中是否能够被氯原子或其他游离氯所取代。如果挥发性甲基硅氧烷在漂白工艺中可被氯化，氯代甲基硅氧烷将进入造纸污水处理过程中，最终可能将会被释放到环境水体中，可能对环境和人体健康构成潜在威胁。

Xu 等[78]使用气相色谱-四极杆飞行时间质谱联用仪（Q-TOF GC/MS）研究了使用（聚）二甲基硅氧烷和氯气分别作为消泡剂和漂白剂的造纸工厂漂白过程中挥发性甲基硅氧烷的氯代反应。结果表明，在研究的造纸工厂生产过程的水样和固体样品中，D4、D5 和 D6 的一氯代产物 D3D（CH_2Cl）、D4D（CH_2Cl）和 D5D（CH_2Cl）的平均总浓度范围分别为 0.0430~287 μg/L 和 0.0329~270 μg/g。如图 6-6（a）、（b）和（c）所示，除母体 D4、D5 和 D6 外，氯代阶段（C）水样的全扫描色谱图中发现其他未知产物。其次，该研究采集了中国山东某造纸工厂的污水处理工艺过程中各单元的进出污水样品和纸浆/污泥混合样品检测到造纸废水处理过程中，D3D（CH_2Cl）、D4D（CH_2Cl）和 D5D（CH_2Cl）在水样中浓度范围为 0.113~8.68 μg/L，固体样品中浓度范围为 0.888~26.2 μg/g，固水分配值（468~3982 L/kg）比它们所对应的母体化合物高 1.08~4.82 倍。D3D（CH_2Cl）~D5D（CH_2Cl）在整个污水处理过程中的去除率为 77.1%~81.6%，其主要的去除途径为吸附于污泥（约占 35.7%~74.1%）及在污水初沉池中的去除（约占 7.19%~32.5%），在好氧污泥处理过程中几乎不降解，如图 6-7 所示。

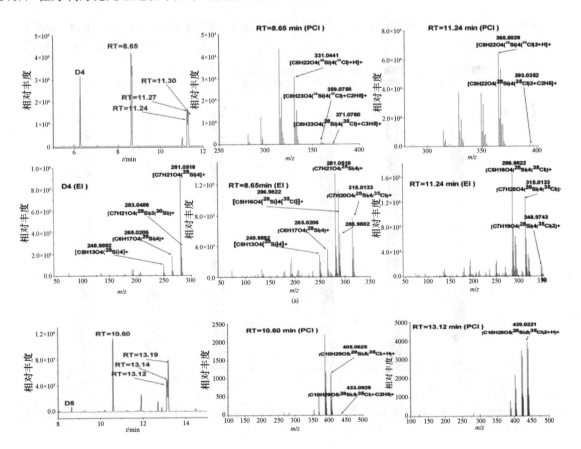

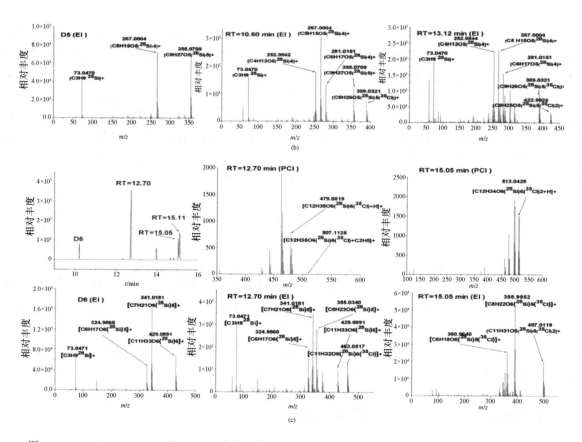

图 6-6　D4（a）、D5（b）和 D6（c）氯代产物的总离子流图（每幅图中第一个小图）和 EI/PCI 质谱图

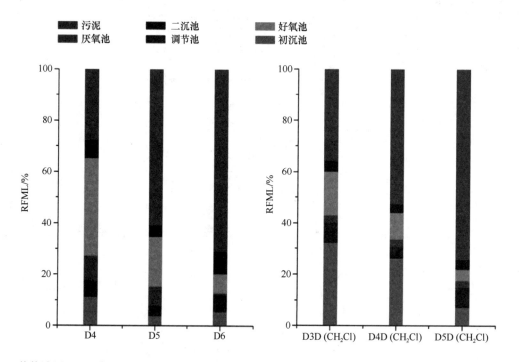

图 6-7　传统造纸工厂（氯漂白）甲基硅氧烷及其一氯代产物在每个工艺流程中的质量损失相对分数（RFML）

4 人群暴露

4.1 硅氧烷生产工厂

鉴于硅氧烷巨大的产量和广泛的使用,其在生产/使用过程中可能存在的职业/普通人群暴露行为应该引起重视,需要结合大气、灰尘、个人护理品等环境样品以及人体血液/组织样本系统评估硅氧烷的暴露途径和人体负荷。然而,目前相关研究十分匮乏。Hanssen 等[79]采集了挪威女性的血液样本研究甲基硅氧烷的浓度水平与个人护理品情况之间的关系,结果发现 D4 的最高浓度为 12.7 ng/mL,D5 和 D6 在绝大多数人的血液中未检出,而且还证实人体血液中甲基硅氧烷的浓度水平与个人护理品使用量之间并无明显关系。

Xu 等[80]以山东省东部地区某硅氧烷生产工厂周边环境以及该厂职工及周边居民为研究对象,研究硅氧烷生产厂周边不同环境介质如空气、灰尘和土壤中甲基硅氧烷的浓度水平及环境行为,同时也研究了该工厂工人血液中甲基硅氧烷浓度水平及其甲基硅氧烷暴露水平和暴露时间的关系。通过对该工厂离职工人血液中甲基硅氧烷浓度水平的分析,估算了人体血液中甲基硅氧烷的衰减速度,这也是国际上第一次对甲基硅氧烷在职业暴露人群血液中的分布、累积及消减行为的研究。

首先,研究了硅氧烷生产工厂周边环境中甲基硅氧烷的浓度水平和环境行为,在该工厂的 6 个车间的室内空气样品中,17 种甲基硅氧烷均有检出,且检出率均为 100%。所有车间室内空气中环型硅氧烷的平均浓度为 34μg/m^3(D6)~2.7 mg/m^3(D4)。在该工厂的下风向,随着距离的增加 B1-B7 大气中挥发性甲基硅氧烷(D4~D6,L3~L6)总浓度以指数形式降低。另外,在 7 种挥发性的甲基硅氧烷中,环型挥发性甲基硅氧烷可能比线型硅氧烷更容易实现大气环境中的长距离迁移。

其次,通过对甲基硅氧烷职业暴露水平进行研究,硅氧烷生产工厂 A 区的室内灰尘中甲基硅氧烷浓度比对照区高 1~5 个数量级,并且该工厂的工人甲基硅氧烷的职业暴露水平要远大于使用日常生活用品的暴露水平。B 区(处于硅氧烷的下风向)的居民区中,在室内空气和灰尘样品中没有检出甲基硅氧烷。由于受硅氧烷工厂的影响,B 区内的室外大气和土壤样品中环型硅氧烷浓度比对照区高 1~3 个数量级。总体来说,硅氧烷周边地区的甲基硅氧烷人群暴露水平比对照区 C 地区高 2~3 个数量级。

同时,研究了三个区域人群血浆中甲基硅氧烷的分布情况,如图 6-8,A 区和 B 区人群血浆中甲基硅氧烷的检出浓度和检出频率高于 C 区:硅氧烷工厂在职工人(n=72)血浆中环型甲基硅氧烷 D4、D5 和 D6 的平均浓度分别为 206 ng/g(df=100%)、215 ng/g(df=100%)和 88.7 ng/g(df=100%)。在职工人血浆中检出除 L4 外所有线型硅氧烷(L3~L16)目标物,平均浓度为 5.62~451 ng/g(df=80.5%~100%)。在 B 区血浆样本(n=14)中,D4、D5 和 D6 可检出,平均浓度分别为 13.5 ng/g、57.8 ng/g 和 4.56 ng/g(df=36%~100%)。

最后,研究血浆中甲基硅氧烷的衰减情况,有 32 个来自于该工厂 A3 车间的离职工人(离职时间 3~320 d,没有居住在 B 区)共 57 个血浆样本。总体来说,离职工人血浆中环型和线型甲基硅氧烷的浓度随离职时间的延长而降低,离职时间大于 85 d 的血浆样本中没有检测到苯基硅氧烷的存在,如图 6-9 所示。目前,没有相关文献报道关于线型甲基硅氧烷的血浆衰减模型。在 Xu 等的研究中采用一室衰减模型研究环型和线型甲基硅氧烷的血浆半衰期。根据拟合结果,甲基硅氧烷在人体血浆中的半衰期要比典型的 POPs 化合物低,说明其在血浆中衰减较快。

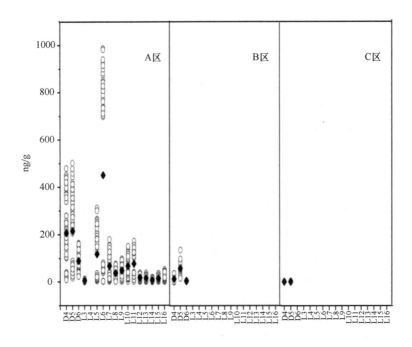

图 6-8　研究人群血浆中甲基硅氧烷分布

作者通过对硅氧烷工厂周边环境中甲基硅氧烷的环境行为及人体暴露情况的研究，证明硅氧烷工厂在生产过程中不可避免地会对周边环境造成一定程度的污染。有机硅生产厂的工人由于职业卫生防护措施的缺乏，其血浆中甲基硅氧烷浓度要明显高于普通人群。并且甲基硅氧烷在人体血浆中不易累积，其血浆半衰期比较短。但由于其亲脂性结构，容易在人体脂肪和器官中积累。

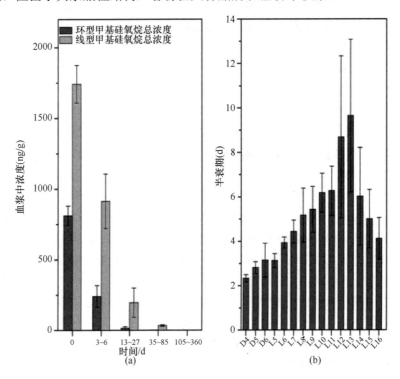

图 6-9　(a) 不同离职时间工人血浆中环型/线型硅氧烷总浓度；(b) 血浆中甲基硅氧烷半衰期

4.2 硅氧烷使用工厂

Xu 等[81]还研究了一类新型 POPs 类污染物-甲基硅氧烷在我国普通人群和建筑、汽车制造和纺织工厂职业人群的暴露水平[38]。通过计算发现，上述三类工业中工人人均甲基硅氧烷使用水平要比我国普通人群高 2~5 个数量级。工业区室内空气和灰尘中甲基硅氧烷浓度比生活区高 1~3 个数量级。环型（D4~D6）和线型硅氧烷（L5~L16）在 528 个工人血液中均有检出，浓度 1.00~252ng/mL，检出率 3.7%~71%，而普通人群血液中仅检出了环形硅氧烷（D4~D6），浓度 1.10~7.50ng/mL，检出率 1.7%~3.7%。在职业暴露过程中，防颗粒呼吸系统可以减少环形硅氧烷 30%的暴露剂量和 74%的线型硅氧烷暴露剂量。另外，PM_{10}提高了线型硅氧烷的暴露剂量。甲基硅氧烷（D4~D6，L6~L11）的脂肪-血液分配系数约为 5.3~241mL/g。与环型硅氧烷相比，线型硅氧烷在人体脂肪内的累积趋势更加明显。其中，L8~L10 在一般人群脂肪中的半衰期约为 1.49~1.80a（图 6-10）。

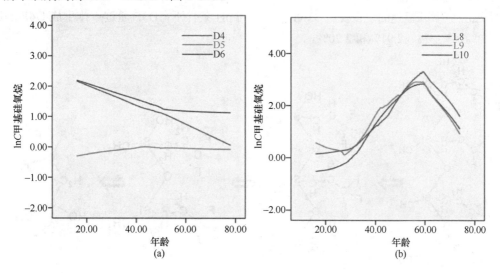

图 6-10 甲基硅氧烷在不同年龄普通人群脂肪中的分布

5 改性硅氧烷环境行为的初步研究

随着有机硅工业的不断发展，为使聚硅氧烷具有化学活性及其他特殊性能从而满足新的工业应用，近年来，大量研究开始关注聚硅氧烷的侧基改性。目前多种改性聚硅氧烷已相继问世并且逐步实现商品化，从而使聚硅氧烷已的用途进一步扩展。

5.1 氟硅氧烷

在聚硅氧烷侧链中引入含氟官能团可得到聚氟烷基硅氧烷。含氟基团的引入，使得聚氟烷基硅氧烷结合了有机硅和有机氟材料的优良特性，具有更高的光、热以及化学稳定性，表现出更优异的耐高、低温能力和超强的疏水疏油性质，使之在各个领域得到新的应用。迄今为止，最常见的商品化的聚氟烷基硅氧烷是聚三氟丙基甲基环三硅氧烷（PMTFPS）。作为重要的氟硅聚合物，PMTFPS 常通过氟硅氧烷单体（如三氟丙基甲基环三硅氧烷）开环聚合的方法合成。相关文献表明，PMTFPS 中通常会残留一定量的氟化硅氧烷单体。由于聚硅氧烷材料以及氟硅聚合物的大量生产和使用，作为合成聚硅氧烷和聚氟硅氧烷的重要单体，甲基硅氧烷和氟硅氧烷在生产和使用过程中有可能释放从而进入环境中。近年来，作为一种新型污染物，甲基硅氧烷已引起科学界的广泛关注。目前，已有多篇文献报道甲基硅氧烷在水体、底泥、大气等多种环境介质中均有检出。此外，部分研究发现甲基硅氧烷会对生物体产生一定的不良影响。而氟硅氧烷的相关研究还未见报道，氟硅氧烷的环境行为及其可能引起的环境健康效应亟待关注。

氟硅氧烷是一种改性硅氧烷，主要包括三氟丙基甲基环三硅氧烷和三氟丙基甲基环四硅氧烷。Zhi 等[82]研究了中国威海某硅氧烷生产工厂周围河水及底泥样品中氟硅氧烷的赋存情况。在河水中，D3F 和 D4F 的浓度范围分别为 2.11～290ng/L 和 7.27～166 ng/L。底泥中 D3F 和 D4F 的浓度范围分别为 11.8～5476 ng/g 和 14.9～6249 ng/g。模拟实验表明，水环境中 D3F 和 D4F 主要衰减方式是开环水解。在不同 pH（5.2、6.4、7.2、8.3 和 9.2）条件下，两者的半衰期范围分别为 80.6～154h 和 267～533h。甲基三氟丙基硅二醇是氟硅氧烷主要的水解产物（图 6-11），其在河水中的浓度为 72.1～182.9 ng/L。此外，水解过程中的重排反应是不可忽视的。在反应 336h 后，不同 pH 条件下，D3F 通过重排生成 D4F 的比例分别为 2.72%、2.93%、3.13%、2.91%和 2.70%。

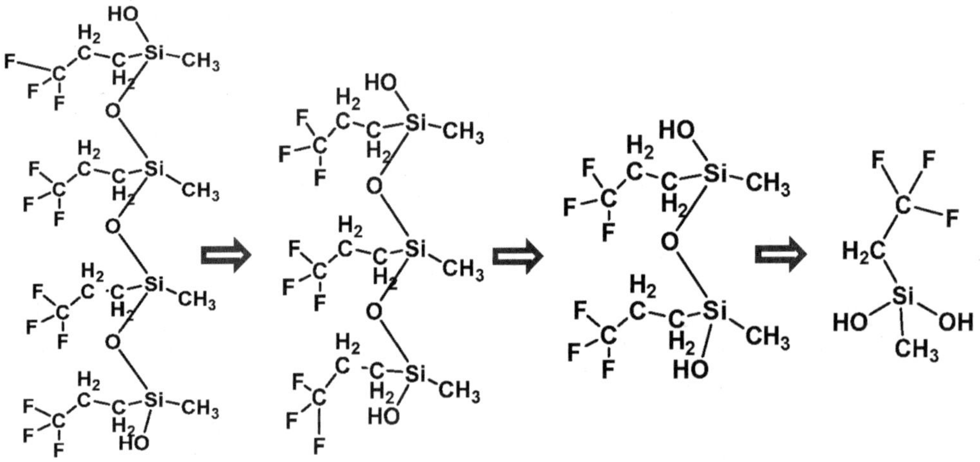

图 6-11 三氟丙基硅氧烷的水解机制

5.2 苯基硅氧烷

由于全球使用量巨大（200 万 t/a），硅氧烷的环境排放和迁移转化行为引起了科研人员的关注，然而目前相关研究主要集中于二甲基硅氧烷化合物（产量占硅氧烷总量的 80%），而对于其他种类的硅氧烷研究极少。Xu 等[83]通过 GC-QTOF/MS 质谱全扫描（full scan）模式，在山东地区某生活污水厂进水中识别出两类含有苯环的环型硅氧烷化合物（三苯基三甲基环三硅氧烷-P3、四苯基四甲基环三硅氧烷-P4），通过分析质谱特征离子和计算特征离子丰度，发现 P3 和 P4 分别含有 2 种（cis-P3、trans-P3）和 4 种（cis-P4、trans-P4a、trans-P4b、trans-P4c）顺反异构体（图 6-12），其响应比分别为 1∶3 和 1∶4∶2∶1。进而通过质量平衡计算发现污水中 95%以上的 P3 和 P4 吸附在活性污泥中，并可随污泥施肥进入土壤，其在土壤中的衰减速度比相同链长的二甲基硅氧烷慢 10 倍以上，且其降解产物之一为高毒性的苯。上述结果表明，虽然苯基硅氧烷的全球产量/使用量仅为二甲基硅氧烷的 1/6，但其环境归趋和生态健康效应需要引起作者的重视。

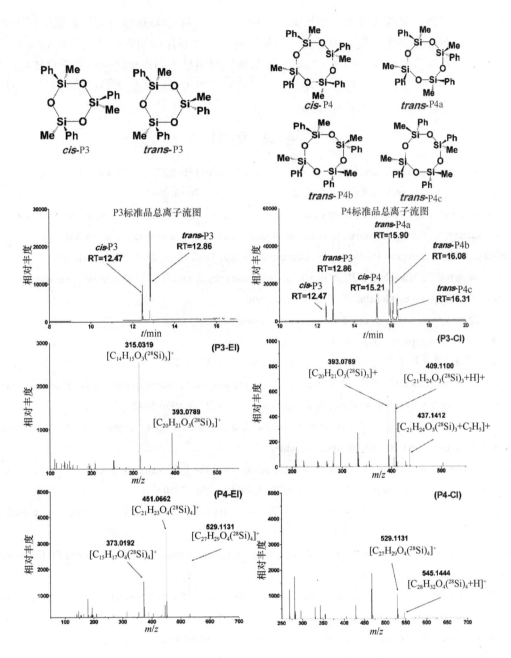

图 6-12 苯基硅氧烷 EI 和 CI 质谱图

6 展　　望

目前，国内外已经比较系统地介绍了甲基硅氧烷在各类环境介质中的分布水平和归趋行为。但是，相关研究工作主要围绕生活区域开展，而极少关注工业区域。实际上，由于甲基硅氧烷在工业区域的使用量占总量 90%以上，其在工业区域的排放、迁移、转化（尤其是某些特殊转化行为）有待系统深入研究。另外，我国在开展硅氧烷的典型区域，特别是硅氧烷生产及使用工业地区的硅氧烷水平、环境行为目前还比较缺乏。另外，暴露人群血液中污染物负荷水平、富集/衰减速度是评价硅氧烷污染物人体健康效应的重要参数，目前还尚未出现相关研究。另外，由于改性基团的引入，侧链改性硅氧烷（如氟硅氧烷、苯基硅氧烷、乙烯基硅氧烷）的物化性质与甲基硅氧烷明显不同，也将导致其迁移转化、生物富集、生物毒性等归趋行为与甲基硅氧烷存在差异，其对环境和生态可能造成的潜在风险需要引起重视。因此，研究典型区域中改性硅氧烷的环境归趋具有重要现实意义。

参 考 文 献

[1] Silicones Environmental, Health and Safety Council of North America (SEHSC). 2011. http://www.sehsc.com/d5.asp.

[2] 卜新平. 国内外有机硅行业市场现状与发展趋势. 化学工业, 2008, 26(6): 39-46.

[3] Silicones Environmental, Health and Safety Council of North America (SEHSC). 2013. http://www.sehsc.com/d5.asp.

[4] Horii Y, Kannan K. Survey of organosilicone compounds, including cyclic and linear siloxanes, in personal-care and household products.Archives of Environmental Contamination and Toxicology, 2008, 55(4): 701-710.

[5] Lassen C, Hansen C L, Mikkelsen S H, et al. Siloxanes: Consumption, toxicity and alternatives. Environmental Project No. 1031 2005. copenhagen: Danish Ministry of the Environment.

[6] Giawin. 2012. http://www.giawin.com/show_news.asp?newsid=46

[7] EPA. 2002. http://www.epa.gov/oppt/iur/tools/data/2002-vol.htm.

[8] 孟宾. 国内外有机硅市场发展现状及趋势. 化工新型材料, 2012, 40(8): 1-4.

[9] Kierkegaard A, Van Egmond R, Mclachlan M S. Cyclic volatile methylsiloxane bioaccumulation in flounder and ragworm in the Humer Esturary. Environmental Science & Technology, 2011, 45(14): 5936-5942.

[10] Kaj L, Andersson J, Cousins A, et al. Results from the Swedish national screening programme 2004 subreport 4: Siloxanes. IVL Swedish Environmental ResearcInstitute, Stockholm, 2005

[11] Warner N A, Evenset A, Christensen G, et al. Volatile siloxanes in the European arctic: Assessment of sources and spatial distribution. Environmental Science and Technology, 2010, 44(19): 7705-7710.

[12] Kierkegaard A, Mclachlan M S. Determination of decamethylcyclopentasiloxane in air using commercial solid phase extractioncartridges. Journal of Chromatography A, 2010, 1217: 3557-3560.

[13] Gobas F A P C, Wolf W D, Burkhard L P, et al. Revisiting bioaccumulation criteria for POPs and PBT assessments. Integr Environ Assess Manag, 2010, 5(4): 624-637.

[14] Fackler P H, Dionne E, Hartley D A, et al. Bioconcentration by fish of a highly volatile silicone compound in a totally enclosed aquatic exposure system. Environmental Toxicology & Chemistry, 1995, 14(10): 1649-1656.

[15] Parrott J L, Alaee M, Wang D, et al. Fathead minnow (*Pimephales promelas*) embryo to adult exposure to decamethylcyclopentasiloxane (D5). Chemosphere, 2013, 93(5): 813-818.

[16] KR D. ^{14}C - dodecamethylcyclohexasiloxane (^{14}C–D6): Bioconcentration in the fathead minnow (*Pimephales promelas*) under FlowThrough test conditions. Dow Corning Corporation, Silicones Environment, Health and Safety Council (SEHSC), 2005, Cited from the Report of the Assessment for D6 by Environment Canada and Health Canada.

[17] Annelin R B, Frye C L. The piscine bioconcentration characteristics of cyclic and linear oligomeric permethylsiloxanes. Science of the Total Environment, 1989, 83(1-2): 1-11.

[18] Borga K, Fjeld E, Kierkegaard A, et al. Consistency in trophic magnification factors of cyclic methyl siloxanes in pelagic freshwater food webs leading to brown trout. Environmental Science & Technology, 2013, 47(24): 14394-14402.

[19] Powell D E, Suganuma N, Kobayashi K, et al. Trophic dilution of cyclic volatile methylsiloxanes (cVMS) in the pelagic marine food web of Tokyo Bay, Japan. Science of The Total Environment, 2016: S0048969716323774.

[20] Lieberman M W, Lykissa E D, Barrios R, et al. Cyclosiloxanes produce fatal liver and lung damage in mice. Environmental

Health Perspectives, 1999, 107(2): 161-165.

[21] Burnsnaas L A, Mast R W, Klykken P C, et al. Toxicology and humoral immunity assessment of decamethylcyclopen tasiloxane (D5) following a 1-month whole body inhalation exposure in Fischer 344 rats. Toxicological Sciences An Official Journal of the Society of Toxicology, 1998, 43(1): 28.

[22] Burns-Naas L A, Mast R W, Meeks R G, et al. Inhalation toxicology of decamethylcyclopentasiloxane (D5) following a 3-month nose-only exposure in fischer 344 rats. Toxicological Sciences, 1998, 43(2): 230-240.

[23] He B, Rhodes-Brower S, Miller M R, et al. Octamethylcyclotetrasiloxane exhibits estrogenic activity in mice via ERα. Toxicology & Applied Pharmacology, 2003, 192(3): 254-261.

[24] Quinn A L, Dalu A, Meeker L S, et al. Effects of octamethylcyclotetrasiloxane (D4) on the luteinizing hormone (LH) surge and levels of various reproductive hormones in female Sprague–Dawley rats. Reproductive Toxicology, 2007, 23(4): 532-540.

[25] Jr M K, Wilga P C, Kolesar G B, et al. Evaluation of octamethylcyclotetrasiloxane (D4) as an inducer of rat hepatic microsomal cytochrome P450, UDP-glucuronosyltransferase, and epoxide hydrolase: A 28-day inhalation study. Toxicological Sciences, 1998, 41(1): 29-41.

[26] McKim, J. M. Potential estrogenic and antiestrogenic activity of the cyclic siloxane octamethylcyclotetrasiloxane (D4) and the linear siloxane hexamethyldisiloxane (HMDS) in immature rats using the uterotrophic assay. Toxicological Sciences, 2001, 63(1): 37-46.

[27] Sousa J V, Mcnamara P C, Putt A E, et al. Effects of octamethylcyclotetrasiloxane (OMCTS) on freshwater and marine organisms. Environmental Toxicology & Chemistry, 2010, 14(10): 1639-1647.

[28] Kent D J, Mcnamara P C, Putt A E, et al. Octamethylcyclotetrasiloxane in aquatic sediments: Toxicity and risk assessment. Ecotoxicology & Environmental Safety, 1994, 29(3): 380-389.

[29] Zhu L Y, Wang C Y, Zhang H, et al. Degradation and Mineralization of Bisphenol A by Mesoporous Bi(2)WO(6) under Simulated Solar Light Irradiation. Environmental Science & Technology, 2010, 44(17): 6843-6848.

[30] Parrott J L, Alaee M, Wang D, et al. Fathead minnow (*Pimephales promelas*) embryo to adult exposure to decamethylcyclopentasiloxane (D5). Chemosphere, 2013, 93(5): 813-818.

[31] Silicones Environmental H A S C. A prolonged sediment toxicity test with Chironomusriparius using spiked sediment. Wildlife International, LTD: Report of the Assessment for D4 by Environmental Canada and Health, 2008, Canada: SEHSC.

[32] Norwood W P, Alaee M, Sverko E, et al. Decamethylcyclopentasiloxane (D5) spiked sediment: Bioaccumulation and toxicity to the benthic invertebrate *Hyalella azteca*. Chemosphere, 2013, 93(5): 805-812.

[33] Velicogna J, Ritchie E, Princz J, et al. Ecotoxicity of siloxane D5 in soil. Chemosphere, 2012, 87(1): 77-83.

[34] McKim J M. Potential estrogenic and antiestrogenic activity of the cyclic siloxane octamethylcyclotetrasiloxane (D4) and the linear siloxane hexamethyldisiloxane (HMDS) in immature rats using the uterotrophic assay. Toxicological Sciences, 2001, 63(1): 37-46.

[35] Quinn AL, Dalu A, Meeker LS, et al. Effects of octamethylcyclotetrasiloxane (D4) on the luteinizing hormone (LH) surge and levels of various reproductive hormones in female Sprague-Dawley rats. Reproductive Toxicology, 2007, 23(4): 532-540.

[36] Quinn A L, Regan J M, Tobin J M, et al. *In vitro* and *in vivo* evaluation of the estrogenic, androgenic, and progestagenic potential of two cyclic siloxanes. Toxicological Sciences, 2007, 96(1): 145-153.

[37] Daly G L, Wania F. Organic contaminants in mountains. Environmental Science and Technology, 2005, 39: 385-398.

[38] Yucuis R A, Stanier C O, Hornbuckle K C. Cyclic siloxanes in air, including identification of high levels in Chicago and distinct diurnal variation. Chemosphere, 2013, 92: 905-910.

[39] Kaj L, Andersson J, Palm Cousins A, et al. Results from the Swedish National Screening Programme 2004, Subreport 4: Siloxanes; IVL: Stockholm, 2005.

[40] Pieri F, Katsoyiannis A, Martellini T, et al. Occurrence of linear and cyclic volatile methyl siloxanes in indoor air samples (UK and Italy) and their isotopic characterization. Environment International, 2013, 59: 363-371.

[41] Tran T M, Abualnaja K O, Asimakopoulos A G, et al. A survey of cyclic and linear siloxanes in indoor dust and their implications for human exposures in twelve countries. Environment International, 2015, 78: 39-44.

[42] Navea J G, Xu S, Stanier C O, et al. Heterogeneous uptake of octamethylcyclotetrasiloxane (D4) and decamethylcyclopentasiloxane (D5) onto mineral dust aerosol under variable RH conditions. Atmospheric Environment, 2009, 43(26): 4060-4069.

[43] Whelan M J, Estrada E, Van Egmond R. A modeling assessment of the atmospheric fate of volatile methyl siloxanes and their reaction products. Chemosphere, 2004, 57: 1427-1437.

[44] Atkinson R. Kinetics of the gas phase reactions of a series of organosilicon compounds with OH and NO_3 radicals and O_3 at $297 \pm 2K$. Environmental Science and Toxicology, 1991, 25: 863-866.

[45] Brooke D N, Crookes M J, Gray D, et al. Environmental Risk Assessment Report: Decamethylcyclopentasiloxane; Environment Agency of England and Wales: Bristol. 2009.

[46] Brooke D N, Crookes M J, Gray D, et al. Environmental Risk Assessment Report: Octamethylcyclotetrasiloxane; Environment Agency of England and Wales: Bristol. 2009.

[47] Brooke D N, Crookes M J, Gray D, et al. Environmental Risk Assessment Report: Diamethylcyclohexasiloxane; Environment Agency of England and Wales: Bristol. 2009.

[48] Krogseth I S, Kierkegaard A, Mclachlan M S, et al. Occurrence and seasonality of cyclic volatile methyl siloxanes in arctic Air. Environmental Science & Technology, 2013, 47(1): 502-509.

[49] Mclachlan M S, Kierkegaard A, Hansen K M, et al. Concentrations and fate of decamethylcyclopentasiloxane (D5) in the atmosphere. Environmental Science and Technology, 2010, 44(14): 5365-5370.

[50] Kozerski G E, Xu S, Miller J, et al. Determination of soil-water sorption coefficients of volatile methylsiloxanes. Environmental Toxicology & Chemistry, 2015, 33(9): 1937-1945.

[51] Xu S, Kozerski G, Mackay D. Critical review and interpretation of environmental data for volatile methylsiloxanes: partition Properties. Environmental Science & Technology, 2014, 48(20): 11748-11759.

[52] Zhang Z, Qi H, Ren N, et al. Survey of Cyclic and linear siloxanes in sediment from the Songhua River and in sewage sludge from wastewater treatment plants, Northeastern China. Archives of Environmental Contamination and Toxicology, 2011, 60(2): 204-211.

[53] Bletsou A A, Asimakopoulos A G, Stasinakis A S, et al. Mass loading and fate of linear and cyclic siloxanes in a wastewater treatment plant in Greece. Environmental Science & Technology, 2013, 47(4): 1824-1832.

[54] Wang D G, Steer H, Tait T, et al. Concentrations of cyclic volatile methylsiloxanes in biosolid amended soil, influent, effluent, receiving water, and sediment of wastewater treatment plants in Canada. Chemosphere, 2013, 93(5): 766-773.

[55] Durham J. Non-regulated study: Method development and preliminary assessment of the hydrolysis kinetics of decamethylcyclopentasiloxane (D5) according to the principles of OECD guideline 111. Draft report. Auburn, MI: Health and Environmental Sciences, Dow Corning Corporation, 2005.

[56] Durham J. Soil-water distribution of decamethylcyclopentasiloxane (D5) using a batch equilibrium method. Auburg, MI: Health and Environmental Science, Dow Corning Corp, 2007.

[57] Bletsou A, Asimakopoulos A, Stasinakis A, et al. Mass loading and fate of linear and cyclic siloxanes in a wastewater treatment plant in Greece. Environmental Science and Technology, 2013, 47: 1824-1832.

[58] 申凯. 有机硅氧烷在市政废水处理厂的归趋研究. 大连: 大连海事大学学位论文, 2014.

[59] Xu L, Shi Y, Cai Y. Occurrence and fate of volatile siloxanes in a municipal Wastewater Treatment Plant of Beijing, China. Water Research, 2013, 47(2): 715-724.

[60] 代晋国, 宋乾武, 王红雨, 等. 我国垃圾渗滤液处理存在问题及对策分析. 环境工程, 2011(S1): 185-188.

[61] 赵德福. 2015 年有机硅行业大数据分析及利润分析. http://www.sci99.com/sdprice/20471787.html.

[62] Schweigkofler M, Niessner R. Determination of siloxanes and VOC in landfill gas and sewage gas by canister sampling and GC-MS/AES analysis. Environmental Science & Technology, 1999, 33(20): 3680-3685.

[63] Paxeus N. Organic compounds in municipal landfill leachates. Water Science & Technology, 2000, 42(7): 323-332.

[64] Xu L, Xu S, Zhi L, et al. Methylsiloxanes release from one landfill through yearly cycle and their removal mechanisms (especially hydroxylation) in leachates. Environmental Science & Technology, 2017: acs.est.7b03624.

[65] Wang D G, Steer H, Tait T, et al. Concentrations of cyclic volatile methylsiloxanes in biosolid amended soil, influent, effluent, receiving water, and sediment of wastewater treatment plants in Canada. Chemosphere, 2013, 93(5): 766-773.

[66] Consuelo Sánchez-Brunete, Miguel E, Albero B, et al. Determination of cyclic and linear siloxanes in soil samples by ultrasonic-assisted extraction and gas chromatography－mass spectrometry. Journal of Chromatography A, 2010, 1217(45): 7024-7030.

[67] Sanchís, Josep, Cabrerizo A, Galbán-Malagón, Cristóbal, et al. Unexpected occurrence of volatile dimethylsiloxanes in Antarctic soils, vegetation, phytoplankton, and krill. Environmental Science & Technology, 2015, 49(7): 4415-4424.

[68] Mackay D, Gobas F, Solomon K, et al. Comment on "Unexpected occurrence of volatile dimethylsiloxanes in Antarctic soils, vegetation, phytoplankton, and krill". Environmental Science & Technology, 2015, 49(12): 7504-7506.

[69] Griessbach E F C, Lehmann R G. Degradation of polydimethylsiloxane fluids in the environment: A review. Chemosphere, 1999, 38(6): 1461-1468.

[70] Lehmann R G, Miller J R, Xu S H, et al. Degradation of silicone polymer at different soil moistures. Environmental Science and Technology, 1998, 32: 1260-1264.

[71] Lehmann RG, Varaparth S, Annelin RB, Arndt J L. Degradation of silicone polymer in a variety of soils. Environmental Toxicology and Chemistry, 1995, 14: 1299-1305.

[72] Xu S H, Chandra G. Fate of cyclic methylsiloxanes in soils. 2. Rates of degradation and volatilization. Environmental Science and Technology, 1999, 33: 4034-4039.

[73] Xu S H, Kozerski G, Mackay D. Critical review and interpretation of environmental data for volatile methylsiloxanes: Partition properties. Environmental Science and Technology, 2014, 48: 11748-11759.

[74] Xu S H, Kropscott B. Challenges in analytical method development and validation for trace level dimethylsilanediol in water. SETAC Europe 25th Annual Meeting held in Barcelona, Catalonia. Spain: 2015.

[75] Xu S H. Fate of cyclic methylsiloxanes in soils. 1. The degradation pathway. Environmental Science and Technology, 1999, 33: 603-608.

[76] Chang, J. Silicones poised for rapid growth. ICIS Chemical Business. 2006. http://www.icis.com/resources/news/2006/06/03/2014560/silicones-poised-for-rapid-growth/.

[77] http://www.giawin.com/show_news.asp?newsid=46. 2012.

[78] Xu L, He X, Zhi L, et al. Chlorinated methylsiloxanes generated in papermaking process and their fate in wastewater treatment processes. Environmental Science & Technology, 2016, 50(23): 435-439.

[79] Hanssen L, Warner N A, Braathen T, et al. Plasma concentrations of cyclic volatile methylsiloxanes (cVMS) in pregnant and postmenopausal Norwegian women and self-reported use of personal care products (PCPs). Environment International, 2013, 51: 82-87.

[80] Xu L, Shi Y, Wang T, et al. Methyl siloxanes in environmental matrices around a siloxane production facility, and their

distribution and elimination in plasma of exposed population. Environmental Science & Technology, 2012, 46(21): 11718-11726.

[81] Xu L, Shi Y, Liu N, et al. Methyl siloxanes in environmental matrices and human plasma/fat from both general industries and residential areas in China. Science of The Total Environment, 2015, 505(1): 454-463.

[82] Zhi L Q, Xu L, Qu Y, et al. Identification and elimination of fluorinated methylsiloxanes in environmental matrices near a manufacturing plant in eastern China. Environmental Science & Technology, 2018, 52: 12235-12243.

[83] Xu L, Xu S H, Zhang Q L, et al. Sources and fate of cyclic phenylmethylsiloxanes in one municipal wastewater treatment plant and biosolids-amended soil. Environmental Science & Technology, 2018, 52: 9835-9844.

作者：徐 琳[1]，蔡亚岐[1]
[1] 中国科学院生态环境研究中心

第 7 章　汞的环境污染研究进展

▶ 1. 人为源大气汞排放研究进展 /156

▶ 2. 水生生态系统汞的迁移转化 /159

▶ 3. 汞在森林生态系统生物地球化学研究进展 /164

▶ 4. 汞的甲基化研究进展 /170

本章导读

汞是通过大气进行长距离跨境传输的全球污染物，受到学术界和国际社会广泛关注。近年来，学术界对汞在地表环境生物地球化学循环规律进行了系统研究，在汞的迁移转化方面获得新的认识。本章重点总结了人类活动大气汞排放的研究在源排放特征和清单等方面取得的进展，发现我国超低排放技术具有非常高的除汞效率，分析了重点源和区域及全球汞排放清单研究进展，提出了未来人为源大气汞排放研究的重点任务；系统总结了水生生态系统汞的迁移转化规律的研究进展，介绍了水体汞的光化学转化、硫化汞形态转化、颗粒态汞形态转化、海洋生态系统汞的迁移转化研究热点及生态系统演变对水环境汞生态健康风险的影响等方面的研究进展；系统总结了汞在森林生态系统生物地球化学循环研究的新进展，介绍了森林生态系统汞生物地球化学循环过程、森林生态系统汞质量平衡和森林生态系统汞同位素分馏规律等方面的研究进展，指出了今后开展工作的方向；系统总结了汞的甲基化研究进展，阐明了环境中甲基汞的来源与分布、汞的微生物甲基化、甲基汞的微生物去甲基化和汞的非生物甲基化等方面的研究进展，并提出了汞甲基化的未来研究方向。

关键词

汞，甲基汞，人为源，水生生态系统，森林生态系统，甲基化

1 人为源大气汞排放研究进展

人为源汞排放是造成全球汞污染负荷增加的根本原因。因此，研究人为源汞排放特征、量化大气汞排放的时空分布，是控制人为源汞污染和研究汞的地球化学循环的重要基础。近年来，人为源汞排放研究在深化大气汞排放特征研究的基础上，在清单分辨率、时间跨度和清单校验等方面都取得了重要进展。

1.1 人为源汞排放特征研究

人为源汞排放特征的研究主要关注汞在工艺过程的形态转化、脱除和同位素分馏等特征。相较于2015年以前的研究，近年来大气汞排放特征研究的进展主要体现在两个方面。其一，研究对象更加广泛，涵盖了清单中的主要行业。其二，引入同位素维度的排放特征研究。

由于燃煤电厂是大气汞排放最重要的管控源也是全球分布最广的排放源，因此，燃煤电厂是汞排放特征研究最为深入的排放源。近年来，国外燃煤电厂的控制技术相对比较稳定。国内燃煤电厂由于空气质量改善的压力，大气污染控制技术的更新速度非常快，目前普遍采用超低排放控制技术。因此，近年来关于燃煤电厂的研究主要关注汞在超低排放控制技术中的形态转化和脱除[1-16]。研究表明，开展超低技术改造的机组，其污染控制设施的大气汞脱除效率可高达99.3%。其中，静电除尘器（ESP）、低低温电除尘器（LTESP）、布袋除尘器（FF）、湿式电除尘器（WESP）的脱除效率分别为19.6%~60%、47.3%~89.3%、9%~92%和6%~71.4%；湿法脱硫塔（WFGD）和干法脱硫塔（DFGD）的脱汞效率分别为22.2%~92.2%和20%左右。烟气汞的形态是影响大气汞在污染控制设施中脱除效果的重要因素。

除燃煤电厂外，有色金属冶炼（铅、锌、铜和工业黄金）、水泥生产、钢铁生产、垃圾焚烧、民用燃煤、污染场地等排放源的大气汞排放特征也陆续引起关注。在有色金属冶炼企业的研究表明，采用双转双吸制酸进行二氧化硫控制的高温火法冶炼企业，金属生产过程污染控制设施组合的烟气汞脱除效率普

遍在 99%以上[17-19]。然而，大量的汞进入了污酸、硫酸等副产物中，这使得有色金属冶炼行业成为重要的水土汞释放源[17]。此外，锌冶炼废物再利用的二次大气汞排放约占冶炼厂总排放的 40.6%～94.6%，成为潜在的大气汞二次排放源[17]。在水泥熟料生产行业，新型干法水泥生产过程的汞循环导致原料中的汞 90%以上进入到大气中[20, 21]。即便这两年水泥行业增加了选择性非催化还原技术的应用，对大气汞排放也基本上没有起到有效协同控制效果[22]。在钢铁生产过程的大气汞排放特征研究表明，焦化和烧结过程是钢铁生产的最主要排放节点[23]。测试企业烧结过程的大气污染控制设施能够去除 81.1%～97.5%的汞[24]。生活垃圾焚烧过程的测试表明[25]，中国现有生活垃圾焚烧设施污染控制设备的脱汞效率在 33.6%～95.2%的范围内波动。污染控制设备的脱汞效率与单位飞灰的活性炭喷射量呈显著正相关。民用燃煤的排放特征测试更新了该排放源烟气汞的形态分布，研究指出民用燃煤烟气中 Hg^0，而非 Hg_p，是烟气汞最主要的形态[26, 27]。在瑞士污染场地的研究表明[28]，污染场地中 Hg^0 和甲基汞的比例非常低，汞主要以 Hg^{2+} 结合有机质等土壤组分的形式存在。然而，污染场地中的汞在中等还原条件就非常容易被还原。这意味着容易被水淹没的污染场地是非常重要的大气汞排放源。

除了对汞形态转化和脱除的研究外，随着汞同位素技术应用的普及，也有一些学者开始研究人为源汞同位素分馏特征。全球范围内煤样中 $\delta^{202}Hg$ 的范围是–3.9‰～0.8‰，$\Delta^{199}Hg$ 的范围是–0.6‰～0.4‰[29]。我国煤中汞质量分馏值为–0.95‰±0.75‰，汞非质量分馏不显著，$\Delta^{199}Hg$ 在 0.01‰±0.10‰范围内[29]。煤燃烧过程中，炉渣中富集汞的重同位素[30]。烟气污控设施的副产物飞灰和脱硫石膏中富集汞的轻同位素，这是由于 Hg^0 氧化过程中轻同位素反应速率较大[30, 31]。采用海水脱硫时，脱硫废水相较于原海水出现汞的重同位素富集[32, 33]。现场观测数据和质量衡算证明，燃煤电厂尾气中的汞富集重同位素。通过比较燃煤电厂附近降水样品和背景点降水样品中的 $\delta^{202}Hg$，发现二者具有显著差别，证明源排放是附近大气汞的主要来源[34]。

1.2 人为源汞排放清单进展

1.2.1 重点源排放清单研究

随着人们对源排放特征认识的深入，不少行业的大气汞排放清单都得到了更新和补充[35-37]，清单的分辨率和不确定性都得到显著的改善。Liu 等[38]建立了中国燃煤电厂机组排放量的分布图，研究同时根据机组发电量的时间变化趋势，将传统的年排放量清单进一步优化到月变化清单，显著提升了清单的时空分辨率[38]。Zhao 等[39]研究了洗煤副产物的流向，指出副产物的二次利用额外贡献了 77.2t 的大气汞排放。Wu 等[36]基于钢铁生产工艺过程的汞流向和排放特征，更新了中国钢铁行业 2000～2015 年的大气汞排放量。研究发现，中国钢铁行业大气汞排放从 2000 年的 21.6t 增长到 2015 年的 94.5t。烧结和炼焦过程的大气汞排放量分别占 35%～46%和 25%～32%。清单的不确定性从以往研究的(–80%, 100%)降低为(–29%, 77%)。Kim 等[35]对韩国牙科汞合金焚烧过程的大气汞排放量进行估算发现，调研的医院和诊所中，采用汞齐的假牙的比例约占 9.9%，单位含有汞齐的假牙中的汞含量为 0.2g，这导致韩国牙科汞合金焚烧每年的大气汞排放量在 42.53～48.86 kg 的范围内。Wu 等[37]的研究表明，2012 年中国铅锌铜金属生产过程的大气汞排放量为 100.4t。对冶炼副产物的流向进行跟踪显示，副产物利用过程将额外贡献 47.8t 的大气汞排放量。Cui 等[40]估算发现贵州省 1990～2016 年民用燃煤大气汞量累计达到 48.9t。Hg^0 是贵州民用燃煤大气汞排放的主要形态。按照《关于汞的水俣公约》第八条"排放"的要求，各缔约方需在公约生效 5 年内提供相关来源的点源排放清单。这就使得基于行政区域的汞排放清单的精度无法达到公约履约的要求。因此，未来重点行业的排放清单仍需进一步向基于点源的排放清单发展。此外，随着研究的深入，物质流向中鉴别出的二次排放源及重点环节的排放仍须评估和定量。

2017年《关于汞的水俣公约》生效后,不少学者开始关注重点源大气汞的减排潜力及控制措施。因此,基于情景分析的未来年大气汞排放量也成为重点源排放研究的热点。Wu 等[41]估算了不同大气汞排放限值情景下,中国燃煤电厂的大气汞变化趋势。研究发现,将中国燃煤电厂排放限值由 30 μg/m³ 降低到 15μg/m³,短期内存在排放总量上升的风险。基于不同情景下排放总量的变化趋势,研究建议我国燃煤电厂在 2025 年之后实施 5μg/m³ 的排放限值。研究同时指出采用多污染物控制技术可达到这个限值的要求并能够有效控制大气汞的排放量。Takiguchi 等[42]基于燃煤电厂的大气汞排放量、原煤汞浓度、未来技术的发展等信息,建议日本新建燃煤电厂的大气汞排放限值为 8μg/m³。Sung 等[43]预测了韩国五个大气汞重点管控源的大气汞排放趋势,发现 2014 年五个重点管控源的大气汞排放量为 4.48t。在不采用控制措施和采用控制措施的情景下,2022 年大气汞的排放量分别为 6.06t 和 2.66t。Wu 等[44]指出,在最严控制情景下,中国五个大气汞排放源的排放量可从 2015 年的 371t 下降到 2030 年的 68t。多污染物控制措施和替代性措施是未来中国大气汞减排的主要措施,但是水泥熟料生产需要额外采用专门的烟气脱汞技术,有色金属冶炼需专门脱汞技术控制硫酸中的汞输出。

1.2.2 区域及全球大气汞排放清单研究

随着重点源清单不确定性的降低,区域和全球大气汞排放清单的定量也更加准确[45, 46]。近年来,人为源排放清单的研究不再局限于某一年份的排放的量化和不确定度性的降低,而更加关注排放的变化趋势,从而评估区域和全球汞污染控制的成效。Zhang 等[45]研究表明,中国人为源大气汞排放总量从 2000 年的 356t 持续增长至 2010 年的 538t,年均增长率为 4.2%;2010 年,工业燃煤、燃煤电厂、有色金属冶炼和水泥生产分别贡献了 22.3%、18.6%、18.1% 和 18.3% 的大气汞排放。Wu 等[46]研究发现,改革开放以来中国大气汞排放累积达到 13294t。燃煤电厂 SO_2 控制措施、工业燃煤行业的颗粒物和 SO_2 控制措施、有色金属冶炼厂的制酸过程都对大气汞排放有显著的协同控制效果,2011 年开始中国大气汞排放总量开始下降[46]。Bourtalas 和 Themelis[47]研究了美国 1989~2014 年大气汞排放,发现 2014 年美国大气汞排放仅为 51.4t。生活垃圾焚烧和能源与废物燃烧等主要源的大气汞排放均得到了有效控制。此外,汞物质流向的研究促进了排放源的识别和清单的发展。相较于单一行业的排放因子模式,物质流分析的方法将各个单一考虑的行业进行了关联。这种方法一方面有助于识别传统清单忽视的大气汞排放源,另一方面有助于将大气汞排放与水土汞释放的控制统筹管理,避免跨介质传输。Hui 等[48]综合考虑了不同排放源的物料传输,从而建立了中国人为源汞流向。研究指出,2010 年中国大气汞排放量为 633t,其中由于副产物/废物再利用额外增加了 102t 大气汞排放。未来大气汞排放的控制可能增加向水土的汞释放量,大气汞排放和水土汞释放的履约应该统筹考虑。

在全球清单方面,《全球汞评估报告 2018》[49]估算 2015 年人为源大气汞排放为 2220t,约占全球大气汞排放总量(含人为源、自然源和再排放源)的 30%。人为源排放中,亚洲仍然是大气汞排放的热点区域,约占 49%;其次是南美和撒哈拉沙漠周边的非洲国家。在所有排放源中,小手工炼金仍然是最主要的排放源,约占 38%,该部门的排放同时也是不确定性和争议最大的;其次是固定燃煤源和有色金属冶炼,分别占 21% 和 15%。Streets 等[50]估算发现全球大气汞排放从 2010 年的 2188t 增长到 2015 年的 2390t。从汞排放的区域分布看,美国、欧洲 OECD 成员以及加拿大的大气汞排放呈现下降趋势,而美洲中部、南亚和东非仍然处于上升范围内。东亚地区近年来的大气汞排放增速显著下降,但是大气汞排放量仍然高达 1012t。未来大气汞排放控制的热点区域仍是亚洲及小手工炼金区域。除历史排放的估算外,未来减排量的预测及相关控制措施的确定也是研究的热点。这些排放结果也得到了观测数据的验证。Zhang 等[51]等基于清单数据和模型模拟认为,1990 年以来人为源大气汞排放的下降是大气汞浓度的下降的重要原因。

1.2.3 排放清单的校验

除了"至下而上"的清单方法学的不断完善外，对清单的校验工作也取得了一定的发展。Song 等[52]基于全球大气汞的观测数据，建立了"至上而下"的排放清单，从而对已有清单进行校验。研究认为，亚洲地区的大气汞排放量在 650~1770t/a 的水平，高于现有全球大气汞排放量的估算。Landis 等[53]和 Weigelt 等[54]对比烟囱排放口和烟羽中汞的形态分布后指出，当前燃煤电厂大气汞排放清单的 Hg^{2+} 的排放比例存在明显的高估。此外，随着汞同位素技术应用的普遍性，有些学者尝试着将源排放同位素特征谱引入排放清单中，从而增加定量全球汞循环的约束指标。Sun 等[55]首次建立了全球包含同位素分馏特征的大气汞排放清单。研究指出，1850~2010 年，全球大气汞排放的 $\delta^{202}Hg$ 从[46]–1.1‰上升到[46]–0.7‰，$\Delta^{199}Hg$ 从–0.02‰下降到–0.04‰。在此基础上，Sun 等[56]开发了包含同位素分馏特征的箱式模型，从同位素的角度评估清单和模式机制中的不确定性。包含同位素的源排放清单为未来清单的校验和大气汞污染的来源解析提供了新的约束条件。

1.3 展望

近几年，人类活动大气汞排放的研究在源排放特征和清单等方面均取得了比较重要的进展。未来大气汞排放研究将主要从以下方面展开。其一，人为源长时间序列的排放特征研究。现有排放特征研究主要采用离线采样的方法开展相关的工作，采样周期往往在数天或者数周内，无法反映人为源长时间序列的变化趋势。未来有必要开展排放特征的长期研究，从而确定人为源排放特征的长期变化趋势，评估污染控制设施脱汞效果的稳定性。此外，源排放特征谱是同位素来源解析的重要依据。当前人为源同位素的研究主要集中在燃煤电厂开展，其他行业鲜见报道。源工艺过程对汞同位素分馏的影响及相关机制仍有待研究。其二，从物质流向角度研究大气汞的排放。当前全球主要采用多污染物协同控制措施进行大气汞排放的控制，这使得大量的汞进入到各种副产物中。由于汞具有较高的挥发性，因此，副产物一旦经历高温过程就极有可能将汞重新释放到烟气中，导致大气汞的二次排放。此外，大气汞排放控制极有可能带来水体和土壤汞释放的增加，并直接影响含汞废物的汞含量和处理。因此，有必要开展区域及全球范围内的汞物质流向研究，识别大气汞二次排放关键环节，并从全流程开展汞污染控制和公约履约，避免汞的跨介质污染。其三，清单校验工作仍需完善。当前空气质量模式模拟出来的 Hg^{2+} 和 Hg_p 与观测数据仍然具有非常大的差距。清单的形态分布是影响大气汞物种模拟结果的重要因素。因此，有必要在清单校验方法上进行研究，通过引入其他观测变量校验排放数据。

2 水生生态系统汞的迁移转化

作为汞研究关注的焦点，汞在水生生态系统中的迁移转化控制着其全球生物地球化学循环及环境和健康效应。水环境是汞污染排放的重要汇之一，排放至大气与土壤中的汞也可通过干湿沉降和侵蚀淋溶等途径进入环境水体[57]。据估计，目前全球通过人为途径释放至大气中的汞约为 2000t/a，而每年大气沉降至环境水体中的汞超过 3700t[58, 59]。更为重要的是，水生生态系统是汞甲基化、生物富集与生物放大的重要场所[60, 61]。进入到环境水体中的汞主要为无机汞，经过一系列的形态转化过程后会形成毒性更强的甲基汞，并通过食物链富集较快速地积累至生物体[62]。通常环境水体中甲基汞浓度低于 1ng/L，但经过食物链放大后鱼体中甲基汞的浓度可达数百甚至数千 μg/kg，这导致世界许多海洋和淡水环境中鱼体内广泛检测到汞超标的现象[63]。沿水生食物链传递的甲基汞还可进一步进入更高营养级的陆生生物（如鸟和爬行及哺乳动物等）体内[64]。因此，水体环境是影响环境汞风险的主要生态系统。

水环境是汞循环的活跃场所，输入至水环境中的汞可发生一系列的形态转化及迁移过程。这些过程对汞的全球传输和环境健康效应有至关重要的影响，是目前水生生态系统汞研究的热点。进入水体的汞可在厌氧底泥中被硫还原菌[65, 66]、铁还原菌[67, 68]、产甲烷菌[69]等具有 *hgcAB* 基因簇的微生物甲基化[60, 70]，并且近来研究表明在淡水湖泊以及海水水柱中亦存在显著的汞甲基化[71, 72]，其对极地海水甲基汞来源的贡献可达 47%[72]；甲基汞可经由微生物或（光）化学途径降解[73]，甲基化与去甲基化之间的平衡控制着水环境中甲基汞的浓度及在食物链中的累积富集。二价汞的还原是水体中零价汞生成的关键过程，在海水[74]和淡水[75]中普遍存在。化学[76]与光化学[75]等非生物反应或微藻与细菌介导的生物过程[77, 78]均可引起二价汞的还原。生成的零价汞可在水体中经由微生物与（光）化学途径再次被氧化[79]。水体中这种同时发生的氧化/还原过程控制着水中零价汞［溶解气态汞（DGM）］的浓度及经由水-气交换挥发进入大气的汞通量[80]。总之，水生生态系统是零价汞-二价汞-甲基汞三者相互转化的主要场所，同时也是这些主要汞形态发生相态转化过程的介质（气相-溶解相-颗粒吸附相）。本节并非是对水环境中汞的这些形态转化和相态转化过程研究进行面面俱到的详述，而是有针对性地对若干相对较"新"并且又至关重要的研究热点进行总结，以期对推进这些领域的汞研究提供思路借鉴。化合物的环境行为和毒理学特征均表现出明显差异。

2.1 水环境中汞的光化学转化

汞在水体中的光化学反应是汞生物地球化学循环非常重要的环节，它调控着许多与汞归趋及生物积累相关的关键过程。其中零价汞和二价汞之间的光氧化还原和甲基汞的光降解是水体汞光化学的主要研究过程。光介导的二价汞还原是水体中可挥发零价汞的主要生成途径[81]，控制着水-气界面的汞交换迁移。零价汞的光氧化可生成二价汞，增加了水环境中可用于甲基汞生成的无机二价汞的量[82]。尽管有极少关于水体中汞光化学甲基化的报道[83-85]，但多项研究表明水体中存在显著的甲基汞光降解。这一过程对于维持水中较低的甲基汞浓度，调控甲基汞的生物累积与放大具有重要的意义。对于不同水体，甲基汞光降解可占表层水体中甲基汞总量的 30%~80%[86, 87]，其中在美国 Everglades 大湿地表层水中约 31.4%的甲基汞经由光降解途径去除[88-90]。

目前对水环境汞的光化学转化（尤其是二价汞光还原与甲基汞光降解）过程已经进行了大量的研究，并探讨了辐射性质（紫外或可见光）及水化学因素［如氯化物和溶解性有机质（DOM）］等对汞光化学反应的影响[91-93]。尤其有较多研究关注了 DOM 在汞光化学转化中的作用[94-97]，把 DOM 的效应归纳为如下三个方面：①DOM 与汞的强亲和力可形成 Hg-DOM 络合物，在光化学反应过程中可发生分子间/分子内的电荷转移；②DOM 可参与很多光化学反应生成或清除自由基及活性氧物种（ROS），例如激发三线态 DOM（^{3}DOM*）、羟基自由基和单线态氧等，这些自由基类物种都能参与到汞的光化学反应中；③DOM 中的发色官能团可吸收光辐射，具有很强的光衰减效应[98, 99]，可通过改变太阳光辐射强度影响汞的光化学反应。

值得注意的是，除了研究水体中溶解相汞光化学过程外，近来有关颗粒物介导的异相光化学反应逐渐引起关注。前期研究一般认为水体悬浮颗粒物（SPM）可吸收/散射部分光照，通过改变辐射性质及强度而影响汞光化学反应过程[100]。研究表明湖水中的悬浮颗粒物越多，DGM 的生成量就越低[101]，与未过滤比，过滤湖水可增加 30%的 DGM 产生量[102]，这些研究结果都肯定了 SPM 的减光作用。相似的效应也可发生在沿海大陆架区域，高 SPM 导致了较低的汞还原速率常数[103]。但令人十分意外的是，也有研究发现颗粒物的存在可导致汞光化学反应增强。例如，保存在透明和黑色瓶中的过滤水样，相对于未过滤的样品，其中的 DGM 浓度分别低 41%和 48%[74]。这暗示着除了影响光辐射之外，颗粒物在汞光化学反应中还起着其他作用，例如颗粒物表面结合的汞可能直接参与光化学反应，但对此尚缺乏实验证据[104,105]。考虑到水体颗粒物通常具有半导体性质，在光照条件下可产生光电子-空穴对[导带（CB）以及价带（VB）]，分别具有较强的还原和氧化能力，因此可能发生电子-空穴对诱导的汞光化学反应（$E^0_{Hg(II)/Hg(I)}$=0.91 V，$E^0_{Hg(I)/Hg(0)}$=0.79 V，$E^0_{Hg(II)/Hg(0)}$=0.85 V）。另外，水体颗粒物受到光照辐射激发会产生自由基和活性氧物种，这些光敏物种可以直接与吸附在颗粒物表面的汞形态反应，也可能会扩散到溶液中，参与溶解相中的汞光化学反应。目前极为缺乏颗粒物介导的异相汞光化学反应方面的研究。

2.2 水体环境硫化汞的形态转化及在汞循环中的作用

硫化汞是水环境中汞重要的汇，在多种环境与生物介质中广泛存在[106]。近年来，不断在底泥[107]、植物叶片[108]、鱼[109]、贻贝[110]及石油相关样品[111,112]等多种介质中发现硫化汞的赋存，对其在环境中的生成及转化过程也不断有新的认识。

在环境中，硫化汞可通过多种化学与生物途径生成。在 Hg^{2+}-DOM-S^{2-} 三元体系中，X 射线吸收光谱揭示二价汞主要以结构无序的 β-HgS 纳米胶体形式存在；β-HgS 粒径或有序性随 S^{2-} 浓度增加而增加，随 DOM 芳香性增加而降低[113]，这可能与 DOM 中芳香性基团对 β-HgS 的稳定作用有关。另一项研究也提示，对于海洋源 DOM，巯基以外的其余 DOM 官能团也在 β-HgS 稳定中发挥作用[114]。环境微生物硫代谢过程生成的 S^{2-} 可与 Hg^{2+} 反应生成 HgS。微生物硫代谢过程可将半胱氨酸转化为 S^{2-}，从而促进 HgS 的生成[115]。相比于指数生长期微生物，处于稳定生长期或凋亡诱导剂处理（导致微生物 ATP 降低）的微生物 HgS 生成显著降低，表明这一硫化过程与微生物活性有关[115]。进一步研究显示，所生成硫化汞为 α-HgS 与 β-HgS，其比例取决于 Hg^{2+} 与 S^{2-} 浓度，但在过量半胱氨酸（1000μmol /L）存在下，仅有 α-HgS 形成[116]。扫描透射电镜显示，对于革兰氏阴性菌（大肠杆菌和硫还原地杆菌），硫化汞纳米颗粒主要位于胞外与细胞表面；但对于革兰氏阳性菌（枯草芽孢杆菌）硫化汞纳米颗粒主要位于细胞质与细胞膜[116]。植物叶片可吸收大气中的零价汞，因此植物凋落物是汞进入湿地等水生态系统的重要途径[117]。近边结构高能分辨 X 射线吸收谱分析显示纳米硫化汞是植物叶片中汞赋存的重要形式[108]。$HgCl_2$ 暴露的小型淡水鱼类唐鱼，其体内约 57%的汞也以 Hg_xS_y 团簇的形式存在[109]。这些结果表明，动植物体内也存在汞的硫化过程，这可能是生物的一种脱毒机制。生物体内汞硫化的途径尚需进一步阐明，其对汞循环的影响（如凋落物汞的再释放与甲基化）也亟需进一步研究。

近些年的研究发现底泥或土壤中硫（硫酸盐/硫代硫酸钠）的添加对硫化汞的循环与汞的生物摄入有重要影响。这一过程涉及硫循环以及 S^{2-}、有机硫导致的汞硫化与溶解，影响因素较为复杂。一些研究显示，硫酸盐的加入促进了巯基结合态汞向硫化汞的转化，从而降低了根际土壤中汞的移动与水稻摄入[118]；但也有研究表明，硫酸盐（50～1000 mg/kg）的引入增加了土壤中硫化汞的溶解，促进了汞的甲基化（提高 28%～61%）与水稻籽粒中的甲基汞累积（提高 22%～55%）[119]。类似地，硫代硫酸钠对汞的移动与植物吸收的影响也较为复杂。对于芥菜（*Brassica juncea*），土壤中硫代硫酸钠的加入可导致其与汞的络合，促进汞的移动、植物吸收与转运[120]，芥菜体内的汞主要以硫化汞纳米颗粒形式存在[120]；但对于油菜（*Brassica napus*）、玉米（*Zea mays*）、甘薯（*Ipomoea batatas*）与水稻（*Oryza sativa*），土壤中硫代硫酸钠的加入则可导致纳米颗粒态 β-HgS 的生成[121,122]，促进土壤中汞的固定，降低植物对汞的吸收[122]。硫代硫酸钠的不同影响可能与不同土壤条件导致硫、汞循环的差异有关。以上结果提示，在未来研究中应进一步加强对环境中硫循环与硫-汞耦合的理解，从而进一步揭示硫对汞循环影响的微观机制。

硫化汞及纳米硫化汞可被微生物利用，参与汞的生物地球化学循环。薄膜扩散梯度被动采样结合多同位素示踪（$^{204}Hg^{2+}$、$^{199}Hg^{2+}$-FeS、$^{196}Hg^{2+}$-DOM、^{200}HgS）对底泥中不同形态无机汞的甲基化研究显示，虽然纳米硫化汞的甲基化速率低于其他无机汞形态，但总汞的甲基化率仍可达 6%以上[123]。另一项底泥微宇宙模拟结合多同位素示踪 [^{201}Hg（NO$_3$）$_2$、^{202}Hg^{2+}-DOM、β-^{198}HgS、CH$_3$^{204}HgCl] 的研究发现，β-HgS 的甲基化率约为其他无机汞的 1/20，主要受控于氧化还原介导的 Hg^{2+}溶解度；β-HgS 转化为零价汞的挥发速率常数约为其他无机汞的 1/10[124]。未来研究应进一步厘清硫化汞生物利用的机制问题（直接摄入还是溶解后摄入）。

2.3 颗粒形态汞参与的水环境汞迁移转化过程

对于大部分淡水[125-128]与海水[129,130]系统，颗粒形态汞（结合在悬浮颗粒物上的无机与甲基汞）是水中汞形态的主要组成部分，是汞在环境水体迁移的重要形式[131]。无机汞与甲基汞在颗粒相与溶解相之间的分配系数主要受颗粒态/溶解态有机质组成与浓度的控制[126,129,132]，受水华[129,133]、森林野火[134]、季节[132,135]等因素的间接影响。由于悬浮颗粒物是底栖动物的重要食物来源，悬浮颗粒物及颗粒态汞浓度的增加会增加大型底栖动物的汞摄入[136,137]。此外，近年来发现的颗粒结合态零价汞在环境水体的存在（可占表层水中总零价汞的 50%以上）表明悬浮颗粒物与零价汞的结合[138]可能抑制水体中溶解性气态汞的形成与挥发，导致悬浮颗粒物与 DGM 的负相关[139]。在传统认识中，通常认为颗粒态汞是惰性的，将其当作汞的汇来处理。近年来一些研究表明汞在水-颗粒微界面可能具有丰富的形态转化，并可被微生物吸收利用，表明颗粒态汞在水环境汞转化过程中的重要作用不容忽视。

实验室模拟研究表明，二价汞可快速吸附在半胱氨酸涂层的赤铁矿颗粒、底泥颗粒物上，且其解吸率较低（0.1%~4%），而二价汞的强络合配体（如：低分子量硫醇和腐植酸）可将其从赤铁矿颗粒（60%）及沉积物颗粒（<6%）上部分解吸下来[140]。模拟甲基化实验进一步显示，厌氧条件下颗粒态二价汞可被脱硫弧菌 ND132 吸收和甲基化，其甲基化率比二价汞的解吸率高 4~10 倍，表明颗粒态汞与微生物的直接接触可促进二价汞的解吸与微生物摄入[140]。现场分析发现，日内瓦湖沉降颗粒物中总汞浓度与底泥相当，而甲基汞浓度则约为底泥的 10 倍[135,141]。沉降颗粒物与底泥中甲基汞的去甲基化速率常数（k_d）相当，但沉降颗粒物中汞甲基化速率常数（k_m）较底泥高一个数量级[141]。因此，沉降颗粒物中甲基汞生成净潜力（k_m/k_d）约为底泥的 10 倍，水体颗粒物中汞原位甲基化可能是控制沉降颗粒中甲基汞水平的关键过程[141]。最近一项研究也表明，湖泊中水华爆发后藻类的沉降和分解可进一步增加甲基化微生物的丰度与活性，从而促进底泥中汞的甲基化[142]。在海洋低氧区及颗粒态有机碳高矿化区，甲基汞占总汞比例较高，也提示颗粒结合态汞在这一区域可能存在微生物甲基化[143]。

对美国弗吉尼亚州 Waynesboro 南河（South River）各介质中汞同位素的分析发现，悬浮颗粒物相和溶解相汞同位素之间存在 δ^{202}Hg 值约 0.28‰的负偏移，表明汞在颗粒相与溶解相之间的分配可能存在同位素分馏[144]。美国田纳西州橡树岭 East Fork Poplar Creek（EFPC）河流各介质中汞同位素的分析发现，水中颗粒态汞同位素组成与工业使用的零价汞类似（δ^{202}Hg=−0.42‰±0.09‰，Δ^{199}Hg≈0）；下游颗粒态汞 δ^{202}Hg 增加 0.53‰，Δ^{199}Hg 降低−0.10‰；而溶解态汞 δ^{202}Hg 从上游的−0.44‰~0.18‰变为下游的约 0.01‰±0.10‰[145]。溶解和颗粒态汞同位素组成的变化并不能完全由流动过程中的稀释和沉降等物理过程解释，其可能经受了微生物与光化学还原等过程[145]，而悬浮颗粒物可能介导了二价汞的光还原[96]或甲基汞的去甲基化[97]过程。

2.4 海洋环境汞迁移转化研究的热点问题

海洋是汞环境地球化学循环的一个重要源和汇。排放到环境中的汞可以通过大气沉降、废水排放和河流流入等多种途径进入海洋。受光照、有机物、微生物以及其他化学活性物质的影响，汞在海洋环境

中会发生复杂的形态转化、生物累积和迁移过程。一方面，可以转变为挥发态的汞而进入大气中，成为二次排放源；另一方面，能够被生物体吸收，随食物链在生物体内累积和放大，或者向沉积物转移。根据联合国环境规划署（UNEP）的"Global Mercury Assessment 2018"，每年大约有3800t汞由大气沉降至海洋，同时有3400t汞蒸发重新进入大气；由陆源排放随河流进入海洋的汞，每年大约300t；此外，大约100t汞会由海洋底部的地质作用进入海洋，同时海水的垂直运动及颗粒物的沉降又会将部分汞带入洋底。由于海洋在全球汞循环中的重要地位，针对海洋环境中汞迁移转化的研究可以更加深入了解汞在海洋不同介质、不同深度及海洋生态系统中的环境行为，从而对汞的全球循环过程以及沿海居民或食用海产品人群的暴露风险有更加深入的了解。

目前，针对海洋各介质，包括海洋大气、海水、沉积物及海洋生物中汞形态分布、时空变化及影响因素是开展最为广泛也相对成熟的研究。由于不同海域的污染来源、水文和环境条件以及生态系统存在较大差异，导致汞的分布特征和演变趋势也不尽相同。对我国近海沉积物及生物样品的分析结果表明，在我国近海区域存在一定程度的汞污染，且呈现明显的时空变化，如珠江口及近岸海域沉积物中汞含量在20世纪70年代后明显增加，与当地经济发展具有密切关系[146]。总体来看，洋流在汞的运输和分布中起重要作用，对汞的空间分布具有明显影响；汞在沉积物中的沉积受总有机碳和pH的影响，其中总有机碳的影响更为显著[147]。在渤海不同营养级生物样品中，营养级对甲基汞的富集起主要作用，脂肪含量的影响并不显著；相反，无机汞含量与营养级负相关，呈现生长稀释效应[148]。近期，Meng等[149]构建了一个组合的SC-UDF（SC, source contribution；UDF, urban distance factor）模型，一方面定量计算了我国四个海域沉积物中不同汞源的贡献百分比，另一方面揭示了沿岸城市对近海汞污染的影响范围，可用于快速判定区域内其他未知采样点的主导汞源，进而针对性地制定汞污染控制对策。

由于海水和表层沉积物中具有活跃的微生物种群，以及富含其他有利于烷基化的化合物，是环境中重要的汞烷基化及去烷基化场所[150]。近期一些研究利用科考船随航采集的海水悬浮颗粒物及沉积物进行现场或实验室模拟试验，通过改变各环境参数如pH、光照等对其中汞的烷基化和去烷基化速率及影响因素进行了深入研究[151-156]，在海洋环境中汞迁移转化过程方面获得了新的认识。但是，大多数研究仍集中在甲基化和去甲基化过程，对海水和沉积物中其他的汞烷基化和去烷基化过程研究较少。

近年来，汞同位素技术迅速发展并展现出巨大的潜力，在海洋汞迁移转化研究中也发挥了重要作用。例如，Yin等[157]通过我国珠江口及南海沉积物中汞的同位素组成特征，证明周边工业和城市排放是珠三角沉积物中汞的主要来源。Xu等[158]使用二元混合模型，发现我国南海水箱养殖渔场中非肉食性鱼中绝大部分的汞主要来自于饲料，而肉食性鱼体中的汞主要来自于饲料与内脏的结合。同时，甲基汞浓度以及甲基汞在鱼体中的比例与$\delta^{202}Hg$值成正比，但是与$\Delta^{199}Hg$值没有关系。可以预见，汞同位素方法在示踪海洋环境中汞的来源和迁移转化过程方面具有广阔的应用前景。

2.5 生态系统演变对水环境汞生态健康风险的影响

近年来一些研究从整体出发，探索汞在水环境中的迁移转化与汞排放控制对生物体汞累积及健康风险的相对作用，这类研究尤其值得关注。作为目前最严重的全球环境问题之一，汞污染造成的水环境中甲基汞的生成和食物链累积富集并进而通过鱼类等水产品消费造成人体汞暴露及健康风险是汞研究关注的焦点问题。鱼体内汞的富集水平一方面受到汞排放（即水生生态系统中的汞输入与输出）的影响，另一方面也受控于水环境中汞的甲基化/去甲基化、氧化/还原和吸附/解吸等一系列迁移转化过程[63]以及各种环境因素（如天然有机质[159, 160]、硫酸盐[161]、氧化/还原电位[159]等）。降低汞的环境健康风险是全世界128个国家签署《关于汞的水俣公约》（Minamata Convention on Mercury，2013年1月19日签署，2017年8月16日正式生效）的主要目的，而这一目的的实施是通过在全球范围内控制和减少汞排放。由于排放和产生毒性的汞化学形态不同并且从排放到健康风险要经过复杂的汞生物地球化学循环过程,因此除汞排放外,水环境中汞迁移转化过程及控制因素也是影响汞生态及健康风险的关键。这使得实施《关于汞的水俣公约》的成效将不仅取决于汞减排量,也受控于水环境中的汞循环过程及控制因素。

自 20 世纪 80 年代以来,世界许多国家地区已开始采取关闭汞矿和限制含汞产品使用等汞减排措施[162]。这些措施显著降低了北美和欧洲等地区汞的排放量显著降低[58],然而在这些区域鱼体内汞风险并未相应减小,报道的鱼体内汞浓度变化趋势并不一致[163-171]。汞排放量与鱼体内汞变化趋势的不同也间接表明一些系统中,相较于汞排放量,生态系统演变对汞生态健康风险的影响可能更大。这一观点也在之前的一些研究中得到证实。例如,研究表明加拿大几个湖泊中鱼体内汞年际变化主要受全球气候变暖导致的温度升高而非汞排放量变化控制[172, 173]。Adirondack 湖过去二十多年的监测数据统计分析结果表明,流域面积、水位、pH 和鱼个体状况是控制鱼体内汞水平的关键因素[174]。此外,浮游生长的"生物稀释"作用也被广泛报道可以降低水环境汞的风险[175, 176]。这些研究结果表明弄清生态系统演变(包括食物链和环境因素变化)对水环境汞生物地球化学的影响是科学评价水环境汞风险的关键。

中国是世界上汞排放量较大的国家,水体和沉积物中汞的含量也较高,汞的潜在生态健康风险较高。然而目前报道的鱼体内汞的含量普遍相对较低[177-179],存在汞实际生态环境风险与排放强度不符的问题。目前对于该现象的解释主要包括几个方面,一是生态环境破坏及污染导致中国大部分鱼类处于较低营养级[177-179],二是过度捕捞导致经济鱼类寿命较短[177],三是富营养化引起的"生物稀释"效应[178]。鉴于我国高汞现存量和排放压力,《关于汞的水俣公约》签署后虽然汞的排放量将逐渐减少,但今后很长一段时间水环境仍将面临汞潜在风险较高的问题,如何合理评价和预测生态系统演变条件下中国水环境汞的环境效应和生态健康风险是水环境汞迁移转化研究应该关注的一个重点。

3 汞在森林生态系统生物地球化学研究进展

如前所述,人为源排放的汞包括:气态元素汞(GEM,Hg^0)、气态氧化汞(GOM,Hg^{II})与颗粒态汞(PBM,HgP)[180, 181]。GOM 与 PBM 由于化学性质活泼,能通过干湿沉降较快速从大气中去除,而进入地表与海洋生态系统[181, 182]。而 GEM 具有较强的化学惰性,在大气中的停留时间约为 0.5~1.5 年,随全球大气运动而进行长距离传输[181, 183]。在太阳辐射、大气活性自由基与大气氧化物的作用下,气态元素汞能够被缓慢地氧化为气态氧化汞,或氧化产物吸附在颗粒物上形成颗粒态汞[181, 183]。除了上述的去除方式外,气态元素汞能通过干沉降的方式进入海洋与陆地生态系统。对于陆地生态系统,最引人注意的是植被的叶片通过气孔吸收大气中的元素汞,并以凋落物的方式进入森林生态系统[184]。相关研究表明,全球森林凋落物汞的沉降总量为 1000~1200 Mg/a[185, 186],这相当于全球大气约 20%的总汞库存量[183]。显然,大部分人为源排放的汞通过大气传输、植被转化、凋落物沉降而最终被积累到森林土壤中。

沉降在森林土壤的汞,可随地表径流进入下游水生生态系统而造成甲基汞的污染。由于森林表层土壤富含对汞具有极强亲和性的有机质,研究表明只有少量(<10%)已沉降的汞会迁移转化进入到水生生态系统[187-191]。除此之外,最引人关注的是沉降累积在土壤的汞在光、微生物及非生物作用下还原成 Hg^0,再次释放到大气环境中[192, 193]。目前对土壤汞向水生生态系统的迁移途径的认识已比较清楚,但是对于沉降汞的再释放过程认识尚有很大的不确定性。从全球尺度看,相比其他陆地生态系统,学术界对森林系统与大气间 Hg^0 交换通量估算的误差最大,其通量交换范围为 −727~+707 Mg/a,这使得森林生态系统到底是大气 Hg^0 的源还是汇的问题成为争论的焦点[194, 195]。因此,研究森林生态系统汞的生物地球化学过程对认识全球汞的生物地球化学循环及评估生态环境的健康风险起着至关重要的作用。

3.1 森林生态系统汞循环过程

森林作为成熟的多介质生态系统,汞在森林生态系统的地球化学循环主要包括以下 4 个过程(图 7-1):①大气-植被叶片界面间汞交换过程(包括了植物通过气孔从大气中吸收 Hg^0[196]、植物气孔向周围大气再释放 Hg^0[184, 197]、叶片表面吸附大气中的 Hg^{II} 和 Hg_p 以及叶片表面的汞的光致还原过程[184, 198]);②大气-森林土壤界面间汞交换(主要包括了大气汞的沉降、表层土壤的还原作用以及深层土壤的还原过程[199-203]);③植被-土壤系统的 Hg 交换(包括植被蒸腾作用从土壤溶液吸收 Hg 并向地上部份传输[197, 204-206]、植被凋落物向土壤输入 Hg[186, 207]、穿透雨洗刷叶片表面的 Hg 和树干茎流洗刷树枝树干表面的 Hg 后,向土壤输入 Hg[187-190])、植被叶片吸收土壤释放的 Hg^0(间接过程);④地表径流-土壤-地下渗流系统的汞交换[208]。

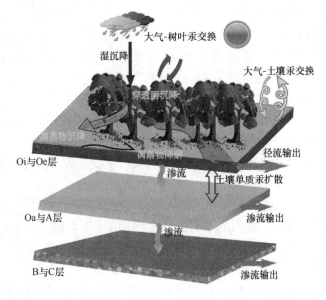

图 7-1 森林生态系统汞的生物地球化学循环过程

目前,森林生态系统汞循环不确定性来源主要是以下两个方面。第一个方面是关于大气-植被叶片界面间汞交换通量的数据匮乏。全球关于大气-树叶间汞交换通量监测的森林站点不到 10 个[194]。该数据的缺乏使得在估算大气-树叶间汞交换通量时,产生了极大不确定性[194]。当前的估算模型研究工作只能将森林凋落物的汞输入来代替叶片生长周期内大气汞在叶片中净积累量[186, 209]。这类近似的替代虽然能够为模拟大气-叶片间汞的通量交换提供一个相对可靠的参考数据集,但是该过程忽略了叶片与大气间双向的汞交换过程,而且植被叶片中的汞并不完全来源于大气零价汞的沉降且存在叶片向植被枝茎等部分的传输,这显然会导致模型估算中的偏差[210]。第二个方面是关于大气-土壤界面汞通量交换动力学研究很少。尽管在全球已经有很多监测站点报道该净交换通量结果[194, 195],然而大气-土壤 Hg^0 交换通量是一个净交换通量的结果,杂糅了诸多的动力学过程,对于汞在土壤中演化的化学动力学研究很少。缺乏足够的关于土壤中汞的还原机制的认识,使得大气-土壤间汞通量交换的模型构建变得十分困难。如何厘清大气-土壤交换过程的各个动力学过程的贡献度,是下一步建立完善的地气交换模型的基础,也是当前的研究难点。

3.2 森林生态系统汞的质量平衡

近 20 年来,前人针对于森林生态系统汞质量平衡开展了大量的研究,研究主要集中在欧洲以及北美的北方森林/温带森林，通过对森林生态系统空旷地降水、森林内穿透雨、沉降的凋落物和地表径流输出中汞含量的长期观测，初步估算了森林生态系统汞输入和输出情况[187-190]。结果表明，森林生态系统从大气中转移的汞绝大部分（>80%）最终被保留在森林土壤中，证实了森林是大气 Hg^0 的汇。Agnan 等[194]估算了全球森林生态系统每年能够从大气中转移 Hg^0 的量为 59Mg（–727～+707 Mg/a），存在极大的不确定性。

我国学者在森林系统中也开展了类似的工作研究，如云南的哀牢山[184]，贵州的雷公山和鹿冲关[191]、重庆的铁山坪[191]、缙云山和四面山[211-213]、四川贡嘎山[200]、吉林长白山[214]以及江西千烟洲和湖南会同[197,215]等森林系统，其结果表明大气汞与森林的交换通量为–215～58μg/（$m^2·a$）。除江西千烟洲和湖南会同亚热带针叶林外，其他站点的结果均认为森林生态系统是大气 Hg^0 的汇。

在森林生态系统中，植被叶片与大气间汞交换是一个双向进行的过程[184, 216-218]。早期的研究表明，植被能够通过叶片的蒸腾作用向大气释放汞[197, 219, 220]，并且有的模型模拟的结果也阐明了森林系统中汞的释放能够占到自然源汞排放的 75%[221, 222]。然而，由于植被根系铁膜能够阻止土壤溶液中的汞进入到植物地上部分，且添加稳定同位素[204]和天然稳定同位素[184, 214, 223, 224]的数据都表明了植物叶片中大部分的汞（可达 99%）来自于大气。植被能够从大气中吸收 Hg^0，并在叶片中逐渐积累，但同时也能够向大气中再释放 Hg^0[184, 225]。相比于利用凋落物估算叶片对大气汞净积累的方法，利用动力学通量袋法能够完成小尺度的枝条-大气间的 Hg^0 的交换测量[184, 197, 226]，能够实现大气-植被叶片间瞬时交换通量的测量。利用汞稳定同位素手段结合动力学通量袋的研究成果表明，植被的再释放作用约能够占到植被初始吸收通量的 30%[184]。然而对于叶片-大气测量的研究还非常有限，仍缺乏大量的数据来建立更加完善全球生态系统汞循环模型。

森林土壤中汞的储量巨大，且在汞的全球地球化学循环中较为活跃[227-229]。土壤中汞的来源主要包括来自于干沉降（包括凋落物的汞输入和雨水冲刷植被过程汞的增加）和湿沉降（空旷降水汞输入）[230]。土壤中汞的归趋包括土壤在微生物、光和有机质暗反应作用下的还原再释放到大气的过程、汞在土壤中的老化累积和地表径流/地下渗流汞流失过程。

对全球森林而言，森林中穿透雨汞输入为 1338Mg/a[189, 191, 214, 231-240]。亚洲地区（数据来源于中国森林的数据结果）的穿透雨的汞输入量是欧美地区的 2～3 倍，主要是由于其穿透雨中汞浓度偏高引起的[186, 230]。中国森林穿透雨中汞含量高于欧洲和北美森林地区可以归因于两点：其一，中国空旷地降水中汞含量偏高；其二，中国高汞环境下，森林地区冠层滞留了更多的 Hg^{II} 和 Hg_P，穿透雨洗刷叶片过程中带入了更多的汞。大量的数据表明了，森林地区穿透雨汞浓度大概是空旷地降水汞浓度的 2 倍以上是[183, 188-191, 231, 234-239, 241-243]。结合全球森林地表接受凋落物汞输入量为 1000～1200Mg/a[185, 186]，综合可知，汞的干沉降是大气中汞沉降进入森林生态系统的主要方式（图 7-2），可占总沉降量的 70%～85%。

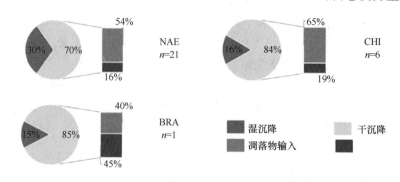

图 7-2 森林生态系统汞沉降所占比例饼形图

NAE 表示北美和欧洲的数据结果，CHI 表示的是中国的数据结果，BRA 表示的是巴西的数据结果（引自参考文献[186]）

森林土壤与大气 Hg^0 间的交换过程是一个双向的过程（图 7-1）[244]，净通量结果表现为大气 Hg^0 的源（表 7-1）。基于动力学通量箱法[202,245]，前人针对森林土壤与大气间 Hg^0 交换开展了大量的测量工作。但是测量结果存在较大的不确定性 [−1.3~81.2 μg/(m²·a)]，进而导致估算全球森林土壤的汞通量时也产生了一定的不确定性（0~500 Mg/a）[194]。估算全球森林土壤汞通量时的误差主要来自于以下几个方面：首先，测量方法的不一致性（包括通量箱的规格、材质、采样流速的不同等）[194,202,246]；其次，森林内部土壤空间异质性大，相近的位置通量存在巨大差异[194]；第三，测量时间较短，通常仅为几天到几周，缺乏长时间多季节的测量结果[195]；最后，前人的研究大多数集中在欧洲和北美的北方/温带森林[估算值为（173±127）Mg/a]，对于热带和亚热带森林的观测数据较少[估算值为（300±450）Mg/a]。而热带和亚热带森林面积占了全球森林面积的一半，用少量的数据进行整体的估计，能够造成较大的估算误差。因此，未来的森林土壤-大气间 Hg^0 交换的研究应该更多地关注热带和亚热带地区的森林。

表 7-1 全球森林系统大气-土壤 Hg^0 交换通量结果及主要影响的环境因素　　　　[单位：ng/(m²·h)]

森林类型	位置	均值	SD	影响因素	参考文献
亚热带森林	亚洲	7.2	5.3	温度,湿度和大气汞浓度	[247], [246], [248], [212], [211], [249], [200]
热带森林	南美洲	1.65	0.7	温度和湿度	[250], [251], [252]
温带/北方森林	亚洲	1.8	1.42	温度	[214], [253]
	北美洲	1.05	1.3	温度和湿度	[254], [255], [256], [257], [258], [259], [260], [261]
	欧洲	0.27	0.49	温度和湿度	[262], [245], [263], [219]

土壤中先前沉降的"老"汞（几十年前或者上百年前沉降到土壤中的汞）对土壤与大气间 Hg^0 交换通量的贡献份额尚不清楚。原因就是，动力学通量箱仅仅测到大气-土壤间 Hg^0 净交换通量，而无法辨别土壤中复杂的循环过程。目前，只有极少数的研究表明次表层土壤（≥5cm）深度的"老"汞也可能参与地表土壤与大气间 Hg^0 的交换过程[264]，且该过程对土壤表面与大气间 Hg 交换通量的贡献可能随着土壤的深度增加而降低[199,201]。然而，Demers 等[265]利用汞同位素示踪了大气-土壤交换界面的汞的循环过程，结果表明森林大气-土壤间 Hg^0 交换过程是 Hg^0 在土壤-大气界面间的一个快速循环过程，即大气 Hg^0 向土壤界面沉降随后这部分汞迅速从土壤界面向大气环境再释放，而不是先前沉降的"老"汞被还原的结果。综上所述，控制大气-森林土壤 Hg^0 交换的机理十分复杂，对于土壤大气界面再释放汞的来源还缺乏统一的认识，目前还无法科学回答大气 Hg^0 沉降到土壤的"老"汞需要多少时间的老化才不再参与大气与土壤界面的汞循环这一科学问题，这也极大制约了科学评估未来执行汞减排政策的效果。

在地表径流-土壤-地下渗流系统汞的交换过程中，由于森林系统表层土壤富含有机质，有机质对于 Hg 具有极强的亲和力，因此地下水中的汞含量相比于降水要低 2~7 倍[228]，造成森林系统地下渗流汞的输出量几乎可以忽略。

3.3　森林生态系统汞稳定同位素研究

自然界中汞具有 7 种稳定同位素，分别是 ^{196}Hg、^{198}Hg、^{199}Hg、^{200}Hg、^{201}Hg、^{202}Hg 和 ^{204}Hg，自然丰度分别为 0.15%、10.0%、16.8%、23.1%、13.2%、29.8%、6.85%。随着多接收电感耦合等离子体质谱仪（MC-ICP-MS）的快速发展，汞同位素的测量精度普遍达到±0.1‰（2σ），极大地促进了汞同位素的发展[266-268]。汞同位素具备了特征性的"三维"同位素分馏体系，分别为汞同位素的质量分馏（MDF，通常用 $\delta^{202}Hg$ 表示），汞的奇数同位素非质量分馏（odd-MIF，通常用 $\Delta^{199}Hg$ 表示）和汞的偶数同位素非质量分馏（even-MIF，通常用 $\Delta^{200}Hg$ 表示）[269,270]。自然界几乎所有过程都能够引起汞同位素的质量分馏，包括化学过程（如光致还原过程、光致氧化过程、有机质的暗还原过程、有机质的暗氧化过程、汞的甲基化和光致去甲基化过程等）[266, 271-275]、物理过程（如吸附、脱吸附、蒸发、挥发和扩散等）[276-281]以及微生物参与的生物地球化学过程（如微生物的还原过程及微生物参与的甲基化过程和去甲基化过程）[266, 282-286]。汞奇数同位素非质量分馏仅仅发生在一些特殊的地球化学过程中，如Hg^{II}的光致还原过程、Hg^0的光致氧化过程、有机质的暗氧化/暗还原过程、甲基化过程和光致去甲基化过程等[266, 271-274, 276-279, 287-289]。较大的奇数汞同位素非质量分馏（$\Delta^{199}Hg>\sim0.4‰$）被发现在液相光还原反应[266, 274, 288, 290]、光致去甲基化过程[266, 285, 291]和气相光氧化过程[275]。较小的奇数汞同位素非质量分馏（$\Delta^{199}Hg<\sim0.4‰$）被报道在有机质的暗还原/暗氧化过程[272, 273]、蒸发过程[276, 277]和Hg^{II}与-SR/-SH络合过程[278]。较小的奇数汞同位素非质量分馏通常是由于核体积效应（NVE）驱动的[292, 293]，并形成$\Delta^{199}Hg/\Delta^{201}Hg$斜率约为1.6[273, 277, 278]。与此相反的是，核磁效应（MIE）驱动下的奇数汞同位素非质量分馏过程通常会导致更大的非质量分馏，并且形成$\Delta^{199}Hg/\Delta^{201}Hg$斜率为$1.0\sim1.2$[266, 274, 288, 291]。在湿沉降样品中[265, 294, 295]及荧光灯管涂层中[296]，科学家们还发现明显的偶数汞同位素非质量分馏（$\Delta^{200}Hg>0.1‰$）。汞同位素的偶数非质量分馏中$\Delta^{200}Hg/\Delta^{204}Hg$比值为$-0.6\sim-0.5$，但是其产生机理并不清楚，很可能是与高层大气光化学氧化过程相关[270]。

汞稳定同位素技术为进一步认识汞在森林系统的地球化学过程（如汞的沉降过程、再释放过程）的研究提供了新的视角。植物叶片、大气、雨水、土壤及母岩通常分别具有特征性汞同位素指纹。就全球而言，大气中的$\delta^{202}Hg$均值为正值（$0.59‰\pm0.29‰$），$\Delta^{199}Hg$值为负值（$0.59‰\pm0.29‰$）[184, 214, 224, 265]；雨水中的$\delta^{202}Hg$值为负值（$-0.57‰\pm0.47‰$），而$\Delta^{199}Hg$值为明显的正值（$0.44‰\pm0.23‰$）[265, 294, 295]；树叶、凋落物、表层土壤中汞的$\delta^{202}Hg$与$\Delta^{199}Hg$均为负值[56, 184, 193, 265, 297]；而母岩中汞的$\delta^{202}Hg$一般为负值，且$\Delta^{199}Hg$的值一般趋近于0。值得一提的是，森林系统中只有降水的中观测到明显的$\Delta^{200}Hg$，大小为$0.25‰\pm0.19‰$[265, 294, 295]。

3.3.1 大气-叶片间汞通量过程汞同位素分馏

大气作为植物叶片中汞的最主要来源，植被吸收大气汞的过程中能够产生明显的汞同位素的质量分馏（MDF），相比于大气成熟叶片中MDF的偏差能够达到$-3‰$[184, 214, 224, 265]。这是由于植物叶片在吸收大气Hg^0时，优先选择较轻的同位素，导致成熟叶片及凋落物保留了偏负的$\delta^{202}Hg$。而研究也表明了，植被叶片存在明显的再释放Hg^0的过程，是导致成熟叶片及凋落物中汞比大气汞存在偏负$\Delta^{199}Hg$的原因（$-0.1‰\sim-0.2‰$）[184]，且再释放汞存在特征性的汞同位素指纹，偏负的$\delta^{202}Hg$和偏正的$\Delta^{199}Hg$。考虑到全球植被吸收大气汞而产生的大气汞汇相当可观，因此，植被吸收大气汞及再释放汞的过程产生的汞同位素的分馏，有可能对全球大气汞同位素的组成产生深远的影响。

3.3.2 大气-土壤间汞通量过程汞同位素分馏

土壤中结合态的汞经过还原作用在土壤空隙中的形成气态单质汞[199]，进而参与大气-土壤界面间的汞通量交换。土壤中汞还原的途径主要分为在微生物[298]、光[299]及非生物质的暗反应[76]等。利用汞稳定同位素示踪土壤中汞的老化过程发现，有机质暗还原过程是其中不可忽视的一个过程[193, 297]。随着土壤剖面的加深，土壤中的汞存在着$\Delta^{199}Hg$信号逐渐变负的特征，主要归因于有机质暗还原过程造成Hg^{II}丢失。在瑞典Histosl的剖面及美国14个森林剖面中$\delta^{202}Hg$随土壤加深逐渐变正[193, 227]，Jiskra等[193]将其归因于微生物还原作用和有机质暗还原过程的综合作用，而Zheng等[272]则强调Hg^0的氧化过程中的平衡分馏是造成$\delta^{202}Hg$变正的可能原因。而在法属圭亚那铁铝土剖面中$\delta^{202}Hg$随土壤加深逐渐变负，主要是由于含铁基团的氧化过程导致$\delta^{202}Hg$偏负[297]。目前森林土壤剖面中汞同位素的变化的原因仍未达成共识，存在着诸多的争议。尤其是其中非质量分馏的变化规律，在其他森林土壤剖面中并未被觉察到[227]。Sun等[56]最新的汞同位素全球模型的研究成果中，没有考虑森林地区土壤过程中$\Delta^{199}Hg$的变化，因此，无法厘清汞在森林土壤内部的地球化学循环过程，限制了全球汞同位素模型的建立，甚至会造成估算的不确定性。

3.3.3 地表径流过程汞同位素分馏

利用汞稳定同位素示踪了地表径流中汞的来源,发现地表径流中汞同位素组成与土壤中汞的同位素组成基本一致,其 $\Delta^{200}Hg$ 的信号特征为$-0.12‰\sim-0.01‰$[208]。与天然降水中汞同位素偶数非质量分馏特征（+）$\Delta^{200}Hg$ 显著不同,说明了降水中汞同位素信号经过穿透雨洗刷叶片-土壤有机质过滤后已经完全被稀释而变得难以察觉到。地表径流表现出与表层土壤相一致的信号特征,这有利于较为准确地构建森林系统汞的同位素质量模型。然而,穿透雨洗刷植被叶片的过程目前研究仍缺乏同位素的证据,难以厘清天然降水中（+）$\Delta^{200}Hg$ 信号消失具体是发生在哪一过程中。因此,未来关于穿透雨过程的汞同位素组成变化的研究为进一步辨别水中 Hg^{II} 在森林生态系统中的源汇关系提供基础数据。

3.4 总结与展望

虽然从 20 世纪 90 年代开始,森林系统汞的生物地球化学过程的研究取得显著的进展,但从全球尺度看,相比其他陆地生态系统,森林系统与大气间 Hg^0 交换通量估算的误差最大,通量交换范围为$-727\sim703$ Mg/a,这使得森林生态系统到底是大气 Hg^0 的源还是汇的问题上成为争论的焦点[194]。为了全面认识森林系统汞的生物地球化学过程,下述的关键科学问题与任务可能是未来研究的重点。

大气-叶片间汞交换过程观测数据匮乏,存在较大不确定性。长久以来,学术界一直将凋落物向土壤的汞输入量来表征叶片从大气中吸收汞的通量[230],而忽略了大气-叶片间汞交换的生物地球化学过程。动力学通量袋的实验则由于通量袋的温室效应和壁效应的影响[184],一直被人诟病。少量的通量袋直接测定大气-叶片汞交换的实验也存在着较大的不确定性。例如,亚热带气候下的哀牢山常绿阔叶林植被表现为大气汞的汇,而同样是亚热带气候下的千烟洲常绿针叶林则表现为大气汞的源[197]。如此截然不同的结果,制约了对于森林大气-叶片间汞交换过程的理解。亟需更多的野外测量的数据结果和一致的测量方法来有助于建立更加完善的大气-叶片汞交换过程模型。

土壤与大气间汞的化学动力学过程认识匮乏。土壤系统除了其自身复杂的生物地球化学过程外,还包括过去沉降的"老"汞（几十或上百年前沉降到土壤的汞）与表层土壤的"新"汞间的复杂交互作用。土壤汞的老化效应可能会降低先前沉降汞的活性,使得汞在土壤不同层位的结合状态发生变化,导致汞在土壤中的可移动性与生物化学活性的变化。目前,只有极少数的研究表明次表层土壤（≥5cm 深度）的"老"汞也可能参与地表土壤与大气间汞的交换过程[264],且该过程对土壤表面与大气间汞交换通量的贡献可能随着土壤的深度增加而降低[199]。由于无法科学回答大气汞沉降到土壤后需要多少时间的老化才不再参与大气与土壤界面的汞循环这一科学问题,这制约了进一步对陆地生态系统汞的生物地球化学循环规律的认识。

森林生态系统汞循环过程中汞稳定同位素的分馏研究仍非常有限。汞稳定同位素作为一种有效的手段可以示踪汞在森林汞生物地球化学循环过程,但是目前的研究还很有限,很多科学问题还认识不清。如森林地气交换过程中汞同位素的分馏是如何进行的,土壤中复杂的生物地球化学过程如何改变汞同位素组成,降水中偏正的偶数汞同位素非质量分馏信号最终归趋到了哪里及穿透雨过程汞同位素组成是如何变化的,土壤老化过程中的有机质暗反应过程在全球森林土壤中是否是普遍存在,等等。上述这些关键科学问题的解决,为进一步认识森林系统在全球汞的生物地球化学循环所扮演的角色奠定了坚实的基础与保障。

4 汞的甲基化研究进展

4.1 甲基汞的性质和危害

据估计,在工业时代,生态系统中汞(Hg)的输入量增加了2~5倍[300]。汞是一种全球性污染物,可以从污染源长距离运输到世界各地,在生态系统和食物网中普遍存在,造成严重的环境和人类健康问题[301]。其中,由于极强的毒性、生物累积性和生物活性,甲基汞(CH_3Hg^+)是一种最受关注的汞存在形式[302]。甲基汞极易随食物链累积放大,生物浓缩因子可达到6个数量级以上,易被人体吸收并蓄积[303, 304]。甲基汞是一种发育神经毒素,可穿过血脑屏障,对中枢神经系统造成不可逆转的损害;也可以通过胎盘屏障进入胎儿体内,且在胎儿大脑和其他组织中蓄积的浓度可超过母体,对婴幼儿的神经发育系统造成严重威胁[305, 306]。

4.2 甲基汞的来源和分布

在自然条件下,汞的甲基化主要发生在缺氧水和沉积物中。在大多数湖泊中,水体甲基汞来自底层缺氧水中甲基汞的扩散和运输。然而当从深水层输送的甲基汞含量可忽略不计时,海洋和湖泊表层水中依然存在大量的甲基汞,这说明在好氧条件下也可以产生甲基汞[302]。在海洋表层及深水区,甲基汞的浓度很低;在中间水层,特别是缺氧层,甲基汞的浓度最高[152, 307-311]。然而深海水层可能受到海底热泉羽流影响,导致甲基汞浓度升高[309]。在少数情况下,大气也可能是表层水体甲基汞的重要来源[302]。

食用海产品是全球人类甲基汞摄入的主要方式之一[305, 312],尤其是食用处于较高营养级的鱼类[313]。例如有研究[314]表明,随着鱼类食用量的增加,血液中甲基汞的浓度也会显著增加。此外,水稻也是人们摄入甲基汞的重要途径[315, 316]。例如,有研究[316]对汞矿开采及冶炼区居民的头发进行测定,发现甲基汞含量较高,主要原因是食用的大米含有较高浓度的甲基汞。

4.3 汞的微生物甲基化

4.3.1 无机汞的生物可利用性

土壤和沉积物间隙水中,甲基汞主要由厌氧微生物如硫酸盐还原菌、铁还原菌等将无机Hg(II)进行甲基化而产生[68, 317]。甲基汞的生成与土壤和底泥中的总无机汞浓度并不成正比,只有部分的无机汞可以被厌氧微生物甲基化[62, 318, 319]。因此,无机汞的生物可利用性是甲基汞合成的关键因素。

硫离子和含巯基的溶解态的天然有机物(NOM)与无机Hg^{2+}具有极强的亲和性,极易与土壤和沉积物间隙水中的无机Hg(II)络合[320]。传统观点[319, 321]认为,只有溶解态的憎水性硫化汞(HgS)络合物可以作为甲基汞的前体物进入微生物体内合成甲基汞。近年来研究发现[322, 323],亲水性含汞化合物也可以通过主动运输的方式进入菌体内部合成甲基汞。

无机汞形态的不同往往会造成生物可用性的差异。首先不同络合态汞的生物甲基化效率可能存在明显差别[324, 325]。例如,分子量低的溶解性有机物能促进甲基化过程,当分子量太高时便不能促进甲基化过程[326]。其次，无机汞的尺寸也会影响其生物可利用性。例如，HgS 纳米颗粒的微生物甲基化速率远高于黑辰砂和朱辰砂晶体的微生物甲基化速率[327, 328]。与黑辰砂晶体相比，HgS 纳米颗粒物更容易溶解，且在巯基络合离子交换过程中具有更强的反应活性。另外，有研究[329]利用汞同位素示踪法，发现河口沉积物中的无机汞的甲基化速率依次为：β-^{201}HgS（s）＜α-^{199}HgS（s）＜≡FeS-^{202}Hg(II)＜NOM-^{196}Hg(II)＜^{198}Hg(NO$_3$)$_2$(aq)。此外，当 Hg^{2+}被还原为 Hg（0）后，也可被 *Desulfovibrio desulfuricans* ND132 甲基化[330]。有研究[331]对 *Geobactersulfurreducens* PCA 和 *Desulfovibrio desulfuricans* ND132 汞甲基化过程中的汞稳定同位素分馏进行分析，发现约 35%的无机汞被甲基化后，甲基汞的 δ^{202}Hg 值比初始的无机汞值高出 4‰，说明微生物优先利用无机 ^{202}Hg 作为甲基化的底物。

4.3.2 甲基化微生物的种类和活性

在自然环境中，甲基汞主要通过厌氧微生物的汞甲基化产生[317]。目前为止，含有汞甲基化基因 *hgcAB* 的微生物约为 140 种[332-334]，经试验证明具有汞甲基化能力的微生物约 50 种。其中，硫酸盐还原菌（SRB）、铁还原菌（FeRB）和产甲烷菌是最主要的甲基化微生物。

SRB 是一类重要的专性厌氧汞甲基化细菌，它以乳酸或丙酮酸等有机物作为电子供体，使硫酸盐和亚硫酸盐等被还原为 H$_2$S。根据电子给体的类型，它分为乙酸型、乳酸型和丙酮酸型。其中与比非乙酸型 SRB 相比，乙酸型 SRB 的汞甲基化率较高或者不变[335, 336]。多种 SRB 具有甲基化能力[333, 337]，包括 *Desulfovibrio*、*Desulfotomaculum* 和 *Desulfobulbus* 属等。例如，*Desulfovibrio desulfuricans* ND132 能够将细胞中的 Hg（0）氧化（通过与含巯基分子反应），并将氧化后的汞转化为甲基汞[338]。

部分 FeRB 也具有汞甲基化能力，如 *Geobactersulfurreducens*。另外，添加低浓度的半胱氨酸可以显著提高 *Geobactersulfurreducens* 的甲基化速率[323]。然而半胱氨酸会抑制 *Geobactersulfurreducens* PCA 突变体（缺失细胞色素 c）的汞甲基化。此外，*Geobactersulfurreducens* PCA 突变体产生的甲基汞是野生型 *Geobactersulfurreducens* PCA 的两倍。最近的研究表明，厌氧条件下铁还原菌 *Geobacterbemidjiensis* Bem 可以还原 Hg(II)并产生甲基汞[339]。实验证明[322]，FeRB *Geobactersulfurreducens* 对 Hg(II)的吸收很大程度上依赖于细胞外部介质中与 Hg(II)结合的巯基化合物的性质，其中一些巯基化合物能够促进 Hg(II)的吸收和甲基化，而另一些则抑制 Hg(II)的吸收和甲基化。SRB *Desulfovibrio desulfuricans* ND132 的 Hg(II)吸收体系比 FeRB *Geobactersulfurreducens* 具有更高的亲和力，在较强巯基络合物存在下更加促进了汞的甲基化[322]。

前期研究表明[340]藻类生物膜上发生的汞甲基化可能与产甲烷微生物有关，并检测到了产甲烷微生物的 16S rRNA 序列，其中包括 *Methanococcales*、*Methanobacteria* 和 *Methanosarcinales*。另外在纯培养实验中，*Methanomethylovoranshollandica*、*Methanolobustindarius* 和 *Methanospirillumhungatei* JF-1 被证明可以将汞甲基化[333, 341]。最新研究[342]发现，*Methanocellapaludicola* SANAE、*Methanocorpusculumbavaricum*、*Methanofollisliminatans* GKZPZ 和 *Methanosphaerulapalustris* E1-9c 也能够产生甲基汞；并且，在低硫化物浓度（<100μmol/L）和 0.5～5mmol/L 半胱氨酸存在时甲基化程度最大。对 *Methanomethylovoranshollandica* 来说，添加高达 5mmol/L 的半胱氨酸可以促进甲基汞的产生和细胞的生长；相反，硫化物抑制了甲基汞的生成[342]。产甲烷菌之间的甲基化速率存在内在的差异，但某些物种的汞甲基化速率与已知的 SRB 和 FeRB 甲基化速率相同[342]。

除物种外，微生物活性在汞甲基化效率中也起着重要作用。例如，在许多缺氧的水环境中，SRB 是主要的汞甲基化物种；然而在受含铁废水排放影响的沉积物和富铁沉积物中，*Geobacteraceae* 具有较高的甲基化活性，是进行汞甲基化的主要微生物[343]。另外，与浮游态相比，当 SRB 形成生物膜后，其汞甲基化效率更高[344, 345]。然而也有实验[346]发现，厌氧水层中浮游态微生物对汞甲基化的贡献可能比生物膜更强。此外，不同的微生物可以协同促进汞的甲基化。如硫氧化菌可将硫氧化为硫酸盐，当硫氧化菌与硫酸盐还原菌共生时，可为其提供电子受体，使硫酸盐还原菌对汞的甲基化作用增强[344]。

4.3.3 汞甲基化的微生物转化机制

前期对微生物汞甲基化机制的研究的主要集中在 SRB 上，具体包括 SRB 完全氧化和非完全氧化[347]。SRB 完全氧化过程主要由甲基钴胺素蛋白复合物提供甲基，通过乙酰辅酶 A 途径对 Hg(II)进行甲基化[348, 349]；而非完全氧化机制与乙酰辅酶 A 进行主要的碳代谢途径相关，并不通过乙酰辅酶 A 途径[335]。所以，在没有乙酰辅酶 A 或者在甲基汞抑制剂存在时，仍有一些 SRB 菌株可以进行汞甲基化[335, 350]。

近期，汞甲基化基因 *hgcA* 和 *hgcB* 的发现进一步阐明了汞微生物甲基化的遗传机制[334, 351]。实验[334]表明，Hg(II)在 *hgcA* 基因编码的卟啉蛋白的作用下产生甲基汞，而类咕啉辅因子在 *hgcB* 基因编码的铁氧化还原蛋白作用下还原。当 *Desulfovibrio desulfuricans* ND132 和 *Geobactersulfurreducens* PCA 菌株中的 *hgcA* 或 *hgcB* 基因被敲除后，便无法进行汞的甲基化[334]；当重新植入 *hgcA* 和 *hgcB* 基因时，细菌可以使汞重新被甲基化。因此，*hgcAB* 基因是微生物汞甲基化的必要条件[333]。另外，*Geobactersulfurreducens* PCA 和 *Desulfovibrio desulfuricans* ND132 汞甲基化基因 *hgcAB* 缺失后，Hg(II)的还原速率增加，但 Hg（0）的氧化速率降低，且细胞的巯基含量降低，说明汞的吸附率和吸收率下降[352]。*hgcAB* 基因几乎存在于所有的厌氧（但不是有氧）环境中，包括开放海洋的含氧层、无脊椎动物消化道、解冻的永冻土、沿海"死区"、土壤、沉积物和极端环境[332]。

4.3.4 环境因子对汞微生物甲基化的影响

环境中有机物的组成在甲基汞的形成中起重要作用。首先，有机物质可与无机汞络合，影响其生物可利用度。例如在含有微摩尔每升浓度的硫化物和低 Hg、溶解性有机物（DOM）摩尔比的缺氧水环境中，微生物汞甲基化程度随 DOM 硫化度的增加而增加[353]。这是因为，DOM 与硫化物的反应（称为硫化）可以显著提高 DOM 中还原硫官能团的含量，有利于高生物利用度的 Hg-DOM 络合物的形成或抑制了 HgS 的生长和聚集，从而提高汞的生物可利用性。有机物可以作为电子供体，影响甲基化微生物的活性和种类。例如添加水稻秸秆可以明显提升汞矿区稻田土中的甲基汞的浓度[354, 355]，这主要是因为秸秆被微生物分解为有机质后，提高了土壤中的可溶性碳，从而增加微生物的甲基化活性。另外，NOM 可以通过介导 Hg^0 和 Hg^{II} 的氧化还原循环影响无机汞的转化和生物可利用性。研究发现[356]，在黑暗缺氧条件下，还原性有机物能够同时还原和氧化汞。Hg^{II} 的还原主要是由还原醌类引起的；Hg^0 的氧化是由巯基官能团通过氧化络合作用控制的，例如低分子量硫醇化合物、谷胱甘肽和巯基乙酸可以在还原条件下氧化 Hg^0。有研究发现[357]，汞与 DOM 的光化学反应会减少可被 Sn^{II} 还原的汞含量，降低汞的生物利用度，导致 *Geobactersulfurreducens* PCA 产生的甲基汞减少 80%。

硫酸盐是影响汞甲基化的重要因素之一。首先，作为电子受体，硫酸盐可显著提高土壤和沉积物中 SRB 的活性，并促进甲基化速率[358]。同时，硫酸盐还原产生的硫离子也可与汞形成多硫化物，导致汞活性和生物利用度增加。例如，在贵州汞矿的稻田中，一些采样点的甲基汞含量较高[359]，可能是因为硫酸盐刺激了由 SRB 介导的汞甲基化，并且高浓度的硫离子促进汞的溶出，提高了无机汞的活性。然而，在一定条件下当硫离子与无机汞生成 HgS 沉淀时，汞的生物可利用性降低[360]。

硒酸盐或亚硒酸盐能够减少水生生物或土壤植物对汞的吸收和积累。例如，在水稻土中施加硒酸钠或亚硒酸钠后，谷物中甲基汞的浓度显著降低[361]，并与土壤甲基汞浓度正相关。此外，在富硒土壤中发现了 HgSe 纳米颗粒。因此，推测土壤中硒和汞的相互作用可能会减少水稻中甲基汞的积累。

Schaefer 等[362]通过研究发现，Zn^{II} 和 Cd^{II} 等特定痕量金属对 Hg^{II} 的生物体吸收和甲基化有抑制作用，而其他金属如 Ni^{II}、Co^{II} 或 Fe^{II} 则不起作用；在不改变 Hg^{II} 形态的情况下，当通过与硝酸氨三乙酸络合降低游离 Zn^{II} 的浓度时，可减轻 Hg^{II} 甲基化的抑制作用。Zn^{II} 的抑制作用也表明，当 Hg-半胱氨酸配合物或

中性 $HgCl_2$ 主导 Hg^{II} 的形态时,荷电物种和中性物种都是通过主动而非被动的方式进入胞质的,因此推测 Hg^{II} 吸收是细胞吸收 Zn^{II} 过程中金属转运体偶然摄取的结果[362]。然而,有研究发现[363],厌氧环境中 Zn^{2+} 可与汞竞争结合巯基化合物,导致 Hg^{II} 更易被 *Geobactersulfurreducens* PCA 还原并降低 Hg^{II} 的甲基化。

除此之外,其他环境因子,如植物、pH、光照、温度等可通过对微生物活性、汞的可利用性、群落结构等造成影响,从而影响汞的微生物甲基化。例如,增加一些种类的植物可以提升盐沼底泥中的微生物活性,产生的甲基汞浓度是不含植物时的几十倍以上[364]。

4.4 甲基汞的微生物去甲基化

绝大多数甲基化微生物也能够去甲基化,其速率与 Hg^{II}、甲基汞浓度、微生物的种类和活性等因素有关[365]。例如,某些甲烷氧化菌,如 *Methylosinustrichosporium* OB3b,能够迅速地吸收和降解甲基汞,*Methylosinustrichosporium* OB3b 的去甲基化作用随着甲基汞浓度的增加而增强[366]。*Geobacterbemidjiensis* Bem 也可以去甲基化[339],并且对数期细胞降解甲基汞的能力比稳定期更强。

微生物去甲基化机制包括还原性去甲基化和氧化去甲基化[367]。在还原性去甲基化过程中,有机汞裂解酶(MerB)在微生物还原下破坏了 C—Hg 键,生成无机 Hg^{II} 和 CH_4,然后汞还原酶(MerA)将 Hg^{II} 还原成 Hg^0[368,369]。在氧化去甲基化过程中,甲基汞被微生物氧化成无机 Hg^{II}、CO_2 和少量 CH_4。

4.5 汞的非生物甲基化

在环境中,汞也可以通过非生物作用甲基化。大量实验发现,在气相和水相中,无机 Hg^{II}、Hg^{I} 和 Hg^0 能够通过化学反应产生甲基汞[370]。如在紫外线或太阳光下,甲烷、甲醇、乙酸等有机小分子可以作为甲基供体,与蒸气 Hg^0 或氯化汞发生气相反应,生成甲基汞[370]。另外,多种生物分子、腐殖质和环境污染物等也可提供甲基,将无机汞甲基化,例如甲基锡、碘甲烷等[371-373]。

甲基供体的浓度、类型和水化学条件可以影响汞的非生物甲基化速率。比如,光化学甲基化过程中,巯基化合物[374]或无机硫化物[375]有助于甲基汞的生成;水体的 pH 和盐类均会对甲钴胺的汞甲基化反应造成影响[376]。

4.6 结论与展望

甲基汞广泛存在于水、土壤和大气中。环境中甲基汞的含量是由无机汞生物可利用性、甲基化微生物的类型和活性、汞甲基化的转化机制、环境因子和去甲基化等多种因素决定的。为了进一步探究甲基汞的环境行为,阐明甲基汞的生成和转化机制,降低其产生的环境和健康风险,有以下几方面需要更深入地研究:①对复杂环境条件作用下甲基汞前体物的化学形态和生物有效性的预测;②不同形态无机汞进入微生物细胞的控制因素;③食物链中甲基汞富集和放大的影响因素和控制方法;④探究更高效、快速的天然水体甲基汞降解和治理措施;⑤人体甲基汞暴露后的摄入机制、健康风险和控制方法。

参 考 文 献

[1] 赵毅, 韩立鹏. 660 MW 超低排放燃煤电站汞分布特征研究. 环境科学学报, 2019, 39(03): 853-858.
[2] 魏绍青, 滕阳, 李晓航, 等. 300 MW 等级燃煤机组煤粉炉与循环流化床锅炉汞排放特性比较. 燃料化学学报, 2017, 45(08): 1009-1016.
[3] 王丽. 超低排放机组中汞、砷和硒等重金属的迁移特性研究. 杭州: 浙江大学学位论文, 2018.

[4] 宋畅, 张翼, 郝剑, 等. 超低排放电厂脱硫及湿式电除尘废水中汞排放分析. 中国电力, 2017, 50(11): 164-167.

[5] 宋畅, 张翼, 郝剑, 等. 燃煤电厂超低排放改造前后汞污染排放特征. 环境科学研究, 2017, 30(05): 672-677.

[6] 宋畅, 刘钊, 汪涛, 等. 超低排放电厂 PM, SO_2, NO_x 及汞污染排放特征. 华北电力大学学报(自然科学版), 2017, 44(06): 93-99.

[7] 任攀杰, 李晓瑞, 高小武, 等. 基于实测的山西省 300 MW 以上燃煤机组烟气汞排放现状分析与研究. 电力科技与环保, 2017, 33(05): 47-48.

[8] 钱莲英, 徐哲明, 李震宇, 等. 燃煤机组超低排放改造对汞的脱除效果研究. 环境科学与管理, 2016, 41(04): 64-67.

[9] 李永生, 许月阳, 薛建明. 630MW 燃煤超低排放机组 SCR 对汞的协同作用研究. 动力工程学报, 2018, 38(11): 914-918.

[10] 华晓宇, 章良利, 宋玉彩, 等. 燃煤机组超低排放改造对汞排放的影响. 热能动力工程, 2016, 31(07): 110-116+39.

[11] 崔立明, 黄志杰, 莫华, 等. 不同超低排放技术路线的协同脱汞实测与研究. 中国电力, 2017, 50(10): 136-139+43.

[12] 陈瑶姬. 燃煤电厂超低排放系统中烟气汞的迁移规律研究. 电力科技与环保, 2017, 33(01): 9-11.

[13] 陈璇. 燃煤机组超低排放改造对汞排放的影响. 北京: 华北电力大学学位论文, 2018.

[14] 陈坤洋, 郭婷婷, 王海刚, 等. 超低排放改造后燃煤烟气净化设备协同脱汞潜力分析. 中国电力, 2018, 51(06): 160-165.

[15] Zhang Y, Yang J, Yu X, et al. Migration and emission characteristics of Hg in coal-fired power plant of China with ultra low emission air pollution control devices. Fuel Process Technol, 2017, 158: 272-280.

[16] Zhao S, Duan Y, Yao T, et al. Study on the mercury emission and transformation in an ultra-low emission coal-fired power plant. Fuel, 2017, 199: 653-661.

[17] Wu Q R, Wang S X, Hui M L, et al. New insight into atmospheric mercury emissions from zinc smelters using mass flow analysis. Environ Sci Technol, 2015, 49(6): 3532-3539.

[18] Wu Q R, Wang S X, Yang M, et al. Mercury flows in large-scale gold production and implications for Hg pollution control. J Environ Sci, 2018, 68: 91-99.

[19] Yang M, Wang S X, Zhang L, et al. Mercury emission and speciation from industrial gold production using roasting process. Journal of Geochemical Exploration, 2016, 170: 72-77.

[20] Wang F Y, Wang S X, Zhang L, et al. Characteristics of mercury cycling in the cement production process. J Hazard Mater, 2016, 302: 27-35.

[21] Li X, Li Z, Wu T, et al. Atmospheric mercury emissions from two pre-calciner cement plants in Southwest China. Atmos Environ, 2019, 199: 177-188.

[22] 王小龙. 水泥生产过程中汞的排放特征及减排潜力研究. 浙江: 浙江大学学位论文, 2017.

[23] Wang F Y, Wang S X, Zhang L, et al. Mercury mass flow in iron and steel production process and its implications for mercury emission control. J Environ Sci, 2016, 43: 293-301.

[24] Xu W Q, Shao M P, Yang Y, et al. Mercury emission from sintering process in the iron and steel industry of China. Fuel Process Technol, 2017, 159: 340-344.

[25] Li G L, Wu Q R, Wang S X, et al. The influence of flue gas components and activated carbon injection on mercury capture of municipal solid waste incineration in China. Chem Eng J, 2017, 326: 561-569.

[26] Cui Z, Li Z, Zhang Y, et al. Atmospheric mercury emissions from residential coal combustion in Guizhou province, southwest China. Energ Fuel, 2019, 33(3): 1937-1943.

[27] Zhao C, Luo K. Household consumption of coal and related sulfur, arsenic, fluorine and mercury emissions in China. Energ Policy, 2018, 112: 221-232.

[28] Gilli R S, Karlen C, Weber M, et al. Speciation and mobility of mercury in soils contaminated by legacy emissions from a

chemical factory in the rhone valley in canton of valais, Switzerland. Soil Syst, 2018, 2(3): 22-28.

[29] Sun R Y, Sonke J E, Heimburger L E, et al. Mercury stable isotope signatures of world coal deposits and historical coal combustion emissions. Environ Sci Technol, 2014, 48(13): 7660-7668.

[30] Sun R Y, Heimburger L E, Sonke J E, et al. Mercury stable isotope fractionation in six utility boilers of two large coal-fired power plants. Chemical Geology, 2013, 336: 103-111.

[31] Tang S L, Feng C H, Feng X B, et al. Stable isotope composition of mercury forms in flue gases from a typical coal-fired power plant, Inner Mongolia, northern China. J Hazard Mater, 2017, 328: 90-97.

[32] Huang S Y, Yuan D X, Lin H Y, et al. Fractionation of mercury stable isotopes during coal combustion and seawater flue gas desulfurization. Appl Geochem, 2017, 76: 159-167.

[33] Lin H Y, Peng J J, Yuan D X, et al. Mercury isotope signatures of seawater discharged from a coal-fired power plant equipped with a seawater flue gas desulfurization system. Environ Pollut, 2016, 214: 822-830.

[34] Sherman L S, Blum J D, Keeler G J, et al. Investigation of local mercury deposition from a coal-fired power plant Using Mercury Isotopes. Environ Sci Technol, 2012, 46(1): 382-390.

[35] Kim H-J, Park J-H, Sakong J. Estimation of mercury emission from incineration of extracted teeth with dental amalgam fillings in South Korea. Int J Env Res Pub He, 2018, 15(7): 469-478.

[36] Wu Q R, Gao W, Wang S X, et al. Updated atmospheric speciated mercury emissions from iron and steel production in China during 2000—2015. Atmos Chem Phys, 2017, 17(17): 10423-10433.

[37] Wu Q R, Wang S X, Zhang L, et al. Flow analysis of the mercury associated with nonferrous ore concentrates: Implications on mercury emissions and recovery in China. Environ Sci Technol, 2016, 50(4): 1796-1803.

[38] Liu K Y, Wang S X, Wu Q R, et al. A highly resolved mercury emission inventory of Chinese coal-fired power plants. Environ Sci Technol, 2018, 52(4): 2400-2408.

[39] Zhao C, Luo K L. Sulfur, arsenic, fluorine and mercury emissions resulting from coal-washing byproducts: A critical component of China's emission inventory. Atmos Environ, 2017, 152: 270-278.

[40] Cui Z K, Li Z G, Zhang Y Z, et al. Atmospheric Mercury Emissions from Residential Coal Combustion in Guizhou Province, Southwest China. Energ Fuel, 2019, 33(3): 1937-1943.

[41] Wu Q, Wang S, Liu K, et al. Emission-limit-oriented strategy to control atmospheric mercury emissions in coal-fired power plants towards the implementation of Minamata Convention . Environ Sci Technol, 2018, 52(19): 11087-11093.

[42] Takiguchi H, Tamura T. Mercury emission control in Japan. Asian Journal of Atmospheric Environment, 2018, 12(1): 37-46.

[43] Sung J-H, Oh J-S, Mojammal A H M, et al. Estimation and future prediction of mercury emissions from anthropogenic sources in South Korea. J Chem Eng Jpn, 2018, 51(9): 800-808.

[44] Wu Q, Li G, Wang S, et al. Mitigation options of atmospheric Hg emissions in China. Environ Sci Technol, 2018, 52(21): 12368-12375.

[45] Zhang L, Wang S X, Wang L, et al. Updated emission inventories for speciated atmospheric mercury from anthropogenic sources in China. Environ Sci Technol, 2015, 49(5): 3185-3194.

[46] Wu Q R, Wang S X, Li G L, et al. Temporal trend and spatial distribution of speciated atmospheric mercury emissions in China during 1978–2014. Environ Sci Technol, 2016, 50(24): 13428-13435.

[47] Bourtsalas A C, Themelis N J. Major sources of mercury emissions to the atmosphere: The US case. Waste Manag, 2019, 85: 90-94.

[48] Hui M L, Wu Q R, Wang S X, et al. Mercury flows in China and global drivers. Environ Sci Technol, 2017, 51: 222-231.

[49] Arctic monitoring and assessment programme and united nations environment programme(AMAP/UNEP). Geneva,

Switzerland: AMAP/UNEP, 2018.

[50] Streets D G, Horowitz H M, Lu Z, et al. Global and regional trends in mercury emissions and concentrations, 2010—2015. Atmos Environ, 2019, 201: 417-427.

[51] Zhang Y X, Jacob D J, Horowitz H M, et al. Observed decrease in atmospheric mercury explained by global decline in anthropogenic emissions. Proc Natl Acad Sci U S A, 2016, 113(3): 526-531.

[52] Song S, Selin N E, Soerensen A L, et al. Top-down constraints on atmospheric mercury emissions and implications for global biogeochemical cycling. Atmos Chem Phys, 2015, 15(12): 7103-7125.

[53] Landis M S, Ryan J V, Ter Schure A F H, et al. Behavior of mercury emissions from a commercial coal-fired power plant: the relationship between stack speciation and near-field plume measurements. Environ Sci Technol, 2014, 48(22): 13540-13548.

[54] Weigelt A, Slemr F, Ebinghaus R, et al. Mercury emissions of a coal-fired power plant in Germany. Atmos Chem Phys, 2016, 16(21): 13653-13668.

[55] Sun R Y, Streets D G, Horowitz H M, et al. Historical(1850-2010)mercury stable isotope inventory from anthropogenic sources to the atmosphere. Elementa-Sci Anthrop, 2016, 4: 1-15.

[56] Sun R Y, Jiskra M, Amos H M, et al. Modelling the mercury stable isotope distribution of Earth surface reservoirs: Implications for global Hg cycling. Geochim Cosmochim Ac, 2019, 246: 156-173.

[57] Streets D G, Horowitz H M, Lu Z F, et al. Global and regional trends in mercury emissions and concentrations, 2010-2015. Atmospheric Environment, 2019, 201: 417-427.

[58] UNEP. Global Mercury Assessment 2013: Sources, Emissions, Releases and Environmental Transport, 2013.

[59] UNEP. Global Mercury Assessment 2018: Sources, Emissions, Releases and Environmental Transport, 2018.

[60] Regnell O, Watras C J. Microbial mercury methylation in aquatic environments: A critical review of published field and laboratory studies. Environmental Science & Technology, 2019, 53(1): 4-19.

[61] La Colla N S, Botte S E, Marcovecchio J E. Mercury cycling and bioaccumulation in a changing coastal system: from water to aquatic organisms. Marine Pollution Bulletin, 2019, 140: 40-50.

[62] Harris R C, Rudd J W M, Amyot M, et al. Whole-ecosystem study shows rapid fish-mercury response to changes in mercury deposition. Proc Natl Acad Sci U S A, 2007, 104(42): 16586-16591.

[63] Morel F M M, Kraepiel A M L, Amyot M. The chemical cycle and bioaccumulation of mercury. Annu Rev Ecol Syst, 1998, 29: 543-566.

[64] Cristol D A, Brasso R L, Condon A M, et al. The movement of aquatic mercury through terrestrial food webs. Science, 2008, 320(5874): 335-335.

[65] Compeau G C, Bartha R. Sulfate-reducing bacteria - principal methylators of mercury in anoxic estuarine sediment. Applied and Environmental Microbiology, 1985, 50(2): 498-502.

[66] Gilmour C C, Henry E A, MITCHELL R. Sulfate stimulation of mercury methylation in fresh-water dediments. Environmental Science & Technology, 1992, 26(11): 2281-2287.

[67] Fleming E J, Mack E E, Green P G, et al. Mercury methylation from unexpected sources: Molybdate-inhibited freshwater sediments and an iron-reducing bacterium. Applied and Environmental Microbiology, 2006, 72(1): 457-464.

[68] Kerin E J, Gilmour C C, Roden E, et al. Mercury methylation by dissimilatory iron-reducing bacteria. Applied and Environmental Microbiology, 2006, 72(12): 7919-7921.

[69] Hamelin S, Amyot M, Barkay T, et al. Methanogens: Principal methylators of mercury in lake periphyto. Environmental Science & Technology, 2011, 45(18): 7693-700.

[70] Parks J M, Johs A, Podar M, et al. The genetic basis for bacterial mercury methylation. Science, 2013, 339(6125): 1332-1335.

[71] Eckley C S, Hintelmann H. Determination of mercury methylation potentials in the water column of lakes across Canada. Science of the Total Environment, 2006, 368(1): 111-125.

[72] Lehnherr I, St Louis V L, Hintelmann H, et al. Methylation of inorganic mercury in polar marine waters. Nat Geosci, 2011, 4(5): 298-302.

[73] Li Y, Cai Y. Progress in the study of mercury methylation and demethylation in aquatic environments. Chinese Science Bulletin, 2013, 58(2): 177-185.

[74] Amyot M, Gill G A, Morel F M M. Production and loss of dissolved gaseous mercury in coastal seawater. Environmental Science & Technology, 1997, 31(12): 3606-3611.

[75] Amyot M, Mierle G, Lean D R S, et al. Sunlight-induced formation of dissolved gaseous mercury in lake waters. Environmental Science & Technology, 1994, 28(13): 2366-2371.

[76] Alberts J J, Schindler J E, Miller R W, et al. Elemental mercury evolution mediated by humic acid. Science, 1974, 184(4139): 895-896.

[77] Poulain A J, Aayot M, Findlay D, et al. Biological and photochemical production of dissolved gaseous mercury in a boreal lake. Limnology and Oceanography, 2004, 49(6): 2265-2275.

[78] Gregoire D S, Poulain A J. A physiological role for Hg-II during phototrophic growth. Nat Geosci, 2016, 9(2): 121-125.

[79] Ci Z J, Zhang X S, Yin Y G, et al. Mercury redox chemistry in waters of the eastern Asian Seas: From polluted coast to clean open ocean. Environmental Science & Technology, 2016, 50(5): 2371-2380.

[80] Mason R P, Fitzgerald W F, Morel F M M. The biogeochemical cycling of elemental mercury - anthropogenic influences. Geochimica Et Cosmochimica Acta, 1994, 58(15): 3191-3198.

[81] St Pierre K A, St Louis V L, Lehnherr I, et al. Drivers of mercury cycling in the rapidly changing glacierized watershed of the high arctic's largest lake by volume(Lake Hazen, Nunavut, Canada). Environmental Science & Technology, 2019, 53(3): 1175-1185.

[82] Pestana I A, Almeida M G, Bastos W R, et al. Total Hg and methylmercury dynamics in a river-floodplain system in the Western Amazon: Influence of seasonality, organic matter and physical and chemical parameters. Science of the Total Environment, 2019, 656: 388-399.

[83] Siciliano S D, O'driscoll N J, Tordon R, et al. Abiotic production of methylmercury by solar radiation. Environ Sci Technol, 2005, 39(4): 1071-1077.

[84] Yin Y, Chen B, Mao Y, et al. Possible alkylation of inorganic Hg(II)by photochemical processes in the environment. Chemosphere, 2012, 88: 8-16.

[85] Yin Y, Li Y, Tai C, et al. Fumigant methyl iodide can methylate inorganic mercury species in natural waters. Nature Communications, 2014, 5.

[86] Sellers P, Kelly C A, Rudd J W M, et al. Photodegradation of methylmercury in lakes. Nature, 1996, 380(6576): 694-697.

[87] Klapstein S J, O'driscoll N J. Methylmercury biogeochemistry in freshwater ecosystems: A review Focusing on DOM and photodemethylation. Bulletin of Environmental Contamination and Toxicology, 2017.

[88] Li Y, Mao Y, Liu G, et al. Degradation of methylmercury and its effects on mercury distribution and cycling in the Florida Everglades. Environmental Science & Technology, 2010, 44(17): 6661-6666.

[89] Zhang D, Yin Y, Li Y, et al. Critical role of natural organic matter in photodegradation of methylmercury in water: Molecular weight and interactive effects with other environmental factors. Science of the Total Environment, 2017, 578: 535-541.

[90] Tai C, Wu H, Li Y, et al. Photodegradation mechanism of methyl mercury in environmental waters. Chinese Science Bulletin, 2017, 62(1): 70-78.

[91] Lescord G L, Emilson E J S, Johnston T A, et al. Optical properties of dissolved organic matter and their relation to mercury concentrations in water and biota across a remote freshwater drainage basin. Environmental Science & Technology, 2018, 52(6): 3344-3353.

[92] Luo H W, Yin X P, Jubb A M, et al. Photochemical reactions between mercury(Hg)and dissolved organic matter decrease Hg bioavailability and methylation. Environmental Pollution, 2017, 220: 1359-1365.

[93] Vudamala K, Chakraborty P, Sailaja B B V. An insight into mercury reduction process by humic substances in aqueous medium under dark condition. Environmental Science and Pollution Research, 2017, 24(16): 14499-14507.

[94] Lee S, Roh Y, Kim K W. Influence of chloride ions on the reduction of mercury species in the presence of dissolved organic matter. Environmental Geochemistry and Health, 2019, 41(1): 71-79.

[95] Asaduzzaman A, Riccardi D, Afaneh A T, et al. Environmental mercury chemistry in-silico. Accounts of Chemical Research, 2019, 52(2): 379-388.

[96] O'driscoll N J, Vost E, Mann E, et al. Mercury photoreduction and photooxidation in lakes: Effects of filtration and dissolved organic carbon concentration. Journal of Environmental Sciences-China, 2018, 68: 151-159.

[97] Munson K M, Lamborg C H, Boitenau R M, et al. Dynamic mercury methylation and demethylation in oligotrophic marine water. Biogeosciences, 2018, 15(21): 6451-6460.

[98] Zhang T, Hsu-Kim H. Photolytic degradation of methylmercury enhanced by binding to natural organic ligands. Nat Geosci, 2010, 3(7): 473-476.

[99] Baughman G L, Gordon J A, Wolfe N L, et al. Chemistry of organomercurials in aquatic systems. Washington, DC: U.S. Environmental Protection Agency, 1973.

[100] Vost E E, Amyot M, O'driscoll N J. Photoreactions of mercury in aquatic systems// Environmental Chemistry and Toxicology of Mercury. John Wiley & Sons, Inc. 2012: 193-218.

[101] Tseng C M, Lamborg C, Fitzgerald W F, et al. Cycling of dissolved elemental mercury in Arctic Alaskan lakes. Geochimica et Cosmochimica Acta, 2004, 68(6): 1173-1184.

[102] Garcia E, Amyot M, Ariya P A. Relationship between DOC photochemistry and mercury redox transformations in temperate lakes and wetlands. Geochimica et Cosmochimica Acta, 2005, 69(8): 1917-1924.

[103] Whalin L, Kim E H, Mason R. Factors influencing the oxidation, reduction, methylation and demethylation of mercury species in coastal waters. Marine Chemistry, 2007, 107(3): 278-294.

[104] Klapstein S J, O'driscoll N J. Methylmercury biogeochemistry in freshwater ecosystems: A review focusing on DOM and photodemethylation. Bulletin of Environmental Contamination and Toxicology, 2018, 100(1): 14-25.

[105] O'driscoll N J, Vost E, Mann E, et al. Mercury photoreduction and photooxidation in lakes: Effects of filtration and dissolved organic carbon concentration. Journal of Environmental Sciences, 2018, 68: 151-159.

[106] Chen Y, Yin Y, Shi J, et al. Analytical methods, formation, and dissolution of cinnabar and its impact on environmental cycle of mercury. Critical Reviews in Environmental Science and Technology, 2017, 47(24): 2415-2447.

[107] Voros D, Diazsomoano M, Gerslova E, et al. Mercury contamination of stream sediments in the North Bohemian Coal District(Czech Republic): Mercury speciation and the role of organic matter. Chemosphere, 2018, 211: 664-673.

[108] Manceau A, Wang J, Rovezzi M, et al. Biogenesis of mercury-sulfur nanoparticles in plant leaves from atmospheric gaseous mercury. Environmental Science & Technology, 2018, 52(7): 3935-3948.

[109] Bourdineaud J P, Gonzalez R M, Rovezzi M, et al. Divalent mercury in dissolved organic matter is bioavailable to fish and

accumulates as dithiolate and tetrathiolate complexes. Environmental Science & Technology, 2019, 4: 282-291.

[110] Manceau A, Bustamante P, Haouz A, et al. Mercury(II)binding to metallothionein in mytilus edulis revealed by high energy-resolution XANES spectroscopy. Chemistry-a European Journal, 2019, 25(4): 997-1009.

[111] Avellan A, Stegemeier J P, Gai K, et al. Speciation of mercury in selected areas of the petroleum value Chain. Environmental Science & Technology, 2018, 52(3): 1655-1664.

[112] Ruhland D, Nwoko K, Perez M, et al. AF4-UV-MALS-ICP-MS/MS, spICP-MS, and STEM-EDX for the characterization of metal-containing nanoparticles in gas condensates from petroleum hydrocarbon samples. Analytical Chemistry, 2019, 91(1): 1164-1170.

[113] Poulin B A, Gerbig C A, Kim C S, et al. Effects of sulfide concentration and dissolved organic matter characteristics on the structure of nanocolloidal metacinnabar. Environmental Science & Technology, 2017, 51(22): 13133-13142.

[114] Mazrui N M, Seelen E, King'ondu C K, et al. The precipitation, growth and stability of mercury sulfide nanoparticles formed in the presence of marine dissolved organic matter. Environmental Science-Processes & Impacts, 2018, 20(4): 642-656.

[115] Thomas S A, Gaillard J-F. Cysteine addition promotes sulfide production and 4-fold Hg(II)-S coordination in actively metabolizing escherichia coli. Environmental Science & Technology, 2017, 51(8): 4642-4651.

[116] Thomas S A, Rodby K E, Roth E W, et al. Spectroscopic and microscopic evidence of biomediated HgS species formation from Hg(II)- cysteine complexes: Implications for Hg(II)bioavailability. Environmental Science & Technology, 2018, 52(17): 10030-10039.

[117] Meng B, Li Y, Cui W, et al. Tracing the uptake, transport, and fate of mercury in sawgrass (*Cladium jamaicense*) in the Florida Everglades using a multi-isotope technique. Environmental Science & Technology, 2018, 52(6): 3384-3391.

[118] Li Y, Zhao J, Guo J, et al. Influence of sulfur on the accumulation of mercury in rice plant (*Oryza sativa L.*) growing in mercury contaminated soils. Chemosphere, 2017, 182: 293-300.

[119] Li Y, Zhao J, Zhong H, et al. Understanding enhanced microbial MeHg production in mining-contaminated paddy soils under sulfate amendment: Changes in Hg mobility or microbial methylators? Environmental Science & Technology, 2019, 53(4): 1844-1852.

[120] Wang J, Anderson C W N, Xing Y, et al. Thiosulphate-induced phytoextraction of mercury in Brassica juncea: Spectroscopic investigations to define a mechanism for Hg uptake. Environmental Pollution, 2018, 242: 986-993.

[121] Liu T, Wang J, Feng X, et al. Spectral insight into thiosulfate-induced mercury speciation transformation in a historically polluted soil. Science of the Total Environment, 2019, 657: 938-944.

[122] Li Y, Li H, Yu Y, et al. Thiosulfate amendment reduces mercury accumulation in rice (*Oryza sativa L*). Plant and Soil, 2018, 430(1-2): 413-422.

[123] Ndu U, Christensen G A, Rivera N A, et al. Quantification of mercury bioavailability for methylation using diffusive gradient in thin-film samplers. Environmental Science & Technology, 2018, 52(15): 8521-8529.

[124] Zhu W, Song Y, Adediran G A, et al. Mercury transformations in resuspended contaminated sediment controlled by redox conditions, chemical speciation and sources of organic matter. Geochimica Et Cosmochimica Acta, 2018, 220: 158-179.

[125] Baptista-Salazar C, Richard J-H, Horf M, et al. Grain-size dependence of mercury speciation in river suspended matter, sediments and soils in a mercury mining area at varying hydrological conditions. Applied geochemistry, 2017, 81: 132-142.

[126] Morway E D, Thodal C E, Marvin-Dipasquale M. Long-term trends of surface-water mercury and methylmercury concentrations downstream of historic mining within the Carson River watershed. Environmental Pollution, 2017, 229: 1006-1018.

[127] LIANG P, FENG X, YOU Q, et al. The effects of aquaculture on mercury distribution, changing speciation, and

bioaccumulation in a reservoir ecosystem. Environmental Science and Pollution Research, 2017, 24(33): 25923-25932.

[128] Sun X, Zhang Q, Kang S, et al. Mercury speciation and distribution in a glacierized mountain environment and their relevance to environmental risks in the inland Tibetan Plateau. Science of the Total Environment, 2018, 631-632: 270-278.

[129] Jedruch A, Kwasigroch U, Beldowska M, et al. Mercury in suspended matter of the Gulf of Gdansk: Origin, distribution and transport at the land-sea interface. Marine Pollution Bulletin, 2017, 118(1-2): 354-367.

[130] Cossa D, De Madron X D, Schafer J, et al. Sources and exchanges of mercury in the waters of the Northwestern Mediterranean margin. Progress in Oceanography, 2018, 163: 172-183.

[131] Kelly C A, Rudd J W M. Transport of mercury on the finest particles results in high sediment concentrations in the absence of significant ongoing sources. Science of the Total Environment, 2018, 637: 1471-1479.

[132] Cesario R, Mota A M, Caetano M, et al. Mercury and methylmercury transport and fate in the water column of Tagus estuary(Portugal). Marine Pollution Bulletin, 2018, 127: 235-250.

[133] Taylor V F, Buckman K L, Seelen E A, et al. Organic carbon content drives methylmercury levels in the water column and in estuarine food webs across latitudes in the Northeast United States. Environmental Pollution, 2019, 246: 639-649.

[134] Jensen A M, Scanlon T M, Riscassi A L. Emerging investigator series: The effect of wildfire on streamwater mercury and organic carbon in a forested watershed in the southeastern United States. Environmental Science-Processes & Impacts, 2017, 19(12): 1505-1517.

[135] Diez E G, Graham N D, Loizeau J L. Total and methyl-mercury seasonal particulate fluxes in the water column of a large lake(Lake Geneva, Switzerland). Environmental Science and Pollution Research, 2018, 25(21): 21086-21096.

[136] Jedruch A, Beldowska M, Graca B. Seasonal variation in accumulation of mercury in the benthic macrofauna in a temperate coastal zone(Gulf of Gdansk). Ecotoxicology and Environmental Safety, 2018, 164: 305-316.

[137] Jedruch A, Beldowska M, Ziolkowska M. The role of benthic macrofauna in the trophic transfer of mercury in a low-diversity temperate coastal ecosystem(Puck Lagoon, southern Baltic Sea). Environmental Monitoring and Assessment, 2019, 191(3):

[138] Wang Y, Li Y, Liu G, et al. Elemental mercury in natural waters: Occurrence and determination of particulate Hg(0). Environmental Science & Technology, 2015, 49(16): 9742-9749.

[139] Cesario R, Poissant L, Pilote M, et al. Dissolved gaseous mercury formation and mercury volatilization in intertidal sediments. Science of the Total Environment, 2017, 603: 279-289.

[140] Zhang L, Wu S, Zhao L, et al. Mercury sorption and desorption on organo-mineral particulates as a source for microbial methylation. Environmental Science & Technology, 2019, 53(5): 2426-2433.

[141] Diez E G, Loizeau J L, Cosio C, et al. Role of settling particles on mercury methylation in the oxic water column of freshwater systems. Environmental Science & Technology, 2016, 50(21): 11672-11679.

[142] Lei P, Nunes L M, Liu Y R, et al. Mechanisms of algal biomass input enhanced microbial Hg methylation in lake sediments. Environment international, 2019, 126: 279-288.

[143] Canario J, Santos E J, Padeiro A, et al. Mercury and methylmercury in the Atlantic sector of the Southern Ocean. Deep-Sea Research Part Ii-Topical Studies in Oceanography, 2017, 138: 52-62.

[144] Washburn S J, Blum J D, Demers J D, et al. Isotopic characterization of mercury downstream of historic industrial contamination in the South River, Virginia. Environmental Science & Technology, 2017, 51(19): 10965-10973.

[145] Demers J D, Blum J D, Brooks S C, et al. Hg isotopes reveal in-stream processing and legacy inputs in East Fork Poplar Creek, Oak Ridge, Tennessee, USA. Environmental Science-Processes & Impacts, 2018, 20(4): 686-707.

[146] Shi J B, Ip C C, Zhang G, et al. Mercury profiles in sediments of the Pearl River Estuary and the surrounding coastal area of

South China. Environ Pollut, 2010, 158(5): 1974-1979.

[147] Meng M, Shi J B, Yun Z J, et al. Distribution of mercury in coastal marine sediments of China: Sources and transport. Mar Pollut Bull, 2014, 88(1-2): 347-353.

[148] Meng M, Shi J B, Liu C B, et al. Biomagnification of mercury in mollusks from coastal areas of the Chinese Bohai Sea. Rsc Adv, 2015, 5(50): 40036-40045.

[149] Meng M, Sun R Y, Liu H W, et al. An integrated model for input and migration of mercury in Chinese coastal sediments. Environ Sci Technol, 2019, 53(5): 2460-2471.

[150] Semeniuk K, Dastoor A. Development of a global ocean mercury model with a methylation cycle: Outstanding issues. Glob Biogeochem Cycle, 2017, 31(2): 400-433.

[151] Mazrui N M, Jonsson S, Thota S, et al. Enhanced availability of mercury bound to dissolved organic matter for methylation in marine sediments. Geochimica Et Cosmochimica Acta, 2016, 194: 153-162.

[152] Soerensen A L, Schartup A T, Gustafsson E, et al. Eutrophication increases phytoplankton methylmercury concentrations in a coastal sea-a baltic sea case study. Environmental Science & Technology, 2016, 50(21): 11787-11796.

[153] Cesario R, Hintelmann H, Mendes R, et al. Evaluation of mercury methylation and methylmercury demethylation rates in vegetated and non-vegetated saltmarsh sediments from two Portuguese estuaries. Environmental Pollution, 2017, 226: 297-307.

[154] Ndungu K, Schaanning M, Braaten H F V. Effects of organic matter addition on methylmercury formation in capped and uncapped marine sediments. Water Res, 2016, 103: 401-407.

[155] Bratkic A, Koron N, Guevara S R, et al. Seasonal variation of mercury methylation potential in pristine coastal marine sediment from the gulf of Trieste(Northern Adriatic Sea). Geomicrobiol J, 2017, 34(7): 587-595.

[156] Liang P, Wu S C, Zhang C, et al. The role of antibiotics in mercury methylation in marine sediments. J Hazard Mater, 2018, 360: 1-5.

[157] Yin R, Feng X, Chen B, et al. Identifying the sources and processes of mercury in subtropical estuarine and ocean sediments using Hg isotopic composition. Environmental Science & Technology, 2015, 49(3): 1347-1355.

[158] Xu X, Wang W X. Mercury exposure and source tracking in distinct marine-caged fish farm in southern China. Environmental Pollution, 2017, 220, Part B: 1138-1146.

[159] Jiang T, Skyllberg U, Bjorn E, et al. Characteristics of dissolved organic matter(DOM)and relationship with dissolved mercury in Xiaoqing River-Laizhou Bay estuary, Bohai Sea, China. Environ Pollut, 2017, 223: 19-30.

[160] Graham A M, Cameron B K T, Hajic H A, et al. Sulfurization of dissolved organic matter increases Hg-sulfide-dissolved organic matter bioavailability to a Hg-methylating bacterium. Environ Sci Technol, 2017, 51(16): 9080-9088.

[161] Myrbo A, Swain E B, Johnson N W, et al. Increase in nutrients, mercury, and methylmercury as a consequence of elevated sulfate reduction to sulfide in experimental wetland mesocosms. J Geophys Res-Biogeosci, 2017, 122(11): 2769-2785.

[162] Mohapatra S P, Nikolova I, Mitchell A. Managing mercury in the great lakes: An analytical review of abatement policies. Journal of Environmental Management, 2007, 83(1): 80-92.

[163] Bhavsar S P, Gewurtz S B, Mcgoldrick D J, et al. Changes in mercury levels in great lakes fish between 1970s and 2007. Environmental Science & Technology, 2010, 44(9): 3273-3279.

[164] Cross F A, Evans D W, Barber R T. Decadal declines of mercury in adult bluefish (1972—2011) from the Mid-Atlantic Coast of the USA. Environmental Science & Technology, 2015, 49(15): 9064-9072.

[165] Azim M E, Kumarappah A, Bhavsar S P, et al. Detection of the spatiotemporal trends of mercury in lake erie fish communities: A bayesian approach. Environmental Science & Technology, 2011, 45(6): 2217-2226.

[166] Gandhi N, Tang R W K, Bhavsar S P, et al. Fish mercury levels appear to be increasing lately: a report from 40 years of monitoring in the province of Ontario, Canada. Environmental Science & Technology, 2014, 48(10): 5404-5414.

[167] Drevnick P E, Lamborg C H, Horgan M J. Increase in mercury in pacific yellowfin tuna. Environmental Toxicology and Chemistry, 2015, 34(4): 931-934.

[168] Wyn B, Kidd K A, Burgess N M, et al. Increasing mercury in yellow perch at a hotspot in Atlantic Canada, kejimkujik national park. Environmental Science & Technology, 2010, 44(23): 9176-9181.

[169] Brigham M E, Sandheinrich M B, Gay D A, et al. Lacustrine responses to decreasing wet mercury deposition rates-results from a case study in Northern Minnesota. Environmental Science & Technology, 2014, 48(11): 6115-6123.

[170] Hutcheson M S, Smith C M, Rose J, et al. Temporal and spatial trends in freshwater fish tissue mercury concentrations associated with mercury emissions reductions. Environmental Science & Technology, 2014, 48(4): 2193-2202.

[171] Zhou C L, Cohen M D, Crimmins B A, et al. Mercury temporal trends in top predator fish of the Laurentian great lakes from 2004 to 2015: Are concentrations still decreasing? Environmental Science & Technology, 2017, 51(13): 7386-7394.

[172] Lucotte M, Paquet S, Moingt M. Climate and physiography predict mercury concentrations in game fish species in Quebec lakes better than anthropogenic disturbances. Archives of Environmental Contamination and Toxicology, 2016, 70(4): 710-723.

[173] Evans M, Muir D, Brua R B, et al. Mercury trends in predatory fish in great slave lake: The influence of temperature and other climate drivers. Environmental Science & Technology, 2013, 47(22): 12793-12801.

[174] Dittman J A, Driscoll C T. Factors influencing changes in mercury concentrations in lake water and yellow perch(Perca flavescens)in Adirondack lakes. Biogeochemistry, 2009, 93(3): 179-196.

[175] Pickhardt P C, Folt C L, Chen C Y, et al. Algal blooms reduce the uptake of toxic methylmercury in freshwater food webs. Proc Natl Acad Sci U S A, 2002, 99(7): 4419-4423.

[176] Chen C Y, Folt C L. High plankton densities reduce mercury biomagnification. Environmental Science & Technology, 2005, 39(1): 115-121.

[177] Pan K, Chan H D, Tam Y K, et al. Low mercury levels in marine fish from estuarine and coastal environments in southern China. Environmental Pollution, 2014, 185: 250-257.

[178] Liu B, Yan H Y, Wang C P, et al. Insights into low fish mercury bioaccumulation in a mercury-contaminated reservoir, Guizhou, China. Environmental Pollution, 2012, 160: 109-117.

[179] Cheng H F, Hu Y N. Understanding the paradox of mercury pollution in China: High concentrations in environmental matrix yet low levels in fish on the market. Environmental Science & Technology, 2012, 46(9): 4695-4696.

[180] 冯新斌, 付学吾, JONAS S, 等. 地表自然过程排汞研究进展及展望. 生态学杂志, 2011, 30(5): 845-856.

[181] Obrist D, Kirk J L, Zhang L, et al. A review of global environmental mercury processes in response to human and natural perturbations: Changes of emissions, climate, and land use. Ambio, 2018, 47(2): 116-140.

[182] Pirrone N, Hedgecock I M, SPROVIERI F. New Directions: Atmospheric mercury, easy to spot and hard to pin down: impasse?. Atmos Environ, 2008, 42(36): 8549-8551.

[183] Lindberg S, Bullock R, EBINGHAUS R, et al. A synthesis of progress and uncertainties in attributing the sources of mercury in deposition. Ambio, 2007, 36(1): 19-32.

[184] Yuan W, Sommar J, Lin C J, et al. Stable isotope evidence shows re-emission of elemental mercury vapor occurring after reductive loss from foliage. Environ Sci Technol, 2019, 53(2): 651-660.

[185] Obrist D. Atmospheric mercury pollution due to losses of terrestrial carbon pools?. Biogeochemistry, 2007, 85(2): 119-123.

[186] Wang X, Bao Z, Lin C J, et al. Assessment of global mercury deposition through litterfall. Environ Sci Technol, 2016,

50(16): 8548-8557.

[187] Scherbatskoy T, Shanley J B, Keeler G J. Factors controlling mercury transport in an upland forested catchment. Water Air & Soil Pollution, 1998, 105(1-2): 427-438.

[188] Schwesig D, Matzner E. Pools and fluxes of mercury and methylmercury in two forested catchments in Germany. Sci Total Environ, 2000, 260(1): 213-223.

[189] St Louis V L, Rudd J W M, Kelly C A, et al. Importance of the forest canopy to fluxes of methyl mercury and total mercury to boreal ecosystems. Environ Sci Technol, 2001, 35(15): 3089-3098.

[190] Larssen T, Wit H A D, Wiker M, et al. Mercury budget of a small forested boreal catchment in southeast Norway. Sci Total Environ, 2008, 404(2): 290-296.

[191] Wang Z, Zhang X, Xiao J, et al. Mercury fluxes and pools in three subtropical forested catchments, southwest China. Environ Pollut, 2009, 157(3): 801-808.

[192] Driscoll C T, Mason R P, Chan H M, et al. Mercury as a global pollutant: Sources, pathways, and effects. Environ Sci Technol, 2013, 47(10): 4967-4983.

[193] Jiskra M, Wiederhold J G, Skyllberg U, et al. Mercury deposition and re-emission pathways in boreal forest soils investigated with Hg isotope signatures. Environ Sci Technol, 2015, 49(12): 7188-7196.

[194] Agnan Y, Le Dantec T, Moore C W, et al. New constraints on terrestrial surface atmosphere fluxes of gaseous elemental mercury using a global database. Environ Sci Technol, 2016, 50(2): 507-524.

[195] Zhu W, Lin C J, Wang X, et al. Global observations and modeling of atmosphere-surface exchange of elemental mercury: A critical review. Atmos Chem Phys, 2016, 16(7): 4451-4480.

[196] Du S H, Fang S C. Uptake of elemental mercury-vapor by C_3-species and C_4-species. Environmental and Experimental Botany, 1982, 22(4): 437-443.

[197] Luo Y, Duan L, Driscoll C T, et al. Foliage/atmosphere exchange of mercury in a subtropical coniferous forest in south China. J Geophys Res-Biogeo, 2016, 121(7): 2006-2016.

[198] Buchanan B B, Gruissem W, Jones R L. Biochemistry and molecular biology of plants. John Wiley & Sons, 2015.

[199] Obrist D, Pokharel A K, Moore C. Vertical profile measurements of soil air suggest immobilization of gaseous elemental mercury in mineral soil. Environ Sci Technol, 2014, 48(4): 2242-2252.

[200] Fu X W, Feng X B, Wang S F. Exchange fluxes of Hg between surfaces and atmosphere in the eastern flank of Mount Gongga, Sichuan province, southwestern China. J Geophys Res-Atmos, 2008, 113(D20): 232.

[201] Sigler I J M, Lee X. Gaseous mercury in background forest soil in the northeastern United States. J Geophys Res-Biogeo, 2006, 111(G2): 18.

[202] Lin C J, Zhu W, Li X C, et al. Novel dynamic flux chamber for measuring air-surface exchange of Hg-o from soils. Environ Sci Technol, 2012, 46(16): 8910-8920.

[203] Gustin M S, Biester H, Kim C S. Investigation of the light-enhanced emission of mercury from naturally enriched substrates. Atmos Environ, 2002, 36(20): 3241-3254.

[204] Cui L, Feng X, Lin C J, et al. Accumulation and translocation of ^{198}Hg in four crop species. Environmental Toxicology & Chemistry, 2014, 33(2): 334-340.

[205] Ericsen J A, Gustin M S. Foliar exchange of mercury as a function of soil and air mercury concentrations. Sci Total Environ, 2004, 324(1): 271-279.

[206] Ericsen J A, Gustin M S, XIN M, et al. Air-soil exchange of mercury from background soils in the United States. Sci Total Environ, 2006, 366(2-3): 851-863.

[207] Wang X, Lin C J, Lu Z Y, et al. Enhanced accumulation and storage of mercury on subtropical evergreen forest floor: Implications on mercury budget in global forest ecosystems. J Geophys Res-Biogeo, 2016, 121(8): 2096-2109.

[208] Jiskra M, Wiederhold J G, Skyllberg U, et al. Source tracing of natural organic matter bound mercury in boreal forest runoff with mercury stable isotopes. Environmental Science Processes & Impacts, 2017, 19(10): 1235.

[209] Risch M R, Dewild J F, Gay D A, et al. Atmospheric mercury deposition to forests in the eastern USA. Environmental pollution, 2017, 228: 8-18.

[210] Siwik E I H, Campbell L M, Mierle G. Distribution and trends of mercury in deciduous tree cores. Environ Pollut, 2010, 158(6): 2067-2073.

[211] Ma M, Sun T, Du H X, et al. A two-year study on mercury fluxes from the soil under different vegetation cover in a subtropical region, South China. Atmosphere, 2018, 9(1): 2156-2169.

[212] Ma M, Wang D Y, Du H X, et al. Mercury dynamics and mass balance in a subtropical forest, southwestern China. Atmos Chem Phys, 2016, 16(7): 4529-4537.

[213] Ma M, Wang D Y, Sun R G, et al. Gaseous mercury emissions from subtropical forested and open field soils in a national nature reserve, southwest China. Atmos Environ, 2013, 64: 116-123.

[214] Fu X W, Zhu W, Zhang H, et al. Depletion of atmospheric gaseous elemental mercury by plant uptake at Mt. Changbai, Northeast China. Atmos Chem Phys, 2016, 16(20): 12861-12873.

[215] Yu Q, Luo Y, Wang S, et al. Gaseous elemental mercury(GEM)fluxes over canopy of two typical subtropical forests in south China. Atmospheric Chemistry & Physics, 2018, 18(1): 1-27.

[216] Converse A D, Riscassi A L. Seasonal variability in gaseous mercury fluxes measured in a high-elevation meadow. Atmos Environ, 2010, 44(18): 2176-2185.

[217] Osterwalder S, Fritsche J, Alewell C, et al. A dual-inlet, single detector relaxed eddy accumulation system for long-term measurement of mercury flux. Atmospheric Measurement Techniques, 2016, 9: 509-524.

[218] Sommar J, Zhu W, Shang L, et al. Seasonal variations in metallic mercury(Hg^0)vapor exchange over biannual wheat - corn rotation cropland in the North China Plain. Biogeosciences, 2016, 13(7): 2029-2049.

[219] Lindberg S E, Hanson P J, Meyers T P, et al. Air/surface exchange of mercury vapor over forests: The need for a reassessment of continental biogenic emissions. Atmos Environ, 1998, 32(5): 895-908.

[220] Lindberg S E, Jackson D R, Huckabee J W, et al. Atmospheric emission and plant uptake of mercury from agricultural soils near the almaden mercury mine. J Environ Qual, 1979, 8(4): 572-578.

[221] Shetty S K, Lin C J, Streets D G, et al. Model estimate of mercury emission from natural sources in East Asia. Atmos Environ, 2008, 42(37): 8674-8685.

[222] Bash J O, Miller D R, Meyer T H, et al. Northeast United States and southeast Canada natural mercury emissions estimated with a surface emission model. Atmos Environ, 2004, 38(33): 5683-5692.

[223] Enrico M, Le Roux G, Marusczak N, et al. Atmospheric mercury transfer to peat bogs dominated by gaseous elemental mercury dry deposition. Environ Sci Technol, 2016, 50(5): 2405-2412.

[224] Yu B, Fu X W, Yin R S, et al. Isotopic Composition of atmospheric mercury in China: New evidence for sources and transformation processes in air and in vegetation. Environ Sci Technol, 2016, 50(17): 9262-9269.

[225] Laacouri A, Nater E A, Kolka R K. Distribution and uptake dynamics of mercury in leaves of common deciduous tree species in Minnesota, U.S.A. Environ Sci Technol, 2013, 47(18): 10462-104670.

[226] Graydon J A, St Louis V L, Lindberg S E, et al. Investigation of mercury exchange between forest canopy vegetation and the atmosphere using a new dynamic chamber. Environ Sci Technol, 2006, 40(15): 4680-4688.

[227] Zheng W, Obrist D, Weis D, et al. Mercury isotope compositions across North American forests. Global Biogeochem Cy, 2016, 30(10): 1475-1492.

[228] Grigal D F. Mercury sequestration in forests and peatlands: A review. J Environ Qual, 2003, 32(2): 393-405.

[229] Obrist D, Johnson D W, Lindberg S E, et al. Mercury distribution across 14 US forests. Part I: Spatial patterns of concentrations in biomass, litter, and soils. Environ Sci Technol, 2011, 45(9): 3974-3981.

[230] 王训, 袁巍, 冯新斌. 森林生态系统汞的生物地球化学过程. 化学进展, 2017, (09): 72-82.

[231] Wan Q, Feng X B, Lu J, et al. Atmospheric mercury in Changbai Mountain area, northeastern China II. The distribution of reactive gaseous mercury and particulate mercury and mercury deposition fluxes. Environmental Research, 2009, 109(6): 721-727.

[232] Fu X, Feng X, Zhu W, et al. Elevated atmospheric deposition and dynamics of mercury in a remote upland forest of southwestern China. Environ Pollut, 2010, 158(6): 2324-2333.

[233] Fu X W, Feng X, Dong Z Q, et al. Atmospheric gaseous elemental mercury(GEM)concentrations and mercury depositions at a high-altitude mountain peak in south China. Atmos Chem Phys, 2010, 10(5): 2425-2437.

[234] Lindberg S E. Forests and the global biogeochemical cycle of mercury: The importance of understanding air/vegetation exchange processes. 1996.

[235] Rea A W, Keeler G J, Scherbatskoy T. The deposition of mercury in throughfall and litterfall in the lake champlain watershed: A short-term study. Atmos Environ, 1996, 30(19): 3257-3263.

[236] Grigal D F, Kolka R K, Fleck J A, et al. Mercury budget of an upland-peatland watershed. Biogeochemistry, 2000, 50(1): 95-109.

[237] Sheehan K D, Fernandez I J, Kahl J S, et al. Litterfall mercury in two forested watersheds at Acadia National Park, Maine, USA. Water Air Soil Poll, 2006, 170(1-4): 249-265.

[238] Choi H D, Holsen T M, Hopke P K. Atmospheric mercury(Hg)in the Adirondacks: Concentrations and sources. Environ Sci Technol, 2008, 42(15): 5644-5653.

[239] Fisher J A, Jacob D J, Soerensen A L, et al. Riverine source of Arctic Ocean mercury inferred from atmospheric observations. Nat Geosci, 2012, 5(7): 499-504.

[240] Fostier A H, Forti M C, Guimaraes J R D, et al. Mercury fluxes in a natural forested Amazonian catchment(Serra do Navio, Amapa State, Brazil). Sci Total Environ, 2000, 260(1-3): 201-211.

[241] Iverfeldt A. Mercury in forest canopy throughfall water and its relation to atmospheric deposition. Water Air Soil Poll, 1991, 56: 553-564.

[242] Munthe J, Hultberg H, Iverfeldt Å. Mechanisms of deposition of methylmercury and mercury to coniferous forests. Water Air & Soil Pollution, 1995, 80(1-4): 363-371.

[243] Lee Y H, Bishop K H, Munthe J. Do concepts about catchment cycling of methylmercury and mercury in boreal catchments stand the test of time? Six years of atmospheric inputs and runoff export at Svartberget, northern Sweden. Sci Total Environ, 2000, 260(1-3): 11-20.

[244] Wang X, Lin C J, Feng X. Sensitivity analysis of an updated bidirectional air-surface exchange model for elemental mercury vapor. Atmos Chem Phys, 2014, 14(12): 6273-6287.

[245] Xiao Z F, Mubthe J, Schroeder W H, et al. Vertical fluxes of volatile mercury over forest soil and lake surfaces in Sweden. Tellus B, 1991, 43(3): 267-279.

[246] Zhou J, Wang Z W, Zhang X S, et al. Investigation of factors affecting mercury emission from subtropical forest soil: A field controlled study in southwestern China. J Geochem Explor, 2017, 176: 128-135.

[247] Du B Y, Wang Q, Luo Y, et al. Field measurement of soil mercury emission in a Masson pine forest in Tieshanping, Chongqing in Southwestern China. Huan Jing Ke Xue, 2014, 35(10): 3830-3835.

[248] Ma J, Hintelmann H, Kirk J L, et al. Mercury concentrations and mercury isotope composition in lake sediment cores from the vicinity of a metal smelting facility in Flin Flon, Manitoba. Chem Geol, 2013, 336: 96-102.

[249] Fu X W, Feng X B, Zhang H, et al. Mercury emissions from natural surfaces highly impacted by human activities in Guangzhou province, South China. Atmos Environ, 2012, 54: 185-193.

[250] Almeida M D, Marins R V, Paraquetti H H M, et al. Mercury degassing from forested and open field soils in Rondonia, Western Amazon, Brazil. Chemosphere, 2009, 77(1): 60-66.

[251] Carpi A, Fostier A H, Orta O R, et al. Gaseous mercury emissions from soil following forest loss and land use changes: Field experiments in the United States and Brazil. Atmos Environ, 2014, 96: 423-429.

[252] Magarelli G, Fostier A H. Influence of deforestation on the mercury air/soil exchange in the Negro River Basin, Amazon. Atmos Environ, 2005, 39(39): 7518-7528.

[253] Han J S, Seo Y S, Kim M K, et al. Total atmospheric mercury deposition in forested areas in South Korea. Atmos Chem Phys, 2016, 16(12): 7653-7662.

[254] Carpi A, Lindberg S E. Application of a Teflon (TM) dynamic flux chamber for quantifying soil mercury flux: Tests and results over background soil. Atmos Environ, 1998, 32(5): 873-882.

[255] Poissant L, Casimir A. Water-air and soil-air exchange rate of total gaseous mercury measured at background sites. Atmos Environ, 1998, 32(5): 883-893.

[256] Zhang H, Lindberg S E, Marsik F J, et al. Mercury air/surface exchange kinetics of background soils of the Tahquamenon River watershed in the Michigan Upper Peninsula. Water Air Soil Poll, 2001, 126(1-2): 151-169.

[257] Nacht D M, Gustin M S. Mercury emissions from background and altered geologic units throughout Nevada. Water Air Soil Poll, 2004, 151(1-4): 179-193.

[258] Schroeder W H, Beauchamp S, Edwards G, et al. Gaseous mercury emissions from natural sources in Canadian landscapes. J Geophys Res-Atmos, 2005, 110(D18): 2322-2332.

[259] Kuiken T, Gustin M, Zhang H, et al. Mercury emission from terrestrial background surfaces in the eastern USA. II: Air/surface exchange of mercury within forests from South Carolina to New England. Appl Geochem, 2008, 23(3): 356-368.

[260] Kuiken T, Zhang H, Gustin M, et al. Mercury emission from terrestrial background surfaces in the eastern USA. Part I: Air/surface exchange of mercury within a southeastern deciduous forest(Tennessee)over one year. Appl Geochem, 2008, 23(3): 345-355.

[261] Choi H D, Holsen T M. Gaseous mercury fluxes from the forest floor of the Adirondacks. Environ Pollut, 2009, 157(2): 592-600.

[262] Schroeder W H, Munthe J, Lindqvist O. Cycling of mercury between water, air, and soil compartments of the environment. Water Air Soil Poll, 1989, 48(3-4): 337-347.

[263] Ferrara R, Maserti B E, Andersson M, et al. Mercury degassing rate from mineralized areas in the Mediterranean basin. Water Air Soil Poll, 1997, 93(1-4): 59-66.

[264] Mazur M E E, Eckley C S, Mitchell C P J. Susceptibility of soil bound mercury to gaseous emission as a function of source depth: An enriched isotope tracer investigation. Environ Sci Technol, 2015, 49(15): 9143-9149.

[265] Demers J D, Blum J D, Zak D R. Mercury isotopes in a forested ecosystem: Implications for air-surface exchange dynamics and the global mercury cycle. Global Biogeochem Cy, 2013, 27(1): 222-238.

[266] Bergquist B A, Blum J D. Mass-dependent and -independent fractionation of Hg isotopes by photoreduction in aquatic

systems . Science, 2007, 318(5849): 417-420.

[267] Yin R S, Feng X B, Foucher D, et al. High precision determination of mercury isotope ratios using online mercury vapor generation system coupled with multi-collector inductively coupled plasma-mass spectrometry . Chinese J Anal Chem, 2010, 38(7): 929-934.

[268] 冯新斌, 尹润生, 俞奔, 等. 汞同位素地球化学概述. 地学前缘, 2015, 22(5): 124-135.

[269] Blum J D, Sherman L S, Johnson M W. Mercury isotopes in earth and environmental sciences //Jeanloz R. Annu Rev Earth Pl Sc. 2014: 249-269.

[270] Blum J D, Johnson M W. Recent developments in mercury stable isotope analysis . Rev Mineral Geochem, 2017, 82: 733-757.

[271] Yang L, Sturgeon R. Isotopic fractionation of mercury induced by reduction and ethylation . Anal Bioanal Chem, 2009, 393(1): 377-385.

[272] Zheng W, Demers J D, Lu X, et al. Mercury stable isotope fractionation during abiotic dark oxidation in the presence of thiols and natural organic matter . Environ Sci Technol, 2019, 53(4): 1853-1862.

[273] Zheng W, Hintelmann H. Nuclear field shift effect in isotope fractionation of mercury during abiotic reduction in the absence of light . J Phys Chem A, 2010, 114(12): 4238-4245.

[274] Zheng W, Hintelmann H. Mercury isotope fractionation during photoreduction in natural water is controlled by its Hg/DOC ratio . Geochim Cosmochim Ac, 2009, 73(22): 6704-6715.

[275] Sun G Y, Sommar J, Feng X B, et al. Mass-dependent and -independent fractionation of mercury isotope during gas-phase oxidation of elemental mercury vapor by atomic Cl and Br . Environ Sci Technol, 2016, 50(17): 9232-9241.

[276] Estrane N, Carignan J, Sonke J E, et al. Mercury isotope fractionation during liquid-vapor evaporation experiments . Geochim Cosmochim Ac, 2009, 73(10): 2693-2711.

[277] Ghosh S, Schauble E A, Couloume G L, et al. Estimation of nuclear volume dependent fractionation of mercury isotopes in equilibrium liquid-vapor evaporation experiments . Chem Geol, 2013, 336: 5-12.

[278] Wiederhold J G, Cramer C J, Daniel K, et al. Equilibrium mercury isotope fractionation between zissolved Hg(II)species and thiol-Bound Hg . Environ Sci Technol, 2010, 44(11): 4191-4197.

[279] Jiskra M, Wiederhold J G, Bourdon B, et al. Solution speciation controls mercury isotope fractionation of Hg(II)sorption to goethite . Environ Sci Technol, 2012, 46(12): 6654-6662.

[280] Koster Van Groos P G, Esser B K, Williams R W, et al. Isotope effect of mercury diffusion in air . Environ Sci Technol, 2014, 48(1): 227-233.

[281] Yin R S, Feng X B, Meng B. Stable mercury isotope variation in rice plants (*Oryza sativa* L.) from the wanshan mercury mining district, SW China . Environ Sci Technol, 2013, 47(5): 2238-2245.

[282] Kritee K, Blum J D, Barkay T. Mercury Stable Isotope Fractionation during Reduction of Hg(II)by Different Microbial Pathways . Environ Sci Technol, 2008, 42(24): 9171-9177.

[283] Kritee K, Blum J D, Johnson M W, et al. Mercury stable isotope fractionation during reduction of Hg(II)to Hg(0)by mercury resistant microorganisms . Environ Sci Technol, 2007, 41(6): 1889-1895.

[284] Kritee K, Blum J D, Reinfelder J R, et al. Microbial stable isotope fractionation of mercury: A synthesis of present understanding and future directions . Chem Geol, 2013, 336: 13-25.

[285] Malinovsky D, Vaneaecke F. Mercury isotope fractionation during abiotic transmethylation reactions . Int J Mass Spectrom, 2011, 307(1-3): 214-224.

[286] Perrot V, Bridou R, Pedrero Z, et al. Identical Hg isotope mass dependent fractionation signature during methylation by

sulfate-reducing bacteria in sulfate and sulfate-free Environment. environ Sci Technol, 2015, 49(3): 1365-1373.

[287] Blum J D, Popp B N, Drazen J C, et al. Methylmercury production below the mixed layer in the North Pacific Ocean. Nat Geosci, 2013, 6(10): 879-884.

[288] Zheng W, Hintelmann H. Isotope fractionation of mercury during its photochemical reduction by low-molecular-weight organic compounds. J Phys Chem A, 2010, 114(12): 4246-4253.

[289] Jimenez M M, Perrot V, Epov V N, et al. Chemical kinetic isotope fractionation of mercury during abiotic methylation of Hg(II) by methylcobalamin in aqueous chloride media. Chem Geol, 2013, 336: 26-36.

[290] Rose C H, Ghose S, Blum J D, et al. Effects of ultraviolet radiation on mercury isotope fractionation during photo-reduction for inorganic and organic mercury species. Chem Geol, 2015, 405: 102-111.

[291] Chandan P, Ghose S, Bergquist B A. Mercury isotope fractionation during aqueous photoreduction of monomethylmercury in the presence of dissolved organic matter. Environ Sci Technol, 2015, 49(1): 259-267.

[292] Bigeleisen J. Nuclear size and shape effects in chemical reactions. Isotope chemistry of the heavy elements. J Am Chem Soc, 1996, 118(15): 3676-3680.

[293] Schauble E A. Role of nuclear volume in driving equilibrium stable isotope fractionation of mercury, thallium, and other very heavy elements. Geochim Cosmochim Ac, 2007, 71(9): 2170-2189.

[294] Chen J B, Hintelmann H, Feng X B, et al. Unusual fractionation of both odd and even mercury isotopes in precipitation from Peterborough, ON, Canada. Geochim Cosmochim Ac, 2012, 90: 33-46.

[295] Gratz L E, Keeler G J, Blum J D, et al. Isotopic composition and fractionation of mercury in Great Lakes precipitation and ambient air. Environ Sci Technol, 2010, 44(20): 7764-7770.

[296] Mead C, Lyons J R, Johnson T M, et al. Unique Hg stable isotope signatures of compact fluorescent lamp-sourced Hg. Environ Sci Technol, 2013, 47(6): 2542-2547.

[297] Guedron S, Arnouroux D, Tessier E, et al. Mercury isotopic fractionation during pedogenesis in a tropical forest soil catena (French Guiana): Deciphering the impact of historical gold mining. Environ Sci Technol, 2018, 52(20): 11573-11582.

[298] Fritsche J, Obrist D, Alewell C. Evidence of microbial control of Hg-0 emissions from uncontaminated terrestrial soils. J Plant Nutr Soil Sc, 2008, 171(2): 200-209.

[299] Carpi A, Lindberg S E. Sunlight-mediated emission of elemental mercury from soil amended with municipal sewage sludge. Environ Sci Technol, 1997, 31(7): 2085-2091.

[300] Jonsson S, Andersson A, Nilsson M B, et al. Terrestrial discharges mediate trophic shifts and enhance methylmercury accumulation in estuarine biota. Science Advances, 2017, 3(1): 679-688.

[301] Fitzgerald W F, Clarkson T W. Mercury and monomethylmercury: Present and future concerns. Environmental health perspectives, 1991, 96: 159-166.

[302] Morel F M M, Kraepiel A M L, Amyot M. The chemical cycle and bioaccumulation of mercury. Annual Review of Ecology, Evolution, and Systematics, 1998, 29(1): 543-566.

[303] Wiener J G, Krabbenhoft D P, Heinz G H, et al. Ecotoxicology of mercury//Hoffman D J, Rattner B A, Burton JG A, et al. Handbook of Ecotoxicology (2nd ed). Boca Raton, FL: Lewis Publishers. 2003: 409-463.

[304] Clarkson T W, Magos L, Myers G J. The toxicology of mercury: Current exposures and clinical manifestations. N Engl J Med, 2003, 349(18): 1731-1737.

[305] Mergler D, Anderson H A, Chan L H M, et al. Methylmercury exposure and health effects in humans: A worldwide concern. Ambio, 2007, 36(1): 3-11.

[306] Clifton J C. Mercury exposure and public health. Pediatr Clin N Am, 2007, 54(2): 237-241.

[307] Canario J, Santos E J, Padeiro A, et al. Mercury and methylmercury in the Atlantic sector of the Southern Ocean. Deep Sea Research Part II: Topical Studies in Oceanography, 2017, 138: 52-62.

[308] Cossa D, De Madron X D, Schafer J, et al. The open sea as the main source of methylmercury in the water column of the Gulf of Lions(Northwestern Mediterranean margin). Geochimica et Cosmochimica Acta, 2017, 199: 222-237.

[309] Bowman K L, Hammerschmidt C R, Lamborg C H, et al. Mercury in the North Atlantic Ocean: The US GEOTRACES zonal and meridional sections. Deep Sea Research Part II: Topical Studies in Oceanography, 2015, 116: 251-261.

[310] Cossa D, Heimburger L E, Lannuzel D, et al. Mercury in the Southern Ocean. Geochimica et Cosmochimica Acta, 2011, 75(14): 4037-4052.

[311] Heimburger L E, Cossa D, Marty J C, et al. Methyl mercury distributions in relation to the presence of nano- and picophytoplankton in an oceanic water column(Ligurian Sea, North-western Mediterranean). Geochimica et Cosmochimica Acta, 2010, 74(19): 5549-5559.

[312] Sunderland E M. Mercury exposure from domestic and imported estuarine and marine fish in the US seafood market. Environmental Health Perspectives, 2007, 115(2): 235-242.

[313] WHO. Environmental health criteria No. 101: Methylmercury. Geneva; World Health Organization(WHO). 1990.

[314] Bjornberg K A, Vahtera M, Grawe K P, et al. Methyl mercury exposure in Swedish women with high fish consumption. Science of the Total Environment, 2005, 341(1-3): 45-52.

[315] Zhang H, Feng X B, Larssen T, et al. In inland China, rice, rather than fish, is the major pathway for methylmercury exposure. Environmental Health Perspectives, 2010, 118(9): 1183-1188.

[316] Feng X B, Li P, Qiu G L, et al. Human exposure to methylmercury through rice intake in mercury mining areas, Guizhou province, China. Environmental Science & Technology, 2008, 42(1): 326-332.

[317] Compeau G C, Bartha R. Sulfate-reducing bacteria: PRINCIPAL METHYLATORS of mercury in anoxic estuarine sediment. Applied and Environmental Microbiology, 1985, 50(2): 498-502.

[318] Munthe J, Bodaly R A, Branfireun B A, et al. Recovery of mercury-contaminated fisheries. Ambio, 2007, 36(1): 33-44.

[319] Benoit J M, Gilmour C C, Mason R P, et al. Sulfide controls on mercury speciation and bioavailability to methylating bacteria in sediment pore waters(vol 33, pg 951, 1999). Environmental Science & Technology, 1999, 33(10): 1780.

[320] Andrews J C. Mercury speciation in the environment using X-ray absorption spectroscopy //ATWOOD D A. Recent Developments in Mercury Science. 2006: 1-35.

[321] Benoit J M, Mason R P, Gilmour C C. Estimation of mercury-sulfide speciation in sediment pore waters using octanol-water partitioning and implications for availability to methylating bacteria. Environmental Toxicology and Chemistry, 1999, 18(10): 2138-2141.

[322] Schaefer J K, Rocka S S, Zheng W, et al. Active transport, substrate specificity, and methylation of Hg(II)in anaerobic bacteria. Proceedings of the National academy of Sciences of the United States of America, 2011, 108(21): 8714-8719.

[323] Schaefer J K, Morel F M. High methylation rates of mercury bound to cysteine by *Geobacter sulfurreducens*. Nature Geoscience, 2009, 2(2): 123-126.

[324] Hammerschmidt C R, Fitzgerald W F. Methylmercury cycling in sediments on the continental shelf of southern New England. Geochimica et Cosmochimica Acta, 2006, 70(4): 918-930.

[325] Hammerschmidt C R, Fitzgerald W F. Geochemical controls on the production and distribution of methylmercury in near-shore marine sediments. Environmental Science & Technology, 2004, 38(5): 1487-1495.

[326] Siciliano S D, O'driscoll N J, Tordon R, et al. Abiotic production of methylmercury by solar radiation. Environmental Science & Technology, 2005, 39(4): 1071-1077.

[327] Zhang T, Kim B, Leyard C, et al. Methylation of mercury by bacteria exposed to dissolved, nanoparticulate, and microparticulate mercuric sulfides. Environmental Science & Technology, 2012, 46(13): 6950-6958.

[328] Zhang T, Kucarzyk K H, Kim B, et al. Net methylation of mercury in estuarine sediment microcosms amended with dissolved, nanoparticulate, and microparticulate mercuric sulfides. Environmental Science & Technology, 2014, 48(16): 9133-9141.

[329] Jonsson S, Skyllberg U, Nilsson M B, et al. Mercury methylation rates for geochemically relevant Hg-II species in sediments. Environmental Science & Technology, 2012, 46(21): 11653-11659.

[330] Hu H Y, Lin H, Zheng W, et al. Oxidation and methylation of dissolved elemental mercury by anaerobic bacteria. Nat Geosci, 2013, 6(9): 751-754.

[331] Janssen S E, Schaefer J K, Barkay T, et al. Fractionation of mercury stable isotopes during microbial methylmercury production by iron- and sulfate-reducing bacteria. Environmental Science & Technology, 2016, 50(15): 8077-8083.

[332] Podar M, Gilmour C C, Brandt C C, et al. Global prevalence and distribution of genes and microorganisms involved in mercury methylation. Science Advances, 2015, 1(9): 1-13.

[333] Gilmour C C, Podar M, Bullock A L, et al. Mercury methylation by novel microorganisms from new environments. Environmental Science & Technology, 2013, 47(20): 11810-11820.

[334] Parks J M, Johs A, Podar M, et al. The genetic basis for bacterial mercury methylation. Science, 2013, 339(6125): 1332-1335.

[335] Ekstrom E B, Morel F M M, Benoit J M. Mercury methylation independent of the acetyl-coenzyme a pathway in sulfate-reducing bacteria. Applied and Environmental Microbiology, 2003, 69(9): 5414-5422.

[336] King J K, Kostka J E, Frischer M E, et al. Sulfate-reducing bacteria methylate mercury at variable rates in pure culture and in marine sediments. Applied and Environmental Microbiology, 2000, 66(6): 2430-247.

[337] Benoit J M, Gi; mour C C, Mason R P. Aspects of bioavailability of mercury for methylation in pure cultures of *Desulfobulbus propionicus*(1pr3). Applied and Environmental Microbiology, 2001, 67(1): 51-58.

[338] Wang Y, Schaefer J K, Mishra B, et al. Intracellular Hg(0)oxidation in *Desulfovibrio desulfuricans* ND132. Environmental Science & Technology, 2016, 50(20): 11049-11056.

[339] Lu X, Liu Y, Johs A, et al. Anaerobic mercury methylation and demethylation by *Geobacter bemidjiensis* Bem. Environmental Science & Technology, 2016, 50(8): 4366-4373.

[340] Hamelin S, Amyot M, Barkay T, et al. Methanogens: Principal methylators of mercury in lake periphyton. Environmental Science & Technology, 2011, 45(18): 7693-7700.

[341] Yu R Q, Reinfelder J R, Hines M E, et al. Mercury methylation by the methanogen *Methanospirillum hungatei*. Applied and Environmental Microbiology, 2013, 79(20): 6325-6330.

[342] Gilmour C C, Billock A L, Mcburney A, et al. Robust mercury methylation across diverse methanogenic archaea. Mbio, 2018, 9(2):

[343] Bravo A G, Zopfi J, Buck M, et al. Geobacteraceae are important members of mercury-methylating microbial communities of sediments impacted by waste water releases. Isme Journal, 2018, 12(3): 802-812.

[344] Lin T Y, Kampalath R A, Lin C-C, et al. Investigation of mercury methylation pathways in biofilm versus planktonic cultures of *Desulfovibrio desulfuricans*. Environmental Science & Technology, 2013, 47(11): 5695-5702.

[345] Lin C C, Jay J A. Mercury methylation by planktonic and biofilm cultures of *Desulfovibrio desulfuricans*. Environmental Science & Technology, 2007, 41(19): 6691-6697.

[346] Huguet L, Castelle S, Schafer J, et al. Mercury methylation rates of biofilm and plankton microorganisms from a

hydroelectric reservoir in French Guiana . Science of the Total Environment, 2010, 408(6): 1338-1348.

[347] Han S, Narasingarao P, Obraztsova A, et al. Mercury speciation in marine sediments under sulfate-limited conditions . Environmental Science & Technology, 2010, 44(10): 3752-3757.

[348] Choi S C, Chase T, Bartha R. Metabolic pathways leading to mercury methylation in *Desulfovibrio desulfuricans* LS . Applied and Environmental Microbiology, 1994, 60(11): 4072-4077.

[349] Choi S C, Chase T, Bartha R. Enzymatic catalysis of mercury methylation by *Desulfovibrio desulfuricans* LS . Applied and Environmental Microbiology, 1994, 60(4): 1342-1346.

[350] Ekstrom E B, Morel F M. Cobalt limitation of growth and mercury methylation in sulfate-reducing bacteria . Environmental Science & Technology, 2007, 42(1): 93-99.

[351] Poulain A J, Barkay T. Cracking the mercury methylation code . Science, 2013, 339(6125): 1280-1281.

[352] Lin H, Hurt R A, Jr, Johs A, et al. Unexpected effects of gene deletion on interactions of mercury with the methylation-deficient mutant delta *hgcAB* . Environmental Science & Technology Letters, 2014, 1(5): 271-276.

[353] Graham A M, Cameron B K T, Hajic H A, et al. Sulfurization of dissolved organic matter increases Hg-sulfide-dissolved organic matter bioavailability to a Hg-Methylating bacterium. Environmental Science & Technology, 2017, 51(16): 9080-9088.

[354] Zhu H, Zhong H, Evans D, et al. Effects of rice residue incorporation on the speciation, potential bioavailability and risk of mercury in a contaminated paddy soil . Journal of Hazardous Materials, 2015, 293(0): 64-71.

[355] Liu Y R, Dong J X, Han L L, et al. Influence of rice straw amendment on mercury methylation and nitrification in paddy soils . Environmental Pollution, 2016, 209: 53-59.

[356] Zheng W, Liang L, Gu B. Mercury reduction and oxidation by reduced natural organic matter in anoxic environments. Environmental Science & Technology, 2012, 46(1): 292-299.

[357] Luo H W, Yin X, Jubb A M, et al. Photochemical reactions between mercury(Hg)and dissolved organic matter decrease Hg bioavailability and methylation . Environmental Pollution, 2017, 220: 1359-1365.

[358] Martin D R C R, Mateo R, Jiemenez-Moreno M. Is gastrointestinal microbiota relevant for endogenous mercury methylation in terrestrial animals? . Environmental Research, 2017, 152: 454-461.

[359] Zhao L, Qiu G, Anderson C W N, et al. Mercury methylation in rice paddies and its possible controlling factors in the Hg mining area, Guizhou province, Southwest China . Environmental Pollution, 2016, 215: 1-9.

[360] Paquette K, Helz G. Solubility of cinnabar(red HgS)and implications for mercury speciation in sulfidic waters . Water, Air, and Soil Pollution, 1995, 80(1): 1053-1056.

[361] Wang Y J, Dang F, Zhao J T, et al. Selenium inhibits sulfate-mediated methylmercury production in rice paddy soil . Environmental Pollution, 2016, 213: 232-239.

[362] Schaefer J K, Szczuka A, Morel F M M. Effect of divalent metals on Hg(II)uptake and methylation by bacteria . Environmental Science & Technology, 2014, 48(5): 3007-3013.

[363] Lin H, Morrell F J L, Rao B, et al. Coupled mercury-cell sorption, reduction, and oxidation on methylmercury production by *Geobacter sulfurreducens* PCA . Environmental Science & Technology, 2014, 48(20): 11969-11976.

[364] Canario J, Caetano M, Vale C, et al. Evidence for elevated production of methylmercury in salt marshes . Environmental Science & Technology, 2007, 41(21): 7376-7382.

[365] Bridou R, Monperrus M, Gonzalez P R, et al. Simultaneous determination of mercury methylation and demethylation capacities of various sulfate‐reducing bacteria using species‐specific isotopic tracers. Environmental Toxicology and Chemistry, 2011, 30(2): 337-344.

[366] Lu X, Gu W, Zhao L, et al. Methylmercury uptake and degradation by methanotrophs . Science Advances, 2017, 3(5): 659-667.

[367] Marvin D M, Agee J, Mcgowan C, et al. Methyl-mercury degradation pathways: A comparison among three mercury-impacted ecosystems . Environmental Science & Technology, 2000, 34(23): 4908-4916.

[368] Schaefer J K, Yagi J, Reinfelder J R, et al. Role of the bacterial organomercury lyase(MerB)in controlling methylmercury accumulation in mercury-contaminated natural waters . Environmental Science & Technology, 2004, 38(16): 4304-4311.

[369] Barkay T, Miller S M, Summers A O. Bacterial mercury resistance from atoms to ecosystems. FEMS Microbiology Review, 2003, 27(2-3): 355-384.

[370] 阴永光, 李雁宾, 蔡勇, 等. 汞的环境光化学. 环境化学, 2011, 30(1): 84-91.

[371] Chen B W, Chen P, He B, et al. Identification of mercury methylation product by *tert*-butyl compounds in aqueous solution under light irradiation. Marine Pollution Bulletin, 2015, 98(1-2): 40-46.

[372] Celo V, Lean D R S, Scott S L. Abiotic methylation of mercury in the aquatic environment. Science of the Total Environment, 2006, 368(1): 126-137.

[373] Falter R. Experimental study on the unintentional abiotic methylation of inorganic mercury during analysis: Part 1: Localisation of the compounds effecting the abiotic mercury methylation. Chemosphere, 1999, 39(7): 1051-1073.

[374] 左跃钢, 庞叔薇. 巯基化合物存在下无机汞的光化学甲基化. 环境科学学报, 1985, 5(2): 239-243.

[375] Akagi H, Fujita Y, Takabatake E. Photochemical methylation of inorganic mercury in the presence of mercuric sulfide. Chemistry Letters, 1975, 4(2): 171-176.

[376] Chen B W, Wang T, Yin Y G, et al. Methylation of inorganic mercury by methylcobalamin in aquatic systems. Applied Organometallic Chemistry, 2007, 21(6): 462-467.

作者：冯新斌[1], 王书肖[2], 蔡 勇[3,6], 张 彤[4], 吴清茹[2], 刘广良[5], 阴永光[6], 袁 巍[1]

[1] 中国科学院地球化学研究所, [2] 清华大学环境学院, [3] 山东大学, [4] 南开大学, [5] 江汉大学, [6] 中科院生态环境研究中心

第8章 铁基材料及其污染控制技术研究进展

- 1. 引言 /194
- 2. 纳米零价铁及其重金属废水治理技术 /194
- 3. 铁矿物及其污染控制技术研究进展 /201
- 4. 基于高铁酸盐的污染控制技术及原理 /214
- 5. 微生物介导亚铁氧化耦合硝酸盐还原机制与环境效应 /220
- 6. 展望 /226

本章导读

铁是地壳中元素丰度排在第四的过渡金属元素，广泛存在大气气溶胶、天然水体、土壤和动植物体内，其多变的价态及多种多样的化学特性，使其在地质演替、生物进化及环境污染物迁移和转化过程中都扮演着极其重要的角色。因此，基于铁基材料发展的环境污染控制技术已成为环境和地球化学领域的研究热点。本章总结了近五年来国内外学者基于零价铁、铁矿物、高铁酸盐发展的环境污染控制原理及技术研究进展，并综述了微生物介导亚铁氧化耦合硝酸盐还原机制与环境效应，为深入研究铁环境化学奠定基础。

关键词

零价铁，铁矿物，高铁酸盐，污染控制，环境修复，微生物耦合

1 引　言

铁是地壳中元素丰度排在第四的过渡金属元素，广泛存在大气气溶胶、天然水体、土壤和动植物体内。由于铁元素价态多变，化学性质活泼，在地质演替、生物进化及环境污染物迁移和转化过程中都扮演着极其重要的角色。因此，基于铁基材料发展的环境污染控制技术已成为环境和地球化学领域的研究热点。从低价态到高价态，常见的铁基材料有零价铁、二价铁矿物、三价铁矿物、高铁酸盐等。它们既是廉价友好的环境污染物吸附剂，又可通过电子传递过程发生氧化还原反应，影响环境污染物的迁移、转化、归趋和生物利用等。本章总结了近五年来国内外学者基于零价铁、铁矿物、高铁酸盐发展的环境污染控制原理及技术研究进展，并综述了微生物介导亚铁氧化耦合硝酸盐还原机制与环境效应，为深入研究铁环境化学及其污染控制技术奠定基础。

2 纳米零价铁及其重金属废水治理技术

土壤及水体重金属污染控制近年得到广泛关注，重金属污水是相关重金属污染重要源头。近年来我国污水处理排放标准不断提高，也对重金属污水处理技术提出更高要求。各种相关技术研究论文及报道爆发式出现。其中，功能纳米材料去除水中重金属近年来得到广泛关注和研究，而获得放大验证和实际应用的并不多见。特别是，在众多功能纳米材料中，纳米零价铁成为为数不多的获得工程化应用的纳米材料之一。

2.1 纳米零价铁的结构与制备

纳米零价铁（nZVI）具有"核-壳"结构，即核心为金属铁，外层包裹一层铁氧化物[1, 2]。图 8-1（c）、(f) 和（i）为 STEM 明场像（bright-field，BF），图 8-1（a）、(d) 和（g）为 STEM 高角度环形暗场像（high-angle annular dark-field，HAADF），纳米零价铁多为球形颗粒，粒径集中分布在 20～100 nm，其聚集成链状。单个纳米零价铁颗粒的内核区域和外壳区域都存在明显的衬度差异，说明内核和外壳的结构

及化学组成存在显著差异，即纳米零价铁具有典型的"核-壳"结构[1]。纳米零价铁的"核-壳"结构是由于零价铁在环境中自发的氧化过程生成的铁氧化物层包裹在零价铁外部而形成的，壳层厚度约为 2～4 nm。XPS 和 EELS 等化学分析手段进一步表明，铁氧化物层是由靠近 Fe^0-O 界面的混合二价/三价铁氧化物和靠近水相界面的三价铁氧化物组成[3]。

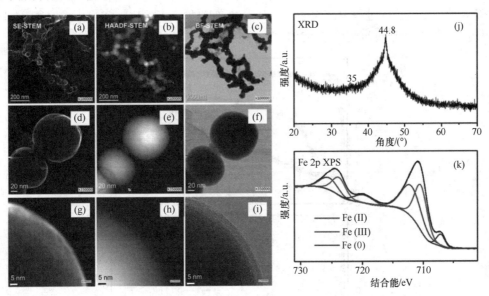

图 8-1 纳米零价铁的表面化学和晶相特征：扫描透射电镜图、X 射线衍射图和 X 射线光电子能谱分析（引自参考文献[1]）
(a)、(d)、(g) 二次电子（SE）图像；(b)、(e)、(h) STEM 高角度环形暗场像（HAADF）；(c)、(f)、(i) STEM 明场像（BF）；(j) X 射线衍射图；(k) Fe 2 p 的 X 射线光电子能谱分析

新鲜纳米零价铁的 XRD 谱图在衍射角 $2\theta=44.8°$ 存在一个弱且宽化的衍射特征峰 [图 8-1 (j)]，该峰为体心立方的 α-Fe（0）(110) 晶面衍射峰，宽化的衍射峰表明其铁核是粒径较小的多晶颗粒组成[1]。XPS 分析结果进一步证明"核-壳"结构。研究结果表明，纳米零价铁表面的氧化层与块状铁表面致密的钝化膜有明显区别，颗粒大小、合成方法和保存条件等对其组成结构存在较大影响。例如，溅射法合成的纳米零价铁氧化层主要为 γ-Fe_2O_3 或未完全氧化的 Fe_3O_4；气体还原法合成的纳米零价铁氧化层主要由 Fe_3O_4 组成；而液相还原法合成的纳米零价铁氧化层化学组成与 FeOOH 相似。

纳米零价铁的合成方法多种多样，包括液相还原法、气相还原法、溅射法、蒸发凝聚法、气相热分解法、机械球磨法等。其中，硼氢化钠还原铁盐制备纳米零价铁设备简单，无需高温高压，是实验室研究最常采用的方法，但应用于大规模实际处理存在成本过高的缺陷；为此张伟贤研究组在 2007 年开始探索采用精细球磨制备纳米零价铁并获得成功[4]。

2.2 纳米零价铁在水环境转化

零价铁标准电极电位 E^0[Fe(II)/Fe（0）] 为-0.44V，具有较强的还原性。纳米零价铁具有更大比表面积（约 30 m^2/g），因此反应活性更高，投加到去离子水中可快速与水分子和溶解氧发生反应[5, 6]

$$2Fe^0_{(s)} + 4H^+_{(aq)} + O_{2(aq)} \rightarrow 2Fe^{2+}_{(aq)} + 2H_2O_{(l)} \tag{8-1}$$

$$Fe^0_{(s)} + 2H_2O_{(l)} \rightarrow Fe^{2+}_{(aq)} + H_{2(g)} + 2OH^-_{(aq)} \tag{8-2}$$

上述反应将消耗水中的氢离子或产生氢氧根离子，使溶液的 pH 上升。此外，随着溶解氧及其他氧化性物质的消耗，溶液的氧化还原电位也将显著下降。图 8-2 为去离子水中投加不同剂量（0.01 g/L，0.05

g/L，0.1 g/L 和 0.75 g/L）纳米零价铁后 pH 和氧化还原电位（E_h）的变化曲线。从图中可以看出，纳米零价铁的加入使 pH 上升了 2~3 个单位，最终稳定在 pH 7.5~9.5 的范围内，整个体系呈碱性环境。pH 的迅速上升使体系维持在碱性范围内，有利于金属阳离子的吸附和沉淀。与此同时，随着纳米零价铁的加入，体系的氧化还原电位也迅速下降，最终维持在–400～–600 mV 的强还原性范围内，有利于重金属的还原去除。此外，在反应过程中产生的一系列腐蚀产物[如 Fe^{2+}、$Fe(OH)_2$、FeOOH、Fe_3O_4、绿锈等]既能吸附重金属，也可通过沉淀/共沉淀作用去除金属。

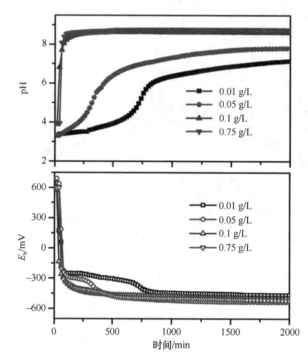

图 8-2　新鲜纳米零价铁投加到去离子水中 pH 和 E_h 随时间变化趋势图（引自参考文献[7]）

纳米零价铁在典型水环境条件下转化如图 8-3 所示[1]。新鲜纳米零价铁在无氧水环境腐蚀 3 天后，少量颗粒氧化成片层状结构，大部分颗粒为球状且核壳界线分明，仅壳层厚度略增加，主要物相仍为无定形结构 α-Fe^0，产生少量 Fe_3O_4。在扰动复氧水相中腐蚀 3 天后形貌结构发生较大变化，由球状颗粒演变片层状结构，再结晶形成 γ-FeOOH。静态复氧条件下纳米零价铁腐蚀速率较慢，反应 30 天颗粒仍存在大量零价铁且具有强还原性，主要的腐蚀产物为 γ-FeOOH 和少量 Fe_3O_4；但此时部分颗粒核心完全被腐蚀产生中空结构铁氧化物，部分颗粒未被完全腐蚀但是核壳界限模糊，壳层结构增厚[7]。扰动复氧条件下腐蚀 1 天的颗粒基本失去还原能力，主要产物为 Fe_3O_4，颗粒由"核-壳"结构腐蚀完全溶解形成镂空结构的氧化物。随着反应时间继续延长至 3 天，腐蚀产物演变为片层状 γ-FeOOH，表明缓慢连续供氧（静态复氧）有利于 γ-FeOOH 的形成；而快速供氧（扰动复氧）首先生成 Fe_3O_4 腐蚀产物，随着氧气的继续供给，Fe_3O_4 转变为 γ-FeOOH[8, 9]。重金属溶液，例如 Cr(VI)溶液中，低浓度 Cr（≤10mg/L）条件下生成的还原产物 Cr(Ⅲ)-Fe(Ⅲ) 沉积物量少不均匀地分布在纳米零价铁内造成更多表面缺陷位点，促进纳米零价铁腐蚀。高浓度 Cr（≥100mg/L）溶液中，在表面形成 Cr(Ⅲ)-Fe(Ⅲ)沉积物保护层分布于核壳之间抑制纳米零价铁腐蚀，颗粒结构基本不发生变化[10]。水相中无机阴离子对纳米零价铁性质演变存在影响：NO_3^- 可竞争纳米零价铁表面活性位点并被还原，同时还能够腐蚀颗粒表面的氧化层，形成 Fe_3O_4 铁的矿物质；SO_4^{2-} 和 HPO_4^{2-} 对纳米零价铁颗粒的物相演变影响较小，除形成少量铁的氧化物外还生成一些铁的矿化物钝化层，阻碍活性位点形成而减弱纳米零价铁腐蚀；HCO_3^- 降低颗粒的腐蚀速率和腐蚀程度。高分子有机电解质 APAM 和 CMC 包覆于颗粒表面起保护作用,阻碍通过表面高活性位点与水分子和溶解氧反应,有效起到防止腐蚀作用,且短时间腐蚀后对颗粒活性影响较小。

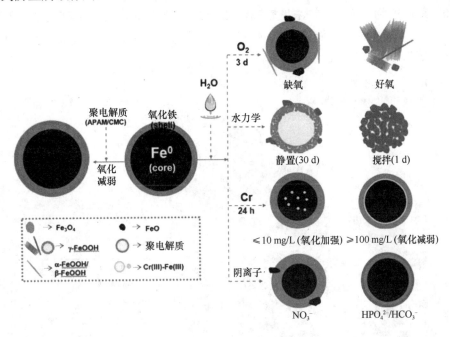

图 8-3　纳米零价铁在水相反应中表面化学和晶相转化模型图(引自参考文献[7])

2.3　纳米零价铁与重金属反应

2.3.1　反应机理

大量实验表明,纳米零价铁能够有效去除铜、砷、铬、镉、铅、镍、锌等多种重金属污染物,并且速率很快。纳米零价铁独特的"核-壳"结构和物理、化学性质使其能够通过多种作用实现重金属的分离富集。纳米零价铁与重金属的作用机理主要包括吸附、还原、沉淀/共沉淀三种[2]。首先,水中的纳米零价铁氧化物壳被羟基官能团(—OH)所覆盖,能够通过静电作用和表面络合作用吸附重金属离子,将其固定于纳米铁表面。其次,纳米零价铁核心部分的 Fe^0 是较强的还原剂,能够作为电子供体,将氧化性强于 Fe(II)的重金属离子还原到较低价态(例如,金属单质等稳定态)。再次,Fe^0 腐蚀使整个体系呈碱性,尤其是纳米零价铁表面,OH 浓度较高,有利于重金属离子的表面沉淀。另外,铁腐蚀产生的 Fe(II)和 Fe(III)还能与 As(III/V)等含氧阴离子发生共沉淀作用,实现污染物的去除。

张伟贤研究组采用 XPS 和 XRD 等研究了多种重金属离子与纳米零价铁反应后在纳米零价铁表面的存在形态,认为当某金属的 E^0 比 Fe^0 的高(正)很多时,该金属主要通过还原作用被纳米铁去除,如 Cu(II)、Hg(II)、Ag(I)等,覆盖在零价铁外部的铁氧化壳层对还原过程中的电子传递有很大的影响,从核心的零价铁到表面的重金属可能存在三种电子传递途径:①电子直接通过颗粒表面缺陷(如空穴、晶界等)进行传递;②铁氧化物层由混合二价/三价铁氧化物构成,可视为半导体,介导电子传递;③纳米零价铁表面的二价铁作为电子供体,为重金属的还原提供电子。当某金属的 E^0 比 Fe^0 的更低(负)或接近时,该金属主要通过吸附作用被去除,如 Zn(II)、Cd(II)等;当某金属的 E^0 比 Fe^0 的稍高(正)时,该金属可通过吸附、还原的双重作用被去除,如 Ni(II)、Co(II)、Pb(II)等[2]。纳米零价铁与几种常见的重金属反应机理图见图 8-4，吸附、还原、沉淀及共沉淀是 nZVI 同步去除水中重金属的主要机理。nZVI 与 Cu(II)反应后出现树枝状产物（SEM），经 XRD 分析主要为 Cu^0 和 Cu_2O；nZVI 与 As(V)反应时：XPS 表明存在 As(III)和 As（0）还原产物；在 As(V)浓度较高（如 500 mg/L）时，可与 nZVI 腐蚀产生的 Fe^{2+} 形成 $Fe_3(AsO_4)_2$ 沉淀[11]；Cu(II)存在亦能显著提高 nZVI 对 As(V)的去除[12]。nZVI 腐蚀产物（Fe^{2+}、OH^-等）可与 Zn(II)、Pb(II)等形成（共）沉淀。

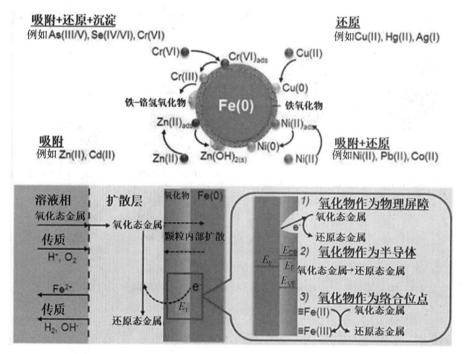

图 8-4　纳米零价铁与重金属反应机理及反应过程示意图（引自参考文献[3]）

纳米零价铁与 Ni(II)反应：Ni(II)可以通过吸附和还原的双重作用被纳米零价铁有效去除，Ni(II)与纳米零价铁的反应为研究纳米零价铁的结构、性质及与重金属的反应过程提供了很好的例证。Li 等的 XPS 研究发现，Ni(II)与纳米零价铁反应时首先在物理吸附的作用下被束缚到纳米零价铁表面，然后通过化学吸附被进一步固定，并逐步还原为 Ni^0。其反应过程如下[13]：

$$\equiv FeOH + Ni^{2+} \rightarrow \equiv FeO\text{-}Ni^{2+} + H^+ \tag{8-3}$$

$$\equiv FeONi^+ + H_2O \rightarrow \equiv FeONi\text{-}OH + H^+ \tag{8-4}$$

$$\equiv FeONi^+ + Fe^0 + H^+ \rightarrow FeO\text{-}Ni + Fe^{2+} \tag{8-5}$$

一般认为，溶解态的重金属离子首先需要扩散穿过液膜到达纳米零价铁表面，通过静电作用/表面络合作用吸附到表面，然后通过沉淀、还原等作用被进一步固定。

2.3.2　复杂体系及其他传统材料比较

以上有关 nZVI 去除重金属研究局限于简单水环境体系，对复杂实际废水缺乏考察。张伟贤研究组随后选取几种常见重金属废水，进行长期考察。现场长期（>150 d）监测表明，有色冶炼、电镀等行业废水往往成分复杂，存在多种重金属，且浓度高、波动大（如 As 可达 1000 mg/L，Pb 可达 2500 mg/L）[12]。

张伟贤研究组通过实验室小试首先考察了 nZVI 处理重金属废水的可行性及作用机理，比较研究常见重金属去除方法处理此类废水可行性。纳米零价铁分离多种重金属的反应速率明显高于普通铁基材料（铁氧化物或普通尺度零价铁粉），处理容量是活性炭、氧化铝、沸石和离子交换树脂等材料的 10～1000 倍。

图 8-5 比较了相同投加量的纳米零价铁和 400 目的微米铁粉在相同时间内对 4 种常见重金属的去除效果。由图 8-5 可知,纳米零价铁对 Pb(II)、Zn(II)、As(V)、Cr(VI)的最大负荷分别可达 790.9 mg/g-Fe、567.1 mg/g-Fe、603.6 mg/g-Fe 及 86.3 mg/g-Fe,远高于微米铁的去除容量。综上,纳米零价铁和普通微米铁粉相比优势在于:①比表面积大,表面活性高,与重金属的反应速率和去除容量都高于微米铁;②颗粒表面铁氧化物膜结构不同,纳米零价铁表面缺陷多,更有利于还原反应中的电子传递,因而与污染物的反应更加彻底;③纳米零价铁表面活性点位多,抗钝化能力优于微米铁;④纳米零价铁颗粒尺寸小,能有效分散悬浮在水中,与污染物接触更加充分,混合条件对传质的影响没有微米铁大[3, 12]。

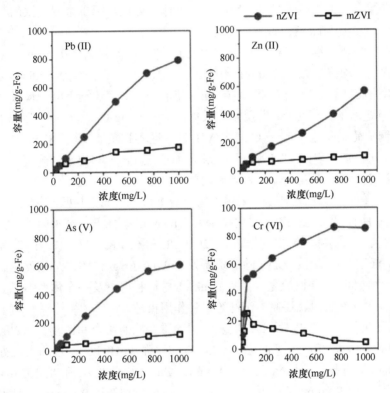

图 8-5　纳米零价铁和微米零价铁(400 目)重金属去除容量比较(引自参考文献[3])

与其他材料相比,nZVI 也存在很多优势。结果表明,沉淀[Ca(OH)$_2$、Na$_2$S]、吸附等方法均难以实现多种重金属同步去除,残余浓度较高(结果在现场研究中得到验证)。nZVI 可实现废水中多种重金属阳离子及含氧阴离子同步有效去除,如 nZVI 可在 1 h 内去除水中 Cu、As、Zn、Ni 四种重金属(初始浓度 6~130 mg/L),残余浓度均低于 0.5 mg/L。纳米零价铁因以上独特的结构和性质、高反应活性及良好的环境相容性,使其脱颖而出,且具有处理效果好等优点。

2.4　纳米零价铁工程应用

2.4.1　重力分离

nZVI 应用于水处理,反应后需固液分离,从而实现 nZVI 回用及保障出水水质。张伟贤研究组分别在沉降柱体(2.5 m)和小型连续流竖流沉淀池考察了 nZVI 重力分离[14]。沉降柱实验表明,nZVI(0.5 g/L)可通过重力沉降实现固液分离,其中 90%以上 nZVI 在 1 h 内完成沉降;竖流式沉淀池沉降实验表明,对于 1 g/L nZVI,当表面负荷由 3.0 m^3/(m^2·h)降至 0.5 m^3/(m^2·h)时,出水中悬浮铁浓度由 0.56 g/L 降至

0.07 g/L。固相分析表明，nZVI 快速沉降可归因于其较低的表面电荷（～4 mV，pH=8.4）及微米级团聚体（D_{50}=70 μm）。实验同时发现，少部分 nZVI 及腐蚀产物沉降性能较差，即使在较低表面负荷 [0.1 m^3/(m^2·h)] 或较长沉降时间（3 h）下，上清液悬浮铁浓度仍可达 30 mg/L。通过投加少量（1%，w/w）阴离子型聚丙烯酰胺（PAM）可有效促进 nZVI 沉降并减少悬浮物浓度。粒径分析表明，PAM 修饰后 nZVI 团聚体粒径显著增大，D_{50} 由 7 μm 增至 355 μm。少量 PAM 加入对 nZVI 除 Ni(II) 反应活性无明显抑制。

2.4.2 中试研究

随后，在中试规模上，张伟贤研究组分别考察了该体系处理实际重金属废水的可行性及关键影响因素，完善了相关理论、方法及配套技术。

（1）考察了 nZVI 处理含铜线路板制造废水可行性[15]。在流量为 1.0 m^3/h，反应区水力停留时间（HRT）为 1.6 h，nZVI 平均投量为 0.23 kg/m^3 条件下，废水中平均铜浓度从 52.5 mg/L 降至 1.7 mg/L，且出水水质稳定。降低反应区 HRT（由 1.6 h 到 0.6 h）使铜处理效率逐渐下降，最低降至 27%。结合反应器中 nZVI 浓度，考察进水铜负荷影响：发现在较低负荷下[<10 mg-Cu/(g-Fe·h)]，系统除铜效率大于 90%。本研究中，nZVI 除铜负荷最高达 343 mg-Cu/g-nZVI；反应区容积负荷高达 1876 g-Cu/(m^3·d)，污泥中铜含量高达 25%（w/w）。研究还发现，nZVI 反应区氧化还原电位（E_h）与出水铜浓度呈显著正相关，即出水铜浓度随 E_h 下降而减小，可作为反应器除铜效果的判别指标。

（2）针对含砷多重金属复杂冶炼废水，采用多级反应体系考察 nZVI 处理该废水可行性[12]。在流量为 0.4 m^3/h，各级反应区 HRT 为 4 h，nZVI 平均投量为 2.1 kg/m^3 条件下，废水中砷、铜、锌、镍等 7 种重金属得到同步去除，其中平均砷、铜浓度分别从 520 mg/L 和 67 mg/L 降至 10 mg/L 和 0.1 mg/L，总体出水中各重金属浓度均低于 0.1 mg/L；nZVI 除砷负荷为 239 mg-As/g-nZVI；一级反应器污泥中砷含量高达 10%。研究还发现，两级间 nZVI 回流是一种提高 nZVI 材料利用率及其在体系中停留时间的可行措施。研究证实 nZVI 除 As(V) 存在吸附、还原、沉淀等多种作用机理。

（3）实验进一步研究考察了 nZVI 和传统石灰法去除重金属的结合和比较[16]。制酸废水中含有高浓度的铅和锌，平均浓度分别达 610 mg/L 和 195 mg/L，且酸性强（pH=2.5），经石灰法处理后出水中仍残留铅 19 mg/L，锌 18 mg/L，远超排放标准（铅：0.1 mg/L，Zn：1.0 mg/L）。在流量为 0.4 m^3/h，反应区 HRT 为 4 h，nZVI 平均投量为 1.2 kg/m^3 条件下，出水中铅、锌浓度均低于 0.1 mg/L，且出水水质未受进水波动影响。小试及反应产物固相分析表明，石灰法难以达标的原因有：水质波动导致石灰投量难以精确控制、铅、锌氢氧化物的复溶（pH=12.5）、沉淀粒径小难沉降等。研究表明，纳米零价铁去除重金属可有效避免高 pH 条件下复溶、难沉降分离等问题，其与 nZVI 相对温和的 pH（8.6）及其自我调节稳定机制有关；此外，nZVI 晶种效应可有效增大反应产物粒径并促进重金属沉淀分离（图 8-6）。

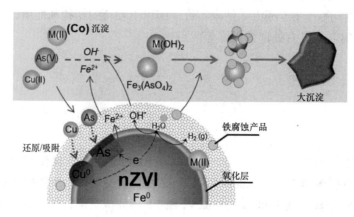

图 8-6　nZVI 处理复杂废水机理（引自参考文献[17]）

2.4.3 工程应用

在以上系统研究基础上,在较大规模(~400 m³/d)上,张伟贤研究组考察了 nZVI 处理高浓度含铜含砷废水可行性,验证小试及中试研究结论并考察相应措施可行性[17]。研究采用两级"反应—分离—回用"强化工艺(图 8-7),采用级间回流提高 nZVI 利用率,采用 E_h 实时监测反馈调节 nZVI 投加,系统出水稳定。长期运行表明,nZVI 平均投量为 0.4~0.5 kg/m³(年均使用 5.3×10^4 kg nZVI 处理 1.2×10^5 m³ 废水),各级反应区 HRT 为 2 h,废水中平均砷、铜浓度分别从 110 mg/L、103 mg/L 降至 0.29 mg/L、0.16 mg/L,其他重金属同步有效分离;其中 nZVI 除砷、除铜负荷分别达 245 mg-As/g-nZVI 和 226 mg-Cu/g-nZVI,总体重金属去除负荷超过 500 mg-重金属/g-nZVI。运行也证明,反应区 E_h 及时反馈调控 nZVI 投加灵敏,表现出较强抗冲击负荷能力。污泥中铜、砷含量分别达 10%、8%,可进一步回收。研究还发现,废水微量的 Au 在 nZVI 中得到富集:系统产生的污泥金可达 100 g/t,平均约 41g/t[18],所有产生的污泥被循环,进行重新资源化处理。nZVI 富集废水中痕量金原理如图 8-8 所示。工程实践表明,nZVI 在处理类似重金属工业废水具备显著特色和应用前景。

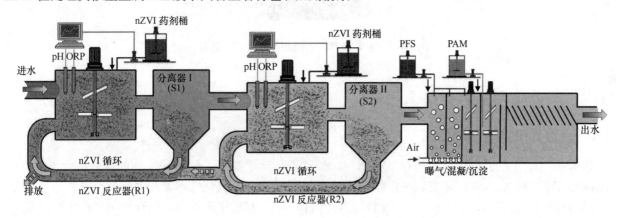

图 8-7 纳米零价铁处理废水工艺流程图(引自参考文献[17])

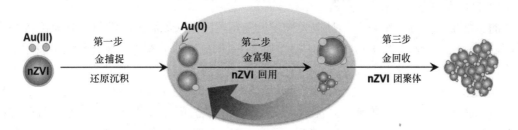

图 8-8 纳米零价铁富集废水中痕量金原理(引自参考文献[18])

3 铁矿物及其污染控制技术研究进展

铁是地壳中元素丰度排在第四的过渡金属元素,但因其性质活泼、价态多变,大多数铁元素以铁矿物存在于自然界。铁矿物种类繁多,其中以赤铁矿、针铁矿和磁铁矿等铁的(氢)氧化物最为常见。铁矿物具有多种多样的化学特性,在地质演替、生物进化及环境污染物迁移和转化过程中都扮演着极其重要的角色。同时,由于其廉价易得,环境友好,表面活性位点丰富等特点,铁矿物已被广泛地用于发展绿色友好的水体及土壤污染控制技术，成为当前环境领域的研究热点。本节将介绍常见铁矿物的形成、制备及表征，并重点综述基于铁矿物的吸附性质和氧化还原性质发展的环境污染控制技术（图8-9）。

图 8-9　铁矿物及其污染控制技术简要示意图

3.1　铁矿物的形成、制备及表征

3.1.1　自然界中铁矿物的形成

铁矿物广泛存在于土壤、水体和沉积物中，驱动着地球上物理、化学和生物的部分或全部过程。Navrotsky 等[19]在 *Science* 综述文章中这样描述道："很难找到一个氧化铁不参与的化学反应和环境过程，从火星表面到地球深处，从废弃工厂到高科技磁性储存材料，从鸽子大脑和趋磁性细菌到药物传输系统，氧化铁都广泛存在。它们是岩石和土壤的组分，是腐蚀和细菌过程的构成，而且是铁质营养元素的来源。"图 8-10 展示了自然界中存在的铁矿物[20]。

铁矿物的自然形成过程非常复杂，且很大程度上受环境因素（如温度、湿度及酸碱度等）的影响。简单来说，以铁氧化物为例，含 Fe(II)原生矿石［如含 Fe(II)硅酸盐］首先在风化作用中发生不可逆转的溶解和氧化，释放的 Fe^{3+} 主要以沉淀的形式存在，即为铁（氢）氧化物。形成的铁氧化物又可在自然环境中被微生物还原溶解为 Fe^{2+}，通过氧化或水解，溶解的 Fe^{2+} 又会生成铁氧化物，形成一个自然铁循环过程。与铁氧化物不同，硫铁矿物通常在无氧的沉积物或沉积岩中形成[20]。首先，溶解性硫酸盐或含硫的有机化合物被硫还原细菌还原或分解成 H_2S。形成的 H_2S 可与不同的铁矿物反应，生成黑色的一硫化铁（FeS）。硫单质存在的条件下，FeS 又可与硫单质反应生成黄铁矿（FeS_2）[21]。自然条件下形成的铁矿物有结晶态，也有无定形态。通常，结晶态的铁矿物主要有赤铁矿（$\alpha\text{-}Fe_2O_3$）、针铁矿（$\alpha\text{-}FeOOH$）、纤铁矿（$\gamma\text{-}FeOOH$）、磁赤铁矿（$\gamma\text{-}Fe_2O_3$）和黄铁矿（FeS_2）等，无定形态的主要有 2-线水铁矿、6-线水铁矿、沃特曼矿和绿绣矿等。

自然界中的铁矿物受周围环境影响会相互转化。几乎所有的铁矿物都可以在适当的条件下通过脱水、脱羟基及氧化还原等方式进行结构重排，转化为另外至少两种铁矿物[22]。在有氧环境中，赤铁矿和针铁矿在热力学上通常是自然界中最稳定的铁矿物，是许多其他铁矿物转化的最终产物。例如，水铁矿通过热脱水、脱羟基作用可转化为赤铁矿和磁赤铁矿，通过再沉淀作用可转化为针铁矿，在 Cl^- 存在的酸性介质中可转化为正方针铁矿；纤铁矿可通过还原作用转化为磁铁矿，而磁铁矿又可通过氧化作用转化为磁赤铁矿或赤铁矿。铁矿物的相互转化往往会影响其表面固定元素的可利用性及环境效应。

图 8-10 自然界中广泛存在的铁矿物（引自参考文献[20]）

3.1.2 铁矿物的人工制备方法

实验室中，铁矿物一般以二价铁盐或三价铁盐为原料制得。总的来说，铁矿物的制备方法有化学法、物理法和生物法三大类[23]。图 8-11 对比展示了这三大类合成方法。其中，化学法以其简单、高效和可控等优点成为合成铁矿物使用最广泛的方法。常见化学合成铁矿物的方法有：沉淀法、水热法和微乳化法等。例如，FeOOH，Fe_3O_4 和 $\gamma\text{-}Fe_2O_3$ 通常是向铁盐溶液里加入强碱，使所得悬浮液沉淀、陈化而制得[24]；赤铁矿和针铁矿可分别在 pH=0.8~2.6 和 pH=8.0~10.0 的条件下 100~200℃水热转化 $Fe(OH)_3$ 得到[25]；Zhang 等[26]以 $FeSO_4 \cdot 7H_2O$、$Na_2S_2O_3 \cdot 5H_2O$ 和 S 单质为原料，通过温度为 200℃、时间为 24h 的一锅水热反应制得了黄铁矿。Cong 等[27]通过微乳化法，将 P123、$EO_{20}PO_{70}EO_{20}$ 和 $Fe(NO_3)_3 \cdot 9H_2O$ 在室温下混合制得 α-FeOOH 纳米棒。

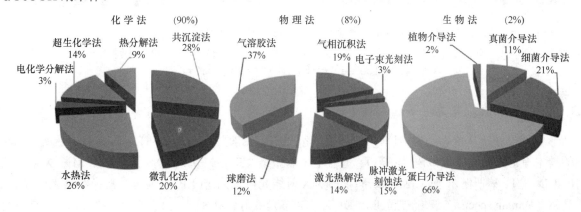

图 8-11 铁矿物合成方法的对比（引自参考文献[23]）

由于铁矿物的性质很大程度上会影响其化学反应特性及应用，因此不同形貌、尺寸及晶面暴露的铁矿物的可控合成一直是材料合成领域的研究热点。Sayed 和 Polshettiwar[28]利用一种微波辅助模板溶剂热合成技术，通过改变前驱体铁盐的类型，用相同的合成步骤即可得到 6 种不同形貌的铁氧化物。如图 8-12 所示，分别以七水合硫酸亚铁、无水草酸亚铁、六水合三氯化铁、无水硝酸铁、无水葡萄糖酸亚铁和五羰基铁为前驱体，可得到纳米棒状的 α-FeOOH、果壳状的 β-FeOOH 四方纤铁矿、扭曲立方块状的 α-Fe_2O_3、纳米立方块状的 α-Fe_2O_3 和 γ-Fe_2O_3、多孔球状的无定形氧化铁和定向自组装花状的 γ-Fe_2O_3。Kakuta 研究组[29]通过水解硝酸铁或氧化 Fe(OH)$_2$ 合成了杆状的、饼状的和阵列状的针铁矿，并探究了针铁矿的形貌与其光催化氧化乙醛的活性之间的相关性。Zhang 等[30]通过控制硝酸铁与氢氧化钠/钾反应的时间、温度及投料比等因素合成了 8.7 nm、10.1 nm、16.6 nm、26.8 nm 和 38.2 nm 五种不同尺寸的针铁矿纳米颗粒。此外，铁氧化物晶面暴露的调控手段也已经趋于成熟。例如，将六水三氯化铁、乙醇和乙酸钠在 180 ℃下反应 12 h，洗涤烘干可获得 {001} 晶面暴露的赤铁矿纳米片；将六水三氯化铁、乙醇、油酸和油酸钠在 180 ℃下反应 12 h，洗涤烘干可获得 {012} 晶面暴露的赤铁矿纳米立方块；将六水三氯化铁、氯化铵和水在 120 ℃下反应 12 h 得到前驱体 β-FeOOH，再将前驱体 β-FeOOH 在马弗炉中以 520 ℃ 煅烧 2 h 可获得 {110} 晶面暴露的赤铁矿纳米棒[31]。不同的暴露晶会直接影响铁氧化物与污染物的吸附配位构型，从而决定其对污染物的固定能力。这些工作为揭示铁矿物结构与性能及其环境效应之间的关系奠定了基础。

图 8-12　六种不同形貌铁氧化物 SEM 图（引自参考文献[28]）

（a）纳米棒状的 α-FeOOH；（b）果壳状的 β-FeOOH 四方纤铁矿；（c）扭曲立方块状的 α-Fe_2O_3；（d）纳米立方块状的 α-Fe_2O_3 和 γ-Fe_2O_3；
（e）多孔球状的无定形氧化铁；（f）定向自组装花状的 γ-Fe_2O_3

3.1.3　铁矿物的表征技术

从形貌到尺寸，从价态到官能团，铁矿物的化学结构与组成普遍复杂多样。常见的用于铁矿物结构表征的技术有：扫描电子显微镜（SEM）、透射电子显微镜（TEM）、原子力显微镜（AFM）、X 射线粉末衍射（XRD）、^{57}Fe 穆斯堡尔谱（Mössbauer spectra）、X 射线光电子能谱（XPS）、傅里叶变换红外光谱（FTIR）、拉曼光谱（Raman spectra）、动态光散射（DLS）、静态光散射（SLS）[23]。表 8-1 详细总结了这些常见铁矿物的性质表征技术。运用这些表征技术，有助于我们探明铁矿物的宏观及微观结构、表面及体相结构，加深对铁矿物本质的认识与了解，对揭示其结构与环境修复性能之间的构效关系具有重大意义。

表 8-1　铁矿物结构、组成与分散状态的表征

特性	表征参数	表征技术
结构	形貌、尺寸	透射电子显微镜（TEM）、扫描电子显微镜（SEM）
	晶体结构	X射线粉末衍射（XRD）、^{57}Fe 穆斯堡尔谱（Mössbauer spectra）
	分子/原子空间排列	原子力显微镜（AFM）
组成	元素组成	X射线荧光光谱（XRF）
	物相组成	^{57}Fe 穆斯堡尔谱（Mössbauer spectra）
	价态组成	X射线光电子能谱（XPS）、^{57}Fe 穆斯堡尔谱（Mössbauer spectra）
	官能团组成	傅里叶变换红外光谱（FTIR）、拉曼光谱（Raman spectra）
分散状态	分散性	动态光散射（DLS）
	团聚度	静态光散射（SLS）

3.2　基于铁矿物吸附性质的环境污染控制技术

污染物吸附是指污染物通过物理作用或化学作用向固体吸附剂发生质量转移的过程。铁矿物由于其尺寸小、比表面积高、表面活性位点丰富及廉价易得等特点，被认为是一种环境友好的并具有工业应用价值的吸附剂[32]。更重要的是，大多数铁矿物具有磁性，使其易于从反应体系中分离，便于回收循环利用，可降低水处理经济成本[33]。因此，基于铁矿物的吸附性质发展环境污染控制技术受到越来越多的关注。譬如，Hu 等[34]研究发现纳米 γ-Fe_2O_3 对水体中 Cr(VI)的吸附具有较高的选择性，最大吸附量可达 19.2 mg/g。通过研究吸附机制，发现在较低 pH 下，γ-Fe_2O_3 吸附 Cr(VI)的作用力主要是静电引力；而在较高 pH 下，CrO_4^{2-} 主要与 γ-Fe_2O_3 表面的—OH 以离子交换的方式吸附。Nassar 等[35]系统地研究了 Pb(II) 在 Fe_3O_4 表面的吸附动力学和热力学，证实了 Pb(II)在 Fe_3O_4 纳米材料上的吸附属于物理吸附，并具有吸热性和自发性。吸附在 30min 内达到平衡，最大吸附量为 36.0 mg/g。

尽管铁矿物吸附水体污染物具有高效性和特异性，但通常以超细颗粒的形式存在。这往往会导致其在流动体系中产生易团聚、难分离和导流系数较低等问题[36]。为了克服上述困难和提高吸附效率，近年来，铁矿物涂覆材料和改性铁矿物材料等多功能材料得到了广泛关注。譬如，Li 等[37]制备合成了一种氨基肟功能化二氧化硅涂层 Fe_3O_4 核壳材料（Fe_3O_4@SiO_2-AO）用于 U(VI)的吸附。其中，Fe_3O_4 核具有超顺磁性，便于材料的分离和收集；SiO_2 壳可避免 Fe_3O_4 核发生团聚，并保护其不被酸溶液氧化或溶解。实验证明，相比于核壳 Fe_3O_4@SiO_2 材料，表面氨基肟功能化的 Fe_3O_4@SiO_2-AO 材料具有更高的 U(VI) 吸附性能。这说明表面胺肟基团具有较强的螯合能力，能增强材料对 U(VI)的吸附作用。批量吸附实验结果表明，Fe_3O_4@SiO_2-AO 对 U(VI)的吸附性能依赖于 pH 大小，而与溶液离子强度大小无关。这意味着 Fe_3O_4@SiO_2-AO 吸附 U(VI)主要是通过形成了内球配位化合物。在 pH=5.0±0.1，T = 298 K 的条件下，最大吸附量可达 0.441 mmol /g，符合 Langmuir 吸附模型。此外，吸附 U(VI)的 Fe_3O_4@SiO_2-AO 可以很容易地被磁铁从水溶液中分离出来，并被 HCl 溶液有效地洗脱回收（图 8-13）。这种铁矿物的表面功能化的核壳结构吸附剂不仅克服了传统铁矿物吸附剂易团聚的难题，同时也提高了其吸附效率及回收率，为将来高效铁基吸附剂的设计提供了思路。

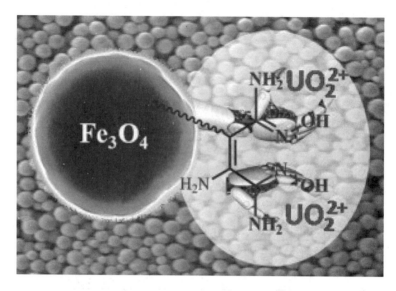

图 8-13　氨基肟功能化 Fe$_3$O$_4$@SiO$_2$ 核壳磁性微球对 U(VI)的吸附示意图（引自参考文献[37]）

基于上述吸附剂的设计思路，Fortner 研究组[38]结合非水相合成法和双层相转移法制备出了一系列高度均匀的双层有机酸包覆的 Fe$_3$O$_4$ 单分散材料用于水体中 U(VI)的吸附和分离。这种材料的尺寸大小和表面涂层性质都是可调控的，可直接评价涂覆材料的尺寸和涂层性质与吸附性能之间的相关性。批量吸附实验表明，尺寸为 8 nm 的油酸（OA）双层包覆（OA-IONPs）和单十二烷基磷酸钠（SDP）双层包覆（SDP-IONPs）Fe$_3$O$_4$ 材料具有最优的 U(VI)吸附性能，最大吸附量分别为 635 mg U/g Fe 和 657 mg U/g Fe。X 射线近边吸收光谱（XANES）表明，在不同双层包覆材料上吸附的 U(VI)被不同程度地还原为 U(IV)。扩展 X 射线吸收精细结构（EXAFS）分析表明，铀离子主要是与 Fe$_3$O$_4$ 表面的双层有机涂层进行配位络合，进一步证实了有机涂层对提高 U(VI)吸附的重要性（图 8-14）。这项工作进一步证明了铁矿物包覆技术在水体重金属污染处理中的应用价值，并为评价涂覆材料性质与吸附性能之间的构效关系奠定了基础。

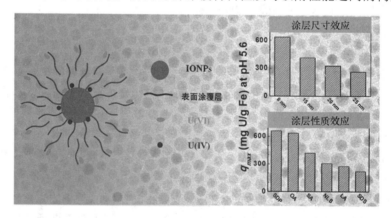

图 8-14　双层有机酸包覆的单分散纳米 Fe$_3$O$_4$ 对 U(VI)的吸附（引自参考文献[38]）

除了上述涂层包覆技术外，元素取代技术也可有效提高铁矿物的吸附性能。陆现彩研究组[39]在针铁矿合成过程中引入锰元素，制备出了部分 Fe(III)被 Mn(III)取代的针铁矿（MnGeo），并发现随着 Mn(III)取代量的增大，针铁矿对 Pb^{2+}的吸附性能也随之提升。在锰含量为 0%、3.4%、5.7%、10.8%、12.9%的情况下，其对 Pb^{2+}的最大吸附量分别为 16.34 mg/g、20.70 mg/g、27.47 mg/g、60.61 mg/g、90.09 mg/g。扩展 X 射线吸收精细结构（EXAFS）分析表明，Pb^{2+}在针铁矿上的吸附为规则的 edge-sharing 构型的配合物，$R_{\text{PB-Fe}}$ 为 3.31 Å；而在 Mn(III)取代的针铁矿中，Pb^{2+}优先地与表面 Mn(III)位点结合形成 edge-sharing

构型的配合物，其 R_{PB-Mn}=3.47 Å。Mn(III)的引入为 Pb^{2+} 的吸附提供了新的优先吸附位点，从而提高针铁矿对 Pb^{2+} 的吸附性能（图 8-15）。由于铁和锰共存于自然界中，铁矿物中铁被锰取代的现象普遍存在。因此，该研究揭示了自然界中含锰铁矿物对重金属污染物吸附与固定的潜能，并为高效铁基吸附剂的合成提供了新契机。

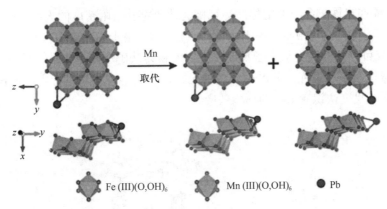

图 8-15　Mn(III)取代的针铁矿对 Pb(II)的吸附构型示意图（引自参考文献[39]）

除了开发基于铁矿物的新型功能吸附材料，从原子水平上探究重金属离子在铁矿物上的吸附机理也受到了研究者们的广泛关注，并对深入揭示该过程的环境效应具有重大意义。2016 年，张礼知研究组[40]初步从原子水平认识了不同特定晶面暴露的赤铁矿纳米晶吸附 Cr(VI)的内在机制。结合原位衰减全反射红外光谱（ATR-FTIR）中赤铁矿纳米晶上表面吸附的 Cr(VI)的 Cr—O 振动频率、密度泛函理论（DFT）计算的振动分析及 Cr(VI)的 K 边 X 射线吸收精细结构光谱（K-edge EXAFS），他们发现 Cr(VI)在赤铁矿表面形成单齿单核和双齿双核两种内球配位模式的铬配合物。其中，单齿单核配合物存在于赤铁矿{001}晶面，双齿双核配合物存在于赤铁矿{110}晶面。这些不同的吸附模式会直接影响赤铁矿晶面铬的吸附量大小。单齿单核对应的 Cr(VI)吸附密度为 5.39 #Cr nm^{-2}，双齿双核对应的吸附量为 10.79 #Cr nm^{-2}。这说明赤铁矿{110}晶面暴露更有利于 Cr(VI)的吸附（图 8-16）。这项工作启迪我们可以通过暴露一些特殊晶面来提高重金属离子的吸附去除，同时也为铁矿物分子尺度修复机理提供了新思路。

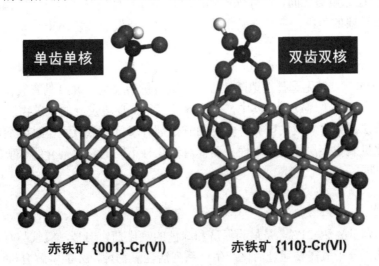

图 8-16　赤铁矿晶面依赖的 Cr(VI)吸附示意图（引自参考文献[40]）

2019 年，张礼知课题组又系统地使用了 DFT、EXAFS 和 ATR-FTIR 等手段从原子水平上研究了铀酰离子在赤铁矿{001}、{012}和{110}晶面的吸附，发现铀酰在赤铁矿三个晶面都形成内球配位，{001}晶面为边缘共享双齿双核构型（^{2}E），{012}和{110}晶面都是角共享双齿双核构型（^{2}C）。吸附动力学实验显示，铀酰在{012}和{110}晶面的吸附密度相近（0.3 #U nm^{-2}），而且大于其在{110}晶面的吸附密度（0.18 #U nm^{-2}）。这说明铀酰在赤铁矿晶面的配位微环境极大地决定其吸附密度的大小，即 ^{2}C 铀酰表面配合物比 ^{2}E 构型更加有利于铀酰在赤铁矿上的吸附（图 8-17）[41]。由于赤铁矿在自然界中主要是暴露出{001}和{012}晶面，该研究可以帮助我们从分子尺度理解铀酰和赤铁矿这些主要暴露晶面的相互作用，从根本上理解铀元素的地球化学循环过程，并准确预测铀元素在长期地球化学过程中的命运、迁移和转化，可以进一步从分子、原子尺度认识赤铁矿的环境效应。

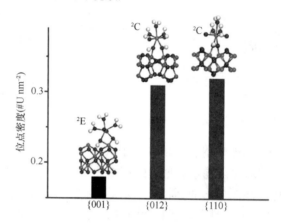

图 8-17　U(VI)被赤铁矿吸附的配位微环境和吸附量的关系（引自参考文献[41]）

3.3　基于铁矿物氧化还原性质的环境污染控制技术

铁矿物中的铁通常有两种价态：Fe(III)和 Fe(II)。由于 Fe(II)/Fe(III)的氧化还原电势处在主要的 C、O、N 和 S 元素物种氧化还原电势之间，因此铁矿物的氧化还原特性使其在地球化学循环和环境污染物的迁移和转化中都扮演着重要的角色，基于铁矿物的氧化还原性质发展环境污染物修复技术成为国际前沿研究热点[42]。

高级氧化技术（advanced oxidation processes，AOPs）是利用一定方法产生具有强氧化性的羟基自由基（·OH）的技术。·OH 可以无选择性地降解绝大多数有机污染物，并且具有氧化效率高、去除污染速度快以及不会造成二次污染等优点[43]。其中，利用廉价易得、环境友好的铁矿物活化氧化剂产生·OH 来实现污染物净化的方法得到了广泛的研究。常见的氧化剂有氧气、双氧水、过硫酸盐等，针对不同的氧化剂，铁矿物的活化机制也不尽相同。下面主要针对这几种不同的氧化剂介绍铁矿物在活化氧化剂降解污染物方面的最新进展。

3.3.1　活化分子氧

氧气在空气中的含量高达 21%，是最为经济和绿色的氧化剂。但是，氧气与有机污染物的直接反应是自旋受阻的。因此，分子氧需要通过活化产生一系列活性氧物种，才能降解有机污染物[44, 45]。低价态铁基材料，包括零价铁和二价铁矿物均表现出还原活性，具有活化分子氧的能力。目前，对于二价铁矿物对有机污染物的直接还原转化已有大量的报道，但对其与氧气相互作用产生活性氧物种的认识却十分有限，而实际上这部分产生的活性氧物种对环境污染物转化和修复起着至关重要的作用。周东美研究组发现，磁铁矿纳米颗粒在有氧条件下可以活化分子氧产生•OH，实现对 2-氯联苯（2-CB）的氧化降解。•OH 的整个产生过程分为两步：首先，磁铁矿中的 Fe^{2+} 与 O_2 反应生成超氧负离子自由基（$•O_2^-$），$•O_2^-$ 结合 H^+ 生成 H_2O_2；其次，溶解的 Fe^{2+} 能迅速将 H_2O_2 分解产生•OH。图 8-18 为该过程的示意图。通过分析鉴定 2-CB 的降解中间产物，发现有羟基化的产物存在，证实•OH 参与了 2-CB 的降解过程[46]。这项工作阐明了低价铁矿物在有氧条件下对污染物的氧化转化途径，为发展基于二价铁矿物活化分子氧修复污染物技术奠定了基础。

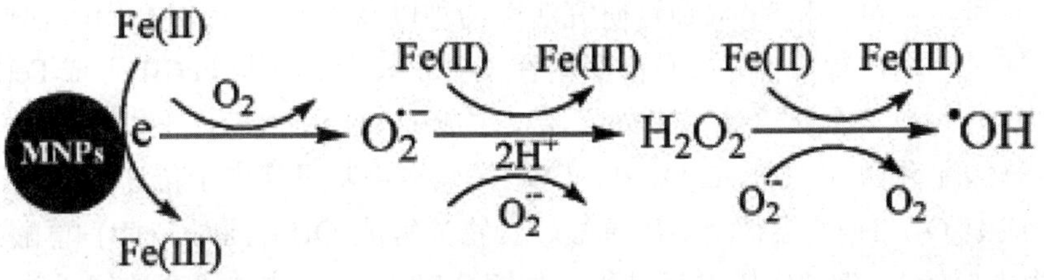

图 8-18　磁铁矿纳米颗粒活化分子氧降解 2-CB 机理（引自参考文献[46]）

受这篇文章启发，研究者们逐渐认识到了二价铁矿物产生的活性氧物种在环境污染物氧化降解中不可忽视的作用，并对其具体的产生机制进行了深入的研究。譬如，2016 年，袁松虎研究组系统研究黄铁矿在有氧条件下和无氧条件下产生•OH 的机制[47]。他们发现在无氧条件下，•OH 是由黄铁矿表面硫缺陷氧化 H_2O 产生的，因此，表层的更新可以使更多的硫缺陷暴露出来，从而产生更多的硫缺陷，提高•OH 的产率；而在有氧条件下，•OH 主要来自于黄铁矿表面二价铁活化氧气，其次来自于表面硫缺陷氧化 H_2O 和中间溶出物质 Fe(II)/硫物种活化氧气。在氧气的活化过程中，表面二价铁首先通过 2 电子途径还原氧气为 H_2O_2，H_2O_2 随后通过芬顿反应被分解为•OH（图 8-19）。不管是有氧条件下还是无氧条件下产生的•OH，均能够氧化重金属 As(III) 和有机污染物磺胺。这些发现阐明了二价铁矿物在不同气氛条件下产生•OH 的机制，启迪我们可以在不同环境条件下利用黄铁矿不同的氧化还原特性进行重金属和有机污染物的转化和降解。

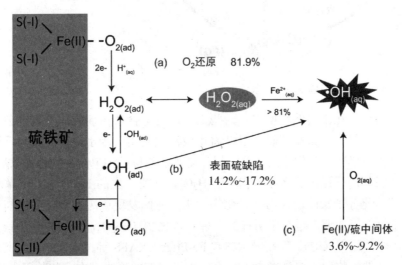

图 8-19　有氧条件和无氧条件下黄铁矿产生•OH 的机理（引自参考文献[47]）

除了二价铁矿物可以活化分子氧外，大多数三价铁矿物具有半导体性质，也可通过光催化作用产生的光生电子实现分子氧的活化。例如，赤铁矿被认为是一种环境友好的 n 型半导体，其禁带宽度较窄，约为 2.0～2.2 eV。因此赤铁矿可吸收到 600 nm 的可见光，收集了高达 40%的太阳光谱能量。然而，赤铁矿的光催化性能却受到电子和空穴复合速率高、空穴扩散长度低（2～4 nm）和电导率差等因素的限制[48, 49]。研究人员已经做了很多尝试来克服赤铁矿的这些不足，如通过形成纳米结构来降低复合速率，通过掺杂合适的金属来提高电导率，以及提高电荷转移能力[48]。其中，引入多元羧酸形成铁矿物/多元羧酸二元体系，通过配体-金属电荷转移过程，增强其光催化性能最为简单高效。譬如，李芳柏研究组分别利用α-FeOOH 和 α-Fe_2O_3 与草酸组成活化分子氧体系，探究了该体系在紫外光照下对五氯酚的降解作用[50]。他们发现在草酸浓度为 1.2 mmol/L 的条件下，光照 1h 后，五氯酚在 α-FeOOH/草酸体系和 α-Fe_2O_3/草酸体系中的降解率分别可达到68%和83%。通过研究活性物种以及铁物种的浓度变化，他们发现Fe(III)/Fe(II)循环驱动着整个反应过程的进行。首先，草酸吸附在铁矿物表面，并与Fe(III)形成 Fe(III)-草酸表面配合物；随后，这些 Fe(III)-草酸表面配合物被紫外光激发生成•$C_2O_4^-$ 和 •CO_2^- 自由基，同时 Fe(III)-草酸配合物被还原为 Fe(II)-草酸配合物；产生的•$C_2O_4^-$ 和 •CO_2^- 等碳自由基可将电子传递给氧气，生成•O_2^- 和•OOH 等自由基，随后生成 H_2O_2；H_2O_2 会被 Fe(II)-草酸配合物分解成•OH，同时 Fe(II)-草酸配合物又被氧化为Fe(III)-草酸配合物，形成 Fe(III)/Fe(II)循环过程，如图 8-20 所示。这项研究评估了铁矿物/多元羧酸体系的环境污染物修复性能，阐明了该体系中铁矿物与多元羧酸的相互作用机制及铁循环机理。

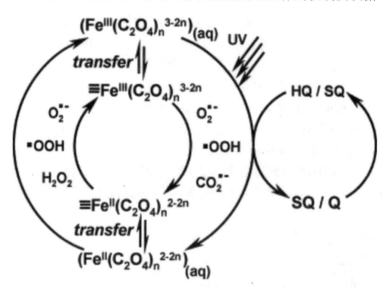

图 8-20　铁氧化物/草酸体系在光照条件下 Fe(III)/Fe(II)循环示意图（引自参考文献[50]）

事实上，铁矿物不仅仅存在于自然环境中，一些工业过程中含铁药剂的加入也将必然导致其排放的废弃污泥中铁物种的存在。最近，张礼知研究组[51]将市政污泥利用一步水热法转化为污泥水热催化剂，并利用其中固有的铁物种与草酸相互作用活化分子氧，发现草酸与污泥中的铁物种可以形成表面配合物，该配合物不仅可以在光照条件下直接活化分子氧产生 H_2O_2，还将产生的 H_2O_2 迅速分解为•OH 用于降解各种有机污染物。将其活化分子氧的活性与铁氧化物对比，他们发现该污泥水热催化剂与草酸相互作用在光照条件下活化分子氧的能力要远远高于所选的三种铁氧化物（Fe_2O_3，FeOOH 和 Fe_3O_4）。为了探明其性能较优越的原因，他们结合 XRD、穆斯堡尔谱和 Fe-Edge EXAFS 等技术系统表征其铁物种。结果显示，该污泥水热催化剂中的铁物种主要是以含铁黏土矿物绿泥石的形式存在，与其周围的 Si 元素和 Al 元素形成 Fe—O—Si 键和 Fe—O—Al 键这种强的相互作用。密度泛函理论证明，Fe—O—Si 这种强的相互作用可以降低铁和草酸配合物的 HOMO-LUMO 轨道能级差，从而提高其反应活性（图 8-21）。该研究对比了含铁黏土矿物与铁氧化物在污染物治理方面的性能及差异，并为以后设计合成高效铁基催化剂指明了方向。

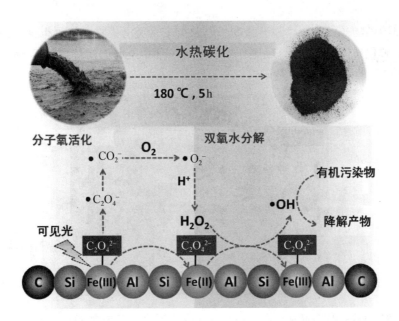

图 8-21 污泥水热催化剂中含铁黏土矿物与草酸相互作用活化分子氧机制（引自参考文献[51]）

3.3.2 活化双氧水

H_2O_2 可在 Fe^{2+} 作用下迅速分解产生 •OH，Fe^{2+} 同时被氧化为 Fe^{3+}，这便是传统的均相芬顿反应，因其成本低廉、安全高效和环境友好等优点常被应用于有机污染水体的处理中。然而，其 Fe^{3+}/Fe^{2+} 循环困难、pH 工作范围窄、易生成铁泥、无法回收利用等缺点也限制其在大规模的实际污水处理中的应用[43]。近年来，利用铁矿物代替 Fe^{2+} 活化分解 H_2O_2 发展异相芬顿（Fenton）氧化技术受到了广泛关注。虽然异相 Fenton 氧化技术能拓宽 pH 工作范围、铁基催化剂可回收再利用，但反应体系中 Fe(III)/Fe(II) 循环效率依旧很低。因此，提高 Fe(III)/Fe(II) 循环效率是异相 Fenton 氧化技术发展面临的关键科学问题。

目前，文献已有大量提高异相 Fenton 反应中 Fe(III)/Fe(II) 循环效率的方法研究报道，例如，与光、电和超声等方法结合发展异相 Fenton 联用技术、将铁矿物与其他材料复合制备高效电子传递异相 Fenton 材料等[52]。近年来，研究者发现，将能络合还原表面 Fe(III) 的有机/无机配体引入异相 Fenton 反应体系能大幅提升 Fe(III)/Fe(II) 循环效率。目前，常见的用于调控 Fe(III)/Fe(II) 循环的配体有乙二胺四乙酸（EDTA）、羧甲酸-β-环糊精（CMCD）和谷氨酸二乙酸（GLDA）等。譬如，Matta 等[53]向磁铁矿异相 Fenton 体系中引入了 CMCD，有效提高了 2,4,6-三硝基苯酚的降解效率。虽然这些配体的加入能有效调控异相 Fenton 反应中铁循环，然而其生物难降解的特性导致了不可避免的二次污染。因此，探索环境友好、成本低廉的配体将对异相 Fenton 氧化技术的发展具有重要意义。

2017 年，张礼知研究组[54]发现羟胺的引入能有效提高针铁矿异相 Fenton 体系中污染物降解效率，而羟胺随污染物一起降解，不会引起二次污染。通过原位 ATR-FTIR 光谱和 DFT 计算研究针铁矿表面铁还原动力学，他们发现羟胺上的氧原子首先与针铁矿表面 Fe(III) 进行配位，形成内球配位模型。通过分析溶液中及表面铁价态的变化，他们发现在 pH=5.0 的条件下溶液中未检测到铁离子的溶出，而在针铁矿表面上观察到了 Fe(III) 向 Fe(II) 转化的现象，这说明羟胺能通过电子传递有效促进针铁矿表面 Fe(III)/Fe(II) 循环。电子自旋共振（ESR）技术证实了体系中有 •OH 产生，同时加入 F 将 •OH 从针铁矿表面分离出来之后，溶液中 •OH 的含量升高，进一步说明表面 •OH 的产生。因此，不同于传统异相 Fenton 反应，该反应体系整个过程中未检测到铁离子的溶出，Fe(III)/Fe(II) 的循环、双氧水的分解以及 •OH 的产生均发生在针铁矿表面，被称之为表面 Fenton 反应。在该表面 Fenton 体系中，染料类污染物（罗丹明 B、甲基橙和亚甲基蓝)、农药类污染物(阿特拉津和五氯酚)以及抗生素类污染物(林可霉素、四环素和氯霉素)均能在 1h 之内被降解矿化(图 8-22)。这项工作不仅发展了环境友好的配体调控 Fenton 铁循环的方法,更为表面 Fenton 化学反应的研究奠定了基础。

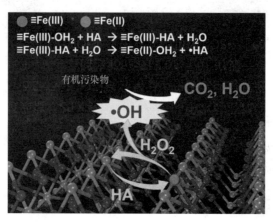

图 8-22　羟胺促进针铁矿表面铁循环及表面 Fenton 体系降解有机污染物示意图(引自参考文献[54])

自然界中存在着大量的天然有机小分子配体,如草酸、抗坏血酸及柠檬酸等。这些天然有机小分子配体兼具络合性与还原性,且环境友好,可望成为调控异相 Fenton 反应铁循环的良好配体。张礼知研究组[55]发现,广泛存在于蔬菜和水果中的抗坏血酸可与赤铁矿表面铁配位络合,促进赤铁矿 Fenton 反应中 Fe(III)/Fe(II)循环降解除草剂甲草胺,并从分子层面探究了抗坏血酸与赤铁矿表面相互作用机制。首先,他们利用形貌控制合成方法合成了{012}和{001}晶面暴露的赤铁矿纳米晶。结合原位 ATR-FTIR、DFT 计算等分析发现,抗坏血酸在赤铁矿{001}晶面形成了双齿单核配位模式,而在{012}晶面上形成单齿单核配位模式。通过对比 Fenton 反应活性,他们发现{001}晶面形成的双齿单核铁-抗坏血酸配位模式更有利于 H_2O_2 的分解以及甲草胺的降解。由于抗坏血酸和赤铁矿在自然环境普遍存在,这项研究可以帮助我们深刻理解铁循环和具有氧化还原活性的元素的地球化学循环以及污染物迁移、转化之间的内在关系,并启示我们利用铁循环实现污染控制和环境修复(图 8-23)。

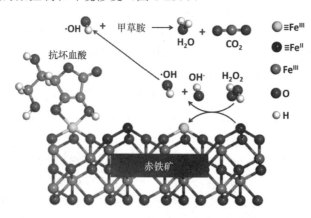

图 8-23　抗坏血酸促进赤铁矿表面铁循环芬顿降解有机污染物示意图(引自参考文献[55])

铁矿物与污染物共存于沉积物及地下水中,铁循环会强烈影响着环境污染物的迁移与转化。如果增加沉积层铁矿物表面 Fe(II)含量,就能增强铁矿物活化氧化剂能力,达到利用沉积物中铁矿物原位修复有机污染场地地下水目的。基于这一思路,张礼知研究组[56]利用抗坏血酸调控含水层沉积物的 Fe(III)/Fe(II)循环,提高其分解 H_2O_2 产生•OH 的效率,进而构建了有机污染地下水原位氧化修复体系。沉积物样品的

X 射线粉末衍射以及穆斯堡尔谱表征显示，沉积物样品含有丰富的铁矿物，包括赤铁矿、针铁矿、绿泥石和蒙脱石等。他们将沉积物与水按 5 g/30 mL 的比例混合来模拟沉积物与地下水共存的状态，发现抗坏血酸具有较强络合能力和还原性，能吸附在沉积物表面与铁离子配位，进而发生由抗坏血酸向表面 Fe(III) 的电子转移过程，使得沉积物表面原位产生更多的表面 Fe(II)，显著提高沉积物分解 H_2O_2 的效率。虽然抗坏血酸的加入将沉积物系统的 pH（8.9）降低至 5.7，但这一过程并没有提高体系中溶解铁浓度，而随着反应的进行，沉积物系统的 pH 逐渐上升到 8.3。这一现象说明这种加入抗坏血酸调控 Fe(III)/Fe(II) 循环的方法对沉积物环境影响较小。该原位修复体系可以有效降解对氯硝基苯、甲草胺、氯霉素以及罗丹明 B 等一系列有机污染物。而模拟柱实验中，双氧水的分解效果和污染物的降解效果都十分理想。更为重要的是，抗坏血酸随对氯硝基苯污染物一起降解，避免了二次污染（图 8-24）。这项工作揭示了铁循环调控在环境修复领域的重要意义，并为地下水原位修复提供了新思路。

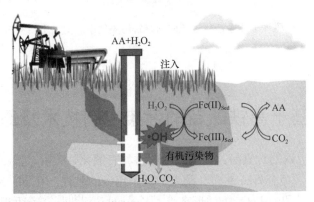

图 8-24 抗坏血酸活化含水层沉积物原位化学氧化修复有机污染地下水示意图（引自参考文献[56]）

3.3.3 活化过硫酸盐

过硫酸盐活化技术属于高级氧化技术中的一种。过硫酸（PS）盐或过一硫酸盐（PMS）可通过光、热或化学方法被活化产生具有强氧化能力的单电子氧化剂 $\cdot SO_4^-$（E^0=2.5～3.1 V）。$\cdot SO_4^-$ 能快速地与有机污染物反应，二级反应速率常数一般在 10^7～10^{10} L·s/mol 之间。通常，Fe^{2+} 是用于化学法活化过硫酸盐产生 $\cdot SO_4^-$ 的常用活化剂[式（8-6）]。$\cdot SO_4^-$ 又可与水反应产生 $\cdot OH$[式（8-7）][57]。$\cdot SO_4^-$ 和 $\cdot OH$ 均能参与污染物的降解。

$$Fe^{2+} + S_2O_8^{2-} \rightarrow Fe^{3+} + SO_4^{2-} + \cdot SO_4^- \tag{8-6}$$

$$\cdot SO_4^- + H_2O \rightarrow HSO_4^- + \cdot OH \tag{8-7}$$

与传统均相 Fenton 反应一样，Fe^{2+} 活化过硫酸盐也面临着铁离子易沉淀、难回收等问题。为了克服这一难题，异相铁矿物催化剂同样被用于代替 Fe^{2+} 活化过硫酸盐，如何提高铁矿物活化过硫酸盐中 Fe(III)/Fe(II) 循环效率仍是面临的关键科学问题。然而，相对于 Fenton 反应，铁矿物活化过硫酸盐中 Fe(III)/Fe(II) 循环调控的研究却较少。Hanna 和 Vione 研究组利用紫外光照射促进磁铁矿 Fe(III)/Fe(II) 循环活化过硫酸盐降解苯酚，发现紫外光的照射确实可以使 Fe_3O_4 表面 Fe(III) 向 Fe(II) 转化，提高苯酚的降解效率（图 8-25）[58]。同时，他们还对比了分别在天然水体基质中和在自由基清除剂存在下 UV/Fe_3O_4 活化过硫酸盐体系与 UV/Fe_3O_4 活化 H_2O_2 体系的性能差异，发现相比于 UV/Fe_3O_4/H_2O_2 体系，UV/Fe_3O_4/PS 体系受天然有机质或自由基捕获剂的影响较小，但 UV/Fe_3O_4/PS 矿化苯酚效率不如 UV/Fe_3O_4/H_2O_2 体系的效率高。该研究借助紫外光实现了 Fe_3O_4/PS 体系中铁的有效循环，并启示我们如何根据水质选择最合适的水处理工艺。

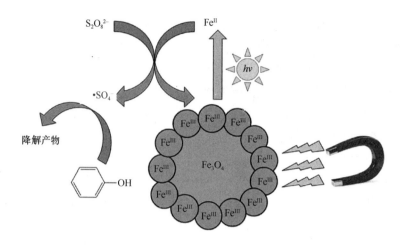

图 8-25　紫外光/磁铁矿体系活化过硫酸盐降解污染物示意图（引自参考文献[58]）

4　基于高铁酸盐的污染控制技术及原理

高铁酸盐[Fe(VI)]是+6 价铁的含氧酸盐，是一种同时具有氧化、吸附、絮凝、杀菌、消毒等多种功能的绿色强氧化剂，因独特的环境友好特性而受到人们越来越多的重视。在 pH 为 4.0～10.0 的水环境中，Fe(VI)主要有 FeO_4^{2-}、$HFeO_4^-$ 和 H_2FeO_4 三种形态[59]，Kamachi 等[60]研究表明相比于去质子化的 FeO_4^{2-}，质子化的 $HFeO_4^-$ 和 H_2FeO_4 含氧配体的自旋密度更高、氧化能力更强，因此 Fe(VI)的氧化能力随 pH 升高逐渐减弱，在酸性条件下氧化还原电位为 2.20 V，而在碱性条件下氧化还原电位为 0.70 V。与传统的氧化剂（O_3、Cl_2、H_2O_2、$KMnO_4$）相比，Fe(VI)具有更高的氧化还原电位，能够快速氧化一些含有不饱和官能团的有机物如内分泌干扰物和药物与个人护理品[61]，以及有害有毒的无机物，如硫化物、硫氰根、亚硝酸根等[62]。Fe(VI)的强氧化性使其在去除水中多种不易降解的有机污染物的同时还具有杀菌功能，其还原产物新生态氢氧化铁具有较强的絮凝作用和吸附作用，可有效去除水中的重金属（如铜、镍）及含氧酸根（如磷酸盐、砷酸根）[63]。近几十年来，高铁酸盐制备方法日渐成熟，产率和成品纯度不断提高，技术上已经逐渐能够满足生产和实际应用的需要。

4.1　Fe(VI)氧化特性及强化氧化研究

4.1.1　Fe(VI)氧化反应动力学及机理

Fe(VI)和大多数有机物和无机物的氧化还原反应遵循二级反应动力学模型，如式（8-8）所示。

$$-\frac{d[Fe(VI)]}{dt}=k_{app}[Fe(VI)]_{tot}[X]_{tot} \tag{8-8}$$

式中，k_{app} 是表观的二级反应动力学速率常数，动力学实验是利用停留光谱仪在假一级动力学条件下（即 $[X]_{tot} \gg [Fe(VI)]_{tot}$）观测 Fe(VI)的浓度变化实现的，$k_{app}$ 即由假一级速率常数计算得来；$[Fe(VI)]_{tot}$ 表示 Fe(VI)的总浓度；$[X]_{tot}$ 表示目标物的总浓度，这些目标物主要是与环境相关的化合物比如氰化物、含氮含硫化合物和新型有机污染物。由于酸性条件下 Fe(VI)的自分解太快，这些动力学实验大多是在中性及碱性条件下进行。Fe(VI)的形态分布与反应体系的pH密切相关。此外，部分有机物和无机物在水溶液中发生解离，因此，k_{app}与pH的关系可以通过Fe(VI)各形态分布与有机污染物的酸碱解离常数来进一步进行模拟，如式（8-9）所示。

$$k_{app}\left[Fe(VI)\right]_{tot}[X]_{tot} = \sum_{j=1,2,\cdots}^{i=1,2,3,4} k_{ij}\alpha_i\beta_j\left[Fe(VI)\right]_{tot}[X]_{tot} \tag{8-9}$$

式中，k_{ij}表示各形态之间的反应速率常数，α_i和β_j表示Fe(VI)和有机污染物对应的比例系数，i和j表示Fe(VI)和有机污染物在水溶液中的形态数目。

氧化还原反应的本质是氧化物和还原物之间的电子转移，Fe(VI)的氧化反应遵循1e或者2e的转移机制，分别产生不稳定的Fe(V)和Fe(IV)，中间价态铁Fe(V)和Fe(IV)也可继续发生1e或者2e转移变成更低价态，这取决于所氧化的目标化合物。目前对高铁酸盐反应机理的假设主要基于以下几种方法：①^{18}O示踪法：根据标记的^{18}O转移反应进行推理。Goff和Murmann[64]最先提出Fe(VI)与无机污染物反应的2e转移机制。他们根据化学计量学及^{18}O转移反应发现Fe(VI)通过直接氧转移，即2e转移机制氧化亚硫酸盐，形成Fe(IV)和硫酸根。但是考虑到高价铁含氧酸盐与水分子之间能够进行氧交换[65]，通过^{18}O示踪法可能会得出错误的结论；②动力学方法：Sharma[66]根据Fe(VI)氧化无机物的速率常数（k）以及无机物的单电子还原电位（$E_{(1)}^o$）或双电子还原电位（$E_{(2)}^o$）之间的关系来鉴别电子转移机理，$\log k$与氰根（CN^-）、超氧自由基（$\cdot O_2^-$）、亚硫酸根（SO_3^{2-}）等无机物的$E_{(1)}^o$呈线性关系则表明Fe(VI)氧化这些无机物发生1e转移产生Fe(V)（图8-26）；$\log k$与连二亚硫酸根（$S_2O_4^{2-}$）、亚砷酸根（AsO_3^{3-}）、亚硒酸根（SeO_3^{2-}）、亚硝酸根（NO_2^-）、H_2O_2等无机物的$E_{(2)}^o$呈线性关系则表明Fe(VI)氧化这些无机物发生2e转移产生Fe(IV)（图8-26）；③产物分析法：利用精密光谱仪器检测Fe(V)、Fe(IV)和自由基等反应产物的光学信号，是一种最直接的方法，但是以往检测Fe(V)/Fe(IV)及研究它们的性质一般都是用脉冲辐解技术来开展的，由于所需的反应条件和设备苛刻，因此到目前为止相关文献较少。Sharma和Cabelli[67]用脉冲辐解法发现Fe(VI)与亚硫酸盐的第一步反应实际上是1e转移生成Fe(V)和亚硫酸根自由基（$\cdot SO_3^-$），而后Fe(V)与亚硫酸盐的反应机制是氧转移即2e转移。

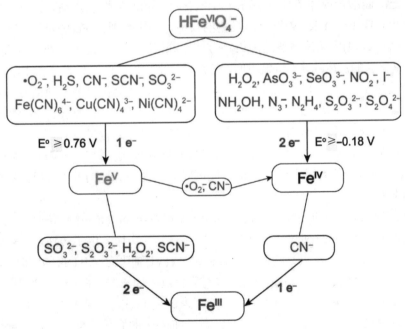

图8-26　Fe(VI)氧化过程中的电子转移及Fe(V)和Fe(IV)生成（引自参考文献[62]）

4.1.2 强化 Fe(VI)氧化新型有机污染物的研究进展

由于传统的水处理工艺无法有效去除内分泌干扰物、药物和个人护理品等新型有机污染物，因此在污水处理厂二级出水中仍能够检测到这些有机污染物的存在。水中的新型有机污染物虽然浓度低，但会通过饮用水和食物链的富集作用对人体和环境生物造成严重的危害，已引起了人们对水质安全的担忧。化学氧化是破坏水中有机污染物结构甚至使其矿化的有效手段之一。Yang 等[68]研究了 Fe(VI)在污水处理厂二级出水中对 68 种新型有机污染物的去除效果，结果表明，Fe(VI)选择性地氧化去除富含供电子基团的新型有机污染物，例如含苯酚结构的雌激素和三氯生、含苯胺结构的抗生素、含胺基结构的酸性药物、含双键的卡马西平、雄激素、孕激素和糖皮质激素，但是 Fe(VI)不与三氯卡班、3 个雄激素（表雄酮、雄酮、5α-二氢睾酮）、7 个酸性药物（氯贝酸、二四滴、二甲四氯苯氧基乙酸、布洛芬、非诺洛芬、吉非罗奇、酮洛芬）、2 个中性药物（扑米酮和环磷酰胺）和脱水红霉素发生反应。此外，在 pH 较低情况下，高铁酸钾自身降解速率快，氧化作用时间短，对有机物去除不完全。探索 Fe(VI)氧化污染物的反应机制并对其氧化过程进行强化是高铁酸盐技术的一个研究热点。传统的强化方法如与臭氧、光催化联用技术虽然在一定程度上可以强化 Fe(VI)的氧化效果（ref），但是仍无法有效氧化对于 Fe(VI)本身难以氧化的有机微污染物。近年来，很多研究者致力于提升 Fe(VI)氧化过程的效率，并取得了显著的进展。

Manoli 等[69]通过向弱碱性、非缓冲的 Fe(VI)与有机污染物溶液中加入少量的酸（硝酸、盐酸或乙酸）以提高 Fe(VI)对有机污染物的氧化效能，作者认为由加酸引起的有机污染物氧化效率的提升可能是由于酸/Fe(VI)混合溶液中产生了 Fe(IV)和 Fe(V)，但是作者没有提供任何 Fe(V)/Fe(IV)产生的直接证据，而且作者也没有区分加酸引起 pH 降低对有机污染物去除的影响。Feng 等[70]的研究结果表明，在 pH8.0 和 9.0 时，氨氮的加入可以使得 Fe(VI)氧化氟甲喹的反应速率提高 5~12 倍，提高的倍数与所加氨氮浓度及 pH 有关。这些作者进一步发现过一硫酸盐和 Fe(VI)对氧化四种氟喹诺酮有协同作用[71]。Manoli 等[72]的研究结果显示，固体 SiO_2 也可以在 pH8.0 左右大大提高 Fe(VI)对咖啡因的氧化速率。但是，以上研究都没有去探明加入氨氮、过一硫酸盐或固体 SiO_2 时 Fe(VI)氧化污染物的活性氧化剂。

研究表明中间价态铁 Fe(V)/Fe(IV)氧化污染物的反应速率常数要比 Fe(VI)高出 2~6 个数量级，而且 Fe(V)/Fe(IV)可以降解 Fe(VI)所无法降解的污染物[61-63]。利用脉冲辐解技术，Sharma 等[73]比较了 Fe(VI)、Fe(V)、Fe(IV)氧化氰化物的反应速率，结果表明，三者与氰化物反应的活性顺序为 Fe(V)> Fe(IV)> Fe(VI)。Terryn 等[74]进一步通过量化计算证实 Fe(VI)、Fe(V)、Fe(IV)氧化氰化物的反应活性与实验数据一致。研究发现 Fe(VI)在实际水体中的除污染效能反而更强，由于实际水体中成分复杂，Fe(VI)能和许多还原剂反应生成大量的中间价态 Fe(V)和 Fe(IV)，因此可以推测，实际水体的背景成分可能对 Fe(VI)除污染起到了强化作用[75]。因此，如果能在 Fe(VI)氧化有机污染物的过程中快速诱导 Fe(V)/Fe(IV)的产生并有效利用 Fe(V)/Fe(IV)来氧化去除水中的有机污染物，将不仅可以大大提升 Fe(VI)氧化有机污染物的速率，而且有望可以降解 Fe(VI)本来所无法氧化的有机污染物。

Dong 等[76]通过加入 2，2′-azino-bis-(3-ethylbenzothiazoline-6-sulfonate)（ABTS）提升了 pH 6.0~10.0 范围内 Fe(VI)氧化双氯芬酸钠的反应速率，作者认为 ABTS 的作用是电子传递，ABTS 被 Fe(VI)氧化为 $ABTS^{•+}$ 后 $ABTS^{•+}$ 可以快速氧化双氯芬酸钠，ABTS 被 Fe(VI)氧化为 $ABTS^{•+}$ 的同时，Fe(VI)会被还原为 Fe(V)[77]，但是 Dong 等[76]并没有考虑 Fe(V)的产生对氧化过程的影响。Feng 等[78]使用多种无机还原剂包括羟胺、AsO_3^{3-}、SeO_3^{2-}、NO_2^-、I^-、SO_3^{2-} 以及 $S_2O_3^{2-}$ 来诱导 Fe(IV)/Fe(V)的产生并促进 Fe(VI)对有机污染物的氧化，作者发现这些还原剂在不同程度上提高了 Fe(VI)的氧化效果。但是，部分还原剂只能在反应初始阶段促进 Fe(VI)的氧化，同时还消耗了大量的 Fe(VI)，可能会降低整体的氧化容量，而且对 Fe(VI)本身难降解的有机污染物的去除效果并不显著。Sun 等[79]使用碳纳米管（CNT）活化 Fe(VI)氧化降解溴苯类有机物，研究发现在 pH 7.0~10.0 范围内 CNT 能缓慢提高 Fe(VI)的氧化效率，通过指针化合物实验确认中间价态 Fe(IV)和 Fe(V)是活性氧化剂，且 CNT 还可以吸附氧化过程中产生的多溴联苯和多溴二苯醚，但是 CNT 作为致癌物，人为地引入到水处理过程中存在潜在的环境风险。因此，应该使用高效、经济、安全的还原剂诱导中间价态铁的产生从而促进污染物的氧化。

亚硫酸盐是优良的还原剂，可用作食品防腐剂、抗菌剂。在水处理领域，亚硫酸钠活化高锰酸钾产生微摩尔/升级的 Mn(III)能非常快速地降解水中的污染物[80-82]。基于 Na_2SO_3 对 $KMnO_4$ 的活化作用机制、K_2FeO_4 [Fe(VI)] 与 $KMnO_4$ 的相似性，亚硫酸钠也已用来活化 Fe(VI)达到快速氧化降解污染物的目的。Zhang 等[83]发现叔丁醇和甲醇都可以大大抑制 Fe(VI)/SO_3^{2-} 体系对磺胺甲噁唑的氧化，他们认为•SO_4^- 和 •OH 都是 Fe(VI)/SO_3^{2-} 体系快速去除有机污染物的活性氧化剂。但是，Sun 等[84]认为只有•SO_4^- 是 Fe(VI)/SO_3^{2-} 体系快速氧化有机污染物的活性氧化剂，因为他们发现过量乙醇会抑制 Fe(VI)/SO_3^{2-} 体系对有机污染物的氧化，但叔丁醇基本没影响。然而，Feng 等[85]报道，在有氧条件下除了•SO_3^-、•SO_4^- 和•OH，Fe(V)和 Fe(IV)也在 Fe(VI)/SO_3^{2-} 体系快速氧化有机污染物过程中起作用，但在厌氧条件下，只有•SO_3^-、Fe(V)和 Fe(IV)起作用，因为在厌氧条件下，•SO_3^- 转化为•SO_4^- 和•OH 的过程被切断了。这三个研究组观察到的不同现象及提出的不同反应机制，可能主要是由于不同研究所采用的反应条件[例如 SO_3^{2-}/Fe(VI)摩尔比]不同导致的。然而，虽然 Feng 等[85]提到 Fe(V)和 Fe(IV)可能在 Fe(VI)/SO_3^{2-} 体系中起作用，但是他们并没有提供直接的证据。跟其他提升 Fe(VI)氧化过程的方法相比，SO_3^{2-} 活化 Fe(VI)的方法效果更显著，而且 SO_3^{2-} 的最终氧化产物是水中常见的阴离子 SO_4^{2-}，无二次污染问题。Shao 等[86]在研究 Na_2SO_3 强化 Fe(VI)去除水中有机污染物过程中发现在其他反应条件固定的情况下，Na_2SO_3 多次少量加入时有机污染物的去除效率比 Na_2SO_3 一次性投加时效果更佳，因此提出利用难溶的 $CaSO_3$ 作为 SO_3^{2-} 缓释剂替代 Na_2SO_3 以活化 Fe(VI)。研究结果表明在不同 $CaSO_3$ 剂量、不同 pH 时，$CaSO_3$ 的加入可以提高 Fe(VI)氧化抗生物、药物、农药等有机污染物的反应速率 6.5~173.7 倍。虽然 Fe(VI)-$CaSO_3$ 体系氧化有机污染物的速度比 Fe(VI)-Na_2SO_3 慢，但是在同样的氧化剂及其他反应条件下，达到平衡的时候，Fe(VI)-$CaSO_3$ 体系比 Fe(VI)-Na_2SO_3 对有机污染物的去除率更高或相当。此外，Fe(VI)-$CaSO_3$ 体系还能氧化 Fe(VI)本身难以氧化的有机污染物如阿特拉津和布洛芬。通过淬灭剂实验、指针化合物实验、竞争动力学实验、产物分析实验等，证实 $CaSO_3$ 能促进 Fe(VI)降解有机污染物主要是由于 $CaSO_3$ 可以诱导 Fe(VI)原位产生 Fe(IV)/Fe(V)（图 8-27）。

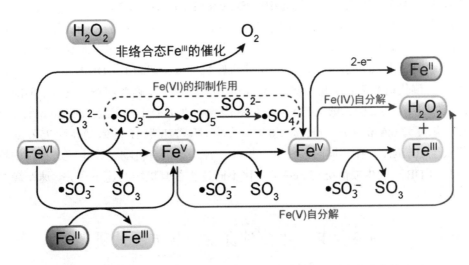

图 8-27　Fe(VI)-$CaSO_3$ 体系中 Fe(V)和 Fe(IV)生成路径（引自参考文献[86]）

4.2 Fe(VI)氧化过程中消毒副产物生成情况

一般认为，Fe(VI)氧化有机污染物过程中不会产生有毒有害的副产物。Sharma[66]根据 Fe(VI)与过量 Br⁻在 pH 为 8.5 时反应的动力学实验得出 Fe(VI)无法氧化 Br⁻的结论，并认为 Fe(VI)克服了臭氧的氧化受限于溴酸盐产生的问题。但是 Jiang 等[87]和 Huang 等[88]却在经 Fe(VI)氧化的含 Br⁻水样中检测到了活性溴及 BrO_3^- 的生成。研究发现，BrO_3^- 的生成和 Br⁻浓度、Fe(VI)投加量、pH、缓冲盐和共存阴离子密切相关。水中 Br⁻浓度越大，pH 越低，Fe(VI)浓度越高，则越有利于 BrO_3^- 的生成。在 pH 为 5.0 的碳酸盐缓冲溶液中，当 Fe(VI)投加量为 10 mg/L、Br⁻浓度为 200~1000 μg/L 时，BrO_3^- 的生成量为 12.5~273.8 μg/L，远大于 10 μg/L 的饮用水标准限值[88]。BrO_3^- 的产生受 pH 的影响很大，当 pH 升至 7.0，相同条件下的 BrO_3^- 生成量大幅降低至 20 μg/L 以下[88]。这是因为 Fe(VI)氧化 Br⁻的二级反应速率常数（k_{Br}）随 pH 的升高而降低，比如，当 pH 从 6.2 升至 8.0，k_{Br} 相应地从 10.9 L·s/mol 降至 0.1 L·s/mol 以下[87-89]。研究发现 Fe(VI)自分解产物 H_2O_2 和 Fe(III)氢氧化物颗粒对 BrO_3^- 生成起到了至关重要的作用[87, 88]。H_2O_2 可以迅速将 Fe(VI)、Fe(V)和 Fe(IV)氧化 Br⁻的中间产物 HOBr 还原为 Br⁻，从而阻碍 BrO_3^- 的生成。而 Fe(III)氢氧化物颗粒能够催化 Fe(VI)快速氧化 H_2O_2，抑制了 H_2O_2 与 HOBr 的反应。然而，在磷酸盐缓冲溶液中，由于磷酸盐络合了 Fe(III)，阻碍 Fe(III)氢氧化物颗粒的生成，反应过程中积累了大量的 H_2O_2，所以磷酸盐的存在则有助于抑制 BrO_3^- 的产生。在天然水体中加入 0.1 mg/L 的 Br⁻，BrO_3^- 的生成量低于 3 μg/L，说明天然水体中的背景成分也起到了抑制作用[87]；由于天然有机物（NOM）广泛存在于实际水体中，Fe(VI)氧化 Br⁻过程中产生的 HOBr 会进一步氧化 NOM 产生溴代消毒副产物（Br-DBPs），如三溴甲烷和溴乙酸，但是经检测 TOBr 的生成量低于 15 μg/L，这样低的生成量不会对饮用水安全构成威胁[87]。

含 I⁻源水经传统氧化技术，如液氯[90]、氯胺[91]、高锰酸钾[92]，氧化后会生成毒性比 Br-DBPs 更强的碘代消毒副产物（I-DBPs）。由于 Fe(VI)具有强氧化性，在近中性条件下氧化 I⁻的二级反应速率常数（k_I）大于 10^4 L·s/mol[66]，Fe(VI)氧化含 I⁻水过程中 I-DBPs 的生成情况也引起了越来越多的重视。Zhang 等[93]使用 Fe(VI)对含 I⁻水样进行预氧化处理，并考察其对液氯、氯胺消毒过程中三碘甲烷（IF）生成的影响。研究发现，Fe(VI)对 IF 生成的影响与 Fe(VI)投量密切相关。Fe(VI)投量大于 2.0 mg-Fe/L 时，已检测不出 IF。Wang 等[94]研究了含 I⁻水体外加腐殖酸（HA）经 Fe(VI)氧化后 I-DBPs 的情况。考察的副产物有 IF 和碘乙酸（IAA），研究发现，在 Fe(VI)投量为 10~80 μmol/L、I⁻浓度为 20 μmol/L、HA 浓度为 5 mg-C/L、pH 为 7.0 时，反应时间 12 h 后检测不出消毒副产物。Shin 等[95]研究发现 Fe(VI)氧化 I⁻遵循两电子转移反应机制生成 Fe(IV)和 HOI，而 HOI 进一步被 Fe(VI)氧化或者在 Fe(VI)的作用下发生歧化，最终产物为无毒的 IO_3^-。由于 Fe(VI)氧化 HOI 速率和 HOI 歧化速率较快，从而抑制了 I-DBPs 的生成。Dong 等[96]则研究了 Fe(VI)氧化 4 种碘代造影剂有机物（碘帕醇、碘海醇、泛影酸、碘普罗胺）过程中 I-DBPs 产生情况，考察的副产物有 IF、IAA 和三碘乙酸（TIAA）。结果发现，在 Fe(VI)投量为 500 μmol/L、碘帕醇浓度为 50 μmol/L、NOM 浓度为 0.5 mg-C/L、pH 为 9.0 时，反应时间 72h 后 IF 的浓度高达 75.6 μg/L，而 IAA 和 TIAA 的浓度却低于 10 μg/L。在相同的条件下，另外 3 种造影剂有机物经氧化后这 3 种消毒副产物的浓度均低于 10 μg/L。I-DBPs 的生成情况与 Fe(VI)氧化不同造影剂有机物的速率有关，速率越大则越有利于消毒副产物的产生。

4.3 Fe(VI)还原产物特性及其在去除砷及重金属方面的研究

Fe(VI)具有强氧化性和不稳定性，易还原、自分解生成新生态的 Fe(III)。由于新生态铁的粒径更小，比表面积更高，颗粒表面更加光滑和圆润，其凝聚、絮凝、吸附、沉降性能和铁盐混凝剂如 $FeCl_3$ 截然不同。Goodwill 等[97]比较了实验室配水和实际水库水中 Fe(VI)还原产生的新生态铁和 $FeCl_3$ 水解生成的 Fe(III)颗粒的各种特性。结果表明，两者具有相似的表面电荷，但它们的粒径大小分布和颗粒数目并不相同。新生态铁的粒径分布呈现双峰模式，在 60~70 nm 和 250~1000 nm 分别出现了一个高峰和一个强度较低的分布峰，平均粒径为 172 nm；而 $FeCl_3$ 水解生成的 Fe(III)颗粒只有一个 200~300 nm 的高峰，平均粒径为 276 nm，高于新生态铁。XPS 分析结果显示新生态铁颗粒中含有赤铁矿（$\alpha\text{-}Fe_2O_3$）和 FeOOH，而铁盐混凝剂中只有 FeOOH。Lv 等[98]发现在 Fe(VI)投量为 3.0 mg-Fe/L 和 pH 为 7.5 时新生态铁由无定形以及晶型的铁氧化物组成，两者含量比例为 2.3∶1，而 $FeCl_3$ 水解生成的 Fe(III)颗粒主要是无定形的氢氧化铁；作者还研究了 NOM 对新生态铁的沉降性能的影响，研究发现较低浓度的 NOM 易被 Fe(VI)氧化，对絮体生长及沉降性能影响较小，而较高浓度的 NOM 会增加新生态铁表面电荷，阻碍了新生态铁的凝聚，从而悬浮在上清液中难以沉降，反而会增加水中浊度及纳米颗粒污染，对出水水质及水环境造成不利影响。将 Fe(VI)预氧化与常规混凝结合可以解决此项缺陷[99]，常规混凝颗粒径大、沉降性能好，Fe(VI)原位生成的新生态铁可以高效吸附水中重金属和其他微污染物，同时很容易被混凝剂形成的絮体去除。

一般认为，新生态铁通过吸附/絮凝的方式去除水中砷及重金属污染物，与 pH 直接相关的 Fe(VI)分解特性和新生态铁表面电荷决定了其去除金属污染物的效率。近年来研究者采用多种手段详细分析表征了 Fe(VI)分解产生的新生态铁去除金属污染物的机理。Prucek 等[100]比较了高铁酸盐原位和异位产生的铁（氢）氧化物除 As(III)效率，发现原位除砷效率大大高于异位除砷；并采用 XRD、高分辨 XPS 和穆斯堡尔谱表征了铁氧化物的物质结构和砷的形态、分布，结果表明，在高铁酸盐原位分解除砷过程中，砷不仅吸附在表面，还有大量的砷嵌入到了 $\gamma\text{-}Fe_2O_3$ 尖晶石结构中，从而提高了除砷效率，而异位除砷过程中，砷仅仅分布在 $\gamma\text{-}Fe_2O_3$ 的表面（图 8-28）。Prucek 等[101]进一步研究了高铁酸盐对多种重金属[Cd(II)、Co(II)、Ni(II)、Cu(II)] 在不同[Fe(VI)]/[M(II)]条件下的去除效果，结果发现，在 pH 为 6.6，[Fe(VI)]/[M(II)] 为 2∶1 时，Co(II)、Ni(II)、Cu(II)几乎完全去除，而对于 Cd(II)，在[Fe(VI)]/[M(II)]为 15∶1 时去除率仅仅达到 75%；进一步的研究发现，Co(II)、Ni(II)、Cu(II)更易于形成相应的 MFe_2O_4 尖晶石，也更易于嵌入到 $\gamma\text{-}Fe_2O_3$ 结构中，而 Cd(II)仅仅吸附在 $\gamma\text{-}Fe_2O_3$ 的表面，推测重金属离子半径和电子结构可能显著影响了重金属去除效率及其化学行为。高铁酸盐去除砷及重金属后的产物主要以磁性的 $\gamma\text{-}Fe_2O_3$ 为主，易于从水中分离，而且金属离子嵌入在 $\gamma\text{-}Fe_2O_3$ 结构中不易于释放分离。Yang 等[102,103]用 Fe(VI)对 2 种有机砷（对氨基苯胂酸和硝羟苯胂酸）进行氧化处理，并研究了氧化过程中 Fe(VI)对游离砷的去除效果。研究发现，Fe(VI)不仅能有效氧化有机砷，在氧化的同时新生态铁吸附去除了氧化过程中释放的 As(V)，FTIR 分析结果显示，As(V)与新生态铁（氢）氧化物络合形成 As—(OFe)键。

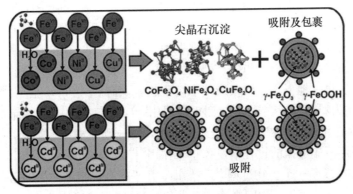

图 8-28　Fe(VI)原位和异位除砷的机理示意图（引自参考文献[100]）

5 微生物介导亚铁氧化耦合硝酸盐还原机制与环境效应

铁的生物地球化学氧化还原过程对铁的矿化、污染物的转化和营养物质的流失有影响。有氧条件下，亚铁被氧气通过生物和化学作用氧化；在厌氧的条件下被亚硝酸盐、二氧化锰等氧化剂氧化。本节主要综述厌氧条件下微生物介导的硝酸盐还原亚铁氧化反应。已有研究报道，中性厌氧条件下，硝酸盐还原亚铁氧化过程在铁、氮元素相互作用中发挥着重要作用。然而，目前这一代谢过程的机制很大程度上是未知的。虽然早期的研究认为 Fe-N 耦合是一个生物作用过程，但近期的研究证实这一过程存在生物和化学两种机制。硝酸盐还原的中间产物亚硝酸盐和一氧化氮可以在该体系中与亚铁发生化学反应。因此，酶促亚铁氧化作用被过高估计。目前，生物和化学作用亚铁氧化的相对贡献仍然不清楚。本节综述旨在全面阐述硝酸盐还原亚铁氧化过程中的生物和化学过程。未来的研究需要进一步回答微生物介导的硝酸盐还原亚铁氧化是酶促作用氧化亚铁耦合硝酸盐还原的生物过程，还是在富亚铁环境中由生物作用还原硝酸盐诱导的产物与亚铁的化学副反应过程。

5.1 概述

铁是地壳中含量第四丰富的元素，也是生物圈中含量最高的氧化还原活性金属[104]。自然铁循环过程包括三价铁还原和亚铁氧化两部分。有氧条件下，氧气可以通过生物和化学作用氧化亚铁[105]。在中性有氧条件下，亚铁与氧气之间发生的 Fenton 反应已被广泛的研究[106]；在中性微氧条件下，亚铁可被微氧型亚铁氧化微生物氧化[107]；在中性厌氧条件下，缺少氧气化学作用氧化亚铁的竞争，硝酸盐还原亚铁氧化微生物在氧化亚铁过程中起着重要作用。硝酸盐还原亚铁氧化微生物是严格厌氧的，在沉积或陆地厌氧环境中的 Fe(II) 到 Fe(III) 氧化过程中发挥重要作用[108]。NRFO 具有参与全球范围内 Fe(II) 氧化的潜力，然而，这种代谢对全球氮循环的意义目前尚不清楚。

硝酸盐是生物圈中重要的环境成分之一，也是全球氮库存中广泛存在的高活性含氮物之一[109]。它是植物和微生物的营养物质，也可作为许多细菌、古菌以及一些真核生物的电子受体[110,111]。硝酸盐还原是一个常见的微生物作用过程[112,113]。尽管异化硝酸盐还原很可能不如同化硝酸盐还原过程重要，但异化硝酸盐还原的中间产物对元素生物地球化学过程有重要的影响，如，在无氧条件下氧化 Fe(II)[109]。

在厌氧条件下，硝酸盐和亚铁不能直接发生化学反应[114]，反硝化作用与亚铁氧化过程的耦合可由硝酸盐还原亚铁氧化微生物介导的，该类微生物已被发现二十多年[115]。硝酸盐还原亚铁氧化微生物不受限于光和氧环境，因此它们可能比光合型和微氧型微生物更丰富[104]。硝酸盐还原亚铁氧化过程已在多种环境中被发现，如海洋、温泉、地下水、微咸或淡水沉积物、深海[116-119]，水稻土[120]，湿地沉积物[121-123]，和热液喷口[124]，甚至在火星可能存在铁-氮代谢也被提出[125]。在深海中，研究者发现硝酸盐还原是依赖于 Fe 的，且 Fe(II) 的存在可以显著提高硝酸盐还原速率，这一现象清楚地阐明在自然的富铁环境中硝酸盐还原亚铁氧化反应在还原硝酸盐的过程中发挥重要作用[119]。

早期的研究认为 NRFO 是一个生物过程[115]，然而，几乎所有已报道的 NRFO 微生物都可以通过反硝化作用直接还原硝酸盐，且反硝化的中间产物（亚硝酸盐和 NO）可以快速化学氧化 Fe(II)[114]。由此产生疑问，Fe(II) 氧化是酶促的（生物作用）还是与亚硝酸盐或 NO 反应（化学作用）的结果，以及两种作用的相对贡献又如何。

5.2 微生物耦合硝酸盐还原和亚铁氧化

第一个硝酸盐还原亚铁氧化群落是从淡水沉积物中分离得到的[126]，此后的二十多年陆续有很多此类微生物被分离得到并进行了准确的分类。几乎所有的已被分离得到的硝酸盐还原亚铁氧化微生物都是兼性营养型，即需要有机底物，如乙酸钠，参与亚铁氧化和硝酸盐还原过程[127]。由此引出疑问，Fe(II)氧化是酶促反应，还是仅仅是由异养硝酸盐还原的中间产物介导的化学反应[115]。此外，研究报道90%的硝酸盐还原微生物具有利用硝酸盐和乙酸钠进一步氧化亚铁的能力[128]。目前也分离得到一些能在没有任何有机基质的自养条件下存活的微生物，这些微生物能利用Fe(II)作为唯一的电子供体，CO_2作为唯一的碳源进行生长[116, 129-132]，它们与有氧条件下广泛存在的微氧型亚铁氧化微生物很相似[117]。大多数微氧型亚铁氧化微生物都是自养的，但NRFO微生物主要是异养的（或兼性营养型）。本节根据微生物介导的硝酸盐还原和亚铁氧化过程的最新假设（图8-29），即NRFO微生物包括兼性营养型、异养型和自养型，进行讨论，具体内容如下。

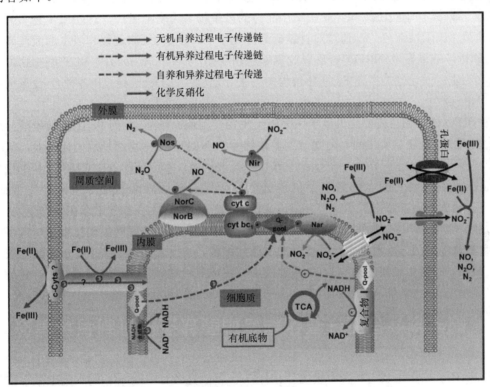

图8-29 微生物介导硝酸盐还原耦合亚铁氧化过程示意图（引自参考文献[239]）

NAD^+：氧化型辅酶I，NADH：还原型辅酶I，TCA：三羧酸循环，Nar：硝酸盐还原酶，Nir：亚硝酸盐还原酶，NorC/NorB：一氧化氮还原酶，Nos：氧化亚氮还原酶，cyt bc_1/cyt c：细胞色素 bc_1/c，c-Cyts：细胞色素 c

兼性营养型的NRFO微生物可以通过异养过程在有机电子供体存在的情况下直接还原硝酸盐，同时，共存的Fe(II)也可以通过自养过程为硝酸盐还原提供电子。研究报道，在3~4mm深的土层中，每克干土含1.6×10^6个兼性营养型的NRFO微生物，这些微生物群落或单菌可以在厌氧条件下以亚铁为唯一的电子供体，或者在乙酸钠存在条件下以亚铁作为额外的电子供体[126, 133]。这一新过程可能对水生沉积物缺氧区Fe(III)的形成有重要贡献。从猪排泄废水中分离得到的 *Dechlorosoma suillum* strain PS[134]和从河流沉积物中发现的 *Dechloromonas* sp. UWNR4[135]均能够在厌氧条件下以硝酸盐为电子受体、乙酸钠为电子供体迅速氧化亚铁。Kappler、Schink 和 Newman 从湖泊沉积物中分离得到一株厌氧 NRFO 细菌（*Acidovorax* sp. strain BoFeN1）。Strain BoFeN1 在以乙酸钠为有机共底物的兼性营养条件下还原硝酸盐氧化亚铁，在细胞周围形成三价铁矿物结壳[136]。同时在淡水/沟渠沉积物和地下水中也发现了其他的菌，如 *Acidovorax* sp.（BrG1，2AN，TPSY）[126, 138-140]。在同一沟渠沉积物中，还发现另外两种菌株（*Aquabacterium* sp. BrG2 和 *Thermomonas* sp. BrG3）能在乙酸盐存在条件下耦合 Fe(II)氧化和硝酸盐还原[126, 137]。有研究者利用一株分离得到的 NRFO 细菌（*Klebsiella oxytoca* FW33AN）研究 Fe(II)的形态和氧化速率对 NRFO 过程产生的三价铁氢氧化物的矿物种类是否有影响[141]。Kumaraswamy 等分离得到一株兼性自养型 NRFO 菌，*Paracoccus ferrooxidans* strain BDN-1，该菌株可以使用[Fe(II)EDTA]$^{2-}$作为电子供体，硝酸盐或亚硝酸盐作为电子受体，但仍然需要乙醇作为共底物[129]。研究者从 Staraya Russa Resort 的含铁、微咸的低温温泉沉积物中分离到嗜中性铁氧化菌 *Hoeflea siderophila* Hf1，该菌能在兼性和有机异养生长条件下氧化亚铁还原硝酸盐[118]。从水稻土中分离得到的 *Rhodocyclaceae* sp. strain paddy-1 在含有 5mmol/L 乙酸钠兼性营养条件下可还原 97%的硝酸盐、氧化 86%的 Fe(II)。上述的兼性营养型 NRFO 微生物在不同环境的微生物群落中被发现。在水稻土中发现亚铁氧化耦合硝酸还原过程，采用分子生物手段分析 NRFO 微生物的多样性结果显示，具有硝酸还原耦合亚铁氧化能力的微生物多为变形菌门，这些微生物多为具有高 GC 含量 DNA 的革兰氏阳性细菌[120]。通过对针铁矿悬浮中 Fe-N 氧化还原过程和淡水河流漫滩沉积物中微生物群落的分析，结果显示 NRFO 群落以脱氯单胞菌（如，*Dechloromonas*）和三价铁还原 δ-变形杆菌（*Geobacter*）为主；而这些类群在 Fe 循环过程中占主导地位，且 Geobacter 和各种 β-变形菌门在循环培养中均参与 NRFO[142]。

异养型 NRFO 微生物也需要有机电子供体用于生物作用还原硝酸盐，但 Fe(II)是被硝酸盐还原中间产物（亚硝酸盐或 NO）化学氧化的，因此，大多数异养型 NRFO 微生物都是反硝化细菌。从不同土壤中分离到 3 株异养型 NRFO 菌 *Paracoccus denitrificans* strains（ATCC17741、ATCC19367 和 Pd 1222），这些微生物被证实可以利用乙酸盐作为电子供体进行反硝化，随后硝酸盐还原中间产物亚硝酸盐氧化 Fe(II)[143, 144]。研究发现从咸水沉积物中分离出来的 strain HidRe2 能在乙酸钠存在条件下，在 pH 约为 7 时还原硝酸盐、氧化 Fe(II)；然而，该菌株也被发现是广泛存在的嗜常温的反硝化细菌[145]。*Pseudomonas stutzeri* LS-2 是从中国南方水稻土中分离得到的反硝化细菌，其反应动力学结果显示硝酸盐被生物作用还原生成亚硝酸盐，产物亚硝酸盐会与 Fe(II)快速反应并生成三价铁氧化物导致细胞结壳，并最终抑制微生物的反硝化作用[146]。另一株从人脊髓液中分离的 *Pseudomonas stutzeri* strain（ATCC 17588）也被发现具有利用乙酸钠为有机底物进行反硝化，并化学作用氧化 Fe(II)的能力[143]。由于反硝化细菌在不同环境中分布广泛，异养型 NRFO 微生物也可能广泛存在于富亚铁的缺氧环境中。考虑到微生物群落中的异养菌（反硝化菌），在滨海淡水湖底泥缺氧顶层，硝酸盐还原和光合型铁氧化微生物竞争 Fe(II)。硝酸盐还原菌在夜晚氧化 Fe(II)，白天主要由光合菌氧化 Fe(II)[147]。有研究者对中性 pH 的水稻土微宇宙体系中厌氧 NRFO 过程中微生物群落组成和多样性的变化进行了研究[148, 149]。结果显示，硝酸盐还原亚铁氧化过程中主要的微生物菌属是 *Azospira*、*Zoogloea* 和 *Dechloromonas*。此外，这些硝酸盐还原微生物生成的亚硝酸盐同样可以化学作用氧化亚铁。这些发现增进了我们对中性 pH 环境中缺乏和存在铁(II)的缺氧环境中硝酸盐还原过程的理解，并扩展了我们对这些过程所涉微生物群落的认识[148]。

自养型 NRFO 微生物不需要有机电子供体作为共底物，但需要 CO_2 为碳源，并以亚铁为唯一电子供体，耦合硝酸盐还原和亚铁氧化过程。富集群落（culture KS）是一个能够完全氧化 Fe(II)耦合反硝化生成 N_2 的自养型 NRFO 群落，其是广泛运用的研究 NRFO 过程的模式群落[150-152]。其他的自养型 NRFO 微生物也被陆续分离和报道。如，从意大利 Vulcano 的一个浅海海底热液系统分离得到的一个球状的厌氧亚铁氧化古菌（*Ferroglobus placidus* DSM 10642）被确认为一株中性超嗜热菌，能在 65~95℃条件下正常生长[116]。Weber 等从厌氧的淡水湖沉积物中分离出一株亚铁氧化细菌——strain 2002[130]。该研究表明，

strain 2002是一株厌氧嗜常温、嗜中性的无机自养型亚铁氧化微生物，但其并不能在无机自养生长条件下传代培养，这些过程中一定还存在未知的代谢路径[153]。Bosch等研究证实，*Thiobacillus denitrificans*可以厌氧氧化纳米级黄铁矿颗粒，并以硝酸盐为电子受体[154]。另一株嗜中性的自养型NRFO微生物（*Azospira bacterium* TR1）从不列颠哥伦比亚Trail的一个生物修复场地分离得到，该细菌在厌氧条件下能够耦合硝酸盐还原亚铁氧化[155]。从深层沉积物中分离到一株以碳酸氢盐为唯一碳源的反硝化细菌（*Microbacterium* sp.），该菌株可以利用Fe(II)-EDTA作为硝酸盐还原的唯一电子供体，此外，反应过程中未出现细胞结壳，也没有补充底物被消耗[156]。分别从土壤和淡水沉积物中分离得到两株微生物 *Pseudomonas* species（W1和SZF 15），研究表明它们能在自养生长条件下还原硝酸盐和氧化亚铁，最终硝酸盐的最大去除量相当于总Fe(II)氧化量的71%[157,158]。尽管这些研究已经分离得到自养型NRFO微生物，也发现了可能的自养过程，但NRFO反应是否真的不需要共底物仍存在疑问。因为，很多研究在自养培养实验中排除了有机电子供体，但在准备接种微生物的过程并没有将细胞从残留有机物中储存的碳排除，这些有机物可能在自养培养过程中进一步被用作有机碳源干扰实验。

5.3 硝酸盐生物作用还原为亚硝酸盐和化学反硝化作用

针对亚铁与硝酸盐反应的研究起始于1966年[159]。虽然有研究显示在酸性和有氧条件下亚铁可以还原硝酸盐，但值得注意的是这个反应只发生在高温条件下[159]。在常温、中性、缺氧条件下，虽然热力学理论显示亚铁与硝酸盐反应是可行的，但实际上该反应并不能发生[114]。Fe(III)/Fe(II)的氧化还原电位在$-314 \sim 14$ mV之间，这均远负于硝酸盐还原过程中所涉及的氧化还原电对（NO_3^-/NO_2^-, +430 mV；NO_2^-/NO, +350 mV；NO/N_2O, +1180 mV；N_2O/N_2, +1350 mV）[104]，但实验结果显示在没有催化剂的情况下硝酸盐与亚铁不能反应[114]。硝酸盐与亚铁的化学反应可以通过许多催化剂加速，如Cu^{2+}、铁氢氧化物，甚至微生物表面[114,160,161]。

能够完成异化硝酸盐还原为亚硝酸盐的微生物在环境中广泛存在，这些微生物主要存在于含硝酸盐的严氧环境中，包括土壤[162,163]、海洋沉积物[164]以及人类的消化系统[165]。该反应可由膜结合硝酸还原酶（NAR）或周质硝酸还原酶（NAP）催化[166]。在海洋沉积物中，许多微生物脱氮的第一步是将硝酸盐异化还原为亚硝酸盐[167,168]。受各种地球化学条件，如温度、有机质浓度、pH和碳源等的影响，亚硝酸盐可以在环境中积累到mmol/L浓度[122,136,169-171]。相比其他氮循环过程，硝酸盐还原是亚硝酸盐的主要来源，且亚硝酸盐能够被亚铁化学还原（该反应称为：化学反硝化[168,172-174]）。有广泛报道称在厌氧富铁环境中，微生物（如*Ferroglobus placidus*）作用下异化硝酸盐还原可与亚铁氧化过程耦合[175]，因此，亚铁与亚硝酸盐的化学反应是理解亚铁与硝酸盐相互作用的基础，此外，处于竞争关系的生物作用亚硝酸盐还原与化学反硝化过程的相对贡献需要评估[115]。

微生物-Fe(II)-硝酸盐体系中亚硝酸盐与亚铁的化学反应已引起人们的广泛关注[114]。实际上，亚硝酸盐与亚铁的化学反应（化学反硝化）早在几十年前就已经发现。均相体系中的亚硝酸盐与亚铁反应在很多研究中被提到[159,176]。有研究调查了土壤矿物和金属阳离子在亚硝酸盐还原过程中的作用，结果显示土壤矿物和金属离子对亚硝酸盐的还原并无影响，而Fe(II)可以促进亚硝酸盐的分解[177]。此外，无论是化学还是生物作用亚铁氧化均会产生Fe(II)-Fe(III)中间矿物，如绿锈和菱铁矿[178]。研究表明，亚硝酸盐能较快速地氧化绿锈和菱铁矿中的结构亚铁[115,179]。也有研究关注配体对Fe(II)氧化过程的影响，结果表明，在微生物-Fe(II)-硝酸盐体系中，柠檬酸盐等强有机配体也能显著加速亚硝酸盐对Fe(II)的化学氧化[128]，此外，亚硝酸盐与Fe(II)配体之间的化学反应动力学、机理和产物的分析结果也支持了这一结果[180-182]。随后，研究者采用动力学模型和表面表征等不同方法考察了亚硝酸盐与亚铁伴生矿物（如，Fe(II)-硅酸盐、Fe(II)吸附在纤铁矿上、绿锈、菱铁矿、含水铁氧化物和Fe(II)）之间的反应[179,182-187]。

尽管对 Fe(II)和亚硝酸盐化学反应的研究取得了上述进展,但化学反硝化作用在 NRFO 过程中的作用直到最近才开始被重视。Picardal 描述了 NRFO 环境下可能涉及的非生物反应[114]。鉴于 Fe(II)与亚硝酸盐的快速非生物反应,在微生物 NRFO 的研究中生物和非生物反应的相对贡献需要进行全面的评价。因此,在各种反硝化细菌参与的 NRFO 系统中,研究者特意对 Fe(II)氧化的非生物过程进行了检测[144, 188]。N_2O 稳定同位素技术被用于区分厌氧环境下的化学反硝化作用。其中,N_2O 的 SP 值被认为是一种定性区分自然系统中非生物和生物 N_2O 释放源的方法[189]。亚硝酸盐的 N、O 同位素被用作一种强有力的表征化学反硝化和生物的 N_2O 产生过程的方法[190, 191]。研究者采用 N 标记的方法测定反硝化作用产生的 N_2O 的速率,结果表明,化学反硝化就足以解释维达湖中 N_2O 含量高的原因[192]。研究者利用了同位素方法和微传感器研究连续流动培养体系,用于解析潮间带沉积物中 N_2O 的生成机制。结果表明,在硝酸盐水平升高的培养过程中,N_2O 通量的增加不受细菌直接活性的调节,而主要受真菌反硝化和/或非生物反应(如,化学反硝化)的催化[193]。

Chen 等通过反应动力学、氮同位素分馏和二次矿物表征等方法研究了 *Pseudogulbenkiania* strain 2002 在没有任何有机共底物的条件下硝酸盐还原耦合 Fe(II)氧化的过程[194]。结果表明,NRFO 过程中同时存在生物和化学 Fe(II)氧化过程。Liu 等利用 strain BoFeN1 和 *Pseudogulbenkiania* sp. 2002 研究 NRFO 过程,并建立了一个动力学模型来量化生物过程和化学反硝化的相对贡献[195]。结果表明,对于亚硝酸盐的还原,生物作用的贡献高于化学反硝化;对于 Fe(II)氧化,亚硝酸盐化学氧化 Fe(II)的贡献高于生物作用。这些发现提供了 NRFO 过程中化学反硝化和生物作用的定量信息,有助于理解地下环境中铁和氮的全球生物地球化学循环机制。

5.4 酶促亚铁氧化耦合硝酸盐还原

关于 Fe(II)和 NO_3^- 的生物反应,目前还不清楚 Fe(II)是被酶促作用氧化还是仅被异养硝酸盐还原反应的 N 中间产物化学氧化[144, 196]。虽然没有直接证据证明 NRFO 过程中存在酶促 Fe(II)氧化[197, 198],但是一些间接的观察证实了生物 Fe(II)氧化过程的存在[199, 200]。在微生物-铁(II)-硝酸盐体系中分离化学和生物过程是一项非常具有挑战性的工作[201],但是,研究者根据动力学和 N_2O 同位素分馏的结果明确的区分开了生物和化学过程。Kopf、Henny 和 Newman 成功地利用动力学模型方法量化了化学 Fe(II)氧化过程,并证明 *Pseudogulbenkiania* sp. strain MAI-1 可以直接生物氧化 Fe(II)[128]。Jamieson 等[202]发现单独的亚硝酸的还原量不足以氧化如此数量的 Fe(II),其中 60%~75%的 Fe(II)氧化是由所研究的微生物酶促作用氧化的。Peng 等[203]发现的络合态的亚铁(如,柠檬酸、EDTA、胡敏酸和富里酸)会抑制微生物硝酸盐还原亚铁氧化,且只有未络合的(无机)亚铁才能引起亚硝酸盐在周质和胞外的积累,而络合态亚铁不能进入周质也不能积累亚硝酸盐。研究者对化学和生物过程相对贡献的计算结果表明,虽然酶促 Fe(II)氧化的相对贡献较低,但其确实在 Fe(II)氧化中发挥了作用,尤其是在硝酸盐或亚硝酸盐浓度很低的条件下[195]。

尽管有上述证据证实生物 Fe(II)氧化的存在,但 NRFO 细菌中哪些蛋白与 Fe(II)氧化有关仍不清楚。研究者提出的 NRFO 过程的可能机制,Fe(II)的电子直接传递给酶,如亚铁氧化酶、硝酸盐还原酶和细胞色素 bc1 复合物等[204-209],最终由硝酸盐或其还原中间产物得到电子[210, 211]。原位监测 Strain 2002 在存在亚铁和硝酸盐的情况下其细胞色素 c 被还原,表明至少有一种细胞色素 c 可能参与了从亚铁到呼吸链的电子传递过程[212, 213]。虽然目前还没分离确定 NRFO 过程中涉及的酶[198, 214],但有报道称氧化铁细菌的 *c* 型细胞色素(*c*-Cyts)很可能参与 Fe(II)与细胞膜之间的电子转移过程[210, 215, 216]。因此,*c*-Cyts 的表征及其在 Fe(II)氧化中的作用对于理解酶促 Fe(II)氧化过程是至关重要的[207, 217, 218]。

基于 *c*-Cyts 的光谱性质,最近科学家开发了一种检测限较低的新型分光光度计,该仪器可以原位研究细胞悬液中 *c*-Cyts 的氧化还原动力学[217, 219, 220]。利用该分光光度计可以成功检测活细胞悬液中 NRFO

菌（如，strain 2002 和 BoFeN1）参与的 Fe(II) 与 c-Cyts 的反应[195]，这一结果证明 c-Cyts 理论上可从 Fe(II) 处获得电子。尽管关键蛋白介导电子转移的潜在机制尚不清楚，但鉴于原位光谱法能够直接观察关键蛋白（c-Cyts），因此它是一种非常有前途的酶学机制识别工具[217, 221]，此外，c-Cyts 氧化还原转化动力学的研究进一步强调了细胞色素在微生物铁呼吸链中的作用。Popovic 等研究表明[222]，土壤细菌（*Paracoccus denitrificans*）的 c-Cyts 氧化酶的结构和热力学指标与光合型亚铁氧化细菌（*Rhodobacter sphaeroides*）的相似[223, 224]，这就说明 NRFO 细菌的细胞色素 c 可能是表征 Fe(II) 氧化能力的重要因素。虽然 c-Cyts 被发现参与 NRFO，但仍不确定通过 c-Cyts 的酶促 Fe(II) 氧化是否与特定电子传递链的硝酸盐还原过程有关。

5.5 环境意义

NRFO 过程在含 Fe 和含 N 物质的化学和微生物介导的氧化还原转化过程中发挥作用。如图 8-30 所示，对铁矿物而言，Fe(II) 和 Fe(III) 的氧化还原循环导致铁矿物的转化，从而助于污染物的迁移和转化，如重金属（As，Sb，U）和氯化有机化合物。对含氮物来说，硝酸盐的还原与许多自然或工程水系统中的 N 污染物直接相关，此外，温室气体 N_2O 的产生对全球气候变化具有潜在的影响。因此，深入了解含氮和含铁物之间可能发生的氧化还原反应，将有助于更好地理解自然环境中 Fe 与 N 循环之间的联系。

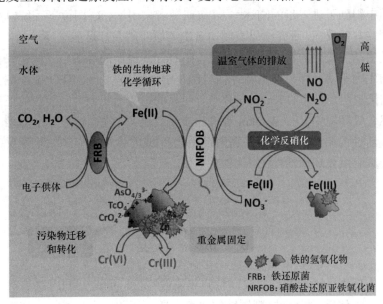

图 8-30　微生物介导亚铁氧化耦合硝酸盐还原过程的环境效应（引自参考文献[239]）

铁和锰氧化物对重金属和放射性核素的吸附长期以来被认为是固定这些污染物的重要机制[225, 226]。基于最近描述的 *Dechlorosoma* 种类微生物介导的厌氧 NRFO 过程，可以开发一种潜在的在还原环境中稳定重金属和放射性核素（HMR）的生物修复策略。结果表明，*Dechlorosoma suillum* 氧化 Fe(II) 是一种稳定和固定环境中 HMR 的新方法[227]。Senn 和 Hemond 报道在美国马萨诸塞州的市内 Upper Mystic 湖[228]，厌氧条件下砷（As）的循环受到硝酸盐还原耦合氧化 Fe(II) 过程的影响，亚铁氧化产生的颗粒状的含水铁氧化物能够吸附 As，同时反应过程使 As(III) 氧化为 As(V)。在美国地质调查局科德角研究点的一个厌氧、铁还原砂含水层区域，当亚铁浓度降低到不可测时硝酸浓度开始增加，而亚铁氧化过程伴随着 As(V) 的去除，污染源处的 As(V) 可通过被含水铁氧化沉淀包裹而去除[229]，此研究再次强调了 NRFO 的 As 固定过程中的重要作用。在水稻土中，添加硝酸盐和亚铁可以富集更多的 NRFO 微生物，导致根际土壤溶液中溶解态 Fe(II) 浓度大大降低。研究表明，NRFO 菌可导致 As 共沉淀或吸附于土壤中 Fe(III) 矿物，从而降低水稻对还原态 As 的吸收[230]。研究者采用连续流动填砂柱实验结果表明，微生物氧化 As(III) 和 Fe(II) 与反硝化作用有关，通过形成 Fe(III) 氢氧化物包裹的沙粒吸附 As(V) 的，增加水中砷在厌氧环境中的固定[231]。厌氧 NRFO 细菌的新陈代谢不受到砷酸的影响，NRFO 细菌在亚铁氧化过程中可以有效地固定砷（>96%），能有效地将溶解态砷浓度降低到接近甚至低于当前饮用水 10μg / L 的限制[232]。对单一的厌氧硝酸盐还原亚铁氧化微生物 Citrobacter freundii strain, PXL1 同步去除水中硝酸盐和亚砷酸盐的潜力进行评估，结果表明厌氧 NRFO 微生物具有原位修复硝酸盐和亚砷酸盐污染地下水的潜力[233]。Wang 等报道向水稻田中同时施用亚铁和硝酸盐[234]，通过增强非生物和生物铁氧化还原转化和矿化作用有效地增强了 As 的氧化和固定，从而有效地降低了 As 在水稻植株中的积累，这一发现增加了对 Fe/N/As 生物地球化学循环的认识，同时对 As 污染水稻田的农业管理和治理至关重要。

硝酸盐是公认的微生物介导的常见过程[112, 113]，Fe(II) 氧化对硝酸盐还原和温室气体排放（即 N_2O）的影响鲜有报道，这主要是人们对氮在生物和非生物过程中的转化认识不足所致。虽然这些非生物 N_2O 的产生途径已被发现近一个世纪，但在现代生物研究中往往被忽视，因此有必要进一步研究非生物反应对 N_2O 产生的重要性[235]。研究者提出一个初步假设（亚铁轮假设），硝酸盐在森林土壤中的非生物固定化是由硝酸盐和亚铁之间的生物和非生物过程造成的[236]。虽然还不清楚非生物和生物反应是如何发生的，但研究结果强调了 NRFO 在公认的 N 循环中的存在和重要性。铁与氮的相互作用在河岸森林沉积物的铁和氮循环动力学中得到了进一步的阐述[237]。有研究者采用 N、O 同位素系统研究了不同环境条件下亚铁非生物还原亚硝酸盐的反应。研究结果表明，Fe(II) 还原亚硝酸盐可能是环境 N_2O 的重要非生物来源，特别是在氧化还原发生动态变化的富铁环境中[190]。在 Doane 最近的综述中，他建议在未来的研究中，应该像评估我们熟悉的生物驱动循环一样重新评估自然化学氮循环[238]。NRFO 过程在含铁的深海中被观察到，模拟培养实验结果表明，东非的 Kabuno 海湾有明显的硝酸盐还原。硝酸盐的还原依赖于铁，表明这种反应常发生在天然含铁环境中[119]。Wankel 等进一步研究表明，在高浓度硝酸盐培养条件下，N_2O 通量的增加不受细菌直接活性的调节，而主要受真菌反硝化和/或非生物反应（如化学反硝化）的催化，尤其是在沿海生态系统中具有氧化还原活性的沉积物中[193]。

6 展　望

铁基材料具有成本低廉、环境友好、表面位点丰富等特性，因此，利用铁基材料发展绿色友好环境治理和修复技术已成为水体/土壤污染控制领域的热点话题，相关研究也取得了突破性的进展。然而，在实际应用过程中仍存在一些问题，例如，零价铁易腐蚀板结、铁矿物 Fe(II)/Fe(III) 循环困难、高铁酸盐制备成本高等。因此，低成本实现零价铁的功能化，发展高效铁矿物铁循环策略，开发绿色高铁酸盐的制备及活化方法，阐明生物和化学过程对 NRFO 过程的相对贡献仍是目前铁环境化学领域亟待解决的关键科学问题。

参 考 文 献

[1] Liu A R, Wang W, Liu J, et al. Nanoencapsulation of arsenate with nanoscale zero-valent iron(nZVI): A 3D perspective. Science Bulletin, 2018, 63(24): 1641-1648.

[2] Li X Q, Zhang W X. Sequestration of Metal cations with zerovalent iron nanoparticles: A study with high resolution 9-ray photoelectron spectroscopy(HR-XPS). Journal of Physical Chemistry C, 2007, 111(19): 6939-6946

[3] Huang X Y, Wang W, Ling L, et al. Heavy metal-nZVI reactions: the Core-shell structure and applications for heavy metal

treatment. Acta Chimica Sinica, 2017, 75: 529-537.

[4] Li S L, Yan W L, Zhang W X. Solvent-free production of nanoscale zero-valent iron(nZVI)with precision milling. Green Chemistry, 2009, 11(10): 1618-1626.

[5] Hua Y L, Liu J, Gu T H, et al. The colorful chemistry of nanoscale zero-valent iron(nZVI). Journal of Environmental Sciences, 2018, 67(5): 1-3.

[6] Chen K F, Li S L, Zhang W X. Renewable hydrogen generation by bimetallic zero valent iron nanoparticles. Chemical Engineering Journal, 2011, 170(2-3): 562-567.

[7] Liu A, Liu J, Han J, et al. Evolution of nanoscale zero-valent iron(nZVI)in water: Microscopic and spectroscopic evidence on the formation of nano- and micro-structured iron oxides. Journal of Hazardous Materials, 2017, 322(Pt A): 129-135.

[8] Liu A R, Liu J, Pan B C, et al. Formation of lepidocrocite(γ-FeOOH)from oxidation of nanoscale zero-valent iron(nZVI)in oxygenated water. RSC Advances, 2014, 4: 57377-57382.

[9] Liu A R, Liu J, Zhang W X. Transformation and composition evolution of nanoscale zero valent iron(nZVI)synthesized by borohydride reduction in static water. Chemosphere, 2015, 119: 1068-1074.

[10] Huang X Y, Ling L, Zhang W X. Nanoencapsulation of hexavalent chromium with nanoscale zero-valent iron: High resolution chemical mapping of the passivation layer. Journal of Environmental Sciences, 2018, 67(5): 7-16.

[11] Yan W L, Ramos M A V, Koel B E, et al. Multi-tiered distributions of arsenic in iron nanoparticles: Observation of dual redox functionality enabled by a core-shell structure. Chemical Communications, 2010, 46(37): 6995-6997.

[12] Li S L, Wang W, Liu Y Y, et al. Zero-valent iron nanoparticles (nZVI) for the treatment of smelting wastewater: A pilot-scale demonstration. Chemical Engineering Journal, 2014, 254: 115-123.

[13] Li X Q, Zhang W X. Iron nanoparticles: the core−shell structure and unique properties for Ni(II)sequestration. Langmuir, 2006, 22(10): 4638-4642.

[14] Wang W, Li S L, Lei H, et al. Enhanced separation of nanoscale zero-valent iron (nZVI) using polyacrylamide: Performance, characterization and implication. Chemical Engineering Journal, 2015, 260: 616-622.

[15] Li S L, Wang W, Yan W L, et al. Nanoscale zero-valent iron (nZVI) for the treatment of concentrated Cu(II)wastewater: A field demonstration. Environmental Science-Processes & Impacts, 2014, 16: 524-533.

[16] Wang W, Hua Y, Li S, et al. Removal of Pb(II)and Zn(II)using lime and nanoscale zero-valent iron(nZVI): A comparative study. Chemical Engineering Journal, 2016, 304: 79-88.

[17] Li S, Wang W, Liang F, et al. Heavy metal removal using nanoscale zero-valent iron(nZVI): Theory and application. Journal of Hazardous Materials, 2017, 322: 163-171.

[18] Li S L, Li J H, Wang W, et al. Recovery of gold from wastewater using nanoscale zero-valent iron. Environmental Science: Nano, 2019, 6: 519-527.

[19] Navrotsky A, Mazeina L, Majzlan J. Size-driven structural and thermodynamic complexity in iron oxides. Science, 2008, 319(5870): 1635-1638.

[20] Tartaj P, Morales M P, Gonzalez-Carreño T, et al. The iron oxides strike back: From biomedical applications to energy storage devices and photoelectrochemical water splitting. Advanced Materials, 2011, 23(44): 5243-5249.

[21] Berner R A. Sedimentary pyrite formation. American Journal of Science, 1970, 268: 1-23.

[22] Cornell R M, Schwertmann U. The iron oxides: Structure, properties, reactions, occurrences and uses. Second Edition. 2004.

[23] Ali A, Zafar H, Zia M, et al. Synthesis, characterization, applications, and challenges of iron oxide nanoparticles. Nanotechnology, Science and Applications, 2016, 9: 49-67.

[24] Mohapatra M, Anand S. Synthesis and applications of nano-structured iron oxides/hydroxides: Review. International Journal of Engineering Science and Technology, 2010, 2(8): 127-146.

[25] Christensen A N. Hydrothermal preparation of goethite and hematite from amorphous Iron(III)hydroxide. Acta Chemica Scandinavica, 1968, 22: 1487-1490.

[26] Liu W, Wang Y, Ai Z, et al. Hydrothermal synthesis of FeS_2 as a high-efficiency Fenton reagent to degrade alachlor via superoxide-mediated Fe(II)/Fe(III)cycle. ACS Applied Materials & Interfaces, 2015, 7(51): 28534-28544.

[27] Geng F, Zhao Z, Cong H, et al. An environment-friendly microemulsion approach to α-FeOOH nanorods at room temperature. Materials Research Bulletin, 2006, 41(12): 2238-2243.

[28] Sayed F N, Polshettiwar V. Facile and sustainable synthesis of shaped iron oxide nanoparticles: Effect of iron precursor salts on the shapes of iron oxides. Scientific Reports, 2015, 5: 9733.

[29] Kakuta S, Numata T, Okayama T. Shape effects of goethite particles on their photocatalytic activity in the decomposition of acetaldehyde. Catalysis Science & Technology, 2014, 4(1): 164-169.

[30] Zhang H, Bayne M, Fernando S, et al. Size-dependent bandgap of nanogoethite. Journal of Physical Chemistry C, 2011, 115(36): 17704-17710.

[31] Zhou X, Lan J, Liu G, et al. Facet-mediated photodegradation of organic dye over hematite architectures by visible light. Angewandte Chemie International Edition, 2012, 51(1): 178-182.

[32] Dave P N, Chopda L V. Application of iron oxide nanomaterials for the removal of heavy metals. Journal of Nanotechnology, 2014, 2014: 1-14.

[33] Lata S, Samadder S R. Removal of arsenic from water using nano adsorbents and challenges: A review. Journal of Environmental Management, 2016, 166: 387-406.

[34] Hu J, Chen G, Lo I M. Removal and recovery of Cr(VI)from wastewater by maghemite nanoparticles. Water Research, 2005, 39(18): 4528-4536.

[35] Nassar N N. Rapid removal and recovery of Pb(II)from wastewater by magnetic nanoadsorbents. Journal of Hazardous Materials, 2010, 184(1-3): 538-546.

[36] Siddiqui S I, Chaudhry S A. Iron oxide and its modified forms as an adsorbent for arsenic removal: A comprehensive recent advancement. Process Safety and Environmental Protection, 2017, 111: 592-626.

[37] Zhao Y, Li J, Zhao L, et al. Synthesis of amidoxime-functionalized Fe_3O_4@SiO_2 core–shell magnetic microspheres for highly efficient sorption of U(VI). Chemical Engineering Journal, 2014, 235: 275-283.

[38] Li W, Troyer L D, Lee S S, et al. Engineering nanoscale iron oxides for uranyl sorption and separation: Optimization of particle core size and bilayer surface coatings. ACS Applied Materials & Interfaces, 2017, 9(15): 13163-13172.

[39] Liu H, Lu X, Li M, et al. Structural incorporation of manganese into goethite and its enhancement of Pb(II) adsorption. Environmental Science & Technology, 2018, 52(8): 4719-4727.

[40] Huang X, Hou X, Song F, et al. Facet-dependent Cr(VI)adsorption of hematite nanocrystals. Environmental Science & Technology 2016, 50(4): 1964-1972.

[41] Huang X, Hou X, Wang F, et al. Molecular-scale structures of uranyl surface complexes on hematite facets. Environmental Science: Nano, 2019, 6(3): 892-903.

[42] Huang X, Hou X, Zhang X, et al. Facet-dependent contaminant removal properties of hematite nanocrystals and their environmental implications. Environmental Science: Nano, 2018, 5(8): 1790-1806.

[43] Qin Y, Song F, Ai Z, et al. Protocatechuic acid promoted alachlor degradation in Fe(III)/H_2O_2 Fenton system. Environmental Science & Technology, 2015, 49(13): 7948-7956.

[44] Zhao K, Zhang L, Wang J, et al. Surface structure-dependent molecular oxygen activation of BiOCl single-crystalline nanosheets. Journal of the American Chemical Society, 2013, 135(42): 15750-15753.

[45] Mu Y, Ai Z, Zhang L. Phosphate shifted oxygen reduction pathway on Fe@Fe_2O_3 core-shell nanowires for enhanced reactive

oxygen species generation and aerobic 4-chlorophenol degradation. Environmental Science & Technology, 2017, 51(14): 8101-8109.

[46] Fang G D, Zhou D M, Dionysiou D D. Superoxide mediated production of hydroxyl radicals by magnetite nanoparticles: Demonstration in the degradation of 2-chlorobiphenyl. Journal of Hazardous Materials, 2013, 250-251: 68-75.

[47] Zhang P, Yuan S, Liao P. Mechanisms of hydroxyl radical production from abiotic oxidation of pyrite under acidic conditions. Geochimica et Cosmochimica Acta, 2016, 172: 444-457.

[48] Mishra M, Chun D-M. α-Fe_2O_3 as a photocatalytic material: A review. Applied Catalysis A: General, 2015, 498: 126-141.

[49] Khedr M H, Abdel Halim K S, Soliman N K. Synthesis and photocatalytic activity of nano-sized iron oxides. Materials Letters, 2009, 63(6-7): 598-601.

[50] Lan Q, Li F B, Sun C X, et al. Heterogeneous photodegradation of pentachlorophenol and iron cycling with goethite, hematite and oxalate under UVA illumination. Journal of Hazardous Materials, 2010, 174(1-3): 64-70.

[51] Chen N, Shang H, Tao S, et al. Visible light driven organic pollutants degradation with hydrothermally carbonized sewage sludge and oxalate via molecular oxygen activation. Environmental Science & Technology, 2018, 52(21): 12656-12666.

[52] Pignatello J J, Oliveros E, MacKay A. Advanced oxidation processes for organic contaminant destruction based on the Fenton reaction and related chemistry. Critical Reviews in Environmental Science and Technology, 2006, 36(1): 1-84.

[53] Matta R, Hanna K, Kone T, et al. Oxidation of 2, 4, 6-trinitrotoluene in the presence of different iron-bearing minerals at neutral pH. Chemical Engineering Journal, 2008, 144(3): 453-458.

[54] Hou X, Huang X, Jia F, et al. Hydroxylamine promoted goethite surface Fenton degradation of organic pollutants. Environmental Science & Technology, 2017, 51(9): 5118-5126.

[55] Huang X, Hou X, Jia F, et al. Ascorbate-promoted surface iron cycle for efficient heterogeneous Fenton alachlor degradation with hematite nanocrystals. ACS Applied Materials & Interfaces, 2017, 9(10): 8751-8758.

[56] Hou X, Huang X, Li M, et al. Fenton oxidation of organic contaminants with aquifer sediment activated by ascorbic acid. Chemical Engineering Journal, 2018, 348: 255-262.

[57] Jaafarzadeh N, Ghanbari F, Ahmadi M. Catalytic degradation of 2, 4-dichlorophenoxyacetic acid(2, 4-D)by nano-Fe_2O_3 activated peroxymonosulfate: Influential factors and mechanism determination. Chemosphere, 2017, 169: 568-576.

[58] Avetta P, Pensato A, Minella M, et al. Activation of persulfate by irradiated magnetite: Implications for the degradation of phenol under heterogeneous photo-Fenton-like conditions. Environmental Science & Technology, 2015, 49(2): 1043-1050.

[59] Rush J D, Zhao Z W, Bielski B H J. Reaction of ferrate(VI)/ferrate(V)with hydrogen peroxide and superoxide anion: A stopped-flow and premix pulse radiolysis study. Free Radical Research, 1996, 24(3): 187-198.

[60] Kamachi T, Kouno T, Yoshizawa K. Participation of multioxidants in the pH dependence of the reactivity of ferrate(VI). The Journal of Organic Chemistry, 2005, 70(11): 4380-4388.

[61] Sharma V K. Ferrate(VI)and ferrate(V)oxidation of organic compounds: Kinetics and mechanism. Coordination Chemistry Reviews, 2013, 257(2): 495-510.

[62] Sharma V K. Oxidation of inorganic contaminants by ferrates(VI, V, and IV)-kinetics and mechanisms: A review. Journal of Environmental Management, 2011, 92(4): 1051-1073.

[63] Sharma V K, Zboril R, Varma R S. Ferrates: Greener oxidants with multimodal action in water treatment technologies. Accounts of Chemical Research, 2015, 48(2): 182-191.

[64] Goff H, Murmann R K. Studies on mechanism of isotopic oxygen exchange and reduction of ferrate(VI)ion(FeO_4^{2-}). Journal of the American Chemical Society, 1971, 93(23): 6058-6065.

[65] Pestovsky O, Bakac A. Aqueous ferryl(IV)ion: Kinetics of oxygen atom transfer to substrates and oxo exchange with solvent water. Inorganic Chemistry, 2006, 45(2): 814-820.

[66] Sharma V K. Oxidation of inorganic compounds by ferrate(VI)and ferrate(V): One-electron and two-electron transfer steps. Environmental Science & Technology, 2010, 44(13): 5148-5152.

[67] Sharma V K, Cabelli D. Reduction of oxyiron(V)by sulfite and thiosulfate in aqueous solution. The Journal of Physical Chemistry A, 2009, 113(31): 8901-8906.

[68] Yang B, Ying G G, Zhao J L, et al. Removal of selected endocrine disrupting chemicals(EDCs)and pharmaceuticals and personal care products(PPCPs)during ferrate(VI)treatment of secondary wastewater effluents. Water Research, 2012, 46(7): 2194-2204.

[69] Manoli K, Nakhla G, Ray A K, et al. Enhanced oxidative transformation of organic contaminants by activation of ferrate(VI): Possible involvement of Fe-V/Fe-IV species. Chemical Engineering Journal, 2017, 307: 513-517.

[70] Feng M, Cizmas L, Wang Z, et al. Activation of ferrate(VI)by ammonia in oxidation of flumequine: Kinetics, transformation products, and antibacterial activity assessment. Chemical Engineering Journal, 2017, 323: 584-591.

[71] Feng M B, Cizmas L, Wang Z Y, et al. Synergistic effect of aqueous removal of fluoroquinolones by a combined use of peroxymonosulfate and ferrate(VI). Chemosphere, 2017, 177: 144-148.

[72] Manoli K, Nakhla G, Feng M B, et al. Silica gel-enhanced oxidation of caffeine by ferrate(VI). Chemical Engineering Journal, 2017, 330: 987-994.

[73] Sharma V K, O'Connor D B, Cabelli D E. Sequential one-electron reduction of Fe(V)to Fe(III)by cyanide in alkaline medium. The Journal of Physical Chemistry B, 2001, 105(46): 11529-11532.

[74] Terryn R J, Huerta-Aguilar C A, Baum J C, et al. Fe-VI, Fe-V, and Fe-IV oxidation of cyanide: Elucidating the mechanism using density functional theory calculations. Chemical Engineering Journal, 2017, 330: 1272-1278.

[75] Lee Y, Yoon J, Von Gunten U. Kinetics of the oxidation of phenols and phenolic endocrine disruptors during water treatment with ferrate(Fe(VI)). Environmental Science & Technology, 2005, 39(22): 8978-8984.

[76] Dong H Y, Qiang Z M, Lian J F, et al. Promoted oxidation of diclofenac with ferrate(Fe(VI)): Role of ABTS as the electron shuttle. Journal of Hazardous Materials, 2017, 336: 65-70.

[77] Lee Y, Kissner R, von Gunten U. Reaction of ferrate(VI)with ABTS and self-decay of ferrate(VI): Kinetics and mechanisms. Environmental Science & Technology, 2014, 48(9): 5154-5162.

[78] Feng M B, Jinadatha C, McDonald T J, et al. Accelerated oxidation of organic contaminants by ferrate(VI): The overlooked role of reducing additives. Environmental Science & Technology, 2018, 52(19): 11319-11327.

[79] Sun S, Jiang J, Qiu L, et al. Activation of ferrate by carbon nanotube for enhanced degradation of bromophenols: Kinetics, products, and involvement of Fe(V)/Fe(IV). Water Research, 2019, 156: 1-8.

[80] Sun B, Guan X H, Fang J Y, et al. Activation of manganese oxidants with bisulfite for enhanced oxidation of organic contaminants: The involvement of Mn(III). Environmental Science & Technology, 2015, 49(20): 12414-12421.

[81] Sun B, Li D, Linghu W S, et al. Degradation of ciprofloxacin by manganese(III)intermediate: Insight into the potential application of permanganate/bisulfite process. Chemical Engineering Journal, 2018, 339: 144-152.

[82] Sun B, Bao Q Q, Guan X H. Critical role of oxygen for rapid degradation of organic contaminants in permanganate/bisulfite process. Journal of Hazardous Materials, 2018, 352: 157-164.

[83] Zhang J, Zhu L, Shi Z Y, et al. Rapid removal of organic pollutants by activation sulfite with ferrate. Chemosphere, 2017, 186: 576-579.

[84] Sun S F, Pang S Y, Jiang J, et al. The combination of ferrate(VI)and sulfite as a novel advanced oxidation process for enhanced degradation of organic contaminants. Chemical Engineering Journal, 2018, 333: 11-19.

[85] Feng M B, Sharma V K. Enhanced oxidation of antibiotics by ferrate(VI)-sulfur(IV)system: Elucidating multi-oxidant mechanism. Chemical Engineering Journal, 2018, 341: 137-145.

[86] Shao B B, Dong H Y, Sun B, et al. Role of ferrate(IV)and ferrate(V)in activating ferrate(VI)by calcium sulfite for enhanced oxidation of organic contaminants. Environmental Science & Technology, 2019, 53, (2): 894-902.

[87] Jiang Y J, Goodwill J E, Tobiason J E, et al. Bromide oxidation by ferrate(VI): The formation of active bromine and bromate. Water Research, 2016, 96: 188-197.

[88] Huang X, Deng Y, Liu S, et al. Formation of bromate during ferrate(VI)oxidation of bromide in water. Chemosphere, 2016, 155: 528-533.

[89] Lee Y, von Gunten U. Oxidative transformation of micropollutants during municipal wastewater treatment: Comparison of kinetic aspects of selective(chlorine, chlorine dioxide, ferrate(VI), and ozone)and non-selective oxidants(hydroxyl radical). Water Research, 2010, 44(2): 555-566.

[90] Criquet J, Allard S, Salhi E, et al. Iodate and iodo-trihalomethane formation during chlorination of iodide-containing waters: Role of bromide. Environmental Science & Technology, 2012, 46(13): 7350-7357.

[91] Liu S G, Li Z L, Dong H Y, et al. Formation of iodo-trihalomethanes, iodo-acetic acids, and iodo-acetamides during chloramination of iodide-containing waters: Factors influencing formation and reaction pathways. Journal of Hazardous Materials, 2017, 321: 28-36.

[92] Zhao X D, Ma J, von Gunten U. Reactions of hypoiodous acid with model compounds and the formation of iodoform in absence/presence of permanganate. Water Research, 2017, 119: 126-135.

[93] Zhang M S, Xu B, Wang Z, et al. Formation of iodinated trihalomethanes after ferrate pre-oxidation during chlorination and chloramination of iodide-containing water. Journal of the Taiwan Institute of Chemical Engineers, 2016, 60: 453-459.

[94] Wang X S, Liu Y L, Huang Z S, et al. Rapid oxidation of iodide and hypoiodous acid with ferrate and no formation of iodoform and monoiodoacetic acid in the ferrate/I-/HA system. Water Research, 2018, 144: 592-602.

[95] Shin J, von Gunten U, Reckhow D A, et al. Reactions of ferrate(VI)with iodide and hypoiodous acid: Kinetics, pathways, and implications for the fate of iodine during water treatment. Environmental Science & Technology, 2018, 52(13): 7458-7467.

[96] Dong H Y, Qiang Z M, Liu S G, et al. Oxidation of iopamidol with ferrate(Fe(VI)): Kinetics and formation of toxic iodinated disinfection by-products. Water Research 2018, 130: 200-207.

[97] Goodwill J E, Jiang Y J, Reckhow D A, et al. Characterization of particles from ferrate preoxidation. Environmental Science & Technology, 2015, 49(8): 4955-4962.

[98] Lv D Y, Zheng L, Zhang H Q, et al. Coagulation of colloidal particles with ferrate(VI). Environmental Science-Water Research & Technology, 2018, 4(5): 701-710.

[99] Ma J, Liu W. Effectiveness of ferrate(VI)preoxidation in enhancing the coagulation of surface waters. Water Research, 2002, 36(20): 4959-4962.

[100] Prucek R, Tucek J, Kolarik J, et al. Ferrate(VI)-induced arsenite and arsenate removal by *in situ* structural incorporation into magnetic iron(III)oxide nanoparticles. Environmental Science & Technology, 2013, 47(7): 3283-3292.

[101] Prucek R, Tucek J, Kolarik J, et al. Ferrate(VI)-prompted removal of metals in aqueous media: Mechanistic delineation of enhanced efficiency via metal entrenchment in magnetic oxides. Environmental Science & Technology, 2015, 49(4): 2319-2327.

[102] Yang T, Wang L, Liu Y L, et al. Removal of organoarsenic with ferrate and ferrate resultant nanoparticles: Oxidation and adsorption. Environmental Science & Technology, 2018, 52(22): 13325-13335.

[103] Yang T, Liu Y L, Wang L, et al. Highly effective oxidation of roxarsone by ferrate and simultaneous arsenic removal with *in situ* formed ferric nanoparticles. Water Research, 2018, 147: 321-330.

[104] Kappler A, Straub K L. Geomicrobiological cycling of iron. Reviews in Mineralogy & Geochemistry, 2005, 59(1): 85-108.

[105] Emerson D, Fleming E J, McBeth J M. Iron-oxidizing bacteria: An environmental and genomic perspective. Annual Review

of Microbiology, 2010, 64(1): 561-583.

[106] Jain B, Singh A K, Kim H, et al. Treatment of organic pollutants by homogeneous and heterogeneous Fenton reaction processes. Environmental Chemistry Letters, 2018, 16(3): 947-967.

[107] Hedrich S, Schlomann M, Johnson D B. The iron-oxidizing proteobacteria. Microbiology, 2011, 157(6): 1551-1564.

[108] Melton E D, Rudolph A, Behrens S, et al. Influence of nutrient concentrations on MPN quantification and enrichment of nitrate-reducing Fe(II)-oxidizing and Fe(III)-reducing bacteria from littoral freshwater lake sediments. Geomicrobiology Journal, 2014, 31(9): 788-801.

[109] Kuypers M M M, Marchant H K, Kartal B. The microbial nitrogen-cycling network. Nature Reviews Microbiology, 2018, 16(5): 263-276.

[110] Zumft W G. Cell biology and molecular basis of denitrification. Microbiology and Molecular Biology Reviews, 1997, 61(4): 533-616.

[111] Hayatsu M, Tago K, Saito M. Various players in the nitrogen cycle: Diversity and functions of the microorganisms involved in nitrification and denitrification. Soil Science and Plant Nutrition, 2008, 54(1): 33-45.

[112] Castresana J, Saraste M. Evolution of energetic metabolism: the respiration-early hypothesis. Trends in Biochemical Sciences, 1995, 20(11): 443-448.

[113] Ducluzeau A L, Lis R V, Duval S, et al. Nitschke W. Was nitric oxide the first deep electron sink? Trends in Biochemical Sciences, 2009, 34(1): 9-15.

[114] Picardal F. Abiotic and microbial interactions during anaerobic transformations of Fe(II) and NO_x. Frontiers in Microbiology, 2012, 3: 112-8.

[115] Klueglein N, Kappler A. Abiotic oxidation of Fe(II) by reactive nitrogen species in cultures of the nitrate-reducing Fe(II)oxidizer *Acidovorax* sp. BoFeN1 - questioning the existence of enzymatic Fe(II)oxidation. Geobiology, 2013, 11(2): 180-190.

[116] Hafenbradl D, Keller M, Dirmeier R, et al. *Ferroglobus placidus gen* nov, sp nov, a novel hyperthermophilic archaeum that oxidizes Fe^{2+} at neutral pH under anoxic conditions. Archives of Microbiology, 1996, 166(5): 308-314.

[117] Emerson D. Biogeochemistry and microbiology of microaerobic Fe(II)oxidation. Biochemical Society Transactions, 2012, 40(6): 1211-1216.

[118] Sorokina A Y, Chernousova E Y, Dubinina G A. *Hoeflea siderophila* sp. nov., a new neutrophilic iron-oxidizing bacterium. Microbiology, 2012, 81(1): 59-66.

[119] Michiels Céline C, Darchambeau F, Roland F A E, et al. Iron-dependent nitrogen cycling in a ferruginous lake and the nutrient status of Proterozoic oceans. Nature Geoscience, 2017, 10(3): 217-221.

[120] Ratering S, Schnell S. Nitrate-dependent iron(II)oxidation in paddy soil. Environmental Microbiology, 2001, 3(2): 100-109.

[121] Weiss J V, Emerson D, Megonigal J P. Geochemical control of microbial Fe(III)reduction potential in wetlands: Comparison of the rhizosphere to non-rhizosphere soil. FEMS Microbiology Ecology, 2004, 48(1): 89-100.

[122] Weber K A, Urrutia M M, Churchill P F, et al. Anaerobic redox cycling of iron by freshwater sediment microorganisms. Environmental Microbiology, 2006, 8(1): 100-113.

[123] Wang H, Hu C, Han L, et al. Effects of microbial cycling of Fe(II)/Fe(III) and Fe/N on cast iron corrosion in simulated drinking water distribution systems. Corrosion Science, 2015, 100: 599-606.

[124] Emerson D, Moyer C L. Neutrophilic Fe-oxidizing bacteria are abundant at the Loihi Seamount Hydrothermal Vents and play a major role in Fe oxide deposition. Vaccine Research, 2002, 68(1): 3085-3093.

[125] Alex P, Pearson V K, Schwenzer S P, et al. Nitrate-dependent iron oxidation: A potential mars metabolism. Frontiers in Microbiology, 2018, 9: 513-527.

[126] Straub K L, Benz M, Schink B, et al. Anaerobic, nitrate-dependent microbial oxidation of ferrous iron. Applied and Environmental Microbiology, 1996, 62(4): 1458-1460.

[127] Li B, Tian C, Zhang D, et al. Anaerobic nitrate-dependent iron(II)oxidation by a novel autotrophic bacterium, *Citrobacter freundii* Strain PXL1. Geomicrobiology Journal, 2014, 31(2): 138-144.

[128] Kopf S H, Henny C, Newman D K. Ligand-enhanced abiotic iron oxidation and the effects of chemical versus biological iron cycling in anoxic environments. Environmental Science & Technology, 2013, 47(6): 2602-2611.

[129] Kumaraswamy R, Sjollema K, Kuenen G, et al. Nitrate-dependent $[Fe(II)EDTA]^{2-}$ oxidation by *Paracoccus ferrooxidans* sp. nov., isolated from a denitrifying bioreactor. Systematic and Applied Microbiology, 2006, 29(4): 276-286.

[130] Achenbach L A, Weber K A, Pollock J, et al. Anaerobic nitrate-dependent iron(II)bio-oxidation by a novel lithoautotrophic betaproteobacterium, strain 2002. Applied and Environmental Microbiology, 2006, 72(1): 686-694.

[131] Zhao L, Dong H, Kukkadapu R, et al. Biological oxidation of Fe(II)in reduced nontronite coupled with nitrate reduction by *Pseudogulbenkiania* sp Strain 2002. Geochimica et Cosmochimica Acta, 2013, 119: 231-247.

[132] Zhao L, Dong H, Edelmann R E, et al. Coupling of Fe(II)oxidation in illite with nitrate reduction and its role in clay mineral transformation. Geochimica et Cosmochimica Acta, 2017, 200: 353-366.

[133] Straub K L, Buchholz-Cleven B E. Enumeration and detection of anaerobic ferrous iron-oxidizing, nitrate-reducing bacteria from diverse European sediments. Applied and Environmental Microbiology, 1998, 64(12): 4846-4856.

[134] Chaudhuri S K, Lack J G, Coates J D. Biogenic magnetite formation through anaerobic biooxidation of Fe(II). Applied and Environmental Microbiology, 2001, 67(6): 2844-2848.

[135] Chakraborty A, Picardal F. Neutrophilic, nitrate-dependent, Fe(II)oxidation by a *Dechloromonas* species. World Journal of Microbiology & Biotechnology, 2013, 29(4): 617-623.

[136] Kappler A, Schink B, Newman D K. Fe(III)mineral formation and cell encrustation by the nitrate-dependent Fe(II)-oxidizer strain BoFeN1. Geobiology, 2005, 3(4): 235-245.

[137] Straub K L, Schber W A, Buchholz-Cleven B E E, Schink B. Diversity of ferrous iron-oxidizing, nitrate-reducing bacteria and their involvement in oxygen-independent iron cycling. Geomicrobiology Journal, 2004, 21(6): 371-378.

[138] Chakraborty A, Picardal F. Induction of nitrate-dependent Fe(II)oxidation by Fe(II)in *Dechloromonas* sp strain UWNR4 and *Acidovorax* sp strain 2AN. Applied Microbiology and Biotechnology, 2013, 79(2): 748-752.

[139] Carlson H K, Clark I C, Blazewicz S J, et al. Fe(II)oxidation is an innate capability of nitrate-reducing bacteria that involves abiotic and biotic reactions. Journal of Bacteriology, 2013, 195(14): 3260-3268.

[140] Byrne-Bailey K G, Weber K A, Chair A H, et al. Completed genome sequence of the anaerobic iron-oxidizing cacterium *Acidovorax ebreus* strain TPSY. Journal of Bacteriology, 2010, 192(5): 1475-1476.

[141] Senko J M, Dewers T A, Krumholz L R. Effect of oxidation rate and Fe(II)state on microbial nitrate-dependent Fe(III)mineral formation. Applied and Environmental Microbiology, 2005, 71(11): 7172-7177.

[142] Coby A J, Picardal F, Shelobolina E, et al. Repeated anaerobic microbial redox cycling of iron. Applied and Environmental Microbiology, 2011, 77(17): 6036-6042.

[143] Muehe E M, Gerhardt S, Schink B, et al. Ecophysiology and the energetic benefit of mixotrophic Fe(II)oxidation by various strains of nitrate-reducing bacteria. FEMS Microbiology Ecology, 2009, 70(3): 335-343.

[144] Nicole K, Fabian Z, York-Dieter S, et al. Potential role of nitrite for abiotic Fe(II)oxidation and cell encrustation during nitrate reduction by denitrifying bacteria. Applied and Environmental Microbiology, 2014, 80(3): 1051-1061.

[145] Benz M, Brune A, Schink B. Anaerobic and aerobic oxidation of ferrous iron at neutral pH by chemoheterotrophic nitrate-reducing bacteria. Archives of Microbiology, 1998, 169(2): 159-165.

[146] Li S, Li X, Li F. Fe(II)oxidation and nitrate reduction by a denitrifying bacterium, Pseudomonas stutzeri LS-2, isolated from

paddy soil. Journal of Soils and Sediments, 2018, 18(4): 1668-1678.

[147] Melton E D, Schmidt C, Kappler A. Microbial iron(II)oxidation in littoral freshwater lake sediment: The potential for competition between phototrophic vs. nitrate-reducing iron(II)-oxidizers. Frontiers in Microbiology, 2012, 3: 197-209.

[148] Li X, Zhang W, Liu T, et al. Changes in the composition and diversity of microbial communities during anaerobic nitrate reduction and Fe(II)oxidation at circumneutral pH in paddy soil. Soil Biology & Biochemistry, 2016, 94: 70-79.

[149] Hu M, Chen P, Sun W, et al. A novel organotrophic nitrate-reducing Fe(II)-oxidizing bacterium isolated from paddy soil and draft genome sequencing indicate its metabolic versatility. Rsc Advances, 2017, 7(89): 56611-56620.

[150] Weber K A, Picardal F W, Roden E E. Microbially catalyzed nitrate-dependent oxidation of biogenic solid-phase Fe(II)compounds. Environmental Science & Technology, 2001, 35(8): 1644-1650.

[151] Blöthe M, Roden E E. Composition and activity of an autotrophic Fe(II)-oxidizing, nitrate-reducing enrichment culture. Applied and Environmental Microbiology, 2009, 75(21): 6937-6940.

[152] He S, Tominski C, Kappler A, et al. Metagenomic analyses of the autotrophic Fe(II)-oxidizing, nitrate-reducing enrichment culture KS. Applied and Environmental Microbiology, 2016, 82(9): 2656-2668.

[153] Kappler A, Emerson D, Gralnick J, et al. Geomicrobiology of iron. Boca Raton, FL, USA: CRC Press, 2015.

[154] Bosch J, Lee K Y, Jordan G, et al. Anaerobic, nitrate-dependent oxidation of pyrite nanoparticles by Thiobacillus denitrificans. Environmental Science & Technology, 2012, 46(4): 2095-2101.

[155] Mattes A, Gould D, Taupp M, et al. A novel autotrophic bacterium isolated from an engineered wetland system links nitrate-coupled iron oxidation to the removal of As, Zn and S. Water, Air, & Soil Pollution, 2013, 224(4): 1490-1505.

[156] Zhang H, Wang H, Yang K, et al. Nitrate removal by a novel autotrophic denitrifier (*Microbacterium* sp.) using Fe(II)as electron donor. Annals of Microbiology, 2014, 65(2): 1069-1078.

[157] Su J, Shao S, Huang T, et al. Anaerobic nitrate-dependent iron(II)oxidation by a novel autotrophic bacterium, *Pseudomonas* sp. SZF15. Journal of Environmental Chemical Engineering, 2015, 3(3): 2187-2193.

[158] Zhang H, Wang H, Yang K, et al. Autotrophic denitrification with anaerobic Fe^{2+} oxidation by a novel *Pseudomonas* sp W1. Water Science and Technology, 2015, 71(7): 1081-1087.

[159] Chao T T, Kroontje W. Inorganic nitrogen transformations through the oxidation and reduction of Iron 1. Soil Science Society of America Journal, 1966, 30(2): 193-196.

[160] Postma D. Kinetics of nitrate reduction by detrital Fe(II)-silicates. Geochimica et Cosmochimica Acta, 1990, 54(3): 903-908.

[161] Hansen H C B, Koch C B, Krogh H N, et al. Abiotic nitrate reduction to ammonium: Key role of green rust. Environmental Science & Technology, 1996, 30(6): 2053-2056.

[162] Cleemput O V, Samater A H. Nitrite in soils: Accumulation and role in the formation of gaseous N compounds. Fertilizer Research, 1996, 45(1): 81-89.

[163] Philippot L, Hallin S, Schloter M. Ecology of denitrifying prokaryotes in agricultural soil. Advances in Agronomy, 2007, 96: 249-305.

[164] Kraft B, Tegetmeyer H E, Sharma R, et al. The environmental controls that govern the end product of bacterial nitrate respiration. Science, 2014, 345(6197): 676-679.

[165] Lundberg J O, Weitzberg E, Gladwin M T. The nitrate-nitrite-nitric oxide pathway in physiology and therapeutics. Nature Reviews Drug Discovery, 2008, 7(2): 156-167.

[166] Moreno-Vivian C, Cabello P, Martinez-Luque M, et al. Prokaryotic nitrate reduction: Molecular properties and functional distinction among bacterial nitrate reductases. Journal of Bacteriology, 1999, 181(21): 6573-6584.

[167] Preisler A, de Beer D, Lichtschlag A, et al, Jørgensen B B. Biological and chemical sulfide oxidation in a Beggiatoa inhabited marine sediment. ISME Journal, 2007, 1(4): 341-353.

[168] Tsementzi D, Wu J, Deutsch S, et al. SAR11 bacteria linked to ocean anoxia and nitrogen loss. Nature, 2016, 536(7615): 179-195.

[169] Betlach M R, Tiedje J M. Kinetic explanation for accumulation of nitrite, nitric oxide, and nitrous oxide during bacterial denitrification. Applied and Environmental Microbiology, 1981, 42(6): 1074-1084.

[170] Constantin H, Fick M. Influence of C-sources on the denitrification rate of a high-nitrate concentrated industrial wastewater. Water Research, 1997, 31(3): 583-589.

[171] Glass C, Silverstein J A. Denitrification kinetics of high nitrate concentration water: pH effect on inhibition and nitrite accumulation. Water Research, 1998, 32(3): 831-839.

[172] Lam P, Lavik G, Jensen M M, et al. Revising the nitrogen cycle in the Peruvian oxygen minimum zone. Proceedings of the National Academy of Sciences of the United States of America, 2009, 106(12): 4752-4757.

[173] Kampschreur M J, Kleerebezem R, de Vet W W, et al. Reduced iron induced nitric oxide and nitrous oxide emission. Water Research, 2011, 45(18): 5945-5952.

[174] Bristow L A, Callbeck C M, Larsen M, et al. N_2 production rates limited by nitrite availability in the Bay of Bengal oxygen minimum zone. Nature Geoscience, 2017, 10(1): 24-29.

[175] Weber K A, Achenbach L A, Coates J D. Microorganisms pumping iron: Anaerobic microbial iron oxidation and reduction. Nature Reviews Microbiology, 2006, 4(10): 752-764.

[176] Moraghan J T, Buresh R J. Chemical reduction of nitrite and nitrous oxide by ferrous iron. Soil Science Society of America Journal, 1976, 41(1): 47-50.

[177] Nelson D W, Bremner J M. Role of soil minerals and metallic cations in nitrite decomposition and chemodenitrification in soils. Soil Biology & Biochemistry, 1970, 2(1): 1-8.

[178] Melton E D, Swanner E D, Behrens S, et al. The interplay of microbially mediated and abiotic reactions in the biogeochemical Fe cycle. Nature Reviews Microbiology, 2014, 12(12): 797-808.

[179] Hansen H C B, Borggaard O K, Sørensen J. Evaluation of the free energy of formation of Fe(II)-Fe(III)hydroxide-sulphate (green rust) and its reduction of nitrite. Geochimica et Cosmochimica Acta, 1994, 58(12): 2599-2608.

[180] Pearsall K A, Bonner F T. Aqueous nitrosyliron(II)chemistry. 1. Reduction of nitrite and nitric oxide by iron(II)and (Trioxodinitrato) iron(II)in acetate buffer. Intermediacy of nitrosyl hydride. Inorganic chemistry, 1982, 21: 1978-1985.

[181] Pearsall K A, Bonner F T. Aqueous nitrosyliron(II)chemistry. 2. Kinetics and mechanism of nitric oxide reduction. The dinitrosyl complex. Chemischer Informationsdienst, 1982, 13(33): 1973-1978.

[182] Cleemput O V, Baert L. Nitrite stability influenced by iron compounds. Soil Biology & Biochemistry, 1983, 15(2): 137-140.

[183] Sørensen J, Thorling L. Stimulation by lepidocrocite (γ-FeOOH) of Fe(II)-dependent nitrite reduction. Geochimica et Cosmochimica Acta, 1991, 55(5): 1289-1294.

[184] Ottley C J, Davison W, Edmunds W M. Chemical catalysis of nitrate reduction by iron(II). Geochimica et Cosmochimica Acta, 1997, 61(9): 1819-1828.

[185] Rakshit S, Matocha C J, Coyne M S. Nitrite reduction by siderite. Soil Science Society of America Journal, 2008, 72(4): 1070-1077.

[186] Tai Y L, Dempsey B A. Nitrite reduction with hydrous ferric oxide and Fe(II): Stoichiometry, rate, and mechanism. Water Research, 2008, 43(2): 546-552.

[187] Dhakal P, Matocha C J, Huggins F E, Vandiviere M M. Nitrite reactivity with magnetite. Environmental Science & Technology 2013, 47(12): 6206-6213.

[188] Etique M, Jorand F P, Zegeye A, et al. Abiotic process for Fe(II)oxidation and green rust mineralization driven by a heterotrophic nitrate reducing bacteria (*Klebsiella mobilis*). Environmental Science & Technology, 2014, 48(7): 3742-3751.

[189] Jones L C, Brian P, Lezama Pacheco J S, et al. Stable isotopes and iron oxide mineral products as markers of chemodenitrification. Environmental Science & Technology, 2015, 49(6): 3444-3452.

[190] Buchwald C, Grabb K, Hansel C M, et al. Constraining the role of iron in environmental nitrogen transformations: Dual stable isotope systematics of abiotic NO_2- reduction by Fe(II) and its production of N_2O. Geochimica et Cosmochimica Acta, 2016, 186: 1-12.

[191] Grabb KC, Buchwald C, Hansel C, et al. A dual nitrite isotopic investigation of chemodenitrification by mineral-associated Fe(II) and its production of nitrous oxide. Geochimica et Cosmochimica Acta, 2017, 196: 388-402.

[192] Ostrom N E, Gandhi H, Trubl G, et al. Chemodenitrification in the cryoecosystem of Lake Vida, Victoria Valley, Antarctica. Geobiology, 2016, 14(6): 575-587.

[193] Wankel S D, Ziebis W, Buchwald C, et al. Evidence for fungal and chemodenitrification based N_2O flux from nitrogen impacted coastal sediments. Nature Communications, 2017, 8: 15595.

[194] Chen D, Liu T, Li X, et al. Biological and chemical processes of microbially mediated nitrate-reducing Fe(II) oxidation by *Pseudogulbenkiania* sp. strain 2002. Chemical Geology, 2018, 476: 59-69.

[195] Liu T, Chen D, Luo X, et al. Microbially mediated nitrate-reducing Fe(II) oxidation: Quantification of chemodenitrification and biological reactions. Geochimica et Cosmochimica Acta, 2018, in press.

[196] Nordhoff M, Tominski C, Halama M, et al. Insights into nitrate-reducing Fe(II) oxidation mechanisms through analysis of cell-mineral associations, cell encrustation, and mineralogy in the chemolithoautotrophic enrichment culture KS. Applied and Environmental Microbiology, 2017, 83(13): 1-19.

[197] Beller H R, Peng Z, Legler T C, et al. Genome-enabled studies of anaerobic, nitrate-dependent iron oxidation in the chemolithoautotrophic bacterium Thiobacillus denitrificans. Frontiers in Microbiology, 2013, 4: 249-265.

[198] Schaedler F, Lockwood C, Lueder U, et al. Microbially mediated coupling of Fe and N cycles by nitrate-reducing Fe(II)-oxidizing bacteria in littoral freshwater sediments. Applied and Environmental Microbiology, 2017, 84(2): 11-21.

[199] Rentz J A, Kraiya C, Luther G W, et al. Control of ferrous iron oxidation within circumneutral microbial iron mats by cellular activity and autocatalysis. Environmental Science & Technology, 2007, 41(17): 6084-6089.

[200] Shelobolina E, Konishi H, Xu H, et al. Isolation of phyllosilicate–iron redox cycling microorganisms from an illite–smectite rich hydromorphic soil. Frontiers in Microbiology, 2012, 3: 134-143.

[201] Schmid G, Zeitvogel F, Hao L, et al. 3D analysis of bacterial cell-(iron)mineral aggregates formed during Fe(II) oxidation by the nitrate-reducing *Acidovorax* sp. strain BoFeN1 using complementary microscopy tomography approaches. Geobiology, 2014, 12(4): 340-361.

[202] Jamieson J, Prommer H, Kaksonen A, et al. Identifying and quantifying the intermediate processes during nitrate-dependent iron(II) oxidation. Environmental Science & Technology, 2018, 52(10): 5771-5781.

[203] Peng C, Sundman A, Bryce C, et al. Oxidation of Fe(II) organic matter complexes in the presence of the mixotrophic nitrate-reducing Fe(II)-oxidizing bacterium Acidovorax sp BoFeN1. Environmental Science & Technology, 2018, 52(10): 5753-5763.

[204] Einsle O, Messerschmidt A, Huber R, et al. Mechanism of the si9-electron reduction of nitrite to ammonia by cytochrome c nitrite reductase. Journal of the American Chemical Society, 2002, 124(39): 11737-11745.

[205] Bird L J, Bonnefoy V, Newman D K. Bioenergetic challenges of microbial iron metabolisms. Trends in microbiology, 2011, 19(7): 330-340.

[206] Carlson H K, Clark I C, Melnyk R A, et al. Toward a mechanistic understanding of anaerobic nitrate-dependent iron oxidation: Balancing electron uptake and detoxification. Frontiers in Microbiology, 2012, 3: 57-62.

[207] Han R, Li F, Liu T, et al. Effects of incubation conditions on Cr(VI) reduction by c-type cytochromes in intact *Shewanella*

oneidensis MR-1 cells. Frontiers in Microbiology, 2016, 7: 746-758.

[208] Han R, Li X, Wu Y, et al. *In situ* spectral kinetics of quinone reduction by c-type cytochromes in intact *Shewanella oneidensis* MR-1 cells. Colloids and Surfaces A, 2017, 520: 505-513.

[209] Liu T, Wang Y, Li X, et al. Redox dynamics and equilibria of c-type cytochromes in the presence of Fe(II)under anoxic conditions: Insights into enzymatic iron oxidation. Chemical Geology, 2017, 468: 97-104.

[210] Liu J, Wang Z, Belchik S M, et al. Identification and characterization of MtoA: *a decaheme c-type* cytochrome of the neutrophilic Fe(II)-oxidizing bacterium *Sideroxydans lithotrophicus* ES-1. Frontiers in Microbiology 2012, 3: 37-48.

[211] He S, Barco R A, Emerson D, et al. Comparative genomic analysis of neutrophilic iron(II)oxidizer genomes for candidate genes in extracellular electron transfer. Frontiers in Microbiology, 2017, 8: 1584-1601.

[212] Ilbert M, Bonnefoy V. Insight into the evolution of the iron oxidation pathways. Biochimica et Biophysica Acta, 2013, 1827(2): 161-175.

[213] Ishii S, Joikai K, Otsuka S, et al. Denitrification and nitrate-dependent Fe(II)oxidation in various Pseudogulbenkiania strains. Microbes and Environments, 2016, 31(3): 293-298.

[214] Laufer K, Røy H, Jørgensen B B, et al. Evidence for the existence of autotrophic nitrate-reducing Fe(II)-oxidizing bacteria in marine coastal sediment. Applied and Environmental Microbiology, 2016, 82(20): 6120-6131.

[215] Weber K A, Hedrick D B, Peacock A D, et al. Physiological and taxonomic description of the novel autotrophic, metal oxidizing bacterium, *Pseudogulbenkiania* sp strain, 2002. Applied and Environmental Microbiology, 2009, 83(3): 555-565.

[216] David E, Field E K, Olga C, et al. Comparative genomics of freshwater Fe-oxidizing bacteria: implications for physiology, ecology, and systematics. Frontiers in Microbiology, 2013, 4: 254-271.

[217] Luo X, Wu Y, Li X, et al. The *in situ* spectral methods for examining redox status of c-type cytochromes in metal-reducing/oxidizing bacteria. Acta Geochimica, 2017, 36(3): 544-547.

[218] Liu T, Wu Y, Li F, et al. Rapid redox processes of c-type cytochromes in a living cell suspension of *Shewanella oneidensis* MR-1. Chemistryselect, 2017, 2(3): 1008-1012.

[219] Blake Ii R C, Griff M N. *In situ* spectroscopy on intact *Leptospirillum ferrooxidans* reveals that reduced cytochrome 579 is an obligatory intermediate in the aerobic iron respiratory chain. Frontiers in Microbiology, 2012, 3: 136-146.

[220] Blake Ii R C, Anthony M D, Bates J D, et al. *In situ* spectroscopy reveals that microorganisms in different phyla use different electron transfer biomolecules to respire aerobically on soluble iron. Frontiers in Microbiology, 2016, 7: 1963-1972.

[221] Luo X, Wu Y, Liu T, et al. Quantifying redox dynamics of c-type cytochromes in a living cell suspension of dissimilatory metal-reducing bacteria. Analytical Sciences, 2018, in press.

[222] Popovic D M, Leontyev I V, Beech D G, Stuchebrukhov A A.On similarity of cytochrome c oxidases in different organisms. Proteins: Structure Function and Bioinformatics, 2010, 78(12): 2691-2698.

[223] Jiao Y, Newman D K. The pio operon is essential for phototrophic Fe(II)oxidation in *Rhodopseudomonas palustris* TIE-1. Journal of Bacteriology, 2007, 189(5): 1765-1773.

[224] Croal L R, Jiao Y, Newman D K. The fox operon from Rhodobacter strain SW2 promotes phototrophic Fe(II)oxidation in *Rhodobacter capsulatus* SB1003. Journal of Bacteriology, 2006, 189(5): 1774-1782.

[225] Means J L, Crerar D A, Duguid J O. Migration of radioactive wastes: radionuclide mobilization by complexing agents. Science, 1978, 200(4349): 1477-1481.

[226] Means J L, Crerar D A, Borcsik M P, et al. Adsorption of Co and selected actinides by Mn and Fe oxides in soils and sediments. Geochimica et Cosmochimica Acta, 1978, 42(12): 1763-1773.

[227] Lack J G, Chaudhuri S K, Kelly S D, et al. Immobilization of radionuclides and heavy metals through anaerobic bio-oxidation of Fe(II). Applied and Environmental Microbiology, 2002, 68(6): 2704-2710.

[228] Senn D B, Hemond H F. Nitrate controls on iron and arsenic in an urban lake. Science, 2002, 296(5577): 2373-2376.

[229] Hoehn R, Isenbeck-Schroeter M, Kent D B, et al. Tracer test with As(V)under variable redox conditions controlling arsenic transport in the presence of elevated ferrous iron concentrations. Journal of Contaminant Hydrology, 2006, 88(1-2): 36-54.

[230] Chen X P, Zhu Y G, Hong M N, et al. Effects of different forms of nitrogen fertilizers on arsenic uptake by rice plants. Environmental Toxicology and Chemistry, 2008, 27(4): 881-887.

[231] Sun W J, Sierraalvarez R, Milner L, et al. Arsenite and ferrous iron oxidation linked to chemolithotrophic denitrification for the immobilization of arsenic in anoxic environments. Environmental Science & Technology, 2009, 43(17): 6585-6591.

[232] Hohmann C, Winkler E, Morin G, et al. Anaerobic Fe(II)-oxidizing bacteria show as resistance and immobilize as during Fe(III)mineral precipitation. Environmental Science & Technology, 2010, 44(1): 94-101.

[233] Li B, Pan X, Zhang D, et al. Anaerobic nitrate reduction with oxidation of Fe(II)by *Citrobacter freundii* strain PXL1: A potential candidate for simultaneous removal of As and nitrate from groundwater. Ecological Engineering, 2015, 77: 196-201.

[234] Wang X, Liu T, Li F, et al. Effects of simultaneous application of ferrous iron and nitrate on arsenic accumulation in rice grown in contaminated paddy soil. ACS Earth and Space Chemistry. 2018, 2(2): 103-111.

[235] Zhu-Barker X, Cavazos A R, Ostrom N E, et al. The importance of abiotic reactions for nitrous oxide production. Biogeochemistry, 2015, 126(3): 251-267.

[236] Davidson E A, Chorover J, Dail D B. A mechanism of abiotic immobilization of nitrate in forest ecosystems: The ferrous wheel hypothesis. Global Change Biology, 2003, 9(2): 228-236.

[237] Clement J C, Shrestha J, Ehrenfeld J G, et al. Ammonium oxidation coupled to dissimilatory reduction of iron under anaerobic conditions in wetland soils. Soil Biology & Biochemistry, 2005, 37(12): 2323-2328.

[238] Doane T A. The Abiotic Nitrogen Cycle. ACS Earth and Space Chemistry, 2017, 1(7): 411-421.

[239] Liu T, Chen D, Li X, et al. Microbially mediated coupling of nitrate reduction and Fe(II)oxidation under anoxic conditions. FEMS Microbiology Ecology 2019, in press.

作者：张礼知[1]，陈　娜[1]，艾智慧[1]，张伟贤[2]，李少林[2]，关小红[2]，邵彬彬[2]，
李芳柏[3]，刘同旭[3]，陈丹丹[3]，李晓敏[3]

[1]华中师范大学，[2]同济大学，[3]广东省生态环境技术研究所

第 9 章　持久性有机污染物植物吸收、迁移与转化的分子机制

- 1. 引言 /240
- 2. POPs的植物吸收与迁移 /240
- 3. POPs的植物转化 /243
- 4. POPs手性物质的植物吸收和转化 /248
- 5. 展望 /249

本章导读

持久性有机污染物（persistent organic pollutants，POPs）已经成为全球性的环境问题，引起人们的广泛关注。作为地球上的重要生命形式，植物对 POPs 的生物地球化学循环起到十分重要的作用。植物对 POPs 的吸收与转化可以降低环境介质中 POPs 的含量和毒性，起到植物修复的作用；植物对 POPs 的吸收与迁移使 POPs 可以通过食物链进入人体，产生较高的环境暴露风险。本章主要探讨了 POPs 的植物吸收、转化与迁移的分子机制。

关键词

持久性有机污染物，植物，吸收，转化，迁移

1 引 言

持久性有机污染物（persistent organic pollutants，POPs）具有持久性、长距离迁移性、生物积累性和生物毒性，成为备受关注的全球性环境污染物。植物是环境介质中 POPs 进入生物圈的重要途径，对 POPs 也具有一定的储存作用。研究植物吸收、转化 POPs 的分子机制，有利于针对环境污染采取相应措施，增强植物吸收作用和降解作用从而增强植物修复的效率，或者有效抑制植物吸收作用从而降低 POPs 进入食物链的风险，也有利于加深对 POPs 环境行为和归趋的认识，为正确评估其持久性和环境毒性效应提供科学数据。

2 POPs 的植物吸收与迁移

除了类激素物质，植物对绝大多数有机物的吸收都是通过被动扩散的方式进行[1]。植物对 POPs 的吸收主要包括：根从土壤或水中吸收，茎和叶从空气中吸收，茎和叶从颗粒物中吸收。不同的吸收过程受到不同因素的影响，机制也不尽相同[2]。

2.1 土壤中 POPs 的植物吸收和迁移

土壤是 POPs 重要的汇，因此土壤中 POPs 的植物吸收和富集备受关注。植物根系从土壤或者水中吸收、富集 POPs，受到 POPs 自身物理化学性质（辛醇-水分配系数 K_{ow} 和水溶性等）、环境介质的性质以及植物种类等多种因素的影响。土壤中的有机质含量、土壤类型以及其他一些因素（土壤 pH 和厌氧有氧条件等）均会显著影响 POPs 的植物吸收和富集。土壤中存在的有机物通常吸附在土壤颗粒上，研究表明，2,2′,4,4′-四溴二苯醚（2,2′,4,4′-tetrabromodiphenyl ether，BDE-47）、2,2′,4,4′,5-五溴二苯醚（2,2′,4,4′,5-pentabromodiphenyl ether，BDE-99）和 2,2′,4,4′,6-五溴二苯醚（2,2′,4,4′,6-pentabromodiphenyl ether，BDE-100）可以强烈地与土壤中的有机质结合，从而降低其可萃取性和生物可利用性[3]。因此土壤中的有机质（soil organic matter，SOM）与植物会竞争吸附/吸收土壤中的 POPs。通过对比白萝卜对水、硅砂和土壤中 4-溴二苯醚（4-bromodiphenyl ether，BDE-3）的吸收与富集，发现土壤中的有机质会结合 BDE-3，延长作物根中富集的 BDE-3 浓度达到平衡的时间[4]。小麦吸收、富集土壤中十溴二苯醚（decabromodiphenyl ether，BDE-209）时，根部富集系数（RCF）与 SOM 含量呈现显著的负相关性[5]。SOM 含量从 1.1%升至 4.5%时，土壤中菲的生物有效性从 3.3%降至 0.7%[6]。向农田中施用农作物秸秆，秸秆的竞争吸附会显著降低 BDE-47 在胡萝卜根中的累积[7]。

POPs 自身的理化性质，尤其是 K_{OW}，对其在土壤/水-植物体系中的分配也起到重要作用。在水培暴露实验中，疏水性的农药更倾向于分配到大麦苗根系上，吸附/吸收作用在很短时间内即可达到平衡，几乎所有的暴露化合物最终都富集在根系上[8]。通常不同性质的有机化合物，其 logRCF 与 logK_{OW} 会表现出显著的正相关性[9]。杨树水培暴露于分子结构类似但具有不同 K_{OW} 的化合物，如 4-氯联苯（4-monochlorobiphenyl，CB-3）、4,4′-二氯联苯（4,4′-dichlorobiphenyl，CB-15）、2,4,4′-三氯联苯（2,4,4′-trichlorobiphenyl，CB-28）、2,2′,5,5′-四氯联苯（2,2′,5,5′-tetrachlorobiphenyl，CB-52）和 3,3′,4,4′-四氯联苯（3,3′,4,4′-tetrachlorobiphenyl，CB-77）后，logRCF 和 logK_{OW} 即呈现显著的正相关关系[10]。这也就意味着 POPs 化合物的疏水性越强，越容易与植物根部的脂类等有机物质结合，从而富集在植物根部。

植物的种类和生长阶段对其根部吸收富集 POPs 也具有重要影响。不同植物以及植物的不同生长阶段根系中的脂含量均存在一定的差异，根中脂含量越高，其对有机污染物的吸收和富集就越快[11]。因此，有机污染物的疏水性越强，植物根部的脂质含量越高，有机污染物与根的结合作用就越强。如果这种结合作用发生在根部的外表面，则污染物不易被吸收进入根的内部，如果这种结合作用发生在根的内部，则化合物不易在植物体内发生迁移，因此在土培暴露中，植物根中的脂含量与根部富集的 BDE-209 的含量表现出显著的正相关性，但根中的高脂肪含量抑制了根中污染物向地上部分的迁移[12]。实际上，以目前的研究手段还很难有效区分植物根部富集的污染物，有多少是吸附在根的表面，又有多少被吸收进入根的内部，尤其是只在植物根部检出有机污染物，而地上部分无法检出目标化合物的情况下，更难以证实污染物是否能够被植物吸收。

植物通过根系吸收到体内的 POPs 可以向地上部分进行迁移，从根部到达茎部和叶部，甚至进入果实[13]。进入叶中的 POPs 还可以通过植物挥发作用进入空气中，这一现象对苯和三氯乙烯这类挥发性较强的物质尤为明显[14, 15]。短链氯化石蜡（SCCPs）暴露后，大量的 SCCPs 也可以迁移至南瓜叶片中，并通过叶片挥发至空气中，而其脱氯产物比母体化合物的挥发性更大，也更易通过植物挥发进入空气中[16]。

对于 POPs 的吸收和迁移机制已经有比较深入的研究。POPs 首先会吸附在根的表面进行富集，然后同水一起进入根部。POPs 穿过根尖无角质覆盖的细胞的未成熟细胞壁，之后沿着细胞间隙（质外体形式）或者穿过细胞（共质体形式）迁移至运输组织——木质部和韧皮部[17-19]。在经过的皮层细胞中，细胞壁为多孔结构，POPs 可以自由穿梭至根部的内皮层，即根部维管束外围排列的致密单层细胞，然后通过渗透作用缓慢进入木质部和韧皮部[20]。不同的物质可能会以不同的形式迁移。二硝基甲苯和二硝基苯在植物体内以共质体的形式进行迁移，而菲和萘则以质外体的形式迁移[21]。有机物的分子大小也会影响其从根部向上的迁移，分子小的 POPs 容易穿透细胞膜，从而容易进出木质部和韧皮部[18]，更易实现从根部向地上部分的迁移。

2.2 空气中 POPs 的植物吸收和迁移

由于 POPs 属于半挥发性物质，广泛存在于空气中，其中 logK_{OW} 值较低的物质更易以气态分子的形式存在，因此，植物通过地上部分（茎和叶）吸收空气中的有机污染物也是植物吸收环境中 POPs 的重要途径。植物地上部分表皮中的角质（聚合脂类）对植物吸附和吸收空气中的有机污染物起到至关重要的作用[22]。例如，叶片吸收空气中的菲时，菲并不是均匀分布在叶片表皮上，而是呈簇状分布，并且菲向叶片中层的聚合脂类进行迁移时也是沿着这样的簇状通道进行[23]。

空气中的 POPs 可以通过吸附作用分配到颗粒物上。这个分配过程受到环境条件、颗粒物性质和 POPs

的物理化学性质等因素的影响。而分配到颗粒物上的有机污染物，也会被植物的地上部分捕集，并与植物发生分配，从而造成植物的POPs暴露，而这一过程则与颗粒物的干、湿沉降行为有关。$\log K_{OA} > 11$的化合物被认为更倾向于以颗粒物吸附的形式沉降到叶片上[24]，而有些颗粒物甚至可以永久性地与植物叶片（地上部分）发生结合[25]。在研究树皮中的全氟类化合物时，发现不同种类的树皮对全氟化合物的富集行为不同，而这种差异与多种因素有关，但其中由于树皮的形貌特征不同而造成的对颗粒物的截留行为的不同是很重要的一个影响因素。杨树、槐树的树皮表面比松树的树皮表面更加粗糙多孔，具有很多深裂隙，树皮样品即便经过流水冲洗，依然有细颗粒被保留在树皮的裂隙中，因此杨树和槐树树皮中的全氟化合物含量和单体分布与大气颗粒物中全氟化合物的组成类似，而松树树皮中全氟化合物的单体分布特征则与大气气相中全氟化合物的组成类似。被截留在树皮中的颗粒物，经过雨水的冲刷、浸润，其上富集的污染物会与植物发生交换，利于植物吸收作用的进行。此外，植物地上部分吸收的POPs类物质还可以在植物体内向根部迁移。在南瓜幼苗的空气暴露中，茎和叶中的六氯丁二烯（hexachlorobutadiene，HCBD）会迅速向根部迁移，导致HCBD在根中富集[26]。

2.3 POPs植物吸收的模型

有研究人员建立了POPs植物吸收的数学模型，以期能够预测不同物质在不同植物和不同环境条件下的吸收特性。相关模型主要分为三类：回归模型、稳态模型和动态模型。其中动态模型是回归模型和稳态模型的综合体，涉及的参数最多，计算也最为复杂，更适合于急性暴露的过程；而回归模型与稳态模型，主要涉及的是平衡状态时植物的吸收，因而更适合于存在平衡态的长期暴露过程[27]。尽管在具体的模型公式上存在差别，但上述三类模型均用到了逸度模型和分配模型的概念[28]。

逸度（f）是有机物从一个部分逃逸到另一个部分的趋势，达到平衡时，不同部位的逸度相同。逸度可以表示为植物某个部位的有机物浓度（C）与逸度容量常数（Z）的比值。两个不同部位Z值的比值（k）为这两个部位的分配系数[29]。根据Mackay和Paterson的研究，在环境介质如空气、土壤和水中，有机物平衡时的逸度均可以用物理量来表示[30]。而有机物在根中的逸度为其在水中的逸度与RCF的乘积；有机物在茎中的逸度为其在水中的逸度与茎富集因子（茎中浓度与水中浓度的比值，SCF）的乘积；有机物在叶中的逸度为其在空气中的逸度与叶富集因子（叶中浓度与空气中浓度的比值，FCF）的乘积。但该逸度模型还存在一定的局限性：在预测中使用的RCF、SCF和FCF是基于少量化合物拟合的公式进行计算的，不具备普适性；同时计算中忽略了不同植物对有机物吸收的差异性[29]；同时这种模型在对叶中的有机物的浓度预测中忽略了植物内部的迁移过程。

在分配模型中，有机污染物的吸收被认为是被动过程，水既是污染物的溶剂也是污染物植物吸收和迁移的载体[31]。土壤中POPs类物质主要是吸附在土壤矿物质，以及分配在土壤有机质（SOM）中，而植物可以吸收的有机污染物主要是分配在SOM中的。分配在SOM中的有机物浓度为土壤中的有机污染物浓度与SOM质量分数的乘积。而土壤间隙水中的有机污染物浓度则为SOM上有机污染物的浓度与SOM-水分配系数的乘积。植物组织是由水、碳水化合物、脂肪和蛋白质等组成，不同组分与水之间具有不同的分配系数。所有组分各自分配系数与其质量分数乘积之和与植物体内溶液中POPs浓度的乘积描述了整个植物或者植物某组分中有机污染物的浓度。有机污染物进入植物体内的方式是主动吸收还是被动吸收，可以用有机污染物在植物体内溶液中的浓度与土壤间隙水中的溶度的比值a加以评估，在主动吸收过程中，一般$a>1$，而在被动吸收中，$a<1$。a的值会受到有机物在土壤中的浓度、暴露时间、有机物理化性质以及植物中脂肪-水比例等因素的影响[32]。通常而言，基于分配模型的计算通常会导致预测值低于实验值[33, 34]。分配模型也存在一定缺陷，其中只考虑了植物从根部的吸收过程以及向上的迁移过程。实际上，对于叶片从空气中的吸收过程，也可以用相似的方法进行计算和预测[35]。虽然还有待于进一步

的完善,这些模型的建立,还是为实际农业生产和污染地区污染状况评估提供了强有力的预测工具。

2.4 POPs 对植物的影响

有机污染物暴露会影响植物的正常生命活动,高浓度的有机污染物如果超出植物能够承载的负荷,会对植物造成严重的毒害作用,甚至导致植物死亡。通常,暴露于一定浓度的有机污染物,植物的生长、叶绿素和可溶性蛋白等含量以及植物体内的各种酶的活性也会随之变化。在对小麦进行浓度为 0.1~10 mg/L 的全氟辛基磺酸(perfluorooctane sulfonate,PFOS)暴露后,小麦的生长受到轻微的促进作用,同时叶片中的叶绿素和可溶性蛋白含量增加,小麦根和叶中超氧化物歧化酶(SOD)和过氧化物酶(POD)的活性显著增强。但当暴露浓度增加到 200 mg/L 时,小麦的生长、叶绿素和可溶性蛋白含量以及酶活性均受到抑制[36]。全氟辛酸(perfluorooctanoic acid,PFOA)同样会引起植物体内氧化还原应激的改变。EC_{50} 的 PFOA 暴露于拟南芥后,会在其累积部位显著增加 H_2O_2 和丙二醛的含量,H_2O_2 可以引起植物的氧化损伤,丙二醛可以破坏大分子结构[37]。暴露全氟辛基磺酰胺(perfluorooctane sulfonamide,FOSA,1.856 nmol/mL)后,南瓜中 SOD、POD、谷胱甘肽转移酶(GSTs)和细胞色素 P450 酶(CYP450 酶)的活性均有所增强[38]。暴露于 2,4-二溴苯酚(2,4-dibromophenol,2,4-DBP)后,植物中糖基转移酶(UGTs)的活性与糖基化代谢产物的生成呈现显著的正相关性[39],表明了糖基转移酶在糖基化代谢过程中的作用。

3 POPs 的植物转化

3.1 POPs 植物转化过程概述

POPs 可以通过吸收和迁移过程进入到植物的不同部位,不同部位 POPs 的代谢过程一直是研究人员关注的热点问题。由于植物具有与动物肝脏类似的酶系和代谢过程,因此植物被视为"绿色肝脏"[40]。植物中有机污染物的代谢过程也可以大致分为三相反应过程,即Ⅰ相、Ⅱ相和Ⅲ相反应过程。Ⅰ相反应过程是有机污染物的活化过程,通过引入某些基团或者脱去某些基团以增加污染物的反应活性,如羟基化和去甲基化等反应过程[41-46];Ⅱ相反应一般为Ⅰ相反应产物在植物体内的进一步代谢。在这个过程中,内源性物质如糖苷和谷胱甘肽等化合物与Ⅰ相产物结合,使污染物水溶性和毒性进一步降低[40];Ⅲ相反应主要涉及Ⅱ相产物的隔离与固定,如进入液泡和细胞壁等过程[40, 47, 48]。不同的代谢和转化过程是由不同的酶介导的。典型的调控Ⅰ相反应的酶主要包括 CYP450 酶、POD、乙醇脱氢酶(ADH)和乙醛脱氢酶(ALDH)和硝基还原酶(NaR)等。Ⅱ相反应的酶主要包括甲基转移酶(MTs)、磺基转移酶(SULTs)、UGTs 以及 GSTs 等。而关于Ⅲ相反应相关酶系的研究还比较匮乏,但一些转运蛋白,如 ATP-binding cassette(ABC 蛋白)可能会参与到Ⅱ相产物进入液泡的过程[49],而参与细胞壁合成的一些酶,如纤维素合成酶等则会参与到Ⅱ相产物进入或者被包裹至细胞壁的过程中。通常,POPs 的植物转化是多种酶同时作用且相互交织的,因此多种反应产物往往同时存在。

3.2 POPs 植物转化中酶的作用

在 CYP450 酶的作用下,有机污染物在植物体中生成具有环氧结构的中间产物,进而发生水解生成羟基化产物[50]。在反应体系中加入CYP450酶活性抑制剂1-氨基苯并三唑（ABT）和17-十八炔酸（ODYA），暴露物CB-3的羟基化产物显著减少，从而证明了CYP450酶在污染物羟基化过程中的作用[51]。对大豆暴露8∶2氟调醇（8∶2 fluorotelomer alcohol，8∶2 FTOH）时，会生成8∶2氟调醛和8∶2氟调羧酸，检测Ⅰ相反应过程中几种酶的活性，发现ADH和ALDH的活性显著增强，从侧面证明了ADH和ALDH在8∶2 FTOH转化过程中的作用[52]。对于含卤原子的有机污染物如PCBs和PBDEs等，在植物体内通常也会发生还原脱卤作用。卤原子的去除也会增加有机污染物的反应活性。将从植物中提取的NaR与2,2′,4,4′,5,5′-六氯联苯（2,2′,4,4′,5,5′-hexachlorobiphenyl，CB-153）体外孵育后，发现CB-153的脱氯过程显著增强[53]；提高NaR的浓度，2,4,4′-三溴二苯醚（2,4,4′-tribrominated diphenyl ether，BDE-28）、BDE-47、BDE-99和BDE-209的脱溴速率也随之增强，这种剂量效应关系说明了NaR在植物体内有机污染物脱卤过程中的重要作用[54]。

MTs可以将植物体内的甲基供体，如S-腺苷甲硫氨酸（SAM）提供的甲基转移至有机污染物分子中，甲基受体原子一般为O、S、N等。SULTs一般利用3′-磷酸腺苷-5′-磷酸硫酸酯作为磺酸基团供体，将磺酸基团转移至含有羟基的污染物上。在GSTs的作用下，还原态的GSH可与有机污染物发生亲核加成或者亲核取代反应[48]。对于卤代有机污染物，例如PCBs、氯乙烯等，通常是在脱氯的基础上与GSH发生亲核取代反应[55, 56]。

3.3　POPs在植物中典型的Ⅰ相反应

羟基化是POPs在植物体内最常见的Ⅰ相反应。POPs类物质引入羟基后反应活性增强，有利于进一步反应，但有时生成的羟基化产物较其母体化合物的毒性效应更强[57-61]。对于含有芳香环的POPs类物质如PCBs和PBDEs的研究相对较多，应用的植物模型主要有杨树、南瓜和玉米等，本节对反应发生的位点及难易程度进行了评估与归纳。杨树水培暴露于单氯取代的CB-3时，联苯的苯环上所有未被氯原子取代的位点均可发生羟基化，即可以生成五种单羟基取代的OH-CB-3。这些羟基化产物中，以对位取代的4′-OH-CB-3的含量最高（0.86%），其次为邻位取代的2′-OH-CB-3（0.14%）[62]。杨树水培暴露于CB-77时，在暴露很短时间内即可检测到羟基化的产物，最先能检测到的羟基化产物为邻位的6-OH-CB-77，随着暴露时间的延长，还可以进一步检测到间位的5-OH-CB-77，但最终的含量仍以6-OH-CB-77为主[63, 64]。杨树暴露于2,2′,3,5′,6-五氯联苯（2,2′,3,5′,6-pentachlorobiphenyl，CB-95）时，对位取代的4′-OB-CB-95也在羟基化产物中占主导[65]。由以上的实验结果可知，在杨树中，PCBs的羟基化位点顺序为：对位>邻位>间位。但是由于羟基化是PCBs在植物体内活化的第一步，后续进行的反应会从一定程度上影响羟基化产物的含量，因此实验结果只能从一定程度上反映PCBs发生羟基化反应位点的优先级。

不同植物体内同一化合物的羟基化位点，以及不同化合物的羟基化反应位点都会有所差异。在生菜和番茄中，CB-77的羟基化产物有别于杨树，检出了2-OH-CB-77[66]。南瓜幼苗暴露于BDE-28时，会生成多种羟基化产物，除了BDE-28的羟基化产物外，还有溴重排以及低溴代PBDEs的羟基化产物[67]。由于缺乏标准品以及2′-OH-BDE-28和3′-OH-BDE-28的共流出，很难判断BDE-28羟基化的规律。在对南瓜幼苗水培暴露BDE-47时，整个暴露体系中以间位的5-OH-BDE-47为主，其次为6-OH-BDE-47，而其他的羟基化产物则是溴原子重排的产物[68]。对玉米进行BDE-28和BDE-47水培暴露时发现，低溴代的羟基化产物占主导[69]。可能的原因是玉米的脱溴作用强于南瓜。

从取代基团对苯环不同位点电子云密度的影响来看，苯环上的卤原子会因为吸电子效应造成苯环的反应活性降低，但同时由于共轭作用，会使邻位和对位的电子云密度的抑制效应降低，因此间位的电子云密度降低的最为明显。当两个苯环以联苯的形式相连或者通过醚键相连，其中的一个（卤代）苯基（苯氧基）可视为另一个苯环的取代基，卤原子与苯基/苯氧基的邻对位定位效应，使得PCBs和PBDEs苯环上不同位点的电子云密度不同,从而羟基化活性不同,电子云密度越大反应活性越强。羟基化反应的发生还会受到酶作用位点的影响。对于不含苯环的物质如六溴环十二烷(1,2,5,6,9,10-hexabromocyclododecane,HBCD)等,在玉米等植物内也会发生羟基化,除了单羟基取代的 HBCD 和五溴环十二烷,还可以检测出双羟基取代的 HBCD 和五溴环十二烷[70],这可能意味着六溴环十二烷的羟基化更易进行。

多数 POPs 物质含有卤原子,脱卤作用也是 POPs 类物质降解过程中的重要步骤。有植物存在时,可以极大地促进土壤中 POPs 脱卤作用的发生。CB-52、CB-77 和 CB-153 在土壤-柳枝稷系统中反应 32 周后,整个反应体系中 PCBs 的降解率分别为 63.1%±0.5%,54.2%±1.1%和 51.5%±0.4%,显著高于纯土壤体系中这几种物质的降解率(29.1%±3.9%,20.1%±2.0%和 30.1%±1.2%),而降解的主要产物即为脱氯产物[71]。BDE-209 在土壤-植物体系中也会发生逐级的脱溴反应,最终检测到 19 种低溴代的 PBDEs,其中也含有被禁止生产使用的五溴和八溴二苯醚[12]。在土壤-植物共存的研究体系中,除了植物根系分泌物和土壤微生物对 POPs 具有降解作用以外,卤代 POPs 物质也可以在植物体内发生脱卤反应。从苜蓿叶片中提取出的硝基还原酶以及从玉米中提取的硝基还原酶的标准品均能够使 PCB-153 发生脱氯反应。向反应溶液中加入钼盐后,脱氯效率大幅增加[72]。在对玉米和南瓜水培暴露 CB-28,2,2',4,4'-四氯联苯(2,2',4,4'-tetratrichlorobiphenyl,CB-47)和 BDE-28,BDE-47 以及 BDE-99 时,均可检测出这些物质的脱卤产物[63, 67, 68, 73-75]。植物和微生物体内的脱卤作用存在一定的差异性。微生物的脱卤作用容易发生在芳香化合物的间位上,而植物介导的反应更倾向于发生在邻位和对位[76, 77]。总体而言,PBDEs 和 PCBs 在植物体内生成的脱卤转化产物占母体化合物的比率通常比较低,一般小于 1%。此外,芳环上的卤原子还会发生重排反应。玉米暴露 BDE-47 后,会生成邻位脱溴的 BDE-28,再发生邻位脱溴生成 4,4'-二溴二苯醚(4,4'-dibromodiphenyl ether,BDE-15),以及溴原子重排的脱溴产物 3,4'-二溴二苯醚(3,4'-dibromodiphenyl ether,BDE-13)[68]。

不含芳香环的化合物 HBCD,也会在植物中发生脱溴作用,生成五溴环十二烷和四溴环十二烷,不同构型的 HBCD 脱溴的比例为 1.3%～3.6%[70]。对于链状卤代烃 SCCPs 的单体化合物 1,2,5,5,6,9,10-七氯癸烷(1,2,5,5,6,9,10-heptachlorodecane),在南瓜中的脱氯产物可以占到总暴露量的 42%,同时还有 4%的氯重排产物[16];8:2 FTOH 在大豆中同样会发生脱氟作用,脱氟后的 7:3 氟调羧酸和 PFOA 浓度随着暴露时间逐渐增加,最终达到 5.6%和 6.0%,成为最主要的代谢产物[48]。

综上所述,脱卤过程既可以发生在苯环上也可以发生脂肪环以及链烃上,从产物量与母体化合物量的比率上来看,苯环上的脱卤作用比脂肪环和链烃上的更弱。但对于 1,1,1-三氯-2,2-双(4-氯苯基)乙烷[1,1,1-trichloro-2,2-bis-(4-chlorophenyl)ethane,DDT]而言,在苯环上和饱和碳上均含有氯原子,在植物作用下,脱氯反应通常会在饱和碳上发生,分别生成 2,2-双(4-氯苯基)-1,1-二氯乙烷(DDD)和 2,2-双(4-氯苯基)-1,1-二氯乙烯(DDE)[78, 79],而苯环上的脱氯反应及其产物则未见报道。

除了这些典型的 I 相反应,还存在其他的由 I 相酶介导的反应,如裂解反应以及聚合反应等。具有 POPs 性质的四溴双酚 A(tetrabromobisphenol A,TBBPA)在水稻细胞中可以检测到 TBBPA 裂解后生成的两种单苯环产物,其中一种为 DBHPA,占到总转化产物的 50%以上(占暴露母体化合物总量的 2.2%),另一种为 DBHP,约占总转化产物含量的 46.4%[80]。在南瓜植株中,TBBPA 也可以转化生成这两种单苯环裂解产物,且总的转化率占到总暴露量的 4%左右[81]。这些实验结果证明裂解反应是 POPs 类物质在植物体内的重要转化过程。与裂解反应相反,聚合反应在植物体内也可能发生。低等植物-海洋藻类体内含有溴过氧化物酶(BPOs),在这种酶的催化下,2,4,6-三溴酚(2,4,6-tribromophenol)先生成氧中心自由基,进而生成聚合产物,包括多种 OH-PBDEs 和 PBDDs[82, 83]。在高等植物中,相关的研究还较为缺乏,但是在动物中,类似的聚合反应也同样得到证实。对动物和人的微粒体以及小鼠暴露三氯生(triclosan,TCS)、苯酚(phenol)、双酚 A(bisphenol A,BPA)以及苯并[a]芘(benzoapyrene)等,不仅可以发现相同物质之间的聚合产物，还可以检出不同物质之间的聚合产物，这表明酚类物质在动物体内的聚合反应是一个普遍的代谢途径，而且这个过程主要是由 I 相酶，如 CYP450 酶介导[84]。由于高等植物的酶系与动物和人体肝微粒体中的酶系一致，因此，可以断定在高等植物体内，类似的聚合反应也有可能发生。

3.4 POPs 在植物中典型的 II 相反应

研究比较普遍的 POPs 的 II 相反应是甲基化过程。在水稻中，4'-羟基-2,3,4,5-四氯联苯（4'-hydroxy-2,3,4,5-tetrachlorobiphenyl，4'-OH-CB-61）和 3'-羟基-2,3,4,5-四氯联苯（3'-hydroxy-2,3,4,5-tetrachlorobiphenyl，3'-OH-CB-61）能生成相应的甲基化产物 4'-甲氧基-2,3,4,5-四氯联苯（4'-methoxy-2,3,4,5-tetrachlorobiphenyl，4'-MeO-CB-61）和 3'-甲氧基-2,3,4,5-四氯联苯（3'-methoxy-2,3,4,5-tetrachlorobiphenyl，3'-MeO-CB-61）[85]。在南瓜中，6 种四溴代的 OH-PBDEs 均可以生成相应甲基化产物 MeO-PBDEs[86]。

在植物暴露 PCBs 或者 PBDEs 之后的代谢研究中，可以同时检测到羟基取代的 PCBs、PBDEs 和甲氧基取代的 PCBs、PBDEs，检测到的羟基化产物中羟基取代位点和甲氧基化合物中甲氧基的取代位点通常具有对应关系，而且取代位点一致的两类产物之间还存在一定的负相关关系。分别检测 BDE-28 和 BDE-47 暴露后植物体内的羟基取代产物和甲氧基取代产物，发现 3'-MeO-BDE-28 和 6-MeO-BDE-47 分别为这两种物质最主要的甲氧基取代产物，而取代位点一致的羟基化产物 3'-OH-BDE-28 和 6-OH-BDE-47 的含量则较低[69]。由于 PCBs 和 PBDEs 本身不具有羟基，一般无法直接发生甲基化反应，因此体系中的甲氧基取代的化合物很可能是 PCBs 或 PBDEs 先发生羟基化反应，再进一步发生甲基化反应从而生成甲氧基取代化合物。但在实际环境样品检测中，发现的羟基取代化合物和甲氧基取代化合物之间，取代位点并非总是完全一致，因此对于 PCBs 和 PBDEs 是羟基化后发生甲基化反应，还是甲氧基直接接合到苯环上发生甲氧基化反应还无法定论。

在植物体内，甲基化与去甲基化通常是可逆的反应过程，即甲基化产物还可以发生去甲基化反应。通常去甲基化反应和甲基化反应的速率也有所不同。对水稻体内 MeO-PCBs 的去甲基化和 OH-PCBs 的甲基化的研究显示，两类母体化合物分别单独暴露 5 天后，18% 的 4'-MeO-CB-61 发生去甲基化反应，生成 4'-OH-CB-61，而只有 1% 的 4'-OH-CB-61 发生甲基化生成 4'-MeO-CB-61，两种反应的速率相差较大[85]。在水稻、玉米和小麦等不同植物中，3'-甲氧基-2,3,5,6-四氯联苯（3'-methoxy-2,3,5,6-tetrachlorobiphenyl，3'-MeO-CB-65）和 4'-甲氧基-2,2',4,5,5'-五氯联苯（4'-Methoxy-2,2',4,5,5'-pentachlorobiphenyl，4'-MeO-CB-101）的去甲基化速率也快于其对应的 OH-PCBs 的甲基化速率[87]，这也就意味着对于 MeO-PCBs 和 OH-PCBs 而言，更倾向于发生去甲基化过程。

对于含有溴原子的芳香化合物 MeO-PBDEs 而言，其去甲基化的速率与含氯的 MeO-PCBs 存在显著的差异。不同取代位点和不同溴原子数的 MeO-PBDEs 暴露于大豆、小麦、水稻、玉米以及南瓜等不同植物时，其去甲基反应产物与母体化合物的含量比率通常都低于其对应结构的 OH-PBDEs 的甲基化产物比率，即在 MeO-PBDEs 和 OH-PBDEs 之间，甲基化反应更容易发生[86, 88]。

对于 TBBPA 而言，在对南瓜幼苗暴露的 15 天时间内，其甲基化反应迅速发生，且在 4 天时其甲基化产物的含量达到最高，随后会迅速下降，并保持较低浓度，直至暴露结束。但是南瓜苗暴露于双甲氧基四溴双酚 A（TBBPA DME）时，去甲基化产物 TBBPA 的含量在开始时变化缓慢，但是在暴露 9 天时，TBBPA 的含量迅速增加，暴露结束时，TBBPA 的含量则相对较高[81]。这与 4'-MeO-CB-61 和 4'-OH-CB-61 的相互转化趋势存在较大差异，意味着忽略生成规律单纯比较暴露结束时刻的甲基化产物或者去甲基化产物的量来判断反应进行的速率可能会存在一定的误差，因为甲基化或者去甲基化反应产物的剩余量不仅与生产反应的速率有关，还与其可能发生的进一步转化反应的去除速率有关。

有机污染物本身具有羟基或者经过Ⅰ相反应加上羟基后，还可以进一步发生硫酸酯化、糖基化以及与谷胱甘肽和氨基酸结合等代谢过程。本身含有羟基的污染物，如 TBBPA、OH-PBDEs 等，其进一步的结合反应较容易发生。对淡水微藻暴露 TBBPA 后，可以检测到硫酸酯和葡萄糖结合的产物[89]。而对本身不含羟基的有机污染物，一般则需要在植物体内先发生羟基化反应，之后才能进行Ⅱ相的结合反应，因此这类有机污染物结合产物的形成依赖于羟基化产物的生成。对杨树水培暴露 CB-3，经过 25 天后，由于暴露浓度（1μg/mL）较高，其羟基化产物含量达到 600 ng/g 以上，进而能检测到 3 种与羟基化产物相对应的硫酸酯化产物[62, 90]。但在杨树水培暴露 CB-77 时，由于暴露浓度相对较低（1ng/mL 和 10ng/mL），没有检测到含硫的代谢物[63]。

由于多数具有 POPs 特性的物质都不含有羟基，这也就意味着，其Ⅱ相转化是受到Ⅰ相反应速率限制的。在研究 DDT 在植物体内的转化途径时，发现 DDT 可以在未灭菌土壤-菠菜和卷心菜体系中转化为 2,2-双-(4-氯苯基)乙酸［2,2-bis-(4-chlorophenyl)acetic acid，DDA］和其他极性代谢产物，但是由于这些极性产物的含量很低，未能对其进行结构鉴定；将 DDT 和 DDA 分别暴露于大豆和小麦的悬浮细胞 2 天后，发现 DDT 转化为 DDA 的量小于暴露总量的 1.4%，而 DDA 转化为极性更强的物质的量为 56.5%，其中包括 DDA 的葡萄糖结合产物[91]。2,4-二溴酚暴露于胡萝卜细胞 5 天后，糖基化的产物（以单葡萄糖基-2,4-二溴酚定量）占到母体化合物总暴露量的 9.3%[39]。

TBBPA 在淡水微藻中会生成与葡萄糖结合的产物[89]，而在人体中会转化为与葡萄糖醛酸结合的产物[92, 93]。这表明植物体内的糖基化过程与动物和人体内的糖基化过程存在一定的差异，前者倾向于利用葡萄糖，后者更倾向于利用葡萄糖醛酸[94, 95]。在高等植物体内，有机污染物的糖基化产物生成后，糖基部分可以进一步被丙二酰基修饰[96]。丙二酰基修饰是植物对外源性有机污染物独特的适应机制[95]。胡萝卜暴露 TCS 后，检测到了较多的与糖基（葡萄糖基和核糖基）以及丙二酰基糖苷结合的产物[97]。在辣根属植物（A. rusticana）的根中，TCS 糖基化的产物糖基上的羟基可以被乙酰基、磺酸基进一步取代[98]。

外源有机污染物与氨基酸和谷胱甘肽相结合，可以极大地增强代谢物的水溶性，是污染物代谢转化的重要过程。氨基酸和谷胱甘肽在植物对外源有机污染物的解毒中也起着至关重要的作用。在拟南芥细胞暴露双氯灭痛（diclofenac）后，检测出了双氯灭痛与谷氨酸结合的产物，占到代谢产物的 70% 以上，证明与谷氨酸结合是双氯灭痛代谢的主要过程[99]。2,4-DBP 进入胡萝卜细胞后，也可以与丙氨酸、乙酰丙氨酸等进行结合[39]。植物中 8∶2 FTOH 的转化产物与 GSH 结合后，还存在可能的裂解产物——半胱氨酰结合产物[48]。

3.5　POPs 在植物中典型的Ⅲ相反应

有机污染物及其糖基化产物会被固定到细胞壁中。将 4-氯苯胺（4-chloroaniline）和 2,4-二氯苯酚（2,4-dichlorophenol）暴露于玉米和土豆细胞后，将两种植物的细胞壁经酶水解后，检测到了母体化合物，占暴露量的比例分别为 27%～31% 和 32%～47%，同时还检测到了与 β-D-葡萄糖苷的结合物，占总暴露量的比例分别为 44%～47% 和 56%～64%[100]，这一结果表明母体化合物以及其糖基结合的产物可能会被细胞壁包裹或者结合到细胞壁上，从一定程度上阐释了Ⅲ相反应的部分机制。

除了与葡萄糖结合进入细胞壁外，有机污染物还可以通过与木质素结合进入细胞壁。五氯苯酚（pentachlorophenol，PCP）暴露于水稻和小麦的细胞后，在细胞壁不可萃取的残留物中检测出了 PCP 羟基化产物 1,2-二羟基-3,4,5,6-四氯苯（1,2-dihydroxy-3,4,5,6-tetrachlorobenzene）与木质素结合生成的产物[101]。对水稻植株进行 3-氯苯胺和 3,4-二氯苯胺的水培暴露后，发现有 40% 的结合产物存在于根部的木质素中[102]。玉米暴露 4-氯苯胺后，细胞壁木质素中的结合态物质含量最高（17.0%），其他结合产物依次是在果胶（13.3%）、蛋白质（4.0%）、淀粉（3.8%）、半纤维素（1.3%）和纤维素（0.6%）中；在 2,4-

二氯苯酚的暴露中，也发现了较高含量的木质素结合产物（10.1%）[100]。这些实验结果都表明，有机污染物与木质素的结合是探究污染物在植物体内Ⅲ相反应中不可忽略的过程。

除了进入细胞壁被固定外，有机污染物与谷胱甘肽等物质结合后，还存在进入液泡隔离的过程。研究认为谷胱甘肽和半胱氨酰等结合产物很可能存在于细胞质或者质外体中，既有可能进一步发生反应，也有可能是稳定的终产物[103, 104]。与谷胱甘肽等结合的两性物质会在ABC蛋白的作用下从细胞质进入液泡，从而使外源有机污染物隔离，达到解毒的目的[105]。

3.6 POPs转化产物的检测与筛查方法

POPs在植物体内的多种转化方式既有可能先后发生，也有可能同时发生，产物性质差异巨大。例如PCBs和PBDEs发生羟基化反应后，水溶性增强，而羟基化产物进一步发生甲基化时，水溶性会减弱，亲脂性增强；而羟基化反应之后经过硫酸酯化和糖基化反应，则会进一步增强水溶性，糖基化的物质还会进一步发生硫酸酯化等反应，因此，若是希望在暴露体系中尽量多地发现不同的代谢产物，一定要针对不同性质代谢产物采用不同的前处理方法。同样地，不同性质的代谢产物的检测也需要用到不同的仪器。比如极性比较强的糖基化产物需要用到液相质谱的检测，而甲基化的产物则需要利用气相质谱进行检测。在检测过程中，目的性地筛查某种或者某几种类型的代谢产物时，往往无法全面理解POPs类物质在植物体内的转化过程。因此用非目的性筛查的方法检测POPs在植物体内的代谢产物，对全面了解污染物的代谢十分必要。由于POPs类物质多含有卤原子，因此可以采用以下非目的性筛选的方式：对于含有溴和氯POPs物质，由于溴和氯均具有天然丰度接近的同位素，因此这些物质在质谱上的（准）分子离子峰簇中含有质量差值为$2u$的质谱峰，同时这些质谱峰的丰度也存在特定的比例。为了能够更加直观地观测到这些含溴和含氯的POPs物质的代谢产物，通常会将质谱的同位素特征与质量缺失两者结合[106-108]。对于含氟的化合物和代谢产物，其质谱中会存在特定的碎片离子，如$C_2F_5^-$和$C_3F_7^-$等。通过提取这些特征碎片离子，可以定性检测含氟代谢产物[109]。

4 POPs手性物质的植物吸收和转化

在POPs化合物中，有一些物质存在手性对映体，如氯丹（trans/cis-chlordane），o,p'-DDT以及多种PCBs，尽管手性对映体的物理化学性质相同，但是其生物活性存在较大的差异，因而在植物体内的吸收、迁移和转化行为也存在较大的差异，有时还会存在对映体选择性。

在手性物质研究中，为了更好地描述手性物质的组成，通常会引入对映体分数（enantiomer fractions，EF）这一指标。EF的计算方式为（+）对映体浓度和总物质浓度的比值或者在色谱柱上首先洗脱的对映体（E_1）浓度与总对映体（E_1+E_2）浓度的比值[110]。手性物质表现为外消旋时，EF= 0.5。植物对手性对映体表现的选择性更倾向于发生在植物体内的迁移和转化过程中。在植物的根、茎和叶中，(−) 2,2',3,4',6-五氯联苯（2,2',3,4',6-pentachlorobiphenyl，CB-91）、CB-95和2,2',3,3',6,6'-六氯联苯（2,2',3,3',6,6'-hexachlorobiphenyl，CB-136）更倾向于富集在莲花的不同组织中，而2,2',3,4',5',6-六氯联苯（2,2',3,4',5',6-hexachlorobiphenyl，CB-149）、2,2',3,3',4,6,6'-七氯联苯（2,2',3,3',4,6,6'-hepta-chlorobiphenyl，CB-176）和2,2',3,4,4',5',6-七氯联苯（2,2',3,4,4',5',6-heptachlorobiphenyl，CB-183）并没有发生显著的手性选择性富集[111]。但是在其他植物、动物中，这三种手性物质的迁移和转化过程均具有手性选择性[112-114]。在杨树的水培暴露实验中，暴露20天后，水培溶液中CB-95的外消旋依然存在，这证明杨树对两种对映体的富集作用不存在选择性,但是(−)CB-95 在植物中的去除更为迅速,在木栓的底部到中部,EF 值逐渐从 0.449 ± 0.012 至 0.307 ± 0.051[115]。CB-95 的羟基化产物 5-OH-CB-95 在杨树中的迁移和转化过程中均具有手性选择性,(−)OH-CB-95 更容易在植物体内积累,在木质部、树皮和茎中的 EF 值从 0.643 ± 0.110 至 0.835 ± 0.087[116]。桉树树叶和松针吸收空气中的 PCBs,如 2,2′,3,3′,6-五氯联苯(2,2′,3,3′,6-pentachlorobiphenyl,CB-84)、2,2′,3,3′,4,6′-六氯联苯(2,2′,3,3′,4,6′-hexachlorobiphenyl,CB-132)、CB-136、CB-149 和 CB-183 等,其中 CB-84、CB-132 和 CB-183 等物质在植物叶片中的手性组成和空气中的相比没有显著差异($p > 0.317$),而(−)CB-95、136 和 149 却更容易在桉树树叶和松针中富集[117],即植物桉树叶和松针吸收空气中不同的手性 PCBs 时表现出了不同的选择性,但这种现象也可能与手性化合物后续代谢过程有关。

手性 PCBs 多数可以发生选择性代谢[118]。通过同源建模和分子对接的方法,发现一组阻旋异构体与 CYP2B1 的结合的亲和力存在差异,这可能是导致阻旋异构体在转化中发生手性选择的原因[119]。物质在植物体内转化的手性选择性会影响到母体物质的手性特征。在 CYP450 酶的作用下,PCBs 的代谢速率通常随着分子量增加而降低,随邻位-对位的氢原子数增加而增加。在桉树树叶中吸附的 PCBs 中,代谢速率快的,其偏离 EF = 0.5 的幅度越大,而代谢速率慢的,其偏离 EF = 0.5 的幅度越小[120],CB-84 的 EF 值为 0.414 ± 0.051,CB-183 的 EF 值为 0.494 ± 0.023[117],即 CB-84 的代谢速率较 CB-183 更快。

5　展　望

POPs 在植物体内吸收、迁移和转化等过程的研究还存在很多不明确的地方,有待进一步探究。①POPs 在植物体内的迁移,尤其是叶片吸收的 POPs 向下迁移的分子机制。②不同结构的 POPs 在植物体内不同的迁移行为以及植物挥发行为。含有脂肪链、脂肪环以及苯环的 POPs 在植物体内的吸收和迁移存在比较明显的差异,但是造成这种差异的分子机制还不明确。③POPs 物质在植物体内新的转化途径。分析检测技术的进步,为全面、系统地探究 POPs 在植物体内的转化途径提供了可能性。④POPs 在植物体内转化的分子机制。不同 POPs 在植物体内的代谢途径和代谢的比例存在较大的差异,但是造成差异的具体原因和规律还有待进一步探究。因此相关分析方法学也是植物与污染物交互作用机制研究的重要技术难点。

参 考 文 献

[1] Bromilow R H, Chamberlain K. Principles governing uptake and transport of chemicals. *In* Trapp S, McFarlane J C, Eds. Plant Contamination: Modelling and Simulation of Organic Chemical Processes. London: Lewis Publishers, 1995: 38-64.

[2] Collins C, Fryer M, Grosso A. Plant uptake of non-ionic organic chemicals. Environmental Science & Technology, 2006, 40(1): 45-52.

[3] Mueller K E, Mueller-Spitz S R, Henry H F, et al. Fate of pentabrominated diphenyl ethers in soil: Abiotic sorption, plant uptake, and the impact of interspecific plant interactions. Environmental Science & Technology 2006, 40(21): 6662-6667.

[4] Yang C Y, Chang M L, Wu S C, et al. Partition uptake of a brominated diphenyl ether by the edible plant root of white radish(*Raphanus sativus* L.). Environmental Pollution, 2017, 223: 178-184.

[5] Li H L, Qu R H, Yan L G, et al. Field study on the uptake and translocation of PBDEs by wheat (*Triticum aestivum* L.) in soils amended with sewage sludge. Chemosphere, 2015, 123: 87-92.

[6] White J C, Kelsey J W, Hatzinger P B, et al. Factors affecting sequestration and bioavailability of phenanthrene in soils. Environmental Toxicology and Chemistry, 1997, 16(10): 2040-2045.

[7] Xiang L L, Sheng H J, Xu M, et al. Reducing plant uptake of a brominated contaminant (2,2′,4,4′-tetrabrominated diphenyl ether) by incorporation of maize straw into horticultural soil. Science of The Total Environment, 2019, 663: 29-37.

[8] Shone M G T, Bartlett B O, Wood, A V. A Comparison of the uptake and translocation of some organic herbicides and a systemic fungicide by barley: II. Relationship between uptake by roots and translocation to shoots. Journal of Experimental Botany, 1974, 25(2): 401-409.

[9] Briggs G G, Bromilow R H, Evans A A. Relationships between lipophilicity and root uptake and translocation of non-ionized chemicals by barley. Pesticide Science, 1982, 13: 495-504.

[10] Liu J Y, Schnoor J L. Uptake and translocation of lesser-chlorinated polychlorinated biphenyls(PCBs)in whole hybrid poplar plants after hydroponic exposure. Chemosphere, 2008, 73: 1608-1616.

[11] Yang C Y, Chang M L, Wu S C, et al. Sorption equilibrium of emerging and traditional organic contaminants in leafy rape, Chinese mustard, lettuce and Chinese cabbage. Chemosphere, 2016, 154: 552-558.

[12] Huang H L, Zhang S Z, Christie P, et al. Behavior of decabromodiphenyl ether(BDE-209)in the soil-plant system: Uptake, translocation, and metabolism in plants and dissipation in soil. Environmental Science & Technology, 2010, 44(2): 663-667.

[13] Yang C Y, Wu S C, Lee C C, et al. Translocation of polybrominated diphenyl ethers from field-contaminated soils to an edible plant. Journal of Hazardous Materials, 2018, 351: 215-223.

[14] Collins C D, Bell J N B, Crews C. Benzene accumulation in horticultural crops. Chemosphere, 2000, 40: 109-114.

[15] Ma X M, Burken J G. TCE diffusion to the atmosphere in phytoremediation applications. Environmental Science & Technology, 2003, 37(11): 2534-2539.

[16] Li Y L, Hou X W, Yu M, er al. Dechlorination and chlorine rearrangement of 1,2,5,5,6,9,10-heptachlorodecane mediated by the whole pumpkin seedlings. Environmental Pollution, 2017, 224: 524-531.

[17] Zhang C, Feng G, Liu Y W, et al. Uptake and translocation of organic pollutants in plants: A review. Journal of Integrative Agriculture, 2017, 16(8): 60345-60357.

[18] Kvesitadze G, Khatisashvili G, Sadunishvili T, et al. Plants for remediation: Uptake, translocation and transformation of organic pollutants. *In* Öztürk M, Ashraf M, Aksoy A, Ahmad M S A, Hakeem K R, Eds. Plants, Pollutants and Remediation. Netherlands, USA: Springer, 2015: 241-305.

[19] Sitte P, Ziegler H, Ehrendorfer F, et al. Lehrbuch der Botanik für Hochschulen (Vol 33). *In* Fischer S, Ed. New York: Gustav Fischer Verlag, 1991.

[20] Trapp M, Mc Farlane C. Modeling and simulation of organic chemical processes. *In* Trapp S, Mcarlane J C, Eds. Plant Contamination. USA: CRC, 1994.

[21] Su Y, Zhu Y. Transport mechanisms for the uptake of organic compounds by rice(*Oryza sativa*)roots. Environmental Pollution, 148: 94-100.

[22] Chen B L, Li Y G, Guo Y T, et al. Role of the extractable lipids and polymeric lipids in sorption of organic contaminants onto plant cuticles. Environmental Science & Technology, 2008, 42(5): 1517-1523.

[23] Li Q Q, Chen B L. Organic pollutant clustered in the plant cuticular membranes: Visualizing the distribution of phenanthrene in leaf cuticle using two-photon confocal scanning laser microscopy. Environmental Science & Technology, 2014, 48(9): 4774-4781.

[24] McLachlan M S. Framework for the interpretation of measurements of SOCs in plants. Environmental Science & Technology, 1999, 33(11): 1799-1804.

[25] Smith K E C, Jones K C. Particles and vegetation: Implications for the transfer of particle-bound organic contaminants to vegetation. Science of The Total Environment, 2000, 246(2-3): 207-236.

[26] Hou X W, Zhang H Y, Li Y L, et al. Bioaccumulation of hexachlorobutadiene in pumpkin seedlings after waterborne

exposure. Environmental Science: Processes & Impacts, 2017, 19: 1327-1335.

[27] Collins C D, Fryer M E. Model intercomparison for the uptake of organic chemicals by plants. Environmental Science & Technology, 2003, 37(8): 1617-1624.

[28] Paterson S, Mackay D, Mcfarlane C. A model of organic chemical uptake by plants from soil and the atmosphere. Environmental Science & Technology, 1994, 28: 2259-2266.

[29] Calamari D, Vighi M, Bacci, E. The use of terrestrial plant biomass as a parameter in the fugacity model. Chemosphere, 1987, 16: 2359-2364.

[30] Mackay D, Paterson S. Calculating fugacity. Environmental Science & Technology, 1981, 15(9): 1006-1014.

[31] Chiou C T, Sheng G Y, Manes M. A partition-limited model for the plant uptake of organic contaminants from soil and water. Environmental Science & Technology, 2001, 35(7): 1437-1444.

[32] Wu X, Zhu L Z. Prediction of organic contaminant uptake by plants: Modified partition-limited model based on a sequential ultrasonic extraction procedure. Environmental Pollution, 2019, 246: 124-130.

[33] Yang Z Y, Zhu L Z. Performance of the partition-limited model on predicting ryegrass uptake of polycyclic aromatic hydrocarbons. Chemosphere, 2007, 67: 402-409.

[34] Zhang M, Zhu L Z. Sorption of polycyclic aromatic hydrocarbons to carbohydrates and lipids of ryegrass root and implications for a sorption prediction model. Environmental Science & Technology, 2009, 43(8): 2740-2745.

[35] Müller J F, Hawker D W, Connell D W. Calculation of bioconcentration factors of persistent hydrophobic compounds in the air/vegetation system. Chemosphere, 1994, 29: 623-640.

[36] Qu B, Zhao H, Zhou J. Toxic effects of perfluorooctane sulfonate (PFOS) on wheat (*Triticum aestivum* L.) plant. Chemosphere, 2010, 79: 555-560.

[37] Yang X, Ye C, Liu Y, et al. Accumulation and phytotoxicity of perfluorooctanoic acid in the model plant species *Arabidopsis thaliana*. Environmental Pollution, 2015, 206: 560-566.

[38] Zhao S, Liang T, Zhou T, et al. Biotransformation and responses of antioxidant enzymes in hydroponically cultured soybean and pumpkin exposed to perfluorooctane sulfonamide(FOSA). Ecotoxicology and Environmental Safety, 2018, 161: 669-675.

[39] Sun J Q, Chen Q, Qian Z, et al. Plant uptake and metabolism of 2,4-dibromophenol in carrot: *In vitro* enzymatic direct conjugation. Journal Agriculture and Food Chemistry, 2018, 66(17): 4328-4335.

[40] Sandermann H. Higher plant metabolism of xenobiotics: The 'green liver' concept. Pharmacogenetics, 1994, 4(5): 225-241.

[41] Bártíková H, Skálová L, Stuchlíková L, et al. Xenobiotic-metabolizing enzymes in plants and their role in uptake and biotransformation of veterinary drugs in the environment. Drug Metabolism Reviews, 2015, 47(3): 374-387.

[42] Durán N, Esposito E. Potential applications of oxidative enzymes and phenoloxidase-like compounds in wastewater and soil treatment: A review. Applied Catalysis B: Environment, 2000, 28(2), 83-99.

[43] Strommer J. The plant *ADH* gene family. The Plant Journal, 2011, 66(1): 128-142.

[44] Brocker C, Vasiliou M, Carpenter S, et al. Aldehyde dehydrogenase(ALDH)superfamily in plants: Gene nomenclature and comparative genomics. Planta, 2013, 237(1): 189-210.

[45] He D, Lei Z, Xing H, et al. Genome-wide identification and analysis of the aldehyde dehydrogenase(ALDH)gene superfamily of Gossypium raimondii. Gene, 2014, 549(1): 123-133.

[46] Siminszky B. Plant cytochrome P450-mediated herbicide metabolism. Phytochemistry Reviews, 2006, 5(2-3): 445-458.

[47] Dietz A C, Schnoor J L. Advances in phytoremediation. Environmental Health Perspectives, 2001, 109: 163-168.

[48] Schröder P, Collins C. Conjugating enzymes involved in xenobiotic metabolism of organic xenobiotics in plants. International Journal of Phytoremediation, 2002, 4(4): 247-265.

[49] Rea P A. Plant ATP-binding cassette transporters. Annu Rev Plant Biol, 2007, 58: 347-375.

[50] Hamdane D, Zhang H, Hollenberg P. Oxygen activation by cytochrome P450 monooxygenase. Photosynthesis Research, 2008, 98: 657-666.

[51] Zhai G, Lehmler H, Schnoor J L. Inhibition of cytochromes P450 and the hydroxylation of 4-monochlorobiphenyl in whole poplar. Environmental Science & Technology, 2013, 47(13): 6829-6835.

[52] Zhang H, Wen B, Hu X, et al. Uptake, translocation, and metabolism of 8: 2 fluorotelomer alcohol in soybean(*Glycine max* L. Merrill). Environmental Science & Technology, 2016, 50(24): 13309-13317.

[53] Magee K D, Michael A, Ullah H, et al. Dechlorination of PCB in the presence of plant nitrate reductase. Environmental Toxicology and Pharmacology, 2008, 25(2): 144-147.

[54] Huang H, Zhang S, Wang S, et al. In vitro biotransformation of PBDEs by root crude enzyme extracts: Potential role of nitrate reductase (NaR) and glutathione S-transferase (GST) in their debromination. Chemosphere, 2013, 90: 1885-1892.

[55] Blanchette B N, Singh B R. An enzyme based dechlorination of a polychlorinated biphenyl (PCB) mixture, Aroclor 1248, using glutathione *S*-transferases from the northern quahog *Mercenaria mercenaria*. Journal of Protein Chemistry, 2003, 22(4): 377-386.

[56] Munter T, Cottrell L, Golding B T, et al. Detoxication pathways involving glutathione and epoxide hydrolase in the *in vitro* metabolism of chloroprene. Chemical Research in Toxicology, 2003, 16(10): 1287-1297.

[57] Maddox C, Wang B, Kirby P A, et al. Mutagenicity of 3-methylcholanthrene, PCB3, and 4-OH-PCB3 in the lung of transgenic BigBlue® rats. Environmental Toxicology and Pharmacology, 2008, 25(2): 260-266.

[58] Boxtel A L, Kamstra J H, Cenijn P H, et al. Microarray analysis reveals a mechanism of phenolic polybrominated diphenylether toxicity in zebrafish. Environmental Science & Technology, 2008, 42(5): 1773-1779.

[59] Ptak A, Ludewig G, Lehmler H J, et al. Comparison of the actions of 4-chlorobiphenyl and its hydroxylated metabolites on estradiol secretion by ovarian follicles in primary cells in culture. Reproductive Toxicology, 2005, 20(1): 57-64.

[60] Machala M, Bláha L, Lehmler H J, et al. Toxicity of hydroxylated and quinoid PCB metabolites: Inhibition of gap junctional intercellular communication and activation of aryl hydrocarbon and estrogen receptors in hepatic and mammary cells. Chemical Research in Toxicology, 2004, 17(3): 340-347.

[61] Wan Y, Liu F, Wiseman S, et al. Interconversion of hydroxylated and methoxylated polybrominated diphenyl ethers in Japanese medaka. Environmental Science & Technology, 2010, 44(22): 8729-8735.

[62] Zhai G, Lehmler H J, Schnoor J L. Hydroxylated metabolites of 4-monochlorobiphenyl and its metabolic pathway in whole poplar plants. Environmental Science & Technology, 2010, 44(10): 3901-3907.

[63] Liu J Y, Hu D F, Jiang G B, et al. *In vivo* biotransformation of 3,3′,4,4′-tetrachlorobiphenyl by whole plants-poplars and switchgrass. Environmental Science & Technology, 2009, 43(19): 7503-7509.

[64] Zhai G, Lehmler H J, Schnoor J L. Identification of hydroxylated metabolites of 3,3′,4,4′-tetrachlorobiphenyl and metabolic pathway in whole poplar plants. Chemosphere, 2010, 81: 523-528.

[65] Ma C, Zhai G, Wu H, et al. Identification of a novel hydroxylated metabolite of 2,2′,3,5′,6-pentachlorobiphenyl formed in whole poplar plants. Environmental Science and Pollution Research, 2016, 23(3): 2089-2098.

[66] Bock C. Untersuchungen des Metabolismus von 3,3′,4,4′-tetrchlorbiphenyl (PCB77) und 2,2′,5-trichlorbiphenyl(PCB 18)in ausgewaehlten pflanzlichen *in-vitro*-systemen. Landbauforschung Volkenrode, 1999, 207 (Sonderheft): 1-106.

[67] Yu M, Liu J, Wang T, et al. Metabolites of 2,4,4′-tribrominated diphenyl ether (BDE-28) in pumpkin after *in vivo* and *in vitro* exposure. Environmental Science & Technology, 2013, 47(23): 13494-13501.

[68] Sun J, Liu J, Yu M, et al. In vivo metabolism of 2,2′,4,4′-tetrabromodiphenyl ether (BDE-47) in young whole pumpkin plant. Environmental Science & Technology, 2013, 47(8): 3701-3707.

[69] Wang S, Zhang S, Huang H, et al. Debrominated, hydroxylated and methoxylated metabolism in maize (*Zea mays* L.) exposed to lesser polybrominated diphenyl ethers (PBDEs). Chemosphere, 2012, 89: 1295-1301.

[70] Huang H, Zhang S, Lv J, et al. Experimental and theoretical evidence for diastereomer-and enantiomer-specific accumulation and biotransformation of HBCD in maize roots. Environmental Science & Technology, 2016, 50(22): 12205-12213.

[71] Meggo R E, Schnoor J L, Hu D. Dechlorination of PCBs in the rhizosphere of switchgrass and poplar. Environmental Pollution, 2013, 178: 312-321.

[72] Magge K D, Michael A, Ullah H, et al. Dechlorination of PCB in the presence of plant nitrate reductase. Environmental Toxicology and Pharmacology, 2008, 25(2): 144-147.

[73] Xu X, Wen B, Huang H, et al. Uptake, translocation and biotransformation kinetics of BDE-47, 6-OH-BDE-47 and 6-MeO-BDE-47 in maize (*Zea mays* L.). Environmental Pollution, 2016, 208: 714-722.

[74] Wang S, Zhang S, Huang H, et al. Uptake, translocation and metabolism of polybrominated diphenyl ethers (PBDEs) and polychlorinated biphenyls (PCBs) in maize (*Zea mays* L.). Chemosphere, 2011, 85: 379-385.

[75] Zhao M, Zhang S, Wang S, et al. Uptake, translocation, and debromination of polybrominated diphenyl ethers in maize. Journal of Environmental Sciences, 2012, 24(3): 402-409.

[76] Gerecke A C, Hartmann P C, Heeb N V, et al. Anaerobic degradation of decabromodiphenyl ether. Environmental Science & Technology, 2005, 39(4): 1078-1083.

[77] Field J A, Sierra-Alvarez R. Microbial transformation and degradation of polychlorinated biphenyls. Environmental Pollution, 2008, 155: 1-12.

[78] Garrison A W, Nzengung V A, Avants J K, et al. Phytodegradation of *p, p′*-DDT and the Enantiomers of *o, p′*-DDT. Environmental Science & Technology, 2000, 34(9): 1663-1670.

[79] Gao J, Garrison A W, Hoehamer C, et al. Uptake and phytotransformation of *o, p′*-DDT and *p, p′*-DDT by axenically cultivated aquatic plants. Journal of Agriculture and Food Chemistry, 2000, 48(12): 6121-6127.

[80] Wang S, Cao S, Wang Y, et al. Fate and metabolism of the brominated flame retardant tetrabromobisphenol A(TBBPA)in rice cell suspension culture. Environmental Pollution, 2016, 214: 299-306.

[81] Hou X, Yu M, Liu A, et al. Biotransformation of tetrabromobisphenol A dimethyl ether back to tetrabromobisphenol A in whole pumpkin plants. Environmental Pollution, 2018, 241: 331-338.

[82] Lin K, Gan J, Liu W. Production of hydroxylated polybrominated diphenyl ethers from bromophenols by bromoperoxidase-catalyzed dimerization. Environmental Science & Technology, 2014, 48(20): 11977-11983.

[83] Arnoldsson K, Andersson P L, Haglund P. Formation of environmentally relevant brominated dioxins from 2,4,6,-tribromophenol via bromoperoxidase-catalyzed dimerization. Environmental Science & Technology, 2012, 46(13): 7239-7244.

[84] Ashrap P, Zheng G, Wan Y, et al. Discovery of a widespread metabolic pathway within and among phenolic xenobiotics. Proceedings of the National Academy Sciences of the United States of America, 2017, 114(23): 6062-6067.

[85] Sun J, Pan L, Su Z, et al. Interconversion between methoxylated and hydroxylated polychlorinated biphenyls in rice plants: An important but overlooked metabolic pathway. Environmental Science & Technology, 2016, 50(7): 3668-3675.

[86] Sun J, Liu J, Liu Y, et al. Reciprocal transformation between hydroxylated and methoxylated polybrominated diphenyl ethers in young whole pumpkin plants. Environmental Science & Technology Letters, 2014, 1(4): 236-241.

[87] Sun J, Pan L, Chen J, et al. Uptake, translocation, and metabolism of hydroxylated and methoxylated polychlorinated biphenyls in maize, wheat, and rice. Environ Science and Pollution Research, 2018, 25(1): 12-17.

[88] Pan L, Sun J, Wu X, et al. Transformation of hydroxylated and methoxylated 2,2′,4,4′,5-brominated diphenyl ether (BDE-99) in plants. Journal of Environmental Sciences, 2016, 49: 197-202.

[89] Peng F Q, Ying G G, Yang B, et al. Biotransformation of the flame retardant tetrabromobisphenol-A (TBBPA) by freshwater microalgae. Environmental Toxicology and Chemistry, 2014, 33(8): 1705-1711.

[90] Zhai G, Lehmler H J, Schnoor J L. Sulfate metabolites of 4-monochlorobiphenyl in whole poplar plants. Environmental Science & Technology, 2013, 47(1): 557-562.

[91] Arjmand M, Sandermann H. Metabolism of DDT and related compounds in cell suspension cultures of soybean (*Glycine max* L.) and wheat (*Triticum aestivum* L.). Pesticide Biochemistry and Physiology, 1985, 23(3): 389-397.

[92] Arbuckle T E, Marro L, Davis K, et al. Exposure to free and conjugated forms of bisphenol A and triclosan among pregnant women in the MIREC cohort. Environmental Health Perspectives, 2014, 123(4): 277-284.

[93] Ho K L, Yuen K K, Yau M S, et al. Glucuronide and sulfate conjugates of tetrabromobisphenol A (TBBPA): Chemical synthesis and correlation between their urinary levels and plasma TBBPA content in voluntary human donors. Environment International, 2017, 98: 46-53.

[94] Rowland A, Miners J O, Mackenzie P I. The UDP-glucuronosyltransferases: Their role in drug metabolism and detoxification. The International Journal of Biochemistry & Cell Biology, 2013, 45(6): 1121-1132.

[95] Meech R, Miners J O, Lewis B C, et al. The glycosidation of xenobiotics and endogenous compounds: versatility and redundancy in the UDP glycosyltransferase superfamily. Pharmacology & Therapeutics, 2012, 134(2): 200-218.

[96] Taguchi G, Ubukata T, Nozue H, et al. Malonylation is a key reaction in the metabolism of xenobiotic phenolic glucosides in Arabidopsis and tobacco. The Plant Journal, 2010, 63(6): 1031-1041.

[97] Macherius A, Eggen T, Lorenz W, et al. Metabolization of the bacteriostatic agent triclosan in edible plants and its consequences for plant uptake assessment. Environmental Science & Technology, 2012, 46(19): 10797-10804.

[98] Macherius A, Seiwert B, Schröder P, et al. Identification of plant metabolites of environmental contaminants by UPLC-QToF-MS: The *in vitro* metabolism of triclosan in horseradish. Journal of Agriculture and Food Chemistry, 2014, 62(5): 1001-1009.

[99] Fu Q, Ye Q, Zhang J, et al. Diclofenac in Arabidopsis cells: Rapid formation of conjugates. Environmental Pollution, 2017, 222: 383-392.

[100] Pogány E, Pawlizki K H, Wallnöfer P R. Formation, distribution and bioavailability of cell wall bound residues of 4-chloroaniline and 2, 4-dichlorophenol. Chemosphere, 1990, 21: 349-358.

[101] Scheel D, Schaefer W, Sandermann H. Metabolism of pentachlorophenol in cell suspension cultures of soybean (*Glycine max* L.)and wheat (*Triticum aestivum* L.). General results and isolation of lignin metabolites. Journal of Agriculture and Food Chemistry, 1984, 32: 1237-1241.

[102] Still G G, Balba H M, Mansager E R. Studies on the nature and identity of bound chloroaniline residues in plants. Journal of Agriculture and Food Chemistry, 1981, 29: 739-746.

[103] Schröder P. Fate of glutathione *S*-conjugates in plants: Cleavage of the glutathione moiety. In Hatzios KK, Ed. Regulation of Enzymatic Systems Detoxifying Xenobiotics in Plants. The Netherlands: Kluwer Acad Publ, 1997: 233-244.

[104] Marrs K A. The functions and regulation of glutathione *S*-transferases in plants. Annual Review of Plant Physiology and Plant Molecular Biology, 1996, 47: 127-158.

[105] Rea P A, Li Z S, Lu Y P, et al. From vacuolar GS-X pumps to multispecific ABC transporters. Annual Review of Plant Physiology and Plant Molecular Biology, 1998, 49: 727-760.

[106] Ballesteros-Gómez A, Ballesteros J, Ortiz X, et al. Identification of novel brominated compounds in flame retarded plastics containing TBBPA by combining isotope pattern and mass defect cluster analysis. Environmental Science & Technology, 2017, 51(3): 1518-1526.

[107] Cariou R, Omer E, Léon A, et al. Screening halogenated environmental contaminants in biota based on isotopic pattern and

mass defect provided by high resolution mass spectrometry profiling. Analytica Chimica Acta, 2016, 936: 130-138.

[108] Jobst K J, Shen L, Reiner E J, et al. The use of mass defect plots for the identification of(novel)halogenated contaminants in the environment. Analytical and Bioanalytical Chemistry, 2013, 405(10): 3289-3297.

[109] Liu Y N, Richardson E S, Derocher A E, et al. Hundreds of unrecognized halogenated contaminants discovered in polar bear serum. Angewandte Chemie International Edition, 2018, 57(50): 16401-16406.

[110] Harner T, Wiberg K, Norstrom R. Enantiomer fractions are preferred to enantiomer ratios for describing chiral signatures in environmental analysis. Environmental Science & Technology, 2000, 34(1): 21-220.

[111] Dai S H, Wong C S, Qiu J, et al. Enantioselective accumulation of chiral polychlorinated biphenyls inlotus plant (*Nelumbonucifera* spp.). Journal of Hazardous Materials, 2014, 280: 612-618.

[112] Wong C S, Garrison A W, Smith P D, et al. Enantiomeric composition of chiral polychlorinated biphenyl atropisomers in aquatic and riparian biota. Environmental Science & Technology, 2001, 35(12): 2448-2454.

[113] Ross M S, Verreault J, Letcher R J, et al. Chiral organochlorine contaminants in blood and eggs of glaucous gulls (*Larus hyperboreus*) from the Norwegian Arctic. Environmental Science & Technology, 2008, 42(19): 7181-7186.

[114] Serrano R, Fernandez M, Rabanal R, et al. Congener-specific determination of polychlorinated biphenyls in shark and grouper livers from the northwest African Atlantic Ocean. Archives of Environmental Contamination and Toxicology, 2000, 38(2): 217-224.

[115] Zhai G S, Hu D F, Lehmler H J, et al. Enantioselective biotransformation of chiral PCBs in whole poplar plants. Environmental Science & Technology, 2011, 45(6): 2308-2316.

[116] Zhai G S, Gutowski S M, Lehmler H J, et al. Enantioselective transport and biotransformation of chiral hydroxylated metabolites of polychlorinated biphenyls in whole poplar plants. Environmental Science & Technology, 2014, 48(20): 12213-12220.

[117] Chen S J, Tian M, Zheng J, et al. Elevated levels of polychlorinated biphenyls in plants, air, and soils at an e-waste site in southern China and enantioselective biotransformation of chiral PCBs in plants. Environmental Science & Technology, 2014, 48(7): 3847-3855.

[118] Warner N A, Martin J W, Wong C S. Chiral polychlorinated biphenyls rre biotransformed enantioselectively by mammalian cytochrome P-450 isozymes to form hydroxylated metabolites. Environmental Science & Technology, 2009, 43(1): 114-121.

[119] Lu Z, Wong C S. Factors affecting phase I stereoselective biotransformation of chiral polychlorinated biphenyls by rat cytochrome P-450 2B1 isozyme. Environmental Science & Technology, 2011, 45(19): 8298-8305.

[120] Borlakoglu J T, Wilkins J P G. Correlations between the molecular structures of polyhalogenated biphenyls and their metabolism by hepatic microsomal monooxygenases. Comparative Biochemistry and Physiology Part C: Comparative Pharmacology, 1993, 105(1): 113-117.

作者：侯兴旺[1]，刘稷燕[1]
[1] 中国科学院生态环境研究中心

第10章 中国大气环境化学研究进展

- 1. 引言 /257
- 2. 大气光化学污染 /257
- 3. 大气成核和新粒子形成机制 /259
- 4. 二次有机气溶胶与大气非均相/多相反应 /262
- 5. 展望 /264

本章导读

大气环境化学是认识空气污染和气候变化问题的重要学科领域。随着我国大气污染防治工作的深入推进，大气中 $PM_{2.5}$ 浓度逐年下降，但仍处于高位且下行压力持续增大，光化学污染开始凸显并呈恶化态势，我国大气复合污染下的大气化学机制研究成为环境科学领域的热点问题。近年来，国内外学者围绕我国区域大气复合污染的关键化学过程开展了大量研究工作，取得了一系列重要进展。本章对国内外学者在大气光化学污染、大气成核和新粒子形成机制、二次有机气溶胶与大气非均相/多相反应等大气环境化学前沿科学问题方面取得的重要研究成果进行介绍，并对未来研究的重点方向进行展望。

关键词

区域大气复合污染，霾，大气光化学污染，大气成核，大气非均相化学，大气多相化学，二次有机气溶胶

1 引 言

大气环境化学是一门专门研究污染物在大气中的来源、迁移、转化、归宿及其对大气环境、人体健康、生态系统和气候影响的学科，是环境化学的重要学科方向，也是化学和地球科学高度交叉的学科领域。该学科方向随全球和区域大气环境问题的出现、演变而不断发展，同时又指导人们控制大气污染，解决大气环境及其影响等问题。"十三五"以来，随着大气污染防治工作的深入推进，我国区域大气复合污染的形势发生了深刻变化：主要区域城市大气中 $PM_{2.5}$ 浓度逐年下降，化学组成发生显著变化（硫酸盐比重下降，硝酸盐和有机物比例上升），$PM_{2.5}$ 污染总体仍处于高位且下行压力持续增大；光化学（臭氧）污染开始凸显并呈恶化态势。在此背景下，国内外学者围绕我国区域大气复合污染的关键化学过程开展了大量研究工作，在大气成核和新粒子形成机制、二次有机气溶胶形成机理、大气颗粒物表面/界面的非均相反应、大气氧化性与自由基化学、$PM_{2.5}$ 与臭氧的协同控制机制、实验室模拟和外场观测技术发展与应用等方面取得了很大进展。本章对近两年来该学科方向，特别是我国科研工作者主要研究进展进行总结和归纳。

2 大气光化学污染

大气光化学污染是发达地区和经济快速发展地区普遍面临的区域性环境问题之一。它以高浓度的臭氧、过氧乙酰硝酸酯（PAN）、醛类等二次污染物为主要标志，对人体健康、农作物生长和生态环境具有严重危害。近年来，随着社会经济的快速发展和能源消耗的迅速增加，我国大气光化学污染形势日益严峻，臭氧作为2013年以来我国大部分地区六项监测指标中唯一"不降反升"的污染物，已经成为我国城市空气质量达标的重要挑战，引起了政府和科研人员的深切关注。近几年，相关学者开始从更长时间跨度和更大空间尺度来研究我国近地面臭氧浓度的变化趋势，从全球尺度以及各类自然源与人为排放层面分析我国臭氧的来源，深入探讨除臭氧前体物（VOCs 和 NO_x）以外的其他活性物种和化学过程对臭氧生成的影响，评估臭氧暴露对人体健康、生态环境和社会经济的不利影响，聚焦"$PM_{2.5}$ 和臭氧协同控制"开展了大量研究工作，取得了很大进展。本节重点对2017年以来关于我国大气光化学污染研究所取得的重要进展进行梳理。

国际全球大气化学计划（IGAC）2018年正式发布了全球对流层臭氧评估项目（TOAR）研究报告，该报告以专刊系列论文的形式发表在 Elementa: Science of the Anthropocene 上。TOAR 搜集了全球范围内 9000 多个站点的臭氧长期监测数据，旨在全面分析全球不同区域对流层臭氧的长期趋势、现状以及对气候变化、人体健康、农作物和生态系统的影响。总体上，2000～2014 年全球对流层臭氧变化趋势呈现出明显的地区性和季节性特征，其中欧美大部分地区夏季日间臭氧浓度呈下降趋势，冬季则呈上升趋势或者无显著变化[1]。与欧美地区相比，我国具有十年以上跨度臭氧监测的站点十分有限，但这些站点所记录的臭氧浓度在过去十几年均呈现显著的升高趋势。2003～2015 年华北地区泰山顶和上甸子区域本底站的地面臭氧浓度年均增幅达 1～2 ppb/a[2, 3]；1994～2013 年青藏高原瓦里关全球大气本底站的地面臭氧增幅亦有 0.2～0.3 ppb/a[4]。2013 年以来，环保部环境空气质量监测网络加强了对臭氧的监测，这为深入分析近年来我国臭氧污染的变化趋势和特征提供了宝贵资料。Wang 等[5]发现 2013～2015 年北京、上海、兰州和成都的臭氧日最大 8 小时平均浓度上升了 12%～34%；Li 等[6]发现 2014～2017 年间北京、上海和成都的臭氧浓度增幅为 1.1～1.7 ppb/a，扣除气象因素的影响后我国东部城市群的夏季臭氧增幅可达 1～3 ppb/a，其中京津冀地区增长最快。Lu 等[7]分析了 TOAR 和我国环境空气质量监测网络的臭氧数据，发现当前中国城市臭氧污染程度要显著高于同纬度的欧美日韩等发达国家，且从历史角度看我国主要城市的臭氧浓度相当于甚至高于美国 20 世纪 80 年代的水平（见图 10-1），足见污染形势不容乐观。未来仍需要对我国不同区域地面臭氧的演变趋势及其驱动因素进行更加全面的分析。

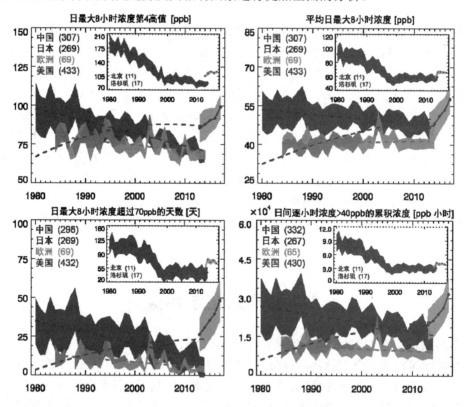

图 10-1① 　1980～2017 年暖季（4～9月）中国（红色），日本（紫色），欧洲（橙色）和美国（蓝色）城市地区地面观测臭氧浓度的变化

小图中显示了北京（红色）和洛杉矶（蓝色）的臭氧变化。对于日本、欧洲和美国，只纳入在 1980～2014 年间有超过 25 年有效记录的城市站点。对于中国，纳入了 2013～2017 年有连续观测的 74 个主要城市的站点。图中阴影区域表示各项指标平均值±50％标准偏差的范围，括号内数值表示站点数量。虚线表示欧洲和美国臭氧浓度变化的线性拟合曲线以及日本臭氧浓度变化的抛物线拟合曲线。（图片来自文献[7]）

---

① 彩图信息请扫描封底二维码查看。

一些学者利用化学传输模式从区域或全球尺度分析了我国对流层臭氧的复杂来源以及人为活动和自然因素的贡献。Sun 等[8]用 GEOS-Chem 模型量化了人为排放和气象条件变化对我国中东部地区臭氧趋势的影响，发现人为污染排放增加是近十多年（2003～2015）区域臭氧持续增长的主要原因，而气象条件则是导致臭氧污染年际变化和空间差异的重要因素。Ni 等[9]利用 GEOS-Chem 双向耦合模型分析了全球 8 个主要源区对我国春季臭氧的贡献，发现国外人为源排放对我国近地面臭氧浓度贡献了 2～11 ppbv，对 2 km 以下和对流层上层的贡献分别高达 40%～60%和 85%，其中日韩、东南亚、南亚地区对我国东部沿海和华南地区具有显著贡献。Lu 等[10]发现臭氧自然来源（不包含全球人为排放时的臭氧浓度）对我国夏季近地面臭氧日最大 8 小时平均浓度的贡献高达 70%，其中生物源 VOCs 排放、闪电 NO_x 排放、土壤 NO_x 排放、生物质燃烧排放和平流层输送分别贡献了 1～10 ppbv，不同地区各类自然来源的贡献有明显差异。总体来看，自然源对我国月平均近地面臭氧浓度起主导作用，而国内人为源排放对臭氧污染事件的贡献更为重要（高达 38%～69%），因此在夏季高温条件下对关键前体物进行控制可以有效避免我国城市臭氧超标现象。

关于臭氧光化学机制的研究最近也取得了一些重要进展，特别是在大气非均相化学和卤素活化等过程对臭氧的影响方面取得了一些新的认识。香港理工大学王韬课题组将 HONO、N_2O_5、$ClNO_2$ 等活性氮氧化物的来源和化学过程加入 WRF-Chem 模式，模拟分析了氮氧化物非均相化学过程对我国不同地区臭氧生成的影响，发现 HONO 和 $ClNO_2$ 可以显著增加华北平原、长三角和珠三角等重点区域的地面臭氧（6%～13%）和边界层臭氧浓度（5%～6%）[11]。更为重要的是，由于 HONO 和 $ClNO_2$ 光解可以同时释放 NO_x 和自由基（OH 和 Cl 原子），考虑新的氮氧化物非均相化学机制后可以显著改变臭氧生成对 NO_x 和 VOCs 敏感性的诊断结果，即加入 HONO 和 $ClNO_2$ 均导致臭氧生成机制更倾向于"NO_x 和 VOCs 共同控制"[12]，该结果对于重新评估臭氧生成机制和制定污染控制对策具有重要的参考价值。2018 年年底，《美国科学院院刊》在线发表了南京信息工程大学廖宏团队与哈佛大学的合作研究成果，该工作利用 2013～2017 年我国环境空气质量监测网络的监测数据和 GEOS-Chem 模型解释了 $PM_{2.5}$ 和 O_3 控制的"跷跷板"效应，认为近年来我国城市大气中 $PM_{2.5}$ 浓度的显著下降导致 HO_2 自由基在颗粒物表面非均相反应汇的减弱，对我国近年来臭氧浓度的升高具有重要贡献[6]。因此，$PM_{2.5}$ 与 O_3 的协同控制已经成为今后我国大气污染防治的重大课题。

此外，一些学者也初步评估了我国臭氧污染对人体健康和生态系统的不利影响。相关研究表明，臭氧暴露可能是慢性阻塞性肺病病人心肺血管功能减弱的主要原因[13]；据估计我国 2016 年由于臭氧暴露导致的死亡人数约为 7 万人，经济损失高达 7.6 亿美元左右[14]。此外，由于臭氧污染，我国生态系统每年净初级生产力减少约 14%（0.6 Pg 碳）[15]。

综上，我国大气光化学污染问题已经凸显并呈逐步恶化态势，臭氧污染控制已经迫在眉睫，国家和各地也加大了对臭氧污染监测和研究的支持力度，未来仍需在臭氧污染的演变态势、化学机制、区域传输、生态和健康影响以及管理决策等方面开展大量系统深入的研究工作，进而支撑臭氧污染防治和空气质量持续改善。2019 年 3 月 29 日，中国环境科学学会臭氧污染控制专业委员会在成都正式成立，未来将大力推进臭氧污染控制相关领域的学术发展与防控进程，推动我国城市和区域大气臭氧污染防治从科学走向实践。

3 大气成核和新粒子形成机制

大气中部分气体分子在随机碰撞中可以形成稳定的分子簇，并逐渐生长为纳米颗粒物，被称为大气成核（atmospheric nucleation）；如果成核过程所形成的分子簇和纳米颗粒物不与大气中已存在颗粒物碰并而损失，则可以继续生长形成大气新颗粒物而成为大气新粒子形成事件（new particle formation）[16]。大气新粒子形成事件是一种重要的二次气溶胶形成途径，国内外学者已经在不同的大气环境中广泛地观测到大气新粒子形成事件：其最直观的体现是成核模态的颗粒物数浓度的急剧增加，并且这些成核模态的纳米颗粒物的粒径会持续增长至爱根模态。目前学术界已经广泛承认了硫酸分子在绝大多数地区大气成核过程中的关键作用[17]。有关大气新粒子生成事件国内研究现状的综述可以参照文献[18, 19]，此处仅介绍其中引发学者较为关注的部分国内外代表性研究工作。

3.1 近期实验室模拟研究进展

有关大气成核与新粒子形成的研究进展建立在外场观测、实验室模拟、理论预测的综合研究中。在实验室模拟方面，以欧盟科学家主导的 CERN-CLOUD 系列实验最为著名，开展了包括硫酸—水二元成核[20]、硫酸—氨气—水三元成核[20]、硫酸—有机胺—水三元成核[21, 22]、硫酸—极低挥发性有机化合物—水三元成核[23, 24]、离子诱导成核[20]、离子诱导下的纯生物源有机化合物反应成核[25]等机制在内的一系列实验室模拟，研究了不同条件下成核关键前体物的浓度与成核速率的对应关系、成核过程中中性和带电分子簇的化学组分和生成顺序、不同类别的前体物对分子簇和纳米颗粒物生长的定量贡献等，形成了多个重大科学成果。

3.2 近期外场观测研究进展

通过将外场观测中所获得的大气成核特征与 CERN-CLOUD 实验的模拟结果进行对比，国外学者已在少数具有相对洁净大气的地点确定了真实大气成核机制，包括芬兰北方森林地区的硫酸—极低挥发性有机化合物—水三元成核[24, 26]、瑞士少女峰自由大气中不同大气条件下的硫酸—极低挥发性有机化合物—水成核或硫酸—氨气—水三元成核[27]、爱尔兰西海岸 Mace Head 地区的碘酸（HIO_3）成核[28]、以及南极海岸地区离子诱导下的硫酸—氨气成核[29]。此外，单萜烯氧化生成的高氧化度有机分子可以在极低的气体硫酸浓度条件下形成仅由生物源高氧化度有机分子构成的离子团簇，可能是工业革命前大气新粒子生成的一种机制[30]。这些研究表明全球各地的大气成核机制可能各不相同，具有地域特征。

上述国外外场观测研究主要集中于相对洁净的大气条件下，而中国城市地区的实际大气条件显著不同，可导致大气新粒子生成和生长的前体物浓度较高，同时可作为新生颗粒物碰并汇的已存颗粒物浓度也较高；同时新粒子生成事件与污染地区灰霾事件是否存在一定关联尚不完全清楚。最近几年，我国学者开展了大量的有关大气成核与新粒子形成及其前体物的外场观测研究，其中北京大学胡敏课题组的研究暗示北京地区的大气颗粒物高污染可能最终源自于空气洁净时期的大气新粒子形成事件，将这两个大气化学领域的前沿难题联系起来：在大气新粒子形成事件中高数浓度的纳米颗粒物得以形成，并在接下来的几天时间尺度内粒径得以连续增长，最终形成大气颗粒物污染乃至灰霾事件（如图 10-2 所示）[31]。

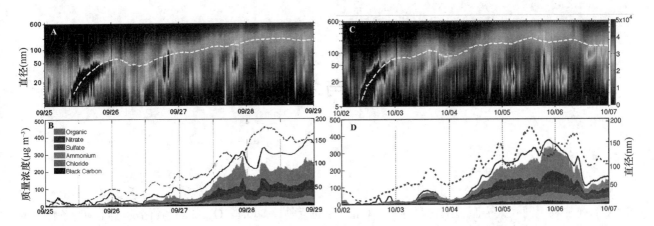

图 10-2　A，C 分别为两次 $PM_{2.5}$ 污染事件过程中的颗粒物粒径分布和平均粒径（白色线）变化情况；B，D 分别为两次 $PM_{2.5}$ 污染事件过程中 $PM_{2.5}$ 质量浓度（实线），平均粒径（虚线），$PM_{1.0}$ 的化学组分的变化情况

复旦大学王琳课题组则应用了大气常压界面—飞行时间质谱和硝酸根化学电离—飞行时间质谱技术对大气成核关键气态前体物及其团簇进行识别（如图 10-3 所示），在上海市大气中首次发现并证实了硫酸—二甲胺—水三元成核机制可以解释所观测到的大气新粒子形成事件[32]。

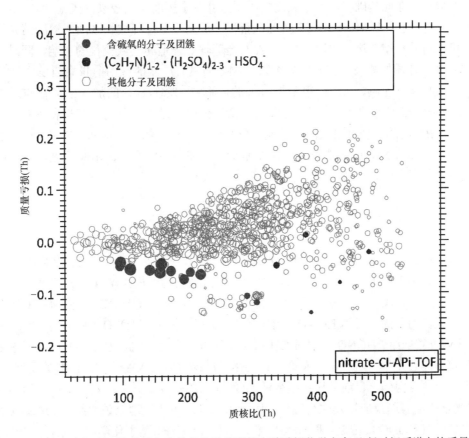

图 10-3　上海市大气中识别到的关键气态前体物及其团簇在硝酸根化学电离-飞行时间质谱上的质量亏损图

4 二次有机气溶胶与大气非均相/多相反应

4.1 二次有机气溶胶

二次有机气溶胶（SOA）在区域污染、大气辐射平衡以及气候变化等方面具有重要作用。由于生成 SOA 的前体物不同（人为源、生物源等）、氧化途径的竞争性（OH、O_3、NO_3 氧化及不同 NO_x 条件下的光氧化等）及大气环境条件的影响（颗粒相、液相、多相等），使得 SOA 光学性质存在显著差异，而实现 SOA 光学特性准确测量，特别是针对 SOA 重要前体物在不同波长下光学特性研究尤为重要。中国科学院化学研究所葛茂发等结合光腔衰荡光谱、光声光谱仪系统研究了苯系物、长链烷烃、柠檬烯在不同条件下氧化生成 SOA 的光学性质[33-35]，并在此基础上开发了多波长光腔衰荡光谱仪。通过研究 SOA 重要潜在源——长链烷烃 SOA 的光学性质发现[36]，正十二烷（C12）、正十五烷（C15）和正十七烷（C17）的光氧化得到的 SOA 的复折射率（RI）在波长 532 nm 处主要以散射为主。低 NO_x 条件，无机种子的存在导致 RI 值的实部减少，而高 NO_x 条件，无种子条件下的 RI 值低于种子条件下的 RI 值，研究结果与化学成分解析相互验证。在此基础上，中科院化学所团队开发了多波长光腔衰荡光谱（MCRD-AES）。通过使用可调光源和平行腔技术，MCRD-AES 实现从紫外—可见—近红外波长区间气溶胶消光特性的在线测量。研究结果可以帮助模型改进，用以评估 SOA 对全球辐射强迫和气候变化的影响，同时高精度和波长依赖性的气溶胶光学性质检测技术在科学研究及监测方面具有广阔的应用前景。Chen 等[37]研究了 TiO_2 对二次有机气溶胶形成的影响，在间二甲苯与 NO_x 的体系中，发现 TiO_2 的加入会抑制 SOA 的形成。TiO_2 对于 SOA 的抑制强弱与 NO_x 的初始浓度以及种子气溶胶的加入有关。通过对气相和颗粒相产物分析发现 TiO_2 对于 SOA 抑制作用的一个直接原因是抑制了气相中羰基化合物的形成；间二甲苯的主要光催化产物是间甲基苯甲醛，而加入 NO_x 后，硝酸-3-甲基苄基酯也会在体系中产生。由此可见，TiO_2 的加入影响了反应动力学及微观机制[37]。

4.2 HONO

羟基自由基（OH）是大气中最重要的氧化剂之一，氧化性极强，可以与有机物发生一系列的光氧化过程，导致臭氧、过氧乙酰硝酸酯（PAN）和大量二次污染物的形成，增强大气的氧化能力。对流层大气中 OH 自由基的主要来源有 O_3、醛类和亚硝酸（HONO）光解，其中 HONO 光解对 OH 自由基的贡献越来越多地受到人们的关注，日间 HONO 光解产生的 OH 自由基占日间 OH 总量的 34%～56%，尤其当清晨 O_3 和甲醛浓度较低时，HONO 光解对 OH 自由基的贡献高达 80%。因此，HONO 严重影响着大气的氧化能力。尤其在最近几年，随着研究人员对大气化学过程逐渐深入的认识，仪器的发展，对 HONO 的研究日益成为一个热点问题。目前普遍被接受的 HONO 来源有以下五个：直接排放、均相反应、非均相反应、光解或光增强的表面反应以及生物过程。由于 HONO 的生成机理众说纷纭，模式模拟得到的 HONO 浓度远低于实际大气 HONO 的浓度。长光程吸收光谱法（LOPAP）基于化学湿法采样和光度测量的原理，采用双通道原理，利用差减法最大限度排除 NO_2 和 PAN 等气体的干扰影响。国内中国科学院化学研究所及北京大学都已经利用该方法搭建了具有自主知识产权的 HONO 自动在线分析仪，与国内外仪器对比，具有优异的准确性和稳定性，尤其针对我国的重污染条件下的观测研究更具有优势，已经应用在多次大型观测中，均取得了良好的成效，也得到了国内外同行的广泛认可。

北京地区 2014 年 2 月 22 日~3 月 2 日及 2016 年 12 月 16~23 日均经历了一个典型的霾污染过程和一个清洁过程，Hou 和 Zhang 等对这两个时期观测数据进行分析，发现 2014 年冬季，由于重污染天气下多种污染物聚集，导致 HONO 的未知来源相比于清洁天更多[38]；在 2016 年的观测中发现，机动车直接排放以及 NO 的均相反应是北京城区夜间 HONO 的直接来源[39]。同时，HONO 与 $PM_{2.5}$ 以及相对湿度之间存在着密切联系。Tong 等对 2014 年冬季北京城区和郊区的观测，发现城区污染期和清洁期 HONO 浓度分别为郊区的 2.63 倍和 1.80 倍。城区 HONO 浓度与 CO，NO_x 的相关性较好，表明直接排放是影响城区 HONO 与 CO 和 NO_x 相关性的主要因素。比较了城区和郊区在污染和清洁时期气相反应和非均相反应的夜间产率。由于城区的高 NO 浓度，气相反应的贡献在污染和清洁时期都是显著的。在郊区，非均相反应是 HONO 的主要来源，其产率在污染和清洁时期分别是气相反应的 11.9 倍和 20.0 倍[40, 41]。

4.3 多相反应

多相化学反应是二次颗粒物爆发性增长的主要驱动力之一，我国尤其是京津冀地区污染物排放强度大，高浓度的大气颗粒物为大气痕量气体及光氧化剂等重要活性物种提供了多相反应的重要平台，这些物种之间的多相化学转化、相互耦合会进一步对区域环境造成影响。目前污染物的多相物理化学过程以及不同相间的迁移转化规律还存在很大的不确定性，灰霾形成过程中污染物的气-液-固多相反应机制还不清楚，极大限制了数值模拟结果的合理性，同时也对我国对准确评估环境政策的影响提出了巨大的挑战。

大气颗粒物通常为包含一次气溶胶如矿尘、海盐等和二次气溶胶如硫酸盐、硝酸盐等的复杂混合物。Tan 等发现 NO_2 与 $(NH_4)_2SO_4$ 并不反应，而当 $CaCO_3$ 与 $(NH_4)_2SO_4$ 混合后，在"干态"下混合体系与 NO_2 的反应主要以 $CaCO_3$ 与 NO_2 的反应为主，且反应后生成的 NO_3^- 浓度与颗粒物中 $CaCO_3$ 含量呈线性关系。当湿度增加后，$(NH_4)_2SO_4$ 不仅参与了反应，甚至促进 NO_3^- 在混合物表面的生成[42]。Na_2SO_4 和 $CaCO_3$ 都能与 NO_2 发生反应，干态条件下 NO_3^- 的生成量和生成速率符合 NO_2 与纯组分反应中 NO_3^- 生成的线性叠加，随着湿度增加，$Ca(NO_3)_2$ 与 Na_2SO_4 在混合颗粒表面的相互作用对 NO_3^- 的生成具有促进作用，使混合颗粒与 NO_2 的反应能力大于纯组分反应活性的加和，存在 1 + 1 > 2 的现象。同时，逐渐有 $CaSO_4 \cdot 0.5H_2O$ 和 $CaSO_4 \cdot 2H_2O$ 晶体的形成[43]。研究结果表明颗粒物混合后 NO_2 发生的多相反应，既影响了二次硝酸盐的生成量，也会影响硫酸盐的存在形式，使城市区域的污染变得更为复杂。

Zhang 等发现相对湿度的增加能够促进 $CaCO_3$ 表面二次硫酸盐的快速形成，SO_4^{2-} 生成浓度最大值（反应 200 min）、初始阶段和稳定阶段生成速率的最大值分别是干态下的 14 倍、1.5 倍和 43 倍[44]。进一步，他们发现减小 $CaCO_3$ 的粒径也会促进硫酸盐的形成。发现 RH 会影响形成的硫酸钙的形貌，而颗粒物的粒径无论对产物的形貌还是大小都没有影响。当 RH≤50%，表面会形成微晶形态的硫酸钙，而当 RH≥60%，形成的液态水层能够通过促进不断生成的硫酸钙水合物的形成和凝聚，从而促进硫酸钙棒状晶体的形成和增长，极限长度大约为 2 μm。结合已有研究中给出的燃煤飞灰和电厂排放出的硫酸钙形貌，为大气中硫酸钙棒状晶体非均相形成路径提供了实验证据[44]。

4.4 环境污染控制

葛茂发等在挥发性有机物（VOCs）的消除方面取得了较为系统的研究成果，重点利用界面效应、形貌效应、缺陷效应、尺寸效应等发展高效消除 VOCs 的催化剂体系，建立了催化剂结构与催化活性间的构效关系。在界面效应研究方面，利用 Pt/ZrO_2 催化剂中混晶相结构形成的相界面调控活性物种，使 Pt 物种优先吸附、稳定在相界面处并与界面产生强的相互作用，诱导电子从 Pt 转移到界面处形成带正电荷的 Pt，从而加快了氧物种在活性位的解离和迁移，实现了甲醛的室温完全转化。此外，基于水滑石特有结构，构建了 Pt-Fe/Ni 活性界面。金属-载体间强相互作用使得金属-载体界面处的 Fe^{3+} 物种容易被还原为 Fe^{2+}，有利于氧气在该活性位点的解离，进而提高了甲醛催化氧化的活性和稳定性。在缺陷效应研究方面，建立了缺陷贡献的半定量评价方法，利用 Mn 掺杂引入氧空位缺陷，有效调控 $Mn_xZr_{1-x}O_2$ 催化剂表面的缺陷浓度，结合活性数据与理论计算，从定量的角度证明氧空位是催化氧化甲苯性能提高的关键性因素。此外，在刻蚀构筑表面缺陷研究方面，利用酸刻蚀效应诱导 Mn_2O_3 表面形成大量缺陷结构，提高了催化剂吸附和活化 O_2 的能力，进而提升了甲苯催化活性。在形貌效应研究方面，我们首次报道了双面纳米梳结构的 Co_3O_4-B 催化剂，其独特的双面纳米梳形貌赋予催化剂更多的缺陷结构、丰富的表面 Co^{3+} 和晶格氧物种，在消除挥发性有机物苯时，该催化剂表现出较好的催化活性和优异的反应稳定性。在尺寸效应研究方面，利用胶体沉积-原位热分解的方法合成 Au/MnO_2 催化剂，低温煅烧的 Au/MnO_2 样品具有高分散的较小尺寸 Au 纳米粒子、丰富的表面 Mn^{4+} 物种、优良的低温还原性和氧物种移动能力，从而有效提高甲苯催化氧化性能。

5 展　　望

党的十九大报告提出"持续实施大气污染防治行动，打赢蓝天保卫战"，对新时代的大气污染防治工作提出了更高的目标和要求。当前，我国大气污染防治已经步入深水区，$PM_{2.5}$ 浓度下行压力持续增加，臭氧污染不断加剧，$PM_{2.5}$ 与臭氧协同控制将成为今后我国大气污染防治工作的主旋律，这对"科学治污"和"精准控制"提出了更高的要求。大气环境化学学科的发展和理论创新无疑将成为我国大气污染防治工作进一步推进的核心驱动力。今后，大气重污染过程的大气化学成因、大气氧化性变化、新粒子形成、单颗粒表界面化学、大气非均相化学、二次气溶胶、云雾化学、天气气候对大气成分的影响、大气污染与健康等仍然是大气环境化学学科的重要研究内容。此外，中美贸易战也为我国科技发展提出了新的课题和挑战，针对我国大气环境化学研究和污染防治的关键技术突破（如污染监测技术、模型算法等）也是今后大气环境化学发展的重要方向。

参 考 文 献

[1] Gaudel A, Cooper O, Ancellet G, et al. Tropospheric Ozone Assessment Report: Present-day distribution and trends of tropospheric ozone relevant to climate and global atmospheric chemistry model evaluation. Elementa Science of Anthropocene, 2018, 6(39): 790-811.

[2] Ma Z, Xu J, Quan W, et al. Significant increase of surface ozone at a rural site, north of eastern China. Atmospheric Chemistry and Physics, 2016, 16(6): 3969-3977.

[3] Sun L, Xue L, Wang T, et al. Significant increase of summertime ozone at Mount Tai in Central Eastern China. Atmospheric Chemistry and Physics, 2016, 16(16): 10637-10650.

[4] Xu W, Xu X, Lin M, et al. Long-term trends of surface ozone and its influencing factors at the Mt Waliguan GAW station, China – Part 2: The roles of anthropogenic emissions and climate variability. Atmospheric Chemistry and Physics, 2018, 18(2): 773-798.

[5] Wang W N, Cheng T H, Gu X F, et al. Assessing spatial and temporal patterns of observed ground-level ozone in China. Scientific reports, 2017, 7(1): 3651.

[6] Li K, Jacob D J, Liao H, et al. Anthropogenic drivers of 2013–2017 trends in summer surface ozone in China. Proceedings of

the National Academy of Sciences, 2019, 116(2): 422-427.

[7] Lu X, Hong J, Zhang L, et al. Severe surface ozone pollution in China: A global perspective. Environmental Science & Technology Letters, 2018, 5(8): 487-494.

[8] Sun L, Xue L, Wang Y, et al. Impacts of meteorology and emissions on summertime surface ozone increases over central eastern China between 2003 and 2015. Atmospheric Chemistry and Physics, 2019, 19(3): 1455-1469.

[9] Ni R, Lin J, Yan Y, et al. Foreign and domestic contributions to springtime ozone over China. Atmos Chem Phys, 2018, 18(15): 11447-11469.

[10] Lu X, Zhang L, Chen Y, et al. Exploring 2016-2017 surface ozone pollution over China: source contributions and meteorological influences. Atmospheric Chemistry and Physics Discussion, 2019, 2019: 1-45.

[11] Zhang L, Li Q, Wang T, et al. Combined impacts of nitrous acid and nitryl chloride on lower-tropospheric ozone: new module development in WRF-Chem and application to China. Atmos Chem Phys, 2017, 17(16): 9733-9750.

[12] Li Q, Zhang L, Wang T, et al. "New" Reactive Nitrogen Chemistry Reshapes the Relationship of Ozone to Its Precursors. Environmental science & technology, 2018, 52(5): 2810-2818.

[13] Li H, Wu S, Pan L, et al. Short-term effects of various ozone metrics on cardiopulmonary function in chronic obstructive pulmonary disease patients: Results from a panel study in Beijing, China. Environmental Pollution, 2018, 232: 358-366.

[14] Maji K J, Ye W F, Arora M, et al. Ozone pollution in Chinese cities: Assessment of seasonal variation, health effects and economic burden. Environmental Pollution, 2019, 247: 792-801.

[15] Yue X, Unger N, Harper K, et al. Ozone and haze pollution weakens net primary productivity in China. Atmospheric Chemistry and Physics, 2017, 17: 6073-6089.

[16] Zhang R, Khalizov A, Wang L, et al. Nucleation and growth of nanoparticles in the atmosphere. Chemical Reviews, 2011, 112(3): 1957-2011.

[17] Sipilä M, Berndt T, Petäjä T, et al. The role of sulfuric acid in atmospheric nucleation. Science, 2010, 327(5970): 1243-1246.

[18] Chu B, Kerminen V M, Bianchi F, et al. Atmospheric new particle formation in China. Atmos Chem Phys, 2019, 19(1): 115-138.

[19] Wang Z, Wu Z, Yue D, et al. New particle formation in China: Current knowledge and further directions. Science of the Total Environment, 2017, 577: 258-266.

[20] Kirkby J, Curtius J, Almeida J, et al. Role of sulphuric acid, ammonia and galactic cosmic rays in atmospheric aerosol nucleation. Nature, 2011, 476(7361): 429.

[21] Almeida J, Schobesberger S, Kürten A, et al. Molecular understanding of sulphuric acid–amine particle nucleation in the atmosphere. Nature, 2013, 502-359.

[22] Kürten A, Jokinen T, Simon M, et al. Neutral molecular cluster formation of sulfuric acid–dimethylamine observed in real time under atmospheric conditions. Proceedings of the National Academy of Sciences, 2014, 111(42): 15019-15024.

[23] Riccobono F, Schobesberger S, Scott C E, et al. Oxidation products of biogenic emissions contribute to nucleation of atmospheric particles. Science, 2014, 344(6185): 717-721.

[24] Schobesberger S, Junninen H, Bianchi F, et al. Molecular understanding of atmospheric particle formation from sulfuric acid and large oxidized organic molecules. Proceedings of the National Academy of Sciences, 2013, 110(43): 17223-17228.

[25] Kirkby J, Duplissy J, Sengupta K, et al. Ion-induced nucleation of pure biogenic particles. Nature, 2016, 533(7604): 521.

[26] Kulmala M, Kontkanen J, Junninen H, et al. Direct observations of atmospheric aerosol nucleation. Science, 2013, 339(6122): 943-946.

[27] Bianchi F, Tröstl J, Junninen H, et al. New particle formation in the free troposphere: A question of chemistry and timing. Science, 2016, 352(6289): 1109-1112.

[28] Sipilä M, Sarnela N, Jokinen T, et al. Molecular-scale evidence of aerosol particle formation via sequential addition of HIO$_3$. Nature, 2016, 537(7621): 532.

[29] Jokinen T, Sipilä M, Kontkanen J, et al. Ion-induced sulfuric acid–ammonia nucleation drives particle formation in coastal Antarctica. Science advances, 2018, 4(11): 9744.

[30] Rose C, Zha Q, Dada L, et al. Observations of biogenic ion-induced cluster formation in the atmosphere. Science advances, 2018, 4(4): eaar5218.

[31] Guo S, Hu M, Zamora M L, et al. Elucidating severe urban haze formation in China. Proceedings of the National Academy of Sciences, 2014, 111(49): 17373-17378.

[32] Yao L, Garmash O, Bianchi F, et al. Atmospheric new particle formation from sulfuric acid and amines in a Chinese megacity. Science, 2018, 361(6399): 278-281.

[33] Li J, Li K, Wang W, et al. Optical properties of secondary organic aerosols derived from long-chain alkanes under various NO$_x$ and seed conditions. Science of the Total Environment, 2017, 579: 1699-1705.

[34] Li K, Li J, Liggio J, et al. Enhanced light scattering of secondary organic aerosols by multiphase reactions. Environmental science & technology, 2017, 51(3): 1285-1292.

[35] Peng C, Wang W, Li K, et al. The Optical Properties of Limonene Secondary Organic Aerosols: The Role of NO$_3$, OH, and O$_3$ in the Oxidation Processes. Journal of Geophysical Research: Atmospheres, 2018, 123(6): 3292-3303.

[36] Li J, Wang W, Li K, et al. Development and application of the multi-wavelength cavity ring-down aerosol extinction spectrometer. Journal of Environmental Sciences, 2019, 76: 227-237.

[37] Chen Y, Tong S, Wang J, et al. Effect of Titanium Dioxide on Secondary Organic Aerosol Formation. Environmental Science & Technology, 2018, 52(20): 11612-11620.

[38] Hou S, Tong S, Ge M, et al. Comparison of atmospheric nitrous acid during severe haze and clean periods in Beijing, China. Atmospheric Environment, 2016, 124: 199-206.

[39] Zhang W, Tong S, Ge M, et al. Variations and sources of nitrous acid(HONO)during a severe pollution episode in Beijing in winter 2016. Science of The Total Environment, 2019, 648: 253-262.

[40] Tong S, Hou S, Zhang Y, et al. Comparisons of measured nitrous acid(HONO)concentrations in a pollution period at urban and suburban Beijing, in autumn of 2014. Science China Chemistry, 2015, 58(9): 1393-1402.

[41] Tong S, Hou S, Zhang Y, et al. Exploring the nitrous acid(HONO)formation mechanism in winter Beijing: direct emissions and heterogeneous production in urban and suburban areas. Faraday discussions, 2016, 189: 213-230.

[42] Tan F, Tong S, Jing B, et al. Heterogeneous reactions of NO$_2$ with CaCO$_3$–(NH$_4$)$_2$SO$_4$ mixtures at different relative humidities. Atmospheric Chemistry and Physics, 2016, 16(13): 8081-8093.

[43] Tan F, Jing B, Tong S, et al. The effects of coexisting Na$_2$SO$_4$ on heterogeneous uptake of NO$_2$ on CaCO$_3$ particles at various RHs. Science of the Total Environment, 2017, 586: 930-938.

[44] Zhang Y, Tong S, Ge M, et al. The influence of relative humidity on the heterogeneous oxidation of sulfur dioxide by ozone on calcium carbonate particles. Science of The Total Environment, 2018, 633: 1253-1262.

作者：陈建民[1]，胡　敏[2]，葛茂发[3]，王　琳[1]，薛丽坤[4]
[1]复旦大学，[2]北京大学，[3]中国科学院化学研究所，[4]山东大学

第11章 环境抗生素与耐药基因污染研究进展

▶ 1. 引言 /268

▶ 2. 抗生素使用 /268

▶ 3. 抗生素环境污染 /269

▶ 4. 耐药基因环境污染 /271

▶ 5. 展望 /276

本章导读

抗生素残留及其引起的细菌耐药性引起了国际社会的高度重视，已成为全球性环境健康问题。本章介绍了近年来环境抗生素与耐药基因污染的重要研究进展，并展望今后的研究方向。中国是抗生素的生产和使用大国，医疗和畜禽养殖业中都存在抗生素的滥用和过度使用问题，而常规污水处理工艺无法有效去除抗生素和耐药基因，造成水、土壤和大气环境介质中抗生素和耐药基因的普遍污染。由此可见，加强环境中抗生素和耐药基因污染特征、迁移传播规律和控制技术研究极其重要。

关键词

抗生素，耐药基因，城市污水处理厂，养殖环境，大气，土壤，环境污染

1 引 言

抗生素是一类重要的药物，用于治疗人类和动物的各种疾病，还用作畜牧养殖饲料的添加剂以促进动物生长与疾病预防[1]。抗生素大量使用的直接后果是造成环境中抗生素残留的广泛存在，对生态环境与人类健康产生严重的影响。其中最受关注的问题就是细菌长期暴露于抗生素所产生的耐药性以及耐药菌与耐药基因的广泛传播[2]。抗生素残留及其引起的细菌耐药性引起了国际社会的高度重视，已成为全球性环境健康问题。世界卫生组织 2000 年报告将细菌耐药性作为 21 世纪人类面临的最严峻的问题，号召全球共同抑制耐药性的发展[3]，并于 2015 年提出了全球行动计划，号召各成员国也制定相应的控制抗生素耐药行动计划 [4]。我国已于 2016 年制定了《中国遏制细菌耐药行动计划（2016～2020 年）》，来加强抗生素药物管理，遏制细菌耐药，维护人民群众健康。

2 抗生素使用

抗生素种类繁多，常用的抗生素有磺胺类、喹诺酮类、四环素类、大环内酯类、β-内酰胺类、氨基糖苷类、碳青霉烯类和头孢菌素类。抗生素的大量生产和使用，导致抗生素从人类和动物等使用源头释放到环境中，造成环境抗生素污染。

我国是世界上最大的抗生素生产国和消费国之一，据估计 2013 年我国抗生素总使用量为 16.2 万 t，其中人用抗生素占总量的 48%，其余为兽用抗生素[5]。我国住院病人和门诊病人使用抗生素的比率很高，抗生素处方约占医院处方药的 50%，而发达国家的医院只有 10%[6]。与人类使用相比，我国畜牧业中抗生素使用量更大，兽用抗生素使用量从 2007 年的 46% 增加到 2013 年的 52%。氟苯尼考、林可霉素、泰乐菌素、恩诺沙星、磺胺喹恶啉、恩诺沙星和泰乐菌素在我国年使用量都超过 1000t[5, 7]。研究表明人类和动物服用和注射抗生素后，50%～90%将以母体或活性代谢产物的形式随粪便和尿液排出体外，最终进入环境。因此，抗生素的大量使用、滥用或过度使用被认为是中国抗生素耐药性增加的主要原因。

3 抗生素环境污染

经使用，抗生素通过多种途径进入环境，包括人类生活污水、畜禽养殖和水产养殖的废水废物、医院废水、制药厂工业废水和农业径流等，造成水环境和土壤环境抗生素污染。抗生素环境污染的源汇过程如图11-1所示，其中城市污水处理厂和畜禽养殖场等是抗生素的主要污染源，而水环境和土壤是重要的汇。

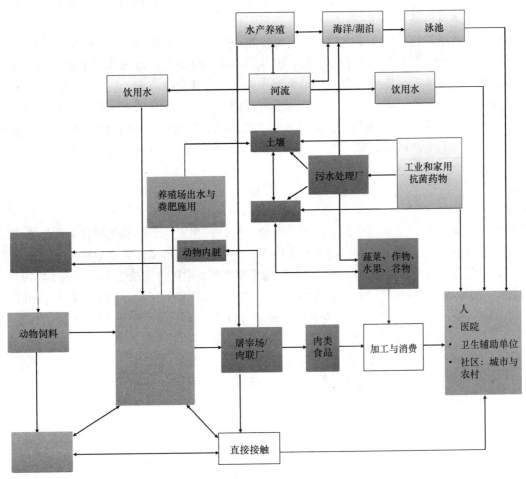

图 11-1 抗生素污染的源汇过程[8]

3.1 城市污水处理厂中抗生素污染

大量研究表明中国污水处理厂的进水和出水可检出各种抗生素[5, 9-28]，检出频率高的抗生素包括：磺胺嘧啶、磺胺甲嗪、磺胺甲噁唑、甲氧苄啶、四环素、土霉素、环丙沙星、恩诺沙星、诺氟沙星、氧氟沙星、罗红霉素-H_2O，检出浓度水平在进水和出水中从几 ng/L 到几十 μg/L 不等，表明这些抗生素在传统污水处理厂工艺中的无法有效去除。事实上，某些抗生素如四环素类和氟喹诺酮类抗生素检测出较高的水相去除率，也主要是通过污泥相的吸附作用，而并不是真正意义上的降解去除[11, 13, 15, 22, 25, 27, 29]。

3.2 水环境中抗生素污染

中国主要河流地表水和沉积物中广泛检出各种抗生素，检出浓度范围从未检测到几 μg/L[15, 30-43]。其中河流地表水检出率高的抗生素有磺胺甲噁唑、土霉素、环丙沙星、诺氟沙星、氧氟沙星、克拉霉素和脱水红霉素，浓度常高达数 μg/L。其中应该注意的是，在城市污水排放口和养殖废水排放口下游河流中往往能检出更高浓度的抗生素。Zhou 等[42]发现中国北方主要河流黄河、海河和辽河沉积物中氟喹诺酮类和四环素类的浓度较高，而磺胺类和大环内酯类的浓度相对较低。Yang 等也发现在中国南方珠江流域沉积物中类似的分布规律[38]。这也说明由于物理化学性质的差异，进入环境后不同类型抗生素会存在不同的相分配、迁移和削减等环境行为。

应光国课题组基于构建的多介质逸度模型模拟发现，我国各流域抗生素排放量存在较大差异[5]，抗生素排放密度最高的是位于华南地区的珠江流域，其次是华北地区的海河流域和华东地区的太湖和钱塘江流域。广州、深圳、北京和上海等人口密集型大城市在抗生素排放量方面的贡献显著。抗生素排放量在地域上的差异与著名的"胡焕庸线"十分吻合，划分成明显的东部和西部两个部分，其中中国东部的抗生素排放量密度是西部流域的 6 倍以上，这反映了人类活动对抗生素分布的巨大影响。这与我国河流地表水和沉积物的抗生素监测结果基本一致。

3.3 养殖环境中抗生素污染

随着集约化畜禽（猪、家禽和牛）养殖业的高速发展，抗生素的生产和使用量也在不断增长。事实上，用以防止畜禽感染和促进生长的兽用抗生素由于动物肠道吸收不良，大多数抗生素最终进入畜禽粪便中。畜禽粪便已成为农业和整体环境中残留兽药的主要来源，中国兽用抗生素总量为 84240t[5]，畜禽粪便产量已超过 2000 万 t。我国学者针对畜禽养殖废物中抗生素污染做了大量研究，在猪、鸡和牛粪样品中测定了不同种类的兽用抗生素[44-57]。其中氟喹诺酮（FQ）、磺胺类（SA）和四环素（TC）类是畜禽废物中检出率最高的抗生素类型。恩诺沙星在鸡粪中最高检出浓度达到 1420.76 mg/kg，是迄今为止中国报道的最高抗生素残留浓度，诺氟沙星在鸡粪中的残留浓度也高达 225.45 mg/kg[54]。土霉素和氯四环素也是畜禽粪便中最常见的两种抗生素，检出浓度分别高达到 416.8 mg/kg（鸡粪）[53]和 764.4 mg/kg（猪粪）[51]。

3.4 土壤中抗生素污染

抗生素可以通过中污水灌溉、污泥回用或填埋，以及畜禽粪便有机肥料使用而引入土壤。不同土壤来源的抗生素浓度差异很大[23, 31, 44, 45, 47, 48, 52, 55-60]，通常在养殖场附近土壤中能检出较高浓度的抗生素。如：在猪场污水排放附近的土壤中检出金霉素浓度高达 12.9 mg/kg[55]，在猪场附近的农田中检出土霉素、磺胺嘧啶和磺胺甲噁唑浓度分别高达 4.24 mg/kg、2.45 mg/kg 和 2.41 mg/kg[48]。此外，由于有机蔬菜生产区畜禽粪肥被广泛用作替代营养源施用，有机蔬菜生产区土壤中抗生素残留问题也备受关注。在中国山东省的一个重要蔬菜种植区，检出环丙沙星和氧氟沙星的最大浓度为 0.652 mg/kg 和 0.288 mg/kg[58, 61]。另外中水灌溉的土壤中也发现了抗生素的积累[62]，虽然浓度远低于畜禽养殖场周边土壤，但也应该引起重视。

4 耐药基因环境污染

环境是耐药基因的重要贮存库。迄今为止，在我国地表水、沉积物、污泥和土壤等多种环境介质中都检出各种耐药菌和耐药基因。环境中抗生素残留可能进一步对环境细菌产生选择性压力，促进耐药性和耐药基因产生。抗生素耐药基因（ARG）已被认为是一种新兴的环境污染物[63]。耐药基因的广泛传播和扩散，尤其是从环境细菌向人类病原体传播等可能严重降低抗生素的疗效，对公众造成严重的健康威胁。

4.1 城市污水处理系统中耐药基因的污染现状及来源

4.1.1 耐药基因在污水处理系统中的污染现状

大量研究发现，经过污水处理系统多级工艺处理后，出水中仍然可以检出多种 ARGs 的存在，并对周围受纳水体造成污染。Auerbach 等[64]分析了不同时间多次采集的美国 Wisconsin 两个污水处理厂的进、出水等样品，发现了多种四环素耐药基因 $tetA-E$、$tetG$、$tetM$、$tetO$、$tetQ$、$tetS$。Munir 等[65]在 USA 密歇根州的污水处理系统进水、二沉池出水、出水中均检出相当高浓度的 $tetW$、$tetO$ 和 $sul1$ 耐药基因。Borjesson 等[66]研究了污水处理系统中甲氧西林耐药基因（$mecA$），这种基因在一整年的样品中均能检出，且季节性变化不明显。Pruden 等[67]在美国科罗拉多州北部地区城市污水处理厂和饮用水厂均能检测到四环素耐药基因（$tetO$ 和 $tetW$）与磺胺耐药基因（$sul1$ 和 $sul2$）。可见，ARGs 在污水处理系统中广泛存在，而且多种耐药基因并存。

与污水相比，好氧、厌氧生物处理单元的活性污泥中集中了更多的耐药基因。这些处理单元往往含有高浓度的抗生素、耐药细菌和耐药基因。Auerbach 等[64]发现活性污泥样品中多种四环素类耐药基因。Munir 等[65]发现污泥样品中耐药基因（$tetW$、$tetO$ 和 $sul1$）含量均高于水中。Zhang 等[68]采集了全球各地 15 个污水处理厂的活性污泥，定性和定量检测 14 种四环素耐药基因，其中六种（$tetA$、$tetC$、$tetG$、$tetM$、$tetS$ 和 $tetX$）在所有水厂全部检出，其中 $tetG$ 浓度最高（$1.75×10^{-2} \sim 2.43×10^{-3}$ copies/16s），$tetC$ 和 $tetA$ 浓度其次。由此可见 ARGs 在污水处理系统的污染已经成为全球性环境问题。

污水处理系统中分离出的细菌常对多种抗生素具有耐药性，即多重耐药性。Huang 等[69]对北京污水处理厂出水中的细菌进行耐药性培养，发现出水中许多异样细菌含有抗青霉素、氯霉素、头孢霉素和利福平的抗性，而且发现一些细菌同时对 5～6 种抗生素耐药。Łuczkiewicz 等[70]选择采自市政污水处理厂厌氧罐和回流活性污泥的样品，在选择培养基上培养粪便大肠菌和球大肠菌，发现这两种细菌分别有 19 和 17 种耐药性，其中包括环丙沙星、氨苄青霉素、四环素等抗药率都在 20%～34%。综上所述，污水处理系统是环境耐药基因污染的重要源头。

4.1.2 污水处理系统中耐药基因的来源分析

污水处理厂的污水包括居民生活污水、医疗废水、制药废水和养殖废水等。其中市政污水处理厂主要位于城市生活区附近，污水主要来自于城市生活污水以及包含医院在内的公共基础设施的污水[71]；而分散在工厂和养殖场附近的污水处理厂，污水来源主要为工业废水和养殖废水等[72,73]，此类来源的污水可能含有大量抗生素及其耐药基因。影响污水处理系统中耐药基因的产生、形成的因素有很多。

第一，由于污水处理系统的进水往往不是单一来源，其组成复杂。城市污水处理系统的进水中的细菌通常接触过居民生活中或是医院中使用的抗生素，其中有相当一部分已经成为抗生素抗性细菌，并且携带抗生素耐药基因。Fuentefria[74]对医院出水中的绿脓杆菌进行检测，结果显示绿脓杆菌具有抗生素抗性。Duong 等[75]在越南河内的医疗废水中检测出抗氟喹诺酮的抗性细菌，并从处理医院废水生物膜中葡萄球菌检测到的 *mecA*[76]。Volkmann 等[77]使用 TaqMan 探针法检测进水中含有医疗废水的市政污水处理厂的样品，结果表明肠球菌的 *vanA*、葡萄球菌的 *mecA* 和肠杆菌的 *ampC* 的检出率分别为 21%、78%和 0。Ghosh 等[78]在城市生活污水处理系统进水水源中检测出了 *tetA*、*tetO*、*tetX*、*Int1* 四种基因型。Chen 等[79]也在北卡罗来纳的一个城市污水处理厂进水中发现了耐药基因 *ermF*、*ermB*、*ermX* 和 *tetG*。总体而言，生活污水是市政污水处理厂 ARGs 的主要来源，而接纳医疗废水和制药废水的污水处理系统会有相当高的耐药细菌和耐药基因。

第二，污水中残留的抗生素等化学成分构成筛选抗性细菌的环境选择压力。环境中残留的抗生素，会刺激微生物的新陈代谢，从而促进环境中耐药基因的产生和选择。目前许多研究发现，环境中的抗生素残留与耐药基因之间有很好的相关性[80, 81]。Li 等[19, 82]在检测北京 7 个养猪场的废水和土壤后发现，喹诺酮耐药基因和喹诺酮抗生素之间存在着显著的相关性。他们还对养猪场废水中的氯霉素及其耐药基因进行研究对比，也发现了相同的结论。此外，研究还发现，低浓度的抗生素也对耐药细菌和 ARGs 增殖扩散有促进作用[76, 77]。研究表明，污水中的抗生素残留以及微生物菌群相互作用是造成污水处理系统耐药基因广泛存在的重要因素。

4.1.3 耐药基因在污水处理系统中的归趋

经过污水处理系统处理后，传统的理化指标（COD、总氮、总磷等）大幅度降低，但耐药基因的去除效果有限。ARGs 可以在不同的细菌之间水平转移，从而在一定条件下加速 ARGs 扩散与传播，污水处理系统恰恰为耐药基因的传播扩散创造了良好的场所[83]。污水处理系统的主体工艺是以去除有机物为主的二级处理，包括传统活性污泥法、生物滤池、生物转盘等。有些水厂还包括三级处理（深度处理），如：臭氧、紫外消毒或加氯消毒。污水处理厂对控制 ARGs 的传播具有重要作用，但其去除效果取决于处理工艺和运行情况。Munir 等[65]在研究了不同的处理工艺对 ARGs 的去除后，发现膜生物反应器对污水 ARGs 有明显的去除效果，出水的 ARGs 浓度可以降低 3～4 个数量级。相对而言，生物滤池的处理效果要差些。Novo 等[84]研究了淹没式曝气滤池和滴滤池的处理效果，发现前者能够使微生物数量大幅减少（约两个数量级），而后者去除率低于 50%。可见不同处理工艺对于 ARGs 的去除效果差异较大。

污水处理厂的生物处理工艺主要包括活性污泥法和生物膜法。常见的活性污泥法导致污水处理厂出水中抗性基因浓度比进水降低。例如，四环素抗性基因通过市政污水处理厂活性污泥法处理后，出水中的浓度比进水低几个数量级[64]。但进一步研究发现耐药基因的相对浓度并没有发生显著的变化，这表明污水中抗性基因浓度降低的原因有可能是微生物总量的减少，这与在最终的脱水污泥中检测到很高浓度的抗性基因的结果一致。也有研究表明其他处理工艺，比如 SBR 工艺、氧化沟等处理方式对抗性基因的影响相对较小，甚至经过这些工艺后出水中的耐药基因浓度不变或者耐药菌的比例有所升高。Fan 等[85]发现在 SBR 工艺中残存的红霉素可以诱导筛选出进水中不存在的 *ereA* 基因。Luczkiewicz 等[70]在采用 A²/O 工艺的污水处理系统的出水中发现肠球菌和大肠杆菌的耐药比率升高。Zhang 等[86]发现 A/O 工艺污水处理系统的出水中微生物氯霉素的耐药比率由 25%提高至 35%。生物膜法对耐药基因的影响目前尚没有明确的定论，Akinbowale 等[87]认为生物膜由于减少了出水中的微生物量可以进而减少抗性基因，而 Gibbs 等[88]则发现生物膜法出水中 *mecA* 基因的浓度反而升高了。研究说明，污水处理系统中微生物间的相互接触有利于耐药基因的水平传播扩散。

污水处理系统中物理化学工艺主要是通过絮凝、沉降和高级氧化等方式。初次沉淀池可以去除较大颗粒的固体物质，但对耐药基因的变化影响不明显。二次沉淀池只是通过将耐药基因从污水混合物中沉降到污泥中而实现出水中耐药基因浓度的降低，但并没有导致耐药基因数量上的真正消除。而化学处理方式则可以破坏耐药基因结构而彻底降低耐药基因排放进入环境的风险。污水处理过程中的高级氧化（氯消毒、臭氧、紫外等）对耐药基因的去除有着明显的作用。高级氧化的作用剂量与作用时间对高级氧化去除耐药基因的效果影响显著[89]。Macauley等[90]发现较低浓度的氯剂量不能明显地去除污水中的耐药基因，低剂量反而会提高污水中的抗性水平，当剂量达到一定量时才能有效减少耐药基因的含量。紫外消毒也是一种消减耐药基因的有效方式[90, 91]，但是这与紫外线的强度有明显关系，低剂量的紫外线照射很难达到去除耐药基因的效果，只有当照射强度达到一定程度时才能有效地破坏耐药基因。Auerbach等[64]发现低剂量的紫外消毒不能减少出水中 *tetQ* 和 *tetG* 的浓度。Zhang等[86]的研究甚至发现经过紫外消毒后出水中单一和多重耐药菌的比率反而有所升高。

4.2 养殖环境耐药基因污染

动物养殖过程中抗生素的使用对动物的肠道微生物群落产生重要影响，进而促进粪便中细菌的耐药性[92-94]，因此动物粪便已成为重要的耐药菌库、耐药基因库和携带耐药基因的可转移质粒库[95-102]。养殖环境中耐药基因的污染特征既有行业特性（畜禽、水产养殖），也有普遍性。

4.2.1 畜禽养殖环境中细菌耐药性

畜禽养殖环境的细菌耐药性主要集中在养猪业、养鸡业、养牛业。研究已证实畜禽养殖是细菌耐药性的重要源头，耐药基因在猪、鸡、牛等养殖场中普遍流行，并通过食物链、空气或直接接触向其他环境或健康人群传播[103-110]。畜禽养殖耐药菌的研究主要集中在与疾病相关的细菌种类，如：糖肽类和链阳菌素耐药的肠球菌属、氟喹诺酮类耐药的弯曲杆菌、多重耐药的大肠杆菌、沙门氏菌、猪链球菌，以及产超广谱β-内酰胺酶的细菌、耐甲氧西林金黄色葡萄球菌[104, 110-112]。Yang等[113]分析了2000年北京和河北省养猪场和养鸡场大肠杆菌菌株的抗菌敏感性，发现大多数大肠杆菌菌株对多种抗生素均具有耐药性。Cheng等[114]调查了中国东部杭州8个畜牧场不同规模养殖场间四环素耐药基因和磺胺耐药基因丰度无显著性差异。研究表明，养殖环境中的耐药菌已出现多种交叉耐药或共耐药现象，克隆传播和质粒等移动元件介导的水平传播是耐药菌在养殖环境中广泛传播的主要分子机制[104, 109, 110, 112, 115]。

畜禽养殖场及周边环境是抗生素耐药菌和耐药基因的重要源。畜禽养殖场及周边环境中的多种环境介质如粪便、土壤、纳污河流及潟湖（水样和沉积物）样品中存在多种耐药菌和耐药基因[116-119]。Zhu等[120]采用高通量qPCR技术对三个大型养猪场从粪便处理到土地处置的三个阶段中的耐药基因类型和浓度进行了评估。与无抗生素肥料或土壤控制相比，检测到149个独特耐药基因。Li等研究了10个典型畜禽养殖粪便样本中耐药基因的广谱分布特征，发现四环素、磺胺、大环内酯和氯霉素耐药基因均显著富集[121]。应光国课题组多年来对我国畜禽养殖环境中耐药菌的传播和耐药基因的污染特征开展了大量研究工作[122-125]。结果表明我国不同规模的集约化养猪场环境中，94%的大肠杆菌菌株为耐药菌，其中98%为多重耐药菌，携带多种耐药基因和基因盒。进一步研究发现，养猪场粪污排放带来的耐药基因污染输出大于养鸡场[123]。猪场比鸡场更易促进耐药基因在受纳土壤中的多样化，猪场废物排放对受纳水环境中耐药基因污染大于养鸡场[122]。养猪废水排放能增加土壤和蔬菜中微生物群落多样性，可检出动物肠道来源的变形菌门、拟杆菌门和厚壁门细菌等。受纳养猪废水浇灌的蔬菜地表层、中层、底层土的耐药基因数量比对照菜地分别高出5.70倍、14.7倍、5.93倍，养猪场粪污的影响对土壤耐药基因组的演化起重要作用。

养殖场常见的污水处理设施、粪便堆肥处理等手段并不能有效去除抗生素和耐药基因[123-125]，从而使耐药基因和耐药菌伴随着废水或者粪肥农田利用进入河流、农田、地下水等受纳环境，对陆地生态系统、水生生态系统形成潜在威胁，对环境健康和人体健康具有一定风险[126-129]。

4.2.2 水产养殖环境中细菌耐药性

水产养殖常用的抗生素类型包括磺胺类、青霉素类、大环内酯类、喹诺酮类、氟苯尼考和四环素类[130, 131]。目前水产养殖环境中细菌耐药性研究主要侧重于水产病原相关的菌株，如沙门氏菌、气单胞菌属、弧菌属、肠杆菌科细菌等[105, 132-137]。研究发现，饲料、鱼和养殖水体环境共有多种耐药菌，携带 $sul1$、$sul2$、$tetS$、$tetL$、bla_{CMY-2}、bla_{CTX} 等耐药基因[138, 139]。应光国课题组研究发现鱼塘复合养殖系统的肠杆菌科细菌耐药严重[140, 141]。98.5%的细菌具有抗生素耐药性，37.9%表现出多重耐药性，83.7%的细菌携带整合子，58.8%携带耐药基因盒[140]。从传统水产养殖和复合水产养殖粪便、养殖水体、沉积物等环境中分离的大肠杆菌65.4%为耐药菌，同时这些菌株携带四环素类 tet 基因、磺胺类 sul 基因、超广谱β-内酰胺酶 ESBLs 等基因。进一步对比研究发现，复合水产养殖模式中抗生素耐药菌、耐药基因污染远远大于传统养殖水产养殖模式，表明复合水产养殖是抗生素耐药菌和耐药基因的重要污染源[141]。

近年来研究也阐明了水产养殖水体、沉积物等环境中多种耐药基因的污染特征，如四环素类 tet 基因、磺胺类 sul 基因、甲氧苄啶类 dfr 基因、氟苯尼考基因 $floR$ 和喹诺酮类 PMQR 基因[142-150]。进一步研究发现了水产养殖环境中耐药基因与多种细菌种群相关[151]，如 γ-变形菌的弧菌属和海单胞菌属可能是 tetB\tetD 的宿主菌[149]。

尽管水产养殖业中抗生素的使用频率和使用方式不同于畜禽养殖，但是抗生素耐药菌依然广泛检出，因此，耐药基因在水产养殖环境中的污染不容忽视。总体而言，水产养殖环境中分离的菌株主要携带的耐药基因主要有：①喹诺酮类耐药基因：DNA 促旋酶基因和拓扑异构酶基因、PMQR 基因；②四环素类 tet 基因等；③氯霉素和氟苯尼考耐药基因；④磺胺类 sul 基因；⑤β-内酰胺类 ESBLs 基因家族；⑥移动元件相关的基因，如整合子和基因盒相关基因等；⑦大环内酯类 erm 基因等。

4.3 大气环境耐药基因污染

环境中耐药基因的研究较多地集中于土壤和水体环境中，对于空气中耐药基因的研究较少。然而，近年来大气环境耐药基因也逐渐引起了关注。首先，空气污染是影响人类健康的重要环境参数[152]，降低空气污染水平可减少中风、心脏病、肺癌以及慢性和急性呼吸道疾病，包括哮喘导致的疾病负担。我国多数城市颗粒物污染严重，如 PM_{10}、$PM_{2.5}$ 等，这些颗粒物上除了含有有毒有害化学污染物外，还附着多种微生物，这些微生物可能携带耐药基因，因此随着人们对空气质量要求的日益提升，人们逐渐关注空气中的生物污染包括耐药基因的状况。

其次，空气是耐药基因在环境中扩散的主要介质之一。空气中的微生物主要来自于地表环境，土壤、水体中的微生物可通过扬尘、泼溅等方式进入大气环境，并进而附着于空气中的颗粒物上，形成生物气溶胶。由于生物气溶胶具有体积小、重量轻等特性，容易受到空气流动的影响，在环境中扩散，甚至受风力影响，进行较长距离的迁移，这些微生物所携带的耐药基因也随之在环境完成扩散和迁移。

最后，大气环境也是人类暴露耐药基因的主要介质。人和动物每时每刻都在进行着呼吸，空气中的耐药基因可通过呼吸进入人或动物的呼吸道内。同时，空气微生物还可通过直接接触，定殖于人或动物体表[153, 154]；此外，空气微生物还可影响植物叶际微生物[155]，人或动物通过进食或与植物接触均有可能导致耐药基因的迁移[156]，从而间接危害人类健康。

4.3.1 大气环境耐药基因的组成和丰度

采用分离培养及分子生物学技术（包括 PCR、定量 PCR 和宏基因组测序等），在不同区域空气样品中均已检测到多种条件致病菌和耐药基因，这些条件致病菌包括大肠杆菌、金黄色葡萄球菌、芽胞杆菌、假单胞菌、肠球菌等[157-160]，其中大肠杆菌和金黄色葡萄球菌是研究最多的条件致病菌。从空气样品中检出的耐药基因基本涵盖了目前常见的抗生素类型，其中四环素类、β内酰胺类、磺胺类和大环内酯类耐药基因研究较多，也是检出频率较高的耐药基因[161-166]。

值得注意的是，从空气样品中还检测到所谓人类治疗细菌感染的"最后防线"抗生素耐药基因，比如万古霉素耐药基因 van[163,166]、碳青霉烯类耐药基因 bla_{CARB}、bla_{IMP}、bla_{OXA} 和超广谱β内酰胺酶 bla_{TEM}、bla_{CTX-M} 等[163,167]。此外，在空气样品中还检测到多种可介导水平基因转移的可移动遗传元件（MGEs），包括转座酶和整合酶基因[163,166,168,169]，其中I类整合酶基因拷贝数可达 10^4 copies/m^3 [164]。这些研究结果表明耐药基因在空气中广泛存在，大气环境可能是耐药基因的潜在储存库，空气样品同时检测到耐药基因和可移动遗传元件暗示着在空气生物气溶胶中可能发生水平基因转移，并且这些基因进入人体后可能转移到人体微生物中。

大气环境中不同类型的耐药基因其拷贝数和相对丰度均有较大差异。空气中耐药基因的拷贝数通常较低，大约在 1～1000 copies/m^3 [164]，这主要是由于空气中微生物的生物量较少。而耐药基因的相对丰度（相对于 16S rRNA 基因）则表现出较大的变异性，其范围可从 10^{-5} 到约 5 copies/16S rRNA[166]。目前关于空气耐药基因丰度与其他环境介质比较的数据较少，从有限的宏基因组测序结果分析来看，空气样品中耐药基因相对丰度跟其他介质中相比较高[163]。同时，耐药基因的结构组成上与其他环境介质相比也有显著差异[165]。

4.3.2 大气环境耐药基因来源

空气微生物的来源是地表环境微生物，土壤、水体、人和动物等均与空气直接接触，从而释放微生物到空气中，其所携带的耐药基因也随之进入大气环境。由于空气与多种环境介质的交互作用，使得部分空气样品中微生物种类较为丰富，同时其耐药基因的种类也较高[163]。因此人们研究了地表环境中耐药基因富集区域对空气中耐药基因的影响，其中特别关注医院、动物养殖场、污水处理厂等区域。

医院是细菌耐药性高发区域，近年来，医院环境耐药性明显升高[170,171]，特别是医院环境的耐药基因通常位于人类致病菌上，其感染人类并导致抗生素治疗失败的风险较大。接触传播和空气传播是医院致病菌感染和耐药基因传播的主要途径[172]。因此医院空气环境耐药基因的研究更多地集中于生物气溶胶中耐药菌的分离鉴定及耐药基因，目前已从医院空气环境中分离到多种病原菌，包括金黄色葡萄球菌、铜绿假单胞菌、鲍曼不动杆菌、嗜麦芽窄食单胞菌、肠杆菌等，并且鉴定出多种耐药基因，其中许多菌株均表现出多重耐药表型[158,159,173-175]。此外，有少量研究报道了医院周边空气环境中细菌群落和耐药基因组成，医院周边空气环境潜在病原菌比例高于其他空气样品[176]，从中分离出的金黄色葡萄球菌具有多重耐药表型[177]。

由于抗生素常作为饲料添加剂用以预防动物细菌感染和促进动物生长，动物养殖业也是耐药基因高发区域[178]。对于动物养殖场空气环境耐药基因的研究主要集中于养猪场、养鸡场和养牛场，其中许多研究也集中于养殖场生物气溶胶中潜在病原菌的分离和耐药表型分析，包括肠球菌属、大肠杆菌属、葡萄球菌属、弯曲杆菌属和产气荚膜梭状芽胞杆菌属等[167,179]，特别关注金黄色葡萄球菌[180,181]。此外，研究者也采用 PCR 和高通量测序技术对养殖场空气样品中的细菌群落结构和耐药基因进行了检测，养殖场空气中耐药基因有很大一部分来自于动物粪便[179]和堆肥处理系统[164]，其中检测到的耐药基因包括四环素类、beta 内酰胺类、大环内酯类和磺胺类耐药基因等，同时还检测到整合酶基因。在养殖场不同区域（如养殖区和办公区）的空气样品中细菌群落结构有显著差异[182,183]，并与耐药基因组成显著相关[164]。动物养殖场内生物气溶胶可向周边环境扩散，有研究表明在养牛场下游空气样品中，牛肠道相关菌群的比例升高，同时四环类耐药基因显著富集[184]。最近的一项研究表明，动物养殖场内携带有质粒编码的扩诺酮类耐药大肠杆菌可扩散的周边空气中[157]。

污水处理厂由于汇集了来自人类和动物的大量抗生素残留、耐药细菌和耐药基因，也是耐药基因的热点区域[185-188]。污水处理工艺中曝气过程和剩余污泥脱水干燥过程均可将污水和污泥中的微生物及其耐药基因释放到空气环境中，并对周边环境产生影响[189,190]。制药污水处理厂气溶胶中可分离培养到多种条件致病菌，超过45%的分离菌株表现出对2种以上的抗生素具有耐药性[191,192]。通过对污水处理厂上风处、下风处及其处理厂内空气样品比较，发现污水处理厂周边样品均可检测到高活性的生物气溶胶，生物反应池和污泥处理室耐药基因水平较高[193]。宏基因组测序结果表明，动物养殖场污水处理厂和城市污水处理厂空气样品中均能检测到多种耐药基因，但其组成有显著差异，其中氨基糖苷类、大环内脂林可霉素链阳杀菌素B类和四环素类是动物养殖场的主要耐药基因类型，而城市污水处理厂中多重耐药基因和杆菌肽耐药基因占优势[194]。

城市是大量人口的聚集点，近年来随着我国城市化和工业的发展，城市人口急速增加，城市居民也越来越关心空气质量问题。全球空气样品耐药基因分析表明城市空气样品中丰度可达100倍以上差异，其中beta内酰胺酶 bla_{TEM} 丰度最高，其次是喹诺酮类耐药基因 $qepA$[166]。在城市空气样品中也分离到多种耐药细菌，包括肠杆菌科细菌、苯甲异噁唑青霉素耐药金黄色葡萄球菌、耐甲氧西林金黄色葡萄球菌等[192,195,196]，并从中检测到四环素类耐药基因 $tetK$，大环内酯类耐药基因 $ermC$ 和万古霉素耐药基因 $vanA$[196]。此外，还有研究表明城市农贸市场活禽交易区耐药基因丰度显著高于周边样品[197]。

4.3.3 大气环境耐药基因的影响因子

大气环境耐药基因的组成和丰度受到空气周边环境、细菌群落组成、气候条件和理化因子（温度、湿度、紫外强度、化学污染物和空气污染）的影响[172]。正如前文所述，由于空气微生物主要来源于地表环境微生物，与空气联系的周边环境介质中耐药基因的组成和丰度将显著影响空气耐药基因的组成，因此较多的研究侧重于陆地上耐药基因高发区域如医院、养殖场、污水处理厂等对空气耐药基因的影响。此外，有研究表明城市灰尘中含有多种耐药基因，并且具有明显的季节差异，这些灰尘中耐药基因可通过空气流动进入大气环境中，影响空气中的耐药基因组成[198,199]。细菌是耐药基因的主要宿主微生物，因此也是影响耐药基因组成的重要因素之一[200]。空气中耐药基因的组成和丰度也显著受到细菌群落组成的影响[165,166,177]。

在众多因子中，空气质量对空气耐药基因的影响受到了广泛的关注，空气污染严重时，尤其是雾霾天气中空气耐药基因的多样性和丰度通常高于正常水平[163,165,166,169,201]。推测其可能原因是首先雾霾天气中，颗粒物含量较高，为微生物附着提供了良好的环境，其微生物生物量通常较高；其次含有较高生物量的颗粒物也可能增加了细菌间的水平基因转移过程；再次空气颗粒物中可能含有较多的重金属和有机化合物，包括抗生素，对耐药细菌起了共选择的作用[184]；最后，雾霾天气可能减少了空气的紫外辐射强度，也有利于微生物的存活。目前，关于空气污染影响耐药基因组成的具体机制还不明，今后应加强这方面研究探明不同因素对耐药基因的贡献。

5 展 望

细菌对抗生素的耐药性是全球医疗、卫生和食品安全关注的重要问题。近年来，细菌耐药性的增强和全球范围内的传播严重威胁了人类健康和经济发展。细菌耐药性不仅仅在医疗领域和动物养殖业中引起了广泛的关注，近年来人们逐渐意识到耐药基因在环境中的持久性残留、传播和扩散要比抗生素本身的危害要大，耐药基因的研究在环境科学领域也日益受到关注，并且提出应该采用"One health"（大健康理念，或同一健康理念）的方法来研究耐药基因和耐药病原菌在人类、动物和环境介质中的扩散及风险。近年来科学家们已针对不同环境中耐药基因的多样性和环境行为开展了广泛的研究，包括养殖水域、污水处理厂、制药厂废水、河流、沉积物和土壤等，在不同环境介质中均检测到多种耐药基因，并且其多样性和丰度受到人类活动的显著影响。环境介质中已广泛检出抗生素和耐药基因污染，环境中抗生素耐药性的流行日益引起学术界和政府的重视。目前国内已经有人、畜耐药性监测网络平台，国家层面应该建立环境耐药性监测网络平台。

抗生素和耐药基因的传播扩散可能对人类、动物和生态系统健康构成潜在的风险，因此有必要开展不同区域尺度抗生素和耐药基因的源汇过程研究，研究环境细菌耐药的各种驱动机制，探讨环境抗生素和耐药基因的长期归趋，建立定量风险评价模型和方法体系。针对污水、粪便入田，开展土壤环境抗生素和耐药基因消减规律和环境效应研究，尤其应该加强植物吸收抗生素和耐药基因的机理研究。

相对土壤和水体环境，大气环境中耐药基因的研究较少。目前已经在大气环境耐药基因的组成和丰度、来源和影响因子等方面取得了一些进展，但仍然存在着许多科学问题亟待解决。目前大气环境耐药基因的研究主要集中在城市环境，特别是医院、养殖场、污水处理厂等热点区域，对大气环境耐药基因组成、丰度、成因、扩散的整体认知还不足，无法有效开展不同区域尺度的定量暴露风险评价。对空气微生物或耐药基因的扩散过程还未有系统研究，应加强空气耐药基因的扩散过程模型研究以预测排放源耐药基因的传播扩散规律，建立空气耐药基因的人群健康风险评估方法。

针对重要污染源，应该研发有针对性的抗生素和耐药基因污染控制技术。除优化现有污水处理工艺外，应加强深度处理技术去除抗生素和耐药基因的机理研究，尤其需要加强医院、制药厂和养殖场等行业废水和废物中抗生素和耐药基因的去除机理和去除技术研究，从源头有效降低抗生素和耐药基因的污染排放量。

参 考 文 献

[1] Sarmah A K, Meyer M T, Boxall A B A. A global perspective on the use, sales, exposure pathways, occurrence, fate and effects of veterinary antibiotics (VAs) in the environment. Chemosphere, 2006, 65(5): 725-759.

[2] Mazel D. Integrons: agents of bacterial evolution. Nature Reviews Microbiology, 2006, 4(8): 608.

[3] Organization W H. WHO annual report on infectious disease: Overcoming antimicrobial resistance. WHO, Geneva, Switzerland. 2000.

[4] Organization W H. Draft global action plan on antimicrobial resistance. World Health Organization, available at< http://www.who.int/drugresistance/AMR_DRAFT_GAP_1_ Oct_2014_for_MS_consultation. pdf>(last visited June 11, 2015), 2014.

[5] Zhang Q Q, Ying G G, Pan C G, et al. Comprehensive evaluation of antibiotics emission and fate in the river basins of China: Source analysis, multimedia modeling, and linkage to bacterial resistance. Environmental Science & Technology, 2015, 49(11): 6772-6782.

[6] Yin X X, Song F J, Gong Y H, et al. A systematic review of antibiotic utilization in China. Journal of Antimicrobial Chemotherapy, 2013, 68(11): 2445-2452.

[7] Van Boeckel T P, Brower C, Gilbert M, et al. Global trends in antimicrobial use in food animals. Proceedings of the National Academy of Sciences of the United States of America, 2015, 112(18): 5649-5654.

[8] Davies J and Davies D. Origins and evolution of antibiotic resistance. Microbiology and Molecular Biology Reviews, 2010, 74(3): 417-33.

[9] Chang H, Hu J Y, Wang L Z, et al. Occurrence of sulfonamide antibiotics in sewage treatment plants. Chinese Science

Bulletin, 2008, 53(4): 514-520.

[10] Chang X S, Meyer M T, Liu X Y, et al. Determination of antibiotics in sewage from hospitals, nursery and slaughter house, wastewater treatment plant and source water in Chongqing region of Three Gorge Reservoir in China. Environmental Pollution, 2010, 158(5): 1444-1450.

[11] Gao L H, Shi Y L, Li W H, et al. Occurrence of antibiotics in eight sewage treatment plants in Beijing, China. Chemosphere, 2012, 86(6): 665-671.

[12] Gulkowska A, Leung H W, So M K, et al. Removal of antibiotics from wastewater by sewage treatment facilities in Hong Kong and Shenzhen, China. Water Research, 2008, 42(1-2): 395-403.

[13] Hou J, Wang C, Mao D, et al. The occurrence and fate of tetracyclines in two pharmaceutical wastewater treatment plants of Northern China. Environmental Science and Pollution Research, 2016, 23(2): 1722-1731.

[14] Hu X G, He K X, Zhou Q X. Occurrence, accumulation, attenuation and priority of typical antibiotics in sediments based on long-term field and modeling studies. Journal of Hazardous Materials, 2012, 225: 91-98.

[15] Jia A, Wan Y, Xiao Y, et al. Occurrence and fate of quinolone and fluoroquinolone antibiotics in a municipal sewage treatment plant. Water Research, 2012, 46(2): 387-394.

[16] Leung H W, Minh T B, Murphy M B, et al. Distribution, fate and risk assessment of antibiotics in sewage treatment plants in Hong Kong, South China. Environment International, 2012, 42: 1-9.

[17] Li B and Zhang T. Mass flows and removal of antibiotics in two municipal wastewater treatment plants. Chemosphere, 2011, 83(9): 1284-1289.

[18] Li B, Zhang T, Xu Z Y, et al. Rapid analysis of 21 antibiotics of multiple classes in municipal wastewater using ultra performance liquid chromatography-tandem mass spectrometry. Analytica Chimica Acta, 2009, 645(1-2): 64-72.

[19] Li J, Shao B, Shen J Z, et al. Occurrence of chloramphenicol-resistance genes as environmental pollutants from swine feedlots. Environmental Science & Technology, 2013, 47(6): 2892-2897.

[20] Peng X Z, Wang Z D, Kuang W X, et al. A preliminary study on the occurrence and behavior of sulfonamides, ofloxacin and chloramphenicol antimicrobials in wastewaters of two sewage treatment plants in Guangzhou, China. Science of the Total Environment, 2006, 371(1-3): 314-322.

[21] Shao B, Chen D, Zhang J, et al. Determination of 76 pharmaceutical drugs by liquid chromatography-tandem mass spectrometry in slaughterhouse wastewater. Journal of Chromatography A, 2009, 1216(47): 8312-8318.

[22] Sun Q, Li M Y, Ma C, et al. Seasonal and spatial variations of PPCP occurrence, removal and mass loading in three wastewater treatment plants located in different urbanization areas in Xiamen, China. Environmental Pollution, 2016, 208: 371-381.

[23] Wang D, Sui Q, Lu S G, et al. Occurrence and removal of six pharmaceuticals and personal care products in a wastewater treatment plant employing anaerobic/anoxic/aerobic and UV processes in Shanghai, China. Environmental Science and Pollution Research, 2014, 21(6): 4276-4285.

[24] Xu J, Xu Y, Wang H M, et al. Occurrence of antibiotics and antibiotic resistance genes in a sewage treatment plant and its effluent-receiving river. Chemosphere, 2015, 119: 1379-1385.

[25] Xu W H, Zhang G, Li X D, et al. Occurrence and elimination of antibiotics at four sewage treatment plants in the Pearl River Delta (PRD), South China. Water Research, 2007, 41(19): 4526-4534.

[26] Yan Q, Gao X, Huang L, et al. Occurrence and fate of pharmaceutically active compounds in the largest municipal wastewater treatment plant in Southwest China: Mass balance analysis and consumption back-calculated model. Chemosphere, 2014, 99: 160-170.

[27] Zhang H M, Liu P X, Feng Y J, et al. Fate of antibiotics during wastewater treatment and antibiotic distribution in the effluent-receiving waters of the Yellow Sea, northern China. Marine Pollution Bulletin, 2013, 73(1): 282-290.

[28] Zhou L J, Ying G G, Liu S, et al. Occurrence and fate of eleven classes of antibiotics in two typical wastewater treatment plants in South China. Science of the Total Environment, 2013, 452: 365-376.

[29] Li W H, Shi Y L, Gao L H, et al. Occurrence and removal of antibiotics in a municipal wastewater reclamation plant in Beijing, China. Chemosphere, 2013, 92(4): 435-444.

[30] Chen B W, Yang Y, Liang X M, et al. Metagenomic profiles of antibiotic resistance genes (ARGs) between human impacted estuary and deep ocean sediments. Environmental Science & Technology, 2013, 47(22): 12753-12760.

[31] Chen K and Zhou J L. Occurrence and behavior of antibiotics in water and sediments from the Huangpu River, Shanghai, China. Chemosphere, 2014, 95: 604-612.

[32] Li N, Zhang X B, Wu W, et al. Occurrence, seasonal variation and risk assessment of antibiotics in the reservoirs in North China. Chemosphere, 2014, 111: 327-335.

[33] Luo Y, Xu L, Rysz M, et al. Occurrence and transport of tetracycline, sulfonamide, quinolone, and macrolide antibiotics in the Haihe River Basin, China. Environmental Science & Technology, 2011, 45(5): 1827-1833.

[34] Tong L, Huang S B, Wang Y X, et al. Occurrence of antibiotics in the aquatic environment of Jianghan Plain, central China. Science of the Total Environment, 2014, 497: 180-187.

[35] Xu W H, Yan W, Li X D, et al. Antibiotics in riverine runoff of the Pearl River Delta and Pearl River Estuary, China: Concentrations, mass loading and ecological risks. Environmental Pollution, 2013, 182: 402-407.

[36] Xue B M, Zhang R J, Wang Y H, et al. Antibiotic contamination in a typical developing city in south China: Occurrence and ecological risks in the Yongjiang River impacted by tributary discharge and anthropogenic activities. Ecotoxicology and Environmental Safety, 2013, 92: 229-236.

[37] Yan C X, Yang Y, Zhou J L, et al. Antibiotics in the surface water of the Yangtze Estuary: Occurrence, distribution and risk assessment. Environmental Pollution, 2013, 175: 22-29.

[38] Yang J F, Ying G G, Zhao J L, et al. Simultaneous determination of four classes of antibiotics in sediments of the Pearl Rivers using RRLC-MS/MS. Science of the Total Environment, 2010, 408(16): 3424-3432.

[39] Yang J F, Ying G G, Zhao J L, et al. Spatial and seasonal distribution of selected antibiotics in surface waters of the Pearl Rivers, China. Journal of Environmental Science and Health Part B-Pesticides Food Contaminants and Agricultural Wastes, 2011, 46(3): 272-280.

[40] Zhang R J, Zhang G, Zheng Q, et al. Occurrence and risks of antibiotics in the Laizhou Bay, China: Impacts of river discharge. Ecotoxicology and Environmental Safety, 2012, 80: 208-215.

[41] Zheng S L, Qiu X Y, Chen B, et al. Antibiotics pollution in Jiulong River estuary: Source, distribution and bacterial resistance. Chemosphere, 2011, 84(11): 1677-1685.

[42] Zhou L J, Ying G G, Zhao J L, et al. Trends in the occurrence of human and veterinary antibiotics in the sediments of the Yellow River, Hai River and Liao River in northern China. Environmental Pollution, 2011, 159(7): 1877-1885.

[43] Zhu S C, Chen H, Li J N. Sources, distribution and potential risks of pharmaceuticals and personal care products in Qingshan Lake basin, Eastern China. Ecotoxicology and Environmental Safety, 2013, 96: 154-159.

[44] Hou J, Wan W, Mao D, et al. Occurrence and distribution of sulfonamides, tetracyclines, quinolones, macrolides, and nitrofurans in livestock manure and amended soils of Northern China. Environmental Science and Pollution Research International, 2015, 22(6): 4545-54.

[45] Hu X G, Zhou Q X, Luo Y. Occurrence and source analysis of typical veterinary antibiotics in manure, soil, vegetables and groundwater from organic vegetable bases, northern China. Environmental Pollution, 2010, 158(9): 2992-2998.

[46] Hu X G, Yi L, Zhou Q X, et al. Determination of thirteen antibiotics residues in manure by solid phase extraction and high performance liquid chromatography. Chinese Journal of Analytical Chemistry, 2008, 36(9): 1162-1166.

[47] Huang Y J, Cheng M M, Li W H, et al. Simultaneous extraction of four classes of antibiotics in soil, manure and sewage

sludge and analysis by liquid chromatography-tandem mass spectrometry with the isotope-labelled internal standard method. Analytical Methods, 2013, 5(15): 3721-3731.

[48] Ji X L, Shen Q H, Liu F, et al. Antibiotic resistance gene abundances associated with antibiotics and heavy metals in animal manures and agricultural soils adjacent to feedlots in Shanghai; China. Journal of Hazardous Materials, 2012, 235: 178-185.

[49] Li Y X, Li W, Zhang X L, et al. Simultaneous determination of fourteen veterinary antibiotics in animal feces by solid phase extraction and high performance liquid chromatography. Chinese Journal of Analytical Chemistry, 2012, 40(2): 213-217.

[50] Li Y X, Zhang X L, Li W, et al. The residues and environmental risks of multiple veterinary antibiotics in animal faeces. Environmental Monitoring and Assessment, 2013, 185(3): 2211-2220.

[51] Pan X, Qiang Z M, Ben W W, et al. Residual veterinary antibiotics in swine manure from concentrated animal feeding operations in Shandong Province, China. Chemosphere, 2011, 84(5): 695-700.

[52] Qiao M, Chen W D, Su J Q, et al. Fate of tetracyclines in swine manure of three selected swine farms in China. Journal of Environmental Sciences, 2012, 24(6): 1047-1052.

[53] Zhang H B, Luo Y M, Wu L H, et al. Residues and potential ecological risks of veterinary antibiotics in manures and composts associated with protected vegetable farming. Environmental Science and Pollution Research, 2015, 22(8): 5908-5918.

[54] Zhao L, Dong Y H, Wang H. Residues of veterinary antibiotics in manures from feedlot livestock in eight provinces of China. Science of the Total Environment, 2010, 408(5): 1069-1075.

[55] Zhou L J, Ying G G, Liu S, et al. Excretion masses and environmental occurrence of antibiotics in typical swine and dairy cattle farms in China. Science of the Total Environment, 2013, 444: 183-195.

[56] Zhou L J, Ying G G, Liu S, et al. Simultaneous determination of human and veterinary antibiotics in various environmental matrices by rapid resolution liquid chromatography-electrospray ionization tandem mass spectrometry. Journal of Chromatography A, 2012, 1244: 123-138.

[57] Zhou L J, Ying G G, Zhang R Q, et al. Use patterns, excretion masses and contamination profiles of antibiotics in a typical swine farm, South China. Environmental Science-Processes & Impacts, 2013, 15(4): 802-813.

[58] Li X W, Xie Y F, Li C L, et al. Investigation of residual fluoroquinolones in a soil-vegetable system in an intensive vegetable cultivation area in Northern China. Science of the Total Environment, 2014, 468: 258-264.

[59] Wu L H, Pan X, Chen L K, et al. Occurrence and distribution of heavy metals and tetracyclines in agricultural soils after typical land use change in east China. Environmental Science and Pollution Research, 2013, 20(12): 8342-8354.

[60] Wu N, Qiao M, Zhang B, et al. Abundance and diversity of tetracycline resistance genes in soils adjacent to representative swine feedlots in China. Environmental Science & Technology, 2010, 44(18): 6933-6939.

[61] Li X W, Xie Y F, Wang J F, et al. Influence of planting patterns on fluoroquinolone residues in the soil of an intensive vegetable cultivation area in northern China. Science of the Total Environment, 2013, 458: 63-69.

[62] Fang H, Wang H, Cai L, et al. Prevalence of antibiotic resistance genes and bacterial pathogens in long-term manured greenhouse soils as revealed by metagenomic survey. Environmental Science and Technology, 2015, 49(2): 1095-1104.

[63] Pruden A, Pei R T, Storteboom H, et al. Antibiotic resistance genes as emerging contaminants: Studies in northern Colorado. Environmental Science & Technology, 2006, 40(23): 7445-7450.

[64] Auerbach E A, Seyfried E E, Mcmahon K D. Tetracycline resistance genes in activated sludge wastewater treatment plants. Water Research, 2007, 41(5): 1143-1151.

[65] Munir M, Wong K, Xagoraraki I. Release of antibiotic resistant bacteria and genes in the effluent and biosolids of five wastewater utilities in Michigan. Water Research, 2011, 45(2): 681-693.

[66] Stefan B R, Sara M, Andreas M, et al. A seasonal study of the *mecA* gene and *Staphylococcus aureus* including methicillin-resistant *S. aureus* in a municipal wastewater treatment plant. Water Research, 2009, 43(4): 925-932.

[67] Amy P, Ruoting P, Heather S, et al. Antibiotic resistance genes as emerging contaminants: studies in northern Colorado. Environmental Science & Technology, 2006, 40(23): 7445.

[68] Xu Xiang Z and Tong Z. Occurrence, abundance, and diversity of tetracycline resistance genes in 15 sewage treatment plants across China and other global locations. Environmental Science & Technology, 2011, 45(7): 2598-2604.

[69] Huang J J, Hu H Y, Lu S Q, et al. Monitoring and evaluation of antibiotic-resistant bacteria at a municipal wastewater treatment plant in China. Environment International, 2012, 42: 31-36.

[70] Łuczkiewicz A, Jankowska K, Fudala-Książek S, et al. Antimicrobial resistance of fecal indicators in municipal wastewater treatment plant. Water Research, 2010, 44(17): 5089-5097.

[71] Yi Fang L, Chih Hung W, Rajendra Prasad J, et al. Presence of plasmid pA15 correlates with prevalence of constitutive MLS(B) resistance in group A streptococcal isolates at a university hospital in southern Taiwan. International Journal of Antimicrobial Agents, 2007, 29(6): 1167-1170.

[72] Yvonne A and Andreas P. The tetracycline resistance determinant Tet 39 and the sulphonamide resistance gene *sulII* are common among resistant *Acinetobacter spp.* isolated from integrated fish farms in Thailand. Journal of Antimicrobial Chemotherapy, 2007, 59(1): 23-27.

[73] Yvonne A and Dorthe S. Class 1 integrons and tetracycline resistance genes in *alcaligenes, arthrobacter,* and *Pseudomonas spp.* isolated from pigsties and manured soil. Applied and Environmental Microbiology, 2005, 71(12): 7941-7947.

[74] Fuentefria D B, Ferreira A E, Corção G. Antibiotic-resistant *Pseudomonas aeruginosa* from hospital wastewater and superficial water: Are they genetically related? Journal of Environmental Management, 2011, 92(1): 250-255.

[75] Anh D H, Ngoc Ha P, Hoang Tung N, et al. Occurrence, fate and antibiotic resistance of fluoroquinolone antibacterials in hospital wastewaters in Hanoi, Vietnam. Chemosphere, 2008, 72(6): 968-973.

[76] Schwartz T, Kohnen W, Jansen B, et al. Detection of antibiotic-resistant bacteria and their resistance genes in wastewater, surface water, and drinking water biofilms. FEMS Microbiology Ecology, 2003, 43(3): 325-335.

[77] Volkmann H, Schwartz T, Bischoff P, et al. Detection of clinically relevant antibiotic-resistance genes in municipal wastewater using real-time PCR(TaqMan). Journal of Microbiological Methods, 2004, 56(2): 277-286.

[78] Das A, Baiswar P, Patel D P, et al. Recycling of on- and off-farm plant biomass in organic rice production at subtropical hill ecosystem of north eastern Indian Himalayas. Progress in Environmental Science and Technology, Vol Ii, Pts a and B, ed. S. C. Li, et al. 2009. 2484-2490.

[79] Chen J, Michel F C, Sreevatsan S, et al. Occurrence and Persistence of Erythromycin Resistance Genes (erm) and Tetracycline Resistance Genes (tet) in Waste Treatment Systems on Swine Farms. Microbial Ecology, 2010, 60(3): 479-486.

[80] Wu N, Qiao M, Zhang B, et al. Abundance and diversity of tetracycline resistance genes in soils adjacent to representative swine feedlots in China. Vol. 44. 2010. 6933-6939.

[81] Yi L, Daqing M, Michal R, et al. Trends in antibiotic resistance genes occurrence in the Haihe River, China. Environmental Science & Technology, 2010, 44(19): 7220.

[82] Juan L, Thanh W, Bing S, et al. Plasmid-mediated quinolone resistance genes and antibiotic residues in wastewater and soil adjacent to swine feedlots: potential transfer to agricultural lands. Environmental Health Perspectives, 2012, 120(8): 1144-1149.

[83] Sungpyo K, Hongkeun P, Kartik C. Propensity of activated sludge to amplify or attenuate tetracycline resistance genes and tetracycline resistant bacteria: a mathematical modeling approach. Chemosphere, 2010, 78(9): 1071-1077.

[84] Novo A and Manaia C M. Factors influencing antibiotic resistance burden in municipal wastewater treatment plants. Appl Microbiol Biotechnol, 2010, 87(3): 1157-1166.

[85] Caian F and Jianzhong H. Proliferation of antibiotic resistance genes in microbial consortia of sequencing batch reactors (SBRs) upon exposure to trace erythromycin or erythromycin-H_2O. Water Research, 2011, 45(10): 3098-3106.

[86] Zhang Y, Marrs C F, Simon C, et al. Wastewater treatment contributes to selective increase of antibiotic resistance among Acinetobacter spp. Science of the Total Environment, 2009, 407(12): 3702-3706.

[87] Akinbowale O L, Peng H, Barton M D. Diversity of tetracycline resistance genes in bacteria from aquaculture sources in Australia. Journal of Applied Microbiology, 2010, 103(5): 2016-2025.

[88] Gibbs S G, Green C F, Tarwater P M, et al. Airborne antibiotic resistant and nonresistant bacteria and fungi recovered from two swine herd confined animal feeding operations. Journal of Occupational & Environmental Hygiene, 2004, 1(11): 699-706.

[89] C Dodd M. Potential impacts of disinfection processes on elimination and deactivation of antibiotic resistance genes during water and wastewater treatment. Vol. 14. 2012. 1754-1771.

[90] Macauley J J, Qiang Z, Adams C D, et al. Disinfection of swine wastewater using chlorine, ultraviolet light and ozone. Water Research, 2006, 40(10): 2017-2026.

[91] Huang J J, Hu H Y, Tang F, et al. Inactivation and reactivation of antibiotic-resistant bacteria by chlorination in secondary effluents of a municipal wastewater treatment plant. Water Research, 2011, 45(9): 2775-2781.

[92] Langford F M, Weary D M, Fisher L. Antibiotic resistance in gut bacteria from dairy calves: A dose response to the level of antibiotics fed in milk. Journal of Dairy Science, 2003, 86(12): 3963-3966.

[93] Looft T, Johnson T A, Allen H K, et al. In-feed antibiotic effects on the swine intestinal microbiome. Proceedings of the National Academy of Sciences of the United States of America, 2012, 109(5): 1691-1696.

[94] Witte W. Selective pressure by antibiotic use in livestock. International Journal of Antimicrobial Agents, 2000, 16: S19-S24.

[95] Bibbal D, Dupouy V, Ferre J P, et al. Impact of three ampicillin dosage regimens on selection of ampicillin resistance in *Enterobacteriaceae* and excretion of *bla* (TEM) genes in swine feces. Applied and Environmental Microbiology, 2007, 73(15): 4785-4790.

[96] Enne V I, Cassar C, Sprigings K, et al. A high prevalence of antimicrobial resistant *Escherichia coli* isolated from pigs and a low prevalence of antimicrobial resistant *E-coli* from cattle and sheep in Great Britain at slaughter. FEMS Microbiology Letters, 2008, 278(2): 193-199.

[97] Schwaiger K, Harms K, Holzel C, et al. Tetracycline in liquid manure selects for co-occurrence of the resistance genes *tet*(M) and *tet*(L) in *Enterococcus faecalis*. Veterinary Microbiology, 2009, 139(3-4): 386-392.

[98] Binh C T, Heuer H, Kaupenjohann M, et al. Piggery manure used for soil fertilization is a reservoir for transferable antibiotic resistance plasmids. FEMS Microbiology Ecology, 2008, 66(1): 25-37.

[99] Binh C T T, Heuer H, Gomes N C M, et al. Similar bacterial community structure and high abundance of sulfonamide resistance genes in field-scale manures. Manure: management, uses and environmental impacts. Nova Science Publishers, Hauppauge, NY, 2010: 141-166.

[100] Duriez P and Topp E. Temporal dynamics and impact of manure storage on antibiotic resistance patterns and population structure of *Escherichia coli* isolates from a commercial swine farm. Applied and Environmental Microbiology, 2007, 73(17): 5486-5493.

[101] Heuer H, Focks A, Lamshoeft M, et al. Fate of sulfadiazine administered to pigs and its quantitative effect on the dynamics of bacterial resistance genes in manure and manured soil. Soil Biology & Biochemistry, 2008, 40(7): 1892-1900.

[102] Heuer H, Kopmann C, Binh C T T, et al. Spreading antibiotic resistance through spread manure: characteristics of a novel plasmid type with low %G plus C content. Environmental Microbiology, 2009, 11(4): 937-949.

[103] Chen L, Chen Z L, Liu J H, et al. Emergence of RmtB methylase-producing *Escherichia coli* and *Enterobacter cloacae* isolates from pigs in China. Journal of Antimicrobial Chemotherapy, 2007, 59(5): 880-885.

[104] Deng Y, Zeng Z, Chen S, et al. Dissemination of IncFII plasmids carrying *rmtB* and *qepA* in *Escherichia coli* from pigs, farm workers and the environment. Clinical Microbiology and Infection, 2011, 17(11): 1740-1745.

[105] Jiang H X, Tang D, Liu Y H, et al. Prevalence and characteristics of beta-lactamase and plasmid-mediated quinolone resistance genes in *Escherichia coli* isolated from farmed fish in China. Journal of Antimicrobial Chemotherapy, 2012, 67(10): 2350-2353.

[106] Liu J H, Wei S Y, Ma J Y, et al. Detection and characterisation of CTX-M and CMY-2 beta-lactamases among *Escherichia coli* isolates from farm animals in Guangdong Province of China. International Journal of Antimicrobial Agents, 2007, 29(5): 576-581.

[107] Wang X M, Jiang H X, Liao X P, et al. Antimicrobial resistance, virulence genes, and phylogenetic background in *Escherichia coli* isolates from diseased pigs. FEMS Microbiology Letters, 2010, 306(1): 15-21.

[108] Yang X, Liu W, Liu Y, et al. F33: A-: B-, IncHI2/ST3, and IncI1/ST71 plasmids drive the dissemination of *fosA3* and *bla*$_{CTX-M-55/-14/-65}$ in *Escherichia coli* from chickens in China. Frontiers in Microbiology, 2014, 5: 688.

[109] Yao Q, Zeng Z, Hou J, et al. Dissemination of the *rmtB* gene carried on IncF and IncN plasmids among *Enterobacteriaceae* in a pig farm and its environment. Journal of Antimicrobial Chemotherapy, 2011, 66(11): 2475-2479.

[110] Zhao J, Chen Z, Chen S, et al. Prevalence and dissemination of oqxAB in Escherichia coli isolates from animals, farmworkers, and the environment. Antimicrobial Agents and Chemotherapy, 2010, 54(10): 4219-4224.

[111] Barton M D. Impact of antibiotic use in the swine industry. Current Opinion in Microbiology, 2014, 19: 9-15.

[112] Wang J, Lin D C, Guo X M, et al. Distribution of the multidrug resistance gene *cfr* in *Staphylococcus* isolates from pigs, workers, and the environment of a hog market and a slaughterhouse in guangzhou, China. Foodborne Pathogens and Disease, 2015, 12(7): 598-605.

[113] Yang H C, Chen S, White D G, et al. Characterization of multiple-antimicrobial-resistant *Escherichia coli* isolates from diseased chickens and swine in China. Journal of Clinical Microbiology, 2004, 42(8): 3483-3489.

[114] Cheng W, Chen H, Su C, et al. Abundance and persistence of antibiotic resistance genes in livestock farms: a comprehensive investigation in eastern China. Environment International, 2013, 61: 1-7.

[115] He L, Partridge S R, Yang X, et al. Complete nucleotide sequence of pHN7A8, an F33: A-: B- type epidemic plasmid carrying *bla*$_{CTX-M-65}$, *fosA3* and *rmtB* from China. Journal of Antimicrobial Chemotherapy, 2013, 68(1): 46-50.

[116] Aarestrup F M, Hasman H, Jensen L B, et al. Antimicrobial resistance among *Enterococci* from pigs in three European countries. Applied and Environmental Microbiology, 2002, 68(8): 4127-4129.

[117] Rodriguez-Mozaz S, Chamorro S, Marti E, et al. Occurrence of antibiotics and antibiotic resistance genes in hospital and urban wastewaters and their impact on the receiving river. Water Research, 2014, 69C: 234-242.

[118] Hartmann M, Frey B, Mayer J, et al. Distinct soil microbial diversity under long-term organic and conventional farming. ISME J, 2015, 9(5): 1177-1194.

[119] Czekalski N, Sigdel R, Birtel J, et al. Does human activity impact the natural antibiotic resistance background? Abundance of antibiotic resistance genes in 21 Swiss lakes. Environment International, 2015, 81: 45-55.

[120] Zhu Y G, Johnson T A, Su J Q, et al. Diverse and abundant antibiotic resistance genes in Chinese swine farms. Proceedings of the National Academy of Sciences of the United States of America, 2013, 110(9): 3435-3440.

[121] Li B, Yang Y, Ma L P, et al. Metagenomic and network analysis reveal wide distribution and co-occurrence of environmental antibiotic resistance genes. Isme Journal, 2015, 9(11): 2490-2502.

[122] He L Y, Liu Y S, Su H C, et al. Dissemination of antibiotic resistance genes in representative broiler feedlots environments: identification of indicator ARGs and correlations with environmental variables. Environmental Science and Technology, 2014, 48(22): 13120-13129.

[123] He L Y, Ying G G, Liu Y S, et al. Discharge of swine wastes risks water quality and food safety: Antibiotics and antibiotic resistance genes from swine sources to the receiving environments. Environment International, 2016, 92–93: 210-219.

[124] Zhang M, He L Y, Liu Y S, et al. Fate of veterinary antibiotics during animal manure composting. Science of the Total

Environment, 2019, 650(Pt 1): 1363-1370.

[125] Zhang M, Liu Y S, Zhao J L, et al. Occurrence, fate and mass loadings of antibiotics in two swine wastewater treatment systems. Science of the Total Environment, 2018, 639: 1421-1431.

[126] Fang H, Han L, Zhang H, et al. Dissemination of antibiotic resistance genes and human pathogenic bacteria from a pig feedlot to the surrounding stream and agricultural soils. Journal of Hazardous Materials, 2018, 357: 53-62.

[127] Jia S, Zhang X X, Miao Y, et al. Fate of antibiotic resistance genes and their associations with bacterial community in livestock breeding wastewater and its receiving river water. Water Research, 2017, 124: 259-268.

[128] Ma L, Xia Y, Li B, et al. Metagenomic assembly reveals hosts of antibiotic resistance genes and the shared resistome in pig, chicken, and human feces. Environmental Science and Technology, 2016, 50(1): 420-427.

[129] Hanna N, Sun P, Sun Q, et al. Presence of antibiotic residues in various environmental compartments of Shandong province in eastern China: Its potential for resistance development and ecological and human risk. Environment International, 2018, 114: 131-142.

[130] Mo W Y, Chen Z, Leung H M, et al. Application of veterinary antibiotics in China's aquaculture industry and their potential human health risks. Environmental Science and Pollution Research International, 2015, 41(3): 40-49.

[131] Sapkota A, Sapkota A R, Kucharski M, et al. Aquaculture practices and potential human health risks: current knowledge and future priorities. Environment International, 2008, 34(8): 1215-1226.

[132] Hatha A A and Lakshmanaperumalsamy P. Antibiotic resistance of Salmonella strains isolated from fish and crustaceans. Letters in Applied Microbiology, 1995, 21(1): 47-49.

[133] Hatha M, Vivekanandhan A A, Joice G J, et al. Antibiotic resistance pattern of motile aeromonads from farm raised fresh water fish. International Journal of Food Microbiology, 2005, 98(2): 131-134.

[134] Schmidt A S, Bruun M S, Dalsgaard I, et al. Incidence, distribution, and spread of tetracycline resistance determinants and integron-associated antibiotic resistance genes among motile aeromonads from a fish farming environment. Applied and Environmental Microbiology, 2001, 67(12): 5675-5682.

[135] Schmidt A S, Bruun M S, Dalsgaard I, et al. Occurrence of antimicrobial resistance in fish-pathogenic and environmental bacteria associated with four danish rainbow trout farms. Applied and Environmental Microbiology, 2000, 66(11): 4908-4915.

[136] Akinbowale O L, Peng H, Barton M D. Antimicrobial resistance in bacteria isolated from aquaculture sources in Australia. Journal of Applied Microbiology, 2006, 100(5): 1103-1113.

[137] Rodriguez-Blanco A, Lemos M L, Osorio C R. Integrating conjugative elements as vectors of antibiotic, mercury, and quaternary ammonium compound resistance in marine aquaculture environments. Antimicrobial Agents and Chemotherapy, 2012, 56(5): 2619-2626.

[138] Huang Y, Zhang L, Tiu L, et al. Characterization of antibiotic resistance in commensal bacteria from an aquaculture ecosystem. Frontiers in Microbiology, 2015, 6: 914.

[139] Yang J, Wang C, Wu J, et al. Characterization of a multiresistant mosaic plasmid from a fish farm Sediment Exiguobacterium sp. isolate reveals aggregation of functional clinic-associated antibiotic resistance genes. Applied and Environmental Microbiology, 2014, 80(4): 1482-1488.

[140] Su H C, Ying G G, Tao R, et al. Occurrence of antibiotic resistance and characterization of resistance genes and integrons in Enterobacteriaceae isolated from integrated fish farms in south China. Journal of Environmental Monitoring, 2011, 13(11): 3229-3236.

[141] Zhang R Q, Ying G G, Su H C, et al. Antibiotic resistance and genetic diversity of *Escherichia coli* isolates from traditional and integrated aquaculture in South China. Journal of Environmental Science and Health Part B-Pesticides Food Contaminants and Agricultural Wastes, 2013, 48(11): 999-1013.

[142] Gao P, Mao D, Luo Y, et al. Occurrence of sulfonamide and tetracycline-resistant bacteria and resistance genes in aquaculture environment. Water Research, 2012, 46(7): 2355-2364.

[143] Liang X M, Nie X P, Shi Z. [Preliminary studies on the occurrence of antibiotic resistance genes in typical aquaculture area of the Pearl River Estuary]. Huan Jing Ke Xue, 2013, 34(10): 4073-4080.

[144] Muziasari W I, Managaki S, Parnanen K, et al. Sulphonamide and trimethoprim resistance genes persist in sediments at Baltic Sea aquaculture farms but are not detected in the surrounding environment. PloS One, 2014, 9(3): e92702.

[145] Seyfried E E, Newton R J, Rubert K F, et al. Occurrence of tetracycline resistance genes in aquaculture facilities with varying use of oxytetracycline. Microbial Ecology, 2010, 59(4): 799-807.

[146] Tamminen M, Karkman A, Lohmus A, et al. Tetracycline resistance genes persist at aquaculture farms in the absence of selection pressure. Environmental Science and Technology, 2011, 45(2): 386-391.

[147] Yuan J, Ni M, Liu M, et al. Occurrence of antibiotics and antibiotic resistance genes in a typical estuary aquaculture region of Hangzhou Bay, China. Marine Pollutution Bulletin, 2019, 138: 376-384.

[148] Ng C, Chen H, Goh S G, et al. Microbial water quality and the detection of multidrug resistant E. coli and antibiotic resistance genes in aquaculture sites of Singapore. Marine Pollutution Bulletin, 2018, 135: 475-480.

[149] Jang H M, Kim Y B, Choi S, et al. Prevalence of antibiotic resistance genes from effluent of coastal aquaculture, South Korea. Environmental Pollution (Barking, Essex: 1987), 2018, 233: 1049-1057.

[150] Su H, Liu S, Hu X, et al. Occurrence and temporal variation of antibiotic resistance genes (ARGs) in shrimp aquaculture: ARGs dissemination from farming source to reared organisms. Science of the Total Environment, 2017, 607-608: 357-366.

[151] Fang H, Huang K, Yu J, et al. Metagenomic analysis of bacterial communities and antibiotic resistance genes in the Eriocheir sinensis freshwater aquaculture environment. Chemosphere, 2019, 224: 202-211.

[152] Heft-Neal S, Burney J, Bendavid E, et al. Robust relationship between air quality and infant mortality in Africa. Nature, 2018, 559(7713): 254-258.

[153] Brunetti A E, Lyra M L, Melo W G P, et al. Symbiotic skin bacteria as a source for sex-specific scents in frogs. Proceedings of the National Academy of Sciences, 2019.

[154] Lehtimäki J, Sinkko H, Hielm-Björkman A, et al. Skin microbiota and allergic symptoms associate with exposure to environmental microbes. Proceedings of the National Academy of Sciences, 2018, 115(19): 4897-4902.

[155] Flies E J, Skelly C, Negi S S, et al. Biodiverse green spaces: a prescription for global urban health. Frontiers in Ecology and the Environment, 2017, 15(9): 510-516.

[156] von Hertzen L. Plant microbiota: implications for human health. British Journal of Nutrition, 2015, 114(9): 1531-1532.

[157] Wu B, Qi Q, Zhang X, et al. Dissemination of *Escherichia coli* carrying plasmid-mediated quinolone resistance (PMQR) genes from swine farms to surroundings. Science of the Total Environment, 2019, 665: 33-40.

[158] Solomon F B, Wadilo F, Tufa E G, et al. Extended spectrum and metalo beta-lactamase producing airborne *Pseudomonas aeruginosa* and *Acinetobacter baumanii* in restricted settings of a referral hospital: a neglected condition. Antimicrobial Resistance and Infection Control, 2017, 6.

[159] Saadoun I, Jaradat Z W, Al Tayyar I A, et al. Airborne methicillin-resistant *Staphylococcus aureus* in the indoor environment of King Abdullah University Hospital, Jordan. Indoor and Built Environment, 2015, 24(3): 315-323.

[160] Sapkota A R, Ojo K K, Roberts M C, et al. Antibiotic resistance genes in multidrug-resistant Enterococcus spp. and Streptococcus spp. recovered from the indoor air of a large-scale swine-feeding operation. Letters In Applied Microbiology, 2006, 43(5): 534-540.

[161] 薛银刚, 刘菲, 王利平, 等. 气溶胶中抗生素抗性基因研究进展: 以养殖场和医院为例. 生态毒理学报, 2017, 12(06): 27-37.

[162] 贺小萌, 曹罡, 邵明非, 等. 空气中抗性基因(ARGs)的研究方法及研究进展. 环境化学, 2014, 33(05): 739-747.

[163] Pal C, Bengtsson-Palme J, Kristiansson E, et al. The structure and diversity of human, animal and environmental resistomes. Microbiome, 2016, 4(1): 54.

[164] Gao M, Qiu T, Sun Y, et al. The abundance and diversity of antibiotic resistance genes in the atmospheric environment of composting plants. Environment International, 2018, 116: 229-238.

[165] Hu J, Zhao F, Zhang X X, et al. Metagenomic profiling of ARGs in airborne particulate matters during a severe smog event. Science of the Total Environment, 2018, 615: 1332-1340.

[166] Li J, Cao J, Zhu Y G, et al. Global survey of antibiotic resistance genes in air. Environmental Science and Technology, 2018, 52(19): 10975-10984.

[167] 高敏, 仇天雷, 秦玉成, 等. 养鸡场空气中抗性基因和条件致病菌污染特征. 环境科学, 2017, 38(02): 510-516.

[168] 刘菲, 许霞, 屠博文, 等. 某集约化肉鸡饲养场 PM2.5 中抗生素抗性基因的分布特征. 环境科学, 2019, 40(02): 567-572.

[169] Xie J W, Jin L, Luo X S, et al. Seasonal disparities in airborne bacteria and associated antibiotic resistance genes in PM2.5 between urban and rural sites. Environmental Science & Technology Letters, 2018, 5(2): 74-79.

[170] Organization W H. Antimicrobial resistance: global report on surveillance. 2014: Available: http://www.who.int/drugresistance/documents/surveillancereport/en/.

[171] Mahnert A, Moissl-Eichinger C, Zojer M, et al. Man-made microbial resistances in built environments. Nature Communications, 2019, 10(1): 968.

[172] Herfst S, Böhringer M, Karo B, et al. Drivers of airborne human-to-human pathogen transmission. Current Opinion in Virology, 2017, 22: 22-29.

[173] Drudge C N, Krajden S, Summerbell R C, et al. Detection of antibiotic resistance genes associated with methicillin-resistant *Staphylococcus aureus* (MRSA) and coagulase-negative *staphylococci* in hospital air filter dust by PCR. Aerobiologia, 2012, 28(2): 285-289.

[174] Gilbert Y, Veillette M, Duchaine C. Airborne bacteria and antibiotic resistance genes in hospital rooms. Aerobiologia, 2010, 26(3): 185-194.

[175] Lis D O, Pacha J Z, Idzik D. Methicillin resistance of airborne coagulase-negative *staphylococci* in homes of persons having contact with a hospital environment. American Journal of Infection Control, 2009, 37(3): 177-82.

[176] Li H, Zhou X Y, Yang X R, et al. Spatial and seasonal variation of the airborne microbiome in a rapidly developing city of China. Science of the Total Environment, 2019, 665: 61-68.

[177] Gao X L, Shao M F, Wang Q, et al. Airborne microbial communities in the atmospheric environment of urban hospitals in China. J Hazard Mater, 2018, 349: 10-17.

[178] Zhu Y G, Johnson T A, Su J Q, et al. Diverse and abundant antibiotic resistance genes in Chinese swine farms. Proceedings of the National Academy of Sciences of the United States of America, 2013, 110(9): 3435-3440.

[179] 刘长利, 郑国砥, 王磊, 等. 养猪场空气中抗性基因和条件致病菌污染特征. 应用生态学报, 2018, 29(08): 2730-2738.

[180] Liu D, Chai T, Xia X, et al. Formation and transmission of Staphylococcus aureus (including MRSA) aerosols carrying antibiotic-resistant genes in a poultry farming environment. Science of the Total Environment, 2012, 426: 139-145.

[181] Lenart-Boron A, Wolny-Koladka K, Stec J, et al. Phenotypic and molecular antibiotic resistance determination of airborne coagulase negative *staphylococcus spp.* strains from healthcare facilities in southern poland. Microbial Drug Resistance, 2016, 22(7): 515-522.

[182] Hong P-Y, Li X, Yang X, et al. Monitoring airborne biotic contaminants in the indoor environment of pig and poultry confinement buildings. Environmental Microbiology, 2012.

[183] Alvarado C S, Gandara A, Flores C, et al. Seasonal changes in airborne fungi and bacteria at a dairy cattle concentrated animal feeding operation in the southwest United States. Journal of Environmental Health, 2009, 71(9): 40-44.

[184] McEachran A D, Blackwell B R, Hanson J D, et al. Antibiotics, bacteria, and antibiotic resistance genes: aerial transport from cattle feed yards via particulate matter. Environmental Health Perspectives, 2015, 123(4): 337-343.

[185] Rizzo L, Manaia C, Merlin C, et al. Urban wastewater treatment plants as hotspots for antibiotic resistant bacteria and genes spread into the environment: A review. Science of the Total Environment, 2013, 447: 345-360.

[186] Michael I, Rizzo L, McArdell C S, et al. Urban wastewater treatment plants as hotspots for the release of antibiotics in the environment: A review. Water Research, 2013, 47(3): 957-995.

[187] An X L, Su J Q, Li B, et al. Tracking antibiotic resistome during wastewater treatment using high throughput quantitative PCR. Environment International, 2018, 117: 146-153.

[188] Su J Q, An X L, Li B, et al. Metagenomics of urban sewage identifies an extensively shared antibiotic resistome in China. Microbiome, 2017, 5(1): 84.

[189] De Luca G, Zanetti F, Perari A C, et al. Airborne coagulase negative *staphylococci* produced by a sewage treatment plant. International Journal of Hygiene and Environmental Health, 2001, 204(4): 231-238.

[190] 高新磊, 邵明非, 贺小萌, 等. 污水处理厂空气介质抗生素抗性基因的分布. 生态毒理学报, 2015, 10(05): 89-94.

[191] Zhang M, Zuo J, Yu X, et al. Quantification of multi-antibiotic resistant opportunistic pathogenic bacteria in bioaerosols in and around a pharmaceutical wastewater treatment plant. Journal of Environmental Sciences (China), 2018, 72: 53-63.

[192] Teixeira J V, Cecilio P, Goncalves D, et al. Multidrug-resistant Enterobacteriaceae from indoor air of an urban wastewater treatment plant. Environmental Monitoring and Assessment, 2016, 188(7): 59-67.

[193] Li J, Zhou L T, Zhang X Y, et al. Bioaerosol emissions and detection of airborne antibiotic resistance genes from a wastewater treatment plant. Atmospheric Environment, 2016, 124: 404-412.

[194] Yang Y, Zhou R, Chen B, et al. Characterization of airborne antibiotic resistance genes from typical bioaerosol emission sources in the urban environment using metagenomic approach. Chemosphere, 2018, 213: 463-471.

[195] Perez H R, Johnson R, Gurian P L, et al. Isolation of Airborne Oxacillin-Resistant Staphylococcus aureus from Culturable Air Samples of Urban Residences. Journal of Occupational and Environmental Hygiene, 2011, 8(2): 80-85.

[196] Gandolfi I, Franzetti A, Bertolini V, et al. Antibiotic resistance in bacteria associated with coarse atmospheric particulate matter in an urban area. Journal of Applied Microbiology, 2011, 110(6): 1612-1620.

[197] 房文艳, 高新磊, 李继, 等. 城市社区农贸市场空气微生物及抗生素抗性基因研究. 生态毒理学报, 2015, 10(05): 95-99.

[198] Zhou H, Wang X L, Li Z H, et al. Occurrence and distribution of urban dust-associated bacterial antibiotic resistance in northern China. Environmental Science & Technology Letters, 2018, 5(2): 50-55.

[199] Mazar Y, Cytryn E, Erel Y, et al. Effect of dust storms on the atmospheric microbiome in the eastern Mediterranean. Environmental Science and Technology, 2016, 50(8): 4194-202.

[200] Forsberg K J, Patel S, Gibson M K, et al. Bacterial phylogeny structures soil resistomes across habitats. Nature, 2014, 509(7502): 612-6.

[201] Xie J, Jin L, He T, et al. Bacteria and antibiotic resistance genes (ARGs) in PM2.5 from China: Implications for human exposure. Environmental Science and Technology, 2018, 53(2): 963-972.

作者：应光国[1]*，汪美贞[2]，罗义[3]，苏建强[4]，何良英[1]，刘有胜[1]，杨凤霞[5]，陈则友[3]

[1]华南师范大学，[2]浙江工商大学，[3]南开大学，[4]中国科学院城市环境研究所，[5]农业农村部环境保护科研监测所

第 12 章　环境微塑料的形成、分析、行为及风险管控研究进展

- 1. 微塑料的环境来源与分布 /289
- 2. 微塑料的检测及表征 /299
- 3. 微塑料的生物毒性及人体暴露 /304
- 4. 微塑料的环境行为 /308
- 5. 微塑料污染的环境风险和管控 /312
- 6. 展望 /314

本章导读

粒径<5 mm细微塑料颗粒的环境污染问题近年来引起了社会各界的广泛关注，其在海洋、淡水、陆地和大气环境中的生态风险以及与环境中其他污染物的复合污染效应已成为环境学科领域的一个新兴研究热点。本文综述了环境中微塑料及纳米塑料的来源、分布、检测方法、毒性、载体作用、以及风险管控等方面的国内外最新研究进展，并对环境微塑料的未来研究方向提出了展望。

关键词

微塑料，毒性效应，分布和检测，复合污染，风险管控

随着高分子合成技术的进步，塑料工业的发展给人类提供了各种各样的塑料制品。塑料以其质量轻、耐腐蚀、易加工成型、成本低、使用方便等优点，被广泛应用于国民经济的多个行业，从工农业生产到衣食住行，塑料制品已深入到社会的每一个角落，进入到人们的生产、生活的各个领域。根据欧洲塑料工业协会（Plastic-Europe）和中国《塑料工业》编辑部的最新统计表明[1, 2]，在2009～2017年，世界塑料年总产量从 2.5×10^4 万t增长到了 3.48×10^4 万t。目前常见的塑料制品材质主要包括聚乙烯（PE），聚苯乙烯（PS），聚氯乙烯（PVC）和聚丙烯（PP）等，过去五十年，塑料产品的生产和使用量急剧增长，据统计，2014年全球塑料制品的产量已经达到3亿t[3]，并且仍然在持续增加。塑料的大量使用以及随意抛弃导致其在环境中大量积累；塑料制品极难降解，在环境中的存留时间可以达到100年甚至更长[4]。残留在环境中的塑料制品在一系列的物理、化学和生物因素的共同作用下，形成粒径小于5mm的微塑料颗粒甚至更小的纳米塑料（粒径小于1μm）[5-7]。由于可以随水力、风力作用而长距离的运输，微塑料广泛分布于环境介质中，主要包括水体、沉积物（沙滩和底泥）以及生物体内[8-10]。微塑料污染已成为生态系统中一个新的安全隐患，并对生态安全甚至人类健康产生潜在威胁[11]。

1 微塑料的环境来源与分布

1.1 环境微塑料的来源

自从20世纪50年代以来，随着塑料工业的兴起和发展，塑料垃圾的污染已成为全球性的环境问题。人类已经生产了83亿t塑料，其中的63亿t被废弃成为垃圾。塑料垃圾中，仅有9%被回收利用，12%进行了焚烧处理，剩余79%进入垃圾填埋场或自然环境[12]。

环境中的微塑料可分为原生微塑料和次生微塑料。原生微塑料是以微小粒径形态直接释放到环境中的塑料颗粒，主要来源有在工业生产过程中最初就被制造成微小粒径的合成树酯颗粒，被人为添加到日用化学品中的塑料微珠，或是塑料制品在制造、使用或维护过程中磨损产生的颗粒；次生微塑料是进入环境中的大尺寸塑料垃圾在紫外线、波浪、风力等作用下，逐步老化破碎分解形成的塑料颗粒[13, 14]。

1.1.1 原生微塑料的环境来源

原生微塑料绝大部分源于陆地（98%），只有2%来自海上活动。合成衣物洗涤、合成橡胶轮胎磨损等过程产生的微塑料约占原生微塑料总量的一半，原生微塑料通过公路径流、废水处理系统、风力等途

径进入环境。估算原生微塑料进入海洋的排放量约为 80 万~250 万 t/a，约占海洋微塑料总量的 15%~31%[15]。

1) 合成纺织纤维

在合成纺织物的洗涤过程中，衣物纤维脱落也是原生微塑料的重要来源。这些纤维经污水处理排放后进入自然水体，并可能最终进入海洋[16, 17]。Boucher 等对家用洗衣机的洗衣废水研究表明，单件衣服每次洗涤可产生大于 1900 根纤维，并估算合成衣物洗涤产生的纤维总量占原生微塑料总量的 34.8%[15]。

2) 合成橡胶轮胎

车辆轮胎外部材料由合成聚合物基质组成，主要成分包括苯乙烯丁二烯橡胶（约占 60%）、其他橡胶和添加剂等。车辆行驶过程中，轮胎摩擦路面会造成磨损，大量合成橡胶颗粒粉尘（主要小于 80 μm）散落在路面上和空气中[18]，一部分进入邻近的土地，另一部分通过降水或人工洒水进入排水沟，从而进入自然水体和海洋。挪威、瑞典和德国的轮胎橡胶颗粒粉尘年排放量估计分别为 0.45 万 t、1 万 t 和 11 万 t[14]。轮胎磨损产生的微塑料排放量约占原生微塑料总量的 28.3%[15]。

3) 道路标记材料

在道路基础设施的建设和维护过程中，道路标记使用的材料中有大量的油漆、热塑性塑料、预聚物和环氧树酯，在道路标记使用过程中，由于风化和磨损，大量微塑料将释放到环境中。有研究表明，道路标记材料向环境释放的微塑料占原生微塑料总量的 7%[15]。

4) 船舶涂料

船舶涂料广泛应用于船舶的各个部分，包括船体、船舱、甲板及设备，船舶涂料包括固体涂料、防腐涂料和防污涂料。用作涂料的塑料材料包括聚氨酯、环氧涂料、乙烯基涂料和油漆[19]。微塑料用在涂料中，可以起到降低涂料密度，提高硬度和耐用性，提供消光效果，增强颜色等作用。在船舶使用、维护、清洁过程中，涂料中的微塑料将可能直接排放进入水体。有研究表明，船舶涂料向环境释放的微塑料占原生微塑料总量的 3.7%[15]。

5) 化妆品和个人护理用品塑料微珠

部分化妆品和个人护理用品，如洗面奶、牙膏、沐浴乳、洗手液等面部和身体清洁产品，可能人为添加塑料微珠，用于去除角质、油脂、死皮等，某些类型的化妆品添加塑料微珠用于增加产品美观度。一些化妆品和个人护理用品中，塑料微珠的重量占产品重量的 10%，每克产品中多达数千个塑料微珠[20, 21]。

据报道，欧洲国家用于化妆品和个人护理用品的塑料微珠使用量为 4073t/a，PE 是用作皮肤清洁产品塑料微珠的主要材料，占总量的 93%[21]。塑料微珠在消费者清洗过程中被冲入污水系统，并进入水体和海洋。化妆品和个人护理用品中的塑料微珠排放量占原生微塑料总量的 2%[15]。

6) 树脂颗粒

树脂颗粒是塑料加工过程使用的原料或半成品，通常为 2~5 mm 的圆柱形、方形等形状的颗粒，在制造、加工、运输和回收过程中，都有可能被泄漏到环境中。批量的树脂颗粒中，由于原料生产中的污染，也可能含有较细的塑料粉尘。挪威 PE 工厂过去十年的数据表明，树脂颗粒泄漏量为 0.04%[14]。Essel 等[22]对德国塑料行业树脂颗粒泄漏量按 0.1%~1.0%计算，得到德国塑料行业的微塑料损失为 2.1 万~21 万 t/a。这两项研究均认为树脂泄漏是环境中原生微塑料的重要来源[21]，约占原生微塑料总量的 0.3%[15]。

7) 含塑料粉尘

各类塑料制品广泛应用于人类的生产和生活中。在塑料制品的全生命周期过程中，因磨损、风化、倾倒释放出的塑料颗粒，可归类到含塑料粉尘[21, 22]，具体包括合成鞋底、合成炊具等塑料制品的磨损，人造草皮的风化，喷砂研磨过程，含橡胶颗粒沥青道路的使用等，均有塑料颗粒向环境的释放。含塑料粉尘排放量约占原生微塑料总量的 24%[15]。

1.1.2 次生微塑料的环境来源

次生微塑料的来源包括陆源和海源的塑料垃圾。大量塑料垃圾通过各种途径进入海洋，模拟研究表明全球192个沿海国家仅2010年向海洋输入的塑料垃圾就达480万~1270万吨[23]。陆地的塑料垃圾通过雨污水系统、河流沟渠、风力、人为等途径进入海洋，据估计，陆源塑料垃圾占海洋垃圾总量的80%[24, 25]；海上来源的塑料垃圾主要来自海上设施、船舶运输、渔业养殖活动等[26]。

1) 陆源塑料垃圾

微塑料垃圾的陆地来源主要包括垃圾填埋场、河流和洪水、排水口、非法倾倒[27]、交通运输、滨海旅游区[14]、农用薄膜等。混合收集的生活垃圾中，含有塑料袋、一次性塑料餐具、塑料包装、快餐盒等大量塑料垃圾。目前人类产生的塑料垃圾中，接近50亿吨进入垃圾填埋场或者自然环境中[12, 28]。垃圾填埋场包括有生活垃圾卫生填埋场、工业和建筑垃圾填埋场、垃圾简易填埋场和垃圾非正规堆放点等。垃圾填埋场底或堆放点的轻质塑料垃圾在风力、降水、生物摄食等作用下极易重新进入环境；此外，在经济发展水平和垃圾管理水平较低的国家，还存在居民向公共场所、路边、河道沿岸随意丢弃垃圾的现象。此外，垃圾运输、塑料原料和制品运输过程中的塑料泄漏，都是次生微塑料的重要来源。

2) 海源塑料垃圾

微塑料垃圾的海洋来源主要包括渔业的破损渔网、绳索、鱼线等废弃渔具，水产养殖的泡沫、塑料网具、绳索、浮标等，客船、货船、渔船、游艇、军用或执法船只的生活垃圾、绳索和维修废弃物，海上作业平台的生活垃圾和工作使用过程中的固体废弃物[26]。渔业活动中，大量使用合成聚合物制作的渔网和绳索，如尼龙、PE、PP等，渔网和绳索破旧后很难回收利用，往往被渔民丢弃在海里或岸边。据统计，约10%的海洋垃圾是废弃的渔具[29]。

目前，水产养殖业的鱼类占全球鱼类产量的一半，随着水产养殖业的迅速发展，塑料制品的使用也在增加[30]。水产养殖业消耗大量塑料浮标，主要由EPS制成，使用寿命仅为2~3年。EPS废弃后极易破碎成碎片，散落在海面和海滩上，成为海洋漂浮垃圾和海滩垃圾的主要成分。

各类船舶和海上作业平台产生的生活垃圾、绳索、维修废弃物等，也是微塑料垃圾的重要来源，由于缺乏监管，大量塑料垃圾被丢弃到海中和码头岸边。

1.2 海洋环境中微塑料

1.2.1 海水

PP、PE以及泡沫塑料（如EPS），由于其密度小于海水，漂浮在海表，而PVC、PTFE等则由于密度大于海水则会下沉（表12-1）。基于塑料本身的特点，一旦塑料进入海洋，就会在在水体中持续存在并聚积，并通过风和表层流的作用在全球海洋输移[31]。自从20世纪70年代首次揭示了微塑料的存在以来[32]，有关河口、近岸、近海和大洋中的微塑料浓度被陆续报道[31]，包括在北冰洋和南大洋等海域也有微塑料的检出[33, 34]。据估算，约有7000~35000t塑料（包括微塑料）漂浮在开阔大洋[35]，超过250000t和超过5万亿片的塑料聚积在海洋中[36]。

全球海水中微塑料平均浓度从赤道东太平洋的4.8×10^{-6}个/m³[37]到瑞典近岸海域的8.6×10^3个/m³[38]，最大差异达9个量级。45%研究的微塑料均值浓度在0.01~10个/m³[39]。太平洋是目前开展相关研究最多的海域，海表微塑料的浓度从27000~448000个/km²，0.004~9200个/m³[40]不等。大西洋中微塑料浓度相对低于太平洋，在污染严重海区为1500个/km²和2.5个/m³[31, 40, 41]。地中海是微塑料污染的热点区，浓度为平均62000个/km²，这可能是由于地中海为城市化地区包围的"封闭海域"[40, 42]。

表 12-1 常见塑料聚合物类型、应用、密度

聚合物	缩写	应用	密度/(g/cm³)	行为 a
发泡聚苯乙烯（膨胀泡沫）	EPS	保温盒、浮子、杯、防震包装材料、快餐食品的包装等	0.01～0.64	
聚丙烯	PP	绳、瓶盖、齿轮、包装带、汽车保险杆和轮壳罩等	0.90～0.94	漂浮
聚乙烯	PE	塑料袋、食品包装材料、注塑制品、管材、建设用材等	0.91～0.95	
聚苯乙烯丁二烯共聚物（丁苯橡胶）	SBR	汽车轮胎、胶带、胶管、电线电缆等	1.04	
聚苯乙烯	PS	日用包装容器、一次性用品、电气、仪表外壳、家用电器等	1.04～1.09	
聚酰胺或尼龙	PA	工程塑料、渔网、工业用布、缆绳、传送带、帐篷、降落伞等	1.02～1.16	
聚丙烯腈（腈纶）	PAN	毛毯和地毯、滑雪衣、船帆、军用帆布、帐篷、医用器具、污水处理和回用等	1.18	下沉
聚甲基丙烯酸甲酯（丙烯酸）	PMMA	信号灯设备、仪表盘、储血容器、灯光散射器、饮料杯、文具等	1.10～1.25	
聚氯乙烯	PVC	建筑材料、工业制品、日用品、地板革、地板砖、人造革、管材、电线电缆、包装膜等	1.16～1.58	
聚碳酸酯	PC	玻璃装配业、汽车工业和电子、电器工业	1.2	
聚氨基甲酸乙酯（聚氨酯）	PU	油漆、涂料、密封胶、粘合剂、屋顶防水保温层、方向盘、保险杆、减震垫、人造革、塑胶运动场等	1.20	
醋酸纤维素	CA	香烟过滤嘴	1.22～1.29	
聚对苯二甲酸丁酯	PBT	电子电器、车门把手、保险杆、分电盘盖、挡泥板、键盘、灯罩等	1.30～1.40	
聚对苯二甲酸乙二醇酯	PET	无菌包装材料、纺织品、电影胶片、电气绝缘材料、电容器膜、包装瓶、微波烘箱烤盘等	1.34～1.39	
人造丝		纺织品、卫生用品	1.50	
聚四氟乙烯	PTFE	绝缘材料、防粘涂层、特氟龙器具等	2.10～2.30	

a 塑料在海中的漂浮和下沉行为根据与海水密度（1.025 g/cm³）比较而得

高浓度的微塑料不只是存在于工业化程度高的区域和人口密度高的地区附近，即使在远离人类的偏远地区，如开阔大洋、北冰洋和南大洋等海域也检出了高浓度微塑料。风和海流是控制表层水体[41]微塑料水平分布的主要因素。在开阔大洋，微塑料会在海洋环流的辐聚区发生聚积，不但模型预测[43]，而且在北太平洋和南太平洋[44, 45]、北大西洋[41]、南大西洋和印度洋[35]的多次现场观测都证明了这一点。在东北太平洋的一项为期12年（2001～2012年）的研究中发现，环流辐聚区漂浮微塑料的浓度（156800 个/km²）比非辐聚区（1864 个/km²）高84倍[44]。在世界大洋的五个亚热带环流的辐聚区，微塑料浓度均比非辐聚区要高[35]。沿海地区表层水中微塑料的空间分布受地表径流和河流输入程度的季节性影响，强降雨通常会引起沿海地区微塑料浓度的升高。有大量报告称在暴雨事件发生后立即出现高浓度微塑料[39, 46, 47]。河口附近羽状区的扩散速度和范围可影响微塑料的短期空间分布和漂浮物斑块的运动。半个多世纪以来，全球塑料需求和产量呈指数增长，预计未来塑料碎片进入海洋环境的数量将会增加。此外，即使塑料碎片向海洋环境的新输入完全停止，预计海洋环境中先前积聚的大型塑料碎片的风化和破碎将持续存在，从而产生大量的微塑料。因此，从长远来看，海洋微塑料的数量可能会增加。

为进一步比较海表微塑料的空间分布，仅针对在海洋环境中使用 0.2～0.5 mm 的 Manta 和 Neuston

网收集表层水体中微塑料的 50 个研究结果进行了比较(图 12-1),其中 50%的研究结果使用 0.3～0.35 mm 网目采样。最高浓度是采用 0.5 mm 网目的瑞典斯泰农松德的聚乙烯港的 102550 个/m³[38],南非的 Durban 港以 1200 个/m³ [48]紧随其后。北太平洋环流辐聚区的海表微塑料浓度为 33 个/m³[49]和 2.23 个/m³[50],南大洋为 0.031 个/m³[34],北冰洋为 0.34 个/m³[31]。韩国的巨济岛近岸微塑料浓度高达 47 个/m³ [51],东亚海日本海域 3.70 个/m³ [52],我国的渤海、黄海、东海和南海分别为 0.33、0.33、0.21 和 0.034 个/m³。最低检测到的采样点浓度是美国 San Gabriel 河口的 0.002 个/m³ [53]。

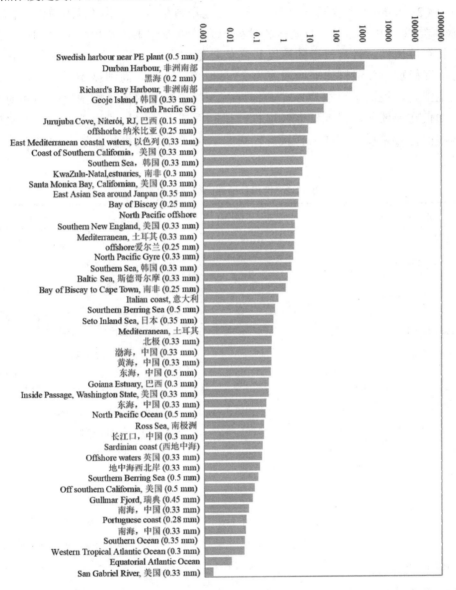

图 12-1　全球海表微塑料浓度,括号中数字表示网筛孔径

1.2.2　沙滩、沉积物

微塑料不仅在海表水体,在次表层水体、海床的全水柱中也持续存留和积累[54]。漂浮微塑料由于风和海流的作用会被冲上岸滩[55],大块塑料和微塑料在海滩上直接暴露于紫外线辐射和高温,为塑料的风化提供了有利条件,可产生更小的微米和纳米级颗粒[54, 56]。重于海水的微塑料(如聚酯类和 PVC)可能沉降在海床,尽管尺寸和形状(例如微塑料的纵横比)、海表张力可能会影响其下沉速率。微生物在微塑

料上的附着,微塑料与浮游生物聚集体的相互作用,以及微塑料随粪便排出,也会导致轻质微塑料沉降进入底栖环境[54]。早期的微塑料监测研究主要集中在大尺寸(1~5 mm)微塑料上,如沙滩上的工业树酯原粒[57]。自 2000 年以来,对海滩和潮下带沉积物(包括深海海床)尺寸<1 mm 的次生小型微塑料也进行了广泛的监测[58-61]。

研究认为海滩和海滩沉积物中微塑料污染主要源于人口稠密地区的排放[62, 63]。20 世纪 70 年代首次报道了海滩微塑料,成为早期研究的主要关注点[64-67];这些在海滩上浓度很高(100000 个 m^{-2})的工业树酯原粒(2~5 mm)主要源于大型港口和当地塑料生产厂。世界各地海滩中的微塑料分布见表 12-2。与水体类似,由于研究中的采样和预处理方法以及结果表征单位的不同,导致报道的海滩微塑料结果不具有可比性。目前报道的浓度最高值是德国东弗里西亚群岛海滩的 62100 个/kg。有研究者统计,海滩沉积物中微塑料的平均浓度为 25~47897 个/m^2,海滩微塑料平均浓度的最大差异在三个数量级的范围内,远低于海水差异[39]。海滩沉积物中微塑料的平均浓度范围相对窄可通过采样方法来解释,如仅在海滩的滨线或残骸线取样,以断面或多线取样的调查很少,多数研究设置单个或多个位于滨线的样方采样。滨线采样较易,可与其他类似研究进行比较,但可能导致过高估计沙滩微塑料的量。

表 12-2 世界各地海滩微塑料的浓度

站点	粒径范围/ mm	浓度
海滩,比利时	0.038~1	92.8 ± 37.2 个/kg
海滩,马耳他共和国	—	>1000 个/m^2
海滩,英国	—	8 个/kg
海滩,葡萄牙	—	6 个/m^2
East Frisian Islands,德国	—	621 个 10 /g
Norderney,德国	<1	1.7~2.3 个/kg
南波罗的海滩	>0.045	39 ± 10 个/kg;外海滩 25 个/kg,高度城市化的湾区海滩 53 个/kg
锦州,中国	0.001~5	102.9 ± 39.9 个/kg
葫芦岛,中国	0.001~5	163.3 ± 37.7 个/kg
葫芦岛,中国	0.001~5	9.25(2.73~39.3)个/kg
东戴河,中国	0.001~5	117.5 ± 23.4 个/kg
滨州,中国	0.001~5	4.87(1.17~9.34)个/kg
日照,中国	0.001~5	12.3(2.34~44.0)个/kg
山东,中国	0.3~5	740 ± 2458(50~1000)个/kg
洋口,如东,中国	0.001~5	2.53(2.34~29.5)个/kg
南汇嘴,上海,中国	0.001~5	53 ± 12 个/kg
金山,上海,中国	0.001~5	7.20(2.34~22.6)个/kg
沙仑,中国台湾	0.038~5	7.20 个/kg
白沙湾,中国台湾	0.038~5	0.47 个/kg
外木山,基隆,中国台湾	0.038~5	28.3 个/kg
福隆,中国台湾	0.038~5	0.93 个/kg
泉州,中国	0.001~5	2.43(0.39~10.1)个/kg

续表

站点	粒径范围/mm	浓度
广东沿岸，中国	0.315～5	121.7±128.1（4.53～319.5）个/kg
大鹏湾，中国	0.001～5	30.7（2.34～102.6）个/kg
珠海，中国	—	5.07±2.99 个/kg
香港，中国	0.315～5	102.1 个/kg，湿季
香港，中国	0.315～5	16.2±36.3 个/kg，旱季
香港，中国	—	0.047±0.01 个/kg
澳门，中国	—	0.19±0.25 个/kg
江门，中国	0.001～5	3.24（2.34～4.67）个/kg
沙扒湾，阳江，中国	0.001～5	5014 个/kg
三娘湾，钦州，中国	0.001～5	3.63（3.11～4.28）个/kg
钦州，中国	—	0.20±0.08 个/kg
北海，中国	—	0.44±0.15 个/kg
北海，中国	0.001～5	6080 个/kg
涠洲岛，北海，中国	—	2.60±2.00 个/kg
海口，中国	0.001～5	7934 个/kg
万宁，中国	0.001～5	8714 个/kg
三亚，中国	0.001～5	6872 个/kg
亚龙湾，三亚，中国	0.001～5	3.02（1.17～5.06）个/kg
Geoje，韩国	0.001～5	2721 个/L

研究表明，人口密度与海滩微塑料的浓度间并无直接相关性[54]。城市和偏远地区沙滩中的微塑料浓度相似[68]，旅游和非旅游海滩间微塑料浓度也无显著空间差异[69]，表明海滩微塑料与附近的陆源输入或人口密度无关，应是通过水动力过程和洋流从开阔大洋输运来的[54]。季节变化等自然因素也会影响海滩上微塑料的浓度和分布。雨季香港海滩微塑料的含量显著高于干季，说明珠江的输送在香港的微塑料数量和分布方面起着重要作用[70]。此外，风和表层流会影响微塑料从海到海滩的输运和沉积，即使附近没有塑料生产厂或捕捞活动，迎风海滩与顺风海滩相比含有更多的微塑料，表明表层流是微塑料输运和沉积到海滩的驱动力[71]。

潮下带微塑料平均浓度数据有限，但结果显示其分布范围较窄（图12-2），为1～8000 个/kg。港口为热点污染区域，如加拿大的Halifax港最高，为8000 个/kg[72]。沿海和陆架沉积物中的微塑料浓度相对较低（15～145 个/kg）。Van Cauwenberghe 等[61]调查了大西洋和地中海深海沉积物中的微塑料，平均浓度为400 个/m^2。根据[73]的研究，在人迹罕至的Kuril-Kamchatka海沟沉积物中发现了2020 个/m^2的微塑料，大多数是高密度聚合物纤维。由于此类聚合物通常在其进入环境处聚积[73]，因此，推测这些微塑料是通过洋流输运到此的[74]。另有研究报道，在大西洋、地中海和印度洋的海底沉积物中存在微纤维，据估计，印度洋海底沉积物中纤维浓度为40亿根/km^2[75]。潮下带沉积物中微塑料的浓度受季节变化的影响。对钦奈海岸表层沉积物淹水前后的微塑料分布和浓度的研究发现，洪水后微塑料的浓度比洪水前高三倍，表明新的微塑料在洪水期间通过河流被冲刷入海[76]。

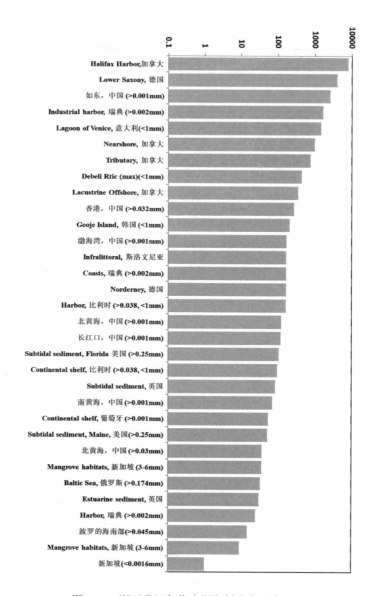

图 12-2 潮下带沉积物中微塑料浓度（个/kg）

1.3 淡水环境中微塑料

河流是海洋微塑料的主要来源之一。经模型估算，每年从河流进入海洋的微塑料约为 1.15～2.41 百万吨[77]。因此，淡水环境（河流、湖泊等）中的微塑料污染受到广泛关注。已报道的世界各地的湖泊水体微塑料污染差异较大（图 12-3），其平均浓度分布从 0.02 个/m^3（北美五大湖的 Lake Huron）[78] 到 2.46 个/m^3（中国太湖）[79]。靠近人类活动区的微塑料浓度远高于湖泊中心区域的浓度[79]，证实了人类活动对微塑料污染的影响。与湖泊相比，河流更易受人类活动影响而含有更多的微塑料，河流的微塑料浓度远高于湖泊（图 12-4）。例如，Faure 等[80]检测到河流和湖泊中微塑料浓度分别为 7 ± 0.2 个/m^3 和 0.51 ± 0.67 个/m^3，究其原因，很有可能是因为河流的流动性及流经人类活动区的面积都远大于湖泊。

在全球范围内，对微塑料在河流水体表面分布的研究相对于湖泊的研究更多。为更好地对比全球各河流的微塑料浓度，本章中仅采用拖网采样数据（表 12-3）。欧洲河流的相关研究报道较多，微塑料在欧洲各河流中的浓度范围跨越三个数量级（0.028～64 个/m^3）。在同一个国家的多条河流中，微塑料浓度范围较为稳定，例如 Rech 等[81]报道了位于智利的四条河流，浓度范围为 0.17～0.74 个/m³。从目前可得到的数据来看，微塑料在世界河流中的浓度分布区域性差异并不明显。

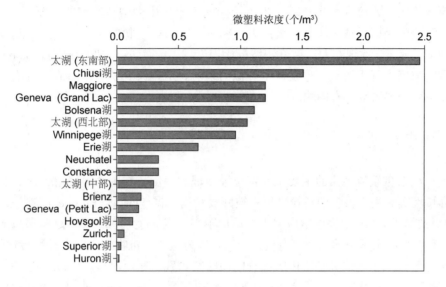

图 12-3　湖泊水体中微塑料的污染浓度（数据来源：文献 [78～80，82，83]）

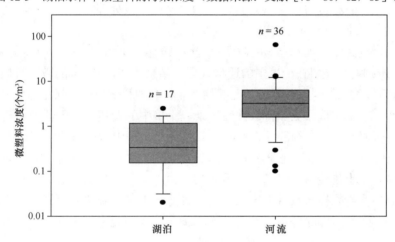

图 12-4　湖泊和河流中微塑料污染浓度对比

箱式图分别展示的是浓度中值、25% 和 75% 的浓度值

表 12-3　全球部分已报道的河流水体表面微塑料浓度（仅采用了拖网采样数据）

河流	国家	文献报道浓度	拖网孔径
Danube	奥地利	0.317±4.66 个/m³	500 μm
Thames	英国	0.028 个/m³	300 μm
Seine	法国	0.28～0.47 个/m³	330 μm
Rhine	瑞士	17 个/m³	300 μm
Po	意大利	2.1，4.3 个/m²	330 μm
Ottawa	加拿大	1.35 个/m³	100 μm
Snake，Lower Columbia	美国	2.57±2.95 个/m³	100 μm
Goiana	巴西	7.13～19 个/100 m³	300 μm
Elqui，Maipo，Manle，BioBio	智利	0.17～0.74 个/m³	1 mm

Hurley 等[85]证实洪水会导致河底沉积微塑料浓度显著下降。通过雨季和旱季河流中微塑料浓度的对比，Faure 等[80]发现 Venoge 河在雨季（均值 64 个/m³）的微塑料浓度远高于旱季（均值 6.5 个/m³）。除此之外，河流流域的人口密度和城市化程度也是影响河流水体中微塑料浓度的重要因素。Kataoka 等[86]分析了日本多条河流中微塑料浓度与其流域的人口密度及城市化现状的相关性，结果发现，微塑料浓度与二者的相关性都很显著，体现在人口密度和城市化程度越高，微塑料浓度也越高。Baldwin 等[87]对美国多条河流中微塑料浓度进行了监测，也发现了类似的趋势。这些研究说明，河流流域范围内的人类活动对河流甚至海洋中微塑料浓度具有影响。

1.4 土壤环境中微塑料

陆地或土壤环境中微塑料污染的主要来源有垃圾填埋和倾倒、农用地膜残留、含有微塑料污泥的再利用、随意丢弃垃圾等。由于很多塑料垃圾很难被回收利用，据 Plastics Europe 报道[88]，在欧洲每年有 30.8%的塑料垃圾被填埋。PE 薄膜通常应用于农用地膜[89]，若未能妥善处理，这些塑料薄膜将会残留在土壤中，并且能有效吸附土壤中的农药等有机化合物，成为这些有机物在土壤中迁移转化的传播介质[90]。除此之外，Carr 等[91]证实了污水处理厂中大部分的微塑料都残留在污泥里。Mahon 等[92]在爱尔兰的污泥样品中也检测到相当高的微塑料浓度（4200～15000 个/kg 干重）。污泥在农业中的再利用使得这些残留的微塑料被转移到土壤中。Nizzetto 等[93]估算通过污泥的直接应用或处理过的固体废弃物排放所进入到陆地环境的微塑料约为 125～850t/（a·百万人口）。

与水体相比，土壤流动性差，进入土壤的微塑料一般只能在土壤生物的作用下来进行重新分配。Huerta Lwanga 等[94]探究了蚯蚓对微塑料在土壤中的迁移过程，结果表明，73.5%的微塑料颗粒都被蚯蚓从土壤表面转移到了土壤内部。Maaβ 等[95]也证实了土壤跳虫能将微塑料颗粒向土壤下方转移，导致微塑料颗粒进入更深的土壤内部，对土壤造成更加持久性的污染。土壤中的微塑料由于其形态与土壤颗粒相似，容易被土壤生物误食，研究指出，蚯蚓的死亡率随着其生活的土壤中微塑料的浓度增大而增大[96]。更为严重的是，微塑料对土壤中的一些有害化学物质具有较强的吸附性，Hodson 等[97]指出，土壤中的微塑料成为了重金属（例如锌）从土壤转移到蚯蚓的传播介质。由于土壤成分复杂造成微塑料难被分离提取，微塑料在土壤中分布的不均给采样布点带来困难，因此目前对土壤中微塑料污染的研究较少，有待进行更加全面深入的研究。在韩国，Jang 等报道约有 44%的海洋微塑料是来源于陆地；从源头上控制微塑料的产生与排放，才能从根本上解决海洋微塑料污染的难题。

1.5 灰尘和大气中微塑料

灰尘是环境污染物的一个重要载体，尤其是室内灰尘。建筑材料、清洁用品、室内活动和室外灰尘的侵入都是室内灰尘中污染物的来源[98]。人们常使用的包装产品，电子产品，医疗产品，织物等经磨损、风化产生的塑料材质的颗粒也可能成为灰尘中微塑料的来源。Dris 等[99]在距法国巴黎城市中心 10 km 左右的公寓和办公室内，采集了室内空气及除尘袋里的灰尘样品。根据显微镜观察结果，公寓室内灰尘中微塑料的浓度在 190～670 个纤维/mg 灰尘。办公室空气中微塑料的浓度（4.0～59.4 纤维/m³）高于公寓环境（1.1～18.2 纤维/m³），且不同公寓内空气样品中的微塑料浓度也存在差异，这可能是和不同的生活方式有关（例如，衣物是晾干还是烘干，是否使用地毯等）。室内灰尘中纤维的尺寸范围（50～4850 μm）大于室内空气中的微塑料（50～3250 μm）。在纤维状微塑料的成分组成上，67%的纤维由天然成分组成（棉花、醋酸纤维、羊毛等），33%的纤维包含人工合成聚合物（尼龙等）。该研究证明了室内环境中人体暴露于微塑料的风险，尽管人体对于该研究尺寸内微塑料的摄入和吸入的概率很小。此外，室外灰尘也可能是水、空气、土壤等环境介质中微塑料的源或汇。Dehghani 等[100]在伊朗德兰黑的市中心地区的 10 个街道样品中，检测到微塑料的浓度在 83～605 个/30 g 灰尘干重。在检测到的微塑料当中，尺寸集中在 250～500 μm 之间，颜色以橘色、黄色、黑色、灰色四种为主，并且形状以球状和纤维状为主。Abbasi 等[101]研究了伊朗布什尔市街道灰尘中微塑料的污染情况，发现微塑料的浓度在 21～165.8 个/g 灰尘，这个结果低于 Dehghani 等[100]报道的结果，可能是由于采样点选取的不同。通过主成分分析，微塑料和微橡胶的产生与重金属元素浓度呈现出相关性。

随着大气污染的持续加重，大气环境中的微塑料也得到关注。建筑材料、人工草皮等产生的颗粒，垃圾填埋场产生的飘尘，都可能成为大气环境中微塑料的来源。土壤表层的微塑料也可能在风的载带下进入大气环境中。周倩等[102]报道了烟台市大气环境中微塑料的形状、浓度、成分和季节性差异。烟台大气环境中的微塑料以纤维状为主，占比 95%，主要由聚酯和 PVC 构成；其他形态的微塑料，如碎片、薄膜和发泡等也有少量发现，成分分别以 PE、PVC 和 PS 为主。根据调查结果估算，研究区域大气沉降的微塑料量约为 2.33×10^{13} 个/a，相当于 0.9～1.4t/a。因此，滨海城市大气环境中的微塑料很可能是海岸和近海海域微塑料污染的重要来源。在我国广东省东莞市的室外降尘中，Cai 等[103]发现了纤维、泡沫、碎片、薄膜等不同形状的微塑料，主要是纤维状，尺寸在 200～700 μm 的最多，占比 30%左右。大气降尘中微塑料浓度为 36 ± 7 个/m²/d，主要由纤维素（73%）、PE（14%）、PP（9%）、PS（4%）构成。Dris 等[104]研究了法国巴黎城市和郊区室外降尘中的微塑料，结果表明城区室外降尘中微塑料的浓度（110 ± 96 个/m²/d）显著高于郊区（53 ± 38 个/m²/d），且几乎全是由纤维组成，其中，50%的纤维成分是天然纤维（棉花或羊毛），21%是天然合成纤维（醋酸纤维和人造丝），17%是纯合成纤维（PET、PA 等），12%是纤维混合物（PET 和 PU、棉花和 PA）。他们猜测衣物是室外降尘中微塑料的主要来源。Dris 等[99]的另一项研究发现，室内空气中的微塑料浓度（0.4～59.4 个纤维/m³）明显高于室外空气中微塑料的浓度（0.3～1.5 个纤维/ m³）。研究结果提示，室内空气通过空气交换可向室外大气中输送微塑料。

2 微塑料的检测及表征

2.1 微米尺度采样与检测

野外环境中微塑料样品的采集主要分为水体样品（表面或水柱）、沉积物样品（沙滩或底泥）、生物体样品（水生生物或陆生生物）。

水体微塑料的样品采集主要形式有拖网采集水体表层样品[105]、水泵或水桶采集大体积水体表层或水柱样品、金属筛网采集水体微表层样品[51]。不同采集方法所对应的水量差异很大，拖网采集滤过水量一般为 10～1000 m³，水泵采集滤过水量一般为 1～100 L[106]。同一水体中，不同采样方法所得到的微塑料浓度具有数量级的差异，微表层水采集>大体积水采集>拖网采集[51]。海洋环境中微塑料研究大多是针对海表漂浮微塑料的，主要采用 Manta 网或 Neuston 网拖网采样，使用网眼为 10～1000 μm 不等。也有通过具不同孔径的大体积过滤装置来采集聚积在海表的漂浮微塑料[39,51,106]，或使用垂直拖网，连续浮游生物记录仪、分层网，或在次表层走航泵水采样[39,107]。随着微塑料粒径的降低，其丰度急剧增加[35,52]。因此，在微塑料丰度比较中考虑采样和检测的微塑料尺寸下限是至关重要的。表 12-4 将不同研究的取样或采样后处理中的网目尺寸进行了分类[39]，67%研究的网目在 280～505 μm 间，6%使用了较大的网目（900～1000 μm），16%使用较小的网目（120～250μm），其余 11%的研究使用了网目为

表 12-4　表层海水中微塑料浓度（个/m³）及所用采样网（改编自参考文献 [39]）

网目/μm	文献数量	变化范围	平均值±标准差	中位值	75%位值	25%位值
10～80	8	0.17～8654	2444 ± 2841	1841	2679	599
120～250	11	0.012～969	93 ± 291	1.15	2.57	0.49
280～350	38	0.00028～7.68	0.96 ± 2.05	0.031	0.28	0.01
450～505	9	0.00002～1.69	0.22 ± 0.55	0.015	0.063	0.012
900～1000	4	0.0000048～0.000341	0.00028 ± 0.00034	0.00018	0.00044	0.00022

10～80 μm 的精细浮游植物网。根据不同的网目范围，海表微塑料的平均（中位值）浓度，10～80 μm 为 2.4（1.8）× 10^3 个/m³，120～250 μm 为 93（1.15）个/m³，280～350 μm 为 9.6（0.31）× 10^{-1} 个/m³，250～505 μm 为 2.2（0.15）× 10^{-1} 个/m³，900～1000 μm 为 2.8（1.8）× 10^{-4} 个/m³。与采样区域和采样时间无关，海表微塑料浓度表现出与网目的负相关关系。

Reisser 等[108]探究了微塑料在水体垂直方向的分布，研究发现绝大部分的微塑料主要分布于表层水体 0.5 m 以内。总的来说，为了使所采集到的样品更具代表性和可比性，大面积水域的水体样品采集建议使用微塑料拖网，采集表层 0.5 m 水体中的样品，在不堵塞网底管的前提下，所过滤的水量建议在 100 m³ 以上。

沉积物中微塑料的采样主要分为沙滩或河岸泥沙采集和底泥采集[59]。与底泥采集相比，泥沙采集的难度相对较低，耗费较少，大部分已报道的沉积物中微塑料的分布都是沙滩或河岸沉积物中浓度[109, 110]，少量研究报道了底泥中微塑料的浓度[111, 112]。沙滩或河岸沉积物采集一般选择其高潮水线，采集深度主要在表层 1～20 cm，单个样品的采集面积主要为 0.0025～1 m² 范围[113]。底泥样品的采集主要采用不锈钢抓泥斗，采集河床或海床表层 1～10 cm 的底泥[61, 112]。

生物体微塑料采样主要采集水生生物（鱼类、贝类、浮游微生物类）[114-116]或陆生生物（鸟类和蚯蚓类）[117, 118]。这些生物的来源分为市场购买和野外或养殖场随机打捞。Davidson 和 Dudas[119]对比了野生和养殖的贝类中微塑料的含量，发现并无显著差异。一般来说，鱼类和鸟类主要取其胃肠道进行分析，贝类取其软组织，浮游生物和蚯蚓类因为体积较小，不适合进行解剖，故而整体分析。

采集到的样品一般需运回实验室进行分离提取和定性定量分析。沉积物样品由于其成分复杂，需要先用比重大的液体进行浮选分离,浮选出来的颗粒物再被进一步的筛分和过滤得到不同粒径范围的微塑料样品[59, 113]。将样品进行消解（常用消解液有过氧化氢、蛋白酶 K、芬顿试剂、强酸等[113]）后得到纯净的塑料样品，根据颗粒物的形态可将微塑料样品分为碎片类、薄膜类、泡沫类、小球类、线类、纤维类等。

2.2　纳米尺度检测方法

粒径小于 100 nm 的塑料污染颗粒被称为纳米塑料[120]，它们可以通过更大的塑料碎片形成。由于其较高的比表面积以及在生物体内迁移的潜力，使得这类颗粒特别引人关注。与此同时，它们细小的粒径尺寸也给进一步的处理和观察增加了难度，在环境样品中很难被分离和识别，这使得我们对环境中纳米塑料的认识也十分有限。

2.2.1　纳米塑料的分离

微塑料在环境外力作用下持续破碎，最终会形成纳米级塑料。由于尺寸的差异，现行微塑料分离方法并不完全适用于纳米塑料。所以在生物、海水、沉积物或土壤等真实环境样品中，开展纳米塑料的检测研究还是相当少。目前为止，为工程纳米颗粒开发的方法被证明是有效的，例如分子排阻色谱技术（size exclusion chromatography，SEC）、凝胶电泳和磁场场流分离技术（magnetic field-flow fractionation，MFFF），这些技术也许可以借鉴应用到纳米塑料的分离中。

分子排阻色谱法的分离原理为凝胶色谱柱的分子筛机制。样品通过一系列具备分子筛效应的多孔凝胶，根据待测组分的分子大小进行分离。目前已有文献报道，利用分子排阻色谱法可以检测塑料的分子量及其分布[121, 122]，然而将 SEC 应用到纳米塑料的研究仍有一定的困难，因为如果两种纳米塑料粒子都能进入凝胶孔隙，即使它们的大小有差别，也不会有好的分离效果。

凝胶电泳法是在电场作用下，纳米粒子穿过凝胶介质，其迁移速率取决于颗粒所带电荷和尺寸[123]。琼脂糖凝胶因为其孔径大、均一度好，制备方便可以被用于更广尺度范围纳米颗粒的分离[124]。然而，使用凝胶电泳进行纳米塑料分离时，由于其粒径过小可能会被嵌入凝胶中，给后续的分析带来挑战，而且凝胶电泳技术并不只对塑料具有特异性，因此也会有其他的干扰。

场流分离（field-flow fractionation，FFF）作为一类分离技术，主要用于分离 1 nm～100 μm 的生物大分子、聚合物分子、胶体以及有机相或水相中的微粒悬浮物[125, 126]。MFFF 是场流分离技术的重要分支，以磁场力为主要作用力，利用磁性粒子磁响应性及所受重力的综合作用对微粒进行区分和筛选[127]。将 MFFF 应用于纳米塑料分离的最大技术难点是需要一种方法去磁化它们，这可能比微塑料的磁化更具挑战性，因为磁性纳米颗粒太大，从而无法有效地与纳米塑料结合。

上述技术尽管在纳米微塑料分离中有所应用，但其目标物的浓度远高于我们所预期分离的环境基质中纳米塑料的浓度。因而，在进一步提高分离效率之前，它们还很难被应用于环境样品的纳米微塑料分离中。

2.2.2 纳米塑料的检测

环境样品中的纳米塑料一旦分离后，应该从视觉上和化学角度上进行表征。然而在现有技术下，难以将它们与没有化学结构信息的天然纳米颗粒区分开。所以目前纳米塑料难以表征，甚至难以检测。目检技术是一种廉价且方便的检测方法，可以反映包括粒子的形貌、大小和颜色等信息，减少后续化学表征的粒子数量，对于粒径大于 500 μm 的颗粒，肉眼或光学显微镜是可行的[128]。然而，即使有较好的光，受到衍射极限的限制，传统光学显微镜成像的分辨率也很难突破 200 nm[129]。有研究者尝试采用红外原子力显微镜（AFM-IR）[130]分析纳米塑料，但是由于纳米塑料比微料塑料更具异质性，该系统的适用性有待进一步探究[131]。扫描电子显微镜（scanning electron microscopy，SEM）可以观察到更小的颗粒及其表面形貌，但由于成本高和复杂的样品制备，所以该方法是不适用的，而且这项技术不能证实塑料的存在[132, 133]。用透射电镜（transmission electron microscopy，TEM）可以检测到纳米塑料，有学者在生物摄食纳米塑料的研究中用 TEM 检测人为加入的荧光纳米塑料。但透射电镜提供不了任何化学结构方面的信息，同时由于纳米塑料的非晶态结构，透射电镜对纳米塑料的可视化效果也不理想，而且需要重金属染色剂[134]。此外，Zhang 等[135]采用双光子激发和定时检测的方法，在非侵入性条件下观察到了 Musa exotica 体内聚集的 12 nm Eu 发光标记的纳米塑料，与电子显微镜相比，该技术具有简化样品制备的优点，然而它的衍射仍然是有限的。

纳米颗粒追踪分析（nanoparticle tracking analysis，NTA）技术，在降解研究中被用来测量悬浮纳米塑料的浓度和尺寸分布，结合布朗运动和光散射特性来识别 30 nm 以下的颗粒。Lambert 等[5]利用纳米粒子追踪分析模型证明了次生纳米塑料的生成，发现样品溶液中纳米粒子浓度可高达 1.26×10^8 个/mL，比空白对照高 3 倍。但从浓度量级来看，微塑料可使用的"逐个粒子"分析的方法，对纳米塑料而言很难实现。除了不能提供可视化的粒子图像，无法确定纳米塑料的化学结构也是 NTA 技术的主要缺点之一[136]。

聚合物颗粒的化学识别对环境样品中纳米塑料的分析具有重要意义，因为它一方面确认了纳米塑料在系统中存在的问题，另一方面它可以提供粒子额外的化学特性，例如提供关于添加剂存在和老化的信息[137]。然而，在微塑料分析中经常使用的振动光谱由于它们的光谱分辨率受到光源波长的限制，仅能识别微米范围内的颗粒，不能实现对纳米塑料的鉴定。除了光谱学，质谱鉴定是微塑料分析中另一种常用的方法。其中，裂解气相色谱质谱联用（Py-GS-MS）具有高灵敏度，使其更适合环境样品中纳米塑料的检测。但存在误判的可能，因为不同的聚合物可能产生相似的热解产物。Mintenig 等[138]提出了一个定量分析纳米塑料的框架，采用非对称流场流分离与 Py-GS-MS 技术联用，实现了对纳米塑料分离、鉴定的可能，但该系统的实用性仍有待证实。

2.3 塑料种类、形态、表面特征分析

2.3.1 塑料的种类分析

合成塑料聚合物目前超过 5000 种[139]，其中 PP、PE、PVC、PET 和 PS[140]共占塑料聚合物的 80%。塑料聚合物材料是为了不同的目的而生产的，因此具有特定的属性。环境中塑料材质的识别对塑料的环境丰度和影响研究至关重要。早期研究将目检与燃烧性能以及塑料极性相结合。例如，在 Carpenter 和 Smith[141]的研究中，根据塑料颗粒的燃烧性能，判定尼龙收集网中一些白色颗粒不是 PS、PVC 或聚丙烯酸酯类。随着技术的进步，人工分选和电感耦合等离子体（ICP）、拉曼光谱、近红外（NIR）光谱、X 射线荧光光谱以及原子吸收光谱（AAS）都被用于识别塑料的种类。通过对比样品的特征峰和谱库中的标准样品特征峰来判定微塑料的结构类型。由于红外光谱仪操作简单，仪器价格相对便宜，多数微塑料定性分析选用傅里叶红外光谱仪[113]，常见微塑料样品的傅里叶红外光谱图如 12-5 所示。各分析方法的比较见表 12-5。

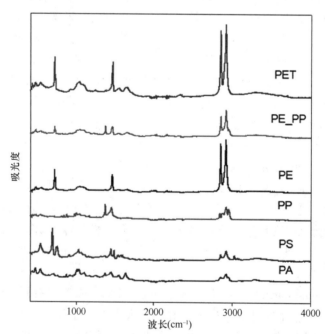

图 12-5 常见微塑料样品的傅里叶红外光谱图

样品来自于珠江入海口表面水体

表 12-5 微塑料常见检测技术

检测技术	优缺点描述	时间
人工分类	是一种基于物理性质差异的传统方法，但它存在一些缺点，如易出错，成本高，耗时	2008[142] 1994[143]
电感耦合等离子体（ICP）	提供有效的雾化，激发和电离，对于高精度液体样品的分析具有优势，在塑料分析之前，这种方法需要对样品进行溶解，耗时	1991[144] 2003[145]
拉曼光谱（RS）	可用于检测分子带，可靠性高，信号弱	1988[146] 2011[147]
傅里叶红外光谱（FTIR）	最广泛使用的方法，准确识别塑料类型，不受荧光干扰，识别塑料颗粒最小粒径为 20 μm，不能分析有色或不透明的塑料	2013[148] 2004[58]
X 射线荧光光谱（XRF）	可准确识别 PVC	2011[149]
原子吸收光谱（AAS）	灵敏度高，样品必须被粉碎并溶解在液体中，耗时	2000[150] 2010[151]
裂解气相色谱-质谱分析技术（Py-GC-MS）	颗粒被分解，对热降解产物进行分析，获得更精细的聚合物成分信息，会对样品本身造成破坏，不利于后续研究	2019[152]
激光诱导击穿光谱（LIBS）	从等离子体的衰变过程发出元素特征线，通过监测它们的位置和强度进行定性和定量分析，样品预处理的需求较少，灵敏度低和严重的基质效应	2019[152]

2.3.2 塑料的形态和表面特征分析

为了获得塑料样品形态的信息，显微操作是最可行的方法，因为它可以直接获得塑料的几何形状和表面特征。显微镜有许多不同的操作模式，其中两组成像技术最为突出：光学显微镜，电子显微镜。使用光学显微镜，可以对塑料的形态进行观察，其中可添加荧光染料尼罗红，它能选择性地对大多数合成聚合物（橡胶除外）进行染色[153]。电子显微镜分为扫描电子显微镜和透射电子显微镜。扫描电子显微镜（SEM）可以获得塑料的高分辨率图像，也可与其他仪器结合使用，如能量色散光谱（EDS）或能量色散 X 射线光谱（XEDS），以测量粒子表面发射的光谱线，对塑料样品进行微观和元素分析[154]。透射电子显微镜（TEM）检测样品下方的发射电子束，需要高达 300 kV 的高电子加速电压和非常薄的样品（塑料薄膜类），由于透射操作模式，TEM 提供关于颗粒内部而不是表面的信息。

结晶度是塑料重要的热力学指标，差示扫描量热法是测量塑料结晶度和玻璃化温度最为普遍的技术手段，计算公式如下：

$$X_c(\%) = \frac{\Delta H_m - \Delta H_c}{\Delta H_m^0} \times 100 \tag{12-1}$$

其中，$X_c(\%)$ 为结晶度含量，ΔH_m 为熔化焓，ΔH_c 为冷结晶焓，ΔH_m^0 为 100%结晶聚合物的理论熔化焓。特别是当塑料经过风化、光照后，结晶度会产生一系列的改变，研究显示经过实验室紫外线处理后的 PE，由于分子量的降低，聚合物的剩余部分因分子重组而结晶度的增高，但是当塑料长期处于缺氧环境中，聚合物由于分子交联，结晶度降低[155]。通过差示扫描量热法表征聚合物结晶行为是在高于晶体熔化范围的高温下进行的，这意味着在测试时不能反映固体半结晶聚合物的形态结构。

热重分析[156,157]是研究煤、塑料和生物质燃烧的普遍方法，通过热重分析获得样品燃烧的动力学特征[158,159]。Cepeliogullar 等[160]研究了塑料—生物质混合物的共热解特性和动力学，发现了 PVC 热解过程的三个步骤。

塑料机械性能也是衡量塑料材质的重要指标。当暴露于恶劣的环境条件下，例如高温、强紫外线照射、机械负荷或腐蚀性环境[161]，塑料制品会失去大部分的机械性能。研究显示，对实验室老化的 PE 材料进行传统的拉伸试验，发现随着暴露时间的延长，断裂伸长率迅速降低，失去韧性[162]。尽管现今有很多塑料表征的技术手段，但这些方法只能分析一种或几种参数，很显然难以满足对塑料特别是微塑料的全方位探究，因此从不同角度丰富微塑料的分析监测方法仍是迫在眉睫的工作。

3 微塑料的生物毒性及人体暴露

3.1 微塑料进入生物体的途径与赋存特征

3.1.1 微塑料进入生物体的途径

大量的野外研究表明，在 220 多种不同营养级的生物体内能检测到微塑料污染[163]。微塑料可以通过摄食、吸附和粪合等三种方式进入生物体（图 12-6）。其中，动物摄食被认为是微塑料进入动物体的主要途径，也是微塑料对动物构成潜在危害的重要原因[164, 165]。

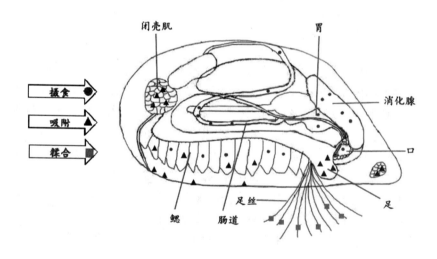

图 12-6 微塑料进入生物体的途径

1）摄食途径

微塑料与小型浮游生物具有类似的尺寸，这导致高营养级的浮游动物在正常摄食过程中会被动摄取微塑料或把它当成自然猎物而摄取[164]。Ory 等[166]在对太平洋亚热带环流区域的调查中发现，琥珀条纹鱼倾向于摄食蓝色的微塑料碎片，与其捕食的浮游动物在颜色和尺寸上极为相似。微塑料不仅可以直接被生物摄食，还可以通过不同营养级动物之间的摄食行为逐级传递，比如通过贻贝传递给滨蟹，通过浮游动物传递给糠虾[167-169]。

生物的摄食机制和生理结构都会影响其对微塑料的摄食，相对于自由游泳和腐食性生物，底栖滤食性生物对微塑料颗粒更为敏感，能够摄取更多的微塑料[170]。比如双壳类动物贻贝，它在摄取食物前会根据颗粒质量对尺寸类似的粒子进行区分，不合适的颗粒会被当成假粪排出[171]。微塑料随着海水从进水管进入贻贝体内后，会被鳃表面的黏液所捕获，其中能够被识别的部分会以假粪的形式被排出体外，而被捕获的微塑料会被鳃上皮细胞同化或被转移到口和消化系统中去。大量研究中，在贻贝的鳃、肠道和假粪中都找到了微塑料，证实了贻贝对微塑料的摄食行为[172]。此外，底栖滤食性鱼类比目鱼摄入微塑料的比例（75%）要远高于浮游捕食性鱼类银白鱼（25%），杂食性鱼类比草食和肉食性鱼类体内的微塑料含量高[173, 174]。

2）吸附途径

除了胃肠道，在生物的其他部位也不断检测出微塑料。野外调查和室内模拟研究都在贻贝的足、闭壳肌、性腺等器官中发现了微塑料[175]。这些器官与贻贝的摄食无关，因此这些研究表明，生物体除了摄食还可以通过吸附的方式对微塑料进行积累，且通过吸附进入贻贝的微塑料大约占到总量的一半[175]。

3）糅合途径

相对于其他与摄食无关的器官，贻贝的足中含有的微塑料含量最高，甚至超过了胃，仅次于肠道，这个发现预示着足对微塑料的积累可能有更为复杂的方式[175]。贻贝足上的腺体通过分泌足丝去附着基底，当它的足在伸出壳外分泌足丝的过程中，会与环境中的微塑料直接接触，新生足丝有可能与微塑料糅合在一起，影响贻贝足丝的完整性和它的附着功能（未发表数据）。已经有研究表明微塑料的暴露导致贻贝新生足丝减少，且足丝韧性减弱[176]。

3.1.2 微塑料在生物体内的赋存特征

目前对微塑料在生物体内赋存特征的研究主要集中于两个方面。一方面，通过野外调查来探索微塑料在生物体不同器官中的分布特征；另一方面，通过室内暴露和清除实验来研究生物体对微塑料的摄入和排出。由于微塑料在生物体内的迁移直接影响到它的生物毒性效应，它在进入消化系统后的转移路径和停留时间一直是当前研究的关注点。微塑料的尺寸、形状以及生物体自身的结构特点等都会影响微塑料在生物体不同器官中的分布[177-179]。

1）微塑料尺寸的影响

微塑料的尺寸是决定其生物可利用性的一个关键因素。例如，小于 250 μm 的微塑料在贻贝胃肠道中的比例高于其他各个器官[175]。虽然大部分摄入的微塑料会随着粪便被排出体外[180, 181]，但是尺寸较小的微塑料（3 或 9.6 μm）可能会通过消化道进入血液循环系统[182]。Lu 等[179]通过对比发现 5 μm 的 PS 颗粒可以进入鱼的肝脏，而较大尺寸的粒子（20 μm）只在鳃和肠道中被检测到。类似的研究也指出大于 20 μm 的微塑料不能转移到鱼的肝脏和肌肉中[178, 183]。然而也有相反的研究报道了鱼肌肉和肝脏中尺寸超过 100 μm 的微塑料的存在[184-186]。有关微塑料，尤其大粒径微塑料进入生物组织中的可能性需进一步研究证实；究竟多大尺寸的微塑料能够穿过肠膜转移到生物体各个部位还需要进一步研究；这个转运过程比较复杂，受到微塑料其他特征以及生物体自身的影响，但毋庸置疑的是尺寸起着至关重要的作用。

2）微塑料形状的影响

除了尺寸外，微塑料的形状也直接影响了它在生物体内的迁移和分布。然而目前的室内暴露实验多使用单一类型的微塑料（多为小球或颗粒），对形状这一因素在微塑料的摄取和排除过程中的影响研究相对较少。Jabeen 等[177]用不同形状的微塑料暴露鱼后，发现仅有纤维状的微塑料被摄取。然而 Qu 等[187]用三种形状的微塑料暴露贻贝后，发现贻贝吸收的球状微塑料最多。一般情况下，较小的微塑料纤维在进入消化道前更容易被嵌入鱼的鳃组织[178]。不同形状的微塑料在消化道中的排除也是存在差异的，它们一旦被摄取，纤维可能缠绕在鳃或其他组织上面不易排出，而肠道中的小球则可以更快地被排出体外[188, 189]。

3）生物因素的影响

生物体自身的摄食机制和生理结构也会影响其对微塑料的摄取和体内分布。野外调查发现杂食性鱼类马面鲀消化道中的微塑料丰度相对较高，肠道和胃部微塑料的赋存量与组织结构的复杂程度有密切关系[190]。Collard 等[185]通过对三种鲱形目鱼类的过滤容积比较后发现，沙丁鱼具有高的滤水面积和更密的鳃耙，因此摄食了更多更小的微塑料。一些等足目动物前胃中的细密塞网结构可以有效阻止 1 μm 以上的微塑料进入消化器官[191]。甲壳类动物的消化道结构相对复杂，它们前肠中有用于研磨的几丁质盘，尺寸较大个体的盘间距也较大，可以让更多的微塑料进入中肠排出体外，而较小个体的消化道中会积累更多的微塑料[192]。

3.2 塑料的毒理学暴露方法

由于野外环境高度复杂，因此很难确定微塑料对水生生物具体的毒害作用。在实验室内模拟野外环境研究微塑料毒性的方法已经被大多数学者所认可[193, 194]。

1）当前存在的问题

野外环境中的微塑料不论是在形状和尺寸上，还是聚合物类型上，都是非常之多和复杂的[195]。在形状上，微塑料可分为纤维、碎片、薄膜、球、泡沫等；在尺寸上，微塑料可被破碎分解成纳米尺寸的塑料；在聚合物类型上，又分为通用塑料（如 PE、PP、PVC、PS 等），工程塑料（如 PA、PC 等）[196]。这些性质使得微塑料在环境中呈现出不均匀的空间分布特征，加之野外环境的不断变化，如塑料风化、生物污损[197-199]，又使得微塑料处在一个动态变化过程中。然而，当前微塑料的室内毒理学暴露实验，大多数却仅采用了形状规则、尺寸精确和聚合物单一的商业微珠或者碎片作为研究对象[40]，并没有考虑到微塑料的异质性和动态变化，例如，PS 荧光微球就常作为受试材料，被用来检测微塑料对生物的动力学过程和毒性效应[200, 201]；另一个常用的受试材料是 PE 颗粒或碎片[202, 203]。

室内实验中的微塑料暴露浓度大多要高于野外真实浓度，甚至高达几个数量级。例如，Phuong 等[40]比较发现当前室内暴露使用的最低浓度大约是野外检测到的最高浓度的 4500 倍。除了目前使用的暴露材料和受试浓度与真实环境不相符以外，许多暴露实验采用的暴露时间、模式生物种类以及评价指标也大相径庭[204, 205]，这些现象不仅导致实验结果千差万别，甚至南辕北辙，而且无法比较不同研究之间的结果。总之，目前的室内暴露实验既不能很好地反映野外真实环境下微塑料对生物的毒理学效应，不适合于微塑料的生态与健康风险评估，也无法获得可靠的科学证据作为制定微塑料污染管控法案的支撑。

2）未来的发展方向

由于目前室内研究与真实环境相去甚远，因此获取有效的微塑料室内毒理学数据将是今后微塑料研究面临的最严峻的挑战之一。要解决这一难题需要我们重新认识微塑料污染物的本质特征，并在此基础上重建适合于微塑料毒理学研究的新方法。首先，在野外调查中，应该尽可能多地获取真实环境中的微塑料信息，包括丰度、大小、形状和颜色等等；其次，建立"微塑料的特征谱"更全面地表征环境中的微塑料，在室内制备具有环境特征的微塑料；最后，使用环境特征微塑料在环境浓度下进行暴露实验（图 12-7）。

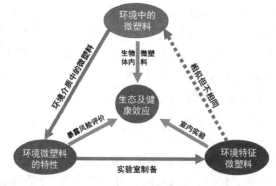

图 12-7 真实环境中的和制备的微塑料的差异及关系

除了尽可能地模拟野外真实环境中的微塑料特性、结合特异性指标检测外，还需要考虑生物对微塑料的真实"摄食（feeding）"途径和行为。例如，鱼类摄入微塑料的最主要途径是依靠摄食，而鱼类的摄食包括摄食行为、摄食量、摄食频率和摄食节律等多个方面的特性，摄食行为又是由食物感知、口味评价、食物吞咽或拒绝等构成的复杂行为链。但目前研究还未揭示知鱼类摄食微塑料的过程和机制。因此，当前迫切需要深入加强这方面的研究，以便更准确地揭示环境中微塑料可能的生态毒性和风险。

3.3 微塑料的毒性效应

如果说微塑料研究的过去十年证明了其"普遍存在"，今后十年的主要目标则是阐明其"生态危害"。目前已有大量研究表明，微塑料对生物存活、生长、生殖、摄食活性、生理生化参数、组织病理学以及基因表达水平等指标均有直接毒性效应[164, 206, 207]。

3.3.1 微球和颗粒的毒性

当前研究最多的是 PS 微球对生物的毒性。例如，PS 暴露对微藻的生长和桡足类 F0 和 F1 代的存活有显著的影响，此外 PS 还会导致贻贝摄食活性和血细胞吞噬活性的降低以及牡蛎子代产量和生长的降低[208-212]。目前也有许多研究使用微塑料粉末或颗粒进行暴露实验。Lei 等[213]选取了五种常见的微塑料颗粒（PE、PP、PVC、PS 和 PA）进行暴露实验，发现斑马鱼和秀丽线虫的存活率均降低，斑马鱼的肠道褶皱破损、上皮细胞破裂并出现炎症，线虫的生长生殖受到影响。

3.3.2 碎片和纤维的毒性

尽管目前在环境中检测到的微塑料中，大部分是纤维和不规则的碎片，但在室内暴露中，对它们的毒性研究还相对有限。例如，PP 纤维暴露 4 周导致滨海螃蟹的摄食量以及维持生长的能量均降低；乙烯醋酸乙烯酯（EVA）纤维暴露 6 周使得红鲫鱼成鱼的健康系数显著降低，鳃丝出现破损和断裂现象；PP 纤维长期暴露（8 个月）降低了海螯虾的营养可利用性[177, 192, 214, 215]。Rochman 等[203, 216]发现暴露于 PE 碎片中的青鳉鱼的肝脏存在糖原耗竭、脂肪空泡化和单细胞坏死等情况，此外碎片暴露还会导致雌鱼的 *ERα*、*Vtg-I* 和 *Chg-H* 这三个基因的显著下调。

3.3.3 纳米塑料的毒性

纳米粒径塑料可经由生物的消化道转移到循环系统、免疫系统乃至组织细胞中，引发集体免疫系统的炎症反应等[217, 218]。例如，Kashiwada 等[219]在青鳉鱼的大脑、睾丸、肝脏和血中均检测到了 39.4 nm 的纳米 PS 塑料，并且纳米塑料能够穿透血脑屏障进入大脑；Tussellino 等[220]发现 50 nm PS 粒子可以进入非洲爪蟾胚胎的肠道细胞中。蓄积在生物体内的塑料可通过食物链传递到高营养级生物甚至人体内[221]。

3.4 微塑料对人体的暴露风险

微塑料在环境介质中广泛存在，其对人体的暴露也被日益关注。通常认为，饮食暴露是微塑料人体暴露的重要途径。鱼类和贝类为人类提供了约 20%的动物蛋白摄入量[222]。室内研究发现，鱼类有很强的微塑料摄入能力[216, 223]，而现场调查也发现，许多经济鱼类体内含有大量微塑料[224]。双壳贝类动物摄食时，将大量的水泵入外壳的套膜腔中，保留了悬浮在鳃上的微塑料颗粒[182, 225, 226]。虽然室内实验得到的水生动物体内微塑料浓度要比真实环境高出许多[40, 227]，但是鱼类和贝类作为食物是人体微塑料暴露的重要来源之一。除水产品外，其他食品中也发现有微塑料。蜂蜜和食糖中检出了合成微纤维（>40 μm）和塑料碎片（多数 10～20 μm）[228]，德国啤酒中也检出了微塑料主要是碎片（109 个/L）[229]，海盐中的微塑料主要是 PET 和 PE[230]。

灰尘的摄入和吸入是微塑料人体暴露的另一重要途径。巴黎人口密集城区大气沉积物中的 30% 纤维是塑料（主要 7～15 μm）[104]。伊朗街道灰尘中含有 250～500 μm 的微塑料颗粒或碎片[100]。中国城市地区室内外灰尘中 PET 微塑料的浓度范围分别为 1550～12000 mg/kg 和 212～9020 mg/kg，PC 微塑料的浓度中位数为 4.6 mg/kg（室内）和 2.0 mg/kg（室外）[231]。室外灰尘中的 PET 微塑料浓度与国民生产总值（GDP）和人口密度（PD）都呈现出正相关关系[231]。对于职业暴露，PVC 和 PA 微塑料的室内暴露水平可分别达到 0.5 和 0.8 个/mL[232]。微塑料的大量暴露会刺激人体的视觉和嗅觉。职业暴露于对位芳纶、聚酯和聚酰胺纤维的荷兰工人，临床表现有咳嗽、呼吸困难、喘息和痰增多等症状[233]。尽管关于微塑料人体暴露风险的研究很少，但越来越多的毒性试验表明，微塑料特别是纳米塑料暴露的健康风险不容忽视。

4 微塑料的环境行为

4.1 微塑料的形成、风化和迁移转运

环境中的次生微塑料主要由残留在环境中的塑料制品通过水流、风力及其他机械作用力以及紫外线、环境微生物等的物理、化学和生物共同作用（所谓"老化过程"）下而形成[5, 6]。此外，微米塑料以及纳米塑料还有可能作为个人护理品中的添加成分随着生活废水直接进入环境[62, 234, 235]。微塑料从陆地传输到海洋的过程中（图 12-8），河流起到重要的运输载体作用[236]。进入河流的密度较小的微塑料颗粒（例如 PE、PP、PS 等密度小于水的塑料）主要漂浮在水体表面，密度较大的颗粒（例如 PVC、PET 等密度大于水的塑料）会渐渐沉入河底，在河床上沉积，部分塑料颗粒在漂浮过程中可能随潮汐涌动滞留在河岸。随着水流量的增大，河岸滞留和河底沉积的微塑料可能会有一部分重新回到水体并随着水流进入海洋。

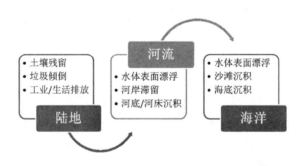

图 12-8 微塑料在环境中由陆地进入海洋的传输过程

海洋环境中微塑料的迁移过程包括漂流、悬浮、沉降（缓慢沉降和快速沉降）、再悬浮、搁浅、再漂浮和埋藏等。这些过程可以循环往复地发生，也可能自发终止。例如，漂浮微塑料由于表面生物附着或泥沙絮凝等原因导致其密度逐渐增大，发生悬浮、沉降过程；沉入海底的微塑料在海流的作用下可能发生再悬浮或漂浮，也可能在海底发生水平方向的移动或者被沉积物埋藏。这些复杂的迁移过程导致微塑料在海洋中呈现明显的垂向分布特征：底层沉积物中的微塑料数量远大于表层水，中层水中微塑料的含

量最少[59]，海岸沉积物中的微塑料含量高于深海沉积物[237]。此外，微塑料可能在洋流和风力的作用下进入大洋，也可能被海浪冲刷上岸发生搁浅，搁浅的微塑料可能发生再漂浮，重新进入海洋中[238]。

微塑料通过机械破碎、化学氧化等过程形成以后，在环境中尤其是河口、海洋潮滩等地区，经过风力、水力和光照等物理因素，以及生物降解和生物扰动等生物因素的作用，其表面发生物理和化学变化，如在微观形貌上出现划痕、皱褶、微孔、裂纹等风化特征，同时产生羧酸、醛、酯或酮等含氧官能团[239]。例如，发泡类微塑料的风化产生油酸腈、芥酸酰胺、α-N-去甲基美沙醇、1,1-二苯基-螺[2,3]-己烷-5-羧酸甲酯、棕榈酸十八酯和棕榈酸十六酯等含氧、含氮化学物质[239]。在河口及附近潮滩沉积物中的树脂颗粒表面有侵蚀和破损现象，存在酯基和酮基，可能与生物作用、光氧化作用有关[240]。我国黄渤海的一些河口及其附近潮滩沉积物或土壤中存在大量的不同形貌类型的微塑料，且多数微塑料表面粗糙，有的还附着铁氧化物、黏土矿物等物质[239]。这些微塑料在环境中长期风化，导致表面结构、化学组成和理化性质发生显著变化，可能在很大程度上影响微塑料自身在环境中的迁移行为、与环境中污染物的相互作用、微生物附着等，进而影响其在海洋和陆地环境中的生态效应和风险。

4.2 微塑料对环境污染物的吸附和解吸

4.2.1 pH、盐度、温度等溶液条件的影响

有研究表明，多氯联苯（PCBs）在PP微塑料的吸附容量随温度（19～27℃）的降低而增大[241]，pH降低使全氟辛磺酸（PFOS）在PE和PS上的吸附更强[242]。盐度对吸附的影响与污染物种类有关。盐度对污染物菲和滴滴涕（DDT）在PVC和PE上的吸附[243]和麝香等在PP上吸附[244]无显著影响；盐度上升可增加PE和PS颗粒上PFOS和PCBs的吸附[242, 245]，但是盐的存在却降低全氟烷基物质（PFASs）和抗生素在微塑料表面的吸附[246, 247]。

4.2.2 微塑料类型、粒径及表面老化的影响

微塑料的分子组成和结构在其吸附有机污染物的过程中起着重要作用[242, 248]。比如，Mueller等认为由于显著降低的玻璃化转变温度，PP比PS显示出更高的吸附性[249]。PE对芘表现出最高的亲和力，其次是PS和PVC[250]，同样，菲在微塑料上的吸附量顺序为PE >PS >PVC[103]。PS和其羧基化产物（PS-COOH）对18种PFASs的亲和力高于高密度聚乙烯（HDPE）[246]。Li等认为因其多孔性结构和氢键，PA对淡水体系中抗生素的吸附能力最强，而其他四种微塑料（PE，PS，PP，PVC）的吸附容量较低[247]。七种脂肪族和芳香族有机吸附质（正己烷、苯等）在四种不同微塑料上的吸附量大小顺序为PA <PE <PVC <PS，表明其他因素（比如π-π相互作用）促进PS吸附芳香族化合物[251]。

有机污染物在微塑料上的吸附容量通常随着微塑料粒径的减小而增加[241, 244]，纳米微塑料的吸附容量高于微米级[245, 252]。研究表明，老化导致显著的表面氧化和微小的局部微裂纹形成，老化后PS微塑料对有机化合物（非极性、单极性、双极性脂肪族化合物；非极性、单极性和双极性芳香族化合物）的吸附系数均有所减低[249, 253]。

4.2.3 污染物性质的影响

当多种污染物同时存在时，可能产生吸附的拮抗作用，例如DDT降低菲在微塑料上的吸附[254]，随着水溶液中芘浓度的增加，菲在微塑料和天然沉积物上的吸收和分配系数（K_d）均下降[103]。大量研究显示，有机污染物在塑料表面的吸附与污染物的Log K_{ow}呈正相关[242, 246-249]，表明疏水作用对吸附的重要性[251]。PAHs与芳香族聚合物PS表面之间的π-π相互作用解释了PAHs在纳米PS塑料上的高吸附和非

线性吸附[252]。吸附强弱还和微塑料的表面积（表面吸附）和微塑料内部的孔径（分配作用）密切相关[255]。Liu 等发现非极性化合物（芘、多溴联苯醚 BDE47）倾向于吸附在 PS 微塑料（玻璃态聚合物结构）的内部基质中（导致吸附质的物理截留），而极性化合物（壬基酚）只在表面吸附[256]。

4.2.4 解吸作用

León 等在地中海的周边地区采集了塑料碎片（>1 mm），发现塑料碎片中吸附的大部分污染物（PAHs、PPCPs、塑料添加剂和农药等）在最初 24h 内解吸，具有较高的转移到海水、沉积物和生物体中的可能[257]。肠道中的表面活性剂会加快污染物的解吸速率[258]。另外，微塑料和污染物性质也会影响解吸；芘和 BDE47 在 PS 上的吸附能力强且解吸迟滞系数大，可能是因为非极性化合物倾向于吸附在 PS 微塑料（玻璃态聚合物结构）的内部基质中，导致吸附物被物理截留，而在结晶程度较高的 PE 塑料中则不会发生迟滞现象[256]。对于具有解吸迟滞现象的非极性和弱极性污染物，会在环境中随着微塑料而发生迁移，扩大其污染范围。

4.3 微塑料对污染物生物富集的影响

微塑料上负载的疏水性有机污染物（HOCs）可随着微塑料的摄入而被生物体吸收。室内研究发现，微塑料和污染物共存时，往往能促进 HOCs 在水生生物体内富集，并改变污染物的毒性效应和作用机制。例如，50 nm 的微塑料可以显著增加菲在大型蚤体内的富集，并和菲体现出加和毒性效应[259]。微塑料还可以增加雌二醇和双酚 A 分别在斑马鱼幼鱼和成鱼体内的生物累积量并影响共存污染物的毒性效应。还有报道指出，微塑料的存在会增加芘在胆汁中代谢产物的浓度，降低虾虎鱼（*Pomatoschistus microps*）幼鱼死亡率以及异柠檬酸脱氢酶的活性等[260]。相反地，模型预测结果显示在受到 HOCs 污染的生态系统中，微塑料的载体作用十分有限，甚至可能降低污染物在生物体内的富集。例如，Gouin 等[261]利用平衡分配概念建立了 HOCs 在大气、水体、沉积物和塑料之间的海岸生态系统分配模型。结果发现由于沉积物和溶解性有机质对 HOCs 的竞争吸附作用，最终分配进入 PE 微塑料的污染物仅占<0.1%。同样地，Koelmans 等研究者通过模拟微塑料存在时开放海岸环境中 HOCs 在沙蚕体内的富集状况，发现微塑料对 PCBs 的稀释和清除作用超过了其载体效应[117, 262, 263]。

图 12-9 微塑料对所负载疏水性有机物（HOCs）的迁移和累积作用的两类冲突观点，即实验室研究发现，微塑料和污染物共存时往往能促进有机污染物在水生生物体内富集，而模型预测结果显示在受到 HOC 污染的生态系统中，微塑料的载体作用十分有限，甚至可能降低污染物在生物体内的富集[117, 208, 259, 261-266]。

这一观点看起来相互冲突，原因实际上是由于两类研究没有在同一个框架体系下开展，也就是说它们检验的假设不同。具体来说，大部分室内暴露研究采用了非平衡状态的实验设计，仅采用经 HOCs 染毒的塑料进行暴露，而没有考虑其他的环境介质，并且采用的测试生物没有或极少有 HOCs 污染，这就迫使 HOCs 从塑料向生物体迁移。因此，此类研究主要证实了微塑料上负载的 HOCs 可以沿着逸度梯度方向往生物体迁移。而在大部分的模型研究中，除了考察微塑料的迁移作用外，还同时考察环境中其他运移 HOCs 途径的贡献，研究表明 HOCs 通过微塑料的摄入在生物（沙蚕、鱼类和海鸟）体内的富集贡献远低于其他自然运移载体（食物颗粒等）的贡献[266]。但这两类研究也有共识，即：当微塑料与污染物共存时，水生生物摄入微塑料既可能增加生物体的污染水平，又可能降低污染物在生物体内的累积，主要取决于污染物在微塑料和生物体之间的逸度梯度[267]。

总体而言，室内暴露实验常常缺乏对复杂环境介质的考量，而微塑料的摄入途径又很难与其他平行共存的 HOCs 自然运移途径（水体、空气、沉积物、溶解性有机质和食源性颗粒物等）区分开来。模型研究虽然考察了多种环境介质，但是通常缺乏模型验证环节，因而只能提供生物富集的间接证据，也就

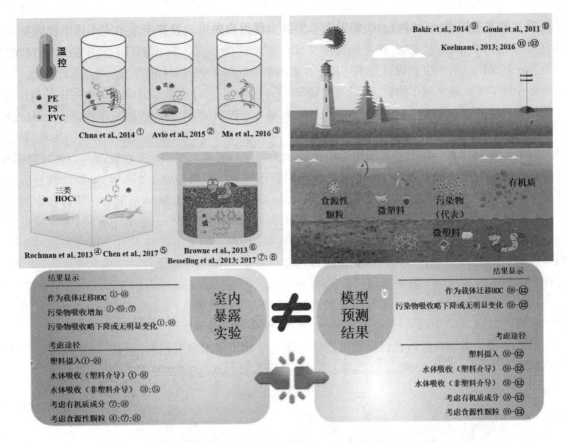

图 12-9　微塑料对所负载疏水性有机物（HOCs）的迁移和累积作用的两类冲突观点

限制了其研究结论的可信度。因此，模拟自然环境的室内暴露和模型研究应相互结合，使所得暴露数据更接近真实环境，以便模型更准确进行机制解析。

4.4　微塑料对污染物环境迁移转化的影响

微塑料对污染物有较强的吸附性能，微塑料上的污染物富集浓度可较水体浓度高出几个数量级[268]。室内和野外研究均发现，微塑料对 PCBs、HOCs 的吸附能力与沉积物或沉积物中的有机组分相当[245, 269]。虽然微塑料对污染物有良好的吸附性能，但大粒径微塑料的迁移能力并不高。当粒径达到纳米级别的时候，微塑料颗粒在砂质土壤中的迁移能力迅速上升，并且能够携带有机污染物迁移[256]。研究者通过建立了海洋中塑料的迁移模型，预测发现微塑料的水平迁移量比其他随着洋流迁移的物质质量低 4~6 个数量级[270]，因此微塑料与共存污染物可能对原位水生生物的生态风险更高。

但值得注意的是，微塑料可以显著改变污染物的垂直迁移和分布。在水环境中，通常高密度的微塑料下沉，而低密度的微塑料上浮、停留在水面表层（约占 46%）[271]。但随着微生物的栖居，微塑料的疏水性下降，颗粒密度增加，浮力逐渐下降，从而影响其在水体中的垂直分布[272]。例如，许多藻类会排出胞外多糖，而胞外多糖可以在湍流等作用下形成黏性颗粒的团聚体[273-275]，微塑料可以被包裹进入这些藻类团聚体中而被垂直运输到水体底部[164]。负载于微塑料上的污染物也将随之发生垂直迁移，从而对不同水层的水生生物产生生态效应。在土壤生态系统中，微塑料对污染物的迁移与污染物极性密切相关。研究发现，低浓度的纳米塑料会显著增强非极性化合物（芘）和弱极性化合物（BDE47）在饱和土壤柱中的迁移，但对三种极性化合物（双酚 A、双酚 F 和壬基酚）的迁移基本没有显著影响[256]。

污染物也可能随着微塑料在食物链中迁移。微塑料自身的食物链迁移已在桡足类/多毛类浮游生物—糠虾[168]，贻贝—滨蟹[167]的二级食物链中证实。纳米级微塑料（24 nm）还被证实可以沿着斜生栅藻—大型蚤—鲫鱼的三级水生食物链发生迁移，并影响鲫鱼的脂质代谢和行为活动[276]。微塑料对污染物的迁移和转化在近年受到越来越多的关注。研究发现苯并[a]芘可以随着 PE 微塑料沿着卤虫向斑马鱼迁移[277]。通用理论模型（MICROWEB）预测发现，微塑料摄入越多，PCBs 的生物放大效应越弱，而 PAHs 的生物放大效应则越强[278]。

微塑料的存在同样有可能改变环境中有机污染物的降解和转化。例如，蒽在微塑料存在时降解变慢，这主要是由于吸附在微塑料上蒽的生物有效性低于吸附于天然沉积颗粒物上[279]。污染物降解变慢也可能导致其被生物富集的机会增加[278]。Ma 等[217]发现，50 nm 的 PS 颗粒显著抑制了菲在培养基中的削减和降解，分别表现在菲的半衰期显著延长及代谢产物和母体的比例显著降低，而 10 μm 的 PS 颗粒对菲的削减和降解无显著影响。这种差异可能是由于菲在 50 nm PS 颗粒上的吸附量高于 10 μm 的 PS 颗粒[217]。Qu 等的结果表明，PVC 微塑料存在可以改变泥鳅体内文拉法辛和 O-去甲基文拉法辛的转化[280]，当 PVC 微塑料存在时，更多的文拉法辛随着微塑料被泥鳅（*Misgurnus anguillicaudatus*）摄入，肝组织中代谢成 O-去甲基文拉法辛并积累[280]。Oliveira 等认为 PE 微塑料增加了虾虎鱼（*Pomatoschistus microps*）胆汁中芘代谢物的浓度[260]。

综上所述，虽然微塑料对有机污染物的迁移转化和生物富集的研究已有报道，但是对相关机理还缺乏深入研究。

5 微塑料污染的环境风险和管控

5.1 微塑料污染的社会、经济、政治效应

5.1.1 微塑料污染的社会效应

微塑料的社会效应，主要表现在微塑料污染对人类社会所造成的负面影响，主要体现在以下几个方面：①微塑料对河流、湖泊、地下水和海水水质和沉积物的影响。微塑料作为广泛存在于淡水和海洋等水体的一种新型污染物，已受到国内外学者的高度关注[28, 281]。②微塑料对水生生物的影响。微塑料容易被浮游动物、底栖生物、鱼类误食，并对生物体产生危害[282]，微塑料能通过食物链从低营养级向高营养级水生生物传递[283]，并可能危害人体健康。由于被摄入的微塑料会在较短时间内随粪便一起排出生物体外[284]，微塑料是否会进入到生物体其他器官尚需进一步研究[285]。③微塑料富集污染物的风险。环境中微塑料由于颗粒小、具有疏水性等特征，是持久性有机污染物等有毒有害化学物质的载体，并可通过各种迁移途径扩散，影响污染物的全球分布[286]，但目前对环境中微塑料在老化或降解过程中的污染物结合与释放机制等关键科学问题仍缺乏了解[287]。④微塑料引起公众的广泛关注，一方面有利于提高社会对塑料污染危害性的认识，减少对塑料制品的使用和丢弃，但另一方面可能引发公众的过度恐慌。

5.1.2 微塑料污染的经济效应

微塑料污染对渔业产生负面经济影响，具体包括直接经济影响和间接经济影响。直接经济影响表现为，由于水生生物误食微塑料，导致生物的摄食、生长、产卵受到危害，并可能导致生物的死亡，从而

造成渔业资源的损失；间接经济影响是水生生物误食微塑料后，通过食物链传递，进一步影响人类健康。在对微塑料进行防治过程中，也将产生大量的经济成本，主要表现在岸滩垃圾、漂浮垃圾、水底垃圾的打捞和清理，以及在河流、排水口设置垃圾拦截和收集设施等。

5.1.3 微塑料污染的政治效应

微塑料不仅对环境和人体健康存在潜在威胁，还涉及跨界污染、产业结构调整和国际治理等问题。目前，国际上对塑料垃圾问题的关注逐渐从科学研究层面向实质性污染管控和全球治理延伸，微塑料污染问题已从单一的环境问题演变为集环境问题、经济问题和政治问题为一体的复杂问题[28]。

在全球层面达成的多边环境协定中，有三个多边环境协议与海洋垃圾密切相关。《联合国海洋法公约》（UNCLOS）为与海洋有关的问题提供了广泛的法律框架。《国际防止船舶造成污染公约》（MARPOL 73/78）附则Ⅴ涉及海洋垃圾污染，禁止将塑料从船上向海洋排放。《伦敦公约》（LC）制定了向海洋环境倾倒废弃物和其他物质的规范，以减少海洋污染。2018年5月，联合国环境署专家咨询小组向第三届联合国环境大会提交了《防治海洋塑料垃圾和微塑料：评估相关国际、区域和次区域治理战略和办法的效力》报告（草案稿），建议用3~4年的时间完成一项新的具有国际法律约束力的文书。

目前来看，在应对海洋塑料垃圾问题上，软法是治理体系发挥作用最为显著的机制，国际上尚无以减少海洋塑料垃圾和微塑料为主要目标的具有约束力的多边环境协议。与海洋垃圾相关的全球性软法有世界粮农组织（FAO）的《负责任渔业行为守则》（CCRF）、《保护海洋环境免受陆上活动影响全球行动纲领》（GPA）、《檀香山战略——海洋垃圾预防和管理全球框架》，联合国环境署（UNEP）在2012年建立的"海洋垃圾全球伙伴关系"（GPML）等。2012年联合国可持续发展大会通过了题为《我们希望的未来》成果文件，要求成员国实施相关公约和计划，在2025年实现"大幅度减少海洋垃圾"的目标。联合国大会通过的第70/1号决议批准了2030年可持续发展目标，目标14要求保护和可持续利用海洋和海洋资源，并明确强调要减少海洋垃圾污染[28]。2017年，二十国集团通过《G20 海洋垃圾行动计划》。2018年6月9日，以加拿大为首G7国家签署《海洋塑料宪章》（美国和日本未签署），承诺到2030年对至少55%的塑料包装进行回收和再利用。由此可以看出，海洋垃圾与微塑料治理已引起国际社会的日益关注。

5.2 微塑料及塑料污染的管控

5.2.1 法律和政策层面

国家和地方在法律和政策上的行动是减轻塑料污染的主要手段，通过使用禁塑（微珠、塑料袋等）、渔具回收激励措施等机制，取得了积极和可衡量的进展。作为国家层面专门的海洋垃圾法案，日本2009年通过的《促进海洋垃圾处置法》（LPMLD），旨在控制和减少海洋垃圾的产生，授权中央政府制定海洋垃圾政策[288]。韩国的海洋环境管理立法中包含了海洋垃圾相关条款，在《韩国海洋环境管理法案》（MEM Act）中授权制定《海洋垃圾管理计划》，并提出了垃圾循环再利用、海洋废弃物流域管理计划、渔具回购计划及渔具标记计划等[288]。美国2006年通过了《海洋垃圾研究、预防和减少法案》（MDRPRA），实施了海洋垃圾计划（MDP）。美国部分城市制定了"零垃圾"计划，部分州实施塑料袋有偿使用制度，并禁止在个人护理品中添加塑料微珠。

中国虽然尚未出台专门的海洋垃圾法案，但近年出台了《水污染防治行动计划》《土壤污染防治行动计划》《渤海综合治理攻坚战行动计划》和《农业农村污染治理攻坚战行动计划》，发布了《国务院办公厅关于限制生产销售使用塑料购物袋的通知》（限塑令）和《禁止洋垃圾入境推进固体废物进口管理制度改革实施方案》（禁废令），开展了"打击固体废物环境违法行为专项行动"（清废行动2018），实施了《生活垃圾分类制度实施方案》，并构建陆海统筹的综合防治体系，加强城镇、农村、河流、船舶港口等关键区域的垃圾管理与防治工作，有效减少了海洋垃圾的陆源和海源输入。

5.2.2 管理和技术层面

垃圾填埋场选址和运营对减少塑料进入环境非常重要。美国、菲律宾、巴西、新西兰等国对垃圾填埋场的选址和运营方式进行了限定[288]，例如禁止建设露天堆放场，不得在滨海地区、洪泛区和湿地建设填埋场等。地震、海啸、台风等自然灾害可以产生大量的海洋垃圾。2009年世界银行关于重点国家灾害风险管理项目的报告列出了20个易受自然灾害影响的高风险国家，其中有14个是沿海国家[289]。美国、日本、海地等国在发生重大灾难后，均采取措施应对产生的大量海洋塑料垃圾[288]。

开展垃圾强制分类收集是实现塑料等材料回收和循环利用的必要途径，在一定程度上可以避免或减少塑料在填埋或堆放过程中进入环境。日本通过开展分类收集制度，2004年的垃圾产量比峰值时1996年减少了32.1%，垃圾回收量则增加了25.1%[290]。德国的垃圾回收率在1990年时只有15%，开展分类收集后，2008年垃圾回收率超过了64%。

开展陆源垃圾清理是有效的海洋垃圾防治手段。美国洛杉矶市和旧金山市使用政府资金用于清理海洋垃圾[291]；韩国政府实施各种计划用于奖励清理海洋垃圾，并制定了回购计划，鼓励并资助渔民和社区清理海洋垃圾[292]。

其他防治海洋垃圾的手段还包括开展渔具标记和回收，加强船舶垃圾管理，打击非法倾倒行为等。

5.2.3 企业和市场层面

联合国环境署甄别了诸如税收、收费、罚款和处罚等市场手段，认为可以应用于塑料污染管控[293]。这些机制背后包含的基本原则包括污染者付费、用户自付和全部成本回收等制度。澳大利亚、美国、德国和韩国采用了一系列市场手段，有效地减少了乱扔垃圾现象，从源头减少海洋垃圾。企业的社会责任感会对塑料制品的研发、销售和全生命周期等产生重要影响。有研究报告表明，许多大公司将包装和塑料可持续利用视为企业社会责任的一部分，为避免品牌负面形象，均加大力度改善包装材料和包装技术。2011年塑料行业协会发布了《海洋垃圾解决方案全球宣言》，已有34个国家的60个行业协会签署了宣言[294]。

5.2.4 社会公众层面

基于社会民间力量采取自下而上减少海洋污染的治理方案，可产生实质性的效果。非政府组织是有影响力的社区治理行动者，绿色和平组织（Greenpeace）、世界自然基金会（WWF）和世界自然保护联盟（IUCN）等大型非政府组织有专门的海洋垃圾计划和资助，开展了大量的净滩活动、公益宣传教育和研究工作。对于海洋塑料垃圾问题，社区团体或非政府组织，对治理所能产生的影响各不相同，进而可能改变政府法规和行业政策[28]。

建立完善的海洋垃圾污染宣传教育体系，引起公众对海洋污染的关注，激发公众的环保意识，进而改变消费行为，自觉养成垃圾收集和分类的习惯，以及促进垃圾的循环利用和无害化处理等，都有助于从源头上解决塑料污染问题[295]。

6 展望

目前关于微塑料的调查研究正逐步从水体环境尤其是海洋转到陆地环境，而大气环境中的微塑料调查研究仍处于起步阶段。研究分析方法仍然局限于粒径相对较大的毫米和微米级塑料颗粒,对粒径更小的亚微米和纳米颗粒在环境中的调查研究方法学仍然需要技术层面上的突破。同时,复杂环境如土壤、沉积物中的塑料分离方法,微塑料表面携带污染物的分离鉴定方法也有待进一步改进。在微塑料的毒理学及生物富集的研究中,应该推进实验方法和实验设置的标准化,并且尽量保持室内实验条件设置与真实环境的一致性,尤其是考虑塑料材质选择的合理性、不同生物的生活习性、构造特征以及在食物链生态位不同而带来的微塑料摄入途径的多样性。微塑料在生物体内示踪定量技术的发展是未来微塑料生物富集研究领域取得突破的关键。大尺度环境中微塑料的分布和迁移规律,其携带污染物在食物链和食物网中的传播规律,以及生物分子水平上的毒性机理是未来需要解决的关键科学问题。考虑微塑料污染目前已经及日后可能产生的社会、经济、政治效应,国家和地方应在法律和政策上采取行动,从管理和技术层面规范化垃圾填埋场地的选址和运营,企业和市场层面制定一系列奖惩措施,从源头控制塑料垃圾的产生和制造,在公众层面大力宣传并进一步开展垃圾强制分类收集等措施是塑料垃圾管控的必经途径。

致谢:"十三五"国家重点研发计划"海洋环境安全保障"重点专项(编号:2016YFC1402200)。

参 考 文 献

[1] 刘朝艳. 2016~2017年世界塑料工业进展(Ⅰ). 塑料工业, 2018, 46: 1-12, 32.

[2] Europe P. Plastics: the facts.2018. https://www.plastics.gl/.

[3] Anderson J C, Park B J, Palace V P. Microplastics in aquatic environments: Implications for Canadian ecosystems. Environmental Pollution, 2016, 218: 269-280.

[4] Hamer J, Gutow L, Kohler A, et al. Fate of microplastics in the marine isopod Idotea emarginata, Environmental Science & Technology. 2014, 48: 13451-13458.

[5] Lambert S, Wagner M. Characterisation of nanoplastics during the degradation of polystyrene. Chemosphere, 2016, 145: 265-268.

[6] Stolte A, Forster S, Gerdts G, et al. Microplastic concentrations in beach sediments along the German Baltic coast. Marine Pollution Bulletin, 2015, 99: 216-229.

[7] Ter Halle A, Jeanneau L, Martignac M, et al. Nanoplastic in the North Atlantic subtropical gyre. Environmental Science & Technology, 2017, 51: 13689-13697.

[8] Auta H S, Emenike C U, Fauziah S H. Distribution and importance of microplastics in the marine environment: A review of the sources, fate, effects, and potential solutions. Environment International, 2017, 102: 165.

[9] Zhang K, Su J, Xiong X X, et al. Microplastic pollution of lakeshore sediments from remote lakes in Tibet plateau, China. Environmental Pollution, 2016, 219: 450-455.

[10] Steer M, Cole M, Thompson R C, et al. Microplastic ingestion in fish larvae in the western English Channel. Environmental Pollution, 2017, 226: 250-259.

[11] Thompson R C, Moore C J, vom Saal F S, et al. Plastics, the environment and human health: current consensus and future trends. Philosophical Transactions of the Royal Society B, 2009, 364.

[12] Geyer R, Jambeck J R, Law K L. Production, use, and fate of all plastics ever made. Science Advances, 2017, 3: 170-782.

[13] Kershaw P J, Rochman C M. Sources, fate and effects of microplastics in the marine environment: part 2 of a global assessment. 2015, 90: 96.

[14] Sundt P, Schulze P E, Syversen F. Sources of microplastics pollution to the marine environment. Norwegian Environment Agency, 2014, 17: 220.

[15] Boucher J, Friot D. Primary Microplastics in the Oceans: A Global Evaluation of Sources. 2017.

[16] Mark Anthony B, Phillip C, Niven S J, et al. Accumulation of microplastic on shorelines woldwide: sources and sinks. Environmental Science & Technology, 2011, 45: 9175-9179.

[17] Magnusson K, Eliasson K, Fråne A, et al. Swedish Sources and Pathways for Microplastics to the Marine Environment, A Review of Existing Data. IVL Swedish Environmental Research Institute, 2016, 183.

[18] Verschoor A, de Poorter L, Roex E, et al. Quick Scan and Prioritization of Microplastic Sources and Emissions. RIVM Letter report, 2014, 156.

[19] OECD. Coating Industry (Paints, Lacquers and Varnishes). 2014.

[20] Leslie H. Plastic in Cosmetics: Are we polluting the environment through our personal care? Plastic ingredients that contribute to marine microplastic litter. Report of UNEP, 2015.

[21] Lassen C, Hansen S F, Magnusson K, et al. Microplastics: occurrence, effects and sources of releases to the environment in Denmark. Copenhagen K: Danish Environmental Protection Agency, 2015.

[22] Essel R, Engel L, Carus M, et al. Sources of microplastics relevant to marine protection in Germany. Texte, 2015, 64

[23] Jambeck J R, Geyer R, Wilcox C, et al. Plastic waste inputs from land into the ocean. Science, 2015, 347 : 768-771.

[24] Sheavly S, Register K. Marine debris & plastics: environmental concerns, sources, impacts and solutions. Journal of Polymers and the Environment, 2007, 15: 301-305.

[25] Gregory M R, Ryan P G. Pelagic plastics and other seaborne persistent synthetic debris: a review of Southern Hemisphere perspectives, in: Marine Debris. Springer, 1997, 49-66.

[26] Niaounakis M. 1 - The problem of marine plastic debris, in: M. Niaounakis (Ed.) Management of Marine Plastic Debris. William Andrew Publishing, 2017, 1-55.

[27] Liffmann M, Boogaerts L. Linkages between land-based sources of pollution and marine debris, in: Marine Debris. Springer, 1997, 359-366.

[28] 王菊英, 林新珍. 应对塑料及微塑料污染的海洋治理体系浅析. 太平洋学报, 2018, 26: 83-91.

[29] Macfadyen G, Huntington T, Cappell R. Abandoned, lost or otherwise discarded fishing gear, Food and Agriculture Organization of the United Nations (FAO). 2009.

[30] Algalita, Plastic pollution and the aquaculture industry, (2017).

[31] Lusher A L, Tirelli V, O'Connor I, et al. Microplastics in Arctic polar waters: the first reported values of particles in surface and sub-surface samples. Scientific Reports, 2015, 5.

[32] Carpente E J, Anderson S J, Miklas H P, et al. Polystyrene spherules in coastal waters. Science, 1972, 178: 749-750.

[33] Cozar A, Marti E, Duarte C M, et al. The Arctic Ocean as a dead end for floating plastics in the North Atlantic branch of the Thermohaline Circulation. Science Advances, 2017, 3.

[34] Isobe A, Uchiyama-Matsumoto K, Uchida K, et al. Microplastics in the Southern Ocean. Marine Pollution Bulletin, 2017, 114: 623-626.

[35] Cozar A, Echevarria F, Ignacio Gonzalez-Gordillo J, et al. Plastic debris in the open ocean. Proceedings of the National Academy of Sciences of the United States of America, 2014, 111: 10239-10244.

[36] Eriksen M, Lebreton L C M, Carson H S, et al. Plastic pollution in the world's oceans: more than 5 trillion plastic pieces weighing over 250, 000 tons afloat at sea. Plos One, 2014, 9.

[37] Spear L B, Ainley D G, Ribic C A. Incidence of plastic in seabirds from the tropical pacific, 1984-91 - relation with distribution of species, sex, age, season, year and body-weight. Marine Environmental Research, 1995, 40 : 123-146.

[38] Noren F. Small plastic particles in coastal Swedish waters. N-Research report, 2007, 11.

[39] Shim W J, Hong S H, Eo S. Marine microplastics: abundance, distribution and composition, in: Microplastic Contamination in Aquatic Environments – an Emerging Matter of Environmental Urgency, 2018, 1-26.

[40] Phuong N N, Zalouk-Vergnoux A, Poirier L, et al. Is there any consistency between the microplastics found in the field and those used in laboratory experiments? Environmental Pollution, 2016, 211: 111-123.

[41] Law K L, Moret-Ferguson S, Maximenko N A, et al. Plastic accumulation in the North Atlantic subtropical gyre. Science, 2010, 329 : 1185-1188.

[42] Collignon A, Hecq J H, Galgani F, et al. Annual variation in neustonic micro- and meso-plastic particles and zooplankton in the Bay of Calvi (Mediterranean-Corsica). Marine Pollution Bulletin, 2014, 79: 293-298.

[43] Lebreton L C M, Greer S D, Borrero J C. Numerical modelling of floating debris in the world's oceans. Marine Pollution Bulletin, 2012, 64: 653-661.

[44] Law K L, Moret-Ferguson S E, Goodwin D S, et al. Distribution of surface plastic debris in the eastern Pacific Ocean from an 11-year data set. Environmental Science & Technology, 2014, 48: 4732-4738.

[45] Eriksen M, Maximenko N, Thiel M, et al. Rifman, Plastic pollution in the South Pacific subtropical gyre. Marine Pollution Bulletin, 2013, 68: 71-76.

[46] Kang J H, Kwon O Y, Shim W J. Potential threat of microplastics to zooplanktivores in the surface waters of the southern sea of Korea. Archives of Environmental Contamination and Toxicology, 2015, 69: 340-351.

[47] Yonkos L Y, Friedel E A, Perez-Reyes A C, et al. Microplastics in four estuarine rivers in the Chesapeake Bay, USA. Environmental Science & Technology, 2014, 48: 14195-14202.

[48] Nel H A, Hean J W, Noundou X S, et al. Do microplastic loads reflect the population demographics along the southern African coastline? Marine Pollution Bulletin, 2017, 115: 115-119.

[49] Goldstein M C, M. Rosenberg, L. Cheng. Increased oceanic microplastic debris enhances oviposition in an endemic pelagic insect, Biology Letters, 8(2012): 817-820.

[50] Moore C J, Moore S L, Leecaster M K, et al. A comparison of plastic and plankton in the North Pacific central gyre. Marine Pollution Bulletin, 2001, 42: 1297-1300.

[51] Song Y K, Hong S H, Jang M, et al. Large accumulation of micro-sized synthetic polymer particles in the sea surface microlayer. Environmental Science & Technology, 2014, 48: 9014-9021.

[52] Isobe A, Uchida K, Tokai T, et al. East Asian seas: a hot spot of pelagic microplastics. Marine Pollution Bulletin, 2015, 101: 618-623.

[53] Moore C J, Moore S L, Weisberg S B, et al. A comparison of neustonic plastic and zooplankton abundance in southern California's coastal waters. Marine Pollution Bulletin, 2002, 44: 1035-1038.

[54] Li W C. The occurrence, fate, and effects of microplastics in the marine environment, microplastic contamination in aquatic environments, in: E. Zeng(Eds)Microplastic Contamination in Aquatic Environments - An Emerging Matter of Environmental Urgency. Elsevier, 2018, 133-173.

[55] Isobe A, Kubo, Tamura Y, et al. Selective transport of microplastics and mesoplastics by drifting in coastal waters. Marine Pollution Bulletin, 2014, 89: 324-330.

[56] Song Y K, Hong S H, Jang M, et al. Combined effects of UV exposure duration and mechanical abrasion on microplastic fragmentation by polymer type. Environmental Science & Technology, 2017, 51: 4368-4376.

[57] Gregory M R. Accumulation and distribution of virgin plastic granules on new-zealand beaches. New Zealand Journal of Marine and Freshwater Research, 1978, 12: 399-414.

[58] Thompson R C, Olsen Y, Mitchell R P, et al. Lost at sea: where is all the plastic? Science, 2004, 304: 838-838.

[59] Hidalgo-Ruz V, Gutow L, Thompson R C, et al. Microplastics in the marine environment: a review of the methods used for identification and quantification. Environmental Science & Technology, 2012, 46: 3060-3075.

[60] Lee J, Hong S, Song Y K, et al. Relationships among the abundances of plastic debris in different size classes on beaches in

South Korea. Marine Pollution Bulletin, 2013, 77: 349-354.

[61] van Cauwenberghe L, Vanreusel A, Mees J, et al. Microplastic pollution in deep-sea sediments. Environmental Pollution, 2013, 182: 495-499.

[62] Browne M A, Crump P, Niven S J, et al. Accumulation of microplastic on shorelines woldwide: sources and sinks. Environmental Science & Technology, 2011, 45: 9175-9179.

[63] Naidoo T, Glassom D, Smit A J. Plastic pollution in five urban estuaries of KwaZulu-Natal, South Africa. Marine Pollution Bulletin, 2015, 101: 473-480.

[64] Gregory M R. Plastic pellets on New Zealand beaches. Marine Pollution Bulletin, 1977, 8: 82-84.

[65] Gregory M R. Virgin plastic granules on some beaches of eastern canada and bermuda. Marine Environmental Research, 1983, 10: 73-92.

[66] Shiber J G. Plastic pellets on the coast of lebanon. Marine Pollution Bulletin, 1979, 10: 28-30.

[67] Shiber J G. Plastic pellets on spains costa-del-sol beaches. Marine Pollution Bulletin, 1982, 13: 409-412.

[68] Reisser J, Shaw J, Wilcox C, et al. Marine plastic pollution in waters around Australia: Characteristics, concentrations, and pathways. Plos One, 2013, 8.

[69] Laglbauer B J L, Franco-Santos R M, Andreu-Cazenave M, et al. Macrodebris and microplastics from beaches in Slovenia. Marine Pollution Bulletin, 2014, 89: 356-366.

[70] Cheung P K, Cheung L T O, Fok L. Seasonal variation in the abundance of marine plastic debris in the estuary of a subtropical macro-scale drainage basin in South China. Science of the Total Environment, 2016, 562: 658-665.

[71] Ivar do Sul J A, Spengler A, Costa M F. Here, there and everywhere: small plastic fragments and pellets on beaches of Fernando de Noronha (Equatorial Western Atlantic). Marine Pollution Bulletin, 2009, 58: 1236-1238.

[72] Mathalon A, Hill P. Microplastic fibers in the intertidal ecosystem surrounding Halifax Harbor, Nova Scotia. Marine Pollution Bulletin, 2014, 81: 69-79.

[73] Fischer V, Elsner N O, Brenke N, et al. Plastic pollution of the Kuril-Kamchatka Trench area (NW pacific), Deep-Sea Research Part II: Topical Studies in Oceanography, 2015, 111: 399-405.

[74] Engler R E. The complex interaction between marine debris and toxic chemicals in the ocean. Environmental Science & Technology, 2012, 46: 12302-12315.

[75] Woodall L C, Sanchez-Vidal A, Canals M, et al. The deep sea is a major sink for microplastic debris. Royal Society Open Science, 2014, 1.

[76] Veerasingam S, Mugilarasan M, Venkatachalapathy R, et al. Influence of 2015 flood on the distribution and occurrence of microplastic pellets along the Chennai coast, India. Marine Pollution Bulletin, 2016, 109: 196-204.

[77] Lebreton L C M, van der Zwet J, Damsteeg J W, et al. River plastic emissions to the world's oceans. Nature Communications, 2017, 8: 15611.

[78] Eriksen M, Mason S, Wilson S, et al. Microplastic pollution in the surface waters of the Laurentian Great Lakes. Marine Pollution Bulletin, 2013, 77: 177-182.

[79] Su L, Xue Y, Li L, et al. Microplastics in Taihu Lake, China. Environmental Pollution, 2016, 216: 711-719.

[80] Faure F, Demars C, Wieser O, et al. Plastic pollution in Swiss surface waters: nature and concentrations, interaction with pollutants. Environmental Chemistry, 2015, 12: 582-591.

[81] Rech S, Macaya-Caquilpán V, Pantoja J F, et al. Sampling of riverine litter with citizen scientists - findings and recommendations. Environmental Monitoring and Assessment, 2015, 187: 187-335.

[82] Fischer E K, Paglialonga L, Czech E, et al. Microplastic pollution in lakes and lake shoreline sediments - a case study on Lake Bolsena and Lake Chiusi (central Italy). Environmental Pollution, 2016, 213: 648-657.

[83] Anderson P J, Warrack S, Langen V, et al. Microplastic contamination in Lake Winnipeg, Canada. Environmental Pollution, 2017, 225: 223-231.

[84] Free C M, Jensen O P, Mason S A, et al. High-levels of microplastic pollution in a large, remote, mountain lake. Marine Pollution Bulletin, 2014, 85: 156-163.

[85] Hurley R, Woodward J, Rothwell J J. Microplastic contamination of river beds significantly reduced by catchment-wide flooding. Nature Geoscience, 2018, 11: 251-257.

[86] Kataoka T, Nihei Y, Kudou K, et al. Assessment of the sources and inflow processes of microplastics in the river environments of Japan. Environmental Pollution, 2019, 244: 958-965.

[87] Baldwin A K, Corsi S R, Mason S A. Plastic debris in 29 Great Lakes tributaries: relations to watershed attributes and hydrology. Environmental Science & Technology, 2016, 50: 10377-10385.

[88] Plastics Europe, Plastics-the facts 2016. An analysis of European plastics production, demand and waste data., PlasticsEurope, 2016.

[89] Hablot E, Dharmalingam S, Hayes D G, et al. Effect of simulated weathering on physicochemical properties and inherent biodegradation of PLA/PHA nonwoven mulches. Journal of Polymers & the Environment, 2014, 22: 417-429.

[90] Ramos L, Berenstein G, Hughes E A, et al. Polyethylene film incorporation into the horticultural soil of small periurban production units in Argentina. Science of the Total Environment, 2015, 523: 74-81.

[91] Carr S A, Liu J, Tesoro A G. Transport and fate of microplastic particles in wastewater treatment plants. Water Research, 2016, 91: 174-182.

[92] Mahon A M, Connell B O, Healy, et al. Microplastics in sewage sludge: effects of treatment, 2017, 51.

[93] Nizzetto L, Futter M, Langaas S. Are agricultural soils dumps for microplastics of urban origin? Environmental Science and Technology, 2016, 50: 10777-10779.

[94] Huerta Lwanga E, Gertsen H, Gooren H, et al. Incorporation of microplastics from litter into burrows of *Lumbricus terrestris*, Environmental Pollution, 2017, 220: 523-531.

[95] Maaß S, Daphi D, Lehmann A, et al. Transport of microplastics by two collembolan species. Environmental Pollution, 2017, 225: 456-459.

[96] Huerta Lwanga E, Gertsen H, Gooren H, et al. Microplastics in the terrestrial ecosystem: implications for Lumbricus terrestris (Oligochaeta, Lumbricidae). Environmental Science & Technology, 2016, 50: 2685-2691.

[97] Hodson M E, Duffus-Hodson C, Clark A, et al. Plastic bag derived-microplastics as a vector for metal exposure in terrestrial invertebrates. Environmental Science & Technology, 2017, 51: 4714-4721.

[98] Mercier F, Glorennec P, Thomas O, et al. Organic contamination of settled house dust: a review for exposure assessment purposes. Environmental Science & Technology, 2011, 6716.

[99] Dris R, Gasperi J, Mirande C, et al. A first overview of textile fibers, including microplastics, in indoor and outdoor environments. Environmental Pollution, 2017, 221: 453-458.

[100] Dehghani S, Moore F, Akhbarizadeh R. Microplastic pollution in deposited urban dust, Tehran metropolis, Iran. Environmental Science and Pollution Research, 2017, 20360.

[101] Abbasi S, Keshavarzi B, Moore F, et al. Investigation of microrubbers, microplastics and heavy metals in street dust: a study in Bushehr city, Iran. Environmental Earth Sciences, 2017, 1.

[102] 周倩, 骆永明. 滨海城市大气环境中发现多种微塑料及其沉降通量差异. 科学通报, 2017, 62: 3902-3909.

[103] Cai L, Wang J, Peng J, et al. Characteristic of microplastics in the atmospheric fallout from Dongguan city, China: preliminary research and first evidence. Environmental Science and Pollution Research, 2017, 24: 24928-24935.

[104] Dris R, Gasperi J, Saad M, et al. Synthetic fibers in atmospheric fallout: A source of microplastics in the environment?

Marine Pollution Bulletin, 2016, 104: 290-293.

[105] Zhang W, Zhang S, Wang J, et al. Microplastic pollution in the surface waters of the Bohai Sea, China. Environmental Pollution, 231(2017)541-548.

[106] Zhao S, Zhu L, Wang T, et al. Suspended microplastics in the surface water of the Yangtze Estuary System, China: first observations on occurrence, distribution. Marine Pollution Bulletin, 2014, 86: 562-568.

[107] Cai M, He H, Liu M, et al. Lost but can't be neglected: huge quantities of small microplastics hide in the South China Sea. Science of the Total Environment, 2018, 633: 1206-1216.

[108] Reisser J, Slat B, Noble K, et al. The vertical distribution of buoyant plastics at sea: an observational study in the North Atlantic Gyre. Biogeosciences, 2015, 12: 1249-1256.

[109] Zhao S, Zhu L, Li D. Characterization of small plastic debris on tourism beaches around the South China Sea. Regional Studies in Marine Science, 2015, 62: 55-62.

[110] Yu X, Peng J, Wang J, et al. Occurrence of microplastics in the beach sand of the Chinese inner sea: the Bohai Sea. Environmental Pollution, 2016, 214: 722-730.

[111] Vianello A, Boldrin A, Guerriero P, et al. Microplastic particles in sediments of Lagoon of Venice, Italy: first observations on occurrence, spatial patterns and identification, Estuarine. Coastal and Shelf Science, 2013, 130: 54-61.

[112] Peng G, Zhu B, Yang D, et al. Microplastics in sediments of the Changjiang Estuary, China. Environmental Pollution, 2017, 225: 283-290.

[113] Mai L, Bao L J, Shi L, et al. A review of methods for measuring microplastics in aquatic environments. Environmental Science and Pollution Research, 2018, 25: 11319-11332.

[114] Silva J D B, Barletta M, Lima A R A, et al. Use of resources and microplastic contamination throughout the life cycle of grunts (*Haemulidae*) in a tropical estuary. Environmental Pollution, 2018, 242: 1010-1021.

[115] Cho Y, Shim W J, Jang M, et al. Abundance and characteristics of microplastics in market bivalves from South Korea. Environmental Pollution, 2019, 245: 1107-1116.

[116] Rist S, Baun A, Hartmann N B. Ingestion of micro- and nanoplastics in *Daphnia magna* - Quantification of body burdens and assessment of feeding rates and reproduction. Environmental Pollution, 2017, 228: 398-407.

[117] Besseling E, Foekema E M, van den Heuvel-Greve M J, et al. The effect of microplastic on the uptake of chemicals by the lugworm *Arenicola marina* (L.) under environmentally relevant exposure conditions. Environmental Science & Technology, 2017, 51: 8795-8804.

[118] Zhao S, Zhu L, Li D. Microscopic anthropogenic litter in terrestrial birds from Shanghai, China: not only plastics but also natural fibers. Science of the Total Environment, 2016, 550: 1110-1115.

[119] Davidson K, Dudas S E. Microplastic ingestion by wild and cultured manila clams (*Venerupis philippinarum*) from Baynes Sound, British Columbia. Archives of Environmental Contamination and Toxicology, 2016, 71: 147-156.

[120] Nguyen B, Claveau-Mallet D, Hernandez L M, et al. Separation and analysis of microplastics and nanoplastics in complex environmental samples. Accounts of Chemical Research, 2019, 52: 858-866.

[121] Pretorius N O, Rode K, Simpson J M, et al. Analysis of complex phthalic acid based polyesters by the combination of size exclusion chromatography and matrix-assisted laser desorption/ionization mass spectrometry. Analytica Chimica Acta, 2014, 808: 94-103.

[122] Busnel J P, Degoulet C. Size exclusion chromatography analysis of block terpolymers using refractometry and dual UV detection. Polymer Testing, 2006, 25: 358-365.

[123] Zanchet D, Micheel C M, Parak W J, et al. Electrophoretic isolation of discrete Au nanocrystal/DNA conjugates. Nano Letters, 2001, 1: 32-35.

[124] Xu X, Caswell K K, Tucker E, et al. Size and shape separation of gold nanoparticles with preparative gel electrophoresis. Journal of Chromatography A, 2007, 1167: 35-41.

[125] Messaud F A, Sanderson R D, Runyon J R, et al. An overview on field-flow fractionation techniques and their applications in the separation and characterization of polymers. Progress in Polymer Science, 2009, 34: 351-368.

[126] Malik M I, Pasch H. Field-flow fractionation: new and exciting perspectives in polymer analysis. Progress in Polymer Science, 2016, 63: 42-85.

[127] 刘昱. 新型磁场场流分离体系的构建和评价. 天津: 天津大学硕士学位论文, 2013.

[128] Rocha-Santos T, Duarte A C. A critical overview of the analytical approaches to the occurrence, the fate and the behavior of microplastics in the environment, TrAC Trends in Analytical Chemistry, 2015, 65: 47-53.

[129] Betzig E, Patterson G H, Sougrat R, et al. Imaging intracellular fluorescent proteins at nanometer resolution. Science, 2006, 313: 1642-1645.

[130] Dazzi A, Prater C B, Hu Q, et al. AFM-IR: Combining atomic force microscopy and infrared spectroscopy for nanoscale chemical characterization. Applied Spectrosopy., 2012, 66: 1365-1384.

[131] Renner G, Schmidt T C, Schram J. Analytical methodologies for monitoring micro (nano) plastics: which are fit for purpose? Current Opinion in Environmental Science & Health, 2018, 1: 55-61.

[132] Silva A B, Bastos A S, Justino C I L, et al. Microplastics in the environment: challenges in analytical chemistry – a review. Analytica Chimica Acta, 2018, 1017: 1-19.

[133] Shim W J, Hong S H, Eo S E. Identification methods in microplastic analysis: a review. Analytical Methods, 2017, 9: 1384-1391.

[134] Gigault J, Pedrono B, Maxit B, et al. Marine plastic litter: the unanalyzed nano-fraction. Environmental Science: Nano, 2016, 3: 346-350.

[135] Zhang T, Wang C, Dong F, et al. Uptake and translocation of styrene maleic anhydride nanoparticles in murraya exotica plants as revealed by noninvasive, real-time optical bioimaging. Environmental Science & Technology, 2019, 53: 1471-1481.

[136] La Rocca, Di Liberto G, Shayler P J, et al. Application of nanoparticle tracking analysis platform for the measurement of soot-in-oil agglomerates from automotive engines. Tribology International, 2014, 70: 142-147.

[137] Schwaferts C, Niessner R, Elsner M, et al. Methods for the analysis of submicrometer- and nanoplastic particles in the environment. TrAC Trends in Analytical Chemistry, 2019, 112: 52-65.

[138] Mintenig S M, Bäuerlein P S, Koelmans A A, et al. Closing the gap between small and smaller: towards a framework to analyse nano- and microplastics in aqueous environmental samples. Environmental Science: Nano, 2018, 5: 1640-1649.

[139] Martin W, Scott L. Freshwater Microplastics: Emerging Environmental Contaminants? Springer, Netherlands, Europe, 2018.

[140] Prata J C, Da Costa J P, Lopes I, et al. Effects of microplastics on microalgae populations: a critical review. Science of The Total Environment, 2019, 665: 400-405.

[141] Edward A, Carpenter J, Smith J A K L. Plastics on the Sargasso Sea Surface. Science, 1972, 1240.

[142] Siddiqui M N, Gondal M A, Redhwi H H. Identification of different type of polymers in plastics waste, Journal of Environmental Science and Health. Part A, Toxic/Hazardous Substances & Environmental Engineering, 2008, 43: 1303-1310.

[143] Burgiel J, Butcher W, Halpern R, et al. Cost evaluation of automated and manual post-consumer plastic bottle sorting systems. Final Report, 1994.

[144] Marshall J, Franks J, Abell I, et al. Determination of trace elements in solid plastic materials by laser ablation-inductively coupled plasma mass spectrometry. Journal of Analytical Atomic Spectrometry, 1991, 6: 145-150.

[145] Stepputat M, Noll R. On-line detection of heavy metals and brominated flame retardants in technical polymers with

laser-induced breakdown spectrometry. Applied Optics, 2003, 6210.

[146] Allen V, Kalivas J H, Rodriguez R G. Post-consumer plastic identification using Raman spectroscopy. Applied Spectroscopy, 1999, 53: 672-681.

[147] Anzano J, Bonilla B, Montull-Ibor, et al. Plastic identification and comparison by multivariate techniques with laser-induced breakdown spectroscopy. Journal of Applied Polymer Science (Print), 2011, 121: 2710-2716.

[148] Fortes F J, Moros J, Lucena P, et al. Laser-induced breakdown spectroscopy. Analytical Chemistry, 2013, 85: 640-669.

[149] Boueri M, Motto-Ros V, Lei W Q, et al. Identification of Polymer Materials Using Laser-Induced Breakdown Spectroscopy Combined with Artificial Neural Networks. Appl. Spectrosc, 2011, 65: 307-314.

[150] Ernst T, Popp R, van Eldik R. Quantification of heavy metals for the recycling of waste plastics from electrotechnical applications. Talanta, 2000, 53: 347-357.

[151] Duarte A T, Dessuy M B, Silva M M, et al. Determination of cadmium and lead in plastic material from waste electronic equipment using solid sampling graphite furnace atomic absorption spectrometry. Microchemical Journal, 2010, 96: 102-107.

[152] Liu K, Tian D, Li C, et al. A review of laser-induced breakdown spectroscopy for plastic analysis. TrAC Trends in Analytical Chemistry, 2019, 110: 327-334.

[153] Maes T, Jessop R, Wellner N, et al. A rapid-screening approach to detect and quantify microplastics based on fluorescent tagging with Nile Red. Scientific Reports, 2017, 7: 44501.

[154] Rios Mendoza L M, Balcer M. Microplastics in freshwater environments: a review of quantification assessment. Trends in Analytical Chemistry, 2019, 113: 402-408.

[155] Briassoulis D, Babou E, Hiskakis M, et al. Degradation in soil behavior of artificially aged polyethylene films with pro-oxidants. Journal of Applied Polymer Science, 2015, 132: 42289.

[156] Pinto F, Franco C, André R N, et al. Effect of experimental conditions on co-gasification of coal, biomass and plastics wastes with air/steam mixtures in a fluidized bed system. Fuel, 2003, 82: 1967-1976.

[157] Wang G, Zhang J, Shao J, et al. Characterisation and model fitting kinetic analysis of coal/biomass co-combustion. Thermochimica Acta, 2014, 591: 68-74.

[158] Burra K G, Gupta A K. Synergistic effects in steam gasification of combined biomass and plastic waste mixtures. Applied Energy, 2018, 211: 230-236.

[159] Sharypov V I, Marin N, Beregovtsova N G, et al. Co-pyrolysis of wood biomass and synthetic polymer mixtures. Part I: influence of experimental conditions on the evolution of solids, liquids and gases. Journal of Analytical and Applied Pyrolysis, 2002, 64: 15-28.

[160] Cepeliogullar O, PÜTÜN A E. Thermal and kinetic behaviors of biomass and plastic wastes in co-pyrolysis. Energy Conversion and Management, 2013, 75: 263-270.

[161] Celina M C. Review article: Review of polymer oxidation and its relationship with materials performance and lifetime prediction. Polymer Degradation and Stability, 2013, 98: 2419-2429.

[162] Selke S, Auras R, Tuan Anh N, et al. Evaluation of Biodegradation-Promoting Additives for Plastics. Environmental science & technology, 2015, 49: 3769-3777.

[163] Lusher A L, Welden N A, Sobral P, et al. Sampling, isolating and identifying microplastics ingested by fish and invertebrates. Analytical Methods, 2017, 9: 1346-1360.

[164] Wright S L, Thompson R C, Galloway T S. The physical impacts of microplastics on marine organisms: a review. Environmental Pollution, 2103, 178: 483-492.

[165] Jovanovic B. Ingestion of microplastics by fish and its potential consequences from a physical perspective. Integrated Environmental Assessment and Management, 2017, 13: 510-515.

[166] Ory N C, Sobral P, Ferreira J L, et al. Amberstripe scad *Decapterus muroadsi* (Carangidae)fish ingest blue microplastics resembling their copepod prey along the coast of Rapa Nui(Easter Island)in the South Pacific subtropical gyre. Science of the Total Environment, 2017, 586: 430-437.

[167] Farrell P, Nelson K. Trophic level transfer of microplastic: *Mytilus edulis* (L.) to *Carcinus maenas* (L.). Environmental Pollution, 2013, 177: 1-3.

[168] Setälä O, Fleming-Lehtinen V, Lehtiniemi M. Ingestion and transfer of microplastics in the planktonic food web. Environmental Pollution, 2014, 1852014.

[169] Santana M F M, Moreira F T, Turra A. Trophic transference of microplastics under a low exposure scenario: insights on the likelihood of particle cascading along marine food-webs. Marine Pollution Bulletin, 2017, 121: 154-159.

[170] Setälä O, Norkko J, Lehtiniemi M. Feeding type affects microplastic ingestion in a coastal invertebrate community. Marine Pollution Bulletin, 2016, 102: 95-101.

[171] Ward J E, Shumway S E. Separating the grain from the chaff: particle selection in suspension- and deposit-feeding bivalves, Journal of Experimental Marine Biology and Ecology, 2004, 300: 83-130.

[172] Li J, Lusher A L, Rotchell J M, et al. Using mussel as a global bioindicator of coastal microplastic pollution. Environmental Pollution, 2019, 244: 522-533.

[173] McGoran A R, Clark P F, Morritt D. Presence of microplastic in the digestive tracts of European flounder, *Platichthys flesus*, and European smelt, *Osmerus eperlanus*, from the River Thames. Environmental Pollution, 2017, 220: 744-751.

[174] Mizraji R, Ahrendt C, Perez-Venegas D, et al. Is the feeding type related with the content of microplastics in intertidal fish gut? Marine Pollution Bulletin, 2017, 116: 498-500.

[175] Kolandhasamy P, Su L, Li J, et al. Adherence of microplastics to soft tissue of mussels: a novel way to uptake microplastics beyond ingestion. Science of the Total Environment, 2018, 610: 635-640.

[176] Green D S, Colgan T J, Thompson R C, et al. Exposure to microplastics reduces attachment strength and alters the haemolymph proteome of blue mussels (*Mytilus edulis*). Environmental Pollution, 2019, 246: 423-434.

[177] Jabeen K, Li B, Chen Q, et al. Effects of virgin microplastics on goldfish (*Carassius auratus*). Chemosphere, 2018, 213: 323-332.

[178] Su L, Deng H, Li B, et al. The occurrence of microplastic in specific organs in commercially caught fishes from coast and estuary area of east China. Journal of Hazardous Materials, 2019, 365: 716-724.

[179] Lu Y, Zhang Y, Deng Y, et al. Uptake and accumulation of polystyrene microplastics in zebrafish (*Danio rerio*) and toxic effects in liver. Environmental Science & Technology, 2016, 50: 4054-4060.

[180] Woods M N, Stack M E, Fields D M, et al. Microplastic fiber uptake, ingestion, and egestion rates in the blue mussel (*Mytilus edulis*). Marine Pollution Bulletin, 2018, 137: 638-645.

[181] Gonçalves C, Martins M, Sobral P, et al. An assessment of the ability to ingest and excrete microplastics by filter-feeders: a case study with the Mediterranean mussel. Environmental Pollution, 2019, 245: 600-606.

[182] Browne M A, Dissanayake A, Galloway T S, et al. Ingested microscopic plastic translocates to the circulatory system of the mussel, *Mytilus edulis* (L.). Environmental Science & Technology, 2008, 42: 5026-5031.

[183] Devriese L I, van der Meulen M D, Maes T, et al. Microplastic contamination in brown shrimp (*Crangon crangon*, Linnaeus 1758) from coastal waters of the Southern North Sea and Channel area. Marine Pollution Bulletin, 2015, 98: 179-187.

[184] Abbasi S, Soltani N, Keshavarzi B, et al. Microplastics in different tissues of fish and prawn from the Musa Estuary, Persian Gulf. Chemosphere, 2018, 205: 80-87.

[185] Collard F, Gilbert B, Compere P, et al. Microplastics in livers of European anchovies (*Engraulis encrasicolus*, L.). Environmental Pollution, 2017, 229: 1000-1005.

[186] Akhbarizadeh R, Moore F, Keshavarzi B. Investigating a probable relationship between microplastics and potentially toxic elements in fish muscles from northeast of Persian Gulf. Environmental Pollution, 2018, 232: 154-163.

[187] Qu X, Su L, Li H, et al. Assessing the relationship between the abundance and properties of microplastics in water and in mussels. Science of the Total Environment, 2018, 621: 679-686.

[188] De Witte B, Devriese L, Bekaert K, et al. Quality assessment of the blue mussel (*Mytilus edulis*): comparison between commercial and wild types. Marine Pollution Bulletin, 2014, 85: 146-155.

[189] Renzi M, Guerranti C, Blaslovic A. Microplastic contents from maricultured and natural mussels. Marine Pollution Bulletin, 2018, 131: 248-251.

[190] Jabeen K, Su L, Li J, et al. Microplastics and mesoplastics in fish from coastal and fresh waters of China. Environmental Pollution, 2017, 221: 141-149.

[191] Hämer J, Gutow L, Koehler A, et al. Fate of microplastics in the marine isopod *Idotea emarginata*. Environmental Science & Technology, 2014, 48: 13451-13458.

[192] Welden N A C, Cowie P R. Environment and gut morphology influence microplastic retention in langoustine, Nephrops norvegicus. Environmental Pollution, 2016, 214: 859-865.

[193] Foley, Feiner Z S, Malinich T D, et al. A meta-analysis of the effects of exposure to microplastics on fish and aquatic invertebrates. Science of the Total Environment, 2018, 631-632: 550-559.

[194] Huffer T, Praetorius A, Wagner S, et al. Microplastic exposure assessment in aquatic environments: learning from similarities and differences to engineered nanoparticles. Environmental Science & Technology, 2017, 51: 2499-2507.

[195] Cole M, Lindeque P, Halsband C, et al. Microplastics as contaminants in the marine environment: a review. Marine Pollution Bulletin, 2011, 62: 2588-2597.

[196] Verschoor A J. Towards a definition of microplastics: considerations for the specification of physico-chemical properties, in: RIVM Letter report 2015-0116. National Institute for Public Health and the Enviroment, 2015.

[197] Oberbeckmann S, Loder M G J, Labrenz M. Marine microplastic-associated biofilms - a review. Environmental Chemistry, 2015, 12: 551-562.

[198] O'Brine T, Thompson R C. Degradation of plastic carrier bags in the marine environment. Marine Pollution Bulletin, 2010, 60: 2279-2283.

[199] Rummel C D, Jahnke A, Gorokhova E, et al. Impacts of biofilm formation on the fate and potential effects of microplastic in the aquatic environment. Environmental Science & Technology Letter, 2017, 4: 258-267.

[200] Cole M, Galloway T S. Ingestion of nanoplastics and microplastics by pacific oyster larvae. Environmental Science & Technology, 2015, 49: 14625-14632.

[201] Hu L, Su L, Xue Y, et al. Uptake, accumulation and elimination of polystyrene microspheres in tadpoles of *Xenopus tropicalis*. Chemosphere, 2016, 164: 611-617.

[202] Karami A, Groman D B, wilson S P, et al. Biomarker responses in zebrafish(*Danio rerio*)larvae exposed to pristine low-density polyethylene fragments. Environmental Pollution, 2017, 223: 466-475.

[203] Rochman C M, Kurobe T, Flores I, et al. Early warning signs of endocrine disruption in adult fish from the ingestion of polyethylene with and without sorbed chemical pollutants from the marine environment. Science of the Total Environment, 2014, 493: 656-661.

[204] Karami A. Gaps in aquatic toxicological studies of microplastics. Chemosphere, 2017, 184: 841-848.

[205] Li L, Su L, Cai H, et al. The uptake of microfibers by freshwater Asian clams (*Corbicula fluminea*) varies based upon physicochemical properties. Chemosphere, 2109, 221; 107-114.

[206] Adam V, Yang T, Nowack B. Toward an ecotoxicological risk assessment of microplastics: comparison of available hazard

and exposure data in freshwaters. Environmental Toxicology and Chemistry, 2019, 38: 436-447.

[207] Anbumani S, Kakkar P. Ecotoxicological effects of microplastics on biota: a review. Environmental Science and Pollution Research, 2018, 25: 14373–14396.

[208] Browne M A, Niven S J, Galloway T S, et al. Microplastic moves pollutants and additives to worms, reducing functions linked to health and biodiversity. Current Biology, 2013, 23: 2388-2392.

[209] Canesi L, Ciacci C, Bergami E, et al. Evidence for immunomodulation and apoptotic processes induced by cationic polystyrene nanoparticles in the hemocytes of the marine bivalve Mytilus. Marine Environmental Research, 2015, 111: 34-40.

[210] Lee K W, Shim W J, Kwon O Y, et al. Size-dependent effects of micro polystyrene particles in the marine copepod *Tigriopus japonicus*. Environmental Science & Technology, 2013, 47: 11278-11283.

[211] Sjollema S B, Redondo-Hasselerharm P, Leslie H A, et al. Do plastic particles affect microalgal photosynthesis and growth? Aquatic Toxicology, 2016, 170: 259-261.

[212] Sussarellu R, Suquet M, Thomas Y, et al. Oyster reproduction is affected by exposure to polystyrene microplastics. Proceedings of the National Academy of Sciences, 2016, 113: 2430-2435.

[213] Lei L, Wu S, Lu S, et al. Microplastic particles cause intestinal damage and other adverse effects in zebrafish *Danio rerio* and nematode *Caenorhabditis elegans*. Science of the Total Environment, 2018, 619-620: 1-8.

[214] Watts A J, Urbina M A, Corr S, et al. Ingestion of plastic microfibers by the crab *Carcinus maenas* and its effect on food consumption and energy balance. Environmental Science & Technology, 2015, 49: 14597-14604.

[215] Murray F, Cowie P R. Plastic contamination in the decapod crustacean *Nephrops norvegicus*(Linnaeus, 1758). Marine Pollution Bulletin, 2011, 62: 1207-1217.

[216] Rochman C M, Hoh E, Kurobe T, et al. Ingested plastic transfers hazardous chemicals to fish and induces hepatic stress. Scientific Reports, 2013, 3: 3263.

[217] Ma Y, Huang A, Cao S, et al. Effects of nanoplastics and microplastics on toxicity, bioaccumulation, and environmental fate of phenanthrene in fresh water. Environmental Pollution, 2016, 219: 166-173.

[218] Chen Q, Gundlach M, Yang S, et al. Quantitative investigation of the mechanisms of microplastics and nanoplastics toward zebrafish larvae locomotor activity, Science of the Total Environment, 2017, 584-585: 1022-1031.

[219] Kashiwada S. Distribution of nanoparticles in the see-through medaka (*Oryzias latipes*). Environmental Health Perspectives, 2006, 114: 1697-1702.

[220] Tussellino M, Ronca R, Formiggini F, et al. Polystyrene nanoparticles affect *Xenopus laevis* development. Journal of Nanoparticle Research, 2015, 17: 70.

[221] Carbery M, O'Connor W, Palanisami T. Trophic transfer of microplastics and mixed contaminants in the marine food web and implications for human health. Environmental International, 2018, 115: 400-409.

[222] Fao R, FI. The state of world fisheries and aquaculture, 2012. State of World Fisheries & Aquaculture, 2010, 4: 40–41.

[223] Mazurais D, Ernande B, Quazuguel P, et al. Evaluation of the impact of polyethylene microbeads ingestion in European sea bass (*Dicentrarchus labrax*) larvae. Marine Environmental Research, 2015, 112: 78-85.

[224] Bessa F, Barría P, Neto J M, et al. Occurrence of microplastics in commercial fish from a natural estuarine environment. Marine Pollution Bulletin, 2018, 128: 575-584.

[225] Evan Ward J, Shumway S E. Separating the grain from the chaff: particle selection in suspension- and deposit-feeding bivalves. Journal of Experimental Marine Biology and Ecology, 2004, 300: 83-130.

[226] Brillant M G S, Macdonald B A. Postingestive selection in the sea scallop, *Placopecten magellanicus* (Gmelin): the role of particle size and density. Journal of Experimental Marine Biology and Ecology , 2000, 253: 211-227.

[227] JCosta J P D, Santos P S M, Duarte A C, et al. (Nano) plastics in the environment – sources, fates and effects. Science of the Total Environment, 216, 566-567: 15-26.

[228] Liebezeit G, Liebezeit E. Non-pollen particulates in honey and sugar, Food Additives & Contaminants. Part A. Chemistry, Analysis, Control, Exposure & Risk Assessment, 2013, 30: 2136-2140.

[229] Liebezeit G, Liebezeit E. Synthetic particles as contaminants in German beers, Food Additives & Contaminants. Part A. Chemistry, Analysis, Control, Exposure & Risk Assessment, 2014, 31: 1574-1578.

[230] Yang D, Shi H, Li L, et al. Microplastic pollution in table salts from China. Environmental Science & Technology, 2015, 22: 13622-13627.

[231] Liu C, Li J, Zhang Y, et al. Widespread distribution of PET and PC microplastics in dust in urban China and their estimated human exposure. Environment International, 2019, in press.

[232] Wright S L, Kelly F J. Plastic and human health: a micro issue? Environmental Science & Technology, 2017, 51: 6634-6647.

[233] Kremer A M, Pal T M, Boleij J S M, et al. Airway hyper-responsiveness and the prevalence of work-related symptoms in workers exposed to irritants, American Journal of Industrial Medicine, 26(1994)655-669.

[234] Wardrop P, Shimeta J, Nugegoda D, et al. Chemical pollutants sorbed to ingested microbeads from personal care products accumulate in fish. Environmental Science & Technology, 2016, 50: 4037-4044.

[235] Hernandez L M, Yousefi N, Tufenkji N. Are There nanoplastics in your personal care products? Environmental Science & Technology Letters, 2017, 4: 280-285.

[236] Schmidt C, Krauth T, Wagner S. Export of plastic debris by rivers into the sea. Environmental Science & Technology, 2017, 51: 12246-12253.

[237] Imhof H K, Ivleva N P, Schmid J, et al. Contamination of beach sediments of a subalpine lake with microplastic particles. Current Biology, 2013, 23: R867-R868.

[238] Edyvane K S, Dalgetty A, Hone P W, et al. Long-term marine litter monitoring in the remote great Australian Bight, South Australia. Marine Pollution Bulletin, 2004, 48: 1060-1075.

[239] 周倩, 章海波, 周阳, 等. 滨海河口潮滩中微塑料的表面风化和成分变化. 科学通报, 2018, 63: 214-224.

[240] Veerasingam S, Saha M, Suneel V, et al. Characteristics, seasonal distribution and surface degradation features of microplastic pellets along the Goa coast, India. Chemosphere, 2016, 159: 496-505.

[241] Zhan Z, Wang J, Peng J, et al. Sorption of 3, 3', 4, 4'-tetrachlorobiphenyl by microplastics: a case study of polypropylene. Marine Pollution Bulletin, 2016, 110: 559-563.

[242] Wang F, Shih K M, Li X Y. The partition behavior of perfluorooctanesulfonate (PFOS) and perfluorooctanesulfonamide (FOSA) on microplastics. Chemosphere, 2015, 119: 841-847.

[243] Bakir A, Rowland S J, Thompson R C. Transport of persistent organic pollutants by microplastics in estuarine conditions. Estuarine Coastal & Shelf Science, 2014, 140: 14-21.

[244] Zhang X, Zheng M, Wang L, et al. Sorption of three synthetic musks by microplastics. Marine Pollution Bulletin, 2017, 23: 318.

[245] Velzeboer I, Kwadijk C J, Koelmans A A. Strong sorption of PCBs to nanoplastics, microplastics, carbon nanotubes, and fullerenes. Environmental Science & Technology, 2014, 48: 4869-4876.

[246] Llorca M, Schirinzi G, Martínez M, et al. Adsorption of perfluoroalkyl substances on microplastics under environmental conditions. Environmental Pollution, 2018, 235: 680-691.

[247] Li J, Zhang K, Zhang H. Adsorption of antibiotics on microplastics. Environmental Pollution, 2018, 237: 460.

[248] Lee H, Shim W J, Kwon J H. Sorption capacity of plastic debris for hydrophobic organic chemicals. Science of The Total Environment, 2014, 470-471: 1545-1552.

[249] Mueller A, Becker R, Dorgerloh U, et al. The effect of polymer aging on the uptake of fuel aromatics and ethers by microplastics. Environmental Pollution, 2018, 240: 639-646.

[250] Wang W, J. Wang J. Comparative evaluation of sorption kinetics and isotherms of pyrene onto microplastics. Chemosphere, 2017, 193: 567-573.

[251] Hüffer T, Hofmann T. Sorption of non-polar organic compounds by micro-sized plastic particles in aqueous solution. Environmental Pollution, 2016, 214: 194-201.

[252] Liu L, Fokkink R, Koelmans A A. Sorption of polycyclic aromatic hydrocarbons to polystyrene nanoplastic. Environmental Toxicology & Chemistry, 2016, 35: 1650-1655.

[253] Hüffer T, Weniger A K, Hofmann T. Data on sorption of organic compounds by aged polystyrene microplastic particles. Data in Brief, 2018, 236: 218-225.

[254] Bakir A, Rowland S J, Thompson R C. Competitive sorption of persistent organic pollutants onto microplastics in the marine environment. Marine Pollution Bulletin, 2012, 64: 2782-2789.

[255] Bayo J A, Martínez A, Guillén M, et al. Microbeads in commercial facial cleansers: threatening the environment. Clean – Soil Air Water, 2017, 45: 1600683.

[256] Liu J, Ma Y, Zhu D, et al. Polystyrene nanoplastics-enhanced contaminant transport: role of irreversible adsorption in glassy polymeric domain. Environmental Science & Technology, 2018, 52: 2677-2685.

[257] León V M, García I, González E, et al. Potential transfer of organic pollutants from littoral plastics debris to the marine environment. Environmental Pollution, 2018, 236: 442-453.

[258] Bakir A, Rowland S J, Thompson R C. Enhanced desorption of persistent organic pollutants from microplastics under simulated physiological conditions. Environmental Pollution, 2014, 185: 16-23.

[259] Ma Y, Huang A, Cao S, et al. Effects of nanoplastics and microplastics on toxicity, bioaccumulation, and environmental fate of phenanthrene in fresh water. Environmental Pollution, 2016, 219: 166-173.

[260] Oliveira M, Ribeiro A, Hylland K, et al. Single and combined effects of microplastics and pyrene on juveniles(0+ group)of the common goby *Pomatoschistus microps* (Teleostei, Gobiidae). Ecological Indicators, 2013, 34: 641-647.

[261] Gouin T, Roche N, Lohmann R, et al. A thermodynamic approach for assessing the environmental exposure of chemicals absorbed to microplastic. Environmental Science & Technology, 2011, 45: 1466-1472.

[262] Koelmans A A, Besseling E, Wegner A, et al. Plastic as a carrier of POPs to aquatic organisms: a model analysis, Environmental Science & Technology, 47 7812-7820.

[263] Besseling E, Wegner A, Foekema E M, et al. Effects of microplastic on fitness and PCB bioaccumulation by the lugworm *Arenicola marina* (L.). Environmental Science & Technology, 2013, 47: 593-600.

[264] Avio G G, Gorbi S, Regoli F. Experimental development of a new protocol for extraction and characterization of microplastics in fish tissues: first observations in commercial species from Adriatic Sea. Marine Environmental Research, 2015, 111: 18-26.

[265] Chua E M, Shimeta J, Nugegoda D, et al. Assimilation of polybrominated diphenyl ethers from microplastics by the marine amphipod, *Allorchestes compressa*. Environmental Science & Technology, 2014, 48: 8127-8134.

[266] Koelmans A A, Bakir A, Burton G A, et al. Microplastic as a vector for chemicals in the aquatic environment: critical review and model-supported reinterpretation of empirical studies. Environmental Science & Technology, 2016, 50: 3315-3326.

[267] Koelmans A A. Modeling the role of microplastics in bioaccumulation of organic chemicals to marine aquatic organisms. A critical review, in: Bergmann M., Gutow L., Klages M.(eds) Marine Anthropogenic Litter, Spinger, Cham, 2015, 309-324.

[268] Law K L, Thompson R C. Microplastics in the seas. Science, 2014, 345: 144-145.

[269] Chen Q, Reisser J, Cunsolo S, et al. Pollutants in plastics within the North Pacific subtropical gyre. Environmental Science &

Technology, 2018, 52: 446-456.

[270] Zarfl C, Matthies M. Are marine plastic particles transport vectors for organic pollutants to the Arctic? Marine Pollution Bulletin, 2010, 60: 1810-1814.

[271] Agency U S E P. Municipal solid waste in the United States: 2005 facts and figures, Office of Solid Waste, EPA; Available through National Technical Information Service, Washington, D.C.Springfield, VA, 2006.

[272] Lobelle D, Cunliffe M. Early microbial biofilm formation on marine plastic debris. Marine Pollution Bulletin, 2011, 62: 197-200.

[273] Moriceau B, Garvey M, Ragueneau O, et al. Evidence for reduced biogenic silica dissolution rates in diatom aggregates. Marine Ecology Progress Series, 2007, 333: 129-142.

[274] Thornton D C O. Diatom aggregation in the sea: mechanisms and ecological implications. European Journal of Phycology, 2002, 37: 149-161.

[275] Turner J T. Zooplankton fecal pellets, marine snow and sinking phytoplankton blooms. Aquatic Microbial Ecology, 2002, 27: 57-102.

[276] Cedervall T, Hansson L A, Lard M, et al. Food chain transport of nanoparticles affects behaviour and fat metabolism in fish. Plos One, 2012, 7: 32254.

[277] Batel A, Linti F, Scherer M, et al. The transfer of benzo[α]pyrene from microplastics to *Artemia nauplii* and further to zebrafish via a trophic food web experiment-CYP1A induction and visual tracking of persistent organic pollutants. Environmental Toxicology & Chemistry, 2016, 35: 1656-1666.

[278] Diepens N J, Koelmans A A. Accumulation of plastic debris and associated contaminants in aquatic food webs. Environmental Science & Technology, 2018, 52: 8510-8520.

[279] Kleinteich J, Seidensticker S, Marggrander N, et al. Microplastics reduce short-term effects of environmental contaminants. Part II: polyethylene particles decrease the effect of polycyclic aromatic hydrocarbons on microorganisms. International Journal of Environmental Research & Public Health, 2018, 15: 287.

[280] Qu H, Ma R, Wang B, et al. Enantiospecific toxicity, distribution and bioaccumulation of chiral antidepressant venlafaxine and its metabolite in loach (*Misgurnus anguillicaudatus*) co-exposed to microplastic and the drugs. Journal of Hazardous Materials, 2018, 370: 203-211.

[281] 丁剑楠, 张闪闪, 邹华, 等. 淡水环境中微塑料的赋存、来源和生态毒理效应研究进展. 生态环境学报, 2017, 26: 1619-1626.

[282] Alford L K, Corcoran P, Driedger A, et al. Microplastics pollution in the Great Lakes ecosystem: summary of presentations at IAGLR 2014, in: 57th Annual Conference on Great Lakes Research(IAGLR 2014), Hamilton. Ontario, 2014.

[283] Nelms S E, Galloway T S, Godley B J, et al. Investigating microplastic trophic transfer in marine top predators, Environmental Pollution, 2018, 238: 999-1007.

[284] Grigorakis S, Mason S A, Drouillard K G. Determination of the gut retention of plastic microbeads and microfibers in goldfish (*Carassius auratus*). Chemosphere, 2017, 169: 233-238.

[285] 吴辰熙, 潘响亮, 施华宏, 等. 我国淡水环境微塑料污染与流域管控策略. 中国科学院院刊, 2018, 33: 1012-1020.

[286] Rios L M, Moore C, Jones P R. Persistent organic pollutants carried by synthetic polymers in the ocean environment. Marine Pollution Bulletin, 2007, 54: 1230-1237.

[287] 周倩, 章海波, 李远, 等. 海岸环境中微塑料污染及其生态效应研究进展. 科学通报, 2015, 33: 3210-3220.

[288] UNEP, Marine Litter Legislation: A Toolkit for Policymakers, 2016.

[289] Desvarieux M, Demmer R T, Jacobs, et al. Periodontal bacteria and hypertension: the oral infections and vascular disease epidemiology study (INVEST). Journal of Hypertension, 2010, 28: 1413-1421.

[290] Sakai S, Ikematsu T, Hirai Y, et al. Unit-charging programs for municipal solid waste in Japan. Waste Management, 2008, 28: 2815-2825.

[291] Stickel B H, Jahn A, Kier B. The cost to West Coast communities of dealing with trash, reducing marine debris. US EPA report, 2012.

[292] Morishige C. Marine debris prevention projects and activities in the Republic of Korea and United States: a compilation of project summary reports. USA NOAA Technical Memorandum NOS-OR&R-36, 2010.

[293] UNEP. Guidelines on the Use of Market-based and Economic Instruments to Address the Problem of Marine Litter, U.N.E. Programme, Nairobi, Kenya, 2009.

[294] vince J, Hardesty B D. Plastic pollution challenges in marine and coastal environments: from local to global governance. Restoration Ecology, 2017, 25: 123-128.

[295] 李道季. 海洋微塑料污染状况及其应对措施建议. 环境科学研究, 2019, 32: 197-202.

作者：马骑骉[1]，陈启晴[2]，张微微[3]，麦 磊[4]，汪 磊[5]，鞠茂伟[3]，施华宏[2]，王菊英[3]，曾永平[4]，季 荣[1]

[1] 南京大学，[2] 华东师范大学，[3] 国家海洋环境监测中心，[4] 暨南大学，[5] 南开大学

第13章 饮用水消毒及消毒副产物研究进展

- 1. 引言 /331
- 2. 消毒方式 /332
- 3. 消毒副产物的识别及生成特征 /334
- 4. 消毒副产物及其前体物的分析检测方法 /342
- 5. 饮用水中消毒副产物的浓度分布 /344
- 6. 消毒副产物的毒理学特征 /346
- 7. 消毒副产物的控制及削减 /348
- 8. 困难及展望 /351

本章导读

消毒是饮用水处理中必不可少的环节,通过采用化学或物理的手段杀灭水中的病原微生物,防止介水传染病的大范围传播和流行。但是,由于饮用水原水中存在天然有机质、溴化物、碘化物等天然无机物,以及人工合成化学品,它们可能与投加的消毒剂如氯、臭氧、二氧化氯、氯胺等发生一系列化学反应,生成一些有毒有害污染物,被称作消毒副产物(DBPs),对人体健康造成长期负面影响。

本章比较全面地综述了国内外饮用水消毒和消毒副产物领域的主要研究进展,包括:饮用水消毒方式、消毒副产物的生成机制、消毒副产物的分析检测方法及其在饮用水中的浓度分布、消毒副产物的毒理学效应和流行病学调查、消毒副产物的削减与控制方式,最后讨论了消毒和消毒副产物研究领域存在的问题,并展望了未来的发展趋势。希望读者能够通过阅读本章,对国内外在饮用水消毒与消毒副产物的研究现状有一个全面的认识。

关键词

饮用水,消毒,消毒副产物,生成机制,分析检测,浓度分布,毒理学,削减

1 引 言

随着人口增长和生活水平的提高,人们对淡水资源的需求量日益增加,对用水品质的要求也越来越高;同时,随着人类活动的加剧引起了水资源的严重污染。因此,为了得到足量、品质合格的饮用水,对水处理的技术和工艺提出了更高的要求。消毒是水处理工艺中一个必要环节,既可以杀灭水中的病原微生物保障水质的卫生学安全,也可以去除传统水处理方法(如凝聚、沉降和过滤)不能完全去除的微量污染物。自从美国在水处理工艺中引入消毒后,霍乱、伤寒、痢疾的发病率分别降低了 80%、50%、90%[1]。然而,由于饮用水原水中存在大量有机和无机化合物,消毒处理中使用的消毒剂在杀灭病原微生物的同时,也可能与水体中的化学污染物发生反应,生成毒性更高的消毒副产物,威胁人体健康。

1974 年,荷兰研究者 Rook 和美国研究者 Bellar 相继在经氯化消毒处理的饮用水中检出了三卤甲烷(THMs)类副产物。此后,各国研究者对消毒副产物的生成机制、分析检测、浓度水平、毒理效应以及控制和削减等很多方面开展了大量的研究工作。截至 2018 年年底,全球共发表了 11068 项研究成果,其中来自美国的研究成果数量最多,约占全球总研究成果的 30%,中国次之,约占 17.6%。1974~2018 年,国内外消毒及消毒副产物研究成果的年度分布如图 13-1 所示。自 2000 年开始该领域的研究快速发展,2010 年后再次迎来更加快速的发展。从研究成果的出版刊物分析,《水研究》(*Water Research*)、《环境科学与技术》(*Environment Science and Technology*)是该领域最重要、最受欢迎的期刊。

水源水中存在一定浓度的天然有机质(如富里酸、腐殖质)、人和动物排泄物(如尿素、尿酸等)、各种无机物(如溴化物、碘化物等)以及多种人工合成化学品等,它们在消毒过程中将发生多种化学反应(如取代、水解、加成、氧化等)生成三卤甲烷(THMs)、卤乙酸(HAAs)、卤代硝基甲烷(HNMs)、卤乙腈(HANs)、卤代乙酰胺(HAcAms)、二甲基亚硝胺(NDMA)等消毒副产物[2]。为了检测并识别水中的消毒副产物,国内外学者对消毒副产物的检测方法和仪器进行了优化和改进,迄今识别的消毒副产物已经超过 700 种,而且还不断有新的消毒副产物被检出。常用的分析检测方法有气相色谱-质谱(GC-MS)、气相色谱-电子捕获检测器(GC-ECD)、气相色谱-质谱-质谱(GC-MS-MS)、液相色谱-质谱(HPLC-MS)、液相色谱-质谱-质谱(LC-MS-MS)等。为了提高检测灵敏度,降低检测限,近年来一些更

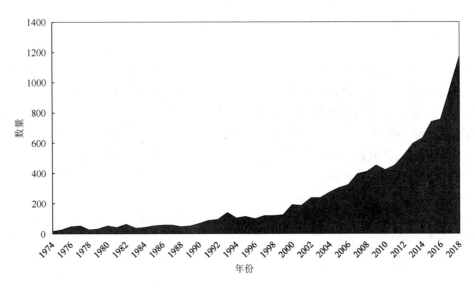

图 13-1　1974～2018 年间国内外饮用水消毒及消毒副产物的研究成果统计

为先进的仪器，如全二维气相色谱-质谱法（GC×GC-MS）、液相色谱-电喷雾电离-傅里叶变换离子回旋共振质谱法（ESI FT-ICR MS）被用于消毒副产物及其前体物的识别与检测。

随着研究的深入，许多学者揭示了消毒副产物的各种毒理学效应。1976 年，美国国家癌症研究所首次证实三氯甲烷在动物试验中能诱发肿瘤，说明该化合物对人体健康存在潜在威胁，如致突变、致畸、致癌等"三致"效应。该研究结果一经报道，引起公众的普遍关注，饮用水氯化消毒处理所引起的水质安全问题也因此成为该领域研究的热点。为了降低消毒副产物带来的潜在危害，国内外学者对消毒副产物的控制和削减技术、方法进行了探索。控制消毒副产物的生成主要从削减前体物和优化消毒方法两个方面考虑。有效削减消毒副产物前体物的方法主要包括混凝沉淀、活性炭吸附、膜过滤、磁性离子交换以及预氧化等；通过优化消毒方法，如改变消毒剂或使用联合消毒方式亦可降低消毒副产物的生成。对于在消毒处理中已经生成的消毒副产物，可以采用煮沸、超滤、纳滤、反渗透、高级氧化、吸附、离子交换、电渗析等多种化学、物理方法以及生物方法去除。

经过四十多年的发展，饮用水消毒及消毒副产物的研究取得了极大进展，但目前仍然面临很多挑战，如开发更加先进的消毒副产物检测识别方法，甄别消毒副产物的潜在危害，以及研发更加高效控制消毒副产物生成的技术等，这需要不同学科的科学家和工程师密切合作，以保证饮用水安全。

2　消毒方式

饮用水的消毒方式已有多种，其中应用时间最长、最普遍的消毒方式是氯化消毒，其他的传统消毒方式还有氯胺消毒、臭氧消毒、紫外线消毒、二氧化氯消毒、电化学消毒等。近年来，为减少消毒副产物的生成，达到更好的消毒效果，许多联合消毒方式得到发展，如氯胺-氯消毒、紫外-氯消毒、预氧化-氯消毒等。

2.1　氯化消毒

氯化消毒是一种应用最广泛的消毒技术，具有价格低廉、操作简便、消毒持久等优点。常用的消毒剂有氯气、液氯、漂白粉[Ca（ClO）Cl]和漂粉精[Ca（ClO）$_2$]等，其主要的有效成分是次氯酸（HClO）。当HClO扩散到带负电荷的细菌表面，穿透细胞膜进入细菌内部，氧化破坏细菌的酶系统而使细菌死亡[3]。同时，由于HClO的反应活性高，也能与水体中的污染物反应生成消毒副产物，如三卤甲烷（THMs）、卤乙酸（HAAs）、卤乙腈（HANs）等。

2.2 氯胺消毒

氯胺消毒是通过氯和氨反应生成一氯胺（NH$_2$Cl）、二氯胺（NHCl$_2$）和三氯胺（NCl$_3$），它们在水中缓慢生成HClO而进行的。氯胺具有稳定性好、持久杀菌等优点，也可大大降低消毒副产物如THMs和HAAs的生成。另外，氯胺的氧化能力比氯弱，会减轻对管网的腐蚀。但是，氯胺消毒也存在缺点，如氯胺的氧化能力弱，往往通过提高氯胺浓度和延长水力停留时间才能达到杀菌目的。近年研究发现氯胺消毒处理中容易生成毒性更高的含氮消毒副产物，如亚硝胺、卤代硝基甲烷、卤代酰胺等[4]。

2.3 臭氧消毒

臭氧消毒作为氯消毒的一种替代方法，在饮用水处理中应用越来越多，尤其是欧洲地区。臭氧可以彻底、广谱、无残留地杀灭细菌繁殖体和芽孢、病毒、真菌等，并可破坏肉毒杆菌毒素，对霉菌也有极强的杀灭作用。臭氧是一种强氧化剂，其灭菌有以下三种形式：第一，臭氧能氧化分解细菌内部葡萄糖所需的酶，导致细菌灭活死亡；第二，直接与细菌、病毒作用，破坏其细胞器和DNA、RNA，使细菌的新陈代谢受到破坏，导致细菌死亡；第三，透过细胞膜组织，侵入细胞内，作用于外膜的脂蛋白和内部的脂多糖，使细菌发生通透性畸变而溶解死亡。臭氧会很快分解为氧气，其对环境的影响小。臭氧消毒的不足是成本高，对管道有腐蚀作用，控制和检测的技术要求高。

2.4 紫外线消毒

紫外线消毒是一种物理消毒方式，通过破坏微生物细胞中的DNA或RNA的分子结构，造成生长性细胞死亡和（或）再生性细胞死亡，进而达到杀菌的效果。紫外线消毒具有诸多优点，如：操作简单，不引入其他化学物质，有望实现"零消毒副产物"；对耐氯微生物隐孢子虫、贾第虫等的杀灭效果好。紫外线消毒的不足在于不能持续杀菌，另外，水中的某些化学物质会吸收紫外线，发生转化，如酚类、芳香化合物以及硝酸盐等。

2.5 电化学消毒

近年来，电化学消毒在海水净化及一些小型饮用水处理中得到重视。电化学消毒主要是依靠电场作用，通过电化学反应装置对水中细菌等进行杀灭去除。电极是影响电化学消毒方法的关键因素。所用的阳极应采用惰性材料，如金属钛、钌钛、石墨等，通电以后阳极不会溶出，具有较小的过电位。近年来，为了提高电化学消毒的效率，研究者研发了多种新型电极材料，特别关注电极对电氧化反应的催化作用，如二氧化钛涂覆的钛电极、银掺杂的二氧化铅电极等，都被证明在直接电氧化消毒中具有催化作用。曾抗美在研究电化学法处理饮用水时发现使用钛板做电极，消毒30 min后，大肠杆菌的去除率达100%[5]。清华大学胡洪营等[6]在电化学消毒中使用氧化铜纳米线作电极，获得了极好的消毒效果，工作电压1 V，每小时能耗25 J/L，7 s内细菌的去除率大于7 log。

2.6 联合消毒

如上所述，各种单项消毒技术都或多或少存在一定的不足，为了减少消毒副产物的生成，减少消毒剂的投加量，提高消毒效率，联合消毒方式快速发展，如氯-氯胺消毒（Cl_2-NH_3Cl）、紫外线-氯（胺）消毒[UV-Cl_2（NH_3Cl）]、预氧化-氯化消毒等。

氯-氯胺联合消毒技术结合了氯的强氧化性及氯胺消毒过程产生消毒副产物少的优点[7]。两者结合既可降低氯胺的浓度、保证消毒能力，又可减少消毒副产物的生成。二氧化氯-氯/氯胺联合消毒利用了二氧化氯的强氧化性优于氯/氯胺与水中有机物发生氧化反应，因此减少了氯/氯胺和有机物反应生成消毒副产物的生成量；此外，水中余氯通过氧化 ClO_2^- 生成 ClO_2 减少了二氧化氯消毒副产物 ClO_2^- 量，增加了残余的 ClO_2 量，增强了联合消毒剂对水中细菌的持续杀菌能力。研究发现二氧化氯-氯/氯胺联合消毒工艺不仅对大肠杆菌和脊髓灰质炎病毒的灭活具有协同作用，对水体中其他微生物的灭活作用也强于单一氯消毒。陈国青等[8, 9]用二氧化氯-氯的混合消毒剂灭活 f_2 噬菌体，结果表明二氧化氯-氯混合消毒剂比氯灭活 f_2 效果好，且受水体酸碱度的影响要小得多，在不同酸碱度均有良好的消毒效果。由于单一氯消毒无法去除饮用水中的耐氯微生物，而紫外消毒属于物理处理，不具有持续消毒能力，因此将紫外消毒与氯（胺）消毒联合使用[10-13]。二者结合既可有效杀灭耐氯微生物，也可达到持续消毒的效果，同时也减少了消毒剂的使用量。吕东明等的研究发现，紫外-氯联合消毒过程中，在紫外剂量为 40 mJ/cm^2 的常规剂量下，即使氯的投加量 CT 值为 5 $mg/(L·min)$（约为单一氯消毒剂量的 1/12）时，联合消毒工艺对病原体的灭活率亦可达 99.99%以上[14]。通过预氧化技术可降低水中消毒副产物前体物的浓度或改变其性质，进而降低消毒副产物的生成潜力。研究发现，与单独氯化消毒相比，高锰酸钾预氧化-活性炭吸附-氯消毒联合处理后，含碳消毒副产物（C-DBPs）和含氮消毒副产物（N-DBPs）的生成潜力显著降低，6 种 C-DBPs（二氯甲烷、三氯甲烷、二氯乙酸、三氯乙酸、1,1-二氯丙酮、1,1,1-三氯丙酮）的生成量减少了 60%～90%，6 种 N-DBPs（二氯乙腈、三氯乙腈、三氯硝基甲烷、二氯乙酰胺、三氯乙酰胺、二甲基亚硝胺）减少了 64%～93%[15]。

3 消毒副产物的识别及生成特征

截至目前已经从消毒处理后的水体中识别出超过 700 种消毒副产物，主要种类包括：三卤甲烷（THMs）、卤代乙酸（HAAs）、卤代乙腈（HANs）、卤代乙酰胺（HAcAms）、卤代硝基甲烷（HNMs）、卤代酮（HKs）、卤代酚以及醛类等[16-17]。

3.1 三卤甲烷（THMs）

自 20 世纪 70 年代首次从氯化消毒处理后的水体中检出三氯甲烷（TCM）以来，国内外加强了对氯消毒过程中 THMs 生成的研究[18-19]。水消毒处理中常见的 THMs 主要有含氯三卤甲烷（Cl-THMs）、含溴三卤甲烷（Br-THMs）以及含碘三卤甲烷（I-THMs），其中检出频率最高的是 TCM。消毒过程中，消毒剂与有机物反应生成 THMs 的机制、生成条件非常复杂，影响因素也多。另外，生成 THMs 的前体物种类也非常多，其中苯酚类化合物是腐殖质的基本结构单元，在氯化消毒中容易生成 THMs。间苯二酚在氯消毒过程中生成 TCM 的转化机制如图 13-2 所示[20]，反应主要包括两个阶段：首先，间苯二酚在次氯酸和次溴酸的作用下发生亲电取代，生成卤代酚；其次，卤代酚类化合物进一步发生水解、脱羧等反应生成 TCM 和三溴甲烷（TBM）。

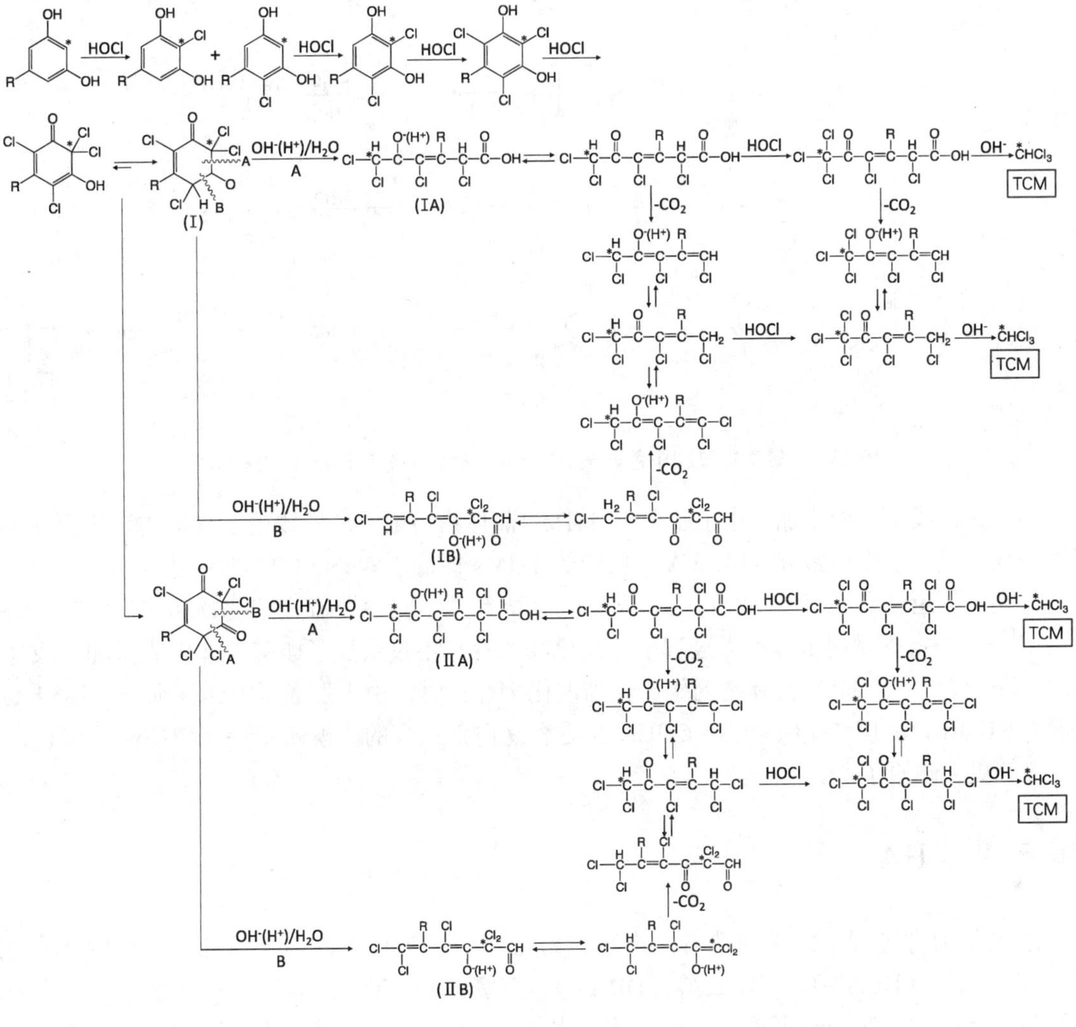

图 13-2　氯消毒处理中间苯二酚转化为 TCM 的路径图

浙江大学 Li 等研究了色氨酸（Tyr）作为前体物在氯消毒过程中的转化行为，并提出了 TCM 和三碘甲烷（TIM）的生成路径[21]，如图 13-3 所示。

不同的消毒方式也会影响 THMs 的生成。由于氯的氧化能力远强于氯胺，与氯胺消毒相比，氯消毒过程能生成更多的 THMs 和 HAAs。若在氯化消毒体系中引入紫外线后，UV-Cl_2 消毒过程比单独氯消毒产生的 THMs 更多，而 UV-Cl_2 消毒过程中 THMs 的生成量是 UV-NH_2Cl 消毒过程中的 6 倍，这可能是由于 UV-Cl_2 会产生更多的活性物质，增加了 THMs 的生成[22, 23]。引入 TiO_2 光催化剂后，光催化处理会使 NOM 部分转化从而减少总三卤甲烷（TTHMs）的产生[24]。同时，增加预氧化过程，如高铁酸盐预氧化，能降低 NOM 的浓度或改变其形态，从而降低后续氯消毒过程中 THMs 的产生。当然前处理方式也会影响 THMs 的生成，楚文海等发现，MBR 处理比 A^2/O 处理后的水生成 THMs 的潜力更高[23]。

此外，消毒条件的改变影响消毒副产物的生成。在氨基酸的氯化过程中，THMs 的生成量随氯剂投加量的增加而增大；在一定氯剂量下，随氯化时间的延长，THMs 的生成量增多；pH 升高在不同程度上促进了氨基酸转化为 THMs[25]。在饮用水 UV-氯消毒过程中，THMs 的生成量随紫外强度的增大而增大。

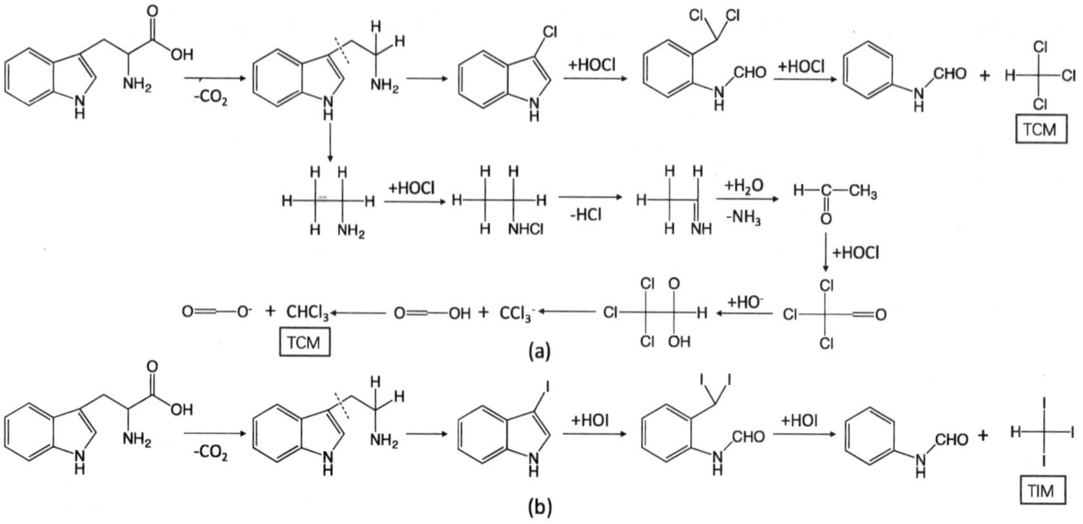

图 13-3　氯消毒过程中色氨酸（Tyr）转化成 TCM 和 TIM 的路径图

这可能是由于紫外线剂量的增加促进了小分子 THMs 前体物的生成，导致更多有机物转化为 THMs，而在相同紫外线剂量下，中压紫外（MP-UV）生成的 THMs 比低压紫外（LP-UV）更多。

水中的无机物也会影响消毒副产物的生成，Li 等发现当饮用水中存在溴化物和碘化物时，THMs 的生成量增多[25]。在富含碘的水体消毒过程中，溴化物的存在不仅增加了碘酸盐的生成量和生成速率，也增加了溴、碘混合取代 THMs 前体物的生成。当溴化物存在时，由于形成了氧化性强的次溴酸（HOBr）和氯溴胺（NHBrCl），使二氯碘甲烷（$CHCl_2I$）更容易向毒性高的二溴碘甲烷（$CHBr_2I$）转化，生成的 THMs 中含有更多碘原子[26]。

3.2　卤乙酸（HAAs）

卤乙酸也是消毒过程中常见的消毒副产物，检出率较高的卤乙酸有一氯乙酸（CAA）、一溴乙酸（BAA）、二氯乙酸（DCAA）、二溴乙酸（DBAA）、三氯乙酸（TCAA）等，水处理厂中含量最多的是 TCAA 和 DCAA[27-33]。Serrano 等在西班牙某饮用水处理厂的不同工艺段取水，并检测各种消毒副产物的浓度，结果发现原水中检出 2 种 HAAs（TCAA 和 DCAA），二氧化氯预氧化后有 5 种新的 HAAs 生成，氯胺消毒后 7 种 HAAs 的浓度均大幅升高，并且生成了碘代乙酸[34]。

Zhang 等发现胞嘧啶在氯消毒过程中会生成 HAAs，并提出了胞嘧啶生成 DCAA 和 TCAA 的路径，如图 13-4 所示。胞嘧啶在氯化消毒过程中发生取代反应、水解反应、消去反应、加成反应及脱氨基后转化成 DCAA 及 TCAA[38]。

刘绍刚等提出了含天然有机物及碘离子的水体在氯化消毒后生成碘乙酸（IAAs）的转化路径，如图 13-5 所示。含羧基的天然有机物在经过烯醇化、取代、加成及氧化反应后生成三碘乙酸（TIAA）；含有氨基的有机物首先生成卤代腈类中间体如碘乙腈、二碘乙腈，继续发生加成反应和水解脱氨基后生成碘乙酸（IAA）和二碘乙酸（DIAA）[39]。

不同消毒方式及消毒条件的改变影响 HAAs 的生成。木质素在氯化过程中会产生较多的 TCAA，在氯胺化过程中产生较多的 DCAA。Krasna 等发现氯胺消毒增加了 IAAs 的生成[35]。刘绍刚等[39]也发现氯胺消毒比氯消毒过程生成的 IAA、DIAA、TIAA 更多。氯胺消毒过程中，随着氯胺剂量的增加，DIAA 和 IAA 的生成量逐渐增大，而 TIAA 的浓度先显著增大后逐渐降低，这主要是由于大量的 I^- 直接被一氯胺水解产生的自由氯氧化为 IO_3^-，降低了 TIAA 的生成。而当消毒体系中 I^- 浓度增大时，TIAA 的浓度先

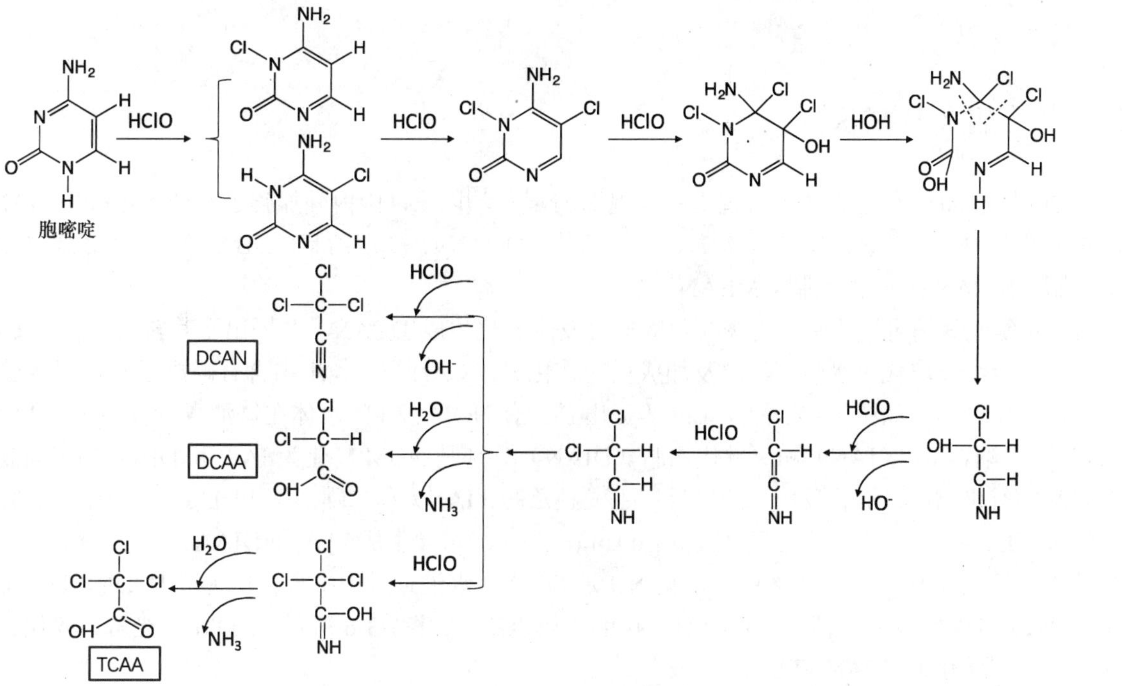

图 13-4　氯消毒过程中胞嘧啶转化为 DCAA 和 HAAs 的路径图

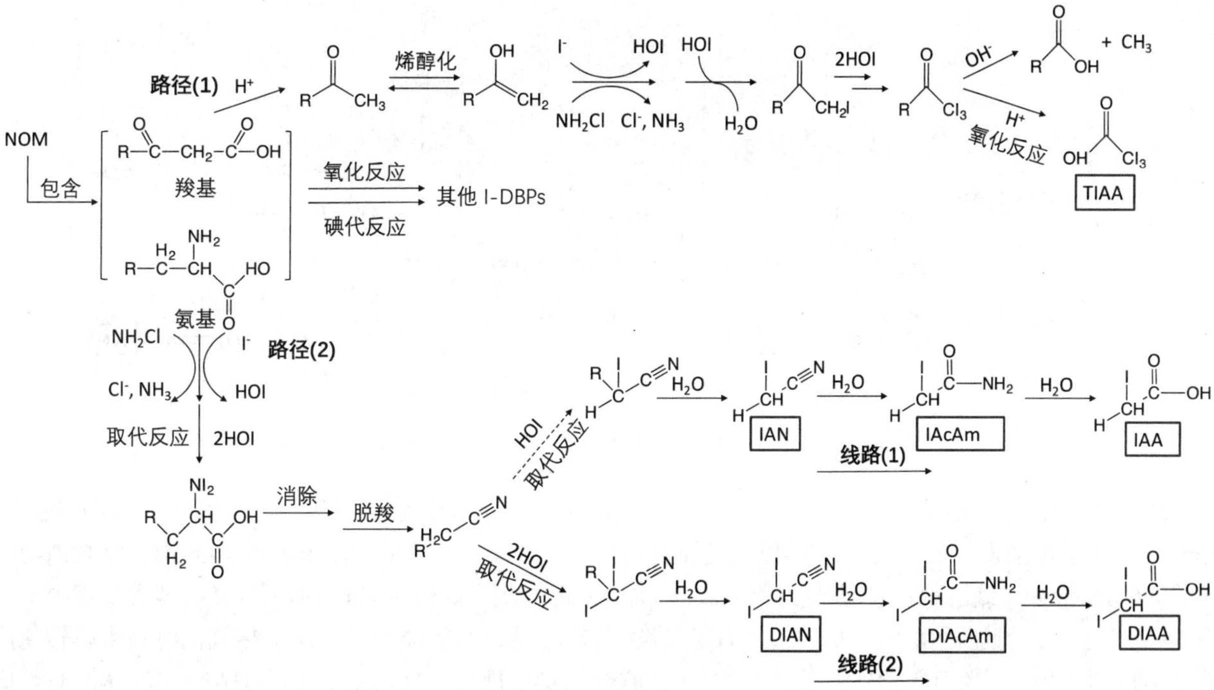

图 13-5　氯消毒过程中 NOM 转化为 HAAs 的路径图

增大后减小，DIAA 的浓度逐渐增大，IAA 的浓度逐渐减小，说明过量的 I⁻ 竞争 NOM 与一氯胺的反应，导致碘代消毒副产物的生成量减少。此外，水中的无机离子（如碘离子、溴离子等）也会改变 HAAs 的生成种类，随着 Br⁻/I⁻ 增大，TIAA 和 DIAA 浓度减小，溴碘乙酸（BrIAA）浓度增大。另外，温度对消毒副产物的生成也有一定的影响，Serrano 等发现 HAAs 的生成受季节的影响，温暖季节生成的 HAAs 明显高于寒冷季节[34]。

3.3 卤乙腈（HANs）

自 20 世纪 80 年代在氯化消毒过程中发现 HANs[36,37]以来，国内外加强了对消毒过程中 HANs 的研究，常见的 HANs 有三氯乙腈（TCAN）、二氯乙腈（DCAN）、二溴乙腈（DBAN）、氯溴乙腈（BCAN）、一氯乙腈（MCAN）、一溴乙腈（MBAN）等。

Zhang 等在氯消毒过程中发现嘧啶类物质会发生转化生成 DCAN，并提出了其转化路径，如图 13-4 所示。胞嘧啶经过取代、水解、消除及加成反应转化成 DCAN[38]。当水中含有碘离子时，消毒过程中生成碘代消毒副产物。如图 13-5 所示，Liu 等提出了含有 NOM 及 I$^-$的水体在氯消毒过程中生成 IAAs 的转化路径[39]。碘代乙腈（IAN）和二碘代乙腈（DIAN）分别是 NOM 转化为 IAA 和 DIAA 的中间副产物，含有氨基的有机物经过取代反应和脱羧反应生成碘乙腈（IAN）和二碘乙腈（DIAN），继续水解后生成碘代乙酰胺（IAcAm）和二碘代乙酰胺（DIAcAm），进一步水解生成 IAA 和 DIAA。

藻类有机质（AOM）是水体中的常见 NOM，在饮用水消毒过程中会转化生成 HANs，De Vera 等提出了 AOM 在 UV-氯（胺）消毒过程中 HANs 的生成机制，如图 13-6 所示，AOM 经光解、取代、消除和脱羧反应最终转化为 TCAN 和 DCAN[40]。

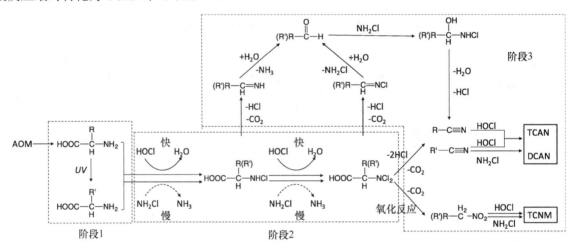

图 13-6 藻类有机质在 UV-氯（胺）消毒过程中转化为 HANs 和 TCNM 的路径图

Huang 等发现，HANs 是氯消毒处理中主要的 N-DBPs，而 HAcAms 是氯胺消毒的主要 N-DBPs[4]。Tian 等对比了氯消毒、氯胺消毒、氯-氯胺消毒过程中，HANs 的生成量，发现消毒 24 h 内，氯胺消毒生成的 DCAN 最少，氯消毒和氯-氯胺消毒生成的 DCAN 浓度相当；消毒 72 h 内，氯-氯胺消毒过程中形成的 DCAN 最多，可能是由于 DCAN 在 NH_2Cl 存在时较稳定，水解能力弱，而单独氯化时易水解转化成其他消毒副产物[7]。当引入紫外线后，在 UV-Cl_2 联合消毒处理比氯化消毒产生的 HANs 多，MP-UV 比 LP-UV 的效果更明显。预氧化能降低水中 NOM 的浓度或改变其形态，从而减少消毒副产物的生成。楚文海等研究了自由态氨基酸和结合态氨基酸在氯化消毒过程中的转化，发现高锰酸钾预氧化可减少 HANs 前体物的生成，进而降低了 HANs 的生成。随着 pH 的升高，DCAN 的生成量先增加后降低，在 pH5.5 时生成量达到最高[14]。De Vera 等人发现 O_3 可与含氮有机物反应生成硝基化合物 R-NO_2，减少了 HANs 前体物的生成，从而降低了 HANs 的生成潜力[40]。

3.4 卤代乙酰胺（HAcAms）

国内外研究者在饮用水氯/氯胺消毒过程中已经检测到了多种类型的HAcAms[41, 42]。如美国[35]、中国[43]、澳大利亚[44]、英国[45]等国家的研究者均在经消毒处理的水中检测到了HAcAms。

Chuang等[45]用4-羟基苯甲酸作为腐殖质的模型化合物，研究了其在氯-氯胺消毒过程中的转化机制，检测到二氯乙酰胺（DCAcAm）和三氯乙酰胺（TCAcAm）等多种副产物，并提出了其转化路径，如图13-7所示。4-羟基苯甲酸经过脱羧、取代、消除、加成及氧化等一系列反应最后转化为TCAcAm和DCAcAm。

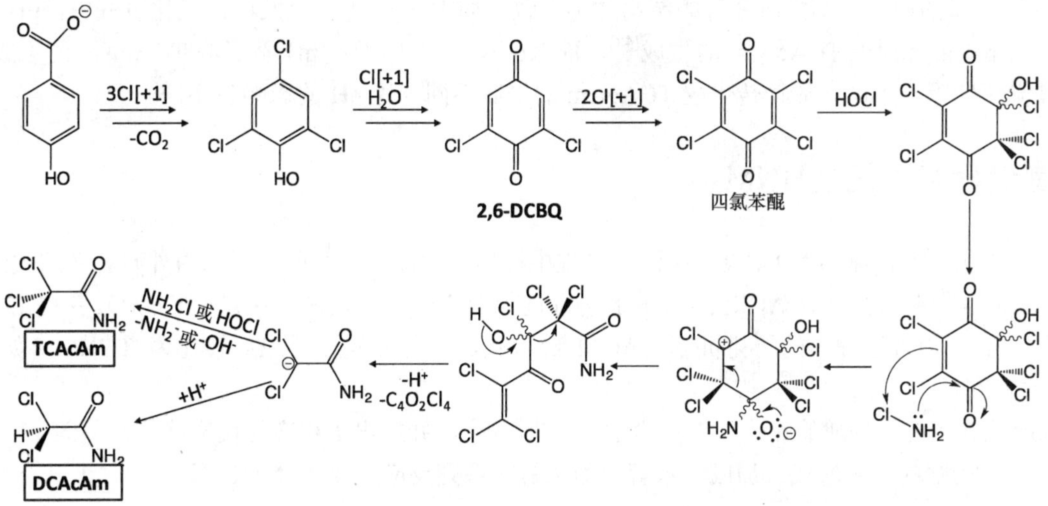

图 13-7　氯-氯胺消毒过程中4-羟基苯甲酸转化为HAcAms的路径图

楚文海等[46]研究了饮用水消毒过程中自由氨基酸和寡肽转化生成HAcAms的路径，如图13-8所示，氨基酸可经过取代、脱羧以及消除反应转化生成DCAcAm。

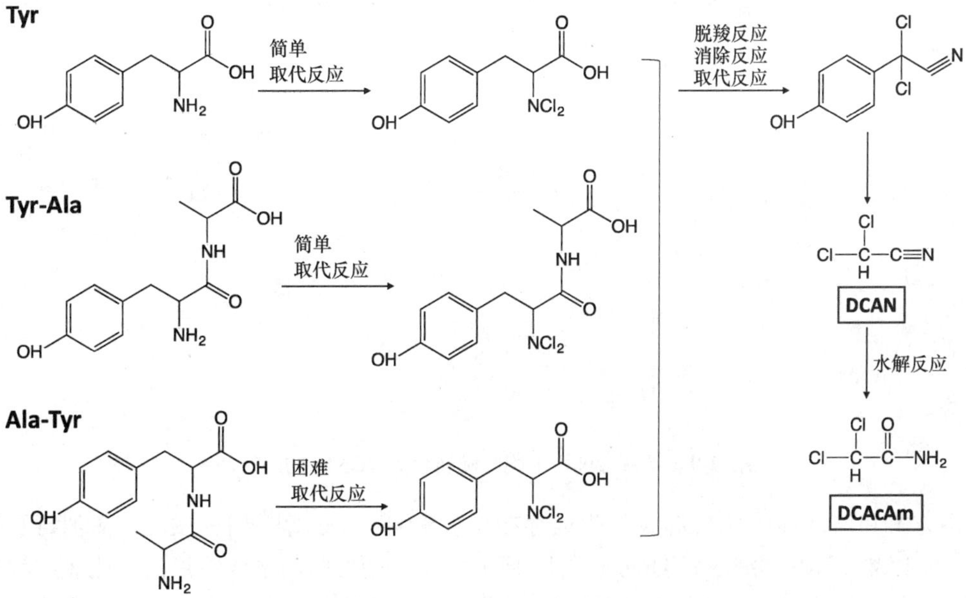

图 13-8　氯消毒过程中氨基酸和寡肽转化成DCAN和DCAcAm的路径图

消毒方式及消毒条件的改变影响 HAcAms 的生成。Huang 等探讨了饮用水氯消毒、氯胺消毒及氯-氯胺联合消毒过程中消毒副产物的生成情况，检出 4 种 HAcAms，其中 DCAcAm 为主要组分。三种不同消毒方式下，氯-氯胺联合消毒中生成的 DCAcAm 最多，氯消毒中生成的 DCAcAm 最少，氯-氯胺联合消毒中延长预氯化时间可减少 HAcAms 的生成[4]。De Vera 等[40]在研究臭氧预氧化-氯化消毒过程中消毒副产物的生成潜力时发现，臭氧预氧化主要是通过影响总卤代乙酰胺（THAcAms）的前体物而降低其生成潜力。刘绍刚等[39]研究了富碘水在氯胺消毒过程中碘代消毒副产物的生成及种类，在消毒过程中检出了二碘代乙酰胺（DIAcAm），当体系中一氯胺浓度增大时，DIAcAm 的生成量先增加后降低，主要原因是，大量的 I$^-$ 被一氯胺水解产生的自由氯氧化为 IO$_3^-$，使 DIAcAm 生成量降低。Kosaka 等[47]发现随着 pH 升高，DCAcAm 生成量升高，主要是由于在较高 pH 时有更多的 DCAN 水解生成 DCAcAm。关于 pH 对 TCAcAm 生成量的影响，不同研究者的结论并不一致。如 Huang 等研究发现，氯化过程中，pH 在 6～9 范围内，低 pH 更有利于 TCAcAm 的生成[48]，而 Kosaka 等[47]发现，pH 不影响 TCAcAm 的生成。这些结果的差异可归因于不同前体物转化成 TCAcAm 的机制不同，受 pH 的影响也不同。

3.5 卤代硝基甲烷（HNMs）

HNMs 是另一类常见的 N-DBPs，在消毒过程中检出率较高，已经引起了国内外研究者的关注[35, 49]。De Vera 等[40]在研究富含 AOM 的饮用水消毒过程中检出了三氯硝基甲烷（TCNM），并提出了 AOM 在 UV-氯（胺）消毒过程中的转化机制，AOM 经取代、消除和脱羧反应后转化为 TCNM（如图 13-6 所示）。

Zhang 等在嘧啶类物质的氯化过程中检出了 TCNM[38]，并提出了相应的反应路径，如图 13-9 所示。在氯化消毒中胞嘧啶发生取代、加成、水解及消去等一系列反应生成 TCNM。

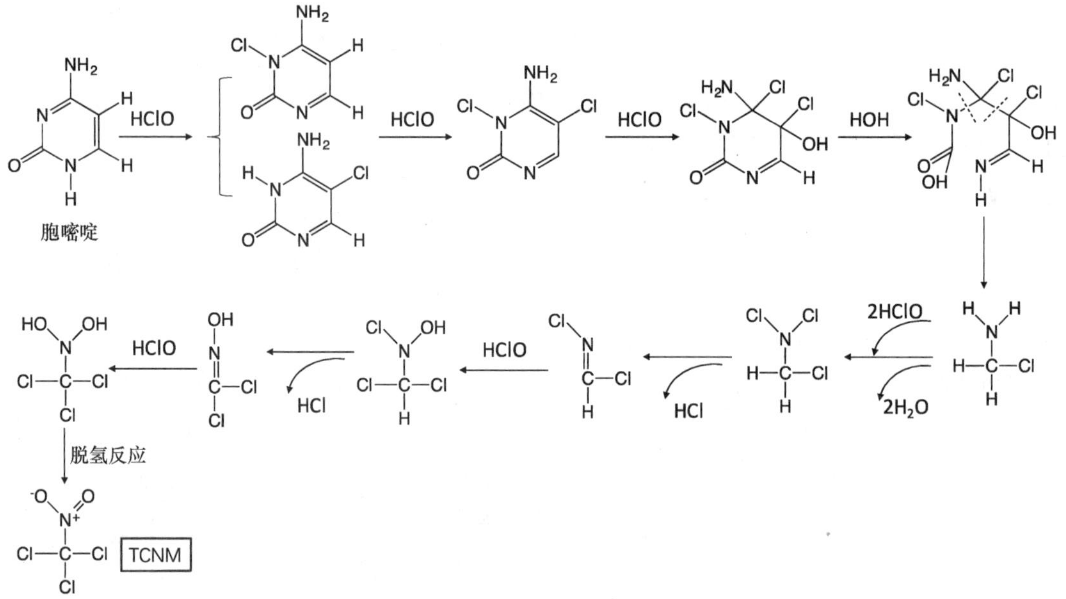

图 13-9 氯消毒过程中胞嘧啶转化为 TCNM 的路径图

不同消毒方式和消毒条件对 HNMs 的生成存在显著影响。Lee 等[50]研究发现，氯消毒和氯胺消毒过程中 TCNM 的生成量基本一致，而 Dotson 等[51]研究发现含硝酸盐的水体经氯消毒比氯胺消毒生成更多 TCNM。同时，若在氯消毒前先进行臭氧预氧化[52]或 UV 辐射[53-54]，均会增加 HNMs 生成量，可能是由于臭氧预氧化或 UV 处理增加了 HNMs 前体物的生成[10]。消毒体系的酸碱性对 HNMs 的生成也有重要影响，研究发现随着 pH（5～9）的升高，氯化过程中 HNMs 的生成量逐渐增大。另外，水中无机离子也会影响 HNMs 的生成量，增加水中硝酸盐[55]和溴离子[37]的浓度，会引起 HNMs 的生成量增多。

3.6 亚硝胺（NMAs）

自从在氯消毒、氯胺消毒的饮用水中发现 N-二甲基亚硝胺（NDMA）以来，亚硝胺类新型消毒副产物引起了国内外研究者的重点关注[35, 56-62]。Schreiber 和 Ferre 等[60, 61]研究了氯胺与二甲胺（DMA）反应生成 NDMA 的转化机制，如图 13-10 所示，DMA 主要与二氯胺反应先生成氯代偏二甲肼（Cl-UDMH），进一步被氧化生成 NDMA。

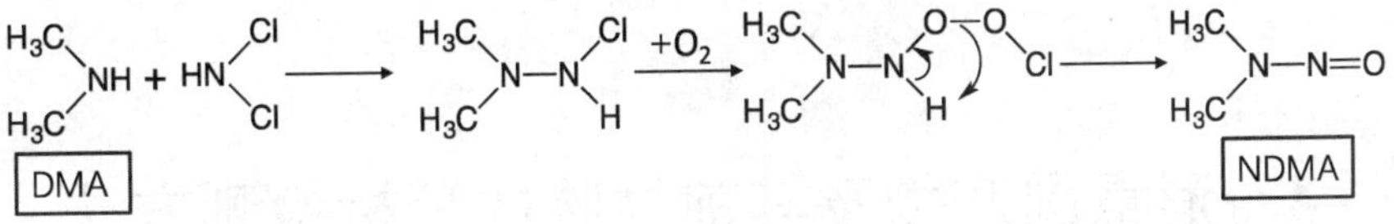

图 13-10 氯胺消毒过程中二甲胺转化为 NDMA 的路径图

臭氧氧化二甲胺生成较低浓度的 NDMA，且随 pH 升高 NDMA 产率也增加。一些含 N-N 键的化合物如偏二甲肼、氨基脲、磺酰胺类能在臭氧作用下生成更多的 NDMA，其摩尔产率可达到 50%以上。Gan 等发现丁酰肼能与二氧化氯反应生成 NDMA，提出了肼类化合物在二氧化氯消毒体系中反应生成偏二甲肼类中间体，而后继续氧化生成 NDMA，如图 13-11 所示[63]。

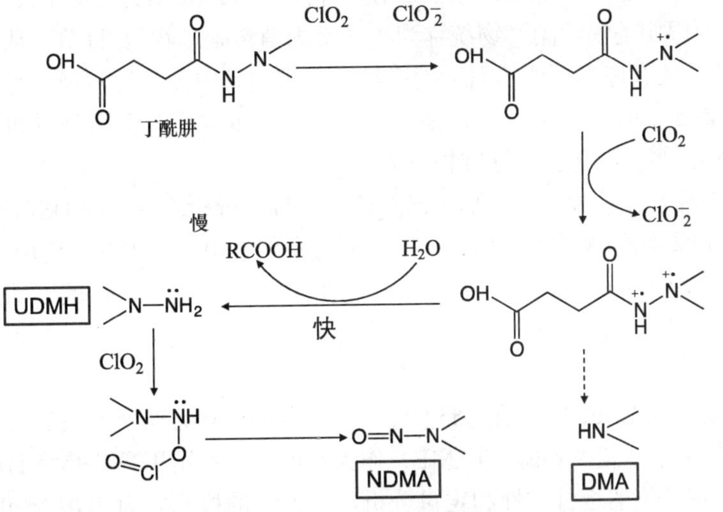

图 13-11 氯胺消毒过程中丁酰肼转化为 NDMA 的路径图

Schreiber 和 Farre[60-61]的研究表明 NDMA 主要在氯胺消毒过程中生成，消毒条件的改变对其生成量有较大影响。Schreiber 等[60]发现，当向含有氨的水中加入氯，且 $Cl_2:NH_3$ 比例较高时，生成更多的二氯胺，更有利于 NDMA 的生成。Farre 等[61]提出在污水二级处理过程中若先投加氨后投加 NaClO，会增加 NDMA 的生成量；若先投加氯后投加氨，并延长氯化时间，可降低 NDMA 的生成量[60, 64-68]。Park 等[69]研究了 NDMA 的生成潜力，发现氯胺消毒前先进行自由氯、二氧化氯预氧化等处理有利于减少 NDMA 的生成，而臭氧预氧化会促进聚二烯丙基二甲基氯化铵（polyDADMAC）生成 NDMA。

3.7 其他消毒副产物

除了以上 5 类常见消毒副产物之外，近年来在饮用水消毒处理中还检测了许多其他类型的消毒副产物。Park 等人[70]在水处理厂的臭氧消毒水体中检出了甲醛和乙醛。Jeong 等[71]在美国水处理厂出水中检测到了多种卤代醛，其中最多的是二氯乙醛。2006 年，Bull 等预测水消毒过程中可能生成卤代苯醌类副产物[72]，2009~2010 年，Zhao 等在饮用水中检测到了二氯苯醌 DCBQ、二溴苯醌 DBBQ、二氯溴苯醌 DCMBQ 以及三氯苯醌 TCBQ[73]。3-氯-4-二氧甲基-5-羟基-2（5 氢）-呋喃（MX）也是一种氯化消毒副产物，Holmbom 等于 1981 年首次在氯化漂白的木浆液中发现，1986 年 Hemming 等首次在氯化消毒的饮用水中检出 MX。此后，美国、英国、日本和中国等相继在氯化消毒饮用水中检到 MX。

4 消毒副产物及其前体物的分析检测方法

4.1 消毒副产物检测

考虑到消毒副产物的毒理学效应，许多国家和国际机构将部分常见的消毒副产物列入水质标准，并限定了其最大允许浓度，称为"常规消毒副产物"。例如，我国在《饮用水水质标准》（GB/T5749-2006）中对 TCM、DCM、一溴二氯甲烷（BDCM）、二溴一氯甲烷（DBCM）、DCAA、TCAA 等进行了限定。而其他未在水质标准中限定的消毒副产物统称为"非常规消毒副产物"。目前，从水体中识别出的消毒副产物已经超过 700 种，还有许多可能的消毒副产物尚未识别。因此，对常规消毒副产物和非常规消毒副产物的检测和识别仍然是一项重要的任务，消毒副产物的检测主要有目标分析和非目标分析。常规消毒副产物如 THMs、HAAs 等已经建立了标准检测方法。

针对 THMs，美国环境保护局（USEPA）已经建立了标准分析方法，如 USEPA 551.1 和 USEPA 534.2 等。其中 USEPA551.1 采用液-液萃取（LLE）技术浓缩富集水中的 THMs，运用气相色谱-电子捕获检测器分析，该方法的检测限为 0.002~0.055 μg/L[74]。除此之外，检测 THMs 的方法还有很多，如使用顶空固相微萃取与气相色谱联用、液相微萃取与气相色谱联用等方法，既可以减少萃取剂对环境的污染，也可以提高检测灵敏度[75-77]。

HAAs 的分析主要采取 USEPA 552.1、552.2、552.3 和 6251B 方法及其改进方法[78-81]，该方法采用甲基叔丁基醚（MTBE）或甲基叔戊基醚（TAME）作为萃取剂，利用甲醇共热将 HAAs 衍生化为甲酯，并通过气相色谱-电子捕获检测器进行定性和定量分析，该方法的检测限为 0.012~0.27 μg/L[80]。为了降低消毒副产物分析的检测限，提高检测的灵敏度，研究者们在标准方法的基础上，通过改良分析仪器或优化实验条件开发了新的分析检测方法。王海鸥等[82]以 MTBE 为萃取剂，采用液-液微萃取-气相色谱法测定水中 THMs 的含量，色谱柱选用柱效高、色谱分辨率高、进样量少、灵敏度高的毛细管柱，检测器为灵敏度高于普通 ECD 检测器的 μ-ECD（微池电子捕获检测器），各消毒副产物 TCM、CCl_4、BDCM、DBCM 和 TBM 的检测限分别为 0.002、0.004、0.008、0.0012 和 0.011 μg/L，该法显著优于国标法。Chen 等[83]开发了冷离子阱-气相色谱-质谱法检测水中 HNMs，该方法降低了 HNMs，尤其是二溴氯硝基甲烷（DBCNM）和三溴硝基甲烷（TBNM）在进样过程中的热降解损失。

对于未知消毒副产物，常用的检测技术主要包括 GC-MS、LC-MS 技术等。从 19 世纪 70 年代早期开始，质谱技术在饮用水质量分析中逐渐发挥关键作用。作为最早用于消毒副产物的检测工具，迄今为止大部分消毒副产物的分析检测仍离不开 GC-MS，主要适合于分析低极性、挥发或半挥发性的消毒副产物，如 THMs、低极性溴代或碘代消毒副产物、部分含氮消毒副产物如亚硝胺类[84,85]、HAmAcs[86-87]等。Plewa 等[88]利用 GC-MS 获取分子离子和碎片离子信息，推测出 5 种 IAAs [碘乙酸、溴代碘乙酸、(E)-2-碘-3-甲基-丁二酸、(Z)-和(E)-3-溴-3-碘-丙烯酸]的分子结构，并利用购买或自行合成的碘乙酸标准品证实了它们的分子结构。Nihemaiti 等[89]利用高分辨 GC-MS 分析间苯二酚经氯胺化处理后的产物，发现 500 μmol/L 间苯二酚在 pH7 条件下加入 5.6 mmol/L NH_2Cl 后不仅生成了多种常见的含氮消毒副产物，如 DCAN、DCAcAm 和 TCAcAm，还检出了 6 种产率很高的未知产物，根据高分辨质谱提供的精确分子量推测出它们含有杂环结构。

对于亲水性强或分子量大且不易挥发的消毒副产物，LC-MS 的检测效果更好。Pereira 等[90]发现 17β-雌二醇和雌酮在臭氧消毒后生成了 4 种主要消毒副产物，其结构通过超高效液相色谱与三重四极杆-飞行时间质谱联用技术（UPLC-QqQ-QTOF-MS）进行识别。魏东斌等[91]利用 UPLC-QTOF-HRMS 识别了二苯甲酮类防晒剂在次氯酸消毒体系中生成的 10 余种未见报道的消毒副产物，发现氯化消毒处理中酮类化合物可能发生 Baeyer-Villiger 氧化生成酯类产物，并经水解生成卤代酚类产物，且该反应具有一定的普适性。张相如等[92]利用 LC-MS 的母离子检测模式可高效识别水体中的高极性含碘副产物，同时模拟研究了烹饪过程中含碘消毒副产物的生成机制。此外，对于某些分子量小但极性高、不易挥发、易水解的化合物，如极性碘代消毒副产物和卤代乙酰胺类化合物，也可运用 LC-MS 检测识别[93-96]。

为了能够检测出更多的消毒副产物，降低检测限，提高检测灵敏度，需要优化分析方法和检测仪器。Li 等[96]采用全二维气相色谱-质谱法，结合 OECD QSAR Toolbox Ver. 3.2，检测饮用水中的挥发、半挥发性消毒副产物。该方法检测了水在氯化消毒、氯胺消毒和臭氧消毒过程中的消毒副产物，每种消毒后的水样中均检测出 500 多种消毒副产物。最近，液相色谱-电喷雾电离傅里叶变换离子回旋共振质谱法（ESI FT-ICR MS）的应用得到了迅速发展，该方法在检测未知消毒副产物及其前体物方面优势显著。已有多项研究报道了饮用水消毒体系中检测到 659 种一氯代产物，348 种二氯代产物，441 种一溴代产物，37 种二溴代产物，178 种一碘代产物以及 13 种二碘代产物，其中仅有 10 种化合物已知[97-100]。这些研究结果说明，ESI FT-ICR MS 可以为未知消毒副产物的组成和结构分析提供帮助。

除化学方法检测之外，生物检测和毒性效应引导的消毒副产物识别方法具有明显优势。显著的毒性效应是消毒副产物备受关注的最主要原因，加之消毒副产物种类繁多，因此在毒性效应引导下可更加快速方便地识别高毒性消毒副产物。陈妙等提出了毒性效应引导的高风险消毒副产物识别方法，可提高识别消毒后水体中高风险消毒副产物的准确性和效率[101]。例如，胡建英等[102]在研究 17β-雌二醇（E2）为前体物的氯化消毒副产物时，通过 ESI-LC/MS 识别出 7 种主要产物，并通过比较次氯酸与雌二醇不同反应时间段反应液的雌激素活性，发现反应 60 min 时消毒副产物的毒性比反应 120 min 时的更高，由于无法将母体及消毒副产物完全分离，因此通过化学合成制备出其中的 4 种产物 4-氯-E2、2,4-二氯-E1、2,4-二氯-E2 和产物 C（雌二醇的氯代产物）。根据这 4 种产物的剂量效应关系结合 LC-MS 定量检测结果，最后确定 2,4-二氯-E1 和 2,4-二氯-E2 是反应时间为 120 min 和 180 min 产生的消毒副产物中主要的雌激素活性物质。肖铭等[91]的研究结果表明，在二苯甲酮-4 的氯化消毒过程中共检出数十种消毒副产物，其中 3-氯代-二苯甲酮-4 显示出较高的遗传毒性。李立平[103]的研究结果显示头孢唑啉在氯化消毒过程中会产生遗传毒性高的亚砜类副产物。

4.2 消毒副产物前体物解析方法

已有大量研究表明水体中的 NOM，尤其是腐殖酸类物质是消毒副产物的主要前体物。由于我国水源普遍受到污染，水体中除 NOM 之外，排入水中的人工合成化学污染物也可能成为消毒副产物的重要前体物。

由于水体中有机污染物种类多、性质差异大，很难分别研究其在消毒过程中副产物的生成特征。目前通常将有机物根据其性质差异进行分馏，然后分别研究各馏分在消毒处理中副产物的生成潜能，从中识别出对消毒副产物贡献显著的前体物组分。陈超等根据 Leenheer 等提出的方法[104]，按照不同化学特性的有机物在不同树脂上的吸附特征，将水中有机物分为疏水性有机酸、疏水性有机碱、疏水中性有机物、亲水性有机酸、亲水性有机碱以及亲水性中性有机物。检测发现疏水性有机酸占原水总溶解性有机物的 36%，亲水性有机酸占 26%，而疏水性有机碱和亲水性有机碱分别只占 4% 左右。其中富含腐殖质的疏水性有机酸对 THMs 和 HAAs 生成潜能的贡献都远远高于其他组分的有机物，是这两类消毒副产物的主要前体物[105]。王成坤等使用简化的树脂富集法对北方某水源水中亚硝胺前体物进行了解析，发现亲水性有机物和小分子有机物对亚硝胺的生成贡献更大[106]。不过，树脂富集解析法需要水样量大、事先纯化树脂、操作复杂且时间长、结果波动大、缺乏有效的质控信息。

美国加州大学洛杉矶分校的 Rosario-Ortiz 等开发了用于解析水中有机物的一种新方法——快速极性分析法（PRAM）[107]。该方法通过比较吸附前后水中有机物浓度（以 UV_{254} 或 DOC 计）的变化，获得吸附在不同种类固相萃取小柱上的有机物量，可定量描述水中天然有机物的极性分布特性。该方法需要的水样量小，测试可以在 2 h 内完成，洗脱也比较简单，工作量小；其次，固相萃取小柱是标准化产品，吸附重现性好，在操作过程中可以运行多个平行样，误差一般在 10% 以内。Chen 等首次尝试将 PRAM 方法用于亚硝胺前体物的解析。通过优选的强阳离子交换树脂小柱（SCX）和非极性十八烷（C18）固相萃取小柱，发现水源水和污水中的亚硝胺前体物普遍具有正电性和非极性的通用结构特征（既可被 SCX 富集，也可被 C18 富集）[108]。该基本结构的提出为亚硝胺前体物在水环境中的分布特性研究和控制技术开发提供了新思路。Bei 等利用亚硝胺前体物在通常 pH 情况下带正电的特点，利用阳离子交换小柱富集水中的痕量亚硝胺前体物，首次报道黑臭水体中的污染物也是亚硝胺的重要前体物，还首次鉴别出一种新的 NDMA 前体物——福美锌[109]。

5　饮用水中消毒副产物的浓度分布

消毒副产物因其显著的生物毒性效应而备受关注，同样，其在饮用水中的浓度也是决定其健康风险的重要因素。在氯消毒过程中，水中 NOM，如腐殖酸、富里酸和藻类等与投加的消毒剂氯发生取代、加成和氧化反应生成氯代消毒副产物（Cl-DBPs）。目前检测到的 Cl-DBPs 已多达数百种，包括 THMs、HAAs、HANs、MX 等，其中 THMs 和 HAAs 两者含量之和约占全部 Cl-DBPs 的 80% 以上。二氧化氯消毒产生的消毒副产物主要为亚氯酸盐（ClO_2^-）和氯酸盐（ClO_3^-）。臭氧可以氧化水中的有机物生成酮类、羧酸和醛类化合物，还可以氧化溴离子为次溴酸，导致 Br-DBPs 的生成。

Kransner 等[35]于 2000~2002 年调查了美国 12 家水厂消毒副产物的浓度分布，发现自来水中 THMs 含量为 4~164 μg/L，HAAs 为 5~130 μg/L。Williams 等人 1997 年对加拿大 35 座氯消毒水厂、10 座氯胺消毒水厂和 6 座臭氧消毒水厂出水中的消毒副产物进行了调查，THM4 在氯消毒水厂出水中浓度范围为 1.6~120.8 μg/L（中位值为 17.2 μg/L），氯胺消毒水厂中浓度范围为 2.9~80.1 μg/L（中位值为 19.7 μg/L），臭氧消毒的水厂中浓度范围为 2.5~74.9 μg/L（中位值为 57.4 μg/L）；DCAN 在氯和氯胺消毒的水厂中浓度中位值均为 1.0 μg/L，臭氧消毒的水厂中浓度中位值为 0.61 μg/L；DCAA 和 TCAA 在氯消毒水厂出水中浓度中位值分别为 9.0 μg/L 和 13.0 μg/L，在氯胺消毒出水中浓度中位值分别为 7.7 μg/L 和 6.9 μg/L，在臭氧消毒出水中浓度中位值分别为 6.4 μg/L 和 1.5 μg/L[110]。Weisel 等[111]在美国新泽西州进行了一项人群氯化消毒副产物暴露评价的研究，在自来水中检出的 TTHMs 平均含量为 33 μg/L(0.03～260 μg/L)，DCAA 为 19 μg/L（0.33～110 μg/L），TCAA 为 18 μg/L（0.25～120 μg/L）。Krasner 等还对美国 23 座污水处理厂出水进行了类似调查，发现 TCNM 在氯消毒和氯胺消毒的污水处理厂出水中浓度分别为 ND～0.7 μg/L 和 ND～0.6 μg/L[112]。

在欧洲，Goslan 等[113]对苏格兰 7 座水厂进行了检测，THM4 在氯消毒和氯胺消毒的水厂出水中浓度中位值分别为 106 μg/L 和 48 μg/L，在氯消毒水厂出水中，冬季 THM4 的浓度达到最高值 419 μg/L；HAAs 在氯消毒和氯胺消毒的水厂出水中浓度中位值分别为 44 μg/L 和 16 μg/L；HANs 在氯消毒和氯胺消毒水厂出水中浓度中位值分别为 1.7 μg/L 和 1.3 μg/L；TCNM 在氯消毒和氯胺消毒水厂出水中浓度中位值均为 0.1 μg/L。希腊研究者 Malliarou 于 1999～2000 年间调查了雅典 4 座饮用水处理厂出水中卤代消毒副产物的含量，结果显示 THM4、HAA9、TCAN、水合氯醛（CH）、1,1-二氯丙酮以及 1,1,1-三氯丙酮的含量分别为 ND～18.9 μg/L、ND～11.03 μg/L、ND～1.06 μg/L、0.32～0.69 μg/L、0.25～0.54 μg/L 和 0.03～0.67 μg/L。Malliarou 等在 2005 年对英国 3 家水务公司辖区内采集 30 份自来水样品，测得 THM4 的浓度均值为 27.6～50.9 μg/L，HAAs 为 35～95 μg/L（最大值 244 μg/L）[114]。Egorov 等[115]测得俄罗斯 Cherepovets 市饮用水中 THMs 浓度为 70～205 μg/L，HAAs 为 30～150 μg/L。在英国、芬兰、美国等国家的自来水中均检出强诱变剂 MX，含量为 2～67 ng/L[116]。Egorov 等[115]2003 年从俄罗斯 Cherepovets 市饮用水中检出 MX 的浓度为 50～160 ng/L。

在亚洲，Lee 等对韩国 35 个饮用水厂出水中的消毒副产物含量进行了调查，HAAs 浓度中位值为 6.96 μg/L，其中 DCAA 和 TCAA 的中位值分别为 2.15 μg/L 和 3.75 μg/L；HANs 浓度中位值为 2.34 μg/L，其中 DCAN 为 2.51 μg/L；TCNM 的中位值为 0.46 μg/L（ND～4.19 μg/L）[117]。日本 5 座城市的 9 份饮用水样中也检出了 MX，含量为 3～9 ng/L[118]。

自 20 世纪 80 年代以来，很多研究报道了我国城市供水中多种消毒副产物的浓度情况。同济医学院杨晓萍等 1983～1984 年调查了氯消毒过程中三氯甲烷的含量，三座水厂出水中 TCM 含量在 7.5～24 μg/L[119]。中国疾病预防与控制中心邓瑛等[120]调查发现大庆、北京、天津、郑州、长沙和深圳 6 个城市饮用水中 THMs 和 HAAs 的浓度分别为 ND～92.8 μg/L 和 ND～40 μg/L。中国科学院生态环境研究中心丁欢欢等[121]调查了中国 31 个重点城市 70 座水厂出水中 28 种消毒副产物的浓度水平和分布特征。结果表明，28 种消毒副产物中有 21 种被检出，其平均检出率为 50%。THM4 和 HAAs 是最主要的消毒副产物，其浓度中位值分别为 10.53 μg/L 和 10.95 μg/L；两种检出的 I-THMs（DCIM 和 BCIM）浓度范围为 ND～5.58 μg/L；HANs 的浓度范围为 ND～39.20 μg/L，中位值为 1.11 μg/L；4 种 HNMs 中只有 TCNM 和 TCNM 被检出，其最高浓度分别为 0.96 μg/L 和 0.28 μg/L。清华大学贝尔等[122]调查了我国 23 省份 44 座城市 164 个水样中 9 种亚硝胺类 NMAs 消毒副产物的含量，发现 NDMA 是最主要的 NMAs 组分，在水厂出水和用户龙头水中平均浓度分别为 11 ng/L 和 13 ng/L。而加利福尼亚对 NDMA 的控制标准为 10 ng/L，调查结果中约 26%的水厂出水和约 29%的龙头水超过该值。NDMA 在所检测水样的检出率为 37%，是美国的 3.6 倍。王成坤对北方某自来水厂进行了一年的连续月度检测，发现水厂出水中亚硝胺浓度由于水源切换而呈现明显的季节性变化[123]。当水源为河水时出水中 NDMA、N-亚硝基吗啉（NMOR）和 N-亚硝基吡咯烷（NPYR）浓度分别为 6.9 ng/L、3.3 ng/L 和 3.1 ng/L，而当采用另一水源时，出水中 NDMA、N-亚硝基甲基乙胺（NMEA）和 NPYR 分别为 10.1 ng/L、4.9 ng/L 和 4.7 ng/L。东北师范大学 Yang 等调查了以长江和黄河为水源的 5 座水厂中 HAcAms 的含量，调查结果显示 HAcAms 的浓度顺序为 Cl-HAcAms（15.3 μg/L）≫ Cl-Br-HAcAms（1.2 μg/L）＞ Br-HAcAms（1.0 μg/L），DCAcAm 的浓度最高（10.7 μg/L）[124]。

6 消毒副产物的毒理学特征

随着毒理学研究的逐渐深入及毒性测试技术的不断进步，研究者们不断揭示着消毒副产物对人体和生态环境的危害，消毒副产物所引起的健康问题也越来越受到重视。特别是近十年来，关于消毒副产物的细胞毒性、遗传毒性、三致效应、内分泌干扰效应的研究报道越来越多。

6.1 细胞毒性

由于操作简单，细胞毒性测试被广泛用于评价消毒副产物的毒性，反映在消毒副产物作用下导致细胞膜溶解破裂或细胞活性下降的程度，这些变化往往可通过对细胞进行染色或荧光信号的变化等予以观察。

李杏放等[125]利用人膀胱癌细胞株 T24 检测了饮用水中 4 种新型卤代苯醌类消毒副产物的细胞毒性，发现 2,6-二氯-1,4-苯醌、2,6-二氯-3-甲基-1,4-苯醌、2,3,6-三氯-1,4-苯醌、2,6-二溴苯醌的半数抑制浓度 IC_{50} 值分别为 95、110、151、142 μmol/L，并且发现卤代苯醌类化合物诱导细胞产生活性氧造成细胞损伤。

Muellner 等[126]利用中华仓鼠卵巢癌细胞（CHO）测试了 7 种 HANs 的慢性细胞毒性，并将这些实验数据与已报道的 CHO 毒性结果进行对比，发现碘代和溴代卤乙腈是其中慢性细胞毒性最高的消毒副产物，且 HNMs 和 HANs 的遗传毒性相当。

Zhou 等[127]用溴化噻唑蓝四氮唑 MTT 法检测消毒副产物吩嗪的细胞毒性，结果表明吩嗪对 T24 细胞和 HepG2 细胞的 IC_{50} 分别为 0.50 mmol/L 和 2.04 mmol/L。Du 等用实时细胞电子分析系统检测卤代苯醌的细胞毒性，检测结果显示 T24 细胞对二氯苯醌的 72 h-IC_{50} 值为 3.1 mmol/L[128]。细胞膜的完整性是反映消毒副产物细胞毒性的重要指标，当细胞释放乳酸脱氢酶时，说明细胞膜的完整性受损。当小鼠脑神经瘤细胞 Neuro-2a 暴露 CAA 时，可检测到胞外乳酸脱氢酶浓度急剧增加，Neuro-2a 细胞的半致死剂量为 1.5 mmol/L，该结果表明一定浓度的 CAA 可导致神经细胞膜损伤[129]。一般情况下，碘代消毒副产物的细胞毒性高于相应溴代副产物，更高于氯代副产物，Plewa 等用 CHO 细胞研究了 HAAs 的 72 h 慢性细胞毒性，并测试了当细胞密度降低 50%时卤乙酸的浓度（%C1/2），研究结果显示，HAAs 的细胞毒性顺序为：BAA > DBAA > CAA > TBAA > DCAA > TCAA，其中 BAA 和 TCAA 的%C1/2 分别为 8.9×10^{-6} 和 1.752×10^{-2} mol/L[130]。Richardson 等用 CHO 细胞检测了 13 种碘代消毒副产物的细胞毒性，其毒性顺序为碘乙酸 >（E）-3-溴-2-碘-丙烯酸 > 三碘甲烷 >（E）-3-溴-3-碘-丙烯酸 >（Z）-3-溴-3 碘丙烯酸 > 二碘乙酸 > 一溴一碘乙酸 >（E）-2-碘-3-甲基丁烯二酸 > 一溴二碘甲烷 > 二溴碘甲烷 > 一溴一氯碘甲烷 ≈ 一氯二碘甲烷 > 二氯碘甲烷，其中碘乙酸和二氯碘甲烷的%C1/2 分别为 2.95×10^{-6} mol/L 和 4.13×10^{-3} mol/L[131]。

HNMs 是饮用水消毒过程中一种常见的消毒副产物，已经被 USEPA 认定为优先控制污染物。Plewa 等用 CHO 细胞研究了 HNMs 的 72 h 慢性细胞毒性，并测试了当细胞密度降低 50%时的浓度（%C1/2），结果显示 HNMs 的细胞毒性顺序为：二溴硝基甲烷 > 二溴一氯硝基甲烷 > 一溴硝基甲烷 > 三溴硝基甲烷 > 一溴二氯硝基甲烷 > 一溴一氯硝基甲烷 > 二氯硝基甲烷 > 一氯硝基甲烷 > 三氯硝基甲烷。其中二溴硝基甲烷和三氯硝基甲烷的%C1/2 分别为 6.09×10^{-6} mol/L 和 5.36×10^{-4} mol/L，溴代硝基甲烷的细胞毒性高于氯代硝基甲烷的细胞毒性[88]。

6.2 遗传毒性

遗传毒性是污染物作用于有机体，导致其遗传物质在染色体水平、分子水平和碱基水平上受到各种损伤，从而造成的毒性作用。由于基因毒性与致癌性密切相关，遗传毒性实验也广泛用于评价消毒副产物的毒性特征。常见的遗传毒性试验有反映DNA损伤/修复的彗星实验和SOS/umu实验，反映基因突变的Ames实验以及表征染色体损伤的微核试验等。

Ragazzo等[132]利用 *Salmonella typhimurium* 细胞的Ames实验结合人体肝脏肿瘤细胞株HepG2的微核试验和彗星实验检测了新型消毒剂过氧甲酸消毒处理后消毒副产物的遗传毒性，并用洋葱（*Allium cepa*）根尖细胞检验消毒出水的致突变性，结果发现过氧甲酸消毒不会导致毒性上升。国际癌症研究机构（IARC）在人类致癌风险评估工作会议中指出氯二溴甲烷可引起染色体畸变以及姊妹染色单体交换，溴仿可引起姊妹染色单体交换以及微核的形成[133]。HepG2细胞的彗星实验结果显示，HAAs对细胞DNA的损伤能力顺序为 I-HAAs > Br-HAAs > Cl-HAAs，其最低可观测效应浓度分别为 0.01 μmol/L，0.1 μmol/L 和 100 μmol/L[134]。

Plewa等测试了卤乙酸对CHO细胞DNA的损伤能力，结果显示卤乙酸对CHO的DNA损伤能力顺序为：一溴乙酸 > 一氯乙酸 > 二溴乙酸 > 三溴乙酸，二氯乙酸和三氯乙酸对DNA的损伤程度与对照组无显著差异，其中一溴乙酸和三溴乙酸的单细胞凝胶电泳遗传毒性潜力分别为：1.7×10^{-6} mol/L 和 2.456×10^{-3} mol/L[130]。Richardson检测了13种碘代消毒副产物的遗传毒性，发现其中7种消毒副产物表现出遗传毒性，毒性顺序为碘乙酸 > 二碘乙酸 > 一氯二碘甲烷 > 一溴碘乙酸 >（*E*）-2-碘-3-甲基丁二酸 >（*E*）-3-溴-3-碘-丙烯酸 >（*E*）-3-溴-2-碘-丙烯酸。其中碘乙酸和（*E*）-3-溴-2-碘-丙烯酸的单细胞凝胶电泳遗传毒性潜力分别为 8.70×10^{-6} mol/L 和 7.58×10^{-3} mol/L[131]。

Wang等的研究显示[135]，当水体中含有较低浓度的 NH_3-N 时，消毒后遗传毒性下降，而当水体中 NH_3-N 浓度较高时，消毒后遗传毒性升高，揭示遗传毒性与N-DBPs有关。Plewa等测试了HNMs对CHO细胞DNA的损伤能力，发现HNMs对DNA的损伤能力顺序为：二溴硝基甲烷 > 一溴二氯硝基甲烷 > 三溴硝基甲烷 > 三氯硝基甲烷 > 一溴硝基甲烷 > 二溴一氯硝基甲烷 > 一溴一氯硝基甲烷 > 二氯硝基甲烷 > 一氯甲烷。其中二溴硝基甲烷和二氯硝基甲烷的单细胞凝胶电泳遗传毒性潜力分别为 2.62×10^{-5} mol/L 和 2.15×10^{-3} mol/L。溴代硝基甲烷的遗传毒性高于氯代硝基甲烷的遗传毒性[136]。

6.3 其他毒性

近年来消毒副产物所引起的内分泌干扰性、发育毒性、生殖毒性、神经毒性等负面生物效应也越来越受到关注。在化学品分类和毒性预测方面，定量结构效应关系（QSAR）发挥重要作用。张相如等[137]利用 *dumerilii* 沙蚕测试了新型芳香族消毒副产物的发育毒性，并通过QSAR拟合出消毒副产物的 log$P$（辛醇-水分配系数）、E_{LUMO}-E_{HOMO}（亲电或亲核性）和 pKa（离子化程度）等参数与其毒性结果EC50之间的相关关系，从实验和理论上证实了芳香族消毒副产物的发育毒性比脂肪族消毒副产物更强。

Narotsky等[138]将妊娠期的F344大鼠分别暴露4种THMs的混合物THM4（TCM、BDCM、CDBM、TNM）、5种HAAs的混合物HAA5（CAA、DCAA、TCAA、BAA、DBAA）及这9种消毒副产物的混合物DBP9，结果表明这3组混合物在暴露剂量 ≥ 613 μmol/kg/d 时都可能导致F344大鼠妊娠丢失，存活的幼崽在HAA5暴露剂量 ≥ 308 μmol/kg/d 时会发生眼部发育畸形。也有报道显示DCAA和DBAA对人体和实验动物具有神经毒性[139-140]。DBAA可导致F344大鼠液体摄入量减少（从17～21 mL/d下降至13.6 mL/d）、体重下降（从163.7～269.4 g下降至151.0～246.4 g）、腹泻、四肢无力、肌张力减退、感觉运动神经衰退、毛发脱落等症状，解剖观察大鼠神经系统，发现暴露组大鼠的脊髓神经纤维发生退化，脊髓灰质和白质发生了一定程度的细胞液泡化[139]。任洪强等[141]利用代谢组学结合组织病理学和氧化应激分析等手段，研究发现 BANs 和 HAcAms 会对小鼠的氨基酸、能量和脂质代谢造成干扰。

6.4 流行病学调查

自 1974 年从氯消毒饮水中检出 THMs 后，世界各地相继开展了大量流行病学研究。1991 年，国际癌症研究机构（IARC）的报告指出，直肠癌、结肠癌和膀胱癌与居民饮用氯化消毒的水有显著相关性，胃、脑、肺和肝等癌症死亡率也与氯消毒副产物有关。一项针对美国新英格兰地区居民经不同途径（游泳、淋浴、饮用）暴露 THMs 与膀胱癌发病率的流行病学调查结果表明，膀胱癌发病率与 TTHMs（日均饮用摄入低暴露量组：TTHMs, OR=1.53, 95%CI: 1.01, 2.32, p-trend=0.16；积累饮用摄入暴露组：OR=1.45, 95%CI: 0.95, 2.2, p-trend=0：13；淋浴高暴露组：OR=1.43, 95% CI: 0.80, 2.42, p-trend=0.10）和溴代三卤甲烷（日均饮用摄入低暴露量组：OR=1.98, 95%CI: 1.19, 3.29, p-trend=0.03；积累饮用摄入暴露组：OR=1.78, 95%CI: 1.05, 3.00, p-trend=0.02）的含量之间存在显著的相关关系，但与游泳时间（OR=0.94, 95%CI: 0.55, 1.59, p-rend=0.09）几乎没有关系[142]。Chowdhury 等[143]利用伤残调整生命年（(DALYs）方法评估了沙特阿拉伯某脱盐水厂中 THMs 的致癌风险，发现其平均 DALY 为每年 25.1。

华中科技大学公共卫生学院长期开展消毒副产物的生殖毒性研究。程英惠等[144]探讨了孕期血中氯化消毒副产物 THMs 浓度与胎儿出生体重之间的关系。研究发现孕期全血中 TBM、Br-THMs 和 TTHMs 浓度升高与新生儿出生体重降低存在剂量-效应关系（趋势性 p 值分别为 0.089、0.079 和 0.067）。与 TBM（<1.67 ng/L）、Br-THMs（<3.20 ng/L）和 TTHMs（<10.42 ng/L）低暴露组的孕妇相比，TBM（>3.43 ng/L）、Br-THMs（>5.74 ng/L）和 TTHMs（>17.80 ng/L）高暴露组孕妇所生新生儿出生体重分别平均降低了 87.60 g（95% CI: –175.33～0.12）、78.21 g（95% CI: –165.84～9.42）和 74.74 g（95%CI: –162.76～13.28）。曾强等在武汉同济医院调查了 401 名男性血样中 THMs 的含量与精子质量参数和血清中总睾酮浓度的关系，结果显示血样中 BDCM（β=–0.13 million；95% CI: –0.22, –0.03）和 DBCM（β=–4.74%；95% CI: –8.07, –1.42）分别与精子数量下降和精子直线运动能力下降均有密切关系。TTHMs 总浓度与精子浓度的下降（p=0.07）存在明显的剂量-效应关系，另外，DBCM 与血清总睾酮浓度的下降（p=0.07）存在显著的剂量-效应关系[145]。

7 消毒副产物的控制及削减

就目前的水消毒技术而言，各种消毒处理中生成消毒副产物已经成为不可避免的事实，如何有效控制、削减消毒副产物的生成，降低人体摄入浓度成为一项亟待解决的问题。从水处理的工艺流程分析，控制及削减消毒副产物的生成可能有以下三条途径：削减消毒副产物的前体物、优化消毒方式、去除消毒副产物[146]。其中消毒处理之前削减消毒副产物的前体物是最有效的方法[147]。

7.1 消毒副产物前体物的去除

目前为止，已经报道过很多去除消毒副产物前体物的方法，如混凝[148-149]、吸附[150-151]、膜过滤[152-153]、磁性离子交换[154-156]、预氧化[157-160]等。

由于消毒副产物前体物的物理化学性质差异大,决定了前处理及去除方法的差异。如混凝可以去除水中分子量大、疏水性强、含芳香环的有机污染物,而对低分子量亲水性有机物去除效果不佳[161-162]。当用混凝去除水体中疏水性酸组分后,THMs 的生成量从 59 μg/L 降低到 39 μg/L,而对于含低分子量前体物的水体(MW<5kDa),混凝对 THMs 的生成量没有显著影响。

活性炭吸附是一种去除污染物的有效方法,由于活性炭具有比表面积大、吸附容量大等诸多优点,因此常作为水处理单元使用。将颗粒活性炭吸附(GAC)与消毒处理结合有利于减少消毒副产物的生成,研究发现若将 GAC 置于消毒之前,THM4、HAA5、HAA9 和 TOX 的去除率分别降低了 29.6%、30.7%、31.2%和 37.6%;若将 GAC 置于消毒之后,THM4、HAA5、HAA9 和 TOX 的去除率分别降低了 96.9%、78.2%、88.6%和 63.2%[163]。由于粉状活性炭(PAC)的比表面积更大,常用于水处理的不同阶段。研究发现与单独的混凝相比,PAC 对 THMs 前体物的去除率从 10.5%提高到 70%,对 N-DBPs 前体物的去除率从 45%提高到 93%[164]。近年来,石墨烯引起了很多研究者的关注,Liu 等在消毒前用磁性石墨烯氧化物(MGO)作为吸附剂处理 4 种水体,发现经 MGO 处理的水,消毒后多种消毒副产物的生成量降低了 7%~98%[165],说明 MGO 对消毒副产物前体物的去除有良好的效果。

预氧化技术在水处理中常用来降解水中的还原性和大分子污染物,常用的预氧化方法有高铁酸盐、臭氧、过硫酸盐、高锰酸盐预氧化以及高级氧化技术(如:UV/H_2O_2 和 UV/过硫酸盐)。研究发现高铁酸盐预氧化可以显著降低 I-THMs 的生成,当预氧化过程中高铁酸盐浓度大于 2 mg/L 时,未检出 I-THMs 的生成[166]。通常情况下,预氧化技术,如臭氧预氧化,可以降低 THMs 的生成潜力[167],但是由于臭氧是一种强氧化剂,可以将有机大分子(20~1000 kDa)转化为小分子(0.3~10 kDa),消毒后副产物的种类增多。

臭氧-(生物)活性炭深度处理工艺可以有效去除 THMs、HAAs 和 NMAs 的前体物,对抗生素等新型污染物也有较好地去除效果,在我国得到了大规模的应用[168-169]。该工艺去除 NMAs 的机理包括:臭氧氧化可以将部分非极性前体物破坏或者转化为极性前体物;活性炭可以吸附 NMAs 前体物的非极性端;微生物则利用自身表面的负电荷吸附 NMAs 前体物中带正电荷的二烷基胺官能团,随后缓慢降解[170-171]。研究证实,二甲胺等 NMAs 前体物可被微生物作为碳源和氮源利用[172]。

Ersan 等[173]研究了纳滤对降低消毒副产物生成的影响,发现纳滤对 NDMA、HNMs、THMs 前体物的去除效率可以达 57%~83%,48%~87%,72%~97%。构建人工湿地可以加强 COD_{Mn},NH_4^+-N,TN,DOC,UV_{254},THMs 和 HAAs 的去除。磁性离子交换(MIEX)是一种去除卤代消毒副产物前体物的有效方法,可以去除带负电荷的 DOC,尤其是低分子亲水性有机物[174]。清华大学陈超课题组首次开发了阳离子交换去除 NMAs 前体物的技术[175]。对典型前体物的研究结果表明,Lagergren 准一级动力学模型和准二级动力学模型对前体物离子交换过程模拟很好。在离子交换柱(床)模式下,流速、硬度离子对交换效果具有显著影响,5 m/h 流速下穿透时交换容量利用达 70%以上,进水中 NMAs 前体物基本全部去除。

由于水体中污染物成分复杂,单项水处理技术难以完全去除消毒副产物前体物,联合使用多项技术将更大程度地去除消毒副产物前体物。由于 MIEX 对不同官能团的选择性不同[161, 176-180],其可去除的污染物类型与混凝相比显著不同,因此将二者联合可更好去除 NOM,降低消毒副产物的生成潜力[181]。膜过滤技术在去除天然有机物和人工合成化学品方面前景广阔,而膜过滤具有易污染和去除低分子量组分效率低的缺点,结合 PAC 和树脂预处理,可以达到防止膜污染、延长膜使用寿命的目的[182]。将 GAC 与氯化铁(吸附-混凝)结合可用于处理反渗透浓缩液,对 DOC、UV_{254} 和 TDS 的去除率分别从 16.9%、18.9%和 39.7%升高到 91.8%、96%和 76.5%[183]。Metcalfe 等[184]将悬浮离子交换、凝固、陶瓷膜过滤结合处理废水,与传统的砂滤相比,DOC、UVA、THMFP、HAAFP、Br-DBPs 的去除率分别提高了 50%、62%、62%、62%和 47%。高锰酸钾及活性炭吸附可以通过改变消毒副产物的前体物,降低某些 C-DBPs、N-DBPs

的生成潜力[66, 146]。楚文海等[185]发现将高锰酸钾氧化与活性炭吸附相结合（PM-PAC）能显著增强 DOC、DON、NH_3-N 及藻类的去除。将臭氧预氧化与 GAC 结合，可有效去除 NMAs 前体物，可将 NDMA 的生成潜力从 194 ng/L 降低到 94 ng/L[122]。

7.2 优化消毒方法

由于消毒方式和消毒条件在很大程度上影响消毒副产物的生成，因此可以通过优化消毒方式、消毒剂用量、接触时间等方式减少消毒副产物的生成。

尽管在过去的几十年中，氯是最常用的消毒剂，但由于氯消毒过程中可能生成多种消毒副产物，促使研究者寻找其他替代消毒剂，如二氧化氯、氯胺、臭氧和高锰酸盐等[186, 187]或者采用组合消毒的方法。游离氯和氯胺是常用的两种氯消毒剂，二者各有优劣。游离氯灭菌效果好，但生成卤代消毒副产物的量比氯胺高得多；氯胺的氧化性比游离氯低，效果持久，生成卤代消毒副产物较少，但生成亚硝胺、碘代副产物的量则比游离氯高。在对乙酰氨基酚的氯胺消毒处理中 TCM、DCAN、DCAcAm 和 TCAcAm 的产率分别为 0.004%、0.005%、0.39%和 0.035%，低于氯消毒的产率 2.25%、0.54%、0.63%、0.2%[188]。但也有研究表明氯胺消毒过程中将生成更多的 HAcAms。在含有 NOM 的二级出水消毒过程中，氯胺消毒和氯消毒过程中产生的 HAcAms 分别为 88.5 nmol/L 和<30 nmol/L[189]。另外，氯胺消毒的主要消毒剂是一氯氨，由于一氯氨分子中含有氮元素，为 N-DBPs 的生成提供了可能的氮源。另外，Wang 等报道了氯胺消毒会促进 I-DBPs 的生成[190]，氯胺消毒的水处理厂中 IAA 和 IF（分别为 15 μg/L 和 11 μg/L）的生成量均高于氯消毒处理（分别为 2 μg/L 和 ND）中的生成量[191]。

清华大学张晓健、陈超等研究开发了一种新型顺序氯化消毒工艺——"短时游离氯后氯胺顺序消毒工艺"（以下简称"顺序氯化"）[192]，可以安全经济地实现细菌灭活、消毒副产物控制和管网生物稳定性等多方面的要求。顺序氯化与单独氯、氯胺消毒相比，对多种指示微生物的灭活效果更佳，显示该工艺中游离氯与氯胺具有协同消毒作用[193]。近期的研究表明，该工艺也能很好地实现 THMs、HAAs 和 NMAs 的综合控制[194]。

消毒剂的用量也会影响消毒副产物的生成，降低氯的投加量将减少消毒副产物的生成。折点加氯时，折点之前氯的投加量与 THMs 和 HAAs 的生成量没有明显的线性关系，折点之后，随着氯的投加量增加，THMs 和 HAAs 的生成量急剧升高[195, 196]。然而，消毒剂投加量的减少是有限的，因为它必须满足杀灭病原体和保持余氯水平的基本要求。近年来有研究者提出了多步加氯的消毒方式，在水处理过程中，与一步加氯相比，在总氯投加量相同的情况下，两步加氯和三步加氯使消毒副产物的生成量分别降低了 16.7%和 23.4%[197, 198]。

除了投加化学消毒剂，物理消毒或者物理化学消毒引起了越来越多的关注。Huo 等研制了一种氧化铜纳米线（CuONW）改性的三维铜泡沫电极，并在 1 V 工作电压下获得了优良的消毒性能[6]。在处理过程的 7s 内，能耗为 25 J/L，由于该方法投加的氯浓度较低，消毒副产物的形成潜力降低。当超声与 ClO_2 结合消毒时，ClO_2^- 和 ClO_3^- 的生成量分别从 1.37 mg/L 和 0.17 mg/L 升高到 4.71 mg/L 和 2.57 mg/L[187]。

UV 消毒具有不引入化学物质的优点，与氯或氯胺联合消毒时既可达到杀灭微生物又可达到持续消毒的目的，常用的紫外消毒剂量为 40 mJ/cm²。但是，高剂量紫外处理（500～1000 mJ/cm²）则会激活水中的多种离子，导致消毒副产物的增加。Liu 等发现 UV 与氯联合消毒时生成的 TCM、DCAA 和 CNCl 分别是单独氯消毒时的 2.12、1.67 和 2.26 倍[11]。

7.3 消毒副产物的去除

对于在消毒过程中已经形成的消毒副产物，可以采用煮沸、超滤、纳滤、反渗透、高级氧化、吸附、离子交换、电渗析等多种化学、物理或生物方法去除。其中，饮用水末端用户可以通过加热的方式去除消毒副产物，减少人体暴露。煮沸是一种有效的"去毒"过程，可以去除水中的一些消毒副产物，减少人体对自来水中消毒副产物的接触。研究发现自来水煮沸 3 min 后，THMs 的去除率超过 92.3%[103]。煮沸 5 min 后，卤代消毒副产物的总含量降低了 62.3%，其中 Br-DBPs 和 Cl-DBPs 的去除率分别为 62.8%和 61.1%[200]。

膜过滤可有效截留部分污染物。反渗透、超滤和纳滤已广泛应用于水中痕量消毒副产物的去除。由于尺寸排斥和电荷排斥的综合效应，反渗透/纳滤可以截留超过 90%的 HAAs[201]。但存在的主要问题是该处理能耗高，且当水中污染物浓度过高时，限制了膜在去除消毒副产物过程中的大量使用。

由于大多数消毒副产物都是卤代有机化合物，通过氧化还原反应脱去消毒副产物中的卤素原子是去除消毒副产物的可能途径。亚硫酸盐/UV$_{254}$ 工艺可以在 15 min 内通过脱氯去除 100%的一氯乙酸，而紫外线照射几乎不能降解一氯乙酸[202]。一些双金属催化剂（铜/铁、钯/铁）可在 60 min 内通过还原脱卤去除 TBM 和 TBAA[203]。在 UV$_{254}$-TiO$_2$ 催化体系中，在甲酸盐（FM）和溶解氧的存在下，TCAA 可快速通过羧基阴离子自由基（CO$_2^-$）机制发生脱氯去除[204]。

吸附法是一种既可用于去除消毒副产物前体物，又可直接去除消毒副产物的方法。近年已研发出多种新型吸附材料（例如纳米复合材料），能够有效去除消毒副产物。由于零价铁具有表面积大、表面活性高、价格低廉等优点，已广泛用于去除水中的氯代烃、硝基苯、氯代酚、多氯联苯、重金属和阴离子等多种污染物[205-207]。有研究者开发了零价铁/活性炭纳米复合材料（NZVI/AC），可协同去除 90%的 THMs。由于石墨烯具有巨大的表面积，被认为是装载无机纳米颗粒的理想支撑材料[208]。石墨烯/Fe^0 纳米复合材料（G-NZVI）可去除几乎 100%的 TCNM[209]。氯-强碱型阴离子交换树脂（D201-Cl）是一种价廉、高效的饮用水溴酸盐脱除剂[210]。石英砂也是一种用于去除饮用水中有机污染物的低成本材料[211]。将石英砂和超声技术结合，可有效去除饮用水中 12 种常规消毒副产物，其去除效率超过 20%[212]。

优化消毒技术，去除消毒副产物的前体物，提高消毒效率和去除已生成的消毒副产物都是降低饮用水中消毒副产物含量的有效措施。此外，保护水源不受污染也同样重要。据报道，叶片有机质和藻类有机质分别是 C-DBPs 和 N-DBPs 的潜在前体物[213]。研究表明，在消毒处理过程中，沉积物可将大量 NOM 释放到水中，并进一步转化为 THMs 和 HAAs[214]。另外，一些无机离子（溴化物、碘化物和亚硝酸盐）是 Br-DBPs、I-DBPs 和 N-DBPs 的前体物，这些无机离子的引入将促进消毒副产物的生成[201, 215-216]。氯胺消毒过程中，HNMs 的生成量与亚硝酸盐和溴化物的含量呈正相关[215]，THMs 和 HANs 的生成量与溴化物的含量呈正相关[216]。此外，在 BP-4 氯化过程中，碘离子的存在增加了 I-DBPs 的生成[217]。

8　困难及展望

截止目前文献中报道的消毒副产物虽然已经超过 700 多种，但饮用水氯化消毒过程中产生的总有机卤素（TOX）还有超过 50%无法确定结构。同样，在饮用水臭氧氧化过程中形成的生物可同化有机碳（AOC）也有超过 50%没有确认结构。如果没有先进的检测仪器和优化的方法，很难识别未知消毒副产物的分子结构。因此，第一，今后饮用水消毒及消毒副产物研究的首要任务是开发有效的样品预处理方法，以及灵敏、快速的消毒副产物检测分析方法；第二，随着人类活动的加剧，水体污染程度日趋严重，水源水中的成分十分复杂，已经发现了数以万计的天然和人工合成化学品，这些都有可能是消毒副产物的潜在前体物。揭示消毒过程中各种前体物生成消毒副产物的转化机制已成为一项艰巨而重要的工作；第三，生物体对消毒副产物的内暴露、外暴露以及各种不良毒理效应更是复杂，需要通过长期、系统的毒理学研究，甚至通过流行病学调查进行揭示；第四，国内外对消毒副产物的控制和管理对策仍然薄弱，应大力发展和推广切实有效的消毒副产物和前体物降解技术，尽快提出有效的评价和管理措施框架；第五，消毒副产物的毒性效应是公众最关注的问题，因此毒性效应引导法可以将所有与消毒副产物相关的研究结合起来，包括高风险消毒副产物的识别、生成机制、消除技术的选择和评估管理。毒性引导的消毒副产物研究将是一种创新和有效的研究方法[101]。总之，消毒和消毒副产物的研究是多学科的，涉及化学、毒理学、流行病学和工程技术等。为预防介水传染疾病的爆发、削减消毒副产物的生成、降低消毒副产物的潜在健康风险，不同学科的科学家和工程师之间需要建立更加密切的合作。

参 考 文 献

[1] Ohanian E, Mullin C, Orme J. Health effects of disinfectants and disinfection by-products: a regulatory perspective. Water Chlorination: Chem. Environ. Impact Health Effects, 1990, 6: 75-86.

[2] Richardson S. Drinking water disinfection by-products. Encyclopedia Environ. Anal. Remed, 1998, 3: 1398-1421.

[3] International programme on chemical safety disinfectants and disinfectant by-products. Environmental Health Criteria 216. World Health Organization, Geneva, 2000.

[4] Huang H, Chen B, Zhu Z. Formation and speciation of haloacetamides and haloacetonitriles for chlorination, chloramination, and chlorination followed by chloramination. Chemosphere, 2017, 166: 126-134.

[5] 曾抗美, 史建福, 刘桂华. 电化学法进行饮用水消毒研究. 中国给排水, 1999, 15: 16-18

[6] Huo Z, Xie X, Yu T, et al. Nanowire- modified three-dimensional electrode enabling low-voltage electroporation for water disinfection. Environ. Sci. Technol., 2016, 50(14): 7641-7649.

[7] Tian C, Liu R, Guo T, et al. Chlorination and chloramination of high-bromide natural water: DBPs species transformation. Sep. Purif. Technol., 2013, 102: 86-93.

[8] 陈国青, 梁增辉. 氯和二氧化氯灭活 f2 噬菌体机理的初步探讨. 中国公共卫生, 1997, 13(6): 356-357.

[9] 陈国青, 梁增辉. 氯和二氧化氯灭活 f2 噬菌体的效果观察. 中国公共卫生, 1998, 14(7): 431-432.

[10] Guo Z, Lin Y, Xu B, et al. Factor affecting THM, HAN and HNM formation during UV-chlor (am) ination of drinking water. Chem. Eng. J., 2016, 306: 1180-1188.

[11] Liu W, Cheung L, Yang X, et al. THM, HAA and CNCl formation from UV irradiation and chlor (am) ination of selected organic waters. Water Res., 2006, 40: 2033-2043.

[12] Dahlén J, Bertilsson S, Pettersson C. Effects of UV-A irradiation on dissolved organic matter in humic surface waters. Environ. Int., 1996, 22: 501-506.

[13] Frimmel F. Impact of light on the properties of aquatic natural organic matter. Environ. Int., 1998, 24: 559–571.

[14] 吕东明. 饮用水多屏障消毒策略及紫外消毒技术的应用. 净水技术, 2019, 38(1): 1-6.

[15] Chu W, Yao D, Gao N, et al. The enhanced removal of carbonaceous and nitrogenous disinfection by-product precursors using integrated permanganate oxidation and powdered activated carbon adsorption pretreatment. Chemosphere, 2015, 141: 1-6.

[16] Boorman G, DellarcoV, Dunnick J, et al. Drinking water disinfection byproducts: review and approach to toxicity evaluation. Environ. Health Perspect., 1999, 107(11): 207-217.

[17] Krasner S, McGuire M, Jacangelo J, et al. The occurrence of disinfection byproducts in U.S. drinking water. J. Am. Water Works Assoc., 1989, 81(8): 41-53.

[18] Bellar T, Lichtenberg J, Kroner R. The occurrence of organohalides in chlorinated drinking waters. J. Am. Water Works Ass., 1974, 66: 703-706.

[19] Rook J. Formation of haloforms during chlorination of natural waters, Water Treat. Exam., 1974, 23: 234-243.

[20] Boyce S, Hornig J. Reaction pathways of trihalomethane formation from the halogenation of dihydroxyaromatic model compounds for humic acid. Environ. Sci. Technol., 1983, 17(4): 202-211.

[21] Li C, Lin Q, Dong F. Formation of iodinated trihalomethanes during chlorination of amino acid in waters. Chemosphere, 2019, 217: 355-363.

[22] Meng Y, Wang Y, Han Q, et al. Trihalomethane (THM) formation from synergic disinfection of biologically treated municipal wastewater: effect of ultraviolet (UV) irradiation and titanium dioxide photocatalysis on dissolve organic matter fractions. Chem. Eng. J., 2016, 303: 252-260.

[23] Chu W, Gao N, Krasner S, et al. Formation of halogenated C-, N-DBPs from chlor(am)ination and UV irradiation of tyrosine in drinking water. Environ. Pollut., 2012, 161: 8-14.

[24] Gerrity D, Mayer B, Ryu H, et al. A comparison of pilot-scale photocatalysis and enhanced coagulation for disinfection byproduct mitigation. Water Res., 2009, 43: 1597-1610.

[25] Li C, Gao N, Chu W, et al. Comparison of THMs and HANs formation potential from the chlorination of free and combined histidine and glycine. Chem. Eng. J., 2017, 307: 487-495.

[26] Allard S, Tan J, Joll C, et al. Mechanistic study on the formation of Cl-/Br-/I-trihalomethanes during chlorination/chloramination combined with a theoretical cytotoxicity evaluation. Environ. Sci. Technol., 2015, 49: 11105-11114.

[27] Cancho B, Ventura F, Galceran M. Behavior of halogenated disinfection by-products in the water treatment plant of Barcelona, Spain. Bull. Environ. Contam. Toxicol., 1999, 63: 610-617.

[28] Golfinopoulos S, Nikolaou A. Formation of DBPs in the drinking water of Athens, Greece: a ten-year study. Global Nest J., 2005, 7: 106-118.

[29] Golfinopoulos S, Nikolaou A. Survey of disinfection by-products in drinking water in Athens, Greece. Desalination, 2005, 176: 13-24.

[30] Lebel G, Benoit F, Williams D. A one-year of halogenated disinfection by-products in the distribution system of treatment plants using three different disinfection processes. Chemosphere, 1997, 34: 2301-2317.

[31] Rodriguez M, Serodes J, Levallois P. Behaviour of trihalomethnes and haloacetic acids in a drinking water distribution system. Water Res., 2004, 38: 4367-4382.

[32] Rodriguez M, Serodes J, Levallois P, et al. Chlorinated disinfection by- products in drinking water according to source, treatment, season, and distribution location. J. Environ. Eng. Sci., 2007, 6: 355-365.

[33] Mercier-Shanks C, Serodes J, Rodriguez M. Spatio-temporal variability of non-regulated disinfection by-products within a drinking water distribution network. Water Res., 2013, 47: 3231-3243.

[34] Serrano M, Montesinos I, Cardador M, et al. Seasonal evaluation of the presence of 46 disinfection by-products throughout a drinking water treatment plant. Sci. Total Environ., 2015, 517: 246-258.

[35] Krasner S, Weinberg H, Richardson S, et al. Occurrence of a new generation of disinfection byproducts. Environ. Sci. Technol., 2006, 40: 7175-7185.

[36] Oliver B. Dihaloacetonitriles in drinking water: algae and fulvic acid as precursors. Environ. Sci. Technol., 1983, 17(2): 80-83.

[37] Trehy M, Yost R, Miles C. Chlorination byproducts of amino acids in natural waters. Environ. Sci. Technol., 1986, 20(11): 1117-1122.

[38] Zhang B, Xian Q, Gong T, et al. DBPs formation and genotoxicity during chlorination of pyrimidines and purines bases. Chem. Eng. J., 2017, 307: 884-890.

[39] Liu S, Li Z, Dong H, et al. Formation of iodo-trihalomethanes, iodo-acetic acids, and iodo-acetamides during chloramination of iodide-containing waters: factors influencing formation and reaction pathways. J. Hazard Mater., 2017, 321: 28-36.

[40] De Vera G, Stalter D, Gernjak W, et al. Towards reducing DBP formation potential of drinking water by favouring direct ozone over hydroxyl radical reactions during ozonation. Water Res., 2015, 87: 49-58.

[41] Krasner S, Lee C, Chinn R, et al. Bromine incorporation in regulated and emerging DBPs and the relative predominance of mono-, di-, and trihalogenated DBPs. In: Water quality technology conference and exposition, Cincinnati, OH, USA, 2008.

[42] Shah A, Mitch W. Halonitroalkanes, halonitriles, haloamides, and N-nitrosamines: a critical review of nitrogenous disinfection byproduct formation pathways. Environ. Sci. Technol., 2012, 46: 119-131.

[43] Chu W, Gao N, Yin D, et al. Trace determination of 13 haloacetamides in drinking water using liquid chromatography triple quadrupole mass spectrometry with atmospheric pressure chemical ionization. J. Chromatogr. A, 2012, 1235: 178-181.

[44] Liew D, Linge K, Joll C, et al. Determination of halonitromethanes and haloacetamides: an evaluation of sample preservation and analyte stability in drinking water. J. Chromatogr. A, 2012, 1241: 117-122.

[45] Chuang Y, McCurry D, Tung H, et al. Formation pathways and trade-offs between haloacetamides and haloacetaldehydes during combined chlorination and chloramination of lignin phenols and natural waters. Environ. Sci. Technol., 2015, 49: 14432-14440.

[46] Chu W, Li D, Gao N, et al. Comparison of free amino acids and short oligopeptides for the formation of trihalomethanes and haloacetonitriles during chlorination: effect of peptide bond and pre-oxidation. Chem. Eng. J., 2015, 281: 623-631.

[47] Kosaka K, Nakai T, Hishida Y, et al. Formation of 2, 6-dichloro-1, 4-benzoquinone from aromatic compounds after chlorination. Water Res., 2016, 110: 48-55.

[48] Huang H, Wu Q, Tang X, et al. Formation of haloacetonitriles and haloacetamides and their precursors during chlorination of secondary effluents. Chemosphere, 2016, 144: 297-303.

[49] Hu J, Song H, Addison J, et al. Halonitromethane formation potentials in drinking waters. Water Res., 2010, 44(1): 105-114.

[50] Lee W, Westerhoff, P, Croue J. Dissolved organic nitrogen as a precursor for chloroform, dichloroacetonitrile, N-nitrosodimethylamine, and trichloronitromethane. Environ. Sci. Technol., 2007, 41(15): 5485-5490.

[51] Dotson A, Westerhoff P, Krasner S. Nitrogen enriched dissolved organic matter (DOM) isolates and their affinity to form emerging disinfection byproducts. Water Sci. Technol., 2009, 60: 135-143.

[52] Hoigne J, Bader H. The formation of trichloronitromethane (chloropicrin) and chloroform in a combined ozonation/ chlorination treatment of drinking water. Water Res., 1988, 22(3): 313-319.

[53] Reckhow D, Linden K, Kim J, et al. Effect of UV treatment on DBP formation. J. Am. Water Works Assoc., 2010, 102(6): 100-113.

[54] Shah A, Dotson A, Linden K, et al. Impact of UV disinfection combined with chlorination/chloramination on the formation of halonitromethanes and haloacetonitriles in drinking water. Environ. Sci. Technol., 2011, 45(8): 3657-3664.

[55] Hong H, Qian L, Xiao Z, et al. Effect of nitrite on the formation of halonitromethanes during chlorination of organic matter from different origin. J. Hydrol., 2015, 531(3): 802-809.

[56] Choi J, Valentine R. Formation of N-nitrosodimethylamine (NDMA) from reaction of monochloramine: a new disinfection by-product. Water Res., 2002, 36: 817-824.

[57] Mitch W, Sharp J, Trussell R, et al. N-nitrosodimethylamine(NDMA)as a drinking water contaminant: a review. Environ. Eng. Sci., 2003, 20: 389-404.

[58] Bond T, Huang J, Templeton M, et al. Occurrence and control of nitrogenous disinfection by-products in drinking water – a review. Water Res., 2011, 45: 4341-4354.

[59] Mitch W, Sedlak D. Formation of N-nitrosodimethylamine (NDMA) from dimethylamine during chlorination. Environ. Sci. Technol., 2002, 36: 588-595.

[60] Schreiber I, Mitch W. Influence of the order of reagent addition on NDMA formation during chloramination. Environ. Sci.

Technol., 2005, 39: 3811-3818.

[61] Farre M, Reungoat J, Argaud F, et al. Fate of N-nitrosodimethylamine, trihalomethane and haloacetic acid precursors in tertiary treatment including biofiltration. Water Res., 2011, 45: 5695-5704.

[62] Schreiber I, Mitch W. Nitrosamine formation pathway revisited: the importance of chloramine speciation and dissolved oxygen. Environ. Sci. Technol., 2006, 40: 6007-6014.

[63] Gan W, Bond T, Yang X, et al. Role of chlorine dioxide in N-Nitrosodimethylamine formation from oxidation of model amines. Environ. Sci. Technol., 2015, 49(19): 11429-11437.

[64] Wilczak A, Assadi-Rad A, Lai H, et al. Formation of NDMA in chloraminated water coagulated with DADMAC cationic polymer. J. Am. Water Works Assoc., 2003, 95: 94-106.

[65] Charrois J, Hrudey S. Breakpoint chlorination and free-chlorine contact time: implications for drinking water N-nitrosodimethylamine concentrations. Water Res., 2007, 41: 674-682.

[66] Chen Z, Valentine R. The influence of the pre-oxidation of natural organic matter on the formation of N-nitrosodimethylamine (NDMA). Environ. Sci. Technol., 2008, 42: 5062-5067.

[67] Krasner S, Mofidi A, Liang S. DBP formation resulting from short-term contact with chlorine, followed by long-term contact with chloramines, Proceeding of Annual Conference of Am. Water Works Assoc., 2003, 1348-1365.

[68] Golfinopoulos S, Nikolaou A, Lekkas T. The occurrence of disinfection by-products in the drinking water of Athenes, Greece. Environ. Sci. Pollut. Res. Int., 2003, 10(6): 368-372.

[69] Park S, Padhye L, Wang P, et al. N-nitrosodimethylamine (NDMA) formation potential of amine-based water treatment polymers: effects of in situ chloramination, breakpoint chlorination, and pre-oxidation. J. Hazard Mater., 2015, 282: 133-140.

[70] Park K, Choi S, Lee S, et al. Comparison of formation of disinfection by-products by chlorination and ozonation of wastewater effluents and their toxicity to Daphnia magna. Environ. Pollut., 2016, 215: 314-321.

[71] Jeong C, Postigo C, Richardson S, et al. Occurrence and comparative toxicity of haloacetaldehyde disinfection byproducts in drinking water. Environ. Sci. Technol., 2015, 49: 13749-13759.

[72] Bull R, Reckhow D, Rotello V, et al. Use of Toxicological and Chemical Models to Prioritize DBP Research, AWWA Research Foundation, Denver, CO, US, 2006.

[73] Zhao Y, Qin F, Boyd J, et al. Characterization and determination of chloro-and bromo-benzoquinones as new chlorination disinfection byproducts in drinking water, Anal. Chem., 2010, 82: 4599-4605.

[74] Environmental Protection Agency (EPA Method 551.1). Determination of clorination disinfection byproducts, chlorinated solvents, and halogenated pesticides/herbicides in drinking water by liquid-liquid extraction and gas chrmatography with electron-capture detection. Cincinnati, USEPA, 1995.

[75] 刘赞明. LC-MS/MS 在消毒副产物分析中的应用. 北京: 北京化工大学硕士学位论文, 2013.

[76] Antoniou C, Koukouraki E, Diamadopoulos E. Determination of chlorinated volatile organic compounds in water and municipal wastewater using headspace-solid phase microextraction-gas chromatography. J. Chromatogr. A, 2006, 1132: 310-314.

[77] Vora-adisak N, Varanusupakul P. A simple supported liquid hollow fiber membrane microextraction for sample preparation of trihalomethanes in water samples. J. Chromatogr. A, 2006, 1121: 236-241.

[78] American Public Health Association (APHA Method 6251B). Standard methods for the examination of water and wastewater, 22th. Washingto DC, 2012.

[79] Environmental Protection Agency (EPA Method 552.2). Determination of haloacetic acids and dalapon in drinking water by liquid-liquid extraction, derivatization and gas chromatography with electron capture detection. Cincinnati, USEPA, 1995.

[80] Environmental Protection Agency (EPA Method 552.3). Determination of haloacetic acids and dalapon in drinking water by

liquid-liquid microextraction, derivatization, and gas chromatography with electron capture detection. Cincinnati, USEPA, 2003.

[81] Environmental Protection Agency (EPA Method 552.1). Determination of haloacetic acids and dalapon in drinking water by ion-exchange liquid-solid extraction and gas chromatography with an electron capture detector. Cincinnati, USEPA, 1992.

[82] 王海鸥, 陈忠林, 张学军. 液液萃取气相色谱法测定水中三卤甲烷. 哈尔滨商业大学学报: 自然科学版, 2011, 27(3): 297-300.

[83] Chen H, Yin J, Zhu M, et al. Cold on-column injection coupled with gas chromatography/mass spectrometry for determining halonitromethanes in drinking water. Anal. Meth., 2016, 8(2): 362-370.

[84] McDonald J, Harden N, Nghiem L, et al. Analysis of N-nitrosamines in water by isotope dilution gas chromatography-electron ionisation tandem mass spectrometry. Talanta, 2012, 99: 146-154.

[85] Zhang H, Ren S, Yu J, et al. Occurrence of selected aliphatic amines in source water of major cities in China. J. Environ. Sci., 2012, 24: 1885-1890.

[86] Le Roux J, Nihemaiti M, Croué J. The role of aromatic precursors in the formation of haloacetamides by chloramination of dissolved organic matter. Water Res., 2016, 88: 371-379.

[87] Zhang N, Liu C, Qi F, et al. The formation of haloacetamides, as an emerging class of N-DBPs, from chlor (am) ination of algal organic matter extracted from Microcystis aeruginosa, Scenedesmus quadricauda and Nitzschia palea. RSC Adv, 2017, 7: 7679-7687.

[88] Plewa M, Wagner E, Richardson S, et al. Halonitromethane drinking water disinfection byproducts: chemical characterization and mammalian cell cytotoxicity and genotoxicity. Environ. Sci. Technol., 2004, 38: 4713-4722.

[89] Nihemaiti M, Le Roux J, Hoppe-Jones C, et al. Formation of haloacetonitriles, haloacetamides, and nitrogenous heterocyclic byproducts by chloramination of phenolic compounds. Environ. Sci. Technol., 2017, 51: 655-663.

[90] Pereira R, De Alda M, Joglar J, et al. Identification of new ozonation disinfection byproducts of 17β-estradiol and estrone in water. Chemosphere, 2011, 84: 1535-1541.

[91] Xiao M, Wei D, Yin J, et al. Transformation mechanism of benzophenone-4 in free chlorine promoted chlorination disinfection. Water Res., 2013, 47: 6223-6233.

[92] Pan Y, Zhang X, Li Y. Identification, toxicity and control of iodinated disinfection byproducts in cooking with simulated chlor(am)inated tap water and iodized table salt. Water Res., 2016, 88: 60-68.

[93] Wang A, Lin Y, Xu B, et al. Degradation of acrylamide during chlorination as a precursor of haloacetonitriles and haloacetamides. Sci Total Environ., 2018, 615: 38-46.

[94] Chai Q, Zhang S, Wang X, et al. Effect of bromide on the transformation and genotoxicity of octyl-dimethyl-p-aminobenzoic acid during chlorination. J. Hazard. Mater., 2017, 324: 626-633.

[95] Pan Y, Li W, An H, et al. Formation and occurrence of new polar iodinated disinfection byproducts in drinking water. Chemosphere, 2016, 144: 2312-2320.

[96] Li C, Wang D, Li N, et al. Identifying unknown by-products in drinking water using comprehensive two-dimensional gas chromatography-quadrupole mass spectrometry and in silico toxicity assessment. Chemosphere, 2016, 163: 535-543.

[97] Wang X, Wang J, Zhang Y, et al. Characterization of unknown iodinated disinfection byproducts during chlorination/chloramination using ultra-high resolution mass spectrometry. Sci. Total Environ., 2016, 554-555: 83-88.

[98] Zhang H, Zhang Y, Shi Q, et al. Study on transformation of natural organic matter in source water during chlorination and its chlorinated products using ultrahigh resolution mass spectrometry. Environ. Sci. Technol., 2012, 46: 4396-4402.

[99] Zhang H, Zhang Y, Shi Q, et al. Characterization of low molecular weight dissolved natural organic matter along the treatment trait of a water-works using Fourier transform ion cyclotron resonance mass spectrometry. Water Res., 2012, 46:

5197-5204.

[100] Houtman C, van Oostveen A, Brouwer A, et al. Identification of estrogenic compounds in fish bile using bioassay-directed fractionation. Environ. Sci. Technol., 2004, 38: 6415-6423.

[101] 陈妙, 魏东斌, 杜宇国. 毒性效应引导的高风险消毒副产物识别方法. 中国科学: 化学, 2018, 10: 1207-1216.

[102] Hu J, Cheng S, Aizawa T, et al. Products of aqueous chlorination of 17β-estradiol and their estrogenic activities. Environ Sci Technol., 2003, 37: 5665-5670.

[103] Li L, Wei D, Wei G, et al. Transformation of cefazolin during chlorination process: products, mechanism and genotoxicity assessment. J. Hazard. Mater., 2013, 262: 48-54.

[104] Leeheer J. Comprehensive approach to reparative isolation and fractionation of dissolved organic carbon from natural waters and wastewaters. Environ. Sci. Technol., 1981, 15(5): 578-587.

[105] Chen C, Zhang X, Zhu L, et al. Disinfection by-products and their precursors in a water treatment plant in north China: seasonal changes and fraction analysis. Sci. Total Environ., 2008, 397: 140-147.

[106] Wang C, Zhang X, Wang J, et al. Effects of organic fractions on the formation and control of N-nitrosamine precursors during conventional drinking water treatment processes. Sci. Total Environ., 2013, 449: 295-301.

[107] Rosario-Ortiz F, Snyder S, Suffet I. Characterization of dissolved organic matter in drinking water sources impacted by multiple tributaries. Water Res., 2007, 41(18): 4115-4128.

[108] Chen C, Leavey S, Krasner S, et al. Applying polarity rapid assessment method and ultrafiltration to characterize NDMA precursors in wastewater effluents. Water Res., 2014. 57: 115-126.

[109] Bei E, Liao X, et al. Identification of nitrosamine precursors from urban drainage during storm events: A case study in southern China. Chemosphere, 2016, 160: 323-331.

[110] Williams D, LeBel G, Benoit F. Disinfection by-products in Canadian drinking water. Chemosphere, 1997, 34: 299-316.

[111] Weisel C, Kim H, Haltmeier P, et al. Exposure estimates to disinfection by-products of chlorinated drinking water. Eviron. Health Perspect, 1999, 107(2): 103-110.

[112] Krasner S, Westerhoff P, et al. Occurrence of disinfection by-products in United States wastewater treatment plant effluent. Environ. Sci. Technol, 2009, 43(21): 8320-8325.

[113] Goslan E, Krasner S, Bower M, et al. A comparison of disinfection by-products found in chlorinated and chloraminated drinking waters in Scotland. Water Res., 2009, 43: 4698-4706.

[114] Malliarou E, Collins C, Graham N, et al. Haloacetic acids in drinking water in the United Kingdom. Water Res., 2005, 39: 2722-2730.

[115] Egorov A, Tereschenko A, Altshul L, et al. Exposures to drinking water chlorination by-products in a Russian city. Int. J. Hyg. Environ. Health., 2003, 206(6): 539-551.

[116] Meier J, Blazak W, Knohl R. Mutagenic and clastogenic properties of 3-chloro-4-(dichloromethyl)-5-hydroxy-2(5H)-furanone: a potent bacterial mutagen in drinking water. Environ. Mol. Mutagen., 1987, 10(4): 411-424.

[117] Lee K, Kim B, Hong J, et al. A study on the distribution of chlorination by-products (CBPs) in treated water in Korea. Water Res., 2001, 35: 2861-2872.

[118] EPA Method 551.1: Determination of chlorinated disinfection by-Products, chlorinated solvents, and halogenated pesticides/herbicides in drinking water by liquid-liquid extraction and gas chromatography with electron capture detection. USEPA, Office of Water, Technical Support Center, Cincinnati, 1998.

[119] 杨晓萍, 蔡宏道. 饮用水中三氯甲烷与致突变物. 环境科学, 1986, 7(3): 6-9.

[120] 邓瑛, 魏建荣, 鄂学礼, 等. 中国六城市饮用水中氯化消毒副产物分布的研究. 卫生研究, 2008, 37(2): 207-210.

[121] Ding H, Meng L, Zhang H, et al. Occurrence, profiling and prioritization of halogenated disinfection by-products in drinking

water of China. Environ. Sci.: Process. Impacts, 2013, 15(7): 1424-1429.

[122] Bei E, Shu Y, Li S, et al. Occurrence of nitrosamines and their precursors in drinking water systems around mainland China. Water Res., 2016, 98: 168-175.

[123] Wang C, Liu S, Wang J, et al. Monthly survey of N-nitrosamines yield in a conventional water treatment plant in North China. J. Environ. Sci., 2015, 38: 142-149.

[124] Yang W, Dong L, Luo Z, et al. Application of ultrasound and quartz sand for the removal of disinfection byproducts from drinking water. Chemosphere. 2014, 101: 34-40.

[125] Du H, Li J, Moe B, et al. Cytotoxicity and oxidative damage induced by halobenzoquinones to T24 bladder cancer cells. Environ Sci. Technol., 2013, 47: 2823-2830.

[126] Muellner M, Wagner E, McCalla K, et al. Haloacetonitriles vs. regulated haloacetic acids: are nitrogen-containing DBPs more toxic? Environ. Sci. Technol., 2007, 41: 645-651.

[127] Zhou W, Lou L, Zhu L, et al. Formation and cytotoxicity of a new disinfection by-product(DBP)phenazine by chloramination of water containing diphenylamine. J. Environ. Sci., 2012, 24(7): 1217-1224.

[128] Du H, Li J, Moe B, et al. A real-time cell- electronic sensing method for comparative analysis of toxicity of water contaminants. Anal. Methods, 2014, 6(7): 2053-2058.

[129] Lu T, Su C, Tang F, et al. Chloroacetic acid triggers apoptosis in neuronal cells via a reactive oxygen species-induced endoplasmic reticulum stress signaling pathway. Chem. Biol. Interact., 2015, 225: 1-12.

[130] Plewa M, Kargalioglu Y, Vankerk D, et al. Mammalian cell cytotoxicity and genotoxicity analysis of drinking water disinfection by-products. Environ. Mol. Mutagen., 2002, 40: 134-142.

[131] Richardson S, Fasano F, Ellington J, et al. Occurrence and mammalian cell toxicity of iodinated disinfection byproducts in drinking water. Environ. Sci. Technol., 2008, 42: 8330-8338.

[132] Ragazzo P, Feretti D, Monarca S, et al. Evaluation of cytotoxicity, genotoxicity, and apoptosis of wastewater before and after disinfection with performic acid. Water Res., 2017, 116: 44-52.

[133] IARC. Re-evaluation of some organic chemicals, hydrazine and hydrogen peroxide, vol. 71, International Agency for Research on Cancer, Lyon, France, 1999.

[134] Zhang L, Xu L, Zeng Q, et al. Comparison of DNA damage in human-derived hepatoma line(HepG2)exposed to the fifteen drinking water disinfection byproducts using the single cell gel electrophoresis assay. Mutat. Res., 2012, 741(1-2): 89-94.

[135] Wang L, Hu H, Wang C. Effect of ammonia nitrogen and dissolved organic matter fractions on the genotoxicity of wastewater effluent during chlorine disinfection. Environ. Sci. Technol., 2007, 41(1): 160-165.

[136] Bond T, Templeton M, Mokhtar Kamal N, et al. Nitrogenous disinfection byproducts in English drinking water supply systems: occurrence, bromine substitution and correlation analysis. Water Res., 2015, 85: 85-94.

[137] Yang M, Zhang X. Comparative developmental toxicity of new aromatic halogenated DBPs in a chlorinated saline sewage effluent to the marine polychaete *Platynereis dumerilii*. Environ. Sci. Technol., 2013, 47: 10868-10876.

[138] Narotsky M, Best D, McDonald A, et al. Pregnancy loss and eye malformations in offspring of F344 rats following gestational exposure to mixtures of regulated trihalomethanes and haloacetic acids. Reproduct. Toxicol., 2011, 31: 59-65.

[139] Moser V, Phillips P, Levine A, et al. Neurotoxicity produced by dibromoacetic acid in drinking water of rats. Toxicol. Sci., 2004, 79: 112-122.

[140] Stacpoole P, Harwood H, Cameron D, et al. Chronic toxicity of dichloroacetate: possible relation to thiamine deficiency in rats. Fundamental Appl. Toxicol., 1990, 14: 327-337.

[141] Deng Y, Zhang Y, Zhang R, et al. Mice In vivo toxicity studies for monohaloacetamides emerging disinfection byproducts based on metabolomic methods. Environ Sci Technol., 2014, 48: 8212-8218.

[142] Freeman L, Cantor K, Baris D, et al. Bladder cancer and water disinfection by-product exposures through multiple routes: a population-based case–control study (New England, USA). Environ. Health Perspec., 2017, 125: 67010.

[143] Chowdhury S, Chowdhury I, Zahir M. Trihalomethanes in desalinated water: Human exposure and risk analysis. Human Ecol. Risk Assess, 2018, 24: 26-48.

[144] 程英惠, 徐红霞, 黄莉莉, 等. 孕期氯化消毒副产物三卤甲烷暴露对新生儿出生体重的影响. 环境与健康杂志, 2017, 34(9): 782-785.

[145] Zeng Q, Li M, Xie S, et al. Baseline blood trihalomethanes, semen parameters and serum total testosterone: a cross-sectional study in China. Environ. Int., 2013, 54: 134-140.

[146] Kristiana I, Joll C, Heitz A. Powdered activated carbon coupled with enhanced coagulation for natural organic matter removal and disinfection by-product control: application in a Western Australian water treatment plant. Chemosphere, 2011, 83: 661-667.

[147] Bond T, Goslan E, Parsons S, et al. Treatment of disinfection by‐product precursors. Environ. Technol., 2011, 32: 1-25.

[148] Matilainen A, Vepsäläinen M, Sillanpää M. Natural organic matter removal by coagulation during drinking water treatment: A review. Adv. Colloid. Interface Sci., 2010, 159: 189-197.

[149] Liu H, Liu R, Tian C, et al. Removal of natural organic matter for controlling disinfection by-products formation by enhanced coagulation: A case study. Sep. Purif. Technol., 2012, 84: 41-45.

[150] Qi S, Schideman L. An overall isotherm for activated carbon adsorption of dissolved natural organic matter in water. Water Res., 2008, 42: 3353-3360.

[151] Velten S, Knappe D, Traber J, et al. Characterization of natural organic matter adsorption in granular activated carbon adsorbers. Water Res., 2011, 45: 3951-3959.

[152] Zhang M, Li C, Benjamin M, et al. Fouling and natural organic matter removal in adsorbent/membrane systems for drinking water treatment. Environ. Sci. Technol., 2003, 37: 1663-1669.

[153] Lamsal R, Montreuil K, Kent F, et al. Characterization and removal of natural organic matter by an integrated membrane system. Desalination, 2012, 303: 12-16.

[154] Boyer T, Singer P. A pilot-scale evaluation of magnetic ion exchange treatment for removal of natural organic material and inorganic anions. Water Res., 2006, 40: 2865-2876.

[155] Kingsbury R, Singer P. Effect of magnetic ion exchange and ozonation on disinfection by-product formation. Water Res., 2013, 47: 1060-1072.

[156] Comstock S, Boyer T. Combined magnetic ion exchange and cation exchange for removal of DOC and hardness. Chem. Eng. J., 2014, 241: 366-375.

[157] Moslemi M, Davies S, Masten S. Hybrid ozonation–ultrafiltration: The formation of bromate in waters containing natural organic matter. Sep. Purif. Technol., 2014, 125: 202-207.

[158] Sarathy S, Mohseni M. Effects of UV/H_2O_2 advanced oxidation on chemical characteristics and chlorine reactivity of surface water natural organic matter. Water Res., 2010, 44: 4087-4096.

[159] Huang X, Leal M, Li Q. Degradation of natural organic matter by TiO_2 photocatalytic oxidation and its effect on fouling of low-pressure membranes. Water Res., 2008, 42: 1142-1150.

[160] Liu S, Lim M, Fabris R, et al. TiO_2 photocatalysis of natural organic matter in surface water: impact on trihalomethane and haloacetic acid formation potential. Environ. Sci. Technol., 2008, 42: 6218-6223.

[161] Drikas M, Chow C, Cook D. The impact of recalcitrant organic character on disinfection stability, trihalomethane formation and bacterial regrowth: an evaluation of magnetic ion exchange resin (MIEX®) and alum coagulation. J. Water Supply Res. Technol. AQUA, 2003, 52(7): 475-487.

[162] Fearing D, Banks J, Guyetand S, et al. Combination of ferric and MIEX® for the treatment of a humic rich water. Water Res., 2004, 38(10): 2551-2558.

[163] Jiang J, Zhang X, Zhu X, et al. Removal of intermediate aromatic halogenated DBPs by activated carbon adsorption: a new approach to controlling halogenated DBPs in chlorinated drinking water. Environ. Sci. Technol., 2017, 51(6): 3435-3444.

[164] Wang F, Gao B, Yue Q, et al. Effects of ozonation, powdered activated carbon adsorption, and coagulation on the removal of disinfection by-product precursors in reservoir water. Environ. Sci. Pollut. Res., 2017, 24(21): 17945-17954.

[165] Liu Z, Wang X, Luo Z, et al. Removing of disinfection by-product precursors from surface water by using magnetic graphene oxide. Plos One, 2015, 10(12): 143819.

[166] Zhang M, Xu B, Wang Z, et al. Formation of iodinated trihalomethanes after ferrate pre-oxidation during chlorination and chloramination of iodide-containing water. J. Taiwan Inst. Chem. Eng., 2016, 60: 453-459.

[167] Hu J, Chu W, Sui M, et al. Comparison of drinking water treatment processes combinations for the minimization of subsequent disinfection by-products formation during chlorination and chloramination. Chem. Eng. J., 2018, 335: 352-361.

[168] Chen C, Zhang X, He W, et al. Comparison of seven kinds of drinking water treatment processes to enhance organic material removal: a pilot test. Sci. Total Environ., 2007, 382: 93-102.

[169] Bei E, Wu X, Qiu Y, et al. A tale of two water supplies in China: finding practical solutions to urban and rural water supply problems. Accounts Chem. Res., 2019, 52(4): 867-875.

[170] Liao X, Bei E, Li S, et al. Applying the polarity rapid assessment method to characterize nitrosamine precursors and to understand their removal by drinking water treatment processes. Water Res., 2015, 87: 292-298.

[171] Liao X, Chen C, Xie S, et al. Nitrosamine precursor removal by BAC: adsorption versus biotreatment case study. J. AWWA., 2015, 107(9): 454-463.

[172] Liao X, Chen C, Zhang J, et al. Dimethylamine biodegradation by mixed culture enriched from drinking water biofilter. Chemosphere, 2015, 119: 935-940.

[173] Ersan M, Ladner D, Karanfil T. The control of N-nitrosodimethylamine, Halonitromethane, and Trihalomethane precursors by Nanofiltration. Water Res., 2016, 105: 274-281.

[174] Yang Y, Lu J, Yu H, et al. Characteristics of disinfection by-products precursors removal from micro-polluted water by constructed wetlands. Ecol. Eng., 2016, 93: 262-268.

[175] Li S, Zhang X, Bei E, et al. Capability of cation exchange technology to remove proven N-nitrosodimethylamine precursors. J. Environmental Sci., 2017. 58: 331-339.

[176] Bolto B, Dixon D, Eldridge R, et al. Removal of natural organic matter by ion exchange. Water Res., 2002, 36(20): 5057-5065.

[177] Allpike B, Heitz A, Joll C, et al. Size exclusion chromatography to characterize DOC removal in drinking water treatment. Environ. Sci. Technol., 2005, 39(7): 2334-2342.

[178] Mergen M, Adams B, Vero G, et al. Characterisation of natural organic matter (NOM) removed by magnetic ion exchange resin (MIEX® Resin). Water Sci. Technol. Water Supply., 2009, 9(2): 199-205.

[179] Bond T, Goslan E, Parsons S, et al. Disinfection by-product formation of natural organic matter surrogates and treatment by coagulation, MIEX and nanofiltration. Water Res., 2010, 44: 1645-1653.

[180] Kristiana I, Allpike B, Joll C, et al. Understanding the behaviour of molecular weight fractions of natural organic matter to improve water treatment processes. Water Sci. Technol. Water Supply., 2010, 10(1): 59-68.

[181] Watson K, Farre M, Knight N. Enhanced coagulation with powdered activated carbon or MIEX® secondary treatment: a comparison of disinfection by-product formation and precursor removal. Water Res., 2015, 68: 454-466.

[182] Huang W, et al. Effects of macro-porous anion exchange and coagulation treatment on organic removal and membrane

fouling reduction in water treatment. Desalination, 2015, 355: 204-216.

[183] Sun Y, Yang Z, Ye T, et al. Evaluation of the treatment of reverse osmosis concentrates from municipal wastewater reclamation by coagulation and granular activated carbon adsorption. Environ. Sci. Pollut. Res., 2016, 23(13): 13543-13553.

[184] Metcalfe D, Rockey C, Jefferson B, et al. Removal of disinfection by-product precursors by coagulation and an innovative suspended ion exchange process. Water Res., 2015, 87: 20-28.

[185] Chu W, Yao D, Gao N, et al. The enhanced removal of carbonaceous and nitrogenous disinfection by-product precursors using integrated permanganate oxidation and powdered activated carbon adsorption pretreatment. Chemosphere, 2015, 141: 1-6.

[186] Fan M, Qiao J, Wang W, et al. Study on water disinfection by-products during 2004 to 2010. J. Med. Pest Control, 2011, 27(3): 200-204.

[187] Zhou X, Zhao J, Li Z, et al. Influence of ultrasound enhancement on chlorine dioxide consumption and disinfection by-products formation for secondary effluents disinfection. Ultrason. Sonochem., 2016, 28: 376-381.

[188] Ding S, Chu W, Bond T, et al. Formation and estimated toxicity of trihalomethanes, haloacetonitriles, and haloacetamides from the chlor (am) ination of acetaminophen. J. Hazard. Mater., 2018, 341: 112-119.

[189] Huang H, Chen B, Zhu Z. Formation and speciation of haloacetamides and haloacetonitriles for chlorination, chloramination, and chlorination followed by chloramination. Chemosphere, 2017, 166: 126-134.

[190] Wang X, Wang J, Zhang Y, et al. Characterization of unknown iodinated disinfection byproducts during chlorination/chloramination using ultra-high resolution mass spectrometry. Sci. Total Environ., 2016, 554-555: 83-88.

[191] Wei X, Chen X, Wang X, et al. Occurrence of regulated and emerging iodinated DBPs in the Shanghai drinking water. PLOS One, 2013, 8(3): 59677.

[192] Chen C, Zhang X, He W, et al. Simultaneous control of microorganisms, disinfection by-products by Sequential Chlorination. Biomed. Environ. Sci., 2007, 20: 119-125.

[193] Zhang X, Chen C, Wang Y. Synergetic inactivation of microorganisms by short-term free chlorination and subsequent monochloramination in drinking water disinfection. Biomed. Environ. Sci., 2007, 20: 373-380.

[194] Liao X, Chen C, Yuan B, et al. Control of nitrosamines, THMs, and HAAs in heavily impacted water with O_3-BAC. J. Am. Water Works Ass., 2017, 109(6): 3-13.

[195] Du Y, Lv X, Wu Q, et al. Formation and control of disinfection byproducts and toxicity during reclaimed water chlorination: a review. J. Environ. Sci., 2017, 58: 51-63.

[196] Yang X, Shang C, Huang J. DBP formation in breakpoint chlorination of wastewater. Water Res., 2015, 39(19): 4755-4767.

[197] Li Y, Zhang X, Yang M, et al. Three-step effluent chlorination increases disinfection efficiency and reduces DBP formation and toxicity. Chemosphere, 2017, 168: 1302-1308.

[198] Li Y, Yang M, Zhang X, et al. Two-step chlorination: a new approach to disinfection of a primary sewage effluent. Water Res., 2017, 108: 339-347.

[199] Huo Z, Xie X, Yu T, et al. Nanowire-modified three-dimensional electrode enabling low-voltage electroporation for water disinfection. Environ. Sci. Technol., 2016, 50(14): 7641-7649.

[200] Pan Y, Zhang X, Wagner E, et al. Boiling of simulated tap water: effect on polar brominated disinfection byproducts, halogen speciation, and cytotoxicity. Environ. Sci. Technol., 2014, 48(1): 149-156.

[201] Yang L, She Q, Wan M, et al. Removal of haloacetic acids from swimming pool water by reverse osmosis and nanofiltration. Water Res., 2017, 116: 116-125.

[202] Li X, Ma J, Liu G, et al. Efficient reductive dechlorination of monochloroacetic acid by sulfite/UV process. Environ. Sci. Technol., 2012, 46(13): 7342-7349.

[203] Zha X, Ma L, Liu Y. Reductive dehalogenation of brominated disinfection byproducts by iron based bimetallic systems. RSC Adv., 2016, 6(20): 16323-16330.

[204] Liu X, Zhong J, Fang L, et al. Trichloroacetic acid reduction by an advanced reduction process based on carboxyl anion radical. Chem. Eng. J., 2016, 303: 56-63.

[205] Fu F, Dionysiou D, Liu H. The use of zero-valent iron for groundwater remediation and wastewater treatment: a review. J. Hazard. Mater., 2014, 267: 194-205.

[206] Liang L, Sun W, Guan X, et al. Weak magnetic field significantly enhances selenite removal kinetics by zero valent iron. Water Res., 2014, 49: 371-380.

[207] Liu Y, Phenrat T, Lowry G. Effect of TCE concentration and dissolved groundwater solutes on NZVI-promoted TCE Dechlorination and H2 evolution. Environ. Sci. Technol., 2007, 41(22): 7881-7887.

[208] Xiao J, Gao B, Yue Q, et al. Removal of trihalomethanes from reclaimed-water by original and modi- fied nanoscale zero-valent iron: characterization, kinetics and mechanism. Chem. Eng. J., 2015, 262: 1226-1236.

[209] Chen H, Cao Y, Wei E, et al. Facile synthesis of graphene nano zero-valent iron composites and their efficient removal of trichloronitromethane from drinking water. Chemosphere, 2016, 146: 32-39.

[210] Chen R, et al. Sorption of trace levels of bromate by macroporous strong base anion exchange resin: influencing factors, equilibrium isotherms and thermodynamic studies. Desalination, 2014, 344: 306-312.

[211] Liu J, Wang X, Fan B. Characteristics of PAHs adsorption on inorganic particles and activated sludge in domestic wastewater treatment. Bioresour. Technol., 2011, 102(9): 5305-5311.

[212] Yang F, Zhang J, Chu W, et al. Haloactamides versus halomethanes formation and toxicity in chloraminated drinking water. J. Hazard. Mater., 2014, 274: 156-163.

[213] Sun H, Song X, Ye T, et al. Formation of disinfection by-products during chlorination of organic matter from phoenix tree leaves and Chlorella vulgaris. Environ. Pollut., 2018, 243: 1887-1893.

[214] Hong H, Huang F, Wang F, et al. Properties of sediment NOM collected from a drinking water reservoir in South China, and its association with THMs and HAAs formation. J. Hydrol., 2013, 476: 274-279.

[215] Hong H, Qian L, Xiong Y, et al. Use of multiple regression models to evaluate the formation of halonitromethane via chlorination/chloramination of water from Tai Lake and the Qiantang River, China. Chemosphere, 2015, 119: 540-546.

[216] Hong H, Song Q, Mazumder A, et al. Using regression models to evaluate the formation of trihalomethanes and haloacetonitriles via chlorination of source water with low SUVA values in the Yangtze River Delta region, China. Environ. Geochem. Health., 2016, 38: 1303-1312.

[217] Yang F, Wei D, Xiao M, et al. The chlorination transformation characteristics of benzophenone-4 in the presence of iodide ions. J. Environ. Sci., 2017, 58: 93-101.

作者：魏东斌[1]，孙雪凤[1]，陈　妙[1]，邱　玉[2]，贝　尔[2]，陈　超[2]，杜宇国[1]

[1] 中国科学院生态环境研究中心，[2] 清华大学

第 14 章 大气污染物毒理学效应研究进展

▶ 1. 引言 /364

▶ 2. 大气污染物暴露对呼吸系统的影响 /364

▶ 3. 大气污染物暴露对心血管系统的影响 /369

▶ 4. 大气污染物暴露对中枢神经系统的影响 /373

▶ 5. 展望 /376

本章导读

随着世界人口的增加、工业生产和交通运输的发展以及煤炭、石油等能源利用的增长，大气受到严重污染。大气污染对人群健康的影响早在1900年就有报道，而随着新能源的出现、燃料的变化及生产过程的革新，特别是大气污染类型的转化，研究污染物对健康的危害，提出对大气污染危害的防治对策一直是学界关注的热点问题。世界卫生组织公布的数据显示，2016年室内外空气污染造成了世界范围内七百万人口的死亡，占到全球总死亡率的1/8，而在死亡率贡献中列前五位的疾病分别为缺血性心脏病、中风、慢性阻塞性呼吸道疾病、肺癌和急性下呼吸道感染。但到目前为止，除传统的呼吸系统损伤外，关于其他损伤靶点的文献报道不一致，"暴露-效应-机制"的毒理学实验证据值得深入探讨。本章将对大气污染物暴露的健康危害、毒理学效应以及分子作用机制等方面的研究进行综合介绍。

关键词

大气污染物，健康风险，毒理学效应，作用机制

1 引 言

大气污染可以导致人群中死亡率和发病率显著升高。早在1900年就有研究报道显示燃煤相关的煤烟与雾形成的烟雾导致城市人口死亡率急剧升高。随后两项公认的美国队列研究表明，暴露于空气中的细颗粒物与死亡率特别是呼吸系统以及心肺系统的死亡率相关[1,2]。20世纪开始，随着生产过程的革新、新能源的出现以及燃料的变化，导致大气污染类型的转化，污染物对健康的危害以及其毒理学效应更加复杂多变。目前有多项研究估计了大气污染对我国居民造成的疾病负担。2018年5月2日，世界卫生组织在日内瓦发布的最新报告显示，包括中国在内的全球许多国家或地区，空气污染仍然处于危险水平，全球90%的人呼吸的空气中含有高浓度的污染物，而空气污染物在2016年直接或者间接导致了全球700万人死亡，其中成人慢性阻塞性肺疾病（COPD）、肺癌、中风以及心脏病引起的死亡分别占到了43%、29%、25%、24%[3]。在我国，大多数城市的研究报告以及实验研究显示$PM_{2.5}$、SO_2、NO_2和O_3等空气污染物浓度增加与心血管疾病、呼吸系统疾病以及总死亡率升高之间存在显著的统计学相关性[4,5]。目前我国正经历大气环境恶化且居民全方位暴露于严重复合污染的时期，而相应的大气污染毒理机制与健康影响研究工作仍处于探索阶段，一是大气环境污染的多元性给毒理-健康研究提出了全新挑战，二是对于大气污染物生物学效应与毒理学机制的认识不足。大气污染物主要通过呼吸进入气管、支气管，对呼吸道具有腐蚀和刺激作用，与呼吸系统损伤和多种疾病的相关性一直是学术界关注的重点。虽然越来越多的证据显示大气污染物还会对心脑等系统产生不良影响，但除传统的肺损伤外，关于其他损伤靶点的文献报道很不一致，毒性效应也多停留在"相关性"水平上，缺乏"暴露-效应"关系的实验证据。在作用机制方面，氧化应激和炎性被认为是损伤发生的重要基础，但分子信号过程并不清楚，是否还存在其他调控环节也不明确。

2 大气污染物暴露对呼吸系统的影响

大气污染物能够通过多种途径威胁人类健康，除少部分污染物能通过皮肤接触的方式外，大部分大

气污染物主要通过呼吸直接作用于呼吸系统从而对人体健康造成影响。对于大气污染引起的呼吸系统疾病，主要是通过引起气道上皮和肺实质受损，造成肺部氧化应激，吞噬作用受阻，炎症细胞浸润，细胞免疫失调等不良反应，最终导致哮喘、慢性非阻塞性疾病以及肺癌等多种肺部疾病的发生。

2.1 大气污染与哮喘

哮喘是肺部以及呼吸道常见的长期炎性疾病。其特征是可变和反复出现的可逆性气流阻塞和支气管痉挛。症状主要包括喘息、咳嗽、胸闷和呼吸短促等。近几十年来，室内外环境的变化导致世界范围内哮喘、鼻炎、湿疹（过敏性和非过敏性）的患病率尤其是儿童患病率显著增加。2010~2012年，通过对中国室内环境与儿童健康展开调研，发现确诊哮喘患病率为1.7%~9.8%（平均为6.8%），相比于1990年的0.91%和2000年的1.50%有大幅增长[6]。哮喘患病率的大幅增长，已经不能完全用遗传因素以及饮食习惯等来解释，越来越多的流行病学研究将原因指向了大气污染。在对10个欧洲城市进行的一项研究中，14%的儿童哮喘病例和15%的儿童哮喘恶化病例与道路交通有关的污染物暴露相关[7]。孕期暴露也是增加儿童哮喘发病率的重要因素。在加拿大安大略省统计了2006~2012年出生的761172个儿童，其中110981个儿童确诊患有哮喘，分析其孕期空气污染物暴露和儿童哮喘的相关关系发现孕中期暴露于$PM_{2.5}$[每四分位数间距（IQR）增加的风险比（HR）=1.07，95%CI：1.06~1.09]与儿童哮喘发生发展相关[8]。除了易感性较强的儿童，有关总体人群的数据也显示了空气污染物与哮喘的相关性。在中国上海，每年$PM_{2.5}$增加10 μg/m³会引起过敏性鼻炎和哮喘的患病率的升高，比值比（OR）分别为1.20（95%CI：1.11~1.29）和1.10（95%CI：1.03~1.18）[9]。然而，关于大气污染物以及哮喘的相关性的研究仍存在不一致的结果。加拿大随访期间共观察到74543例COPD事件，87141例哮喘和12908例肺癌。在单个污染物模型中，环境超细颗粒物中每个四分位数增加与事件性COPD（HR=1.06，95%CI：1.05~1.09）相关，但与哮喘（HR=1.00，95%CI：1.00~1.01）或肺癌（HR=1.00，95%CI：0.97~1.03）不相关[10]。因此，颗粒物与哮喘发病率之间的相关性还需进一步研究。对于NO_2与哮喘的研究发现，在对10个欧洲城市进行的一项研究中，14%的儿童哮喘病例和15%的儿童哮喘恶化病例与道路交通有关的污染物暴露相关[11]。同时Hansel等指出，NO_2浓度的升高与学龄前儿童所表现出的说话障碍、咳嗽以及一些夜间哮喘症状的增加相关，而采取降低室内NO_2浓度等一些保护措施后，这些易感人群的哮喘发病率显著降低[12]。

哮喘模型的建立有国际上公认的标准化的建模方法，因此关于空气污染物和哮喘的动物实验研究较为广泛。有研究发现$PM_{2.5}$增加了哮喘小鼠嗜酸性粒细胞和中性粒细胞的数量，以及肺泡灌洗液（BALF）中促炎因子TNF-α和Th2细胞因子IL-4和IL-10的水平，同时加重了炎性细胞浸润，引起了杯状细胞增生和肺组织微观结构的改变[13]。孕期暴露于机动车尾气污染物能够增加子代小鼠血清OVA-IgE、气道炎症、气道高反应性、Th2和Th17相关细胞因子[14]。PM暴露能够通过增强幼年期小鼠Th2免疫反应，氧化应激和降低DNA甲基转移酶（DNMT）的表达来增加小鼠成年期罹患哮喘的可能性[15]。$PM_{2.5}$的提取物能够增加气道上皮细胞细胞毒性，激活抗原呈递细胞和T细胞，从而加重哮喘等呼吸系统疾病[16]。暴露于PM会增加中性粒细胞浸润以及TNF-α和IFN-γ的表达，并增强过敏性免疫反应，包括嗜酸性粒细胞浸润和Th2相关细胞因子（IL-5和IL-13）水平上升[17]。孕期超细颗粒物暴露能够抑制子代小鼠的免疫应激反应[18]。而将妊娠期小鼠暴露于二手烟后发现，孕期二手烟暴露显著影响了子代小鼠的肺泡发育，而雄性小鼠在肺部发育过程中比雌性小鼠更为敏感[19]。类似的研究还发现，孕期二手烟暴露引起子代miR-130a的表达可能下调，而miR-16和miR-221表达可能上调，并可能调控 *Hif1α* 介导的凋亡、血管生成和免疫相关信号通路，最终可能引起肺发育不良的发生[20]。此外，孕期香烟烟雾暴露还能够减弱肺部NK细胞功能，刺激肺部气道炎症反应发生，最终可能加重子代小鼠哮喘症状[21]。对小鼠进行孕期NO_2暴露后发现，孕期NO_2暴露能够引起子代小鼠支气管和血管周围炎性细胞浸润和支气管周围胶原蛋白沉积,子代小鼠肺部Ⅱ型细胞因子IL-4和IL-13表达上升,Ⅰ型细胞因子IFN-γ表达下降,但是随着子代小鼠的发育而逐渐恢复到正常水平;进一步进行过敏原刺激和哮喘建模发现,孕期NO_2暴露能够增加子代小鼠对过敏原的易感性以及哮喘易感性;而哮喘关键基因*Il4*的启动子区甲基化可能参与了上述病症反应过程[22,23]。当雌性小鼠在怀孕早期暴露于*Aspergillus fumigatus*后,子2代小鼠嗜酸性气道炎症会加重而且*Il4* CpG^{-408}和CpG^{-393}位点的甲基化水平降低[24]。更有研究证实孕期NO_2暴露与线粒体抗氧化相关基因甲基化状态的改变相关[25]。综上,大气污染物本体暴露以及孕期暴露可能通过表观遗传机制导致Th1/Th2细胞分化失衡,从而造成过敏性免疫反应等哮喘症状。

2.2 大气污染与慢性阻塞性肺疾病

慢性阻塞性肺疾病(COPD)是全球范围内发病和死亡率领先的疾病之一,最新的慢性阻塞性肺疾病全球倡议(GOLD)指南将COPD定义为"可预防和治疗的疾病",其特征在于"持续的气流受限逐渐加重,并且与气道和肺部吸入毒性颗粒和气体所产生的慢性炎性相关"[26]。预计到2020年将成为第三大死因和第五大致残因素[27]。流行病学研究表明COPD影响了5%~19% 40岁以上的成年人。吸烟(主动或被动)被认为是造成COPD发生发展最重要的原因。此外,其他风险因素也可能在COPD发生发展中起重要作用,这些危险因素包括职业性灰尘暴露,户外和室内空气污染、燃料燃烧,社会经济地位较低等。另外,有证据表明空气污染也是COPD发生发展的重要危险因素。

Ko等研究表明香港居民COPD的住院风险与$PM_{2.5}$浓度呈现显著的正相关,$PM_{2.5}$浓度每升高10 μg/m^3,COPD的住院率增加3.1%(95%CI:2.6%~3.6%)[28]。法国农民吸烟人数比一般人群少,但是COPD的发病率高。通过调查50个农场的75个农民发现,高浓度$PM_{2.5}$暴露与血清中细胞因子IL-13和IL-8水平的改变相关,这可能是造成COPD发病率升高的原因[29]。然而,一项在伯明翰的研究呈现相反的趋势,即$PM_{2.5}$每升高15 μg/m^3,COPD的住院率下降3.9%(95%CI:-9.0%~-1.6%)[30]。然而,关于罗马的一项研究并没有发现COPD住院率与$PM_{2.5}$正向的相关关系[31]。同样,关于$PM_{2.5}$与COPD死亡率相关关系的研究也存在不一致的结果。Schwartz等研究了美国6个社区$PM_{2.5}$与COPD死亡率之间的关系,结果显示$PM_{2.5}$浓度每升高10 μg/m^3,COPD的死亡率增加了3.3%(95%CI:1.0%~5.7%)[32]。相反地,在奥地利,Neuberger发现$PM_{2.5}$浓度与COPD死亡率之间并没有明显的相关关系,其他一些研究虽然观察到了$PM_{2.5}$与COPD住院率与死亡率之间正向的相关关系,但是没有显著性,这两者之间关系的研究仍要继续[33]。关于其他大气污染物的研究发现,COPD发病率与35年平均NO_2水平(风险比:1.08,95%CI:1.04~1.14,每四分位数范围为5.8 μg/m^3)有关[34]。此外,主要生活区域NO_2浓度升高与呼吸困难症相关,而卧室NO_2浓度的增加与夜间症状的增加和COPD严重恶化风险有关[35],CO与COPD也有类似的相关关系[36]。有研究发现主要生活区域NO_2浓度升高与呼吸困难症相关,而卧室NO_2浓度的增加与夜间症状的增加和COPD严重恶化风险有关[37]。另外,对北京地区23例慢性阻塞性肺疾病患者进行的一项定组研究发现,$PM_{2.5}$、PM_{10}和SO_2浓度的增加分别会导致呼出气NO增加13.6%(95%CI:4.8%~23.2%),9.2%(95%CI:2.1%~16.8%)和34.2%(95%CI:17.3%~53.4%)[38]。此外,还有学者探讨了生命早期空气污染物暴露与COPD发病的相关性。Martinez在*The New England Journal of Medicine*杂志上详细介绍了COPD的生命早期起源假说,他认为尽管香烟烟雾仍然是COPD发病一个主要的病因,但与多种生物机制相关的遗传、环境和发育因素,以及在生长发育时期发挥重要作用的因素,都可以能够降低儿童1s最大用力呼气量(FEV1),并进一步降低成年人的FEV1,从而增加COPD的患病风险[39]。

COPD模型的建立主要包括香烟吸入型、内毒素诱导型、细菌感染型以及基因调控型等。然而,COPD是多因素诱导的临床综合征,其发病机制十分复杂,至今国内外还没有完全建立符合人类COPD标准的动物模型[40]。目前关于空气污染物与COPD建模的动物实验研究十分缺乏,现有的研究仅从COPD典型症状方面分析污染物与 COPD 的相关关系。人群实验研究发现暴露于低水平的室内炭黑导致 COPD 患者的脂质过氧化和氧化性 DNA 损伤[41]。有一些研究发现附着于空气污染物上的有机碳，元素碳，NO_3^- 和 NH_4^+ 可能是 $PM_{2.5}$ 导致 COPD 患者 *Nos2a* 基因 DNA 甲基化降低和 FeNO 升高的主要原因[42]。通过对香烟烟雾暴露诱导的 COPD 模型鼠进行 $PM_{2.5}$ 暴露，发现 $PM_{2.5}$ 暴露能够导致 Notch 信号通路过度活化，从而导致 COPD 模型鼠的免疫功能更加紊乱[43]。对 COPD 患者支气管上皮细胞进行机动车尾气暴露后发现，污染物暴露能够引起上皮细胞 IL-8 以及抗氧化因子的分泌和释放、细胞色素 p450 表达上升以及抑制超氧化物歧化酶-1（SOD-1）的表达[44]。同样地，$PM_{2.5}$ 暴露能够造成健康和 COPD 患者支气管上皮细胞的线粒体功能受损，导致线粒体活性氧簇（ROS）水平上升，腺苷三磷酸（ATP）产量下降[45]。颗粒物能够直接作用于 T 细胞中的多环芳烃受体，从而增加 Th17 细胞的分化，并增加 IL-17 的表达，导致肺组织炎症的增加[46]。20 ppm NO_2 暴露能够诱导小鼠发生进行性气道炎性反应以及肺实质局灶性炎症，表现为中性粒细胞和巨噬细胞浸润，杯状细胞和黏液增生[47]。此外，通过对小鼠进行 $PM_{2.5}$ 和 SO_2、NO_2 复合暴露，发现空气污染物复合暴露能可以通过下调 miR-338-5p 的表达进而增加其靶基因 *Hif1α* 的表达，从而作用于 *Hif1α/Fhl-1* 通路，造成 COPD 典型症状肺动脉高压相关的肺血管内皮细胞和呼吸功能的损伤[48]。$PM_{2.5}$ 暴露能够对不同生命阶段的小鼠造成不同程度的肺损伤，包括：呼吸功能受损、肺组织病理学改变、氧化应激以及炎性反应等，其中 10 月龄小鼠最为敏感。

同时，$PM_{2.5}$ 暴露 4 周可以通过 H3K27ac 富集的 *Stat2* 和 *Bcar1* 降低肺功能并诱导肺部炎症；另外，这些不良影响在暴露结束后 2 周恢复到正常水平。然而，在恢复 2 周后仍然可以在肺泡中观察到持续性的颗粒负载，这可能是引起肺部疾病的长期潜在风险的因素之一[49]。而 miR-181c 能够通过调控其靶标因子 *Ccn1* 抑制香烟烟雾诱导的 COPD 病症的发生[50]。有关孕期空气污染物暴露与 COPD 相关性的研究也有报道。Drummond 对怀孕小鼠进行孕期和出生后成年期香烟烟雾暴露后发现，孕期香烟烟雾暴露能显著降低子鼠的肺功能，而成年期暴露仅能影响受到孕期暴露的小鼠的肺功能，在成年后，处理组与对照组相关基因表达没有显著差异[51]。目前，基因调控型的动物模型受到国内外研究人员的重视，这有助于进一步阐述大气污染物诱发和加重 COPD 的发病机制。

2.3 大气污染与肺癌

肺癌是全球最常见的癌症之一，具有极高的死亡率，据世界卫生组织（WHO）估计每年死于肺癌的患者约占全球总癌症死亡病例的 1/4[52]。2013 年 WHO 的国际癌症研究机构（IARC）将将大气污染列为确定的人类致癌物。研究结果显示，在排除了吸烟和室内空气污染等混杂因素后，发现大气重污染区肺癌发病率是轻污染区的 1.3 倍。因此，大气污染成为诱发肺癌的又一不可忽视的重要因素，严重威胁公众健康。

大气颗粒物与肺癌风险的增加最密切相关，研究表明人为源的可入肺颗粒物（$PM_{2.5}$）暴露导致的全球肺癌死亡率为 12.8%（5.9%~18.5%）[53]。欧洲 17 国队列研究对 2095 例肺癌病例进行跟踪调查，结果显示肺癌风险与 PM_{10} 之间有统计学意义[（HR=1.22，95%CI：1.03~1.45）/（10 μg/m³）][54]。2002 年，这一研究小组重新证明了颗粒物和肺癌死亡率的相关关系即长期暴露于 $PM_{2.5}$ 时，$PM_{2.5}$ 浓度每升高 10 μg/m³，肺癌的死亡率的风险上升 8%[55]。关于一项针对终身不吸烟者的队列研究表明 $PM_{2.5}$ 浓度每增加 10 μg/m³，男性和女性的肺癌死亡率均会增加 15%~27%[56]。一项从 1986~2003 年跟踪调查了 3355 例事件病例的队列研究发现，$PM_{2.5}$ 每增加 10 μg/m³，肺癌发病风险升高 1.17 倍（95%CI：0.93~1.47）[57]。

为数不多的流行病学研究结果显示 SO_2 暴露也与肺癌致死率显著相关。陈士杰等利用灰色关联度模型，对整体人群的肺癌死亡率资料与大气污染物年均浓度资料进行测算，结果显示，肺癌死亡率与 8 年前 SO_2 的关联度最大，提示 SO_2 对肺癌影响的潜伏期为 7 年[58]。曹杰等通过对 70947 名病患十年跟踪调查研究,结果发现,大气 SO_2 浓度每增加 10 μg/m³,人群心肺疾病和肺癌的总死亡率显著增加 3.2%(95%CI:2.3%~4.0%)[59]。日本研究者曾收集了 1974~1983 年大气污染数据资料,并对随后 9 年内 6687 例死亡病例中 518 例死于肺癌的病例进行分析,结果发现,SO_2 暴露可显著增加肺癌死亡率[60]。我国台湾学者最新发表的研究结果显示,SO_2 暴露可增加女性肺癌的发病率,且 SO_2 暴露与肺鳞状细胞癌发病的相关性要高于肺腺癌[61]。上述流行病学研究结果提示 SO_2 暴露与肺癌致死率显著相关。

交通源空气污染环境暴露与肺癌的发病率和死亡率也存在显著相关关系。研究发现常年暴露于空气污染的职业司机肺癌的发病率和死亡率明显增加[62]。2016 年,在一项双污染物模型的研究中发现 PM_{10} 和 SO_2 联合 NO_2 对肺癌死亡率的影响效果显著,其中 NO_2 联合 PM_{10} 的 HRs 达到 1.051[63]。一项队列研究分析了 1988~2009 年在加州癌症登记处确定的 352053 名肺癌患者及其空气污染物平均浓度,发现 SO_2、NO_2、PM_{10}、$PM_{2.5}$ 对患者诊断的 HRs 分别为 1.30(95% CI:1.28~1.32)、1.04(95% CI:1.02~1.05)、1.26(95% CI:1.25~1.28)和 1.38(95% CI:1.35~1.41)[64]。对全球范围内 21 项队列研究进行 meta 分析并量化长期暴露在室外空气污染中导致肺癌的风险后,研究发现肺癌死亡率或发病率的风险分别增加 7.23%(95% CI:1.48~13.31)/10 μg/m³ $PM_{2.5}$,13.17%(95% CI:5.57~21.30)/10 ppb NO_2 和 14.76%(95% CI:1.04~30.34)/10 ppb SO_2。Yang 等对 48 项队列研究 meta 分析发现 NO_2 与所有病因之间存在正相关,对肺癌死亡率的 HRs 为 1.05(95% CI:1.02~1.08;PI:0.94~1.17)[65]。对韩国人群为基础的病例对照研究发现,暴露于 PM_{10} 和 NO_2 与肺癌风险呈正相关,HRs 分别为 1.010(95%CI:1.001~1.020)/10 μg/m³ 和 1.008(95%CI:0.999~1.016)/10 μg/m³[66]。

流行病学研究的结果表明大气污染暴露与肺癌死亡率之间有一定的相关性,但是也存在争议。由于不同地区/国家大气污染的情况不同,以及收集和研究的方法不同,结果方面存在一定的争议。因此,针对以上的问题,需要进行更大规模的流行病学研究,特别是针对不同区域/国家之间的比较研究。同时,大气污染暴露对肺癌的影响存在显著的延迟性,因此长期的流行病学研究至关重要。

尽管大气颗粒物已经是肺癌确认的病因之一,但是其致肺癌的确切分子机制还不是很清楚。前期研究通过氨基甲酸酯诱发肺癌,随后进行 $PM_{2.5}$ 暴露,发现 $PM_{2.5}$ 能够增加氨基甲酸酯处理的肺组织中结节的数量,从而加重肺癌的症状[67]。暴露于交通源污染物导致小鼠肺组织中炎症基因(Cxcl11 和 Tnfs4),过敏性哮喘基因(Clca3 和 Prg2)以及肺癌基因(Agr2、Col11a1 和 Sostdc1)表达上升[68]。与生理盐水组相比,$PM_{2.5}$ 暴露组荷瘤小鼠肿瘤结节数量增加,肺泡灌洗液中蛋白水平升高,MMP1、IL-1β 和 VEGF 表达升高,且 12 种血管生成因子的水平增加[69]。

然而,不同区域大气颗粒物的成分存在较大差异,其对健康效应的影响也不一致。桑楠课题组近年来针对 $PM_{2.5}$ 及其成分诱导肺癌发生发展的确切分子机制开展了系统的研究。多环芳烃类化合物被公认为 $PM_{2.5}$ 致癌主要成分。颗粒物中的致癌性成分 PAHs 呈现出较强的季节分布特征,Yue 等在太原的研究发现我国北方城市冬季颗粒物中的 PAHs 最高,且主要为燃煤来源,致癌风险较其他季节也最高。并且进一步通过细胞实验证实 PM 及其负载的 PAHs 能够通过氧化应激诱导上皮间质转化和胞外基质断裂从而促进肺癌细胞的侵袭转移[70]。

而附着于颗粒物上的 PAHs 与 NO_x 反应形成的硝基多环芳烃(N-PAHs)作为二次气溶胶的重要成分及其具有较强的生物学毒性效应,受到了公众的极大关注[71,72]。研究表明,全球各地区均发现了 N-PAHs 的污染,尽管环境中 NPAHs 的浓度很低,只有其母体 PAHs 的 1/10~1/100,但实验数据已证实 N-PAHs 比 PAHs 具有更强的直接致突性和致癌性,并把 NPAHs 归为"可能的人类致癌物"[73-75]。该课题组最近的研究发现,采暖期霾天采集 $PM_{2.5}$ 附着的 N-PAHs 总含量显著高于其他天气状况下采集的 N-PAHs,致癌风险明显高于其他天气,且柴油机废气颗粒物标记物 1-NP 在霾天的致癌风险为 2.75×10^{-8},远高于其他 N-PAHs。为了进一步研究 N-PAHs 对肺癌肿瘤转移的促进作用,使用不同浓度 N-PAHs 处理肺癌细胞,发现 N-PAHs 可通过介导 MST/Hippo 信号传导功能失调,抑制 YAP 磷酸化,导致 YAP/TAZ 的核转运,激活增强子，最终调控癌靶基因的表达，驱动肺癌-血管内皮细胞黏附效应的发生，并进一步通过动物实验证实 nitro-PAH 可以通过激活 *Hippo-Yap* 信号通路显著促进肺癌裸鼠肿瘤发生血行性转移。

此外，中国北方冬季雾霾的主要特征包括混合高度低，相对湿度高，一次污染物大量排放和二次无机气溶胶（特别是硫酸盐）的快速生成，雾霾期间高水平硫酸盐气溶胶促进肺癌的发生发展也已引起广泛的关注。云洋等研究发现二次硫酸盐气溶胶既可通过氧化应激介导核转录因子异常表达，也可通过改变 DNA 甲基化水平，调控 EMT 机制，进而促进小鼠肺癌细胞发生侵袭转移效应。也有研究表明 $PM_{2.5}$ 能够通过改变 microRNA（miRNA）的表达诱导 EMT 机制，同时诱导肺癌细胞的肿瘤干细胞特性[76]。以上结果提示表观遗传机制可能是 PM 致肺癌发生发展的又一关键分子机制[77]。

综上，流行病学研究中有关大气细颗粒物暴露与肺癌死亡率之间的关系仍存有争议，毒理学研究则表明大气细颗粒物能够通过 EMT 以及表观遗传等方式加重肺癌症状。

3 大气污染物暴露对心血管系统的影响

大气污染物通过引起肺部炎症、全身炎症、氧化应激、内皮细胞功能失调和血小板减少、血栓形成前病变、动脉粥样硬化、冠状动脉疾病和充血性心力衰竭等对心脏产生不利影响。大量研究表明 PM 空气污染以及其他几种空气污染物（CO、NO_2 和 SO_2）与高血压患病率、动脉粥样硬化、心脏衰竭和心肌纤维化等相关。

3.1 大气污染物与高血压

据世界卫生组织（WHO）统计资料显示，心血管疾病居全球死因的首位，全球心血管病死亡人数占总死亡人数的 1/3。美国心脏协会发布的《心脏病与卒中统计数据（2017 版）》显示，美国人口的约 34% 患有高血压。我国 2017 年度《中国心血管病报告》显示，我国心血管病现患人数 2.9 亿，其中高血压 2.7 亿。血压是人体心血管健康的重要指标，血压升高是动脉粥样硬化和冠心病的重要危险因素，也是心力衰竭的重要原因，严重威胁人类健康和生命安全。

目前，关于高血压与大气污染的研究主要分为两类：短期暴露研究和长期暴露研究。前者研究大气污染物的急性效应，主要包括短期的时间序列分析；后者研究大气污染物的慢性效应，例如多年暴露的队列研究。目前，关于大气污染物与高血压发生关系的已知的样本量最大的研究是在加拿大安大略省进行的人群队列研究[78]，该研究从加拿大全国人口调查和社区健康调查人群中筛选年龄≥35 岁、无高血压、无心血管疾病共 35303 名居民进行追踪，随访时间是 1996~2010 年，使用安大略省高血压数据库确定随访期间高血压发生情况，通过卫星遥感获得地面 $PM_{2.5}$ 浓度数据，结果发现，研究期间共有 8649 人发生高血压，2296 人死亡，调整年龄和性别后，$PM_{2.5}$ 每增加 10 μg/m³，高血压发生的风险增加 10%。Kateryna 等[79]研究了欧洲 9 个国家 15 个队列人群血压变化与交通相关空气污染的关系，这是关于大气污染物与血压之间关系研究的最大规模队列研究，结果显示，距住宅 100 m 以内的主要道路上的交通负荷与非药物参与者的收缩压和舒张压升高具有正相关关系。总的来说，对于急性暴露研究，Meta 分析结果显示，SO_2、$PM_{2.5}$ 和 PM_{10} 的浓度每增加 10 μg/m³，高血压的患病风险与 SO_2、$PM_{2.5}$ 和 PM_{10} 的短期暴露显著相关[80-86]。而 NO_2 和 PM_{10} 的浓度每增加 10 μg/m³，高血压的患病风险与 NO_2 和 PM_{10} 的长期暴露显著相关[78, 86-95]。除流行病学研究证据外，实验室毒理实验的结果也证明了大气污染物与高血压之间的相关关系。Bartoli 等[96]将 13 只雌性混种犬暴露于大气颗粒物浓度为环境空气颗粒物浓度 30 倍的哈佛大学颗粒物浓缩实验室（HAPC）中［粒径 0.15～2.5 μm，浓度（358.1±306.7）μg/m³］，结果显示，实验组收缩压、舒张压和平均脉压均显著升高。此外，通过对 3 个英文数据库（PubMed、Embase 和 ISI Web of Science）和 4 个中文数据库（CNKI、维普、中国生物医学文献数据库、万方）2017 年 5 月 25 日前的所有文献进行系统检索发现，大气污染与血压升高和高血压呈正相关关系，且地理和社会人口因素也会影响大气污染对高血压的影响，例如，男性、亚洲人、北美人和大气污染水平较高的地区人群，该相关关系表现的更为显著[97]。

大气污染物导致高血压发生的毒性作用机制在于系统性炎症反应和氧化压力，从而引发动脉周围交感神经的重构[98, 99]。Schins 等研究发现，暴露于 $PM_{2.5}$ 后，大鼠血液内毒素含量、IL-8 和 TNF-α 水平显著升高[100]。氧化应激反应也会增加炎性因子的循环，继而导致内皮功能紊乱，血管稳态失衡，氮氧化物的生物利用率降低，致使外周血管收缩，从而最终导致高血压的发生[99, 101]。Lu 等[102]通过体内外实验，证实 $PM_{2.5}$ 暴露导致活性氧水平升高，损害肾脏 D1 受体介导的钠排泄，从而引起高血压的发生。此外，大气颗粒物可使自主神经系统失衡，交感神经的敏感性增加，神经调控血管收缩作用增强，从而导致血压升高[90]。

3.2　大气污染物与动脉粥样硬化

动脉粥样硬化（atherosclerosis，AS）是心血管疾病最主要的病理基础[103, 104]。美国心脏协会发布的《心脏病与卒中统计数据（2017 版）》显示，心血管疾病所致死亡中冠状动脉粥样硬化性心脏病排第一位。我国 2017 年度《中国心血管病报告》显示，我国居民心血管疾病占总死亡原因的 40%以上，其中冠状动脉粥样硬化性心脏病居心血管病患病率和死亡率的第一位，并且其患病率和死亡率仍处于持续上升阶段。

Koken 等[105]收集了美国科罗拉多州丹佛市 1993 年 7 月～1997 年 8 月的 PM_{10} 和气态污染物（O_3、NO_2、SO_2 和 CO）的数据，将其与同时期 65 岁以上男性女性因心血管系统疾病住院人数进行比较分析，结果显示，O_3 与冠状动脉粥样硬化住院风险增加有关。

Rivera 等[106]分析了 2007～2010 年西班牙赫罗纳省 2780 名参与者交通相关空气污染（主要是 NO_2）与亚临床动脉粥样硬化［颈动脉内膜-中层厚度（IMT）和踝臂厚度指数（ABI）］之间的关系，这些参与者分别来自 12 个城镇，代表了该地区地理多样性，且环境空气污染水平也相差较大。结果显示，长期交通相关暴露于动脉粥样硬化亚临床标志物之间表现出了显著的相关性，且在受教育程度高的人群和 60 岁以上男性人群中表现出了更强的关联性。

2005 年，Künzli 等[107]对 798 名临床患者的居住区进行地理编码，以确定环境 $PM_{2.5}$ 的年平均浓度，通过检测患者 IMT 的变化来揭示 $PM_{2.5}$ 长期暴露与 AS 之间的关系，结果显示，IMT 与 $PM_{2.5}$ 存在一定的相关性，并且这一关系在老年女性患者中更为显著，该研究是 AS 与大气细颗粒物污染之间联系的第一个流行病学证据。Hoffmann 等[108]通过德国人口稠密和高度工业化地区的 4494 位中老年人的队列研究发现，$PM_{2.5}$ 长期暴露与冠状动脉钙化形成具有相关性。Allen 等[109]对美国 5 个城市的 1147 名人群进行了队列研究，结果发现腹主动脉钙化风险与 $PM_{2.5}$ 的长期暴露显著相关。因此，大气 $PM_{2.5}$ 污染与动脉粥样硬化之间存在着明确的相关性。

血管内皮细胞损伤被认为是 AS 发病的早期关键性环节[110]，而血管内皮作为体内的芳香烃受体，对二噁英、多氯联苯、PAHs 等环境污染物造成的损伤具有高度的敏感性[111]。Wang 等[112]研究发现人脐静脉内皮细胞（HUVECs）暴露于柴油排放 $PM_{2.5}$ 后，被细胞吞噬的颗粒物不能被细胞自噬所消除，胞内 ROS 不断积累，最终导致内皮功能障碍。Rui 等[113]发现 $PM_{2.5}$ 暴露会引发内皮细胞 ROS 的产生，进而导致细胞活力显著下降，内皮功能紊乱，在使用 N-乙酰半胱氨酸（NAC）作为 ROS 的清除剂后，可抑制 ROS 的增加，对内皮细胞起到保护作用。Bo 等[114]研究发现大气 $PM_{2.5}$ 对内皮细胞造成氧化压力，引起 HUVECs 的 ROS 水平升高、超氧化物歧化酶（SOD）水平降低，进而通过检测到 IL-6 和 TNF-α 的分泌发现内皮细胞引起了严重的炎性反应。综上所述，$PM_{2.5}$对HUVECs的毒性及作用机制，主要在于通过氧化压力产生ROS，继而引发内皮细胞功能紊乱与炎症反应发生，而这些不利影响与AS的发生发展密切相关。

体外实验通常影响因素少，可控性好，但是体内的细胞需要经过内环境才能与外界进行物质交换，因此，需要通过选择合适的动物模型来验证污染物在体内的实际毒性作用情况。Araujo等[115]将载脂蛋白E基因敲除（$ApoE^{-/-}$）小鼠暴露于洛杉矶高速公路附近，研究大气颗粒物促AS的作用，结果显示，$PM_{2.5}$暴露小鼠会产生全身氧化应激反应，其早期AS病变较暴露于洁净空气的小鼠更为严重。Ying等[116]在使用高脂饲料喂食$ApoE^{-/-}$小鼠的同时进行$PM_{2.5}$暴露实验，结果显示小鼠主动脉诱导型一氧化氮合成酶（iNOS）表达明显增加，同时超氧化物歧化酶（SOD）生成增加，蛋白质大量硝化，复合斑块面积明显增加，这就证实了$PM_{2.5}$通过诱导血管活性氧和反应性氮物种而导致AS的发生。

3.3 大气污染物与心力衰竭

心力衰竭是指心脏的舒张功能和收缩功能发生障碍，不能将静脉回心血量充分排出心脏，从而导致动脉系统灌注不足，无法为周围组织提供血液和氧气以满足其代谢需求。在病理生理学上，心输出量（CO）绝对或相对量降低。心力衰竭是心脏疾病发展的终末阶段，主要危险因素包括冠状动脉疾病、高血压、糖尿病、心脏病家族史、肥胖、慢性肺病或使用心脏毒素等，最终可导致心脏性猝死或全身灌注不足导致的慢性多器官衰竭[117]。我国统计数据得出，广州每日大气污染物PM_{10}、SO_2或NO_2每升高10 $\mu g/m^3$，由心力衰竭发病导致的每日紧急救护车调度分别增加3.54%、5.29%和4.34%[118]。在美国宾夕法尼亚州阿勒格尼县，研究者通过病例交叉方法评估环境空气污染与居住地医疗保险接受者（年龄≥65岁）的充血性心力衰竭住院率之间的关系，揭示PM_{10}、CO、NO_2和SO_2与住院率显著正相关，其中CO和NO_2的关联性最强；这些结果表明，来自交通相关来源的空气污染的短期升高可能引发老年心力衰竭患者的急性心脏代偿反应，并且具有某些合并症的人可能更容易受到这些影响[119]。

纤维化是细胞自我强化的一种状态，伴随着成纤维细胞的增殖和活化成更活跃的肌成纤维细胞、胶原蛋白沉积和细胞外基质的净积累、胶原交联变化和胶原蛋白周转失衡，是一种常见的组织对慢性损伤的病理生理反应。一方面，伤口愈合和组织重塑等修复保护机制响应细胞压力和损伤，以保持器官的功能完整性系统。另一方面，正常愈合部位放松管制并继续暴露于慢性损伤最终导致组织纤维化、细胞外基质的大量沉积、瘢痕形成和器官衰竭[120]。纤维化几乎存在于所有心脏疾病的病理学过程，是导致心肌僵硬度增加、心室舒张和收缩功能紊乱以及继发心力衰竭的重要因素[121-123]。纤维化可由一系列信号激活，包括：①上皮/内皮屏障急性损伤；②纤维化细胞因子的释放，主要是TGFβ；③肌成纤维细胞的基质活化；④炎症细胞的招募；⑤ROS的诱导；⑥胶原生成细胞的激活[120]。研究表明，当心肌胶原蛋白含量升高2~3倍时可导致心室舒张期的硬度增加以及充盈异常，升高4倍以上可导致心室收缩功能下降[124]。

心脏舒缩功能一般由多普勒超声心动图法测定，其中表征左心室收缩功能的参数是射血分数EF和缩短分数FS，表征左心室舒张功能的参数有E/A、E/E'和E'/A'。EF指每搏输出量占心室舒张末期容积量的百分比，与心脏的收缩功能呈正相关，即左室收缩能力越强，EF值越大。FS值过低则表示左室收缩功能发生障碍。

气态污染物对机体的纤维化效应主要集中在肺部，对心肌纤维化研究甚少。秦国华等对SD大鼠进行了7 mg/m^3 SO_2染毒30天，随后通过超声检测大鼠心脏功能，结果发现SO_2暴露显著降低表征左心室收缩功能的指标EF和FS，而NALC预处理可缓解SO_2诱导的左心室收缩功能抑制[125]。这表明，当大鼠暴露于SO_2时，心脏喷射血液效率受损，心脏没有将足够的血液泵送到身体的其他部位。

$PM_{2.5}$损害心血管的途径可以简单地概括为3条主要途径[126]：①颗粒吸入肺泡，激发局部炎症反应，随后又继发全身炎症反应，并通过全身炎症反应影响心血管系统；②极细颗粒可以直接跨越肺泡膜转位到血液中，从而直接影响内皮功能，或损害线粒体等细胞器[127]；③颗粒物质通过激活肺泡表面的敏感受体来改变自主神经活动，从而间接影响心血管系统。上述三种过程均可促使一系列与 CVD 相关的病理学过程的发生，如血管收缩、内皮紊乱、动脉粥样硬化、血压升高等血管方面的改变，或血栓形成、血小板凝集、血脂异常等血液方面的变化[128, 129]。近几年研究者发现，除上述血管和血液异常外，PM 还可引发心肌肥大与纤维化，而心肌纤维化的发生可进一步导致心室舒缩功能改变。

吸入 $PM_{2.5}$ 与不良的心室重塑和心肌肥大以及心肌纤维化的恶化相关[130, 131]。大气 $PM_{2.5}$ 的化学组成非常复杂，与其来源、形成方式、粒径、自然气候条件等多种因素密切相关，因此 $PM_{2.5}$ 的组成成分及特性会随着季节的更替而发生变化，由此引起的人体健康效应也不尽相同。另外，研究表明易感人群如老年人、女性、患有冠状动脉疾病和糖尿病的患者等对于 $PM_{2.5}$ 诱导引发的 CVD 发病率和死亡率承担更大的风险[128, 132]。

采集太原市春夏秋冬四季的 $PM_{2.5}$ 分别对 H9C2 大鼠心肌细胞进行染毒，发现四季 $PM_{2.5}$ 暴露均可引起纤维化标志基因 *Col1a1* 的表达升高，且冬季 $PM_{2.5}$ 暴露后升高最为显著。而太原市冬季 $PM_{2.5}$ 处理不同年龄（4 周龄、4 月龄和 10 月龄）小鼠 4 周，结果显示 $PM_{2.5}$ 暴露可诱导幼年和老年组小鼠胶原沉积及纤维化标志物表达的升高，其中老年组小鼠变化最为明显[123]。其他针对浓缩的大气 PM 和炭黑暴露的研究也得到了相似的结果[133-135]。交通源 $PM_{2.5}$ 主要来源于柴油机废气，有研究报道，与过滤空气对照组或成年后暴露柴油废气的小鼠相比，孕期暴露于柴油废气的小鼠在主动脉缩减术（TAC）后，对心肌纤维化的易感性显著增加[131]。总而言之，由于北方城市冬季燃煤影响，四季中冬季 $PM_{2.5}$ 所诱导的纤维化效应最为显著；而从纤维化相关基因的表达情况来看，老年小鼠和胎儿阶段及幼年时期小鼠对纤维化更为敏感。

当涉及 $PM_{2.5}$ 诱导心肌纤维化分子机制时，大多数研究往往集中在炎症反应、氧化应激及 TGFβ 的释放。研究发现心肌细胞中 ROS 形成的增加与老年人心脏的舒张功能降低等不利变化有关[136]。炭黑颗粒急性暴露 4 天能够降低老年小鼠的心脏收缩性，并且这种功能下降与 ROS 产生增加相关[135]。秦国华等发现冬季 $PM_{2.5}$ 暴露 4 周后，与对照组相比，老年小鼠心脏和肺部 MDA 含量、ROS、炎性因子 IL-6 显著上调，这一结果表明 $PM_{2.5}$ 暴露可以诱导心脏和肺部氧化损伤及炎症反应，揭示 $PM_{2.5}$ 通过氧化应激和炎症反应诱导心脏损伤的可能性[123]。此外，在胎儿和幼儿阶段暴露柴油废气，明显改变了小鼠成年后肺中的炎性细胞因子，也间接支持了这一观点[131]。TGFβ1 是纤维化发生过程中的重要介导因子，可通过与其受体 TGFβR1 及 TGFβR2 结合激活典型的 *TGFβ/Smad* 信号通路，磷酸化 SMAD2/3，从而诱导胶原含量的增加，也可能通过不依赖于 *Smad* 的信号通路，如激活 NOX4 氧化还原通路使胶原产生和分泌增加，进而调控纤维母细胞的表型和功能。$PM_{2.5}$ 暴露可导致 4 周龄和 10 月龄小鼠心脏组织中通路蛋白 NOX-4 和 TGFβ1 表达量增加，并激活了 SMAD，提示心肌纤维化的发生可能与 NOX4-TGFβ1-Smad 信号通路相关[123]。

3.4 大气污染物与心肌纤维化

心肌纤维化的发生可进一步导致心室舒缩功能改变。PM 急性和慢性暴露都可能导致心脏功能紊乱，而老年小鼠和胎儿阶段及幼年时期小鼠更为敏感。Clarke 等研究了 18 月龄和 28 月龄小鼠炭黑急性暴露 3 h 后，第四天心脏功能的变化，结果发现 28 月龄暴露组小鼠的收缩末期直径（LVEds）和舒张末期左室直径（LVEdd）显著增加，FS、舒张期后壁厚度（RWT）和 EF 明显降低；而这些指标在 18 月龄组的小鼠中则未观察到明显变化，上述数据揭示急性炭黑暴露更易导致衰老小鼠心脏功能受损[137]。为了进一步确定衰老小鼠的易感性，对不同年龄小鼠（4 周龄、4 月龄以及 10 月龄）亚慢性暴露于太原市冬季 $PM_{2.5}$

4周，发现 $PM_{2.5}$ 可显著升高 10 月龄小鼠的心率和血压，且有显著的时间效应关系；4 周龄小鼠心率只与暴露前相比有明显差异，收缩压无明显趋势；4 月龄小鼠心率血压均无明显变化。由组织多普勒测量可知，$PM_{2.5}$ 暴露可显著改变所有年龄组小鼠的心脏舒张功能，但仅有老年组小鼠的收缩功能会受 $PM_{2.5}$ 暴露的影响[123]。以上结果进一步证实，$PM_{2.5}$ 暴露可诱导小鼠心脏功能紊乱，主要表现在舒张功能上，且老年组小鼠更敏感。

然而，对成年小鼠长期暴露 PM 也可能导致心脏功能障碍：Wold 及其同事研究发现，浓缩 $PM_{2.5}$ 慢性暴露 9 个月可引起雄性成年小鼠心脏重塑并引起心脏舒缩功能障碍，与对照组相比，LVEds 和 LVEdd 均增加，这些变化进而导致左室收缩功能降低；而暴露组表征舒张功能的指标 E/A 值低于对照组[134]。除此之外，胎儿阶段及幼年时期暴露于柴油尾气或浓缩的 $PM_{2.5}$ 导致在成年后表现出明显的心脏重构，EF 和 FS 减少，LVEds、LVEdd、收缩末期容积（LVEsv）、舒张末期容积（LVEdv）均增加，这些结构改变与 $PM_{2.5}$ 暴露小鼠的左室收缩性降低相关，最终导致明显的舒张功能障碍，证明早期暴露于 $PM_{2.5}$ 导致成年后心脏功能障碍和心力衰竭的易感性增加[138, 139]。另外，$PM_{2.5}$ 暴露 3 周可通过诱导肺部氧化应激，从而加剧心衰模型小鼠（横向主动脉缩窄术 TAC）的肺部炎症、血管重塑和右心室肥大，导致心衰发病率增加[140]。

实验和临床证据表明心肌纤维化的发生可能是可逆的[141]。有研究表明慢性暴露浓缩 $PM_{2.5}$ 15 周可诱发高血压模型大鼠（5 周龄）血压升高，显著降低心脏搏出量和输出量；而在停止暴露 2 周后，这些指标均得到了恢复，表明长期暴露于 $PM_{2.5}$ 可导致可逆性心脏功能障碍和心肌肥厚[142]。对 10 月龄敏感小鼠 $PM_{2.5}$ 暴露 4 周后进行为期 2 周的恢复实验，结果发现，停止暴露后各项指标，如心率、收缩压、心脏收缩和舒张功能在两周内可以恢复至正常水平，收缩压早于心率发生恢复，收缩功能早于舒张功能发生恢复。此外，沉积的胶原蛋白在停止暴露后可以被分解。停止暴露 1 周和 2 周后，与暴露 4 周 $PM_{2.5}$ 组相比，心脏和肺部 MDA 均显著降低。以上结果表示 $PM_{2.5}$ 对心肺氧化损伤、心脏功能紊乱及心肌纤维化效应均是可逆的[123]。

为了全面地考察 $PM_{2.5}$ 暴露对老年小鼠的心脏损伤，张英英等基于 GC-MS 的代谢组学分析技术对小鼠心脏代谢物进行非靶向分析，共发现了 40 种差异代谢物，与对照组相比，$PM_{2.5}$ 组中有 24 种代谢物上调，16 种代谢物下调，差异代谢物主要富集在碳水化合物代谢（延胡索酸、苹果酸、核糖 5-磷酸盐、柠檬酸、琥珀酸、苏糖酸、草酸、肌醇、磷酸盐、胆固醇、丙酮酸、乳酸、2-羟丁酸和丙三醇）、脂肪酸代谢（壬二酸、十二烷酸、壬酸、戊二酸和花生四烯酸）、氨基酸代谢（天门冬氨酸、戊二酸、焦谷氨酸、丙氨酸、4-羟脯氨酸、甲酸、赖氨酸、巯基乙胺、异亮氨酸、亮氨酸和缬氨酸）、核苷酸代谢（核糖 5-磷酸盐、腺嘌呤、腺苷酸和尿苷）以及烟碱酰胺代谢（烟碱酰胺）[143]。

4 大气污染物暴露对中枢神经系统的影响

大气污染是世界范围内一个主要的环境健康威胁。大气污染物来源广泛，成分复杂，流行病学和实验研究提示污染物暴露不仅会对心肺系统造成损害，还会通过多种途径对中枢神经系统产生不良影响，引发中风、认知功能障碍和神经退行性疾病［如帕金森（PD）、阿尔兹海默症（AD）、多发性硬化（MS）］等多种疾病发生。目前研究认为，持久的氧化应激、长期神经炎性以及内皮功能障碍是大气污染物直接诱导神经系统损伤的主要机制。

4.1 大气污染物与脑缺血性损伤

缺血性脑中风占中风发病率的 80%，与大脑血液供应不足有关，也称缺血性脑血管病，是指由于血栓对脑动脉的堵塞导致脑组织血液供应不足，神经元死亡造成偏瘫和意识障碍，基本病理反应主要表现为血液流变学改变、血脑屏障破坏、脑水肿以及脑梗死等[144]。内皮功能障碍和炎症反应是缺血性脑中风发生发展的主要分子机制[145, 146]。一方面，内皮功能损伤引起内皮依赖性的血管扩张减弱，促进脂质和细胞的渗透性、细胞外基质沉积或溶解、脂质蛋白氧化、血小板激活和血栓形成；另一方面，炎症反应可破坏血脑脊液屏障，损伤内皮细胞，炎性细胞在毛细血管中聚集和浸润，凝血级联反应和炎症恶性循环被激活，血栓形成加快将动脉管腔部分或完全堵塞，严重影响脑部血流的供应。由于脑组织对氧气和能量的需求量极大，缺血后会立刻发生自由基的呼吸爆破、钙离子超载和谷氨酸堆积，继而梗塞周边组织去极化和细胞凋亡加重进一步损伤脑组织。

中风是导致死亡的第二大原因，也是全球成人残疾的第三大原因，而大气污染已成为全球中风负担的第三大因素，占中风负担的 29.2%[147]。最早关于大气污染与中风的相关研究发生在 20 世纪 80 年代早期，英格兰和威尔士的一项研究调查了气象变量波动和脑血管死亡率之间的关系，发现与大气颗粒物空气污染水平强烈相关[148]。几年后，一项在中国的研究发现室内煤烟是中风的危险因素，与年龄，血压和吸烟无关[149]。最近研究表明，气态和颗粒物大气污染暴露与中风入院或中风死亡之间存在密切的时间关联[150]。因为中风是一种异质性疾病，其疾病类型依赖于潜在的血管危险因素和急性期触发，大气污染对中风的影响可能与触发因子的不同而有所差异。Lisabeth 等研究发现即使在污染水平相对较低的社区，$PM_{2.5}$ 和 O_3 暴露仍与缺血性卒中风明显相关[151]。Chen 等在中国进行了第一个全国医院的中风前瞻性登记队列，其中包括 2007~2008 年访问医院的 12291 名缺血性中风患者。所有患者均被随访 1 年并记录随访期间的死亡情况。参与者通过使用卫星遥控器估算患病前 3 年所接触环境中 PM_1，$PM_{2.5}$，PM_{10} 和 NO_2 的浓度。发现 PM_1 和 $PM_{2.5}$ 的前驱暴露与缺血性中风后的死亡率增加显著相关[152]。美国学者研究发现在 1999~2010 年间发生的 727 例中风中，539 例是缺血性的，122 例是出血性的，且 PM_{10} 与缺血性中风之间存在统计学显著相关性[153]。然而，流行病学结果受样本量、调查范围和评定方法等方面的局限，且混杂因素较多，使得其结果可靠性较低，无法给出定性而统一的结论，更不能阐明相关分子机制。

实验研究表明，短期暴露于 SO_2、NO_2 及 PM_{10} 等空气污染物可能通过心源性栓塞诱导缺血性中风发生[154]，且吸入的纳米颗粒可以在血管疾病部位累积[155]。另一研究发现利用不同浓度的 SO_2 处理 Wistar 大鼠，SO_2 浓度依赖性地升高 ET-1，iNOS，COX-2 和 Icam-1 mRNA 和蛋白表达水平。采用大脑中动脉线栓（MCAO）技术建立缺血性脑中风模型，建模成功后吸入过滤空气的模型大鼠大脑皮层内皮因子（*ET-1* 和 *eNOS*）和炎性因子（iNOS，COX-2 和 Icam-1）表达发生变化，神经元凋亡数目增多，并伴随有局灶性脑缺血坏死；而 SO_2 吸入暴露则会使这一变化显著增强，并抑制模型大鼠术后再灌注过程脑缺血局灶区域以及神经行为学的恢复，加重损伤效应[156]。在体和体外 SO_2 暴露实验证实，COX-2 催化花生四烯酸（AA）和内源性大麻素（eCBs）代谢参与海马神经元突触功能和缺血性损伤过程可作为损伤发生的效应标记，其介导的 AA 和 eCBs（主要是单酰基甘油类 eCBs，2-AG）脂质代谢反应是 SO_2 吸入诱导神经功能损伤的重要的调控环节：自由基攻击→COX-2 表达上调造成前列腺素 E2（PGE2）释放增加并作用于 EP2/EP4 受体（主要是 EP2 受体）→激活 cAMP/PKA 途径→神经元损伤[157]。

此外，利用 PM 暴露神经元，星形胶质细胞和小胶质细胞的体外研究发现，PM 暴露会导致突触功能的改变[158]和炎性细胞因子[159]的上调。采集自不同季节（春、夏、秋、冬）的 PM_{10} 样品对原代培养皮层神经元、神经元缺糖缺氧（OGD）模型和清洁级雄性 Wistar 大鼠进行染毒，发现不同季节 PM_{10} 染毒可诱导原代培养皮层神经元炎性标记物（iNOS、COX-2 和 ICAM-1）表达；增加磷酸化 Ca^{2+}/钙调蛋白依赖性蛋白激酶Ⅱα（p-CaMKⅡα）和p-CREB的蛋白水平；并诱发即早基因（*C-JUN*和*C-FOS*）的高表达。同时，增加原代培养皮层神经元OGD模型的易感性，并呈现明显的季节效应。另外，不同季节PM_{10}样品也能导致大鼠脑组织病理学损伤，诱导神经元凋亡加剧，引发内皮功能障碍、炎症反应，以及随后的突触损伤，诱导神经功能障碍[160]。低浓度$PM_{2.5}$暴露原代海马神经元，可诱导细胞COX-2的过度表达，并且$PM_{2.5}$处理海马脑片20 min即观察到PGE2的释放和增强的兴奋性突触传递，研究还发现ROS参与$PM_{2.5}$诱导的COX-2/PGE2过表达和随后的兴奋性突触传递，表明$PM_{2.5}$通过ROS-NF-κB途径激活COX-2表达，增加PGE2释放，增强兴奋性突触传递并诱发神经炎性的发生[161]。

4.2 大气污染物与神经退行性损伤

神经退行性疾病是一类严重威胁人类健康的疾病，其致病机理主要包括氧化应激机理、神经炎症机理、细胞凋亡机理等。长期暴露或急性暴露在大气污染物中可以直接损伤中枢神经系统，或污染物引起呼吸系统和免疫系统等产生有害因子，通过外周循环到达大脑，导致大脑的神经炎症、神经毒性、氧化应激等反应，最终产生神经退行性病变，如AD、PD、认知障碍和神经发育障碍等。AD是典型的神经退行性疾病之一，临床表现为渐进性记忆衰退和认知功能损伤等，病理学特征主要包括β-淀粉样蛋白（Aβ）沉积和TAU蛋白高度磷酸化，线粒体功能障碍是其早期症状。

国内外研究显示，交通源污染物或大气环境污染物接触可导致人群脑组织TAU蛋白过度磷酸化、Aβ斑块沉积、炎性因子表达升高、神经退行性病变以及学习记忆功能损伤。Jung等对我国台湾95690个人（年龄≥65岁）进行历时9年的队列随访研究显示，O_3浓度每增加10.91 ppb，AD患病率风险增加211%，$PM_{2.5}$浓度每增加4.34 μg/m³，AD患病风险会增加138%，该结果提示人群长期高于EPA标准暴露于O_3和$PM_{2.5}$与AD的患病风险增加显著相关[162]。Calderón-Garcidueñas团队对清洁区和污染区域儿童、成人和犬类额叶皮层的解剖样品进行分析，发现空气污染可导致脑组织基因异常表达、氧化应激、DNA损伤、神经炎性及AD型神经退行性病理学改变[163-165]；且污染区域的暴露人群中40%出现tau蛋白过度磷酸化，51%出现Aβ弥漫性斑块[166]。然而以上结果缺乏大气污染物暴露与疾病发生的实验证据，更缺少对其致病机理的深入解析。

现有研究证实，$PM_{2.5}$，SO_2和NO_2复合暴露能够造成小鼠空间学习记忆能力的衰退和引起细胞凋亡相关基因（*p53*，*BAX*和*BCL-2*）的异常表达。此外，这些改变与线粒体的形态变化，ATP的减少，线粒体分裂蛋白的升高和融合蛋白的下降相关[167]。SO_2吸入暴露诱导神经炎性和退行性变发生，改变突触结构和功能可塑性，损伤空间记忆能力，亚慢性暴露条件下（6 h/d，28 d）损伤发生的最低效应浓度为7 mg/m³[168,169]。这一过程与MAGL调控2-AG通路密切相关，可能涉及CB1R依赖的PPARγ信号途径[170]。$PM_{2.5}$作为全球疾病负担的主要环境风险因素之一，动物实验研究发现，$PM_{2.5}$与甲醛共暴露比单独暴露更有可能诱发AD样病变，其机制涉及氧化应激和炎症机制[171]。不同浓度的$PM_{2.5}$暴露可通过系统炎性造成小鼠海马组织突触结构和功能的改变，破坏空间学习记忆能力。在这一过程中，*Bace1*激活具有重要作用。在此基础上利用特异性干扰技术证实，神经炎性因子激活NF-κB p65，NF-κB p65与miR-574-5p启动子区的结合下调miR-574-5p表达，进而miR-574-5p与*Bace1* 3′UTR区结合反向调控*Bace1*表达，而miR-337-5p会通过与*Tau*（*Mapt*）非编码区3′端结合调控*Tau*的表达，进而影响空间学习记忆能力并最终导致突触可塑性破坏和神经功能损伤[172,173]。此外，在$PM_{2.5}$暴露后的小鼠海马组织中观察到显著的代谢组学改变，包括9个降低的代谢物和11个增加的代谢物。代谢变化主要涉及能量代谢，胆固醇代谢，花生四烯酸代谢，肌醇磷酸代谢和天冬氨酸代谢[174]。

同时，多项研究表明NO_2相关大气污染物或交通源污染物暴露与认知功能损伤及神经退行性病变相关。NO_2暴露显著上调大鼠皮层组织ROS和脂质过氧化产物MDA水平，引发氧化应激反应，导致线粒体膜损伤、嵴断裂和线粒体溶解等超微结构改变，并浓度依赖性地抑制线粒体呼吸链复合酶 II、IV、V 活性及 ATP 的合成，致使线粒体功能紊乱。此外，高浓度 NO_2 可显著降低线粒体生物合成调控因子（PGC-1α、NRF1 和 TFAM）的蛋白表达，抑制线粒体的自我修复[175]。进一步研究发现，NO_2 吸入暴露导致正常 C57BL/6J 和 AD 模型鼠空间学习记忆功能受损，增加 Aβ42 的合成及聚积。此过程伴随小胶质细胞、星形胶质细胞异常活化和神经元退行性改变。其间，NO_2 导致 APP/PS1 小鼠皮层 5487 个基因异常表达，KEGG 通路分析显示，差异基因主要富集于突触功能、学习记忆、Aβ 产生及花生四烯酸（AA）代谢等路径。在 NO_2 染毒期间注射 JZL184，可通过抑制单酰基甘油脂肪酶（MAGL）活性提高内源性 2-花生四烯酸甘油酯（2-AG）含量并降低 AA 代谢产物，缓解因暴露所致神经炎症、神经元退行、Aβ42 沉积及学习记忆功能障碍2)[176]。另一研究发现，NO_2 暴露还可导致脑组织胰岛素信号传导路径（IRS-1-AKT）紊乱，干扰下游糖原合成酶激酶-3β（GSK-3β）活性，引起 TAU 蛋白过度磷酸化，降低谷氨酸受体及突触后致密物表达。与此同时，NO_2 激活 MAPK/JNK 通路导致脑、肝脏、骨骼肌 IRS-1-AKT 路径异常，引发系统性胰岛素抵抗，并代偿性增加血清胰岛素水平，进一步恶化脑内胰岛素信号损伤，导致 tau 蛋白磷酸化增加[177]。

5 展 望

我国大气污染与健康的研究起步较晚，但近年来在大气污染毒理学和流行病学研究方法、理论和成果上均有显著的进步。我国大气污染毒理学研究尽管在发表论文数量上已跃居全球第二，但与欧美发达国家相比，仍存在诸多不足，与国外先进水平仍有不小的差距，未来应加强以下几个方面的工作。

（1）缺少亚慢性和慢性实验。慢性和亚慢性实验对于探索空气污染的长期毒性或未知毒性具有重要的作用。通过慢性和亚慢性动物实验，可以发现一些空气污染所致损伤的趋势性并观察其毒性作用的机制，同时探索机体细胞或组织在长期暴露中的改变，从而为预防或治疗空气污染所致损伤提供依据，但是目前由于实验仪器设备的限制，我国对于空气污染的慢性和亚慢性实验相对较少。

（2）我国目前还没有商品化的真实环境动态吸入染毒装置，难以开展对外界真实暴露情况的体内实验。环境与气候暴露的动态装置对于研究空气污染和气候的毒性具有重要的作用。该装置可模拟人体实际长期低剂量接触空气污染物的方式进行动物暴露；可通过此装置探索颗粒物对机体的急性、慢性毒性作用；暴露时长可以几天也可以数月。通过长期试验研究，观察颗粒物对机体各系统的毒性效应，发现早期生物学效应，便于今后对颗粒物毒性进行预防和治疗。我国学者在真实环境暴露研究方面中进行了探索[178]，发表了在华北重污染地区开展的长期染毒的研究结果，为深入认识空气污染的多器官毒性效应和机制提供重要科学证据。

（3）缺乏对一些特定疾病的动物模型研究，如高血压等心血管疾病、糖尿病、肿瘤等疾病的动物模型，以深入探讨大气污染物对这些慢性人类疾病的致病机理。

经济的持续高速增长、机动车保有量的快速增加和以煤炭为主的能源结构，使得我国当前正面临着严峻的大气污染形势，尤其是近年来我国中东部多次出现的大范围雾霾天气，更加重了人们对大气污染健康危害的忧虑。国内外多项权威研究均报道了 $PM_{2.5}$ 为代表的大气污染已成为我国最主要的环境问题和公共卫生问题之一。在当前的大气污染形势下，一方面应积极开展大气污染毒理学研究，可为深入和全面理解大气污染物的致病机制提供科学依据，也可为据此探索相应的效应阻断或防护措施提供一些思路。另一方面，应开展更多高质量的流行病学研究，为我国大气污染的健康风险提供更精确的估计，也为我国制修订环境空气质量标准、开展健康风险评估提供更充分的依据。

参 考 文 献

[1] Dockery D W, Pope C A, Xu X, et al. An association between air pollution and mortality in six U.S. cities. The New England Journal of Medicine, 1993, 329(24): 1753-1759.

[2] Lipfert F W, Perry H M, Jr, Miller J P, et al. The Washington University-EPRI Veterans' Cohort Mortality Study: Preliminary results. Inhal Toxicol, 2000, 12(Suppl 4): 41-73.

[3] WHO. WHO issues latest global air quality report: Some progress, but more attention needed to avoid dangerously high levels of air pollution. http://www.wpro.who.int/china/mediacentre/releases/2018/20180502-WHO-Issues-Latest-Air-Quality- Report/en/.Access Date: 2019/02/01.2018.

[4] Lin H, Wang X, Liu T, et al. Air pollution and mortality in China. Advances in Experimental Medicine and Biology, 2017, 1017: 103-121.

[5] Ding R, Jin Y, Liu X, et al. Characteristics of DNA methylation changes induced by traffic-related air pollution. Mutation Research-Genetic Toxicology and Environmental Mutagenesis, 2016, 796: 46-53.

[6] Zhang Y P, Li B Z, Huang C, et al. Ten cities cross-sectional questionnaire survey of children asthma and other allergies in China. Chinese Science Bulletin, 2013, 58(34): 4182-4189.

[7] Perez L, Declercq C, Iñiguez C, et al. Chronic burden of near-roadway traffic pollution in 10 European cities (APHEKOM network). European Respiratory Journal, 2013, 42(3): 594-605.

[8] Lavigne E, Belair M A, Rodriguez Duque D, et al. Effect modification of perinatal exposure to air pollution and childhood asthma incidence. European Respiratory Journal, 2018, epub ahead of print, doi: 10.1183/13993003.01884-2017.

[9] Chen F, Lin Z, Chen R, et al. The effects of $PM_{2.5}$ on asthmatic and allergic diseases or symptoms in preschool children of six Chinese cities, based on China, Children, Homes and Health (CCHH) project. Environmental Pollution, 2018, 232: 329-337.

[10] Weichenthal S, Bai L, Hatzopoulou M, et al. Long-term exposure to ambient ultrafine particles and respiratory disease incidence in Toronto, Canada: A cohort study. Environmental Health, 2017, 16(1): 64.

[11] Perez L, Declercq C, Iniguez C, et al. Chronic burden of near-roadway traffic pollution in 10 European cities (APHEKOM network). European Respiratory Journal, 2013, 42(3): 594-605.

[12] Hansel N N, Breysse P N, Mccormack M C, et al. A longitudinal study of indoor nitrogen dioxide levels and respiratory symptoms in inner-city children with asthma. Environmental Health Perspectives, 2008, 116(10): 1428-1432.

[13] Zhang X, Zhong W, Meng Q, et al. Ambient $PM_{2.5}$ exposure exacerbates severity of allergic asthma in previously sensitized mice. Journal of Asthma, 2015, 52(8): 785-794.

[14] Manners S, Alam R, Schwartz D A, et al. A mouse model links asthma susceptibility to prenatal exposure to diesel exhaust. Journal of Allergy and Clinical Immunology, 2014, 134(1): 63-72.

[15] Mei M, Song H, Chen L, et al. Early-life exposure to three size-fractionated ultrafine and fine atmospheric particulates in Beijing exacerbates asthma development in mature mice. Particle and Fibre Toxicology, 2018, 15(1): 13.

[16] Chowdhury P H, Okano H, Honda A, et al. Aqueous and organic extract of $PM_{2.5}$ collected in different seasons and cities of Japan differently affect respiratory and immune systems. Environmental Pollution, 2018, 235: 223-234.

[17] Huang K L, Liu S Y, Chou C C, et al. The effect of size-segregated ambient particulate matter on Th1/Th2-like immune responses in mice. PLoS One, 2017, 12(2): e0173158.

[18] Rychlik K A, Secrest J R, Lau C, et al. In utero ultrafine particulate matter exposure causes offspring pulmonary immunosuppression. Proceedings of the National Academy of Sciences of the United States of America, 2019, 116(9): 3443-3448.

[19] Noël A, Penn A, Zaman H, et al. Sex-specific lung functional changes in adult mice exposed only to second-hand smoke in utero. Respiratory Research, 2017, 18(1): 104.

[20] Singh S P, Chand H S, Langley R J, et al. Gestational exposure to sidestream (secondhand) cigarette smoke promotes transgenerational epigenetic transmission of exacerbated allergic asthma and bronchopulmonary dysplasia. The Journal of Immunology, 2017, 198(10): 3815-3822.

[21] Ferrini M, Carvalho S, Cho Y H, et al. Prenatal tobacco smoke exposure predisposes offspring mice to exacerbated allergic airway inflammation associated with altered innate effector function. Particle and Fibre Toxicology, 2017, 14(1): 30.

[22] Yue H, Yan W, Ji X, et al. Maternal exposure of BALB/c mice to indoor NO_2 and allergic asthma syndrome in offspring at adulthood with evaluation of DNA methylation associated Th2 polarization. Environmental Health Perspecttives, 2017, 125(9): 097011.

[23] Yue H, Yan W, Ji X, et al. Maternal exposure to NO_2 enhances airway sensitivity to allergens in BALB/c mice through the JAK-STAT6 pathway. Chemosphere, 2018, 200: 455-463.

[24] Niedzwiecki M, Zhu H, Corson L, et al. Prenatal exposure to allergen, DNA methylation, and allergy in grandoffspring mice. Allergy, 2012, 67(7): 904-910.

[25] Gruzieva O, Xu C J, Breton C V, et al. Epigenome-wide meta-analysis of methylation in children related to prenatal NO_2 air pollution exposure. Environmental Health Perspecttives, 2017, 125(1): 104-110.

[26] Vogelmeier C F, Criner G J, Martinez F J, et al. Global Strategy for the diagnosis, management, and prevention of chronic obstructive lung disease 2017 report. GOLD Executive Summary. American Journal of Respiratory and Critical Care Medicine, 2017, 195(5): 557-582.

[27] Viegi G, Maio S, Pistelli F, et al. Epidemiology of chronic obstructive pulmonary disease: Health effects of air pollution. Respirology, 2006, 11(5): 523-532.

[28] Ko F W, Tam W, Wong T W, et al. Temporal relationship between air pollutants and hospital admissions for chronic obstructive pulmonary disease in Hong Kong. Thorax, 2007, 62(9): 780-785.

[29] Audi C, Baiz N, Maesano C N, et al. Serum cytokine levels related to exposure to volatile organic compounds and $PM_{2.5}$ in dwellings and workplaces in French farmers - a mechanism to explain nonsmoking COPD. International Journal of Chronic Obstructive Pulmonary Disease, 2017, 12: 1363-1374.

[30] Anderson H R, Bremner S A, Atkinson R W, et al. Particulate matter and daily mortality and hospital admissions in the west midlands conurbation of the United Kingdom: Associations with fine and coarse particles, black smoke and sulphate. Occupational and Environmental Medicine, 2001, 58(8): 504-510.

[31] Belleudi V, Faustini A, Stafoggia M, et al. Impact of fine and ultrafine particles on emergency hospital admissions for cardiac and respiratory diseases. Epidemiology, 2010, 21(3): 414-423.

[32] Schwartz J, Dockery D W, Neas L M. Is daily mortality associated specifically with fine particles? Journal of the Air & Waste Management Association, 1996, 46(10): 927-939.

[33] Neuberger M, Moshammer H, Rabczenko D. Acute and subacute effects of urban air pollution on cardiopulmonary emergencies and mortality: Time series studies in Austrian cities. International Journal of Environmental Research and Public Health, 2013, 10(10): 4728-4751.

[34] Andersen Z J, Hvidberg M, Jensen S S, et al. Chronic obstructive pulmonary disease and long-term exposure to traffic-related air pollution: A cohort study. American Journal of Respiratory and Critical Care Medicine, 2011, 183(4): 455-461.

[35] Hansel N N, Mccormack M C, Belli A J, et al. In-home air pollution is linked to respiratory morbidity in former smokers with chronic obstructive pulmonary disease. American Journal of Respiratory and Critical Care Medicine, 2013, 187(10): 1085-1090.

[36] Van Vliet EDS, Kinney P L, Owusu-Agyei S, et al. Current respiratory symptoms and risk factors in pregnant women cooking with biomass fuels in rural Ghana. Environment International, 2019, 124: 533-540.

[37] Hansel N N, Mccormack M C, Belli A J, et al. In-home air pollution is linked to respiratory morbidity in former smokers

with chronic obstructive pulmonary disease. American Journal of Respiratory and Critical Care Medicine, 2013, 187(10): 1085-1090.

[38] Wu S, Ni Y, Li H, et al. Short-term exposure to high ambient air pollution increases airway inflammation and respiratory symptoms in chronic obstructive pulmonary disease patients in Beijing, China. Environment International, 2016, 94: 76-82.

[39] Martinez F D. Early-life origins of chronic obstructive pulmonary disease. The New England Journal of Medicine, 2016, 375(9): 871-878.

[40] 彭青和, 李泽庚, 刘向国. 等. 慢性阻塞性肺疾病动物模型建立研究进展. 辽宁中医药大学学报, 2013, (9): 116-119.

[41] Grady S T, Koutrakis P, Hart J E, et al. Indoor black carbon of outdoor origin and oxidative stress biomarkers in patients with chronic obstructive pulmonary disease. Environment International, 2018, 115: 188-195.

[42] Chen C Y, Chen C P, Lin K H. Biological functions of thyroid hormone in placenta. International Journal of Molecular Sciences, 2015, 16(2): 4161-4179.

[43] Gu X Y, Chu X, Zeng X L, et al. Effects of $PM_{2.5}$ exposure on the Notch signaling pathway and immune imbalance in chronic obstructive pulmonary disease. Environmental Pollution, 2017, 226: 163-173.

[44] Vaughan A, Stevanovic S, Jafari M, et al. The effect of diesel emission exposure on primary human bronchial epithelial cells from a COPD cohort: N-acetylcysteine as a potential protective intervention. Environmental Research, 2019, 170: 194-202.

[45] Leclercq B, Kluza J, Antherieu S, et al. Air pollution-derived $PM_{2.5}$ impairs mitochondrial function in healthy and chronic obstructive pulmonary diseased human bronchial epithelial cells. Environmental Pollution, 2018, 243(Pt B): 1434-1449.

[46] Van Voorhis M, Knopp S, Julliard W, et al. Exposure to atmospheric particulate matter enhances Th17 polarization through the aryl hydrocarbon receptor. PLoS One, 2013, 8(12): e82545.

[47] Wegmann M, Fehrenbach A, Heimann S, et al. NO_2-induced airway inflammation is associated with progressive airflow limitation and development of emphysema-like lesions in C57BL/6 mice. Experimental and Toxicologic Pathology, 2005, 56(6): 341-350.

[48] Ji X, Zhang Y, Ku T, et al. MicroRNA-338-5p modulates pulmonary hypertension-like injuries caused by SO_2, NO_2 and $PM_{2.5}$ co-exposure through targeting the HIF-1alpha/Fhl-1 pathway. Toxicology Research (Camb), 2016, 5(6): 1548-1560.

[49] Ji X, Yue H, Ku T, et al. Histone modification in the lung injury and recovery of mice in response to $PM_{2.5}$ exposure. Chemosphere, 2019, 220: 127-136.

[50] Du Y, Ding Y, Chen X, et al. MicroRNA-181c inhibits cigarette smoke-induced chronic obstructive pulmonary disease by regulating CCN1 expression. Respiratory Research, 2017, 18(1): 155.

[51] Drummond D, Baravalle E M, Lezmi G, et al. Combined effects of in utero and adolescent tobacco smoke exposure on lung function in C57Bl/6J mice. Environmental Health Perspecttives, 2017, 125(3): 392-399.

[52] Loomis D, Grosse Y, Lauby S B, et al. The carcinogenicity of outdoor air pollution. Lancet Oncology, 2013, 14(13): 1262-1263.

[53] Evans J, Donkelaar A V, Martin R V, et al. Estimates of global mortality attributable to particulate air pollution using satellite imagery. Environmental Research, 2013, 120(1): 33-42.

[54] Haidong K. Globalisation and environmental health in China. Lancet, 2014, 384(9945): 721-723.

[55] Pope C A, Burnett R T, Thun M J, et al. Lung cancer, cardiopulmonary mortality, and long-term exposure to fine particulate air pollution. JAMA, 2002, 287(9): 1132-1141.

[56] Turner M C, Krewski D, Pope C A, et al. Long-term ambient fine particulate matter air pollution and lung cancer in a large cohort of never-smokers. American Journal of Respiratory and Critical Care Medicine, 2011, 184(12): 1374-1381.

[57] Ru J H, Yanlin Z, Carlo B, et al. High secondary aerosol contribution to particulate pollution during haze events in China. Nature, 2014, 514(7521): 218-222.

[58] 陈士杰, 李秀央, 周连芳. 大气污染物致肺癌的潜伏期灰色定量分析. 中华流行病学杂志, 2003, 24(3): 233-235.

[59] Cao J, Yang C, Li J, et al. Association between long-term exposure to outdoor air pollution and mortality in China: A cohort study. Journal of Hazardous Materials, 2011, 186(2-3): 1594-1600.

[60] Turkan S, Kalkan M, Ahin C K. An Association between long-term exposure to ambient air pollution and mortality from lung cancer and respiratory diseases in Japan. Journal of Epidemiology, 2011, 21(2): 132.

[61] Tseng C Y. Cell type specificity of female lung cancer associated with sulfur dioxide from air pollutants in Taiwan: An ecological study. BMC Public Health, 2012, 12(1): 4.

[62] Chen G, Wan X, Yang G, et al. Traffic-related air pollution and lung cancer: A meta-analysis. Thorac Cancer, 2015, 6(3): 307-318.

[63] Chen X, Zhang L W, Huang J J, et al. Long-term exposure to urban air pollution and lung cancer mortality: A 12-year cohort study in Northern China. Science of the Total Environment, 2016, 571: 855-861.

[64] Eckel S P, Cockburn M, Shu Y H, et al. Air pollution affects lung cancer survival. Thorax, 2016, 71(10): 891-898.

[65] Atkinson R W, Butland B K, Anderson H R, et al. Long-term concentrations of nitrogen dioxide and mortality: A meta-analysis of cohort studies. Epidemiology, 2018, 29(4): 460-472.

[66] Lamichhane D K, Kim H C, Choi C M, et al. Lung cancer risk and residential exposure to air pollution: a Korean population-based case-control study. Yonsei Medical Journal, 2017, 58(6): 1111-1118.

[67] Pereira F A, Lemos M, Mauad T, et al. Urban, traffic- related particles and lung tumors in urethane treated mice. Clinics (Sao Paulo), 2011, 66(6): 1051-1054.

[68] Yang J, Chen Y, Yu Z, et al. Changes in gene expression in lungs of mice exposed to traffic-related air pollution. Molecular and Cellular Probes, 2018, 39: 33-40.

[69] Yang B, Xiao C. $PM_{2.5}$ exposure significantly improves the exacerbation of A549 tumor-bearing CB17-SCID mice. Environmental Toxicology and Pharmacology, 2018, 60: 169-175.

[70] Yue H, Yun Y, Gao R, et al. Winter polycyclic aromatic hydrocarbon-bound particulate matter from peri-urban north china promotes lung cancer cell metastasis. Environmental Science & Technology, 2015, 49(24): 14484-14493.

[71] Kamens R M, Guo J, Guo Z, et al. Polynuclear aromatic hydrocarbon degradation by heterogeneous reactions with N_2O_5 on atmospheric particles. Atmospheric Environmentpart Ageneral Topics, 1990, 24(5): 1161-1173.

[72] Nielsen T. Reactivity of polycyclic aromatic hydrocarbons towards nitrating species. Environmental Science & Technology, 1984, 18(3): 157.

[73] Durant J L, Jr W F B, Lafleur A L, et al. Human cell mutagenicity of oxygenated, nitrated and unsubstituted polycyclic aromatic hydrocarbons associated with urban aerosols. Mutation Research, 1996, 371(3-4): 123.

[74] Reisen F, Arey J. Atmospheric reactions influence seasonal PAH and nitro-PAH concentrations in the Los Angeles basin. Environmental Science & Technology, 2005, 39(1): 64-73.

[75] Wang W, Jariyasopit N, Schrlau J, et al. Concentration and photochemistry of PAHs, NPAHs, and OPAHs and toxicity of $PM_{2.5}$ during the Beijing Olympic Games. Environmental Science & Technology, 2011, 45(16): 6887-6895.

[76] Wei H, Liang F, Cheng W, et al. The mechanisms for lung cancer risk of $PM_{2.5}$: Induction of epithelial-mesenchymal transition and cancer stem cell properties in human non-small cell lung cancer cells. Environmental Toxicology, 2017, 32(11): 2341-2351.

[77] Yue H, Yun Y, Gao R, et al. Winter polycyclic aromatic hydrocarbon-bound particulate matter from Peri-urban North China promotes lung cancer cell metastasis. Environmental Science & Technology, 2015, 49(24): 14484-14493.

[78] Chen H, Burnett R T, Kwong J C, et al. Spatial association between ambient fine particulate matter and incident hypertension. Circulation, 2014, 129(5): 562-569.

[79] Fuks K B, Weinmayr G, Foraster M, et al. Arterial blood pressure and long-term exposure to traffic-related air pollution: an analysis in the European Study of Cohorts for Air Pollution Effects (ESCAPE). Environmental Health Perspecttives, 2014,

122(9): 896-905.

[80] Guo Y, Tong S, Li S, et al. Gaseous air pollution and emergency hospital visits for hypertension in Beijing, China: A time-stratified case-crossover study. Environmental Health, 2010, 9: 57.

[81] Guo Y, Tong S, Zhang Y, et al. The relationship between particulate air pollution and emergency hospital visits for hypertension in Beijing, China. Science of the Total Environment, 2010, 408(20): 4446-4450.

[82] Szyszkowicz M, Rowe B H, Brook R D. Even low levels of ambient air pollutants are associated with increased emergency department visits for hypertension. Canadian Journal of Cardiology, 2012, 28(3): 360-366.

[83] Brook R D, Kousha T. Air pollution and emergency department visits for hypertension in edmonton and calgary, Canada: A case-crossover study. American Journal of Hypertension, 2015, 28(9): 1121-1126.

[84] Nascimento L F, Francisco J B. Particulate matter and hospital admission due to arterial hypertension in a medium-sized Brazilian city. Cadernos de Saude Publica, 2013, 29(8): 1565-1571.

[85] Qorbani M, Yunesian M, Fotouhi A, et al. Effect of air pollution on onset of acute coronary syndrome in susceptible subgroups. Eastern Mediterranean Health Journal, 2012, 18(6): 550-555.

[86] Cai Y, Zhang B, Ke W, et al. Associations of short-term and long-term exposure to ambient air pollutants with hypertension: A systematic review and meta-analysis. Hypertension, 2016, 68(1): 62-70.

[87] Oudin A, Stromberg U, Jakobsson K, et al. Hospital admissions for ischemic stroke: does long-term exposure to air pollution interact with major risk factors? Cerebrovascular Diseases, 2011, 31(3): 284-293.

[88] Chen S Y, Wu C F, Lee J H, et al. Associations between long-term air pollutant exposures and blood pressure in elderly residents of Taipei city: A cross-sectional study. Environmental Health Perspecttives, 2015, 123(8): 779-784.

[89] Babisch W, Wolf K, Petz M, et al. Associations between traffic noise, particulate air pollution, hypertension, and isolated systolic hypertension in adults: the KORA study. Environmental Health Perspecttives, 2014, 122(5): 492-498.

[90] Coogan P F, White L F, Jerrett M, et al. Air pollution and incidence of hypertension and diabetes mellitus in black women living in Los Angeles. Circulation, 2012, 125(6): 767-772.

[91] Dong G H, Qian Z M, Xaverius P K, et al. Association between long-term air pollution and increased blood pressure and hypertension in China. Hypertension, 2013, 61(3): 578-584.

[92] Foraster M, Kunzli N, Aguilera I, et al. High blood pressure and long-term exposure to indoor noise and air pollution from road traffic. Environmental Health Perspecttives, 2014, 122(11): 1193-1200.

[93] Johnson D, Parker J D. Air pollution exposure and self-reported cardiovascular disease. Environmental Research, 2009, 109(5): 582-589.

[94] Levinsson A, Olin A C, Modig L, et al. Interaction effects of long-term air pollution exposure and variants in the GSTP1, GSTT1 and GSTCD genes on risk of acute myocardial infarction and hypertension: A case-control study. PLoS One, 2014, 9(6): e99043.

[95] Sorensen M, Hoffmann B, Hvidberg M, et al. Long-term exposure to traffic-related air pollution associated with blood pressure and self-reported hypertension in a Danish cohort. Environmental Health Perspecttives, 2012, 120(3): 418-424.

[96] Bartoli C R, Wellenius G A, Diaz E A, et al. Mechanisms of inhaled fine particulate air pollution-induced arterial blood pressure changes. Environmental Health Perspecttives, 2009, 117(3): 361-366.

[97] Yang B Y, Qian Z, Howard S W, et al. Global association between ambient air pollution and blood pressure: A systematic review and meta-analysis. Environmental Pollution, 2018, 235: 576-588.

[98] Franklin B A, Brook R, Arden Pope C. Air pollution and cardiovascular disease. Current Problems in Cardiology, 2015, 40(5): 207-238.

[99] Brook R D, Rajagopalan S. Particulate matter, air pollution, and blood pressure. Journal of the American Society of Hypertension, 2009, 3(5): 332-350.

[100] Schins R P, Lightbody J H, Borm P J, et al. Inflammatory effects of coarse and fine particulate matter in relation to chemical and biological constituents. Toxicology and Applied Pharmacology, 2004, 195(1): 1-11.

[101] Fuks K, Moebus S, Hertel S, et al. Long-term urban particulate air pollution, traffic noise, and arterial blood pressure. Environmental Health Perspecttives, 2011, 119(12): 1706-1711.

[102] Lu X, Ye Z, Zheng S, et al. Long-term exposure of fine particulate matter causes hypertension by impaired renal d1 receptor-mediated sodium excretion via upregulation of g-protein-coupled receptor kinase type 4 expression in sprague-dawley rats. Journal of the American Heart Association, 2018, 7(1): pii: e007185.

[103] Bai Y, Sun Q. Fine particulate matter air pollution and atherosclerosis: Mechanistic insights. Biochimica et Biophysica Acta, 2016, 1860(12): 2863-2868.

[104] Lusis A J. Atherosclerosis. Nature, 2000, 407(6801): 233-241.

[105] Koken P J, Piver W T, Ye F, et al. Temperature, air pollution, and hospitalization for cardiovascular diseases among elderly people in Denver. Environmental Health Perspecttives, 2003, 111(10): 1312-1317.

[106] Rivera M, Basagana X, Aguilera I, et al. Association between long-term exposure to traffic-related air pollution and subclinical atherosclerosis: The REGICOR study. Environmental Health Perspecttives, 2013, 121(2): 223-230.

[107] Kunzli N, Jerrett M, Mack W J, et al. Ambient air pollution and atherosclerosis in Los Angeles. Environmental Health Perspecttives, 2005, 113(2): 201-206.

[108] Hoffmann B, Moebus S, Mohlenkamp S, et al. Residential exposure to traffic is associated with coronary atherosclerosis. Circulation, 2007, 116(5): 489-496.

[109] Allen R W, Criqui M H, Diez Roux A V, et al. Fine particulate matter air pollution, proximity to traffic, and aortic atherosclerosis. Epidemiology, 2009, 20(2): 254-264.

[110] Gimbrone M A, Jr, Garcia C G. Endothelial cell dysfunction and the pathobiology of atherosclerosis. Circulation Research, 2016, 118(4): 620-636.

[111] Gdula A J, Czepiel J, Toton Z J, et al. Docosahexaenoic acid regulates gene expression in HUVEC cells treated with polycyclic aromatic hydrocarbons. Toxicology Letters, 2015, 236(2): 75-81.

[112] Wang J S, Tseng C Y, Chao M W. Diesel exhaust particles contribute to endothelia apoptosis via autophagy pathway. Toxicological Sciences, 2017, 156(1): 72-83.

[113] Rui W, Guan L, Zhang F, et al. $PM_{2.5}$-induced oxidative stress increases adhesion molecules expression in human endothelial cells through the ERK/AKT/NF-kappaB-dependent pathway. Journal of Applied Toxicology, 2016, 36(1): 48-59.

[114] Bo L, Jiang S, Xie Y, et al. Effect of vitamin E and omega-3 fatty acids on protecting ambient $PM_{2.5}$-Induced inflammatory response and oxidative stress in vascular endothelial cells. PLoS One, 2016, 11(3): e0152216.

[115] Araujo J A, Barajas B, Kleinman M, et al. Ambient particulate pollutants in the ultrafine range promote early atherosclerosis and systemic oxidative stress. Circulation Research, 2008, 102(5): 589-596.

[116] Ying Z, Kampfrath T, Thurston G, et al. Ambient particulates alter vascular function through induction of reactive oxygen and nitrogen species. Toxicological Sciences, 2009, 111(1): 80-88.

[117] Tanai E, Frantz S. Pathophysiology of heart failure. Comprehensive Physiology, 2015, 6(1): 187-214.

[118] Yang C, Chen A, Chen R, et al. Acute effect of ambient air pollution on heart failure in Guangzhou, China. International Journal of Cardiology, 2014, 177(2): 436-441.

[119] Wellenius G A, Bateson T F, Mittleman M A, et al. Particulate air pollution and the rate of hospitalization for congestive heart failure among medicare beneficiaries in Pittsburgh, Pennsylvania. American Journal of Epidemiology, 2005, 161(11): 1030-1036.

[120] Kisseleva T, Brenner D A. Mechanisms of fibrogenesis. Experimental Biology and Medicine (Maywood), 2008, 233(2): 109-122.

[121] Andersen S, Nielsen K J E, Vonk Noordegraaf A, et al. Right ventricular fibrosis. Circulation, 2019, 139(2): 269-285.

[122] Goh K Y, He L, Song J, et al. Mitoquinone ameliorates pressure overload-induced cardiac fibrosis and left ventricular dysfunction in mice. Redox Biology, 2019, 21: 101100.

[123] Qin G, Xia J, Zhang Y, et al. Ambient fine particulate matter exposure induces reversible cardiac dysfunction and fibrosis in juvenile and older female mice. Particle and Fibre Toxicology, 2018, 15(1): 27.

[124] Weber K T, Sun Y, Campbell S E. Structural remodelling of the heart by fibrous tissue: Role of circulating hormones and locally produced peptides. European Heart Journal, 1995, 16 Suppl N: 12-18.

[125] Qin G, Wu M, Wang J, et al. Sulfur dioxide contributes to the cardiac and mitochondrial dysfunction in rats. Toxicological Sciences, 2016, 151(2): 334-346.

[126] Mills N, Donaldson K, Pw, Boon N, et al. Adverse cardiovascular effects of air pollution. Nature Clinical Practice Cardiovascular Medicine, 2009, 115(6): 36-44.

[127] Geiser M, Kreyling W G. Deposition and biokinetics of inhaled nanoparticles. Particle and Fibre Toxicology, 2010, 7(1): 2.

[128] Brook R D, Rajagopalan S, Pope C A, et al. Particulate matter air pollution and cardiovascular disease an update to the scientific statement from the American Heart Association. Circulation, 2010, 121(21): 2331-2378.

[129] Chin M T. Basic mechanisms for adverse cardiovascular events associated with airpollution. Heart, 2015, 101(4): 253-256.

[130] Wold L E, Ying Z, Hutchinson K R, et al. Cardiovascular remodeling in response to long-term exposure to fine particulate matter air pollution. Circulation-Heart Failure, 2012, 5(4): 452-461.

[131] Weldy C S, Liu Y, Chang Y C, et al. In utero and early life exposure to diesel exhaust air pollution increases adult susceptibility to heart failure in mice. Particle and Fibre Toxicology, 2013, 10(1): 59.

[132] Lee B J, Kim B, Lee K. Air Poll ution exposure and cardiovascular disease. Toxicological Research, 2014, 30(2): 71.

[133] Sancini G, Farina F, Battaglia C, et al. Health risk assessment for air pollutants: alterations in lung and cardiac gene expression in mice exposed to Milano winter fine particulate matter ($PM_{2.5}$). PLoS One, 2014, 9(10): e109685.

[134] Wold L E, Ying Z, Hutchinson K R, et al. Cardiovascular remodeling in response to long-term exposure to fine particulate matter air pollution. Circulation Heart Failure, 2012, 5(4): 452.

[135] Tankersley C G, Champion H C, Takimoto E, et al. Exposure to inhaled particulate matter impairs cardiac function in senescent mice. Ajp Regulatory Integrative & Comparative Physiology, 2008, 295(1): R252-263.

[136] Yang X, Doser T A, Fang C X, et al. Metallothionein prolongs survival and antagonizes senescence-associated cardiomyocyte diastolic dysfunction: Role of oxidative stress. Faseb Journal Official Publication of the Federation of American Societies for Experimental Biology, 2006, 20(7): 1024-1026.

[137] Tankersley C G, Champion H C, Takimoto E, et al. Exposure to inhaled particulate matter impairs cardiac function in senescent mice. American Journal of Physiology-Regulatory Integrative and Comparative Physiology, 2008, 295(1): R252-263.

[138] Gorr M W, Velten M, Nelin T D, et al. Early life exposure to air pollution induces adult cardiac dysfunction. American Journal of Physiology Heart & Circulatory Physiology, 2014, 307(9): 1353-1360.

[139] Weldy C S, Liu Y, Chang Y C, et al. In uteroand early life exposure to diesel exhaust air pollution increases adult susceptibility to heart failure in mice. Particle and Fibre Toxicology, 2013, 10(1): 59.

[140] Yue W, Tong L, Liu X, et al. Short term $PM_{2.5}$ exposure caused a robust lung inflammation, vascular remodeling, and exacerbated transition from left ventricular failure to right ventricular hypertrophy. Redox Biology, 2019, 22: 101161.

[141] Kong P, Christia P, Frangogiannis N G. The pathogenesis of cardiac fibrosis. Cellular and Molecular Life Sciences, 2014, 71(4): 549-574.

[142] Ying Z, Xie X, Bai Y, et al. Exposure to concentrated ambient particulate matter induces reversible increase of heart weight in spontaneously hypertensive rats. Particle and Fibre Toxicology, 2015, 12(1): 15.

[143] Zhang Y, Ji X, Ku T, et al. Ambient fine particulate matter exposure induces cardiac functional injury and metabolite

alterations in middle-aged female mice. Environmental Pollution, 2019, 248: 121-132.

[144] 杨牧祥, 于文涛, 丁宁, 等. 中风康对局灶性脑缺血大鼠脑病理形态学的影响. 中医研究, 2006, 19(11): 7-10.

[145] Sommer C J. Ischemic stroke: experimental models and reality. Acta Neuropathologica, 2017, 133(2): 245-261.

[146] Esenwa C C, Elkind M S. Inflammatory risk factors, biomarkers and associated therapy in ischaemic stroke. Nature Reviews Neurology, 2016, 12(10): 594-604.

[147] Global, regional, and national age-sex specific all-cause and cause-specific mortality for 240 causes of death, 1990—2013: A systematic analysis for the Global Burden of Disease Study 2013. Lancet, 2015, 385(9963): 117-171.

[148] Knox E G. Meteorological associations of cerebrovascular disease mortality in England and Wales. Journal of Epidemiology and Community Health, 1981, 35(3): 220-223.

[149] Zhang Z F, Yu S Z, Zhou G D. Indoor air pollution of coal fumes as a risk factor of stroke, Shanghai. American Journal of Public Health, 1988, 78(8): 975-977.

[150] Shah A S, Lee K K, Mcallister D A, et al. Short term exposure to air pollution and stroke: Systematic review and meta-analysis. BMJ, 2015, 350: h1295.

[151] Lisabeth L D, Escobar J D, Dvonch J T, et al. Ambient air pollution and risk for ischemic stroke and transient ischemic attack. Annals of Neurology, 2008, 64(1): 53-59.

[152] Chen G, Wang A, Li S, et al. Long-term exposure to air pollution and survival after ischemic stroke. Stroke, 2019, 50(3): 563-570.

[153] Fisher J A, Puett R C, Laden F, et al. Case-crossover analysis of short-term particulate matter exposures and stroke in the health professionals follow-up study. Environment International, 2019, 124: 153-160.

[154] Chung J W, Bang O Y, Ahn K, et al. Air pollution is associated with ischemic stroke via cardiogenic embolism. Stroke, 2017, 48(1): 17-23.

[155] Miller M R, Raftis J B, Langrish J P, et al. Inhaled nanoparticles accumulate at sites of vascular disease. ACS Nano, 2017, 11(5): 4542-4552.

[156] Sang N, Yun Y, Li H, et al. SO_2 inhalation contributes to the development and progression of ischemic stroke in the brain. Toxicological Sciences, 2010, 114(2): 226-236.

[157] Sang N, Yun Y, Yao G Y, et al. SO_2-induced neurotoxicity is mediated by cyclooxygenases-2-derived prostaglandin E_2 and its downstream signaling pathway in rat hippocampal neurons. Toxicological Sciences, 2011, 124(2): 400-413.

[158] Davis D A, Akopian G, Walsh J P, et al. Urban air pollutants reduce synaptic function of CA1 neurons via an NMDA/NO pathway in vitro. Journal of Neurochemistry, 2013, 127(4): 509-519.

[159] Morgan T E, Davis D A, Iwata N, et al. Glutamatergic neurons in rodent models respond to nanoscale particulate urban air pollutants in vivo and in vitro. Environmental Health Perspecttives, 2011, 119(7): 1003-1009.

[160] Guo L, Li B, Miao J J, et al. Seasonal variation in air particulate matter (PM_{10}) exposure-induced ischemia-like injuries in the rat brain. Chemical Research in Toxicology, 2015, 28(3): 431-439.

[161] Li B, Guo L, Ku T, et al. $PM_{2.5}$ exposure stimulates COX-2-mediated excitatory synaptic transmission via ROS-NF-κB pathway. Chemosphere, 2018, 190: 124-134.

[162] Jung C R, Lin Y T, Hwang B F. Ozone, particulate matter, and newly diagnosed Alzheimer's disease: A population-based cohort study in Taiwan. Journal of Alzheimer's Disease, 2015, 44(2): 573-584.

[163] Calderon G L, Solt A C, Henriquez-Roldan C, et al. Long-term air pollution exposure is associated with neuroinflammation, an altered innate immune response, disruption of the blood-brain barrier, ultrafine particulate deposition, and accumulation of amyloid beta-42 and alpha-synuclein in children and young adults. Toxicologic Pathology, 2008, 36(2): 289-310.

[164] Calderon G L, Reed W, Maronpot R R, et al. Brain inflammation and Alzheimer's-like pathology in individuals exposed to severe air pollution. Toxicologic Pathology, 2004, 32(6): 650-658.

[165] Calderon G L, Franco L M, Torres J R, et al. Pediatric respiratory and systemic effects of chronic air pollution exposure: nose, lung, heart, and brain pathology. Toxicologic Pathology, 2007, 35(1): 154-162.

[166] Calderon G L, Kavanaugh M, Block M, et al. Neuroinflammation, hyperphosphorylated tau, diffuse amyloid plaques, and down-regulation of the cellular prion protein in air pollution exposed children and young adults. Journal of Alzheimer's Disease, 2012, 28(1): 93-107.

[167] Ku T, Ji X, Zhang Y, et al. $PM_{2.5}$, SO_2 and NO_2 co-exposure impairs neurobehavior and induces mitochondrial injuries in the mouse brain. Chemosphere, 2016, 163: 27-34.

[168] Yao G, Yue H, Yun Y, et al. Chronic SO_2 inhalation above environmental standard impairs neuronal behavior and represses glutamate receptor gene expression and memory-related kinase activation via neuroinflammation in rats. Environmental Research, 2015, 137: 85-93.

[169] Yao G, Yun Y, Sang N. Differential effects between one week and four weeks exposure to same mass of SO_2 on synaptic plasticity in rat hippocampus. Environmental Toxicology, 2016, 31(7): 820-829.

[170] Li B, Chen M, Guo L, et al. Endogenous 2-arachidonoylglycerol alleviates cyclooxygenases-2 elevation-mediated neuronal injury from SO_2 inhalation via PPARgamma pathway. Toxicological Sciences, 2015, 147(2): 535-548.

[171] Liu X, Zhang Y, Luo C, et al. At seeming safe concentrations, synergistic effects of $PM_{2.5}$ and formaldehyde co-exposure induces Alzheimer-like changes in mouse brain. Oncotarget, 2017, 8(58): 98567-98579.

[172] Ku T, Li B, Gao R, et al. NF-kappaB-regulated microRNA-574-5p underlies synaptic and cognitive impairment in response to atmospheric $PM_{2.5}$ aspiration. Particle and Fibre Toxicology, 2017, 14(1): 34.

[173] Ku T, Chen M, Li B, et al. Synergistic effects of particulate matter ($PM_{2.5}$) and sulfur dioxide (SO_2) on neurodegeneration via the microRNA-mediated regulation of tau phosphorylation. Toxicology Research (Camb), 2017, 6(1): 7-16.

[174] Ning X, Li B, Ku T, et al. Comprehensive hippocampal metabolite responses to $PM_{2.5}$ in young mice. Ecotoxicology and Environmental Safety, 2018, 165: 36-43.

[175] Yan W, Ji X, Shi J, et al. Acute nitrogen dioxide inhalation induces mitochondrial dysfunction in rat brain. Environmental Research, 2015, 138: 416-424.

[176] Yan W, Yun Y, Ku T, et al. NO_2 inhalation promotes Alzheimer's disease-like progression: cyclooxygenase-2-derived prostaglandin E2 modulation and monoacylglycerol lipase inhibition-targeted medication. Scientific Reports, 2016, 6: 22429.

[177] Yan W, Ku T, Yue H, et al. NO_2 inhalation causes tauopathy by disturbing the insulin signaling pathway. Chemosphere, 2016, 165: 248-256.

[178] Li D, Zhang R, Cui L, et al. Multiple organ injury in male C57BL/6J mice exposed to ambient particulate matter in a real-ambient PM exposure system in Shijiazhuang, China. Environmental Pollution, 2019, 248: 874-887.

作者：阚海东[1]，郑玉新[2]，桑楠[3]，王婷[4]，岳慧峰[3]
[1]复旦大学，[2]青岛大学，[3]山西大学，[4]南开大学

第15章　环境毒理组学研究进展

▶ 1. 引言 /387

▶ 2. 暴露组的概念形成及研究进展 /388

▶ 3. 转录组学分析技术及其研究进展 /392

▶ 4. 蛋白组学分析技术及其研究进展 /396

▶ 5. 代谢组学分析技术及其研究进展 /400

▶ 6. 展望 /403

本章导读

环境毒理学主要研究环境污染物对生物有机体,尤其是对人体的影响及作用机制,"组学"技术的发展为环境毒理学的研究提供了新的视角。21世纪毒理学测试新方法指出应将新的毒理学评价方法纳入未来的毒性测试系统中去,用毒性通路的理念从系统生物学方向来阐述毒物的毒性作用。基因组学、转录组学、蛋白组学和代谢组学作为系统生物学的重要组成部分,其高通量筛选特征使其在毒物的毒性作用效应和机制研究上具备很大的优势。本章将针对研究全面可测量环境暴露数据的暴露组,研究机体所有基因表达的RNA(如mRNA)进行全面分析的转录组学,对机体在不同环境条件下蛋白质表达进行定性和定量分析的蛋白质组,研究小分子代谢物和代谢通路变化的代谢组,对近年来其在环境毒理学领域的研究进展进行综述。

关键词

环境毒理,暴露组,转录组,蛋白组,代谢组

1 引 言

20世纪80年代开启的"人类基因组计划"是生命科学史上的里程碑,在研究过程中建立起来的策略、思想与技术,构成了生命科学领域的新学科——基因组学。基因测序技术的快速发展让人们对基因组学在疾病发生、诊断和治疗方面所起的作用抱有极高的期望。然而,科学家逐渐意识到,人类只有很少的疾病是单独由基因决定的[1,2],70%~90%的疾病是基因与环境多种因素共同作用的结果[2-4]。诸多的环境健康研究观察到了疾病发生与单一或几种环境因素的关联关系,但绝大多数关联关系的内在科学证据是有限或不足的。时至今日,科学家们仍然难以回答环境暴露究竟如何并在多大程度上影响着人体健康。环境暴露是多因子的组合并具有时空变异的特点,要揭示诸多与外界环境相关疾病的发病机制,需要具备全局和系统的研究理念,由此"暴露组"的概念逐渐形成并不断发展。由于基因组序列本身并不能提供生物的基因功能信息,因此,人们将眼光聚焦到了基因表达过程中很重要的转录环节,逐渐认识到开展转录组研究的重要性以及必要性。基因的表达分为转录和翻译过程,对同一生物体而言,虽然每个细胞具有相同的基因序列,但不同细胞在特定的时空条件下对基因的表达是不同的,这种不同与转录环节有密不可分的关系。随着生物信息学跨入后基因组时代,以及高通量测序技术的出现,大规模的基因表达研究的序幕已经拉开,转录组学正作为一门新兴学科在生物学前沿研究中发挥关键作用,并逐步成为生命科学研究的热点。蛋白质作为基因翻译的产物,是细胞内的活性分子和具体细胞功能的执行者。蛋白组学作为基因组学的重要补充,可以帮助我们揭示生命活动的本质。因此,致力于实现对生物体所有蛋白质的定性和定量分析蛋白组学应运而生[5]。蛋白组学是研究细胞或机体内所有蛋白质的组成及其变化规律的学科,已经成为后基因组学时代的重要研究内容之一,基于生物质谱技术的分析策略已成为蛋白组学研究的核心技术[6]。代谢组是基因组和蛋白组的下游,基因和蛋白质表达的微小变化会在代谢物上得到放大。代谢物的变化是机体表型或功能改变的直接体现,能够为生物标志物的发现提供新的思路和平台。目前,多组学技术在环境毒理学领域已经有了广泛的应用。

2 暴露组的概念形成及研究进展

2.1 暴露组概念的形成

暴露组（exposome）术语最早由法国流行病学家 Christopher Wild 博士在 2005 年发表的一篇文章中提出，最初的定义为：生活方式、饮食和环境所构成的全部暴露。在随后发表的文章中，Christopher Wild 博士又对暴露组的概念加以补充和完善，将暴露组定义为：内部（代谢、内源激素、体形、肠道微生物、炎症、脂质过氧化、氧化胁迫、衰老等）、特定的外环境部（当地环境、辐射、饮食、生活方式、污染、职业、医疗干预等）和一般的外环境（社会制度、教育、经济状况、心理和生理压力、城市-农村环境差别、气候等）暴露的总和[7]。Christopher Wild 博士提出暴露组概念是希望找到能够帮助流行病学家确定疾病发生原因的手段和工具。一种可测量的暴露组对流行病学研究将是巨大利好，并且有一个可行的暴露组的潜在价值和意义要超越暴露评估和流行病学本身。

美国的 Gary W. Miller 博士和 Dean P. Jones 博士发现 Christopher Wild 博士提出的暴露组概念并不是教科书式的定义，因而建议修改[8]。2014 年，Gary W. Miller 博士和 Dean P. Jones 博士共同提出了修订的暴露组概念[9]：全生命过程中环境影响及其相关生物学响应的累积量度，包括来自于环境、饮食和行为的外暴露和内源生物学过程。此定义关注了暴露组的量化，并且强调了身体对全生命期暴露的响应的重要性。要理解这一修订的暴露组概念，需掌握四个关键词：累积量度（cumulative measure）、相关生物学响应（associated biological response）、行为（behavior）和内源过程（endogenous processes）。累积量度是指构成暴露组的复合因子是可测量的，并且随时间逐渐累积。如果一个因子没有测量手段，那么就不能作为暴露组的一部分。相关生物学响应是指我们身体对不同外部作用的响应方式。例如，一个环境促发因素导致了一个基因的沉默或甲基化，此基因改变可在任意时间测量。相应地，相关生物学响应的累积量度可简单表述为累积生物学响应（cumulative biological response），是指不断发生的身体对所有外部因素刺激的适应和不良适应。行为是指那些自发的、或由外向内对我们身体产生作用的行动，包括影响健康的社会决定因素、由社会关系和习惯引起的压力、精神活动、冒险活动和积极活动（如锻炼身体）。内源过程是指身体内时刻进行的生化反应。人体内在发生上千种反应，以促使营养物质分解和细胞组织的形成，这些反应可产生影响身体健康的副产物，如自由基。在 Escher 等发表的综述文章中[10]，将暴露组的定义用如下直观明了的简图（图 15-1）表示。

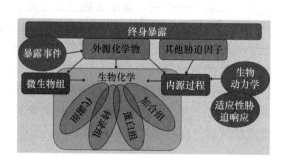

图 15-1 暴露组定义（引自参考文献[10]）

暴露组研究的核心挑战是全面可测量的环境暴露数据的获得。一旦获得这些数据，流行病学家可将它们代入现存的用于疾病预防和管理的公共卫生模型，用于解释基因和环境如何相互作用影响健康和疾病发展。随着暴露组研究的深入，每个人的暴露组会与疾病人群和健康人群形成对照，也会与每个人不同生命阶段的个人行为形成对比，找出造成疾病的原因，从而确认、减少或消除有害暴露。

2.2 暴露组学研究方法学

顾名思义，暴露组学（exposomics）是指研究暴露组的方法及理论体系。它的目标是将专一外部暴露、一般外部暴露和内源生物化学响应有效整合起来，建立起"污染过程-人体暴露-基因表达-人体响应"的研究路线[11]。暴露组学是一庞大的学科体系，如何有效开展暴露组学研究是公共卫生学家和毒理学家面临的现实问题。美国加州大学伯克利分校的Rappaport教授[12]对暴露组学的定义为：暴露组学是研究暴露组以及暴露组对人类疾病过程影响的科学。他与同校的Smith教授[4]共同提出了研究暴露组学的两种策略，一种是"自下而上"（bottom-up）策略，即测量人体在任意时刻每一外暴露源的化学物质，包括空气、水、饮食等，用加和的方式估算个体的暴露水平，检验病例组和对照组是否存在暴露差别，用于寻找疾病发生的原因。这一方法需要耗费大量的精力来检测庞大未知的外部化学物质，可能忽略了内暴露环境中化学物质的信息，而内暴露环境中的化学物质经常随着性别、肥胖、炎症和压力等变化而发生变化。另一种是"自上而下"（up-bottom）策略，即测量个体血液中目标物质（污染物、生物标志物、激素等）的种类和含量，确定导致疾病的物质及暴露来源，并反推暴露和剂量的关系。

"自下而上"与"自上而下"两种策略各有优缺点：前者可以分析有害物质的来源，可以进行大规模的人群研究，但难以与个体的内暴露相关联，无法确定疾病发生的具体原因；后者首先测定人体内暴露环境中的污染物、暴露标志物和效应标志物，可为确定暴露与疾病的关系提供有力证据，但难以进行大规模的人群研究，并且也无法确定有害因素的来源。因此，将两者有机结合，各取其优势，是暴露组学研究中应该采取的策略[13]。

2.3 外暴露组的测量

外暴露组包括了专一外部暴露和一般外部暴露，其组成因素极其复杂。在传统的流行病学研究中，问卷调查是最常用的获取外暴露信息的手段，调查项目通常包括：年龄、性别、体重、职业、饮食、生活习惯、居住环境、工作环境和心理状态等。通过查询当地的地理和气象信息，也可以获得例如辐射、气候、空气污染、水污染和土壤污染等外暴露信息。但人体暴露化学物质的种类和水平无法通过问卷调查和地理气象资料查询获得，并且由于个体在不同地点和不同场景间流动，其外暴露水平难以准确估计。因此，要准确测量个体的外暴露组，需要将问卷调查、资料查询、动态监测技术、高通量污染物检测技术以及病毒和致病微生物检测技术相结合。

借助3S技术可实现个体外暴露组的实时动态监测。3S技术是遥感技术、地理信息系统和全球定位系统的统称，在近20年，3S技术已广泛用于绘制人体对特定污染物的暴露地图，记录、计算并评估人体的外暴露水平[3]。手机及其网络的发展以及各种便携式小型设备（各类便携式环境监测仪器、人体活动和体征记录仪以及可穿戴设备等）的研发，使得通过3S技术获得空间暴露数据更为准确和快捷。手机已成为人们现代生活的必需品，在全球，拥有配置移动、音频、可视和定位系统的手机用户已超过10亿，如果在手机内部配置相应的环境检测系统（如大气颗粒物和挥发性有机污染物的监测装置）及其软件和网络，每位用户的暴露信息传输至终端后可形成巨大的环境暴露数据库。借助3S的动态监测技术能全面逼真地评价个体面对的外暴露。带有遥感式空间参照技术和模型的装置能持续、时效、真实地评价因地理位置、活动以及生活方式等因子造成个人暴露的差异[14]。例如一个人的活动强度、类型及地理位置能反映空气呼吸暴露、接触与摄入有害污染物。但是，借助 3S 的动态监测技术仅适用于特定人群和监测特定污染物，仍无法完全获知人体外暴露的全部信息，亦无法适用于公共人群的健康管理中[11]。要系统、无偏差、准确地获知外暴露化学污染物的种类和水平，需要发展高通量污染物检测技术。

环境中的污染物有数百万种之多，如何快速分析并鉴定出关键有害污染物是外暴露组测量的一大难题。目前已确定的在环境中广泛存在且毒性较高的污染物包括：重金属，挥发性有机污染物（甲醛、二氯甲烷、石油烃等），持久性有机污染物如二噁英（PCDD/Fs）、多氯联苯（PCBs）、有机氯农药（OCPs）、多溴联苯醚（PBDEs）和全氟辛烷磺酸（PFOS）及其盐类等，内分泌干扰物，抗生素，农药，兽药，医药，消毒副产物，含卤阻燃剂和增塑剂。在不同的环境介质和食品中这些污染物的种类分布和水平也会发生变化；同时，这些污染物也存在较大的时空变异。电感耦合等离子质谱仪（ICP-MS）的普及，使得重金属的高通量分析成为可能。通过一次进样，可以分析出样品中几乎全部重金属的水平。而有机污染物的物理化学性质差异较大，仍然难以做到一次分析获得大量污染物的浓度数据。各种高灵敏的气相色谱-质谱仪（GC-MS）和液相色谱-质谱仪（LC-MS）已普遍用于环境有机污染物的检测，相应地也发展出多种多样的有机污染物检测方法。但目前仍然缺乏环境污染物的高通量无偏差的分析技术。

病毒和致病微生物引起的疾病爆发，会对社会安定产生重大影响，并引起公众的巨大恐慌。如何快速有效地鉴别出食源性致病微生物是预防与控制食源性疾病的关键环节。菌培养法是病毒和致病微生物常规检测方法。随着技术的不断进步，免疫分析、代谢组分析、核酸分析、生物芯片技术、生物传感器和表面增强拉曼光谱已用于病毒和致病微生物的快速检测，极大提高了检测的灵敏度和效率。

2.4 体内外源污染物的分析

内暴露组分析包括血液中外源污染物的分析及内源化学物质的分析。在欧洲，英国、法国、西班牙、挪威、希腊、立陶宛 6 个国家基于 6 个母婴队列联合开展了人类生命早期暴露组（HELIX）研究。通过高灵敏度分析手段，在 1301 个血液和尿液样本中共检出 45 种外源污染物或其代谢物（暴露标志物），包括有机氯化合物（多氯联苯、有机氯农药）、多溴联苯醚、环境酚类、有机磷酸盐代谢产物、邻苯二甲酸酯类代谢产物、全氟辛磺酸、有毒元素（砷、镉、铜、汞、铅、锌、钛和钴等）[15, 16]。推荐用非靶向代谢组分析方法来检测体液中的污染物及其代谢产物，通过数据库比对、未知化合物鉴定和数据的可视化处理，可以广谱无偏差地筛查暴露标志物。美国哈佛大学于 2005~2009 年在密西根和得克萨斯州的 16 个县开展了生殖力和环境关系的纵向调查（LIFE）队列研究，共征集 501 对准备生育的夫妇；综合运用 ICP-MS、高分辨气质联用仪和高效液相色谱-串联质谱分析了采集血清和尿液样本中的 128 种内分泌干扰物，包括：多氯联苯、有机氯农药、有机溴代阻燃剂、全氟辛磺酸、金属元素、酚类、邻苯二甲酸酯类代谢产物、杀菌剂、扑热息痛等；发现夫妻双方血清中持久性难降解内分泌干扰物的分布模式趋于相似，而非持久性内分泌干扰物的分布模式在夫妻间差别较大[17]。西班牙环境流行病研究中心组织开展了 INMA-Sabadell 出生队列研究，采用多种分析测试手段检测了 728 名怀孕妇女血清和尿液样品中 45 种化学污染物，包括 OCPs、PBDEs、PFOS、双酚 A（BPA）、金属元素及邻苯二甲酸酯类代谢产物[18]。美国的 Pleil 和 Stiegel[14]发展了一种通过监测呼吸其中生物标志物来揭示人体暴露组的方法，他们让测试者将呼吸气吹过过滤管，肺部水汽中的挥发物和气溶胶被管壁捕获，然后采用气相色谱质谱联用仪靶向分析其中的化学成分，以发现其中的生物标志物，从而推测人体的内暴露情况。

2.5 生物组学技术在暴露组研究中的应用

暴露组研究必须借助高通量生物组学技术,包括:基因组、表观基因组、转录组、蛋白组、代谢组和微生物组。污染物进入人体细胞后会在各个层面改变细胞的生理生化功能,一些污染物可与DNA形成加合物或改变DNA的复制和表达,从而引起转录组和代谢组的变化,转录组学可以更清楚地记录环境因子对机体基因表达的影响。机体的细微变化都会引起蛋白质谱的差异,同时,一些污染物可直接与蛋白酶相结合,改变酶的功能与活性,蛋白质结构、分布、功能和活性的改变可导致疾病的发生;通过开展蛋白组学的研究记录一个组织或细胞的全部蛋白质表达水平、氨基酸序列和翻译后加工以及蛋白质相互作用等信息,在此基础上可以了解细胞的各种生物化学过程以及病理反应。代谢组位于基因组、转录组和蛋白组下游,因此,外源污染物或胁迫因子的刺激可灵敏地反映在代谢物组的变化上。反过来,代谢组的变化也会引起蛋白组和转录组发生相应改变。代谢组的分析策略与体内外源污染物及其生物标志物的分析策略相同,因此,美国 Warth 等[19]基于液质联用技术发展了代谢组、外源污染物及其生物标志物的全分析方法。另外,外源污染物或胁迫因子的长期刺激也可引起表观基因组的改变,例如 DNA 甲基化,并进而引起遗传改变或疾病发生[8]。

人体微生物组(microbiome)也是暴露组学研究的重要内容。人体微生物主要分布在消化道、泌尿生殖道、呼吸道和皮肤等部位,据估计人体内微生物数量是人体细胞数量的 10 倍[8]。人体微生物组在消化、代谢、调解免疫功能、预防和治疗疾病,合成维生素、氨基酸、营养因子、免疫因子、传递物质、前驱体等方面扮演着重要角色,在人体内环境层面上反映人体健康程度,特别是慢性疾病的状况。人体内微生物的种类和定居部位如果发生改变,将导致"微生态失调",如葡萄糖耐受不良、肥胖、II型糖尿病、老龄化相关疾病和非酒精性脂肪性肝病等。肠道微生物菌群在帮助人体代谢营养物质的同时也会降解掉一些有害物质;同时,药物的使用,尤其是抗生素,可以显著改变微生物群落结构和数量,并进而引起肠道功能的改变。

2.6 暴露组学的研究进展

暴露组学的复杂性注定了暴露组学的研究是一系统庞大的工程,需要科学家的顶层设计、政府的大力支持和民众的积极参与。欧美日发达国家已开始进行暴露组研究。暴露组学研究通常基于队列研究来开展。欧洲的 HELIX(human early-life exposome)项目由 6 个国家 13 家合作机构共同承担,测量 32000 多对母婴的环境暴露,以及对儿童成长、发育、健康的后续影响,期望描述欧洲人群的早期暴露和儿童时期健康的关系。已有 10 个欧洲出生队列,BAMSE(瑞典)、GASPII(意大利)、GINIplus 和 LISAplus(德国)、MAAS(英国)、PIAMA(荷兰)和 4 个 INMA 队列(西班牙),研究了空气污染与肺炎、哮喘及中耳炎之间的相关性。塞浦路斯提出了城市环境暴露组的研究框架[20]。2001 年,法国成立了国家职业病监测与预防网络(RNV3P),开展职业病暴露组研究,期望获得职业暴露和患病的关系[21]。比利时于 2017 年启动 EXPOsOMICS 计划,企图通过对空气污染和水污染健康效应的研究将外暴露组和内暴露组整合起来,以发展和建立暴露组学研究方法[22]。

为了推动全环境关联研究或全暴露组关联研究以及人类暴露组计划,美国已发起成立暴露组联盟,由美国加州大学伯克利分校、英国伦敦帝国学院和国王学院、国际癌症研究机构以及美国国家环境健康科学研究所等单位组成。2012 年,美国政府发布《生物监测国家战略》,开展国家生物监测项目(NBP)研究,旨在评估人群营养状态及美国人口对环境化学和有毒有害物质的暴露水平,该项目测量人体中 300 余种与环境相关的化学物质和营养学指标。2013 年,美国成立了第一个暴露组学研究中心 HERCULES,

以提供强大的暴露组学研究方法为己任[23]。日本环境和儿童研究于 2011 年［The Japan Environmentand Children's Study（JECS）］开展了一个出生队列研究，包括 10 万对父母和儿童，拟评估一系列环境因素对儿童健康和发育的影响[24]。

我国的暴露组学研究起步较晚。2013 年发布的《中国人群暴露参数手册（成人卷）》，填补了我国在暴露参数方面的空白，有利于暴露组学研究的开展。同年，吴永宁提出我国食品中化学危害暴露组与毒理学测试新技术的路线图。最近两年，我国国家重点研发计划已开始部署有关暴露组学的研究。

3　转录组学分析技术及其研究进展

3.1　转录组及转录组学的概念形成

转录组（transcriptome）这一概念是由 Velculescu 等在 1997 年首先提出的[25]，指的是在相同环境（或生理条件）下，一个细胞或一种组织中基因所能转录出的所有 RNA 的总和[26]。转录组学（transcriptomics）作为功能基因组学的重要组成部分，通过探索基因的功能和结构，揭示特定生物学过程中的分子机理，是研究生物体转录组的发生和变化规律的学科[27, 28]。当前新型高通量测序技术的运用使转录组学的数据量呈爆炸式增长，极大拓宽了转录组学的范围。因此转录组学便在生物学前沿研究中逐渐成为生命科学研究的热点，为人类探索生命进程、寻找疾病成因、开发新型药物和应对环境气候变化对人类的消极影响提供新方法。转录组学经典研究策略如图 15-2 所示。

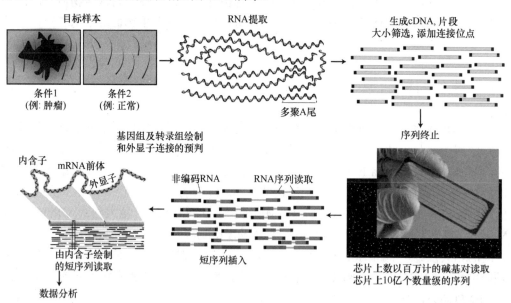

图 15-2　转录组学经典研究策略（引自参考文献[29]）.

3.2　转录组学分析技术

伴随分子生物学技术迅猛飞速发展，我们迎来了后基因组时代，使得高通量分析成为不可或缺的工具，为真正意义上的转录组学研究提供了夯实的技术支撑。现阶段的高通量研究方法主要可以分为三大类：一是基于杂交技术开发的方法，主要是指抑制差减杂交技术、微阵列技术等；二是基于 Sanger 测序开发出的方法，主要包括表达序列标签技术、基因表达系列分析技术、大规模平行标签测序技术；三是基于下一代高通量测序技术的全转录组 RNA 测序技术。

3.2.1 抑制差减杂交技术

抑制差减杂交技术（suppression subtractive hybridization，SSH）是 Diatchenko 等[30]于 1996 年发明的一种快速分离两个不同样本中差异表达基因的技术。SSH 是以抑制 PCR 和差减杂交技术为基础，通过一次差减杂交可使低丰度的 mRNA 序列得以高于 1000 倍的富集，在此过程中通过选择性地抑制在 PCR 过程中非目标序列的扩增，将获得的差异表达基因片段作为分子标记，进行标记辅助选择，实现目标序列的大量富集。SSH 技术适用于许多分子遗传和定位克隆研究，如基因的协同表达以及发现差异表达的基因等。但 SSH 也有自身的局限性：一般只用于两个样品的差异比较分析，而对于多个样品则无能为力；对取材的时期要求严格；所需的起始 RNA 量较大，对于一些来之不易的 RNA 并不适用。

3.2.2 表达序列标签技术

表达序列标签技术（expression sequence tags technology，EST）最初用来作为全基因组测序的低成本替代方案而出现。最早利用 EST 技术的是 1991 年 Adms 等用人脑组织 cDNA 得到的表达序列标签[31]。由 Velculescu 等[32]在 1995 提出的 EST 是通过从 cDNA 文库中随机挑取克隆对其进行大规模测序，获得部分 cDNA 的 5′端或 3′端序列，每条序列均可特异性地代表生物体的器官、组织或细胞在某一特定时期的一个表达基因，长度一般为 200～800 碱基对（bp），每个基因的表达次数越多能够测到的相应 EST 也越多，通过分析即可了解基因表达情况和表达丰度。随着准确性和覆盖率的提高，EST 开始应用于转录分析和蛋白组学等领域。但目前 EST 研究还存在诸多问题，首先，随着 EST 数据的不断增加亟需建立高通量、自动化的 EST 数据分析平台；其次，由于生物基因组极其庞大，所以在获取新 EST 的方面效率较低且耗资巨大；最后，如何避免全世界各研究机构对同一物种进行重复测序也是一个需要解决的问题。

3.2.3 基因表达系列分析技术

基因表达系列分析技术（serial analysis of gene expression，SAGE）由 Velculescu 等在 1995 提出[32]，是一项以 Sanger 测序为基础，用来分析基因群体表达状态的技术，是一种可以快速定量分析大量转录本的方法。SAGE 有两个重要原则：首先需要一个与转录文本位置隔离的、包含足够信息且可唯一识别转录的简短核苷酸序列标签（9～10 个 bp）；其次，短序列标签的连接，以在单个克隆内对多个标签进行测序的方式被有效分析转录，这样可以获得多个连续的短序列 SAGE 标签，而这些标签详细展示了对应基因的表达内容。SAGE 能应用于寻找新基因以及定量比较不同状态下的组织细胞的特异基因表达等。虽然 SAGE 技术有快速、大量以及节省时间等优点，但其不能完整检测到丰度比较低的 mRNA、花费巨大且容易产生假阳性。

3.2.4 大规模平行标签测序技术

大规模平行标签测序技术（massively parallel signature sequencing，MPSS）是 Brenner 等在 2000 年建立的[33]，可以看作是 SAGE 的改进版。是以 DNA 测序为基础建立的大规模、自动化、高通量基因分析技术，通过标签（一般为 10～20 bp）库的建立、微珠与标签的连接、酶切连接反应和生物信息分析等步骤，获得每一标签序列的拷贝数，该拷贝数代表了与该标签序列相应的基因表达水平，即可以在基因序列未知的情况下找出表达水平较低、差异较小的基因并得到相应的 cDNA 序列。MPSS 对于功能基因组研究非常有效,能在短时间内捕获细胞或组织内全部基因的表达特征;MPSS 可提供某一 cDNA 在体内特定发育阶段的拷贝数以及相应 cDNA 的序列,这就为在转录水平上进行基因表达分析提供了强有力的定性和定量手段[34];MPSS 还可以应用于不同丰度基因的差异表达分析,制作基因转录图谱,对新发现的基因加速其克隆和功能分析。MPSS 的优点是可以分析未知序列的基因、覆盖度高、能测得低表达丰度的基因、操作便捷耗时少;不足之处在于需要选择合适的标签序列,如果出现基因和标签之间的非特异性,将容易产生分析错误且成本较高。

3.2.5 微阵列技术

微阵列技术(microarray)是分子生物学领域里程碑式的重大突破,1995 年关于该技术首篇文献报道成为当年 *Science* 最有影响的文章之一[35]。Lipshutz 等在 1997 年最早详细报道了微阵列技术相关的照相平板印刷技术、激光共聚焦扫描、固相表面合成寡核苷酸以及核酸分子杂交技术等,随后其所在的 Affymetrix 公司将这些技术结合起来,研制出最初的基因芯片[36],这也是 Microarray 最被人们熟知的名称。Microarray 通过微加工技术将数以万(百万)计特定序列的 DNA 片段(基因探针)由阵列器有规律地排列固定于 2 cm^2 的硅片或者玻璃片等支持物上故又称基因芯片,这些"基因探针"可与放射标记物 ^{32}P 或荧光素等标记的样品分子进行杂交,通过激光共聚焦显微镜等扫描杂交信号的强度,获得相关基因表达谱,解答生物学领域的问题。最常被应用的 Microarray 技术有三种:玻璃斑点阵列、就地合成阵列和自组装阵列。可按固相载体上的探针不同,其又分为 cDNA 芯片和寡核苷酸芯片。微阵列技术的优点是自动化、小型、平行和数据规模大。其局限性在于 Microarray 只能提供相对浓度,对于复杂的哺乳动物基因组通常很难设计阵列,基于杂交技术的芯片只适用于检测已知序列无法捕获新的 mRNA。

3.2.6 高通量直接全转录组 RNA 测序技术

高通量直接全转录组 RNA 测序技术(RNA-seq)是近年来开发的一种革命性的转录组分析方法,是探索 RNA 的最灵敏方法。RNA-seq 能够全面快速地确定生物样本中几乎所有 RNA(mRNA 和非编码 RNA)序列的特征和丰度,生成的基因组级转录图谱由每个基因的转录结构和(或)表达水平组成[37]。RNA-seq 有诸多优势:RNA-seq 能以单基分辨率揭示转录边界的精确位置,结果也显示出高水平的技术和生物重复性;还可以揭示转录区域中的序列变化;RNA-seq 背景信号很低且无量化上限,有很大的动态表达水平范围,在一项研究中,研究人员分析了酿酒酵母中 1600 万个映射读数,其估计了超过 5 个数量级的范围;RNA-seq 需要较少的 RNA 样品。综上,RNA-Seq 成为当今转录组学研究中最有效的手段。与此同时,RNA-seq 也面临了许多局限与挑战,首先,由于 RNA 不稳定,容易降解,因此给样品处理处理增加的难度,使研究人员对样品纯度、数量和质量的控制仍然十分不易;其次,与 DNA 序列分析相比,RNA-seq 对于数据的读取比数据的比对更具挑战性;此外,与基因组测序相比,RNA 的预期相对丰度变化很大,以往许多研究表明具有最低和最高表达的基因之间相差不小于 10^5~10^7 个数量级,然而,由于 RNA-seq 读取数据的方式类似随机抽样,所以一小部分高表达基因便可以消耗大部分读数,因此对文库的建立提出了更高的要求。

3.3 转录组学技术的研究进展

转录组学技术飞速进入 RNA-seq 时代,基于 RNA-seq 开发出了许多新的技术。2010 年 Jason G Underwood 等[38]在 *Nature Methods* 描述了一种通过转录组高通量测序方法测定 RNA 结构的方法 (Fragmentation sequencing, Fragseq),这是一种高通量 RNA 结构探测方法,通过利用核酸酶 P1 特异性切割单链核酸,对产生的片段进行高通量 RNA 测序,在探测整个小鼠核转录组的过程中,同时准确地映射了具有已知结构的多个 ncRNA 中的单链 RNA 区域，且在 ncRNAs 中鉴定并实验验证了结构化区域，并且探测了两种细胞类型以验证重现性。对于单个细胞来说，转录组的随机变异性往往可以决定细胞的结局，因此由于不同细胞的转录组表型具有异质性，按照理论推测，转录组分析应该以单细胞为研究模型。2014 年北京大学学者在 PNAS 上撰文[39]，利用微流控芯片技术实现了高质量单细胞的 RNA-seq，全面提高了单细胞全转录组分析的准确性和可靠性。该技术将大大提高人们对单个细胞在哺乳动物发育中转录复杂性的理解，以往基于多细胞研究所得的许多研究结论并无法正确反映复杂生物体全面而真实的信息，严重掩盖了生命进程中种种事件中存在的大量随机行为；而该技术是细胞生命分析技术所追求的极限状态，在对传统技术提出极大的挑战同时也指明了未来的研究方向。2016 年 Jonas Frisén 等[40]开发出一种新的被称作空间转录组学（spatial tranomics）的高分辨率方法，该方法主要用于探究在一种组织中哪些基因是有活性的，该文章发表在 Nature Methods 上，它允许在单个组织切片中以空间分辨率对转录组进行可视化和定量分析，通过在具有独特位置条形码的阵列逆转录引物上定位组织切片，展示高质量的 RNA 测序数据。空间转录组学实现了定量基因表达数据和组织切片内 mRNA 分布的可视化，这在基础研究和诊断中有很大应用前景。同年，Nature Methods 推出了年度技术：表观转录组分析（epitranscriptome analysis），主要研究动态 RNA 的分布，调节和功能[41]，发现细胞可以在 mRNA 的生命周期内通过添加/移除一些化学修饰，进行更加直接瞬时的基因表达调控，实现对外界环境变化的快速反应。该技术虽尚处在起步阶段，但可以看到广阔前景。通常，基因组的转录产生稳定和瞬时 RNA，大多数 ncRNAs 都是迅速降解且难以检测到，如今知道它们也发挥着至关重要的功能，Patrick Cramer 等发表报道[42]，瞬时转录组测序（TT-seq）是一种统一绘制整个 RNA 产生单位范围，并估计 RNA 合成和降解速率的方案；TT-seq 还可以映射多聚腺苷酸化位点下游的瞬时 RNA 并揭示转录终止站点；该技术可以在多种疾病中发挥着重要作用。我国在转录组学的研究方面，紧跟国际前沿的脚步，为该领域的研究作出了极大的贡献。

3.4 转录组学在环境毒理学研究中的应用

环境是导致人类疾病发生的主要因素之一，利用转录组学技术识别影响人类健康的关键环境危害因素，深入探讨环境胁迫与基因的交互作用，揭示外源化合物对机体在基因层面造成损伤和突变的过程和机制，探究外界环境恶化对基因的功能改变及其与疾病的关系等是现如今相关领域科研工作者面临的重大挑战。研究人员等用 SAGE 技术通过转录组测序从淡水蟹中获得了六万多个单基因，并通过基因表达谱鉴定，发现多个与 Cd 胁迫相关的基因信号通路（VEGF/Toll/Jak-STAT/MAPK）。另有报道，运用 Microarray 通过基因表达谱的分析鉴定了一组受调节的基因，初步确定环境空气污染可能通过 Nod1-NF-κB 轴对雌性 BALB/c 小鼠的免疫系统产生影响。Yiling Li 等利用高通量 mRNA 测序（RNA-seq）和蛋白组学技术，发现甲基汞暴露可以通过剪接体干扰 RNA 剪接，在甲基汞暴露后观察到总共六百多个异常 RNA 选择性剪接事件，揭露了甲基汞神经毒性的新分子机制。Emmanuel Francois 等利用 RNA-seq 筛选鉴定环境内分泌干扰物与雌激素调控相关的基因中有 55 个基因的表达水平显著增强，38 个基因的表达水平显著降低，证实了利用 RNA-seq 能够更深入了解内分泌干扰物对细胞的调控机制。C. Vega-Retter 等运用 RNA-seq 技术鉴定了栖息于两个不同污染区域的鱼类中的差异表达基因，在一个受污染区域主要上调与细胞增殖以及肿瘤抑制有关的基因，而在另一个污染区域则发现与细胞凋亡过程相关的基因过表达；且栖息于两个污染区域的鱼，鸟氨酸脱羧酶基因（与肿瘤促进和进展相关）均过表达，这些结果表明，在人为污染对表型变化与基因型选择之间的相互作用的影响。Sarah L. O'Beirne 等运用 Microarray 检测了 $PM_{2.5}$ 暴露下吸烟者小气道上皮细胞转录组与非吸烟者的差异，将颗粒物污染和小气道上皮细胞生物学与易感个体的呼吸道疾病（肺功能受损、哮喘、COPD 和肺癌等）发展风险联系起来，在暴露于细颗粒物 30 天后，研究人员检测了 >54000 个全基因组转录本，发现与细胞分裂/细胞凋亡以及氧化应激相关的基因 CDC27/AIMP/2、与组蛋白转录抑制相关的 RBBP7 以及与宿主细胞因子行管的 HCFC1 等表达均有较大差异。Lucie Baillon 等在法国和加拿大不同污染梯度设置八个采样点，利用 RNA-seq 检测野生大西洋鳗的全局肝转录组，确定了 8 种金属和 25 种有机污染物对鱼类的影响，研究发现砷（As）、镉（Cd）、林丹（γ-HCH）是影响鳗鱼转录组的主要因素，与 As 暴露相关的基因涉及血管毒性，与 Cd 相关的基因参与细胞周期和能量代谢，对于 γ-HCH 则影响基因参与脂类代谢和细胞生长。Kelsa Gabehart 团队利用 Microarray 寻找新生小鼠在暴露于臭氧 6h 及 24h 肺组织的基因微阵列表达，发现 543 个基因下调，323 个基因下调，其中鉴定出有显著差异的 50 个基因中只有 RORC/GRP/VREB3/CYP2B6 显著上调，其他均下调，而这些基因涉及细胞分裂/增殖与细胞周期调节。以上结果表明急性臭氧暴露通过全面抑制细胞周期功能改变发育中的新生儿肺中的细胞增殖。Almudena Espín-Perez 等运用 Microarray 转录组学技术结合流行病学队列研究，发现颗粒物（$PM_{10}/PM_{2.5}$）、超细颗粒（UFPC）、氮氧化物（$NO_2/NO/NO_x$）、碳黑（BC）和碳氧化物（CO/CO_2）的个人暴露血液样本中基因的差异表达，发现 *Hsa-miR-197-3p*、*hsa-miR-29a-3p*、*hsa-miR-15a-5p*、*hsa-miR-16-5p* 和 *hsa-miR-92a-3p*（与肺癌和阿尔茨海默症等相关）与暴露相关地显著表达，表明它们作为空气污染的组学-生物标志物的潜在相关性。

4 蛋白组学分析技术及其研究进展

4.1 蛋白组学概述

蛋白组学（proteomics）的定义最早起源于 1995 年，是 Wilkins 等根据"蛋白质"（protein）和"基因组学"（genomics）组合提出[43]。蛋白组学以整体、动态和定量的原则来研究各种蛋白质的功能，包含生物体内全部蛋白质的鉴定与定量，系统地分析生物体内蛋白质表达，细胞定位，蛋白质相互作用，翻译后修饰，不同时间、空间以及细胞类型间的蛋白质周转等。蛋白组学已经成为后基因组时代的重要学科，与基因组学的研究结果相比，其实验结果更能从细胞和个体乃至物种的整体水平来解释生命现象的本质和规律，因此，比基因组学的研究更具有广度和挑战性。

4.2 蛋白组学分析技术

传统的蛋白组学研究主要是利用二维凝胶电泳对细胞或组织中的蛋白质进行分离[44]，该方法的优点是可以实现分子量 10000～300000 Da 蛋白质的覆盖，在一次实验中就能得到上千个蛋白质的斑点，但对分离得到的蛋白质无法进行高灵敏并且快速的鉴定，限制了该技术的进一步发展；20 世纪 80 年代末，质谱仪器发生了两大技术上的突破，使得蛋白质分析从根本上发生了革命性的改变。这两种技术就是软电离技术：电喷雾离子化（ESI）技术[45]和基质辅助激光解吸离子化（MALDI）技术[46]，这两种离子化技术解决了蛋白质或者肽段等不易挥发的生物大分子难以实现高灵敏度离子化的难题，发明这两种技术的科学家美国的芬恩（John Fenn）和日本的田中耕一（Koichi Tanaka），也因此被授予了 2002 年的诺贝尔化学奖。这两种软电离技术与质谱技术相结合，已经成为蛋白组学研究的核心技术之一[47]。

质谱技术用于蛋白组学分析的基本原理是，蛋白质和肽段分子离子化后形成分子离子峰，质谱仪器准确地检测其质荷比（m/z）以及相应的价态，得到其分子量信息并进一步结合母离子丰富的多级碎片离子峰，确定蛋白质或者肽段的氨基酸序列。MALDI 和 ESI 这两种离子化方式可以实现生物分子的高效离子化，而不会使其结构破坏。MALDI 离子化技术是将待测样品与基质分子形成共结晶薄层，在激光的照射下，样品分子吸收基质的能量和质子，快速升华形成气态离子实现离子化。ESI 离子化技术则采用高压电场将分析物溶液形成高电荷密度的微小雾滴，随着雾滴中溶剂分子的进一步挥发，实现样品溶液的离子化，因此，ESI 技术可以实现液相色谱仪与质谱仪联用，用于分析复杂的生物样品。MALDI-MS 技术通常用来分析相对简单的肽段混合物。

质量分析器是质谱仪中的核心技术。目前，常用于蛋白组学研究的生物质谱其质量分析器主要类别有：飞行时间（TOF）、四极杆（Q）、离子阱（IT）、傅里叶变换离子回旋共振（FT-ICR）和静电场轨道阱（Orbitrap）[48,49]。它们可以单独使用，也可以互相组合形成功能更强大的仪器，每一种都有自己的长处与不足[50]。ESI 通常与离子阱或三重四极杆质谱联用；MALDI 则常与 TOF 联用用于分析肽段的精确质量。离子阱质谱具备多级质谱能力，灵敏度较高、扫描速度快、性能稳定，缺点是质量精度较低；FT-ICR 目前是世界上质量准确度最高、分辨率最高的质谱仪，但造价昂贵且操作复杂限制了它在蛋白组学领域的应用。近年推出的 Orbitrap 则较好地平衡了质量精度、分辨率以及仪器占地等因素，同时又具备多级质谱能力，故而得到广泛使用。为了实现蛋白质及相关多肽的高通量准确鉴定，商品化质谱仪采用串联或并联的模式将多个或多种质量分析器进行组合，例如三重四极杆（QqQ）、四极杆-飞行时间联用（Q-TOF）、三重四极杆-飞行时间联用（QqQ-TOF）和飞行时间联用（TOF-TOF）等。一系列基于静电场轨道阱（Orbitrap）的质谱仪，线性离子阱-静电场轨道阱联用（LTQ-Orbitrap）、四极杆-静电场轨道阱联用（Q-Exactive）及 Q-OT-qIT "三合一" 质谱系列（Orbitrap Fusion）等[51]，则因质量分辨率和准确度的大大改善，尤其是仪器尺寸上的减小、运行成本降低等优点，已经成为了蛋白组学研究领域广泛使用的质谱仪。

4.3 蛋白组学的研究策略

基于生物质谱技术的蛋白组学的研究策略主要可以分为两种：自上而下（top-down）和自下而上（bottom-up）策略[52]。自上而下的蛋白组学分析技术，不经过蛋白质酶解的过程，直接对完整蛋白质进行质谱鉴定。完整蛋白首先通过 PAGE 或免疫富集法从复杂的生物样本中分离，然后直接用 ESI 或 MALDI 技术生成离子。生成的离子经碰撞诱导解离（CID）、高能量碰撞诱导解离（HCD）、电子捕获解离（ECD）或电子转移解离（ETD）等方式碎裂，并在串联质谱中分析。"top-down" 研究策略能提供完整蛋白质的生物学信息，保留多种翻译后修饰之间的关联信息，所鉴定到的蛋白质更接近于其在生物体内的真实状态[53]，在蛋白鉴定、分析、序列解析及翻译后修饰表征方面具有潜在的研究优势。但该策略面临着很多的技术挑战，如蛋白质混合物的分离困难、离子化效率低、质谱谱图的数据处理困难，以及对质谱仪器的要求极高等。

"bottom-up" 的策略首先是从生物样品中高效地提取蛋白质并进行酶解，然后进行质谱分析的蛋白鉴定，最后通过数据库检索或者从头测序法对质谱谱图进行解析，从而实现完整蛋白质的间接鉴定。其中，质谱对肽段的有效鉴定主要是通过获得肽段的相对分子质量和碎片离子信息来实现。首先，通过一级质谱扫描，获得肽段母离子的质荷比，再通过串联质谱分析，获得一系列肽段碎片离子信息。二级串联质谱得到的图谱能够为我们提供目标肽段的序列信息和翻译后的修饰信息。"自下而上" 的研究策略分为基于多维液相色谱分离技术平台的 "鸟枪法" 蛋白组学（shotgun proteomics）和基于特定目标物为分析对象的靶标蛋白组学（target proteomics）[54]。前者的实验目的是鉴定越多的蛋白质越好，而后者仅对一小部分目标肽段进行高质量精度高准确度的定性及定量研究。目前以 "鸟枪法" 蛋白组学的应用最为广泛，又被称为发现蛋白组学（discovery proteomics）。

4.4 蛋白组学在环境毒理学研究中的应用

环境毒理学是一门研究外源性环境污染物毒性作用的学科。其主要任务是描述环境污染物的毒性作用、阐明毒性作用机制、开展化学物质危险度评价等。蛋白组学技术具有高通量的特点，能实现对蛋白质的高效快捷定量分析。将蛋白组学技术和环境毒理学相结合，可从蛋白水平上研究外源性化合物对机体的毒作用机制，并从中筛选出具有较高特异性和灵敏度的蛋白标志物，目前已广泛应用于环境污染物的健康风险评价。

4.4.1 重金属毒性机制研究

重金属通常指对生物体生长发育有负面影响的有毒金属，如镉（Cd）、铜（Cu）、铬（Cr）、铅（Pb）、镍（Ni）、锰（Mn）、锌（Zn）以及危险金属，如砷（As）、汞（Hg）。重金属污染已经给生态环境和健康效应带来了严重威胁，本节综述了近年来基于蛋白组学对重金属在水生生物、动物和植物毒性研究中的应用。短期 Cd（100 μg/L）暴露对贻贝蛋白质组的影响较小，采用二维凝胶电泳（2D-DIGE）分离蛋白，进一步利用 MALDI-TOF 鉴定出 12 种差异蛋白，包括结构蛋白、代谢蛋白和应激反应蛋白[55]。而不同贝类对 Cd 暴露的生物学效应也不同，白纹蛤和斑纹蛤对 Cd 暴露（200 μg/L）的蛋白组学研究表明，差异表达蛋白（DEPs）白蛤多于斑马蛤。Cd 暴露共同诱导了两种蛤的关键生物学过程，包括免疫反应和代谢。白蛤蛋白质水解和能量产生有关的过程得到了增强。斑纹蛤的蛋白质水解和能量产生的某些过程耗竭[56]。采用蛋白组学研究了雄性牡蛎和雌性牡蛎对锌暴露的不同反应。在 50 μg/L 或 500 μg/L 锌暴露 30 d 后，雌性牡蛎性腺的锌累积量大于雄性牡蛎且性腺发育加快。雌牡蛎生殖腺中表达的半胱氨酸、组氨酸脱氢酶等具有锌转运和储存功能的蛋白和多功能蛋白明显高于雄牡蛎[57]。以水蚤为模式生物研究了铅（Pb^{2+}）和阿特拉津对水生生物的影响。通过比较对照和胁迫条件下的蛋白图谱，发现 Pb 和阿特拉津对水蚤的影响几乎相反；Pb 抑制了大部分 DEPs，而阿特拉津激活了大部分 DEPs[58]。

大鼠连续 5 周暴露于 Mn（200 mg/L）后，基于 iTRAQ 对 3012 种肠道黏膜细胞锰含量修饰蛋白进行了定性和定量分析，共发现 175 种肠黏膜蛋白在锰暴露下差异表达。这些蛋白与胰腺分泌、蛋白质消化吸收、脂肪消化吸收、氨基酸生物合成、甘油脂代谢、阿尔茨海默症、帕金森病、矿物质吸收相关[59]。Hg[$HgCl_2$：0.1 mg/(kg·d)、1 mg/(kg·d)、5 mg/(kg·d)]暴露后 SD 大鼠肾脏组织的蛋白组学分析表明，硒结合蛋白 1（SBP1）是 Hg 给药后大鼠肾皮质中最明显的上调蛋白。尿中 SBP1 的排泄量呈剂量依赖性显著增加。采用 $HgCl_2$、$CdCl_2$ 或顺铂处理正常肾近端小管细胞 24h，发现 SBP1 在化学诱导肾毒性的病理过程中发挥重要作用，尿中 SBP1 的排泄可作为早期肾损伤的生物标志物[60]。A549 细胞对外源 Cd 暴露的蛋白质组研究发现，鉴定的 53 个差异蛋白根据其生物学过程进行分类，主要涉及代谢过程、细胞过程、发育过程和细胞成分；根据蛋白的分子功能进行分类，主要涉及催化活性和结合活性[61]。对 Pb、As 和甲基汞（MeHg）混合暴露在海马细胞蛋白质水平上的相互作用进行研究。蛋白质组数据的基因功能和通路分析证实了与线粒体功能障碍、氧化应激、m-RNA 剪接和泛素系统功能障碍相关的神经退行性改变发生[62]。重金属对海马细胞的影响顺序为 Pb < As < MeHg。长期接触铅、砷和甲氧基汞等重金属会增加患神经退行性疾病的风险[63]。采用蛋白组学技术研究了跳虫体内铜、锰和镍毒性，结果表明跳虫的免疫系统、神经元生长和金属离子结合中涉及的几种蛋白均上调[64]。

蛋白组学研究能够反映植物应对重金属污染的调控过程，展示主要参与细胞解毒和耐受机制的蛋白质网络和代谢途径。大豆植株分别暴露于 Pb（30 μg/L）和 Hg（0.5 μg/L）后，叶片和结节的氧化应激水平均增加，抗坏血酸过氧化物酶、谷胱甘肽还原酶和过氧化氢酶的活性也被调节。铅和汞 76 种蛋白质有显著影响，P 和 Hg 分别对 33 种和 43 种蛋白质有影响。差异蛋白在功能上与多种细胞功能相关[65]。紫花苜蓿长期暴露于Cd（10 mg/kg）后，179个细胞壁蛋白和30个可溶性部分蛋白的丰度发生了变化。这些蛋白参与细胞壁重构、防御反应、碳水化合物代谢和促进木质化过程[66]。基于蛋白组学研究了菌株Q2-8诱导小麦植株Cd和As吸收减少的分子机制：菌株Q2-8通过提高Cd和As胁迫下根系能量代谢、防御和细胞壁生物合成的效率，减轻Cd和As对小麦幼苗的毒害，小麦根中差异表达的细胞壁生物合成相关蛋白仅存在于Cd胁迫下小麦植株中[67]。采用蛋白质组学对土壤中Ni、Cu和Zn对植物（Ocimum basilicum L.）的影响研究发现：Cu可导致致敏蛋白浓度升高，严重的Cu胁迫还会导致与蒸腾和光合作用相关的特定蛋白质的积累，并得出铜对该植株的危害最大，产生的过敏原最多的结论[68]。水稻幼苗对六价铬（Cr^{6+}）胁迫的蛋白组学反应发现：鉴定出的64种蛋白质参与一系列的细胞过程，包括细胞壁合成、能量生产、初级代谢、电子传递和解毒。细胞壁是水稻抵抗铬胁迫的重要屏障，水稻植株应对Cr胁迫的策略包括将铬离子固定在细胞壁中，减少其易位；激活抗氧化防御，减轻Cr诱导的氧化应激[69]。对镉（Cd^{2+}）、干旱及其复合胁迫下短柄草幼苗叶片进行了蛋白组学分析。采用2D-DIGE检测到117个差异蛋白，参与光合作用/呼吸、能量和碳代谢、压力/防御/解毒、蛋白质折叠和降解、氨基酸代谢。Cd^{2+}和复合胁迫对叶片蛋白质组的影响大于单独的渗透胁迫[70]。

4.4.2 有机污染物毒性机制研究

有机污染物种类繁多，且毒性较大，本节选择了两种典型有机物如多环芳烃（PAHs）和全氟辛烷磺酸（PFOS），对其基于蛋白组学的毒理学研究进行了综述。PAHs是一种高致癌性的污染物，往往在环境中长期存在。为探讨拟南芥对PAHs中菲的氧化应激反应，采用蛋白组学技术鉴定出30种差异表达蛋白，其中核苷二磷酸激酶3（NDPK-3）显著上调，进一步研究了NDPK-3过表达、敲除和野生型植物中应激反应酶的水平。NDPK-3过表达体中APX、CAT、POD、SOD活性均高于野生型；在NDPK-3敲除序列中，这些酶的活性降低。这些数据证实NDPK-3是拟南芥对多环芳烃胁迫响应的正向调节因子[71]。采用二维凝胶电泳结合MALDI-TOF/TOF-MS对菲暴露的小麦根蛋白进行分析和鉴定。菲诱导上调的蛋白与植物的防御反应、抗氧化系统和糖酵解有关；下调的蛋白质涉及高能量化合物的代谢和植物生长[72]。另外一个研究对小麦幼苗进行菲暴露的亚细胞观察表明，叶绿体发生逆转，失去了结构的完整性。采用iTRAQ分析了叶绿体蛋白质谱的变化，共鉴定出517种蛋白，8个与类囊体相关的蛋白被下调，相关基因的表达通过RT-PCR验证，证实了类囊体破坏是叶绿体变形的原因[73]。采用"鸟枪"蛋白组学方法对大西洋鳕鱼（Gadus morhua）的血浆样品进行了分析，共鉴定了369种蛋白质，在暴露于PAHs的鱼类中发现了12种蛋白水平的显著变化，11种上调蛋白主要是免疫球蛋白，表明在暴露的鱼体内触发了免疫反应[74]。绵羊长期饲养在污水污泥施肥牧场上，暴露后检测其肝脏中的PAHs和PCBs，同时采用二维差分凝胶电泳法对肝脏蛋白进行分离，采用液相色谱/串联质谱法对差异表达的蛋白进行鉴定。发现环境化学品暴露影响了外源化合物脱毒反应过程以及雄性和雌性绵阳的肝蛋白组，包括主要的血浆分泌蛋白和代谢酶。通路分析预测了癌症相关通路的失调和脂质代谢的改变[75]。用5种PAHs（芴、蒽、菲、荧蒽和芘）对蚯蚓（Eisenia fetida）进行暴露，并对小于20 kDa的蛋白进行分析，鉴定出54个差异蛋白。其中，三个差异蛋白在暴露于蒽和芘的蚯蚓中表达；一个蛋白选择性地表达于暴露于芴、菲和荧蒽的蚯蚓中[76]。

蛋白组学为PFOS毒性效应的分子机制和生物标志物研究提供了方法。斑马鱼胚胎暴露于PFOS（0.5 mg/L）192 h后，采用二维凝胶电泳进行蛋白组学分析发现了69个差异蛋白，进一步采用MALDI-TOF-MS对目标蛋白进行鉴定。这些蛋白的功能包括解毒、能量代谢、脂质转运/类固醇代谢过程、细胞结构、信号转导和凋亡[77]。对欧洲杜父鱼（Cottus gobio）进行短期PFOS（0.1 mg/L或1 mg/L，96 h）暴露，然后采用二维凝胶电泳和nano-LC-MS/MS相结合的方法，鉴定出的差异表达蛋白包含了应激反应、泛素-蛋白酶体系统、能量代谢、肌动蛋白细胞骨架等功能[78]。欧洲鳗鱼外周血单核细胞（PBMC）体外暴露于PFOS（10 μg/L和1 mg/L，48 h），共鉴定出48个不同功能类群的差异蛋白，涉及细胞骨架、蛋白折叠、细胞信号、蛋白水解途径、碳水化合物和能量代谢[79]。采用胚胎干细胞试验来评估 PFOS 发育毒性（40 μmol/L，10 d），蛋白组学分析共鉴定出 176 个差异蛋白，主要与催化活性、细胞核定位或细胞成分有关[80]。iTRAQ 标记定量蛋白质组技术也被应用于 PFOS 的毒理学研究。PFOS 暴露[1.0 mg/(kg·d)、2.5 mg/(kg·d) 和 5.0 mg/(kg·d)] 7d 后小鼠肝脏的差异蛋白主要参与脂质代谢、转运、生物合成过程以及对刺激的反应[81]。人类肝细胞 LO2 暴露于 PFOS（25～50 mg/L）72 h 后，蛋白组学的结果表明 PFOS 能够通过抑制 HNRNPC、HUWE1、UBQLN1，以及诱导 PAF1 参与 p53 和原癌基因的信号通路激活从而触发凋亡过程[82]。另一个人肝细胞株（HL-7702）暴露于 PFOS（50 μmol/L 48 h 和 96 h）后[83]，差异表达蛋白质与细胞增殖有关，包括肝癌衍生生长因子（Hdgf）和增殖标志物 Mk167 和 Top2α。蛋白质组学也可以应用于发现导致毒性效应的靶点蛋白。应用化学探针标记的方法，在 PFOA 暴露的小鼠肝脏中鉴定到了两种 PFOA 的直接作用靶点蛋白（Acaca 和 Acacb），并说明了其对脂肪代谢的毒理效应[84]。

5 代谢组学分析技术及其研究进展

5.1 代谢组学概念

代谢组学是对机体所有小分子代谢物进行定性和定量分析的一门学科，其研究对象主要是分子量小于 1000 的代谢物。代谢组学研究采用先进的分析检测技术，结合模式识别和大数据统计方法，研究生物体内代谢物随内源以及外源环境变化而产生的动态变化规律，帮助人们更好地了解生物系统对环境和遗传因素变化的反应。其基本原理是外源化合物能够通过破坏细胞正常的结构，改变细胞内小分子代谢物的稳态，从而通过直接或间接生物学效应改变靶组织中的代谢物组成及代谢通路，而代谢物中的生物标志物的动态变化可以指示外源化合物对机体的代谢损伤[85-88]。

5.2 代谢组学分析技术

代谢组学分析流程包括样品采集和预处理、数据的采集和分析。而数据采集是代谢组学流程中最重要的环节，需要借助各种技术平台对具有不同性质的代谢物进行无偏向的分析，核磁共振（NMR）和质谱（MS）技术以它们独特的优势成为目前代谢组学研究中最常用的技术手段[98]。NMR 技术分析能够深入物质内部而不具有破坏性，样品不需要繁琐的前处理步骤，减少了处理过程中的损失，且对含氢化合物均有响应具有普适性，还具有迅速、准确、分辨率高和重现性好等优点；但 NMR 技术也存在灵敏度较低、检测成本较高等不足，限制了其在代谢组学领域的应用。而质谱技术有较高灵敏度，可以实现对多个化合物的同时分析和鉴定，且检测成本相对较低，是代谢组学研究的重要技术之一。近些年质谱及色谱-质谱联用技术快速发展，在代谢组学领域发挥越来越大的作用[90, 91]。

质谱联用技术主要包括 GC-MS 和 LC-MS。GC-MS 主要适用于检测挥发性物质，具有较高的分辨率和灵敏度，而且有标准谱图库，对化合物的定性能力比较强。但是 GC-MS 不能直接用于分析难挥发、极性强、热稳定性差的化合物，需要通过硅烷化、甲基化、酰基化等衍生化方法改善其挥发性、峰形、分离度以及灵敏度，预处理过程较为繁琐。LC-MS 将分离能力强的液相色谱和灵敏度高、准确性好、分辨率高的质谱相结合，对样本制备要求不高，且灵敏度高、线性范围宽，能够实现对复杂生物样本中多种化合物同时定性和定量，成为代谢组学的研究中应用越来越多的技术平台[92, 93]。

靶向代谢组学和非靶向代谢组学（发现代谢组学）已成为代谢组学应用的两种主要模式[94]。靶向代谢组学分析常采用色谱与QqQ或Q-Trap等质谱串联，用多反应监测器（MRM）进行检测。但靶向代谢组学往往只关注部分代谢物信息，不具有整体性，过于依赖基于先验知识所预设的目标分析物。非靶向代谢组学分析中，色谱可与TOF、Orbitrap、IT-TOF、Q-TOF等质谱联用，可观察到整体无偏向的代谢变化，可以发现未知生物标志物。虽然非靶向方法具有分析通量高、数据信息丰富等优点，但存在数据稳定性、重复性差等缺点，且代谢物鉴定需要花费大量的时间[95, 96]。而靶向分析方法虽然能够弥补非靶向方法在定量线性范围、稳定性、重复性等方面的不足，但其分析通量远远比不上非靶向分析，得到的信息量十分有限，并不适用于总体的代谢轮廓分析。因此，拟靶向分析方法被提出，该方法结合了非靶向和靶向分析方法的优点，从而能精确定量多种代谢物，满足代谢分析中大通量、宽线性范围、稳定性重复性好的要求[97, 98]。

5.3 代谢组学相关软件的应用

高通量代谢组学研究的一大难点在于数据的处理和分析。为分析和处理此类数据，各种生物信息学工具应运而生。质谱和核磁分析产生的海量数据通常具有高维度、高噪声、高缺失值、高变异性以及复杂的相关性和冗余性等特点。需要对这些复杂高维的数据集进行预处理、物质鉴定、统计分析和结果解释等。代谢组学的相关软件目的在于：①描述数据集中代谢组学数据的整体结构；②解释不同代谢特征物质与样本表型数据的相关性。代谢组学数据分析的流程主要分为预处理（图谱预处理、数据预处理）、物质鉴定、统计分析及结果解释。每个步骤中所用的方法根据具体的数据类型、数据来源以及用户的实验设计方案有所不同。各种全功能软件工具，其开发目标和侧重点也有差异，用户可根据自己的需求进行合理选择[99]。现在最常用的软件有十几种，针对代谢组学数据处理和分析的若干核心环节，以下四款软件应用最为广泛，它们分别为：

（1）MAVEN是基于现有的开源软件建立的LC-MS数据分析软件，可处理来自三重四极杆和全谱仪器分析的数据，为数据的检验和可视化提供了交互式的工具。该软件具有从特征值提取到代谢物途径分析及数据可视化的整个数据分析流程。同时为便于验证数据，MAVEN运用机器学习算法来自动评估峰值质量[100]。

（2）MZmine是一款开源软件，其主要目标是设计一款用户友好型、灵活性强、易扩展的软件。该软件主要处理LC-MS平台数据，可用于靶标和非靶标代谢组学数据分析。其中数据处理模块采用嵌入式可视化工具，实现立即预览分析结果的功能。软件新增功能包括基于随机抽样一致算法对齐峰列表、在线数据库鉴别峰、改进的同位素模式识别以及实现数据可视化。项目管理是该软件新增的另一核心模块，用户可随时追踪并储存临时的数据分析结果[101]。

（3）MetaboAnalyst是一款完全免费的web平台代谢组学数据分析软件。它是一款结合统计分析、代谢物功能与生物学注释及可视化为一体的在线工具。该软件包含8个功能模块，可归纳成3大类：①探索性数据分析，包含"统计分析"和"时间序列"模块；②功能分析，包含"富集分析""通路分析"和"整合通路分析"模块；③高级分析方法，包含"生物标志物分析""样本量的估计"和"效能分析"模块。此外，它还包含"其他实用程序"模块，该模块具有脂质组学数据分析的特殊功能和化合物身份标识号码（ID）转换工具[102]。

（4）XCMS Online软件是一款基于云计算的数据处理平台，提供高质量的代谢组学数据分析，是一款用户友好型的非靶标代谢组学数据处理软件。XCMS Online继承了XCMS强大的数据预处理功能，同时新增了单因素和多因素统计分析方法、代谢物特征注释和代谢物鉴定，为非靶向代谢组学提供了完整的工作流程方案。另外它将XCMS的命令行界面改为图形用户界面，降低了操作难度[103]。

在完整的数据分析流程中，MAVEN 和 MZmine 软件的优势在于数据的预处理，而 MetaboAnalyst 的优势则在于强大的数据统计分析、高级分析以及功能分析等功能，这对代谢组学数据的解释具有重要意义。MAVEN 软件的显著优势在于能对峰质量进行良好的评估，能提供可信度较高的代谢组学数据，同时能实现数据在代谢途径中的可视化绘图。MZmine 软件除具有全面的数据预处理功能外，还支持用户开发新的算法，所以它既适合于无编程基础的组学工作者也适合一些高级用户。XCMS Online 软件是经典 XCMS 函数包的扩展，使用简便，是一款良好的用户友好型软件。但它不具有良好的扩展性，用户不能自定义算法[99]。随着样本检测技术的不断改进，高通量代谢组学已迎来大数据时代，除以上介绍的四款软件外，还有 statTarget[104]、SMART[105]等新兴软件也是代谢组学研究的有力工具。但当前全功能软件的数量和性能普遍滞后于使用需求，期望更多此类软件被报道，为代谢组学研究提供支持。这也将是未来代谢组学数据分析研究的一大方向。

5.4 代谢组学在典型有机污染物毒性评估中的应用

环境污染物特别是持久性有机污染物（POPs）对机体的暴露一般具有以下特点：①接触剂量通常较小，有些污染物接触剂量远低于无可见有害作用水平；②长时间内反复接触甚至终生接触，如职业暴露；③多种环境污染物同时作用于机体，环境化学污染物种类繁多，往往多以低浓度混合物形式存在；④接触的人群易感性差异大。传统的毒性测试通量低、周期长、敏感度低，显然不能满足日益增长的污染物毒性测试的需求。代谢组学的发展则为低剂量典型有机污染物毒性效应和毒性作用机制评估提供了新视角。代谢组学是基因组学和蛋白组学的下游学科，代谢物的变化是机体对环境因素影响的最终应答，由于机体代谢物的变化可灵敏地指示和确证外来干扰在组织和器官水平的毒性效应以及毒性作用靶位点，因此借助代谢组学技术来评价环境污染物暴露带来的毒性效应，并进而推断其毒性作用的分子机制，具有快速、灵敏度高、选择性强等特点，特别是在低剂量或环境剂量污染物的毒性效应评估方面具有很大的优势。

近年来，代谢组学逐渐开始被用于 POPs 及其他典型有机污染物类物质毒性效应的研究。Lankadurai 等[106]研究了全氟辛酸（PFOA）和 PFOS 对蚯蚓（*Eisenia fetida*）的代谢干扰，发现 PFOA 和 PFOS 都能引起蚯蚓体内代谢物的显著变化，亮氨酸、精氨酸、谷氨酸、麦芽糖和三磷酸腺苷（ATP）是 PFOA 和 PFOS 暴露的潜在标志物。另外 PFOA 和 PFOS 暴露还会损伤线粒体内膜结构，从而加速脂肪酸氧化，干扰 ATP 合成。Ji 等[107]研究了多溴联苯醚（PBDEs）对蚯蚓（*Eisenia fetida*）的毒性作用，结果表明 2,2′,4,4′-四溴联苯醚（BDE-47）主要引起蚯蚓体内甜菜碱、氨基酸、ATP、葡萄糖、麦芽糖和琥珀酸，标志着 BDE-47 干扰蚯蚓体内能量代谢，引起渗透压改变。Zeng 等[108]研究了 BPA 暴露后大鼠尿液中代谢物的变化，结果表明氨基酸、多胺、核苷、有机酸等 42 种代谢物发生显著变化，缬氨酸、亮氨酸和异亮氨酸生物合成、谷氨酰胺和谷氨酸代谢等代谢通路受到严重干扰，另外显著变化代谢物中包括多种神经递质以及与神经传递相关代谢物，表明 BPA 能够干扰大鼠神经系统。Li 等[109]研究发现大鼠口服 BPA 后可引起半乳糖代谢紊乱。Wei 等[110]采用 UHPLC-Orbitrap 高分辨质谱研究了 BDE-47 的毒性机制，结果表明 BDE-47 可以通过抑制戊糖磷酸途径使人乳腺癌细胞 MCF-7 的代谢紊乱，并引起氧化应激。

环境化学污染物种类繁多，低剂量长期暴露是环境污染暴露的显著特征。O'Kane 等[111]对 SD 大鼠通过口服暴露低剂量多氯联苯（PCBs）和二噁英（2,3,7,8-TCDD）（0.1 μg/kg TEQ），借助 UHPLC/Q-TOF MS 对大鼠血样中小分子代谢物进行分析，结果表明 PCBs 和 2,3,7,8-TCDD 即使在低剂量暴露下也会引起大鼠体内多个代谢物的显著性变化，鉴定出的显著变化代谢物主要包括氨基酸和脂类代谢物。耿柠波等[112]以 HepG2 细胞为研究对象，探讨了环境浓度水平短链氯化石蜡（SCCPs）对细胞小分子代谢物的干扰作用。结果表明 SCCPs 暴露后，HepG2 细胞小分子代谢物与对照组相比有明显的变化，细胞在糖代谢、氨基酸代谢和脂肪酸代谢方面发生不同程度的紊乱。王菲迪等[97]的研究结果表明，低剂量（0.05 mg/L）的六溴环十二烷（HBCD）暴露即可诱发氧化胁迫效应和代谢紊乱，结合多重比较分析、代谢通路富集分析以及代谢通路限速酶活性分析，研究了 HBCD 的细胞毒性作用机制。

另外，环境污染物多以低浓度混合物形式存在，因此不可避免地对人类造成复杂多样的健康危害，即潜在联合毒性效应，代谢组学技术的发展大大提高了系统揭示机体对化学混合物的生物应答和确认作用机制相关生物标志物的可能性[113-116]。Song 等[117, 118]用代谢组学研究了苯并芘（BaP）和滴滴涕（DDT）单独和联合暴露分别对翡翠贻贝性腺和鳃的代谢干扰，发现 BaP 并没有引起翡翠贻贝性腺的代谢物变化，但是 BaP 引起翡翠贻贝鳃中支链氨基酸的显著下降从而干扰渗透压调节，DDT 则同时引起翡翠贻贝性腺和鳃渗透压调节和能量代谢的扰动，而当 BaP 和 DDT 联合暴露时，翡翠贻贝性腺的渗透压调节和能量代谢受到干扰，但是翡翠贻贝鳃的代谢物则未受到任何扰动。另外，代表单个代谢特征随污染物暴露变化总合的代谢影响水平指数（MELI）的计算可以用来探讨代谢组层面的剂量-效应关系。同时，MELI 的计算还能够定量描述化学混合物暴露干扰的相关代谢通路的联合毒性效应类型，即协同、加和或者拮抗效应。王菲迪等[119]采用代谢组学分析策略，结合常规毒性测试手段，研究了 PAHs 和 SCCPs 的联合毒性作用，基于拟靶向代谢组分析得到无偏且精确的代谢物信息，通过计算代谢影响水平指数来定量描述联合暴露对整体代谢和专一代谢通路的联合暴露效应类型。发现 PAHs 和 SCCPs 联合暴露对 HepG2 细胞总代谢水平的影响表现为加和效应。

6 展　　望

近年来，组学技术的飞速发展为环境毒理学研究提供了新的思路和强有力的技术支持。但是，各个组学技术在其发展道路上仍然面临着巨大的挑战。暴露组的概念得到了不断的完善和更新，但研究技术手段和组织方式的滞后仍限制着这一庞大计划的有效开展。暴露组研究人体从出生至生命结束整个生命过程中每时每刻暴露于外界环境（空气、水、饮食），以及社会环境和个人发展阶段及行为；由于诸多因素处在动态变化之中，给暴露组的研究增加许多不确定性。暴露组涉及的内外暴露因素众多，目前仍缺乏强大的数学算法和数学模型能将这些复杂因素关联起来，并从中解构出暴露和疾病的关联关系。另外，研究和报告的标准化和研究数据的共享方式也是当前急需解决的问题。转录组及转录组学的的研究技术近几年来依托于 RNA-Seq 得到快速发展，如全长转录组测序、数字基因表达谱分析、小 RNA 测序、降解组测序、长链非编码 RNA 测序、单细胞转录组测序等等。RNA-Seq 与转录组学的下一个重大挑战是靶向更复杂的转录组，识别和跟踪来自所有基因的稀有 RNA 同种型的表达变化。推动实现这一目标的技术包括配对末端测序，链特异性测序以及使用更长的读数以增加覆盖率和深度等。而诸如空间转录组学以及表观转录组学等正处于起步阶段的新技术，也亟需在解决特异性问题的前提下，加快依赖于抗体富集的转录组与蛋白组相结合的新技术的开发。蛋白组学及代谢组学技术在高通量、高灵敏度的分析特征生物标志物发现、环境污染物低剂量暴露、复合污染研究等领域发挥了独特的优势。但是二者无论是现有的分析仪器、分析技术，还是数据采集和数据分析上都需要进一步提高。近年来蛋白组学及代谢组学高效样品预处理技术，高准确度高覆盖定性与定量技术和数据解析技术的发展推动了其分析的覆盖度，将这些新技术与新方法引入毒理学研究，有助于全景式地揭示机体暴露于污染物的调控规律。

挑战伴随着机遇，环境毒理组学技术目前遇到的挑战也恰恰是未来发展的机遇。相信随着多组学研究方法的不断优化和完善，必将使环境安全性评价更准确、高效及全面。环境毒理组学技术研究环境毒理学需要各个学科的融合和交叉，以及各个领域科学家的紧密合作，逐步建立多仪器分析相结合，整体表征与局部表征相结合，定性与定量分析相结合，组学分析与临床生化分析相结合，构建多学科参与、多分析方法整合、大数据分析的多组学平台。逐步完善暴露组学研究方法并建立健全人类暴露组信息数据库，有望在不久的将来揭开人体暴露、作用机制和疾病成因的本质，为人类健康保护和疾病预防发挥重要作用。在转录组方面，更应专注于新的测序技术，随着测序成本持续下降，预计 RNA-Seq 将取代微阵列用于涉及确定转录组结构和动力学的许多应用。而不断发展高通量、高灵敏度与高精确度的现代分析技术，建立可利用的该研究领域的相应数据库及专家系统将是蛋白组学及代谢组学未发展的关键。毒性作用通路是一个复杂的网络，也是深入揭示毒性作用机制的有效方法，暴露于外源化学物是如何破坏这些路径并最终导致了负的健康效应，暴露组、转录组、蛋白组和代谢组相结合才能为毒理学评估提供综合视角，系统揭示污染物的毒性作用机制。因此，多组学融合的环境毒理组学技术将会成为毒理学发展的必然趋势。

参 考 文 献

[1] Paul L, Niels V H, Pia K V, et al. Environmental and heritable factors in the causation of cancer: Analyses of cohorts of twins from Sweden, Denmark, and Finland. New England Journal of Medicine, 2000, 343(2): 78-85.

[2] Teri A M, Joan B W, Francis S C. Genes, environment and the value of prospective cohort studies. Nat Rev Genet, 2006, 7(10): 812-820.

[3] Thea M E, John P M. Environmental exposures and gene regulation in disease etiology. Environ Health Perspect, 2007, 115(9): 1264-1270.

[4] Stephen M R, Martyn T S. Epidemiology. Environment and disease risks. Science, 2010, 330 (6003): 461-461.

[5] Pandey A, Mann M. Proteomics to study genes and genomes. Nature, 2000, 405(6788): 837-846.

[6] Parag M, Bernhard K. Proteomics: A pragmatic perspective. Nature Biotechnology, 2010, 28: 695-709.

[7] Christopher P W. The exposome: From concept to utility. Int J Epidemiol, 2012, 41(1): 24-3.2

[8] Gary W M, Dean P J. The nature of nurture: refining the definition of the exposome. Toxicol Sci, 2014, 137(1): 1-2.

[9] Martine V, Rémy S, Oliver R, et al. The human early-life exposome (HELIX): Project rationale and design. Environ Health Perspect, 2014, 122(6): 535-544.

[10] Beate I E, Jörg H, Tobias P, et al. From the exposome to mechanistic understanding of chemical-induced adverse effects. Environ Int, 2017, 99: 97-106.

[11] 孙路遥, 王继忠, 彭书传, 等. 暴露组及其研究方法进展. 环境科学学报, 2016, 36 (1): 27-37.

[12] Rappaport S M. Discovering environmental causes of disease. J Epidemiol Community Health, 2012, 66(2): 99-102.

[13] 冷曙光, 郑玉新, 等. 基于生物标志物和暴露组学的环境与健康研究. 中华疾病控制杂志, 2017, 21 (11): 1079-1095.

[14] Joachim D P, Matthew A S. Stiegel. Evolution of Environmental Exposure Science: Using Breath-Borne Biomarkers for "Discovery" of the Human Exposome. Analytical Chemistry, 2013, 85(21): 9984-9990.

[15] Line S H, Amrit K S, Enrique C, et al. In-utero and childhood chemical exposome in six European mother-child cohorts. Environ Int, 2018, 121(Pt 1):751-763.

[16] Clayton S B, Oliver F. Using untargeted metabolomics for detecting exposome compounds. Current Opinion in Toxicology, 018(8): 87-92.

[17] Ming Kei C, Kurunthachalam K, Germaine M L, et al. Toward capturing the exposome: Exposure biomarker variability and co-exposure patterns in the shared environment. Environ Sci Technol, 2018, 52(15): 8801-8810.

[18] Oliver R, Xavier B, Lydiane A, et al. The Pregnancy exposome: Multiple environmental exposures in the INMA-sabadell birth cohort. Environ Sci Technol, 2015 ,49(17):10632-10641.

[19] Benedikt W, Scott S, Mingliang F, et al. Exposome-scale investigations guided by global metabolomics, pathway analysis,

and cognitive computing. Anal Chem, 2017, 89(21): 11505-11513.

[20] Xanthi D A., Konstantinos C M. The framework of urban exposome: Application of the exposome concept in urban health studies. Sci Total Environ, 2018, 636: 963-967.

[21] Laurie F, Vincent B, Regis de G, et al. Occupational exposome: A network-based approach for characterizing Occupational Health Problems. J Biomed Inform, 2011, 44(4): 545-552.

[22] Michelle C T, Paolo V, Eduardo S, et al. Exposomics: final policy workshop and stakeholder consultation. BMC Public Health, 2018, 18(1): 260-271.

[23] 王琼, 董小燕, 高圣华, 等. 环境污染暴露评估的研究进展. 环境与健康杂志, 2016, 33 (11):1025-1029.

[24] 白志鹏, 陈莉, 韩斌. 暴露组学的概念与应用. 环境与健康杂志, 2015, 32 (1): 1-9.

[25] Victor E V, Lin Z, Wei Z, et al. Characterization of the Yeast Transcriptome. Cell, 1997, 88(2): 243–251.

[26] Lockhart D J, Winzeler E A. Genomics, gene expression and DNA arrays. Nature, 2000, 405 (6788): 827-836.

[27] Matthias E F, Wolfgang K, Reinhold S, et al. The human transcriptome: Implications for the understanding of human disease. 2009, 123-149.

[28] Jeffrey A M, Zhong W. Next-generation transcriptome assembly. Nat Rev Genet, 2011, 12(10): 671-682.

[29] Malachi G, Jason R W, Nicholas C S, et al. Informatics for RNA sequencing: A web resource for analysis on the cloud. Plos Computational Biology, 2015, 11(8): 1004393-1004393.

[30] Malachi G, Yun F C L, Aaron P C, et al. Suppression subtractive hybridization: A method for generating differentially regulated or tissue-specific cDNA probes and libraries. Proceedings of the National Academy of Sciences of the United States of America, 1996, 93(12): 6025-6030.

[31] Mark D A, Jenny M K, Gocayne J D, et al. Complementary DNA sequencing: expressed sequence tags and human genome project. Science, 1991, 252(5013): 1651-1656.

[32] Victor E V, Lin Z, Bert V, et al. Serial Analysis of Gene Expression. Science, 1995, 270(5235): 484-487.

[33] Sydney B, Maria J, John B, et al. Gene expression analysis by massively parallel signature sequencing (MPSS) on microbead arrays.Vol.18. 2000, 630-634.

[34] 陈杰. 大规模平行测序技术(MPSS)研究进展. 生物化学与生物物理进展, 2004, 1(8):761-765.

[35] Schena M, Shalon D, Davis R W, et al. Quantitative monitoring of gene expression patterns with a complementary DNA microarray. Science, 1995, 270(5235): 467-470.

[36] Tom S, Marc A, Duncan D, et al. A new dimension for he human genome project: Towards comprehensive expression maps. Nat Genet, 1997, 16(2): 126-132.

[37] Wang Z, Gerstein M, Snyder M. RNA-Seq: A revolutionary tool for transcriptomics. Nat Rev Genet, 2009, 10(1): 57-63.

[38] Jason G U, Andrew V U, Sol K, et al. Frag Seq: Transcriptome-wide RNA structure probing using high-throughput sequencing. Nature Methods, 2010, 7: 995-1001.

[39] Aaron M S, Xiannian Z, Chen C, et al. Microfluidic single-cell whole-transcriptome sequencing. Proceedings of the National Academy of Sciences of the United States of America, 2014, 111(19):7048-7053.

[40] Ståhl P L, Salmén F, Vickovic S, et al. Visualization and analysis of gene expression in tissue sections by spatial transcriptomics. Science, 2016, 353(6294): 78-82.

[41] Xiaoyu L, Xushen X, Chengqi Y. Epitranscriptome sequencing technologies: decoding RNA modifications. Nature Methods, 2016, 14: 23-31.

[42] Björn S, Margaux M, Benedikt Z, et al. TT-seq maps the human transient transcriptome. Science, 2016, 352(6290): 1225-1228.

[43] Wilkins M R, Pasquali C, Appel R D, et al. From proteins to proteomes: Large scale protein identification by

two-dimensional electrophoresis and amino acid analysis. Biotechnology (N Y), 1996, 14(1): 61-65.

[44] Gorg A, Weiss W, Dunn M J. Current two-dimensional electrophoresis technology for proteomics. Proteomics, 2004, 4(12): 3665-3685.

[45] Fenn J B, Mann M, Meng C K, et al. Electrospray ionization for mass spectrometry of large biomolecules. Science, 1989, 246(4926): 64-71.

[46] Hillenkamp F, Karas M. Mass spectrometry of peptides and proteins by matrix-assisted ultraviolet laser desorption/ionization, in Methods in Enzymology, Academic Press, 1990, 280-295.

[47] Steen H, Mann M. The abc's (and xyz's) of peptide sequencing. Nature Reviews Molecular Cell Biology, 2004, 5: 699- 571.

[48] Aebersold R, Mann M. Mass spectrometry-based proteomics. Nature, 2003, 422(6928): 198-207.

[49] Yates J R, Ruse C I, Nakorchevsky A. Proteomics by mass spectrometry: Approaches, advances, and applications. Annual review of biomedical engineering, 2009, 11: 49-79.

[50] Han X, Aslanian A, Yates J R. Mass spectrometry for proteomics. Current Opinion in Chemical Biology, 2008, 12(5): 483-490.

[51] Gallien S, Duriez E, Crone C, et al. Targeted proteomic quantification on quadrupole-orbitrap mass spectrometer. Molecular & amp; Cellular Proteomics, 2012, 11(12): 1709-1723.

[52] Chait B T. Mass Spectrometry: Bottom-Up or Top-Down? Science, 2006, 314(5796): 65-66.

[53] Catherman A D, Skinner O S, Kelleher N L. Top Down proteomics: Facts and perspectives. Biochemical and Biophysical Research Communications, 2014, 445(4): 683-693.

[54] Doerr A. Mass spectrometry–based targeted proteomics. Nature Methods, 2012, 10: 23.

[55] Company R, Antúnez O, Cosson R P, et al. Protein expression profiles in Bathymodiolus azoricus exposed to cadmium. Ecotoxicology and Environmental Safety, 2019, 171: 621-630.

[56] Lu Z, Wang S, Shan X, et al. Differential biological effects in two pedigrees of clam *Ruditapes philippinarum* exposed to cadmium using iTRAQ-based proteomics. Environmental Toxicology and Pharmacology, 2019, 65: 66-72.

[57] Luo L, Zhang Q, Kong X, et al. Differential effects of zinc exposure on male and female oysters (*Crassostrea angulata*) as revealed by label-free quantitative proteomics. Environmental Toxicology and Chemistry, 2017, 36(10): 2602-2613.

[58] Le V Q A, Ahn J Y, Heo M Y, et al. Proteomic profiles of Daphnia magna exposed to lead (II) acetate trihydrate and atrazine. Genes & Genomics, 2017, 39(8): 887-895.

[59] Wang H, Wang S, Cui D, et al. iTRAQ-based proteomic technology revealed protein perturbations in intestinal mucosa from manganese exposure in rat models. RSC Advances, 2017, 7(50): 31745-31758.

[60] Lee E K, Shin Y J, Park E Y, et al. Selenium-binding protein 1: A sensitive urinary biomarker to detect heavy metal-induced nephrotoxicity. Archives of Toxicology, 2017, 91(4): 1635-1648.

[61] Zhao W J, Zhang Z J, Zhu Z Y, et al. Time-dependent response of A549 cells upon exposure to cadmium. Journal of Applied Toxicology, 2018, 38(11): 1437-1446.

[62] Kumar V, Karri V, Edwin M, et al. A system-based comparative proteomics approach to investigate heavy metals mixtures toxicity mechanism relates to the neurodegeneration on hippocampal cell line. Toxicology Letters, 2018, 295(1): S203-S203.

[63] Karri V, Ramos D, Martinez J B, et al. Differential protein expression of hippocampal cells associated with heavy metals (Pb, As, and MeHg) neurotoxicity: Deepening into the molecular mechanism of neurodegenerative diseases. Journal of Proteomics, 2018, 187: 106-125.

[64] Son J, Lee Y S, Lee S E, et al. Bioavailability and toxicity of copper, manganese, and nickel in Paronychiurus kimi (Collembola), and biomarker discovery for their exposure. Archives of Environmental Contamination and Toxicology, 2017, 72(1): 142-152.

[65] Baig M A, Ahmad J, Bagheri R, et al. Proteomic and ecophysiological responses of soybean (*Glycine max* L.) root nodules to Pb and hg stress. BMC Plant Biology, 2018, 18(1): 1-21.

[66] Gutsch A, Keunen E, Guerriero G, et al. Long-term cadmium exposure influences the abundance of proteins that impact the cell wall structure in Medicago sativa stems. Plant Biology, 2018, 20(6): 1023-1035.

[67] Wang X H, Wang Q, Nie Z W, et al. Ralstonia eutropha Q2-8 reduces wheat plant above-ground tissue cadmium and arsenic uptake and increases the expression of the plant root cell wall organization and biosynthesis-related proteins. Environmental Pollution, 2018, 242: 1488-1499.

[68] Georgiadou E C, Kowalska E, Patla K, et al. Influence of heavy metals (Ni, Cu and Zn) on nitro-oxidative stress responses, proteome regulation and allergen production in basil (*Ocimum basilicum* L.) plants. Frontiers in plant science, 2018, 9: 1-21.

[69] Zeng F, Wu X, Qiu B, et al. Physiological and proteomic alterations in rice (*Oryza sativa* L.) seedlings under hexavalent chromium stress. Planta, 2014, 240(2): 291-308.

[70] Cheng Z W, Chen Z Y, Yan X, et al. Integrated physiological and proteomic analysis reveals underlying response and defense mechanisms of Brachypodium distachyon seedling leaves under osmotic stress, cadmium and their combined stresses. Journal of Proteomics, 2018, 170: 1-13.

[71] Liu H, Weisman D, Tang L, et al. Stress signaling in response to polycyclic aromatic hydrocarbon exposure in *Arabidopsis thaliana* involves a nucleoside diphosphate kinase, NDPK-3. Planta, 2015, 241(1): 95-107.

[72] Shen Y, Du J, Yue L, et al. Proteomic analysis of plasma membrane proteins in wheat roots exposed to phenanthrene. Environmental Science and Pollution Research, 2016, 23(11): 10863-10871.

[73] Shen Y, Li J, Gu R, et al. Proteomic analysis for phenanthrene-elicited wheat chloroplast deformation. Environment International, 2019, 123: 273-281.

[74] Enerstvedt K S, Sydnes M O, Pampanin D M. Study of the plasma proteome of Atlantic cod (*Gadus morhua*): Effect of exposure to two PAHs and their corresponding diols. Chemosphere, 2017, 183: 294-304.

[75] Filis P, Walker N, Robertson L, et al. Long-term exposure to chemicals in sewage sludge fertilizer alters liver lipid content in females and cancer marker expression in males. Environment International, 2019, 124: 98-108.

[76] Nam T H, Jeon H J, Mo H, et al. Determination of biomarkers for polycyclic aromatic hydrocarbons (PAHs) toxicity to earthworm (*Eisenia fetida*). Environmental Geochemistry and Health, 2015, 37(6): 943-951.

[77] Shi X, Yeung L W Y, Lam P K S, et al. Protein profiles in zebrafish (*Danio rerio*) embryos exposed to perfluorooctane sulfonate. Toxicological Sciences, 2009, 110(2): 334-340.

[78] Dorts J, Kestemont P, Marchand P A, et al. Ecotoxicoproteomics in gills of the sentinel fish species, Cottus gobio, exposed to perfluorooctane sulfonate (PFOS). Aquatic Toxicology, 2011, 103(1): 1-8.

[79] Roland K, Kestemont P, Hénuset L, et al. Proteomic responses of peripheral blood mononuclear cells in the European eel (*Anguilla anguilla*) after perfluorooctane sulfonate exposure. Aquatic Toxicology, 2013, 128-129: 43-52.

[80] Zhang Y Y, Tang L L, Zheng B, et al. Protein profiles of cardiomyocyte differentiation in murine embryonic stem cells exposed to perfluorooctane sulfonate. Journal of Applied Toxicology, 2016, 36(5): 726-740.

[81] Tan F, Jin Y, Liu W, et al. Global liver proteome analysis using iTRAQ labeling quantitative proteomic technology to reveal biomarkers in mice exposed to perfluorooctane sulfonate (PFOS). Environmental Science & Technology, 2012, 46(21): 12170-12177.

[82] Huang Q, Zhang J, Peng S, et al. Proteomic analysis of perfluorooctane sulfonate-induced apoptosis in human hepatic cells using the iTRAQ technique. Journal of Applied Toxicology, 2014, 34(12): 1342-1351.

[83] Cui R, Zhang H, Guo X, et al. Proteomic analysis of cell proliferation in a human hepatic cell line (HL-7702) induced by perfluorooctane sulfonate using iTRAQ. Journal of Hazardous Materials, 2015, 299: 361-370.

[84] Shao X, Ji F, Wang Y, et al. Integrative chemical proteomics-metabolomics approach reveals Acaca/Acacb as direct molecular targets of PFOA. Analytical chemistry, 2018, 90(18): 11092-11098.

[85] Skiles G. Use of metabonomics for investigative toxicology. Chemical Research in Toxicology, 2002, 15: 1657-1657.

[86] Robertson D G. Metabonomics in toxicology: A review. Toxicological Sciences, 2005, 85(2): 809-822.

[87] Pognan F. Genomics, proteomics and metabonomics in toxicology: Hopefully not 'fashionomics'. Pharmacogenomics, 2004, 5(7): 879-893.

[88] Lindon J C, Nicholson J K, Holmes E, et al. Contemporary issues in toxicology: The role of metabonomics in toxicology and its evaluation by the COMET project. Toxicology and Applied Pharmacology, 2003, 187(3): 37-146.

[89] Marshall, D D, R. Powers. Beyond the paradigm: Combining mass spectrometry and nuclear magnetic resonance for metabolomics. Progress in Nuclear Magnetic Resonance Spectroscopy, 2017, 100: 1-16.

[90] Want E J, Wilson I D, Gika H, et al, Global metabolic profiling procedures for urine using UPLC-MS. Nature Protocols, 2010, 5(6): 1005-1018.

[91] Halket J M, Waterman D, Przyborowska A M, et al. Chemical derivatization and mass spectral libraries in metabolic profiling by GC/MS and LC/MS/MS. Journal of Experimental Botany, 2005, 56(410): 219-243.

[92] Chen Y H, Xu J, Zhang R P, et al. Methods used to increase the comprehensive coverage of urinary and plasma metabolomes by MS. Bioanalysis, 2016, 8(9): 981-997.

[93] Zhao H Z, Li H, Chung, A C K, et al. Large-scale longitudinal metabolomics study reveals different trimester-specific alterations of metabolites in relation to gestational diabetes mellitus. Journal of Proteome Research, 2018, 18: 292-300.

[94] 任向楠, 梁琼麟. 基于质谱分析的代谢组学研究进展. 分析测试学报, 2017, 36(2): 161-169.

[95] Zhou Y, Song R X, Zhang Z S, et al. The development of plasma pseudotargeted GC-MS metabolic profiling and its application in bladder cancer. Analytical and Bioanalytical Chemistry, 2016, 408(24): 6741-6749.

[96] Chen S L, Kong H W, Lu X, et al. Pseudotargeted metabolomics method and its application in serum biomarker discovery for hepatocellular carcinoma based on ultra high-performance liquid chromatography/triple quadrupole mass spectrometry. Analytical Chemistry, 2013, 85(17): 8326-8333.

[97] Wang F D, Zhang H J, Geng N B, et al. New Insights into the cytotoxic mechanism of hexabromocyclododecane from a metabolomic approach. Environmental Science & Technology, 2016, 50(6): 3145-3153.

[98] Rubert J, Zachariasova M, Hajslova J. Advances in high-resolution mass spectrometry based metabolomics studies for food - a review. Food Additives and Contaminants Part a-Chemistry Analysis Control Exposure & Risk Assessment, 2015, 32(10): 1685-1708.

[99] 梁丹丹, 李忆涛, 郑晓皎. 代谢组学全功能软件研究进展. 上海交通大学学报(医学版), 2018, 38(7): 805-810.

[100] Melamud E, Vastag L, Rabinowitz J D. Metabolomic analysis and visualization engine for LC-MS data. Analytical Chemistry, 2010, 82(23): 9818-9826.

[101] Pluskal T, Castillo S, Villar B A, et al. MZmine 2: Modular framework for processing, visualizing, and analyzing mass spectrometry-based molecular profile data. BMC Bioinformatics, 2010, 11: 395-406.

[102] Xia J, Sinelnikov I V, Han B, et al. MetaboAnalyst 3.0-making metabolomics more meaningful. Nucleic Acids Research, 2015, 43(W1): W251-W 257.

[103] Gowda H, Ivanisevic J, Johnson C H, et al. Interactive XCMS Online: Simplifying advanced metabolomic data processing and subsequent statistical analyses. Analytical Chemistry, 2014, 86(14): 6931-6939.

[104] Luan H, Ji F, Chen Y, et al. statTarget: A streamlined tool for signal drift correction and interpretations of quantitative mass spectrometry-based omics data. Analytica Chimica Acta, 2018, 1036: 66-72.

[105] Liang Y J, Lin Y T, Chen C W, et al. SMART: Statistical metabolomics analysis: An R tool. Analytical Chemistry, 2016,

88(12): 6334-6341.

[106] Lankadurai B P, Simpson A J, Simpson M J. H-1 NMR metabolomics of Eisenia fetida responses after sub-lethal exposure to perfluorooctanoic acid and perfluorooctane sulfonate. Environmental Chemistry, 2012, 9(6): 502-511.

[107] Ji C L, Wu H F, Wei L, et al. Proteomic and metabolomic analysis of earthworm *Eisenia fetida* exposed to different concentrations of 2,2′,4,4′-tetrabromodiphenyl ether. Journal of Proteomics, 2013, 91: 405-416.

[108] Zeng J, Kuang H, Hu C X, et al. Effect of bisphenol A on rat metabolic profiling studied by using capillary electrophoresis time-of-flight mass spectrometry. Environmental Science & Technology, 2013, 47(13): 7457-7465.

[109] Li S, Jin Y, Wang J, et al. Urinary profiling of cis-diol-containing metabolites in rats with bisphenol A exposure by liquid chromatography-mass spectrometry and isotope labeling. Analyst, 2016, 141(3): 1144-1153.

[110] Wei J, Xiang L, Yuan Z, et al. Metabolic profiling on the effect of 2,2′,4,4′-tetrabromodiphenyl ether (BDE-47) in MCF-7 cells. Chemosphere, 2018, 192: 297-304.

[111] O'Kane A A, Chevglier O P, Graham S F, et al. Metabolomic profiling of in vivo plasma responses to dioxin-associated dietary contaminant exposure in rats: implications for identification of sources of animal and human Exposure. Environmental Science & Technology, 2013, 47(10): 5409-5418.

[112] Geng N B, Zhang H J, Zhang B Q, et al. Effects of short-chain chlorinated paraffins exposure on the viability and metabolism of human hepatoma HepG2 cells. Environmental Science & Technology, 2015, 49(5): 3076-3083.

[113] Beyer J, Petersen K, Song Y, et al. Environmental risk assessment of combined effects in aquatic ecotoxicology: A discussion paper. Marine Environmental Research, 2014, 96: 81-91.

[114] Ji J, Zhu P, Cui F, et al. The antagonistic effect of mycotoxins deoxynivalenol and zearalenone on metabolic profiling in serum and liver of mice. Toxins (Basel), 2017, 9(28): 1-13.

[115] Ji J, Zhu P, Pi F, et al. GC-TOF/MS-based metabolomic strategy for combined toxicity effects of deoxynivalenol and zearalenone on murine macrophage ANA-1 cells. Toxicon, 2016, 120: 175-184.

[116] Athersuch T. Metabolome analyses in exposome studies: Profiling methods for a vast chemical space. Archives of Biochemistry and Biophysics, 2016, 589: 177-186.

[117] Song Q Q, Chen H, Li Y H, et al. Toxicological effects of benzo(a)pyrene, DDT and their mixture on the green mussel Perna viridis revealed by proteomic and metabolomic approaches. Chemosphere, 2016, 144: 214-224.

[118] Song Q Q, Zheng P F, Qiu L G, et al. Toxic effects of male Perna viridis gonad exposed to BaP, DDT and their mixture: A metabolomic and proteomic study of the underlying mechanism. Toxicology Letters, 2016, 240(1): 185-195.

[119] Wang F D, Zhang H J, Geng N B, et al. A metabolomics strategy to assess the combined toxicity of polycyclic aromatic hydrocarbons (PAHs) and short-chain chlorinated paraffins (SCCPs). Environmental Pollution, 2018, 234: 572-580.

作者：耿柠波[1]，张保琴[1]，宋肖垚[1]，张海军[1]，蔡宗苇[2]，陈吉平[1]
[1]中国科学院大连化学物理研究所，[2]香港浸会大学

第 16 章 环境计算化学与计算毒理学

▶ 1. 引言 /411
▶ 2. 碳纳米材料吸附化学品的模拟预测研究 /413
▶ 3. 化学品环境降解转化行为的模拟预测研究 /416
▶ 4. 化学品毒性机制及效应的模拟预测研究 /428
▶ 5. 展望 /433

本章导读

环境计算化学与计算毒理学属于新兴的交叉学科领域,在化学物质的环境暴露、危害性与风险性评价与预测中,具有重要应用前景。本章重点介绍2017年以来,我国学者在环境计算化学与计算毒理学研究方面取得的一些代表性创新成果,主要涉及碳纳米材料吸附化学品、化学物质降解转化行为、化学物质毒理效应的模拟预测等方面的成果,最后提出对该领域发展的展望。

关键词

环境计算化学,计算毒理学,化学品风险评价,化学品,毒理效应

1 引 言

1828年,德国化学家维勒(F. Wöhler,1800~1882年)将氰酸铵水溶液加热得到了尿素(NH_2CONH_2),开创了人工合成化学品的纪元。1908年,德国化学家哈伯(F. Haber,1868~1934年)申请了合成氨工艺的专利,开启了人类大规模合成化学品的时代。100多年来,人类合成了化肥、农药、药物、各种日用和工业化学品,其中许多化学品是地球生态圈中原本不存在(或者即使存在,浓度也特别低)的物质,属于人为的(anthropogenic)化学品。这些化学品在促进人类社会发展、提升人类生活质量方面,发挥了巨大作用。化学这个学科(尤其是有机化学)在化学品的合成和大规模生产应用方面,也发挥了重要作用。

20世纪60年代,人类大规模使用各种化学品导致对生态和人体健康的不利效应逐渐显现。1962年,Rachel Carson(1907~1964)在《寂静的春天》中,指出了一些人工合成化学品(如有机氯农药、多氯联苯)污染对地球生态环境造成的危害,这本书促使了人类环保意识的觉醒。1972年,联合国在瑞典斯德哥尔摩召开人类首次环境大会。其后近50年间,人类开展了大量的控制、削减化学品污染的工作,然而至今仍未能彻底解决化学品的环境污染问题。例如,综合各种文献资料可以发现,阻燃剂多溴联苯醚(PBDEs)在全球环境以及包括人体在内的生物体内广泛存在(图16-1)。

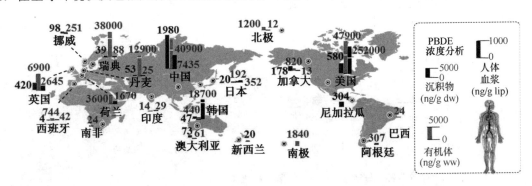

图16-1 阻燃剂多溴联苯醚(PBDEs)在全球水体沉积物以及人体内的浓度水平

美国前副总统戈尔(Al Gore)在为《我们被偷走的未来》所作的序言中指出,"持久性有毒物质污染了整个地球""即使偏远地区的人体中也能检出持久性的工业化学物质""这些化学物质积累在我们的身体内""更严重的是,在子宫中以及通过乳汁,母亲可以将这些化学物质传递给下一代""初步动物和流行病学研究表明,这些化学物质可以引起许多不利效应,如精子数降低、不孕不育、生殖发育畸形、激

素紊乱导致的癌症……"。因此，虽然人类进入了 21 世纪，但是合成化学品仍是人体与生态健康的重要风险源，是人类可持续发展面临的一个重大挑战。

为了应对化学品污染问题，联合国环境署（UNEP）于 2006 年发布了全球化学品管理战略指针（SAICM），希望"到2020年，通过透明和科学的风险评价与风险管理程序，并考虑预先防范措施原则以及向发展中国家提供技术和资金等能力支援，实现化学品生产、使用以及危险废物符合可持续发展原则的良好管理，以最大限度地减少化学品对人体健康和环境的不利影响"。为实现此目标，UNEP 在 2013 年发布了《全球化学品展望》（Global Chemicals Outlook），倡导通过科学合理的化学品管理，控制化学品的全球污染。2019 年 3 月，UNEP 又发布了《全球化学品展望》（第二版），并指出："最大限度减少化学品和废物的不利影响这一全球目标无法在 2020 年实现"。不仅如此，《全球化学品展望》（第二版）还预测，2017～2030 年全球化学品（不包括药品）销售额的增长将要翻一番（图 16-2）[1]，而且预测从 1990 到 2030 年，全球基本化学品产能的增长远高于人口的增长（图 16-3）[1]。

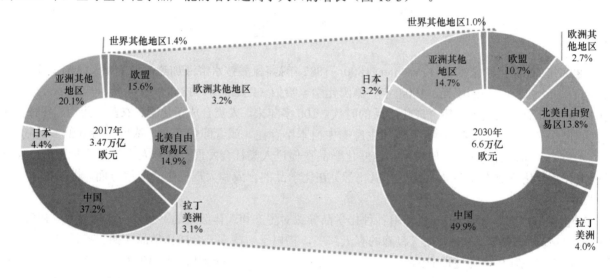

图 16-2　欧洲化学工业理事会预测 2017～2030 年全球化学品（不包括药品）销售额的增长（引自参考文献[1]）

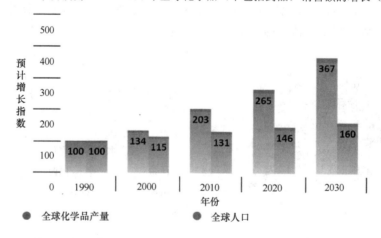

图 16-3　全球基本化学品产能增长与人口增长的预测及比较（引自参考文献[1]）

《全球化学品展望》（第二版）指出，为实现人类社会可持续发展，迫切需要所有利益攸关方在全球采取更加积极进取的行动，尤其需要进一步加强科技对化学品管理的支撑作用，填补包括化学品的危害性在内的数据空白和未知领域。在填补化学品的环境暴露、危害性在内的数据空白和未知领域方面，环境计算化学与计算毒理学大有作为。原因在于，化学品种类众多，仅仅依赖实验测试的技术评价化学品的环境暴露、危害性和风险性，效率低、耗时长而滞后于化学品环境管理的实际需求。另一方面，人类所生存的地球环境，是一个多介质/界面、空间异质、时空连续的系统，有限时间-地点的采样方案以及仪器分析测试的手段也不足以表征地球生态环境的整体和系统性特征，因此需要模型模拟的技术手段。

环境计算化学与计算毒理学属于新兴的交叉学科研究领域，二者在概念的内涵和外延上均有交汇。广义地看，凡是环境化学中涉及的计算模拟研究，或都可归于环境计算化学，也可以认为环境计算化学是环境化学与大数据分析、化学信息学（化学计量学）、计算化学（包括量子化学、分子力学）等学科的交叉。环境计算化学采用计算模拟的方法，揭示和表征化学污染物的形成机制、源解析、多介质迁移转化归趋（环境分布）、毒性效应及生态与健康风险等。计算毒理学主要基于计算化学、化学/生物信息学和系统生物学原理，通过构建计算机（*in silico*）模型，来实现化学品环境暴露、危害性与风险的高效模拟预测（图16-4）[2]。

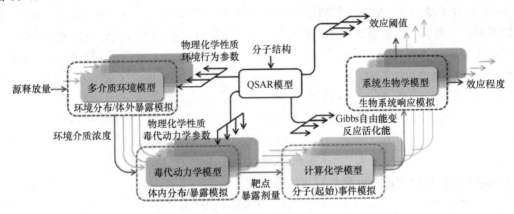

图 16-4　面向化学品风险预测与评价的计算毒理学模型格局（引自参考文献[2]）

2017 年，我们在《环境化学前沿》中撰写了"环境计算化学与预测毒理学"一章。2 年过去了，我国的学者在环境计算化学与计算毒理学研究方面，又取得了丰富的创新性成果。限于篇幅和时间，本章择部分成果加以介绍，以期抛砖引玉，希望为促进我国在环境计算化学与计算毒理学的学科发展发挥正能量。

本章重点介绍近 2 年来我国学者在化学品环境分配行为、降解转化行为、毒理效应的模拟预测方面的一些成果，最后提出我们对该领域发展的展望。

2　碳纳米材料吸附化学品的模拟预测研究

纳米材料是指至少有一维尺度在 1 nm 至 100 nm 范围内的材料，根据其化学组成不同可分为碳纳米材料、金属纳米材料、金属氧化物纳米材料以及量子点等多种不同的类别。纳米材料的大量生产与使用，使其难免被释放进入环境。据估算，仅 2010 年，全球范围内经各种方式释放进入大气、土壤和水中的纳米材料就分别有 0.8 万吨、5.2 万吨和 6.9 万吨[3]。进入环境中的纳米材料会发生一系列的迁移转化过程，同时也会对生物体产生毒性效应。研究纳米材料的环境行为和毒性效应是评价其生态风险的重要内容。因而，发展纳米材料环境行为和毒性效应的模拟预测方法对于评价其生态风险具有重要意义。

纳米材料中的碳纳米材料（包括 C_{60}、碳纳米管、石墨烯和氧化石墨烯等），由于其具有较大的疏水性表面而对有机物产生较强的吸附作用。这种吸附作用不仅会影响碳纳米材料和化学品的环境行为，还会影响它们的毒性效应。因此，模拟预测碳纳米材料对化学品的吸附行为对于全面准确地了解碳纳米材料和化学品的环境归趋、评价其生态风险显得十分必要。

2.1 C_{60}吸附化学品的预测

作为代表性零维碳纳米材料，C_{60}对化学品的吸附预测已有一些研究报道。Yang等[4]发展了基于C_{60}比表面积和微孔体积预测菲在C_{60}表面吸附量的方法。Wang等[5]选用量子力学和分子力学的方法模拟预测了C_{60}对7种溶解性有机质替代物（DOM_R、D-葡萄糖醛酸、对苯醌、五倍子酸、香草醛、对苯二酚、苯甲酸和腺嘌呤）的吸附，分析了吸附作用的机理。在吸附过程中，C_{60}作为电子供体转移电子给DOM_R，C_{60}吸附五倍子酸的主要驱动力为静电作用。还发现C_{60}吸附DOM_R后，表观水溶解度增大。

Sun等[6]采用分子动力学（MD）的方法，模拟了C_{60}对9种不同的低分子量有机酸（DOM的重要组成成分）的吸附，并考察了不同解离形态的低分子量有机酸与C_{60}的相互作用，发现C_{60}对芳香族有机酸的吸附作用强于对脂肪族有机酸的吸附作用，且吸附作用随有机酸分子量的增大而增强；C_{60}对低分子量有机酸中性形态的吸附作用比其对阴离子形态作用强。还发展了基于低分子量有机酸的分子结构参数预测其在C_{60}表面的相互作用能（E_{int}）的模型：

真空中C_{60}对中性低分子量有机酸的相互作用能：

$$E_{int} = -3.737 - 0.033\alpha - 0.307\log K_{OW} \quad (16.1)$$
$$n = 9, \ r = 0.960, \ RMSE = 0.494, \ Q^2_{CUM} = 0.867$$

水相中C_{60}对中性低分子量有机酸的相互作用能：

$$E_{int} = -1.070 - 0.018\alpha - 0.031V_M \quad (16.2)$$
$$n = 9, \ r = 0.925, \ RMSE = 0.560, \ Q^2_{CUM} = 0.867$$

水相中C_{60}对阴离子形态低分子量有机酸的相互作用能：

$$E_{int} = 0.002 - 0.028\alpha - 0.154\mu \quad (16.3)$$
$$n = 9, \ r = 0.949, \ RMSE = 0.706, \ Q^2_{CUM} = 0.834$$

式中，α为分子极化率，$\log K_{OW}$为辛醇/水分配系数的对数值，V_M为分子摩尔体积，μ为偶极矩，n为低分子量有机酸的个数，r为模型的相关系数，RMSE为均方根误差，Q^2_{CUM}为交叉验证系数。

2.2 碳纳米管（CNTs）吸附化学品的预测

碳纳米管（CNTs）依据其管壁层数的不同，可分为单壁碳纳米管（SWNTs）和多壁碳纳米管（MWNTs）。CNTs吸附化学品的模拟预测研究，包含MWNTs和SWNTs吸附预测两方面。目前，可预测化学品在MWNTs表面吸附的模型已有近50个[4, 7-18]，而适用于SWNTs吸附预测的模型仅有十几个[4, 7, 10, 18, 19]。可用于预测有机物在MWNTs表面吸附的模型中，既有可预测吸附量的模型，也有可预测Dubinin-Ashtakhov模型中与吸附能相关的参数E_a的模型，还有可用于预测吸附平衡常数（K）的模型。这些预测模型所使用的描述符既有通过实验方法测定的分子描述参数，也有基于理论计算获取的参数；既有采用多元线性回归构建的线性预测模型，也有使用支持向量机、最小二乘支持向量机等构建的非线性预测模型。模型的应用域，也从最初的单一类型有机物扩大到了脂肪族和芳香族有机物。

可用于预测化学品在SWNTs表面吸附的模型中，所有的模型均是基于实验测定的分子描述参数构建的。具有这些实验测定的分子结构描述符值的有机物仅有三千余种，仍有大量的化学品缺乏相应的分子结构描述符值而无法进行预测。目前SWNTs预测模型的构建方法单一，多为线性回归算法，预测SWNTs

吸附化学品的非线性模型研究尚属空白。Wang等[20]基于文献中搜集的61种有机物在水中SWNTs表面的log$K$值以及三个量子化学描述符和五个分子结构描述符,采用多元线性回归(MLR)和支持向量机(SVM)算法构建了线性模型和非线性模型,模型均具有良好的拟合优度、稳健性和预测能力。MLR模型和SVM模型预测的log$K$值与实验测定的log$K$值拟合效果如图16-5所示,模型的应用域覆盖多种有机化合物,它们可含有如下官能团:>C=C<,—C≡C—,—C_6H_5,>C=O,—COOH,—C(O)O—,—OH,—O—,—F,—Cl,—Br,—NH_2,—NH—,>N—,>N—N<,—NO_2,>N—C(O)—NH_2,>N—C(O)—NH—,—S—和—S(O)(O)—。机理分析表明,在SWNTs吸附有机化合物的过程中,范德瓦耳斯相互作用和疏水作用为主要驱动力。

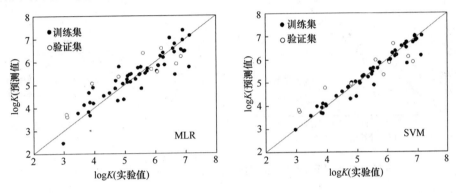

图16-5 模型预测的log$K$值与实验测定的log$K$值拟合图

量子化学计算方法的快速发展和大型服务器运算效率的提升,使得通过量子化学计算而预测SWNTs对有机物的吸附行为成为可能。Zou等[21]采用密度泛函理论(DFT)方法,计算了环己烷、苯及其衍生物、多环芳烃在SWNTs(8,0)表面的吸附能以及吸附过程的自由能变,发现苯、甲苯等12种有机化合物在SWNTs(8,0)表面的吸附能与有机化合物的logK_{OW}值呈线性相关,且有机物的logK_{OW}值越大,越容易吸附在SWNTs表面。此外,还估算了π-π相互作用对吸附能的贡献近似为24%,—NO_2对吸附能的贡献也约为24%。同时,还发现计算得到的吉布斯自由能变(ΔG_{cal})与实验测定的自由能变(ΔG_{exp})具有显著的线性相关性($r=0.97$);ΔG_{cal}与E_a也存在显著的线性相关性($r=0.93$)。该研究为发展碳纳米材料吸附有机物的量子化学模拟预测方法奠定了基础,并可为SWNTs和有机物的环境行为评价提供基础数据。

实际生产的CNTs往往含有掺杂、官能团、缺陷等问题,使其与不含缺陷的CNTs的吸附性能有所差异。以掺氮碳纳米管为例,张馨元等[22]选用DFT的方法,模拟预测了芳香型有机物在水中掺氮SWNTs表面的吸附,并考察了掺氮浓度和掺氮形态对吸附的影响。结果发现,氮掺杂能够增强SWNTs对芳香型有机物的吸附,浓度较高的吡啶氮掺杂的SWNTs具有较强的吸附性能。掺氮SWNTs吸附有机物的过程中疏水作用和π-π作用起主要贡献。

2.3 石墨烯和氧化石墨烯吸附化学品的预测

以往的研究指出,碳纳米材料吸附化学品的过程中,静电相互作用、色散作用、π-π相互作用、疏水作用和氢键作用会同时存在;但是,不同作用对总吸附的贡献仍有待进一步定量区分。并且,气相和水相中,各种作用的贡献差异也仍待进一步揭示。鉴于此,Wang等[23]采用DFT方法,计算预测了38种有机化合物在气相和水相中石墨烯表面的吸附能(E_a),并构建了预测有机化合物在石墨烯表面吸附能的多参数线性自由能关系(pp-LFER)模型,模型具有良好的拟合度和稳健性。模型预测的与DFT计算的E_a的绝对值($|E_a|$)具有较好的拟合效果(图16-6),模型应用域覆盖脂肪烃、苯及其衍生物和多环芳烃。

基于 pp-LFER 模型进一步定量区分了不同作用对总吸附的贡献,发现在气相吸附中,表征色散作用的 lL 项和表征偶极/极化率作用的 sS 项对总吸附能的贡献最显著,贡献范围分别为 32%～74%和 2%～23%;在水相吸附中,代表色散作用和疏水作用的 vV 项贡献最显著,为 72%～90%,其中疏水作用对总吸附的贡献为 1%～14%。此外,还发现有机化合物在石墨烯表面的吸附能比其在 SWNTs 表面的吸附能强,并构建了可预测环己烷、苯、苯胺、甲苯、苯酚和硝基苯在不同管径和手性的 SWNTs 表面吸附能的模型。

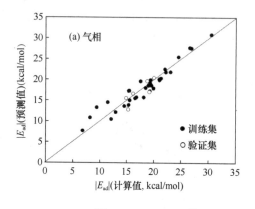
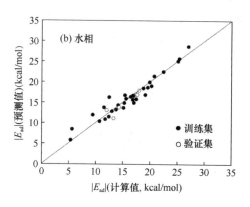

图 16-6　pp-LFER 模型预测的$|E_{ad}|$值与 DFT 计算的$|E_{ad}|$值的拟合图

由于含有氧化官能团的氧化石墨烯是石墨烯类纳米材料的主要组成成员,且含氧官能团也会对有机化合物在氧化石墨烯表面的吸附产生影响[24, 25]。但是不同含氧官能团对吸附影响的机理尚不清楚,石墨烯和氧化石墨烯对有机化合物吸附作用的差异也有待进一步研究。Wang 等[26]采用分子动力学模拟,预测了 43 种有机化合物在水中石墨烯和含有不同氧化官能团如羟基、环氧基和酮羧基的氧化石墨烯表面的 $\log K$ 值,发现石墨烯比氧化石墨烯具有更强的吸附能力;含有羟基或羧基官能团的氧化石墨烯与有机化合物的羟基官能团之间形成了氢键,而氧化石墨烯中的环氧基则没有与所考察的有机化合物形成氢键。进一步构建了可预测有机化合物在水相中石墨烯和氧化石墨烯表面 $\log K$ 值的理论线性溶解能关系(TLSER)模型。所建的 TLSER 模型具有较好的拟合度、稳健性和预测能力,适用于预测不同的芳香类化合物如苯、苯乙醇、苯酚、苯胺、硝基苯、丁腈、卤代苯、酮、酯、联苯及它们的衍生物、多环芳烃和多溴联苯醚在水中石墨烯和氧化石墨烯表面的 $\log K$ 值。机理分析表明,有机物在水中石墨烯表面的 $\log K$ 值主要受氢键酸度(ε_α)和色散/疏水作用(V)影响,有机物在水中氧化石墨烯表面的 $\log K$ 值受 ε_α 的影响最显著。

3　化学品环境降解转化行为的模拟预测研究

3.1　大气中的化学品的降解转化行为

3.1.1　有机胺的降解转化

有机胺是一类重要的挥发性有机化合物,人类活动和自然过程中均可产生[27]。目前大气中检出的有机胺已达 160 余种,浓度可达几百 ppt[27]。有机胺的大气氧化具有生成致癌性亚硝胺/硝胺的潜力[27]。此外，有机胺通过酸碱反应能够增强 H_2SO_4 的成核能力[28, 29]。近年来，有机胺的气相转化得到越来越多的关注。

谢宏彬等重点关注了基于胺溶液的燃烧后捕捉 CO_2（PCCC）过程释放的有机胺的大气转化机制和动力学，系统研究了 PCCC 技术使用的典型溶剂乙醇胺（MEA）的大气氧化及参与 H_2SO_4 成核的机制与动力学[30-32]。2018 年，他们将研究的焦点转到乙醇胺的替代品哌嗪（PZ）的大气转化机制和动力学，运用量子化学及动力学模拟相结合的方法，重点关注 PZ 大气氧化生成致癌性亚硝胺的风险，并与 MEA 进行对比[33]（图 16-7）。研究发现，·Cl 由于与 PZ 的 N 形成独特的两中心三电子键使得其夺取 PZ 的 N 上的 H 形成 N 中心自由基（PZ-N）是最可行的，这与·OH 引发的反应机制不同。·Cl 引发 PZ 的反应速率常数与·OH 引发的反应速率常数相当。结合·OH 引发反应，发现即使在高·Cl（[·Cl]）条件下，PZ 氧化生成 N 中心自由基的产率仍低于 MEA 对应的转化过程，这与仅考虑·OH 引发的反应结果一致。然而，与以前推测不同的是，PZ 氧化生成亚硝胺的产率高于乙醇胺，这主要是由哌嗪的 N 中心自由基生成亚硝胺的产率远远高于乙醇胺的 N 中心自由基所致。因此，该研究阐明 PZ 氧化生成亚硝胺的风险高于 MEA，指出 N 中心自由基产率不能够用以衡量有机胺生成致癌性亚硝胺的产率。因此，为了评价有机胺氧化生成致癌性亚硝胺的环境风险，将来更多的研究应该关注 N 中心自由基化学。

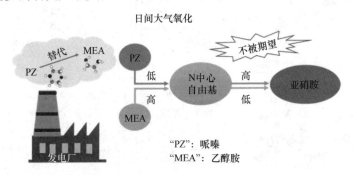

图 16-7 哌嗪和乙醇胺的大气氧化生成亚硝胺关键过程对比示意图

3.1.2 多环芳烃大气氧化形成硝基多环芳烃的机理

硝基多环芳烃（NPAHs）可以诱导产生与大气毒性有关的活性氧化物质（ROS），其还与人类罹患肺癌有关[34, 35]。NPAHs 通常比 PAHs 有更高的致癌性和致突变性。NPAHs 可以通过不完全燃烧过程，如垃圾焚烧等直接排放到大气中，也可以通过 PAHs 在大气中的氧化转化二次生成[36]。在大气中 PAHs 可以被·OH、NO_3 等氧化形成 NPAHs，通常认为反应机理为·OH 和 NO_3 加成到多环芳烃分子中的 C=C 双键，生成具有较高化学活性的 OH-PAHs 或 NO_3-PAHs 加合物，该类加合物能够很快地与大气中的 NO_2 继续发生加成反应，并通过单分子解离直接脱去一分子水或一分子硝酸生成 NPAHs[37]。此外，PAHs 的大气氧化过程以及一些有毒氧化产物（如 NPAHs）的生成与大气中的痕量气体密切相关。近期越来越多的计算化学研究表明甲醇、乙醇、硫酸、硝酸、甲酸、乙酸和水等大气痕量气体可以通过降低甚至完全消除表观势垒以促进气相的 H 迁移反应[38, 39]。

近年来张庆竹课题组采用量子化学计算在分子水平上研究了萘、苊烯、苊、芴、蒽、菲、荧蒽和芘等多环芳烃在大气中由·OH 和 NO_3 引发，经水分子催化转化形成硝基多环芳烃的过程[40-42]，并在此基础上进一步研究了甲醇、乙醇、硫酸、硝酸、甲酸、乙酸等大气痕量气体催化作用，分析了这些催化剂影响 NPAHs 生成过程的具体因素，比较了水、酸和醇气体分子对·OH 和 NO_3 引发 PAHs 的大气氧化生成 NPAHs 过程的影响的异同（图 16-8）。研究发现脱水或脱硝酸过程是 PAHs 和·OH 或 NO_3 自由基发生气相反应生成 NPAHs 的速控步骤。在没有催化剂的情况下，OH-NO_2-PAHs 加合物通过单分子解离经脱去一分子水的反应经过张力较大的四元环过渡态，NO_3-NO_2-PAHs 加合物通过单分子解离脱去一分子硝酸的反应经过张力较小的六元环过渡态，脱水和脱硝酸过程的势垒超过 45 kcal/mol，在通常大气环境中很难发生。甲醇、乙醇、硫酸、硝酸、甲酸、乙酸和水可以通过降低 PAH-OH-NO_2 中间体脱水的活化能以促进 NPAHs 的生成，催化剂的具体催化效果取决于反应复合物催化脱水的反应速率常数、反应复合物生成的平衡常数和催化剂在大气中的浓度。在对流层大气条件下，甲醇、乙醇、硫酸、硝酸、甲酸、乙酸和水都有促进 PAHs 氧化生成 NPAHs 过程的作用，其中水促进·OH 和 NO_3 引发的 PAHs 氧化生成 NPAHs 的效果最为显著，而在醇和酸的浓度有显著提升的情况下，如重污染天气或在接近排放源处，甲醇、乙醇、硫酸、硝酸、甲酸和乙酸对·OH 和 NO_3 引发的 NPAHs 大气转化生成的促进效果比水更显著。该研究提出了大气中新的 NPAHs 形成机理，阐明了水、酸和醇分子的催化作用，发现了大气中新的硝基萘生成源——以往的研究认为大气中的 9-硝基萘来自于直接排放及颗粒物状态萘的非均相反应，解释了野外现场观测结果，为多环芳烃及其衍生物的有效污染控制及治理大气污染和环境立法、环境规划提供理论依据和科技支撑。

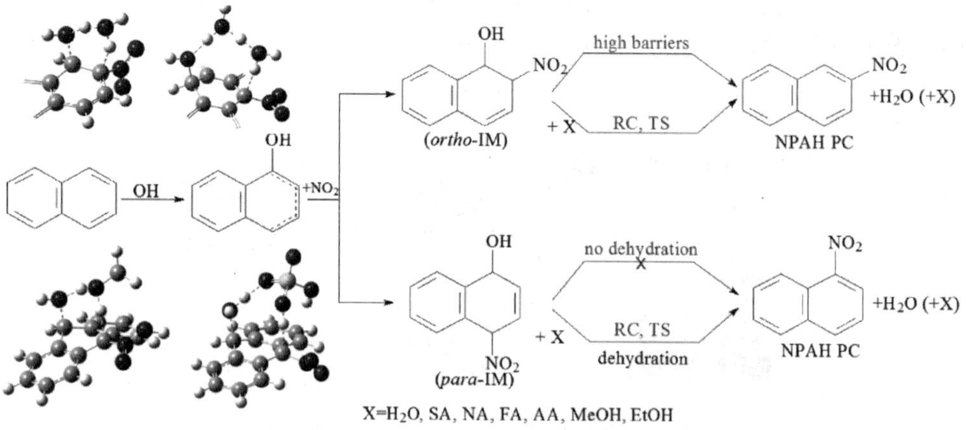

图 16-8　多环芳烃在大气中可被·OH/NO_2 及 NO_3/NO_2 氧化生成硝基多环芳烃的机理

3.1.3　HONO 的降解转化行为

HONO 是大气·OH 的重要来源，对大气氧化性、空气质量、人体健康以及生态环境有显著影响，因此厘清 HONO 的生成与消耗机理（简称生消机理）有着重要的科学意义和社会意义。尽管 HONO 在大气化学中的重要性得到了广泛的共识，关于它的研究也是科学界的关注热点。但是，迄今为止，HONO 的多相形成机理，或者说 HONO 的来源还不是十分的清楚，仍然需要进一步探索[43]。

孙孝敏课题组与合作团队采用理论计算化学和烟雾箱模拟实验分别对 HONO 的形成进行了研究[44]。用 DFT 研究了 NaCl（100）表面上 NO_2 分子的吸附及水解，解释了 HONO 形成机理和 Cl 损耗现象。另外，揭示了 NO_2 在 TiO_2 表面转化生成 HONO 的微观机理（图 16-9）。NO_x，SO_2，VOCs 等污染物可以在 TiO_2 表面上光催化分解。该结果有助于解释工程应用中大规模使用 TiO_2 涂料对局部大气氧化性的影响，进而对空气质量、人体健康和环境产生的影响。在烟雾箱实验中，氮氧化物在臭氧、水存在下可以转化形成 HONO 且反应现象明显；而丙烯和三甲苯在 NO 存在下光反应亦可形成 HONO。HONO 的形成，无疑对大气氧化性产生深刻影响。该研究旨在揭示 HONO 的生消规律，厘清 HONO 生成源对大气氧化性的影响，对二次有机气溶胶的形成及风险评价具有一定的参考意义。

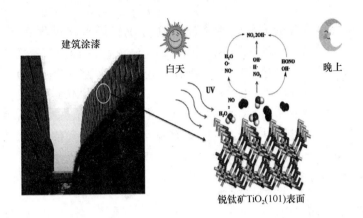

图 16-9　NO_2 在 TiO_2 表面产生 HONO 和 ·OH 的微观转化机理

3.1.4　苯系物和异戊二烯大气氧化形成高含氧低挥发性化合物的机理

过氧自由基（RO_2）是挥发性有机化合物大气光氧化过程中一类重要的自由基中间体。在对流层中，RO_2 自由基与 NO 和 $HO_2/R'O_2$ 的双分子反应与臭氧和二次有机气溶胶的生成密切相关[45]。最近的研究发现，在 NO_x 浓度较低的相对洁净的大气环境中，RO_2 的单分子反应也具有重要的意义。RO_2 的分子内单分子氢迁移以及随后的 O_2 加成，将挥发性有机化合物（VOCs）逐步氧化为高含氧的、含多官能团的产物（HOMs）（图 16-10）[46, 47]。这些产物通常具有极低的挥发性，对二次有机气溶胶（SOA）的生成甚至新颗粒的形成均有重要的贡献。RO_2 的单分子过程是 VOCs 大气自氧化的重要途径，部分过程中生成的 ·OH 也影响着大气氧化能力[48]。

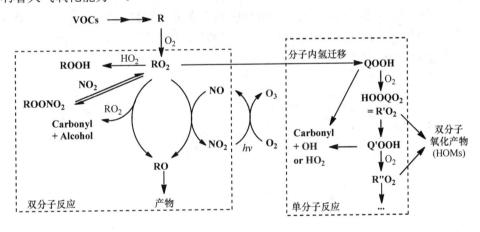

图 16-10　过氧自由基（RO_2）的双分子和单分子反应示意图

王赛男等研究了几类 VOCs 大气氧化中 RO_2 的单分子反应动力学，系统研究了有机醚和二甲基硫醚的大气氧化机理，特别是其中 RO_2 的单分子反应动力学[49, 50]。2017 年，他们发现苯系物大气氧化中生成的双环过氧自由基中间体（BPR）能够从取代烷基上夺取氢原子，发生快速的分子内氢迁移[51]。计算表明，甲苯、乙苯和异丙苯对应的 BPRs 中发生第一次氢迁移反应的有效速率分别为 2.6×10^{-2} s^{-1}、7.0 s^{-1} 和（8.8~14）s^{-1}。随后加 O_2 得到的新过氧自由基还能够发生第二次，甚至第三次氢迁移反应。这一理论预测得到了实验证实。与德国 Berndt 小组合作，利用 CI-APi-TOF（化学电离-大气压力界面-飞行时间）质谱仪，在气相流动管反应器实验中观测到多次氢迁移后的过氧自由基，例如在异丙苯反应中检测到 C_9H_{12}—OH—$(O_2)_n$（$n = 2,3,4$）自由基和 C_9H_{12}—OH—$(O_2)_3$ 脱去 OH 后的羰基化合物 $C_9H_{12}O_6$。2018 年，通过理论预测和实验测量相结合的方法,他们发现在异戊二烯氧化过程中形成的初代过氧自由基 Z-δ-ISOPO$_2$ 也可以发生连续的氢迁移和加 O$_2$,生成 C$_5$H$_9$O$_7$ 和 C$_5$H$_9$O$_9$ 等高含氧的过氧自由基,以及 C$_5$H$_{10}$O$_7$、C$_5$H$_{10}$O$_8$ 和 C$_5$H$_8$O$_6$ 等闭壳层产物[52]。在森林地区典型大气条件下(NO 浓度约为 50 ppt),通过粗略地估算异戊二烯初代氧化过程形成 HOMs 的产率,发现生成 HOMs 通道的比例可达~11%。异戊二烯氧化生成的 HOMs 一方面可能构成 SOA 的重要前体物,但另一方面,反应生成的大量过氧自由基在大气环境中可能与萜烯氧化生成的过氧自由基反应,降低萜烯的 SOA 生成能力[53]。因此,为了更准确地评价 VOCs 大气氧化对 O$_3$、SOA 和大气氧化能力的影响,需要继续研究 RO$_2$ 的单分子和双分子反应动力学,特别是 RO$_2$ 的自反应和交叉反应。

3.1.5 三氧化硫的转化行为

新粒子形成(NPF)是大气中颗粒物和云凝结核的重要来源,对区域环境质量及气候和人体健康均有重大影响。尤其在我国高浓度气态污染物及其强氧化性背景下,NPF 总是频繁发生[54, 55]。成核阶段是 NPF 的关键步骤,但由于实验技术的限制,成核阶段分子水平的物理化学机制,尤其是污染地区 NPF 的机制仍然是一个未解之谜[29, 56]。因此,揭示 NPF 成核阶段的潜在机制,对有效评估 NPF 对区域特别是污染地区的环境和健康效应以及对大气环境的科学研究都有重要意义。

三氧化硫(SO$_3$)是一种重要的空气污染物并具有很高的反应活性[57-60]。通常情况下,其与水(H$_2$O)的反应被认为是大气中 SO$_3$ 主要的消耗路径。然而,在高度污染地区,氨气(NH$_3$)的浓度可高达 160 ppb[61]。基于此,张秀辉课题组与合作团队采用量子化学计算和大气团簇动力学模型相结合的方案,研究了高污染地区重要气态污染物 SO$_3$ 和 NH$_3$ 在反应物 NH$_3$ 催化下的反应(即自催化反应)(图 16-11),发现在 NH$_3$ 浓度较高的干燥地区,此路径是 SO$_3$ 除与 H$_2$O 反应之外的另一条重要消耗路径。同时发现反应产物氨基磺酸(SFA)可不同程度增强城市地区大气关键成核团簇[硫酸-二甲胺团簇(SA-DMA)]的形成速率,并进一步提出高度污染地区 NH$_3$-SO$_3$ 的自催化反应引发的气溶胶新粒子形成的新机制,为我国复合大气污染条件下新粒子形成机制研究提供了新的研究思路和理论指导。

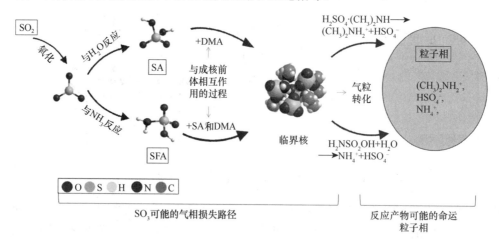

图 16-11　污染物三氧化硫(SO$_3$)可能的大气命运

3.1.6 含氮自由基的降解转化行为

大气气溶胶对地球辐射平衡、全球气候变化和人类健康有重要的影响[62]。要评估气溶胶的各种影响需要了解其在大气中的来源及转化过程。如前所述,NPF 是大气气溶胶和云凝结核的重要来源[29]。研究表明,除了硫酸和水之外,有机物也是 NPF 的重要参与者[28, 63]。由于大气环境中有机物结构的多样性,研究 NPF 的形成机制的主要挑战在于研究不同有机物对颗粒物形成和生长的贡献。很多研究已经探讨了不同类型的有机物，如有机胺，有机酸等的作用[64, 65]。然而，同一类有机物也可能会因取代基不同而对 NPF 过程有不同的影响。

罗一等研究了以丝氨酸（Ser）/苏氨酸（Thr）为代表的氨基酸和以尿素（Urea）/甲酰胺（Formamide）为代表的酰胺化合物与硫酸（SA）及 1~4 个水分子（W）的相互作用，重点关注了这两类含氮化合物中的—OH 及—NH$_2$ 取代基在成核过程中的影响（图 16-12）[66, 67]。结果表明，有机物的取代基不同，所形成的团簇具有不同的结构。在团簇(Thr)(SA)和(Ser)(SA)中，丝氨酸和苏氨酸分别通过羟基和氨基基团与硫酸形成环状结构。尿素团簇与甲酰胺团簇的结构也不尽相同。另外，有机物中的取代基也会影响团簇内的质子转移情况。在没有 W 参与的情况下，(Thr)(SA)和(Ser)(SA)团簇就会发生质子转移。尿素团簇中存在 SA 到 C=O 基团或者 W 的质子转移，而甲酰胺团簇中则没有发生。除了团簇的结构，有机物的取代基还会影响其与 SA 分子的相互作用。与苏氨酸、甘氨酸和丙氨酸相比，丝氨酸与硫酸的相互作用最强。(Formamide)(SA)团簇的分子间相互作用弱于(Urea)(SA)团簇。团簇的吸湿性也会受有机物取代基的影响。即使在低相对湿度下，团簇(Thr)(SA)(W)$_n$（$n = 1$，2）和(Ser)(SA)(W)$_n$（$n = 1$，2）仍然是主要存在类型。与尿素团簇相比，甲酰胺团簇对水分子的亲和性较强。以上结果表明，取代基不同会对含氮化合物参与的成核过程具有一定的影响。因此，为了更好地模拟有机物参与的新粒子生成过程，将来更多的研究应该关注有机物的取代基效应对新粒子生成过程的影响。

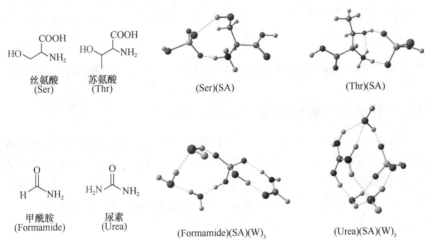

图 16-12　不同取代基的氨基酸和酰胺化合物对硫酸团簇的形成及其水化过程的影响

3.1.7　磷系阻燃剂和增塑剂的大气转化行为

大气和水环境中普遍存在着·OH，它会影响很多化学品的转化和归趋[68, 69]。转化动力学参数即·OH 反应速率常数（k_{OH}）可用于评价化学品的环境持久性，水相 k_{OH} 还可用于评价高级氧化处理体系中化学品的去除效率。此外，化学品与·OH 反应的转化产物又可能具有比母体更大的环境风险。综合这两方面的因素，开展新型有机污染物磷系阻燃剂和增塑剂（OPs）与·OH 反应机制和动力学的研究对其生态风险评价具有重要意义。

李超等通过实验测定、定量构效活性关系（QSAR）和量子化学计算相结合的方法对新型有机物 OPs 与·OH 水相和气相反应机制和动力学进行了研究[70-72]。通过竞争动力学实验测定了 18 种 OPs 与·OH 水相 k_{OH}；基于测定的 k_{OH} 值构建了可用于预测更多 OPs 的 k_{OH} 值的 QSAR 模型（图 16-13），研究揭示了分子中重原子的数目（n_{Hm}）、信息内容指数（第 5 级邻近对称性，IC5）以及分子中最负碳原子电荷（Q_{C^-}）

是影响水中 OPs 与·OH 反应活性的重要参数[70]。此外，基于量子化学计算，筛选了适合不同类别的 OPs 与·OH 反应速率常数的计算方法和溶剂化模型。李超等还针对目前有关芳基取代的 OPs 的大气氧化机制和动力学研究比较少的问题，选择了对甲苯酰磷酸酯（TpCP）作为模型化合物，系统开展了·OH 引发 TpCP 大气氧化机制和动力学的研究[71]。研究发现，整体上摘取—CH_3 上氢原子的反应通道快于·OH 加成到甲苯环上的反应通道，初级反应生成的产物有 TpCP 自由基、TpCP-OH 加合物和二对甲苯基磷酸酯。TpCP 自由基和 TpCP-OH 加合物可以进一步与大气中的 O_2/NO 反应生成苯甲醛取代的磷酸酯、甲基酚取代磷酸酯和双环化自由基等。基于过渡态理论计算出 TpCP 的 k_{OH} 值为 1.9×10^{-12} $cm^3 \cdot mol^{-1} \cdot s^{-1}$，大气半减期为 4.2 d，表明 TpCP 具有潜在的大气持久性，因此未来该化合物需要得到更多的关注。

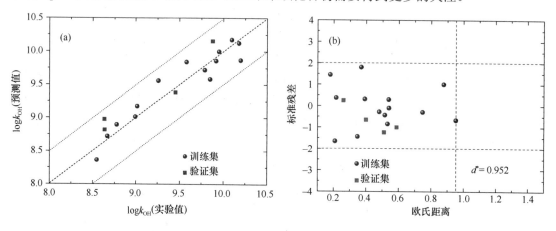

图 16-13　OPs 的 $\log k_{OH}$ 实验值和预测值的拟合图（a）及模型的应用域表征图（b）

3.1.8　木质素热解产物的大气转化机制

木质素作为构成天然木材的基本要素之一，是自然界中含有芳香基团最丰富的物质[73]。木质素的热解可产生高浓度的苯甲氧基类化合物。它们与 SOA[74]及光氧化剂的形成[75]有关，并对人类健康构成健康风险[76]。苯甲氧基类化合物不仅存在于气相中，也存在于大气液相环境中[77]，并随水循环过程进入水体。在不同的地理位置，水溶性有机化合物（WSOCs）可能超过典型的城市污染源[78]。

何茂霞等重点关注了含氧挥发性不饱和化合物、溴系阻燃剂及邻苯二甲酸酯等化学品在大气中的氧化过程及其动力学[79-83]。在此基础上，2018 年，他们将研究的焦点转到基于高级氧化降解技术的环境污染物降解的主要作用机理及其降解产物的水生毒性预测，系统研究了木质素燃烧产物-苯甲氧基类化合物的气/液降解机制和动力学，并运用量子化学及动力学模拟相结合的方法，评估了该类化合物及其转化产物的生态毒理学特征（图 16-14）[84-87]。研究结果表明，NO_3 和·OH 引发苯甲氧基类化合物的反应机制相同，一般通过加成和抽提两种反应类型进行氧化降解，并且这两种反应途径之间有竞争趋势，加成反应对总速率常数的贡献相对较大。其中，·OH 引发的降解速率最高，愈创木酚的气相速率大约为 5.56×10^{-12} $mol^{-1} \cdot s^{-1}$，液相速率是 1.35×10^{10} $mol \cdot L^{-1} \cdot s^{-1}$。$NO_3$ 引发的气相总包反应速率常数约为 10^{-13} $cm^3 \cdot mol^{-1} \cdot s^{-1}$。与 NO_3 和·OH 引发反应不同，O_3 引发的降解反应则是通过进攻苯环上的 C=C 破坏污染物的结构，生成克里奇中间体。此外，O_3 引发的降解速率常数比与·OH 和 NO_3 引发的速率慢，其气相速率常数约为 $10^{-21} \sim 10^{-17}$ $cm^3 \cdot mol^{-1} \cdot s^{-1}$。综上可以看出，在苯甲氧基类污染物环境氧化去除中，·OH 和 NO_3 的贡献率比 O_3 大。在气相研究基础上，经溶剂效应计算得到的热力学及动力学数据发现，该类化合物在水溶液中的降解比气相中的容易，水分子的存在对环境污染物的降解起到了催进作用。另外，通过生态毒性急慢性数值计算表明，经过·OH 和 NO_3 降解后，生成的酚类和醛类产物的 EC_{50} 值小于 10 $mg \cdot L^{-1}$ 甚至更低，对绿藻具有急性毒性，双酚类或硝基苯类化合物的 ChV 值小于 1 $mg \cdot L^{-1}$，对绿藻具有慢性毒性，而绝大部分的 O_3 转化产物的 LC_{50}，EC_{50} 及 ChV 均大于 1000 mg·L^{-1}，不具有急性及慢性毒性。可见，O_3 在苯甲氧基类化合物的氧化降解过程中基本实现了绿色降解，其转化产物均不再具有较强的毒性。通过对一系列不同取代基的单环芳烃与氧化剂的反应及毒性评价发现，具有吸电子基团的化合物反应活性更强，R 基反应活性的顺序大致为—NO_2 > —CF > —SO_3H > —COOH > —CHO > —Cl > —CH_3 > —OCH_3 > —OH > —NH_2。该研究旨在给出这类化合物在典型环境中的降解机制，找到最优反应路径，并总结规律，给出这类污染物及其降解产物的毒性，对更好地理解和认识环境污染物的去除及其风险评价具有重要意义。期望通过这些研究，能给出以解决实验化学中的具体问题为目标和具有广泛应用前景的普适性理论。

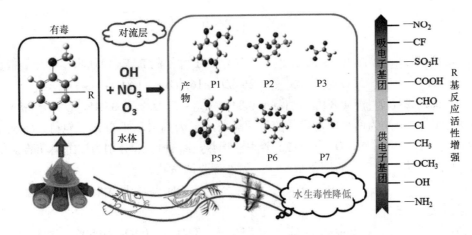

图 16-14　大气/水体中苯甲氧基类化合物和不同氧化剂的主要作用过程及其水生毒性示意图

3.2　水相中的化学品降解转化行为

3.2.1　化学品水解反应的计算模拟与预测

水解反应是化学品在水环境中降解的重要途径之一，化学品的水解半减期由数秒到几年不等[88, 89]。在美国环保局统计的两个数据库（the Federal Insecticide, Fungicide, and Rodenticide Act 和 Collaborative Estrogen Receptor Activity Prediction Project）中，约有超过 40%的化学品包含一个或多个水解官能团[89]。因此，研究化学品的水解反应对评价它们的环境持久性和生态风险十分重要。关于化学品的水解反应研究，主要分为水解路径和水解速率常数两方面。现阶段，对于多水解官能团、水解路径复杂的化学品，鲜有研究。化学品的水解速率常数仍十分缺乏，难以满足其环境持久性和生态风险评价的需求。

张海勤等[90]基于传统水解实验和密度泛函理论（DFT）计算，研究了模型化合物头孢拉定的水解反应（图 16-15）。研究发现，头孢拉定能够发生 β-内酰胺键的水解、分子内氨解和支链酰胺键的碱催化水解。在 β-内酰胺键的水解过程中，有直接水解和间接水解两种机理。酸性条件下，水分子辅助水解为主要路径，中性和碱性条件下，碱催化水解为主要路径。并且，头孢拉定分子中羧基能促进其水解反应。

徐童等[91]以邻苯二甲酸酯（PAEs）为模型化合物，通过与实验值对比，确定最佳 DFT 计算方法，研究了水解反应路径、水解动力学。并且，针对不同反应机理，构建了可高通量预测 PAEs 碱催化水解二级速率常数（k_B）的定量构效活性关系（QSAR）模型（图 16-15）。DFT 计算表明，PAEs 水解有两种限速步骤：OH 亲核进攻酯键 C 原子；离去基团离去。对于大多数 PAEs，离去基团离去为限速步骤，离去基团离去后发生质子转移。具有环状侧链的 PAEs 水解最快，其次为具有直链烷基侧链的 PAEs，含有支链烷基侧链的 PAEs 水解最慢。

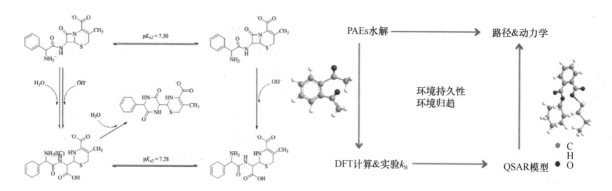

图 16-15 头孢拉定和邻苯二甲酸酯的水解反应

为针对不同反应机理,预测 PAEs 的 k_B 值,首先建立了可预测 PAEs 限速步骤的分类模型:

$$\Delta G^{\ddagger}_{TS2} - \Delta G^{\ddagger}_{TS1} = 26.4D_1 - 16.1D_2 + 11.8D_3 - 8.43 \quad (16.4)$$

$$n_{tra} = 23,\ R^2_{tra} = 0.891,\ Q^2_{LOO} = 0.841;\ n_{ext} = 5,\ R^2_{ext} = 0.945,\ Q^2_{ext} = 0.848$$

式中,$\Delta G^{\ddagger}_{TS1}$ 和 $\Delta G^{\ddagger}_{TS2}$ 分别为两种限速步骤的反应能垒,D_1 为离去基团 O 原子电荷,D_2 为 PAEs 子结构(R_{20}_H)的最高占据分子轨道能,D_3 为 PAEs 芳香性指数,n_{tra} 和 n_{ext} 分别为训练集和验证集数据集,R^2 为决定系数,Q^2_{LOO} 为去一法交叉验证系数,Q^2_{ext} 为外部验证系数。

$\Delta G^{\ddagger}_{TS2} - \Delta G^{\ddagger}_{TS1} > 0$,离去基团离去为限速步骤;反之,$OH^-$ 亲核进攻酯键 C 原子为限速步骤。对于离去基团离去为限速步骤:

$$\log k_{B\_side\ chain} = -4.55D_4 + 2.32D_5 - 1.00D_6 + 0.806 \quad (16.5)$$

$$n_{tra} = 19,\ R^2_{tra} = 0.865,\ Q^2_{LOO} = 0.801,\ RMSE_{tra} = 0.389;$$
$$n_{ext} = 4,\ R^2_{ext} = 0.925,\ Q^2_{ext} = 0.840,\ RMSE_{ext} = 0.311$$

式中,$k_{B\_side\ chain}$ 为 PAEs 单个侧链的 k_B 值,D_4 为 PAEs 的化学硬度,D_5 为离去基团中 sp^3 杂化 C 原子数量,D_6 为 PAEs 侧链中所含支链数量,RMSE 为均方根误差。

对于 OH^- 亲核进攻酯键 C 原子为限速步骤:

$$\log k_{B\_side\ chain} = 5.29D_7 - 1.64 \quad (16.6)$$

$$n_{tra} = 5,\ R^2_{tra} = 0.975,\ Q^2_{LOO} = 0.914,\ RMSE_{tra} = 0.276$$

式中,D_7 为 PAEs 分子极化率。

仅通过传统的实验方法研究种类众多的化学品的水解反应难以满足其环境持久性和生态风险评估的需求。DFT 计算和 QSAR 模型相结合的方法,为研究化学品的水解反应提供了新的思路和方法。

3.2.2 有机化学品与水合电子 e_{aq}^- 的反应速率常数的 QSAR 模型

水合电子(e_{aq}^-)是一种极强的还原剂,还原电势为 –2.9 V[92]。它广泛存在于一些基于高级还原技术处理废水的体系中以及在自然水体中,因此对化学品的迁移转化产生影响[93, 94]。化学品与水合电子反应的速率常数($k_{e_{aq}^-}$)是表征化学品与 e_{aq}^- 反应强度与能力的重要参数,可用于评价水中化学品的去除效率。然而化学品种类繁多,通过实验方法一一获取 $k_{e_{aq}^-}$ 是不现实的。因此,需要建立一种能够快速预测 $k_{e_{aq}^-}$ 的方法,填补现有数据的缺失。

李超等[95]通过 QSAR 模拟和量子化学计算结合的方法,分别建立了预测脂肪族和基于苯环构建的化合物的 $k_{e_{aq}^-}$ 的 QSAR 模型,并探讨了有机物的分子结构与 e_{aq}^- 反应活性的关系。所构建的 QSAR 预测模型均具有良好的拟合度、稳健性和预测能力(图 16-16)。研究发现,分子最低未占据轨道能(E_{LUMO})、单电子还原电势(E_{RED})和分子极化率(α)是影响有机物与 e_{aq}^- 反应活性的最重要的参数。此外,采用量子化学计算研究了 e_{aq}^- 诱导的单电子转移反应（SET）的机制、可行性和卤代有机物的脱卤情况。热动力学的计算结果表明，具有吸电子官能团的化合物通常比具有供电子基团的化合物拥有较高的 $k_{e_{aq}^-}$ 值和较低的反应活化自由能（$\Delta^{\ddagger}G^{\circ}_{SET}$）、反应自由能（$\Delta G_{SET}$）和反应焓变（$\Delta H_{SET}$），表明其 SET 反应更容易发生（图 16-17）。这是因为 e_{aq}^- 是亲核试剂，它倾向于与电子密度低的位点反应，而不易与电子密度高的位点反应。研究还发现 e_{aq}^- 诱导的 SET 反应能使难降解的卤代化合物实现脱卤，表明基于 e_{aq}^- 的高级还原技术是一种有效去除难降解卤代化合物的技术。本研究所构建的模型可为预测污染物在城市污水、饮用水等处理过程中的转化速率提供基本工具，并为筛选及评价污染物是否可以通过 e_{aq}^- 还原降解从水中去除提供理论依据。

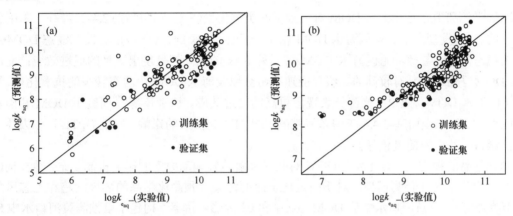

图 16-16 脂肪族化合物（a）与基于苯环构建的化合物（b）的 $k_{e_{aq}^-}$ 的实验值与预测值的拟合图

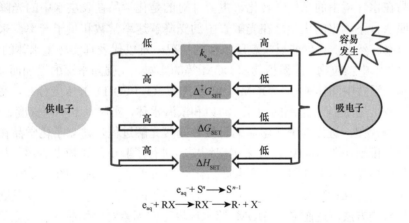

图 16-17 含不同官能团的有机物与 e_{aq}^- 反应的热动力学可行性对比

3.2.3 水中化学品环境光化学行为的模拟预测研究

光化学转化是化学品在水环境中的重要消减途径，且受水中众多溶解性物质[如：溶解性有机质（DOM）、溶解氧、NO_3^-/NO_2^-、CO_3^{2-}/HCO_3^- 以及卤素离子]的影响。各溶解性物质可通过生成的光活性中间体影响化学品的光降解，如图 16-18 所示。目前，针对化学品的水环境光降解行为开展了一些研究。李英杰等[96]研究了 Cl^- 和 Br^- 对磺胺二甲基嘧啶光化学行为的影响，发现激发三线态磺胺二甲基嘧啶可以氧化 Cl^- 和 Br^- 生成相应的卤素自由基，从而导致磺胺二甲基嘧啶发生光致卤化反应。在海水中，磺胺二甲基嘧啶可发生光致卤化反应，并通过飞行时间高分辨质谱检测到了生成的氯代和溴代磺胺二甲基嘧啶中间体。

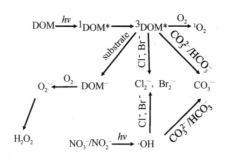

图 16-18　水中主要溶解性物质光致生成光活性中间体的途径

王杰琼等[97, 98]研究了近岸海水 DOM 对三种磺胺类抗生素(SAs)和防晒剂 2,4-二羟基二苯甲酮(BP-1)环境光降解的影响,发现与非养殖区海水 DOM 相比,养殖区海水 DOM 生成的激发三线态 DOM(^{3}DOM*)具有较高的稳态浓度,导致养殖区海水 DOM 对三种 SAs 的光降解呈现出更高的促进效应;而研究发现近岸海水 ^{3}DOM*与 BP-1 具有较高的二级反应速率常数以及海水 DOM 中含有较少的抗氧化剂,导致近岸海水 DOM 比淡水 DOM 对 BP-1 光降解表现出更高的促进效应。张亚南等[99]研究了 DOM 和卤素离子对新型溴代阻燃剂 2,3-二溴丙基-2,4,6-三溴苯氧基醚(DPTE)光降解的影响,发现 DPTE 可以发生 ^{3}DOM*、·OH 以及卤素自由基引发的氧化降解。

考虑到化学品的种类众多,且水中 DOM 对其光降解途径和动力学的影响机制复杂。需要构建化学品在水环境中的光化学转化预测模型,对于防范化学品的污染、预测化学品的环境归趋和生态风险具有重要意义。李英杰等[100]以黄河口水中的 DOM、NO_3^- 和 Cl^- 为影响因素,构建了可预测黄河口水中 SAs 的光降解动力学模型,揭示了黄河口水中溶解性组分对 SAs 光降解动力学影响的内在机理。

以往关于化学品在水环境中的光化学转化,主要针对的是化学品在表层水中的光降解。通常采用在实验室模拟太阳光照条件下测定的化学品在光解管中的光降解速率常数和量子产率,来推测环境水体中的化学品的光降解速率。然而,这仅能反映表层水体的情况,难以代表真实环境水体的情形。

在自然水体中,受水中颗粒物、浮游植物和 DOM 等的影响,光强随水深的增加而衰减。而水下光强的衰减会进一步影响光活性中间体(如,^{3}DOM*、单线态氧 $^1O_2$ 和·OH)的浓度分布,影响有机污染物的降解动力学。周成智等[101]在黄河三角洲的河流、河口和滨海水体,实测水下不同深度、不同时刻的光强,建立了水下光强的预测模型;进而考虑化学品的直接和间接光解过程,建立了化学品在水体不同深度和时刻的光解速率的预测模型(图 16-19),并采用现场实验进行了验证。该模型表明,如果直接采用实验室测定的数据预测其环境光降解半减期,会严重低估化学品的环境持久性(图 16-20)。

除了通过模拟实验的方法构建化学品的光降解动力学参数的预测模型,还可以发展化学品的环境光化学降解途径的计算预测方法,进而揭示 DOM 等水环境影响因素对化学品的环境光降解影响的微观机制。张思玉等[102]选取 2-苯基苯并咪唑-5-磺酸(PBSA)和对氨基苯甲酸(PABA)两种有机防晒剂为模型化合物,采用 DFT 计算和实验模拟的方法,研究了溶解氧、DOM、卤素离子(Cl^-、Br^- 和 I^-)和 HCO_3^-/CO_3^{2-} 对 PBSA 光降解的影响;计算了 $^1O_2$ 与 PBSA 和 PABA 的反应热力学参数,揭示其与 $^1O_2$ 的反应机理,并预测了可能的光解产物。该研究为发展化学品的水环境光降解途径的预测方法奠定了基础。

3.2.4　维生素 B12 催化卤代有机物还原脱卤机理及双元素稳定同位素分馏的计算方法

大多数厌氧微生物体内的还原脱卤酶,通过研究发现是以金属辅酶-维生素 B12(Vitamin B12)的形式存在。由于维生素 B12 分子中起还原脱卤作用的还原态[Co(I)]中的钴原子同时具有很强的还原性和亲核性,导致其催化还原脱卤反应路径的多样性[103-106]。尤其是部分卤代有机化合物在还原过程中生成的中间体分子相比较于母体化合物可能会表现出更强的毒性。而评估维生素 B12 分子催化污染物脱卤过程产生的脱毒效应,需要全面揭示脱卤反应的微观反应机理。

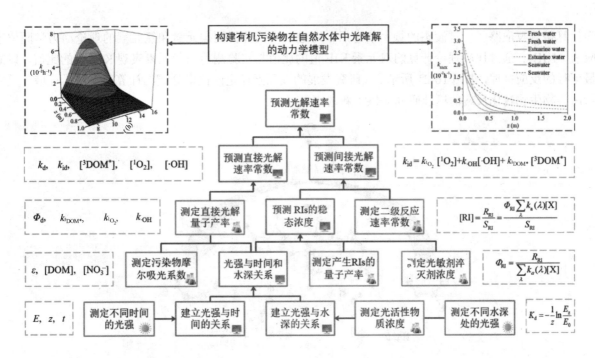

图 16-19　考虑水下光强衰减及直接光解、间接光解过程，预测污染物在水体中光降解动力学的模型

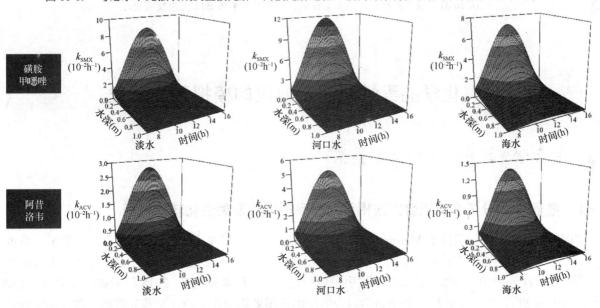

图 16-20　黄河三角洲不同水体中两种污染物光降解速率常数（k）随水深和时间的变化，预测值与实测值具有较好的一致性，均方根误差为 0.31～0.76

在维生素 B12 催化还原脱卤反应机制的研究中，地下水中常见化学品（氯代乙烯类化合物）是研究最多的底物种类。根据大量实验文献报道，维生素 B12 催化还原氯代乙烯类化合物的反应机理存在较大争议，可能存在的反应机理有：电子转移、亲核取代及亲核加成[107]。季力等使用量子化学计算，首次系统地解析了维生素 B12 催化还原氯代乙烯类化合物的反应机理：四氯乙烯和三氯乙烯主要通过亲核取代的反应路径被维生素 B12 催化脱卤，而维生素 B12 催化还原二氯乙烯和氯乙烯主要通过亲核加成的反应路径，并基于反应机理建立了各个反应路径的表征分子结构和反应性构效关系的线性自由能关系模型[108]。季力等在该研究中提出了双元素稳定同位素分馏用于研究反应机理的计算新方法：即采用量子

化学计算得到反应路径的过渡态的频率信息后，可进一步计算得到碳元素和氯元素的理论动力学同位素效应值（KIE）；通过比较理论计算的双元素 KIE 值的比值与实验测得的双元素表观 KIE 值的比值，以确定最可行的反应机理，如图 16-21 所示。该研究发展的双元素稳定同位素分馏的计算新方法为环境体系中痕量污染物生物转化反应的机理研究提供了新的工具。

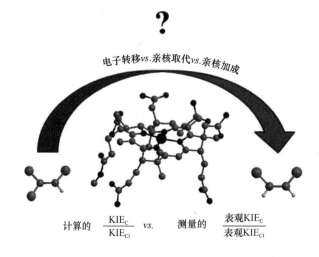

图 16-21　双元素稳定同位素分馏的计算新方法解析 B12 脱卤反应机理

4　化学品毒性机制及效应的模拟预测研究

4.1　酶代谢外源化合物的计算模拟

4.1.1　量子力学/分子力学组合方法揭示化学品的毒理及酶催化转化机理

2013 年诺贝尔化学奖授予 Martin Karplus、Michael Levitt 以及 Arieh Warshel 三位科学家，显示出量子力学/分子力学组合方法在模拟复杂大分子体系尤其是在模拟生物酶参与的催化反应领域的重要地位[109]。量子力学/分子力学组合方法的年被引用量一直呈现一种攀升趋势，2018 年的被引量达到了 13000 余次（图 16-22）。然而量子力学/分子力学组合方法的应用多集中在生命、医药等领域，在环境领域的应用仍然较少[110]。

张庆竹、李延伟等成功将量子力学/分子力学组合方法应用于环境领域[1120-114]，揭示了滴滴涕、多氯联苯、含氟有机污染物等化学品酶催化转化机理。他们与加州大学洛杉矶分校合作，发展了一种环境扰动过渡态采样方法（EPTSS），该方法能够嵌入量子力学/分子力学组合方法，并用于准确描述酶催化自由能变和动力学同位素效应[115]。基于该方法成功揭示了酶 SpnF 在多杀菌素 A 合成中采用一种分叉的过渡态（同一个过渡态分叉成两种产物），证明了酶 SpnF 能够在亚皮秒时间尺度上影响催化反应的选择性。2019 年，张庆竹、李延伟等进一步采用量子力学/分子力学组合方法阐明了 P450 酶作用下萘转化为环氧化萘的反应机理，发现环氧化萘在细胞质环境中 $OH^-/OH\cdot/H_2O/H_3O^+$ 等作用下可以进一步转化生成萘醌等致癌物[116]。

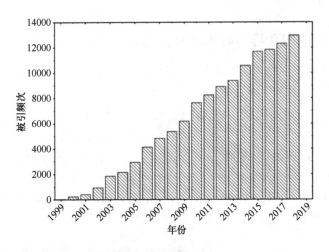

图 16-22　量子力学/分子力学组合方法近年来的年被引次数

4.1.2　双酚 S 类物质甲状腺激素受体 β 拮抗效应分子机制

双酚 S（BPS）及其类似物（BPSs）作为双酚 A（BPA）的类似物，近几年来全球生产与使用量逐年上升[117]，该类化学品在环境中的残留被频繁检测出，目前在人体尿液、乳汁、血液样本中均有不同程度的检出[117-120]。BPSs 在一定程度上具有 BPA 相似的潜在毒性效应，包括细胞毒性、基因毒性、免疫毒性和内分泌干扰效应[117, 121]等，其安全性越来越受到公众的广泛关注。

鲁莉萍和庄树林等采用分子动力学模拟及体外筛选实验评估了环境相关浓度下 BPSs 潜在的甲状腺激素受体（TRβ）干扰效应（图 16-23）[122]。表达人 TRβ 配体结合域（LBD）蛋白并进行了荧光光谱体外结合实验，发现具有不同基团的 BPSs 诱导 TRβ LBD 荧光猝灭并发生红移，显示 TRβ LBD 构象的改变，且四溴双酚 S（TBBPS）和四溴双酚 A（TBBPA）诱导的构象变化更为显著。

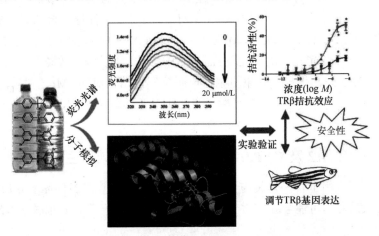

图 16-23　BPSs 诱导甲状腺激素受体 β 干扰效应

100 ns 的分子动力学模拟发现 BPSs 结合 TRβ LBD 诱导人 TRβ LBD 的 Cα 原子 RMSD 较大变化，这与荧光光谱体外结合实验结果一致。模拟发现 TRβ LBD 螺旋 H11-H12 间距发生较大变化，说明了局部构象的显著改变。TRβ LBD 构象的破坏可能影响 TRβ 的转录活性，干扰 TR 异二聚化以及下游配体信号传导。进一步通过重组人 TRβ 基因酵母实验，验证了 TBBPS 和 TBBPA 比 BPS 和 BPA 具有更强的 TRβ 拮抗效应。BPSs 暴露 72 小时后会诱导斑马鱼胚胎中 TRβ 在 mRNA 表达水平的显著改变。BPS 和 TBBPS 的 TRβ 受体干扰效应显示 BPS 不一定是安全的 BPA 替代品[121, 123]。

4.1.3 细胞色素 CYP3A4 酶介导的叶菌唑立体选择性代谢

细胞色素 CYP3A4 酶参与生物体内许多外源性化学品的生物转化，介导化学品潜在的毒性效应[124]。化学品与 P450 酶的交互作用会影响底物特异性结合[125]，引起酶蛋白分子活性位点构象变化。考察 CYP3A4 与化学品的交互作用对阐明化学品代谢的分子机制至关重要。

庄树林、刘维屏和周如鸿等采用分子动力学模拟（MD）、拉伸动力学模拟并结合酶体外代谢实验系统研究了人 CYP3A4 酶介导的顺式叶菌唑（cis-MEZ）立体选择性代谢机制研究[126]。利用含时密度泛函理论（TDDFT）和圆二色光谱仪（CD）构建了叶菌唑绝对构型的手性表征方法。鉴于 cis-MEZ 与 CYP3A4 酶活性位点 Fe 原子成键问题，采用拉伸动力学模拟方法进行构建，采用经典分子动力学模拟方法优化构象，并进一步采用自由能微扰方法评估了结合自由能（图 16-24）。

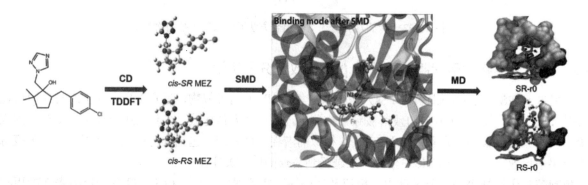

图 16-24　cis-MEZ 与人 CYP3A4 酶的相互作用

对 CYP3A4 与 cis-MEZ 的复合物开展了总共 2.2 μs 的分子动力学模拟，发现 cis-MEZ 两种异构体与 CYP3A4 酶结合模式存在显著差异。cis-RS MEZ 在活性口袋中具有更加伸展的结构，而 cis-SR MEZ 处于活性口袋的疏水核心，导致较多自由能的损失。cis-MEZ 的对映选择性结合诱导 CYP3A4 中活性位点氨基酸构象的变化。自由能微扰计算发现 cis-MEZ 异构体和 CYP3A4 间的范德瓦耳斯力作用是导致相互作用力差异的主要原因。体外荧光光谱试验发现 cis-SR MEZ 引起的荧光强度增加，而 cis-RS MEZ 降低，并且 cis-SR MEZ 对 CYP3A4 酶活性抑制效果更强，说明特异性结合对 cis-MEZ 对映体选择性代谢的影响。揭示了特异性结合对 cis-MEZ 的对映体选择性代谢的影响及其分子机制，为基于结构和代谢机理的相关化学品评估筛选提供了必要信息。

4.1.4 人细胞色素 CYP1A2 及 CYP3A4 酶介导的 2-氯噻吨酮生物转化

噻吨酮（TXs）广泛应用于食品包装与紫外线固化树脂[127]，其在多种环境介质与人体中均被检测到残留[128, 129]。TXs 的频繁暴露可能会导致对人体健康具有潜在危害[130]，目前关于 TXs 的代谢和毒性作用信息比较缺乏。

詹婷洁和庄树林等考察了由人细胞色素 CYP1A2 及 CYP3A4 酶介导的 2-氯噻吨酮（2-Cl-TX）的体外代谢，并分析了 2-Cl-TX 对 CYP1A2 和 CYP3A4 酶活性、mRNA 和蛋白表达的毒性效应[131]。采用四极杆飞行时间质谱仪（Q-TOF-MS/MS）鉴定发现，2-Cl-TX 经人肝微粒体孵育后的两种代谢产物的荷质比比母本化合物的增加了 16 Da，说明 2-Cl-TX 代谢后增加了氧原子。进一步利用 SMARTCyp 预测了 2-Cl-TX 的代谢位点，发现 2-Cl-TX 中硫原子 S.8 与苯环上 C12 需要相对较低的活化能，由此推测 2-Cl-TX 代谢后主要生成磺化氧化产物及羟化产物（图 16-25）。

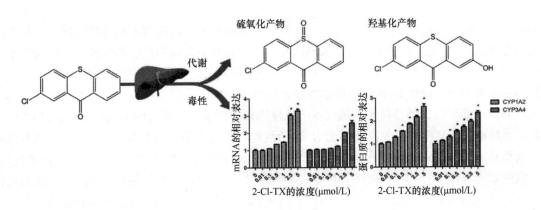

图 16-25　2-氯噻吨酮的体外代谢与毒性作用

进一步将 CYP1A2 抑制剂 α-萘黄酮或 CYP3A4 抑制剂酮康唑加入 2-Cl-TX 的代谢体系，发现 α-萘黄酮的加入导致两种主要代谢产物的含量发生明显变化，而酮康唑的加入使得磺化氧化产物完全消失，说明 CYP1A2 与 CYP3A4 酶在 2-Cl-TX 代谢中起主要作用。CYP 酶抑制实验发现 2-Cl-TX 对 CYP1A2 与 CYP3A4 酶活性具有抑制作用，IC_{50} 分别为 8.36 μmol/L 与 0.86 μmol/L。2-Cl-TX 短期暴露（24 h、48 h）上调 CYP1A2 和 CYP3A4 在 mRNA 水平和蛋白水平的表达，并呈现浓度依赖关系。该研究为噻吨酮的人体健康与安全风险评估提供数据支持。

4.2　环境内分泌干扰效应的毒性通路与模拟预测

4.2.1　个人护理品的毒性兴奋（Hormetic）效应及准确定量表征

个人护理品（PCPs）作为新型污染物 PPCPs 的一大类污染物近年来受到广泛关注和研究。众所周知，所有 PCPs 产品实际上都是由多个化学品构成的混合物。目前关于 PCPs 相关生物效应的研究多是 PCPs 中一些活性成分比如三氯生等的单个物质或是 2~3 个活性成分构成的单个浓度比混合物，这显然不能真正表征各种 PCPs 产品的特性。

为了解 PCPs 产品混合物的毒性的真实情况，刘树深等选择了全球知名品牌旗下畅销的 23 种 PCPs 产品，包括 13 种爽肤水、7 种柔肤水和 3 种化妆水，作为测试污染物（简称 TSM），采用短长期微板毒性分析方法[132]，测定了 TSM 暴露 0.25 h 和 12 h 对青海弧菌的发光抑制毒性，结果发现这些 TSM 混合物具有不完全一致的剂量-效应关系（CRC），很多具有所谓的低量刺激和高剂量-抑制的 Hormesis 现象或 Hormetic 效应。23 种 TSMs 中有 6 种其抑制毒性无论在 0.25 h 还是在 12 h 皆呈单调增加的剂量-效应关系。另外 17 种 TSMs 在 0.25 h 时呈单调增加的剂量-效应关系，但在 12 h 却出现出明显的低浓度刺激高浓度抑制的 Hormesis 现象[133]。刘树深等应用 APTox[134]和 JSFit[135]分别对单调增加和 Hormesis 两种剂量-效应关系进行了最佳拟合，从而准确预测效应下的浓度或不同浓度下的效应，并提出准确表征 Hormesis 现象的 5 个特征参数半数最大刺激效应浓度、最大刺激效应及浓度、零效应点以及半数效应。该研究直接以 PCPs 产品为试验对象，明显有别于目前大多是以其中的几个组分为对象的；JSFit 方法将 Hormetic 曲线看成是一个上升的 Hill 函数和下降的 Hill 函数的乘积并通过初始回归参数的自动化以实现特征参数的准确求解，目前没有其他文献报道。

4.2.2　化合物毒性分类预测模型构建及警示子结构识别

计算毒理学是一个多学科交叉的前沿领域，至少包括三个方面的研究内容：预测模型构建、警示子结构（Structural Alert）识别和计算系统毒理学。随着大数据和人工智能技术的快速发展，计算毒理学也获得了快速发展，目前 QSAR 技术已被学术界、各类化学品制造公司和政府监管机构广泛应用于化学品风险评估。近年来，唐赟教授实验室致力于从事相关领域的研究，并已取得重要进展。

唐赟等主要采用分子指纹和/或分子描述符来表达分子结构，应用多种机器学习方法来构建毒性分类预测模型；同时，采用信息增益技术，识别各类毒性端点的警示子结构（图 16-26）[136]。2017 年 7 月以来，发表了多篇毒性预测模型构建论文[137-142]，包括预测农药对鱼类的水生毒性、环境化学品对淡水甲壳类生物如大型溞（Daphnia magna）的急性毒性[138]和对海洋甲壳类生物如糠虾（Mysidae）的急性毒性[139]以及化合物的基因毒性[140]、生殖毒性[141]和皮肤敏感性[142]。在构建分类预测模型的同时，他们还识别了各类毒性端点（endpoint）的警示子结构，可用于直观判断一个新化合物是否具有某种潜在毒性。当然，这种直观判断还比较粗糙，存在一定比例的假阳性率。因此，他们对警示子结构做了进一步研究，发现某些警示子结构与另一些子结构以某种方式共同存在时，能够变成无毒子结构，这样就为逆转化合物毒性提供了一种新的策略[143]。所发展各类毒性预测模型，已整合入该实验室发展的在线预测服务系统 admetSAR（http：//lmmd.ecust.edu.cn/admetsar2/）[144]。

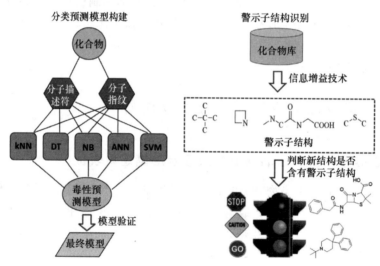

图 16-26　化合物毒性分类预测模型构建及警示子结构识别示意图

4.2.3　基于机器学习对纳米颗粒生殖毒性的关键因素筛选及预测

人工纳米材料由于其优异的理化特性，在生物医疗、化学化工、电子机械和能源环保等诸多领域引起了广泛应用[145]。随之而来的是，人工纳米材料的生态环境风险也引起了普遍关注[146, 147]。不同于普通的重金属或持久性有机污染物，人工纳米材料或颗粒的属性特征是高维的，引起的生物不良效应也是非常复杂的[148]。在化学品的风险评估中，找到纳米颗粒属性特征与生物毒性效应之间的对应关系非常重要，但是二者之间并非简单的一一对应关系，现有的理论知识很难整体解释纳米颗粒属性特征与生物毒性效应之间的关系。随着人工智能的迅速发展，机器学习体现了计算速度快、灵活度高、泛化能力强、自我训练学习等优势，通过其与实验数据相结合，有望准确地预测和挖掘纳米颗粒属性特征与生物毒性效应之间的复杂关系。

胡献刚团队收集了纳米颗粒引起动物生殖毒性的相关文献，提取数据，使用随机森林模型，从高异质性数据中成功筛选出了主导生殖毒性的关键因素（图 16-27）[149]。首先通过 Meta 分析思路，经过文献筛选，以常见的 18 种人工纳米颗粒为研究对象，经过数据提取和整理后，提取了 10 种可能影响纳米颗粒生殖毒性的属性特征及暴露方式。依据模型中衡量因素重要性的参数，即预测的均方差增大，发现纳米颗粒的种类和暴露方式是显著影响纳米颗粒在小鼠睾丸中积累的两个重要因素；而纳米颗粒种类和毒性指标是影响纳米颗粒生殖毒性最重要的两个因素。模型结果表明，动物体内的大量元素（例如，铁和锌）更倾向于积累在小鼠的生殖器官，但其诱发的生殖毒性要远低于贵金属元素构成的纳米颗粒所引发的生殖毒性。模型进一步筛选得出，腹腔注射引起小鼠生殖毒性的风险要高于静脉注射、气管滴注和腹部皮下注射。随后，研究通过结合相似网络和层次聚类模型，识别了本文数据异质性的来源，可视化了影响生殖毒性的关键因素，及其与毒性效应的关系。最后，依据筛选出的关键因素，实现了纳米颗粒引起小鼠生殖毒性的预测。上述研究表明，通过机器学习与实验数据结合，有望为纳米颗粒的生态环境风险的解释说明及预测提供新的思路。

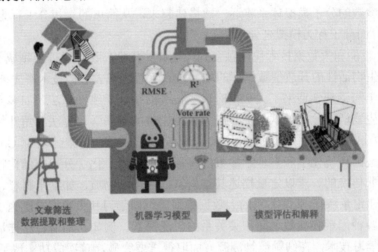

图 16-27 基于机器学习对纳米颗粒生殖毒性的关键因素筛选及预测的示意图

5 展 望

为了研究各种化学品在生态环境中的源和汇，及其对生态和人体健康的不利效应，环境化学、毒理学等主要构建于实验测试基础之上的交叉学科发展起来。长久以来，环境化学、毒理学等学科成功探索了化学物质在环境介质及生物相的迁移转化规律，形成了较为系统的知识理论，积累了一定规模的化学品环境行为参数、毒性测试数据，为化学品的风险评价作出了贡献。

但是人们逐渐意识到，单纯的实验测试体系无力填补化学品风险预测与管理所面临的数据鸿沟。这一基本事实，促使人们依据 QSAR 发展数学/计算机模型快速预测风险评价体系所涉及的危害性、环境行为参数以及毒性终点指标。另一方面，单纯的环境采样和分析测试，也无法对化学物质的环境分布规律有整体性的认知。因此，基于唯象的经验规律和物理化学公式，人们发展了多介质环境模型以描述一种化学物质在不同环境介质中的迁移转化行为。唯象模型依赖于大量的参数，其中与化学物质紧密相关的环境行为参数、物理化学性质等也可以诉诸 QSAR 模型加以预测。这些早期的研究，构成了环境计算化学与计算毒理学发展的萌芽。

然而，QSAR 模型本质上是在化学物质结构特征与其特定的性质之间挖掘相关关系，它难以清晰阐释环境化学和毒理学现象的机制。而机制不明晰，人们就无法安心地将模型用于预测所关心的环境现象。因此，QSAR 模型的应用必然囿于特定的范围——应用域。而 QSAR 模型应用域的拓展又不得不依赖于新的实验测试数据集。早期的计算方法跳不出这个"怪圈",因此相对于"正统"的实验测试体系而言,往往处于"从属"的地位。

近10年来,基于量子力学、经典力学的分子模拟技术逐渐渗透到环境研究领域,成为环境计算化学与计算毒理学家的重要工具。分子模拟能够以近乎无限的分辨率揭示出化学物质在自然环境现象以及污染控制过程中行为的分子机制。与QSAR不同,分子模拟所揭示的分子机制与其计算模拟的结果,即表征特定环境化学或毒理学现象的关键参数之间互为因果。同时,分子模拟是基于构成分子的元素类型、电子结构特征及原子局部成键特点进行参数化的,具有极强的泛化能力,其唯一的局限就是所能模拟体系的大小。而随着高性能计算技术的快速发展,这个局限亦有望得到快速突破。因此,可以预料在未来,基于分子模拟研究环境化学与毒理学现象的研究将会越来越多。具体而言,围绕成熟的唯象模型所需的、与化学物质结构有关参数的分子模拟体系将逐一被搭建起来,从简单的气相分子行为模拟,向水相及气-固、气-液、液-固相界面分子行为模拟延伸。

近5年来,人工智能和大数据技术快速融合到环境计算化学与计算毒理学领域中来。得益于围绕化学品物理化学性质和生物活性的开放型数据库的构建,人工智能和大数据技术率先被应用于QSAR研究领域——环境计算化学与毒理学迎来了大数据时代。传统的统计学线性算法,如多元线性回归和偏最小二乘回归所不擅长的生物活性预测领域,俨然成为各种先进的机器学习算法的角斗场。其中,深度神经网络算法,似乎又成为这个角斗场中的明星。不仅如此,大数据方法学跳出了经典QSAR概念的束缚——用于预测人们所感兴趣的环境化学和毒理学参数的特征不再局限于化学分子的结构特征。突破了这个思维桎梏,人们得以为非传统的、难以定量描述其结构特征的化学物质,如纳米材料、金属、金属氧化物等物质的毒性预测构建定量特征-活性关系的大数据模型。机器学习和大数据热潮也引爆了人们对环境化学与毒理学现象之中各种因素和参数之间相关关系的广泛探索。例如,刘倩、江桂斌课题组利用硅、氧双同位素指纹和机器学习技术,在细颗粒物溯源方面取得重要突破,首次成功实现了SiO_2颗粒的来源区分[150]。可以预见,在未来还会出现更多的对化学物质的结构特征、环境行为特征、健康效应和流行病学特征、功能性材料特性等各种因素之间的相关关系的分析和讨论,并逐渐形成一系列重要科学猜想和新的唯象理论,构建起化学品、人类社会经济活动、生态环境健康以及其他因素之间的定量关系脉络。

当然,计算毒理学本身已经搭建起来的"大厦"(参见图16-4)仍需进一步修缮和添砖加瓦。不忘初心,方得始终。计算毒理学的初心就是要辅助化学品风险预测和评价工作,侧重解决三方面的科学问题:一是如何模拟化学品从源释放到导致不利生态效应的连续过程;二是如何评价与预测种类众多且数目不断增加的化学品的风险;三是如何预测化学品对生态系统不同物种的风险,也就是如何能做到跨物种外推。

解决第一个问题,需要将从源释放到导致不利生态效应的连续过程(source to adverse outcome pathways, S2AOP)分解为适合科学研究的"子过程",并发展唯象理论或找到适用的物理化学规律来逐一描述这些过程,最终通过这些子过程之间数据的无缝对接,还原出完整的S2AOP。目前,多介质环境归趋模型被广泛用于描述化学物质在多种环境介质中的迁移转化过程。而基于生理的毒代动力学(PBTK)模型则被用于描述化学品在生物体内的浓度分布和转化过程。二者的数据对接还需要耦合人群暴露模式和场景参数。如此,化学品的"暴露"过程大致能够得到较为完整的描述。然而,现有多介质环境模型和PBTK模型的空间分辨率与预测准确性均有待提升,这就需要在二者的模型结构上做出大刀阔斧的调整,增加模型内部的复杂性,增添其与外部环境数据交换的接口,让模型更加逼真。

此外,评价化学物质的风险不仅要考虑暴露,更要考虑化学品不利效应。长久以来,化学品的不利效应被规范化为所谓的危害性,成为贴在化学品上的一类属性标签。例如,化学品急性毒性指标(如半数致死浓度和半数效应浓度)、内分泌干扰性、致癌性等均属于化学品危害性标签。危害性标签便于化学品风险管理,也为大数据研究提供了较为规整的数据结构。但是,危害性标签也将原本连续的剂量-效应关系和时间-效应关系按照人为规定的剂量和时间阈限固化为静态指标——人们无法从这些标签中对不利效应机制有更深的理解，而没有明晰的机制，计算毒理学就难以茁壮地成长。

2010年以来，不利结局路径（AOP）的概念体系被多次阐释和讨论，围绕AOP的实用性框架和研发导则也陆续问世。基于体外测试（in vitro）的高通量筛选技术（HTS）、组学技术和基于高分辨细胞/亚细胞影像的高内涵测试技术的发展，为AOP的发现和验证提供了大量的实验证据。近年来，一些顶层终点/不利结局，如皮肤刺激性的AOP已经较为清晰，且能够追溯到具体的关键事件和分子起始事件（MIE）上。AOP框架为计算模拟化学物质的不利效应提供了依据。例如，可构建体现MIE的关键分子体系加以分子模拟。而针对关联AOP中关键事件的一系列过程，都可以发展唯象的理论与模型，如细胞信号通路模型、虚拟组织模型，从而将整个AOP串联起来。

考虑化学品的AOP、代谢及其在生物体内的浓度分布，人们就有希望把整体动物（in vivo）的毒性数据和基于细胞的体外测试毒性数据联系起来，从而进一步对现有的HTS数据的实用性加以验证。可以预见，一些零星的唯象模型最初会以"案例研究"的形式搭建起来，真正棘手的问题仍然是将此类模型应用于其他化学品时不可逃避的参数化过程——每个化学品需要一套参数来驱动关乎其自身的唯象模型。而这恰恰就与第二个问题产生了紧密的联系。

从方法的实用性来看，QSAR、机器学习和大数据技术构成了解决第二个问题的核心方法。具体而言，在化学品多介质环境分配行为参数的模拟预测原理与技术方面，如今针对非离解性有机物的正辛醇/水分配系数K_{OW}、正辛醇/空气分配系数K_{OA}、亨利定律常数K_H、过冷液体蒸气压P_L、水溶解度S_W、土壤（沉积物）吸附系数K_{OC}等风险行为参数，已经有一些预测模型，模型的预测准确性较高，应用域较大。但是，针对可离解性有机物的分配行为参数，仍需要很多探索性研究。体现一定机理性的多参数线性自由能关系（pp-LFER）模型应用于环境分配行为参数的预测，具有较好的预测精准度。但是传统上的pp-LFER模型所应用的分子结构参数（溶质参数）值需要色谱等实验手段测定，很有必要发展pp-LFER溶质参数值的理论预测方法。

相较于化学品环境分配行为，有机化学品在环境中的降解转化行为模拟预测的难度更大。总体上，化学品环境降解转化行为参数的数值都较少。约有1000种化学品具有气相羟基自由基反应二级速率常数（k_{OH}）的数值，500种化学物质具有水相k_{OH}的数值。不到1000种化学品具有微生物降解动力学参数的数值。太阳光照射地球的表面，环境光降解可以发生于大气、水体的表层、土壤和植物的表面，然而，只有百种数量的化学品具有直接光解速率常数（含量子产率）的数值，且是在各种环境条件下测定的数值。因此，发展大气、水体、土壤环境中化学品的降解动力学行为参数、降解途径与产物的预测技术，仍是计算毒理学领域有待攻克的难题，尤其是机理清晰、应用域较广泛的模型。数据挖掘及大数据分析相关技术在此方面大有用武之地。

基于分子模拟来直接预测化学品的环境分配与降解转化行为，具有对实验数据依赖性弱、模型适用范围广的特点。随着数据挖掘、大数据处理、深度学习等技术与计算毒理学的交叉融合，有望开发出高效、准确的化学品毒理风险预测与评价模型。值得注意的是，纳米材料作为一类特殊的化学品，其环境行为和毒理效应的模拟预测仍然具有很大的挑战性。这主要是由于纳米材料对应的分子体系一般都较大，受限于计算机硬件性能，而现有的基于经典力场的方法对一些金属和类金属元素并不支持，导致无法对无机纳米材料开展较大尺度的模拟研究。

回答第三个问题，即跨物种外推的问题，则需要借鉴人类不利效应的模拟框架，即率先为实验研究较充分的、为我们所熟知的模式生物构建相应的AOP，具体过程与为人类构建"效应"模拟的过程较为类似，此处不再赘述。

时至今日，计算模拟方法已经成为了环境化学与毒理学研究不可或缺的要素。计算机模型也成为集理论阐释与实际应用为一体的多面手，加速着科学研究向产业化和化学品管理决策支持工具的转化。计算环境化学与计算毒理学具有鲜明的数字化、网络化、智能化特性，其发展无疑是数字中国建设在生态环境保护与化学品管理领域的一个关键环节，在推动经济社会发展、促进国家治理体系和治理能力现代化、满足人民日益增长的美好生活需要方面将发挥着越来越重要的作用。

参 考 文 献

[1] Global chemicals outlook II: From legacies to innovative solutions. United Nations Environment Programme, 2019.

[2] 王中钰, 陈景文, 乔显亮, 等. 面向化学品风险评价的计算(预测)毒理学. 中国科学: 化学, 2016, 46(2): 222.

[3] Keller A A, Lazareva A. Predicted releases of engineered nanomaterials: From global to regional to local. Environmental Science & Technology Letters, 2014, 1(1): 65-70.

[4] Yang K, Zhu L Z, Xing B S. Adsorption of polycyclic aromatic hydrocarbons by carbon nanomaterials. Environmental Science & Technology, 2006, 40(6): 1855-1861.

[5] Wang Z, Chen J W, Sun Q, et al. C_{60}-DOM interactions and effects on C_{60} apparent solubility: A molecular mechanics and density functional theory study. Environment International, 2011, 37(6): 1078-1082.

[6] Sun Q, Xie H B, Chen J W, et al. Molecular dynamics simulations on the interactions of low molecular weight natural organic acids with C_{60}. Chemosphere, 2013, 92(4): 429-434.

[7] Yang K, Xing B S. Adsorption of organic compounds by carbon nanomaterials in aqueous phase: Polanyi theory and its application. Chemical Reviews, 2010, 110(10): 5989-6008.

[8] Yang K, Wu W H, Jing Q F, et al. Aqueous adsorption of aniline, phenol, and their substitutes by multi-walled carbon manotubes. Environmental Science & Technology, 2008, 42(21): 7931-7936.

[9] Ersan G, Apul O G, Karanfil T. Linear solvation energy relationships (LSER) for adsorption of organic compounds by carbon nanotubes. Water Research, 2016, 98: 28-38.

[10] Xia X R, Monteiro R N A, Riviere J E. An index for characterization of nanomaterials in biological systems. Nature Nanotechnology, 2010, 5(9): 671-675.

[11] Xia X R, Monteiro R N A, Mathur S, et al. Mapping the surface adsorption forces of nanomaterials in biological systems. ACS Nano, 2011, 5(11): 9074-9081.

[12] Apul O G, Wang Q L, Shao T, et al. Predictive model development for adsorption of aromatic contaminants by multi-walled carbon nanotubes. Environmental Science & Technology, 2013, 47(5): 2295-2303.

[13] Wang Q L, Apul O G, Xuan P, et al. Development of a 3D QSPR model for adsorption of aromatic compounds by carbon nanotubes: Comparison of multiple linear regression, artificial neural network and support vector machine. RSC Advances, 2013, 3(46): 23924.

[14] Zhao Q, Yang K, Li W, et al. Concentration-dependent polyparameter linear free energy relationships to predict organic compound sorption on carbon nanotubes. Science Report, 2014, 4: 3888.

[15] Huffer T, Endo S, Metzelder F, et al. Prediction of sorption of aromatic and aliphatic organic compounds by carbon nanotubes using poly-parameter linear free-energy relationships. Water Research, 2014, 59: 295-303.

[16] Liu F, Zou J W, Hu G X, et al. Quantitative structure-property relationship studies on the adsorption of aromatic contaminants by carbon nanotubes. Acta Physico-Chimica Sinica, 2014, 30(9): 1616-1624.

[17] Apul O G, Zhou Y, Karanfil T. Mechanisms and modeling of halogenated aliphatic contaminant adsorption by carbon nanotubes. Journal of Hazardous Materials, 2015, 295: 138-144.

[18] Yu X Q, Sun W L, Ni J R. LSER model for organic compounds adsorption by single-walled carbon nanotubes: Comparison with multi-walled carbon nanotubes and activated carbon. Environmental Pollution, 2015, 206: 652-660.

[19] Ding H, Chen C, Zhang X. Linear solvation energy relationship for the adsorption of synthetic organic compounds on single-walled carbon nanotubes in water. SAR and QSAR in Environmental Research, 2016, 27(1): 31-45.

[20] Wang Y, Chen J W, Tang W H, et al. Modeling adsorption of organic pollutants onto single-walled carbon nanotubes with theoretical molecular descriptors using MLR and SVM algorithms. Chemosphere, 2019, 214: 79-84.

[21] Zou M Y, Zhang J D, Chen J W, et al. Simulating adsorption of organic pollutants on finite (8, 0) single-walled carbon nanotubes in water. Environmental Science & Technology, 2012, 46(16): 8887-8894.

[22] Zhang X, Xie H B, Wei X, et al. Simulating adsorption of organic pollutants on n-doped single-walled carbon nanotubes in water. Chinese Science Bulletin, 2015, 60(19): 1796-1803.

[23] Wang Y, Chen J W, Wei X X, et al. Unveiling adsorption mechanisms of organic pollutants onto carbon nanomaterials by density functional theory computations and linear free energy relationship modeling. Environmental Science & Technology, 2017, 51(20): 11820-11828.

[24] Georgakilas V, Tiwari J N, Kemp K C, et al. Noncovalent functionalization of graphene and graphene oxide for energy materials, biosensing, catalytic, and biomedical applications. Chemical Reviews, 2016, 116(9): 5464-5519.

[25] Dimiev A M, Alemany L B, Tour J M. Graphene oxide. Origin of acidity, its instability in water, and a new dynamic structural model. ACS Nano, 2013, 7(1): 576-588.

[26] Wang Y, Comer J, Chen Z F, et al. Exploring adsorption of neutral aromatic pollutants onto graphene nanomaterials via molecular dynamics simulations and theoretical linear solvation energy relationships. Environmental Science-Nano, 2018, 5(9): 2117-2128.

[27] Ge X L, Wexler A S, Clegg S L. Atmospheric amines - part I. A review. Atmospheric Environment, 2011, 45(3): 524-546.

[28] Zhang R Y, Khalizov A, Wang L, et al. Nucleation and growth of nanoparticles in the atmosphere. Chemical Reviews, 2012, 112(3): 1957-2011.

[29] Zhang R Y, Wang G H, Guo S, et al. Formation of urban fine particulate matter. Chemical Reviews, 2015, 115(10): 3803-3855.

[30] Xie H B, Li C, He N, et al. Atmospheric chemical reactions of monoethanolamine initiated by OH radical: Mechanistic and kinetic study. Environmental Science & Technology, 2014, 48(3): 1700-1706.

[31] Xie H B, Ma F F, Wang Y F, et al. Quantum chemical study on ·Cl-initiated atmospheric degradation of monoethanolamine. Environmental Science & Technology, 2015, 49(22): 13246-13255.

[32] Xie H B, Elm J, Halonen R, et al. Atmospheric fate of monoethanolamine: Enhancing new particle formation of sulfuric acid as an important removal process. Environmental Science & Technology, 2017, 51(15): 8422-8431.

[33] Ma F F, Ding Z Z, Elm J, et al. Atmospheric oxidation of piperazine initiated by ·Cl: Unexpected high nitrosamine yield. Environmental Science & Technology, 2018, 52(17): 9801-9809.

[34] Finlaysonpitts B J, Pitts J N. Tropospheric air pollution: Ozone, airborne toxics, polycyclic aromatic hydrocarbons, and particles. Science, 1997, 276(5315): 1045-1052.

[35] Landvik N E, Gorria M, Arlt V M, et al. Effects of nitrated-polycyclic aromatic hydrocarbons and diesel exhaust particle extracts on cell signalling related to apoptosis: Possible implications for their mutagenic and carcinogenic effects. Toxicology, 2007, 231(2-3): 159-174.

[36] Jariyasopit N, McIntosh M, Zimmermann K, et al. Novel Nitro-PAH formation from heterogeneous reactions of PAHs with NO_2, NO_3/N_2O_5, and OH radicals: Prediction, laboratory studies, and mutagenicity. Environmental Science & Technology, 2014, 48(1): 412-419.

[37] Tsapakis M, Stephanou E G. Diurnal cycle of PAHS, Nitro-PAHS, and Oxy-PAHS in a high oxidation capacity marine background atmosphere. Environmental Science & Technology, 2007, 41(23): 8011-8017.

[38] Kumar M, Busch D H, Subramaniam B, et al. Organic acids tunably catalyze carbonic acid decomposition. Journal of Physical Chemistry A, 2014, 118(27): 5020-5028.

[39] Da S G. Carboxylic acid catalyzed Keto-Enol tautomerizations in the gas phase. Angewandte Chemie-International Edition, 2010, 49(41): 7523-7525.

[40] Zhang Q Z, Gao R, Xu F, et al. Role of water molecule in the gas-phase formation process of nitrated polycyclic aromatic hydrocarbons in the atmosphere: A computational study. Environmental Science & Technology, 2014, 48(9): 5051-5057.

[41] Zhao N, Qingzhu Z Z, Wang W X. Atmospheric oxidation of phenanthrene initiated by OH radicals in the presence of O_2 and NO_x—A theoretical study. Science of the Total Environment, 2016, 563: 1008-1015.

[42] Huang Z X, Zhang Q Z, Wang W X. Mechanical and kinetic study on gas-phase formation of dinitronaphthalene from 1-and 2-nitronaphthalene. Chemosphere, 2016, 156: 101-110.

[43] Zhang C X, Zhang X, Kang L Y, et al. Adsorption and transformation mechanism of NO_2 on NaCl(100) surface: A density functional theory study. Science of the Total Environment, 2015, 524: 195-200.

[44] Lv G C, Sun X M, Zhang C X, et al. Understanding the catalytic role of oxalic acid in SO_3 hydration to form H_2SO_4 in the atmosphere. Atmospheric Chemistry and Physics, 2019, 19(5): 2833-2844.

[45] Orlando J J, Tyndall G S. Laboratory studies of organic peroxy radical chemistry: An overview with emphasis on recent issues of atmospheric significance. Chemical Society Reviews, 2012, 41(19): 6294-6317.

[46] Jokinen T, Sipila M, Richters S, et al. Rapid autoxidation forms highly oxidized RO_2 radicals in the atmosphere. Angewandte Chemie-International Edition, 2014, 53(52): 14596-14600.

[47] Bianchi F, Trostl J, Junninen H, et al. New particle formation in the free troposphere: A question of chemistry and timing. Science, 2016, 352(6289): 1109-1112.

[48] Praske E, Otkjaer R V, Crounse J D, et al. Atmospheric autoxidation is increasingly important in urban and suburban North America. Proceedings of the National Academy of Sciences of the United States of America, 2018, 115(1): 64-69.

[49] Wu R R, Wang S N, Wang L M. New mechanism for the atmospheric oxidation of dimethyl sulfide. The importance of intramolecular hydrogen shift in a CH_3SCH_2OO radical. Journal of Physical Chemistry A, 2015, 119(1): 112-117.

[50] Wang S N, Wang L M. The atmospheric oxidation of dimethyl, diethyl, and diisopropyl ethers. The role of the intramolecular hydrogen shift in peroxy radicals. Physical Chemistry Chemical Physics, 2016, 18(11): 7707-7714.

[51] Wang S N, Wu R R, Berndt T, et al. Formation of highly oxidized radicals and multifunctional products from the atmospheric oxidation of alkylbenzenes. Environmental Science & Technology, 2017, 51(15): 8442-8449.

[52] Wang S N, Riva M, Yan C, et al. Primary formation of highly oxidized multifunctional products in the ·OH-initiated oxidation of isoprene: A combined theoretical and experimental study. Environmental Science & Technology, 2018, 52(21): 12255-12264.

[53] McFiggans G, Mentel T F, Wildt J, et al. Secondary organic aerosol reduced by mixture of atmospheric vapours. Nature, 2019, 565(7741): 587-593.

[54] Schlesinger R B, Kunzli N, Hidy G M, et al. The health relevance of ambient particulate matter characteristics: Coherence of toxicological and epidemiological inferences. Inhalation Toxicology, 2006, 18(2): 95-125.

[55] Haywood J, Boucher O. Estimates of the direct and indirect radiative forcing due to tropospheric aerosols: A review. Reviews of Geophysics, 2000, 38(4): 513-543.

[56] Kulmala M, Petaja T, Ehn M, et al., Chemistry of atmospheric nucleation: On the recent advances on precursor characterization and atmospheric cluster composition in connection with atmospheric new particle formation. Annual Review of Physical Chemistry, 2014, 65: 21-37.

[57] Zhuang Y, Pavlish J H. Fate of hazardous air pollutants in oxygen-fired coal combustion with different flue gas recycling. Environmental Science & Technology, 2012, 46(8): 4657-4665.

[58] Chen L G, Bhattacharya S. Sulfur emission from victorian brown coal under pyrolysis, oxy-fuel combustion and gasification

conditions. Environmental Science & Technology, 2013, 47(3): 1729-1734.

[59] Cao Y, Zhou H C, Jiang W, et al. Studies of the fate of sulfur trioxide in coal-fired utility boilers based on modified selected condensation methods. Environmental Science & Technology, 2010, 44(9): 3429-3434.

[60] Fleig D, Vainio E, Andersson K, et al. Evaluation of SO_3 measurement techniques in air and oxy-fuel combustion. Energy & Fuels, 2012, 26(9): 5537-5549.

[61] Bray C D, Battye W, Aneja V P, et al. Evaluating ammonia (NH_3) predictions in the noaa national air quality forecast capability (NAQFC) using *in-situ* aircraft and satellite measurements from the CALNEX-2010 campaign. Atmospheric Environment, 2017, 163: 65-76.

[62] Von-Schneidemesser E, Monks P S, Allan J D, et al. Chemistry and the linkages between air quality and climate change. Chemical Reviews, 2015, 115(10): 3856-3897.

[63] Kulmala M, Kontkanen J, Junninen H, et al. Direct observations of atmospheric aerosol nucleation. Science, 2013, 339(6122): 943-946.

[64] Zhang R Y, Suh I, Zhao J, et al. Atmospheric new particle formation enhanced by organic acids. Science, 2004, 304(5676): 1487-1490.

[65] Almeida J, Schobesberger S, Kurten A, et al. Molecular understanding of sulphuric acid-amine particle nucleation in the atmosphere. Nature, 2013, 502(7471): 359.

[66] Ge P, Luo G, Luo Y, et al. A molecular-scale study on the hydration of sulfuric acid-amide complexes and the atmospheric implication. Chemosphere, 2018, 213: 453-462.

[67] Ge P, Luo G, Luo Y, et al. Molecular understanding of the interaction of amino acids with sulfuric acid in the presence of water and the atmospheric implication. Chemosphere, 2018, 210: 215-223.

[68] Comes F J. Recycling in the earths atmosphere - the OH radical - its importance for the chemistry of the atmosphere and the determination of its concentration. Angewandte Chemie-International Edition in English, 1994, 33(18): 1816-1826.

[69] Takeda K, Takedoi H, Yamaji S, et al. Determination of hydroxyl radical photoproduction rates in natural waters. Analytical Sciences, 2004, 20(1): 153-158.

[70] Li C, Wei G L, Chen J W, et al. Aqueous OH radical reaction rate constants for organophosphorus flame retardants and plasticizers: Experimental and modeling studies. Environmental Science & Technology, 2018, 52(5): 2790-2799.

[71] Li C, Zheng S S, Chen J W, et al. Kinetics and mechanism of center dot ·OH-initiated atmospheric oxidation of organophosphorus plasticizers: A computational study on tri-p-cresyl phosphate. Chemosphere, 2018, 201: 557-563.

[72] Li C, Chen J W, Xie H B, et al. Effects of atmospheric water on center dot ·OH-initiated oxidation of organophosphate flame retardants: A DFT investigation on TCPP. Environmental Science & Technology, 2017, 51(9): 5043-5051.

[73] Pettersen R C. The chemistry of solid wood. Wood Science Technology, 1984, 207: 57-126.

[74] Coeur T C, Tomas A, Guilloteau A, et al. Aerosol formation yields from the reaction of catechol with ozone. Atmospheric Environment, 2009, 43(14): 2360-2365.

[75] Derwent R G, Jenkin M E, Saunders S M. Photochemical ozone creation potentials for a large number of reactive hydrocarbons under European conditions. Atmospheric Environment, 1996, 30(2): 181-199.

[76] Payton F, Bose R, Alworth W L, et al. 4-methylcatechol-induced oxidative stress induces intrinsic apoptotic pathway in metastatic melanoma cells. Biochemical Pharmacology, 2011, 81(10): 1211-1218.

[77] Pillar E A, Zhou R X, Guzman M I. Heterogeneous oxidation of catechol. Journal of Physical Chemistry A, 2015, 119(41): 10349-10359.

[78] Kawamura K, Tachibana E, Okuzawa K, et al. High abundances of water-soluble dicarboxylic acids, ketocarboxylic acids and alpha-dicarbonyls in the mountaintop aerosols over the north China plain during wheat burning season. Atmospheric

[79] Li J, Cao H J, Han D D, et al. Computational study on the mechanism and kinetics of ·Cl-initiated oxidation of vinyl acetate. Atmospheric Environment, 2014, 94: 63-73.

[80] Han D D, Cao H J, Li J, et al. Computational study on the mechanisms and rate constants of the ·OH-initiated oxidation of ethyl vinyl ether in atmosphere. Chemosphere, 2014, 111: 61-69.

[81] Cao H J, He M X, Han D D, et al. Theoretical study on the mechanism and kinetics of the reaction of 2,2',4,4'-Tetrabrominated diphenyl ether (BDE-47) with OH radicals. Atmospheric Environment, 2011, 45(8): 1525-1531.

[82] Cao H J, He M X, Han D D, et al. ·OH-initiated oxidation mechanisms and kinetics of 2, 4, 4'-tribrominated diphenyl ether. Environmental Science & Technology, 2013, 47(15): 8238-8247.

[83] Han D D, Li J, Cao H J, et al. Theoretical investigation on the mechanisms and kinetics of ·OH-initiated photooxidation of dimethyl phthalate (DMP) in atmosphere. Chemosphere, 2014, 95: 50-57.

[84] Sun J F, Wei B, Mei Q, et al. Ozonation of 3-methylcatechol and 4-methylcatechol in the atmosphere and aqueous particles: Mechanism, kinetics and ecotoxicity assessment. Chemical Engineering Journal, 2019, 358: 456-466.

[85] Wei B, Sun J F, Mei Q, et al. Theoretical study on gas-phase reactions of nitrate radicals with methoxyphenols: Mechanism, kinetic and toxicity assessment. Environmental Pollution, 2018, 243: 1772-1780.

[86] Sun J F, Mei Q, Wei B, et al. Mechanisms for ozone-initiated removal of biomass burning products from the atmosphere. Environmental Chemistry, 2018, 15(1-2): 83-91.

[87] Wei B, Sun J F, Mei Q, et al. Mechanism and kinetic of nitrate radical-initiated atmospheric reactions of guaiacol (2-methoxyphenol). Computational and Theoretical Chemistry, 2018, 1129: 1-8.

[88] Schwarzenbach R P, Gschwend P M, Imboden D M. Environmental organic chemistry, 2nd ed. New Jersey: John Wiley & Sons, 2003.

[89] Tebes S C, Patel J M, Jones W J, et al. Prediction of hydrolysis products of organic chemicals under environmental pH conditions. Environmental Science & Technology, 2017, 51(9): 5008-5016.

[90] Zhang H Q, Xie H B, Chen J W, et al. Prediction of hydrolysis pathways and kinetics for antibiotics under environmental pH conditions: A quantum chemical study on cephradine. Environmental Science & Technology, 2015, 49(3): 1552-1558.

[91] Xu T, Chen J W, Wang Z, et al. Development of prediction models on base-catalyzed hydrolysis kinetics of phthalate esters with density functional theory calculation. Environmental Science & Technology, 2019, DOI: 10.1021/acs.est.9b00574.

[92] Siefermann K R, Abel B. The hydrated electron: A seemingly familiar chemical and biological transient. Angewandte Chemie-International Edition, 2011, 50(23): 5264-5272.

[93] Li X C, Ma J, Liu G F, et al. Efficient reductive dechlorination of monochloroacetic acid by sulfite/UV process. Environmental Science & Technology, 2012, 46(13): 7342-7349.

[94] Thomas S T E, Blough N V. Photoproduction of hydrated electron from constituents of natural waters. Environmental Science & Technology, 2001, 35(13): 2721-2726.

[95] Li C, Zheng S S, Li T T, et al. Quantitative structure-activity relationship models for predicting reaction rate constants of organic contaminants with hydrated electrons and their mechanistic pathways. Water Research, 2019, 151: 468-477.

[96] Li Y J, Qiao X L, Zhang Y N, et al. Effects of halide ions on photodegradation of sulfonamide antibiotics: Formation of halogenated intermediates. Water Research, 2016, 102: 405-412.

[97] Wang J Q, Chen J W, Qiao X L, et al. DOM from mariculture ponds exhibits higher reactivity on photodegradation of sulfonamide antibiotics than from offshore seawaters. Water Research, 2018, 144: 365-372.

[98] Wang J Q, Chen J W, Qiao X L, et al. Disparate effects of DOM extracted from coastal seawaters and freshwaters on photodegradation of 2, 4-dihydroxybenzophenone. Water Research, 2019, 151: 280-287.

[99] Zhang Y N, Wang J Q, Chen J W, et al. Phototransformation of 2,3-dibromopropyl-2,4,6-tribromophenyl ether (DPTE) in natural waters: Important roles of dissolved organic matter and chloride ion. Environmental Science & Technology, 2018, 52(18): 10490-10499.

[100] 李英杰. 河口水中溶解性物质对磺胺类抗生素光降解行为的影响. 大连: 大连理工大学, 2016.

[101] Zhou C Z, Chen J W, Xie H J, et al. Modeling photodegradation kinetics of organic micropollutants in water bodies: A case of the Yellow River estuary. Journal of Hazardous Materials, 2018, 349: 60-67.

[102] Zhang S Y, Chen J W, Qiao X L, et al. Quantum chemical investigation and experimental verification on the aquatic photochemistry of the sunscreen 2-phenylbenzimidazole-5-sulfonic acid. Environmental Science & Technology, 2010, 44(19): 7484-7490.

[103] Stich T A, Brooks A J, Buan N R, et al. Spectroscopic and computational studies of CO^{3+}-corrinoids: Spectral and electronic properties of the B-12 cofactors and biologically relevant precursors. Journal of the American Chemical Society, 2003, 125(19): 5897-5914.

[104] Liptak M D, Brunold T C. Spectroscopic and computational studies of CO^{1+} cobalamin: Spectral and electronic properties of the "superreduced" B-12 cofactor. Journal of the American Chemical Society, 2006, 128(28): 9144-9156.

[105] Cretnik S, Thoreson K A, Bernstein A, et al. Reductive dechlorination of TCE by chemical model systems in comparison to dehalogenating bacteria: Insights from dual element isotope analysis ($^{13}C/^{12}C$, $^{37}Cl/^{35}Cl$). Environmental Science & Technology, 2013, 47(13): 6855-6863.

[106] Giedyk M, Goliszewska K, Gryko D. Vitamin B-12 catalysed reactions. Chemical Society Reviews, 2015, 44(11): 3391-3404.

[107] Kliegman S, McNeill K. Dechlorination of chloroethylenes by COB(I) alamin and cobalamin model complexes. Dalton Transactions, 2008, (32): 4191-4201.

[108] Ji L, Wang C C, Ji S J, et al. Mechanism of cobalamin-mediated reductive dehalogenation of chloroethylenes. ACS Catalysis, 2017, 7(8): 5294-5307.

[109] Thiel W, Hummer G. Chemistry methods for computational chemistry. Nature, 2013, 504(7478): 96-97.

[110] Tratnyek P G, Bylaska E J, Weber E J. In silico environmental chemical science: Properties and processes from statistical and computational modelling. Environmental Science-Processes & Impacts, 2017, 19(3): 188-202.

[111] Li Y W, Shi X L, Zhang Q Z, et al. Computational evidence for the detoxifying mechanism of epsilon class glutathione transferase toward the insecticide DDT. Environmental Science & Technology, 2014, 48(9): 5008-5016.

[112] Li Y W, Zhang R M, Du L K, et al. Catalytic mechanism of C—F bond cleavage: Insights from QM/MM analysis of fluoroacetate dehalogenase. Catalysis Science & Technology, 2016, 6(1): 73-80.

[113] Wang J J, Tang X W, Li Y W, et al. Computational evidence for the degradation mechanism of haloalkane dehalogenase LinB and mutants of Leu248 to 1-chlorobutane. Physical Chemistry Chemical Physics, 2018, 20(31): 20540-20547.

[114] Wang J J, Chen J F, Tang X W, et al. Catalytic mechanism for 2,3-dihydroxybiphenyl ring cleavage by nonheme extradiol dioxygenases bphc: Insights from QM/MM analysis. Journal of Physical Chemistry B, 2019, 123(10): 2244-2253.

[115] Yang Z Y, Yang S, Yu P Y, et al. Influence of water and enzyme spnf on the dynamics and energetics of the ambimodal [6+4]/[4+2] cycloaddition. Proceedings of the National Academy of Sciences of the United States of America, 2018, 115(5): E848-E855.

[116] Bao L, Liu W, Li Y, et al. Carcinogenic metabolic activation process of naphthalene by the cytochrome P450 enzyme 1b1: A computational study. Chemical research in toxicology, 2019, 32(4): 603-612.

[117] Chen D, Kannan K, Tan H L, et al. Bisphenol analogues other than BPA: Environmental occurrence, human exposure, and toxicity-a review. Environmental Science & Technology, 2016, 50(11): 5438-5453.

[118] Qu G B, Liu A F, Hu L G, et al. Recent advances in the analysis of TBBPA/TBBPS, TBBPA/TBBPS derivatives and their transformation products. TrAC-Trends in Analytical Chemistry, 2016, 83: 14-24.

[119] Jin H B, Zhu J, Chen Z J, et al. Occurrence and partitioning of bisphenol analogues in adults' blood from China. Environmental Science & Technology, 2018, 52(2): 812-820.

[120] Morris S, Allchin C R, Zegers B N, et al. Distributon and fate of HBCD and TBBPA brominated flame retardants in North Sea estuaries and aquatic food webs. Environmental Science & Technology, 2004, 38(21): 5497-5504.

[121] Rochester J R, Bolden A L. Bisphenol S and F: A systematic review and comparison of the hormonal activity of bisphenol a substitutes. Environmental Health Perspectives, 2015, 123(7): 643-650.

[122] Lu L P, Zhan T J, Ma M, et al. Thyroid disruption by bisphenol s analogues via thyroid hormone receptor beta: *In vitro, in vivo*, and molecular dynamics simulation study. Environmental Science & Technology, 2018, 52(11): 6617-6625.

[123] Eladak S, Grisin T, Moison D, et al. A new chapter in the bisphenol a story: Bisphenols and bisphenolf are not safe alternatives to this compound. Fertility and Sterility, 2015, 103(1): 11-21.

[124] Rosch A, Anliker S, Hollender J. How biotransformation influences toxicokinetics of azole fungicides in the aquatic invertebrate gammarus pulex. Environmental Science & Technology, 2016, 50(13): 7175-7188.

[125] Sevrioukova I F, Poulos T L. Structure and mechanism of the complex between cytochrome P4503A4 and Ritonavir. Proceedings of the National Academy of Sciences of the United States of America, 2010, 107(43): 18422-8427.

[126] Zhuang S L, Zhang L L, Zhan T J, et al. Binding specificity determines the cytochrome P4503A4 mediated enantioselective metabolism of metconazole. Journal of Physical Chemistry B, 2018, 122(3): 1176-1184.

[127] Matsubara H, Ohtani H. Rapid and sensitive determination of the conversion of UV-cured acrylic ester resins by pyrolysis-gas chromatography in the presence of an organic alkali. Analytical Sciences, 2007, 23(5): 513-516.

[128] Liu R Z, Lin Y F, Hu F B, et al. Observation of emerging photoinitiator additives in household environment and sewage sludge in China. Environmental Science & Technology, 2016, 50(1): 97-104.

[129] Liu R Z, Mabury S A. First detection of photoinitiators and metabolites in human sera from united states donors. Environmental Science & Technology, 2018, 52(17): 10089-10096.

[130] Reitsma M, Bovee T F H, Peijnenburg A, et al. Endocrine-disrupting effects of thioxanthone photoinitiators. Toxicological Sciences, 2013, 132(1): 64-74.

[131] Zhan T J, Pan L M, Liu Z F, et al. Metabolic susceptibility of 2-chlorothioxanthone and its toxic effects on MRNA and protein expression and activities of human CYP1A2 and CYP3A4 enzymes. Environmental Science & Technology, 2018, 52(20): 11904-11912.

[132] Zhu X W, Liu S S, Ge H L, et al. Comparison between the short-term and the long-term toxicity of six triazine herbicides on photobacteria Q67. Water Research, 2009, 43(6): 1731-1739.

[133] Xu Y Q, Liu S S, Wang Z J, et al. Commercial personal care product mixtures exhibit hormetic concentration-responses to *Vibrio qinghaiensis* sp.-Q67. Ecotoxicology and Environmental Safety, 2018, 162: 304-311.

[134] Liu S S, Zhang J, Zhang Y H, et al. Aptox: Assessment and prediction on toxicity of chemical mixtures. Acta Chimica Sinica, 2012, 70(14): 1511-1517.

[135] Wang Z J, Liu S S, Qu R. Jsfit: A method for the fitting and prediction of J- and S-shaped concentration-response curves. RSC Advances, 2018, 8(12): 6572-6580.

[136] Yang H B, Sun L X, Li W H, et al. *In silico* prediction of chemical toxicity for drug design using machine learning methods and structural alerts. Frontiers in Chemistry, 2018, 6: 12.

[137] Li F X, Fan D F, Wang H, et al. *In silico* prediction of pesticide aquatic toxicity with chemical category approaches. Toxicology Research, 2017, 6(6): 831-842.

[138] Cao Q Q, Liu L, Yang H B, et al. *In silico* estimation of chemical aquatic toxicity on crustaceans using chemical category methods. Environmental Science-Processes & Impacts, 2018, 20(9): 1234-1243.

[139] Liu L, Yang H, Cai Y, et al. *In silico* prediction of chemical aquatic toxicity for marine crustaceans via machine learning. Toxicology Research, 2019, in press. DOI: 10.1039/C8TX00331A.

[140] Fan D F, Yang H B, Li F X, et al. *In silico* prediction of chemical genotoxicity using machine learning methods and structural alerts. Toxicology Research, 2018, 7(2): 211-220.

[141] Jiang C, Yang H, Di P, et al. *In silico* prediction of chemical reproductive toxicity using machine learning. Journal of Applied Toxicology, 2019, in press. DOI: 10.1002/jat.3772.

[142] Di P W Y Y M, Jiang C S, Cai Y C, et al. Prediction of the skin sensitizing potential and potency of compounds via mechanism-based binary and ternary classification models. Toxicology In Vitro, 2019, in press. DOI: 10.1016/j.tiv.2019.01.004.

[143] Yang H, Sun L, Li W, et al. Identification of nontoxic substructures: A new strategy to avoid potential toxicity risk. Toxicological Sciences, 2018, 165(2): 396-407.

[144] Yang H B, Lou C F, Sun L X, et al. Admetsar 2.0: Web-service for prediction and optimization of chemical admet properties. Bioinformatics, 2019, 35(6): 1067-1069.

[145] Zhang G Y, Du S X, Wu K H, et al. Two-dimensional materials research. Science, 2018, 360(6389): 15-18.

[146] Wang Z Y, Zhu W P, Qiu Y, et al. Biological and environmental interactions of emerging two-dimensional nanomaterials. Chemical Society Reviews, 2016, 45(6): 1750-1780.

[147] Hu X G, Li D D, Gao Y, et al. Knowledge gaps between nanotoxicological research and nanomaterial safety. Environment International, 2016, 94: 8-23.

[148] Hu X G, Zhou Q X. Health and ecosystem risks of graphene. Chemical Reviews, 2013, 113(5): 3815-3835.

[149] Ban Z, Zhou Q X, Sun A Q, et al. Screening priority factors determining and predicting the reproductive toxicity of various nanoparticles. Environmental Science & Technology, 2018, 52(17): 9666-9676.

[150] Yang X, Liu X, Zhang A, et al. Distinguishing the sources of silica nanoparticles by dual isotopic fingerprinting and machine learning. Nature Communcations, 2019, 10(1): 1620.

作者：陈景文[1]，王中钰[1]，肖子君[1]，王　雅[1]，马芳芳[1]，谢宏彬[1]，张庆竹[2]，徐　菲[2]，孙孝敏[2]，王赛男[3]，王黎明[3]，罗　一[1]，李　超[4]，何茂霞[2]，徐　童[1]，王杰琼[1]，季　力[5]，李延伟[2]，庄树林[5]，刘树深[6]，唐　赞[7]，胡献刚[8]

[1]大连理工大学，[2]山东大学，[3]华南理工大学，[4]东北师范大学，[5]浙江大学，[6]同济大学，[7]华东理工大学，[8]南开大学

第17章　新型环境污染物的生态毒理研究进展

▶ 1. 全氟化合物的生态毒理研究 /445

▶ 2. 新型全氟聚醚羧酸的生态毒理研究 /449

▶ 3. 氯化石蜡的生态毒理研究 /457

▶ 4. 典型有机污染物的内分泌干扰和神经发育毒性效应 /467

本章导读

本章针对全氟及多氟烷基化合物（PFASs）、新型 PFASs 替代品全氟聚醚羧酸（PFECAs）、氯化石蜡（CPs）、新型溴代阻燃剂（如 PBDEs）和有机磷阻燃剂（OPFRs）等典型新型热点污染物，介绍了其环境分布、生物累积和毒性效应，重点综述了毒性效应与机理研究进展，并对未来的研究方向进行了讨论和展望。

关键词

全氟及多氟烷基化合物，全氟聚醚羧酸，氯化石蜡，溴代阻燃剂，有机磷阻燃剂，毒性效应

1 全氟化合物的生态毒理研究

全氟及多氟烷基化合物（per- and polyfluoroalkyl substances，PFASs）是指脂肪烃碳链上连接的氢原子被氟原子全部或部分取代的一类有机化合物，包含 $C_nF_{2n+1}^-$ 基团。根据官能团的不同，主要分为全氟烷基羧酸类（perfluoroalkyl carboxylic acids，PFCAs）、全氟烷基磺酸类（perfluoroalkyl sulfonic acids，PFSAs）、氟化调聚醇（fluorotelomer alcohols，FTOHs）等。其中，具有八个碳原子的全氟辛基羧酸（PFOA）和全氟辛基磺酸（PFOS）的应用最广，也是其他全氟化合物在环境中的转化产物。由于全氟化合物具有优良的热稳定性、化学稳定性及高表面活性及其大规模的生产使用[1]，已经成为全球性的环境污染物，在水、大气、土壤、沉积物等不同的环境介质、野生动物及人体内均检测到不同浓度的 PFASs[2]。生物毒理学研究发现 PFASs 会影响生物体遗传、生殖、内分泌等，部分 PFASs 甚至具有疑似致癌性[3]。因此，2009 年 5 月 PFOS 及其相关产品被《斯德哥尔摩公约》第四次缔约方大会列为新增的持久性有机污染物[1]。在国际上一些大型厂商基本停止 PFASs 生产的情况下，我国的产量仍然呈现缓慢增加的趋势[4]。

由于 PFASs 具有疏水疏油的特点，导致其主要与蛋白质结合，易在动物体的血液和肝脏中蓄积，影响机体的免疫机制、生殖和神经发育系统，造成生物毒性效应。PFCAs 和 PFSAs 在生物体内极难降解，具有较长的半衰期，其半衰期与碳链长度呈正相关。PFASs 在环境中的暴露浓度和时间会影响其对生物体的毒性效应，因此，尽管生物体内的 PFASs 浓度较低，但随着暴露时间延长，PFASs 所引起的毒性效应不容忽视。

1.1 全氟化合物的生物富集和人体暴露

1.1.1 全氟化合物的生物富集

PFASs 作为一种新型持久性有机污染物已经被证实具有生物富集作用，研究主要包括对一些典型区域的野外调查以及代表性物种的实验室模拟实验，并关注 PFASs 化学结构特征如碳链长度、异构体、物种差异等对生物富集效应的影响。Zhong 等[5]研究了全氟烷基酸类物质在锦鲤体内的生物富集机制，发现 PFSAs 和 PFCAs 的吸收速率常数（K_u）都随碳链的增长而增大，然而清除速率常数（K_e）表现出相反的趋势，并且 PFCAs 和 PFSAs 的 BAF 值都与它们的链长成正相关，说明长链 PFSAs 和 PFCAs 更容易在锦鲤体内富集。

PFASs 在生产过程中会产生空间异构体，它们的物理化学性质如 pK_a、溶解度、极性、表面活性剂等

的微小差异导致其不同的环境行为以及在生物富集过程中发生选择性吸收、排泄或降解。Chen 等[6]通过体内实验与体外模拟相结合的方法，研究了全氟磺酰胺（PFOSA）在锦鲤体内生物转化的异构体选择性，结果显示支链 PFOSA 更容易转化为支链 PFOS。而支链 PFOSA 比其直链异构体具有更快的排出速率，导致直链 PFOSA 在锦鲤体内的比例升高。而支链 PFOSA 更高的转化率使锦鲤体内支链 PFOS 的比例增大（图 17-1），这有助于解释在野生动物中观测到的较高支链 PFOS 异构体比例。进一步的与锦鲤肝脏和肾脏的体外模拟实验表明，代谢主要发生在锦鲤肝脏而不是肾脏中。

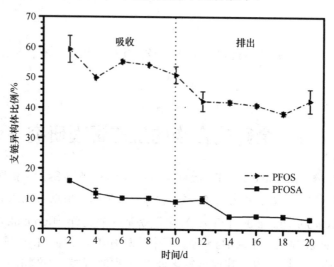

图 17-1　支链 PFOSA 和支链 PFOS 在全鱼体内的比例趋势

研究表明 PFOS 等不仅可以在生物体中富集，还可以通过食物链传递并放大。Fang 等[7]通过对太湖食物网的研究发现，对于 PFOA、PFOS 和 PFOSA 来说，其直链异构体比支链更容易在生物体内富集，导致其直链异构体的比例较高。PFOS 和长碳链的 PFCAs，如全氟癸烷羧酸（PFDA）、全氟十二烷羧酸（PFDoA），能在淡水系统食物链上发生营养等级放大作用，其营养等级放大系数（trophic magnification factor，TMF）分别为 2.43、2.68 和 3.46。

由于施用污泥、污水灌溉等，土壤中也存在一定程度的 PFASs 污染。Zhao 等[8]研究了 PFOSA 在土壤中的生物降解以及在蚯蚓和小麦体内的吸收和代谢。结果显示，PFOSA 可以在土壤微生物的作用下降解为 PFOS。PFOSA 可以被小麦根和蚯蚓从土壤中富集，并且富集能力比 PFOS 强。虽然蚯蚓和小麦都可以将 PFOSA 降解为 PFOS，但是在小麦体内还发现了包括全氟己基磺酸（PFHxS）和全氟丁基磺酸（PFBS）在内的一些短链 PFSAs，这些物质在蚯蚓和土壤中并没有检出，说明小麦降解 PFOSA 的机制与蚯蚓和微生物不同。

1.1.2　全氟化合物的人体暴露

PFASs 在多种人体组织中被检出，比如血液、尿液、胎盘和羊水等，食物、灰尘（或空气）、饮用水等是 PFASs 人体暴露的主要途径。许多研究显示 PFASs 对人体特别是儿童造成了潜在的健康风险，比如干扰甲状腺激素、产生免疫毒性、提高血清脂联素浓度等。婴儿由于器官发育和防御机能不完整，往往对污染物更加敏感，受到的损伤也更大。特别是孕妇在孕期暴露于 PFASs 时，会对胎儿生长和发育产生不良后果。

多项研究发现 PFASs 可以跨越胎盘屏障，通过脐带血由母亲转移给胎儿，因此母亲暴露于 PFASs 会对胎儿的正常生长发育产生潜在危害[9]。对于 PFOS 和 PFOA 异构体，支链异构体的转移比其直链异构体更有效（图 17-2）。Zhang 等[10]通过计算肾清除率（clrenal）发现，PFHxS 和 PFOS 的肾清除率比全氟庚基羧酸（PFHpA）、PFOA 和全氟壬酸（PFNA）低，说明与相同氟化碳链长度的 PFCAs 相比，PFSAs 通过尿排泄的效率更低，且支链 PFOS 和 PFOA 异构体比相应的直链异构体更易排出。分子对接模型结果显示，直链 PFOS 与人血清蛋白（HSA）的结合能力要强于支链 PFOS，这可能是其难以排出的原因。尿排泄是短链 PFCAs（$C_{\leqslant 8}$）的主要清除途径，但是对于长链 PFCAs、PFOS 和 PFHxS 来说，可能还有其他重要的排泄途径起作用。Shi 等[11]发现氯代多氟烷基醚磺酸（Cl-PFESAs）在鱼类消费者和金属电镀工人血清中含量比在对照组中高。含有 8 个碳的 Cl-PFESA 是在人体内最具生物持久性的 PFAS，其肾清除半衰期和总清除半衰期相差 20 倍，说明 C_8 Cl-PFES 存在其他比尿排泄更重要的清除途径。Gao 等[12]研究发现室内灰尘和总悬浮颗粒物是氟化物制造厂工人最重要的暴露途径。Beesoon 等[13]也通过实验证实了直链 PFOS 比支链 PFOS 结合 HSA 的能力更强。这解释了与 PFOA 相比，PFOS 在人体内具有更长的生物半衰期、更高的胎盘转移率和支链异构体更高的肾清除率。Shan 等[14]的调查研究发现，食物是成年人暴露 PFASs 的主要途径。结合 PFASs 在人体中的肾清除率，通过一室模型预测的直链 PFOA 在人体血清中的比值（预测值：98.2%）与实测值（99.7%）接近但略低，说明人体中 PFOA 主要来源于直接暴露，但不排除少量的间接来源。对于 PFOS 来说，模型预测的直链异构体在人血清中所占百分比（69.3%）高于实测值（59.2%），说明 PFOS 除了直接暴露来源外，还存在一定的间接暴露，即人体暴露前体物质，这些前体物质在人体内代谢转化为 PFOS。

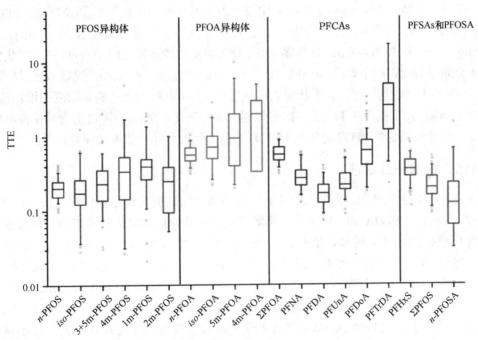

图 17-2 不同链长的 PFCAs、PFSAs 和 PFOSA 的 TTE 值分布

1.2 全氟化合物的毒性效应与机制

1.2.1 全氟化合物对斑马鱼早期的发育毒性

大量研究表明 PFASs 对斑马鱼的早期发育具有一定的影响。Shi 等[15]研究发现 PFOS（1 mg/L、3 mg/L 和 5 mg/L）可导致斑马鱼的孵化延迟，孵化率和幼鱼的存活率显著下降，幼鱼出现严重的畸形包括外胚层畸形、色素减退、卵黄囊水肿、尾部和心脏畸形以及脊柱弯曲。3 mg/L 和 5 mg/L 的 PFOS 会显著抑制

斑马鱼的生长发育，导致体长显著降低。Zheng 等[16]报道 PFOA 对斑马鱼胚胎具有急性毒性，在 48 hpf 时出现水肿和一系列畸形，其 96 hpf 脊柱弯曲畸形的 EC_{50} 是 198 mg/L。在 200 mg/L 或者更高浓度可致斑马鱼的幼鱼孵化延迟和脊柱弯曲在 72 hpf。PFNA 延迟孵化且不透明胚胎的比例与浓度呈正比，在 8 hpf 时一些胚胎开始停止分裂并变得不透明，在 72 hpf 时 LC_{50} 为 84 mg/L。PFOS 在 24 hpf、48 hpf 和 72 hpf 时的 LC_{50} 值分别为 69 mg/L、68 mg/L 和 68 mg/L，在 72 hpf 时出现脊柱弯曲。Zhang 等[17]研究了 PFDoA 对斑马鱼幼鱼的发育毒性，PFDoA（1.2 mg/L 和 6.0 mg/L）使斑马鱼幼鱼体长有所降低，心跳和孵化率随着暴露浓度的增加而降低，畸形率和死亡率则显著增加。

1.2.2 全氟化合物对斑马鱼的甲状腺毒性

全氟化合物作为内分泌干扰污染物，对生物的甲状腺具有一定的干扰作用。斑马鱼胚胎从 2 hpf 暴露在不同浓度的 PFDoA（0 mg/L、0.24 mg/L、1.2 mg/L 和 6 mg/L）直到 96 hpf，在暴露浓度为 6 mg/L 时，甲状腺激素 T_3 和 T_4 的浓度与空白相比显著下降。下丘脑-垂体-甲状腺（HPT）轴上的有关甲状腺激素合成、管理和活性的基因都发生了变化，这些激素和基因的变化说明了 PFDoA 对斑马鱼的甲状腺具有干扰作用[18]。斑马鱼胚胎经过 PFOA 和 PFBA 的亚慢性（0~6 dpf）和慢性（0~28 dpf）暴露，在 6 dpf 时，PFOA 和 PFBA 均可导致后鱼鳔变小且甲状腺过氧化物酶基因 tpo 以及编码鱼鳔表面活性蛋白的基因 sp-a 和 sp-c 的表达均上升。暴露 28 dpf 时，50%的鱼中没有前鱼鳔并且体型也较小。总的来说，PFOA 和 PFBA 能通过鱼鳔表面活性系统产生甲状腺破坏[18]。也有研究报道了斑马鱼胚胎暴露在不同浓度的 PFOS（0 ug/L、100 ug/L、200 ug/L 和 400 ug/L）暴露下，促肾上腺皮质激素释放因子（crf）和促甲状腺激素（tsh）基因表达水平分别显著上调和下调，钠/碘同向转运酶（nis）和脱碘酶 I（dio1）基因表达显著上调，而甲状腺球蛋白（tg）基因的表达则下调，甲状腺素运载蛋白（ttr）基因的表达随着浓度的增加而下调，甲状腺受体 α 和 β（trα 和 trβ）基因在 PFOS 暴露后分别上调和下调；全鱼体中 T_4 的含量保持不变，但 T_3 的水平显著增加。这些结果说明 PFOS 能干扰 HPT 轴上的基因表达并产生甲状腺毒性效应[19]。

1.2.3 全氟化合物的肝脏毒性

全氟化合物容易在生物体的肝脏富集，从而引起学者们对其肝脏毒性的广泛关注。有研究表明一些 PFASs（PFOA、PFNA、PFDA 和 PFOS）能够增强虹鳟鱼肝脏内的氧化应激能力，诱导脂质过氧化反应，从而引起肝脏损伤和肿大现象，增强肝脏肿瘤的发生[20]。Cheng 等[21]将 8 hpf 的斑马鱼胚胎在 0.5 μmol/L 的 PFOS 溶液中暴露 5 个月，发现 PFOS 可通过干扰脂质的合成，脂肪酸氧化和排泄低/极低密度脂蛋白等途径诱导斑马鱼肝脏脂肪变性。通过口服强饲法将雌性 Harlan Sprague-Dawley 大鼠每天暴露于 0~2.0 mg PFDA/kg，持续 28 天，雌性 B6C3F1/N 小鼠一次/周暴露 0~5.0 mg PFDA/kg，持续 4 周，结果发现在用 0.5 mg PFDA/(kg·d) 处理的大鼠中观察到肝细胞坏死和肝肿大；在小鼠中，暴露于 0.625 mg PFDA/(kg·周)后观察到肝肿大（26%~89%），同时在 5.0 mg PFDA/(kg·周)时观察到脾脏萎缩（20%），这些现象均表明 PFDA 慢性长期暴露可导致肝脏毒性[22]。

1.2.4 全氟化合物的生殖毒性

通过口服强饲法给予成年小鼠 0、0.5 mg/(kg·d) 和 10 mg/(kg·d) 的 PFOS 五周，在 10 mg/kg 组中血清睾酮水平显著降低，精子数量下降，在 0.5 mg/kg 和 10 mg/kg 组中观察到睾丸雌激素受体（ER）水平的增加，而仅在 10 mg/kg 组中观察到 ER 表达的减少，这些结果均说明了 PFOS 能诱导睾丸毒性，改变睾丸 ERs 表达，减少生殖细胞的增殖和增加细胞的凋亡[23]。另外，也有许多研究发现 PFASs 能作为激素受体激动剂导致睾酮水平减少。流行病学资料显示 PFOS 暴露引起的雄性生殖力下降可能与精子数量减少有关[24]。

1.2.5 与蛋白质相互作用

研究表明 PFASs 更容易富集在生物体血液、肝脏和肾脏等蛋白含量丰富的组织中，PFASs 与蛋白质的相互作用取决于 PFASs 的化学性质，其疏水性的全氟碳链及酸性的羧酸或磺酸官能团与脂肪酸的结构很类似，能够促进与蛋白的疏水作用和离子相互作用，因此 PFASs 能与多种脂肪酸运输蛋白或结合蛋白受体相互作用。Chen 等[25]通过紫外光谱法发现 PFASs 与牛血清蛋白（BSA）的相互作用导致 BSA 的构象改变并且 BSA 络氨酸残基和色氨酸残基周围微环境的极性发生变化。利用荧光指示剂发现 PFOS 在低浓度时，首要的结合位点是药物位点Ⅱ，而在高浓度时，药物位点Ⅱ达到饱和，PFOS 将会在药物位点Ⅰ有很强的结合。祝凌燕课题组通过荧光光谱法研究 PFASs 与血清白蛋白（BSA 和 HSA）的结合能力，利用荧光探针探究 PFASs 与血清白蛋白的结合位点，结果发现 PFASs 能与血清白蛋白疏水槽上的氨基酸残基发生疏水作用，从而导致血清白蛋白静态猝灭。PFSAs 倾向于在血清白蛋白的色氨酸位点上结合，而 PFCAs 则更倾向于在药物位点上结合。PFCAs 与 BSA 在三个位点上的结合常数之和随着碳链长度的增加而增大，而与 HSA 的结合常数则没有这样的变化趋势。PFASs 会导致蛋白的二级结构和三级结构发生变化，从而可能对蛋白的各种生命活动造成影响。Ren 等[26]利用荧光竞争结合实验调查了 16 个结构不同 PFASs 与人类甲状腺激素受体（TR）相互作用，发现大多数 PFASs 都能与 TR 结合，与 T_3 相比相对结合能为 0.0003~0.05，另外也观察到氟化烷基链长度大于 10 且带有酸官能团的 PFASs 与 TR 的结合最佳。分子对接分析显示，与 T_3 类似，大多数测试的 PFASs 有效地融入 TR 中的 T_3 结合口袋，并与精氨酸 228 形成氢键，这些结果均表明 PFASs 能通过直接结合 TR 来破坏 TR 通路的正常活性。Weiss 等[27]利用放射性配体结合测定实验测试了 24 种 PFASs 以及 6 种结构相似的天然脂肪酸与甲状腺运载蛋白（TTR）的结合能力，结合能依次是 PFHxS > PFOS/PFOA > PFHpA > 全氟-1-辛基亚磺酸钠 > PFNA，且它们与 TTR 的结合能比天然配体 T_4 低 12.5~50 倍。这些蛋白结合研究对认识 PFASs 的组织分布、毒性作用和机理提供了重要的信息。

1.3 展望

作为新型持久性有机污染物，PFASs 可以通过生物富集和生物放大效应在生物体内积累，积累能力受化合物理化性质、生物体自身生活习性及代谢能力、环境条件等多种因素影响。目前对于 PFASs 的研究主要集中在 PFOS 和 PFOA，对其异构体以及全氟化合物前体的研究较少。对于这些物质的研究很大程度上受限于样品前处理技术和仪器分析方法的不完善。所以我们要利用联用技术、高分辨色谱-质谱等技术实现分析方法的完善和优化，为全氟化合物的生态毒性研究提供技术支持。另外，对全氟化合物积累机制的研究还比较缺乏，亟需利用代谢组学、蛋白组学等多种组学的方法从分子水平阐明其生物积累效应与机理。

2 新型全氟聚醚羧酸的生态毒理研究

为减少并最终禁用 PFOS、PFOA 等长链 PFASs，国内外工业界已展开相关替代品研究并取得了实质进展。目前，PFASs 的主要替代策略包括："短链"替代"长链"，如使用全氟碳链更短的 PFBA、PFHxA、PFBS 等；"多氟"替代"全氟"，如使用全氟聚醚（PFPEs），以插入氧杂原子的方式减少长链全氟碳链的"有效长度"[28]。全氟聚醚羧酸（PFECAs）即是一类以后者为替代思路，在聚合物加工和织物整理等领域替代 PFOA 的新型 PFASs。目前已知的 PFECAs 包括 3M 公司生产的 4, 8-二氧杂-3-氢-全氟壬酸（商品名 ADONA）、Asahi 公司生产的 EEA、科慕（杜邦）公司的六氟环氧丙烷（HFPO）二聚体羧酸（HFPO-DA，商品名 GenX）、我国东岳公司的三聚体、四聚体羧酸（HFPO-TA、HFPO-TeA）[28, 29]。此外，环境化学家利用高分辨质谱进行非靶向筛查，在氟化学工业园区下游水体鉴定出多种新型 PFECAs[30]，分子结构亦见图 17-3。部分化合物如 PFMOAA、PFO2HxA 和 PFO3OA，已在美国河流（Cape Fear River）及周边水源水中检出，但因缺少标准品无法准确定量，通过比较峰面积，估算 PFMOAA、PFO2HxA 和 PFO3OA 的含量是 HFPO-DA 的 2~113 倍，暗示上述三种物质可能是该地区含量更高、更为重要的新型替代品[30]。

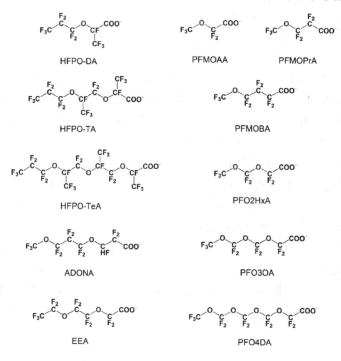

图 17-3　已知 PFECAs 的分子结构（醚氧原子用红色标记）

总之，PFECAs 因在烷基骨架不同位置含有氧杂原子，且部分物质存在支链，结构相比全氟羧酸趋于复杂。本节综述了以 HFPO-TA 为代表的聚醚羧酸替代品的环境分布、生物累积、人群暴露水平及毒性效应，围绕目前存在的问题及未来的研究方向进行了讨论和展望，以期为 PFASs 替代品的环境污染及风险评估提供科学依据。

2.1　PFECAs 替代品的分析方法及环境分布

我们首先利用精馏、碱解、重结晶等方法，从工业级原料中分离并纯化 HFPO-TA，获得高纯度标准品。随后，利用液相色谱-四极杆-飞行时间串联质谱（LC-QTOFMS），通过比较精确分子量和二级碎片峰，定性鉴定出 HFPO-TA 的分子结构，并发现其在我国自然水体、野生鲤鱼体内赋存（图 17-4）；最终基于超高效液相色谱串联三重四极杆质谱（UHPLC-MS/MS），建立了可在不同环境或生物介质中（自然水体、血清、生物组织等）定量检测 33 种传统 PFASs 和 HFPO-DA、HFPO-TA 等新型替代品的分析方法。该方法在各介质中回收率稳定（73%~109%）、灵敏度高（定量限 0.01~0.10 ng/mL），可充分满足 PFASs 替代品的检测[31]。

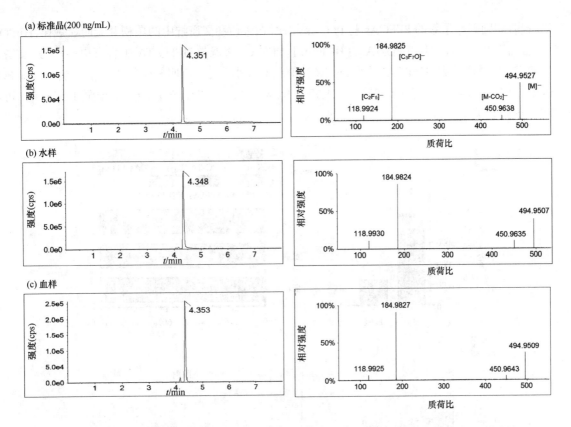

图 17-4　在标准品（a）、小清河水（b）、鲤鱼血液样品（c）的提取离子流图和二级碎片峰
（仪器：Shimadzu LC-20A 串联 AB SCIEX X-500R Q-TOF，ESI- Mode）

小清河自西向东经山东济南、淄博、东营等市流入莱州湾，其中，位于河段中部的山东东岳化工有限公司是亚洲最大的氟聚物生产商。通过采集并检测小清河（图 17-5）及野生鲤鱼（图 17-6）、黑斑蛙体内的 PFASs，我们发现污染点源附近水体中 HFPO-TA 浓度高达 68500 ng/L，经下游稀释后浓度均值 7107 ng/L，占所有检出 PFASs 总浓度的 24%±12%，是仅次于 PFOA（60%±18%）的第二大污染物[31]。下游采集的野生鲤鱼血清中 HFPO-TA 平均浓度为 1527 ng/mL，其生物累积因子 BCF = 186，高于 PFOA 的生物累积因子（BCF = 89）[31]，HFPO-TA 在黑斑蛙体内 $BCF_{whole\ body}$ 值亦显著高于 PFOA[32]。进一步研究 HFPO-TA 在鲤鱼、黑斑蛙体内的组织分布，发现 HFPO-TA 比 PFOA 更易于在肝脏等器官累积[31, 32]。

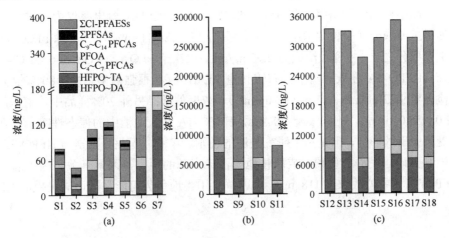

图 17-5　HFPO-TA 等在小清河上游（a）、点源附近（b）、下游（c）的分布组成（根据参考文献[31]修改）

结果表明：相同碳原子数的 PFECAs 与 PFCAs 相比，在鲤鱼和黑斑蛙中的累积并无明显规律，HFPO-DA 的 BCF 显著高于 PFHxA，但 HFPO-TA 的 BCF 低于 PFNA，可能与 HFPO-TA 中含有更多—CF_3 支链，亲水性增加有关。聚醚羧酸中，HFPO-DA 的 log BCF 值相对较小（<1），表明其生物蓄积能力弱，但 HFPO-TA 的 log BCF（2.18 ± 0.44）显著高于 PFOA（1.93 ± 0.34），暗示其生物蓄积能力甚至强于 PFOA，可能与蛋白亲和力更强有关[31, 32]。

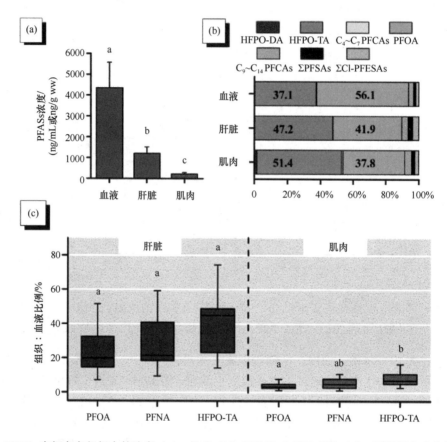

图 17-6　PFASs 在鲤鱼各组织中的浓度（a）、组成（b）及脏器/血液浓度比（c）（图根据参考文献[31]修改）

东岳氟化学工厂周边（8 km 内）居民血清中 HFPO-TA 浓度分布特征见图 17-7。结果显示，HFPO-TA 在几乎全部血清样品中检出（检出率 > 97.9%），但短链的 HFPO-DA 检出率仅为 16.7%~39.6%。PFOA 是当地居民中的最主要成分（中位数浓度），占全部已测 PFASs 的 86% ± 9%。除 PFOA 以外，HFPO-TA 亦是人血清中主要成分之一，中位数浓度为 2.93 ng/mL，仅次于 PFOA、PFOS 和 6∶2 Cl-PFESA（F-53B）排名第四。另外，HFPO-TA 在人血清中分布较其他 PFASs 存在很强的偏态性：80%的样本中 HFPO-TA 浓度分布在未检出至 9.23 ng/mL（均值 2.63 ng/mL），浓度偏低；另外 20%样品中均值浓度 36.8 ng/mL，分布范围为 12.0~55.0 ng/mL。相较其他 PFASs，HFPO-TA 在人血清中的偏态性暗示其在当地居民中的暴露水平可能受到因素的影响，但是，本研究关注的性别、年龄、职业等因素与 HFPO-TA 浓度并无显著性关联。这可能同当地居民的食鱼频率或居住地与污染点源的距离等因素有关。因小清河鲤鱼肌肉中含有高浓度的 HFPO-TA（中位数浓度 118 ng/g ww），食用当地鱼类可能是周边居民摄取 HFPO-TA 的潜在因素[31]。

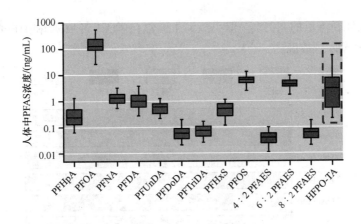

图 17-7　HFPO-TA 等替代品在工厂周边区域人群中分布（图根据参考文献[31]修改）

2.2　新型全氟替代品在全球部分自然水体中的分布特征

在上述工作基础上，进一步研究了中国、美国、英国、瑞典、德国、荷兰、韩国共 7 个国家主要河流或湖泊表层水样 160 份，探讨了新型 PFASs 的分布特征，并通过分析各国水体中 PFASs 结构组成上的异同了解国内外 PFASs 替代品的使用现状和差异[33]。

研究发现 $C_4 \sim C_9$ PFCAs、PFBS、PFHxS 和 PFOS 在所有水样中均有检出，$C_{10} \sim C_{14}$ PFCAs 检出率为 10%~98%。我国巢湖 ΣPFCA+ΣPFSA 均值浓度最高（263 ng/L），其次为太湖、泰晤士河、莱茵河等，瑞典梅拉伦湖浓度最低（16.6 ng/L）（图 17-8）。国外水体短链 PFCAs（$C_4 \sim C_7$）所占比例已显著高于 PFOA。另外，短链 PFBS 在西方国家（6%~34%）水体中所占比重显著高于我国（1%~11%）。此外，我们发现 HFPO-DA、HFPO-TA 和 ADONA 3 种 PFECAs 都有检出，其中 HFPO-DA 在全部水样中的检出率高达 96%，各水体中平均浓度范围为 0.7 ng/L（长江）~14 ng/L（太湖）。HFPO-TA 检出率为 83%，平均浓度范围为 0.1 ng/L（泰晤士河）~5 ng/L（太湖）。ADONA 仅在莱茵河 15 份水样中检出[33]。

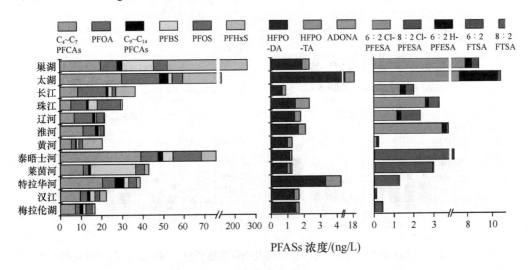

图 17-8　PFASs 在国内外水体中的均值浓度（引自参考文献[33]）

进一步结合水体 PFASs 浓度和年径流量估算出我国河流 PFASs 入海通量（图 17-9）。结果表明，绝大多数的 PFASs 经长江、珠江排放入海，其中长江是绝大多数 PFASs 的主要排放水体（占入海总量的 64%~98%），而珠江和小清河分别为 HFPO-DA（65%）和 HFPO-TA（77%）的主要排放河。尽管小清河的径流量极小，仅为长江径流量的 0.07%，但小清河上游氟化学工厂 PFOA 和 HFPO-TA 的大量排放使之成为我国最为重要的 PFASs 排放源之一[33]。

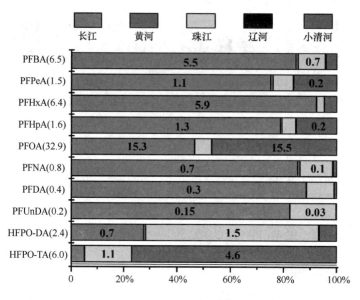

图 17-9 我国主要河流 PFASs 年入海通量及河流相对贡献（t/a）（引自参考文献[33]）

2.3 新型 PFECAs 替代品的肝脏毒性效应与机制

低浓度 HFPO-TA 0.02 mg/(kg·d)和 0.1 mg/(kg·d)暴露 28 天后，小鼠血清中 HFPO-TA 水平与污染区野生鲤鱼血清中含量一致。最低剂量 HFPO-TA 暴露导致小鼠相对肝重增加、肝脏肿大、出现急性肝损伤（图 17-10），并通过激活 PPAR 通路干扰脂质代谢。而 PFOA 在 1.25 mg/(kg·d)浓度下不引起肝损伤；HFPO-TA 肝脏毒性强于 PFOA[34]。

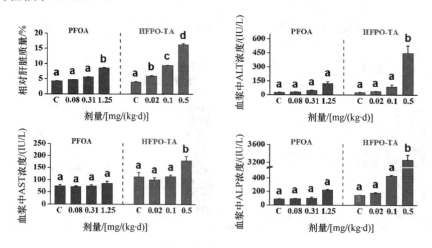

图 17-10 比较 PFOA 和 HFPO-TA 对小鼠脏器及生化指标影响（图根据参考文献[34]修改）

PFOA 和 HFPO-TA 暴露 HL7702 细胞 24 h 后对细胞活力抑制率的浓度-效应曲线（CRC）、拟合模型和相关参数见图 17-11。PFOA 和 PFOS 对 HL7702 细胞活力出现低浓度促进高浓度抑制现象，该现象与本研究组已发表研究一致。HFPO-TA 则表现为 S 型浓度-效应曲线。HFPO-TA 对细胞活力抑制效应的 IC_{50} 较 PFOA 更低，表明其肝细胞毒性更强[35]。

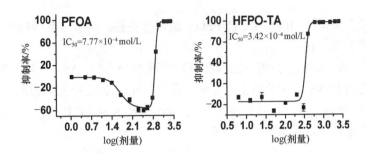

图 17-11　PFOA 和 HFPO-TA 暴露 HL7702 细胞 24 h 后细胞活力抑制效果后
拟合所得浓度-效应关系曲线及 IC_{50} 值（引自参考文献[35]）

尽管本研究中未发现肝脏肿瘤的产生，PPAR 通路、化学品致癌通路和/或肿瘤生成相关基因/蛋白的显著变化暗示 HFPO-TA 或具有致癌潜能。我们通过 Western Blotting 对肝脏中癌症相关蛋白表达水平进行检测，发现 0.1 mg/(kg·d)和 0.5 mg/(kg·d)暴露组小鼠肝脏 AFP 水平较对照组显著升高。另外，Sirt1 水平较对照组显著上调，p21、MTA2 在三个剂量组小鼠肝脏中均有不同程度下调。C-MYC 和 MDM4 水平在 0.1 mg/(kg·d)和 0.5 mg/(kg·d)暴露组小鼠肝脏中显著上调。抑癌因子 p53 蛋白表达水平较对照组无明显改变。暗示 HFPO-TA 暴露可导致细胞非正常增殖，进而造成肝脏肿瘤的发生。基于上述结果，本研究认为 HFPO-TA 并非是安全合适的 PFOA 替代品[34]。

2.4　新型 PFECAs 替代品同蛋白质相互作用

我们的前期研究发现 PFOS 和 PFOA 等可结合人肝脏型脂肪酸结合蛋白（hL-FABP），随着碳链长度的增长，其结合能力增强。PFOS 和 PFOA 以中等强度结合 hL-FABP，其结合摩尔比为 2：1。我们进一步通过多种技术研究了几种 PFOA 替代品与 hL-FABP 及其他蛋白的相互作用。

hL-FABP 蛋白与不同 PFAS 亲和力由解离常数表征，通过竞争取代与蛋白结合的 1,8-ANS 探针，计算出 PFAS 与蛋白解离常数。HFPO-TA 解离常数最低，HFPO-DA 的解离常数则高于 PFOA，表明 HFPO-DA 与蛋白的结合能力较 PFOA 明显减弱，而 HFPO-TA 与蛋白结合能力则强于 PFOA，提示 HFPO-TA 或具有更强的肝脏毒性。骨架中插入 O 原子的 HFPO 同系物在与蛋白结合过程中并非如 PFOA 一般以直链形式结合，相反在 O 原子处发生一定程度的结构扭转，其结合模式亦随之发生改变，与更多氨基酸残基发生相互作用。HFPO-DA 羧酸头部除 R122 残基外，亦可与 S124 和 S39 产生氢键结合，FPO-TA 头部同样与上述氨基酸作用，其尾部与 F50 残基间发生疏水作用[35]（图 17-12）。

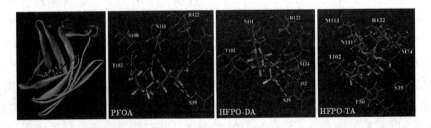

图 17-12　PFOA、HFPO-DA 和 HFPO-TA 与 hL-FABP 蛋白结合模式（引自参考文献[35]）

平衡透析实验是目前公认的测定小分子与蛋白结合参数的标准方法，进一步研究了 PFASs 同血清白蛋白（HSA）的结合能力。研究表明：PFOA 与 HSA 结合位点数为 5.471，HFPO-TA 为 3.18（图 17-13）。HAS 可能的结合位点有 7 个，平衡透析实验测定的亲和力为 HSA 所有位点共同作用的亲和力，K_d 值越小，则亲和力越大。HFPO-TA 与 HSA 的解离常数为 7.6×10^{-5} mol/L，较 PFOA 更小，表明其与蛋白亲和力更高。结合其他实验结果说明 HFPO-TA 较 PFOA 疏水性更强,可能与其碳链中插入 O 原子及 CF_3 支链的存在,导致分子体积变大有关。

HSA 与 PFASs 孵育后,经胰蛋白酶有限酶解后,发现 HSA 同 PFASs 结合后能部分抵抗胰蛋白酶水解作用,且其保护作用随 PFAS 含量的增加而增强。此外,比较不同 PFAS 发现,HFPO-TA 较 PFOA 对蛋白保护效应更强(图 17-14)。进一步通过质谱分子量检测及 N 末端测序对上述核心片段进行分析,序列比对后,发现该片段包含位点 3 及部分位点 6。

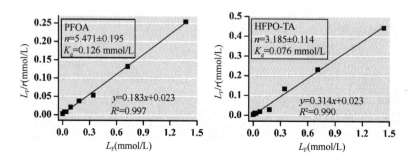

图 17-13　PFOA、HFPO-TA 与 HSA 结合的解离常数及结合位点数拟合图

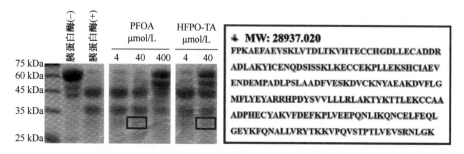

图 17-14　胰蛋白酶酶切 PFOA/HFPO-TA-HSA 电泳图及序列鉴定结果

体外表达纯化两段结合核心区 Domain Ⅰ 和 Domain Ⅱ 用于后续 PFAS 与核心区相互作用研究。通过 ITC 方法测定,获得 PFAS 同两个核心结合区结合的亲和常数(K_a)、结合位点数(n)、结合熵(ΔS)及结合焓(ΔH)(图 17-15)。结果表明,PFOA 和 HFPO-TA 与两段核心区均可结合,亲和常数均表现为 K_a(Domain Ⅰ)< K_a(Domain Ⅱ),即与 Domain Ⅱ 亲和力强于 Domain Ⅰ,该结果与荧光取代结果一致。PFOA 与两个结合核心区结合位点数分别为:0.97 ± 0.07 和 0.99 ± 0.02,表明其结合摩尔比为 1∶1;HFPO-TA 与两个结合核心区结合摩尔比与 PFOA 一致,均为 1∶1。PFOA 与两个结合核心区的结合反应中 $\Delta S \approx 0$ 表明其可通过疏水作用结合,HFPO-TA 与两个结合核心区结合过程中 $\Delta H < 0$、$\Delta S < 0$,表明其主要通过氢键及范德华力结合,其与 Domain Ⅰ 结合过程中的 ΔS 接近 0,表明疏水作用也在其中发挥作用。

综上所述,HFPO-DA、HFPO-TA 不仅存在于污染点源(氟化工园区)周边,已在全球七个国家的河流与湖泊水体中稳定检出。HFPO-TA 可蓄积于我国野生鲤鱼和青蛙的血液、肝脏和肌肉中,且生物累积因子(BCF)显著高于 PFOA,暗示 HFPO-TA 具有更强的生物蓄积性。此外,毒理学实验发现 HFPO-TA 较 PFOA 具有更高的肝脏和细胞毒性,这些研究结果为评估其健康风险提供科学依据。

2.5　展望

结构类似的化合物具有类似的性能,其对环境及生物的毒性效应也类似。一个"有毒"化合物被另一个结构类似的化合物替代,往往并不能真正解决毒害问题,相反可能会陷入"化学品污染—替代—再

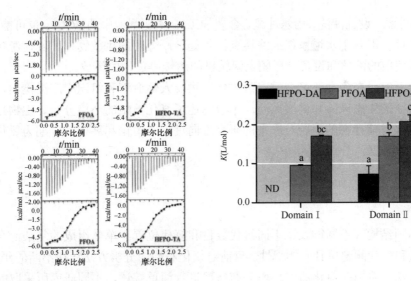

图 17-15　PFASs 与 HSA 两个核心蛋白区域结合力分析

污染"循环往复的怪圈。我们已有数据表明：部分 PFASs 替代品如 HFPO-TA 等的生物蓄积性及毒性效应较传统 PFOA 更强。已有报道仅针对少数已知替代品，其他类型的替代品的环境行为及毒性效应仍有待研究。目前研究新型 PFECAs 面临的问题主要包括：缺少标准品，难以精确定量。McCord 等曾利用 HFPO-DA 等 5 种现有聚醚羧酸标准品，建立混合标准曲线，对缺乏标准品的 PFMOAA、PFO2HxA 和 PFO3OA 进行半定量检测，但测定值与真实值尚存在较大误差，误差可达一个数量级以上[36]；缺乏精确、灵敏的定量分析方法。PFECAs 因含有醚氧原子，热稳定性差，在 LC-MS/MS 检测过程中存在源内裂解现象[31]，母离子在离子源（100～150℃）即裂解为碎片离子，难以获得稳定的[M-H]⁻信号，导致灵敏度差；背景污染高。初步研究已发现液相色谱管路、仪器、耗材中广泛存在此类物质，致使仪器基线和方法空白升高，灵敏度下降。

今后应开展如下工作：①鼓励公司公开替代品的信息，如替代品结构、性质、产量等；②建立此类污染物精确、灵敏的定量分析方法；③鉴定环境中未知结构的 PFASs 替代品，研究其在典型环境介质中的分布特征、生物蓄积性等环境行为和毒性效应；④通过计算毒理学手段，构建替代品结构与毒性之间关系；开展低剂量、长期毒性和复合毒性研究，整合多种组学，从分子、基因等水平研究其毒性机制，通过系统性研究为研发"绿色、低毒、低蓄积性"的新型 PFASs 替代品提供理论基础和技术支撑。

3　氯化石蜡的生态毒理研究

氯化石蜡（chlorinated paraffins，CPs）是正构烷烃的氯代衍生物，可分为短链（$C_{10\sim13}$）、中链（$C_{14\sim17}$）、长链（$C_{>17}$）氯化石蜡，作为阻燃剂、增塑剂广泛应用于塑料、油漆、橡胶、纺织品等工业产品。近年来氯化石蜡在我国经济发达地区环境样品和生物样品中被广泛检出，其相应的毒性作用和生态风险受到极大关注。短链氯化石蜡（SCCPs）在 2017 年被正式列入 POPs 公约受控物质清单。

中国每年都在全球范围内生产大量的 CPs，其中中国所占份额最大，2013 年报告为 105 万吨[37]。生产 CPs 的工厂主要分布在人口密度高、社会经济发达的沿海地区，包括长三角、珠三角地区以及山东、广东、辽宁等省份，而部分产品中 SCCPs 含量占到 80%以上。氯化石蜡属于人类工业产品，生产、储存、运输、使用、废弃物处理过程都会释放氯化石蜡导致其广泛分布[38]。氯化石蜡向水中释放的来源可能出现在制造过程，包括外溢、设施冲洗、雨水冲刷。金属加工/金属切削液中的 SCCPs 也可能由于液桶丢弃、黏附和废液进入水环境，汇入下水道系统最终转到污水处理厂处理的排放物中。其他源包括采矿设备、石油和天然气钻探所使用的液体和设备、船舶上涡轮机的运转等过程的释放。

研究表明，氯化石蜡已在多种环境介质、野生动植物及人体血液和组织中广泛检出，表明具有持久性和蓄积性，存在潜在的生态风险和健康风险。本节综述了氯化石蜡的环境分布、生物累积、人群暴露水平及毒性效应，围绕目前存在的问题及未来的研究方向进行了讨论和展望，以期为氯化石蜡的环境污染及风险评估提供科学依据。

3.1 环境分布与迁移

氯化石蜡是由不同结构（不同链长、不同氯代数目和氯代位置）单体组成的复杂混合物，这一特点使得其精准成分分析和含量测定更具有挑战性。目前对氯化石蜡的分析方法还在不断的研究和优化当中，尚未有统一的标准方法。本节对氯化石蜡环境分布情况进行简单综述，不同研究所采取的分析方法有所不同，在此不再分别介绍。

3.1.1 环境介质中氯化石蜡的分布

Diefenbacher[39]等研究了苏黎世地区大气中 SCCPs 的含量水平和分布特征，2013 年春季调查结果其含量在 1.1～42 ng/m^3；而我国（北京地区）大气含量是该地 30～50 倍之多[40]。局地特征和季节对大气中氯化石蜡含量影响很大。Huang[41]等发现室内空气中 SCCPs 含量远高于室外，因为室内有装饰用品、油画、油漆、皮革等含有 SCCPs 成分，可能是室内释放源。而这一结果与在德国巴伐利亚地区[42]的结果相反，该地区公共场所灰尘浓度远高于家居室内浓度。可能与不同地区人口分布、生活习惯有关。另外，CPs 容易吸附于 $PM_{2.5}$ 中，存在潜在的健康风险。

CPs 具有较强的疏水性，所以在水体中含量较低，一般为 ng/L 级，局部达到 μg/L 级。渤海中[43]的 CPs 同系物分布与工业污水中分布特征相似；距离人类生产区越近的水体含量越高[44]，表明人类排放是自然海水中 CPs 的重要来源。

辽东湾海域[45]、东海海域[46]沉积物中 SCCPs 与总有机碳（TOC）呈现显著线性相关，说明有机质含量会显著影响 SCCPs 迁移、分布，但也存在例外[47]。英国和挪威地区调查[48]显示土壤中 SCCPs 含量远高于其他 POPs，且发现随着纬度增高 SCCPs 含量明显下降，推测可能是受大气传输能力的限制。

我国海洋生物体 SCCPs 含量明显高于发达国家，不同生物体中 SCCPs 含量与体脂呈线性相关，说明脂肪含量会显著影响赋存和积累。K_{ow} 是影响生物体富集能力的关键因素，该值较低的同系物容易在水中迁移，较高的容易赋存在底部沉积物中，这也导致底栖生物含量高于其他生物。北极地区生物体内也有检出[49]，其中 C_{10}-CPs 是含量最多的同系物，表明其具有高挥发性、低颗粒物附着性和长距离迁移能力[50]。

氯化石蜡具有较大挥发性，容易进入大气进行迁移，可以远距离迁移至全球各地区包括两极地区，短链同系物更容易长距离迁移[51]，生物体内分析结果[52]也说明了这一点。另外其具有高疏水性（log K_{ow} 在 4.39～8.69），可生物富集、放大。在生物体内的富集会受到碳链长度和氯含量的影响，研究证明其与 log K_{ow} 有正相关性[53]。

3.1.2 长三角地区淀山湖水生生物体内 SCCPs 含量

淀山湖位于黄浦江上游，是上海市水源地之一，因此了解 CPs 生产地区水生生态系统中的 CPs 分布特性十分重要。Zhou 等[54]以 GC-LRMS-NICI 的方法检测了淀山湖湖水中 14 种水生生物体内 24 种目标 SCCPs 含量以及同系物存在形式；以碳氮稳定同位素的方法进行营养位的计算。

结果表明淀山湖水生生物体内 SCCPs 含量远高于挪威海湾、密歇根湖、安大略湖、西班牙埃布罗河等地的报道,但与我国其他区域(辽东湾、渤海海域)的检出结果一致。目前仍然存在缺乏统一分析方法的问题,所以进行地区间的横向比较结果并不十分可靠。体内含量最高的是鲤鱼、鲫鱼和蜗牛;而调查范围内底栖生物平均体内浓度高于非底栖生物体,证明沉积物是 SCCPs 的源和汇。

对同系物存在形式进行分析(图 17-16、图 17-17)发现,碳链数 $C_{10\sim11}$ 和氯代数 $Cl_{6\sim7}$ 在生物体内平均含量最高,与其他地区的结果不尽一致,可能是由于生产情况不同所导致的。前期研究认为碳链长、氯代数高的 SCCPs 更容易生物累积、难以从生物体内排出[55],在非底栖生物体内可得以验证。

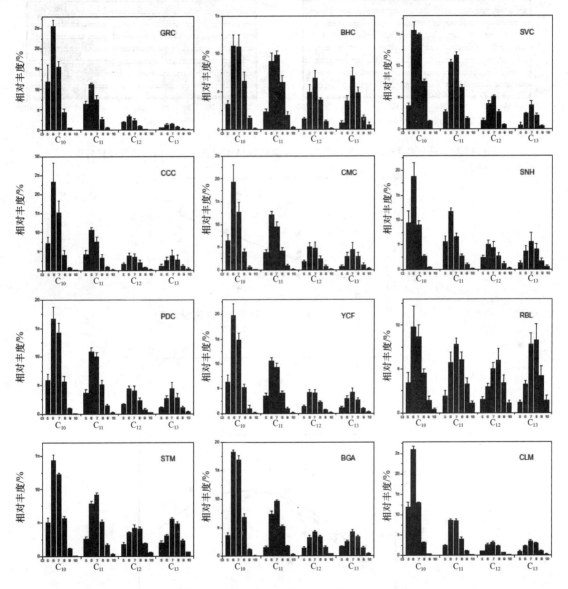

图 17-16 不同生物体内 SCCPs 同系物碳链长特征(引自参考文献[54])

发现每种生物体内 SCCPs 与营养位均呈现显著正相关,表明 SCCPs 随着食物链出现生物放大情况。对脂肪归一化的 SCCPs 同系物浓度和营养位进行线性拟合发现存在显著正相关(23 种 SCCPs),且营养放大因子(TMF)在 1.19~1.57,表明淀山湖水生食物网中存在生物放大效应。发现 TMF 与碳链长存在显著正相关(图 17-18),表明不同碳链长的同系物的生物放大潜力不同;另外 TMF 和 $\log K_{ow}$ 出现二次相关关系(图 17-19),与其他 POPs 较为一致。综上,本研究表明淀山湖水生生物体内 SCCPs 含量处于世界范围的中上水平，底栖生物含量最高；在不同同系物中，短链同系物含量最高；而所有同系物均出现了生物放大特征。

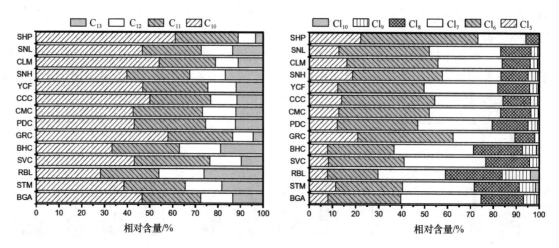

图 17-17　不同生物体内 SCCPs 的链长、氯代数目相对含量（引自参考文献[54]）

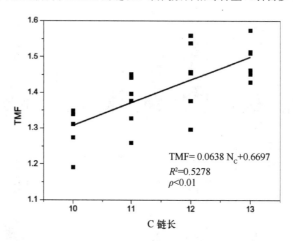

图 17-18　营养级放大因子与碳链长之间关系（引自参考文献[54]）

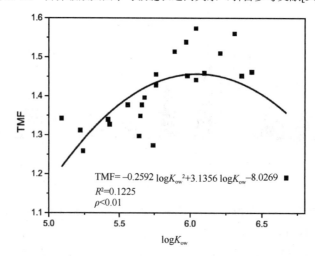

图 17-19　营养级放大因子与 $\log K_{ow}$ 之间关系（引自参考文献[54]）

3.2 毒性效应与机制

3.2.1 内分泌干扰效应

Gong 等[56]对 Sprague Dawley 雄性大鼠进行食物相 SCCPs 暴露，研究血液和肝脏中 TH 稳态和甲状腺形态的变化。研究了 TH 合成、调控、转运、代谢、分解的基因表达情况；同时用分子对接的方法模拟不同 CP 分子与本构雄烷的结合情况。研究结果表明，SCCPs 暴露组显著降低了游离甲状腺素 T_4 和三碘甲状腺氨酸 T_3 的水平，通过反馈机制使得促甲状腺素 TSH 水平显著升高。肝尿苷二磷酸葡萄糖醛酸转移酶和有机阴离子载体的基因表达水平出现显著升高，表明鼠肝脏中甲状腺素代谢受到刺激。SCCPs 对甲状腺腺体组织（整体质量、组织结构、甲状腺合成基因表达）没有明显的改变。说明 TH 变化与 HPT 轴基因无关；但是发现Ⅱ相代谢酶 *Ugt1a1* 等蛋白水平显著升高，说明 T_4 代谢受到影响（甲状腺激素代谢受到促进作用）；T_3 流入肝脏过程受到激发（转运基因上调），流出过程无影响，说明 T_3 在肝脏中出现积累；同时Ⅰ相代谢酶（P450 2B1，即 CAR 通路的生物标志物）表达水平显著升高，表明暴露于 SCCPs 后大鼠的本构型雄激素受体（CAR）信号出现上调。分子对接的结果表明，SCCPs 通过疏水性作用于 CAR 结合，亲和作用与氯代度数有关。文章认为，CAR 通路是甲状腺激素 TH 流入肝脏比例增加的内在机制（图 17-20）；而雄性大鼠 TH 缺乏的主要原因可能是酶异常导致过量分解。SCCPs 与其他很多污染物作用方式类似，并非与 TH 转运蛋白结合，而是诱导Ⅰ、Ⅱ相代谢酶发生干扰作用。

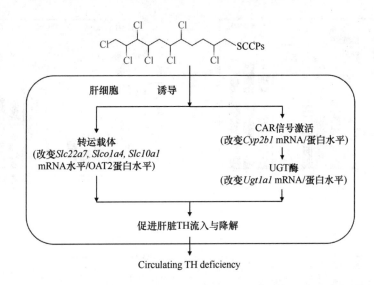

图 17-20 可能的 SCCPs 诱导甲状腺紊乱机制（图根据参考文献[56]修改）

Liu 等[57]报道称低浓度暴露组（0.5 μg/L）SCCPs 暴露会导致甲状腺激素稳态失调，斑马鱼体内甲状腺激素转运蛋白（TTR）、Ⅱ型脱碘酶（ID2）、Ⅲ型脱碘酶（ID3）表达显著降低，并呈现剂量效应关系。

Zhang 等[58]在离体条件下研究了三种 SCCPs 的内分泌干扰效应及可能的途径（受体介导、非受体介导途径），发现雌激素受体 α 是介导的雌激素效应是主要的内分泌干扰途径，两种 SCCPs 产生了明显的拮抗作用；而第三种 SCCPs 产生了显著的糖皮质激素受体 GR 拮抗作用。另外，发现几种类固醇合成相关基因表达显著上调，说明 SCCPs 也可以通过非核受体介导的方式发挥内分泌干扰效应。

3.2.2 神经毒性

前期研究表明，腹腔注射 CPs 会对小鼠基本运动行为产生干扰[59]；会影响突触正常生理功能，即钠依赖胆碱摄取明显减少[60]，表明其可能存在潜在的神经毒性；又有研究发现了不同链长、不同氯取代程度的氯化石蜡所具有的物化特性[61]、细胞毒性[62]、肝脏损伤[63]、在动物体内代谢以及分布情况[64]有所不同，于是我们猜想氯化石蜡结构的差异可能同样会导致神经毒性的差异。同时，已有的毒性研究大多将实验室合成的纯品进行简单混合进行生物暴露实验，然而实际环境中存在的氯化石蜡多来自于工业产品的泄露污染。因此采用商用氯化石蜡产品进行毒性测试更能模拟真实环境暴露情况。于是我们利用斑马鱼作为模式生物利用行为学实验对猜想进行验证。

利用碳骨架法和脱氯加氘法对四种商用氯化石蜡（CP-42、CP-52a、CP-52b、CP-70）进行结构分析，获得四种 CPs 碳链长排序为 CP-52b < CP-70 < CP-42 < CP-52a，氯化程度排序 CP-42 < CP-52a < CP-70。

以斑马鱼为模式动物进行氯化石蜡暴露，利用 viewpoint zebrabox 测试平台，在光暗交替刺激条件下测试仔鱼（受精后第 5~6 天）在单位光照周期内的运动距离、行进角度和社交行为以量化表征仔鱼行为能力（方法优化自参考文献[65]）。

挑选发育正常的幼鱼进行行为学测试，发现仔鱼运动行为受到显著影响。光照期、黑暗期及全测试期内的游动总距离明显下降表明正常行为能力受到干扰（图 17-21），主要出现在中高浓度暴露组（100 μg/L、1000 μg/L）。除了长链低氯代的 CP-42 组未出现显著干扰之外，对比其他组结果发现，随着氯代度数的升高，抑制作用加强；随着碳链长度变短，抑制作用也会加强。

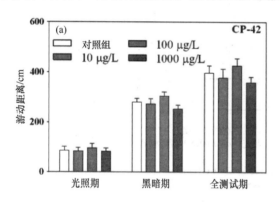
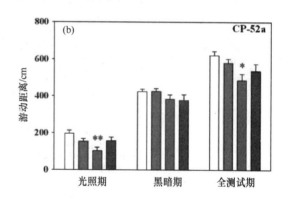
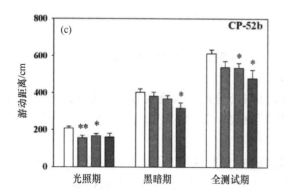
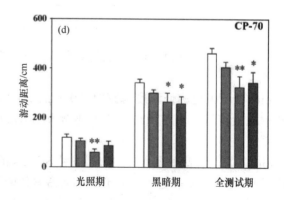

图 17-21　不同氯代度数、碳链长度的商用氯化石蜡对斑马鱼的游动距离的影响
（结果以平均值 ± SEM 表示，$n=24$，$*p<0.05$，$**p<0.01$）

氯代程度越高、碳链长度越长的氯化石蜡毒性效应越强。比较仔鱼在不同角度范围的转动次数（图 17-22）发现链长较短的氯化石蜡（CP-52b 和 CP-70）暴露组在中幅度（±10°~±90°）转角次数显著降低，而长链（CP-52a）暴露组出现低中浓度刺激、高浓度抑制的现象。CP-42 暴露组基本无效应，而 CP-70 组呈现了最强烈的干扰效应；CP-52b 组出现显著转角次数抑制效应，而 CP-52a 低浓度暴露组则相反，在–90°~0°~+90°区间转动次数显著增加。与该暴露组行进距离结果结合分析发现，仔鱼出现了更多定点转动而未向前运动；转角偏好与仔鱼方向感、平衡、认知有关，表明可能是 CPs 暴露导致方向感或是其他方面出现了紊乱。

以两条仔鱼间平均单次接触时长表示社交行为能力，发现高浓度暴露显著干扰了仔鱼正常社交能力（图 17-23）。氯代数相同的情况下，长链（CP-52a）使单次社交行为持续时间延长，而短链（CP-52b）抑制社交行为；与转角相似，这两者又呈现出相反结果，表明确实可能存在不同的干扰机制。

综上所述，商用氯化石蜡暴露具有典型的神经行为效应，可显著干扰斑马鱼正常游泳行为，斑马鱼仔鱼出现游动能力下降、转角偏好异常等行为效应，且成剂量相关性，认为此结果可代表中枢神经系统的异常。氯化程度越高、碳链长度越短的氯化石蜡毒性效应越强，这与内分泌干扰效应的研究结果一致。未来将继续从分子蛋白层面入手研究氯化石蜡干扰中枢神经系统的内在机制。行为效应从表观上证明了氯化石蜡对斑马鱼存在神经发育毒性，研究结果可为后期毒性机制研究及化学品分类管控提供依据。

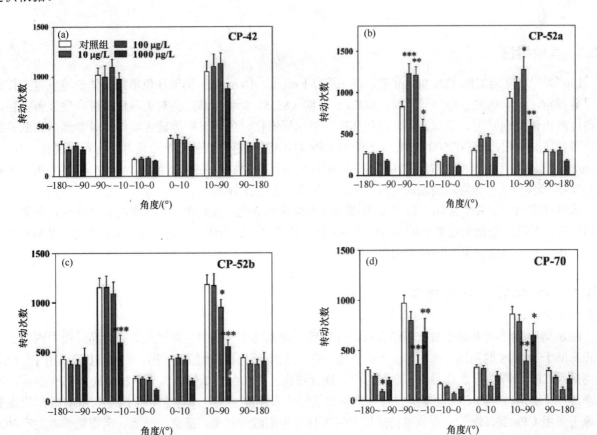

图 17-22 不同氯代度数、碳链长度的商用氯化石蜡对斑马鱼的行进转角情况的影响（结果以平均值± SEM 表示，$n=24$，$*p<0.05$，$**p<0.01$，$***p<0.001$）

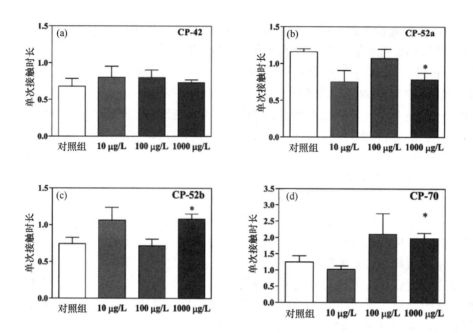

图 17-23 不同氯代度数、碳链长度的商用氯化石蜡对斑马鱼的社交行为的影响（结果以平均值± SEM 表示，$n=24$，$*p<0.05$）

3.2.3 其他毒性

Liu 等[57]发现 SCCPs 暴露 96 hpf 后，高浓度（1 mg/L、10 mg/L）出现仔鱼形态、体长等发育毒性终点显著异常。组织病理学检测[63]发现，SCCPs 长期（85 d）食物相暴露会对斑马鱼肝脏产生显著损害，使得肝脏出现空泡减少、局部炎症、点状坏死、组织纤维化、色素沉积等显著损伤。对非洲爪蟾胚胎的发育毒性研究表明，浓度在 500 mg/L 以下的 SCCPs 对爪蟾无致死效应，但亚致死毒性终点出现较多异常：5 mg/L 组出现发育延迟和形态异常（水肿、肠道发育异常、脊柱弯曲等）；低浓度（0.5 mg/L）组出现谷胱甘肽-S-转移酶活性异常增高，表明斑马鱼体内氧化还原水平异常。

体外代谢组学研究[62]表明，SCCPs 暴露会改变细胞内氧化还原状态，干扰细胞正常代谢，刺激不饱和脂肪酸和长链脂肪酸的过氧化酶β-内酰胺酶氧化，且干扰氨基酸代谢和糖酵解，相关机制与 PPAR 途径有关。斑马鱼的研究结果[66]同样表明存在代谢干扰。

3.3 人体内暴露与健康风险

Gao 等[67]对不同食品中氯化石蜡含量研究发现，肉制品中含量明显高于蔬菜类食品。而中国式烹饪方式有助于 SCCPs 的消除；日本喜欢食用生鱼，因此虽然鱼体含量远低于中国，但实际摄入与中国持平。食物摄入 CPs 量取决于饮食习惯、烹调习惯、饮食结构、人体参数等等。肉类食物是人类 CPs 暴露的重要来源，动物饲料中的 CPs 会富集在畜养禽畜间接导致人体健康风险。动物饲料中的 S/MCCPs 浓度主要来源于商用 CPs 混合物[68]。在日本的研究发现多种食物来源（谷类、蔬菜、糖类、肉蛋奶等农副产品）中都发现了 SCCPs 的存在。认为食物可能是主要暴露途径，但目前暂无显著健康风险[53]。

Qiao[69]等采用 GC×GC-ECNI-TOFMS 分析方法对北京地区 21 对匹配的母亲和胎儿脐带血中的 SCCPs 和 MCCPs 进行了检测，并且对血液中五种甲状腺激素含量进行了分析。研究结果显示母血中 SCCPs 和 MCCPs 的浓度范围在 21.7～373 ng/g ww 和 3.76～31.8 ng/g ww，胎儿脐带血中 SCCPs 和 MCCPs 的浓度范围在 8.51～107 ng/g ww 和 1.33～12.9 ng/g ww（图 17-24）。MCCPs 含量少于 SCCPs，可能是由环境分布特性以及生物累积特性所决定；而母体血液中 SCCPs 与 MCCPs 的绝对含量以及脂肪归一化后的相对含量都存在显著相关性（$p<0.01$）说明二者在人体中具有相似的累积、转运、迁移路径；然而在脐带血样中却没有发现类似的结果，说明不同的同系物存在不同的胎盘转运效率。S/MCCPs 会穿过胎盘屏障进入胎儿体内，即存在代际传递。

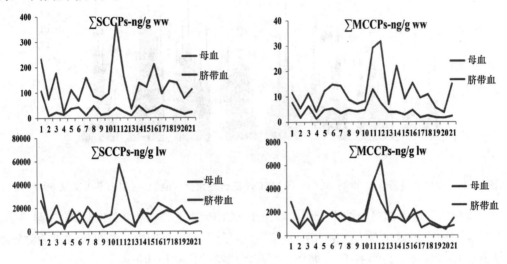

图 17-24　母血和脐带血样品中短链和中链氯化石蜡的含量（引自参考文献[69]）

对母血和脐带血中氯化石蜡同系物结构特征进行了分析（图 17-25），发现 C_{10} 同系物、$Cl_{6\sim7}$ 同系物含量最为丰富，而这一特征可能与地区有关。氯化石蜡同系物的 $\log K_{ow}$ 随着碳链长的增加而增加，这会导致疏水性强的长链同系物偏向于在高脂肪含量的组织器官积累，这也解释了为什么氯化石蜡在血液中与乳汁中（其他报道）的分布特征不同。配对的脐带血和母血中的 CPs 同系物结构组成相似但比例有轻微差异，也说明不同同系物有着不同的代际传递效率。

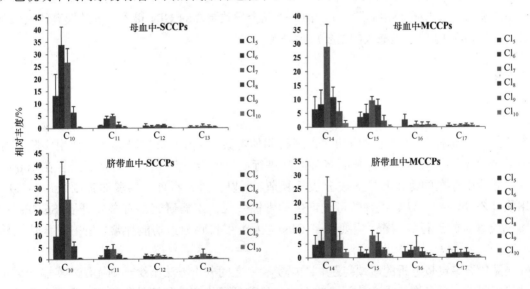

图 17-25　母血和脐带血中氯化石蜡同系物丰富度情况（引自参考文献[69]）

对不同链长和氯代度数的 CPs 在子代/母代间的传递比例进行对比（图 17-26），发现胎盘转移能力与碳链长度、氯代度数呈正比。即物化结构参数会影响 CPs 脂溶性，进而影响同系物的胎盘转运效率和人体内分布特征；高脂溶性、分子质量轻的 CPs 更容易通过胎盘转移，且胎盘转运形式是被动转运（自由扩散）。

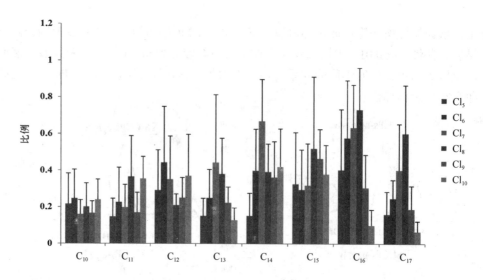

图17-26　母血到脐带血中48种SCCPs和MCCPs的传递比例（$n=21$）（引自参考文献[69]）

对血样中五种甲状腺激素的检测结果表明，母血中SCCPs质量分布和TSH激素水平呈显著正相关。综上，本研究结果表明中链氯化石蜡（MCCPs）和短链氯化石蜡（SCCPs）都可以通过胎盘转运进入子代体内，且暴露在CPs环境中会影响TSH循环，存在污染物的代际传递风险。

Gao等[70]调查了食物、室内空气、室内灰尘、饮用水样品中S/MCCPs含量，以这些外暴露数据对人体健康风险进行了综合评价。北京室内灰尘中S/MCCPs含量为92/82 μg/g，室内空气中S/MCCPs含量83/3.4 ng/m^3，而食物中S/MCCPs含量83/56 ng/g。最重要的暴露途径是食物相暴露和室内灰尘吸入，室内灰尘和空气对婴儿具有更高健康风险。除婴儿外，综合评价的结果表明人体健康风险不显著。本研究的结果（北京）是瑞典的10~100倍（大气浓度）；与中国一家购物中心室内空气中浓度含量持平。短链、中链氯化石蜡的监测结果是其他POPs的100~1000倍。室外大气可稀释氯化石蜡浓度，比室内含量降低10倍之多，推测室内有CPs污染源；室内空气SCCPs含量与建筑类型有关，其中居民楼含量最高，认为来源于特有的装修、室内循环系统等特征。油炸食品会导致更高的CPs摄入，导致较高的摄入风险；饮食摄入量高于日韩约10倍，是成人的主要风险来源。

3.4　展望

目前氯化石蜡的毒性研究仍处于起步阶段，尤其是免疫毒性、神经毒性等的研究数据极为缺乏；下一步建议在关注毒性效应的同时增加对毒理机制的考察。目前的毒性研究大多使用实验室合成的氯代烷烃纯品或将不同氯含量、碳链长的多种纯品进行简单混合，但实际上氯化石蜡存在10000种以上同系物，在环境中存在的氯化石蜡的成分极其复杂。所以在进行毒理学实验时需要采用与实际环境相符的商用氯化石蜡产品，以真正判断可能产生的毒性效应。疾病研究：有必要开展SCCPs对人体潜在的健康风险评价的流行病调查，需要了解CPs与人类甲状腺激素功能异常、生殖发育功能异常有无相关。

要解决氯化石蜡定量分析的瓶颈问题。不同实验室之间分析方法不统一则无法将实验结果进行横向对比。寻找可行的定量参数和更高效便捷的方法（目前是氯化度数和解卷积算法）用来分析鉴定氯化石蜡混合物中的不同同系物。

尽管目前很多国家和地区已经对SCCPs的生产和使用加以管控，但在国内缺乏肉制品及其他食物中SCCPs的含量。推荐进入水质、食品加工行业标准。目前加拿大、瑞典等国家已经设立CPs的推荐日最大容许摄入浓度以便加以管控。很多组织机构（ECHA、UNEP、IARC 等）已经将 CPs 列入多种管控清单，但目前中国还没有出台控制政策，建议加快落实政策的制定。

鉴于 SCCPs 已经被纳入 POPs 清单受到使用限制，目前 M/LCCPs 作为 SCCPs 的替代品仍在广泛应用，目前已有很多研究单位开始关注 MCCPs 的安全性问题，发现其环境分布含量、生物富集特性以及毒性都并不像我们想象得那么安全。故今后要继续深入开展对 M/LCCPs 的安全性调查及风险评估工作。

4 典型有机污染物的内分泌干扰和神经发育毒性效应

4.1 增塑剂类污染物对生殖内分泌系统的影响

塑料在现代生活中的用途涉及方方面面，塑料产品中的双酚 A（bisphenol A，BPA）和邻苯二甲酸酯类（phthalate esters，PAEs）物质也可以通过各种途径释放到日常生活环境当中。大量研究已经证实，BPA 和 PAEs 均可对生物体生殖内分泌系统产生影响，因此受到广泛关注。

4.1.1 双酚 A 及其替代物

前期研究表明，BPA 具有弱雌激素效应。越来越多的研究显示 BPA 在分子水平上的作用机制十分复杂，可能涉及：①经典受体途径，即通过细胞核内的雌激素受体（estrogen receptors，ERs），雌激素相关受体（estrogen related receptors，ERRs）等发生作用，表现出促进或抑制效应；②非基因组途径，即通过细胞膜上少量膜雌激素受体（membrane ERs，mERs）或 G 蛋白耦联受体 30（G protein-coupled receptor 30，GPR30）等激活细胞内信号通路；③激素合成与代谢途径（如各种关键限速酶）；④表观遗传学途径（如 DNA 甲基化和组蛋白修饰）等[71]。BPA 可能通过上述途径干扰下丘脑-垂体-性腺（hypothalamus-pituitary-gonadal，HPG）轴的多个环节，从而对生殖内分泌系统的中枢调控、激素合成、性腺发育和繁殖功能等不同层次产生有害效应（图 17-27）。

图 17-27 BPA 对生殖内分泌系统的影响机制

❋表示可能的作用位点；KiSS：吻素；GnRH：促性腺激素释放激素；LH：黄体生成素；FSH：卵泡刺激素；GPR30/54：G 蛋白耦联受体 30/54；ER：雌激素受体；VTG：卵黄蛋白原；STAR：类固醇合成急性调节蛋白；CYPs：细胞色素 P450；PPARγ：过氧化物酶体增生物激活受体 γ

BPA可能通过上述途径中的一种或几种干扰生物体的生殖内分泌系统和相关功能。以哺乳动物为对象的研究表明，对发情前期雌性小鼠侧脑室注射BPA（20 μg/kg），能通过前腹侧室旁核吻素（kisspeptin，KiSS）及其受体G蛋白耦联受体54（G protein-coupled receptor 54，GPR54）介导的途径，影响视前区促性腺激素释放激素（gonadotropin-releasing hormone，GnRH）mRNA，以及促黄体生成素（luteinizing hormone，LH）和雌二醇（17β-estradiol，E_2）的水平[72]。在以鱼类为对象的研究中，研究人员以200 μg/L BPA暴露斑马鱼后，发现雌雄鱼血浆中雌激素、雄激素、卵泡刺激素（FSH）和促黄体生成素（LH）含量显著下降，导致产卵量下降，但性腺组织结构没有明显变化[73]。同时，对鱼类的研究还表明，BPA在很低剂量下（1 μg/L）即可抑制雄性鲤鱼精巢的发育，导致性腺结构改变[74]。类似地，将雄性稀有鮈鲫暴露于15 μg/L BPA后，会对构成血精屏障的塞尔托利氏细胞的基因表达有显著影响[75]。然而尚没有证据显示BPA会影响细胞核内雄激素受体的转录活性或者由细胞膜AR介导的细胞信号通路。目前认为BPA对雄性生殖系统的影响可能是由于ERβ受到抑制所致[71]，关于这一点还要进一步深入研究。

近期的研究揭示了表观遗传学在BPA内分泌干扰效应中的作用。例如，将稀有鮈鲫雌鱼暴露BPA 35天后卵巢的总DNA甲基化量显著升高，卵巢中*CYP19a1a*基因甲基化含量在7 d和35 d暴露时间组均显著抑制，而且*CYP19a1a* mRNA表达水平与卵巢中CpGs位点甲基化含量呈负相关[76]。在另一项研究中，以BPA（13.75 μg/L）暴露雌性稀有鮈鲫7d和14d后，发现卵巢中细胞色素P450（Cytochrome P450）*CYP17a1*和*CYP11a1*的DNA甲基化水平和转录水平显著相关[77]。此外，研究人员还发现BPA暴露雄性斑马鱼会损伤其精子形成并且改变睾丸细胞的表观形态，进一步研究结果表明，在BPA浓度为2000 μg/L时，会引发精巢中组蛋白H3K9ac、H3K14ac、H4K12ac高度乙酰化和DNA超甲基化[78]。这些结果说明BPA暴露可通过影响基因甲基化水平干扰类固醇合成基因的转录。

越来越多的流行病学调查发现，BPA暴露与女性生殖功能异常和疾病如多囊卵巢综合征、子宫内膜病变、乳腺癌和流产早产等具有相关性[79]。研究表明，不育女性尿液中BPA含量较正常生育女性更高[80]，且接受体外受精女性的尿液中BPA含量越高，其子宫反应性越差，导致体外受精的成功率降低[81]。因此，女性体内BPA的含量可能与其不育有很大关系。同时流行病学调查也发现，BPA暴露与男性生殖功能异常如精子质量降低、隐睾、尿道下裂等疾病风险增加也具有相关性[71]。还有报道显示，BPA暴露会导致年轻男性睾丸间质细胞容量减少和精子数量降低[82]。相对于BPA暴露对成年人的内分泌干扰效应以及与疾病发生的关系，BPA对婴幼儿和儿童的潜在影响受到极大关注。研究人员通过对学龄前儿童的尿液样本检测发现，BPA占所有检测的溴酚类化合物的94%，位居第二[83]。而近几年的研究显示，BPA暴露会影响学龄儿童和婴幼儿体内的激素水平，具有内分泌干扰效应。例如，根据一项针对中国学龄儿童调查，结果表明儿童暴露BPA与特发性中枢性性早熟（ICPP）呈现显著相关性[84]。因此，需要高度关注BPA对儿童生长发育等潜在影响。

值得注意的是，BPA的替代物，双酚S（bisphenol S，BPS）和双酚F（bisphenol F，BPF）由于结构与BPA非常相似，也具有雌激素受体干扰活性，并且BPS能够激活细胞膜上的ERα诱导的信号通路 [如丝裂原活化蛋白激酶（mitogen-activated protein kinase，MAPK）和caspase-8等]。总之，BPF和BPS的雌激素活性与BPA几乎处于同一水平[85]。BPF和BPS在环境介质和人体内的含量也与BPA相当，并且随之使用量增加还会更高，因此它们对生态环境和人类健康的潜在危害也需要进一步密切关注。

4.1.2 DEHP及其代谢产物

邻苯二甲酸二辛酯 [di(2-ethylhexyl)phthalate，DEHP] 是用量最大，环境分布最广泛的邻苯二甲酸酯类物质，主要用作塑料产品的增塑剂及添加剂。离体研究结果表明，DEHP可能更倾向于干扰雄激素受体途径：在受体报告基因研究中，DEHP在未表现出对雌激素受体的干扰效应，但对雄激素受体同时表现出激动效应和拮抗效应（EC_{50}>100μmol/L）[86]。活体研究结果也证实了，DEHP可以干扰雄性生殖系统，特别是胚胎阶段子代的精巢发育过程[87]。例如，在哺乳动物研究中，以较低剂量［0.05 mg/(kg·d) 和 5 mg/(kg·d)］DEHP 暴露妊娠及哺乳期母代小鼠，其雄性后代精巢中 FSHR 和 LHR 的表达量均出现显著下降，精囊重量比对照组轻 20%～25%，精液浓度下降 50%，精子活力降低 20%[88]。在以鱼类为对象的研究中，以 50 mg/(kg·d) 剂量给雄性斑马鱼注射 DEHP 10 天后，其精巢内成熟精子比例下降，精母细胞比例增加，进一步研究发现，这可能是由于精巢中过氧化物酶体增生物激活受体（peroxisome proliferator-activated receptor，PPAR）信号通路受到影响[89]。因此，DEHP 也可能通过多个受体通路对雄性生殖系统产生有害效应。

最近的研究证实，母代小鼠暴露于 DEHP 可能会导致子代精巢 DNA 甲基化水平和 DNA 甲基转移酶表达水平的变化[90]。以 300 mg/(kg·d) 的 DEHP 暴露雄性小鼠，观察到雄性小鼠后代会表现出睾丸发育不全等症状；结合精子转录组和甲基组数据，发现 DEHP 能够持续诱导精囊内相关内分泌蛋白和抗原基因簇启动子甲基化的沉默，这些基因在精子生理学中发挥着基本的作用，同时 DEHP 诱导与启动子去甲基化相关的三个基因表达水平上调，另外，DEHP 导致 mir-615 的表达水平上升，micro RNA 启动子甲基化降低[91]。上述研究均表明，DEHP 对雄性后代生殖系统的影响也可能与表观遗传修饰有关。

值得注意的是，哺乳动物研究显示，DEHP 的主要代谢产物邻苯二甲酸单乙基己基酯［mono (2-ethylhexyl) phthalate，MEHP］也可以影响子代胚胎精巢中固醇类激素合成而引起生殖毒性，并且毒性效应强于 DEHP[92]。流行病学调查结果也表明男性尿液中 DEHP 及 MEHP 含量均与精子畸形率具有相关性[93]。因此，DEHP 对雄性生殖内分泌系统的影响在某种程度上与其代谢产物密切相关，在后续研究中需充分考虑到体内代谢过程对内分泌干扰效应的影响。

此外，DEHP 及其代谢产物对雌性动物生殖内分泌干扰效应的研究报道逐渐增加。例如针对雌性大鼠卵巢细胞的研究表明，DEHP 可通过降低排卵前期卵泡颗粒细胞对 E_2 的生成从而抑制排卵，而 MEHP 也可以降低 E_2 的合成，进一步研究发现可能是由于 PPAR 介导的信号通路受到干扰从而抑制了 P450 芳香化酶的活性[87]。研究表明，DEHP 能够诱导子宫内膜异位症的发生，进一步研究发现可能是由于 DEHP 上调了与阿尔多酮还原酶（AKRs）相关基因 *AKR1C3* 表达量，进而增加 AKRs 活性，最终导致子宫内膜异位症的发生[94]。另一项实验研究了 DEHP 对断奶期至成熟期之间小鼠卵泡的发育情况，结果表明 DEHP 可以抑制此时期卵泡发育因子 mRNA 的基因表达，但是却导致卵巢 miRNA 的表达量上升[95]。流行病学研究也发现孕期 DEHP 和 MEHP 暴露可能增加妊娠早期流产风险[96]。因此，需要进一步关注 DEHP 及其代谢产物对雌性生殖系统的影响。

4.2 溴代阻燃剂对甲状腺内分泌系统的影响

4.2.1 多溴二苯醚

PBDEs 及其羟基化代谢产物（HO-PBDEs）的结构与生物体内甲状腺激素（T_3，T_4）十分相似，因此成为近年来最受关注的一类甲状腺内分泌干扰物。

一些来自对哺乳动物或者鱼类的实验表明，PBDEs 的暴露可以引起甲状腺激素水平的异常（主要是 T_4 水平降低），并具有母源传递效应[97, 98]。例如，BDE-99 水相暴露成年斑马鱼 28 天后，雌鱼和雄鱼 T_4 水平均显著下降，且雄鱼 T_3 水平显著升高；BDE-99 染毒成年斑马鱼可显著干扰子一代仔鱼体内 T_4 水平，并影响 HPTL（hypothalamic-pituitary-thyroid-liver，HPTL）轴[97]。我们前期研究表明，BDE-209 可降低斑马鱼雌鱼血清中 T_3 和 T_4 水平，并可干扰子一代仔鱼体内 T_3 水平[98]。流行病学研究发现，PBDEs 可能影响人类甲状腺激素并增加甲状腺疾病风险，但相关性可能与 PBDEs 的含量以及年龄、性别等因素有关[99]。中国山东 deca-BDE 生产厂的 72 名职业工人血样检测表明，BDE-209 含量与 T_3 总含量（total T_3,

tT_3)呈较弱的正相关,但与T_4总含量(total T_4,tT_4)具有显著正相关关系,且 BDE-209 含量增加 10 倍,tT_4 和 tT_3 分别增加了 7.8% 和 5.4%[100]。中国广东省一家医院对清远电子垃圾拆解区及附近 76 名男性居民分别抽取血样和精液,检测发现血液中 $\sum$PBDEs(BDE-28,99,100,153,154,183,209)水平与游离 T_3(free T_3,fT_3)水平具有显著负相关关系;男性精液中 $\sum$PBDEs 水平与 T_3、T_4 的水平无关,但与促甲状腺激素(thyroid stimulating hormone,TSH)具有显著负相关关系[101]。很多研究也证实,母亲体内的 PBDEs 可通过胎盘、脐带血以及乳汁等途径传递给后代,影响婴幼儿的甲状腺系统。加拿大魁北克省一家医院对 380 名妇女在孕早期和分娩时分别抽取血样,检测发现孕早期 PBDEs 水平与血清甲状腺素的水平无关,但是到分娩时 PBDEs 水平与甲状腺素水平呈负相关[102]。美国俄亥俄州 162 对母子的血样抽取检测表明,16±3 周妊娠时孕妇血清中 $\sum$PBDEs(BDE-28,47,99,100,153)与 3 周岁儿童血清中 fT_3 水平具有显著正相关关系;孕妇血清中 $\sum$PBDEs 含量增加,男孩血清中游离 T_4(free T_4,fT_4)水平显著下降,但女孩血清中 fT_4 水平升高;此外,孕妇血清 $\sum$PBDEs 与儿童血清中 TSH 呈负相关,且孕妇血清 $\sum$PBDEs 增加 10 倍,儿童血清中 TSH 下降了 27.6%[103]。最近美国一项关于人类胎盘的研究表明,样本中 PBDEs 总量与基因组 DNA 甲基化水平具有正相关关系,这种影响可能传递子代,从而对后代的甲状腺系统及生长发育产生更为深远的影响[104]。

很多研究证实,HO-PBDEs 对甲状腺激素的干扰能力高于其母体化合物,因此研究人员对其作用机制十分感兴趣。通过离体和计算机模拟等方法证实,HO-PBDEs 的羟基可与甲状腺激素受体(thyroid hormone receptor,TR)、甲状腺激素转运蛋白(transthyretin,TTR)、甲状腺素结合球蛋白(thyroid binding globulin,TBG)以及甲状腺素磺酸基转移酶(thyroxylsulfonate transferase 1A1,SULT1A1)的残基形成氢键作用,表现出比母体 PBDEs 更强的结合能力[105]。例如,BDE-47 对人胎盘绒毛膜癌细胞(BeWo)的 SULT 酶活性没有影响,但是 3-OH-BDE-47 和 6-OH-BDE-47 对 3,3′-T2 SULT 酶活性均表现出具有较强的抑制作用(IC_{50} 分别为 48.9 nmol/L 和 395.5 nmol/L)[106]。6-OH-BDE-47 可下调斑马鱼胚胎甲状腺激素受体 β(TRβ)的 mRNA 表达,诱导脑部细胞凋亡,从而影响斑马鱼胚胎的神经行为,即蜷缩频率显著增加,但通过显微注射 TRβ 的 mRNA 可以进行部分的挽救,因此 6-OH-BDE-47 可通过甲状腺激素受体 β(TRβ)影响斑马鱼的神经系统[107]。此外,最近的研究表明 HO-PBDEs 还可能通过阻止共调节因子的募集来抑制甲状腺激素的正常功能,从而表现出抗甲状腺激素效应[108]。关于 HO-PBDEs 与人类甲状腺系统疾病的流行病学调查也取得了一定的进展。Liu 等[109]对甲状腺癌患者的研究表明,患者体内 PBDEs 含量与甲状腺激素水平没有显著相关性,但 HO-BDE-47 与游离 T_4 具有显著负相关关系,而 HO-PBDEs 总量与 TSH 具有显著正相关关系。这项研究首次揭示了 HO-PBDEs 与癌症患者甲状腺功能异常的相关性,同时也表明 HO-PBDEs 可能较母体化合物对人体甲状腺系统的影响更显著。

4.2.2 新型溴代阻燃剂

近年来的研究表明,作为 PBDEs 替代品的新型溴代阻燃剂在世界范围的多种非生物介质和野生动物体内广泛存在,而它们是否能够引起甲状腺干扰效应也成为关注的热点问题。目前的证据显示,与其他替代品相比,五溴二苯醚和十溴二苯醚的替代品,四溴邻苯二甲酸双(2-乙基己基)酯[bis(2-ethylhexyl)2,3,4,5-tetrabromophthalate,TBPH]和十溴二苯乙烷(decabromodiphenyl Eher,DBDPE)表现出较强的甲状腺内分泌干扰能力。例如,报告基因实验表明,TBPH 和它的代谢产物四溴邻苯二甲酸单-2-乙基己基酯[mono(2-ethylhexyl)tetrabromophthalate,TBMEPH]均表现出具有较强的抗甲状腺激素活性(IC_{50} 分别为 0.1 μmol/L 和 32.3 μmol/L),但对 PXR 均表现出激动效应(EC_{50} = 5.5 μmol/L 和 2.0 μmol/L),提示 TBPH 及其代谢产物可能通过多种途径影响甲状腺内分泌系统[110]。活体实验也发现,将围产期大鼠灌喂 TBMEHP 2 天后,母鼠出现明显的低甲状腺症状,血清中 T_3 显著下降,T_4 水平未发生变化,但脱碘酶活性显著降低[111]。雄性大鼠灌喂 500mg/kg bw/day DBDPE 28 天后,对 tT_4、tT_3 和 fT_4 没有影响,但是 fT_3

显著下降，TSH 和促甲状腺素释放激素（thyrotropin-releasing hormone，TRH）显著升高，并且诱导甲状腺氧化损害，干扰 HPT 轴，提示 DBDPE 可能通过氧化应激损伤和 HPT 轴影响甲状腺内分泌系统[112]。此外，在流行病学的调查研究中还发现人类血液中 TBPH 的含量与 T_3 的含量与呈正相关的关系[113]。以上证据均显示 TBPH 及其代谢产物可能具有甲状腺干扰效应。作为 PBDEs 替代品，TBPH 及其他新型溴代阻燃剂对生态系统和人类健康的安全性需要进一步的评估。

4.3 有机磷阻燃剂的神经发育毒性效应

近年来，有机磷阻燃剂（organophosphorus flame retardant，OPFRs）作为 PBDEs 类的替代品大量生产和使用，在各种环境介质及生物体中均有不同程度的检出。由于 OPFRs 的化学结构与有机磷农药具有相似的磷酸二酯键，而有机磷农药被广泛证实具有神经毒性，因此研究人员十分关心 OPFRs 是否可以通过类似的作用方式引起神经毒性效应。

4.3.1 TDCIPP

国外学者将未分化的 PC12 细胞暴露于 TDCIPP（0～50 μmol/L）及典型有机磷农药毒死蜱（chlorpyrifos，CPF，50 μmol/L）24 小时后，发现 TDCIPP 对 PC12 细胞表现出显著的神经毒性，且具有剂量-效应依赖关系，同时最高剂量组的毒性效应与 CPF 的毒性作用相当[114]。中国学者以 SH-SY5Y 细胞为对象，研究了 TDCIPP 诱导神经毒性的作用机制。他们发现以较高剂量的 TDCIPP（25～100 μmol/L）暴露 SH-SY5Y 细胞 24 h 后，可诱导细胞内的活性氧（Reactive oxygen species，ROS）升高，线粒体膜电势下降，促进细胞凋亡的发生[115]；而以较低剂量的 TDCIPP（0～5 μmol/L）处理未分化的 SH-SY5Y 细胞 3～5 d 后，发现 0～2.5 μmol/L TDCIPP 处理组细胞的存活率并没有受到影响，但是 2.5 μmol/L TDCIPP 处理组细胞表面的轴突显著增多，而且含有长轴突的神经元数量也显著增加，同时神经元分化的生物标志物 MAP2 蛋白的表达显著升高，表明 TDCIPP 能够诱导 SH-SY5Y 细胞分化为成熟的神经元[116]。以上研究结果表明，TDCIPP 在较低剂量下能促进神经细胞的分化，而在较高剂量下可诱导神经细胞内氧化应激，诱导细胞凋亡。

由于在离体研究中，TDCIPP 表现出了较强的神经毒性潜力，因此研究人员对其在活体当中的神经毒性效应也十分关注。中国学者将斑马鱼胚胎暴露于 TDCIPP（0 μg/L、4 μg/L、20 μg/L 和 100 μg/L）5 d 后，在仔鱼体内检测到很高含量的 TDCIPP 及其代谢产物 BDCIPP，表明 TDCIPP 很容易在生物体内富集并发生代谢，但低剂量下的短期 TDCIPP 暴露未造成明显神经发育毒性[117]。当以更高剂量（500 μg/L）暴露斑马鱼胚胎时，即使短期（5 d）暴露也可以显著诱导神经发育毒性，包括孵化率和存活率降低，畸形率增加（如脊柱弯曲），神经元发育相关的基因和蛋白（MBP、SYN2A 和 α1-tubulin）显著降低，同时运动行为发生改变，这与典型的神经毒物毒死蜱（CPF，100 μg/L）暴露导致的神经毒性类似。但 CPF 暴露显著抑制了斑马鱼幼鱼体内乙酰胆碱酯酶（acetylcholinesterase，AChE）和丁酰胆碱酯酶（butyrylcholinesterase，BChE）的活性，TDCIPP 暴露则对这两种酶的活性未产生显著影响[118]。上述结果表明，TDCIPP 暴露可在生物体内引起神经发育毒性，但其作用机制可能与 CPF 有所不同。

4.3.2 其他 OPFRs

除 TDCIPP 外，其他 OPFRs 也被证实可以在生物内诱导神经发育毒性。国外学者将未分化的 PC12 细胞分别暴露于 50 μmol/L 的 CPF、TCEP、TCIPP 及三-（2,3-二溴丙基）-磷酸酯[tris（2,3-dibromopropyl）phosphate，TDBPP] 6 d 后，发现它们均可以减少 PC12 细胞的数量，但都不能改变未分化的 PC12 细胞的分化状态。在加入 NGF 的条件下，CPF 可以促进未分化的 PC12 细胞朝向多巴胺能神经元细胞分化，抑制其向胆碱能神经元细胞的分化；TDBPP 可以同时促进 PC12 细胞朝向胆碱能及多巴胺能神经元细胞分化的过程；而 TCEP 和 TCPP 仅可促进 PC12 细胞向胆碱能神经元细胞的分化过程[114]。上述结果表明，OPFRs 具有潜在的神经毒性，但其作用机制可能与 CPF 有所区别，且不同的 OPFRs 对神经元细胞分化表现出不同的作用。

在活体研究中也得到了类似的结论。研究人员分别以磷酸三丁酯（tributyl phosphate，TNBP，3125 μg/L）、磷酸三（2-丁氧基）乙酯［(tris（2-butoxyethyl) phosphate，TBOEP，6250 μg/L）］及 TPhP（625 μg/L）暴露日本青鳉胚胎，发现它们均可以抑制仔鱼在持续光照条件下和明暗交替光照条件下的运动速度，表明这 3 种 OPFRs 都具有潜在的神经毒性。进一步研究发现，TNBP 暴露可显著诱导仔鱼体内的 AChE 酶活性和 *ache* 基因的表达；TPhP 暴露则显著抑制仔鱼体内 AChE 酶活性和 *ache* 的基因表达；而 TBOEP 暴露对该酶的活性和基因表达没有显著影响[119]。类似地，我国学者将受精后 2 小时斑马鱼胚胎暴露于 TCIPP 或 TCEP（0 mg/L、100 mg/L、500 mg/L、2500 mg/L）或毒死蜱（CPF，100 mg/L）至 120 小时后，发现 CPF 显著抑制仔鱼体内 AChE 活性，但 TCIPP 和 TCEP 对该酶的活性无影响；但是 TCIPP 和 TCEP 暴露可以导致神经元发育相关的基因和蛋白显著降低，包括 MBP、SYN2A、α1-tubulin、SHHa 和 GAP43，同时导致幼鱼的运动行为发生改变，表明 TCIPP 和 TCEP 的作用方式与 CPF 不同，但仍可能通过影响中枢神经神经系统的发育影响其相关功能[120]。

上述研究结果提示了 OPFRs 神经毒性作用机制的复杂性，科学家们仍需进行深入的研究。尽管如此，以上这些证据已经显示 OPFRs 具有一定的神经毒性，提示进一步评估 OPFRs 的生态环境风险具有重要的意义。

4.4 展望

通过上述内容可以看出，有机污染物对内分泌系统和神经发育过程的影响具有特殊性和复杂性。首先，它们为何在很低的剂量下即可产生长期甚至跨代的影响，即使是针对我们上述提到的这些典型有机污染物，这一点都尚未完全研究清楚，因此后续需要对非基因组途径、表观遗传学以及其他可能的机制开展更深入的研究。其次，虽然我们只介绍了这些污染物的主要干扰效应，但任何一种污染物的作用靶点都不是单一的，它们可能同时干扰生殖、甲状腺内分泌系统，以及神经、免疫和代谢系统等等。一方面我们对这些毒性效应的认识还不全面，需要进一步完善，另一方面如何利用已有的证据，建立高效便捷的毒性评估和风险预测体系，更好地为保护生态环境和人类健康提供支持，也是我们面临的亟待解决的问题之一。更为复杂的是，不管是野生生物还是人类，所接触的环境中往往是同时存在很多种污染物，它们很可能作用于相同的靶点，产生协同或者拮抗作用，因此，因此机体内的变化是多种污染物共同作用的结果。对此，目前尚没有合理有效的评估方法，这也将是环境毒理学领域研究的长期目标。

参 考 文 献

[1] Lindstrom A B, Strynar M J, Libelo E L. Polyfluorinated compounds: Past, present, and future. Environmental Science & Technology, 2011, 45(19): 7954-7961.

[2] Houde M, De Silva A O, Muir D C, et al. Monitoring of perfluorinated compounds in aquatic biota: an updated review. Environmental Science & Technology, 2011, 45(19): 7962-7973.

[3] Andersen M E, Butenhoff J L, Chang S C, et al. Perfluoroalkyl acids and related chemistries--toxicokinetics and modes of action. Toxicological Sciences : An official journal of the Society of Toxicology, 2008, 102(1): 3-14.

[4] Greaves A K, Letcher R J, Sonne C, et al. Tissue-specific concentrations and patterns of perfluoroalkyl carboxylates and sulfonates in East Greenland polar bears. Environmental Science & Technology, 2012, 46(21): 11575-11583.

[5] Zhong W, Zhang L, Cui Y, et al. Probing mechanisms for bioaccumulation of perfluoroalkyl acids in carp (*Cyprinus carpio*):

Impacts of protein binding affinities and elimination pathways. Science of the Total Environment, 2019, 64(7): 992-999.

[6] Chen M, Qiang L, Pan X, et al. *In Vivo* and *in Vitro* isomer-specific biotransformation of perfluorooctane sulfonamide in common carp (*Cyprinus carpio*). Environmental Science & Technology, 2015, 49(23): 13817-13824.

[7] Fang S, Chen X, Zhao S, et al. Trophic magnification and isomer fractionation of perfluoroalkyl substances in the food web of Taihu Lake, China. Environmental Science & Technology 2014, 48(4): 2173-2182.

[8] Zhao S, Zhou T, Wang B, et al. Different biotransformation behaviors of perfluorooctane sulfonamide in wheat (*Triticum aestivum* L.)from earthworms (*Eisenia fetida*). Journal of Hazardous Materials, 2018, 346: 191-198.

[9] Zhao L, Zhang Y, Zhu L, et al. Isomer-specific transplacental efficiencies of perfluoroalkyl substances in human whole blood. Environmental Science & Technology Letters, 2017, 4(10): 391-398.

[10] Zhang Y, Beesoon S, Zhu L, et al. Biomonitoring of perfluoroalkyl acids in human urine and estimates of biological half-life. Environmental Science & Technology, 2013, 47(18): 10619-10627.

[11] Shi Y, Vestergren R, Xu L, et al. Human exposure and elimination kinetics of chlorinated polyfluoroalkyl ether sulfonic acids (Cl-PFESAs). Environmental Science & Technology, 2016, 50(5): 2396-2340.

[12] Gao Y, Fu J, Cao H, et al. Differential accumulation and elimination behavior of perfluoroalkyl acid isomers in occupational workers in a manufactory in China. Environmental Science & Technology, 2015, 49(11): 6953-6962.

[13] Beesoon S, Martin J W. Isomer-specific binding affinity of perfluorooctanesulfonate (PFOS) and perfluorooctanoate (PFOA) to serum proteins. Environmental Science & Technology, 2015, 49(9): 5722-5731.

[14] Shan G, Wang Z, Zhou L, et al. Impacts of daily intakes on the isomeric profiles of perfluoroalkyl substances (PFASs) in human serum. Environment International, 2016, 89-90: 62-70.

[15] Shi X, Du Y, Lam P K S, et al. Developmental toxicity and alteration of gene expression in zebrafish embryos exposed to PFOS. Toxicology and Applied Pharmacology, 2008, 230(1): 23-32.

[16] Zheng X M, Liu H L, Shi W, et al. Effects of perfluorinated compounds on development of zebrafish embryos. Environmental Science and Pollution Research, 2011, 19(7): 2498-2505.

[17] Zhang S, Guo X, Lu S, et al. Exposure to PFDoA causes disruption of the hypothalamus-pituitary-thyroid axis in zebrafish larvae. Environmental Pollution, 2018, 235: 974-982.

[18] Godfrey A, Hooser B, Abdelmoneim A, et al. Thyroid disrupting effects of halogenated and next generation chemicals on the swim bladder development of zebrafish. Aquatic Toxicology, 2017, 193: 228-235.

[19] Shi X, Liu C, Wu G, et al. Waterborne exposure to PFOS causes disruption of the hypothalamus-pituitary-thyroid axis in zebrafish larvae. Chemosphere, 2009, 77(7): 1010-1018.

[20] Benninghoff A D, Orner G A, Buchner C H, et al. Promotion of hepatocarcinogenesis by perfluoroalkyl acids in rainbow trout. Toxicological Sciences : An official journal of the Society of Toxicology, 2012, 125(1): 69-78.

[21] Cheng J, Lv S, Nie S, et al. Chronic perfluorooctane sulfonate (PFOS) exposure induces hepatic steatosis in zebrafish. Aquatic Toxicology, 2016, 176: 45-52.

[22] Frawley R P, Smith M, Cesta M F, et al. Immunotoxic and hepatotoxic effects of perfluoro-*n*-decanoic acid (PFDA) on female Harlan Sprague-Dawley rats and B6C3F1/N mice when administered by oral gavage for 28 days. Journal of Immunotoxicology, 2018, 15(1): 41-52.

[23] Qu J H, Lu C C, Xu C, et al. Perfluorooctane sulfonate-induced testicular toxicity and differential testicular expression of estrogen receptor in male mice. Environmental Toxicology and Pharmacology, 2016, 45: 150-157.

[24] Toft G, Jonsson B A, Lindh C H, et al. Exposure to perfluorinated compounds and human semen quality in Arctic and European populations. Human Reproduction, 2012, 27(8): 2532-2540.

[25] Chen H, He P, Rao H, et al. Systematic investigation of the toxic mechanism of PFOA and PFOS on bovine serum albumin

by spectroscopic and molecular modeling. Chemosphere, 2015, 129: 217-224.

[26] Ren X M, Zhang Y F, Guo L H, et al. Structure-activity relations in binding of perfluoroalkyl compounds to human thyroid hormone T_3 receptor. Archives of Toxicology, 2015, 89(2): 233-242.

[27] Weiss J M, Andersson P L, Lamoree M H, et al. Competitive binding of poly- and perfluorinated compounds to the thyroid hormone transport protein transthyretin. Toxicological Sciences, 2009, 109(2): 206-216.

[28] Wang Z, Cousins I T, Scheringer M, et al. Fluorinated alternatives to long-chain perfluoroalkyl carboxylic acids (PFCAs), perfluoroalkane sulfonic acids (PFSAs) and their potential precursors. Environmental International, 2013, 60: 242-248.

[29] Song X, Vestergren R, Shi Y, et al. Emissions, transport, and fate of emerging per- and polyfluoroalkyl substances from one of the major fluoropolymer manufacturing facilities in China. Environmental Science & Technology, 2018, 52(17): 9694-9703.

[30] Sun M, Arevalo E, Strynar M, et al. Legacy and emerging perfluoroalkyl substances are important drinking water contaminants in the Cape Fear River watershed of North Carolina. Environmental Science & Technology Letters, 2016, 3(12): 415-419.

[31] Pan Y, Zhang H, Cui Q, et al. First report on the occurrence and bioaccumulation of hexafluoropropylene oxide trimer acid: An emerging concern. Environmental Science & Technology, 2017, 51(17): 9553-9560.

[32] Cui Q, Pan Y, Zhang H, et al. Occurrence and tissue distribution of novel perfluoroether carboxylic and sulfonic acids and legacy per/polyfluoroalkyl substances in black-spotted frog (*Pelophylax nigromaculatus*). Environmental Science & Technology, 2018, 52(3): 982-990.

[33] Pan Y, Zhang H, Cui Q, et al. Worldwide distribution of novel perfluoroether carboxylic and sulfonic acids in surface water. Environmental Science &Technology, 2018, 52(14): 7621-7629.

[34] Sheng N, Pan Y, Guo Y, et al. Hepatotoxic effects of hexafluoropropylene oxide trimer acid (HFPO-TA), a novel perfluorooctanoic acid (PFOA) alternative, on mice. Environmental Science & Technology, 2018, 52(14): 8005-8015.

[35] Sheng N, Cui R, Wang J, et al. Cytotoxicity of novel fluorinated alternatives to long-chain perfluoroalkyl substances to human liver cell line and their binding capacity to human liver fatty acid binding protein. Archives of Toxicology, 2018, 92(1): 359-369.

[36] Mccord J, Newton S, Strynar M. Validation of quantitative measurements and semi-quantitative estimates of emerging perfluoroethercarboxylic acids (PFECAs) and hexfluoroprolyene oxide acids (HFPOAs). Journal of Chromatography A, 2018, 1551: 52-58.

[37] UNEP/POPS/POPRC.12/11/Add3. Report of the Persistent Organic Pollutants Review Committee on the Work of its Twelfth Meeting: Risk Management Evaluation on Short-chain Chlorinated Paraffins, 2017.

[38] 王亚韡, 傅建捷, 江桂斌. 短链氯化石蜡及其环境污染现状与毒性效应研究. 环境化学, 2009, 28(1): 1-9.

[39] Diefenbacher P S, Bogdal C, Gerecke A C, et al. Short-chain chlorinated paraffins in zurich, switzerland—Atmospheric concentrations and emissions. Environmental Science & Technology, 2015, 49(16): 9778-9786.

[40] Wang T, Han S, Yuan B, et al. Summer–winter concentrations and gas-particle partitioning of short chain chlorinated paraffins in the atmosphere of an urban setting. Environmental Pollution, 2012, 171: 38-45.

[41] Huang H, Gao L, Xia D, et al. Characterization of short- and medium-chain chlorinated paraffins in outdoor/indoor PM_{10}/$PM_{2.5}$/$PM_{1.0}$, in Beijing, China. Environmental Pollution, 2017, 225: 674-680.

[42] Hilger B, Fromme H, VLkel W, et al. Occurrence of chlorinated paraffins in house dust samples from Bavaria, Germany. Environmental Pollution, 2013, 175: 16-21.

[43] Zhao N, Cui Y, Wang P W, et al. Short-chain chlorinated paraffins in soil, sediment, and seawater in the intertidal zone of Shandong Peninsula, China: Distribution and composition. Chemosphere, 2019, 220: 452-458.

[44] Gao Y, Zhang H, Su F, et al. Environmental occurrence and distribution of short chain chlorinated paraffins in sediments and soils from the Liaohe River Basin, P. R. China. Environmental Science & Technology, 2012, 46(7): 3771-3778.

[45] Ma X, Zhang H, Wang Z, et al. Bioaccumulation and trophic transfer of short chain chlorinated paraffins in a marine food web from Liaodong Bay, North China. Environmental Science & Technology, 2014, 48(10): 5964-5971.

[46] Zhao Z, Li H, Wang Y, et al. Source and migration of short-chain chlorinated Paraffins in the Coastal East China Sea using multiproxies of marine organic geochemistry. environmental Science & Technology, 2013, 47(10): 5013-5022.

[47] Qiao L, Xia D, Gao L, et al. Occurrences, sources and risk assessment of short- and medium-chain chlorinated paraffins in sediments from the middle reaches of the Yellow River, China. Environmental Pollution, 2016, 219: 483-489.

[48] Halse A K, Schlabach M, Schuster J K, et al. Endosulfan, pentachlorobenzene and short-chain chlorinated paraffins in background soils from Western Europe. Environmental Pollution, 2015, 196: 21-28.

[49] Li H, Fu J, Pan W, et al. Environmental behaviour of short-chain chlorinated paraffins in aquatic and terrestrial ecosystems of Ny-Alesund and London Island, Svalbard, in the Arctic. Science of The Total Environment, 2017, 590-591: 163-170.

[50] 国家环境分析测试中心. 持久性有机污染物(POPs)区域污染现状和演变趋势. 北京: 中国环境出版社, 2015.

[51] Li H, Fu J, Zhang A, et al. Occurrence, bioaccumulation and long-range transport of short-chain chlorinated paraffins on the Fildes Peninsula at King George Island, Antarctica. Environment International, 2016, 94: 408-414.

[52] Reth M, Ciric A, Christensen G N, et al. Short- and medium-chain chlorinated paraffins in biota from the European Arctic — Differences in homologue group patterns. Science of the Total Environment, 2006, 367(1): 252-260.

[53] Fisk A T, Cymbalisty C D, Tomy G T, et al. Dietary accumulation and depuration of individual C10-, C11- and C14-polychlorinated alkanes by juvenile rainbow trout (*Oncorhynchus mykiss*). Aquatic Toxicology, 1998, 43(2): 209-221.

[54] Zhou Y, Yin G, Du X, et al. Short-chain chlorinated paraffins (SCCPs) in a freshwater food web from Dianshan Lake: Occurrence level, congener pattern and trophic transfer. Science of The Total Environment, 2018, 615: 1010-1018.

[55] Van Mourik L M, Gaus C, Leonards P E G, et al. Chlorinated paraffins in the environment: A review on their production, fate, levels and trends between 2010 and 2015. Chemosphere, 2016, 155: 415-428.

[56] Gong Y, Zhang H, Geng N, et al. Short-chain chlorinated paraffins (SCCPs) induced thyroid disruption by enhancement of hepatic thyroid hormone influx and degradation in male Sprague Dawley rats. Science of The Total Environment, 2018, 625: 657-666.

[57] Liu L, Li Y, Coelhan M, et al. Relative developmental toxicity of short-chain chlorinated paraffins in Zebrafish (*Danio rerio*)embryos. Environmental Pollution, 2016, 219: 1122-1130.

[58] Zhang Q, Wang J, Zhu J, et al. Assessment of the endocrine-disrupting effects of short-chain chlorinated paraffins in *in vitro* models. Environment International, 2016, 94: 43-50.

[59] Eriksson P, Jan -Erik Kihlström. Disturbance of motor performance and thermoregulation in mice given two commercial chlorinated paraffins. Bulletin of Environmental Contamination and Toxicology, 1985, 34(2): 205-209.

[60] Eriksson P, Nordberg A. The effects of DDT, DDOH-palmitic acid, and a chlorinated paraffin on muscarinic receptors and the sodium-dependent choline uptake in the central nervous system of immature mice. Toxicology & Applied Pharmacology, 1986, 85(2): 121-127.

[61] Bettina H, Hermann F, Wolfgang V, et al. Effects of chain length, chlorination degree, and structure on the octanol-water partition coefficients of polychlorinated *n*-Alkanes. Environmental Science & Technology, 2011, 45(7): 2842-2849.

[62] Geng N, Zhang H, Zhang B, et al. Effects of short-chain chlorinated paraffins exposure on the viability and metabolism of human hepatoma HepG2 cells. Environmental Science & Technology, 2015, 49(5): 3076-3083.

[63] Cooley H M, Fisk A T, Wiens S C, et al. Examination of the behavior and liver and thyroid histology of juvenile rainbow trout (*Oncorhynchus mykiss*) exposed to high dietary concentrations of C_{10}-, C_{11}-, C_{12}- and C_{14}-polychlorinated *n*-alkanes.

Aquatic Toxicology, 2001, 54(1-2): 81-99.

[64] Geng N B, Zhang H J, Xing L G, et al. Toxicokinetics of short-chain chlorinated paraffins in Sprague-Dawley rats following single oral administration. Chemosphere, 2016, 145: 106-111.

[65] Zhang B, Xu T, Huang G, et al. Neurobehavioral effects of two metabolites of BDE-47 (6-OH-BDE-47 and 6-MeO-BDE-47) on zebrafish larvae. Chemosphere, 2018, 200: 30-35.

[66] Ren X, Zhang H, Geng N, et al. Developmental and metabolic responses of zebrafish (*Danio rerio*) embryos and larvae to short-chain chlorinated paraffins (SCCPs) exposure. Science of The Total Environment, 2018, 622-623: 214-221.

[67] Gao W, Cao D, Lv K, et al. Elimination of short-chain chlorinated paraffins in diet after Chinese traditional cooking: A cooking case study. Environment International, 2019, 122: 340-345.

[68] Shujun D, Xiaomin L, Xiaoou S, et al. Concentrations and congener group profiles of short- and medium-chain chlorinated paraffins in animal feed materials. Science of The Total Environment, 2019, 647: 676-681.

[69] Lin Q, Lirong G, Minghui Z, et al. Mass fractions, congener group patterns, and placental transfer of short- and medium-chain chlorinated paraffins in paired maternal and cord serum. Environmental Science & Technology, 2018, 52(17): 10097-10103.

[70] Gao W, Cao D, Wang Y, et al. External exposure to short-and medium-chain chlorinated paraffins for the general population in Beijing, China. Environmental Science & Technology, 2018, 52(1): 32-39.

[71] Acconcia F, Pallottini V, Marino. Molecular mechanisms of action of BPA. Dose-Response, 2015, 13: 1-9.

[72] Wang X, Chang F, Bai Y, et al. Bisphenol a enhances kisspeptin neurons in anteroventral periventricular nucleus of female mice. Journal of Endocrinology, 2014, 221(2): 201-213.

[73] Fang Q, Shi Q, Guo Y, et al. Enhanced bioconcentration of bisphenol A in the presence of nano-TiO_2 can lead to adverse reproductive outcomes in zebrafish. Environmental Science and Technology, 2016, 50: 1005-1013.

[74] Mandich A, Bottero S, Benfenati E, et al. *In vivo* exposure of carp to graded concentrations of bisphenol A. General and Comparative Endocrinology, 2007, 153: 15-24.

[75] Tao S, Wang L, Zhu Z, et al. Adverse effects of bisphenol A on Sertoli cell blood-testis barrier in rare minnow *Gobiocypris rarus*. Ecotoxicology and Environmental Safety, 2019, 171: 475–483.

[76] Liu Y, Yuan C, Chen S, et al. Global and *cyp19a1a* gene specific DNA methylation in gonads ofadult rare minnow Gobiocypris rarus under bisphenol A exposure. Aquatic Toxicology, 2014, 156: 10-16.

[77] Zhang T, Liu Y, Chen H, et al. The DNA methylation status alteration of two steroidogenic genes in gonads of rare minnow after bisphenol A exposure. Comparative Biochemistry and Physiology Part C: Toxicology Pharmacology, 2017, 198: 9-18.

[78] González R S, Lambó M, Fernández D C, et al. Male exposure to bisphenol A impairs spermatogenesis and triggers histone hyperacetylation in zebrafish testes. Environmental Pollution, 2019, 248: 368-379.

[79] Rochester J R. Bisphenol A and human health: A review of the literature. Reprodutive Toxicology, 2013, 42: 132-155.

[80] Caserta D, Bordi G, Ciardo F, et al. The influenceof endocrine disruptors in a selected population of infertile women. Gynecolodgical Endocrinology, 2013, 29: 444-447.

[81] Ehrlich S, Williams P, Missmer S, et al. Urinarybisphenol A concentrations and early reproductive health outcomes among women undergoing IVF. Human Reproduction, 2012, 27: 3583-3592.

[82] Adoamnei E, Mendiola J, Vela S F, et al. Urinary bisphenol A concentrations are associated with reproductive parameters in young men. Environmental Research, 2018, 161: 122-128.

[83] Liu Y, Yan Z, Zhang Q, et al. Urinary levels, composition profile and cumulative risk of bisphenols in preschool-aged children from Nanjing suburb, China. Ecotoxicology and Environmental Safety, 2019, 172: 444–450.

[84] Chen Y, Wang Y, Ding G, et al. Association between bisphenol A exposure and idiopathic central precocious puberty (ICPP)

among school-aged girls in Shanghai, China. Environment International, 2018a, 115: 410–416.

[85] Rochester J R, Bolden A L. Bisphenol S and F: A systematic review and comparison of the hormonal activity of bisphenol A substitutes. Environmental Health Perspectives, 2015, 123: 643-650.

[86] Shen O, Du G, Sun H, et al. Comparison of *in vitro* hormone activities of selected phthalates using reporter gene assays. Toxicology Letters, 2009, 191: 9-14.

[87] Zarean M, Keikha M, Poursafa P, et al. A systematic review on the adverse health effects of di-2-ethylhexyl phthalate. Environmental Science and Pollution Research, 2016, 23: 24642-24693

[88] Pocar P, Fiandanese N, Secchi C, et al. Exposure to di (2-ethyl-hexyl) phthalate (DEHP) *in utero* and during lactation causes long-term pituitary-gonadal axis disruption in male and female mouse offspring. Endocrinology, 2012, 153: 937-948.

[89] Uren W T M, Lewis C, Filby A L, et al. Mechanisms of toxicity of di(2-ethylhexyl)phthalate on the reproductive health of male zebrafish. Aquatic Toxicology, 2010, 99: 360-369.

[90] Wu S, Zhu J, Li Y, et al. Dynamic effect of di-2-(ethylhexyl)phtalate on testicular toxicity: Epigenetic changes and their impact on gene expression. International Journal of Toxicology, 2010, 29: 193-200.

[91] Stenz L, Escoffier J, Rahban R, et al. Testicular dysgenesis syndrome and long lasting epigenetic silencing of mouse sperm genes involved in the geproductive system after prenatal exposure to DEHP. PLoS One, 2017, 12: eo170441.

[92] Magdouli S, Daghrir R, Brar S K, et al. Diethylhexylphtalate in the aquatic and terrestrial environment: A critical review. Journal of Environmental Management, 2013, 127: 36-49.

[93] Wang Y X, You L, Zeng Q, et al. Phthalate exposure and human semen quality: Results from an infertility clinic in China. Environmental Research, 2015, 142: 1-9.

[94] Kim L Y, Kim, M R, Kim J H, et al. Aldo-keto reductase activity after diethylhexyl phthalate exposure in eutopic and ectopic endometrial cells. European Journal of Obstetrics & Gynecology and Reproductive Biology, 2017, 215: 215-219.

[95] Liu J, Wang W X, Zhu J L, et al. Di (2-ethylhexyl) phthalate (DEHP) influences follicular development in mice between the weaning period and maturity by interfering with ovarian development factors and microRNAs. Environmental Toxicology, 2018, 33: 535-544.

[96] Arbuckle T E, Davis K, Marro L, et al. Phthalate and bisphenol A exposure among pregnant women in Canada-results from the MIREC study. Environment International, 2014, 68: 55-65.

[97] Wu L Y, Li Y F, Ru H J, et al. Parental exposure to 2, 2', 4, 4'5 - pentain polybrominated diphenyl ethers (BDE-99) causes thyroid disruption and developmental toxicity in zebrafish. Toxicology and Applied Pharmacology, 2019, 372: 11-18.

[98] Chen L G, Wang X F, Zhang X H, et al. Transgenerational endocrine disruption and neurotoxicity in zebrafish larvae after parental exposure to binary mixtures of decabromodiphenyl ether (BDE-209) and lead. Environmental Pollution, 2017, 230: 96-106.

[99] Hoffman K, Sosa J A, Stapleton H M. Do flame retardant chemicals increase the risk for thyroid dysregulation and cancer? Current Opinion in Oncology, 2017, 29: 7-13.

[100] Chen T, Niu P Y, Kong F L, et al. Disruption of thyroid hormone levels by decabrominated diphenyl ethers (BDE-209) in occupational workers from a deca-BDE manufacturing plant. Environment International, 2018, 120: 505-515.

[101] Yu Y J, Lin B G, Chen X C, et al. Polybrominated diphenyl ethers in human serum, semen and indoor dust: Effects on hormones balance and semen quality. Science of the Total Environment, 2019, 671: 1017-1025.

[102] Abdelouahab N, Langlois M F, Lavoie L, et al. Maternal and cord-blood thyroid hormone levels and exposure to polybrominated diphenyl ethers and polychlorinated biphenyls during early pregnancy. American Journal of Epidemiology, 2013, 178: 701-713.

[103] Vuong A M, Braun J M, Webster G M, et al. Polybrominated diphenyl ether (PBDE) exposures and thyroid hormones in

children at age 3 years. Environment International, 2018, 117: 339-347.

[104] Kappil M A, Li Q, Li A, et al. *In utero* exposures to environmental organic pollutants disrupt epigenetic marks linked to fetoplacental development. Environmental Epigenetics, 2016, 2(1): dw013.

[105] Butt C M, Stapleton H M. Inhibition of thyroid hormone sulfotransferase activity by brominated flame retardants and halogenated phenolics. Chemical Research in Toxicology, 2013, 26: 1692–1702.

[106] Leonetti C P, Butt C M, Stapleton H M. Disruption of thyroid hormone sulfotransferase activity by brominated flame retardant chemicals in the human choriocarcinoma placenta cell line, BeWo. Chemosphere, 2018, 197: 81-88.

[107] Wang F, Fang M L, Hinton D E, et al. Increased coiling frequency linked to apoptosis in the brain and altered thyroid signaling in zebrafish embryos (Danio rerio) exposed to the PBDE metabolite 6-OH-BDE-47. Chemosphere, 2018, 198: 342-350.

[108] Chen Q, Wang X, Shi W, et al. Identification of thyroid hormone disruptors among HO-PBDEs: *In vitro* investigations and co-regulator involved simulations. Environmental Science and Technology, 2016, 50(22): 12429-12438.

[109] Liu S, Zhao G, Li J, et al. Association of polybrominated diphenylethers (PBDEs) and hydroxylated metabolites (OH-PBDEs) serum levels with thyroid function in thyroid cancer patients. Environmental Research, 2017, 159: 1-8.

[110] Skledar D G, Tihomir T, Carino A, et al. New brominated flame retardants and their metabolites as activators of the pregnane x receptor. Toxicology Letters, 2016, 259: 116-123.

[111] Springer C, Dere E, Hall S J, et al. Rodent thyroid, liver and fetal testis toxicity of the monoestermetabolite of bis-(2-ethylhexyl) tetrabromonophtalate (TBPH), a novel brominated flame retardant present in indoor dust. Environmental Health Perspectives, 2012, 120: 1711-1719.

[112] Wang Y, Chen T, Sun Y, et al. A comparison of the thyroid disruption induced by decabrominated diphenyl ethers (BDE-209) and decabromodiphenyl ethane (DBDPE) in rats. Ecotoxicology and Environmental Safety, 2019, 174: 224-235.

[113] Johnson P I, Stapleton H M, Mukherjee B, et al. Associations between brominated flame retardants in house dust and hormone levels in men. Science of the Total Environment, 2013, 445: 177-184.

[114] Dishaw L V, Powers C M, Ryde I T, et al. Is the pentaBDE replacement, tris (1,3-dichloropropyl) phosphate (TDCPP), a developmental neurotoxicant? Studies in PC12 cells. Toxicology and Applied Pharmacology, 2011, 256: 281-289.

[115] Li R W, Zhou P J, Guo Y Y, et al. Tris (1,3-dichloro-2-propyl) phosphate-induced apoptotic signaling pathways in SH-SY5Y neuroblastoma cells. Neurotoxicology, 2017, 58: 1-10.

[116] Li R W, Zhou P J, Guo Y Y, et al. The involvement of autophagy and cytoskeletal regulation in TDCIPP-induced SH-SY5Y cell differentiation. Neurotoxicology, 2017, 62: 14-23.

[117] Wang Q, Lai N L, Wang X, et al. Bioconcentration and transfer of the organophorous flame retardant 1, 3-dichloro-2-propyl phosphate causes thyroid endocrine disruption and developmental neurotoxicity in zebrafish larvae. Environmental Science and Technology, 2015, 49(8): 5123-5132.

[118] Li R, Zhang L, Shi Q, et al. A protective role for autophagy in TDCIPP-induced developmental neurotoxicity in zebrafish larvae. Aquatic Toxicology, 2018, 199: 46-54.

[119] Sun L W, Tan H N, Peng T, et al. Developmental neurotoxicity of organophosphate flame retardants in early life stage of Japanese medaka (*Oryzias latipes*). Environmental Toxicology and Chemistry, 2017, 35(12): 2931-2940.

[120] Li R, Wang H, Mi C, et al. The adverse effect of TCIPP and TCEP on neurodevelopment of zebrafish embryos/larvae. Chemosphere , 2019, 220: 811-817.

作者：尹大强[1]，祝凌燕[2]，周炳升[3]，戴家银[4]，徐挺[1]，杨心悦[1]

[1]同济大学，[2]南开大学，[3]中国科学院水生生物研究所，[4]中国科学院动物研究所

第 18 章　环境砷污染与健康研究进展

- 1. 引言 /480
- 2. 环境砷污染来源 /480
- 3. 环境砷污染健康危害 /485
- 4. 环境砷污染致病机制 /488
- 5. 环境砷污染健康风险评估 /495
- 6. 展望 /499

本章导读

随着经济社会的快速发展，由砷污染引发的环境问题和健康问题日益突出，其不仅影响社会的可持续发展，同时也影响民众的身体健康和生活质量。砷作为自然界普遍存在并被广泛使用的类金属元素，能够通过特定的自然过程或人为活动实现环境中的富集，地球化学元素的分布不均和工农业用途均会造成环境污染，导致环境中居民发生急慢性健康损害甚至癌症。随着环境砷污染与健康危害事件的频繁发生，其污染来源、健康损害及其机制、健康风险评估也越来越受到国内外学者的重视。为反映国内外环境砷污染与健康的研究动态，梳理其研究成果及经验，促进我国砷污染与健康研究领域的交流，本章较系统全面地介绍了环境砷污染来源、砷污染健康危害、致病机制及健康风险评估相关研究进展。

关键词

砷，环境砷污染，健康损害，分子机制，健康风险评估

1 引 言

砷作为自然界普遍存在并被广泛使用的类金属元素，能够通过特定的自然过程或人为活动形成环境中的富集和污染，并通过直接吸食或生物地球化学循环进入到人体造成危害。美国毒物和疾病登记署将砷列在"致病毒物名单"的首位，国际癌症研究机构将其列为Ⅰ类致癌物。长期砷暴露造成的健康危害现已成为世界范围内的重大公共卫生问题。据世界卫生组织（WHO）报告的数据显示：全球范围内长期暴露于饮水砷＞10 μg/L（WHO推荐标准）的人口超过两亿[1]；我国现状亦不容乐观，研究发现，可能近2000万人正面临长期砷暴露的威胁[2]。因此，长期砷暴露对健康的危害及防控研究任重道远。

本章重点从环境砷污染的来源、环境砷污染的健康危害、环境砷污染致病机制、环境砷污染的健康风险评估等方面对环境砷污染与健康做了较系统全面的介绍。

2 环境砷污染来源

随着国内外环境砷污染健康危害事件的频繁发生，其污染来源及生物地球化学循环过程也越来越受到国内外研究人员的重视，进一步理清环境砷污染的来源及其生物地球化学循环过程对控制环境砷污染对人类健康的损害具有重要作用。

2.1 环境砷污染概况

砷（arsenic，As）是自然环境中普遍存在的一种类金属元素，在地壳中含量较丰富（约为1.5 mg/kg）。砷及其化合物可在岩石、煤炭、土壤、沉积物、地下水和地表水中富集，人类通过不同途径摄入和吸收上述富集的砷及其化合物可对其机体造成不可逆性的健康损害。砷进入自然环境主要有两种途径：一种是生物地球化学循环的自然过程，包括自然的区域性分布不均衡、雨水侵蚀、微生物对砷及其化合物的代谢、森林燃烧、火山爆发等；另一种是人为因素所造成的砷污染，包括各类矿产的开采冶炼，含砷化

学物（如农药、杀虫剂、涂料和木材防腐剂）的大量使用，工业生产（如玻璃、制药、电子和半导体）过程以及饲料添加剂的使用等[3]。

自然环境砷污染现象在全球范围均有发生，如在印度、孟加拉国、智利、中国、巴基斯坦、美国、墨西哥、阿根廷、泰国、越南、匈牙利、尼泊尔等多个国家都存在地下水高砷污染问题。据不完全估计，全球约有1.4亿人受饮用水砷污染的影响，至少有近1亿人面临因砷暴露所致的各类疾病的威胁[4]。此外，中国贵州省和陕西省尚有世界上特有的燃煤型砷中毒病区[5]。

随着现代工农业的不断发展，人类在生产生活过程中遭受砷污染导致的健康损害问题亦越来越突出。其中最早的案例出现在1890年波兰某矿区，当地居民集中出现的健康问题是由于其水源被含砷硫化矿物污染所致[3]。随后，人类生活生产过程中出现的砷污染事件层出不穷，如1955年日本森永奶粉因使用了受砷污染的添加剂而导致当地130名儿童死亡的"日本毒奶粉事件"[6]；近年来，我国因人为活动而造成的砷污染事件也时有发生，如2006年，湖南岳阳某化工厂因违规排放砷污染的工业废水而导致岳阳新墙河受到严重砷污染，其水砷浓度高达0.31~0.62 mg/L，远远超过国家水砷标准，导致周边8万多居民饮水困难，健康受到严重威胁[7]；2007年，贵州省独山县某企业违规排放砷污染废水，使都柳江和麻球河流域水砷含量严重超标，造成17人中毒，约2万多居民生活用水受到威胁[8]；2008年，云南省阳宗海发生严重砷污染现象，造成2.6万居民的饮水安全受到威胁，严重危害当地生态系统[9]。目前国际上对大尺度、多受体的砷排放来源研究十分有限。20世纪80年代，Jerome等[10]估算了全球范围内排放到水体、土壤和大气的微量元素总量，发现有色金属开采与冶炼、石油、煤炭、木材燃烧、钢铁制造、水泥制造、肥料播撒、废物处理等是主要的人为砷污染排放来源。Han等[11]研究发现，从工业时代的1850年到2000年的150年间，人为活动造成的砷污染排放逐年攀升，全球累计砷排放总量约453万t，其中矿业作业产生的砷量占比72.6%，是人为活动造成环境砷污染的首要来源。陈琴琴等[12]对我国31个省（直辖市、自治区）2010年度砷污染排放进行了较全面的估算，结果显示我国2010年度直接排放砷约5.75万t，其中排入土壤的约3.37万t，排入水体的约1.98万t，排入大气的约0.40万t。

2.2 环境砷污染来源

如前所述，环境砷污染来源主要有自然来源和人为来源两种方式。自然来源主要指源于岩石或矿物的自然风化、大气和雨水的侵蚀以及微生物活动、地热活动、火山爆发等，砷由原来难以与土壤、岩石等介质产生化学反应、溶解、迁移的状态，转变为活泼、易溶解、易随流动介质迁移而进入自然环境而造成砷污染。

人为来源则主要来自于人类工农业生产和生活过程。如人类农业活动中使用大量含砷化肥（如砷酸铅、亚砷酸钠、乙酰亚砷酸铜和砷酸钙等无机砷；稻宁、稻脚青和巴黎绿等有机砷）、农药（主要含一甲基胂酸和二甲基亚胂酸）、木材防腐剂（铬砷合剂、砷酸锌、砷酸钠等）、饲料添加剂（苯砷酸化合物）等，造成广泛的环境砷污染与累积。此外，人类工业生产过程中亦会产生不同程度的砷污染，如冶金和半导体工业、电子工业、化学工业、冶炼工业等广泛使用含砷化合物（如砷化镓、砷化铜等），也成为砷污染的主要来源之一[3]。

2.2.1 大气砷污染来源

大气中的砷有自然来源，也有人为来源。研究估算，大气圈中砷的保有量约为$1.74×10^6$ kg，其中多数存在于北半球（约为$1.48×10^6$ kg）[13]。另有研究显示城市区域大气中砷含量约为3~180 ng/m^3，而远离人类聚集地的偏远区域大气中砷的浓度仅为0.01~1 ng/m^3 [14]。

1）自然来源

远离人类聚集地的偏远区域大气中的砷主要为自然来源，如火山爆发释放到大气中的砷含量约为 17150 t/a，自然发生的森林大火、木材燃烧所释放砷的量约为 125～3345 t/a，土壤微生物低温活动释放砷的量约为 160～26200 t/a，海洋循环流动释放到大气中的砷约为 27 t/a[15, 16]。此外，含砷矿物的风化侵蚀所释放的砷也是大气砷污染的重要自然来源。自然环境中形成的高砷矿物主要与硫结合形成硫化物，如雄黄（AsS）、雌黄（As_2S_3）、毒砂（FeAsS）、硫砷铜矿（Cu_3AsS_4）、辉钴矿（CoAsS）、砷黝铜矿（$Cu_{12}As_4S_{13}$）和砷镍矿（NiAs）等，这些自然矿物受风化侵蚀作用，每年可向大气释放约 1980 t 砷[17]。

2）人为来源

人类农业、工业生产过程中所造成的砷污染是目前人类聚集地区域空气砷污染的主要来源。Pacyna 等[18]统计了全球范围内不同年份大气中砷污染的主要人为排放源及排放量，发现有色金属冶炼是大气中砷的主要人为排放源，化工石油燃料燃烧次之。

（1）有色金属冶炼。由于含铜、锡、铅、镍、钴、银、金等元素的天然矿物多数含砷，因此砷成为上述矿物开采和冶炼的主要副产品。研究显示，1983 年全球初级铅和铜冶炼的砷排放因子分别是 200～400 g/t 和 1000～1500 g/t[19]。到 1995 年，新的开采和冶炼技术大大降低了冶炼过程中含砷污染物的排放，初级铅和铜的砷排放因子分别降到了 3～5 g/t 和 100～500 g/t[17]。到 21 世纪初，欧盟对含砷污染物排放控制更加严格，规定铅和铜冶炼的砷排放因子限值分别为 3 g/t 和 50 g/t。而到 2009 年，欧盟再次调整大气含砷污染物排放限值，初级铅和铜冶炼的砷排放因子进一步降低到 2.1 g/t 和 39 g/t，仅占 1983 年砷排放因子下限值的 1.05% 和 3.9%[20]。

（2）煤的燃烧。砷是一种典型的亲煤元素，与煤中有机或无机物均有很强的亲和力。1985 年，Yudovich 等[21]报道了烟煤和褐煤的平均砷含量分别为（20±3）mg/kg 和（14±4）mg/kg，2004 年的进一步评估中，该值下降为（9.0±0.8）mg/kg 和（7.4±1.4）mg/kg。但部分国家或地区如俄罗斯阿尔泰山脉褐煤[20]、英国南威尔士烟煤[22]、土耳其西部烟煤[23]、美国阿拉巴马州烟煤[20]的砷含量远超出世界平均含量。值得注意的是，20 世纪 70 年代发现，我国贵州部分地区（如黔西南州兴仁县）烟煤砷含量严重超标，当地居民因长期使用高砷污染的煤而导致的燃煤污染型地方性砷中毒至今仍为我国特有的地方病病种[24]。燃煤作为世界主要能源之一，其年均消耗量呈逐步增加趋势，中国能源统计年鉴数据表明，1995～2010 年国内燃煤消耗量增加了 126.79%，其中工业、电力和其他燃煤消耗量分别上升 242.09%、96.60% 和 7.15%，而家庭燃煤消耗量则下降 32.31%[25]。煤在燃烧过程中，部分砷经高温挥发并冷凝富集在颗粒物表面，最终以砷氧化物蒸汽或颗粒物的形式释放到大气中[16]。煤完全燃烧后（>400℃），其中的砷便会形成三氧化二砷蒸气，在空气中冷凝形成三氧化二砷微粒而悬浮；而不完全燃烧时，部分砷随燃煤烟尘进入空气，两者共同组成砷飘尘而污染空气[21]。因高砷煤燃烧污染的室内空气中，飘尘砷含量远较煤中的砷含量高，可达煤中砷含量的数百倍。

除煤燃烧外，石油和天然气的燃烧也可向大气排放一定量的砷。有报道显示，2009 年欧洲石油和天然气燃烧的砷排放因子分别为 1～4.3 μg/MJ 和 0.09 μg/MJ，2000 年和 2010 年欧洲石油燃烧的砷排放量分别为 122.86 t 和 97.32 t[26]。

（3）含砷化学物的使用。砷在过去被广泛应用于生活生产当中（如农药、杀虫剂、防腐剂等），尤其从 19 世纪开始，包括砷酸铅（$PbAsO_4$）、砷酸镁（$MgAsO_4$）、砷酸钙（$CaAsO_4$）和砷酸锌（$ZnAsO_4$）等以含砷无机盐为主的农药被多个国家普遍用于果园害虫防治。到 20 世纪 60 年代 DDT 出现之前，全世界砷产量的 80% 被用于杀虫剂的生产[27]。到 20 世纪末，杀虫剂中砷的消耗量下降了 50%，其中有机砷杀虫剂逐渐替代了过去单一的无机砷盐杀虫剂。在杀虫剂的使用过程中，大气是其长距离扩散的重要媒介，杀虫剂的空中喷洒、农作物和土壤的挥发以及土壤的风蚀作用等都会导致空气砷含量的增加。Mukai 等[28]研究发现，有机砷杀虫剂的使用，导致大气中 MMA 和 DMA 的含量高达 485 pg/m^3 和 53 pg/m^3，而在距杀虫剂使用点 20~30 km 区域的大气 MMA 和 DMA 平均浓度仅为 1.4 pg/m^3 和 0.43 pg/m^3。Matschullat 等[16]研究发现，全球大气砷的流通量中，含砷农药和杀虫剂的释放量（3440 t/a）不容忽视。

砷作为多数防腐剂的主要成分之一（主要的含砷木材防腐剂有加铬砷酸铜和氨铜砷酸锌），曾被广泛应用于木材防腐工业，据报道，20 世纪末防腐剂的砷用量约占世界砷产量的 30%。Helsen 等[29]研究发现，在使用加铬砷酸铜热处理木材过程中，约 8%~95%的砷可挥发进入大气环境中，每年约有 150 t 砷经由防腐剂的使用而进入到大气环境中。

（4）垃圾焚烧。垃圾焚烧过程中，大量的砷可因高温挥发到烟气中，最终富集于飞灰中，通过各种途径污染大气、水体和土壤等环境。1983 年全球城市污泥、垃圾焚烧的砷排放因子和大气砷排放量分别为 5.0~10 g/t、1.1~2.8 g/t 和 15~60 t、154~392 t，1990 年分别为 5.0 g/t、1.1 g/t 和 37 t、87 t[29]。Alvim-Ferraz 等[30]对葡萄牙某大型医院医疗废物焚烧的研究表明，砷的排放因子为 0.0599~0.364 g/t，而美国环保署规定的医疗废物砷排放因子为 0.121 g/t。Sullivan 等[31]对澳大利亚污泥焚烧的研究发现，污泥中砷的平均含量为 5.6 g/t（干重），焚烧的砷排放因子为 4.7 g/t。2009 年，欧洲城市垃圾、工业垃圾、农业垃圾以及医疗废物焚烧的砷排放因子分别为 0.01 g/t、0.016 g/t、0.058 g/t 和 1.3 g/t[32]。

（5）其他人为来源。除上述主要人为来源外，玻璃生产、添加剂、干燥剂、染料、催化剂、合金和半导体工业以及动物医药等工业生产所消耗砷约占全球砷总产量的 20%，在上述生产加工、使用和后续处理等过程均可产生大气砷排放。

2.2.2 水体砷污染来源

天然水体中砷含量一般处于较低水平。通常情况下，海水砷浓度在 1~8 μg/L 范围内，其化学形态以砷酸根离子为主。地表或地下淡水资源中，正常情况下砷浓度通常为 1~10 μg/L，而在砷污染区域（如硫化物矿区），其浓度可达 100~5000 μg/L，富氧化性的水体中以砷酸盐为主，而还原性水体中则主要富集亚砷酸盐[3]。此外，在天然水体中尚存在一些有机砷化合物，如 AsBe、MMA 和 DMA。同样地，水体中砷的来源也主要是自然来源和人为来源。

1）自然来源

据不完全估计，全球范围内每年通过自然作用释放到水环境中的砷约为 2.2 万 t[33]。地壳中的砷含量约为 1.5~3 mg/kg，而大多数岩石中砷含量约为 0.5~2.5 mg/kg[34]。在自然风化作用，每年岩石风化释放进入生物圈的砷含量约 4.5 万 t，其中部分砷可通过活化迁移进入湖泊、河流、海洋等自然水体[35]。

自然环境中的许多地热水中砷含量普遍较高，因此高砷地热水是水体砷污染的重要来源。在日本、新西兰、美国阿拉斯加、俄罗斯西伯利亚堪察加半岛、中国云南和西藏等国家和地区地均有砷含量非常丰富的地热水资源。研究显示日本地热水砷含量可高达 1.8~6.4 mg/L，新西兰地热水砷含量甚至达到 8.5 mg/L，其高砷形成原因与当地黑色页岩的地质结构密切相关[36]。

除地热水之外，因特殊的水文地质环境条件及生物地球化学作用，全球范围内存在许多严重危害人体健康的高砷地下水区域，其主要类型分为冲击平原和三角洲型和干旱半干旱地区的内陆盆地型。上述地区人群因长期暴露于高砷地下水而导致的饮水型地方性砷中毒已成为严重危害人类健康的全球性公共卫生问题[3]。此外，其他一些硫化物矿区或矿化区，由于含砷硫化物矿物的氧化也可能在其所在区域形成高砷地下水。

2）人为来源

水体中砷的主要人为来源有农业生产活动和工业生产过程中含砷产品使用所产生的"三废"排放。据统计，全世界每年因人类生活生产活动而排入水体中的砷约有 120 万 t，其中工农业生产中砷的排放是造成水体砷污染的主要原因[33]。

在农业生产过程中含砷化肥和农药的使用，是造成河流湖泊水体砷污染的重要人为来源。含砷化肥如磷肥中的砷含量通常在 20～50 mg/kg，有时甚至高达到数百 mg/kg。含砷农药中主要含有砷酸铅、砷酸钙、亚砷酸钠、甲基胂、甲基胂酸二钠和砷酸铜等，这些不同价态的砷在施用后约有 0.1%～10%可转化为可溶性的砷而释放进入河流湖泊[37]。

含砷工业"三废"的排放，特别是矿业开采活动，是导致水体砷污染的另一个重要原因。砷是百余种矿物的重要组成成分，常以金属硫化物矿石或金属砷酸盐形式存在，因此在有色金属矿的开采及冶炼过程中往往会造成含砷废水的集中排放。其次是硫酸工业等的排放，硫酸工业使用了高砷硫铁矿作原料，其引起的砷污染事件已多次发生。若在未经处理或处理不达标的情况下排放工业生产废水，或含砷矿区受到雨水冲刷则会使大量砷进入河流湖泊[38]。

2.2.3 土壤砷污染来源

同大气和水体砷污染来源相似，土壤砷的来源也包括自然成因和人为来源两个方面。

未受污染土壤中的砷主要来自母岩的风化，土壤中砷的含量通常要高于母岩。不同土壤类型砷含量不一样，未被污染的土壤一般含有砷 1～40 mg/kg，由花岗岩风化形成的砂质土中砷含量水平最低，但冲积土和有机土的含量则较高[3]。

影响土壤中砷浓度的自然因素除母岩外，还包括气候、土壤有机质和无机质组成以及氧化还原电势。许多相关的研究发现母岩的类型在影响土壤金属含量方面起决定作用。花岗岩、石灰岩和基性岩风化形成的棕壤，其砷含量分别为 7.48 mg/kg、11.9 mg/kg 和 8.47 mg/kg[39]。

土壤中自然产生的砷主要为无机砷类，其也可与土壤中的有机质相结合。在氧化性和有氧条件下，五价砷酸盐类（As^{5+}）是一类稳定的形态，其可被吸附于黏土、铁和锰氧化物或氢氧化物和一些有机质中。在富含铁的地层中，砷可反应生成砷酸铁而沉淀下来。在还原性条件下，砷化物主要是以亚砷酸盐（As^{3+}）存在。此外，无机砷化物还可被一些微生物甲基化，在氧化性条件下，生成 MMA、DMA 和三甲基胂氧化物（TMAsO）。在厌氧条件下，它们可被还原成具挥发性和易被氧化的甲基胂。砷酸铁（$FeAsO_4$）和砷酸铝（$AlAsO_4$）是酸性土中主要砷化物，它们比砷酸钙（Ca_3AsO_4）更难溶于水，后者是盐性土和石灰质土中的主要化学形式。土壤 pH 和氧化还原电势对被吸附的砷酸有显著的影响。在相同的 pH 条件下，它也可受到土壤类型的影响，即从灰碳土到棕碳土再到栗钙土依次增加[39]。

未被污染的土壤砷浓度普遍在 5～15 mg/kg。Boyle 和 Jonasson[40]提出世界土壤中砷平均浓度为 7.2 mg/kg，Shacklett 等[41]提出美国土壤砷平均浓度为 7.4 mg/kg。Ure 和 Berrow[42]给出砷平均值为 11.3 mg/kg。泥炭和沼泽土壤的砷浓度可能更高（平均值 13 mg/kg），这主要是因为在还原条件下硫化物矿物不断增加的结果。Shotyk 等[43]在瑞士两个泥炭剖面中发现的最大砷浓度为 9 mg/kg，而在剖面里也发现了微量矿物成分的存在。在诸如含黄铁矿的页岩、金属矿脉以及脱水的红树林沼泽等富含硫化物地带，由于黄铁矿的氧化而形成的酸性硫酸盐土壤也可以富集砷。Dudas[44]发现加拿大富含黄铁矿页岩风化产生的酸性硫酸盐土壤中的砷浓度达到 45mg/kg。

诸如当地工业中的冶炼、化石燃料燃烧的产物和农业中的杀虫剂、磷肥等都可将额外的砷引入土壤中。Ure 和 Berrow[42]指出在为果树长期使用含砷的农药后，果园土壤中砷浓度范围达到 366～732 mg/kg。

矿业活动产生的尾矿渣和废水污染的沉积物和土壤砷浓度可比自然条件下的砷浓度值高几个数量级。在尾矿堆和尾矿污染的土壤中砷可高达数千 mg/kg。如此高的砷浓度不仅反映出原生富砷硫化物矿物的大量存在，而且反映出原生富砷矿物经过次生变化形成的砷酸铁盐和铁的氧化物的增加。尾矿堆中原生的硫化物矿物易于氧化，而次生矿物在氧化条件下的地表水和地下水中具有不同的溶解度。臭葱石（$FeAsO_4·2H_2O$）是一种常见的硫化物氧化产物，其溶解度决定了周围环境中砷的浓度，特别是在氧化条件下，砷经常被铁的氧化物紧密结合，此时的砷相对稳定，易于在土壤中形成累积[3]。

2.2.4 食品砷污染来源

总体而言，食品中的砷主要来源于食品生产、加工和储存过程。通常情况下，动物机体、植物中均含有微量的砷。食物中的砷主要来源于其生长过程中土壤、大气和水体的吸收与富集。人体摄取的砷大部分来源于饮用水和食物，通过呼吸途径进入体内的砷不及1%。在墨西哥和印度部分地方性砷中毒病区人体从食物中摄入的砷量占总砷摄入量的20%。Schoof等[45]对美国和欧洲普通人群的食物砷摄入量调查显示，稻米中砷摄入量的贡献率仅次于鱼类产品，位居第2位。在印度饮用水受到轻度砷污染的地区，Uchino等[46]发现蔬菜和谷类作物的砷输入是人群摄取砷的主要途径，其摄入量占总砷摄入量的73.3%以上。李筱薇等[47]对中国12个省、市、自治区的膳食调查表明，成年男子从蔬菜和粮食类（包括谷类、豆类、薯类）中摄入的砷占膳食砷摄入总量的75.6%，其中粮食类占60.4%。王振刚等[48]对湖南石门雄黄矿区附近人群砷暴露（主要途径为水和食物）的调查显示，未受砷污染地区的人群中食物砷的贡献率占83.6%。由此可见，土壤—食物—人体暴露是普通人群摄入砷的最重要途径，蔬菜和粮食中的砷往往是非职业性暴露的主要摄入途径。肖细元等[49]对蔬菜和粮油作物中砷的累积特点和富集能力进行了统计分析，结果表明，土壤砷浓度直接影响粮油作物的砷含量，土壤中砷含量与粮油作物的砷含量呈极显著的正相关，清洁区和污染区蔬菜砷含量变幅分别为 0.001~1.07 mg/kg 和 0.001~8.51 mg/kg（鲜重），均值分别为 0.035 mg/kg 和 0.068 mg/kg。不同种类蔬菜的砷含量由大到小依次为：叶菜类>根茎类>茄果类>鲜豆类；清洁区和污染区粮油作物的砷含量变幅分别为 0.001~2.20 mg/kg 和 0.007~6.83 mg/kg（干重），均值分别为 0.081 mg/kg 和 0.294 mg/kg，其中水稻的砷含量显著高于小麦和玉米。从富集系数来看，叶菜类蔬菜的砷富集系数最高，芹菜、蕹菜、茼蒿、芥菜等蔬菜的抗砷污染能力较弱，而粮食作物玉米的抗砷污染能力较强。

动物因其所处环境不同，摄入量存在差异而导致它们之间砷积蓄量存在较大不同。海洋中腔肠动物、某些软体动物和部分甲壳动物砷含量为 0.005~0.3 mg/kg，淡水鱼平均砷含量为 0.54 mg/kg（湿基），但某些鱼肝油砷含量可达到 77 mg/kg（湿基）[3]。

在食品原料、辅料、食品加工、储存、运输和销售过程中使用和接触的机械、管道、容器、包装材料以及因工艺需要加入的食品添加剂中，均可造成砷对食品的污染。特定的环境条件也可造成严重的食品砷污染，例如贵州、陕西等地区高砷煤燃烧所产生的烟气也是食物砷污染的重要污染源，居民燃用高砷煤做饭取暖，炉灶无烟囱，玉米、辣椒等放于炉灶上层烘烤，使食物受到室内煤烟污染，农民通过长期食入与吸入途径摄取大量的砷造成了严重的健康损害问题。

3 环境砷污染健康危害

环境中砷可通过经口摄入、呼吸道吸入和皮肤吸收等途径进入人体，相对稳定地分布于皮肤、肝、肺、肾等组织器官中。肝脏是砷的主要代谢器官，最终约有70%的砷（无机砷和/或有机砷）经肾脏通过尿液排出，无机砷（iAs）在体内的存留时间比有机砷长，且 iAs 的排泄过程更长[50]。在整个吸收、分布、代谢和排泄的过程中，砷及其代谢产物可与机体多个器官系统产生相互作用，目前研究认为，砷可对全身几乎所有器官造成危害，被美国毒物和疾病登记署（ATSDR）认定为最高级优先管理化学毒物[51]。本部分综述了砷暴露对人体多器官、多系统的损害作用。

3.1 砷对皮肤系统的影响

皮肤作为机体最大的器官,已被证实为最容易受到砷中毒影响的器官。由于皮肤损伤是慢性砷中毒的早期表现,因此通常被当作慢性砷中毒的典型标志[52]。慢性砷暴露引起的皮肤损害特征为皮肤色素改变(色素脱失、色素沉着)、皮肤角化和皮肤癌,这些关键特征现已被包括中国在内的多个国家作为慢性砷中毒的诊断标准。

(1) 皮肤色素改变:慢性砷中毒病人皮肤色素改变包括色素沉着和色素脱失两种,通常表现为细小的斑点、雨滴状的色素沉着或脱色,也有局限性或斑片状色素沉着,其中在躯干和四肢明显,一般不涉及面部皮肤[53]。

(2) 皮肤角化过度:慢性砷中毒皮肤角化过度主要发生在手掌和脚底。早期阶段,受累皮肤可能有一种硬化的、砂砾状特征,通常进展形成易观察到的隆起、点状、疣状角化灶。严重情况下,手掌和脚底出现弥漫性疣状病变,脚底可能出现严重的裂纹和裂缝。角化皮肤的病理组织学检查显示广泛的角化过度和角化不全,常可见不同程度的鳞状细胞增生和真皮无嗜碱性变性[54]。

(3) 皮肤恶性病变:慢性砷中毒皮肤病变严重者可出现鲍文氏病。一般认为该疾病从发病时起就是一种皮内癌,其皮肤病变通常为红斑、色素沉着、结痂、裂隙和角化,有些可能是结节状、溃疡或侵蚀。2012 年 IARC 基于中国台湾西南高水砷暴露地区的生态学研究,确认了饮水砷暴露与皮肤癌之间的因果关系。到目前为止,几乎所有已发表的关于砷暴露与皮肤癌之间关系的研究,都为砷暴露与非黑色素瘤皮肤癌(基底细胞癌和鳞状细胞癌)间具有关联性提供了证据[55, 56]。

3.2 砷对消化系统的影响

急性砷中毒消化系统症状明显,如呕吐、腹痛、腹泻等,严重者可发生昏迷并伴发中毒性肝病等,若抢救不及时可在短期内死亡。慢性砷暴露对消化系统的影响主要表现为食欲减退、恶心、腹痛、腹胀、腹泻、便秘、肝区疼痛及肝脾肿大等[57]。砷所致肝病的临床体征包括食管静脉曲张出血、腹水、黄疸或肝肿大,血清肝功能指标异常等;在严重毒性的晚期,可能发生肝纤维化、非肝硬化门静脉纤维化、肝硬化甚至肝衰竭[57]。研究显示,印度西孟加拉邦的砷暴露(水砷 0.05~3.20 mg/L)人群中,血清球蛋白、碱性磷酸酶、丙氨酸氨基转移酶和天门冬氨酸氨基转移酶水平均明显升高,并伴随肝脏肿大[58]。中国贵州燃煤污染型砷中毒病区的研究发现,砷中毒患者可出现不同程度的肝肿大和肝损害,肝硬化腹水是当地砷中毒人群的主要死因之一[59]。砷暴露与肝癌的相关性至今仍无定论,但一项纳入了 7 项研究的 Meta 分析结果显示,长期的饮水砷暴露可导致肝癌的死亡率增加[60]。

3.3 砷对呼吸系统的影响

长期从环境中接触过量的砷,可能导致肺癌的高发和多种非恶性呼吸系统疾病,中国台湾、日本、智利、阿根廷和美国的高砷接触人群中,均观察到了这种关联。在采矿或磨矿过程中吸入砷尘或烟尘常会发生呼吸系统并发症,如呼吸道症状、气道上皮损伤、肺功能受损、慢性阻塞性肺疾病等[54]。队列研究中观察到,一些常见临床呼吸道症状包括慢性咳嗽、胸声、呼吸急促、痰中带血等呼吸问题均与当地人群砷暴露明显相关[61]。前瞻性队列研究发现,孟加拉国人群低至中等水平的砷暴露会导致肺功能受损和结核病的高发[62]。Dauphine 及其同事发现,在子宫内和儿童时期通过饮水接触砷与人类长期肺功能和非恶性肺疾病有关[63]。在子宫内和生命早期暴露于高水平饮水砷(平均 152.13 μg/L)的儿童,其肺功能较非砷暴露儿童明显降低[64]。

3.4 砷对神经系统的影响

脑是砷中毒的关键靶器官，不同形态的砷可分布在大脑的各个部位，其中在脑垂体的蓄积量相对最高[65]。急性砷中毒引起的神经系统症状发展相当快，常见的症状包括头痛、幻觉、四肢无力、麻痹、癫痫和昏迷等[66]。越来越多的人群和动物研究表明，长期砷暴露与神经功能缺陷有关，常见的临床症状包括足底感觉异常、疼痛和麻木，感觉神经比运动神经对砷更为敏感，其机制与氧化应激诱导的细胞凋亡及对神经递质的影响有关[67, 68]。神经心理学方面的研究证实砷暴露可对儿童记忆和语言学习能力造成严重损害[69]。此外，亦有因长期低水平砷暴露而引起的阿尔茨海默症及其相关疾病的报道[70]。

3.5 砷对免疫系统的影响

近年来，砷对机体免疫系统的影响受到了国内外研究人员的广泛关注。在儿童和成人的多项横断面研究结果均表明，砷可能具有免疫抑制作用[71]。巨噬细胞是免疫球蛋白免疫毒性的主要靶细胞，砷作用于巨噬细胞后，会导致其黏附能力丧失及表面标志物表达的改变，进而影响内吞和吞噬作用。除巨噬细胞外，砷暴露还会损害淋巴细胞的发育、活化、增殖和功能[72, 73]。中国贵州的研究表明，燃煤砷暴露可导致患者免疫功能出现持久性损害，包括引起人体免疫球蛋白、补体含量、细胞因子的改变，抑制T淋巴细胞的增殖和功能[74]。此外，砷在体内可促进炎症反应，增加体内炎症分子的表达。最近的研究结果表明，砷在生命早期具有免疫毒性，在子宫内接触砷会影响免疫介导细胞和胸腺发育[75, 76]。孟加拉国的一项研究发现，长期接触砷可引起学龄前儿童体内的 Th1 细胞因子水平降低，进而导致细胞介导的免疫功能下降，但 Th2 细胞因子未见变化[77]。

3.6 砷对心血管系统的影响

大量研究显示，高砷暴露与心血管疾病（CVD）及其协同危险因素（如动脉粥样硬化、高血压、心律失常和糖尿病）之间存在强关联性，并呈一定剂量依赖关系，同时与某些重要的心脏并发症如心肌损伤、心律失常和心肌病有直接关系[78, 79]。来自中国台湾西南地区、美国犹他州和孟加拉国等的队列研究显示，高砷饮水暴露可明显增加当地人群心血管疾病发病风险和死亡率[80]。来自西班牙的研究发现，低至中等水平的城市饮水砷暴露（< 1～118 μg/L）也与人群心血管疾病死亡率上升明显相关[81]。此外，砷暴露与高血压的发病关联在孟加拉国、伊朗、印度和中国等多个国家均有报道[54]。然而基于 Meta 分析的结果表明，由于证据有限，砷暴露与高血压之间的关系仍然存在争议和不确定性[82]。心电图 QT 间期代表心室的缩小和复极化，QT 间期长短的变化与恶性室性心律失常的发生密切相关。最近的许多研究表明，饮水中砷的暴露与 QT 间期延长有很强的相关性，且低砷暴露水平即可明显改变 QT 间期持续时间[83, 84]。乌脚病（BFD）是一种典型的外周血管疾病，临床表现为严重的动脉硬化、进行性动脉闭塞和随后的下肢坏疽。中国台湾西南部地区的流行病学研究证实，饮水中高砷暴露和无机砷甲基化代谢能力降低是 BFD 发展的主要原因[85]。

3.7 砷对内分泌系统的影响

众所周知，砷是一种环境内分泌干扰物，其可影响包括胰腺、甲状腺、性腺等多个内分泌相关组织器官[54]。胰腺产生胰岛素、胰高血糖素并调节机体的血糖水平。研究表明，砷可在胰腺中积累，进而降低胰岛素的分泌和细胞活力[86]。来自塞浦路斯、孟加拉国、墨西哥、瑞典、中国与美国等多个国家的人群流行病学调查发现，高砷暴露或长期低水平砷暴露均与 2 型糖尿病（T2DM）的发病风险密切相关[54]。甲状腺在调节人体代谢方面发挥着重要作用，负责三碘甲状腺素（T3）和甲状腺素（T4）的分泌，砷与甲状腺之间的相互作用尚未被充分证实。然而最近有研究报道，发现户外工作人员职业砷暴露与其对甲状腺激素和甲状腺的影响存在明显相关关系[87]。来自美国的研究亦显示通过地下水接触低水平砷（2～22 μg/L）与居住在得克萨斯州西部乡村的人群甲状腺功能减退明显相关[88]。

3.8 砷对泌尿系统的影响

砷对泌尿系统的损伤主要集中在肾脏清除砷及其代谢产物的过程中。越来越多的证据表明，砷暴露可导致人类和小鼠肾内尿路上皮组织和肾细胞癌[54]。在智利等地的水砷浓度降低后，肾癌（尿路上皮癌和肾细胞癌）、肾疾病死亡率和蛋白尿发生率的也发生下降，这为两者间的因果关系提供了证据[89]。砷中毒引起肾损伤（急性肾小管坏死）的主要临床表现有硫脲水平降低，血清肌酐、血尿素氮和蛋白尿水平升高等[90]；此外，砷还可能对近曲小管、肾脏毛细血管和肾小球造成损害[91]。近年的研究表明，慢性肾病是砷暴露引起机体损伤的一种重要并发症，来自美国、智利、孟加拉国等砷中毒病区人群流行病学的研究均发现暴露在高砷水平的环境中会显著增加当地人群肾脏疾病的死亡率[54]；斯里兰卡学者成功建立了农药砷暴露与慢性病因不明性肾病（CKDu）流行之间的可能联系，组织病理学研究表明，砷暴露与肾小管间质性肾炎、肾小球萎缩和肾小球丢失之间存在明显相关性；CKDu 患者尿砷、管状蛋白尿和中性粒细胞明胶酶相关脂质运载蛋白（NGal）水平明显升高（>300 μg/mg 肌酐/dL）[92]。

3.9 砷对生殖发育系统的影响

研究表明，无机砷暴露对生殖和发育均有明显毒性。砷可影响男性和女性的性器官，并可能导致两性的生育问题。在男性中，砷可能通过引起睾丸激素合成下降、凋亡和坏死诱导性腺功能障碍[93]。实验研究和人群调查均表明，无机砷及其代谢物很容易通过胎盘进入胎儿体内，引起胎儿生长发育迟缓甚至死亡[94, 95]。最近用斑马鱼动物模型进行的一项研究表明，砷暴露通过下调 Dvr1 的表达来影响胚胎发育[96]。包括前瞻性队列研究在内的许多流行病学研究均已提供证据表明孕期饮水中砷暴露可能会对胎儿和婴儿的生长及存活造成剂量依赖性损害[97, 98]。来自孟加拉国的横断面研究发现，饮水砷暴露（范围为 0～1710 μg/L）与当地人群的自然流产、死产和新生儿死亡有很强的相关性[99]；在中国台湾东北部的砷中毒流行地区，饮用砷污染（范围 0.15～3585 μg/L）井水与和胎儿早产以及低出生体重之间也存在一定关联[100]。

4 环境砷污染致病机制

自然环境中的砷及其化合物通过多种途径进入机体后，经过一系列复杂的生物转运转化过程，几乎可以影响人体的每一个器官。这些过程受到砷及其化合物的理化性质、机体状态、环境等多种因素的影响，同时，砷与机体的作用机制十分复杂，无机砷至今尚未成功复制出动物致癌模型，其确切的致病致癌机制目前仍未明确。近年来，随着检测分析方法、研究手段等的不断发展，在环境砷污染致病机制的研究上取得了一系列新进展，主要涉及氧化损伤、遗传损伤、信号通路表达改变、表观遗传调控、免疫损伤及炎症反应等方面。本节重点综述了近年来环境砷污染致病机制的主要研究进展。

4.1 砷致氧化损伤

氧化应激（oxidative stress，OS）指机体在遭受有害刺激时，体内活性氧自由基（reactive oxygen species，ROS）和活性氮自由基（reactive nitrogen species，RNS）等高活性分子产生过多，氧化程度超出抗氧化物的清除能力，造成氧化、抗氧化系统失衡，进而导致组织损伤的过程。多年来研究已证实，组织或细胞氧化应激是砷对机体的主要损害效应[101]。砷在体内的代谢主要是一系列氧化-还原反应和甲基化反应，无机砷化合物（As^{III}/As^{V}）进入机体后，通过氧化还原和甲基化反应生成一甲基胂酸（MMA^{V}）或一甲基亚胂酸（MMA^{III}），在此基础上进一步进行氧化还原反应生成二甲基胂酸（DMA^{V}）或二甲基亚胂酸（DMA^{III}）；在长期摄入海产品的人群体内还可检测到一定量的三甲基胂氧化物（$TMAO^{V}$）[51]。在上述代谢过程中可产生不同类型自由基，砷及其甲基化代谢产物亦可直接攻击细胞造成损伤并产生活性氧自由基从而诱发氧化应激，进一步可造成机体内多种细胞和分子信号异常、细胞周期改变、凋亡紊乱等结局。

4.1.1 砷对ROS的影响

如上所述，砷在细胞内的代谢过程中可产生多种类型的ROS。目前研究认为砷暴露诱导ROS升高的主要途径有亚砷酸盐氧化为砷酸盐过程中产生ROS、砷在体内代谢过程中产生甲基胂而介导ROS产生、通过线粒体电子传递链的复合物Ⅰ和复合物Ⅲ和甲基化砷使铁蛋白释放铁而引发Haber-Weiss反应等。实验研究表明，砷可诱导多种体外细胞系产生氧自由基和过氧化氢（H_2O_2），也可增加细胞内与ROS密切相关的酶的表达，如超氧化物歧化酶、苏/丝氨酸激酶和苯醌氧化还原酶等[102]。DMA^{III}与氧分子反应形成二甲基胂过氧化物，通过氧化应激而导致DNA氧化损伤，始动砷的致癌过程。近年来众多研究发现，砷可诱导体内最重要的抗氧化信号通路——Keap1-Nrf2/ARE信号通路改变，也证实了砷可介导ROS生成[103, 104]。8-羟基脱氧苷（8-OHdG）是一种敏感的DNA氧化损伤标记，在砷诱导的皮肤癌组织中可检测出高浓度8-OHdG，进一步研究认为，8-OHdG的升高既可能是ROS的作用，也可能是直接的电子转移引起DNA损伤所致[105]。

4.1.2 砷对抗氧化酶的影响

大量研究证实，砷在诱导ROS的同时，亦可影响多种抗氧化酶的活性，如超氧化物歧化酶、谷胱甘肽-S-转移酶、谷胱甘肽过氧化物酶、谷胱甘肽还原酶、血红素加氧酶、硫氧蛋白还原酶和NADPH氧化酶和细胞色素P450酶等[106]。砷对这些抗氧化酶活性的变化，与砷暴露剂量、暴露时间和作用靶器官的不同均密切相关。一般情况下，低水平短期砷暴露可诱导上述抗氧化酶的活性升高，而慢性砷暴露则通常导致其活性降低。研究已证实，砷可诱导NADPH氧化酶的关键亚基高表达、磷酸化活化或膜转位[107]。

GSH是组织细胞中巯基最丰富的非蛋白，其可直接与外来亲电子基团结合，也可作为谷胱甘肽-S-转移酶和谷胱甘肽过氧化酶等酶的辅助因子，还可在砷的氧化还原及甲基化代谢中作为重要的电子供体，在抵抗砷所诱发的氧化应激中发挥重要作用。此外，GSH还可参与调节半胱氨酸蛋白酶、应激激酶和转录因子等砷毒作用的靶蛋白特定巯基残基的氧化还原反应过程。因此，大剂量急性砷暴露和慢性砷暴露均可消耗大量GSH，甚至引起GSH耗竭，低水平短期砷暴露可导致肝组织GSH水平适应性或代偿性增加，而长期低水平砷暴露，则会导致GSH明显减少[108]。

4.1.3 砷对关键抗氧化通路的影响

Keap1-Nrf2/ARE作为体内最重要的抗氧化信号通路，近年来成为砷致氧化应激机制的研究热点。其中 Nrf2 和它的胞浆抑制蛋白 Keap1 是细胞抗氧化反应的中枢调节分子，多种氧化剂和亲电子剂均能促使 Nrf2 与 Keap1 解离，进而核转位激活并与抗氧化反应原件（ARE）结合而启动下游一系抗氧化应激蛋白和 II 相解毒酶的基因转录。砷及其化合物已被证实是常见的 Nrf2 激活剂，近年来大量研究已证实，砷可激活多种人类细胞系中的 Keap1-Nrf2/ARE 通路，包括皮肤角质形成细胞、成骨细胞、膀胱上皮细胞、胎盘绒毛膜癌细胞、骨髓瘤细胞等，进而上调 HO-1、NQO1、过氧化物氧化还原酶 1 和 γ-谷氨酰半胱氨酸合成酶等[109]。而在长期砷暴露人群或砷暴露导致的体外恶性转化细胞中，Nrf2 或其调控的抗氧化基因/蛋白的表达水平也保持升高，如在对燃煤型砷中毒人群的研究中发现，长期砷暴露可导致砷中毒人群外周血细胞中 Nrf2 表达增加、Nrf2-ARE 结合能力增强，诱导下游基因表达，并与砷中毒患者的肝损伤程度相关[110]；在低剂量亚砷酸钠长期染毒所致恶性转化的 HaCaT 细胞中，也存在 Nrf2 的高表达，提示砷诱导细胞恶性转化可能与 Nrf2 介导的抗氧化反应紊乱有关[111]。总之，Keap1-Nrf2/ARE 抗氧化通路失调参与砷致肿瘤的发生发展，其具体的机制尚待进一步研究。

4.2 砷致遗传损伤

砷是一种遗传毒物，其对细胞 DNA 损伤效应和 DNA 修复抑制是其致病致癌作用的重要分子机制。砷可引起哺乳动物和人细胞 DNA 单链断裂、微核、姊妹染色单体互换、染色体畸变的发生率增加，且还可增强紫外线等其他 DNA 损伤物质的致突变能力，诱导细胞增殖异常、凋亡从而产生相关损害效应。

4.2.1 砷致 DNA 损伤

研究显示，低水平 As^{3+} 能诱导人肝细胞的 DNA 蛋白质交联物形成，而且 DNA 蛋白质交联物随砷浓度增加而升高，具有剂量-效应关系[112]。DNA 链断裂常由于氧化 DNA 加合物和 DNA-蛋白质交联的切除产生，采用单细胞凝胶电泳（SCGE）对砷中毒人群血细胞进行检测，发现 DNA 单链断裂发生早于其临床表现[113]。关于砷暴露对机体基因突变的影响，研究较多的是 $p53$ 基因，$p53$ 基因与染色体畸变、氧化应激、生长因子分泌异常，细胞增殖加快、肿瘤进展等过程关系密切，在 P53 功能丧失的情况下，砷暴露可通过诱导 DNA 修复缺陷，细胞周期蛋白 D1 的表达和 P53 依赖性 P21 表达升高，引起 DNA 损伤[114]。此外，砷可抑制编码人端粒酶的逆转录酶亚基 hTERT 基因表达水平，引起端粒酶活性降低，进而促进 DNA 损伤、基因组不稳定及肿瘤的发生[115]。砷及其代谢产物均可引起染色体畸变。对芬兰饮水砷暴露的居民中尿砷浓度与淋巴细胞染色体畸变之间的关联研究发现，在多重回归模型中排除性别、年龄和吸烟等因素后，总尿砷含量与染色体畸变率成正相关，且尿中 MMA/总砷比值与染色体畸变率也呈明显正相关[116]。研究表明，不同有机和无机砷化合物作用于人成纤维细胞，可诱导染色体缺失和断裂，其诱变能力由弱到强依次为 TMA<MMA<DMA<砷酸<亚砷酸[117]。砷及其主要代谢产物 DMA 均可明显诱发非整倍体的产生，但并不抑制纺锤体的形成，提示砷化物可抑制有丝分裂、干扰细胞周期中纺锤体的动力、加速纤维微管的聚合作用[118]。

4.2.2 砷致 DNA 修复障碍

机体对 DNA 损伤有多种修复系统，以保证遗传信息的完整性和稳定性，研究表明，砷暴露诱导 DNA 损伤的同时还可干扰 DNA 损伤修复过程，并且由于 DNA 修复酶对砷的抑制较敏感，砷对 DNA 修复有关的酶的抑制比 DNA 修复酶本身的作用更强。在贵州省燃煤型砷中毒病区的研究发现，砷可引起砷中毒患者 DNA 损伤修复基因（MGMT、hMLH1、hMSH2、ERCC1、ERCC2）启动子区高甲基化，从而抑制其表达[119, 120]。体外研究发现，砷可增加成纤维细胞对紫外线的敏感性，其机制可能是砷与 DNA 修复有关的 DNA 连接酶或修复酶的巯基相结合，从而抑制 DNA 的损伤修复。另有研究发现，三价砷可以结合到范可尼贫血互补 L 基因泛素连接酶的环指结构区而诱导 DNA 损伤修复蛋白范可尼贫血 D2 基因的泛素化降解,从而抑制 DNA 修复[121]。砷的三价甲基化代谢产物 MMA^{III} 和 DMA^{III} 可抑制在 DNA 修复、基因组稳定性维持和细胞周期控制中均发挥重要作用的 ADP-核糖基生成,还可在致癌过程中对几种基础核苷酸切除修复(BER)的关键分子产生明显影响[122]。

4.3 砷致信号通路异常

细胞所有的生命活动均受到细胞信号通路的调控,砷引起细胞信号通路异常是当前砷中毒机制研究热点之一。砷可通过多种途径影响细胞代谢、炎性反应、氧化应激、细胞凋亡等相关通路,诱导一系列细胞功能相关基因表达水平异常,参与砷所致的多器官、多系统损伤。

4.3.1 砷与 NF-κB 信号通路

核因子-κB(nuclear factor kappa B,NF-κB)是一种多向性、多功能的核转录因子,可调控一大类基因的表达,如免疫相关受体、细胞因子、炎症因子、黏附分子等,在免疫细胞的激活、细胞凋亡和增殖、炎症反应等过程中发挥重要作用。在静息状态时,NF-κB 与其抑制因子 IκB 结合以失活形式存在于细胞胞浆中,上游信号的刺激可使 NF-κB 激酶抑制剂(IKK)通过磷酸化或形成分子内二硫键而被激活,进而使 IκB 发生磷酸化并从 NF-κB/IκB 复合物中脱落下来,被激活 NF-κB 转入细胞核中,结合到靶基因的启动子区,从而启动靶基因表达。

不同浓度和时间砷暴露对不同细胞引起 NF-κB 信号通路改变,可产生不同效应[123]。在贵州省燃煤型砷中毒人群中的研究发现,砷暴露可以通过激活 NF-κB 改变炎症相关基因的表达、促进促炎症介质的过量产生,参与砷中毒的发生发展。在低剂量亚砷酸钠(1 μmol/L)致 HaCaT 细胞恶性转化过程中,砷可以激活 NF-κB,进入核内的 NF-κB 结合到 mot-2 启动子区,从而调控 mot-2 转录表达水平进而阻止 p53 的核转位,最终导致 P53 功能失活[124]。砷通过 ROS 激活 NF-κB 通路,可调控一些抗氧化或保护基因(超氧化物歧化酶、过氧化氢酶、金属硫蛋白、血红素氧合酶-1 和谷胱甘肽过氧化物酶 1)和促氧化基因(黄嘌呤氧化还原酶、氮氧合酶、花生四烯酸 5-脂氧合酶、细胞色素 P450 酶)表达,这些基因产物又可以反过来对细胞内的 ROS 进行负反馈或正反馈调控,形成一个复杂调控网络[125]。

4.3.2 砷与 MAPK 信号通路

丝裂原活化蛋白激酶(mitogen-activated protein kinase,MAPK)是哺乳动物体内广泛存在的一类丝/苏氨酸蛋白酶,可被一系列的细胞外信号或刺激激活,对细胞增殖、分化、凋亡、转化有重要调控作用,并与炎症、肿瘤等的发生密切相关。MAPK 通路主要包括 3 条亚信号通路:细胞外信号转导激酶(ERK)信号通路、c-Jun N-端激酶(JNK)信号通路和 p38 信号通路。ERK 通路的主要功能是促进细胞增殖与分化,而 JNK 和 p38 信号通路主要是促进细胞凋亡与死亡。

砷化物能够在多种类型细胞中激活 MAPK 通路,且与其所致细胞增殖和恶性转化有关。亚砷酸钠激活不同 MAPK 亚信号通路主要取决于其浓度、作用时间和靶细胞类型。亚砷酸钠所致肝 TRL1215 细胞恶性转化过程中抑制 JNK 信号通路、激活 ERK 信号通路,从而促进细胞增殖,抑制其凋亡[126]。不同浓度砷可激活不同 MAPK 亚信号通路,对细胞的增殖和凋亡产生双向作用,大鼠肺支气管上皮细胞暴露于低剂量亚砷酸钠(2 μmol/L)可激活 ERK 信号通路,提高细胞增殖率,此作用可被 ERK 抑制剂所阻断;而在高剂量(20 μmol/L)亚砷酸钠暴露下,可激活 JNK 信号通路,提高细胞凋亡率,此作用同样可被 JNK 抑制剂所阻断[127]。

4.3.3 砷与 PI3K/Akt 信号通路

磷脂酰肌醇 3-激酶（phosphatidylinositol 3-hydroxy kinase，PI3K）是细胞内重要的信号转导分子，响应于外源性生长因子或细胞因子刺激，其可在细胞膜上催化 PtdIns4,5-二磷酸（PIP2）的磷酸化以产生脂质第二信使 PtdIns 3,4,5-三磷酸（PIP3），进一步活化 Akt，通过磷酸化并调控下游多种与细胞代谢、凋亡、增殖和分化有关的蛋白。

越来越多的研究表明，砷可通过激活 PI3K/Akt 通路，在砷致细胞损伤和恶性转化中发挥作用。采用 As_2O_3 处理胶质瘤细胞后，磷酸化 Akt 蛋白表达下降，凋亡相关蛋白表达升高，从而促进了细胞凋亡[128]。砷处理人类支气管上皮细胞（BEAS-2B）和原代人支气管上皮（HBEC）后，可激活 AKT 和 mTOR 活性，细胞表现出较高的增殖率、存活率和锚定独立生长等恶性细胞转化标志[129]。亚砷酸盐处理皮肤 HaCaT 细胞可以激活 PI3K/Akt 通路，进而引起其下游细胞周期素 D1 高表达和细胞增殖加快，用 PI3K/Akt 通路抑制剂预处理细胞或敲除 Akt 后，再用亚砷酸盐处理 HaCaT 细胞，则不引起上述改变，提示 PI3K/Akt 通路在砷所致 HaCaT 细胞恶性转化过程中发挥重要作用[130]。

4.3.4 砷与 HIFs 信号通路

缺氧诱导因子（hypoxia-inducible factors，HIFs）是细胞感知和适应氧气水平变化的中枢调控因子，其不仅是缺氧条件下维持氧稳态的关键因子，对 ROS 和 NO 也很敏感。目前已发现的 HIF 家族包括 HIF-1、HIF-2 和 HIF-3，其均由 HIFα 和 HIFβ 两个亚单位构成的异源二聚体复合物。HIFα 既是调节亚基，又是活性亚基，其蛋白质稳定性和活性均受细胞内氧浓度的调节；HIFβ 是许多转录因子的共同亚基，其对氧不敏感，可存在于任何氧浓度下。

研究表明，砷可产生 ROS，也可使细胞利用氧障碍而产生缺氧，从而使细胞 HIFs 泛素化分解途径受阻而高表达，进一步与调控基因的缺氧反应元件（hypoxia response element，HRE）结合而引起调控基因的高表达。HIFs 调控基因的功能相当广泛，涉及细胞的能量代谢、增殖和生存、凋亡、血管生成、浸润与转移等多种细胞生物学效应。研究表明，亚砷酸可激活 ERK 信号通路，抑制 HIF-2α 的泛素蛋白酶体降解，从而诱导转录因子 HIF-2α 水平升高，升高的 HIF-2α 一方面通过调控抑癌基因 *p53*，使其失活，从而刺激细胞过度增殖、促进细胞发生恶性转化；另一方面通过直接作用于 Twist1 和 Bmi1 而促进细胞发生上皮-间质（epithelial-to-mesenchymal transition，EMT）和肿瘤干细胞（cancer stem cells，CSCs）特性获得，还可通过间接调控亚砷酸盐诱导的炎症反应而进一步促进细胞发生 EMT，最终在细胞恶性转化中发挥重要作用[131]。

4.4 砷致表观遗传改变

随分子生物学的发展，表观遗传学在砷中毒及砷致癌机制中的作用成为研究热点。目前，相关研究主要集中在砷对机体 DNA 甲基化、组蛋白修饰非编码 RNA（non-coding RNA）表达及功能的影响上。这些表观遗传在调节基因表达、维持细胞正常功能、胚胎发育中起重要作用，其调控模式的失常参与了多种重大疾病的发生，如肿瘤、自身免疫性疾病、代谢性疾病等，众多的研究结果表明，表观遗传机制广泛参与了砷的致病致癌过程[132-134]。

4.4.1 砷致 DNA 甲基化改变

DNA 甲基化模式对于哺乳动物的发育至关重要，是研究最多的表观遗传改变。DNA 甲基化是在 DNA 甲基转移酶（DNMT）的作用下，以 *S*-腺苷甲硫氨酸（SAM）为甲基供体，将甲基基团结合到 5′-CpG-3′

中胞嘧啶的第 5 位碳原子上，形成 5-甲基胞嘧啶（5-MeC）的 DNA 序列修饰，DNA 甲基化状态可调节基因转录、影响基因表达调控和基因组稳定性等，在疾病和肿瘤的发生和演进过程中扮演重要角色。

砷在环境和生物体中均有甲基化代谢的转化过程，进入机体的无机砷以 SAM 为甲基供体，经过两次氧化还原后分别生产一甲基胂酸和二甲基胂酸。无机砷在人体的甲基化代谢与 DNA 甲基化可能存在对甲基供体 SAM 的竞争，也可能直接影响 DNMTs 的活力，从而影响机体的 DNA 甲基化状态。细胞和动物实验及流行病学研究表明，砷暴露可影响全基因组 DNA 甲基化水平，如导致小鼠肝组织、人皮肤角质形成细胞、恶性转化的人前列腺上皮细胞、人外周血淋巴细胞等全基因组低甲基化[135]。同时砷暴露还可影响特定基因（如癌基因、抑癌基因、DNA 损伤修复基因等）的甲基化水平，如无机砷可引起叙利亚猴胚胎细胞发生癌变，在癌变细胞中发现 *c-myc*、*c-fos* 和 *c-Ha-ras* 等原癌基因发生低甲基化，进而引起基因表达水平升高而产生细胞周期紊乱[136]。另有研究发现摄食缺乏甲基饲料的小鼠接触无机砷，可以加剧 DNA 低甲基化，如引起癌基因 *c-Ha-ras* 基因 5'端调控区的低甲基化[137]。在饮水和燃煤型砷中毒研究中发现，砷可导致砷中毒患者 $p15^{INK4\beta}$、*p16* 基因、DNA 损伤修复基因（*MGMT*、*hMLH1*、*hMSH2*、*XPD*、*ERCC1*、*ERCC2*）高甲基化，进而影响其表达，参与砷中毒发生发展[119, 120]。

此外，妊娠期砷暴露可能通过影响表观基因组，导致胎儿的出生缺陷、发育障碍和生命晚期疾病[138]。研究者检测了英国新罕布什尔州的一个出生队列的 134 名婴儿的脐带血 DNA 的甲基化情况，结果显示，基于砷暴露水平的不同，CpG 的甲基化存在差异，在 44 个 CpG 岛中，与于低剂量砷暴露组相比，高剂量砷暴露组中 75%的 CpG 表现出更高的甲基化水平[139]。墨西哥的一项妊娠队列研究（饮用水砷浓度为 0.456～236 μg/L）筛查了 38 个脐带血样本中超过 40 万个 CpG 位点的甲基化变化情况，并重点研究了 16 个与砷相关的甲基化状态改变的基因，16 个基因中有 7 个基因的 DNA 甲基化水平与胎龄和头围的差异有关，且这 16 个基因被富集到特定转录因子的结合位点，这些转录因子被证实可被砷暴露改变并影响下游信号通路[140]。

4.4.2 砷调控组蛋白修饰

在细胞中，核小体是染色质的基本组成单位，其由核心组蛋白八聚体（2 个拷贝的 H2A，H2B，H3 和 H4）及缠绕在外的 DNA 组成，组蛋白 N 末端氨基酸残基可发生由特异性酶引起的多种共价修饰，包括甲基化、乙酰化、磷酸化、糖基化、瓜氨酸化和泛素化，从而改变染色质结构而影响基因的转录水平。

组蛋白修饰与砷毒作用密切相关，H3K4、H3K9、H3K18、H3K27、H3K36、H4K20 等组蛋白修饰参与了砷中毒发生发展，三价无机砷处理哺乳动物细胞可引起 H3K9me2、H3K4me2、H3K4me3 增加及 H3K27me3 减少[141]。亚砷酸钠作用于构建的组蛋白 H3 赖氨酸修饰位点突变的 HBE 细胞株，发现 H3K4 甲基化修饰在亚砷酸钠诱导的细胞毒性和遗传毒性中起重要作用[142]。砷暴露人群外周血淋巴细胞 H3K18ac 和 H3K36me3 水平改变与尿砷、发砷水平及尿 8-OHdG 水平相关，表明其参与了砷致氧化损伤[143]。在砷中毒患者中，淋巴细胞组蛋白 H3K36me3、H4K20me1、H4K20me2 修饰水平的改变与砷诱导的 DNA 损伤密切相关[144]。

4.4.3 砷与非编码 RNA

近来的研究鉴定了大量非编码 RNA，包括微小 RNA（microRNA，miRNA）、长链非编码 RNA（long non-coding RNA，lncRNA）和环状 RNA（circular RNA，circRNA）等。这些调节性的 RNA 可以在基因转录、RNA 成熟和蛋白质翻译等各个水平调控基因表达，参与发育、分化和新陈代谢等多种生物过程，也参与许多人类疾病的发生发展。

miRNA 为长度在 20～22 个核苷酸的单链 RNA 分子，miRNA 可以通过与靶 RNA 分子的 3'非翻译区结合而阻滞其翻译，从而抑制其蛋白表达水平。采用 miRNA 芯片对燃煤砷暴露人群进行外周血 miRNA

检测，发现 miR-21、miR-145、miR-155 和 miR191 在外周血中水平升高，进一步的研究显示其分别介导了砷对 DNA 损伤、氧化损伤及凋亡等多种生物学效应的影响[145, 146]。用 2 μmol/L 砷酸钠处理人淋巴母细胞 TK6 细胞 6 天，观察到有 5 种 miRNA（miR-210，miR-22，miR-34a，miR-221 和 miR-222）表达水平显著改变[147]。研究表明，砷诱导氧化应激的同时可激活多种 miRNA，如 miR-9、miR-21、miR-125b、miR-121 等的表达；miR-200b 和 miR-21 的表达改变可能参与砷诱导的细胞恶性转化[148]。亚砷酸盐可通过增加 miR-21 表达水平而促进 ERK/NF-κB 信号通路激活，引起人胚肺成纤维细胞发生恶性转化[149]。

lncRNA 通常是指长度大于 200 个核苷酸的非编码 RNA 转录本。目前 lncRNA 在砷的毒作用机制研究中尚在起步阶段，有研究发现亚砷酸盐可以通过上调 lncRNA MALAT1 抑制 HIF-1 的泛素化降解途径，进而引起 HIF-1α 水平升高，HIF-1α 高表达进一步诱导细胞发生瓦伯格（Warburg）效应，从而导致其终产物乳酸水平升高，促进肝上皮细胞发生恶性转化[150]。

4.5 砷与免疫功能损伤和炎症反应

免疫系统作为机体抵抗外界有害物质损害的重要防御体系，在各器官系统维持正常功能中发挥着不可替代的作用。经饮用水慢性砷暴露与心血管系统、神经系统疾病、糖尿病以及多组织器官恶性肿瘤的发生密切相关，在一定程度上都与机体的免疫功能受损和炎症反应有关。越来越多的证据表明，砷可引起免疫器官组织结构改变、免疫细胞亚群异常、相关免疫分子表达改变等免疫功能异常[151]。

来自孟加拉、印度、墨西哥等国家砷污染地区的人群流行病学研究发现，砷暴露可影响机体免疫器官的功能，尤其孕妇砷暴露可导致胎儿胸腺发育异常[76]。小鼠实验发现亚砷酸盐可引起胸腺细胞脂质过氧化，同时砷暴露后处于细胞周期 G1 期的胸腺细胞大量增加，细胞凋亡加重，进而影响胸腺发挥正常免疫功能[152]。长期自由饮水砷暴露可诱导小鼠脾细胞凋亡，从而影响脾脏发挥免疫功能[153]。在印度西孟加拉饮用无机砷污染井水的慢性砷中毒病人和有皮肤损伤的病人中，70%以上发现了肝大和（或）脾大的严重炎性反应[154]。

T 细胞亚群之间的平衡是维持免疫系统内部环境稳定的中心环节，CD_4^+/CD_8^+ 比值保持动态平衡，对机体的细胞免疫和体液免疫具有重要调节作用。研究显示，砷暴露可降低患者外周血中 CD_3^+，CD_4^+ 细胞及 CD_4^+/CD_8^+ 比值，导致 T 细胞刺激指数及增殖能力抑制，降低 T 淋巴细胞和 B 淋巴细胞的百分比[74, 155]。不同水平砷暴露可引起大鼠调节性 T 细胞（Treg）和辅助性 T 细胞-17（Th17）分化失衡，进而诱导其 Treg、Th17 细胞相关抑炎因子（如 IL-10、TGF-β1 与促炎因子（IL-17、IL-6）分泌异常，最终促进砷致大鼠肝脏炎性损伤[156]。长期砷暴露不仅使小鼠脾脏的重量减少，还使脾脏的 CD_4^+ 数量减少及 CD_4^+/CD_8^+ 比例下降，抑制脾单核细胞的增殖及 IL-2、IL-6、IL-12 和 γ-干扰素等重要细胞因子的分泌。无机砷对巨噬细胞亦有很强的细胞毒作用，亚砷酸盐和砷酸盐的半数致死亡（IC_{50}）浓度分别为 5 μmol/L 和 500 μmol/L，其主要导致细胞坏死（80%）和部分细胞凋亡（20%），并且在细胞毒性剂量范围内明显诱发炎性细胞因子的释放[72]。

多种人群流行病学调查及动物实验发现，砷暴露可引起多种免疫相关基因/蛋白以及部分炎症因子的表达水平异常，包括主要组织相容性复合体 II，CD69 等免疫调节相关基因和蛋白的表达水平下调，巨噬细胞抑制因子单核细胞化学引诱物蛋白 1 和其他细胞因子（如 IL-1α、IL-6、IL-8）水平升高，以及 IL-1β 和 TNF-α 等炎症相关因子的失调[157]。上皮细胞（UROtsa）暴露于 50 nmol/L MMA^{III}，可诱导 ROS 产生和 IL-6、IL-8、IL-1、巨噬细胞游走抑制因子等细胞炎症因子的过度积累，以及不同的信号通路和与炎症反应相关的转录因子（NF-kB、AP-1、c-Jun、ERK、p38 和 AKT）的活化[158]。砷暴露可引起炎性细胞因子 IL-6 分泌与释放增加，进而激活信号转导和转录激活因子 3（STAT3），STAT3 的激活通过诱导 miR-21 的高表达而促进肺支气管上皮细胞发生上皮间质转化，从而导致细胞恶性转化[159]。

5 环境砷污染健康风险评估

环境健康风险评估是处理各类环境污染引起的健康危害事件，制定公共卫生相关政策、法规与标准，进而采取可行的健康干预措施，并与媒体及公众进行风险交流，开展公众环境健康服务的必要工具和手段。环境健康风险评估的核心是在识别环境污染物健康危害的基础上，定量评估不同污染物暴露总量及其生物有效性，建立环境污染物暴露与健康危害效应的剂量-反应关系，同时控制风险评估中伴随的不确定性，最大程度减少风险管理的成本。早在20世纪30年代，国际上便开始环境健康风险评估研究，此阶段多采用毒物鉴定法进行环境污染物对健康影响的定性分析，随后于50年代提出了安全系数法，用于估算环境污染物人群暴露的可接受剂量，到80年代环境健康风险评估体系基本形成。1983年，美国国家科学院出版《联邦政府的风险评估：管理程序》[160]，将其评估程序概括为危害识别、剂量-反应关系评估、暴露评估和风险特征分析，该科学体系至今被包括我国在内的多个国家和组织采用。目前有关砷污染的环境和健康风险评估方面资料相对较少，本部分主要介绍砷暴露健康风险评估体系及目前主要的评估方法。

5.1 环境砷污染的危害识别

急性或亚急性高水平无机砷暴露可引起明显的胃肠功能紊乱，包括从轻微的腹部绞痛和腹泻到严重的危及生命的出血性肠胃炎均可出现。而在长期暴露砷的情况下，上述症状往往不明显，但如本章第二节所述，长期暴露无机砷对机体多个组织器官均有诸多不良影响。在癌症研究方面：IARC于1987年将砷确定为人类致癌物[161]，在2012年的评估中亦指出，已有充分证据表明无机砷暴露可引起肺癌、膀胱癌和皮肤癌，另与肝癌、前列腺癌、肾癌等存在明显正相关关系[162]。神经系统疾病方面：急性无机砷中毒最初可引起胃肠道或心血管等症状，之后伴随有中枢神经系统和周围神经系统相关症状；慢性砷暴露中直接出现周围神经病变较少。除急性神经毒性和周围神经病变外，大脑发育阶段暴露无机砷也可能会导致不良健康结局[69]。孟加拉国、印度和中国饮水型砷中毒病区进行的流行病学研究均显示砷暴露儿童智力水平明显低于正常儿童[163,164]。生殖发育毒性方面：研究证实，无机砷或其甲基化代谢物可通过胎盘屏障对胎儿造成多种损伤，包括发育迟缓甚至死亡。孟加拉国、印度、智利和中国台湾东北部的人群流行病学研究均报道高砷暴露地区胎儿、新生儿及产后死亡率、低出生体重、子痫前症和先天畸形等发生率明显高于较非砷暴露地区[100,165-169]。免疫毒性方面：来自印度、孟加拉国、墨西哥等国家的人群流行病学研究均发现砷暴露可影响机体免疫器官的正常功能[170]。值得关注的是，研究还发现高砷暴露地区孕妇的免疫功能明显低下，可进一步影响胎儿免疫器官的正常发育[171]。此外，国内学者从不同层面探讨了我国特有的燃煤型砷中毒人群免疫功能改变情况，发现燃煤砷暴露可造成当地人群出现明显免疫毒性，如改变T淋巴细胞亚群比例及补体水平、造成免疫炎症因子分泌异常等[172]。砷暴露对糖尿病发病率的影响方面：现有的研究数据提供了足够的证据证明处于较高水平饮水砷（≥150 μg/L）暴露的人群糖尿病发病率与砷暴露存在明显关联，但是较低水平砷暴露（<150μg/L）证据尚不充分[173]。砷暴露对心血管病的影响方面：许多流行病学研究表明慢性饮水砷暴露与各种心血管疾病之间存在剂量-反应关系，包括微循环障碍、动脉硬化、高血压、冠心病和脑梗塞等[174-177]。

对有机砷而言，研究发现长期接触MMA的狗会出现腹泻和呕吐症状。在啮齿动物的生物测定中，

并没有发现暴露于 MMA^{5+} 后致癌的证据。美国环保署农药项目办公室认为没有证据证明 MMA 是一种致癌物质[178]。甲状腺和膀胱已被确定为 DMA 的靶器官，在亚慢性和慢性 DMA 暴露大鼠中均观察到甲状腺病变，主要体现在甲状腺滤泡内立方细胞转变为柱状上皮细胞的发生率[179]；长期暴露于 DMA^{5+} 还可导致雄性大鼠发生膀胱肿瘤。IARC 认为目前有足够的证据表明 DMA^{5+} 在实验动物中具有致癌性。2009 年，IARC 亦将 DMA 列为人类可能致癌物质[162]。

5.2 环境砷污染剂量-反应关系评估

如前所述，无机砷或有机砷暴露可导致严重的全身性损害，因此环境砷暴露健康风险评估一直是当前研究的重点和难点。在健康风险评估中，剂量-反应关系评估是重中之重，而砷的剂量-反应关系模式目前尚存在争议。既往研究多采用无阈线性外推模式来评定砷暴露剂量-反应关系，该模式假定砷从零剂量开始即可能对机体产生不良影响。来自中国台湾地区的砷暴露剂量-反应关系相关研究显示，即使饮用水中砷浓度为 50 μg/L 以下，个体终生暴露致癌风险概率依然高达千分之一，远远高出公众可接受标准（百万分之一）[180]。因此，研究人员建议将饮水中砷含量标准改为 10 μg/L，甚至更低。但另有证据表明，砷暴露对机体健康的影响并非完全按线性模式变化，而是有可能遵循有阈或非线性剂量-反应模式。一方面表现在人群和动物层面的相关研究均提示砷元素本身可能具有某些生理功能；另一方面一些分子毒理学的研究和人群代谢资料均对无阈线性模式提出质疑[181]。

虽然目前无法明确砷暴露的具体剂量-反应模式，但正因为砷暴露风险评估过程中所暴露出关键性数据的缺乏，也为今后的研究指明了方向。考虑到饮用水的安全性，近年来多个国家均在饮用水砷标准限值调整上作出努力。人群研究证实，即使饮水砷浓度控制在小于最大污染水平（maximum contamination level，MCL）50 μg/L，长期砷暴露人群依然存在较高的致癌风险。为此，多个国家均已将饮用水砷的 MCL 限值从 50 μg/L 进行下调。加拿大提出了 25 μg/L 的过渡性标准；日本、欧洲一些国家将饮用水砷限值下调至 10 μg/L；美国环保署（EPA）经过综合研究与分析，亦将饮用水砷卫生标准从 50 μg/L 修订为 10 μg/L；我国目前则结合国情分阶段进行，当前修订颁布的饮用水砷卫生标准城市集中供水为 10 μg/L，农村分散式供水依然为 50 μg/L[3]。目前，究竟砷暴露于多少浓度或接触多长时间亦或伴随何种条件以及受哪些因素影响才能引发疾病，依然是国内外学者研究关注的焦点。因此迫切需要对低剂量（0～50 μg/L）砷暴露与健康效应的关系进行系统研究。来自比利时的研究表明，机体暴露于低砷水平（水中 20～50 μg/L；空气中无机砷 0.3 μg/m³）时，对肿瘤发生率没有明显的影响[182]。由于人群研究存在诸多影响因素，低浓度砷暴露单一指标的亚临床变化有时难以确认为砷毒性效应。因此阐明该问题，对不同国家采取切实有效和经济的砷暴露防控措施均有重大的理论和实际意义。目前，包括我国在内的世界各国学者都在关注和进行相关研究。

5.3 环境砷污染暴露评估

暴露评估作为健康风险评估的一个重要环节，其不仅要给出个体或群体砷暴露的准确估计值，还要提供如暴露时间、暴露方式和频度等砷暴露特征，以用于后续砷暴露风险特征分析。目前，对水体、燃煤和大气等不同介质中的砷暴露评估已有一些报道，尤其近年来数学模型模拟以及剂量重建等方法的进展使得砷暴露评估的科学性有了大幅度提升。如 Mann 等[183]采用经口、气管滴注和静脉注射等方法率先构建了家兔和仓鼠的无机砷暴露生理药代动力学（PBPK）评估模型，通过该模型得到了五价砷、三价砷、一甲基胂酸（MMA）和二甲基胂酸（DMA）四种代谢产物在尿液和组织中的分布水平。另通过该模型结合组织、器官对砷的亲和度设置了皮肤、肺脏、肝脏、肾脏以及其他组织器官五个隔室，并通过体重换算得到相应生理参数。当机体组织吸收率、代谢速率、亲和度等某些生理参数缺乏时，可通过 PBPK 模型拟合进行估计，可较准确地预测砷代谢产物的排泄速率。随后 Mann 等[184]又对模型中的体重、吸收速率和代谢速率等参数进行了校正，并外推到人群，结果发现该模型对上述四种代谢产物在人体内吸收、分布、代谢和排泄拟合效果良好。EI-Masri 等[185]在人体中构建了 PBPK 模型，用于推断经口摄入无机砷后，组织、尿液中砷及其代谢产物的浓度。该模型包括无机砷（三价砷、五价砷等）及其代谢产物（MMA、DMA 等）两个亚模型，每个亚模型均采用流量极限隔室的方法来构建。模型中的组织、器官选定依据基于砷化物的理化性质、暴露途径、靶组织和代谢位点，能够更客观地对人体砷暴露水平进行综合估计。

在人体砷暴露量估算方面，美国环保署给出的经口砷暴露参考剂量（RfD）为 0.0003 mg/(kg·d)，相当于 70 kg 体重的个体每天饮用 2 L 砷含量为 0.01 mg/L 的水所摄入砷总量[186]。该参考剂量是基于中国台湾地区砷暴露地区人群研究的结果而提出的，该研究显示暴露于饮水砷含量为 0.17 mg/L 的人群皮肤病变及心血管疾病的发生风险明显高于暴露于 0.009 mg/L 饮水砷的人群。该研究进一步根据当地人群生活习惯、砷暴露浓度、暴露频率等因素而推算出的未观察到有害作用水平（NOAEL）和观察到有害作用最低水平（LOAEL）分别为 0.0008 mg/(kg·d)和 0.014 mg/(kg·d)[187, 188]。因此，在具体推算 RfD 时，美国环保署综合考虑各种不确定因素，最终以 NOAEL 的近 3 倍为不确定系数，进而获得 0.0003 mg/(kg·d)的经口砷暴露 RfD。目前国内外尚没有明确的经呼吸砷暴露参考浓度（RfC）。

5.4 环境砷污染风险特征分析

风险特征分析是健康风险评估的最后总结阶段，即通过综合分析前三阶段的相关信息，定性或定量估算砷暴露在机体、系统或人群中引起健康损害的概率，并阐述其伴随的不确定性以及可能引起的公众健康问题。目前国际上有关砷化物暴露健康风险特征分析资料主要来自于美国环保署（EPA）。

在经口暴露无机砷化合物方面，美国 EPA 根据中国台湾饮水砷污染地区大样本人群队列研究结果推断出 1.5 mg/(kg·d)砷暴露量与皮肤癌发生风险的癌症斜率因子（cancer slope factors, CSF，常用以评估环境致癌物暴露与癌症发生风险的关联性），得到每 0.02 μg/L 饮水砷暴露的癌症发生风险概率为百万分之一；在经呼吸道暴露无机砷方面，美国 EPA 给出的结果为每 0.0002 μg/m³ 空气砷暴露的癌症发生风险概率为百万分之一[189]；每日耐受摄入量（TDI）是指一种化学物长期暴露或终生暴露而不引起健康损害效应的最高剂量。美国国家公共卫生和环境研究所研究人员通过大量实验得到长期经口或经呼吸暴露无机砷的 TDI 分别为 1.0μg /(kg·d)和 1.0 μg/m³[190]。

有机砷化合物暴露方面，美国 EPA 农药项目组基于 MMA 暴露后大鼠体重变化、食物消耗、胃肠道病理组织学及甲状腺功能改变等指标，得到长期 MMA 暴露的 RfD 为 0.03 mg/(kg·d)；基于大鼠膀胱上皮组织增殖能力改变得到长期 DMA 暴露的 RfD 为 0.014 mg/(kg·d)[191]。在有机砷暴露癌症风险特征分析方面，美国 EPA 农药项目组经过多项测定认为未见明显证据证明 MMA 暴露具有致癌性，而 DMA 在不导致细胞增殖的剂量下不具有致癌性。而 IARC 2009 年的报道则将 MMA 和 DMA 认定为"具有潜在致癌风险的化学物"[192]。

在气体砷化物方面，20 世纪 90 年代美国 EPA 综合风险信息系统（IRIS）结合气体砷化物暴露动物模型中溶血效应及脾脏变化情况得到长期吸入气体砷化合物的 RfD 为 0.05 μg/m³。WHO 基于气体砷暴露对动物血液学指标的影响给出的气体砷化物暴露健康风险指导标准同样为 0.05 μg/m³[193]。2008 年，美国加利福尼亚州环境保护署（Cal EPA）给出急性和慢性气体砷化物暴露健康风险参考值分别为 0.2 μg/m³ 和 0.015 μg/m³[194]。

5.5 环境砷污染健康风险评估的不确定性

5.5.1 砷暴露剂量和评估模型的不确定性

砷暴露和健康损害作用之间的关系可通过不同的剂量范围进行计算，包括砷累积剂量、终身日均摄入量、砷暴露峰值以及多种不同统计学模型的使用。剂量和模型的使用能影响风险评估的计算，风险评估或多或少存在不确定性，其依赖于如何用剂量准确测定出暴露水平以及如何使剂量更接近观察终点（如癌症）。在风险影响评估中，美国 EPA 指出：某些情况下，如果从短期暴露方式到长期暴露方式来评估毒性，增加日摄入量能较好地反映儿童非致癌效应。然而，致癌效应可用终身暴露来估算，其中需考虑到儿童剂量的增加。剂量反应模型反映了终身暴露（从童年到成年），可预期终身致癌风险，但该模型是否恰当，目前尚存在的不确定性。与短期高剂量暴露比较，对于相对低剂量长期暴露作用的了解比较少，这些数据的缺乏是砷暴露健康风险评估中不确定性的主要来源。当用动物实验数据外推至人群研究来选择合适的剂量反应模型和剂量时就增加了不确定性。

目前，砷作用方式和砷暴露致癌作用最相关的剂量还不清楚。剂量影响着流行病学研究结果的解释，由于使用不适当的剂量研究使砷与癌症之间缺乏联系，例如，如果暴露持续时间和剂量对砷诱发癌症都重要，那么累积暴露可能与癌症相关，而峰值暴露则可能与癌症无相关。对于流行病学研究结果的解释和比较不同的研究必须考虑到复杂的剂量问题，因此，剂量的选择也给砷健康风险评估增添了不确定性。

5.5.2 食物砷暴露的不确定性

对砷暴露人群，除了调整假定的高暴露剂量，还应调整膳食砷摄入量，因为饮食也是砷暴露的重要来源。在生态流行病学研究中观测到的剂量-反应数据必须考虑饮食因素，其有助于计算总砷摄入量，并估计人群从食物摄入砷量占砷总摄入量的比例，食物砷是不确定性的一个重要来源。调查发现，对于人群食物砷摄入量，总砷浓度最高是海鲜（从淡水鱼 160 mg/kg 到深水鱼 2360 mg/kg），但海鲜中无机砷含量较低，平均为 1~2 mg/kg；大米中无机砷含量最高（74 mg/kg）、其次是面粉（11 mg/kg）、葡萄汁（9 mg/kg）、熟菠菜（6 mg/kg）[3]。因此，就无机砷摄入量而言，膳食中谷类和农产品所占比例较大。但是，目前关于不同个体和群体消耗食物中砷水平和品种的研究资料较少，其食物中砷的作用仍不确定。虽然有研究用较明确的数据估算美国人群总砷暴露与健康风险，但由于饮食、生活习惯及人种差异，对其他地区如中国台湾地区砷暴露人群食物中砷摄入量并不适用。这些数据都需要合适的剂量反应模型来计算总砷摄入量。

在分析砷的数据资料时要考虑通过饮食摄入的潜在影响，因为暴露程度的准确测量决定暴露与结果之间的关联强度。暴露误差的存在往往导致风险评估的无效，虽然食物对癌症发病率影响的偏倚程度仍然有争论，但饮食对砷流行病学研究的影响却是可能的。饮食和营养的 3 种方式可能会影响砷暴露与健康危害（包括癌症）之间关系：①混杂饮食；②低质量饮食所致易感性增强；③由于不同的饮食习惯，难以用某一地区调查结果外推到其他群体。

虽然大多数砷与癌症的流行病学研究中饮食残留无法估算，但在其他研究比如动物实验中已被测量，而且并不与其他易测量的暴露因素相混淆。尽管如此，由于没有明确证据证明砷诱发癌变的分子机制，故仍然存在不可测量却能产生轻微影响的混杂因素。除了混杂因素外，因素之间的交互作用或作用改变也可能使暴露与结果之间的关联增强。例如，吸烟者患某特定疾病的危险性可能增加 4 倍，暴露于辐射其患病危险性可增加一倍。如果吸烟并同时暴露于辐射，则患病危险性增加 20 倍，表明存在暴露和其他因素的交互作用。

6 展望

砷及其化合物因其广泛分布于环境，对生物体具有慢性累积性损害乃至癌变的严重后果，已成为全球性公共卫生问题，并受到各国政府、学界、社会的高度关注。其中地方性砷中毒波及至少22个国家约2亿人口，且至今存在影响因素复杂、致病机制不清、无特效治疗药物等瓶颈问题，致其健康损害问题难以得到彻底解决。因此，多学科融合深化开展研究对推进环境砷污染与健康问题的解决具有重要的科学价值和深远的社会意义。

6.1 砷与环境多因素交互作用对健康的影响研究有待加强

毒物间的交互作用是一个极为复杂的问题，可呈相加作用或拮抗作用，亦可能是毒性剧增的协同作用。砷及其化合物与许多重金属和有毒化学物共存于自然界，而现有研究多集中研究单一砷化物所致健康损害，其结果存在局限性。因此加强开展砷与其他有害因素对健康的影响研究具有重要实际意义。

6.2 砷中毒机制与转化应用研究有待深入

目前，砷致健康损害的机制研究众多，但存在较多不足，如多数机制研究为小作坊操作模式，存在研究团队各自为政、缺乏疾病整体观的单兵作战研究模式；各实验室缺乏统一的标准操作规范及质量控制体系，限制了不同实验室研究结果的比较和利用；多仅针对某一具体机制开展研究，各假说间缺乏相互联系；多以体外细胞及体内动物实验为主，来自砷暴露人群研究的直接证据相对偏少，且有限的人群研究尚存在人群数量有限、不同人群间可比性差、研究结果不能互证互用等缺陷，导致机制研究与实际应用需求脱节，转化应用价值有限。因此，深化砷中毒机制及其转化应用研究，不仅可为砷中毒患者的临床救治提供新思路和指导，尚可为完善砷中毒的诊断标准和建立防治效果评价体系等提供依据。

6.3 建立或完善我国慢性砷中毒人群研究队列有待重视

目前，关于砷暴露与不良妊娠结局、发育障碍、成年期慢性疾病（心脑血管疾病、高血压、糖尿病等）有关已是不争的事实，但对其间关系的认识十分有限，需要更多前瞻性人群研究结果予以证实；地方性砷中毒防控后期管理亦需对人群暴露水平、健康情况等进行长期追踪与监测。因此，利用我国不同暴露途径和暴露水平的砷中毒人群资源建立研究队列意义重大。其不仅可为砷的远期毒作用研究提供支撑，更重要的是可防患于未然发挥健康预警作用，早期发现问题早期干预。

6.4 加强易感人群和敏感生命阶段的识别与干预研究意义重大

环境-遗传和多种其他因素在砷致病致癌中发挥作用，这些因素可在不同生命阶段和不同个体中产生各种影响，但对于砷在整个生命周期中如何影响这些过程，相关研究尚处于起步阶段。因此，将分子水平的微观研究与人群流行病学的宏观研究相结合，科学识别各种易感因素和生命阶段，并通过干预易感个体的早期生命阶段，达到预防或减少甚至消除砷致健康损害的目标。

6.5 环境砷污染健康风险评估研究须与时俱进

环境健康风险评估是进行环境污染物科学管理的重要环节，在国家科技部、生态环境部的多项规划中被列为重点推进任务，但其任重而道远。目前，环境砷暴露健康风险评估的研究有限，多基于单一暴露途径（如饮水），以明确的健康损害效应（如组织器官功能改变、致畸、致癌、死亡等）作为结局终点，而多途径、不同水平砷暴露对机体分子层面的改变或远期健康效应等方面的风险评估研究十分匮乏。随着互联网技术的不断发展和完善，人类已进入"大数据"时代，传统的描述/机制毒理学研究亦逐步向计算/预测毒理学转变。大数据最核心的应用是预测，即通过高级数学算法处理海量数据，通过一系列分析得出预测结果。基于大数据的环境健康风险评估体系可实时、系统地对环境有害因素与人群健康损害效应大数据进行收集、分析和整合，已成为当前环境健康风险评估研究的风向标，亦为环境砷污染健康风险评估提供了新的策略。

近年来，国际上提出了 21 世纪风险评估（RISK21）等的新理念和新方法，其中 RISK21 是一种基于问题形成、暴露驱动的渐进式风险评估方法，可充分利用现有数据用视图的方式展示暴露信息及其不确定性，优化各种已知信息，达成风险或安全决策。2017 年，美国国家科学院、美国国家工程院和美国国家医学科学院联合发布了《应用 21 世纪科学进展改进风险相关评价》的报告，全面回顾了如何应用暴露科学、毒理科学及流行病学新进展来改进风险相关评价，并提出了研究与发展的方向和策略。这些新理念和新方法为我国风险评估研究与应用实践提供了借鉴和参考。

6.6 砷中毒靶向药物研发亟待加强

对砷中毒现症患者尽早治疗、减少病残、提高生命质量，是现阶段我国地方性砷中毒防控的重点内容之一，然而无针对性特效治疗药物和干预手段等限制了相关工作的进展。生物信息学结合大数据分析和药效验证具有提高药物研发效益的优势。使用动物或细胞疾病模型进行表型筛选可提供更好的性能，但也有其自身的缺点，包括费用高、通量低、机制不明确以及人类疾病谱覆盖度有限等。大数据驱动的策略正逐渐用于解决这些困难。由于砷中毒目前尚无特效治疗药物，现症患者主要采用对症治疗。因此，利用大数据信息寻找潜在药物靶点，研究老药新用途具有事半功倍的现实意义。

参 考 文 献

[1] WHO Arsenic in drinking-water. Background document for preparation of WHO Guidelines for drinkingwater quality. Geneva, Switzerland: World Health Organization, 2003.

[2] Rodriguez L L, Sun G, Berg M, et al. Groundwater Arsenic Contamination Throughout China. Science, 2013, 341(6148): 866-868.

[3] 张爱华, 郑宝山, 王杰, 等. 砷与健康. 北京. 科学出版社, 2008.

[4] Ravenscroft P, Brammer H, Richards K. Arsenic Pollution: A Global Synthesis. United Kingdom: Wiley, 2009.

[5] Yu G Q, Sun D J, Yan Z. Health effects of exposure to natural arsenic in groundwater and coal in China: An overview of occurrence. Environmental Health Perspectives, 2007, 115(4): 636-642.

[6] Hideo T, Hideaki T, Akira O. Long-term prospective study of 6104 survivors of arsenic poisoning during infancy due to contaminated milk powder in 1955. Journal of Epidemiology, 2010, 20(6): 439-445.

[7] 何华先, 程朝辉, 姚海珊, 等. 一起水源水砷污染事件的调查与处理. 中国热带医学, 2007, 7(11): 2158-2159.

[8] 陈明. 采取合理措施避免跨界污染——贵州独山瑞丰矿业砷污染事件解析. 环境保护, 2009(3): 63-64.

[9] 都里. 阳宗海污染事件的深刻警示. 民主与法制, 2008(22): 1-1.

[10] Nriagu, Jerome O. Arsenic in the environment. Part 2: Human health and ecosystem effects. Advances in Environmental Science & Technology, 1994.

[11] Han F X, Yi S, Monts D L, et al. Assessment of global industrial-age anthropogenic arsenic contaminatio. Die Naturwissenschaften, 2003, 90(9): 395-401.

[12] 陈琴琴. 中国砷污染排放清单研究. 南京: 南京大学, 2013.

[13] 龚仓, 徐殿斗, 马玲玲. 大气颗粒物中砷及其形态的研究进展. 化学通报. 2014, 77(6): 502-509.

[14] 胡立刚, 蔡勇. 砷的生物地球化学. 化学进展, 2009, 21(2): 458-466.

[15] Bissen M, Frimmel F H. Arsenic: A review. Part I: Occurrence, toxicity, speciation, mobility. Acta Hydrochimica Et Hydrobiologica, 2003, 31(1): 9-18.

[16] Matschullat J. Arsenic in the geosphere: A review. Science of the Total Environment, 2000, 249(1): 297-312.

[17] Nriagu J O. Arsenic in the environment, Part 1: Cycling and characterization. New York: John Wiley, 1994.

[18] Pacyna J M, Pacyna E G. An assessment of global and regional emissions of trace metals to the atmosphere from anthropogenic sources worldwide. Environmental Reviews, 2001, 9(4): 269-298.

[19] Nriagu J O, Pacyna J M. Quantitative assessment of worldwide contamination of air, water and soils by trace metals. Nature, 1988, 333(6169): 134-139.

[20] EEA (European Environment Agency). European Union emission inventory report 1990—2009 under the UNECE Convention on Long-range Transboundary Air Pollution (LRTAP). Akusherstvo I Ginekologiia, 2010, 42(1): 22-23.

[21] Yudovich Y E, Ketris M P. Arsenic in coal: A review. International Journal of Coal Geology, 2005, 61(3): 141-196.

[22] Gayer R A, Rose M, Dehmer R, et al. Impact of sulphur and trace element geochemistry on the utilization of a marine-influenced coal-case study from the South Wales Variscan foreland basin. International Journal of Coal Geology, 1999, 40(2): 151-174.

[23] Karayigit A I, Spears D A, Booth C A. Antimony and arsenic anomalies in the coal seams from the Gokler coalfield, Gediz, Turkey. International Journal of Coal Geology, 2000, 44(1): 1-17.

[24] Zheng B, Ding Z, Huang R, et al. Issues of health and disease relating to coal use in southwestern China. International Journal of Coal Geology, 1999, 40(2–3): 119-132.

[25] 国家统计局能源统计司, 国家能源综合司. 中国能源统计年鉴 2011. 北京. 中国统计出版社, 2011.

[26] Espreme. EU-Project espreme-integrated assessment of heavy metal releases in Europe. 2012.08.01.

[27] Garelick H, Jones H, Dybowska A, et al. Arsenic Pollution Sources. 2008, 197: 17-60.

[28] Mukai H, Ambe Y. Detection of monomethylarsenic compounds originating from pesticide in airborne particulate matter sampled in an agricultural area in Japan. Atmospheric Environment, 1987, 21(1): 185-189.

[29] Helsen L. Sampling technologies and air pollution control devices for gaseous and particulate arsenic: A review. Environmental Pollution, 2005, 137(2): 305-315.

[30] Alvim M C M, Afonso S A V. Incineration of different types of medical wastes: emission factors for particulate matter and heavy metals. Environmental Science & Technology, 2003, 37: 3152-3157.

[31] Sullivan R, Woods I. Using emission factors to characterise heavy metal emissions from sewage sludge incinerators in Australia. Atmospheric Environment, 2000, 34(26): 4571-4577.

[32] European Environment Agency (EEA). EMEP/EEA Air pollutant emission inventory guidebook-2009. [2012-08-01].

[33] 赵金艳, 王金生, 郑骥. 含砷废渣的处理处置技术现状. 资源再生, 2011(11): 58-59.

[34] 魏大成. 环境中砷的来源. 国外医学(医学地理分册), 2003, 24(4): 173-175.

[35] 环境保护部应急办. 砷污染应急处置技术. 北京: 中国环境科学出版社, 2010.

[36] 郭华明, 杨素珍, 沈照理. 富砷地下水研究进展. 地球科学进展, 2007, 22(11): 1109-1117.

[37] 陈怀满, 郑春荣, 周东美, 等. 土壤中化学物质的行为与环境质量. 北京: 科学出版社, 2002.

[38] 吴万富, 徐艳, 史德强, 等. 我国河流湖泊砷污染现状及除砷技术研究进展. 环境科学与技术, 38(S1): 190-197.

[39] 翁焕新, 张霄宇, 邹乐君. 中国土壤中砷的自然存在状况及其成因分析. 浙江大学学报(工学版), 2000, 34(1): 88-92.

[40] Boyle R W, Jonasson I R. The geochemistry of arsenic and its use as an indicator element in geochemical prospecting. Journal of Geochemical Exploration, 1984, 20(3): 223-302.

[41] Shacklette H T, Boerngen J G, Keith J R. Selenium, fluorine, and arsenic in surficial materials of the conterminous United States. Circular, 1974.

[42] Ure A M, Berrow M L. The elemental constituents of soils. *In*: Environmental Chemistry. London: Royal Society of Chemistry, 1982, 94-204.

[43] Shotyk W, Cheburkin A K, Appleby P G, et al. Two thousand years of atmospheric arsenic, antimony, and lead deposition recorded in an ombrotrophic peat bog profile, Jura Mountains, Switzerland. Earth & Planetary Science Letters, 1996, 145(1-4): E1-E7.

[44] Dudas M J. Accumulation of native arsenic in acid sulphate soils in Alberta. Canadian Journal of Soil Science, 1987, 67(2): 317-331.

[45] Schoof R A, Yost L J, Eickhoff J, et al. A market basket survey of inorganic arsenic in food. Food & Chemical Toxicology, 1999, 37(8): 839-846.

[46] Uchino T, Roychowdhury T, Ando M, et al. Intake of arsenic from water, food composites and excretion through urine, hair from a studied population in West Bengal, India. Food & Chemical Toxicology, 2006, 44(4): 455-461.

[47] 李筱薇, 高俊全, 王永芳, 等. 2000 年中国总膳食研究——膳食砷摄入量. 卫生研究, 2006, (1): 66-69.

[48] 王振刚, 何海燕, 严于伦, 等. 石门雄黄矿地区居民砷暴露研究. 卫生研究, 1999, 28(1): 12-14.

[49] 肖细元, 陈同斌, 廖晓勇, 等. 我国主要蔬菜和粮油作物的砷含量与砷富集能力比较. 环境科学学报, 2009, 29(2): 291-296.

[50] Goyer R A, Clarkson T W. Toxic effects of metals. *In*: Klaassen C D (Ed.). Casarett & Doull's Toxicology. The Basic Science of Poisons., 5th ed. McGraw-Hill Health Professions Division, 1996.

[51] ATSDR. Public Health Statement for Arsenic. Available at: http: //www.atsdr.cdc.gov/phs/phs.asp?id=18&tid=3.

[52] Rahman M M, Ng J C, Naidu R. Chronic exposure of arsenic via drinking water and its adverse health impacts on humans. Environmental Geochemistry & Health, 2009, 31(1 Supplement): 189-200.

[53] Guha Mazumder D N. Chronic arsenic toxicity & human health. Indian Journal of Medical Research, 2008, 128(4): 436.

[54] Flora S J S. Handbook of Arsenic Toxicology. Chapter 6, Health effects chronic arsenic toxicity, Academic Pr Inc, 2015: 1-41.

[55] Tseng W P. Effects and dose-response relationships of skin cancer and blackfoot disease with arsenic.Environmental Health Perspectives 1977; 19: 109-119.

[56] Liao W T, Yu C H, Lam C E, et al. Differential effects of arsenic on cutaneous and systemic abnormality: Focusing on CD_4^+ cell apoptosis in patients with arsenic induced Bowen's disease. Carcinogenesis, 2009, 30: 1064-1072.

[57] Kapaj S, Peterson H, Liber K, et al. Human health effects from chronic arsenic poisoning: A review. Journal of Environmental Science and Health, Part A, 2006, 41(10): 2399-2428.

[58] Mazumder D N, Das G J, Santra A, et al. Chronic arsenic toxicity in west Bengal—The worst calamity in the world. Journal of the Indian Medical Association, 1998, 96(1): 4-7.

[59] Liu J, Zheng B, Aposhian H V, et al. Chronic arsenic poisoning from burning high-arsenic-containing coal in Guizhou, China. Environmental Health Perspectives, 2002, 110(2): 119-122.

[60] Wang W, Cheng S, Zhang D. Association of inorganic arsenic exposure with liver cancer mortality: A meta-analysis. Environmental Research, 2014, 135: 120-125.

[61] Parvez F, Chen Y, Brandtrauf P W, et al. A prospective study of respiratory symptoms associated with chronic arsenic exposure in Bangladesh: Findings from the Health Effects of Arsenic Longitudinal Study (HEALS). Thorax, 2015, 65(6): 528.

[62] Parvez F, Chen Y, Yunus M, et al. Arsenic exposure and impaired lung function. Findings from a large population-based prospective cohort study. American Journal of Respiratory and Critical Care Medicine, 2013, 188(7): 813-819.

[63] Dauphiné D C, Ferreccio C, Guntur S, et al. Lung function in adults following in utero and childhood exposure to arsenic in drinking water: Preliminary findings. International Archives of Occupational & Environmental Health, 2011, 84(6): 591-600.

[64] Reciovega R, Gonzalezcortes T, Olivascalderon E, et al. *In utero* and early childhood exposure to arsenic decreases lung function in children. Journal of Applied Toxicology, 2015, 35(4): 358-366.

[65] Sánchezpeña L C, Petrosyan P, Morales M, et al. Arsenic species, AS3MT amount, and AS3MT gen expression in different brain regions of mouse exposed to arsenite. Environmental Research, 2010, 110(5): 428-434.

[66] Bartolomé B, Córdoba S, Nieto S, et al. Acute arsenic poisoning: Clinical and histopathological features. British Journal of Dermatology, 2015, 141(6): 1106-1109.

[67] Kannan G M, Tripathi N, Dube S N, et al. Toxic effects of arsenic(III) on some hematopoietic and central nervous system variables in rats and guinea pigs. Journal of Toxicology: Clinical Toxicology, 2001, 39: 675-682.

[68] Mathew L, Vale A, Adcock J E. Arsenical peripheral neuropathy. Practice Neurology, 2010, 10: 34-38.

[69] Vahidnia A, Vand V G B, De Wolff F A. Arsenic neurotoxicity: A review. Human & Experimental Toxicology, 2007, 26(10): 823-832.

[70] O'Bryant S E, Edwards M, Menon C V, et al. Long-term low-level arsenic exposure is associated with poorer neuropsychological functioning: A project FRONTIER study. International Journal of Environmental Research and Public Health, 2011, 8(3): 861-874.

[71] Selgrade M K. Immunotoxicity the risk is real. Toxicological Sciences, 2007, 100(2): 328-332.

[72] Lemarie A, Morzadec C, Bourdonnay E, et al. Human macrophages constitute targets for immunotoxic inorganic arsenic. The Journal of Immunology, 2006, 177(5): 3019-3027.

[73] Srivastava R K, Li C, Chaudhary S C, et al. Unfolded protein response (UPR) signaling regulates arsenic trioxide-mediated macrophage innate immune function disruption. Toxicology and Applied Pharmacology, 2013, 272(3): 879-887.

[74] Zeng Q B, Luo P, Gu J Y, et al. PKCθ-mediated Ca^{2+}/NF-AT signalling pathway may be involved in T-cell immunosuppression in coal-burning arsenic-poisoned population. Environmental Toxicology and Pharmacology, 2017, 55: 44-50.

[75] Moore S E, Prentice A M, Wagatsuma Y, et al. Early-life nutritional and environmental determinants of thymic size in infants born in rural Bangladesh. Acta Pãdiatrica, 2009, 98(7): 1168-1175.

[76] Raqib R, Ahmed S, Sultana R, et al. Effects of *in utero* arsenic exposure on child immunity and morbidity in rural Bangladesh. Toxicology Letters, 2009, 185(3): 197-202.

[77] Ahmed S, Moore S E, Kippler M, et al. Arsenic exposure and cell-mediated immunity in pre-school children in rural bangladesh. Toxicological Sciences, 2014, 141(1): 166-175.

[78] Benowitz N L. Cardiotoxicity in the workplace. Occupational Medicine, 1992, 7(3): 465-478.

[79] Manna P, Sinha M, Sil P C. Arsenic-induced oxidative myocardial injury: Protective role of arjunolic acid. Archives of Toxicology, 2008, 82(3): 137-149.

[80] Chen Y, Karagas M R. Arsenic and cardiovascular disease: New evidence from the United States. Annals of Internal Medicine, 2013. 159: 713-714.

[81] Ma José M, Boix R, Pastor B R, et al. Arsenic in public water supplies and cardiovascular mortality in Spain. Environmental Research, 2010, 110(5): 454.

[82] Abir T, Rahman B, D'Este C, et al. The association between chronic arsenic exposure and hypertension: A meta-analysis. Journal of Toxicology, 2012, 2012: 1-13.

[83] Mordukhovich I, Wright R O, Amarasiriwardena C, et al. Association between low-level environmental arsenic exposure and QT interval duration in a general population study. American Journal of Epidemiology, 2009, 170(6): 739-746.

[84] Wang C H, Chen C L, Hsiao C K, et al. Increased risk of Qt prolongation associated with atherosclerotic diseases in arseniasis-endemic area in southwestern coast of Taiwan. Toxicology and Applied Pharmacology, 2009, 239(3): 320-324.

[85] Tseng C H, Chong C K, Chen C J, et al. Dose-response relationship between peripheral vascular disease and ingested inorganic arsenic among residents in blackfoot disease endemic villages in Taiwan. Atherosclerosis, 1996, 120: 125-133.

[86] Lu T H, Su C C, Chen Y W, et al. Arsenic induces pancreatic β-cell apoptosis via the oxidative stress-regulated mitochondria-dependent and endoplasmic reticulum stress-triggered signaling pathways. Toxicology Letters, 2011, 201(1): 15-26.

[87] Ciarrocca M, Tomei F, Caciari T, et al. Exposure to arsenic in urban and rural areas and effects on thyroid hormones. Inhalation Toxicology, 2012, 24(9): 589-598.

[88] Gong G, Basom J, Mattevada S, et al. Association of hypothyroidism with low-level arsenic exposure in rural West Texas. Environmental Research, 2015, 138: 154-160.

[89] Smith A H, Marshall G, Liaw J, et al. Mortality in young adults following *in utero* and childhood exposure to arsenic in drinking water. Environmental Health Perspectives, 2012, 120(11): 1527-1531.

[90] Sasaki A, Oshima Y, Fujimura A. An approach to elucidate potential mechanism of renal toxicity of arsenic trioxide. Experimental Hematology (New York), 2007, 35(2): 252-262.

[91] Rahman M M, Ng J C, Naidu R. Chronic exposure of arsenic via drinking water and its adverse health impacts on humans. Environmental Geochemistry & Health, 2009, 31(1 Supplement): 189-200.

[92] Jayasumana M, Paranagama P, Amarasinghe M, et al. Possible link of chronic arsenic toxicity with chronic kidney disease of unknown etiology in Sri Lanka. Journal of Nature Science Research, 2013. 3: 64-73.

[93] Shen H, Xu W, Zhang J, et al. Urinary metabolic biomarkers link oxidative stress indicators associated with general arsenic exposure to male infertility in a Han Chinese Population. Environmental Science & Technology, 2013, 47(15): 8843-8851.

[94] Golub M S, Macintosh M S, Baumrind N. Developmental and reproductive toxicity of inorganic arsenic: Animal studies and human concerns. Journal of Toxicology and Environmental Health, Part B, 1998, 1(3): 199-237.

[95] Tabacova S, Iii E S H, Gladen B C. Developmental toxicity of inorganic arsenic in whole embryo culture: Oxidation state, dose, time, and gestational age dependence. Toxicology and Applied Pharmacology, 1996, 138(2): 298-307.

[96] Li X, Ma Y, Li D, et al. Arsenic impairs embryo development via down-regulating Dvr1 expression in zebrafish. Toxicology Letters, 2012, 212(2): 161-168.

[97] Hopenhayn R C, Browning S R, Hertz P I, et al. Chronic arsenic exposure and risk of infant mortality in two areas of Chile. Environmental Health Perspectives, 2000, 108(7): 667-673.

[98] Guo J X, Hu L, Yand P Z, et al. Chronic arsenic poisoning in drinking water in Inner Mongolia and its associated health effects. Journal of Environmental Science and Health, Part A, 2007, 42(12): 1853-1858.

[99] Milton A H, Smith W, Rahman B, et al. Chronic arsenic exposure and adverse pregnancy outcomes in bangladesh. Epidemiology, 2005, 16(1): 82-86.

[100] Yang C Y, Chang C C, Tsai S S, et al. Arsenic in drinking water and adverse pregnancy outcome in an arseniasis-endemic area in northeastern Taiwan. Environmental Research, 2003, 91(1): 29-34.

[101] Flora S J. Arsenic-induced oxidative stress and its reversibility. Free Radical Biology and Medicine 2011, 51(2): 257-281.

[102] Jomova K, Jenisova Z, Feszterova M, et al. Arsenic: Toxicity, oxidative stress and human disease. Journal of Applied Toxicology, 2011, 31(2): 95-107.

[103] Wang H, Zhu J, Li L, et al. Effects of Nrf2 deficiency on arsenic metabolism in mice. Toxicology and Applied Pharmacology, 2017, 337: 111-119.

[104] Wang C, Niu Q, Ma R, et al. The variable regulatory effect of arsenic on Nrf2 signaling pathway in mouse: A systematic review and meta-analysis. Biological Trace Element Research, 2018, 1-22.

[105] Matsui M, Nishigori C, Toyokuni S. The role of oxidative DNA damage in human arsenic carcinogenesis: Detection of 8-hydroxy-2'-deoxyguanosine in arsenic-related Bowen's disease. The Journal of Investigative Dermatology. 1999, 113(1): 26-31.

[106] Xu M, Rui D, Yan Y, et al. Oxidative damage induced by arsenic in mice or rats: A systematic review and meta-analysis. Biological Trace Element Research. 2017, 176(1): 154–175.

[107] Hughes M F, Beck B D, Chen Y, et al. Arsenic exposure and toxicology: A historical perspective. Toxicological Sciences, 2011, 123(2): 305-332.

[108] 安艳, 李春春, 邓晗依. 砷诱发氧化应激研究现状. 国外医学医学地理杂志, 2015, 36(03): 165-173.

[109] Lau A, Whitman S A, Jaramillo M C, et al. Arsenic-mediated activation of the Nrf2-Keap1 antioxidant pathway. Journal of Biochemical and Molecular Toxicology, 2013, 27(2): 99-105.

[110] 王祺, 张爱华, 李军, 等. 核因子 E2 相关因子 2-抗氧化反应元件结合能力及其下游基因表达与燃煤型砷中毒肝损伤关系探讨. 中华地方病学杂志. 2015, 34(6): 401-405.

[111] Wang D P, Ma Y, Yang X, et al. Hypermethylation of the Keap1 gene inactivates its function, promotes Nrf2 nuclear accumulation, and is involved in arsenite-induced human keratinocyte transformation. Free Radical Biology and Medicine. 2015, 89: 209-219.

[112] Ramirez P, Del Razo L M, Gutierrez R M C, et al. Arsenite induces DNA-protein crosslinks and cytokeratin expression in the WRL-68 human hepatic cell line. Carcinogenesis, 2000, 21(4): 701-706.

[113] 张爱华, 黄晓欣, 李军, 等. 用 SCGE 法检测燃煤污染型砷中毒患者血细胞 DNA 的损伤作用. 中国地方病学杂志, 2000, 19(1): 7-9.

[114] Zhang A H, Hong F, Yang G H, et al. Unventilated indoor coal-fired stoves in Guizhou Province, China: Cellular and genetic damage in villagers exposed to arsenic in food and air. Environmental Health Perspectives. 2007, 115(4): 653–658.

[115] Ferrario D, Collotta A, Carfi M, et al. Arsenic induces telomerase expression and maintains telomere length in human cord blood cells. Toxicology, 2009, 260(1-3): 132-141.

[116] Mäki P J, Kurttio P, Paldy A, et al. Association between the clastogenic effect in peripheral lymphocytes and human exposure to arsenic through drinking water. Environmental & Molecular Mutagenesis, 2015, 32(4): 301-313.

[117] Yamanaka K, Hayashi H, Kato K, et al. Involvement of preferential formation of apurinic/apyrimidinic sites in dimethylarsenic-induced DNA strand breaks and DNA-protein crosslinks in cultured alveolar epithelial cells. Biochem Biophys Res Commun, 1995, 207(1): 244-249.

[118] Mass M J, Wang L. Arsenic alters cytosine methylation patterns of the promoter of the tumor suppressor gene *p53* in human lung cells: A model for a mechanism of carcinogenesis. Mutation Research, 1997, 386: 263-277.

[119] 潘雪莉, 张爱华, 黄晓欣. 燃煤污染型砷中毒人群 MGMT 基因甲基化、转录及表达的研究. 环境与健康杂志, 2010, 27(4): 283-287.

[120] Zhang A H, Li H Y, Xiao Y, et al. Aberrant methylation of nucleotide excision repair genes is associated with chronic arsenic poisoning. Biomarkers. 2016, 12: 1-10.

[121] Jiang J, Bellani M, Li L, et al. Arsenite binds to the RING finger domain of FANCL E3 ubiquitin ligase and inhibits DNA interstrand cross-link repair. ACS Chemical Biology, 2017, 12: 1858-1866.

[122] Shen S, Wang C, Weinfeld M, et al. Inhibition of nucleotide excision repair by arsenic. Chinese Science Bulletin, 2013, 58(2): 214-221.

[123] Wei M, Liu J M, Xu M C, et al. Divergent effects of arsenic on NF-κB signaling in different cells or tissues: A systematic review and meta-analysis. International Journal of Environmental Research and Public Health, 2016, 13(2): 163.

[124] Li Y, Ling M, Xu Y, et al. The repressive effect of NF-kappaB on p53 by mot-2 is involved in human keratinocyte transformation induced by low levels of arsenite. Toxicological Sciences, 2010, 116(1): 174-182.

[125] Gong X, Ivanov V N, Hei TK. 2, 3, 5, 6-Tetramethylpyrazine (TMP) down-regulated arsenic-induced heme oxygenase-1 and ARS2 expression by inhibiting Nrf2, NF-κB, AP-1 and MAPK pathways in human proximal tubular cells. Archives of Toxicology, 2016, 90(9): 2187-2200.

[126] Qu W, Fuquay R, Sakurai T, et al. Acquisition of apoptotic resistance in cadmium-induced malignant transformation: Specific perturbation of JNK signal transduction pathway and associated metallothionein overexpression. Molecular carcinogenesis, 2006, 45(8): 561-571.

[127] Lau A T, Li M, Xie R, et al. Opposed arsenite-induced signaling pathways promote cell proliferation or apoptosis in cultured lung cells. Carcinogenesis, 2004, 25(1): 21-28.

[128] 李洋, 杨东波, 张俊和, 等. LY294002通过抑制胶质瘤的PI3K/Akt通路增强三氧化二砷毒性作用. 哈尔滨医科大学学报, 2014, (1): 13-17.

[129] Chen Q Y, Costa M. PI3K/Akt/mTOR signaling pathway and the biphasic effect of arsenic in carcinogenesis. Molecular Pharmacology, 2018, 94: 784-792.

[130] Hamann I, Klotz L O. Arsenite-induced stress signaling: Modulation of the phosphoinositide 3'-kinase/Akt/FoxO signaling cascade. Redox Biology, 2013, 1(1): 104-109.

[131] Xu W C, Luo F, Sun B F, et al. HIF-2α, acting via miR-191, is involved in angiogenesis and metastasis of arsenite-transformed HBE cells. Toxicology Research (Camb). 2016, 5(1): 66-78.

[132] Ren X, Mchale C M, Skibola C F, et al. An emerging role for epigenetic dysregulation in arsenic toxicity and carcinogenesis. Environmental Health Perspectives, 2011, 119(1): 11-19.

[133] 张爱华, 王大朋. 重视地方性砷中毒机制假说间相互作用, 提升机制研究及其转化应用价值. 中华地方病学杂志, 2016, 35(1): 1-3.

[134] Bustaffa E, Stoccoro A, Bianchi F, et al. Genotoxic and epigenetic mechanisms in arsenic carcinogenicity. Archives of Toxicology, 2014, 88(5): 1043-1067.

[135] Tellezplaza M, Tang W Y, Shang Y, et al. Association of global DNA methylation and global DNA hydroxymethylation with metals and other exposures in human blood DNA samples. Environmental Health Perspectives, 2016, 122(9): 229-239.

[136] Takahashi M, Barrett J C, Tsutsui T. Transformation by inorganic arsenic compounds of normal Syrian hamster embryo cells into a neoplastic state in which they become anchorage-independent and cause tumors in newborn hamsters. International Journal of Cancer, 2002, 99(5): 629-634.

[137] Chen H, Liu J, Zhao C Q, et al. Association of c-myc overexpression and hyperproliferation with arsenite-induced malignant transformation. Toxicol Appl Pharmacol, 2001, 175(3): 260-268.

[138] Green B B, Karagas M R, Punshon T, et al. Epigenome-wide assessment of DNA methylation in the placenta and arsenic exposure in the New Hampshire Birth Cohort Study (USA). Environmental Health Perspectives, 2016, 124(8): 1253-1260.

[139] Koestler D C, Avissar Whiting M, Houseman E A, et al. Differential DNA methylation in umbilical cord blood of infants exposed to low levels of arsenic *in utero*. Environmental Health Perspectives, 2013, 121(8): 971-977.

[140] Rojas D, Rager J E, Smeester L, et al. Prenatal arsenic exposure and the epigenome: identifying sites of 5-methylcytosine alterations that predict functional changes in gene expression in newborn cord blood and subsequent birth outcomes. Toxicological Science. 2015, 143(1): 97-106.

[141] Roy R V, Son Y O, Pratheeshkumar P, et al. Epigenetic targets of arsenic: Emphasis on epigenetic modifications during carcinogenesis. Journal of Environmental Pathology Toxicology & Oncology, 2015, 34(1): 63-84.

[142] 牛林梅, 章征保, 曾晓雯, 等. 组蛋白修饰改变在亚砷酸钠毒性效应中的作用研究. 癌变畸变突变, 2014, 26(2): 88-93.

[143] Ma L, Li J, Zhang Z B, et al. Specific histone modification responds to arsenic-induced oxidative stress. Toxicology and Applied Pharmacology. 2016, 302: 52-61.

[144] Li J, Ma L, Wang X L, et al. Modifications of H3K9me2, H3K36me3 and H4K20me2 may be involved in arsenic-induced genetic damage. Toxicology Research. 2016, 5(5): 1380-1387.

[145] Sun B F, Xue J C, Li J, et al. Circulating miRNAs and their target genes associated with arsenism caused by coal-burning. Toxicology Research. 2017, 6(2): 162-172.

[146] Zeng Q B, Zou Z L, Wang Q L, et al. Association and risk of five miRNAs with arsenic-induced multiorgan damage. Science of the Total Environment, 2019, 680: 1-9.

[147] Marsit C J, Eddy K, Kelsey K T. MicroRNA responses to cellular stress. Cancer Research, 2006, 66: 10843-10848.

[148] Rao C V, Sanya P, Altaf M, et al. Biological effects and epidemiological consequences of arsenic exposure, and reagents that can ameliorate arsenic damage *in vivo*. Oncotarget, 2017, 8(34): 57605-57621.

[149] Ling M, Li Y, Xu Y, et al. Regulation of miRNA-21 by reactive oxygen species-activated ERK/NF-κB in arsenite-induced cell transformation. Free Radical Biology and Medicine, 2012, 52(9): 1508-1518.

[150] Luo F, Liu X L, Ling M, et al. The lncRNA MALAT1, acting through HIF-1α stabilization, enhances arsenite-induced glycolysis in human hepatic L-02 cells. Biochimica et Biophysica Acta. 2016, 1862(9): 1685-1695.

[151] Dangleben N L, Skibola C F, Smith M T. Arsenic immunotoxicity: A review. Environmental Health 2013, 12(1): 73.

[152] Singh Manish K, Yadav Suraj S, Gupta V, et al. Immunomodulatory role of *Emblica officinalis* in arsenic induced oxidative damage and apoptosis in thymocytes of mice. BMC Complementary and Alternative Medicine, 2013, 13: 193.

[153] Stepnik M, Stańczyk M, Arkusz J, et al. Assessment of apoptosis in thymocytes and splenocytes from mice exposed to arsenate in drinking water: Cytotoxic effects of arsenate on the cells *in vitro*. Environmental Letters, 2005, 40(2): 16.

[154] Guha M D N. Effect of chronic intake of arsenic-contaminated water on liver. Toxicology and Applied Pharmacology. 2005, 206(2): 169-175.

[155] Zhao L, Yang S, Guo Y Y, et al. Chronic arsenic exposure in drinking water interferes with the balances of T lymphocyte subpopulations as well as stimulates the functions of dendritic cells in vivo. International Immunopharmacology, 2019, 71: 115-131.

[156] 刘永莲, 张爱华, 王大朋, 等. 调节性 T 细胞、辅助性 T 细胞-17 相关免疫因子在砷暴露大鼠外周血中的差异表达及意义. 中华地方病学杂志.2017, 36(1): 11-14.

[157] Zhang J, Zhang Y, Wang W, et al. Double-sided personality: Effects of arsenic trioxide on inflammation. Inflammation, 2018, 41: 1128-1134.

[158] Escudero Lourdes C, Medeiros M K, Cárdenas G M C, et al. Low level exposure to monomethyl arsonous acid-induced the over-production of inflammation-related cytokines and the activation of cell signals associated with tumor progression in a urothelial cell model. Toxicology and Applied Pharmacology, 2010, 244(2): 162-173.

[159] Pratheeshkumar P, Son Y O, Divya S P, et al. Oncogenic transformation of human lung bronchial epithelial cells induced by arsenic involves ROS-dependent activation of STAT3-miR-21-PDCD4 mechanism. Scientific Reports, 2016, 6: 37227.

[160] National Research Council (NRC). Risk assessment in the federal government: Managing the process. Washington, DC:

National Academy Press.1983

[161] Grund S C, Kunibert H, Uwe W H. Arsenic and arsenic compounds. Iarc Monographs on Evaluation of Carcinogenic Risks to Man, 1987, 7: 39.

[162] IARC Working Group on the Evaluation of Carcinogenic Risks to Humans. Arsenic, metals, fibres, and dusts. Iarc Monogr Eval Carcinog Risks Hum, 2012, 100(Pt C): 11-465.

[163] Wasserman G A, Liu X, Parvez F, et al. Water arsenic exposure and children's intellectual function in Araihazar, Bangladesh. Environmental Health Perspectives, 2004, 112: 1329-33.

[164] von Ehrenstein O S, Poddar S, Yuan Y, et al. Children's intellectual function in relation to arsenic exposure. Epidemiology, 2007, 18: 44-51.

[165] Jin Y, Xi S, Li X, et al. Arsenic speciation transported through the placenta from mother mice to their newborn pups. Environ Res, 2006, 101: 349-55.

[166] Vahter M. Health effects of early life exposure to arsenic. Basic Clin Pharmacol Toxicol, 2008, 102: 204-211.

[167] Ahmad S A, Sayed M H, Barua S, et al. Arsenic in drinking water and pregnancy outcomes. Environ Health Perspect, 2001, 109: 629-631.

[168] Hopenhayn C, Ferreccio C, Browning S R, et al. Arsenic exposure from drinking water and birth weight. Epidemiology, 2003, 14: 593-602.

[169] Huyck K L, Kile M L, Mahiuddin G, et al. Maternal arsenic exposure associated with low birth weight in Bangladesh. J Occup Environ Med, 2007, 49: 1097-104.

[170] NRC. Critical Aspects of EPA's IRIS Assessment of Inorganic Arsenic: Interim Report. Washington, DC: The National Academies Press, 2014.

[171] Das N, Paul S, Chatterjee D, et al. Arsenic exposure through drinking water increases the risk of liver and cardiovascular diseases in the population of West Bengal, India. BMC Public Health, 2012, 12: 639.

[172] 张爱华, 王大朋. 从免疫学角度深化认识砷所致全身性损害机制及其防治策略. 中华地方病学杂志, 2017, 36(1): 7-10.

[173] Maull E A, Ahsan H, Edwards J, et al. Evaluation of the association between arsenic and diabetes: A National Toxicology Program workshop review. Environmental Health Perspectives, 2012, 120: 1658-1670.

[174] Wang C H, Hsiao C K, Chen C L, et al. A review of the epidemiologic literature on the role of environmental arsenic exposure and cardiovascular diseases. Toxicology and Applied Pharmacology, 2007, 222: 315-326.

[175] Navas A A, Sharrett A R, Silbergeld E K, et al. Arsenic exposure and cardiovascular disease: A systematic review of the epidemiologic evidence. American Journal of Epidemiology, 2005, 162: 1037-1049.

[176] Moon K, Guallar E, Navas A A. Arsenic exposure and cardiovascular disease: An updated systematic review. Current Atherosclerosis Reports, 2012, 14: 542-555.

[177] Abhyankar L N, Jones M R, Guallar E, Navas-Acien A. Arsenic exposure and hypertension: A systematic review. Environmental Health Perspectives, 2012, 120: 494-500.

[178] US EPA. Revised Reregistration Eligibility Decision for MSMA, DSMA, CAMA, and cacodylic acid. The Office of Pesticide Programs. USEPA, Washington, DC, 2006.

[179] Arnold L L, Eldan M, Nyska A, et al. Dimethylarsinic acid: Results of chronic toxicity/oncogenicity studies in F344 rats and in B6C3F1 mice. Toxicology 2006, 223: 82-100.

[180] Brown K G, Chen C J. Significance of exposure assessment to analysis of cancer risk from inorganic arsenic in drinking water in Taiwan. Risk Analysis, 1995, 15475-15484.

[181] Clewell H J, Gentry P R, Yager J W. Considerations for a Biologically Based Risk Assessment for Arsenic// Arsenic: Exposure Sources, Health Risks, and Mechanisms of Toxicity. John Wiley & Sons, Inc, 2015.

[182] Buchet J P, Lison D. Mortality by cancer in groups of the Belgian population with a moderately increased intake of arsenic. International Archives of Occupational & Environmental Health, 1998, 71(2): 125-130.

[183] Mann S, Droz P O, Vahter M. A physiologically based pharmacokinetic model for arsenic exposure. Ⅰ. Development in hamsters and rabbits. Toxicology and Applied Pharmacology, 1996, 137: 8-22.

[184] Mann S, Droz P O, Vahter M. A physiologically based pharmacokinetic model for arsenic exposure. Ⅱ Validation and application in humans. Toxicology and Applied Pharmacology, 1996, 140: 471-486.

[185] Elmasri H A, Kenyon E M. Development of a human physiologically based pharmacokinetic (PBPK) model for inorganic arsenic and its mono- and di-methylated metabolites. Journal of Pharmacokinetics & Pharmacodynamics, 2008, 35(1): 31-68.

[186] U.S. Environmental Protection Agency (USEPA). Integrated Risk Information System (IRIS), Arsenic. CASRN 7440-38-2. 1998. Washington, DC, U.S. Environmental Protection Agency (USEPA).

[187] Tseng W P. Effects and dose-response relationships of skin cancer and blackfoot disease with arsenic. Environmental Health Perspectives, 1977, 19(19): 109-119.

[188] Tseng W P, Chu H M, How S W, et al. Prevalence of skin cancer in an endemic area of chronic arsenicism in Taiwan. Journal of the National Cancer Institute, 1968, 40(3): 453-463.

[189] US EPA. Toxicological review of inorganic arsenic. EPA IRIS. Washington, DC, 1993.

[190] Baars A. Re-evaluation of human-toxicological maximum permissible risk levels.(The National Institute of Public Health and the Environment), Bilthoven, the Netherlands; 2001, 25-29.

[191] Arnold L L, Eldan M, van Gemert M, et al. Chronic studies evaluating the carcinogenicity of monomethylarsonic acid in rats and mice. Toxicology 2003, 190: 197-219.

[192] Straif K, Benbrahim T L, Baan R, et al. WHO International Agency for Research on Cancer Monograph Working Group. A review of human carcinogens-part C: metals, arsenic, dusts, and fibres. Lancet Oncology, 2009, 10: 453-454.

[193] US EPA. Toxicological review of arsine. EPA IRIS. Washington, DC, 1994.

[194] Cal EPA. Inorganic arsenic reference exposure levels Sacramento: State of California Office of Environmental Health Hazard Assessment, 2008.

作者：张爱华[1]，王大朋[1]，姚茂琳[1]
[1] 贵州医科大学

第 19 章 邻苯二甲酸酯增塑剂的人体暴露与健康风险

▶ 1. 引言 /511
▶ 2. 邻苯二甲酸酯增塑剂的环境污染及人体暴露现状 /513
▶ 3. 邻苯二甲酸酯增塑剂的体内代谢及其与慢性疾病的关联 /520
▶ 4. 邻苯二甲酸酯增塑剂的健康效应与毒理机制 /525
▶ 5. 展望 /530

本章导读

邻苯二甲酸酯（phthalate esters，PAEs）又称为酞酸酯，由邻苯二甲酸酐与醇反应生成，是一类使用最广泛的增塑剂。近年来，随着对 PAEs 毒性的认识，其环境污染、人体暴露及健康风险受到广泛关注。调查发现水体、土壤及大气中都存在不同程度的 PAEs 污染，在饮用水、食品、化妆品及医疗用品中都有 PAEs 大量检出。PAEs 可通过饮食、呼吸、皮肤等途径进入人体，并在体内快速转化为相应的单酯或葡萄糖醛酸结合物等代谢产物，并经尿液、粪便或汗液排出体外。流行病学调查发现体内 PAEs 暴露与生殖疾病、肥胖、胰岛素抵抗、糖尿病等多种疾病的发病率具有正相关性。目前已证实邻苯二甲酸(2-乙基)已酯（DEHP）等多种 PAEs 及其代谢产物具有内分泌干扰效应，一些 PAEs 在某些产品中已被禁用或限制使用。本章将针对 PAEs 增塑剂在环境介质中分布、人体暴露、健康效应及其毒性机制等方面的研究进展进行综合介绍。

关键词

邻苯二甲酸酯，增塑剂，环境介质，人体暴露，健康风险，毒性机制

1 引　言

邻苯二甲酸酯（phthalate esters，PAEs）是邻苯二甲酸和相应的醇发生费歇尔酯化反应形成的酯的统称，具有无色无味、电性能好、挥发性低和耐低温等特点，可有效增加产品的可塑性、柔韧性或膨胀性，因而大量应用于塑料、涂料、橡胶的合成工艺，还广泛应用于起泡剂、农药载体、化妆品、驱虫剂等的生产。各种邻苯二甲酸酯的用途主要取决于它们的分子量，高分子量的 PAEs，如邻苯二甲酸二(2-乙基)已酯（DEHP）、邻苯二甲酸二异壬酯（DINP）等主要用作制造聚氯乙烯塑料的增塑剂；而低分子量的 PAEs，如邻苯二甲酸二乙酯（DEP）、邻苯二甲酸二甲酯（DMP）、邻苯二甲酸二异丁酯（DIBP）等主要用于个人护理产品、化妆品、杀虫剂和药物等（表 19-1）。PAEs 是目前世界上生产量大、应用面广的人工合成有机化合物，特别是在塑料工业中的应用量很大，其中聚氯乙烯（PVC）塑料制品中 PAEs 的含量可高达 20%~50%。全球每年 PAEs 的使用量在 800 万 t 以上。PAEs 与塑料产品分子之间并不形成共价键，而是以氢键或范德华力相连接，彼此保留相对独立的化学性质。因此，随着时间的推移，在生产、使用、废弃及后处理过程中，PAEs 很容易释放到外界环境中，导致 PAEs 普遍性污染。PAEs 已成为目前最主要的新型污染物之一[1]。图 19-1 列出了 PAEs 的结构通式及目前环境中常见几种 PAEs 的结构式，PAEs 分子中烷基侧链的碳原子数通常为 1~13，由于碳原子数的不同，其物理化学性质也存在一定的差别。随着 PAEs 烷基侧链中碳原子数的增加，正辛醇-水分配系数 K_{ow} 显著增加，在水中的溶解度 S_w 则明显降低。工业上常用的 PAEs 有 15 种之多，有超过 30 余种的 PAEs 在全球不同区域的水体、沉积物、土壤、空气、玩具以及食品和饮用水中检测出来，最常见且含量较高的 PAEs 有 DEHP、DINP、邻苯二甲酸二异癸酯（DIDP）、邻苯二甲酸二正辛酯（DNOP）、邻苯二甲酸丁苄酯（BBP）、邻苯二甲酸二丁酯（DBP）等。其中高分量子的 DEHP 和 DINP 是全球使用量最大的两种 PAEs，分别占总 PAEs 产量的 50%和 25%，且生产量逐年增加，2019 年产量预计可达 667 万 t。

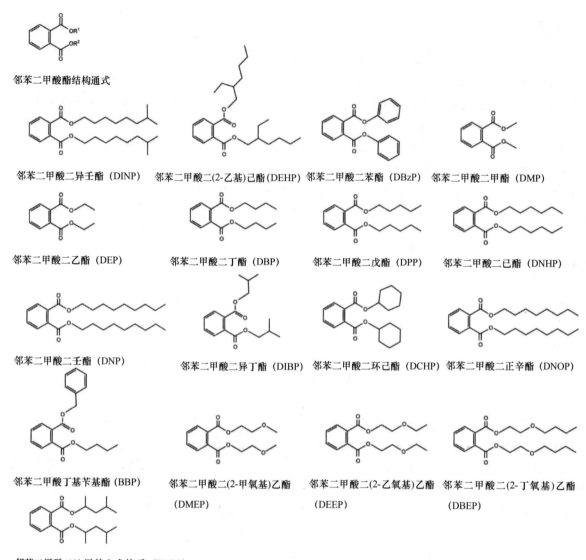

图 19-1 PAEs 结构通式及环境中常见的几种 PAEs 结构式

表 19-1 常见 PAEs 的主要用途[2]

PAEs	主要用途
邻苯二甲酸二异癸酯	聚氯乙烯增塑剂/软化剂、涂层用塑料膏（塑料溶胶）（家具、炊具、油布、合成革和墙壁）覆盖物）、旋转成型/旋转铸造（制造中空塑料产品，如玩具、游戏和运动球、漏斗）；药丸、食品包装、耐热电线、PVC 地板材料；防腐、防污油漆、密封剂和纺织油墨等
邻苯二甲酸二异壬酯	电线电缆，柔性聚氯乙烯薄膜，涂层织物，汽车零件，建筑和建筑（防水）材料、乙烯基地板、鞋类、密封件、层压膜和含 PVC 的学校用品（如香味橡皮擦和铅笔盒）；非聚合物应用，如黏合剂、油漆、表面活性剂和 T 恤印刷油墨；塑料、玩具（塑料书、球、娃娃和卡通人物）和婴儿用品（垫子、衣服、连指手套、奶嘴的覆盖物、肥皂包装和淋浴垫）、层压板、树脂、表面活性剂、丝网印刷油墨以及软质 PVC 制品（垫圈、胶靴等）
邻苯二甲酸二辛酯	PVC、建筑和汽车、地板、医疗和卫生产品（例如，血液/透析袋、透析设备、注射器、植入物、导管）、地砖和家具装饰、墙面覆盖物；玩具；食品包装；涂料；液压/液电容器中的介电流体；光棒、塑料薄膜、手套、电缆和电线地板、浴帘中的溶剂、乙基纤维素树脂等

续表

PAEs	主要用途
邻苯二甲酸二异庚酯	汽车变速器润滑油、转向液中的润滑油及作为增塑剂、橡胶等塑料制品
邻苯二甲酸二苯酯	硝基和乙基纤维素、聚苯乙烯、苯酚和乙烯树脂中的增塑剂
邻苯二甲酸丁苄酯	PVC 产品中的增塑剂，如聚乙烯地砖、黏合剂和密封剂、汽车护理产品、玩具、食品包装、合成材料皮革，工业溶剂，个人护理产品
邻苯二甲酸二戊酯	玻璃纤维和橡胶、黏合剂、油漆/清漆、灌浆剂中作为溶剂
邻苯二甲酸二丁酯	醋酸纤维素塑料、个人护理产品（例如，指甲油和溶剂、化妆品）、油漆、清漆、涂层（如药品）；许多其他化妆品中的香味成分等
邻苯二甲酸二丙酯	聚氯乙烯树脂，黏合剂黏合剂，纤维素薄膜
邻苯二甲酸二乙酯	个人护理产品（如香水、指甲油）、表层涂料（如药品）、染料、杀虫剂；在香水中用作固定剂和溶剂，驱蚊剂、樟脑替代品、食品表面润滑剂和药品包装等
邻苯二甲酸二甲酯	蚊蝇驱虫剂；外寄生虫药；芳香成分化妆品，家庭和个人护理产品；儿童玩具和儿童保育用品、油漆、油漆、塑料和橡胶

2 邻苯二甲酸酯增塑剂的环境污染及人体暴露现状

由于 PAEs 与塑料分子之间是以氢键或范德华力连接，而不是紧密的共价键结合，因此，在塑料制品的生产与使用过程中，PAEs 极易从塑料制品中浸出而进入环境。近年来，研究人员对 PAEs 在环境多种介质中的分布及赋存状态进行了广泛的调查。结果表明，在全球不同区域的水体及沉积物、土壤、室内空气及颗粒物等多种环境介质中均存在不同程度的 PAEs 污染；同时，在饮用水、食品、个人护理品（化妆品、香水、指甲油等）及医疗用品等人体暴露途径中都有 PAEs 的检出。PAEs 可以通过食品、呼吸、皮肤接触等多种途径进入人体，在人体中也有广泛检出。

2.1 水体及沉积物中的 PAEs

与其他合成化合物类似，PAEs 可通过各种途径进入水生系统，如直接或间接废水排放、固体废弃物的堆放、干湿沉降、雨水淋洗以及 PVC 塑料的缓慢释放等，大气中的 PAEs 也通过干沉降或雨水淋洗而转入水环境。由于 PAEs 的溶解度较低，因此水环境中的 PAEs 大部分会与颗粒紧密相连并最终积聚在底部沉积物中。在过去几年中，在大型河流的水体及沉积物中都检测到 PAEs 污染。

Zhang 等[3]研究了长江中下游段的 PAEs 分布情况，共检测到 16 种 PAEs 同系物，其浓度范围与长江江苏段地表水（Σ6PAEs 为 0.178~1.474 μg/L）基本一致[4]，但低于北京湖泊中的 PAEs 浓度（0.386~3.184 μg/L）[5]，这可能是因为北京地区的人口密度较大及生活水平较高[6]。该研究还发现，深层水体的 PAEs 含量要高于表层水体。DEHP、DIBP 和 DBP 是三种检出率最高的 PAEs，同时该研究指出海水中的 DEHP、沉积物中的 DEHP 及 DNBP（邻苯二甲酸二丁基酯）都超过了环境风险水平。16 种 PAEs 的浓度分布随空间位置变化很大，在长江入海口处检测到 Σ16PAEs 的浓度最高（3.421 μg/L），认为可能是河海边界区域存在一个边缘过滤（marginal filter）效应，由于水动力学和生物地球化学过程，使河海边界区域成为一个"栅栏"，污染在此处发生沉淀、团聚和吸附[7]。He 等[8]检测了巢湖水体及悬浮颗粒物（suspended particulate matter，SPM）中 6 种 PAEs 的浓度，结果表明巢湖中普遍存在 PAEs 污染，在溶解相和颗粒相中 Σ6PAEs 的浓度分别为 0.370～13.2 μg/L 及 14.4～7129 μg/L，其中 DIBP 和 DBP 是巢湖中含量最高的两种 PAEs；从时间分布上来看，水相中 PAEs 浓度呈现出季节性变化趋势，而 SPM 中的 PAEs 浓度变化规律不明显；从空间分布来看，湖泊和河口之间存在着明显的空间差异。同时该研究指出 PAEs 的疏水性对其在复杂环境系统中的分配影响不大，K_{ow} 模型可能不适合天然湖泊等复杂多介质环境中 PAEs 的分配预测。Cheng 等[9]研究了珠江三角洲地区淡水鱼塘中 13 种 PAEs 的浓度及分布情况，发现 DEHP 是水相及沉积物中的主要 PAEs，分别占 ΣPAEs 的 70.1%和 66.1%，有 3 处采样点的 DEHP 浓度超过了中国《地表水环境质量标准》（8.0 μg/L），沉积物中 ΣPAEs 的浓度要高于珠江三角洲地区河流和河口沉积物中的浓度，DEHP 和 DBP 的平均浓度超过了推荐的环境风险域值，对珠江三角洲的水产养殖鱼塘环境构成了潜在风险。总体上，我国大部分水体中都存在不同程度的 PAEs 污染，有些水域的 PAEs 已临近或超过环境风险域值，既对水生生物构成了潜在的威胁，同时 PAEs 在水生生物中的积累，能够通过食物链传递到人体，对人体健康造成危害。

2.2 土壤中的 PAEs

工业烟尘沉降、污水灌溉、施用肥料、堆积的农田塑料薄膜及塑料废品中的 PAEs 在自然力的长期作用下溶出而进入土壤，造成土壤污染，其中农用塑料薄膜的应用是农田土壤中最重要的 PAEs 来源。2011 年中国农业使用的塑料薄膜数量约为 229 万 t，覆盖面积达到 1980 万 hm^2 [10,11]。此外，土壤中 PAEs 的浓度也会受到大气沉降、土壤群落和气象条件的影响[12,13]。Niu 等[14]于 2014 年对中国农田土壤进行普查，发现所检测的 15 种 PAEs 的总浓度变化很大，范围为 75.0～6369 μg/kg，平均值为 1088 μg/kg。所有土壤样品中均检测 DMP、DEP、DIBP、DNBP 和 DMEP。在分析的 PAEs 同系物中，DEHP 的浓度最高，其次是 DIBP，DNBP 和 DNP，其平均残留水平分别为 821 μg/kg、74.7 μg/kg、65.8 μg/kg 和 69.6 μg/kg。福建省、广东省和新疆维吾尔自治区土壤中 ΣPAEs 含量最高。重庆市、湖北省和河南省的土壤中 PAEs 的含量也较高。然而，东北地区的 PAEs 总水平相对较低。

Wang 等[15]分析了西安市市区土壤中的 6 种被美国环境保护署（EPA）列为优先控制 PAEs 的含量，结果发现城市土壤中 6 种 PAEs（Σ6PAEs）的总浓度为 193.0～19146 μg/kg，平均为 1369 μg/kg，其中 DBP 和 DEHP 是城市土壤中最重要的两种 PAEs。工业区、交通区和居民区的 PAEs 水平相对较高。城市土壤中的 PAEs 主要来源于增塑剂或添加剂的应用、化妆品和个人护理产品的使用、建筑材料和家居用品的排放、工业过程和大气沉降。城市土壤中的某些 PAEs 浓度超过了土壤容许浓度和环境风险水平。可见，无论是农田土壤还是城市土壤都受到不同程度的 PAEs 污染。

2.3 室内空气及颗粒物中的 PAEs

室内空气中 PAEs 主要来源于室内建筑材料、装修、装饰材料及生活用品、个人护理品中掺杂的 PAEs 的挥发。室内空气中 PAEs 主要有三种赋存状态：空气中游离存在、附着在空气中悬浮的颗粒物表面、吸附在室内灰尘表面，其中 PAEs 主要以结合在悬浮颗粒物或灰尘中的状态存在。Papadopoulos 等[16]分析了室内灰尘中 194 种有机物，发现 PAEs 是最优先污染物。Hu 等[17]对我国 6 个城市的室内主要空气污染物进行测量，发现 DEHP 和 DBP 是室内灰尘中最常见的半挥发性有机化合物（semi-volatile organic compounds，SVOCs）。Blanchard 等[18]发现与室内悬浮颗粒结合是室内 DEHP 的主要赋存状态，约占 DEHP

总量的 50%，另有约 30%结合在灰尘中。Zhu 等[19]检测了中国 6 个地区采集的 120 份室内灰尘样品，研究了 9 种常见 PAEs 的浓度、分布及人体暴露情况，结果发现 PAEs 总浓度（∑9PAEs）为 2.31～1590 μg/g 变化（平均 150 μg/g），其中东北地区室内灰尘中 PAEs 浓度最高，平均为 394 μg/g，而西南地区室内灰尘中的 PAEs 浓度较低，为 52.1 μg/g。DEHP 是室内颗粒物含量最高的 PAEs，图 19-2 比较了美国（均值 340 μg/g，2003 年）[20]、丹麦（均值 858 μg/g，2003 年）[21]、德国（均值 703 μg/g，2004 年）[22]、瑞典（均值 770 μg/g，2005 年）[23]、美国（均值 304 μg/g，2011 年）[24]、中国（均值 228 μg/g，2010 年）[24]、中国（均值 1190 μg/g，2012 年）[25]、日本（均值 147 μg/g，2010 年）[26]、美国（均值 107 μg/g，2017 年）[27]等不同国家室内灰尘中 DEHP 的浓度。可见尽管各地区空气中检测到 PAEs 浓度有所差别，但各个国家和地区都存在 PAEs 污染。

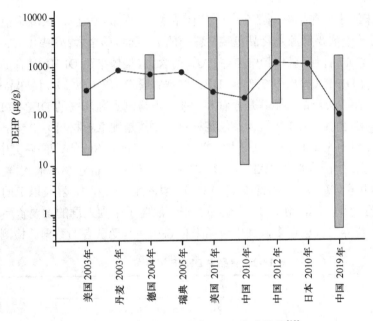

图 19-2　各国室内灰尘中 PAEs 浓度比较[19]

2.4　PAEs 在饮用水及食品中的污染现状

饮用水中存在 PAEs 的广泛污染，日常生活中所使用的瓶装水，运送生活所需的日用水塑料管道都可以成为 PAEs 的外暴露来源。Domínguez-Morueco 等[28]检测了西班牙马德里地区自来水中的 PAEs 含量，结果显示自来水中检出了 DMP、DEP 及 DBP 三种 PAEs，DBP 的含量最高，可达 633 ng/L。几项研究都报告饮用水源中 DEHP 具有较高浓度[29-31]，其中中国台湾地区自来水中最高浓度可达 172 ng/L，瓶装水中的 DEHP 浓度可达 300 ng/L[32]。DEHP 的高浓度可能是与其在日常生活用品中的广泛应用有关。鉴于饮用水源存在 PAEs 污染情况，各国都对饮用水中的 PAEs 进行了严格控制，中国《生活饮用水卫生标准》（GB 5749—2006）规定，DEP、DBP 和 DEHP 在饮用水中浓度限值为 300 μg/L、3 μg/L 和 8 μg/L；美国 EPA 规定饮用水中 DEHP 的上限浓度为 6 μg/L，世界卫生组织和欧盟限定饮用水中 DEHP 的浓度为 8 μg/L。饮用水中 PAEs 的种类及浓度取决于水源、季节和储存类型[33]。Shi 等[30]调查了饮用水供应的不同阶段 PAEs 的去除情况，发现过滤、混凝、好氧生物降解、氯化、臭氧氧化等处理过程，可去除 95%的 DMP、83%的 DEP、79%的 BBP 和 99%的 DEHP，该研究还报道煮沸法可去除 90%以上的 DEHP。Liu 等[34]在全国范围内调查了自来水厂饮用水中 6 种 PAEs（DEP、DMP、DBP、BBP、DEHP、DNOP）的污染情况，在6种目标PAEs中,DBP和DEHP含量最高,中值(±四分位间距)分别为0.18±0.47μg/L和0.18±0.97μg/L,均未超过中国饮用水水质标准的限值。中国北方地区饮用水中的PAEs普遍高于南方和东部地区。依据调查的浓度,对终生暴露风险进行了评估,表明即使在保守情况下(95%的风险),饮用水中的PAEs对中国居民也不会造成致癌风险。Zaki等[35]调查了埃及市场上瓶装水（PET塑料瓶）中的6种常见PAEs,在购买后立即分析的水样中只检测到DEHP和DBP,检测频率为50%和58%,平均浓度分别为0.104 μg/L和0.082 μg/L,并且发现瓶装水的储存时间、温度和瓶装水中检测到的PAEs的浓度之间存在正相关,表明PAEs可从PET塑料材料中渗漏到饮用水中。Salazar-Beltran等[36]检测了墨西哥蒙特雷城市中的10个商业品牌的聚对苯二甲酸乙二醇酯（PET）瓶装饮用水中的PAEs,结果发现在瓶装水样品中检测到的PAEs主要是DBP,其浓度为20.5~82.8 μg/L。迁移试验表明塑料容器中这些污染物的存在可能取决于原材料及其生产过程中的条件。

食品中的PAEs来源主要包括有两部分,一是食品本身含有的PAEs,来源于植物生长过程中蓄积在植物体中的PAEs；另一个重要来源是食品包装材料中的PAEs转移至食品中[37]。Fierens等[38]检测了比利时市场上400种食品（含12种包装材料）中8种PAEs的含量,整体来说DEHP是食品中含量较多的PAEs,检出率为81%,其次为DIBP（检出率75%）、DBP（69%）和BBP（58%）。调味品中DMP的含量最高,达4238.0 μg/kg鲜重,在谷类产品（尤其是面食和大米）中检测到高浓度的DEP和DIBP,植物油中BBP含量最高,达1127 μg/kg鲜重。DIBP是食品包装材料（尤其是硬纸板材料）中检出率最高的PAE,在硬纸板材料中其浓度为24.0~523.0 ng/cm^2（平均值为162.0 ng/cm^2）。Sui等[39]的研究中,对食物中的DEHP浓度进行了分析,确定了中国人群的PAEs膳食暴露程度,研究人群包括一般人群（2~100岁）和四个年龄组,即2~6岁的儿童,7~12岁的青少年,13~17岁的年轻人和18岁及以上的成年人。并在2011~2012年收集了1704份食品样本,分为12个食品类别,涵盖了中国人群的主要食品,食物消费数据来自于2002年进行的中国营养与健康调查,其中包括来自68959名受试者的数据。检测结果如表19-2所示。

表19-2 食品中DEHP浓度[39]　　　　　　　　　　（单位：mg/kg）

序号	食品	样本数	<LOD	平均值	中位值	范围
1	谷物	167	82	0.13（MB）	0.04	ND-1.07
2	肉	153	69	0.23（MB）	0.04	ND-3.41
3	牛奶	242	181	0.05（UB）	0.03	ND-0.51
4	蔬菜根	116	63	0.18（MB）	0.02	ND-2.03
5	蔬菜叶	125	52	0.05（MB）	0.03	ND-0.22
6	水产	169	93	0.08（MB）	0.02	ND-0.70
7	饮料	205	186	0.06（UB）	0.05	ND-0.50
8	饮用水	165	123	0.01（UB）	0	ND-0.03
9	植物油	140	98	0.21（UB）	0.2	ND-1.32
10	方便面	121	95	0.18（UB）	0.1	ND-0.39
11	果冻	54	39	0.08（UB）	0.05	ND-0.49
12	果酱	47	38	0.08（UB）	0.05	ND-0.96
总计		1704	1201	—	—	—

注：ND,未检出；MB,中界；UB,上界

将DEHP的平均浓度与个体食物消耗数据组合可以来估计膳食暴露。结果发现,食品中的DEHP含量范围为0~3.41 mg/kg；肉类（0.23 mg/kg）和植物油（0.21 mg/kg）的平均值最高。一般人群、儿童和成人的每日DEHP平均膳食摄入量分别为2.34 μg/kg、4.51 μg/kg和2.03 μg/kg体重,97.5%摄入量分别为每天

5.22 µg/kg、8.43 µg/kg 和 3.64 µg/ kg 体重。儿童群体的 DEHP 膳食摄入的主要食物来源是谷物（39.44%）、饮用水（16.94%）和肉类（15.81%）；成人的主要食物来源是谷物（44.57%），肉类（15.70%）和饮用水（12.28%）。该小组还根据以下公式评估了 DEHP 的每日摄入量：

$$y_i = \frac{\sum_{k=1}^{p} X_i k C_k}{\text{bw}_i} \tag{19-1}$$

式中，y_i 表示饮食中消费者的 DEHP 摄入量 [µg/(kg 体重·d)]；$X_i k$ 表示消费者 i 来自食物的消费量 k（g）；C_k 是食物中 DEHP 的平均浓度（mg/kg）；bw_i 是指消费者 i 的体重（kg）；p 是消费者 i 消费的食物数量。经过计算发现，一般人群 DEHP 的平均膳食暴露量约为每天 2.34 µg/kg 体重，2~6 岁儿童每天 4.51 µg/kg 体重，成人每天 2.03 µg/kg 体重，说明 DEHP 在居民日常饮食中广泛存在[39]。

2.5 个人护理品中及医疗用品中的 PAEs

PAEs 除常用作塑料的增塑剂外，还广泛应用于个人护理产品中，如 PAEs 应用于指甲油中以降低产品脆性，应用于头发护理品中，使发胶在头发表面形成柔韧的膜而避免头发僵硬，应用于护肤产品中增加皮肤的柔顺感。同时，PAEs 还可作为一些产品的溶剂和香水的抗挥发剂。因此，PAEs 在个人护理产品中广泛检出。2007 年，Koniecki 等[40]检测了加拿大多个省份的零售商店中 252 种个人护理产品（表 19-3），包括香水、护发产品（头发喷雾剂，摩丝和凝胶）、除臭剂（包括止汗剂）、指甲油、乳液、皮肤清洁剂和婴儿用品及 98 种婴儿护理产品。发现在关注的 18 种 PAEs 中，检测到 DEP、DMP、DIBP、DBP 和 DEHP。检出频率为 DEP（103/252，252 种产品中有 103 种产品检出）> DBP（15/252）> DIBP（9/252）> DEHP（8/252）> DMP（1/252）。在几乎所有类型的调查产品中都检出 DEP，这与 Hubinger 等[41]

表 19-3 化妆品和个人护理产品中 DEHP 的每日皮肤接触剂量的估计[40]

产品类型	产品应用（g/次）	使用频率（次/d）	保留系数	暴露估计值[µg/（kg 体重·d）]				
				DMP 最大值	DEP 中值	DEP 最大值	DnBP 最大值	DEHP 最大值
香水	0.61	3	1		2.6	39		0.8
洗剂	8	1	1			37		
头发护理	5	1	0.1			0.5	0.015	
除臭剂	0.5	1	1	0.03		1.5		
指甲油	0.25	0.28	1				0.34	0.01
皮肤清洁剂	2.5	2	0.1			0.12		0.01
总计（成年女性）				0.03	2.6	78	0.36	0.082
婴儿润肤露	1.4	0.14	1			0.37		0.01
婴儿洗发水	0.51	0.27	0.01			0.001		
防尿布疹护肤膏	1.4	1.72	1			20		
婴儿油	1.3	1.57	1			0.1		
总计 0.5~4 岁						20		0.01
婴儿润肤露	1.4	0.14	1			0.75		0.02
婴儿洗发水	0.51	0.27	0.01			0.003		
防尿布疹护肤膏	1.4	1.72	1			41		
婴儿油	1.3	1.57	1			0.21		
总计 0~6 个月						42		0.02

的报道是一致的，其中香水的检测水平最高（25542 μg/g，相当于 2.6%）；而 DBP 主要存在于指甲油产品中，最高浓度为 24304 μg/g（2.4%）。在其他产品中也发现了 DBP，如头发喷雾剂、头发摩丝、皮肤清洁剂和婴儿洗发水，但浓度低得多（36 μg/g 或更低）。该研究小组还估算了三个年龄组人群（成年女性，以 60 kg 计），幼儿（0.5～4 岁）和婴儿（0～6 个月）经过使用化妆品和个人护理产品，所产生的 PAEs 的每日暴露量（表 19-3）。但是，这一暴露估计是基于现有的产品使用模式数据，而不是基于概率模型的人口使用分布。对于成年女性，确定 DEP 的最大日暴露量为 78 μg/(kg 体重·d)，而其他 3 种 PAEs 的最大日暴露量较低 [DEHP：0.82 μg/(kg 体重·d)；DBP：0.36 μg/(kg 体重·d)；DMP：0.03 μg/(kg 体重·d)]。

Bao 等[42]检测了上海 198 个个人护理品中的 11 种 PAEs 含量，结果发现 DEP 是最常见的 PAEs（检出率 29.8%），其次是 DIBP（检出率 6.6%），并估算了人体通过个人护理品的暴露剂量，女性成人每日暴露于 DEP、DMEP、DIBP、DBP、DPP 和 DEHP 的平均值分别为 0.018 μg/(kg 体重·d)、0.012 μg/(kg 体重·d)、0.002 μg/(kg 体重·d)、0.001 μg/(kg 体重·d)、0.003 μg/(kg 体重·d)和 0.002 μg/(kg 体重·d)。

除饮用水、食品、空气等常见暴露方式外，PAEs 还可通过医疗途径进入人体。目前常用的医用塑料为 PVC，主要用于血袋、静脉输液袋、输血器、输液器、肠道和肠道外营养袋、腹膜透析袋、鼻饲管及血液透析管路等。患者在接触 PVC 医用塑料时，PAEs 便可能溶出而进入人体。在医疗途径中与 PAEs 接触的风险等级，取决于患者病程种类、频率和周期等因素。Malarvannan 等[43]最近检测了比利时和荷兰两所医院每天使用的各种医疗器械和基本附件中的 PAEs 含量，结果表明，在多种医疗器械中都有 DEHP 的检出。

2.6 PAEs 人体暴露现状

明确 PAEs 在人体内的含量及分布是准确评估其对人体健康危害的基础，因此近年来，多个课题组都开展了 PAEs 在体内含量的人群调查研究。由于环境中的 PAEs 无处不在，生物样本（如血液、尿液等）在收集、运输、储存以及分析测定的系列过程中都可能会遭受 PAEs 的污染。鉴于此，对于 PAEs 暴露的测量，国内外学者通常采取以检测尿液中代谢产物（如邻苯二甲酸单酯，mPAEs）的方法来进行暴露定量分析，毕竟尿液代谢产物不存在遭受污染的情况。所以，对 PAEs 的暴露评估一般不是测定它的母体形式，而是测量其特有的代谢产物如单酯和次级氧化产物（即生物标志物）。

近年来，研究人员测定了不同国家或地区人体内的 PAEs 代谢产物的含量。美国疾病控制和预防中心（CDC）在 1999～2000 年首次揭示了国家尺度下美国居民尿液中 PAEs 代谢物的水平[44]。Colacino 等对 2003～2004 年国家健康和营养检查调查（NHANES）收集的 2350 份居民尿液进行分析，再次报道了美国居民的 PAEs 暴露情况，发现 DEHP 及其代谢产物与家禽类的消费有关，而尿液中 MEP（邻苯二甲酸单乙酯，DEP 的代谢产物）与蔬菜，特别是番茄、马铃薯的消费有关[45]。Wittassek 等[46]发现，在 1988～2003 年，德国学生至少暴露于 DNBP、DIBP、BBzP、DEHP 及 DINP 5 种 PAEs 中。Gao 等[47]于 2010 年在中国大陆共采集了 108 名青年（18～22 岁）的尿液样本，检测了尿液中的 14 种 mPAEs 含量，结果发现 MBP（邻苯二甲酸单丁酯，DMP 的代谢产物）和 MIBP（邻苯二甲酸单异丁酯，DIBP 的代谢产物）是含量最高的两种 mPAEs，其几何平均数分别为 67.0 ng/mL 和 57.2 ng/mL，这一结论与 Guo 等[48]及 Wang 等[49]的结论相近[50, 51]。该研究还比较了多个国家或地区人群尿样中 mPAEs 的含量，结果如图 19-3 所示。结果显示，日本（中位数：84.3 ng/mL）[52]、科威特（中位数：94.1 ng/mL）[53]和墨西哥（中位数：82.1 ng/mL）[54]人群中的 MBP 浓度高于中国人群。德国人群的 MBP 浓度最高（中位数：181 ng/mL）[55]。研究还发现，中国男青年尿液中 MIBP 的含量要高于其他国家的报道值（除丹麦男性外）[56]，表明中国男青年对 DIBP 的接触较高。

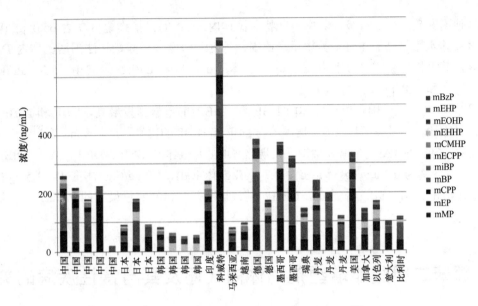

图 19-3 不同国家人群体内 mPAEs 浓度比较[47]

人体尿液中 mPAEs 的浓度主要用于描述 PAEs 相对暴露水平，而不能直接反映此类物质在人体内的暴露情况。因此，研究人员以尿液中 mPAEs 的浓度为基础，通过尿液中此类物质的浓度来计算人体对于污染物的暴露量，可以更加准确地评价人体对于污染物的暴露情况。研究人员建立了多种计算模型，可以通过尿液中 mPAEs 的浓度来计算人体 PAEs 的暴露量，如常用的 EDI (estimated daily intake，估计每日摄入量) 模型[57]：

$$\mathrm{EDI} = CV \times \frac{M_1}{M_2} \times \frac{1}{f} \times \frac{1}{W} \tag{19-2}$$

式中，EDI 是个人每日 PAEs 摄入量 [μg/(kg 体重·d)]，C 是尿液中 PAEs 代谢物浓度 (μg/L)，V 是人体每日尿液排泄量 (L/d)，M_1 和 M_2 分别为 PAEs 母体及其代谢物的分子量 (g/mol)，W 是体重 (kg)，f 是相对于排出的 PAEs 代谢物的摩尔分数。

表 19-4 列出了几个不同国家和地区人群体内 PAEs 的暴露情况。可以看出，DEP、DBP、DIBP 和 DEHP 是人体暴露程度较高的 PAEs 物质。印度、科威特、马来西亚居民体内 DEP 的暴露水平相对较高，平均

表 19-4 不同国家人群体内 PAEs 日暴露量　　[单位：μg/(kg 体重·d)]

国家	采样年份	样品数/个	年龄/岁	DMP	DEP	DBP	DIBP	DEHP	BBzP
美国[59]	1988～1994	289	20～60		12	1.5		0.71	0.88
美国[59]	1999～2000	2536	6～20		5.4			0.7	
中国[48]	2010	183	37±19	0.6	1.1	8.5		3.4	
中国[58]	2012	782	8～11	0.3	0.7	1.9	1.5	3.7	0.01
德国[59]	2001/2003	119	20～29			2.2	1.5	6.4	0.22
丹麦[60]	2006～2008	129	6～21			4.29		4.04	0.62
加拿大[61]	2007～2009	3236	6～49		1.09				
日本[59]	2004	35	20～70		2	1.3		1.8	
韩国[53]	2006～2007	60	35±11		177 (2.95)	110 (1.83)		102 (1.70)	
印度[53]	2010	22	46±17		1228 (20.5)	178 (2.97)		339 (5.65)	
科威特[53]	2010	46	21±15		3900 (65.0)	822 (13.7)		435 (7.25)	
马来西亚[53]	2010	29	30±9		693 (11.6)	124 (2.07)		97.7 (1.63)	
越南[53]	2010	30	49±18		64.0 (1.07)	173 (2.88)		112 (1.87)	
法国[62]	2007	279			1	1.5	2.2	5.8	0.4
比利时[63]	2013	52	1～12		1.47	2.38	0.48	3.37	0.42

日暴露量比其他国家高一个数量级，其中科威特居民体内 DEP 的日暴露量可高达 65.0 μg/(kg 体重·d)。DEHP 是人体暴露水平最高的 PAEs 类物质，德国、丹麦、印度、科威特和法国居民体内 DEHP 的暴露水平相对较高，其平均日暴露量分别为 6.4 μg/(kg 体重·d)、4.04 μg/(kg 体重·d)、5.65 μg/(kg 体重·d)、7.25 μg/(kg 体重·d)和 5.8 μg/(kg 体重·d)。

Guo 等[53]的研究发现中国青年男性体内 DBP 和 DIBP 的日暴露量最高，DBP 和 DIBP 的总暴露量达 8.5 μg/(kg 体重·d)，明显高于其他国家居民体内 DBP 和 DIBP 的日暴露量。Wang 等[58]调查了上海、江苏和浙江地区儿童体内多种 PAEs 的暴露情况，发现这些地区儿童体内 DMP、DEP、DBP、DIBP、DEHP 和 DBzP 的日暴露量分别为 0.3 μg/(kg 体重·d)、0.7 μg/(kg 体重·d)、1.9 μg/(kg 体重·d)、1.5 μg/(kg 体重·d)、3.7 μg/(kg 体重·d)和 0.01 μg/(kg 体重·d)，DEHP 是该地区儿童日暴露量最高的 PAEs。

3 邻苯二甲酸酯增塑剂的体内代谢及其与慢性疾病的关联

3.1 PAEs 体内代谢

不同环境介质中的 PAEs，如玩具、灰尘、饮用水、食物等，可通过不同的暴露路径到达人体表面，最后主要通过口入、皮肤、呼吸三种途径被吸入人体，构成人体总暴露。其中，经口暴露是 PAEs 进入人体的主要方式，其次是通过灰尘和室内空气的吸入[64, 65]。经口暴露包括直接摄入食物和饮用水中的 PAEs，以及偶尔摄入灰尘、土壤或手表面的 PAEs 等路径；呼吸暴露包括吸入气相及颗粒物中的 PAEs；而皮肤暴露过程则较为复杂，首先 PAEs 通过气相传递、颗粒物沉降、衣服传递、日用化妆品等的皮肤接触而到达皮肤表面，然后通过皮肤渗透进入血液。PAEs 在血液中一般会与人血清白蛋白（human serum albumin，HSA）或脂肪酸结合蛋白（fatty acid-binding protein，FABP）等结合而被运输到全身各个器官[66, 67]。

PAEs 进入人体后会被代谢并通过尿液、汗液或粪便等排出体外[68, 69]，肝脏是 PAEs 代谢的主要器官。PAEs 在排泄出人体之前会发生两个阶段的化学转变：第一阶段，PAEs 被酯酶或脂肪酶转化为相应的单酯，即 mPAEs[70, 71]；在第二阶段，在葡萄糖醛酸基转移酶（glucuronyl transferase，GT）作用下，这些亲水性代谢物将转化为葡萄糖醛酸结合物，并经尿液或汗液等排泄出体外[72]。如 DEHP 经口摄食进入肠道后，一部分以母体形式直接吸收，另一部分则在胰腺酶和肠道脂肪酶的作用下迅速发生 I 相代谢，水解为邻苯二甲酸单(2-乙基己基)酯（monoethylhexyl phthalate，MEHP），MEHP 一部分在肠道内被吸收，一部分经过多步氧化继续代谢为带有亲脂性脂肪侧链的氧化物（5OH-MEHP，5cx-MEPP，5oxo-MEHP，2cx-MMHP）。这些单酯和氧化产物可以在葡萄糖醛酸基转移酶作用下发生 II 相代谢，与葡萄糖醛酸苷发生偶联。DEHP 在人体内的代谢如图 19-4 所示。

由于与 PAEs 降解有关的酶类在体内各组织中的分布不同，PAEs 母体化合物进入不同组织或器官后，其代谢速度也有区别。DEHP 及其代谢产物通过血液分布于肝脏、肾脏、胃肠道、脂肪组织、肌肉、皮肤等全身各个组织中，其中肝脏和脂肪组织中的含量较高[64]。大鼠实验发现，大多数 DEHP 可在 24 h 内完全代谢，在肺脏内的半衰期为 1.5 h，肝脏中为 28.4 h，而储存于脂肪组织中的 DEHP 长时间不能完全代谢，半衰期长达 156 h[75, 76]。通过男性人体试验志愿者的研究发现，一次性口服 48.1 mg DEHP 后，监测其 3 种代谢产物，发现血液和尿液中 MEHP 的最高峰出现在口服后 2 h，而二级代谢产物的氧化形式 5OH-MEHP 及 5oxo-MEHP 的最大浓度出现在口服后 4 h。血液中的主要代谢产物是 MEHP，而尿液中 5OH-MEHP 占主导地位。44 h 后，约有 47%的 DEHP 以 MEHP(7.3%)、5OH-MEHP(24.7%)及 5oxo-MEHP

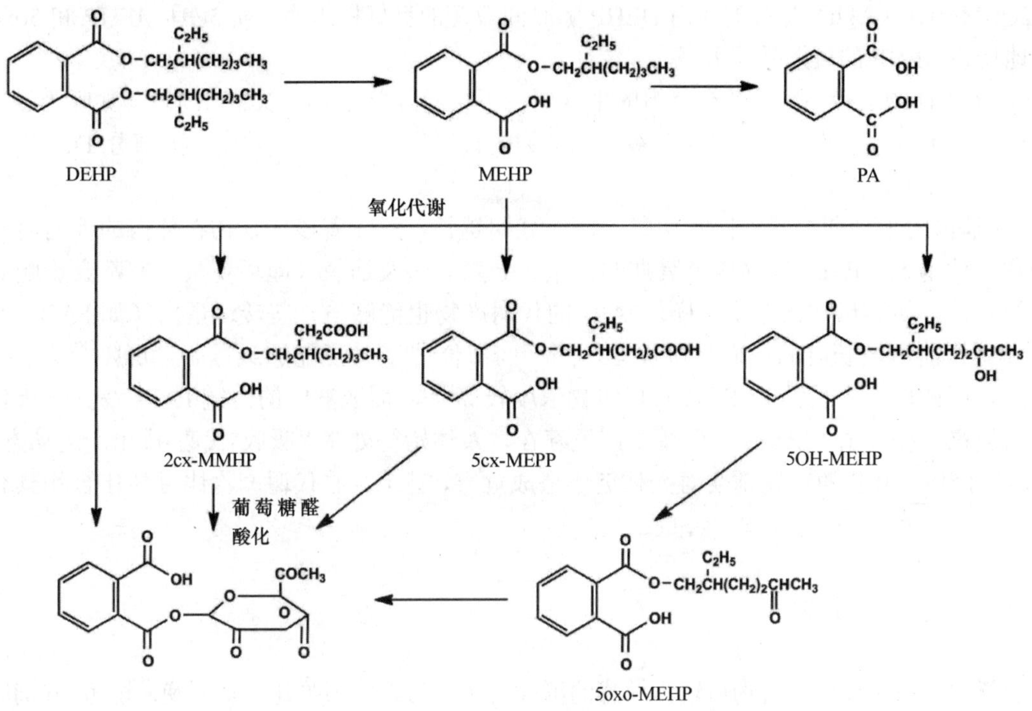

图 19-4 PAEs 在人体内的主要代谢途径示意图（根据参考文献[73, 74]绘制）

（14.9%）的形式从尿液中排出，因而可以推断 DEHP 在人体内的半衰期约为 48 h[77]。Koch 等研究发现，DINP 在 1.27 mg/kg bw 的暴露下，48 h 后，约有 43.6% 的 DINP 以代谢物的形式排出体外，包括 OH-MINP（20.2%）、COO-MINP（10.7%）和 oxo-MINP（10.6%）及少量的 MINP（2.2%）[78]。这些研究表明 DEHP、DINP 这两种环境中最常见的 PAEs 在体内停留了大约 2 d 时间。Mittermeier 等[79]研究了 4 名男性志愿者（23~58 岁）单次摄入 D_4-MEHP 或 D_4-MNBP（邻苯二甲酸单丁基酯，DNBP 的代谢产物）后检测其代谢物情况，结果发现 90% 以上的代谢物在前 22 h 内出现在尿液中，D_4-MEHP 及其 4 种次级代谢物（D_4-5OH-MEHP、D_4-5oxo-MEHP、D_4-2cx-MMHP 及 D_4-5cx-MEPP）的平均排泄量为给药剂量的 62%，其中 D_4-5cx-MEPP 的最高值为 15%。D_4-MEHP 的平均半衰期为 3.5 h，但其代谢物的半衰期要更长一些，D_4-5OH-MEHP、D_4-5oxo-MEHP、D_4-5OH-MEHP 及 D_4-5cx-MEPP 的半衰期分别为 6.5 h、6.6 h、21.4 h 和 8.9 h。在 D_4-MNBP 研究中，D_4-MNBP 及其次级代谢产物的总回收率在 52%~130%，半衰期为 (1.9±0.5) h 的单酯占总摄入剂量的大部分（92%），而次代谢物 D_4-3OH-MNBP 和 D_4-3cx-MPP 分别占摄入剂量的 7.1% 和 1.0%。总的来说，口服后邻苯二甲酸单酯 MEHP 和 MNBP 的动力学与其双酯的性质是相似的。

PAEs 及其代谢产物可以通过尿液、粪便（胆汁排泄物等）及汗液等形式排出体外，少数 PAEs 则会长期滞留在脂肪或分泌至乳汁中，其中尿液是最主要的排出途径。动物及人体实验发现，PAEs 的代谢物主要是通过尿液排出，而未代谢的母体化合物则主要通过粪便排出体外。一般来说，长链 PAEs（如 DIDP、DINP、DEHP、DNHP 等）主要通过尿液、汗液及粪便以葡萄糖醛酸结合物的形式排出体外，而短链的 PAEs（如 DMP、DEP、DBP 等）通常以其相应单酯的形式排出体外，如 DEHP 进入体内 24 h 后，仅有不到 10% 的 DEHP 原液经尿液排出，而约 67% 的 DEHP 转化为代谢产物经尿液排出，所以人们习惯将 MEHP 作为 DEHP 的生物标记物。但有文献报道，尿液中 MEHP 所代表的 DEHP 小于原剂量的 10%，且与 DEHP 的其他代谢产物相比，MEHP 的半衰期最短（5 h），因此，半衰期相对较长的 5cx-MEPP（12~

15 h）和 2cx-MMHP（24 h）是检测体内 DEHP 随时间改变的良好标志物，而 5OH-MEHP 和 5oxo-MEHP 则能更好地反映 DEHP 的短期暴露水平。

PAEs 进入人体的途径不同，其在体内的半衰期也不相同。当 DEHP 以非口途径（如呼吸、皮肤吸收等途径）进入人体后，其转化为单酯及二级氧化产物的代谢反应不会很快发生，大部分 DEHP 会以母体化合物形式直接排出体外。

PAEs 在体内的半衰期与其毒性密切相关，半衰期越长，意味着该化合物在体内的停留时间越长，对健康的危害就越大。PAEs 在体内半衰期的长短、暴露时间及途径（即环境条件）都会影响其在人体内的代谢及毒性。如其母体化合物一样，PAEs 的代谢产物也能够与一些转运蛋白（如 FABPs）结合，从而可以在体内滞留较长时间，而且可被运送到不同部位[70]，甚至能够透过胎盘屏障[80]。由于 PAEs 的疏水性质，因此母乳中长链 PAEs 及其代谢物浓度较高[81]，而尿液中的短链 PAEs 及其一级代谢产物的含量相对较高一些。由于 PAEs 在环境中广泛存在，人体始终处在"吸收-代谢-排出"的动态平衡中，因此，母体化合物及其代谢产物都会对人体健康造成危害，甚至某些代谢产物比母体化合物具有更强的毒性作用。

3.2 PAEs 与生殖健康

孕妇及新生儿是对环境有害物质较为敏感的群体。从胚胎着床到胎儿成功分娩，短短 40 周时间将完成人类数百万年的进化史，期间任一时段暴露于环境有害物质都将有极大可能损害孕产结局及新生儿健康。PAEs 及其代谢产物对孕产妇及新生儿的暴露及其不良孕产结局的研究在近十年得到较多关注。Snijder 等[82]对荷兰 4680 名孕妇开展前瞻性队列研究，发现孕期接触 PAEs 可对新生儿出生体重、头围和身长产生不利影响；此外，PAEs 暴露与胎盘重量减少显著相关。Swan 等[83]研究表明，孕期暴露于 PAEs 可致出生男婴肛殖距（AGD）缩短；孕妇尿液中 MEP、MBP、MBzP 浓度与肛殖指数（AGI：AGD/体重）呈显著负相关，该研究结果表明孕期 PAEs 暴露可致子代男婴生殖系统发育不良。进一步研究发现，PAEs 的代谢产物不但可通过胎盘屏障进入胎儿体内[84]，亦可通过母乳增加新生儿暴露，从而影响子代发育。Huang 等[85]研究发现，婴儿体内性激素水平与母乳中 PAEs 水平有一定关联，而早产可降低婴儿体内 DEHP 的清除率。总的来说，当前研究多认为，孕产妇 PAEs 及其代谢产物暴露可对子代发育造成影响，但尚不明确其影响的程度及作用机理。

近年来，PAEs 与女性不孕之间的关系引起人们的关注。Messerlian 等[86]在 2016 年调查研究发现，尿液中高浓度 DEHP 可致不孕女性窦卵泡数减少，表明 PAEs 暴露可影响女性卵巢储备。Hauser 等[87]发现，不孕女性尿液中 DEHP 代谢产物浓度与体外授-胚胎移植（*in vitro* fertilization and embryo transfer，IVF-ET）治疗周期的获卵数、卵子成熟率、临床妊娠率及胎儿出生率呈负相关，与胚胎质量无明显关联。Wu 等[88]指出，不孕夫妇 PAEs 暴露情况与卵裂期胚胎质量无明显关联，然而男性患者尿液中 PAEs 及其代谢产物浓度与囊胚质量成负相关，由于胚胎自身基因组激活正是发生在这一阶段，提示孕前 PAEs 暴露可影响人类生殖潜能。Du 等[89]对 112 名接受 IVF-ET 治疗的 110 份尿液及配对卵泡液进行检测，未观察到 PAEs 代谢产物与 IVF-ET 妊娠结局之间存在明显关联的证据。作者认为，样本量较小或许是无法寻找到两者之间关联的原因。2018 年，同一团队指出，PAEs 尿液浓度与血清抑制素 B（inhibin B，INHB）之间存在负相关，表明 PAEs 暴露对窦卵泡发育可能产生不良影响[90]。

3.3 PAEs 与肥胖

一项基于美国健康与营养调查（NHANES），1999～2002 年的横断面研究表明美国成年男性尿液中几种 mPAEs（MBzP、MEHHP、MEOHP 和 MEP）的浓度与腰围呈现显著的正相关[91]。类似地，Hatch 等[92]也发现年龄为 20～59 岁的男性尿液中 MBzP、MEOHP、MEHHP、MEP 和 MBP 的浓度与腰围和身高体重指数（BMI）呈显著正相关，而对于年龄 20～59 岁的女性，其尿液中 MEHP 的浓度与 BMI 呈显著负相关。在一项针对美国护士进行的为期 10 年的前瞻性研究中，尿液中几种 mPAEs（MBzP 和 MBP）的浓度与体重的增加量呈显著的剂量-效应关系，而 MEP 和 MEHP 的浓度与体重的改变量呈非单调的剂量-效应关系[93]。另外，流行病学研究也调查了 PAEs 暴露与几种脂质指标的相关性。Lin 等[94]针对 1016 名老年人的研究发现，仅仅 MIBP 在女性血清中的浓度与总脂肪含量、躯干脂肪含量以及皮下脂肪组织呈现正相关性，而在男性血清中未观察到相关性。

除此之外，几项研究也报道了儿童 PAEs 暴露与其肥胖的相关性，研究结果具有性别和种族差异。基于 NHANES 2007～2010 年数据的横断面研究表明男孩尿液中低分子量 mPAEs（包括 MBP、MEP 和 MIBP）的浓度与较高的肥胖风险有关[95]。类似地，Zhang 等[96]针对 493 名儿童的研究表明尿液中 MBP 和总的低分子量 mPAEs 的浓度与男孩肥胖呈正相关，而 MEHP、MEHHP 和 ∑DEHP 的浓度与女孩肥胖呈负相关。Trasande 等[97]针对 NHANES 2003～2008 年数据的研究表明仅仅对于非西班牙裔黑人，尿液中增加的低分子量的 mPAEs 的浓度与较高的超重和肥胖患病风险有关。另外，Teitelbaum 等[98]在一项前瞻性研究中调查了尿液中 mPAEs 的浓度与一年后 BMI 和腰围等指标的关系，结果表明超重儿童尿液中 MEP 和总的低分子量 mPAEs 的浓度与 BMI、腰围有显著的剂量-效应关系。特别地，几项研究也调查了产前 PAEs 暴露与后代肥胖的相关性，然而研究结果存在差异。一项针对 89 对母婴配对人群的前瞻性研究表明脐带血中 DEHP 代谢产物 mEOHP 的浓度与男孩出生 11 个月后的 BMI 呈显著负相关[99]。同时，Valvi 等[100]的研究表明孕妇妊娠期尿液中低分子量 mPAEs 的浓度与男婴体重、BMI 呈现负相关趋势，而与女婴体重、BMI 呈现正相关趋势。

3.4 PAEs 与糖尿病

糖尿病是一种由于机体内胰岛素缺乏或胰岛素在肝脏、骨骼肌、脂肪等靶细胞不能正常发挥生理作用而引起的体内糖、蛋白以及脂肪代谢发生紊乱的综合征，主要可以分为 3 种类型——Ⅰ 型糖尿病、Ⅱ 型糖尿病（T2DM）及妊娠糖尿病。尽管遗传、肥胖、不良饮食习惯和久坐的生活方式等是糖尿病患病的危险因素，然而，近年来，越来越多的流行病学研究发现 PAEs 暴露也与糖尿病患病风险有关。许多横断面研究已经表明尿液（或血清）中几种 mPAEs（MEHP、MEHHP、MEOHP、MMP、MEP、MCPP、MBzP、MIBP 和 MBP）的浓度与 T2DM 患病风险有正相关性[101-105]。而一项针对 123 位肥胖人群的横断面研究表明尿液中 mPAEs 浓度与 T2DM 患病风险无关[106]，这可能与较少的样本数量有关。特别地，Sun 等[107]针对美国护士进行的前瞻性研究表明仅仅在平均年龄为 45.6 岁的亚组人群中尿液中 mPAEs 的浓度与 T2DM 存在显著正相关性，而在平均年龄为 65.6 岁人群中却不存在相关性，不一致的结果可能与研究人群所处更年期状态不同有关。

同时，一些研究也调查了 PAEs 暴露与胰岛素抵抗、空腹血糖以及糖化血红蛋白等指标的相关性。绝大多数研究表明尿液（或血清）中 mPAEs 的浓度与胰岛素抵抗存在显著的正相关性。而针对空腹血糖和糖化血红蛋白，研究结果存在差异。一项针对加拿大人群的横断面研究表明尿液中几种 mPAEs（MIBP、MEHP、MEHHP、MBZP、MCPP 和 ∑DEHP）与糖化血红蛋白的水平呈正相关[102, 108-111]。而在另一项针对肥胖人群的研究表明尿液中 MCPP 和 MEHP 的浓度与糖化血红蛋白的浓度呈显著负相关[106]。类似地，James-Todd 等[102]基于 NHANES 2001～2008 年数据的研究也表明高浓度的 mCPP 浓度与糖化血红蛋白呈负相关。几项流行病学研究已经表明尿液（或血清）中 MIBP、∑DEHP、MBP、MCPP、MBzP 和 MEHHP 的浓度与空腹血糖呈显著正相关[102, 108, 112]。相反，一项基于 NHANES 2001～2008 年的数据表明尿液中 MBzP 的浓度与空腹血糖水平呈负相关[102]。而 Chen 等[111]针对我国台湾的 12～30 岁人群的研究未发现 PAEs 的暴露与空腹血糖存在相关性。目前，大部分研究采用的横断面研究，因此需要进一步采用队列研究来探索 PAEs 暴露与糖尿病及其相关指标的因果关系。此外，研究人群中糖尿病前期患者或糖尿病人群降糖药物的服用或生活方式的改变可能会影响 PAEs 暴露与糖尿病及相关指标的相关性。

3.5　PAEs 与呼吸系统疾病

目前，多项研究已经表明职业暴露于 PAEs 能够增加人群患呼吸道系统疾病的风险。一项针对于消防员的病例-对照研究表明相比于未暴露人群，消防员暴露于燃烧的 PVC 塑料后，无论是短期内还是长期的情况下，均表现出更为频繁和严重的呼吸系统病症，同时，暴露者更易得哮喘或支气管炎[113]。另外，对于来自 PVC 加工厂的工人，暴露于 PVC 热降解产物和 PAEs 的工人与未暴露者相比表现出更多的上呼吸道病症[114]。对于儿童，一项前瞻性病例-对照研究表明，儿童在出生两年之内患支气管梗阻的风险与 PVC 地板以及纺织墙体材料的使用有显著相关性[115]。这与几项横断面研究的结果也是一致的[116, 117]。同时，研究也表明室内灰尘中邻苯二甲酸丁基苄基酯（BBP）的浓度与儿童鼻炎的得病风险有关，灰尘中 DEHP 的浓度与哮喘有显著相关性[23, 118]。此外，几项研究也调查了产前 PAEs 的暴露与婴幼儿患呼吸系统疾病的关系。一项队列研究表明母亲孕期尿液中∑DEHP 和 MBzP 的浓度与儿童喘息、支气管炎或哮喘的患病风险存在显著正相关[119]。同时，Whyatt 等[120]和 Ku 等[121]发现除了 MBzP，母亲尿液中 MBP 或 MEHP 的浓度也与儿童哮喘的患病风险存在显著正相关性。

3.6　PAEs 与其他疾病的相关性调查

有少量文献报道了 PAEs 暴露与患癌风险的关系。López-Carrillo 等[54]针对墨西哥女性进行的一项病例-对照研究表明尿液中 MEP 的浓度与乳腺癌患病风险呈显著正相关，而且对于未绝经女性的风险更大，相反，MBzP 和 MCPP 的浓度与乳腺癌患病风险存在负相关关系。而在另一项针对美国人群的病例-对照研究表明尿液中较高浓度的 MEHP 与高的乳腺癌患病风险有关，而其他的 mPAEs（MBP、MBzP、MEHHP、MEOHP、MEP 和 MMP）浓度与乳腺癌患病风险无关[122]。同时，一项基于 NHANES 1999～2004 年数据的研究表明 20 岁以上女性尿液中 7 种 mPAEs 的浓度（MBP、MIBP、MEP、MCPP、MBzP、MEHP 和 MEHHP）与乳腺癌患病风险无显著相关性[123]。

流行病学研究已经调查了孕期 PAEs 的暴露与婴幼儿神经行为的关系。Yolton 等[124]的研究表明对于 5 周大的婴儿，其母亲尿液中较高浓度的 MBP 与婴儿改善的行为组织、增强的自我控制等有关；对于男婴，较高浓度 DEHP 的代谢产物与更多非最优的反射（nonoptimal reflexes）有关。此外，孕期 MEHHP、MEOHP 和 MBP 的暴露也与 6 个月婴儿的智力发育指数和运动发展指数呈现显著的负相关[125]。同时，研究也表明母亲孕期尿液中低分子量 mPAEs 的浓度与 4～9 岁儿童的行为问题有显著正相关[126]，MBP、MEOHP 和 MEHP 的浓度与儿童外显行为问题（externalizing problem behaviors，EPB）存在显著正相关性[127]。除此之外，横断面研究也表明儿童尿液中 DEHP 代谢产物的浓度与其词汇测试得分存在显著负相关性[128]。低分子量 mPAEs 的浓度也与较多的社会缺陷以及较差的社会认知、社会交往和社会意识有关[129]。

近年来，PAEs 与高血压、肝脏疾病等的发病率进行了相关性调查，但总得来说，目前 PAEs 与这些疾病发生率之间的流行病学调查较少，且有些结论相互矛盾，尚难以证明 PAEs 暴露与人群疾病之间的关系。

4 邻苯二甲酸酯增塑剂的健康效应与毒理机制

4.1 生殖毒性

卵巢具有排卵和分泌甾体激素的功能。越来越多的动物及体外实验表明，卵巢是PAEs及其代谢产物的作用靶点之一。PAEs，特别是其代谢产物暴露可致卵泡发育动力学改变。一项研究表明孕期环境水平DEHP暴露可致成年F1代雌鼠卵巢中始基卵泡减少，窦前卵泡增加，卵泡池耗竭加速，诱发卵巢早衰[130]。一项体外实验证实，大鼠窦前卵泡经MEHP（10～80 μg/mL）体外培养10天后，发育至窦卵泡阶段的比例明显下降[131]。PAEs及其代谢产物可诱导排卵异常、干扰黄体形成。另一项研究将斑马鱼暴露于含0.02～40 μg/L DEHP 的水溶液中，可抑制其卵母细胞生发泡的破裂，后者是排卵前减数分裂恢复的必经过程[132]。Lovekamp等[133]以DEHP对雌性大鼠灌胃染毒，发现与对照组相比，实验组自然排卵周期延长、动情周期延长，出现排卵障碍及多囊样改变。PAEs及其代谢产物可影响卵巢甾体激素的生成。Lovekamp等[133]的实验结果显示，MEHP可降低芳香化酶mRNA的表达水平，进而影响雌二醇（estradiol, E2）的合成，证实MEHP可通过抑制芳香化酶mRNA转录来抑制睾酮（testosterone, T）向E2的转化。Svechnikova等[134]研究发现，MEHP可抑制颗粒细胞中胆固醇向线粒体的转运，从而使孕酮分泌量下降。胆固醇进入线粒体是孕酮合成的关键步骤，这个过程受到类固醇合成急性调节蛋白（steroidogenic acute regulatory protein, StAR）的调节。有研究报道DEHP可降低雄性大鼠睾丸间质细胞StAR mRNA的表达，认为PAEs可能通过降低StAR的表达进而影响雌性大鼠颗粒细胞孕酮的合成。由于PAEs及其代谢产物种类繁多、结构不同，其可能的毒理作用及机制亦不同。总的来说，PAEs及其代谢产物产生生殖毒性的机制主要但不限于以下几个方面：①干扰性激素的合成。PAEs可通过干扰性激素合成过程中的关键酶类（如芳香化酶）的活性从而影响性激素的合成。②对雌/雄激素受体信号通路的直接干扰。最近Park等[135]研究发现，DEHP等多种PAEs能够增强雌激素受体（estrogen receptor, ER）或抑制雄激素受体（androgen receptor, AR）介导的转录活性。③对雌/雄激素信号通路的间接干扰。PAEs作为一种内分泌干扰污染物，对PPAR、TR和AhR等信号通路都具有干扰作用，这些通路又会与雌/雄激素信号通路之间形成复杂的"交叉对话（cross-talk）"，最终干扰雌/雄激素受体信号通路。

4.2 肥胖、胰岛素抵抗及糖尿病

PAEs暴露能够诱发肥胖，目前认为这主要是与PAEs能够促进脂肪生成及前脂肪细胞分化有关，而后者主要是PAEs能够激活过氧化物酶体增殖剂激活受体（peroxisome proliferators-activated receptors, PPARs）受体信号通路而引起的[136-139]。PAEs是亲脂性的，容易在脂肪组织中积累，且在脂肪组织中难以代谢，脂肪细胞可能是PAEs长期暴露的主要靶点之一。在小鼠3T3-L1前脂肪细胞中研究发现，MEHP暴露可以明显促进前脂肪细胞的分化，应用RT-qPCR检测发现，MEHP可以促进脂肪分化的关键调控基因PPARγ的上升表达；MEHP对脂肪分解、葡萄糖摄取/糖酵解和线粒体呼吸都有干扰作用[140]。Yin等[141]报道BBP暴露也能够导致转录因子C/EBPα（CCAAT-enhancer-binding protein α）和PPARγ的表达增强，它们的下游基因的表达也明显升高，且具有时间-依赖性，应用GC/MC检测发现BBP暴露干扰了细胞甘油单体生成和脂肪酸合成相关的代谢图谱。Hao等[142]应用小鼠体内实验也表明围产期DEHP暴露可能会

增加后代肥胖的发生率。可见，PPARγ 受体在 PAEs 诱导脂肪生成及细胞分化中起了重要作用。有研究报道，在分化成熟的 3T3-L1 脂肪细胞中，MEHP 暴露也可以激活 PPARγ 通路而诱发炎症。除 PPARγ 信号通路外，也有文献报道 PAEs 可通过其他机制来促进脂肪积累。如最近 Qi 等[143]发现，在 3T3-L1 细胞中，MEHP 可以增加细胞内 STAT-3 的转录及蛋白表达及磷酸化，通过 TYK-2/STAT-3 通路而促进脂肪生成。

胰岛素抵抗（insulin resistance）是指各种原因导致的胰岛素促进葡萄糖摄取和利用的效率下降，从而导致正常数量的胰岛素不足以产生对脂肪细胞、肌肉细胞和肝细胞正常的胰岛素响应，是产生 II-型糖尿病的根本原因之一。研究人员对 PAEs 诱发胰岛素抵抗的分子机制进行了一定的探索。Deng 等[144]将 DBP 以 0.5 mg/(kg·d)、5 mg/(kg·d)、50 mg/(kg·d)的剂量口服暴露小鼠 7 周，发现单独暴露于 50 mg/(kg·d) DBP 可显著降低胰岛素分泌和葡萄糖耐受，但对胰岛素抵抗无影响。然而，结合高脂肪饮食和链脲佐菌素（STZ）治疗，DBP 暴露明显加重了葡萄糖不耐症、胰岛素耐受性和胰岛素抵抗，并导致胰腺和肾脏病变。该小组还同时研究了 DBP 在胰岛素信号传导途径中的作用，发现 DBP 暴露可破坏 PI3K 表达和 Akt 磷酸化，降低胰腺 GLUT-2 水平，结果表明，DBP 通过阻断胰岛素信号通路和损害胰岛素分泌而加重了 II-型糖尿病。Viswanathan 等[145]应用 50 μmol/L 和 100 μmol/L DEHP 和 MEHP 暴露大鼠 L6 肌管细胞 24 h，然后用胰岛素刺激 20 min，结果发现溶胶和质膜组分中的 GLUT-4 水平显著降低。此外，DEHP 和 MEHP 暴露显著改变了 GLUT-4 易位相关的胰岛素信号分子（Rab8a）和相关蛋白质（IRAP、SNAP23、Syntaxin4、Munc18c 等）的表达，影响 GLUT-4 从胞质溶胶到质膜的移位，从而导致葡萄糖摄取量减少。在生物体内，胰岛素信号通路与炎症反应、JNK 通路、NF-κB 通路、PPAR 通路及 Nrf2 通路等发生"交叉对话"，PAEs 也可通过这些通路而诱发胰岛素抵抗及糖尿病的发生[146]。

4.3 内分泌干扰效应

内分泌系统是腺体（位于大脑、胃肠系统、肾脏、胰腺、卵巢、睾丸、甲状腺等）将激素直接分泌到血液中，经血液循环进入靶器官后发挥生物学作用。这些激素包括氨基酸类（肾上腺素、褪黑激素、甲状腺素等）、类花生酸（前列腺素、白三烯等）、肽类（脂联素、降钙素、胰岛素、胰高血糖素、催产素、促甲状腺素等）和类固醇类（雄激素、雌激素、孕酮、盐皮质激素、糖皮质激素等）。多数 PAEs 都是亲脂性的，很容易进入人体血液（或其他体液）并很快地代谢为 I 相及 II 相产物。如环境中最常见的 DEHP 及 DINP 自进入人体到完全排出体外大约需要 2 天时间，在被排出人体之前，一些 PAEs 的母体化合物及代谢产物能够与一些内分泌信号系统发生作用，产生内分泌干扰效应，属于内分泌干扰物（endocrine disrupting chemicals，EDCs）[147]。

有关 PAEs 类的内分泌干扰效应的研究数据主要来源于动物及细胞实验，但是流行病学调查也发现 PAEs 与性异常（性反转、子宫内膜异位症、生殖发育改变、青春期过早和生育能力下降）及过敏和哮喘、肥胖、胰岛素抵抗和 II 型糖尿病、注意力下降、多动症、自闭症等都有相关性，这些都是内分泌干扰效应的典型特征。

PAEs 可通过多种方式产生内分泌干扰效应，其中对内分泌激素信号通路的干扰是其发挥内分泌毒性的重要路径。多项研究表明，雌激素受体（estrogen receptor，ER）、甲状腺受体（thyroid receptor，TR）及信号通路介导了 PAEs 及其代谢物产生内分泌毒性的[148-150]。Engel 等[151]利用报告基因分析法检测了 PAEs 及其代谢产物的内分泌效应，发现在研究测试的 15 种 PAEs 中，BzBP、DIDP 和 DTDP 能够显著激活 ERα 和 ERβ，而 DEP、DNBP、DIBP、DNPP 和 DIPP 等对 ERα 具有弱激活作用，而对 ERβ 无明显效应。其中 BzBP 在 10 μmol/L 浓度下达到 ERα 和 ERβ 的最大激活，与 E2 诱导的激活效应相比，BzBP 的 ERα 和 ERβ 激活率分别为 76%和 12%（分别以 0.1 nmol/L E2 对 ERα 及 1 nmol/L E2 对 ERβ 的激活效应为 100%）。与 PAEs 相比，该研究中测试的 19 种代谢物均未激活 ERα 或 ERβ。有趣的是，PAEs 对 ER

受体的作用是双重的,当单独暴露时,PAEs 表现有弱的激活效应,当与 E2 共暴露时,PAEs 则表现出拮抗效应。同时,作者还发现,除 DINP 和 DIDP 外,其余检测的 13 种 PAEs 均具有雄激素受体的抑制活性。Ghisari 等[152]应用雌激素受体荧光素酶报告载体稳定转染的 MVLN 细胞,发现 BBP、DBP 等 PAEs 能够激活 ER。同时作者应用 GH3 细胞增殖实验研究 PAEs 的甲状腺受体激活效应,发现 BBP、DEHP、DBP 等均具有甲状腺受体激活效应。Kim 等[148]的研究也证实 DEHP 具有激活 TR 受体的能力,并可通过 TR 受体激活而进一步诱发甲状腺细胞增殖、DNA 损伤等毒性。

除可直接与这些核受体结合而诱发内分泌干扰效应外,PAEs 还可通过多种途径直接或间接的干扰激素的合成,影响激素在体内的稳态。Li 等[153]研究了 MEHP 对大鼠卵巢颗粒细胞(GCS)增殖和分泌的影响作用,发现 MEHP 暴露能够显著抑制 GCS 的增殖,并增加性激素受体和孕酮合成中的关键酶(StAR、cP450scc 及 3β-HSD)的表达,刺激类固醇激素的分泌。除对 AR 受体的拮抗作用外,PAEs 还可通过抑制睾丸 3β-羟基甾体脱氢酶(3β-hydroxysteroid dehydrogenase, 3β-HSD)的活性而产生抗雄激素效应。3β-羟基甾体脱氢酶是一种甾体生成酶,催化孕烯醇酮生成孕酮、17α-羟孕酮、雄烯二醇生成睾酮等多步化学反应,在甾体激素代谢中起着重要作用。Yuan 等[154]考察了几种 PAEs 对 3β-HSD 活性的影响作用,结果发现对侧链为直链的 PAEs 来说,C_1、C_2、C_7 和 C_8 的 PAEs 只在大于 1000 μmol/L 时对大鼠的 3β-HSD 显示出较弱的抑制活性,而对人 3β-HSD 的活性几乎没有影响;C3、C4 和 C5 的 PAEs 对大鼠 3β-HSD 的具有抑制作用,其 IC_{50} 值为 62.7 μmol/L、30.3 μmol/L 和 33.8 μmol/L,对人 3β-HSD 的 IC_{50} 值为 123.0 μmol/L、24.1 μmol/L 和 25.5 μmol/L。Ye 等[155]给大鼠进行 DEHP 灌胃处理 30 天,结果表明 DEHP 导致大鼠甲状腺和肝脏组织病理改变,包括甲状腺滤泡腔直径减小、肝细胞水肿等,三碘甲状腺原氨酸(T3)、甲状腺素(T4)和促甲状腺激素释放素(thyrotropin releasing hormone, TRH)降低,同时研究发现,DEHP 暴露后,肝脏酶(包括 Ugt1a1,CYP2b1,Sult1e1,Sult2b1)被显著诱导。总之,DEHP 暴露能够扰乱甲状腺激素的体内平衡,降低甲状腺素水平,而 Ras/Akt/TRHr 通路及肝脏酶在 DEHP 的甲状腺干扰中起了重要作用。文献报道 PAEs 对组成性雄甾烷受体(constitutive androstane receptor, CAR)、孕烷 X 受体(pregnane X receptor, PXR)都有激活作用[156],这些受体的激活,也会引发机体产生内分泌干扰毒性。

4.4 表观遗传毒性

表观遗传(epigenetics)是指基因的 DNA 序列没有发生改变的情况下,基因功能发生了可遗传的变化,并最终导致表型的变化。研究发现,PAEs 具有表观遗传毒性,能够在不引起 DNA 序列改变的情况下产生可遗传的毒性作用。近年,PAEs 的表观遗传毒性受到人们的关注,发现 PAEs 可通过改变基因启动子区域 DNA 的甲基化水平及 microRNA(miRNA)而产生表观遗传毒性(图 19-5)。

DNA 甲基化是表观遗传修饰的主要形式之一,DNA 甲基化能够引起染色质结构、DNA 构象、DNA 稳定及 DNA 与蛋白质相互作用方式的改变,从而参与调控许多重要的生物学现象和过程。在人类或其他哺乳动物体内,DNA 的甲基化主要发生在胞嘧啶上,形成 5-甲基胞嘧啶(5-methylcytosine, 5mC),整个基因组中 5mC 占总胞嘧啶的 2%~7%,而其中 70% 的 5mC 存在于 CpG 二核苷酸中。分散于 DNA 中的 CpG 约有 70%~90% 通常是甲基化的,而位于结构基因启动子区域的 CpG 岛通常处于非甲基化状态。如果基因启动子区域的 CpG 岛被甲基化,则会抑制该基因的表达,相反,如果结构基因启动子区域的 CpG 岛发生异常的去甲基化,则会导致该基因的大量表达。在 Wistar 孕鼠中,DEHP 暴露可以降低葡萄糖转运体(GLUT-4)启动子区域的 DNA 甲基化水平,从而上调 GLUT4 的表达,严重影响葡萄糖稳态和胰岛素信号传导[157]。Wu 等报道孕期暴露 DEHP 可以诱导小鼠睾丸 DNA 总甲基化水平上升 10%,DNA 甲基化水平的变化可能在母体 DEHP 暴露引起的睾丸功能异常中起重要作用[158]。大鼠实验表明 DEHP 和 DBP 通过表观遗传对成年发病疾病的影响很大,包括青春期异常、卵巢疾病(多囊卵巢和原发性卵巢功能不

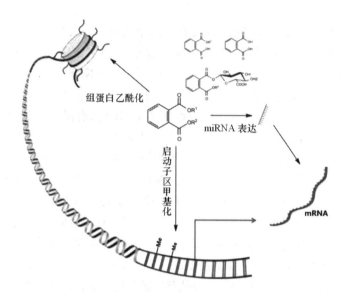

图 19-5　PAEs 及其代谢产物诱发表观遗传毒性的可能机制

全)、睾丸功能障碍和肥胖等[159]。这是由表观遗传印记基因启动子区的 DNA 甲基化介导的，在父系遗传的胰岛素样生长因子 2 受体（IGF2R）基因和 PEG3（Paternally Expressed Gene 3）尤为明显[159, 160]。有研究发现 DBP 通过 PPARs 降低大鼠肝脏中 c-myc 基因 CpG 岛的甲基化水平，从而促进癌症的发生[161]。流行病学调查也发现，在人体内 DEHP 的产前暴露与父系表达的 $IGF2$ 基因（该基因在胚胎和胎盘的生长中起着重要作用）的甲基化水平降低有明显的相关性[162]。一些流行病学调查也发现 PAEs 暴露水平与人体基因的甲基化异常有关，Wang 等[163]对 256 名儿童进行调查，评估了 PAEs 暴露对表观遗传变化的影响以及对哮喘的作用，结果发现当儿童尿液中 5OH-MEHP 含量较高时，TNF-α 基因启动子的甲基化程度较低。

MicroRNA（miRNA）是一类短的单链内源性非编码 RNA 分子，大约由 21～25 个核苷酸组成，可以通过与互补的 mRNA 选择性结合而抑制该基因的表达。miRNA 广泛存在于动植物中，是一类重要的基因表达调控手段。研究发现，PAEs 暴露可以改变 miRNA 的表达图谱，从而改变机体的正常生理功能。Meruvu 等[164]应用 MEHP 暴露人绒毛膜滋养层细胞（HTR8/SVneo），发现 MEHP 暴露能够导致 3 种 miRNA（miR-17-5p，miR-155-5p，miR-126-3p）的上升，说明胎盘暴露于 MEHP 可能导致异常 miRNA 表达，从而引发妊娠并发症；在另一项研究中，该研究组还发现 MEHP 能够诱导 miR-16 表达上升，并进一步导致细胞凋亡[165]。最近 Martinez 等[166]在一项横断面研究中发现，尿中 PAEs 代谢产物的浓度与滤泡液中 miRNA 表达的改变有关联。

组蛋白修饰的改变也是引起表观遗传的重要机制之一。组蛋白是染色体基本结构蛋白，组蛋白与 DNA 紧密结合形成核小体结构。组蛋白的乙酰化（通常发生在 Lys 或 Arg 残基上）或甲基化，可以改变组蛋白与 DNA 之间的亲和力，从而改变基因的表达，但目前关于 PAEs 暴露与组蛋白修饰异常的报道较少。

4.5　代谢组学研究

代谢组学是继基因组学和蛋白组学之后发展起来的一种新兴的组学技术，其研究对象主要是分子量 1000 Da 以内的内源性代谢物，包括氨基酸、多肽、有机酸、维生素、多酚、生物碱、核酸、碳水化合物、脂质和其他小分子。相比于其他组学，代谢组学具有一些优势。第一，相对于基因水平和蛋白质水平，代谢物水平上发生的变化更容易检测；第二，基因组学和蛋白质组学需要进行全基因组测序和建立含有大量表达标签的数据库，相比之下，代谢组学的分析手段更简易；第三，与基因和蛋白的数量相比，代谢物的数量相对较少，便于后续分析和验证；第四，根据生物体内小分子代谢物和代谢通路的变化，可以寻找代谢物与生理病理变化的相对关系，系统揭示疾病发生发展的机制。

代谢组学可分为非靶向和靶向两种。非靶向代谢组学是对生物体内源性代谢物进行的一种全面系统的无偏向分析，主要用于差异代谢物的筛选；而靶向代谢组学则是针对特定的一种或几种途径的代谢产物进行定量分析，是一种有偏向的代谢组学分析[167]。

目前，代谢组学检测平台主要包括核磁共振（NMR）、气相色谱-质谱（GC-MS）、液相色谱-质谱（LC-MS）等（表19-5）。除了用于药物研发、药物筛选以及药物作用机制和临床评价外，代谢组学也被用于化学品暴露研究，通过对生物体内小分子代谢物进行定性定量分析，探索化学品暴露下代谢物与生理病理变化的关系，系统揭示化学品暴露对疾病发生发展的机制。

表19-5 三种代谢组学检测平台的优缺点

检测平台	优点	缺点
NMR	高通量；普适性；良好的客观性和重现性；快速（2~3 min/样）；预处理简单，不需要衍生化和分离；无损伤性；能检测样品中大多数有机类化合物；仪器使用寿命长	检测动态范围窄；灵敏度和分辨率较MS低；样品用量相对大（0.1~0.5 mL）；仪器成本高
GC-MS	高通量；高精密度、灵敏度及重现性；具有可参考的标准谱图数据库，易于定性；样品用量适中（0.1~0.2 mL）；可检测样品中大多数有机和部分无机分子	需要衍生化；需要分离；分析速度较慢（20~40 min/样）；较难鉴定新化合物；不适用于难挥发性、热不稳定性物质的分析
LC-MS	高通量；高分别率、灵敏度；检测动态范围宽；样品处理简单，不需要衍生化；样品用量少（10~100 μL）；适用于热不稳定性、不易挥发、不易衍生化和分子量较大的物质	分析速度较慢（15~40 min/样）；缺少可以参考的标准谱图数据库；较难鉴定新化合物；成本较高；仪器寿命较短

目前，已经有研究利用代谢组学分析了PAEs暴露对啮齿动物体内代谢产物的影响。基于GC-MS平台，Xia等[168]分析了怀孕小鼠暴露于DBP后母鼠血清、胎盘和胎儿脑组织中的代谢产物，发现与对照组相比，暴露组中碳水化合物（如果糖和乳酸）、氨基酸（如异亮氨酸、2-氧基-3-甲基戊酸酯、缬氨酸）、脂质（如油酸、硬脂酸、花生四烯酸、顺式5,8,11,14,17-二十碳五烯酸）和嘌呤（如黄嘌呤、尿囊素和尿素）等代谢存在显著差异。此外，利用LC-MS平台，Zhang等[169]发现大鼠在DEHP和多氯联苯（PCB1254）联合暴露12 d后能够引起磷脂代谢差异，如磷脂酰胆碱（18：4/18：1）、溶血磷脂酰乙醇胺（18：2/0：0）和14种溶血磷脂，同时，色氨酸和苯丙氨酸浓度也显著降低。

Banerjee等[170]利用NMR平台，发现大鼠孕期暴露于BBzP后，与对照组相比，高剂量暴露组雄性后代大脑中谷氨酸、磷酸乙醇胺和肉毒碱有显著差异，同时，高暴露组雄性后代睾丸中肌酐、肌醇、磷胆碱、磷乙醇胺、谷氨酸、谷胱甘肽、甜菜碱、抗坏血酸、柠檬酸盐、腺嘌呤、甘油、组胺、三甲胺、肌苷、天冬氨酸、乙酸酯、酪氨酸、N-乙酰半胱氨酸和O-乙酰肉碱的含量存在显著差异，而在雌性后代大脑和子宫中代谢物没有表现出显著差异，表明了BBzP的抗雄激素效应。同时，另一项研究也表明雄性大鼠暴露于PAEs的混合物15周后，代谢组学分析显示出下调的类固醇蛋白以及雄激素代谢的改变[171]。

离体研究也探索了PAEs暴露对脂肪代谢的影响。Xu等[172]的研究表明大鼠的HRP-1滋养层细胞暴露于DEHP或MEHP后，10类脂质代谢物的水平（包括胆固醇酯、二酰基甘油、三酰基甘油、磷脂酰胆碱、磷脂酰乙醇胺、磷脂酰丝氨酸、溶血磷脂酰胆碱、心磷脂和鞘磷脂）显著增高，表明DEHP或其代谢物MEHP的暴露能够改变大鼠胎盘细胞系的脂质代谢，这可能会导致胎盘中的必需脂肪酸或脂质稳态发生异常，从而影响胎儿的正常发育。

除此之外，代谢组学也被用于人群流行病学研究，以期发现与PAEs暴露相关的生物标志物和毒性机制。Zhou等[173]通过对115名孕妇血浆和尿液中的代谢物进行研究，发现较高浓度的PAEs暴露与增加的脂质生成、炎症以及核酸代谢的改变有关。在另一项针对 750 名孕妇的研究表明，尿液中 mPAEs 的浓度与三羧酸循环（TCA cycle）的代谢产物（包括琥珀酸、柠檬酸和乙酸）呈显著负相关[174]。此外，一项横断面研究调查了男性 PAEs 暴露与尿液中代谢产物的相关性，结果显示在 PAEs 高暴露组中乙酰神经氨酸、肉毒碱 C8∶1、肉毒碱 C18∶0、胱氨酸、苯甘氨酸、苯丙酮酸和谷氨苯丙氨酸的水平显著增加，而肉毒碱 C16∶2、双乙酰精胺、丙氨酸、牛磺酸、色氨酸、鸟氨酸、甲基葡聚糖酸、羟基-前列腺素 E2、酮基-前列腺素 E2 的水平显著降低，研究表明，低水平 PAEs 的暴露与氧化应激和脂肪酸氧化的增加以及前列腺素代谢的降低有关[175]。

综上所述，目前利用代谢组学手段来研究 PAEs 暴露对生物体生理病理影响的研究还相对较少，因此，今后需要进一步开展相关研究，通过运用代谢组学在内的新兴技术和手段，探索与 PAEs 暴露相关的生物标志物和潜在机理[102]。

5 展 望

随着流行病学调查及毒理学研究的深入，PAEs 类增塑剂的健康危害已引起了人们的广泛关注。各国针对几种毒性危害较大的 PAEs 进行了严格控制，如美国环境保护署（EPA）将 DMP、DEP、DBP、DOP、BBP 和 DEHP 列为优先控制的污染物；美国《消费品安全改进法案》（CPSIA）规定儿童玩具及护理用品中的 DBP、BBP、DEHP、DNOP、DINP、DIDP 含量不得超过 0.1%。我国也将 DEP、DMP、DOP 3 种 PAEs 确定为环境优先控制污染物，将 17 种 PAEs 类物质列入"食品中可能违法添加的非食用物质和易滥用的食品添加剂名单"，并规定了食品、食品添加剂中的 DEHP、DINP、DBP 最大残留量。但是目前这些化合物并没有完全禁止使用，考虑到我国是 PAEs 生产和使用的大国，因此，PAEs 的人体健康风险将值得长期关注。下一阶段，仍需加大对普通人群及职业暴露人群的流行病学调查，用更科学的统计方法来明确 PAEs 暴露与人体健康的因果关系。其次，仍需进一步明确 PAEs 的毒性效应及分子机制，包括 PAEs 经肠道或皮肤进入人体血液的方式，在体内转运的方式以及产生内分泌干扰效应的分子机制等。将表观遗传学、代谢组学等新兴技术方法用于 PAEs 的毒理学研究，在更微观的水平上研究 PAEs 产生毒性作用的分子机理。考虑到 PAEs 具有生殖毒性及表观遗传毒性，因此，需要进一步研究 PAEs 是否存在跨代遗传的毒性效应及其可能的分子机制。最后，随着 DEHP 等传统 PAEs 的限制使用，近年来有一些新的替代品不断出现，如结构与 DINP 极其类似的环己烷-1,2-二羧酸二异壬酯（DINCH）等，这些新型替代品的毒性如何，介导其毒性作用的分子机制是什么，仍是环境毒理学家下一步需要详细研究的课题。

参 考 文 献

[1] Halden R U. Plastics and health risks. Annual Review of Public Health, 2010, 31: 179-194.

[2] Benjamin S, Masai E, Kamimura N, et al. Phthalates impact human health: Epidemiological evidences and plausible mechanism of action. Journal of Hazardous Materials, 2017, 340: 360-383.

[3] Zhang Z M, Zhang H H, Zhang J, et al. Occurrence, distribution, and ecological risks of phthalate esters in the seawater and sediment of Changjiang River Estuary and its adjacent area. Science of the Total Environment, 2018, 619-620: 93-102.

[4] He H, Hu G J, Sun C, et al. Trace analysis of persistent toxic substances in the main stream of Jiangsu section of the Yangtze River, China. Environmental Science and Pollution Research, 2011, 18(4): 638-648.

[5] Zheng X, Zhang B T, Teng Y. Distribution of phthalate acid esters in lakes of beijing and its relationship with anthropogenic activities. Science of the Total Environment, 2014, 476-477: 107-113.

[6] Zhang M, Xu J, Xie P. Metals in surface sediments of large shallow eutrophic Lake Chaohu, China. Bulletin of Environmental Contamination and Toxicology, 2007, 79(2): 242-245.

[7] Yang H, Zhuo S, Xue B, et al. Distribution, historical trends and inventories of polychlorinated biphenyls in sediments from Yangtze River Estuary and adjacent East China Sea. Environmental Pollution, 2012, 169: 20-26.

[8] He Y, Wang Q, He W, et al. The occurrence, composition and partitioning of phthalate esters (PAEs) in the water-suspended particulate matter (SPM) system of Lake Chaohu, China. Science of the Total Environment, 2019, 661: 285-293.

[9] Cheng Z, Liu J B, Gao M, et al. Occurrence and distribution of phthalate esters in freshwater aquaculture fish ponds in Pearl River Delta, China. Environmental Pollution, 2019, 245: 883-888.

[10] Cai Q Y, Mo C H, Wu Q T, et al. Occurrence of organic contaminants in sewage sludges from eleven wastewater treatment plants, China. Chemosphere, 2007, 68(9): 1751-1762.

[11] Mo C H, Cai Q Y, Li Y H, et al. Occurrence of priority organic pollutants in the fertilizers, China. Journal of Hazardous Materials, 2008, 152(3): 1208-1213.

[12] Zeng F, Cui K, Xie Z, et al. Occurrence of phthalate esters in water and sediment of urban lakes in a subtropical city, Guangzhou, South China. Environment International, 2008, 34(3): 372-380.

[13] Zeng F, Cui K, Xie Z, et al. Phthalate esters (PAEs): emerging organic contaminants in agricultural soils in peri-urban areas around Guangzhou, China. Environmental Pollution, 2008, 156(2): 425-434.

[14] Niu L, Xu Y, Xu C, et al. Status of phthalate esters contamination in agricultural soils across China and associated health risks. Environmental Pollution, 2014, 195: 16-23.

[15] Wang L, Liu M, Tao W, et al. Pollution characteristics and health risk assessment of phthalate esters in urban soil in the typical semi-arid city of Xi'an, Northwest China. Chemosphere, 2018, 191: 467-476.

[16] Papadopoulos A, Vlachogiannis D, Maggos T, et al. A semi-quantitative approach for analysing low-volatile organic compounds in house dust using an SFE method: Significant common features and particular differences of the extracts. Journal of Supercritical Fluids, 2013, 82: 268-281.

[17] Hu J, Li N, Lv Y, et al. Investigation on indoor air pollution and childhood allergies in households in six chinese cities by subjective survey and field measurements. International Journal of Environmental Research and Public Health, 2017, 14(9): 979.

[18] Blanchard O, Glorennec P, Mercier F, et al. Semivolatile organic compounds in indoor air and settled dust in 30 French dwellings. Environmental Science & Technology, 2014, 48(7): 3959-3969.

[19] Zhu Q, Jia J, Zhang K, et al. Phthalate esters in indoor dust from several regions, China and their implications for human exposure. Science of the Total Environment, 2019, 652: 1187-1194.

[20] Rudel R A, Camann D E, Spengler J D, et al. Phthalates, alkylphenols, pesticides, polybrominated diphenyl ethers, and other endocrine-disrupting compounds in indoor air and dust. Environmental Science & Technology, 2003, 37(20): 4543-4553.

[21] Clausen P A, Lindeberg Bille R L, Nilsson T, et al. Simultaneous extraction of di(2-ethylhexyl)phthalate and nonionic surfactants from house dust. Concentrations in floor dust from 15 Danish schools. Journal of Chromatography. A, 2003, 986(2): 179-190.

[22] Fromme H, Lahrz T, Piloty M, et al. Occurrence of phthalates and musk fragrances in indoor air and dust from apartments and kindergartens in Berlin (Germany). Indoor Air, 2004, 14(3): 188-195.

[23] Bornehag C G, Sundell J, Weschler C J, et al. The association between asthma and allergic symptoms in children and phthalates in house dust: a nested case-control study. Environmental Health Perspectives, 2004, 112(14): 1393-1397.

[24] Guo Y, Kannan K. Comparative assessment of human exposure to phthalate esters from house dust in China and the United States. Environmental Science & Technology, 2011, 45(8): 3788-3794.

[25] Kang Y, Man Y B, Cheung K C, et al. Risk assessment of human exposure to bioaccessible phthalate esters via indoor dust around the Pearl River Delta. Environmental Science & Technology, 2012, 46(15): 8422-8430.

[26] Ait Bamai Y, Araki A, Kawai T, et al. Associations of phthalate concentrations in floor dust and multi-surface dust with the interior materials in Japanese dwellings. Science of the Total Environment, 2014, 468-469: 147-157.

[27] Subedi B, Sullivan K D, Dhungana B. Phthalate and non-phthalate plasticizers in indoor dust from childcare facilities, salons, and homes across the USA. Environmental Pollution, 2017, 230: 701-708.

[28] Domínguez-Morueco N, González-Alonso S, Valcárcel Y. Phthalate occurrence in rivers and tap water from central Spain. Science of the Total Environment, 2014, 500-501: 139-146.

[29] Hashizume K, Nanya J, Toda C, et al. Phthalate esters detected in various water samples and biodegradation of the phthalates by microbes isolated from river water. Biological & Pharmaceutical Bulletin, 2002, 25(2): 209-214.

[30] Shi W, Hu X, Zhang F, et al. Occurrence of thyroid hormone activities in drinking water from eastern China: contributions of phthalate esters. Environmental Science & Technology, 2012, 46(3): 1811-1818.

[31] Liu Y, Chen Z, Shen J. Occurrence and removal characteristics of phthalate esters from typical water sources in northeast china. Journal of Analytical Methods in Chemistry, 2013, 2013: 419-439.

[32] Liou S H, Yang G C, Wang C L, et al. Monitoring of PAEMs and beta-agonists in urine for a small group of experimental subjects and PAEs and beta-agonists in drinking water consumed by the same subjects. Journal of Hazardous Materials, 2014, 277: 169-179.

[33] Dominguez-Morueco N, Gonzalez-Alonso S, Valcarcel Y. Phthalate occurrence in rivers and tap water from central Spain. Science of the Total Environment, 2014, 500-501: 139-146.

[34] Liu X, Shi J, Bo T, et al. Occurrence and risk assessment of selected phthalates in drinking water from waterworks in China. Environmental Science and Pollution Research International, 2015, 22(14): 10690-10698.

[35] Zaki G, Shoeib T. Concentrations of several phthalates contaminants in Egyptian bottled water: Effects of storage conditions and estimate of human exposure. Science of the Total Environment, 2018, 618: 142-150.

[36] Salazar-Beltran D, Hinojosa-Reyes L, Ruiz-Ruiz E, et al. Determination of phthalates in bottled water by automated on-line solid phase extraction coupled to liquid chromatography with UV detection. Talanta, 2017, 168: 291-297.

[37] Yang J, Song W, Wang X, et al. Migration of phthalates from plastic packages to convenience foods and its cumulative health risk assessments. Food Additives & Contaminants: Part B, 2019: 1-8.

[38] Fierens T, Servaes K, Van Holderbeke M, et al. Analysis of phthalates in food products and packaging materials sold on the Belgian market. Food and Chemical Toxicology, 2012, 50(7): 2575-2583.

[39] Sui H X, Zhang L, Wu P G, et al. Concentration of di(2-ethylhexyl) phthalate (DEHP) in foods and its dietary exposure in China. International Journal of Hygiene and Environmental Health, 2014, 217(6): 695-701.

[40] Koniecki D, Wang R, Moody RP, et al. Phthalates in cosmetic and personal care products: Concentrations and possible dermal exposure. Environmental Research, 2011, 111(3): 329-336.

[41] Hubinger J C, Havery D C. Analysis of consumer cosmetic products for phthalate esters. Journal of Cosmetic Science, 2006, 57(2): 127-137.

[42] Bao J, Wang M, Ning X, et al. Phthalate concentrations in personal care products and the cumulative exposure to female adults and infants in Shanghai. Journal of Toxicology and Environmental Health. Part A, 2015, 78(5): 325-341.

[43] Malarvannan G, Onghena M, Verstraete S, et al. Phthalate and alternative plasticizers in indwelling medical devices in pediatric intensive care units. Journal of Hazardous Materials, 2019, 363: 64-72.

[44] Silva M J, Barr D B, Reidy J A, et al. Urinary levels of seven phthalate metabolites in the U.S. population from the National Health and Nutrition Examination Survey (NHANES) 1999—2000. Environmental Health Perspectives, 2004, 112(3):

331-338.

[45] Colacino J A, Harris T R, Schecter A. Dietary intake is associated with phthalate body burden in a nationally representative sample. Environmental Health Perspectives, 2010, 118(7): 998-1003.

[46] Wittassek M, Wiesmuller G A, Koch H M, et al. Internal phthalate exposure over the last two decades—A retrospective human biomonitoring study. International Journal of Hygiene and Environmental Health, 2007, 210(3-4): 319-333.

[47] Gao C J, Liu L Y, Ma W L, et al. Phthalate metabolites in urine of Chinese young adults: Concentration, profile, exposure and cumulative risk assessment. Science of the Total Environment, 2016, 543(Pt A): 19-27.

[48] Guo Y, Wu Q, Kannan K. Phthalate metabolites in urine from China, and implications for human exposures. Environment International, 2011, 37(5): 893-898.

[49] Wang H, Zhou Y, Tang C, et al. Urinary phthalate metabolites are associated with body mass index and waist circumference in Chinese school children. PloS One, 2013, 8(2): e56800.

[50] Han X, Cui Z, Zhou N, et al. Urinary phthalate metabolites and male reproductive function parameters in Chongqing general population, China. International Journal of Hygiene and Environmental Health, 2014, 217(2-3): 271-278.

[51] Liu L, Bao H, Liu F, et al. Phthalates exposure of Chinese reproductive age couples and its effect on male semen quality, a primary study. Environment International, 2012, 42: 78-83.

[52] Itoh H, Iwasaki M, Hanaoka T, et al. Urinary phthalate monoesters and endometriosis in infertile Japanese women. Science of the Total Environment, 2009, 408(1): 37-42.

[53] Guo Y, Alomirah H, Cho H S, et al. Occurrence of phthalate metabolites in human urine from several Asian countries. Environmental Science & Technology, 2011, 45(7): 3138-3144.

[54] Lopez-Carrillo L, Hernandez-Ramirez R U, Calafat A M, et al. Exposure to phthalates and breast cancer risk in northern Mexico. Environmental Health Perspectives, 2010, 118(4): 539-544.

[55] Koch H M, Rossbach B, Drexler H, et al. Internal exposure of the general population to DEHP and other phthalates—Determination of secondary and primary phthalate monoester metabolites in urine. Environmental Research, 2003, 93(2): 177-185.

[56] Joensen U N, Frederiksen H, Blomberg Jensen M, et al. Phthalate excretion pattern and testicular function: A study of 881 healthy Danish men. Environmental Health Perspectives, 2012, 120(10): 1397-1403.

[57] Guo Y, Wu Q, Kannan K. Phthalate metabolites in urine from China, and implications for human exposures. Environment International, 2011, 37(5): 893-898.

[58] Wang B, Wang H, Zhou W, et al. Urinary excretion of phthalate metabolites in school children of China: Implication for cumulative risk assessment of phthalate exposure. Environmental Science & Technology, 2015, 49(2): 1120-1129.

[59] Wittassek M, Koch H M, Angerer J, et al. Assessing exposure to phthalates - the human biomonitoring approach. Molecular Nutrition & Food Research, 2011, 55(1): 7-31.

[60] Frederiksen H, Aksglaede L, Sorensen K, et al. Urinary excretion of phthalate metabolites in 129 healthy Danish children and adolescents: Estimation of daily phthalate intake. Environmental Research, 2011, 111(5): 656-663.

[61] Saravanabhavan G, Walker M, Guay M, et al. Urinary excretion and daily intake rates of diethyl phthalate in the general Canadian population. Science of the Total Environment, 2014, 500-501: 191-198.

[62] Zeman F A, Boudet C, Tack K, et al. Exposure assessment of phthalates in French pregnant women: results of the ELFE pilot study. International Journal of Hygiene and Environmental Health, 2013, 216(3): 271-279.

[63] Dewalque L, Charlier C, Pirard C. Estimated daily intake and cumulative risk assessment of phthalate diesters in a Belgian general population. Toxicology Letters, 2014, 231(2): 161-168.

[64] Rusyn I, Peters J M, Cunningham M L. Modes of action and species-specific effects of di-(2-ethylhexyl)phthalate in the liver.

Critical Reviews in Toxicology, 2006, 36(5): 459-479.

[65] Fromme H, Gruber L, Schlummer M, et al. Intake of phthalates and di(2-ethylhexyl)adipate: results of the integrated exposure assessment survey based on duplicate diet samples and biomonitoring data. Environment International, 2007, 33(8): 1012-1020.

[66] Xie X, Wang Z, Zhou X, et al. Study on the interaction of phthalate esters to human serum albumin by steady-state and time-resolved fluorescence and circular dichroism spectroscopy. Journal of Hazardous Materials, 2011, 192(3): 1291-1298.

[67] Carbone V, Velkov T. Interaction of phthalates and phenoxy acid herbicide environmental pollutants with intestinal intracellular lipid binding proteins. Chemical Research in Toxicology, 2013, 26(8): 1240-1250.

[68] Hines E P, Calafat A M, Silva M J, et al. Concentrations of phthalate metabolites in milk, urine, saliva, and serum of lactating North Carolina women. Environmental Health Perspectives, 2009, 117(1): 86-92.

[69] Silva M J, Reidy J A, Samandar E, et al. Detection of phthalate metabolites in human saliva. Archives of Toxicology, 2005, 79(11): 647-652.

[70] Frederiksen H, Skakkebaek N E, Andersson A M. Metabolism of phthalates in humans. Molecular Nutrition & Food Research, 2007, 51(7): 899-911.

[71] Niino T, Ishibashi T, Ishiwata H, et al. Characterization of human salivary esterase in enzymatic hydrolysis of phthalate esters. Journal of Health Science, 2003, 49(1): 76-81

[72] Axelsson J, Rylander L, Rignell-Hydbom A, et al. Phthalate exposure and reproductive parameters in young men from the general Swedish population. Environment International, 2015, 85: 54-60.

[73] Koch H M, Bolt H M, Preuss R, et al. New metabolites of di (2-ethylhexyl) phthalate (DEHP) in human urine and serum after single oral doses of deuterium-labelled DEHP. Archives of Toxicology, 2005, 79(7): 367-376.

[74] Koch H M, Preuss R, Angerer J. Di (2-ethylhexyl) phthalate (DEHP): Human metabolism and internal exposure—An update and latest results. International Journal of Andrology, 2006, 29(1): 155-165; discussion 181-155.

[75] Oishi S, Hiraga K. Distribution and elimination of di-2-ethylhexyl phthalate (DEHP) and mono-2-ethylhexyl phthalate (MEHP) after a single oral administration of DEHP in rats. Archives of Toxicology, 1982, 51(2): 149-155.

[76] Oishi S. Effects of phthalic acid ester on testicular mitochondrial function in the rat. Archives of Toxicology, 1990, 64(2): 143-147.

[77] Koch H M, Bolt H M, Angerer J. Di (2-ethylhexyl) phthalate (DEHP) metabolites in human urine and serum after a single oral dose of deuterium-labelled DEHP. Archives of Toxicology, 2004, 78(3): 123-130.

[78] Koch H M, Angerer J. Di-iso-nonylphthalate (DINP) metabolites in human urine after a single oral dose of deuterium-labelled DINP. International Journal of Hygiene and Environmental Health, 2007, 210(1): 9-19.

[79] Mittermeier A, Volkel W, Fromme H. Kinetics of the phthalate metabolites mono-2-ethylhexyl phthalate (MEHP) and mono-n-butyl phthalate (MnBP) in male subjects after a single oral dose. Toxicology Letters, 2016, 252: 22-28.

[80] Li X, Sun H. Distribution of phthalate metabolites between paired maternal-fetal samples. Environmental Science & Technology, 2018, 52(11): 6626-6635.

[81] Hogberg J, Hanberg A, Berglund M, et al. Phthalate diesters and their metabolites in human breast milk, blood or serum, and urine as biomarkers of exposure in vulnerable populations. Environmental Health Perspectives, 2008, 116(3): 334-339.

[82] Snijder C A, Roeleveld N, Te Velde E, et al. Occupational exposure to chemicals and fetal growth: the Generation R Study. Human Reproduction, 2012, 27(3): 910-920.

[83] Swan S H, Main K M, Liu F, et al. Decrease in anogenital distance among male infants with prenatal phthalate exposure. Environmental Health Perspectives, 2005, 113(8): 1056-1061.

[84] Li X, Sun H, Yao Y, et al. Distribution of phthalate metabolites between paired maternal-fetal samples. Environmental

Science & Technology, 2018, 52(11): 6626-6635.

[85] Huang P C, Kuo P L, Chou Y Y, et al. Association between prenatal exposure to phthalates and the health of newborns. Environment International, 2009, 35(1): 14-20.

[86] Messerlian C, Wylie B J, Minguez-Alarcon L, et al. Urinary concentrations of phthalate metabolites and pregnancy loss among women conceiving with medically assisted reproduction. Epidemiology, 2016, 27(6): 879-888.

[87] Hauser R, Gaskins A J, Souter I, et al. Urinary Phthalate metabolite concentrations and reproductive outcomes among women undergoing in vitro fertilization: Results from the EARTH study. Environmental Health Perspectives, 2016, 124(6): 831-839.

[88] Wu H, Ashcraft L, Whitcomb B W, et al. Parental contributions to early embryo development: Influences of urinary phthalate and phthalate alternatives among couples undergoing IVF treatment. Human Reproduction, 2017, 32(1): 65-75.

[89] Du Y Y, Fang Y L, Wang Y X, et al. Follicular fluid and urinary concentrations of phthalate metabolites among infertile women and associations with in vitro fertilization parameters. Reproductive Toxicology, 2016, 61: 142-150.

[90] Du Y Y, Guo N, Wang Y X, et al. Urinary phthalate metabolites in relation to serum anti-Müllerian hormone and inhibin B levels among women from a fertility center: A retrospective analysis. Reproductive Health, 2018, 15(1): 33.

[91] Stahlhut R W, van Wijngaarden E, Dye T D, et al. Concentrations of urinary phthalate metabolites are associated with increased waist circumference and insulin resistance in adult U.S. males. Environmental Health Perspectives, 2007, 115(6): 876-882.

[92] Hatch E E, Nelson J W, Qureshi M M, et al. Association of urinary phthalate metabolite concentrations with body mass index and waist circumference: a cross-sectional study of NHANES data, 1999—2002. Environmental Health, 2008, 7: 27.

[93] Song Y, Hauser R, Hu F B, et al. Urinary concentrations of bisphenol A and phthalate metabolites and weight change: A prospective investigation in US women. International Journal of Obesity, 2014, 38(12): 1532-1537.

[94] Lind P M, Roos V, Ronn M, et al. Serum concentrations of phthalate metabolites are related to abdominal fat distribution two years later in elderly women. Environmental Health, 2012, 11: 21.

[95] Buser M C, Murray H E, Scinicariello F. Age and sex differences in childhood and adulthood obesity association with phthalates: analyses of NHANES 2007—2010. International Journal of Hygiene and Environmental Health, 2014, 217(6): 687-694.

[96] Zhang Y, Meng X, Chen L, et al. Age and sex-specific relationships between phthalate exposures and obesity in Chinese children at puberty. PloS One, 2014, 9(8): e104852.

[97] Trasande L, Attina T M, Sathyanarayana S, et al. Race/ethnicity-specific associations of urinary phthalates with childhood body mass in a nationally representative sample. Environmental Health Perspectives, 2013, 121(4): 501-506.

[98] Teitelbaum S L, Mervish N, Moshier E L, et al. Associations between phthalate metabolite urinary concentrations and body size measures in New York City children. Environmental Research, 2012, 112: 186-193.

[99] de Cock M, de Boer M R, Lamoree M, et al. First year growth in relation to prenatal exposure to endocrine disruptors—A Dutch prospective cohort study. International Journal of Environmental Research and Public Health, 2014, 11(7): 7001-7021.

[100] Valvi D, Casas M, Romaguera D, et al. Prenatal phthalate exposure and childhood growth and blood pressure: evidence from the spanish INMA-sabadell birth cohort study. Environmental Health Perspectives, 2015, 123(10): 1022-1029.

[101] Svensson K, Hernandez-Ramirez R U, Burguete-Garcia A, et al. Phthalate exposure associated with self-reported diabetes among Mexican women. Environmental Research, 2011, 111(6): 792-796.

[102] James-Todd T, Stahlhut R, Meeker J D, et al. Urinary phthalate metabolite concentrations and diabetes among women in the National Health and Nutrition Examination Survey (NHANES) 2001—2008. Environmental Health Perspectives, 2012, 120(9): 1307-1313.

[103] Piecha R, Svacina S, Maly M, et al. Urine levels of phthalate metabolites and bisphenol A in relation to main metabolic

syndrome components: dyslipidemia, hypertension and type 2 diabetes. a pilot study. Central European Journal of Public Health, 2016, 24(4): 297-301.

[104] Dong R, Zhao S, Zhang H, et al. Sex differences in the association of urinary concentrations of phthalates metabolites with self-reported diabetes and cardiovascular diseases in Shanghai adults. International Journal of Environmental Research and Public Health, 2017, 14(6): E598.

[105] Lind P M, Zethelius B, Lind L. Circulating levels of phthalate metabolites are associated with prevalent diabetes in the elderly. Diabetes Care, 2012, 35(7): 1519-1524.

[106] Dirinck E, Dirtu A C, Geens T, et al. Urinary phthalate metabolites are associated with insulin resistance in obese subjects. Environmental Research, 2015, 137: 419-423.

[107] Sun Q, Cornelis M C, Townsend M K, et al. Association of urinary concentrations of bisphenol A and phthalate metabolites with risk of type 2 diabetes: A prospective investigation in the Nurses' Health Study (NHS) and NHSII cohorts. Environmental Health Perspectives, 2014, 122(6): 616-623.

[108] Dales R E, Kauri L M, Cakmak S. The associations between phthalate exposure and insulin resistance, beta-cell function and blood glucose control in a population-based sample. Science of the Total Environment, 2018, 612: 1287-1292.

[109] Stahlhut R W, van Wijngaarden E, Dye T D, et al. Concentrations of urinary phthalate metabolites are associated with increased waist circumference and insulin resistance in adult U.S. males. Environmental Health Perspectives, 2007, 115(6): 876-882.

[110] Kim J H, Park H Y, Bae S, et al. Diethylhexyl phthalates is associated with insulin resistance via oxidative stress in the elderly: A panel study. PloS One, 2013, 8(8): e71392.

[111] Chen S Y, Hwang J S, Sung F C, et al. Mono-2-ethylhexyl phthalate associated with insulin resistance and lower testosterone levels in a young population. Environmental Pollution, 2017, 225: 112-117.

[112] Olsen L, Lind L, Lind P M. Associations between circulating levels of bisphenol A and phthalate metabolites and coronary risk in the elderly. Ecotoxicology and Environmental Safety, 2012, 80: 179-183.

[113] Markowitz J S. Self-reported short- and long-term respiratory effects among PVC-exposed firefighters. Archives of Environmental Health, 1989, 44(1): 30-33.

[114] Nielsen J, Fahraeus C, Bensryd I, et al. Small airways function in workers processing polyvinylchloride. International Archives of Occupational and Environmental Health, 1989, 61(7): 427-430.

[115] Jaakkola J J K, Oie L, Nafstad P, et al. Interior surface materials in the home and the development of bronchial obstruction in young children in Oslo, Norway. American Journal of Public Health, 1999, 89(2): 188-192.

[116] Jaakkola J J, Verkasalo P K, Jaakkola N. Plastic wall materials in the home and respiratory health in young children. American Journal of Public Health, 2000, 90(5): 797-799.

[117] Jaakkola J J, Parise H, Kislitsin V, et al. Asthma, wheezing, and allergies in Russian schoolchildren in relation to new surface materials in the home. American Journal of Public Health, 2004, 94(4): 560-562.

[118] Kolarik B, Naydenov K, Larsson M, et al. The association between phthalates in dust and allergic diseases among Bulgarian children. Environmental Health Perspectives, 2008, 116(1): 98-103.

[119] Gascon M, Casas M, Morales E, et al. Prenatal exposure to bisphenol A and phthalates and childhood respiratory tract infections and allergy. Journal of Allergy and Clinical Immunology, 2015, 135(2): 370-378.

[120] Whyatt R M, Perzanowski M S, Just A C, et al. Asthma in inner-city children at 5~11 years of age and prenatal exposure to phthalates: the columbia center for children's environmental health cohort. Environmental Health Perspectives, 2014, 122(10): 1141-1146.

[121] Ku H Y, Su P H, Wen H J, et al. Prenatal and postnatal exposure to phthalate esters and asthma: A 9-year follow-up study of

a taiwanese birth cohort. PloS One, 2015, 10(4): e0123309.

[122] Holmes A K, Koller K R, Kieszak S M, et al. Case-control study of breast cancer and exposure to synthetic environmental chemicals among Alaska native women. International Journal of Circumpolar Health, 2014, 73(1): 25760.

[123] Morgan M, Deoraj A, Felty Q, et al. Environmental estrogen-like endocrine disrupting chemicals and breast cancer. Molecular and Cellular Endocrinology, 2017, 457: 89-102.

[124] Yolton K, Xu Y, Strauss D, et al. Prenatal exposure to bisphenol A and phthalates and infant neurobehavior. Neurotoxicology and Teratology, 2011, 33(5): 558-566.

[125] Kim Y, Ha E H, Kim E J, et al. Prenatal exposure to phthalates and infant development at 6 months: Prospective mothers and children's Environmental Health (MOCEH) study. Environmental Health Perspectives, 2011, 119(10): 1495-1500.

[126] Engel S M, Miodovnik A, Canfield R L, et al. Prenatal phthalate exposure is associated with childhood behavior and executive functioning. Environmental Health Perspectives, 2010, 118(4): 565-571.

[127] Lien Y J, Ku H Y, Su P H, et al. Prenatal exposure to phthalate esters and behavioral syndromes in children at 8 years of age: Taiwan maternal and infant cohort study. Environmental Health Perspectives, 2015, 123(1): 95-100.

[128] Cho S C, Bhang S Y, Hong Y C, et al. Relationship between environmental phthalate exposure and the intelligence of school-age children. Environmental Health Perspectives, 2010, 118(7): 1027-1032.

[129] Miodovnik A, Engel S M, Zhu C, et al. Endocrine disruptors and childhood social impairment. Neurotoxicology, 2011, 32(2): 261-267.

[130] Pocar P, Fiandanese N, Berrini A, et al. Maternal exposure to di(2-ethylhexyl)phthalate (DEHP) promotes the transgenerational inheritance of adult-onset reproductive dysfunctions through the female germline in mice. Toxicology and Applied Pharmacology, 2017, 322: 113-121.

[131] Wan X, Zhu Y, Xili M A, et al. Effect of DEHP and its metabolite MEHP on in vitro rat follicular development. Journal of Hygiene Research, 2010, 39(3): 268-270, 274.

[132] Oliana C, Luca T, Claudia S, et al. DEHP impairs zebrafish reproduction by affecting critical factors in oogenesis. PLoS One, 2012, 5(4): e10201.

[133] Tara L S, Davis B J. Mechanisms of phthalate ester toxicity in the female reproductive system. Environmental Health Perspectives, 2003, 111(2): 139-145.

[134] Svechnikova I, Svechnikov K, Soder O. The influence of di-(2-ethylhexyl) phthalate on steroidogenesis by the ovarian granulosa cells of immature female rats. Journal of Endocrinology, 2007, 194(3): 603-609.

[135] Park C, Lee J, Kong B, et al. The effects of bisphenol A, benzyl butyl phthalate, and di(2-ethylhexyl)phthalate on estrogen receptor alpha in estrogen receptor-positive cells under hypoxia. Environmental Pollution, 2019, 248: 774-781.

[136] Pomatto V, Cottone E, Cocci P, et al. Plasticizers used in food-contact materials affect adipogenesis in 3T3-L1 cells. Journal of Steroid Biochemistry and Molecular Biology, 2018, 178: 322-332.

[137] Sakuma S, Sumida M, Endoh Y, et al. Curcumin inhibits adipogenesis induced by benzyl butyl phthalate in 3T3-L1 cells. Toxicology and Applied Pharmacology, 2017, 329: 158-164.

[138] Chiang H C, Wang C H, Yeh S C, et al. Comparative microarray analyses of mono(2-ethylhexyl)phthalate impacts on fat cell bioenergetics and adipokine network. Cell Biology and Toxicology, 2017, 33(6): 511-526.

[139] Hurst C H, Waxman D J. Activation of PPARalpha and PPARgamma by environmental phthalate monoesters. Toxicological Sciences, 2003, 74(2): 297-308.

[140] Chiang H C, Kuo Y T, Shen C C, et al. Mono (2-ethylhexyl) phthalate accumulation disturbs energy metabolism of fat cells. Archives of Toxicology, 2016, 90(3): 589-601.

[141] Yin L, Yu KS, Lu K, et al. Benzyl butyl phthalate promotes adipogenesis in 3T3-L1 preadipocytes: A high content cellomics

and metabolomic analysis. Toxicology in Vitro, 2016, 32: 297-309.

[142] Hao C, Cheng X, Xia H, et al. The endocrine disruptor mono-(2-ethylhexyl)phthalate promotes adipocyte differentiation and induces obesity in mice. Bioscience Reports, 2012, 32(6): 619-629.

[143] Qi W, Zhou L, Zhao T, et al. Effect of the TYK-2/STAT-3 pathway on lipid accumulation induced by mono-2-ethylhexyl phthalate. Molecular and Cellular Endocrinology, 2019, 484: 52-58.

[144] Deng T, Zhang Y, Wu Y, et al. Dibutyl phthalate exposure aggravates type 2 diabetes by disrupting the insulin-mediated PI3K/AKT signaling pathway. Toxicology Letters, 2018, 290: 1-9.

[145] Viswanathan M P, Mullainadhan V, Chinnaiyan M, et al. Effects of DEHP and its metabolite MEHP on insulin signalling and proteins involved in GLUT4 translocation in cultured L6 myotubes. Toxicology, 2017, 386: 60-71.

[146] Zhang W, Shen X Y, Zhang W W, et al. Di-(2-ethylhexyl)phthalate could disrupt the insulin signaling pathway in liver of SD rats and L02 cells via PPARgamma. Toxicology and Applied Pharmacology, 2017, 316: 17-26.

[147] Barakat R, Seymore T, Lin P P, et al. Prenatal exposure to an environmentally relevant phthalate mixture disrupts testicular steroidogenesis in adult male mice. Environmental Research, 2019, 172: 194-201.

[148] Kim S, Park G Y, Yoo Y J, et al. Di-2-ethylhexylphthalate promotes thyroid cell proliferation and DNA damage through activating thyrotropin-receptor-mediated pathways *in vitro* and *in vivo*. Food and Chemical Toxicology, 2019, 124: 265-272.

[149] Yu Z, Han Y, Shen R, et al. Gestational di-(2-ethylhexyl)phthalate exposure causes fetal intrauterine growth restriction through disturbing placental thyroid hormone receptor signaling. Toxicology Letters, 2018, 294: 1-10.

[150] Park C, Lee J, Kong B, et al. The effects of bisphenol A, benzyl butyl phthalate, and di(2-ethylhexyl)phthalate on estrogen receptor alpha in estrogen receptor-positive cells under hypoxia. Environmental Pollution, 2019, 248: 774-781.

[151] Engel A, Buhrke T, Imber F, et al. Agonistic and antagonistic effects of phthalates and their urinary metabolites on the steroid hormone receptors ERalpha, ERbeta, and AR. Toxicology Letters, 2017, 277: 54-63.

[152] Ghisari M, Bonefeld-Jorgensen E C. Effects of plasticizers and their mixtures on estrogen receptor and thyroid hormone functions. Toxicology Letters, 2009, 189(1): 67-77.

[153] Li N, Liu T, Guo K, et al. Effect of mono-(2-ethylhexyl)phthalate (MEHP) on proliferation of and steroid hormone synthesis in rat ovarian granulosa cells in vitro. Journal of Cellular Physiology, 2018, 233(4): 3629-3637.

[154] Yuan K, Zhao B, Li X W, et al. Effects of phthalates on 3beta-hydroxysteroid dehydrogenase and 17beta-hydroxysteroid dehydrogenase 3 activities in human and rat testes. Chemico-Biological Interactions, 2012, 195(3): 180-188.

[155] Ye H, Ha M, Yang M, et al. Di2-ethylhexyl phthalate disrupts thyroid hormone homeostasis through activating the Ras/Akt/TRHr pathway and inducing hepatic enzymes. Scientific Reports, 2017, 7: 40153.

[156] Laurenzana E M, Coslo D M, Vigilar M V, et al. Activation of the constitutive androstane receptor by monophthalates. Chemical Research in Toxicology, 2016, 29(10): 1651-1661.

[157] Rajesh P, Balasubramanian K. Phthalate exposure in utero causes epigenetic changes and impairs insulin signalling. Journal of Endocrinology, 2014, 223(1): 47-66.

[158] Wu S, Zhu J, Li Y, et al. Dynamic effect of di-2-(ethylhexyl)phthalate on testicular toxicity: Epigenetic changes and their impact on gene expression. International Journal of Toxicology, 2010, 29(2): 193-200.

[159] Manikkam M, Tracey R, Guerrero-Bosagna C, et al. Plastics derived endocrine disruptors (BPA, DEHP and DBP) induce epigenetic transgenerational inheritance of obesity, reproductive disease and sperm epimutations. PloS One, 2013, 8(1): e55387.

[160] Li L, Zhang T, Qin X S, et al. Exposure to diethylhexyl phthalate (DEHP) results in a heritable modification of imprint genes DNA methylation in mouse oocytes. Molecular Biology Reports, 2014, 41(3): 1227-1235.

[161] Kostka G, Urbanek-Olejnik K, Wiadrowska B. Di-butyl phthalate-induced hypomethylation of the c-myc gene in rat liver.

Toxicology and Industrial Health, 2010, 26(7): 407-416.

[162] Becker K, Goen T, Seiwert M, et al. GerES Ⅳ: phthalate metabolites and bisphenol A in urine of German children. International Journal of Hygiene and Environmental Health, 2009, 212(6): 685-692.

[163] Wang I J, Karmaus W J, Chen S L, et al. Effects of phthalate exposure on asthma may be mediated through alterations in DNA methylation. Clinical Epigenetics, 2015, 7: 27.

[164] Meruvu S, Zhang J, Choudhury M. Mono-(2-ethylhexyl) phthalate increases oxidative stress responsive mirnas in first trimester placental cell line HTR8/SVneo. Chemical Research in Toxicology, 2016, 29(3): 430-435.

[165] Meruvu S, Zhang J, Bedi Y S, et al. Mono-(2-ethylhexyl)phthalate induces apoptosis through miR-16 in human first trimester placental cell line HTR-8/SVneo. Toxicology in Vitro, 2016, 31: 35-42.

[166] Martinez R M, Hauser R, Liang L, et al. Urinary concentrations of phenols and phthalate metabolites reflect extracellular vesicle microRNA expression in follicular fluid. Environment International, 2019, 123: 20-28.

[167] Gómez C, Gallart-Ayala H. Metabolomics: A tool to characterize the effect of phthalates and bisphenol A. Environmental Reviews, 2018, 26(4): 351-357.

[168] Xia H, Chi Y, Qi X, et al. Metabolomic evaluation of di-n-butyl phthalate-induced teratogenesis in mice. Metabolomics, 2011, 7(4): 559-571.

[169] Zhang J, Yan L, Tian M, et al. The metabonomics of combined dietary exposure to phthalates and polychlorinated biphenyls in mice. Journal of Pharmaceutical and Biomedical Analysis, 2012, 66: 287-297.

[170] Banerjee R, Pathmasiri W, Snyder R, et al. Metabolomics of brain and reproductive organs: Characterizing the impact of gestational exposure to butylbenzyl phthalate on dams and resultant offspring. Metabolomics, 2012, 8(6): 1012-1025.

[171] Gao H T, Xu R, Cao W X, et al. Combined effects of simultaneous exposure to six phthalates and emulsifier glycerol monosterate on male reproductive system in rats. Toxicology and Applied Pharmacology, 2018, 341: 87-97.

[172] Xu Y, Knipp G T, Cook T J. Effects of di-(2-ethylhexyl)-phthalate and its metabolites on the lipid profiling in rat HRP-1 trophoblast cells. Archives of Toxicology, 2006, 80(5): 293-298.

[173] Zhou M, Ford B, Lee D, et al. Metabolomic markers of phthalate exposure in plasma and urine of pregnant women. Frontiers in Public Health, 2018, 6: 298.

[174] Maitre L, Robinson O, Martinez D, et al. Urine Metabolic Signatures of multiple environmental pollutants in pregnant women: an exposome approach. Environmental Science & Technology, 2018, 52(22): 13469-13480.

[175] Zhang J, Liu L, Wang X, et al. Low-level environmental phthalate exposure associates with urine metabolome alteration in a chinese male cohort. Environmental Science & Technology, 2016, 50(11): 5953-5960.

作者：孙红文[1]，张连营[1]，段义爽[1]

[1] 南开大学

第 20 章　低浓度化学品 Hormesis 效应的机制与应用研究进展

▶ 1. 引言 /541

▶ 2. 低浓度化学品Hormesis效应的广谱性 /542

▶ 3. 低浓度化学品Hormesis效应的发生条件 /543

▶ 4. 低浓度化学品Hormesis效应的机制研究 /546

▶ 5. 低浓度化学品Hormesis效应的应用研究 /550

▶ 6. 展望 /556

本章导读

低浓度暴露是真实环境中化学品污染的一个普遍规律。与常规实验室研究的高剂量暴露导致的抑制作用不同,低浓度暴露往往呈现出促进作用,这种"低促高抑"称为 Hormesis 效应。寻找到一个能引起广谱 Hormesis 效应的代表性化合物,是验证 Hormesis 效应普遍性的一个有力补充;洞悉 Hormesis 效应的产生条件,是研究 Hormesis 效应机理及其应用的前期基础;揭示基于不同信号分子和相关通路以及随时间变化的 Hormesis 效应机理,是探寻 Hormesis 效应本质调控者的关键,为未来确定 Hormesis 效应公认机制指明方向;拓展 Hormesis 效应在不同领域的应用,不仅是 Hormesis 效应在实际环境中具有普遍性的有力验证,更有望于推动真实环境中以化学品低浓度暴露为特征的环境问题等的有效解决。本章将针对低浓度化学品 Hormesis 效应的广谱性、发生条件、分子机制和不同领域应用进行较全面的综述。

关键词

低浓度化学品的促进作用,Hormesis 效应的广谱性,Hormesis 效应发生条件,基于信号分子的 Hormesis 机制,不同领域应用

1 引 言

自从 16 世纪毒理学家 Paracelsus 提出"剂量决定毒性(the dose makes the poison)"的理论以来,传统毒理学家一直是以阈值模型[图 20-1(a)]和线性非阈值模型[图 20-1(b)]来分别评估非致癌物和致癌物的危险程度。但是,1887 年 Schulz 在对酵母菌的研究中首次发现,许多有毒物质在高剂量对酵母菌表现出抑制作用但在低剂量却能刺激二氧化碳的释放。这被认为是毒物兴奋效应(Hormesis)现象的首次发现。Southam 等在研究红雪松提取物对真菌的作用时也观察到这一现象,并首次将这一现象命名为"Hormesis"[1]。早期研究中,Hormesis 效应虽常被人们发现,但由于刺激作用比较弱(对照组的 130%~160%),而仍被当作实验误差,受到多方面的质疑[2]。一直到 2003 年,美国毒理学家 Calabrese 等在 Nature 杂志上发表题为 Toxicology rethinks its central belief 的评论文章[3],质疑了传统的阈值模型和线性非阈值模型,重新提出了 Hormesis 剂量-效应关系模型[hormetic dose-response model,图 20-1(c)]。所谓的 Hormesis 是指毒物或污染物对生物体的剂量-效应关系表现为在高剂量时产生抑制作用,而在低剂量时产生刺激效应的特殊现象[4]。随后,系列杂志上也发表了多篇关于 Hormesis 的评论,使毒物刺激效应成为近年来的关注热点。

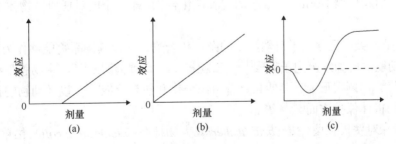

图 20-1 三种剂量-效应关系模型示意图

(a)阈值模型;(b)线性非阈值模型;(c)Hormesis 模型

越来越多的研究表明，Hormesis 效应可能具有普遍性，它在不同的生物模型、测试终点以及化合物类别下都普遍存在[5]。目前，在各类生物（包括动物、植物、微生物）、各类毒物（包括致癌物和非致癌物）及各类生命现象（包括肿瘤形成、生殖、生长、寿命及代谢等）中都发现了 Hormesis 现象。其范围几乎涵盖了包括重金属化合物、氰化物、多环芳烃、多氯联苯、有机砷化物以及农药和一些抗生素在内的大量有毒物质。Calabrese 由此认为 Hormesis 效应是一种普遍存在的生物学现象，并呼吁将 Hormesis 模型替代阈值模型和线性非阈值模型作为剂量-效应关系评价的缺省模型[6]。

因此，Hormesis 的概念刚一提出，即在多个领域内产生了轰动性的影响。

（1）其中首当其冲的是风险评价和风险管理。我们生活中所接触的每日允许摄入量（ADI）、最大残留限量（MRL）等概念，大多是根据毒理学实验获得的数据，采用阈值模型或线性非阈值模型进行风险评价而估算得出的安全阈值。因为阈值模型和线性非阈值模型的本质均为剂量-效应单调相关，因而只要低于某一特定临界剂量（EPA 称之为 RfD，即"参考剂量"），暴露就被认为是安全的。而 Hormesis 是非单调相关的，因此，有毒理学家认为根据 Hormesis 的规律，毒物在低剂量存在时实际上不是一件坏事，似乎没有必要如此浪费财力去处理那些低剂量无害的化学品[7]。因此，它的研究对化学品管理及公共卫生机构制定政策是非常重要的，而且与广大群众健康、国家经济也密切相关。

（2）Hormesis 的剂量-效应关系代表了整个生物学领域中剂量-效应概念的模式转变，它影响了毒理学家如何选择生物模型、观测终点、设计研究、评估危险，甚至如何提出他们要研究或实验的问题与假说。

（3）在医学和药理学领域，Hormesis 带来的冲击亦不小。研究表明，现阶段使用的大多抗生素、抗病毒剂和抗肿瘤制剂以及大量的其他药物都表现出 Hormesis 效应的非单调剂量-效应关系：一个剂量可能是临床有效的，但另一剂量则可能是有害的。如一些抗肿瘤药物（如苏拉明）在高剂量下抑制细胞增殖而具有临床疗效，而在低剂量条件下又成为一种局部激动剂，可以促进细胞增殖[8]。

可见，Hormesis 向传统的认识提出了挑战，将从根本上改变整个危险度评价模式，成为危险度评价进程中的一个里程碑[3]。因此，Hormesis 效应是否具有广谱性，Hormesis 效应在什么条件下产生，为什么会有 Hormesis 效应，Hormesis 效应对我们的生活和生产实践具有怎么样的指导意义？这些问题的回答及其相关研究极其重要。

2 低浓度化学品 Hormesis 效应的广谱性

Hormesis 效应已经在广泛的模式生物和宽泛的受试化合物上被报道出来[9]，然而，对于 Hormesis 的普遍性却一直存有质疑，这不仅源于目前尚无统一的机制来解释 Hormesis 效应，更源于目前尚未寻找到一个非常合适的能够引起广谱 Hormesis 效应的代表性化合物，即一个能对任何生物体都具有 Hormesis 效应的化合物。

吲哚（indole）是一类广泛存在于自然界中的杂环化合物，它及其同系物能够作为信号分子，调控生物的多方面的生理功能，比如细菌孢子形成[10]、质粒稳定[11]、细菌耐药[12]、生物膜形成[13]以及毒力因子释放[14]；植物顶端优势、侧根形成、茎伸长、管组织分化和热带反应[15]；以及动物的神经系统[16, 17]、消化系统[10]和癌症后的机体保护方面[18, 19]等。

研究发现[20]，吲哚对费氏弧菌（*Aliivibrio fischeri*）、大肠杆菌（*Escherichia coli*）、枯草芽孢杆菌（*Bacillus subtilis*）、铜绿微囊藻（*Microcystis aeruginosa*）、羊角月牙藻（*Selenastrum capricornutum*）、人皮肤成纤维细胞（human skin fibroblasts）和人宫颈癌细胞（human cervical cancer cells）均具有 Hormesis 效应（图 20-2）。通过比较吲哚、吲哚同系物（二氢吲哚与吲哚乙酸）和吲哚结构类似物［(*R*)-3-吡咯烷醇

和 N-环乙基吡咯烷醇]对费氏弧菌生物荧光随时间变化 Hormesis 效应的机理，证明了吲哚环是吲哚能够作用于细菌的群体效应（QS）系统从而引起费氏弧菌生物荧光随时间变化 Hormesis 效应的关键结构。原核生物（费氏弧菌、大肠杆菌、枯草芽孢杆菌和铜绿微囊藻）通过群体感应信号分子来进行交流，而真核生物（羊角月牙藻、人皮肤成纤维细胞和人宫颈癌细胞）则可以通过第一信使以及第二信使完成细胞间交流[20]。因此，推测吲哚可能通过作用于受试生物的细胞间交流，从而影响细胞的代谢或生长，最终导致了广谱性 Hormesis 效应的产生。

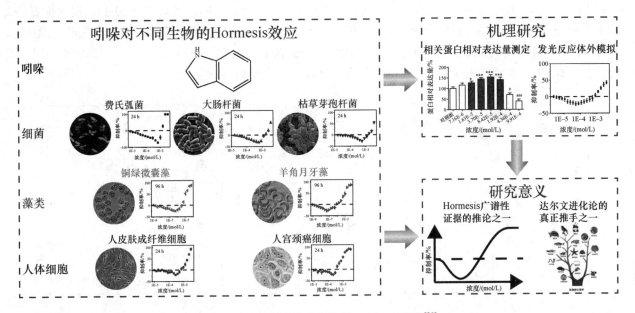

图 20-2　吲哚对不同模式生物的 Hormesis 效应[20]

虽然乙醇、甲醛、氨等小分子化合物被频繁报道具有 Hormesis 效应[9]，但是它们不是生物体的信号分子，因而不具有可能对所有生物都产生 Hormesis 效应的前提条件。吲哚可能是目前发现的第一个具有广谱 Hormesis 效应的化合物，这对于"目前在环境毒理学领域可能具有颠覆意义，但却一直备受质疑"的 Hormesis 从此被广泛接受起到非常积极的推动作用。另一方面，结合"只有具有 tnaA 基因的细菌才能产生吲哚，真核生物没办法产生吲哚"的前人研究[22]和"吲哚对不同生物具有不同的 Hormesis 效应"的本研究成果，同济大学环境学院化学品环境效应与人体健康课题组（即林志芬课题组）推测，环境驱动细菌产生吲哚，而不同生物由于对吲哚的 Hormesis 响应范围不同从而处于 Hormesis 效应的被促进或被抑制阶段，前者因为适应而适者生存于大量细菌存在的环境，而后者因为被抑制增加了被细菌生存的环境所淘汰的可能。这便是吲哚具有广谱 Hormesis 效应的重要生态学意义：细菌通过吲哚对其他生物的 Hormesis 效应可能是"物竞天择，适者生存"的达尔文进化论的真正推手之一。

3　低浓度化学品 Hormesis 效应的发生条件

物质和能量是影响生物体生长和代谢的基本因素，比如化学信号分子通过与之相应的生物受体蛋白相互作用从而调控生物效应，而生物生长环境中的能源物质种类和含量对生物生长都起到了关键作用。因此，外源污染物进入生物体内通过与生物体内信号分子竞争结合生物蛋白进而会表现出异常的生物效应。可见，外源污染物的结构、外源污染物对生物体暴露时间以及生物体的能量供给等都是决定 Hormesis

效应发生的关键因素。

　　Hormesis 效应产生的条件之一：结构上的相似性——外源污染物必须是内源化合物的拟似物。以费氏弧菌为受试生物，呋喃酮和吡咯酮类化合物为受试化合物，采取 0～24 h 发光毒性测定和分子对接技术（molecular docking）开展了化合物结构对费氏弧菌 Hormesis 效应影响的研究。研究结果发现：在对费氏弧菌 0～24 h 的发光毒性测定中，呋喃酮和吡咯酮类的化合物对费氏弧菌的发光均产生了 Hormesis 效应。通过分子对接的结果表明，呋喃酮和吡咯酮类的化合物均可以与费氏弧菌群体感应内源信号分子 C6 的靶蛋白 LuxR 结合，且结合位点与 C6 结合 LuxR 的位点相同，结合自由能也较高（图 20-3）。据此推测，呋喃酮和吡咯酮对费氏弧菌的 Hormesis 效应与费氏弧菌的群体感应有关。当费氏弧菌没有产生群体感应时，此时 C6 浓度较低，呋喃酮和吡咯酮由于其结构与费氏弧菌的内源信号分子 C6 类似，因此它们可以在费氏弧菌自身 C6 产生不足的情况下与 LuxR 结合，从而促进费氏弧菌的发光。

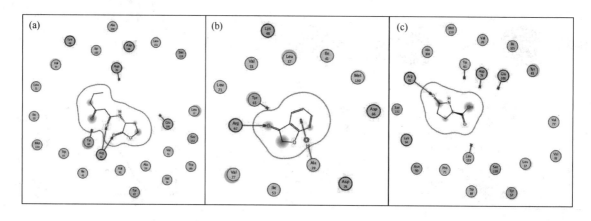

图 20-3　吲哚对不同模式生物的 Hormesis 效应

（a）内源信号分子 C6 与 LuxR 蛋白的对接结果；（b）外源污染物呋喃酮（3-苯并呋喃酮）与 LuxR 蛋白的对接结果；（c）外源污染物吡咯酮（2-吡咯酮-5-羧酸）与 LuxR 蛋白的对接结果

　　Hormesis 效应产生的条件之二：时间上的前置性——外源污染物在内源信号分子尚未分泌或分泌不足的情况下与相应蛋白的结合，即"在错误的时间遇到相似的化合物"。测定了 0～24 h 内磺胺（磺胺邻二甲氧嘧啶，sulfadoxine，SDX）作用下费氏弧菌的生长和发光的剂量-效应曲线，结合费氏弧菌 0～24 h 的生长和发光曲线，探究了磺胺 SDX 作用于费氏弧菌产生发光的 Hormesis 效应与暴露时间之间的联系。结果表明，由于在 0～24 h 内费氏弧菌的群体感应分为三个阶段，因此磺胺 SDX 对费氏弧菌的 Hormesis 也分为三个阶段（图 20-4）。第一阶段发生在费氏弧菌生长的迟缓期，此时费氏弧菌还没有产生群体感应，费氏弧菌内源信号分子 C6 的浓度较低，磺胺 SDX 可以代替 C6 结合于 LuxR 蛋白，进而促进费氏弧菌的发光，产生 Hormesis 效应；第二阶段发生于费氏弧菌生长的对数期，此时费氏弧菌开始产生群体感应，C6 的浓度逐渐增大，越来越多的 C6 开始占据其靶蛋白 LuxR 上的结合位点，这导致磺胺 SDX 与 LuxR 的结合量减少，从而 Hormesis 减弱；第三阶段的 Hormesis 效应发生于费氏弧菌的稳定期，此时群体感应逐渐减弱，C6 的量逐渐减小，可以与磺胺 SDX 结合的 LuxR 的量相对增加，Hormesis 效应相比第二阶段逐渐增强。

　　Hormesis 效应产生的条件之三：能源上的补偿性——在培养基内 C 源不足时，生物将加入的含碳化合物作为其碳源物质从而维持自身生长。测定了 0～24 h 内不同混合碳源体系（不同的组成和各组分浓度）中磺胺 SDX 作用下费氏弧菌的生长和发光的剂量-效应关系，结合不同混合碳源体系下费氏弧菌的生长和发光曲线，探讨了磺胺作用于费氏弧菌产生发光的 Hormesis 效应与费氏弧菌生长能源之间的联系。研究结果表明，在培养基中加入葡萄糖的情况下，费氏弧菌会优先利用葡萄糖作为能源物质，在这个过程中，

第20章 低浓度化学品 Hormesis 效应的机制与应用研究进展

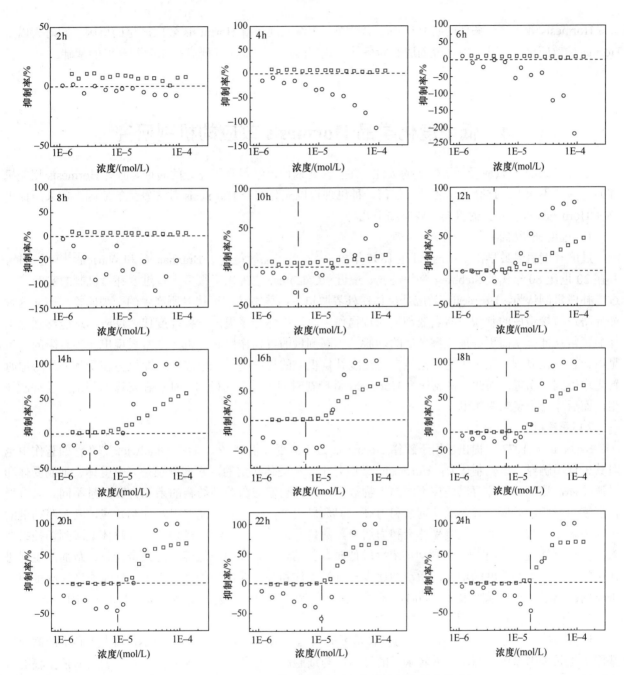

图 20-4 磺胺 SDX 对费氏弧菌生长和发光的剂量-效应关系

图中○为磺胺 SDX 对费氏弧菌发光的剂量-效应关系；□为磺胺 SDX 对费氏弧菌生长的剂量-效应关系；竖直方向的灰色虚线标记的是磺胺 SDX 对费氏弧菌发光 Hormesis 效应的最大促进浓度

因为葡萄糖更容易被利用，因此费氏弧菌更早地进入了第一个对数生长期。但在费氏弧菌利用葡萄糖的过程中，葡萄糖可能会通过影响费氏弧菌合成 C6 从而推迟的群体感应的产生从而抑制了费氏弧菌的发光总量，并进一步对磺胺 SDX 引起费氏弧菌发光 Hormesis 效应产生影响。而在培养基中加入葡萄糖对磺胺 SDX 引起费氏弧菌发光的 Hormesis 效应最明显的影响是会导致 Hormesis 效应在费氏弧菌进入二次生长后逐渐减小或消失。推测这是因为费氏弧菌利用原始碳源进行生长时产生了 C6，随着 C6 浓度的增大，费氏弧菌产生群体感应。因此随着 C6 浓度的继续增大，磺胺 SDX 与 LuxR 的结合会减少，从而导致费氏弧菌 Hormesis 效应减弱或消失。

Hormesis 效应产生条件的研究有利于更微观、更深入地了解 Hormesis 效应产生的原因，同时为以后 Hormesis 效应机制及其应用研究创造实验条件，是保障这些后期深入研究得以顺利开展的基础。

4 低浓度化学品 Hormesis 效应的机制研究

尽管 Hormesis 效应受到了广泛的关注，并且其普遍性已经得到了充分的验证，但对 Hormesis 相关机理的研究并不深入，迄今为止还没有任何一种机理可以充分解释 Hormesis 效应发生的原因。前人所提出解释 Hormesis 效应发生的机制主要有如下几种。

1) 过度补偿机制

过度补偿机制最早由 Townsend 在 1897 年提出[23]，后来陆续得到了 Branham[24]和 Warren[25]等的支持，并在 20 世纪 80 年代被 Stebbing[26, 27]发展。最近，Calabrese 也提出了很多支持过度补偿机制的实例[28]。过度补偿机制认为，Hormesis 效应是机体在外界胁迫下内稳态受到干扰时所产生的一种应答，而非这种外界胁迫直接造成的刺激。即当低剂量的毒物作用时，机体会表现出一种轻度代偿反应，这能够强化机体的正常功能，从而更好地抵御之后的刺激；当高剂量的毒物作用时，机体则常表现出一种有限的、不足以让自身恢复到对照水平的补偿能力。过度补偿机制的提出基于一个前提，即机体在受到外界胁迫时首先遭受到了伤害，因此该机制认为 Hormesis 效应在时间上具有这样的特征：在暴露的初期呈现抑制作用，而后才会出现刺激作用。

2) 受体机制

Szabadi 于 1977 年提出了用于解释 Hormesis 效应的受体机制[29]，后来 Chadwick[30]在他的著作中也对其进行了解释。该机制认为，Hormesis 效应是由三种受体决定的，它们分别是兴奋受体、抑制受体和过渡受体，这三种受体可以调控不同的生物效应，并且它们与激动剂结合的亲和力和数量不同。兴奋受体能够引发刺激效应，其对激动剂亲和力高，但数量较少；抑制受体能够引发抑制效应，其对激动剂的亲和力较低，数量较多；过渡受体亲和力居中，数量不定，只是单纯与毒物结合，整体上起到调控兴奋效应向抑制转变的作用[31]。当激动剂浓度较低时，其结合高亲和力但低容量的兴奋受体，从而产生促进作用；随着激动剂浓度的增加，激动剂逐步与过渡受体和抑制受体相结合，产生的抑制作用逐渐抵消促进作用；当激动剂的暴露浓度较高时，激动剂结合低亲和力但高容量的受体亚型，产生抑制作用。

3) 基因表达与调控作用机制

研究表明，毒物低剂量刺激效应与多种基因的表达和调控密切相关[32]，如 DNA 修复基因、应激蛋白基因、生长调节基因、启动子或转录子的激活、与细胞凋亡有关的基因等。一些研究表明低剂量辐射可通过激活多种信号转导通路而激活 AP-1、SRF、ATFZ 和 NF-KB 等转录因子[33]。

4) 氧化应激机制

需氧细胞在代谢过程中会产生活性氧（ROS），当机体内的 ROS 得不到清理而大量累积时，其会对细胞产生毒害作用。通常情况下，机体内 ROS 的产生和清除处于动态平衡的状态。已有研究表明，ROS 也是对细胞具有广泛调节作用的信息分子，参与细胞信号的转导、激活转录因子，影响基因的表达，从而促使细胞增殖和分化。所以在适当的浓度下，ROS 对机体有一定的保护作用。

5) DNA 损伤修复机制

一些有毒物质在低剂量时可以通过修复 DNA 损伤的方式对机体表现出刺激效应。已有研究证明[34]，一些物质在低浓度时可以促进机体内某些细胞 DNA 的合成，或通过合成相关蛋白的方式减少有毒物质对 DNA 的损害。

6）免疫功能增强机制

有研究表明，低剂量的丝裂霉素 C 可激活机体的免疫系统，产生白细胞介素-2，诱导淋巴因子激活的杀伤细胞，增强机体对外源性化合物的抵抗力[35]。

然而，这些理论一方面缺乏足够的实验证据支持，另一方面都仅能从某一水平或某一效应入手，未能抓住 Hormesis 效应机制的本质调控者这一难点，因而不能完全解释 Hormesis 现象的产生。

4.1 个体水平的基于 ROS 和 NO 的氧化应激机制

在前期研究的基础上，同济大学环境学院化学品环境效应与人体健康课题组（即林志芬课题组）认为，在个体水平上，ROS 和 NO 可能是 Hormesis 效应的本质调控者，这基于以下这样几个事实：①生物体在受到外界胁迫时会产生 ROS 和 NO[36, 37]，它们与 Hormesis 效应类似，是一种普遍存在的生物学现象，没有生物与污染物种类的区别。②大量研究表明，ROS 和 NO 作为第二信使，可能是胁迫信号转导的中心环节[38, 39]。③ROS 和 NO 本身对生物体的作用具有两面性特征，例如低剂量的 H_2O_2 和 NO 对生物体有刺激作用，而高剂量的 H_2O_2 和 NO 对生物体有损伤作用[40-42]，因此它们可能是化合物 Hormesis 效应的本质贡献者[43]。④在氧化应激机制中，H_2O_2 和 NO 不仅作为自由基相互之间发生复杂的化学反应，而且作为信号分子也存在着复杂的信号交互作用（cross-talk）[44, 45]。H_2O_2 和 NO 存在的这种复杂的关系，可能是阻碍 Hormesis 效应机理得以揭示的根本原因。

因此，同济大学林志芬课题组选取磺胺类抗生素作为研究对象，采用大肠杆菌和枯草芽孢杆菌作为模式生物，测定了细菌在磺胺类抗生素作用下的 Hormesis 效应，并采用荧光探针方法检测了细菌在不同浓度抗生素作用下产生的胞内 ROS 的浓度，分析了 ROS 和 Hormesis 的关系；构建了一氧化氮合酶（nos）缺失的枯草芽孢杆菌和表达 nos 的大肠杆菌，考察了 NO 对 Hormesis 效应产生的影响。结果发现，磺胺在引起促进作用的浓度范围内，使细菌胞内的 ROS 水平轻微的升高；而在引起抑制作用的浓度范围内，胞内的 ROS 水平明显的升高。据此提出了 ROS 可能是磺胺引起细菌 Hormesis 效应的一个原因，即 ROS 水平在轻微升高时，扮演了信号分子的角色，调控细菌的相关通路从而引起了生长促进作用。此外，磺胺对表达了 nos 的大肠杆菌没有 Hormesis 效应，而对敲除了 nos 基因的枯草芽孢杆菌呈现了 Hormesis 效应。可见，细菌胞内的 NO 可能是磺胺对细菌生长没有呈现 Hormesis 效应的原因，即 NO 可能会消除由 ROS 引起的 Hormesis 效应。因此，该课题组提出了 ROS 和 NO 在抗生素引起的细菌 Hormesis 效应中的作用，指出了 ROS 是 Hormesis 效应的本质调控者这一机制，并表明了 NO 和 ROS 的交互作用会消除 ROS 所引起的 Hormesis 效应（图 20-5）。

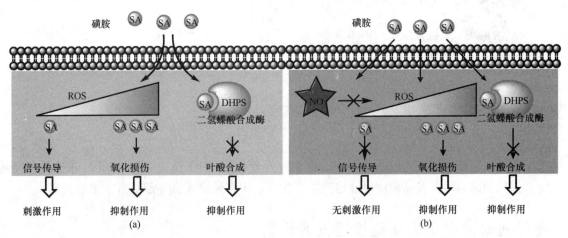

图 20-5　磺胺对大肠杆菌和枯草芽孢杆菌产生毒性效应的机制图

(a) 大肠杆菌，Hormesis 效应；(b) 枯草芽孢杆菌，无 Hormesis 效应

4.2 群体水平的基于群体感应 QS 信号分子的机制

从个体水平开展基于 ROS 和 NO 的氧化应激 Hormesis 机制研究的同时,同济大学林志芬课题组从群体水平还开展了基于群体感应 QS 信号分子的机制探讨。群体感应(quoum sensing,QS)是细菌通过分泌和感应一种称作自诱导分子(autoinducer,AI)的化学信号分子进行交流,协调群体行为的一种现象,被认为是细菌之间的"语言"[46]。其中,C6(3-oxo-C6-HSL)是目前研究最多也最为成熟的细菌群体效应信号分子之一。

该课题组选取了磺胺类抗生素(磺胺氯哒嗪,sulfachloropyridazine,SCP)和细菌群体效应信号分子 C6 为研究对象和费氏弧菌为受试生物,测定了以磺胺 SCP 对费氏弧菌的 Hormesis 效应,探究了外源信号分子 C6 对以磺胺 SCP 为代表的磺胺类抗生素毒性的调控效应,测量了在 Hormesis 曲线不同浓度下的相关基因(*luxI* 和 *luxR*)表达,测定了当费氏弧菌体内 LitR 蛋白的量改变时,在磺胺 SCP 的作用下的生物效应以及相关基因表达的改变,阐明了磺胺类抗生素的 Hormesis 机制:磺胺 SCP 进入细菌体内后,除了传统的与 DHPS 蛋白作用而抑制生长影响发光菌发光等生理活性外,还可作用于 LitR 蛋白通路(图 20-6);由于 LitR 蛋白存在两种活性状态,一是促进 *luxR* 的表达,另外一种是促进 *dhps* 的表达,磺胺 SCP 作用于 LitR 蛋白能够促使 LitR 蛋白促进 *dhps* 表达的活性状态向促进 *luxR* 的活性状态转换,促进 *luxR* 的表达,翻译出更多的 LuxR 蛋白,促进细菌发光,从而产生 Hormesis 作用[47]。

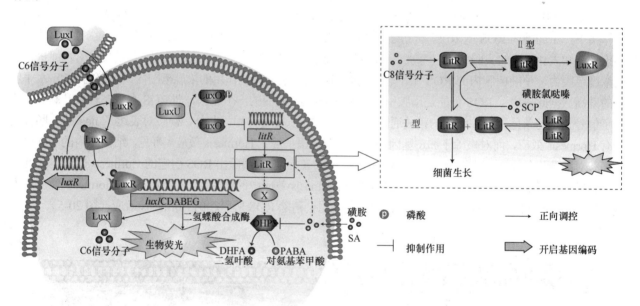

图 20-6 磺胺对费氏弧菌 Hormesis 效应的机制示意图[47]

此基于群体效应信号通路的机制是 Hormesis 的一个新机制,对于抗生素抗性基因、未来抗菌新药(特别是抗生素替代药品)的开发、应用以及这些新药的环境生态风险评价等都具有重要的意义。

4.3 基于随时间变化的跷跷板理论的机制

目前有关于 Hormesis 机制的探究大部分研究并没有说明时间对 Hormesis 效应的影响,因而提出的机制仅仅只能在某个特殊的时间点解释 Hormesis 效应；同时，许多研究者将刺激作用和抑制作用孤立开，没有意识到一个化合物可以在测试终点上同时引起刺激作用和抑制作用。因此，以往提出的机制尚未能解释 Hormesis 效应随时间变化的维度特性。

因此，同济大学林志芬课题组测定了五种抗菌剂磺胺吡啶（sulfapyridine，SPY）、磺胺增效剂（trimethoprim，TMP）、盐酸四环素（tetracycline hydrochloride，TH）、群体感应抑制 3-苯并呋喃酮[benzofuran-3(2H)-one，B3O]和群体感应抑制剂 4-溴代-5 亚甲基溴代-呋喃酮[(Z)-4-bromo-5-(bromomethylene)-2(5H)-furanone，C30]对费氏弧菌生物荧光从 1～24 h 的毒性效应。结果表明，SPY、TMP、TH、B3O 和 C30 均引起了费氏弧菌生物荧光随时间变化的 Hormesis 效应。

同时，根据课题组前期研究[48]发现的 Hormesis 剂量效应曲线模型是两个 Logistic 方程（分别反映刺激作用和抑制作用）的数学加和[图 20-7（b）]，将上述五种抗菌剂随浓度变化的 Hormesis 曲线都分别分解成刺激作用和抑制作用[图 20-7（a）]，并结合相关蛋白的相对表达量测定、模拟发光反应的探究等结果，提出了一个新的模型——跷跷板模型对受试化合物导致费氏弧菌生物荧光随时间变化的 Hormesis 效应进行了解释[49]：认为化合物刺激作用和抑制作用之间的相关影响可以由一个摆动的跷跷板恰当而清晰地反映出来，其中跷跷板的一边代表刺激作用（S），而另一边代表抑制作用（I）。如图 20-7（c）所示：当跷跷板向刺激端倾斜，表明 S 大于 I，此时化合物对生物荧光表现出刺激效应；当跷跷板处于平衡状态，表明 S 等于 I；当跷跷板向抑制端倾斜，表明 S 小于 I，此时化合物对生物荧光表现出抑制效应。众所周知，跷跷板的摆动来源于两端的向下力的不同。在本研究中，跷跷板刺激端的向下力等同于刺激曲线上确定点的倾斜度（K_S）的绝对值，这反映的是刺激作用的增长率；跷跷板抑制端的向下力等同于抑制曲线上确定点的倾斜度（K_I）的绝对值，这反映的是抑制作用的增长率。比较 K_S 和 K_I 的大小能够帮助判别跷跷板将要摆动的方向：当 K_S 大于 K_I，跷跷板有向刺激端摆动的趋势；当 K_S 等于 K_I，跷跷板将会保持稳定；K_S 小于 K_I，跷跷板有向抑制端摆动的趋势。

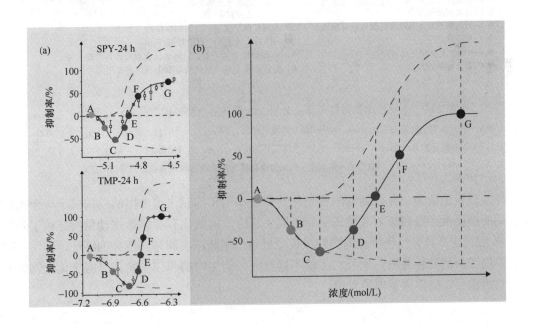

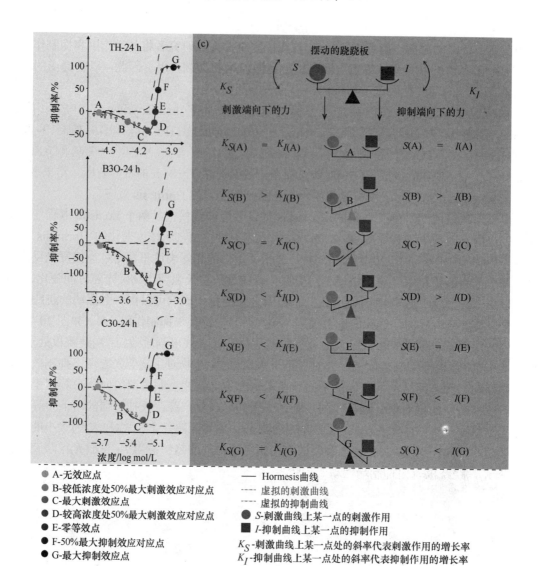

图 20-7 五种抗菌剂对费氏弧菌 Hormesis 效应的跷跷板模型机制示意图[49]

（a）受试化合物在 24 h 对费氏弧菌生物荧光的 Hormesis 效应的拟合曲线，包括七个特殊且重要的点：A，B，C，D，E，F 和 G，以及其分解的刺激曲线和抑制曲线；（b）24 h 典型的 Hormesis 效应曲线，包括七个特殊且重要的点：A，B，C，D，E，F 和 G，以及其分解的刺激曲线和抑制曲线；（c）在 24 h 刺激作用和抑制作用构成的跷跷板从 A 点状态到 G 点状态有规律地摆动

总之，跷跷板模型认为，化合物随浓度变化的刺激作用和抑制作用占据两端构成了一个随化合物浓度增加而摆动的跷跷板；跷跷板随着费氏弧菌不同生长阶段的变化有规律地摆动最终导致了随时间变化的 Hormesis 效应。这个具有生物特性的跷跷板模型展现出了随浓度和时间变化的特征，不仅对 Hormesis 的潜在机制提供了新的独特见解，而且对了解不同效应间的生物加和在生物体暴露于化合物时产生进化型适应性策略的重要作用提供了理论支撑。

5 低浓度化学品 Hormesis 效应的应用研究

由于 Hormesis 效应的普遍性存在，且 Hormesis 向传统的认识提出了挑战，因此，Hormesis 在多个领域内产生了轰动性的影响，比如，化学品管理及公共卫生机构中的风险评价和风险管理，毒理学领域中所有相关研究，医学和药理学领域中药物用量等等。因此，掌握和应用 Hormesis 效应对人类生产和生活有着重要意义。

5.1 中药领域

中药因其神奇疗效而被广泛应用，是我国几千年一直传承下来的主要治病方式。但是，由于中药用药的复杂性，导致目前中药方剂还是按照"君臣佐使"的配药思路，剂量上按照药典大致范围和医生个人经验进行确定。这种剂量上的粗略、定性和笼统的"剂量确定"的依据，使得中药方的传承更多基于经验基础，缺乏科学的理论指导和客观标准。

Hormesis 效应在中医药领域非常普遍。许多草药的萃取物在单一或混合作用于动物或人体细胞等体外模型时，都会引起 Hormesis 效应。例如，骨碎补、地鳖虫、红花和广地龙等几种中药在低浓度时能够促进人体成骨细胞的增殖，因此被用于治疗骨折等损伤[50-52]。中药引起的毒物兴奋效应，也会以预处理或预暴露（preconditioning）的形式出现。例如，用少量的蒿本内酯（提取自川芎和当归）预处理 P12 细胞后能够使细胞免受缺氧缺糖造成的死亡。中药中各化合物处于低剂量下能够对各种生理功能起到刺激的作用，即 Hormesis 效应中的"低剂量促进作用"，从而达到中医"调"的目的。另一方面，中医中常用以毒攻毒的原理。比如喜用蝎子、蜈蚣、蟾蜍等治疗癌症。又如，雷公藤被认为是有毒的草药，但是可以被用来祛风除湿治疗关节炎。这些草药的剂量应该定在高剂量，即发挥毒物兴奋效应中的"高剂量抑制作用"。可见，中医"调"和"治"对应 Hormesis 效应的"低促"和"高抑"。

因此，同济大学林志芬课题组提出，Hormesis 效应可提供科学定量中药剂量的依据[53]。中药用药一般是先采用高剂量"以毒攻毒"的"治"，后再加低剂量加强免疫机能的"调"，这不仅在常见病多发病中，在恶性肿瘤临床治疗和疑难病症或者是治疗效果欠佳的疾病，也都是利用中药之 Hormesis 效应，往往能起到良好作用。在"调"的过程中，药物的剂量通常是相对较低的。但是，很多草药中的活性成分的含量是很低的，很多情况下，甚至太低以至于不能产生最大的促进作用。如图 20-8（b）所示，药物的

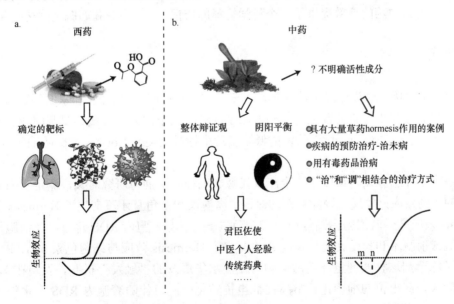

图 20-8　中药和西药在临床用药剂量上的对比[53]

（a）药物的活性成分和作用靶标通常是明确的，药物的剂量根据剂量效应关系来确定；（b）中医采用整体和辩证的角度来看待疾病，试图寻求机体的阴阳平衡。中药中的活性成分通常是不确定的。用药剂量往往通过医生的个人经验和药典推荐的剂量

剂量在 m 时能够通过毒物兴奋效应产生有益作用，但是 m 剂量并非产生最大促进的剂量。为了使药物"调"的作用发挥到最大，我们应当按照剂量 n 来给药。这个最优的剂量 n 在中医中常被成为超大剂量。中药中早已有很多超大剂量用药而产生奇特功效的例子，比如明代医家张景岳以擅长于超大剂量应用熟地黄，而有"张熟地"之雅称；民国年间名医陆仲安以擅用黄芪著称，而且用量颇大，至 8~10 两，有"陆黄芪"之称。

依据 Hormesis 效应进行中药剂量定量，在对待不同的疾病时，应充分考虑因病而异。某种中药在某个剂量下，对一种疾病可能是利用 Hormesis 效应中的抑制作用，而在对另一种疾病的时候，可能发挥的是 Hormesis 效应中的刺激作用。即同一种中药在对不同的疾病时扮演的角色不一样，这可能源于某主药中的某一主要成分具有 Hormesis 效应的特性，因此在不同剂量档次，表现出不同的药效作用。另外，中药的每个药方都是由几种甚至十几种草药构成的复合药方。其中活性比较高或含量比较高的成分，可能会在草药剂量相对较小时就发挥作用，但是对于那些活性较小或含量较小的成分，只有在草药剂量较高的时候才能发挥作用。因此，在某一草药的剂量下，并非每一种活性成分都能够有效发挥作用。此外，单味药或药方中的某些成分之间会发生协同的相互作用，也会对 Hormesis 效应所对应的剂量区域产生影响。因此在对待不同的疾病时，要充分考虑中药的剂量选择。

同样地，Hormesis 效应也依赖于不同患者的个体差异情况。同一化合物对同种生物的效应与生物个体差异密切相关，因此，西医往往通过体重、体表面积、患者年龄等参数计算和调整用药剂量，而目前中医往往是结合"望闻问切"来确定病人的身体状况，最终根据经验综合确定方剂中各味中药的用量。由于毒物兴奋效应的浓度区间范围往往比较大，是传统毒性阈值的 1/20 到 1/1000[54]。因此，通过"望闻问切"尚无法准确确定不同病人的达到 Hormesis 效应最大促进时中药的剂量，建议在今后大量科学实验基础上，结合现代先进科技，提出一些人体个体差异的评价方法，从而能有效地预测出达到 Hormesis 效应最大促进下所需中药浓度。

总之，中医经常被诟病的一个问题是用药定量化缺少确凿的科学证据，目前对于中医的研究仍充满挑战性。基于中医的治疗的复杂性，以及中西医理论具有本质的不同。同济大学林志芬课题组通过对中药中普遍的 Hormesis 现象的归纳和分析，提出 Hormesis 效应可能是中药起药理作用的本质机制（如图 20-8 所示）。这为中药的疗效提供了一个科学解释的依据，并在中医和西医之间构建起了一架桥梁。

5.2　水华领域

关于水华爆发的原因，学界已经争论多年，主要是围绕着氮磷含量、pH 和水温等这些藻类生长的外部条件展开的。然而近些年有研究发现，一些低浓度的环境因子例如重金属、氨基酸和农药在水华形成过程中扮演着重要角色，并认为环境中残留的低浓度污染物对藻类的 Hormesis 效应可能是水华爆发的一个诱因。

因此，同济大学林志芬课题组以三嗪类和取代脲类七种除草剂为研究对象，测定它们对铜绿微囊藻和羊角月牙藻的生物效应，发现七种除草剂均对铜绿微囊藻和羊角月牙藻产生了 Hormesis 效应。同时，测定了 Hormesis 效应产生时细胞内高低电位细胞色素 b_{559} 含量、胞内 ROS 含量以及细胞外 NO 含量的变化，并通过流式细胞术和拉曼光谱检测等实验探究了 Hormesis 效应与细胞色素 b_{559}、ROS、NO 之间的关系，探讨了除草剂这种低剂量环境污染物对蓝藻水华爆发的"触发"机制，发现在 Hormesis 效应刺激逐渐增强阶段，高电位的细胞色素 b_{559} 向低电位转化，同时伴随着胞内 ROS 含量的上升。据此推断（如图 20-9 所示），藻细胞通过将高电位的细胞色素 b_{559} 向低电位转化，从而缓解除草剂造成的光抑制；但与此同时，这个过程造成了电子向氧气的传递，从而形成 ROS。ROS 含量的一定水平的上升造成了刺激作用产生。在 Hormesis 效应刺激逐渐消失和抑制增强阶段，观察到了细胞色素 b_{559} 重新从低电位转化为高电位；同时发现了胞内 ROS 含量下降以及胞内 NO 含量下降。据此论证了三者产生变化的相互关系，即 NO 能够导致细胞色素 b_{559} 从低电位转向高电位，而这一转化能够引起胞内 ROS 含量的下降。ROS 的下降和由此引起的级联反应是除草剂抑制作用产生的根本原因。

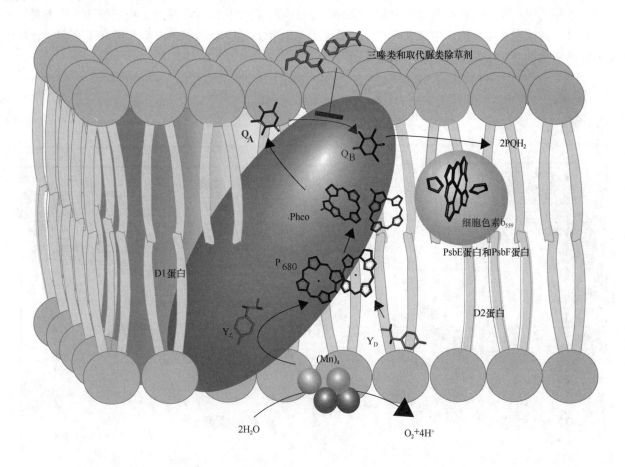

图 20-9　除草剂对铜绿微囊藻和羊角月牙藻产生 Hormesis 效应的机制图

总之，基于除草剂对藻类的 Hormesis 效应中的刺激作用，以及任何化合物对任何生物在适宜条件下都可导致 Hormesis 效应的普遍性，同济大学林志芬课题组推测目前环境中残留的大量污染物可对藻类产生 Hormesis 效应，而且由于实际环境中污染物的低浓度污染特征导致了这些效应往往呈现出 Hormesis 的刺激阶段，一旦外界条件适宜（即氮磷含量、pH 和水温等），藻类就可大量生长。可见，在水华藻类爆发的过程中，低浓度环境污染物对水华藻的刺激作用能够起到一个最初始的激发作用，是造成水华爆发的重要诱因之一。

5.3　食品添加剂领域

基于食品添加剂因其使用量大，使用范围广泛，以及每天低剂量经口进入人体的特点，同济大学林志芬课题组选择了较为经常使用的呋喃酮和吡咯酮类食品添加剂，以人宫颈癌细胞（human cervical cancer cells，HeLa 细胞）和人皮肤成纤维细胞（human skin fibroblasts，HSF 细胞）为受试体，测定了呋喃酮和吡咯酮对 HeLa 细胞和 HSF 细胞的毒性作用，发现有 5 种呋喃酮对 HeLa 细胞具有明显的 Hormesis 作用（刺激效应最大达到 37%），所测试的 5 种吡咯酮对 HeLa 细胞均无明显的 Hormesis 作用；而 6 种呋喃酮对 HSF 细胞产生了 Hormesis 效应（刺激效应最大达到 33%），3 种吡咯酮对 HSF 细胞产生了 Hormesis 效应（刺激效应最大达到 46%）。

同时，测定了细胞内 ROS、NO 以及细胞外 NO 含量，细胞分裂指标（细胞周期、端粒长度等）和细胞凋亡指标［线粒体膜电位、胱天蛋白酶（Caspase）3 等］的变化情况，探讨了呋喃酮和吡咯酮类食品添加剂对正常细胞和癌细胞作用结果的差异，提出了基于 ROS 和 NO 调控的呋喃酮类和吡咯酮类食品添加剂对癌细胞的 Hormesis 效应机制（图 20-10）。

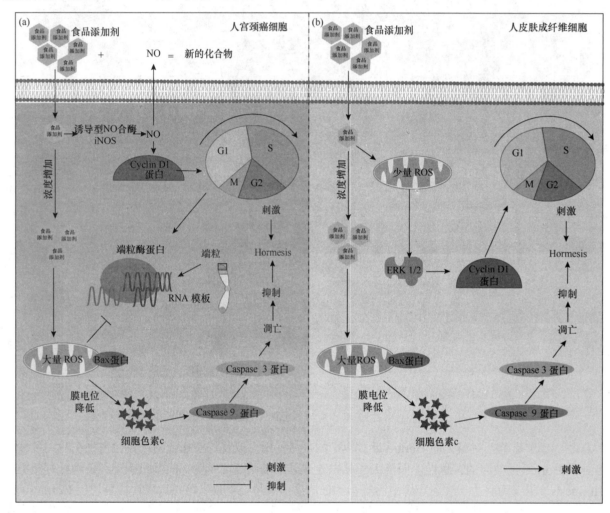

图 20-10　呋喃酮类和吡咯酮类食品添加剂对癌细胞的 Hormesis 效应机制
（a）呋喃酮类食品添加剂对 HeLa 细胞；（b）呋喃酮类和吡咯酮类食品添加剂对 HSF 细胞

呋喃酮类食品添加剂对 HeLa 细胞基于 ROS 和 NO 的 Hormesis 效应机制如图 20-10（a）所示，低剂量的呋喃酮类食品添加剂暴露时通过负反馈调节机制降低了胞外 NO 的含量，低浓度的 NO 促进细胞 Cyclin D1 蛋白的表达，进而加快了 HeLa 细胞从 G0/G1 期向 S 期的转换速度，S 期的细胞通过端粒酶的调节使得端粒长度变长，产生了刺激作用。而当呋喃酮类和吡咯酮类食品添加剂的浓度进一步增加时，细胞内产生大量的 ROS 启动线粒体凋亡通路，开启 Caspase 级联反应最终引起细胞凋亡，从而呈现出抑制作用。

呋喃酮类和吡咯酮类食品添加剂对 HSF 细胞基于 ROS 调控的 Hormesis 效应作用机制如图 20-10（b）所示，低浓度化合物暴露引起 HSF 细胞产生少量的 ROS，激活 ERK1/2 蛋白，上调 Cyclin D1 蛋白表达量，加快细胞从 G1 期向 S 期的转换，产生了刺激作用。当呋喃酮和吡咯酮类化合物的暴露浓度进一步增加时，HSF 细胞内的 ROS 含量持续增加，此时大量的 ROS 能够启动线粒体凋亡通路，从而引发细胞凋亡，表现出抑制作用。

从 Hormesis 作用机制看，呋喃酮和吡咯酮类食品添加剂对正常细胞（HSF 细胞）和癌细胞（HeLa 细胞）主要区别在于后者中 NO 的调控作用。低浓度呋喃酮和吡咯酮类化合物暴露时，HeLa 细胞主要在细胞浆与细胞膜上高表达 iNOS，iNOS 在被激活后能够长时间地合成 NO。并且 HeLa 细胞的细胞膜本身通透性更强[55]，成纤维细胞存在着生长因子（FGF）家族，这些生长因子能够刺激内皮细胞迁徙、增殖，从而促进伤口新生血管形成[56, 57]，所以 HeLa 细胞相较于 HSF 细胞受到外界刺激后由于细胞膜维持完整性的能力更差，从而导致 HeLa 细胞产生大量 NO 后更易转移至胞外，使 HeLa 细胞产生 Hormesis 效应。

因此，基于 NO 和呋喃酮的相互作用导致 HeLa 细胞产生 Hormesis 效应的原因，同济大学林志芬课题组已经开展了 NO 和呋喃酮的相互作用，希望结合定量结构-活性相关的理论，探讨化合物结构对其 Hormesis 效应产生的影响，从而为未来设计出能对癌细胞不产生（或难产生）Hormesis 效应的食品添加剂的结构提供一个理论支撑。

5.4 混合化合物联合毒性领域

环境中化合物的复合暴露这一普遍规律使得混合化合物联合毒性的研究迫在眉睫，然而混合化合物联合毒性作用的复杂性使得相关研究裹足不前，其中混合物联合毒性的"交叉现象"便是阻碍研究进展的重大因素之一。

混合物联合毒性的"交叉现象"是指混合物联合毒性作用随混合物浓度变化而发生改变的现象，近年来一直被频繁报道出来[58-62]。例如，Zhang 等[60]利用浓度加和（concentration addition，CA）模型对离子液体和杀虫剂的联合毒性进行探究时发现混合物的联合毒性作用展现了低浓度拮抗、中浓度相加和高浓度协同的交叉现象。González-Pleiter 等[61]发现红霉素和左氧氟沙星对蓝藻 *Anabaena* CPB4337 生长联合毒性的实际剂量效应曲线（CRC）和相应的 CA 曲线发生了交叉，联合毒性作用表现为低浓度区域拮抗作用、中浓度区域相加作用、高浓度区域协同作用。由于这种现象是在使用模型法判别联合毒性作用时由实际的混合物剂量效应曲线和模型剂量效应曲线发生交叉导致的，所以这种现象被称为"交叉现象"[63]。交叉现象中多种形式的联合毒性作用揭示出混合物组分浓度对于判别混合物联合毒性作用的重要性，这一发现对于今后混合物潜在毒性的预测以及相应环保政策的制定都具有深远影响。然而相关机制研究一片空白。

同济大学林志芬课题组在研究磺胺类抗生素和红霉素对大肠杆菌（*Escherichia coli*，*E. coli*）的联合毒性时，发现 CRC 与 CA 模型剂量效应曲线交叉现象的产生与磺胺类抗生素对 *E. coli* 的 Hormesis 效应密切相关，因为提出了基于磺胺 Hormeis 作用的磺胺与红霉素对大肠杆菌交叉现象的机制（图 20-11）[63]：低浓度磺胺对 *E. coli* 生长的刺激作用导致了外排泵的表达，外排泵会泵出 *E. coli* 体内的红霉素，因而混合物在低浓度时联合毒性作用表现为拮抗；高浓度磺胺和红霉素的抑制作用会对 *E. coli* 的生长形成双阻断，因而混合物在高浓度时联合毒性作用表现为协同。可见，混合物中单一组分的 Hormesis 效应可能是混合物联合毒性作用随混合物浓度变化的主要原因。

总之，深入探究混合物中各单一组分的 Hormesis 效应与机制可能是明晰交叉现象的有效途径，可望促进混合物联合毒性研究的发展。

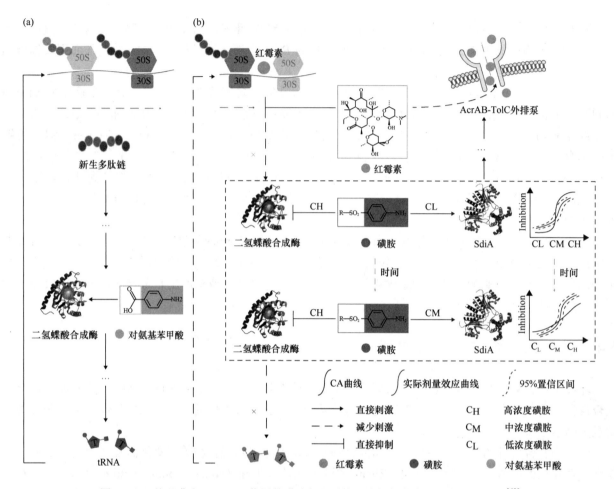

图 20-11　基于磺胺 Hormesis 作用的磺胺与红霉素对大肠杆菌交叉现象的机制图[63]

6　展　　望

基于 Hormsis 现象具有普遍性，环境中污染物具有低浓度暴露的普遍规律，以及人体主动和被动暴露于低浓度化合物的特征等，开展 Hormsis 的研究极其必要。未来有关研究可从下面几方面深入：

（1）深化 Hormesis 机制的研究。采用基因组学、蛋白质组学和代谢组学等，从分子水平、通路水平进一步更深入探讨 Hormesis 现象发生的原因。

（2）探索各类典型化合物的生物效应特征与其 Hormsis 效应的相关关系。比如，抗生素既可导致细菌产生 Hormesis 效应，又可诱导细菌产生耐药性，都可看作细菌应激的表现形式。那么，其细菌耐药性与 Hormesis 效应之间存在着怎么样的内在本质联系？

（3）寻找人类各种疾病与化合物 Hormesis 效应的可能相关关系。化合物污染影响着人类的健康已经是不争的事实，但却一直无法获得它们之间的相关关系的全面实验证据。从低浓度化合物的 Hormesis 效应角度研究，可能是对常规从高剂量角度进行研究的一个有力补充，可望带来全新的研究视界和全新的相关理论。

（4）推动化合物 Hormesis 效应在人类生活和生产实践各个领域的应用研究。比如，如何在 Hormesis 效应模型指导下，进一步真正定量化和规范化中医药科学用药；如何在 Hormesis 效应机制的作用下，挖掘出能"治标也治本"的治理水华新方法。

参 考 文 献

[1] Southam C M. Effects of extract of western red-cedar heartwood on certain wood-decaying fungi in culture. Phytopathology, 1943, 33: 517-524.

[2] Roberts S M. Another view of the scientific foundations of hormesis. Critical Reviews in Toxicology, 2001, 31(4-5): 631-635.

[3] Calabrese E J, Baldwin L A. Toxicology rethinks its central belief. Nature, 2003, 421(6924): 691-692.

[4] Calabrese E J, Baldwin L A. Defining hormesis. Human & Experimental Toxicology, 2016, 21(2): 91-97.

[5] Calabrese E J, Blain R. The occurrence of hormetic dose responses in the toxicological literature, the hormesis database: An overview. Toxicology and Applied Pharmacology, 2005, 202(3): 289-301.

[6] Calabrese E J. Hormesis is central to toxicology, pharmacology and risk assessment. Human & Experimental Toxicology, 2010, 29(4): 249-261.

[7] Calabrese E J, Baldwin L A. Applications of hormesis in toxicology, risk assessment and chemotherapeutics. Trends in Pharmacological Sciences, 2002, 23(7): 331-337.

[8] Boylan M, van den Berg H W, Lynch M. The antiproliferative effect of suramin towards tamoxifen-sensitive and resistant human breast cancer cell lines in relation to expression of receptors for epidermal growth factor and insulin-like growth factor-I: Growth stimulation in the presence of tamoxifen. Annals of Oncology, 1998, 9: 205-211.

[9] Calabrese E J, Blain R B. The hormesis database: the occurrence of hormetic dose responses in the toxicological literature. Regulatory Toxicology and Pharmacology, 2011, 61(1): 73-81.

[10] Stamm I, Lottspeich F, Plaga W. The pyruvate kinase of *Stigmatella aurantiaca* is an indole binding protein and essential for development. Molecular Microbiology, 2005, 56(5): 1386-1395.

[11] Chant E L, Summers D K. Indole signalling contributes to the stable maintenance of *Escherichia coli* multicopy plasmids. Molecular Microbiology, 2007, 63(1): 35-43.

[12] Hirakawa H, Inazumi Y, Masaki T, et al. Indole induces the expression of multidrug exporter genes in *Escherichia coli*. Molecular Microbiology, 2005, 55(4): 1113-1126.

[13] Di Martino P, Fursy R, Bret L, et al. Indole can act as an extracellular signal to regulate biofilm formation of *Escherichia coli* and other indole-producing bacteria. Canadian Journal of Microbiology, 2003, 49(7): 443-449.

[14] Anyanful A, Dolan‐Livengood J M, Lewis T, et al. Paralysis and killing of *Caenorhabditis elegans* by enteropathogenic *Escherichia coli* requires the bacterial tryptophanase gene. Molecular Microbiology, 2005, 57(4): 988-1007.

[15] Davies P J. Plant hormones: Physiology, biochemistry and molecular biology. Berlin: Springer Science & Business Media, 2013.

[16] Hoyer D, Clarke D E, Fozard J R, et al. International Union of Pharmacology classification of receptors for 5-hydroxytryptamine(Serotonin). Pharmacological Reviews, 1994, 46(2): 157-203.

[17] Zhang L S, Davies S S. Microbial metabolism of dietary components to bioactive metabolites: Opportunities for new therapeutic interventions. Genome Medicine, 2016, 8(1): 1-18.

[18] Aggarwal B B, Ichikawa H. Molecular targets and anticancer potential of indole-3-carbinol and its derivatives. Cell Cycle, 2005, 4(9): 1201-1215.

[19] Eisenbrand G, Hippe F, Jakobs S, et al. Molecular mechanisms of indirubin and its derivatives: Novel anticancer molecules with their origin in traditional Chinese phytomedicine. Journal of Cancer Research & Clinical Oncology, 2004, 130(11): 627-35.

[20] Sun H, Zheng M, Song J, et al. Multiple-species hormetic phenomena induced by indole: A case study on the toxicity of indole to bacteria, algae and human cells. Science of the Total Environment, 2019, 657: 46-55.

[21] Berridge M J. The molecular basis of communication within the cell. Scientific American, 1985, 253(4): 142-152.

[22] Lee J H, Lee J. Indole as an intercellular signal in microbial communities. FEMS Microbiology Reviews, 2010, 34(4): 426-444.

[23] Townsend C O. The correlation of growth under the influence of injuries. Annals of Botany, 1897, 11(44): 509-532.

[24] Branham S E. The effects of certain chemical compounds upon the course of gas production by baker's yeast. Journal of Bacteriology, 1929, 18(4): 247-264.

[25] Warren S. The histopathology of radiation lesions. Physiological Reviews, 1944, 24(2): 225-238.

[26] Stebbing A R D. Hormesis—The stimulation of growth by low levels of inhibitors. Science of The Total Environment, 1982, 22(3): 213-234.

[27] Stebbing A R D. A theory for growth hormesis. Mutation Research/Fundamental and Molecular Mechanisms of Mutagenesis, 1998, 403(1): 249-258.

[28] Calabrese E J. Overcompensation stimulation: A mechanism for hormetic effects. Critical Reviews in Toxicology, 2001, 31(4-5): 425-470.

[29] Szabadi E. A model of two functionally antagonistic receptor populations activated by the same agonist. Journal of Theoretical Biology, 1977, 69(1): 101-112.

[30] Chadwick W, Maudsley S. The Devil is in the Dose: Complexity of Receptor Systems and Responses. *In*: Mattson M P, Calabrese E J, (eds). Hormesis. Humana Press, 2010, 95-108.

[31] 范萌知, 张一倩, 刘幸俊. Hormesis 现象及其同工 G-蛋白耦合受体机制. 毒理学杂志, 2014, 2(28): 159-163.

[32] Sun H, Pan Y, Gu Y, et al. Mechanistic explanation of time-dependent cross-phenomenon based on quorum sensing: A case study of the mixture of sulfonamide and quorum sensing inhibitor to bioluminescence of *Aliivibrio fischeri*. Science of The Total Environment, 2018, 630: 11-19.

[33] 高常安, 张爱茜, 蔺远, 等. 酚类化合物非单调剂量-效应毒理学机制的 QSAR 研究. 科学通报, 2009, 54(2): 161-170.

[34] von Zglinicki T, Edwall C, Ostlund E, et al. Very low cadmium concentrations stimulate DNA synthesis and cell growth. Journal of Cell Science, 1992, 103(4): 1073-1081.

[35] Aringa S, Karimine N, Takamuku K, et al. Correlation of eosinophilia with clinical response in patients with advanced carcinoma treated with low-dose recombinant interleukin-2 and mitomycin C. Cancer Immunology, Immunotherapy, 1992, 35(4): 246-250.

[36] Kohanski M A, Dwyer D J, Hayete B, et al. A common mechanism of cellular death induced by bactericidal antibiotics. Cell, 2007, 130(5): 797-810.

[37] Laskin J D, Heck D E, Laskin D L. Nitric oxide pathways in toxic responses. John Wiley & Sons, Ltd, 2009.

[38] Finkel T. Signal transduction by reactive oxygen species. The Journal of Cell Biology, 2011, 194(1): 7-15.

[39] Nunoshiba T, Wishnok J S, Tannenbaum S R, et al. Activation by nitric oxide of an oxidative-stress response that defends *Escherichia coli* against activated macrophages. Proceedings of the National Academy of Sciences, 1993, 90(21): 9993-9997.

[40] Semchyshyn H M. Hormetic concentrations of hydrogen peroxide but not ethanol induce cross-adaptation to different stresses in budding yeast. International journal of microbiology, 2014, 485792.

[41] Luna López A, González-Puertos V Y, López-Diazguerrero N E, et al. New considerations on hormetic response against oxidative stress. Journal of cell communication and signaling, 2014, 8(4): 323-331.

[42] Brugmann W B, Firmani M A. Low concentrations of nitric oxide exert a hormetic effect on Mycobacterium tuberculosis *in vitro*. Journal of clinical microbiology. 2005, 43(9): 4844-4846.

[43] Ludovico P, Burhans W C. Reactive oxygen species, ageing and the hormesis police. FEMS yeast research, 2014, 14(1): 33-39.

[44] Husain M, Bourret T J, Mccollister B D, et al. Nitric oxide evokes an adaptive response to oxidative stress by arresting respiration. Journal of Biological Chemistry, 2008, 283(12): 7682-7689.

[45] Gusarov I, Nudler E. NO-mediated cytoprotection: Instant adaptation to oxidative stress in bacteria. Proceedings of the National Academy of Sciences of the United States of America, 2005, 102(39): 13855-13860.

[46] Miller M B, Bassler B L. Quorum sensing in bacteria. Annual Reviews in Microbiology, 2001, 55(1): 165-199.

[47] Yao Z, Wang D, Wu X, et al. Hormetic mechanism of sulfonamides on *Aliivibrio fischeri* luminescence based on a bacterial cell-cell communication. Chemosphere, 2019, 215: 793-799.

[48] Deng Z, Lin Z, Zou X, et al. Model of hormesis and its toxicity mechanism based on quorum sensing: A case study on the toxicity of sulfonamides to *Photobacterium phosphoreum*. Environmental Science & Technology, 2012, 46(14): 7746-7754.

[49] Sun H, Calabrese E J, Zheng M, et al. A swinging seesaw as a novel model mechanism for time-dependent hormesis under dose-dependent stimulatory and inhibitory effects: A case study on the toxicity of antibacterial chemicals to *Aliivibrio fischeri*. Chemosphere, 2018, 205: 15-23.

[50] Wang W, Sheu S, Chen Y, et al. Evaluating the bone tissue regeneration capability of the chinese herbal decoction danggui buxue tang from a molecular Biology perspective. BioMed Research International, 2014, 853234.

[51] Fu Y, Chen K, Chen Y, et al. Earthworm (*Pheretima aspergillum*) extract stimulates osteoblast activity and inhibits osteoclast differentiation. BMC complementary and alternative medicine, 2014, 14(1): 440.

[52] Lam P Y, Ko K M. Schisandrin B as a hormetic agent for preventing age-related neurodegenerative diseases. Oxidative Medicine and Cellular Longevity, 2012, 250825.

[53] Wang D, Calabrese E J, Lian B, et al. Hormesis as a mechanistic approach to understanding herbal treatments in traditional Chinese medicine. Pharmacology and Therapeutics, 2018, 184: 42-50.

[54] Fu Y, Sheu S, Chen Y, et al. Porous gelatin/tricalcium phosphate/genipin composites containing lumbrokinase for bone repair. Bone, 2015, 78: 15-22.

[55] Ramos D M, But M, Regezi, et al. Expression of integrin β6 enhances invasive behavior in oral squamous cell carcinoma. Matrix Biology, 2002, 21(3): 297-307.

[56] Lawrence W T. Physiology of the acute wound. Clinics in Plastic Surgery, 1998, 25(3): 321-340.

[57] Herndon D N. Growth hormones and factors in surgical patients. Advances in surgery, 1992, 25: 65-97.

[58] Payne J, Scholze M, Kortenkamp A. Mixtures of four organochlorines enhance human breast cancer cell proliferation. Environmental Health Perspectives, 2001, 109(4): 391-397.

[59] Richter M, Escher B I. Mixture toxicity of reactive chemicals by using two bacterial growth assays as indicators of protein and DNA damage. Environmental Science & Technology, 2005, 39(22): 8753-8761.

[60] Silva E, Rajapakse N, Scholze M, et al. Joint effects of heterogeneous estrogenic chemicals in the E-screen: Exploring the applicability of concentration addition. Toxicological Sciences, 2011, 122(2): 383-394.

[61] Zhang J, Liu S S, Liu H L. Effect of ionic liquid on the toxicity of pesticide to *Vibrio qinghaiensis* sp.-Q67. Journal of Hazardous Materials, 2009, 170(2-3): 920-927.

[62] González-Pleiter M, Gonzalo S, Rodea-Palomares I, et al. Toxicity of five antibiotics and their mixtures towards photosynthetic aquatic organisms: Implications for environmental risk assessment. Water Research, 2013, 47(6): 2050-2064.

[63] Sun H, Ge H, Zheng M, et al. Mechanism underlying time-dependent cross-phenomenon between concentration-response curves and concentration addition curves: A case study of sulfonamides-erythromycin mixtures on *Escherichia coli*. Scientific Reports, 2016, 6: 33718.

作者：林志芬[1]，孙昊宇[1]

[1] 同济大学

第21章 环境纳米材料与技术研究进展

▶ 1. 引言 /561

▶ 2. 催化臭氧氧化耦合膜分离技术原理与应用 /561

▶ 3. 多相催化高级氧化水中难降解有机物研究进展 /563

▶ 4. 天然氢氧化镁纳米材料环境应用的基础理论研究进展 /566

▶ 5. 有机-无机协同作用对有机重金属微污染物的强化吸附机制研究进展 /569

▶ 6. 限域结构复合纳米材料的构效调控与水处理应用研究进展 /570

▶ 7. 展望 /572

本章导读

纳米技术在污染控制和环境修复领域有重大应用前景，环境纳米技术及其原理是环境化学的热点研究方向之一，而高性能纳米材料的理性设计和可控制备是环境纳米技术工程化应用的关键。近期，基于新型纳米材料的吸附、膜分离、高级氧化及其耦合集成技术取得了重要研究进展，部分已开展工程化应用示范。本章将介绍新型环境纳米材料与技术研究的最新亮点工作，并指出环境纳米材料和技术研究存在的不足以及未来的重点发展方向。

关键词

纳米反应器，催化臭氧氧化，类芬顿反应，过硫酸盐活化，纳米限域结构

1 引 言

纳米技术在污染控制和环境修复领域有重大应用前景，高性能纳米材料的理性设计和可控制备是环境纳米技术工程化应用的关键。近期，基于新型纳米材料的吸附分离、高级氧化及其耦合集成技术取得了重要研究进展，部分已开展工程化示范。例如，基于纳米反应器的催化臭氧氧化耦合膜分离技术，可以实现饮用水源水以及废水的高效处理；表面络合增强的金属单反应位芬顿催化、双反应中心芬顿催化、氮化碳活化过硫酸盐等多相催化高级氧化技术能高效处理水中的难降解有机污染物；对天然氢氧化镁纳米材料的表面缺陷构筑和物相调控循环，可实现水中重金属离子的选择性提取和回收利用；有机-无机复合纳米材料的协同作用有助于水中有机-重金属复合微污染物的强化去除；限域结构复合纳米材料在吸附、高级氧化、膜分离等技术中都展现出较大应用前景。本章将简要介绍新型环境纳米材料与技术研究的最新亮点工作。

2 催化臭氧氧化耦合膜分离技术原理与应用

臭氧氧化与膜分离是水处理领域广泛应用的技术。膜分离技术依靠机械截留作用分离水中的污染物，但如果不能将其分解，被截留的污染物容易造成膜污染，而通过膜孔的小分子有可能使滤出水具有生物毒性。在膜分离过程中添加臭氧（O_3）的主要目的是利用臭氧氧化或催化氧化分解截留在膜面的污染物、能够通过膜孔的小分子以及浓缩液中的污染物，抑制膜污染和二次污染，提高出水水质[1]。

臭氧添加位置不同，其作用也不同。在膜单元之前设置臭氧氧化单元作为预处理，可将大分子有机物分解成能通过膜孔的小分子，从而减缓膜污染[2]；在膜单元中添加 O_3，并在进水中添加臭氧催化剂[3]或在分离膜表面负载催化剂能提高 O_3 转化成羟基自由基（·OH）的效率，分解膜面截留的污染物，从而减缓膜污染[4]；固定床催化臭氧氧化还可作为后处理单元分解膜单元浓缩液[5, 6]中的污染物。上述形式中，在分离膜表面负载催化剂的催化臭氧氧化耦合分离膜是研究重点，已报道了催化剂负载于微滤[7]或超滤膜[8]、板式[9]、管式[10]或中空纤维膜[11]等耦合形式，其高效去除污染物、减缓膜污染、提高通量等多种功效得到认可。目前主要的关注点包括高效催化剂[12-15]、在分离膜上负载催化剂的方法[16, 17]和对不同目标物的降解效果[18, 19]等方面。

随着纳米技术的发展和我国环保技术的市场需求增强，催化臭氧氧化耦合膜分离技术的研究已经不再局限于抑制膜污染和二次污染，一些工作开始着眼于该技术的微观工作机制和实际应用，本小节概述了这两方面的代表性工作。

2.1 纳米反应器原理

催化臭氧氧化过程依靠 O_3 在催化剂表面转化生成的·OH 降解被膜截留的污染物。虽然·OH 与有机物反应的速度非常快 $[k > 6.0 \times 10^8$ L/(mol×s)$]^{[20]}$，但其寿命很短（3.15×10^{-6} s）[21]，作用范围只在催化剂表面附近的 10 nm 以内[22]。常见的催化臭氧氧化采用固定床形式，将催化剂颗粒装填在塔式反应器中使用，催化剂颗粒间距在亚毫米尺度，比 10 nm 高 4 个数量级，因此·OH 利用率有较大的提高空间。超滤（10 nm < 孔径 < 100 nm）或纳滤（1 nm < 孔径 < 10 nm）膜孔的尺寸与·OH 的有效扩散距离相符，因此能充分利用·OH 提高难降解性污染物的分解效率。

Zhang 等以孔径 50 nm 的 MnO_2/Al_2O_3 超滤膜为研究对象，发现 O_3 在膜孔内的分解速率是在体相溶液中的 428 倍。根据膜面材料等电点、溶液 pH、溶液中荷电离子对 O_3 分解速率的影响规律，他们认为纳米空间限制带来的双电层重叠和水分子簇定向排列是导致膜孔内 O_3 分解成·OH 加速的原因[23]。

Quan 等开发了·OH 的定量表征方法[24]，率先对 O_3 在膜孔中转换成·OH 的效率以及·OH 在膜孔中降解污染物的效率做了较为系统的研究[25]。他们以氧化锌（ZnO）纳米管阵列作为分离膜，将每根纳米管视为反应器，通过·OH 探针实验和动力学模型分析发现纳米反应空间内 O_3 的传质系数是催化 O_3 分解产生·OH 的动力学系数的 720 倍，而相同材质的微孔结构中 O_3 传质是限速步骤。纳米尺度空间还对内部流场具有约束作用，使·OH 在固-液相界面的富集与暴露强度（暴露强度参数 $R_{CT} > 3.2 \times 10^{-6}$）远超过报道的大部分多孔催化剂（介于 $10^{-9} \sim 10^{-6}$ 之间）。这种纳米结构效应随着反应器尺寸的缩小而增强，当纳米通道平均直径下降到 10 nm 时，R_{CT} 增大到 8.0×10^{-5}（图 21-1）。对应的污染物降解速率相对于多孔纳米催化剂提高 350 倍。根据电子顺磁共振（EPR）分析结果，O_3 直接或间接（通过单线态氧）与 ZnO 作用生成超氧自由基阴离子（$O_2^{\cdot-}$），$O_2^{\cdot-}$ 易与 O_3 迅速反应最终产生·OH。

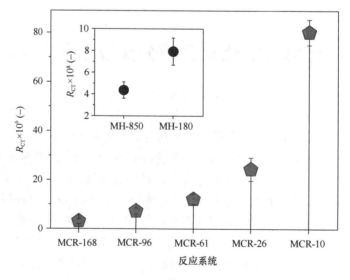

图 21-1　微通道催化反应器（MCR）通道直径（MCR 后面的数字代表孔径，单位 nm）对催化臭氧氧化过程产生的·OH 在固-液相界面的暴露强度（R_{CT}）的影响。插图为直径 180 nm 和 850 nm 的通道中的 R_{CT}（引自参考文献[25]）

2.2 中试设备及其水处理性能

面向应用的中试工艺设备从 2013 年开始出现，最早的报道应用臭氧氧化耦合膜分离技术处理微污染地表水，目标是得到饮用水[26]。该设备采用孔径 60 nm 左右的平板陶瓷超滤膜，在膜单元通入 O_3，使膜单元反洗间隔时间延长了 2 倍，滤出水经过生物活性炭柱，分解残留 O_3 并去除氨氮。该设备的处理能力 120 m^3/d，连续 20 天实验结果表明氨氮去除率 90%（初始浓度 3.6 mg/L）、TOC 去除率 76%（初始浓度 3.8 mg/L）、浊度去除率 99%（出水浊度 0.14 NTU）。

随后，Quan 等研发了两套催化臭氧氧化耦合膜分离工艺设备。第一套设备纯水通量 12 m^3/d[27]，膜单元包括两个膜组件，每个组件内放置 7 根 19 通道管式 Al_2O_3 陶瓷膜，陶瓷膜直径 30 mm，长 1016 mm，孔隙率 40%，孔径 2 μm，有效面积 0.7 m^2。通过浸渍提拉后高温煅烧的方法在陶瓷膜基底的表面上负载了 Ti-Mn 氧化物功能层，负载功能层后平均孔径 80 nm 左右。在膜单元前设置砂滤罐去除较大颗粒物、在膜单元后设置活性炭柱分解残余 O_3 和小分子污染物。中试实验在大连某海产养殖场完成，设备每运行 4 h 反洗一次，持续运行 4 h 约生产 2 m^3 回用水，排出浓缩液 88 L，废水回收率为 95.8%。

在小型设备的基础上增加膜组件数量制成纯水通量 100 m^3/d 的催化臭氧氧化膜分离装置[28]，该装置由三部分构成，聚丙烯纤维（PP 棉）粗滤部分用于过滤污水处理厂沉淀池出水中的悬浮颗粒物，保护膜组件；核心单元是催化臭氧氧化功能膜组件，用于分离并分解污染物；活性炭罐，用于分解过量 O_3 和残余的小分子污染物。三部分设备整合到一个集装箱内，形成撬装设备。该设备在鞍山某印染废水处理厂连续运行 30 天，通量稳定在 110 L/(m×h)，过膜 COD 低于 50 mg/L，过活性炭后 COD 低于 4 mg/L，出水色度为 1，总悬浮颗粒物和大肠杆菌（*Escherichia coli*）在出水中无检出，满足中水回用要求。另外在辽宁某市政污水厂进行 40 天的现场中试，设备进出水 COD 分别为 31 mg/L 和 18 mg/L，进出水悬浮固体（SS）分别为 26 mg/L 和 7 mg/L，进出水大肠杆菌分别为 2.4×10^6 MPN/L 和 50 MPN/L，出水水质能够稳定的满足《水污染物综合排放标准》（DB 11/307-2013），运行动力成本约为 0.13 元/t，表明该设备具有良好的污水处理效果，达到实用化水平。

2.3 小结

催化臭氧氧化耦合膜分离技术具有实用化前景，系统和深入地理解其耦合机制有利于设计出最佳的催化功能膜、组件和工艺设备，不断改善臭氧传质并提高·OH 产生及利用效率、降低投资及运行成本。

3 多相催化高级氧化水中难降解有机物研究进展

工业废水以及城市污水中难降解有机污染物的去除，特别是工业废水中高浓度难降解有机物的去除，是上述水体净化达标排放的关键难题。芬顿（Fenton）反应以及过硫酸盐活化技术在该方面具有很大的潜力，成为国际研究热点。但在实际应用中，经典的 Fe^{2+} 与过氧化氢（H_2O_2）反应体系存在三个主要的问题：①反应要求较强酸性（pH 2~3）；②产生大量铁泥；③H_2O_2 的利用率低。目前发展的多相类芬顿催化剂虽然一定程度上解决了反应酸性问题，但是仍然涉及固定化的金属离子的氧化还原，难以解决反应速率低和 H_2O_2 无效分解的问题[29]。过硫酸盐（PS）可以看作是 H_2O_2 的衍生物，得益于其无二次污染的优点，非金属催化剂在活化过硫酸盐消除水中有机污染物方面的研究已引起环境研究者的关注[30]。

针对本领域的热点与难点，胡春课题组基于界面微观过程，构建材料表面的微电场，形成贫富电子反应微区，深入研究在水-固微界面污染物、H_2O_2、PS在反应微区的作用过程，揭示了污染物高效转化的机制，并基于材料表面微电场构建了双反应中心固相类Fenton催化体系，以及氮化碳表面电子极化的PS活化催化体系，解决了相应技术在水处理中的瓶颈问题。本节对相关研究进展进行简要介绍。

3.1 表面络合促进金属单反应位芬顿催化过程

金属单反应位Fenton催化过程是指Fenton催化剂依靠某一种或多种金属物种通过自身单一位点的反复氧化还原来实现H_2O_2的活化。简而言之，指其氧化还原过程发生在同一金属位点，只是价态反复变化。严格地说，经典Fenton反应以及基于经典Fenton反应原理开发出的多相Fenton催化体系都属于金属单反应位Fenton过程。胡春课题组针对难生物降解的芳环类污染物的催化降解，开发出系列金属单反应位多相Fenton催化剂，包括铜掺杂介孔γ型氧化铝（γ-Cu-Al$_2$O$_3$）[31]、铜掺杂介孔二氧化硅（SiO$_2$）纳米微球（Cu-MSMs）[32]和铜掺杂铝强化的蒲公英状SiO$_2$纳米纤维球（DCASNs）[33]。在此系列研究中，胡春课题组发现了多相Fenton反应络合促进机制（图21-2）：Fenton反应时芳环类污染物和芳环类中间产物的酚羟基易与催化剂表面的Cu物种发生脱质子型络合作用，以铜为反应中心形成σ-Cu^{2+}-配体络合物。H_2O_2除了在其传统意义的还原过程中生成一个·OH外，还可以直接攻击络合于催化剂表面的有机自由基加合物进而生成一个羟基化产物和另一个·OH，同时抑制HO$_2^·$/O$_2^{·-}$和O$_2$的生成，避免了H_2O_2的无效分解，极大提升了H_2O_2的有效利用率。这一过程中，配体上的电子迅速传递给Cu^{2+}，使之快速还原为Cu$^+$，打破了Fenton反应的速率限制步骤，加快整个循环反应的进行，显著提升催化剂的催化活性和效率。利用这样的Fenton催化体系，双酚A（BPA）、2-氯酚以及苯妥英等多种含芳环的物质能够在很短时间内被完全去除，且在原始污染物完全消失之前，H_2O_2的利用率维持在90%左右。

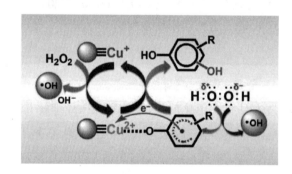

图21-2 表面络合促进金属单反应位Fenton催化过程（引自参考文献[31]）

3.2 双反应中心芬顿催化体系构建与界面反应过程

基于金属单反应位催化机制的Fenton催化体系虽然显著提升了催化剂活性和H_2O_2利用率，然而这一机制仍存在不足之处，即σ-Cu^{2+}-配体络合物的形成强烈依赖于具有酚羟基的芳环类物质。基于此机制的催化剂对于不含芳环的物质的降解，其催化效率则会明显下降。究其本质原因在于其金属单反应位的构建仍然没有实质性脱离经典Fenton反应中金属离子自身的氧化还原[29]。

据此，胡春课题组首次提出并成功构建出电子极性分布的双反应中心Fenton催化体系。目前构建双反应中心催化体系的方法主要有晶格掺杂取代和表面有机络合-聚合[29]，例如利用铜铝钛三金属晶格

掺杂 SiO_2 蒲公英辐射状纳米球而形成的 d-TiCuAl-SiO_2 Ns 双反应中心 Fenton 催化体系[34]。由于掺杂的三种金属原子的电负性差异，造成晶格氧周围的电子密度分布不均匀，形成了以晶格 Cu 为中心的高电子密度区和以晶格 Al、Ti 为中心的低电子密度区。Fenton 反应时，H_2O_2 在高电子密度区（Cu 中心）被还原，而反应过程中由污染物降解产生的有机自由基中间体代替 H_2O_2 被吸附到低电子密度区（Al/Ti 中心）并作为电子供体释放电子而自身被持续氧化降解，由于铜的相对较高的电负性，释放的电子被迅速转移到铜中心。整个过程避免了 H_2O_2 的氧化，使得 H_2O_2 被最大限度地用于降解有机污染物。而 OH-CCN/CuCo-Al_2O_3[35]、CN-Cu(II)-$CuAlO_2$[36]、Cu-MP NCs[37]和 CMS-RGO NSs[38]等新型高效双反应中心 Fenton 催化体系则是利用表面有机络合-聚合的方式构建而成的。有机配体与金属基底之间通过 C-O-M（M 指代过渡金属）键桥呈 σ 型键连，从而诱发了富电子金属中心和大 π 键上缺电子中心的形成。一系列实验证据揭示了其高效产生·OH 的反应机制（图 21-3）[35, 36]：在 H_2O_2 存在的情况下，富电子金属中心大量的电子传递给 H_2O_2，使 H_2O_2 主要发生还原性反应，转化为·OH；同时，H_2O 的电子被有机配体上的缺电子中心剥夺，使其氧化，也转化为·OH。缺电子中心获得的电子经 C-O-M 键桥传递到富电子金属中心，以维持整个系统电子得失平衡。因此，这些双反应中心 Fenton 催化体系体现出很高的·OH 转化率。该催化体系对于多种有机污染物的降解也表现出极佳的催化活性、稳定性以及高的 H_2O_2 利用率。

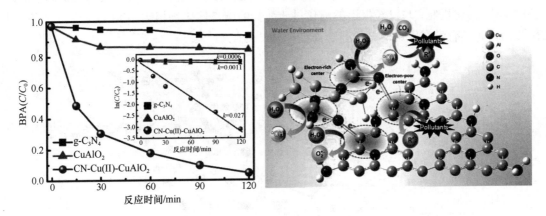

图 21-3　CN-Cu(II)-$CuAlO_2$ 芬顿催化活性评价及其双反应中心反应过程（引自参考文献[36]）

3.3　氮化碳表面电子结构调控与强化过硫酸盐活化降解有机污染物

石墨相氮化碳（g-C_3N_4）是一种稳定的有机聚合非金属材料，近年来在光催化领域中得到广泛的研究[39]。但是受限于 g-C_3N_4 较慢的电子迁移速率，其在活化过硫酸盐降解有机物污染物方面的研究报道并不多见。基于 g-C_3N_4 电子结构的可调控性，胡春课题组将基于表面微电场的双反应中心催化理念引入到过硫酸盐活化领域，并对 g-C_3N_4 进行氧掺杂在其表面构建微电场，增强氧掺杂 g-C_3N_4（O-CN）与过硫酸盐之间的电子转移，强化过硫酸盐活化，实现在无外加辅助条件下有机污染物的高效降解[40]。具体来说，由于氧原子比碳原子具有更高的电负性，氧原子取代 g-C_3N_4 中 sp^2 杂化的氮原子后，氧原子周围会聚集更多的电子，形成高电子密度区，而氧原子邻近的碳原子电荷密度相应降低，形成低电子密度区；理论计算结果证实了 O-CN 中不同电子密度区域的形成。与原始 g-C_3N_4 相比，O-CN 展现出更高活化过一硫酸盐（PMS）的能力，在 45 min 内可以将 0.05 mmol/L 的 BPA 完全降解。同时，O-CN 具有良好的稳定性，重复利用 5 次后仍然可以氧化降解 94%的 BPA（图 21-4）。

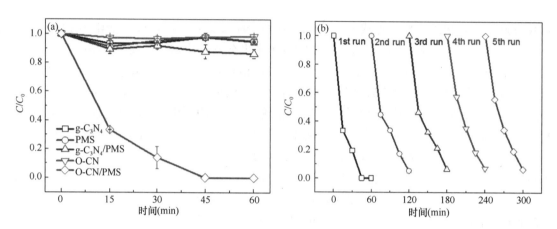

图 21-4　氧掺杂 g-C$_3$N$_4$（O-CN）活化 PMS 降解有机污染物的高活性（a）及其稳定性（b）（引自参考文献[40]）

O-CN 与 PMS 反应产生两种电子转移途径，即 PMS 在富电子的氧原子处发生还原反应，生成·OH 和硫酸根自由基（SO$_4^{·-}$），而在缺电子的碳原子周围发生氧化反应，形成过一硫酸根自由基（SO$_5^{·-}$），SO$_5^{·-}$ 进而与水分子反应产生单线态氧（^{1}O$_2$）。基于此，胡春课题组首次提出了 PMS 同时氧化还原的新的活化机理（图 21-5），在·OH、SO$_4^{·-}$ 和 ^{1}O$_2$ 的共同作用下实现 BPA、2-氯酚等有机污染物的快速降解。得益于这种独特的 PMS 活化机制，与已经报道的活化过硫酸盐的碳基催化剂相比，O-CN 表现出更好的稳定性，使 O-CN 成为一种具有潜在应用前景的非金属过硫酸盐活化剂[40]。

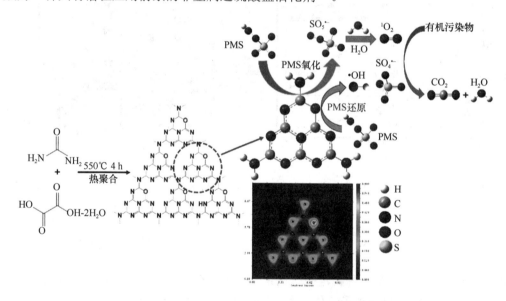

图 21-5　氧掺杂 g-C$_3$N$_4$ 活化 PMS 降解有机污染物机理（引自参考文献[40]）

4　天然氢氧化镁纳米材料环境应用的基础理论研究进展

氢氧化镁［Mg(OH)$_2$］是一种环境友好、成本低廉、广谱性的天然环境纳米材料，主要优势如下：①Mg(OH)$_2$ 是具有 pH 缓冲性的固态弱碱，其在使用中最高 pH 不超过 9，符合美国环保署《清洁水法案》所规定的限值；②Mg(OH)$_2$ 是一种动力学上非常稳定的纳米材料，人工合成的 Mg(OH)$_2$ 通常会天然地稳定在纳米相,因而其活性高、吸附和反应能力均很强;③安全、无毒、无害;④成本低。我国是盐湖资源大国,有储量丰富、成本低廉的镁盐可作为纳米 $Mg(OH)_2$ 生产的原材料。在过去十年中,全球每年 $Mg(OH)_2$ 在污染治理方面的消耗量超过 48 万吨,主要应用于酸性废水处理和烟气脱硫,在重金属脱除、印染废水脱色、养殖废水脱磷、脱铵等方面也有实际应用尝试。$Mg(OH)_2$ 虽然由于其自身的优势,在工业界得到了一定的认可,但其基础研究与实际应用需求的匹配度上存在着比较大的脱节。林璋教授团队围绕着水中低浓度重金属离子选择性提取和高浓度重金属的循环富集回收利用问题,以广谱性的 $Mg(OH)_2$ 作为研究对象深入探讨了表面缺陷的构筑以及材料自身的物相调控循环在重金属离子选择性提取和回收利用上的重要意义。

4.1 表面缺陷 $Mg(OH)_2$ 的调控合成及对低浓度重金属氧阴离子的选择性吸附

纳米 $Mg(OH)_2$ 对许多阴离子型重金属离子和分子(如铬酸盐、磷酸盐、阴离子染料分子)具有较高的吸附亲和力,可以用于富集各种阴离子污染物。为了优化回收策略,进一步提高对重金属氧阴离子在低浓度下的吸附亲和力和选择性十分必要。此外,已有研究表明,MgO 羟基化得到的 $Mg(OH)_2$ 可以与 As(V) 形成稳定的内球络合物,从而具有较高的 As(V) 吸附效率。而 MgO 在羟基化过程中不可避免的会出现 OH 缺陷,表面缺陷的存在是否可以进一步增强纳米 $Mg(OH)_2$ 对重金属氧阴离子的吸附能力?林璋教授团队结合理论计算探索了 $Mg(OH)_2$ 表面的缺陷调控如何实现对砷酸根[As(V)]、铬酸根[Cr(VI)]等阴离子型重金属污染物的选择性吸附[41]。首先,如图 21-6 所示,将重金属离子置于 $Mg(OH)_2$ 表面氢氧根点缺陷上方,通过 DFT 计算优化后发现,As(V)、Cr(VI) 都可牢固地吸附在 $Mg(OH)_2$ 表面缺陷处,并且两者的氧原子替代了原 $Mg(OH)_2$ 表面氢氧根中的氧原子。这说明羟基缺陷的 $Mg(OH)_2$ 极大提升了对 As(V)、Cr(VI) 的吸附能。对吸附在 $Mg(OH)_2$(001)表面-OH 缺陷位的 As(V) 的局域密度态和电子密度分析表明,$Mg(OH)_2$ 的缺陷位点对 As(V) 具有类似化学吸附的强化学结合相互作用,这也导致

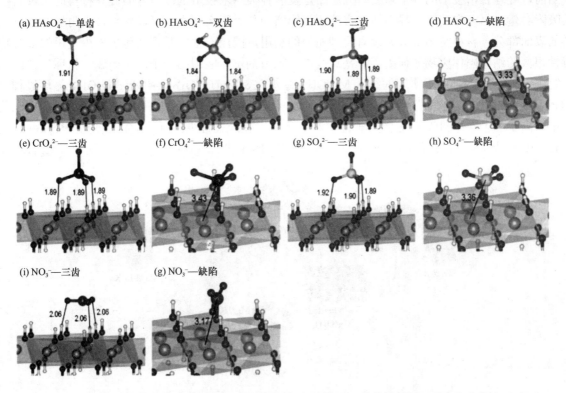

图 21-6 一系列含氧阴离子吸附在 $Mg(OH)_2$(001)表面的 DFT 优化结构图(引自参考文献[41])

了缺陷态的 $Mg(OH)_2$ 对 As(V)具有选择性吸附能力。随后通过强还原和退火处理成功制备了具有表面羟基缺陷的 $Mg(OH)_2$，如图 21-7 所示，缺陷丰富的 $Mg(OH)_2$ 对 As(V)、Cr(VI)的吸附量远高于缺陷贫乏的 $Mg(OH)_2$；比表面积标化后，前者对 As(V)的吸附性能更优异，这应归因于表面缺陷的强亲和力。该成果结合理论与实验系统地解释了 $Mg(OH)_2$ 表面羟基缺陷对于砷酸根和铬酸根吸附性能提升的根本原因，为吸附剂的缺陷调控提供了新的思路。

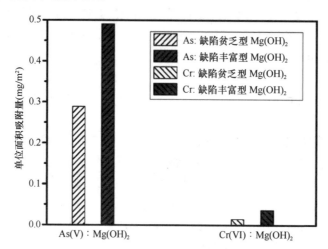

图 21-7　缺陷丰富型和缺陷贫乏型 $Mg(OH)_2$ 对 As(V)和 Cr(VI)的单位表面积吸附量（引自参考文献[41]）

4.2　利用 CO_2 加压调控 $Mg(OH)_2$ 相循环提取回收含铅废水中 Pb^{2+}

当前，处理含铅废水的一个重要研究思路是发展各类纳米吸附剂：一方面提高材料的比表面积以增加其吸附容量，另一方面通过界面活性官能团修饰增加其吸附亲和力。可以看到，吸附机制仅仅利用到材料的表面部分，材料主体部分并没有得到充分的利用，因而产生的是表面包裹着少量铅的混合物。这些毒性很大的纳米固体废物如何处置，是这些纳米吸附剂实际应用的一个瓶颈问题。林璋教授团队设计和研究了一种在 CO_2 辅助下实现的纳米 $Mg(OH)_2$→离子态 $Mg(HCO_3)_2$ →纳米 $Mg(OH)_2$ 循环过程对水中 Pb(II)分离回收的效果和机制[42]。通过纳米花状 $Mg(OH)_2$ 与可溶性 $Mg(HCO_3)_2$ 之间的可逆相变工艺实现废水中铅离子的去除和回收。技术原理如图 21-8 所示，在此过程中，Pb^{2+} 首先通过离子交换和沉淀固定在 $Mg(OH)_2$ 表面。然后经 CO_2 处理，将负载 Pb 的纳米 $Mg(OH)_2$ 转化为可溶性 $Mg(HCO_3)_2$，

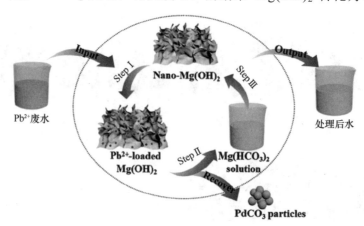

图 21-8　利用 CO_2 加压调控 $Mg(OH)_2$ 相循环提取回收含铅废水中 Pb^{2+} 技术原理图（引自参考文献[42]）

含 Pb 物相转化为不溶性 PbCO$_3$，实现 Pb^{2+} 提纯回收利用。浓缩的 Mg(HCO$_3$)$_2$ 进行热处理，转化为不溶性 Mg$_3$(OH)$_2$(CO$_3$)$_4$·8H$_2$O，再进行煅烧得到 MgO，进一步水化得到 Mg(OH)$_2$，再次用于废水中 Pb^{2+} 的去除和回收。本工作不仅对含铅废水的处理有非常大的意义，而且也为使用 Mg(OH)$_2$ 来处理和回收含其他重金属离子（如 Hg^{2+}、Cu^{2+}、Zn^{2+}、Co^{2+}、Ni^{2+}）的废水提供了很好的范例。

5 有机-无机协同作用对有机重金属微污染物的强化吸附机制研究进展

全球范围内由人类合成和地质原因形成了数以千计的有机微污染物。它们在水中的浓度虽然很低，但某些有机为污染物会在生物体内累积，引起负面生态效应以及基因变异、癌症等严重健康效应，对人类健康造成极大危害，因此其去除问题亟待解决。有机重金属化合物（OMC）是其中一类新型的微污染物，其特点是结构中不仅含有有机基团，同时连接着重金属元素，形成了一种有机-无机复合的结构，如有机砷、有机锡、有机锑等。对于这类污染物，采用降解的方法会产生同样高毒性的重金属离子，如砷酸根、锑酸根等，可能会对水环境造成二次污染。因此，急需在不改变 OMC 结构的基础上发展能将其从水中直接快速去除的深度处理方法。

利用材料与污染物间的界面化学作用有可能通过直接捕获的方法实现水中微污染物的深度去除。但是，由于 OMC 在水中的浓度很低，且受大量共存的杂离子影响，因此对材料的亲和力提出了很高的要求。由于 OMC 结构中除重金属基团外，一般还存在苯环及氢键位点（羟基、氨基、硝基等），可能与吸附材料间产生重金属配位、π-π 及氢键等作用。因此，若能设计材料与 OMC 间同时发生以上两种或多种作用，有望通过这些作用力的协同效应提高材料对 OMC 的亲和力。

为了实现这类污染物的高效去除，林璋教授课题组以有机砷为例进行了系列研究。有机砷是环境中一类典型的 OMC 微污染物，其中，洛克沙胂（roxarsone，ROX）和阿散酸（p-arsanilic acid，p-ASA）由于具有广谱抗菌性，被广泛应用于动物饲料添加剂中。为了提高吸附材料对有机砷的亲和力，林璋教授团队首先以四氧化三铁（Fe$_3$O$_4$）纳米颗粒为活性中心，以石墨烯为基体并提供有机官能团，设计了一种三维复合纳米材料 Fe$_3$O$_4$@RGO。其对 ROX 的吸附表现出明显的协同作用，吸附容量及亲和力与 Fe$_3$O$_4$ 和石墨烯的混合物相比，分别提高了 4 倍和 11 倍，并且可以在 90 min 内将废水中的 As 浓度降至 50 ppb 以下[43]。机理研究表明，Fe$_3$O$_4$@RGO 在吸附 ROX 过程中存在 As-Fe 内球配位、π-π 堆积和氢键三种作用力。其对 ROX 的协同增强吸附主要是由于负载在石墨烯表面使 Fe$_3$O$_4$ 暴露出更多的（400）晶面，也使其表面羟基含量增加，因此缩短了 As-Fe 原子间距，同时增强了 As-Fe 配位和氢键作用。此外，石墨烯基体自身还可以通过 π-π 作用直接吸附 ROX。复合后，材料与 ROX 间的弱静电引力使 π-π 作用得到增强（图 21-9）。

在此基础上，林璋教授团队根据阿散酸的结构特点，设计了一种表面带缺陷的锆（Zr）基金属有机框架（MOF）材料（UiO-67），并对其进行氨基改性，希望利用其 Zr 基核心对砷的强配位作用，协同含氨基的配体与阿散酸间的 π-π 和氢键作用，增强材料对阿散酸的吸附能力[44]。从吸附结果来看，氨基改性后，UiO-67 对阿散酸的亲和力有明显增强，是原始 UiO-67 的 3 倍左右。其中，As-Zr 配位是 UiO-67 吸附阿散酸的主要作用力。氨基改性后，As-Zr 配位及 π-π 堆积的结合能均有所增强，并且在吸附过程中形成了新的氢键作用力（图 21-10）。这种协同作用提高了 UiO-67 对阿散酸的亲和力，可实现废水中有机砷的高效提取。同时用量仅为活性炭的 1/40，因此在实际环境污染治理中有极大的应用潜力。

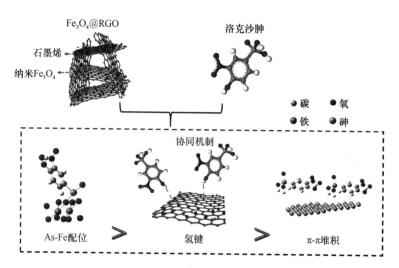

图 21-9　Fe$_3$O$_4$@RGO 纳米复合材料对 ROX 的协同增强吸附机制（引自参考文献[43]）

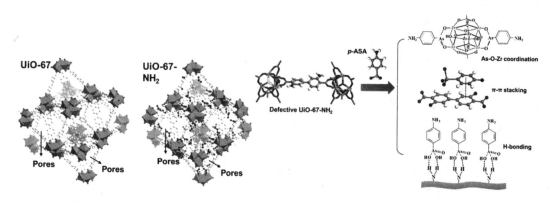

图 21-10　UiO-67-NH$_2$ 结构及其对 p-ASA 的协同吸附机制（引自参考文献[44]）

鉴于重金属配位是吸附有机砷过程中最强的作用力之一，且可通过氢键作用得到协同增强，受天然蛋白质结构中存在的氢键网络可增强其对重金属特异性识别的现象启发，林璋教授团队进一步以纳米纤维素为基质，结合 ZrO(OH)$_2$ 纳米颗粒，设计了一种仿生复合材料 ZNC。其中由纳米纤维素引入的氢键网络，可通过调整 ZrO(OH)$_2$ 配位核心的局部结构、改变 As-Zr 重金属配位构型，并强化 As-Zr 内球配位络合物的稳定性的多重协同机制，来增强有机砷与 Zr 之间的重金属配位，从而进一步强化材料对有机砷的亲和力。与现有其他吸附材料相比，ZNC 对低浓度有机砷的亲和力提高了一个数量级以上。其对实际废水中有机砷的吸附选择性明显增强，对 p-ASA 和 ROX 的吸附量与纯水中相比仅下降了 2.8% 及 2.6%，可实现实际废水中低浓度有机砷的深度去除。同时，基于此机制，发现仿生材料对低浓度有机锡及有机磷的亲和力也可提高 3 倍以上。因此，基于 OMC 结构特点提出的有机-无机协同增强机制，有望实现对水中低浓度有机态重金属的深度去除。

6　限域结构复合纳米材料的构效调控与水处理应用研究进展

通过将纳米材料固定化制备具有限域结构的大颗粒复合材料是克服其易团聚失活、难操作、潜在环境风险等规模化水处理应用瓶颈最为有效的策略之一，研究限域条件下纳米材料对污染物的去除转化机制对于推动纳米水处理化学与技术的发展具有重要意义。

6.1 树脂基纳米复合材料

树脂基纳米复合材料作为极少数已实现商品化生产的水处理纳米材料,有机结合了纳米颗粒高除污活性与树脂载体大颗粒、高机械强度、低流体阻力等特性,且可高效再生与循环利用,目前已应用于各类规模化水处理工程中。最新研究表明,纳米孔树脂载体的网孔限域效应还会赋予复合纳米材料一些特殊的性质,提高其去除复杂污废水体系中磷、砷、氟、铬等污染物的性能。

铁系氧化物因对磷酸盐吸附容量高、选择性好、环境友好、廉价易得等特点,被广泛应用于水体中磷酸盐的深度去除。在实际含磷污水中往往含有大量的 Ca^{2+},在一定 pH 和浓度条件下,Ca^{2+} 易与磷酸盐形成磷酸钙并在铁氧化物表面发生沉积,严重影响铁氧化物对磷酸盐的去除与循环使用。近期,潘丙才课题组将水合氧化铁负载于具有纳米孔结构的毫米尺寸交联聚苯乙烯球内,制备出可规模化生产与应用的复合纳米材料,并探究了 Ca^{2+} 对此类铁系复合材料除磷性能的影响[45]。研究首次发现,在这样特殊的纳米限域体系内,Ca^{2+} 对纳米水合氧化铁除磷不仅没有传统的抑制作用,反而具有显著的增强作用。进一步研究表明,这主要是由于在纳米限域体系内,氧化铁吸附除磷、磷酸钙及羟基磷灰石形成这两个过程实现了分区独立进行,从而减缓甚至消除了 Ca^{2+} 在氧化铁除磷过程中的抑制作用;另一方面,Ca^{2+} 可通过形成多元复合物强化磷的去除。这一工作对于研究纳米限域体系中污染物净化过程中的特殊物理化学效应具有重要的启示意义。

2016 年,美国湿地基金会、美国国家鱼类和野生动物基金会、加拿大安大略省环境部等机构联合发起了全球最大的水处理挑战赛——乔治·巴利水奖(George Barley Water Prize),总奖金额 1000 万美金,确定以天然水中磷的深度处理为目标,开展从实验方案、小试、中试到工程化验证的全球技术挑战赛。潘丙才课题组主持开发的基于纳米复合材料的地表水深度除磷技术从 208 个国际方案中脱颖而出,入围全球前 10 强,并在中试阶段比赛(2018 年 2~6 月,多伦多)中获得亚军,入围 2019 年在美国佛罗里达州进行的工程示范总决赛(5000~10000 t/d)。这是获奖队伍中唯一采用纳米技术的团队,吸引了国际知名媒体采访报道,提升了我国环境纳米技术的国际影响力。

树脂基铁氧化物复合纳米材料对水中砷的深度去除也有重要应用前景,但限域结构中铁氧化物纳米颗粒的尺寸调控与性能提升仍是一个难点。潘丙才课题组用闪冻(flash freezing)技术制备了外径为毫米尺寸的介孔聚苯乙烯小球,并在其网孔内原位负载粒径小于 10 nm 且均一可调的 α-羟基氧化铁(α-FeOOH)纳米颗粒,发现该复合纳米材料的除砷能力远高于尺寸 18×60 nm 的 α-FeOOH 纳米棒(数量级提升)[46],这主要是由于聚苯乙烯网孔中的亚 10 nm 的 α-FeOOH 纳米颗粒有更多表面羟基,对 As(V)的亲和力更强。

此外,潘丙才课题组将水合氧化锆(HZO)纳米颗粒通过叔胺基($R-NR_2$)连接负载到超高交联聚苯乙烯阴离子交换树脂(HCA)的网孔中,制备了 HZO@HCA 复合纳米材料,并对比了该材料与商品化锆基复合纳米吸附剂 HZO@D201 除氟性能[47]。相较于 HZO@D201 的大孔-介孔结构,HZO@HCA 具有大量的微孔结构,并使大分子 NOM 分子难以接近纳米 HZO 的活性位点,水中高达 20 mg-C/L 的腐殖酸对 HZO@HCA 的吸附除氟性能基本没有影响;此外,$R-NR_2$ 基团有利于强化吸附在复合材料外表面的 NOM 的脱附过程,从而消除了 NOM 对传统材料除氟的抑制效应。

近期,高冠道课题组利用大孔聚苯乙烯小球对苯胺优良的吸附能力,采用一步快速聚合的方法,原位制备了结构完好并便于实际应用的聚苯胺@聚苯乙烯(PANI@PS)复合纳米材料。研究表明,球状的 PANI@PS 复合材料能在近中性(pH 6.0)条件下高效去除 Cr(VI),相比于粉末态聚苯胺,其对 Cr(VI)的去除效率提高了 5.4 倍,且高于文献报道的其他多种吸附材料[48]。PANI@PS 复合材料高效去除 Cr(VI)的性能主要是由于其丰富的作用位点、多孔结构以及由此产生的限域效应。更有意义的是,受纳米孔道限域效应的影响,PANI@PS 的表面电位在中性水溶液条件下变为负值,可以有效捕获带正电的 Cr(III),

从而同步实现了 Cr(VI)的吸附/还原和 Cr(III)的原位捕获。

6.2 纳米限域结构强化类 Fenton 催化反应及膜分离过程

最近,潘丙才课题组以经典的类 Fenton 催化体系为研究对象,选用多壁碳纳米管(MWCNT,内径约 7 nm)为 Fe_2O_3(粒径约 2 nm)的模板载体,分别制备出管内与管外负载 Fe_2O_3 的复合催化剂 Fe_2O_3@CNT 与 Fe_2O_3/CNT,比较研究了两种催化剂介导的类 Fenton 反应降解有机污染物的性能与机制[49]。研究发现,Fe_2O_3/CNT-H_2O_2 类芬顿体系降解有机污染物的活性物种为经典的·OH,而 Fe_2O_3@CNT-H_2O_2 产生的活性物种仅为 $^1O_2$,且相同条件下后者对亚甲基蓝的降解速率是前者的 22.5 倍。不仅如此,在 pH 值 5~9 的范围内,Fe_2O_3@CNT-H_2O_2 均可保持高效稳定的污染物降解性能,这有望拓展类 Fenton 催化剂在碱性条件下的应用;另外,限域催化体系可选择性氧化易被吸附的有机污染物,氧化速率与吸附亲和力成线性关系。作者推测,碳纳米管(CNTs)的限域结构从动力学与热力学两方面影响了类 Fenton 催化的反应路径。这项工作有望为限域条件下水处理纳米新技术的发展提供重要参考。

近年来,非均相催化臭氧氧化技术快速发展,这一技术旨在通过强化臭氧转化为·OH 自由基提高 O_3 降解有机污染物的能力。现有研究报道的臭氧催化剂多为微米或纳米尺度,易流失、难分离,在规模化水处理应用中受到限制。潘丙才课题组通过浸渍法制备了一种面向气泡柱臭氧氧化的 CeTiZr 复合氧化物介孔毫米球催化剂 CTZO[50],直径在 0.8~1.0 mm,孔径分布主要集中在 4~7 nm,比表面积为 180 m^2/g,Ce 元素均匀分布于载体中形成固溶体。CTZO 表现出优于氧化铈(CeO_2)和 TZO 载体的催化臭氧分解的性能。选取草酸作为难降解小分子羧酸的代表,在气泡柱反应器中,CZTO 与 TZO 载体对草酸的吸附性能接近,但前者显著促进了草酸的臭氧催化氧化过程,草酸在臭氧氧化后被彻底矿化为 CO_2。CTZO 在 8 次循环实验中未观察到明显的催化活性降低,同时无显著的 Ce 溶出,证明 CTZO 具有较高的催化活性和良好的稳定性。此外,CTZO 催化臭氧氧化降解水中对氯间苯二酚、布洛芬、双氯芬酸、诺氟沙星、磺胺甲恶唑等污染物均不同程度(20%~60%)提高了污染物矿化率,展现出良好的应用前景。

新型膜基材料是水污染深度处理技术的重要发展方向。张炜铭课题组研制了一种基于 ZIF-8 的 MOF 杂化膜[51],使用聚偏氟乙烯(PVDF)超滤膜作为基底,通过原位界面聚合等过程,使负载的 ZIF-8 功能层可牢牢稳固在膜表面,有助于改善传统共混法制膜导致的膜水通量小、缺陷多、机械强度低、有效位点被覆盖的缺点,为新型膜材料制备方法改良提供了新思路。相比于传统的吸附材料,ZIF-8 具有高水稳定性,其网状结构的咪唑骨架及内含不饱和 Zn 金属中心在水中生成的大量羟基官能团可实现对高浓度 NaCl 溶液中微量 Ni^{2+} 的高选择性富集,达到对含镍废水的深度处理标准。该课题组近期通过颗粒预分散、蒸发溶剂、热致相转化等一系列手段,制备了功能颗粒[MIL-53(Fe)颗粒]负载量超过 60%的超高 MOF 负载量的 MIL-PVDF 多功能超滤杂化膜,用于水中不同污染物的同步吸附、催化及过滤高效去除[52]。基于其超高的颗粒负载量及其同步吸附催化性能,该杂化膜可实现对水中亚甲基蓝的高效去除。同时,由于制备过程中的孔结构控制,该杂化膜同时具有较好的超滤性能,可以实现对牛血清蛋白的高效截留,同时保持了较高的水通量。

7 展 望

用于污染控制和环境修复的纳米技术及其原理是环境化学的热点研究方向之一,大量的纳米材料,如富勒烯、碳纳米管、纳米金属氧化物、石墨烯、金属有机框架等材料已广泛应用于水、大气、土壤等污染控制与治理方面的研究,我国也已成为国际上最为重要的环境纳米技术与理论研究高地之一。总体而言，目前环境纳米技术研究仍主要停留在实验室研究，如何在保持相关纳米技术与理论创新的同时，推动这类技术从实验室走向工程化应用，应是今后需要重点关注的课题；一些关键瓶颈如工程化应用纳米材料的理性设计与制备、纳米处理单元与传统污染治理工艺的耦合、复杂环境下材料工作性能的影响与机制、环境纳米技术的长期应用效能与安全效应等，应该得到更多的关注。

参 考 文 献

[1] Chen C J, Fang P Y, Chen K C. Permeate flux recovery of ceramic membrane using TiO_2 with catalytic ozonation. Ceramics International, 2017, 43: S758-S764.

[2] Lehman S G, Liu L. Application of ceramic membranes with pre-ozonation for treatment of secondary wastewater effluent. Water Research, 2009, 43: 2020-2028.

[3] Zhang Y, Zhao P, Li J, et al. A hybrid process combining homogeneous catalytic ozonation and membrane distillation for wastewater treatment. Chemosphere, 2016, 160: 134-140.

[4] Wang Y H, Chen K C, Chen C R. Combined catalytic ozonation and membrane system for trihalomethane control. Catalysis Today, 2013, 216: 261-267.

[5] Wang H, Wang Y, Lou Z, et al. The degradation processes of refractory substances in nanofiltration concentrated leachate using micro-ozonation. Waste Management, 2017, 69: 274-280.

[6] Liu P, Zhang H, Feng Y, et al. Removal of trace antibiotics from wastewater: A systematic study of nanofiltration combined with ozone-based advanced oxidation processes. Chemical Engineering Journal, 2014, 240: 211-220.

[7] Wang Z, Chen Z, Chang J, et al. Fabrication of a low-cost cementitious catalytic membrane for p-chloronitrobenzene degradation using a hybrid ozonation-membrane filtration system. Chemical Engineering Journal, 2015, 262: 904-912.

[8] Zhu Y, Zhang H, Zhang X. Study on catalytic ozone oxidation with nano-tiO_2 modified membrane for treatment of municipal wastewater. Asian Journal of Chemistry, 2014, 26: 3871-3874.

[9] He Z, Xia D, Huang Y, et al. 3D MnO_2 hollow microspheres ozone-catalysis coupled with flat-plate membrane filtration for continuous removal of organic pollutants: Efficient heterogeneous catalytic system and membrane fouling control. Journal of Hazardous Materials, 2018, 344: 1198-1208.

[10] Zhu B, Hu Y, Kennedy S, et al. Dual function filtration and catalytic breakdown of organic pollutants in wastewater using ozonation with titania and alumina membranes. Journal of Membrane Science, 2011, 378: 61-72.

[11] Yu W, Brown M, Graham N J D. Prevention of PVDF ultrafiltration membrane fouling by coating MnO_2 nanoparticles with ozonation. Scientific Reports, 2016, 6: 30144.

[12] Guo Y, Xu B, Qi F. A novel ceramic membrane coated with MnO_2–Co_3O_4 nanoparticles catalytic ozonation for benzophenone-3 degradation in aqueous solution: Fabrication, characterization and performance. Chemical Engineering Journal, 2016, 287: 381-389.

[13] Zhu Y, Quan X, Chen F, et al. CeO_2-TiO_2 coated ceramic membrane with catalytic ozonation capability for treatment of tetracycline in drinking water. Science of Advanced Materials, 2012, 4: 1191-1199.

[14] Guo Y, Song Z, Xu B, et al. A novel catalytic ceramic membrane fabricated with $CuMn_2O_4$ particles for emerging UV absorbers degradation from aqueous and membrane fouling elimination. Journal of Hazardous Materials, 2018, 344: 1229-1239.

[15] Zhu Y, Chen S, Quan X, et al. Hierarchical porous ceramic membrane with energetic ozonation capability for enhancing water treatment. Journal of Membrane Science, 2013, 431: 197-204.

[16] Byun S, Davies S H, Alpatova A L, et al. Mn oxide coated catalytic membranes for a hybrid ozonation–membrane filtration: Comparison of Ti, Fe and Mn oxide coated membranes for water quality. Water Research, 2011, 45: 163-170.

[17] Cheng X, Liang H, Qu F, et al. Fabrication of Mn oxide incorporated ceramic membranes for membrane fouling control and

enhanced catalytic ozonation of *p*-chloronitrobenzene. Chemical Engineering Journal, 2017, 308: 1010-1020.

[18] Alpatova A L, Davies S H, Masten S J. Hybrid ozonation-ceramic membrane filtration of surface waters: The effect of water characteristics on permeate flux and the removal of DBP precursors, dicloxacillin and ceftazidime. Separation and Purification Technology, 2013, 107: 179-186.

[19] Park H, Choi H. As(III)removal by hybrid reactive membrane process combined with ozonation. Water Research, 2011, 45: 1933-1940.

[20] Buxton G V, Greenstock C L, Helman W P, et al. Critical Review of rate constants for reactions of hydrated electrons, hydrogen atoms and hydroxyl radicals(·OH/·O⁻)in Aqueous Solution. Journal of Physical and Chemical Reference Data, 1988, 17: 513.

[21] Attri P, Kim Y H, Park D H, et al. Generation mechanism of hydroxyl radical species and its lifetime prediction during the plasma-initiated ultraviolet (UV) photolysis. Scientific Reports, 2015, 5: 9332.

[22] Carretero-Genevrier A, Boissiere C, Nicole L, et al. Distance dependence of the photocatalytic efficiency of TiO_2 revealed by in situ ellipsometry. Journal of the American Chemical Society, 2012, 134: 10761-10764.

[23] Fan X, Zhang X. Characteristics of ozone decomposition inside ceramic membrane pores as nano-reactors. Water Science and Technology: Water Supply, 2014, 14: 421-428.

[24] Zhang S, Quan X, Wang D. Fluorescence microscopy image-analysis(FMI)for the characterization of interphase HO· production originated by heterogeneous catalysis. Chemical Communications, 2017, 53: 2575-2577.

[25] Zhang S, Quan X, Wang D. Catalytic ozonation in arrayed zinc oxide nanotubes as highly efficient minicolumn catalyst reactors (MCRs): Augmentation of hydroxyl radical exposure. Environmental Science & Technology, 2018, 52: 8701-8711.

[26] Guo J, Wang L, Zhu J, et al. Highly integrated hybrid process with ceramic ultrafiltration-membrane for advanced treatment of drinking water: A pilot study. Journal of Environmental Science and Health, Part A, 2013, 48: 1413-1419.

[27] Chen S, Yu J, Wang H, et al. A pilot-scale coupling catalytic ozonation–membrane filtration system for recirculating aquaculture wastewater treatment. Desalination, 2015, 363: 37-43.

[28] Zhang J, Yu H, Quan X, et al. Ceramic membrane separation coupled with catalytic ozonation for tertiary treatment of dyestuff wastewater in a pilot-scale study. Chemical Engineering Journal, 2016, 301: 19-26.

[29] Lyu L, Hu C. Heterogeneous fenton catalytic water treatment technology and mechanism. Progress in Chemistry, 2017, 29: 981-999.

[30] Duan X G, Sun H Q, Wang S B. Metal-Free carbocatalysis in advanced oxidation reactions. Accounts of Chemical Research, 2018, 51: 678-687.

[31] Lyu L, Zhang L L, Wang Q Y, et al. Enhanced fenton catalytic efficiency of gamma-Cu-Al_2O_3 by sigma-Cu^{2+}-ligand complexes from aromatic pollutant degradation. Environmental Science & Technology, 2015, 49: 8639-8647.

[32] Lyu L, Zhang L L, Hu C. Enhanced Fenton-like degradation of pharmaceuticals over framework copper species in copper-doped mesoporous silica microspheres. Chemical Engineering Journal, 2015, 274: 298-306.

[33] Lyu L, Zhang L L, Hu C, et al. Enhanced Fenton-catalytic efficiency by highly accessible active sites on dandelion-like copper-aluminum-silica nanospheres for water purification. Journal of Materials Chemistry A, 2016, 4: 8610-8619.

[34] Lyu L, Zhang L L, Hu C. Galvanic-like cells produced by negative charge nonuniformity of lattice oxygen on d-TiCuAl-SiO_2 nanospheres for enhancement of Fenton-catalytic efficiency. Environmental Science-Nano, 2016, 3: 1483-1492.

[35] Lyu L, Zhang L L, He G Z, et al. Selective H_2O_2 conversion to hydroxyl radicals in the electron-rich area of hydroxylated C-g-C_3N_4/ CuCo-Al_2O_3. Journal of Materials Chemistry A, 2017, 5: 7153-7164.

[36] Lyu L, Yan D B, Yu G F, et al. Efficient destruction of pollutants in water by a dual-reaction center fenton-like process over carbon nitride compounds-complexed Cu(II)-$CuAlO_2$. Environmental Science & Technology, 2018, 52: 4294-4304.

[37] Lyu L, Han M E, Cao W R, et al. Efficient fenton-like process for organic pollutant degradation on Cu-doped mesoporous polyimide nanocomposites. Environmental Science-Nano, 2019, 6: 798-808.

[38] Han M, Lyu L, Huang Y, et al. In situ generation and efficient activation of H_2O_2 for pollutant degradation over $CoMoS_2$ nanosphere-embedded rGO nanosheets and its interfacial reaction mechanism. Journal of Colloid and Interface Science, 2019, 543: 214-224.

[39] Ding F, Yang D, Tong Z W, et al. Graphitic carbon nitride-based nanocomposites as visible-light driven photocatalysts for environmental purification. Environmental Science-Nano, 2017, 4: 1455-1469.

[40] Gao Y W, Zhu Y, Lyu L, et al. Electronic structure modulation of graphitic carbon nitride by oxygen doping for enhanced catalytic degradation of organic pollutants through peroxymonosulfate activation. Environmental Science & Technology, 2018, 52: 14371-14380.

[41] Ou X, Liu X, Liu W, et al. Surface defects enhance the adsorption affinity and selectivity of $Mg(OH)_2$ towards As(V) and Cr(VI) oxyanions: A combined theoretical and experimental study. Environmental Science: Nano, 2018, 5: 2570-2578.

[42] Liu X, Song K, Liu W, et al. Removal and recovery of Pb from wastewater through a reversible phase transformation process between nano-flower-like $Mg(OH)_2$ and soluble $Mg(HCO_3)_2$. Environmental Science: Nano, 2019, 6: 467-477.

[43] Tian C, Zhao J, Zhang J, et al. Enhanced removal of roxarsone by Fe_3O_4@3D graphene nanocomposites: Synergistic adsorption and mechanism. Environmental Science: Nano, 2017, 4: 2134-2143.

[44] Tian C, Zhao J, Ou X, et al. Enhanced adsorption of *p*-arsanilic acid from water by amine-modified UiO-67 as examined using extended X-ray absorption fine structure, X-ray photoelectron spectroscopy, and density functional theory calculations. Environmental Science & Technology, 2018, 52: 3466-3475.

[45] Zhang Y Y, She X W, Gao X, et al. Unexpected favorable role of Ca^{2+} in phosphate removal by using nanosized ferric oxides confined in porous polystyrene beads. Environmental Science & Technology, 2019, 53: 365-372.

[46] Zhang X L, Cheng C, Qian J S, et al. Highly efficient water decontamination by using sub-10 nm FeOOH confined within millimeter-sized mesoporous polystyrene beads. Environmental Science & Technology, 2017, 51: 9210-9218.

[47] Zhang X L, Zhang L, Li Z X, et al. Rational design of antifouling polymeric nanocomposite for sustainable fluoride removal from NOM-rich water. Environmental Science & Technology, 2017, 51: 13363-13371.

[48] Ding J, Pu L T, Wang Y F, et al. Adsorption and reduction of Cr(VI) together with Cr(III) sequestration by polyaniline confined in pores of polystyrene beads. Environmental Science & Technology, 2018, 52: 12602-12611.

[49] Yang Z C, Qian J S, Yu A Q, et al. Singlet oxygen mediated iron-based Fenton-like catalysis under nanoconfinement. Proceedings of National Academy of Sciences of the United States of America, 2019, 116: 6659-6664.

[50] Shan C, Xu Y, Hua M, et al. Mesoporous Ce-Ti-Zr ternary oxide millispheres for efficient catalytic ozonation in bubble column. Chemical Engineering Journal, 2018, 338: 261-270.

[51] Li T, Zhang W M, Zhai S, et al. Efficient removal of nickel(II) from high salinity wastewater by a novel PAA/ZIF-8/PVDF hybrid ultrafiltration membrane. Water Research, 2018, 143: 87-98.

[52] Ren Y, Li T, Zhang W M, et al. MIL-PVDF blend ultrafiltration membranes with ultrahigh MOF loading for simultaneous adsorption and catalytic oxidation of methylene blue. Journal of Hazardous Materials, 2019, 365: 312-321.

作者：全燮[1]，潘丙才[2]，陈威[3]，胡春[4]，林璋[5]，闫兵[4,6]

协稿：于洪涛[1]，单超[2]，吕来[4]，高耀文[4]

统稿：姜传佳[3]

[1] 大连理工大学，[2] 南京大学，[3] 南开大学，[4] 广州大学，[5] 华南理工大学，[6] 山东大学

第22章 纳米材料的环境转化与归趋研究进展

▶ 1. 引言 /577

▶ 2. 金属基纳米材料的环境转化与归趋 /578

▶ 3. 碳纳米材料的环境转化与归趋 /581

▶ 4. 复合纳米材料及多元金属基纳米材料的环境转化与归趋 /584

▶ 5. 复杂环境中纳米材料转化与归趋研究的新方法 /585

▶ 6. 展望 /587

本章导读

纳米材料释放到环境中后，其存在形态会受到环境因素的作用而改变，进而影响其环境归趋和生物效应。本章总结了纳米材料环境转化与归趋的最新研究进展，包括：天然有机质等环境因素对金属基纳米材料的胶体稳定性、多孔介质迁移特性、溶解和化学转化（如硫化）过程的影响；金属基纳米材料在土壤环境（尤其是植物根区）等实际环境或接近真实环境中的溶解与转化过程；阳离子种类、天然有机质等水化学条件对碳纳米材料的胶体稳定性和多孔介质迁移特性的影响；碳纳米材料的水环境转化过程（如化学还原、光化学转化）对其理化性质和胶体稳定性的影响；以及复合纳米材料和多元金属基纳米材料的环境行为。此外，本章还总结了近年来研究复杂环境中纳米材料转化与归趋的新实验技术和模型方法，并指出纳米材料环境转化与归趋研究领域未来的重点方向。

关键词

人工纳米材料，环境转化，胶体稳定性，溶解，天然有机质

1 引　言

纳米材料具有诸多优良特性，使其在很多领域都有广泛的应用，如工业催化、电子电气、生物医学、航天航空、化妆品、环境和能源等领域。伴随着纳米产业的快速发展和人工纳米材料的大量使用，纳米材料的环境释放、环境累积、环境行为和生物效应引起人们的关注[1-4]。纳米尺度赋予材料很多无与伦比的奇特性能，然而这些奇特的理化、光电性质也恰恰使得纳米材料具有很多未知的、难以预测的健康和环境风险。对人工纳米材料的环境和健康效应的研究几乎是伴随着纳米技术而诞生和发展。2008年，美国先后成立了两个研究纳米材料环境影响的联合研究中心，分别是加州大学洛杉矶分校领衔成立的加州大学纳米技术环境影响中心（UC CEIN）和杜克大学领衔成立的纳米技术环境影响中心（CEINT），都旨在探明纳米材料的环境行为和生物效应，为纳米技术的安全及可持续应用提供科学依据。经济合作与发展组织（OECD）也于2007年成立了人造纳米材料工作组，指导开展了人造纳米材料测试项目，以考察用于研究化学品毒理的OECD试验指南方法是否适用于纳米毒理学研究[5-7]。与此同时，中国许多学者也较早地关注了该领域，并且开展了前瞻性的研究工作。

纳米零价金属、金属氧化物等金属基纳米材料以及碳纳米材料进入环境后，环境条件（如光照、温度、pH、离子强度、天然胶体等）会显著地改变原始态纳米材料的存在形态[8-13]。同时，水环境中的有机大分子（如腐殖酸、蛋白质、脂质等）也会与纳米材料发生作用，并影响纳米材料的团聚、多孔介质迁移、溶解等环境过程[14-16]。经过环境转化的纳米材料可以进入水生生物（如细菌、藻类、甲壳类、鱼类、哺乳动物等）和植物体内，继而发生体内累积并导致毒性效应[17]。最终，水生生物以及植物摄入的纳米材料可能会通过食物链传递进入人体，带来未知的健康风险[18]。关于纳米材料的环境转化与归趋已有大量研究，本章将对该研究领域的最新进展进行简要归纳，以期为后续的相关研究提供参考。关于环境中纳米材料生物效应与分子机制的最新研究进展将在第23章中进行介绍。

2 金属基纳米材料的环境转化与归趋

2.1 金属基纳米材料的胶体稳定性和多孔介质迁移性

腐殖质、胞外多聚物（EPS）、蛋白质等天然有机质（NOM）能显著影响纳米尺寸的零价金属、金属氧化物等金属基纳米材料的表面电荷、亲疏水性和空间位阻效应，从而影响金属基纳米材料在水环境中的胶体稳定性。最近的研究进一步证实了天然有机质对金属基纳米材料胶体稳定性的重要影响。例如，Jung 等研究了 Pony 湖富里酸（PLFA）对 2 种表面配体修饰的银（Ag）纳米颗粒（NPs）在水中分散性的影响[19]，发现：当水体中无 PLFA 时，分枝状聚乙烯亚胺（BPEI）修饰的 Ag 纳米颗粒（BPEI-AgNPs）的分散性高于柠檬酸根修饰的 Ag 纳米颗粒（Cit-AgNPs）；当 PLFA 存在时，Cit-AgNPs 会保持高分散性和均一的粒度分布，但 BPEI-AgNPs 则迅速发生团聚；经过 3 天后，Cit-AgNPs 仍保持很好的分散性，而 BPEI-AgNPs 的团聚程度则与 PLFA 的浓度有关。他们的发现证实，低浓度 PLFA 吸附到表面带正电荷的 BPEI-AgNPs 后，表面电荷的中和导致了 BPEI-AgNPs 团聚的发生。而高浓度 PLFA 条件下，PLFA 的过量吸附使得表面电荷呈现负电性，增加了静电排斥作用和空间位阻，提高了纳米颗粒的分散性[19]。而 Lin 等研究了在 Na^+ 和 Ca^{2+} 存在时，EPS 对二氧化钛（TiO_2）纳米颗粒稳定性的影响。他们发现，在高浓度 Ca^{2+}（10～40 mmol/L）条件下，EPS 会增加 TiO_2 纳米颗粒的团聚速率[20]。最近的一项研究则发现，亚硒酸盐还原菌释放的 EPS 可以吸附到其产生的零价硒纳米颗粒表面，提高零价硒纳米颗粒的胶体稳定性及其固定零价汞的能力[21]。Levak 等发现，在海水中，蛋白质的存在会使 AgNPs 保持长期的分散[16]。Sheng 等则发现，除了离子强度、pH 和蛋白质浓度外，蛋白质分子的种类与大小也会影响赤铁矿（Fe_2O_3）纳米颗粒的分散性，并证明分子的空间位阻比静电作用更重要[22]。

纳米材料表面的包覆层对其胶体稳定性也有重要影响。例如，最近的一项研究报道了腐殖酸和硫离子共存体系对 Ag 纳米颗粒在缺氧条件下胶体稳定性的影响，发现其影响效果与 Ag 纳米颗粒的包覆层有关：聚乙烯吡咯烷酮（PVP）包被的 Ag 纳米颗粒，其胶体稳定性和粒径分布几乎不受影响；而对于柠檬酸根包被的 Ag 纳米颗粒，腐殖酸和硫离子都会引起其粒径的快速增加，而且腐殖酸的效果更为显著[23]。最近的另一项研究比较了 4 种不同表面修饰的氧化铜（CuO）纳米颗粒在生物和环境介质中的胶体稳定性，发现生物介质中的蛋白质和氨基酸对 4 种 CuO 纳米颗粒的胶体稳定性都有重要影响，而且影响方式和程度与 CuO 纳米颗粒的表面修饰有关[24]。若包覆层因环境因素而发生变化，则会改变纳米材料的胶体稳定性。例如，金（Au）纳米颗粒表面的 PVP 包覆层在紫外光照下会发生快速氧化并部分降解，未降解的部分贴合到 Au 纳米颗粒的表面，位阻效应变弱，因而大幅降低了 Au 纳米颗粒的胶体稳定性[25]。也有研究发现，NOM 能显著影响 Au 纳米颗粒在土壤或沉积物胶体颗粒表面的黏附和异相团聚过程，不论 Au 纳米颗粒的表面修饰如何[26]。除 NOM 外，自然和人工水环境中还存在人为排放的表面活性剂，这些表面活性物质会对纳米材料的胶体稳定性产生显著影响。例如，带负电的十二烷基硫酸盐（SDS）会吸附到在近中性 pH 下带正电的 TiO_2 纳米颗粒表面；在一定的 SDS 浓度下，TiO_2 纳米颗粒表面的正电荷被中和，进而发生团聚形成大的团聚体，而 SDS 分子间的疏水作用也能促进 TiO_2 纳米颗粒的团聚[27]。

金属基纳米材料在土壤等多孔介质中的附着沉积和迁移行为受多孔介质理化性质的显著影响[28]。例如，纳米 TiO_2 在水体沉积物中的迁移受到 pH、有机质以及铁/铝氧化物的影响[29]。而土壤的化学组分也能显著影响氧化铈（CeO_2）纳米颗粒的迁移特性及植物可利用性：土壤中的黏土组分能促进 CeO_2 纳米颗粒的截留，从而降低植物对 Ce 元素的吸收，而土壤有机质组分则能够促进植物对 Ce 元素的吸收；在低有机质含量的土壤中，柠檬酸根包被的 CeO_2 纳米颗粒比无包被的 CeO_2 纳米颗粒在土壤中的迁移性强，更容易被植物吸收[30]。腐殖酸盐对聚丙烯酸（PAA）包被的纳米零价铁（nZVI）在氧化硅多孔介质中迁移性的影响与多孔介质的表面性质有关：对于表面性质均一的多孔介质（如标准石英砂柱），腐殖酸盐能大幅提高 PAA-nZVI 的迁移性，而对于表面不均一的多孔介质（如污染场地采样得到的实际砂土），腐殖酸盐对 PAA-nZVI 迁移性的影响则不明显[31]。在氧化硅或铁氧化物涂覆的氧化硅饱和多孔介质中，大肠杆菌（Escherichia coli）能促进 PVP 包覆的 Ag 纳米颗粒的沉积；而在实际的地下含水层多孔介质中，则没有显著影响[32]。土壤微生物分泌的纤维素酶（一种典型的胞外酶）则能显著降低 TiO_2 纳米颗粒在氧化硅表面的沉积速率[33]。此外，最近的一项研究通过批式黏附和沉降实验，考察了不同粒径和表面修饰的 Ag 纳米颗粒在污水处理过程中的去除，发现 Ag 纳米颗粒的去除效率主要与其粒径有关，与表面包覆层无关，而且受污泥浓度和流体的剪切速率影响[34]。

关于金属基纳米材料的多孔介质迁移行为已有大量研究，最近的一项研究通过对以往研究结果的元分析（meta-analysis），提出了影响人工纳米材料在饱和多孔介质中迁移滞留行为的关键因素，并首次明确了现有的颗粒传输模型不适用的情况[35]。以往大部分研究是在连续流动的饱和柱实验中进行，而最近一项研究考察了不饱和柱实验条件下流动中断以及水溶液离子强度对 Ag 纳米颗粒在壤质砂土中迁移的影响，并进行了数值模拟，发现：流动中断过程中土壤溶液离子强度的变化以及 Ag 纳米颗粒在空气-水界面的黏附对 Ag 纳米颗粒的迁移性有重要影响[36]。另一项研究则发现：相较于不饱和砂柱，Ag 纳米颗粒在不饱和土柱中的迁移更慢，穿透曲线的形状也不同[37]，说明今后的研究应该更关注实际土壤中复杂组分对纳米材料迁移行为的影响。

2.2 金属基纳米材料的溶解与转化

纳米 Ag、纳米氧化锌（ZnO）、纳米 CuO 等金属基纳米材料在环境水体中会发生溶解以及硫化等化学转化过程，进而显著影响其环境归趋、生物可利用性以及毒性效应[38-43]。金属基纳米材料在水环境中的溶解过程已有系统全面的研究[44,45]，但其在土壤环境（尤其是植物根区）中的溶解与转化行为研究较少。

最近的研究表明，NOM 对金属基纳米材料的溶解过程有较为复杂的影响。例如，Jiang 等比较了 16 种 NOM 组分对 ZnO 纳米颗粒溶解性的影响，发现 ZnO 纳米颗粒的溶解速率常数（k_{obs}）与 NOM 的比紫外吸光度（SUVA）、芳香族碳和羧基碳含量呈现正相关，而与氢/碳比和脂族碳的含量呈现负相关[46]；他们的发现证明，芳香碳的含量是决定 NOM 促进 ZnO 纳米颗粒溶解的关键影响因素。Jung 等发现，PLFA 对 Cit-AgNPs 和 BPEI-AgNPs 的溶解速率体现出不同的影响：在低浓度（5 mg/L）时，两种 AgNPs 的溶解均会降低。在高浓度（30 mg/L）时，BPEI-AgNPs 的溶解速率会增加，而 Cit-AgNPs 的溶解并无显著变化[19]。Gunsolus 等比较了 Suwannee 河腐殖酸（SRHA）、Suwannee 河富里酸（SRFA）和 PLFA 对 Cit-AgNPs 溶解性的影响，发现 SRHA、SRFA 对 Cit-AgNPs 的溶解几乎没有影响，但 PLFA 可以抑制 Cit-AgNPs 的溶解。他们推测上述差异主要是由于 PLFA 富含含硫官能团，特别是巯基官能团。含硫官能团对 AgNPs 具有高亲和力，会限制分子氧与 AgNPs 表面活性位点的接触，从而抑制 Cit-AgNPs 的氧化与溶解[47]。最近的另一项研究则发现，谷胱甘肽、植物螯合肽等含硫生物分子促进 AgNPs 溶解的能力与分子中巯基的数量呈正相关[48]。Collin 等通过研究 SRFA、PLFA 和 Pahokee peat 富里酸（PPFA）对硫化纳米银（sAgNPs）溶解的影响，发现 SRFA 和 PPFA 可以减缓 sAgNPs 的溶解，而 PLFA 会促进 sAgNPs 的溶解[49]。Levak 等考察了 Ag 纳米颗粒在 15 天时间内的离子释放动力学过程，发现 Cit-AgNPs 的 Ag^+ 释放会在第一天迅速增加，之后达到相对稳定的数值（约为银总含量的 85%），而牛血清白蛋白（BSA）的存在会显著地抑制 Ag^+ 的释放[16]。Liu 等则发现，BSA 对 AgNPs 溶解的影响较为复杂：BSA 最初能促进 AgNPs 的溶解，但长期来看却抑制其溶解[50]。Miao 等发现，相较于藻酸盐和 BSA，EPS 具备更丰富的化学官能团和金属离子络合能力，因此会更有效地促进 CuO 纳米颗粒的离子释放与溶解[51]。最近的研究表明，NOM 也会影响硫化汞纳米颗粒的粒径、结晶度以及溶解度和生物可利用性[52]。

最近，Gao 等研究了 CuO 纳米颗粒在土壤中的溶解动力学以及铜元素的可提取性随时间的变化，发现在含水率 16%的土壤中，CuO 纳米颗粒的溶解过程仍然符合一级动力学模型；而且，与添加硝酸铜的对照组不同，添加 CuO 纳米颗粒的土壤中，铜元素的可提取性随时间而增加[53]。Gao 等进一步发现，小麦（$Triticum\ aestivum$）根系对 CuO 纳米颗粒溶解的影响较为复杂：一方面，根能将根区土壤的 pH 提高 0.4～0.6 个 pH 单位，有助于减缓 CuO 纳米颗粒的溶解；另一方面，根系分泌物却能提高孔隙水中溶解态铜的浓度[54]。大麦（$Hordeum\ vulgare$ L.）根部的表面可以引发 CeO_2 纳米颗粒中 Ce(IV)的还原并促进 CeO_2 纳米颗粒的植物吸收[55]。Peng 等则研究了 CuO 纳米颗粒在土壤-水稻（$Oryza\ sativa$）系统中的转化与归趋，发现 CuO 纳米颗粒能降低系统的氧化还原电位并提高土壤溶液的电导率，同时，CuO 纳米材料转化为硫化亚铜（Cu_2S）和单质 Cu；在淹水条件下，CuO 纳米颗粒的生物可利用性大幅降低，但在干-湿循环过程中，其生物可利用性却大幅提高[56]。在水稻田中，Ag 纳米颗粒的转化与归趋也受土壤 NOM 和氧化还原条件的影响[57]。此外，土壤中的铁氧化物胶体矿物能与 Ag 纳米颗粒发生异相团聚并抑制其溶解[58]。

单质金属以及金属氧化物纳米颗粒在富含硫离子的水环境中容易发生硫化转化过程[59-62]。最近，关于金属基纳米材料的硫化过程有了更深入的认识。例如，李灵香玉课题组研究了有氧条件下，代表性金属硫化物矿物——硫化铜（CuS）纳米颗粒对 Ag 纳米颗粒的硫化作用，并揭示了硫化过程的机制：Ag 纳米颗粒发生氧化溶解释放的 Ag^+离子与 CuS 纳米颗粒表面发生阳离子交换，生成硫化银（Ag_2S）纳米颗粒[63]。NOM 会抑制 Ag 纳米颗粒的硫化[63]，而 Ag 纳米颗粒的高分子表面修饰[64]以及水中的表面活性剂[65]也会影响其硫化过程。例如，巯基化的聚乙烯醇（PEG）可以促进 Ag 纳米颗粒的硫化，并抑制硫化后纳米颗粒在 SiO_2 表面的沉降；相反，PVP 则会抑制 Ag 纳米颗粒的硫化，并促进其在 SiO_2 表面的沉降[64]。祝凌燕课题组最近研究了不同性质的表面活性剂对 Ag 纳米线硫化过程的影响，发现：阴离子型、两性离子型（zwitterionic）以及非离子型表面活性剂对 Ag 纳米线的溶解与硫化没有影响，阳离子型表面活性剂则能显著促进 Ag 纳米线的溶解与硫化[65]。Gogos 等最近则研究了 BSA、藻酸盐、腐殖酸等不同种类的 NOM 对 CuO 纳米颗粒硫化过程的影响，发现：当水中不存在 NOM 时，CuO 先转化形成无定形的 Cu_xS 纳米颗粒（粒径<10 nm），然后转化为晶态 CuS 纳米颗粒，NOM 则会显著减缓 Cu_xS 转化为 CuS 的过程，而且，当存在高浓度（1 g/L）的 BSA 时，反应结束时仍有 30%的 CuO 没有发生转化[66]。纳米零价铁[67, 68]和铁氧化物纳米颗粒[69]在硫离子存在的条件下也会发生硫化转化，而且腐殖酸等有机大分子会影响 nZVI 的硫化过程[68]。

如上所述，单质金属和金属氧化物等金属基纳米材料在厌氧环境中易发生硫化而转化成纳米金属硫化物[70]，但最新的研究表明，硫化金属纳米材料并不是金属基纳米材料的最终稳定形态[71]，在某些条件下也会发生进一步转化和生物吸收。例如，李灵香玉等最近发现，在次氯酸钠存在时，Ag_2S 纳米颗粒会发生活性氧物种（ROS）介导的快速溶解；NOM 会加速这一过程，而溶解氧的存在会抑制 Ag_2S 纳米颗粒的溶解[72]。而且，虽然 Ag_2S 纳米颗粒比 Ag 纳米颗粒溶解释放的银离子低一个数量级，但水生植物浮萍（$Landoltia\ punctata$）暴露于两种纳米颗粒后，其根部累积的银含量基本相同；在浮萍根部中，Ag 纳米颗粒主要转化为 Ag_2S 以及银-硫基络合物，而 Ag_2S 纳米颗粒的组分不变，仍为 Ag_2S[73]。此外，Ag 纳米颗粒以及 Ag_2S 纳米颗粒在小麦的根表皮及根内也能发生化学转化，并引起不同的毒性效应[74]。

最近有研究人员考察了纳米 CuO、CuS、Ag、Ag_2S 等不同的金属基纳米材料在中型淡水湿地生态系统中的空间分布、化学转化、长期稳定性以及生物可利用性，发现：暴露 1～9 个月后，无论初始形态如何，上述纳米材料都有大约 50%存在于水蕴草（$Egeria\ densa$）组织内，其余主要存在于表层沉积物中，水层中含量很少；在表层沉积物中，纳米 Ag 和 CuO 快速发生化学转化，而在水蕴草组织中，仍能观测到未转化的纳米 Ag、CuO 以及 CuS；纳米 CuO 和 CuS 在表层沉积物中都转化为 Cu-有机络合物，另有少量单质 Cu；纳米 Ag 和 Ag_2S 经过 6~9 个月的转化后，化学形态比较接近，在表层沉积物和水蕴草组织中都主要（70%~90%）以 Ag_2S 的形式存在，其余 10%~30%以 Ag-巯基络合物的形式存在[75]。最近的一项研究发现，纳米 Ag 的投加方式（长期少量投加或脉冲式投加）对其在中型淡水湿地生态系统中的归趋及生态效应有重要影响[76, 77]；而另一项研究表明，CeO_2 纳米颗粒在中型淡水湿地生态系统中的转化和归趋与其粒径有很大关系[78]。

微生物也能影响金属基纳米材料的环境转化与溶解。例如，模式土壤微生物枯草芽孢杆菌（*Bacillus subtilis*）及其分泌物可以影响 Ag 纳米颗粒的硫化过程：枯草芽孢杆菌可以将 Ag 纳米颗粒完全硫化，生成晶态 Ag_2S、无定形态 Ag_2S 以及 Ag-巯基络合物，而细菌分泌物则可将 Ag 纳米颗粒部分硫化，生成 $Ag@Ag_2S$ 核壳结构[79]。枯草芽孢杆菌也能促进 CeO_2 纳米颗粒的还原溶解[80]。零价的纳米金在水中热力学稳定，但最近的研究表明，水蕴草表面的生物膜可以诱发 Au 纳米颗粒的氧化溶解；其中，促进 Au 纳米颗粒溶解能力最强的生物膜中含有能大量释放氰根离子的微生物，其释放的氰根离子能高效促进零价金的氧化以及金离子的络合[81]。此外，土壤微生物紫色色杆菌（*Chromobacterium violaceum*）释放的氰根离子也能促进 Au 纳米颗粒在土壤环境中的氧化溶解[82]。环境中重要的产电微生物奥奈达希瓦氏菌（*Shewanella oneidensis* MR-1）能将赤铁矿（Fe_2O_3）纳米颗粒还原转化为磁铁矿（Fe_3O_4）纳米颗粒[83]，而 Fe_3O_4 纳米颗粒在富氧水体中则能被快速氧化成磁赤铁矿（γ-Fe_2O_3），比表面积和表面电荷发生改变，胶体稳定性下降，形成微米级的团聚体[84]。

2.3 小结

最近的研究证实，金属基纳米材料在水环境中的胶体稳定性受其表面包覆层、天然有机质以及人工释放的表面活性物质等因素的显著影响；其在土壤等多孔介质中的迁移特性不但取决于多孔介质的复杂组分和表面性质，也与水力学条件密切相关。天然有机质和微生物也能显著影响金属基纳米材料的溶解和化学转化过程，进而影响其生物可利用性。最近，研究人员开始关注金属基纳米材料在土壤环境（尤其是植物根区）以及中型淡水湿地生态系统等实际环境或接近真实环境中的溶解与转化行为，并取得一些初步的重要发现。

3 碳纳米材料的环境转化与归趋

3.1 碳纳米材料的胶体稳定性及多孔介质迁移

已有研究表明，氧化石墨烯（GO）、柴油车油烟纳米颗粒[85]、表面羧基化的炭黑纳米颗粒[86]等表面带负电的纳米材料的胶体稳定性[85, 87, 88]和在多孔介质中的迁移特性[89]受水环境中阳离子的显著影响。最近，陈威课题组发现，一价和二价阳离子能显著抑制 GO 以及还原氧化石墨烯（RGO）在饱和石英砂柱中的迁移，而且不同阳离子对石墨烯材料迁移的抑制作用强弱遵循 Hofmeister 序列（图 22-1）[90]：对于碱金属离子，顺序为 $Na^+ < K^+ < Cs^+$；对于碱土金属，其顺序为 $Mg^{2+} < Ca^{2+} < Ba^{2+}$。这一现象主要是因为：离子水合半径较大的阳离子（如 Cs^+ 和 Ba^{2+}）与离子水合半径较小的阳离子（如 Na^+ 和 Mg^{2+}）相比水合

作用更弱，与石英砂以及石墨烯材料的作用更强。而且，Cs^+、Ca^{2+}、Ba^{2+}等离子能与石墨烯材料表面的官能团形成内圈配位络合物，能够更有效地中和石墨烯材料表面的负电荷，并通过"桥键"作用（以及可能存在的截留作用）加剧石墨烯材料在石英砂表面的沉积。

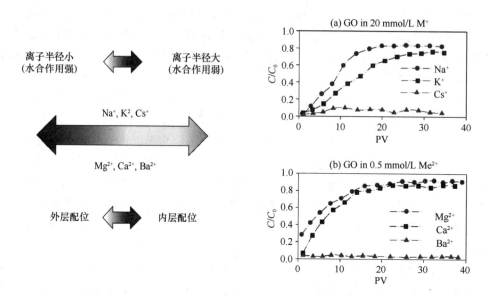

图 22-1　金属阳离子抑制石墨烯在饱和多孔介质中迁移的 Hofmeister 效应（引自参考文献[90]）

碳纳米材料的胶体稳定性也受到 NOM 的显著影响。例如，SRHA、SRFA 等 NOM 能显著提高 GO 在水环境中的胶体稳定性，其增强机制主要是 NOM 吸附引起的空间位阻排斥，而静电排斥的贡献较小；芳香碳含量越高的 NOM，其稳定 GO 的能力越强，说明 π-π 作用是 GO-NOM 相互作用的重要机制[91]。最近，祝凌燕课题组发现，BSA 对 GO 在水中胶体稳定性的影响与 BSA 的浓度有关：在水溶液离子强度一定的条件下，随着 BSA 浓度的升高，GO 的胶体稳定性先降低，后升高[92]。当 BSA 浓度较低时，BSA 中带正电的赖氨酸残基会与 GO 表面带负电的官能团通过静电作用结合，从而促进 GO 的团聚，而当 BSA 浓度较高时，吸附到 GO 表面的 BSA 会通过位阻作用增强 GO 的胶体稳定性。

土壤及地下含水层等多孔介质中的矿物组分也能影响石墨烯材料的迁移行为。例如，高岭土、蒙脱石、伊利石等黏土矿物虽然整体带负电荷，但却能抑制表面带负电的 GO 在饱和石英砂柱中的迁移；这主要是因为黏土层板的边缘带正电荷，引起石墨烯在黏土颗粒表面的沉积；其中高岭土的边缘面积比例最高，因此抑制石墨烯迁移的效果最显著[93]。此外，地下含水层中的主要矿物成分铁氧化物也可显著影响石墨烯的迁移行为，原因在于带正电荷的铁氧化物为带负电的 GO 提供了结合的点位，且包覆铁氧化物的石英砂表面粗糙程度较高，利于 GO 的沉积[94]。其中，针铁矿和 Fe_2O_3 对 GO 迁移的抑制作用要大于水铁矿，这是由于在相同铁含量条件下，针铁矿和赤铁矿对多孔介质表面电荷的影响比水铁矿更为显著，更不利于 GO 的迁移。同种价态不同种类阳离子对 GO 在包覆针铁矿的石英砂中迁移的抑制作用也遵循 Hofmeister 序列，这是由于离子水合半径既影响其压缩双电层的能力，也决定了其作为"桥键"促进 GO 在多孔介质上吸附的能力。而 GO 在赤铁矿纳米胶体中则会发生异相团聚形成纳米杂化体[95]。

石墨烯材料在人工和自然水环境中发生的转化过程能显著影响其物理化学性质及胶体稳定性。例如，GO 释放进入环境后可在环境中的还原性物质（如 S^{2-} 等）作用下发生化学还原转化。最近，陈威课题组研究了化学还原过程对石墨烯材料胶体稳定性的影响，发现 GO 以及不同还原方法（水合肼、硼氢化钠或抗坏血酸还原）制备的 RGO 在水中的团聚行为受石墨烯材料表面性质和水中阳离子的影响[96]。研究发现，石墨烯材料在氯化钠水溶液中的临界聚沉浓度（CCC）随其表面碳氧比（C/O）的升高而降低，即还原程度越高的石墨烯材料越容易发生团聚；但石墨烯材料团聚行为的上述变化主要是由于表面疏水性的改变，而非表面电荷的变化。二价阳离子（如 Ca^{2+} 和 Mg^{2+}）比一价阳离子（如 K^+ 和 Na^+）促进石墨烯团聚的能力更强，而且无论是一价或者二价阳离子，水合密度更低的阳离子（如 K^+ 和 Ca^{2+}）比水合密度更高的阳离子（如 Na^+ 和 Mg^{2+}）能更有效地促进石墨烯的团聚，这主要是由于前者能更有效地中和石墨烯的表面电荷。此外，Ca^{2+} 还可以通过"桥键"作用加剧石墨烯的团聚。然而，在钙、镁离子浓度相对较高（如 0.68 mmol/L 的 Ca^{2+} 和 0.24 mmol/L 的 Mg^{2+}）的模拟地下水中，石墨烯材料发生团聚的难易程度却与其还原程度无显著相关性，而是与表面官能团的种类和含量有关。

石墨烯的化学还原转化也能影响其在多孔介质中的迁移特性。例如，陈威课题组最近研究了经还原处理的 GO 表面理化性质与水化学因素交互作用对其迁移能力的影响机制，发现 GO 和 3 种 RGO（分别用水合肼、硼氢化钠和抗坏血酸还原 GO 制得）在饱和多孔介质中迁移行为的差别并不简单地取决于材料的理化性质，而是取决于材料理化性质与水化学条件的共同作用[97]。例如，在 Na^+、K^+、Mg^{2+} 溶液中，GO 和 RGO 的迁移能力与材料表面的 C/O 比有较好的相关性，符合传统的 DLVO 理论。但在 Ca^{2+} 溶液中，还原程度最低的 VC-RGO 迁移能力反而低于另外两种 RGO，其本质原因在于 VC-RGO 表面的羧基（一种具有金属络合能力的含氧官能团）数量最多，能够通过 Ca^{2+} 的"桥键"作用与石英砂结合，导致其在石英砂上沉积。

3.2 碳纳米材料在水环境中的光化学转化

石墨烯材料进入自然水体环境中后可在光照条件下发生转化，但其在不同波长的自然光照射下发生转化的路径、机制和效应研究较少。最近研究发现：自然光照，尤其是紫外波段，可以导致 GO 表面含氧官能团明显减少且 GO 片层发生明显的破碎[98]。模拟自然光和紫外波段光照后，GO 的 C/O 比分别增加了 41%和 31%，相比之下，可见波段光照射后 C/O 比只增加了 5.5%，且形态只发生褶皱，并未产生显著的破碎。在紫外和可见光照射下均观察到 GO 形成羟基自由基（·OH）、超氧自由基阴离子（$O_2^{·-}$）和单线态氧（$^1O_2$）三种 ROS，但比例不同。$^1O_2$ 主要在可见光照射下形成，$O_2^{·-}$ 和·OH 主要在紫外光照射下形成。羟基自由基对 GO 的羟基化作用是导致 GO 分解破碎的主要原因。光照导致 GO 表面含氧官能团的减少而影响了水环境中 GO 的胶体稳定性，同时 GO 形态的变化及疏水性的增强使其对疏水性有机污染物的吸附能力显著增强。

紫外线照射是水和废水处理的常用技术，GO 在紫外光照射下的光化学转化对其在环境中的迁移、归趋以及生物效应[99]至关重要。考虑到实际水体中广泛存在硝酸盐，陈威课题组研究了硝酸盐对 GO 光化学转化的影响及作用机制[100]。研究发现，不同浓度的硝酸盐对 GO 在紫外光照射下的光化学转化机制有重要影响：在不含硝酸盐或者硝酸盐浓度较低（≤0.1 mmol/L）时，经紫外光照后 GO 主要发生直接光解，即含氧官能团被光还原；而当硝酸盐浓度较高（≥1 mmol/L）时，则主要发生间接光解，大片层的 GO 破碎成细小的碎片（图 22-2）。原因在于硝酸盐光照产生的具有强氧化性的·OH 是引起 GO 光化学转化的最主要成分。结合 X 射线电子能谱（XPS）、拉曼光谱和 UV-vis 光谱等谱学研究结果推测：·OH 不仅能攻击 GO 边缘或缺陷处的羟基和羧基等含氧官能团，还能以亲电加成反应等途径来攻击 GO 的碳骨架，导致大片的 GO 最终被撕裂为碎片。

紫外线照射也能影响富勒烯（C_{60}）纳米材料的化学转化和胶体稳定性。例如，波长 351 nm 的长波紫外线（UVA）能显著加快水中游离态氯对 C_{60} 纳米团聚体的氧化过程，且产物的氧化程度与暗态氧化产物相比更高；光氧化产物的表面亲水性和胶体稳定性都有所提高[101]。同样地，UVA 也能促进水中游离态氯对富勒醇的氧化转化，并改变氯氧化产物的表面亲水性及其在憎水性表面的沉积特性[102]。

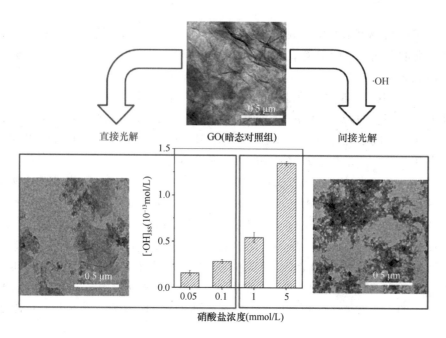

图 22-2　不同浓度的硝酸盐对紫外光照射下 GO 光化学转化的影响（引自参考文献[100]）

3.3　小结

最近的研究证实，碳纳米材料在水环境中的胶体稳定性以及在多孔介质中的迁移特性受水中阳离子种类、天然有机质等水化学条件的显著影响。石墨烯等碳纳米材料在人工和自然水环境中发生的化学转化过程（如化学还原、光化学转化）也能显著影响其物理化学性质及胶体稳定性。

4　复合纳米材料及多元金属基纳米材料的环境转化与归趋

过去十余年间，关于纳米材料环境行为与影响的研究大都选择单组分的纳米材料（如纳米 TiO_2、纳米 Ag、碳纳米管等）为研究对象，系统研究了这些模型纳米材料的环境转化、归趋和生物效应。但最近的纳米技术工业化应用和实验室研究越来越关注含有多种组分的复合纳米材料[尤其是两种以上纳米材料紧密结合形成的纳米杂化体（nanohybrid）[103]]以及含有两种以上金属元素而具有单一物相的多元金属基纳米材料。相较于单组分的碳纳米材料或含有一种金属元素的金属基纳米材料，复合纳米材料以及多元金属基纳米材料具有独特的理化性质；因此，其环境行为和生态效应与单组分纳米材料以及几种单组分纳米材料的混合共存体系[104-115]相比，可能具有很大差异，不能用以前基于单组分纳米材料研究获得的规律来预测。研究典型复合纳米材料以及多元金属基纳米材料的环境转化与归趋，对于完善纳米技术环境影响研究体系、提高纳米材料环境风险预测模型的可靠度，有重要的理论和实际意义。

最近有研究报道了碳纳米材料与纳米金属氧化物形成的纳米杂化体的团聚及多孔介质迁移规律。例如，Wang 等研究了碳纳米管-Fe_3O_4 纳米杂化体[116]以及 RGO-金属氧化物纳米杂化体[117]在饱和多孔介质中的迁移，发现大部分实验现象仍能用传统的 DLVO 理论以及胶体过滤理论解释，但也观察到一些用传统理论无法解释的现象。而最近的另一项研究则发现，多壁碳纳米管（MWCNT）与 TiO_2 纳米颗粒形成的 MWCNT-TiO_2 纳米杂化体在水中的团聚行为不能用传统的电动力学规律解释，而必须考虑纳米杂化体的特殊性质（如分形维度、表面粗糙度和电荷不均一性等）[118]。

多元金属基纳米材料的环境行为和生物效应，最近也引起研究人员的关注。例如，有研究报道，锂离子电池中常用的正极材料钴酸锂镍锰（$Li_xNi_yMn_zCo_{1-y-z}O_2$）在水环境中会溶解释放 Ni^{2+}、Co^{2+} 等金属离子，并对奥奈达希瓦氏菌（革兰氏阴性）[119]以及枯草芽孢杆菌（革兰氏阳性）[120]产生毒性。在水溶液中，磷酸根会吸附到钴酸锂（$Li_xCo_{1-x}O_2$）纳米颗粒表面，改变其表面电荷，并使其在水溶液中长时间保持稳定分散[121]；而且，钴酸锂和钴酸锂镍锰纳米颗粒会对水生生物造成不同毒性效应[122]。也有研究报道了碳化钨钴（WCCo）纳米颗粒在土壤环境中的归趋和生物效应[123, 124]。胡焱弟课题组的研究发现，相较于纯相水合铁纳米颗粒，含有 Mn、Al 杂质离子的水合铁纳米颗粒在 Al_2O_3 表面的沉积更慢；而在 SiO_2 表面，Mn 离子掺杂的水合铁纳米颗粒沉积更快，Al 离子掺杂则抑制了沉积过程[125]。范文宏课题组最近则报道了钙钛矿纳米材料（如 $LaFeO_3$、$YFeO_3$、$BiFeO_3$、$LaMnO_3$、$LaCoO_3$ 等）的环境转化及其对大型溞（Daphnia magna）的毒性效应[126]。

需要注意的是，复合纳米材料以及多元金属基纳米材料的种类无穷无尽，没有必要也无法对其逐一进行研究。拟开展的研究课题应该具有充分的理论价值、实际意义和环境相关性，并且能预期得到一些共性规律，从而利于修正和完善纳米技术环境影响研究体系、提高纳米材料环境风险预测模型的可靠度。

5 复杂环境中纳米材料转化与归趋研究的新方法

5.1 实验技术

最近，一些新的技术手段被用于研究复杂环境中纳米材料的转化与归趋过程，如团聚、多孔介质迁移、化学转化、溶解、吸附和生物吸收等。例如，改进的纳米颗粒跟踪分析（nanoparticle tracking analysis，NTA）技术被成功用于研究垃圾填埋场渗滤液中人工纳米材料的均相及异相团聚过程[127]。纳米材料在环境界面的黏附效率（attachment efficiency）一般采用静态柱实验测定，而 Wiesner 课题组最近发展出一种混合法用于测定复杂水环境中纳米材料的黏附效率[128]。微区拉曼成像技术最近被用于研究 Ag 纳米颗粒在水合矿物表面附着的分子机制[129]，而频谱激发极化法（spectral induced polarization，SIP）则被用于研究超顺磁铁氧化物纳米颗粒在砂柱中的迁移[130]。最近，倾斜柱状薄膜（SCTF）与各向异性相差光学显微镜（ACOM）联用技术被用于可视化 TiO_2 纳米颗粒的沉积过程[131]。长周期 X 射线驻波荧光产率谱被用于研究 Ag 纳米颗粒在奥奈达希瓦氏菌生物膜-Al_2O_3-水界面的动态分配过程[132]。微流体器件最近被用于模拟河口条件下盐度梯度对 C_{60} 纳米团聚体胶体稳定性的影响，发现盐度梯度的存在会提高纳米材料的胶体稳定性[133]。

同位素分馏和同位素示踪技术在纳米材料水环境转化与生物吸收研究中有了新的应用。例如，刘倩课题组用同位素分馏技术研究了腐殖酸对 Ag 纳米颗粒在光照条件下的水环境转化及持久性的影响[134]。毛亮课题组用放射性同位素 ^{14}C 标记研究了石墨烯沿水生食物链的营养级传递[135]，而张智勇课题组用放射性同位素 ^{141}Ce 标记研究了 CeO_2 纳米颗粒沿陆生食物链的营养级传递[136]。多同位素示踪技术最近也被用于研究纳米 $ZnO^{[137]}$、纳米 $Ag^{[138]}$ 以及 CdSe/ZnS 量子点[139]的环境分布、溶解转化及生物吸收与累积等过程。

电感耦合等离子体质谱（ICP-MS）与新型分离技术的的联用以及具有特殊测定模式的 ICP-MS 最近也被用于研究复杂环境中纳米材料的转化与归趋。例如，刘景富课题组将中空纤维流场流分离（hollow fiber flow field-flow fractionation，HF5）与 ICP-MS 联用技术用于研究低浓度（如 10 ng/mL）的 Ag 纳米颗粒在复杂水环境中的团聚过程，以及 Ag^+ 离子生成 Ag 纳米颗粒的过程[140]。最近，有研究报道以 Au@Ag 核壳结构纳米颗粒为示踪剂，用单颗粒模式的 ICP-MS（spICP-MS）研究纳米银在复杂水环境中的化学转化[141]，而单细胞模式的 ICP-MS（scICP-MS）最近被用于量化单个藻类细胞中金离子和金纳米颗粒的吸收与分布[142]。激光剥蚀 ICP-MS 最近也被用于研究斑马鱼（Danio rerio）胚胎对 Ag、Au、CuO、ZnO 等金属基纳米颗粒的吸收以及纳米颗粒在斑马鱼胚胎表面和内部的分布[143]。快速采集技术（FAST）spICP-MS 可以测定粒径更小（5~10 nm）的 Ag 纳米颗粒，被用于比较不同粒径的 Ag 纳米颗粒在模拟污水处理厂中的去除效率[144]。

碳纳米材料的吸附过程一般采用批实验进行研究，而柱色谱（column chromatography）技术最近被用于研究 MWCNT 对硝酸根、溴离子等无机离子的吸附[145]。热点归一化的表面增强拉曼光谱（SERS）最近被用于实时监测柠檬酸根包被的金纳米颗粒表面的有机配体交换动态过程[146]，而衰减全反射傅里叶变换红外光谱则首次被用于原位研究 TiO_2 纳米颗粒表面抗坏血酸、柠檬酸、腐殖酸、BSA 等有机配体的置换反应[147]。高分辨扫描透射电子显微镜（HR-STEM）最近也被用于研究 CeO_2 以及 Ce_2O_3 纳米颗粒在复杂环境介质中的化学转化[148, 149]。无梯度能斯特平衡溶出法（AGNES）这一电化学分析技术被用于研究人工唾液环境中 ZnO 纳米颗粒的溶解及磷酸根诱导的转化过程[150]。此外，有研究人员探索了用短期高温实验（25~80℃ 下处理 4 周）模拟长期老化过程对金属氧化物纳米材料粒径、表面电荷以及胶体稳定性的影响[151]。

王文雄课题组近期将聚集诱导发光技术用于研究单细胞浮游藻类对 Ag 纳米颗粒的吸收，发现 Ag 纳米颗粒不能进入细小裸藻（Euglena gracilis）细胞内部[152]。此外，除了大量实验室研究以及中型淡水湿地生态系统中进行的研究，最近也有研究人员在真实湖泊中投加 Ag 纳米颗粒以研究不同营养级的鱼类的 Ag 吸收和体内分布[153]。

5.2 模型方法

近期，有多种数值和概念模型被应用于模拟纳米材料在环境中的迁移、团聚、吸附、转化、生物吸收等环境行为。例如，Walker 课题组用 COMSOL 模拟软件建立了基于 DLVO 理论和水力学规律的数值模型，用于模拟荧光乳胶纳米颗粒在多孔介质中的迁移，并用荧光显微镜获得的可视化数据进行验证[154]。也有研究者将群体平衡模型（population-balance model）与碰撞频率和沉降速率模型结合，模拟研究了羟基磷灰石纳米颗粒的粒径和浓度的时空变化，发现团聚过程发生的阶段对羟基磷灰石纳米颗粒的环境归趋有重要影响[155]。群体平衡模型最近也被用于研究 GO 在复杂流体条件下的团聚和沉积过程[156]。分子动力学模拟（molecular dynamics simulations）计算最近被成功用于研究 GO 的团聚过程[157]，单壁碳纳米管（SWCNT）的酶促降解过程[158]，GO 对有机化合物的吸附[159-162]，以及富勒烯、纳米 CeO_2 等人工纳米材料与有机农药分子的吸附作用机制[163]。

由于纳米材料的特殊理化性质，传统的污染物归趋模型不能准确模拟纳米材料的环境行为，而美国环保署（US EPA）的 Bouchard 等最近在常用的水质分析模拟程序（water quality analysis simulation program，WASP）模型中引入颗粒碰撞速率和黏附效率等参数，并将改进的 WASP 模型用于模拟 MWCNT 在地表水环境中的迁移与归趋[164]以及 GO 的光转化与迁移过程[165]。需要注意的是，传统的基于憎水性有机污染物热力学平衡分配的生物吸收模型不适用于 TiO_2 等金属基纳米材料，而应采用生物动力学模型来模拟金属基纳米材料的生物吸收和累积过程[166]，同时需要考虑纳米材料可能发生的溶解、转化、团聚、迁移过程[167, 168]。Keller 课题组最近开发了 nanoFate 模型，并将其示范用于模拟纳米 CeO_2、CuO、TiO_2、ZnO 等纳米金属氧化物在旧金山湾区的归趋[169]。Domercq 课题组最近则提出了一种描述城市水环境系统中纳米材料释放与归趋的综合模型框架，该模型有很高的时间和空间分辨率[170]。而 Wiesner 课题组最近则将基于功能分析（functional assay）的模型用于研究 5 种不同的纳米材料在淡水湿地环境中的迁移，并用中型淡水湿地生态系统的实测数据进行验证，发现除了与悬浮固体的异相团聚，纳米材料自身的团聚以及在植物表面的黏附对于模型预测结果有重要影响，在以后的模型研究中需要加以考虑[171]。

6 展　望

关于纳米材料在水环境中的转化与归趋机制，在过去十余年的研究中已经取得了较为系统、深入的认识。纳米材料进入环境后，可在多种环境因素的作用下发生转化，使其化学组成、粒径分布、表面官能团、表面电荷等理化性质发生显著改变，进而影响其环境行为与生物效应。其中，环境中大量存在的天然有机质能吸附到纳米材料表面，并显著影响纳米材料的胶体稳定性、溶解性和生物可利用性。除纳米零价金属、金属氧化物等金属基纳米材料以及碳纳米材料外，复合纳米材料以及多元金属基纳米材料的环境归趋和生物效应也开始受到关注。此外，近年来一些新的实验技术和模型方法被应用于研究纳米材料的环境转化、归趋和生物效应，加深了我们对这些问题的认识。

但需要注意的是，前期研究多为单因素或多因素控制变量实验研究，以考察纳米材料的性质以及各种环境因素对纳米材料环境转化、迁移等过程的影响规律与机制，在更为复杂的、更接近实际环境条件的模拟实验系统中进行的研究较少。而且，相较于纳米材料在水环境中的溶解与转化过程，纳米材料在土壤环境中的转化与归趋过程研究较少。近期在中型淡水湿地生态系统中进行的研究，有助于阐明纳米材料在复杂真实环境中的转化与归趋机制和规律。随着纳米材料分析检测技术的发展完善，以及新的模型方法的建立，今后应量化研究真实环境或接近真实环境的模拟环境中，多种因素共同作用对纳米材料转化与归趋过程的影响规律和机制。

参 考 文 献

[1] Sun T Y, Mitrano D M, Bornhoft N A, et al. Envisioning nano release dynamics in a changing world: Using dynamic probabilistic modeling to assess future environmental emissions of engineered nanomaterials. Environmental Science & Technology, 2017, 51: 2854-2863.

[2] Pourzahedi L, Vance M, Eckelman M J. Life cycle assessment and release studies for 15 nanosilver-enabled consumer products: Investigating hotspots and patterns of contribution. Environmental Science & Technology, 2017, 51: 7148-7158.

[3] Laux P, Riebeling C, Booth A M, et al. Challenges in characterizing the environmental fate and effects of carbon nanotubes and inorganic nanomaterials in aquatic systems. Environmental Science-Nano, 2018, 5: 48-63.

[4] Arvidsson R, Baun A, Furberg A, et al. Proxy measures for simplified environmental assessment of manufactured nanomaterials. Environmental Science & Technology, 2018, 52: 13670-13680.

[5] Schmutz M, Som C, Krug H F, et al. Digging below the surface: the hidden quality of the OECD nanosilver dossier. Environmental Science-Nano, 2017, 4: 1209-1215.

[6] Hansen S F, Hjorth R, Skjolding L M, et al. A critical analysis of the environmental dossiers from the OECD sponsorship programme for the testing of manufactured nanomaterials. Environmental Science-Nano, 2017, 4: 282-291.

[7] Noordhoek J W, Verweij R A, van Gestel C A M, et al. No effect of selected engineered nanomaterials on reproduction and survival of the springtail *Folsomia candida*. Environmental Science-Nano, 2018, 5: 564-571.

[8] Wagner S, Gondikas A, Neubauer E, et al. Spot the difference: Engineered and natural nanoparticles in the environment—

Release, behavior, and fate. Angewandte Chemie-International Edition, 2014, 53: 12398-12419.

[9] Lead J R, Batley G E, Alvarez P J J, et al. Nanomaterials in the environment: Behavior, fate, bioavailability, and effects—An updated review. Environmental Toxicology and Chemistry, 2018, 37: 2029-2063.

[10] Espinasse B P, Geitner N K, Schierz A, et al. Comparative persistence of engineered nanoparticles in a complex aquatic ecosystem. Environmental Science & Technology, 2018, 52: 4072-4078.

[11] Li Y, Zhao J, Shang E X, et al. Effects of chloride ions on dissolution, ROS generation, and toxicity of silver nanoparticles under UV irradiation. Environmental Science & Technology, 2018, 52: 4842-4849.

[12] Guo X R, Yi Y G, Tan Z Q, et al. Environmentally relevant freeze-thaw cycles enhance the redox mediated morphological changes of silver nanoparticles. Environmental Science & Technology, 2018, 52: 6928-6935.

[13] Wu X H, Neil C W, Kim D, et al. Co-effects of UV/H_2O_2 and natural organic matter on the surface chemistry of cerium oxide nanoparticles. Environmental Science-Nano, 2018, 5: 2382-2393.

[14] Philippe A, Schaumann G E. Interactions of dissolved organic matter with natural and engineered inorganic colloids: A review. Environmental Science & Technology, 2014, 48: 8946-8962.

[15] Mudunkotuwa I A, Grassian V H. Biological and environmental media control oxide nanoparticle surface composition: the roles of biological components (proteins and amino acids), inorganic oxyanions and humic acid. Environmental Science-Nano, 2015, 2: 429-439.

[16] Levak M, Buric P, Sikiric M D, et al. Effect of protein corona on silver nanoparticle stabilization and ion release kinetics in Artificial seawater. Environmental Science & Technology, 2017, 51: 1259-1266.

[17] Zhang J, Guo W, Li Q, et al. The effects and the potential mechanism of environmental transformation of metal nanoparticles on their toxicity in organisms. Environmental Science-Nano, 2018, 5: 2482-2499.

[18] Yin C Y, Zhao W L, Liu R, et al. TiO_2 particles in seafood and surimi products: Attention should be paid to their exposure and uptake through foods. Chemosphere, 2017, 188: 541-547.

[19] Jung Y J, Metreveli G, Park C B, et al. Implications of Pony Lake fulvic acid for the aggregation and dissolution of oppositely charged surface-coated silver nanoparticles and their ecotoxicological effects on *Daphnia magna*. Environmental Science & Technology, 2018, 52: 436-445.

[20] Lin D, Story S D, Walker S L, et al. Influence of extracellular polymeric substances on the aggregation kinetics of TiO_2 nanoparticles. Water Research, 2016, 104: 381-388.

[21] Wang X N, Song W J, Qian H F, et al. Stabilizing interaction of exopolymers with nano-Se and impact on mercury immobilization in soil and groundwater. Environmental Science-Nano, 2018, 5: 456-466.

[22] Sheng A X, Liu F, Xie N, et al. Impact of proteins on aggregation kinetics and adsorption ability of hematite nanoparticles in aqueous dispersions. Environmental Science & Technology, 2016, 50: 2228-2235.

[23] Milne C J, Lapworth D J, Gooddy D C, et al. Role of humic acid in the stability of ag Nanoparticles in suboxic conditions. Environmental Science & Technology, 2017, 51: 6063-6070.

[24] Ortelli S, Costa A L, Blosi M, et al. Colloidal characterization of CuO nanoparticles in biological and environmental media. Environmental Science-Nano, 2017, 4: 1264-1272.

[25] Louie S M, Gorham J M, Tan J J, et al. Ultraviolet photo-oxidation of polyvinylpyrrolidone(PVP)coatings on gold nanoparticles. Environmental Science-Nano, 2017, 4: 1866-1875.

[26] El Hadri H, Louie S M, Hackley V A. Assessing the interactions of metal nanoparticles in soil and sediment matrices—A quantitative analytical multi-technique approach. Environmental Science-Nano, 2018, 5: 203-214.

[27] Loosli F, Stoll S. Effect of surfactants, pH and water hardness on the surface properties and agglomeration behavior of engineered TiO_2 nanoparticles. Environmental Science-Nano, 2017, 4: 203-211.

[28] Wang Z W, Wang X Y, Zhang J Y, et al. Influence of surface functional groups on deposition and release of TiO_2 nanoparticles. Environmental Science & Technology, 2017, 51: 7467-7475.

[29] Fisher-Power L M, Cheng T. Nanoscale titanium dioxide ($nTiO_2$) transport in natural sediments: Importance of soil organic matter and Fe/Al oxyhydroxides. Environmental Science & Technology, 2018, 52: 2668-2676.

[30] Layet C, Auffan M, Santaella C, et al. Evidence that soil properties and organic coating drive the phytoavailability of cerium oxide nanoparticles. Environmental Science & Technology, 2017, 51: 9756-9764.

[31] Micic V, Schmid D, Bossa N, et al. Impact of sodium humate coating on collector surfaces on deposition of polymer-coated nanoiron particles. Environmental Science & Technology, 2017, 51: 9202-9209.

[32] Chen F M, Yuan X M, Song Z F, et al. Gram-negative *Escherichia coli* promotes deposition of polymer-capped silver nanoparticles in saturated porous media. Environmental Science-Nano, 2018, 5: 1495-1505.

[33] Akanbi M O, Hernandez L M, Mobarok M H, et al. QCM-D and NanoTweezer measurements to characterize the effect of soil cellulase on the deposition of PEG-coated TiO_2 nanoparticles in model subsurface environments. Environmental Science-Nano, 2018, 5: 2172-2183.

[34] Cornelis G, Forsberg-Grivogiannis A M, Skold N P, et al. Sludge concentration, shear rate and nanoparticle size determine silver nanoparticle removal during wastewater treatment. Environmental Science-Nano, 2017, 4: 2225-2234.

[35] Goldberg E, McNew C, Scheringer M, et al. What factors determine the retention behavior of engineered nanomaterials in saturated porous media? Environmental Science & Technology, 2017, 51: 2729-2737.

[36] Makselon J, Zhou D, Engelhardt I, et al. Experimental and numerical investigations of silver nanoparticle transport under variable flow and ionic strength in soil. Environmental Science & Technology, 2017, 51: 2096-2104.

[37] Yecheskel Y, Dror I, Berkowitz B. Silver nanoparticle (Ag-NP) retention and release in partially saturated soil: column experiments and modelling. Environmental Science-Nano, 2018, 5: 422-435.

[38] Mitzel M R, Lin N, Whalen J K, et al. *Chlamydomonas reinhardtii* displays aversive swimming response to silver nanoparticles. Environmental Science-Nano, 2017, 4: 1328-1338.

[39] Azimzada A, Tufenkji N, Wilkinson K J. Transformations of silver nanoparticles in wastewater effluents: Links to Ag bioavailability. Environmental Science-Nano, 2017, 4: 1339-1349.

[40] Koser J, Engelke M, Hoppe M, et al. Predictability of silver nanoparticle speciation and toxicity in ecotoxicological media. Environmental Science-Nano, 2017, 4: 1470-1483.

[41] Baccaro M, Undas A K, de Vriendt J, et al. Ageing, dissolution and biogenic formation of nanoparticles: How do these factors affect the uptake kinetics of silver nanoparticles in earthworms? Environmental Science-Nano, 2018, 5: 1107-1116.

[42] Schultz C L, Gray J, Verweij R A, et al. Aging reduces the toxicity of pristine but not sulphidised silver nanoparticles to soil bacteria. Environmental Science-Nano, 2018, 5: 2618-2630.

[43] Poynton H C, Chen C, Alexander S L, et al. Enhanced toxicity of environmentally transformed ZnO nanoparticles relative to Zn ions in the epibenthic amphipod *Hyalella azteca*. Environmental Science-Nano, 2019, 6: 325-340.

[44] Molleman B, Hiemstra T. Time, pH, and size dependency of silver nanoparticle dissolution: The road to equilibrium. Environmental Science-Nano, 2017, 4: 1314-1327.

[45] Hedberg J, Blomberg E, Wallinder I O. In the search for nanospecific effects of dissolution of metallic nanoparticles at freshwater-Like conditions: A critical review. Environmental Science & Technology, 2019, 53: 4030-4044.

[46] Jiang C J, Aiken G R, Hsu-Kim H. Effects of natural organic matter properties on the dissolution kinetics of zinc oxide nanoparticles. Environmental Science & Technology, 2015, 49: 11476-11484.

[47] Gunsolus I L, Mousavi M P S, Hussein K, et al. Effects of humic and fulvic acids on silver nanoparticle stability, dissolution, and toxicity. Environmental Science & Technology, 2015, 49: 8078-8086.

[48] Marchioni M, Gallon T, Worms I, et al. Insights into polythiol-assisted AgNP dissolution induced by bio-relevant molecules. Environmental Science-Nano, 2018, 5: 1911-1920.

[49] Collin B, Tsyusko O V, Starnes D L, et al. Effect of natural organic matter on dissolution and toxicity of sulfidized silver nanoparticles to *Caenorhabditis elegans*. Environmental Science-Nano, 2016, 3: 728-736.

[50] Liu C, Leng W N, Vikesland P J. Controlled evaluation of the impacts of surface coatings on silver nanoparticle dissolution rates. Environmental Science & Technology, 2018, 52: 2726-2734.

[51] Miao L Z, Wang C, Hou J, et al. Enhanced stability and dissolution of CuO nanoparticles by extracellular polymeric substances in aqueous environment. Journal of Nanoparticle Research, 2015, 17: 404.

[52] Poulin B A, Gerbig C A, Kim C S, et al. Effects of sulfide concentration and dissolved organic matter characteristics on the structure of nanocolloidal metacinnabar. Environmental Science & Technology, 2017, 51: 13133-13142.

[53] Gao X Y, Spielman-Sun E, Rodrigues S M, et al. Time and nanoparticle concentration affect the extractability of Cu from CuO NP-amended soil. Environmental Science & Technology, 2017, 51: 2226-2234.

[54] Gao X, Avellan A, Laughton S, et al. CuO nanoparticle dissolution and toxicity to wheat (*Triticum aestivum*) in rhizosphere soil. Environmental Science & Technology, 2018, 52: 2888-2897.

[55] Rico C M, Johnson M G, Marcus M A. Cerium oxide nanoparticles transformation at the root-soil interface of barley (*Hordeum vulgare* L.). Environmental Science-Nano, 2018, 5: 1807-1812.

[56] Peng C, Xu C, Liu Q L, et al. Fate and transformation of cuO nanoparticles in the soil-rice system during the life cycle of rice plants. Environmental Science & Technology, 2017, 51: 4907-4917.

[57] Li M, Wang P, Dang F, et al. The transformation and fate of silver nanoparticles in paddy soil: Effects of soil organic matter and redox conditions. Environmental Science-Nano, 2017, 4: 919-928.

[58] Wang R, Dang F, Liu C, et al. Heteroaggregation and dissolution of silver nanoparticles by iron oxide colloids under environmentally relevant conditions. Environmental Science-Nano, 2019, 6: 195-206.

[59] Levard C, Reinsch B C, Michel F M, et al. Sulfidation processes of pVP-coated silver nanoparticles in aqueous solution: Impact on dissolution rate. Environmental Science & Technology, 2011, 45: 5260-5266.

[60] Ma R, Levard C, Michel F M, et al. Sulfidation mechanism for zinc oxide nanoparticles and the effect of sulfidation on their solubility. Environmental Science & Technology, 2013, 47: 2527-2534.

[61] Ma R, Stegemeier J, Levard C, et al. Sulfidation of copper oxide nanoparticles and properties of resulting copper sulfide. Environmental Science-Nano, 2014, 1: 347-357.

[62] Fan D M, Lan Y, Tratnyek P G, et al. Sulfidation of iron-based materials: A review of processes and implications for water treatment and remediation. Environmental Science & Technology, 2017, 51: 13070-13085.

[63] Zhang X X, Xu Z L, Wimmer A, et al. Mechanism for sulfidation of silver nanoparticles by copper sulfide in water under aerobic conditions. Environmental Science-Nano, 2018, 5: 2819-2829.

[64] Nguyen M L, Murphy J A, Hamlet L C, et al. Ligand-dependent Ag_2S formation: Changes in deposition of silver nanoparticles with sulfidation. Environmental Science-Nano, 2018, 5: 1090-1095.

[65] Zhang Y Q, Xia J C, Xu J L, et al. Impacts of surfactants on dissolution and sulfidation of silver nanowires in aquatic environments. Environmental Science-Nano, 2018, 5: 2452-2460.

[66] Gogos A, Voegelin A, Kaegi R. Influence of organic compounds on the sulfidation of copper oxide nanoparticles. Environmental Science-Nano, 2018, 5: 2560-2569.

[67] Qin H J, Guan X H, Bandstra J Z, et al. Modeling the kinetics of hydrogen formation by zerovalent iron: Effects of sulfidation on micro- and nano-scale particles. Environmental Science & Technology, 2018, 52: 13887-13896.

[68] Bhattacharjee S, Ghoshal S. Sulfidation of nanoscale zerovalent iron in the presence of two organic macromolecules and its

effects on trichloroethene degradation. Environmental Science-Nano, 2018, 5: 782-791.

[69] Kumar N, Pacheco J L, Noel V, et al. Sulfidation mechanisms of Fe(III)-(oxyhydr)oxide nanoparticles: A spectroscopic study. Environmental Science-Nano, 2018, 5: 1012-1026.

[70] Gogos A, Thalmann B, Voegelin A, et al. Sulfidation kinetics of copper oxide nanoparticles. Environmental Science-Nano, 2017, 4: 1733-1741.

[71] Le Bars M, Legros S, Levard C, et al. Drastic change in zinc speciation during anaerobic digestion and composting: Instability of nanosized zinc sulfide. Environmental Science & Technology, 2018, 52: 12987-12996.

[72] Li L X Y, Xu Z L, Wimmer A, et al. New insights into the stability of silver sulfide nanoparticles in surface water: Dissolution through hypochlorite oxidation. Environmental Science & Technology, 2017, 51: 7920-7927.

[73] Stegemeier J P, Colman B P, Schwab F, et al. Uptake and distribution of silver in the aquatic plant *Landoltia punctata* (Duckweed) exposed to silver and silver sulfide nanoparticles. Environmental Science & Technology, 2017, 51: 4936-4943.

[74] Pradas del Real A E, Vidal V, Carriere M, et al. Silver nanoparticles and wheat roots: A complex interplay. Environmental Science & Technology, 2017, 51: 5774-5782.

[75] Stegemeier J P, Avellan A, Lowry G V. Effect of initial speciation of copper- and silver-based nanoparticles on their long-term fate and phytoavailability in freshwater wetland mesocosms. Environmental Science & Technology, 2017, 51: 12114-12122.

[76] Colman B P, Baker L F, King R S, et al. Dosing, not the dose: Comparing chronic and pulsed silver nanoparticle exposures. Environmental Science & Technology, 2018, 52: 10048-10056.

[77] Ward C S, Pan J F, Colman B P, et al. Conserved microbial toxicity responses for acute and chronic silver nanoparticle treatments in wetland mesocosms. Environmental Science & Technology, 2019, 53: 3268-3276.

[78] Geitner N K, Cooper J L, Avellan A, et al. Size-based differential transport, uptake, and mass distribution of ceria (CeO_2) nanoparticles in wetland mesocosms. Environmental Science & Technology, 2018, 52: 9768-9776.

[79] Eymard-Vernain E, Lelong C, Pradas del Real A E, et al. Impact of a model soil microorganism and of Its secretome on the fate of silver nanoparticles. Environmental Science & Technology, 2018, 52: 71-78.

[80] Xie C J, Zhang J Z, Ma Y H, et al. *Bacillus subtilis* causes dissolution of ceria nanoparticles at the nano-bio interface. Environmental Science-Nano, 2019, 6: 216-223.

[81] Avellan A, Simonin M, McGivney E, et al. Gold nanoparticle biodissolution by a freshwater macrophyte and its associated microbiome. Nature Nanotechnology, 2018, 13: 1072-1077.

[82] McGivney E, Gao X Y, Liu Y J, et al. Biogenic cyanide production promotes dissolution of gold nanoparticles in soil. Environmental Science & Technology, 2019, 53: 1287-1295.

[83] Luo H W, Zhang X, Chen J J, et al. Probing the biotransformation of hematite nanoparticles and magnetite formation mediated by *Shewanella oneidensis* MR-1 at the molecular scale. Environmental Science-Nano, 2017, 4: 2395-2404.

[84] Demangeat E, Pedrot M, Dia A, et al. Colloidal and chemical stabilities of iron oxide nanoparticles in aqueous solutions: The interplay of structural, chemical and environmental drivers. Environmental Science-Nano, 2018, 5: 992-1001.

[85] Chen C Y, Huang W L. Aggregation kinetics of diesel soot nanoparticles in wet environments. Environmental Science & Technology, 2017, 51: 2077-2086.

[86] Han Y, Hwang G, Park S, et al. Stability of carboxyl-functionalized carbon black nanoparticles: the role of solution chemistry and humic acid. Environmental Science-Nano, 2017, 4: 800-810.

[87] Li Q Q, Chen B L, Xing B S. Aggregation kinetics and self-assembly mechanisms of graphene quantum dots in aqueous solutions: Cooperative effects of pH and electrolytes. Environmental Science & Technology, 2017, 51: 1364-1376.

[88] Gao Y, Chen K, Ren X M, et al. Exploring the aggregation mechanism of graphene oxide in the presence of radioactive

elements: Experimental and theoretical studies. Environmental Science & Technology, 2018, 52: 12208-12215.

[89] Xia T J, Fortner J D, Zhu D Q, et al. Transport of sulfide-reduced graphene oxide in saturated quartz sand: Cation-dependent retention mechanisms. Environmental Science & Technology, 2015, 49: 11468-11475.

[90] Xia T J, Qi Y, Liu J, et al. Cation-inhibited transport of graphene oxide nanomaterials in saturated porous media: The hofmeister effects. Environmental Science & Technology, 2017, 51: 828-837.

[91] Jiang Y, Raliya R, Liao P, et al. Graphene oxides in water: Assessing stability as a function of material and natural organic matter properties. Environmental Science-Nano, 2017, 4: 1484-1493.

[92] Sun B B, Zhang Y Q, Chen W, et al. Concentration dependent effects of bovine serum albumin on graphene oxide colloidal stability in aquatic environment. Environmental Science & Technology, 2018, 52: 7212-7219.

[93] Lu T T, Xia T J, Qi Y, et al. Effects of clay minerals on transport of graphene oxide in saturated porous media. Environmental Toxicology and Chemistry, 2017, 36: 655-660.

[94] Qi Z C, Du T T, Ma P K, et al. Transport of graphene oxide in saturated quartz sand containing iron oxides. Science of the Total Environment, 2019, 657: 1450-1459.

[95] Feng Y P, Liu X T, Huynh K A, et al. Heteroaggregation of graphene oxide with nanometer- and micrometer-sized hematite colloids: Influence on nanohybrid aggregation and microparticle sedimentation. Environmental Science & Technology, 2017, 51: 6821-6828.

[96] Qi Y, Xia T J, Li Y, et al. Colloidal stability of reduced graphene oxide materials prepared using different reducing agents. Environmental Science-Nano, 2016, 3: 1062-1071.

[97] Xia T J, Ma P K, Qi Y, et al. Transport and retention of reduced graphene oxide materials in saturated porous media: Synergistic effects of enhanced attachment and particle aggregation. Environmental Pollution, 2019, 247: 383-391.

[98] Du T T, Adeleye A S, Zhang T, et al. Influence of light wavelength on the photoactivity, physicochemical transformation, and fate of graphene oxide in aqueous media. Environmental Science-Nano, 2018, 5: 2590-2603.

[99] Hou W C, Lee P L, Chou Y C, et al. Antibacterial property of graphene oxide: The role of phototransformation. Environmental Science-Nano, 2017, 4: 647-657.

[100] Duan L, Zhang T, Song W H, et al. Photolysis of graphene oxide in the presence of nitrate: Implications for graphene oxide integrity in water and wastewater treatment. Environmental Science-Nano, 2019, 6: 136-145.

[101] Wu J W, Li W, Fortner J D. Photoenhanced oxidation of C_{60} aggregates (nC_{60}) by free chlorine in water. Environmental Science-Nano, 2017, 4: 117-126.

[102] Wu J W, Alemany L B, Li W L, et al. Photoenhanced transformation of hydroxylated fullerene (fullerol) by free chlorine in water. Environmental Science-Nano, 2017, 4: 470-479.

[103] Saleh N B, Aich N, Plazas-Tuttle J, et al. Research strategy to determine when novel nanohybrids pose unique environmental risks. Environmental Science-Nano, 2015, 2: 11-18.

[104] Tong T Z, Wilke C M, Wu J S, et al. Combined toxicity of nano-ZnO and nano-TiO_2: From single- to multinanomaterial systems. Environmental Science & Technology, 2015, 49: 8113-8123.

[105] Wilke C M, Tong T Z, Gaillard J F, et al. Attenuation of microbial stress due to nano-ag and nano-TiO_2 interactions under dark conditions. Environmental Science & Technology, 2016, 50: 11302-11310.

[106] Pagano L, Pasquali F, Majumdar S, et al. Exposure of *Cucurbita pepo* to binary combinations of engineered nanomaterials: physiological and molecular response. Environmental Science-Nano, 2017, 4: 1579-1590.

[107] Josko I, Oleszczuk P, Skwarek E. Toxicity of combined mixtures of nanoparticles to plants. Journal of Hazardous Materials, 2017, 331: 200-209.

[108] Wilke C M, Wunderlich B, Gaillard J F, et al. Synergistic bacterial stress results from exposure to nano-Ag and nano-TiO_2

mixtures under light in environmental media. Environmental Science & Technology, 2018, 52: 3185-3194.

[109] Zhang Y Q, Qiang L W, Yuan Y T, et al. Impacts of titanium dioxide nanoparticles on transformation of silver nanoparticles in aquatic environments. Environmental Science-Nano, 2018, 5: 1191-1199.

[110] Kryuchkova M, Fakhrullin R. Kaolin alleviates graphene oxide toxicity. Environmental Science & Technology Letters, 2018, 5: 295-300.

[111] Wilke C M, Gaillard J F, Gray K A. The critical role of light in moderating microbial stress due to mixtures of engineered nanomaterials. Environmental Science-Nano, 2018, 5: 96-102.

[112] Wilke C M, Petersen C, Alsina M A, et al. Photochemical interactions between n-Ag_2S and n-TiO_2 amplify their bacterial stress response. Environmental Science-Nano, 2019, 6: 115-126.

[113] Wu B, Wu J L, Liu S, et al. Combined effects of graphene oxide and zinc oxide nanoparticle on human A549 cells: bioavailability, toxicity and mechanisms. Environmental Science-Nano, 2019, 6: 635-645.

[114] Lian F, Yu W C, Wang Z Y, et al. New insights into black carbon nanoparticle-induced dispersibility of goethite colloids and configuration-dependent sorption for phenanthrene. Environmental Science & Technology, 2019, 53: 661-670.

[115] Huang B, Wei Z B, Yang L Y, et al. Combined toxicity of silver nanoparticles with hematite or plastic nanoparticles toward two freshwater algae. Environmental Science & Technology, 2019, 53: 3871-3879.

[116] Wang D J, Park C M, Masud A, et al. Carboxymethy lcellulose mediates the transport of carbon nanotube-magnetite nanohybrid aggregates in water-saturated porous media. Environmental Science & Technology, 2017, 51: 12405-12415.

[117] Wang D J, Jin Y, Park C M, et al. Modeling the transport of the "new-horizon" reduced graphene oxide-metal oxide nanohybrids in water-saturated porous media. Environmental Science & Technology, 2018, 52: 4610-4622.

[118] Das D, Sabaraya I V, Zhu T R, et al. Aggregation behavior of multiwalled carbon nanotube-titanium dioxide nanohybrids: probing the part-whole question. Environmental Science & Technology, 2018, 52: 8233-8241.

[119] Gunsolus I L, Hang M N, Hudson-Smith N V, et al. Influence of nickel manganese cobalt oxide nanoparticle composition on toxicity toward *Shewanella oneidensis* MR-1: Redesigning for reduced biological impact. Environmental Science-Nano, 2017, 4: 636-646.

[120] Feng Z V, Miller B R, Linn T G, et al. Biological impact of nanoscale lithium intercalating complex metal oxides to model bacterium *B-subtilis*. Environmental Science-Nano, 2019, 6: 305-314.

[121] Laudadio E D, Bennett J W, Green C M, et al. Impact of phosphate adsorption on complex cobalt oxide nanoparticle dispersibility in aqueous media. Environmental Science & Technology, 2018, 52: 10186-10195.

[122] Niemuth N J, Curtis B J, Hang M N, et al. Next-generation complex metal oxide nanomaterials negatively impact growth and development in the benthic invertebrate *Chironomus riparius* upon settling. Environmental Science & Technology, 2019, 53: 3860-3870.

[123] Ribeiro M J, Maria V L, Soares A, et al. Fate and effect of nano tungsten carbide cobalt (WCCo) in the soil environment: Observing a nanoparticle specific toxicity in *Enchytraeus crypticus*. Environmental Science & Technology, 2018, 52: 11394-11401.

[124] Noordhoek J W, Pipicelli F, Barone I, et al. Phenotypic and transcriptional responses associated with multi-generation exposure of *Folsomia candida* to engineered nanomaterials. Environmental Science-Nano, 2018, 5: 2426-2439.

[125] Dai C, Liu J J, Hu Y D. Impurity-bearing ferrihydrite nanoparticle precipitation/deposition on quartz and corundum. Environmental Science-Nano, 2018, 5: 141-149.

[126] Zhou T T, Fan W H, Liu Y Y, et al. Comparative assessment of the chronic effects o five nano-perovskites on *Daphnia magna*: A structure-based toxicity mechanism. Environmental Science-Nano, 2018, 5: 708-719.

[127] Mehrabi K, Nowack B, Dasilya Y A R, et al. Improvements in nanoparticle tracking analysis to measure particle aggregation

and mass distribution: A case study on engineered nanomaterial stability in incineration landfill leachates. Environmental Science & Technology, 2017, 51: 5611-5621.

[128] Geitner N K, O'Brien N J, Turner A A, et al. Measuring nanoparticle attachment efficiency in complex systems. Environmental Science & Technology, 2017, 51: 13288-13294.

[129] Brittle S W, Foose D P, O'Neil K A, et al. A raman-based imaging method for characterizing the molecular adsorption and spatial distribution of silver nanoparticles on hydrated mineral surfaces. Environmental Science & Technology, 2018, 52: 2854-2862.

[130] Mellage A, Holmes A B, Linley S, et al. Sensing coated iron-oxide nanoparticles with spectral induced polarization (SIP): Experiments in natural sand packed flow through columns. Environmental Science & Technology, 2018, 52: 14256-14265.

[131] Kananizadeh N, Peev D, Delon T, et al. Visualization of label-free titanium dioxide nanoparticle deposition on surfaces with nanoscale roughness. Environmental Science-Nano, 2019, 6: 248-260.

[132] Desmau M, Gelabert A, Levard C, et al. Dynamics of silver nanoparticles at the solution/biofilm/mineral interface. Environmental Science-Nano, 2018, 5: 2394-2405.

[133] Gigault J, Balaresque M, Tabuteau H. Estuary-on-a-chip: Unexpected results for the fate and transport of nanoparticles. Environmental Science-Nano, 2018, 5: 1231-1236.

[134] Zhang T Y, Lu D W, Zeng L X, et al. Role of secondary particle formation in the persistence of silver nanoparticles in humic acid containing water under light irradiation. Environmental Science & Technology, 2017, 51: 14164-14172.

[135] Dong S P, Xia T, Yang Y, et al. Bioaccumulation of ^{14}C-labeled graphene in an aquatic food chain through direct uptake or trophic transfer. Environmental Science & Technology, 2018, 52: 541-549.

[136] Ma Y H, Yao Y, Yang J, et al. Trophic transfer and transformation of CeO_2 nanoparticles along a terrestrial Food. Chain: Influence of Exposure Routes. Environmental Science & Technology, 2018, 52: 7921-7927.

[137] Laycock A, Romero-Freire A, Najorka J, et al. Novel multi-isotope tracer approach to test ZnO nanoparticle and soluble Zn bioavailability in joint soil exposures. Environmental Science & Technology, 2017, 51: 12756-12763.

[138] Yang Q Q, Shan W Y, Hu L G, et al. Uptake and transformation of silver nanoparticles and ions by rice plants revealed by dual stable isotope tracing. Environmental Science & Technology, 2019, 53: 625-633.

[139] Supiandi N I, Charron G, Tharaud M, et al. Isotopically labeled nanoparticles at relevant concentrations: How low can we go? The case of CdSe/ZnS QDs in surface waters. Environmental Science & Technology, 2019, 53: 2586-2594.

[140] Tan Z Q, Yin Y G, Guo X R, et al. Tracking the transformation of nanoparticulate and ionic silver at environmentally relevant concentration levels by hollow fiber flow field-flow fractionation coupled to ICPMS. Environmental Science & Technology, 2017, 51: 12369-12376.

[141] Merrifield R C, Stephan C, Lead J. Determining the concentration dependent transformations of Ag nanoparticles in complex media: Using SP-ICP-MS and Au@Ag core shell nanoparticles as tracers. Environmental Science & Technology, 2017, 51: 3206-3213.

[142] Merrifield R C, Stephan C, Lead J R. Quantification of Au nanoparticle biouptake and freshwater algae using single Cell-ICP-MS. Environmental Science & Technology, 2018, 52: 2271-2277.

[143] Bohme S, Baccaro M, Schmidt M, et al. Metal uptake and distribution in the zebrafish (*Danio rerio*) embryo: Differences between nanoparticles and metal ions. Environmental Science-Nano, 2017, 4: 1005-1015.

[144] Tuoriniemi J, Jurgens M D, Hassellov M, et al. Size dependence of silver nanoparticle removal in a wastewater treatment plant mesocosm measured by FAST single particle ICP-MS. Environmental Science-Nano, 2017, 4: 1189-1197.

[145] Metzelder F, Schmidt T C. Environmental conditions influencing sorption of inorganic anions to multiwalled carbon nanotubes studied by column chromatography. Environmental Science & Technology, 2017, 51: 4928-4935.

[146] Wei H R, Leng W N, Song J, et al. Real-time monitoring of ligand exchange kinetics on gold nanoparticle surfaces enabled by hot spot-normalized surface-enhanced raman scattering. Environmental Science & Technology, 2019, 53: 575-585.

[147] Wu H B, Gonzalez-Pech N I, Grassian V H. Displacement reactions between environmentally and biologically relevant ligands on TiO_2 nanoparticles: Insights into the aging of nanoparticles in the environment. Environmental Science-Nano, 2019, 6: 489-504.

[148] Merrifield R C, Arkill K P, Palme R E, et al. A high resolution study of dynamic changes of Ce_2O_3 and CeO_2 nanoparticles in complex environmental media. Environmental Science & Technology, 2017, 51: 8010-8016.

[149] Su Y, Tong X, Huang C, et al. Green algae as carriers enhance the bioavailability of ^{14}C-labeled few-layer graphene to freshwater snails. Environmental Science & Technology, 2018, 52: 1591-1601.

[150] David C A, Galceran J, Quattrini F, et al. Dissolution and phosphate-induced transformation of ZnO nanoparticles in synthetic saliva probed by AGNES without previous solid–liquid separation. Comparison with UF-ICP-MS. Environmental Science & Technology, 2019, 53: 3823-3831.

[151] Briffa S M, Lynch I, Trouillet V, et al. Thermal transformations of manufactured nanomaterials as a proposed proxy for ageing. Environmental Science-Nano, 2018, 5: 1618-1627.

[152] Zhang L Q, Wang W X. Dominant role of silver ions in silver nanoparticle toxicity to a unicellular alga: Evidence from luminogen imaging. Environmental Science & Technology, 2019, 53: 494-502.

[153] Martin J D, Frost P C, Hintelmann H, et al. Accumulation of silver in Yellow Perch (*Perca flavescens*) and Northern Pike (*Esox lucius*) from a lake dosed with nanosilver. Environmental Science & Technology, 2018, 52: 11114-11122.

[154] Chen C, Waller T, Walker S L. Visualization of transport and fate of nano and micro-scale particles in porous media: modeling coupled effects of ionic strength and size. Environmental Science-Nano, 2017, 4: 1025-1036.

[155] Babakhani P, Doong R A, Bridge J. Significance of early and late stages of coupled aggregation and sedimentation in the fate of nanoparticles: Measurement and modeling. Environmental Science & Technology, 2018, 52: 8419-8428.

[156] Babakhani P, Bridge J, Phenrat T, et al. Aggregation and sedimentation of shattered graphene oxide nanoparticles in dynamic environments: A solid-body rotational approach. Environmental Science-Nano, 2018, 5: 1859-1872.

[157] Tang H, Zhao Y, Yang X N, et al. New insight into the aggregation of graphene oxide using molecular dynamics simulations and extended derjaguin-landau-verwey-overbeek theory. Environmental Science & Technology, 2017, 51: 9674-9682.

[158] Chen M, Zeng G M, Xu P A, et al. Understanding enzymatic degradation of single-walled carbon nanotubes triggered by functionalization using molecular dynamics simulation. Environmental Science-Nano, 2017, 4: 720-727.

[159] Tang H, Zhao Y, Yang X N, et al. Understanding the pH-dependent adsorption of ionizable compounds on graphene oxide using molecular dynamics simulations. Environmental Science-Nano, 2017, 4: 1935-1943.

[160] Tang H, Zhao Y, Shan S J, et al. Wrinkle- and edge-adsorption of aromatic compounds on graphene oxide as revealed by atomic force microscopy, molecular dynamics simulation, and density functional theory. Environmental Science & Technology, 2018, 52: 7689-7697.

[161] Wang Y, Comer J, Chen Z F, et al. Exploring adsorption of neutral aromatic pollutants onto graphene nanomaterials via molecular dynamics simulations and theoretical linear solvation energy relationships. Environmental Science-Nano, 2018, 5: 2117-2128.

[162] Tang H, Zhao Y, Shan S J, et al. Theoretical insight into the adsorption of aromatic compounds on graphene oxide. Environmental Science-Nano, 2018, 5: 2357-2367.

[163] Geitner N K, Zhao W L, Ding F, et al. Mechanistic insights from discrete molecular dynamics simulations of pesticide-nanoparticle interactions. Environmental Science & Technology, 2017, 51: 8396-8404.

[164] Bouchard D, Knightes C, Chang X J, et al. Simulating multiwalled carbon nanotube transport in surface water systems using

the water quality analysis simulation program (WASP). Environmental Science & Technology, 2017, 51: 11174-11184.

[165] Han Y L, Knightes C D, Bouchard D, et al. Simulating graphene oxide nanomaterial phototransformation and transport in surface water. Environmental Science-Nano, 2019, 6: 180-194.

[166] Isaacson C W, Sigg L, Ammann A A, et al. Interactions of TiO$_2$ nanoparticles and the freshwater nematode *Plectus aquatilis*: Particle properties, kinetic parameters and bioconcentration factors. Environmental Science-Nano, 2017, 4: 712-719.

[167] Jiang C J, Castellon B T, Matson C W, et al. Relative contributions of copper oxide nanoparticles and dissolved copper to Cu uptake kinetics of gulf killifish (*Fundulus grandis*) embryos. Environmental Science & Technology, 2017, 51: 1395-1404.

[168] Vijver M G, Zhai Y J, Wang Z, et al. Emerging investigator series: The dynamics of particle size distributions need to be accounted for in bioavailability modelling of nanoparticles. Environmental Science-Nano, 2018, 5: 2473-2481.

[169] Garner K L, Suh S, Keller A A. Assessing the risk of engineered nanomaterials in the environment: development and application of the nano fate model. Environmental Science & Technology, 2017, 51: 5541-5551.

[170] Domercq P, Praetorius A, Boxall A B A. Emission and fate modelling framework for engineered nanoparticles in urban aquatic systems at high spatial and temporal resolution. Environmental Science-Nano, 2018, 5: 533-543.

[171] Geitner N K, Bossa N, Wiesner M R. Formulation and validation of a functional assay-driven model of nanoparticle aquatic transport. Environmental Science & Technology, 2019, 53: 3104-3109.

作者：陈　威[1]，刘思金[2]，潘丙才[3]，全　燮[4]，闫　兵[5,6]

协稿：姜传佳[1]，徐　明[2]，徐立宁[2]，戚　豫[2]

统稿：姜传佳[1]

[1]南开大学，[2]中国科学院生态环境研究中心，[3]南京大学，[4]大连理工大学，[5]广州大学，[6]山东大学

第 23 章 环境中纳米材料的生物效应与分子机制研究进展

- 1. 引言 /598
- 2. 环境冠对纳米材料生物效应的影响 /598
- 3. 纳米材料不同层次的生物效应与关键调控因子 /604
- 4. 纳米材料的植物毒性与分子机制 /611
- 5. 纳米材料的水生毒理效应 /614
- 6. 纳米材料对超重群体的毒性效应 /617
- 7. 纳米材料与污染物的复合生物效应 /618
- 8. 雾霾颗粒物关键致毒组分研究 /622
- 9. 展望 /625

本章导读

纳米材料释放到环境中后，可能对生态环境和人类健康造成潜在危害。本章聚焦于环境中纳米材料生物效应与分子机制的最新研究进展，从纳米材料在复杂环境中的转化（尤其是纳米材料表面环境冠的形成及其对纳米材料毒性效应和毒性机制的影响），纳米材料在不同层次（生物分子、细胞、个体、生态系统）的生物效应，纳米材料的植物和水生毒理效应，纳米材料-无机/有机污染物复合污染的毒性效应与毒性机制，纳米材料对超重群体的毒性效应，以及雾霾颗粒物关键致毒组分研究等方面总结了关于纳米材料生态与健康风险的最新发现，指出当前研究存在的主要问题及未来的研究方向。

关键词

人工纳米材料，环境转化，环境冠，生物效应，毒性机制

1 引 言

纳米合成技术的高速发展，使越来越多微纳尺度的新型材料得以有效开发。由于限域空间的影响，纳米材料可具有量子效应、小尺寸效应、自组装效应等诸多优点，因而目前在很多领域得到了广泛的应用。据统计，目前有超过1892种纳米产品在39个国家和地区销售。纳米材料释放到环境中后，可能对生态环境和人类健康造成潜在危害。纳米材料在环境中以及生物体内会发生复杂的转化过程，进而影响其环境归趋、毒性效应与毒性机制[1]。在众多影响因素中，环境和生物体内大量存在的有机质能吸附到纳米材料表面形成冠层，并显著影响纳米材料的胶体稳定性、溶解性、生物可利用性和毒性效应。另一方面，纳米材料与环境中的有机污染物、无机污染物共存，可能发生协同或拮抗效应，造成纳米材料-污染物复合暴露风险。其中，大气细颗粒物是典型的纳米尺寸的核载体吸附多种有机污染物、无机污染物形成的复杂纳米材料——污染物复合物，其环境健康风险也受到了广泛的关注。

本章聚焦纳米材料的生物效应与分子机制，从纳米材料的环境转化（尤其是纳米材料表面"环境冠"的形成），不同层次的生物效应，纳米材料的植物和水生毒理效应及纳米材料——污染物复合污染的环境健康风险等方面综述纳米材料的生态与健康风险的最新研究进展，指出当前研究主要存在的问题及未来的研究方向。

2 环境冠对纳米材料生物效应的影响

纳米材料在实际环境水体中会发生复杂的转化过程，进而影响其环境归趋、毒性效应与毒性机制[2]。特别地，真实水环境中广泛存在的有机大分子，如腐殖质等天然有机质（NOM）、胞外聚合物（extracellular polymeric substances，EPS）、蛋白质等[3]，会在原始态纳米材料表面形成"环境冠（environmental corona）"或"生态冠（eco-corona）"[3, 4]。与原始态纳米材料的理化特性不同，环境冠的存在会赋予纳米材料新的生物学特性（biological identity）[5-7]，影响纳米材料的水环境行为（如团聚、沉降、转化、溶解等），并介导纳米材料与生物的作用，进而导致不同的毒性效应[8, 9]。因此，有必要针对纳米材料的环境冠进行系统的研究，才能最终阐明纳米材料的生态环境与健康风险。本节将概述水环境中纳米材料环境冠的形成与特征，以及环境冠对纳米材料毒性效应和机制的影响。

2.1 纳米材料环境冠的形成与特征

当纳米材料进入生物体液（如血液、淋巴液、肺泡液）后，体液内的生物大分子会与纳米材料发生作用，在其表面形成"生物冠（biological corona）"[10, 11]。典型的生物冠，如蛋白冠（protein corona）和脂冠（lipid corona），会通过屏蔽效应、配体-受体作用、免疫识别作用等机制，影响纳米材料的细胞毒性、体内分布、组织代谢、免疫反应等过程[12-14]。因此，生物冠的研究对于揭示纳米材料的人体健康效应和发展疾病诊疗技术非常重要，相关研究也发展迅速。

相较于生物冠，环境冠主要指人工纳米材料进入环境后，在自然环境条件下，其表面形成的一层或多层由环境大分子所组成的结构，如"NOM冠"、"EPS冠"和蛋白冠（图23-1）[15, 16]。例如，在真实水环境中，由于纳米材料具有很高的比表面积与表面自由能，容易吸附周围环境中的有机大分子（如腐殖质、蛋白质、多糖、表面活性剂、脂质等），降低纳米材料自身的表面自由能并提高稳定性[15, 17]。在纳米材料的表界面，环境大分子可以通过静电作用、范德瓦耳斯力、疏水作用、氢键等方式被吸附[18]。环境大分子与纳米材料的表界面作用强弱受纳米材料自身物理化学性质的影响，如化学组成、形状、粒径、表面性质等[3]。同时，理论上，环境冠并非静态的，而是会受水环境条件（如光照、温度、pH、离子强度等）变化的影响而产生动态的变化[19]。例如，当NOM和蛋白质共存时，在二氧化钛（TiO_2）纳米颗粒（NPs）表面形成的环境冠，其组成随NOM和蛋白质引入的顺序而动态变化[20]。

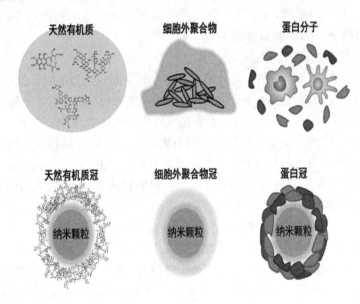

图 23-1 人工纳米材料的典型环境冠

然而，目前针对纳米材料环境冠的毒理研究相对比较缺乏，亟待被系统地阐述。与生物体液相比，真实水环境更为复杂，特别是不同水体（如河流、湖泊和海洋等）的时间和空间差异，难以系统地比较和归类。同时，真实水体的环境条件与化学组成复杂，需要综合运用不同的技术和方法对环境冠进行表征与解析。针对纳米材料环境冠的毒理研究，需要多学科交叉研究的背景与策略。

依据目前已有研究报道，人工纳米材料在水环境中形成的典型环境冠可归为以下几类。

1) 天然有机质冠（NOM corona）

天然有机质是水环境中广泛存在的各种有机化合物的总称，包括腐殖质和多种生物分子（如有机酸、多肽、脂质、糖、胺、醇等）。不同有机分子与人工纳米材料的作用强弱和机制存在明显差异，因此也决定了其形成的 NOM 冠具有复杂的成分和形成机制。例如，一项研究发现，磁赤铁矿（γ-Fe_2O_3）纳米颗粒表面的腐殖酸长链分子构象会受颗粒表面性质和 pH 的影响；在 pH 为 9 时，γ-Fe_2O_3 纳米颗粒表面的腐殖酸长链分子呈现出拉伸构象，并提高了 γ-Fe_2O_3 纳米颗粒的胶体稳定性[21]。Pallem 等则发现，腐殖酸在金（Au）纳米颗粒表面的吸附作用依赖于其表面的配体：当 Au 纳米颗粒的表面配体为β-D-葡萄糖时，腐殖酸会与其发生配体交换；而当 Au 纳米颗粒的表面配体为柠檬酸时，腐殖酸会重叠覆盖在其表面[22]。Chang 等也发现，腐殖酸吸附会造成多壁碳纳米管（MWCNT）表面的十二烷基硫酸钠（SDS）解吸附，并改变其表面性质[23]。

不同 NOM 组分在纳米材料表面的吸附特性不同。例如，Baalousha 等研究了银（Ag）纳米颗粒表面 NOM 冠的化学特征，发现其通常富含氮、硫化合物，但 NOM 的分子量、不饱和键和含氧基团与不同水源中的 NOM 紧密相关[15]。Ghosh 等比较了三种化学结构不同的腐殖酸组分与氧化铝（Al_2O_3）纳米颗粒作用的差异[24]。他们发现，在 Ca^{2+} 存在的条件下，腐殖酸的表面吸附会影响 Al_2O_3 纳米颗粒的胶体稳定性，并且与三种腐殖酸组分的极性和链长存在明显的关联性。而二维相关尺寸排除色谱（2D-CoSEC）和三维荧光-平行因子分析（EEM-PARAFAC）研究则发现，高分子量且具有荧光性的腐殖质组分更优先吸附到 TiO_2、氧化锌（ZnO）等金属氧化物纳米材料的表面[25]。碳纳米管（CNTs）对不同 NOM 组分的吸附与 NOM 的化学性质（如芳香性、氧化程度、羧基酸性等）密切相关：相较于水体中微生物源的 NOM，CNTs 优先吸附陆地源的、经过氧化分解作用产生的 NOM[26]。而且，不同分子量的 NOM 组分在 CNTs 上的吸附能力也不相同：对于腐殖酸，CNTs 主要吸附分子量在 0.5～2 kDa 的组分；对于富里酸，主要吸附分子量在 1～3 kDa 的组分；< 0.4 kDa 的组分则不吸附[26]。

2）胞外聚合物冠（EPS corona）

在水环境中，微生物（如细菌、真菌）会分泌出 EPS，其主要成分为多糖和蛋白质（二者在 EPS 中的总含量约为 70%～90%）[27]，此外还包括部分脂质、核酸和无机物质。EPS 的分子量主要介于 10～30 kDa。对于微生物，EPS 具有重要作用：一方面 EPS 能够在水环境中固定微生物，保证其生存环境的稳定；另一方面，EPS 可以保证微生物细胞免受水环境中有毒有害物质（如重金属和人工纳米材料）的损害[28]。

在真实水环境中，纳米材料可以与 EPS 发生作用，并形成 EPS 冠。Verneuil 等发现，谷皮菱形藻（*Nitzschia palea*）分泌的 EPS 会被双层碳纳米管吸附，且吸附成分主要含有类蛋白聚合物（10 kDa 和 174 kDa）[29]。Jain 等则发现，在废水中，生物合成的硒纳米颗粒（BioSe NPs）表面存在 EPS 吸附层；EPS 吸附层的组分包括羧酸、磷酸基、磺酸基、亚磺酸基、硫醇基以及氨基等多种官能团，因此会影响 BioSe NPs 的表面电荷。同时，他们证实 EPS 吸附成分的主要化学元素包括碳、氮、氧、磷、硫、钙和铁。进一步，他们还解析了 EPS 冠的主要成分，包括多糖 [（313.8±3.5）mg/g]、蛋白质 [（144.1±2.1）mg/g]、腐殖质样物质 [（158.2±2.3）mg/g] 和 DNA [（4.6±0.8）mg/g] 等[30]。Liu 等发现砷还原菌株（*Pantoea sp.* IMH）通过还原反应产生的 AuNPs 表面存在约 3 nm 厚的 EPS 冠，主要含有膜蛋白、脂蛋白和磷脂[31]。而近期的另一项研究则发现，不同生长状态的大肠杆菌（*Escherichia coli* K-12 MG1655）所分泌的 EPS 在 Ag 纳米颗粒表面形成的 EPS 冠的成分不同[32]。Wang 等研究了 ZnO 以及氧化硅（SiO_2）纳米颗粒与 EPS 的作用，发现 EPS 中类蛋白物质的酰胺（N—H 和 C—N）和羧酸类物质的羰基（C=O）是纳米颗粒的重要键合位点[33]。Sheng 等重点研究了奥奈达希瓦氏菌（*Shewanella oneidensis* MR-1）分泌的 EPS 中外膜细胞色素 c（OmcA）与赤铁矿（Fe_2O_3）纳米颗粒的作用[34]。他们发现，在 pH 为 5.7 和低盐浓度条件下，Fe_2O_3 纳米颗粒（9 nm、36 nm 和 112 nm）均会吸附 OmcA，但吸附效率随 OmcA 浓度增加先升高后降低。但在高盐浓度条件下，吸附效率随 OmcA 浓度增加而逐渐降低。并且，OmcA 的吸附能力与 Fe_2O_3 纳米颗粒的粒径呈现一定的相关性。除了纳米颗粒的粒径，Nevius 等证明聚苯乙烯（PS）纳米颗粒的表面官能团和电荷（SO_4^-、—COO^- 和—NH_2）会直接影响纳米颗粒与 EPS 的作用强度[35]。Adeleye 等则发现，TiO_2 纳米颗粒对 EPS 的吸附取决于颗粒的表面积、电荷和亲疏水性等因素，且 EPS 的—COOH 会通过去质子化的方式与钛离子结合[18]。

3）蛋白冠（protein corona）

目前，纳米材料在哺乳动物体液（如血液）中形成的蛋白冠研究相对比较深入。然而，对纳米材料在水生生物体内形成的蛋白冠的研究还非常有限。纳米材料进入水生生物体后，会与其体液或黏液接触，并在纳米颗粒的表面迅速形成蛋白冠。之后，纳米材料可以被细胞摄入，进而与细胞内的蛋白质作用并形成蛋白冠。在蛋白冠形成的同时，蛋白质的分子构象或生物功能都可能发生变化。相较于 NOM 冠和 EPS 冠，蛋白冠具有更重要的生理意义，它会介导纳米材料与细胞的作用。

目前，越来越多的研究开始关注纳米材料在水生生物体内形成的蛋白冠。例如，Yue 等研究了 Ag 纳米颗粒在虹鳟鱼（Oncorhynchus mykiss）鳃细胞（RTgill-W1）内的蛋白冠组分，并识别出与细胞膜黏附、摄入和囊泡运输、应激响应等功能相关的蛋白质。他们的发现证明，对于纳米材料，蛋白冠具有重要的生物功能[36]。Gao 等研究了聚乙烯吡咯烷酮（PVP）包被的银纳米颗粒（PVP-AgNPs）在小口黑鲈（Micropterus dolomieu）血清中形成的蛋白冠，发现蛋白冠的厚度与表面电荷受鱼性别的影响，具有性别特异性。蛋白质组学分析结果显示，在雌鱼血清中，蛋白冠含有卵黄特异性蛋白（vitellogenin 和 zona pellucida），这意味着 PVP-AgNPs 可能会在鱼卵细胞内累积并具有生殖毒性[37]。最近另一项研究发现，具有性别特异性的蛋白冠也影响斑马鱼（Danio rerio）免疫细胞对 SiO_2 纳米颗粒的响应[38]。Canesi 等发现在紫贻贝（Mytilus galloprovincialis）血淋巴血清中，不同纳米材料，如氨基化 PS NPs、氧化铈（CeO_2）NPs 和 TiO_2 NPs 等形成的蛋白冠具有特异性[39, 40]。他们的研究结果显示，MgC1q6 蛋白是氨基化 PS NPs 的蛋白冠的主要成分，而表面带负电荷的 CeO_2 NPs 则主要吸附 Cu/Zn-SOD 蛋白[41]。此外，水生生物的分泌物中的蛋白质也可以与纳米材料作用，形成蛋白冠。例如，Mu 等发现斑马鱼分泌物中的蛋白质会被氧化石墨烯（GO）纳米片吸附，并影响 GO 纳米片的形貌和毒性[42]。Bourgeault 等研究了 SiO_2 NPs 和 TiO_2 NPs 与紫贻贝鳃黏液蛋白的相互作用，发现高丰度的外套膜外蛋白（extrapallial protein, EP）几乎没有被吸附，其归因于 EP 中组氨酸的高含量，使得蛋白质分子不易于在纳米颗粒的表面展开[16, 43]。他们还发现主穹窿蛋白（Major Vault Protein）只会在 SiO_2 NPs 表面吸附，而没有在 TiO_2 NPs 的蛋白冠中检测到[16]。最近，Wheeler 课题组开发了一种基于机器学习的模型，用于根据蛋白质和纳米材料的理化性质以及水化学条件来预测纳米 Ag 表面形成的蛋白冠的指纹特征[44]。

2.2 环境冠对纳米材料毒性效应的影响

目前，关于人工纳米材料对水生生物的毒性效应已有不少科学发现，然而大部分研究只关注了原始态纳米材料的毒性，而且暴露实验通常在"理想"环境条件下开展，很难反映真实环境中的暴露状态。因此，越来越多的研究开始关注生态环境大分子（如 NOM、EPS 和蛋白质）在纳米材料表面形成的环境冠对纳米材料生物毒性的影响。

研究表明，NOM 冠的形成对金属基纳米材料的生物毒性效应有显著影响。例如，Li 等发现纳米零价铁（nZVI）在吸附 NOM 后，对大肠杆菌的毒性会显著降低[45]。Ostermeyer 等研究了柠檬酸根包被的 Ag 纳米颗粒（Cit-AgNPs）对氨氧化细菌（Nitrosomonas europaea）的毒性，发现牛血清白蛋白（BSA）和藻酸盐均可降低 Cit-AgNPs 的毒性[46]。然而，Wirth 等却发现，腐殖酸可以提高 AgNPs 的稳定性，而不会增加 AgNPs 对生物膜中荧光假单胞菌（Pseudomonas fluorescens）的毒性[47]。纳米 Ag 还能抑制微生物形成生物膜，但最近的研究发现，腐殖酸能够减轻纳米 Ag 对荧光假单胞菌形成生物膜的抑制作用[48]。Cupi 等考察了 NOM 和老化过程分别对纳米 Ag 等纳米材料毒性的影响，发现老化过程并没有降低纳米 Ag 对大型溞（Daphnia magna）的毒性，但 20 mg/L 的 NOM 却可以完全消除纳米 Ag 的毒性[49]。Jung 等研究了Pony Lake富里酸（PLFA）对柠檬酸根修饰的Ag纳米颗粒（Cit-AgNPs）和分枝状聚乙烯亚胺（BPEI）修饰的Ag纳米颗粒（BPEI-AgNPs）环境行为和毒性的影响，发现高浓度的PLFA可以增强Cit-AgNPs与BPEI-AgNPs的胶体稳定性，并且其溶解性会随PLFA浓度而变化。基于上述发现，他们证实PLFA可以降低AgNPs对大型溞超过70%的毒性[50]。Gao等研究了Suwannee River腐殖酸（SRHA）对AgNPs毒性的影响。结果发现，当AgNPs的暴露浓度为50 μg/L时，网纹溞（Ceriodaphnia dubia）的死亡率随SRHA浓度（0～20 mg/L）的增加而降低[51]。

纳米金属氧化物、金属硫化物量子点等纳米材料的毒性效应也受到NOM冠等环境冠的影响。例如，NOM降低了ZnO纳米颗粒的团聚，同时减少了ZnO纳米颗粒释放出的Zn^{2+}的生物可利用性，以及由ZnO纳米颗粒诱导的氧化应激损伤；相比之下，NOM并未影响MnO_2纳米颗粒的团聚行为及毒性[52]。Ong等发现，腐殖酸可以减轻ZnO纳米颗粒等多种纳米材料对斑马鱼的发育毒性[53]。大型溞分泌的蛋白质等生物分子能在纳米ZnO、纳米CeO_2等金属氧化物纳米材料的表面形成环境冠，影响其团聚状态和溶解性，进而影响其被大型溞的摄取以及造成的毒性效应[54]。最近的一项研究通过比较硫化镉（CdS）量子点对野生型酿酒酵母（Saccharomyces cerevisiae）以及经过基因编辑的变异菌株的细胞毒性发现：相较于野生型酵母菌株，删除编码某些蛋白基因的变异酵母菌株对CdS量子点的毒性有更高的耐受度，说明蛋白冠在CdS量子点与酵母细胞的相互作用中发挥重要功能[55]。

近期，也有研究者考察不同分子和化学性质的NOM组分对Ag纳米颗粒等纳米材料毒性效应的影响。例如，Gunsolus等发现，PLFA会降低Cit-Ag纳米颗粒对奥奈达希瓦氏菌的毒性，而Suwannee River富里酸（SRFA）和SRHA影响并不显著[56]。Yang等发现，PLFA和SRFA会降低Cit-AgNPs对秀丽隐杆线虫（Caenorhabditis elegans）肠上皮组织的损伤和致死率[57]。并且，PLFA降低纳米Ag毒性的效果更为显著，主要归因于PLFA比SRFA具有更高的硫含量，可以更有效地结合银离子[58]。Unrine课题组则比较了不同来源的NOM影响硫化纳米银（sAgNPs）对秀丽隐杆线虫的毒性。他们发现，PLFA可以增强sAgNPs的溶解性并体现出更强的毒性，但SRFA和PPFA却降低了sAgNPs的溶解性和毒性效应[59]。Kteeba等比较了不同的NOM存在时，粒径10～30 nm的ZnO纳米颗粒对斑马鱼胚胎孵化率的影响，结果显示，不同NOM存在时，ZnO纳米颗粒毒性降低顺序依次为：腐殖酸≥不同来源的NOM（既含有腐殖酸也含有富里酸）>藻酸≫BSA[60]。周丛生物膜的EPS会降低Fe_2O_3纳米颗粒对微生物的毒性，而且EPS的不同组分（溶解态、表面松散连接、表面紧密连接）在其中起的作用不同[61]。林道辉课题组最近也发现，不同的EPS组分（溶解态或表面连接）对纳米TiO_2与小球藻（Chlorella pyrenoidosa）的相互作用有不同的影响[62]。

2.3 环境冠对纳米材料毒性机制的影响

与传统环境污染物（如重金属）不同，作为一类新型环境污染物，纳米材料的毒性效应和机制更加复杂[1]。在水环境中，环境冠可以改变纳米材料的环境行为，如溶解性、胶体稳定性等，进而影响其毒性效应[50, 59, 63]。此外，环境冠还可以介导纳米材料与水生生物的作用，并展现出与原始态纳米材料不同的毒性作用机制。例如，在鱼的血清中，纳米材料可以形成蛋白冠，而蛋白冠中的特定蛋白质组分可以影响纳米材料与细胞的相互作用，如细胞膜黏附、细胞摄入、囊泡运输、应激响应等[36]。目前已知，原始态纳米材料可以通过诱发氧化应激、破坏遗传物质、激活炎症反应、干扰代谢平衡、诱导细胞死亡等机制导致毒性损伤[64-66]。然而，环境冠的形成是否会改变原始态纳米材料的毒性机制仍不清楚。因此，为了明确纳米材料与生物体之间的相互作用和毒性效应，必须揭示环境冠介导纳米材料细胞毒性的作用机制[67, 68]。

2.3.1 生物摄入与体内累积

环境冠对纳米材料生物摄入与体内累积的影响因纳米材料和生物种类而异。腐殖酸等NOM通常会降低纳米材料及其释放的金属离子的生物摄取。例如，Ha等发现，SRHA和胎牛血清（FBS）均可显著影响富勒烯C_{60}纳米颗粒的粒径和表面电荷，并降低Caco-2细胞对C_{60}纳米颗粒的摄入[69]。Chen等同样发现，腐殖酸可以影响C_{60}纳米颗粒的粒径与表面电荷，并降低大型溞和斑马鱼对C_{60}纳米颗粒的摄入[70]。Jiang等发现，SRHA、SRFA等NOM可以降低大底鳉（*Fundulus grandis*）胚胎对氧化铜（CuO）纳米颗粒以及溶解态铜离子的吸收速率常数，而且芳香碳含量越高的NOM，降低铜吸收的效果越明显[71]。Collin等发现腐殖酸可以影响CeO_2纳米颗粒在秀丽隐杆线虫的体内累积；当腐殖酸浓度升高时，CeO_2纳米颗粒在秀丽隐杆线虫体内的累积减少，毒性也降低[72]。腐殖酸也能抑制恶臭假单胞菌（*Pseudomonas putida*）对Fe_2O_3纳米颗粒的黏附和吸收，并减少活性氧物种（ROS）的产生，从而降低其毒性[73]。然而，Yang等研究了PLFA影响秀丽隐杆线虫对AgNPs的摄取行为，发现PLFA会减少秀丽隐杆线虫对Ag^+的摄取，但不会影响其对Cit-AgNPs的摄取[57]。同样的，Oliver等研究了腐殖酸对静水椎实螺（*Lymnaea stagnalis*）通过食物摄入AgNPs的影响，发现腐殖酸对AgNPs的体内累积与毒性并无显著性的影响[74]。最近，Li等证实，NOM会影响AgNPs的胶体稳定性和溶解性，并降低其对水稻（*Oryza sativa*）幼苗的生物有效性；而且，不同NOM组分的影响程度不同，巯基含量高的NOM组分能更有效地降低AgNPs的生物有效性[75]。

但也有研究发现，NOM会促进纳米材料的生物摄入与体内累积。例如，Wang等发现溶解态有机质（DOM）可以增强铜绿微囊藻（*Microcystis aeruginosa*）对CuO纳米颗粒的细胞摄入[76]。Hayashi等则发现，AgNPs会吸附蚯蚓（*Eisenia fetida*）的体腔蛋白（EfCP），并可以促进蚯蚓体腔细胞对AgNPs的摄入[77]。Nasser等研究了大型溞分泌的蛋白质对PS NPs的影响，发现大型溞的分泌蛋白会快速在PS NPs表面形成环境冠，并导致纳米颗粒分散性变差。相较于原始态的PS NPs，环境冠导致大型溞体内累积的PS NPs更难被代谢出体外，并对其进食行为产生了影响[78]。

2.3.2 膜黏附与膜损伤

Kang等发现，EPS中的多糖可以还原Au^{3+}离子形成粒径2～15 nm的金纳米颗粒并黏附于大肠杆菌的表面[79]。Mensch等发现NOM可以改变金刚石纳米颗粒的粒径和ζ电位，并减少金刚石NPs在革兰氏阴性的奥奈达希瓦氏菌细胞膜表面的附着，从而降低金刚石NPs对奥奈达希瓦氏菌的膜损伤[80]。另外一项研究也证明，NOM可以降低nZVI纳米颗粒在大肠杆菌表面的黏附，从而降低其毒性效应[45]。Liu等发现，4种有机大分子（SRFA、SRHA、藻酸盐和BSA）在硫化银（Ag_2S）纳米颗粒表面吸附的厚度不同，依次为BSA＞SRHA＞藻酸盐＞SRFA；而且，四种有机大分子吸附后的Ag_2S NPs对大肠杆菌细胞膜损伤的程度刚好与上述次序相反[81]。另有研究发现，紫贻贝的血淋巴血清可以增加CeO_2 NPs和PS NPs对原代血细胞的毒性，主要表现为溶酶体膜的稳定性降低、吞噬能力下降以及胞外ROS水平的改变[40, 41]。Mu等则发现，在水中，GO纳米片会与斑马鱼的分泌物结合，并表现出对斑马鱼胚胎更高的毒性，如死亡率、致畸性、β-半乳糖苷酶上调和线粒体膜电位异常[42]。腐殖酸与不同官能化的石墨烯纳米材料共暴露后，腐殖酸能显著降低石墨烯纳米材料对斜生栅藻（*Scenedesmus obliquus*）的毒性，其作用机制除了通过改变石墨烯纳米材料的结构和表面电荷从而减少其与藻类细胞的接触以及作为抗氧化剂消耗ROS外，腐殖酸还可以减轻石墨烯纳米材料对藻类细胞的物理穿透和损伤[82]。

2.4 小结

综上所述，在真实水环境中，环境冠对纳米材料的环境行为与毒性效应存在重要的影响。目前已有不少研究关注了纳米材料的环境冠，但依然有许多重要的科学问题需要在未来进一步探索：①目前研究主要集中于环境冠对纳米材料环境行为的研究，忽视了不同环境条件下，环境冠成分与化学特征的变化，以及对纳米材料毒性效应的影响；②很大比例的研究关注了 NOM 冠与纳米材料的作用，但有关 NOM 冠成分解析与形成机制的研究还比较有限，没有从分子水平和化学机制上得出具体的作用规律；③相较于 NOM 冠，有关水环境以及水生生物的 EPS 冠和蛋白冠的研究难度大，因此相应的研究还比较缺乏；④对环境冠影响纳米材料的毒性效应规律还没有形成统一的认识，依然存在许多矛盾的观点和结果，需要更多的系统研究才能得出明确的结论；⑤环境冠介导纳米材料与细胞的表界面作用过程以及毒性机制还处于比较初步的研究阶段，除了环境冠影响纳米材料的细胞摄入、体内累积、细胞膜黏附和膜损伤，更具体的分子损伤机制还有待于揭示；⑥目前已报道工作更多关注的是静态的环境冠，缺乏对水环境中环境冠的动态变化进行研究。只有通过对以上科学问题的探索，才能提供最为关键的科学依据，并最终回答纳米材料的生态环境和健康风险这一重大问题。

3 纳米材料不同层次的生物效应与关键调控因子

纳米材料在被广泛应用的同时，也造成了不可避免的环境暴露与人体接触风险。考虑到纳米材料本身非常复杂，其环境赋存状态与生物作用效应区别于传统的小分子污染物，系统探讨调控纳米效应的理化性质，解析影响其环境过程、生物毒性与健康效应的关键因子，对于指导新型纳米材料的设计与合成，或开展纳米风险评估具有重要的科学意义（图 23-2）。

依据调控纳米材料表面理化性质的不同手段，可分为物理吸附调控、化学共价连接调控和介于两者之间的以范德瓦耳斯力和疏水作用为基础的强相互作用调控。不同调控手段均可对纳米材料的表面性质产生影响，进而影响或改变纳米材料的表面理化性质。与此同时，纳米材料的材质、尺寸、形状等固有性质在上述表面调控过程中可保持相对稳定。因此，表面理化性质与内核属性可共同决定纳米材料的生物反应活性。

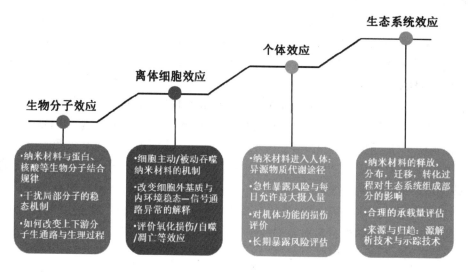

图 23-2 纳米材料不同层次的生物学效应评价

3.1 生物分子效应

纳米材料产生生物效应的基础是与生物分子发生接触和相互作用。"纳米-生物"界面过程是指发生在纳米材料表面及生物成分的表面（如蛋白质、膜、磷脂、内吞小泡、细胞器、DNA 和生物液体）之间的动态的物理化学相互作用、动力学以及热力学交换过程[83, 84]。研究纳米-生物界面效应对于评估纳米技术生物安全性，拓展其在生物领域的应用范围有重要意义。关于该方面的研究多侧重以下三个方面：纳米颗粒理化性质（大小、形状、表面电荷、官能团种类、团聚状态等）、液-固界面作用（表面能、离子吸附、双电层作用等）、颗粒-生物界面作用（蛋白质、细胞膜、受体-配体结合等），其中关键性的假设是纳米-生物界面过程是相对稳态的，而真实界面过程是不均匀、动态/瞬态的。这种动态过程既来自于纳米颗粒性状的改变，也体现在生物调控导致的不均一/非稳态（如细胞分泌物、细胞膜结构改变、细胞活性等）。动态情况下的纳米-生物界面效应尤其值得关注，因其具有纳米颗粒暴露剂量低、纳米尺度接触、考虑生物应激调控等特点，更能反映真实的体内（in vivo）暴露下的生物效应。如何综合考虑细胞动态调控（如信号网络）在纳米-生物界面中的作用是预测界面效应、颗粒物摄入及生物效应的重要方面。

张承东课题组的研究主要关注动态过程中的纳米-生物界面效应，具有生物介导、只发生在近距离纳米尺度及多重界面反应同时进行等特点。①界面电子传递[85]。发现低剂量下 GO 可与大肠杆菌发生界面电子传递作用，通过干扰呼吸链导致胞内电子通过膜上细胞色素 c 传递至细胞外，产生大量胞外 ROS，低剂量（~20 mg/L）即可高效杀菌，而同时 GO 被还原进一步加速了电子传递过程。该项研究提出了一种新的 GO 微生物毒性机理，而这一现象只发生在 GO 不对细胞膜蛋白造成显著结构改变、微生物具有高度呼吸活性情况下，即细胞活性介导的界面电子传递过程。②界面络合[86]。CNTs 表面羧基官能团能与催化过程的关键离子锰离子发生络合，干扰锰过氧化物酶的催化反应、抑制酶的反转。这一界面络合作用仅发生在酶活性中心、酶的锰离子结合位点、CNTs 三者处于纳米尺度空间的情况下。所以不应当只考虑基于纳米颗粒理化性质与生物界面性质的界面驱动力，而需额外关注当纳米颗粒与生物不断接近处于纳米尺度时才发生的界面反应。③界面氧化还原反应[87]。揭示还原氧化石墨烯（RGO）与辣根过氧化物酶之间同时发生的界面氧化还原反应（自由基湮灭）、酶与 RGO 之间电子传递作用，可显著稳定酶结构、保持酶活性 10 倍以上。这种动态反应同时改变了 RGO 的理化性状。

生物体内富含大量的蛋白质、核酸等有机分子，纳米材料与生物大分子结合形成相对稳定的生物分子蛋白冠，可能会扰乱体内的分子平衡，表现在含纳米材料的局部区域因吸附导致特异或非特异的分子浓度升高，进而改变生物分子结构功能与代谢稳态。当纳米材料暴露在生理环境下，其表面双电层会因离子强度的骤增而压缩，导致不同颗粒之间静电斥力减小，并进一步导致纳米材料的不稳定性聚集趋势。一般来讲，生物大分子例如蛋白质的快速吸附会增加一层保护屏障，增大颗粒间的斥力，保持胶体的稳定并进一步达到吸附-解吸平衡。生物分子在颗粒表面的吸附-解析平衡受到相互作用力的影响，通常用吸附常数 K_a 或解离常数 K_d 表示。当前的研究显示，纳米材料的多种性质均可以影响相互作用力和吸附平衡。例如，Jiang 等[88]研究发现表面羧基修饰密度越大越有利于碳纳米管吸附蛋白质。相较于带负电荷的表面，带正电的修饰层倾向于吸附更多的蛋白质。Qi 等[89]通过研究不同还原程度及表面官能团的石墨烯材料对 FBS 蛋白质的吸附作用发现，石墨烯材料会选择性吸附高分子量的蛋白组分，还原程度更高的石墨烯材料对蛋白质的吸附亲和力更高，主要的吸附机制是疏水性作用；然而，还原程度更高的（即表面疏水性更强的）石墨烯材料对蛋白质的吸附容量反而更低，这主要是因为还原程度更高的石墨烯材料团聚程度更高。最近 Xu 等[90]利用聚类分析的手段系统性探究了三类共 21 种金纳米材料的形状、粒径、Zeta 电位等对蛋白质吸附的影响，发现水合粒径和 Zeta 电位起主要作用，但是其他的物理化学性质也起了一定的作用。多种因素对吸附量的影响实际上是吸附作用类型起着主要调控作用，其中疏水作用、范德瓦耳斯力、静电作用等单独或联合作用是最为常见的类型。

吸附-解吸平衡建立后,蛋白质-纳米材料的相互作用域和吸附构型也基本稳定。在稳定的共轭物体系中,蛋白质与纳米颗粒接触的作用域对于揭示蛋白质的吸附构型有着重要的意义。举例来讲,不同表面修饰的纳米金球与BSA的相互作用域有显著差异,导致其吸附构型不同[91]。另外,材质也是影响相互作用域的重要因素。目前探究相互作用域的手段主要是二维核磁共振和红外光谱,但面对较为复杂的蛋白质,当前技术仍难以完全胜任,因此新一代以氢氘交换质谱和交联质谱为核心的研究技术正在不断得到发展与完善。纳米材料对生物分子的吸附一方面增强了自身的稳定性,另一方面也会改变大分子本身的结构与功能。例如纳米材料对过氧化氢酶[92]、白蛋白[93]、凝血因子XII[94]等可造成不同程度的影响。总之,纳米材料对生物大分子的效应主要集中在固-液界面上,可受到以表面修饰分子为基础的Zeta电位、包覆层种类以及以粒径为基础的水合粒径、表面积等多重因素的综合作用。

3.2 离体细胞效应

目前诸多研究证实,纳米材料的理化性质对其生物效应影响巨大。以金纳米颗粒作为核材料,闫兵课题组[95]在多种人源细胞系中系统研究了纳米颗粒的理化性质对细胞氧化应激的调控。研究发现,金纳米颗粒的疏水性和表面正电荷密度可以调控细胞氧化应激水平,而表面负电荷密度不能有效调控细胞氧化应激水平。进一步的机制研究表明,疏水和表面带正电的金纳米颗粒可以分别通过激活NADPH氧化酶和线粒体产生氧化应激。从纳米医药的应用角度,疏水和表面带正电的金纳米颗粒导致的氧化应激对紫杉醇具有化疗增敏效果,在癌症治疗中具有潜在的应用价值(图23-3)。

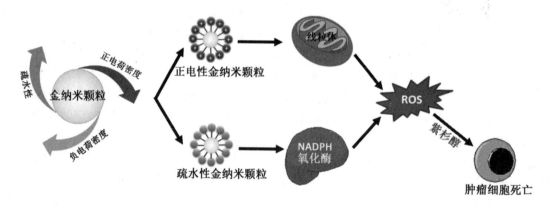

图23-3 疏水和表面带正电的金纳米颗粒可以通过不同的机制调控细胞氧化应激,并对紫杉醇具有化疗增敏作用[95]

二维纳米材料会对哺乳动物细胞产生毒性效应[96, 97],而厚度是二维纳米材料至关重要的理化性质之一。闫兵课题组近期的一项研究[98]结果显示,多层二硫化钼(MoS$_2$)纳米片可以被细胞广泛摄入,而少层MoS$_2$纳米片则更倾向于贴附在细胞表面,极少被细胞摄入。生物效应评价结果显示,在不引起显著细胞毒性的剂量条件下,尽管两种不同厚度的二维MoS$_2$纳米片的细胞定位有很大差异,它们能诱导细胞产生类似水平的自噬。进一步的机制分析结果表明,多层MoS$_2$纳米片诱导摄入依赖的细胞自噬,而贴附在细胞膜表面的少层MoS$_2$纳米片可通过与细胞膜表面受体相互作用,诱导细胞发生mTOR依赖的自噬。

富集生物分子的纳米材料可能会被免疫系统识别并清除,也可能通过配体-受体途径以及非特异性途径("特洛伊木马"形式[99])进入细胞。基于离体实验的研究结果显示,巨噬细胞易于吞噬纳米颗粒[100],但聚乙烯醇(PEG)等高分子表面修饰可以改变这一趋势[101, 102];对于特定的纳米材料,癌细胞相较于正常细胞吞噬能力更强[103-105],且二者的内吞(endocytosis)机制并不相同[106]。不同表面修饰、形状和粒径的纳米颗粒的内吞行为也存在较大差异[107]。例如，表面带正电的纳米颗粒更容易进入细胞，而中性表面的纳米颗粒难以进入细胞；球状的纳米颗粒比棒状的纳米颗粒更易进入细胞；而对于尺寸因素来说，吞噬量与粒径在一定范围内呈现倒 U 型曲线，即存在最适吞噬粒径，实验和理论计算的结果一般认为是 50 nm[107, 108]。在吞噬机制研究中，网格蛋白依赖性、小窝蛋白依赖性或非依赖性通路是公认的三条主要吞噬路径。除超小粒径的纳米颗粒外，大多数纳米颗粒的内吞方式均是网格蛋白和/或小窝蛋白通路。基于抑制剂的研究认为，较小的颗粒通过小窝蛋白的协助进入细胞，而较大的颗粒通过网格蛋白的协助进入细胞[109]。在吞噬的基础上，相应的细胞毒性也不容忽视。例如稀土金属纳米材料对磷酸根的攫取会导致细胞膜结构的损伤[110, 111]，纳米银会对大鼠脑部神经元产生氧化应激损伤[112, 113]，而超小粒径的量子点会对细胞产生炎性和自噬效应[114]。

目前纳米材料对细胞的毒性效应主要集中在以 ROS 为中心的氧化应激通路，以及由此引发的炎性效应和自噬作用上，而对于其他效应的研究，尤其是关于细胞形态结构、跨膜过程、内在能量传输与表观遗传等并不充分。现有的一些探索显示，SiO_2 纳米颗粒暴露内皮细胞后导致细胞间的黏附连接受损，间隙打开引起渗漏效应和形态改变，该过程受到纳米颗粒密度的调控[115]。纳米银暴露胚胎干细胞后可以抑制细胞分化，干扰 X 染色体失活进程，但并不产生明显的 ROS 变化[116]。纳米银和 GO 的暴露会引起细胞基因组甲基化水平变化，进而影响膜骨架结构、能量代谢和离子通道相关蛋白质[117]。综合而言，纳米材料的细胞效应主要以相互接触作用后产生内吞行为为基础，因此受到化学组成、表面修饰、形状、粒径等因子的调控。

此外，纳米材料在环境或生物介质中发生的转化过程也会对其生物效应产生显著影响。例如，刘思金课题组最近研究了 GO 在环境中的化学还原过程对其毒性效应的影响[118]：首先模拟环境中硫还原过程对 GO 进行还原，得到两种不同还原程度的 RGO；相较于原始态 GO，两种 RGO 的氧含量、羟基和环氧基含量均有明显降低，同时材料边缘发生卷曲。随后，选取两种巨噬细胞——永生化的巨噬细胞系 J774A.1 和小鼠骨髓原代巨噬细胞 BMDM 进行暴露，结果表明，两种 RGO 能够诱导巨噬细胞产生更强的炎症反应，而这种毒性效应的增强可能是由于官能团较少、疏水性更强的 RGO 在巨噬细胞内更高的团聚程度所导致的。由此得出结论，环境过程会改变 GO 的理化性质，通过影响其在细胞内的赋存状态影响其毒性效应。陈威课题组最近则研究了 GO 在模拟肺泡液中的转化过程对其生物效应的影响[119]。研究发现，GO 进入模拟肺泡液（Gamble 溶液）及模拟溶酶体液（ALF）后均会发生还原反应，还原程度 Gamble-GO < ALF-GO；两种转化过后的材料在形貌、表面官能团种类和数量上都与原始 GO 有很大的差别。通过单一组分分析及模型分子的实验发现，有机酸小分子在模拟肺泡液中起到了主要的还原作用（即将环氧基和羧基打开形成羟基）。进一步的研究发现，Gamble-GO 和 ALF-GO 在培养基中团聚的形态不同，导致了两种材料被巨噬细胞吞噬和排出的效应有明显差别。原因是 ALF-GO 表面含有大量的羧基，使其更容易形成"边缘相连"（edge-to-edge）的团聚体，而 Gamble-GO 则会通过疏水作用形成"层层堆叠"（layer-to-layer）的团聚体，二者团聚状态的差别最终导致了对巨噬细胞作用上的差异。

3.3 个体效应

纳米材料的体外研究并不能完全满足其生物学效应评价的需要，模式动植物则可在个体层次为安全性评估提供重要参考。基于纳米材料可能的暴露途径与环境赋存状态，常选用的动植物有鼠类、鱼类、南瓜幼苗等。作为广泛使用的纳米材料之一，关于纳米银的动物实验研究相对较多。通过对孕鼠暴露不同粒径的纳米银，分析乳汁中纳米颗粒含量可以得出小粒径纳米银具有世代传递的特征[120]。新生大鼠的滴鼻实验显示，纳米银可靶向脑组织累积且难以消除[121]。当纳米材料伴随血液流动时，高密度纳米材料贴近血管壁移动[122]。纳米材料的体内分布与表面修饰和粒径密切相关。例如，羧基修饰的纳米颗粒易于和中性粒细胞相互作用进而被肾脏清除[123]。粒径在 10~20 nm 的颗粒主要在肝脏中被迅速清除[124]，而粒径 200 nm 以上的颗粒主要在脾脏中清除[125]，小于 8 nm 的颗粒则在肾脏中迅速代谢[126]。虽然在基于生物体整体、器官或组织层面，已积累了一系列纳米毒理学的研究数据，然而人们对单细胞水平上纳米材料分布规律的认识仍不清晰，因此该方向研究成为当前纳米毒理学的新兴领域。

纳米材料的环境释放必然造成其对环境生物的影响；而且，纳米材料的环境转化过程影响其生态毒性效应。在可见光照射下，某些半导体金属氧化物纳米材料可以产生羟基自由基（·OH）和超氧自由基阴离子（$O_2^{\cdot -}$）等 ROS[127]，进而引起生物的氧化应激水平上升和细胞结构的损伤。Hu 等近期研究发现[128, 129]，单层 MoS_2 和单层二硫化钨（WS_2）在光照作用下能缓慢地向水体中释放金属离子，并发生相的转变，但是释放出的金属离子并没有引起明显的生物毒性效应，生物毒性效应主要是由层状的纳米材料引起。环境中的腐殖酸能与单层 MoS_2 非共价键结合，从而改变 MoS_2 纳米片的表面形貌、层状结构和胶体稳定性。此外，电子从腐殖酸转移至 MoS_2 纳米片上，能显著提高其类过氧化物酶催化活性并促进其化学溶解，释放出硫离子和钼离子。不同相的单层 WS_2 能导致不同的生物效应，包括影响细胞对纳米颗粒的摄取、氧化应激、膜损伤、脂质氧化和代谢，并且金属 1T 相的 WS_2 比半导体 2H 相的 WS_2 表现出更强的毒性，这种差异可能主要源于自由基产生能力的不同（图 23-4）。上述研究表明，二维过渡金属硫化物的环境转化复杂，并且与生物之间有密切的相互作用，在应用和处置二维过渡金属硫化物的过程中需要充分考虑其生态环境风险。

基于植物模型，当前研究主要探讨了纳米材料的吸收[130]、分布转化[131]与毒性效应[132]，已有研究证实，水环境的离子强度、有机质浓度和成分、植物的种类、纳米材料粒径等多因素共同影响了纳米材料在植物体内的行为，详见本章第 4 节。另外，纳米材料对模式生物暴露途径的差异，也可造成其产生不同的毒性作用，这也体现了纳米材料环境与生物效应的复杂性。

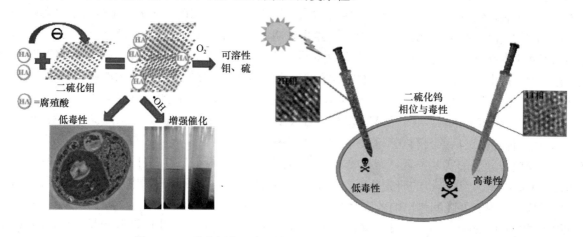

图 23-4　二硫化钼和二硫化钨的环境转化及其生物效应[128, 129]

3.4　生态系统效应

生态系统中纳米材料的迁移转化过程，近来也逐渐引起人们的关注；尤其，纳米材料是否具有类似于持久性有机污染物（POPs）的跃迁传递规律与食物链富集效应，是研究者探讨的重点问题。在水生食物链传递中，针对纳米 CeO_2[133]、纳米金[134]、纳米 TiO_2[135]的研究表明，纳米材料可沿食物链传递，这一过程会受到纳米尺寸的调控，但并不表现出生物放大效应。然而，陆生食物链的研究结果显示，量子点[136]、纳米 CeO_2[137]、纳米金[138]等纳米材料具有生物放大效应。基于当前不同的研究发现，考虑到实验室内构建的食物链与选取的纳米材料种类、粒径、表面修饰上的差异性，针对纳米材料与受试食物链，建立统一的实验规范尤为重要。

最近，吴李君课题组研究了 Ag 纳米颗粒在大肠杆菌-秀丽隐杆线虫食物链中的传递过程和毒性效应。他们使用不同粒径的 Ag 纳米颗粒喂食大肠杆菌，电镜下可以观察到不同粒径的纳米颗粒均能被大肠杆菌所携带；秀丽隐杆线虫摄食了携带 Ag 纳米颗粒的大肠杆菌后，在其皮下组织、生殖腺细胞内及消化道壁内均发现有 Ag 纳米颗粒分布，且线虫体内 Ag 的含量与其喂食的大肠杆菌携带量高度相关[139]。进一步对线虫的子代数目、种群增长、存活率和寿命进行检测，结果显示，大肠杆菌携带的纳米 Ag 粒径越小，对线虫的毒性效应越明显。他们还发现，环境因素也会对纳米材料在食物链中的传递过程有所影响：使用不同离子强度溶液处理的纳米银暴露大肠杆菌后再将大肠杆菌喂食线虫，纳米银经食物链传递后对线虫的毒性增加，包括诱导线虫体内氧化应激水平上升、生殖腺细胞凋亡数目增加、子代数减少、寿命缩短、运动能力受损等[140]。与此同时，高离子强度溶液处理后的纳米银在大肠杆菌及线虫体内的生物累积量也显著增加。

王震宇课题组最近以环境暴露浓度最高的人工纳米材料——纳米 TiO_2 为代表，研究了纳米材料进入海洋底栖环境后，在底栖食物链中的营养转移、累积和毒性。选用的海洋食物链为沙蚕—大菱鲆（Scophthalmus maximus）和藻类—栉孔扇贝（Chlamys farreri）。研究发现，纳米 TiO_2 可以沿着食物链从双齿围沙蚕（Perinereis aibuhitensis）传递到大菱鲆幼鱼，但是未发生生物放大效应；大菱鲆幼鱼通过摄食纳米 TiO_2 处理过的沙蚕和直接的水体暴露后，Ti 主要累积在鳃和消化道，其次是皮肤、肝和肌肉，并且降低了幼鱼的生长速率，增加了致病率；通过幼鱼的组织学和营养成分分析，发现肝和脾发生组织病变，影响了营养代谢，造成鱼体营养成分改变（粗蛋白降低，粗脂肪升高），表明鱼体的营养品质降低[141]。进一步研究发现，在藻类—扇贝食物链中有明显的生物放大效应，其原因为纳米 TiO_2 能够吸附在新月菱形藻（Nitzschia closterium）细胞表面并能够被藻细胞所吸收[142]。此外，在沙蚕、大菱鲆和扇贝的排泄物中均检出大量的纳米 TiO_2。纳米 TiO_2 随粪便的排出能够使其环境风险和归趋具有更大的不确定性[141, 142]。最近，有研究发现，纳米 TiO_2 在小球藻—网纹溞食物链中的营养级传递潜势与 TiO_2 的晶相有关[143]；林道辉课题组最近则发现，纳米 TiO_2 在小球藻—大型溞食物链中的生物累积和生物放大效应受纳米 TiO_2 粒径和晶相结构（如金红石相比例）显著影响[144]。最近的一项研究还发现，铜离子可以影响 MWCNT 在大型溞以及黑头呆鱼（Pimephales promelas）体内的累积和营养级传递[145]。

除营养级传递外，纳米材料对环境中生物种群结构和生态功能的影响也受到关注。例如，长期低剂量纳米 Ag 暴露能改变溪流环境中真菌和细菌的种群结构及其分解落叶的活性[146]，也能改变溪流环境中周丛生物[147]以及海洋环境中浮游生物[148]的微生物群落结构和生态功能。不同粒径和表面修饰的 Au 纳米颗粒则会影响土壤微生物群落结构以及与营养元素循环有关的酶活性[149]。此外，将土壤跳虫（Folsomia candida）暴露于 Ag 纳米颗粒后，其消化道中的微生物群落会发生改变，且抗生素抗性基因减少[150]。最近的一项微宇宙（microcosm）实验研究则表明，增加生态系统中的营养级复杂度可以降低纳米 Ag 对生物的影响，因此用单一物种进行的纳米毒性实验可能高估纳米材料的环境风险[151]。

金属氧化物、金属硫化物以及碳纳米材料也能影响环境中微生物的种群结构与生态功能。例如，纳米 ZnO 能影响淡水生态系统中真菌的群落结构以及分解落叶的活性[152]。Fe_2O_3 纳米颗粒会显著影响周丛生物的微生物群落组成、生物多样性和代谢活性[153]，而纳米 Fe_3O_4 则会在短时间内改变湿地土壤中细菌和古细菌的组成以及电产甲烷活性[154]。La_2O_3、Nd_2O_3、Gd_2O_3 等稀土金属氧化物则会富集土壤微生物的抗生素抗性基因，提高其对四环素等抗生素的抗性[155]。Ag_2S 纳米颗粒能影响淡水生物膜的微生物群落组成和生物活性[156]。石墨烯[157]和纳米 CuO[158] 会影响污水处理系统中活性污泥的细菌群落结构及其去除氮、磷营养元素等水体污染物的功能，而纳米 ZnO 和纳米 CuO 则会引起污泥厌氧消化过程中抗生素抗性基因的扩散[159]。最近有研究发现，金属基纳米材料施用于农田后，也会影响农田土壤中微生物的群落结构和酶活性：其中纳米 Ag 会显著抑制某些酶的活性，纳米 TiO_2 可能轻微降低酶活性，纳米 ZnO 或

CuO 没有影响或增强酶活性[160]。也有研究人员用毒代动力学模块耦合的能量收支模型方法考察纳米 CeO_2 对大豆及其共生菌的生长和生态功能的影响[161]。而炭黑、MWCNT、石墨烯等碳纳米材料对大豆根区原核微生物种群结构的影响则与大豆植物的生长阶段有关：在生殖生长期的影响大于营养生长期[162]。

3.5 纳米-生物效应的构效关系及预测模型

目前关于纳米材料生物效应的研究大多基于实验手段合成特定理化性质的纳米材料，以及后续在细胞水平和动物水平的实验研究。纳米材料的理化性质是多种多样的，而且每种性质可在较大的范围内发生变化。因此，仅仅通过实验手段难以揭示纳米材料的理化性质与生物活性之间的关系。实验研究结合计算化学方法为探索纳米材料对生物效应的调控提供了新的思路。闫兵课题组[163]采用组合化学方法合成了 7 种表面性质（疏水性、表面正电荷密度、表面负电荷密度、表面氢键受体密度、表面氢键给体密度、表面π键密度、表面配体分子几何结构）变化的金纳米系列，并在 A549 和 HEK293 两个细胞系中研究了各种纳米材料对 HO-1 的表达水平及细胞摄入水平的调控。闫兵课题组综合运用实验和计算化学手段，建立了基于理化性质的纳米材料生物效应构效关系及预测模型，并将该模型用于纳米材料生物效应的预测及具有特定生物效应的纳米材料的辅助设计。通过实验验证，使用该模型计算得到的纳米材料生物效应的预测值与实验值有很好的一致性（图 23-5）。这一发现为高效、低成本的纳米材料合理设计和安全应用提供了重要指导。在此基础上，Zhu 等[164]通过构建虚拟纳米材料库，评估纳米材料的溶剂可及表面，建立了纳米材料亲疏水性的预测模型。通过对比纳米材料的实验 LogP 和使用该预测模型得出的预测 LogP，证实了建立的纳米材料亲疏水性预测模型的可靠性。

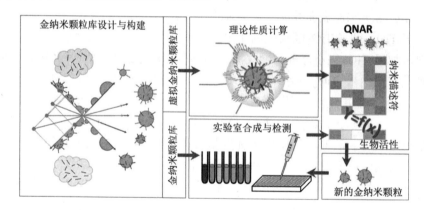

图 23-5 基于 QNAR（Quantitative Nanostructure-Activity Relationships）模型的纳米材料设计、生物效应预测和实验验证[163]
基于实验中的纳米材料结构，通过计算化学手段可以获得各种纳米材料的描述符数值。将这些描述符数值与实验获得的生物效应值进行关联，从而得到对纳米材料引起下游生物效应的预测模型。基于预测模型，可以预测未知纳米材料的生物效应，并可以通过实验对预测的准确性进行验证

3.6 小结

虽然现有研究积累了纳米材料不同理化性质调控其生物学效应的大量实验数据，但仍有诸多科学问题亟待深入研究。探讨纳米材料全生命周期过程的完整的代谢流研究还非常缺乏；大范围环境尺度的研究，可参考 POPs 的系统研究与同位素示踪技术对源解析和转化归趋的研究；具体的生物学通路，例如外排通路、生物分子冠的形成规律与变化机制仍然需要进一步的深入探索。

4 纳米材料的植物毒性与分子机制

土壤是环境中多数人工纳米材料的主要归宿，污泥回用、污水灌溉、大气沉降等途径都有可能导致人工纳米材料进入土壤，特别是农用土壤。作为陆地生态系统的初级生产者，植物（尤其是农作物）势必有越来越多的机会暴露于纳米材料这种新型污染物。此外，近年来纳米技术在农业上的应用也展现出巨大的前景，纳米材料在种子处理、植物生长与发育、病虫害控制与预警、有毒农业化学品的监测等方面都有着优异的表现[165]，以纳米农药、纳米肥料[166]、纳米传感器为代表的纳米农业化学品制剂越来越多的得到开发并且进入市场。因此，无论是无意排放还是有意应用，人工纳米材料与植物的接触不可避免，研究纳米材料与植物的相互作用关乎粮食安全、食品安全和人类健康。明晰纳米材料在植物体内的吸收、富集行为和毒性效应对评估其生态及健康风险、保障纳米技术在农业的可持续应用都具有重要意义。我国在这一领域的研究开展的较早，中国科学院高能物理所的赵宇亮、张智勇课题组，浙江大学林道辉课题组，无锡大学王震宇课题组，中国科学院生态环境研究中心的张淑贞课题组，都是世界上比较早的在纳米-植物领域开展前瞻性研究的小组。近年来，有关纳米材料植物毒性效应的研究大量涌现。据 Lv 等[167]最近的评述文章统计，近十年发表的有关"纳米-植物毒性"的论文共约 430 篇。本节将阐述在纳米材料植物毒性效应领域中过往的研究结果、研究现状和进展，着重阐述纳米材料的生物积累以及毒性效应。

4.1 纳米材料的植物吸收路径

纳米材料植物毒性效应研究的关键科学问题之一是纳米材料能否进入植物体内以及怎样进入植物体内（吸收途径），即纳米材料在植物体内的吸收、运转和生物积累。明晰这些问题直接关系到食品安全和人类健康等重大问题[167]。纳米材料区别于传统污染物最主要的特征是其超细颗粒物的本质，这一特征决定了其不能用常规的分析无机和有机化合物的方法（如 ICP-OES、ICP-MS、GC-MS、LC-MS）来定量和定位其在植物体内的分布。近年来，研究者不断尝试各种不同的技术手段来分析纳米材料在植物体内的吸收、分布和生物转化。例如，电镜（SEM、TEM）技术、同步辐射微聚焦 X 射线荧光（μXRF）分析技术[168]、X 射线吸收近边结构（XANES）[169,170]、荧光/激光共聚焦显微镜、双光子显微镜[171,172]、拉曼光谱[173]、电感耦合等离子体质谱（ICP-MS）的单颗粒检测[174]、激光剥蚀电感耦合等离子体质谱（LA-ICP-MS）和同位素示踪技术[175-177]等。尽管这些技术存在各自的优缺点，但纵览过去十几年在此领域的研究结果，利用现有技术在纳米材料植物吸收途径和机理方面已经取得一些进展。

例如，Zhang 等[178]将水培的黄瓜（Cucumis sativus）幼苗从根部暴露于 2000 mg/L 的纳米 CeO_2 悬浮液 21 天，试图探寻植物对纳米 CeO_2 的吸收和植物体内生物转化情况。通过透射电镜和 XANES 等技术，Zhang 等发现 CeO_2 纳米颗粒可以穿透植物根部的表皮，进入细胞间隙并进入地上部分（茎）。作者们对于纳米 CeO_2 进入植物体内的生物转化也进行了研究，在黄瓜根和茎分别发现了磷酸铈和碳酸铈；作者推测在根表面部分 CeO_2 被植物分泌的有机酸以及具有还原作用的代谢产物所溶解，以 Ce^{3+} 离子形态进入植物体内，在植物体内又与磷酸盐、碳酸盐等发生化学反应或者形成螯合物。同年，Wang 等学者[179]发现，CuO 纳米颗粒在玉米幼苗内，既可以通过根部由导管（xylem）向上运输，也可以经由筛管（phloem）从地上部位运输到地下部位。在同一年，Servin 等[180]利用 μXRF 和 μXAS 发现了黄瓜幼苗根部吸收和运转纳米 TiO_2 的直接证据，并发现纳米 TiO_2 在植物体内没有进行生物转化。Zhao 等[181]将玉米幼苗暴露于 FITC 荧光标记的纳米 CeO_2，并用激光共聚焦显微镜观察到很多纳米颗粒分布在根部表皮、皮质的细胞壁上，因此推测纳米材料可能的质外体运输路径。这些证据都表明,纳米颗粒能够被植物吸收、进入植物体内。最近,Ma 等用分根系统研究了纳米 CeO_2 在水培的黄瓜植株体内的运输和生物转化,发现:纳米 CeO_2 在植物体内会发生生物转化而被部分还原为 Ce(III),而且 Ce(III)和 Ce(IV)都能经由导管从根部向上运输,但只能以含 Ce(IV)的纳米 CeO_2 形式经由筛管从芽向根部运输[182]。而且,CeO_2 纳米颗粒在水培黄瓜植株体内的运输和生物转化过程与纳米颗粒的微观形貌有关[183]。最近另一项研究发现,表面修饰的 CeO_2 纳米颗粒被水培小麦(*Triticum aestivum*)幼苗根部吸收的强弱与纳米颗粒的表面电荷有关——表面带正电的 CeO_2 纳米颗粒与小麦根部的结合更强,但较长时间(34 h)暴露后,表面带负电及不带电的 CeO_2 纳米颗粒运输到小麦叶片的效率更高;该研究还发现,无论 CeO_2 纳米颗粒表面带何种电荷,在小麦根部和叶片都检测到了 Ce(III),说明 CeO_2 纳米颗粒发生了还原转化[184]。最近,张智勇课题组系统比较了 4 种不同植物对 CeO_2 纳米颗粒的转化和体内运输,并揭示了造成不同的原因[185]。

但同时也有很多实验结果显示了纳米材料不能被植物根部吸收的证据。例如,Taylor 等[186]发现金纳米颗粒(7~108 nm)并不能进入拟南芥(*Arabidopsis thaliana*)根部。造成这些看似矛盾的结果背后,是因为纳米材料能否进入植物体内取决于诸多因素,例如,纳米颗粒的化学组成、晶体结构、表面电性、尺寸、形貌、表面修饰、植物种类等。植物细胞壁上的孔道直径大约在 5~20 nm,所以普遍认为的纳米材料可以被根吸收的尺寸是小于 20 nm。此外,植物的细胞壁结构、蒸腾速率、导管尺寸等都是影响植物吸收纳米材料难易程度的因素[187]。因此,不同研究的结果之间很难进行比较。但可以肯定的是,某些拥有适当尺寸和理化特性的纳米颗粒能够通过根表面的孔道穿透外表皮,进入皮质、内表皮,从而进入导管,甚至向地上部位运输。而有关吸收途径,根据目前的研究结果,植物吸收纳米材料的两大路径分别是质外体途径和共质体途径。对于单个植物细胞来说,内吞作用是主要的吸收路径。目前,细胞穿透(cellular penetration)是普遍接受的植物吸收纳米材料的途径,尽管确切的机理尚不十分明确[187, 188]。以往多数研究只设置苗期实验,集中于纳米材料被根、叶等营养器官的吸收和富集,而开展整个生命周期的实验,探究纳米材料在果实中的积累方面的研究尚十分缺乏。公众普遍关心的问题是纳米材料是否能够通过植物果实积累进入食物链。遗憾的是,受研究手段的限制,玉米、水稻、小麦等重要作物果实中纳米材料积累的证据尚未见报道。

到目前为止,极少有研究探寻纳米颗粒从叶片进入植物的途径。而在实际环境中,部分颗粒物由大气沉降进入到陆地生态系统,直接与植物叶片接触的情况并不少见。Zhao 等[189, 190]通过电镜技术观测到生菜和菠菜叶片气孔直径分别为 15~17 μm 和 10~15 μm。因此,粒径为几十纳米甚至几百纳米的纳米颗粒,都有可能通过气孔进入叶片。Hong 等[191]将黄瓜幼苗叶片暴露于 CeO_2 纳米颗粒,发现 CeO_2 纳米颗粒可以通过叶片吸收进入黄瓜幼苗,进而输送到地下部分(根)。当然也有一些矛盾的结果,例如,Birbaum 等[192]发现纳米 CeO_2 可以被叶片吸收,但并没有被转运到根部。Kranjc 等[193]比较了芸芥(arugula)和茅菜(escarole)幼苗叶片对 Pt 纳米颗粒的吸收及向根部的转移,发现叶片的表面能(亲疏水性)显著影响 Pt 纳米颗粒在叶片的黏附和团聚状态:Pt 纳米颗粒在低表面能的叶片(芸芥幼苗叶片)上更易发生团聚和黏附;对于两种植物,Pt 纳米颗粒都难以从叶片向根部的转移。最近,Zhao 等研究了 CuO 纳米颗粒在水生植物凤眼莲(*Eichhornia crassipes*)植株体内的吸收、分布和存在形态,发现:凤眼莲的根和沉水叶均是 CuO 纳米颗粒进入植物体的途径,其中沉水叶片的吸收贡献率约为根部的 3 倍,是 CuO 纳米颗粒进入凤眼莲体内的主要途径;进入植物体后,CuO 纳米颗粒主要存在于细胞壁、细胞质和大液泡中,一部分以 CuS 和 $Cu_3(PO_4)_2$ 的形式存在[194]。而 Xiong 等则发现,CuO 纳米颗粒的叶面暴露会增加莴苣(*Lactuca sativa* L.)和甘蓝(*Brassica oleracea* L.)等蔬菜的叶面铜吸收[195]。如前所述,纳米材料的物理化学特性(如尺寸、表面电性)、植物类型、叶片表面的形貌和亲/疏水性,都是影响和决定植物叶片能否吸收纳米材料的因素。但目前普遍接受的纳米材料通过叶片被吸收的途径有蜡质层(cuticle)和气孔(stomatal)。

有关纳米颗粒在植物体内被吸收、转运和转化的更多最新成果和技术，请参见Gardea-Torresdey等[196]、Miralles等[197]、Wiesner等[187]和我国学者Lv等[167]的综述文章。综上所述，此领域经过多年的努力，取得了一些进展，对纳米颗粒进入植物的途径和生物转化过程取得了一定的认识，但更多机理尚不十分明确，尤其是纳米颗粒在植物亚细胞（叶绿体、线粒体）水平上的定位方面的研究十分缺乏。此外，多数研究都是在水培条件下完成的，虽然回答了一些问题，但与真实土壤中纳米材料的行为相差甚远：在真实土壤环境中，纳米材料在根际附近会进行复杂的物理、化学反应，发生团聚、溶解、与土壤组分（土壤矿物、有机质）的吸附等过程，这些都会影响纳米材料以及相应的金属离子的吸收[198-200]。受到研究手段的限制，这个领域的研究具有一些挑战，必须不断地开发新技术，并借助生物、物理、化学领域的新技术来寻求突破。同时，需要植物生理学、毒理学、材料学、分析化学等学科交叉协作来解决这些具有挑战的问题。

4.2 纳米材料的植物毒性效应及其机理

纳米材料植物毒性效应的关键科学问题之二是纳米材料的植物毒性效应，或者说植物对纳米材料的响应。纳米颗粒进入植物体后，很可能参与植物某些生理和代谢过程，对植物的生理过程造成影响，如影响基因表达[201]，干扰植物体内的抗氧化代谢途径[190, 202]，从而造成毒性效应，甚至能跨世代影响子代植物的生理和营养状态[203]。2007年，Lin和Xing[204]率先研究了5种人工纳米材料（MWCNT、Al、Al_2O_3、Zn、ZnO）对6种作物的毒性效应。继而，各国学者针对更多纳米材料与不同植物的相互作用开展了广泛的研究。如前所述，纳米材料植物毒性效应领域近年来涌现出许多研究论文。Rico等[132]、Miralles等[197]、Schwab等[187]、Du等[205]、Marmiroli等[206]的综述文章对这些研究结果已经进行总结和概括。总体来看，目前研究结果表明，在低剂量暴露条件下，纳米材料对植物的毒性效应较低[207]。此外，纳米材料的可移动性、被植物吸收的可能性都比较低[187]。总之，纳米材料的植物毒性取决于其化学组成、粒径、表面电性、植物类型[208]等种种因素，因为篇幅所限，在此不一一赘述。本小节仅对纳米植物毒理学存在的一些问题和可能被忽视的方面进行讨论和探索。

纳米材料植物毒性效应方面的研究存在的一些共识性问题包括：暴露期短、剂量高、培养介质（营养液、砂培）并非真实的土壤[207]。尽管这一类的研究在早期评估纳米材料的潜在风险是必要的，但是所得到的数据并不能充分评估现实中因素复杂的环境暴露风险。由于纳米颗粒与土壤组分的相互作用，纳米材料在真实土壤中的毒性，要远远小于在营养液和基质中的毒性。此外，纳米材料的对植物的效应存在剂量-效应关系，高浓度下往往表现出毒性，然而，某些纳米材料在低剂量下表现出促进作物生长的趋势。更多研究应该在低浓度下开展，充分利用其正面效应。

其次，评价纳米材料对植物的影响多集中在测定与生长有关的指标，如生物量、根长、叶面积等[209-211]，或者测定与光作用相关的生理指标，如光合作用效率、呼吸和蒸腾速率等。这些表型指标是基因、蛋白质和代谢产物调控的下游表现，往往不能判断植物的早期响应，更重要的是无法揭示致毒机理[132]。有些研究试图从分子水平上探究纳米材料的毒性效应机理，但只关注少数的、有限的基因/蛋白质/代谢产物，很难全面揭示生命体的整体响应[212]。组学技术（基因组学、转录组学、蛋白质组学、代谢组学）很可能是解决这一具有挑战性的研究课题的有效工具。Majumda等[213]利用蛋白质组学，研究了纳米材料和植物的相互作用。Zhao等[214]利用基于核磁共振和质谱的非靶向代谢组学平台，研究了多种植物对不同纳米材料（纳米铜、纳米银）的代谢组学响应，揭示了毒性效应机理和植物自身的解毒机理。Soria等则将ICP-MS与液相色谱-四极飞行时间质谱（LC-QToF/MS）技术相结合，一方面测定拟南芥对纳米银的吸收，另一方面测定其代谢物谱，两方面结合分析有望阐明纳米材料生物效应的机制[215]。

再次，以往的很多研究只关注纳米材料对植物某一组织的影响，由植物根系分泌物和微生物组成的根际对于纳米材料的响应，极少被关注。植物和土壤微生物有着密不可分的关系。植物将光合作用固定碳的 25%～40%以根系分泌物的形式释放到根际，这些化合物（糖、氨基酸、有机酸、蛋白质等）不但能直接影响纳米 Au、纳米 Cu 等纳米材料及相应金属离子的植物吸收[216, 217]，还是根际微生物生长必不可少的碳源[218]。Zhao 等的研究发现，纳米 Cu 胁迫下，黄瓜根系分泌物的组成被改变，植物大量上调各种氨基酸以及与抗氧化有关的酚类物质[214]。可以猜测，纳米材料很可能通过改变植物根系分泌物，即土壤微生物的食物供给结构，间接影响土壤微生物的基因多样性和丰度。反之，微生物群落结构的改变也可能影响植物的生长。例如，土壤中有益微生物群落的改变，如生物防治微生物、固氮微生物，会影响植物对营养元素的吸收和抗病能力[219]。因此，只考虑纳米材料与植物的二元关系，而忽略土壤微生物的变化和响应的研究方法，不能够全面评估纳米材料的环境效应。

4.3 小结

随着纳米技术越来越广泛的应用，种类繁多的纳米材料将释放到环境中，研究纳米材料对植物的毒性效应对评估其生态及健康风险具有重要意义。另一方面，随着纳米技术在农业上展现出巨大的潜力，人工纳米材料促进植物生长、提高植物抗逆性和作物产量等正面效应[220-222]的报道越来越多，因此，研究纳米农药对植物的毒性效应是推广纳米技术在农业中的安全、可持续应用以及规避潜在生态风险的重要前提。纳米材料与植物相互作用领域下一阶段的目标是机理方面的深入研究，需要借助更多先进的手段以及多学科的交叉。

5 纳米材料的水生毒理效应

目前，应用较广泛的人工纳米材料主要包括金属纳米材料、金属氧化物纳米材料和碳纳米材料等。随着纳米材料的广泛应用，大量的纳米材料不可避免地进入水环境中，并在水生生物体内累积，对水生生物产生毒性效应，进而对水生态环境产生潜在危害。本节对人工纳米材料水生毒理效应的研究进展进行简要介绍。

5.1 金属纳米材料

金属纳米材料主要包括纳米 Ag、纳米 Cu 等。其中纳米 Ag 是金属纳米材料中使用量最大的一种；目前，关于纳米 Ag 生态毒性效应的研究也最多。研究表明，水体中存在的低浓度纳米 Ag 会导致稀有鮈鲫肝脏和腮组织产生明显损伤[223]。而且，水体途径暴露纳米 Ag 对鱼类胚胎发育具有明显毒性，可引起日本青鳉（*Oryzias latipes*）胚胎表面绒毛膜破裂、胚胎及内容物释出[224]。此外，纳米 Ag 可穿过斑马鱼和黑头呆鱼胚胎表面的绒毛膜孔道进入体内[225, 226]。纳米 Ag 长期低剂量暴露具有生殖毒性效应，可显著诱导斑马鱼卵巢和精巢组织凋亡细胞增多，并降低雌鱼的生殖能力[227]。纳米 Ag 还能影响斑马鱼消化道中的菌群结构，而且这种效应因斑马鱼性别而异[228]。特别需要指出的是，Ag 纳米颗粒在水环境中会溶解释放 Ag^+ 离子，并引起毒性。例如，研究发现，纳米 Ag 对原生动物四膜虫（*Tetrahymena thermophila*）的毒性效应主要是由 Ag^+ 引起，而非纳米 Ag 诱导产生的 ROS 所致。经纳米 Ag 暴露后，四膜虫金属硫蛋白基因表达上调量显著高于氧化应激相关基因表达，意味着 Ag^+ 起重要作用。同样，通过比较纳米 Ag 和 Ag^+，也发现 Ag^+ 在纳米 Ag 对四膜虫的毒性效应中起主要作用[229]。纳米 Ag 可影响大型溞的 RNA 聚合酶、蛋白质消化吸收等；粒径越小毒性越强，但与粒径相比，表面包覆层是决定其毒性的主要因素[230]。表面包覆层的配体组成也显著影响 Au 纳米颗粒对斑马鱼胚胎的毒性[231]。

与纳米 Ag 相似，纳米 Cu 的水体暴露也引起对鱼类胚胎发育的急性毒性。一项研究比较了不同粒径（25 nm、50 nm、100 nm）的纳米 Cu 对斑马鱼胚胎的毒性效应，发现各种粒径的纳米 Cu 均影响斑马鱼胚胎的孵化率、增加畸形率，抑制仔鱼运动行为，且粒径越小毒性效应越强。此外，研究也发现纳米 Cu 的毒性效应主要是由颗粒状的纳米 Cu 转化产生的铜离子引起[232]。最近，Brun 等首次用活体（in vivo）实验证据表明，Cu 纳米颗粒等纳米材料可以引发斑马鱼胚胎皮肤和肠道细胞中与免疫系统调控相关的基因改变[233]。

5.2 金属氧化物纳米材料

常见的金属氧化物纳米材料主要包括纳米 TiO_2、纳米 ZnO、纳米 CuO 等。纳米 TiO_2 是应用最广泛的纳米材料之一，用于化妆品、防晒霜以及建筑材料表面涂层等，可以直接排放到水环境中。早期的研究认为，与其他金属氧化物纳米材料（如纳米 ZnO、纳米 CuO）相比，纳米 TiO_2 的毒性较低，在无紫外线照射的条件下，只有浓度达到 100 mg/L 以上时才对生物具有毒性效应[234]；甚至在紫外线照射下，某些高分子包被的 TiO_2 纳米颗粒对发育期斑马鱼等水生生物的毒性也很低[235]。因此纳米 TiO_2 被认为是低毒或"无毒"的环境友好纳米材料。近期的一项研究指出，纳米 TiO_2 暴露会导致泥蚶（Tegillarca granosa）体内多巴胺、氨基丁酸（GABA）、乙酰胆碱（ACh）三种神经递质浓度显著升高，同时，抑制乙酰胆碱酶（AChE）活性，与神经递质编码调制酶（AChE、GABAT、MAO）和受体（mAChR3、GABAD、DRD3）相关的基因表达显著下调，对泥蚶产生神经毒性效应[236]。但总体而言，纳米 TiO_2 的急性暴露对水生生物的毒性较低，而长期暴露可对受试生物产生毒性效应。而且，纳米 TiO_2 在发生化学转化或老化后，其对水生生物的生物可利用性以及生态毒理效应也会发生变化[237, 238]。

纳米 ZnO、纳米 CuO 等溶解性较强的金属氧化物纳米材料的水生毒理效应更为复杂，需要考虑其在环境介质中的溶解和离子释放[239]。一项研究指出，纳米 ZnO 暴露会导致青鳉胚胎的孵化率降低，孵化延迟，畸形率、死亡率和心率显著增加；而且，纳米 ZnO 对青鳉死亡率和心率的影响显著大于 Zn^{2+} 离子[240]。纳米二氧化铅（PbO_2）会在日本青鳉肠道内发生还原溶解，释放 Pb^{2+} 离子，并造成铅的生物累积和毒性效应[241]。与重金属对鱼类暴露作用的主要靶器官相似，纳米 ZnO 暴露后，在鲤（Rutilus rutilus caspicus）鳃中累积含量最高，其次由高到低依次为肝脏、肾脏和肌肉。纳米 ZnO 暴露可引起生物标志物金属硫蛋白显著表达，说明释放的 Zn^{2+} 起了主要作用。此外，纳米 ZnO 的暴露显著增加对鳃、肝和肾等组织的损伤，并与组织中 Zn 累积量相关[242]。一项研究比较了纳米 CuO、纳米 ZnO 和纳米 NiO 对斑马鱼的毒性，发现纳米 CuO 毒性最强，且纳米金属氧化物的毒性可能与抑制 DNA 的合成和修复，以及诱导 ROS 生成有关[243]。通过对比纳米 CuO 和纳米 Cu 对斑马鱼胚胎的毒性效应，Harper 课题组最近发现，团聚状态、表面电荷以及 Cu^{2+} 离子释放都不能充分解释两种纳米材料的毒性差异，而 ROS 产生能力的不同是其毒性差异的主要原因[244]。而且，金属氧化物纳米材料对不同发育阶段的斑马鱼（胚胎或仔鱼）的毒性有所不同[245]。

最近，Zhao 等人研究了纳米 CuO 对水生植物凤眼莲的毒性效应[194]，发现：①纳米 CuO 对凤眼莲的毒性效应明显高于 CuO 大颗粒，其释放的 Cu^{2+} 离子对植物的生长也有一定的抑制作用，但没有纳米颗粒的作用显著，表明纳米 CuO 的生物毒性效应主要是其纳米的粒径效应本身所致；②纳米 CuO 暴露下，凤眼莲根冠和分生区受到严重损伤，细胞排列疏松无序，根冠有脱离根尖的趋势，其保护功能受到严重损坏，根尖伸长区出现断根，但对成熟区影响不显著；③纳米 CuO 对叶片的损伤主要表现为通过诱导表皮保卫细胞产生 H_2O_2，打通了细胞膜上的 Ca^{2+} 离子通道，使得胞质 Ca^{2+} 离子浓度升高，诱导气孔关闭，气体交换受阻,从而抑制了光合作用和呼吸作用,导致沉水叶片失绿变黄,乃至腐烂。

5.3 碳纳米材料

随着碳纳米管和石墨烯等碳纳米材料越来越广泛的应用,其对水生生物的毒性效应备受关注。近期的一项研究指出,MWCNT 暴露导致菲律宾蛤仔(Ruditapes philippinarum)代谢活力增强,糖原及蛋白质浓度降低;同时,显著诱导了菲律宾蛤仔的氧化应激反应,并抑制了胆碱酯酶的活性,具有神经毒性效应[246]。GO 暴露可引起斑马鱼鳃细胞凋亡和坏死数量增加,并诱导 ROS 大量生成,引起鳃和肝组织损伤[247]。而且,GO 会引起斑马鱼胚胎的发育毒性,其毒性机制包括 DNA 修饰、蛋白质糖基化以及 ROS 的过度生成[248]。最近,赵建等研究比较了三种不同石墨烯材料,即 GO、RGO 与多层石墨烯(MG)对小球藻的毒性效应和可能的毒性机制[249]。研究表明,三种石墨烯材料对藻细胞的半数效应浓度(EC_{50})分别为:GO(37 mg/L)、RGO(34 mg/L)、MG(62 mg/L),悬浮性良好的 GO 会通过遮蔽效应抑制藻细胞生长,且遮蔽效应产生的毒性占 GO 总毒性的 17%。通过团聚效应实验及光学显微镜观察发现,疏水性的 RGO、MG 比 GO 更容易与藻细胞发生异相团聚,进而增加 RGO、MG 与藻细胞相互接触的概率。进一步研究表明,三种石墨烯材料均会破坏藻细胞的膜完整性,且 RGO 导致膜损伤程度最强,氧化胁迫和物理穿刺/提取的综合作用是三种石墨烯材料产生膜损伤的主要原因(图 23-6)。另外,三种石墨烯材料均会吸附培养基中的宏量元素(N、P、Mg、Ca)导致培养基中养分耗竭,进而间接对藻细胞产生毒性。表面含有大量官能团的 GO 引起的养分耗竭最强,养分耗竭产生的毒性占 GO 总毒性的 53%。林道辉课题组最近则发现,MWCNTs 发生氧化后更易与藻细胞发生团聚,对小球藻的急性毒性增强[250]。

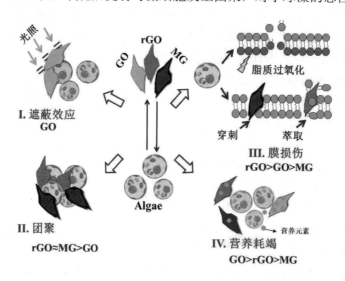

图 23-6 不同类型石墨烯材料对藻类的致毒机制[249]

赵建等还进一步研究了天然矿物颗粒和 NOM 如何影响石墨烯材料对藻类的毒性。他们以 Al_2O_3 为模式矿物颗粒,研究了不同形貌的 Al_2O_3 颗粒对 GO 藻毒性的影响机制[251]。结果表明,GO 可通过与藻细胞的直接接触对藻细胞产生毒性,主要的致毒机制包括遮蔽效应、氧化胁迫和物理损伤(图 23-7)。不同类型的 Al_2O_3 与 GO 的结合均可缓解 GO 的毒性效应,其中纤维状的 Al_2O_3 对毒性缓解的程度最强,其次是大颗粒 Al_2O_3 和小球形 Al_2O_3。主要的毒性缓解机制是 GO 可以与 Al_2O_3 发生异相团聚,并且 Al_2O_3 可以包裹在细胞表面从而减小了藻细胞与 GO 的直接接触,进而降低了 GO 的毒性效应。不同形态的 Al_2O_3 均能显著缓解由 GO 造成的物理损伤、遮蔽效应和氧化胁迫。同时,NOM 能够缓解 GO 对藻类的毒性效应,其主要机制是 NOM 能够包裹在 GO-Al$_2$O$_3$ 异相团聚体的表面,通过空间位阻作用减少 GO 与藻细胞的直接物理接触。这些研究结果表明,客观评价纳米材料的生物安全性时,需充分考虑 NOM 和矿物颗粒等环境因素的影响。

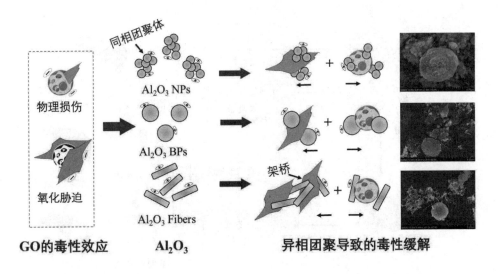

图 23-7 Al$_2$O$_3$ 颗粒与 GO 的异相团聚及其对 GO 藻毒性的影响[251]

5.4 小结

随着纳米材料的大量生产和使用,释放到水环境的纳米材料可能对水生生物产生潜在环境生态风险。目前开展的研究主要关注急性暴露对受试生物的毒性效应。基于纳米材料的独特物理化学性质,释放到环境中的纳米材料会发生团聚、分散、与其他物质结合、转化等过程。因此,研究纳米材料对生物的潜在毒性需要考虑其在环境中的行为、生物可利用性、转化等诸多因素。

6 纳米材料对超重群体的毒性效应

近年来,随着人们饮食结构和生活方式的改变,超重、肥胖作为一种慢性代谢相关疾病,已在全球范围内迅速蔓延。超重群体的易感性不止表现在其显著增加了个体的死亡风险和脂肪肝、糖尿病、胰岛素抵抗综合征、心血管疾病甚至某些癌症的发生风险,更表现在其对环境有害因素(如 POPs、大气颗粒物等)反应更加敏感。因此,在超重群体中开展纳米材料及其与污染物复合物暴露的健康效应研究具有重要的理论和现实意义。

纳米银是在日常生活用品中应用最为广泛的纳米材料之一,其环境健康风险,尤其是对超重群体等环境有害刺激的易感群体的健康风险也更应受到大家的关注。闫兵课题组[252]在近期的一项研究中,对比评价了纳米银在正常及高脂饮食诱导的超重小鼠体内生物分布和毒性效应的差异,并对其中的关键分子机制作了阐述。结果显示,连续 14 天口服暴露纳米银(300 mg/kg)在正常小鼠中没有表现出显著的毒性效应,但在超重小鼠肝脏中引起显著的炎症和肝细胞水样变性,并加剧了肝细胞脂肪变性程度,促进超重小鼠非酒精性脂肪肝发病进程。通过设立银离子对照,对肝脏中银离子和银纳米颗粒进行了定性、定量的分析,文章作者发现银离子可以在超重小鼠肝脏中被还原为银纳米颗粒,上述纳米银暴露造成的毒性损伤来源于银纳米颗粒，而不是银离子。进一步的机制研究结果显示，银纳米颗粒暴露引起超重小鼠肝脏 Kupffer 细胞的活化及相关炎症信号通路（NF-κB、JNK 以及 p38 MAPK）的激活（图 23-8）。同时，银纳米颗粒暴露显著抑制了脂肪酸氧化（*Ppard* 及其靶基因）相关基因的表达，加剧了肝脏的脂肪积累（图 23-8）。

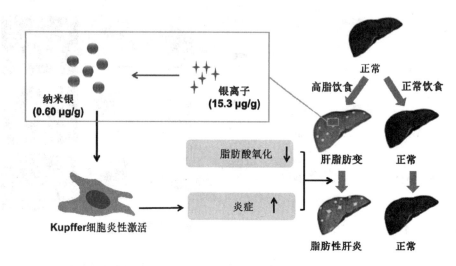

图 23-8 纳米银在超重小鼠中诱导脂肪肝变加重的机制[252]

在超重小鼠肝脏中积累或还原的银纳米颗粒诱导肝脏 Kupffer 细胞炎性激活，引起肝脏的炎症反应。同时，纳米银可显著下调脂肪酸氧化相关基因的表达，引起肝脏中脂肪积累。纳米银诱导的超重小鼠肝脏炎症和脂肪酸氧化能力下降共同作用，加剧高脂饮食诱导的脂肪肝疾病进程

纳米材料与环境污染物共暴露可显著改变污染物在细胞、低等水生生物以及高等哺乳动物中的吸收分布，引起不同于污染物单独暴露的毒性效应。然而，纳米材料-环境污染物复合体系在超重群体中的生物分布行为及毒性效应研究鲜有报道。闫兵课题组[253]的研究选用纳米 ZnO 为纳米材料的代表，选用铅离子为典型重金属污染物的代表，评价了 ZnO 纳米材料和铅离子共暴露在超重小鼠中的生物分布行为及毒性效应。结果显示，经口服暴露后，纳米 ZnO 显著增加了铅离子在正常及超重小鼠肾脏、脾脏、肝脏等主要脏器的累积量。这一方面可能是由于纳米 ZnO 作为重金属离子的载体，促进了铅离子的吸收，另一方面，游离的铅离子比纳米 ZnO-铅离子复合物更容易被各脏器清除。高脂饮食饲喂引起超重小鼠小肠通透性的增加，进一步加剧了铅离子在超重小鼠中的积累。毒性评价结果显示，纳米 ZnO 与铅离子共暴露引起了超重小鼠体重的显著性下降。此外，铅离子在主要脏器的积累对超重小鼠的肾脏和脾脏没有造成显著的病理学改变，但加重了超重小鼠的肝脏损伤。

7 纳米材料与污染物的复合生物效应

进入环境中的纳米材料凭借较高的比表面积和独特的表面化学性质，可以与环境中多种有机污染物、无机污染物发生吸附、解吸、氧化、还原等一系列复杂的相互作用，这一过程不仅会影响纳米材料自身的理化性质、赋存状态和生物效应，同时也会对环境污染物的毒性效应产生显著影响[254, 255]。这是当前纳米材料环境效应研究领域的一个热点，相关报道正逐渐增加（图 23-9）。纳米材料-污染物复合生物效应研究方法主要有三类[255]：第一类以污染物为主要对象，探究纳米材料对其生物效应的影响；第二类以纳米材料为主，探究污染物对纳米材料生物效应的影响；第三类将两者等同，进行析因实验，即将两个或

多个因素的各处理水平正交分组实验,它不仅可以检验各因素内部不同水平间的差异,还可检验两个或多个因素间是否存在交互作用。三种方法中,析因实验最为系统可靠,但由于总实验组数等于各因素处理水平数的乘积,工作量大,相关研究较少。

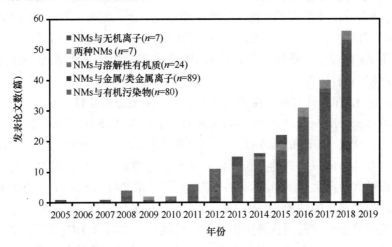

图 23-9　纳米材料(NMs)和污染物复合生物效应研究文献统计

数据来自 Web of Science,以图例中关键词检索得到所有文献后进一步分析确定相关文献数量。检索时间:2019 年 3 月 6 日

纳米材料影响污染物毒性的作用机制主要包括两个方面[256]:①理化机制方面,纳米材料通过吸附/解吸,氧化/还原等理化作用影响污染物的环境行为及毒性效应;②生物学机制方面,纳米材料通过影响污染物在生物体内的累积、分布、代谢,进而影响污染物毒性。此外,纳米材料还可以通过上调生物体的自我防御机制来抵抗污染物的毒性效应[256]。因此,纳米材料与污染物间可能发生复杂的拮抗、协同、加和、独立等复合生物效应。例如,有研究发现 TiO_2 纳米颗粒与阿特拉津能协同抑制小球藻生长,但其与五氯苯间的复合效应为加和作用,而与六氯苯和多氯联苯(PCB-77)为拮抗作用[257]。纳米材料与污染物间拮抗效应的原因主要包括[255]:纳米材料通过吸附作用降低污染物在暴露介质或生物体内的有效浓度,或通过氧化还原作用将高毒性污染物转化为低毒性物质;污染物在纳米材料上的吸附钝化了纳米材料的生物活性,消耗其产生的 ROS,抑制其生物毒性;纳米材料与污染物竞争生物膜上的结合位点,减弱各自的生物毒性。纳米材料与污染物间协同或加和作用则主要归因为"木马效应",即纳米材料吸附载运污染物进入生物体,增强其毒性[255]。此外,纳米材料和污染物也可能通过影响生物的细胞膜结构、基因表达、生物代谢等,改变生物的耐受性,从而增强或减弱共存污染物的毒性效应[255, 258]。

7.1　金属基纳米材料与污染物复合作用

目前,纳米 TiO_2 与共存污染物复合暴露的研究较多。与其他纳米颗粒一样,纳米 TiO_2 具有大的比表面积以及表面活性基团,因此能通过物理或者化学方式吸附环境中共存的其他物质[256, 259]。一些学者研究了纳米 TiO_2 与多种重金属、典型有机污染物以及其他纳米材料(如纳米 Ag[260])复合暴露的环境行为和毒性效应。其中一些研究指出,纳米 TiO_2 可以作为污染物的载体,增加重金属[如 Cd、Zn、As(V)、Pb 等]在大型溞[261-263]、四膜虫[264]等模式生物体内的累积、生物可利用性及其毒性。例如,许安课题组最近发现,低浓度的 TiO_2 纳米颗粒可以吸附并且携带砷通过不同的胞吞途径进入细胞,诱导线粒体损伤,提高氧化应激水平,进而协同增强砷在哺乳动物细胞中诱导的 DNA 双链断裂、微核产生以及基因突变等遗传毒性效应[265]。而且,As、Cd、Ni 等重金属离子在 TiO_2 纳米颗粒表面的吸附会影响纳米颗粒在水中的沉降行为及其沿深度方向的分布,进而导致重金属自身在水环境中的垂直分布发生变化,并最终引起纳米颗粒以及重金属对不同水层水生生物的暴露水平的差异。例如，许安课题组以底栖生物秀丽隐杆线虫为模型，发现 TiO_2 纳米颗粒可以加剧重金属在线虫体内的累积以及诱导的生殖和发育毒性[266]。但也有研究指出，纳米 TiO_2 与重金属［Cu、Cd、As(V)、Pb］对铜绿微囊藻、大型溞和卤虫无节幼体（Artemia salina nauplii）复合暴露后，降低了其生物可利用性和毒性[267-270]。最近，Yan 等研究了 TiO_2 纳米颗粒对 As 在卤虫无节幼体体内累积、分布、排出以及毒性效应的影响，发现：TiO_2 纳米颗粒促进了 As 的生物累积，但同时加速了 As 的排出；此外，TiO_2 纳米颗粒降低了高生物活性态 As 的比例，因此总体上降低了 As 对卤虫无节幼体的毒性[271]。

也有研究表明，纳米 TiO_2 与共存污染物复合暴露后，能通过食物链传递。例如，将纳米 TiO_2 与 As 复合暴露于海洋微拟球藻（Nannochloropsis maritima），纳米 TiO_2 增加了 As 的积累，并可经过食物链传递给海洋浮游动物盐水卤虫（Artemia salina）[270]。而将纳米 TiO_2 与 Cd^{2+} 离子复合暴露于盐水卤虫，尽管高剂量的纳米 TiO_2（如 400 mg/L）对盐水卤虫没有明显的急性毒性效应，但不同剂量的纳米 TiO_2 与镉复合暴露时，对镉的毒性效应产生了不同的影响，即低剂量纳米 TiO_2（5 mg/L）增加了 Cd（5 mg/L）对盐水卤虫的急性毒性，而高剂量的纳米 TiO_2（400 mg/L）则降低了镉的毒性效应。显然，复合暴露的毒性效应与纳米材料以及污染物的暴露剂量密切相关，这可能是因为纳米 TiO_2 能影响镉的吸附和生物可利用性[272]。因此，由于暴露剂量、时间的差异，纳米 TiO_2 与重金属复合暴露后表现出毒性效应的复杂性。目前，纳米 TiO_2 与重金属复合暴露对鱼类毒性效应的研究较少，早期的一篇文献发现，纳米 TiO_2 与镉复合暴露，增加了镉在鲤鱼体内的生物可利用性[273]。而近期的研究表明，纳米 TiO_2 不仅增加了铅在斑马鱼仔鱼体内的累积，同时显著地增强了铅对斑马鱼的甲状腺内分泌干扰效应和神经发育毒性[274]。此外，TiO_2 纳米颗粒与重金属间的协同毒性还具有跨世代传递效应。例如，许安课题组使用 TiO_2 纳米颗粒与镉共同暴露线虫，发现低剂量 TiO_2 纳米颗粒即可通过促进镉在生殖细胞中的累积，进而增强镉对亲代和子代(F1、F2 代)线虫的毒性效应[275]。

金属氧化物纳米材料与有机污染物复合暴露对水生生物毒性效应也受到了广泛的关注。一项研究指出，纳米 TiO_2 与三丁基锡（TBT）复合暴露，与单独暴露相比引起软体动物鲍鱼胚胎氧化损伤，导致孵化率下降和畸形率增加[276]。另一项研究则发现，纳米 TiO_2 单独暴露后，对盐水卤虫没有产生显著的急性毒性，但存在纳米 TiO_2 时显著改变了菲对盐水卤虫的急性毒性效应。与菲单独暴露相比，低浓度（5 mg/L）的纳米 TiO_2 暴露增加了菲的毒性效应，而高浓度（400 mg/L）的纳米 TiO_2 暴露降低了菲的毒性效应[272]。PVP 包被的 Fe_3O_4 纳米颗粒则能减轻原油水溶组分对河口桡足类生物（Amphiascus tenuiremis）的毒性[277]。

对鱼类复合暴露的研究发现，当斑马鱼胚胎共暴露于十溴联苯醚（BDE-209）和低剂量纳米 TiO_2（100 μg/L）时，纳米 TiO_2 可增加 BDE-209 的生物可利用性，同时也增加 BDE-209 的代谢产物的含量，引起甲状腺内分泌干扰效应和发育神经毒性效应[278]。当斑马鱼胚胎与纳米 TiO_2 和五氯酚（PCP）共暴露后，并没有增加 PCP 对斑马鱼仔鱼的生物可利用性，但是增强了仔鱼对 PCP 的代谢能力，表现为中间代谢产物四氯氢醌显著升高以及相关的 ROS 生成，并引发脂质过氧化和 DNA 损伤（8-OHDG）[279]。而当纳米 TiO_2 与双酚 A（BPA）复合暴露后，能增加斑马鱼成鱼对纳米 TiO_2 和 BPA 的生物可利用性，主要分布在肝脏、性腺、大脑，表现出"木马效应"，诱导明显的繁殖毒性，表现为抑制产卵[280]。此外，复合暴露纳米 TiO_2 与 BPA，积累在母代的纳米 TiO_2 与 BPA 可传递给子代，引起子代的甲状腺内分泌干扰效应和神经发育毒性[280]，也改变斑马鱼肠道中的微生物群落结构，并引起诸如斑马鱼生长和肠道健康（如神经传递、上皮屏障渗透性、炎症和氧化应激）相关的响应，且表现出性别和浓度依赖关系[281]。纳米 TiO_2 也能增加微囊藻毒素 LR（microcystin-LR）在斑马鱼体内的生物累积，并引起生殖毒性[282]以及子代的发育毒性[283]。纳米 TiO_2 与二噁英（TCDD）复合暴露引起欧洲鲈鱼拮抗免疫响应基因的表达和 DNA 损伤[284]。

但是也有研究显示，纳米材料与有机污染物复合暴露表现出效应的复杂性。如上文提及的，4 种有机氯污染物（阿特拉津、六氯苯、PCB77、五氯苯）与纳米 TiO_2 对小球藻的复合暴露表现出多种不同的效应,包括协同、加和、拮抗[257],这与暴露的剂量、时间以及污染物的特性有关。最近,Martin-de-Lucia等研究了纳米 SiO$_2$、TiO$_2$、Fe$_3$O$_4$ 等无机纳米材料与含有 46 种有机微污染物的废水共暴露对蓝细菌(*Anabaena* sp. PCC 7120 CPB4337)的毒性效应,发现:低浓度的纳米 SiO$_2$ 和 TiO$_2$ 与废水中的微污染物产生拮抗作用,从而降低毒性;而纳米 Fe$_3$O$_4$ 以及高浓度的纳米 SiO$_2$ 和 TiO$_2$ 则与微污染物产生协同作用,增加了毒性[285]。锌铝层状双氢氧化物(LDH)纳米材料与传统防锈剂 2-巯基苯并噻唑(MBT)复合得到的新型防锈剂对菲律宾帘蛤的急性毒性远低于单独的 MBT,因此,LDH 纳米材料降低了 MBT 防锈剂的生态风险[286]。

此外,金属基纳米材料也能影响有机污染物以及(类)金属的植物吸收和毒性效应。例如,纳米 TiO$_2$ 能显著降低水稻对四环素的吸收以及四环素对水稻造成的毒性[287]。纳米 ZnO 能显著降低水稻对 As(V)和 As(III)的吸收,而纳米 CeO$_2$ 则对水稻的 As 吸收没有影响[288]。纳米 CuO 能通过缩短水稻的生长周期而降低水稻籽实中 As 的累积[20]。纳米羟基磷灰石能降低 Pb 在水稻体内的转移以及对根部细胞的毒性[289]。nZVI 能降低拟南芥对 As(V)的吸收,从而减轻 As(V)对其的毒性[290]。Wu 等最近研究了 nZVI 对丛枝菌根真菌(*arbuscular mycorrhizal*)-玉米(*Zea mays* L.)共生系统中重金属生物吸收的影响,发现:nZVI 降低了玉米对 Zn、Pb 等重金属的吸收,但同时抑制了丛枝菌根真菌的生长及其对重金属的稳定作用[291]。纳米 CeO$_2$ 与 Cd^{2+} 离子的共暴露能改变水培大豆幼苗根区的物理、化学、生物过程(如促进根系分泌物的释放),进而影响植物对 Cd 和 Ce 的吸收和累积:纳米 CeO$_2$ 对大豆幼苗根部的 Cd 含量没有影响,却能显著降低 Cd 从根向芽的运输;Cd^{2+} 离子暴露降低了大豆根部 Ce 的浓度,却显著提高了芽中 Ce 的浓度[292]。

以上研究显示,纳米 TiO$_2$ 等金属基纳米材料能作为污染物的载体,增加污染物对受试生物的可利用性乃至毒性。但也有研究表明,纳米材料与污染物复合暴露可引起包括协同、拮抗、加和等多种效应,表现出作用的复杂性,这与暴露剂量、时间、污染物的性质、作用方式等因素密切相关。

7.2 碳纳米材料与污染物复合作用

碳纳米材料由于其独特的物理化学特性,可吸附与之共存的污染物[293-296],因此可影响污染物的环境行为、生物可利用性和毒性[297,298]。碳纳米材料可以作为有机污染物的载体,通过"木马效应"增加有机污染物的生物有效性和毒性。例如,将碳纳米粉与代表性多环芳烃(PAHs)苯并芘 BaP 复合暴露于斑马鱼胚胎,碳纳米粉促进了 BaP 在斑马鱼胚胎体内的累积,并激活机体的应激反应通路,诱导更强的细胞毒性效应[299]。同样地,将 C$_{60}$ 纳米颗粒与 BaP 共同暴露于鱼鳃细胞,也会增强 BaP 的细胞毒性[300]。最近的一项研究发现,SWCNT 吸附的炔雌醇(EE2)能在鱼类肠道环境条件下发生解吸并保持生物活性,与雌激素受体结合并将其激活[301]。而 GO 也能增加水稻对 PAHs 的吸收以及 PAHs 对水稻的生物毒性[302]。

但在某些情况下,碳纳米材料能降低有机污染物的生物有效性和毒性。例如,GO 纳米片和炭黑纳米颗粒能降低 BaP 对卤虫(*Artemia franciscana*)和鱼肝脏细胞的生物有效性[303]。当 SWCNT 与全氟辛烷磺酸盐(PFOS)复合暴露后,在斑马鱼肝脏、肠、鳃和脑组织中 PFOS 的含量随 SWCNT 浓度的增加而降低,而在斑马鱼表皮中却增加,这可能是由于 SWCNT 吸附 PFOS,形成较大的团聚体,降低了斑马鱼对 PFOS 的生物可利用性。但是,复合暴露能诱导产生更多的 ROS,并显著抑制肝组织中超氧化物歧化酶(SOD)、过氧化氢酶(CAT)和乙酰胆碱酯酶(AChE)的活性[304]。而当衣藻(*Chlamydomonas reinhardtii*)共暴露于 GO 和含有多种有机微污染物的废水时,有机微污染物在 GO 表面的吸附以及对 GO 团聚的促进作用会引起拮抗效应从而降低对衣藻的毒性[305]。许安课题组近期的研究结果显示,GO 可与哺乳动物细胞发生相互作用,黏附在细胞膜表面或被部分内吞入胞质中,诱导细胞内氧化应激水平升高,并进一步促使 GO 在细胞中诱导下游降解通路顺畅的完整的细胞自噬效应；作为细胞内广泛存在的一种抵御外界恶劣环境的重要的自我保护机制，自噬帮助细胞有效拮抗了有机污染物多氯联苯诱导的细胞毒性和遗传毒性[306]。由于碳纳米材料的比表面积、形状、尺寸等物理化学性质存在极大差异，而且毒性效应与暴露的剂量、时间等因素密切相关，因此，碳纳米材料与污染物复合作用的环境行为、生物可利用性也表现出多样性[307]。

除有机污染物外，GO 还可以影响无机污染物的毒性效应。许安课题组以秀丽隐杆线虫为模型，发现 GO 可以显著降低无机砷诱导的生殖毒性。究其机制，主要包括两个方面：一方面，与 GO 和多氯联苯的作用类似，GO 也可在线虫体内诱导具有保护作用的细胞自噬效应，进而降低砷引起的线粒体损伤以及氧化应激；另一方面，GO 还可以下调线虫体内砷特异性结合蛋白 LEC-1 的表达，促使砷排出体外，降低砷在线虫体内的累积，进而拮抗砷的毒性[308]。最近，刘思金课题组发现了低剂量 GO 提升重金属物细胞毒性的一个新机制：GO 可以作用于巨噬细胞并且改变细胞膜的骨架和完整度，并进一步改变细胞膜的通透性从而导致细胞对 Cd^{2+} 等重金属离子的摄入增加[309]。同时发现，如果改变暴露方式，改为先使用低剂量暴露 GO 前处理，随即暴露低于致死量的 Cd^{2+} 离子，也会刺激细胞内高 ROS 表达且导致细胞凋亡。研究显示，GO 暴露可以改变巨噬细胞的激活状态进而影响重金属离子对巨噬细胞的毒性。该研究发现为科学评价人工纳米材料的环境和生物学效应提供了新的视角：即使人工纳米材料自身的毒性等效应不是非常显著，但仍可通过影响环境中重金属等污染物的环境行为和毒性而产生显著的负面效应。另一方面，需要注意的是，某些无机离子是生物必需的营养元素，GO 对其的吸附作用可能影响生物体对这些无机营养元素的摄取，进而造成毒性效应。例如，Yu 等发现，GO 能通过络合捕获铁离子而使酿酒酵母、白念珠菌（*Candida albicans*）等真核微生物以及荧光假单胞菌等原核微生物处于胞外缺铁的营养状态，扰乱了其与铁相关的生理和代谢过程，进而抑制了微生物细胞的生长[310]。

7.3 小结

目前，关于纳米颗粒与环境中污染物复合暴露对生物毒性效应的研究受到重视，但是研究结论往往相互矛盾。鉴于真实环境的复杂性，特别是当存在天然有机质时，研究纳米材料与污染物复合暴露毒性效应的生态环境风险，首先应充分了解纳米材料与有机质以及污染物的环境行为、物理化学作用特征，进而可有效揭示对生物有效性、代谢转化和毒性的影响机制。此外，有关纳米材料与污染物复合生物效应的研究主要在水环境介质中展开（90.6%），土壤介质中的相关研究十分缺乏（6.1%）。土壤是环境中纳米材料与污染物的一个重要的汇，研究土壤中纳米材料与污染物的复合生物效应具有显著现实意义，但由于土壤介质的复杂性，相关研究更具挑战性。另外，当前对纳米材料与污染物复合生物效应的研究一般旨在为评估纳米材料的环境风险提供科学支持，但是一定组合与配比的纳米材料-生物互作体系可以在保证生物安全的基础上协同去除污染物。这是一个崭新的研究领域，可拓展纳米材料与污染物复合效应研究的方向，并为治理环境污染提供新技术、新方法。

8 雾霾颗粒物关键致毒组分研究

大气颗粒物严重危害人类健康，其中细颗粒物（$PM_{2.5}$）污染及其对人体健康的危害备受公众关注。$PM_{2.5}$ 中的超细颗粒物（<100 nm）比表面积最大，容易成为很多有毒有害成分（如多环芳烃、重金属、病原微生物等）的载体，或成为反应体。目前关于大气中雾霾颗粒物的毒性报道多集中于流行病学调查

的相关性研究[311]，以及对 PM$_{2.5}$ 中水溶性成分、脂溶性组分和不可溶物的细胞或动物实验研究[312]。这些报道中雾霾颗粒物的组分分类仍然相当复杂，难以解释其生物效应产生的原因及机制。

PM$_{2.5}$ 细颗粒物的组成结构可简单描述为存在一个惰性吸附核，在核表面吸附有无机盐离子、重金属、有机物和微生物等多种成分。为了在多样性混合组分中找到对 PM$_{2.5}$ 整体毒性贡献较大的因子，闫兵课题组采用还原论研究方法尝试解决雾霾致毒机制问题，假设颗粒物的基质核材料或者其表面吸附的重金属、有机物等成分可能在 PM$_{2.5}$ 的毒性效应中起着重要的作用，分别设计、合成了核材料理化性质多样和吸附多种有机物、重金属及类金属污染物的模型细颗粒库，系统探究了细颗粒物的关键致毒组分及毒性机制。

8.1 细颗粒物核材料理化性质对其细胞毒性的影响

细颗粒物经呼吸系统进入肺，随后进入血液，循环至全身各器官，包括肾脏、肝脏等。闫兵课题组[313]合成了表面配体结构多样化的纳米尺度的细颗粒物系列（FUP1-7），系统评价了其与多种典型环境污染物的毒性效应。结果显示，细颗粒物表面的配体结构显著影响其对人肺支气管上皮细胞 16HBE 和人胚肾细胞 HEK293 的毒性效应，多污染物与细颗粒物复合暴露引起增强的细胞毒性，但不影响表面配体结构改变引起的细颗粒物细胞毒性的变化趋势。

进一步的机制分析结果显示，细颗粒物暴露诱导细胞产生氧化应激的水平与细胞毒性水平一致，多种污染物与细颗粒物复合暴露显著增强细颗粒物诱导细胞产生 ROS 的能力，但不影响细颗粒物表面配体结构改变引起的细胞 ROS 水平的变化趋势。这些结果也说明细颗粒物及其与多种污染物的复合物诱导 16HBE 和 HEK293 细胞产生 ROS 是其引起细胞毒性的可能机制。

8.2 细颗粒物吸附污染物对其健康效应的影响

考虑碳是真实雾霾颗粒物的主要核组分，闫兵课题组[314]选取碳纳米颗粒作为核材料，设计、合成了基于碳纳米颗粒的细颗粒吸附多种有机物、重金属及类金属污染物的模型细颗粒库，用于模拟雾霾颗粒，研究其毒性效应及机制。定量分析确定两种细颗粒所吸附的关键污染物［Cr(VI)、As(III)、Pb^{2+} 和 BaP］组分与大气中测定的这类污染物浓度具有高度的相似性（图 23-10）。在此基础上，从细颗粒物形态、在水环境中的分散状态及细胞毒性等方面，系统评价了基于碳纳米颗粒的模型细颗粒物与实际大气细颗粒物的异同。透射电镜表征结果显示，未吸附污染物及吸附多种关键污染物的模型细颗粒物在形态上与济南地区收集的雾霾颗粒具有类似的粒径和形态。基于碳纳米颗粒的系列模型细颗粒物在水或培养基中的水合粒径及 Zeta 电位也与济南雾霾颗粒物类似。此外，随着模型细颗粒物表面吸附污染物种类的增加，其对人正常支气管上皮细胞 16HBE 的细胞毒性逐渐增强，同时吸附了 4 种不同污染物的模型细颗粒物与济南雾霾颗粒物表现出类似的毒性效应。这些结果充分显示，采用还原理论建立的模型细颗粒物与实际大气雾霾颗粒具有类似的理化性质和生物效应，可用于模拟雾霾颗粒物，探究其健康效应及致毒的分子机制。

经呼吸进入肺脏的细颗粒物可引起显著的肺部炎症，而细颗粒物造成肺部炎症的关键致毒组分及致毒机制尚不清楚。应用模型细颗粒物，闫兵课题组系统探究雾霾颗粒的免疫毒性及机制。结果显示，相对于其他模型细颗粒物，吸附了 Pb^{2+} 的模型细颗粒物表现出更强的细胞免疫毒性，表明 Pb^{2+} 在细颗粒物诱导的 16HBE 细胞免疫响应中起关键作用。进一步的长链非编码 RNA 的芯片分析结果显示，含 Pb^{2+} 模型细颗粒物和济南雾霾颗粒物显著下调新型长链非编码 RNA lnc-PCK1-2：1 的表达，而 lnc-PCK1-2：1 基因过表达则显著缓解含 Pb^{2+} 模型细颗粒物诱导的细胞免疫响应。上述结果充分说明，细颗粒物复合 Pb^{2+} 是雾霾颗粒诱导细胞产生炎症响应的关键组分，含 Pb^{2+} 细颗粒物通过抑制 lnc-PCK1-2：1 的表达对 16HBE 细胞产生免疫毒性。

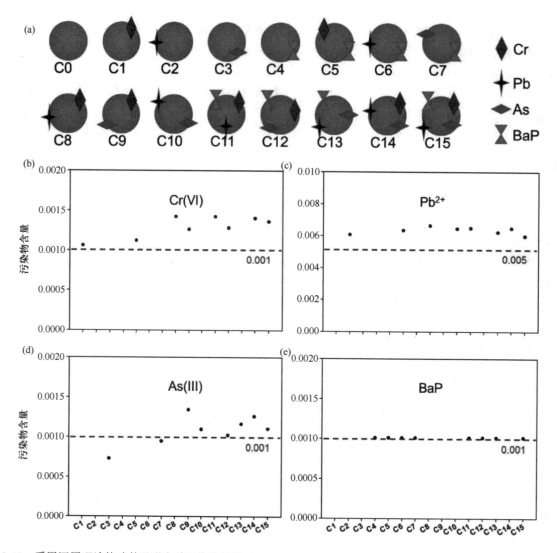

图 23-10 采用还原理论构建的吸附多种污染物的模型细颗粒物模式图（a）和模型细颗粒物中各污染物含量（b～e）

图中虚线标注显示广州雾霾中各污染物组分的含量（引自参考文献[314]）

细颗粒物经呼吸道进入肺脏后，首先与肺表面活性剂接触，并发生相互作用。考虑肺表面活性剂主要存在于肺泡中，为更加真实地模拟细颗粒物人体暴露过程，闫兵课题组[314]选用人肺腺癌细胞 A549 为细胞模型，探究了肺表面活性剂对模型细颗粒物及广州雾霾细胞毒性的影响。研究结果显示，细颗粒物复合 Cr(VI)/Pb^{2+}是其细胞毒性的主要致毒组分，肺表面活性剂 Curosurf$^®$可显著降低含 Cr(VI)/Pb^{2+}细颗粒物及广州雾霾的细胞毒性。进一步的机制分析结果显示，Curosurf$^®$可诱导细胞发生自噬，进而缓解细颗粒物诱导的细胞毒性。

8.3 小结

PM$_{2.5}$严重危害人类健康。然而，由于 PM$_{2.5}$颗粒物组分的复杂性和时空的多变性，鉴别大气细颗粒物中的关键致毒组分，揭示细颗粒物致毒的关键分子机制是目前细颗粒物健康效应研究的瓶颈。经验证，采用还原理论构建模型细颗粒物库的方法为明确细颗粒物各组分对其毒性效应的贡献，揭示细颗粒物健康风险的内在机制提供了新思路。然而，运用模型细颗粒物库探索 PM$_{2.5}$健康风险，会得到海量的细胞毒性大数据，如何综合分析相关毒性大数据，得出细颗粒物健康风险的一般性规律，在后续的研究中尤为重要。因此，结合生物信息学和计算化学的大数据运算的优势，构建 $PM_{2.5}$ 潜在健康危害的预测模型，将为 $PM_{2.5}$ 健康风险的提早预防提供坚实的理论基础。

9 展　望

综上所述，纳米材料可以在多个层次上与生物分子、细胞、动植物个体乃至生态系统发生复杂的相互作用，导致生态毒性、生物毒性和健康风险。纳米材料自身的理化性质（组成、形貌、尺寸、表面修饰等），经环境、生物转化之后被赋予的新的理化性质等均会对其生物行为和健康效应产生影响，是纳米材料生物效应的关键调控因子。考虑释放到环境中的纳米材料势必与环境中的有机污染物、无机污染物共存，造成纳米材料与多种污染物复合暴露的风险，越来越多的研究开始关注真实环境状态下多种污染物复合暴露的生态效应和健康风险。大气细颗粒物是典型的纳米材料-有机污染物、无机污染物的复合结构，采用还原理论构建模型细颗粒物库的方法为复杂大气细颗粒物关键致毒组分和毒性分子机制的研究提供了新思路。

在纳米材料生物效应和健康风险研究中仍然存在着许多不足，无法一一列举，以下只是其中几个问题：

（1）现有研究多聚焦纳米材料自身理化性质与其环境健康风险的关联性分析，经环境、生物转化后的纳米材料的详细表征和生物效应的研究尚有欠缺。

（2）现有的水质标准通常针对单一污染物，真实环境中纳米材料及其与多种污染物复合暴露对人体的安全剂量范围尚不明确。

（3）对现有环境复合纳米污染物生物毒性和健康效应的综合分析方法、手段，尤其是结合生物信息学和计算化学的大数据分析优势开发纳米材料及其与多种污染物复合暴露的环境健康风险预测模型，尚有很大的提升空间。

针对尚未解决的科学问题，我们应该建立行之有效的研究方案，发展针对纳米材料环境、生物转化过程检测的技术手段，并开展多学科交叉研究，为纳米材料安全应用、生态系统安全和人类健康提供强有力的理论和技术支撑。

参 考 文 献

[1] Zhang J, Guo W, Li Q, et al. The effects and the potential mechanism of environmental transformation of metal nanoparticles on their toxicity in organisms. Environmental Science-Nano, 2018, 5: 2482-2499.

[2] Zhang Q R, Liu X L, Meng H Y, et al. Reduction pathway-dependent cytotoxicity of reduced graphene oxide. Environmental Science-Nano, 2018, 5: 1361-1371.

[3] Pulido R G, Leganes F, Fernandez P F, et al. Bio-nano interface and environment: A critical review. Environmental Toxicology and Chemistry, 2017, 36: 3181-3193.

[4] Malysheva A, Lombi E, Voelcker N H. Bridging the divide between human and environmental nanotoxicology. Nature Nanotechnology, 2015, 10: 835-844.

[5] Albanese A, Walkey C D, Olsen J B, et al. Secreted biomolecules alter the biological identity and cellular interactions of nanoparticles. ACS Nano, 2014, 8: 5515-5526.

[6] Monopoli M P, Aberg C, Salvati A, et al. Biomolecular coronas provide the biological identity of nanosized materials. Nature Nanotechnology, 2012, 7: 779-786.

[7] Markiewicz M, Kumirska J, Lynch I, et al. Changing environments and biomolecule coronas: consequences and challenges

for the design of environmentally acceptable engineered nanoparticles. Green Chemistry, 2018, 20: 4133-4168.

[8] Wang Z Y, Zhang L, Zhao J, et al. Environmental processes and toxicity of metallic nanoparticles in aquatic systems as affected by natural organic matter. Environmental Science-Nano, 2016, 3: 240-255.

[9] Surette M C, Nason J A. Nanoparticle aggregation in a freshwater river: The role of engineered surface coatings. Environmental Science-Nano, 2019, 6: 540-553.

[10] Kelly P M, Aberg C, Polo E, et al. Mapping protein binding sites on the biomolecular corona of nanoparticles. Nature Nanotechnology, 2015, 10: 472-479.

[11] Wu W H, Zhang R J, McClements D J, et al. Transformation and speciation analysis of silver nanoparticles of dietary supplement in simulated human gastrointestinal tract. Environmental Science & Technology, 2018, 52: 8792-8800.

[12] Caracciolo G, Farokhzad O C, Mahmoudi M. Biological identity of nanoparticles in vivo: Clinical implications of the protein corona. Trends in Biotechnology, 2017, 35: 257-264.

[13] Olenick L L, Troiano J M, Vartanian A, et al. Lipid corona formation from nanoparticle interactions with bilayers. Chem, 2018, 4: 2709-2723.

[14] Cai R, Chen C. The crown and the scepter: roles of the protein corona in nanomedicine. Advanced Materials, 2019, in press.

[15] Baalousha M, Afshinnia K, Guo L D. Natural organic matter composition determines the molecular nature of silver nanomaterial-NOM corona. Environmental Science-Nano, 2018, 5: 868-881.

[16] Bourgeault A, Legros V, Gonnet F, et al. Interaction of TiO_2 nanoparticles with proteins from aquatic organisms: the case of gill mucus from blue mussel. Environmental Science and Pollution Research, 2017, 24: 13474-13483.

[17] Jayalath S, Larsen S C, Grassian V H. Surface adsorption of nordic aquatic fulvic acid on amine-functionalized and non-functionalized mesoporous silica nanoparticles. Environmental Science-Nano, 2018, 5: 2162-2171.

[18] Adeleye A S, Keller A A. Interactions between algal extracellular polymeric substances and commercial TiO_2 nanoparticles in aqueous media. Environmental Science & Technology, 2016, 50: 12258-12265.

[19] Omar F M, Aziz H A, Stoll S. Aggregation and disaggregation of ZnO nanoparticles: Influence of pH and adsorption of Suwannee River humic acid. Science of the Total Environment, 2014, 468: 195-201.

[20] Liu J, Simms M, Song S, et al. Physiological effects of copper oxide nanoparticles and arsenic on the growth and life cycle of rice (Oryza sativa japonica 'Koshihikari'). Environmental Science & Technology, 2018, 52: 13728-13737.

[21] Ghosh S, Jiang W, McClements D J, et al. Colloidal stability of magnetic iron oxide nanoparticles: Influence of natural organic matter and synthetic polyelectrolytes. Langmuir, 2011, 27: 8036-8043.

[22] Pallem V L, Stretz H A, Wells M J M. Evaluating aggregation of gold nanoparticles and humic substances using fluorescence spectroscopy. Environmental Science & Technology, 2009, 43: 7531-7535.

[23] Chang X J, Bouchard D C. Surfactant-wrapped multiwalled carbon nanotubes in aquatic systems: Surfactant displacement in the presence of humic acid. Environmental Science & Technology, 2016, 50: 9214-9222.

[24] Ghosh S, Mashayekhi H, Bhowmik P, et al. Colloidal stability of Al_2O_3 nanoparticles as affected by coating of structurally different humic acids. Langmuir, 2010, 26: 873-879.

[25] Phong D D, Hur J. Using two-dimensional correlation size exclusion chromatography (2D-CoSEC) and EEM-PARAFAC to explore the heterogeneous adsorption behavior of humic substances on nanoparticles with respect to molecular sizes. Environmental Science & Technology, 2018, 52: 427-435.

[26] Ateia M, Apul O G, Shimizu Y, et al. Elucidating adsorptive fractions of natural organic matter on carbon nanotubes. Environmental Science & Technology, 2017, 51: 7101-7110.

[27] Lecloux A J. Discussion about the use of the volume-specific surface area (VSSA) as criteria to identify nanomaterials according to the EU definition. Journal of Nanopart Research, 2015, 17: 447.

[28] Ikuma K, Decho A W, Lau B L T. When nanoparticles meet biofilms-interactions guiding the environmental fate and accumulation of nanoparticles. Frontiers in Microbiology, 2015, 6: 591.

[29] Verneuil L, Silvestre J, Randrianjatovo I, et al. Double walled carbon nanotubes promote the overproduction of extracellular protein-like polymers in *Nitzschia palea*: An adhesive response for an adaptive issue. Carbon, 2015, 88: 113-125.

[30] Jain R, Jordan N, Weiss S, et al. Extracellular polymeric substances govern the surface charge of biogenic elemental selenium nanoparticles. Environmental Science & Technology, 2015, 49: 1713-1720.

[31] Liu W J, Wang L Y, Wang J, et al. New insights into microbial-mediated synthesis of Aubiolayer nanoparticles. Environmental Science-Nano, 2018, 5: 1757-1763.

[32] Faghihzadeh F, Anaya N M, Astudillo C C, et al. Kinetic, metabolic and macromolecular response of bacteria to chronic nanoparticle exposure in continuous culture. Environmental Science-Nano, 2018, 5: 1386-1396.

[33] Wang Q, Kang F X, Gao Y Z, et al. Sequestration of nanoparticles by an EPS matrix reduces the particle-specific bactericidal activity. Scientific Reports, 2016, 6: 21379.

[34] Sheng A X, Liu F, Shi L, et al. Aggregation kinetics of hematite particles in the presence of outer membrane cytochrome OmcA of *Shewanella oneidensis* MR-1. Environmental Science & Technology, 2016, 50: 11016-11024.

[35] Nevius B A, Chen Y P, Ferry J L, et al. Surface-functionalization effects on uptake of fluorescent polystyrene nanoparticles by model biofilms. Ecotoxicology, 2012, 21: 2205-2213.

[36] Yue Y, Behra R, Sigg L, et al. Silver nanoparticle-protein interactions in intact rainbow trout gill cells. Environmental Science-Nano, 2016, 3: 1174-1185.

[37] Gao J J, Lin L, Wei A, et al. Protein corona analysis of silver nanoparticles exposed to fish plasma. Environmental Science & Technology Letters, 2017, 4: 174-179.

[38] Hayashi Y, Miclaus T, Murugadoss S, et al. Female versus male biological identities of nanoparticles determine the interaction with immune cells in fish. Environmental Science-Nano, 2017, 4: 895-906.

[39] Canesi L, Balbi T, Fabbri R, et al. Biomolecular coronas in invertebrate species: Implications in the environmental impact of nanoparticles. NanoImpact, 2017, 8: 89-98.

[40] Canesi L, Ciacci C, Fabbri R, et al. Interactions of cationic polystyrene nanoparticles with marine bivalve hemocytes in a physiological environment: Role of soluble hemolymph proteins. Environmental Research, 2016, 150: 73-81.

[41] Sendra M, Volland M, Balbi T, et al. Cytotoxicity of CeO_2 nanoparticles using *in vitro* assay with *Mytilus galloprovincialis* hemocytes: Relevance of zeta potential, shape and biocorona formation. Aquatic Toxicology, 2018, 200: 13-20.

[42] Mu L, Gao Y, Hu X G. Characterization of biological secretions binding to graphene oxide in water and the specific toxicological mechanisms. Environmental Science & Technology, 2016, 50: 8530-8537.

[43] Mathe C, Devineau S, Aude J C, et al. Structural determinants for protein adsorption/non-adsorption to silica surface. PLoS One, 2013, 8: 81346.

[44] Findlay M R, Freitas D N, Mobed M M, et al. Machine learning provides predictive analysis into silver nanoparticle protein corona formation from physicochemical properties. Environmental Science-Nano, 2018, 5: 64-71.

[45] Li Z Q, Greden K, Alvarez P J J, et al. Adsorbed polymer and nom limits adhesion and toxicity of nano scale zerovalent iron to *E. coli*. Environmental Science & Technology, 2010, 44: 3462-3467.

[46] Ostermeyer A K, Mumuper C K, Semprini L, et al. Influence of bovine serum albumin and alginate on silver nanoparticle dissolution and toxicity to *nitrosomonas europaea*. Environmental Science & Technology, 2013, 47: 14403-14410.

[47] Wirth S M, Lowry G V, Tilton R D. Natural organic matter alters biofilm tolerance to silver nanoparticles and dissolved silver. Environmental Science & Technology, 2012, 46: 12687-12696.

[48] Bertuccio A J, Tilton R D. Silver sink effect of humic acid on bacterial surface colonization in the presence of silver ions and

nanoparticles. Environmental Science & Technology, 2017, 51: 1754-1763.

[49] Cupi D, Hartmann N B, Baun A. The influence of natural organic matter and aging on suspension stability in guideline toxicity testing of silver, zinc oxide, and titanium dioxide nanoparticles with *Daphnia Magna*. Environmental Toxicology and Chemistry, 2015, 34: 497-506.

[50] Jung Y J, Metreveli G, Park C B, et al. Implications of pony lake fulvic acid for the aggregation and dissolution of oppositely charged surface-coated silver nanoparticles and their ecotoxicological effects on *Daphnia magna*. Environmental Science & Technology, 2018, 52: 436-445.

[51] Gao J, Powers K, Wang Y, et al. Influence of Suwannee River humic acid on particle properties and toxicity of silver nanoparticles. Chemosphere, 2012, 89: 96-101.

[52] Noventa S, Rowe D, Galloway T. Mitigating effect of organic matter on the *in vivo* toxicity of metal oxide nanoparticles in the marine environment. Environmental Science-Nano, 2018, 5: 1764-1777.

[53] Ong K J, Felix L C, Boyle D, et al. Humic acid ameliorates nanoparticle-induced developmental toxicity in zebrafish. Environmental Science-Nano, 2017, 4: 127-137.

[54] Briffa S M, Nasser F, Valsami J E, et al. Uptake and impacts of polyvinylpyrrolidone (PVP) capped metal oxide nanoparticles on *daphnia magna*: role of core composition and acquired corona. Environmental Science-Nano, 2018, 5: 1745-1756.

[55] Ruotolo R, Pira G, Villani M, et al. Ring-shaped corona proteins influence the toxicity of engineered nanoparticles to yeast. Environmental Science-Nano, 2018, 5: 1428-1440.

[56] Gunsolus I L, Mousavi M P S, Hussein K, et al. Effects of humic and fulvic acids on silver nanoparticle stability, dissolution, and toxicity. Environmental Science & Technology, 2015, 49: 8078-8086.

[57] Yang X Y, Jiang C J, Hsu K H, et al. Silver nanoparticle behavior, uptake, and toxicity in *caenorhabditis elegans*: Effects of natural organic matter. Environmental Science & Technology, 2014, 48: 3486-3495.

[58] Manceau A, Nagy K L. Quantitative analysis of sulfur functional groups in natural organic matter by XANES spectroscopy. Geochimica et Cosmochimica Acta, 2012, 99: 206-223.

[59] Collin B, Tsyusko O V, Starnes D L, et al. Effect of natural organic matter on dissolution and toxicity of sulfidized silver nanoparticles to *Caenorhabditis elegans*. Environmental Science-Nano, 2016, 3: 728-736.

[60] Kteeba S M, Adawi H I, Rayis O A, et al. Zinc oxide nanoparticle toxicity in embryonic zebrafish: Mitigation with different natural organic matter. Environmental Pollution, 2017, 230: 1125-1140.

[61] Tang J, Zhu N Y, Zhu Y, et al. Distinguishing the roles of different extracellular polymeric substance fractions of a periphytic biofilm in defending against Fe_2O_3 nanoparticle toxicity. Environmental Science-Nano, 2017, 4: 1682-1691.

[62] Gao X, Zhou K J, Zhang L Q, et al. Distinct effects of soluble and bound exopolymeric substances on algal bioaccumulation and toxicity of anatase and rutile TiO_2 nanoparticles. Environmental Science-Nano, 2018, 5: 720-729.

[63] Huang T D, Sui M H, Yan X, et al. Anti-algae efficacy of silver nanoparticles to Microcystis aeruginosa: Influence of NOM, divalent cations, and pH. Colloids and Surfaces A-Physicochemical and Engineering Aspects, 2016, 509: 492-503.

[64] Xia T, Kovochich M, Liong M, et al. Comparison of the mechanism of toxicity of zinc oxide and cerium oxide nanoparticles based on dissolution and oxidative stress properties. ACS Nano, 2008, 2: 2121-2134.

[65] Heinlaan M, Ivask A, Blinova I, et al. Toxicity of nanosized and bulk ZnO, CuO and TiO_2 to bacteria *vibrio fischeri* and crustaceans *Daphnia magna* and *Thamnocephalus platyurus*. Chemosphere, 2008, 71: 1308-1316.

[66] Yan K, Liu Y H, Yang Q L, et al. Evaluation of the novel nanoparticle material - CdSe quantum dots on *Chlorella pyrenoidosa* and *Scenedesmus obliquus*: Concentration-time-dependent responses. Ecotoxicology and Environmental Safety, 2019, 171: 728-736.

[67] Nel A E, Madler L, Velegol D, et al. Understanding biophysicochemical interactions at the nano-bio interface. Nature

Materials, 2009, 8: 543-557.

[68] Cheng L C, Jiang X M, Wang J, et al. Nano-bio effects: interaction of nanomaterials with cells. Nanoscale, 2013, 5: 3547-3569.

[69] Ha Y, Wang X Z, Liljestrand H M, et al. Bioavailability of fullerene under environmentally relevant conditions: Effects of humic acid and fetal bovine serum on accumulation in lipid bilayers and cellular uptake. Environmental Science & Technology, 2016, 50: 6717-6727.

[70] Chen Q Q, Yin D Q, Li J, et al. The Effects of humic acid on the uptake and depuration of fullerene aqueous suspensions in two aquatic organisms. Environmental Toxicology and Chemistry, 2014, 33: 1090-1097.

[71] Jiang C J, Castellon B T, Matson C W, et al. Relative contributions of copper oxide nanoparticles and dissolved copper to cu uptake kinetics of gulf killifish (*Fundulus grandis*) embryos. Environmental Science & Technology, 2017, 51: 1395-1404.

[72] Collin B, Oostveen E, Tsyusko O V, et al. Influence of natural organic matter and surface charge on the toxicity and bioaccumulation of functionalized ceria nanoparticles in *Caenorhabditis elegans*. Environmental Science & Technology, 2014, 48: 1280-1289.

[73] Ouyang K, Walker S L, Yu X Y, et al. Metabolism, survival, and gene expression of *Pseudomonas putida* to hematite nanoparticles mediated by surface-bound humic acid. Environmental Science-Nano, 2018, 5: 682-695.

[74] Oliver A L S, Croteau M N, Stoiber T L, et al. Does water chemistry affect the dietary uptake and toxicity of silver nanoparticles by the freshwater snail *Lymnaea stagnalis*? Environmental Pollution, 2014, 189: 87-91.

[75] Li M, Dang F, Fu Q L, et al. Effects of molecular weight-fractionated natural organic matter on the phytoavailability of silver nanoparticles. Environmental Science-Nano, 2018, 5: 969-979.

[76] Wang Z Y, Li J, Zhao J, et al. Toxicity and internalization of CuO nanoparticles to prokaryotic alga *Microcystis aeruginosa* as affected by dissolved organic matter. Environmental Science & Technology, 2011, 45: 6032-6040.

[77] Hayashi Y, Miclaus T, Scavenius C, et al. Species differences take shape at nanoparticles: Protein corona made of the native repertoire assists cellular interaction. Environmental Science & Technology, 2013, 47: 14367-14375.

[78] Nasser F, Lynch I. Secreted protein eco-corona mediates uptake and impacts of polystyrene nanoparticles on *Daphnia magna*. Journal of Proteomics, 2016, 137: 45-51.

[79] Kang F X, Qu X L, Alvarez P J J, et al. Extracellular saccharide-mediated reduction of Au^{3+} to gold nanoparticles: New insights for heavy metals biomineralization on microbial surfaces. Environmental Science & Technology, 2017, 51: 2776-2785.

[80] Mensch A C, Hernandez R T, Kuether J E, et al. Natural organic matter concentration impacts the interaction of functionalized diamond nanoparticles with model and actual bacterial membranes. Environmental Science & Technology, 2017, 51: 11075-11084.

[81] Liu Y L, Yang T, Wang L, et al. Interpreting the effects of natural organic matter on antimicrobial activity of Ag_2S nanoparticles with soft particle theory. Water Research, 2018, 145: 12-20.

[82] Zhang Y, Meng T, Guo X, et al. Humic acid alleviates the ecotoxicity of graphene-family materials on the freshwater microalgae *Scenedesmus obliquus*. Chemosphere, 2018, 197: 749-758.

[83] Nel A E, Madler L, Velegol D, et al. Understanding biophysicochemical interactions at the nano-bio interface. Nature Materials, 2009, 8: 543-557.

[84] Mu Q, Jiang G, Chen L, et al. Chemical basis of interactions between engineered nanoparticles and biological systems. Chemical Reviews, 2014, 114: 7740-7781.

[85] Zhao H R, Zhang C D, Wang Y Q, et al. Self-damaging aerobic reduction of graphene oxide by *Escherichia coli*: Role of go-mediated extracellular superoxide formation. Environmental Science & Technology, 2018, 52: 12783-12791.

[86] Zhang C, Chen W, Alvarez P J. Manganese peroxidase degrades pristine but not surface-oxidized (carboxylated) single-walled carbon nanotubes. Environmental Science & Technology, 2014, 48: 7918-7923.

[87] Zhang C, Chen S, Alvarez P J J, et al. Reduced graphene oxide enhances horseradish peroxidase stability by serving as radical scavenger and redox mediator. Carbon, 2015, 94: 531-538.

[88] Zhao X, Hao F, Lu D, et al. Influence of the surface functional group density on the carbon-nanotube-induced α-chymotrypsin structure and activity alterations. ACS Applied Materials & Interfaces, 2015, 7: 18880-18890.

[89] Qi Y, Chen W, Liu F, et al. Aggregation morphology is a key factor determining protein adsorption on graphene oxide and reduced graphene oxide nanomaterials. Environmental Science-Nano, 2019, in press.

[90] Xu M, Soliman M G, Sun X, et al. How entanglement of different physicochemical properties complicates the prediction of *in vitro* and *in vivo* interactions of gold nanoparticles. ACS Nano, 2018, 12: 10104-10113.

[91] Huang R X, Carney R R, Ikuma K, et al. Effects of surface compositional and structural heterogeneity on nanoparticle-protein interactions: Different protein configurations. ACS Nano, 2014, 8: 5402-5412.

[92] Zhang R, Jia C, Zhao L, et al. Characterization of the interaction between carbon black and three important antioxidant proteins using multi spectroscopy and modeling simulations. Chemosphere, 2019, 222: 823-830.

[93] Zhao X, Lu D, Hao F, et al. Exploring the diameter and surface dependent conformational changes in carbon nanotube-protein corona and the related cytotoxicity. Journal of Hazardous Materials, 2015, 292: 98-107.

[94] Hao F, Liu Q S, Chen X, et al. Exploring the heterogeneity of nanoparticles in their interactions with plasma coagulation factor XII. ACS Nano, 2019, 13: 1990-2003.

[95] Sun H, Liu Y, Bai X, et al. Induction of oxidative stress and sensitization of cancer cells to paclitaxel by gold nanoparticles with different charge densities and hydrophobicities. Journal of Materials Chemistry B, 2018, 6: 1633-1639.

[96] Liu S, Shen Z Y, Wu B, et al. Cytotoxicity and efflux pump inhibition induced by molybdenum disulfide and boron nitride nanomaterials with sheetlike structure. Environmental Science & Technology, 2017, 51: 10834-10842.

[97] Gray E P, Browning C L, Wang M J, et al. Biodissolution and cellular response to MoO_3 nanoribbons and a new framework for early hazard screening for 2D materials. Environmental Science-Nano, 2018, 5: 2545-2559.

[98] Zhou X, Jia J, Luo Z, et al. Remote induction of cell autophagy by 2D MoS_2 nanosheets via perturbing cell surface receptors and mTOR pathway from outside of cells. ACS Applied Materials & Interfaces, 2019, 11: 6829-6839.

[99] Limbach L K, Wick P, Manser P, et al. Exposure of engineered nanoparticles to human lung epithelial cells: influence of chemical composition and catalytic activity on oxidative stress. Environmental Science & Technology, 2007, 41: 4158-4163.

[100] Moore A, Weissleder R, Bogdanov Jr A. Uptake of dextran-coated monocrystalline iron oxides in tumor cells and macrophages. Journal of Magnetic Resonance Imaging, 1997, 7: 1140-1145.

[101] Lunov O, Syrovets T, Loos C, et al. Differential uptake of functionalized polystyrene nanoparticles by human macrophages and a monocytic cell line. ACS Nano, 2011, 5: 1657-1669.

[102] Weissleder R, Kelly K, Sun E Y, et al. Cell-specific targeting of nanoparticles by multivalent attachment of small molecules. Nature Biotechnology, 2005, 23: 1418.

[103] Wang L, Liu Y, Li W, et al. Selective targeting of gold nanorods at the mitochondria of cancer cells: implications for cancer therapy. Nano Lettersers, 2011, 11: 772-780.

[104] Zhou W, Shao J, Jin Q, et al. Zwitterionic phosphorylcholine as a better ligand for gold nanorods cell uptake and selective photothermal ablation of cancer cells. Chemical Communications, 2010, 46: 1479-1481.

[105] Zhang Y, Yang M, Portney N G, et al. Zeta potential: A surface electrical characteristic to probe the interaction of nanoparticles with normal and cancer human breast epithelial cells. Biomedical Microdevices, 2008, 10: 321-328.

[106] Saha K, Kim S T, Yan B, et al. Surface functionality of nanoparticles determines cellular uptake mechanisms in mammalian

cells. Small, 2013, 9: 300-305.

[107] Chithrani B D, Ghazani A A, Chan W C W. Determining the size and shape dependence of gold nanoparticle uptake into mammalian cells. Nano Letters, 2006, 6: 662-668.

[108] Zhang S, Gao H, Bao G. Physical principles of nanoparticle cellular endocytosis. ACS Nano, 2015, 9: 8655-8671.

[109] Conner S D, Schmid S L. Regulated portals of entry into the cell. Nature, 2003, 422: 37.

[110] Li R, Ji Z, Chang C H, et al. Surface interactions with compartmentalized cellular phosphates explain rare earth oxide nanoparticle hazard and provide opportunities for safer design. ACS Nano, 2014, 8: 1771-1783.

[111] Mirshafiee V, Sun B, Chang C H, et al. Toxicological profiling of metal oxide nanoparticles in liver context reveals pyroptosis in kupffer cells and macrophages versus apoptosis in hepatocytes. ACS Nano, 2018, 12: 3836-3852.

[112] Sun C, Yin N, Wen R, et al. Silver nanoparticles induced neurotoxicity through oxidative stress in rat cerebral astrocytes is distinct from the effects of silver ions. Neurotoxicology, 2016, 52: 210-221.

[113] Yin N, Liu Q, Liu J, et al. Silver nanoparticle exposure attenuates the viability of rat cerebellum granule cells through apoptosis coupled to oxidative stress. Small, 2013, 9: 1831-1841.

[114] Qin Y, Zhou Z W, Pan S T, et al. Graphene quantum dots induce apoptosis, autophagy, and inflammatory response via p38 mitogen-activated protein kinase and nuclear factor-κB mediated signaling pathways in activated THP-1 macrophages. Toxicology, 2015, 327: 62-76.

[115] Tay C Y, Setyawati M I, Leong D T. Nanoparticle density: A critical biophysical regulator of endothelial permeability. ACS Nano, 2017, 11: 2764-2772.

[116] Zhang J, Chen Y J, Gao M, et al. Silver nanoparticles compromise female embryonic stem cell differentiation through disturbing X chromosome inactivation. ACS Nano, 2019, 13: 2050-2061.

[117] Chen Y, Xu M, Zhang J, et al. Genome-wide dna methylation variations upon exposure to engineered nanomaterials and their implications in nanosafety assessment. Advanced Materials, 2017, 29: 1604580.

[118] Wu Y K, Wang F F, Wang S H, et al. Reduction of graphene oxide alters its cyto-compatibility towards primary and immortalized macrophages. Nanoscale, 2018, 10: 14637-14650.

[119] Qi Y, Liu Y, Xia T, et al. The biotransformation of graphene oxide in lung fluids significantly alters its inherent properties and bioactivities toward immune cells. NPG Asia Materials, 2018, 10: 385-396.

[120] Morishita Y, Yoshioka Y, Takimura Y, et al. Distribution of silver nanoparticles to breast milk and their biological effects on breast-fed offspring mice. ACS Nano, 2016, 10: 8180-8191.

[121] Wen R, Yang X, Hu L, et al. Brain-targeted distribution and high retention of silver by chronic intranasal instillation of silver nanoparticles and ions in Sprague–Dawley rats. Journal of Applied Toxicology, 2016, 36: 445-453.

[122] Tang S, Peng C, Xu J, et al. Tailoring renal clearance and tumor targeting of ultrasmall metal nanoparticles with particle density. Angewandte Chemie International Edition, 2016, 128: 16273-16277.

[123] Fromen C A, Kelley W J, Fish M B, et al. Neutrophil–particle interactions in blood circulation drive particle clearance and alter neutrophil responses in acute inflammation. ACS Nano, 2017, 11: 10797-10807.

[124] Sun T, Zhang Y S, Pang B, et al. Engineered nanoparticles for drug delivery in cancer therapy. Angewandte Chemie International Edition, 2014, 53: 12320-12364.

[125] Moghimi S M, Hunter A C, Murray J C. Long-circulating and target-specific nanoparticles: Theory to practice. Pharmacological Reviews, 2001, 53: 283-318.

[126] Deen W M, Lazzara M J, Myers B D. Structural determinants of glomerular permeability. American Journal of Physiology-Renal Physiology, 2001, 281: F579-F596.

[127] Wang D, Zhao L X, Ma H Y, et al. Quantitative analysis of reactive oxygen species photogenerated on metal oxide

nanoparticles and their bacteria toxicity: The role of superoxide radicals. Environmental Science & Technology, 2017, 51: 10137-10145.

[128] Zou W, Zhou Q, Zhang X, et al. Environmental transformations and algal toxicity of single-layer molybdenum disulfide regulated by humic acid. Environmental Science & Technology, 2018, 52: 2638-2648.

[129] Yuan P, Zhou Q, Hu X. The phases of WS2 nanosheets influence uptake, oxidative stress, lipid peroxidation, membrane damage, and metabolism in algae. Environmental Science & Technology, 2018, 52: 13543-13552.

[130] Zhu H, Han J, Xiao J Q, et al. Uptake, translocation, and accumulation of manufactured iron oxide nanoparticles by pumpkin plants. Journal of Environmental Monitoring, 2008, 10: 713-717.

[131] Dan Y, Zhang W, Xue R, et al. Characterization of gold nanoparticle uptake by tomato plants using enzymatic extraction followed by single-particle inductively coupled plasma–mass spectrometry analysis. Environmental Science & Technology, 2015, 49: 3007-3014.

[132] Rico C M, Majumdar S, Duarte G M, et al. Interaction of nanoparticles with edible plants and their possible implications in the food chain. Journal of Agricultural and Food Chemistry, 2011, 59: 3485-3498.

[133] Zhao X C, Yu M, Xu D, et al. Distribution, bioaccumulation, trophic transfer, and influences of CeO_2 nanoparticles in a constructed aquatic food web. Environmental Science & Technology, 2017, 51: 5205-5214.

[134] Ferry J L, Craig P, Hexel C, et al. Transfer of gold nanoparticles from the water column to the estuarine food web. Nature Nanotechnology, 2009, 4: 441-444.

[135] Zhu X, Wang J, Zhang X, et al. Trophic transfer of TiO_2 nanoparticles from daphnia to zebrafish in a simplified freshwater food chain. Chemosphere, 2010, 79: 928-933.

[136] Werlin R, Priester J H, Mielke R E, et al. Biomagnification of cadmium selenide quantum dots in a simple experimental microbial food chain. Nature Nanotechnology, 2010, 6: 65.

[137] Majumdar S, Trujillo R J, Hernandez V J A, et al. Cerium biomagnification in a terrestrial food chain: Influence of particle size and growth stage. Environmental Science & Technology, 2016, 50: 6782-6792.

[138] Judy J D, Unrine J M, Bertsch P M. Evidence for biomagnification of gold nanoparticles within a terrestrial food chain. Environmental Science & Technology, 2011, 45: 776-781.

[139] Luo X, Xu S M, Yang Y N, et al. Insights into the ecotoxicity of silver nanoparticles transferred from *Escherichia coli* to *caenorhabditis elegans*. Scientific Reports, 2016, 6: 36465.

[140] Yang Y N, Xu G M, Xu S M, et al. Effect of ionic strength on bioaccumulation and toxicity of silver nanoparticles in *caenorhabditis elegans*. Ecotoxicology and Environmental Safety, 2018, 165: 291-298.

[141] Wang Z Y, Yin L Y, Zhao J, et al. Trophic transfer and accumulation of TiO_2 nanoparticles from clamworm (*Perinereis aibuhitensis*) to juvenile turbot (*Scophthalmus maximus*) along a marine benthic food chain. Water Research, 2016, 95: 250-259.

[142] Wang Z Y, Xia B, Chen B J, et al. Trophic transfer of TiO_2 nanoparticles from marine microalga (*Nitzschia closterium*) to scallop (*Chlamys farreri*) and related toxicity. Environmental Science-Nano, 2017, 4: 415-424.

[143] Iswarya V, Bhuvaneshwari M, Chandrasekaran N, et al. Trophic transfer potential of two different crystalline phases of TiO_2 NPs from *Chlorella* sp to *Ceriodaphnia dubia*. Aquatic Toxicology, 2018, 197: 89-97.

[144] Chen X J, Zhu Y, Yang K, et al. Nanoparticle TiO_2 size and rutile content impact bioconcentration and biomagnification from algae to daphnia. Environmental Pollution, 2019, 247: 421-430.

[145] Cano A M, Maul J D, Saed M, et al. Trophic transfer and accumulation of multiwalled carbon nanotubes in the presence of copper ions in *Daphnia magna* and fathead minnow (*Pimephales promelas*). Environmental Science & Technology, 2018, 52: 794-800.

[146] Tlili A, Jabiol J, Behra R, et al. Chronic exposure effects of silver nanoparticles on stream microbial decomposer communities and ecosystem functions. Environmental Science & Technology, 2017, 51: 2447-2455.

[147] Gil A C, Tlili A, Schirmer K, et al. Long-term exposure to silver nanoparticles affects periphyton community structure and function. Environmental Science-Nano, 2018, 5: 1397-1407.

[148] Tsiola A, Toncelli C, Fodelianakis S, et al. Low-dose addition of silver nanoparticles stresses marine plankton communities. Environmental Science-Nano, 2018, 5: 1965-1980.

[149] Asadishad B, Chahal S, Cianciarelli V, et al. Effect of gold nanoparticles on extracellular nutrient-cycling enzyme activity and bacterial community in soil slurries: Role of nanoparticle size and surface coating. Environmental Science-Nano, 2017, 4: 907-918.

[150] Zhu D, Zheng F, Chen Q L, et al. Exposure of a soil collembolan to Ag nanoparticles and $AgNO_3$ Disturbs its associated microbiota and lowers the incidence of antibiotic resistance genes in the gut. Environmental Science & Technology, 2018, 52: 12748-12756.

[151] Wu F, Harper B J, Harper S L. Differential dissolution and toxicity of surface functionalized silver nanoparticles in small-scale microcosms: Impacts of community complexity. Environmental Science-Nano, 2017, 4: 359-372.

[152] Du J J, Zhang Y Y, Cui M H, et al. Evidence for negative effects of ZnO nanoparticles on leaf litter decomposition in freshwater ecosystems. Environmental Science-Nano, 2017, 4: 2377-2387.

[153] Tang J, Zhu N Y, Zhu Y, et al. Responses of periphyton to Fe_2O_3 nanoparticles: A physiological and ecological basis for defending nanotoxicity. Environmental Science & Technology, 2017, 51: 10797-10805.

[154] Xiao L L, Liu F H, Liu J C, et al. Nano-Fe_3O_4 particles accelerating electromethanogenesis on an hour-long timescale in wetland soil. Environmental Science-Nano, 2018, 5: 436-445.

[155] Qi L, Ge Y, Xia T, et al. Rare earth oxide nanoparticles promote soil microbial antibiotic resistance by selectively enriching antibiotic resistance genes. Environmental Science-Nano, 2019, 6: 456-466.

[156] Liu S Q, Wang C, Hou J, et al. Effects of silver sulfide nanoparticles on the microbial community structure and biological activity of freshwater biofilms. Environmental Science-Nano, 2018, 5: 2899-2908.

[157] Nguyen H N, Castro W S L, Rodrigues D F. Acute toxicity of graphene nanoplatelets on biological wastewater treatment process. Environmental Science-Nano, 2017, 4: 160-169.

[158] Wang X H, Li J, Liu R, et al. Responses of bacterial communities to CuO nanoparticles in activated sludge system. Environmental Science & Technology, 2017, 51: 5368-5376.

[159] Huang H N, Chen Y G, Yang S Y, et al. CuO and ZnO nanoparticles drive the propagation of antibiotic resistance genes during sludge anaerobic digestion: Possible role of stimulated signal transduction. Environmental Science-Nano, 2019, 6: 528-539.

[160] Asadishad B, Chahal S, Akbari A, et al. Amendment of agricultural soil with metal nanoparticles: effects on soil enzyme activity and microbial community composition. Environmental Science & Technology, 2018, 52: 1908-1918.

[161] Klanjscek T, Muller E B, Holden P A, et al. Host-symbiont interaction model explains non-monotonic response of soybean growth and seed production to nano-CeO_2 exposure. Environmental Science & Technology, 2017, 51: 4944-4950.

[162] Ge Y, Shen C C, Wang Y, et al. Carbonaceous nanomaterials have higher effects on soybean rhizosphere prokaryotic communities during the reproductive growth phase than during vegetative growth. Environmental Science & Technology, 2018, 52: 6636-6646.

[163] Wang W, Sedykh A, Sun H, et al. Predicting nano–bio interactions by integrating nanoparticle libraries and quantitative nanostructure activity relationship modeling. ACS Nano, 2017, 11: 12641-12649.

[164] Wang W, Yan X, Zhao L, et al. Universal nanohydrophobicity predictions using virtual nanoparticle library. Journal of

Cheminformatics, 2019, 11: 6.

[165] Nuruzzaman M, Rahman M M, Liu Y, et al. Nanoencapsulation, nano-guard for pesticides: A new window for safe application. Journal of Agricultural and Food Chemistry, 2016, 64: 1447-1483.

[166] Elmer W H, White J C. The use of metallic oxide nanoparticles to enhance growth of tomatoes and eggplants in disease infested soil or soilless medium. Environmental Science-Nano, 2016, 3: 1072-1079.

[167] Lv J, Christie P, Zhang S. Uptake, translocation, and transformation of metal-based nanoparticles in plants: recent advances and methodological challenges. Environmental Science-Nano, 2019, 6: 41-59.

[168] Wang P, Lombi E, Sun S K, et al. Characterizing the uptake, accumulation and toxicity of silver sulfide nanoparticles in plants. Environmental Science-Nano, 2017, 4: 448-460.

[169] Spielman S E, Lombi E, Donner E, et al. Temporal evolution of copper distribution and speciation in roots of *Triticum aestivum* exposed to CuO, Cu(OH)$_2$, and CuS nanoparticles. Environmental Science & Technology, 2018, 52: 9777-9784.

[170] Dwivedi A D, Yoon H, Singh J P, et al. Uptake, Distribution, and transformation of zerovalent iron nanoparticles in the edible plant *Cucumis sativus*. Environmental Science & Technology, 2018, 52: 10057-10066.

[171] Bonilla B N J, Paez A, Reyes A, et al. Two-photon microscopy and spectroscopy studies to determine the mechanism of copper oxide nanoparticle uptake by sweetpotato roots during postharvest treatment. Environmental Science & Technology, 2018, 52: 9954-9963.

[172] Zhang T R, Wang C X, Dong F Q, et al. Uptake and translocation of styrene maleic anhydride nanoparticles in *Murraya exotica* plants as revealed by noninvasive, real-time optical bioimaging. Environmental Science & Technology, 2019, 53: 1471-1481.

[173] Das K K, You Y Q, Torres M, et al. Development and application of a digestion-Raman analysis approach for studying multiwall carbon nanotube uptake in lettuce. Environmental Science-Nano, 2018, 5: 659-668.

[174] Deng Y Q, Petersen E J, Challis K E, et al. Multiple method analysis of TiO$_2$ nanoparticle uptake in rice (*Oryza sativa* L.) Plants. Environmental Science & Technology, 2017, 51: 10615-10623.

[175] Chen L Y, Wang C L, Li H L, et al. Bioaccumulation and toxicity of ^{13}C-skeleton labeled graphene oxide in wheat. Environmental Science & Technology, 2017, 51: 10146-10153.

[176] Davis R A, Rippner D A, Hausner S H, et al. *In vivo* tracking of copper-64 radiolabeled nanoparticles in *lactuca sativa*. Environmental Science & Technology, 2017, 51: 12537-12546.

[177] Dang F, Chen Y Z, Huang Y N, et al. Discerning the sources of silver nanoparticle in a terrestrial food chain by stable isotope tracer technique. Environmental Science & Technology, 2019, 53: 3802-3810.

[178] Zhang P, Ma Y, Zhang Z, et al. Biotransformation of ceria nanoparticles in cucumber plants. ACS Nano, 2012, 6: 9943-9950.

[179] Wang Z, Xie X, Zhao J, et al. Xylem- and phloem-based transport of CuO nanoparticles in maize (*Zea mays* L.). Environmental Science & Technology, 2012, 46: 4434-4441.

[180] Servin A D, Castillo M H, Hernandez V J A, et al. Synchrotron micro-XRF and micro-XANES confirmation of the uptake and translocation of TiO$_2$ nanoparticles in cucumber (*Cucumis sativus*) plants. Environmental Science & Technology, 2012, 46: 7637-7643.

[181] Zhao L, Peralta V J R, Ren M, et al. Transport of Zn in a sandy loam soil treated with ZnO NPs and uptake by corn plants: electron microprobe and confocal microscopy studies. Chemical Engineering Journal, 2012, 184: 1-8.

[182] Ma Y H, He X, Zhang P, et al. Xylem and phloem based transport of CeO$_2$ nanoparticles in hydroponic cucumber plants. Environmental Science & Technology, 2017, 51: 5215-5221.

[183] Zhang P, Xie C J, Ma Y H, et al. Shape-dependent transformation and translocation of ceria nanoparticles in cucumber plants. Environmental Science & Technology Letters, 2017, 4: 380-385.

[184] Spielman S E, Lombi E, Donner E, et al. Impact of surface charge on cerium oxide nanoparticle uptake and translocation by wheat (*Triticum aestivum*). Environmental Science & Technology, 2017, 51: 7361-7368.

[185] Zhang P, Ma Y H, Xie C J, et al. Plant species-dependent transformation and translocation of ceria nanoparticles. Environmental Science-Nano, 2019, 6: 60-67.

[186] Taylor A F, Rylott E L, Anderson C W N, et al. Investigating the toxicity, uptake, nanoparticle formation and genetic response of plants to gold. PLoS One, 2014, 9: e93793.

[187] Schwab F, Zhai G, Kern M, et al. Barriers, pathways and processes for uptake, translocation and accumulation of nanomaterials in plants—Critical review. Nanotoxicology, 2016, 10: 257-278.

[188] Lin S, Reppert J, Hu Q, et al. Uptake, translocation, and transmission of carbon nanomaterials in rice plants. Small, 2009, 5: 1128-1132.

[189] Zhao L, Ortiz C, Adeleye A S, et al. Metabolomics to detect response of lettuce (*Lactuca sativa*) to $Cu(OH)_2$ nanopesticides: oxidative stress response and detoxification mechanisms. Environmental Science & Technology, 2016, 50: 9697-9707.

[190] Zhao L, Huang Y, Adeleye A S, et al. Metabolomics reveals $Cu(OH)_2$ nanopesticide-activated anti-oxidative pathways and decreased beneficial antioxidants in spinach leaves. Environmental Science & Technology, 2017, 51: 10184-10194.

[191] Hong J, Peralta V J R, Rico C, et al. Evidence of translocation and physiological impacts of foliar applied CeO_2 nanoparticles on cucumber (*Cucumis sativus*) plants. Environmental Science & Technology, 2014, 48: 4376-4385.

[192] Birbaum K, Brogioli R, Schellenberg M, et al. No Evidence for cerium dioxide nanoparticle translocation in maize plants. Environmental Science & Technology, 2010, 44: 8718-8723.

[193] Kranjc E, Mazej D, Regvar M, et al. Foliar surface free energy affects platinum nanoparticle adhesion, uptake, and translocation from leaves to roots in arugula and escarole. Environmental Science-Nano, 2018, 5: 520-532.

[194] Zhao J, Ren W T, Dai Y H, et al. Uptake, distribution, and transformation of CuO NPs in a floating plant *Eichhornia crassipes* and related stomatal responses. Environmental Science & Technology, 2017, 51: 7686-7695.

[195] Xiong T T, Dumat C, Dappe V, et al. Copper oxide nanoparticle foliar uptake, phytotoxicity, and consequences for sustainable urban agriculture. Environmental Science & Technology, 2017, 51: 5242-5251.

[196] Gardea T J L, Rico C M, White J C. Trophic transfer, transformation, and impact of engineered nanomaterials in terrestrial environments. Environmental Science & Technology, 2014, 48: 2526-2540.

[197] Miralles P, Church T L, Harris A T. Toxicity, uptake, and translocation of engineered nanomaterials in vascular plants. Environmental Science & Technology, 2012, 46: 9224-9239.

[198] Qiu H, Smolders E. Nanospecific phytotoxicity of CuO nanoparticles in soils disappeared when bioavailability factors were considered. Environmental Science & Technology, 2017, 51: 11976-11985.

[199] Peng C, Chen S, Shen C S, et al. Iron plaque: A barrier layer to the uptake and translocation of copper oxide nanoparticles by rice plants. Environmental Science & Technology, 2018, 52: 12244-12254.

[200] Layet C, Santaella C, Auffan M, et al. Phytoavailability of silver at predicted environmental concentrations: does the initial ionic or nanoparticulate form matter? Environmental Science-Nano, 2019, 6: 127-135.

[201] Landa P, Dytrych P, Prerostova S, et al. Transcriptomic response of *arabidopsis thaliana* exposed to CuO nanoparticles, bulk material, and ionic copper. Environmental Science & Technology, 2017, 51: 10814-10824.

[202] Zhang H L, Du W C, Peralta V J R, et al. Metabolomics reveals how cucumber (*Cucumis sativus*) reprograms metabolites to cope with silver ions and silver nanoparticle-induced oxidative stress. Environmental Science & Technology, 2018, 52: 8016-8026.

[203] Rico C M, Johnson M G, Marcus M A, et al. Intergenerational responses of wheat (*Triticum aestivum* L.) to cerium oxide nanoparticles exposure. Environmental Science-Nano, 2017, 4: 700-711.

[204] Lin D, Xing B. Phytotoxicity of nanoparticles: Inhibition of seed germination and root growth. Environmental Pollution, 2007, 150: 243-250.

[205] Du W, Tan W, Peralta V J R, et al. Interaction of metal oxide nanoparticles with higher terrestrial plants: Physiological and biochemical aspects. Plant Physiology and Biochemistry, 2017, 110: 210-225.

[206] Ruotolo R, Maestri E, Pagano L, et al. Plant response to metal-containing engineered nanomaterials: An omics-based perspective. Environmental Science & Technology, 2018, 52: 2451-2467.

[207] Servin A D, White J C. Nanotechnology in agriculture: Next steps for understanding engineered nanoparticle exposure and risk. NanoImpact, 2016, 1: 9-12.

[208] Yuan L, Richardson C J, Ho M, et al. Stress responses of aquatic plants to silver nanoparticles. Environmental Science & Technology, 2018, 52: 2558-2565.

[209] Rawat S, Pullagurala V L R, Hernandez M M, et al. Impacts of copper oxide nanoparticles on bell pepper (*Capsicum annum* L.) plants: A full life cycle study. Environmental Science-Nano, 2018, 5: 83-95.

[210] Lee S M, Raja P M V, Esquenazi G L, et al. Effect of raw and purified carbon nanotubes and iron oxide nanoparticles on the growth of wheatgrass prepared from the cotyledons of common wheat (*Triticum aestivum*). Environmental Science-Nano, 2018, 5: 103-114.

[211] Rui M M, Ma C X, White J C, et al. Metal oxide nanoparticles alter peanut (*Arachis hypogaea* L.) physiological response and reduce nutritional quality: A life cycle study. Environmental Science-Nano, 2018, 5: 2088-2102.

[212] Zhao L, Peng B, Hernandez V J A, et al. Stress response and tolerance of *Zea mays* to CeO_2 nanoparticles: cross talk among H_2O_2, heat shock protein, and lipid peroxidation. ACS Nano, 2012, 6: 9615-9622.

[213] Majumdar S, Almeida I C, Arigi E A, et al. Environmental effects of nanoceria on seed production of common bean (*Phaseolus vulgaris*): A proteomic analysis. Environmental Science & Technology, 2015, 49: 13283-13293.

[214] Zhao L, Huang Y, Hu J, et al. ^{1}H NMR and GC-MS based metabolomics reveal defense and detoxification mechanism of cucumber plant under nano-Cu stress. Environmental Science & Technology, 2016, 50: 2000-2010.

[215] Soria N G C, Montes A, Bisson M A, et al. Mass spectrometry-based metabolomics to assess uptake of silver nanoparticles by *Arabidopsis thaliana*. Environmental Science-Nano, 2017, 4: 1944-1953.

[216] Avellan A, Schwab F, Masion A, et al. Nanoparticle uptake in plants: gold nanomaterial localized in roots of *Arabidopsis thaliana* by X-ray computed nanotomography and hyperspectral imaging. Environmental Science & Technology, 2017, 51: 8682-8691.

[217] Huang Y X, Zhao L J, Keller A A. Interactions, transformations, and bioavailability of nano-copper exposed to root exudates. Environmental Science & Technology, 2017, 51: 9774-9783.

[218] Grayston S J, Vaughan D, Jones D. Rhizosphere carbon flow in trees, in comparison with annual plants: The importance of root exudation and its impact on microbial activity and nutrient availability. Applied Soil Ecology, 1997, 5: 29-56.

[219] Vessey J K. Plant growth promoting rhizobacteria as biofertilizers. Plant Soil, 2003, 255: 571-586.

[220] Wu H H, Shabala L, Shabala S, et al. Hydroxyl radical scavenging by cerium oxide nanoparticles improves *Arabidopsis salinity* tolerance by enhancing leaf mesophyll potassium retention. Environmental Science-Nano, 2018, 5: 1567-1583.

[221] Hao Y, Yuan W, Ma C X, et al. Engineered nanomaterials suppress Turnip mosaic virus infection in tobacco (*Nicotiana benthamiana*). Environmental Science-Nano, 2018, 5: 1685-1693.

[222] Xiong L, Wang P, Hunter M N, et al. Bioavailability and movement of hydroxyapatite nanoparticles (HA-NPs) applied as a phosphorus fertiliser in soils. Environmental Science-Nano, 2018, 5: 2888-2898.

[223] 周群芳, 孙成, 刘伟, 等. 基于稀有鮈鲫模型研究水环境中纳米银的毒理学效应. 科学通报, 2015, 60: 645-653.

[224] Wu Y, Zhou Q. Dose- and time-related changes in aerobic metabolism, chorionic disruption, and oxidative stress in

embryonic medaka (*Oryzias latipes*): Underlying mechanisms for silver nanoparticle developmental toxicity. Aquatic Toxicology, 2012, 124-125: 238-246.

[225] Lee K J, Nallathamby P D, Browning L M, et al. *In vivo* imaging of transport and biocompatibility of single silver nanoparticles in early development of zebrafish embryos. ACS Nano, 2007, 1: 133-143.

[226] Laban G, Nies L F, Turco R F, et al. The effects of silver nanoparticles on fathead minnow (*Pimephales promelas*) embryos. Ecotoxicology, 2010, 19: 185-195.

[227] Ma Y B, Lu C J, Junaid M, et al. Potential adverse outcome pathway (AOP) of silver nanoparticles mediated reproductive toxicity in zebrafish. Chemosphere, 2018, 207: 320-328.

[228] Ma Y B, Song L Y, Lei Y, et al. Sex dependent effects of silver nanoparticles on the zebrafish gut microbiota. Environmental Science-Nano, 2018, 5: 740-751.

[229] Juganson K, Mortimer M, Ivask A, et al. Mechanisms of toxic action of silver nanoparticles in the protozoan *Tetrahymena thermophila*: From gene expression to phenotypic events. Environmental Pollution, 2017, 225: 481-489.

[230] Hou J, Zhou Y, Wang C, et al. Toxic effects and molecular mechanism of different types of silver nanoparticles to the aquatic crustacean *Daphnia magna*. Environmental Science & Technology, 2017, 51: 12868-12878.

[231] Truong L, Zaikova T, Schaeublin N M, et al. Residual weakly bound ligands influence biological compatibility of mixed ligand shell, thiol-stabilized gold nanoparticles. Environmental Science-Nano, 2017, 4: 1634-1646.

[232] Hua J, Vijver M G, Ahmad F, et al. Toxicity of different-sized copper nano- and submicron particles and their shed copper ions to zebrafish embryos. Environmental Toxicology and Chemistry, 2014, 33: 1774-1782.

[233] Brun N R, Koch B E V, Varela M, et al. Nanoparticles induce dermal and intestinal innate immune system responses in zebrafish embryos. Environmental Science-Nano, 2018, 5: 904-916.

[234] Sharma V K. Aggregation and toxicity of titanium dioxide nanoparticles in aquatic environment—A Review. Journal of Environmental Science and Health, Part A, 2009, 44: 1485-1495.

[235] Felix L C, Folkerts E J, He Y, et al. Poly(acrylic acid)-coated titanium dioxide nanoparticle and ultraviolet light co-exposure has minimal effect on developing zebrafish (*Danio rerio*). Environmental Science-Nano, 2017, 4: 658-669.

[236] Guan X, Shi W, Zha S, et al. Neurotoxic impact of acute TiO_2 nanoparticle exposure on a benthic marine bivalve mollusk, *Tegillarca granosa*. Aquatic Toxicology, 2018, 200: 241-246.

[237] Fan W H, Lu H T, Wang W X. Aging influences on the biokinetics of functional TiO_2 nanoparticles with different surface chemistries in *Daphnia magna*. Environmental Science & Technology, 2018, 52: 7901-7909.

[238] Georgantzopoulou A, Carvalho P A, Vogelsang C, et al. Ecotoxicological effects of transformed silver and titanium dioxide nanoparticles in the effluent from a lab-scale wastewater treatment system. Environmental Science & Technology, 2018, 52: 9431-9441.

[239] Moore J D, Avellan A, Noack C W, et al. Time-dependent bacterial transcriptional response to CuO nanoparticles differs from that of Cu^{2+} and provides insights into CuO nanoparticle toxicity mechanisms. Environmental Science-Nano, 2017, 4: 2321-2335.

[240] Cong Y, Jin F, Wang J, et al. The embryotoxicity of ZnO nanoparticles to marine medaka, *Oryzias melastigma*. Aquatic Toxicology, 2017, 185: 11-18.

[241] Ng D Q, Chu Y, Tan S W, et al. In vivo evidence of intestinal lead dissolution from lead dioxide (PbO_2) nanoparticles and resulting bioaccumulation and toxicity in medaka fish. Environmental Science-Nano, 2019, 6: 580-591.

[242] Khosravi K K, Lofrano G, Pak N H, et al. Effects of ZnO nanoparticles in the caspian roach (*Rutilus rutilus caspicus*). Science of the Total Environment, 2018, 626: 30-41.

[243] Hou J, Liu H, Wang L, et al. Molecular toxicity of metal oxide nanoparticles in *Danio rerio*. Environmental Science &

Technology, 2018, 52: 7996-8004.

[244] Denluck L, Wu F, Crandon L E, et al. Reactive oxygen species generation is likely a driver of copper based nanomaterial toxicity. Environmental Science-Nano, 2018, 5: 1473-1481.

[245] Peng G T, He Y, Zhao M, et al. Differential effects of metal oxide nanoparticles on zebrafish embryos and developing larvae. Environmental Science-Nano, 2018, 5: 1200-1207.

[246] De Marchi L, Neto V, Pretti C, et al. The impacts of emergent pollutants on *Ruditapes philippinarum*: Biochemical responses to carbon nanoparticles exposure. Aquatic Toxicology, 2017, 187: 38-47.

[247] Souza J P, Baretta J F, Santos F, et al. Toxicological effects of graphene oxide on adult zebrafish (*Danio rerio*). Aquatic Toxicology, 2017, 186: 11-18.

[248] Zhang X L, Zhou Q X, Zou W, et al. Molecular mechanisms of developmental toxicity induced by graphene oxide at predicted environmental concentrations. Environmental Science & Technology, 2017, 51: 7861-7871.

[249] Zhao J, Cao X, Wang Z, et al. Mechanistic understanding toward the toxicity of graphene-family materials to freshwater algae. Water Research, 2017, 111: 18-27.

[250] Zhang L Q, Lei C, Yang K, et al. Cellular response of *Chlorella pyrenoidosa* to oxidized multi-walled carbon nanotubes. Environmental Science-Nano, 2018, 5: 2415-2425.

[251] Zhao J, Dai Y H, Wang Z Y, et al. Toxicity of GO to freshwater algae in the presence of Al_2O_3 particles with different morphologies: importance of heteroaggregation. Environmental Science & Technology, 2018, 52: 13448-13456.

[252] Jia J, Li F, Zhou H, et al. Oral exposure to silver nanoparticles or silver ions may aggravate fatty liver disease in overweight mice. Environmental Science & Technology, 2017, 51: 9334-9343.

[253] Jia J, Li F, Zhai S, et al. Susceptibility of overweight mice to liver injury as a result of the ZnO Nanoparticle-enhanced liver deposition of Pb^{2+}. Environmental Science & Technology, 2017, 51: 1775-1784.

[254] Canesi L, Ciacci C, Balbi T. Interactive effects of nanoparticles with other contaminants in aquatic organisms: Friend or foe? Marine Environmental Research, 2015, 111: 128-134.

[255] Deng R, Lin D, Zhu L, et al. Nanoparticle interactions with co-existing contaminants: joint toxicity, bioaccumulation and risk. Nanotoxicology, 2017, 11: 591-612.

[256] Liu Y, Nie Y, Wang J, et al. Mechanisms involved in the impact of engineered nanomaterials on the joint toxicity with environmental pollutants. Ecotoxicology and Environmental Safety, 2018, 162: 92-102.

[257] Zhang S, Deng R, Lin D, et al. Distinct toxic interactions of TiO_2 nanoparticles with four coexisting organochlorine contaminants on algae. Nanotoxicology, 2017, 11: 1115-1126.

[258] Deng R, Zhu Y, Hou J, et al. Antagonistic toxicity of carbon nanotubes and pentachlorophenol to *Escherichia coli*: Physiological and transcriptional responses. Carbon, 2019, 145: 658-667.

[259] Naasz S, Altenburger R, Kühnel D. Environmental mixtures of nanomaterials and chemicals: The Trojan-horse phenomenon and its relevance for ecotoxicity. Science of the Total Environment, 2018, 635: 1170-1181.

[260] Wilke C M, Wunderlich B, Gaillard J F, et al. Synergistic bacterial stress results from exposure to nano-ag and nano-TiO_2 mixtures under light in environmental media. Environmental Science & Technology, 2018, 52: 3185-3194.

[261] Tan C, Fan W H, Wang W X. Role of titanium dioxide nanoparticles in the elevated uptake and retention of cadmium and zinc in *Daphnia magna*. Environmental Science & Technology, 2012, 46: 469-476.

[262] Li M, Luo Z, Yan Y, et al. Arsenate accumulation, distribution, and toxicity associated with titanium dioxide nanoparticles in *Daphnia magna*. Environmental Science & Technology, 2016, 50: 9636-9643.

[263] Li L, Sillanpää M, Schultz E. Influence of titanium dioxide nanoparticles on cadmium and lead bioaccumulations and toxicities to *Daphnia magna*. Journal of Nanoparticle Research, 2017, 19: 223.

[264] Yang W W, Wang Y, Huang B, et al. TiO$_2$ nanoparticles act as a carrier of Cd bioaccumulation in the ciliate *Tetrahymena thermophila*. Environmental Science & Technology, 2014, 48: 7568-7575.

[265] Wang X N, Liu Y, Wang J, et al. Amplification of arsenic genotoxicity by TiO$_2$ nanoparticles in mammalian cells: New insights from physicochemical interactions and mitochondria. Nanotoxicology, 2017, 11: 978-995.

[266] Wang J J, Dai H, Nie Y G, et al. TiO$_2$ nanoparticles enhance bioaccumulation and toxicity of heavy metals in *Caenorhabditis elegans* via modification of local concentrations during the sedimentation process. Ecotoxicology and Environmental Safety, 2018, 162: 160-169.

[267] Chen J, Qian Y, Li H, et al. The reduced bioavailability of copper by nano-TiO$_2$ attenuates the toxicity to *Microcystis aeruginosa*. Environmental Science and Pollution Research, 2015, 22: 12407-12414.

[268] Fan W, Peng R, Li X, et al. Effect of titanium dioxide nanoparticles on copper toxicity to *Daphnia magna* in water: Role of organic matter. Water Research, 2016, 105: 129-137.

[269] Tan L Y, Huang B, Xu S, et al. Aggregation reverses the carrier effects of TiO$_2$ nanoparticles on cadmium accumulation in the waterflea *Daphnia magna*. Environmental Science & Technology, 2017, 51: 932-939.

[270] Yang F, Zeng L, Luo Z, et al. Complex role of titanium dioxide nanoparticles in the trophic transfer of arsenic from *Nannochloropsis maritima* to *Artemia salina nauplii*. Aquatic Toxicology, 2018, 198: 231-239.

[271] Yan C Z, Yang F, Wang Z S, et al. Changes in arsenate bioaccumulation, subcellular distribution, depuration, and toxicity in *Artemia salina nauplii* in the presence of titanium dioxide nanoparticles. Environmental Science-Nano, 2017, 4: 1365-1376.

[272] Lu J, Tian S, Lv X, et al. TiO$_2$ nanoparticles in the marine environment: Impact on the toxicity of phenanthrene and Cd^{2+} to marine zooplankton *Artemia salina*. Science of The Total Environment, 2018, 615: 375-380.

[273] Zhang X, Sun H, Zhang Z, et al. Enhanced bioaccumulation of cadmium in carp in the presence of titanium dioxide nanoparticles. Chemosphere, 2007, 67: 160-166.

[274] Miao W, Zhu B, Xiao X, et al. Effects of titanium dioxide nanoparticles on lead bioconcentration and toxicity on thyroid endocrine system and neuronal development in zebrafish larvae. Aquatic Toxicology, 2015, 161: 117-126.

[275] Wang J, Nie Y, Dai H, et al. Parental exposure to TiO$_2$ NPs promotes the multigenerational reproductive toxicity of Cd in *Caenorhabditis elegans* via bioaccumulation of Cd in germ cells. Environmental Science-Nano, 2019, 6: 1332-1342.

[276] Zhu X, Zhou J, Cai Z. TiO$_2$ Nanoparticles in the marine environment: impact on the toxicity of tributyltin to abalone (*Haliotis diversicolor supertexta*) embryos. Environmental Science & Technology, 2011, 45: 3753-3758.

[277] Alabresm A, Mirshahghassemi S, Chandler G T, et al. Use of PVP-coated magnetite nanoparticles to ameliorate oil toxicity to an estuarine meiobenthic copepod and stimulate the growth of oil-degrading bacteria. Environmental Science-Nano, 2017, 4: 1859-1865.

[278] Wang Q, Chen Q, Zhou P, et al. Bioconcentration and metabolism of BDE-209 in the presence of titanium dioxide nanoparticles and impact on the thyroid endocrine system and neuronal development in zebrafish larvae. Nanotoxicology, 2014, 8: 196-207.

[279] Fang Q, Shi X, Zhang L, et al. Effect of titanium dioxide nanoparticles on the bioavailability, metabolism, and toxicity of pentachlorophenol in zebrafish larvae. Journal of Hazardous Materials, 2015, 283: 897-904.

[280] Fang Q, Shi Q, Guo Y, et al. Enhanced bioconcentration of bisphenol a in the presence of nano-TiO$_2$ can lead to adverse reproductive outcomes in zebrafish. Environmental Science & Technology, 2016, 50: 1005-1013.

[281] Chen L, Guo Y, Hu C, et al. Dysbiosis of gut microbiota by chronic coexposure to titanium dioxide nanoparticles and bisphenol a: implications for host health in zebrafish. Environmental Pollution, 2018, 234: 307-317.

[282] Cheng H C, Yan W, Wu Q, et al. Adverse reproductive performance in zebrafish with increased bioconcentration of microcystin-LR in the presence of titanium dioxide nanoparticles. Environmental Science-Nano, 2018, 5: 1208-1217.

[283] Wu Q, Yan W, Liu C S, et al. Parental transfer of titanium dioxide nanoparticle aggravated MCLR- induced developmental toxicity in zebrafish offspring. Environmental Science-Nano, 2018, 5: 2952-2965.

[284] Della T C, Buonocore F, Frenzilli G, et al. Influence of titanium dioxide nanoparticles on 2,3,7,8-tetrachlorodibenzo-*p*-dioxin bioconcentration and toxicity in the marine fish European sea bass (*Dicentrarchus labrax*). Environmental Pollution, 2015, 196: 185-193.

[285] Martin D L I, Campos M M C, Aguera A, et al. Reverse Trojan-horse effect decreased wastewater toxicity in the presence of inorganic nanoparticles. Environmental Science-Nano, 2017, 4: 1273-1282.

[286] Martins R, Oliveira T, Santos C, et al. Effects of a novel anticorrosion engineered nanomaterial on the bivalve *Ruditapes philippinarum*. Environmental Science-Nano, 2017, 4: 1064-1076.

[287] Ma C X, Liu H, Chen G C, et al. Effects of titanium oxide nanoparticles on tetracycline accumulation and toxicity in *Oryza sativa* (L.). Environmental Science-Nano, 2017, 4: 1827-1839.

[288] Wang X X, Sun W J, Zhang S, et al. Elucidating the effects of cerium oxide nanoparticles and zinc oxide nanoparticles on arsenic uptake and speciation in rice (*Oryza sativa*) in a hydroponic system. Environmental Science & Technology, 2018, 52: 10040-10047.

[289] Ye X X, Wang G Z, Zhang Y X, et al. Hydroxyapatite nanoparticles in root cells: reducing the mobility and toxicity of Pb in rice. Environmental Science-Nano, 2018, 5: 398-407.

[290] Zhang W L, Lo I M C, Hu L M, et al. Environmental risks of nano zerovalent iron for arsenate remediation: Impacts on cytosolic levels of inorganic phosphate and MgATP^{2-} in *arabidopsis thaliana*. Environmental Science & Technology, 2018, 52: 4385-4392.

[291] Wu S L, Vosatka M, Vogel M K, et al. Nano zero-valent iron mediated metal(loid) uptake and translocation by arbuscular mycorrhizal symbioses. Environmental Science & Technology, 2018, 52: 7640-7651.

[292] Rossi L, Sharifan H, Zhang W L, et al. Mutual effects and in planta accumulation of co-existing cerium oxide nanoparticles and cadmium in hydroponically grown soybean (*Glycine max* (L.) Merr.). Environmental Science-Nano, 2018, 5: 150-157.

[293] Hou L, Fortner J D, Wang X M, et al. Complex interplay between formation routes and natural organic matter modification controls capabilities of C_{60} nanoparticles (nC_{60}) to accumulate organic contaminants. Journal of Environmental Sciences, 2017, 51: 315-323.

[294] Wang L L, Zhu D Q, Chen J W, et al. Enhanced adsorption of aromatic chemicals on boron and nitrogen co-doped single-walled carbon nanotubes. Environmental Science-Nano, 2017, 4: 558-564.

[295] Metzelder F, Funck M, Schmidt T C. Sorption of heterocyclic organic compounds to multiwalled carbon nanotubes. Environmental Science & Technology, 2018, 52: 628-637.

[296] Liu J, Ma Y, Zhu D Q, et al. Polystyrene nanoplastics-enhanced contaminant transport: role of irreversible adsorption in glassy polymeric domain. Environmental Science & Technology, 2018, 52: 2677-2685.

[297] Linard E N, Apul O G, Karanfil T, et al. Bioavailability of carbon nanomaterial-adsorbed polycyclic aromatic hydrocarbons to pimphales promelas: Influence of adsorbate molecular size and configuration. Environmental Science & Technology, 2017, 51: 9288-9296.

[298] Deng Y Q, Eitzer B, White J C, et al. Impact of multiwall carbon nanotubes on the accumulation and distribution of carbamazepine in collard greens (*Brassica oleracea*). Environmental Science-Nano, 2017, 4: 149-159.

[299] Della T C, Parolini M, Del G L, et al. Adsorption of B(α)P on carbon nanopowder affects accumulation and toxicity in zebrafish (*Danio rerio*) embryos. Environmental Science-Nano, 2017, 4: 1132-1146.

[300] Li J Y, Hu L X, Ying G G, et al. Co-exposure of C_{60} fullerene with benzo[a]pyrene results in enhanced biological effects in cells as determined by Fourier-transform infrared spectroscopy. Environmental Science-Nano, 2017, 4: 1404-1418.

[301] Bisesi J H, Robinson S E, Lavelle C M, et al. Influence of the gastrointestinal environment on the bioavailability of ethinyl estradiol sorbed to single-walled carbon nanotubes. Environmental Science & Technology, 2017, 51: 948-957.

[302] Li X K, Mu L, Hu X G. Integrating proteomics, metabolomics and typical analysis to investigate the uptake and oxidative stress of graphene oxide and polycyclic aromatic hydrocarbons. Environmental Science-Nano, 2018, 5: 115-129.

[303] Rodd A L, Castilho C J, Chaparro C E F, et al. Impact of emerging, high-production-volume graphene-based materials on the bioavailability of benzo(a)pyrene to brine shrimp and fish liver cells. Environmental Science-Nano, 2018, 5: 2144-2161.

[304] Li Y, Men B, He Y, et al. Effect of single-wall carbon nanotubes on bioconcentration and toxicity of perfluorooctane sulfonate in zebrafish (*Danio rerio*). Science of The Total Environment, 2017, 607-608: 509-518.

[305] Martin L I, Campos M M C, Aguera A, et al. Combined toxicity of graphene oxide and wastewater to the green alga *Chlamydomonas reinhardtii*. Environmental Science-Nano, 2018, 5: 1729-1744.

[306] Liu Y, Wang X N, Wang J, et al. Graphene oxide attenuates the cytotoxicity and mutagenicity of PCB 52 via activation of genuine autophagy. Environmental Science & Technology, 2016, 50: 3154-3164.

[307] Zhang H Y, Chen W X, Shen X F, et al. Influence of multi-walled carbon nanotubes and fullerenes on the bioaccumulation and elimination kinetics of phenanthrene in geophagous earthworms (*Metaphire guillelmi*). Environmental Science-Nano, 2017, 4: 1887-1899.

[308] Dai H, Liu Y, Wang J J, et al. Graphene oxide antagonizes the toxic response to arsenic via activation of protective autophagy and suppression of the arsenic-binding protein LEC-1 in *Caenorhabditis elegans*. Environmental Science-Nano, 2018, 5: 1711-1728.

[309] Zhu J Q, Xu M, Wang F F, et al. Low-dose exposure to graphene oxide significantly increases the metal toxicity to macrophages by altering their cellular priming state. Nano Research, 2018, 11: 4111-4122.

[310] Yu Q L, Zhang B, Li J R, et al. Graphene oxide significantly inhibits cell growth at sublethal concentrations by causing extracellular iron deficiency. Nanotoxicology, 2017, 11: 1102-1114.

[311] Guo C, Zhang Z, Lau A K H, et al. Effect of long-term exposure to fine particulate matter on lung function decline and risk of chronic obstructive pulmonary disease in Taiwan: A longitudinal, cohort study. The Lancet Planetary Health, 2018, 2: e114-e125.

[312] Xu F, Qiu X, Hu X, et al. Effects on IL-1β signaling activation induced by water and organic extracts of fine particulate matter ($PM_{2.5}$) *in vitro*. Environmental Pollution, 2018, 237: 592-600.

[313] Bai X, Liu Y, Wang S, et al. Ultrafine particle libraries for exploring mechanisms of $PM_{2.5}$-induced toxicity in human cells. Ecotoxicology and Environmental Safety, 2018, 157: 380-387.

[314] Jia J, Yuan X, Peng X, et al. Cr(VI)/Pb^{2+} are responsible for $PM_{2.5}$-induced cytotoxicity in A549 cells while pulmonary surfactant alleviates such toxicity. Ecotoxicology and Environmental Safety, 2019, 172: 152-158.

作者：闫 兵[1,2]，陈 威[3]，胡献刚[3]，林道辉[4]，刘思金[5]，潘丙才[6]，全 燮[7]，王震宇[8]，吴李君[9]，许 安[9]，张承东[10]，赵 建[11]，赵丽娟[6]，周炳升[12]，周群芳[5]

协稿：贾建博[1]，孙海南[2]，郝 放[5]，戚 豫[5]，徐立宁[5]，徐 明[5]，刘 赟[9]，黄 青[9]

统稿：姜传佳[3]

[1]广州大学，[2]山东大学，[3]南开大学，[4]浙江大学，[5]中国科学院生态环境研究中心，[6]南京大学，[7]大连理工大学，[8]江南大学，[9]中国科学院合肥物质科学研究院，[10]北京师范大学，[11]中国海洋大学，[12]中国科学院水生生物研究所

第24章 水污染与控制技术研究

- 1. 引言 /643
- 2. 物化水处理方法与技术 /644
- 3. 生化水处理方法与技术 /663
- 4. 废水处理新技术系统 /676
- 5. 废水资源/能源新技术 /688

本章导读

本章从四个方面对国内外废水处理研究工作深入分析：①物化水处理方法与技术（高级氧化技术：电催化氧化、自由基氧化、类芬顿、零价铁还原；富集分离技术：膜分离、混凝、吸附、消毒）；②生化水处理方法与技术（脱氮除磷技术：Anammox脱氮技术、深度脱氮技术、生物除磷技术；废水处理实践：城市污水处理、工业废水处理）；③废水处理新技术系统（生物电化学技术、膜生物反应器技术、化工废水处理技术、厌氧处理新技术、物化-生物耦合处理技术）；④废水资源/能源新技术（有机碳回收技术、磷盐回用技术、重金属回收技术）。结合作者自身的科研工作，系统总结了各研究方向的研究现状和重要进展。同时，介绍了废水处理新技术原理和进展，分析了废水处理理论和技术的发展趋势、应用情况，并展望了未来的发展前景。

关键词

废水处理，物化技术，生化处理，废水资源化，废水能源化

1 引 言

近几十年，随着我国城镇化和工业化进程的加快、经济迅猛发展，导致了工业废水和生活污水的大量排放，造成了严重的水环境污染，不仅限制了社会的发展，而且威胁到了人类的生存环境和人体健康。因此，针对水环境污染，研发高效、绿色、经济的废水处理方法迫在眉睫，也成为国内外环境界关注的热点。

废水处理技术可分为物理处理法、化学处理法和生物处理法三大类。物理处理法（如重力分离、浮选、沉淀、吸附等）主要利用物理作用分离废水中的污染物质，但对污染物不能彻底降解或转化，多用于组分回收，具有设备简单、操作方便、去除效果好等优点，应用广泛；化学处理法（如混凝、膜分离、高级氧化法、还原法等）通过化学作用或化学反应彻底转化或分解废水中的污染物，降低或去除其对人体/动植物的危害，具有快速高效、设备操作性高、容易实现自动化等优点，对多种毒性较大污染物具有较好的处理效果；生物处理法利用微生物的新陈代谢将废水中的有机物和部分无机物去除或转化的一种处理方法，根据微生物的作用环境分为好氧生物处理和厌氧生物处理。该法具有投资少、运行费用低廉、效果显著等优点，在工业废水和生活污水的处理中得到了极为广泛的应用，成为占主导地位的水处理技术。

当前废水处理理念逐渐从污染物的去除向资源化、能源化转变，随着我国废水排放标准的不断提高以及废水中新型污染物种类的不断发现，迫切需要提升废水处理效能和水平，需要将化学、生物学、材料学乃至人工智能等学科的理念和技术不断应用到废水处理中。一方面，这些交叉和应用为传统废水处理技术的发展带来崭新的活力，新的分析方法、表征手段有利于深入研究微观机理和反应理论，不断揭示复杂的水环境体系；另一方面，交叉融合带来的新材料、新技术、新工艺的发展将使得水污染控制技术更高效、更经济，从而促进其在实际废水处理中的转化应用。

2 物化水处理方法与技术

2.1 高级氧化/还原处理技术

2.1.1 电催化氧化

难降解工业废水处理一直是世界性难题，传统的处理方法难以满足提标减排的新要求。基于解决实际问题的应用需求和学科发展的必然选择，电化学方法正在受到广泛关注。电化学法以电子为反应物驱动有机物降解，具有高效、简单、环境友好、易于实现自动化等特点，尤其是废水中的盐离子能够作为电化学反应的电解质，为解决常规方法不能解决的问题提供了重要途径[1,2]。电催化氧化是最为常见的电化学处理方法，它利用在电极表面发生直接或间接氧化反应去除有机污染物。在直接氧化法中，有机污染物直接在阳极表面失去电子而被氧化，一般在较低电位下发生；间接氧化法是指有机污染物被阳极产生的强氧化性中间产物（如·OH、Cl_2、$SO_4^{·-}$、H_2O_2 等）介导氧化的过程，需要在较高电位下才能够完成（图 24-1）。近年来，电催化氧化法在废水处理，尤其是分散式废水处理领域展现出巨大的发展潜力。然而，仍有很多原理性和技术性的问题尚未解决，尤其是如何突破电极材料和能耗对系统放大性能的制约。因此，在推进电催化氧化法工程化应用的进程中，人们一直力图从基础科学角度解决原理性问题，从实际应用角度解决技术性问题。

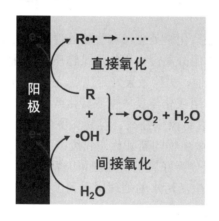

图 24-1 电催化氧化中的直接氧化和间接氧化

1）电极材料与电催化氧化原理

无论是直接氧化还是间接氧化，电催化反应都在电极材料与电解质溶液的界面上发生，电极材料的性质决定了界面反应机制、废水处理效率与能耗，以及处理成本，是最重要的影响因素。理想的电极材料应具备以下特征：①良好的导电性；②较高的物理化学稳定性和寿命；③对电催化反应具有高活性和高选择性；④廉价易得，绿色环保[3]。迄今为止，人们尚未开发出满足所有上述标准的电极材料，尤其是电极的活性和稳定性本身就存在矛盾，这为电极材料和反应机理的研究提出了挑战。尽管如此，人们在长期实践探索过程中仍然获得了一些重要的发现和有价值的研究成果。

（1）直接电催化氧化。

对于大部分有机污染物（如染料、有机农药、持久性有机物、抗生素等）来说，直接氧化一般效率较低，有机物及其氧化中间产物（如苯酚类、苯胺类物质）很可能在电极表面发生电聚合反应造成电极污染和钝化，覆盖活性位点降低处理效率，但对于全氟类有机污染物来说，直接氧化却是最有效的方式。比如，大部分高级氧化法对全氟辛酸（PFOA，$C_7F_{15}COOH$）氧化降解几乎没有效果，这是因为 PFOA 中的 F 元素具有极强的电负性，长链—CF_2—键键能很高[4]，但电催化氧化是个例外。在这一研究领域，牛军峰等做了大量的开创性工作，他们发现在 Ce 掺杂纳米晶体 PbO_2 薄膜电极上施加足够高的电位可大幅降低 PFOA 分解的活化能，支链末端的—COOH 基团在电极表面失去一个电子生成—COO•，随后在 O_2 和·OH 的作用下逐个将—CF_2—键断裂，最终仅在 2 h 的时间内便可将其矿化为 CO_2 和 HF[5, 6]。虽然后续断链反应过程·OH 也十分重要，但第一步直接单电子转移反应最为关键。最近的研究发现，当废水中含有硫酸根时，电极孔内的强酸性微环境能够促使硫酸根氧化为硫酸根自由基（$SO_4^{·-}$），在电极直接氧化和 $SO_4^{·-}$ 单电子转移氧化的协同作用下，PFOA 的氧化效率得到了显著提升[7]。可以说，全氟类有机污染物的高效去除是直接电催化氧化在污染物去除方面最具特色的应用之一，是其他传统方法不可替代的。

（2）间接电催化氧化。

高效间接氧化需要电极材料具有足够高的析氧电位，此外·OH 在电极表面弱吸附以保证具有更高的反应活性（惰性电极）。人们发现，找到这样的惰性电极并非易事，目前报道最多的电极有金属氧化物涂层电极（也叫形稳电极，即 DSA）[8]、掺硼金刚石（BDD）[9]和亚氧化钛[10]。从图 24-2 可以看出：金属氧化物涂层电极存在电导率低、涂层脱落和重金属离子溶出风险等问题；BDD 是效果最好也是研究最多的电极材料，但价格十分昂贵，工程价值被大大削弱。近年来，亚氧化钛作为一种新型电极材料在电催化氧化废水处理中很好地平衡了各个要素之间的矛盾，各方面的性能更加均衡，正在引起人们广泛的兴趣。

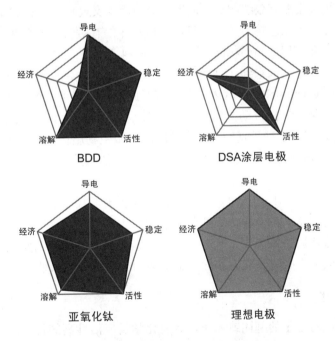

图 24-2 电催化氧化电极材料的比较

研究证实，BDD 和亚氧化钛能够产生准自由态的·OH，这种特性使得它们对各类持久性有机物、有机农药、抗生素等污染物有着很高的去除效率，对某些难降解实际工业废水（如制革废水、石化废水、乳品废水、造纸废水、垃圾渗沥液等）的处理效果也优于其他的高级氧化方法[11, 12]。这些废水通常含有高浓度的难降解有机物和氯离子，因此 COD 和氨氮的去除主要通过产生的活性 Cl_2 和 HClO 实现。同时，在阳极高电位的条件下，还可能生成氯酚类物质和高氯酸盐等副产物，因此，可以在电氧化后端使用活性炭对这些污染物进行吸附处理以达到降低水体二次污染风险的目的[13]。此外，电化学氧化的能耗也是人们关心的问题。在实际废水处理中，通常需要在高电流密度工况下运行，这不仅降低库伦效率而且增加吨水处理能耗和成本。比如，当使用 BDD 阳极处理垃圾渗沥液时，低电流密度工况（反应限制，库伦效率较高）的能耗为 22 kW·h/kgCOD；而高电流密度工况（传质限制，库伦效率较低）下的能耗上升到 95 kW·h/kgCOD，增加了 4 倍[14]。为了降低能耗，人们提出梯级降流的对策，以保证有机污染物的氧化尽可能在极限电流密度下发生[15]。最近，人们发现惰性电极能够氧化硫酸根生成 $SO_4^{\cdot-}$ 等，这些强氧化性中间产物均能够通过非法拉第过程氧化水中的有机污染物[7, 16]。氧化性自由基与大多数有机物的反应速率很快（$10^6 \sim 10^9$ L·mol^{-1}·s^{-1}），这意味着间接电催化氧化反应主要受传质过程的限制，因此强化污染物向电极界面的传质过程是提升有机污染物降解性能的重要途径。

2）电催化氧化系统的放大与应用

一个电化学过程的形成大致需要经历以下几个阶段：基本构想—实验室研究—工艺研究—放大试验—工业化应用。电催化氧化废水处理工程应用的重要前提是合理设计电解槽反应器，需要考虑的因素包括传质、传热、反应动力学、电极表面电流密度和电势分布等。这些因素在系统放大后可能对综合处理效能产生重要的影响。虽然处理效率的小幅度提升看似微不足道，但可能对节省吨水处理成本产生重要意义。遗憾的是，目前绝大部分关于电催化氧化的研究仍停留在实验室水平上，中试甚至小试规模的研究仍然凤毛麟角。事实上，有许多在小规模试验中看起来无关痛痒的问题都有可能导致系统放大失败。这些问题可能体现在：①随着电极尺寸的增大，电流和电势在电极表面的分布将不再均匀，进而直接影响到有机物的降解效率和能耗；②伴随着阳极析氧反应，阴极析出氢气，氢气和氧气在同一个系统中存在爆炸的可能，使用离子交换膜能有效缓解这一问题，但膜污染又会增加运行成本；③电极反应产生热量，这些热量在小试研究中几乎可以忽略不计，一旦放大后对热的有效管理就会变得非常重要。

电催化氧化技术的发展进程遵循 Gartner 周期律（图 24-3）[17]。自 20 世纪 70 年代 Dabrowski 首次提出电化学废水处理的概念和方法以来到 21 世纪初，电催化氧化技术经历了技术萌芽期、期望膨胀期。以 Compton、Krik、Kotz 和 Comninellis 为代表的科学家在此领域开展了大量的研究工作，但由于电极材料和高能耗的限制，这些成果并未能推动电催化氧化技术的工程化应用。在此之后该技术进入了泡沫破裂低谷期，2000~2010 年之间几乎处于真空期（在此阶段，微生物燃料电池技术正在从萌芽期走向期望膨胀期）。随着近年来工业废水问题的日益突出，材料科学和新能源技术的快速更新和发展，电催化氧化技术又重新回到了人们的视野中，大量高水平论文再次出现在国际顶尖级期刊上，中试和生产规模的反应工艺也如雨后春笋般的出现；尤其是风电、水电、核电等新能源技术的进步带来的电能成本降低将使电化学法在废水处理领域具有更大的竞争力。相信随着更多人的关注和参与，电催化氧化废水处理技术在不久的将来会迎来它的稳步爬升恢复期，一旦突破了电极材料和高能耗的瓶颈，将更加明确它的商业化发展途径，相关的研究成果也必将在理论和应用上发展到一个更高的水平。

2.1.2 自由基活化氧化

自由基是指含有非成对电子的基团或原子，具有化学反应活性高的特点，从而在生物、化学、环境等领域发挥重要的作用。在水污染控制领域，1976 年，Hoigné 和 Bader 最先提出基于自由基的高级氧化技术（AOPs）的概念：通过不同途径和方式产生·OH 的过程。随着 AOPs 的发展，体系中依赖的自由基从·OH 扩展到其他类型自由基，如硫酸根自由基（$SO_4^{\cdot-}$）、卤素自由基（HRs）等新型自由基。这些新型自由基具有与·OH 不同的氧化还原电位，如图 24-4 所示，氧化还原电位排序：·OH ≈ $SO_4^{\cdot-}$ > $Cl^{\cdot}$ > $Br^{\cdot}$ > $I^{\cdot}$。

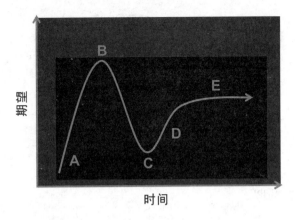

图 24-3 电催化氧化技术的 Gartner 周期律
A-技术萌芽期；B-期望膨胀期；C-泡沫破裂低谷期；D-稳步爬升恢复期；E-应用成熟期

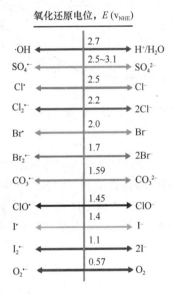

图 24-4 不同自由基的氧化还原电位

1）水中自由基的产生方式

水中自由基的产生方式有原生和次生两种方式。一般来讲，AOPs 的目的是产生强氧化性的原生自由基，如·OH、$SO_4^{\cdot-}$、$Cl^{\cdot}$等，但由于这些自由基的化学活性很高，在水中会发生一系列副反应，生成次生自由基。自由基与一般的氧化不同之处在于，自由基氧化速率高，体系受水质本底影响极大，导致特定 AOP 体系中包含原生自由基和多种次生自由基的综合作用。

如图 24-5（a）所示，水处理 AOPs 中原生自由基的产生方式主要有通过光、电、热、过渡金属、非金属等方式激发氧化剂（如臭氧、过氧化氢、过硫酸盐、氯、氧气等）产生·OH、$SO_4^{\cdot-}$或$Cl^{\cdot}$，不同激活方式采用的反应器形式不同。水处理中常见 AOPs 体系有：紫外/H_2O_2、紫外/过硫酸盐（PS）、紫外/氯、O_3/H_2O_2、热活化 PS、过渡金属离子或氧化物催化 O_3、H_2O_2 和 PS，还有近年来发现的非金属活化 H_2O_2 或 PS 体系[18, 19]。其中，非金属活化产生自由基是近年来的研究热点。朱本占课题组 2007 年最早发现卤苯醌可以活化过氧化氢产生·OH[20]，马军课题组、周东美课题组发现了盐酸羟胺、对苯醌、天然有机物、苯酚可以活化 H_2O_2 或 PS，产生·OH 或 $SO_4^{\cdot-}$[21]，Curtin 大学的 Sun 课题组研究了碳基材料（如活性炭、石墨烯、碳纳米管）可以活化 PS 产生 $SO_4^{\cdot-}$[22]。

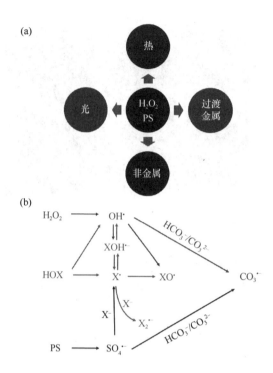

图 24-5 水处理中原生自由基（a）和次生自由基（b）的产生方式

2）次生自由基的生成机理

原生自由基除了与目标污染物反应,还会与水及水中的背景物质（如 DOM、H_2O/HO^-、HCO_3^-/CO_3^{2-}、Cl^-/Br^- 等）反应,产生次生自由基,如碳酸根自由基（$CO_3^{\cdot-}$）、卤素自由基（$Cl^{\cdot}$、$Br^{\cdot}$、$Cl_2^{\cdot-}$、$Br_2^{\cdot-}$ 等）。如图 24-5（b）所示,$SO_4^{\cdot-}$ 可与 H_2O/HO^- 反应生成 $\cdot OH$,$SO_4^{\cdot-}$ 可与 HCO_3^-/CO_3^{2-} 反应生成 $CO_3^{\cdot-}$,$SO_4^{\cdot-}$ 可与 Cl^-/Br^- 反应生成 $Cl^{\cdot}$、$Br^{\cdot}$,并进一步转化为 $Cl_2^{\cdot-}$、$Br_2^{\cdot-}$ 等,$Cl^{\cdot}$ 可以与 HO^- 反应生成 $ClOH^{\cdot-}$,并可进一步转化为 $\cdot OH$。

由于实际水体的复杂性,AOPs 体系中原生自由基产生之后,会发生非常复杂的自由基转化反应,造成实际水处理中原生自由基浓度的大幅度下降,但在此过程中生成的次生自由基,也可以对目标污染物起到一定程度的降解作用。例如,在实际水体中,紫外/氯体系中原生自由基 $\cdot OH$ 和 $Cl^{\cdot}$ 与水质背景物质反应,转化产生一系列次生自由基如 $ClO^{\cdot}$、$CO_3^{\cdot-}$、$ClBr^{\cdot-}$,对特定结构的 PPCPs 的降解起到重要作用[23]。

3）自由基氧化水中有机污染物的动力学

对于水中常见的有机污染物,不同自由基的氧化特性存在差异,如图 24-6 所示,$\cdot OH$、$SO_4^{\cdot-}$、$Cl_2^{\cdot-}$、$ClO^{\cdot}$ 和 $CO_3^{\cdot-}$ 与水中 40 种 PPCPs 的二级反应速率常数的范围差别较大,其中 $\cdot OH$ 的速率常数在 $10^9 \sim 10^{10}$ L·mol^{-1}·s^{-1} 之间,而 $SO_4^{\cdot-}$ 的速率常数在 $10^7 \sim 10^{10}$ L·mol^{-1}·s^{-1} 之间,其他三种自由基（$Cl_2^{\cdot-}$、$ClO^{\cdot}$ 和 $CO_3^{\cdot-}$）的速率常数范围分布范围更宽。这说明不同的自由基有不同的氧化特性,相对 $\cdot OH$,其他自由基具有更强的选择性。例如,$SO_4^{\cdot-}$ 对含有卤素的化合物去除效率明显高于 $\cdot OH$,可以有效降解含氟化合物和含氯化合物[24, 25]。含氯自由基（如 $Cl^{\cdot}$、$Cl_2^{\cdot-}$、$ClO^{\cdot}$）对含有氨基、羟基或甲氧基、萘环、碳碳双键等供电子基团污染物的选择性较强[26, 27]。$Cl^{\cdot}$ 和咖啡因以及卡马西平的二级反应速率常数分别是 5×10^{10} L·mol^{-1}·s^{-1} 和 5.6×10^{10} L·mol^{-1}·s^{-1},高于 $\cdot OH$ 和 $SO_4^{\cdot-}$ 的速率常数一个数量级[28, 29];$Cl_2^{\cdot-}$ 和 $ClO^{\cdot}$ 与含有供电子基团的 PPCPs 的二级反应速率常数大于 10^8 L·mol^{-1}·s^{-1},反应活性与 PPCPs 结构上的取代基供电性呈正相关关系[23]。$CO_3^{\cdot-}$ 对苯胺类、苯酚类以及磺胺类物质具有较强的活性,其二级反应速率常数都大于 10^8 L·mol^{-1}·s^{-1}[23]。

自由基与水中有机物的反应速率,取决于二级反应速率常数与自由基浓度的乘积,某些氧化性较弱的自由基，如 ClO·、CO$_3^{·-}$，虽然其二级反应速率常数较低，但由于稳定性较强，在水中的稳态浓度远远高于氧化性强的自由基（如·OH、SO$_4^{·-}$、Cl·），导致其对特定污染物的降解起到更加重要的作用[26]。

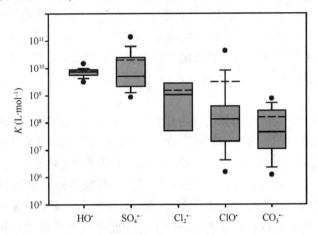

图 24-6　不同自由基（·OH，SO$_4^{·-}$，Cl$_2^{·-}$，ClO·，CO$_3^{·-}$）与水中常见 40 种 PPCPs 反应的二级反应速率常数，其中实线表示中位值

4）自由基氧化有机物的机理和路径

自由基氧化有机物的机理有电子转移、夺氢、加成反应。·OH 和 SO$_4^{·-}$氧化的路径包含羟基化、脱烷基、脱羧基等过程[30]，Cl·氧化还可导致氯取代[31]。同时，对有机物的不同官能团，自由基的作用机制不完全相同[30, 32-34]。大部分的有机污染物的母体物质可被自由基有效降解，但并不能完全矿化，会伴随各种中间产物的生成。

自由基氧化产生的一些中间产物可能具有毒性，因此，AOPs 氧化过程对水质毒性的影响也值得关注。自由基氧化对体系毒性的影响主要取决于中间产物和母体物质毒性的相对差异。关于自由基氧化对水质毒性的影响的报道，有毒性增加、减少和变化不大三种情况。例如，Sedlak 课题组在 PNAS 上报道了·OH 在氧化苯酚的过程中对小鼠的毒性增加，主要是由于生成 2-丁烯-1,4-二醛的中间产物，此物质可选择性与蛋白质中的半胱氨酸反应，影响蛋白质的活性[35]；UV/氯会增加体系中的氯代中间产物生成，从而增强体系的细胞毒性[36]；·OH 氧化诺贝酸显著降低体系对水蚤的毒性[37]；UV/H$_2$O$_2$ 和 UV/PS 均可氧化抗生素类物质甲氧苄啶和磺胺甲噁唑，降低其抗生素活性，但增加了体系的急性毒性[38]。目前的毒性评估建立在不同的模式生物的基础上，并且有较大差异。以上结果的表明，自由基氧化对有机物毒性的影响，需要进行更加深入的评估，单一的模式生物并不能全面反应毒性，应根据母体物质的特性选择合适的毒性评估方法[39]，需要在不同层面对毒性进行系统评价。

2.1.3 类芬顿

芬顿反应最早由法国科学家 Fenton 在 1893 年发现[40]，并因其能产生强氧化性的羟基自由基等活性氧物种（ROS）被广泛应用于水中有机污染物的去除。虽然传统芬顿反应具有反应快、氧化活性高等特点，但因其易生成铁泥、H$_2$O$_2$ 利用率不高、需要酸性环境等缺点，使得芬顿技术很难被工业推广。而类芬顿是近年来发展起来的高效水处理技术之一，具有反应效率高、铁泥生成量少、使用 pH 范围宽等优点，被认为是处理工业废水的理想手段之一。

1）光-芬顿

光-芬顿法是指在传统芬顿反应的基础上引入紫外光或可见光照射，从而进一步促进 H$_2$O$_2$ 或过一硫酸盐（PMS）等的分解。紫外光照射不仅可以提高 H_2O_2 的分解效率，还能促进 Fe（III）向 Fe（II）的转化，降低铁的用量。Ma 等通过在 α-羟基酸(AHA)/Fe(II)/PMS 芬顿体系中引入 UV-A 紫外光，实现了中性条件下对苯酚的高效降解，10 分钟内降解率可达 81.7%[41]。Rizzo 等采用亚氨基二琥珀酸(IDS)-Cu 复合物作为催化剂，并以 UV-C 紫外光照射作为辅助手段构建了 UV-C/H_2O_2/IDS-Cu 类芬顿体系，该体系不仅可以快速矿化苯酚，还能使废水中的大肠杆菌快速失活[42]。目前，光芬顿技术仍然存在光能利用率低、反应器能耗大、成本高等缺点。

2）光催化协同芬顿

利用半导体光催化剂产生的光生电子促进芬顿反应中铁离子的循环，是光催化协同芬顿反应的基本原理。Malato 等利用 Fe/TiO_2 基光催化协同芬顿技术对四种水溶性农药的降解进行了中试研究[43]。研究发现当吡虫啉、伐虫脒、灭多威的初始浓度为 50 mg/L，敌草隆的初始浓度为 30 mg/L，TiO_2 的浓度为 200 mg/L，铁的初始浓度为 0.05 mmol/L 时，该体系可实现对农药分子 90%的矿化率。Xing 等开发了一系列铁氧化物负载石墨烯基光催化协同芬顿体系[44, 45]。石墨烯表面聚集大量光生电子，可促进铁离子的吸附，提高铁离子的循环效率。该体系在中性条件下可实现对罗丹明 B 等有机污染物分子的高效去除。目前，光催化剂本身量子产率低、光生电子寿命短、光催化剂与铁氧化物之间作用力较弱，电子传输效率低等问题是限制该技术走向工业化的关键。

3）电-芬顿

电-芬顿技术的特点是电解池阴极还原 O_2 产生 H_2O_2，阳极氧化产生 Fe（II）或外加 Fe（II），Fe（II）分解 H_2O_2 产生大量·OH 和 Fe（III），而 Fe（III）又可在阴极得电子还原成 Fe（II），从而实现芬顿反应的循环。Zhao 等以金属-树脂为前驱体，采用一步溶胶凝胶法合成 FeCuC 气溶胶，并采用 CO_2 和 N_2 对其进行造孔活化，以增加其比表面积，并以此气溶胶为电极材料构建了电芬顿体系。该电极上高分散的 Fe^0 和 Cu^0 促进了界面电子的迁移，提高了 H_2O_2 的产量。该体系反应 1h 后，对工业染料废水的 COD 去除率可达到 83%[46]。Zhan 等利用有序介孔 $NiMn_2O_4$ 修饰作后的碳毡作为阴极构建了电芬顿体系。修饰过后的阴极不仅比表面积增大，其表面暴露的 Mn^{4+}/Mn^{3+} 和 Ni^{3+}/Ni^{2+} 氧化还原电对，还提高了电子的传导率，促进了 ROS 的生成。该体系在 1.5h 内实现对环丙沙星 100%的降解率[47]。目前电芬顿法的改进主要集中于提高阴极的比表面积，以提高 H_2O_2 的生成量。

4）超声-芬顿

超声-芬顿法的原理是产生了膨胀与压缩循环，该循环会产生空化气泡。气泡破裂会造成局部高温高压，一方面会使部分有机污染物直接热解，另一方面也会产生羟基自由基。Sun 等采用超声辅助芬顿反应处理酸性黑 1（AB1）废水，试验证实当超声频率为 40 kHz（能量密度：50 W/L）时，可显著提高芬顿反应降解 AB1 的效率[48]。Wang 等研究发现超声不仅可以促进斯沃特曼铁矿表面铁离子的溶出，还能提高 H_2O_2 的分解效率，从而实现对双酚 A 的快速降解[49]。目前超声-芬顿体系普遍存在能耗较高的问题。

5）微波-芬顿

微波是利用介质分子间的相互摩擦，来提高芬顿体系的温度，从而促进 H_2O_2 分解生成羟基自由基。Wang 等在传统芬顿体系中引入微波辅助，减少了污泥的生成，提高了高浓度制药废水的可生物降解性[50]。微波-芬顿法的温度易控制，处理效率高，反应时间短，但是目前该法还未有大规模工业化应用的实例。

6）零价铁-芬顿

零价铁-芬顿法就是将铁单质作为芬顿反应的主催化剂，由于其还具有很强的还原能力，所以零价铁可以参与后续的还原反应，为芬顿反应提供大量的 Fe（II）。零价铁-芬顿法具有原料廉价易得，处理效率高，pH 适用范围广等优点。Kallel 等利用 Fe^0/H_2O_2 体系处理橄榄油厂废水，研究发现 pH 等于 2~4 时，反应 1h 后，废水的颜色消失且酚类污染物的含量显著减少[51]。与微米级零价铁相比，纳米零价铁具有更大的比表面积，可提供更多的活性位点。Wang 等构建了纳米零价铁（nZVI）类芬顿体系，在初始 pH 为

6.1，4-氯-3-甲基苯酚（CMP）的初始浓度为 0.7 mmol/L，Fe^0 添加量为 0.5 g/L，H_2O_2 浓度为 3.0 mmol/L 时，该体系在 15min 内可实现对 CMP 的完全降解[52]。但是零价铁易氧化失活，循环稳定性差，利用率低等是限制其工业应用的主要问题。

7）单原子-芬顿

单原子-芬顿是近年来新兴的一种非均相类芬顿体系，单原子催化剂具有均匀分布的金属活性中心，其独特的电子性质可以最大化原子效率，同时单原子催化剂具有较低的金属离子溶出，可提高催化剂的稳定性。Liu 等将单原子钴锚定在氮掺杂石墨烯上，得到具有双反应活性中心的 FeCo@NC 催化剂，氮掺杂石墨烯不仅对有机污染物具有较强的吸附能力，而且其表面的 CoN_4 位点可降低 PMS 的吸附能，促进 PMS 分解生成单线态氧，从而实现对双酚 A 的高效降解。目前单原子-芬顿还处于探索阶段，其制备方法复杂，催化剂产量低，运行成本高，不适合工业推广。

8）助催化芬顿

（1）有机酸助催化芬顿。

在芬顿反应中加入具有还原性和络合能力的有机分子作为助催化剂，既可以促进铁离子的循环又可以抑制铁离子的流失。Zhang 等研究发现原儿茶酸、抗坏血酸等利用其还原性和络合能力显著提高了芬顿反应中 Fe（III）向 Fe（II）的转化效率。相比于传统芬顿体系，在 pH 为 3.6 时，$Fe(III)/PCA/H_2O_2$ 类芬顿体系对甲草胺的降解率提高了近 10000 倍[53, 54]。有机酸助催化剂在芬顿反应中容易被 ROS 分解，因此反应中需要加入大量助催化剂，这不仅会增加成本，还会降低有机分子的矿化率，因此寻找高效稳定的无机助催化剂是解决问题的关键。

（2）硫化钼助催化芬顿。

硫化钼助催化芬顿技术是最近由 Xing 等开发的一种高效类芬顿反应体系[55-57]。该技术主要利用硫化钼等金属硫化物表面暴露的高活性金属中心，实现对 Fe(III) 向 Fe(II) 的快速、高效转化。研究发现，当 H_2O_2 和 Fe^{2+} 浓度分别控制在 0.4 mmol/L 和 0.07 mmol/L 时，硫化钼助催化作用可将 H_2O_2 的分解效率从 28 % 提高至 75%。该体系对染料废水、苯系物废水、垃圾渗滤液等工业废水具有显著的 COD 去除效果。目前如何实现助催化剂的模块化生产，提高硫化钼的利用率，进一步降低成本，是推动该技术工业化的关键。

综上所述，近年来类芬顿技术在水污染物控制领域取得了一系列的研究进展，但仍然存在很多科学问题和技术难题，如铁离子循环效率低、易生成铁泥、稳定性差、运行成本高等。理论上，通过引入无机助催化剂既能提高芬顿反应的活性和稳定性，又能促进铁的再生，有望开发出效率高、投资小的类芬顿技术用于水中污染物的去除。

2.1.4 零价铁还原

铁是地壳含量位列第四的元素，是较活泼金属（标准氧化还原电位 E^0=−0.44 V）。零价铁（zero-valent iron，ZVI）是以铁单质为主要成分外表面包裹少量铁氧化物的材料。1994 年，Gillham 等报道 ZVI 能高效降解 1,4 二氯乙烷[58]。近 20 年来，ZVI 无毒、丰富、廉价、易于生产，且还原过程易于维护，通过对污染物电子定向转移，将污染物转化为无毒或毒性较小的物质，已成功用于含氯化有机化合物（COCs）[59]、硝基芳香化合物（NACs）[60]、砷[61-63]、重金属[64-66]、硝酸盐[67]、染料[68]和苯酚污染的地下水和废水的修复/处理（表 24-1）。

1）降解有机污染物

ZVI 已被证明能够对环境中存在的 COCs 包括三氯乙烯、三氯乙烷、氯乙烯、四氯甲烷以及六六六等起到有效降解作用，并在原位 COCs 处理过程中展现出良好的去除效果。在这些应用中，ZVI 主要作为电子供体，将电子传递给有机物。脂族氯化化合物的半反应在中性 pH 下具有+ 0.5 V～+ 1.5 V 的标准

表 24-1　可被 ZVI 去除的污染物种类及其主要来源

类型	举例	主要来源
重（类）金属	铬、汞、铅、锌、镍、镉 钴、砷、锑、铜、银、铀	矿采、冶炼、电镀、皮革等行业废水、废渣
非金属	硝酸盐/亚硝酸盐、磷酸盐 高氯酸盐、硒酸盐/亚硒酸盐	酸洗废水、农业灌溉、市政污水等
卤代有机物	氯甲烷、溴甲烷、氯乙烯 氯苯、氯酚、四氯化碳、多氯联苯	工业清洗水、香料浸出/萃取废水、灭火剂等
含氮类有机物	硝基苯、偶氮类染料	医药、印染等化工废水

还原电位。因此，零价铁材料还原反应具有强热力学驱动力[69]。

$$Fe^0 \longrightarrow Fe^{2+} + 2e^-,\ E^0 = -0.44\ V \quad (24\text{-}1)$$

$$RCl + 2e^- + H^+ \longrightarrow RH + Cl^-,\ E^0 = 0.5 \sim 1.5\ V\ (pH=7) \quad (24\text{-}2)$$

1996 年，文献首次报道了纳米尺寸的零价铁（nanoscale zero-valent iron，nZVI）的合成，并能够大幅度提升 COCs 的降解效率[70]。同时，nZVI 也存在易氧化、易团聚、稳定性差等问题。众多研究者采用表面修饰、负载、包裹 nZVI，通过颗粒表面静电排斥作用防止颗粒团聚，有效提升稳定性和降解效率[71]。双金属 nZVI，如 Pd-nZVI 也可产生活性氢物质并裂解碳-卤键，促进反应速率大幅提高，减少中间产物的积累，生成更加饱和的最终产物。图 24-7 阐明了 nZVI 和 Pd-nZVI 还原 TCE 的反应机理。

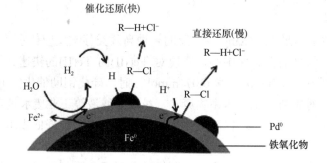

图 24-7　nZVI 和 Pd-nZVI 还原脱氯机理

近年研究发现用硫化物或二硫化物等低价硫对 ZVI 进行表面硫化处理能够大幅度提高还原脱氯速率，同时抑制产氢、减缓 ZVI 腐蚀，在有氧条件下也能够促进多相催化降解有机物，成为研究热点[72]。

潜在致癌物质 2,4,6-三硝基甲苯、2,4-二硝基甲苯和 2,6-二硝基甲苯等硝基芳族化合物被广泛应用于各类工业领域。由于硝基会抑制硝基芳族化合物的生物降解，因此传统的生物工艺无法处理含硝基芳族化合物的废水。研究表明零价铁能有效地降解废水中的硝基芳族化合物，将其转化成芳族胺化合物。而且，纳米零价铁能够大幅提高硝基芳族化合物降解速率[73]。另外，缺氧条件下零价铁能够还原偶氮和蒽醌染料脱色而得到广泛关注。

2）处理无机污染物

（1）As(III) 和 As(V)。

自然界的砷，常见于地下水中，主要是亚砷酸盐 [As(III)] 和砷酸盐 [As(V)] 的形式存在，是毒性最大、致癌化学元素之一，被世界卫生组织定为优先污染物。ZVI 去除砷的机理相当复杂，主要涉及吸附、还原、表面沉淀以及与各种铁腐蚀产物（如铁/三价铁）的共沉淀[74, 75]。运行条件 [如 pH、DO、硬度和腐殖酸（HA）等] 对于总砷的去除效率具有重要影响。

HR-XPS 多线分析和原位 XAS 研究了砷（III）与 nZVI 的反应，在砷（III）固定过程中发现了吸附、氧化、还原和 nZVI 内迁移的现象[76, 77]。砷（III）的氧化发生在氧化铁壳层上，而还原是一个多步骤的过程，伴随着砷通过氧化层扩散，形成靠近单质 Fe 核的砷-铁金属间相，砷以多层方式分布在纳米颗粒的氧化物壳上（图 24-8）。

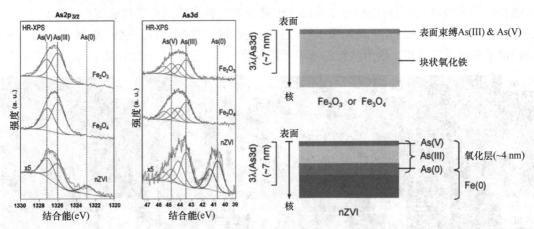

图 24-8　HR-XPS 多线分析得出不同价态 As 在 nZVI 表面的分层分布

球差校正扫描透射电镜（Cs-STEM）结合能量色散谱（XEDS）聚焦单个 nZVI 几个平方纳米范围元素分布成像和定量分析，得到 As、O 和 Fe 0.32 nm 精细步长的分布规律，观察到 As 在纳米零价铁核-壳间形成～1 nm 厚度的富集物，与 HR-XPS 的结果相吻合（图 24-9）。

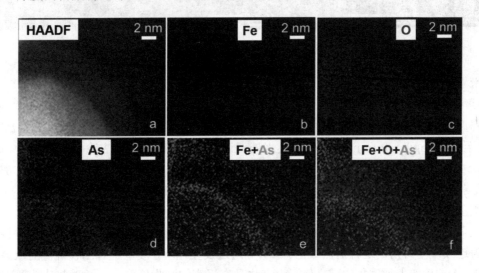

图 24-9　As 在纳米颗粒上的分布规律

（2）金属阳离子。

ZVI 与不同电化学性能的金属阳离子的反应中，金属种类的最终状态取决于金属相对于铁的氧化还原电位。如氧化还原电位较高的 Cu(II) 和 Ag(I) 被快速还原而去除。对于其他具有接近铁的还原电位的物种，在氧化铁外壳上的吸附、沉淀和共沉淀是它们有效去除的重要机制。此外，由于广泛腐蚀，ZVI 对溶液 pH 的调节作用利于金属沉淀的形成和去除。

近年来，利用球差校正扫描透射电镜研究了单个纳米零价铁与类/重金属的微观反应。研究表明纳米零价铁去除重金属污染物的界面反应很大程度上取决于类/重金属的离子电位、氧化还原电位（E^0）、反应速率和它们对 pH 的不同响应。对于强氧化性含氧离子（如 $Cr_2O_7^{2-}$、UO_2^{2+} 等）在界面处断裂金属–O 键后在还原驱动和纵向晶界增多的条件下扩散并被包裹在纳米颗粒内；E^0 略高于铁的中等氧化剂（如 AsO_4^{3-}、SeO_3^{2-} 等）界面断键后扩散并稳定在纳米颗粒的核-壳间；E^0 比铁高且反应快的阳离子（如 Ag^+、Cu^{2+} 等）在颗粒表面还原富集；E^0 接近铁且反应慢的阳离子（如 Ni^{2+} 等），在氧化还原驱动下，这些重金属与表面能较低的 Fe 原子的逆向移动速度差促使内部形成空腔；E^0 远低于铁的重金属阳离子（如 Cs^+、Zn^{2+}）则因对 pH 不同的响应而发生均匀吸附或成簇沉淀吸附（图 24-10）。nZVI 为捕获和/或封装这些无机污染物提供了多种反应途径，这为传统的基于吸附的修复方法提供了一种有吸引力的替代方法[78]。

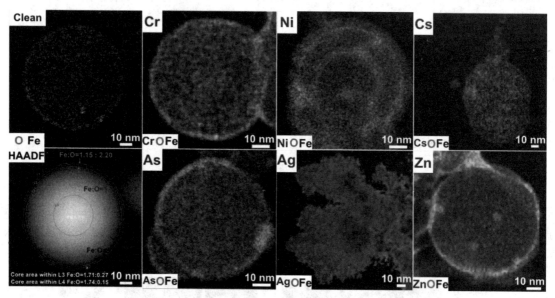

图 24-10　单个纳米零价铁与类/重金属微观反应定性定量研究

（3）纳米零价铁还原工程化应用。

采用"两级纳米零价铁并联＋曝气混凝沉淀"工艺组合（图 24-11）将纳米零价铁用于大规模重金属废水处理。工程调试期间（127 d）的监测结果表明：废水中主要污染物砷、铜平均浓度分别从 110 mg/L、103 mg/L 降至 0.29 mg/L、0.16 mg/L，低于废水处理设计排放标准，平均去除率均超过 99.7%；其他重金属（如 Co、Cr、Ni、Pb、Zn 等）浓度均降至 0.01 mg/L 左右，远低于相应排放标准；废水中 Se、Sb、Au 等元素也几乎被完全去除[79]。结合调试期间纳米零价铁的消耗量及废水中重金属离子的累积截留量，本工程条件下纳米零价铁除砷、除铜负荷分别达 245 mg-As/g-nZVI 和 226 mg-Cu/g-nZVI，总体重金属去除负荷超过 500 mg-重金属/g-nZVI，优于多数重金属离子吸附材料或离子交换材料。图 24-11（b）和（c）为连续一个月内的运行监测数据。此外，通过反复循环，废水中有回收价值的重金属离子[如 Cu(II)、Ag(I)、Au(III) 等]被富集浓缩，使污泥产物具有极高的回收价值，可抵消部分处理成本，甚至可能成为企业的另一种创收方式。以本工程为例，处理过程产生的污泥作为富金矿被再次返回该冶炼厂冶炼过程的最前端，真正实现了废水中有价金属的回收再利用。

零价铁还原各种污染物是一个多反应复合过程，尤其在复杂的环境体系中，微观机理和反应理论还有待于更深入的研究。同时，将其应用于实际重金属污染物处理和重金属污染修复也是研究的重点。零价铁技术实际应用中涉及的问题主要有：①材料的大规模制备及成本控制；②针对不同环境介质（如废水、土壤、底泥等）中重金属污染物的可行性工艺研发及配套设备开发；③工艺运行过程中的调控参数及稳态控制；④纳米材料潜在生态风险及规避方法。

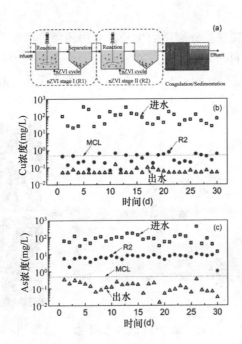

图 24-11 纳米零价铁处理冶炼废水工艺流程（a），工程运行期间进出水中 As、Cu 浓度变化（b，c）

2.2 富集分离技术

2.2.1 膜分离

水是生命之源，优质水资源是保障生态稳定、社会发展和人类存续的根本。污/废水治理及资源化是目前全球环境领域的焦点问题。膜分离技术由于其分离精度高、产水水质好、无二次污染、占地面积小等优势被广泛应用于各工业领域的废水处理、回用及资源化，是解决当前水污染等问题的关键技术[80]。微滤、超滤、纳滤、反渗透等传统膜分离技术已被广泛用于工业生产，但也遇到了一些瓶颈问题，例如，水通量、选择性与服役寿命无法兼得等。这些问题的产生与膜材料本身的性质密切相关，膜材料的精细设计与制备是突破现有膜性能的主要方法[81, 82]。与此同时，一些新兴的膜技术，如正渗透、膜接触器等，也取得了长足进展，在一些特殊领域展现出了充分的潜力。本小节针对各膜过程的共性难题或突出问题，从膜材料设计与制备和新型膜工艺开发两个角度简要介绍近年来水处理用膜技术的研究进展。

1）膜材料的设计与制备

理想状态下的膜分离过程应具有三个特点，包括：高选择性、高通量和长服役寿命。但目前的膜材料往往面临选择性与通量此消彼长（Trade-Off 效应）、服役寿命不足（由膜污染、膜材料损伤或老化造成）等问题。针对这些问题，研究人员在膜材料的设计与制备领域，开展了大量研究工作。

分离膜孔径均一化是提升膜分离过程精度，突破选择性和通量 Trade-Off 效应的关键，均孔膜因此应运而生。均孔膜不仅是一种新结构的分离膜，更代表着分离膜，尤其是超微滤膜的发展方向。利用嵌段共聚物微相分离的特性，可获取规整排列、孔径在 10～50 nm 范围内连续可调的均孔结构。但在嵌段共聚物基均孔膜走向工业化之前，须首先解决嵌段共聚物原材料的成本和强度的瓶颈问题[83]。

利用多元胺的水溶液与多元酰氯的有机溶液之间的界面聚合反应，可在支撑基膜表面形成致密的聚酰胺分离层，是当前最便捷高效、应用最广泛的纳滤膜、反渗透膜、正渗透膜制备方法。由于多元胺与多元酰氯之间的反应速率快，对界面聚合反应进程的控制被认为是精确调控聚酰胺分离层结构的关键[84]。在多孔基膜表面构筑亲水中间层可优化水相单体在膜表面的分布均匀程度，同时可减缓水相单体向有机相反应区的扩散速率，进而控制聚酰胺分离层的生长，获得超薄分离层[85-87]。但中间层的大面积便捷制备是需要解决的难题[88]。最近，科研人员发现，在水相溶液中添加亲水单分子可获得具有管状和点状纳米图灵结构的高通量高选择性纳滤膜（图 24-12），这为界面聚合反应原位调控以及界面聚合成膜理论研究提供了新的灵感[89]。

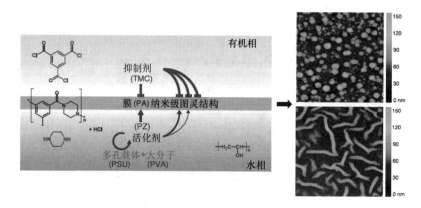

图 24-12　图灵结构聚酰胺膜生成示意图[89]

通过向成膜溶液中添加水通道蛋白或其衍生物、碳纳米管、石墨烯、金属-有机骨架、共价有机骨架等构筑混合基质膜可突破膜选择性和通量之间 Trade-Off 效应，成为目前分离膜领域的研究热点[90]。纳米材料表界面性质的合理设计以及开发巧妙的有机-无机复合方法是增加纳米材料在混合基质膜中分散性和负载量的关键。此外，关于混合基质膜构效关系的认知有待提高。分离性能的提升是源于纳米材料本身的孔道还是纳米材料与本体材料之间的相界面目前尚无定论。

膜污染是所有膜过程不可避免的问题，不可逆膜污染影响了滤膜的服役寿命。形成不可逆膜污染的原因是污染物与膜表面形成强相互作用，充分了解污染物的性质及膜-污染物相互作用机制能够更有针对性地制备抗污染滤膜材料。通过表面偏析、涂敷和表面接枝等方法在膜表面引入亲水分子链可有效减缓膜污染。但单一的亲疏水性调控往往很难提供广泛的抗污染效果，因此动态两亲（亲水/疏水）滤膜的研制正在引起关注。例如，通过在聚偏氟乙烯多孔膜表面上引入氟岛封端的 pH 响应高分子链段构筑具有清洗响应的抗污染表面，提升了滤膜的抗污染性能和清洗效率[91]。对于芳香聚酰胺反渗透膜来说，粗糙的膜表面是膜污染的重要原因。通过设计新型成膜过程（如分子层层自组装）或调控界面聚合反应来构建光滑膜表面是研制抗污染反渗透膜的热点[92]。

具有高强度、良好化学稳定的膜材料是保障服役寿命所必须。近年来，研究人员通过编制管增强、多孔膜复合、筋线增强等方式有效抑制了中空纤维超微滤膜断丝的问题[93]。通过活性单体分子结构设计或聚酰胺表面化学修饰等方法显著提升了纳滤膜和反渗透膜的耐氯性能[94, 95]。

2）新型膜工艺开发

由渗透压驱动的膜分离过程近年来正在引起人们越来越多的关注，因为它们在水处理和新能源领域都展现了巨大的应用潜力[96]。利用渗透压进行水处理的正渗透工艺（FO）[97]以及利用渗透压进行发电的压力阻尼渗透工艺（PRO）[98]是目前研究的热点。FO 无需外加压力驱动，所以理论上无须消耗能量。但是，FO 工艺的主要限制因素是驱动溶液的分离和回用。因此，研发易于分离和回用的新型驱动溶液是 FO 工艺目前主要的挑战。而在 PRO 工艺中，限制其应用的主要因素是膜表面的浓差极化和膜污染。

膜蒸馏（MD）技术具有截留率高、产水水质好、浓缩倍数高等优势，在海水淡化、废水零排放方面展现出良好的应用前景。但疏水膜的润湿问题一直制约着膜蒸馏的产业化。除了研制抗润湿疏水膜，基于膜润湿机理开发便捷的预处理工艺也是保障膜蒸馏稳定运行的有效方法。最近，研究人员开发了气液界面萃取-MD 耦合工艺，在料液进入疏水膜组件前去除其中的表面活性物质，显著延长了膜蒸馏稳定运行时间[99]。

厌氧膜生物反应器（AnMBR）具有高负荷、低能耗、污泥量少、产水水质好、产甲烷等优势，在强调可持续发展理念的今天，这种膜工艺逐步赢得了人们的青睐。但膜污染、溶解性甲烷丢失等问题制约着该技术的工业应用。如何低能耗地控制膜污染以及回收溶解性甲烷是推进 AnMBR 技术在水处理领域应用的主要挑战[100]。

膜曝气生物反应器（MABR）是利用透气膜进行曝气供氧的污水生物处理新工艺，具有无泡曝气、异向传质和分层结构的特点，可实现同步脱氮除磷。最近，MABR 技术被成功应用于城市黑臭河治理（图 24-13），已建立多处示范工程，取得了良好的社会反响[101]。

图 24-13 （a）和（b）分别是水草式 MABR 曝气膜组件和浮动式 MABR 河道水体修复系统的照片；（c）和（d）分别是修复前后河道照片[101]

综上所述，与传统的水处理技术相比膜分离具有显著的优势，但膜分离技术也面临诸多问题和挑战。高选择性、高通量和长服役寿命是理想膜材料应该具备的三个特点。通过优化成膜过程、科学合理设计膜表界面以获得具有高选择性、高通量、高强度、抗污染的膜材料是分离膜领域的研究热点。针对瓶颈问题，定向设计及优化膜材料与膜过程，是新型膜工艺走向应用的必经之路。

2.2.2 混凝

混凝是指通过投加絮凝剂，使污染物脱稳、聚集、沉降的水处理技术。"混凝"和"絮凝"同义，目前研究者更习惯使用"絮凝"的称谓。絮凝技术具有操作简便、成本低廉、改造容易等优点，是水处理过程的最基本单元之一。近两年，关于絮凝技术的研究进展主要集中在以下四个方面：新型絮凝剂的研发、絮凝剂去除污染物的特征和机理研究、絮凝工艺的研究以及利用絮凝进行污泥脱水的探究。

1）新型絮凝剂的研发

研发新型絮凝剂强化絮凝单元对污染物的去除、扩展絮凝的应用范围一直都是研究的热点。近两年，新型絮凝剂的研发主要集中在无机盐类絮凝剂和无机-有机共价键型絮凝剂上。对于无机盐类絮凝剂，2018年，针对水华问题，有研究使用溶胶-凝胶法发明了新型钛基干凝胶絮凝剂，制备过程中使用乙酰丙酮作为水解引发剂，一方面可促进钛凝胶的形成，从而有效去除水体的蓝藻细菌，另一方面，乙酰丙酮可作为光活化剂有效降解蓝藻细菌产生的微囊藻毒素；该絮凝剂较传统的四氯化钛和聚合四氯化钛有明显的优势[102]。同时该课题组详细探究了该新型钛凝胶絮凝剂对于有机物的去除效率高的机理[103]。该钛基絮凝剂的研发为传统无机絮凝剂的改良提供了参考。

无机盐类絮凝剂去除污染物的能力不断提升，但去除污染物类型仍局限在浊度、藻类等大分子物质，而近年来新型的无机-有机共价键型絮凝剂拓宽了絮凝去除污染物的范围[104]。但与此同时，共价键型絮凝剂也面临着无机、有机组分之间极易相互作用、交联、失效的问题，一方面使絮凝性能下降，另一方面絮凝剂也难以稳定储存。针对这一问题，2018年，研究者从仿生学角度出发，通过模仿海洋生物海葵的身体结构及捕食特征，研发了一种仿生纳米絮凝剂[105]。絮凝剂具有独特的核壳胶束结构，解决了常规絮凝剂易失稳的难题。在絮凝过程中，絮凝剂能发生构型反转，类似海葵的捕食过程：由铝硅复合物组成的"壳"水解相变，去除悬浮物和胶体；由有机官能团组成的"核"外翻并环绕固定于"壳"上，像海葵的触手捕捉小分子污染物。絮凝剂对污水处理厂二级出水中颗粒物、胶体以及溶解性小分子的去除率均达到90%以上。仿海葵纳米絮凝剂对于水深度处理工艺的简化具有重要意义。但目前该絮凝剂制备方法复杂，原材料成本较高，距离真正的工业化应用仍有较大的距离。

2）絮凝剂去除污染物的特征和机理研究

目前工业应用最为广泛的絮凝剂仍是传统的无机、有机絮凝剂；传统絮凝剂虽然应用时间久，但其对某些新型污染物的去除效果、特征和机理仍不明确。近两年，一些学者研究了传统无机盐类和有机高分子类絮凝剂对一些新型污染物的去除特征和机理。研究的污染物对象主要包括备受关注的纳米颗粒、重金属以及溶解性有机物等。例如，目前具有较好应用前景的超精细炭黑颗粒吸附剂，极易在水体中残留，已有研究详细探究了23种不同参数的氯化铝絮凝剂和絮凝方式对于超精细炭黑颗粒的去除特征，为降低超纳米颗粒的环境风险提供了理论依据[106]。针对亲水性生物高分子多糖、蛋白质等极易造成膜污染的难题，有研究者考察了絮凝过程对色氨酸、酪氨酸以及腐殖酸三种类型生物聚合物的去除规律，指出将絮凝应用为膜滤过程的前处理手段时，需要依据水体中生物高分子的类型选择絮凝剂和絮凝条件[107]。另外，一些研究详细探究了絮凝对重金属的去除特征，并发现水体中呈现颗粒态、胶体态和溶解态的天然有机物对重金属的去除有很大影响[108]。以上研究通过揭示絮凝原理对水体污染物的去除特征，为实际工程中絮凝剂的选取以及絮凝参数的调整提供了依据。

3）絮凝工艺的研究

近两年，关于絮凝工艺研究的方向与以往没有太大变化，仍主要集中在絮凝-膜工艺和絮凝-高级氧化工艺上，絮凝-膜工艺的研究热点主要是通过絮凝形成的絮体修饰膜组件，减少膜污染现象的发生；而絮凝-高级氧化工艺的研究热点则在于强化藻类的去除，降低水华风险。

在絮凝-膜工艺的研究中，有研究充分利用絮凝剂水解絮体的松散性和强吸附性，针对不同污染物构建了系列絮凝-膜耦合工艺；针对饮用水源中普遍存在的有机污染物，研究将混凝剂水解絮体分批次注入膜池，在滤膜表面构建了动态的"三明治"式絮体保护层，有效减缓了膜污染，且形成的滤饼层松散易清洗[109]。此外，针对水体中重金属粒径小容易透过超滤膜的问题，有报道研究了以地表水中近年来备受关注的 Sb(V)为目标污染物，考察了絮体-超滤耦合工艺对其去除能力。研究发现膜池内注入铁盐絮体使Sb(V)具有更高的去除效率并可有效降低膜污染[110]。

利用絮凝-高级氧化工艺去除藻类具有很大的应用优势；该过程中，高级氧化可有效去除微生物的胞外物,从而使藻类更易被絮凝过程去除。近两年,文献报道最多的高级氧化方法主要包括芬顿法和光催化两种,絮凝过程中产生的铁离子既参与了絮凝过程,又可催化自由基的生成,使高级氧化过程无需外加化学试剂,例如,利用絮凝-UV/H_2O_2强化铁絮凝技术去除藻青菌[111],利用絮凝-O_3去除藻类[112]等。

4) 絮凝用于污泥脱水的研究

污泥处置是废水处理的一部分,其中污泥脱水是污泥处置过程中最重要的环节。污泥是由微生物、微生物胞外聚合物(EPS)等组成的高度异质化的聚集体,含水率一般在90%以上,传统的絮凝方法主要是利用污泥的胶体特性,使细小的污泥形成大絮体与水分离;但是由于EPS具有极高的结合水的能力,因此传统絮凝只能加快污泥脱水,但在降低污泥含水率方面效果并不明显。

针对以上问题,近两年利用絮凝进行污泥深度脱水取得系列进展,主要是通过新的絮凝药剂或通过新工艺,破坏微生物的EPS,促进颗粒内部水的释放。例如,有研究开发了两种基于骨架助滤和有机高分子絮凝的厌氧消化污泥强化脱水新技术,采用诱导原位结晶助滤耦合高分子絮凝,发现原位结晶形成的磷酸铵镁结晶(MAP)可以通过吸附EPS促使絮体的致密化,改善污泥的过滤行为[113]。还提出了甲醇和无机混凝剂耦合调理强化污泥脱水性能的新型技术,发现了在絮凝过程中添加甲醇,可破坏蛋白质的水化层并诱导凝聚,实现污泥结合水的脱除[114]。

另外一些研究则利用絮凝耦合高级氧化工艺强化污泥脱水。高级氧化过程中,氧化剂分解生成高活性的自由基,攻击和破坏EPS聚合物主链的断裂,从而促进污泥脱水的效率。2018年,一些研究报道了一种无需外加过氧化氢的新的污泥调理方式[115],在纳米零价铁和空气的激活下氧化降解污泥胞外聚合物,将结合水转化为自由水;同时加入的聚二甲基二烯丙基氯化铵(PDMDAAC)药剂,可以起到络合和絮凝的作用。目前,利用絮凝进行污泥脱水正在向着更加经济、高效的方向进行。

综上所述,近年来,絮凝技术通过与其他技术的结合,其应用范围越来越广,但絮凝技术也面临问题和挑战。共价键型絮凝剂去除小分子的新功能为新型絮凝剂的研发指明了方向,扩展絮凝剂去除污染物的范围是新型絮凝剂研发的热点;另外,絮凝-高级氧化、絮凝-膜技术的耦合工艺可强化水体污染物的去除以及污泥脱水,但面临着能耗高的问题,开发经济、环保的絮凝工艺也成为研究的热点。

2.2.3 吸附

吸附法由于其适应范围广、处理效果好、可重复使用等优点而成为去除水体中污染物最为高效的技术之一。一直以来,吸附材料的制备一直是吸附领域的研究重点,但如何低成本地设计并开发出高效能的吸附剂已成为吸附研究的核心[116]。针对污染物处理及同步资源化的最终目标,吸附剂的再生及吸附工艺的搭建已成为近年来的研究热点[117]。下面分别针对污/废水中的重金属离子和典型的新型有机污染物的去除以及在此基础上的吸附工艺的搭建,总结近年来吸附技术的研究进展。

1) 重金属的吸附去除

水体中重金属的残留已经严重危害到生态系统及人类健康。同时,重金属废水本身就是一个巨大的宝贵资源库[118]。如何通过吸附技术修复重金属污染进而实现资源化显得尤为重要。根据具体的废水特征,合理设计吸附材料可以提升重金属靶向吸附能力[119],优化吸附工艺可以进一步实现资源的高值转化[120]。通过源头定向设计与制备吸附材料,可以极大限度地从废水中选择性捕获金属资源。Liu等[121]通过硫代内酯化功能化丙烯酰胺的自由基聚合一步得到PAM-TL吸附剂,该吸附剂对Ag^+展现出较高的吸附容量(145.2 mg/g),是对应离子印迹材料(IIP)的5.8倍,且其选择性分配系数则是IIP的5.1倍。Yu等[122]将2-羟乙基-甲基丙烯酸酯高分子(PHEMA)接枝到Pd离子印迹表面,系统研究了不同接枝度的高分子刷对吸附容量与抗共存离子干扰性能的影响,结果表明接枝了高分子刷的Pd离子印迹材料的选择性吸附能力得到进一步提升。吸附剂对重金属选择能力的提升对于重金属废水(特别是含Pt、Pd、Ag、Cu等稀贵金属的废水)的低价处理及同步资源化具有重要的应用前景。对于吸附机理研究,Luo等[123]通过DFT

计算确定出单宁酸基吸附剂（TA@Zr）的具体吸附位点位于 TA 分子苯环上的邻位酚羟基，且通过不同金属吸附能的 DFT 计算，解释 TA@Zr 对 Pb^{2+} 的选择性靶向吸附，进而阐明邻位酚羟基与 Pb^{2+} 发生空间匹配配位进而发生离子交换的吸附机理。

在极大限度追求吸附容量和选择性的基础上，吸附剂的再生已成为新的研究热点。传统的再生方式主要是加药洗脱，但加入的大量药剂会增加再生成本及环境不友好。对于重金属吸附，新型的再生技术有光还原、电还原、光-电还原等，甚至还可以通过纳米技术实现原有吸附物的包埋以构建新鲜吸附界面。Wu 等[124]利用弱电介入将吸附在氨肟功能化碳布表面位点上的 Cu^{2+}、Pb^{2+} 和 Cd^{2+} 还原脱附去除，位点的更新进一步提升了吸附容量，高达 2300 mg/g，是表面吸附基商业过滤器的 2~3 个数量级。Kim 等[125]构建出导电性的聚苯胺纳米纤维电极，通过电化学调控聚苯胺的氧化还原态来实现 Hg^{2+} 的吸附与快速脱附。

2）典型新型有机污染物的吸附去除

随着人口的急剧增长，污/废水中药物与个人护理品（PPCPs）和全氟有机化合物（PFOS）等新型有机污染物也急剧增多。这类新型有机污染物结构稳定，难以生化降解，且大部分分子含有效价官能团，对生态环境中的微生物存在较大的毒害及抗性基因的形成及传播[126]。吸附就成为一种非常适宜的处理手段。对于有机污染物的吸附，主要的吸附机理包括疏水、静电、离子交换、氢键作用等[127]。传统的吸附剂（活性炭、离子交换树脂、高分子聚合物等）存在吸附容量低、易释放等缺点[128]。对传统材料的改性及制备新材料是两种常用的手段来进一步提升对有机污染物的吸附处理能力。对碳材料实现孔结构、表面官能团、比表面积、亲疏水性等物理特性的可控制备，可以有望得到价廉且高吸附容量的吸附剂。Javed 等[129]将廉价的沥青通过煅烧和 KOH 高温活化得到超高比表面多孔的活性炭（AS），其比表面积高达 3851 m^2/g 是商业活性炭的 4.7 倍，这导致 AS 对双酚 A 的吸附容量高达 1113 mg/g。Li 等[130]制备出阳离子型钍基 MOF（SCU-8），具有较大的内部半径（约 2.2 nm）及很高的比表面积，其对 PFOS 污染物具有高的吸附容量及吸附速率。邢宝山等[131]首次报道少层黑磷对阳离子型亚甲基蓝和阴离子型刚果红具备高的吸附容量，分别高达（1232±283）mg/g 和（230±9）mg/g，通过 TEM 与 AFM 及 DFT 理论计算手段阐明出褶皱诱导的吸附机理。对于有机物的复合污染（硝态氮、磷等），Elimelech 等[105]开发出核壳结构的海葵状胶束纳米吸附剂（AMC）实现多种污染物的一步去除，这有望简化水处理过程且减少处理时间、用地需求及处理成本。另一方面，有机污染物脱附的主要方式是氧化降解。Xiao 等[132]通过电芬顿技术在活性炭-PTFE 电极表面实现污染物的氧化去除，实现再生。Ren 等[133]制备出超疏水表面活性剂掺杂的导电高分子琥珀辛酯磺酸钠掺杂的聚吡咯 [Ppy(AOT)]，系统研究了电化学可逆的吸附与再生。

3）吸附工艺的搭建

在吸附材料高效研发的基础上，如何根据实际的环境条件合理地设计吸附工艺是走向实际应用中不可或缺的阶段。目前实际应用过程中最为常见的当属固定床工艺，但固定床的成本高、占地大且吸附剂使用量大等问题限制了其实际应用。将合适的吸附材料组装到水处理膜的表面，可以进一步提升吸附材料的潜力，且可以更好地糅合进已有的污水膜处理装置。Bolisetty 等[134]开发出类淀粉蛋白-碳复合膜，各种不同的金属离子（Hg^{2+}、Pb^{2+}、Au^{3+}、Pd^{2+}）都可以被快速截留在膜表面，如 Na_2PdCl_4 溶液的浓度可以从 12.2 mg/L 降低到<0.019 mg/L，溶液由浅黄色变成无色（图 24-14）。对于含有大量醇类的抗生素制药废水，如何靶向实现抗生素分子的吸附进而氧化去除具有现实意义。Luo 等[135]首次制备出表面具有分子印迹空腔的 TiO_2/WO_3 光催化剂，该催化剂可以从复杂废水中选择性吸附抗生素类污染物，进而光催化去除。Wu 等[124]构建出高分子修饰的碳毡电极一体化反应器，能够快速地纯化污染的水体，去除速率高达 3000 L/(h·m^2)，且在极短的时间内（仅 3s）可以使重金属的浓度降低到小于 5 mg/L（图 24-15）。这种小型的电化学装置在解决极端条件下饮用水方面有着广阔的应用前景。

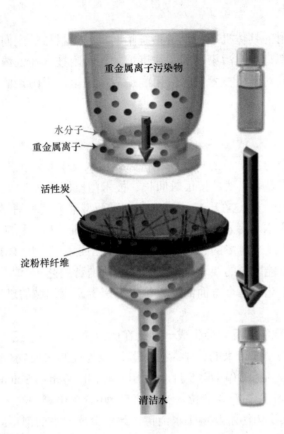

图 24-14　淀粉样蛋白-碳复合水处理膜的示意图及吸附前后 Na_2PdCl_4 溶液颜色变化[134]

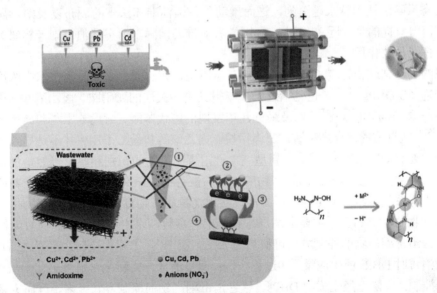

图 24-15　高分子修饰的碳毡电极一体化反应器及重金属吸附去除的微观机理示意图[124]

4）总结与展望

吸附技术可以有效地去除污废水中各种不同类型的污染物，特别是重金属离子和典型的新型有机污染物。面向资源化的重金属离子的吸附，开发靶向的功能吸附材料一直是研究侧重点，且相关的吸附技术与工艺也得到了广泛的工程应用。但对于具体的选择性吸附机理还需要进一步深入探究[136]，且实际的工程应用成本还尚高于市场现有的吸附工艺。另一方面，废水中有机污染物的吸附尚处于实验室研究阶段，对于不同有机污染物吸附的具体机理还尚未系统阐明，吸附材料的定向设计还处于初步探索阶段。随着废水排放标准的提高及新型有机污染物的关注度增加，吸附技术在低浓度有机污染物的深度去除方面将展现较好的应用前景。针对吸附技术实际应用的最后一公里，吸附剂的高效再生及吸附工艺的设计与优化将成为新的研究重点。

2.2.4 消毒

消毒是供水和污水处理中至关重要的一环，是保障饮用水安全和水生态安全的关键措施[80]。由于氯系消毒剂价格低廉、消毒效果显著、消毒持续时间长、技术门槛低，从20世纪开始至今，其便被广泛用于给水厂和污水厂的消毒处理[137]。这在一定程度上抑制了新型消毒剂和消毒技术的发展。然而，近些年也出现了一些因素驱动了饮用水和污水消毒在理论研究、技术研发和工程应用领域的发展。譬如，①饮水健康和水生态风险问题得到越来越多的关注，新型高毒性消毒副产物（DBPs）不断被发现；②为满足更加严格的标准规范或实现节能降耗的要求，需要寻求新的消毒工艺等[138]。本节尝试围绕消毒副产物、氯替代消毒方法和组合消毒工艺等三个方面简述近年来水消毒研究领域的进展。

1) 消毒副产物

自从1974年研究人员发现氯消毒后会生成一种潜在的致癌物质——氯仿以来，权衡微生物污染带来的介水疾病风险和消毒副产物暴露带来的化学毒性风险即成为控制水质安全的关键核心问题之一，消毒副产物逐渐得到更多关注[137]。截至2019年5月，经国际通用的Web of Science数据库检索，涉及消毒副产物的文章9000余篇，相关作者20000余位，文章来自90余个国家和地区，所出版文章中涉及的消毒副产物多来自饮用水消毒，其次为污水及再生水消毒，此外游泳池水的消毒副产物问题也逐渐得到关注。45年来，环境、水处理和公共卫生等领域专业人员的不懈努力使得我们对消毒副产物的重要性、生成机制以及健康和生态风险的认识越来越深刻，进而推动了各国标准和法规的颁布以及消毒副产物分析和控制技术的发展。值得一提的是，近30年来，我国学者对推动国际水消毒和消毒副产物研究领域的理论发展和技术进步发挥了重要作用[139]。

目前已发现的消毒副产物有700多种，包括三卤甲烷、卤乙酸、卤代醛等在水中浓度相对较高的典型含碳消毒副产物（C-DBPs）[140]，也包括近十余年得到更多关注的亚硝胺、卤乙酰胺、卤乙腈等毒性更高的含氮消毒副产物（N-DBPs）[141]，以及最近几年发现的卤代苯醌、卤代苯甲醛和卤代苯乙腈等芳香性消毒副产物[142-145]。从目前的研究结果看，三卤甲烷、卤乙酸等传统C-DBPs并不是导致饮用水和污水消毒后毒性升高的主要驱动因素[146]。因此，新型高毒性消毒副产物的识别和鉴定仍然是消毒副产物领域未来研究的重点之一。并且，更多的消毒副产物应被纳入流行病学研究，建立能够精准表征消毒副产物暴露水平的检测技术体系与成组生物毒性分析方法，实现消毒副产物毒性风险的精确评估。

水中消毒副产物的种类和含量主要由消毒方式、工艺参数和水中存在的有机物与无机物等前体物的组成所决定的。故消毒副产物的控制方式可概括为如下三类[147]：①源头控制：消毒之前通过去除DBPs的前体物来降低消毒时DBPs的生成量；②过程控制：调整消毒方式和优化消毒工艺参数来降低DBPs的生成量；③末端控制：去除已经生成的DBPs。尽管消毒副产物的控制方式很多，但研究多集中在单一方法对单类消毒副产物的控制上，未来需从源头控制和过程控制角度加强对多种类消毒副产物的综合控制研究。并且新型消毒副产物的生成机制、在供水管网和二次供水中的迁移转化规律等研究仍然存在大量空白。其毒性的替代指标，包括出厂水中的总有机卤素、综合毒性和前体物总量等指标应用仍处在实验室研究阶段[138]，有待进一步加强研究与应用。相对于饮用水，污水成分更为复杂，污染物种类和数量相对更多[148]，消毒处理时消毒剂的投加量更高，进而产生的消毒副产物浓度可能更高，种类可能更加复杂[149]。因此，未来需加强污水及其再生水消毒后高毒性副产物的识别，以及考察其对受纳水体的生态影响，并结合消毒前污水处理技术与工艺，实现病原微生物和消毒副产物的综合控制，保障水质安全。

2）氯替代消毒方法

考虑到消毒副产物、耐氯菌以及消毒剂运输和储存的问题，次氯酸钠、紫外线、二氧化氯等替代液氯消毒的消毒剂或消毒方法已较为成熟并得到规模化应用[150-152]。目前，过氧化物在污水消毒领域也逐渐得到应用，过氧乙酸可能是目前这类物质中最具有潜力的消毒剂，然而其在环境中的残留毒性需要进一步评估[153]；其他替代消毒剂在供水和污水消毒中应用的研究也有报道，如过甲酸、银/过氧化氢以及纳米银等。尽管它们尚不具备规模化应用的潜力，但这些技术仍可能被用于某些特定情况。譬如用银质材料来控制输水管网中的生物膜或者减少医院等小型水循环系统中存在的顽固病原体等。任何一种消毒技术都有一定的局限性，在消毒应用中，需要综合考虑原水水质、出水水质要求、消毒剂来源与安全性、消毒副产物生成潜力、水厂工艺特点等因素，并通过技术经济比较确定所需采用的消毒剂及运行工艺参数。

3）组合消毒工艺

由于将两种及以上的消毒手段组合使用可以弥补单一消毒方式存在的局限性，近年来组合消毒工艺逐渐得到更多的关注。目前研究较多的是将紫外线、臭氧以及基于它们的高级氧化技术（如紫外线/过氧化氢、臭氧/过氧化氢、紫外线/过硫酸盐、紫外线/二氧化钛等）作为第一步的消毒工艺，有效的杀灭水中的各种病原微生物以及消减消毒副产物前体物。其中利用紫外消毒可以灭活水中的贾第鞭毛虫和隐孢子虫等原生动物[154]，利用紫外线/过氧化氢可以有效破坏含氮消毒副产物前体物[155]，然后再投加液氯、氯胺、次氯酸钠或二氧化氯等不易分解的消毒剂来维持持续消毒效果，通过不同阶段各种消毒剂的相互协同作用，取长补短，扩大微生物控制的覆盖面，进而取得较好的消毒效果；同时可以大大减少消毒副产物的生成量并协同去除其他微量污染物，达到安全消毒和提升水质的效果。

3 生化水处理方法与技术

3.1 脱氮除磷技术

3.1.1 Anammox/Canon 脱氮技术

总氮是引起水生态问题的主要物质，"十三五"以来已引起高度重视。目前污水 C/N 低，难满足传统硝化反硝化工艺脱总氮需求，生物脱氮工艺向着高效、节能和低碳的方向发展。厌氧氨氧化（anaerobic ammonium oxidization，Anammox）是指在微生物作用下氨氮和亚硝酸氮直接转化为氮气的过程，是目前已知的最为短程的水体氮素去除形式。和所有的新技术一样，Anammox 技术应用也有难点和挑战：①厌氧氨氧化功能菌增殖缓慢、代谢条件严格；②前序短程硝化难控制、多菌群协作难调控等。针对此问题，深入识别功能微生物的生化行为及其分子机制，对于解决 Anammox 技术瓶颈问题具有重要的现实意义。

1）Anammox 菌代谢分子机制

Anammox 菌生长速率慢与其氮素和碳素代谢途径有着密切的联系。Anammox 菌将氨氮和亚硝酸盐氮转化为氮气和硝酸盐氮的同时，将 CO_2 或碳酸氢盐固定为有机碳［式（24-3）］[156, 157]，所以 Anammox 菌的生长伴随着氮素代谢过程［式（24-4）］。由于约 4 mol 的亚硝酸盐氮被氧化才会生成 1 mol 的有机碳［式（24-4）］，即氨氮与亚硝氮和碳酸氢盐的反应进行 15 次才可固定 1 mol 的无机碳［式（24-3）］，这是导致 Anammox 菌生长速率缓慢的重要原因[158]。

$$1NH_4^+ + 1.32NO_2^- + 0.066HCO_3^- + 0.13H^+ \longrightarrow 1.02N_2 + 0.26NO_3^- + 0.066CH_2O_{0.5}N_{0.15} + 2.03H_2O \quad (24\text{-}3)$$

$$0.26NO_2^- + 0.066HCO_3^- \longrightarrow 0.26NO_3^- + 0.066CH_2O_{0.5}N_{0.15} \qquad (24\text{-}4)$$

虽然 Anammox 菌氮素与碳素的反应过程得到广泛认可，但是对于氨氮与亚硝酸盐氮反应生成氮气的具体过程及其分子机制的认知不断被修正 [式（24-5）]，且这一过程因菌种而异。在 2015 年之前，公认的反应包括如下三个步骤：①亚硝酸盐氮在 nirK 或 nirS 型亚硝酸盐还原酶的作用下被还原为一氧化氮 [式（24-6）][159-161]；②一氧化氮与氨氮在联氨合酶（hzsBCA）的作用下发生归中反应生成肼 [式（24-7）][157, 162, 163]；③肼在联氨脱氢酶（hdh）的作用下被氧化为氮气 [式（24-8）][164, 165]，典型菌属包括 Ca. Jettenia 和 Ca. Scalindua。而 Ca. Kuenenia 生成肼的方式不同于上述过程，研究学者通过对 hzsBCA 基因进行蛋白提纯和晶型结构的分析发现，该蛋白复合体具有两个独特的血红素 c 类型的活性位点供氧化还原反应进行，因此生成肼的过程 [式（24-7）] 包括一氧化氮被还原为羟胺及羟胺与氨氮反应生成肼 [式（24-9）和式（24-10）] 两个步骤，该成果于 2015 年被刊登于 Nature[166] 杂志。同年，研究学者利用宏基因组测序和多种生物信息学技术拼接了 Ca. Brocadia sinica 草图，且在其基因组上并未找到 nirK 或 nirS 基因[167]。2016 年，Oshiki 等利用同位素示踪技术探明在 Ca. Brocadia 体内亚硝酸盐氮直接被还原为羟胺 [式（24-11）]，之后羟胺与氨氮反应生成肼 [式（24-10）]，而且催化亚硝酸盐氮还原的酶为细胞色素 c 型 hao 类蛋白[168]。次年，Lawson 等利用转录组测序技术发现该基因在 Ca. Brocadia 体内可被大量表达[165]，再次证明了 Ca. Brocadia 菌氮素代谢途径的特殊性。由于该代谢途径不产生一氧化氮这种中间代谢产物，不但降低了氮素代谢过程对 Anammox 细胞的毒性，提高其脱氮效率；而且进一步减少脱氮过程中释放的温室气体，真正实现高效减排。

$$1NH_4^+ + 1NO_2^- \longrightarrow 1N_2 + 2H_2O \qquad (24\text{-}5)$$

$$NO_2^- + 2H^+ + e^- \longrightarrow NO + H_2O \qquad (24\text{-}6)$$

$$NO + NH_4^+ + 2H^+ + 3e^- \longrightarrow N_2H_4 + H_2O \qquad (24\text{-}7)$$

$$N_2H_4 \longrightarrow N_2 + 4H^+ + 4e^- \qquad (24\text{-}8)$$

$$NO + 3H^+ + 3e^- \longrightarrow NH_2OH \qquad (24\text{-}9)$$

$$NH_2OH + NH_4^+ \longrightarrow N_2H_4 + H_2O + H^+ \qquad (24\text{-}10)$$

$$NO_2^- + 5H^+ + 4e^- \longrightarrow NH_2OH + H_2O \qquad (24\text{-}11)$$

联氨脱氢酶 hdh 蛋白的空间结构对于 Anammox 菌能够维持高效的氮素代谢具有重要意义。在 Anammox 菌氮素代谢过程中，肼氧化生成氮气的反应是释放低电位电子（−750 mV）和产生质子梯度及生成 ATP 的过程，对于 Anammox 菌的生长至关重要。由于催化该反应的 hdh 蛋白与 Anammox 菌小体在空间上间隔较远[169]，因此 hdh 蛋白不仅需要将该反应释放的电子捕获、储存，还需将电子精确地运送至 Anammox 菌小体膜上的醌池中以完成后续的氮素代谢。2019 年，Akram 等结合冷冻电子显微镜和晶体结构检测技术探明 hdh 蛋白复合体由 192 个细胞色素基团构成了横跨整个复合体的电子传输网络，这有利于增加电子在储存和传输过程中的间距，防止产生静电作用和电子损失；同时复合体上具有选择性的"过滤"结构，有利于电子被精确地传输至受体，该成果被刊登于 Science Advances[170] 杂志。这种巧夺天工的蛋白结构对于提高 Anammox 工艺的氮素污染物去除效率具有深远的意义。

2）Anammox 菌团聚机理

Anammox 菌具有强烈的聚集倾向[171]，这对于增强微生物的环境适应能力、降低其敏感性具有重要作用。胞外聚合物在这一团聚过程中起到了核心作用，主要体现在以下 3 个方面[172-174]：①藻酸盐类胞外多糖（polysaccharide, PS）呈凝胶状，形成有利于污泥团聚的交叉网络结构，触发污泥颗粒化进程；②胞外蛋白（protein, PN）具有大量电负性基团（羧基、羟基、偶氮基等）及疏水性官能团，促进污泥团聚并稳定团聚体结构；③高 PN/PS 比增强污泥的疏水性及絮凝能力。与其他污泥类型相比，Anammox

污泥的 PN 具有更大量的疏水性氨基酸及更松散的二级结构[175]，使污泥呈现出更为强烈的聚集趋势。

Anammox 菌经碳素代谢生成的有机碳一部分用于生长繁殖，另一部分被用于合成胞外聚合物。胞外聚合物的分泌对于 Anammox 菌群的团聚和 Anammox 菌的截留具有重要作用，因此对于 Anammox 菌群胞外聚合物的研究备受研究学者关注[173]。最初，关于 Anammox 菌群胞外聚合物的研究多侧重于其组分和相应的官能团与占比的检测[172, 173]，而对于不同组分的合成及其在胞外的代谢途径关注较少。2018 年，Zhao 等利用宏代谢组技术检测表明，Anammox 菌群胞外蛋白质主要由一些疏水性氨基酸的代谢形成，如苯丙氨酸、丙氨酸和亮氨酸等；胞外多糖的主要合成通路包括尿嘧啶核苷-N-乙酰葡糖胺和尿嘧啶核苷-N-半乳糖胺等核苷糖代谢途径；胞外脂质则主要由磷脂酰乙醇胺代谢形成[174]。同时，这些胞外聚合物的合成途径随着反应器启动阶段和氮素负荷变化及外源添加有机碳源或酰化高丝氨酸内酯类信号分子等操作条件的变化而发生改变[176, 177]。因此，探究 Anammox 菌胞外聚合物的合成途径及其在不同操作条件下的变化情况，可为 Anammox 工艺运行提供切实可行的操作策略，以提高菌群团聚和脱氮效率。

细菌利用信号分子来调节整个菌群的各种行为，而这些行为通常对整个菌群有利，并在一定条件下能够提高菌群的存活能力[178]。研究表明，面对不利环境时（低温、溶解氧、低 pH、底物浓度波动等），厌氧氨氧化菌胞内环二鸟苷酸含量的上调增强了其团聚性能[176]。近来，利用宏基因组学和多种生物信息学联用的技术手段分析发现，Anammox 菌群的胞外多聚物的形成是由多种细菌合作完成的，如 Anammox 菌代谢分泌的核苷糖在绿弯菌体内聚合为高聚糖并转运至胞外形成胞外多糖[177]。这些物质被分泌至胞外后作为伴生异养菌的碳源和电子供体，促进体系中硝酸盐氮被还原为亚硝酸盐氮，进一步被 Anammox 菌利用，这也有利于提高 Anammox 工艺的脱氮性能。

3）Anammox 技术工程应用

欧洲、亚洲和北美已有超过 200 个基于 Anammox 的污水处理工艺的工程案例[179]。到目前为止 Anammox 仍主要应用于污泥消化液和含高氨氮工业废水的处理，且技术已经比较成熟，在美国和欧洲有较多的实际工程案例，例如，奥地利 Strass 污水处理厂（污泥消化液）、荷兰 Olburgen 污水厂（马铃薯加工厂废水）等[180]。在我国，Anammox 工艺的工程应用也主要集中在高氨氮废水的处理，例如通辽梅花味精厂废水的处理和山东滨州安琪酵母废水的处理。

尽管基于 Anammox 的主流城镇污水处理工艺可在降低成本的同时为总氮排放提标提供可能性，但是城镇污水低氨氮以及冬季低温的特点仍是该工艺应用的主要挑战[178, 181]。新加坡樟宜污水厂水温常年在 30℃左右，所以难点只在进水低氨氮；而我国西安第四污水厂同时存在着低温（10~20℃）、低氨氮两个难点。因此，就主流 Anammox 工艺的应用来说，后者的运行机理及改造经验更值得探讨和研究。

综上所述，工业废水的处理采用 Anammox 工艺较为适宜，且在我国仍有较大的发展空间；主流 Anammox 工艺的应用仍需要进一步的探索和研究，新型水处理材料的开发、反应器构型的设计等可能是解决应用瓶颈的一些途径。

3.1.2 深度脱氮技术

过多的氮素排放到环境水体中，会引起水体的富营养化，对生态环境造成危害。随我国经济社会的发展，社会用水量逐年飙升，采用更严格的氮素排放标准有助于控制向环境中排放的氮素总量。《水污染防治行动计划》明确规定要强化城镇生活污染治理，加快城镇污水设施建设与改造。京津冀和太湖等区域也制定了更为严格的地方标准以控制污染物排放总量。目前城镇污水处理厂的污水经二级生化处理后，其一级 A 达标排放水中仍含有较高含量的总氮（TN），远远超出地表水 IV 类水体水质标准。鉴于生态环境的要求与排放标准的趋严，污水处理单位需要强化现有脱氮技术或对二级出水进行深度脱氮[182, 183]。

1）深度脱氮的瓶颈

污水深度脱氮通常指对城镇污水处理厂二级出水进一步处理，使出水中的总氮进一步降低以满足更严格的环境要求。污水处理厂二级出水具有低 COD、低总氮、低碳氮比和低浓度等特点，若利用传统生物技术对其进行深度脱氮则需要投加大量外加碳源，增加了操作成本，且易造成二次污染。新型深度脱氮技术的发展势在必行。

2）深度脱氮研究进展

（1）反硝化滤池。

反硝化滤池（DNBF）可以用于对污水处理厂二级出水进行生物脱氮，其脱氮容积负荷一般为 0.8～3.0 kg N/(m³·d)[184]。因其占地面积小、出水水质好、产污泥量少、有机物容积负荷高、处理效率高等特点成为近年来污水再生回用的研究热点。由于二级生化出水中底物的浓度较低，可生化性较差，一般需向 DNBF 中投加碳源。不同的碳源会导致 DNBF 中出现不同的优势菌属，进而影响整体脱氮效果。多数研究发现甲醇作为碳源时，滤池的挂膜效果和硝氮去除率较高，同时反冲洗时间更短、运行周期长，且投加量较少[185, 186]。亦有研究发现以乙酸盐为外加碳源的 DNBF 中，除磷效果更好[187]。

（2）固相反硝化。

固相反硝化是指向污水中投加不溶于水的固相碳源，使其作为生物膜载体并为反硝化细菌提供电子供体。适合处理碳氮比较低的二级出水。由于固态碳源需先被胞外酶分解为溶于水的小分子，而后才可被用生物膜中的反硝化菌用作电子供体，故可避免向水体中释放过量碳源，防止出水 COD 超标。固相反硝化的速率受主要受碳源类型、水力条件和操作参数的影响。目前常见的固态碳源主要有可生物降解聚合物和天然纤维素[188]。木屑作为固态碳源脱氮速率可达 1～10 g N/(m³·d)，且具有较长的生命周期（5～15 年），成本较低，是较优的固态碳源[189]。固相反硝化中的微生物大多数来自于 β 变形菌门中的 Comamonadaceae 科[190]。固相反硝化同时也存在 DOC 释放、氨氮生成、亚硝酸盐积累和温室气体排放等缺点。其对 PPCPs 的去除也是今后研究的一个方向。

（3）自养反硝化。

自养反硝化细菌利用无机物作为电子供体，以无机碳作为碳源进行反硝化过程。如图 24-16 所示，与异养反硝化相比，自养反硝化不需要提供有机电子供体，降低了操作成本，同时污泥产量低，具有广阔的应能用前景[191]。

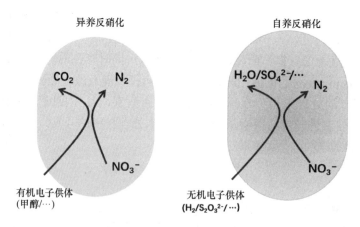

图 24-16　异养反硝化与自养反硝化

在缺氧条件下，氢自养反硝化微生物能够以氢气为电子供体进行反硝化过程。相对于直接向污水中鼓泡曝氢气，氢基质膜生物膜反应器（MBfR）利用中空纤维膜，通过膜曝气的方式向水中输送氢气，氢气被附着在纤维膜外表面的生物膜所利用驱动反硝化进行。MBfR 的氢气传输效率高且过程更为安全[192]。目前已有应用氢基质 MBfR 对地下水进行深度脱氮处理的工程案例，处理效果与效益分析均优于传统的离子交换法[193]。利用 MBfR，赵和平团队进一步研究了基于甲烷的反硝化技术。研究发现以甲烷为外加碳源时，MBfR不仅可以高效去除硝酸盐，还可以有效还原废水中的氧化态重金属离子（铬酸盐、硒酸盐等），使其转化为更易沉淀的形式从污水中去除[194,195]。

硫细菌可以利用无机碳源，以低价态含硫化合物进行自养反硝化。袁莹等[196]发现采用不同形式硫化物对低浓度硝酸盐废水进行反硝化时，$Na_2S_2O_3$系统的脱氮效果最优，Na_2S系统的脱氮效果最差。硫自养反硝化的速率和异养反硝化相当，但硫化物溶解度低，反硝化过程产生硫酸盐污染以及碱度消耗等问题限制了硫自养反硝化在实际工程中的应用。除硫细菌外，铁型反硝化菌也可以在厌氧条件下，利用亚铁离子或零价铁作为电子供体，将硝酸盐或亚硝酸盐转化为气态氮化物。工业废水中常含有亚铁离子，利用铁型反硝化菌进行脱氮可以实现"以废治废"，但该技术现在尚处于研发阶段，需要解决菌源与运行参数的优化、三价铁表面堆积影响微生物活性等问题[197]。

微藻以光合自养方式生长，过程无需有机碳源，适合二级出水的脱氮处理。微藻深度脱氮面临的主要挑战是固液分离问题。当微藻以悬浮方式生长时，反应器内的藻体浓度往往较低，整体的沉降性能较差，造成出水的SS偏高，且微藻死亡后仍会向水中释放大量氨氮，造成二次污染。此外，使用微藻进行深度脱氮还面临占地面积大、操作成本高等问题。这些问题有望通过发展出更为高效和紧凑的反应器如膜光生物反应器等加以解决[198,199]。

（4）新兴技术。

反渗透（RO）：工业污水及垃圾渗滤液处理中，反渗透与前端生物处理工艺相结合时，可将生化处理出水中的氨氮进一步降低。反渗透产水可以回流至前段工艺进行处理。万金保等[200]使用氧化沟-反渗透工艺处理垃圾渗滤液，发现末端反渗透处理系统可以将氨氮浓度从二沉池出水中的 77 mg/L 降至 9.1 mg/L。经RO工艺处理过的出水基本可以满足中水回用标准，能够产生一定的经济效益。

厌氧氨氧化：短程硝化-厌氧氨氧化工艺中，氨氧化细菌（AOB）在好氧环境下将氨氮氧化为亚硝酸盐。厌氧环境下，厌氧氨氧化菌以二氧化碳为碳源，以亚硝酸盐为电子受体，氨为电子供体生成氮气，进而将氮素从污水中清除。由于采用短程硝化，运行费用可降低90%，节省占地50%，且整个过程无需额外投加碳源，污泥产量也大大降低[201]。该工艺的核心步骤是实现亚硝酸盐在反应器内的累积，目前主要通过淘汰系统中亚硝酸盐氧化菌（NOB）和富集氨氧化菌（AOB）的方式实现。但厌氧氨氧化应用于二级出水的深度处理会面临氨氮负荷较低，导致厌氧氨氧化菌难以富集的问题。

3）小结

总体上看，污水处理厂出水的总氮低、碳氮比也较低，对这些对深度脱氮均产生了不利影响。异养反硝化速率较快，但外加碳源的投入大大增加了操作成本。自养反硝化不需要有机碳源，但较低的总氮浓度往往会使微生物难以富集。从以上两个角度看，强化现有二级处理的脱氮工艺，将深度脱氮处理放置于前端或许是成本上较优的选择。

3.1.3　生物除磷技术

强化生物除磷（EBPR）是目前城市污水处理厂中广泛采用的除磷工艺。作为微生物驱动的过程，EBPR可以很好的融入城市污水生物处理工艺中，利用城市污水自身含有的有机物作为碳源实现除磷过程；聚磷菌（PAOs）对污水中磷酸盐的富集作用，也为后续的磷回收创造了条件[202,203]。本小节对EBPR系统中关键功能微生物、碳源需求以及最近发展的一些新工艺模式进行简要综述。

1）EBPR系统中的关键功能微生物

Ca. Accumulibacter是公认的、在EBPR过程中起关键作用的PAOs[204,205]。*Ca.* Accumulibacter以小分子挥发性脂肪酸（VFAs）作为优先碳源，厌氧条件下通过分解胞内多聚磷酸盐摄取VFAs并将其转化为聚羟基脂肪酸酯（PHA）储于胞内，好氧条件下降解PHA，过量摄取磷酸盐合成多聚磷酸盐并完成细胞增殖。目前，对EBPR过程机制的认识很大程度上建立在对*Ca.* Accumulibacter代谢特征研究的基础

上[205]。迄今发现的 *Ca.* Accumulibacter 依据其聚磷酸盐激酶基因（*ppk1*）序列不同，划分为两个主进化枝（clade I 和 II）和 14 个分进化枝（IA-IE 以及 IIA-III），不同进化枝的 *Ca.* Accumulibacter 表现出碳源利用及反硝化能力上的差异[206, 207]。

近年来，*Tetrasphaera* 作为新型 PAOs 被发现在丹麦等地的污水处理厂中具有比 *Ca.* Accumulibacter 更高的丰度[204, 209]。与 *Ca.* Accumulibacter 不同，*Tetrasphaera* 可以利用氨基酸和糖类，不合成 PHA，并且可以在厌氧条件下发酵氨基酸和糖类，并可通过发酵作用获得能量厌氧吸磷[209]。最近的研究显示，在丹麦 6 个污水处理厂中 *Tetrasphaera* 对于除磷的贡献超过了 *Ca.* Accumulibacter[208]。其在世界其他地区 EBPR 系统中的普遍性和关键作用有待于验证。

聚糖菌（GAOs）是 EBPR 系统中另一类关键微生物，它们在厌氧条件下分解糖原产生能量，摄取 VFAs，将其转化为 PHA 储存于胞内，好氧条件下降解 PHA 用于细胞增殖及糖原再生。GAOs 不聚磷，因而对除磷无贡献，但它们与 PAOs 竞争碳源，是 EBPR 系统中需要进行控制的微生物类群[202, 203]。目前，对于 GAOs 的研究多集中在 *Ca.* Competibacter 和 *Defluviicoccus*，它们也是实际污水处理系统中常见的 GAOs。两类 GAOs 存在碳源利用与摄取机制上的差异，*Ca.* Competibacter 不能高效摄取丙酸，*Defluviicoccus* 则具有较弱的乙酸摄取能力。*Ca.* Competibacter 借助氢离子电势进行 VFAs 的摄取，而 *Defluviicoccus* 更多的依靠自由扩散。通过优化碳源供应，可以实现对 *Ca.* Competibacter 和 *Defluviicoccus* 的控制[202]。在实际污水处理系统中 GAOs 与 PAOs 可以长期共存，真实环境下 PAOs 和 GAOs 之间的相互作用关系有待于更深入的研究[209]。

现在分子生物技术的进步为 EBPR 工艺中的关键功能微生物的研究提供了强大的工具，以 *Ca.* Accumulibacter 为例，通过高通量测序，目前已获得 1 个全基因组图谱和 10 余个全基因草图，为从基因层面更好地认识它们的代谢潜能提供了依据。然而，最近的研究[208]发现，在实际污水处理系统中，高达 70%的除磷作用可能源于除 *Ca.* Accumulibacter 和 *Tetrasphaera* 之外的 PAOs 的贡献。这也暗示了对于实际 EBPR 系统中未知 PAOs 的识别鉴定及其生理生化特征的研究方面还有很多的工作需要开展。

2）EBPR 的碳源需求

有机碳源是影响 EBPR 工艺稳定性的重要因素，VFAs 作为 *Ca.* Accumulibacter 的优先碳源在 EBPR 系统中受到重点关注。根据 *Ca.* Accumulibacter 的生化定量模型，每去除 1 mg 的磷，需要消耗 24 mg 乙酸。为保证 EBPR 除磷效果，一般要求进水 COD/P>30，具体数值依据污水组成以及工艺布局和运行方式的不同而存在差异。在污水碳源不足的条件下，通过补充乙酸等 VFAs 是增强 EBPR 除磷效果的有效方式。但从经济性考虑，需要寻找更加廉价易得的碳源，通过剩余污泥或餐厨垃圾等的可控发酵来生产 VFAs 作为 EBPR 的碳源补充是非常具有的应用前景的方式[210]。近年来，利用乙醇、葡萄糖、氨基酸、甘油等作为碳源的 EBPR 系统的可行性在实验室中也得到了证实，显示了这些非 VFAs 类的有机物作为 EBPR 补充碳源的潜力[211]。此外，对实际污水处理厂污泥的研究显示其中的 PAOs 可以利用除 VFAs 之外的多种碳源，即使 *Ca.* Accumulibacter 也表现出碳源利用方面高度的多样性[212]。污水中除了 VFAs 之外的有机物对于 EBPR 的贡献应予以更多的关注。

3）EBPR 的新工艺形式

随着污水处理系统节能降耗和物质能源回收越来越受到重视，近年来，国内外研究者也提出了 EBPR 的一些新工艺形式，为降低 EBPR 过程能源、碳源消耗，更好的耦合其他污染物的去除，以及现有 EBPR 工艺的升级改造提供了思路。

最近，Carvalho 等[213]报道了一种光自养辅助的 EBPR 系统（photo-EBPR），通过光合细菌和聚磷菌的复合，在光照条件下，无需曝气实现了除磷，此外发现，磷酸盐摄取不完全依赖于 PHA 的降解，其中的 PAOs 可能直接从光照中获取了能量。Oyserman 等[213]也报道了类似的光合 EBPR 系统，利用藻类的光合作用为聚磷和硝化过程提供氧气，这些系统有潜力节省曝气的能源消耗，而且有望减少碳源需求。

针对高硫酸盐污水，陈光浩课题组近年开发了一种硫循环协同生物除磷系统（S-EBPR），厌氧条件下硫酸盐被还原至单质硫/多聚硫，好氧条件下还原性硫单质氧化产生能量供给磷酸盐摄取，区别于常规

EBPR系统，S-EBPR系统的磷循环过程很可能是由一些新型的硫代谢相关PAOs所介导[215]。Rubio-Rincon等[15]在高进水硫离子浓度条件下富集了一种混合营养PAO：*Thiothrix caldifontis*，该菌可以在厌氧条件下聚硫，并在好氧条件下将硫氧化为硫酸盐获取能量实现聚磷，显示了利用无机硫取代有机碳源实现EBPR的潜力。

针对低碳源条件下，彭永臻课题组构建一系列的反硝化除磷新系统，其中包括基于双污泥系统的A^2/O-BAF[217]，可以实现反硝化除磷率97%以上；以及基于后置内源反硝化除磷的AOA系统[218]，可以在C/N=4.4的条件下实现出水总无机氮和磷酸盐分别低于3 mg/L和0.2 mg/L。这些研究为实现低碳源条件下高效的同步脱氮除磷提供了方案。

与废水能源回收的结合方面，近年来，一种利用反硝化产N_2O用于能量回收的CANDO工艺被提出，Gao等[219]利用反硝化除磷菌群来驱动亚硝酸盐还原，将亚硝酸盐还原产N_2O与EBPR过程结合实现同步除磷。除了优势PAOs：*Ca.* Accumulibacter clade IA，菌群中的其他具有反硝化基因缺陷的微生物对N_2O积累具有贡献。

此外，在运行模式方面，郝晓地等[220]提出通过在厌氧池提取富磷上清液进行磷酸盐的沉淀回收的方式，不但可以削减后续好氧过程中磷酸盐负荷，而且有利于节省碳源消耗，对于低碳源污水的除磷，以及污水厂的升级改造都是值得考虑的方案。通过减少好氧过程的曝气也可实现对能耗直接削减，Keene等[221]在中试级别的UCT工艺中考察了低溶氧条件下的除磷效率，在平均溶解氧低至0.33 mg/L条件下，除磷效率保持在了90%以上，这与*Ca.* Accumulibacter高的氧亲和能力有关。

4）展望

随着我国污水排放标准的不断提高，城市污水厂面临越来越严格的出水总磷和磷酸盐的排放标准。这对EBPR的工艺稳定性及除磷效能的保障提出了更高的要求。对于实际EBPR系统中关键功能菌特别是目前未知的PAOs和GAOs的识别及其生化特征的研究将为更好地把握EBPR过程机制及其关键控制因素提供依据。我国城市污水有机物浓度普遍偏低，有限碳源条件下的高效生物强化除磷技术的开发是研究重点。此外，随着污水处理理念逐渐从污染物的去除向资源、能源回收转变，如何平衡污水碳源回收与EBPR碳源需求之间的矛盾也是重要的研究课题。

3.2 废水处理实践

3.2.1 城市污水

随着我国环境保护事业的深入开展，国家对城市污水处理工作的总体要求逐步提高，排放标准日益严格，管理水平稳步提升。面向生态文明建设重大战略要求，我国水处理领域的专家、学者和工业界人士通力合作，取得了丰硕的研究成果。本节仅就主流厌氧氨氧化、污水处理系统微生物组、基于人工智能的污水处理技术三个热点研究方向做简要论述，以期为污水处理工程实践提供一定的思考。

1）主流厌氧氨氧化

厌氧氨氧化技术是一种新型污水脱氮技术。厌氧氨氧化细菌能够在厌氧条件下以铵盐为电子供体、以亚硝酸盐为电子受体进行生长和代谢。由于厌氧氨氧化细菌的化能自养性和厌氧性，厌氧氨氧化工艺相比传统生物脱氮技术更为经济。全球大型厌氧氨氧化工程已经超过100座，其中75%用于处理城市污水。目前这一领域的代表性研究机构和商业公司包括：荷兰代尔夫特理工大学、新加坡公用事业局、瑞士EAWAG公司、哈尔滨工业大学、北京工业大学、浙江大学、美国DC Water公司、荷兰帕克公司、威立雅公司、瑞士EssDe公司等。

近年来，主流厌氧氨氧化概念被广泛讨论和实践。该技术能够在较低进水氮浓度（20~75mg/L）和较低温度（10～28℃）等不利条件下在污水厂的主流实现碳、氮同步脱除[222,223]。技术的核心功能源于厌氧氨氧化细菌同氨氧化细菌、异养菌和除磷菌等功能菌群的微生态平衡和协同作用[224]。目前国外至少有六家城市污水处理厂在积极尝试升级到主流厌氧氨氧化，包括荷兰 Dokhaven 污水处理厂、奥地利 Strass 污水处理厂、瑞士 Glarnerland 污水处理厂、新加坡樟宜污水处理厂、美国 Chesapeake-Elizabeth 污水处理厂、美国 Blue Plains 污水处理厂。

如何长期稳定运行主流厌氧氨氧化是亟待解决的难题。从微生物生理生态学角度，由于厌氧氨氧化细菌生长非常缓慢且对温度、DO 等环境因子敏感，在复杂群落中难以与异养细菌、亚硝酸盐氧化细菌等竞争，从而导致菌群"适配性"不强，这是主流厌氧氨氧化难以成功的根本原因。针对这一问题，我国学者已经开展了一系列尝试。例如，提出并实现了短程反硝化-厌氧氨氧化协同脱氮[225]，发现了某些厌氧氨氧化细菌能够氧化小分子有机物[226]，借助 MBfR 反应器实现厌氧氨氧化细菌-反硝化厌氧甲烷氧化细菌协同脱氮[227]。

未来的研究可以从解决菌群的"适配性"入手，开发新型厌氧氨氧化工艺并建立相应的运行操作方式：①发现或改造厌氧氨氧化菌的新型代谢途径；②利用新技术实现反应器内微环境的精准调控（pH、温度、SRT、DO 等）；③结合生物学和电化学原理，开创全新的工艺型式实现协同脱氮除碳除磷。

2）污水处理微生物组

城市污水处理厂的活性污泥微生物组研究对探索微生物生态学原理和加深对污水处理工艺的认知具有重要意义。随着基因测序技术的不断升级，该方法越来越多地应用于分析城市污水厂活性污泥的微生物组，为环境化学、环境工程、微生物生态学提供了崭新的研究思路。

近年来主要的研究进展包括三个方面。第一，好氧工艺的微生物组特征。由我国学者主导的全球城市污水处理厂微生物组研究揭示了全球污水处理厂的微生物群落多样性，挖掘了背后的群落构建机制。该研究通过系统抽样方法分析了全球六大洲 23 个国家 269 个污水厂近 1200 个活性污泥样品的 16S rDNA 序列。研究发现，组成细菌群落的 28 个核心 OTU 与活性污泥的功能密切相关。研究证实，活性污泥的细菌群落未表现出纬度梯度差异性，其空间周转率与研究的尺度有关[228]。第二，厌氧工艺的微生物组特征。由我国学者参与的全球厌氧消化微生物组研究分析了来自 5 个国家 51 个城市污水处理厂 90 座大型厌氧消化工程的 148 个污泥样本。研究发现，塑造厌氧消化微生物组结构的关键驱动因素是预处理、温度和盐度等运行条件，而地理位置并不是主要驱动因素[229]。第三，抗生素耐药细菌和抗性基因分布特征。污水处理厂是抗生素耐药细菌和抗生素抗性基因的热点。污水处理厂作为一个连接人类生活和自然系统的关键节点，具有丰富的微生物遗传多样性。研究证实，污水中残留的抗生素及其他共选择筛选压力对抗生素抗性基因的水平转移起到了推动作用[230]。

随着基因组学、转录组学、蛋白质组学和代谢组学领域技术的不断突破，预期在未来这些新型研究手段将大量投入到污水处理厂微生物组研究中，帮助人们加深对生物处理过程的理解，改善工艺的性能。

3）基于人工智能的污水处理技术

活性污泥法生物处理技术是世界上最普遍应用的污水处理技术[231]。随着我国现代化进程的不断加快，我国城市污水处理正在以前所未有的速度发展和扩大。据统计，2004～2014 年，全国废水排放总量由 482.4 亿 t 增至 716.18 亿 t，增幅高达 48.46%。2009～2015 年，我国城镇污水处理厂数量由 1878 座增至 3542 座，年复合增长率达到 11.25%。按照总投资统计，国内外污水厂对污泥的处理处置费用占据了总运行费用的 60%～70%[232,233]。随着城市生活污水及工业废水处理量的增加、处理排放标准的进一步提高，随之产生的城市剩余污泥量也越来越大[234,235]。预计到 2020 年，我国的市政污泥产量将达到 6000～9000 万 t。由此可见，我国城市污水的处理处置的问题已十分突出，开发并建立稳定高效的城市污水处理技术势必成为未来污水厂新建及改造的重点和难点。开发新型或改进的强化污水处理的智能优化控制生物污水处理系统，对未来污水生物处理技术的发展应用具有重要的指导意义和巨大的工程应用价值。

机器学习（machine learning）从本质上是一个多学科的领域。它吸取了人工智能、概率统计、计算复杂性理论、控制论、信息论等多学科的成果[236]。机器学习这门学科所关注的问题是：计算机程序如何随着经验积累自动提高性能？近年来，从智能机器人、自驾驶汽车，到检测信用卡交易欺诈的数据挖掘程序等领域都体现了机器学习的成功应用[237]。在未来，运用机器学习的方法，计算机能从医疗记录中学习，获取治疗新疾病的最有效方法；住宅管理系统分析住户的用电模式，以降低能源消耗；个人软件助理跟踪用户的兴趣，并为其选择最感兴趣的在线新闻。由此可见，运用机器学习的方法，计算机便可以同时从污水厂几十年以及实时在线的数据中，获得自动控制的最佳控制策略，从而真正实现污水处理过程及污水厂运行的智能化调控。近年来，机器学习相关算法已成功应用于如传统厌氧-缺氧-好氧（A^2O）工艺过程参数优化设置[238]、基于机器学习的污水厂能耗评估[239]、基于人工智能的污水厂效能预测[240]等方面的研究。可见，基于机器学习算法的优点，开展污水处理系统的智能化控制研究，可以有效解决具有非线性、多元耦合影响、以及控制效果的非时效性为特征系统的问题。基于机器学习的污水处理智能调控可以有效实现新型/改进污水处理工艺在实际污水处理厂的稳定运行，从而实现如不同反应工艺、不同规模、以及不同污水处理工况下的稳定、高效运行，旨在实现污水处理过程的实时优化控制、预测准备、应对突发危机、自我修复等智能化调控能力，以期全面实现现有污水处理过程的智能化控制。

3.2.2 难降解工业废水

根据国家统计局和环境保护部出版的《中国环境统计年鉴2016》[241]统计数据，我国污水及废水排放总量达到735.3亿t，工业废水排放量199.5亿t，占排放总量的27.1%，其中，化学需氧量（COD）排放总量为1046.5万t，NH_4^+-N排放总量为141.8万t。污染不仅降低水的使用功能价值，还改变生态系统，使人类付出代价。未来，水资源稀缺将是我国化工、制药、纺织、采矿和能源等行业面临的最大挑战之一。工业领域普遍存在工艺流程长、所用原材料种类多、多介质转化/交换的过程，所排放出的废水具有成分复杂、污染物浓度高、毒害性大以及可生化性差等特点，表现为含有潜热、稀相/多相共存及具有回收价值的特征。

燃料化工、造纸化工、印染化工及制药化工行业的废水排放量分别占工业废水排放总量的18.0%、13.1%、12.5%、3.0%。四个行业的排放量加和占我国工业废水总排放量的46.6%，其COD和NH_4^+-N的排放量分别占工业废水总排放量的37.7%和32.7%。表24-2从COD、BOD、B/C值（BOD_5/COD）、NH_4^+-N、TP等方面统计了制革、造纸等10个行业的废水基本水质数据，表明行业之间的废水差异明显。

表24-2 高强度排放一些行业的废水特征[242-247]

废水类型	COD (mg/L)	BOD_5 (mg/L)	B/C	NH_4^+-N (mg/L)	SS (mg/L)	PO_4^{3-} (mg/L)
造纸废水	2468~3580	516~720	0.201~0.211	35~46	500~1500	—
纺织废水	2000~4000	200~800	0.125~0.2	4~8	—	2
印染废水	400~3000	100~800	0.2~0.26	100	100~200	0.2
焦化废水	1000~8000	230~2050	0.23~0.250	250~500	—	0.1~0.3
制药废水	12000~15000	4700	0.512	—	1700	
制革废水	8000~16000	2000~5000	0.313	450		
淀粉废水	35000	16000	0.457		13300	
乳品废水	3500	2200	0.629	120	—	
饮料废水	1800	1000	0.556	—	—	

工业废水承载的经济负荷大，工业门类多以及工业废水组成的复杂性，面临着资源、能耗与风险之间的多元矛盾[248, 249]。我国每年用于废水处理的含氮、硫、磷化合物及酸碱量非常巨大，能耗总量达到

320亿～588亿 kW·h，吨纸生产用水量为100～150 t，吨钢生产用水量为5～20 t，单位GDP的废水排放量达到9.6 m³/万元。

(1) 燃料化工废水。燃料化工主要包括煤制焦、煤制油、煤制气、油制气的化工工艺过程。由于选用的基本原材料主要来自于煤和石油等，因而众多元素存在于生产过程中，且普遍的燃料化工过程均涉及高温反应，多种金属元素共存尤其是富于价态变化的过渡金属的催化作用，使燃料化工废水呈现出组分复杂、浓度高、难处理的特征[250]。普遍的燃料化工废水中以酚类、苯类为代表的有机物其COD值处于3000～30000 mg/L范围内[251,252]，废水中含有很高浓度的含氮化合物（NH_3、SCN^-、CN^-、NO_2^-、NO_3^-、有机氮），普遍能检测到杂环芳烃、多环芳烃之类的持久性有机污染物（POPs），使废水表现出综合毒性效应。燃料化工过程从原材料复配、催化转化/合成裂解到产品纯化的生产过程中都涉及各种化工原料的投加，废水中盐分的含量很高。

(2) 造纸化工废水。主要包括制浆工段、废纸脱墨、纤维合成、染色漂白等生产工艺排放出来的液体/废液。其中的主要污染物包括木素、半纤维素、细小纤维、悬浮物、油墨、染料、溶解有机污染物和无机污染物等[253,254]。废水的COD值处在300～2000 mg/L的普遍范围内，TN达到30～50 mg/L，TP为1.0～17.4 mg/L，SS为100～500 mg/L。造纸行业的废水表现为排放量大、SS高、B/C值低、色度大、缺氮缺磷、消毒前驱物丰富等特征。

(3) 印染化工废水。包括染料化工生产过程中产生的染料废水和印染过程中产生的染整废水。随着科技进步，染料的合成工艺突飞猛进，目前染料种类超过千种，中国已经是世界上第一位的染料生产大国，其中合成有机染料出口量达到25.9万 t（国家统计局2016年数据）。为了加强染料分子的显色基团如—N=N—、—N=O、—SO_3Na、—OH、—NH_2等与纤维分子之间的反应，提高上染率，除了要高温条件下进行外，还需要用到各种助剂和盐分，从而使得染料化工废水表现出浓度高、组成复杂、显色度、高盐分、难降解以及有毒有害的特征，并且混有各类卤代物、苯胺、酚类、氯酚（CPs）、单环苯烃（MAHs）、多环芳烃（PAHs）等各种持久性有机污染物。

(4) 制药化工废水。制药工业包括医药、农药、兽药等行业的生产过程当中排放的废水，涉及化工制药、生物制药、中药合成、分离提取不同的领域与工段。因药物产品丰富、生产工艺条件而差异性较大，具有污染物浓度高，COD达到5000～20000 mg/L，BOD达到2000～10000 mg/L，SS达到5000～23000 mg/L，TN达到600～1000 mg/L，还有氰化物、抗生素、部分重金属等有害物质。因此，制药化工废水[255,256]也表现为有机物浓度的巨大差异、分子结构复杂、高盐分、环境影响敏感等特征。

(5) 冶炼废水。冶炼行业主要包括铅锌冶炼、稀土冶炼及少量的稀贵金属回收冶炼。铅锌火法冶炼废水主要是烧结、熔炼、制酸、煤气炉等生产工艺产生的废水以及循环冷却水，湿法冶炼废水主要是浸出、萃取分离、电解等生产工艺产生的废水。铅锌冶炼废水的产量约为金属产品量的5.5～6.1倍，即每生产1 t金属，约产生废水5.5～6.1 m³。表24-3为典型铅锌冶炼废水的水质，铅锌冶炼废水重金属含量高，Zn^{2+}、Cd^{2+}的含量分别达到152 mg/L及60 mg/L以上；含盐量高，Na^+、Cl^-、SO_4^{2-}的含量分别达到3700 mg/L、5500 mg/L和1000 mg/L；铅锌冶炼废水的有机物浓度相对较低，COD及BOD分别为460 mg/L和140 mg/L[257]。

表24-3 典型铅锌冶炼厂废水水质

	pH	Cond. (mS/cm)	Ca^{2+} (g/L)	Cl^- (g/L)	CO_3^{2-} (g/L)	SO_4^{2-} (g/L)	F^- (mg/L)	Zn (mg/L)	Cd (mg/L)	Pb (mg/L)	COD (mg/L)
	8.72	7.33	0.31	0.51	0.026	3.05	11.2	10.59	56.48	6.51	112
铅锌冶炼废水	6.13	7.82	nd	0.11	nd	2.75	nd	1500	0.51	0.33	65
	6.9	13	1.0	5.5	nd	0.9		152	60	1.0	460

在稀土元素分离和纯化的过程中所排放的废水,除了重金属成分以外,还含有氟化物、氨氮、放射性元素等污染物。据统计,每吨稀土矿的生产工序排放废水均超过 10 m³ [258]。稀土生产废水主要包括稀土皂废水、碱皂废水、碳铵沉淀母液、碳铵沉淀洗涤废水、草酸沉淀母液和草酸沉淀洗涤废水等,各工艺段所产生的废水性质差异大,需对各工艺段废水的水质进行单独分析,通过特征的识别寻求合理的处理工艺。表 24-4 为典型稀土生产企业的工艺废水水质。

表 24-4 典型稀土工艺废水主要水质 (单位:mg/L,pH 除外)

废水种类	稀土皂废水	碱皂废水	碳铵沉淀母液	碳铵沉淀洗涤废水	草酸沉淀母液	草酸沉淀洗涤废水
水量(m³/d)	226.0	82.5	80.5	112.7	71.0	236.0
pH	0.48	6.48	6.34	6.67	0.17	1.46
TDS	356000	44100	101000	35000	375000	14100
Cl^-	96800	10998	27730	8930	37130	1739
NH_3-N	50500	5424	13152	3269	356	1.2
TOC	173	494	8.6	3.4	1430	43.9
COD	287	1032	14.2	nd	258	nd
F^-	0.14	nd	0.68	nd	0.05	nd
Cu	3.68	5.20	2.74	0.38	26.45	0.75
Zn	16.55	24.40	55.94	1.44	152.1	1.86
Pb	67.82	1.22	0.13	0.04	1.31	nd
Mn	23.91	2.34	0.54	0.61	4.19	0.051
Cd	0.19	0.14	0.062	nd	nd	nd
Ni	2.12	3.72	2.07	0.22	18.44	0.35
Cr	1.64	nd	2.05	0.16	16.74	0.22
As	0.15	0.085	nd	nd	nd	nd
Hg	0.03	nd	0.006	nd	nd	nd

稀贵金属冶炼废水多以强酸性、高盐度以及含重金属离子等特点为主,废水 COD 浓度约为 200 mg/L,对 COD 有贡献的物质主要包括胺和酰胺类等难降解的有机物[259]。认为冶炼废水普遍具有重金属含量高,毒性效应普遍,有机物难以降解,氮素与盐分多态化等的特征。

(6)矿山废水。矿山废水主要来源于采掘、破碎、球磨、选矿等生产工段,废水性质与矿种有关,表 24-5 为典型铅锌矿山企业生产废水的水质。可见,废水以高重金属含量、高盐度为主要特征,废水中的 COD 范围为 314~642 mg/L。

表 24-5 典型铅锌矿生产工艺废水水质 (单位:mg/L)

	pH	COD	TOC	Na	K	Ca	Fe	Pb	Zn
硫尾水	6.85	642.24	15.63	32.17	122.98	809.4	0.353	4.45	7.27
铅锌废水	12.19	367.24	19.65	13.39	43.96	578.3	0.197	1.44	nd
硫精废水	8.04	314.65	35.98	15.32	51.91	664.8	0.119	0.86	nd
	Cu	Cr	Cd	Mn	F^-	Cl^-	SO_4^{2-}	NO_3^-	NO_2^-
硫尾水	nd	nd	nd	2.17	0.46	13.23	684.70	30.80	nd
铅锌废水	nd	nd	nd	0.66	3.95	65.19	17.50	nd	
硫精废水	nd	nd	nd	0.26	nd	2.51	551.56	13.75	nd

矿种不同，选矿废水性质差异较大。目前，大部分的矿山废水仍主要随尾矿输送至尾矿库中，部分废水在尾矿库澄清后回用至主生产流程，其余部分在处理达标后随地表径流排放出尾矿库库区。尾矿库外排水即从矿山尾矿库中经地表径流排出库区的废水，由于尾矿在堆积过程中常伴有自然风化及微生物氧化的过程，尾矿中的元素硫被氧化为 SO_4^{2-}，从而导致外排水的酸性增强。

典型工业废水普遍具有有机物浓度高、营养元素失衡、毒性效应普遍、氮素与盐分多态化等特征，除此之外，工业废水中还隐藏着技术、资源、能量、污染、管理等信息，其处理技术应该考虑各种信息的相互作用。针对上述典型工业废水，我国目前缺乏成熟实用的综合性水质指标/指数的指导与应用，如废水中高能量物质的判断、营养成分/组成的判断、可资源利用盐分化合物途径的判断等，使废水原水的资源属性、废水处理方法的有效性以及废水处理工艺技术的合理性/经济性缺乏科学依据，困扰着我国走生态型工业发展的道路。

1) 工业废水处理生化技术

(1) 生化处理技术。

目前，要对废水生化处理技术下一个明确定义是非常困难的。根据电子供/受体的不同，生物处理工艺可以通过厌氧、缺氧/水解、好氧等若干单元组成，通过单元组合与条件控制可以实现除碳、脱氮、氨化、硝化、反硝化、脱硫、除磷、产酸、产碱、减毒等目标。典型工业废水中含有 100～1000 mg/L（总氮计）浓度范围的含氮化合物，多以 NH_4^+、SCN^-、CN^-、NO_2^-、NO_3^- 形式存在，如焦化废水的组成约为 COD：TN：TP=4000：400：0.1（不同企业生产工艺，数据会在一定范围内变化，不作具体分析）；如果 B/C(BOD_5/COD) 取值 0.5（包含了还原性无机物氧化与脱氮的耗氧量），则 BOD：TN：TP=100：20：0.005。由此认为，这类废水表现出富氮缺磷的水质特征，硫化物也普遍存在。因此，不仅显示出达标的生物脱氮技术具有挑战性，在资源/能源提供以及盐分的生成/归趋方面也必须同步考虑。

(2) 生物脱氮方法原理。

生物脱氮的方法学研究是目前废水处理研究中最活跃的领域之一，原因是氮元素进入环境后所发生的水体富营养化将会带来严重的后果[260]，氮元素又是农业生产过程中的营养元素，氮元素可与氢、碳、氧、硫等构成各种含氮化合物。含氮化合物的氮素一般以还原价态（-3）表现出最强还原性，这类化合物有 NH_4^+、CN^-、SCN^- 及溶解性有机氮（DON）等，普遍存在酸碱平衡，表现为弱酸或弱碱的分子形式，在适宜的 pH 条件以及热作用的存在下，分子化合物可以从水相中转移到气相实现相分离。

氮的生物转化是一个高度复杂的过程，展现出参与的微生物众多，酶的类型众多，产物及中间产物富于变化，受控于电子转移的不同表现形式。空气中的 N_2 通过固氮菌的作用将气态氮转化为 NH_4^+-N，继而被生物同化吸收转化为 ON；ON 经微生物的氨化作用转化为 NH_4^+-N；氨氮经氨氧化菌的作用转化为 NO_2^-，NO_2^- 经硝化菌的作用转化为 NO_3^-；NO_3^-/NO_2^- 再经反硝化菌或厌氧氨氧化菌的作用转化为 N_2 或 N_2O。N_2、NO、N_2O、NH_3 从水相释放到大气中实现了氮素的地球化学循环[261]。

图 24-17 展示了环境中微生物对氮循环转化反应的丰富性。在工业废水处理中，微生物对氮元素的转化和脱除通过氨氧化菌、亚硝酸盐氧化菌、硝酸盐还原菌，厌氧氨氧化菌等的共同作用来实现。

最近发现的一些氮循环中的微生物，如可以使氨氧化到亚硝酸盐的古生菌和彻底氨氧化为硝酸盐的全程硝化菌，在废水处理系统中被发现，但它们在这些系统中所扮演的角色还有待于更进一步的明确。在氧气受限制的脱氮系统中可能存在更加丰富的微生物劳动分工[263]。在过去的十年里，许多由微生物引起的新的氮转化反应和途径被发现，然而，仍然存在许多反应不能由已知的酶催化来解释，尽管宏观热力学上可行，对此，可能存在迄今为止未被发现的生物化学催化剂。例如，依赖硝酸盐的氨氧化作用不能通过厌氧氨氧化途径来进行，因为氨盐需要首先被氧化成中间产物羟胺或者类似的含氧物质。

$$5NH_4^+ + 3NO_3^- \longrightarrow 4N_2 + 9H_2O + 2H^+ \quad (\Delta G^0 = -297 \text{ kJ·mol}^{-1} \text{ } NH_4^+) \qquad (24\text{-}12)$$

相似地，激活氨盐新的生化途径依赖于铁和锰的生物催化作用：

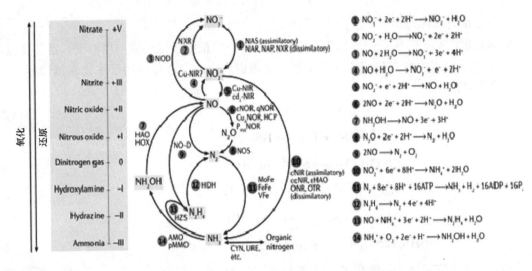

图 24-17 环境中微生物对氮循环转化的反应[262]

$$NH_4^+ + 6Fe^{3+} + 2H_2O \longrightarrow NO_2^- + 6Fe^{2+} + 8H^+ \quad (\Delta G^0 = -247 \text{ kJ·mol}^{-1} \text{ NH}_4^+) \quad (24\text{-}13)$$

$$2NH_4^+ + 6Fe^{3+} \longrightarrow N_2 + 6Fe^{2+} + 8H^+ \quad (\Delta G^0 = -303 \text{ kJ·mol}^{-1} \text{ NH}_4^+) \quad (24\text{-}14)$$

目前与上述反应过程有关的酶及其催化机理还不清楚。将这些未被发现反应的微生物鉴定出来的唯一方法就是在条件可控的实验室内培养它们。很显然，对这些氮转化微生物的生理生化条件需要与废水的环境即溶液性质关联起来。

通过上述分析，我们发现氮循环微生物基于其生理和生化功能的多样性，结合反应器的条件控制，所构建的水处理技术更加趋向于低能量与自清洁的方向。对此，亟需从废水性质出发，基于氮的生物转化路径的多样性，转变传统视角，探索适应绿色发展理念、寻求经济有效的典型工业废水脱氮新方法原理。

（3）生物脱氮工艺。

硝化反硝化工艺：氨氮经好氧氨氧化菌与亚硝酸氧化菌转化为硝酸盐氮，硝酸盐氮在无氧条件下结合电子供体经反硝化菌转化为氮气。硝化反硝化工艺需要供给电子受体（氧气）以及电子供体（含硫或碳化合物），同时伴随生物质的产生及温室气体的释放（CO_2 和 N_2O）；为了防止亚硝酸的积累以及减少氧化亚氮的形成，在反硝化过程中往往需要提供过量碳源，结果导致出水 COD 值增高，造成化学物质以及动力的浪费。

亚硝化反硝化工艺：氨氮经好氧氨氧化菌的作用转化为亚硝酸盐氮，亚硝酸盐氮结合电子供体经反硝化菌的作用转化为分子氮。因此该工艺成为一种较为节能减耗的短程生物脱氮技术。与硝化反硝化工艺相比，氧气需求量节省 33%左右，碳源需求量节省约 35%。此工艺的难点是 DO、温度、碱度、污泥龄（SRT）等方面的精准控制，将氨氧化为亚硝酸根[264]。未来防止具有毒性的亚硝酸根直接排放进入水体，需要通过后续的硝化工艺来深度处理含氮化合物，此过程需要补充一定量的碳源。

短程硝化厌氧氨氧化工艺：氨氮经好氧氨氧化菌作用转化为亚硝酸盐氮，亚硝酸盐氮结合氨氮经厌氧氨氧化菌的作用转化为分子氮。其化学计量式为式（24-15）～式（24-16）：

$$NH_4^+ + 1.24O_2 + 0.16CO_2 + 0.04HCO_3^- \longrightarrow 0.04C_5H_7O_2N + 0.96NO_2^- + 0.94H_2O + 1.9H^+ \quad (24\text{-}15)$$

$$NH_4^+ + 1.32NO_2^- + 0.066HCO_3^- + 0.13H^+ \longrightarrow 1.02N_2 + 0.26NO_3^- + 0.066CH_2O_{0.5}N_{0.15} + 2.03H_2O \quad (24\text{-}16)$$

此过程仅需要消耗少量的氧气，无需外加电子供体；相比于硝化反硝化工艺，氧气需求节省60%左右，无需有机碳源。工艺的关键之一是短程硝化阶段 NO_2^-/NH_4^+ 比例的稳定实现，反应不彻底以及反应过程中产生 NO_3^-，需要后续反硝化工艺的配合才能完全消除 TN 的污染[265]。另外，由于厌氧氨氧化微生物为自养型微生物，其生长缓慢，世代周期长，受其他化合物的抑制，系统容易崩溃，给工艺的技术实践和管理带来很大的挑战[266]。

（4）重金属废水资源综合利用技术。

在重金属硫化沉淀后，为保证处理后出水中重金属离子含量低于排放或回用标准，通常需保证硫离子的过量，即处理后出水中含有一定量的硫化物，可考虑作为电子供体应用于生物脱氮过程[267]，如式（24-17）～式（24-22）所示：

$$S^{2-} \longrightarrow S+2e^- \tag{24-17}$$

$$S^{2-}+H^+ \longrightarrow HS^- \tag{24-18}$$

$$S+4H_2O \longrightarrow SO_4^{2-}+8H^++6e^- \tag{24-19}$$

$$NO_3^-+2H^++2e^- \longrightarrow NO_2^-+H_2O \tag{24-20}$$

$$NO_2^-+4H^++3e^- \longrightarrow 0.5N_2+2H_2O \tag{24-21}$$

$$5HS^-+8NO_3^-+3H^+ \longrightarrow 5SO_4^{2-}+4N_2+4H_2O \tag{24-22}$$

脱氮过程中，硫化物被用作能源和电子供体去除亚硝酸盐或硝酸盐[268]。与异养反硝化脱氮相比，使用硫化物的自脱氮过程是处理有机碳缺乏废水的有效替代方案，不仅消除了脱氮过程对碳源的需求[269]，并且产生的污泥少，也间接地降低了污泥的处理成本。

2）工业废水生化处理技术的未来思考

废水处理生物过程中，除了脱氮技术之外，还存在广泛的其他研究领域，如生物反应器理论、生物分子工程学理论、同步除磷除硫的生物学原理等。针对污染控制方法原理，把控制过程分解为产品相分离、能量元素分离、营养元素去除、有机物矿化、微污染物转化和盐分分离纯化等不同的阶段。

废水中氮的生物去除涉及众多的氮转化反应如氨化、亚硝化、硝化、反硝化和厌氧氨氧化等，这些氮转化反应的组合形成了复杂且多样的脱氮路径，这些路径包括传统的硝化反硝化、节能低耗的亚硝化反硝化和无需有机碳源最为节能的短程硝化厌氧氨氧化。然而，目前应用的废水处理工艺都未能实现这三种及以上脱氮路径的整合，这是工艺的结构配置等造成的结果，无法通过简单的操作在同一个工艺中实现不同的脱氮路径，给废水处理的升级改造带来极大的困难，总氮去除面临着技术方面的挑战。

工业废水普遍表现出有机污染物浓度高、营养元素失衡、生物毒性抑制、氮素与盐分呈多态化等的特征，其处理往往需要从相分离、污染物转化等多角度来考虑。废水处理工艺中，如果要解决好总氮去除-碳源利用-能源消耗三者之间的矛盾，通过在线回收或循环回用其内部资源，实现自净化作用。利用产生的污泥制备污泥炭吸附剂，再投加到工业废水中吸附污染物，实现污泥的资源化短程循环回用，表现出较高的 COD 去除能力及很好的节能效果[270, 271]。

针对原理与工艺的结合，探索基础科学问题，在物质-能量-电子流动层面上分析组合工艺过程中污染物的转化，考虑能量物质的分离与质能效应、总氮浓度趋零途径、重金属风险归趋与高风险污染物去除的协同机制，对此，需要不断发现循环理论，开创新的水处理工艺，追求风险最小化的污染控制过程。

4　废水处理新技术系统

4.1　生物电化学技术

4.1.1　微生物电化学技术的简要原理

微生物是环境中最古老、最高效的分解者。在生命起源的初期，微生物对地球元素的赋存形态的变化扮演着重要角色。早期地表的氧被以铁为主的矿物消耗,很快进入了无氧的环境。为了生存,在这种厌氧环境下微生物演化出了诸多生存手段,充分利用环境中的电子受体。从生物电化学的角度来看,微生物在此电子传递过程中,可获取电子供体(底物,如单质硫)和最终电子受体之间的电势能差合成 ATP,作为生命活动所需的能量来源。这部分可获取的能量可由公式 $\Delta G=-n\Delta EF$ 计算。这里,ΔG 代表电子由底物传递到最终电子受体的吉布斯自由能,n 代表反应的转移电子数,ΔE 代表底物和最终电子受体之间的电势差。可见,二者的电势差越大,微生物可获取的吉布斯自由能就越多。微生物在厌氧环境中"想尽办法",最大程度获取氧化还原电位较高的电子受体。其中的一些在厌氧环境中演化出了还原铁锰氧化物的特殊能力。它们将代谢产生的电子通过内嵌于细胞膜和细胞周质中的细胞色素传导至胞外,最终实现了电子的跨膜直接传递。这类微生物与环境中的其他细菌共存形成了导电生物膜,电子可跨过几十微米厚的生物膜直接传递[272]。基于此,以电活性微生物氧化污染物的反应为阳极,高电位电子受体还原反应(如氧气、三价铁、六价铬等)为阴极,以直接获取或部分利用二者的电化学势能差为目标,科学家和工程师构建了大量的微生物电化学系统(microbial electrochemical system, MES)。二十年来科学家搭建了不同构型的 MES,实现了废水中的有机污染物向电能的直接转化[273]。

4.1.2 微生物电化学技术用于废水能源回收

早期的 MES 研究集中于多种有机废水和生物质的高效电能转化,最终目标是实现污水处理能量自持,甚至能量正输出。随着系统结构设计优化和低成本高性能活性炭电极材料的突破,实验室水平使用低成本纯碳电极材料制备的 MES 最高功率密度可接近 3000 mW/m^2 [274]。然而在其放大应用于实际废水处理中,存在功率密度大幅度衰减的难题。使用相似的反应器构型,以实际生活污水为底物,在实验室 28 mL 空气阴极反应器中获得的功率密度为 304 mW/m^2,而放大至 85 L 后功率密度仅为 83 mW/m^2 [275]。插槽式的阴阳极模块设计使 MES 实现了进一步放大,在 1.5 m^3 的反应器中插入了 336 对阴阳极模块,获得了 (406±30) mW/m^3 的功率密度,即 2.03×10^{-3} $kW\cdot h/m^3$ 的污水电能产出[276]。值得注意的是,该 MES 反应器的出水水质达到了我国一级 A 的标准,停留时间 5 h 即可实现化学需氧量(COD)去除率 91%,总氮去除率 64% 和氨氮去除率 91%,成本仅为 1135 美元/m^3。以上表明,MES 作为一种水处理技术是可行的,产生的电能作为可原位应用的"产物",有望进一步得到提升,未来或许能发挥更大的作用。由此认为,MES 性能提升的核心是基于对微生物间的电子传递规律和电活性生物膜形成过程的科学认知。

4.1.3 微生物胞外电子传递和成膜过程

近年来,科学家在微生物胞外电子传递领域的研究进展不断刷新人类的认知。Lovley 教授在 2005 年发现 *Geobacter* 的 IV 型伞毛"纳米导线"可作为胞外电子传递通路[277],其可能的导电机理是在伞毛上定向排列了大量的细胞色素 OmcS[278]。而近期 Wang 等利用冷冻电镜发现了"纳米导线"的本质并非 IV 型伞毛,而只是由细胞色素 OmcS 定向排列组成的微结构,相邻 OmcS 亚基间通过共用一个血红素共价键及 π-π 堆积作用强化亚基间的联系[279]。另一类模式电活性细菌 *Shewanella oneidensis* MR-1 的"纳米导线"则是由细胞外膜突起形成的,是细胞膜的一部分[280]。尽管在胞外电子传递机制上依然存在诸多争议,但这代表了自然界中的一种普遍现象,包括古菌 *Methanospirillum hungatei* 在内,均发现了这种类似的纳米导线结构[281]。借助这种导电结构,不同微生物间可以共享代谢产生的电子,实现种间直接电子传递(DIET)。先前的研究证实了模式电活性细菌 *G. sulfurreducens* 和 *G. metallireducens* 可直接通过胞外电子的传递进行种间互营共生作用[282],而近期又发现了其不依赖伞毛共生的证据[283]。这也从另一个侧面反应了伞毛在细菌间电子传递的作用是存在争议的。研究还发现,电活性微生物与产甲烷菌间同样可进行 DIET,*Geobacter* 代谢产生的电子可传递给产甲烷菌(例如 *Methanosaeta*)[284],而基于 *G. sulfurreducens* 和 *P. aestuarii* 光合 DIET 的发现将其拓展到有光的环境[285]。甲烷氧化古菌与 *Geobacter* 间同样具有潜在的 DIET

过程[286]。

而面对未来的水处理工程应用，混菌生物膜的成膜过程也得到了广泛关注。基于双光源明场成像系统，发现了污水中电活性细菌的成膜过程与感应电场强度密切相关[287]。处于强电场核心区的生物膜形成了直径约 30 μm 的红色团聚体，而团聚体向弱电场边缘区域逐渐减少。一些信号类物质对电活性微生物的选择也是有贡献的，例如 AHL 等传统群体感应信号物质[288]或者亚抑制浓度（sub-MIC）的妥布霉素[289]。基于 G. sulfurreducens PCA 的转录组学发现，亚抑制浓度妥布霉素显著上调了与产电相关的细胞色素和 IV 型伞毛的表达。这可能是 Geobacter 对环境胁迫的一种应激反应，低浓度的其他对 Geobacter 有毒性的物质可能具有同样的作用。

4.1.4　基于微生物电化学原理的废水处理和污泥处置

MES 在厌氧废水/污泥处理提效领域具有更加广阔的应用前景。以偶氮染料微生物脱色为例，传统的研究认为偶氮染料还原酶系统是脱色的主要途径。近期，Liu 等发现了模式电活性微生物 G. sulfurreducens 可依靠强大的胞外电子传递系统快速脱色，细胞色素 OmcB、OmcC 和 OmcE 均参与其中[290]。研究者还发现电活性微生物的另一项功能——加快反硝化速率。不可利用硝酸盐的电活性细菌 G. sulfurreducens 可加速反硝化微生物群落形成、消除迟滞期，减缓亚硝氮积累，使混菌中反硝化功能基因 nirS 表达量显著提升，最大反硝化速率提升了 51%[291]。当 MES 中加入 2 g/L 硝酸盐时，反硝化效率可达（12.2±0.6）kg 硝态氮/(m^3·d)，群落中 Thauera 得到强化[292]。

基于 DIET 原理，MES 可用于加速厌氧废水处理，提升甲烷产率。在处理啤酒废水的升流式厌氧污泥床（UASB）反应器中，发现了高度导电的 Geobacter 和产甲烷 Methanosaeta 的微生物聚集体[293]，而在厌氧反应器中添加诸如碳布、碳毡和活性炭等材料可加速产甲烷过程[294]。甚至是在高固含量的污泥中，加入导电的磁铁矿（Fe$_3$O$_4$）可显著抑制短链脂肪酸的积累，甲烷产量提升了 27%[295]。而发酵产生的甲烷也可在 MES 中作为唯一底物进行微生物电化学氧化，产出电流密度高达 1130 mA/m^2 [296]。

4.1.5　基于微生物电化学传感器的水质监测

电活性生物膜建立了微生物活性与电流信号的直接联系，基于 MES 的生化需氧量（BOD）传感器和水质生物毒性预警传感器逐渐走向商业化应用[297]。近年来，研究发现自养的生物阴极可作为探测原件，极大拓展了 MES 传感器的应用场合[298]。例如，已有研究者将其用于饮用水消毒副产物毒性的检测[299]。与植物根际结合，开发出了一种植物根际 MES 传感器用于预警酸雨[300]。当酸雨滴到水稻叶片后，2 min 内信号通过富马酸、半乳糖和葡萄糖通路由叶片经根际传导至微生物，并触发电流衰减预警。该传感器可进一步拓展至植物/作物的健康监测，未来有望在智慧农业领域拓展应用范围。

综上所述，近年来 MES 的新功能不断涌现，潜在的应用领域也得到了大范围拓展。未来的研究需要在混菌电子传递理论上进一步取得突破，并基于这些科学发现指导新技术研发，实现产业化应用，使微生物电化学技术早日造福人类。

4.2　膜生物反应器技术

4.2.1　概述

膜生物反应器（membrane bioreactor，MBR）是一种基于传统活性污泥法（CAS）与膜过滤过程耦合的污水深度处理与回用技术[301, 302]。膜过滤具有高效的固液分离效果，因此 MBR 通常可维持较高污泥浓度（MLSS = 8～15 g/L），保证长世代周期微生物的保育、富集，具有占地面积小、剩余污泥产量低、出水水质好且易于自动化管理等技术优势。根据《中国水处理行业可持续发展战略研究报告（膜工业卷）》统计结果，MBR 目前已在全球超过 200 个国家及地区建设并投产运行，污水处理量超过 1 万 t/a 的工程总处理规模已达 2000 万 t/d[303]。据美国 Grand View Research 咨询公司预测，全球 MBR 市场将于 2025 年达到 82.7 亿美元规模。同时，英国 Cranfield 大学 Judd 教授通过在 MBR 网站（www.thembrsite.com）开展的定向问卷调查结果显示（如图 24-18 所示），MBR 用户急切希望科研界在新型膜材料制备、膜过滤与污染过程解析、反应器效能优化（预处理工艺设备研发、过程控制与优化）以及工艺创新研发等方面开展相关应用基础研究，从而实现 MBR 整体节能降耗与稳定运行[304]。

近年来，我国高校和科研机构在 MBR 基础研究领域处于国际一流水平，我国学者在 MBR 膜污染机理解析、抗污染膜材料制备、新型 MBR 工艺等方面取得了一系列研究成果。根据 Web of Science 数据库检索，全球范围内发表 MBR 相关论文 9947 篇（截至 2019 年 5 月），其中中国学者发表 2565 篇（含合作发表），占比超过 25%，发文数量超过美国，位居全球第一。在 MBR 应用领域，涌现了以碧水源、膜天、斯纳普等为代表的具有国际竞争力的 MBR 膜供应商。目前，我国 MBR 工程累计处理规模已超过 1400 万 t/d，位居全球第一。

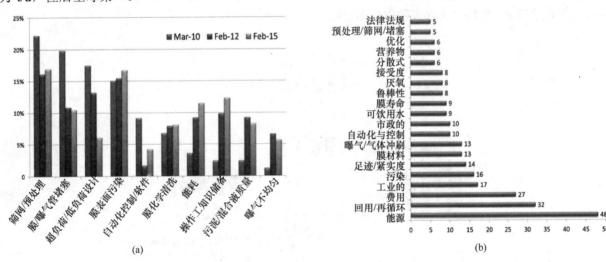

图 24-18　MBR 在线问卷调查结果

(a) 问题"根据你的经验，哪些技术因素或缺陷制约了 MBR 的运行效果"的答案反馈；
(b) 问题"未来 MBR 技术该如何发展"答案中的关键词词频[304]

4.2.2　研究进展与动态

1）新型膜材料制备

新型膜材料的研发一直是 MBR 领域的重要研究方向之一。相比于传统膜过滤过程，由于 MBR 活性污泥混合液具有浓度高、溶解性有机物成分复杂等特征，如何制备抗有机与抗生物污染的新型膜材料得到研究者的持续关注。清华大学、天津工业大学、同济大学等研究团队开展了聚偏氟乙烯（PVDF）、聚醚砜（PES）等膜材料的改性研究，构建了系列基于膜表面官能团负载以及纳米颗粒共混的新型膜材料[305, 306]。针对 MBR 的生物污染问题，同济大学研究团队制备了以季铵盐共混和表面接枝的系列抗生物污染膜材料[307]，解析了其直接接触杀菌与间接杀菌共同作用的抗污染机理[308]。此外，大连理工大学和同济大学等团队在新型功能膜（如电催化活性功能膜、导电膜）开展了相关研究[309-311]等，揭示了电场诱发强氧化剂（如过氧化氢、羟基自由基）生成的膜界面行为以及抗污染机理。总体而言，我国在 MBR 新型膜材料制备方面的研究处于蓬勃发展状态。

2）膜污染过程解析与控制

在次临界通量条件过滤过程中，溶解性有机物与微生物在膜表面及孔道内吸附沉积进而形成凝胶层污染仍是 MBR 存在的一个技术挑战。我国学者在 MBR 膜过滤与污染过程解析方面表现十分活跃，在膜面污染物系统性表征、膜面污染物形成机制等方面取得了系列研究成果[305, 312]（膜污染形成过程如图 24-19 所示）。近年来随着组学（omics）技术的突破性发展，研究人员利用 Hiseq 高通量测序发现 MBR 反应器中微生物分泌多糖与胞外聚合物的基因水平与传统活性污泥法没有显著差异，但膜污染物中涉及胞外聚合物运输的基因丰度较低[313]。中山大学研究团队利用蛋白质组学对膜面滤饼层中蛋白质多样性进行了系统性分析[314]。随着多种分析手段的交互式应用，可以从分子水平上进一步解析膜污染的科学原理。

针对 MBR 实际运行过程中的膜污染问题，我国学者近期在物理及化学清洗方面也进行了系统性研究（膜清洗过程如图 24-19 所示）。以 NaOCl 清洗剂为例，研究发现 NaOCl 在污染层中扩散具有 pH 依附性，在高 pH 条件下，去质子化的次氯酸根传质速率更高且呈现非均匀扩散特征，可以有效松散污染层的基体结构。但需要注意的是，微生物长期暴露于清洗药剂中可能导致菌群演替，如清华大学团队研究发现经 15 mg Cl_2/L 预处理后，残留细菌可能分泌分子量更高的胞外聚合物从而导致更严重的膜污染[315]。因此，膜清洗研究过程中需要综合考虑多种因素的影响。

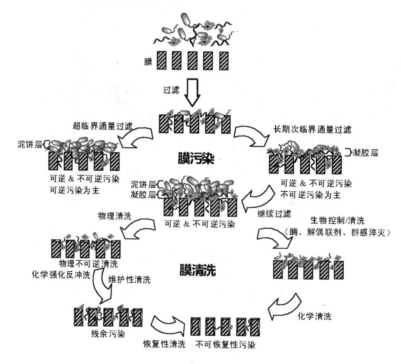

图 24-19 膜生物反应器膜污染与膜清洗过程示意图

3）反应器效能优化及工艺创新

膜格栅是保证 MBR 稳定运行的重要预处理单元。由于膜组件运行过程中需气液两相扰动以减缓滤饼层形成，为避免污水中发丝等杂质对组件的缠绕，MBR 需要比传统污水处理工艺更为精细的格栅分离设施（例如间隙 0.1～0.5 mm 的超细格栅）。同济大学团队针对现有格栅单元的局限性，研发了针对 MBR 的杂质分离设施[316]，可以有效分离传统膜格栅不能拦截去除的杂质，保障 MBR 的稳定运行。

在 MBR 实际操作过程中，保证曝气均匀并减少水力死区是降低 MBR 运行能耗的关键。近年来，借助计算流体力学（computational fluid dynamics，CFD）商业软件包，我国学者针对 MBR 反应器曝气条件（例如曝气量、气泡大小、曝气位置）和结构优化研究取得重要研究进展。中国科学院生态环境研究中心研究团队利用 Ansys Design Modeler 和 Fluent 对大型膜生物反应器的设计及运行进行了模拟优化，模拟结果建议可减小反应器下部断面面积，调整角区角度或者增加搅拌泵，从而减小死区占比，增大下部流速[317]。此外，生化反应动力学与 CFD 模拟单元可以进行耦合，从而实现 MBR 反应器内动力学与流体力学多尺度层面的模拟与工艺优化。清华大学团队构建了基于出水氨氮浓度反馈的自动曝气方法（如图 24-20 所示），用于大型 MBR 污水处理厂的节能降耗实践，实现了曝气能耗降低 20%[318]。

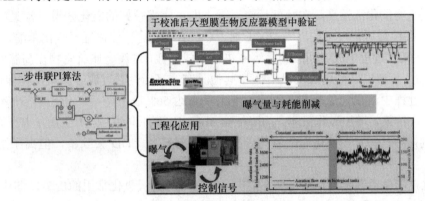

图 24-20 某大型 MBR 污水处理厂基于出水氨氮浓度反馈的自动曝气节能降耗实践[317]

MBR 新工艺的研发也是国内外污水处理领域的研究热点，中国科学技术大学、哈尔滨工业大学、清华大学、同济大学研究员开展了厌氧 MBR（包括厌氧动态膜 MBR）、正渗透 MBR[302]、微生物燃料电池 MFC-MBR 等新型组合工艺研究[319,320]。

4.2.3 展望

近年来，我国在 MBR 研究和应用方面取得了巨大进步，MBR 技术在我国污水处理与资源化领域发挥了重要作用。在 MBR 应用方面，我国已建设了大量 MBR 污水处理工程，总处理能力位居全球第一。在应用基础研究方面，我国在膜污染机制与膜污染控制技术研究、抗污染膜材料制备、新型 MBR 工艺研发等方面处于国际一流水平。未来，新型膜材料、膜污染控制、工艺节能降耗、污水资源化能源化等是 MBR 领域重要的研究与发展方向。

4.3 化工废水处理技术

化工行业是我国重要支撑产业，生产过程涉及大量物质转换和"三废"排放，尤其是废水具有组成复杂、污染物浓度高、水质波动大等特点，是典型的难处理工业废水。以煤化工行业的湿熄焦废水为例，内含数百种有机污染物、高浓度氰化物、氨氮和无机盐等，生物毒性强、难降解，废水组成还随煤品质、生产参数发生显著变化，达标处理难度极大。当前，化工废水一般需采用组合处理工艺。由于细分行业的化工废水组成差异大，治理需求各不相同，本节主要介绍几种化工废水难降解有机污染物的治理技术，也是本领域的研究热点和难点。

4.3.1 高浓度有机废水湿式氧化预处理技术

生化法是去除废水中有机污染物的主要工艺，但化工废水直接进行生化处理的可能出现效率低、系统易崩溃，预处理通常可以减毒和提高可生化性。湿式氧化法是在高温（140～300℃）、高压（0.5～20.0 MPa）、过量氧气的条件下降解有机物，设备要求和运行成本远低于超临界氧化，并且废水 COD 大于 20 g/L 时可实现系统自热，是一种绿色节能的高浓度有机废水预处理技术。

加入非均相催化剂或氧化试剂可降低湿式氧化所需的反应温度，提高氧化效率。非均相催化剂包括碳材料、过渡金属和贵金属。碳材料成本低，酸碱条件下均可使用，受到广泛关注。Stuber课题组[321]在碳材料催化湿式氧化领域开展了大量研究，在140～160℃时，碳材料催化湿式氧化可完全去除酚及取代物，但TOC去除率非常低，且碳材料自身氧化消耗严重。即使作为载体，碳材料同样易被氧化，使催化剂稳定性降低[322]，因此，碳材料在高温、高含氧的湿式氧化体系不太具有应用前景。

过渡金属氧化物催化活性高，在较低反应温度下即可实现较好的氧化效果。常见的过渡金属氧化物中，CuO的催化活性最强，反应温度为200℃时，Cu/MCM-41[323]催化湿式氧化苯酚，TOC去除率达到90%以上。但湿式氧化过程生成中间产物使溶液快速变成酸性，过渡金属氧化物在高温和酸性溶液中极易变成离子溶出，如Cu溶出率可达60%以上[323]，其他具有较高活性的过渡金属氧化物催化剂如FeO、MnO等，在湿式氧化过程中溶出现象也非常严重，溶出率高达50%以上[324, 325]。金属溶出不仅影响催化剂活性，同时带来二次污染，增加操作成本。为降低金属溶出，孙承林课题组[326]开发出一系列$ZnFe_{2-x}Al_xO_4$尖晶石相催化剂，在160℃下催化湿式氧化苯酚能获得较高的COD去除率，但离子溶出问题仍未彻底解决。

贵金属催化剂催化活性更高，且几乎不溶出。Pt、Ru是湿式氧化常用的贵金属催化剂，Pt/TiO_2-CeO_2催化氧化苯酚，160℃下反应2 h时COD去除率达到96%[327]。Ru催化活性高，且氧化生成CO_2的选择性高[328]。虽然贵金属催化剂解决了离子溶出问题，但处理杂环有机污染物时易失活：Pt催化剂处理含氮废水时活性降低明显，第二轮活性比第一轮降低20%以上[322]，处理含硫有机物时，杂环上硫原子与活性金属形成不可逆的金属-硫键，使贵金属中毒失活[329]，因此，贵金属催化剂在湿式氧化处理含杂原子污染物的报道极少。

与其他高级氧化过程联用是提高湿式氧化效率的另一种策略。Quintanilla等[330]采用H_2O_2-催化湿式氧化组合工艺处理苯酚溶液，可明显提高低温条件（127℃）下TOC去除率。曹宏斌课题组[326]开发出过硫酸盐-湿式氧化工艺处理高浓度苯并噻唑废水，比H_2O_2耦合工艺效果更佳，并且可回收高分子聚合物（产率60%以上）。曾光明[331]课题组也发现在催化湿式氧化过程加入过硫酸盐，能显著提高腐殖酸的处理效果。

4.3.2 低浓度有机废水催化臭氧氧化深度处理技术

经过预处理和生化法组合处理之后，化工废水COD可降至200～400 mg/L，还需要深度处理以满足要求越来越高的废水排放标准。臭氧氧化能力较强，且可在常温常压下操作，是有机物降解的重要技术选择。但目前高压放电产臭氧的效率偏低，处理废水的运行成本相对较高，一般用于生化出水的深度处理单元，或作为膜脱盐回用的预处理单元。

为提高臭氧氧化效率，一般加入非均相催化剂，利用反应中产生大量具有强氧化性的羟基自由基来矿化废水中有机污染物，能实现更高的臭氧利用率与污染物去除效率。非均相臭氧氧化催化剂主要分为金属和非金属两类，其中金属催化剂已在去除水中的有机污染物方面展现出很好的能力。如Yang等[332]合成MnO_x/Al_2O_3催化剂，能有效催化臭氧氧化并矿化水体中的医药类污染物，如苯妥英、苯海拉明、双氯芬酸钠和安替比林等。Martins等[333]合成$MnCeO_x$催化剂，催化臭氧氧化包括6种酚酸的模拟含酚废水，酚酸的TOC去除率大大提高。金属催化剂在臭氧氧化过程同样存在金属溶出问题，而非金属催化剂其环境友好，成本较低，近年来受到越来越多的关注[334]。

在催化机理研究方面，马军课题组[335]、胡春课题组[336]分别制备蜂窝陶瓷负载催化剂、SBA-15负载铁/铝催化剂催化臭氧处理低浓度有机物，并分别提出表面羟基、表面Lewis酸性位催化臭氧分解活化的反应机理。曹宏斌课题组[337, 338]发现碳材料表面含氧官能团、缺陷结构均可催化臭氧分解，并基于DFT理论计算和电子顺磁共振光谱发现，超氧自由基、单线态氧等活性物种可能在催化臭氧降解酚类污染物过程起作用。基于大比表面积和富表面基团碳材料与过渡金属复合策略，开发出商业化臭氧氧化催化剂，活性显著高于活性炭和锰砂，并可稳定使用三年以上。通过优化反应器设计提高臭氧利用效率并解决其他工程难题，将该技术率先应用于焦化废水深度处理工程，出水 COD 可降至 50 mg/L，稳定满足焦化行业和辽宁省的最新排放标准，目前已在鞍钢、武钢、邯钢、攀钢等大型钢铁企业和煤化工企业建立 10 余套示范工程，单套处理规模 100～400 m^3/h。

4.3.3 含盐有机废水电化学-臭氧耦合处理技术

含盐有机废水是一种典型化工废水，可能来源于生产过程或膜处理过程的浓水。含盐高浓度有机废水可以采用铁碳内电解、湿式氧化技术处理后进入生化处理系统，低浓度有机废水则大多采用化学氧化法深度处理。

电化学氧化处理含盐废水时，可合理利用废水中无机盐作为电解质导电。电化学氧化一般分为直接氧化和间接氧化，直接氧化是水体中有机污染物直接在电极表面被氧化，间接氧化是通过电极生成活性自由基降解有机污染物。目前电极主要有活性电极（RuO_2、IrO_2 等）和非活性电极（PbO_2、掺硼金刚石电极等），这两种氧化途径均对电极要求高，因为电极腐蚀及尺寸较小，通常存在处理效率低，能耗大的问题。采用电化学与臭氧耦合的方式，通过阳极氧化和阴极还原 O_3 可显著提升活性氧自由基的生成速率，强化有机物处理。Kishimoto 课题组[339]采用 Ti/Pt 阳极和 Ti 阴极结合臭氧的方式处理新兴有机化工试剂 1,4-二氧六环，可将底物降解率由电化学过程的 1%、臭氧氧化的 13% 提升到 54%。为进一步提升电化学与臭氧耦合效果，王玉珏课题组[340]提出电化学-过臭氧化耦合技术，利用炭黑阴极还原 O_2 原位生成 H_2O_2，与 O_3 发生过臭氧化反应产生羟基自由基。该技术具有装置简单易操作，绿色无二次污染的优点。通过优化臭氧浓度、电流强度、电极材料和电解质种类等参数，在染料、医药、化学原料等化工废水处理中取得良好效果[341-344]。例如处理 Orange Ⅱ 染料废水时，电化学-过臭氧化技术可在 45min 去除 95.7% 的 TOC，而单独臭氧氧化和电化学氧化 90 min 也仅分别为 55.6% 和 15.3%。处理 1,4-二氧六环时，电化学-过臭氧化技术在 120 min 实现 96.6% 的 TOC 去除率，远远优于臭氧氧化（6.1%）和掺硼金刚石电极的电化学过程（26.9%），并且耦合过程能耗仅为臭氧氧化技术的 15% 和掺硼金刚石电极的电化学技术的 67%。曹宏斌课题组[345]采用三电极体系研究该技术，发现 O_3 电解和过臭氧化反应协同产生羟基自由基，共同作用于有机污染物去除。由于处理过程中溶液 pH 降低，有机污染物的处理效率大幅下降，还研发出 C_3N_4-Mn/CNT 非均相催化剂应用于电化学-过臭氧化过程[346]，将草酸的降解效率提升 57 倍，极大地提高了该技术去除酸性废水中有机物的效率。此外，将掺硼金刚石电极引入电化学-过臭氧化技术，通过阳极氧化，O_3 电解和过臭氧化技术的三重结合，可进一步提高处理效果。

4.3.4 展望

催化氧化法与生化法联用是解决有机废水污染的重要措施，为提高有机物氧化效率，可开发新材料催化的传统氧化技术，或新型催化氧化过程。目前国内正在开展的氧化技术多且杂，应深入梳理各种氧化技术的优劣及更具体的适用范围，例如湿式氧化处理后废水变成强酸性，更适于处理高浓度碱性有机废水，西门子公司开发的 Zimpro@湿式空气氧化技术就成功应用于石化行业碱液处理。另外，在开发新型高效的氧化技术时，应注重与其他学科交叉，借鉴材料、催化领域的最新进展，例如通过简单方法制备单原子催化剂，可显著提高不同氧化、还原过程中有机物催化降解速率。

4.4 厌氧处理新技术

当前，以水解酸化-产甲烷原理建立的传统厌氧消化工艺面临以下技术难题：①复杂有机物的水解酸化效率较低，导致有机固废厌氧消化时间过长（20天以上）；②高浓度有机物产生大量有机酸，破坏厌氧系统的pH平衡，导致产甲烷受抑制乃至失败，致使厌氧系统操作不稳定，产气效率较低。

4.4.1 异化铁还原强化厌氧消化技术

异化铁还原（dissimilatory iron reduction）是铁还原菌氧化底物产生电子、并将电子传递给Fe(III)（氢）氧化物使之还原的过程。在厌氧系统中，以Fe(III)为末端电子受体进行的呼吸可获得更多能量，使异化铁还原在分解难降解物和复杂有机物方面更具优势。研究表明，铁还原菌对大部分苯类化合物、多糖和腐植酸具有分解能力[347-349]。Baek等报道，将水铁矿置入污泥厌氧消化反应器内可显著加速复杂有机物降解，同时Fe(II)浓度显著增高，表明厌氧水解酸化效率与异化铁还原密切相关[350]。然而在大多数情况下，异化铁还原速率并不高，这与铁还原菌的胞外电子传递受限相关。加入蒽醌类、核黄素等电子穿梭体物质可以加速电子在异化铁还原菌与铁氧化物间的传递，从而提高异化铁还原对有机底物的利用效果，达到加速大分子物质厌氧分解的目的[351]。值得注意的是，异化铁还原速率还与铁的晶型、氧化还原电位有关。无定形的铁氧化物还原速率优于晶体结构的铁氧化物；氧化还原电势高的铁氧化物对有机底物的利用效率更高（水铁矿> α-FeOOH > α-Fe$_2$O$_3$> Fe$_3$O$_4$）[352]。在厌氧处理时，应针对性地选择铁氧化物类型，为复杂有机物的强化厌氧分解提供合适的末端电子受体。图24-21为异化铁强化厌氧产甲烷示意图。

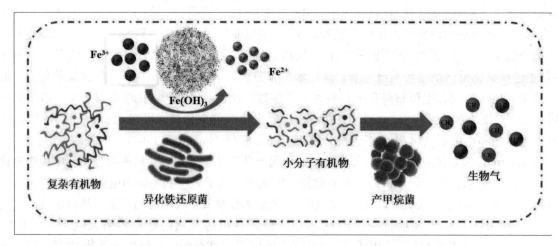

图24-21 异化铁强化厌氧产甲烷原理图

4.4.2 零价铁强化厌氧产甲烷技术

目前，产甲烷菌可利用的底物仅限于乙酸和一元碳，其他有机物只有经产酸发酵为乙酸或一元碳后，才能被甲烷菌利用[353]。然而，厌氧产甲烷却受限于电子受体H$^+$的低电势（H$^+$+2e===H$_2$，E_{H+/H_2} = −414 mV，NHE/pH=7）。为此，作为H$^+$接受有机电子的产物，H$_2$必须维持在很低的分压以下（10^{-4}~10^{-5}atm），才能在热力学上保证有机物厌氧代谢的持续进行[354]。而嗜氢产甲烷菌作为消耗H$_2$的主要微生物，却在大部分厌氧系统中不是主流（以嗜乙酸甲烷菌为主）[355]。因此在操作条件波动时很容易导致H$_2$积累，无法完成种间氢传递（interspecies hydrogen transfer，IHT，某些系统为甲酸传递），致使产甲烷停顿乃至失败。张耀斌等发现，零价铁可以作为电子供体直接被耗氢产甲烷菌利用（4Fe^0+8H$^+$+CO$_2$===4Fe^{2+}+CH$_4$+2H$_2$O），这种代谢活动显著促进耗氢产甲烷菌的生长，其丰度从8%~9%提高到20%以上[355]。计算表明，零价铁强化的IHT系统里，产氢产乙酸的吉布斯自由能降低10%~20%，突破IHT的热力学屏障，推动厌氧呼吸的右移[357]。零价铁可以提高铁氧化还原酶、乙酸激酶等重要酶的活性2~34倍，促进大分子有机物的水解酸化，复杂有机物如剩余污泥的水解酸化效率和产甲烷效率分别提高20%~50%以上。

4.4.3 生物电化学强化厌氧技术

氢在厌氧中的转化，还可以通过外加电势的方法突破其热力学壁垒（E_{H^+/H_2} = −414 mV），如阴极产甲烷的生物电解池（MEC）。在该系统里，H^+在阴极电势下，被氢型甲烷菌转化为甲烷（$CO_2+8H^++8e^-$ ══ CH_4+2H_2O，$E^0 \approx$ −440 mV，NHE/pH=7）或甲酸等，从而改善有机物的厌氧氧化和底物分解。Cheng[358]等研究表明，在双室 MEC 利用嗜氢产甲烷菌作为生物阴极还原 CO_2 可以提高产甲烷速率和程度，同时在 MEC 阳极能有效缓解酸性积累和底物分解。阳极氧化效率与阳极电活性微生物的富集有直接关系。研究显示[359]，MEC 系统微生物测序显示，电极上的地杆菌丰度为 25.8%，而电极上的甲烷菌以 *Methanosaeta* 为主。这使人联想到近期被发现的另一种新的产甲烷模式——地杆菌和 *Methanosaeta* 之间的直接种间电子传递产甲烷（direct interspecies electron transfer，DIET）。

4.4.4 DIET 厌氧消化及其强化方法

2010 年，Lovley 等发现，共培养的两种地杆菌（*Geobacter*）能通过其导电的 pili 传递电子，以乙醇为电子供体将延胡索酸还原[281,360]。随后发现，这种直接电子传递方式，也可以在地杆菌和产甲烷菌（如 *Methanosaeta*）之间进行，将乙醇氧化产生的电子，通过两种微生物之间的电子交换直接还原 CO_2 为甲烷（图 24-22）。DIET 产甲烷的这一发现，颠覆了对厌氧消化的认知：有机物无需经历产酸过程，可能直接通过两类微生物之间的电子传递，将有机物转化为甲烷。相比依靠电子载体 H_2 在两类微生物之间扩散的 IHT 方式，两类微生物之间的 DIET 电子传递更高效，其电子传递速率是 IHT 的 10^6 倍。要构建 DIET 代谢，首先需要富集地杆菌。在通常的厌氧反应器内，地杆菌的丰度并不高，因此大多数常规厌氧反应器的 DIET 产甲烷不明显。在厌氧系统内投加 Fe_2O_3、$Fe(OH)_3$ 等高价铁（氢）氧化物[361,362]，可从电子终端上富集具有地杆菌，从而强化 DIET。另一方面，导体碳材料如碳棒等均被报道可替代地杆菌的 pili[363]，进行微生物间厘米级的电子传输，即从电子连接通路上加快 DIET 产甲烷。DIET 的这一设计思路，已经在高浓度有机废水处理、有机固废厌氧消化产甲烷（城市污泥、畜禽粪便、餐厨垃圾等）等的工程应用中获得验证，有效提高了甲烷产率和底物分解效率。

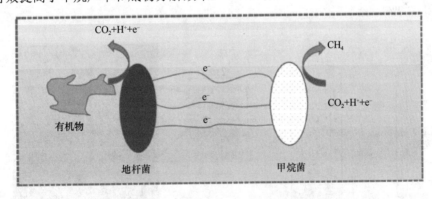

图 24-22　DIET 强化厌氧消化原理图

4.4.5 展望

厌氧产甲烷是高负荷有机废物能源化处理的最经济、有效的手段之一，但受限于传统原理的局限，越来越不能满足实际需要。随着有机废物产量的增加和国家生态化建设的要求，对高效厌氧技术的需求越来越迫切。建立在厌氧代谢新原理上的 DIET 产甲烷，可以有效提高有机底物的厌氧消化速率，大幅增加沼气中的甲烷含量并减排二氧化碳，有效提高抗冲击性能。可以预料，该技术将在高负荷有机废物（废水）处理中发挥越来越重要的作用。为此，还需深入研究该种路径产甲烷的技术特点、运行原理与调控机制，为该技术的大规模工程应用提供理论和技术支撑。

4.5 光催化-生物降解直接耦合技术

各种物理、化学和生物的水处理方法日新月异，然而任何技术都无法独自解决复杂的水处理问题。特别是随着有毒/难降解有机污染物排放量的增加，单一处理方法的局限性更为凸显。物化-生物耦合处理技术能够互作互补，强化有机污染物的高效加速去除、矿化与毒性削减。光催化-生物降解直接耦合技术（intimate coupling of photocatalysis and biodegradation，ICPB）是传统耦合技术的革新，在处理生物抑制性、高毒性污染物方面具有显著优势。

4.5.1 光催化与生物降解直接耦合技术的基本结构

2008 年，美国工程院院士 Bruce E. Rittmann 教授课题组首次构建了 ICPB 体系[364, 365]。他们把 TiO_2 负载于毫米级的多孔载体骨架表面，并将生物膜培养于载体空隙内部。载体外表面的催化剂受紫外光激发产生光生空穴电子对，攻击废水中的生物抑制性/毒性污染物，产生的可生物降解中间体被孔隙中的微生物迅速利用并矿化。由于生物膜附着于载体内部，避开了紫外辐射以及光生羟基自由基的攻击，保障了体系中微生物的活性和稳定性。ICPB 突出的特点即为：光催化反应与生物降解作用在时间上和空间上同步完成，由此被命名为"直接耦合反应"[366, 367]。上海师范大学张永明教授与 Rittmann 教授合作，从电子传递角度研究了 ICPB 中催化氧化对加速污染物降解的贡献[368, 369]。ICPB 技术解决了分段物化-生物降解组合技术的不足，避免了高级氧化控制不当导致污染物降解不完全或氧化过度的问题[370-372]。2015 年，周丹丹等以光强为 5.38×10^{-5} einstein/(L·s) 的 LED 作为光源，将可见光催化引入至 ICPB 体系（图 24-23）[373]。这一改变解决了传统 ICPB 的两个瓶颈问题：一是使 ICPB 直接利用太阳光成为可能；二是避免了可溶性微生物产物（SMP）释放引起的溶解性有机碳（DOC）升高的问题[374]。可见光响应的 ICPB 体系中，载体上的生物膜量与生物活性具有更好的稳定性，模式污染物苯酚的去除率与矿化度分别提高了 32.6%和 42.2%。

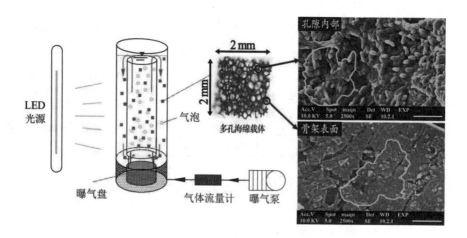

图 24-23 可见光 ICPB 构造（左）与多孔海绵载体上光催化材料与生物膜分布（右）[373]

4.5.2 ICPB 技术应用与机制

近年来，基于 ICPB 体系已经开展较为系统和深入的工作，包括以优势活性物种影响为背景的催化剂筛选[375, 376]、催化剂自组装负载方法建立与优化[377]、直接耦合反应动力学模拟与验证[378]，以及直接耦合机制[364-367, 373]等，使 ICPB 技术神秘面纱日渐揭开，在多种典型污染物降解与矿化方面的优势愈发凸显。

1) ICPB 技术处理氯酚废水

氯酚的芳环结构和氯代原子使其具有很强的毒性和抗降解能力，是污/废水中典型的生物抑制性污染物。李国政等[365, 368]解析了三氯酚（TCP）的去除与矿化机制。单一光催化对 TCP（初始浓度为 14 μmol/L）的去除率为 93%，但对 DOC 的去除没有显著的贡献。与之相比，ICPB 反应不仅进一步提升了 TCP 的去除效率，出水 DOC 的浓度较单独的光催化反应降低了 90%。但是，当光催化反应产生大量半衰期较长的光生 H_2O_2 时，ICPB 体系中活细胞比率将显著下降。有共基质存在时，活细胞比率可提高 20%，且生物多样性指数显著提升，氯酚的去除率与矿化率分别了提高 27%和 23%[373]。

2) ICPB 技术处理抗生素废水

抗生素类物质对生物活性具有强烈的抑制作用（对活性污泥半数效应浓度低至 2.2 mg/L）[378]，因而传统的生物处理法几乎不能降解废水中的抗生素[379]。利用可见光响应 ICPB 技术处理初始浓度为 30 mg/L 的盐酸四环素（TCH）时发现，ICPB 体系中生物膜活性良好，富集了对 TCH 有较强耐受性的菌属，包括 *Methylibium*、*Runella*、*Comamonas* 和 *Pseudomonas* 等。ICPB 体系内微生物降解作用的存在，降低了光催化中间产物对自由基的竞争，使得 TCH 母体的降解效率高达 90%。特别是，与单独光催化反应相比，COD 去除率提高了 20%以上[380]。采用外加基质的手段，生物膜中的活菌比例从 56%提高至 86%，TCH 的降解效率进一步提高至 95%，降解速率常数提高了 40%。同时，生物群落中富集了 *Thauera*、*Pseudomonas*、*Runella* 等与 TCH 或其中间产物降解有关的菌属，强化了 TCH 光催化降解中间产物的生物矿化[381]。UPLC/MS/MS 分析进一步揭示，投加共基质后 ICPB 中产生的一些含 π-π 共轭体系的小分子中间产物能被微生物进一步降解。

3) ICPB 技术处理其他废水

ICPB 技术在处理染料、苯酚、硝基苯、DNT、吡啶等其他生物抑制性污染物时，也展现出了上述类似的优势。

4.5.3 ICPB-阳极电解池及电能转化机制

将生物电化学电池（MxC）和光电化学电池（PEC）耦合，用于含生物抑制性有机物类废水处理是一种极具潜力的新型技术。Li 课题组和 Qian 课题组在 2014 年均实现光催化与生物耦合阳极的概念[382, 383]。然而，在这种三明治结构（由内至外：电极—赤铁矿—生物膜）的电极中，半导体会直接与生物膜接触，光照后其产生的活性物种对生物膜产生一定的伤害作用。ICPB 的概念近年被引入 MxC 中，以具有导电功能的泡沫碳为电极基底构建了光催化-生物直接耦合阳极[384, 385]。当 4-氯酚初始浓度为 25 mg/L 时，ICPB 阳极相较于光阳极和生物阳极对污染物的降解分别提高了 31%和 13%。光生电子的产生使得 ICPB 阳极产电能力也相较于生物阳极提高了 50%。在 ICPB 阳极运行时，阳极呼吸菌（ARB）的细胞色素 C 的氧化还原反应活性增强，表明 ARB 生物膜对光生电子传导起到关键作用。

4.5.4 展望

ICPB 作为一种新型的物化-生物耦合处理技术，对废水中毒性/难降解有机物的去除与矿化有着独特优势。通过材料能级匹配设计，能够与生物化学电池进一步耦合，提高生物抑制性有机物的处理与能源化效率。未来需要定向设计与生物降解耦合的物化技术，拓展研发其他物化技术与生物降解直接耦合的方法，并在实际废水处理中得以应用。

5 废水资源/能源新技术

5.1 有机碳回收技术

污水中有巨大的有机碳资源（包括污水自身的有机物及生物处理产生的污泥有机质）。未经处理的城市污水中含有的能量是传统活性污泥法所需的 9 倍[386]，其中的有机碳可提供约 1.9 kW·h·m^{-3} 的能量，然而传统的活性污泥法不仅耗费相当大量的能量用于曝气，而且还使大部分有机碳被内源呼吸消耗掉，这导致污水有机资源的浪费。因此，开发废水中有机碳的回收技术是实现可持续污水处理理念行之有效的方法，它包括如何在降低污水处理工艺能耗的同时最大化回收污水中的有机碳（即有机碳的分离富集）及有机碳高效资源化利用。

5.1.1 有机碳分离富集

1）膜技术

与其他分离方式相比，膜分离技术具有高效、设备体积小、易于自动化控制等优点。Wang 等[387]采用改进的自形成动态膜生物反应器（SF-DMBR）在连续运行条件下能回收废水中 80%的有机物。但是，膜分离技术普遍具有膜污染严重并导致膜组件使用周期短、运行成本升高等不足，它阻碍了膜技术的进一步发展。Katsuki 等[388]发现采用柠檬酸化学强化反洗的物理清洗方式能够有效控制膜污染，并且 75%的有机物被成功回收。

2）混凝-微滤

混凝-过滤是通过混凝作用产生微絮物吸附污水中有机物的一种过程，得到的污泥将进一步资源化利用，实现有机碳的回收和再利用。Jin 等[389]提出了一种利用混凝-微滤（CCM）组合系统进行污水预浓缩，反应器连续运行 295 h，平均净流量 13.3 L/(m^2·h)，浓缩后是有机物浓度达到约 16000 mg COD/L，有机物回收率接近 70%，但吸附剂滤饼层存在一定负面影响。Gong 等[390]研究了吸附剂滤饼层对实际污水的影响，结果表明吸附剂滤饼层控制了膜污染的过程，表明混合混凝/吸附微滤是一种很有前途的污水有机质回收工艺。

5.1.2 有机碳厌氧发酵制甲烷

目前，厌氧仍然是回收污水中有机碳最行之有效的技术。有机物在厌氧条件下经过微生物的水解、酸化、产氢、产乙酸及产甲烷等反应，被转化成沼气而回收。传统厌氧消化技术采用中温（35℃）、长水力停留时间（HRT=15~20 d）的操作条件，最大的缺点是处理时间长，导致处理设备体积大，基础设施投入高，并增加了维护成本。有研究表明，污泥中大颗粒有机物的初始水解成为传统厌氧消化技术的限速步骤[391]。因此，通过投加碱液、高温预处理、超声波预处理或者在高温（55~70℃）下厌氧消化等方法可以提高水解速度，从而提高甲烷产量[392-395]。例如，Li[396]等将过氧化钙和超声波相结合对污水处理产生的污泥进行预处理，不仅提高了污泥水解效果，还可以增加污泥中可生物降解有机物的比例，产甲烷效率提高了 36%。

5.1.3 微生物发酵制聚羟基烷基酸酯

聚羟基烷基酸酯（polyhydroxyalkanoates，PHA）具有良好的生物相容、生物可降解及塑料的热加工性能，可作为生物医用材料和生物可降解包装材料。利用混合菌群进行开放式发酵，将废水中有机质生物转化为小分子挥发性脂肪酸（VFAs）；再通过活性污泥中具有较高PHA合成能力的菌群进行富集，实现废水有机质转化为PHA；最后可通过分离提纯工艺，得到高纯度的PHA产品。但混菌发酵工艺有机负荷较低，导致PHA产量较低，同时还会受到菌群絮体性状失稳（污泥膨胀）的影响，使得该工艺不具备竞争优势，因此如何优化混菌发酵工艺工况条件，提升其PHA的产量成了目前所关注的热点。Xu[397]等采用游离亚硝氮处理硝化污泥，应用于A^2O-SBR和N-SBR组成的双污泥系统中，在满足营养物质去除的同时，将PHA含量从（1.46±0.10）mmol-C/g VSS提高到了（2.43±0.12）mmol-C/g VSS。Li[398]等根据相关性分析，提出了提高污泥浓度、促进胞外聚合物的生成、减小污泥絮体的粒径可以进一步促进PHA累积的方法。

5.1.4 合成生物柴油类物质

生物柴油是一种碳中性能源，可部分替代化石燃料，是典型的"绿色能源"，具有环保性能好、发动机启动性能好、燃料性能好，原料来源广泛、可再生等特性。一些微生物（包括微藻、真菌、酵母和细菌）能够积累超过细胞干重20%的油类物质。用微藻处理乳制品废水时，不仅可合成生物柴油（它由38%碳水化合物、15%蛋白质和22%脂质组成），而且去油微藻中提取的还原糖含量最高（达到54.12%），进一步将所得还原糖用酿酒酵母发酵生产生物乙醇，其产量高达116.2 mg/g[399]。但大部分研究还处在实验室规模阶段，与大规模生产存在一定差异，因此利用微藻制备生物柴油的效果及经济效益需要进行全过程的分析。

5.1.5 生产蛋白

单细胞蛋白是指富含高蛋白的食用微生物。使用微生物（藻类和细菌），将废水中的有机质及营养盐同化合成为单细胞蛋白，一直废水资源化的研究热点。例如Saejung与Thammaratana[400]报道了利用光合细菌进行废水处理并生产单细胞蛋白，发现当初始pH为7、光照强度为4000勒克斯时，生物产量及生物量生产力达到3.2 g/g COD和2.1 g/(L·d)，生物量蛋白质含量达到60.1%，有较好的蛋白质回收效果。然而，在营养物质和基质处于气态之前，细菌和藻类就停止生长，有机碳被消耗后，藻类生长受到抑制[401]，因此如何提高废水转化为蛋白质产量值得受到更多关注。此外，污水生物处理产生的剩余污泥中含有大量蛋白类物质，如何高效回收蛋白质是提高污泥资源化利用价值的一个重要方面。García等[402]首次评估了水热水解污泥作为蛋白质回收的可再生资源的潜力，并从蛋白质回收率和选择性两个方面，对水解污泥中蛋白质的分离方法进行了评价，这些分离方法通常应用于工业规模。结果表明，硫酸铵的加入是最佳的分离方法，可使蛋白质回收率达到86%以上，选择性最高。

5.1.6 制备碳材料

污泥中富含碳元素，因此将其碳化后制备成吸附材料是简单易行的资源化利用方式。一般而言，碳材料对污染物的吸附能力强弱主要由其孔径分布和总孔容积所决定，利用氮气吸附和脱附量计算得到的BET表面积是表征该项指标最常用的方法。通常，污泥直接碳化而不经过任何活化之后得到的材料，其BET最高可达到511.3 m^2/g[403]。除此之外，污泥经高温裂解得到的碳材料还可以应用于能源领域。例如，Yin[404]等将利用污泥高温热解制成的生物炭应用于微生物电解池（MEC）中，促进了污泥水解和产酸过程，增强了有机物氧化和阳极上的电子传递，使该系统的甲烷产量提高了24.7%。

5.2 磷回用技术

磷,是地球生命活动的基础元素,也是人类生存与发展的重要资源。人类活动使氮磷等营养元素过量排放进入江河湖海,超过水体自净能力时,会导致水体富营养化,对生态环境与人类健康造成不良影响[405]。另一方面,磷是一种自然资源,磷矿石不可再生,国土资源部在2012年已经将磷矿石列为"2010年后不能满足国民经济发展需求的20个矿种"之一。我国是当今世界上最大的磷肥生产国及消耗国[406],实现磷资源的可持续利用对农业与社会发展至关重要。因此,实现污水中磷资源的回收及再利用,将是行之有效的解决方法之一。

5.2.1 磷回用技术的总体发展趋势

目前,常用的废水除磷工艺包括化学沉淀法、结晶法、生物法等,研究表明,通过现有除磷工艺处理后,污水中90%的磷以化学或生物的形式存留在剩余污泥中[407]。近几年,研究者尝试开发新型磷回收技术:Li 等利用膜生物反应器强化化学除磷-厌氧消化过程,通过微生物还原反应释放污泥中的化学磷[408];Zhang 等采用聚苯乙烯基纳米水合氧化铁复合材料进行磷吸附,同时发现 Ca^{2+} 可通过形成多元复合物强化磷的去除[409];Buliauskaitė 等将纤发菌属和嘉利翁氏菌属形成的生物氧化铁用于磷回收过程,结果表明其具有较强的磷去除与回收能力[410]。

在磷回用技术发展的过程中,以微生物燃料电池(MFC)为基础的生物电化学系统(BES)能够在净化污水的同时,回收能源与氮磷资源[411],具有良好的发展潜力。BES 在磷的回收技术研究方面包括两大类:①基于富集/沉淀过程的磷回收,通过电场驱动 NH_4^+ 和 PO_4^{3-} 迁移得到较高浓度的氮磷回收液,再以鸟粪石等形式回收磷[412];②基于光合过程的磷去除/回收。Zhang 等构建了由沉积物 MFC 和微藻组成的光合生物电化学系统去除了约70%的磷[413];Jiang 等将藻类光合反应器与 MFC 系统耦合,污水依次经过阳极、阴极和藻类光合反应器,实现了92%的磷去除[414]。生物电化学系统在城镇污水及尿液中磷的回收方面,取得了一系列重要进展,为废水资源/能源新技术的发展提供了新理论与技术支撑。

5.2.2 生物电化学系统回收城镇污水中的磷

Cusick 与 Logan 在2012年提出单室型微生物电解池(MEC)同步产氢和鸟粪石,利用阴极的高 pH 使磷以鸟粪石形式结晶沉淀,磷酸盐去除率为20%~40%,氢生产率为 $0.7\sim2.3\ m^3$-$H_2/(m^3\cdot d)$,MEC 生产鸟粪石的能量消耗可被氢回收能量折减,以降低鸟粪石回收的成本[415]。Zhang 等构建了同步去除氮磷的 R^2-BES 系统,利用外加电场使得 NH_4^+ 透过阳离子交换膜,PO_4^{3-} 通过与 OH^- 置换透过阴离子交换膜,实现废水中氮磷的同步去除,在0.8 V 的外加电压下,R^2-BES 除去了83%的 NH_4^+ 和52%的 PO_4^{3-} [416]。随后,其团队提出利用微生物电解池-正渗透(MEC-FO)复合系统富集磷,以鸟粪石沉淀形式回收。模拟污泥厌氧消化液先进入 MEC 阳极室,电场作用下 NH_4^+ 通过阳离子交换膜迁移至阴极室,同时有机物降解。MEC 阳极出水进入 FO 系统,实现水回收和磷浓缩,磷回收率达到79.5%[417]。Ye 等构造了以阳离子交换膜为分隔材料的 MFC,在阴极的高 pH 下,通过化学沉淀法去除与回收污水中的磷,去除率94.9%[418]。

清华大学黄霞教授团队构建了污水净化—产电—氮磷浓缩一体化的微生物氮磷回收电池(microbial nutrient recovery cell, MNRC)(图24-24)[419]。在该团队先前研发的微生物脱盐电池(microbial desalination cell, MDC)的基础上,通过改变两电极间阴、阳离子交换膜的排列方式。运行过程中,污水依次流经阳极和阴极,MNRC 利用阳极产电微生物氧化有机物产生的电能,驱动污水中的 NH_4^+ 和 PO_4^{3-} 穿过离子交换膜向氮磷回收室迁移,从而在净化污水的同时产生电能并回收氮磷。经 MNRC 处理后,模拟污水中 COD、NH_4^+-N 和 PO_4^{3-}-P 的去除率可达90%、97%和64%,回收液中的 NH_4^+-N 和 PO_4^{3-}-P 分别可达污水中原浓度的 1.5 倍和 2.2 倍，为后续鸟粪石的生产提供了有利基础。MNRC 处理实际生活污水，氨氮和总磷的平均去除率分别为 84%和 72%，回收液中 NH_4^+-N 和总磷的浓缩倍数分别在 1.50～1.71 倍及 2.01～2.91 倍[420]。

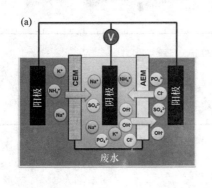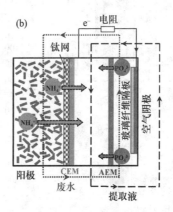

图 24-24 （a）R^2-BES[416]和（b）MNRC[419]原理示意图

在此基础上，团队构建了综合净化 BES 系统——新型的污水自持能源驱动水深度净化和氮磷回收集成系统[421]。采用堆叠构型，以多层离子交换膜构建氮磷分离中间腔室，显著提高了电流的利用效率。污水依次经过阳极、阴极和脱盐膜堆淡水室，可实现污水深度净化与同步氮磷回收。经该系统处理后，实际生活污水出水优于国家一级 A 排放标准。系统连续运行，回收液中总氮、氨氮、总磷的浓度可提升至污水中原浓度的 1.8 倍、1.4 倍、2.6 倍，并利用鸟粪石法从富集液中回收氮磷元素。

5.2.3 生物电化学系统回收尿液中的磷

尿液是重要的氮磷污染源，贡献了市政污水中 80%以上的氮元素和 50%以上的磷元素[422]，同时，尿液中的 COD 浓度为 8000～10000 mg/L，蕴含着丰富的化学能，适合作为 BES 的处理对象。Ledezma 等利用微生物电解/电渗析池回收尿液中的氮、磷、钾，可产生含有 1.87 mol/L NH_4^+-N、0.29 mol/L PO_4^{3-}-P 和 0.18 mol/L K^+ 的浓缩液，有助于促进尿液营养素在肥料或蛋白质生产行业中的再利用[423]。

清华大学先后构建了污水驱动的尿液氮磷分离回收系统[424]及尿液自驱动氮磷回收系统 U-Power（图 24-25）[421]，分别将污水及尿液中的化学能转化为电能，驱动尿液中 NH_4^+ 和 PO_4^{3-} 的跨膜迁移及浓缩回收，同时完成污水的深度净化。前者从尿液中回收了 76%～87%的 N 及 72%～93%的 P；后者对尿液中 COD、总氮、总磷的去除率分别为 93.8%、73.1%、86.2%，浓水中的氮磷回收浓度分别为 1234 mg/L 和 101 mg/L，净化及回收效率较高，不需要额外能量输入，且能稳定产电（21.3 W/cm^3）。

5.2.4 展望

废水中磷资源的回收与再利用是废水资源化发展的重要方向之一。废水磷回用技术原理包括结晶法、生物法、化学沉淀法等，以此为基础，研究者又尝试结合吸附法、离子交换法、膜技术等废水处理方法，开发新型磷回收技术。其中，BES 系统在城市污水及尿液磷回收方面的研究成果，为废水中氮磷资源的回收提供了新的发展方向和理论与技术基础。

BES 氮磷回收技术的后续研究工作可聚焦在技术原理及实际应用两方面。建议未来深入研究 BES 回收氮磷的机理与调控策略，进一步分析包括 NH_4^+ 和 PO_4^{3-} 在内的各类离子的竞争迁移规律，定向强化氮磷迁移回收，减少电荷损失，以期提升 MNRC 及综合净化生物电化学系统运行效率。实际应用方面，在已开展研究基础上，进行材料优选，降低装备成本，开展中试研究与小型示范工程。

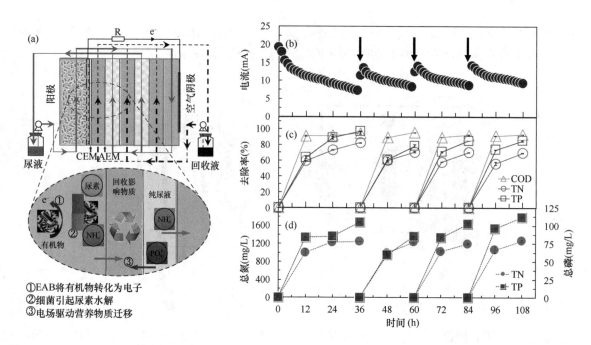

图 24-25 （a）U-Power 系统原理示意图及其（b）输出电流，（c）尿液 COD、TN、TP 去除率，（d）氮磷回收情况

5.3 重金属回收技术

矿冶、石油、化工、电镀等工业生产中，都会产生大量含有重金属例如铜、锌、镍、铬等元素的废水，并已经成为全球关注的重要环境问题[425]。由于重金属难以由微生物降解为无害物质，游离的重金属离子易与水环境中各种有机或者无机配体等结合生成络合物[426]，使重金属的迁移能力增强，促进了生物吸收，进而经食物链迁移转化和富集，对人类健康造成了严重危害。然而，这些重金属废水在造成严重环境污染问题的同时，也导致了重金属资源本身的极大浪费。因此，在实现这些重金属废水达标排放的同时，通过技术创新使重金属得以资源化回收，成为当前的研究热点之一。

重金属回收技术主要是依据化学作用和物理作用，将废水中重金属离子转化可再利用形态。目前，废水中重金属回收技术主要有两类：第一类是化学处理法，是指通过外加驱动或者投加药剂，发生化学反应去除废水中重金属离子的方法，达到将有毒物质转化为可利用新物质的目的。具体方法有电解法、光电催化法、化学沉淀法等。第二类是物理化学法，是指废水中的重金属在不改变其化学形态的条件下进行吸附、浓缩、分离的方法，这类方法的特点是重金属可以原来的形态加以回收利用。具体方法有电解法、光电催化法、化学沉淀法、膜分离法、离子交换法、吸附法、蒸发浓缩法等。

5.3.1 电解法

电解法是在电解质溶液中通以直流电流，阴离子迁移向阳极并在阳极上发生氧化反应，阳离子迁移向阴极并在阴极上发生还原反应，电解质溶液中的金属阳离子在阴极被还原并沉积在阴极板上，从而降低废水中重金属含量，阴极得以回收重金属。电解法处理废液效率高，装置紧凑、用地少、产生污泥少，便于控制管理，在国内外得到广泛应用。Colantonio 等已经证明了电解可以实现废水中重金属镉的回收，并探讨了不同条件下回收不同形式镉的机理[427]。Collivignarelli 等采用电解法成功地回收了酸洗废液中的金属铜和镍，并计算了能耗效益，为工业应用提供了一定基础[428]。电解法主要用于电镀废水等含有较高重金属浓度的处理，不适于处理较低浓度的含重金属离子的废水。

5.3.2 光电催化法

电还原将金属离子以零价态金属形式沉积是回收重金属的有效方法，但对络合态重金属回收速率慢，效率较低。光催化体系中产生的活性物种可以有效将金属络合物氧化破络合。在光催化体系中光生空穴和光生电子较容易复合，因此考虑将光催化技术与电化学技术协同作用，即光电催化法，通过在光阳极和阴极之间施加一定的偏压，将光生电子引导到阴极上，从而实现了光生电子和空穴的分离，极大地降低了光生空穴和光生电子的复合率。

赵旭等研究报道了光电催化过程可协同氧化 Cu-EDTA 络合物与同步实现电还原阴极回收铜，揭示了羟基自由基为主导的氧化破络合机理，阐明了 EDTA 氧化中间产物与二价铜形成络合物被电还原为零价铜的微观过程（图 24-26 左图）[429]。催化氧化铜氰络合物过程中，随着氰根被氧化，释放的铜在强碱性条件下水解易形成铜氧化物沉积在光阳极抑制氧化反应进行。申请者通过外加焦磷酸盐或有机络合物与释放的重金属离子快速形成络合物，抑制金属氧化物形成与在阳极沉积，促进了阴极回收，实现氰根在强碱性条件下的快速氧化与同步电还原金属回收，获得了高纯度零价态的铜（纯度>95%），申请者揭示了光生空穴与羟基自由基共同氧化氰根的作用机理[430]。研究还发现，银氰络合物氧化过程中绝大部分的单质形态的银沉积到阴极，沉积在光阳极表面的微量 Ag_2O 与光阳极半导体产生异质结，促进了光生电荷分离，提高了络合物氧化效率[431]。进一步将过硫酸盐引入到电化学系统，有效提高了重金属络合物的氧化破络合效率；研发发现过硫酸盐与铜氰络合物发生类芬顿反应生成的硫酸根自由基是强化破氰的关键物种，而沉积在阳极的铜氧化物也可活化过硫酸盐产生硫酸根自由基。进一步发现硫酸根自由基转换为羟基自由基过程中可降低体系 pH，促进电还原回收金属（图 24-26 右图）[432]。外加的过硫酸盐解决了常规电氧化污染物效率较低的问题，为高效处理重金属络合物废水与强化回收金属提供了新方法。

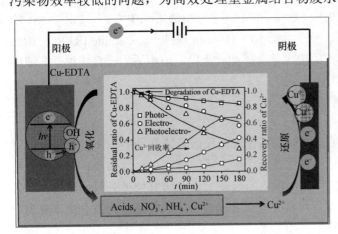 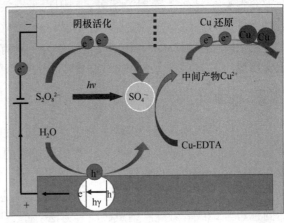

图 24-26　Cu-EDTA 光电氧化与 Cu 电还原回收（左）；过硫酸盐强化 Cu-EDTA 氧化与铜阴极还原（右）

5.3.3 化学沉淀法

化学沉淀法是在运行过程中，通过加入沉淀剂，使得金属离子与沉淀剂中的某些基团结合，产生沉淀，实现重金属的去除。按照加入药剂的不同，化学沉淀法可以分为：碱性沉淀法、硫化物沉淀法、铁盐沉淀法和螯合沉淀法等。化学沉淀法较为成熟实用，成本也低，可以用于处理大批量的重金属废水。经过这种工艺处理后得到的含重金属污泥，可制成各种工业材料，也可以将其掺入煤渣、黏土等烧制成砖的方法，达到资源化和综合利用[433]。例如，Özverdi 和 Erdem 等利用硫化物沉淀法实现了 Cu^{2+}、Cd^{2+} 和 Pb^{2+} 的有效去除[434]。化学沉淀方法对络合态重金属的沉淀去除能力有限。废水中的组分通常较复杂，化学沉淀方法难以得到高纯度的金属沉淀物。

5.3.4 膜分离法

膜分离法是利用薄膜的选择透过性，在外力作用下溶液中的溶质和溶剂分离 以达到提纯、浓缩和净化[435]。而重金属废水中的溶质成份主要是无机盐，利用膜截留方式使废水中的重金属等盐类与水体分开，得到含低盐量的水，又能获得浓度较高的金属浓缩液。在预处理工艺作用下，将废水中影响分离膜性能和寿命的大分子杂质去除，透过膜的脱盐水返回清洗槽中回用，浓缩废水返回循环水箱，以达到水质净化和重金属回用的双重目的。通过膜分离技术处理重金属废水，可以提高废水的高回收率，有效降低成本，无需引入其他化学物质，具有很好的经济和环境效益[436]。例如，我国杭州海洋所等单位开发了多种膜技术组合应用以及膜技术，在提高电镀废水中重金属离子回收方面取得了一定成果[437]。

5.3.5 离子交换法

离子交换法是目前废水中重金属回收技术中常用的方法之一。主要作用原理是利用树脂表面活性基团与废水中金属离子进行离子交换，将金属离子吸附于树脂上，吸附饱和后，可通过再生剂（酸/碱）处理后的洗脱液对重金属进行回收，而树脂通过再生剂再生进行重复使用[438]。该技术关键是：①选用吸附选择性好，交换容量大，容易再生的离子交换树脂；②回用水的质量能够满足生产需要；③回收所得的金属有资源化的渠道；④设计开发投资省、占地小、操作简单、便于推广的设备；⑤废水回用处理过程中不会引起新的污染物。其中，离子交换树脂的性能及运行条件对重金属离子的去除影响显著[439]。因此，目前国内外关于离子交换法去除水中重金属的研究主要集中在对树脂的改性和运行条件优化等方面[440-442]。例如有研究表明，利用聚乙烯亚胺改性的离子交换树脂对含 Cu^{2+} 废水进行处理，改性后的树脂交换速率、交换容量有了显著提高[443]；另外，对阳离子交换树脂 Ceralite IR120 去除电镀废水中 Cu^{2+}、Zn^{2+}、Ni^{2+} 的运行条件进行系统研究后发现，在室温、pH 4～6、反应时间 80～110 min 条件下，Ceralite IR120 树脂对污染物去除效率最高[444]。离子交换法作为一种有效的废水重金属回收手段，但不适用于直接处理高浓度重金属废水，且对于含络合态重金属废水的处理效率低。

5.3.6 吸附法

吸附法可以用来处理废水中的重金属。通过吸附材料的高比表面积或者特殊功能基团，对废水中重金属进行物理吸附或者化学吸附。目前所采用的材料主要有活性炭、活性污泥、活性焦炭、壳聚糖、粉煤灰、蟹壳、褐煤、风化煤、陶粒、膨润土等，也有利用丰富的硅藻土资源研究处理吸附金属离子，效果较好[445]。吸附法简单易操作，经济性较好，处理后可将含有重金属的吸附材料进行固液分离，分离后的固体可焚烧或者煅烧制备成砖块等方法对重金属进行回收。

5.3.7 蒸发浓缩法

蒸发浓缩法，是指将高浓度的重金属废液进行蒸发，使溶液浓缩，回收重金属的方法。可用于处理含铬、铜、银及镍离子废液。该方法的工艺简单、节约生产用水、占地小、操作简单；但因能耗大、操作费用高、一般只作为一种辅助处理手段。它对高浓度重金属废液和膜浓缩液处理是一个经济、可行的处理方法，在实际应用中有很高的推广价值。

5.3.8 结论与展望

综上所述，针对废水中重金属的回收技术，高效益大规模在工程上应用是其技术成熟的标志之一。因此，后续研究工作应主要集中围绕如何在实际废水处理中获得好的处理效果、方便的使用方式、低的使用价格、高的经济收益和低的维护成本，同时与其他工艺更好的组合等方面开展，这些工作具有重要的现实意义与价值。

参 考 文 献

[1] 曲久辉, 刘会娟. 水处理电化学原理与技术. 北京: 科学出版社, 2017.

[2] Comninellis C, Chen G. Electrochemistry for the Environment. New York: Springer, 2010.

[3] Pletcher D, Walsh F C. Industrial Electrochemistry. London: Chapman and Hall, 1982.

[4] Key B D, Howell R D, Criddle C S. Fluorinated organics in the biosphere. Environmental Science & Technology, 1997, 31: 2445-2454.

[5] Niu J, Lin H, Xu J, et al. Electrochemical mineralization of perfluorocarboxylic acids (PFCAs) by Ce-doped modified porous nanocrystalline PbO_2 film electrode. Environmental Science & Technology, 2012, 46: 10191-10198.

[6] Niu J, Lin H, Gong C, et al. Theoretical and experimental insights into the electrochemical mineralization mechanism of perfluorooctanoic acid. Environmental Science & Technology, 2013, 47: 14341-14349.

[7] Liu G, Zhou H, Teng J, et al. Electrochemical degradation of perfluorooctanoic acid by macro-porous titanium suboxide anode in the presence of sulfate. Chemical Engineering Journal, 2019, 371: 7-14.

[8] Martínez-Huitle C A, Rodrigo M A, Sires I, et al. Single and coupled electrochemical processes and reactors for the abatement of organic water pollutants: A critical review. Chemical Reviews, 2015, 115: 13362-13407.

[9] Comninellis C. Electrocatalysis in the electrochemical conversion/combustion of organic pollutants for waste water treatment. Electrochimica Acta, 1994, 39: 1857-1862.

[10] Hayfield P C S. Development of a new material-monolithic Ti_4O_7 Ebonex® ceramic. Cambridge, U.K: Royal Society of Chemistry. 2002.

[11] Radjenovic J, Sedlak D L. Challenges and opportunities for electrochemical processes as next- generation technologies for the treatment of contaminated water. Environmental Science & Technology, 2015, 49: 11292-11302.

[12] You S, Liu B, Gao Y, et al. Monolithic porous Magnéli-phase Ti_4O_7 for electro-oxidation treatment of industrial wastewater. Electrochimica Acta, 2016, 214: 326-335.

[13] Arevalo E, Calmano W. Studies on electrochemical treatment of wastewater contaminated with organotin compounds. Journal of Hazardous Materials, 2007, 146: 540-545.

[14] Woisetschlager D, Humpl B, Koncar M, et al. Electrochemical oxidation of wastewater—Opportunities and drawbacks. Water Science & Technology, 2013, 68: 1173-1179.

[15] Urtiaga A, Gomez P, Arruti A, et al. Electrochemical removal of tetrahydrofuran from industrial wastewaters: Anode selection and process scale-up. Journal of Chemical Technology & Biotechnology, 2014, 89: 1243-1250.

[16] Farhat A, Keller J, Tait S, et al. Removal of persistent organic contaminants by electrochemically activated sulfate. Environmental Science & Technology, 2015, 49: 14326-14333.

[17] McCutcheon J R. Avoiding the hype in developing commercially viable desalination technologies. Joule, 2019.

[18] Miklos David B, Remy C, Jekel M, et al. Evaluation of advanced oxidation processes for water and wastewater treatment—A critical review. Water Research, 2018, 139: 118-131.

[19] Wang J, Wang S. Activation of persulfate (PS) and peroxymonosulfate (PMS) and application for the degradation of emerging contaminants. Chemical Engineering Journal, 2018, 334: 1502-1517.

[20] Zhu B, Kalyanaraman B, Jiang G. Molecular mechanism for metal-independent production of hydroxyl radicals by hydrogen peroxide and halogenated quinones. Proceedings of the National Academy of Sciences of the United States of America, 2007,

104(45): 17575-17578.

[21] Fang G, Gao J, Dionysios D, et al. Activation of persulfate by quinones: Free radical reactions and implication for the degradation of PCBs. Environmental Science & Technology, 2013, 47(9): 4605-4611.

[22] Jian K, Duan X, Chen W, et al. Nitrogen-doped bamboo-like carbon nanotubes with Ni encapsulation for persulfate activation to remove emerging contaminants with excellent catalytic stability. Chemical Engineering Journal, 2018, 332: 98-408.

[23] Guo K, Wu Z, Yan S, et al. Comparison of the UV/chlorine and UV/H_2O_2 processes in the degradation of PPCPs in simulated drinking water and wastewater: Kinetics, radical mechanism and energy requirements. Water Research, 2018, 147: 184-194.

[24] Hou S, Li L, Shang C, et al. Degradation kinetics and pathways of haloacetonitriles by the UV/persulfate process. Chemical Engineering Journal, 2017, 320: 478-484.

[25] Lutze Holger V, Brekenfeld J, Naumov S, et al. Degradation of perfluorinated compounds by sulfate radicals—New mechanistic aspects and economical considerations. Water Research, 2017, 129: 509-519.

[26] Guo K, Wu Z, Shang C, et al. Radical chemistry and structural relationships of PPCP degradation by UV/chlorine treatment in simulated drinking water. Environmental Science & Technology, 2017, 51(18): 10431-10439.

[27] Hua Z, Guo K, Kong X, et al. PPCP degradation and DBP formation in the solar/free chlorine system: Effects of pH and dissolved oxygen. Water Research, 2019, 150: 77-85.

[28] Wang W, Wu Q Huang Nan, Wang Ting, et al. Synergistic effect between UV and chlorine (UV/chlorine) on the degradation of carbamazepine: Influence factors and radical species. Water Research, 2016, 98: 190-198.

[29] Sun P, Lee W, Zhang R, et al. Degradation of DEET and caffeine under UV/chlorine and simulated sunlight/chlorine conditions. Environmental Science & Technology, 2016, 50(24): 13265-13273.

[30] Luo X, Zheng Z, Greaves J, et al. Trimethoprim: Kinetic and mechanistic considerations in photochemical environmental fate and AOP treatment. Water Research, 2012, 46(4): 1327-1336.

[31] Wu Z, Fang J, Xiang Y, et al. Roles of reactive chlorine species in trimethoprim degradation in the UV/chlorine process: Kinetics and transformation pathways. Water Research, 2016, 104: 272-282.

[32] Anipsitakis George P, Dionysiou Dionysios D, Gonzalez Michael A. Cobalt-mediated activation of peroxymonosulfate and sulfate radical attack on phenolic compounds implications of chloride ions. Environmental Science & Technology, 2006, 40(3): 1000-1007.

[33] Mártire Daniel O, Rosso Janina A, Bertolotti Sonia, et al. Kinetic Study of the Reactions of Chlorine Atoms and $Cl_2^{\cdot-}$ Radical Anions in Aqueous Solutions. II. Toluene, Benzoic Acid, and Chlorobenzene. Journal of Physical Chemistry A, 2001, 105(22): 5385-5392.

[34] Aravindakumar Charuvila T, Nien Schuchmann Man, Rao Balijepalli S, et al. The reactions of cytidine and 2′-deoxycytidine with $SO_4^{\cdot-}$ revisited. Pulse radiolysis and product studies. Organic & Biomolecular Chemistry, 2003, 1(2): 401-408.

[35] Prasse C, Ford B, Nomura D K, et al. Unexpected transformation of dissolved phenols to toxic dicarbonyls by hydroxyl radicals and UV light. Proceedings of the National Academy of Sciences of the United States of America, 2018, 115(10): 2311-2316.

[36] Gao Z, Lin Y, Xu B, et al. Effect of UV wavelength on humic acid degradation and disinfection by-product formation during the UV/chlorine process. Water Research, 2019, 154: 199-209.

[37] Candido João P, Andrade Sandro J, Fonseca Ana L, et al. Ibuprofen removal by heterogeneous photocatalysis and ecotoxicological evaluation of the treated solutions. Environmental Science and Pollution Research, 2016, 23(19): 19911-19920.

[38] Zhang R, Yang Y, Huang C, et al. UV/H$_2$O$_2$ and UV/PDS treatment of trimethoprim and sulfamethoxazole in synthetic human urine: Transformation products and toxicity. Environmental Science & Technology, 2016, 50(5): 2573-2583.

[39] Rizzo L. Bioassays as a tool for evaluating advanced oxidation processes in water and wastewater treatment. Water Research, 2011, 45(15): 4311-4340.

[40] Fenton H J H. Oxidation of tartaric acid in presence of iron. Journal of the Chemical Society, 1894, 65: 899-910.

[41] Zhang J, Song H, Liu Y, et al. Remarkable enhancement of a photochemical Fenton-like system (UV-A/Fe(II)/PMS) at near-neutral pH and low Fe (II)/peroxymonosulfate ratio by three alpha hydroxy acids: Mechanisms and influencing factors. Separation and Purification Technology, 2019, 224: 142-151.

[42] Fiorentino A, Cucciniello R, Di Cesare A, et al. Disinfection of urban wastewater by a new photo-Fenton like process using Cu-iminodisuccinic acid complex as catalyst at neutral pH. Water Research, 2018, 146: 206-215.

[43] Malato S, Blanco J, Cáceres J, et al. Photocatalytic treatment of water-soluble pesticides by photo-Fenton and TiO$_2$ using solar energy. Catalysis Today, 2002, 76: 209-220.

[44] Dong C, Lu J, Qiu B, et al. Developing stretchable and graphene-oxide-based hydrogel for the removal of organic pollutants and metal ions. Applied Catalysis B: Environmental, 2018, 222: 146-156.

[45] Qiu B, Xing M, Zhang J. Stöber-like method to synthesize ultralight, porous, stretchable Fe$_2$O$_3$/graphene aerogels for excellent performance in photo-Fenton reaction and electrochemical capacitors. Journal of Materials Chemistry A, 2015, 3: 12820-12827.

[46] Zhao H, Qian L, Guan X, et al. Continuous bulk FeCuC aerogel with ultradispersed metal vanoparticles: An efficient 3D heterogeneous electro-Fenton cathode over a wide range of pH 3~9. Environmental Science & Technology, 2016, 50: 5225-5233.

[47] Sun Y, Li Y, Mi X, et al. Evaluation of ciprofloxacin destruction between ordered mesoporous and bulk NiMn$_2$O$_4$/CF cathode: efficient mineralization in a heterogeneous electro-Fenton-like process. Environmental Science-Nano, 2019, 6: 661-671.

[48] Sun J, Sun S, Sun J, et al. Degradation of azo dye Acid black 1 using low concentration iron of Fenton process facilitated by ultrasonic irradiation. Ultrasonics Sonochemistry, 2007, 14: 761-766.

[49] Li X, Zhang Y, Xie Y, et al. Ultrasonic-enhanced Fenton-like degradation of bisphenol A using a bio-synthesized schwertmannite catalyst. Journal of Hazardous Materials, 2018, 344: 689-697.

[50] Yang Y, Wang P, Shi S, et al. Microwave enhanced Fenton-like process for the treatment of high concentration pharmaceutical wastewater. Journal of Hazardous Materials, 2009, 168: 238-245.

[51] Kallel M, Belaid C, Boussahel R, et al. Olive mill wastewater degradation by Fenton oxidation with zero-valent iron and hydrogen peroxide. Journal of Hazardous Materials, 2009, 163: 550-554.

[52] Xu L, Wang J. A heterogeneous Fenton-like system with nanoparticulate zero-valent iron for removal of 4-chloro-3-methyl phenol. Journal of Hazardous Materials, 2011, 186: 256-264.

[53] Qin Y, Song F, Ai Z, et al. Protocatechuic acid promoted alachlor degradation in Fe(III)/H$_2$O$_2$ Fenton system. Environmental Science & Technology, 2015, 49: 7948-7956.

[54] Hou X, Huang X, Ai Z, et al. Ascorbic acid/Fe@Fe$_2$O$_3$: A highly efficient combined Fenton reagent to remove organic contaminants. Journal of Hazardous Materials, 2016, 310: 170-178.

[55] Xing M, Xu W, Dong C, et al. Metal sulfides as excellent co-catalysts for H$_2$O$_2$ decomposition in Advanced Oxidation Processes. Chem, 2018, 4: 1359-1372.

[56] Dong C, Ji J, Shen B, et al. Enhancement of H$_2$O$_2$ decomposition by the co-catalytic effect of WS$_2$ on the Fenton reaction for

the synchronous reduction of Cr(VI)and remediation of phenol. Environmental Science & Technology, 2018, 52: 11297-11308.

[57] Liu J, Dong C, Deng Y, et al. Molybdenum sulfide Co-catalytic Fenton reaction for rapid and efficient inactivation of Escherichia coli. Water Research, 2018, 145: 312-320.

[58] Gillham R W, O'Hannesin S F. Enhanced degradation of halogenated aliphatics by zero-valent iron. Groundwater, 1994, 32(6): 958-967.

[59] Su C, Puls R W, Krug T A, et al. A two and half-year-performance evaluation of a field test on treatment of source zone tetrachloroethene and its chlorinated daughter products using emulsified zero valent iron nanoparticles. Water Research, 2012, 46(16): 5071-5084.

[60] Gu C, Jia H, Li H, et al. Synthesis of highly reactive subnano-sized zero-valent iron using smectite clay templates. Environmental Science & Technology, 2010, 44(11): 4258-4263.

[61] Neumann A, Kaegi R, Voegelin A, et al. Arsenic removal with composite iron matrix filters in Bangladesh: a field and laboratory study. Environmental Science & Technology, 2013, 47(9): 4544-4554.

[62] Yan W, Ramos M A, Koel B E, et al. As (III) sequestration by iron nanoparticles: study of solid-phase redox transformations with X-ray photoelectron spectroscopy. Journal of Physical Chemistry C, 2012, 116(9): 5303-5311.

[63] Ling L, Zhang W. Visualizing arsenate reactions and encapsulation in a single zero-valent iron nanoparticle. Environmental Science & Technology, 2017, 51(4): 2288-2294.

[64] Li X, Zhang W. Sequestration of metal cations with zero valent iron nanoparticles a study with high resolution X-ray photoelectron spectroscopy (HR-XPS). Journal of Physical Chemistry C, 2007, 111(19): 6939-6946.

[65] Ling L, Zhang W. Enrichment and encapsulation of uranium with iron nanoparticle. Journal of the American Chemical Society, 2015, 137(8): 2788-2791.

[66] Ling L, Huang X Y, Zhang W X. Enrichment of precious metals from wastewater with core-shell nanoparticles of iron. Advanced Materials, 2018, 30(17): 1705703.

[67] Jiang Z, Lv L, Zhang W, et al. Nitrate reduction using nanosized zero-valent iron supported by polystyrene resins: role of surface functional groups. Water Research, 2011, 45(6): 2191-2198.

[68] Shirin S, Balakrishnan V K. Using chemical reactivity to provide insights into environmental transformations of priority organic substances: The Fe^0-mediated reduction of Acid Blue 129. Environmental Science & Technology, 2011, 45(24): 10369-10377.

[69] Luo S, Qin P, Shao J, et al. Synthesis of reactive nanoscale zero valent iron using rectorite supports and its application for Orange II removal. Chemical Engineering Journal, 2013, 223: 1-7.

[70] Wang C, Zhang W. Synthesizing nanoscale iron particles for rapid and complete dechlorination of TCE and PCBs. Environmental Science & Technology, 1997, 31(7): 2154-2156.

[71] Johnson R L, Nurmi J T, O'Brien Johnson G S, et al. Field-scale transport and transformation of carboxymethylcellulose-stabilized nano zero-valent iron. Environmental Science & Technology, 2013, 47(3): 1573-1580.

[72] Han Y, Yan W. Reductive dechlorination of trichloroethene by zero-valent iron nanoparticles: Reactivity enhancement through sulfidation treatment. Environmental Science & Technology, 2016, 50(23): 12992-13001.

[73] Zhang X, Lin Y, Shan X, et al. Degradation of 2,4,6-trinitrotoluene (TNT) from explosive wastewater using nanoscale zero-valent iron. Chemical Engineering Journal, 2010, 158(3): 566-570.

[74] Mak M S H, Rao P, Lo I M C. Effects of hardness and alkalinity on the removal of arsenic(V) from humic acid-deficient and humic acid-rich groundwater by zero-valent iron. Water Research, 2009, 43(17): 4296-4304.

[75] Bang S, Johnson M D, Korfiatis G P, et al. Chemical reactions between arsenic and zero-valent iron in water. Water Research, 2005, 39(5): 763-770.

[76] Yan W, Vasic R, Frenkel A I, et al. Intraparticle reduction of arsenite (As(III)) by nanoscale zerovalent iron (nZVI) investigated with in situ X-ray absorption spectroscopy. Environmental Science & Technology, 2012, 46(13): 7018-7026.

[77] Ling L, Zhang W. Sequestration of arsenate in zero-valent iron nanoparticles: Visualization of intraparticle reactions at angstrom resolution. Environmental Science & Technology Letters, 2014, 1(7): 305-309.

[78] Ling L, Huang X, Li M, et al. Mapping the reactions in a single zero-valent iron nanoparticle. Environmental Science & Technology, 2017, 51(24): 14293-14300.

[79] Li S, Wang W, Liang F, et al. Heavy metal removal using nanoscale zero-valent iron (nZVI): Theory and application. Journal of Hazardous Materials, 2017, 322: 163-171.

[80] Shannon M A, Bohn P W, Elimelech M, et al. Science and technology for water purification in the coming decades. Nature, 2008, 452(7185): 301-310.

[81] Fane A G, Wang R, Hu M. Synthetic membranes for water purification: status and future. Angewandte Chemie International Edition, 2015, 54: 3368-3386.

[82] Werber J R, Osuji C O, Elimelech M. Materials for next-generation desalination and water purification membranes. Nature Reviews Materials, 2016, 1: 1-15.

[83] 汪勇, 邢卫红, 徐南平. 均孔膜. 化工学报, 2016, 67(1): 27-40.

[84] Liu S, Wu C, Hung W, et al. One-step constructed ultrathin Janus polyamide nanofilms with opposite charges for highly efficient nanofiltration. Journal of Materials Chemistry A, 2017, 5(44): 22988-22996.

[85] Karan S, Jiang Z, Livingston A G. Sub-10 nm polyamide nanofilms with ultrafast solvent transport for molecular separation. Science, 2015, 348(6241): 1347-1351.

[86] Wu M, Lv Y, Yang H, et al. Thin film composite membranes combining carbon nanotube intermediate layer and microfiltration support for high nanofiltration performances. Journal of Membrane Science, 2016, 515: 238-244.

[87] Zhu Y, Xie W, Gao S, et al. Single-walled carbon nanotube film supported nanofiltration membrane with a nearly 10 nm thick polyamide selective layer for high-flux and high-rejection desalination. Small, 2016, 12(36): 5034-5041.

[88] Gao S, Zhu Y, Gong Y, et al. Ultrathin polyamide nanofiltration membrane fabricated on brush-painted single-walled carbon nanotube network support for ion sieving, ACS Nano, 2019, 13(5): 5278-5290.

[89] Tan Z, Chen S, Peng X, et al. Polyamide membranes with nanoscale Turing structures for water purification. Science, 2018, 360: 518-520.

[90] Cheng Y, Ying Y, Japip S, et al. Advanced porous materials in mixed matrix membranes. Advanced Materials, 2018, 30(47): 1802401-1802420.

[91] Ma Z, Lu X, Wu C, et al. Functional surface modification of PVDF membrane for chemical pulse cleaning. Journal of Membrane Science, 2017, 524: 389-399.

[92] Gu J, Lee S, Stafford C, et al. Molecular layer-by-layer assembled thin-film composite membranes for water desalination. Advanced Materials, 2013, 25(34): 4778-4782.

[93] 吴松. 复合增强三孔中空纤维膜的制备. 天津: 天津工业大学, 2017.

[94] Wei X, Wang Z, Zhang Z, et al. Surface modification of commercial aromatic polyamide reverse osmosis membranes by graft polymerization of 3-allyl-5, 5-dimethylhydantoin. Journal of Membrane Science, 2010, 351(1-2): 222-233.

[95] Cho K L, Hill A J, Caruso F, et al. Chlorine resistant glutaraldehyde crosslinked polyelectrolyte multilayer membranes for desalination. Advanced Materials, 2015, 27(17): 2791-2796.

[96] Klaysom C, Cath T Y, Depuydt T, et al. Forward and pressure retarded osmosis: potential solutions for global challenges in energy and water supply. Chemistry Society Review, 2013, 42(16): 6959-6989.

[97] Cath T Y, Childress A E, Elimelech M. Forward osmosis: Principles, applications, and recent developments. Journal of Membrane Science, 2006, 281(1-2): 70-87.

[98] Logan B E, Elimelech M. Membrane-based processes for sustainable power generation using water. Nature, 2012, 488(7411): 313-319.

[99] Zhao X, Lu X, Liu Z, et al. Gas-liquid interface extraction: An effective pretreatment approach to retard pore channel wetting in hydrophobic membrane application processes. Journal of Membrane Science, 2019, 574: 174-180.

[100] Li W, Yu H. Advances in energy-producing anaerobic biotechnologies for municipal wastewater treatment. Engineering, 2016, 2(4): 438-446.

[101] 孙临泉. MABR 技术在城市受污染河道修复中的应用研究. 天津: 天津大学, 2015.

[102] Wang X, Wang X, Wei Z, et al. Potent removal of cyanobacteria with controlled release of toxic secondary metabolites by a titanium xerogel coagulant. Water Research, 2018, 128: 341-349.

[103] Wang X, Gan Y, Guo S, et al. Advantages of titanium xerogel over titanium tetrachloride and polytitanium tetrachloride in coagulation: A mechanism analysis. Water Research, 2018, 132: 350-360.

[104] Zhao H, Wang L, Hanigan D, et al. Novel ion-exchange coagulants remove more low molecular weight organics than traditional coagulants. Environmental Science & Technology, 2016, 50: 3897-3904.

[105] Liu J, Cheng S, Cao N, et al. Actinia-like multifunctional nanocoagulant for single-step removal of water contaminants. Nature Nanotechnology, 2019, 14(1): 64.

[106] Nakazawa Y, Matsui Y, Hanamura Y, et al. Minimizing residual black particles in sand filtrate when applying super-fine powdered activated carbon: Coagulants and coagulation conditions. Water Research, 2018, 147: 311-320.

[107] Ding Q, Yamamura H, Yonekawa H, et al. Differences in behaviour of three biopolymer constituents in coagulation with polyaluminium chloride: Implications for the optimisation of a coagulation- membrane filtration process. Water Research, 2018, 133: 255-263.

[108] Hargreaves A J, Vale P, Whelan J, et al. Impacts of coagulation- flocculation treatment on the size distribution and bioavailability of trace metals (Cu, Pb, Ni, Zn) in municipal wastewater. Water Research, 2018, 128: 120-128.

[109] Ma B W, Li W, Liu R, et al. Multiple dynamic Al-based floc layers on ultrafiltration membrane surfaces for humic acid and reservoir water fouling reduction. Water Research, 2018, 139: 291-300.

[110] Ma B W, Wang X, Liu R, et al. Enhanced antimony(V) removal using synergistic effects of Fe hydrolytic flocs and ultrafiltration membrane with sludge discharge evaluation. Water Research, 2017, 121: 171-177.

[111] Jia P, Zhou Y, Zhang X, et al. Cyanobacterium removal and control of algal organic matter (AOM) release by UV/H_2O_2 pre-oxidation enhanced Fe(II) coagulation. Water Research, 2018, 131: 122-130.

[112] Liu B, Qu F, Yu H, et al. Membrane fouling and rejection of organics during algae-laden water treatment using ultrafiltration: A comparison between in situ pretreatment with Fe(II)/persulfate and ozone. Environmental Science & Technology, 2018, 52: 765-774.

[113] Wang Q, Zhang W, Yang Z, et al. Enhancement of anaerobic digestion sludge dewatering performance using in-situ crystallization in combination with cationic organic polymers flocculation. Water Research, 2018, 146: 19-29.

[114] Xu Q, Wang Q, Zhang W, et al. Highly effective enhancement of waste activated sludge dewaterability by altering proteins properties using methanol solution coupled with inorganic coagulants. Water Research, 2018, 138: 181-191.

[115] Xiao K, Pei K, Wang H, et al. Citric acid assisted Fenton-like process for enhanced dewaterability of waste activated sludge

with in-situ generation of hydrogen peroxide. Water Research, 140: 232-242.

[116] Peng Y, Huang H, Zhang Y, et al. A versatile MOF-based trap for heavy metal ion capture and dispersion. Nature Communications, 2018, 9(1): 187.

[117] Su X, Kushima A, Halliday C, et al. Electrochemically-mediated selective capture of heavy metal chromium and arsenic oxyanions from water. Nature Communications, 2018, 9(1): 4701.

[118] Yin X, Long J, Xi Y, et al. Recovery of silver from wastewater using a new magnetic photocatalytic ion-imprinted polymer. ACS Sustainable Chemistry & Engineering, 2017, 5(3): 2090-2097.

[119] Huang N, Zhai L, Xu H, et al. Stable covalent organic frameworks for exceptional mercury removal from aqueous solutions. Journal of the American Chemical Society, 2017, 139(6): 2428-2434.

[120] Luo X, Zhang K, Luo J, et al. Capturing lithium from wastewater using a fixed bed packed with 3-D MnO_2 ion cages. Environmental Science & Technology, 2016, 50(23): 13002-13012.

[121] Liu X, Yang L, Luo X, et al. A novel non-imprinted adsorbent with superior selectivity towards high-performance capture of Ag(I). Chemical Engineering Journal, 2018, 348: 224-231.

[122] Yu H, Shao P, Fang L, et al. Palladium ion-imprinted polymers with PHEMA polymer brushes: Role of grafting polymerization degree in anti-interference. Chemical Engineering Journal, 2019, 359: 176-185.

[123] Zou L, Shao P, Zhang K, et al. Tannic acid-based adsorbent with superior selectivity for lead(II)capture: Adsorption site and selective mechanism. Chemical Engineering Journal, 2019, 364: 160-166.

[124] Wu T, Liu C, Kong B, et al. Amidoxime-functionalized macroporous carbon self-fefreshed electrode materials for rapid and high-capacity removal of heavy metal from water. ACS Central Science, 2019, 5(4): 719-726.

[125] Kim Y, Lin Z, Jeon I, et al. Polyaniline nanofiber electrodes for reversible capture and release of mercury(II)from water. Journal of the American Chemical Society, 2018, 140(43): 14413-14420.

[126] Zhang Q, Ying G, Pan C, et al. Comprehensive evaluation of antibiotics emission and fate in the river basins of China: Source analysis, multimedia modeling, and linkage to bacterial resistance. Environmental Science & Technology, 2015, 49(11): 6772-6782.

[127] Cheng C, Cai Y, Guan G, et al. Hydrophobic‐force‐driven removal of organic compounds from water by reduced graphene oxides generated in agarose hydrogels. Angewandte Chemie, 2018, 130(35): 11347-11351.

[128] Dhaka S, Kumar R, Deep A, et al. Metal–organic frameworks (MOFs) for the removal of emerging contaminants from aquatic environments. Coordination Chemistry Reviews, 2019, 380: 330-352.

[129] Javed H, Luong D, Lee C, et al. Efficient removal of bisphenol-A by ultra-high surface area porous activated carbon derived from asphalt. Carbon, 2018, 140: 441-448.

[130] Li Y, Yang Z, Wang Y, et al. A mesoporous cationic thorium-organic framework that rapidly traps anionic persistent organic pollutants. Nature Communications, 2017, 8(1): 1354.

[131] Zhao Q, Ma W, Pan B, et al. Wrinkle-induced high sorption makes few-layered black phosphorus a superior adsorbent for ionic organic compounds. Environmental Science: Nano, 2018, 5(6): 1454-1465.

[132] Xiao Y, Hill J. Benefit of hydrophilicity for adsorption of methyl orange and electro-fenton regeneration of activated carbon-polytetrafluoroethylene electrodes. Environmental Science & Technology, 2018, 52(20): 11760-11768.

[133] Ren Y, Lin Z, Mao X, et al. Superhydrophobic, surfactant-doped, conducting polymers for electrochemically reversible adsorption of organic contaminants. Advanced Functional Materials, 2018, 28(32): 1801466.

[134] Bolisetty S, Mezzenga R. Amyloid-carbon hybrid membranes for universal water purification. Nature Nanotechnology, 2016, 11(4): 365-371.

[135] Luo X, Deng F, Min L, et al. Facile one-step synthesis of inorganic-framework molecularly imprinted TiO_2/WO_3 nanocomposite and its molecular recognitive photocatalytic degradation of target contaminant. Environmental Science & Technology, 2013, 47(13): 7404-7412.

[136] Alvarez P J J, Chan C K, Elimelech M, et al. Emerging opportunities for nanotechnology to enhance water security. Nature Nanotechnology, 2018, 13(8): 634.

[137] Sedlak D L, von Gunten U. The Chlorine Dilemma. Science, 2011, 331: 42-43.

[138] Blanca J, Chen C, Chu W H, et al. Global Trends & Challenges in Water Science, Research and Management (Disinfection section), International Water Association, 2016.

[139] Sun X F, Chen M, Wei D B, et al. Research progress of disinfection and disinfection by-products in China. Journal of Environmental Sciences, 2019, 81: 52-67.

[140] Richardson S D, Plewa M J, Wagner E D, et al. Occurrence, genotoxicity, and carcinogenicity of regulated and emerging disinfection by-products in drinking water: A review and roadmap for research. Mutation Research/reviews in Mutation Research, 2007, 636: 178-242.

[141] Shah A D, Mitch W A. Halonitroalkanes, halonitriles, haloamides, and N-nitrosamines: A critical review of nitrogenous disinfection byproduct formation pathways. Environmental Science & Technology, 2012, 46: 119-131.

[142] Zhao Y, Qin F, Boyd J M, et al. Characterization and determination of chloro and bromo-benzoquinones as new chlorination disinfection byproducts in drinking water. Analytical Chemistry, 2010, 82: 4599-4605.

[143] Pan Y, Zhang X R. Four groups of new aromatic halogenated disinfection byproducts: Effect of bromide concentration on their formation and speciation in chlorinated drinking water. Environmental Science & Technology, 2013, 47: 1265-1273.

[144] Yang M, Zhang X. Comparative developmental toxicity of new aromatic halogenated DBPs in a chlorinated saline sewage effluent to the marine polychaete *Platynereis dumerilii*. Environmental Science & Technology, 2013, 47: 10868-10876.

[145] Zhang D, Chu W, Yu Y, et al. Occurrence and stability of chlorophenylacetonitriles: A new class of nitrogenous aromatic DBPs in chlorinated and chloraminated drinking waters. Environmental Science & Technology Letters, 2018, 5: 394-399.

[146] Li X, Mitch W A. Drinking water disinfection byproducts (DBPs) and human health effects: multidisciplinary challenges and opportunities. Environmental Science & Technology, 2018, 52: 1681-1689.

[147] 高乃云, 楚文海, 严敏, 等. 饮用水消毒副产物形成与控制研究. 北京: 中国建筑工业出版社, 2011.

[148] Li Y, Li A, Xu J, et al. Formation of soluble microbial products (SMP) by activated sludge at various salinities. Biodegradation, 2013, 24: 69-78.

[149] 胡洪营, 王丽莎, 魏东斌. 污水消毒面临的技术挑战及其对策. 世界科技研究与发展, 2005, 27: 36-41.

[150] 李梦凯, 强志民, 史彦伟, 等. 紫外消毒系统有效辐射剂量测试方法研究进展. 环境科学学报, 2012, 32: 513-520.

[151] Dong H, Qiang M, Lian J, et al. Degradation of nitro-based pharmaceuticals by UV photolysis: Kinetics and simultaneous reduction on halonitromethanes formation potential. Water Research, 2017, 119: 83-90.

[152] Yang X, Guo W, Zhang X, et al. Formation of disinfection by-products after pre-oxidation with chlorine dioxide or ferrate. Water Research, 2013, 47: 5856-5864.

[153] Antonelli M, Mezzanotte V, Panouillères M. Assessment of peracetic acid disinfected effluents by microbiotests. Environmental Science & Technology, 2009, 43: 6579-6584.

[154] Linden K G, Shin G A, Faubert G, et al. UV disinfection of Giardia lamblia cysts in water Environmental Science & Technology, 2002, 36: 2519-2522.

[155] Chu W, Gao N, Yin D, et al. Impact of UV/H_2O_2 pre-oxidation on the formation of haloacetamides and other nitrogenous disinfection byproducts during chlorination. Environmental Science & Technology, 2014, 48: 12190-12198.

[156] Strous M, Heijnen J J, Kuenen J G, et al. The sequencing batch reactor as a powerful tool for the study of slowly growing anaerobic ammonium-wxidizing microorganisms. Applied Microbiology and Biotechnology, 1998, 50: 589-596.

[157] Strous M, Pelletier E, Mangenot S, et al. Deciphering the evolution and metabolism of an anammox bacterium from a community genome. Nature, 2006, 440(7085): 790-794.

[158] Kartal B, van Niftrik L, Keltjens J T, et al. Anammox-Growth Physiology, Cell Biology, and Metabolism. Advances in Microbial Physiology, 2012, 60.

[159] Van de Vossenberg J, Woebken D, Maalcke W J, et al. The metagenome of the marine anammox bacterium "*Candidatus Scalindua profunda*" illustrates the versatility of this globally important nitrogen cycle bacterium. Environmental Microbiology, 2013, 15(5): 1275-1289.

[160] Hira D, Toh H, Migita C T, et al. Anammox organism KSU-1 expresses a NirK-type copper-containing nitrite reductase instead of a NirS-type with cytochrome cd1. FEBS Letters, 2012, 586(11): 1658-1663.

[161] Ali M, Oshiki M, Okabe S. Simple, rapid and effective preservation and reactivation of anaerobic ammonium oxidizing bacterium "Candidatus Brocadia sinica". Water Research, 2014, 57(0): 215-222.

[162] Kartal B, Maalcke W J, de Almeida N M, et al. Molecular mechanism of anaerobic ammonium oxidation. Nature, 2011, 479(7371): 127-130.

[163] Kartal B, De Almeida N M, Maalcke W J, et al. How to make a living from anaerobic ammonium oxidation. FEMS Microbiology Reviews, 2013, 37(3): 428-461.

[164] Speth D R, in't Zandt M H, Guerrero-Cruz S, et al. Genome-based microbial ecology of anammox granules in a full-scale wastewater treatment system. Nature Communications, 2016, 7: 11172.

[165] Lawson C E, Wu S, Bhattacharjee A S, et al. Nature Publishing Group, 2017. Metabolic network analysis reveals microbial community interactions in anammox granules. Nature Communications, 2017, 8: 15416.

[166] Dietl A, Ferousi C, Maalcke W J, et al. The inner workings of the hydrazine synthase multiprotein complex. Nature, 2015, 527(7578): 394–397.

[167] Oshiki M, Shinyako-Hata K, Satoh H, et al. Draft genome sequence of an anaerobic ammonium-oxidizing bacterium, "*Candidatus Brocadia sinica*". Genome Announcements, 2015, 3(2): 3-4.

[168] Oshiki M, Ali M, Shinyako-hata K, et al. Hydroxylamine-dependent anaerobic ammonium oxidation (anammox) by "*Candidatus Brocadia sinica*". Environmental Microbiology, 2016, 18: 3133-3143.

[169] de Almeida N M, Neumann S, Mesman R J, et al. Immunogold localization of key metabolic enzymes in the anammoxosome and on the tubule-like structures of *Kuenenia stuttgartiensis*. Journal of Bacteriology, 2015, 197(14): 2432-2441.

[170] Akram M, Dietl A, Mersdorf U, et al. A 192-heme electron transfer network in the hydrazine dehydrogenase complex. Science Advances, 2019, 5(4): 4310.

[171] Chen T, Zheng P, Shen L. Growth and metabolism characteristics of anaerobic ammonium-oxidizing bacteria aggregates. Applied Microbiology and Biotechnology, 2013, 97(12): 5575-5583.

[172] Hou X, Liu S, Zhang Z. Role of extracellular polymeric substance in determining the high aggregation ability of anammox sludge. Water Research, 2015, 75(5): 51-62.

[173] Jia F, Yang Q, Liu X. Stratification of extracellular polymeric substances (EPS) for aggregated anammox microorganisms. Environmental Science & Technology, 2017, 51(6): 3260-3268.

[174] Zhao Y, Feng Y, Li J. Insight into the aggregation capacity of Anammox consortia during reactor start-up. Environmental Science & Technology, 2018, 52(6): 3685-3695.

[175] Yin C, Meng F, Chen G. Spectroscopic characterization of extracellular polymeric substances from a mixed culture

dominated by ammonia-oxidizing bacteria. Water Research, 2015, 68: 740-749.

[176] Guo Y, Liu S, Tang X, et al. Insight into c-di-GMP regulation in anammox aggregation in response to alternating feed loadings. Environmental Science & Technology, 2017, 51(16): 9155-9164.

[177] Zhao Y, Liu S, Jiang B, et al. 2018. Genome-centered metagenomics analysis reveals the symbiotic organisms possessing ability to cross-feed with anammox bacteria in anammox consortia. Environmental Science & Technology, 2018, 52(19): 11285-11296.

[178] Waters C M, Bassler B L. Quorum sensing: Cell-to-cell communication in bacteria. Annual Review of Cell and Developmental Biology, 2005, 21: 319-346.

[179] Cao Y, van Loosdrecht M C M, Daigger G T. Mainstream partial nitrition–anammox in municipal wastewater treatment: status, bottlenecks, and further studies. Applied Microbiology and Biotechnology, 2017, 101(4): 1365-1383.

[180] Lackner S, Gilbert E M, Vlaeminck S E, et al. Full-scale partial nitritation/anammox experiences—An application survey. Water Research, 2014, 55: 292-303.

[181] Laureni M, Fal S P, Robin O, et al. Mainstream partial nitritation and anammox: Long-term process stability and effluent quality at low temperatures. Water Research, 2016, 101: 628-639.

[182] Zhang Q, Yang W, Ngo H, et al. Current status of urban wastewater treatment plants in China. Environment International, 2016, 92-93: 11-22.

[183] Han D, Currell M, Cao G. Deep challenges for China's war on water pollution. Environmental Pollution, 2016, 218(218): 1222-1233.

[184] Chaali M, Naghdi M, Brar SK, et al. A review on the advances in nitrifying biofilm reactors and their removal rates in wastewater treatment. Journal of Chemical Technology and Biotechnology, 2018, 93(11): 3113-3124.

[185] Wei N, Shi Y, Wu G. Removal of nitrogen and phosphorus from the secondary effluent in tertiary denitrifying biofilters combined with micro-coagulation. Journal of the International Association on Water Pollution Research, 2016, 73(11): 2754.

[186] Wei N, Shi Y, Wu G, et al. Tertiary Denitrification of the secondary effluent by denitrifying biofilters packed with different sizes of quartz sand. Water, 2014, 6(5): 1300-1311.

[187] Debarbadillo C. Tertiary Denitrification and very low phosphorus limits: A practical look at phosphorus limitations on denitrification filters. Proceedings of the Water Environment Federation, 2006, 123(6): 3454-3465.

[188] Wang J, Chu L. Biological nitrate removal from water and wastewater by solid-phase denitrification process. Biotechnology Advances, 2016, 34(6): 1103-1112.

[189] Robertson W D. Nitrate removal rates in woodchip media of varying age. Ecological Engineering, 2010, 36(11): 1581-1587.

[190] Khan S T, Horiba Y, Yamamoto M, et al. Members of the family Comamonadaceae as primary poly (3-hydroxybutyrate-co-3-hydroxyvalerate)-degrading denitrifiers in activated sludge as revealed by a polyphasic approach. Applied and Environmental Microbiology, 2002, 68(7): 3206-14.

[191] Di Capua F, Papirio S, Lens P N L, et al. Chemolithotrophic denitrification in biofilm reactors. Chemical Engineering Journal, 2015, 280: 643-657.

[192] Martin K J, Nerenberg R. The membrane biofilm reactor (MBfR) for water and wastewater treatment: Principles, applications, and recent developments. Bioresource Technology, 2012, 122(10): 83-94.

[193] Evans P, Smith J, Singh T, et al. Perchlorate Destruction and Potable Water Production Using Membrane Biofilm Reduction and Membrane Filtration. ESPCT Final Report, 2014.

[194] Zhong L, Lai C, Shi L. Nitrate effects on chromate reduction in a methane-based biofilm. Water Research, 2017, 115: 130-137.

[195] Lai C, Wen L, Shi L. Selenate and Nitrate Bioreductions Using Methane as the Electron Donor in a Membrane Biofilm Reactor. Environmental Science & Technology, 2016, 50: 10179-10186.

[196] Katsuki K, Masahiko N, Yoshimasa W. Nitrate removal by a combination of elemental sulfur-based denitrification and membrane filtration. Water Research, 2002, 36(7): 1758-1766.

[197] 王茹, 赵治国, 郑平. 铁型反硝化: 一种新型废水生物脱氮技术. 化工进展, 2019, 38(4): 2003-2010.

[198] Luo Y, P Le-Clech, Henderson R K. Simultaneous microalgae cultivation and wastewater treatment in submerged membrane photobioreactors: A review. Algal Research, 2017, 24: 425-437.

[199] Gonçalves A L, Pires J C M, Simões M. A review on the use of microalgal consortia for wastewater treatment. Algal Research, 2017: 403-415.

[200] 万金保, 余晓玲, 吴永明, 等. UASB-氨吹脱-氧化沟-反渗透处理垃圾渗滤液. 水处理技术, 2019(05): 135-138.

[201] Star W R L, Abma W R, Blommers D, et al. Startup of reactors for anoxic ammonium oxidation: Experiences from the first full-scale anammox reactor in Rotterdam. Water Research, 2007, 41(18): 4149-4163.

[202] Oehmen A, Lemos P C, Carvalho G, et al. Advances in enhanced biological phosphorus removal: From micro to macro scale. Water Research, 2007, 41: 2271-2300.

[203] Yuan Z, Pratt S, Batstone D J. Phosphorus recovery from wastewater through microbial processes. Current Opinion in Biotechnology, 2012, 23(6): 878-883.

[204] Stokholm-Bjerregaard M, McIlroy S J, Nierychlo M, et al. A critical assessment of the microorganisms proposed to be important to enhanced biological phosphorus removal in full-scale wastewater treatment systems. Frontiers in Microbiology, 2017, 8: 718.

[205] He S, McMahon K D. Microbiology of 'Candidatus Accumulibacter' in activated sludge. Microbial Biotechnology, 2011, 4(5): 603-619.

[206] Skennerton C T, Barr J J, Slater F R, et al. Expanding our view of genomic diversity in Candidatus Accumulibacter clades. Environmental Microbiology, 2015, 17: 1574-1585.

[207] Mao Y, Graham D W, Tamaki H, et al. Dominant and novel clades of Candidatus Accumulibacter phosphatis in 18 globally distributed full-scale wastewater treatment plants. Scientific Reports, 2015, 5: 11857.

[208] Fernando E Y, McIlroy S J, Nierychlo M, et al. Resolving the individual contribution of key microbial populations to enhanced biological phosphorus removal with Raman–FISH. The ISME Journal, 2019, DOI: 10.1101/387795.

[209] Nielsen P H, McIlroy S J, Albertsen M, et al. Re-evaluating the microbiology of the enhanced biological phosphorus removal process. Current Opinion in Biotechnology, 2019, 57: 111-118.

[210] Zheng X, Zhou W, Wan R, et al. Increasing municipal wastewater BNR by using the preferred carbon source derived from kitchen wastewater to enhance phosphorus uptake and short-cut nitrification-denitrification. Chemical Engineering Journal, 2018, 344: 556-564.

[211] Shen N, Zhou Y. Enhanced biological phosphorus removal with different carbon sources. Applied Microbiology and Biotechnology, 2016, 100: 4735-4745.

[212] Qiu G, Zuniga-Montanez R, Law Y, et al. Polyphosphate-accumulating organisms in full-scale tropical wastewater treatment plants use diverse carbon sources. Water Research, 2019, 149: 496-510.

[213] Carvalho V C F, Freitas E B, Silva P J, et al. Oehmen A The impact of operational strategies on the performance of a photo-EBPR system. Water Research, 2018, 129: 190-198.

[214] Oyserman B O, Martirano J M, Wipperfurth S, et al. Community assembly and ecology of activated sludge under photosynthetic feast−famine conditions. Environmental Science & Technology, 2017, 51: 3165-3175.

[215] Wu D, Ekama G A, Lu H, et al. A new biological phosphorus removal process in association with sulfur cycle. Water Research, 2013, 47: 3057-3069.

[216] Rubio-Rincon F J, Welles L, Lopez-Vazquez C M, et al. Long-term effects of sulphide on the enhanced biological removal of phosphorus: The symbiotic role of Thiothrix caldifontis. Water Research, 2017, 116: 53-64.

[217] Zhang M, Wang C, Peng Y, et al. Organic substrate transformation and sludge characteristics in the integrated anaerobic anoxic oxic–biological contact oxidation (A^2/O-BCO) system treating wastewater with low carbon/nitrogen ratio. Chemical Engineering Journal, 2016, 283: 47-57.

[218] Zhao W, Huang Y, Wang M, et al. Post-endogenous denitrification and phosphorus removal in an alternating anaerobic/oxic/anoxic (AOA) system treating low carbon/nitrogen (C/N) domestic wastewater. Chemical Engineering Journal, 2018, 339: 450-458.

[219] Gao H, Liu M, Griffin J S, et al. Complete nutrient removal coupled to nitrous oxide production as a bioenergy source by denitrifying polyphosphate-accumulating organisms. Environmental Science & Technology, 2017, 51: 4531-4540.

[220] 郝晓地, 方晓敏, 李天宇, 等. 污水处理厂升级改造中的认识误区. 中国给水排水, 2018, 34(4): 11-15.

[221] Keene N A, Reusser S R, Scarborough M J, et al. Pilot plant demonstration of stable and efficient high rate biological nutrient removal with low dissolved oxygen conditions. Water Research, 2017, 121: 72-85.

[222] Cao Y, van Loosdrecht M C, Daigger G T. Mainstream partial nitrition-anammox in municipal wastewater treatment: status, bottlenecks, and further studies. Applied Microbiology and Biotechnology, 2017, 101(4): 1365-1383.

[223] Agrawal S, Seuntjens D, Cocker P D. Success of mainstream partial nitritation/anammox demands integration of engineering, microbiome and modeling insights. Current Opinion in Biotechnology, 2018, 50: 214-221.

[224] Lauren M, Weissbrodt D, Villez K, et al. Biomass segregation between biofilm and flocs improves the control of nitrite-oxidizing bacteria in mainstream partial nitritation and anammox processes. Water Research, 2019, 154: 104-116.

[225] Ma B, Qian W, Yuan C. Achieving mainstream nitrogen removal through coupling Anammox with denitratation. Environmental Science & Technology, 2017, 51(15): 8405-8413.

[226] Tao Y, Huang X, Gao D, et al. NanoSIMS reveals unusual enrichment of acetate and propionate by an anammox consortium dominated by Jettenia asiatica. Water Research, 2019, 159: 223-232.

[227] Xie G J, Liu T, Cai C. Achieving high-level nitrogen removal in mainstream by coupling anammox with denitrifying anaerobic methane oxidation in a membrane biofilm reactor. Water Research, 2018, 131: 196-204.

[228] Wu L, Ning D, Zhang B, et al. Global diversity and biogeography of bacterial communities in wastewater treatment plants. Nature Microbiology, 2019, 4(7): 1183-1195.

[229] Mei R, Nobu M K, Narihiro T. Operation-driven heterogeneity and overlooked feed-associated populations in global anaerobic digester microbiome. Water Research, 2017, 124: 77-84.

[230] Manaia C M, Rocha J, Scaccia N. Antibiotic resistance in wastewater treatment plants: Tackling the black box. Environment International, 2018, 115: 312-324.

[231] Lin S, Jin Y, Fu L. Microbial community variation and functions to excess sludge reduction in a novel gravel contact oxidation reactor. Journal of Hazardous Materials, 2009, 165(1-3): 1083-1090.

[232] 田红. 超声波在城市剩余活性污泥中的传输特性的模拟及实验研究. 重庆: 重庆大学, 2010.

[233] Liu Y. Bioenergetic interpretation on the So/X(o) ratio in substrate-sufficient batch culture. Water Research, 1996, 30(11): 2766-2770.

[234] 戴晓虎. 我国城镇污泥处理处置现状及思考. 给水排水, 2012, 38(2): 1-5.

[235] 王少坡, 孙力平, 于静洁, 等. 原位剩余污泥减量技术的研究进展. 天津城市建设学院学报, 2009, 15(2): 99-106.

[236] Mitchell T M. 机器学习. 北京: 机械工业出版社, 2003.

[237] Oonk S, Spijker J. A supervise machine-learning approach towards geochemical predictive modelling in archaeology. Journal of Archaeological Science, 2015, 59: 80-88.

[238] Pang J W, Yang S S, He L, et al. Intelligent control/operational strategies in WWTPs through an integrated Q-learning algorithm with ASM2d-guided reward. Water, 2019, 11(5): 927.

[239] Torregrossa D, Leopold U, Hernández-Sancho F, et al. Machine learning for energy cost modelling in wastewater treatment plants. Journal of Environmental Management, 2018, 223: 1061-1067.

[240] Nourani V, Elkiran G, Abba S I. Wastewater treatment plant performance analysis using artificial intelligence-an ensemble approach. Water Science & Technology, 2018, 78: 2064-2076.

[241] 国家统计局, 环境保护部. 中国环境统计年鉴 2016. 北京: 中国统计出版社, 2016.

[242] 王元月, 魏源送, 张树军. 厌氧氨氧化技术处理高浓度氨氮工业废水的可行性分析. 环境科学学报, 2013, 33(9): 2359-2368.

[243] 韦朝海, 朱家亮, 吴超飞, 等. 焦化行业废水水质变化影响因素及污染控制. 化工进展, 2011, 30(1): 225-232.

[244] 李伟成, 田哲, 戚伟康, 等. 制药废水的厌氧生化性评价及 UASB 处理. 环境工程学报, 2014, (12): 5156-5160.

[245] 贾永强, 李伟, 贾立庄, 等. Fenton 氧化深度处理高浓度造纸废水的中试实验. 环境工程学报, 2014, (1): 215-221.

[246] 李湘溪, 吴超飞, 韦朝海, 等. 焦化废水处理过程中盐分变化及其影响因素. 化工进展, 2016, (11): 3690-3700.

[247] Mutamim N S A, Noor Z Z, Hassan M A A, et al. Membrane bioreactor: Applications and limitations in treating high strength industrial wastewater. Chemical Engineering Journal, 2013, 225: 109-119.

[248] 任南琪. 高浓度难降解有机工业废水生物处理技术关键. 给水排水, 2010, 36(9): 1-3.

[249] 杨敏, 高迎新, 张昱, 等. 我国环境工程领域工业废水治理技术取得新突破——基于污染物特征的工业废水处理与资源化技术推动区域可持续发展. 科技促进发展, 2016, 12(02): 211-216.

[250] Patra S, Roy E, Madhuri R, et al. Fast and selective preconcentration of europium from wastewater and coal soil by graphene oxide/silane@Fe_3O_4 dendritic nanostructure. Environmental Science & Technology, 2015, 49(10): 6117-6126.

[251] Sun M, Chen J, Dai X M, et al. Controlled separation of low temperature coal tar based on solvent extraction-column chromatography. Fuel Process Technology, 2015, 136: 41-49.

[252] Cui W, Zheng H, Niu M, et al. Product compositions from catalytic hydroprocessing of low temperature coal tar distillate over three commercial catalysts. React Kinet Mech Catal, 2016, 119(2): 491-509.

[253] Huber P, Nivelon S, Ottenio P, et al. Coupling a chemical reaction engine with a mass flow balance process simulation for scaling management in papermaking process waters. Industrial & Engineering Chemistry Research, 2012, 52(1): 421-429.

[254] Xu L, He X, Zhi L, et al. Chlorinated methylsiloxanes generated in the papermaking process and their fate in wastewater treatment processes. Environmental Science & Technology, 2016, 50(23): 12732-12741.

[255] Shi X, Lefebvre O, Ng K K, et al. Sequential anaerobic-aerobic treatment of pharmaceutical wastewater with high salinity. Bioresource Technology, 2014, 153: 79-86.

[256] Hu H, Jiang C, Ma H, et al. Removal characteristics of DON in pharmaceutical wastewater and its influence on the N-nitrosodimethylamine formation potential and acute toxicity of DOM. Water Research, 2017, 109: 114-121.

[257] 解晓杰. 改良型微曝氧化沟/微电解/絮凝技术处理金属冶炼废水. 中国给水排水, 2018, 34(16): 65-68.

[258] 陈涛, 李宁, 晏波, 等. 稀土湿法冶炼废水污染治理技术与对策. 化工进展, 2014, 33(5): 1306-1311.

[259] Wang H M, Ren Z Y. Bioelectrochemical metal recovery from wastewater: A review. Water Research, 2014, 66: 219-232.

[260] Xu Z, Jiang Y, Zhou G. Nitrogen cycles in terrestrial ecosystems: Climate change impacts and mitigation. Environmental Reviews, 2016, 24(2): 132-143.

[261] Galloway J N, Winiwarter W, Leip A, et al. Nitrogen footprints: Past, present and future. Environment Research Letters, 2014, 9(11): 115003.

[262] Kuypers M M M, Marchant H K, Kartal B. The microbial nitrogen-cycling network. Nature Reviews Microbiology, 2018, 16(5): 263-276.

[263] Pan J, Ma J, Wu H, et al. Application of metabolic division of labor in simultaneous removal of nitrogen and thiocyanate from wastewater. Water Research, 2019, 150: 216-224.

[264] Perera M K, Englehardt J D, Tchobanoglous G, et al. Control of nitrification/denitrification in an onsite two-chamber intermittently aerated membrane bioreactor with alkalinity and carbon addition: Model and experiment. Water Research, 2017, 115: 94-110.

[265] Azari M, Walter U, Rekers V, et al. More than a decade of experience of landfill leachate treatment with a full-scale anammox plant combining activated sludge and activated carbon biofilm. Chemosphere, 2017, 174: 117-126.

[266] 冯玉杰, 张照韩, 于艳玲, 等. 基于资源和能源回收的城市污水可持续处理技术研究进展. 化学工业与工程, 2015, (5): 20-28.

[267] Moraes B S, Orrú J G T, Foresti E. Nitrogen and sulfide removal from effluent of UASB reactor in a sequencing fed-batch biofilm reactor under intermittent aeration. Journal of Biotechnology, 2013, 164(3): 378-385.

[268] Chung J, Amin K, Kim S, et al. Autotrophic denitrification of nitrate and nitrite using thiosulfate as an electron donor. Water Research, 2014, 58(1): 169-178.

[269] Fajardo C, Mosquera-Corral A, Campos J L, et al. Autotrophic denitrification with sulphide in a sequencing batch reactor. Journal of Environmental Management, 2012, 113, 552-556.

[270] Zhou H, Wei C, Zhang F, et al. Energy-saving optimization of coking wastewater treated by aerobic bio-treatment integrating two-stage activated carbon adsorption. Journal of Cleaner Production, 2018, 467-476.

[271] Zhou H, Wei C, Zhang F, et al. A comprehensive evaluation method for sludge pyrolysis and adsorption process in the treatment of coking wastewater. Journal of Environmental Management, 2019, 235: 423-431.

[272] Lovley D R. Syntrophy goes electric: Direct interspecies electron transfer. Annual Review Microbiology, 2017, 71: 643-664.

[273] Logan B E, Wallack M J, Kim K-Y, et al. Assessment of microbial fuel cell configurations and power densities. Environmental Science & Technology Letters, 2015 2: 206-214.

[274] Wang X, Feng C, Ding N, et al. Accelerated OH– transport in activated carbon air cathode by modification of quaternary ammonium for microbial fuel cells. Environmental Science & Technology, 2014, 48: 4191-4198.

[275] Rossi R, Jones D, Myung J, et al. Evaluating a multi-panel air cathode through electrochemical and biotic tests. Water Research, 2019, 148: 51-59.

[276] He W, Dong Y, Li C, et al. Field tests of cubic-meter scale microbial electrochemical system in a municipal wastewater treatment plant. Water Research, 2019, 155: 372-380.

[277] Reguera G, McCarthy K D, Mehta T, et al. Extracellular electron transfer via microbial nanowires. Nature, 2005, 435: 1098-1101.

[278] Leang C, Qian X, Mester T, et al. Alignment of the c-Type Cytochrome OmcS along Pili of Geobacter sulfurreducens. Applied and Environmental Microbiology, 2010, 76: 4080-4084.

[279] Wang F, Gu Y, O'Brien J P, et al. Structure of microbial nanowires reveals stacked hemes that transport electrons over micrometers. Cell, 2019, 177: 361-369.e10.

[280] Pirbadian S, Barchinger S E, Leung K M, et al. Shewanella oneidensis MR-1 nanowires are outer membrane and periplasmic extensions of the extracellular electron transport components. PNAS, 2014, 111: 12883-12888.

[281] Walker D J F, Martz E, Holmes D E, et al. The archaellum of methanospirillum hungatei is electrically conductive. mBio, 2019, 10: e00579-19.

[282] Summers Z M, Fogarty H E, Leang C, et al. Direct exchange of electrons within aggregates of an evolved syntrophic coculture of anaerobic bacteria. Science, 2010, 330: 1413-1415.

[283] Liu X, Zhuo S, Rensing C, et al. Syntrophic growth with direct interspecies electron transfer between pili-free Geobacter species. The ISME journal, 2018, 12: 2142-2151.

[284] Rotaru A-E, Shrestha P M, Liu F, et al. A new model for electron flow during anaerobic digestion: direct interspecies electron transfer to Methanosaeta for the reduction of carbon dioxide to methane. Energy & Environmental Science, 2014, 7: 408-415.

[285] Ha P T, Lindemann S R, Shi L, et al. Syntrophic anaerobic photosynthesis via direct interspecies electron transfer. Nature Communications, 2017, 8: 13924.

[286] McGlynn S E, Chadwick G L, Kempes C P, et al. Single cell activity reveals direct electron transfer in methanotrophic consortia. Nature, 2015, 526: 531.

[287] Du Q, Mu Q, Cheng T, et al. Real-time imaging revealed that exoelectrogens from wastewater are selected at the center of a gradient electric field. Environmental Science & Technology, 2018, 52: 8939-8946.

[288] Liu W, Cai W, Ma A, et al. Improvement of bioelectrochemical property and energy recovery by acylhomoserine lactones (AHLs) in microbial electrolysis cells (MECs). Journal of Power Sources, 2015, 284: 56-59.

[289] Zhou L, Li T, An J, et al. Subminimal inhibitory concentration (sub-MIC) of antibiotic induces electroactive biofilm formation in bioelectrochemical systems. Water Research, 2017, 125: 280-287.

[290] Liu Y, Zhang F, Li J, et al. Exclusive extracellular bioreduction of methyl orange by azo reductase-free Geobacter sulfurreducens. Environmental Science & Technology, 2017, 51: 8616-8623.

[291] Wan Y, Zhou L, Wang S, et al. Syntrophic growth of *Geobacter sulfurreducens* accelerates anaerobic denitrification. Front Microb, 2018, 9.

[292] Huang H, Cheng S, Li F, et al. Enhancement of the denitrification activity by exoelectrogens in single-chamber air cathode microbial fuel cells. Chemosphere, 2019, 225: 548-556.

[293] Morita M, Malvankar N S, Franks A E, et al. Potential for direct interspecies electron transfer in methanogenic wastewater digester aggregates. mBio, 2011, 2: e00159-11.

[294] Dang Y, Holmes D E, Zhao Z, et al. Enhancing anaerobic digestion of complex organic waste with carbon-based conductive materials. Bioresource Technology, 2016, 220: 516-522.

[295] Wang T, Zhang D, Dai L, et al. Magnetite triggering enhanced direct interspecies electron transfer: A scavenger for the blockage of electron transfer in anaerobic digestion of high-solids sewage sludge. Environmental Science & Technology, 2018, 52: 7160-7169.

[296] Yu L, Yang Z, He Q, et al. Novel gas diffusion cloth bioanodes for high-performance methane-powered microbial fuel cells. Environmental Science & Technology, 2019, 53: 530-538.

[297] Prévoteau A, Rabaey K. Electroactive biofilms for sensing: Reflections and perspectives. ACS Sensors, 2017, 2: 1072-1085.

[298] Jiang Y, Liang P, Liu P, et al. A novel microbial fuel cell sensor with biocathode sensing element. Biosensors & Bioelectronics, 2017, 94: 344-350.

[299] Prévoteau A, Clauwaert P, Kerckhof F-M, et al. Oxygen-reducing microbial cathodes monitoring toxic shocks in tap water. Biosensors & Bioelectronics, 2019, 132: 115-121.

[300] Li T, Wang X, Zhou Q, et al. Swift acid rain sensing by synergistic rhizospheric bioelectrochemical responses. ACS Sensors,

2018, 3: 1424-1430.

[301] Xiao K, Liang S, Wang X. Current state and challenges of full-scale membrane bioreactor applications: a critical review. Bioresource Technology, 2019, 271: 473-481.

[302] Wang X, Zhang J, Chang V. Removal of cytostatic drugs from wastewater by an anaerobic osmotic membrane bioreactor. Chemical Engineering Journal, 2018, 339: 153-161.

[303] 郑祥, 魏源送, 王志伟, 等. 中国水处理行业可持续发展战略研究报告(膜工业卷Ⅱ). 北京: 中国人民大学出版社, 2016.

[304] Judd S. The status of industrial and municipal effluent treatment with membrane bioreactor technology. Chemical Engineering Journal, 2016, 305: 37-45.

[305] Meng F, Zhang S, Oh Y. Fouling in membrane bioreactors: An updated review. Water Research, 2017, 114: 151-180.

[306] Liang S, Xiao Kang, Zhang S. A facile approach to fabrication of superhydrophilic ultrafiltration membranes with surface-tailored nanoparticles. Separation and Purification Technology, 2018, 203: 251-259.

[307] Ping M, Zhang X, Liu M. Surface modification of polyvinylidene fluoride membrane by atom-transfer radical-polymerization of quaternary ammonium compound for mitigating biofouling. Journal of Membrane Science, 2019, 570-571: 286-293.

[308] Zhang X, Ma J, Tang C Y. Antibiofouling polyvinylidene fluoride membrane modified by quaternary ammonium compound: direct contact-killing versus induced indirect contact-killing. Environmental Science & Technology, 2016, 50(10): 5086-5093.

[309] Huang J, Wang Z, Zhang J. A novel composite conductive microfiltration membrane and its anti-fouling performance with an external electric field in membrane bioreactors. Scientific Reports, 2015, 5: 9268.

[310] Liu Y, Cheng C, Bai S. The performance of Pd-rGO electro-deposited PVDF/carbon fiber cloth composite membrane in MBR/MFC coupled system. Chemical Engineering Journal, 2019, 365: 317-324.

[311] Zheng J, Wang Z, Ma J. Development of an electrochemical ceramic membrane filtration system for efficient contaminant removal from waters. Environmental Science & Technology, 2018, 52(7): 4117-4126.

[312] Chen W, Qian C, Hong W L. Evolution of membrane fouling revealed by label-free vibrational spectroscopic imaging. Environmental Science & Technology, 2017, 51: 9580-9587.

[313] Ma J, Wang Z, Li H. Metagenomes reveal microbial structures, functional potentials, and biofouling-related genes in a membrane bioreactor. Applied Microbiology and Biotechnology, 2016, 100(11): 5109-5121.

[314] Zhou Z, Meng F, He X. Metaproteomic analysis of biocake proteins to understand membrane fouling in a submerged membrane bioreactor. Environmental Science & Technology, 2015, 49(2): 1068-1077.

[315] Wang Y, Wu Y, Tong X. Chlorine disinfection significantly aggravated the biofouling of reverse osmosis membrane used for municipal wastewater reclamation. Water Research, 2019, 154: 246-257.

[316] 刘简, 李汉冲, 梅晓洁, 等. 基于微网分离技术的污水杂质分离工艺优化. 净水技术, 2017, 41(10): 105-110.

[317] 柳蒙蒙, 陈梅, 杨敏, 等. 基于CFD的大型膜生物反应器的设计及运行优化. 环境工程学报, 2018, 12(2): 552-558.

[318] Sun J, Liang P, Yan X. Reducing aeration energy consumption in a large-scale membrane bioreactor: Process simulation and engineering application. Water Research, 2016, 93: 205-213.

[319] Wang Y, Sheng G, Li W. Development of a novel bioelectrochemical membrane reactor for wastewater treatment. Environmental Science & Technology, 2011, 45(21): 9256-9261.

[320] Ma J, Wang Z, He D. Long-term investigation of a novel electrochemical membrane bioreactor for low-strength municipal wastewater treatment. Water Research, 2015, 78: 98-110.

[321] Suárez-Ojeda M E, Fabregat A, Stüber F, et al. Catalytic wet air oxidation of substituted phenols: Temperature and pressure

effect on the pollutant removal, the catalyst preservation and the biodegradability enhancement. Chemical Engineering Journal, 2007, 132: 105-115.

[322] Cao S, Chen G, Hu X, et al. Catalytic wet air oxidation of wastewater containing ammonia and phenol over activated carbon supported Pt catalysts. Catalysis Today, 2003, 88: 37-47.

[323] Wu Q, Hu X, Yue P, et al. Copper/MCM-41 as catalyst for the wet oxidation of phenol. Applied Catalysis B-Environmental, 2001, 32: 151-156.

[324] Quintanilla A, Fraile A F, Casas J A, et al. Phenol oxidation by a sequential CWPO-CWAO treatment with a Fe/AC catalyst. Journal of Hazardous Materials, 2007, 146: 582-588.

[325] Fortuny A, Bengoa C, Font J, et al. Bimetallic catalysts for continuous catalytic wet air oxidation of phenol. Journal of Hazardous Materials, 1999, 64: 181-193.

[326] Xu A, Yang M, Qiao R, et al. Activity and leaching features of zinc-aluminum ferrites in catalytic wet oxidation of phenol. Journal of Hazardous Materials, 2007, 147: 449-456.

[327] Rocha M A L, Del Ángel G, Torres-Torres G, et al. Effect of the Pt oxidation state and Ce^{3+}/Ce^{4+} ratio on the Pt/TiO_2-CeO_2 catalysts in the phenol degradation by catalytic wet air oxidation (CWAO). Catalysis Today, 2015, 250: 145-154.

[328] Duprez D, Delanoe F, Barbier Jr J, et al. Catalytic oxidation of organic compounds in aqueous media. Catalysis Today, 1996, 29: 317-322.

[329] 肖天存, 安立敦, 刘恒宁. 负载型贵金属催化剂的硫中毒机理——工业催化剂寿命估价方法的研究, 工业催化, 1992, 48: 53-61.

[330] Quintanilla A, Casas J A, Rodriguez J J. Hydrogen peroxide-promoted-CWAO of phenol with activated carbon. Applied Catalysis B: Environmental, 2010, 93: 339-345.

[331] Xu X, Zeng G, Peng Y, et al. Potassium persulfate promoted catalytic wet oxidation of fulvic acid as a model organic compound in landfill leachate with activated carbon. Chemical Engineering Journal, 2012, 200: 25-31.

[332] Yang L, Hu C, Nie Y, et al. Catalytic ozonation of selected pharmaceuticals over mesoporous alumina-supported manganese oxide. Environmental Science & Technology, 2009, 43(7): 2525-2529.

[333] Martins R C, Quinta-Ferreira R M. Catalytic ozonation of phenolic acids over a Mn-Ce-O catalyst. Applied Catalysis B: Environmental, 2009, 90(1-2): 268-277.

[334] Wang Y, Cao H, Chen C, et al. Metal-free catalytic ozonation on surface-engineered graphene: Microwave reduction and heteroatom doping. Chemical Engineering Journal, 2019, 355: 118-129.

[335] Ma J, Cao Z. Quantifying the perturbations of persistent organic pollutants induced by climate change. Environmental Science & Technology, 2010, 44(22): 8567-8573.

[336] Bing J, Hu C, Nie Y, et al. Mechanism of catalytic ozonation in Fe_2O_3/Al_2O_3@SBA-15 aqueous suspension for destruction of ibuprofen. Environmental Science & Technology, 2015, 49(3): 1690-1697.

[337] Wang Y, Cao H, Chen L. Tailored synthesis of active reduced graphene oxides from waste graphite: Structural defects and pollutant-dependent reactive radicals in aqueous organics decontamination. Applied Catalysis B: Environmental, 2018, 229: 71-80.

[338] Wang Y, Xie Y, Sun H. Efficient catalytic ozonation over reduced graphene oxide for *p*-hydroxylbenzoic acid (PHBA) destruction: Active site and mechanism, ACS Applied Materials & Interfaces, 2016, 8(15): 9710-9720.

[339] Kishimoto N, Nakagawa T, Asano M, et al. Ozonation combined with electrolysis of 1,4-dioxane using a two-compartment electrolytic flow cell with solid electrolyte. Water Research, 2008, 42: 379-385.

[340] Yuan S, Li Z, Wang Y. Effective degradation of methylene blue by a novel electrochemically driven process.

Electrochemistry Communications, 2013, 29: 48-51.

[341] Bakheet B, Yuan S, Li Z, et al. Electro-peroxone treatment of Orange II dye wastewater. Water Research, 2013, 47: 6234-6243.

[342] Li X, Wang Y, Zhao J, et al. Electro-peroxone treatment of the antidepressant venlafaxine: Operational parameters and mechanism. Journal of Hazardous Materials, 2015, 300: 298-306.

[343] Frangos P, Wang H, Shen W, et al. A novel photoelectro-peroxone process for the degradation and mineralization of substituted benzenes in water. Chemical Engineering Journal, 2016, 286: 239-248.

[344] Wang H, Bakheet B, Yuan S, et al. Kinetics and energy efficiency for the degradation of 1, 4-dioxane by electro-peroxone process. Journal of Hazardous Materials, 2015, 294: 90-98.

[345] Guo Z, Xie Y, Wang Y, et al. Towards a better understanding of the synergistic effect in the electro-peroxone process using a three electrode system. Chemical Engineering Journal, 2018, 337: 733-740.

[346] Guo Z, Zhou L, Cao H, et al. C_3N_4–Mn/CNT composite as a heterogeneous catalyst in the electro-peroxone process for promoting the reaction between O_3 and H_2O_2 in acid solution. Catalysis Science & Technology, 2018, 8: 6241-6251.

[347] Lovley D R, Phillips E J P. Novel mode of microbial energy metabolism: Organic carbon oxidation coupled to dissimilatory reduction of iron or manganese. Applied and Environmental Microbiology, 1988, 54, (6): 1472 - 1480.

[348] Tern N, Mejia J, He S, et al. Dual Role of Humic substances as electron donor and shuttle for dissimilatory iron reduction.Environmental Science and Technology, 2018, 52(10): 5691-5699.

[349] Podosokorskaya O A, Kadnikov V V, Gavrilov S N, et al. Characterization of *Melioribacter roseus* gen. nov., sp nov., a novel facultatively anaerobic thermophilic cellulolytic bacterium from the class Ignavibacteria, and a proposal of a novel bacterial phylum Ignavibacteriae. Environmental Microbiology, 2013, 15, (6): 1759-1771.

[350] Baek G, Kim J, Lee C. Influence of ferric oxyhydroxide addition on biomethanation of waste activated sludge in a continuous reactor. Bioresource Technology, 2014, 166: 596-601.

[351] Royer R A, Burgos W D, Fisher A S, et al. Enhancement of biological reduction of hematite by electron shuttling and Fe(II) complexation. Environmental Science & Technology, 2002, 36(9): 1939-1946.

[352] Bird L J, Bonnefoy V, Newman D K. Bioenergetic challenges of microbial iron metabolisms. Trends in Microbiology, 2011, 19(7): 330-340.

[353] Demirel B, Scherer P. The roles of acetotrophic and hydrogenotrophic methanogens during anaerobic conversion of biomass to methane: a review. Reviews in Environmental Science and Bio/Technology, 2008, 7(2): 173-190.

[354] Stams A J M, Plugge C M. Electron transfer in syntrophic communities of anaerobic bacteria and archaea. Nature Reviews Microbiology, 2009, 7(8): 568-577.

[355] De Vrieze J, Hennebel T, Boon N, et al. Methanosarcina: The rediscovered methanogen for heavy duty biomethanation. Bioresource Technology, 2012, 112: 1-9.

[356] Liu Y, Zhang Y, Quan X, et al. Optimization of anaerobic acidogenesis by adding Fe-0 powder to enhance anaerobic wastewater treatment. Chemical Engineering Journal, 2012, 192: 179-185.

[357] Meng X, Zhang Y, Li Q, et al. Adding Fe-0 powder to enhance the anaerobic conversion of propionate to acetate. Biochemical Engineering Journal, 2013, 73: 80-85.

[358] Cheng S, Xing D, Call D F, et al. Direct biological conversion of electrical current into methane by electromethanogenesis. Environmental Science & Technology, 2009, 43, (10): 3953-3958.

[359] Zhao Z, Zhang Y, Quan X, et al. Evaluation on direct interspecies electron transfer in anaerobic sludge digestion of microbial electrolysis cell. Bioresource Technology, 2016, 200: 235-244.

[360] Shrestha P M, Rotaru A, Summers Z M, et al. Transcriptomic and genetic analysis of direct interspecies electron transfer. Applied and Environmental Microbiology, 2013, 79(7): 2397-2404.

[361] Zhao Z, Li Y, Yu Q, et al. Ferroferric oxide triggered possible direct interspecies electron transfer between Syntrophomonas and Methanosaeta to enhance waste activated sludge anaerobic digestion. Bioresource Technology, 2018, 250: 79-85.

[362] Wang M, Zhao Z, Niu J, et al. Potential of crystalline and amorphous ferric oxides for biostimulation of anaerobic digestion. ACS Sustainable Chemistry & Engineering, 2019, 7, (1): 697-708.

[363] Zhao Z, Zhang Y, Li Y, et al. Potentially shifting from interspecies hydrogen transfer to direct interspecies electron transfer for syntrophic metabolism to resist acidic impact with conductive carbon cloth. Chemical Engineering Journal, 2017, 313: 10-18.

[364] Zhang Y, Sun X, Chen L, et al. Integrated photocatalytic-biological reactor for accelerated 2,4,6-trichlorophenol degradation and mineralization. Biodegradation, 2012, 23: 189-198.

[365] Li G, Park S, Rittmann B E. Developing an efficient TiO_2-coated biofilm carrier for intimate coupling of photocatalysis and biodegradation. Water Research, 2012, 46: 6489-6496.

[366] Marsolek M D, Torres C I, Hausner M, et al. Intimate coupling of photocatalysis and biodegradation in a photocatalytic circulating-bed biofilm reactor. Biotechnology and Bioengineering, 2008, 101: 83-92.

[367] Li G, Park S, Kang D, et al. 2, 4, 5-trichlorophenol degradation using a novel TiO_2-coated biofilm carrier: roles of adsorption, photocatalysis, and biodegradation. Environmental Science & Technology, 2011, 45: 8359-8367.

[368] Bai Q, Yang L, Li R, et al. Accelerating quinoline biodegradation and oxidation with endogenous electron donors. Environmental Science & Technology, 2015, 49, 11536-11542.

[369] Tang Y, Zhang Y, Yan N. The role of election donors generated from UV photolysis for accelerating pyridine biodegradation. Biotechnology and Bioengineering, 2015, 112, 1792-1800.

[370] Essam T, Amin M A, El Tayeb O, et al. Sequential photochemical-biological degradation of chlorophenols. Chemosphere, 2007, 66: 2201-2209.

[371] Botia D C, Rodriguez M S, Sarria V M. Evaluation of UV/TiO_2 and UV/ZnO photocatalytic systems coupled to a biological process for the treatment of bleaching pulp mill effluent. Chemosphere, 2012, 89: 732-736.

[372] Suryaman D, Hasegawa K, Kagaya S. Combined biological and photocatalytic treatment for the mineralization of phenol in water. Chemosphere, 2006, 65: 2502–2506.

[373] Zhou D, Xu Z, Dong S, et al. Intimate coupling of photocatalysis and biodegradation for degrading phenol using different light types: visible light vs. UV light. Environmental Science & Technology, 2015, 49: 7776-7783.

[374] Wen D, Li G, Xing R, et al. 2, 4-DNT removal in intimately coupled photobiocatalysis the roles of adsorption, photolysis, photocatalysis, and biotransformation. Environment Biotechnology, 2012, 95: 263-272.

[375] Zhao M, Shi J, Zhao Z, et al. Enhancing chlorophenol biodegradation: Using a co-substrate strategy to resist photo-H_2O_2 stress in a photocatalytic-biological reactor. Chemical Engineering Journal. 2018, 352: 255-261.

[376] Ma D, Zou D, Zhou D, et al. Phenol removal and biofilm response in coupling of visible-light driven photocatalysis and biodegradation: Effect of hydrothermal treatment temperature. Inter Biodeter Biodegrad, 2015, 104: 178-185.

[377] Dong S, Dong S, Tian X, et al. Role of self-assembly coated Er^{3+}: $YAlO_3/TiO_2$ in intimate coupling of visible-light-responsive photocatalysis and biodegradation reactions. Journal of Hazardous Materials, 2016, 302: 386-394.

[378] Halling-Sørensen B. Inhibition of aerobic growth and nitrification of bacteria in sewage sludge by antibacterial agents. Archives of Environmental Contamination and Toxicology, 2001, 40: 451-460.

[379] Li B, Zhang T. Biodegradation and adsorption of antibiotics in the activated sludge process. Environmental Science &

Technology. 2010, 44: 3468-3473.

[380] Xiong H, Zou D, Zhou D, et al. Enhancing degradation and mineralization of tetracycline using intimately coupled photocatalysis and biodegradation (ICPB). Chemical Engineering Journal, 2017, 316: 7-14.

[381] Xiong H, Dong S, Zhang J. Roles of an easily biodegradable co-substrate in enhancing tetracycline treatment in an intimately coupled photocatalytic-biological reactor. Water Research, 2018, 136: 75-83.

[382] Li D, Cheng Y, Li L, et al. Light-driven microbial dissimilatory electron transfer to hematite. Physical Chemistry Chemical Physics, 2014, 16: 23003-23011.

[383] Qian F, Wang H, Ling Y, et al. Photoenhanced electrochemical interaction between shewanella and a hematite nanowire photoanode. Nano Letters, 2014, 14: 3688-3693.

[384] Zhou D, Dong S, Shi J, et al. Intimate coupling of an N-doped TiO_2 photocatalyst and anode respiring bacteria for enhancing 4-chlorophenol degradation and current generation. Chemical Engineering Journal, 2017, 317: 882-889.

[385] Zhou D, Dong S, Ki D, et al. Photocatalytic-induced electron transfer via anode-respiring bacteria (ARB) at an anode that intimately couples ARB and a TiO_2 photocatalyst. Chemical Engineering Journal, 2018, 338: 745-751.

[386] Scherson Y D, Criddle C S. Recovery of freshwater from wastewater: Upgrading process configurations to maximize energy recovery and minimize residuals. Environmental Science & Technology, 2014, 48(15): 8420-8432.

[387] Wang L, Liu H, Zhang W, et al. Recovery of organic matters in wastewater by self-forming dynamic membrane bioreactor: Performance and membrane fouling. Chemosphere, 2018, 203: 123-131.

[388] Kimura K, Honoki D, Sato T. Effective physical cleaning and adequate membrane flux for direct membrane filtration (DMF) of municipal wastewater: Up-concentration of organic matter for efficient energy recovery. Separation and Purification Technology, 2017, 181: 37-43.

[389] Jin Z, Gong H, Temmink H, et al. Efficient sewage pre-concentration with combined coagulation microfiltration for organic matter recovery. Chemical Engineering Journal, 2016, 292(0): 130-138.

[390] Gong H, Jin Z, Wang Q, et al. Effects of adsorbent cake layer on membrane fouling during hybrid coagulation/adsorption microfiltration for sewage organic recovery. Chemical Engineering Journal, 2017, 317: 751-757.

[391] Schievano A. Can two-stage instead of one-stage anaerobic digestion really increase energy recovery from biomass? Applied Energy, 2014, 124(1): 335-342.

[392] Yang X, Wen L, Liu X, et al. Bio-augmentative volatile fatty acid production from waste activated sludge hydrolyzed at pH 12. RSC Advances, 2015, 5(62): 50033-50039.

[393] Liu X, Xu Q, Wang D, et al. Thermal-alkaline pretreatment of polyacrylamide flocculated waste activated sludge: Process optimization and effects on anaerobic digestion and polyacrylamide degradation. Bioresource Technology, 2019, 281: 158-167.

[394] Cheng C, Zhou Z, Qiu Z, et al. Enhancement of sludge reduction by ultrasonic pretreatment and packing carriers in the anaerobic side-stream reactor: Performance, sludge characteristics and microbial community structure. Bioresource Technology, 2018, 249: 298-306.

[395] Li Y, Chen Y, Wu J. Enhancement of methane production in anaerobic digestion process: A review. Applied Energy, 2019, 240: 120-137.

[396] Li X, Liu Y, Xu Q, et al. Enhanced methane production from waste activated sludge by combining calcium peroxide with ultrasonic: Performance, mechanism, and implication. Bioresource Technology, 2019, 279: 108-116.

[397] Xu Q, Liu X, Yang G, et al. Free nitrous acid-based nitrifying sludge treatment in a two-sludge system obtains high polyhydroxyalkanoates accumulation and satisfied biological nutrients removal. Bioresource Technology, 2019, 284: 16-24.

[398] Li H, Zhang J, Shen L, et al. Production of polyhydroxyalkanoates by activated sludge: Correlation with extracellular polymeric substances and characteristics of activated sludge. Chemical Engineering Journal, 2019, 361: 219-226.

[399] Hemalatha M, Sravan J S, Min B, et al. Microalgae-biorefinery with cascading resource recovery design associated to dairy wastewater treatment. Bioresource Technology, 2019, 284: 424-429.

[400] Saejung C, Thammaratana T. Biomass recovery during municipal wastewater treatment using photosynthetic bacteria and prospect of production of single cell protein for feedstuff. Environmental Technology, 2016, 23: 3055-3061.

[401] Rasouli Z, Valverde-Perez B, D'Este M, et al. Nutrient recovery from industrial wastewater as single cell protein by a co-culture of green microalgae and methanotrophs. Biochemical Engineering Journal, 2018, 134: 129-135.

[402] García M, Urrea J L, Collado S, et al. Protein recovery from solubilized sludge by hydrothermal treatments. Waste Management, 2017, 67: 278-287.

[403] Wang G, Li Q, Dzakpasu M, et al. Impacts of different biochar types on hydrogen production promotion during fermentative co-digestion of food wastes and dewatered sewage sludge. Waste Management, 2018, 80: 73-80.

[404] Yin C, Shen Y, Yuan R, et al. Sludge-based biochar-assisted thermophilic anaerobic digestion of waste-activated sludge in microbial electrolysis cell for methane production. Bioresource Technology, 2019, 284: 315-324.

[405] Muenich R L, Kalcic M, Scavia D. Evaluating the impact of legacy P and agricultural conservation practices on nutrient loads from the Maumee river watershed. Environmental Science & Technology, 2016, 50(15): 8146-8154.

[406] Yuan Z, Jiang S, Sheng H, et al. Human perturbation of the global phosphorus cycle: Changes and consequences. Environmental Science & Technology, 2018, 52(5): 2438-2450.

[407] Cornel P, Schaum C. Phosphorus recovery from wastewater: needs, technologies and costs. Water Science and Technology, 2009, 59(6): 1069-1076.

[408] Li R, Wang X, Li X. A membrane bioreactor with iron dosing and acidogenic co-fermentation for enhanced phosphorus removal and recovery in wastewater treatment. Water Research, 2018, 129: 402-412.

[409] Zhang Y, She X, Gao X, et al. Unexpected favorable role of Ca^{2+} in phosphate removal by using nanosized ferric oxides confined in porous polystyrene beads. Environmental Science & Technology, 2019, 53(1): 365-372.

[410] Buliauskaitė R, Wilfert P, Suresh Kumar P, et al. Biogenic iron oxides for phosphate removal. Environmental Technology, 2018: 1-7.

[411] Sun M, Zhai L, Li W, et al. Harvest and utilization of chemical energy in wastes by microbial fuel cells. Chemical Society Reviews, 2016, 45(10): 2847-2870.

[412] Jain A, He Z. "NEW" resource recovery from wastewater using bioelectrochemical systems: Moving forward with functions. Frontiers of Environmental Science & Engineering, 2018, 12(4): 1.

[413] Zhang Y, Noori J S, Angelidaki I. Simultaneous organic carbon, nutrients removal and energy production in a photomicrobial fuel cell (PFC). Energy & Environmental Science, 2011, 4(10): 4340-4346.

[414] Jiang H, Luo S, Shi X, et al. A novel microbial fuel cell and photobioreactor system for continuous domestic wastewater treatment and bioelectricity generation. Biotechnology Letters, 2012, 34(7): 1269-1274.

[415] Cusick R D, Logan B E. Phosphate recovery as struvite within a single chamber microbial electrolysis cell. Bioresource Technology, 2012, 107: 110-115.

[416] Zhang F, Li J, He Z. A new method for nutrients removal and recovery from wastewater using a bioelectrochemical system. Bioresource Technology, 2014, 166: 630-634.

[417] Zou S, Qin M, Moreau Y, et al. Nutrient-energy-water recovery from synthetic sidestream centrate using a microbial electrolysis cell - forward osmosis hybrid system. Journal of Cleaner Production, 2017, 154: 16-25.

[418] Ye Y, Ngo H, Guo W, et al. Feasibility study on a double chamber microbial fuel cell for nutrient recovery from municipal wastewater. Chemical Engineering Journal, 2019, 358: 236-242.

[419] Chen X, Sun D, Zhang X, et al. Novel self-driven microbial nutrient recovery cell with simultaneous wastewater purification. Scientific Reports, 2015, 5: 15744.

[420] Gao Y, Sun D, Wang H, et al. Urine-powered synergy of nutrient recovery and urine purification in a microbial electrochemical system. Environmental Science: Water Research & Technology, 2018, 4(10): 1427-1438.

[421] Zuo K, Cai J, Liang S, et al. A ten liter stacked microbial desalination cell packed with mixed ion-exchange resins for secondary effluent desalination. Environmental Science & Technology, 2014, 48(16): 9917-9924.

[422] Larsen T A, Gujer W. Separate management of anthropogenic nutrient solutions (human urine). Water Science and Technology, 1996, 34(3): 87-94.

[423] Ledezma P, Jermakka J, Keller J, et al. Recovering ntrogen as a solid without chemical dosing: Bio-electroconcentration for recovery of nutrients from urine. Environmental Science & Technology Letters, 2017, 4(3): 119-124.

[424] Chen X, Gao Y, Hou D, et al. The microbial electrochemical current accelerates urea hydrolysis for recovery of nutrients from source-separated urine. Environmental Science & Technology Letters, 2017, 4(7): 305-310.

[425] Fu F, Wang Q. Removal of heavy metal ions from wastewaters: A review. Journal of Environmental Management, 2011, 92: 407-418.

[426] Fu F, Xie L, Tang B, et al. Application of a novel strategy—Advanced Fenton-chemical precipitation to the treatment of strong stability chelated heavy metal containing wastewater. Chemical Engineering Journal, 2012, 189-190: 283-287.

[427] Colantonio N, Kim Y. Cadmium(II)removal mechanisms in microbial electrolysis cells. Journal of Hazardous Materials, 2016, 311: 134-141.

[428] Collivignarelli M C, Abbà A, Bestetti M, et al. Electrolytic recovery of nickel and copper from acid pickling solutions used to treat metal surfaces. Water, Air, & Soil Pollution, 2019, 230: 101.

[429] Zhao X, Guo L B, Zhang B F, et al. Photoelectrocatalytic oxidation of CuII-EDTA at the TiO_2 electrode and simultaneous recovery of CuII by electrodeposition. Environmental Science & Technology, 2013, 47: 4480-4488.

[430] Zhao X, Zhang J J, Qiao M, et al. Enhanced photoelectrocatalytic decomposition of copper cyanide complexes and simultaneous recovery of copper with a Bi_2MoO_6 electrode under visible light by $EDTA/K_4P_2O_7$. Environmental Science & Technology, 2015, 49: 4567-4574.

[431] Qi F J, Yang B, Wang Y B, et al. H_2O_2 Assisted photoelectrocatalytic oxidation of Ag-Cyanide complexes at metal-free g-C_3N_4 photoanode with simultaneous Ag recovery. ACS Sustainable Chemistry & Engineering, 2017, 5: 5001-5007.

[432] Zeng H, Liu S, Chai B, et al. Enhanced photoelectrocatalytic decomplexation of Cu-EDTA and Cu recovery by persulfate activated by UV and cathodic reduction. Environmental Science & Technology, 2016, 50: 6459-6466.

[433] 吴长淋. 电镀污泥的性质及资源化研究进展. 资源节约与环保, 2018, 4: 94+123.

[434] Ozverdi A, Erdem M. Cu^{2+}, Cd^{2+} and Pb^{2+} adsorption from aqueous solutions by pyrite and synthetic iron sulphide. Journal of Hazardous Materials, 2006, 137: 626-632.

[435] Vaneeckhaute C, Darveau O, Meers E. Fate of micronutrients and heavy metals in digestate processing using vibrating reversed osmosis as resource recovery technology. Separation and Purification Technology, 2019, 223: 81-87.

[436] 王乐译. 膜法处理电镀废水. 水处理技术, 2012, 3: 131-132.

[437] 周大众, 韩锡荣, 黄浩, 等. 电镀废水中重金属处理技术研究现状与发展. 广州化工, 2014, 42: 16-18.

[438] Prelot B A I, Marchandeau F, Zajac J. On the real performance of cation exchange resins in wastewater treatment under conditions of cation competition: the case of heavy metal pollution. Environmental Science and Pollution Research, 2014, 21:

9334-9343.

[439] Alyuz B, Veli S. Kinetics and equilibrium studies for the removal of nickel and zinc from aqueous solutions by ion exchange resins. Journal of Hazardous Materials, 2009, 167: 482-488.

[440] Asci Y, Kaya S. Sorption of cobalt(II) from an aqueous medium using Amberlite 200C and Dowex 88 resins: Equilibrium and kinetic studies. Desalination and Water Treatment, 2016, 57: 13091-13105.

[441] Fu Y, Zhao J, Wang Q. Wan L, Adsorption efficiency of lignin-based ion exchange resin on heavy metal ions. Chinese Journal of Environmental Engineering, 2016, 10: 4314-4318.

[442] Pal P, Banat F. Comparison of heavy metal ions removal from industrial lean amine solvent using ion exchange resins and sand coated with chitosan. Journal of Natural Gas Science and Engineering, 2014, 18: 227-236.

[443] Revathi M, Basha C A, Velan M. Removal of copper(II) ions from synthetic electroplating rinse water using polyethyleneimine modified ion-exchange resin. Desalination and Water Treatment, 2016, 57: 20350-20367.

[444] Meyyappan, Saravanan M, Chiya A B, et al. Removal of copper, nickel, and zinc ions from electroplating rinse water. Clean-Soil Air Water, 2012, 40: 66-79.

[445] Wang Y, Wang X, Li J, et al. Coadsorption of tetracycline and copper(II) onto struvite loaded zeolite—An environmentally friendly product recovered from swine biogas slurry. Chemical Engineering Journal, 2019, 371: 366-377.

作者：尤世界[1]，方晶云[2]，邢明阳[3]，凌岚[4]，武春瑞[5]，赵华章[6]，罗旭彪[7]，楚文海[4]，刘思彤[6]，赵和平[8]，邱光磊[9]，王爱杰[10]，韦朝海[9]，王鑫[11]，王志伟[4]，曹宏斌[12]，谢勇冰[12]，赵赫[12]，张耀斌[13]，周丹丹[14]，陈银广[4]，张潇源[15]，赵旭[10]，张英[11]，周明华[11]，黄霞[15]，韦朝海[9]，俞汉青[16]

[1]哈尔滨工业大学，[2]中山大学，[3]华东理工大学，[4]同济大学，[5]天津工业大学，[6]北京大学，[7]南昌航空大学，[8]浙江大学，[9]华南理工大学，[10]中国科学院生态环境研究中心，[11]南开大学，[12]中国科学院过程工程研究所，[13]大连理工大学，[14]东北师范大学，[15]清华大学，[16]中国科学技术大学

第25章 矿山环境污染控制研究进展

- 1. 引言 /719
- 2. 源头控制——矿山尾矿中重金属释放抑制技术 /720
- 3. 过程控制——酸性矿山废水中重金属去除研究 /726
- 4. 次生矿物——酸性矿山废水中重金属自然归宿 /731
- 5. 生态恢复——矿山废弃地治理技术研究及应用 /736
- 6. 展望 /738

本章导读

酸性矿山废水（AMD）主要由尾矿中的金属硫化物矿物（如黄铁矿、磁黄铁矿、黄铜矿、闪锌矿等）在微生物、Fe^{3+}、O_2等的综合作用下形成，对 AMD 形成机制及治理的研究不仅有利于矿山的可持续发展，而且对矿区环境保护与生态安全有着重要意义。矿山环境污染控制是一项复杂的系统工程，为了减少 AMD 的产生，人们开发了多种尾矿综合利用和处理处置技术；然而这仍不可避免会产生一些 AMD，因此人们研发了系列的 AMD 处理技术，旨在降低 AMD 对下游环境和生态安全的影响。本章将针对 AMD 产生的源头控制、迁移的过程控制进行较全面的综述，并对 AMD 中次生矿物的形成与转化及其对重金属环境行为的影响进行较系统的总结，对矿山废弃地极端环境微生物生态、生态恢复技术研究与应用实践现状进行较深入的探讨。

关键词

矿山污染，源头控制，吸附去除，次生矿物，生态恢复，重金属

1 引　言

矿山环境是一种重要的原生与次生环境的复合体，特别是金属矿山，既是资源的集中地，又是生态环境的污染源。矿业活动给矿山地球化学环境及周围生态系统带来了巨大的影响，如废矿与尾矿的排放堆积、酸性矿山排水、河流与土壤重金属污染、破坏地下水及生态平衡等。在硫化物金属矿山开采过程中这些影响极为明显，金属硫化物的氧化会释放出大量的酸性矿山废水（acid mine drainage，AMD）。AMD 往往具有较强的酸性，严重的地方 pH 可低至 2~3，有时甚至低于 2，AMD 水体中富含 SO_4^{2-}、Fe^{3+} 以及大量的生物可利用形态的有毒有害重金属离子（如铅、镉、砷、铜、铬等）[1]。AMD 具有很大的危害性，它不但会增强有毒有害重金属元素的化学活动性及生物可利用性，同时还会造成土壤和水体的酸化，使污染环境的修复变得更加困难[2]。以往的研究表明，在矿山关闭后几十年甚至更长时间内，采选冶集中区及尾矿堆积区产生的污染都很难自然消除，其周边的生态环境也很难再恢复到最原始的状态[3, 4]。

我国作为矿业大国，长期以来的矿山开采及利用，造成了大量的尾矿堆积。而且由于目前我国涉及矿山开采方面的环保法规尚不够完善，采矿业造成的环境污染现象较为普遍，其后果也非常严重。如广东韶关著名的"癌症村"上坝村，由于村民的饮用水源和农业灌溉水源——横石河上游接纳了多金属硫化物矿山"大宝山"矿排出的矿山废水，使得耕地中铅和镉的浓度超标十几到几十倍，村民因癌症死亡的人数占总死亡人数的 80%以上，此外皮肤病、肝病也是常见多发病。矿山的开采与矿石的冶炼已经成为我国环境介质（水体、土壤、大气和生态系统）中重金属的最主要来源之一。

近几十年来，环境学家们在如何消除 AMD 的污染方面做了大量的研究。如采用中和沉淀、生物吸收、生物转化、生物络合、生物沉淀及湿地处理等方法对矿山酸性废水进行达标处理；利用植物修复等方法对已被矿山酸性废水污染的土壤进行修复等。这些工作对矿山环境的治理及生态环境修复等方面做出的贡献是显著的。然而，不管采用任何处理方法，AMD 的处理都是一项技术复杂、费用高昂的末端治理工程。通过技术手段从源头上防止 AMD 的产生才是现在 AMD 污染治理的优先选择。AMD 污染治理和其他很多环境污染治理情况一样，应该"防重于治"。源头控制方法是 AMD 的前期控制技术，无论在技术上还是在费用上都要比 AMD 的后期处理具有优势。

2 源头控制——矿山尾矿中重金属释放抑制技术

金属硫化物矿山在开采过程中会产生大量的尾矿，其中所包含的重金属通常以金属硫化物矿物（如黄铁矿、磁黄铁矿、黄铜矿、方铅矿、闪锌矿、砷黄铁矿等）的形式存在于尾矿之中。这些金属硫化物矿物在自然风化的作用下，容易通过化学和生物氧化作用形成 AMD。AMD 是采矿活动带来的最严重的环境问题之一。在自然条件下，AMD 的形成主要受制于尾矿或废石堆中金属硫化物矿的氧化反应。因此，为了对金属硫化物矿物氧化所引起的环境问题有一个科学的认识，也为了对由此引起的污染进行有效控制，近几十年来，人们针对金属硫化物矿物的氧化过程做了大量的研究工作，对金属硫化物矿物的氧化机理有了较为深入的探讨。以黄铁矿为例，黄铁矿在氧化过程中会生成硫酸，使得黄铁矿周围的环境成酸性。在反应初期，黄铁矿的氧化主要以空气中的氧气为主，活性三价铁离子含量较低，因此黄铁矿的氧化速度较慢。随着黄铁矿氧化反应的持续进行，反应中产生的酸使得黄铁矿周围环境的 pH 逐渐降低，当环境中的 pH 降为 2.5 以下时，Fe^{3+} 的水解产物大量的溶解，使得 Fe^{3+} 的反应活性迅速增加，因而，黄铁矿的氧化变得更为迅速。Mazumder 等[5]发现，当溶液的 pH 低于 3.0 时，Fe^{3+} 引起的黄铁矿氧化速率要比由 O_2 引起的氧化反应速率高出 3~4 个数量级。另外，有研究显示微生物（如氧化亚铁硫杆菌）的存在，能够极大增大黄铁矿的氧化速率。在微生物的催化作用下，黄铁矿的氧化速率可提高 10^4 倍以上[6]。因此要想从源头上控制 AMD 的形成，必须阻碍黄铁矿等硫化物矿物与环境中的空气、水和其他氧化剂发生接触，抑制或减缓硫化物矿物的氧化反应速率，从而在根本上减少 AMD 的产生。

2.1 物理屏蔽法

物理屏蔽法是阻隔金属硫化物尾矿与外界环境接触的较为传统的方法之一，其主要是通过在尾矿顶部设置覆盖层以减少氧气和降水的进入，降低硫化矿物的氧化速率。物理屏蔽一般可根据上部覆盖层性质分为干式覆盖法和湿式覆盖法两大类。

2.1.1 干式覆盖法

干式覆盖法是利用一层具有极低渗透系数的覆盖材料将尾矿密封，从而减少氧气和水分向尾矿的渗透。土壤是干式覆盖法中最常用的覆盖材料，Payant 等[7]在一个基于实验室的研究中表明，与未覆盖的尾矿相比，土壤覆盖法能使硫化矿尾矿的产酸速率降低三个数量级。而 Adu-Wusu 等[8]通过 3 年的现场监测发现，采用黏土作为覆盖层能使进入尾矿库的降水量减少 93%。然而，采用土壤覆盖法防止黄铁矿氧化的效果会随着时间推移而变化。如研究者对澳大利亚北部地区 RumJungle 矿区废弃岩堆表面土覆盖的有效性进行了评价，发现尽管保护层最初降低了硫化物脱氧化率，但其氧化速率随着时间的推移而迅速增加[9]。同样地，在安大略省贝塞尔白矿，尽管有土壤覆盖，黄铁矿的氧化几年后仍会迅速发生[10]。而 Swanson 等[11]发现土壤覆盖在经历了急性雨季和旱季的地区是无效的，因为在干燥过程中容易出现裂缝，这种裂缝降低了土壤覆盖的效果。除了使用土壤作为干式覆盖材料外，还可以使用土工膜等人工合成材料对尾矿堆进行覆盖，Meek 等[12]研究了美国西弗吉尼亚州的一个采用土工膜进行覆盖的尾矿库工程，监测发现，在进行覆盖后，尾矿库表面尾矿砂的氧化程度降低了 70%。但用这类材料覆盖大量废物需要的成本过高，这也妨碍了其常规的现场使用[13]。此外，塑料或聚合物衬垫容易开裂，维修费用非常高。

而作为人工合成材料的替代物，一些固体废弃物也曾被用作硫化物尾矿堆的覆盖材料。如粉煤灰等固体废弃物就曾被用于采矿废弃物处置场上部覆盖层的制作材料[14]。粉煤灰中含有大量的碱性物质，这些碱性物质可以与酸性矿山废水中的硫酸根反应生成石膏类物质，它们可以不断填堵粉煤灰之间的孔隙，使其渗透系数降低，最终可以有效隔绝采矿废弃物与空气和水的接触。木屑、树皮、城市污泥和泥炭等有机废物同样可为硫化物尾矿提供一个 pH 缓冲区，中和硫化矿氧化过程中产生的酸并消耗氧气，在尾矿中创造缺氧条件，从而抑制氧化亚铁硫杆菌的生长。Yanful 等[15]就曾对以树皮作为覆盖材料的有效性进行了评估，他们认为，树皮提供的有机质可以导致废石中微生物产生的可溶性生长产物与游离 Fe^{3+} 发生络合作用，而这种络合降低了三价铁作为氧化剂的有效性。同样，Reardon 等[16]也曾评估过木材废物在限制硫化物尾矿氧化的潜力，发现它提供了一个有效的氧气屏障，然而这种方法的有效期却很短。另外，Backes 等[17]研究了在粪肥和污泥等有机废物存在下黄铁矿的氧化。他们发现这些有机物质对黄铁矿氧化有抑制作用。当然，尽管这些有机残留物可以为 AMD 提供短期的解决方案，而且具有成本效益的优势，但它们还存在一些其他问题，如从污泥中提取的有机物质含有有机酸和重金属等有害物质，这些有害物质有可能污染矿区周围的环境。例如，Pond 等[18]曾报道，尽管有机污泥等能控制 AMD 的产生，减少铜矿尾矿中 SO_4^{2-} 的浓度，但却增加了有毒金属如砷、铜、镍和锌的浓度。另外，使用植被覆盖可以获得与其他有机材料相似的 AMD 控制结果。然而，植被也只能提供短期的解决办法。如 Fugill 等[19]曾报道，在煤炭上建立植被可以减少酸的产生，但只能维持不到两个生长季。

2.1.2 湿式覆盖法

湿式覆盖法是另一种应用广泛的酸性矿山废水源头控制方法。对于已经产生的采矿废弃物，当地形条件能够满足要求时，可以通过灌水对其进行覆盖，或直接将硫化物尾矿存于湖泊底部。通过在水中处理这些废物，为尾矿创造了高度缺氧的条件，从而抑制了硫化物矿物的氧化。实验室和实地研究都曾表明，水下处理是稳定硫化物尾矿的有效方法之一。Woikersdorfer 等[20]通过对挪威的一个尾矿库研究发现，经过湿式覆盖后，尾矿库周围地下水的 pH 由原来的 2.0 升高到 4.0，铜离子的浓度由原来的 530 mg/L 降低到 1 mg/L。

对于大型矿山，水覆盖法曾是他们减少氧气与含硫化合物废物接触的常用方法之一，因为与其他选择相比，水覆盖法的成本相对较低。但是，对于多数地表开采区水覆盖法可能不是一个切实可行的替代办法。这项技术仅限于可永久改变地下水位以完全覆盖硫化物废物的地点，或者可以将硫化物矿物放入湖中的地点。Johnson 等[6]认为，在矿井只是部分被淹没地点，水覆盖是有问题的。这是由于地下水位的季节性升降，会使得黄铁矿暴露在大气中时会发生氧化。而矿山废弃物在湖泊中的倾倒则需要严格的工程设计和适当的维护，以最大限度地减少堤坝破坏的风险，这往往是不符合成本效益的。

2.2 表面钝化法

钝化法是利用化学反应在金属硫化物矿物颗粒的表面形成一层不溶的、惰性的和致密的保护膜，从而使外界的氧化剂无法与硫化物矿物本体接触。与传统覆盖法相比，钝化处理是从微观（单个颗粒）角度来保护金属硫化物矿物。该法具有操作简单、成本低的优点，是最具发展前景的方法之一。近几十年来研究者已陆续发现了多种有效的金属硫化物矿物表面钝化剂。

2.2.1 无机钝化剂

1）磷酸盐

将磷酸盐应用于金属硫化物尾矿中可在其矿物表面形成稳定的磷酸铁钝化层，物理上抑制氧化剂与硫化矿本体的接触，从而限制硫化物矿物的进一步氧化[21-23]。磷酸盐钝化层的形成是通过将硫化物矿物与含氧化剂（H_2O_2）、缓冲液和磷酸盐（KH_2PO_4）的溶液相接触而形成的[24]。在此过程中，金属硫化矿在过氧化氢的存在下被氧化，在矿物表面产生高价金属离子，而高价金属离子则可与磷酸根反应生成耐酸沉淀，包覆在金属硫化物矿物表面[25]。

Huang 等[26,27]报道，用 0.5% H_2O_2+0.02 mol/L KH_2PO_4 淋洗草霉状黄铁矿 1000 min 后，黄铁矿的氧化得到有效控制。与 0.5% H_2O_2（对照）和 0.5% H_2O_2 + 0.013 mol/L EDTA 溶液淋洗的相比较，磷酸盐处理的黄铁矿氧化程度分别下降了约 70%和 50%。他们用 X 射线衍射、扫描电子显微镜技术和化学分析证实了磷酸盐处理后的黄铁矿表面有 $FePO_4$ 膜的存在。最近，Choi 等[28]也证实无论是在批量试验还是实地试验中，KH_2PO_4 的应用显著降低了 II-Gwang 矿样品的硫化物氧化。然而，磷酸盐钝化剂的长效性却不强。Evangelou[29]进行了一项长期的现场试验，结果表明，磷酸盐处理煤矸石的效果仅为前 15 周，之后矿石的氧化速度迅速增加。而 Mauric 等[30]也证明，对含金属硫化矿尾矿进行磷酸盐处理仅在短期内有效。此外，磷酸盐钝化剂在防止硫化矿生物氧化方面的效果亦不佳，因为磷酸盐是促进微生物生长的重要营养物质。Vandiviere 等[31]也表明，在硫化物氧化细菌存在的情况下，磷酸盐钝化层不能抑制黄铁矿氧化。而且，磷酸盐钝化剂只有在 pH 较高（>4）时才能稳定。因此需要对矿堆连续监测以保持其所需的 pH，而这无形增加了实际操作的难度。

2）硅酸盐

Evangelou[32]通过发现黄铁矿的硅酸盐表面钝化剂而获得专利，这是一种新的减少金属硫化物矿物氧化的技术。使用这种技术，将硫化矿尾矿用含有缓冲剂和低浓度氧化剂（如 H_2O_2）的硅酸盐溶液进行处理，可在金属硫化物矿物表面形成一层金属硅酸盐钝化层。在硅酸盐钝化金属硫化物矿物的过程中因在矿物表面形成了两层保护膜——羟基铁层和硅氧层，因此这种钝化膜具有很强的抵抗外界氧化物的能力。而且与磷酸盐钝化剂相比，硅酸盐钝化剂所显示的最大优越性在于它们能抵抗酸性的攻击（低至 pH 2~4）。

许多研究均在实验室条件下证实了硅酸盐钝化剂在防止金属硫化物矿物氧化方面的有效性[33-36]。另外 Vandiviere 等[37]对硅酸盐钝化硫化物矿物的有效性进行了实地验证，他们在加拿大对尾矿进行了硅酸盐钝化处理，发现在尾矿表面形成的硅酸盐钝化层能显著阻止硫化矿的氧化。而且，硅酸盐钝化层的有效性比磷酸盐钝化层更长，也不会像磷酸盐那样容易引起富营养化。当然该类钝化剂在使用过程中仍需使用 H_2O_2 对硫化物尾矿进行预处理。

3）碱性物质

在硫化物残渣中添加碱性材料是预防 AMD 的一种相对常用的方法。最常用的碱性材料包括氢氧化钠（NaOH）、石灰［CaO 或 $Ca(OH)_2$］、石灰石（$CaCO_3$）、碳酸钠（Na_2CO_3）和碳酸氢钠（$NaHCO_3$）等。在硫化物尾矿中加入碱性物质，可提高尾矿的 pH，pH 的升高一方面可降低微生物的活性，从而使硫化物矿物的生物氧化受到抑制。另一方面，还可在硫化物矿物表面形成一层 $Fe(OH)_3$ 等难溶性物质，从而阻隔被包裹矿物的氧化。Huminicki 等[38]就研究了在 $NaHCO_3$ 溶液条件下，黄铁矿能自发形成的 FeO(OH) 膜对硫铁矿实现保护。而且他们发现随着反应进行，矿物表面的保护膜会增厚变密，可以有效地阻止氧化剂和硫化物矿物的接触，从而抑制了 AMD 的形成。虽然用石灰等碱性物质对硫化物尾矿进行处理，可在减少酸性废水的排放，然而为使其长期保持效果，需要定期添加碱性物质，这无形中增加了运营的成本。

近年来，由于粉煤灰具有成本低、就地利用率高、自愈能力强等优点，被认为是一种合适的碱性替代材料[39]。粉煤灰可在两个方面阻止 AMD 的产生，一方面，粉煤灰能在短时间内使尾矿堆 pH 升高，导致沉淀氢氧化铁胶体包覆着黄铁矿颗粒。而另一方面，粉煤灰中富含硅和氧化铝，会产生一种硅酸钙凝胶，这种凝胶能够结合惰性固体颗粒，实现对硫化物矿物的保护。这些产物均能对氧化剂和水的扩散产生强大的屏障。许多实验室研究的作者报告了使用粉煤灰稳定富含黄铁矿的尾矿和防止 AMD 的情况。如

Xenidis 等[40]通过室内土柱试验发现，只要投加少量的粉煤灰即可明显提高尾矿砂渗出液的 pH。当投加量增多后，还可以降低尾矿砂的渗透系数。来自安大略省 Sudbury 矿的实地研究也表明，粉煤灰可以有效地稳定产生酸的尾矿[41]。另一项针对瑞典法伦硫化矿尾矿的研究表明，粉煤灰显著降低了硫化矿的氧化速率[42]。Yeheyis[43]通过室内静态和动态两种试验方法也证明，粉煤灰的加入可以有效阻止酸性矿山废水的产生。当然，这种方法也存在其弊端，如果粉煤灰与矿物混合不均匀，粉煤灰过多的地方呈碱性，没有的地方硫化矿氧化继续进行。

4）高锰酸盐

杜邦公司开发了硫化物材料的高锰酸盐钝化技术[44]。在这项技术中，首先用高 pH（>12）的碱性溶液冲洗硫化物矿物表面，然后用高锰酸钾和氧化镁对矿石进行处理，在硫化物矿物表面形成一层惰性的铁锰氧化物层，它可以抑制硫化物矿物的氧化，从而减少酸的生成。

相比许多其他钝化剂，高锰酸盐稳定性较好，且具有较强的耐酸性能，而且高锰酸盐可以很容易地在野外施用而不会干扰目标尾矿。McCloskey[45]在蒙大拿州的金日光矿进行了高锰酸盐处理的野外实验研究。结果表明，41 周后，尾矿的产酸量、SO_4^{2-} 和金属元素均有所下降，中试试验表明，高锰酸盐钝化法能显著降低污染物排放达 5 年以上。Ji 等[46]也发现，在存在 $KMnO_4$ 的情况下，Young-Dong 煤矿样品在 8 天内产生的 SO_4^{2-} 浓度相比没有 $KMnO_4$ 条件下得到了显著降低。但高锰酸盐钝化法对风化矿物样品进行钝化处理面临的最大问题是必须维持较高的 pH。

2.2.2 有机钝化剂

1）8-羟基喹啉

在硫化物表面覆盖一层 8-羟基喹啉可以抑制黄铁矿的氧化。Lan 等[47]在实验室使用 0.0034 mol/L 8-羟基喹啉对黄铁矿进行包膜试验，发现它对黄铁矿的化学氧化和生物氧化都有明显的抑制作用。研究表明，钝化液的 pH 和温度对其成膜效果均会产生影响[48,49]。钝化剂在 pH 为 5 时形成的钝化膜效果最好。而成膜溶液温度由25℃提高到40℃时有利于 8-羟基喹啉铁膜的抑制氧化效果的提高。同时通过相关实验，他们还提出的 8-羟基喹啉铁膜技术，不仅具有 $FePO_4$ 包膜抑制化学氧化的优点，还具有杀菌剂抑制 T.F 菌的功能，因此，无论在接菌湿润条件下，还是在干燥条件下，均表现出显著的抑制黄铁矿氧化的效果。这一结果为综合包膜和杀菌技术于源头治理 AMD 污染技术以及寻找以克服季铵盐类阳离子表面活性杀菌剂易被雨水淋失、短效等缺点的固体杀菌剂开发与利用提供了一条全新的思路和科学依据。但 8-羟基喹啉钝化剂的致命缺点在于它对水生生物有毒，且价格昂贵，可能不适合现场应用。

2）腐殖酸

腐殖酸对于钝化黄铁矿具有潜在的吸引力，因为腐殖酸对氧化物表面具有高亲和力[50]，导致黄铁矿的电化学活性降低[51]。Lalvani 等[52]发现，采用腐殖酸作为钝化剂可以使黄铁矿氧化率的降低程度高达 98%。Ačai 等[53]的工作也支持腐殖酸钝化黄铁矿的潜力。而且腐殖酸广泛分布于土壤和水中，对环境不会构成太大威胁。但腐植酸的缺点是腐植酸在 pH 低于 2 时是不溶的，导致它无法在强酸性环境中有效使用。此外，腐殖酸对细菌氧化的影响及其长期有效性仍需进一步深入研究。

3）脂肪酸

Nyavor 等[54]在实验室研究了脂肪酸的有效性，发现它能明显抑制黄铁矿的化学和生物氧化。另外 Jiang 等[55]利用一种单饱和脂肪酸——油酸钠在黄铁矿的表面形成一层疏水型的保护膜以阻隔黄铁矿与外界氧化剂的接触，从而保护黄铁矿免受氧化。在他们的研究中，他们对用油酸钠包膜和未包膜的黄铁矿进行了电化学和表面接触角的测量。从电化学测量结果来看，未钝化处理的黄铁矿表现出极高的电化学活性，在电位大于 100 mV 时，出现了很大的氧化峰，说明这时黄铁矿经历了很强的氧化过程。而当用 10mg/L 的油酸钠钝化黄铁矿后，钝化黄铁矿在很大的测量范围内（小于 700 mV）都没有发现时显的电流信号，只有当电压大于 800 mV 时，钝化黄铁矿才开始被氧化。而从上述两种矿物的表面接触角测量结果来看，钝化之前，黄铁矿在不同 pH 接触角的变化范围为 30°～50°。钝化处理后，黄铁矿的接触角增加得相当明显，当溶液 pH 为 6 时，其接触角达到了 100°。这说明油酸钠的钝化大大增加了黄铁矿表面的疏水性。这种疏水性的增加有利于减少水分子对黄铁矿表面的攻击，从而减少黄铁矿的表面氧化速率。但这类钝化剂有毒且价格昂贵，因此在现场规模应用这种化学物质可能不可行[56]。

4）乙酰丙酮

乙酸丙酮具有一定程度的抑制黄铁矿氧化的能力，Belzile 等[57]通过比较乙酰丙酮、腐殖酸、木质素磺酸铵、草酸及硅酸钠对两种不同来源的黄铁矿的钝化性能发现，上述几种化合物都在一定程度上能抑制黄铁矿的氧化，尤其是用双氧水对黄铁矿进行预氧化处理后，其钝化效果更好。而相比而言，乙酰丙酮和硅酸钠在常温下对黄铁矿进行钝化处理后，其钝化效果最佳。但乙酰丙酮的长效性目前还未得到证实。另外，此研究也表明草酸也是一种效果极佳的黄铁矿钝化剂，其对 Kidd Creek 矿区的黄铁矿进行钝化处理后，能使其氧化速率减少 58%，但草酸钝化最大的缺点是钝化条件比较苛刻，其温度需要达到 65℃，导致这种钝化剂使用起来比较困难。

5）聚乙烯多胺

聚乙烯多胺因其良好的缓冲能力和还原特性而可用于阻止硫化物的氧化。Cai 等[58]研究表明三乙烯四胺（TETA）对磁黄铁矿氧化具有明显的抑制作用。同样，Chen 等[59]研究了 TETA 和二乙烯三胺（DETA）对黄铁矿氧化的影响，并发现这两种钝化剂都能有效地阻止黄铁矿的氧化。而两者相比 TETA 比 DETA 更有效，这主要是因为 TETA 的疏水性更强，可以在矿物表面形成更好的疏水膜。而且 TETA 可以很好地抑制氧化亚铁硫杆菌的生长，从而实现对黄铁矿生物氧化的抵制。尽管如此，聚乙烯多胺因其对环境有毒而限制了它的使用。

6）邻苯二酚类

Satur 等[60,61]于 2007 年提出载体微胶囊（CME）作为一种抑制黄铁矿氧化的技术。在 CME 中，使用一种水溶性有机物作为载体，将钛转移到黄铁体表面。然后有机载体分解并在黄铁矿表面形成一层薄薄的 TiO_2 或 $Ti(OH)_4$ 钝化层。他们以邻苯二酚为载体，二氧化钛中的钛为金属离子，研究了这种技术。结果发现以此法可在黄铁矿表面形成稳定的耐酸和氧化剂钝化层。然而，他们的实验是在 pH 在 5～6 的 N_2 气氛下进行的，且实验只持续了很短的时间（7 d）。另外 Jha 等[62]使用硅-邻苯二酚络合物 $[Si(cat)_3^{2-}]$ 来钝化处理煤中的黄铁矿和金属硫化物，可在矿物表面形成一层薄薄的氧化硅或氢氧化硅钝化膜，此钝化膜即使在氧化亚铁硫杆菌的作用下，也能显著抑制黄铁矿的氧化。但此法的长效性也还未得到证实，且此法需要大量的邻苯二酚，导致处理成本较高。

7）有机硅烷

近年来，有机硅烷钝化剂因其独特的优势而成为人们研究的热点。有机硅烷具有安全可靠、无毒、无污染、耐氧化、防水性强的特点而被广泛应用于金属材料的防腐[63,64]及建筑材料的防护[65]，而最近研究者们也开始将有机硅烷引入金属硫化物矿物的钝化处理当中。如 Khummalai 等[66]研究了一系列有机硅烷，包括甲基三甲氧基硅烷（MTMOS）、硅酸四甲酯、四乙氧基硅烷和 *N*-(2-氨基乙基)-3-氨丙基三甲氧基硅烷对砷黄铁矿氧化的抑制性能。他们观察到 MTMOS 所形成的钝化层比其他钝化层能更好地抑制砷黄铁矿的生物和化学氧化。You 等[67]也发现聚硅氧烷能在黄铁矿表面形成一层疏水保护膜，这层保护膜能大大地降低黄铁矿的化学及电化学氧化速率，且随聚硅氧烷浓度的增加其钝化性能随之增强。Diao 等[68]研究了正硅酸乙酯（TEOS）和 *N*-丙基三甲氧基硅烷（NPS）对黄铁矿氧化速率的抵制效果。他们将一定浓度的两种钝化剂对未经预氧化处理的黄铁矿样品进行包膜处理后，分别测量钝化前后的黄铁矿在 H_2O_2 和 *Acidithiobacillus ferrooxidans* 菌液的侵蚀下铁离子和硫酸根离子的释放速率。结果表明，两种钝化剂分别能使黄铁矿的化学氧化速率降低 59%和 96%，而使其生物氧化速率降低 69%和 95%。笔者也考察了三种有机硅烷：巯丙基三甲氧基硅烷（PropS-SH）、乙烯基三甲氧基硅烷（VTMS）及氨丙基三甲氧基硅烷（APS）对黄铁矿化学氧化的钝化效果[69]。实验结果表明，经 APS、VTMS 和 PropS-SH 钝化后的黄铁矿在化学浸取实验中的总铁浸出抑制率分别为 49.4%、71.4%和 89.2%；硫酸根离子的浸出抑制率分别达到 45.1%、69.1%和 89.6%。这表明此三种有机硅烷钝化剂均能不同程度地实现黄铁矿的表面钝化。有机硅烷能实现对硫化物矿物的表面钝化，主要是通过硅烷分子间 Si—O—Si 键的交联聚合而形成一层网状膜，然后通过 Si—O—Me 键（Me 为金属硫化物矿物表面的金属元素）牢固地吸附在金属硫化物矿物表面[68, 69]。因此，为使处理过程中有机硅烷钝化膜足够致密，所添加的有机硅烷浓度往往较高。从目前已有的研究报道及笔者的前期研究结果来看，有机硅烷只有在浓度相对较高的条件下才能实现对金属硫化物矿物的有效保护。但因有机硅烷价格本身较高，用高浓度的有机硅烷作为尾矿中金属硫化物矿物的钝化剂势必会带来昂贵的处理费用。

因此，为在较低有机硅烷浓度的条件下实现对金属硫化物矿物的良好钝化，笔者课题组还将纳米 SO_2 作为"填充剂"引入 PropS-SH 钝化膜矩阵中，制备了 PropS-SH/纳米粒子复合钝化剂[70]。发现纳米 SO_2 的加入，可在较少 PropS-SH 用量的条件下，大幅度提高涂层致密度、厚度及机械强度，同时，纳米 SO_2 的填充还能大大增加腐蚀介质渗透涂层的路径曲折性，阻碍腐蚀介质渗透，减缓其在涂层中的扩散（相关示意图见图 25-1），从而可在不损害 PropS-SH 钝化性能的条件下实现有机硅烷用量的减量化。

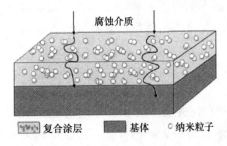

图 25-1　腐蚀介质在有纳米粒子作填充剂的复合钝化膜中的扩散示意图

以上研究表明，有机硅烷类钝化剂在金属硫化物矿物的表面钝化处理中具有较好的应用前景，然而目前制约该类钝化剂推广应用的最大瓶颈依然是其长效性问题。经此类钝化剂包膜后的硫化物矿物在外界环境氧化介质（如 O_2、Fe^{3+} 和微生物等）的长期侵蚀下，不可避免地会使表面钝化膜受到破坏而产生裂缝，因钝化膜不能实现自我修复，所以氧化介质容易以这些裂缝为突破点，逐渐侵蚀硫化物矿石，最终导致钝化膜失效。正是由于以上原因，才使得经这些有机硅烷类钝化剂的有效期往往只有几个月时间，超过这个时间，有机硅烷类钝化剂的钝化效果将降低。因此如何提高钝化剂的自修复性能，使其在金属硫化物矿物表面形成的钝化膜在局部受到损坏时能及时对膜层缺陷进行弥补修复，也就成了该技术急需解决的一个难点。

2.3　生物抑制法

天然矿区环境中广泛分布着大量的浸矿微生物，这些微生物的存在使硫化物矿物的氧化速率增加了成千上万倍，从而产生大量的 AMD，造成矿区及周边生态环境的重金属及酸污染。为了有效地降低矿物的风化速度，减轻其带来的环境危害，就必须抑制其中浸矿微生物的生长。以氧化亚铁硫杆菌为例，研究表明，杀菌剂十二烷基磺酸钠（SDS）和十六烷基三甲基溴化胺（CTAB）能有效地抑制金属硫化物尾矿的生物氧化和酸化，添加 SDS 缓释滴丸处理的新尾矿在浸矿溶液中 30d 总 Fe 溶出率比对照低 84.4%，浸矿溶液 pH、E_h 基本保持初始值不变，是一种可用于尾矿污染原位控制的有效途径[1]。

3 过程控制——酸性矿山废水中重金属去除研究

近几十年来，各国政府和科学家在如何消除矿山废水污染方面做了大量的工作。为了减少 AMD 的产生，开发了多种尾矿综合利用和处理处置技术。例如，采用钝化剂对尾矿进行钝化，在尾矿表面形成致密的惰性钝化膜，将尾矿与水、空气、微生物和氧化剂隔离，从而抑制尾矿的氧化及溶蚀速率，减少 AMD 的产生。然而，仅仅依靠对尾矿进行处理尚不能完全解决现有的污染问题，仍不可避免地产生 AMD，因此，还需要通过合理的技术手段对所排放的 AMD 进行治理，去除其中的 H^+、SO_4^{2-} 和重金属离子，确保矿区下游灌溉水及饮用水的安全[71]。

3.1 酸性矿山废水中重金属的去除方法

AMD 中重金属的去除有化学沉淀、离子交换、氧化还原、混凝、吸附等方法，其中吸附法被广泛研究，人们试图开发高吸附容量、低成本的吸附材料。近年来，农业废弃物作为一类生物量大、可再生循环、再生周期短、可生物降解、环境友好的天然纤维材料受到青睐，并在矿区环境污染控制中得到成功应用。

3.1.1 化学法

化学法是向 AMD 中投加化学物质，提高水的 pH，使水中的金属离子和硫酸根离子与所投加的物质发生化学反应产生沉淀，从而使重金属得以去除的方法。按投加药剂的种类不同，分为中和法和硫化物沉淀法。中和法已成为处理 AMD 中最为普遍的方法，这与该方法具有操作简单、管理方便、工作环境好和处理费用低等优点，并有较多可供使用的化学药剂有关，比如有石灰、石灰石、电石渣、碳酸钠以及氢氧化钠等。中和法可明显地提高 AMD 的 pH，同时去除重金属和硫酸根离子。一般来说，用石灰和石灰石处理 AMD 适应性强。例如，Charerntanyarak 采用石灰处理化学混凝沉淀方法可处理浓度为 150 mg/L 的 Cd^{2+} 废水。若 pH 大于 9.5，或持续加药可进一步降低 Cd^{2+} 的浓度[72]。但是，处理后生成的硫酸钙渣量大，且脱水难，不易堆存处理。若投加硫化物，形成重金属硫化物的溶解度比氢氧化物的溶解度低，效果更好。而且水质处理后的 pH 不高，一般不需要再用酸进行中和处理。通常使用的硫化剂有硫化氢、硫化钠等。但是，硫化剂有毒性，在与重金属沉淀时，可能会有部分硫化物在水中残留而对水体造成污染。此外，有可能会产生 H_2S，形成二次污染，故限制了硫化物沉淀法的广泛应用。

3.1.2 浮选法

该方法先通过化学法将重金属离子形成氢氧化物或硫化物沉淀，再用鼓气上浮法去除。也可采用电解上浮法，在电解过程将重金属络合物氧化分解生成重金属氢氧化物，使其被铝（或铁）阳极溶解形成的氢氧化铝或氢氧化铁吸附，形成共沉淀；另外，还有离子浮选法是通过投加阴离子表面活性剂，与水体重金属离子形成具有表面活性剂的络合物或者螯合物[73]。

3.1.3 微生物法

在 AMD 处理中，微生物可起到，①吸附作用：细菌、霉菌、活性污泥、藻类及某些高等植物可通过吸附和离子交换等物理化学机制，将环境中的重金属吸收进体内。研究表明：在铅矿污水中利用氰细菌、藻类和高等植物的吸附作用,可使 Pb、Cu、Zn 和 Cd 等重金属的去除率可达 99%[74]。②生物转化作用:例如,某些抗汞的假单胞杆菌可产生金属汞离释酶,可将 Hg^{2+} 还原为金属汞[75]。③络合作用:动胶菌、蓝细菌、硫酸盐还原菌以及某些藻类,能够产生胞外聚合物,如多糖、糖蛋白、脂多糖等,具有大量的阴离子基团,能与金属离子结合。例如,有些细菌、真菌、蓝细菌和藻类在限定铁条件下,产生一些称作铁载体(siderophpres)的低分子量化合物,能与 Fe、Cu 等金属发生络合反应[76]。④沉淀作用:微生物在其生长过程中会释放出许多代谢产物,如硫化氢和有机物等,这些产物能与金属反应生成沉淀从而固定重金属。例如,柠檬酸细菌具有一种抗镉的酸性磷酸酯酶,可分解有机的 2-磷酸甘油,并产生的 HPO_4^{2-} 与 Cd^{2+} 形成 $CdHPO_4$ 沉淀[76]。在众多微生物中,利用 SRB 菌进行生物去除 AMD 中重金属被认为最有前途的修复技术[77]。为了发挥 SRB 菌的高效作用,基质混合添加物的配比非常关键[78],这也是微生物法使用过程中不稳定的原因。

3.1.4 人工湿地和渗透性反应屏障法

AMD 在人工湿地的土壤(填料)中可以发生各种化学反应,如化学沉淀、离子交换和拮抗反应;同时,水生植物吸收积累和微生物的氧化还原可增强对重金属的去除效果[79]。可以说,人工湿地是物理化学及微生物共同作用的结果。但是,人工湿地需要的实施范围很大,因此,对渗透性反应屏障(PRB)进行了新的改进,尤其是治理 AMD 对地下水体重金属污染效果更好[80],它由埋在窄沟中的活性物质组成。在大多数情况下,活性材料的组成包括零价铁(ZVI),以及常用的改性沸石、石灰石,活性炭、含 SRB 的有机基质等。其中,填料类型、流量和停留时间是影响 PRB 去除能力的主要因素[81, 82]。

3.1.5 膜分离技术

膜技术主要是通过利用一种特殊的半透膜,在外界给予的压力作用和不改变溶液中化学形态的前提下,会在膜的两侧将溶剂和溶质进行分离或浓缩的方法。膜分离技术包括超滤、反渗透、纳滤和电渗析等方法。这些膜分离技术在 AMD 应用非常广泛,例如:NF(陶瓷和聚合物 NF270 膜)(预处理通过曝气和微滤)可去除 96%AMD 中包括重金属离子在内的多种离子[83];单通道 RO 用于煤矿排出的 AMD(预处理经过氧气炉渣和石灰/苏打灰软化),处理后的废水会可达到饮用水的标准[84]。

3.1.6 吸附法

以上的各种技术和方法各有优缺点。相比于这些方法,吸附法是一种更为流行和有效的处理技术[85]。吸附法主要包括物理吸附和化学吸附。其中,物理吸附指吸附剂与吸附质之间由于分子间力(范德瓦耳斯力)和静电力而产生的吸附[86]。其特点是吸附强度受吸附质性质影响大,分子间作用力没有选择性,吸附质一般不稳定在吸附剂表面的固定位置上,可以在界面范围内自由移动。因此,这种吸附牢固程度很弱。化学吸附则指的是吸附剂与吸附质发生化学反应。在反应过程中,反应热较大,约为 40~420 kJ/mol,所以牢固性能更好[87]。而且一种吸附剂只能与某一种或几种吸附质发生反应,因此,化学吸附更具有吸附专属性能。通常地,吸附法的相关技术和理论基本都是围绕着吸附剂展开的,与此同时,各种类型吸附材料的研制又会对吸附法的理论和技术起到促进作用。近些年来,新材料和新技术层出不穷,加快了高效吸附剂的研制。尤其是研制吸附能力良好的低成本吸附材料更是成为当今吸附领域研究的热点和重点。目前,常见的低成本吸附材料主要集中在矿物材料、工业固体废物和农林废弃物三大类。

1)矿物吸附材料

矿物吸附材料主要有以下几种。①膨润土:以蒙脱石为主要矿物成分的黏土岩。经过改性后,吸附性能可显著提高。特别是钠基型膨润土具有更大的膨胀性和阳离子交换容量,适合作为重金属吸附剂[88]。②海泡石:一种纤维状的含水硅酸镁黏土矿物。其内部结构表面积丰富,含有多孔性矿物质,网状孔径大，表面拥有众多羟基。实验表明，海泡石经过氧化氢、硝酸钠改性后可以提升对 Pb^{2+} 的吸附容量，其吸附机理解释为海泡石硅氧四面体的外缘有大量的 Si—OH，可以与 Pb^{2+} 形成表面络合作用；另外，通过其表面大量的—OH 和溶液中—OH 共同的作用，将 Pb^{2+} 沉淀下来[89]。③凹凸棒石：一种具层链状结构的含水富镁硅酸盐黏土矿物，含有蒙脱石、高岭石、水云母和海泡石等矿物，呈土块状形貌，遇水不膨胀，吸附性能强。例如，改性粒状凹凸棒石对 Cu、Ni、Cd 的最大理论吸附量分别为 66.8 mg/g、71.7 mg/g、54.5 mg/g[90]。④沸石：一类架状构造的含水铝硅酸盐矿物，其晶体结构具很多空腔空间，常见的有辉沸石、钠沸石、菱沸石、方沸石、浊沸石、片沸石和钙沸石等。其结构和晶体特性决定沸石具有良好的吸附性。例如，原生的斜发沸石（clinoptilolite）和氯化钠预处理的斜发沸石，对水中镉离子的吸附能力分别为 0.18 meq/g 和 0.12 meq/g[91]。⑤黏土：含水铝矿物，这些矿物构成了胶体分数小于 2 μ 土壤、沉积物、岩石和水，也可能是混合物组成细粒黏土矿物和其他黏土大小的晶体矿物，如石英、碳酸盐和金属氧化物[92]。研究发现：在 pH 为 6 时，利用生黏土和破碎黏土砖粉对 Pb、Cd、Zn 进行吸附，前者去除率分别为 69.0%、22.7%和 12.9%，后者则为 77.3%、29.5%和 13.5%[93]。

2）工业固体废物

工业固体废弃物是指人类在工业过程中取用目的成分后，弃去的固体物质和泥浆状物质，虽然这部分物质已经失去自身的主要特性，但是其特殊的产生环境和物化性质，可以应用在处理 AMD 重金属污染中。这样的材料有污泥、粉煤矿、水淬渣及煤矸石等。研究人员从两种植物（Aboño 和 Cangas del Narcea）燃烧后得到粉煤灰，不但可以中和酸性废水，将 Cd^{2+} 沉淀，而且还可以吸附氢氧化镉，吸附量为 8.0 mg/g[94]；还有利用电镀污泥作为吸附剂处理水中 Cd^{2+}，常温下吸附容量达到 40 mg/g[95]；也有工程试验发现，尾气污泥颗粒可吸附去除 AMD 中 100%的 Cu 和 Fe、97%的 Zn 以及 94%的 Mn，效果十分显著[96]。

3）农林废弃物

农林废弃物是指农业和林业生产与加工过程中产生的剩余物（淘汰物），如秸秆、皮壳、浆液、渣沫及屑粉。它们的特点是数量巨大、价格低廉、可再生循环、可生物降解。而且农业废弃物结构中的孔隙度高、比表面积大，容易进行物理吸附，还含有一些活性物质，例如单宁、黄酮醇以及果胶质等有效活性基团，极易产生化学吸附。因此，农业废弃物非常适合作为 AMD 重金属离子的吸附材料。总的来看，这些农林废弃物是由纤维素、半纤维素和木质素三种成分组成。纤维素的基本结构单元是 D-吡喃式葡萄糖基（即失水葡萄糖），属于线形高分子化合物；葡萄糖基之间的联结都是 β-苷键联结；每个基环上均具有三个醇羟基，即在 C_2、C_3 位置的是仲醇羟基，在 C_6 位置的是伯醇羟基（图 25-2）。这些羟基使得纤维素可以发生氧化、醚化、润胀、接枝共聚等化学反应。半纤维素是一种共聚物的总称。在半纤维素分子中，一般出现由某种单糖作为单体，通过 β-1,4 键相互联结形成的组成部分，即木聚糖、甘露聚糖以及半乳聚糖，这些单体以 C—C 键、C—O—C 键等形式连接而成。木质素是苯基丙烷结构单元，通过醚键和碳-碳键联结而成的芳香族高分子化合物[97]。构成木质素三种主要单体分别是紫丁香基丙烷、愈创木基丙烷和对羟基苯丙烷。这三者之间的关系是：纤维素组成微细纤维，进而构成纤维细胞壁的网状骨架，其分子排列是规则有序、聚集成束的；而半纤维素主要结合在纤维素微纤维的表面，并且相互连接，构成一个坚硬的细胞相互连接的网络；木质素则起到使植物细胞壁上的细胞相连的作用，并具有增强细胞壁和黏合纤维的作用。在这些组分里，纤维素有两个非常值得关注的特点：一是纤维素的 3 个—OH 都具有化学活性，尤其是 C_6 位的—OH 化学活性最高；二是纤维素聚合结构中存在大量的氢键。这些氢键既可以发生在同一个纤维素链相邻两个葡萄糖结构中，如 C_6 位上的氧与 C_2 或 C_3 的—OH 之间，也可以形成在相互靠近的不同纤维素链中，从而形成超分子氢键结构[98]，并使得纤维素上这些活性的—OH 被束缚起来。因此，未改性的纤维素的吸附能力是很低的，需要通过必要的化学方法来增加其吸附的性能。实际上，化学改性的实质上就是纤维素的—OH 被其他更为有效的功能基团所取代，使得纤维素的功能更强、应用范围更广[99]。

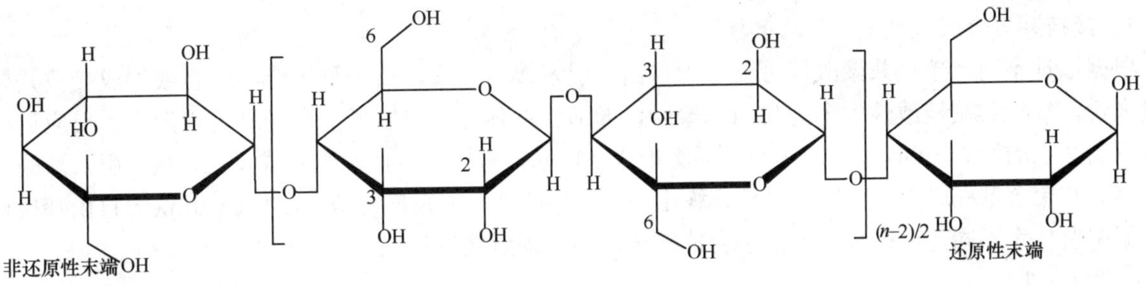

图 25-2　纤维素分子结构

3.2　农林废弃物的改性及其对重金属的吸附去除

我国农业废弃物的种类非常丰富，其中玉米秸秆、稻草秸秆和花生壳所占比重很大。然而绝大部分秸秆作为生活能源的燃料后还田，或者就地燃烧还田。实际上，秸秆焚烧后还田后增产比直接还田还低，不但破坏了生态平衡，使土壤肥力衰竭，而且焚烧时容易造成空气污染和存在火灾隐患；同时，秸秆作为燃料的热能利用率极低，对资源利用也是极大的浪费[100]。为了赋予秸秆更高的使用价值和更多的利用途径，研究者利用丰富的农林废弃物资源，改性制备出去除水体重金属离子的吸附剂。

3.2.1　农林废弃物的改性方法

1）酸改性与碱改性

直接用酸或碱对材料进行化学浸泡，可以破坏其木质素的致密结构，提高其孔隙度，暴露出更多的活性基团。其中，碱润胀是应用最早、最广以及最为有效的纤维素化学预处理方法。按作用次序来看，碱液最先与最外层的木质素表面接触，木质素的酸性醇羟基对碱产生吸附作用，会使其表面与碱液处于饱和平衡状态；接着木质素与碱发生化学作用；最后发生化学水解作用，木质素从表面脱离出来；然后，碱液与半纤维素产生剥离作用和分解反应，导致半纤维素也被分离出来。最后碱液与纤维素作用，生成碱纤维素，纤维素大分子被分解成小分子，纤维素的葡萄糖分子也被脱离出来，直至到达一个相对稳定的状态[101]。碱润胀的结果使农林废弃物纤维素组分占到了较大比例，不但提高了纤维素的可及度，而且更有利于纤维素的进一步化学改性。

2）衍生化改性

衍生化改性也称为直接功能化改性，其实质就是纤维素的定点取代。依据纤维素的葡萄糖基环的第2、3、6个碳原子上三个极性游离—OH 的反应活性，发生被个别取代或者全部取代的现象，从而成为各种结构、性能以及作用不同的纤维素化合物，或者称为纤维素衍生物。这种改性还可分为：①酯化改性：将纤维素内的羟基转变为酯基，氢键减少或者消失，分子间相互作用减弱，纤维素成为一种纤维素酯[102]。②卤化改性：在改性过程中，在纤维素化合物分子中引入卤素原子以及生产卤化物的反应过程。③氧化改性：根据发生在不同的羟基的氧化反应，主要有以下几种氧化方式，一是伯羟基氧化成羧基；二是链末端环节中的还原性基团氧化成羧基；三是葡萄糖酐环节中第2、3位碳上的仲羟基氧化成羧基，或是仲羟基在环不破裂的前提下氧化成酮基；四是第1、5位碳发生破裂，第1位碳上发生氧化；五是第1、2位碳发生破裂，第1位碳形成碳酸酯基团，第2位碳上氧化成羧基；六是在纤维素大分子环节间"氧桥"氧化形成过氧化物[103]。④醚化改性：醚化反应的生成产物是纤维素高分子中羟基上的氢被烃基取代的生成物。生成物根据其离子性分为四类：纤维素烷基醚（非离子纤维素醚）、阴离子纤维素醚、阳离子纤维素醚以及两性离子纤维素醚[104]。

3）接枝共聚合成改性

根据原理不同，接枝共聚改性可以分为自由基引发接枝共聚、离子型聚合、缩合或加成接枝共聚。一般来说，纤维素接枝改性大多为自由基聚合，即活性单体为带单电子的自由基连锁聚合。反应过程是在葡萄糖基上的游离—OH 上（通过断环后）形成自由基，与接枝单体发生链式反应，从而引入新的功能性支链。其聚合过程通常由引发、增长、终止和链转移四个基元反应构成。O'Connell 认为自由基接枝共聚根据引发体系不同，可以分为光化学合成接枝、高能辐射接枝和化学方法引发三种类型[105]。

4）酶改性

酶改性使得纤维素纤维的长度减小，而宽度和平均扭结指数有所增加。酶的使用还可以使纤维素细胞壁得到破坏，导致其结构上变形，这种变形有可能会暴露出更多的羟基，增加纤维素结构对重金属离子的吸附能力。

3.2.2 农林废弃物改性材料的制备

1）秸秆纤维素的醚化和接枝化改性

醚化改性是以丙烯腈作为醚化剂，先在碱润胀处理下，秸秆纤维素成为碱纤维素。加入丙烯腈后，丙烯腈的氰基（—CN）与纤维素的伯—OH 发生取代反应，成为纤维素新的官能团（Cell-CORCN，Cell 为纤维素），即为醚化改性玉米秸秆吸附剂[106, 107]；接枝化改性则以丙烯腈作为接枝改性的单体，先在高锰酸钾引发下，纤维素分子链上葡萄糖基 C_2 和 C_3 之间断开，而且这两个位置的仲—OH 首先被氧化为醛基，醛基很容易发生重排，转变成烯醇结构。烯醇进一步与锰离子反应，在纤维素大分子上产生自由基，进而诱发单体进行接枝共聚反应。—CN 就以支链的形式接入到纤维素的大分子链上，成为接枝改性玉米秸秆吸附剂[108]。

2）秸秆纤维素的螯合改性

螯合改性是在醚化或者接枝化引入—CN 的基础上，进行胺化处理，使得纤维素上不同位置的—CN 发生不同的作用：一部分的—CN 被胺基（—NH$_2$）等取代，发生交联作用，使得相邻纤维素主链之间连接起来。另一部分的—CN 依然保持不变，它们会在 pH 的调节下，与硫化钠反应制备出硫基螯合纤维材料（—H$_2$N—C=S 基团）[109]；或者浸入氯乙酸盐溶液，将 pH 调至弱碱性，对这部分的—CN 进行羧基化处理，研制出羧基化螯合纤维材料（—H$_2$N—C=OOH 基团）[110]。

3）秸秆纤维素的季铵化改性

这种方法是先碱液润胀稻草秸秆纤维素，然后环氧氯丙烷与纤维素进行交联反应，生成环氧纤维素醚，导致氧环开环，氧原子存在一对未成对的电子，可与三甲胺中心氮原子所携带的孤对电子配对成键。碳氮单键形成后，氮原子相连的四个碳，形成季铵结构。季铵结构的中心氮原子呈现正电性，具有类似盐的性质。最后通过稀 HCl 作用，将 Cl$^-$ 负载季铵结构上，制备出季铵化改性稻草秸秆吸附剂[111]。

4）花生壳的化学和酶改性

以花生壳为研究对象，通过经济实用、步骤简便的氧化改性方法，大批量制备出可便于现场使用，去除效果显著的吸附剂。氧化改性以高锰酸钾作为氧化剂，氧化作用使得花生壳纤维素结构的第 1、2 位碳发生破裂，第 1 位碳形成碳酸酯基团，第 2 位碳上氧化成羧基。在改性后，锰元素在吸附剂表面主要是以 MnO_2 的形态[112]。而生物改性是通过构建漆酶-ABTS [2,2'-连氮基-双-(3-乙基苯并二氢噻唑啉-6-磺酸)，$C_{18}H_{24}N_6O_6S_4$] 的体系进行生物改性。在这个体系下，花生壳中含量偏高的木质素被很好地降解，纤维素上的羟基可脱离氢键的束缚和木质素的包裹，发挥其吸附性能[113]。

3.2.3 改性农林废弃物吸附去除水中重金属的应用实例

从目前来看，农业废弃物改性吸附材料去除水中重金属离子的研究基本停留在实验室里试验阶段，

AMD重金属离子的实地去除应用案例非常少。笔者团队针对东江源区钨矿造成的水体重金属污染情况，开展农业废弃物吸附去除AMD中重金属的应用示范工程。

示范地岿美山钨矿地处江西省定南县岿美山镇，属于东江源区。调查发现，盖矿区的AMD重金属污染情况严重，其枯水期和丰水期的重金属（Zn、Cd、Pb、Cu、Cr和As）含量均超出我国《地表水环境质量标准》（GB 3838—2002）中的三类标准。以出水的特征污染物Cd^{2+}为例，枯水期和丰水期的浓度分别为0.009 mg/L和0.259 mg/L，是三类标准的1.8倍和51.8倍。

结合岿美山钨矿尾矿库的运行及渗出水水质现状，矿区重金属污染控制示范工程的AMD重金属污染控制目标为：渗出水中典型重金属镉的去除率维持30%以上。在前期的调查中发现定南的农业废弃物非常丰富，如板栗壳、花生壳、玉米秸秆和杉树枝叶等。花生壳产量最为可观，更容易收集处理。综合比较改性玉米秸秆、稻草秸秆和花生壳的制备方法和吸附性能，高锰酸钾氧化花生壳的改性方法操作简便、实用性强、经济成本较低、吸附容量更大，适合现场大量制备。

在经过连续6个月的现场吸附测试后，第三方监测数据显示：吸附拦截装置对尾矿库渗出水中重金属的去除具有较好的效果，对镉的去除效果较稳定，去除率均在30%以上。

4 次生矿物——酸性矿山废水中重金属自然归宿

AMD污染的流域环境中，由于较强的酸度和较高浓度的硫酸根、铁与重金属离子，易形成大量特征性的含铁硫酸盐次生矿物[114]，例如：pH<3时易形成黄铁矾（ihleite）；pH在2.8～4时形成施氏矿物（schwertmannite）。这些次生矿物的形成和转化过程，对流域内的重金属起到吸附和共沉淀的作用[115-117]，影响着水体中重金属的环境行为与归宿[118,119]。

4.1 次生矿物的形成

在酸性矿山废水这种低pH、高硫酸根和高铁离子浓度的环境中，极易形成特征性的富铁次生矿物如施氏矿物、黄铁矾类矿物，另外还有水铁矿、针铁矿和纤铁矿等。次生矿物的形成主要取决于环境条件，当pH<3、有高浓度硫酸根并伴随存在一价阳离子时，可形成黄铁矾类矿物；随pH升高，可形成施氏矿物、水铁矿和针铁矿等[120]。Chen等[121]对广东省韶关市大宝山的横石河流域进行矿物学组分的调查研究结果表明，AMD水样中铁离子活度均在溶解曲线之上，且沉积物中的含铁羟基化合物随河水pH缓慢上升沿河呈现演替现象：pH<2.7的河段，以针铁矿、黄钾铁矾为主；pH为2.8～3.9，以施氏矿物为主；pH为3.9～5.2的河段以施氏矿物和水铁矿为主；pH>5.2，以水铁矿和针铁矿为主。

4.1.1 施氏矿物

施氏矿物是一种赭色的铁系含氧硫酸盐矿物[119]，于1990年由德国学者Schwertmann和Bigham等在酸性矿井水中发现[122]，其形成的最佳pH范围为2.8～4，化学式为$Fe_8O_8(OH)_{8-2x}(SO_4)_x$（$1 \leqslant x \leqslant 1.75$），一般自然界中形成的施氏矿物呈典型的球形海胆与针状毛刺结构，结晶度较低，粒径约为400 nm，比表面积可达250 m^2/g，其晶体内部结构与β-FeOOH极相似，为双层孔道结构，结构内的硫酸根以二齿双核形式与Fe结合。施氏矿物的内部结构图25-3所示。

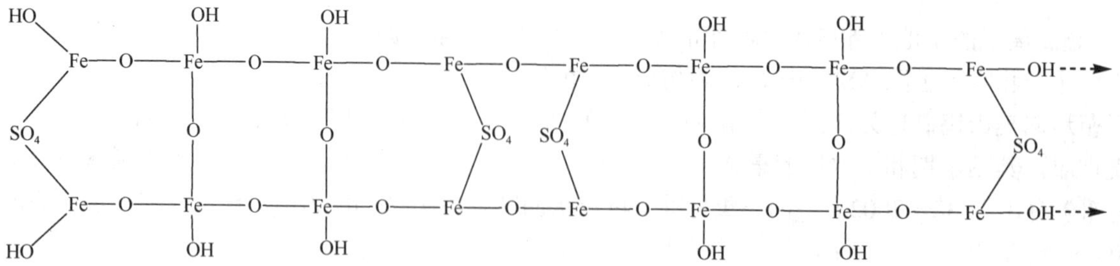

图 25-3 施氏矿物的内部结构[123]

4.1.2 黄铁矾类矿物

黄铁矾类矿物（M[Fe$_3$(OH)$_6$(SO$_4$)$_2$]）也是 AMD 中常见的羟基硫酸盐次生矿物，其化学式中 M 可以是 Na$^+$、K$^+$、NH$_4^+$或者 H$_3$O$^+$等一价阳离子。黄铁矾类矿物表面相对粗糙，为三方晶系，常呈类似立方体的菱面体。图 25-4 为黄铁矾的晶体模型，Fe^{3+}位于略微扭曲的八面体中，SO$_4^{2-}$构成四面体[124, 125]。

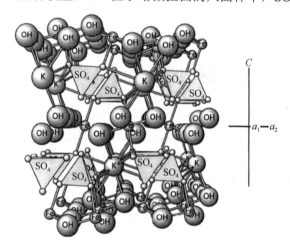

图 25-4 黄铁矾的晶体模型[125]

4.2 次生矿物的稳定性

环境因子如 pH、E_h、温度、共存离子、还原性物质，溶解性有机质等的改变都会引起铁硫酸盐次生矿物的溶解和相转变[115, 126-130]。pH 是铁硫酸盐矿物相转化最关键的影响因子，影响矿物转化反应速率和重金属的分配[131]。在 pH 升高时，施氏矿物最终转变为更稳定的针铁矿，转化速率随 pH 增高而增大[132-133]，并释放 H$^+$：

$$\text{Fe}_8\text{O}_8(\text{OH})_{8-2x}(\text{SO}_4)_x(s) + 2x\text{H}_2\text{O} \rightleftharpoons 8\text{FeOOH}(s) + 2x\text{H}^+ + x\text{SO}_4^{2-} \quad (1 \leq x \leq 1.75) \tag{25-1}$$

而当 pH 降低（pH<3），且环境中存在一价阳离子时，施氏矿物也会转变为结晶度较高的黄铁矾矿物[125]：

$$\text{Fe}_8\text{O}_8(\text{OH})_{8-2x}(\text{SO}_4)_x(s) + (\text{K}^+, \text{Na}^+, \text{NH}_4^+, \text{H}_3\text{O}^+) + \text{SO}_4^{2-} + \text{H}^+ \longrightarrow (\text{K}, \text{Na}, \text{NH}_4, \text{H}_3\text{O})\text{Fe}_3(\text{SO}_4)_2(\text{OH})_6 + \text{Fe}^{3+} \tag{25-2}$$

温度对施氏矿物转化也起着非常重要的作用。4℃条件下，施氏矿物能够在 5 年内稳定存在[134]；当温度从 10℃升高至 20℃，施氏矿物向更稳定的晶相转化的速率提升 3.8 倍[126]。水环境中共存离子 Fe^{2+}等能起到催化施氏矿物相转变的作用，Burton 等[135]发现当 pH 大于 5 时，二价铁对施氏矿物的催化相转变速率是在无添加二价铁条件下的几个数量级倍数，并且能在3 h内完成相转变。

4.3 次生矿物的相转化

按照诱导次生矿物相转变原因的不同，可以将次生矿物相转化分为化学转化和生物转化两类。化学转化主要是指环境条件的改变，如环境因子pH、温度、Fe^{2+}、共存离子等发生变化时，容易引起矿物产生溶解重结晶或催化溶解重结晶，形成在相应环境条件下矿物相较为稳定的矿物。生物转化主要依赖于厌氧微生物如异化铁还原菌对含铁矿物中Fe(III)有很强的还原作用，不仅能够还原溶解次生矿物，释放相应的亚铁等离子，进而重新沉淀结晶生成新的矿物。

4.3.1 化学转化

目前研究的最多的关于次生矿物化学转化的因素是二价铁的催化。Fe^{2+}能够作为一种有效的催化剂，催化亚稳态的铁氧化物向稳定的矿物相快速转化，Fe^{2+}催化转化的速率与Fe^{2+}吸附程度有关。结晶度较低的氧化铁矿物更容易发生亚铁催化溶解重结晶行为，如水铁矿和纤铁矿，在亚铁作用下易形成赤铁矿和针铁矿等次生矿物[116,136]。铁氧化物表面外源二价铁离子与内部结构态Fe(III)的反应机理是亚铁催化铁氧化物相转化的关键问题，通常认为此过程如下：Fe^{2+}吸附于Fe(III)氧化物表面，Fe(II)与结构性Fe(III)发生电子传递，氧化物还原性溶解耦合原子交换，矿物发生相转化[137]。在铁硫化物形成过程中，pH>5时，Fe^{2+}催化施氏矿物转化的速率比无Fe^{2+}时速率快几个数量级。在pH>6、$Fe^{2+}_{(aq)}$≥5mmol/L时，只需3～5 h施氏矿物就能完全转化[138]。

溶解性有机质由于其含有各种活性官能团（如羧基、羟基和酚基），可以改变铁矿物的表面化学性质和胶体稳定性，从而可能影响环境中微量金属污染物的迁移和转化。因此，DOM与矿物之间的相互作用已得到广泛研究。研究结果表明，高的pH有利于含铬施氏矿物中铁、硫酸根和铬离子的释放。掺杂的Cr抑制了施氏矿物向针铁矿的转变，Cr的释放动力学主要受控于实验过程中新形成的二次矿物层。而二次矿物主要为针铁矿。在此过程中，首先假设不存在还原作用，DOM-Fe或DOM-M（M为重金属）形成复合体是施氏矿物结构崩塌的主要原因，紧接着DOM与金属的络合物脱离矿物表面，促使反应向右进行，矿物溶解；其次，DOM对施氏矿物作用机制还取决于自身是否对次生矿物及掺杂的重（类）金属有还原作用，所以在还原的过程中，施氏矿物溶解随之释放重金属和硫酸根，而DOM被氧化后可能还存在络合能力，因此也存在配体促溶过程。因此，一般可以将作用机制概括为还原和配体促进溶解[115,139,140]。

共存离子(如变价金属离子As和Cu)通过吸附Fe^{2+}或与Fe^{2+}发生氧化还原反应影响Fe^{2+}催化效率[117]。如As的存在使Fe^{2+}对矿物催化效率下降，As(V)与施氏矿物的结合能力强于As(III)，其抑制作用相较于As(III)更加明显。同时，施氏矿物转化过程中针铁矿的形成可降低砷的迁移性，有助于砷在固相中的稳定存在并阻止其向环境中释放[141]。利用化学法合成负载含氧阴离子（CrO_4^{2-}、MoO_4^{2-}、AsO_4^{3-}）的施氏矿物，探究其在不同Cu^{2+}浓度与pH条件下施氏矿物的溶解和相转变。在pH约为5时Cu^{2+}能够大量吸附在不同施氏矿物的表面，覆盖矿物表面的反应位点，阻碍不同施氏矿物中Fe^{3+}和SO_4^{2-}的溶解且随着溶液中Cu^{2+}浓度的升高而增强。三种含氧阴离子对施氏矿物稳定性均具有较强的促进作用，作用大小顺序为AsO_4^{3-}≈MoO_4^{2-}>CrO_4^{2-} [142]。含氧阴离子CrO_4^{2-}和AsO_4^{3-}与Fe(III)表面基团之间的亲和力大于SO_4^{2-}与Fe(III)表面基团之间的吸引力，通过减缓矿物的溶解和转化速率使得矿物结构更稳定[143]。

4.3.2 生物转化

在铁硫酸盐次生矿物的形成和转化过程中，微生物起到非常重要的作用[144]。它们能够还原构次生矿物，释放亚铁离子或者单质硫，致使重新沉淀结晶生成新的矿物[145-147]。同时土壤或沉积物中铁的生物氧化还原作用对其他元素的生物地球化学过程有着重要的影响，会影响大多数矿物的分解从而影响有毒重金属的迁移与固定[148-153]。数十年来，诸多研究表明异化铁还原菌（dissimilatory iron reducing bacteria，DIRB）能够还原大多数三价铁氧化物及氢氧化物，如水铁矿、针铁矿、无定形铁氧化物、磁铁矿等含三价铁的黏土矿物[136, 154-156]。刘邓[157]发现典型异化铁还原菌 Shewanella 能够以绿脱石的 Fe(III) 为电子受体；硫酸盐还原菌不仅能以硫酸盐为电子受体，而且能利用黏土矿物的 Fe^{3+} 进行铁呼吸；产甲烷古菌（Methanosarcma barkeri）在特定底物条件下也能够还原黏土矿物中的 Fe^{3+}。

厌氧微生物可参与含铁硫酸盐次生矿物的相转变。研究多集中在铁还原菌参与次生矿物的相变过程。Ouyang 等[158]利用 Shewanella oneidensis MR-1 还原黄铁矾，黄铁矾还原后生成中间产物绿锈和针铁矿及终产物磁铁矿和菱铁矿等含铁矿物；Smeaton 等[159]用 Shewanella putrefaciens CN32 厌氧还原负载铅的黄钾铁矾，推测相变产物包括针铁矿、氢氧化铁等。由环境条件改变所导致的次生矿物相转变极为缓慢，有时可达数月甚至数年之久[126, 160]。而 Bertel 等[161]发现一株具有三价铁还原能力的硫酸盐还原菌 Desulfosporosinus sp. Strain GBSRB4.2 可加速施氏矿物向针铁矿转变。同时也有研究认为，Fe^{2+} 能刺激 SRB 的活性，土壤中 Fe^{2+} 的增加会促使 SRB 的代谢活动更为旺盛，使其生长高峰期延长进而提高硫酸盐的还原速率和硫化物的累积量[162]。Bao 等[163]采用实验室模拟的方法对 AMD 污染环境中微生物对铁硫酸盐次生矿物相转变的影响进行研究，结果表明微生物作用下施氏矿物和黄钾铁矾发生相转变的关键是矿物中 Fe^{3+} 的还原（图 25-5）。施氏矿物和黄钾铁矾中的矿物相 Fe(III) 可被 IRB 以直接的、酶促作用还原，或被 SRB 产生的硫化氢间接还原，从而导致矿物溶解并发生相转变。铁硫酸盐次生矿物在向针铁矿转变的过程中释放 H^+、Fe^{2+} 和 SO_4^{2-}，铁硫酸盐次生矿物的相转变对 AMD 的水化学性质有重要影响。

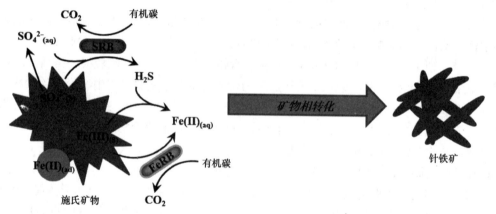

图 25-5　微生物作用下施氏矿物相转变示意图[163]

影响微生物还原含铁矿物的主要因素除了微生物、矿物类型及混合比例以外，水化学条件（溶液 pH、溶液化学成分及有无"电子穿梭体"）也十分重要。探讨电子如何从细胞膜传递至结构 Fe(III) 是铁还原菌与含铁矿物相互作用的核心热点问题。已有研究表明，Shewanella oneidensis MR-1 的完整电子途径包括 4 种细胞色素 C：CymA、MtrA、MtrC 和 MtrB，将电子从内膜经由细胞质和外膜传递至细胞表面。MtrA 和 MtrB 常出现于外膜和细胞质中，用来将电子由 CymA 传递至 MtrC，再将电子传递至细胞外部电子受体[152, 164]。Shane 等[165]发现，当使用针铁矿作为 Shewanella oneidensis MR-1 电子受体时，内膜基因和外膜基因的表达量都要远小于用 Fe^{3+} 作为电子受体时的表达量，并且当针铁矿浓度变化时，内膜基因和外膜基因的表达量也会发生一定的变化。Zeng 等[166]以施氏矿物为模型矿物，研究矿区中常见的羟基硫酸盐高铁次生矿物在微生物群落介导的还原、溶解与转化过程，对不同电子传递模式下的施氏矿物与硫酸盐还原菌群共培养过程中理化性质的变化、矿相转化过程、铁硫元素形态变化进行了系统的分析，研究结果表明在直接的电子传递模式中，添加了外源电子穿梭体 AQDS 后会加速还原性溶解过程；而间接组则主要由 Enterobacteriaceae 科中的未知属构成一个相对丰度超过 90%的单一优势群落，该单一优势群落说明前期 Enterobacteriaceae 科微生物具备产生内源氧化还原有机质去间接还原施氏矿物，且外源的电子穿梭体 AQDS 的加入也不会显著改变其群落结构。

4.4 次生矿物与酸性矿山废水中重金属的归趋

AMD 中含有大量重金属是其成为环境威胁的重要原因之一。AMD 复杂化学条件下大量次生铁氧化物与（类）重金属共存，共存的金属离子会以共沉淀或吸附形式与铁氧化物结合。当金属与铁氧化物结合后将改变其晶体结构、比表面积、稳定性等性质，从而提高铁氧化物的吸附能力或阻碍晶体的生长。针铁矿中 Fe(III)部分被 Cu^{2+}、Ni^{2+}、Co^{2+} 取代，将导致铁氧化物表面积的增加和对砷吸附能力的增强[167]。水铁矿中掺杂的 Al 会抑制 Fe(II)催化水铁矿向高结晶态铁氧化物转化[168]；同时 Al 取代会抑制 Fe^{2+} 与针铁矿间的 Fe 原子交换[169]。铁氧化物中含有过渡态金属时不易被 Fe^{2+} 催化进行晶相转化，与纯铁氧化物相比具有更强的微量元素固定能力[137]。

据报道，施氏矿物和黄铁矾类矿物均对 AMD 环境中的重（类）金属有一定的吸持作用，并且两者有着类似的吸持机制。首先，两者均容易发生类质同象现象。如黄钾铁矾类矿物结构［$AB_3(TO_4)_2(OH)_6$］的形成过程对很多重金属起到"清道夫"的作用，即通过置换和吸附的方式固定重金属，其中 A、B 和 T 点位可以被 Pb^{2+}、Hg^+、Cu^{2+}、Cr^{3+}、Al^{3+}、CrO_4^{2-}、PO_4^{3-}、AsO_4^{3-} 等离子部分置换进入黄钾铁矾结构之中[170, 171]。而施氏矿物结构中的 Fe^{3+} 和 SO_4^{2-} 点位也常存在广泛的类质同象现象，其结构中的 Fe^{3+} 常被 Cu^{2+}、Pb^{2+}、Cd^{2+} 等取代，隧道结构内 SO_4^{2-} 可被离子半径相当、电荷相当、与 Fe^{3+} 配位能力强的含氧阴离子 CrO_4^{2-}、AsO_4^{3-}、SeO_4^{2-} 等取代，从而使得 AMD 中的 Cr、Cd、As 等毒害元素随施氏矿物析出而钝化到固相中[120, 143, 172-174]。其次，两者均具有较好的吸附性能，具有较大的比表面积，对重金属具有良好的吸附能力，可将重金属离子吸附固定于其表面[174]。因此在两种矿物的形成过程中 Cr、Cd、As 等有毒有害元素以吸附或置换的方式进入矿物中[174-176]。如 Michael 等[177]发现在受重金属污染的河流环境中黄铁矾对 As、Cu、Pb 和 Zn 等重金属具有共沉淀作用，降低了下游重金属浓度。

施氏矿物和黄铁矾多以复合物形式沉积于 AMD 河床或受 AMD 影响的河流中，是酸性水体中有毒有害元素的重要沉淀库[178-180]。野外采集的施氏矿物共沉淀了许多重金属杂质，如 As(III)、As(V)均能够大量吸附于矿物表面或以砷酸盐形式进入施氏矿物晶格结构[141, 181]。共沉淀的重金属不但本身迁移性降低，同时从一定程度上影响矿物相转化。对比预先吸附 As(III)与纯施氏矿物的转化过程，发现砷在释放后又快速回到新矿物相表面，并且 As 对新矿物的成型具有抑制作用[181]。根据文献的报道，有报道自然形成的施氏矿物，在 pH 为 2~3 的 AMD 中，反应 323 天时已检测不到施氏矿物的存在，且产物为针铁矿和黄钾铁矾，在相转变的过程中 Zn^{2+}、Cu^{2+}、Pb^{2+}、Cd^{2+}、Ni^{2+} 和 Co^{2+} 等离子也随着释放到溶液中[114]。

铁硫酸盐次生矿物对 AMD 中的有毒重金属通过吸附和共沉淀作用产生了"自然钝化"或"清除"作用[143, 182]。亚稳态次生矿物氧化还原作用和溶解-再沉淀反应控制了重金属向周围环境的迁移。矿物转化过程中晶体中结构态重金属被释放，形成游离态重金属[137]；同时，AMD 中游离态重金属在矿物重结晶过程中被固定在次生矿物结构中[169]。Xie 等[139]探索了富里酸作用下含铬施氏矿物的相转变和铬的释放，经过对 Cr 形态的萃取分析表明，在 pH 为 3.2 时，矿物表面吸附态（F2）和无定型矿物结合态（F3）的铬百分比含量从原来的 87.5%下降到 54.5%；而当 pH=6.5 时，从 86.4%下降为 69.7%。这说明随着二次矿物的形成，铬重新吸附到矿物的表面或者随着矿物的成矿共沉淀到矿物的晶格中。因此矿区中，次生矿物的成矿以及相转变的过程中会伴随着重金属的固定、释放和再固定等环境行为。

5 生态恢复——矿山废弃地治理技术研究及应用

目前，我国10余万家矿山企业，每天采掘出矿产品3000万t，每年向国家输送矿产品近100亿t，为国家经济发展作出了重大贡献。在矿山开采过程中，大量土地因失去利用价值而成为矿业废弃地，如排土场、尾矿、废石场、采矿区和塌陷地等。据统计，全国矿山开采占用和损坏的土地面积已达165.8万hm^2。在自然条件下，矿山废弃地的生态系统经过自然演替可以恢复，但需要100年以上，因此，通过人工干预在相对较短的时间内尽快恢复矿山废弃地的生态环境显得尤为必要[183]。为了合理利用土地和严格保护耕地，加强矿山环境恢复，2012年国土资源部发布了《关于开展工矿废弃地复垦利用试点工作的通知》，开展历史遗留矿山废弃地复垦试点工作，经过3年实践，2015年国土资源部根据生态文明建设等新形势、新要求制定了《历史遗留工矿废弃地复垦利用试点管理办法》。矿山废弃地作为一种以重金属毒害为主要特征的极端裸地，其生态系统自然演替过程与机理研究是原生演替理论的重要组成部分；同时，耐性种或生态型的形成与进化规律研究也将深刻揭示重金属污染的生态效应[184]。

5.1 矿山废弃地极端环境微生物生态

大多有色金属矿业废弃地土壤都含有较高的金属硫化物，而这些金属硫化物被氧化后能够生成硫酸，从而使得土壤酸化，加剧重金属的溶出迁移，导致植物难以生长，携带这些毒性物质的AMD也可随流水排入河流、港湾而影响水生生物的生存。

Huang等[185]发现尾矿酸化过程中嗜酸古细菌的丰度远高于AMD，且酸化早期和晚期阶段的优势微生物类群分别属于真细菌与古细菌，提出嗜酸古细菌 *Ferroplasma acidophilum* 可能是尾矿酸化后期阶段起主导作用的关键催化微生物，这是对AMD形成机制的一个新认识。Kuang等[186]对来自我国华南地区14个矿区的理化差异较大的AMD样本进行16S rRNA的V4高变区测序，结合理化因素的分析揭示了环境因素的变化是极端生境中微生物群落组成差异的主导因素，其中pH影响最为显著；结合由全球其他课题组对该类生境研究所发表的数据分析，更加确证了影响AMD微生物群落多样性变化的主导因素是环境差异而非地域差异。Chen等[187]通过对广东韶关凡口铅锌尾矿酸化过程的地球化学过程与微生物群落关系的研究指出不同酸化阶段是由不同微生物类群所主导的，调控pH可有效控制微生物群落结构、酸化进程与重金属释放。Chen等[188]基于温室模拟研究，建立了矿业废弃物酸化的生物地球化学模型，富集培养了嗜酸铁硫氧化微生物群落，构建了好氧反应器并成功用于AMD的预处理。研究表明，物种数据与环境因子的相关性低于群落宏基因组和宏转录组，极端生境的群落进化速率显著高于普通生境，而且极端微生物能通过高突变速率、基因交流和松弛的选择压力获得更高的基因组可塑性和代谢适应潜力，进一步揭示了极端微生物群落的适应性进化机制[189, 190]。

5.2 矿山废弃地生态恢复技术研究

在强酸环境下重金属溶解性增加，土壤板结严重，熟化受阻，致使氮、磷、钾等营养元素供应不足，植物无法生长。这种裸露的矿山废弃地常常因缺乏植被覆盖成为持久的污染源。然而，在极端酸性重金属矿业废弃地上建立一个安全、稳定、自维持的生态系统是一个巨大的挑战[191]。矿山废弃地的生态恢复是指减少人为因素对矿山生态的不良干扰，降低环境承受的压力，依靠自然生态系统的自我调节功能，再加上部分科学合理的人工干预，恢复遭受破坏的生态环境，促使生态系统稳定健康地发展。自然状态下，生态自然的恢复是一个漫长的自然演替过程，而借助人工调控植被组成和进行一些辅助管理措施等，可以使生态环境迅速改善，加速生态恢复进程。

矿业废弃地属于退化生态系统，生态恢复除了自然净化，通常有物理修复、化学修复和生物修复。矿山废弃土地本身拥有自然净化的能力。自然净化是指将生态系统恢复至原动植物群落能够正常繁殖的状态的自然过程。但是，自然净化是一个非常缓慢的过程。物理修复就是利用物理措施对于废弃地的表面进行固定处理。化学修复就是在废弃地表面施用化学物质，固定表层尾矿。物理和化学修复技术因废弃地中毒性物质的大量存在而受到限制，这就使得高效的生物修复技术受到青睐。生物修复技术包括植物修复技术和微生物处理技术。

Gu 等[192]利用农业废弃物芦竹根茎制备的生物炭对矿山废弃土进行修复。实践证明，该材料能够很好地降低矿区废弃土的毒性。并通过 90 d 的实验确定了该改性材料的最佳投加量为 3%。过多的投加虽然能更大程度的降低 Cd、Cu、Pb 等金属阳离子的毒性，却会增加 As 和 Sb 的阴离子氧化物的迁移性。Penido 等[193]利用污水厂的污泥制备生物炭材料，这种碳材料不仅能够改良矿山废弃土壤的性质，同时对土壤中的 Cd、Pb、Zn 能起到固定作用。

植物修复技术是利用植物吸收、聚集、降解、固定环境中的污染物的能力，从而减少或减轻污染物毒性的技术。超积累植物的运用作为植物修复技术中的核心，近 20 年来，得到了充分的发展。超积累植物具有高效吸收重金属，且在富集重金属后不会表现出明显中毒特征而存活并能繁殖的特征。几种常见的超累积植物包括：壶瓶碎米荠（Se）、芒萁（稀土元素）、海州香薷（Cu）、美洲商陆（Mn）、蜈蚣草（As）、东南景天和伴矿景天（Cd/Zn）[194]。但是，植物修复也存在一些明显的弊端，例如，植物生长条件的限制以及修复耗时较长。

从矿区分离微生物并用以微生物修复技术，从而减缓重金属污染的方法现今广为热议。重金属耐受型微生物种属可作为生物吸附剂用以高效去除重金属离子[195]。重金属能够与微生物细胞表面的官能团或是细胞质内的物质结合，在微生物细胞内部富集。但是，当细胞破裂时或是在某些特殊条件下，重金属将重新释放出来。生物诱导碳酸盐沉淀法是另一种新型实用的生物修复技术，可用以治理矿山废弃土中的重金属污染。微生物诱导沉淀是生物地球化学循环的基础过程，即离子或者化合物与微生物代谢分泌物的反应并使之随即产生矿物颗粒沉淀的过程。生物诱导碳酸盐沉淀是其中重要的一种。微生物释放脲酶，尿素在脲酶的催化下产生碳酸盐，碳酸盐便能与金属离子结合，降低其迁移性。碳酸盐生物矿化可以将可迁移形态的重金属固定为稳定的结晶态。Zhao 等[196]从矿山废弃地中分离得到了 GZ-22 菌，这种菌属不仅能通过生物吸附降低土壤中可迁移性镉的浓度，并且能够通过生物矿化作用，将 Cd 转化为 $CdCO_3$。

5.3 矿山废弃地生态恢复技术应用

矿山废弃地实现生态恢复的难点在于，其含有大量的金属和类金属，土质疏松且土壤颗粒细小。大量的矿山废弃地若不能得到有效的处理，将成为环境潜在的巨大风险。通过生态恢复降低矿山废弃土地的潜在毒性元素的浓度，人类及生态系统的健康将免于危害。

德国和美国最早于 20 世纪 20 年代开展矿山废弃地生态恢复工作，主要包括沉陷区植被恢复、废弃地复垦技术、固废物利用、重金属去除等。德国的莱茵露天煤矿和鲁尔井工煤矿通过农业复垦进行土地恢复，将采坑回填并于其上方覆盖 1.0 m 厚的剥离的黏土，通过施肥和种植作物以改良土壤，恢复土壤的生产力，改善环境。美国于20世纪30年代先后制定关于露天开采以及矿区土地复垦工作法规，边开采边开展复垦工作，使得复垦率达到85%。以采矿业为支柱产业的澳大利亚，在防止地表侵蚀和稳定地表技术、处置扰动土方面位于世界前列，其采用效用-效率优化原则来设计生态修复方案并通过先进的设备观测生态恢复过程，不仅注重土地的恢复，还考虑到动物栖息地的恢复，目前已经形成高科技指导、多专业联合、综合修复等技术手段[197]。王美仙等[198]对美国5个矿山废弃地的生态修复案例进行了深入研究，从土壤改良、水体净化、植被重建3个方面总结归纳案例的生态修复技术要点和特色，为我国矿山废弃地的生态修复工作提供了借鉴。

黄艳红[199]通过开展尾矿库的植物适生性研究，分析尾矿库不同恢复模式下的植物多样性、扬尘产生量、地表径流量、地表径流金属元素含量，表明尾矿库生态恢复可以明显控制重金属的迁移扩散，减少地表径流和扬尘，而尾矿库闭库过程的土壤铺覆并不能起到重金属污染控制作用。杨胜香[200]通过对一个极端酸化的重金属矿业废弃地-大宝山多金属矿排土场进行详尽的野外调查、理化分析和净产酸潜力分析，确定了该废弃地生态恢复的限制因素，采取原位基质改良模式在废弃地上直接改土，添加石灰和鸡粪作为改良剂，采用营养袋育苗并引入土壤种子库，进行合理的生态配置，在排土场废弃地上建立起新的植被。侯晓龙等[201]研究了不同植被措施对福建省紫金山金铜矿废弃地的恢复效果，发现马尾松+胡枝子+香根草+本地河滩草、枫香+本地河滩草和桉树+本地河滩草三种植被配置模式的植被恢复效果较好。曾宪坤等[202]报道采取无土植被恢复技术和原位改良修复技术对城门山铜矿强酸性尾矿库、排土场进行的生态恢复取得了较好的效果。薛生国[203]对湘潭锰矿矿业废弃地进行了植物和土壤调查，在深入了解了尾矿废弃地的基本理化性质和毒性特征后，采用基质改良和抗性植物筛选方法，探讨不同处理方式下矿业废弃地对植物生长状况、吸收养分、重金属毒害的影响，结果表明：基质改良有利于植物生长；施肥能抑制重金属的吸收量，减轻重金属毒害；适于尾矿废弃地生长的抗性植物有冬青卫矛、荷花玉兰、棕榈和海桐，适于矿渣废弃地生长的植物有栾树、千头柏和棕榈。江西德兴铜矿采用"原位基质改良+直接植被"生态恢复技术，集成酸化预测控制技术、土壤重金属毒性控制技术、微生物群落调控技术、先锋植物与野生植物群落演替技术、土壤原位基质改良与熟化控制技术、水土保持与控制技术以及土壤种子库技术等技术体系，恢复排土场、采坑边坡等矿业废弃地总面积达到18.9万m^2，实施的工程项目均取得良好效果，形成了自维持、不退化的植被系统，植被覆盖度均达90%以上，植物种类数目10种以上，且涵盖了乔灌草等三种类型，显著降低了对周边环境的重金属污染，对于稳定边坡、保持水土具有十分明显的作用[204, 205]。

6 展望

矿区环境污染控制是一项复杂的系统工程，通过大量的研究工作，笔者认为：首先，需要提高矿石和尾矿的利用水平，减少矿物开采和冶炼过程中废弃物的排放及其所带来的污染；其次，对尾矿进行必要的钝化处理，抑制其露天堆放期间的化学和生物氧化，从源头上减少尾矿中重金属的释放。然而，这样仍不可避免酸性矿山废水的产生，因此，还需要通过合理的技术手段对所排放的酸性矿山废水进行治理，去除其中的H^+、SO_4^{2-}和重金属离子，确保矿区下游灌溉水及饮用水的安全。

钝化法是是从微观（单个颗粒）角度来保护金属硫化物矿物，是较具发展前景的方法之一。从目前已发现的各种钝化剂来看，尽管它们在实验室条件下被证明是有效的，但它们往往在野外条件下容易失效或需要大量的试剂来确保有效性。因此，目前如何提高金属硫化物矿物的野外长效性及降低运营成本仍是金属硫化物矿物表面钝化剂研究中亟待解决的问题。当然如果一次施加钝化剂后，能在几年甚至更长时间保证金属硫化物矿物的钝化效果，则在长期来看也是降低钝化处理成本的一种方式，但遗憾的是，目前所发现的这些硫化物矿物表面钝化膜在遭到破坏后均无法实现自修复，从而其有效期都不长。因此，提高金属硫化矿物表面钝化膜的自修复能力和长效性也是将来此研究的一个主要方向。另外，目前源头控制 AMD 的技术还没有哪种单独的方法在技术是完成成熟的。所以，在某些情况下，各种技术的组合可能对实现 AMD 遏制有所帮助。例如，将干覆盖法（如土壤）与钝化材料结合使用，或将无机钝化剂与有机钝化剂一起施用，可能比单独使用任何一种技术更有效。当然，这些组合技术的有效性还有待将来的实验验证。

吸附技术是去除 AMD 中重金属行之有效的方法，而农林废弃物是潜在高效的吸附材料。研究表明，农林废弃物的吸附性能还可以通过适当的改性设计还可以实现更多的提升。但是，可能仍存在一个明显的问题，即改性的步骤过多会导致制备过程复杂化，进而使得工艺的费用影响材料的成本问题。这会对农林废弃物的工程化应用产生不利的影响。因此，权衡改性方法和成本之间的关系很重要。另外，到目前为止，农林废弃物在 AMD 中重金属污染修复的应用还是很少。尽管笔者团队已经成功地开展农林废弃物吸附去除 AMD 中重金属的示范工程，但是，这些吸附材料在应用过程中的性能持久性、使用稳定性和过程抗干扰性还可以进一步提高。而且，工程中的吸附拦截装置安装、运行以及与吸附材料的契合等，还需要进一步去检验和优化。

污染物在实际环境中的存在形态决定了其环境与生物地球化学行为。对有毒元素而言，污染物在环境中的迁移、转化、控制和消除过程，其实质是元素在不同环境介质发生的时空改变和形态转化。因此，从大尺度环境归趋、中尺度环境行为到微观界面作用等不同层面的环境过程研究而言，对污染物存在形态进行准确鉴定和分析均是揭示其本质的关键。在 AMD 环境中，存在次生矿物与重金属间动态的相互作用，金属硫化物矿区重金属的环境行为不仅受控于次生矿物的形成，还与次生矿物的溶解相转变以及产物的生成密切相关。因此在环境修复中应考虑次生矿物相转变引起重金属二次释放的环境问题。

参 考 文 献

[1] 党志，卢桂宁，杨琛，等. 金属硫化物矿区环境污染的源头控制与修复技术. 华南理工大学学报(自然科学版), 2012, 40(10): 83-89.

[2] Bhattacharya A, Routh J, Jacks G, et al. Environmental assessment of abandoned mine tailings in Adak, Västerbotten district (northern Sweden). Applied Geochemistry, 2006, 21(10): 1760-1780.

[3] Gunsinger M R, Ptacek C J, Blowes D W, et al. Evaluation of long-term sulfide oxidation processes within pyrrhotite-rich tailings, Lynn Lake, Manitoba. Journal of Contaminant Hydrology, 2006, 83(3): 149-170.

[4] Schippers A, Kock D, Schwartz M, et al. Geomicrobiological and geochemical investigation of a pyrrhotite-containing mine waste tailings dam near Selebi-Phikwe in Botswana. Journal of Geochemical Exploration, 2007, 92(2): 151-158.

[5] Mazumdar A, Goldberg T, Strauss H. Abiotic oxidation of pyrite by Fe(III) in acidic media and its implications for sulfur isotope measurements of lattice-bound sulfate in sediments. Chemical Geology, 2008, 253(1-2): 30-37.

[6] Johnson D B, Hallberg K B. Acid mine drainage remediation options: A review. Science of the Total Environment, 2005, 338(1): 3-14.

[7] Payant S, St-Arnaud L C, Yanful E K. Evaluation of techniques for preventing acidic rock drainage. Sudbury. Conference on Mining and the Enviroment, 1995.

[8] Adu-Wusu C, Yanful E K. Performance of engineered test covers on acid-generating waste rock at Whistle mine, Ontario. Canadian Geotechnical Journal, 2006, 43(1): 1-18.

[9] Timms G P, Bennett J W. The effectiveness of covers at Rum Jungle after fifteen years. Proceedings 5th international conference on acid rock drainage. Society for mining, Metallurgyand Exploration, 2000, 2: 813-818.

[10] Wang H L, Shang J Q, Kovac V, et al. Utilization of Atikokan coal fly ash in acid rockdrainage from musselwhite mine tailings. Canadian Geotechnical Journal, 2006, 43: 229-243.

[11] Swanson D A, Barbour S L, Wilson G W. Dry-site versus wet-site cover design. In: Proceedings of the 4th international conference on acid rock drainage. Vancouver, 1997, IV: 1595-1610.

[12] Meek F. Evaluation of acid prevention techniques used in surface mining. International Land Reclamation and Mine Drainage Conference. Pittsburgh, PA.: Bureau of Mines SP 06B-94, 1994: 24-29.

[13] Skousen J, Foreman J. Water management techniques for acid mine drainage control. Green Lands, 2000, 30: 44-53.

[14] 赵玲, 王荣锌, 李官, 等. 矿山酸性废水处理及源头控制技术展望. 金属矿山, 2009, 7: 131-135.

[15] Yanful E, Orlandea M. Controlling acid drainage in a pyritic mine waste rock. Part II: Geochemistryof drainage. Water Air Soil Pollut, 2000, 124: 259-284.

[16] Reardon E J, Poscente P J. A study of gas compositions in sawmill waste deposits: An evaluationof the use of wood waste in close-out of pyrite tailings. Reclamation and Revegetation Research, 1984, 3: 109-128.

[17] Backes C A, Pulford I D, Duncan H J. Studies on the oxidation of pyrite in colliery spoil: Inhibition of the oxidation by amendment treatments. Reclamation and Revegetation Research, 1987, 6: 1-11.

[18] Pond A P, White S A, Milczarek M, et al. Accelerated weathering of biosolidamendedcopper mine tailings. Journal of Environmental Quality, 2005, 34: 1293-1301.

[19] Fugill R J, Sencindiver J C. Effect of topsoil and vegetation on the generation of acid minedrainage from coal refuse. Charleston, Proceedings seventh annual WV surface mine drainage taskforce symposium, West Virginia mining and reclamation association, 1986.

[20] Wolkersdorfer C, Bowell R. Contemporary reviews of mine water studies in Europe. Mine Water and the Environment, 2004, 23(4): 161.

[21] Evangelou V P. Microencapsulation of pyrite by artificial inducement of $FePO_4$ coatings. Pittsburgh, Proceedings of second international conference on the abatement of acid drainage, 1994.

[22] Evangelou V P. Potential microencapsulation of pyrite by artificial inducement of ferric phosphate coatings. Journal of Environmental Quality, 1995, 24: 535-542.

[23] Georgopoulou Z J, Fytas K, Soto H, et al. Feasibility and cost of creating an ironphosphate coating on pyrrhotite to prevent oxidation. Environmental Geology, 1996, 28: 61-69.

[24] Huang X, Evangelou V P. Suppression of pyrite oxidation rate by phosphate addition. Environmental Geochemistry of Sulfide Oxidation, 1994, 550, 562-573.

[25] 党志, 刘云, 卢桂宁, 等, 金属矿山尾矿钝化技术与原理. 北京: 科学出版社, 2014.

[26] Huang X, Evangelou V. Iron phosphate coating: A novel approach to controlling pyrite oxidation. Pedosphere, 1997, 7(2): 103-110.

[27] Huang X. Suppressing pyrite oxidation via iron phosphate coating. Tailings and Mine Waste, 2004.

[28] Choi J, Ji M K, Yun H S, et al. Inhibition of sulfide mineral oxidation by surface coating agents: Batch and field studies. Journal of Hazardous Materials, 2012, 229: 298-306.

[29] Evangelou V P. Pyrite microencapsulation technologies: Principles and potential field application. Ecological Engineering, 2001, 17(2-3): 165-178.

[30] Mauric A, Lottermoser B G. Phosphate amendment of metalliferous waste rocks, Century Pb-Zn mine, Australia: laboratory and field trials. Applied Geochemistry, 2011, 26(1): 45-56.

[31] Vandiviere M M, Evangelou V P. Comparative testing between conventional and microencapsulation approaches in controlling pyrite oxidation. Journal of Geochemical Exploration, 1998, 64(1-3): 161-176.

[32] Evangelou V P. Oxidation proof silica surface coating iron sulfides. US Patent, No. 5, 494, 703, 1996.

[33] Zhang Y L, Evangelou V P. Formation of ferric hydroxide-silica coatings on pyrite and its oxidation behavior. Soil Science, 1998, 163(1): 53-62.

[34] Takashi N, Takeshi H, Masami Y, et al. Preventing the escape of harmful elements using silica coating. Journal of the Japan Society of Engineering Geology, 2003, 43(6): 390-395.

[35] Crdenes V, Eynde V D, Paradelo R, et al. Passivation techniques to prevent corrosion of iron sulfides in roofing slates. Corrosion Science, 2009, 51(10): 2387-2392.

[36] Bessho M, Wajima T, Ida T, et al. Experimental study on prevention of acid mine drainage by silica coating of pyrite waste rocks with amorphous silica solution. Environmental Earth Sciences, 2011, 64(2): 311-318.

[37] Vandiviere M M, Evangelou V P. Comparative testing between conventional and microencapsulation approaches in controlling pyrite oxidation. Journal of Geochemical Exploration, 1998, 64(1-3): 161-176.

[38] Huminicki, Danielle M C, Rimstidt J D. Iron oxyhydroxide coating of pyrite for acid mine drainage control. Applied Geochemistry, 2009, 24(9): 1626-1634.

[39] Pérez-López R, Cama J, Nieto J M, et al. The iron-coating role on the oxidation kinetics of a pyritic sludge doped with fly ash. Geochimica et Cosmochimica Acta, 2007, 71(8): 1921-1934.

[40] Xenidis A, Mylona E, Paspaliaris I. Potential use of lignite fly ash for the control of acid generation from sulphidic wastes. Waste Management, 2002, 22(6): 631-641.

[41] Shang J Q, Wang H L, Kovac V, et al. Site-specific study stabilization of acid generating mine tailing using coal fly ash. Journal of Materials in Civil Engineering, 2006, 18(2): 140-151.

[42] Hallberg R O, Granhagen J R, Liljemark A. A fly ash/biosludge dry cover for the mitigation of AMD at the falun mine. Chemie Der Erde-Geochemistry, 2005, 65: 43-63.

[43] Yeheyis M B, Shang J Q, Yanful E K. Long-term evaluation of coal fly ash and mine tailing co-placement: A site specific study. Journal of environmental management, 2009, 91(1): 237-244.

[44] Thompson J S. Process for treating iron-containing sulfi de rocks and ores: US Patent, No. 6, 086, 847, 2000.

[45] McCloskey A L. Prevention of acid mine drainage generation from open-pit highwalls-final report. Mine Waste Technology Program Activity III, Project 26, 2005.

[46] Ji S W, Cheong Y W, Yim G J, et al. ARD generation and corrosion potential of exposed roadside rockmass at Boeun and Mujoo, South Korea. Environmental Geology, 2007, 52(6): 1033-1043.

[47] Lan Y, Huang X, Deng B. Suppression of pyrite oxidation by iron 8-hydroxyquinoline. Archives of Environmental Contamination and Toxicology, 2002, 43(2): 168-174.

[48] 兰叶青, 黄骁, 胡霭堂. 有机难溶盐膜抑制黄铁矿氧化的研究. 环境科学学报, 1999, 19(4): 405-409.

[49] 兰叶青, 周钢, 刘正华, 等. 用电化学研究表面膜抑制黄铁矿氧化效果. 南京农业大学学报, 2000, 23(3): 93-96.

[50] Sposito G. The surface chemistry of soils. New York: Oxford University Press, 1984.

[51] Duval J F L, Sorrenti E, Waldvogel Y, et al. On the use of electro kinetic phenomena of the second kind for probing electrode kinetic properties of modified electron conducting surfaces. Physical Chemistry Chemical Physics, 2007, 9(14): 1713-1729.

[52] Lalvani S B, Deneve B A, Weston A. Prevention of pyrite dissolution in acid media. Corrosion, 1991, 47(1): 55-61.

[53] Ačai P, Sorrenti E, Polakovič M, et al. Pyrite passivation by humic acid investigated by inverse liquid chromatography. Colloids and Surfaces A: Physicochemical and Engineering Aspects, 2009, 337(1-3): 39-46.

[54] Nyavor K, Egiebor N O, Fedorak P M. Suppression of microbial pyrite oxidation by fatty acid amine treatment. Science of the total environment, 1996, 182(1-3): 75-83.

[55] Jiang C L, Wang X H, Parekh B K. Effect of sodium oleate on inhibiting pyrite oxidation. International Journal of Mineral

Processing, 2000, 58(1-4): 305-318.

[56] Sahoo P K, Kim K, Equeenuddin S M, et al. Current approaches for mitigating acid mine drainage. Reviews of Environmental Contamination and Toxicology, 2013, 226: 1-32.

[57] Belzile N, Maki S, Chen Y W, et al. Inhibition of pyrite oxidation by surface treatment. Science of the Total Environment, 1997, 196(2): 177-186.

[58] Cai M F, Dang Z, Chen Y W, et al. The passivation of pyrrhotite by surface coating. Chemosphere, 2005, 61: 659-667.

[59] Chen Y W, Yuerong L, Cai M F, et al. Preventing oxidation of iron sulfide minerals by polyethylene polyamines. Minerals Engineering, 2006, 19(1): 19-27.

[60] Satur J, Hiroyoshi N, Tsunekawa M, et al. Carrier-microencapsulation for preventing pyrite oxidation. International Journal of Mineral Processing, 2007, 83(3-4): 116-124.

[61] Satur J, Hiroyoshi N, Ito M, et al. Carrier-microencapsulation for suppressing floatability and oxidation of pyrite in copper mineral processing. Proceedings of the COM2007—46th conference of metallurgists hosting Cu 2007 the sixth international coppercobre conference. Toronto, Canada: Mineral Processing, 2007, 2: 25-30.

[62] Jha R K T. Carrier microencapsulation using Si and catechol to suppress pyrite flotation and oxidation. Sapporo Hokkaido University, Ph.D. thesis 2010.

[63] Flis J, Kanoza M. Electrochemical and surface analytical study of vinyl-triethoxy silane films on iron after exposure to air. Electrochimica Acta, 2006, 51(11): 2338-2345.

[64] Batan A, Mine N, Douhard B, et al. Evidence of covalent bond formation at the silane–metal interface during plasma polymerization of bis-1, 2-(triethoxysilyl) ethane (BTSE) on aluminium. Chemical Physics Letters, 2010, 493: 107-112.

[65] Hagiwara Y, Shimojima A, Kuroda K. Alkoxysilylated-derivatives of double-four-ring silicate as novel building blocks of silica-based materials. Chemistry of Materials, 2007, 20(3): 1147-1153.

[66] Khummalai N, Boonamnuayvitaya V. Suppression of arsenopyrite surface oxidation by solgel coatings. Journal of Bioscience and Bioengineering, 2005, 99(3): 277-284.

[67] You G X, Yu C C, Lu Y, et al. Evaluation of the protective effect of polysiloxane coating on pyrite with electrochemical techniques. Electrochimica Acta, 2013, 93: 65-71.

[68] Diao Z H, Shi T H, Wang S Z, et al. Silane-based coatings on the pyrite for remediation of acid mine drainage. Water Research, 2013, 47(13): 4391-4402.

[69] Ouyang Y T, Liu Y, Zhu R L, et al. Pyrite oxidation inhibition by organosilane coatings for acid mine drainage control. Minerals Engineering, 2015, 72: 57-64.

[70] Liu Y, Hu X, Xu Y. PropS-SH/SiO_2 nanocomposite coatings for pyrite oxidation inhibition to control acid mine drainage at the source. Journal of Hazardous Materials, 2017, 338: 313-322.

[71] 党志, 郑刘春, 卢桂宁, 等. 矿区污染源头控制: 矿山废水中重金属的吸附去除. 北京: 科学出版社, 2015.

[72] Charerntanyarak L. Heavy metals removal by chemical coagulation and precipitation. Water Science and Technology, 1999, 39: 135-138.

[73] Rubio J, Souza M L, Smith R W. Overview of flotation as a wastewater treatment technique. Minerals Engineering, 2002, 15: 139-155.

[74] Gale N L. Biotechnology for the Mining, Metal refining and Fossil Fuel Industries. USA: John Witey and Sons, 1985.

[75] 胡稳奇, 张志光. 微生物方法在重金属污染处理中的应用现状及展望. 大自然探索, 1995(2): 58-62.

[76] 王保军, 杨惠芳. 微生物与重金属的相互作用. 重庆环境科学, 1996, 18(1): 35-38.

[77] Zhang M, Wang H. Preparation of immobilized sulfate reducing bacteria (SRB) granules for effective bioremediation of acid mine drainage and bacterial community analysis. Minerals Engineering, 2016, 92: 63-71.

[78] Muhammad S N, Kusin F M, Muhammad S N, et al. Passive treatment of acid mine drainage using mixed substrates: batch experiments. Procedia Environmetal Science, 2015, 30: 157-161.

[79] Ayora C, Macías F, Torres E, et al. Recovery of rare earth elements and yttrium from passive-remediation systems of acid mine drainage. Environmental Science & Technology, 2016, 50: 8255-8262.

[80] Wang Y, Pleasant S, Jain P, et al. Calcium carbonatebased permeable reactive barriers for iron and manganese groundwater remediation at landfills. Waste Management, 2016, 53: 128-135.

[81] Gibert O, R€otting T, Cortina J L, et al. In-situ remediation of acid mine drainage using a permeable reactive barrier in Aznalcollar (Sw Spain). Journal of Hazardous Materials, 2011, 191: 287-295.

[82] Kefeni K K, Msagati T A M, Mamba B B. Acid mine drainage: Prevention, treatment options, and resource recovery: A review. Journal of Cleaner Production, 2017, 151: 475-493.

[83] Wadekar S S, Vidic R D. Comparison of ceramic and polymeric nanofiltration membranes for treatment of abandoned coal mine drainage. Desalination, 2018, 440: 135-145.

[84] Masindi V. Recovery of drinking water and valuable minerals from acid mine drainage using an integration of magnesite, lime, soda ash, CO_2 and reverse osmosis treatment processes. Journal of Environmental Chemical Engineering, 2017, 5: 3136-3142.

[85] Zheng L C, Dang Z, Yi X Y, et al. Equilibrium and kinetic studies of adsorption of Cd(II) from aqueous solution using modified corn stalk. Journal of Hazardous Materials, 2010, 176: 650-656.

[86] 陈国华. 应用物理化学. 北京: 化学工业出版社, 2008.

[87] 黎松强. 水污染控制与资源化工程. 武汉: 武汉理工大学出版社, 2009.

[88] 张巍. 膨润土在水污染治理中吸附无机污染物的应用进展. 工业水处理, 2018, 38(11): 10-16.

[89] 刘崇敏, 黄益宗, 于方明, 等. 改性海泡石对 Pb 吸附特性的影响. 环境化学, 2013, 32(11): 2024-2029.

[90] 赵超凡, 黄晓岚, 张言, 等. 改性粒状凹凸棒石对 Cu、Ni、Cd 离子的吸附性能研究. 环境工程, 2014, (7): 67-72.

[91] Gedik K, Imamoglu I. Affinity of clinoptilolite-based zeolites towards removal of Cd from aqueous solutions. Sep. Sci. Technol, 2008, 43: 1191-1207.

[92] Bhattacharyya K G, Gupta S S. Adsorption of a few heavy metals on natural and modified kaolinite and montmorillonite: A review. Adv. Colloid. Interf. Sci, 2008, 140: 114-131.

[93] El-Shahat M F, Shehata A M A. Adsorption of lead, cadmium and zinc ions from industrial wastewater by using raw clay and broken clay-brick waste. Asian Journal of Chemistry, 2013, 25(8): 4284-4288.

[94] Ayala J, Blanco F, García P, et al. Asturian fly ash as a heavy metals removal material. Fuel, 1998, 77(11): 1147-1154.

[95] Bhatnagar A, Minocha A K. Utilization of industrial waste for cadmium removal from water and immobilization in cement. Chemical Engerring Journal, 2009, 150: 145-151.

[96] Jafaripour A, Rowson N A, Ghataora G S. Utilisation of residue gas sludge (BOS sludge) for removal of heavy metals from acid mine drainage (AMD). Internaiona Journal of Mining Process 2015, 144: 90-96.

[97] 杨淑惠. 植物纤维化学. 北京: 中国轻工业出版社, 2005.

[98] Nishiyama Y, Langan P, Chanzy H. Crystal structure and hydrogen bonding system in cellulose IB from synchrotron X-ray and neutron fiber diffraction. Journal of American Chemical Society, 2002, 124: 9074-9082.

[99] Liu C F, Sun R C, Zhang A P, et al. Preparation of sugarcane bagasse cellulosic phthalate using an ionic liquid as reaction medium. Carbohydrate Polymers, 2007, 68: 17-25.

[100] 郑刘春, 党志, 曹威, 等. 基于改性农业废弃物的矿山废水中重金属吸附去除技术及应用. 华南师范大学学报(自然科学版), 2015, 47(1): 1-12.

[101] 王德翼. 苎麻纤维素化学与工艺学-脱胶和改性. 北京: 科学出版社, 2001.

[102] 邵自强, 李志强, 刘建华. 纤维素酯在涂料中的研究与应用. 纤维素科学与技术, 2005, 13(3): 46-55.

[103] 李琳, 赵帅, 胡红旗. 纤维素氧化体系的研究进展. 纤维素科学与技术, 2009, 17(3): 59-64.

[104] 张光华, 朱军风, 徐晓凤. 纤维素醚的特点、制备及在工业中的应用. 纤维素科学与技术, 2006, 14(1): 59-65.

[105] O'Connell D W, Birkinshaw C, O'Dwyer T F. Heavy metal adsorbents prepared from the modification of cellulose: A review. Bioresoure Technology, 2008, 99: 6709-6724.

[106] Zheng L C, Dang Z, Zhu C F, et al. Removal of cadmium(II) from aqueous solution by corn stalk graft copolymers. Bioresource Technology, 2010, 101: 5820-5826.

[107] 陈德翼, 郑刘春, 党志, 等. Cu^{2+}和Pb^{2+}存在下改性玉米秸秆Cd^{2+}的吸附. 环境化学, 2009, 28(3): 379-382.

[108] Zheng L C, Zhu C F, Dang Z, et al. Preparation of cellulose derived from corn stalk and its application for cadmium ion adsorption from aqueous solution. Carbohydrate Polymers, 2012, 90: 1008-1015.

[109] Zheng L C, Yang Y B, Meng P P, et al. Absorption of cadmium(II)via sulfur-chelating based cellulose: Characterization, isotherm models and their error analysis. Carbohydrate Polymers, 2019, 209: 38-50.

[110] Zheng L C, Peng D, Meng P, et al. Promotion effects of nitrogenous and oxygenic functional groups on cadmium(II) removal by carboxylated corn stalk, Journal of Cleaner Product, 2018, 201: 609-623.

[111] Cao W, Dang Z, Zhou X Q, et al. Removal of sulphate from aqueous solution using modified rice straw: Preparation, characterization and adsorption performance. Carbohydrate Polymers, 2011, 85: 571-577.

[112] 林芳芳, 易筱筠, 党志, 等. 改性花生壳对水中Cd^{2+}和Pb^{2+}的吸附研究. 农业环境科学学报, 2011, 30(7): 1404-1408.

[113] 赵雅兰, 易筱筠, 雷娟, 等. 基于镉吸附的花生壳酶改性研究. 矿物岩石地球化学通报, 2014, 33(2): 208-213.

[114] Equeenuddin S M, Tripathy S, Sahoo P K, et al. Metal behavior in sediment associated with acid mine drainage stream: Role of pH. Journal of Geochemical Exploration, 2013, 124: 230-237.

[115] Xie Y Y, Yi X Y, Shah K J, et al. Elucidation of desferrioxamine B on the liberation of chromium from schwertmannite. Chemical Geology, 2019, 513: 133-142.

[116] Maillot F, Morin G, Juillot F, et al. Structure and reactivity of As(III)-and As(V)-rich schwertmannites and amorphous ferric arsenate sulfate from the Carnoulès acid mine drainage, France: Comparison with biotic and abiotic model compounds and implications for As remediation. Geochimica et Cosmochimica Acta, 2013, 104: 310-329.

[117] Zhang S L, Jia S Y, Yu B, et al. Sulfidization of As(V)-containing schwertmannite and its impact on arsenic mobilization. Chemical Geology, 2016, 420: 270-279.

[118] 周立祥. 酸性矿山废水中生物成因次生铁矿物的形成及环境工程意义. 地学前缘, 2008, 15(6): 74-82.

[119] 周立祥. 生物矿化: 构建酸性矿山废水新型被动处理系统的新方法. 化学学报, 2017, 75: 552-559.

[120] Lee J S, Chon H T. Hydrogeochemical characteristics of acid mine drainage in the vicinity of an abandoned mine, Daduk Creek, Korea. Journal of Geochemical Exploration, 2006, 88(1): 37-40.

[121] Chen M Q, L u GN, Guo C L, et al. Sulfate migration in a river affected by acid mine drainage from the Dabaoshan mining area, South China. Chemosphere, 2015, 119: 734-743.

[122] Bigham J M, Schwertmann U, Carlson L, et al. A poorly crystallized oxyhydroxysulfate of iron formed by bacterial oxidation of Fe(II)in acid mine waters. Geochimica et Cosmochimica Acta, 1990, 54 : 2743-2758.

[123] Bigham J M, Schwertmann U, Traina S J, et al. Schwertmannite and the chemical modeling of iron in acid sulfate waters. Geochimica et Cosmochimica Acta, 1996, 60(12): 2111-2121.

[124] Gasharova B, Göttlicher J, Becker U. Dissolution at the surface of jarosite: An in situ AFM study. Chemical Geology, 2005, 215(1): 499-516.

[125] Savage K S, Bird D K, O'Day P A. Arsenic speciation in synthetic jarosite. Chemical Geology, 2005, 215(1): 473-498.

[126] Knorr K H, Blodau C. Controls on schwertmannite transformation rates and products. Applied Geochemistry, 2007, 22(9):

2006-2015.

[127] Kumpulainen S, von der Kammer F, Hofmann T. Humic acid adsorption and surface charge effects on schwertmannite and goethite in acid sulphate waters. Water Research, 2008, 42(8): 2051-2060.

[128] Jones A M, Collins R N, Rose J, et al. The effect of silica and natural organic matter on the Fe(II)-catalysed transformation and reactivity of Fe(III)minerals. Geochimica et Cosmochimica Acta, 2009, 73(15): 4409-4422.

[129] Caraballo M A, Rimstidt J D, Macías F, et al. Metastability, nanocrystallinity and pseudo-solid solution effects on the understanding of schwertmannite solubility. Chemical Geology, 2013, 360: 22-31.

[130] Vithana C L, Sullivan L A, Burton E D, et al. Stability of schwertmannite and jarosite in an acidic landscape: Prolonged field incubation. Geoderma, 2015, 239: 47-57.

[131] Regenspurg S, Brand A, Peiffer S. Formation and stability of schwertmannite in acidic mining lakes. Geochimica et Cosmochimica Acta, 2004, 68(6): 1185-1197.

[132] Davidson L E, Shaw S, Benning L G. The kinetics and mechanisms of schwertmannite transformation to goethite and hematite under alkaline conditions. American Mineralogist, 2008, 93(8-9): 1326-1337.

[133] Schwertmann U, Carlson L. The pH-dependent transformation of schwertmannite to goethite at 25℃. Clay Minerals, 2005, 40(1): 63-66.

[134] Jönsson J, Persson P, Sjöberg S, et al. Schwertmannite precipitated from acid mine drainage: phase transformation, sulphate release and surface properties. Applied Geochemistry, 2005, 20(1): 179-191.

[135] Burton E D, Johnston S G. Impact of silica on the reductive transformation of schwertmannite and the mobilization of arsenic. Geochimica et Cosmochimica Acta, 2012, 96: 134-153.

[136] Hansel C M, Benner S G, Neiss J, et al. Secondary mineralization pathways induced by dissimilatory iron reduction of ferrihydrite under advective flow. Geochimica et Cosmochimica Acta, 2003, 67(16): 2977-2992.

[137] Frierdich A J, Catalano J G. Controls on Fe(II)-activated trace element release from goethite and hematite. Environmental Science & Technology, 2012, 46(3): 1519-1526.

[138] Burton E D, Bush R T, Sullivan L A, et al. Schwertmannite transformation to goethite via the Fe(II)pathway: Reaction rates and implications for iron-sulfide formation. Geochimica et Cosmochimica Acta, 2008, 72(18): 4551-4564.

[139] Xie Y Y, Lu G N, Ye H, et al. Fulvic acid induced the liberation of chromium from CrO_4^{2-}-substituted schwertmannite. Chemical Geology, 2017, 475: 52-61.

[140] Xie Y Y, Lu G N, Ye H, et al. Role of dissolved organic matter in the release of chromium from schwertmannite: kinetics, repartition, and mechanisms. Journal of Environmental Quality, 2017, 46: 1088-1097.

[141] Burton E D, Bush R T, Johnston S G, et al. Sorption of arsenic(V) and arsenic(III) to schwertmannite. Environmental Science & Technology, 2009, 43(24): 9202-9207.

[142] Li J F, Xie Y Y, Lu G N, et al. Effect of Cu(II) on the stability of oxyanion-substituted schwertmannite. Environmental Science and Pollution Research, 2018, 25(16): 15492-15506.

[143] Regenspurg S, Peiffer S. Arsenate and chromate incorporation in schwertmannite. Applied Geochemistry, 2005, 20(6): 1226-1239.

[144] Sun W, Xiao T, Sun M, et al. Diversity of the sediment microbial community in the Aha watershed(Southwest China)in response to acid mine drainage pollution gradients. Applied and Environmental Microbiology, 2015, 81(15): 4874-4884.

[145] Lovley D R, Phillips E J P. Organic matter mineralization with reduction of ferric iron in anaerobic sediments. Applied and Environmental Microbiology, 1986, 51(4): 683-689.

[146] Roden E E, Kappler A, Bauer I, et al. Extracellular electron transfer through microbial reduction of solid-phase humic substances. Nature Geoscience, 2010, 3(6): 417-421.

[147] Flynn T M, O'Loughlin E J, Mishra B, et al. Sulfur-mediated electron shuttling during bacterial iron reduction. Science, 2014, 344(6187): 1039-1042.

[148] Stewart B D, Amos R T, Nico P S, et al. Influence of uranyl speciation and iron oxides on uranium biogeochemical redox reactions. Geomicrobiology Journal, 2011, 28(5-6): 444-456.

[149] Masue S Y, Loeppert R H, Fendorf S. Alteration of ferrihydrite reductive dissolution and transformation by adsorbed As and structural Al: Implications for As retention. Geochimica et Cosmochimica Acta, 2011, 75(3): 870-886.

[150] Lu X, Wang H. Microbial oxidation of sulfide tailings and the environmental consequences. Elements, 2012, 8(2): 119-124.

[151] Ziegler B A, McGuire J T, Cozzarelli I M. Rates of As and Trace-Element Mobilization Caused by Fe Reduction in Mixed BTEX–Ethanol Experimental Plumes. Environmental Sscience & Technology, 2015, 49(22): 13179-13189.

[152] Shi L, Dong H, Reguera G, et al. Extracellular electron transfer mechanisms between microorganisms and minerals. Nature Reviews Microbiology, 2016. 14, 651-662.

[153] Li J, Lu J, Lu X, et al. Sulfur Transformation in microbially mediated pyrite oxidation by *Acidithiobacillus ferrooxidans*: Insights from X-ray photoelectron spectroscopy-based quantitative depth profiling. Geomicrobiology Journal, 2016, 33(2): 118-134.

[154] Dong H, Fredrickson J K, Kennedy D W, et al. Mineral transformations associated with the microbial reduction of magnetite. Chemical Geology, 2000, 169(3): 299-318.

[155] Kukkadapu R K, Zachara J M, Smith S C, et al. Dissimilatory bacterial reduction of Al-substituted goethite in subsurface sediments. Geochimica et Cosmochimica Acta, 2001, 65(17): 2913-2924.

[156] Dippon U, Schmidt C, Behrens S, et al. Secondary mineral formation during ferrihydrite reduction by Shewanella oneidensis MR-1 depends on incubation vessel orientation and resulting gradients of cells, Fe^{2+} and Fe minerals. Geomicrobiology Journal, 2015, 10(32): 878-889.

[157] 刘邓. 不同厌氧微生物功能群对粘土矿物结构的还原作用及其矿物转变. 武汉: 中国地质大学, 2012.

[158] Ouyang B J, Lu X C, Liu H, et al. Reduction of jarosite by *Shewanella oneidensis* MR-1 and secondary mineralization. Geochimica et Cosmochimica Acta, 2014, 124: 54-71.

[159] Smeaton C M, Fryer B J, Weisener C G. Intracellular precipitation of Pb by *Shewanella putrefaciens* CN32 during the reductive dissolution of Pb-jarosite. Environmental Science & Technology, 2009, 43(21): 8086-8091.

[160] Wang H, Bigham J M, Tuovinen O H. Formation of schwertmannite and its transformation to jarosite in the presence of acidophilic iron-oxidizing microorganisms. Materials Science and Engineering: C, 2006, 26(4): 588-592.

[161] Bertel D, Peck J, Quick T J, et al. Iron transformations induced by an acid-tolerant *Desulfosporosinus* species. Applied and Environmental Microbiology, 2012, 78(1): 81-88.

[162] Zhao H, Xia B, Qin J, et al. Hydrogeochemical and mineralogical characteristics related to heavy metal attenuation in a stream polluted by acid mine drainage: A case study in Dabaoshan Mine, China. Journal of Environmental Sciences, 2012, 24(6): 979-989.

[163] Bao Y P, Guo C L, Lu G N, et al. Role of microbial activity in Fe(III) hydroxysulfate mineral transformations in an acid mine drainage-impacted site from the Dabaoshan Mine. Science of the Total Environment, 2018, 616-617: 647-657.

[164] Shi L, Rosso K M, Clarke T A, et al. Molecular underpinnings of Fe(III) oxide reduction by *Shewanella oneidensis* MR-1. Frontiers in Microbiology, 2012, 3: 50.

[165] Shane S R, Susan L B, Ming T. Reduction of soluble and insoluble iron froms by membrane fractions of *Shewanella oneidensis* grown under aerobic and anaerobic condititons. Applied and Environmental Microbiology, 2006, 72(4): 2925-2935.

[166] Zeng Y F, Wang H, Guo C L, et al. Schwertmannite transformation via direct or indirect electron transfer by a sulfate

reducing enrichment culture. Environmental Pollution, 2018, 242: 738-748.

[167] Mohapatra M, Sahoo S K, Anand S, et al. Removal of As(V) by Cu(II)-, Ni(II)-, or Co(II)-doped goethite samples. Journal of Colloid and Interface Science, 2006, 298(1): 6-12.

[168] Hansel C M, Learman D R, Lentini C J, et al. Effect of adsorbed and substituted Al on Fe(II)-induced mineralization pathways of ferrihydrite. Geochimica et Cosmochimica Acta, 2011, 75(16): 4653-4666.

[169] Latta D E, Gorski C A, Scherer M M. Influence of Fe^{2+}-catalysed iron oxide recrystallization on metal cycling. Biochemical Society Transactions, 2012, 40: 1191-1197.

[170] Alarcón R, Gaviria J, Dold B. Liberation of adsorbed and co-precipitated arsenic from jarosite, schwertmannite, ferrihydrite, and goethite in seawater. Minerals, 2014, 4(3): 603-620.

[171] Kendall M R, Madden A S, Madden M E E, et al. Effects of arsenic incorporation on jarosite dissolution rates and reaction products. Geochimica et Cosmochimica Acta, 2013, 112(3): 192-207.

[172] Casiot C, Lebrun S, Morin G, et al. Sorption and redox processes controlling arsenic fate and transport in a stream impacted by acid mine drainage. Science of the Total Environment, 2005, 347(1): 122-130.

[173] Antelo J, Fiol S, Gondar D, et al. Cu(II)incorporation to schwertmannite: Effect on stability and reactivity under AMD conditions. Geochimica et Cosmochimica Acta, 2013, 119: 149-163.

[174] Liao Y H, Liang J R, Zhou L X. Adsorptive removal of As(III) by biogenic schwertmarmite from simulated As-contaminated groundwater. Chemosphere, 2011, 83: 295-301.

[175] Baron D, Palmer C D. Solid-solution aqueous-solution reactions between jarosite ($KFe_3(SO_4)_2(OH)_6$) and its chromate analog. Geochimica et Cosmochimica Acta, 2002, 66: 2841-2853.

[176] Courtin N A, Bril H, Neel C, et al. Arsenic in iron cements developed within tailings of a former metalliferous mine -Enguialès, Aveyron, France. Applied Geochemistry, 2003, 18(3): 395-408.

[177] Michael F H, Johnnie N M, Christine V P, et al. Direct observation of heavy metal-mineral association from the Clark Fork River Superfund Complex: Implications for metal transport and bioavailability. Geochimica et Cosmochimica Acta, 2005, 69(7): 1651-1663.

[178] Lee J E, Kim Y. A quantitative estimation of the factors affecting pH changes using simple geochemical data from acid mine drainage. Environmental Geology, 2008, 55(1): 65-75.

[179] 陈福星, 周立祥. 生物催化合成的施氏矿物对废水中Cr(VI)的吸附. 中国环境科学, 2006, 26(1): 11-15.

[180] 王长秋, 马生凤, 鲁安怀, 等. 黄钾铁矾的形成条件研究及其环境意义. 岩石矿物学杂志, 2006, 24(6): 607-611.

[181] Paikaray S, Göttlicher J, Peiffer S. Removal of As(III) from acidic waters using schwertmannite: surface speciation and effect of synthesis pathway. Chemical Geology, 2011, 283(3): 134-142.

[182] Gagliano W B, Brill M R, Bigham J M, et al. Chemistry and mineralogy of ochreous sediments in a constructed mine drainage wetland. Geochimica et Cosmochimica Acta, 2004, 68(9): 2119-2128.

[183] 陈伟, 宁平, 黎慧娟, 等. 矿山废弃地生态环境恢复治理进展. 环境工程技术学报, 2017, 7(1): 78-87.

[184] 束文圣, 叶志鸿, 张志权, 等. 华南铅锌尾矿生态恢复的理论与实践. 生态学报, 2003, 23(8): 1629-1639.

[185] Huang L N, Zhou W H, Hallberg K B, et al. Spatial and temporal analysis of the microbial community in the tailings of a Pb-Zn mine generating acidic drainage. Applied & Environmental Microbiology, 2011, 77(15): 5540-5544.

[186] Kuang J L, Huang L N, Chen L X, et al. Contemporary environmental variation determines microbial diversity patterns in acid mine drainage. The ISME Journal, 2013, 7(5): 1038-1050.

[187] Chen L X, Li J T, Chen Y T, et al. Shifts in microbial community composition and function in the acidification of a lead/zinc mine tailings. Environmental Microbiology, 2013, 15(9): 2431-2444.

[188] Chen Y T, Li J T, Chen L X, et al. Biogeochemical processes governing natural pyrite oxidation and release of acid

metalliferous drainage. Environmental Science & Technology, 2014, 48(10): 5537-5545.

[189] Hua Z S, Han Y J, Chen L X, et al. Ecological roles of dominant and rare prokaryotes in acid mine drainage revealed by metagenomics and metatranscriptomics. The ISME Journal, 2015, 9(6): 1280-1294.

[190] Chen L X, Hu M, Huang L N, et al. Comparative metagenomic and metatranscriptomic analyses of microbial communities in acid mine drainage. The ISME Journal, 2014, 9(7): 1579-1592.

[191] 关军洪, 郝培尧, 董丽, 等. 矿山废弃地生态修复研究进展. 生态科学, 2017, 36(2): 193-200.

[192] Gu J H, Yao J, Jordan G, et al. Arundo donax L. stem-derived biochar increases As and Sb toxicities from nonferrous metal mine tailings. Environmental Science and Pollution Research, 2018, DOI: 10.1007/s11356-018-2780-x.

[193] Penido E S, Martins G C, Mendes T B M, et al. Combining biochar and sewage sludge for immobilization of heavy metals in mining soils. Ecotoxicology and Environmental Safety, 2019, 172: 326-333.

[194] Li J T, Hanumanth K G, Wu L H, et al. Hyperaccumulator plants from China: A synthesis of the current state of knowledge. Environmental Science & Technology, 2018, 52(21): 11980-11994.

[195] Sulaymon A H, Ebrahim S E, Mohammed R M J. Equilibrium, kinetic, and thermodynamic biosorption of Pb(Ⅱ), Cr(Ⅲ), and Cd(Ⅱ) ions by dead anaerobic biomass from synthetic wastewater. Environmental Science and Pollution Research, 2013, 20(1): 175-187.

[196] Zhao Y, Yao J, Yuan Z M, et al. Bioremediation of Cd by strain GZ-22 isolated from mine soil based on biosorption and microbially induced carbonate precipitation. Environmental Science and Pollution Research, 2017, 24(1): 372-380.

[197] 胡晓萧, 李小英. 矿山废弃地生态修复中土壤基质改良技术研究综述. 现代农业科技, 2018, (1): 184-186, 190.

[198] 王美仙, 贺然, 董丽, 等. 美国矿山废弃地生态修复案例研究. 建筑与文化, 2015, (12): 99-101.

[199] 黄艳红. 锰尾矿库植物适生性研究及生态恢复的环境效益. 长沙: 中南大学, 2013.

[200] 杨胜香. 广东大宝山多金属矿排土场生态恢复. 广州: 中山大学, 2010.

[201] 侯晓龙, 庄凯, 刘爱琴, 等. 福建省紫金山金铜矿废弃地不同植被配置模式的恢复效果分析. 水土保持通报, 2012, 32(6): 147-151+157.

[202] 曾宪坤, 熊挺宇. 生态重建技术在城门山铜矿的应用. 有色冶金设计与研究, 2018, 39(5): 4-7.

[203] 薛生国. 湘潭锰矿矿业废弃地生态恢复技术试验研究. 长沙: 中南林学院, 2002.

[204] 陈波. 矿山排土场生态恢复实践——以德兴铜矿水龙山为例. 江西建材, 2017, (22): 278-279.

[205] 吴启明. 重金属矿山废弃地生态恢复新技术的应用. 铜业工程, 2018, (6): 17-19+40.

作者：党　志[1], 刘　云[2], 郑刘春[3], 谢莹莹[4], 叶　翰[1], 卢桂宁[1], 束文圣[5], 周立祥[6], 姚　俊[7]

[1]华南理工大学，[2]湘潭大学，[3]华南师范大学，[4]韩山师范学院，[5]华南师范大学，[6]南京农业大学，[7]中国地质大学（北京）

第 26 章 危险废物处理与资源化研究进展

▶ 1. 引言 /750

▶ 2. 危险废物中重金属的处置与资源化 /751

▶ 3. 危险废物中持久性有机污染物的污染控制技术 /762

▶ 4. 危险废物能源利用与清洁工艺途径 /770

▶ 5. 展望 /780

本章导读

我国危险废物产生量大、污染问题严重，无害化处置和综合利用技术处于发展阶段，开发无害化处置和资源化技术是目前国家与行业的迫切需求。危险废物中所含有的重金属，毒性高难处理，但资源化利用附加值高。有效分离提取重金属是重金属危险废物无毒化处理和资源化利用的重要途径，通过水热调控关键物相的转变、生长及重金属形态变化实现重金属深度提取新策略，可以为消除重金属环境危害提供新途径；采用选冶联合分的工艺，可为铁酸锌类废渣中有价金属高效回收提供短流程新方法。另一方面，近年随着我国垃圾焚烧设施的不断增加，垃圾焚烧规模也都在不断增大。但是垃圾焚烧烟气的污染物控制技术、焚烧飞灰这种危险废物的处理处置与资源化利用成为亟待解决的课题。此外，随着便携式电子设备与电动汽车的广泛使用，电池已成为人们生活中不可缺少的部分，电子废物处理处置与资源化也成为固体废物处置与利用的重要方向之一。"绿色全量资源化"的清洁工艺途径是源头控制危险废物产生的新思路。本章将针对危险废物中的重金属回收、持久性有机污染物原位分解、电池的处理与资源化研究和应用以及绿色全量资源化新工艺途径等内容进行综述。

关键词

危险废物，综合处理与利用，重金属，多氯联苯，电池

1 引　言

2018 年中国统计年鉴表明，我国产生危险废物 6936 万 t，目前综合处理与利用率依然较低。危险废物具有毒性（包括浸出毒性、急性毒性、生物毒性等）、腐蚀性、易燃性、感染性、化学反应性等五种重要特征，环境与生态危害大。危险废物产生于工业企业领域如矿山、制药、煤油化工、电镀及部分金属冶炼等，并涵盖农业领域（废弃的农药瓶）、商业与办公领域（镍镉电池、氧化汞电池、含油废物、飞灰、部分工业污泥、油漆、电路板等），与生活及社会发展息息相关。2017 年，环境保护部发布《"十三五"全国危险废物规范化管理督查考核工作方案》，危险废物的处理处置与资源化已成为国家与行业的重大需求。

目前，焚烧、热解、安全填埋、固化处理以及物理、化学与生化处理等是危险废物的主要处理方式。危险废物的综合利用科学技术研究还有待深入，提高综合环境资源的可持续利用及可能存在二次环境污染等难题急待解决。此外，生活新产品的使用也给危险废物综合处理与利用带来新挑战。例如，从固体废物中提取重金属，普遍的问题是提取效率低，尤其是废物中包夹态重金属是提取的难点与挑战；含铁酸锌的废渣中回收铁锌，均存在工艺冗长、工艺成本高、二次污染等问题；高危铀矿多重金属修复体系复杂。此外，焚烧飞灰这种危险废物成为环境风险控制重要对象，原位催化分解技术是焚烧烟气污染控制的新研究方向；生活汽车启停电池、动力电池、备用电源等带来废铅酸蓄电池环境危害大、回收难等问题；便携式电子设备与电动汽车的广泛使用使得锂极材料回收利用成为危险废物的综合治理重要课题。同时，能够实现源头控制危险废物的"绿色全量资源化"的清洁工艺也成为新的探索命题。

总之，危险废物综合处理与利用技术的开发是环境化学重要研究方向之一，并随着国家的高度重视成为国家与行业的重大需求，也逐渐发展形成了较为完善的方法体系和研究主题。作为现代环境化学学科的重要组成部分，危险废物综合处理与利用不仅为其他相关学科如冶金学科、化工科学、农业科学等提供了重要的支撑，更进一步解决相关工业企业的瓶颈问题，解决生活及社会发展的关键环境瓶颈问题。

随着环境科学深入发展、多学科交叉融合、新技术理论提出和新产品的出现,危险废物综合处理与利用研究不断推进。目前,我国危险废物的综合利用还处于发展阶段,危险废物综合处理与利用研究急须进一步夯实新理论、开发新技术,依然存在诸多挑战。

2 危险废物中重金属的处置与资源化

矿物开采及工业生产中会产生大量重金属(如铬、砷)的危险废物,如处理不当易暴露在环境中,并在大气、水体、土壤、生物体中扩散,而这些重金属生物毒性高,具有生物累积性,可以直接威胁人类的健康。重金属固化稳定化及分离提取是危险废物处理与资源化的典型处置手段;植物-微生物-半导体矿化是尾矿危险废渣的综合有效手段。由于含重金属危险废物的成分、物相复杂,往往有多种金属共存,重金属危险废物的处理存在工艺冗长、处理成本高、二次污染等问题,需要在科学机理的深入探索和认识的基础上发展新的关键技术。

2.1 基于物相调控的重金属提取技术

从工业固体废物中提取重金属、减少排放,无论是对于环境保护还是资源回收来说都非常必要。我国近年来重金属污染严重,提取回收重金属尤为重要。过去几十年,研究者尝试了很多从固体废物中提取重金属的方法,但普遍的问题是提取效率不足,尤其是包夹态重金属是提取的难点与挑战。华南理工大学林璋等近期在调控晶体生长实现固体废物中重金属深度提取方面取得重要进展。课题组在前期研究中成功利用水热调控纳米晶快速生长实现吸附态重金属的完全脱附分离[1]以及独立成相的重金属的分离提取[2,3],在近期工作中首次提出通过水热调控关键物相的转变、生长及重金属形态变化,实现重金属深度提取的策略,以永久消除其环境危害[3]。

以含 Cr(VI)硫酸钙渣这一典型危险工业固体废物为例,作者对原渣进行系统分析时发现(如图 26-1 所示),Cr(VI)虽难以洗脱分离但具有明显的缓释性,主要原因是 Cr(VI)包夹在硫酸钙中。由于硫酸钙由三种形式:$CaSO_4·2H_2O$(DH)、$CaSO_4·0.5H_2O$(HH)和 $CaSO_4$(AH),如图 26-2 所示,它们的结构完全不同。其中二水硫酸钙是低温稳定相(<40℃),无水硫酸钙是高温稳定相(>100℃),半水硫酸钙是亚稳相。作者提出通过水热调控硫酸钙各物相间的转变,在溶解再结晶过程中,使吸附或包夹在矿物上的 Cr(VI)完全释放,此外特定的矿化剂可能对关键物相转变和生长有重要的影响。

通过比较在不同矿化剂作用下水热处理后的 Cr(VI)提取率,发现同样在 120℃水热下,过硫酸盐效果最好,Cr(VI)提取率大于 99%,处理后渣浸出达到一般固体废物标准。如图 26-3 所示,处理前后的实物照片和 SEM 形貌有明显变化。原渣有 Cr(VI)的特征黄色,微观上由很多几微米小颗粒聚集而成;无矿化剂水热处理后,虽然样品颗粒明显增粗,但仍有 Cr(VI)的特征黄色,微观上是由很多细针状晶体自组装成微米级晶簇;用过硫酸钾作为矿化剂水热处理后的白色粉末,晶粒长大上 10 倍且表面光滑、分散性好。

作者以提取效果最佳的过硫酸钾为矿化剂,进一步研究了水热温度和水热时间等因素对 Cr(VI)提取率的影响。如图 26-4 所示,水热温度是 Cr(VI)提取关键因素,只有温度达到 120℃以上才能实现 Cr(VI)完全提取。如图 26-5 所示,温度高于 120℃时,发生了二水硫酸钙转变成无水硫酸钙的过程。如图 26-6 所示,生成的无水硫酸钙与原渣中二水硫酸钙的形貌完全不同,说明二水硫酸钙与无水硫酸钙之间的物相转变是结构完全重整的过程,是溶解再结晶过程。其中溶解过程导致包夹态 Cr(VI)完全释放,而再结晶过程中,水热条件有利于晶体生长和 Cr(VI)脱除。

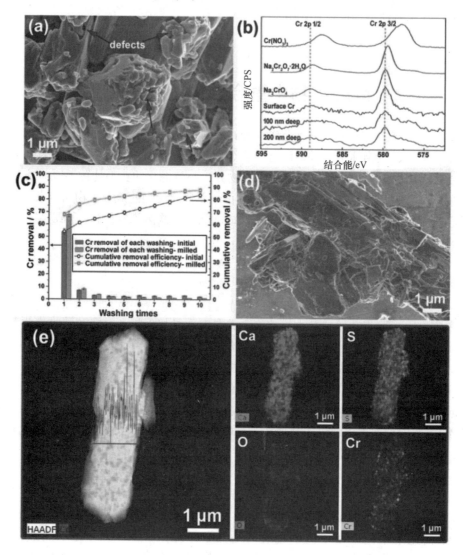

图 26-1　含 Cr(VI)硫酸钙渣原渣表征

(a) SEM 形貌;(b) 不同深度的 XPS 图谱;(c) 原渣中 Cr(VI)的水洗效率,黑色为原渣,红色为研磨后渣;(d) 冲洗后原渣 SEM 形貌;
(e) 冲洗后原渣的球差电镜元素分布

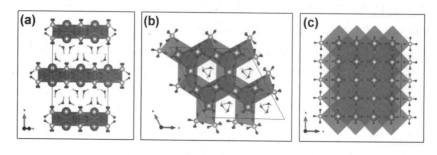

图 26-2　硫酸钙不同形式的晶体结构图

(a) 二水硫酸钙 DH(单斜,空间群 $C2/c$);(b) 半水硫酸钙 HH(单斜,I2);(c) 无水硫酸钙 AH(正交,$Bmmb$)

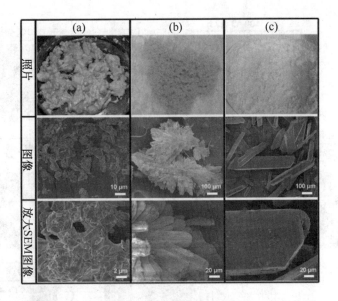

图 26-3 原渣和处理后渣的实物照片和 SEM 形貌图

(a) 原渣；(b) 无矿化剂水热处理后；(c) 过硫酸钾作为矿化剂水热处理后

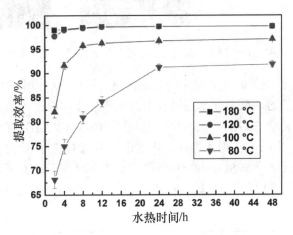

图 26-4 不同水热温度和水热时间条件下 Cr(VI)的提取效率（过硫酸钾 0.5 mol/L）

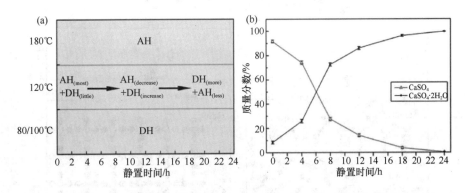

图 26-5 （a）经过不同水热温度处理后随冷却静置时间的 XRD 分析，(b) 水热温度为 120℃，水热 12 h 后，在冷却静置过程中二水硫酸钙与无水硫酸钙物相含量变化（通过 XRD 精修全谱拟合测得）

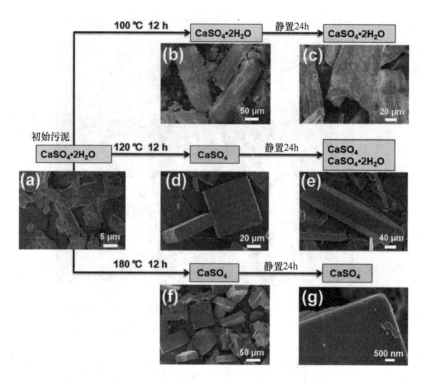

图 26-6　过硫酸盐矿化剂作用下，经过不同水热条件处理后样品的物相和 SEM 形貌演变

作者进一步系统研究并验证了矿化剂过硫酸盐的作用。以过硫酸盐作为矿化剂处理后，体系中上清液的 pH 明显下降，这是因为过硫酸盐在水热过程中发生热活化，通过一系列复杂的反应，最终产生 H^+ 和 SO_4^{2-} 离子。作者通过 Cr K 边 EXAFS 分析证实了原渣中 Cr(VI) 以 CrO_4^{2-} 的形式存在，而经过过硫酸盐水热处理后上清液中 Cr(VI) 以 $Cr_2O_7^{2-}$ 形式存在。作者进一步用 DFT 计算得到 CrO_4^{2-}、$Cr_2O_7^{2-}$、SO_4^{2-} 离子与硫酸钙的结合能排序为 $CrO_4^{2-} \approx SO_4^{2-} \gg Cr_2O_7^{2-}$，如图 26-7 所示，说明 CrO_4^{2-} 最容易与硫酸钙结合，而 $Cr_2O_7^{2-}$ 最难与硫酸钙结合。这从理论基础上验证了 CrO_4^{2-} 容易包夹在硫酸钙中，而 $Cr_2O_7^{2-}$ 不易与硫酸钙结合，容易脱除。

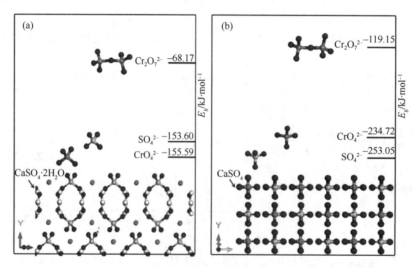

图 26-7　DFT 计算不同离子（CrO_4^{2-}、$Cr_2O_7^{2-}$、SO_4^{2-}）与硫酸钙的结合能
（a）二水硫酸钙；（b）无水硫酸钙

因此，如图26-8所示，硫酸钙渣中包夹态Cr(VI)的全提取机理可以总结为：

（1）在合适的水热温度下，二水硫酸钙转变为无水硫酸钙，是溶解再沉积过程，导致包夹态CrO_4^{2-}的释放。

（2）过硫酸盐作为矿化剂，提供H^+和SO_4^{2-}，其中H^+导致释放的CrO_4^{2-}转变成$Cr_2O_7^{2-}$离子，由于不易与硫酸钙结合，避免其再回到渣中；SO_4^{2-}促进硫酸钙晶体的生长和Cr(VI)脱除。

（3）过硫酸盐热活化产生·OH和SO_4^-自由基，虽有利于氧化工业固体废物中的微量有机质，利于Cr(VI)脱除，但并不是最主要原因。

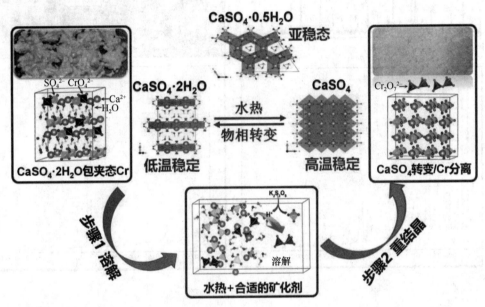

图26-8　过硫酸盐水热作用下包夹态Cr(VI)的提取机理

综上所述，研究独创性地提出水热矿化剂协同调控Cr(VI)赋存关键矿物相和重金属形态变化策略实现包夹态Cr(VI)深度提取，为固体废物中重金属的提取提供了新思路。公斤级实验表明达到了完全脱毒的目标，说明这个策略是可行的。由于水热法简单、成熟、低成本，通过优化矿化剂配方，该策略目前已建立1000 t/a工业线。鉴于石膏危险废物在环境中大量的存在，不仅含铬、砷等阴离子，而且有些含汞、铅等阳离子，利用上述晶体调控协同重金属形态稳定策略，很有希望面向其他石膏危险废物的处理在工业中得到广泛应用。

2.2　选冶联合分离新工艺

冶金提取重金属过程往往存在工艺冗长、成本高、二次污染等问题[5]。铁酸锌是一种普遍存在于冶炼废渣中的尖晶石型化合物，由于晶体结构稳定，铁锌分离回收难，给此类废渣的高效资源化带来困难。过去几十年，对上述问题，中南大学柴立元等提出了选冶联合分离铁锌的新工艺，为铁酸锌类废渣中有价金属高效回收奠定了基础[6]。工艺的关键在于调控还原条件，使铁酸锌选择性分解为磁性铁氧化物与氧化锌，研究明确了铁酸锌选择性还原关键调控因子及调控机理，在近期研究中借助现代热分析动力学方法阐述了铁酸锌选择性还原动力学机制[7-10]。

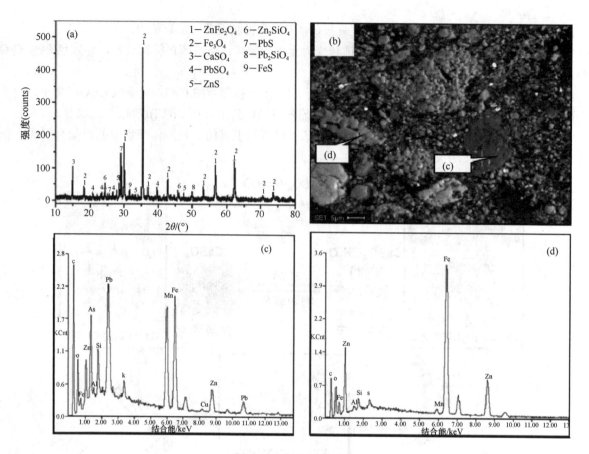

图 26-9 锌浸出渣表征

(a) XRD 图谱;(b) SEM 图谱;(c),(d)EDS 图谱

以锌浸渣这一典型危险废物为例,作者对锌浸出渣进行物相分析确认渣中铁锌主要物相为铁酸锌(图 26-9)。扫描电镜图谱证实渣中不同物相相互嵌布,结合能谱分析可知,铁、锌等分布较为分散,且与铅、砷、硅、锰化合物相互夹杂,颗粒内部微观结构复杂。

由于温度超过 900℃会使物料烧结,影响后续铁锌分离,作者通过相平衡计算考察了 600~900℃下,铁酸锌还原焙烧相变历程。Fe-Zn-O 三元相图(图 26-10)分析表明铁酸锌还原失氧过程可分为四个阶段,各阶段产物组成依次为 Fe_3O_4+ZnO、$Fe_3O_4+FeO+ZnO$、$FeO+Fe+ZnO$、$Fe+ZnO$,其中,理想的铁酸锌选择性还原是使铁酸锌的还原过程终止在第一阶段。

对于 C-CO 还原体系,定义 V_{CO} 为还原强度,其表达式如式(26-1):

$$V_{CO} = P_{CO}/(P_{CO} + P_{CO_2}) \tag{26-1}$$

通过铁酸锌与一氧化碳还原反应平衡计算,确定了还原体系为 $CO-CO_2$ 混合气体,并进一步确定了不同还原强度下,铁酸锌还原产物组成,如图 26-11 所示。

作者进一步通过单因素热重实验考察了不同 CO 强度、CO 浓度及焙烧温度下铁酸锌等温还原规律,结果如图 26-12 所示,可知,CO 强度对铁酸锌的还原速率影响较小,主要影响反应到达还原减速期后还原度高低,进一步证明 CO 强度是决定铁酸锌还原程度的关键因素。当 CO 强度低于 30%时,随着焙烧时间的延长,样品还原度在 40 min 左右到达减速期,之后随时间的延长,还原度上升不明显,且 CO 强度升高虽可使减速期对应还原度提高,但其值始终低于 100%,从还原度角度出发可推断四氧化三铁并未进一步还原为氧化亚铁。对还原样品 XRD 分析证实,当还原强度低于 30%时,焙烧样品 XRD 图谱中未检测到氧化亚铁特征峰,与图 26-11 理论计算结果相符。

第 26 章 危险废物处理与资源化研究进展

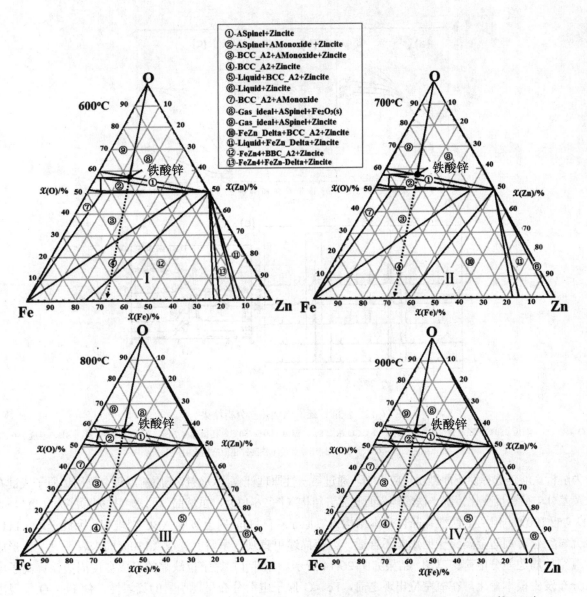

图 26-10　600℃～900℃下 Fe-Zn-O 系相平衡图（600℃～Ⅰ，700℃～Ⅱ，800℃～Ⅲ，900℃～Ⅳ）

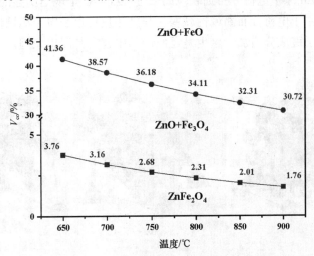

图 26-11　不同还原强度下铁酸锌还原产物组成

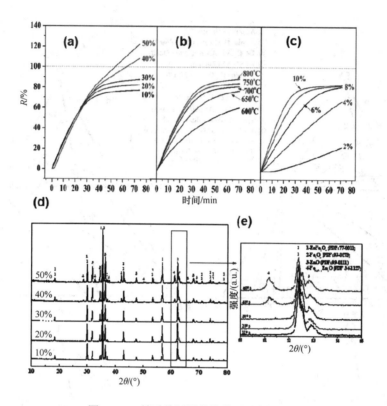

图 26-12　铁酸锌还原焙烧热重分析结果

（a）CO 浓度 8%，温度 750℃，还原强度 10%～50%；（b）CO 浓度 8%，温度 600～750℃，还原强度 20%；（c）CO 浓度 2%～8%，温度 750℃，还原强度 20%；（d）、（e）不同还原强度下铁酸锌还原产物 XRD 图谱

为解析铁酸锌选择性还原微观反应机制，通过第一性原理计算了铁酸锌还原前后氧空位周边原子成键布居数及差分电荷密度变化情况。铁酸锌为面心立方结构（图 26-13），虽然晶体面密度排列顺序为（111）＞（110）＞（100），但由于铁酸锌晶体基本结构为氧原子在[111]晶向上按照 ABC 密堆积，因此选取（111）面研究氧空位对周边金属原子的影响不合适。为使描述更直观，本研究选取（110）面来表现氧空位形成前后氧空位周边差分电荷密度图，结果如图 26-14 所示。由图可知氧空位缺陷的形成会使周边铁锌离子电子云分布发生根本变化，在氧空位出现之前，Fe、O 原子电荷分布呈现明显的离域性，使 Fe—O 呈现出较明显的离子键性质；而 Zn 原子电荷分布则表现局域性，Zn—O 呈现出明显的共价键性质。而形成氧空位以后，氧空位周边的铁原子电荷分布离域性减弱，与之相反锌原子电子离域性增强，导致 Zn—O 键共价键成分减少，对应的离子键成分增加，这也是还原产品中锌容易浸出原因。

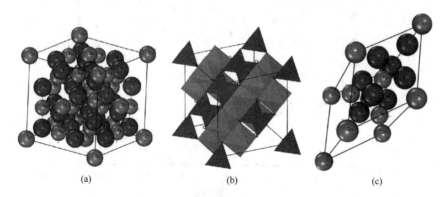

图 26-13　铁酸锌晶胞（a）及晶体结构（b）、原胞（c）图示

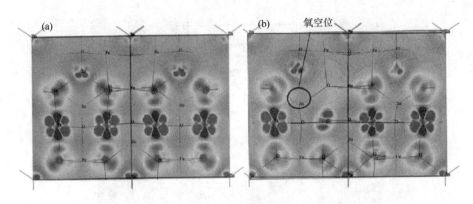

图 26-14 铁酸锌电荷密度图

成键的重叠布居数（表 26-1）分析结果表明，氧空位缺陷形成后，周边 Zn 原子三个 Zn—O 键布居数由 0.39 升至 0.46，键长由 1.98598 Å 降至 1.92391 Å，表明 Zn—O 强度增强。而 Fe 原子对应的五个 Fe—O 变化情况较为复杂，其中两个 Fe—O 键强度显著增强，两个键次之，一个 Fe—O 键则大为削弱，布居数降至 0.26，长度增长至 2.21106 Å。

表 26-1 铁酸锌还原前后氧空位周边原子成键布居数变化情况

	键类型	布居数	键长/Å
	Zn—O	0.39	1.98598
	Fe—O	0.33	2.02616
还原前	Fe，1-O，1；Fe，1-O，5	0.40	1.92582
	Fe，1-O，6～7	0.34	1.92775
	Fe，1-O，8	0.21	2.21106

由上述计算可知，铁酸锌在还原失氧历程中，与氧空位缺陷相连的 Zn—O 键增强，Zn 位于某顶点位置的中间态四面体结构，由于该结构具有较大偶极矩，为重新达到稳定结构，中间态四面体结构必将发生迁移，重新与 O 原子成键形成 Zn，O 配位数为 4 的稳定四面体结构。而与氧空位缺陷相连的 Fe,1-O,8 键在失氧过程中将优先断裂，Fe 原子与其余四个 O 原子重新成键组成四面体结构，形成反式尖晶石结构，即四氧化三铁。因此，四氧化三铁的形成是一个原位生成的过程，而氧化锌的生成过程为异位成核-核生长过程。

综上所述，研究独创性地提出铁酸锌选择性还原分解新思路，为含铁酸锌废渣中有价金属分离回收提供了新技术工艺。基于铁酸锌选择性还原热力学及微观反应机制研究，得出铁酸锌选择性还原关键调控因子以及还原过程微观速率控制步骤。由于该方法具有工艺流程简单、铁锌源头分离等优势，该方案目前已建立 5000 t/a 工业示范线，未来有望在工业生产中广泛推广应用。

2.3 含铀多金属修复处置

本节以位于川西北地区的碳硅泥岩型铀矿为例，阐述多金属尾矿修复相关研究。随着核工业的迅速发展，人们对铀矿进行开采与利用的同时，将产生尾矿废渣，其所含放射性核素和重金属会通过风化、剥蚀、滴流和渗透而迁移和扩散，对土壤、水、动植物和人类造成危害。

西南科技大学董发群等研究重金属赋存状态与微生物多样性之间的关系，发现对于重金属的有效性分析和重金属污染的微生物修复具有重要意义。为此，作者研究了不同污染区中微生物的分布、数量、组成特征及其与周围环境中重金属元素特别是铀的相关性。调研区域如图 26-15 所示。

图 26-15 （a）地质环境调查区域图；（b）尾矿处理后样貌图

通过对×××铀矿床的岩性分析，其含矿岩石主要为炭质硅质岩，并含有少量的方解石、白云母、钠长石等。矿石中的铀主要以独立矿物（沥青铀矿）的形式产出，吸附状态的铀则赋存于炭质、泥质物中，并与多种重金属共存[11]。

此外，细菌、放线菌、真菌作为土壤生态系统中微生物群落的重要组成种类，是实现生态功能、标记环境变化的重要因素[12]。中长沟和羊肠沟微生物量较苏里木塘微生物的含量多，说明重金属含量丰富的土壤中微生物含量较多（图 26-16）。为了考察铀对矿区土壤微生物群落生长和代谢的影响，作者设计向铀矿区污染土壤样品添加外源铀实验。结果表明，土壤生物学指标在一定程度上可以表示铀胁迫下的土壤环境质量演变过程[13]。

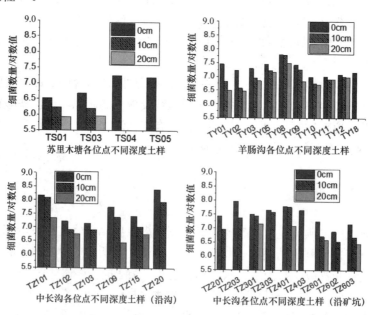

图 26-16 同一采样点不同深度土样中细菌分布图

由γ射线等高线图可知（图26-17），采矿区域为辐射危险区域，最高辐射值达到52.0×10⁻⁸ Gy/h[14]。重金属污染负荷指数法（PLI）分析表明该矿区土壤处于中等污染程度，铀浓度在矿坑处最高达到232.70 mg/kg，重金属砷浓度的平均值为47.26 mg/kg，超出了土壤质量三级标准。由于温泉、取水口及居住区离矿区较近，具有较大的潜在风险。

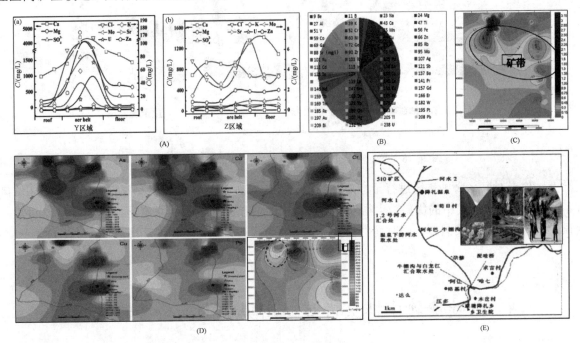

图26-17 川西某铀矿区域内环境污染评价

(A) 水环境测试分布图 [(a) Y区域水样的离子分布，(b) Z区域水样的离子分布]；(B) 水样的全元素检测分析分布图；(C) γ射线辐射场填充彩色等高线图；(D) 矿区土壤中重金属（As，Cd，Cr，Cu，Pb，U）的空间分布特征图；(E) 矿区动植物环境潜在风险

通过对采矿区土壤及岩石进行分析发现，该区域具有光催化活性较强的TiO_2存在，并且从地表水优势菌种中分离并鉴定出考克氏菌这一耐辐射的光电能微生物。作者建立了半导体矿物-微生物-铀这一实验体系，成功实现了利用光电子协同微生物对铀价态的调控。此外，微生物与反应后的产物共同附着电极上，这表明微生物是通过附着在电极上直接接受电子[15]，而铀则可能与微生物直接接触得电子被还原或者进入胞内发生矿化作用。且光电子能促进微生物的生长，并延长其稳定期。

因此，矿区铀污染修复机制可总结如下[16]（图26-18）：

（1）直接作用：半导体在太阳光的照射下产生光电子，重金属通过与矿物相接触得到电子直接被还原。

（2）间接作用：光催化产生电子，通过矿物与微生物的直接接触传递到微生物细胞膜上的蛋白类物质或进入胞内参与三羧酸循环，促进其生长代谢，再由微生物产生源电子与吸附在其表面或者胞内的铀相互作用。

为使生物复垦技术在该矿区得到充分利用，并进一步解决重金属污染。作者提出：利用该矿区的优势植物、优势微生物以及存在的大量TiO_2，将植物修复、微生物还原以及半导体光催化三种方法联合起来，使生态恢复更快捷、更有效。为此，作者对该矿区优势植物进行了调研，并筛选出了两种转移系数高且富集效果明显的植物——尼泊尔酸模和珠芽蓼。

因此可对采矿区和浸矿区进行矿渣回填，然后进行工程加固，最后再利用筛选得到的重金属富集植物进行修复。矿区水经植物修复、半导体矿物光催化、矿化池、泥碳自然净化系统等不同阶段的联合修复作用，可最终达到该矿区水达标排放的目的。

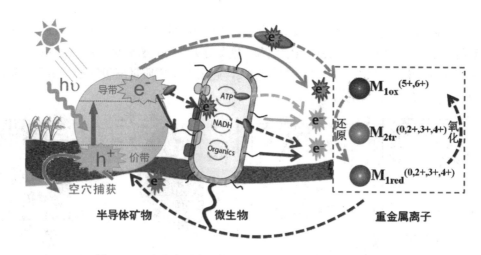

图 26-18 光生电子介导微生物对 U（VI）价态转变的机理

综上，作者对×××铀矿区进行了地质、环境调查，对典型污染区重金属离子赋存状态与环境风险进行了评价，并筛选出其中的优势微生物以及优势植物，提出了联合植物-微生物-半导体矿物法以进一步解决该矿区的重金属污染问题，同时使生物复垦效果显著提高，保护其生态环境。

3 危险废物中持久性有机污染物的污染控制技术

危险废物在焚烧处置过程中，焚烧烟气中的持久性有机污染物的控制是需要解决的重要问题。《关于持久性有机污染物的斯德哥尔摩公约》，将多氯代二苯并-p-二噁英（PCDDs）、多氯代二苯并呋喃（PCDFs）和多氯联苯（PCBs）等列入到首批需要削减和控制的 12 种 POPs 中。另一方面，焚烧过程中产生的焚烧飞灰（以下简称飞灰），又成为环境风险控制的主要对象。因此，针对传统焚烧技术，开发 POPs 的原位分解技术、飞灰安全处置与利用措施成为有机危险废物处置与利用的重要内容。

3.1 催化分解技术

危险废物在焚烧过程中产生的焚烧烟气组成极其复杂，主要污染物包括粉尘（飞灰）、酸性气体、重金属和 POPs 等。2001 年 5 月签署的《关于持久性有机污染物的斯德哥尔摩公约》，将 PCDDs、PCDFs 和 PCBs 列入到首批需要削减和控制的 12 种 POPs 中，并提出在 2028 年前全球范围内消除 POPs。在焚烧烟气处置工艺中，粉尘脱除主要采用除尘器，我国危险废物焚烧的烟气净化系统通常设置袋式除尘器，在袋式除尘器前端喷射活性炭，能够吸附氮氧化物、PCDD/Fs 等，从而脱除烟气中的污染物。活性炭不仅是一种吸附材料，也被用来作为许多催化剂反应的载体。在活性炭上负载金属，对低浓度 PCBs 废气进行吸附的同时利用活性金属对 PCBs 进行催化分解，即采用物理吸附和化学脱氯还原结合的方法，实现对低浓度 PCBs 的去除，大大延长了活性炭的使用周期，且一定程度上避免了二噁英的生成。孙轶斐课题组在前期研究中选取廉价的过渡金属（Fe、Ni、Cu、Zn）代替传统活性炭负载金属使用的贵金属，开发了利用离子交换法制备高负载量的活性炭负载过渡金属催化剂，获得了理想的 PCBs 分解效率[17-19]。

通过离子交换法制备的离子交换型催化剂（IRM-C）中过渡金属的负载率高达 50%以上，与传统浸渍法制备的活性炭负载过渡金属催化剂（LaM-C）相比较，金属负载率高出近 10 倍。催化剂的能谱分析结果如图 26-19 所示，离子交换法制备的活性炭负载过渡金属催化剂在外型上保持原离子交换树脂的球形形貌，在催化剂内部过渡金属 Fe、Ni、Cu 和 Zn 分布均匀，密度高于活性炭。相反，对于浸渍法制备的活性炭负载过渡金属催化剂，过渡金属仅分布在活性炭表面，而且过渡金属的分布密度也低于活性炭。因此，离子交换型活性炭催化剂不仅在金属负载量上高于浸渍型催化剂，金属在催化剂中的分布也优于浸渍型催化剂[20]。

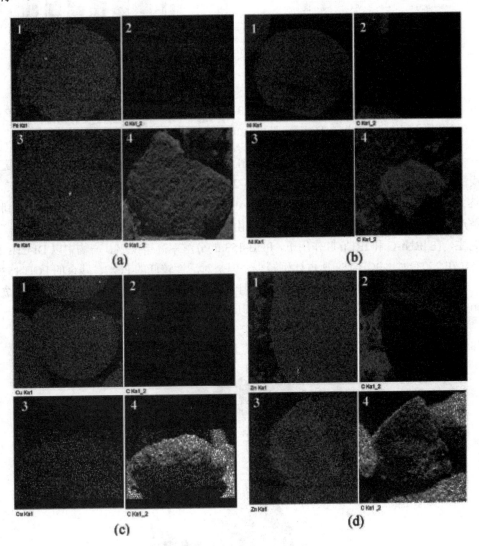

图 26-19 活性炭负载过渡金属催化剂的 SEM 谱图

照片 1 和 2 显示的是 IRM-C 中金属和活性炭的分布，照片 3 和 4 显示的是 LaM-C 中金属和活性炭的分布；
(a) Fe-C，(b) Ni-C，(c) Cu-C，(d) Zn-C

作者研究了八种活性炭负载过渡金属催化剂在不同反应温度下催化分解 PCB-153 的效果[21]，如图 26-20（a）所示，当分解温度为 250℃时，以 IRFe-C、LaFe-C、IRNi-C、LaNi-C、IRCu-C、LaCu-C、IRZn-C 和 LaZn-C 为催化剂的 PCB-153 分解率分别为 77.2%、49.8%、99.6%、66.9%、99.0%、84.7%、90.5% 和 62.7%，随着温度的升高，所有催化剂的活性均增强，当温度达到 350℃时，除了 LaZn-C 的催化效率为 94.7%外，其余催化剂的反应活性均在 97%以上。在所有催化剂中，活性炭负载镍、铜的催化活性最强，活性炭负载锌、铁次之，其中离子交换法制备的活性炭负载过渡金属活性优于浸渍法制备的活性炭负载过渡金属催化剂。

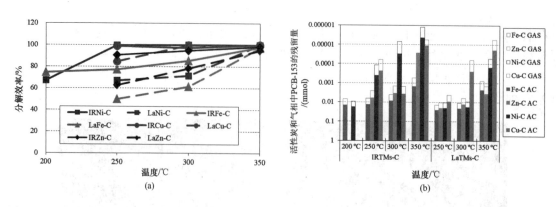

图 26-20 （a）不同活性炭负载过渡金属催化分解 PCB-153 的效率与温度的关系，（b）不同活性炭负载过渡金属催化剂催化分解 PCB-153 后残留 PCB-153 在活性炭和气相中的分布与温度的关系

通过进一步分析催化反应结束后残留 PCB-153 在活性炭和气相中的分布情况，发现温度为 250℃时，95%以上的残留 PCB-153 分布在活性炭上，如图 26-20（b）所示，随着温度升高，PCB-153 残留向气相转移，温度达到 350℃时，PCB-153 在 IRNi-C、IRCu-C、IRZn-C 和 LaCu-C 中的残留均降到 60%以下。PCB-153 分解产物分布如图 26-21 所示，在 250℃时由于 PCB-153 的分解率普遍较低，因此产物中以残余的 PCB-153 为主，在 IRNi-C 和 IRCu-C 作用下，PCB-153 的分解率高于 99%，产物中以 Di-CBs 和联苯为主。随着温度升高，在 250℃时，PCB-153 所占比例减少，但是对活性较低的 Fe-C，残余的 PCB-153 仍然占产物的绝大部分，当温度升高到 350℃时，除了 LaZn-C，由于其催化 PCB-153 分解的效率仅为 94.7%，PCB-153 残留高于 0.005 μmol，其余产物的量都在 0.005 μmol 以下，产物中联苯和 Di-CBs 占主要部分。

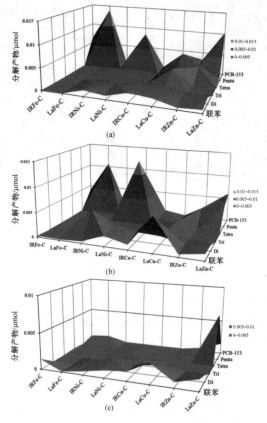

图 26-21 不同催化剂作用下 PCB-153 的分解产物分布
（a）250℃；（b）300℃；（c）350℃

研究为了进一步探究活性炭负载过渡金属催化剂催化分解多氯联苯的性能，开发了活性炭负载双元过渡金属催化剂（Ni/Cu-C，Ni/Zn-C 和 Ni/Pd-C），以 Aroclor 1254 为反应目标物质，系统地研究了不同催化剂对 Aroclor 1254 催化分解的效率，并探讨了反应过程随时间和反应温度变化的趋势[22]。如图 26-22 所示，由反应物浓度和时间的关系揭示了该反应过程是准一级反应的动力学作用机理，且不同催化剂对 PCBs 的分解效率均随反应温度的升高而增加；通过对比廉价金属催化剂 Ni/Zn-C、Ni/Cu-C 和贵金属催化剂 Ni/Pd-C 对 Aroclor 1254 的分解效率发现，在 300℃条件下负载廉价过渡金属催化剂和负载贵金属催化剂的分解效率均达到 99%以上，因此可以实现由过渡金属替代贵金属负载活性炭分解多氯联苯，降低催化剂成本。

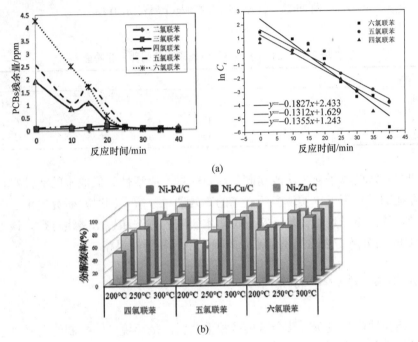

图 26-22 反应条件对活性炭负载双元金属催化分解多氯联苯性能的影响
(a) 反应时间；(b) 反应温度

同时研究了不同金属配比的催化剂对 Aroclor 1254 分解效率的影响，并与单一金属负载型催化剂对 Aroclor 1254 的分解效率进行对比。如图 26-23 所示，研究证明了双元金属负载型催化剂优于单一金属催化剂，该结论通过密度泛函理论计算证实。作者发现双元金属 Ni/Cu 比单一金属 Ni 和 Cu 对多氯联苯中 Cl 的吸附能均大（图 26-24，表 26-2），说明双元金属中 Ni 和 Cu 的协同作用较单一金属对 Cl 具有更强的吸附作用，从而提高了对多氯联苯的分解能力[23]。

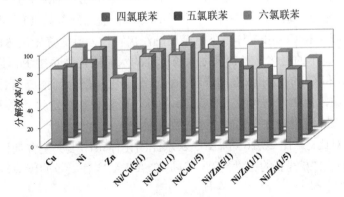

图 26-23 催化剂种类对活性炭负载双元金属催化分解多氯联苯性能的影响

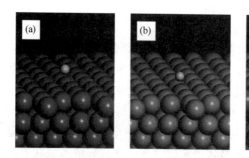

图 26-24 Cl 原子吸附在不同金属表面的构型

(a) Ni；(b) Cu；(c) Ni/Cu

表 26-2 Cl 原子与不同金属表面的间距和吸附能

金属表面	间距/Å	吸附能/eV
Ni	2.11	−1.83
Cu	2.17	−1.72
Ni-Cu	2.14	−1.94

综上所述，研究开发的离子交换法成功制备高负载量、高活性、低成本的活性炭负载过渡金属催化剂，在无氢供体存在的条件下可以实现多氯联苯的高效催化分解。同时，研究提出了多氯联苯的脱氯反应机理和反应路径，依据量子理论计算证实，说明本研究的方法和技术路线可行。该工作为活性炭喷射工艺中 PCBs 等 POPs 的分解与去除提供了重要的理论指导和新的思路。

3.2 焚烧飞灰固化/稳定化技术

焚烧过程不可避免地产生飞灰，其会富集多种重金属，是环境风险控制的主要对象。世界各国飞灰处置的主流技术是将飞灰固化/稳定化以后进入填埋场，水泥固化和螯合剂稳定化是最为常用的固化/稳定化技术。两种技术对重金属的固定机理不同，在填埋场特殊的地球化学条件下，重金属固定的长效性可能存在较大差异，目前国内外相关研究甚为缺乏。针对上述问题，清华大学刘建国教授团队开展了较为深入系统的研究并取得重要进展。该团队在前期研究中重点分析了飞灰水泥固化、自然老化 6 年后的理化特性[24]，进而采用地球化学模拟方法，对比自然老化条件下飞灰水泥固化体系与螯合剂稳定化体系中重金属化学形态的差异[25]，为飞灰固化/稳定化材料选择、工艺设计和环境风险管理提供科学依据。

在相同自然老化条件及老化时间（6 年）下，首先对比了水泥固化飞灰（Ce-6-FA）和螯合剂稳定化飞灰（Ch-6-FA）微观形貌差异。由图 26-25 可知，Ch-6-FA 和 Ce-6-FA 样品中碳酸盐呈现出不同的形貌。在 Ch-6-FA 中碳酸盐发育较充分，微观形貌多种多样。不同微观形貌的碳酸盐具有不同的重金属浸出特性。

进一步分析了 Ch-6-FA 和 Ce-6-FA 体系下重金属的浸出特性差异。首先是达标性批次浸出试验结果。如表 26-3 所示，在经历 6 年的老化，当初满足入场标准的水泥固化体系中 Pb 浸出浓度已超标，同样，螯合剂稳定化体系对 Pb 和 Cd 的浸出浓度同样超标。水泥固化和螯合剂稳定化对于飞灰中重金属的长效固定可能都存在一定问题，但水泥固化技术中仅 Pb 超标，而螯合剂稳定化体系下超标的重金属为 Pb 和 Cd，其他重金属中 Cu、Zn 浸出浓度也远高于水泥固化体系。

其次是酸中和能力对比。从图 26-26 可知，飞灰水泥固化体系比飞灰螯合剂稳定化体系具有更高的酸中和能力。虽然 Ch-6-FA 中方解石的含量更高，但是 Ce-6-FA 含有更多的碳酸盐，尤其是有无定型的碳酸盐。

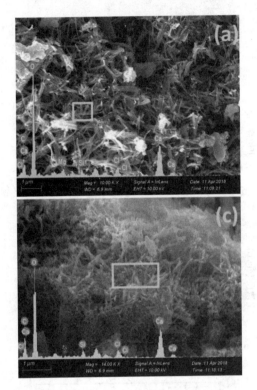

图 26-25　老化飞灰体系中碳酸盐形貌

(a, b, c) Ch-6-FA；(d) Ce-6-FA

表 26-3　达标性批次试验浸出结果　　　　　　　　　　　　（单位：mg/L）

	Pb	Cd	Cu	Cr	Zn	pH
Ce-6-FA	4.01	0.03	0.07	0.25	0.25	12.2
Ch-6-FA	3.00	1.46	3.45	0.01	6.89	6.22
GB 16889—2008	0.25	0.15	40	4.5	100	—

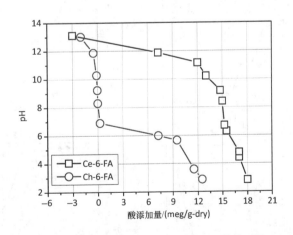

图 26-26　Ce-6-FA 与 Ch-6-FA 酸中和能力对比

采用 pH 相关性浸出方法进一步研究了不同重金属浸出随 pH 变化规律，评价实际场景中重金属释放的最大风险。如图 26-27 所示，除了 Pb 和 Zn，Ce-6-FA 体系下 Cd、Cu 和 Cr 的最大浸出量均低于 Ch-6-FA，表明水泥固化体系对这三种重金属具有较强的固定效果。通过浸出液溶解度模型拟合，发现在 Ce-6-FA

中 Cu、Pb 和 Zn 的溶解度预测曲线很好地符合了实验点，表明它们的浸出是受到溶解度控制的，从而为地球化学模拟提供了可靠的基础。

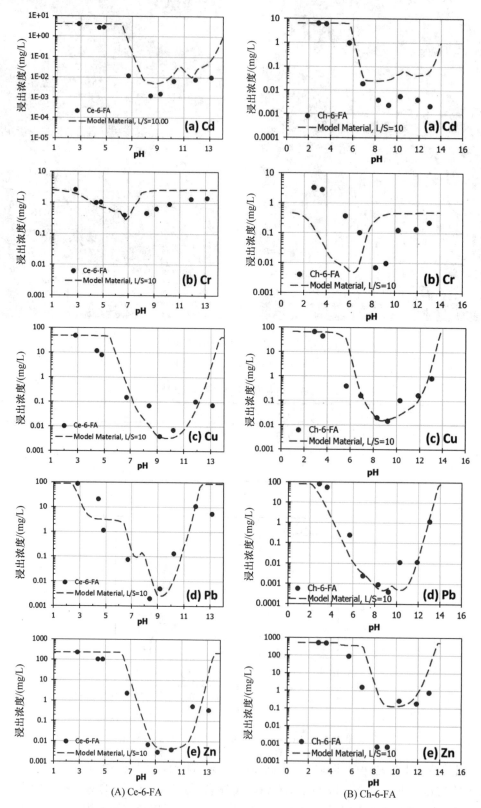

图 26-27　Ce-6-FA（A）和 Ch-6-FA（B）体系下重金属浸出浓度随 pH 变化及溶解度模拟预测曲线

在 pH 相关性浸出试验的基础上进行地球化学模拟。将样品中：①主要和微量元素最大可释放量（1<pH<14）；②主要矿物质组成和化学反应方程式；③样品中主要活性吸附表面（包括固相 HA 和 HFO/HAO）含量；④DOC 和 DIC 含量，输入地球化学模拟软件 LeachXS^{TM}，得到重金属化学形态分布。如图 26-28 所示，对于 Cd，在飞灰本身 pH 条件下，Ce-6-FA 中 Cd 浸出控制相为 CdSiO$_3$，Ch-6-FA 中控

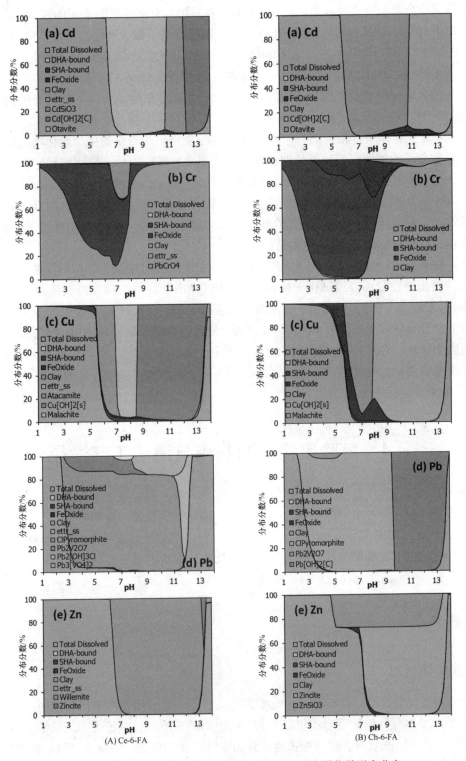

图 26-28　Ce-6-FA（A）和 Ch-6-FA（B）中重金属化学形态分布

制相为 $CdCO_3$。镉的硅酸盐具有较好的化学和物理稳定性，水泥固化体系中 $CdSiO_3$ 存在 Cd 比螯合剂稳定化体系中 $CdCO_3$ 更稳定。对于 Pb，在自身 pH 条件下，Ce-6-FA 中 Pb 浸出控制相主要是 $Pb_2(OH)_3Cl$，其次是 ClPyromorphite[$Pb_5(PO_4)_3Cl$]，Ch-6-FA 中 Pb 浸出控制相是 ClPyromorphite。$\log K$（ClPyromorphite）=84.34，表明该 pH 条件下螯合剂稳定化体系中 Pb 的固定具有较好的长期稳定性。

此外，吸附同样会影响重金属浸出，高 pH 下 Pb^{2+} 容易被具有活性吸附表面的物质吸附，例如 HFO、HAO。Ch-6-FA 中 HFO、DOC 吸附一定量的 Pb，所以同样导致了螯合剂稳定化体系中 Pb 释放量较高。对于 Cu、Cu 的浸出与 HFO 紧密相关，在 Ch-6-FA 中 HFO、DOC 含量高于 Ce-6-FA，所以 HFO、DOC 结合态重金属含量较高。所以，Ch-6-FA 对 Cu 固定的长效性弱于 Ce-6-FA。相比 Ch-6-FA，Ce-6-FA 体系中水泥水化产物的风化降解途径复杂，形成的多种二次矿物和非矿物组分，尤其是非矿物组分具有更高的吸附，从而使得 Pb、Cr 和 Cd 的化学形态更复杂，稳定性也强于螯合剂稳定化体系。

对于螯合剂稳定化体系中，螯合剂的加入获得足够的 S 以沉淀重金属生成硫化物沉淀。初始阶段金属硫化物控制着重金属的溶解度。经历长期的自然风化和紫外线作用后，螯合剂稳定化体系中有机硫螯合物发生降解，造成了重金属的释放，以及螯合物降解产生的 DOC 也增加了 Cr、Cd、Cu 的浸出。因此，Ch-6-FA 比 Ce-6-FA 中重金属的浸出量会更高。含有二硫代氨基甲酸盐（DTCR）的螯合剂具有硫化物的功能，能够高效固定重金属，但该重金属螯合物形成和解体过程中会产生大量盐分，飞灰中 Pb 的固定是基于无机络合作用，高盐分会导致 Pb 的大量释放。

综上所述，如图 26-28 所示，Ce-6-FA 和 Ch-6-FA 体系中矿物组成及非矿物相含量差异显著，从而影响其浸出行为。浸出特性分析及地球化学模拟表明，Pb 浸出特征与溶解度控制预测曲线吻合较好。在 1<pH<14 范围内，Ce-6-FA 中 $Pb_5(PO_4)_3Cl$ 所占的比例远高于 Ch-6-FA。Ch-6-FA 中重金属螯合物的解体过程中产生大量的盐分会导致无机络合 Pb 的释放，从而导致螯合剂稳定化体系中 Pb 的释放量高于水泥固化体系。水泥固化体系中 $CdSiO_3$ 比螯合剂稳定化体系中 $CdCO_3$ 更稳定，表明水泥固化体系对 Pb、Cd 固定的长效性优于螯合剂稳定化体系。

4　危险废物能源利用与清洁工艺途径

随着便携式电子设备与电动汽车的广泛使用，电池已成为人们生活中不可缺少的部分，但废旧电池存在严重环境污染风险。如：废铅酸蓄电池中铅是一种强毒性重金属，废铅酸蓄电池已列入我国《国家危险废物名录》；电镀污泥中含有大量有色金属，而目前常用的处理方法是将其直接填埋处理，并未将其中含有的金属利用。电子垃圾能源利用成为危险废物处置与利用的重要方向。此外，从源头控制，梯级消减并利用矿产资源的清洁工艺途径也为危险废物控制提供了新思路。

4.1　废电池能源化

4.1.1　铅膏制能源材料

铅酸蓄电池具备安全性高、价格低、回收率高等优点，广泛应用于汽车启停电池、动力电池、备用电源等领域。截至 2018 年，铅酸蓄电池仍占全部二次电池的 55%，拥有最大市场份额。2017 年，全球精铅产量 1155 万 t，再生铅占 60%。废铅酸蓄电池作为主要再生铅资源，被誉为"城市矿产"。铅也是一种强毒性重金属，废铅酸蓄电池已列入我国《国家危险废物名录》。因此，废铅酸蓄电池具有资源性与污染性的双重特性。废铅酸蓄电池中废铅膏为电池正负极板失效的活性物质，其主要成分包括 $PbSO_4$（~50 wt%）、PbO_2（~28 wt%）、PbO（~9 wt%）及单质 Pb（~4 wt%）等多价态的含铅组分及其他微量杂质元素，是废铅酸蓄电池中含量最高、最难回收的部分。目前，国内外广泛采用火法冶炼的废铅膏回收再生工艺，温度高达 1000℃以上，产生大量挥发性铅尘和 SO_x。传统火法再生铅引发的"血铅"等环境污染风险受到广泛关注。

国际铅锌组织及许多研究机构研发积极探索替代传统火法再生铅工艺的新途径，湿法回收工艺被认为是实现常温废铅膏清洁回收的最佳途径。华中科技大学杨家宽等与剑桥大学合作率先提出有机酸浸出-低温焙烧工艺，通过柠檬酸等有机酸与废铅膏含铅组分进行络合反应，获得的柠檬酸铅前驱物，再通过 300~500℃低温热解制备出新型铅粉（PbO/Pb）。该新型铅粉可以直接用于制备铅酸蓄电池的极板活性物质，相对于其他湿法工艺，包括酸浸-电解联合工艺、碱溶-结晶工艺等，进一步缩短了再生铅的循环利用过程（图 26-29）。这种工艺消除了高温熔炼带来的铅尘及 SO_x 排放的环境风险，省去了高温熔炼、精炼、球磨氧化等高耗能工艺环节。

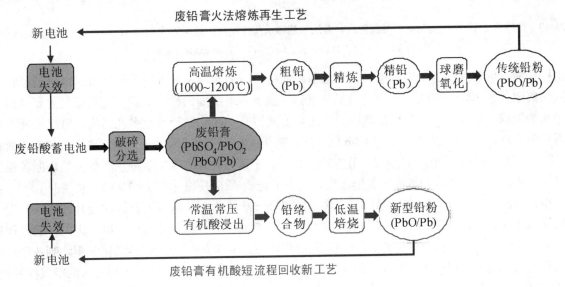

图 26-29　传统废铅酸蓄电池火法再生循环流程与有机酸短流程回收新工艺对比

该团队系统研究了柠檬酸-柠檬酸钠、乙酸-柠檬酸钠两种有机酸浸出工艺。在浸出过程中，柠檬酸钠用于硫酸铅的脱硫，双氧水用于二氧化铅的还原转化，柠檬酸根离子参与结晶反应。在乙酸-柠檬酸钠浸出过程中，乙酸用于调节浸出体系的 pH，从而促进柠檬酸铅的结晶、减少柠檬酸的投加量。研究发现两种浸出体系的最佳 pH 不同（柠檬酸-柠檬酸钠浸出体系的 pH 为 3~4，乙酸-柠檬酸钠浸出体系的pH 为 5~6），还发现 pH 是柠檬酸铅前驱体晶体构型的关键调控因子[26]，柠檬酸-柠檬酸钠、乙酸-柠檬酸钠浸出体系下的柠檬酸铅的分子式分别为 $Pb(C_6H_6O_7)·H_2O$，$Pb_3(C_6H_5O_7)_2·3H_2O$[27]。$Pb(C_6H_6O_7)·H_2O$ 呈现方形片状［图 26-30（a）］，而 $Pb_3(C_6H_5O_7)_2·3H_2O$ 呈现棒状［图 26-30（d）］；其分子结构与晶胞堆积图也有显著差异［图 26-30（b）、（c）、（e）、（f）］[28]。比较两种浸出体系，在乙酸-柠檬酸钠的浸出体系下，柠檬酸铅中铅与柠檬酸根离子的摩尔比为 3∶2，而在柠檬酸-柠檬酸钠浸出体系下，柠檬酸铅中铅与柠檬酸根离子的摩尔比为 1∶1，因此，乙酸-柠檬酸钠浸出体系消耗更少的柠檬酸。

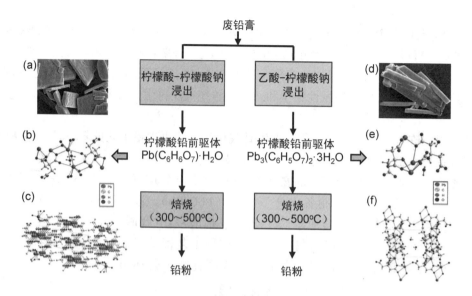

图 26-30 柠檬酸-柠檬酸钠、乙酸-柠檬酸钠两种湿法工艺体系及柠檬酸铅前驱体的 SEM 图、分子结构及晶胞堆积图
(a), (b), (c) $Pb(C_6H_6O_7)·H_2O$; (d), (e), (f) $Pb_3(C_6H_5O_7)_2·3H_2O$

由柠檬酸铅前驱体低温焙烧直接制备电池用超细铅粉，是实现短流程的关键环节。浸出结晶得到的柠檬酸铅前驱体在 300~500℃、空气或者氮气气氛下制备新型超细铅粉[29]。新型超细铅粉主要是由氧化铅、金属铅组成，还含有一定量柠檬酸根裂解之后的残留碳。超细铅粉具有多孔结构，氧化铅或者金属铅附着在多孔碳表面 [图 26-31（a）和（b）]。由于铅粉作为活性物质应用于铅酸蓄电池时，铅粉氧化度（氧化铅占铅粉质量的百分比）需要控制在 75%~85%，申请人发现在不同氧分压下低温焙烧制备铅粉中 C 残留及转化是影响铅粉物相组成的关键因素，提出了传质-反应模型 [图 26-31（c）和（d）][30]。在焙烧过程中，裂解产物的多孔结构使得焙烧产物中始终处于缺氧状态，随着氧分压的提升，氧气与裂解产生的炭反应生成还原性气体 CO，CO 与氧化铅反应生成金属铅，导致氧化度的下降。因此，通过调控合理气氛，可以制备所需氧化度的铅粉。基于传质-反应模型，申请人提出了裂解-氧化两步法制备氧化度可控的铅粉，用于指导大规模铅粉制备，即先在缺氧条件下进行裂解反应，然后再与通入的空气接触反应，制备所需氧化度的铅粉。

将制备的新型铅粉应用于铅酸电池电化学体系。通过搭建了三电极体系对铅粉的电化学特性进行了研究，循环伏安法测试表明，焙烧温度为 370℃、焙烧时间为 20 min 的铅粉表现了良好的可逆性和循环稳定性。作者将新型超细铅粉直接作为电池制备负极活性物质的原料，制备 2V/2Ah 铅酸蓄电池，发现新型超细含碳铅粉较球磨铅粉制备的电池具有更高的初始容量及活性物质利用率[31]。

此外，阐明了新型铅粉中 C 骨架材料与纳米 PbO 晶体原位形成 PbO@C 复合材料微观结构特征 [图 26-32（a）]，提出了超细含碳铅粉制备高能量密度的铅炭电池的方法。将不同温度下裂解的 PbO@C 材料应用于铅炭电池负极，研究发现 PbO@C 复合材料中的炭材料在铅炭电池负极中可提升电解液渗透传输效率，从而提高铅炭电池的充电接受效率。PbO@C 复合材料中 Pb 与 C 的结合使借助炭材料传输的电解液能够得到原位高效利用 [图 26-32（b）]，电池 20h 率下活性物质利用率提高了将近 50%（与不添加炭材料的空白样对比）。同时，Pb 与 C 的结合提供了更多的电化学反应活性位点，可有效抑制极板的不可逆硫酸盐化。添加 P600 的电池性能更优，高倍率部分荷电状态（HRPSoC）的循环寿命从空白样的 11000 圈增加到 40000 圈 [图 26-32（c）和（d）][32]。

图 26-31 （a）370℃焙烧温度下的铅粉 SEM 图；（b）铅粉中多孔碳 SEM 图；（c）和（d）铅粉焙烧过程中传质-反应模型

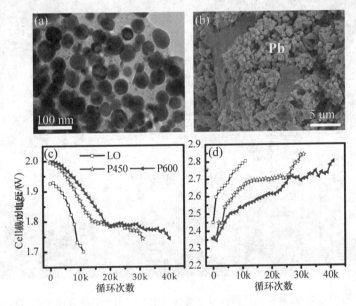

图 26-32 （a）600℃裂解铅粉（P600）的 TEM 图；（b）添加 P600 的铅炭电池负极活性物质的 SEM 图；（c）添加不同特性铅粉的铅炭电池 HRPSoC 循环放电截止电压；（d）充电截止电压

综上所述，研究针对目前国内外普遍关注的废铅酸蓄电池高温火法再生中面临的环境污染的风险，提出了有机酸柔性浸提短流程回收工艺新方法及再生铅的高值化利用的新思路。研究团队与湖北金洋冶金股份有限公司共建了一条年处理量 200t 级别的废铅酸蓄电池湿法回收中试生产线，已经初步实现了废铅膏有机酸湿法短流程回收方法。该工作为电子废弃物通过有机酸络合晶型调控实现杂质高效分离及高值化转化提供了重要的理论指导。

4.1.2 电镀污泥制储锂材料

随着便携式电子设备与电动汽车的广泛使用，人们对于高能量、高功率密度和长循环稳定的锂离子电池的需要越来越迫切。锡基负极材料因为其高的储锂容量（$Li_{4.4}Sn$：992 mA h/g）受到了越来越多的关注，被视为可以取代石墨成为下一代商用锂离子电池的负极材料。然而锡基材料在充放电过程中会产生巨大的体积膨胀，导致其循环稳定性与倍率性能较差，限制了锡基材料的规模化生产应用。如何解决锡基材料的体积膨胀问题，提高其循环稳定性与倍率性能仍然是一个挑战。

另一方面电镀污泥中含有大量有色金属，而目前常用的处理方法是将其直接填埋处理，并没有将其中含有的金属利用起来[2,3]。因此若能实现利用电镀锡泥制备高性能锡基锂离子电池负极材料，不仅能够解决危险废弃物的环境污染问题，也能同时制备高性能锂离子负极材料，这对于锂电负极材料的设计与危险废弃物的综合治理具有重要意义。

华南理工大学林璋、邓洪等在前期研究中成功利用枯草芽孢杆菌实现吸附态重金属铬的吸附分离[33]，在近期工作中首先通过枯草芽孢杆菌吸附电镀锡泥中的锡，再进一步制备高效锂电负极材料（图 26-33）。该工作发展了一种实现从废弃物到高值化储能材料的废物高值化转化策略[34]。

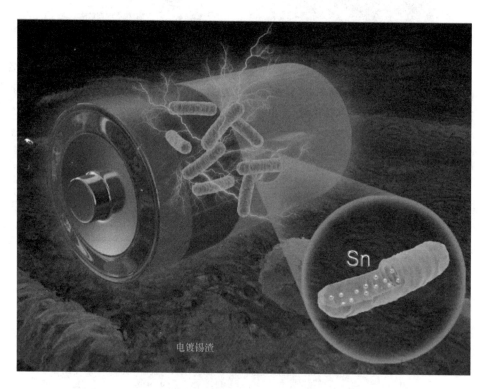

图 26-33　枯草芽孢杆菌提取电镀锡渣中的锡用于制备锂电负极材料

作者利用枯草芽孢杆菌将电镀污泥酸浸液中的锡回收提取，再结合冷冻干燥，惰性气体氛围煅烧，成功制备了具有超小纳米锡颗粒的 Sn@C 复合材料（图 26-34）。如图 26-35 所示，结合 XRD 精修结果，拉曼和 XPS 数据进一步证明了 Sn@C 复合材料之中的锡存在形态主要有锡单质和少量未完全还原的四氧化三锡。Sn@C 复合材料的形貌和能谱图见图 26-36。该研究为电子危险废物处理提供了新途径。

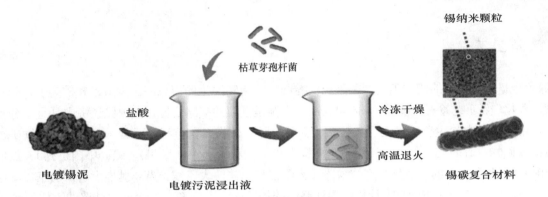

图 26-34　Sn@C 复合材料的制备过程

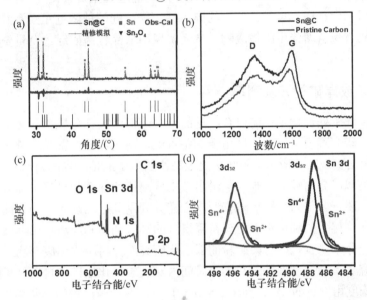

图 26-35　Sn@C 复合材料的 XRD 精修、Raman、XPS 表征结果

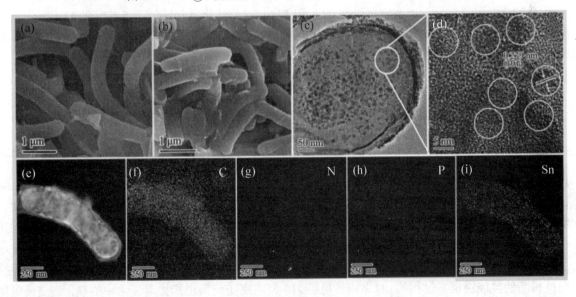

图 26-36　Sn@C 复合材料的形貌和能谱表征

4.2 清洁工艺途径

"绿色全量资源化"的清洁工艺途径是源头控制危险废物产生的新思路。本节以磷矿为例,阐述"绿色全量资源化"新工艺途径。磷矿(phosphorite)资源是重要的战略资源,在国民经济中具有十分重要的地位和作用。磷矿中除主要矿物氟磷灰石外,主要的脉石矿物如白云石、石英、方解石、褐铁矿、黏土和硅酸铝盐矿物;有的矿区的磷矿还伴生具有提取价值的稀土(中国贵州)、碘(中国贵州)、放射性元素(美国)。工业上成熟的产业化生产技术路线为湿法和热法两大路线。当今世界上磷矿加工以湿法路线为主,以磷矿中的 P_2O_5 折成100%计算,湿法路线的加工能力占到90%以上,热法路线占比不到10%。废弃物包括固体废弃物尾矿,加工环节产生的废弃物等。"绿色全量资源化",首先以提高中低品位磷矿适应性及磷的综合利用率为目标,再者将伴生的硅、钙、镁、硅等种微量元素资源化利用,同时解决加工过程清洁生产问题或避免新增伴生固体废物,从而实现中低品位磷矿绿色全量资源化利用技术,推动磷产业绿色健康发展。

4.2.1 中低品位磷矿或尾矿的开发利率

中低品位磷矿和磷尾矿处理技术目前已有广泛的研究,但因大部分技术存在工艺复杂、成本高等问题,使其仍处于实验室研究阶段。如磷尾矿用于制备水泥、建筑用砖、微晶玻璃等技术虽然在实验室取得较好的成果,但仍无法实现工业应用。用酸法处理高镁磷尾矿的技术也因白云石中的 Mg 和 Ca 同属于碱土类,化学性质相类似,分离较为困难,从浸出液中分离镁的化合物只限于实验阶段,还没形成大规模的工业生产。同时,磷资源作为稀缺战略资源,应充分利用其中的磷元素。因此,从磷资源综合利用和技术经济层面出发,磷尾矿的利用重点应该在磷资源的高效利用和经济使用。磷企业主要从豆科禾本科植物根系分泌物中低分子量有机酸对磷矿粉的磷活化研究,探讨了固体废弃物通过生物活化达到资源农业利用的途径。研究了不同浓度柠檬酸在不同时间内对安宁风化原矿和晋宁尾矿矿粉磷的活化释磷结果显示,当柠檬酸的浓度由 0.1 mol/L 升高至 2.5 mol/L 时,两种磷矿粉的释磷累积量随着酸浓度的增加而增加,但当柠檬酸的浓度由 0.1 mol/L 升高至 0.5 mol/L 时,安宁原矿在 24 h 和 144 h 内的活化量则只分别提高了 22.22%和 2.91%、而晋宁尾矿在 24 h 和 144 h 内的活化量却降低了 5.47%和 23.18%。此外,苹果酸对安宁原矿和晋宁尾矿矿粉中磷的释磷累积量随时间的变化显示,苹果酸的浓度为 0.1 mol/L 时对两种磷矿粉的释磷累积量显著高于浓度 0.5 mol/L 和 2.5 mol/L,而 2.5 mol/L 苹果酸与 0.5 mol/L 苹果酸相比,对安宁原矿和晋宁尾矿在 24 h 和 144 h 内的活化量分别提高了 225.47%、-3.77%、254.19%和 108.08%。

难溶性磷转化为可供植物利用形态的磷,从而促进中低品位磷矿粉中磷的利用。因此生物活化不失为磷尾矿资源化利用的一种有效方法和短流程利用途径。

4.2.2 磷矿伴生资源的回收利用探讨

1) 碘的利用技术

针对磷矿的热法加工和湿法加工环节工艺特征,碘的回收工艺主要有离子交换法、空气吹出法、溶剂萃取法。离子交换法是最常用的富集方法,由于原料中碘含量低,不足以析出碘,因此需要对碘进行富集。该法是先加酸酸化,通过加氧化剂氧化成单质碘再通过离子交换柱吸附碘,然后用溶剂碱洗涤从树脂中解析出碘,解析液中的碘经酸化处理可析出单质碘,最后经分离和精制得到成品碘。在此方法中,离子交换树脂的性能决定了过程的 pH 范围。利用 717 型阴离子交换树脂提取废水中的碘,碘交换吸附率高达 96%,总收率达 85%。离子交换法(流程图见图 26-37)是最常用的方法,此方法可以提高碘的浓度,但易受杂质干扰,操作烦琐,成本高。

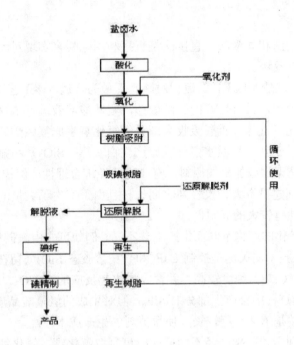

图 26-37　离子交换法工艺路线图

空气吹出法首先被日本伊西化学工业公司所采用，目前是一种优越的工艺，主要是从天然气卤水中制碘。该法是将含碘的母液用硫酸酸化，再加入氧化剂氧化，使碘盐转化为单质碘，同时吹入空气，把碘吹出，吹出后的含碘空气再经过吸收、结晶和精制工序制成粗碘（图 26-38）。从稀磷酸中回收碘采用过此方法。此法虽然成本低，设备占地面积小，但对含碘品位要求比较高，要求稀磷酸中 $\rho(I)$ 一般在 50～100 g/L，回收率在 65%～75%，因此，目前国内很少采用此方法制碘。

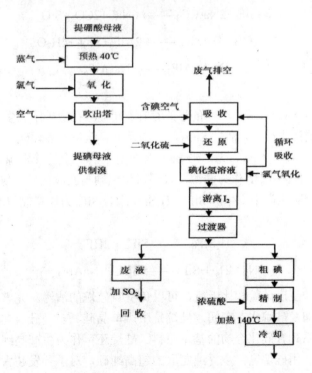

图 26-38　空气吹出法工艺路线图

2）氟的利用技术

氟资源的利用分为直接法和间接法。直接法是直接利用硫酸来实现对氟硅酸的分解。间接法主要有利用氟硅酸来生产氟化盐以及铵法。

直接法：用硫酸分解氟硅酸制取 HF 的流程对磷肥厂最为合适。该工艺过程不消耗辅助原料，没有热转化。用过的硫酸可循环使用，具有良好的经济效益。氟硅酸溶液进行浓缩后，加入硫酸溶液，可将氟硅酸分解成四氟化硅和氟化氢气体，硫酸吸收大部分的氟化氢生成氟磺酸。四氟化硅不易被吸收，呈气体逸出。将四氟化硅纯化，使其与高温水蒸气反应产生白炭黑（SiO_2）和氟化氢气体。上述硫酸通过解吸过程释放氟化氢。该工艺无需添加附加原料，有一定的经济合理性，但控制比较复杂、物料循环量大、氟单程转化效率低、设备基建投资大、投资回收期长。且由于产生了大量的稀硫酸，只能应用于有普钙生产装置的磷肥企业，因此有较大的局限性。

间接法：因由氟硅酸转化的产物不同而生产工艺不同。例如可先由氟硅酸制取 CaF_2，再按 CaF_2 制取氢氟酸的工艺生产 HF。该技术的关键是控制 CaF_2 和介稳溶胶态 SiO_2（留在溶液中）的分离条件，可加入适量絮凝剂。所得产品（CaF_2）纯度较高，并含少量未起反应的 $CaCO_3$。所得的 CaF_2 按常规方法制取 HF。在 H_2SO_4 分解 CaF_2 的过程中会产生部分 H_2SiF_6，但对生成的氢氟酸品质影响不大。目前该工艺技术已经非常成熟，唯一的弊端是工艺流程较长，使得消耗的生产成本较高。

氟硅酸生产氟化盐：利用磷肥副产经不同反应，可制取氟化钙、氟化钾、氟化铝等氟化盐。氟化铝是工业生产的助熔剂，常为无色或白色结晶，性质很稳定，加热的情况下可水解，主要用于炼铝。氟化铝由氟硅酸和氢氧化铝反应、分离制得，同时联产白炭黑（SiO_2）。工艺流程如下：

$$H_2SiF_6+2Al(OH)_3+(n-4)H_2O \longrightarrow 2AlF_3+SiO_2 \cdot nH_2O \downarrow$$

铵化法：工艺过程包括晶种生成、氨解、偏铝酸钠的制备和冰晶石合成 4 个方面。碳酸氢铵或氨水提供氨根，NH_4F 为中间产物，冰晶石中的 Al^{3+}、Na^+ 离子由硫酸盐提供，同是副产硫酸铵。反应流程如下：

$$Na_2SiF_6+2Na_2CO_3 \longrightarrow 6NaF+2CO_2+SiO_2$$

$$2Al(OH)_3+H_2SiF_6 \longrightarrow 2AlF_3+SiO_2+4H_2O$$

$$3NaF+AlF_3 \longrightarrow Na_3AlF_6$$

3）硅的利用技术

硅资源在磷矿中大多以 SiO_2 等酸不溶物的形式存在。硅资源在磷矿中大多以 SiO_2 等酸不溶物的形式存在，在磷肥生产和湿法磷酸生产中，活性 SiO_2 可参与反应，但是不消耗硫酸，在此过程中生成了 SiF_4、H_2SiF_6、Na_2SiF_6、K_2SiF_6、$(NH_4)_2SiF_6$、无定形 SiO_2 等。对于磷矿中的硅资源有两种利用方法。

（1）以 SiF_4、H_2SiF_6、Na_2SiF_6 等作为硅源，采用氢化铝钠与四氟化硅无氯工艺进行反应制备硅烷。其原理为以在湿法磷酸、磷肥生产中产生的 SiF_4、H_2SiF_6、Na_2SiF_6 为原料制得 SiF_4，使其与 $NaAlH_4$ 反应生产 SiH_4。反应式为

$$H_2SiF_6 \longrightarrow SiF_4 + 2HF$$

$$NaAlH_4 + SiF_4 \longrightarrow SiH_4 + NaAlF_4$$

该法生产过程与传统法生产硅烷相比无氯，可以免受氯硅烷的沾污。完整的生产线包括四氟化硅生产车间、氢化铝钠生产车间及硅烷生产车间。硅烷是生产单晶硅、多晶硅、非晶硅、金属硅化物、氮化硅、碳化硅、氧化硅等一系列含硅化合物的基本原料。硅烷还可作为生产高纯硅的原料，可大幅度提高生产能力和降低生产成本。用硅烷进一步提纯制取的超高纯硅，对于开发超大规模集成电路和红外探测器都有重要的用途。近几年，非晶硅太阳能电池发展也十分迅速。

(2) 以无定形 SiO_2 为硅源制备纯硅。无定形 SiO_2 属磷肥副产固体废弃物，性能稳定。以无定形 SiO_2 为硅源制备纯硅，可采用热还原法，包括金属热还原法、非金属热还原法、耦合热还原法，较常用的是非金属热还原法中的氢热还原法和碳热还原法。前者的优点是硅收率高，无中间产物 SiC 生成，且在此方法中还原剂是氢气，其氧化产物无污染，无需纯化过程，成本低，环境效益和经济效益较好；后者工艺相对成熟，成本低，原料易得，产品的纯度高，多用于工业生产。其主要反应为

$$SiO_2 + 2C \longrightarrow Si + 2CO$$

目前该技术还在实验室阶段。

4) 钙的利用技术

磷石膏的综合利用在国内已取得了可喜的成果，如水泥缓凝剂和建筑石膏板、石膏粉等工艺（图 26-39），但是磷石膏资源化利用仍任重道远。

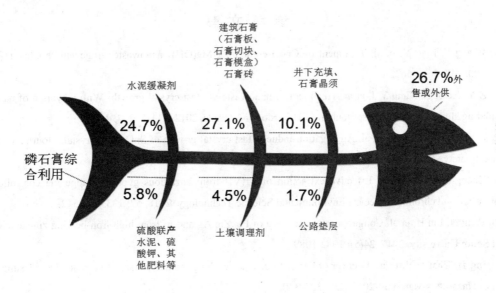

图 26-39　较为成熟的磷石膏利用的比例

5) 镁的利用技术

磷矿中镁主要以白云石形式存在，利用磷矿中（特别是反浮选尾矿）酸镁、碳酸钙等分解温度不同，运用热法煅烧工艺进行分解，得到磷灰石、碳酸钙及氧化镁等产品。从经济合理性方面考虑，选用反浮选尾矿经烘干、煅烧等高能耗工艺，产品且为磷灰石、碳酸钙经济价值不高的产品，且产品运输半径小，目前对公司而言经济性不合理。

肥料级一水硫酸镁工艺路线较为简单，技术成熟。先将硫酸和含氧化镁的矿粉（白云石、菱苦土、蛇纹石）或其煅烧粉按比例计量；在中和罐中加入水或洗水、母液，在搅拌下徐徐加入矿粉，再加入硫酸进行中和反应，控制浆液 pH；将中和液保持一定温度反应一段时间，经叶片过滤机过滤，将清液打入结晶器；冷却，进行离心分离为七水硫酸镁，将湿料送入脱水干燥机，于 200℃ 以上脱水干燥制得一水硫酸镁。目前该技术比较符合公司现状，机油产业化价值。

综上所述，磷矿是不可再生资源，云磷集团针对我国磷资源传统生产工艺过程中资源能源利用率低、能耗高、固体废物排放量大等产业难题，基于对传统生产工艺过程及资源利用情况的系统梳理，提出中低品位磷矿"绿色全量资源化"技术路线思路，构建了清洁工艺途径，提出了源头减量危险废物产生的新思路。

5 展　　望

危险废物处置涉及工业、市政与医疗行业等，与生活及社会发展息息相关，是国家与行业的重大需求。危险废物涵盖工业废物、市政废物等，面临重金属、高毒性有机物等污染问题，其理论研究、技术手段及产业化应用等均涉及复杂环境化学问题，并与材料、化工、冶金、生物、热工等多学科交叉。总体上，危险废物处理和资源化利用在环境化学领域依然处于发展阶段，无论在基础科学研究、技术开发及产业运用方面，均具有良好前景。目前，我国危险废物资源化利用水平与国际依然有较大差距，急需进一步发展，机遇与挑战并存。

参 考 文 献

[1] Liu W, Huang F, Liao Y, et al. Treatment of Cr-VI-containing Mg(OH)$_2$ nanowaste. Angewandte Chemie- international Edition, 2008, 47: 5619-5622.

[2] Zhuang Z Y, Xu X J, Wang Y J, et al. Treatment of nanowaste via fast crystal growth: With recycling of nano-SnO$_2$ from electroplating sludge as a study case. Journal of Hazardous Materials, 2012, 211-212: 414-419.

[3] Zhuang Z Y, Huang F, Lin Z, et al. Aggregation-induced fast crystal growth of SnO$_2$ nanocrystals. Journal of the American Chemical Society, 2012, 134: 16228-16234.

[4] Liu W, Zheng J, Qu X, et al. Effective extraction of Cr(VI) from hazardous gypsum sludge via controlling the phase transformation and chromium species. Environmental Science Technology, 2018, 52: 13336-13342.

[5] Peng B, Peng N, Liu H, et al. Comprehensive recovery of Fe, Zn, Ag and In from high iron-bearing zinc calcine. Journal of Central South University, 2017, 24(5): 1082-1089.

[6] Li M, Peng B, Chai L Y, et al. Recovery of iron from zinc leaching residue by selective reduction roasting with carbon. Journal of Hazardous Materials, 2012, 237: 323-330.

[7] Peng N, Peng B, Liu H, et al. Leaching kinetics modelling of reductively roasted zinc calcine. Canadian Metallurgical Quarterly, 2017, 56(3): 301-307.

[8] Peng N, Peng B, Liu H, et al. Reductive roasting and ammonia leaching of high iron-bearing zinc calcines. Mineral Processing & Extractive Metallurgy Imm Transactions, 2018, 127(1): 1-9.

[9] Yan H, Chai L Y, Peng B, et al. Reduction roasting of high iron-bearing zinc calcine under a CO-CO$_2$ gas: An investigation of the chemical and mineralogical transformations. Jom, 2013, 65(11): 1589-1596.

[10] Yan H, Chai L Y, Peng B, et al. A novel method to recover zinc and iron from zinc leaching residue. Minerals Engineering, 2014, 55: 103-110.

[11] 宗美荣, 董发勤, 刘明学, 等. 西部某铀矿采矿坑口辐射及重金属分布. 吉林大学学报(地球科学版), 2015, (S1).

[12] Che Y L, Wang H, Hu H Y, et al. Research progresses on analytical technologies used in microbial community structure and diversity. Ecology and Environment, 2005, 14(1): 127-133.

[13] 张倩. 若尔盖铀矿区土壤的微生物群落分布及对外源铀的响应. 绵阳: 西南科技大学, 2013.

[14] Zong M R, Dong F Q, Liu M X, et al. A Detailed Investigation on the Environmental Effect of an Uranium Mine in Western China Including γ-ray Radiation Formation and Microbe Distribution. Journal of Nanoscience and Nanotechnology, 2017, 17(9): 6614-6619.

[15] Zhou L, Dong F Q. Surface interaction between metazeunerite and an indigenous microorganism Kocuria rosea: Implications

for bioremediation of As-U tailings. Chemical Engineering Journal, 2019, 359: 393-401.

[16] 董发勤, 刘明学, 郝瑞霞, 等. 矿物光电子-微生物体系重金属离子价态调控及其环境效应研究进展. 矿物岩石地球化学通报, 2018, 37(1): 28-38.

[17] Sun Y F, Liu X Y, Kainuma M, et al. Dechlorination of polychlorinated biphenyls by iron and its oxides, Chemosphere, 2015, 137: 78-86.

[18] Sun Y F, Fu X, Qiao W, et al. Dechlorination of 2, 2', 4, 4', 5, 5'-hexachlorobiphenyl by thermal reaction with activated carbon-supported copper and zinc. Frontiers of Environmental Science and Engineering, 2013, 7: 827-832.

[19] Sun Y F, Takaoka M, Takeda N, et al. Decomposition of 2,2',4,4',5,5'-hexachlorobiphenyl with iron supported on an activated carbon from an ion-exchange resin. Chemosphere, 2012, 88: 895-902.

[20] Sun Y F, Liu L N, Oshita K, et al. Influence of activated-carbon-supported transition metals on the decomposition of polychlorobiphenyls. Part II: Chemical and physical characterization and mechanistic study. Chemosphere, 2016, 159: 668-675.

[21] Sun Y F, Tao F, Liu L N, et al. Influence of activated-carbon-supported transition metals on the decomposition of polychlorobiphenyls. Part I: Catalytic decomposition and kinetic analysis. Chemosphere, 2016, 159: 659-667.

[22] Xu J N, Liu Y W, Tao F, et al. Kinetics and reaction pathway of Aroclor 1254 removal by novel bimetallic catalysts supported on activated carbon. Science of The Total Environment, 2019, 651: 749-755.

[23] Liu Y W, Diao X G, Tao F, et al, Takaoka M and Sun Y. Insight into the low-temperature decomposition of Aroclor 1254 over activated carbon-supported bimetallic catalysts obtained with XANES and DFT calculations. Journal of Hazardous Materials, 2019, 366: 538-544.

[24] Du B, Li J, Fang W, et al. Characterization of naturally aged cement-solidified MSWI fly ash. Waste Management, 2018, 80: 101-110.

[25] Du B, Li J, Fang W, et al. Comparison of long-term stability under natural ageing between cement solidified and chelator-stabilised MSWI fly ash. Environmental Pollution, 2019, 250: 68-78.

[26] Zhang W, Yang J K, Hu Y C, et al. Effect of pH on desulphurization of spent lead paste via hydrometallurgical process. Hydrometallurgy, 2016, 164: 83-89.

[27] Zhu X F, He X, Yang J K, et al. Leaching of spent lead acid battery paste components by sodium citrate and acetic acid. Journal of Hazardous Materials, 2013, 250-251(8): 387-396.

[28] Zhang W, Yang J K, Zhu X F, et al. Structural study of a lead (II) organic complex—A key precursor in a green recovery route for spent lead-acid battery paste. Journal of Chemical Technology and Biotechnology, 2016, 91(3): 672-679.

[29] Li L, Zhu X F, Yang D N, et al. Preparation and characterization of nano-structured lead oxide from spent lead acid battery paste. Journal of Hazardous Materials, 2012, 203-204(4): 274-282.

[30] Hu Y C, Yang J K, Zhang W, et al. A novel leady oxide combined with porous carbon skeleton synthesized from lead citrate precursor recovered from spent lead-acid battery paste. Journal of Power Sources, 2016, 304: 128-135.

[31] Yang D N, Liu J W, Wang Q, et al. A novel ultrafine leady oxide prepared from spent lead pastes for application as cathode of lead acid battery. Journal of Power Sources, 2014, 257: 27-36.

[32] Hu Y C, Yang J K, Hu J P, et al. Synthesis of nanostructured PbO@C composite derived from spent lead-acid battery for next-generation lead-carbon battery. Advanced Functional Materials, 2018, 28(9): 1705294.

[33] Pan X H, Liu Z J, Chen Z, et al. Investigation of Cr(VI) reduction and Cr(III) immobilization mechanism by planktonic cells and biofilms of Bacillus subtilis ATCC-6633. Water Research, 2014, 55: 21-29.

[34] Ye X C, Lin Z H, Liang S J, et al. Upcycling of electroplating sludge into ultrafine Sn@C nanorods with highly stable lithium storage performance. Nano Letters, 2019, 19(3): 1860-1866.

作者：林 璋[1]，宁 平[2]，孙轶斐[3]，柴立元[4]，闵小波[4]，董发勤[5]，刘建国[6]，杨家宽[7]，李耀基[8]，邓 洪[1]，张 静[9]，关清卿[2]

[1]华南理工大学，[2]昆明理工大学，[3]北京航空航天大学，[4]中南大学，[5]西南科技大学，[6]清华大学，[7]华中科技大学，[8]云南磷化集团有限公司，[9]中国科学院生态环境研究中心

第27章 全球气候变化与环境健康研究进展

- 1. 引言 /784
- 2. 全球气候变化及其发展趋势 /785
- 3. 典型污染物环境过程与气候变化 /787
- 4. 其他气候变化与环境健康问题 /794
- 5. 展望 /796

本章导读

全球气候变化和环境健康是当前人类社会面临的重大问题。二氧化碳、臭氧等温室气体浓度升高、全球变暖、极端天气事件频发、海洋酸化等将在很大程度上影响生态系统结构与生态服务功能，并通过调控环境介质及生物响应，影响重金属、持久性有机物、新型污染物如纳米材料、抗生素等的迁移、转化过程和毒性效应以及病原微生物的传播，进而通过食物链传递和直接暴露等途径，影响环境健康。本章将在概括分析全球气候变化趋势的基础上，从环境化学的角度，重点对气候变化影响下典型污染物环境过程和毒性效应以及可能导致的生态安全和环境健康问题进行较全面的综述。

关键词

气候变化，环境污染，生态安全，粮食安全，环境健康

1 引 言

全球环境变化，特别是全球气候变化是当今国际社会普遍关注的全球性问题。随着人口增长、社会经济发展和科学技术进步，人类活动对全球气候变化的影响越来越大[1]。人类活动，尤其是矿物燃料燃烧，释放了大量二氧化碳和其他温室气体，使更多的热量滞留在大气层低层并影响了全球气候[2]。过去130年中，全球气温上升约0.85℃，海平面上升、冰川融化、极端气候事件变得更加剧烈和频繁[3]。全球变暖对局部区域会有一定益处，例如在温和的气候中冬季死亡减少以及在某些地区提高粮食产量，但气候变化的整体环境影响很可能主要是负面的[4]。

全球气候变化被认为是21世纪最大的全球健康威胁和挑战[5]。到21世纪后期，气候变化可能会在区域和全球层面上增加干旱、洪涝发生的频率和严重程度，可能会影响淡水供应及个人卫生，增大腹泻病的风险，并影响到水源性疾病和通过昆虫、蜗牛或其他冷血动物传播的疾病。气温升高和降水变化很可能会减少许多最贫穷地区的主要粮食品种生产，使营养不良和营养不足的问题加剧。超常高温及其导致的花粉及其他气源性致敏原水平、臭氧和空气中其他污染物水平升高，加剧了心血管和哮喘等呼吸道疾病的发作。世界人口有一半以上生活在距海洋60 km以内的地区，不断上升的海平面以及越来越极端的气候事件将破坏家园、医疗设施及其他必要的服务设施，人们可能被迫迁移，转而使包括从精神障碍到传染病等一系列健康影响的风险升高。此外，气候条件改变还会影响到污染物在环境中的迁移、转化过程和生物有效性，也会改变生物和人类对污染物的暴露浓度和适应能力，从而带来难以预测的生态和环境健康问题。在仅考虑到小部分可能的健康影响，并假设持续的经济增长和卫生进展的前提下，世界卫生组织的一次评估结论是，气候变化预计将在2030～2050年，每年多造成约25万人死亡；其中预计有3.8万老年人死于气温过高，4.8万人死于腹泻，6万人死于疟疾，以及9.5万人死于儿童营养不良[6]。

2 全球气候变化及其发展趋势

2.1 全球气候变化的定义和缘起

气候变化通常是指气候平均值和气候离差值（单项数值与平均值之间的差）出现了统计意义上的显著变化，如平均气温、平均降水量、最高气温、最低气温，以及极端天气事件等的变化。人们常说的全球变暖就是气候变化的重要表现之一[7]。

《联合国气候变化框架公约》将"气候变化"定义为："经过相当一段时间的观察，在自然气候变化之外由人类活动直接或间接地改变全球大气组成所导致的气候改变"。人类活动，特别是工业革命以来的人类活动，是造成目前全球变暖的主要原因，其中包括人类生产、生活所造成的二氧化碳、甲烷等温室气体的排放、土地利用变化、城市化等等。数据显示，工业革命以来，发达国家大量消费化石能源所产生的二氧化碳累积排放占全球同期总排放量的70%，导致大气中温室气体浓度显著增加，加剧了以变暖为主要特征的全球气候变化[8]。

2014年11月经初审议通过的联合国政府间气候变化专门委员会（IPCC）第五次评估报告指出，2000～2010年，人为温室气体排放量平均每年增长2.2%，高于此前30年1.3%的年均增长率。大量温室气体排放对全球自然生态系统产生显著影响。IPCC报告指出，在1880～2012年，全球海陆表面平均温度升高了0.85℃。在北半球，1983～2012年可能是过去1400年中最暖的30年。自1971年以来，全球冰川普遍出现退缩现象，格陵兰冰盖和南极冰盖的冰储量减少，北极海冰面积以每十年3.5%～4.1%的速率缩小。20世纪以来，全球海平面上升19 cm，平均每年上升1.7 mm。海平面上升，冰川融化，降水规律变化，极端气候事件变得更加剧烈和频繁[9]。

2.2 全球气候变化的发展趋势

对地球气候的科学诊断，要依赖全球平均温度、海平面高度、南极冰川、山地冰川等数据指标。从《中国气候变化蓝皮书（2019）》中的数据可以看出，2018年全球气候系统变暖趋势进一步持续，中国是全球气候变化的敏感区之一。

世界气象组织（WMO）发布的全球气候状况声明，2018年全球平均温度较工业化前高约1℃，过去四年（2015年、2016年、2017年和2018年）是有完整气象观测记录以来最暖的四个年份。2018年成为有现代海洋观测记录以来海洋最暖的年份，全球海洋热含量打破了2017年刚刚创下的纪录；全球平均海平面继续加速上升，2018年全球平均海平面比2017年上升约3.7 mm；全球山地冰川仍处于物质高亏损状态，北极和南极海冰范围均较常年同期明显偏小。假如按照目前的趋势继续下去，到2100年，海平面便会上升15～95 cm，造成低洼地区和岛屿普遍淹没。近年来几次破记录的飓风、水灾和热浪已引起人们关注，热带海洋变暖也会使热带气旋变得更加频繁、更加严重。

此外，由于人口的增长、人类活动频繁和社会经济的迅速发展，大气中的二氧化碳浓度正在逐年递增，二氧化碳浓度已由工业革命前的约280 µmol/mol增加到目前的约400 µmol/mol。IPCC（联合国政府间气候变化专门委员会）第五次评估报告估计到2100年大气CO_2浓度将增加到538 µmol/mol（RCP4.5）或670 µmol/mol（RCP6.0）[10]。

海洋酸化的概念最早由 Caldeira 等[11]于 2003 年提出，海洋酸化（ocean acidification）是指由于海洋吸收、释放大气中过量的 CO_2，使海水正在逐渐变酸。工业革命以来，海洋吸收人类排放的 CO_2，导致地球表层海水 pH 下降了 0.1，预计到 2100 年海水的 pH 还将下降 0.3。海水酸性的增加，将改变海水化学平衡，使依赖于化学环境稳定性的多种海洋生物乃至生态系统面临巨大威胁[12, 13]。目前，全球海洋酸化正以前所未有的速度快速发展（图 27-1）。据英国卡迪夫大学近日发布的研究结果，目前海洋酸化达到 1400 万年来未见的水平[14]。

图 27-1　海洋酸化过程及趋势图（引自参考文献[12]）

全球气候变化对自然生态系统和社会经济的影响正在加速。2018 年全球气候相关灾害发生次数为 1980 年以来最多的一年，所造成的损失超过全球自然灾害经济损失总量的 90%，其中最主要是由于降水的改变所导致的。伴随着全球气候的变化，极端天气事件如强降水额干旱时间频率可能发生变化，未来全球气候变化可能会改变大气降水的空间分布和时间变异特性，进而改变水资源的空间配置状态，加剧部分流域的水资源供给压力直接影响到水资源稀缺地区的可持续发展（图 27-2）。高温、干旱、强降水等极端天气气候事件对粮食安全、人体健康及自然生态系统服务功能带来重大威胁。

随着全球现代工业化的蓬勃发展，大量化石燃料燃烧、含氮肥料使用和机动车尾气排放，大气中一氧化碳（CO）、氮氧化物（NO_x）和挥发性有机污染物（VOCs）含量急剧增加，这些气体在太阳光照射下会发生光化学氧化反应产生臭氧，从而导致近地层大气臭氧浓度逐年升高。全球平均近地层 O_3 浓度以每年 0.5%～2.0%的增幅由最初的仅 10 nmol/mol 增加到 50 nmol/mol，已超过部分敏感植物的伤害阈值 40 nmol/mol，预计在 2050 年 O_3 浓度还将增加 20%～25%，平均臭氧浓度可达 65 nmol/mol，在 2100 年将继续增加 40%～60%，浓度将达到 80 nmol/mol[15, 16]。而这一系列的气候变化问题也势必会对生态系统以及人类健康带来极大的威胁和难以预估的影响。

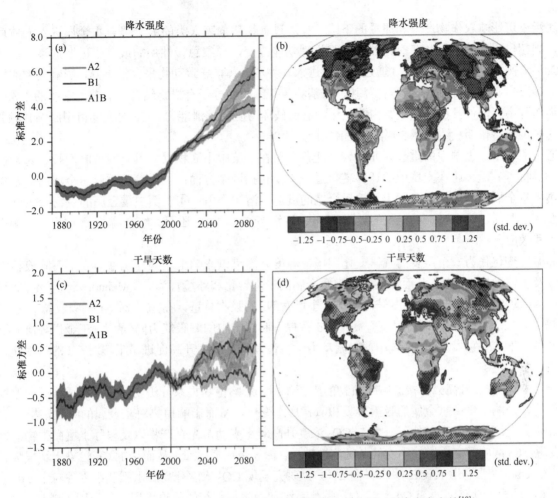

图 27-2　IPCC 预测的全球降水强度与干旱天数变化趋势图（引自参考文献[10]）

3　典型污染物环境过程与气候变化

污染物环境过程与机制研究是环境化学研究的重点任务之一。环境中有毒化学品的迁移、转化过程及生态毒理、环境健康效应受环境介质条件如温度、湿度、酸度等影响很大[17-20]，气候及其相关的大气组分变化、温度变化、降水过程变化、土壤水分变化、土壤理化性质变化等，除了改变生物的生物性状从而影响其适应能力之外，也将对污染物的迁移、转化过程及其生物可利用性和生态效应产生影响，对于过去已经释放到环境中并存在于土壤和沉积物中的遗留污染物，例如汞，全球气候变化可以改变外界环境，使得物质被再活化或释放得更快。温度升高会增加许多生物的代谢率，从而增加某些污染物的生物累积和生物放大的可能性[21]。全球气候变化可能与其他自然和人为因素结合，影响自然环境中有毒化学品的运输和转化[22-25]，进而影响生态安全和环境健康。

3.1　重金属的环境化学过程与气候变化

重金属作为一类典型污染物，进入土壤、水体以及大气环境后，将造成不同程度的环境污染。土壤重金属污染可影响农作物产量和质量的下降，并通过食物链危害人类的健康[26]。重金属进入人体后不易排泄，当超过人体的生理负荷时，就会引起生理功能改变，导致急慢性疾病或产生远期危害。以重金属镉（Cd）为例，污染土壤中的 Cd 通过食物链进入动物和人体，导致动物采食量下降，影响动物的繁殖性能[27]。Cd 进入动物体内后，能与含羟基、氨基和巯基等的高分子有机物结合，影响酶系统的正常功能，还可损伤肾小管，使肾小管发生结构性变化，进而破坏肾脏的排泄能力。Cd 对人体可引起"痛痛病"和肾功能的失调，潜伏期长达 10～30 年，且治疗极为困难。

通常情况下，土壤中的重金属元素较难迁移，可在土壤中不断积累，具有残留时间长、隐蔽性强、毒害大等特点。重金属在土壤中的迁移转化过程主要包括四个方面：①土壤胶体对重金属的吸附作用；②土壤中重金属的配位作用；③土壤中重金属的沉淀和溶解作用；④土壤中重金属的生物转化过程[28]。气候变化主要通过对土壤中重金属的吸附-解吸、溶解-沉淀、氧化-还原、络合-解离等过程的影响来影响其行为和效应。

如温度变化作为一个重要的气候变化因子，温度升高可以影响土壤的理化性质、土壤固液相的表面反应以及土壤微生物活性，从而影响土壤中重金属的形态转化和环境行为[29]。Mehadi 研究了不同温度下，土壤对 Ni^+ 的吸附，结果表明温度升高增加了土壤对 Ni^+ 的吸收总量，在 30℃时 Ni^+ 的吸收总量是 15℃时的 2 倍[30]。Li 等的研究表明，温度升高使 Cd 在棕壤和黑土中的解吸能力降低[31]。王金贵等研究了在不同温度条件下 Cd 在典型农田土壤中的吸附动力学特性，发现温度升高促进了土壤对 Cd 的吸附速率和吸附容量[32]。

大气 CO_2 浓度升高会直接影响到植物的光合作用，影响植物根系分泌物的量和组成成分，改变植物根际环境，影响土壤中重金属的吸附解吸和赋存形态变化，从而影响植物对重金属的吸收富集[33-36]，Jia 等通过研究土壤 Cd 污染条件下，大气 CO_2 浓度升高对小麦幼苗光合产物以及根际土壤微生物活性的影响，发现 CO_2 浓度升高会增加小麦幼苗可溶性糖，总糖和蛋白质含量，从而增加 Cd 污染土壤根际有机酸含量[33]。Guo 等[36]利用 FACE 平台（图 27-3），研究发现 CO_2 浓度升高后水稻和小麦谷粒中 Cd 的含量增加，Cu 的含量降低，改变 Cd 污染对土壤酶系统和微生物系统的胁迫作用，其原因可能与土壤 pH 变化及重金属的形态转化有关，该研究表明，CO_2 浓度升高会改变土壤和植物中污染元素的分布，从而可能影响农产品质量和安全性。Duval 等[37]在有机质含量低的砂壤土中进行了 11 年的 CO_2 浓度升高实验，研究发现 CO_2 浓度升高促进了土壤中污染元素在植株体内的积累，从而导致表层土壤中可提取的污染元素含量下降。

图 27-3 中国 FACE 平台

大气臭氧浓度升高目前也是国际上广为关注的问题，Guo 等[38]在 O_3-FACE 平台系统的研究发现，O_3 浓度升高提高了污染土壤处理组小麦各组织（茎叶、稻壳和谷粒）中的 Cd 浓度，O_3 浓度升高和 Cd 污染联合胁迫强烈影响了抗氧化酶（POD、APX 和 CAT）的活性，从而加速对小麦叶片中的氧化胁迫。王亚波等利用 OTC 平台研究了大气臭氧浓度升高对模拟稻田污染土壤中 Cd 形态转化和植物吸收的影响，研究表明臭氧浓度升高促进了土壤中 Cd 向生物有效性更高的形态转化，提高了水稻籽粒中 Cd 的含量，从而加剧了重金属 Cd 进入食物链的风险[39]。尹微琴等在麦季土壤中发现类似的结果[40]。

尽管重金属污染问题比较普遍，但其对气候变化的响应研究才刚刚起步，少量的报道也是针对常见的 Cd、Cu 等元素的典型案例研究和趋势分析，且研究工作多在模拟污染土壤条件下开展，对于真实环境下 Cd、As、Hg 等更多元素的定量评估尚显不足，对于复合大气二氧化碳、臭氧、温度等联合升高胁迫的综合效应研究尤其缺少，对于水环境中重金属环境行为、生态效应对气候变化响应的研究也不多见，加强气候变化影响下不同区域特征污染物差异性响应规律与机制研究十分必要。

3.2 持久性有机污染物的环境化学过程与气候变化

持久性有机污染物（persistent organic pollutants，POPs）具有长期残留性、生物蓄积性、半挥发性和高毒性，能够在大气环境中长距离迁移。土壤作为 POPs 的蓄积库，全球各地土壤都不同程度地受到了 POPs 的污染，甚至包括南北极地区[41]。POPs 由于特殊的物理化学性质，具有一定的生物累积和生物放大作用[42-45]。气候变化会影响持久性有机污染物的迁移和环境归趋，影响其在大气-土壤系统中的浓度，特别是在对气候变化十分敏感的北极地区[51,52]。在过去的几十年，由于各国对 POPs 生产和使用的限制，经大气传输到北极大气中的 POPs 的含量有所减少，但气候变暖会使沉积在水体、冰川等库中的 POPs 再次挥发到大气中[53]，从而影响到 POPs 在环境中的迁移转化过程。

多环芳烃（PAHs）作为一类典型的 POPs，具有强的致癌性、致畸性、致突变性和生物难降解性[46]。释放到大气中的 PAHs 主要通过大气干湿沉降进入土壤并在土壤中长久积累[47,48]。进入土壤的 PAHs 通过挥发进入大气，或吸附于土壤有机质，也可向下淋溶或渗滤到地下水，也可被植物、土壤动物等吸收富集，或被土壤微生物降解[49]。PAHs 在土壤中的损失过程与损失程度与许多因素有关，包括土壤类型（矿物质含量及有机质含量）、多环芳烃的物理化学性质、土壤温度及土壤湿度、氧化还原电位、养分有效性、降解微生物的存在及其活性以及多环芳烃对降解微生物的生物有效性等[50]。Friedman 等[54]利用 GEOS-Chem 模型模拟 2000~2050 年气候变化对全球大气中多环芳烃含量的影响，发现到 2050 年，北半球大气中常见多环芳烃菲和芘的含量会小幅度上升。相应地，土壤中多环芳烃的含量也将会受到影响，进而影响植物的生长和发育。

大气温度及 CO_2 浓度升高会直接或间接地影响植物根际环境，如改变根系分泌物的组成和数量，影响土壤微生物群落[55,56]。温度及 CO_2 浓度的改变也可能会影响土壤中多环芳烃的降解。Iqbal 等[57]发现高温能够促进土壤中可降解多环芳烃的土著微生物的生长，增强其生存能力，还能降低土壤颗粒对多环芳烃的吸附性，使更多的有机物能够被微生物降解。Ai 等[58]利用 OTC 平台，研究了大气 CO_2/O_3 浓度升高、植物多样性对模拟草地生态系统 PAHs 残留的影响及机制，发现大气 CO_2、O_3 浓度升高、植物多样性对土壤中微生物群落组成和数量有显著影响。该研究中，革兰氏阳性菌是降解土壤中 PAHs 的主要菌群，CO_2 浓度升高降低了革兰氏阳性菌的生物量，O_3 浓度升高同时降低了革兰氏阳性菌和革兰氏阴性菌的生物量，植物多样性对革兰氏阳性菌无显著影响，CO_2、O_3 浓度升高抑制了土壤中 PAHs 的降解（图 27-4）。Ai 等[59]还利用 FACE（Free-Air Concentration Enrichment）平台进行的关于大气 CO_2 浓度升高对稻田土壤多环芳烃残留影响的研究也发现了类似的结果，CO_2 浓度升高改变了土壤微生物群落的组成，特别是降低了部分降解 PAHs 的微生物群落的生物量，如放线菌和真菌。此外，CO_2 浓度升高还减少了作用于

PAHs 降解的功能基因，从而提高了土壤中 PAHs 的累积，该研究还通过采集 FACE 条件下和正常大气条件下的土壤，进行了 ^{14}C-菲的矿化试验，发现 FACE 条件下新鲜土样对 ^{14}C-菲的矿化速率明显低于正常大气条件下的土样，这也进一步证明了 CO_2 浓度升高将导致土壤中 PAHs 累积作用增强。Zhang 等[60]通过温室盆栽实验，研究了大气温度、CO_2 浓度单独及共同升高条件下，土壤芘污染对雌性及雄性欧洲山杨幼苗生长和酚类物质积累的影响，发现土壤芘污染（50 mg/kg）分别使两种性别欧洲云杉幼苗的茎生物量积累和叶面积在正常大气条件下降低了 9% 和 6%，并且在大气温度升高（38%）、CO_2 浓度升高（37%）以及温度和 CO_2 浓度共同升高条件下芘对叶面积的降幅更大，且对雄性山杨叶面积的负面效应比雌性山杨更大。大气温度升高对两种性别欧洲山杨的株高、基径、生物量、比叶面积都有促进作用，并且雌性山杨受益更大，而 CO_2 浓度升高对山杨生长的影响较小。正常大气条件下，芘对两种性别山杨叶中的总酚类物质含量没有显著影响，但在温度升高、CO_2 浓度升高以及温度和 CO_2 浓度共同升高条件下叶中总酚类物质含量分别增大了 16%、1% 和 20%。土壤中芘的残留量在大气 CO_2 浓度升高条件下更高，而大气温度升高对其没有显著影响。该结果表明，欧洲山杨对于大气温度、CO_2 浓度升高以及土壤芘污染的复合胁迫都具有很好的化学防御能力，但在气候变化条件下芘对欧洲山杨的风险依然存在，对雄性欧洲山杨的风险要大于雌性欧洲山杨。

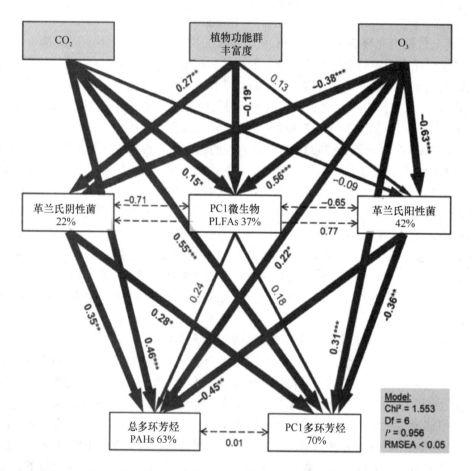

图 27-4　CO_2 浓度升高、O_3 浓度升高和植物功能群丰富度对土壤中微生物和多环芳烃残留影响的结构方程模型结果
（红色箭头：负相关，蓝色箭头：正相关，数字上的星号表示影响显著）①（引自参考文献[58]）

---

① 彩图信息可通过扫描封底二维码获取。

3.3 纳米材料的环境化学过程与气候变化

近年来由于纳米技术的快速发展，纳米材料已经广泛应用于化妆品、食品、药品等领域，纳米材料能够通过多种途径进入到人类的生活和生存环境当中[61, 62]。纳米材料包括碳纳米材料、零价金属NPs、金属氧化物NPs、量子点和纳米聚合物等多种类型[63]。其中金属氧化物NPs（MO_x NPs）是极其重要的一类纳米材料，包括ZnO、Fe_3O_4、CuO、Al_2O_3、TiO_2、CeO_2等。随着大量生产及广泛应用，MO_x NPs进入土壤、水环境中的量也逐渐增多。Keller等[64]对2010年全球$nTiO_2$的生产和使用量进行了市场调查，并估算了排放量，2010年全球生产的约8.8万t的$nTiO_2$中有超过3.8万t被排放后最终进入土壤。Sun等[65]通过动态概率模型模拟，预测到2030年欧洲$nTiO_2$的生产总量将达到170万t，最终排放进入土壤的$nTiO_2$预计将达40万t。$nTiO_2$进入土壤的途径主要通过污泥的农用、工业生产过程中的泄露等[66]。由于MO_x NPs具有金属毒性和纳米毒性的双重效应，进入环境后可能产生的环境风险逐渐受到人们的关注[67]。

随着MO_x NPs的广泛使用，它必然会不断地向环境中释放，使其成为一种潜在的环境污染物。释放到环境中的MO_x NPs很大一部分进入土壤，MO_x NPs在土壤中不可溶，且主要以纳米颗粒的形式存在，它能够被植物吸收而累积在植物体内，从而可能对陆地生态系统造成影响。研究气候变化对其环境过程与效应的研究也有少许报道，如Qi等[68]通过水培实验发现$nTiO_2$（50～200 mg/L）的加入能够缓解高温（35℃）对番茄幼苗叶片光合作用的不利影响。Du等[69]利用FACE平台研究大气CO_2浓度升高（570 μmol/mol和370 μmol/mol）条件下，$nTiO_2$（50 mg/kg和200 mg/kg）对土壤-水稻系统的影响，发现在正常大气条件下，土壤中两种浓度的$nTiO_2$对水稻地上部、地下部及总生物量以及谷粒产量都没有显著影响，但在大气CO_2浓度升高条件下，$nTiO_2$使水稻生物量及谷粒产量均显著降低；大气CO_2升高及$nTiO_2$联合作用还会影响土壤微生物群落组成，改变特定细菌类群及原生生物的多样性和丰度，进而影响其功能的发挥；在高CO_2浓度下，200 mg/kg的$nTiO_2$使植株不同组织中Ti的积累量增加（图27-5）。Zhang等[70]研究了大气CO_2浓度升高和温度升高条件下不同浓度的$nTiO_2$污染在土壤—植物系统中的迁移转化过程，研究发现在高浓度$nTiO_2$处理组，大气CO_2浓度升高对根系$nTiO_2$的吸收没有显著的影响，但大气温度升高显著促进了植物根系对$nTiO_2$的吸收，从而加剧了$nTiO_2$进入生物体的环境风险（图27-6）。除此之外，两种性别欧洲白杨叶中总酚类物质含量在大气温度升高条件下显著降低，在CO_2浓度升高条件下显著增大。

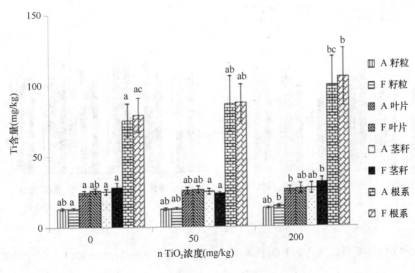

图27-5 CO_2浓度升高对水稻各组织中Ti浓度的影响（引自参考文献[69]）

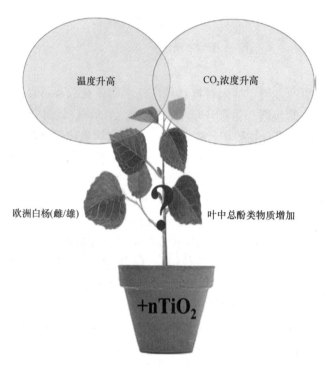

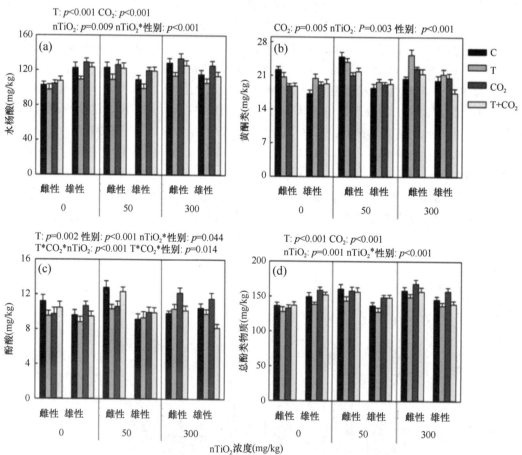

图 27-6　温度和 CO_2 浓度升高条件下 TiO_2 纳米颗粒对欧洲白杨（*Populus tremula* L.）复合胁迫的影响研究（引自参考文献[70]）

在 50 mg/kg 和 300 mg/kg 的 $nTiO_2$ 污染处理中，大气温度升高和/或 CO_2 浓度升高均增加了雌性白杨叶中总酚类物质的含量，表明在未来气候变化条件下，雌性欧洲白杨对土壤 $nTiO_2$ 污染具有更强的化学防御能力，而雄性白杨更易受食草动物、土壤 $nTiO_2$ 污染以及气候变化复合胁迫的影响。在纳米材料和气候变化复合胁迫下，两种性别欧洲白杨的竞争能力可能会发生改变，它们在自然界的性别比例以及遗传多样性将会受到影响，最终可能影响其物种演替。

气候变化对水环境纳米材料毒性效应影响的研究也引起了一定程度的关注。Yin 等[71]研究了 CO_2 浓度升高条件下水-沉积物生态系统中 ZnO 纳米颗粒对金鱼（Carassius auratus）毒性效应，发现 CO_2 浓度升高通过降低水的 pH 提高了悬浮液中的 Zn 含量，从而提高 nZnO 的生物利用度和毒性。与正常大气环境相比，CO_2 浓度升高导致了鱼类组织中锌的积累量增加（在肝脏、大脑和肌肉中分别增加 43.3%、86.4%和 22.5%）。CO_2 浓度升高也加剧了 nZnO 对鱼类的氧化损伤，导致更高的 ROS 强度，更高的 MDA 和 MT 含量以及肝脏和脑中 GSH 含量的降低。

纳米材料作为一类新型污染物，种类多、性质差异大，在环境中检测困难，现有的研究报道尚不足以对气候变化影响下纳米材料的环境风险给出较为全面的认识。加强更多种类的纳米材料以及纳米材料与其他污染物复合作用对气候变化响应的研究，系统揭示和评价未来环境条件下的潜在环境效应和风险，对于更好地发展纳米科技十分重要。

3.4 农药的环境化学过程与气候变化

农药是现代农业生产中的重要工具，包括杀虫剂、除草剂等，在使用过程中除了发挥正常的杀虫除草功能的同时，农药的残留问题同样引起了人们的广泛关注。气候变化可能对植物和牲畜害虫（杂草、昆虫和微生物）的生物学特性产生广泛影响，大气 CO_2 浓度升高、冬季最低气温的升高和生长季节的延长极有可能改变害虫的分布和种群数量[72-79]，许多害虫的生长活动与温度升高密切相关，随着全球气候变暖，预计世界上许多地方的农药使用率会增加[80-82]。在气候变化的背景下，农药的效率可能会降低，进而影响某些杂草的除草剂耐受性[83-87]，害虫压力的增加和杀虫剂有效成分的减少可能导致杀虫剂使用量的增加，田间污染和食物链暴露风险变大[88]。

农药残留的增加，可能导致更多的农药进入到水生生态系统，对水产品的安全、分配和消费以及环境健康产生影响[89-91]。另外，温度增加，一些疾病对于水产养殖的感染也会增加（如鱼腥草、黄杆菌等），从而使得水生养殖需要使用的药物量增加[92]。Beulke 等[93]量化了直接或间接由全球气候变化对地表水和地下水农药使用所产生的浓度、归宿和运输变化的影响，发现在全球气候变化下，地表水和地下水中农药的浓度可能会增加，对于某些农药，峰值浓度可能会增加几个数量级，全球气候变化的间接影响（即对施用农药的量和施用时间的影响）对地表水暴露的影响要大于单独气候变化对农药化学归趋和运输的影响。图 27-7 示意了农药等污染物的来源、归宿和运输的变化将对食品、饮用水、空气等产生多方面的影响，进而影响人体暴露和人体健康[94]。深入理解气候变化影响下农药等有机物的环境过程与效应机制，并耦合与地下水过程的交互作用，建立科学的系统评估方法和预测模型，对于深入揭示气候变化对环境健康的影响十分重要。

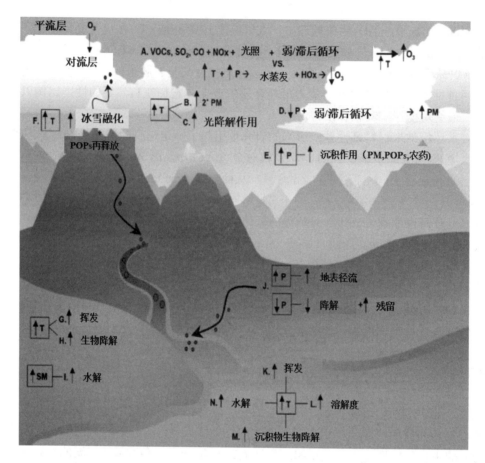

图 27-7　气候变化对农药等各种污染物环境分布的影响（引自参考文献[94]）

A. 温度升高加上臭氧前体、阳光和弱/停滞循环会增加臭氧的形成速度，而水蒸气会增加臭氧的破坏；B. 温度升高可能促进次生 PM 的形成；C. 温度升高可能导致持久性有机污染物的光降解增加；D. 气候变化导致降水增加将增加 PM，持久性有机污染物和农药的湿沉降；E. 降水减少加上弱/停滞循环可能会在区域内增加 PM；F. 融化的雪、冰和冰川将释放并重新混合在这些曾经冷冻过的基质中隔离的持久性有机污染物；G. 温度升高会增加持久性有机污染物和其他农药从土壤到大气的挥发性；H. 温度升高会增加土壤和沉积物中持久性有机污染物和农药的微生物降解；I. 增强的土壤水分会增加农药的水解降解，但可能不会影响持久性有机污染物，因为它们相对耐水解；J. 降水增加将增加农药和 POPs 径流进入水生系统的可能性，而降水量的减少将改善化学物质径流，但可能会增加持留性；K. 水温升高会增加持久性有机污染物和其他农药从水到大气的挥发性；L. 水温升高会增加持久性有机污染物的溶解度，使它们更容易保留在水中；M. 水温升高会增加微生物活性，增加土壤和沉积物中持久性有机污染物和农药的降解；N. 水温升高会使农药水解增加到更少或更多的生物活性降解物；2°PM＝次级 PM；CO＝一氧化碳；HO_x＝HO_2 + OH；NO_x＝氮氧化物；O_3＝臭氧；P＝降水；PM＝颗粒物质；POPs＝持久性有机污染物；SM＝土壤水分；T＝温度；VOCs＝挥发性有机化合物

4　其他气候变化与环境健康问题

4.1　营养元素问题

随着大气中二氧化碳浓度升高至 540～960 ppm 之间（21 世纪末预期浓度范围），主要粮食作物中碳水化合物（比如淀粉和糖类）含量增加，蛋白质含量有可能会显著下降[95-97]。当生长在 2100 年预期大气二氧化碳浓度水平（540～958 ppm）条件时，大麦、小麦、水稻和马铃薯，比在正常大气二氧化碳浓度

(315～400 ppm) 条件下蛋白质浓度降低了 6%～15%，但对玉米和高粱的蛋白质含量没有显著影响[95]。大气二氧化碳浓度升高，植物体内的矿物质和痕量元素含量将会降低，比如铁、锌、钙、镁、铜、硫、磷的浓度都将降低[98-103]，其原因一方面是升高的二氧化碳浓度会增加植物组织中碳水化合物积累，进而稀释包括矿物质在内的其他营养素的含量；另一方面，升高的二氧化碳浓度降低了植物对水分的需求，从而减少了植物根系吸收营养物质的数量[104, 105]。Zhu 等通过 FACE 研究发现，二氧化碳浓度升高稻米中维生素 B1、B2、B5 和 B9 皆有所下降，从而对农作物营养价值和农产品品质产生影响[106]，越来越多的证据表明，饮食中碳水化合物和蛋白质的比例增加，不利于人类代谢和体质，影响身体健康[107-110]。

4.2 花粉问题

大气 CO_2 浓度升高一方面造成全球变暖，另一方面作为光合作用的原料也促进了植物生长[111]。在更高温度和更高 CO_2 浓度的环境中，植物生长更加茂盛，也会产生更多的花粉。温暖的气候让植物的生长季变得更长，从而增加了空气中的花粉数量和存在时间。有研究发现，在大气 CO_2 浓度升高的作用下，植物花粉暴露量有可能增加 202%[112]，该研究模拟了温室气体升高对梯牧草（Timothy grass）花粉产量的影响，梯牧草是欧洲和美国常见的一种禾本科植物，它的花粉是一种重要的人类过敏源，在温室气体当中，CO_2 会刺激植物生长和繁育，一定浓度的 CO_2 可以刺激植物产生更多花粉，这会增加空气中的花粉浓度，并且增加过敏原暴露和患病的人数。花粉会导致哮喘和过敏性鼻炎等危害[113-115]。可预计的全球气候变化将延长植物花粉季节，并增加花粉和孢子的破裂，同时花粉的应变性会增加，从而加剧花粉过敏症和哮喘病[116, 117]。气候变化在延长空气中花粉存留时间的同时，也将扩大花粉的分布范围。一些原本生活在南方的植物由于气候变暖开始向北拓展，如由于气候变化引起的豚草属植物的入侵和繁殖，已导致欧美地区季节性过敏症患者的数量或将急剧增多，过去不太受到花粉困扰的北方地区今后可能就会面临严重的花粉过敏问题。此外，有研究发现，气候变化将导致海河流域更加干燥，使该地区内花粉过敏症患者增加，未来海河流域花粉过敏症将有可能向普遍化、严重化发展[118]。

4.3 天然毒素问题

天然毒素由藻类（例如，微囊藻毒素）、细菌（例如，肉毒杆菌）、植物（例如，配糖生物碱、茴香醚）和真菌（例如，黄曲霉毒素、赭曲霉毒素、玉米赤霉烯酮）产生，并且包括一些剧毒化学品，这些化合物会对人类健康产生一系列影响，包括死亡、肝癌、肝硬化、胃肠道疾病、神经毒性和严重的皮肤毒性[119-122]。天然毒素的生产对环境因素如温度和湿度非常敏感，因此气候变化将直接导致天然毒素产生量的变化。此外，全球气候变化也会引起昆虫载体的分布和活性的变化进而增加植物对霉菌毒素的暴露和脆弱性，农业系统地表径流和面源污染的变化可能会增加或减少藻类繁殖的发生。因此，全球气候变化可能会在一些地区增加人类接触植物毒素和真菌毒素的机会[123]，并导致目标器官毒性和癌症的增加，特别是在真菌感染流行的地区，人群最易受伤害[124]。天然毒素的增加也可能危害某些地区的食品和饮用水安全，从而对人类健康和福祉产生负面影响。

4.4 病原微生物问题

能使人、畜禽和植物致病的微生物统称为病原体微生物（pathogenic microorganisms）。病原微生物可以通过空气、水体、土壤和食物在环境中不断传播[125]。大气中病原微生物主要有八叠球菌、细球菌、枯草杆菌和酵母菌孢子等。病原微生物在大气环境中分布的主要影响因素包括微生物数量与人口密度动植物数量、土壤和地面覆盖情况，以及气温、湿度、气流和日照等因素。水体中的病原微生物主要来源于地面的微生物、大气中漂浮的微生物。引起水体病原微生物污染的主要菌种有肠道细菌（大肠菌群、粪链球菌和梭状芽孢杆菌等）、病原菌（沙门氏菌属、志贺氏菌属和霍乱弧菌等）。病原微生物进入水体，会直接或间接地危害人类健康和影响渔业生产[126]。土壤也会受到病原微生物的污染。土壤中病原微生物污染的来源包括：在土壤中施用未经彻底无害化处理的人畜粪便，未经处理的生活污水、医院污水、屠宰场和饲养场污水以及含有病原体的工业废水进入农田灌溉或污泥农用等。病原体微生物进入土壤后，一般能在土壤中存活一定的时间，在条件适宜的情况下，可以通过不同的途径使人感染发病，给人类健康带来极大的影响。如在被污染的土壤上种植蔬菜瓜果，人生吃这些瓜果蔬菜就有可能感染致病。自然土壤中已存在的致病菌，人体与土壤接触也易得病，如破伤风、肉毒中毒及某些霉菌病等。病原体微生物也会通过动植物的介导作用，被动植物摄取，从而进入到食物链中，最终威胁到人类健康[127]。气候变化可能会通过改变疾病媒介生物进化速度或动物宿主健康的环境变化，增加人畜共患病率。昆虫媒介会增加新鲜农产品的食源性病原体污染的概率，气候改变将会影响传播细菌病原体的害虫和其他微生物的范围和分布，这些微生物可以将细菌病原体（如沙门氏菌）传播到新鲜农产品中。此外，当前抗生素使用普遍，释放到环境中将导致更多抗生素抗性病原体的产生[128]，气候条件的变化很可能导致"新兴"或"新"病原体的出现。目前，关于全球气候变化对病原微生物环境过程影响的报导还比较缺乏，这方面的信息还不够全面，还有待研究者的进一步深入研究。

4.5 大气臭氧浓度升高问题

随着当前世界工业化和城市化进程的不断加快，大部分国家和地区近地层大气臭氧平均浓度不断增加，大气臭氧污染问题也日趋严重[129]。臭氧作为当前大气中最重要的污染物之一，臭氧浓度过高则是无形杀手[130]。首先，臭氧会危害人体健康，过高浓度的臭氧气体吸入人体内后，会迅速转化为活性氧自由基，主要氧化人体中的不饱和脂肪酸，从而造成细胞损伤，严重者可形成癌变。高浓度臭氧对眼睛也有轻度刺激性，会损害人体中枢神经，导致思维紊乱。也会影响甲状腺功能以及人体细胞的新陈代谢过程，加速机体的衰老速度，造成骨骼早期钙质化等[131]。

大气中臭氧浓度不断升高不仅会威胁人体和动物体的健康，其对植物的生长发育也会造成严重的伤害[132, 133]。许多研究表明，臭氧浓度升高抑制了作物的生长发育，会改变植株碳代谢库，导致植株叶片的与光合作用相关的色素浓度降低，叶绿素的结构和组成也会发生改变，丙二醛的含量增加，膜脂过氧化加剧[134]。与此同时，臭氧胁迫降低了植株气孔导度、蒸腾速率以及呼吸强度等相关气体交换的参数[135, 136]，还降低了最大光合速率和光合作用等相关参数，进而使得净光合速率显著降低[137]，最终导致作物减产[138, 139]。还有研究发现，长时间暴露于高浓度的臭氧后，作物的生育期也会发生变化，植株生物量、叶面积、株高、分蘖数等指标都会发生显著降低[140]。另外，臭氧对植物气孔导度、抗氧化系统、光合系统、呼吸作用、膜透性等均会造成不良的影响[141, 142]，导致植株的光合产物分配不均、植株的凋落物以及植物根系的分泌物种类和数量的改变，最终会对整个生态系统造成不良影响[143]。此外，高臭氧浓度会显著增加小麦籽粒中氨基酸和蛋白质含量，增加了籽粒中的 K、Ca、Mg、P、Mn、Cu 和 Zn 元素含量以及水稻中的 Mg、K、Mn 和 Cu 含量，但显著降低籽粒中淀粉含量[144]，从而影响到农产品的品质。

5 展望

气候变化主要通过两种方式影响人类健康：第一，通过改变已经受到气候或天气因素影响的健康问题的严重程度或频率；第二，在之前未发生的地方产生前所未有的或意料之外的健康问题或威胁。例如，气候变化会影响携带莱姆病等疾病的蚊子的分布和行为，也会影响某些地区的水和食物质量，对人类健康产生危害。此外，全球气候变化对精神健康和幸福的影响也是其对人类健康影响的一部分。针对当前全球气候变化与环境健康的研究现状，从环境化学的角度看，下面几个方面的研究有待进一步加强：

（1）加强污染物在地球关键带系统中迁移转化及其对气候变化的响应的机理研究。污染物的迁移转化过程是环境化学研究的重点内容之一，环境条件的改变将引起污染物环境行为和生态效应的变化，在目前已有的研究中多关注单一介质内部或两个介质之间的污染物环境行为对气候变化的响应，从地球关键带的角度，系统研究气候变化影响下污染物在关键带系统的迁移转化及其再平衡过程，深入分析不同气候条件下人体和生物暴露的环境风险，对于综合防控气候变化下的环境健康问题十分重要。

（2）加强抗生素抗性基因等新型污染物环境过程对气候变化的响应机制研究。抗生素的使用导致的环境风险问题目前已引起国内外的广泛关注，基于疾病防控、畜禽养殖等领域对抗生素的依赖性，抗生素今后将持续释放进入环境，抗生素与抗性基因对生态系统和人体健康的潜在威胁当前国内外已有大量的研究报道，但在气候变化的影响下，抗生素的迁移转化与环境归趋、抗性基因的产生与迁移过程会发生怎样的变化，这些变化过程又将对环境健康产生何种性质的影响，目前还不得而知。

（3）加强污染物环境过程与环境健康及气候变化的耦合关系研究。污染物对环境健康的影响不仅取决于污染物暴露浓度的本身，受体的适应性对于其健康响应也十分重要。在未来气候变化条件下，不同区域、不同年龄、不同性别的人群对污染物的敏感性是不同的，尤其是气候变化对不同群体造成的心理的影响，也会导致其对暴露的适应性产生差异，从而造成不同的环境健康响应，如何在全球尺度评估气候变化对环境健康的影响，同时关注其区域的差异性响应，对于更好地应对气候变化十分重要。

（4）加强气候变化影响下环境健康风险防控对策与机制研究。气候变化影响下的环境健康风险影响因素众多，既有传统的污染物的环境暴露问题，又有病原菌与病源微生物问题，还有花粉及其过敏问题及粮食品质及粮食安全问题，甚至还包括气候变化对心理健康的影响问题，众多的问题交织在一起，再加上区域发展的不平衡性和差异性，使得气候变化与环境健康的问题尤其复杂，今后的研究既要关心气候变化影响下的科学机制问题，还要将其与社会经济发展耦合，增强对全球气候变化可能带来的各种环境健康问题的预警预报能力，并对环境健康风险干预措施的结果进行科学模拟和评估，提升环境健康对气候变化适应能力。

参 考 文 献

[1] 叶笃正. 人类活动引起的全球性气候变化及其对我国自然、生态、经济和社会发展的可能影响. 中国科学院院刊, 1986(2): 18-26.

[2] 李军. 气候变化影响人类健康. 中华环境, 2016(Z1): 37-39.

[3] Oliver J E. Intergovernmental Panel in Climate Change(IPCC). Encyclopedia of Energy Natural Resource & Environmental Economics, 2013, 26(2): 48-56.

[4] 科学网. 气候变暖或危及全球粮食产量. 科技创新导报, 2014(26): 2-2.

[5] Seneviratne S I, Nicholls N, Easterling D, et al. Changes in climate extremes and their impacts on the natural physical environment: An overview of the IPCC SREX report. Managing the Risks of Extreme Events and Disasters to Advance Climate Change Adaptation, 2012.

[6] 谈建国, 黄家鑫. 热浪对人体健康的影响及其研究方法. 气候与环境研究, 2004, 9(4): 680-686.

[7] IPCC. Climate Change 2013: The Physical Science Basis. //Stocker T F, Qin D, Plattner G K, Tignor M, Allen S K, Boschung J, Nauels A, Xia Y, Bex V, Midgley P M (Eds). Contribution of Working Group I to the Fifth Assessment Report of the Intergovernmental Panel on Climate Change. Cambridge University Press, Cambridge, UK and New York, NY. 2013:

1535.

[8] Rieder H E, Fiore A M, Horowitz L W, et al. Projecting policy-relevant metrics for high summertime ozone pollution events over the eastern United States due to climate and emission changes during the 21st century. Journal of Geophysical Research: Atmospheres, 2015, 120(2): 784-800.

[9] Pryor S C, Scavia D, Downer C, et al. Midwest. Climate change impacts in the United States: The third national climate assessment, 2014, 418-440.

[10] IPCC. Summary for policymakers. // Edenhofer O, Pichs-Madruga R, Sokona Y, Farahani E, Kadner S, Seyboth K, Adler A, Baum I, Brunner S, Eickemeier P, Kriemann B, Savolainen J, Schlömer S, von Stechow C, Zwickel T, Minx J C (Eds). Climate Change 2014: Mitigation of Climate Change. Contribution of Working Group III to the Fifth Assessment Report of the Intergovernmental Panel on Climate Change, Cambridge University Press, Cambridge, United Kingdom and New York, NY, USA, 2014.

[11] Caldeira K, Wickett M E. Oceanography: anthropogenic carbon and ocean pH. Nature, 2003, 425(6956): 365.

[12] Raven J, Caldeira K, Elderfield H, et al. Ocean acidification due to increasing atmospheric carbon dioxide. Science, 2005, 215(2): 1-60.

[13] Hall S J M, Rodolfo R, Martin S, et al. Volcanic carbon dioxide vents show ecosystem effects of ocean acidification. Nature, 2008, 454(7200): 96-99.

[14] Fabry V J, Seibel B A, Feely R A, et al. Impacts of ocean acidification on marine fauna and ecosystem processes. Trends in Ecology & Evolution, 2013, 28(3): 178-186.

[15] Karlsson P E, Klingberg J, Engardt M, et al. Past, present and future concentrations of ground-level ozone and potential impacts on ecosystems and human health in northern Europe. Science of The Total Environment, 2017, 576: 22-35.

[16] Wang X, Manning W, Feng Z, et al. Ground-level ozone in China: Distribution and effects on crop yields. Environmental Pollution, 2007, 147(2): 394-400.

[17] Baker A C, Wright M S, Stepanauskas R, et al. Co-selection of antibiotic and metal resistance. Trends in Microbiology, 2006, 14(4): 176-182.

[18] Davis J A, Hetzel F, Oram J J, et al. Polychlorinated biphenyls(PCBs)in San Francisco Bay. Environmental Research, 2007, 105(1): 67-86.

[19] Brammer H, Ravenscroft P. Arsenic in groundwater: A threat to sustainable agriculture in South and South-east Asia. Environment International, 2009, 35(3): 647-654.

[20] Gaze W H, Zhang L, Abdouslam N A, et al. Impacts of anthropogenic activity on the ecology of class 1 integrons and integron-associated genes in the environment. Isme Journal, 2011, 5(8): 1253-1261.

[21] Moore M V. Potential effects of climate change on freshwater ecosystems of the New England/Mid-Atlantic Region. Freshwater Ecosystems and Climate Change in North America, 1997, 11(8): 925-947.

[22] Gouin T, Armitage J M, Cousins I T, et al. Influence of global climate change on chemical fate and bioaccumulation: The role of multimedia models. Environmental Toxicology and Chemistry, 2013, 32(1): 20-31.

[23] Boxall A, Hardy A, Beulke S, et al. Impacts of Climate Change on Indirect Human Exposure to Pathogens and Chemicals from Agriculture. Environmental Health Perspectives, 2009, 117(4): 508-514.

[24] Bloomfield J P, Williams R J, Gooddy D C, et al. Impacts of climate change on the fate and behaviour of pesticides in surface and groundwater—A UK perspective. Science of the Total Environment, 2006, 369(1-3): 163-177.

[25] Armitage J M, Quinn C L, Wania F. Global climate change and contaminants—an overview of opportunities and priorities for modelling the potential implications for long-term human exposure to organic compounds in the Arctic. Journal of Environmental Monitoring, 2011, 13(6): 1532.

[26] 崔斌, 王凌, 张国印, 等. 土壤重金属污染现状与危害及修复技术研究进展. 安徽农业科学, 2012, 40(1): 373-375.

[27] 赵中秋, 朱永官, 蔡运龙. 镉在土壤-植物系统中的迁移转化及其影响因素. 生态环境, 2005, 1(2): 282-286.

[28] 房存金. 土壤中主要重金属污染物的迁移转化及治理. 当代化工, 2010, 39(4).

[29] 徐笠, 陆安祥, 王纪华. 温度变化对重金属植物有效性影响的研究进展. 江苏农业科学, 2016, 44(10).

[30] Mehadi. Reaction of Nickel with soils and goethite: equilibrium and kinetic studies. 1993. University of New Hampshire.

[31] Li X H, Zhou Q X, Wei S H, et al. Adsorption and desorption of carbendazim and cadmium in typical soils in northeastern China as affected by temperature. Geoderma, 2011, 160(3/4): 347-354.

[32] 王金贵, 吕家珑, 张瑞龙, 等. 不同温度下镉在典型农田土壤中的吸附动力学特征. 农业环境科学学报, 2012, 31(6): 1118-1123.

[33] Jia X, Zhang C, Zhao Y, et al. Three years of exposure to lead and elevated CO_2, affects lead accumulation and leaf defenses in, Robinia pseudoacacia, L. seedlings. Journal of Hazardous Materials, 2018, 349: 215-223.

[34] Jia X, Zhao Y H, Liu T, et al. Leaf defense system of, Robinia pseudoacacia, L. seedlings exposed to 3, years of elevated atmospheric CO_2, and Cd-contaminated soils. Science of the Total Environment, 2017, 605-606: 48-57.

[35] Li Z, Tang S, Deng X, et al. Contrasting effects of elevated CO_2 on Cu and Cd uptake by different rice varieties grown on contaminated soils with two levels of metals: Implication for phytoextraction and food safety. Journal of Hazardous Materials, 2010, 177(1-3): 352-361.

[36] Guo H, Zhu J, Zhou H, et al. Elevated CO_2 levels affects the concentrations of copper and cadmium in crops grown in soil contaminated with heavy metals under fully open-air field conditions. Environmental Science & Technology, 2011, 45(16): 6997-7003.

[37] Duval B D, Dijkstra P, Natali S M, et al. Plant-Soil distribution of potentially toxic elements in response to elevated atmospheric CO_2. Environmental Science & Technology, 2011, 45(7): 2570-2574.

[38] Guo H, Tian R, Zhu J, et al. Combined cadmium and elevated ozone affect concentrations of cadmium and antioxidant systems in wheat under fully open-air conditions. Journal of Hazardous Materials, 2012, 209-210: 27-33.

[39] 王亚波, 魏思雨, 孙月, 等. 大气臭氧胁迫对稻季土壤 Cd 生物有效性的影响. 农业环境科学学报, 2017(8): 1462-1467.

[40] 尹微琴, 朱红, 周宇澄, 等. 麦季土壤镉生物有效性对大气臭氧浓度升高的响应. 扬州大学学报(农业与生命科学版), 2018(2): 35-42.

[41] Jana K, Nina M, Zdeněk M, et al. Persistent organic pollutants in soils and sediments from James Ross Island, Antarctica. Environmental Pollution, 2008, 152(2): 416-423.

[42] Hites, Ronald A. Polybrominated diphenyl ethers in the environment and in people: A meta-analysis of concentrations. Environmental Science & Technology, 2004, 38(4): 945-956.

[43] Menzie C A, Potocki B B, Santodonato J. Exposure to carcinogenic PAHs in the environment. Environmental Science & Technology, 1992, 26(7): 1278-1284.

[44] Phillips D H. Polycyclic aromatic hydrocarbons in the diet. Mutation Research, 1999, 443(1-2): 139.

[45] Mueller K E, Mueller S S R, Henry H F, et al. Fate of pentabrominated diphenyl ethers in soil: Abiotic sorption, plant uptake, and the impact of interspecific plant interactions. Environmental Science & Technology, 2006, 40(21): 6662-6667.

[46] Alcock R E, Bacon J, Bardget R D, et al. Persistence and fate of polychlorinated biphenyls (PCBs) in sewage sludge-amended agricultural soils. Environmental Pollution, 1996, 93(1): 83-92.

[47] Holoubek I, DuEk L, Milan S, et al. Soil burdens of persistent organic pollutants—Their levels, fate and risk. Part I. Variation of concentration ranges according to different soil uses and locations. Environmental Pollution, 2009, 157(12): 3207-3217.

[48] Bernhard A, Bernd M. B, Peta L, et al. Regionalized concentrations and fingerprints of polycyclic aromatic hydrocarbons (PAHs) in German forest soils. Environmental Pollution, 2015, 203: 31-39.

[49] Kuppusamy S, Thavamani P, Venkateswarlu K, et al. Remediation approaches for polycyclic aromatic hydrocarbons (PAHs) contaminated soils: Technological constraints, emerging trends and future directions. Chemosphere, 2017, 168, 944-968.

[50] Bueno M M, Springael D, Ortega C J J. Effect of a nonionic surfactant on biodegradation of slowly desorbing PAHs in contaminated soils. Environmental Science & Technology, 2011, 45(7): 3019-3026.

[51] Henry W, Macleod M, Konrad H. Emissions, fate and transport of persistent organic pollutants to the Arctic in a changing global climate. Environmental Science & Technology, 2013, 47(5), 2323-2330.

[52] Ma J, Cao Z. Quantifying the perturbations of persistent organic pollutants induced by climate change. Environmental Science & Technology, 2010, 44(22): 8567-8573.

[53] Ma J, Hung H, Tian C, et al. Revolatilization of persistent organic pollutants in the Arctic induced by climate change. Nature Climate Change, 2011, 1(5): 255-260.

[54] Friedman C L, Zhang Y, Selin N E. Climate change and emissions impacts on atmospheric PAH transport to the Arctic. Environmental Science & Technology, 2014, 48(1): 429-437.

[55] Tscherko D, Kandeler E, Jones T H. Effect of temperature on below-ground N-dynamics in a weedy model ecosystem at ambient and elevated atmospheric CO_2 levels. Soil Biology & Biochemistry, 2001, 33(4): 491-501.

[56] Rajkumara M, Prasadb M N V, Swaminathana S, et al. Climate change driven plant–metal–microbe interactions. Environment International, 2013, 53: 74-86.

[57] Iqbal J, Metosh D C, Portier R J. Temperature effects on bioremediation of PAHs and PCP contaminated south Louisiana soils: A laboratory mesocosm study. Journal of Soils and Sediments, 2007, 7(3): 153-158.

[58] Ai F, Eisenhauer N, Jousset A, et al. Elevated tropospheric CO_2 and O_3 concentrations impair organic pollutant removal from grassland soil. Scientific reports, 2018, 8(1): 5519.

[59] Ai F, Eisenhauer N, Xie Y, et al. Elevated CO_2 accelerates polycyclic aromatic hydrocarbon accumulation in a paddy soil grown with rice. PLOS ONE, 2018, 13(4): e0196439.

[60] Zhang Y, Virjamo V, Sobuj N, et al. Elevated temperature and CO_2 affect responses of European aspen (*Populus tremula*) to soil pyrene contamination. Science of the Total Environment, 2018, 634: 150-157.

[61] Warheit D B, Sayes C M, Reed K L, et al. Health effects related to nanoparticle exposures: Environmental, health and safety considerations for assessing hazards and risks. Pharmacology and Therapeutics, 2008, 120(1): 35-42.

[62] Oberdorster G, Oberdorster E, Oberdorster J, et al. Nanotoxicology: An emerging discipline evolving from studies of ultrafine particles. Environmental Health Perspectives, 2005, 113(7): 823-839.

[63] Ju-Nam Y, Lead J R. Manufactured nanoparticles: An overview of their chemistry, interactions and potential environmental implications. Science of the Total Environment, 2008, 400(1-3): 396-414.

[64] Keller A, Mcferran S, Lazareva A, et al. Global life cycle releases of engineered nanomaterials. Journal of Nanoparticle Research, 2013, 15(6): 1692.

[65] Sun T Y, Mitrano D M, BornhFt N A, et al. Envisioning nano release dynamics in a changing world: Using dynamic probabilistic modeling to assess future environmental emissions of engineered nanomaterials. Environmental Science & Technology, 2017, 51(5): 2854-2863.

[66] Simonin M, Guyonnet J P, Martins J M F, et al. Influence of soil properties on the toxicity of TiO_2 nanoparticles on carbon mineralization and bacterial abundance. Journal of Hazardous Materials, 2015, 283: 529-535.

[67] Schrand A M, Rahman M F, Hussain S M, et al. Metal-based nanoparticles and their toxicity assessment. Wiley Interdisciplinary Reviews Nanomedicine & Nanobiotechnology, 2010, 2(5): 544-568.

[68] Qi M, Liu Y, Li T. Nano-TiO_2 improve the photosynthesis of tomato leaves under mild heat stress. Biological Trace Element Research, 2013, 156(1-3): 323-328.

[69] Du W, Gardea T J L, Xie Y, et al. Elevated CO_2 levels modify TiO_2 nanoparticle effects on rice and soil microbial communities. Science of the Total Environment, 2016, 578: 408-416.

[70] Zhang Y, Virjamo V, Sobuj N, et al. Sex-related responses of European aspen (*Populus tremula* L.) to combined stress: TiO_2 nanoparticles, elevated temperature and CO_2 concentration. Journal of Hazardous Materials, 2018, 352: 130-138.

[71] Yin Y, Hu Z, Du W, et al. Elevated CO_2 levels increase the toxicity of ZnO nanoparticles to goldfish (*Carassius auratus*) in a water-sediment ecosystem. Journal of Hazardous Materials, 2016, 327: 64-70.

[72] Bebber, Patrick D. Range-expanding pests and pathogens in a warming world. Annual Review of Phytopathology, 2015, 53(1): 150605182533006.

[73] Rosa A. Sánchez G, Alex C A, Hansson B, et al. Evolutionary consequences of climate‐induced range shifts in insects. Biological Reviews, 2016, 91(4): 1050-1064.

[74] Bale J S, Hayward S A. Insect overwintering in a changing climate. Journal of Experimental Biology, 2010, 213(6): 980.

[75] Zavala J A, Casteel C L, Delucia E H, et al. Anthropogenic increase in carbon dioxide compromises plant defense against invasive insects. Proceedings of the National Academy of Sciences, 2008, 105(13): 5129-5133.

[76] Bridget F. O'Neill, Zangerl A R, Delucia E H, et al. Leaf temperature of soybean grown under elevated CO_2 increases Aphis glycines (Hemiptera: Aphididae) population growth. Insect Science, 2011, 18(4): 419-425.

[77] Rosenzweig C, Iglesias A, Yang X B, et al. Climate change and extreme weather events; Implications for food production, plant diseases, and pests. Global Change and Human Health, 2001, 2(2): 90-104.

[78] Ziska, L H, Runion G B. Future weed, pest, and disease problems for plants. Agroecosystems in a Changing Climate. Newton, P.C.D., R.A. Carran, G.R. Edwards, and P.A. Niklaus, Eds. CRC Press, Boca Raton, FL, 2007, 261-287.

[79] Bailey S W. Climate change and decreasing herbicide persistence. Pest Management Science, 2004, 60(2): 158-162.

[80] Boxall A, Hardy A, Beulke S, et al. Impacts of climate change on indirect human exposure to pathogens and chemicals from agriculture. Environmental Health Perspectives, 2009, 117(4): 508-514.

[81] Chen C C, Mccarl B A. An investigation of the relationship between pesticide usage and climate change. Climatic Change, 2001, 50(4): 475-487.

[82] Ziska L H. Increasing minimum daily temperatures are associated with enhanced pesticide use in cultivated soybean along a latitudinal gradient in the Mid-Western United States. PLoS ONE, 2014, 9(2): e98516.

[83] Ziska L H, Bunce J A. Future atmospheric carbon dioxide may increase tolerance to glyphosate. Weed Science, 1999, 47(5): 608-615.

[84] Ziska L H, Faulkner S, Lydon J. Changes in biomass and root: shoot ratio of field-grown Canada thistle(Cirsium arvense), a noxious, invasive weed, with elevated CO_2: Implications for control with glyphosate. Weed Science, 2004, 52(4): 584-588.

[85] Manea A, Leishman M R, Downey P O. Exotic C4 grasses have increased tolerance to glyphosate under elevated carbon dioxide. Weed Science, 2011, 59(January-March 2011): 28-36.

[86] Ziska, L H. Climate change, carbon dioxide and global crop production: Food security and uncertainty. Handbook on Climate Change and Agriculture. Dinar A, Mendelsohn R, Eds. Cheltenham, United Kingdom: Edward Elgar Publishing, 2011, 9-31.

[87] Ziska L H, Mcconnell L L. Climate change, carbon dioxide, and pest biology: Monitor, mitigate, manage. Journal of Agricultural and Food Chemistry, 2015, 64(1): 6-12.

[88] Delcour I, Spanoghe P, Uyttendaele M. Literature review: Impact of climate change on pesticide use. Food Research International, 2015, 68: 7-15.

[89] Organization W H. Antimicrobial resistance: Global report on surveillance. Australasian Medical Journal, 2014, 7(4): 237.

[90] Cooper K M, Mcmahon C, Fairweather I, et al. Potential impacts of climate change on veterinary medicinal residues in livestock produce: An island of Ireland perspective. Trends in Food Science & Technology, 2015, 44(1): 21-35.

[91] Gormaz, J G, Fry J P, Erazo D C. Love, Public health perspectives on aquaculture. Current Environmental Health Reports, 2014, 1(3): 227-238.

[92] Lafferty K D, Porter J W, Ford S E. Are diseases increasing in the ocean? Annual Review of Ecology, Evolution, and Systematics, 2004, 35(1): 31-54.

[93] Beulke S, Boxall A, Brown C, et al. Assessing and Managing the Impacts of Climate Change on the Environmental Risks of Agricultural Pathogens and Contaminants: Pathways for Transport of Contaminants in Agricultural Systems and Potential Effects of Climate Change. Final Report to Defra for Project SD0441. Fera, Sand Hutton, York, UK. 2007.

[94] Noyes P D, Mcelwee M K, Miller H D, et al. The toxicology of climate change: Environmental contaminants in a warming world. Environment International, 2009, 35(6): 971-986.

[95] Ziska L H, Morris C F, Goins E W. Quantitative and qualitative evaluation of selected wheat varieties released since 1903 to increasing atmospheric carbon dioxide: Can yield sensitivity to carbon dioxide be a factor in wheat performance. Global Change Biology, 2010, 10(10): 1810-1819.

[96] Taub D R, Miller B, Allen H. Effects of elevated CO_2 on the protein concentration of food crops: A meta-analysis. Global Change Biology, 2008, 14, 565-575.

[97] Myers, S S, Zanobetti A, Kloog I, et al. Increasing CO_2 threatens human nutrition. Nature, 2014, 510(7503): 139-142.

[98] Jablonski L M, Curtis W P S. Plant reproduction under elevated CO_2 conditions: A meta-analysis of reports on 79 crop and wild species. New Phytologist, 2002, 156(1): 9-26.

[99] Irakli L. Hidden shift of the ionome of plants exposed to elevated CO_2 depletes minerals at the base of human nutrition. ELife, 2014, 3.

[100] Manderscheid R, Bender J, Jager H J, et al. Effects of season long CO_2 enrichment on cereals. II. Nutrient concentrations and grain quality. Agriculture Ecosystems & Environment, 1995, 54(3): 175-185.

[101] Fernando N, Panozzo J, Tausz M, et al. Elevated CO_2 alters grain quality of two bread wheat cultivars grown under different environmental conditions. Agriculture, Ecosystems & Environment, 2014, 185: 24-33.

[102] Loladze I. Rising atmospheric CO_2 and human nutrition: Toward globally imbalanced plant stoichiometry? Trends in Ecology & Evolution, 2002(10): 457-461.

[103] Cotrufo M F, Ineson P, Scott A. Elevated CO_2 reduces the nitrogen concentration of plant tissues. Global Change Biology, 2010, 4(1): 43-54.

[104] Grifferty A, Barrington S. Zinc uptake by young wheat plants under two transpiration regimes. Journal of Environmental Quality, 2000, 29(2): 443-446.

[105] Mcgrath J M, Lobell D B. Reduction of transpiration and altered nutrient allocation contribute to nutrient decline of crops grown in elevated CO_2, concentrations. Plant, Cell & Environment, 2013, 36(3): 697-705.

[106] Zhu C, Kobayashi K, Loladze I. Carbon dioxide (CO_2) levels this century will alter the protein, micronutrients, and vitamin content of rice grains with potential health consequences for the poorest rice-dependent countries. Science Advances, 2018, 4(5): eaaq1012.

[107] Taub D R, Wang X. Why are Nitrogen concentrations in plant tissues lower under elevated CO_2? A critical examination of the hypotheses. Journal of Integrative Plant Biology, 2008, 50(11): 1365-1374.

[108] Raubenheimer D, Machovsky C G E, Gosby A K, et al. Nutritional ecology of obesity: From humans to companion animals. The British journal of nutrition, 2014, 113(S1).

[109] Layman D K, Boileau R A, Erickson D J, et al. A reduced ratio of dietary carbohydrate to protein improves body composition and blood lipid profiles during weight loss in adult women. Journal of Nutrition, 2003, 133(2): 411-417.

[110] Lake I R, Hooper L, Abdelhamid A, et al. Climate change and food security: Health impacts in developed countries.

Environmental Health Perspectives, 2012, 120(11): 1520-1526.

[111] Rogers C A, Wayne P M, Macklin E A, et al. Interaction of the onset of spring and elevated atmospheric CO_2 on Ragweed(Ambrosia artemisiifolia L.)Pollen Production. Environmental Health Perspectives, 2006, 114(6): 865-869.

[112] Albertine J M, Manning W J, Michelle D C, et al. Projected carbon dioxide to increase grass pollen and allergen exposure despite higher ozone levels. PLoS ONE, 2014, 9(11): e111712.

[113] Sheffield P E, Weinberger K R, Kinney P L. Climate change, aeroallergens, and pediatric allergic disease. Mount Sinai Journal of Medicine A Journal of Translational & Personalized Medicine, 2011, 78(1): 78-84.

[114] Beggs P J. Adaptation to impacts of climate change on aeroallergens and allergic respiratory diseases. International Journal of Environmental Research and Public Health, 2010, 7(8): 3006-3021.

[115] D'Amato G, Cecchi L, D'Amato M, et al. Urban air pollution and climate change as environmental risk factors of respiratory allergy: An Update. Journal of Investigational Allergology & Clinical Immunology, 2010, 20(2): 95-102.

[116] Cotty P J, Jaime G R. Influences of climate on aflatoxin producing fungi and aflatoxin contamination. International Journal of Food Microbiology, 2007, 119(1-2): 109-115.

[117] D"Amato G, Cecchi L. Effects of climate change on environmental factors in respiratory allergic diseases. Clinical and Experimental Allergy: Journal of the British Society for Allergy and Clinical Immunology, 2008, 38(8): 1264-1274.

[118] 闫峰, 安冬, 王娜, 等. 海河流域气候变化对花粉过敏性疾病的影响及适应性对策// 中国气象学会年会, 2012.

[119] Paterson R R M, Lima N. Toxicology of mycotoxins. EXS, 2010, 100: 31-63.

[120] Egmond H. Natural toxins: risks, regulations and the analytical situation in Europe. Analytical and Bioanalytical Chemistry, 2004, 378(5): 1152-1160.

[121] Hawkes C, Ruel M. The links between agriculture and health: An intersectoral opportunity to improve the health and livelihoods of the poor. Bulletin of the World Health Organisation, 2007, 84(12): 984-990.

[122] Miraglia M, Marvin H J P, Kleter G A, et al. Climate change and food safety: An emerging issue with special focus on Europe. Food and Chemical Toxicology, 2009, 47(5): 1009-1021.

[123] Litchfield M H. Estimates of acute pesticide poisoning in agricultural workers in less developed countries. Toxicological Reviews, 2005, 24(4): 271-278.

[124] Magan N, Medina A, Aldred D. Possible climate-change effects on mycotoxin contamination of food crops pre- and postharvest. Plant Pathology(Oxford), 2011, 60(1): 150-163.

[125] 李建标, 朱美财. 环境病原微生物污染与健康. 空军总医院学报, 1999(4): 43-45.

[126] 施闽涛, 何国富. 水环境病原体污染及其健康风险评价研究. 安徽农业科学, 2014, 42(5): 1482-1485.

[127] 蔡祖聪, 黄新琦. 土壤学不应忽视对作物土传病原微生物的研究. 土壤学报, 2016, 53(2): 305-310.

[128] 黄曦, 许兰兰, 黄荣韶, 等. 枯草芽孢杆菌在抑制植物病原菌中的研究进展. 生物技术通报, 2010(1): 24-29.

[129] 刘峰, 朱永官, 王效科. 我国地面臭氧污染及其生态环境效应. 生态环境学报, 2008, 17(4): 1674-1679.

[130] 殷永泉, 李昌梅, 马桂霞, 等. 城市臭氧浓度分布特征. 环境科学, 2004, 25(6): 16-20.

[131] 孔琴心, 刘广仁, 李桂忱. 近地面臭氧浓度变化及其对人体健康的可能影响. 气候与环境研究, 1999, 4(1): 61-66.

[132] Rai R, Agrawal M. Impact of tropospheric ozone on crop plants. Proceedings of the National Academy of Sciences, India Section B: Biological Sciences, 2012, 82(2): 241-257.

[133] Feng Z Z, Hu E Z, Wang X K, et al. Ground-level O_3 pollution and its impacts on food crops in China: A review. Environmental Pollution, 2015, 199: 42-48.

[134] 杨连新, 王余龙, 石广跃, 等. 近地层高臭氧浓度对水稻生长发育影响研究进展. 应用生态学报, 2008, 19(4): 901-910.

[135] 刘晓聪, 董家华, 欧英娟. 大气中 CO_2 与 O_3 浓度升高对植物光合作用影响研究. 环境科学与管理, 2016, 41(3): 152-155.

[136] 徐胜, 何兴元, 陈玮, 等. 高浓度 O_3 对树木生理生态的影响. 生态学报, 2009, 29(1): 368-377.

[137] 列淦文, 叶龙华, 薛立. 臭氧胁迫对植物主要生理功能的影响. 生态学报, 2014, 34(2): 294-306.

[138] Broberg M C, Feng Z, Xin Y, et al. Ozone effects on wheat grain quality—A summary. Environmental Pollution, 2015, 197(1): 203.

[139] Ainsworth E A. Rice production in a changing climate: A meta-analysis of responses to elevated carbon dioxide and elevated ozone concentration. Global Change Biology, 2010, 14(7): 1642-1650.

[140] Feng Z Z, Kobayashi K. Assessing the impacts of current and future concentrations of surface ozone on crop yield with meta-analysis. Atmospheric Environment, 2009, 43(8): 1510-1519.

[141] 黄益宗, 隋立华. 臭氧污染胁迫下植物的抗氧化系统调节机制. 生态毒理学报, 2013, 8(4): 456-464.

[142] 郑有飞, 胡程达, 吴荣军, 等. 臭氧胁迫对冬小麦光合作用、膜脂过氧化和抗氧化系统的影响. 环境科学, 2010, 31(7): 1643-1651.

[143] 平晓燕, 周广胜, 孙敬松. 植物光合产物分配及其影响因子研究进展. 植物生态学报, 2010, 34(1): 100-111.

[144] 冯兆忠, 李品, 袁相洋, 等. 我国地表臭氧生态环境效应研究进展. 生态学报, 2018, 38(5): 1530-1541.

作者：郭红岩[1]，朱永官[2]，冯兆忠[3]，尹　颖[1]，王亚波[1]
[1] 南京大学，[2] 中国科学院城市环境研究所，[3] 南京信息工程大学

第 28 章　冰冻圈环境化学

- 1. 引言 /806
- 2. 冰冻圈简介 /806
- 3. 冰雪及其界面的化学过程 /808
- 4. 北极地区冰冻圈环境化学 /810
- 5. 南极地区冰冻圈环境化学 /815
- 6. 冰川、冻土和高山地区冰冻圈环境化学 /818
- 7. 冰冻圈环境化学的全球效应和展望 /822

本章导读

冰冻圈环境化学是环境化学中相对较新的研究领域，关注化学物质在冰冻圈的来源、分布、迁移归趋和生物效应，及其与气候变化的相互作用。虽然冰冻圈在地表的分布有极大的时间和地域性，冰冻圈环境化学过程对化学污染物的影响则远远超出冰冻圈的范围。本章首先介绍冰冻圈的分布和环境变化，讨论冰、雪界面化学的基本过程，然后以北极、南极和高山地区为例，讨论冰冻圈环境化学的最新研究进展。

关键词

冰冻圈，极地地区，高山地区，汞，持久性有机污染物，有机气溶胶，气候变化

1 引　言

环境化学关注地表环境中的化学物质、化学过程及其对生态系统和人类健康的影响。在环境化学研究中，通常所关心的地表环境包括大气圈、水圈和土壤圈。虽然冰、雪在大气物理和化学过程中的作用早有共识，冰冻圈环境化学、尤其是陆地和海洋冰冻圈环境化学，直到近 20 年才逐渐开始引起关注。与其他地表环境相比，冰冻圈有着特殊的地理、气候和环境特征。作为冻结的固态水体，冰冻圈对气候变化具有高度敏感性；同时，冰冻圈与其他圈层发生直接或间接的物质和能量交换，对全球气候、环境变化和污染物传输有着关键性响应和反馈。

冰冻圈环境化学研究首先是关注化学物质在冰冻圈内的赋存和行为，关注其来源、分布、迁移、归趋和生物效应。冰冻圈（特别是冰川和积雪）是大气降水和干湿沉降的累积，可以反映大气环境变化并用于评估人类活动对大气环境的影响；另外，冰冻圈变化特别是冻结水体的相变和输移会对其他圈层环境化学的要素和过程产生影响。近百年来全球冰冻圈普遍萎缩[1]，对全球和区域气候、水资源、生态环境和人类福祉产生了重大影响并日益受到关注。在全球变化与人类活动相互作用日益增强的背景下，冰冻圈环境化学所关注和研究的问题逐渐成为冰冻圈科学和环境科学研究领域的热点和交叉前沿。

2 冰冻圈简介

2.1 冰冻圈

冰冻圈（cryosphere）亦称冰雪圈、冰圈或冷圈，是指地球表层连续分布且具有一定厚度的、温度在摄氏零度或以下的负温圈层。一般而言，冰冻圈内的水体应处于自然冻结状态；地表水和其他物质混合而成的冻结体也属于冰冻圈的范畴。此外，冰冻圈还包括负温条件下一些微观尺度上的"未冻结水"，例如冰晶表面存在的"准液态"水，以及冻土内部因毛细作用和土壤颗粒吸附作用而形成的未冻结水等。

冰冻圈的主要组成要素包括冰川、冰盖、冻土、积雪、海冰、河湖冰以及固态降水等，这些要素在地球表面水平方向上的分布很不均匀，在高、中纬度地区分布最为密集（图 28-1）。根据冰冻圈要素形成

发育的条件和地理分布，地球冰冻圈可划分为三类[2, 3]：

（1）陆地冰冻圈：主要包括冰川、陆地冻土和积雪。约占全球陆地面积的52%～55%，其中冰川（包括南、北极冰盖和山地冰川等）约占全球陆地面积的10%，多年冻土区约占全球陆地面积的24%。季节性冻土（包括多年冻土活动层）和季节性积雪覆盖范围更大，多年平均最大范围可占北半球陆地面积的一半以上。

（2）海洋冰冻圈：主要包括海冰、冰架、冰山和海底多年冻土。

（3）大气冰冻圈：主要包括大气圈内处于冻结状态的水体，如雪花和冰晶等。

中国是中、低纬度地区冰冻圈发育最为丰富的国家。以冰川、冻土为例，根据第2次冰川编目的最新统计，中国共有冰川48000多条，面积约为$5.2\times10^4\,\mathrm{km^2}$；多年冻土区面积约为$220\times10^4\,\mathrm{km^2}$，其中高海拔多年冻土面积居世界之首。

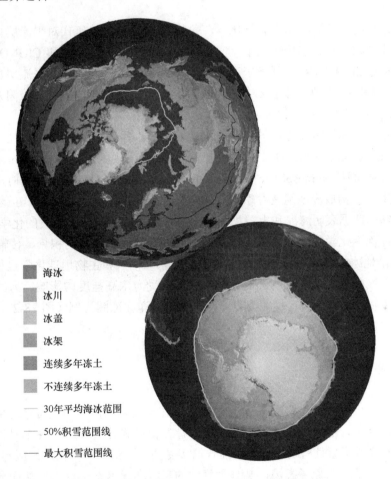

图 28-1　全球冰冻圈分布示意图（引用并修改自参考文献[1]）

2.2 冰冻圈的消融和环境变化

冰冻圈对气候变化最为敏感。由于工业革命以来的全球气候变暖，北半球的陆地和海洋冰冻圈在最近几个世纪、尤其是近几十年来正在经历着快速消融。北冰洋的多年海冰覆盖面积在过去40年里减少了近40%，中部海盆海冰的年均厚度降低了近65%[4]。南大洋的海冰覆盖范围因受洋流变化和其他因素的影响，对气候变暖的响应相对不明显；1979～2012年，南大洋的海冰覆盖范围反而有所增加，但随着周期性深海洋流的变化，这一趋势在未来极有可能发生逆转[5]。全球冰川近几十年全面萎缩，特别是 21 世纪最初 10 年的冰川减少率达到每年 0.81 m 水当量，创 1850 年以来的最高值[6]。格林兰岛的冰盖消融至少已持续 350 年，尤其是近 20 年（1994~2013 年）的消融强度已急增到工业革命前的 250%~575%[7]。南极洲的冰盖在过去 20 年里也有消融，但消融强度低于格陵兰岛[8]。有"第三极"之称的青藏高原地区的高山冰川也基本呈退缩趋势，尤其以高原东南部最为显著，其冰川长度在 1970~2010 年期间每年平均减少为 48.2 m，面积每年减少 0.57%[9]。

3 冰雪及其界面的化学过程

对冰冻圈化学过程的早期研究始于大气化学，尤其是 20 世纪 80 年代对平流层臭氧化学和臭氧空洞的研究。氟利昂类化合物从对流层传输到平流层后，其主要降解产物 HCl 和 $ClONO_2$ 在极地平流层云中发生化学反应，生成的 Cl_2 和 HOCl 经光解产生原子态 Cl，成为南极地区春季平流层臭氧洞形成的主要原因[10]。在这一系列过程中，极地平流层云中的超低温冰晶对气态 HCl 的溶解和吸附及其与 $ClONO_2$ 在冰晶表面的非均相反应起了关键的作用[10]。

几乎在同一时期，北冰洋海岸地区的观测数据显示每年春季海洋边界层的大气中经常出现显著的臭氧亏损事件（ozone depletion event, ODE）[11, 12]。该事件和随后发现的溴剧增事件（bromie explosion event, BEE）[13]和大气汞亏损事件（atmospheric mercury depletion event, AMDE）[14]同时出现。这些事件在南极地区春季亦有发生[15]，构成极地地区春季特有的边界层大气光化学现象[16, 17]。虽然详尽的反应机制尚不完全明确，但足够的证据表明海冰和上覆积雪至少在一定程度上参与了相关的化学反应[18]。

对陆地和海洋冰冻圈化学的早期研究主要集中于冰冻圈中化学物质的积累、传输和释放[19]。但过去 20 年的研究表明，陆地冰雪[20]和海洋冰雪[19]中会发生一系列化学和生物地球化学过程，改变化学物质的形态和反应速率，影响化学物质在生物圈的累积和效应。受海水高盐度的影响，海冰中固、液、气三相共存，卤水中更有生物聚栖，化学过程尤为复杂。以下以海洋冰冻圈为例（图 28-2），对海洋冰雪及其界面的主要化学过程作简单介绍。

3.1 光化学反应

在经历极夜之后，极地地区有长达 6 个月的极昼。在极昼期间，太阳光可穿透部分甚至整个海洋表层冰冻圈，直至表层海水，引发光化学反应[19]。海洋冰冻圈的光化学反应与春季时极地地区边界层中的溴剧增、臭氧亏损和汞亏损事件密切相关。目前普遍接受的观点认为，在极地地区春季的大气边界层内，原本相对惰性的溴离子活化为原子态溴，其与大气中的臭氧和元素态汞反应，从而导致溴剧增、臭氧亏损和汞亏损事件的发生[18]。这其中的关键步骤是溴离子的活化过程，目前认为是通过海水-冰雪-大气界面的多相、链式光化学反应进行[18]：

$$HOBr + Br^- + H^+ \longrightarrow Br_2 + H_2O \tag{28-1}$$

$$Br_2 + h\nu \longrightarrow 2Br \tag{28-2}$$

$$Br + O_3 \longrightarrow BrO + O_2 \tag{28-3}$$

$$BrO + O_3 \longrightarrow HOBr + O_2 \tag{28-4}$$

$$净反应：Br^- + H^+ + 2O_3 \longrightarrow Br + 2O_2 + H_2O \tag{28-5}$$

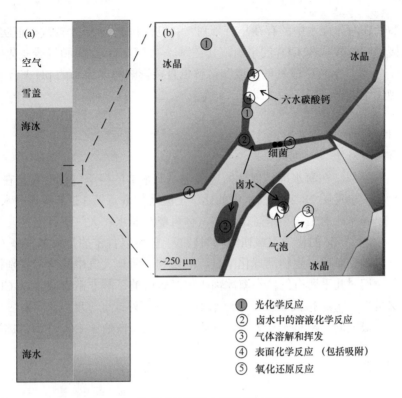

图 28-2 海洋冰冻圈的主要化学过程示意图

(a) 海洋冰冻圈垂直剖面 (左侧：极夜；右侧：极昼)；(b) 海冰的主要化学过程，类似于海冰中的主要化学过程也发生于其上覆雪盖以及陆地冰雪中，只是卤水的体积和盐度都远小于海冰中

早期的研究认为溴离子光化学活化发生于溴离子浓度相对较高的海冰或霜花 (frost flowers) 表面，但最近的研究表明这个活化过程更有可能发生于呈微酸性海冰上的飘雪或雪盖[21, 22]。溴离子光化学活化过程的强度也受制于海冰特征 (盐度、厚度、冰隙等) 和气象条件 (温度、风速等)[23]。

除原子溴之外，冰冻圈光化学反应还可以产生其他高活性物质，如 Cl、H_2O_2、·OH 自由基、NO_x 等。这些物质不仅可以消耗大气中的臭氧和元素态汞，还可以加速大气中有机物的氧化和降解[24]。实验室研究表明，有些发生于冰表面的光化学反应速率远远超过水溶液中的反应速率，反应的产物也会有所不同，这其中的机理目前尚不完全清楚[24, 25]。

3.2 卤水溶液反应

在海水冻结成冰的过程中，海水中的部分盐份会下渗到冰下的海水，其余的则会富集在海冰的表面或冰晶之间，形成盐度远高于海水的卤水 (brine)。受海冰温度以及生长和消融过程影响，卤水在不同季节里可以分布在完全不联通的卤水胞 (brine pockets) 里，也可以联通成卤水通道 (brine channels) (参见图 28-2)。由于体积有限，卤水中化学物质的浓度高度富集，化学反应速率和产物都有可能迥异于淡水或海水中的化学反应[26]。

这类反应中最有代表性的当属海冰中水合碳酸钙的发现。在稀溶液或海水中，当 Ca^{2+} 和 HCO_3^- 达到一定浓度时，会生成沉淀无水碳酸钙，最终成为方解石 (calcite)：

$$Ca^{2+} + 2HCO_3^{2-} \rightleftharpoons CaCO_3 + CO_2 + H_2O \qquad (28-6)$$

而在海冰的卤水中，同样的反应物却生成水合碳酸钙：

$$Ca^{2+} + 2HCO_3^{2-} + 5H_2O \rightleftharpoons CaCO_3 \cdot 6H_2O + CO_2 \qquad (28-7)$$

虽然对水合碳酸钙在海冰卤水中的存在早有预测[27]，但直到 2008 年才首次被证实[28]。由于水合碳酸钙的生成和溶解直接影响碳酸盐和 CO_2 的平衡，其在海洋-大气 CO_2 交换和气候变化中的作用在过去 10 年中引起了广泛关注[29, 30]。水合碳酸钙的生成和溶解也直接影响海冰卤水的 pH [31]，从而影响海冰环境中一系列化学和生物化学过程。

3.3 气体溶解和挥发

与卤水的形成过程类似，在海水冻结成冰的过程中，海水中的溶解气体会富集在卤水中，或挥发为气泡（gas bubbles）[32]（参见图 28-2）。卤水中的化学反应（如本章第 3.2 节所述的水合碳酸钙的沉淀过程）也会产生气体。随着海冰的生长和消融，这些气体可能与海水和大气进行交换，从而改变海-气交换的通量。虽然科学家对海冰气体的化学组成研究相对较少，但可以肯定的是很多气体（如 O_2、CO_2、CH_4、二甲基硫）的相对含量因受海冰中生物地球化学过程的影响，而与大气中的含量差别很大。

值得关注的是，我们对几乎所有气体在海冰卤水中溶解度的了解非常有限。以 CO_2 为例，已有的地球化学和气候模型都是将 CO_2 在摄氏零度以上海水中的溶解度外推至零度以下，来估算 CO_2 在卤水和冰-水界面海水的溶解度。最近的实验室研究表明，这种外推法显著低估了 CO_2 在海洋冰冻圈的溶解度[33]。

3.4 表面化学反应

目前对冰晶表面在冰冻圈化学过程中作用的了解非常有限。倾角激光诱导荧光光谱（glancing-angle laser-induced fluorescence spectroscopy）和拉曼光谱研究发现，很多化合物（如 HCl、HNO_3 和芳香族有机物）在纯冰和淡水冰表面发生的反应与水溶液中显著不同，表明冰晶表面或者是其准液态水层直接参与了化学反应[25]。由于海冰表面通常覆盖一层厚度远超过准液态水层的卤水层，其表面发生的反应可能更接近于卤水溶液中的反应[25]。

3.5 氧化还原反应

海冰中的卤水及其营养物为很多微生物提供了有效的生境，而微生物的活动会导致海冰中溶解氧在特定情况下大幅下降，出现季节性的还原性厌氧环境[34, 35]，进而为一些需要还原性环境的反应创造条件。目前，对海冰中的硝化和反硝化过程已有报道[34]。同时，汞的甲基化过程亦有可能发生[36]。

4 北极地区冰冻圈环境化学

北极地区和南极地区位于地球的两端，其冰冻圈环境化学有诸多相同之处，也有很多不同之处。南极地区是被南大洋包围的、远离人类活动的"海中地"，而北极地区则是被人口稠密、工业发达的北半球陆地所包围的"地中海"。北冰洋和南大洋的海冰环境，以及格陵兰岛和南极洲的冰盖环境也有很多不同，加上大气、海洋环流和生态系统迥异，使两极地区的冰冻圈环境呈现出各自独特的生物地球化学循环。

虽然两极地区几乎没有直接的人为污染源，但污染物可以通过至少以下四种长距离传输方式从低、中纬度地区进入极地地区。

（1）大气传输：由于低、中纬度地区气温较高，挥发性或半挥发性污染物（例如很多有机污染物、元素态汞）会挥发进入大气，随大气环流进入高纬寒冷地区时，则会遇冷沉降下来，这种从低纬向高纬迁移过程中由于冷暖交替出现的一系列距离较短的跳跃过程，被称为"蚱蜢效应"（grasshopper effect）[37]。

（2）海洋传输：进入低、中纬度近海地区的污染物，尤其是水溶态的污染物，可以随着全球海洋洋流远程传输至极地地区。

（3）河流传输：直接汇入极地地区的河流（例如由俄罗斯和加拿大北部注入北冰洋的大河）可以将流域内的污染物经河水和泥沙传输至极地地区。

（4）生物传输：低、中纬度地区的污染物亦可以通过迁徙的鸟类和水生生物传输至极地地区。

对北极地区环境污染物的关注约始于20世纪70年代。与想象中的洁净、未污染环境相反，对北冰洋Amundsen Gulf地区海豹的监测发现，其肝脏内汞的含量超过100 μg/g湿重[38]。大批持久性有机污染物（persistent organic pollutants，POPs）也开始在北冰洋生物体内被监测到[39]。意识到北极地区污染控制需要跨边界合作后，主要环北极国家于1991年签署北极环境保护战略（Arctic Environmental Protection Strategy，AEPS），成立北极监测和评估委员会（Arctic Monitoring and Assessment Programme，AMAP），并在1996年成立北极理事会（Arctic Council）。在过去近30年里，AMAP发布了多份有关北极地区环境污染物的评估报告，并为联合国《关于持久性有机污染物的斯德哥尔摩公约》和《关于汞的水俣公约》提供了有力的支持。

4.1 北冰洋海冰环境中的污染物

北极地区与中、低纬度地区的一个主要环境差异是海冰的存在。作为北冰洋大气、海洋和陆地之间的动态界面，海冰中污染物的来源包括海水（包括输入海水的淡水，如河流和冰川融水）、浅海沉积物以及大气沉降[19]。这些污染物包括汞等金属及其化合物、持久性有机污染物、放射性元素，也包括由新兴工业排放的所谓"新型污染物"（emerging contaminants）。在北极地区，微塑料、原油或燃油泄漏正成为广受关注的新型污染物。随着海冰的消融，北极地区油气开采工业以及海上运输业都有快速发展的趋势和潜力。如何减少原油和燃油泄漏，监测冰冻圈原油和燃油的浓度、化学成分、迁移、降解和生物效应，以及清理和排除泄漏至海冰、海水中的原油、燃油是环北极地区、国家甚至全球性的挑战。

当海水冻结成冰时，与海盐类似，海水中的溶解态污染物会富集于海冰表层和卤水中，因此如果这些污染物在卤水中没有进行后续的化学变化（"保守型污染物"），它们在海冰中的分布应与海冰盐度的分布规律类似。例如，水溶解度相对较大的α-HCH在海冰中的浓度与盐度的相关关系远高于易吸附于颗粒物的γ-HCH以及汞[19]。

当海冰在浅海（水深小于50 m左右）地区生成时，向表层升起的冰晶可加速水体中的湍流和混合，这一过程会促使海水悬浮颗粒物嵌入冰中，从而将浅海沉积物中的污染物转移至海冰中，导致海冰中的污染物浓度远高于海水中的浓度[40, 41]。

大气中的污染物也可以通过干湿沉降进入海冰或其上覆雪盖，这一过程尤其影响北冰洋春季大气中汞向海冰中的沉降。如前文所述，在春季汞亏损过程中，大气中的元素汞被溴原子氧化，生成的二价汞会随干湿沉降有效地进入海冰环境中；而冰雪中的二价汞亦可经光化学还原为元素汞而重新释放到大气中。此外，每年夏季，多年冰表层及其上覆雪盖的消融会形成融池（melt ponds），能非常有效地富集大气沉降污染物[42]。

北冰洋的海冰不是一个扣在海水上静止不动的冰盖，其在时间和空间上都在不停地发生动态变化。从时间上，海冰在不同的季节经历生成、增长、和部分（多年冰，multi-year ice）/完全（当年冰，first-year ice）消融；在空间上，除附着在海岸陆地上的陆源固定冰（landfast ice）外，漂浮在海水上的浮冰随表层洋流而漂移。海冰中的污染物也因此在时间和空间上得以传输（图28-3）。随着海冰的形成、增厚和漂移，污染物既在海水—海冰—大气界面发生垂直传输,也在北冰洋不同的地理位置之间水平传输。比如,冬季冻结在喀拉海海冰中的污染物经历大约 2~5 年的漂流后,抵达弗拉姆海峡,随海冰消融后转移至北大西洋的海水中。

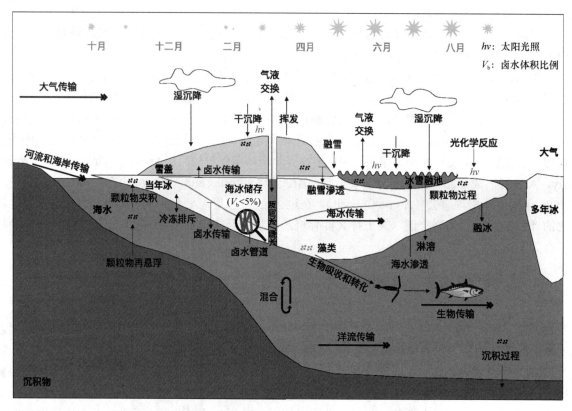

图 28-3　北极地区海洋冰冻圈的主要物理和化学过程示意(引用并修改自参考文献[19])

除了物理传输过程外,海冰中的污染物也会发生化学反应,导致其化学形态、生物有效性和潜在的生物效应发生变化。本章第 3 节所述的化学过程在海冰环境中均有发生(参见图 28-3)。海冰中的很多化学过程受光化学诱导,或者受生物过程影响。

海冰环境中的污染物可以通过多种途径进入海洋食物链。首先,海冰卤水中的微生物、小型浮游生物可以直接摄取卤水中的污染物。春季海冰底部的冰藻以及夏季海冰表层融池的藻类也可以直接吸收卤水及海冰融水中的污染物。此外,夏季海冰的部分或全部消融,也导致海冰中污染物向表层海水中的释放,从而被海洋食物链所吸收和富集。

4.2　北极地区的汞污染

北极监测和评估委员会(AMAP)于 1998 年和 2002 年对北极地区金属污染物汞、铅和镉进行了初次综合评估[39, 43]。结果表明汞是北极地区最值得关注的金属污染物,其在北极地区食物链,尤其是海洋哺乳动物体内的含量很高,而且有上升趋势。由于这些海洋哺乳动物是北极地区原住民(主要是因纽特人)的主要食品来源,因纽特人也成为了世界上受汞污染影响风险最大的人群之一[44]。基于此,汞是北极地区近二十年来研究得最多的污染物之一,发现了很多独特的现象和过程,主要结果总结于 AMAP 2011 年发布的关于北极地区汞的综合评估报告[45]。以下仅对部分结果作一简洁的介绍。

大气中的汞通常分为三种形态:气态零价汞(Hg^0,GEM)、气态氧化态汞(Hg^{2+},GOM)、颗粒态汞（PBM）。大气中的汞主要以 GEM 为主（>95%），GOM 和 PBM 所占比例较少。汞的单质和化合物都有毒性，尤其是有机汞（如甲基汞等）的生物毒性最强。大气中的汞经干湿沉降进入到水体和土壤后，其中的无机汞可以被转化为甲基汞，后者经过食物链生物富集和放大，可以在高营养级生物体内达到较高浓度，进而危害生态环境以及人体健康。汞作为有毒性的全球污染物，一旦从自然源或人为源释放到大气中，会随着气团进行长距离传输，到达北极地区。低纬度地区排放的汞也可以通过海洋、河流和生物传输途径进入北极地区。

大量研究表明，汞在海冰环境中会发生一系列传输和转化过程，影响其在海-气之间的交换以及在海洋生态系统中的富集。Schroeder 等[14]最早在加拿大北极地区 Alert 站点观测到春季时的大气汞亏损事件。后续研究证实汞亏损的发生与溴剧增和臭氧亏损［见反应式（28-1）至式（28-5）］紧密相关，GEM 的氧化是因为大气中卤素自由基（尤其是 Br，BrO）浓度的大量增加（图 28-4）。氧化产物 GOM 和 PBM 经干、湿沉降进入海洋冰冻圈，导致春季冰冻圈、尤其是雪盖和霜花中汞的含量可以高出海水 3~4 个数量级［图 28-5（a）］。雪盖中的氧化汞亦可以发生快速的光还原，生成的 GEM 重新释放到大气，因此只有少量的氧化汞得以留存在冰冻圈中，在夏季冰雪消融期间富集于融池，或淋溶入海水［图 28-5（a）］。汞在北冰洋当年冰和多年冰中的分布和转化（包括甲基化）是近几年的一个研究热点[19, 36, 40]。图 28-5（a）示意汞在北冰洋海水-海冰-大气界面的主要浓度分布[19]。

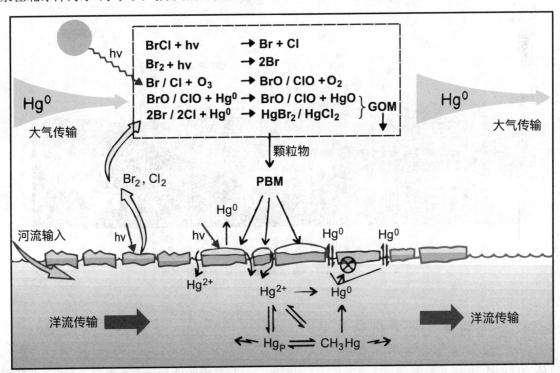

图 28-4　北极地区春季大气汞亏损事件期间汞循环示意图（引用并修改自参考文献[50]）。
GOM：气态氧化态汞；PBM：颗粒态汞

北冰洋海洋哺乳动物体内高含量的汞是甲基汞在海洋食物链中生物富集和放大的结果。然而，它们在时间和空间上的分布显示出极大的差异[45]，其成因尚不完全清楚。早期研究认为这种差异是由于不同海域受到长距离汞传输的不同影响，但越来越多的证据显示这种差异至少在一定程度上是受气候变化的影响，因为气候变化会直接影响汞在大气和海洋之间的交换，在北冰洋的形态分布和甲基化，以及在海洋食物链中的富集和放大[46-49]。

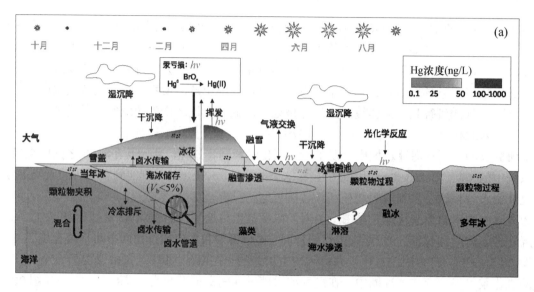

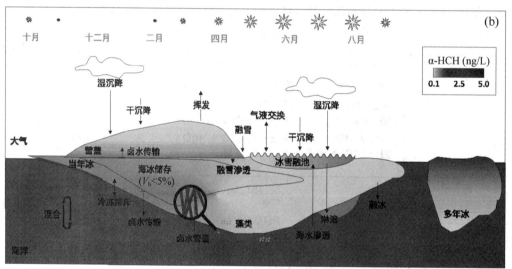

图 28-5 北极地区海洋冰冻圈代表性污染物分布示意图（根据参考文献[19]绘制）

（a）汞；（b）α-HCH。hv：太阳光照；V_b：卤水体积比例

4.3 北极地区的持久性有机污染物

自从 20 世纪 70 年代有监测数据以来，主要 POPs（尤其是挥发性和半挥发性的 POPs）在北极地区的环境和生物体内均有检出[39, 43, 51]。这些 POPs 既包括诸如六氯环己烷（HCH）和多氯联苯（PCB）的"传统"氯代碳氢化合物，也包括新型的溴代和氟代碳氢化合物。它们在高脂肪哺乳动物体内的富集以及对原住民健康的潜在风险在很大程度上促成了联合国《关于持久性有机污染物的斯德哥尔摩公约》的签订、修订和执行。随着全球对 POPs 使用和排放的控制，北极地区海洋和淡水生物体内大部分的 POPs 含量在过去二三十年已呈现明显的下降趋势，其中尤以 α-HCH 的降低趋势最为显著[52]。

与其他污染物类似，POPs 在北冰洋海洋冰冻圈中也呈现一系列复杂的传输和转化过程（参见图 28-3）。以 HCH 为例，其不同异构体因物理化学性质的区别在冰冻圈的来源和分布方面差异很大。水溶性较强的 α-HCH 主要通过洋流和大气沉降进入北冰洋，而 γ-HCH（林丹）则主要由河流和洋流汇入[19]。图 28-5（b）

示意α-HCH 在北冰洋海水-海冰-大气界面的主要浓度分布[19]。对 HCH 在冰冻圈中行为的详细讨论可参见文献[19]。

5 南极地区冰冻圈环境化学

南极地区冰冻圈和北极地区冰冻圈有很多相同的地方，存在着和北极地区相同的人为污染物，比如汞等重金属、POPs、微塑料等随着海水和大气的长距离传输在南极地区都有发现。另外，发生在冰雪表面的传输和转化过程也有相似的地方。但与北极地区不同的是南极地区无冰区的存在，导致南极地区冰冻圈环境化学包括了冰、雪、海水、无冰区、大气和生物的多圈层相互作用。人为排放的 POPs 和金属污染物在南极地区已被多次报道，本节主要介绍多圈层都存在的金属污染物、持久性有机污染物、和对全球气候变化有重要影响的海洋边界层有机气溶胶在南极的分布规律和传输过程。

5.1 南极地区金属污染物

5.1.1 南极大气金属污染物——汞

汞在南极和北极的分布有相同之处也有差异。相同的是南极地区和北极地区的沿海区域春季都会发生大气汞沉降事件，并且夏季随着海冰融化，海水和冰下的汞会重新释放到大气中。不同的是由于地理位置的差异，北极的汞受到北半球中低纬度地区汞排放的影响较大（尤其在冬季）；而南极沿海地区由于受下降风的影响，汞浓度则会受到南极东部冰盖上反应产生的大气汞的显著影响[53]。

对南极地区对流层大气气态汞的研究最早始于 20 世纪 80 年代。早期的研究结果发现在南极海岸附近气态汞浓度非常低（0.23 ng/m³）[54]，然而后续的研究显示大气气态汞的浓度范围在 0.5～1.0 ng/m³ [55]，且不同形态汞的浓度分布在时间和空间上均有较大差异。GOM 在南极点的浓度异常高（0.1～1 ng/m³），占气态总汞（TGM）浓度的 40%以上[56]。GEM 在各站点的分布也有差异，在 Troll 站点，GEM 在冬季的浓度 [（0.98 ± 0.06）ng/m³] 高于夏季 [（0.89 ± 0.29）ng/m³]；在 Dumont d'Urville 站点，GEM 浓度则是夏季 [（0.88 ± 0.32）ng/m³] 略高于冬季 [（0.84 ± 0.11）ng/m³]；在 Dome C 站点发现了出乎预料的高浓度 GEM（0.2～2.3 ng/m³，平均值为 0.85 ng/m³），最大值出现在秋季 [（1.45 ± 0.27）ng/m³]，最小值出现在夏季 [（0.78 ± 0.46）ng/m³]，并且有明显的动态循环过程[53, 55]。研究者对南极沿岸到内陆断面气态总汞（TGM）的分布特征进行了调查，发现内陆地区 TGM（海拔>3000 m）呈现出较高值，昆仑站 TGM 浓度呈现出显著的昼夜差异[57]。南大洋西南海域大气中 GEM 和海水中溶解性气态汞（DGM）的分布在时间和空间上也存在着很大的差异[58]。这些汞形态分布特征的差异可能受到多种因素的影响，比如雪中气态汞的再释放，边界层 OH、O_3、NO_x、RO_2 等自由基含量、氧化损耗以及混合层的对流过程等。

在北极地区发生的大气汞亏损事件，在南极地区的春季也有发生[53, 59]。除了在南极大陆观测到了汞亏损事件，通过对南大洋大气 TGM 和 GEM 以及表层海水中 DGM 的研究发现，南大洋夏季也可以发生汞亏损事件[60, 61]。研究指出，Br 可能在汞亏损事件中发挥重要作用[62]。

南极地区的汞除了来源于大气的远距离传输，海洋本身也可以作为汞的释放源。南大洋海冰下面的海水中有大量的气态汞富集。研究发现，夏季海冰融化将导致更多的汞进入海水并被重新释放到大气中。另外，南极内陆的夏季，雪面汞还存在释放再氧化沉降的动态循环过程，但目前机制尚不明确[55, 56]。极地地区海冰和上覆积雪中汞浓度一般很高，会阻断汞的海气交换过程。当海冰和上覆积雪融化，一方面其自身所含汞会释放出来，另一方面其阻断作用就会消失，导致大气汞浓度上升。而生物活动，比如动物粪便的排放也会导致土壤汞的氧化/沉积/再排放的局部动态循环。

5.1.2 南极无冰区金属污染物

环南极洲无冰区的出现是由气候变暖和冰川后退引起的，虽然无冰区的面积只占南极洲总面积的2%左右，但却是冰、水、岩、土、气和生物相互作用最为活跃的场所。无冰区沉积物中的化学组分可以记录和反映历史时期人类活动对气候、环境和生物等的影响。通过对无冰区粪土沉积物的分析，可以恢复过去千年来南极无冰区污染物含量的变化，比如 Pb、Hg 等[63]。由于远离人类工业区，规避了局地人为排放的影响，因此这些污染物含量的变化能够很好地反映尺度范围内污染物的排放规律。无冰区污染物主要来自于人类活动、大气环流和生物传输富集作用。在 James Ross 岛的地衣中发现了浓度非常高的元素汞和甲基汞，这可能与大气环流有密切关系[64]。另外，有研究者发现生物（如企鹅）对南极地区的金属元素有富集放大作用[65-67]。随南极科考活动的日益增多，人类活动也可能导致南极无冰区的金属污染。但在某些特定区域［如阿德莱德岛（Adelaide Island）］，通过食物链的富集放大作用，企鹅活动对南极环境中金属（铜、锌、铅等）的影响已经超过了人类活动的影响[68]。因此，也需要关注南极动物活动对南极环境质量造成的影响。

5.2 南极地区持久性有机污染物

和北极地区一样，南极地区也发现了POPs的存在。借助"雪龙号"科考平台和南极的科考站点，中国科技工作者开展了南极周边海域大气、海水中POPs的观测。观测发现，不仅传统的POPs，例如有机氯农药、多氯联苯等会传输到南极地区，一些新型POPs也会传输到南极。在南大洋及南极周边海域大气颗粒中检测到了有机磷酸酯类阻燃剂，据推测这种传输不仅仅和大气相关，也和洋流输送相关[69]。除了来源于大气的长距离传输，在南极长城站及邻近的乔治王岛的研究发现土壤、冰川的再释放以及科考活动（如科考站）也是南极大气中POPs不容忽视的来源[70-72]。另外，国外研究者通过对海鸟聚集区的采样观测证明了生物迁移也会向南极环境输送POPs[73]。

由于温度较低，南极环境介质中的POPs半衰期往往会延长，残留时间及相应的环境风险也随之增加。虽然南极生态系统较为简单，食物链较短，但是目前有充足的证据表明，脂溶性较强的POPs，如多氯联苯、DDT 等，会在南极水生生态系统中随着食物链产生生物富集和放大。

5.3 南极海洋边界层有机气溶胶

大气气溶胶是指悬浮在大气中的固态和（或）液态微粒与大气共同组成的多相体系[74]。由于气溶胶颗粒对太阳辐射具有直接的吸收和散射作用，还可以作为云凝结核（CCN）影响成云和降雨，从而改变地表的辐射收支平衡，具有明显的气候效应。南、北极地区作为全球的冷源，局地的气候变化对全球大气环流和气候演变有重要的作用。另外，气溶胶颗粒可以作为污染物的载体，随着大气环流进行远距离传输，对生态环境造成影响。伴随着全球变暖，海冰范围急剧减小，厚度急剧减薄，海气界面发生强烈的物质交换过程，大气成分受到影响，从而影响气候和生态系统。

有机气溶胶是大气颗粒物的主要组分之一（约占其总质量的20%～90%），其来源、老化过程及归趋研究是当今环境化学、地球科学领域的前沿研究课题。有机气溶胶可以分为自然以及人为直接排放的一次有机气溶胶（POA）和在大气中经过一系列反应生成的二次有机气溶胶（SOA）。海洋有机气溶胶的化学组成极其复杂，其中水溶性成分由于其吸湿能力强，对CCN的贡献显著，是海洋有机气溶胶研究中的重点。海水飞沫产生的 POA 是海洋有机气溶胶的主要部分,主要为海盐气溶胶,可以影响云的微物理性质,参与一系列的化学转换过程。另外,研究发现生物成因的 SOA 在远洋对气候有显著的影响。

生物质燃烧是 POA 的最主要来源之一。生物质燃烧气溶胶主要分布在亚微米尺度,在全球尺度上,生物质燃烧气溶胶的净辐射强迫(直接效应和间接效应总和)与硫酸盐气溶胶相当。虽然南极大陆远离人类活动,但已有不少研究在南极地区也发现了生物质燃烧标志物左旋葡萄糖(图 28-6)。南半球海洋上空左旋葡萄糖的含量与北半球相当,说明生物质燃烧对南极地区也有影响,且是南极内陆冰雪有机碳的一个潜在来源,还会影响南极海洋上空气溶胶中的汞[75]。海洋 SOA 的前体物质包括 DMS 和异戊二烯等挥发性有机气体。海洋浮游植物可以排放大量的 DMS,由其氧化生成的甲磺酸(MSA)和非海盐硫酸盐($nss\text{-}SO_4^{2-}$)是远洋地区边界层含硫气溶胶的主要成分,对气溶胶颗粒数量浓度的贡献可以高达 90%[76]。南大洋特别是南极海冰区,被认为是大气 DMS 巨大的源,对全球硫的收支有着重要的影响。对南极含硫气溶胶浓度的主要影响因素有两个方面:一是浮游植物活动,二是大气和海冰物理过程。在南极海冰中,浮游植物量较高时,DMS 浓度也较高。研究发现,南半球春夏季期间,冰间湖海冰动态变化引发浮游植物大量繁殖,从而间接影响含硫气溶胶的浓度[77]。远洋大气中异戊二烯主要来自浮游植物排放和海洋微表层的光化学反应,南极地区由异戊二烯氧化产生的 SOA 含量通常较低,但在藻类爆发时异戊二烯来源的 SOA 浓度会显著上升并能改变 CCN 的物理化学性质[75, 78]。

另外,羧酸和含氮有机物也是海洋有机气溶胶的重要成分,对水溶性有机碳(WSOC)有重要的贡献。在南极沿岸的昭和站、西北太平洋、北大西洋、印度洋、南大洋都检出了显著含量的羧酸。其中草酸是海洋边界层含量最高的羧酸。含氮有机物主要包括胺类和氨基酸以及一些难降解的化合物(如类腐殖质)。但是氨基酸对于海洋气溶胶 WSOC 的贡献并不高,仅为 2%左右[79]。而烷基铵盐,如甲基铵(MMA^+)、二甲基铵(DMA^+)、三甲基铵(TMA^+)等,则是海洋有机气溶胶重要成分,在亚微米尺度对 WSOC 的贡献仅次于 MSA,并且是海洋气溶胶水溶性有机氮(WSON)的主要存在形式(35%)[80]。这些烷基铵盐由海洋生物排放的前导气体在大气中与酸反应产生。此外,海洋有机气溶胶还包括含碘化合物。在南大洋和南极沿岸海域都有含碘的无机和有机气溶胶检出,其前导气体可能为碘代烃(如 CH_2I_2)。

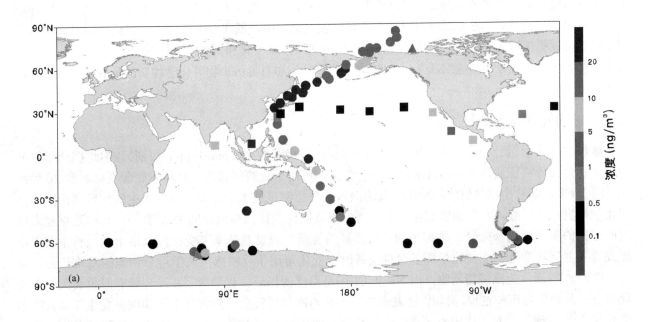

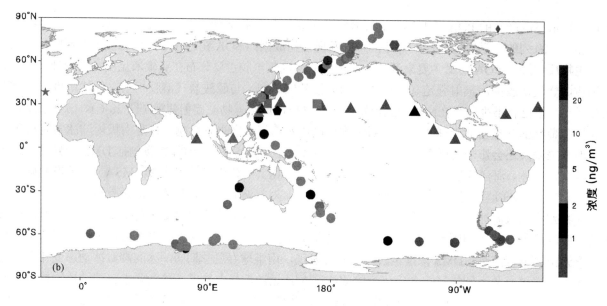

图 28-6 异戊二烯 SOA 标志物[75]（a）和左旋葡萄糖（b）[81]空间分布图

6 冰川、冻土和高山地区冰冻圈环境化学

6.1 冰川环境化学

冰川中的化学组分来自不同时期的大气干湿沉降，是反映大气环境变化的天然档案库。雪冰化学记录研究可以为气候变化、人类活动、生物地球化学循环、地质和宇宙事件等在不同空间和时间尺度上提供直接或间接证据。两极冰盖和山地冰川雪冰化学研究有两方面基本内容，一是通过雪冰化学反映环境状况和过程，二是利用冰芯记录恢复重建古气候环境变化。中、低纬度山地冰川较极地冰川具有较高的积累量且位置更接近人类活动区，在记录和反映区域大气环境变化和人类活动影响方面更具优势。另外，近几十年来全球山地冰川加速缩减[6]，由此释放的营养盐类和有毒污染物等化学物质的区域环境效应也逐渐成为研究热点。

6.1.1 冰川化学组分特征

冰川中的化学组分种类繁多，在雪冰中的含量一般极其微小（ppm～ppq 级别），但这些化学组分的含量和分布等特征通常具有显著的环境指示意义，可用于反映区域大气环境本底和变化。自 20 世纪 60 年代以来，随着全球气候环境变化研究的深入和分析测试技术的进步，极地和中低纬度高山区的冰川化学研究发展迅猛。早期的研究建立了雪冰中氢氧同位素比例与气温的关系[82, 83]，是冰芯重建温度时间序列的重要理论基础[84]。雪冰中的不溶微粒的含量、粒度及其季节变化可以指示大气粉尘和火山爆发等重要环境变迁。主要化学离子含量及其时空变化可用于反映冰川化学组分物源的基本信息，如 Ca^{2+} 和 Mg^{2+} 等通常为陆源指示物，而 Na^+ 和 Cl^- 则是海洋源指示。化学离子在冰川赋存时也会因气温变化引发一系列的后沉积过程，例如雪冰融水的下渗和再冻结过程会导致雪层中化学成分发生迁移，即"淋溶作用"[85]，理解其对雪冰中不同离子含量和分布的影响是深入揭示离子环境意义的基础[86-88]。高山冰川中金属元素的含量和分布除与自然源的沙尘相关,还可以反映人为源排放进入大气的金属的输送和沉降过程。早期的高山冰川金属研究集中在采矿和冶金等人类工业活动密切相关的 Cu、Zn、Pb、Cr、Ni 等[89, 90],近期对毒性强且环境行为复杂的 Hg 较为关注[91-93]。汞在雪气界面和雪冰中的迁移转化活跃,如前所述的极地"大气汞亏损"现象引发了科学界对汞在雪冰中行为和归趋的兴趣。高山地区雪冰中的汞通常与颗粒物相关[94],反映了高山地区高粉尘沉降量的特征,可能指示了大气 Hg 干沉降的重要性及颗粒态汞在冰川中的稳定赋存状态。高山冰川中有机污染物的检出是人类活动影响全球偏远地区的重要标志,有机物的检出种类和含量与冰川海拔高度及其与人类排放源区的距离等有关。一些持久性有机污染物如 PCBs 和 DDT 等在珠峰地区极高海拔冰川中有检出报道[95, 96],但受限于高海拔采样条件和实验测试种类及分析手段差异,数据可比性和精度还有待提高[97]。近期,冰川表面因其具有吸光能力的杂质(特别是黑碳等)可降低冰川表面反照率进而加速冰川消融,成为高山冰川化学研究的焦点[98, 99]。全球范围内山地冰川的黑碳含量在几十到几百 ng/g[2, 100],高于极地 1~2 个量级。冰川中黑碳含量和来源具有显著区域差异,如利用黑碳的 ^{14}C 同位素分析青藏高原大气和冰川中黑碳的化石燃料和生物质燃料相对贡献,发现高原南部冰川中一半以上的黑碳来源于生物质燃烧,与南亚排放相关,而高原北部冰川中黑碳的化石燃料贡献较高[101]。冰川表层雪冰中的矿物粉尘和污染物在物理、化学和生物过程的共同作用下聚集形成"冰尘",具有显著的污染物"聚集"效应,例如在瑞士境内阿尔卑斯山区冰川的冰尘样品分析显示,部分金属 Zn、Sb、Se 和 Hg 在冰尘中的含量显著高于周边冰碛物,甚至比雪冰中的金属含量高 5 个数量级[102]。近年来的强烈增温造成全球很多山地冰川的消融区扩大,冰尘的环境化学特征及其气候环境效应值得深入探讨。

6.1.2 冰芯历史环境记录

冰芯是钻取自冰川内部的连续芯柱,分析其中的物理化学等指标并结合冰芯定年,可以重建过去气候环境变化。高山地区的冰芯历史环境记录特别是工业革命以来的序列为评估人类活动对区域环境的影响提供了重要依据。自 20 世纪末开始,阿尔卑斯山区的冰芯记录陆续报道了数百年来 Pb、Zn、Cu、Cd、Ag、Au 等多种金属元素的历史变化,与区域人类矿业冶金和工业生产活动排放的历史变化趋势相关[90]。其他山地冰川冰芯记录也有类似报道,南美 Quelccaya 冰帽冰芯恢复的过去 1200 年金属 Pb、Zn、Cu、Cd、Ag 的含量及其富集因子历史变化显示,始自 14 世纪中叶,殖民地时期的采矿业和冶金活动就对安第斯山区的大气环境产生了显著干扰[103],在 Illimani Glacier 钻取的冰芯研究则表明 20 世纪燃油的 Pb 排放对大气的破坏程度超过以往任何时期冶金活动的 Pb 排放[104],这为不同时期人类活动对大气环境影响的比对提供了依据。亚洲青藏高原为主体的高山地区冰芯金属历史记录也有类似报道,但与欧洲和南美不同,冰芯记录的金属含量一般并未在 20 世纪前表现为本底变化,而在 20 世纪出现升高,且一直持续到 20 世纪末,这显示亚洲地区工农业的发展进程[92, 105-107]。山地冰川冰芯重建的有机污染物和黑碳等历史记录也为认识区域人为活动排放提供了视角[108-111]。当前,山地冰芯环境记录的研究较前期相比发展放缓,原因包括:冰川消融强烈,高质量冰芯获取更为困难[112, 113];山地冰川较高的粉尘通量限制了连续分样测试技术在高山冰芯化学组分测试中的应用,难以获取更高分辨率的记录;山地冰芯的定年准确度仍需检验和提高,例如 Schuster 等[114]在 2002 年发表了北美 Upper Fremont Glacier 冰芯过去 270 年来汞含量历史变化,被广泛用于评估历史时期人为汞排放对区域环境的影响。但近期的研究利用树轮记录和多参数定年约束修正了冰芯的定年序列,更新了对人为源汞影响大气环境的历史序列[115]。最近也有研究质疑青藏高原内陆古里雅冰帽冰芯的年代范围[116],另外一些报道的冰芯序列近百年序列的精度和准确度也值得再评估。当前,进一步开展冰川雪-气和雪冰内部化学组分累积和变化的现代过程研究、精确解读冰芯中化学指标的环境意义、开发适用于山地冰芯的高分辨测试分析技术、综合利用多参数指标提高冰芯定年的准确性,是未来高山冰芯历史环境记录研究实现突破和发展不可或缺的环节。

6.1.3 冰川消融的区域环境影响

冰川可以作为大气沉降污染物的临时储库，沉降到消融区以及季节性积雪区的部分大气污染物会在当年气温回升时随雪冰消融而释放。在气候变暖背景下，冰川消融区扩大，前期沉降并保存在雪冰中的历史污染物也会释放[49]。在冰川加速消减背景下，冰川消融致微量物质输出（特别是气候和环境意义较强的物质元素如碳、营养离子和持久性污染物等）对大气和被补给陆表和海洋生态系统的影响，成为冰冻圈变化影响的关注点。研究指出，南极和格陵兰岛冰盖融水是近海水域中生物可利用性较高的有机碳[117]和微量元素铁[118]等营养物质的重要来源[119]，其贡献甚至与传统认为处于主导地位的大气沉降相当。同时有研究强调，一些"封存"在雪冰中的污染物会"重新活化"（remobilization）而输出，例如 Ma 等[120]指出，气候变化背景下，北极地区历史时期积累的持久性有机污染物和一些挥发性物质会再次释放进入大气带来新的环境污染。在山地地区，Blais 等[121, 122]在加拿大北部通过一系列研究指出，较其他非冰川融水补给水体，冰川补给湖泊中沉积物和生物体的有机氯农药负荷较高，且主要来自冰川融水输入。随后，Bogdal 等[123, 124]在阿尔卑斯山区通过一系列研究，发现在大气污染物含量近十几年持续下降的背景下，一些高山湖泊表层沉积物的污染物含量却出现上升。通过对比研究和模拟，认为近期冰川加速消融导致污染物输出正成为高山湖泊和其他下游生态系统中污染物的一个重要来源，这一现象陆续在全球其他高山地区被揭示[125-127]。近期，对冰川消融致污染物输出的研究朝向精细化过程描述和定量计算通量方向[128-130]发展，例如 Sun 等[131]对青藏高原内陆典型冰川作用区和补给区水体中汞的形态分布开展了系统观测研究，指出冰川消融直接释放汞的环境风险较低，但运移至下游湿地后甲基化程度升高，对补给区的环境影响风险增加。作者以此建立了高山冰川影响区汞的分布和环境风险分区模型（图 28-7）。山地冰川接近人类生产生活区，其融水径流是一些地区动植物和人类赖以生存的重要水资源，而其消融盛期往往适逢生物活动旺盛期和敏感期[132]。山地冰川较极地冰川对气候更为敏感，并在最近几十年快速萎缩[6]。在山地冰川加速缩减背景下，研究冰川消融输出污染物及其对山区生态环境的影响具有重要实际意义和科学价值，其正成为高山地区冰冻圈环境化学的研究热点。

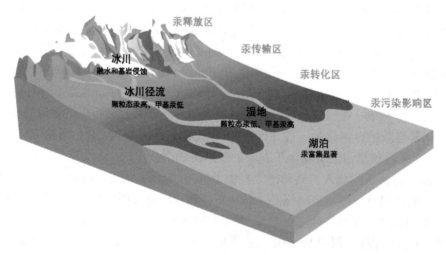

图 28-7 高山冰川影响区汞的分布和环境风险分区模型图（引用并修改自参考文献[131]）

6.2 冻土环境化学

冻土是冰和各种岩石或土等组成的多成分和多相体物质，其中水的状态是影响冻土化学过程的重要因素。冻土并非天然存在并一成不变的，在冻融循环过程及冻结状态会发生一系列化学反应，如溶解反应、水化反应、替代反应、氧化还原反应和离子交换等。低温环境下冻土中的溶解性物质和水分子间的化学反应和产物非常关键[2]。冻结状态的冻土几乎不存在液态水，极大限制了其中的化学反应和物质迁移转化。南极 Jubany 观测站附近土壤剖面中的多环芳烃（PAHs）最高值出现在活动层和多年冻结层界面略下方，随着深度增加浓度逐渐降低，表明冻土层的低渗透性对污染物向下迁移起到"屏障作用"[133]。基于北极地区冻土层总汞含量调查及其与总碳的关系，估算得出北半球冻土中的汞储量为（1656±962）Gg，是一个巨大的汞库[134]。高山地区冻土中化学组分的研究目前还集中在碳、氮等元素的分布、储量和循环过程，目前仅有少量针对环境污染物的研究，如 Sun 等[135]调查了青藏高原东北部典型冻土发育地区土壤垂直剖面中的总汞含量，发现总汞在近地表层位最高，主要与土壤中的有机碳的分布趋势相关，土壤总汞含量与海拔高度呈现负相关趋势，主要是由于高海拔地区土壤以粉质土为主要组分，有机质含量偏低。

气温升高冻土退化可导致土壤中的污染物发生迁移，可进入多年冻土深层、大气或排入地表水甚至地下水。Ci 等[136]在青藏高原内陆 4700 m 海拔典型多年冻土区对土壤剖面中的气态单质汞观测表明，土壤不同层位中的气态单质汞与土壤温度呈显著的指数正相关关系，其中表层 5 cm 土壤中气态单质汞含量对土壤温度上升的响应最为敏感，冻土土壤是显著的大气汞源。气候变暖导致土壤升温与冻土退化，可能增强冻土土壤向大气的汞释放。另外，输油管道工程在冻土区的季节性冻融作用下可能发生泄油，进而对土壤造成污染[137]，目前也有相关研究从污染防控角度探讨石油烃类衍生物在冻土中的迁移及利用微生物治理污染土壤的可行性[137, 138]。

6.3 其他冰冻圈要素环境化学

高山地区积雪是冰冻圈的重要组成部分，与冰川相比，积雪的覆盖区域广且存在季节性，其中的化学组分在春季消融期可以短期内进入大气或下垫面环境。阿尔卑斯山区的研究表明，积雪中的总汞和甲基汞的含量自冬季开始到春末随着雪温的上升而增高，这可能是大气沉降汞在积雪中的持续累积[139]，春末积雪消融是流域水体中污染物如 POPs 的主要来源[140]。高山地区河、湖冰分布和基本物理化学性质的观测很少，且较海冰对区域气候和环境的影响相比作用较小，因此相关的环境化学研究还非常匮乏。青藏高原青海湖和纳木错湖泊沉积物的汞同位素序列研究表明，20 世纪以来 $\Delta^{199}Hg$ 趋向升高，可能是由于气候变暖湖泊封冻期减少，导致 Hg^{2+} 的光化学反应增加[141]，这说明河湖冰物候变化可改变水体自然环境影响水体环境化学。

6.4 青藏高原冰冻圈环境化学

高山地区冰冻圈通常远离直接人为污染源，污染物的传输主要通过大气传输。污染物可以在中、高对流层随大气环流进行水平向的传输。此外，挥发性或半挥发性污染物也可以通过类似于从低、中纬度地区到极地地区的"蚱蜢效应"而从山脚沿山脊传输到山顶。本节以青藏高原为例，简介高山地区冰冻圈环境化学研究趋势。

以青藏高原为主体及周边山系构成的"第三极"地区是全球中、低纬度冰冻圈最发育的地区。这一地区冰冻圈环境化学源于 20 世纪 60 年代开始对冰川雪冰化学组分的测试分析，早期的研究多集中于探讨雪冰离子含量和分布特征[142,143]；20 世纪末到 21 世纪初，随着野外考察的深入和测试技术的进步，雪冰中痕量金属和放射性元素的报道和研究增多[144-148]；21 世纪以来，环境危害大的持久性有毒污染物（如汞和 POPs 等）及气候意义和环境指示意义显著的黑碳和有机物等成为研究关注点[149-152]。

青藏高原是多圈层体现最全且相互作用最为强烈的地区之一，而冰冻圈又是其中变化非常剧烈和敏感的圈层。多圈层相互作用背景下的污染物赋存和迁移转化及其生态环境影响正成为这一地区冰冻圈环境化学研究的前沿。Kang 等[153]对第三极地区大气、冰川和积雪中黑碳等污染物的赋存和环境效应做了详尽综述,认为应加强大气和雪冰协同观测,研究雪冰中污染物的来源及其对气候和冰冻圈变化的影响。Zhang 等[154]系统总结了高原冰川中汞的形态含量及其在冰川径流中的传输和对下游环境的潜在影响,指出冰川消融是促进区域汞循环的活跃因子。冰川消融和冻土退化对区域水体环境的影响也越来越多地被揭示[126, 155, 156]。当前,高原冰冻圈环境化学研究正经历从单一指标到多指标、从散点报道到多站点空间差异研究、从简单测试结果描述向过程机制揭示的转变。高原地区整体气候背景趋向暖湿化[157],高海拔地区的降水形态可能从固态转向液态,而降雪和降水对大气中污染物的清除效率有所差异;高原冰冻圈主要组成要素和分布特征与极地和其他山区存在区别,例如高原地区冰川的海拔高度和跨度大,冰川消融和冰川水文的未来情景与其他高山地区不同。因此,应在考虑区域气候要素变化以及区域冰冻圈特点和变化趋势的基础上开展冰冻圈环境化学研究。此外,虽然高原作为"第三极"与南北极相距甚远,但其环境化学研究的基本技术手段、关注对象和基本规律有相似之处,如 Wang 等[158]对"三极"地区持久性有机污染物的研究现状做了归纳比对和总结。未来基于"三极"冰冻圈环境化学的比对和集成研究也将成为新趋势。

7 冰冻圈环境化学的全球效应和展望

虽然冰冻圈在时间和空间的分布具有一定季节性和地域性差异,但冰冻圈的环境地球化学过程对污染物在区域和全球尺度上均具有重要的影响。从化学机理上,南极地区平流层春季的臭氧空洞、极地地区海洋边界层春季的溴剧增、臭氧和汞亏损现象都是环境化学重要的发现。从污染物传输上,由低纬度到高纬度和由山脚到山顶的"蚱蜢效应",是区域或全球污染物源汇和质量平衡模拟的重要过程。从污染历史和污染源示踪上,冰川提供了为数不多的高分辨率历史环境记录。随着全球变暖,储存在冰冻圈的化学物随冰雪消融而重新释放,成为新的污染源。由于冰冻圈对全球气候环境变化有着关键性响应和反馈,对冰冻圈环境化学的研究也有助于揭示气候变化与环境质量的关系。

需要强调的是,我们目前对于大部分冰冻圈环境化学的研究尚处于监测和数据积累阶段,而对于寒冷温度下的化学反应机理、传输过程以其与微生物和生物的相互作用了解甚微,很多全球环境模型完全忽略冰冻圈过程,或者对其极度简化。此外,随着北冰洋地区海冰的消融,油气开采和海上运输活动将更趋频繁,油气泄漏、微塑料和其他新型污染物对北冰洋脆弱生态系统和原住民的影响也会更趋显著。因此,我们希望有更多的环境化学工作者加入冰冻圈环境化学研究,为预测在全球变化与人类活动相互作用日益增强的背景下污染物的环境行为提供基础。

参 考 文 献

[1] IPCC. Climate Change 2013: The Physical Science Basis. Contribution of Working Group I to the Fifth Assessment Report of the Intergovernmental Panel on Climate Change.Cambridge, United Kingdom and New York, NY, USA, 2013, 1535.

[2] 秦大河, 姚檀栋, 丁永建, 等. 冰冻圈科学概论. 北京: 科学出版社,2017.

[3] Qin D H, Ding Y J, Xiao C D, et al. Cryospheric Science: research framework and disciplinary system. Natl. Sci. Rev, 2018, 5, 255-268.

[4] Lindsay R, Schweiger A. Arctic sea ice thickness loss determined using subsurface, aircraft, and satellite observations. Cryosphere,2015, 9: 269-283.

[5] Zhang L, Delworth T L, Cooke w, et al. Natural variability of Southern Ocean convection as a driver of observed climate trends. Nature Climate Change, 2019, 9: 59-65.

[6] Zemp M, Frey H, Gartner R I, et al. Historically unprecedented global glacier decline in the early 21st century. J. Glaciol, 2015, 61: 745-762.

[7] Trusel L D, Das S B, Osman M B, et al. Nonlinear rise in Greenland runoff in response to post-industrial Arctic warming. Nature, 2018, 564: 104-108.

[8] Hanna E, Navarro F J, Pattyn F, et al. Ice-sheet mass balance and climate change. Nature, 2013, 498: 51-59.

[9] Yao T, Thompson L, Yang W, et al. Different glacier status with atmospheric circulations in Tibetan Plateau and surroundings. Nature Clim. Change 2012, Doi: 10.1038/NCLIMATE1580.

[10] Molina M J, Tso T L, Molina L T, et al. Antarctic stratospheric chemistry of chlorine nitrate, Hydrogen chloride, and ice: Release of active chlorine. Science, 1987, 238: 1253-1257.

[11] Oltmans S J, Komhyr W D. Surface ozone distributions and variations from 1973~1984 measurements at the NOAA geophysical monitoring for climatic-change base-line observatories. J. Geophys. Res. Atmos, 1986, 91: 5229-5236.

[12] Bottenheim J W, Gallant A G, Brice K A. Measurements of NO_y species and O_3 at 82°N latitude. Geophys. Res. Lett, 1986, 13: 113-116.

[13] Barrie L A, Bottenheim J W, Schnell R C, et al. Ozone destruction and photochemical reactions at polar sunrise in the lower Arctic atmopshere. Nature, 1988, 334: 138-141.

[14] Schroeder W H, Anlauf K G, Barrie L A, et al. Arctic springtime depletion of mercury. Nature, 1998, 394: 331-332.

[15] Kreher K, Johnston P V, Wood S W, et al. Ground-basedmeasurements of tropospheric and stratospheric BrO at Arrival Heights (78°S), Antarctica. Geophys. Res. Lett., 1997, 24: 3021-3024.

[16] Barrie L, Platt U. Arctic tropospheric chemistry: An overview. Tellus B, 1997, 49: 450-454.

[17] Frieß U, Hollwedel J, Konig L G, et al. Dynamics and chemistry of tropospheric bromine explosion events in the Antarctic coastal region. J. Geophys. Res, 2004, 109, D06305, doi:10.1029/2003JD004133.

[18] Simpson W R, von Glasow R, Riedel K, et al. Halogens and their role in polar boundary-layer ozone depletion. Atmos. Chem. Phys, 2007, 7: 4375-4418.

[19] Wang F, Pućko M, Stern G. Transport and transformation of contaminants in sea ice. // Thomas D N, Ed. Sea Ice. 3rd ed. Oxford, UK: Wiley-Blackwell, 2017: 472-491.

[20] Bartels R T, Jacobi H W, Kahan T F, et al. A review of air-ice chemical and physical interactions (AICI): liquids, quasi-liquids, and solids in snow. Atmos. Chem. Phys, 2014, 14: 1587-1633.

[21] Pratt K A, Custard K D, Shepson P B, et al. Photochemical production of molecular bromine in Arctic surface snowpacks. Nature Geoscience, 2013, 6: 351-356.

[22] Xu W, Tenuta M, Wang F. Bromide and chloride distribution across the snow-sea ice-ocean interface: A comparative study between an Arctic coastal marine site and an experimental sea ice mesocosm. J. Geophys. Res. Oceans, 2016, doi: 10.1002/2015JC011409.

[23] Moore C W, Obrist D, Steffen A, et al. Convective forcing of mercury and ozone in the Arctic boundary layer induced by leads in sea ice. Nature, 2014, 506: 81-84.

[24] Grannas A M, Bogdal C, Hageman K J, et al. The role of the global cryosphere in the fate of organic contaminants. Atmos. Chem. Phys, 2013, 13: 3271-3305.

[25] Kahan T F, Wren S N, Donaldson D J. A pinch of salt Is all it takes: Chemistry at the frozen water surface. Acc. Chem. Res, 2014, 47: 1587-1594.

[26] O'Concubhair R, Sodeau J R. The Effect of Freezing on Reactions with Environmental Impact. Acc. Chem. Res. 2013, 46:

2716-2724.

[27] Assur A. Composition of sea ice and its tensile strength, U.S. Army Snow Ice and Permafrost Research Establishment: Wilmette, IL, USA, 1960.

[28] Dieckmann G S, Nehrke G, Papadimitriou S, et al. Calcium carbonate as ikaite crystals in Antarctic sea ice. Geophys. Res. Lett, 2008, 35, L08501. doi: 10.1029/2008GL033540.

[29] Rysgaard S, Glud R N, Lennert K, et al. Ikaite crystals in melting sea ice - implications for pCO_2 and pH levels in Arctic surface waters. Cryosphere, 2012, 6: 901-908.

[30] Rysgaard S, Sogaard D H, Cooper M, et al. Ikaite crystal distribution in winter sea ice and implications for CO_2 system dynamics. Cryosphere, 2013, 7, 707-718.

[31] Hare A A, Wang F, Barber D, et al. pH evolution in sea ice grown at an outdoor experimental facility. Mar. Chem, 2013, 154: 46-54.

[32] Tsurikov V L. The formation and composition of the gas content of sea ice. J. Glaciol. 1979, 22: 67-81.

[33] Bailey N, Papakyriakoua T N, Bartels C, Wang F Henry's Law constant for CO_2 in aqueous sodium chloride solutions at 1 atm and sub-zero (Celsius) temperatures. Mar. Chem, 2018, 207: 26-32.

[34] Rysgaard S, Glud R N. Anaerobic N_2 production in Arctic sea ice. Limnol. Oceanogr, 2004, 49: 86-94.

[35] Thomas D N, Dieckmann G S. Sea Ice 2nd Ed. Chichester, UK: Wiley-Blackwell, 2010, 621.

[36] Beattie S A, Armstrong D, Chaulk A, et al. Total and methylated mercury in Arctic multiyear sea ice. Environ. Sci. Technol, 2014, 48: 5575-5582.

[37] Wania F, Mackay D. Tracking the distribution of persistent organic pollutants. Environ. Sci. Technol, 1996, 30: 390A-396A.

[38] Smith T G, Armstrong F A J. Mercury in seals, terrestrial carnivores and principal food items of the Inuit from Holman, N.W.T. J. Fish. Res. Bd Can, 1975, 32: 795-801.

[39] AMAP AMAP Assessment Report: Arctic Pollution Issues; Arctic Monitoring and Assessment Programme: Oslo, Norway, 1998, 859.

[40] Chaulk A, Stern G A, Armstrong D, et al. Mercury distribution and transport across the ocean-sea-ice- atmosphere interface in the Arctic Ocean. Environ. Sci. Technol, 2011, 45: 1866-1872.

[41] Pfirman S, Eicken H, Bauch D, et al. The potential transport of pollutants by Arctic sea ice. Sci. Total Environ, 1995, 159: 129-146.

[42] Pućko M, Stern G A, Barber D G, et al. Mechanisms and implications of alpha-HCH enrichment in melt pond water on Arctic sea ice. Environ. Sci. Technol, 2012, 46: 11862-11869.

[43] AMAP, Arctic Pollution 2002: Persistent Organic Pollutants, Heavy Metals, Radioactivity, Human Health, Changing Pathways. In Arctic Monitoring and Assessment Programme: Oslo, 2002, 112.

[44] Basu N, Horvat M, Evers D C, et al. A state-of-the-science review of mercury biomarkers in human populations worldwide between 2000 and 2018. Environ. Health Persp, 2018, 126, doi.:10.1289/EHP3904

[45] AMAP AMAP Assessment 2011: Mercury in the Arctic, Arctic Monitoring and Assessment Program: Oslo, Norway, 2011, 193.

[46] Wang F, Macdonald R W, Stern G A, et al. When noise becomes the signal: Chemical contamination of aquatic ecosystems under a changing climate. Mar. Pollut. Bull, 2010, 60: 1633-1635.

[47] Wang F, Outridge P M, Feng X B, et al. How closely do mercury trends in fish and other aquatic wildlife track those in the atmosphere? - Implications for evaluating the effectiveness of the Minamata Convention. Sci. Total Environ, 2019, 674: 58-70.

[48] Wang K, Munson K M, Beaupré-Laperrière A, et al. Subsurface seawater methylmercury maximum explains biotic mercury

concentrations in the Canadian Arctic. Sci. Rep, 2018, 8: 14465 doi: 10.1038/s 41598-018-32760-0.

[49] Stern G A, Macdonald R W, Outridge P M, et al. How does climate change influence arctic mercury? Sci. Total Environ, 2012, 414: 22-42.

[50] Macdonald R W, Harner T, Fyfe J. Recent climate change in the Arctic and its impact on contaminant pathways and interpretation of temporal trend data. Sci. Total Environ, 2005, 342: 5-86.

[51] AMAP, AMAP Assessment 2009 - Persistent organic pollutants (POPs) in the Arctic. Sci. Total Environ, 2010, 408: 2851-3051.

[52] Rigét F, Bignert A, Braune B, et al. Temporal trends of persistent organic pollutants in Arctic marine and freshwater biota. Sci. Total Environ, 2018, 649: 99-110.

[53] Angot H, Dastoor A, De Simone F, et al. Chemical cycling and deposition of atmospheric mercury in polar regions: Review of recent measurements and comparison with models. Atmos. Chem. Phys, 2016, 16: 10735-10763.

[54] de Mora S J, Patterson J E, Bibby D M. Baseline atmospheric mercury studies at Ross Island, Antarctica. Antarctic Sci, 2004, 5: 323-326.

[55] Dommergue A, Barret M, Courteaud J, et al. Dynamic recycling of gaseous elemental mercury in the boundary layer of the Antarctic Plateau. Atmos. Chem. Phys, 2012, 12: 11027-11036.

[56] Brooks S, Arimoto R, Lindberg S, et al. Antarctic polar plateau snow surface conversion of deposited oxidized mercury to gaseous elemental mercury with fractional long-term burial. Atmos. Environ, 2008, 42: 2877-2884.

[57] Wang C, Ci Z, Wang Z, et al. Air-sea exchange of gaseous mercury in the East China Sea. Environ. Pollut, 2016, 212: 535-543.

[58] Nerentorp M M G, Gårdfeldt K, Langer S, et al. Seasonal study of mercury species in the Antarctic Sea ice environment. Environ. Sci. Technol, 2016, 50: 12705-12712.

[59] Temme C, Einax J W, Ebinghaus R, et al. Measurements of atmospheric mercury species at a coastal site in the Antarctic and over the south Atlantic Ocean during polar summer. Environ. Sci. Technol, 2003, 37: 22-31.

[60] Xia C, Xie Z, Sun L. Atmospheric mercury in the marine boundary layer along a cruise path from Shanghai, China to Prydz Bay, Antarctica. Atmos. Environ, 2010, 44:1815-1821.

[61] Wang J, Xie Z, Wang F, et al. Gaseous elemental mercury in the marine boundary layer and air-sea flux in the Southern Ocean in austral summer. Sci. Total Environ, 2017, 603-604: 510-518.

[62] Spolaor A, Angot H, Roman M, et al. Feedback mechanisms between snow and atmospheric mercury: Results and observations from field campaigns on the Antarctic plateau. Chemosphere, 2018, 197: 306-317.

[63] Sun L, Xie Z. Changes in lead concentration in Antarctic penguin droppings during the past 3,000 years. Environ. Geol, 2001, 40: 1205-1208.

[64] Zvěřina O, Láska K, Červenka R, et al. Analysis of mercury and other heavy metals accumulated in lichen *Usnea antarctica* from James Ross Island, Antarctica. Environ. Monitor. Assess, 2014, 186: 9089-9100.

[65] Trevizani T H, Figueira R C L, Ribeiro A P, et al. Bioaccumulation of heavy metals in marine organisms and sediments from Admiralty Bay, King George Island, Antarctica. Mar. Pollut. Bull, 2016, 106: 366-371.

[66] 何依芳, 黄清辉, 陈玲, 等. 南极菲尔德斯半岛近岸海洋生物体有机锡污染状况. 环境科学学报, 2018, 38:1256-1262.

[67] Huang T, Sun L, Wang Y, et al. Transport of nutrients and contaminants from ocean to island by emperor penguins from Amanda Bay, East Antarctic. Sci. Total Environ, 2014, 468:578-583.

[68] Chu Z, Yang Z, Wang Y, et al. Assessment of heavy metal contamination from penguins and anthropogenic activities on fildes peninsula and ardley island, Antarctic. Sci. Total Environ, 2019, 646: 951-957.

[69] Cheng W, Xie Z, Blais J M, et al. Organophosphorus esters in the oceans and possible relation with ocean gyres. Environ.

Pollut, 2013, 180: 159-164.

[70] Wang P, Meng W, Li Y, et al. Temporal variation (2011—2014) of atmospheric OCPs at King George Island, west Antarctica. Atmos. Environ, 2018, 191: 432-439.

[71] Li H, Fu J, Zhang A, et al. Occurrence, bioaccumulation and long-range transport of short-chain chlorinated paraffins on the Fildes Peninsula at King George Island, Antarctica. Environ. Int. 2016, 94:408-414.

[72] Chen D, Hale R C, La Guardia M J, et al. Hexabromocyclododecane flame retardant in Antarctica: Research stations as sources. Environ. Pollut, 2015, 206: 611-618.

[73] Cipro C, Bustamante P, Taniguchi S, et al. Seabird colonies as relevant sources of pollutants in Antarctic ecosystems: Part 2—Persistent organic pollutants. Chemosphere, 2019, 214: 866-876.

[74] 王明星. 大气化学概论. 北京: 气象出版社: 2005.

[75] Hu Q H, Xie Z Q, Wang X M, et al. Secondary organic aerosols over oceans via oxidation of isoprene and monoterpenes from Arctic to Antarctic. Sci. Rep, 2013, 3, 2280, doi: 10.1038/srep02280.

[76] Fitzgerald J W. Marine aerosols: a review. Atmos. Environ, 1991, 25: 533-545.

[77] Zhang M, Chen L, Xu G, et al. Linking phytoplankton activity in polynyas and sulfur aerosols over Zhongshan Station, East Antarctica. J. Atmos. Sci, 2015, 72: 4629-4642.

[78] Meskhidze N, Nenes A. Phytoplankton and cloudiness in the Southern Ocean. Science, 2006, 314: 1419-1423.

[79] Mandalakis M, Apostolaki M, Tziaras T, et al. Free and combined amino acids in marine background atmospheric aerosols over the Eastern Mediterranean. Atmos. Environ, 2011, 45: 1003-1009.

[80] Facchini M C, Rinaldi M, Decesari S, et al. Primary submicron marine aerosol dominated by insoluble organic colloids and aggregates. Geophy. Res. Lett, 2008, 35: L17814, doi: 10.1029/2008GL034210.

[81] Hu Q H, Xie Z Q, Wang X M, et al. Levoglucosan indicates high levels of biomass burning aerosols over oceans from the Arctic to Antarctic. Sci. Rep, 2013, 3: 3119, Doi: 10.1038/srep03119.

[82] Dansgaard W. The O^{18}-abundance in fresh water. Geochim. Cosmochim. Acta 1954, 6: 241-260.

[83] Epstein S, Mayeda T. Variation of O^{18} content of waters from natural sources. Geochim. Cosmochim. Acta, 1953, 4: 213-224.

[84] Thompson L G, Yao T, Davis M E, et al. Tropical climate instability: The last glacial cycle from a Qinghai-Tibetan ice core. Science, 1997, 276: 1821-1825.

[85] Davies T D, Vincent C E, Brimblecombe P. Preferential elution of strong acids from a Norwegian ice cap. Nature, 1982, 300: 161-163.

[86] Eichler A, Schwikowski M, Gaggeler H W. Meltwater-induced relocation of chemical species in Alpine firn. Tellus B, 2001, 53: 192-203.

[87] Kang S C, Mayewski P A, Qin D H, et al. Seasonal differences in snow chemistry from the vicinity of Mt. Everest, central Himalayas. Atmos. Environ, 2004, 38: 2819-2829.

[88] Hou S G, Qin D H. The effect of postdepositional process on the chemical profiles of snow pits in the percolation zone. Cold Regions Science and Technology, 2002, 34: 111-116.

[89] Van de Velde K, Ferrari C, Barbante C, et al. A 200 year record of atmospheric cobalt, chromium, molybdenum, and antimony in high altitude alpine firn and ice. Environ. Sci. Technol, 1999, 33: 3495-3501.

[90] Barbante C, Schwikowski M, Doring T, et al. Historical record of European emissions of heavy metals to the atmosphere since the 1650s from Alpine snow/ice cores drilled near Monte Rosa. Environ. Sci. Technol, 2004, 38: 4085-4090.

[91] Eyrikh S, Eichler A, Tobler L, et al. A 320 year ice-core record of atmospheric Hg pollution in the Altai Central Asia. Environ. Sci. Technol, 2017, 51: 11597-11606.

[92] Kang S C, Huang J, Wang F Y, et al. Atmospheric mercury depositional chronology reconstructed from lake sediments and ice core in the Himalayas and Tibetan Plateau. Environ. Sci. Technol, 2016, 50: 2859-2869.

[93] Beal S A, Osterberg E C, Zdanowicz C M, et al. Ice core perspective on mercury pollution during the past 600 years. Environ. Sci. Technol, 2015, 49: 7641-7647.

[94] Zhang Q G, Huang J, Wang F, et al. Mercury distribution and deposition in glacier snow over Western China. Environ. Sci. Technol, 2012, 46: 5404-5413.

[95] Wang X P, Yao T D, Wang P L, et al. The recent deposition of persistent organic pollutants and mercury to the Dasuopu glacier, Mt. Xixiabangma, central Himalayas. Sci. Total Environ, 2008, 394: 134-143.

[96] Kang J H, Choi S D, Park H, et al. Atmospheric deposition of persistent organic pollutants to the East Rongbuk Glacier in the Himalayas. Sci. Total Environ, 2009, 408: 57-63.

[97] Wang X P, Gong P, Wang C F, et al. A review of current knowledge and future prospects regarding persistent organic pollutants over the Tibetan Plateau. Sci. Total Environ, 2016, 573: 139-154.

[98] Xu B Q, Cao J J, Hansen J, et al. Black soot and the survival of Tibetan glaciers. Proc. Nat. Acad. Sci. USA, 2009, 106: 22114-22118.

[99] Lau W K M, Kim M K, Kim K M, et al. Enhanced surface warming and accelerated snow melt in the Himalayas and Tibetan Plateau induced by absorbing aerosols. Environ. Res. Lett, 2010, 5, 025204, doi: 10.1088/1748-9326/5/2/025204.

[100] Ming J, Xiao C D, Cachier H, et al. Black Carbon (BC) in the snow of glaciers in west China and its potential effects on albedos. Atmos. Res, 2009, 92: 114-123.

[101] Li C L, Bosch C, Kang S C, et al. Sources of black carbon to the Himalayan-Tibetan Plateau glaciers. Nature Comm, 2016, 7: 12574, doi: 10.1038/ncomms12574.

[102] Baccolo G, Di Mauro B, Massabo D, et al. Cryoconite as a temporary sink for anthropogenic species stored in glaciers. Sci. Rep, 2017, 7: 9623, doi: 10.1038/s41598-017-10220-5.

[103] Uglietti C, Gabrielli P, Cooke C A, et al. Widespread pollution of the South American atmosphere predates the industrial revolution by 240 y. Proc. Nat. Acad. Sci. USA, 2015, 112: 2349-2354.

[104] Eichler A, Gramlich G, Kellerhals T, et al. Pb pollution from leaded gasoline in South America in the context of a 2000-year metallurgical history. Sci. Adv, 2015, 1: e1400196, doi: 1:e1400196.

[105] Beaudon E, Gabrielli P, Sierra H M R, et al. Central Tibetan Plateau atmospheric trace metals contamination: A 500-year record from the Puruogangri ice core. Sci. Total Environ, 2017, 601: 1349-1363.

[106] Eichler A, Tobler L, Eyrikh S, et al. Ice-core based assessment of historical anthropogenic heavy metal (Cd, Cu, Sb, Zn) emissions in the Soviet Union. Environ. Sci. Technol, 2014, 48: 2635-2642.

[107] Kaspari S, Mayewski P A, Handley M, et al. Recent increases in atmospheric concentrations of Bi, U, Cs, S and Ca from a 350-year Mount Everest ice core record. J. Geophys. Res. Atmos, 2009, 114, D04302, doi:10.1029/2008JD011088.

[108] Bauer S E, Bausch A, Nazarenko L, et al. Historical and future black carbon deposition on the three ice caps: Ice core measurements and model simulations from 1850 to 2100. J. Geophys. Res. Atmos, 2013, 118: 7948-7961.

[109] Pavlova P A, Schmid P, Bogdal C, et al. Polychlorinated biphenyls in glaciers. 1. Deposition history from an alpine ice core. Environ. Sci. Technol, 2014, 48: 7842-7848.

[110] Wang X P, Halsall C, Codling G, et al. Accumulation of perfluoroalkyl compounds in Tibetan Mountain snow: Temporal patterns from 1980 to 2010. Environ. Sci. Technol, 2014, 48: 173-181.

[111] Wang M, Xu B Q, Kaspari S D, et al. Century-long record of black carbon in an ice core from the Eastern Pamirs: Estimated contributions from biomass burning. Atmos. Environ, 2015, 115: 79-88.

[112] Thompson L G, Mosley T E, Davis M E, et al. Tropical glaciers, recorders and indicators of climate change, are disappearing

globally. Annals of Glaciology, 2011, 52: 23-34.

[113] Zhang Q G, Kang S C, Gabrielli P, et al. Vanishing high mountain glacial archives: Challenges and perspectives. Environ. Sci. Technol, 2015, 49: 9499-9500.

[114] Schuster P F, Krabbenhoft D P, Naftz D L, et al. Atmospheric mercury deposition during the last 270 years: A glacial ice core record of natural and anthropogenic sources. Environ. Sci. Technol, 2002, 36: 2303-2310.

[115] Chellman N, McConnell J R, Arienzo M, et al. Reassessment of the Upper Fremont Glacier ice-core chronologies by synchronizing of ice-core-water isotopes to a nearby tree-ring chronology. Environ. Sci. Technol, 2017, 51: 4230-4238.

[116] Hou S G, Jenk T M, Zhang W B, et al. Age ranges of the Tibetan ice cores with emphasis on the Chongce ice cores, western Kunlun Mountains. Cryosphere, 2018, 12: 2341-2348.

[117] Bhatia M P, Das S B, Xu L, et al. Organic carbon export from the Greenland ice sheet. Geochim. Cosmochim. Acta, 2013, 109: 329-344.

[118] Bhatia M P, Kujawinski E B, Das S B, et al. Greenland meltwater as a significant and potentially bioavailable source of iron to the ocean. Nature Geosci, 2013, 6: 274-278.

[119] Hawkings J R, Wadham J L, Tranter M, et al. Ice sheets as a significant source of highly reactive nanoparticulate iron to the oceans. Nature Comm, 2014, 5: 3929.

[120] Ma J, Hung H, Tian C, et al. Revolatilization of persistent organic pollutants in the Arctic induced by climate change. Nature Climate Change, 2011, 1: 255-260.

[121] Blais J M, Schindler D W, Muir D C G, et al. Melting glaciers: A major source of persistent organochlorines to subalpine Bow Lake in Banff National Park, Canada. Ambio, 2001, 30: 410-415.

[122] Blais J M, Schindler D W, Sharp M, et al. Fluxes of semivolatile organochlorine compounds in Bow Lake, a high-altitude, glacier-fed, subalpine lake in the Canadian Rocky Mountains. Limnol. Oceanogr, 2001, 46: 2019-2031.

[123] Bogdal C, Nikolic D, Luthi M P, et al. Release of legacy pollutants from melting glaciers: Model evidence and conceptual understanding. Environ. Sci. Technol, 2010, 44: 4063-4069.

[124] Bogdal C, Schmid P, Zennegg M, et al. Blast from the past: Melting glaciers as a relevant source for persistent organic pollutants. Environ. Sci. Technol, 2009, 43: 8173-8177.

[125] Miner K R, Blais J, Bogdal C, et al. Legacy organochlorine pollutants in glacial watersheds: a review. Environ. Sci.-Process Impacts, 2017, 19: 1474-1483.

[126] Li J, Yuan G L, Wu M Z, et al. Evidence for persistent organic pollutants released from melting glacier in the central Tibetan Plateau, China. Environ. Pollut, 2017, 220: 178-185.

[127] Guzzella L, Salerno F, Freppaz M, et al. POP and PAH contamination in the southern slopes of Mt. Everest (Himalaya, Nepal): Long-range atmospheric transport, glacier shrinkage, or local impact of tourism? Sci. Total Environ, 2016, 544: 382-390.

[128] Vermilyea A W, Nagorski S A, Lamborg C H, et al. Continuous proxy measurements reveal large mercury fluxes from glacial and forested watersheds in Alaska. Sci. Total Environ, 2017, 599: 145-155.

[129] Sun X, Wang K, Kang S, et al. The role of melting alpine glaciers in mercury export and transport: An intensive sampling campaign in the Qugaqie Basin, inland Tibetan Plateau. Environ. Pollut, 2017, 220: 936-945.

[130] Zdanowicz C, Karlsson P, Beckholmen I, et al. Snowmelt glacial and atmospheric sources of mercury to a subarctic mountain lake catchment Yukon Canada. Geochim. Cosmochim. Acta, 2018, 238: 374-393.

[131] Sun X J, Zhang Q G, Kang S C, et al. Mercury speciation and distribution in a glacierized mountain environment and their relevance to environmental risks in the inland Tibetan Plateau. Sci. Total Environ, 2018, 631-632: 270-278.

[132] Meyer T, Wania F. Organic contaminant amplification during snowmelt. Water Res, 2008, 42: 1847-1865.

[133] Curtosi A, Pelletier E, Vodopivez C L, et al. Polycyclic aromatic hydrocarbons in soil and surface marine sediment near Jubany Station (Antarctica). Role of permafrost as a low-permeability barrier. Sci. Total Environ, 2007, 383: 193-204.

[134] Schuster P F, Schaefer K M, Aiken G R, et al. Permafrost stores a globally significant amount of mercury. Geophys. Res. Lett. 2018, 45, 10.1002/2017GL075571.

[135] Sun S W, Kang S C, Huang J, et al. Distribution and variation of mercury in frozen soils of a high-altitude permafrost region on the northeastern margin of the Tibetan Plateau. Environ. Sci. Pollut. Res，2017, 24: 15078-15088.

[136] Ci Z J, Peng F, Xue X, et al. Temperature sensitivity of gaseous elemental mercury in the active layer of the Qinghai-Tibet Plateau permafrost. Environ. Pollut, 2018, 238: 508-515.

[137] 王宝山, 温成成, 孙秦川, 等. 石油烃类污染对青藏高原北麓河地区冻区土壤微生物多样性的影响. 环境工程学报, 2018, 12, (10)：12.

[138] Yang S Z, Wen X, Zhao L, et al. Crude oil treatment leads to shift of bacterial communities in soils from the deep active layer and upper permafrost along the China-Russia crude oil pipeline route. Plos One 2014, 9: e96552. doi: 10.1371/journal.pone.0096552.

[139] Marusczak N, Larose C, Dommergue A, et al. Total mercury and methylmercury in high altitude surface snow from the French Alps. Sci. Total Environ，2011, 409: 3949-3954.

[140] Bizzotto E C, Villa S, Vaj C, et al. Comparison of glacial and non-glacial-fed streams to evaluate the loading of persistent organic pollutants through seasonal snow/ice melt. Chemosphere 2009, 74: 924-930.

[141] Yin R S, Feng X B, Hurley J P, et al. Historical records of mercury stable isotopes in sediments of Tibetan lakes. Sci. Rep. 2016, 6, 23332, doi: 10.1038/srep23332.

[142] Wake C P, Mayewski P A, Xie Z C, et al. Regional distribution of monsoon and desert dust signals recorded in Asian glaciers. Geophys/ Res. Lett, 1993, 20: 1411-1414.

[143] Marinoni A, Polesello S, Smiraglia C, et al. Chemical composition of freshsnow samples from the southern slope of Mt. Everest region (Khumbu-Himal region, Nepal). Atmos. Environ. 2001, 35: 3183-3190.

[144] Huo W M, Yao T D, Li Y F. Increasing atmospheric pollution revealed by Pb record of a 7 000-m ice core. Chin. Sci. Bull，1999, 44: 1309-1312.

[145] Xiao C D, Qin D H, Yao T D, et al. Global pollution shown by lead and cadmium contents in precipitation of polar regions and Qinghai-Tibetan Plateau. Chin. Sci. Bull. 2000, 45: 847-853.

[146] Kang S C, Zhang Q G, Kaspari S, et al. Spatial and seasonal variations of elemental composition in Mt. Everest (Qomolangma) snow/firn. Atmos. Environ，2007, 41: 7208-7218.

[147] Huang J, Kang S C, Zhang Q G, et al. Atmospheric deposition of trace elements recorded in snow from the Mt. Nyainqentanglha region, southern Tibetan Plateau. Chemosphere, 2013, 92: 871-881.

[148] Dong Z W, Shao Y P, Qin D H, et al. Insight into radio-isotope I-129 deposition in fresh snow at a remote glacier basin of Northeast Tibetan Plateau, China. Geophys. Res. Lett, 2018, 45: 6726-6733.

[149] Feng L, Xu J Z, Kang S C, et al. Chemical composition of microbe-derived dissolved organic matter in cryoconite in Tibetan Plateau glaciers: Insights from fourier transform ion cyclotron resonance mass spectrometry analysis. Environ. Sc. Technol, 2016, 50: 13215-13223.

[150] You C, Xu C, Xu B Q, et al. Levoglucosan evidence for biomass burning records over Tibetan glaciers. Environ. Pollut, 2016, 216: 173-181.

[151] Cong Z Y, Gao S P, Zhao W C, et al. Iron oxides in the cryoconite of glaciers on the Tibetan Plateau: Abundance, speciation and implications. Cryosphere, 2018, 12: 3177-3186.

[152] Zhang Y L, Kang S C, Cong Z Y, et al. Light-absorbing impurities enhance glacier albedo reduction in the southeastern

Tibetan plateau. J. Geophys. Res. Atmos, 2017, 122: 6915-6933.

[153] Kang S C, Zhang Q G, Qian Y, et al. Linking atmospheric pollution to cryospheric change in the Third Pole region: Current progresses and future prospects. Natl. Sci. Rev. 2019, doi: 10.1093/nsr/nwz031.

[154] Zhang Q, Sun X, Sun S, et al. Understanding mercury cycling in Tibetan glacierized mountain environment: Recent progress and remaining gaps. Bull. Environ. Contam. Toxicol., 2019, 102: 672-678.

[155] Sharma B M, Nizzetto L, Bharat G K, et al. Melting Himalayan glaciers contaminated by legacy atmospheric depositions are important sources of PCBs and high-molecular-weight PAHs for the Ganges floodplain during dry periods. Environ. Pollut, 2015, 206: 588-596.

[156] Ren J, Wang X P, Gong P, et al. Characterization of Tibetan soil as a source or sink of atmospheric persistent organic pollutants: Seasonal shift and impact of global warming. Environ. Sci. Technol, 2019, 53: 3589-3598.

[157] Chen D, Xu B, Yao T, et al. Assessment of past, present and future environmental changes on the Tibetan Plateau. Chin. Sci. Bull, 2015, 60: 3025-3035.

[158] Wang X, Wang C, Zhu T, et al. Persistent organic pollutants in the polar regions and the Tibetan Plateau: A review of current knowledge and future prospects. Environ. Pollut, 2019, 248: 191-208.

作者：王飞越[1]，谢周清[2]，张强弓[3]

[1] University of Manitoba，[2] 中国科学技术大学，[3] 中国科学院青藏高原研究所

第29章 食物-能源-水：从单一领域到系统关联

- 1. 引言 /833
- 2. FEWS基本概念 /834
- 3. FEWS模型和方法论 /839
- 4. FEWS 关联网络的工程技术 /846
- 5. FEWS的相关政策 /851
- 6. 展望 /853

本章导读

食物、能源和水是人类生存与发展不可或缺的基础性资源，且三者之间存在密切的关联关系，即其中任何一项资源的生产与供给均依赖于另外两项资源。在现阶段有关食物、能源和水的供应政策、规划和管理决策通常是相互孤立的，各要素之间多方面相互作用常常被认为是一个系统对另一个系统的威胁或压力。为了使所有资源供应链更加安全，有必要从整体上（以及每个环节）考虑和解决问题，并评估其对整个食物-能源-水系统（FEWS）稳定性的影响。本章将首先介绍FEWS的发展历程，通过对关联网络的概念表述，对相关的研究方法和方向进行总结概括，另外也归纳了几类国内外学者报道的FEWS研究模型，综述了现有研究成果、技术和应用，并结合国内外相关政策的鼓励提出了对未来FEWS研究发展方向的意见和建议。

关键词

FEWS，食物，能源，水，关联网络

1 引 言

当代经济社会的高速发展带来了人口激增、城市化进程加快和能源消耗模式变化，世界人口预计到2050年将增长到90亿，这种变化和预期加剧了世界对能源和食物的需求，进一步加大了水资源匮乏的压力。从全球角度来看，食物-能源-水系统（FEWS）关系包含了紧密相连的粮食生产、能源供应和水资源系统：粮食代表着农业贸易和消费，是一个国家或地区经济的重要组成部分；能源是农业机械化生产活动、工业和市政用水及污水处理所必需的；水用于人类生活和工业消费、作物灌溉和能源生产。为了人类的健康、繁荣和福利，以及应对不可预测的土地利用变化、人口变化、气候变化等，我们需要加强粮食、能源和水的生产或供应。食物、能源和水的供给不能孤立实现，供应链在每个阶段都是密切相连的，并受一系列共同因素（市场、法律、政策）的影响。为了使所有资源供应链更加安全，有必要从整体上（以及每个环节）综合考虑和解决问题，并评估其对整个系统稳定性的影响。这些复杂的供应系统具有相互依赖性和脆弱性，一旦确定这些相互依赖性和脆弱性，就可以确定若干行动，从而减少浪费、提高效率、增强弹性和增强安全性。FEWS关联关系作为应对全球人口增长、资源短缺和生态环境恶化等问题的系统性管理理念，已引起国际学术界与实践管理领域的广泛关注。

FEWS关联关系理念的萌芽和研究最早可追溯到20世纪70年代，Meadows等撰写的《增长的极限》一书中对全球人口、经济和FEWS等之间的供需与安全问题进行了初步探讨，引发了全球环境学家的广泛关注。中国政府一向重视环境保护和食物供需安全问题，本着对国际环境与资源保护积极负责的态度，参加或缔结了环境与资源保护国际公约和条约三十多件，签约后付出了大量努力——淘汰落后产能、鼓励生态农业及绿色矿山建设等。另外，中国还积极支持了有关国际环境与资源保护的许多重要文件，并把这些国际法文件的精神引入到中国的法律和政策之中。这些文件包括1972年在瑞典斯德哥尔摩发表的《联合国人类环境宣言》、1980年发表的《世界自然资源保护大纲》、1982年在肯尼亚内罗毕发表的《内罗毕宣言》和1992年在巴西里约热内卢发表的《关于环境与发展的里约热内卢宣言》等。而后在中国共产党第十六届中央委员会第五次全体会议上提出了建设"资源节约型社会、环境友好型社会"的伟大目标，力图通过节约资源，通过采取技术和管理等综合措施，不断提高资源利用效率，达到人与自然和谐共生的社会形态，实现尽可能的减少资源消耗和环境代价、满足人们日益增长的物质文化需求的发展模式。其核心内涵是人类的生产和消费活动与自然生态系统协调可持续发展。这一发展理念的提出极大契合了FEWS关联关系的核心思想,也极大推动了相关学术和应用研究在中国的快速落地和迅猛发展。

尽管在FEWS几个关系中存在已知的相互关系,但有关食物、能源和水的政策、规划和管理决策通常是相互孤立的,并没有充分考虑部门之间的权衡。这主要是因为每个独立系统的复杂性,使得理解系统之间的相互关联变得困难,更不用说以综合的方式进行评估。在未来的几十年里,人口增长、生活水平提高和气候变化将使向所有人提供可持续的食物、水和能源变得更加困难。我们需要通过创新研究来增加供应、改善分配、减少浪费、提高效率和减少需求。由于食物-水-能源的关系是如此紧密地交织在一起,一个领域潜在解决方案往往会影响另一个领域。当我们努力满足不断增长的人口的基本需求时,开发一种以系统为导向的整体解决方案对于平衡资源需求至关重要。目前,诸多科研团队投入大量人力、物力、财力开展了很多富有成效的研究工作,研究内容从宏观到微观,从理论研究到技术应用以及管理政策全方位拓展。重要并已开展的研究方向有污水氮磷营养物质回用技术、污水无害化处理技术、可再生能源、微生物燃料电池等。此外还有废水流中营养素循环能力强化、剖析食物(选择氮磷流作为表征指标)-能源-水之间的协同、拮抗等复杂的关联关系;构建涵盖城市氮/磷回收、减少能源消耗与水回用一体化的耦合模型与网络模拟方法,力求建立一套解构城市FEWS的方法体系等。

尽管目前对FEWS关联关系仍然缺乏统一和清晰的界定,现有研究大多是基于不同角度理解与量化"关联关系",但我们必须承认相关治理问题是被嵌入到全球、区域、国家和地方等不同层级的政策、体制、技术和融资选择中,如果深陷在缺乏战略明确性的情况下,在利益相关者中会出现权力、声音和获取信息、资源的渠道分布不均,最终将会产生不公平和不可持续的结果。通常来讲,个体利益相关者的短期胜利与人类长期整体解决方案之间存在巨大的冲突。因此我们迫切需要了解FEWS几个关系中的联系、协同作用和冲突,以便为每个部门最有可能在其他部门产生积极影响的政策提供决策的依据。为了实现这一整合,环境工程研究者需要与农业、能源、卫生、生态学、生物学、社会科学、政策管理等学科的专家协同合作,决策者需要将有关食物、能源、水相互作用的国家、区域和地方相关信息纳入未来一系列条件下制定强有力的政策和规划中。通过这种方式,可以制定政策,确保FEWS几个关系在未来多变的气候条件和不可预测事件发生的情况下,仍然可以确保人类福祉和可持续发展。

2 FEWS基本概念

1983年,联合国大学发起了一个"食物-能源"关联的提案,承认粮食和能源问题之间的重要联系[1]。第二年,联合国大学在巴西巴西利亚举行了"食物、能源和生态系统"会议。1986年,联合国大学又在印度新德里举行了第二届"食物-能源关系和生态系统"国际专题讨论会。20世纪80年代中期美国科学家从研究的角度关注了水和电之间的相互联系,20世纪90年代,世界银行使用"关联系统"(nexus)一词来关注"水、食物和贸易"之间的联系[2]。20世纪90年代中后期至21世纪初,哥伦比亚大学地球研究所哥伦比亚水资源中心对印度的"水-能源-农业"的关联进行了研究,随后,Scott等[3]将"水-能源-发电"的关联应用于墨西哥。随着越来越多的该类主题国际研讨会的召开,世界各地的学者和从业者承认有必要将能源作为关联的核心[4]。随后,2011年波恩"nexus"大会正式宣布了水、能源和食物三大系统的重要性[5],根据Hoff为会议准备的文件,国际社会为应对气候变化和包括人口增长、全球化、经济增长和城市化在内的社会变化而推出了"水-能量-食物"系统(water-energy-food nexus)的概念[6],如图29-1

所示。在波恩"nexus"大会之后,"水、能源和食物安全"关联资源平台由德国联邦政府出台。2012 年,联合国大学综合管理材料和资源学院(United Nations University Institute for Integrated Management of Material Fluxes and of Resources,UNU-FLORES)确定了"综合管理环境资源:水、废弃物和土壤"的概念,2013 年"亚洲及环太平洋地区水-能源-食物关联现状"会议由联合国亚洲及太平洋经济和社会委员会(协议)筹备,并"改革创新了食物-能源-水的关联——应用多尺度综合分析社会和生态系统的新陈代谢(Multi-Scale Integrated Analysis of Societal and Ecosystem Metabolism,MuSIASEM)方法来研究三个领域之间的关系"。同时,联合国粮农组织(Food and Agriculture Organization,FAO)和国际可持续发展研究所(International Institute for Sustainable Development,IISD)在 2013 年提出"水-能源-粮食安全"的关联,为环境投资和风险管理提供切实的规划和决策[5]。联合国粮农组织 2014 年明确了"食物-能源-水关联"的概念,用来描述和处理全球资源系统的复杂和相互关联的特性,人类依靠它来实现不同的社会、经济和环境目标,它也可以平衡不同资源用户的目标和利益,同时保持生态系统的完整性。Allan 指出,"nexus"强调促进与各个部门的合作,并提供了解决学科分歧的机会[5, 7]。

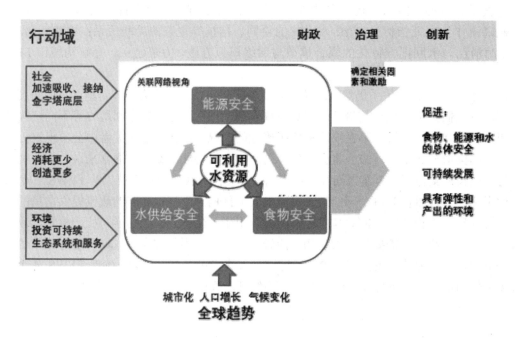

图 29-1 "食物-能源-水"框架图[6]

根据 Hoff 的 nexus 框架[6],我们以可用水资源为中心,同时将全球的社会、经济、环境等其他行为集成在一起进行评估和评价。对于制定其他核心目标,如可持续性或公平性,也同样有效。事实上,考虑到评估整个世界经济论坛关系的复杂性所面临的挑战,开发令人信服的可视化"nexus"方法是一项复杂的任务[8]。

总的来说,"nexus"并没有固定的概念,国际上把"nexus"理解为将不同部门、不同层次的利益相关者的想法和行动联系起来,实现可持续发展的过程[5]。如图 29-2 所示,水系统有"淡水可利用性、综合利用、水循环和流量管理"等自身的循环,粮食系统也涉及"农业生产、消费和废弃物、分配/贸易"等自身的循环,能量系统也有"能量生产、主要原料及分类、能量评估及可靠性"等自身的循环,而三个系统之间又存在着无限的关联和循环,并且在能量和粮食的生产过程中对水有着强烈的需求竞争。同时,从"水-能源-粮食安全"的角度,三者又都有很大的弹性空间。

第29章 食物-能源-水：从单一领域到系统关联

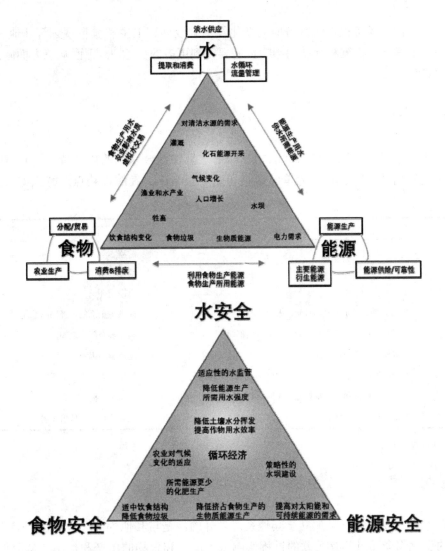

图 29-2 粮食-能源-水的关系图[9]

关于"食物-能源-水"关联系统的研究，一方面认为系统侧重于不同部门之间相互协调，旨在通过各部门之间的相互联系来把握复杂系统的整体特征[10]。由于这三个部门的安全问题都很重要，其中某一个部门的失利都可能对其他两个部门造成压力，因此需要对这些部门进行全面管理。虽然不同的研究者和管理者对这三者的联系都有不同的解释，但可看出，其不同于传统的单学科决策实践，这种关联是为了呼吁通过跨部门协调，进行综合管理，以便减少更多的部门权衡，促进每个部门的可持续发展[11]。

从另外一个方面的研究则认为三者关联的网络系统更重要，系统本身就是一种分析方法，用来量化系统节点之间的链接（即食物、水和能源）。对于这种新兴的方法有许多解释，例如，联合国粮农组织强调，这个关联系统研究方法的功能是系统地分析人与自然的耦合，并通过建立协同作用和管理权衡产生跨部门和规模的自然资源综合管理。Scott等则认为"食物-能源-水"系统方法的本质是产生资源回收[12]，其副产品是资源使用效率的提高。与此同时，Keskinen等认为，"食物-能源-水"系统方法的内容非常丰富，不能从单一的角度来解释[13]。

目前通常从三个方面来阐述这三者之间的联系：内部关联过程（物理化学过程）、对外部影响的响应、系统综合性能。Bizikova等认为[14]，世界经济论坛的研究要求我们提高对三个核心领域的认识：一般通过建模方法（如投入产出分析等）来研究三个或更多资源元素之间联系的性质；考虑到在气候变化和城市化压力的背景下,世界经济论坛研究方向的变化及基础设施、卫生和城市规划等其他领域的变化,对这三种相互关联的不确定性政策和行动的影响也需要得到更多的关注[8]。下面从三方面阐述"食物-能源-水"关联系统的研究进展。

2.1 内部关联分析

内部关联分析通过研究不同领域之间的相互作用来呈现耦合系统的内部特征。为了明确内部关联分析的轮廓,将不同部门之间的交互分为单向影响和交互影响,二者的优缺点及特点如表 29-1 所示[10]。

表 29-1 内部关联分析[10]

	单向影响分析	交互影响分析
关联	单方面关系	多边关系
		双边关系
		反馈闭环
优点	需求资源少(如数据、时间、人力、经济等)	提供更全面的神经网络系统评估
	显示一个特定部门的变化如何影响其他部门	确定多个部门之间的控制或依赖关系
	促进权衡利弊的初步评估	确定驱动因素或过程
	综合管理的起点	从整体的角度为关联问题提供新的解决方案
缺点	无法反映反馈和第二轮效应,该网络分析能力有限	时间、数据、人力和资金方面的资源密集
特点	简单的和有针对性的	复杂的和从整体出发的

2.2 外部影响分析

"食物-能源-水"关联系统不是孤立于其他系统之外的,是与其他外部环境相互影响和相互作用的。任何外部环境的变化都会通过相互关联的过程影响水、能源和食物的生产和使用,从而使关联系统的性能复杂化。根据其来源,外部因素可以分为两组——物理因素和社会因素。

从全球尺度看,20 世纪包括温度与降水在内的主要气候特征值发生了变化,有的地区还呈现出趋势性变化,这种变化已在世界上,特别是在北半球广大地区观测到。美国、中国、俄罗斯、加拿大、英国、澳大利亚等主要国家或地区,几十年乃至近百年的观测资料显示出气候变化的总趋势:即温度升高、降水增多、极端天气情况频繁发生;但地区的差异很大,气候变化的强烈程度以北半球最甚[15]。气候变化、极端天气和自然灾害等物理因素可能通过影响其供应链和生产过程而改变水、能源和粮食的供应。例如,随着气候变化,水体更高的蒸发量会影响水力发电用水和灌溉用水直接业已存在的平衡,从而同时影响能源和农业生产[16]。极端天气,例如干旱和洪水,可能对农业生产力产生显著的负面影响[17];同时水力发电也会受到显著影响[18]。Berardy 和 Chester 发现[19],由于地球近期以来的气候变化导致的气温升高,美国亚利桑那州为了满足灌溉需求,采用农业高度依赖能源密集型水资源供应形式,这将导致更多的水资源和能源消耗。研究表明,由于 2004~2005 年的干旱,伊比利亚地区的水力发电量减少了 40%[20]。为了度过干旱时期,伊比利亚半岛国家转向大量进口化石燃料发电以减少水的用量[20]。

另一方面,社会因素也是重要的外部影响。用户行为和感知等社会因素可能会将资源管理的重点从供给侧转移到需求侧。例如,人类的饮食习惯可以改变整个粮食供应链的能源使用和蓝水足迹。从美国目前的粮食消费模式转向粮食混合饮食可增加 38%的能源消耗和 10%的蓝水足迹[21]。此外,策略和技术可以通过不同资源之间的交互或交换而影响"食物-能源-水"关联系统。Housh 等最近的一项研究表明，在密西西比河流域实施纤维素生物燃料的生产（例如，芒草）有可能减少 NO_3-N 负荷[22]，但芒草种植会增加水的消耗，从而对水流和相关的生态系统造成压力。同时，人口增长带来更大的粮食需求，需要进一步强化土地和水的使用，气候变化可能导致减少地表水和地下水资源的可利用性。

2.3 系统综合性能

除了内部和外部影响分析，对整个系统性能的评估，如弹性[23]、可持续性[24]和适应性，是目前研究的另一个新兴学科领域，并且是对"食物-能源-水"关联系统非常必要的研究领域。

弹性是指系统在外部冲击或扰动期间和之后维持和恢复其基本功能的能力[23]。关联系统的弹性评价对于复杂系统的综合管理和资源的合理开发利用具有重要意义。例如，Suweis 等[25]发现，由于粮食安全日益依赖于全球贸易，人口-粮食耦合体系的弹性正在丧失，且政策变化对系统弹性会产生负面影响。如通过倡导节约用水（如提高灌溉效率）给予节水补贴，但这种政策又会导致扩大灌溉面积而增加用水量，从而加剧水资源短缺，恶化水质，破坏流域的弹性[12]。

系统可持续性对于耦合系统也很重要，可持续性表示系统能够在有限的资源下无限期地继续其功能，因为对系统可持续性的忽视可能导致短期有效的策略，但从长远来看可能会失败。Nguyen 等[24]建立了一个由供水、生物质生产和生物燃料三个子系统组成的生物燃料生产系统，证明了提高供水对生物燃料生产的依赖可能会降低系统的可持续性，这意味着耦合系统的质量可能会随着时间的推移而下降。然而，由于对系统可持续性的定义模糊，对这种性能度量的量化仍然是有限的。

适应性这一概念最初起源于自然科学，特别是进化生态学的研究。尽管适应一词的含义在自然科学中仍然存在争论，但一般而言适应是指个体或者系统通过改善遗传、行为特征适应变化，并通过遗传保留下相应的适应性特征。这一定义涵盖了从生物个体到某一特定物种的种群，乃至整个生态系统的尺度[26]。适应是不同尺度系统中（家庭、社区、群体、区域、国家）的一个过程、一种行动或者结果，当面对气候变化、压力、灾害时，系统能更好地应对、管理或调整。目前适应性的核心概念有以下 5 个。

（1）敏感性：在全球变化领域，IPCC 在 1995 年的报告中提出了气候变化下敏感性的含义"指一个系统对气候条件变化的响应程度，这种响应可能有害，也可能是有益。"Moss 等在评估各个国家对气候变化的脆弱性时，将气候因素当作一种输入，定义敏感性是指系统输出或系统特性响应输入变化而改变的程度。

（2）暴露：人类-环境系统所面临的环境变化的特征及其变化程度，是指系统接触外在变化的特征、强度和速率。

（3）脆弱性：Adger 等对社会科学和自然科学中的脆弱性概念及研究方法的变化进行了梳理，总结了脆弱性研究在不同领域内的研究对象及源流。

（4）适应能力：从个人到全人类的人类系统在面临一系列环境变化时增加（或保持）其个体成员的生活质量的能力。IPCC 早在 2001 年就定义适应能力为"系统调整自身以适应气候变化和极端事件和趋利避害的能力"。

（5）弹性：系统在承受变化压力的过程中吸收干扰、进行结构重组，以保持系统的基本结构、功能、关键识别特征以及反馈机制不发生根本性变化的一种能力。

3 FEWS 模型和方法论

作为至关重要的自然资源，粮食、能源和水是解决饥饿、改善健康和建设可持续经济等全球问题的关键所在。众所周知，生产能源需要水，水的提取、分配和处理又需要能源。此外，粮食生产既需要水又需要能源。资源之间的相互依赖构成食物-能源-水（FEW）的关联关系的基础。基于对这一点认识，管理这些资源的政策也应是相互关联的。然而，在许多政策制定中管理这些相互联系的资源往往是由单独机构来处理以期促进决策，却忽视了资源间的相互依赖和相互联系。过去十年，国际社会在强调这些联系并提高决策者对这些问题的认识上已取得了一些进展，在理论和应用方面也提出了若干模型和框架。值得指出的是，大多数模型都能够以各种各样的方式将上述关联关系恰当地集成到相关自然资源政策中去。

由于这些资源之间的复杂联系，需要建立更复杂的模型以综合的方式来管理自然资源。换句话说，迫切需要综合规划和系统思考，而不只是优化一种资源对另一种资源的使用。通过对当前关于水-能源-粮食间关联关系文献的回顾，我们能清楚地了解已知的、缺失的，并确定 WEF 设计和建模所面临的关键机遇和挑战。为此，我们期望洞悉建立水-能源-食物关系模型的主要挑战和关键考虑，了解模型和框架的研究现状，从而确定是什么阻碍了它们从理论到实践的决策过程，并讨论基于当前模型和框架所形成的政策的含义。

3.1 模型案例

从关联网络的概念出发，国内外学者从不同领域和视角，对 FEWS 概念模型提出了各自的见解和表述。起初，学术界对"关联关系"理解基于该词的直译，将其界定为 FEWS 之间的相互依存关系，即 FEWS 三种资源中任意一方的生产活动和供给过程对其他两方的影响。至今，部分学者仍然沿用这一解释，并在此基础上做出了一些改变，从三者之间的相互影响和相互制约方面进行了界定。

在概念模型的探索和建立过程中，由于研究重点不同，学者们对 FEWS 概念模型的理解角度也呈现多样化。以 FEW 三者中任意一方作为切入点，建立概念模型的研究居多，称之为"中心论"。如斯德哥尔摩环境研究院亚洲开发银行以水资源为核心，认为"可用水量"关系到 FEWS 关联关系的安全；而联合国粮食及农业组织则从粮食安全角度出发，挖掘影响 FEWS 关联关系的关键因素；国际可再生能源机构建议以能源为中心，运用可再生能源技术处理 FEWS 之间的权衡关系以建立概念模型。除此之外，部分学者从资源供给安全的角度解析"关联关系"，并建立概念模型。本节介绍几个典型的概念模型框架，从宏观尺度上指明长期规划的研究方向和方法。

3.1.1 环境生活安全

这个理论框架是为了评估东南亚和海洋国家的生计环境安全而开发的。该框架的重点是应对全球挑战，以实现粮食安全、淡水和能源安全。此外，它将水-能源-粮食-气候关系的概念与现有的自然资源相结合，以实现可持续生存和发展[27]。由于这一框架具有全球性，它的重点是确保可持续利用资源，促进包容性经济增长，并在不同气候条件下维持重要的环境系统功能。

3.1.2 全球政策对话模型

该模型是由国际水资源管理研究所开发的，可作为一种互动的政策规划和情景分析工具，探讨国家范围内水资源的权衡和未来需求。该模型为多部门水规划和分析提供了依据。它的目标是提高人们对国家水资源以及水资源短缺、粮食安全和环境需求之间复杂相互作用的认识，以适应人口增长和国家饮食的变化[12]。这种模型的优点之一在于它不仅给经济带来好处，而且从长远来看也给社会带来好处，为促进社会群体的创建和 WEF 资源管理的决策过程做出贡献。

3.1.3　水、能源和粮食的安全关系

这是为了建立一个面向 WEF 关联关系的新办法，以促进可持续地利用资源和确保安全地获得基本服务。食物、能源和水资源的安全性、可用性是该框架的核心，以反映其作为其他资源的关键驱动力的重要性。该框架旨在加深对水、能源和食物各子系统间的关联关系的理解。该框架的目标可以通过将城市化、人口增长和气候变化等大趋势纳入相关政策决策来实现，目标是在不增加使用 WEF 资源和投资来维持生态系统服务的情况下提高经济增长[6]。

3.1.4　WEF 关系框架

这一框架是在 2011 年世界经济论坛上面世的，主要目标是帮助决策者更好地了解与提前管理资源相关的风险，实现在危机时刻积极地应对。这种方法认识到，水-能源-粮食的关系受到诸如全球治理失败、经济差距和地缘政治冲突等外部因素的影响（而且还受到这些因素的推动）。也就是说，这是水-能源-食物关系的全球风险区域[28]。该框架还显示了在能源短缺和潜在社会稳定影响的背景下，能源与经济安全之间的联系。它强调了粮食生产和水的利用中所涉及的能量需求。与波恩 2011 年 nexus 框架类似，世界经济论坛认为人口增长和环境压力是 nexus 的驱动因素[28]。

3.1.5　WEF 框架

国际山地综合开发中心（ICIMOD）开发了的一个 nexus 框架，主要关注喜马拉雅和南亚地区的生态系统服务。它帮助决策者了解权衡和最大化有效资源使用的利益[29]。该框架强调了生态系统产品的重要性，认为它们必须要受到保护以确保水、能源和粮食部门的恢复能力。

3.1.6　社会和生态系统代谢框架的多尺度综合分析

这一框架是由粮农组织制定的，其目的是通过在社会经济活动和生态限制的不同层次和规模上提供相关特征来研究社会经济系统的新陈代谢模式[30]。该框架被用来分析人口动态、温室气体排放和国家层面土地利用变化等异质性因素之间的关系。该方法采用了复杂系统理论的主要概念，建立在坚实的理论基础上。它允许在构建多层次的社会经济结构时具有特定的灵活性。最后，该框架同时描述了能源、食物和水的流动及其与环境相互作用的复杂系统（社会）之间的相互关系[31]。

在这些研究工具中，定量和定性方法的结合有助于更好地理解相互联系。然而，对于上述模型而言，所需数据的缺乏及其复杂性往往被公认为是一个不可避免的挑战。有限的数据导致只有少量详细的研究工作能够在跨越三个或更多部门面向不同的规模和不同的背景提供真正的解决方案。例如，在目前所研究的模型中几乎看不到空间和时间等基本因素，然而正如我们在上述环节中看到的，这些因素往往在塑造自然资源的可用性和可访问性等方面发挥着举足轻重的作用。

建立全面反映自然资源随时间和空间动态变化的综合框架，可为政府和其他利益相关者提供激励，以提高金融系统的效率和有效性，并通过权衡将相关风险最小化[32]。此外，鉴于这些框架捕捉了供求的时空变化，并突出了自然资源枯竭或滥用的潜在风险和处理，所以应该将其作为长期规划工具[33]。

3.2 模型构建方法

"食物-能源-水"关联系统的研究方法有多种,但是,研究的优先级、目标、规模和数据可用性对于决定应该使用哪种方法进行 FEWS 关联研究至关重要,没有一种方法可以适用于所有情况[34]。从经济学和生态学等学科的基本理论方法出发,FEWS 关联关系的模型构建也就有了基础的手段[35]。本节通过对几种常见模型构建方法的总结以及各自的应用例子和应用范围,为读者提供相关模型构建方法论的概况。

根据模型具体的应用场景、地理尺度和目的,FEW 相关的模型可以分为几种不同的类型。从研究方法来讲,模型可以是基础方法论的、量化的、或者是基于概念的;从地理尺度来讲,模型可以是处理城市尺度的、地区尺度的、国家尺度的、或者是全球尺度的;从模型的构建目的来讲,模型可以通过数据计算体现系统关系,或者为管理机构提供资源管理方法,或是能够为政策和技术发展提供指导。本节着重从研究方法上将文献中的模型进行分类。在每类模型中,地理尺度和构建目的的区别也将会讨论。

表 29-2 常见模型构建方法及应用举例

方法名称	方法简介	应用举例	适合的地理尺度
调查与统计方法	通过实地调查、专家小组、地方机构和政府发布的公共数据的收集以及相关文献来说明关联问题的调查和数理统计是研究关联部门之间相互作用最广泛使用的方法。通过对某一资源生产过程中相关资源消耗的追踪,研究人员可以通过数据分析对水、能源、粮食等资源之间的相互作用进行定量评估	Machell 等[36]发现,发电约占全球淡水总取水量的15%,约70%全球的水被用于粮食生产 Ramirez 等人的研究表明[37],为了粮食安全四个欧洲国家出台一个强制性的卫生法规,要求用于清洗和灭菌的水的温度从1990 年的 60℃ 在增加到 2001 年的 82℃,这导致整个肉类产业总能源的消耗增加了 1/3~2/3 Alimagham 等通过实地调查发现,传统的大豆生产过程比机械化过程更环保,因为前者能耗低,温室气体排放少[38]	适合各种尺度的研究
数据包络分析法	一种非参数模型,度量一系列决策单位将数据输入转化成输出的相对效率来建立模型分析的基础。决策单位的运行效率和资源利用的趋势也可以得到评估	Kourtit 等人[39](Data Envelopment Analysis,DEA)基于全球 39 个城市的效率绩效指标,对其可持续发展状况进行了整体表述,表明欧洲城市在人类环境方面的可持续发展效率得分更高。除此之外,该方法还可以为政策规划提供有价值的信息,关联系统之间的问题进行全面概述[40]	城市尺度 地区尺度
计量经济学分析法	通过分析非实验经济数据,运用统计方法为经济关系提供实证内容[41]。多元回归分析方法为计量经济学分析奠定了基础。然而,计量经济学分析不同于一般统计方法,它通过一组数学方程来表示关联系统,并基于经济学理论来阐明关联问题	计量经济学技术和程序被广泛用于推断因果关系[42]和检验经济理论[43]。例如,Zaman 等人的回归结果[44]揭示了巴基斯坦农业增长和能源消费之间的双向因果关系,证明促进能源利用效率的行动和能源多元化在巴基斯坦确实会带来持续的经济增长,并且在这个关联系统中农业在经济中起着主导作用	国家尺度 城市尺度
生态网络分析(ENA)	由投入产出分析发展而来的[45],是目前评价经济成分与自然成分相互作用的主要方法之一。ENA 将系统视为像食物链一样连接的多个实体的集成,可以对生产和消费交织链中所体现的直接和间接流动进行统一分析,显示出研究多个元素之间权衡的潜力[46]	它被用于分析城市新陈代谢中的代谢流(如水、能源、碳排放)[47]、全球能源贸易[48]和虚拟水[49]。结合其他指标,如链接分析[50]、循环指数等[51],ENA 能检测到一个部门的角色转换,并量化系统性能(可循环性、弹性)。例如,Chen 和 Chen 以北京为例[46],发现大部分城市部门依赖于"能量-水"系统中的制造业,而整个系统对能源系统的影响(±200%)远大于对水系统的影响(±30%)。此外,与自然生态系统相比,"能量-水"系统的平均稳定性更低,因为它有更密集的物质交换[46]	城市尺度 全球尺度
系统动力学模型	一种基于因果机制的自上向下建模方法,它遵循了系统行为由其结构决定的假设。它通过在某一系统内的各要素之间建立因果反馈循环,在宏观和微观两级全面分析多部门系统。这使得它非常适合多学科和多角色问题的评估和分析[52]	Chhipi-Shrestha 等[53]采用这种方法来确定限制加拿大 Penticton 市水系统的可持续性发展的关键因素,并指出城市水系统的运行阶段热水使用是导致能源使用和碳排放的主要诱因。这种方法也广泛应用于政策评价 Newell 等建立了"气候-能源-水"关系的主要因果循图[54],发现技术创新、配电和输电网络冗余的增加以及决策中的动态系统思维有助于提高澳大利亚国家电力市场的弹性。El-Gafy 研究表明[55],与 2010 年基线情景相比,通过小麦贸易模式政策提高埃及小麦自给自足水平,预计 2010~2050 年埃及小麦种植面积和小麦生产水足迹将增加	城市尺度

方法名称	方法简介	应用举例	续表 适合的地理尺度
生命周期分析法（LCA）	对给定产品或过程在整个生命周期内的环境影响进行量化的最广泛使用的方法之一。它可以准确地显示任何单元在其生命周期中的量化，并容易导出其计算过程，具有识别所有可能对环境产生重大影响的输入或输出的特征[56]。基于以上优点，LCA 的方法已被广泛应用于评估"水-能源-粮食"关联系统对环境的影响并指导他们的生产和消费过程，旨在寻求有效的方法来应对目前的资源短缺和全球气候变化	AlAnsari 等人[57]运用生命周期方法（LCA）对能源、水和粮食供应链进行了综合环境评价，结果表明粮食系统对全球变暖的影响最为显著。 Feng 等人则证明[58]，与其他发电技术相比，风力发电在整个生命周期内会显著降低碳排放和用水量。 Foley 等人用 LCA 评价城市供水设施建设和运行对环境的影响并证明[59]，在污水处理厂中，运行能源和温室气体排放与脱氮呈正相关，与除磷无关。这种方法通过比较不同选择的影响，对决策也很有用	区域尺度 国家尺度
（生态）足迹分析	一般指生态足迹，是最为常见的可持续性指标。具体计算方法是衡量满足一定人口活动和物质消耗需要所需要的生物生产力和生产面积	Irabien 和 Darton 针对西班牙温室番茄生产，利用 LCA 和生态足迹分析，通过水足迹、碳足迹以及化学药品足迹的计算，成功地评估了在这个半干旱地区推进温室番茄生产的可持续性。结果显示了该生产模式在历史上对水资源利用的强烈依赖。 Daccache 等利用生态足迹方法研究了地中海地区灌溉农业生产中重要的淡水消费者。结合地理信息系统等工具，研究发现现代化灌溉小麦等作物能减少水的生态足迹，但增加了 CO_2 的排放。因此如何寻找优化的政策和技术策略仍然是这一地区农业生产的重要议题	区域尺度 国家尺度
智能体基模型（ABM）	一种"自下而上（bottom-up）"的方法，本质是一种数值模拟。其中每个"元素"（agent）都是一个独立的实体，在特定的社会环境中具有不同的目标和行为。通过对 agent 进行单独建模和微观模拟，可以在重建和预测整个系统在宏观上的表现和特点[60]。ABM 模型也可以方便地加入如博弈论、计算社会学等数值和理论方法来模拟和控制不同的参数	Smajgl 等利用 ABM 模型方法，针对云南橡胶种植园的扩张和保护当地生物多样性所产生的政策紧张，模拟了政策制定者、当地居民和种植园企业的行为策略。结果表明，贯行的生态系统服务补偿政策可能产生与预期相反的结果从而不利于实现协调经济发展与生态保护的目标	区域尺度
综合指数法	综合指数法通常通过多个指标反映"水-能源-粮食"关联系统的各种社会和环境特征。在这些指标的帮助下，研究人员能够捕捉到"水-能源-粮食"关联系统的整体特性，而不考虑其复杂性[61]	Flörke 等人[40]以城市地表水资源短缺为指标，指出气候变化和人口增长会加剧城市和农业部门之间的水资源竞争，并指出提高灌溉效率成为增强城市水资源安全的重要适应举措	全球尺度 城市尺度

从表（29-2）所总结的方法可以看出，适当的方法根据特定网络系统的规模和研究优先级而有所不同。具体来说，随着系统规模的扩大，可能需要更高程度的数据聚合。相反，当系统缩小时，应该更详细地描述连接系统的内部机制，包括社会和物理机制等。因此，在大规模的关联系统研究中，如可计算的一般均衡模型、计量经济分析及生态网络分析，当系统需要考虑人为关系（如全球供应链的粮食和能源[62]及其探讨社会经济之间的交互关系[63]）时，模型是必要的手段。然而，调查和数学方法可以用于说明几乎所有规模的关联问题，因为它们在克服数据缺乏和在初始阶段提供有价值的见解方面具有优势。

具体来说，随着系统规模的扩大，可能需要更高程度的数据聚合。不同的方法均有它们本身的优缺点及应用范围和条件。例如，智能体基（agent-base）模型更适合揭示个体的决策影响，考虑个体的异质性和生命周期分析，在评估资源消耗和环境相关的政策和技术转移方面具有优势。此外，综合指数法在政策规划中具有很强的社会和环境因素。

3.3 量化分析模型

在构建模型的基本方法基础上，为进一步填补 FEWS 关联关系理论研究与实践管理之间的缺口，学者们试图通过系统集成的方法，深入探究 FEWS 在具体应用模式下的模型，提出优化管理方案，包括长期能源规划模型、水资源评价与规划模型、一般均衡模型以及改进的土壤-资源评估模型等，对 FEWS 演化进行模拟与分析。一些学者和机构也针对 FEWS 模型研究成果开发了一些开源的新工具，为学者们和决策者评估与确定资源分配方案提供公共平台。

本节介绍了几种常见的基于量化分析的模型框架和实例。它们侧重于开发集成的相互关联模型，而且都提供了相应的分析工具。这些模型或框架要么是通过了严格的同行评审，要么是由联合国、国际水资源管理所和粮农组织等权威的非政府组织发布的工具和方法。

3.3.1 可计算一般均衡模型

可计算一般均衡（CGE）模型是一种经济模型，广泛应用于与经济相关的政策分析[63]。CGE 模型通过把握与市场行为和变化相关的相互联系，通过价格机制来评估政策对"水-能源-粮食"关联系统的影响[64]。各种 CGE 模型在关联量化方面取得了相当大的进展，如 GTAPW 模型[65]、GEM-E3 模型[66]和 IMPACT 模型[67]等。Calzadilla 等人利用 GTAPW 模型证明[65]，贸易自由化对农业生产和用水的影响是温和的。Wianwiwat 和 Asafu-Adjaye 发现[68]，促进新政策将增加生产粮食的价格在短期内减少粮食生产，但从长远来看其影响很小，表明这些政策对粮食安全的不利影响是有限的。

3.3.2 家庭尺度 FEWS 模型

这是关注小尺度的，综合了食物、能源和水三者关联关系的模型[69]。通过对 419 个家庭的涵盖两个季度的问卷调查分析，Wa'el 等收集了大量数据，将家庭的生活方式、家庭成员数量、收入、家用电器的效率，产生的固体废弃物和生活污水等细节囊括进了模型细节之中；同时，模型也考虑了大尺度的影响因素，例如气候变化（节气的变化）。该模型关注的尺度非常特别，因而能够给政策和管理提供家庭尺度的关联网络协同作用的关系。该模型可以很容易地扩展到城市尺度为政策制定者提供指导。

3.3.3 多部门系统分析

多部门系统分析模型是 Walker 等研究者基于 Matlab 语言编写对 FEWS 相关的物质流和能量流的城市尺度的模型[70]。该模型还综合了人类活动的影响。模型构建方法偏重基础物理化学过程，如定量分析能量、水、氮磷、碳等物质/能量在关联网络各个要素间的流动。经济相关的影响和温室气体排放也是模型关注的重要参数。由于模型需要精确的物理化学过程建模，所需要的实验或测试的数据量也是非常大的，对模型的应用造成局限。但是该模型还是成功地应用于大伦敦地区并为政策制定者提供政策和投资的指导。

3.3.4 水评估及规划模型

水评价与规划 WEAP 模型是水资源综合规划评价的一种方法。它评估了不同水文和政策情景下的水需求、供应、径流、入渗、作物需求、流量、储存和污染产生、处理、排放和流入河流的水质[71]。WEAP 的优势之一在于它是一个可伸缩的资源规划工具，可以比较供水和需求，并提供预测需求的功能。该建模工具不仅可以对水资源进行建模，还可以对生物量、能源和气候进行建模。

3.3.5 气候、土地利用、能源和水资源战略

该模型是斯德哥尔摩皇家理工学院开发的气候、土地利用、能源和水资源战略（CLEWS）的集成模型框架，应用在国家至跨地区尺度。其中集成了 WEAP-水模型、LEAP-能源模型和 AEZ-土地生产工具，以量化资源使用、温室气体排放以及与实现能源、水和粮食安全目标相关的成本。工具适用于为当前和未来的气候条件下评估依靠雨水和灌溉的作物的适宜性，生物燃料原料作物的潜力，作物变化的实用性

和影响,和在一个观察和预测降雨减少的趋势下确保足够水供应的措施[72]。该模型确定了关联网络各个子系统间正向和负向的反馈回路,并且考虑了气候变化和社会经济变化等因素。模型旨在提供兼顾关联网络三个要素安全的方法,以及利用这些资源的方式和程度对气候变化的影响,从而改变我们在未来获取这些资源的方式。

Welch 等[72]在毛里求斯应用了 CLEWS 模型来分析该岛国的气候、土地、水和能源等因素的相互作用和对政策的影响。毛里求斯有多样的气候,存在水资源匮乏的压力,并且政策偏向减少化石能源的进口且改变现存的农业生产方式,因而是应用 CLEWS 模型较理想的案例。模型结果指出,使用生物能源替代传统能源对环境资源的正向作用被高估了。

3.3.6 水、能源和食物关联网络工具 2.0

Daher 和 Mohtar 在 2015 发表了该在线模型工具[73]。这个工具提供了在 FEWS 中现存的关联联系的量化模型,涵盖了网络中的重要元素,并向科学界和政策制定者提供了一个分析、处理不同情况下国家可持续资源分配的策略平台。模型中的多个参数,例如食物的自足性(某国生产的某类型粮食作物能够满足国内消费的百分比)、水资源情况、能源资源禀赋和国家的进口数量等,都可以进行调整来模拟不同的案例。

模型着重分析了三者之间的物质或能量流动。如图 29-3 所示,在食物自足的情况下,以国家为单位,粮食生产可分为内产内销(DPC)、内产外销(DPE)或是进口(IMP)。进口粮食安全能够以国际市场的贸易来得到保障;内产内销和内产外销则需要消耗本地资源。这样,根据作物产量可以计算出需要的作

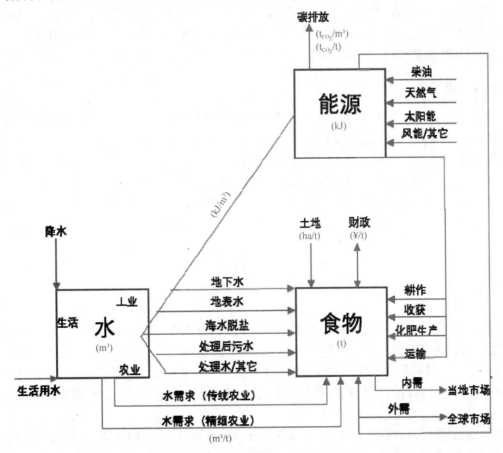

图 29-3 食物-能源-水关联模型构成框架[73]

物面积。又根据作物类型和经济价值的不同，基于市场价格，可以得到生产单位粮食所需要的金钱和卖出得到的利润。另外，也可以根据作物类型不同，计算所需要的水量。具体的农业技术也是要考虑的因素。比如粗放性灌溉和温室作物所需要的水量就非常不同。灌溉方式也是重要的影响因素。为了保证供水安全，地下水、地表水、海水淡化、处理后污水等不同来源以及相关的能量消耗也可以计算出来。农业生产过程中的能量消耗，例如粮食收割、施用化肥等所需能源和碳足迹也可以囊括进该模型。能源安全可以由柴油、天然气、清洁能源等保证。粮食进出口运输的能量和碳足迹也被考虑。

3.3.7 关联网络评估 1.0

这个工具是粮农组织（FAO）开发的[74]，是一个在线的快速评估工具，针对地区中相关食物、能量和水相关介入方所产生的生物经济（bio-economic）压力。该工具结合了量化分析、输入-输出工具和介入方比较模块三个方面的功能。所利用的数据来自于公共或公开的数据源。评估的流程包括内容分析、量化评估、反馈选项和利益相关方对话等模块，可以直接对政策和利益相关方提供专业建议。

3.3.8 农业和能源用水投资诊断工具

这个工具也是粮农组织开发的，它为粮食和能源生产项目用水资源的投资计划提供了评估服务。该工具实际上综合了 FAO 现有的在农业和能源领域的水资源投资进行分析的三个工具：①帮助理解针对食物生产（传统农业和水培农业）中水资源的管理利用潜力和需要的指标体系；②能够反映机构、立法和政策实体的运行特点，提供政策和投资的设计和应用的工具；③对现有和计划中的，在农业和能源领域的水资源投资进行项目评估，在长中短不同时间维度、国家内部或国际间等不同尺度的案例进行分析的工具。该模型分析了水电和灌溉项目对人类、社会和环境，贫困和粮食安全，健康和营养等三个方面的影响，从而确定反映国家体制、法律和政策现实的可行方案[75]。

3.3.9 WEFO 模型

WEFO 模型是一个多时段社会经济学模型[76]。模型使用编程语言 Julia，并使用了线性规划方法，对 FEWS 方面资源管理和决策提供数理支持。基于该模型，我们可以预测不同时段的 FEWS 各要素的生产成本、资源可利用率、社会经济需求和温室气体排放控制等参数。技术上讲，线性规划的解是由单纯形法给出的。优化的目标函数是系统总消耗。系统总消耗可以由系统各个部分的能量消耗、水消耗减去产生的电力、食物产出和减少的温室气体排放得出。由于精确的数学模型细节，使用者（政策制定方）可以通过调节电力和食物的社会经济需求，水资源的分布以及关注的环境压力限制条件，得到不同的优化方案。

3.4 模型小结

由此可见，由于 FEWS 关联关系的内涵较为丰富，其概念模型建立的关注点，从三者之间的相互依存扩展到 FEWS 内部及其作为一个整体与外部环境之间的关系。尤其是 FEW 三者之间客观存在的"关联或耦合关系"，已被相关研究者和政策制定者广泛接受。但整体而言，国际社会对 FEWS 概念模型的建立多种多样，对其内涵和框架的构建未能一致，相关表述也较为宽泛和模糊，缺乏可操作的框架与步骤。

尽管当前 WEF 模型面对着巨大挑战、知识欠缺和实际需要变化，但不应忽视它们的贡献和努力，以便更好地理解 WEF 的相互关联并量化和权衡各个部门，从而面向综合资源建模和管理迈出重要一步。显然，当前所有的研究方案在方法和结构上都是有前途和创新的。目前的模型不仅研究了技术含义，而且超越了社会、气候和政治环境，通过数据共享和时空驱动的整合向高度集成的模型建立迈出关键一步。此外，为比较这三种资源，非常需要建立一个共同的等价基础，当然这也是产生有效的政策建议所亟需的。

总而言之，从相关模型的数量中可以明显看出人们有很大的动力去理解 WEF 各部门之间的相互联系。与此同时，需要基于深入研究而将各种联系纳入决策过程。目前的方法面临着严重的理论匮乏，阻碍了科学家和研究人员开发实用框架。改进跨部门、跨流域、跨国家和跨区域的数据共享，是改进综合连接模型、包括空间和时间等要素的第一步。综合数据系统将产生更强有力的建模工具，对决策过程产生积极影响，并有助于有效管理这些相互关联的资源而改变相关机构和方法。通过对现有水-能源-粮食模型和框架的分析，我们认为决策者在使用这些方法时需要考虑如下 5 个关键问题：

（1）建立能够监督资源之间相互联系的机构，并向它们提供分析工具，以评估资源之间的此消彼长；
（2）鼓励多部门涉众参与，从而使新兴解决方案在各个部门中得到顺利接受；
（3）通过设定多个目标，针对多个部门的资源，设计成本效益政策；
（4）管理替代能源，对水和食物价格产生积极影响；
（5）促进技术创新，以确定更好的方案用于探索要素之间相互联系，从而实现预期的政策效果。

目前的模型和框架为更好地理解 WEF 之间的联系以及量化各部门之间的权衡关系铺平了道路，从而为集成资源建模和管理做出了关键的初步努力。然而，一些可用的模型和框架遇到了方法上的不足，并且可能被限制在特定学科的特定地理规模或范围内。这种限制可能导致由于各种方法的不足而不能充分利用所有资源。相互联结方案优化了对所有这三种资源的使用，然而执行这些策略的机构通常不会以一种综合的方式去执行，而往往偏爱其中一种。这些机构利用自己的员工、数据和工具来制定自己的方案。因此，大多数建议只适用于给定位置和时间点的特定应用领域。

显然非常需要更复杂的建模方法来评估跨部门和参与者之间的 WEF 关联关系。这些方法将帮助决策者制定基于实际状况的政策，以实现可持续性和有效的资源利用。此外，由于许多相互关联的参数随时间和空间而变化，因此有必要从空间和时间上研究资源系统之间的相互联系。在国家层面对水和粮食系统的能源消耗进行系统和定量的评估，使决策者有机会探讨对环境和未来规划的更广泛影响，并提出更好协调的政策选择。

4 FEWS 关联网络的工程技术

食物-能源-水（FEW）是具有全球意义的社会生态关联系统，包括社会、经济和环境等的相互作用。为了将其创建为可持续系统，需要进行跨学科和多个部门的协作研究以及多层面的分析。实际上，关于综合利用食物-能源-水网络的关联关系的工程技术一直是各个学科的研究热点。例如，农业生产中灌溉技术的进步就体现了对食物生产、能源消耗和水资源利用的综合考量；水处理技术中对资源的农业回用也是综合考虑水、能源和农业生产的热门领域；生物质能源的生产也应该综合考量对农业、能源和水的影响。

同样值得注意的是，在世界各国政府和官方实体的鼓励下，重新利用处理过的废水，特别是用于农业灌溉，是一种日益普遍的做法。废水灌溉可能在两个不同的层次上产生影响：改变土壤的理化和微生物特性和/或在土壤中引入和促成化学和生物污染物的积累。前者可能影响土壤生产力和肥力；第二种可能对人类和环境健康构成严重威胁。农业废水的可持续再利用应防止这两种影响，需要进行全面和综合的风险评估。本节将介绍几类典型的综合考虑 FEW 系统的工程应用和技术创新。

4.1 污水中营养物质的回用

生活污水、化肥、食物等工业废水和农田排水等含有大量的氮、磷和无机盐等。这些废水如果不经过处理就直接排放入湖泊、河流，不仅会使天然水体中的营养物质增多，促使自养型生物旺盛生长，还会使得营养元素流失。如果能够发展有效的工程技术，合理处理这些污水，回用其中的水和营养物质作为农业肥料，不仅能够节约农业用水的使用，还能节约在污水处理、化肥生产等过程中消耗的能源。因此，污水回用设计 FEWS 的全部三个因素，成为近年来环境技术的重要研究领域。

4.1.1 鸟粪石沉淀法回用污水中的氮磷营养物质

鸟粪石（MAP）沉淀法是一种高效的处理高氨氮废水的方法，其基本原理是通过向氨氮废水中投加 Mg^{2+} 和 PO_4^{3-}，与 NH_4^+ 生成难溶于水的磷酸铵镁沉淀（$MgNH_4PO_4 \cdot 6H_2O$），从而将氨氮从废水中去除[77]。鸟粪石沉淀法不仅可以将废水中的氨氮以沉淀物的形式固定下来，还能将磷元素固定下来，并且鸟粪石颗粒在形成过程中基本不吸收有毒物质、重金属及有机物，因此，会得到杂质含量较少的化学物，可作为一种高效缓释肥应用于农林业中。因此，很多研究者们着力于通过鸟粪石沉淀法处理废水。例如，Liu 等[78]通过鸟粪石沉淀法从猪场沼液中同时回收氮磷元素，并考察了曝气率和镁磷比对鸟粪石结晶的影响，且对鸟粪石结晶进行了物理、化学分析。此外，Gyutae 等[79]使用污泥灰的酸浸出液与猪场沼液反应，反应产物以鸟粪石的形式沉淀出来，同时去除了废液中的氮磷元素，结果显示氨氮和磷的去除效率分别达到 91.95%和 99.65%。Claudio 等[80]往垃圾渗滤液中分别加磷酸和氧化镁，外加的磷元素和镁元素与垃圾渗滤液中的氮元素反应生成鸟粪石，从而去除了氮元素，结果显示氮元素的去除率最高可达 95%。Ayla 等[81]通过鸟粪石沉淀法从乳制品工业的厌氧废水中回收氮磷元素，他们使用了曲面响应法来优化实验操作条件，结果显示在最佳实验操作条件下，氮、磷元素的去除率分别达 88.4%和 79.3%。Edson 等[82]将可乐饮料废水作为磷元素的来源，往其中加入 $MgCl_2 \cdot 6H_2O$ 和 NH_4Cl 生成鸟粪石，从而达到从饮料废水回收磷元素的目的。

鸟粪石沉淀法不仅用来在实验室规模处理高氨氮废水，还用在了实际工厂应用中。例如，Míriam Cerrillo 等[83]在基于鸟粪石沉淀法批次和连续操作处理猪场沼液的小试实验结果之上，在西班牙的加泰罗利亚地区建立了一个 3.10 m^3 半工业规模的鸟粪石沉淀法装置，且耦合了沼气厌氧消化和堆肥工艺，结果表明用得到的鸟粪石种植的生菜长势很好，未见任何缺乏营养的症状；此外，生菜产量与用磷酸氢二铵种植的生菜相比没有明显差异；经济衡算的结果显示这种处理猪场沼液的成本为 7.45 €/m^3 沼液。Haiming 等[84]使用氧化镁皂化的废水作为镁元素的来源，使用鸟粪石沉淀法对猪场沼液进行处理，首先，通过小试试验考察各操作条件（pH，镁元素的添加量等）对鸟粪石沉淀过程的影响，然后，基于小试试验的数据在内蒙古包头市的近郊区域的一个猪场（每天的废水排放量为 50 m^3）进行了为期 45 天的半工业规模的连续操作实验，结果显示氨氮的平均去除率高达 93%，经济衡算的结果显示：在不考虑人力成本和鸟粪石的商业价值的情况下，用该法处理成本为 6.03 \$/m^3 的沼液。

除此之外，鸟粪石沉淀法还可以与其他方法集成处理废水。例如，张正红等[85]将鸟粪石沉淀法和絮凝法集成起来同步处理猪场沼液：先通过鸟粪石沉淀法实现氮、磷等营养物质的回收，再用絮凝剂对沼液中的悬浮固体、有机质和鸟粪石晶粒进行絮凝作用，同时依靠反应体系中产生的絮体作为晶种诱导鸟粪石结晶反应，充分富集回收沼液中氮、磷及各种有机质。此外，他们还将鸟粪石沉淀与光合细菌复合序批式生物膜反应器集成起来处理猪场沼液[86]：先通过鸟粪石沉淀法去除猪场沼液中的氮、磷和有机物和降低沼液中的悬浮物，提高沼液的碳氮比；然后，通过同时接种光合细菌和活性污泥构建光合细菌复合序批式生物膜反应器进一步净化沼液。冯莹莹等[87]采用鸟粪石沉淀-协同氧化预处理阻燃剂废水（总磷、

总氮中 PO_4^{3-} 和 NH_4^+ 含量较高）。他们先采用鸟粪石沉淀工艺去除废水中的悬浮物、非溶解性有机物和大部分 PO_4^{3-} 和 NH_4^+，后续采用 Fe/C 微电解-Fenton 组合工艺降低废水中 COD，提高废水的可生化性。施毅君等[88]将鸟粪石沉淀法与海水电池相结合处理养殖废水中的 NH_4^+-N 和 P，具体地说：利用海水电池镁阳极腐蚀产生 Mg^{2+}，与虾池养殖废水中的 NH_4^+-和 P 元素形成了鸟粪石，实现了资源化利用，同时，海水电池还用来解决了养殖场的供电问题。

随着鸟粪石沉淀法在处理废水方面取得了不错的效果和成果，研究者们对于该过程的影响因素也做了一些总结：①悬浮固体的含量。一方面，悬浮固体可以提供晶种从而加快鸟粪石的沉淀和结晶速度；另一方面，悬浮固体阻断鸟粪石晶体的活性增长区域，从而能阻碍晶体生长、减缓晶体生长速度[89]。②杂质离子和细菌等的干扰。例如钙离子、锌离子、铜离子、重金属离子（铬离子、砷离子等）、碳酸根离子、硫酸根离子等会影响鸟粪石的反应时间、回收效率、晶体形貌和纯度等。③可溶性磷元素的不足。废水体系中可溶性磷元素占总磷量的比例较低使得鸟粪石沉淀的进程困难。因此，在进行鸟粪石沉淀法处理废水之前，通常通过有酸化、微波辐射等技术预处理磷元素将磷元素转换为可溶性磷元素的形式[90]。④废水中氮、镁和磷元素的比例不协调。理论上，氮、镁和磷元素的摩尔比例为 1：1：1 刚好生成鸟粪石，但实际废水中这三者摩尔比并不是 1：1：1[91]。为了使得元素的去除率达到最大、鸟粪石的生成率最大，需要往废水中外加为 $MgCl_2$、$MgSO_4$、MgO、卤水、海水、水镁石和骨粉废水等[92, 93]添加剂来弥补其他含量不足的元素。⑤合适的废水体系 pH 的调节。废水体系的 pH 会影响体系中的可溶性氮、镁和磷元素的比例，可以通过往废液中通入二氧化碳气体提高溶液 pH[94]，或往废液中添加炉渣使其中的重金属离子发生水解产生 OH^- 而调节 pH[95]。

如前所述，鸟粪石沉淀法之所以引起众多研究者的注意，是因为得到的鸟粪石可以在农业上作为缓释肥进行施用。鸟粪石作为缓释肥施用一方面可以补充作物生长所需要的营养元素，另一方面还能提高营养元素的利用效率，例如 Liu 等[96]研究发现使用鸟粪石的 NO_x 排放量远低于使用普通氮肥的排放量，排放的 NO_x 不仅造成了营养元素的流失，还污染了环境。目前，商业鸟粪石已经成功应用于蔬菜、盆栽植物、树苗等的生长[97]，尤其是对于绿叶子植物更有效，因为镁离子是合成叶绿素必不可少的元素之一。Ponce 等[98]通过研究发现施用鸟粪石（从尿液中获取）的生菜的产量和含磷量要高于施用单一的过磷酸盐的生菜，主要是因为鸟粪石中镁元素的存在对磷元素的溶解和吸收存在着协同作用。Rasul 等[99]对两组玉米分别施用常规化肥（镁含量为 80 kg/ha）和鸟粪石，结果发现这两组玉米的株高、叶面积和干物质产量都很相似。Li 等[97]使用由垃圾渗滤液制备的鸟粪石种植卷心菜，结果发现与化肥（硝酸铵、磷酸钙）种植的卷心菜相比，该卷心菜的长势类似，但镁、磷元素的含量要高。尽管鸟粪石的施用有利于植物的生产，但有研究发现长时间的施用鸟粪石可使得土壤中的钙镁比发生变化，从而改变土壤孔隙度、团聚稳定性和导水率[100]，因此鸟粪石应该与其他化肥共同施用，以减少对土壤的影响。

4.1.2 电化学法回用氮磷

电化学法被认为是从污水中去除和回收氮磷等营养元素的有效手段，简单分为发生在阳极的氧化反应和发生在阴极的还原反应，利用电化学方法在阴极沉淀回收鸟粪石的原理是 H^+ 在阴极附近得到电子生产 H_2，生产的 OH^- 使得电极附近 pH 升高，达到适宜鸟粪石生成的碱性条件，从而诱发鸟粪石的生成。Mariakakis 等[101]将电化学法用于德国的污水处理厂，使用镁棒作为阳极，进水为强化生物除磷工艺的污泥厌氧发酵消化液，成功运行 3 个月后，出水 P 的平均去除率约为 84.2%，出水浓度约为 20.8 mg/L。Alexandra Hug 等[102]将电化学法用于处理尿液，他们使用一个体积为 1L 的单式反应器，阳极采用镁板，阴极采用不锈钢板，结果表明：阳极电位大于 –0.9V，反应器运行 2h 时，95%以上的尿液可以被处理完，且回收得到鸟粪石。

电渗析技术也是一种常见的电化学技术。阴离子交换膜和阳离子交换膜交替排列构成膜堆，再外加电势差即可实现离子的定向迁移。向膜堆中提供合适的电解质溶液，电渗析技术能够实现对电解质溶液的离子分离和浓缩。王晓林利用电渗析技术和鸟粪石反应器集成的技术对水处理系统中厌氧消解后的污泥中的氮磷同时进行资源化，以鸟粪石的形式来回收氮磷盐。在单独连续操作 CED 处理废水时，氮磷盐回收率能够达到 90%～100%；在联合了电渗析池和鸟粪石反应器后，浓缩室中氮磷盐可以降低 60%～70%[103]。

4.1.3 厌氧反应器法回用氮磷

与污水灌溉不同，生活污水的回用技术重点关注对污水的分流、转化和营养物质的回用上。回用的污水可以用作城市分散农业的水源。图 29-4 展示了一个回用生活污水并回用营养物质进行无土农业生产的工艺流程概念[104, 105]。低固体含量的生活污水首先经过厌氧消解膜反应器（AnMBR）。经过生化反应，污水中的氮磷等物质能够以可溶状态被植物利用。同时，回用的水也能够减少灌溉用水需求；厌氧过程产生的沼气也能够作为能源加以利用。对于高固体物质含量的生活废水/废渣（食物残渣等），需要进行热解等预处理来提高后续消解和产气的效率。同时，整个工艺所需要的电力（包括加热、制冷等）可以利用清洁能源。回收并富集的氮磷等营养物质可以用来直接用作无土温室的灌溉用水；或者用来生产鸟粪石后用作肥料[106]。

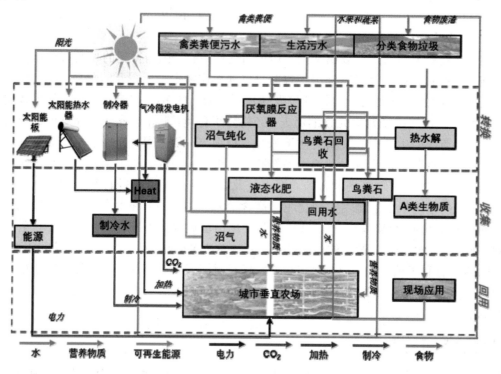

图 29-4　污水回用进行农业生产设计的"食物-能源-水"关系

4.2　可持续能源技术

可持续能源的研究是一种体现能源-食物-水系统关联的重要领域。例如，利用农业生物质作为能源能够降低温室气体排放。比较常见的例子包括生物乙醇和生物柴油。生物乙醇的主要原料是产糖类和产淀粉类作物。生物乙醇可以混合汽油作为燃料。生物柴油的主要原料是有机脂肪酸酯和植物油脂，可以混合入柴油作为燃料。本质上讲，生物质能源就是利用短期的碳循环提供能源，而不是像化石燃料那样依靠地质时间尺度的碳循环[107]。生物质燃料能够直接降低温室气体排放，提高能源供给的安全性，增加单位土地作物的经济价值。已有丰富的文献对相关领域进行研究[108, 109]，本节不再赘述。本节主要从FEW系统的角度，分析生物质燃料技术对FEW系统整体的影响。

从系统关联的角度来看，生物质能源作物的种植可能占用耕地，消耗更多水资源，需要更多的化肥。一些模型模拟的结果就对生物柴油对温室气体排放的负面结果做出过预测。Marta等学者[110]研究了玉米（*Zea mais*）和葵花（*Helianthus annuus*）在Tuscany地区五十年历史的种植模拟（CropSyst模型）。通过整合五十年的气候数据，考虑能耗和水消耗后，能源作物的种植被认为消耗两倍的水，能源净产出也在大部分情况下是负值。Welsch等[72]对毛里求斯玉米种植应用CLEWS模型分析指出，生物柴油的生产在正常年份是有经济收益和环境友好的。但是问题出现在降雨量下降的情况下，温室气体的排放就会显著增加。为了维持淡水的供应和粮食作物的生产，海水淡化和泵取地下水灌溉的能源消耗显著增加。这样，从应用生物柴油中减少的碳排放也不足以弥补因生产能源而增加的排放。

实际的例子也有报道。例如，2003~2007年，全球玉米产量增加的三分之二是用于生物柴油的生产，这间接导致了2008年粮食价格的骤升[73]。在30年的应用历史中，玉米生物乙醇实际上让全球温室气体的排放翻倍，也由于耕种土地的转变和扩张，给灌溉用水产生了更大的压力[111]。

因此，未来对可持续能源相关的新技术有必要先从FEWS系统综合的角度做出全面的资源和政策的分析，以免局限于各要素内部或两两组合的视角，推导出从全局看并不准确的结论或政策建议。

4.3 其他综合技术

即使在传统领域，以食物-能源-水系统的角度进行生产流程优化也是环境领域的重要课题。例如在污水处理厂，合理地利用技术就能够实现能源回用、营养物质回收和水回用。从能量的角度来讲，厌氧消解产生的甲烷气可以作为燃料在合并热能系统（CHPs）中得到利用；沉淀污泥作为生物固废可以用来焚烧发电；或是利用废水提供微生物燃料电池的进水来发电；甚至利用处理厂地形的水头差进行水利发电也曾见报道[112]。在营养物质回收方面，脱水后的污泥可以作为土壤添加剂和酸碱调节剂；如果能够控制市政污水中尿液的分离，大量的氮磷营养物质也可以较方便地回收；活性污泥醇解液因其富含磷、氨和镁，是结晶提取鸟粪石的原料。针对水的回用，污水的农业灌溉是比较传统的回用技术；在工业冷却用水等所需水质不高的情况下，再生水即可满足要求；洗车、冲厕等用水需求也能通过再生水来满足。

农业灌溉用水的调控是一个典型的FEWS课题。在伊朗，每年用于泵取地下水灌溉的能量达到205亿kW·h，消耗20亿升柴油，大概占伊朗总碳排放的3.6%。Karimi的案例分析指出[113]，在伊朗的Gamasiab河谷地区，地下水灌溉是主要的农业灌溉形式。如果能够进行灌溉时间和灌溉水量的模型（SWAP）优化，农业效率、用水效率都能得到提高。同时，由于抽走了较少的地下水，泵水的能量消耗和碳足迹都能得到降低。综合考虑关于食物-能源-水的网络，优化灌溉时间和水量能够对整个关联系统产生正向的促进作用。理想情况下，能耗和碳足迹能够降低40%。这里提到的优化方法主要是使用基于对地下水和土壤类型的Richard's方程。灌溉作业需要综合考虑降水量、土壤性质、农作物耗水量、生长周期等因素来决定合适的灌溉时间安排[113]。在肯尼亚，利用太阳能的区域网灌溉系统也取得了成功。由于不使用柴油作为地下水泵的能源，灌溉的成本和环境足迹得到了显著的减低。平均每户农户可节省灌溉成本400美元，降低CO_2排放521kg[114]。

5 FEWS 的相关政策

尽管近年来在食物-能源-水（FEWS）关系方面的研究出现了重大的、政策驱动性的增长，甚至 INFEWS 各工程技术的进展捷报声此起彼伏，但针对有效、综合治理 FEWS 关系及资源协调过程中所需政策结构和过程变化的研究仍处于起步阶段。在研究过程中一些模型和框架作为区域管理综合方法的一种手段而被开发出来，已经开发出的这些框架和模型不同程度上帮助了决策者了解 FEWS 关系的复杂性，并协助规划和管理这些资源。大多数国家和政府通过沿用或者设置机构来管理这些自然资源，而这些机构因为有着自己的使命和目标，并有自己独有的工作人员、数据、措施和工具，这些情况导致了模型设计中的方法论方法、使用的数据类型以及最终结果解释千差万别，最终使得政策设计变得无所适从[115]。以国家资源政策为例，它优化了粮食生产，却没有考虑到对水、能源和土地的连锁反应，以及这些动态输入成分如何随时间和空间变化。在印度西北部，计划不周的自然资源政策导致农民耗尽了该地区的自然资源，不得已需要使用增加农业投入（化肥和过量的地下水用于灌溉）的方式来支持国家粮食安全[116]。这项政策会导致当地生态系统的资源枯竭，无法达到提高粮食安全水平所期望的政策结果。因此我们发现 FEWS 管理问题是被嵌入到全球、区域、国家和地方不同等级主管机构的政策、体制、技术和融资选择中的，此外各层级之间的强烈互动促使政策对特定事件和结果来作出反应，在特定时期和阶段，不同层级的利益方向并不相同，不同利益相关者的短期获利与人类文明长期整体解决方案之间存在巨大的权衡和鸿沟。站在决策制定者的角度来看，FEW 不仅仅是资源和技术问题，尽管在小范围和小区域内某些政策能发挥作用、提高效率，但是到达一定高度之后，传统的政府组织方式就体现出各部门各自为政的劣势。因此，FEW 问题的解决和相关政策的推出一定是政治的妥协和利益的交换。除此之外，不同收入水平的国家所面临的问题也不在一个层次上。在高收入国家，20 世纪初至中期开发的水处理和分配系统导致公共卫生的显著改善。但是在许多地方，水资源基础设施已经超过了预期的使用寿命，而且基础设施的局限性越来越明显。旧的配水系统管道泄漏，需要修复或更换，以确保水的可靠性和质量。比如美国一些较旧的配水系统和许多住宅管道系统含有铅，在一定的水质和流量条件下，铅可以进入配水，并使居民面临着不健康生活状态的风险。这些国家迫切需要更新和替换这些老化系统。而在低收入国家则面临着一系列不同的配水挑战。在许多地区，废水未经适当处理就排入地表水，污染了水体，使人们无法获得安全的饮用水。他们更迫切需求收集、运输和处理水和废水的集中基础设施，以及采用先进水回用技术的可以提高供水量并回收嵌入的能源的分散式废水处理系统。

5.1 世界各国和国际组织高度重视食物、水及能源安全

联合国粮食及农业组织（FAO）将粮食安全定义为"当所有人在任何时候都从物质、社会和经济上获得足够、安全和营养的食物，满足他们的饮食需求和对积极健康生活的食物偏好。"（世界粮食首脑会议，1996 年）。粮食安全的多维性包括粮食的可得性、稳定性和使用性。粮农组织将粮食安全管理描述为"代表社会成员制定和执行正式和非正式的规则和程序，通过这些规则和程序阐明利益，并作出与一国粮食安全有关的决定。"[117]。粮食安全治理框架通常从全球层面到地方层面，涉及全球供应和需求，由世界贸易组织来监督国际贸易行为，食物安全规则和食品援助。

全球水资源伙伴关系（GWP）将水安全定义为："确保提供质量可接受的充足可靠的水资源，以环境可持续的方式为所有社会和经济活动提供水供应服务；减轻与水有关的风险，如洪水、干旱和污染；解

决共享水域争端可能产生的冲突,特别是在压力不断增加的情况下,并将其转化为双赢的解决方案。"[118]。Hussey将国家级别的水治理定义为"社会系统以协调一致的方式调动能源以实现水资源可持续发展的能力"[4]。这一概念包括能够设计社会接受的公共政策(并动员社会资源支持这些政策),这些政策的目标是可持续发展和利用水资源,并使参与这一过程的不同行动者/利益相关者能够有效地实施这些政策。

国际能源安全主要还是以可承受的价格不间断地提供能源,这是1974年建立国际能源机构的主要目标。石油安全仍然是国际能源署的基石。长期能源安全主要与及时投资相联系,以便根据经济发展和环境需要提供能源。短期能源安全的重点是能源系统对供需平衡的突然变化作出迅速反应的能力。围绕促进能源部门的多样性、效率和灵活性,有一个不断演变的议程,以便它们能够应对能源紧急情况。然而,87%的能源供应来自石油、煤炭和天然气。基于化石能源的不可持续性,世界主要国家和经济体目前都在大力发展新能源和生物能源,来缓解或者彻底解决自己国家所面临的能源短缺问题。

5.2 在国际层级进行 FEW 管理和政策制定的复杂性

人口增长、城市化和气候变化的影响等因素对我们地球的自然资源,特别是水、食物和能源资源造成了巨大的压力。然而,获取这些资源的机会不足往往是一个管理范畴的问题,而不是一个技术或者可用性方面的问题[119]。联合国可持续发展目标或绿色经济的关键概念的通过,是科学和政治界关于处理全球变化和解决地球上资源短缺问题的持续辩论的结果,食物-能源-水(FEW)关系的出现也可以看作是这场辩论的结果之一[120]。它的主要目标是揭示这三种资源之间的联系,并以综合的方式管理它们。

然而想要实现有效的管理是困难的,这里我们以水资源管理为例,到目前为止水是管理最复杂的自然资源。它没有边界,因此没有受政治或行政限制的可能。它要求各个层级都能妥善对待它,这需要所有利益相关者的参与,不管参与者的级别和行政类别。

目前国际水资源冲突不断加剧[121]。超过40%的世界人口生活在跨越国际边界的263个流域[122],这些盆地中只有三分之一在跨界协议下发挥作用。1997年,联合国通过了一项关于国际水道不可通航使用的相对温和的公约,其中包含一项原则声明,即各国应确保"共享河流的可持续和公平使用"。中国是发源于青藏高原的5条主要河流[狮泉河(境外部分为印度河)、雅鲁藏布江、独龙江(境外部分为伊洛瓦底江)、怒江(境外部分为萨尔温江)和沧澜江(境外部分为湄公河)]的源头。在湄公河三角洲地区,国家主权与河岸国家利益之间的紧张关系日益加剧。目前还没有对跨界争端具有管辖权的超国家机构来应对这些紧张情况,根据《联合国宪章》第七章,未经有关国家同意,联合国机构不得进行干预,除非是对和平的威胁。此外,规模不均、政治权力和内部治理实力不同的国家在让国际组织参与达成水资源协定方面有不同的利益[123]。1960年印度和巴基斯坦在世界银行的斡旋下达成的印度和巴基斯坦的水资源共享协议,即使在两国冲突期间,仍然坚持执行。但鉴于国际组织的影响力不断削弱,以及国家实力不断增强,达成此类有效国际协议的前景如今似乎很渺茫。这导致各国政府不愿将争端或谈判国际化。尽管印度河协议得到了遵守,但大坝的盐度、盐水入侵和其他环境影响问题却十分严重,并引发了新的问题。鉴于目前28个联合国各机构、区域委员会和非联合国组织致力于制修订13个不同的重叠水资源相关方案,粮农组织负责可持续发展的助理总干事Alexander Mueller指出,在这些方案之间还有很多未涉及事项,需要加强协调。自从1992年在里约通过了四项都柏林原则,并设立了《21世纪议程》以来,人们对解决问题的理解有了很大的提升。解决方案在政治、社会文化和环境方面变得越来越具体,因此也越来越针对具体的国家和地区。这种方法将要求政治和行政精英在解决方案上达成共识。共识必须以强有力的分析工作为基础,它需要勇于担当的集体领导和为人类谋未来的思想,并更加注重管理。技术对于提高资源利用效率至关重要,包括新的INFEWS技术。但技术不能取代真正的决策人,也不能取代健全的政策和制度。

5.3 夯实工程技术，推动 FEW 政策的革新和发展

尽管在国际层级上 FEW 政策的推动会受到利益相关国家政治、社会文化和环境等方面的制约和相互影响，但是在小范围国家集团和国家内部是有条件进行 FEW 相关政策的革新和发展的。目前世界主要国家和经济体都在大力发展 FEWS 工程技术和革新和发展 FEW 相关政策。FEWS 工程技术的研究开发依赖于若干因素，如技术选择、燃料选择、资源可用性和市场因素，这些因素都可能受到国家资源政策的影响。除了自身的复杂性之外，这些资源也相互关联：水需要产生能源，能源需要提取、分配和处理水，食物生产需要水和能源。这些资源的相互依赖性必然导致管理这些资源的政策也是相互关联的。在许多政策对话中，管理这些相互关联的资源是由单独的机构来处理的，以便于决策，而决策过程往往忽略了资源的相互依赖性和相互关联性。在过去的十年中，在强调这种联系和提高决策者对这些问题的认识方面取得了一些进展。在理论和应用上开发出来多种模型和框架。这些模型和框架中的大多数在许多方面都将有助于将 FEWS 关系正确整合到自然资源政策中。

大多数这些模型和框架的方法论都是建立在确定问题、描述相互联系并最终促进向可持续性过渡的基础上的。不恰当的能源政策会对导致严重的后果。如在印度西北部，计划不周的自然资源政策导致农民使用增加的农业投入（超量化肥和过量的用于灌溉的地下水）以支持国家粮食安全[124]。这一政策的影响是得到一个资源枯竭的当地生态系统，未能实现更高级别粮食安全所需的政策成果。再看看在南非的政策格局中，这种联系是如何发挥的，我们就会发现更多问题和不平衡。南非目前正在评估其能源选择，并制定政策，将适当的"胡萝卜和大棒"应用于各种能源技术，以鼓励可持续能源生产。虽然成本、碳和能源安全方面得到了高度重视，但水需求并不是这一过程的一部分。同样，迄今为止，能源定价还没有成为南非水价战略的一部分[125]。由于这些资源之间的复杂联系，水-能源-粮食关系也是一个决策挑战。虽然已经提出了改进资源政策规划的模型和框架，但需要更复杂的模型来确定综合管理自然资源的实用方法。换句话说，迫切需要综合规划和系统思考，而不是优化一种资源对另一种资源的使用。

5.4 项目资助情况

近年来，各国政府对 FEWS 关联项目加大了资助力度。美国对该类项目研究起步较早，重视程度较高。美国自然基金会在近五年已先后批准了 88 项相关项目，累计资助金额约 1.5 亿美金。此外，美国农业部、食物与农业研究院也对 FEWS 项目进行了大力支持。我国国家自然科学基金委员会（NSFC）与美国国家科学基金会达成合作共识，于 2017 年双方在"食物-能源-水"系统关联领域共同征集资助合作研究项目，最终对"基于耦合水文模型的中美流域尺度"食物-能源-水"系统可持续性比较研究"等 8 个项目予以资助，资助金额超过 2000 万元。项目研究内容涵盖了"食物-能源-水"系统关联所涉及的流域尺度及气候变化背景下的可持续性、系统耦合模型、关联经济的耦合平台构建模拟等，将极大地推动"食物-能源-水"系统关联领域的研究。

6 展　望

现有的研究从不同的方面为"食物-能源-水"的关系提供了一个很好的前景，它可以支持评估权衡，预测潜在的（非预期的）影响，为经常有利益冲突的不同利益相关者提供一个协同优化的基础，并可视

化的影响决策[11]。

然而，目前的研究仍然存在局限性。例如，模型和框架的建立是基于问题的确定、关联关系的描述，并最终向可持续性过渡。由于这需要在增强建模的集成性方面迈上一个大台阶，因此目前仍然存在着相关知识匮乏。例如，空间和时间的动态关系以及如何优化某一目标函数中某一资源对其他资源的影响等就极具挑战性。其中，对于可用建模工具的缺陷往往就认为是由于资源间的高度互连和/或相关数据的缺乏而造成的。优化一个目标函数通常会将问题从一个部门转移到另一个部门，或者将解决方案从短期推迟到长期，进而导致这些问题的解决更具挑战性，成本也更高。鉴于自然资源的动态特点和供求关系，所以揭示一种资源的优化利用对周围环境和其他资源的影响是至关重要的。

综上所述，模型的建立和应用，应该从系统边界、数据不确定性及建模、连接机制、系统评价等方面确定未来的研究方向。未来"水-能源-粮食"系统模型的研究方向需要考虑：①系统边界的定义。②数据的不确定性和模型的不确定性。③涉及与资源流动及其供应链有关的物理、化学过程的基本机制。例如，在水力-水电-农业系统内设置联结问题，其内在机制是指水流状态（如速度、温度）、输沙量、藻类生长、生物化学反应（如通过水沙界面硝化反硝化）、水力发电和粮食生产以及决策机制。④评价耦合联结系统，包括评价指标和定量评价方法。

通过对现有 nexus 工具的了解和评估，我们认为：①能够同时处理水-能源-粮食的模型和框架数量十分有限，而在分析 nexus 时能够拥有多种方法和跨学科方法的模型和框架数量更少。由于缺乏数据共享和可用性，其中一些模型未能有效捕捉到 nexus 组件之间的相互作用。当我们设计出模型工具来分析这些关联时，为了达到关联思维并用于解决多部门资源的复杂性时，我们必须纳入水、能源和粮食以外的几个维度，例如政治、社会和经济等。②受到数据量或是国家层面政策的影响，对关联网络的分析往往局限于地区和国家尺度[126]。③农业部门需要大量能源作为生产的投入，但是很少有水-能源-粮食资源规划方法能够结合能源需求和水需求来将空间种植模式和土地利用纳入其中。④尽管这些模型或框架可为决策者提供有效选择，并利用宏观驱动力在总体水平上进行关联关系的分析，但这些方法往往忽略了较小规模资源管理的复杂性，而在较小规模上的气候和地理等其他因素对供求却呈现出巨大影响。⑤在纳入时空驱动以及资源可用性或可接受性的时空动态方面仍然存在着显著的知识欠缺。然而，鉴于自然资源随时间和空间所发生巨大的变化，上述环节已成为 WEF 框架设计的一个重要组成部分。

参 考 文 献

[1] Sachs I, Silk D. Food and energy: Strategies for sustainable development. Tokyo: United Nations University Press, 1990.

[2] McCalla A. The water, food, and trade nexus. Paper delivered at MENA-MED Conference convened by the World Bank in Marrakesh. 1997.

[3] Scott C A. The water-energy-climate nexus: Resources and policy outlook for aquifers in Mexico. Water Resources Research, 2011, 47(6): 45-51

[4] Hussey K, Pittock J. The energy-water nexus: Managing the links between energy and water for a sustainable future. Ecology and Society, 2012, 17(1): 31.

[5] Endo A, Tsurita I, Burnett K, et al. A review of the current state of research on the water, energy, and food nexus. Journal of Hydrology: Regional Studies, 2017, 11: 20-30.

[6] Hoff H. Understanding the nexus. Background paper for the Bonn 2011 Conference: The water, energy and food security nexus. Stockholm Environment Institute, Stockholm, 2011.

[7] Allan J A. Virtual water-the water, food, and trade nexus. Useful concept or misleading metaphor? Water international, 2003, 28(1): 106-113.

[8] Leck H, Conway D, Bradshaw M, et al. Tracing the water-energy-food nexus: Description, theory and practice. Geography

Compass, 2015, 9(8): 445-460.

[9] D'Odorico P, Davis K F, Rosa L, et al. The global food‐energy‐water nexus. Reviews of geophysics, 2018, 56(3): 456-531.

[10] Zhang C, Chen X, Li Y, et al. Water-energy-food nexus: Concepts, questions and methodologies. Journal of Cleaner Production, 2018, 195: 625-639.

[11] Liu J, Mooney H, Hull V, et al. Systems integration for global sustainability. Science, 2015, 347(6225): 1258832.

[12] Scott C A, Vicuña S, Blanco Gutiérrez I, et al. Irrigation efficiency and water-policy implications for river-basin resilience. Hydrology and Earth System Sciences, 2014, 18(4): 1339-1348.

[13] Keskinen M, Guillaume J, Kattelus M, et al. The water-energy-food nexus and the transboundary context: Insights from large Asian rivers. Water, 2016, 8(5): 193.

[14] Bizikova L, Roy D, Swanson D, et al. The water-energy-food security nexus: Towards a practical planning and decision-support framework for landscape investment and risk management. International Institute for Sustainable Development Winnipeg, Manitoba, 2013.

[15] Wen L, Lin A. 全球气候变化及其影响. 水科学进展, 2003, 14(5): 667-674.

[16] Zeng R, Cai X, Ringler C, et al. Hydropower versus irrigation—An analysis of global patterns. Environmental Research Letters, 2017, 12(3): 034.

[17] Bandara J S, Cai Y. The impact of climate change on food crop productivity, food prices and food security in South Asia. Economic Analysis and Policy, 2014, 44(4): 451-465.

[18] Famiglietti J S. The global groundwater crisis. Nature Climate Change, 2014, 4: 945.

[19] Berardy A, Chester M V. Climate change vulnerability in the food, energy, and water nexus: concerns for agricultural production in Arizona and its urban export supply. Environmental Research Letters, 2017, 12(3): 035.

[20] García-Herrera R, Hernández E, Barriopedro D, et al. The Outstanding 2004/05 drought in the Iberian Peninsula: Associated atmospheric circulation. Journal of Hydrometeorology, 2007, 8(3): 483-498.

[21] Tom M S, Fischbeck P S, Hendrickson C T. Energy use, blue water footprint, and greenhouse gas emissions for current food consumption patterns and dietary recommendations in the US. Environment Systems and Decisions, 2016, 36(1): 92-103.

[22] Housh M, Yaeger M A, Cai X, et al. Managing multiple mandates: A system of systems model to analyze strategies for producing cellulosic ethanol and reducing riverine nitrate loads in the Upper Mississippi River Basin. Environmental Science & Technology, 2015, 49(19): 11932-11940.

[23] Hosseini S, Barker K, Ramirez-Marquez J E. A review of definitions and measures of system resilience. Reliability Engineering & System Safety, 2016, 145: 47-61.

[24] Nguyen T D, Cai X, Ouyang Y, et al. Modeling infrastructure interdependencies, resiliency and sustainability. International Journal of Critical Infrastructures, Special Issue on Resilient and Sustainable Coupled Critical Infrastructures, 2014.

[25] Suweis S, Carr J A, Maritan A, et al. Resilience and reactivity of global food security. Proceedings of the National Academy of Sciences, 2015, 112(22): 6902.

[26] 崔胜辉, 李旋旗, 李扬, 等. 全球变化背景下的适应性研究综述. 地理科学进展, 2011, 30(9): 1088-1098.

[27] Biggs E M, Bruce E, Boruff B, et al. Sustainable development and the water–energy–food nexus: A perspective on livelihoods. Environmental Science & Policy, 2015, 54: 389-397.

[28] Initiative WEFW. Water security: The water-food-energy-climate nexus. Island Press, 2012.

[29] Rasul G. Food, water, and energy security in South Asia: A nexus perspective from the Hindu Kush Himalayan region☆. Environmental Science & Policy, 2014, 39: 35-48.

[30] Giampietro M, Mayumi K, Ramos-Martin J. Multi-scale integrated analysis of societal and ecosystem metabolism

(MuSIASEM): Theoretical concepts and basic rationale. Energy, 2009, 34(3): 313-322.

[31] Giampietro M, Aspinall R J, Ramos-Martin J, et al. Resource accounting for sustainability assessment: The nexus between energy, food, water and land use. Routledge, 2014.

[32] Leck H, Conway D, Bradshaw M, et al. Tracing the water–energy–food nexus: Description, theory and practice. Geography Compass, 2015, 9(8): 445-460.

[33] Bazilian M, Rogner H, Howells M, et al. Considering the energy, water and food nexus: Towards an integrated modelling approach. Energy Policy, 2011, 39(12): 7896-7906.

[34] Liu J, Yang H, Cudennec C, et al. Challenges in operationalizing the water–energy–food nexus. Hydrological Sciences Journal, 2017, 62(11): 1714-1720.

[35] Dai J, Wu S, Han G, et al. Water-energy nexus: A review of methods and tools for macro-assessment. Applied Energy, 2018, 210: 393-408.

[36] Machell J, Prior K, Allan R, et al. The water energy food nexus—Challenges and emerging solutions. Environmental Science: Water Research & Technology, 2015, 1(1): 15-16.

[37] Ramı́rez C A, Patel M, Blok K. How much energy to process one pound of meat? A comparison of energy use and specific energy consumption in the meat industry of four European countries. Energy, 2006, 31(12): 2047-2063.

[38] Alimagham S M, Soltani A, Zeinali E, et al. Energy flow analysis and estimation of greenhouse gases (GHG) emissions in different scenarios of soybean production (Case study: Gorgan region, Iran). Journal of Cleaner Production, 2017, 149: 621-628.

[39] Kourtit K, Suzuki S, Nijkamp P. Tracing high-sustainability performers among world cities - design and application of a multi-temporal data envelopment analysis. Habitat International, 2017, 68: 43-54.

[40] Flörke M, Schneider C, McDonald R I. Water competition between cities and agriculture driven by climate change and urban growth. Nature Sustainability, 2018, 1(1): 51-58.

[41] Wooldridge J M. Introductory econometrics: A modern approach. Toronto: Nelson Education, 2015.

[42] Huang B N, Hwang M J, Yang C W. Causal relationship between energy consumption and GDP growth revisited: A dynamic panel data approach. Ecological Economics, 2008, 67(1): 41-54.

[43] Allard A, Takman J, Uddin G S, et al. The N-shaped environmental Kuznets curve: An empirical evaluation using a panel quantile regression approach. Environmental Science and Pollution Research, 2018, 25(6): 5848-5861.

[44] Zaman K, Awan U, Islam T, et al. Econometric applications for measuring the environmental impacts of biofuel production in the panel of worlds' largest region. International Journal of Hydrogen Energy, 2016, 41(7): 4305-4325.

[45] Chen S, Chen B. Urban energy consumption: Different insights from energy flow analysis, input–output analysis and ecological network analysis. Applied Energy, 2015, 138: 99-107.

[46] Chen S, Chen B. Urban energy-water nexus: A network perspective. Applied Energy, 2016, 184: 905-914.

[47] Chen S, Chen B. Changing urban carbon metabolism over time: Historical trajectory and future pathway. Environmental Science & Technology, 2017, 51(13): 7560-7571.

[48] Duan C, Chen B. Energy-water nexus of international energy trade of China. Applied Energy, 2017, 194: 725-734.

[49] Yang Z, Mao X, Zhao X, et al. Ecological network analysis on global virtual water trade. Environmental Science & Technology, 2012, 46(3): 1796-1803.

[50] Lenzen M. Environmentally important paths, linkages and key sectors in the Australian economy. Structural Change and Economic Dynamics, 2003, 14(1): 1-34.

[51] Finn J T. Measures of ecosystem structure and function derived from analysis of flows. Journal of Theoretical Biology, 1976, 56(2): 363-380.

[52] Stave K A. A system dynamics model to facilitate public understanding of water management options in Las Vegas, Nevada. Journal of Environmental Management, 2003, 67(4): 303-313.

[53] Chhipi-Shrestha G, Hewage K, Sadiq R. Water–energy–carbon nexus modeling for urban water systems: System dynamics approach. Journal of Water Resources Planning and Management, 2017, 143(6): 04017016.

[54] Newell E B, Marsh D M, Sharma D. Enhancing the resilience of the Australian national electricity market: Taking a systems approach in policy development. 2011.

[55] K. El-Gafy I. System dynamic model for crop production, water footprint, and virtual water nexus. Water Resources Management, 2014, 28(13): 4467-4490.

[56] Nair S, George B, Malano H M, et al. Water-energy-greenhouse gas nexus of urban water systems: Review of concepts, state-of-art and methods. Resources, Conservation and Recycling, 2014, 89: 1-10.

[57] Al-Ansari T, Korre A, Nie Z, et al. Development of a life cycle assessment tool for the assessment of food production systems within the energy, water and food nexus. Sustainable Production and Consumption, 2015, 2: 52-66.

[58] Feng K, Hubacek K, Siu Y L, et al. The energy and water nexus in Chinese electricity production: A hybrid life cycle analysis. Renewable and Sustainable Energy Reviews, 2014, 39: 342-355.

[59] Foley J, de Haas D, Hartley K, et al. Comprehensive life cycle inventories of alternative wastewater treatment systems. Water Research, 2010, 44(5): 1654-1666.

[60] Macal C M, North M J. Tutorial on agent-based modeling and simulation. Gothenburg: Winter Simulation Conference (WSC), 2018.

[61] Endo A, Burnett K, Orencio P, et al. Methods of the water-energy-food nexus. Water, 2015, 7(10): 5806-5830.

[62] Ringler C, Bhaduri A, Lawford R. The nexus across water, energy, land and food(WELF): potential for improved resource use efficiency? Current Opinion in Environmental Sustainability, 2013, 5(6): 617-624.

[63] Silviu D A. Simulating the economic impact of resources depletion using a computable general equilibrium model for Romania. Procedia Economics and Finance, 2015, 22: 618-626.

[64] Ge J, Lei Y, Tokunaga S. Non-grain fuel ethanol expansion and its effects on food security: A computable general equilibrium analysis for China. Energy, 2014, 65: 346-356.

[65] Calzadilla A, Rehdanz K, Tol R S. Trade liberalization and climate change: A computable general equilibrium analysis of the impacts on global agriculture. Water, 2011, 3(2): 526-550.

[66] Németh G, Szabó L, Ciscar J C. Estimation of Armington elasticities in a CGE economy–energy–environment model for Europe. Economic Modelling, 2011, 28(4): 1993-1999.

[67] Rosegrant M W, Cai X, Cline S A. World water and food to 2025: dealing with scarcity. Intl Food Policy Res Inst, 2002.

[68] Wianwiwat S, Asafu-Adjaye J. Is there a role for biofuels in promoting energy self sufficiency and security? A CGE analysis of biofuel policy in Thailand. Energy Policy, 2013, 55: 543-555.

[69] Wa'el A H, Memon F A, Savic D A. An integrated model to evaluate water-energy-food nexus at a household scale. Environmental modelling & software, 2017, 93: 366-380.

[70] Walker R V, Beck M B, Hall J W, et al. The energy-water-food nexus: Strategic analysis of technologies for transforming the urban metabolism. Journal of Environmental Management, 2014, 141: 104-115.

[71] WEAP, Water Evaluation and Planning System, 2016 http://www.weap21.org/ index.asp?action=200(verified 04/18 2019).

[72] Welsch M, Hermann S, Howells M, et al. Adding value with CLEWS–Modelling the energy system and its interdependencies for Mauritius. Applied Energy, 2014, 113: 1429-1445.

[73] Daher B T, Mohtar R H. Water–energy–food (WEF) Nexus Tool 2.0: Guiding integrative resource planning and decision-making. Water international, 2015, 40(5-6): 748-771.

[74] FAO. Walking the Nexus Talk: Assessing the Water-Energy-Food Nexus in the Context of the Sustainable Energy for All Initiative, 2014. http://www.fao.org/3/a-i3959e.pdf(verified 4/18 2019).

[75] FAO. Diagnostic tools for investments in water for agriculture and energy, http://www.fao.org/land-water/databases-and-software/diagnostic-tools-for-investment/en/(verified 4/18 2019).

[76] Zhang X, Vesselinov V V. Integrated modeling approach for optimal management of water, energy and food security nexus. Advances in water resources, 2017, 101: 1-10.

[77] 陈静霞, 李咏梅. 鸟粪石沉淀法预处理高氨氮废水的镁盐研究. 环境工程学报, 2011, 5(12): 6.

[78] Liu Y, Kwag J H, Kim J H, et al. Recovery of nitrogen and phosphorus by struvite crystallization from swine wastewater. Desalination, 2011, 277(1): 364-369.

[79] Kwon G, Kang J, Nam J H, et al. Recovery of ammonia through struvite production using anaerobic digestate of piggery wastewater and leachate of sewage sludge ash. Environmental Technology, 2018, 39(7): 831-842.

[80] Di Iaconi C, Pagano M, Ramadori R, et al. Nitrogen recovery from a stabilized municipal landfill leachate. Bioresource Technology, 2010, 101(6): 1732-1736.

[81] Uysal A, Kuru B. Examination of nutrient removal from anaerobic effluent of the dairy processing industry by struvite precipitation using the response surface methodology. Fresenius Environmental Bulletin, 2013, 22(5): 1380-1387.

[82] Foletto E L, Santos WRBd, Mazutti M A, et al. Production of struvite from beverage waste as phosphorus source. Materials Research, 2013, 16: 242-245.

[83] Cerrillo M, Palatsi J, Comas J, et al. Struvite precipitation as a technology to be integrated in a manure anaerobic digestion treatment plant – removal efficiency, crystal characterization and agricultural assessment. Journal of Chemical Technology & Biotechnology, 2015, 90(6): 1135-1143.

[84] Huang H, Chen Y, Jiang Y, et al. Treatment of swine wastewater combined with MgO-saponification wastewater by struvite precipitation technology. Chemical Engineering Journal, 2014, 254: 418-425.

[85] 张正红, 向天勇, 单胜道. 鸟粪石结晶-絮凝同步处理沼液实验研究. 中国沼气, 2014, 32(4): 51-59.

[86] 张正红, 何文辉, 向天勇, 等. 鸟粪石沉淀-光合细菌复合序批式生物膜反应器协同处理猪场沼液. 环境污染与防治, 2018, 40(4): 5.

[87] 冯莹莹, 崔康平, 王郑, 等. 鸟粪石沉淀-协同氧化预处理阻燃剂废水. 环境工程学报, 2017, 11(7): 6.

[88] 施毅君, 胡润芝, 付建灵, 等. 海水电池资源化利用虾池养殖废水的实验研究. 福建师范大学学报(自然科学版): 2019, 35(1): 7.

[89] Le Corre K S, Valsami-Jones E, Hobbs P, et al. Impact of calcium on struvite crystal size, shape and purity. Journal of Crystal Growth, 2005, 283(3): 514-522.

[90] Pan S H, Lo K V, Liao P H, et al. Microwave pretreatment for enhancement of phosphorus release from dairy manure. J Environ Sci Health B, 2006, 41(4): 451-458.

[91] Zhang T, Ding L, Ren H, et al. Ammonium nitrogen removal from coking wastewater by chemical precipitation recycle technology. Water Research, 2009, 43(20): 5209-5215.

[92] Liu Y, Kumar S, Kwag J H, et al. Magnesium ammonium phosphate formation, recovery and its application as valuable resources: A review. Journal of Chemical Technology & Biotechnology, 2013, 88(2): 181-189.

[93] Siciliano A, Rosa S D. Recovery of ammonia in digestates of calf manure through a struvite precipitation process using unconventional reagents. Environmental Technology, 2014, 35(7): 841-850.

[94] Fattah K P, Mavinic D S, Koch F A. Influence of process parameters on the characteristics of struvite pellets. Journal of Environmental Engineering-Asce, 2012, 138(12): 1200-1209.

[95] Wen J, Tao W, Wang Z, et al. Enhancing simultaneous nitritation and anammox in recirculating biofilters: Effects of

unsaturated zone depth and alkalinity dissolution of packing materials. Journal of Hazardous Materials, 2013, 244-245: 671-680.

[96] Liu Y, Gao L, Xue Q, et al. Voltage-dependent anion channel involved in the mitochondrial calcium cycle of cell lines carrying the mitochondrial DNA A4263G mutation. Biochemical and Biophysical Research Communications, 2011, 404(1): 364-369.

[97] Li X Z, Zhao Q L. Recovery of ammonium-nitrogen from landfill leachate as a multi-nutrient fertilizer. Ecological Engineering, 2003, 20(2): 171-181.

[98] González-Ponce R, López-de-Sá E G, Plaza C. Lettuce response to phosphorus fertilization with struvite recovered from municipal wastewater. Ecological Engineering, 2009, 44(2): 426.

[99] Rasul G A M, Esmail A O, Mekha R J. The role of magnesium in increasing of phosphorus fertilizer efficiency and wheat yield. Mesopot J Agric, 2011, 39: 33-39.

[100] Zhang X C, Norton L D. Effect of exchangeable Mg on saturated hydraulic conductivity, disaggregation and clay dispersion of disturbed soils. Journal of Hydrology, 2002, 260(1): 194-205.

[101] Mariakakis I, Bilbao J, Egner S, et al. PilotTesting of Struvite Recovery from Centrate of a German Municipal WWTP Through Electrochemical Precipitation (ePhos Technology). Proceedings of the Water Environment Federation, 2015.

[102] Hug A, Udert K M. Struvite precipitation from urine with electrochemical magnesium dosage. Water Research, 2013, 47(1): 289-299.

[103] 王晓林. 电渗析技术在有机酸生产和剩余污泥氮磷资源化中的应用研究. 合肥: 中国科学技术大学学位论文, 2014.

[104] Shin C, Bae J. Current status of the pilot-scale anaerobic membrane bioreactor treatments of domestic wastewaters: A critical review. Bioresource Technology, 2018, 247: 1038-1046.

[105] Shin C, McCarty P L, Kim J, et al. Pilot-scale temperate-climate treatment of domestic wastewater with a staged anaerobic fluidized membrane bioreactor (SAF-MBR). Bioresource Technology, 2014, 159: 95-103.

[106] Kim Y, Ha E H, Kim E J, et al. Prenatal exposure to phthalates and infant development at 6 months: prospective Mothers and Children's Environmental Health (MOCEH) study. Environmental Health Perspectives. 2011, 119(10): 1495-1500.

[107] Rulli M C, Bellomi D, Cazzoli A, et al. The water-land-food nexus of first-generation biofuels. Scientific reports, 2016, 6: 22521.

[108] McKendry P. Energy production from biomass(part 1): Overview of biomass. Bioresource Technology, 2002, 83(1): 37-46.

[109] Cheng J. Biomass to renewable energy processes. USA: CRC press, 2017.

[110] Dalla Marta A, Natali F, Mancini M, et al. Energy and water use related to the cultivation of energy crops: a case study in the Tuscany Region. Ecology and Society, 2011, 16(2): 53-67.

[111] Taheripour F, Hertel T W, Liu J. The role of irrigation in determining the global land use impacts of biofuels. Energy, Sustainability and Society, 2013, 3(1): 4.

[112] Mo W W, Zhang Q. Energy-nutrients-water nexus: Integrated resource recovery in municipal wastewater treatment plants. Journal of Environmental Management, 2013, 127: 255-267.

[113] Karimi P, Qureshi A S, Bahramloo R, et al. Reducing carbon emissions through improved irrigation and groundwater management: A case study from Iran. Agricultural Water Management, 2012, 108: 52-60.

[114] McCord P, Dell'Angelo J, Baldwin E, et al. Polycentric transformation in Kenyan water governance: A dynamic analysis of institutional and social‐ecological change. Policy Studies Journal, 2017, 45(4): 633-658.

[115] Mabrey D, Vittorio M. Moving from theory to practice in the water–energy–food nexus: An evaluation of existing models and frameworks. Water-Energy Nexus, 2018, 1(1): 17-25.

[116] Dubois O, Faurès J, Felix E, et al. The Water-Energy-Food Nexus: A new approach in support of food security and

sustainable agriculture. Rome, Food and Agriculture Organization of the United Nations, 2014.

[117] Workshop E. Good Food Security Governance: The Crucial Premise to the Twin-Track Approach. ESA Workshop. Rome: Springer Press, 2011.

[118] Ait-Kadi M, Arriens W L. Increasing Water Security: A Development Imperative. Global Water Partnership(GWP): 2012, 1-16.

[119] Pahl-Wostl C. A conceptual framework for analysing adaptive capacity and multi-level learning processes in resource governance regimes. Global Environmental Change, 2009, 19(3): 354-365.

[120] Al-Saidi M, Elagib N A. Towards understanding the integrative approach of the water, energy and food nexus. Science of The Total Environment, 2017, 574: 1131-1139.

[121] Chellaney B. Water: Asia's new battleground. Washington. DC: Georgetown University Press, 2011.

[122] Pearce F. A global treaty on rivers: Key to true water security. Yale Environment, 2012, 360.

[123] Lele U, Klousia-Marquis M, Goswami S. Good governance for food, water and energy security. Aquatic Procedia, 2013, 1: 44-63.

[124] Aggarwal P K, Joshi P K, Ingram J S, et al. Adapting food systems of the Indo-Gangetic plains to global environmental change: key information needs to improve policy formulation. Environmental Science & Policy, 2004, 7(6): 487-498.

[125] Gulati M, Jacobs I, Jooste A, et al. The water–energy–food security nexus: Challenges and opportunities for food security in South Africa. Aquatic Procedia, 2013, 1: 150-164.

[126] Scott C A, Kurian M, Wescoat J L. in Governing the Nexus: Water, Soil and Waste Resources Considering Global Change. Cham: Springer International Publishing, 2015.

作者：陈勇生[1]，左振宇[1,2]，张君枝[1,3]，汪锰[1,4]，张旭[1,5]，钟爽[1,6]，王振兴[1,7]，张伯鹏[1]

[1] 美国佐治亚理工学院，[2] 武汉科技大学，[3] 北京建筑大学，[4] 中国海洋大学，[5] 合肥工业大学，[6] 吉林建筑大学，[7] 生态环境部华南环境科学研究所